Mathematics *for* Polytechnics

As per the latest curriculae of diploma courses prescribed by All India Council of Technical Education (AICTE) and State Boards of Technical Education

- Delhi
- Uttar Pradesh
- Bihar
- Rajasthan
- Maharashtra
- Karnataka
- West Bengal
- Tamil Nadu
- Uttarakhand
- Haryana
- Andhra Pradesh
- Gujarat
- Jharkhand
- Madhya Pradesh

Universities

- Jamia Millia Islamia
- OPJS University, Rajasthan
- Integral University, Lucknow
- Gujarat Technological University

1577 Examples
3174 Exercises
838 Questions

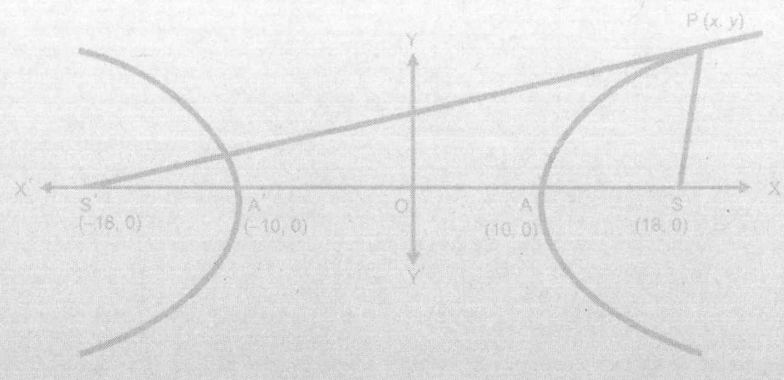

Mathematics *for* Polytechnics

As per the latest curriculae of diploma courses prescribed by All India Council of Technical Education (AICTE) and State Boards of Technical Education

HK Dass MSc
Diploma in Specialist Studies (Maths)
University of Hull, UK
Author of over 50 titles in the fields of engineering and mathematics; popular faculty among students of polytechnics and AMIE and IETE courses

Rama Verma MSc, PhD
Associate Professor
Mata Sundri College for Women
University of Delhi

Rajnish Verma BE, PhD (P), Fellow IETE, MBA
Consultant (Retd.), TCS Ltd.
Ex. DGM, CMC Ltd.

CBS Publishers & Distributors Pvt Ltd

New Delhi • Bengaluru • Chennai • Kochi • Kolkata • Lucknow • Mumbai
Hyderabad • Jharkhand • Nagpur • Patna • Pune • Uttarakhand

Disclaimer

Science and technology are constantly changing fields. New research and experience broaden the scope of information and knowledge. The authors have tried their best in giving information available to them while preparing the material for this book. Although all efforts have been made to ensure optimum accuracy of the material, yet it is quite possible some errors might have been left uncorrected. The publisher, the printer and the authors will not be held responsible for any inadvertent errors or inaccuracies.

Mathematics *for* Polytechnics

ISBN: 978-93-90709-24-3

First Edition: 2022

Published by Satish Kumar Jain and produced by Varun Jain for

CBS Publishers & Distributors Pvt Ltd
4819/XI Prahlad Street, 24 Ansari Road, Daryaganj, New Delhi 110 002, India
Ph: 011-23289259, 23266861, 23266867 Fax: 011-23243014 Website: www.cbspd.com
e-mail: delhi@cbspd.com; cbspubs@airtelmail.in.

Corporate Office: 204 FIE, Industrial Area, Patparganj, Delhi 110 092, India
Ph: 011-4934 4934 Fax: 011-4934 4935 e-mail: publishing@cbspd.com; publicity@cbspd.com

Branches

- **Bengaluru:** Seema House 2975, 17th Cross, K.R. Road, Banasankari 2nd Stage, Bengaluru 560 070, Karnataka, India
  Ph: +91-80-26771678/79 Fax: +91-80-26771680 e-mail: bangalore@cbspd.com
- **Chennai:** 7, Subbaraya Street, Shenoy Nagar, Chennai 600 030, Tamil Nadu, India
  Ph: +91-44-26680620, 26681266 Fax: +91-44-42032115 e-mail: chennai@cbspd.com
- **Kochi:** 42/1325, 36, Power House Road, Opposite KSEB, Kochi-682 018, Kerala, India
  Ph: +91-484-4059061-67 Fax: +91-484-4059065 e-mail: kochi@cbspd.com
- **Kolkata:** 147, Hind Ceramics Compound, 1st Floor, Nilgunj Road, Belghoria, Kolkata-700 056, West Bengal, India
  Ph: 9096713055, 7798394118, 9836841399 e-mail: kolkata@cbspd.com
- **Lucknow:** Basement, Khushnuma Complex, 7-Meerabai Marg (Behind Jawahar Bhawan), Lucknow-226 001, Uttar Pradesh, India
  Ph: +0552-4000032 e-mail:tiwari.lucknowi@cbspd.com
- **Mumbai:** PWD Shed, Gala No. 25/26, Ramchandra Bhatt Marg, Next to JJ Hospital Gate No. 2, Opp Union Bank of India, Noorbaug, Mumbai-400009, Maharashtra, India
  Ph: +91-22-66661880, 66661889 e-mail: mumbai@cbspd.com

Representatives

- Hyderabad 0-9885175004
- Jharkhand 0-9811541605
- Nagpur 0-9421945513
- Patna 0-9334159340
- Pune 0-9623451994
- Uttarakhand 0-9716462459

Printed at Rashtriya Printers, Dilshad Garden, Delhi, India

Foreword

This is one of the best available books on mathematics for the students of polytechnics in India. The three authors of this book are highly competent and have teaching experiences of over three decades. The book covers topics which are consistent with the latest changes in the curriculum. The book explains the basic concepts of mathematics very clearly. It includes a significant number of solved questions which have also appeared in the previous examinations. There are also a large number of unsolved exercises for practice. At the top of this, the authors have made efforts to append latest examination papers of various Boards of Technical Education of previous years.

I congratulate the authors for writing such a useful book with comprehensive explanations of the concepts and analytics of mathematics. I am sure the current and future generation of students interested in polytechnic education in India will greatly benefit from the contents of this book.

Satya Paul
Honorary Professor of Economics
Centre for Social Research and Methods
College of Arts and Social Sciences
Australian National University, Canberra
Adjunct Professor, Amrita University, Kerala, India
Former Professor of Economics
Western Sydney University
and Professor and Head
School of Economics
University of the South Pacific, Suva, Fiji

Preface

The book is based on the rich and varied experience of teaching mathematics for more than 40 years. We have created a fantastic book for diploma students of polytechnics in India, as well as the teaching faculty in these institutions. The text chapters have been arranged according to the latest syllabi (2021-2020-2019) of All India Council for Technical Education (AICTE) and all State Boards of Technical Education and Training. All diploma students, lecturers and other academicians from thousands of polytechnics and institutes in India will benefit greatly from this book.

Earnest effort has been made to make the contents lucid and easy to understand as much as possible. Emphasis has been laid on making the concepts very clear to the students to enable them to score higher marks and better grades. Each and every important step, along with the formulae, is mentioned so that students can follow the subject even without any guidance. Solved examples and exercises included in the book are as per the new trends of the examination papers. The book has a large number of solved examples from the latest examination papers from different State Boards of Technical Education. The students who have studied the subject using this book will be fully confident in solving the problems asked in their examinations.

Salient Features

- Contents specifically selected and presented according to the latest examination pattern to score higher marks and better grades
- Very Short Answer Questions, Long Answer Questions, MCQs at the end of selected chapters
- Easy methodology with step-by-step solutions supplemented with detailed illustrations to clear mathematical concepts
- More than 1577 examples
- More than 3174 exercises and 838 questions given for self-evaluation and practice
- Latest examination papers of various state boards of technical education.

We take this opportunity to express our deep sense of gratitude to Mr SK Jain CMD, Mr Varun Jain Director, Mr YN Arjuna Senior Vice President—Publishing, Editorial and Publicity, and his entire motivated team from CBS Publishers & Distributors, for taking special interest in this book.

We will welcome all kinds of suggestions, corrections and constructive feedbacks to help us make the book more useful and reader friendly. Readers are welcome to point out improvements, mistakes, errors in the book and will be duly recognized too.

HK Dass
Rama Verma
Rajnish Verma

Contents

Unit IV: COORDINATE GEOMETRY

Unit V: FUNCTIONS, LIMITS, AND CONTINUITY

Unit VI: DIFFERENTIAL CALCULUS

Unit VII: INTEGRAL CALCULUS

Unit VIII: DIFFERENTIAL EQUATIONS

Unit IX: LAPLACE TRANSFORMS

Unit X: PROBABILITY AND STATISTICS

AICTE MODEL CURRICULUM FOR DIPLOMA COURSES IN ENGINEERING AND TECHNOLOGY

Detailed First Year Curriculum Contents

SEMESTER I

Course Code	BS101
Course Title	Mathematics I
Number of Credits	3 (L:2,T:1,P:0)
Prerequisites	NIL
Course Category	BS

Course Objectives

This course is designed to give a comprehensive coverage at an introductory level to the subject of trigonometry, differential calculus and basic elements of algebra.

Course Content

Unit I: Trigonometry

Concept of angles, measurement of angles in degrees, grades and radians and their conversions, T-ratios of allied angles (without proof), sum, difference formulae and their applications (without proof). Product formulae (transformation of product to sum, difference and vice versa). T-ratios of multiple angles, sub-multiple angles (2A, 3A, A/2). Graphs of sin x, cos x, tan x and e^x.

Unit II: Differential Calculus

Definition of function; Concept of limits. Four standard limits $\lim\limits_{x\to a}\dfrac{x^n - a^n}{x-a}$, $\lim\limits_{x\to 0}\dfrac{\sin x}{x}$, $\lim\limits_{x\to a}\left(\dfrac{a^x-1}{x}\right)$ and $\lim\limits_{x\to a}(1+x)^{1/x}$.

Differentiation by definition of x^n, $\sin x$, $\cos x$, $\tan x$, e^x and $\log_a x$. Differentiation of sum, product and quotient of functions. Differentiation of function of a function. Differentiation of trigonometric and inverse trigonometric functions, logarithmic differentiation, exponential functions.

Unit III: Algebra

Complex numbers: Definition, real and imaginary parts of a complex number, polar and Cartesian, representation of a complex number and its conversion from one form to other, conjugate of a complex number, modulus and amplitude of a complex number addition, subtraction, multiplication and division of a complex number. De-movier's theorem and its application.

Partial fractions: Definition of polynomial fraction; proper and improper fractions and definition of partial fractions. To resolve proper fraction into partial fraction with denominator containing non-repeated linear factors, repeated linear factors and irreducible non-repeated quadratic factors. To resolve improper fraction into partial fraction.

Permutations and combinations: Value of ${}^nP_r$ and ${}^nC_r$.

Binomial theorem: Binomial theorem (without proof) for positive integral index (expansion and general form); binomial theorem for any index (expansion without proof), first and second binomial approximation with applications to engineering problems.

AICTE MODEL CURRICULUM FOR DIPLOMA COURSES IN ENGINEERING AND TECHNOLOGY

Detailed First Year Curriculum Contents
SEMESTER II

Course Code	BS102
Course Title	Mathematics II
Number of Credits	4 (L:3,T:1,P:0)
Prerequisites	NIL
Course Category	BS

Course Objectives

This course is designed to give a comprehensive coverage at an introductory level to the subject of matrices, integral calculus, coordinate geometry, basic elements of vector algebra and first order differential equations.

Course Content

Unit I: Determinants and Matrices

Elementary properties of determinants upto 3rd order, consistency of equations, Crammer's rule. Algebra of matrices, inverse of a matrix, matrix inverse method to solve a system of linear equations in 3 variables.

Unit II: Integral Calculus

Integration as inverse operation of differentiation. Simple integration by substitution, by parts and by partial fractions (for linear factors only). Use of formulas $\int_0^{\pi/2} \sin^n x\,dx, \int_0^{\pi/2} \cos^n x\,dx$, and $\int_0^{\pi/2} \sin^m x \cos^n x\,dx$ for solving problems where m and n are positive integers.

Applications of integration for (i) Simple problem on evaluation of area bounded by a curve and axes. (ii) Calculation of Volume of a solid formed by revolution of an area about axes. (Simple problems).

Unit III: Co-ordinate Geometry

Equation of straight line in various standard forms (without proof), intersection of two straight lines, angle between two lines. Parallel and perpendicular lines, perpendicular distance formula.

General equation of a circle and its characteristics. To find the equation of a circle, given:

i. Centre and radius
ii. Three points lying on it and
iii. Coordinates of end points of a diameter

Definition of conics (parabola, ellipse, hyperbola) their standard equations without proof. Problems on conics when their foci, directories or vertices are given.

Unit IV: Vector Algebra

Definition notation and rectangular resolution of a vector. Addition and subtraction of vectors. Scalar and vector products of 2 vectors. Simple problems related to work, moment and angular velocity.

Unit V: Differential Equations

Solution of first order and first degree differential equation by variable separation method (simple problems). MATLAB – Simple introduction.

Unit - I : ALGEBRA

Laws of Indices, Algebra and Logarithm

1.1 INTRODUCTION

Welcome dear students and aspirant engineers to the study of mathematics. Let's begin with the basic laws of indices and algebra. As you have already studied exponents, irrational numbers, surds, and much more in your school.

Nevertheless, the concepts have been discussed here to make you acquainted with basics of engineering mathematics, which would be highly useful in coming chapters too. The key to understand any topic is to go through that in detail and please make sure to practice examples.

1.2 EXPONENT

We know that x^4 is a short way of writing $x \cdot x \cdot x \cdot x$. In general, if n is a natural number or a positive integer, then

$x^n = x \cdot x \cdot x \cdot x$ ... to n factors

Here x^n is called the nth *power* of x and x is called the *base*.

The exponent n is placed at the right and above the base x.

Thus $\quad 3^4 = 3 \cdot 3 \cdot 3 \cdot 3 = 81$

$$\left(\frac{2}{3}\right)^5 = \frac{2}{3} \cdot \frac{2}{3} \cdot \frac{2}{3} \cdot \frac{2}{3} \cdot \frac{2}{3} = \frac{32}{243}$$

1.3 LAWS OF EXPONENT (FOR REAL NUMBERS)

(1) $a^m \cdot a^n = a^{m+n}$ (2) $(a^m)^n = a^{mn}$

(3) $\dfrac{a^m}{a^n} = a^{m-n}, (m > n)$ (4) $a^m \, b^m = (ab)^m$

Note. $a^0 = 1$ and $\dfrac{1}{a^m} = a^{-m}$

Example 1. *Simplify:*

(*i*) $2^4 \cdot 2^5$ (*ii*) $3^{\frac{1}{2}} \cdot 3^{\frac{5}{3}}$ (*iii*) $\left(3^{\frac{1}{2}}\right)^8$ (*iv*) $\dfrac{7^{\frac{1}{2}}}{7^{\frac{1}{3}}}$ (*v*) $5^{\frac{1}{4}} \cdot 7^{\frac{1}{4}}$

Solution. (*i*) $2^4 \cdot 2^5 = 2^{4+5} = 2^9$ $\quad [\because \; a^m \cdot a^n = a^{m+n}]$

(*ii*) $3^{\frac{1}{2}} \cdot 3^{\frac{5}{4}} = 3^{\frac{1}{2}+\frac{5}{3}} = 3^{\frac{3+10}{6}} = 3^{\frac{13}{6}}$ $\quad [\because \; a^m \cdot a^n = a^{m+n}]$

(*iii*) $\left(3^{\frac{1}{2}}\right)^8 = 3^{\frac{1}{2}\times 8} = 3^4$ $\qquad [(a^m)^n = a^{mn}]$

(*iv*) $\dfrac{7^{\frac{1}{2}}}{7^{\frac{1}{3}}} = 7^{\frac{1}{2}-\frac{1}{3}} = 7^{\frac{3-2}{6}} = 7^{\frac{1}{6}}$ $\qquad \left[\dfrac{a^m}{a^n} = a^{m-n}\right]$

(*v*) $5^{\frac{1}{4}} \cdot 7^{\frac{1}{4}} = (5\times 7)^{\frac{1}{4}} = 35^{\frac{1}{4}}$ $\qquad [\because\ a^m \,.\, b^m = (ab)^m]$ **Ans.**

Example 2. *Find* (*i*) $(49)^{\frac{1}{2}}$ (*ii*) $(125)^{\frac{1}{3}}$ (*iii*) $(32)^{\frac{3}{5}}$ (*iv*) $(16)^{\frac{3}{2}}$ (*v*) $\dfrac{(13)^{\frac{1}{2}}}{(13)^{\frac{1}{4}}}$

Solution. (*i*) $(49)^{\frac{1}{2}} = (7^2)^{\frac{1}{2}} = 7^{2\times\frac{1}{2}} = 7^1 = 7$

(*ii*) $(125)^{\frac{1}{3}} = (5^3)^{\frac{1}{3}} = 5^{3\times\frac{1}{3}} = 5^1 = 5$

(*iii*) $(32)^{\frac{3}{5}} = (2^5)^{\frac{3}{5}} = 2^{5\times\frac{3}{5}} = 2^3 = 8$

(*iv*) $(16)^{\frac{3}{2}} = (4^2)^{\frac{3}{2}} = 4^{2\times\frac{3}{2}} = 4^3 = 64$

(*v*) $\dfrac{(13)^{\frac{1}{2}}}{(13)^{\frac{1}{4}}} = (13)^{\frac{1}{2}-\frac{1}{4}} = 13^{\frac{1}{4}}$ **Ans.**

Example 3. *Simplify:*

(*i*) $(64)^{\frac{1}{2}} \cdot (8)^{\frac{1}{3}}$ (*ii*) $(27)^{\frac{2}{3}} \cdot (81)^{\frac{1}{4}}$ (*iii*) $(125)^{\frac{1}{3}} \cdot (16)^{\frac{1}{2}}$ (*iv*) $(36)^{\frac{1}{3}} \cdot (8)^{\frac{1}{2}} \cdot (25)^{\frac{1}{2}}$

Solution. (*i*) $(64)^{\frac{1}{2}} \cdot (8)^{\frac{1}{3}} = (2^6)^{\frac{1}{2}} \cdot (2^3)^{\frac{1}{3}} = 2^{6\times\frac{1}{2}} \cdot 2^{3\times\frac{1}{3}} = 2^3 \cdot 2^1 = 2^{3+1} = 2^4$

(*ii*) $(27)^{\frac{2}{3}} \cdot (81)^{\frac{1}{4}} = (3^3)^{\frac{2}{3}} \cdot (3^4)^{\frac{1}{4}} = 3^{3\times\frac{2}{3}} \cdot 3^{4\times\frac{1}{4}} = 3^2 \cdot 3^1 = 3^3$

(*iii*) $(125)^{\frac{1}{3}} \cdot (16)^{\frac{1}{2}} = (5^3)^{\frac{1}{3}} \cdot (2^4)^{\frac{1}{2}} = 5^{3\times\frac{1}{3}} \cdot 2^{4\times\frac{1}{2}} = 5^1 \cdot 2^2 = 5 \cdot 4 = 20$

(*iv*) $(16)^{\frac{1}{3}} \cdot (8)^{\frac{1}{2}} \cdot (25)^{\frac{1}{2}} = (2^4)^{\frac{1}{3}} \cdot (2^3)^{\frac{1}{2}} \cdot (5^2)^{\frac{1}{2}} = 2^{4\times\frac{1}{3}} \cdot 2^{3\times\frac{1}{2}} \cdot 5^{2\times\frac{1}{2}}$

$= 2^{\frac{4}{3}} \cdot 2^{\frac{3}{2}} \cdot 5^1 = 2^{\frac{4}{3}+\frac{3}{2}} \cdot 5 = 2^{\frac{17}{6}} \cdot 5 = 5 \cdot 2^{\frac{17}{6}}$ **Ans.**

Example 4. *Simplify:* (*i*) $\dfrac{(16)^{\frac{1}{3}}}{(2)^{\frac{2}{3}}}$ (*ii*) $\dfrac{(21)^2}{(441)^{\frac{1}{3}}}$ (*iii*) $\dfrac{(35)^{\frac{2}{3}}}{(1225)}$ (*iv*) $\dfrac{(343)^{\frac{1}{3}}}{(49)^{\frac{1}{2}}}$

Solution. (*i*) $\dfrac{(16)^{\frac{1}{3}}}{(2)^{\frac{2}{3}}} = \dfrac{(2^4)^{\frac{1}{3}}}{(2)^{\frac{2}{3}}} = \dfrac{2^{4\times\frac{1}{3}}}{2^{\frac{2}{3}}} = \dfrac{2^{\frac{4}{3}}}{2^{\frac{2}{3}}} = 2^{\frac{4}{3}-\frac{2}{3}} = 2^{\frac{2}{3}}$

(*ii*) $\dfrac{(21)^2}{(441)^{\frac{1}{3}}} = \dfrac{(21)^2}{(21^2)^{\frac{1}{3}}} = \dfrac{(21)^2}{(21)^{2\times\frac{1}{3}}} = \dfrac{(21)^2}{(21)^{\frac{2}{3}}} = (21)^{2-\frac{2}{3}} = (21)^{\frac{4}{3}}$

(*iii*) $\dfrac{(35)^{\frac{2}{3}}}{(1225)} = \dfrac{(35)^{\frac{2}{3}}}{(35)^2} = (35)^{\frac{2}{3}-2} = (35)^{-\frac{4}{3}} = \dfrac{1}{(35)^{\frac{4}{3}}}$

(*iv*) $\dfrac{(343)^{\frac{1}{3}}}{(49)^{\frac{1}{2}}} = \dfrac{(7^3)^{\frac{1}{3}}}{(7^2)^{\frac{1}{2}}} = \dfrac{7^{3\times\frac{1}{3}}}{7^{2\times\frac{1}{2}}} = \dfrac{7^1}{7^1} = 7^{1-1} = 7^0 = 1$ **Ans.**

Example 5. *Solve* $4^{x+1} \,.\, 2^x = 32.$

Solution.

$$4^{x+1} \,.\, 2^x = 32$$

$$(2^2)^{x+1} \,.\, 2^x = 2^5 \quad \Rightarrow \quad 2^{2x+2} \,.\, 2^x = 2^5$$

$$2^{2x+2+x} = 2^5 \quad \Rightarrow \quad 2^{3x+2} = 2^5$$

$$3x + 2 = 5$$

$$3x = 3 \quad \text{or} \quad x = 1$$ **Ans.**

Example 6. *Solve* $2^{x+1} \,.\, 3^{y+2} = \dfrac{1}{6}; \; 2^{2x+1} \,.\, 3^{3y+5} = \dfrac{1}{648}.$

Solution.

$$2^{x+1} \,.\, 3^{y+2} = \frac{1}{6} \Rightarrow 2^x \,.\, 2 \,.\, 3^y \,.\, 3^2 = \frac{1}{6}$$

$$\Rightarrow \quad 2^x \,.\, 3^y \,.\, 2^1 \,.\, 3^2 = \frac{1}{2.3} \quad \Rightarrow \quad 2^x \,.\, 3^y = \frac{1}{2^1 .3.2^1 .3^2}$$

$$\Rightarrow \quad 2^x \,.\, 3^y = \frac{1}{2^2 .3^3} \quad \Rightarrow \quad 2^x \,.\, 3^y = 2^{-2} \,.\, 3^{-3}$$

$$\Rightarrow \quad (2^x \,.\, 3^y)^3 = (2^{-2} \,.\, 3^{-3})^3$$

$$\Rightarrow \quad 2^{3x} \,.\, 3^{3y} = 2^{-6} \,.\, 3^{-9} \quad \text{... (1)}$$

and

$$2^{2x+1} \,.\, 3^{3y+5} = \frac{1}{648}$$

$$\Rightarrow \quad 2^{2x} \,.\, 2^1 \,.\, 3^{3y} \,.\, 3^5 = \frac{1}{8 \times 81}$$

$$\Rightarrow \quad 2^{2x} \,.\, 3^{3y} \,.\, 2^1 \,.\, 3^5 = 2^{-3} \times 3^{-4}$$

$$\Rightarrow \quad 2^{2x} \,.\, 3^{3y} = 2^{-4} \times 3^{-9} \quad \text{...(2)}$$

Dividing (1) by (2), we get

$$\frac{2^{3x} \,.\, 3^{3y}}{2^{2x} \,.\, 3^{3y}} = \frac{2^{-6} \,.\, 3^{-9}}{2^{-4} \,.\, 3^{-9}}$$

$$\Rightarrow \quad 2^x = 2^{-2} \quad \Rightarrow \quad x = -2$$

Substituting $x = -2$ in equation (2), we get

$$2^{-4} \,.\, 3^{3y} = 2^{-4} \,.\, 3^{-9}$$

$$\Rightarrow \quad 3^{3y} = 3^{-9}$$

$$\Rightarrow \quad 3y = -9$$

$$\Rightarrow \quad y = -3, \text{ and } x = -2$$ **Ans.**

1.4 IRRATIONAL NUMBERS

The numbers which can not be expressed in the form $\frac{p}{q}$ are called *irrational number.* Ex : $\sqrt{2}, \sqrt{3}, \sqrt[3]{5}$, etc.

1.5 RADICAL AND RADICAL SIGN

The symbols $\sqrt{\ }, \sqrt[3]{\ }, \sqrt[4]{\ }$ are *radical signs.* Thus $\sqrt{2}, \sqrt{3}, \sqrt[3]{5}$ are also called *radicals.*

The index of a root is the small number written above and to the left of the radical sign $\sqrt{\ }$. Thus, the index number of $\sqrt[3]{6}$ is 3. In square roots, the index 2 is not indicated but understood.

1.6 SURDS

Quantities of the type $\sqrt[n]{a}$, where a is a positive rational number and it is not possible to find exactly the nth root of a, are called surds of order n.

It is an irrational root of a rational number.

Ex : $\sqrt{2}, \sqrt{3}, \sqrt{5}, \sqrt{8}$ and so on.

Note: $\sqrt{2+\sqrt{3}}$ is not a surd because $2 + \sqrt{3}$ is not a rational number.

Remark : $\sqrt[n]{a}$ fails to be a surd if

either (*i*) $a < 0$, *e.g.* $\sqrt{-5}$ is not a surd.

or (*ii*) a is not a rational number.

or (*iii*) the nth root of a is found exactly.

A surd of order 2 is called a quadratic surd and a surd of order 3 is called a cubic surd. Thus $\sqrt{3}$ is a quadratic surd and $\sqrt[3]{4}$ is a cubic surd.

Entire surd: If a surd does not contain a rational factor or term, *e.g.* $\sqrt{2}$ or $\sqrt{2}+\sqrt{3}$.

Mixed surd: If it contains a rational factor or term, *e.g.* $3\sqrt{2}$ or $2+\sqrt{3}$.

Pure surd: If it contains each term a surd, *e.g.* $\sqrt{2}+3\sqrt{5}$.

Binomial surd: If it is a binomial and contains at least one surd, *e.g.* $3+\sqrt{2}$.

Similar surds: Two or more surds are said to be similar when they can be reduced so as to have the same irrational factor. Thus $\sqrt{12}$ and $\sqrt{75}$ are similar surds, being respectively equivalent to $2\sqrt{3}$ and $5\sqrt{3}$..

Note: Surds of the same order can be multiplied, divided and compared.

Example 7. *Which is greater* $\sqrt[3]{5}$ or $\sqrt[4]{4}$?

Solution. $\sqrt[3]{5} = 5^{\frac{1}{3}}$ and $\sqrt[4]{4} = 4^{\frac{1}{4}}$

L.C.M of the denominators of indices 3 and 4 = 12.

$$\sqrt[3]{5} = 5^{\frac{1}{3}} = (5)^{\frac{4}{12}} = (5^4)^{\frac{1}{12}} = (625)^{\frac{1}{12}}$$

$$\Rightarrow \sqrt[4]{4} = 4^{\frac{1}{4}} = (4)^{\frac{3}{12}} = (4^3)^{\frac{1}{12}} = (64)^{\frac{1}{12}}$$

$$\Rightarrow (625)^{\frac{1}{12}} > (64)^{\frac{1}{12}}$$

$$\therefore \quad \sqrt[3]{5} > \sqrt[4]{4}$$

Ans.

EXERCISE 1.1

1. *Find the value of the following:*

(i) $(128)^{\frac{3}{7}}$ **Ans.** 8 (ii) $(243)^{-\frac{2}{5}}$ **Ans.** $\frac{1}{9}$

(iii) $\frac{1}{(216)^{-\frac{2}{3}}}$ **Ans.** 36 (iv) $\left(\frac{8}{27}\right)^{-\frac{4}{3}}$ **Ans.** $\frac{81}{16}$

2. *Solve the following:*

(i) $3^2 . 3^4$ **Ans.** 3^6 (ii) $4^2 . 8^3$ **Ans.** 2^{13}

(iii) $5^3 . 25^2$ **Ans.** 5^7 (iv) $49^2 . 7^3$ **Ans.** 7^7

(v) $(49)^{\frac{1}{2}} \cdot (16)^{\frac{1}{2}}$ **Ans.** 28 (vi) $(125)^{\frac{1}{3}} \cdot (25)^{\frac{1}{2}}$ **Ans.** 5^2

(vii) $(216)^{\frac{1}{3}} \cdot (36)^{\frac{1}{2}}$ **Ans.** 6^2 (viii) $(256)^{\frac{1}{4}} \cdot (32)^{\frac{1}{5}}$ **Ans.** 8

3. *Simplify:*

(i) $\frac{(49)^{\frac{1}{2}}}{(343)^{\frac{1}{3}}}$ **Ans.** 7^0 (ii) $\frac{(27)^2}{(81)^{\frac{1}{3}}}$ **Ans.** $3^{\frac{14}{3}}$

(iii) $\frac{(125)^{\frac{1}{3}}}{(5)^{\frac{4}{3}}}$ **Ans.** $5^{-\frac{1}{3}}$ (iv) $\frac{(121)^{\frac{1}{2}}}{(11)^{\frac{3}{2}}}$ **Ans.** $11^{-\frac{1}{2}}$

(v) $\frac{(16)^{\frac{1}{3}}}{(8)^{\frac{1}{2}}} \times \frac{(4)^{\frac{1}{3}}}{(32)^{\frac{1}{5}}}$ **Ans.** $2^{-\frac{1}{2}}$ (vi) $\frac{(6)^{\frac{1}{3}} \times (9)^{\frac{2}{5}}}{(2) \times (3)^{\frac{2}{3}}}$ **Ans.** $\frac{3^{\frac{7}{15}}}{2^{\frac{2}{3}}}$

(vii) $\frac{(15)^{\frac{2}{3}} (3)^{\frac{1}{2}}}{(5)^2 (9)^{\frac{1}{3}}}$ **Ans.** $\frac{3^{\frac{1}{2}}}{5^{\frac{4}{3}}}$ (viii) $\frac{(30)^{\frac{3}{4}} \cdot (2)^{\frac{1}{3}}}{(25)^{\frac{3}{8}} \cdot (6)^{\frac{13}{12}}}$ **Ans.** $3^{-\frac{1}{3}}$

4. *Simplify the following:*

(i) $\left\{(64)^{\frac{3}{5}}\right\}^2 \times (8^3)^{-2}$ **Ans.** 4

(ii) $\left(\frac{a^x}{a^y}\right)^{x+y} \times \left(\frac{a^y}{a^z}\right)^{y+z} \times \left(\frac{a^z}{a^x}\right)^{z+x}$ **Ans.** 1

(iii) $\left(\frac{x^b}{x^c}\right)^{\frac{1}{bc}} \times \left(\frac{x^c}{x^a}\right)^{\frac{1}{ca}} \times \left(\frac{x^a}{x^b}\right)^{\frac{1}{ab}}$ **Ans.** 1

(iv) $\left(\frac{x^a}{x^b}\right)^{a^2+ab+b^2} \times \left(\frac{x^b}{x^c}\right)^{b^2+bc+c^2} \times \left(\frac{x^c}{x^a}\right)^{c^2+ca+a^2}$ **Ans.** 1

(v) $\left[x^{\frac{b+c}{c-a}}\right]^{\frac{1}{a-b}} \cdot \left[x^{\frac{c+a}{a-b}}\right]^{\frac{1}{b-c}} \cdot \left[x^{\frac{a+b}{b-c}}\right]^{\frac{1}{c-a}}$ **Ans.** 1

(vi) $\left(x^{\frac{1}{2}}+y^{\frac{1}{2}}\right)\cdot\left(x^{\frac{1}{4}}+y^{\frac{1}{4}}\right)\cdot\left(x^{\frac{1}{4}}-y^{\frac{1}{4}}\right)$ **Ans.** $x-y$

(vii) $\dfrac{2^{m+3}\cdot 3^{2m-n}\cdot 5^{m+n+3}\cdot 6^{n+1}}{6^{m+1}\cdot 10^{n+3}\cdot 15^{m}}$ **Ans.** 1

(viii) $\dfrac{2^{n}+2^{n-1}}{2^{n+1}-2^{n}}$ **Ans.** $\dfrac{3}{2}$

(ix) $\dfrac{3^{a}\cdot 3^{a^2-a}}{3^{a+1}3^{a-1}}\times\left[\dfrac{(3^3)^{\frac{\alpha}{3}}}{3^2}\right]^{-\alpha}$ **Ans.** 1

(x) $\left[\dfrac{(a^2b^3)^{\frac{2}{3}}\cdot(a^{-2}\cdot b^{-2}c)^{\frac{3}{2}}}{(a\cdot b)^{-\frac{2}{3}}\cdot c^{\frac{1}{2}}}\right]$ **Ans.** $\dfrac{c^3}{a\cdot b}$

(xi) $\left\{(x^p)^{1-\frac{1}{p}}\right\}^{p^2+p+1}$ **Ans.** x^{p^3-1}

(xii) $\dfrac{1}{1+x^{a-b}+x^{a-c}}+\dfrac{1}{1+x^{b-c}+x^{b-a}}+\dfrac{1}{1+x^{c-a}+x^{c-b}}$ **Ans.** 1

(xiii) Show that $\dfrac{y^{-1}}{x^{-1}+y^{-1}}+\dfrac{y^{-1}}{x^{-1}-y^{-1}}=\dfrac{2xy}{y^2-x^2}$.

(xiv) If $a = xy^{p-1}$, $b = xy^{q-1}$ and $c = xy^{r-1}$, prove that $a^{q-r}\,b^{r-p}\,c^{p-q} = 1$.

(xv) If $a = \sqrt[3]{3}+\left(\sqrt[3]{3}\right)^{-1}$, prove that: $3a^3 - 9a = 10$.

(xvi) If $m = a^x$, $n = a^y$ and $a^z = (m^y\,.\,n^x)^z$, show that: $x\,y\,z = 1$.

(xvii) If $a^x = b^y = c^z$ and $b^2 = ac$, then show that: $\dfrac{1}{x}+\dfrac{1}{z}=\dfrac{2}{y}$.

5. ***Solve the following equations:***

(i) $7^{x+7} = 49^{4x-7}$ **Ans.** $x = 3$

(ii) $2^{x+4} = 2^{x+3} + 4$ **Ans.** $x = -1$

(iii) $\left(\dfrac{a}{b}\right)^{4x-1}=\left(\dfrac{b}{a}\right)^{2x-5}$ **Ans.** $x = 1$

(iv) $9^{x-y} = 81$, $9^{x+y} = 729$ **Ans.** $x=\dfrac{5}{2}, y=\dfrac{1}{2}$

1.7 ALGEBRAIC IDENTITIES

1. $(a + b)^2 = a^2 + 2ab + b^2$
2. $(a - b)^2 = a^2 - 2ab + b^2$
3. $a^2 - b^2 = (a + b)\,(a - b)$
4. $(a + b + c)^2 = a^2 + b^2 + c^2 + 2ab + 2bc + 2ac$
5. $(a + b)^3 = a^3 + b^3 + 3ab\,(a + b)$
   $= a^3 + b^3 + 3a^2b + 3ab^2$
6. $(a - b)^3 = a^3 - b^3 - 3ab\,(a - b)$
   $= a^3 - b^3 - 3a^2b + 3ab^2$

Type I. Factorisation of the perfect square polynomials

Formulae: (*i*) $a^2 + 2ab + b^2 = (a + b)^2$ (*ii*) $a^2 - 2ab + b^2 = (a - b)^2$

(*iii*) $(a + b + c)^2 = a^2 + b^2 + c^2 + 2ab + 2bc + 2ca$

(*a*) Here the first and third terms are perfect squares.

(*b*) The middle term = 2 (product of square roots of first and third terms). This method is illustrated by the following examples.

Example 8. *Factorise:*

(*i*) $4a^2 + 12ab + 9b^2$ (*ii*) $x^2 + 10x + 25$

(*iii*) $9x^2 + 24xy + 16y^2$ (*iv*) $16a^2 - 40ab + 25b^2$

(*v*) $x^2 + 5x + \frac{25}{4}$ (*vi*) $49x^4 - 168x^2y^2 + 144y^4$

Solution. (*i*) $4a^2 + 12ab + 9b^2 = (2a)^2 + (3b)^2 + 2(2a)(3b) = (2a + 3b)^2$ **Ans.**

(*ii*) $x^2 + 10x + 25 = (x)^2 + (5)^2 + 2(x)(5) = (x + 5)^2$ **Ans.**

(*iii*) $9x^2 + 24xy + 16y^2 = (3x)^2 + (4y)^2 + 2(3x)(4y) = (3x + 4y)^2$ **Ans.**

(*iv*) $16a^2 - 40ab + 25b^2 = (4a)^2 + (5b)^2 - 2(4a)(5b) = (4a - 5b)^2$ **Ans.**

(*v*) $x^2 + 5x + \frac{25}{4} = (x)^2 + \left(\frac{5}{2}\right)^2 + 2(x)\left(\frac{5}{2}\right) = \left(x + \frac{5}{2}\right)^2$ **Ans.**

(*vi*) $49x^4 - 168x^2y^2 + 144y^4 = (7x^2)^2 + (12y^2)^2 - 2(7x^2)(12y^2) = (7x^2 - 12y^2)^2$ **Ans.**

EXERCISE 1.2

Factorise the following:

1. $a^2 + 4ab + 4b^2$ **Ans.** $(a + 2b)^2$

2. $x^2 + 12x + 36$ **Ans.** $(x + 6)^2$

3. $16x^2 + 40xy + 25y^2$ **Ans.** $(4x + 5y)^2$

4. $4x^2 + 6x + \frac{9}{4}$ **Ans.** $\left(2x + \frac{3}{2}\right)^2$

5. $9x^4 + 24x^2y^2 + 16y^4$ **Ans.** $(3x^2 + 4y^2)^2$

6. $1 - 8ax + 16a^2x^2$ **Ans.** $(1 - 4ax)^2$

7. $a^2 + a + \frac{1}{4}$ **Ans.** $\left(a + \frac{1}{2}\right)^2$

Type II. Factorising the difference of two squares

If the given polynomial is in the form of the difference of two squares, then its two factors are

(*i*) Sum of the two square roots.

(*ii*) Difference of the two square roots.

$$a^2 - b^2 = (a + b)(a - b)$$

The method will be more clear by the following solved examples.

Example 9. *Factorise:*

(*i*) $x^2 - 4y^2$ (*ii*) $4a^2x^2 - 25b^2y^2$

(*iii*) $a^2 - x^2 - 2xy - y^2$ (*iv*) $4a^2 - 4b^2 + 4a + 1$

Solution. (*i*) $x^2 - 4y^2 = (x)^2 - (2y)^2 = (x + 2y)(x - 2y)$ **Ans.**

(*ii*) $4a^2x^2 - 25b^2y^2 = (2ax)^2 - (5by)^2$

$= (2ax + 5by)\,(2ax - 5by)$ **Ans.**

(iii) $a^2 - x^2 - 2xy - y^2 = a^2 - (x^2 + 2xy + y^2)$

$= a^2 - (x + y)^2 = (a + x + y)(a - x - y)$ **Ans.**

(iv) $4a^2 - 4b^2 + 4a + 1 = (4a^2 + 4a + 1) - 4b^2$

$= (2a + 1)^2 - (2b)^2 = (2a + 2b + 1)(2a - 2b + 1)$ **Ans.**

Example 10. *Factorise:*

(i) $81 - 16x^2$ *(ii)* $5 - 20x^2$ *(iii)* $81x^4 - y^4$

Solution. *(i)* $81 - 16x^2 = (9)^2 - (4x)^2 = (9 + 4x)(9 - 4x)$ **Ans.**

(ii) $5 - 20x^2 = 5(1 - 4x^2) = 5[(1)^2 - (2x)^2] = 5(1 + 2x)(1 - 2x)$ **Ans.**

(iii) $81x^4 - y^4 = (9x^2)^2 - (y^2)^2 = (9x^2 + y^2)(9x^2 - y^2)$

$= (9x^2 + y^2)[(3x)^2 - (y)^2]$

$= (9x^2 + y^2)(3x + y)(3x - y)$ **Ans.**

EXERCISE 1.3

Factorise the following:

1. $x^2 - 4y^2$ **Ans.** $(x + 2y)(x - 2y)$ **2.** $100 - 9z^2$ **Ans.** $(10 + 3z)(10 - 3z)$

3. $x^2 - y^2 + 2x + 1$ **Ans.** $(x + 1 + y)(x + 1 - y)$

4. $x^3 - x$ **Ans.** $x(x + 1)(x - 1)$ **5.** $x^2 - y^2 + 6y - 9$ **Ans.** $(x + y - 3)(x - y + 3)$

6. $9a^2 + 6a + 1 - 36z^2$ **Ans.** $(3a + 1 + 6z)(3a + 1 - 6z)$

7. $x^2 - 1 - 2a - a^2$ **Ans.** $(x + 1 + a)(x - 1 - a)$

8. $4a^2 - (2b - c)^2$ **Ans.** $(2a + 2b - c)(2a - 2b + c)$

9. $18a^2x^2 - 32$ **Ans.** $2(3ax + 4)(3ax - 4)$

(i) Expansion and Factorisation

We will use the formula *(i)* $(a + b + c)^2 = a^2 + b^2 + c^2 + 2ab + 2bc + 2ac$

(ii) $(a + b)^3 = a^3 + b^3 + 3ab(a + b) = a^3 + b^3 + 3a^2b + 3ab^2$

(iii) $(a - b)^3 = a^3 - b^3 - 3ab(a - b) = a^3 - b^3 - 3a^2b + 3ab^2$

(ii) Factorisation of sum and difference of cubes

We will use the following formulae for factorisation of the sum and difference of two cubes:

(i) $a^3 + b^3 = (a + b)(a^2 - ab + b^2)$ *(ii)* $a^3 - b^3 = (a - b)(a^2 + ab + b^2)$

Example 11. *Factorise:* $216a^3 - 125$

Solution. $216a^3 - 125 = (6a)^3 - (5)^3$

$= (6a - 5)[(6a)^2 + (6a)5 + (5)^2]$

$= (6a - 5)(36a^2 + 30a + 25)$ **Ans.**

Example 12. *Factorise:* $a^6 - b^6$

Solution. $a^6 - b^6 = (a^3)^2 - (b^3)^2 = (a^3 + b^3)(a^3 - b^3)$

$= (a + b)(a^2 - ab + b^2)(a - b)(a^2 + ab + b^2)$ **Ans.**

Example 13. *Factorise:* $a^3 - b^3 - a + b$

Solution. $a^3 - b^3 - a + b = (a^3 - b^3) - (a - b)$

$= (a - b)(a^2 + ab + b^2) - 1(a - b)$

$= (a - b)(a^2 + ab + b^2 - 1)$ **Ans.**

Example 14. Prove that: $\dfrac{0.96 \times 0.96 \times 0.96 + 0.04 \times 0.04 \times 0.04}{0.96 \times 0.96 - 0.96 \times 0.04 + 0.04 \times 0.04} = 1$

Solution.
$$\frac{0.96 \times 0.96 \times 0.96 + 0.04 \times 0.04 \times 0.04}{0.96 \times 0.96 - 0.96 \times 0.04 + 0.04 \times 0.04} = \frac{(0.96)^3 + (0.04)^3}{(0.96)^2 - (0.96)(0.04) + (0.04)^2}$$
$$= \frac{a^3 + b^3}{a^2 - ab + b^2} \text{ (where } a = 0.96 \text{ and } b = 0.04)$$
$$= \frac{(a+b)(a^2 - ab + b^2)}{a^2 - ab + b^2}$$
$$= a + b = 0.96 + 0.04 = 1$$
Proved.

EXERCISE 1.4

Factorise the following:

1. $a^3 + 27$ **Ans.** $(a + 3)(a^2 - 3a + 9)$
2. $x^3 + 64$ **Ans.** $(x + 4)(x^2 - 4x + 16)$
3. $1 - 125y^3$ **Ans.** $(1 - 5y)(1 + 5y + 25y^2)$
4. $125x^3 - 343y^3$ **Ans.** $(5x - 7y)(25x^2 + 35xy + 49y^2)$
5. $\dfrac{p^3}{343} + 8q^3$ **Ans.** $\left(\dfrac{p}{7} + 2q\right)\left(\dfrac{p^2}{49} - \dfrac{2pq}{7} + 4q^2\right)$
6. $128x^3y^3 - 250z^3$ **Ans.** $2(4xy - 5z)(16x^2y^2 + 20xyz + 25z^2)$
7. $\dfrac{991 \times 991 \times 991 + 9 \times 9 \times 9}{991 \times 991 - 991 \times 9 \times 9 \times 9}$ **Ans.** 1000
8. $\dfrac{1.03 \times 1.03 \times 1.03 - 0.03 \times 0.03 \times 0.03}{1.03 \times 1.03 + 1.03 \times 0.03 + 0.03 \times 0.03}$ **Ans.** 1
9. $\dfrac{5.01 \times 5.01 \times 5.01 - 0.01 \times 0.01 \times 0.01}{5.01 \times 5.01 + 5.01 \times 0.01 + 0.01 \times 0.01}$ **Ans.** 5

(iii) Factorisation of $a^3 + b^3 + c^3 - 3abc$

Formula: $a^3 + b^3 + c^3 - 3abc = (a + b + c)(a^2 + b^2 + c^2 - ab - bc - ca)$

Proof. $a^3 + b^3 + c^3 - 3abc = (a^3 + b^3) + c^3 - 3abc = [(a + b)^3 - 3ab(a + b)] + c^3 - 3abc$
$$= x^3 - 3abx + c^3 - 3abc \quad \text{[where } a + b = x]$$
$$= (x^3 + c^3) - 3abx - 3abc = (x + c)(x^2 - xc + c^2) - 3ab(x + c)$$
$$= (x + c)(x^2 - xc + c^2 - 3ab) \quad ...(1)$$

Replacing 'x' by $(a + b)$ in (1), we get
$$= (a + b + c)[(a + b)^2 - (a + b)c + c^2 - 3ab]$$
$$= (a + b + c)[a^2 + 2ab + b^2 - ac - bc + c^2 - 3ab]$$
$$= (a + b + c)(a^2 + b^2 + c^2 - ab - bc - ca)$$
Proved.

Example 15. *Factorise:* $a^3 + b^3 + 8c^3 - 6abc$

Solution. $a^3 + b^3 + 8c^3 - 6abc = (a)^3 + (b)^3 + (2c)^3 - 3(a)(b)(2c)$
$$= (a + b + 2c)[a^2 + b^2 + (2c)^2 - ab - b(2c) - a(2c)]$$
$$= (a + b + 2c)(a^2 + b^2 + 4c^2 - ab - 2bc - 2ac)$$
Ans.

Example 16. *Factorise:* $8a^3 + 27b^3 + 64c^3 - 72abc$

Solution. $8a^3 + 27b^3 + 64c^3 - 72abc = (2a)^3 + (3b^3) + (4c)^3 - 3 \times 2a \times 3b \times 4c$

$= (2a + 3b + 4c)\{(2a)^2 + (3b)^2 + (4c)^2 - 2a \times 3b - 3b \times 4c - 4c \times 2a\}$

$= (2a + 3b + 4c)(4a^2 + 9b^2 + 16c^2 - 6ab - 12bc - 8ca)$ **Ans.**

Example 17. *Prove that:*

$$(a + b)^3 + (b + c)^3 + (c + a)^3 - 3(a + b)(b + c)(c + a) = 2(a^3 + b^3 + c^3 - 3abc)$$

Solution. We know that $x^3 + y^3 + z^3 - 3xyz = (x + y + z)(x^2 + y^2 + z^2 - xy - yz - zx)$

$\therefore (a + b)^3 + (b + c)^3 + (c - a)^3 - 3(a + b)(b + c)(c + a)$

$= [(a + b) + (b + c) + (c + a)][(a + b)^2 + (b + c)^2 + (c + a)^2 - (a + b)(b + c)$

$- (b + c)(c + a) - (c + a)(a + b)]$

$= (2a + 2b + 2c)[(a^2 + 2ab + b^2) + (b^2 + 2bc + c^2) + (c^2 + 2ca + a^2) - (ab + ac + b^2 + bc)$

$- (bc + ba + c^2 + ca) - (ca + cb + a^2 + ab)]$

$= 2(a + b + c)(2a^2 + 2b^2 + 2ab + 2bc + 2ca - ab - ac - b^2 - bc - bc - ba - c^2 - ca - ca - a^2 - ab)$

$= 2(a + b + c)(a^2 + b^2 + c^2 - ab - bc - ca)$

$= 2(a^3 + b^3 + c^3 - 3abc)$ **Proved.**

EXERCISE 1.5

Factorise the following:

1. $8a^3 + 27b^3 + c^3 - 18abc$ **Ans.** $(2a + 3b + c)(4a^2 + 9b^2 + c^2 - 6ab - 3bc - 2ac)$
2. $a^3 + 64b^3 - c^3 + 12abc$ **Ans.** $(a + 4b - c)(a^2 + 16b^2 + c^2 - 4ab + 4bc + ac)$
3. $1 - a^3 - b^3 - 3ab$ **Ans.** $(1 - a - b)(1 + a^2 + b^2 + a + b - ab)$
4. $p^3 - 125q^3 + 1 + 15pq$ **Ans.** $(p - 5q + 1)(p^2 + 25q^2 + 1 + 5pq - p + 5q)$
5. $(l + m)^3 + (2m + 3n)^3 + (3n + 4l)^3 - 3(l + m)(2m + 3n)(3n + 4l)$
   Ans. $(5l + 3m + 6n)(13l^2 + 3m^2 + 9n^2 - 12lm + 6ln)$
6. $(x + y)^3 + (y - z)^3 + (z - x)^3 - 3(x + y)(y - z)(z - x)$
   Ans. $2y(3x^2 + y^2 + 3z^2 + 3xy - 3yz - 3zx)$
7. $(p + q)^3 - (q + r)^3 + (r + p)^3 + 3(p + q)(q + r)(r + p)$
   Ans. $2p(p^2 + 3q^2 + 3r^2 + 3pr + 3qr + 3pq)$

(*iv*) Factorisation of $a^3 + b^3 + c^3$ when $a + b + c = 0$

Formula: $a^3 + b^3 + c^3 = 3abc$, when $a + b + c = 0$

Proof. We know that $a^3 + b^3 + c^3 - 3abc = (a + b + c)(a^2 + b^2 + c^2 - ab - bc - ca)$

If $a + b + c = 0$, then

$a^3 + b^3 + c^3 - 3abc = 0 \times (a^2 + b^2 + c^2 - ab - bc - ca)$

or $a^3 + b^3 + c^3 - 3abc = 0$

$a^3 + b^3 + c^3 = 3abc$ **Proved.**

We will use this formula for factorisation.

Example 18. *Factorise:* $(p - q)^3 + (q - r)^3 + (r - p)^3$

Solution. $(p - q)^3 + (q - r)^3 + (r - p)^3$ ...(1)

Put $p - q = a, q - r = b, r - p = c$ in (1), we get

$(p - q)^3 + (q - r)^3 + (r - p)^3 = a^3 + b^3 + c^3$

Now $a + b + c = p - q + q - r + r - p = 0$

$\therefore$ $a^3 + b^3 + c^3 = 3abc$ ... (2)

Putting the value of a, b, c in (2), we get

$(p-q)^3+(q-r)^3+(r-p)^3=3(p-q)(q-r)(r-p)$ **Ans.**

Example 19. *Factorise:* $(x^2-y^2)^3+(y^2-z^2)^3+(z^2-x^2)^3$

Solution. $(x^2-y^2)^3+(y^2-z^2)^3+(z^2-x^2)^3$ ... (1)

Put $x^2-y^2=a,\ y^2-z^2=b,\ z^2-x^2=c$ in (1), we get

$(x^2-y^2)^3+(y^2-z^2)^3+(z^2-x^2)^3=a^3+b^3+c^3$

Now $a+b+c=x^2-y^2+y^2-z^2+z^2-x^2=0$

$\therefore$ $a^3+b^3+c^3=3abc$ ... (2)

Putting the values of a, b, c in (2), we get

$(x^2-y^2)^3+(y^2-z^2)^3+(z^2-x^2)^3=3(x^2-y^2)(y^2-z^2)(z^2-x^2)$ **Ans.**

EXERCISE 1.6

Factorise the following:

1. $(a-2b)^3+(2b-3c)^3+(3c-a)^3$ **Ans.** $3(a-2b)(2b-3c)(3c-a)$
2. $(2x-4y)^3+(4y-3z)^3+(3z-2x)^3$ **Ans.** $3(2x-4y)(4y-3z)(3z-2x)$
3. $(3p-q)^3+(q+2r)^3-(2r+3p)^3$ **Ans.** $-3(3p-q)(q+2r)(3p+2r)$
4. $(l+3m)^3-(3m-4n)^3-(4n+l)^3$ **Ans.** $3(l+3m)(3m-4n)(4n+l)$

1.8 LOGARITHM

There is a great relation between logarithms and exponents.

If $a^x=n$ ...(1)

(where a is a positive real number, $a \neq 1$ and n is also real number)

Then, we rewrite the above statement (1) as $x=\log_a n$ ... (2)

In words, we say 'x is the logarithm of n to the base a' or x equals log n to the base a.

Relation (1) and (2) are equivalent relations.

Relation (1) is in the index form. Relation (2) is in the log form.

Note. (1) log is the short form of logarithm.

(2) Negative numbers and zero have no logarithms, log (–2) and log 0 are meaningless.

Illustration. Consider the following table

$2^4=16$	log of 16 to the base 2 = 4	$\log_2 16=4$
$3^3=27$	log of 27 to the base 3 = 3	$\log_3 27=3$
$4^{-3}=-\frac{1}{64}$	log of $y^2\frac{1}{64}$ to the base 4 = –3	$\log_4\frac{1}{64}=-3$
$10^{-1}=\frac{1}{10}=0.1$	log of 0.1 to the base 10 = –1	$\log_{10}0.1=-1$
$a^0=1$	log of 1 to the base a = 0	$\log_a 1=0$
$a^1=a$	log of a to the base a = 1	$\log_a a=1$

We find two important result in the above illustrations:

(*i*) The logarithm of 1 to any base in '0', *i.e.* $\log_a 1=0$

(*ii*) The logarithm of any quantity to the same base in, '1' *i.e.* $\log_a a=1$

Example 20. *Write the following in the form of logarithms:*

(*i*) $2^7 = 128$ (*ii*) $10^2 = 100$ (*iii*) $3^5 = 243$

Solution. (*i*) The logarithmic form of 2^7 ($2^7 = 128$) is $\log_2 128 = 7$

(*ii*) The logarithmic form of 10^2 ($10^2 = 100$) is $\log_{10} 100 = 2$

(*iii*) The logarithmic form of 3^5 ($3^5 = 243$) is $\log_3 243 = 5$

Example 21. *Write the following in the form of logarithms.*

(*i*) $7^4 = 2401$ (*ii*) $10^{-3} = 0.001$ (*iii*) $11^2 = 121$

Solution. (*i*) The logarithmic form of ($7^4 = 2401$) is

$$\log_7 2401 = 4$$

(*ii*) The logarithmic form of ($10^{-3} = 0.001$) is

$$\log_{10} 0.001 = -3$$

(*iii*) The logarithmic form of ($11^2 = 121$) is **Ans.**

$$\log_{11} 121 = 2.$$

Example 22. *Express each of the following in exponential form*

(i) $\log_5 125 = 3$ (*ii*) $\log_3 81 = 4$ (*iii*) $\log_{10} 10000 = 4$

(*iv*) $\log_2 256 = 8$ (*v*) $\log_6 36 = 2$ (*vi*) $\log_{10} 0.1 = -1$

Solution. We know that $\log_a b = x$ can be written in exponential from as $a^x = b$. Then

(*i*) $\log_5 125 = 3 \Rightarrow 5^3 = 125$

(*ii*) $\log_3 81 = 4 \Rightarrow 3^4 = 81$

(*iii*) $\log_{10} 10000 = 4 \Rightarrow 10^4 = 10000$

(*iv*) $\log_2 256 = 8 \Rightarrow 2^8 = 256$

(*v*) $\log_6 36 = 2 \Rightarrow 6^2 = 36$

(*vi*) $\log_{10} 0.1 = -1 \Rightarrow 10^{-1} = 0.1$ **Ans.**

Example 23. *Find the value of each of the following:*

(*i*) $\log_3 27$ (*ii*) $\log_7 \sqrt[3]{7}$

Solution. (*i*) We know from the definition that

$$\log_a b = x \Leftrightarrow a^x = b,\ a > 0,\ a \neq 1$$

Now, let $x = \log_3 27 \Leftrightarrow 3^x = 27 \Rightarrow 3^x = (3)^3 \Rightarrow x = 3$

Therefore, $x = 3$. Hence, $\log_3 27 = 3$ **Ans.**

(*ii*) We know from the definition that

$$\log_a b = x \Leftrightarrow a^x = b,\ a > 0,\ a \neq 1$$

Now, let $\log_7 \sqrt[3]{7} = x \Leftrightarrow 7^x = \sqrt[3]{7} \Rightarrow 7^x = (7)^{\frac{1}{3}} \Rightarrow x = \dfrac{1}{3}$

Therefore, $\log_7 \sqrt[3]{7} = \dfrac{1}{3}$. **Ans.**

Example 24. *If* $\log_{81} x = \dfrac{3}{2}$, *then find the value of x.*

Solution. We have, $\log_{81} x = \dfrac{3}{2} \Rightarrow (81)^{\frac{3}{2}} = x \Rightarrow (9^2)^{\frac{3}{2}} = x \Rightarrow 9^3 = x$

$\Rightarrow$ $x = 9 \times 9 \times 9 = 729$ **Ans.**

EXERCISE 1.7

Write the following in the form of logarithms:

1. $2^6 = 64$ **Ans.** $\log_2 64 = 6$ **2.** $4^5 = 1024$ **Ans.** $\log_4 1024 = 5$

3. $3^4 = 81$ **Ans.** $\log_3 81 = 4$ **4.** $4^3 = 64$ **Ans.** $\log_4 64 = 3$

5. $7^2 = 49$ **Ans.** $\log_7 49 = 2$ **6.** $8^3 = 512$ **Ans.** $\log_8 512 = 3$

7. $9^{\frac{5}{2}} = 243$ **Ans.** $\log_9 243 = \frac{5}{2}$ **8.** $10^0 = 1$ **Ans.** $\log_{10} 1 = 0$

Express each of the following in exponential form:

9. $\log_5 1 = 0$ **Ans.** $5^0 = 1$ **10.** $\log_{10} 1000 = 3$ **Ans.** $10^3 = 1000$

11. $\log_4 64 = 3$ **Ans.** $4^3 = 64$ **12.** $\log_7 343 = 3$ **Ans.** $7^3 = 343$

13. $\log_{10} 0.001 = -3$ **Ans.** $10^{-3} = 0.001$ **14.** $\log_3 \frac{1}{9} = -2$ **Ans.** $3^{-2} = \frac{1}{9}$

15. $\log_8 4 = \frac{2}{3}$ **Ans.** $8^{\frac{2}{3}} = 4$ **16.** $\log_9 6561 = 4$ **Ans.** $9^4 = 6561$

Find the value of each of the following by the definition of logarithm:

17. $\log_2 16$ **Ans.** 4 **18.** $\log_2 \sqrt{32}$ **Ans.** $\frac{5}{2}$

19. $\log_{10} 10^5$ **Ans.** 5 **20.** $\log_n 1$ **Ans.** 0

1.9 LAWS OF LOGARITHM

In this section, we shall learn the following laws of logarithm. These laws hold for any base $a(a > 0$ and $a \neq 1)$.

(i) First Law (Product Law) $\log_a(mn) = \log_a m + \log_a n$.

The logarithm of the product of two numbers is equal to the sum of their logarithms with reference to the same base.

Proof: Let $\log_a m = x$ and $\log_a n = y$. Then,

$$\log_a m = x \Rightarrow a^x = m \quad \text{... (1)}$$

and $$\log_a n = y \Rightarrow a^y = n \quad \text{... (2) [By definition]}$$

On multiplication of (1) and (2), we get

$$\therefore \quad m \cdot n = a^x \cdot a^y \Rightarrow m \cdot n = a^{x+y} \quad \text{[By laws of indices]}$$

$$\Rightarrow \log_a mn = x + y \quad \text{... (3) [By definition of log]}$$

On putting values of x and y in (3), we get

$$\log_a mn = \log_a m + \log_a n$$

or $$\log 2 \times 3 = \log 2 + \log 3$$

(ii) Second Low (Quotient Law) $\log_a \left(\frac{m}{n}\right) = \log_a m - \log_a n$

The logarithm of quotient of two numbers is equal to the difference of logarithm of the numerator and the logarithm of the denominator.

Proof. Let $\log_a m = x$ and $\log_a n = y$. Then,

$$\log_a m = x \Rightarrow a^x = m \quad \text{... (1)}$$

and $$\log_a n = x \Rightarrow a^y = n \quad \text{... (2) [By definition of log]}$$

On dividing (1) by (2), we get

$$\therefore \quad \frac{a^x}{a^y} = \frac{m}{n} \Rightarrow \frac{m}{n} = a^{x-y} \quad \text{[By laws of indices]}$$

$$\log_e\left(\frac{m}{n}\right) = x - y \quad \text{... (3)}$$

On putting value of $x = \log_a m$ and $y = \log_a n$ in (3), we get

$$\log a\left(\frac{m}{n}\right) = \log_a m - \log_a n \quad \textbf{Proved.}$$

$$\text{or} \quad \log\left(\frac{3}{4}\right) = \log 3 - \log 4$$

(iii) Third Law (Power Law) $\boxed{\log_a m^n = n\log_a m}$

The logarithm of a number raised to a power n is n times the logarithm of the number.

Proof. Let $\log_a m = x$. Then, $a^x = m$

Now, $a^x = m \Rightarrow (a^x)^n = m^n \Rightarrow a^{xn} = m^n$

$\Rightarrow m^n = a^{nx} \Rightarrow \log_a m^n = nx$ [By definition of log]

$\Rightarrow \log_a m^n = n \log_a m$ $(\because x = \log_a m)$

Hence, $\log_a m^n = n\log_a m$ **Proved.**

For example, $\log m^2 = 2\log m$

Example 25. *Find the value:* $\log_3 27\sqrt{729}$

Solution. We have,

$$\log_3 27\sqrt{729} = \log_3 27 + \log_3\sqrt{729} = \log_3(3)^3 + \log_3(729)^{\frac{1}{2}} \quad [\because \sqrt{a} = a^{\frac{1}{2}}]$$

$$= 3\log_3 3 + \frac{1}{2}\log_3(729) \quad [\because \log_a m^n = n\log_a m]$$

$$= 3(1) + \frac{1}{2}\log_3(3^6) = 3 + \frac{1}{2}\cdot 6\log_3 3$$

$$= 3 + \frac{1}{2}\cdot 6(1) = 3 + 3 = 6 \quad [\because \log_3 3 = 1]$$

Example 26. *Show that:* $3\log 4 - 2\log 6 + \log(18)^{\frac{3}{2}} = \log(96\sqrt{2})$

Solution.

$$\text{L.H.S.} = 3\log 4 - 2\log 6 + \log(18)^{\frac{3}{2}}$$

$$= \log 4^3 - \log 6^2 + \log(18)^{\frac{3}{2}} = \log 64 - \log 36 + \log(2\times 3^2)^{\frac{3}{2}}$$

$$= \log\frac{64\times(2\times 3^2)^{\frac{3}{2}}}{36} = \log\frac{64\times 2^{\frac{3}{2}}\times 3^{2\times\frac{3}{2}}}{36} = \log\frac{64\times 2\sqrt{2}\times 27}{36}$$

$$= \log(16\times 2\sqrt{2}\times 3) = \log(96\sqrt{2}) = \text{R.H.S.} \quad \textbf{Proved.}$$

Example 27. *Prove that:*

(*i*) $\log(1 + 2 + 3) = \log 1 + \log 2 + \log 3$ (*ii*) $\log\frac{a^2}{bc} + \log\frac{b^2}{ac} + \log\frac{c^2}{ab} = 0$

(*iii*) $\log_b a \times \log_c b \times \log_a c = 1$

Solution. (*i*) L.H.S. = log (1 + 2 + 3) = log 6 = log (1 × 2 × 3)

= log 1 + log 2 + log 3 = R.H.S.

(*ii*)
$$\text{L.H.S.} = \log\frac{a^2}{bc} + \log\frac{b^2}{ac} + \log\frac{c^2}{ab} = \log\left(\frac{a^2}{bc}\times\frac{b^2}{ac}\times\frac{c^2}{ab}\right)$$
$$= \log\left(\frac{a^2\times b^2\times c^2}{a^2\times b^2\times c^2}\right) = \log 1 = 0 = \text{R.H.S.} \qquad [\because \log 1 = 0]$$

(*iii*) L.H.S. = $\log_b a \times \log_c b \times \log_a c$
$$= \frac{\log_a a}{\log_a b}\times\frac{\log_a b}{\log_a c}\times\frac{\log_a c}{\log_a a} = 1 = \text{R.H.S.}$$ **Proved.**

(Expressing each logarithm to the same base, say *a*, by using base changing formula).

Example 28. *Show that:* $\log\frac{2}{3}+\log\frac{25}{49}+\log\frac{21}{2}+\log\frac{7}{5}=\log 5$ *by two methods.*

Solution. First method
$$\log\frac{2}{3}+\log\frac{25}{49}+\log\frac{21}{2}+\log\frac{7}{5}=\log\frac{2}{3}\times\frac{25}{49}\times\frac{21}{2}\times\frac{7}{5}$$
$$=\log\frac{2\times5\times5\times7\times3\times7}{3\times7\times7\times2\times5}=\log 5$$ **Proved.**

Second method
$$\log\frac{2}{3}+\log\frac{25}{49}+\log\frac{21}{2}+\log\frac{7}{5}$$
$= \log 2 - \log 3 + \log 25 - \log 49 + \log 21 - \log 2 + \log 7 - \log 5$

$= \log 2 - \log 3 + \log 5^2 - \log 7^2 + \log 3 \times 7 - \log 2 + \log 7 - \log 5$

$= \log 2 - \log 3 + 2\log 5 - 2\log 7 + \log 3 + \log 7 - \log 2 + \log 7 - \log 5$

$= \log 5$ **Proved.**

Example 29. *Solve* $\log(x + 4) - \log(3x - 2) = 0$

Solution. $\log(x + 4) - \log(3x - 2) = 0$
$$\log\left(\frac{x+4}{3x-2}\right) = 0$$
$$\log\left(\frac{x+4}{3x-2}\right) = \log 1$$
$$\frac{x+4}{3x-2} = 1 \quad\text{or}\quad x + 4 = 3x - 2$$
$2x = 6$ or $x = 3$ **Ans.**

Example 30. *If* $\frac{\log x}{\log 3} = \frac{\log 81}{\log 9}$, *find x*.

Solution.
$$\frac{\log x}{\log 3} = \frac{\log 3^4}{\log 3^2}$$
$$= \frac{4\log 3}{2\log 3} = 2$$
$\log x = 2\log 3 = \log 3^2 = \log 9$

$x = 9$ **Ans.**

Example 31. *If* $a = \log_x yz$, $b = \log_y zx$, $c = \log_z xy$, *show that*

$$abc = a + b + c + 2$$

Solution.

$$\begin{aligned} abc &= \log_x yz \cdot \log_y zx \cdot \log_z xy \\ &= (\log_x y + \log_x z)(\log_y z + \log_y x)(\log_z x + \log_z y) \\ &= (\log_x y \cdot \log_y z + \log_x y \cdot \log_y x + \log_x z \cdot \log_y z + \log_x z \log_y x)(\log_z x + \log_z y) \\ &= (\log_x z + 1 + \log_x z \log_y z + \log_y z)(\log_z x + \log_z y) \\ &= \log_x z \log_z x + \log_x z \log_z y + \log_z x + \log_z y + \log_x z \log_y z \log_z x \\ &\quad + \log_x z \log_y z \log_z y + \log_y z \log_z x + \log_y z \log_z y \\ &= 1 + \log_x y + \log_z x + \log_z y + \log_y z + \log_x z + \log_y x + 1 \\ &= 2 + (\log_x y + \log_x z) + (\log_y z + \log_y x) + (\log_z x + \log_z y) \\ &= 2 + (\log_x yz + \log_y zx + \log_z xy) \\ &= 2 + a + b + c \end{aligned}$$

Proved.

Example 32. $5^{2x}.7^{2x-1} = 175$

Solution.

$$\log 5^{2x}.7^{2x-1} = \log 175$$

$$\Rightarrow \quad \log 5^{2x} + \log 7^{2x-1} = \log (5^2 \times 7)$$

$$\Rightarrow \quad 2x \log 5 + (2x - 1)\log 7 = 2 \log 5 + \log 7$$

$$\Rightarrow \quad 2x \log 5 - 2 \log 5 + (2x - 1)\log 7 - \log 7 = 0$$

$$\Rightarrow \quad (2x - 2) \log 5 + (2x - 2) \log 7 = 0$$

$$\Rightarrow \quad (2x - 2) (\log 5 + \log 7) = 0$$

$$\Rightarrow \quad (2x - 2) \log 35 = 0$$

$$\therefore \quad 2x - 2 = 0 \quad \text{or} \quad x = 1.$$

Ans.

1.10 COMMON LOGARITHM

Logarithm of numbers to the base 10 are called *common logarithm*. In numerical calculations, the common logarithms are used. Whenever the base is not written, we shall assume the base to be 10.

$10^3 = 1000 \quad \Rightarrow \quad \log (1000) = 3 \quad$ or $\quad \log 10^3 = 3$

$10^2 = 100 \quad \Rightarrow \quad \log (100) = 2 \quad$ or $\quad \log 10^2 = 2$

$10^1 = 10 \quad \Rightarrow \quad \log (10) = 1 \quad$ or $\quad \log 10^1 = 1$

$10^0 = 1 \quad \Rightarrow \quad \log (1) = 0 \quad$ or $\quad \log 10^0 = 0$

$10^{-1} = 0.1 \quad \Rightarrow \quad \log (0.1) = -1 \quad$ or $\quad \log 10^{-1} = -1$

$10^{-2} = 0.01 \quad \Rightarrow \quad \log (0.01) = -2 \quad$ or $\quad \log 10^{-2} = -2$

$10^{-3} = 0.001 \quad \Rightarrow \quad \log (0.001) = -3 \quad$ or $\quad \log 10^{-3} = -3$

and so on.

We conclude that if n is an integer, then $\log_{10} 10^n = n$.

Logarithm to the base e are called natural or Naperian logarithms. Naperian logarithms are used in differential calculus, integral calculus, etc.

1.11 CHARACTERISTIC AND MANTISSA

We know

$10^3 = 1000 \quad \Rightarrow \quad \log (1000) = 3$

$10^2 = 100 \quad \Rightarrow \quad \log (100) = 2$

$10^1 = 10 \Rightarrow \log(10) = 1$

$10^0 = 1 \Rightarrow \log(1) = 0$

$10^{-1} = 0.1 \Rightarrow \log(0.1) = -1$

$10^{-2} = 0.01 \Rightarrow \log(0.01) = -2$

$10^{-3} = 0.001 \Rightarrow \log(0.001) = -3$

Logarithms of a number between 1000 and 100 lies between 3 and 2 (2 + a decimal).

Logarithm of a number between 100 and 10 lies between 2 and 1 (1 + a decimal), and so on.

Numbers	Logarithm
Between 1000 and 100	$2 < \log < 3$ or 2 + a decimal
Between 100 and 10	$1 < \log < 2$ or 1 + a decimal
Between 10 and 1	$0 < \log < 1$ or 0 + a decimal
Between 1 and 0.1	$-1 < \log < 0$ or –1 + a decimal
Between 0.1 and 0.01	$-2 < \log < -1$ or –2 + a decimal
Between 0.01 and 0.001	$-3 < \log < -2$ or –3 + a decimal

Thus logarithm of any number consists of two parts:

(*i*) Integral part (*ii*) Decimal part

Characteristic: The integral part of logarithm is called the *characteristic*. It may be negative or positive.

Mantissa: The decimal part of logarithm is known as the *mantissa*. It is always positive.

$$\log(300) = 2.3010$$

Characteristic (2) Mantissa (.3010)

Number	Characteristic
825	2
36	1
4	0
0.3	–1
0.05	–2
0.002	–3

Rule to Know the Characteristic

Characteristic of the numbers greater than 1.

Number of	Characteristic
3 digits	2
2 digits	1
1 digit	0

EXERCISE 1.8

Find the value of each of the following:

1. $\log_2 16\sqrt{8}$ **Ans.** $\dfrac{11}{2}$

2. $\log_{10} \sqrt[3]{100}$ **Ans.** $\dfrac{2}{3}$

3. $\log_5 \dfrac{\sqrt[4]{25}}{625}$ **Ans.** $-\dfrac{7}{2}$

Prove that the following:

4. $\log 12 = \log 3 + 2\log 2$

5. $\log 360 = 3\log 2 + 2\log 3 + \log 5$

6. $3\log 2 + \log 5 = \log 40$

7. $5\log 3 + \log 9 = \log 2187$

8. $\log\left(\dfrac{9}{14}\right) + \log\left(\dfrac{35}{24}\right) - \log\left(\dfrac{15}{16}\right) = 0$

9. $\log\left(\dfrac{8}{49}\right) + \log\left(\dfrac{28}{3}\right) + \log\left(\dfrac{21}{16}\right) = \log 2$

10. $2\log\left(\dfrac{16}{15}\right) - \log\left(\dfrac{24}{25}\right) + \log\left(\dfrac{27}{32}\right) = 0$

11. If $\log 2 = 0.3010$ and $\log 3 = 0.4771$, find $\log 24$. **Ans.** 1.3801

12. Simplify $\dfrac{\log_2 11}{\log_2 7} - \dfrac{\log_5 11}{\log_5 7}$ **Ans.** 0

13. $\dfrac{1}{\log_a(ab)} + \dfrac{1}{\log_b(ab)} = 1$

14. $\log\left(\dfrac{4}{5}\right) + \log\left(\dfrac{5}{7}\right) + \log\left(\dfrac{7}{4}\right) = 0$

15. $\log\left(\dfrac{a^2}{bc}\right) + \log\left(\dfrac{b^2}{ca}\right) + \log\left(\dfrac{c^2}{ab}\right) = 0$

16. $\log\left(\dfrac{8}{49}\right) + \log\left(\dfrac{28}{3}\right) + \log\left(\dfrac{21}{16}\right) = \log 2$

17. $2\log\left(\dfrac{16}{15}\right) - \log\left(\dfrac{24}{25}\right) + \log\left(\dfrac{27}{32}\right) = 0$

18. $\log\left(\dfrac{75}{16}\right) - 2\log\left(\dfrac{5}{6}\right) + \log\left(\dfrac{8}{27}\right) - \log 2$

19. $\log(3 + 3) \neq \log 3 + \log 3$

20. $\log(1 + 2 + 3) = \log 1 + \log 2 + \log 3$

21. $\log 2 + \log 4 > \log(2 + 4)$

22. $\log 6 - \log 3 < \log(6 - 3)$

23. If $x = \log_7 27$, $y = \log_5 7$, $z = \log_3 5$, show that $xyx = 3$.

24. If $a^2 + b^2 = 7ab$, Show that $\log\left(\dfrac{a+b}{3}\right) = \dfrac{1}{2}[\log a + \log b]$.

25. If $x^2 + y^2 = 8xy$, show that $2\log(x + y) = \log 5 + \log 2 + \log x + \log y$.

26. If $x = \log_a bc$, $y = \log_b ca$, $z = \log_c ab$, show that $\dfrac{1}{1+x} + \dfrac{1}{1+y} + \dfrac{1}{1+z} = 1$.

27. Prove that $x^{\log y - \log z} . y^{\log z - \log x} . z^{\log x - \log y} = 1$

28. Solve: $\log_3 2x + \log_3 7 = \log_3 5$ **Ans.** $x = \dfrac{5}{14}$

Arithmetic Progression and Their Applications

2.1 INTRODUCTION

Frequently in life, we come across many puzzles in which we have been given a succession of numbers of which one number is designated as the first and another as second, yet another as the third and so on. This pattern is known as sequence. For example: population of bacteria at different times form a sequence. The amount of money in a fixed deposit increases in a sequence. The radioactive materials decay in a sequence.

2.2 SEQUENCE

It is a collection of objects ordered in such a way that its first, second, third members, ... can be identified.

A sequence is a succession of numbers or terms formed according to some definite rule.

The nth term in a sequence is denoted by T_n.

For example If $T_n = 2n + 1$; by giving different values of n in T_n, we will get different terms of the sequence.

Thus, $T_1 = (2 \times 1 + 1) = 3, T_2 = (2 \times 2 + 1) = 5, T_3 = (2 \times 3 + 1) = 7.$

Terms

The different numbers which form the sequence are called terms of the sequence, denoted by the symbols $T_1, T_2, T_3, \ldots,$ etc. here the subscripts denote the position of the term.

For example Consider 1, 7, 13, 19, 25, ..., as a sequence,

here, $T_1 = 1, T_2 = 7, T_3 = 13, T_4 = 19, T_5 = 25, \ldots$

2.3 TYPES OF SEQUENCES

1. Finite Sequence

A sequence containing finite number of terms is called a finite sequence.

For example: (*i*) 1, 7, 13, 19, 25 is a finite sequence as it contains only 5 terms.

(*ii*) 2, 12, 22, 32, 43, 52 is a finite sequence as it contains only six terms.

2. Infinite Sequence

A sequence, having unlimited number of terms is known as an infinite sequence.

For example:

(*i*) 1, 7, 13, 19, 25, ..., is an infinite sequence as it contains infinite number of terms.

(*ii*) 6, 10, 14, 18, ..., is an infinite sequence, as it contains infinite number of terms.

2.4 GENERAL TERMS

T_1 represents the first term and T_n represents the nth term in the sequence $T_1, T_2, T_3, ..., T_n$. The nth term of a sequence is denoted by a_n or U_n or T_n is called a general term.

For example third term of the sequence 1, 7, 13, 19, 25, ..., is given by $T_3 = 13$.

Notes: 1. The nth term of the sequence 2, 4, 6, ..., is given by $T_n = 2n$, where n is a natural number.

2. In the sequence of odd numbers 1, 3, 5, ..., the nth term is given by the formula $T_n = 2n - 1$, where n is a natural number.

3. A sequence can be regarded as a function, whose domain is a set of natural numbers.

4. A series is generally represented in a compact form called sigma with notation Σ.

$$\sum_{n=1}^{\infty} T_n = T_1 + T_2 + T_3 + ... + T_n$$

Example 1. *Write down the first five terms of each of the following sequences, whose nth terms are*

(*i*) $T_n = 2n + 5$ (*ii*) $T_n = (-1)^{n-1} (5)^{n+1}$

(*iii*) $T_n = \dfrac{n(n^2 + 5)}{4}$

Solution. (*i*) Here, $T_n = 2n + 5$ ...(1)

Substituting n = 1, 2, 3, 4, 5 in (1), we get

$$T_1 = 2 \times 1 + 5 = 2 + 5 = 7$$
$$T_2 = 2 \times 2 + 5 = 4 + 5 = 9$$
$$T_3 = 2 \times 3 + 5 = 6 + 5 = 11$$
$$T_4 = 2 \times 4 + 5 = 8 + 5 = 13$$
$$T_5 = 2 \times 5 + 5 = 10 + 5 = 15$$

Thus, the required terms are 7, 9, 11, 13, 15 **Ans.**

(*ii*) Here, $T_n = (-1)^{n-1} (5)^{n+1}$

Substituting n = 1, 2, 3, 4, 5 in (1), we get ...(1)

$$T_1 = (-1)^{1-1} (5)^{1+1} = 25$$
$$T_2 = (-1)^{2-1} (5)^{2+1} = -125$$
$$T_3 = (-1)^{3-1} (5)^{3+1} = 625$$
$$T_4 = (-1)^{4-1} (5)^{4+1} = -3125$$
$$T_5 = (-1)^{5-1} (5)^{5+1} = 15625$$

Thus, the required terms are 25, – 125, 625, – 3125, 15625 **Ans.**

(*iii*) Here, $T_n = \dfrac{n(n^2+5)}{4}$ ...(1)

Substituting n = 1, 2, 3, 4, 5 in (1), we get

$$T_1 = \frac{1\,(1^2 + 5)}{4} = \frac{6}{4} = \frac{3}{2}$$

$$T_2 = \frac{2\,(2^2+5)}{4} = \frac{18}{4} = \frac{9}{2}$$

$$T_3 = \frac{3\,(3^2+5)}{4} = \frac{42}{4} = \frac{21}{2}$$

$$T_4 = \frac{4\,(4^2+5)}{4} = \frac{84}{4} = 21$$

$$T_5 = \frac{5\,(5^2+5)}{4} = \frac{150}{4} = \frac{75}{2}$$

Thus, the required terms are $\frac{3}{2}, \frac{9}{2}, \frac{21}{2}, 21, \frac{75}{2}$ **Ans.**

Example 2. *Find the indicated terms in each of the following sequences, whose nth terms are*

(*i*) $T_n = (-1)^{n-1}.n^3$; T_9 (*ii*) $T_n = (n-1)(2-n)(3+n); T_1, T_2, T_3$

(*iii*) $T_n = \begin{cases} \dfrac{n}{\dfrac{n}{96}-1}, & \textit{if n is not the square of a natural number} \\ \dfrac{13}{2}, & \textit{If n is the square of a natural number} \end{cases}$

Find the 440th and 441st terms of the above sequence.

Solution. (*i*) *Here,* $T_n = (-1)^{n-1}.n^3$ ... (1)

Substituting $n = 9$ in (1), we get

$$T_9 = (-1)^{9-1}(9)^3 = (-1)^8\, 729 = 729$$ **Ans.**

(*ii*) Here, $T_n = (n-1)(2-n)(3+n)$ ... (2)

Substituting $n = 1, 2, 3$ in (2), we get

$$T_1 = (1-1)(2-1)(3+1) = 0$$

$$T_2 = (2-1)(2-2)(3+2) = 0$$

$$T_3 = (3-1)(2-3)(3+3) = 2 \times (-1) \times 6 = -12$$ **Ans.**

(*iii*) Since, $n = 440$ is not a perfect square, therefore

$$T_n = \frac{n}{\frac{n}{96}-1} \Rightarrow T_{440} = \frac{440}{\frac{440}{96}-1} = \frac{440}{\frac{344}{96}} = \frac{440 \times 96}{344} = \frac{5280}{43}$$

Again, since $n = 441 = (21)^2$, *i.e.* 441 is a perfect square of 21.

Therefore, $T_{441} = \frac{13}{2}$ **Ans.**

Example 3. *Find the 8th term in the sequence, whose first three terms are 3, 3, 6 and each term after the second is the sum of two terms preceding it.*

Solution. Let T_n be the nth term of the sequence.

Here, $T_1 = 3,\quad T_2 = 3,\quad T_3 = 6$

$$T_n = T_{n-1} + T_{n-2},\ n > 2$$ [given]

$T_4 = T_3 + T_2 = 6 + 3 = 9; \qquad T_5 = T_4 + T_3 = 9 + 6 = 15$

$T_6 = T_5 + T_4 = 15 + 9 = 24; \qquad T_7 = T_6 + T_5 = 24 + 15 = 39$

$T_8 = T_7 + T_6 = 39 + 24 = 63$ **Ans.**

Example 4. *Write the next five terms of the following sequences given by :*

(*i*) $a_1 = a_2 = 2, a_n = a_{n-1} - 1, \quad (n > 2)$

(*ii*) $a_1 = 1 = a_2, a_n = a_{n-1} + a_{n-2}, (n > 2)$. *Find* $\frac{a_{n+1}}{a_n}$ for $n = 1, 2, 3, 4, 5$.

Solution. (*i*) We have, $a_1 = a_2 = 2; \quad a_n = a_{n-1} - 1$

$a_3 = a_{3-1} - 1 = a_2 - 1 = 2 - 1 = 1$

$a_4 = a_{4-1} - 1 = a_3 - 1 = 1 - 1 = 0$

$a_5 = a_{5-1} - 1 = a_4 - 1 = 0 - 1 = -1$

$a_6 = a_{6-1} - 1 = a_5 - 1 = -1 - 1 = -2$

$a_7 = a_{7-1} - 1 = a_6 - 1 = -2 - 1 = -3$ **Ans.**

(*ii*) We have, $a_1 = 1 = a_2, a_n = a_{n-1} + a_{n-2}$

For $n = 1, \frac{a_{n+1}}{a_n} = \frac{a_2}{a_1} = \frac{1}{1} = 1, \qquad a_2 = a_1 + a_0 = 1 + 0 = 1$

For $n = 2, \frac{a_{n+1}}{a_n} = \frac{a_3}{a_2} = \frac{2}{1} = 2, \qquad a_3 = a_2 + a_1 = 1 + 1 = 2$

For $n = 3, \frac{a_{n+1}}{a_n} = \frac{a_4}{a_3} = \frac{3}{2}, \qquad a_4 = a_3 + a_2 = 2 + 1 = 3$

For $n = 4, \frac{a_{n+1}}{a_n} = \frac{a_5}{a_4} = \frac{5}{3}, \qquad a_5 = a_4 + a_3 = 3 + 2 = 5$

For $n = 5, \frac{a_{n+1}}{a_n} = \frac{a_6}{a_5} = \frac{8}{5}, \qquad a_6 = a_5 + a_4 = 5 + 3 = 8$ **Ans.**

EXERCISE 2.1

1. Write the first five terms of the sequence, using the given rule. In each case, the initial value of the index is 1.

   (*i*) $T_n = 2n + 3$ **Ans.** 5, 7, 9, 11, 13

   (*ii*) $T_n = \frac{4n+1}{n+7}$ **Ans.** $\frac{5}{8}, 1, \frac{13}{10}, \frac{17}{11}, \frac{21}{12}$

   (*iii*) $T_n = 2n^2 - n + 1$ **Ans.** 2, 7, 16, 29, 46

   (*iv*) $T_n = \frac{n^2}{3^n}$ **Ans.** $\frac{1}{3}, \frac{4}{9}, \frac{9}{27}, \frac{16}{81}, \frac{25}{243}$

(*v*) $T_n = \dfrac{(-1)^{n-1}}{n^3}$ **Ans.** $1, -\dfrac{1}{8}, \dfrac{1}{27}, -\dfrac{1}{64}, \dfrac{1}{125}$

(*vi*) $T_n = \left(\dfrac{1}{2}\right)^{2n+1} + \left(-\dfrac{1}{2}\right)^{2n}$ **Ans.** $\dfrac{3}{2^3}, \dfrac{3}{2^5}, \dfrac{3}{2^7}, \dfrac{3}{2^9}, \dfrac{3}{2^{11}}$

(*vii*) $T_n = \dfrac{(n+1)^2}{n}$ **Ans.** $4, \dfrac{9}{2}, \dfrac{16}{3}, \dfrac{25}{4}, \dfrac{36}{5}$

(*viii*) $T_n = \dfrac{n(n+1)(2n+1)}{6}$ **Ans.** 1, 5, 14, 30, 55

(*ix*) $T_n = \dfrac{3n^2}{\left(n-\dfrac{1}{2}\right)(n+1)}$ **Ans.** $3, \dfrac{8}{3}, \dfrac{27}{10}, \dfrac{96}{35}, \dfrac{75}{27}$

2. Write the first four terms of the sequence, whose nth term is given as:

(*i*) $(-1)^n \sin\dfrac{n\pi}{2}$ **Ans.** – 1, 0, 1, 0 (*ii*) $(-1)^{n-1} \cos\dfrac{n\pi}{4}$ **Ans.** $\dfrac{\sqrt{2}}{2}, 0, \dfrac{\sqrt{2}}{2}, 1$

3. Find an expression for the nth term of an infinite sequence, whose first four terms are:

(*i*) $1, \dfrac{1}{4}, \dfrac{1}{9}, \dfrac{1}{16}$ **Ans.** $\dfrac{1}{n^2}$ (*ii*) $-\dfrac{1}{2}, \dfrac{1}{6}, -\dfrac{1}{12}, \dfrac{1}{20}$ **Ans.** $\dfrac{(-1)^n}{n(n+1)}$

4. In each case, the first term of a sequence and a recursion formula for succeeding terms are given. If $n \in \{1, 2, 3, 4, 5, 6\}$, write the sequence in each case:

(*i*) $T_1 = 3, \quad T_{n+1} = (T_n - 2)^2$ **Ans.** 3, 1, 1, 1, 1, 1

(*ii*) $T_1 = 2, \quad T_2 = 3 \quad T_n = \dfrac{1}{3}(T_{n-1} + T_{n-2}).$ **Ans.** $2, 3, \dfrac{5}{3}, \dfrac{14}{9}, \dfrac{29}{27}, \dfrac{71}{81}$

5. Find the 18th and 25th terms of the sequence defined by

$$T_n = \begin{cases} n(n+2), & \text{if } n \text{ is an even natural number} \\ \dfrac{4n}{n^2+1} & \text{if } n \text{ is an odd natural number} \end{cases}$$

Ans. $T_{18} = 360, \ T_{25} = \dfrac{50}{313}$

6. Consider the sequence defined by $T_1 = an^2 + bn + c$. If $T_2 = 3, T_4 = 13$ and $T_7 = 113$, show that $3T_n = 17n^2 - 87n + 11$.

7. Find the 19th and 20th terms of the sequence defined by

$$T_n = \begin{cases} n^2, & \text{when } n \text{ is even} \\ n^2+1, & \text{when } n \text{ is odd} \end{cases}$$

Ans. 362, 400

8. Find the first five terms of the sequence $\{T_n\}$. If $T_1 = -5, T_n = \dfrac{T_{n-1}}{n+1}, n \geq 2.$

Ans. $-\dfrac{5}{3}, -\dfrac{5}{12}, -\dfrac{1}{12}, -\dfrac{1}{72}, -\dfrac{1}{504}$

9. Find the first twelve terms of a sequence of primes.

Ans. 2, 3, 5, 7, 11, 13, 17, 19, 23, 29, 31 and 37

10. Find 'a' and 'b', such that 12, $a + b$, $2a$, b are in A.P. **Ans.** $a = 4, b = 6$.

2.5 SERIES

By adding or subtracting the terms of a sequence, we get a series.

Sequence	*Series*
(*i*) 2, 5, 8, 11, 14	2 + 5 + 8 + 11 + 14 is a finite series
(*ii*) 3, 1, – 1, – 3, – 5	3 + 1 – 1 – 3 – 5 is a finite series
(*iii*) 1, 5, 9, 13, 17, ...	1 + 5 + 9 + 13 + 17 + ... is an infinite series

Progression

Sequences following certain patterns are called progressions. In a progression, each term except the first progresses, *e.g.* (*i*) 1, 3, 5, 7, 9, 0, ... (*ii*) 6, 3, 0, –3, –6, ..., etc.

Here, we will discuss arithmetic progression.

2.6 ARITHMETIC PROGRESSION (A.P.)

Consider the sequences which are given below:

(*i*) 1, 3, 5, 7, 9, ... (*ii*) 2, 5, 8, 11, 14, ...

(*iii*) 25, 30, 35, 40, 45, ... (*iv*) $(a - 4b), (a - b), (a + 2b), (a + 5b), (a + 8b)$

Now, we observe that

in example (*i*) $a_1 = 1, a_2 = a_1 + 2 = 1 + 2 = 3, a_3 = a_2 + 2, a_4 = a_3 + 2$, and so on.

(*ii*) $a_1 = 2, a_2 = a_1 + 3, a_3 = a_2 + 3, a_4 = a_3 + 3, a_5 = a_4 + 3$ and so on.

(*iii*) $a_1 = 25, a_2 = a_1 + 5, a_3 = a_2 + 5, a_4 = a_3 + 5, a_5 = a_4 + 5$ and so on.

(*iv*) $a_1 = a - 4b, a_2 = a_1 + 3b, a_3 = a_2 + 3b, a_4 = a_3 + 3b, a_5 = a_4 + 3b$ and so on.

Notes:

In example (*i*) the successive term is obtained by adding 2 to the preceding term, *i.e.* $a_2 = a_1 + 2$.

(*ii*) the successive term is obtained by adding 3 to the preceding term, *i.e.* $a_2 = a_1 + 3$.

(*iii*) the successive term is obtained by adding 5 to the preceding term, *i.e.* $a_2 = a_1 + 5$.

(*iv*) the successive term is obtained by adding $3b$ to the preceding term, *i.e.* $a_2 = a_1 + 3b$.

Here, in all sequences, we observe that successive terms are obtained by adding a fixed number to the preceding terms.

Such sequences are called arithmetic sequences or arithmetic progressions (A.P.)

Definition

A sequence is said to be an arithmetic progression (A.P.), if the difference between every two consecutive terms is the same. A sequence, $T_1, T_2, T_3, ..., T_n$ is called an Arithmetic Progression, if

$$T_1 = a, T_2 = a + d, T_3 = T_2 + d, T_4 = T_3 + d, ..., a_n = a_{n-1} + d, T_{n+1} = T_n + d$$

a is called the first term and the fixed number d is called the common difference of the A.P. nth term is denoted by T_n.

*n*th Term of an A.P. or General Term of an A.P.

Consider the following A.P.

$$a, a + d, a + 2d, a + 3d, \ldots$$

$$T_2 = a + d = a + (2 - 1)d$$

$$T_3 = a + 2d = a + (3 - 1)d$$

$$T_4 = a + 3d = a + (4 - 1)d$$

....

....

$$T_n = a + (n - 1)d$$

$$\boxed{n\text{th term} = T_n = a + (n - 1)d}$$

Choice of Terms in an A.P.

Number of terms	*Terms*	*Common difference*
3	$a - d, a, a + d$	d
4	$a - 3d, a - d, a + d, a + 3d$	$2d$
5	$a - 2d, a - d, a, a + d, a + 2d$	d
6	$a - 5d, a - 3d, a - d, a + d, a + 3d, a + 5d$	$2d$

Example 5. *The nth term of an A.P. is 4n + 1. Write down the first 4 terms and the 18th term of an A.P.*

Solution. Here, $T_n = 4n + 1$ ... (1)

Putting $n = 1, 2, 3, 4$, in (1), we get

$T_1 = 4 \times 1 + 1 = 4 + 1 = 5$, $T_2 = 4 \times 2 + 1 = 8 + 1 = 9$

$T_3 = 4 \times 3 + 1 = 12 + 1 = 13$, $T_4 = 4 \times 4 + 1 = 16 + 1 = 17$

Putting $n = 18$ in (1), we get

$$T_{18} = 4 \times 18 + 1 = 72 + 1 = 73$$

Hence, the first four terms of the A.P. are 5, 9, 13, 17 and the 18th term is 73. **Ans.**

Example 6. *Find the common difference and write the next four terms of the following A.P.*

(*i*) $-1, \frac{1}{4}, \frac{3}{2}, \ldots$ (*ii*) $-8, -2, 4, \ldots$

Solution. We have $-1, \frac{1}{4}, \frac{3}{2}, \ldots$ ... (1)

Here, $T_2 - T_1 = \frac{1}{4} - (-1) = \frac{1}{4} + 1 = \frac{5}{4}$

$$T_3 - T_2 = \frac{3}{2} - \frac{1}{4} = \frac{5}{4}$$

Therefore, (1) is an A.P. with $a = -1, d = \frac{5}{4}$

Now, we know that $T_n = a + (n - 1)d$

$$T_4 = -1 + (4 - 1)\frac{5}{4} = -1 + \frac{15}{4} = \frac{11}{4}$$

$$T_5 = T_4 + d = \frac{11}{4} + \frac{5}{4} = 4, \qquad T_6 = T_5 + d = 4 + \frac{5}{4} = \frac{21}{4}$$

$$T_7 = T_6 + d = \frac{21}{4} + \frac{5}{4} = \frac{13}{2}$$

(*ii*) The given A.P. is – 8, – 2, 4, ... (1)

Here, $T_2 - T_1 = -2 - (-8) = 6$

$T_3 - T_2 = 4 - (-2) = 6$

Therefore (1) is an A.P. with $a = -8, d = 6$

Also, we know that

$$T_n = a + (n-1)d$$

$\therefore$ $T_4 = -8 + (4-1)\,6 = -8 + 18 = 10$

$T_5 = T_4 + d = 10 + 6 = 16$

$T_6 = T_5 + d = 16 + 6 = 22$

$T_7 = T_6 + d = 22 + 6 = 28$ **Ans.**

Example 7. *Which term of the progression* $19, 18\frac{1}{5}, 17\frac{2}{5}, \ldots$ *is the first negative term.*

Solution. The given expression is $19, 18\frac{1}{5}, 17\frac{2}{5}, \ldots$ ... (1)

Here $T_2 - T_1 = \frac{91}{5} - 19 = \frac{91-95}{5} = -\frac{4}{5}$

$T_3 - T_2 = \frac{87}{5} - \frac{91}{5} = -\frac{4}{5}$

Therefore, (1) is an A.P. with $a = 19$, $d = -\frac{4}{5}$

Let the nth term of the given A.P. be the first negative term. Then, nth term < 0

$\Rightarrow$ $T_n < 0$ $\Rightarrow$ $\left[19 + (n-1)\left(-\frac{4}{5}\right)\right] < 0$

$\Rightarrow$ $(99 - 4n) < 0$ $\Rightarrow$ $4n > 99$

$\Rightarrow$ $n > 24\frac{3}{4}$

$\therefore$ $n = 25$, *i.e.* 25th term is the first negative term in the given A.P. **Ans.**

Example 8. *Which term of the sequence 72, 70, 68, 66, ... is 40?* **[S.B.T.E. Jan. 2014]**

Solution. Here, we have

Sequence 72, 70, 68, 66, ... is 40.

$a = 72, d = 70 - 72 = -2, T_n = 40$

$T_n = a + (n-1)d$

$40 = 72 + (n-1)(-2)$

$40 = 72 - 2n + 2$ $\Rightarrow$ $2n = 72 - 40 + 2$

$\Rightarrow$ $2n = 34$

$\therefore$ $n = \frac{34}{2} = 17$

Hence 17th term will be 40. **Ans.**

Example 9. *Find three numbers in A.P. whose sum is 21 and their product is 315.*

[D.I.P.I.E.T.E. June 2018]

Solution. Here, we have

Let the three numbers be:

$(a - d)$, a, and $(a + d)$, then

$$a - d + a + a + d = 21$$

$$3a = 21 \Rightarrow a = 7$$

and $$(a - d) \cdot (a) \cdot (a + d) = 315$$

$$(a^2 - d^2)(a) = 315$$

$$a^3 - ad^2 = 315$$

$$\Rightarrow \quad (7)^3 - 7d = 315$$

$$7d = 343 - 315 = 28$$

$$\therefore \quad d = \frac{28}{7} = 4$$

Hence the required three numbers are

$(a - d), a, (a + d) = 3, 7, 11$ **Ans.**

Example 10. *If the sum of the first n terms of A.P. 85 + 90 + 95 + ... is equal to the sum of first 3n terms of A.P. 9 + 11 + 13 + ..., then find the value of n.*

[S.B.T.E. 2017, Delhi, 2016]

Solution. Here, we have sum of n terms

$$S_n = \frac{n}{2}[2a + (n-1)d]$$

$$= \frac{n}{2}[2 \times 85 + (n-1)(5)] = \frac{n}{2}[5n + 165] \quad \text{... (1)}$$

$$= \frac{3n}{2}[2 \times 9 + (3n-1)(2)] = \frac{3n}{2}[6n + 16] \quad \text{... (2)}$$

$$= \frac{n}{2}[5n + 165] = \frac{3n}{2}[6n + 16] \quad \text{[Given]}$$

$$\Rightarrow \quad 5n + 165 = 18n + 48$$

$$\Rightarrow \quad 13n = 117$$

$$\therefore \quad n = 9$$ **Ans.**

Example 11. *Find the value of K such that the sequence K – 1, K + 3 and 3K – 1 are in A.P.*

[S.B.T.E. 2017]

Solution. Given sequence is:

$K - 1, K + 3, 3K - 1$ are in A.P.

$$\Rightarrow \quad (K + 3) - (K - 1) = (3K - 1) - (K + 3)$$

$$4 = 2K - 4$$

$$\therefore \quad K = 4$$ **Ans.**

Example 12. *Find the sum of 13th terms of the A.P. 3, 8, 13, 18, ...*

Solution. Here, we have

$$a = 3, d = 8 - 3 = 5, n = 13$$

$$T_n = a + (n - 1)d$$
$$T_{13} = 3 + (13 - 1)(5) = 3 + 60 = 63$$ **Ans.**

Example 13. *How many terms are there in the given series 3, 6, 9, 12, ..., 36*

[S.B.T.E., Dec. 2017]

Solution. Given series: 3, 6, 9, 12, ..., 36

$$a = 3,\ d = 6 - 3 = 3,\ T_n = 36$$
$$T_n = a + (n - 1)d$$
$$36 = 3 + (n - 1)(3) = 3 + 3n - 3$$
$$\therefore \quad n = 12$$ **Ans.**

Example 14. *The 2nd, 31st and the last term of an A.P. are $7\frac{3}{4}, \frac{1}{2}$ and $-6\frac{1}{2}$, respectively. Find the first term and the number of terms.*

Solution. Let a be the first term and d the common difference of the A.P.

Given, $$T_2 = 7\frac{3}{4} \quad \Rightarrow \quad a + d = \frac{31}{4} \qquad ... (1)$$

and $$T_{31} = \frac{1}{2} \quad \Rightarrow \quad a + 30d = \frac{1}{2} \qquad ... (2)$$

Subtracting (1) from (2), we get

$$29d = \frac{1}{2} - \frac{31}{4} = -\frac{29}{4} \quad \Rightarrow \quad d = -\frac{1}{4}$$

Putting the value of d in (1), we get

$$a - \frac{1}{4} = \frac{31}{4} \quad \Rightarrow \quad a = \frac{31}{4} + \frac{1}{4} = \frac{32}{4} = 8$$

Let the number of terms be n, so that $T_n = -\frac{13}{2}$

i.e. $$a + (n-1)d = -\frac{13}{2} \quad \Rightarrow \quad 8 + (n-1)\left(-\frac{1}{4}\right) = -\frac{13}{2}$$

$$\Rightarrow \quad 8 - \frac{n}{4} + \frac{1}{4} = -\frac{13}{2} \quad \Rightarrow \quad 32 - n + 1 = -26$$

$$\Rightarrow \quad n = 59$$

Example 15. *Find the middle term(s) in the A.P. 20, 16, 12, ..., – 176.*

Solution. Here the first term, $a = 20$, the common difference, $d = -4$ and the nth term $= -176$.

We know that, $$T_n = a + (n - 1)d$$

$$\Rightarrow \quad -176 = 20 + (n - 1)(-4)$$

$$\Rightarrow \quad (n - 1) = \frac{-176 - 20}{-4} = \frac{196}{4}$$

$$\Rightarrow \quad n - 1 = 49 \quad \Rightarrow \quad n = 50$$

Hence, the middle terms are 25th and 26th.

$$T_{25} = 20 + (25 - 1)(-4) = 20 - 96 = -76$$
$$T_{26} = 20 + (26 - 1)(-4) = 20 - 100 = -80$$ **Ans.**

Example 16. *Split 69 into three parts such that they are in A.P. and the product of two smaller parts is 483.*

Solution. Let the three parts of 69 be $a - d$, a and $a + d$.

Sum of three terms = 69 $\Rightarrow$ $(a - d) + a + (a + d) = 69$

$\Rightarrow$ $3a = 69$ $\Rightarrow$ $a = 23$ ... (1)

Product of two smaller parts = 483 $\Rightarrow$ $a(a - d) = 483$ [using Eq. (1)]

$\Rightarrow$ $23(23 - d) = 483$ $\Rightarrow$ $23 - d = 21$ $\Rightarrow$ $d = 23 - 21 = 2$

Part I = $a - d = 23 - 2 = 21$

Part II = $a = 23$

Part III = $a + d = 23 + 2 = 25$

Hence, three parts are 21, 23, 25. **Ans.**

Example 17. *Divide 32 into four parts which are in A.P., such that the product of extremes to the product of means is 7 : 15.*

Solution. Let the four parts be $(a - 3d)$, $(a - d)$, $(a + d)$, $(a + 3d)$. Then,

sum of four parts = 32

$\Rightarrow$ $(a - 3d) + (a - d) + (a + d) + (a + 3d) = 32$

$\Rightarrow$ $4a = 32$ $\Rightarrow$ $a = 8$

and $$\frac{(a-3d)(a+3d)}{(a-d)(a+d)} = \frac{7}{15}$$

$\Rightarrow$ $$\frac{a^2 - 9d^2}{a^2 - d^2} = \frac{7}{15} \Rightarrow \frac{64 - 9d^2}{64 - d^2} = \frac{7}{15}$$

$\Rightarrow$ $128d^2 = 512$ $\Rightarrow$ $d^2 = 4$ $\Rightarrow$ $d = \pm 2$

Hence, the required parts are 2, 6, 10, 14 **Ans.**

Example 18. *If the nth term of a progression is a linear expression in n, then show that it is an A.P.*

Solution. Let $T_n = an + b$, where a and b are constants.

Then $T_{n-1} = a(n - 1) + b$

$\therefore$ $T_n - T_{n-1} = (an + b) - [a(n - 1) + b]$

$= an - an + b + a - b$

$= a$, which is a constant.

Thus, the difference between any two consecutive terms of the given progression is constant.

Hence, the given progression is an A.P. **Proved.**

Example 19. *If m times the mth term of an A.P. is equal to n times its nth term, show that the (m + n)th term of the A.P. is zero.*

Solution. Let a be the first term and d the common difference, then

$mT_m = nT_n$ [Given]

$\Rightarrow$ $m[a + (m - 1)d] = n[a + (n - 1)d]$

$\Rightarrow$ $[(m^2 - n^2) - (m - n)]d = (n - m)a$

$\Rightarrow$ $[(m - n)(m + n) - (m - n)]d = (n - m)a$ $\Rightarrow$ $(m - n)(m + n - 1)d = (n - m)a$

$\Rightarrow$ $(m + n - 1)d = -a$ $\Rightarrow$ $a + (m + n - 1)d = 0$

$\Rightarrow$ $T_{m+n} = 0$ **Proved.**

Example 20. *If the mth term of an A.P. be* $\left(\frac{1}{n}\right)$ *and nth term be* $\left(\frac{1}{m}\right)$*, then show that its (mn)th term is 1.*

Solution. Let a be the first term and d the common difference

Then $\quad T_m = \frac{1}{n} \quad \Rightarrow \quad a + (m-1)d = \frac{1}{n}$ ... (1)

and $\quad T_n = \frac{1}{m} \quad \Rightarrow \quad a + (n-1)d = \frac{1}{m}$ ... (2)

Subtracting (2) from (1), we get

$$(m-n)d = \left(\frac{1}{n} - \frac{1}{m}\right) \quad \Rightarrow \quad d = \frac{1}{mn}$$

Putting $\quad d = \frac{1}{mn}$ in (1), we get $a = \left[\frac{1}{n} - \frac{(m-1)}{mn}\right] = \frac{1}{mn}$.

$$T_{mn} = a + (mn-1)d = \frac{1}{mn} + (mn-1)\cdot\frac{1}{mn} = \frac{1}{mn} + 1 - \frac{1}{mn} = 1$$ **Proved.**

Example 21. *Let a, b and c be respectively the pth, qth and rth terms of an A.P. prove that*

(*i*) $a(q-r) + b(r-p) + c(p-q) = 0$

(*ii*) $(a-b)r + (b-c)p + (c-a)q = 0$

Solution. Let A be the first term and d the common difference.

Since, $\quad T_p = a \quad \Rightarrow \quad A + (p-1)d = a$ ... (1)

$T_q = b \quad \Rightarrow \quad A + (q-1)d = b$ ... (2)

$T_r = c \quad \Rightarrow \quad A + (r-1)d = c$ ... (3)

(*i*) Multiplying (1) by $(q-r)$, (2) by $(r-p)$ and (3) by $(p-q)$, we get

$(q-r)A + (p-q)(q-r)d = a(q-r)$ ... (4)

$(r-p)A + (a-1)(r-p)d = b(r-p)$ ... (5)

$(p-q)A + (r-1)(p-q)d = c(p-q)$ ... (6)

On adding (4), (5) and (6), we get

$$(q-r)A + (p-1)(q-r)d + (r-p)A + (q-1)(r-p)d + (p-q)A + (r-1)(p-q)d$$
$$= a(q-r) + b(r-p) + c(p-q)$$

$$\Rightarrow \quad A[(q-r) + (r-p) + (p-q)] + [(p-1)(q-r) + (q-1)(r-p) + (r-1)(p-q)]d$$
$$= a(q-r) + b(r-p) + c(p-q)$$

$$\Rightarrow \quad A(0) + (0)d = a(q-r) + b(r-p) + c(p-q)$$

$$\Rightarrow \quad a(q-r) + b(r-p) + c(p-q) = 0$$ **Proved.**

(*ii*) Subtracting (2) from (1), (3) from (2) and (1) from (3), we get

$a - b = (p-q)d$ ... (7)

$b - c = (q-r)d$ ... (8)

$c - a = (r-p)d$ ... (9)

Multiplying (7) by r, (8) by p and (9) by q, and adding them, we get

$$r(a-b) + p(b-c) + q(c-a) = (p-q)dr + (q-r)dp + (r-p)dq$$
$$= d[(p-q)r + (q-r)p + (r-p)q]$$
$$= d \times (0) = 0$$ **Proved.**

Example 22. *In an A.P. of n terms, prove that the sum of kth term from the beginning and kth term from the end is independent of k and equals the sum of the first and the last terms.*

Solution. Let a be the first term and d the common difference of the A.P.

$\therefore$ kth term from the beginning $= T_{k_1} = a + (k - 1)d$ ... (1)

Let l be the last term of the A.P. and $l = a + (n - 1)d$

$\therefore$ The A.P. is $a, a + d, a + 2d, ..., l - 2d, l - d, l$

The A.P. in reverse order can be written as

$$= l, l - d, l - 2d, ..., a + 2d, a + d, a.$$

$\therefore$ kth term from the end of the given A.P.

$$T_{k_2} = [l + (k - 1)(-d)] = [a + (n - 1)d] - (k - 1)d$$
$$= a + (n - 1 - k + 1)d = a + (n - k)d \quad ...(2)$$

Adding (1) and (2), we get

$$\therefore \text{Required sum} = T_{k_1} + T_{k_2} = [a + (k - 1)d] + [a + (n - k)d]$$
$$= 2a + (k - 1 + n - k)d$$
$$= 2a + (n - 1)d \quad \text{(independent of } k) \text{ ... (3)}$$

Also, sum of first and last term $= a + l = a + [a + (n - 1)d] = 2a + (n - 1)d$ ... (4)

Hence, the sum of the first and last terms is independent, of k and (3) = (4) **Proved.**

EXERCISE 2.2

1. Write the first six terms of an A.P. in which

(*i*) $a = 5, d = 4$ **Ans.** 5, 9, 13, 17, 21, 25

(*ii*) $a = 98, d = -3$ **Ans.** 98, 95, 92, 89, 86, 83

(*iii*) $a = 7\frac{1}{2}, d = 1\frac{1}{2}$ **Ans.** $7\frac{1}{2}, 9, 10\frac{1}{2}, 12, 13\frac{1}{2}, 15$

(*iv*) $a = x, d = 3x + 2$ **Ans.** $x, 4x + 2, 7x + 4, 10x + 6, 13x + 8, 16x + 10$

Find the common difference and write the next four terms in each of the following:

2. 2, 5, 8, 11, ... **Ans.** $d = 3$, next four terms are 14, 17, 20, 23

3. $1, \frac{5}{6}, \frac{2}{3}, \frac{1}{2}, ...$ **Ans.** $d = -\frac{1}{6}$, next four terms are $\frac{1}{3}, \frac{1}{6}, 0, -\frac{1}{6}$

4. $2a - 3b, a - 4b, -5b, -a - 6b$ ...

Ans. $d = -(a + b)$, next four terms are $-2a - 7b, -3a - 8b, -4a - 9b, -5a - 10b$

5. $\sqrt{2}, 3\sqrt{2}, 5\sqrt{2}, 7\sqrt{2}$ ...

Ans. $d = 2\sqrt{2}$, next four terms are $9\sqrt{2}, 11\sqrt{2}, 13\sqrt{2}, 15\sqrt{2}$.

How many terms are there in A.P. in each of the following:

6. 10, 13, 16, ..., 52. **Ans.** 15

7. 7, 13, 19, ..., 205. **Ans.** 34

8. $\frac{29}{2}, 12, \frac{19}{2}, ..., -38.$ **Ans.** 22

9. Which term of an A.P.

(*i*) 5, 8, 11, ... is 320? **Ans.** 106

(*ii*) 4, 9, 14, ... is 114? **Ans.** 23

(*iii*) 20, 16, 12, ... is – 96? **Ans.** 30

10. Find the 15th term from the end of the A.P. 3, 5, 7, 9, ..., 201. **Ans.** 173

11. The nth term of a progression is $2n + 3$, prove that it is an A.P. find its 10th term. **Ans.** 23

12. How many three digit numbers are divisible by 7. **Ans.** 128

13. The 5th term of an A.P. is 11 and the 9th term is 7. Find the 16th term. **Ans.** 0

14. The fourth term of an A.P. is ten times the first. Prove that the sixth term is four times the second term.

15. The fourth term of an A.P. is equal to 3 times the first and the seventh term exceeds twice the third term by 1. Find the first term and the common difference. **Ans.** $a = 3, d = 2$

16. If 7 times the 7th term of an A.P. is equal to 11 times its 11th term, show that the 18th term of an A.P. is zero.

17. If the pth term of an A.P. is q and qth term is p, then show that its nth term is $p + q - n$.

18. If $(p + 1)$th term of an A.P. is twice the $(q + 1)$th term, prove that the $(3p + 1)$th term is twice the $(p + q + 1)$th term.

19. Three numbers are in A.P. if the sum of these numbers be 27 and the product 648, find the numbers. **Ans.** 6, 9, 12

20. The sum of three consecutive terms of an A.P. is 21; if the sum of the squares of these terms be 165, find these terms. **Ans.** 4, 7, 10

21. The sum of four numbers in A.P. is 20 and the sum of their squares is 120. Find the numbers. **Ans.** 2, 4, 6, 8

22. The angles of a quadrilateral are in A.P., whose common difference is 10°. Find the angles. **Ans.** 75°, 85°, 95°, 105°

23. In an A.P. if $\frac{a_4}{a_7} = \frac{2}{3}$, find $\frac{a_6}{a_8}$. **Ans.** $\frac{4}{5}$

24. If $a_1, a_2, a_3, ...,$ an be an A.P. of non-zero term, prove that

$$\frac{1}{a_1a_2} + \frac{1}{a_2a_3} + ... + \frac{1}{a_{n-1}a_n} = \frac{n-1}{a_1a_n}.$$

EXERCISE 2.3

1. Find the 15th, 22nd and nth term of the A.P. given by 2, 8, 14, 20, ... **Ans.** 86, 128; $6n - 4$

2. Find the general term of the following A.P.

(*i*) $\frac{1}{2}, \frac{1}{4}, 0, -\frac{1}{4}, ...$ **Ans.** $\frac{3-n}{4}$

(*ii*) $x + b, x + 3b, x + 5b, ...$ **Ans.** $x + (2n - 1)b$

(*iii*) 1, 6, 11, 16, ... **Ans.** $5n - 4$

(*iv*) 54, 50, 46, 42, ... **Ans.** $58 - 4n$

3. (*i*) Find the 15th term from the end of the A.P. : 3, 5, 7, 9, ..., 201. **Ans.** 173

(*ii*) Find the 8th term from the end of the A.P. : 10, 13, ..., 184. **Ans.** 163

4. Which term of the A.P. : 12, 7, 2, –3, ... is – 98? **Ans.** 23

5. How many terms are there in the A.P. $\frac{5}{6}, 1, 1\frac{1}{6}, ..., 3\frac{1}{3}$? **Ans.** 16

6. The 1st term of an A.P. is –7. The common difference is 6 and the last number is 293. Find the number of terms. **Ans.** 51

7. An A.P. consists of 60 terms. If the first and the last terms be 7 and 125 respectively, find the 34th term. **Ans.** 73

8. The third term of an A.P. is p and fourth term is q, find the tenth term and the general term. **Ans.** $7q - 6p$, $4p - 3q + (q - p)n$

9. The 8th term of an A.P. is zero, prove that its 38th term is triple the 18th term.

10. The 1st term of a A.P. is –54 and sixteenth term is 21. Find the 12th term. **Ans.** 1

11. The 8th and 24th term of an A.P. are 45 and 157 respectively, find the 33rd term. **Ans.** 220

12. The 7th term of an A.P. is 20 and its 13th term is 32. Find the A.P. **Ans.** 8, 10, 12, 14, ...

13. The 7th term of an A.P. is –4 and its 13th term is –16. Find the A.P. **Ans.** 8, 6, 4, 2, ...

14. If the nth term of the A.P. : 24, 20, 16, 12, ... is same as nth term of the A.P. : –11, –8, –5, –2, ..., find n. **Ans.** 6

15. Find $a_{30} - a_{20}$ for the A.P. : –9, –14, –19, –24, ... **Ans.** –50

16. The fifth term of an A.P. is thrice the second term and the twelfth term exceeds twice the 6th term by 1. Find the 16th term. **Ans.** 31

17. The sum of the 4th and 8th terms of an A.P. is 24 and the sum of the 6th and 10th terms is 34. Find the first four terms of the A.P. **Ans.** $-\frac{1}{2}, 2, \frac{9}{2}, 7$

18. In a A.P., if the mth term is n and nth term is m, show that rth term is $(m + n - r)$.

19. In a certain A.P., the 32nd term is twice the 12th term. Prove that 70th term is twice the 31st term.

20. Two A.P. have the same common difference. The first term of one of these is 3 and that of the other is 8. What is the difference between their

(*i*) 2nd term? **Ans.** 5

(*ii*) 4th term? **Ans.** 5

(*iii*) 10th term? **Ans.** 5

(*iv*) 30th term? **Ans.** 5

21. Two A.P.s have the same common difference. The difference between their 100th terms is 111222333. What is the difference between their millionth term? **Ans.** 111222333

22. For what value of n, the nth terms of the A.Ps 63, 65, 67, ... and 3, 10, 17, ... are equal? **Ans.** 13

23. Divide 40 into four parts, which are in A.P. such that the ratio of product of extremes to the product of means is 3? **Ans.** 4, 8, 12, 16

24. If in an A.P. ratio of fourth term and 9th term is 1 : 3, find the ratio of 12th term and 5th term. **Ans.** 3 : 1

25. The angles of a quadrilateral are in A.P. whose common difference is 10°. Find the angles. **Ans.** 75, 85, 95, 105
26. How many multiples of 4 lie between 10 and 250? **Ans.** 60
27. How many numbers of two digits are divisible by 6? **Ans.** 15
28. How many numbers of two digits are divisible by 9? **Ans.** 10
29. Divide 32 into four parts, which are in A.P. such that the ratio of product of extremes to the product of means is 7 : 15. **Ans.** 2, 6, 10, 14
30. Find four numbers in A.P. whose sum is 20 and the sum of whose squares is 120. **Ans.** 2, 4, 6, 8
31. Which term of the A.P. 3, 15, 27, 39, ..., will be 120 more than its 64th term? **Ans.** 74
32. Priya was appointed as a lecturer. She was offered a monthly salary of ₹15,000 with annual increment of ₹500. In the tenth year what would be her monthly salary? **Ans.** ₹19500
33. Bhagwati joined a bank on the initial salary of ₹5000 per month with annual increment ₹400. In 20th year what will be her monthly salary. **Ans.** ₹11600
34. Determine the number of term in the A.P. 3, 7, 11, ..., 407. Also, find its 20 term from the end. **Ans.** 102, 331
35. Find the 15th term from the end of an A.P. 6, 10, 14, 18, ..., 102. **Ans.** 46
36. Find the number of term in the A.P. 10, 4, –2, –8. ..., –290. Also, find the 30th term from the end. **Ans.** 51116
37. Find the number of terms in A.P. : 8.5, 6.9, 5.3, ..., 151.5. Also, find its 50th term from the end. **Ans.** 101, –73.1
38. Which term of A.P. : 163, 160, 157, ... is the first negative term. **Ans.** 56
39. Check whether 137 is a term of A.P. 3, 11, 19, 27, ...? **Ans.** No

2.7 PROPERTIES OF ARITHMETIC PROGRESSION

(*i*) If a constant is added to each term of an A.P., then the resulting sequence is also in A.P.

(*ii*) If a constant is subtracted from each term of an A.P. then the resulting sequence is also in A.P.

(*iii*) If each term of an A.P. is multiplied by a constant, then the resulting sequence is also in A.P.

(*iv*) If each term of an A.P. is divided by a nonzero constant, then the resulting sequence is also in A.P.

Examples based on Properties of A.P.

Example 23. *If a, b, c are in A.P., prove that $b + c, c + a, a + b$ are also in A.P.*

Solution. $b + c, c + a, a + b$ will be in A.P. if $(c + a) - (b + c) = (a + b) - (c + a)$

if $\quad a - b = b - c \quad \Rightarrow \quad$ if $2b = a + c$

i.e. if a, b, c are in A.P.

Thus, a, b, c are in A.P. implies $b + c, c + a, a + b$ are in A.P. **Proved.**

Example 24. *If $a + b + c \neq 0$ and $\frac{b+c}{a}, \frac{c+a}{b}, \frac{a+b}{c}$ are in A.P.*

Prove that $\frac{1}{a}, \frac{1}{b}, \frac{1}{c}$ are also in A.P.

Solution. We have, $\dfrac{b+c}{a}, \dfrac{c+a}{b}, \dfrac{a+b}{c}$ are in A.P.

Adding 1 to each term, we get

$$\left(\frac{b+c}{a}+1\right), \left(\frac{c+a}{b}+1\right), \left(\frac{a+b}{c}+1\right) \text{ are in A.P.}$$

i.e.
$$\frac{a+b+c}{a}, \frac{c+a+b}{b}, \frac{a+b+c}{c} \text{ are in A.P.}$$

Dividing each term by $a + b + c$, we get

$$\frac{1}{a}, \frac{1}{b}, \frac{1}{c} \text{ are in A.P.}$$

Proved.

Example 25. *If* $a\left(\dfrac{1}{b}+\dfrac{1}{c}\right), b\left(\dfrac{1}{a}+\dfrac{1}{c}\right), c\left(\dfrac{1}{a}+\dfrac{1}{b}\right)$ *are in A.P., prove that a, b, c, are in A.P.*

Solution. We have, $a\left(\dfrac{1}{b}+\dfrac{1}{c}\right), b\left(\dfrac{1}{a}+\dfrac{1}{c}\right), c\left(\dfrac{1}{a}+\dfrac{1}{b}\right)$ are in A.P.

$$\Rightarrow \quad a\left(\frac{b+c}{bc}\right), b\left(\frac{a+c}{ac}\right), c\left(\frac{b+a}{ab}\right) \text{ are in A.P.}$$

Adding 1 we get
$$\frac{ab+ac}{bc}+1, \frac{ab+bc}{ac}+1, \frac{bc+ac}{ab}+1 \text{ are in A.P.}$$

$$\Rightarrow \quad \frac{ab+ac+bc}{bc}, \frac{ab+bc+ac}{ac}, \frac{bc+ac+ab}{ab} \text{ are in A.P.}$$

Multiplying by $\left(\dfrac{abc}{ab+bc+ac}\right)$ to all the terms, we get

a, b, c are in A.P.

Proved.

Example 26. *If a, b, c are in A.P., prove that*

(*i*) $\dfrac{1}{\sqrt{b}+\sqrt{c}}, \dfrac{1}{\sqrt{c}+\sqrt{a}}, \dfrac{1}{\sqrt{a}+\sqrt{b}}$ *are also in A.P.*

(*ii*) $(b + c)^2 - a^2, (c + a)^2 - b^2, (a + b)^2 - c^2$ are also in A.P.

Solution. If a, b, c are in A.P. [Given]

then $\quad b - a = c - b \quad$ ... (1)

(*i*) Now, if $\dfrac{1}{\sqrt{b}+\sqrt{c}}, \dfrac{1}{\sqrt{c}+\sqrt{a}}, \dfrac{1}{\sqrt{a}+\sqrt{b}}$ are in A.P.

then,
$$\frac{1}{\sqrt{c}+\sqrt{a}} - \frac{1}{\sqrt{b}+\sqrt{c}} = \frac{1}{\sqrt{a}+\sqrt{b}} - \frac{1}{\sqrt{c}+\sqrt{a}}$$

$$\Rightarrow \quad \frac{\sqrt{b}+\sqrt{c}-\sqrt{c}-\sqrt{a}}{(\sqrt{c}+\sqrt{a})(\sqrt{b}+\sqrt{c})} = \frac{\sqrt{c}+\sqrt{a}-\sqrt{a}-\sqrt{b}}{(\sqrt{a}+\sqrt{b})(\sqrt{c}+\sqrt{a})}$$

$$\Rightarrow \quad \frac{\sqrt{b}-\sqrt{a}}{\sqrt{b}+\sqrt{c}} = \frac{\sqrt{c}-\sqrt{b}}{\sqrt{a}+\sqrt{b}}$$

$\Rightarrow \quad (\sqrt{b}-\sqrt{a})(\sqrt{a}+\sqrt{b}) = (\sqrt{b}+\sqrt{c})(\sqrt{c}-\sqrt{b})$

$\Rightarrow \quad b - a = c - b$, which is true as (1).

Hence, $\frac{1}{\sqrt{b}+\sqrt{c}}, \frac{1}{\sqrt{c}+\sqrt{a}}, \frac{1}{\sqrt{a}+\sqrt{b}}$ are in A.P. **Proved.**

(ii) If $(b + c)^2 - a^2, (c + a)^2 - b^2, (a + b)^2 - c^2$ are in A.P., then

$(b + c + a)(b + c - a), (c + a + b)(c + a - b), (a + b + c)(a + b - c)$ are in A.P.

Dividing by $(a + b + c)$, we get

$(b + c - a), (c + a - b), (a + b - c)$ are in A.P.

Subtracting $(a + b + c)$ from each term, we get

$(b + c - a) - (a + b + c), (c + a - b) - (a + b + c), (a + b - c) - (a + b + c)$ are in A.P.

$\Rightarrow \; -2a, -2b, -2c$ are in A.P.

$\Rightarrow \; a, b, c$ are in A.P., which is true.

Hence, $(b + c)^2 - a^2, (c + a)^2 - b^2, (a + b)^2 - c^2$ are in A.P. **Proved.**

EXERCISE 2.4

1. If a, b, c are in A.P., prove that $\frac{ab+ac}{bc}, \frac{bc+ba}{ca}, \frac{ca+bc}{ab}$ are in A.P.
2. Prove that a, b, c are in A.P., if and only if $\frac{1}{bc}, \frac{1}{ca}, \frac{1}{ab}$ are in A.P.
3. If $\frac{b+c-a}{a}, \frac{c+a-b}{b}, \frac{a+b-c}{c}$ are in A.P., prove that $\frac{1}{a}, \frac{1}{b}, \frac{1}{c}$ are also in A.P.
4. If a^2, b^2, c^2 are in A.P., prove that
   (i) $\frac{1}{b+c}, \frac{1}{c+a}, \frac{1}{a+b}$ are also in A.P. *(ii)* $\frac{a}{b+c}, \frac{b}{c+a}, \frac{c}{a+b}$ are also in A.P.
5. If $\frac{1}{b+c}, \frac{1}{c+a}, \frac{1}{a+b}$ are in A.P., prove that $\frac{a}{b+c}, \frac{a}{c+a}, \frac{a}{a+b}$ are also in A.P.
6. If $a^2(b + c), b^2(c + a), c^2(a + b)$ are in A.P., show that either a, b, c are in A.P. or $ab + bc + ca = 0$
7. If $(b - c)^2, (c - a)^2, (a - b)^2$ are in A.P., prove that $\frac{1}{b-c}, \frac{1}{c-a}, \frac{1}{a-b}$ are in A.P.

2.8 SUM OF FIRST *n*-TERMS OF AN A.P.

Let the first term of an A.P. be a and the common difference be d. Let us denote the sum of n terms of an A.P. by S_n, then

$$S_n = a + (a + d) + (a + 2d) + \ldots + [a + (n - 3)d] + [a + (n - 2)d] + [a + (n - 1)d] \quad \ldots (1)$$
$$= t_1 + t_2 + t_3 + \ldots + t_{n-2} + t_{n-1} + t_n$$

By reversing the order of A.P. we get

$$S_n = [a + (n - 1)d] + [a + (n - 2)d] + [a + (n - 3)d] + \ldots + (a + 2d) + (a + d) + a \quad \ldots (2)$$
$$= T_1 + T_2 + T_3 + \ldots + T_{n-2} + T_{n-1} + T_n$$

Adding (1) and (2), term by term, we get

Sum of 1st term of (1) and 1st term of (2) = $t_1 + T_1 = a + a + (n-1)d = 2a + (n-1)d$

Sum of 2nd term of (1) and 2nd term of (2) = $(a + d) + [a + (n-2)d] = 2a + (n-1)d$

Sum of 3rd term of (1) and 3rd term of (2) = $(a + 2d) + [a + (n-3)d] = 2a + (n-1)d$

....

....

Sum of last term of (1) and last term of (2) = $[a + (n-1)d] + a = 2a + (n-1)d$

We know that each S_n of (1) and S_n of (2) has n terms, therefore

$$2S_n = n[2a + (n-1)d] \quad \Rightarrow \quad \boxed{S_n = \frac{n}{2}[2a + (n-1)d]}$$

$$\Rightarrow S_n = \frac{n}{2}[a + a + (n-1)d] \quad \Rightarrow \quad \boxed{S_n = \frac{n}{2}[a + l]} \qquad [\text{last term} = l = a + (n-1)d]$$

Example 27. *Find the sum of indicated number of terms in each of the following arithmetic progressions:*

(*i*) 16, 11, 6, ...; 23 terms (*ii*) – 0.5, – 1.0, – 1.5, ...; 10 terms

(*iii*) $-1, \frac{1}{4}, \frac{3}{2}, ...$; 81 terms (*iv*) $x + y, x - y, x - 3y$ ...; 22 terms

Solution. We know that the sum of n terms of an A.P. is given by

$$S_n = \frac{n}{2}[2a + (n-1)d]$$

where, a is first term, d is common difference and n is number of terms.

(*i*) We have, $a = 16, n = 23, d = 11 - 16 = 6 - 11 = -5$

$$\therefore \quad S_{23} = \frac{23}{2}[2 \times 16 + (23-1)(-5)] = \frac{23}{2}[32 - 110]$$

$$= \frac{23 \times (-78)}{2} = -23 \times 39 = -897$$

(*ii*) We have, $a = -0.5, n = 10, d = -1.0 + 0.5 = -1.5 + 1.0 = -0.5$

$$\therefore \quad S_{10} = \frac{10}{2}[2 \times (-0.5) + (10-1)(-0.5)]$$

$$= 5[-1.0 - 4.5] = 5 \times (-5.5) = -27.5$$

(*iii*) We have, $a = -1, n = 81, d = \frac{1}{4} + 1 = \frac{3}{2} - \frac{1}{4} = \frac{5}{4}$

$$\therefore \quad S_{81} = \frac{81}{2}\left[2 \times (-1) + (81-1)\left(\frac{5}{4}\right)\right]$$

$$= \frac{81}{2}[-2 + 100] = \frac{81 \times 98}{2} = 3969$$

(*iv*) We have, $a = x + y, n = 22, d = (x - y) - (x + y) = (x - 3y) - (x - y) = -2y$

$$\therefore \quad S_{22} = \frac{22}{2}[2(x + y) + (22-1)(-2y)]$$

$$= 11[2x + 2y - 42y] = 11[2x - 40y] = 22(x - 20y)$$ **Ans.**

Example 28. *Sum the series* $1+\frac{1}{3}+\frac{1}{3^2}+\frac{1}{3^3}+\ldots$ *to* ∞ **[S.B.T.E. Dec. 2016]**

Solution. Here we have $1+\frac{1}{3}+\frac{1}{3^2}+\frac{1}{3^3}+\ldots$ to ∞

$$a = 1,\ r = \frac{1}{3}$$

$$S_\infty = \frac{a}{1-r} = \frac{1}{1-\frac{1}{3}} = \frac{1}{\frac{3-1}{3}} = \frac{3}{2}$$ **Ans.**

Example 29. Find the sum of n terms of the series 0.7 + 0.77 + 0.777 + ... **[S.B.T.E. 2015, 2014]**

Solution. Here we have:

$$S_n = 0.7 + 0.77 + 0.777 + \ldots$$
$$= 7\,[0.1 + 0.11 + 0.111 + \ldots]$$
$$= \frac{7}{9}\,[0.9 + 0.99 + 0.999 + \ldots]$$
$$= \frac{7}{9}\left[\left(1-\frac{1}{10}\right)+\left(1-\frac{1}{100}\right)+\left(1-\frac{1}{1000}\right)+\ldots\right]$$
$$= \frac{7}{9}\left[(1+1+1+\ldots)-\left(\frac{1}{10}+\frac{1}{100}+\frac{1}{1000}+\ldots\right)\right]$$
$$= \frac{7}{9}\left[n-\frac{\frac{1}{10}\left(1-\frac{1}{10^n}\right)}{1-\frac{1}{10}}\right]$$
$$= \frac{7}{9}\left[n-\frac{1}{9}\left(1-\frac{1}{10^n}\right)\right]$$

$$\therefore \quad S_n = \frac{7}{81}\left[9n-1+\frac{1}{10^n}\right]$$ **Ans.**

Example 30. *Find the sum of all natural numbers between 100 and 1000 which are multiple of 5.*

Solution. The natural numbers between 100 and 1000, multiple of 5 are 105, 110, ..., 995.

$\therefore$ Required sum = (105 + 110 + ... + 995)

This is an arithmetic series with first term 105 and common difference 5.

Let n be the number to terms in this series.

Then, $T_n = 995 \Rightarrow 105 + (n-1)5 = 995$

$\Rightarrow 995 - 105 = (n-1)5 \Rightarrow 890 = (n-1)5$

$\Rightarrow (n-1) = \frac{890}{5} = 178 \Rightarrow n = 178 + 1 = 179$

$$S_n = \frac{n}{2}[2a+(n-1)d] = \frac{179}{2}[2\times 105+(179-1)5]$$

$$= \frac{179}{2}[210+890] = \frac{179}{2}[1100] = 179 \times 550 = 98450 \quad \textbf{Ans.}$$

Example 31. *How many terms of the A.P.* $-6, -\frac{11}{2}, -5, \ldots$ *are needed to give the sum* -25?

Solution. The given sequence is $-6, -\frac{11}{2}, -5, \ldots$

Here, $\quad T_2 - T_1 = -\frac{11}{2} + 6 = \frac{1}{2};\ T_3 - T_2 = -5 + \frac{11}{2} = \frac{1}{2}$

The given sequence is an A.P. with $a = -6$ and $d = \frac{1}{2}$. ... (1)

Let -25 be the sum of the first n terms $\Rightarrow S_n = -25$

$\Rightarrow \quad \frac{n}{2}[2a+(n-1)d] = -25 \quad \Rightarrow \quad n\left[2(-6)+(n-1)\frac{1}{2}\right] = -50 \quad$ [using Eq. (1)]

$\Rightarrow \quad n\left[-12+\frac{(n-1)}{2}\right] = -50 \quad \Rightarrow \quad n[-24+(n-1)] = -100$

$\Rightarrow \quad -24n + n^2 - n = -100 \quad \Rightarrow \quad n^2 - 25n + 100 = 0$

$\Rightarrow \quad n^2 - 20n - 5n + 100 = 0 \quad \Rightarrow \quad n(n-20) - 5(n-20) = 0$

$\Rightarrow \quad (n-5)(n-20) = 0 \quad \Rightarrow \quad n = 5, 20$ **Ans.**

Example 32. *The sum of n terms of an A.P. is* $4n^2 + 5n$. *Find the series.*

Solution. We have, $\quad S_n = 4n^2 + 5n$

$S_1 = 4(1)^2 + 5(1) = 4 + 5 = 9$

$S_2 = 4(2)^2 + 5(2) = 16 + 10 = 26$

$S_3 = 4(3)^2 + 5(3) - 36 + 15 = 51$

Now, we know that $\quad S_{n-1} + T_n = S_n$

$\Rightarrow \quad T_n = S_n - S_{n-1}$

$\therefore \quad T_1 = S_1 = 9,\ T_2 = S_2 - S_1 = 26 - 9 = 17$

$T_3 = S_3 - S_2 = 51 - 26 = 25$

Hence, the series is 9, 17, 25, ... **Ans.**

Example 33. *Determine the sum of the first 35 terms of an A.P., if its second term is 2 and seventh term is 22.*

Solution. Let the first term of A.P. is '*a*' and common difference is '*d*'. Then,

$T_2 = [a + (2-1)d] = 2$

$\Rightarrow \quad a + d = 2$ ... (1)

And $\quad T_7 = [a + (7-1)d] = 22$

$\Rightarrow \quad a + 6d = 22$ ... (2)

Subtracting (1) from (2), we get $\quad 5d = 20 \quad \Rightarrow \quad d = 4$

Putting $d = 4$ in (1), we get $\quad a + 4 = 2 \quad \Rightarrow \quad a = 2 - 4 = -2$

$\therefore \quad S_{35} = \frac{35}{2}[2(-2)+(35-1)4] = \frac{35}{2}[-4+136] = \frac{35}{2}\times 132 = 2310$ **Ans.**

Example 34. *If the first term of an A.P., is 2 and the sum of first five terms is equal to one fourth of the sum of the next five terms, show that the 20th term is – 112.*

Solution. Let the first term of given A.P. is 'a' and common difference is 'd'.

We have, $T_1 = a = 2$

$$T_1 + T_2 + T_3 + T_4 + T_5 = \frac{1}{4}[T_6 + T_7 + T_8 + T_9 + T_{10}]$$

Sum of 5 terms, where first term is $a = \frac{1}{4}$ sum of 5 terms, where first term is $(a + 5d)$

$$\Rightarrow \quad \frac{5}{2}[2a + (5-1)d] = \frac{1}{4}\times\frac{5}{2}[2(a+5d) + (5-1)d]$$

$$\Rightarrow \quad \frac{5}{2}[2a + 4d] = \frac{1}{4}\times\frac{5}{2}[2a + 14d]$$

$$\Rightarrow \quad (2a + 4d) = \frac{1}{4}[2a + 14d] \quad \Rightarrow \quad 2\times 2 + 4d = \frac{1}{4}[2\times 2 + 14d], \ (a = 2)$$

$$\Rightarrow \quad 8 + 8d = 2 + 7d \quad \Rightarrow \quad d = -6$$

$$\therefore \quad a = 2 \quad \text{and} \quad d = -6$$

Now, $T_{20} = a + (20-1)d = 2 + 19(-6) = 2 - 114 = -112$ **Proved.**

Example 35. *The sum of p, q, r terms of an A.P. are a, b, c, respectively. Show that*

$$\frac{a}{p}(q-r) + \frac{b}{q}(r-p) + \frac{c}{r}(p-q) = 0$$

Solution. Given that $S_p = a, S_q = b, S_r = c$

Let A be the first term and d be the common difference, then

$$\Rightarrow \quad S_p = \frac{p}{2}[2A + (p-1)d] = a \quad \Rightarrow \quad 2A + (p-1)d = \frac{2a}{p} \quad \text{... (1)}$$

$$\Rightarrow \quad S_q = \frac{q}{2}[2A + (q-1)d] = b \quad \Rightarrow \quad 2A + (q-1)d = \frac{2b}{q} \quad \text{... (2)}$$

$$\text{and} \quad S_r = \frac{r}{2}[2A + (r-1)d] = c \quad \Rightarrow \quad 2A + (r-1)d = \frac{2c}{r} \quad \text{...(3)}$$

Multiplying (1) by $(q - r)$, (2) by $(r - p)$ and (3) by $(p - q)$, we get

$$\Rightarrow \quad [2A + (p-1)d](q-r) = \frac{2a}{p}(q-r) \quad \text{... (4)}$$

$$\Rightarrow \quad [2A + (q-1)d](r-p) = \frac{2b}{q}(r-p) \quad \text{... (5)}$$

$$\Rightarrow \quad [2A + (r-1)d](p-q) = \frac{2c}{r}(p-q) \quad \text{... (6)}$$

Adding (4), (5) and (6), we get

$$\frac{2a}{p}(q-r) + \frac{2b}{q}(r-p) + \frac{2c}{r}(p-q)$$

$$= [2A + (p-1)d](q-r) + [2A + (q-1)d](r-p) + [2A + (r-1)d](p-q)$$

$$= 2A(q - r + r - p + p - q) + d[(p-1)(q-r) + (q-1)(r-p) + (r-1)(p-q)]$$

$= 2A(0) + d\,[pq - q - rp + r + qr + p - pq - r + pr - p - qr + q]$
$= 0 + 0 = 0$

$\therefore \quad \frac{a}{p}(q-r)+\frac{b}{q}(r-p)+\frac{c}{r}(p-q)=0$ **Proved.**

Example 36. *The ratio of the sum of m and n terms of an A.P. is $m^2 : n^2$. Show that the ratio of mth and nth term is $2m - 1 : 2n - 1$.*

Solution. Let the first term be 'a' and common difference is d.

$$S_m = \frac{m}{2}\,[2a + (m-1)d] \text{ and } S_n = \frac{n}{2}\,[2a + (n-1)d]$$

Given that, $\frac{S_m}{S_n} = \frac{m^2}{n^2}$

$$\Rightarrow \quad \frac{\frac{m}{2}[2a+(m-1)d]}{\frac{n}{2}[2a+(n-1)d]} = \frac{m^2}{n^2} \quad \Rightarrow \quad \frac{2a+(m-1)d}{2a+(n-1)d} = \frac{m}{n}$$

$\Rightarrow \quad 2an + (mn - n)d = 2am + (mn - m)d$

$\Rightarrow \quad 2an - 2am = (mn - m - mn + n)d$

$\Rightarrow \quad 2a(n - m) = d(n - m) \quad \Rightarrow \quad d = 2a$

$\therefore \quad T_m = a + (m - 1)d = a + (m - 1)2a$ $[\because d = 2a]$

$= a + 2am - 2a = 2am - a$

$\Rightarrow \quad T_m = a(2m - 1)$ ... (1)

and so $\quad T_n = a(2n - 1)$ ... (2)

Dividing (1) by (2), we get

$$\frac{T_m}{T_n} = \frac{a(2m-1)}{a(2n-1)} \quad \Rightarrow \quad \frac{T_m}{T_n} = \frac{2m-1}{2n-1}$$ **Proved.**

Example 37. *The sums of n terms of two arithmetic series are in the ratio of $2n + 1 : 2n - 1$. Find the ratio of their 10th terms.*

Solution. Let the first arithmetic series be $a, a + d, a + 2d, \ldots$ and second arithmetic series be $A, A + D, A + 2D, \ldots$

Given that, $\frac{\frac{n}{2}[2a+(n-1)d]}{\frac{n}{2}[2A+(n-1)D]} = \frac{2n+1}{2n-1} \quad \Rightarrow \quad \frac{2a+(n-1)d}{2A+(n-1)D} = \frac{2n+1}{2n-1}$ ... (1)

We have to find the ratio, $\frac{t_{10}}{T_{10}} = \frac{a+9d}{A+9D} = \frac{2a+18d}{2A+18D} = \frac{2a+(19-1)d}{2A+(19-1)D}$ ... (2)

Putting $n = 19$ in (1), we get

$$\frac{t_{10}}{T_{10}} = \frac{2a+(19-1)d}{2A+(19-1)D} = \frac{2n+1}{2n-1} = \frac{2\times19+1}{2\times19-1} = \frac{39}{37}$$ ... (3)

$$\Rightarrow \quad \frac{t_{10}}{T_{10}} = \frac{39}{37}$$ **Ans.**

Example 38. *In an A.P., if the pth term is* $\frac{1}{q}$ *and the qth term is* $\frac{1}{p}$*, prove that the sum of the first pq terms must be* $\frac{1}{2}$ *(pq + 1).*

Solution. Let the first term of an A.P. be 'a' and common difference be d. Then,

$$T_p = a + (p-1)d = \frac{1}{q} \quad \Rightarrow \quad [(a + (p-1)d] = \frac{1}{q} \qquad \text{... (1)}$$

and $$T_q = a + (q-1)d = \frac{1}{p} \quad \Rightarrow \quad [a + (q-1)d] = \frac{1}{p} \qquad \text{... (2)}$$

Subtracting (2) from (1), we get

$$(p-q)d = \frac{1}{q} - \frac{1}{p} = \frac{p-q}{pq} \quad \Rightarrow \quad d = \frac{1}{pq} \qquad \text{... (3)}$$

Putting $d = \frac{1}{pq}$ in (1), we get $a + (p-1)\cdot\frac{1}{pq} = \frac{1}{q}$

$$\Rightarrow \quad a = \frac{1}{q} - \frac{1}{q} + \frac{1}{pq} \quad \Rightarrow \quad d = \frac{1}{pq} \qquad \text{... (4)}$$

Now, $$S_{pq} = \frac{pq}{2}[2a + (pq-1)d] = \frac{pq}{2}\left[2\cdot\frac{1}{pq} + (pq-1)\frac{1}{pq}\right] \qquad \text{[From (3) and (4)]}$$

$$\Rightarrow \quad S_{pq} = \frac{1}{2}(pq+1)$$

Proved.

Example 39. *The pth term of an A.P. is a and qth term is b. Prove that the sum of its (p + q)th term is* $\frac{p+q}{2}\left[a + b + \frac{a-b}{p-q}\right]$.

Solution. Let A be the first term and D be the common difference of the given A.P. Then,

$$T_p = a \quad \Rightarrow \quad A + (p-1)D = a \qquad \text{... (1)}$$

and $$T_q = b \quad \Rightarrow \quad A + (q-1)D = b \qquad \text{... (2)}$$

Subtracting (2) from (1), we get

$$(p-q)D = a - b \quad \Rightarrow \quad D = \frac{a-b}{p-q} \qquad \text{... (3)}$$

Adding (1) and (2), we get

$$2A + (p+q-2)D = a + b \quad \Rightarrow \quad 2A + (p+q-1)D - D = a + b$$

$$\Rightarrow \quad 2A + (p+q-1)D = a + b + D$$

$$\Rightarrow \quad 2A + (p+q-1)D = a + b + \left(\frac{a-b}{p-q}\right) \qquad \text{... (4) [using (3)]}$$

Now, $$S_{p+q} = \frac{p+q}{2}[2A + (p+q-1)D] \qquad \text{[using (4)]}$$

$$= \frac{p+q}{2}\left[a + b + \frac{a-b}{p-q}\right]$$

Proved.

Example 40. *The sum of n, 2n, 3n terms of an A.P. are S_1, S_2, S_3 respectively. Prove that $S_3 = 3(S_2 - S_1)$*

Solution. Let a be the first term and d be the common difference of the given A.P. Then,

$$\text{Sum of } n \text{ terms} = S_1 = \frac{n}{2}[2a+(n-1)d] \quad \text{... (1)}$$

$$2n \text{ terms} = S_2 = \frac{2n}{2}[2a+(2n-1)d] \quad \text{... (2)}$$

$$3n \text{ terms} = S_3 = \frac{3n}{2}[2a+(3n-1)d] \quad \text{... (3)}$$

Now,
$$S_2 - S_1 = \frac{2n}{2}[2a+(2n-1)d] - \frac{n}{2}[2a+(n-1)d]$$

$$= \frac{n}{2}[4a+2(2n-1)d-2a-(n-1)d]$$

$$= \frac{n}{2}[2a+(4n-2-n+1)d] = \frac{n}{2}[2a+(3n-1)d]$$

$$\Rightarrow \quad 3(S_2 - S_1) = \frac{3n}{2}[2a+(3n-1)d]$$

$$\Rightarrow \quad 3(S_2 - S_1) = S_3$$

[From (3)]. **Proved.**

Example 41. *A man saved Rs. 16500 in 10 years. In each year after the first he saved Rs. 100 more than he did in the preceding year. How much did he save in the first year?*

Solution. The yearly saving forms an A.P. whose common difference is 100 and the sum of whose 10 terms is 16500. If a denotes (in rupees) the saving of the first year, then

$$S_n = \frac{n}{2}[2a+(n-1)d] \quad \Rightarrow \quad 16500 = \frac{10}{2}(2a + 9 \times 100)$$

$$\Rightarrow \quad 16500 = 10a + 4{,}500$$

$$\Rightarrow \quad 10a = 12000 \quad \Rightarrow a = 1200$$

$\therefore$ The first year's saving = ₹1200 **Ans.**

Example 42. *A thief runs away from a police station with a uniform speed of 100 m/min. After one minute a policeman runs behind the thief to catch him. He goes at speed of 100 m/min in first minute and increases his speed 10 m each succeeding minute. After how many minutes, the policeman will catch the thief?*

Solution. Let the policeman catch the thief in n minutes.

Since the thief ran one minute before the police, thus, the time taken by thief before being caught = $(n + 1)$ minutes

Distance travelled by the thief in $(n + 1)$ minutes = $100(n + 1)$ metres

In first minute, speed of policeman = 100 m/minute.

In second minute, speed of policeman = 110 m/minute.

In third minute, speed of policeman = 120 m/minute and so on.

$\therefore$ Speed of 100, 110, 120, ... form an A.P.

Total distance travelled by the policeman in n minutes = $\frac{n}{2}[2 \times 100 + (n-1)10]$

On catching the thief by policeman.

Distance travelled by the thief = Distance travelled by the policeman.

$$100(n + 1) = \frac{n}{2}\,[2 \times 100 + (n - 1)10]$$

$\Rightarrow\quad 100n + 100 = 100n + \frac{n}{2}(n - 1)10 \qquad \Rightarrow\quad 100 = n(n - 1)\,5$

$\Rightarrow\quad n^2 - n - 20 = 0 \qquad \Rightarrow\quad (n - 5)(n + 4) = 0$

$\Rightarrow\quad n - 5 = 0 \qquad \Rightarrow\quad n = 5$

$\Rightarrow\quad n + 4 = 0 \qquad \Rightarrow\quad n = -4$ not possible

Time taken by the policeman to catch the thief = 5 minutes. **Ans.**

EXERCISE 2.5

Find the sum of indicated number of terms in each of the following arithmetic progression:

1. 2, 4, 6, 8, ...; 100 terms. **Ans.** 10, 100

2. 5, 2, – 1, – 4, – 7, ..., n terms **Ans.** $\frac{n}{2}(13 - 3n)$

3. 0.7, 0.71, 0.72, ..., 100 terms. **Ans.** 119.5

4. $15, 14\frac{1}{3}, 13\frac{2}{3}, \ldots$, 40 terms. **Ans.** 80

5. $(a - b)^2; a^2 + b^2, (a + b)^2, \ldots$, n terms. **Ans.** $n[(a - b)^2 + (n - 1)ab]$

6. $\frac{x-y}{x+y}, \frac{3x-2y}{x+y}, \frac{5x-3y}{x+y}, \ldots$, n terms. **Ans.** $\frac{n}{2(x+y)}[n(2x-y)-y]$

7. Find the sum of first n natural numbers. **Ans.** $\frac{n(n+1)}{2}$

8. Find the sum of all natural numbers between 1 and 100, which are divisible by 3. **Ans.** 1683

9. Find the sum of all odd numbers between 100 and 200. **Ans.** 7500

10. Find the sum of 32 terms of an A.P., whose third term is 1 and the 6th term is – 11. **Ans.** – 1696

11. The sum of series of terms in A.P. is 128. If the first term is 2 and the last term is 14, find the common difference. **Ans.** $\frac{4}{5}$

12. Find the number of terms of sequence 21, 18, 15, 12, ..., which must be taken to give a sum of zero. **Ans.** 15

13. The nth term of an A.P. is p and the sum of the first n terms is S. Prove that the first term is $\frac{2S - pn}{n}$.

14. In an arithmetic progression, the sum of p terms is m and the sum of q terms is also m. Find the sum of $(p + q)$ terms. **Ans.** 0

15. The ratio between the sum of n terms of two A.P. is $(7n + 1) : (4n + 27)$. Find the ratio of their 11th terms. **Ans.** 148 : 111

16. The first term of an A.P. is a, the second term is b and the last term is c. Show that the sum of A.P. is $\frac{(b + c - 2a)(c + a)}{2(b - a)}$.

17. If S_n denotes the sum of n terms of an A.P., show that $S_{30} = 3(S_{20} - S_{10})$.

18. The ratio of the sum of n terms of two A.P. is $(2n + 1) : (2n + 5)$; find the ratio of their 5th terms. **Ans.** 11 : 15

EXERCISE 2.6

1. Find the sum of:

(*i*) 2 + 4 + 6 + 8 + ... to 50 terms. **Ans.** 2550

(*ii*) 1 + 7 + 13 + 19 + ... to 40 terms. **Ans.** 4720

(*iii*) – 12 – 8 – 4 + 0 + 4 ... to 20 terms. **Ans.** 520

(*iv*) $2+\frac{7}{2}+5+\frac{13}{2}+...$ to 10 terms. **Ans.** $\frac{175}{2}$

(*v*) 5 + 13 + 21 + 29 + ... + 181. **Ans.** 213

2. Evaluate:

(*i*) 4 + 3 + 8 + 5 + 12 + 7 + ... to 32 terms. **Ans.** 832

(*ii*) 3 + 5 + 7 + 6 + 9 + 12 + 9 + 13 + 17 + ... to 30 terms. **Ans.** 690

3. Find the sum of all numbers, which are divisible by 7 and lying between 50 and 500. **Ans.** 17696

4. Find the sum of all integers between 92 and 786, which are multiples of 9. **Ans.** 33957

5. Find the sum of all natural numbers between 100 and 1000, which are multiples of 5. **Ans.** 98450

6. Find the sum of all odd integers between 300 and 498. **Ans.** 39501

7. Find the sum of all integers between 400 and 579, which are divisible by 10. **Ans.** 8330

8. Find the sum of all 3-digit numbers, which leave the remainder 1, when divided by 4. **Ans.** 123525

9. How many terms of the sequence – 12, – 9, – 6, – 3, ... must be taken to make the sum 54? **Ans.** 12

10. Find the sum of first n natural numbers. **Ans.** $\frac{n(n+1)}{2}$

11. Find the sum of first n even natural numbers. **Ans.** $n(n + 1)$

12. Find the sum of all even numbers between 200 to 500. **Ans.** 52150

13. Show that the sum of first n even natural numbers is equal to $\left(1+\frac{1}{n}\right)$ times the sum of first n odd natural numbers.

14. Find the sum of all four digit numbers, which when divided by 25 leaves 5 as a remainder. **Ans.** 1977300

15. In an A.P., if the first term is 22, the common difference is – 4 and the sum to n terms is 64, find n. Explain double answer. **Ans.** 4 or 8

16. How many terms of the sequence 18, 16, 14, ... should be taken, so that their sum is 78? Explain double answer. **Ans.** 6 or 13

17. How many terms of the A.P. $-6, -\frac{11}{2}, -5, ...$ are needed to give the sum – 25? Explain double answer. **Ans.** 5 or 20

18. Find the number of terms in each of the following:

(*i*) $2 + 4 + 6 + 8 + \dots = 10100$ **Ans.** 100

(*ii*) $-1.0 - 1.5 - 2.0 - 2.5 - \dots = -27$ **Ans.** 9

(*iii*) $-1 + \frac{1}{4} + \frac{3}{2} + \dots = 3969$ **Ans.** 81

(*iv*) $(x + y) + (x - y) + (x - 3y) + \dots = 22\,(x - 20y)$ **Ans.** 22

19. The fourth term of an A.P. is 11 and the eighth term exceeds twice the fourth term by 5. Find the A.P. and sum of first 20 terms. **Ans.** –1, 3, 7, 11, ...; 740

20. Determine, the sum of first 35 terms of an A.P., if second term is 2 and the seventh term is 22. **Ans.** 2310

21. The sum of first 7 terms of an A.P. is 10 and that of next 7 terms is 17. Find the progression. **Ans.** $1, 1\frac{1}{7}, 1\frac{2}{7}, 1\frac{3}{7}, \dots$

22. If the first term of an A.P. is 2 and the sum of first five terms is equal to one-fourth of the sum of the next five terms, find the sum of first 30 terms. **Ans.** – 2550

23. How many terms are there in the A.P. whose first and fifth terms are – 14 and 2 respectively and the sum of the terms is 40? **Ans.** 10

24. The sum of first n terms of an A.P., is zero, show that the sum of next m terms is $\frac{-am(m+n)}{n-1}$, a being the first term.

25. If first term of an A.P. be 100 and the sum of first six terms is five times the sum of next six terms. Find the sum of first eleven terms. **Ans.** 550

26. The sum of first 13 terms of an A.P. is 21 and the sum of first 21 terms is 13. Determine, the sum of first 34 terms. **Ans.** – 34

27. Find the sum of n terms of the A.P., whose nth term is given by $t_n = (5 - 6n)$, $n \in N$. **Ans.** $n(2 - 3n)$

28. The ratio of the sum of n terms of two A.P. is $(3n + 4)$: $(5n + 6)$. Find the ratio of their 7th terms. **Ans.** 43 : 71

29. The sum of the first n terms of two A.P. are in the ratio $(7n + 2) : (n + 4)$. Find the ratio of their 5th terms. **Ans.** 5:1

30. If the sum of the first n terms of two A.P. are in the ratio $(7n - 5) : (5n + 17)$, show that the 6th terms of the two progressions are equal.

31. The ratio between the sum of n terms of two arithmetic progressions is $(7n + 1) : (4n + 27)$. Find the ratio of their 11th terms. **Ans.** 148 : 111

32. The sum of n terms of an A.P. is $(2n + 3n^2)$. Determine, the A.P. and find its rth term. **Ans.** 5, 11, 17, 23, ...; $6r - 1$

33. If the mth term of an A.P. is a and the nth term is b, show that the sum of $(m + n)$ terms is $\left(\frac{m+n}{2}\right)\left[a + b + \left(\frac{a-b}{m-n}\right)\right]$

34. The sum of n terms of a progression is $3n^2 + 4n$. Is this progression an A.P.? If so, find the A.P. and the sum of its rth term. **Ans.** Yes, 7, 13, 19, 25, ...; $(3r^2 + 4r)$

35. If the roots of the equation $(b - c)x^2 + (c - a)x + (a - b) = 0$ are equal, then show that a, b, c are in A.P.

2.9 ARITHMETIC MEAN BETWEEN ANY TWO GIVEN NUMBERS

Let a, b be any two numbers. Let A, be the arithmetic mean between a and b.

Therefore, a, A, b are in A.P.

$\Rightarrow \quad A - a = b - A$ [common difference]

$\Rightarrow \quad 2A = a + b \Rightarrow A = \dfrac{a+b}{2}$

$\therefore$ Arithmetic mean between two given numbers is equal to half their sum.

$$\boxed{\textbf{A. M.} = \frac{a+b}{2}}$$

2.10 *n*th A.M. BETWEEN TWO GIVEN NUMBERS

Let a, b be any two numbers. Let $A_1, A_2, A_3, \ldots, A_n$ be the n A.M. between a and b. Then, $a, A_1, A_2, \ldots, A_n, b$ are in A.P.

Here, b is the $(n + 2)$th term *i.e.*, $T_{n+2} = a + (n + 2 - 1)d \quad \Rightarrow \quad b = a + (n + 1)d$

$\Rightarrow \quad d = \dfrac{b-a}{n+1}$

Thus, the n arithmetic means are given below

$$A_1 = a + d = a + \frac{b-a}{n+1}$$

$$A_2 = a + 2d = a + 2\left(\frac{b-a}{n+1}\right)$$

$$A_3 = a + 3d = a + 3\left(\frac{b-a}{n+1}\right)$$

$$\ldots\ldots \quad \ldots\ldots \quad \ldots\ldots \quad \ldots\ldots$$

$$\ldots\ldots \quad \ldots\ldots \quad \ldots\ldots \quad \ldots\ldots$$

$$A_n = a + nd = a + n\left(\frac{b-a}{n+1}\right)$$

Example 43. *Find the A.M. between 11 and 89.*

Solution. A.M. $= \dfrac{11+89}{2} = \dfrac{100}{2} = 50$ **Ans.**

Example 44. *If* $\dfrac{a^n + b^n}{a^{n-1} + b^{n-1}}$ *is the A.M. between a and b, then find the value of n.*

Solution. We know that A.M. between a and $b = \dfrac{a+b}{2}$

$\Rightarrow \quad \dfrac{a^n + b^n}{a^{n-1} + b^{n-1}} = \dfrac{a+b}{2}$ [Given] $\quad \Rightarrow \quad 2a^n + 2b^n = a^n + ab^{n-1} + a^{n-1}b + b^n$

$\Rightarrow \quad a^n - a^{n-1}b = ab^{n-1} - b^n \quad \Rightarrow \quad a^{n-1}(a-b) = b^{n-1}(a-b)$

$\Rightarrow \quad a^{n-1} = b^{n-1} \; [\because a \neq b] \quad \Rightarrow \quad \left(\frac{a}{b}\right)^{n-1} = 1 = \left(\frac{a}{b}\right)^0, \; \left(\because \left(\frac{a}{b}\right)^0 = 1\right)$

$\Rightarrow \quad n - 1 = 0 \quad \Rightarrow \quad n = 1$ **Ans.**

Example 45. *Insert 6 arithmetic means between 3 and 24.*

Solution. Let $A_1, A_2, A_3, A_4, A_5, A_6$ be six A.M. between 3 and 24.

$\therefore$ 3, $A_1, A_2, A_3, A_4, A_5, A_6$, 24 are in A.P.

We have $T_8 = 24 \quad \Rightarrow \quad 3 + (8 - 1)d = 24 \quad [\because T_n = a + (n - 1)d]$

$\Rightarrow \quad 3 + 7d = 24 \quad \Rightarrow \quad d = 3 \quad (a = 3)$

$\Rightarrow \quad A_1 = a + d = 3 + 3 = 6$

$A_2 = A_1 + d = 6 + 3 = 9$

$A_3 = A_2 + d = 9 + 3 = 12$

$A_4 = A_3 + d = 12 + 3 = 15$

$A_5 = A_4 + d = 15 + 3 = 18$

$A_6 = A_5 + d = 18 + 3 = 21$ **Ans.**

Example 46. *Between 1 and 31, m arithmetic means have been inserted in such a way that the ratio of the 7th and $(m - 1)^{th}$ means is 5 : 9. Find the value of m.*

Solution. Let $A_1, A_2, A_3, A_4, \ldots A_m$ be m A.M. between 1 and 31.

Therefore, 1 $A_1, A_2, \ldots A_m$, 31 are in A.P.

Let d be the common difference of A.P. Here, the total number of terms is $m + 2$ and $T_{m+2} = 31$.

$$\Rightarrow \quad 1 + (m + 2 - 1)d = 31 \quad \Rightarrow \quad (m + 1)d = 30 \quad \Rightarrow \quad d = \frac{30}{m+1}$$

$$A_7 = T_8 = a + 7d = 1 + 7 \times \frac{30}{m+1} = \frac{m+1+210}{m+1} = \frac{m+211}{m+1}$$

....

....

$$A_{m-1} = T_m = 1 + (m - 1)d$$

$$= 1 + (m-1) \times \frac{30}{m+1} = \frac{m+1+30m-30}{m+1} = \frac{31m-29}{m+1}$$

$$\frac{A_7}{A_{m-1}} = \frac{(m+211)/(m+1)}{(31m-29)/(m+1)} = \frac{m+211}{31m-29} = \frac{5}{9} \quad \text{[given]}$$

$$\Rightarrow \quad \frac{m+211}{31m-29} = \frac{5}{9} \quad \Rightarrow \quad 9m + 1899 = 155m - 145$$

$$\Rightarrow \quad 146m = 2044 \quad \Rightarrow \quad m = \frac{2044}{146} = 14$$

Hence, $m = 14$ **Ans.**

Example 47. *Show that the sum of n arithmetic mean between two numbers is n times the single arithmetic mean between them.*

Solution. Let A be the single A.M. between a and b, then $A = \dfrac{a+b}{2}$ ... (1)

Let $A_1, A_2, ..., A_n$ be the n A.M. between a and b. Then,

$a, A_1, A_2, ... A_n, b$ are in A.P.

Let d be the common difference of A.P. Here,

$$T_{n+2} = b$$

$\Rightarrow$ $[a + (n + 2 - 1)d] = b$ $\quad [\because T_n = a + (n - 1)d]$

$\Rightarrow$ $[a + (n + 1)d] = b \Rightarrow d = \dfrac{b-a}{n+1}$

$$A_1 = T_2 = [a + (2 - 1)d] = a + d = a + \frac{b-a}{n+1} \quad ... (2)$$

$$A_2 = T_3 = a + 2d = a + 2\left(\frac{b-a}{n+1}\right)$$

....

....

$$A_n = T_{n+1} = a + nd = a + n\left(\frac{b-a}{n+1}\right) \quad ... (3)$$

Sum of n A.M. $= A_1 + A_2 + A_3 + ... + A_n$

$= \dfrac{n}{2}(A_1 + A_n)$ $\quad \left[\because S_n = \dfrac{n}{2}(a+l)\right]$

$= \dfrac{n}{2}\left[a + \dfrac{b-a}{n+1} + a + n\left(\dfrac{b-a}{n+1}\right)\right]$ $\quad$ [from (2) and (3)]

$= \dfrac{n}{2}\left[2a + (n+1)\left(\dfrac{b-a}{n+1}\right)\right]$

$= \dfrac{n}{2}[2a + b - a] = \dfrac{n}{2}[a + b] = n\left(\dfrac{a+b}{2}\right)$

$\Rightarrow$ Sum of n A.M. $= nA$ $\quad$ [from (1)]

Hence, the sum of n A.M. between a and b is equal to n times the A.M. between a and b. **Proved.**

Example 48. *The sum of two numbers is* $\dfrac{13}{6}$. *An even number of A.M. are being inserted between them. The sum of mean inserted exceeds the number of mean by 1. Find the number of A.M. inserted.*

Solution. Let the numbers be a and b, whose sum is $\dfrac{13}{6}$.

$\Rightarrow$ $a + b = \dfrac{13}{6}$ $\quad ... (1)$

Let $A_1, A_2, ..., A_{2n}$ be $2n$ (even number) A.M. between a and b.

$\Rightarrow$ $A_1 + A_2 + ... + A_{2n} = 2n$ (A.M. between a and b)

$= 2n\left(\dfrac{a+b}{2}\right) = n(a+b) = n\left(\dfrac{13}{6}\right)$ $\quad$ ... (2) [using (1)]

Also, $A_1 + A_2 + ... + A_{2n} = 2n + 1$ $\quad$ ... (3) (given)

From (2) and (3), we get

$$\frac{13}{6}n = 2n + 1 \quad \Rightarrow \quad 13n = 12n + 6 \quad \Rightarrow \quad n = 6$$

$\therefore$ Number of A.Ms. inserted = $2n = 2(6) = 12$ Ans.

Example 49. *If the A.M. between pth and qth terms of an A.P. be equal to the A.M. between rth and sth terms of the A.P., then show that $p + q = r + s$.*

Solution. Let a be the first term and d be the common difference of the given A.P. Then,

$$T_p = p\text{th term} = a + (p-1)d; \qquad T_q = q\text{th term} = a + (q-1)d;$$
$$T_r = r\text{th term} = a + (r-1)d; \qquad T_s = s\text{th term} = a + (s-1)d;$$

It is given that A.M. between T_p and T_q = A.M. between T_r and T_s.

$$\Rightarrow \quad \frac{1}{2}[T_p + T_q] = \frac{1}{2}[T_r + T_s] \quad \Rightarrow \quad T_p + T_q = T_r + T_s$$

$$\Rightarrow \quad \{a + (p-1)d\} + (a + (q-1)d\} = \{a + (r-1)d\} + \{a + (s-1)d\}$$

$$\Rightarrow \quad (p + q - 2)d = (r + s - 2)d \quad \Rightarrow \quad p + q = r + s.$$

Proved.

EXERCISE 2.7

1. Find the A.M. between
   (*i*) 6 and 12 **Ans.** 9
   (*ii*) 5 and 22 **Ans.** 13.5
   (*iii*) $(\cos\theta + \sin\theta)^2$ and $(\cos\theta - \sin\theta)^2$ **Ans.** 1
   (*iv*) $(x + y)^2$ and $(x - y)^2$ **Ans.** $x^2 + y^2$
2. Insert 5 arithmetic means between 4 and 22. **Ans.** 7, 10, ... 19
3. Insert 8 arithmetic means between 17 and 53. **Ans.** 21, 25, ..., 49
4. Find two numbers whose product is 91 and whose A.M. is 10. **Ans.** 7, 13
5. If p arithmetic means are inserted between a and b, prove that $d = \dfrac{b-a}{p+1}$.
6. If n arithmetic means are inserted between 20 and 80 such that the ratio of first mean to the last mean is 1 : 3, then find the value of n. **Ans.** 11
7. Insert 8 A.M. between 2 and 29. Also, verify that the sum of these 8 A.M. is equal to 8 times the A.M. between 2 and 29. **Ans.** 5, 8, 11, 14, 17, 20, 23, 26
8. Find three numbers in A.P. such that their sum is 27 and their product is 648. **[S.B.T.E. Dec. 2016]**
   Ans. 6, 9, 12 or 12, 9, 6
9. How many terms are there in the series 7, 10, 13, ..., 43. **[S.B.T.E. Dec. 2016]**
   Ans. $n = 13$
10. Find the sum of first 10 terms for the sequence 4, 7, 10, 13, ... **[S.B.T.E. Dec. 2016]**
    Ans. 190
11. How many two digit numbers are divisible by 7? **[S.B.T.E. Dec. 2016]**
    Ans. $n = 13$
12. Solve for x, $1 + 6 + 11 + 16 + ... + x = 148$ **[S.B.T.E. Dec. 2016]**
    Ans. $x = 36$

13. Find the sum of first 1000 odd numbers. **[S.B.T.E Dec. 2016]**

Ans. $S_{1000} = 10{,}00000$

14. n A.M. are inserted between 5 and 86, such that the ratio of the first and the last mean is 2:11. Find n. **Ans.** 8

15. There are n A.M. between 3 and 17. The ratio of the last mean to the first mean is 3 : 1. Find the value of n. **Ans. 6**

2.11 APPLICATIONS

Now, we will solve problems based on the applications of A.P.

Example 50. *On the first day strike of physicians in a hospital, the attendance of the O.P.D. was 1500 patients. As the strike continued the attendance declined by 100 patients every day. Find from which day of the strike the O.P.D. would have no patient.*

Solution. We have the attendance of patients on first day (a) = 1500

Attendance declined (d) = – 100 per day

Let there be no patient on the nth day; *i.e.* $T_n = 0$

We know that $T_n = a + (n - 1)d \Rightarrow 0 = 1500 + (n - 1)(-100)$

$\Rightarrow 100n - 100 = 1500 \Rightarrow 100n = 1600$

$\Rightarrow n = 16$

Hence, on the 16th day of strike, the O.P.D. will have no patient. **Ans.**

Example 51. *The ages of the students of a class form an A.P. whose common difference is 4 months. If the youngest student is 8 years old and the sum of the ages of all students of the class is 168 years, find the number of student in the class.*

Solution. The age of the youngest student (a) = 8 years.

Common difference of their ages (d) = 4 months = $\frac{1}{3}$ years.

Sum of ages of all students (S_n) = 168 years.

Let n be the number of students in the class = $\frac{1}{3}$ years.

We know that $S_n = \frac{n}{2}[2a + (n-1)d] \Rightarrow 168 = \frac{n}{2}\left[2 \times 8 + (n-1) \cdot \frac{1}{3}\right]$

$\Rightarrow 2 \times 168 = n\left[16 + \frac{(n-1)}{3}\right] \Rightarrow 336 = 16n + \frac{n(n-1)}{3}$

$\Rightarrow 1008 = 48n + n^2 - n \Rightarrow n^2 + 47n - 1008 = 0$

$\Rightarrow n^2 + 63n - 16n - 1008 = 0 \Rightarrow n(n + 63) - 16(n + 63) = 0$

$\Rightarrow (n + 63)(n - 16) = 0$

Either $n + 63 = 0$ or $n - 16 = 0 \Rightarrow n = -63$ or $n = 16$ [neglect – 63]

Hence, the number of students are 16. **Ans.**

Example 52. *The gate receipts at the show of "Baghbaan" amounted to Rs. 9500 on the first night and showed a drop of Rs. 250 every succeeding night. If the operational expenses of the show are Rs. 2,000 a day, find on which night the show ceases to be profitable?*

Solution. We have cost of gate receipt on the first night $(a) = 9500$

$\therefore$ common difference $(d) = -250$

The show will cease to be profitable on the night when the receipts are just Rs. 2000. Let it happen on the nth night, then

$$2000 = 9500 + (n-1)(-250) \Rightarrow 2000 - 9500 = -250n + 250$$

$$\Rightarrow -7500 - 250 = -250n \Rightarrow -7750 = -250n$$

$$\Rightarrow n = \frac{7750}{250} = 31$$

Hence, on the 31st night the show will cease to be profitable. **Ans.**

Example 53. *Two cars start together in the same direction from the same place. The first goes with uniform speed of 10 km/h. The second goes at a speed of 8 km/h in the first hour and increases the speed by 1/2 km/h each succeeding hour. After how many hours will the second car overtake the first car, if both cars go non-stop.*

Solution. Let the second car overtake the first car after n hours. Then, the two cars travel the same distance in n hours.

Distance travelled by the first car in n hours $= 10n$ km

Distance travelled by the second car in n hours = sum of n terms of an A.P. with first term 8 and common difference $\frac{1}{2}$.

$$= \frac{n}{2}\left[2 \times 8 + (n-1) \times \frac{1}{2}\right] = \frac{n(n+31)}{4}$$

When the second car overtakes the first car, we have

$$10n = \frac{n(n+31)}{4} \Rightarrow 40n = n^2 + 31n \Rightarrow n(n-9) = 0 \Rightarrow n = 9 \quad [\because n \neq 0]$$

Thus, the second car will overtake the first car in 9 hours. **Ans.**

Example 54. *A person buys National Savings Certificates every year of value exceeding the last years purchase by Rs. 100. After 5 years he find that the total value of the certificate is Rs. 5250. Find the value of the certificates purchased by him:*

(i) in the first year *(ii) in the ninth year*

Solution. Let a be the value of the certificates purchased by him in the first year. Common difference is Rs. 100. Then,

$$S_5 = 5250$$

We know that $S_5 = \frac{5}{2}[2a + (5-1)d] \Rightarrow 5250 = \frac{5}{2}[2a + 4 \times 100]$

$$\Rightarrow \frac{5250 \times 2}{5} = 2a + 400 \Rightarrow 2100 = 2a + 400$$

$$\Rightarrow 2a = 2100 - 400 \Rightarrow a = \frac{1700}{2} = 850$$

Hence, the value of certificate in the first year = Rs. 850

$$\therefore T_9 = a + (9-1)d$$

$$\Rightarrow T_9 = 850 + 8 \times 100 = 850 + 800 = 1650$$

which is the value of certificates purchased in the ninth years. **Ans.**

EXERCISE 2.8

1. On the first day of strike of physicians in a hospital, the attendance of the O.P.D. was 750 patients. As the strike continued the attendance declined by 50 patients every day. Find from which day of the strike the O.P.D. would have no patient.
Ans. 16th day

2. A manufacturer of radio-sets produced 600 units in the third year and 700 units in the seventh year. Assuming that the production uniformly increases by a fixed number every year, find
   (*i*) the production in the first year. **Ans.** 550 units
   (*ii*) the total production in 7 years. **Ans.** 4375 units
   (*iii*) the production in the 10th years. **Ans.** 775 units

3. A man saved ₹16500 in ten years. In each year after the first he saved ₹100 more than he did in the preceding year. How much did he saved in the first year?
Ans. ₹1200

4. The sides of a right-angled triangle are in A.P., show that the sides are in the ratio 3 : 4 : 5.

5. A man deposited ₹10000 in a bank at the rate of 5% simple interest annually. Find the amount in 15th year (after 14 years since he deposited the amount) and also calculate the total amount after 20 years. **Ans.** ₹17000; ₹20000

6. A man buys a car for ₹300000, he pays ₹150000 in cash and agrees to pay the balance in yearly instalments of ₹10000 plus 10% interest on the unpaid amount. Find the total amount he paid for the car. **Ans.** ₹420000

7. Rasika buys a D.D.A. flat for ₹1000000. She pays ₹500000 in cash and agrees to pay the balance in yearly instalments of ₹50000 plus 10% interest on the unpaid amount, find the total amount she paid for the flat. **Ans.** ₹1275000

8. The gate receipt at the show of "rain-coat" amounted to ₹6500 on the first night and showed a drop of ₹110 every succeeding night. If the operational expenses of the show are ₹1000 a day, find on which night the show ceases to be profitable.
[Hint : T_n = 1000] **Ans.** 51

9. A tree in each year grows 4 cm less than it did in the previous year. If it grows 1 metre in the first year, in how many years will it have ceased growing?
Ans. At the end of 25 years (26th year), there will be no growth.

10. A man saves ₹32 during the first year, ₹36 in the second year, ₹40 in the third year and so on. In how many years will he save ₹2000. **Ans.** 25 years.

11. 150 workers were engaged to finish a job in a certain number of days. 4 workers dropped out on second day, 4 more workers dropped out on third day and so on. It takes 8 more days to finish the work. Find the number of days in which the work was completed.

EXERCISE 2.9

1. Ruchi saves ₹120 during first month, ₹50 in the next month, ₹180 in the third month. If she continues her savings in this pattern, in how many months, will she save ₹1800? **Ans.** 8 months

2. Shrikant saves ₹32000 during first year, ₹36000 in the next year and ₹40000 in the 3rd year. If he continues his savings in this pattern, in how many years will he save ₹200000? **Ans.** 5 years

3. A manufacturer of T.V. sets, produced 250 set units in the fourth month and 400 units in the seventh month. Assuming that the production uniformly increases by a fixed number every month, find

   (*i*) the production in the 6th month. **Ans.** 350

   (*ii*) total production in 10 months. **Ans.** 3250

   (*iii*) production in 10th month. **Ans.** 550

   (*iv*) production in 8 month. **Ans.** 450

4. The interior angles of a polygon are in A.P. The smallest angle is 120° and the common difference is 5°. Find the number of sides of the polygon.

   [**Hint:** Sum of angles of a polygon of n sides = $(2n - 4)$ rt. angles]

   Ans. 9

5. Find the sum of all 3-digit numbers which leave the remainder 2, when divided by 3. **Ans.** 164850

6. There are 20 trees at equal distance of 5 meters in a line with a well, the distance of the well from the nearest tree being 10 metres. A gardener waters all the trees separately starting from the well and he returns to the well after watering each tree to get water from the next. After watering the last tree he does not go to the well. Find the total distance the gardener will cover in order to water all the trees. **Ans.** 2195 m

7. Two cars start together in the same direction from the same place. The first goes with uniform speed of 10 km/hour. The second goes at a speed of 8 km/hour in the first hour and increases the speed by 1/2 km each succeeding hours. After how many hours will the second car overtake the first, if both cars go nonstop? **Ans.** 9 hours

8. Kamal buys a washing machine for ₹10000. He pays ₹2000 in cash and agrees to pay the balance in instalments of ₹500 plus 10% interest on the unpaid amount. Find the total amount paid for the washing machine. **Ans.** ₹16800

9. A car costs ₹600000. If it depreciates in value, 15% in the first year, 13.5% in the next year and 12% in the third year, and so on. What will be its value at the end of 10 years, all percentage applying to the original cost? **Ans.** ₹105000

10. Surendra purchases a scooter for ₹25000. He pays ₹5000 cash and agrees to pay the balance in annual instalments of ₹2000 plus 10% interest on the unpaid amount. Find the total payment made for the scooter. **Ans.** ₹36000

11. Ritu takes a contract to construct a dispensary upto 31st December 2005. Beyond 31 Dec. 2005 a penalty for delay of construction as follows: ₹100 for first day, ₹150 for second day, ₹200 for the third day etc. How much does a delay of 20 days cost Ritu? **Ans.** ₹11500

12. Krishna Mohan buys a shop for ₹1500000. He pays ₹1000000 in case 4 and agrees to pay balance in 10 annual intalments of ₹50000 each. If the rate of interest is 10% and he pays the instalment with the interest due on the unpaid amount. Find the total cost of the shop. **Ans.** ₹1775000

REVISION EXERCISE

SECTION A

Multiple Choice Questions (MCQs)

1. What is the sum of all natural numbers from 1 to 100?

(*a*) 5050 (*b*) 55 (*c*) 4550 (*d*) 5150 **Ans.** (*a*)

2. The 4th term from the end of A.P. – 11, – 8, – 5, ... 49 is

(*a*) 37 (*b*) 40 (*c*) 43 (*d*) 58 **Ans.** (*b*)

3. Which term of the A.P. 24, 21, 18, ... is the first negative term?

(*a*) 8th (*b*) 9th (*c*) 10th (*d*) 12th **Ans.** (*c*)

4. Which term of an A.P.: 21, 42, 43, ... is 210?

(*a*) 8th (*b*) 9th (*c*) 10th (*d*) 12th **Ans.** (*b*)

5. What is the common difference of the A.P. in which $a_{18} - a_{14} = 32$?

(*a*) 8 (*b*) – 8 (*c*) 4 (*d*) – 4 **Ans.** (*a*)

6. The sequence of numbers – 10, – 6, – 2, 2, ... is

(*a*) an A.P. with $d = -16$ (*b*) an A.P. with $d = 4$

(*c*) an A.P. with $d = -4$ (*d*) not an A.P. **Ans.** (*b*)

7. For what value of p are $2p + 1$, 13, $5p - 3$, three consecutive terms of an A.P.?

(*a*) 2 (*b*) – 2 (*c*) 4 (*d*) – 4 **Ans.** (*c*)

8. Which term of the A.P. 113, 108, 103, ... is the first negative term?

(*a*) 22nd term (*b*) 24th term (*c*) 26th term (*d*) 28th term **Ans.** (*b*)

9. 15th term of the A.P. $x - 7, x - 2, x + 3$, ... is:

(*a*) $x + 63$ (*b*) $x + 73$ (*c*) $x + 83$ (*d*) $x + 53$ **Ans.** (*a*)

10. If $p - 1, p + 3, 3p - 1$ are in A.P. then p is equal to:

(*a*) 4 (*b*) – 4 (*c*) 2 (*d*) – 2 **Ans.** (*a*)

11. The next term of the A.P. $\sqrt{27}, \sqrt{48}, \sqrt{75}, \ldots$ is:

(*a*) $\sqrt{105}$ (*b*) $\sqrt{107}$ (*c*) $\sqrt{108}$ (*d*) $\sqrt{147}$ **Ans.** (*c*)

12. Which term of the A.P. 100, 90, 80, ... is zero?

(*a*) 5th (*b*) 6th (*c*) 10th (*d*) 11th **Ans.** (*d*)

13. Which of the following is not an A.P.?

(*a*) 13, 8, 3, – 2, – 7, – 12 (*b*) 10.8, 11.2, 11.6, 12, 12.4

(*c*) $8\frac{1}{7}, 18\frac{2}{7}, 28\frac{3}{7}, 48\frac{4}{7}, 58\frac{5}{7}$ (*d*) $8\frac{3}{23}, 11\frac{6}{23}, 14\frac{9}{23}, 17\frac{12}{23}$ **Ans.** (*c*)

14. Which term of the A.P. 92, 88, 84, 80, ... is 0?

(*a*) 23 (*b*) 32 (*c*) 22 (*d*) 24 **Ans.** (*d*)

15. Which term of the A.P. 1, 4, 7 ... is 88?

(*a*) 26 (*b*) 27 (*c*) 30 (*d*) 35 **Ans.** (*c*)

SECTION B

Very Short Answer Questions (VSAQs)

1. Find the 20th term from the last term (end) of the A.P. : 3, 8 , 13 ... 253. **Ans.** 158
2. If 6th term of an A.P. is – 10 and its 10th term is – 26, then find the 15th term of the A.P. **Ans.** $a_{15} = -46$
3. Find the value of p, if the numbers x, $2x + p$, $3x + 6$ are three consecutive terms of an A.P. **Ans.** $p = 3$
4. Find the 10th term from the end of the A.P. 8, 10, 12, ..., 126 **Ans.** Term = 108
5. If 8th term of an A.P. is 31st and 15th term is 16 more than 11th term, find the A.P. **Ans.** A.P. = 3, 7, 11, 15 ...
6. For what value of p are $2p$, $p + 10$ and $3p + 2$ in A.P.? **Ans.** $p = 6$
7. Find the number of terms of the series: – 7 + (– 8) + (– 11) + ... + (– 230) **Ans.** 76
8. Find the sum of the first 25 terms of an AP whose nth term is given by $t_n = 7 - 3n$ **Ans.** – 800
9. If S_n denotes the sum of n terms of an A.P. whose common difference is d and first term is a, find $S_n - 2, S_{n-1} + S_{n-2}$. **Ans.** 0
10. Find the 10th term from the end of the A.P. 4, 9, 14, ... 254. **Ans.** 209
11. Find the 12th term of the A.P. $\sqrt{2}, 3\sqrt{2}, 5\sqrt{2}, ...$ **Ans.** $23\sqrt{2}$
12. Find the sum of the first 50 odd natural numbers. **Ans.** 2500
13. 8th term of an A.P. is 37 and its 12th term is 57. Find the A.P. **Ans.** 2, 7, 12
14. Find the sum of first twelve multiples of 7. **Ans.** 546
15. Find the common difference of an A.P. whose first term is $\frac{1}{2}$ and the 8th term is $\frac{17}{6}$. Also write its 4th term. **Ans.** $\frac{1}{3}$ and $\frac{3}{2}$

SECTION C

Short Answer Questions (SAQs)

1. The sum of the 5th and 7th terms of an A.P. is 52 and its 10th term is 46. Find the A.P. **Ans.** $a = 1$, $d = 5$
2. The sum of first-three terms of an A.P. is 33. If the product of the first and third terms exceeds the second term by 29, find the A.P. **Ans.** A.P. = 20, 11, 2, ... or 2, 11, 20, ...
3. The sum of the first n terms of an A.P. is $5n^2 - 3n$. Find the A.P. and hence find its 12th term. **Ans.** 2, 12, 22, 32, ...; $a_{12} = 112$
4. The angles of a triangle are in A.P. The greatest angle is twice the least. Find all angles of the triangle. **Ans.** 40°, 60°, 80°
5. Find the sum of all natural numbers between 200 and 1000 exactly divisible by 6. **Ans.** $S = 79800$
6. Which term of the A.P. : 3, 15, 27, 39, ... will be 120 more than its 21st term? **Ans.** 2, 12, 22, 32 ...; $a_{31} = a_{21} + 120$
7. Find the sum of all two-digits odd positive numbers. **Ans.** $S_n = 2475$

8. Which term is the first negative term in the given A.P. : 23, $21\frac{1}{2}$, 20, ...? **Ans.** $n = 17$

9. If the sum of first n terms of an A.P. is $4n^2 - n$, find the 12th term. **Ans.** $a_{12} = 91$

10. If the sum of first fourteen terms of an AP is 1050 and its first term is 10, find its 20th term. **Ans.** 200

11. How many terms of the A.P. 78, 71, 64, ... are needed to give the sum 465? Also find the last term of this A.P. **Ans.** Number of Terms = 10, 15

12. The sum of the first n terms of an A.P. is given by $S_n = 3n^2 - n$. Determine the A.P. and its 25th term. **Ans.** 2, 8, 14 ...; 146

13. Find three numbers in A.P. whose sum is 15 and whose product is 105. **Ans.** $d = \pm 2$

14. Sum of the first n terms of an A.P. is $5n^2 - 3n$. Find the A.P. and also find its 16th term. **Ans.** 2, 12, 22, 32, ..., 152

15. Find the sum of all three digit numbers which leave the same remainder 2 when divided by 5. **Ans.** 98910

16. Which term of the A.P. 3, 15, 27, 39, ... will be 132 more than its 60th term? **Ans.** 71

17. Find the sum of the first 31 terms of an A.P. whose n^{th} term is given by $3 + \frac{2}{3}n$.

Ans. $\frac{1271}{3}$

18. The sum of the third and seventh term of an A.P. is 6 and their product is 8. Find the sum of the first sixteen terms of the A.P. **Ans.** 76

19. Determine 'a' so that $2a + 1$, $a^2 + a + 1$ and $3a^2 - 3a + 3$ are consecutive terms of an A.P. **Ans.** 1, 2

20. How many terms of the A.P. 9, 17, 25, ..., must be taken to get a sum of 450? **Ans.** 10

21. Find three numbers in A.P. whose sum is 15 and the product is 80. **Ans.** 8, 5, 2

SECTION D

Long Answer Questions (LAQs)

1. Jaipal Singh repays the total loan of ₹118000 by paying every month starting with the first instalment of ₹1000. If he increases the instalment by ₹100 every month. What amount will be paid by him in the 30th instalment? What amount of loan does he still have to pay after 30th instalment? **Ans.** ₹73500, ₹44500

2. Find the sum of all natural numbers less than 100 and divisible by 6. **Ans.** 816

3. If m times the mth term of an A.P. is same as n times the nth term, find its $(m + n)$th term. **Ans.** Zero

4. A manufacturer of T.V. sets produced 600 sets in the third year and 700 sets in the seventh years. Assuming that the production increase uniformly by a fixed number every year, find

(*a*) The production in the first year (b) The production in the 10th year

(*c*) The total production in first 7 years **Ans.** (*a*) 550 (*b*) 775 (*c*) 4375

5. Show that the sum of all odd integers between 1 and 1000 which are divisible by 3 is 166833.

6. The houses of a row are numbered consecutively from 1 to 49. There is a value of x such that the sum of the numbers of the houses preceding the house numbered x is equal to the sum of the numbers of the houses following it. Find this value of x.
Ans. 35
7. A sum of ₹700 is to be used for giving 7 cash prizes to students of a school for their academic performance. If each prize is ₹20 less than its preceding prize, find the value of each of the prizes. **Ans.** 160, 140, 120, 100, 80, 60, 40
8. In an A.P., prove that $a_{m+n} + a_{m-n} = 2a_m$, where n denotes nth term of the A.P.
9. For what value of n, the nth terms of the A.P. 63, 65, 67, ... and 3, 10, 17, ... are equal? **Ans.** 13
10. In November 2009, the number of visitors to a zoo increased daily by 20. If a total of 12300 people visited the zoo in that month, find the number of visitors on 1st November 2009. **Ans.** 120
11. The 4th term of an A.P. is equal to 3 times the first term and the 7th term exceeds twice the 3rd term by 1. Find the A.P. **Ans.** 3, 5, 7
12. Find the sum of all multiple of 9 lying between 300 and 700. **Ans.** 21978

Geometric Progression and Their Applications

3.1 GEOMETRIC PROGRESSION (G.P.)

Let us consider the following sequences:

(*i*) 1, 3, 9, 27, ... (*ii*) $\frac{1}{2}, \frac{1}{4}, \frac{1}{8}, \frac{1}{16}, ...$ (*iii*) $-\frac{1}{3}, \frac{1}{9}, -\frac{1}{27}, \frac{1}{81}, ...$

(*iv*) 0.2, 0.08, 0.032, 0.0128, ... (*v*) $a, ar, ar^2, ar^3, ...$

Here, we observe that

(*i*) $a_1 = 1, \quad \frac{a_2}{a_1} = \frac{3}{1} = 3, \quad \frac{a_3}{a_2} = \frac{9}{3} = 3, \quad \frac{a_4}{a_3} = \frac{27}{9} = 3$ and so on.

(*ii*) $a_1 = \frac{1}{2}, \quad \frac{a_2}{a_1} = \frac{1/4}{1/2} = \frac{1}{2}, \quad \frac{a_3}{a_2} = \frac{1/8}{1/4} = \frac{1}{2}, \quad \frac{a_4}{a_3} = \frac{1/16}{1/8} = \frac{1}{2}$ and so on

(*iii*) $a_1 = -\frac{1}{3}, \frac{a_2}{a_1} = \frac{1/9}{-1/3} = -\frac{1}{3}, \frac{a_3}{a_2} = \frac{-1/27}{1/9} = -\frac{1}{3}, \frac{a_4}{a_3} = \frac{1/81}{-1/27} = -\frac{1}{3}$ and so on

(*iv*) $a_1 = 0.2, \frac{a_2}{a_1} = \frac{0.08}{0.2} = 0.4, \frac{a_3}{a_2} = \frac{0.032}{0.08} = 0.4, \frac{a_4}{a_3} = \frac{0.0128}{0.032} = 0.4$ and so on

(*v*) $a_1 = a, \quad \frac{a_2}{a_1} = \frac{ar}{a} = r, \quad \frac{a_3}{a_2} = \frac{ar^2}{ar} = r, \quad \frac{a_4}{a_3} = \frac{ar^3}{ar^2} = r$ and so on

Thus, we find that every term except the first term bears a constant ratio to the term immediately preceding it. In (*i*) the constant ratio is 3, (*ii*) it is $\frac{1}{2}$, (*iii*) it is $\frac{1}{3}$, (*iv*) the ratio is 0.4 and in (*v*) the constant ratio is r.

Such sequences are called geometric sequences or geometric progressions.

DEFINITION

A sequence of nonzero numbers is said to be a geometric progression, if the ratio of each term, except the first one, by its preceding term is always the same.

This ratio is called the common ratio of a G.P. and is generally denoted by r, the first term by a and n^{th} term is denoted by T_n.

Standard geometric progression is $a, ar, ar^2, ar^3, ..., ar^{n-1}$

3.2 *n*th TERM OF A G.P. (T_n)

If a is the first term and common ratio is r, then $T_n = ar^{n-1}$

Proof

1st term = a

2nd term = $T_2 = ar = ar^{2-1}$

3rd term = $T_3 = ar^2 = ar^{3-1}$

4th term = $T_4 = ar^3 = ar^{4-1}$

5th term = $T_5 = ar^4 = ar^{5-1}$

... ...

n^{th} term = $T_n = ar^{n-1}$

$$\boxed{T_n = ar^{n-1}}$$

Example 1. *Find the 10th term of the geometric series 5 + 25 + 125 + ... Also, find its nth term.*

Solution. We have, 5 + 25 + 125 + ... is a geometric series.

Here, $a = 5$ and $r = \frac{25}{5} = \frac{125}{25} = 5$

We know that $T_n = ar^{n-1} = 5(5)^{n-1} = 5^n$

and $T_{10} = ar^{10-1} = 5(5)^{10-1} = 5^{10}$ **Ans.**

Example 2. *Find 12th term of a G.P. whose 8th term is 192 and the common ratio is 2.*

Solution. We have, $T_8 = 192,\ r = 2$ ... (1)

We also know that $T_8 = ar^7 \Rightarrow 192 = a(2)^7$ [using (1)]

$$\Rightarrow \frac{192}{2^7} = a \Rightarrow \frac{192}{128} = a \Rightarrow a = \frac{3}{2}$$

Now, $T_{12} = ar^{12-1} = ar^{11} = \left(\frac{3}{2}\right)(2)^{11} = 3 \times 2^{10} = 3072$ **Ans.**

Example 3. *The 5th, 8th and 11th terms of a G.P. are p, q and s respectively. Show that $q^2 = ps$.*

Solution. We have, $T_5 = p,\ T_8 = q,\ T_{11} = s$ ... (1)

Now, $T_5 = ar^{5-1} = ar^4 \Rightarrow ar^4 = p$ ... (2) [using (1)]

$T_8 = ar^{8-1} = ar^7 \Rightarrow ar^7 = q$ ... (3) [using (1)]

And $T_{11} = ar^{11-1} = ar^{10} \Rightarrow ar^{10} = s$ ... (4) [using (1)]

On squaring (3), we get

$q^2 = a^2r^{14} = a \cdot a \cdot r^4 \cdot r^{10} = (ar^4)\,(ar^{10})$ [using (2) and (4)]

$\therefore\ q^2 = ps$ **Proved.**

Example 4. *Which term of the geometric sequences:*

(*i*) 2, 8, 32, ... is 131072 ? (*ii*) $\sqrt{3}, 3, 3\sqrt{3}, \ldots$ is 729 ?

(*iii*) $\frac{1}{3}, \frac{1}{9}, \frac{1}{27}, \ldots$ is $\frac{1}{19683}$?

Solution. (*i*) We have, 2, 8, 32, ... is a geometric sequence.

Here, $\frac{T_2}{T_1} = \frac{8}{2} = 4$, *i.e.* $r = 4$ and $a = 2$

Let the nth term of the given sequence be 131072. We know that $T_n = ar^{n-1}$

$\Rightarrow \quad 131072 = (2)\,(4)^{n-1}$

$\Rightarrow \quad 65536 = (4)^{n-1} \Rightarrow (4)^8 = (4)^{n-1} \Rightarrow n - 1 = 8 \Rightarrow n = 9$

Hence, 9th term of the given G.P. is 131072. **Ans.**

(*ii*) We have, $\sqrt{3}, 3, 3\sqrt{3}, \ldots$ is a G.P.

Here, $$r = \frac{3}{\sqrt{3}} = \frac{3\sqrt{3}}{3} = \sqrt{3}, a = \sqrt{3}$$

Let the nth term of the given sequence be 729. Then, $T_n = 729$.

Also, $$T_n = ar^{n-1} = (\sqrt{3})(\sqrt{3})^{n-1+1} = (\sqrt{3})^n$$

$$\Rightarrow \quad 729 = (\sqrt{3})^n \Rightarrow (\sqrt{3})^{12} = (\sqrt{3})^n \Rightarrow n = 12$$

Hence, 12th term of the given G.P. is 729. **Ans.**

(*iii*) We have, $\frac{1}{3}, \frac{1}{9}, \frac{1}{27}, \ldots$ is a G.P. Here, $a = \frac{1}{3}$, $r = \dfrac{\frac{1}{9}}{\frac{1}{3}} = \frac{1}{3}$

Let the nth term of the given sequence be $\frac{1}{19683}$. Then, $T_n = \frac{1}{19683}$.

Also, $T_n = ar^{n-1} \Rightarrow \frac{1}{19683} = \left(\frac{1}{3}\right)\left(\frac{1}{3}\right)^{n-1} = \left(\frac{1}{3}\right)^n \Rightarrow \left(\frac{1}{3}\right)^9 = \left(\frac{1}{3}\right)^n \Rightarrow n = 9$

Hence, 9th term of given G.P. is $\frac{1}{19683}$. **Ans.**

Example 5. *If the 4th and 9th terms of a G.P. be 54 and 13122 respectively. Find the G.P.*

Solution. Let a be the first term and r be the common ratio of the given G.P. Then,

$$T_4 = 4 \quad \text{and} \quad T_9 = 13122$$

$$\Rightarrow \quad ar^3 = 54 \quad \text{and} \quad ar^8 = 13122$$

$$\Rightarrow \quad \frac{ar^8}{ar^3} = \frac{13122}{54} \Rightarrow r^5 = 243 \Rightarrow r^5 = 3^5 \Rightarrow r = 3$$

Putting $r = 3$ in $ar^3 = 54$, we get

$$a(3)^3 = 54 \Rightarrow a = \frac{54}{27} = 2$$

Thus the given G.P. is 2, 6, 18, 54, ... **Ans.**

Example 6. *If the first and nth term of a G.P. are x and y respectively, and if p is the product of the first n terms, prove that $p^2 = (xy)^n$.*

Solution. Let r be the common ratio of the given G.P. Then,

$$y = n\text{th term} = xr^{n-1} \qquad [\because x \text{ is the first term}]$$

$$\Rightarrow \quad r^{n-1} = \frac{y}{x} \Rightarrow r = \left(\frac{y}{x}\right)^{\frac{1}{n-1}}$$

Now, p = product of first n terms

$= x\,(xr)\,(xr^2)\,(xr^3)\,\ldots\,xr^{n-1}$

$= x^n r^{1+2+3+\ldots+(n-1)}$

$= x^n r^{\frac{n(n-1)}{2}}$ $\left[\because 1 + 2 + 3 + \ldots + (n-1) = \frac{n-1}{2}[1+(n-1)] = \frac{n(n-1)}{2}\right]$

$$= x^n \left\{\left(\frac{y}{x}\right)^{\frac{1}{n-1}}\right\}^{\frac{n(n-1)}{2}} = x^n\left(\frac{y}{x}\right)^{\frac{n}{2}} = x^{\frac{n}{2}} \cdot y^{\frac{n}{2}} = (xy)^{\frac{n}{2}}$$

$\Rightarrow \quad p^3 = \left[(xy)^{\frac{n}{2}}\right]^2 \quad \Rightarrow \quad p^2 = (xy)^n.$ **Proved.**

Example 7. *The first term of a G.P. is 1. The sum of the third and fifth terms is 90. Find the common ratio of the G.P.*

Solution. Let r be the common ratio of the G.P. Then,

$T_3 = ar^2 \Rightarrow T_3 = (1)\,r^2 = r^2$ $[\because a = 1]$

and $T_5 = ar^4 \Rightarrow T_5 = (1)\,(r)^4 = r^4$ $[\because a = 1]$

Given, $T_3 + T_5 = 90 \Rightarrow r^2 + r^4 = 90 \Rightarrow r^4 + r^2 - 90 = 0$

$\Rightarrow r^4 + 10r^2 - 9r^2 - 90 = 0 \Rightarrow r^2(r^2 + 10) - 9(r^2 + 10) = 0$

$\Rightarrow (r^2 + 10)(r^2 - 9) = 0 \Rightarrow r^2 - 9 = 0$ $[\because r^2 + 10 \neq 0]$

$\Rightarrow r^2 = 9 \Rightarrow r = \pm 3$

Hence, the common ratio is ± 3. **Ans.**

Example 8. *If the pth, q^{th} and r^{th} terms of a G.P. are a, b and c respectively, prove that*

$$(a^{q-r})\,(b^{r-p})\,(c^{p-q}) = 1$$

Solution. Let A be the first term and R be the common ratio of the given G.P. Then,

p^{th} term $= a \Rightarrow a = AR^{p-1}$ ... (1)

q^{th} term $= b \Rightarrow b = AR^{q-1}$ ... (2)

r^{th} term $= c \Rightarrow c = AR^{r-1}$ ... (3)

Now, $(a^{q-r})\,(b^{r-p})\,(c^{p-q}) = (AR^{p-1})^{q-r}\,(AR^{q-1})^{r-p}\,(AR^{r-1})^{p-q}$

$= (A^{q-r+r-p+p-q})\,(R^{(p-q)(q-r)+(q-1)(r-p)+(r-1)(p-q)})$

[using (1), (2) and (3)]

$= A^0\,(R^{pq-pr-q+r+qr-pq-r+p+pr-rq-p+q}) = A^0 \cdot R^0 = 1$

Hence, $(a^{q-r})\,(b^{r-p})\,(c^{p-q}) = 1.$ **Proved.**

Example 9. *Find all the sequences which are simultaneously arithmetic and geometric progressions.*

Solution. Let $T_1, T_2, T_3, \ldots$ be a sequence which is A.P. as well as G.P.

Let $T_n = a + (n-1)d \quad \forall\, n \in N$

$\therefore$ The sequence is $a, a + d, a + 2d, \ldots$, this is also a G.P.

$\therefore \quad \frac{T_{n+1}}{T_n} = \frac{T_{n+2}}{T_{n+1}} \quad \forall\, n \in N$

$$\Rightarrow \quad \frac{a+nd}{a+(n-1)d} = \frac{a+(n+1)d}{a+nd}$$

$$\Rightarrow \quad (a + nd)^2 = [(a + nd) + d]\,[(a + nd) - d]$$

$$\Rightarrow \quad (a + nd)^2 = (a + nd)^2 - d^2$$

$$\Rightarrow \quad d^2 = 0, \quad i.e. \quad d = 0$$

$\therefore$ The sequence is $a, a + 0, a + 2(0), \ldots$

i.e. $\quad a, a, a, \ldots$ **Ans.**

Example 10. *Find the value of K so that the sequence 2K – 5, K – 4, 10 – 3K forms a G.P.* **[S.B.T.E Dec. 2017]**

Solution. Here we have, $2K - 5, K - 4, 10 - 3K$ are in G.P.

$$\frac{K-4}{2K-5} = \frac{10-3K}{K-4}$$

$$\Rightarrow \quad 7K^2 - 43K + 66 = 0$$

$$K = \frac{-(-43) \pm \sqrt{(43)^2 - 4(7)(66)}}{2 \times 7}$$

$$= \frac{43 \pm \sqrt{1849 - 1848}}{14} = \frac{43 \pm 1}{4}$$

$$\therefore \quad K = \frac{27}{7}, 3$$ **Ans.**

Example 11. *If the 4th, 10th and 16th terms of a G.P. are x, y and z respectively. Prove that x, y, z are in G.P.*

Solution. Let the G.P. be $a, ar, ar^2, \ldots$

Now, $\quad T_4 = ar^3 = x$

$$T_{10} = ar^9 = y$$

$$\Rightarrow \quad T_{16} = ar^{15} = z$$

$$\Rightarrow \quad \frac{y}{x} = \frac{ar^9}{ar^3} = r^6 \qquad \ldots (1)$$

$$\Rightarrow \quad \frac{z}{y} = \frac{ar^{15}}{ar^9} = r^6 \qquad \ldots (2)$$

From (1) and (2), we find that the ratio of cosecutive terms x, y, z is the same *i.e.*, r^6.
Hence, x, y, z are in G.P. **Proved.**

Example 12. *If a, b, c are in A.P. and x, y, z are in G.P., prove that* $x^{b-c}\, y^{c-a}\, z^{a-b} = 1$.

Solution. Since a, b, c are in A.P.

Therefore, $\quad b = a + d \quad \ldots. (1) \quad$ and $\quad c = a + 2d \quad \ldots (2)$

Where, d is the common difference and since x, y, z are in G.P.

$$\therefore \quad \frac{y}{x} = \frac{z}{y} \Rightarrow y^2 = xz \qquad \ldots (3)$$

Now, $\quad x^{b-c} \cdot y^{c-a} \cdot z^{a-b} = x^{(a+d)-(a+2d)} \cdot y^{(a+2d)-a} \cdot z^{a-(a+d)}$ [using (1) and (2)]

$$= x^{-d} \cdot y^{2d}\, z^{-d} = (x^{-1}\, y^2\, z^{-1})d$$

$$= \left(\frac{y^2}{xz}\right)^d = \left(\frac{xz}{xz}\right)^d = (1)^d \qquad \text{[Using (3)]}$$

$$\Rightarrow \qquad x^{b-c}\, y^{c-a}\, z^{a-b} = 1 \qquad [\because (1)^d = 1] \qquad \textbf{Proved.}$$

Example 13. *If the pth and q^{th} terms of a G.P. are q and p respectively, show that $(p + q)^{th}$ term is $\left(\frac{q^p}{p^q}\right)^{\frac{1}{p-q}}$*

Solution. Let a be the first term and r be the common ratio of the G.P. Then,

$$T_p = ar^{p-1} = q \qquad \text{... (1)} \quad \text{and} \quad T_q = ar^{q-1} = p \qquad \text{... (2)}$$

Dividing (1) by (2), we get

$$\frac{ar^{p-1}}{ar^{q-1}} = \frac{q}{p} \Rightarrow r^{p-q} = \frac{q}{p} \Rightarrow r = \left(\frac{q}{p}\right)^{\frac{1}{p-q}}$$

Now,
$$T_{p+q} = ar^{p+q-1} = ar^{p-1} \cdot r^q = qr^q$$

$$= q\left[\left(\frac{q}{p}\right)^{\frac{1}{p-q}}\right]^q = q\left(\frac{q}{p}\right)^{\frac{q}{p-q}} = q\frac{q^{\frac{q}{p-q}}}{p^{\frac{q}{p-q}}} = \frac{q^{1+\frac{q}{p-q}}}{p^{\frac{q}{p-q}}} = \frac{q^{\frac{p}{p-q}}}{p^{\frac{q}{p-q}}} = \frac{(q^p)^{\frac{1}{p-q}}}{(p^q)^{\frac{1}{p-q}}}$$

$$\Rightarrow \qquad T_{p+q} = \left(\frac{q^p}{p^q}\right)^{\frac{1}{p-q}}. \qquad \textbf{Proved.}$$

Example 14. *If a, b, c and d are in G.P., show that*

$$(a^2 + b^2 + c^2)(b^2 + c^2 + d^2) = (ab + bc + cd)^2.$$

Solution. Let r be the common ratio of the G.P., a, b, c, d. Then, $b = ar$, $c = ar^2$ and $d = ar^3$.

$$\text{R.H.S.} = (ab + bc + cd)^2 = (a.ar + ar.\, ar^2 + ar^2.ar^3)^2$$

$$= a^4r^2\, [1 + r^2 + r^4]^2$$

$$\text{L.H.S.} = (a^2 + b^2 + c^2)\, (b^2 + c^2 + d^2)$$

$$= (a^2 + a^2r^2 + a^2r^4)(a^2r^2 + a^2r^4 + a^2r^6)$$

$$= a^2(1 + r^2 + r^4)\, a^2r^2\, (1 + r^2 + r^4)$$

$$= a^4r^2\, (1 + r^2 + r^4)^2$$

L.H.S. = R.H.S. **Proved.**

Example 15. *If* $\frac{a+bx}{a-bx} = \frac{b+cx}{b-cx} = \frac{c+dx}{c-dx}\ (x \neq 0)$ *then show that a, b, c and d are in G.P.*

Solution. Here, $\frac{a+bx}{a-bx} = \frac{b+cx}{b-cx}$

$$\frac{(a+bx)+(a-bx)}{(a+bx)-(a-bx)} = \frac{(b+cx)+(b-cx)}{(b+cx)-(b-cx)} \qquad \text{(componendo and dividendo)}$$

$$\frac{2a}{2bx} = \frac{2b}{2cx} \Rightarrow b^2 = ac$$

So a, b, c are in G.P. ... (1)

Again, $\frac{b+cx}{b-cx} = \frac{c+dx}{c-dx}$

$$\frac{(b+cx)+(b-cx)}{(b+cx)-(b-cx)} = \frac{(c+dx)+(c-dx)}{(c+dx)-(c-dx)} \quad \text{(componendo and dividendo)}$$

$$\frac{2b}{2cx} = \frac{2c}{2dx} \Rightarrow \frac{b}{c} = \frac{c}{d} \Rightarrow c^2 = bd \quad \text{... (2)}$$

So b, c, d are in G.P.

From (1) and (2), we have a, b, c, d are in G.P. **Proved.**

EXERCISE 3.1

Find the common ratio and indicated terms(s) in each of the following sequences:

1. 3, 6, 12, 24, ..., 9th term. **Ans.** $r = 2$, $T_9 = 768$
2. – 6, 18, – 54, ..., 12th term. **Ans.** $r = -3$, $T_{12} = 2(3)^{12}$
3. $2, 2\sqrt{2}, 4$ ...; 13th term. **Ans.** $r = \sqrt{2}$, $T_{13} = 128$
4. $2, -\frac{1}{4}, \frac{1}{32}, \ldots$, 15th term. **Ans.** $r = -\frac{1}{8}$, $T_{15} = (2)^{-41}$
5. $\sqrt{3}, \frac{1}{\sqrt{3}}, \frac{1}{3\sqrt{3}}, \ldots$; 15th term. **Ans.** $r = \frac{1}{3}$, $T_{15} = (3)^{-\frac{27}{2}}$

Which term of the sequence:

6. 5, 10, 20, 40, ... is 5120 ? **Ans.** 11th
7. $2, 1, \frac{1}{2}, \frac{1}{4}, \ldots$ is $\frac{1}{128}$? **Ans.** 9th
8. $1, -\frac{1}{3}, \frac{1}{9}, -\frac{1}{27}, \ldots$ is $-\frac{1}{243}$? **Ans.** 6th
9. $\sqrt{2}, \frac{1}{\sqrt{2}}, \frac{1}{2\sqrt{2}}, \frac{1}{4\sqrt{2}}, \ldots$ is $\frac{1}{512\sqrt{2}}$? **Ans.** 11th
10. The fourth term of a G.P. is 27 and the 7th term is 729. Find the G.P. **Ans.** 1, 3, 9, ...
11. The seventh term of a G.P. is 8 times the fourth term and 5th term is 48. Find the G.P. **Ans.** 3, 6, 12, ...
12. The 5th term of a G.P. is 48 and the 8th term is 384. Find the 12th term. **Ans.** 6144
13. The 4th term of a G.P. is square of its 2nd term and the first term is – 3. Determine its 7th term. **Ans.** – 2187
14. If p^{th} term of a G.P. is P and its qth term is Q, prove that the nth term is $\left(\frac{P^{n-q}}{Q^{n-p}}\right)^{\frac{1}{p-q}}$.
15. Find the number of term of a G.P. whose first is $\frac{3}{4}$ common ratio is 2 and the last term is 384. **Ans.** 10

16. For what value of k, the numbers $1+k, \frac{5}{6}+k, \frac{13}{18}+k$ are in G.P. ? **Ans.** $-\frac{1}{2}$

17. For what value of k, the numbers $-\frac{2}{7}, k, -\frac{7}{2}$ are in G.P.? **Ans.** ± 1

18. If the third term of a G.P. is $6\frac{1}{4}$ and the 7th term is the reciprocal of the third, which term of this G.P. is unity? **Ans.** 5th

19. In any G.P., prove that $T_{n-r} \cdot T_{n+r} = (T_n)^2$.

20. If the G.Ps., 5, 10, 20 ... and 1280, 640, 320, ... have their nth terms equal, find the value of n. **Ans.** 5

21. If in a G.P., the $(p + q)$th term is a and the $(p - q)$th term is b, prove that the pth term is $\sqrt{ab}$.

22. Determine the 8 terms $T_1, T_2, ..., T_8$ of a G.P. given that $\frac{T_3}{T_6} = \frac{8}{27}$ and $T_1 + T_2 = 160$.

Ans. $a = 64, r = \frac{3}{2}$; terms are 64, 96, 144, 216, 324, 486, 729, $1093\frac{1}{2}$.

23. If pth, qth and rth terms of a G.P. are themselves in G.P., show that p, q, r are in A.P.

24. If $\frac{1}{a^x} = \frac{1}{b^y} = \frac{1}{c^z}$ and a, b, c are in G.P., prove that x, y, z are in A.P.

25. If a and b are the roots of $x^2 - 3x + p = 0$ and c, d, are roots of $x^2 - 12x + q = 0$, where a, b, c, d form an G.P. Prove that $(q + p) : (q - p) = 17 : 15$.

26. Find the number of term in G.P. 6, 12, 24, ..., 1536 **[S.B.T.E Dec. 2015]** **Ans.** $n = 9$

3.3 SELECTION OF TERMS IN G.P.

For solving problems on G.P. it is always convenient, if we select the terms as follows when the product of the term is given.

Number of terms	*Terms*	*Common ratio*
3	$\frac{a}{r}, a, ar$	r
4	$\frac{a}{r^3}, \frac{a}{r}, ar, ar^3$	r^2
5	$\frac{a}{r^2}, \frac{a}{r}, a, ar, ar^2$	r

Note: If the product of terms of the G.P. is not given, the terms are chosen as a, ar, ar^2, ar^3.

Example 16. *Find three numbers in G.P. whose sum is 21 and whose product is 216.*

[B.T.E. Delhi Mathematics I Jan. 2009]

Solution. Let the three number in G.P. be $\frac{a}{r}, a, ar$

$$\Rightarrow \quad \frac{a}{r} + a + ar = 21 \qquad \text{[sum = 21]}$$

$$\Rightarrow \qquad a\left(\frac{1}{r}+1+r\right)=21 \qquad \text{... (1)}$$

Product = 216 $\Rightarrow$ $\frac{a}{r}\cdot a\cdot ar=216$

$\Rightarrow$ $a^3 = 216$ $\Rightarrow$ $a=6$ ... (2)

Putting the value of a from (2) in (1), we get

$$6\left(\frac{1}{r}+1+r\right)=21$$

$$\Rightarrow \qquad \frac{1+r+r^2}{r}=\frac{7}{2}$$

$$\Rightarrow \qquad 2+2r+2r^2=7r$$

$$\Rightarrow \qquad 2r^2-5r+2=0$$

$$\Rightarrow \qquad 2r^2-4r-r+2=0$$

$$\Rightarrow \qquad 2r(r-2)-1(r-2)=0$$

$$\Rightarrow \qquad (r-2)(2r-1)=0 \quad \Rightarrow \quad r-2=0 \Rightarrow \quad r=2$$

$$\Rightarrow \qquad 2r-1=0 \quad \Rightarrow \quad r=\frac{1}{2}$$

Case I. When $r = 2$, then the numbers are $\frac{6}{2}$, 6, 6 × 2, *i.e.* 3, 6, 12

Case II. When $r = \frac{1}{2}$, then the numbers are $\frac{6}{\frac{1}{2}}, 6, 6\left(\frac{1}{2}\right)$, *i.e* 12, 6, 3. **Ans.**

Example 17. Find the 4th term from the end of the G.P. 3, 6, 12, 24, ..., 3072

[S.B.T.E Dec. 2015]

Solution. Here we have

$$a=3,\ r=2>1,\ t_n=3072=l$$

$\therefore$ n^{th} term from end $= \frac{l}{r^{n-1}}$

4th term from end $= \frac{3072}{(2)^{4-1}} = 384$ **Ans.**

Example 18. *The first term of G.P. is 4, find the product of first five terms.*

[S.B.T.E Dec., 2017]

Solution. *Here we have* $a = 4$

product of first five terms $= \left(\frac{a}{r^2}\right)\cdot\left(\frac{a}{r}\right)\cdot(a)\cdot(ar)\cdot(ar^2)$

$$= \frac{a^5}{1} = a^5 = 4^5 = 1024 \qquad \textbf{Ans.}$$

Example 19. *Find three numbers in G.P. whose sum is 52 and the sum of whose products in pairs is 624.*

Solution. Let the three numbers in G.P. be $\frac{a}{r}, a, ar$.

$$\text{Sum} = 52 \qquad \Rightarrow \quad \frac{a}{r} + a + ar = 52$$

$$\Rightarrow \qquad a\left(\frac{1}{r} + 1 + r\right) = 52 \qquad \text{... (1)}$$

Sum of products in pairs = 624 $\Rightarrow \quad \frac{a}{r} \cdot a + a \cdot ar + \frac{a}{r} \cdot ar = 624$

$$\Rightarrow \qquad a^2\left(\frac{1}{r} + r + 1\right) = 624 \qquad \text{... (2)}$$

Dividing (2) by (1), we get $a = \frac{624}{52} \Rightarrow a = 12$

Putting $a = 12$ in (1), we get $12\left(\frac{1}{r} + r + 1\right) = 52$

$$\Rightarrow \qquad \frac{r^2 + r + 1}{r} = \frac{52}{12} \qquad \Rightarrow \qquad \frac{r^2 + r + 1}{r} = \frac{13}{3}$$

$$\Rightarrow \qquad 3r^2 + 3r + 3 = 13r \qquad \Rightarrow \qquad 3r^2 - 10r + 3 = 0$$

$$\Rightarrow \qquad (3r - 1)(r - 3) = 0 \qquad \Rightarrow \qquad r = \frac{1}{3} \text{ or } r = 3.$$

Case I: When $r = \frac{1}{3}$, then the numbers are $\frac{12}{\frac{1}{3}}, 12, 12 \times \frac{1}{3},$ *i.e.* 36, 12, 4

Case II: When $r = 3$, then the numbers are $\frac{12}{3}, 12, 12 \times 3,$ *i.e.* 4, 12, 36 **Ans.**

Example 20. *Find four numbers in G.P., whose sum is 85 and product is 4096.*

Solution. Let the four numbers in G.P. be $\frac{a}{r^3}, \frac{a}{r}, ar, ar^3$. ... (1)

$$\text{Product} = 4096 \qquad \Rightarrow \qquad \left(\frac{a}{r^3}\right)\left(\frac{a}{r}\right)(ar)(ar^3) = 4096$$

$$\Rightarrow \qquad a^4 = 4096 \qquad \Rightarrow \qquad a^4 = 8^4 \;\Rightarrow\; a = 8$$

$$\text{Sum} = 85 \qquad \Rightarrow \qquad \left(\frac{a}{r^3} + \frac{a}{r} + ar + ar^3\right) = 85$$

$$\Rightarrow \qquad a\left(\frac{1}{r^3} + \frac{1}{r} + r + r^3\right) = 85$$

$$\Rightarrow \qquad 8\left(r^3 + \frac{1}{r^3}\right) + 8\left(r + \frac{1}{r}\right) = 85 \qquad [\because a = 8]$$

$$\Rightarrow \quad 8\left[\left(r + \frac{1}{r}\right)^3 - 3\left(r + \frac{1}{r}\right)\right] + 8\left(r + \frac{1}{r}\right) = 85$$

$$\Rightarrow \qquad 8\left(r + \frac{1}{r}\right)^3 - 16\left(r + \frac{1}{r}\right) - 85 = 0 \qquad \text{... (2)}$$

Putting $r + \frac{1}{r} = x$ in (2), we get $8x^3 - 16x - 85 = 0$

$\Rightarrow \quad (2x - 5)(4x^2 + 10x + 17) = 0$

$\Rightarrow \quad 2x - 5 = 0 \quad [\because 4x^2 + 10x + 17 = 0$ has imaginary roots$]$

$\Rightarrow \quad 2x - 5 = 0 \quad \Rightarrow \quad x = \frac{5}{2}$

$\Rightarrow \quad r + \frac{1}{r} = \frac{5}{2} \quad \Rightarrow \quad 2r^2 - 5r + 2 = 0$

$\Rightarrow \quad (r - 2)(2r - 1) = 0 \quad \Rightarrow \quad r = 2$ or $r = \frac{1}{2}$

Putting $a = 8$ and $r = 2$ or $r = \frac{1}{2}$ in (1), we obtain the four numbers as 1, 4, 16, 64 or 64, 16, 4, 1. **Ans.**

EXERCISE 3.2

1. If the sum of three numbers in G.P. is 38 and their product is 1728, find them.
   Ans. 8, 12, 18 or 18, 12, 8
2. The product of three numbers in G.P. is 125 and the sum of their products taken in pair is $87\frac{1}{2}$. Find them. **Ans.** 10, 5, $\frac{5}{2}$ or $\frac{5}{2}$, 5, 10
3. The sum of three numbers in G.P. is 21 and the sum of their squares is 189. Find the numbers. **Ans.** 3, 6, 12 or 12, 6, 3
4. The product of first three terms of a G.P. is 1000. If 6 is added to its second term and 7 added to its third term, the terms form an A.P. Find the G.P.
   Ans. 5, 10, 20 or 20, 10, 5
5. The sum of three numbers in G.P. is 56. If we subtract 1, 7, 21 from these numbers in that order, we obtain an arithmetic progression. Find the numbers.
   Ans. 8, 16, 32 or 32, 16, 8
6. The sum of four numbers in G.P. is 820 and their product is 531441. Find the numbers. **Ans.** 1, 9, 81, 729
7. The sum of four numbers is 80. If we subtract 1, 2, 11, 44 from these numbers in that order. We obtain an arithmetic progression. Find the numbers.
   Ans. 2, 6, 18, 54 or 54, 18, 6, 2
8. If the sum of three numbers in G.P. is 38 and their product is 1728 find them.
   [S.B.T.E, Dec. 2015] **Ans.** 8, 12, 18, or 18, 12, 8

3.4 SUM OF FIRST *n* TERMS OF A G.P.

If a and r be the first term and common ratio of a G.P. respectively, then prove that the sum of first n terms of this G.P. is given by $S_n = \frac{a(1-r^n)}{1-r}$, if $r \neq 1$

Proof. Let T_n be the nth term of the given G.P. and S_n be the sum of n terms of the G.P.

$\therefore \quad T_n = ar^{n-1}$

Also $\quad S_n = T_1 + T_2 + ... + T_{n-1} + T_n$

$\Rightarrow \quad S_n = a + ar + ar^2 + ...+ar^{n-2} + ar^{n-1}$ ... (1)

Multiplying both sides by common ratio (r), we get

$\Rightarrow \quad rS_n = ar + ar^2 + ar^3 + ...+ar^{n-1} + ar^n$ ... (2)

Subtracting (2) from (1), we get

$$S_n - rS_n = a + 0 + ... + 0 - ar^n \quad \Rightarrow \quad S_n(1-r) = a - ar^n$$

	$S_n = \dfrac{a(1-r^n)}{1-r}$, **where $r < 1$**
and	$S_n = \dfrac{a(r^n-1)}{r-1}$, **where $r < 1$, $(r \neq 1)$**

Note: If $r = 1$, $S_n = a + a + ...+ n$ terms $= na$.

Example 21. *Find the sum of indicated terms of each of the following geometric progression:*

(*i*) $1, \frac{2}{3}, \frac{4}{9}, ...;$ 10 terms. (*ii*) 0.15, 0.015, 0.0015, ...; 20 terms.

(*iii*) $\sqrt{7}, \sqrt{21}, 3\sqrt{7}, ...; n$ terms.

Solution. (*i*) Here, $a = 1,\ r = \dfrac{\frac{2}{3}}{1} = \dfrac{2}{3}\ (< 1)$ and $n = 10$

We know that, $S_n = \dfrac{a(1-r^n)}{1-r}$, for $r < 1$

$\Rightarrow \quad S_{10} = 1\dfrac{\left(1-\left(\frac{2}{3}\right)^{10}\right)}{1-\frac{2}{3}} = 3\left[1-\left(\frac{2}{3}\right)^{10}\right]$ **Ans.**

(*ii*) $a = 0.15,\ r = \dfrac{0.015}{0.15} = 0.10\ (< 1)$ and $n = 20$. Therefore,

$$S_n = a\left(\frac{1-r^n}{1-r}\right) \Rightarrow S_{20} = 0.15\left(\frac{1-(0.10)^{20}}{1-0.10}\right)$$

$\Rightarrow \quad S_{20} = \dfrac{0.15}{0.90}[1-(0.1)^{20}] = \dfrac{1}{6}[1-(0.1)^{20}]$ **Ans.**

(*iii*) Here, $a = \sqrt{7},\ r = \dfrac{\sqrt{21}}{\sqrt{7}} = \dfrac{\sqrt{7}\times\sqrt{3}}{\sqrt{7}} = \sqrt{3}\ (> 1)$

and number of terms $= n$

Therefore, $S_n = a\left(\dfrac{r^n-1}{r-1}\right)$

$\Rightarrow \quad S_n = \sqrt{7}\left(\dfrac{(\sqrt{3})^n-1}{\sqrt{3}-1}\right) = \dfrac{\sqrt{7}}{2}(\sqrt{3}+1)\left[(3)^{\frac{n}{2}}-1\right]$ **Ans.**

Example 22. *How many terms of the G.P.* $3, \frac{3}{2}, \frac{3}{4}, ...$ *are needed to give the sum* $\frac{3069}{512}$*?*

Solution. Let the sum of n terms of the G.P. $3, \frac{3}{2}, \frac{3}{4}, ...;$ be $\frac{3069}{512}$.

Here, $a = 3$, $r = \frac{\frac{3}{2}}{3} = \frac{1}{2}$ (< 1). Therefore, $S_n = a\left(\frac{1-r^n}{1-r}\right) \Rightarrow \frac{3069}{512} = 3\left(\frac{1-\left(\frac{1}{2}\right)^n}{1-\frac{1}{2}}\right)$

$$\Rightarrow \quad \frac{3069}{512} = 6\left[1-\left(\frac{1}{2}\right)^n\right] \Rightarrow \frac{3069}{512 \times 6} = \left[1-\left(\frac{1}{2}\right)^n\right]$$

$$\Rightarrow \quad \left(\frac{1}{2}\right)^n = 1 - \frac{3069}{3072} = \frac{3072-3069}{3072} = \frac{3}{3072} = \frac{1}{1024}$$

$$\Rightarrow \quad \left(\frac{1}{2}\right)^n = \left(\frac{1}{2}\right)^{10} \Rightarrow n = 10$$

Hence, 10 terms of the given G.P. is needed to give the sum $\frac{3069}{512}$. **Ans.**

Example 23. *Evaluate:* $\sum_{k=1}^{11}(2+3^k)$

Solution.

$$\sum_{k=1}^{11}(2+3^k) = [(2 + 3^1) + (2 + 3^2) + (2 + 3^3) + ... + (2 + 3^{11})]$$
$$= [(2 + 2 + 2 + ... \text{ 11 terms}] + [3 + 9 + 27 + ... \text{ 11 terms}]$$
$$= [22 + (3 + 9 + 27 + ... \text{ 11 terms})] = [22 + S_{11}] \quad ... (1)$$

where, $S_{11} = 3 + 9 + 27 + ... + 11\text{th term}$

It is *a* G.P. with $a = 3$, $r = 3$ (> 1) and $n = 11$

$$\therefore \quad S_n = a\left(\frac{r^n-1}{r-1}\right) \Rightarrow S_{11} = 3\left(\frac{3^{11}-1}{3-1}\right) = 3\left(\frac{3^{11}-1}{2}\right) \quad ... (2)$$

Putting value of S_{11} in (1) from (2), we get

$$\sum_{k=1}^{11}(2+3^k) = \left[22 + 3\left(\frac{3^{11}-1}{2}\right)\right]$$ **Ans.**

Example 24. *The sum of the first three terms of a G.P. is 16 and the sum of the next three terms is 128. Determine the first term, common ratio and the sum to n terms of the G.P.*

Soluion. Let a be the first term and r be the common ratio.

We have, $S_3 = 16 \Rightarrow \frac{a(1-r^3)}{1-r} = 16$ ... (1)

Also, sum of the next three terms = 128 $\Rightarrow$ Sum of first six terms = 128 + 16 = 144

i.e. $S_6 = 144 \Rightarrow \frac{a(1-r^6)}{1-r} = 144$ ... (2)

Dividing (2) by (1), we get

$$\frac{1-r^6}{1-r^3} = \frac{144}{16} \Rightarrow \frac{(1+r^3)(1-r^3)}{(1-r^3)} = 9$$

$$\Rightarrow \quad 1 + r^3 = 9 \Rightarrow r^3 = 8 \Rightarrow r = 2$$

Putting $r = 2$ in (1), we get

$$\frac{a(1-2^3)}{1-2} = 16 \Rightarrow 7a = 16 \Rightarrow a = \frac{16}{7}$$

Here, $a = \frac{16}{7}$ and $r = 2$ (> 1). Therefore,

$$S_n = a\left(\frac{r^n - 1}{r-1}\right) \Rightarrow S_n = \frac{16}{7}\left(\frac{2^n - 1}{2-1}\right) \Rightarrow S_n = \frac{16}{7}(2^n - 1) \quad \textbf{Ans.}$$

Example 25. *In a geometric progression, $\{a_n\}$, if $T_1 = 3$, $T_n = 96$ and $S_n = 189$. Find n.*

Solution. Here, $T_1 = a = 3$, $T_n = 96$, $S_n = 189$

We know that $T_n = ar^{n-1} = 96$ ... (1)

$$S_n = \frac{a(r^n - 1)}{r-1} = \frac{ar^n - a}{r-1}$$

$$\Rightarrow \quad S_n = \frac{a(r^{n-1})r - a}{r-1} \Rightarrow S_n = \frac{96r - a}{r-1} \quad \text{[using (1)]}$$

$$189 = \frac{96r - 3}{(r-1)} \Rightarrow 189r - 189 = 96r - 3 \Rightarrow r = 2, [\because S_n = 189]$$

Now, $T_n = ar^{n-1} \Rightarrow 96 = 3 \times 2^{n-1}$

$$\Rightarrow \quad \frac{96}{3} = 2^{n-1} \Rightarrow 2^{n-1} = 32$$

$$\Rightarrow \quad 2^{n-1} = (2)^5 \Rightarrow n - 1 = 5 \Rightarrow n = 6 \quad \textbf{Ans.}$$

Example 26. *Sum the series:*

$$(x+y) + (x^2 + xy + y^2) + (x^3 + x^2y + xy^2 + y^3) + \ldots \text{ up to } n \text{ terms.}$$

Solution. Required sum

$$S = (x+y) + (x^2 + xy + y^2) + (x^3 + x^2y + xy^2 + y^3) + \ldots \text{ up to } n \text{ terms.}$$

Multiplying and dividing by $(x - y)$, we get

$$S = \frac{1}{x-y}\left[(x^2 - y^2) + (x^3 - y^3) + (x^4 - y^4) + \ldots \text{ up to } n \text{ terms}\right]$$

$$\Rightarrow \quad S = \frac{1}{x-y}\left[(x^2 + x^3 + x^4 + \ldots \text{ upto } n \text{ terms}) - (y^2 + y^3 + y^4 + \ldots \text{ upto } n \text{ terms}\right]$$

$$\Rightarrow \quad S = \frac{1}{x-y}\left[\frac{x^2(1-x^n)}{1-x} - \frac{y^2(1-y^n)}{1-y}\right] \quad \textbf{Ans.}$$

Example 27. *The sum of some terms of G.P. is 315 whose first term and the common ratio are 5 and 2 respectively. Find the last term and the number of terms.*

Solution. Here first term = 5, common ratio = 2

Sum of n terms = 315

$$\Rightarrow \quad \frac{a(r^n - 1)}{r - 1} = 315$$

$$\Rightarrow \quad \frac{5(2^n - 1)}{2 - 1} = 315$$

$$\Rightarrow \quad 2^n - 1 = 63$$

$$\Rightarrow \quad 2^n = 64 = 2^6 \quad \Rightarrow \quad n = 6$$

$$\Rightarrow \quad T_6 = ar^5 = 5(2)^5 = 5 \times 32 = 160$$

Number of terms = 6, last term = 160 **Ans.**

Example 28. *A G.P. consists of an even number of terms. If the sum of all the terms is 5 times the sum of terms occupying odd places. Then find its common ratio.*

Solution. Let number of terms = $2n$

Let the G.P. be $a, ar, ar^2, ar^3, ar^4, \ldots ar^{2n-1}$

$$\text{Sum of all the terms} = \frac{a(r^{2n} - 1)}{r - 1}$$

Number of odd terms = n, Common ratio = r^2

$$\text{Sum of all the odd terms} = \frac{a[(r^2)^n - 1]}{r^2 - 1} = \frac{a[r^{2n} - 1]}{r^2 - 1}$$

But sum of all terms = 5 times (sum of the terms occupying odd places)

$$\Rightarrow \quad \frac{a(r^{2n} - 1)}{r - 1} = \frac{5a(r^{2n} - 1)}{r^2 - 1}$$

$$\Rightarrow \quad 1 = \frac{5}{r + 1} \quad \Rightarrow \quad r + 1 = 5 \quad \Rightarrow \quad r = 4$$

Common ratio = 4 **Ans.**

Example 29. *Let S be the sum, P the product and R the sum of reciprocals of n terms in a G.P. Prove that $P^2R^n = S^n$*

Solution. Let a be the first term and r the common ratio of the G.P. Then,

$$S = a + ar + ar^2 + \ldots + ar^{n-1} = a\left(\frac{r^n - 1}{r - 1}\right) \quad \ldots (1)$$

$$P = a \cdot ar \cdot ar^2 \ldots ar^{n-1} = a^n r^{1+2+3+\ldots+(n-1)} \Rightarrow P = a^n r^{\frac{n(n-1)}{2}} \quad \ldots (2)$$

and
$$R = \frac{1}{a} + \frac{1}{ar} + \frac{1}{ar^2} + \ldots + \frac{1}{ar^{n-1}} \Rightarrow R = \frac{1}{a}\left\{\frac{\left(\frac{1}{r}\right)^n - 1}{\left(\frac{1}{r}\right) - 1}\right\} = \frac{1}{a}\left(\frac{1 - r^n}{1 - r}\right)\frac{1}{r^{n-1}}$$

$$\Rightarrow \qquad R = \frac{1}{a}\left(\frac{r^n - 1}{r - 1}\right)\frac{1}{r^{n-1}} \qquad \text{... (3)}$$

Dividing (1) by (3), we get

$$\frac{S}{R} = a\left(\frac{r^n - 1}{r - 1}\right)\cdot a\left(\frac{r - 1}{r^n - 1}\right)r^{n-1} = a^2 r^{n-1}$$

$$\Rightarrow \qquad \left(\frac{S}{R}\right)^n = (a^2 r^{n-1})^n = a^{2n} r^{n(n-1)} = \left[a^n r^{\frac{n(n-1)}{2}}\right]^2 \quad \Rightarrow \left(\frac{S}{R}\right)^n = P^2 \qquad \text{[using (2)]}$$

$$\therefore \qquad P^2R^n = S^n$$

Proved.

Example 30. *If S_1, S_2 and S_3 are the sums of n, 2n and 3n terms of a G.P., show that*

$$S_1(S_3 - S_2) = (S_2 - S_1)^2$$

Solution. Let a be the first term and r (< 1) the common ratio of the given G.P.

$$\therefore \qquad S_1 = \text{Sum of } n \text{ terms} = \frac{a(1-r^n)}{1-r}, \; S_2 = \text{Sum of } 2n \text{ terms} = \frac{a(1-r^{2n})}{1-r}$$

And $\qquad S_3 = \text{Sum of } 3n \text{ terms} = \dfrac{a(1-r^{3n})}{1-r}$

$$\Rightarrow S_1(S_3 - S_2) = \frac{a(1-r^n)}{1-r}\left[\frac{a(1-r^{3n})}{1-r} - \frac{a(1-r^{2n})}{1-r}\right]$$

$$= \frac{a^2(1-r^n)}{(1-r)^2}[(1-r^{3n}) - (1-r^{2n})] = \frac{a^2(1-r^n)}{(1-r)^2}(r^{2n} - r^{3n})$$

$$= \frac{a^2(1-r^n)}{(1-r)^2}r^{2n}(1-r^n) = \frac{a^2r^{2n}(1-r^n)^2}{(1-r)^2} \qquad \text{... (1)}$$

$$\Rightarrow (S_3 - S_1)^2 = \left[\frac{a(1-r^{2n})}{1-r} - \frac{a(1-r^n)}{1-r}\right]^2$$

$$= \frac{a^2}{(1-r)^2}[1 - r^{2n} - 1 + r^n]^n = \frac{a^2(r^n - r^{2n})^2}{(1-r)^2} = \frac{a^2r^{2n}(1-r^n)^2}{(1-r)^2} \qquad \text{... (2)}$$

From (1) and (2), we get

$$S_1(S_3 - S_2) = (S_2 - S_1)^2$$

Proved.

Recurring Numbers as Geometric Series

Example 31. *Find the sum of the following series upto n terms:*

(*i*) 5 + 55 + 555 + ... (*ii*) 0.6 + 0.66 + 0.666 + ...

Solution. (*i*) We have, 5 + 55 + 555 + ... to n terms

$$= 5[1 + 11 + 111 + \text{... to } n \text{ terms}]$$

$$= \frac{5}{9}[9 + 99 + 999 + \text{... to } n \text{ terms}]$$

$$= \frac{5}{9}[(10 - 1) + (10^2 - 1) + (10^3 - 1) + \ldots + (10^n - 1)]$$

$$= \frac{5}{9}[(10 + 10^2 + 10^3 + \ldots + 10^n) - (1 + 1 + 1 + \ldots + 1)\ n \text{ times}]$$

$$= \frac{5}{9}\left[10 \times \frac{10^n - 1}{10 - 1} - n\right] = \frac{5}{9}\left[\frac{10}{9}(10^n - 1) - n\right] = \frac{50}{81}(10^n - 1) - \frac{5n}{9} \qquad \textbf{Ans.}$$

(*ii*) We have, $0.6 + 0.66 + 0.666 + \ldots$ to n terms

$$= 6 \times 0.1 + 6 \times 0.11 + 6 \times 0.111 + \ldots \text{ to } n \text{ terms}$$

$$= 6[0.1 + 0.11 + 0.111 + \ldots \text{ to } n \text{ terms}]$$

$$= \frac{6}{9}[0.9 + 0.99 + 0.999 + \ldots \text{ to } n \text{ terms}]$$

$$= \frac{2}{3}\left[\frac{9}{10} + \frac{99}{100} + \frac{999}{1000} + \ldots \text{ to } n \text{ terms}\right]$$

$$= \frac{2}{3}\left[\left(1 - \frac{1}{10}\right) + \left(1 - \frac{1}{100}\right) + \left(1 - \frac{1}{1000}\right) + \ldots \text{ to } n \text{ terms}\right]$$

$$= \frac{2}{3}\left[(1 + 1 + 1 + \ldots n \text{ terms}) - \left(\frac{1}{10} + \frac{1}{10^2} + \frac{1}{10^3} + \ldots \text{ to } n \text{ terms}\right)\right]$$

$$= \frac{2}{3}\left[n - \frac{1}{10}\left(\frac{1 - \left(\frac{1}{10}\right)^n}{1 - \frac{1}{10}}\right)\right] = \frac{2}{3}\left[n - \frac{1}{9}\left\{1 - \left(\frac{1}{10}\right)^n\right\}\right]$$

$$= \frac{2n}{3} - \frac{2}{27}(1 - 10^{-n}) \qquad \textbf{Ans.}$$

EXERCISE 3.3

Find the sum of the following G.P:

1. 1, 3, 9, 27, ... to 12 terms. **Ans.** 265720

2. $\frac{1}{2}, \frac{3}{2}, \frac{9}{2}$, ... to 10 terms. **Ans.** $\frac{1}{4}(3^{10} - 1)$

3. $1, -\frac{1}{2}, \frac{1}{4}, -\frac{1}{8}$, ... to 9 terms. **Ans.** $\frac{171}{256}$

4. $2, \frac{-1}{2}, \frac{1}{8}$, ... to 12 terms. **Ans.** $\frac{8}{5}\left[1 - \frac{1}{4^{12}}\right]$

5. x^3, x^5, x^7, ... to n terms. **Ans.** $\frac{x^3(1 - x^{2n})}{1 - x^2}$

6. $1, -a, a^2, -a^3$, ... to n terms. **Ans.** $\left[\frac{1 - (-a)^n}{1 - (-a)}\right]$

7. Given a G.P. with $a = 729$ and 7th term = 64, determine S_7. **Ans.** 2059

8. Find the sum of first n terms and the sum of first 5 terms of the geometric series $1+\frac{1}{3}+\frac{1}{9}+......+\frac{1}{3^{n-1}}+......$ **Ans.** $\frac{3}{2}\left(1-\frac{1}{3^n}\right), \frac{121}{81}$

9. How many terms of the G.P. 3, 3^2, 3^3, ... are needed to give the sum 120? **Ans.** 4

10. Find the sum to n terms of following sequences:

(*i*) 9, 99, 999, 9999, ... **Ans.** $\frac{10}{9}\left[10^n-1\right]-n$

(*ii*) 7, 77, 777, 7777, ... **Ans.** $\frac{7}{9}\left[\frac{10}{9}(10^n-1)\right]$

(*iii*) 7, 7.7, 7.77, 7.777, ... **Ans.** $\frac{7}{81}(4490+10^{-49})$

11. Prove that the sum to n terms of the series 11 + 103 + 1005 + ... is $\frac{10}{9}(10^n-1)+n^2$.

12. Find the sum of the series 2 + 6 + 18 + ... + 4374. **Ans.** 6560

13. Sum the series: $x(x+y)+x^2(x^2+y^2)+x^3(x^3+y^3)+......$ to n terms.

Ans. $x^2\left(\frac{x^{2n}-1}{x^2-1}\right)+xy\left(\frac{(xy)^n-1}{xy-1}\right)$

14. Evaluate the following:

(*i*) $\sum_{k=1}^{n}(2^k+3^{k-1})$ **Ans.** $\frac{1}{2}(2^{n+2}+3^n-5)$ (*ii*) $\sum_{k=2}^{10}4^k$ **Ans.** $\frac{16}{3}(4^9-1)$

15. How many terms of the sequence $\sqrt{3}, 3, 3\sqrt{3}, ...$ must be taken to make the sum $39+13\sqrt{3}$. **Ans.** 6

16. The common ratio of the G.P. is 3 and the last term is 486. If the sum of these terms be 728, find the first term. **Ans.** 2

17. The 4th and 7th terms of a G.P. are $\frac{1}{27}$ and $\frac{1}{729}$ respectively. Find the sum of n terms of the G.P.

Ans. $\frac{3}{2}\left(1-\frac{1}{3^n}\right)$

18. The sum of n terms of a progression is ($2n$ – 1). Show that it is a G.P. Find its common ratio. **Ans.** 2

19. The ratio of the sum of first 3 terms to that of first 6 terms is 125 : 152. Find the common ratio.

$\left[\textbf{Hint. } \frac{a(r^3-1)}{(r-1)}\times\frac{(r-1)}{a(r^6-1)}=\frac{125}{152}, \text{ find } r\right]$ **Ans.** $\frac{3}{5}$

20. If $S_1, S_2, ..., S_n$ are the sums of n terms of n G.Ps. whose first term is 1 and common ratios are 1, 2, 3, ..., n respectively, show that

$S_1 + S_2 + 2S_3 + 3S_4 + ... + (n-1)\,S_n = 1^n + 2^n + 3^n + ... + n^n$.

21. In an increasing G.P., sum of the first and the last term is 66, the product of the second and the last one is 128 and the sum of the terms is 126. How many terms are there in the progression? **Ans.** $n = 6$

22. If S_n denotes the sum to 50 terms of a G.P., show that $(S_{10} - S_{20})^2 = S_{10}(S_{30} - S_{20})$.

23. Let a_n be the nth term of the G.P. of positive numbers. Let $\sum_{n=1}^{100} a_{2n} = \alpha$ and $\sum_{n=1}^{100} a_{2n-1} = \beta$, such that $\alpha \neq \beta$. Show that the common ratio of the G.P. is $\frac{\alpha}{\beta}$.

Hint: Let the G.P. be $a, ar, ar^2, \ldots$

$$\therefore \quad \alpha = a_2 + a_4 + a_6 + \ldots + a_{200} = ar + ar^3 + ar^5 + \ldots + ar^{199} = \frac{ar - (ar^{199})r^2}{1-r}$$

$$= \frac{ar(1-r^{200})}{1-r^2} \text{ etc.}$$

and $\beta = a_1 + a_3 + a_5 + \ldots + a_{199} = a + ar^2 + ar^4 + \ldots + ar^{198}$

$$= \frac{a - (ar^{198})r^2}{1-r^2} = \frac{a(1-r^{200})}{1-r^2}$$

24. Find the sum of all even positive integers less than 200 which are not divisible by 6. **[S.B.T.E Dec. 2016] Ans.** 6534

3.5 SUM OF AN INFINITE G.P.

If a be the first term and r is common ratio of a G.P. such that $|r| < 1$, then the sum to infinity of the G.P. is given by

$$S = \frac{a}{1-r}$$

Proof. We know that, $S_n = \frac{a(1-r^n)}{1-r} = \frac{a}{1-r} - \frac{ar^n}{1-r}$

Since, $|r| < 1$, $r^n \to 0$ as $n \to \infty$, therefore $S = \frac{a}{1-r} - 0$ as $n \to \infty$

Sum to infinity is denoted by S and $S = \frac{a}{1-r}$

$$\Rightarrow \quad \boxed{S_n = \frac{a}{1-r}, \text{ when } n \to \infty}$$

Example 32. *Find the sum to infinity in each of the following geometric progression:*

(*i*) $1, \frac{1}{3}, \frac{1}{9}, \ldots, \infty$ (*ii*) 6, 1.2, 0.24, ..., ∞ (*iii*) 10, – 9, 8.1, ..., ∞

Solution. (*i*) The given G.P. is $1, \frac{1}{3}, \frac{1}{9}, \ldots, \infty$

Here, $a = 1$, and $r = \frac{T_2}{T_1} = \frac{\frac{1}{3}}{1} = \frac{1}{3}$. Since, $\left|\frac{1}{3}\right| = \frac{1}{3} < 1$,

Hence, sum to infinity, $S_\infty = \dfrac{1}{1-\dfrac{1}{3}} = \dfrac{1}{2} = \dfrac{3}{2} = 1.5$ $\left[\because S_\infty = \dfrac{a}{1-r}\right]$

(*ii*) The given G.P. is 6, 1.2, 0.24, ..., ∞

Here, $a = 6$ and $r = \dfrac{T_2}{T_1} = \dfrac{1.2}{6} = 0.2$

Since, $|0.2| = 0.2 < 1.$

Hence, sum to infinity, $S_\infty = \dfrac{6}{1-0.2} = \dfrac{6}{0.8} = 7.5$

(*iii*) The given G.P. is 10, – 9, 8.1, ..., ∞

Here, $a = 10,\ \dfrac{T_2}{T_1} = \dfrac{-9}{10} = -0.9$ Since, $|-0.9| = 0.9 < 1,$

Hence, sum to inifinity, $S_\infty = \dfrac{10}{1-(-0.9)} = \dfrac{10}{1+0.9} = \dfrac{10}{1.9} = \dfrac{100}{19}$ **Ans.**

Example 33. *Prove that*

$$9^{\frac{1}{3}} \cdot 9^{\frac{1}{9}} \cdot 9^{\frac{1}{27}} \cdot \ldots \text{ to } \infty = 3.$$

Solution. We have, $9^{\frac{1}{3}} \cdot 9^{\frac{1}{9}} \cdot 9^{\frac{1}{27}} \cdot \ldots \text{ to } \infty = 9^{\frac{1}{3}+\frac{1}{9}+\frac{1}{27}+\ldots\infty} = 9^S$ ... (1)

where, $S = \dfrac{1}{3} + \dfrac{1}{9} + \dfrac{1}{27} + \ldots + \infty$

This is an inifinite G.S. with $a = \dfrac{1}{3},\ r = \dfrac{T_2}{T_1} = \dfrac{\frac{1}{9}}{\frac{1}{3}} = \dfrac{1}{3}$

Since, $|r| = \left|\dfrac{1}{3}\right| = \dfrac{1}{3} < 1,$ therefore sum to infinity is as follows.

$$S_\infty = \frac{\frac{1}{3}}{1-\frac{1}{3}} = \frac{\frac{1}{3}}{\frac{2}{3}} = \frac{1}{2} \quad \ldots (2) \quad \left[\because S_\infty = \frac{a}{1-r}\right]$$

Putting this value of S in (1), we get

$$9^{\frac{1}{3}} \cdot 9^{\frac{1}{9}} \cdot 9^{\frac{1}{27}} \ldots \text{ to } \infty = 9^{\frac{1}{2}} = (3^2)^{\frac{1}{2}} = 3$$ **Proved.**

Example 34. *Find the sum to infinity of the series*

$$\frac{2}{5} + \frac{3}{5^2} + \frac{2}{5^3} + \frac{3}{5^4} + \ldots + \infty$$

Solution. We have, $\dfrac{2}{5} + \dfrac{3}{5^2} + \dfrac{2}{5^3} + \dfrac{3}{5^4} + \ldots + \infty$

$$= \left(\frac{2}{5} + \frac{2}{5^3} + \frac{2}{5^5} + \ldots + \infty\right) + \left(\frac{3}{5^2} + \frac{3}{5^4} + \frac{3}{5^6} + \ldots + \infty\right)$$

$$= \left(\text{Sum to infinity of G.S. with } a = \frac{2}{5} \text{ and } r = \left(\frac{1}{5}\right)^2\right)$$

$$+ \left(\text{sum to infinity of a G.S. with } a = \frac{3}{25} \text{ and } r = \left(\frac{1}{5}\right)^2\right)$$

$$= \frac{\frac{2}{5}}{1-\left(\frac{1}{5}\right)^2} + \frac{\frac{3}{25}}{1-\left(\frac{1}{5}\right)^2} = \frac{2}{5}\cdot\frac{25}{25-1} + \frac{3}{25}\times\frac{25}{25-1} = \frac{2}{5}\times\frac{25}{24} + \frac{3}{25}\times\frac{25}{24} \qquad \left[\because S_\infty = \frac{a}{1-r}\right]$$

$$= \frac{5}{12} + \frac{3}{24} = \frac{10+3}{24} = \frac{13}{24}$$ **Ans.**

Example 35. *Represent the following as rational numbers:*

(*i*) $0.1\overline{5}$ (*ii*) $0.\overline{712}$

Solution. (*i*) We write

$$0.1\overline{5} = 0.15555 \ldots = 0.1 + 0.05 + 0.005 + 0.0005 + \ldots + \infty$$

$$= 0.1 + \left(\frac{5}{100} + \frac{5}{1000} + \frac{5}{10000} + \cdots + \infty\right)$$

$$= 0.1 + \left(\text{Sum of infinite G.P. with } a = \frac{5}{100} \text{ and } r = \frac{1}{10}\right)$$

$$= 0.1 + \left(\frac{\frac{5}{100}}{1-\frac{1}{10}}\right) = 0.1 + \frac{\frac{5}{100}}{\frac{9}{10}} = 0.1 + \frac{1}{18} = \frac{7}{45}$$ **Ans.**

(*ii*) We write

$$0.\overline{712} = 0.712\ 712\ 712\ 712 \ldots + \infty$$

$$= 0.712 + 0.000712 + 0.000000712 + \ldots + \infty$$

$$= \frac{712}{1000} + \frac{712}{1000000} + \frac{712}{1000000000} + \cdots + \infty$$

$$= \left(\text{Sum of infinite G.P. with } a = \frac{712}{1000} \text{ and } r = \frac{1}{1000}\right)$$

$$= \frac{\frac{712}{1000}}{1-\frac{1}{1000}} = \frac{712}{1000} \times \frac{1000}{999} = \frac{712}{999} \qquad \left[\because S_\infty = \frac{a}{1-r}\right]$$ **Ans.**

Example 36. *If* $x = 1 + a + a^2 + \cdots + \infty$, $y = 1 + b + b^2 + \cdots + \infty$ *and* $|a| < 1$, $|b| < 1$.

Prove that

$$1 + ab + a^2b^2 + \cdots = \frac{xy}{x+y-1}$$

Solution. We have, $x = 1 + a + a^2 + \cdots \text{ to } \infty = \frac{1}{1-a}$ $\quad (\because |a| < 1)$

$$\Rightarrow \quad 1 - a = \frac{1}{x} \Rightarrow a = 1 - \frac{1}{x} \Rightarrow a = \frac{x-1}{x} \qquad \ldots (1)$$

Also, $y = 1 + b + b^2 + \ldots \text{ to } \infty = \dfrac{1}{1-b}$ $(\because |a| < 1)$

$$\Rightarrow \quad 1 - b = \frac{1}{y} \Rightarrow b = 1 - \frac{1}{y} \Rightarrow b = \frac{y-1}{y} \quad \ldots (2)$$

$$\therefore \quad 1 + ab + ab^2 + \ldots \text{ to } \infty = \frac{1}{1-ab} \quad (\because |a| < 1, |b| < 1 \Rightarrow |ab| < 1)$$

$$= \frac{1}{1 - \dfrac{x-1}{x} \cdot \dfrac{y-1}{y}} \text{ [using (1) and (2)]} = \frac{xy}{xy - xy + x + y - 1} = \frac{xy}{x + y - 1}$$ **Proved.**

Example 37. *Find the sum of G.P. series* $1 + \frac{1}{2} + \frac{1}{2^2} + \frac{1}{2^3} + \cdots + \infty$ **[S.B.T.E Dec. 2016]**

Solution. Here, we have $1 + \frac{1}{2} + \frac{1}{2^2} + \frac{1}{2^3} + \cdots + \infty$

$$a = \frac{1}{2}, \quad r = \frac{1}{2} < 1$$

$$S_\infty = \frac{a}{1-r} = \frac{\frac{1}{2}}{1 - \frac{1}{2}} = 1$$ **Ans.**

Example 38. *The sum of first three terms of a G.P. is 16 and the sum of the next three terms is 128. Find the sum of n terms of the G.P.* **[S.B.T.E Dec., 2016]**

Solution. Given: $a + ar + ar^2 = 16$ ... (1)

and $ar^3 + ar^4 + ar^5 = 128$ ... (2)

Dividing (2) by (1), we get $r = 2 > 1$

putting $r = 2$ in (1), we get $a = \frac{16}{7}$

$$S_n = \frac{a(r^n - 1)}{r - 1}, r > 1 = \frac{16(2^n - 1)}{7(2-1)}$$

$$\therefore \quad S_n = \frac{16}{7}(2^n - 1)$$ **Ans.**

Example 39. *The first term of G.P. is 2 and the sum to infinity is 6, find the common ratio.*

Solution. Here, we have

$$a = 2, \quad S_\infty = 6$$

$$S_\infty = \frac{a}{1-r}, \quad r < 1$$

$$\Rightarrow \quad 6 = \frac{2}{1-r} \Rightarrow 6 - 6r = 2$$

$$\therefore \quad 4 = 6r \Rightarrow r = \frac{4}{6} = \frac{2}{3}$$ **Ans.**

Example 40. *The common ratio of a G.P. is* $-\frac{4}{5}$ *and the sum to infinity is* $\frac{80}{9}$. *Find the first term.*

Solution. We have, $r = -\frac{4}{5}$, $S_\infty = \frac{80}{9}$.

We know that, $S_\infty = \frac{a}{1-r} \Rightarrow \frac{80}{9} = \frac{a}{1-\left(-\frac{4}{5}\right)} \Rightarrow \frac{80}{9} = \frac{a}{1+\frac{4}{5}} \Rightarrow \frac{80}{9} \times \frac{9}{5} = a \Rightarrow a = 16$

Hence, the first term of the G.P. is 16. **Ans.**

Example 41. *The sum of an infinite geometric series is 15 and the sum of the squares of these terms is 45. Find the series.*

Solution. Let a be the first term and r the common ratio of the infinite geometric series. Then, it is given that

$$a + ar + ar^2 + \cdots + \infty = 15 \Rightarrow \frac{a}{1-r} = 15 \quad \text{... (1)}$$

$$\text{and} a + a^2r^2 + a^2r^4 + \cdots + \infty = 45 \Rightarrow \frac{a^2}{1-r^2} = 45 \quad \text{... (2)}$$

Dividing (2) by the square of (1), we get

$$\frac{a^2}{1-r^2} \times \frac{(1-r)^2}{a^2} = \frac{45}{15 \times 15} \Rightarrow \frac{1-r}{1+r} = \frac{1}{5}$$

$$\Rightarrow 5 - 5r = 1 + r \Rightarrow 6r = 4 \Rightarrow r = \frac{2}{3}$$

Putting $r = \frac{2}{3}$ in (1), we get $\frac{a}{1-\frac{2}{3}} = 15 \Rightarrow 3a = 15 \Rightarrow a = 5$

Therefore, $T_1 = 5$, $T_2 = a \times r = 5 \times \frac{2}{3} = \frac{10}{3}$, $T_3 = ar^2 = 5 \times \frac{4}{9} = \frac{20}{9}$

Hence, the series is $5 + \frac{10}{3} + \frac{20}{9} + \cdots + \infty$ **Ans.**

Example 42. *If* $S_1, S_2, S_3, ..., S_p$ *denote the sum of infinite G.S. whose first terms are 1, 2, 3, ..., p respectively and whose common ratios are* $\frac{1}{2}, \frac{1}{3}, \frac{1}{4}, \ldots\ldots, \frac{1}{(p+1)}$ *respectively. Show that* $S_1 + S_2 + S_3 + \cdots + S_p = \frac{p(p+3)}{2}$

Solution. Here, for S_1, we have $a = 1$, $r = \frac{1}{2}$ $\therefore$ $S_1 = \frac{1}{1-\frac{1}{2}} = 2$

For S_2, we have $a = 2$, $r = \frac{1}{3}$ $\therefore$ $S_2 = \frac{2}{1-\frac{1}{3}} = 3$

For S_3, we have $a = 3$, $r = \frac{1}{4}$

$$\therefore \quad S_3 = \frac{3}{1-\frac{1}{4}} = 4$$

... ...

... ...

For S_p, we have $a = p$, $r = \frac{1}{p+1}$ $\therefore$ $S_p = \frac{p}{1-\frac{1}{p+1}} = p+1$

Adding all these, we get

$$S_1 + S_2 + S_3 + \cdots + S_p = 2 + 3 + 4 + \cdots + (p+1)$$

$$= \frac{p}{2}[2(2)+(p-1)1] = \frac{p}{2}[p+3] \qquad [\because a = 2, d = 1, n = p]$$

$$= \frac{p(p+3)}{2}$$

Proved.

EXERCISE 3.4

Find the sum to infinity in each of the following geometric progressions:

1. $1, \frac{1}{2}, \frac{1}{4}, \frac{1}{8}, \ldots$ **Ans.** 2

2. $-\frac{5}{4}, \frac{5}{16}, -\frac{5}{64}, \ldots$ **Ans.** – 1

3. 50, 42.5, 36.125, ... **Ans.** $\frac{1000}{3}$

4. 0.3, 0.18, 0.108, ... **Ans.** 0.75

5. $8, 4\sqrt{2}, 4, \ldots$ **Ans.** $8(2+\sqrt{2})$

Prove that

6. $6^{\frac{1}{2}} \cdot 6^{\frac{1}{4}} \cdot 6^{\frac{1}{8}} \cdots \infty = 6$

7. $3^{\frac{1}{2}} \cdot 3^{\frac{1}{4}} \cdot 3^{\frac{1}{8}} \cdots \infty = 3$

For each of the following, find the rational number which will have as its expansion:

8. $0.\bar{3}$ **Ans.** $\frac{1}{3}$

9. $1.\overline{56}$ **Ans.** $\frac{155}{99}$

10. $0.6\bar{8}$ **Ans.** $\frac{31}{45}$

11. $22.\overline{3782}378$ **Ans.** $\frac{223760}{9999}$

12. Sum the given series to infinity: $(\sqrt{2}+1)+1+(\sqrt{2}-1)+\cdots$ **Ans.** $\frac{4+3\sqrt{2}}{2}$

13. The sum of first two terms of an infinite geometric series is 15 and each term is equal to the sum of all the terms following it. Find the the series.

Ans. $10+5+\frac{5}{2}+\frac{5}{4}+\frac{5}{8}+\cdots$

14. If S denotes the sum of an infinite G.P., S_1 denotes the sum of squares of its terms, then prove that the first term and common ratio are respectively $\frac{2SS_1}{S^2+S_1}$ and $\frac{S^2-S_1}{S+S_1}$.

15. If $A = 1 + r^a + r^{2a} + \cdots + \infty$ and $B = 1 + r^b + r^{2b} + \cdots + \infty$, show that $r = \left(\frac{A-1}{A}\right)^{\frac{1}{a}} = \left(\frac{B-1}{B}\right)^{\frac{1}{b}}$.

[Hint: $A = \frac{1}{1-r^a} \Rightarrow r^a = \frac{A-1}{A}$, *etc.*]

16. If the first term of a G.P. is 729 and the seventh term is 64, find S_∞.

Ans. 2187 or $\frac{2187}{5}$

17. The sum of an infinite G.P. is 57 and the sum of their cubes is 9747, find the G.P.

Ans. 19, $\frac{38}{3}, \frac{76}{9}$

18. Find the sum to infinity of the series:

$1, \frac{1}{3}, \frac{1}{2}, \frac{1}{3^2}, \frac{1}{2^2}, \frac{1}{3^3}, \ldots$

Ans. $\frac{5}{2}$

3.6 PROPERTIES OF GEOMETRIC PROGRESSION

In this section, we shall discuss some important properties of geometric progressions.

Property 1. *If all the terms of a G.P. be multiplied or divided by the same quantity, the resulting sequence is also a G.P.*

Proof: Let $T_1, T_2, T_3, \ldots, T_n$ be a G.P. with common ratio r.

Then, $$\frac{T_{n+1}}{T_n} = r, \text{ for all } n \in N \qquad \ldots (1)$$

Let k be a nonzero constant. Multiplying each term of given G.P. by k, we get the new sequence

$$kT_1, kT_2, kT_3 \ldots, kT_n, \ldots$$

Clearly, $$\frac{kT_{n+1}}{kT_n} = \frac{T_{n+1}}{T_n} = r \qquad \text{[using (1)]}$$

Hence, the new sequence also forms a G.P. with common ratio r. **Proved.**

Property 2. *The reciprocals of the terms of a given G.P. form a G.P.*

Proof: Let $T_1, T_2, T_3, \ldots, T_n, \ldots$ be a G.P. with common ratio r. Then,

$$\frac{T_{n+1}}{T_n} = r, \text{ for all } n \in N \qquad \ldots (1)$$

The sequence formed by the reciprocals of the terms of the given G.P. is

$$\frac{1}{T_1}, \frac{1}{T_2}, \frac{1}{T_3}, \ldots, \frac{1}{T_n}, \ldots$$

We have, $$\frac{\frac{1}{T_{n+1}}}{\frac{1}{T_n}} = \frac{T_n}{T_{n+1}} = \frac{1}{r} \qquad \text{[using (1)]}$$

So, the new sequence is a G.P. common ratio $\frac{1}{r}$. **Proved.**

Property 3. *If each term of a G.P. be raised to the same power, the resulting sequence is also a G.P.*

Proof: Let $T_1, T_2, T_3, ..., T_n, ...,$ be a G.P. with common ratio r. Then,

$$\frac{T_{n+1}}{T_n} = r, \text{ for all } n \in N \qquad ... (1)$$

Let k be the non zero real number;

Consider the sequence

$$T_1^k, T_2^k, T_3^k, ..., T_n^k, ...$$

We have, $$\frac{T_{n+1}^k}{T_n^k} = \left(\frac{T_{n+1}}{T_n}\right)^k = r^k, \text{ for all } n \in N. \qquad \text{[using (1)]}$$

Hence, $T_1^k, T_2^k, T_3^k, ..., T_n^k, ...$ is a G.P. with common ratio r^k. **Proved.**

Property 4. *In a finite a G.P., the product of the terms equidistance from the beginning and the end is always same and is equal to the product of the first and the last term.*

Property 5. *The resulting sequence formed by taking the product of the corresponding terms of two G.P. is also a G.P.*

Proof: Let the two G.P. be: $T_1, T_2, ..., T_n, ...$ with common ratio $R \Rightarrow \dfrac{T_{n+1}}{T_n} = R$ ... (1)

and $t_1, t_2, ..., t_n, ...$ with common ratio r. $\Rightarrow \dfrac{t_{n+1}}{t_n} = r$ ... (2)

Multiplying each term of sequence (1) by corresponding term of (2), we get

$$\left(\frac{T_{n+1}}{T_n}\right)\left(\frac{t_{n+1}}{t_n}\right) = Rr$$

Hence, the resulting sequence is also in G.P. with common ratio (Rr). **Proved.**

Property 6. *The resulting sequence, formed by dividing the terms of a G.P. by the corresponding terms of another G.P. is also a G.P.*

Proof: Let the two G.P. be: $T_1, T_2, ..., T_n, ...$ with common ratio R

$$\Rightarrow \quad \frac{T_{n+1}}{T_n} = R \qquad ... (1)$$

and $t_1, t_2, ..., t_n ...$ with common ratio r. $\Rightarrow \dfrac{t_{n+1}}{t_n} = r$ ... (2)

Dividing each term of sequence (1) by corresponding term of (2), we get

$$\frac{\dfrac{T_{n+1}}{T_n}}{\dfrac{t_{n+1}}{t_n}} = \left(\frac{R}{r}\right) \qquad \text{[using (1) and (2)]}$$

Hence, the resulting sequence is also in G.P. with common ratio $\left(\dfrac{R}{r}\right)$. **Proved.**

Example 43. *Three numbers whose sum is 15 are in A.P., if 1, 4, 19 be added to them respectively then they are in G.P. Find the numbers.*

Solution. Let the three numbers be $a - d, a, a + d$. Then, their sum = 15

$\Rightarrow \quad a - d + a + a + d = 15 \;\Rightarrow\; 3a = 15 \;\Rightarrow\; a = 5$

So, the numbers are $5 - d, 5, 5 + d$. Adding 1, 4, 19 respectively to these numbers, we get $6 - d, 9, 24 + d$. These numbers are in G.P.

$\therefore \quad 9^2 = (6 - d)(24 + d) \Rightarrow 81 = 144 - 18d - d^2$

$\Rightarrow \quad d^2 + 18d - 63 = 0 \Rightarrow d^2 + 21d - 3d - 63 = 0 \Rightarrow d(d + 21) - 3(d + 21) = 0$

$\Rightarrow \quad (d + 21)(d - 3) = 0 \Rightarrow d = -21, d = 3$

Hence, the numbers are 26, 5, –16 or 2, 5, 8 **Ans.**

Example 44. *If a, b, c, d are in G.P., show that following are also in G.P.*

(*i*) $a + b, b + c$ and $c + d$

(*ii*) $(a^n + b^n), (b^n + c^n)$ and $(c^n + d^n)$.

Solution. Let r be the common ratio of G.P., then we have $b = ar$, $c = ar^2$, and $d = ar^3$.

(*i*) We have

$$a + b = a + ar = a(1 + r)$$

$$b + c = ar + ar^2 = ar(1 + r)$$

And $$c + d = ar^2 + ar^3 = ar^2(1 + r)$$

Therefore, $\dfrac{b+c}{a+b} = \dfrac{ar(1+r)}{a(1+r)} = r \quad \ldots (1)$ and $\dfrac{c+d}{b+c} = \dfrac{ar^2(1+r)}{ar(1+r)} = r \quad \ldots (2)$

From (1) and (2), we have

$$\frac{b+c}{a+b} = \frac{c+d}{b+c}$$

Therefore, $(a + b), (b + c), (c + d)$ are in G.P. **Proved.**

(*ii*) We have $a^n + b^n = a^n + a^n r^n = a^n(1 + r^n)$

$$b^n + c^n = a^n r^n + a^n r^{2n} = a^n r^n(1 + r^n)$$

and $$c^n + d^n = a^n r^{2n} + a^n r^{3n} = a^n r^{2n}(1 + r^n)$$

Therefore, $$\frac{b^n + c^n}{a^n + b^n} = \frac{a^n r^n(1+r^n)}{a^n(1+r^n)} = r^n \quad \ldots (3)$$

Similarly, $$\frac{c^n + d^n}{b^n + c^n} = \frac{a^n r^{2n}(1+r^n)}{a^n r^n(1+r^n)} = r^n \quad \ldots (4)$$

From (3) and (4), we have

$$\frac{b^n + c^n}{a^n + b^n} = \frac{c^n + d^n}{b^n + c^n}$$

Therefore, $(a^n + b^n), (b^n + c^n), (c^n + d^n)$ are in G.P. **Proved.**

Example 45. *If a, b, c, are in G.P., show that the following are also in G.P.*

$$(i)\ a^2 - b^2, b^2 - c^2, c^2 - d^2 \quad (ii)\ a^2 + b^2 + c^2, ab + bc + cd, b^2 + c^2 + d^2$$

Solution. Since, a, b, c, d are in G.P. Let r be the common ratio of this G.P.

$\therefore \quad b = ar, \quad c = ar^2$ and $d = ar^3$

(*i*) $a^2 - b^2, b^2 - c^2, c^2 - d^2$ are in G.P., If $\dfrac{b^2 - c^2}{a^2 - b^2} = \dfrac{c^2 - d^2}{b^2 - c^2}$

i.e. if $$\frac{(ar)^2 - (ar^2)^2}{a^2 - (ar)^2} = \frac{(ar^2)^2 - (ar^3)^2}{(ar)^2 - (ar^2)^2}$$

$\Rightarrow$ if $\quad \dfrac{a^2r^2(1-r^2)}{a^2(1-r^2)} = \dfrac{a^2r^4(1-r^2)}{a^2r^2(1-r^2)}$

$\Rightarrow$ if $\quad r^2 = r^2$, which is true (since a, b, c, d are in G.P., therefore $a \neq 0$ and $r \neq 1$)

$\therefore \quad a^2 - b^2, b^2 - c^2, c^2 - d^2$ are in G.P. **Proved.**

(ii) $a^2 + b^2 + c^2, ab + bc + cd, b^2 + c^2 + d^2$ are in G.P., if $\dfrac{ab + bc + cd}{a^2 + b^2 + c^2} = \dfrac{b^2 + c^2 + d^2}{ab + bc + cd}$

i.e. if $\quad \dfrac{a(ar) + (ar)(ar^2) + (ar^2)(ar^3)}{a^2 + (ar)^2 + (ar^2)^2} = \dfrac{(ar)^2 + (ar^2)^2 + (ar^3)^2 + (ar^3)^2}{a(ar) + (ar)(ar^2) + (ar^2)(ar^3)}$

$\Rightarrow$ if $\quad \dfrac{a^2r(1 + r^2 + r^4)}{a^2(1 + r^2 + r^4)} = \dfrac{a^2r^2(1 + r^2 + r^4)}{a^2r(1 + r^2 + r^4)}$

$\Rightarrow$ if $\quad r = r$, which is true

$\therefore \quad a^2 + b^2 + c^2, ab + bc + cd, b^2 + c^2 + d^2$ are in G.P. **Proved.**

Example 46. *If $a^2 + b^2$, $ab + bc$ and $b^2 + c^2$ are in G.P., prove that a, b, c are also in G.P.*

Solution. Given that $a^2 + b^2$, $ab + bc$, $b^2 + c^2$ are in G.P.

$\Rightarrow \quad (ab + bc)^2 = (a^2 + b^2)(b^2 + c^2)$

$\Rightarrow \quad a^2b^2 + b^2c^2 + 2ab^2c = a^2b^2 + a^2c^2 + b^2c^2 + b^4 \;\Rightarrow\; b^4 + a^2c^2 - 2ab^2c = 0$

$\Rightarrow \quad (b^2 - ac)^2 = 0 \;\Rightarrow\; b^2 = ac \;\Rightarrow\;$ a, b, c are in G.P. **Proved.**

Example 47. *If a, b, c, d are in G.P., show that*

$$(ab + bc + cd)^2 = (a^2 + b^2 + c^2)(b^2 + c^2 + d^2).$$

Solution. Let r be the common ratio of the given G.P. a, b, c, d. Then,

$$b = ar;\; c = ar^2,\; d = ar^3$$

$$\text{L.H.S.} = (ab + bc + cd)^2 = (a\cdot ar + ar\cdot ar^2 + ar^2\cdot ar^3)^2$$
$$= (a^2r + a^2r^3 + a^2r^5)^2$$
$$= a^4r^2(1 + r^2 + r^4)^2$$
$$\text{R.H.S.} = (a^2 + b^2 + c^2)(b^2 + c^2 + d^2)$$
$$= (a^2 + a^2r^2 + a^2r^4)(a^2r^2 + a^2r^4 + a^2r^6)$$
$$= a^2(1 + r^2 + r^4)a^2r^2(1 + r^2 + r^4)$$
$$= a^4r^2(1 + r^2 + r^4)^2$$

$\Rightarrow \quad$ L.H.S. = R.H.S. **Proved.**

Example 48. *The pth, qth and rth terms of an A.P. as well as those of a G.P. are a, b, c respectively, prove that*

$$a^{b-c}\cdot b^{c-a}\cdot c^{a-b} = 1.$$

Solution. Let x be the first term and d the common difference of the A.P. Then,

$$x + (p - 1)d = a \quad \text{... (1)}$$
$$x + (q - 1)d = b \quad \text{... (2)}$$
$$x + (r - 1)d = c \quad \text{... (3)}$$

On subtracting (2) from (1), and (3) from (2), we get

$$a - b = (p - q)d \qquad \text{... (4)}$$

and $$b - c = (q - r)d \qquad \text{... (5)}$$

Now, let A be the first term and R be the common ratio of the G.P. Then,

$$AR^{p-1} = a \qquad \text{... (6)}$$

$$AR^{q-1} = b \qquad \text{... (7)}$$

$$AR^{r-1} = c \qquad \text{... (8)}$$

On dividing (6) by (7), and (7) by (8), we get

$$\frac{a}{r} = R^{(p-q)} \quad \text{and} \quad \frac{b}{c} = R^{(q-r)}$$

$$\Rightarrow \quad \left(\frac{a}{b}\right)^{\frac{1}{p-q}} = R \quad \text{and} \quad \left(\frac{b}{c}\right)^{\frac{1}{q-r}} = R \Rightarrow \left(\frac{a}{b}\right)^{\frac{1}{p-q}} = \left(\frac{b}{c}\right)^{\frac{1}{q-r}}$$

$$\Rightarrow \quad \left(\frac{a}{b}\right)^{\frac{d}{a-b}} = \left(\frac{b}{c}\right)^{\frac{d}{b-c}} \qquad \text{[using (4) and (5)]}$$

$$\Rightarrow \quad \left(\frac{a}{b}\right)^{b-c} = \left(\frac{b}{c}\right)^{a-b} \qquad \left[\text{Raising each to the power } \frac{(a-b)(b-c)}{d}\right]$$

$$\Rightarrow \quad \frac{a^{b-c}}{b^{b-c}} = \frac{b^{a-b}}{c^{a-b}} \Rightarrow a^{b-c} \cdot c^{a-b} \cdot b^{-(c-a)}$$

$$\Rightarrow \quad a^{b-c} \cdot b^{c-a} \cdot c^{a-b} = 1$$ **Proved.**

EXERCISE 3.5

1. Three numbers are in A.P. and their sum is 15. If 1, 3, 9 be added to them respectively, they form a G.P. Find the numbers. **Ans.** 15, 5, – 5 or 3, 5, 7
2. The sum of three numbers in G.P. is 56. If we subtract 1, 7, 21 from these numbers in that order, we obtain an A.P. Find the numbers. **Ans.** 8, 16, 32
3. If a, b, c are in G.P. Show that the following are also in G.P.:

   (*i*) a^2, b^2, c^2 (*ii*) b^2c^2, c^2a^2, a^2b^2 (*iii*) $\frac{1}{a}, \frac{1}{b}, \frac{1}{c}$ (*iv*) $\frac{1}{a^2}, \frac{1}{b^2}, \frac{1}{c^2},$
4. If a, b, c, d are in G.P. show that $a^2+b^2, b^2+c^2, c^2+d^2$ are also in G.P.
5. If a, b, c, d are in G.P. prove that:

   (*i*) $(b+c)(b+d) = (c+a)(c+d)$ (*ii*) $(a-d)^2 = (b-c)^2 + (c-a)^2 + (d-b)^2$
6. If $\frac{1}{x+y}, \frac{1}{2y}, \frac{1}{y+z}$ are the three consecutive terms of an A.P. prove that x, y, z are the three consecutive terms of a G.P.
7. If a, b, c, d are in G.P., show that $\frac{1}{a^2+b^2}, \frac{1}{b^2+c^2}, \frac{1}{c^2+d^2}$ are in G.P.
8. If a, b, c are in A.P., b, c, d are in G.P. and $\frac{1}{c}, \frac{1}{d}, \frac{1}{e}$ are in A.P., prove that a, c, e are in G.P.

[**Hint:** $b = b = \dfrac{a+c}{2}, c^2 = bd$ and $d = \dfrac{2ce}{c+e} \Rightarrow c^2 = bd = \left(\dfrac{a+c}{2}\right)\left(\dfrac{2ce}{c+e}\right)$

$\Rightarrow c(c + e) = (a + c)e \Rightarrow c^2 = ae$]

9. If a, b, c are in G.P., prove that:

(*i*) $a(b^2 + c^2) = c(a^2 + b^2)$ (*ii*) $a^2b^2c^2\left(\dfrac{1}{a^3} + \dfrac{1}{b^3} + \dfrac{1}{c^3}\right) = a^3 + b^3 + c^3$

(*iii*) $\dfrac{(a+b+c)^2}{a^2+b^2+c^2} = \dfrac{a+b+c}{a-b+c}$ (*iv*) $(a + 2b + 2c)(a - 2b + 2c) = a^2 + 4c^2$

10. If a, b, c, d are in G.P., prove that:

(*i*) $\dfrac{ab - cd}{b^2 - c^2} = \dfrac{a+c}{b}$ (*ii*) $\dfrac{a^2 + ab + b^2}{bc + ca + ab} = \dfrac{b+a}{c+b}$

(*iii*) $(a + b + c + d)^2 = (a + b)^2 + 2(b + c)^2 + (c + d)^2$

11. If a, b, c are in A.P. and a, x, b and b, y, c are in G.P., show that x^2, b^2, y^2 are in A.P.

12. If a, b, c are three distinct real numbers in G.P. and $a + b + c = xb$, then prove that either $x < -1$ or $x > 3$.

[**Hint:** $a + b + c = xb \Rightarrow a + ar + ar^2 = xar \Rightarrow r^2 + (1 - x) r + 1 = 0$.

As r is real $\therefore$ Distinct $\geq 0 \Rightarrow x^2 - 2x - 3 \geq 0 \Rightarrow x < -1$ or $x > 3$]

3.7 GEOMETRIC MEAN BETWEEN TWO GIVEN NUMBERS

1. Single G.M. between two positive numbers

Let a and b be two positive numbers and G be the G.M. between them. Then, a, G, b are in G.P.

$$\Rightarrow \quad \frac{G}{a} = \frac{b}{G}$$

$$\Rightarrow \quad G^2 = ab \Rightarrow \boxed{G = \sqrt{ab}}$$

2. *n*, G.Ms. between two numbers

Let a and b be two numbers and $G_1, G_2, G_3, G_4, \ldots G_n$, be the n G.M. between a and b

$\therefore$ $a, G_1, G_2, G_3, G_4, \ldots\ldots, G_n, b$ are in G.P. with common ratio r.

$$\Rightarrow \quad b = T_{n+2} = ar^{n+1} \Rightarrow r = \left(\frac{b}{a}\right)^{\frac{1}{n+1}}$$

$$\Rightarrow \quad G_1 = ar = a\left(\frac{b}{a}\right)^{\frac{1}{n+1}}, G_2 = ar^2 = a\left(\frac{b}{a}\right)^{\frac{2}{n+1}}, \ldots\ldots, G_n = ar^n = a\left(\frac{b}{a}\right)^{\frac{n}{n+1}}$$

Then, n geometric means are

$$a\left(\frac{b}{a}\right)^{\frac{1}{n+1}}, a\left(\frac{b}{a}\right)^{\frac{2}{n+1}}, a\left(\frac{b}{a}\right)^{\frac{3}{n+1}}, \ldots\ldots, a\left(\frac{b}{a}\right)^{\frac{n}{n+1}}$$

Example 49. *Insert three geometric means between 1 and 256.*

Solution. Let G_1, G_2, G_3 be the three G.Ms. between 1 and 256.

Then, $1, G_1, G_2, G_3, 256$ are in G.P.

Let r be the common ratio and 256 is the fifth term. Then,

$$256 = T_5 = 1(r)^{5-1} = r^4 \Rightarrow r^2 = 16 \Rightarrow r = \pm 4$$

If $r = 4$, then $G_1 = (1) \times 4 = 4, G_2 = (1)(4)^2 = 16, G_3 = (1)(4)^3 = 64$

If $r = -4$, then $G_1 = (1)(-4) = -4, G_2 = (1)(-4)^2 = 16, G_3 = (1)(-4)^3 = -64$

$\therefore$ 4, 16, 64 or – 4, 16, – 64 are the three geometric means **Ans.**

Example 50. *If G.M. between two numbers a and b is G and the two A.M. between them are p and q, then prove that:* $G^2 = (2p - q)(2q - p)$

Solution. We have, $G^2 = ab$ ... (1) ($\because$ G is G.M. between a and b)

Also, p and q are two arithmetic means between a and b.

$\therefore$ a, p, q, b are in A.P.

$\Rightarrow$ $2p = a + q$ and $2q = p + b \Rightarrow 2p - q = a$ and $2q - p = b$

Putting $a = 2p - q$ and $b = 2q - p$ in (1), we get

$G^2 = (2p - q)(2q - p)$ **Proved.**

Example 51. *Find the value of n so that* $\dfrac{a^{n+1} + b^{n+1}}{a^n + b^n}$ *may be the geometric mean between a and b.*

Solution. We know that, G.M. between a and $b = \sqrt{ab}$

$$\Rightarrow \frac{a^{n+1} + b^{n+1}}{a^n + b^n} = \sqrt{ab} \Rightarrow a^{n+1} + b^{n+1} = (a^n + b^n)(ab)^{\frac{1}{2}}$$

$$\Rightarrow a^{n+1} + b^{n+1} = a^{n+\frac{1}{2}} \cdot b^{\frac{1}{2}} + a^{\frac{1}{2}} b^{n+\frac{1}{2}} \Rightarrow a^{n+1} - a^{n+\frac{1}{2}} \cdot b^{\frac{1}{2}} = a^{\frac{1}{2}} b^{n+\frac{1}{2}} - b^{n+1}$$

$$\Rightarrow a^{n+\frac{1}{2}}\left(a^{\frac{1}{2}} - b^{\frac{1}{2}}\right) = b^{n+\frac{1}{2}}\left(a^{\frac{1}{2}} - b^{\frac{1}{2}}\right) \Rightarrow a^{n+\frac{1}{2}} = b^{n+\frac{1}{2}} \qquad \left[\because a \neq b \Rightarrow a^{\frac{1}{2}} - b^{\frac{1}{2}}\ G \neq 0\right]$$

$$\Rightarrow \left(\frac{a}{b}\right)^{n+\frac{1}{2}} = 1 = \left(\frac{a}{b}\right)^0 \Rightarrow n + \frac{1}{2} = 0 \Rightarrow n = -\frac{1}{2}$$ **Ans.**

Example 52. *If* G_1 *and* G_2 *are the two geometric means between b and c and a is their arithmetic mean, then show that:* $G_1^3 + G_2^3 = 2abc$

Solution. We have, a is arithmetic mean between b and c.

$$a = \frac{b + c}{2} \Rightarrow 2a = b + c \qquad \text{... (1)}$$

Also, G_1 and G_2 are geometric means between b and c.

$\therefore$ b, G_1, G_2, c are in G.P.

Let r be the common ratio of above G.P. Then,

$$T_4 = b(r)^{4-1} \Rightarrow c = br^3 \Rightarrow r^3 = \frac{c}{b} \Rightarrow r = \left(\frac{c}{b}\right)^{\frac{1}{3}} \qquad \text{... (2)}$$

Now, $$G_1 = T_2 = b(r)^{2-1} \Rightarrow G_1 = br = b\left(\frac{c}{b}\right)^{\frac{1}{3}} = c^{\frac{1}{3}} b^{\frac{2}{3}} \qquad \text{[using (2)]}$$

$$\Rightarrow G_1^3 = cb^2 \qquad \text{... (3)}$$

And $\quad G_2 = T_3 = b(r)^{3-1} = br^2 = b\left(\frac{c}{b}\right)^{\frac{2}{3}} = b^{\frac{1}{3}} \cdot c^{\frac{2}{3}} \Rightarrow G_2^3 = bc^2$ ... (4)

Adding (3) and (4), we get

$$G_1^3 + G_2^3 = b^2c + bc^2 = bc(b+c) \Rightarrow G_1^3 + G_2^3 = bc(2a) \quad \text{[using (1)]}$$

$$\Rightarrow \quad G_1^3 + G_2^3 = 2abc$$

Proved.

Example 53. *The ratio of the A.M. and G.M. of two positive numbers a and b is m : n. Show that:*

$$a : b = \left(m + \sqrt{m^2 - n^2}\right) : \left(m - \sqrt{m^2 - n^2}\right)$$

Solution. Here, $\frac{A.M.}{G.M.} = \frac{m}{n} \Rightarrow \frac{\frac{a+b}{2}}{\sqrt{ab}} = \frac{m}{n}$

$$\Rightarrow \quad \frac{a+b}{2\sqrt{ab}} = \frac{m}{n}$$

$$\Rightarrow \quad \frac{a+b+2\sqrt{ab}}{a+b-2\sqrt{ab}} = \frac{m+n}{m-n} \quad \text{[Componendo and dividendo]}$$

$$\Rightarrow \quad \frac{(\sqrt{a}+\sqrt{b})^2}{(\sqrt{a}-\sqrt{b})^2} = \frac{m+n}{m-n} \quad \Rightarrow \quad \frac{\sqrt{a}+\sqrt{b}}{\sqrt{a}-\sqrt{b}} = \frac{\sqrt{m+n}}{\sqrt{m-n}}$$

$$\Rightarrow \quad \frac{(\sqrt{a}+\sqrt{b})+(\sqrt{a}-\sqrt{b})}{(\sqrt{a}+\sqrt{b})-(\sqrt{a}-\sqrt{b})} = \frac{\sqrt{m+n}+\sqrt{m-n}}{\sqrt{m+n}-\sqrt{m-n}} \quad \text{[Componendo and dividendo]}$$

$$\Rightarrow \quad \frac{2\sqrt{a}}{2\sqrt{b}} = \frac{\sqrt{m+n}+\sqrt{m-n}}{\sqrt{m+n}-\sqrt{m-n}}$$

$$\Rightarrow \quad \frac{a}{b} = \frac{(m+n)+(m-n)+2\sqrt{(m+n)(m-n)}}{(m+n)+(m-n)-2\sqrt{(m+n)(m-n)}} = \frac{2m+2\sqrt{m^2-n^2}}{2m-2\sqrt{m^2-n^2}}$$

$$\Rightarrow \quad \frac{a}{b} = \frac{m+\sqrt{m^2-n^2}}{m-\sqrt{m^2-n^2}}$$

Proved.

Example 54. *The sum of two numbers is 6 times their geometric mean. Show that the numbers are in the ratio* $3+2\sqrt{2} : 3-2\sqrt{2}$.

Solution. Let two numbers be *a* and *b*. Then, G.M. between *a* and $b = \sqrt{ab}$

Given that sum of two numbers a and $b = 6 \times$ (G.M. between *a* and *b*)

$$\Rightarrow \quad a + b = 6(\sqrt{ab}) \Rightarrow \frac{a+b}{2\sqrt{ab}} = \frac{3}{1}$$

Applying componendo and dividendo, we get

$$\frac{a+b+2\sqrt{ab}}{a+b-2\sqrt{ab}} = \frac{3+1}{3-1} \Rightarrow \frac{(\sqrt{a}+\sqrt{b})^2}{(\sqrt{a}-\sqrt{b})^2} = \frac{4}{2} \Rightarrow \frac{\sqrt{a}+\sqrt{b}}{\sqrt{a}-\sqrt{b}} = \frac{\sqrt{2}}{1}$$

Again, applying componendo and dividendo, we get

$$\frac{\sqrt{a}+\sqrt{b}+\sqrt{a}-\sqrt{b}}{\sqrt{a}+\sqrt{b}-\sqrt{a}+\sqrt{b}}=\frac{\sqrt{2}+1}{\sqrt{2}-1} \Rightarrow \frac{2\sqrt{a}}{2\sqrt{b}}=\frac{\sqrt{2}+1}{\sqrt{2}-1} \Rightarrow \frac{\sqrt{a}}{\sqrt{b}}=\frac{\sqrt{2}+1}{\sqrt{2}-1}$$

Squaring both sides, $\frac{a}{b}=\left(\frac{\sqrt{2}+1}{\sqrt{2}-1}\right)^2 \Rightarrow \frac{a}{b}=\frac{2+1+2\sqrt{2}}{2+1-2\sqrt{2}} \Rightarrow \frac{a}{b}=\frac{3+2\sqrt{2}}{3-2\sqrt{2}}$ **Proved.**

Example 55. *If a, b, c are in G.P. and the equations $ax^2 + 2bx + c = 0$ and $dx^2 + 2ex + f = 0$ have a common root, then show that $\frac{d}{a}, \frac{e}{b}, \frac{f}{c}$ are in A.P.*

Solution. Since a, b, c are in G.P. Therefore,

$$b^2 = ac$$

Now, $ax^2 + 2bx + c = 0 \Rightarrow ax^2 + 2\sqrt{ac} + x + c = 0$

$\Rightarrow (\sqrt{a}x+\sqrt{c})^2 = 0 \Rightarrow \sqrt{a}x+\sqrt{c}=0 \Rightarrow x=-\sqrt{\frac{c}{a}}$ ... (1)

It is given that the equations $ax^2 + 2bx + c = 0$ and $dx^2 + 2ex + f = 0$ have a common root. Also from (1), the equation $ax^2 + 2bx + c = 0$ has equal roots both equal to $-\sqrt{\frac{c}{a}}$.

$\therefore -\sqrt{\frac{c}{a}}$ is a root of the equation $dx^2 + 2ex + f = 0$

$\Rightarrow d\cdot\frac{c}{a}-2e\sqrt{\frac{c}{a}}+f=0$

$\Rightarrow \frac{d}{a}-2e\sqrt{\frac{1}{ac}}+\frac{f}{c}=0$ [Dividing through out by c]

$\Rightarrow \frac{d}{a}-\frac{2e}{b}+\frac{f}{c}=0$ [$\because b^2 = ac$]

$\Rightarrow 2\left(\frac{e}{b}\right)=\frac{d}{a}+\frac{f}{c} \Rightarrow \frac{d}{a},\frac{e}{b},\frac{f}{c}$ are in A.P. **Proved.**

Example 56. *The product of n G.Ms. between any two positive numbers is equal to nth power of the G.M. between them.*

Solution. Let $G_1, G_2, ..., G_n$ be the n G.Ms. between positive numbers a and b.

$\therefore a, G_1, G_2, G_3, ..., G_n, b$ are in G.P.

Let r be the common ratio of this G.P. and b is $(n + 2)^{\text{nd}}$ term.

Now, $b = T_n + 2 = ar^{n+1} \Rightarrow r = \left(\frac{b}{a}\right)^{\frac{1}{n+1}}$

Product of n G.Ms. between a and $b = G_1 \cdot G_2 ....G_n = ar\cdot ar^2 ....ar^n$

$$= a^n r^{1+2+...+n} = a^n \cdot r^{\frac{n}{2}[2(1)+(n-1)1]} = a^n r^{\frac{n(n+1)}{2}}$$

$$= a^n\left[\left(\frac{b}{a}\right)^{\frac{1}{n+1}}\right]^{\frac{n(n+1)}{2}} = a^n\left(\frac{b}{a}\right)^{\frac{n}{2}} = a^{n-\frac{n}{2}}.b^{\frac{n}{2}} = a^{\frac{n}{2}}b^{\frac{n}{2}} = (\sqrt{ab})^n$$

$\therefore$ Product of n G.Ms. between a and b = (G.M. between a and b)^n **Proved.**

EXERCISE 3.6

Find the single geometric mean between each pair of numbers:

1. 1, 25 **Ans.** 5 **2.** 2, 50 **Ans.** 5 **3.** $\frac{1}{2}, 128$ **Ans.** 8

4. 0.027, 7.5 **Ans.** 0.45 **5.** 72, 882 **Ans.** 252 **6.** $c, c^3 ; c > 0$ **Ans.** c^2

Insert the stated number of geometric means between the given numbers:

7. Two between 4 and 500. **Ans.** 20, 100

8. Three between 3 and $\frac{3}{16}$. **Ans.** $\frac{3}{2}, \frac{3}{4}, \frac{3}{8}$

9. Four between $\frac{1}{2}$ and 512. **Ans.** 2, 8, 32, 128

10. Five between $\frac{1}{3}$ and 243. **Ans.** 1, 3, 9, 27, 81

11. Six between 1 and 2187. **Ans.** 3, 9, 27, 81, 243, 729

12. If $k - 1$ is the G.M. between $k - 2$ and $k + 1$, then find the value of k. **Ans.** $k = 3$

13. Insert 5 G.M. between $\frac{1}{3}$ and 9 and verify that their product is the 5th power of the G.M. between $\frac{1}{3}$ and 9. **Ans.** $\frac{1}{\sqrt{3}}, 1, \sqrt{3}, 3, 3\sqrt{3}$

14. Find two positive numbers, whose difference is 12 and whose A.M. exceeds the G.M. by 2. **Ans.** 16, 4

15. Find two numbers, whose arithmetic mean is 34 and the geometric mean is 16. **Ans.** 64, 4

16. Find two numbers, whose sum is 100 and the ratio between A.M. and G.M. is 5 : 4. **Ans.** 20, 80

17. If the A.M. between two positive numbers a and b is twice their G.M., then show that $a : b = 2 + \sqrt{3} : 2 - \sqrt{3}$.

18. If a, b, c be in G.P. and x, y be the A.M. between a, b and b, c respectively, show that

(*i*) $\frac{1}{x} + \frac{1}{y} = \frac{2}{b}$ (*ii*) $\frac{a}{x} + \frac{c}{y} = 2$

19. If G_1 is the first of n G.Ms. between two positive numbers a and b, then show that $G_1^{n+1} = a^n b$.

20. If the A.M. and G.M. between two numbers are in the ratio $m : n$, then prove that the numbers are in ratio $m + \sqrt{m^2 - n^2} : m - \sqrt{m^2 - n^2}$.

3.8 APPLICATIONS OF GEOMETRICAL PROGRESSION

In this section, we shall discuss some problems based upon the applications of geometrical progression.

Example 57. *A person writes a letter to four of his friends. He asks each one of them to copy the letter and mail to four different persons with instruction that they move the chain similarly. Assuming that the chain is not broken and that it costs 50 paisa to mail one letter, determine the amount spent on the postage when the 8th set of letters is mailed.*

Solution. Number of letters in 1st set = 4

Number of letters in the IInd set = 4 + 4 + 4 + 4 = 16

Number of letters in the IIIrd set = 4 + 4 + ... 16 terms = 64

... ...

... ...

The number of letters in the 1st, IInd, IIIrd set, ... are respectively 4, 16, 64, ... which form a G.P.

Here, $$a = 4, \quad r = \frac{T_2}{T_1} = \frac{16}{4} = 4$$

$\therefore$ Number of letters mailed up to 8th set of letters

$$= S_8 = \frac{a(r^8-1)}{r-1} = \frac{4(4^8-1)}{4-1} = \frac{4}{3}(65536-1)$$

$$= \frac{4}{3} \times 65535 = 87380$$

Postage spent @ 50 paisa per letter $= ₹\left(\frac{50}{100} \times 87380\right)$

$= ₹\ 43690$ **Ans.**

Example 58. *The height of a plant at a certain data is 1.6 metre. If it increases by 5 cm. in the following year and if the increase in each year is half of that in the preceding year, show that the plant will never be 1.7 metres high.*

Solution. According to the question, increase in the height of plant in the first, second, third, ..., years are $5, \frac{5}{2}, \frac{5}{4}, \ldots$cm respectively.

Let it reach the height 1.7 m (*i.e.* increase 1.7 m – 1.6 m = 0.1 m = 10 cm) in n years.

Then, the sum of $5, \frac{5}{2}, \frac{5}{4}, \ldots$ to n terms = 10

$$\Rightarrow \quad \frac{5\left(1-\frac{1}{2^n}\right)}{1-\frac{1}{2}} = 10 \qquad \left[\because a = 5, r = \frac{1}{2}, S_n = \frac{a(1-r^n)}{1-r}\right]$$

$$\Rightarrow \quad 10\left(1-\frac{1}{2^n}\right) = 10 \Rightarrow 1-\frac{1}{2^n} = 1 \Rightarrow \frac{1}{2^n} = 0$$

which does not hold for any n. Hence, the plant will never reach a height of 1.7 m. **Proved.**

Example 59. *One side of an equilateral triangle is 24 cm. The mid points of its sides are joined to form another triangle whose mid points are joined to form still another triangle. This process continues indefinitely. Find the sum of the perimeter of all the triangles.*

Solution. The perimeter of the equilateral triangle $ABC = 3 \times 24 = 72$ cm.

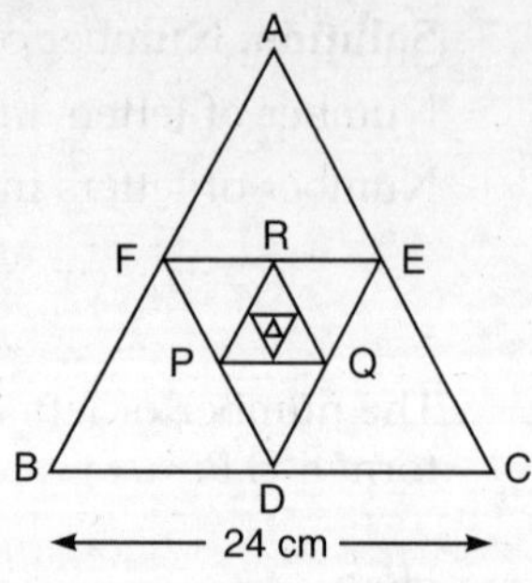

Let DEF be the triangle formed by joining the mid-points of the sides of ΔABC. Then, each side of ΔDEF = 12 cm.

$\therefore$ Perimeter (ΔDEF) = 3 × 12 = 36 cm.

Similarly, each side of third ΔPQR = 6 cm and its perimeter = 3 × 6 = 18 cm.

Since, we can go on forming equilateral triangles by joining the mid points of the sides of the subsequent triangles, therefore, number of such triangles is inifinite and the sum of their perimeters.

$$= 72 + 36 + 18 + \ldots \text{ to } \infty$$

$$= \frac{72}{1-\frac{1}{2}} \qquad \left[\because S_\infty = \frac{a}{1-r}, a = 72, r = \frac{36}{72} = \frac{1}{2}\right]$$

$$= \frac{72}{\frac{1}{2}} = 72 \times 2 = 144 \text{ cm}. \qquad \textbf{Ans.}$$

Example 60. *An insect starts from a point and travels in a straight path 1 mm in the 1st second spell and half of the distance covered in the previous second in the succeeding seconds. In how much time would it reach a point 3 mm away from its starting point?*

Solution. Let the time taken in reaching a point 3 mm away from the starting point by the insect be n seconds. Then,

$$3 = 1 + \frac{1}{2} + \frac{1}{2^2} + \frac{1}{2^3} + \cdots + \text{to } n \text{ terms}$$

$$= \text{Sum of a G.P. with first term } (a) = 1 \text{ and common ratio } (r) = \frac{1}{2}$$

$$= \frac{1\left\{1-\left(\frac{1}{2}\right)^n\right\}}{1-\frac{1}{2}}, \text{ as } r < 1 = \frac{1-\left(\frac{1}{2}\right)^n}{\frac{1}{2}} \qquad \left[\because S_n = \frac{a(1-r^n)}{1-r}\right]$$

$$\Rightarrow \frac{3}{2} = 1-\left(\frac{1}{2}\right)^n \Rightarrow \left(\frac{1}{2}\right)^n = 1-\frac{3}{2} = -\frac{1}{2} \Rightarrow \left(\frac{1}{2}\right)^n = -\frac{1}{2}$$

But, there is no value of n for which $\left(\frac{1}{2}\right)^n = -\frac{1}{2}$

Hence, the insect will never reach a point 3 mm away from its starting point under the given conditions. **Ans.**

Example 61. *A tennis ball when dropped to the ground rebounds to half of the height from which it falls. If it is dropped from a height of 16 metres, find the total distance travelled by the ball till it comes to rest.*

Solution. First time on dropping the distance travelled by the ball = 16 m ↓

Distance travelled by the ball on rebounding = 8 m

↑ 8 ↓

Distance travelled by the ball on falling = 8 m

On second rebound, Distance travelled by the ball = 4 m $\uparrow 4 \downarrow$

Distance travelled by the ball on falling = 4 m

... ...

... ...

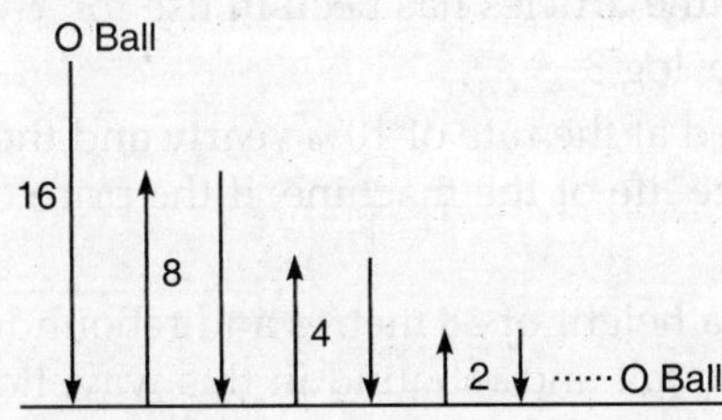

Total distance travelled by the ball

$= 16 + (8 + 8) + (4 + 4) + (2 + 2) + ...$

$= 16 + 2\ [8 + 4 + 2 + ...]$

$= 16 + 2\left(\dfrac{8}{1-\dfrac{1}{2}}\right) \qquad \left(\because S = \dfrac{a}{1-r}\right)$

$= 16 + 2 \times 2 \times 8 = 48$ m. **Ans.**

Example 62. *A manufacturer reckons that the value of a machine, which costs him ₹ 15625 will depreciate each year by 20%. Find the estimated value at the end of 5 years.*

Solution. The present value of a machine = ₹ 15625

The value of the machine next year will be = ₹ $15625 \times \dfrac{80}{100}$

The value of the machine after 2 years will be = ₹ $15625 \times \dfrac{80}{100} \times \dfrac{80}{100}$

The values of the machine at present year, after one year and after 2 years are ₹ 15625, ₹ $15625 \times \dfrac{80}{100}$ and ₹ $15625 \times \dfrac{80}{100} \times \dfrac{80}{100}$.

These value form a G.P.

Here, first term is ₹ 15625 and the common ratio is $\dfrac{80}{100}$, *i.e.* $\dfrac{4}{5}$

The value of the machine after 5 years = ar^5

= ₹ $15625 \times \left(\dfrac{4}{5}\right)^5 = \dfrac{15625 \times 1024}{625 \times 5} = 1024 \times 5$

= ₹ 5120 **Ans.**

EXERCISE 3.7

1. Mr. Raj writes letters to three of his friends. He asks each of them to copy the letter and mail to three different persons with the request that they continue the chain similarly. Assuming that the chain is not broken and that it costs ₹ 2.50 to mail one letter, find the total money spent on postage till the 5th set of letters is mailed.

Ans. ₹ 907.5

2. A square is drawn by joining the mid-points of th sides of a given square. A third square is drawn inside the second square in the same way and this process continues indefinitely. If the side of the first square is 16 cm, determine the sum of the areas of all the squares. **Ans.** 512 sq. cm.
3. An article costs ₹ 512 when new, but by usage, it loses one-fourth of its value yearly. If ₹ y be the value after the articles has been in use for x years, show that
$\log y = (9 - 2x) \log 2 + x \log 3$
4. A machine is depreciated at the rate of 10% yearly and the ultimate scrap value was ₹ 6561. Find the effective life of the machine, if the price of machine is ₹ 10,000. **Ans.** 4 years
5. A ball is dropped from a height of 48 metres and rebounds two third of the distance it falls. It is continued to fall and rebound in this way, how far will it travel before coming to rest. **Ans.** 240 m
6. A bicycle is worth ₹2187 when new, but by usage, it loses one-third of its value yearly. If ₹y be the value after the bicycle has been in use for x years, show that:
$\log y = x\log 2 + (7 - x) \log 3$.
7. The number of bacteria in a culture doubles every hour. If there were 30 bacteria present in the culture originally, how many bacteria will be present at the end of
(*i*) 2nd hour (*ii*) 4th hour (*iii*) nth hour?
Ans. $30(2)^2, 30(2)^4, 30(2)^n$
8. A man deposited ₹10000 in a bank at the rate 5% simple interest annually. Find the amount in 15th year, since he deposited the amount and also calculate the total amount after 20 years. **Ans.** ₹17000, ₹295000

REVISION EXERCISE

SECTION – A

Very Short Answer Questions (VSAQs)

Answer each of the following questions in one word or one sentence or as per exact requirement of the questions:

1. If the fifth term of a G.P. is 2, then write the product of its 9 terms. **Ans.** 512
2. If $(p + q)^{th}$ and $(p - q)^{th}$ terms of a G.P. are m and n respectively, then write its pth term. **Ans.** $\sqrt{mn}$
3. If $\log_x a, a^{x/2}$ and $\log_b x$ are in G.P. then write the value of x. **Ans.** $\log_a (\log_b a)$
4. If the sum of an infinite decreasing G.P. is 3 and the sum of the squares of its term is $\frac{9}{2}$, then write its first term and common difference. **Ans.** $a = 2, r = \frac{1}{3}$
5. If p^{th}, q^{th} and r^{th} terms of a G.P. are x, y, z respectively, then write the value of $x^{q-r} y^{r-p} z^{p-q}$. **Ans.** 1
6. If A_1, A_2 be two AM and G_1, G_2 be two GM between a and b, then find the value of $\frac{A_1 + A_2}{G_1 G_2}$. **Ans.** $\frac{a+b}{ab}$
7. If second, third and sixth terms of an A.P. are consecutive terms of a G.P., write the common ratio of the G.P. **Ans.** 3
8. Write the quadratic equation the arithmetic and geometric means of whose roots are A and G respectively. **Ans.** $x^2 - 2Ax + G^2 = 0$

9. Write the product of n geometric means between two numbers a and b. **Ans.** $(ab)^{n/2}$

10. If $a = 1 + b + b^2 + b^3 + \ldots$ to ∞, then write b in terms of a. **Ans.** $\frac{a-1}{a}$

Long Answer Questions (LAQs)

1. If a, b, c are in G.P., prove that $\log a, \log b, \log c$ are in A.P.
2. If a, b, c are in G.P., prove that
$$\frac{1}{\log_a m}, \frac{1}{\log_b m}, \frac{1}{\log_c m} \text{ are in A.P.}$$
3. If a, b, c are in A.P. and a, b, d, are in G.P., then prove that $a, a - b, d - c$ are in G.P.
4. If pth, qth, rth and sth terms of an A.P. be in G.P., then prove that $p - q, q - r, r - s$ are in G.P.
5. If $\frac{1}{a+b}, \frac{1}{2b}, \frac{1}{b+c}$ are three consecutive terms of an A.P., prove that a, b, c are the three consecutive terms of a G.P.
6. If $x^a = x^{b/2}z^{b/2} = z^c$, then prove that $\frac{1}{a}, \frac{1}{b}, \frac{1}{c}$ are in A.P.
7. Find k such that $k + 9, k - 6$ and 4 form three consecutive terms of a G.P.
8. Three numbers are in A.P. and their sum is 15. If 1, 3, 9 be added to them respectively, they form a G.P. Find the numbers.
9. The sum of three numbers which are consecutive terms of an A.P. in 21. If the second number is reduced by 1 and the third is increased by 1, we obtain three consecutive terms of a G.P. Find the numbers.
10. The sum of three numbers a, b, c in A.P. is 18. If a and b are each increased by 4 and c is increased by 36, the new numbers form a G.P. Find a, b, c.
11. The sum of three numbers in G.P. is 56. If we subtract, 1, 7, 21 from these numbers in that order, we obtain an A.P. Find the numbers.
12. If a, b, c are in G.P., prove that:

    (i) $a(b^2 + c^2) = c(a^2 + b^2)$ (ii) $a^2b^2c^2\left(\frac{1}{a^3}+\frac{1}{b^3}+\frac{1}{c^3}\right) = a^3 + b^3 + c^3$

    (iii) $\frac{(a+b+c)^2}{a^2+b^2+c^2} = \frac{a+b+c}{a-b+c}$ (iv) $\frac{1}{a^2-b^2}+\frac{1}{b^2}=\frac{1}{b^2-c^2}$

    (v) $(a + 2b + 2c)(a - 2b + 2c) = a^2 + 4c^2$.
13. If a, b, c, d are in G.P., prove that:

    (i) $\frac{ab-cd}{b^2-c^2} = \frac{a+c}{b}$ (ii) $(a + b + c + d)^2 = (a + b)^2 + 2(b + c)^2 + (c + d)^2$

    (iii) $(b + c)(b + d) = (c + a)(c + d)$
14. If a, b, c are in G.P., prove that the following are also in G.P.:

    (i) a^2, b^2, c^2 (ii) a^3, b^3, c^3 (iii) $a^2 + b^2, ab + bc, b^2 + c^2$
15. If a, b, c, d are in G.P., prove that

    (i) $(a^2 + b^2), (b^2 + c^2) (c^2 + d^2)$ are in G.P.

    (ii) $(a^2 - b^2), (b^2 - c^2) (c^2 - d^2)$ are in G.P.

    (iii) $\frac{1}{a^2+b^2}, \frac{1}{b^2+c^2}, \frac{1}{c^2+d^2}$ are in G.P.

(*iv*) $(a^2 + b^2 + c^2)$, $(ab + bc + cd)$, $(b^2 + c^2 + d^2)$ are in G.P.

16. If $(a - b)$, $(b - c)$, $(c - a)$ are in G.P., then prove that $(a + b + c)^2 = 3\,(ab + bc + ca)$

17. If a, b, c are in A.P., b, c, d are in G.P. and $\frac{1}{c}, \frac{1}{d}, \frac{1}{e}$ are in A.P., prove that a, c, e are in G.P.

18. If a, b, c are in A.P. and a, x, b and b, y, c are in G.P., show that x^2, b^2, y^2 are in A.P.

19. If a, b, c are in A.P. and a, b, d are in G.P., show that $a, (a - b), (d - c)$ are in G.P.

20. If a, b, c are in G.P., then prove that:

$$\frac{a^2 + ab + b^2}{bc + ca + ab} = \frac{b + a}{c + b}$$

21. If a, b, c are three distinct real numbers in G.P. and $a + b + c = xb$, then prove that either $x < -1$ or $x > 3$.

Multiple Choice Questions (MCQs)

Mark the correct alternative in each of the following:

1. If in an infinite G.P., first term is equal to 10 times the sum of all successive terms, then its common ratio is

(*a*) 1/10 (*b*) 1/11 (*c*) 1/9 (*d*) 1/20 **Ans.** (*b*)

2. If the first term of a G.P. $a_1, a_2, a_3, \ldots$ is unity such that $4a_2 + 5a_3$ is least, then the common ratio of G.P. is

(*a*) – 2/5 (*b*) – 3/5 (*c*) 2/5 (*d*) none of these **Ans.** (*a*)

3. If a, b, c are in A.P. and x, y, z are in G.P., then the value of $x^{b-c}\, y^{c-a}\, z^{a-b}$ is

(*a*) 0 (*b*) 1 (*c*) xyz (*d*) $x^a y^b z^c$ **Ans.** (*b*)

4. The first three of four given numbers are in G.P. and their last three are in A.P. with common difference 6. If first and fourth numbers are equal, then the first number is

(*a*) 2 (*b*) 4 (*c*) 6 (*d*) 8 **Ans.** (*d*)

5. If a, b, c are in G.P. and $a^{1/x} = b^{1/y} = c^{1/z}$, then xyz are in

(*a*) AP (*b*) GP (*c*) HP (*d*) none of these **Ans.** (*a*)

6. If S be the sum, P the product and R be the sum of the reciprocals of n terms of a GP, then P^2 is equal to

(*a*) S/R (*b*) R/S (*c*) $(R/S)^n$ (*d*) $(S/R)^n$ **Ans.** (*d*)

7. The fractional value of 2.357 is

(*a*) 2355/1001 (*b*) 2379/997 (*c*) 2355/999 (*d*) none of these **Ans.** (*c*)

8. If pth, qth and rth terms of an A.P. are in G.P., then the common ratio of this G.P. is

(*a*) $\frac{p-q}{q-r}$ (*b*) $\frac{q-r}{p-q}$ (*c*) pqr (*d*) none of these **Ans.** (*b*)

9. The value of $9^{1/3}.9^{1/9}.9^{1/27} \ldots$ upto ∞, is

(*a*) 1 (*b*) 3 (*c*) 9 (*d*) none of these **Ans.** (*b*)

10. The sum of an infinite G.P. is 4 and the sum of the cubes of its terms is 92. The common ratio of the original G.P. is

(*a*) 1/2 (*b*) 2/3 (*c*) 1/3 (*d*) –1/2 **Ans.** (*a*)

11. If the sum of first two terms of an infinite GP is 1 and every term is twice the sum of all the successive terms, then its first term is

(*a*) 1/3 (*b*) 2/3 (*c*) 1/4 (*d*) 3/4 **Ans.** (*d*)

12. The nth term of a G.P. is 128 and the sum of its n terms is 225. If its common ratio is 2, then its first term is

(*a*) 1 (*b*) 3 (*c*) 8 (*d*) none of these **Ans.** (*a*)

13. If second term of a G.P. is 2 and the sum of its infinite terms is 8, then first term is

(*a*) 1/4 (*b*) 1/2 (*c*) 2 (*d*) 4 **Ans.** (*d*)

14. If a, b, c are in G.P. and x, y are AM between a, b and b, c respectively, then

(*a*) $\frac{1}{x}+\frac{1}{y}=2$ (*b*) $\frac{1}{x}+\frac{1}{y}=\frac{1}{2}$ (*c*) $\frac{1}{x}+\frac{1}{y}=\frac{2}{a}$ (*d*) $\frac{1}{x}+\frac{1}{y}=\frac{2}{b}$ **Ans.** (*d*)

15. If A be one A.M. and p, q be two G.M. between two numbers, then 2A is equal to

(*a*) $\frac{p^3+q^3}{pq}$ (*b*) $\frac{p^3-q^3}{pq}$ (*c*) $\frac{p^2+q^2}{2}$ (*d*) $\frac{pq}{2}$ **Ans.** (*a*)

16. If p, q be two A.M. and G be one G.M. between two numbers, then $G^2 =$

(*a*) $(2p-q)(p-2q)$ (*b*) $(2p-q)(2q-p)$

(*c*) $(2p-q)(p+2q)$ (*d*) none of these **Ans.** (*a*)

17. If x is positive, the sum to infinity of the series

$$\frac{1}{1+x}-\frac{1-x}{(1+x)^2}+\frac{(1-x)^2}{(1+x)^3}-\frac{(1-x)^4}{(1+x)^4}+\ldots \text{ is}$$

(*a*) 1/2 (*b*) 3/4 (*c*) 1 (*d*) none of these **Ans.** (*a*)

18. If $x = (4^3)(4^6)(4^6)(4^9)\ldots(4^{3x}) = (0.0625)^{-54}$, the value of x is

(*a*) 7 (*b*) 8 (*c*) 9 (*d*) 10 **Ans.** (*b*)

19. Given that $x > 0$, the sum $\sum_{n=1}^{\infty}\left(\frac{x}{x+1}\right)^{n-1}$ equals

(*a*) x (*b*) $x+1$ (*c*) $\frac{x}{2x+1}$ (*d*) $\frac{x+1}{2x+1}$ **Ans.** (*b*)

20. In a G.P. of even number of terms, the sum of all terms is five times the sum of the odd terms. The common ratio of the G.P. is

(*a*) $-\frac{4}{5}$ (*b*) $\frac{1}{5}$ (*c*) 4 (*d*) none of these **Ans.** (*c*)

21. Let x be the A.M. and y, z be two G.M. between two positive numbers. Then, $\frac{y^3+z^3}{xyz}$ is equal to

(*a*) 1 (*b*) 2 (*c*) $\frac{1}{2}$ (*d*) none of these **Ans.** (*b*)

22. The product $(32), (32)^{1/6}(32)^{1/36} \ldots$ to ∞ is equal to

(*a*) 64 (*b*) 16 (*c*) 32 (*d*) 0 **Ans.** (*a*)

23. The two geometric means between the numbers 1 and 64 are

(*a*) 1 and 64 (*b*) 4 and 16 (*c*) 2 and 16 (*d*) 8 and 16

(*e*) 3 and 16 **Ans.** (*b*)

24. In a G.P. if the $(m+n)$th term is p and $(m-n)$th term is q, then its mth term is

(*a*) 0 (*b*) pq (*c*) $\sqrt{pq}$ (*d*) $\frac{1}{2}(p+q)$ **Ans.** (*c*)

Some Special Series and Harmonic Progression

4.1 INTRODUCTION

In this chapter, we will discuss the sum to n terms of a series of natural numbers, series of square of natural numbers and sum of cubes of natural numbers.

$\sum$ (Sigma) Notation

It is a Greek capital letter denotes the summation sign. If we put sigma before the general term it means the sum of the series.

$\sum T_n$ = sum of the series whose general term is T_n.

(PI) Notation

It is used for the product of the series.

$\prod T_n$ = Product of series, whose general term is T_n.

Example 1. *Write* $\sum_{k=1}^{6}(5k-2)$ *in expanded form.*

Solution. Replace k by 1, 2, 3, 4, 5, 6 in turn and write the sum. Thus,

$$\sum_{k=1}^{6}(5k-2) = (5\times 1-2)+(5\times 2-2)+(5\times 3-2)+(5\times 4-2)+(5\times 5-2)+(5\times 6-2)$$

$$= 3+8+13+18+23+28 = 93$$ **Ans.**

Example 2. *Write* $\sum_{k=1}^{10} k$ *in expanded form.*

Solution. $\sum_{k=1}^{10} k = 1+2+3+4+5+6+7+8+9+10 = 55.$ **Ans.**

4.2 SUM OF *n* TERMS OF SOME SPECIAL SERIES

I. ***Sum of first n natural numbers*** $(\sum n)$, *i.e.* $1+2+3+\ldots+n = \dfrac{n(n+1)}{2}$

Proof. Let $S_n = 1+2+3+\ldots+n$

This is an A.P. whose first term a is 1 and common difference $d = 1$

$$S_n = \frac{n}{2}[2a+(n-1)d] = \frac{n}{2}[2\times 1+(n-1)\cdot 1] = \frac{n}{2}(n+1)$$ **Proved.**

$$\Rightarrow \qquad \boxed{\sum n = \frac{n(n+1)}{2}}$$

II. ***Sum of squares of first n natural numbers*** $\left(\sum n^2\right)$

i.e. $$1^2 + 2^2 + 3^2 + \ldots + n^2 = \frac{n(n+1)(2n+1)}{6}$$

Proof. We know that, $(x + 1)^3 = x^3 + 3x^2 + 3x + 1$

$$\Rightarrow \qquad (x + 1)^3 - x^3 = 3x^2 + 3x + 1$$

Putting $x = 1, 2, 3, \ldots, (n - 1), n$ successively, we get

$$2^3 - 1^3 = 3 \cdot 1^2 + 3 \cdot 1 + 1$$
$$3^3 - 2^3 = 3 \cdot 2^2 + 3 \cdot 2 + 1$$
$$4^3 - 3^3 = 3 \cdot 3^2 + 3 \cdot 3 + 1$$
$$\ldots \ldots \ldots \ldots \ldots \ldots \ldots \ldots \ldots \ldots$$
$$\ldots \ldots \ldots \ldots \ldots \ldots \ldots \ldots \ldots \ldots$$
$$n^3 - (n - 1)^3 = 3 \cdot (n - 1)^2 + 3 \cdot (n - 1) + 1$$
$$(n + 1)^3 - n^3 = 3 \cdot n^2 + 3 \cdot n + 1$$

Adding all the terms, we get

$(n + 1)^3 - 1^3 = 3\,(1^2 + 2^2 + \ldots + n^2) + 3(1 + 2 + 3 + \ldots + n) + (1 + 1 + 1 + \ldots$ upto n terms)

$$\Rightarrow \qquad (n + 1)^3 - 1^3 = 3\left(\sum_{k=1}^{n} k^2\right) + 3\left(\sum_{k=1}^{n} k\right) + n$$

$$\Rightarrow \qquad n^3 + 3n^2 + 3n = 3\left(\sum_{k=1}^{n} k^2\right) + 3\,\frac{n(n+1)}{2} + n \qquad \left[\because \sum_{k=1}^{n} k = \frac{n(n+1)}{2}\right]$$

$$\Rightarrow \qquad 3\left(\sum_{k=1}^{n} k^2\right) = n^3 + 3n^2 + 3n - \frac{3n(n+1)}{2} - n$$

$$\Rightarrow \qquad 3\left(\sum_{k=1}^{n} k^2\right) = \frac{2n^3 + 3n^2 + n}{2} = \frac{n(n+1)(2n+1)}{2}$$

$$\Rightarrow \qquad \sum_{k=1}^{n} k^2 = \frac{n(n+1)(2n+1)}{6} \Rightarrow \sum n^2 = \frac{n(n+1)(2n+1)}{6}$$

Hence, $$\sum_{k=1}^{n} k^2 = 1^2 + 2^2 + \ldots + n^2 = \frac{n(n+1)(2n+1)}{6}$$

III. ***Sum of cubes of first n natural numbers*** $\left(\sum n^3\right)$

i.e. $$1^3 + 2^3 + 3^3 + \ldots + n^3 = \left\{\frac{n(n+1)}{2}\right\}^2$$

Proof. We know that $(x + 1)^4 = x^4 + 4x^3 + 6x^2 + 4x + 1$

$$\Rightarrow \qquad (x + 1)^4 - x^4 = 4x^3 + 6x^2 + 4x + 1 \qquad \ldots (1)$$

Putting $x = 1, 2, 3, \dots (n-1), n$ successively, we get

$$2^4 - 1^4 = 4 \cdot 1^3 + 6 \cdot 1^2 + 4 \cdot 1 + 1$$
$$3^4 - 2^4 = 4 \cdot 2^3 + 6 \cdot 2^2 + 4 \cdot 2 + 1$$
$$4^4 - 3^4 = 4 \cdot 3^3 + 6 \cdot 3^2 + 4 \cdot 3 + 1$$
$$\dots \dots \dots \dots \dots \dots \dots \dots \dots \dots \dots \dots$$
$$\dots \dots \dots \dots \dots \dots \dots \dots \dots \dots \dots \dots$$
$$n^4 - (n-1)^4 = 4\,(n-1)^3 + 6\,(n-1)^2 + 4\,(n-1) + 1$$

and
$$(n+1)^4 - n^4 = 4 \cdot n^3 + 6 \cdot n^2 + 4 \cdot n + 1$$

Adding all the terms, we get

$$(n+1)^4 - 1^4 = 4(1^3 + 2^3 + \dots\dots + n^3) + 6\,(1^2 + 2^2 + 3^2 + \dots\dots + n^2)$$
$$+ 4\,(1 + 2 + 3 + \dots\dots + n) + (1 + 1 + \dots\dots \text{ to } n \text{ terms})$$

$$\Rightarrow \quad n^4 + 4n^3 + 6n^2 + 4n = 4\left(\sum_{k=1}^{n} k^3\right) + 6\left(\sum_{k=1}^{n} k^2\right) + 4\left(\sum_{k=1}^{n} k\right) + n$$

$$\Rightarrow \quad n^4 + 4n^3 + 6n^2 + 4n = 4\left(\sum_{k=1}^{n} k^3\right) + 6\left(\frac{n(n+1)(2n+1)}{6}\right) + 4\left(\frac{n(n+1)}{2}\right) + n$$

$$\Rightarrow \quad 4\left(\sum_{k=1}^{n} k^3\right) = n^4 + 4n^3 + 6n^2 + 4n - n\,(n+1)\,(2n+1) - 2n\,(n+1) - n$$

$$\Rightarrow \quad 4\left(\sum_{k=1}^{n} k^3\right) = n^4 + 2n^3 + n^2 = n^2\,(n+1)^2$$

$$\Rightarrow \quad \sum_{k=1}^{n} k^3 = \frac{n^2(n+1)^2}{4} \quad \Rightarrow \quad \sum_{k=1}^{n} k^3 = \left(\frac{n(n+1)}{2}\right)^2 = \left(\sum_{k=1}^{n} k\right)^2$$

Hence,
$$\sum_{k=1}^{n} k^3 = 1^3 + 2^3 + \dots + n^3 = \left(\frac{n(n+1)}{2}\right)^2 = \left(\sum_{k=1}^{n} k\right)^2$$

$$\Rightarrow \quad \boxed{\sum n^3 = \left[\frac{n(n+1)}{2}\right]^2} \quad \text{or} \quad \boxed{\sum n^3 = \left(\sum n\right)^2}$$

IV. ***Sum of cubes of first n odd natural numbers:***

$$1^3 + 3^3 + 5^3 + \dots + (2n-1)^3 = n^2\,(2n^2 - 1).$$

Solution. Let $S_n = \sum (\text{odd numbers})^3$

$$\Rightarrow \quad S_n = \sum_{k=1}^{n} (2k-1)^3, \qquad [\because S_n = \text{sum of } n \text{ odd natural numbers}]$$

$$= \sum_{k=1}^{n} (8k^3 - 12k^2 + 6k - 1)$$

$$= 8\left(\sum_{k=1}^{n} k^3\right) - 12\left(\sum_{k=1}^{n} k^2\right) + 6\left(\sum_{k=1}^{n} k\right) - n \qquad [\because 1 + 1 + \dots \text{ to } n \text{ terms} = n]$$

$$= 8 \cdot \frac{1}{4} n^2 (n+1)^2 - 12 \cdot \frac{1}{6} n(n+1)(2n+1) + 6 \cdot \frac{1}{2} n(n+1) - n$$

$$= 2n^2 (n + 1)^2 - 2n (n + 1) (2n + 1) + 3n (n + 1) - n$$

$$= n(n + 1) [2n (n + 1) - 2 (2n + 1) + 3] - n = n(n + 1) (2n^2 - 2n + 1) - n$$

$$= n[(n + 1) (2n^2 - 2n + 1) - 1] = n[2n^3 - 2n^2 + n + 2n^2 - 2n + 1 - 1]$$

$$= n(2n^3 - n) = n^2 (2n^2 - 1)$$

$$\Rightarrow \quad S_n = n^2 (2n^2 - 1) \quad \Rightarrow \quad 1^3 + 3^3 + 5^3 + ... + (2n - 1)^3 = n^2 (2n^2 - 1)$$

Theorem

For a sequence $T_n = an^3 + bn^2 + cn + d$, *prove that* $S_n = a \sum n^3 + b \sum n^2 + c \sum n + nd.$

Proof. We have, $T_n = an^3 + bn^2 + cn + d$

Putting $n = 1, 2, 3, 4, ... (n - 1), n$ successively, we get

$$T_1 = a \cdot 1^3 + b \cdot 1^2 + c \cdot 1 + d$$

$$T_2 = a(2)^3 + b(2)^2 + c(2) + d$$

$$T_3 = a(3)^3 + b(3)^2 + c(3) + d$$

... ...

... ...

$$T_{n-1} = a(n - 1)^3 + b(n - 1)^2 + c(n - 1) + d$$

$$T_n = an^3 + bn^2 + cn + d$$

$$S_n = T_1 + T_2 + T_3 + ... + T_{n-1} + T_n \qquad \text{[on adding all terms]}$$

$$= a(1^3 + 2^3 + 3^3 + .. + n^3) + b(1^2 + 2^2 + 3^2 + ... + n^2) + c(1 + 2 + 3 + ... + n) + nd$$

$$\Rightarrow \quad S_n = a \sum n^3 + b \sum n^2 + c \sum n + nd$$ **Proved.**

Example 3. Find the sum to *n* terms of the series:

$$3 \cdot 8 + 6 \cdot 11 + 9 \cdot 14 + ...$$

Solution. We have, $3 \cdot 8 + 6 \cdot 11 + 9 \cdot 14 + ...$

Each term of this series is formed by the multiplication of the corresponding terms of the following two series.

1st series : $3, 6, 9, ..., 3n$

2nd series : $8, 11, 14, ..., 3n + 5$ $[\because T_n = a + (n - 1)d]$

nth term of the given series = (nth term of 1st series) × (nth term of the 2nd series)

$$\Rightarrow \quad T_n = 3n(3n + 5) = 9n^2 + 15n$$

$$\therefore \quad S_n = \sum T_n = \sum (9n^2 + 15n) = 9 \sum n^2 + 15 \sum n$$

$$= 9 \left\{ \frac{n(n+1)(2n+1)}{6} \right\} + 15 \left\{ \frac{n(n+1)}{2} \right\} = \frac{3}{2} n(n+1) \{(2n+1) + 5\}$$

Hence, $S_n = 3n(n + 1) (n + 3)$ **Ans.**

Example 4. *Find the sum to n terms of the series* $1 \cdot 3 \cdot 6 + 2 \cdot 5 \cdot 9 + 3 \cdot 7 \cdot 12 + ...$

Solution. Each term of the given series is formed by the multiplication of the corresponding terms of the following three series.

1st series : $1, 2, 3, ..., n$ $[T_n = 1 + (n - 1)1]$

2nd series : 3, 5, 7, ..., $(2n + 1)$ $[T_n = 3 + (n - 1)2]$
3rd series : 6, 9, 12, ..., $3(n + 1)$ $[T_n = 6 + (n - 1)3]$
nth term of the given series = (nth term of the 1st series) × (nth term of the 2nd series) × (nth term of the 3rd series)

$$\Rightarrow \quad T_n = n(2n + 1)\,3(n + 1) = 6n^3 + 9n^2 + 3n$$

Now,
$$S_n = \sum T_n = \sum (6n^3 + 9n^2 + 3n) = 6\sum n^3 + 9\sum n^2 + 3\sum n$$

$$= 6\left[\frac{n(n+1)}{2}\right]^2 + 9\left(\frac{n(n+1)(2n+1)}{6}\right) + 3\left(\frac{n(n+1)}{2}\right)$$

$$= \frac{3}{2}[n^4 + 2n^3 + n^2] + \frac{3}{2}[2n^3 + 3n^2 + n] + \frac{3}{2}[n^2 + n]$$

$$= \frac{3}{2}[n^4 + 2n^3 + n^2 + 2n^3 + 3n^2 + n + n^2 + n]$$

$$= \frac{3}{2}[n^4 + 4n^3 + 5n^2 + 2n] = \frac{3}{2}n(n^3 + 4n^2 + 5n + 2) \qquad \textbf{Ans.}$$

Example 5. *Find the sum:*

(*i*) $5^2 + 6^2 + 7^2 + \ldots + 20^2$ (*ii*) $2^3 + 4^3 + 6^3 + \ldots + 18^3$

Solution. (*i*) Let $S = 5^2 + 6^2 + 7^2 + \ldots + 20^2$ ...(1)

Adding and subtracting $1^2 + 2^2 + 3^2 + 4^2$ in given series, we get

$$S = (1^2 + 2^2 + 3^2 + 4^2 + 5^2 + 6^2 + 7^2 + \ldots + 20^2) - (1^2 + 2^2 + 3^2 + 4^2)$$

$$\Rightarrow \quad S = \left(\sum n^2\right)_{n=20} - \left(\sum n^2\right)_{n=4}$$

$$= \frac{20(20+1)(2\cdot 20+1)}{6} - \frac{4(4+1)(2\cdot 4+1)}{6} \qquad \left[\because \sum n^2 = \frac{n(n+1)(2n+1)}{6}\right]$$

$$= \frac{20\times 21\times 41}{6} - \frac{4\times 5\times 9}{6}$$

$$\Rightarrow \quad S = 2870 - 30 = 2840$$

(ii) Let
$$S = 2^3 + 4^3 + 6^3 + \ldots + 18^3 = (2\times 1)^3 + (2\times 2)^3 + (2\times 3)^3 + \ldots + (2\times 9)^3$$

$$= 2^3\,(1^3 + 2^3 + 3^3 + \ldots + 9^3) = 2^3\left(\sum n^3\right)_{n=9} = 2^3 \times \left\{\frac{n(n+1)}{2}\right\}^2_{n=9}$$

$$= 8\times\left(\frac{9(9+1)}{2}\right)^2 = 8 \times (45)^2 = 16200 \qquad \textbf{Ans.}$$

Example 6. *Find the nth term and the sum to n terms of the series:*

$1^2 + (1^2 + 2^2) + (1^2 + 2^2 + 3^2) + \ldots$

Solution. The given series is $1^2 + (1^2 + 2^2) + (1^2 + 2^2 + 3^2) + \ldots$ ... (1)

Let T_n be the nth term of (1). Then,

$$T_n = 1^2 + 2^2 + 3^2 + 4^2 + \ldots n^2 = \sum n^2 = \frac{n(n+1)(2n+1)}{6} = \frac{n^3}{3} + \frac{n^2}{2} + \frac{n}{6}$$

Sum to n terms $= S_n = \sum T_n = \frac{1}{3}\sum n^3 + \frac{1}{2}\sum n^2 + \frac{1}{6}\sum n$

$$= \frac{1}{3}\left(\frac{n(n+1)}{2}\right)^2 + \frac{1}{2}\left(\frac{n(n+1)(2n+1)}{6}\right) + \frac{1}{6}\frac{n(n+1)}{2}$$

$$\Rightarrow \qquad S_n = \frac{n(n+1)}{12}[n(n+1)+(2n+1)+1] = \frac{n(n+1)(n^2+3n+2)}{12}$$

$$= \frac{n(n+1)(n+1)(n+2)}{12} \qquad \Rightarrow \qquad S_n = \frac{n(n+1)^2(n+2)}{12}$$

Ans.

Example 7. *If S_1, S_2, S_3 are the sums of first n natural numbers, their squares, their cubes respectively, show that $9S_2^2 = S_3(1 + 8S_1)$*

Solution. We have, S_1 = Sum of first n natural numbers $= \sum n = \frac{1}{2}n(n+1)$

S_2 = Sum of squares of natural numbers $= \sum n^2 = \frac{n(n+1)(2n+1)}{6}$

and S_3 = Sum of cubes of natural numbers $= \sum n^3 = \left(\frac{n(n+1)}{2}\right)^2$

Now, $$9S_2^2 = 9\left[\frac{n(n+1)(2n+1)}{6}\right]^2 = \frac{1}{4}n^2(n+1)^2(2n+1)^2 \qquad \text{... (1)}$$

and $$S_3(1 + 8S_1) = \frac{1}{4}n^2(n+1)^2\left[1+8\cdot\frac{1}{2}n(n+1)\right] = \frac{1}{4}n^2(n+1)^2(2n+1)^2 \qquad \text{... (2)}$$

From (1) and (2), we get $9S_2^2 = S_3(1 + 8S_1)$ **Proved.**

4.3 HARMONIC PROGRESSION

If $x_1, x_2, x_3, \ldots x_n$ are in arithmetic progression,

then $\frac{1}{x_1}, \frac{1}{x_2}, \frac{1}{x_3}, \ldots, \frac{1}{x_n}$ will be in harmonic progression.

As $\frac{1}{2}, \frac{1}{4}, \frac{1}{6}, \frac{1}{8}, \ldots$ are in H.P. because 2, 4, 6, 8, ... are in A.P.

Example 8. Find 12th term of $\frac{1}{3}, \frac{1}{5}, \frac{1}{7}, \frac{1}{9}, \ldots$

Solution. Consider A.P.

3, 5, 7, 9, ...

$a = 3$

$d = 5 - 3 = 2$

nth term $\quad t_n = [a + (n-1)d]$

12th term, $\quad t_{12} = [3 + (12-1)2]$

$= 3 + 11 \times 2$

$t_{12} = 25$

So, 12th term of given H.P. is $\frac{1}{t_{12}} = \frac{1}{25}$ **Ans.**

Criteria

1. A series of quantities is said to be in a harmonic progression when their reciprocals are in arithmetic progression.

   e.g. $\frac{1}{3}, \frac{1}{5}, \frac{1}{7}$, and $\frac{1}{a}, \frac{1}{a+d}, \frac{1}{a+2d}, \ldots$ are in H.P. as their reciprocals, 3, 5, 7, ..., and $a, a + d, a + 2d$ ... are in A.P.

*n*th term of H.P.

1. Find the nth term of the corresponding A.P. and then take its reciprocal.
2. If the H.P. be $\frac{1}{a}, \frac{1}{a+d}, \frac{1}{a+2d}, \ldots$ then the corresponding A.P. is $a, a + d, a + 2d, \ldots$.
3. T_n of the A.P. is $a + (n - 1)d$ then T_{nth} of the H.P. is ... $\frac{1}{a+(n-1)d}$

4.4 RELATION AMONG A.M., G.M. AND H.M.

For two positive numbers a and b

A.M. = Arithmetic mean = $\frac{a+b}{2}$

G.M. = Geometric mean = $\sqrt{ab}$

H.M. = Harmonic mean = $\frac{2ab}{a+b}$

Multiplying A.M. and H.M., we get A.M. × H.M. = $\frac{a+b}{2} \times \frac{2ab}{a+b} = ab = \text{G.M.}^2$

$$\sum n = 1 + 2 + 3 + \ldots + n = \frac{n(n+1)}{2}$$

$$\sum n^2 = 1^2 + 2^2 + 3^2 + \ldots + n^2 = \frac{n(n+1)(2n+1)}{6}$$

$$\sum n^3 = 1^3 + 2^3 + 3^3 + \ldots + n^3 = \left(\frac{n(n+1)}{2}\right)^2$$

$$\sum (\text{odd numbers})^3 = 1^3 + 3^3 + 5^3 + \ldots + (2n-1)^3 = n^2(2n^2 - 1)$$

$$\sum (\text{even numbers})^3 = 2^3 + 4^3 + 6^3 + \ldots + (2n)^3 = 2^3 \times \left[\frac{n(n+1)}{2}\right]^2$$

4.5 HARMONIC MEAN (H.M.)

The harmonic mean is based on the reciprocals of the numbers averaged. It is defined as the reciprocal of the arithmetic mean of the reciprocal of the individual observation. Thus, by definition

$$\text{H.M.} = \frac{N}{\left(\frac{1}{X_1} + \frac{1}{X_2} + \frac{1}{X_3} + \ldots + \frac{1}{X_n}\right)}$$

where the number of items is large, the computation of harmonic mean in the above manner becomes tedious. To simplify calculations, we obtain reciprocal of various items from the table (given) and apply the following formulae:

In individual observations, $\text{H.M.} = \frac{N}{\sum(1/X)}$

In discrete series, $\text{H.M.} = \frac{N}{\sum\left(f \times \frac{1}{X}\right)}$

in continuous series $\text{H.M.} = \frac{N}{\sum\left(f \times \frac{1}{m}\right)} = \frac{N}{\sum(f/m)}$

4.6 CALCULATION OF HARMONIC MEAN: INDIVIDUAL SERIES

In an individual series, harmonic mean is computed by applying the following formula:

$$\text{H.M.} = \frac{N}{\left(\frac{1}{X_1} + \frac{1}{X_2} + \frac{1}{X_3} + \ldots + \frac{1}{X_n}\right)} = \frac{N}{\sum\left(\frac{1}{X}\right)}$$

X_1, X_2, X_3, *etc.*, refer to the various items of the variable.

Example 9. *Find the harmonic mean from the following:*

2574, 475, 75, 5, 0.8, 0.08, 0.005, 0.0009.

Solution.

Calculation of harmonic mean

X	(I/X)	X	(I/X)
2574	0.0004	0.8	1.2500
475	0.0021	0.08	12.5000
75	0.0133	0.005	200.0000
5	0.2000	0.0009	1111.1111
			Σ (1/X) = 1325.0769

$$\text{H.M.} = \frac{N}{\sum(1/X)} = \frac{8}{1325.0769} = 0.006 \qquad \textbf{Ans.}$$

4.7 CALCULATION OF HARMONIC MEAN: DISCRETE SERIES

In a discrete series, harmonic mean is computed by applying the following formula:

$$\text{H.M.} = \frac{N}{\sum\left(f \times \frac{1}{X}\right)} = \frac{N}{\sum(f/X)}$$

Steps: (*i*) Take the reciprocal of various items of the variable X.

(*ii*) Multiply the reciprocal by respective frequencies and obtain the total, *i.e.*

$$\sum\left(f\times\frac{1}{X}\right)$$

(*iii*) Substitute the values of *N* and $\sum\left(f\times\frac{1}{X}\right)$ in the above formula.

Note: Instead of finding out the reciprocals first and then multiplying them by frequencies it will be far more easier to divide each frequency by the respective value of their variable.

Example 10. *From the following data compute the value of harmonic mean:*

Marks	*10*	*20*	*25*	*40*	*50*
No. of students	*20*	*30*	*50*	*15*	*5*

Solution.

Calculation of harmonic mean

Marks X	*f*	*(f/X)*
10	20	2.000
20	30	1.500
25	50	2.000
40	15	0.375
50	5	0.100
	N = 120	Σ (*f*/X) = 5.975

$$\text{H.M.} = \frac{N}{\sum(f/X)} = \frac{120}{5.975} = 20.08$$ **Ans.**

Example 11. *Calculate Harmonic mean of the following values:*

15, 250, 15.7, 157, 1.57, 105.7, 10.5, 1.06, 25.7 and 0.257.

Solution. We shall find out the reciprocals of the above values from the mathematical tables which are available, instead of manual calculation.

Calculation of harmonic mean

Values (X)	*Reciprocal (1/X)*
15	0.06667
250	0.00400
15.7	0.06369
157	0.00637
1.57	0.63690
105.7	0.00946
10.5	0.09524
1.06	0.94340
25.7	0.03891
0.257	3.89100
N = 10	Σ (1/X) = 5.75564

$$\text{H.M.} = \frac{N}{\sum\left(\frac{1}{X}\right)} = \frac{10}{5.75564} = 1.737 \qquad \textbf{Ans.}$$

Example 12. *Calculate the harmonic mean of the following series:*

Values :	2	6	10	14	18
Frequency :	4	12	20	9	5

Solution. *Calculation of H.M. in a discrete series*

Calculation of harmonic mean

Values (X)	*Frequency (f)*	$\left(f \times \frac{1}{X}\right)$ *or* $\left(\frac{f}{X}\right)$
2	4	2.0000
6	12	2.0000
10	20	2.0000
14	9	0.6429
18	5	0.2778
	$N = 50$	$\Sigma\left(\frac{f}{X}\right) = 6.9207$

$$\text{H.M.} = \frac{N}{\sum\left(\frac{f}{X}\right)} = \frac{50}{6.9207} = 7.25 \qquad \textbf{Ans.}$$

We have used the simplified formula for the calculation of harmonic mean, otherwise we would have to find the reciprocal of each value and multiply it by the respective frequency and then total the products and then divide it by the number of items. It would then have been a tedious process.

4.8 CALCULATION OF HARMONIC MEAN: CONTINUOUS SERIES

For calculating harmonic mean in continuous series the procedure is the same as applied to discrete series. The only difference is that here we take the reciprocal of the mid-points.

Example 13. *From the following data compute the value of harmonic mean:*

Class interval :	10–20	20–30	30–40	40–50	50–60
Frequency :	4	6	10	7	3

Solution. Calculation of harmonic mean

Class interval	*Mid-points (m)*	*Frequency (f)*	f/m
10–20	15	4	0.267
20–30	25	6	0.240
30–40	35	10	0.286
40–50	45	7	0.156
50–60	55	3	0.055
		$N = 30$	$\Sigma (f/m) = 1.004$

$$\text{H.M.} = \frac{N}{\sum(f/m)} = \frac{30}{1.004} = 29.88 \qquad \textbf{Ans.}$$

Example 14. *From the following data, calculate harmonic mean:*

Class interval :	*10–20*	*20–30*	*30–40*	*40–50*	*50–60*
Frequency :	*30*	*75*	*70*	*135*	*220*

Solution.

Calculation of harmonic mean

Class interval (X)	*Frequency (f)*	*Mid-value (m.v.) = m*	*f/m*
10–20	30	15	2
20–30	75	25	3
30–40	70	35	2
40–50	135	45	3
50–60	220	55	4
	$N = 530$		$\Sigma\,(f/m) = 14$

$$\text{H.M.} = \frac{N}{\sum\left(\frac{f}{m}\right)} = \frac{530}{14} = 37.86$$ **Ans.**

Example 15. *From the data given below, if average output per worker by using harmonic mean is 84.48 then find unknown variable f.*

Output :	*70–74*	*75–79*	*80–84*	*85–89*	*90–94*	*95–99*	*100–104*
No. of workers :	*3*	*f*	*15*	*12*	*7*	*6*	*2*

Solution.

Calculation of harmonic mean

Output (X)	*No. of workers (f)*	*Mid-value (m.v.)*	*f/m*
70–74	3	72	0.0416
75–79	f	77	$f/77$
80–84	15	82	0.1829
85–89	12	87	0.1379
90–94	7	92	0.0761
95–99	6	97	0.0619
100–104	2	102	0.0196
	$N = 45 + f$		$\Sigma\left(\frac{f}{m}\right) = 0.52 + \frac{f}{77}$

$$\text{H.M.} = \frac{N}{\sum(f/m)} = \frac{45+f}{0.52+f/77} = 85.48 \quad \Rightarrow \quad f = 5$$ **Ans.**

4.9 USES OF HARMONIC MEAN

The relationship between harmonic mean and arithmetic mean (average) is of great significance. We have seen that *when prices are expressed in quantities* (units per rupee) harmonic mean should be calculated for correct interpretation. If the same prices are expressed in money values (rupees per unit) arithmetic mean gives the correct value of the average.

The above logic can be generalised, if we have to average the ratios involving, price, quantity, speed, time and distance, etc. as follows:

If the given ratios are stated as x unit per y and if values of xs' are given, use harmonic mean, and if values of ys, are given use arithmetic mean. For example, if we have to find out the average speed in *kilometres per hour*, use harmonic mean if *kilometre* (distance travelled) is given. when *hours* (time of journey) are given, use arithmetic mean (In the above case, Kilometre is represented x and hour by y).

The following examples would make the above points clear:

Example 16. *A person drives for 200 kilometres at a speed of 30 kph. He drives another 200 kilometres at a speed of 20 kph. What was the average speed per hour?*

Solution. We have to find out the average speed in kph and we have given the distance travelled, so we should find the harmonic mean.

Calculation of harmonic mean

Speed (X)	*Kilometres travelled (f)*	*f/X*
30	200	200/30
20	200	200/20
	N = 400	Σ (f/X) = 100/6

$$\text{H.M.} = \frac{N}{\sum(f/X)} = \frac{400}{100/6} = \frac{400 \times 6}{100} = 24 \text{ kph}$$

That this is the correct speed can be verified as follows:

First 200 km at a speed of 30 kph would take 6 hrs 40 minutes. Second 200 km at a speed of 20 kph would take 10 h.

Thus, total time taken is 16 h 40 min to cover 400 km

The average speed per hour would be $\frac{400}{16 \text{ h } 40 \text{ min}}$ or $400 \div \frac{50}{3} = 24$ kph **Ans.**

Example 17. *Find the Geometric mean of:*

5, 7, 0, 10, 8 ***(DU, B.Com. 2009)***

Solution. Since one value is zero G.M. is undefined

Example 18. *If four typists take 15, 10, 6 and 5 minutes respectively to type a letter. Than determine the average time required to type a letter if*

(*i*) Four letters are to be typed by each typist

(*ii*) Each typist works for two hours. ***(DU, B.Com. 2008)***

Solution. (*i*) Total time taken to type 16 letters

$$= (15 \times 4) + (10 \times 4) + (6 \times 4) + (5 \times 4) = 144 \text{ minutes}$$

Average time taken to type one letter $= \frac{144}{16} = 9$ minutes

(*ii*) Number of letters typed by A in two hours $= \frac{120}{15}$

Number of letters typed by B, C, D in two hours $= \frac{120}{10}, \frac{120}{6}$ and $\frac{120}{5}$

$$\text{Average time taken to type one letter} = \frac{\text{Total time taken}}{\text{Total number of letters typed}}$$

$$= \frac{120 \times 4}{120\left(\frac{1}{15}+\frac{1}{10}+\frac{1}{6}+\frac{1}{5}\right)} = \frac{60 \times 4}{4+6+10+12} = \frac{240}{32} = 7.5 \text{ minutes} \qquad \textbf{Ans.}$$

Example 19. *A man travelled by car for 3 days. He covered 480 miles each day. On the first day he drove for 10 hours at 48 miles an hour, on the second day he drove for 12 hours at 40 miles an hour and on the last day he drove for 15 hours at 32 miles per hour. What was his average speed.*

Solution. In this problem, the speeds are shown in m.p.h. but we are given the hours travelled (not miles travelled), therefore, arithmetic mean should be calculated.

Calculation of arithmetic mean

Speed (m.p.h.) X	*Hours travelled (f)*	*(fX)*
48	10	480
40	12	480
32	15	480
	$N = 37$	$\Sigma\,(fX) = 1440$

$$\text{Arithmetic mean} = \frac{\sum fX}{N} = \frac{1440}{37} = 38.92 \text{ m.p.h.}$$

However, in this case, harmonic mean could also give us the same result, as the distance travelled is the same. If the distance travelled is the same, there would be no difference between the H.M. and the arithmetic mean. We shall solve the above example by using harmonic mean.

Harmonic mean would be:

$$\frac{N}{\frac{1}{X_1}+\frac{1}{X_2}+\frac{1}{X_3}} = \frac{3}{\frac{1}{48}+\frac{1}{40}+\frac{1}{32}} = \frac{3}{\frac{37}{480}} = \frac{3 \times 480}{37} = \frac{1440}{37} = 38.92 \text{ miles per hour} \quad \textbf{Ans.}$$

It should be observed that the H.M. and the A.M. have the same value. Here, simple H.M. has been used as the distance travelled at various speeds is the same.

4.10 WEIGHTED HARMONIC MEAN

In the last example, though the speed of travelling varied but still the distance travelled at three different speed was the same. However, there might be cases where not only the speed of travelling varies but the distances travelled by various speeds also vary. In such cases, weighted harmonic mean is calculated. The different distances travelled are taken as weights.

However, there is no basic difference in the method by which weighted harmonic mean is calculated and the method by which simple harmonic mean is calculated except that the frequencies (*f*) are supposed to be weight (*w*) in case of the calculation of the weighted harmonic mean.

The formula for the calculation of weighted harmonic mean

$$= \frac{\sum(w)}{\sum(w/X)}, \text{ where } w \text{ is the weight and } X \text{ the value of the variable.}$$

In the calculation of the simple harmonic mean, we had used the formula:

$$\text{H.M.} = \frac{\sum(f)}{\sum(f/X)} = \frac{N}{\sum(f/X)}$$

The following examples would illustrate the calculation of weighted harmonic mean:

Example 20. *A man travels first 900 km of his journey by train at an average speed of 80 kph, next 2000 km by plane at an average speed of 300 kph and, finally, 20 km by taxi at an average speed of 30 km per hour. What is his average speed for the entire journey.*

Solution.

Calculation of weighted harmonic mean

Speed in km per h (X)	*Distance travelled (w)*	*w/X*
80	900	11.25
300	2000	6.66
30	20	0.66
Total	Σ (*w*) = 2920	Σ (*w/X*) = 18.57

$$\text{H.M.}_w = \text{Weighted H.M.} = \frac{\sum(w)}{\sum(w/X)} = \frac{2920}{18.57} = 157.24 \text{ km}$$

Here $\frac{w}{X}$ = Time taken

$$\text{Average} = \frac{\text{Total distance travelled}}{\text{Total time taken}}$$

Thus, the average speed is 157.24 kph. The weighted H.M. has been used as distance travelled at different speeds differ. **Ans.**

Example 21. *A distance of 100 kph is covered by a man as follows:*

(*i*) 80 kph in a car at a speed of 40 kph.

(*ii*) 15 kph on a cycle at a speed of 20 kph.

(*iii*) 5 kph on foot at a speed of 5 kph.

What was his average speed?

Solution.

Calculation of weighted harmonic mean

Speed in km per hour (X)	*Distance travelled (w)*	*w/X*
40	80	2.00
20	15	0.75
5	5	1.00
Total	Σ (*w*) = 100	Σ (*w/X*) = 3.75

$$\text{Weighted H.M.} = \frac{\sum(w)}{\sum(w/X)} = \frac{100}{3.75} = 26.66 \text{ kph}$$ **Ans.**

4.11 ADVANTAGES OF HARMONIC MEAN

(*i*) It satisfies the test of *rigid definition*. Its definition is precise and its value is always definite.

(*ii*) Like arithmetic mean (average) and geometric mean, this average is *also based on all the observations of the series*. It cannot be calculated in the absence of even a single figure.

(*iii*) Harmonic mean is *capable of further algebraic treatment*.

(*iv*) Like geometric mean, H.M. (average) is also *not affected very much by fluctuations of sampling*.

(*v*) *It gives a greater importance to small items* and, as such, a single big item cannot push up its value.

(*vi*) *It measures relative changes* and is extremely useful in averaging certain types of ratios and rates.

4.12 DISADVANTAGES OF HARMONIC MEAN

1. Harmonic mean *is not readily understood* nor can it be calculated with ease.
2. *It gives a very high weightage to small items* and for analysis of economic data, it is not very useful.
3. *It is usually a value which does not exist in a series.*
4. *Generally, it is not a good representative of a statistical series*, unless the phenomenon is such where small items have to be given a very high weightage.

4.13 RELATIONSHIP AMONG ARITHMETIC MEAN, GEOMETRIC MEAN AND HARMONIC MEAN

In any distribution unless all items are equal, there would be a difference in the value of arithmetic mean $(\overline{X})$, geometric mean (G.M.) and harmonic mean (H.M.). The condition would be

$$\overline{X} > \text{G.M.} > \text{H.M.}$$

which means that arithmetic mean would be more than geometric mean which would be more than harmonic mean.

If all the items in a distribution have the same value, then $\overline{X} = \text{G.M.} = \text{H.M.}$ Thus, we can say that in any case,

$$\overline{X} \geq \text{G.M.} \geq \text{H.M.}$$

which means that arithmetic average is equal to or greater than geometric mean which is equal to or greater than harmonic mean.

Example 22. *The arithmetic mean of two observations is 127.5, and their geometric mean is 60. Find their harmonic mean. Also obtain the observations.*

Solution. Let the two observations be a and b,

then $$\text{A.M.} = \frac{a+b}{2}, \quad \text{G.M.} = \sqrt{ab}$$

$$\text{H.M.} = \frac{2ab}{a+b} = \frac{ab}{\frac{a+b}{2}} = \frac{(\text{G.M.})^2}{\text{A.M.}} = \frac{(60)^2}{127.5} = \frac{60 \times 60}{127.5} = 28.23$$

$$\text{A.M.} = \frac{a+b}{2} = 127.5 \quad \Rightarrow \quad a + b = 255$$

$$\text{G.M.} = \sqrt{ab} = 60 \quad \Rightarrow \quad ab = (60)^2 = 3600$$

$$(a-b)^2 = (a+b)^2 - 4ab = (255)^2 - 4 \times 3600 = 65025 - 14400 = 50625$$

$\Rightarrow$ $a - b = \pm 225$

Solving $a + b = 255$ and $a - b = 225$, we get

$a = 240$ and $b = 15$,

and solving $a + b = 225$ and $a - b = -225$ we get

$a = 15$, $b = 240$

Hence, the numbers are 240 and 15. **Ans.**

Example 23. *Prove that between two positive quantities, say a and b.*

$$\text{A.M.} \geq \text{G.M.} \geq \text{H.M.}$$

Solution. $\text{A.M.} = \dfrac{a+b}{2}$, $\text{G.M.} = \sqrt{ab}$, $\text{H.M.} = \dfrac{2ab}{a+b}$

$$\text{A.M.} - \text{G.M.} = \frac{a+b-2\sqrt{ab}}{2} = \frac{\left(\sqrt{a}\right)^2 + \left(\sqrt{b}\right)^2 - 2\sqrt{a}\sqrt{b}}{2} = \left(\sqrt{a}-\sqrt{b}\right)^2 > 0$$

[$\because$ The square of a quantity is always positive]

A.M. – G.M. = a positive quantity $\Rightarrow$ A.M. > G.M.

$$\text{A.M.} \times \text{H.M.} = ab = (\text{G.M.})^2$$

$\Rightarrow$ $\dfrac{\text{A.M.}}{\text{G.M.}} = \dfrac{\text{G.M.}}{\text{H.M.}} \Rightarrow$ G.M. > H.M. [$\because$ A.M. > G.M.]

Hence, A.M. > G.M. > H.M.

If $a = b$, then A.M. = G.M. = H.M.

$\therefore$ A.M. $\geq$ G.M. $\geq$ H.M. **Proved.**

EXERCISE 4.1

Short Answer Questions (SAQs)

1. The first term of an A.P. is a, and the sum of the first p terms is zero, show that the sum of its next q terms is $\dfrac{-a(p+q)q}{p-1}$. [**Hint:** Required sum = $S_{p+q} - S_p$]

2. A man saved Rs. 66000 in 20 years. In each succeeding year after the first year he saved Rs. 200 more than what he saved in the previous year. How much did he save in the first year? **Ans.** Rs. 1400

3. A man accepts a position with an initial salary of Rs. 5200 per month. It is understood that he will receive an automatic increase of Rs. 320 in the very next month and each month thereafter.

   (*a*) Find his salary for the tenth month. **Ans.** Rs. 8080

   (*b*) What is his total earnings during the first year? **Ans.** Rs. 83520

4. If the pth and qth terms of a G.P. are q and p respectively, show that its $(p + q)$th term is $\left(\frac{p^p}{p^q}\right)^{\frac{1}{p-q}}$.

5. A carpenter was hired to build 192 window frames. On first day he made five frames and each day, thereafter he made two more frames than he made the day before. How many days did it take him to finish the job? **Ans.** 12 days

6. We know the sum of the interior angles of a triangle is 180°. Show that the sum of the interior angles of polygons with 3, 4, 5, 6, ... sides form an arithmetic progression. Find the sum of the interior angles for a 21 sided polygon. **Ans.** 3420°

7. A side of an equilateral triangle is 20 cm long. A second equilateral triangle is inscribed in it by joining the mid points of the sides of the first triangle. Find the perimeter of the sixth inscribed equilateral triangle. **Ans.** $\frac{15}{8}$ cm

8. In a potato race, 20 potatoes are placed in a line at intervals of 4 metres with the first potato 24 metres from the starting point. A contestant is required to bring the potatoes back to the starting place one at a time. How far would he run in bringing back all the potatoes? **Ans.** 2480 m

9. In a cricket tournament 16 school teams participated. A sum of Rs. 8000 is to be awarded among themselves as prize money. If the last placed team is awarded Rs. 275 in prize money and the award increases by the same amount for successive finishing places, how much amount will the first place team receive? **Ans.** Rs. 725

10. If $a_1, a_2, a_3, ..., a_n$ are in A.P., where $a_i > 0$ for all i, show that

$$\frac{1}{\sqrt{a_1}+\sqrt{a_2}}+\frac{1}{\sqrt{a_2}+\sqrt{a_3}}+...+\frac{1}{\sqrt{a_{n-1}}+\sqrt{a_n}}=\frac{n-1}{\sqrt{a_1}+\sqrt{a_n}}$$

11. Find the sum of the series
$(3^3 - 2^3) + (5^3 - 4^3) + (7^3 - 6^3) + ...$ to (*i*) n terms (*ii*) 10 terms
Ans. (*i*) $4n^3 + 9n^2 + 6n$ (*ii*) 4960

12. Find the rth term of an A.P., sum of whose first n terms in $2n + 3n^2$.
[Hint: $a_n = S_n - S_{n-1}$] **Ans.** $T_r = 6r - 1$

Long Answer Questions (LAQs)

13. If A is the arithmetic mean and G_1, G_2 be two geometric means between any two numbers, then prove that $2A = \frac{G_1^2}{G_2} + \frac{G_2^2}{G_1}$.

14. If $\theta_1, \theta_2, \theta_3, ..., \theta_n$ are in A.P., whose common difference is d, show that

$$\sec\theta_1 \sec\theta_2 + \sec\theta_2 \sec\theta_3 + ... + \sec\theta_{n-1} \sec\theta_n = \frac{\tan\theta_n - \tan\theta_1}{\sin d}.$$

15. If the sum of p terms of an A.P. is q and the sum of q terms is p, show that the sum of $p + q$ terms is $-(p + q)$. Also, find the sum of first $p - q$ terms $(p > q)$.

16. If pth, qth, and rth terms of an A.P. and G.P. are both a, b and c respectively, show that

$$a^{b-c} . b^{c-a} . c^{a-b} = 1$$

Multiple Choice Questions (MCQs)

Choose the correct answer out of the four given options in each of the Exercises 17 to 26 (MCQ)

17. If the sum of n terms of an A.P. is given by $S_n = 3n + 2n^2$, then the common difference of the A.P. is

(*a*) 3 (*b*) 2 (*c*) 6 (*d*) 4 **Ans.** (*d*)

18. The third term of G.P. is 4. The product of its first 5 terms is:

(*a*) 4^3 (*b*) 4^4 (*c*) 4^3 (*d*) none of these **Ans.** (*c*)

19. If 9 times the 9th term of an A.P. is equal to 13 times the 13th term, then the 22nd term of the A.P. is

(*a*) 0 (*b*) 22 (*c*) 220 (*d*) 198 **Ans.** (*a*)

20. If x, $2y$, $3z$ are in A.P., where the distinct number x, y, z are in G.P., then the common ratio of the G.P. is

(*a*) 3 (*b*) $\frac{1}{3}$ (*c*) 2 (*d*) $\frac{1}{2}$ **Ans.** (*b*)

21. If in an A.P., $S_n = qn^2$ and $S_m = qm^2$, where S, denotes the sum of r terms of the A.P., then S_q equals

(*a*) $\frac{q^3}{2}$ (*b*) mnq (*c*) q^3 (*d*) $(m + n)q^2$ **Ans.** (*c*)

22. Let S_n denote the sum of the first n terms of an A.P. If $S_{2n} = 3S_n$ then $S_{3n} : S_n$ is equal to

(*a*) 4 (*b*) 6 (*c*) 8 (*d*) 10 **Ans.** (*b*)

23. The minimum value of $4^x + 4^{1-x}$, $x \in R$, is

(*a*) 2 (*b*) 4 (*c*) 1 (*d*) 0 **Ans.** (*b*)

24. Let S_n denote the sum of the cubes of the first n natural numbers and s_n denote the sum of the first n natural numbers. Then $\sum_{r=1}^{n}\left(\frac{S_n}{s_n}\right)$ equals

(*a*) $\frac{n(n+1)(n+2)}{6}$ (*b*) $\frac{n(n+1)}{2}$ (*c*) $\frac{n^2+3n+2}{2}$ (*d*) none of these **Ans.** (*a*)

25. If t_n denotes the nth term of the series 2 + 3 + 6 + 11 + 18 + ... then t_{50} is

(*a*) $49^2 - 1$ (*b*) 49^2 (*c*) $50^2 + 1$ (*d*) $49^2 + 2$ **Ans.** (*d*)

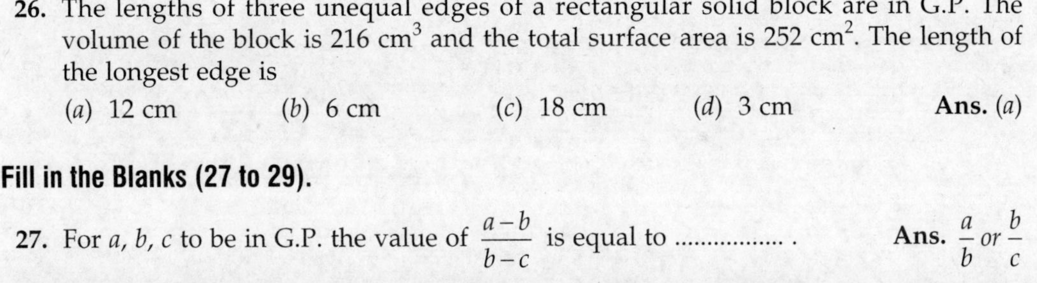

26. The lengths of three unequal edges of a rectangular solid block are in G.P. The volume of the block is 216 cm^3 and the total surface area is 252 cm^2. The length of the longest edge is

(*a*) 12 cm (*b*) 6 cm (*c*) 18 cm (*d*) 3 cm **Ans.** (*a*)

Fill in the Blanks (27 to 29).

27. For a, b, c to be in G.P. the value of $\frac{a-b}{b-c}$ is equal to .... **Ans.** $\frac{a}{b}$ *or* $\frac{b}{c}$

28. The sum of terms equidistant from the beginning and end in an A.P. is equal to .... **Ans.** First term + last term

29. The third term of a G.P. is 4, the product of the first five terms is ....**Ans.** 4^5

State whether the statements in questions 30 to 34 are True or False.

30. Two sequences cannot be in both A.P. and G.P. together. **Ans.** F

31. Every progression is a sequence but the converse, i.e. every sequence is also a progression need not necessarily be true. **Ans.** T

32. Any term of an A.P. (except first) is equal to half the sum of terms which are equidistant from it. **Ans.** T

33. The sum or difference of two G.P., is again a G.P. **Ans.** F

34. If the sum of n terms of a sequence is a quadratic expression then it always represents an A.P. **Ans.** F

Match the questions given under column I with their appropriate answers given under the column II.

35.

Column I	Column II
(*a*) $4, 1, \frac{1}{4}, \frac{1}{16}$	(*i*) A.P.
(*b*) 1, 3, 5, 7	(*ii*) sequence
(*c*) 13, 8, 3, –2, –7	(*iii*) G.P.

Ans. (*a*) ↔ (*iii*), (*b*) ↔ (*i*), (*c*) ↔ (*ii*)

36.

Column I	Column II
(*a*) $1^2 + 2^2 + 3^2 + ... + n^2$	(*i*) $\left(\frac{n(n+1)}{2}\right)^2$
(*b*) $1^3 + 2^3 + 3^3 + ... + n^3$	(*ii*) $n(n + 1)$
(*c*) $2 + 4 + 6 + ... + 2n$	(*iii*) $\frac{n(n+1)(2n+1)}{6}$
(*d*) $1 + 2 + 3 + ... + n$	(*iv*) $\frac{n(n+1)}{2}$

Ans. (*a*) ↔ (*iii*), (*b*) ↔ (*i*), (*c*) ↔ (*ii*) (*d*) ↔ (*iv*)

Partial Fractions

5.1 INTRODUCTION

We know that

$$\frac{1}{x+1}+\frac{1}{x+2}=\frac{2x+3}{(x+1)(x+2)} \qquad \text{... (1)}$$

$$\frac{2x+1}{x-5}-\frac{x}{x+4}=\frac{x^2+14x+4}{(x-5)(x+4)} \qquad \text{... (2)}$$

$$\frac{1}{x+3}-\frac{1}{x-3}+\frac{x}{x^2+2}=\frac{x^3-6x^2-9x-12}{(x+3)(x-3)(x^2+2)} \qquad \text{... (3)}$$

In (1), $\frac{1}{x+1}$ and $\frac{1}{x+2}$ **are the partial fractions of** $\frac{2x+3}{(x+1)(x+2)}$

In (2), $\frac{2x+1}{x-5}$ and $\frac{1}{x+2}$ **are the partial fractions of** $\frac{x^2+14x+4}{(x-5)(x+4)}$

In (3), $\frac{1}{x+3}, \frac{1}{x-3}$ and $\frac{x}{x^2+2}$ **are the partial fractions of** $\frac{x^3-6x^2-9x-12}{(x-3)(x-3)(x^2+2)}$

In this chapter, we will learn to resolve a given fraction into a group of partial fractions.

5.2 PARTIAL FRACTIONS

To express a single rational fraction into the sum of two or more single rational fractions is called **partial fraction resolution.**

For example,

$$\frac{2x+x^2-1}{x(x^2-1)}=\frac{1}{x}+\frac{1}{x-1}-\frac{1}{x+1}$$

$\frac{2x+x^2-1}{x(x^2-1)}$ is the resultant fraction and $\frac{1}{x}+\frac{1}{x-1}-\frac{1}{x+1}$ are its partial fractions.

5.3 POLYNOMIAL

Any expression of the form $P(x) = a_n x^n + a_{n-1} x^{n-1} + ... + a_2 x^2 + a_1 x + a_0$, where $a_n, a_{n-1}, ..., a_2, a_1, a_0$ are real constants, if $a_n \neq 0$ then $P(x)$ is called polynomial of degree n.

5.4 RATIONAL FRACTION

We know that $\frac{p}{q}, q \neq 0$ is called a rational number. Similarly, the quotient of two polynomials is $\frac{N(x)}{D(x)}$, where $D(x) \neq 0$, with no common factors, is called a rational fraction, it is of two types proper fraction and improper fraction:

Proper Fraction

A rational fraction $\frac{N(x)}{D(x)}$ is called a proper fraction if the degree of numerator $N(x)$ is less than the degree of denominator $D(x)$.

Examples:

(*i*) $\dfrac{9x^2 - 9x + 6}{(x-1)(2x-1)(x+2)}$ (*ii*) $\dfrac{6x + 27}{3x^3 - 9x}$

Improper Fraction

A rational fraction $\frac{N(x)}{D(x)}$ is called an improper fraction if the degree of the numerator $N(x)$ is greater than or equal to the degree of the denominator $D(x)$.

Examples:

(*i*) $\dfrac{2x^3 - 5x^2 - 3x - 10}{x^2 - 1}$ (*ii*) $\dfrac{6x^3 - 5x^2 - 7}{3x^2 - 2x - 1}$ (*iii*) $\dfrac{(2x+4)(x-1)}{3x^2 - 5}$

Note: An improper fraction can be expressed by division, as the sum of a polynomial and a proper fraction.

For example, $\dfrac{6x^3 + 5x^2 - 7}{3x^2 - 2x - 1} = (2x + 3) + \dfrac{8x - 4}{x^2 - 2x - 1}$ which is obtained by dividing $6x^3 + 5x^2 - 7$ by $3x^2 - 2x - 1$, then we get a polynomial $(2x + 3)$ and a proper fraction $\dfrac{8x - 4}{x^2 - 2x - 1}$

Steps:

Before resolving a given fraction into simpler partial fractions two steps are important.

Step I. The degree of numerator should be of lower degree than that of the denominator. If the degree of the numerator is greater or equal to the degree of denominator, then the numerator is divided by the denominator until the remainder of lower degree than the denominator.

$$\frac{\text{Numerator}}{\text{Denominator}} = \text{Quotient} + \frac{\text{remainder}}{\text{denominator}}$$

Step II. The denominator is factorised into simplest factors. Repeated factors should always be combined at one place.

Types of original fraction to be resolved into partial fractions

While solving questions on partial fractions, we would deal with the following **five cases**:

(*i*) The denominator has linear nonrepeated factors only.

(*ii*) The denominator has linear factors one or more of these being repeated.

(*iii*) The denominator has nonrepeated irreducible quadratic factors.

(*iv*) The denominator has repeated irreducible quadratic factors.

(*v*) The denominator has only even powers of variable both in the numerator as well as denominator.

(i) Type I

When the denominator contains nonrepeated linear factors only

The following steps need to be remembered:

Step 1. Let the given expression $= \dfrac{A}{I \text{ linear factor}} + \dfrac{B}{II \text{ linear factor}} + \ldots$

Step 2. Multiply both sides by L.C.M.

Step 3. Find the values of $A, B, \ldots$

Step 4. Substitute the values of $A, B, \ldots$ in step (1).

Notes:

1. Values of A, B and C can be found by putting the value of x.
2. Values of A, B and C can also be found by comparing the coefficients of like powers of x as has been done in Example 6.
3. **Short rule.** Suppress the factor $x - 2$ from the denominator of the given fraction and put $x = 2$ [$\because x - 2 = 0$] in the remaining part and thus value of A is obtained.

   Similarly, suppressing the factors $x + 2$ and $x + 4$ term by term, the values of B and C are obtained.

Example 1. *Resolve* $\dfrac{5x+12}{(x+2)(x+3)}$ *into partial fractions.* **[B.T.E, Delhi, 2016]**

Solution. Let $$\frac{5x+12}{(x+2)(x+3)} \equiv \frac{A}{x+2} + \frac{B}{x+3} \qquad \ldots (1)$$

$$\Rightarrow \quad \frac{5x+12}{(x+2)(x+3)} \equiv \frac{A(x+3)+B(x+2)}{(x+2)(x+3)}$$

$$\Rightarrow \quad 5x + 12 \equiv A\,(x+3) + B\,(x+2) \qquad \ldots (2)$$

[This is an identity. It is true for all values of x]

On putting $x = -3$, ($x + 3 = 0 \;\Rightarrow\; x = -3$) in (2), we get

$$-15 + 12 = A \times 0 + B\,(-3+2) \;\Rightarrow\; -3 = -B \;\Rightarrow\; B = 3$$

On putting $x = -2$, ($x + 2 = 0 \;\Rightarrow\; x = -2$) in (2), we get

$$-10 + 12 = A\,(-2+3) + B \times 0 \;\Rightarrow\; 2 = A$$

Substituting the values of A and B is (1), we have $\dfrac{5x+12}{(x+2)(x+3)} = \dfrac{2}{x+2} + \dfrac{3}{x+3}$

These are the required partial fractions. **Ans.**

Example 2. *Express* $\dfrac{3x+8}{6x^2-23x+20}$ *into partial fractions.*

Solution. The factors of $6x^2 - 23x + 20$ are $(3x - 4)$ and $(2x - 5)$.

Let $$\frac{3x+8}{6x^2-23x+20} \equiv \frac{A}{3x-4} + \frac{B}{2x-5} \quad \text{[A and B constants]} \quad ...(1)$$

$$\Rightarrow \quad \frac{3x+8}{6x^2-23x+20} \equiv \frac{A(2x-5)+B(3x-4)}{(3x-4)(2x-5)}$$

$$\Rightarrow \quad 3x + 8 \equiv A(2x-5) + B(3x-4) \quad ...(2)$$

Since the equation is identically true for all the values of x.

On putting $x = \dfrac{4}{3}$ in (2), we get $\left(3x-4=0,\ x=\dfrac{4}{3}\right)$

$$4 + 8 = A\left(\frac{8}{3} - 5\right) + B \times 0 \Rightarrow 12 = \frac{-7}{3}A \Rightarrow A = \frac{-36}{7}$$

On putting $x = \dfrac{5}{2}$ in (2), we get $\left(2x-5=0,\ x=\dfrac{5}{2}\right)$

$$\frac{15}{2} + 8 = A \times 0 + B\left(\frac{15}{2} - 4\right) \Rightarrow \frac{31}{2} = \frac{7}{2}B \Rightarrow B = \frac{31}{7}$$

On substituting the values of A and B is (1), we have

$$\frac{3x+8}{6x^2-23x+20} = \frac{-\frac{36}{7}}{3x-4} + \frac{\frac{31}{7}}{2x-5}$$

Hence, the partial fractional are $-\dfrac{36}{7(3x-4)} + \dfrac{31}{7(2x-5)}$ **Ans.**

Example 3. *If* $\dfrac{2x-1}{(x-2)(x+1)} = \dfrac{A}{x-2} + \dfrac{B}{x+1}$*; find A and B.* **[S.B.T.E. Dec. 2017]**

Solution. Here we have:

$$\frac{2x-1}{(x-2)(x+1)} = \frac{A}{x-2} + \frac{B}{x+1}$$

$$A = \left.\frac{2x-1}{x+1}\right|_{x=2} = \frac{4-1}{2+1}$$

$$\therefore \quad A = 1$$

$$B = \left.\frac{2x-1}{x-2}\right|_{x=-1} = \frac{-2-1}{-3}$$

$$\therefore \quad B = 1$$

Hence $A = 1, B = 1$ **Ans.**

Example 4. *If* $\frac{x+1}{(x-a)(x-3)} = \frac{2}{x-a} + \frac{b}{x-3}$, *find the value of 'a' and 'b'.*

[S.B.T.E. Delhi 2016, 2015]

Solution. Here we have

$$\frac{x+1}{(x-a)(x-3)} = \frac{2}{x-a} + \frac{b}{x-3}$$

$$(x+1) = 2(x-3) + b(x-a) \quad \text{... (1)}$$

put $x = a$ in (1), we get

$a = 7$

Again putting $x = 3$ in (1), we get

$b = -1$

Hence $a = 7, b = -1$ **Ans.**

Example 5. *Resolve* $\frac{x^3+4x^2+2x+2}{x^2-5x+6}$ *into partial fractions.* **[S.B.T.E. Dec. 2015]**

Solution. Given fraction is an improper fraction, the degree of numerator is greater than the denominator.

By division, we get

$$\frac{x^3+4x^2+2x+2}{x^2-5x+6} = x+1+\frac{x-4}{x^2-5x+6} \quad \text{... (1)}$$

Let $\frac{x-4}{x^2-5x+6} = \frac{x-4}{(x-3)(x-2)} = \frac{A}{x-3} + \frac{B}{x-2}$ ... (2)

$$x-4 = A(x-2) + B(x-3) \quad \text{... (3)}$$

putting $x = 3$ in (3), we get

$A = -1$

Again putting $x = 2$ in (3), we get $B = 2$

putting the value of A and B in (2), we get

$$\frac{x-4}{x^2-5x+6} = \frac{-1}{x-3} + \frac{2}{x-2}$$

$$\therefore \quad \frac{x^3+4x^2+2x+2}{x^2-5x+6} = x+1-\frac{1}{x-3}+\frac{2}{x-2}$$ **Ans.**

Example 6. *Resolve into partial fractions of* $\frac{x^2}{(x^2-1)(x+3)}$.

Solution. Let $\frac{x^2}{(x+1)(x-1)(x+3)} \equiv \frac{A}{x+1} + \frac{B}{x-1} + \frac{C}{x+3}$ ... (1)

$$\Rightarrow \quad \frac{x^2}{(x+1)(x-1)(x+3)} \equiv \frac{A(x-1)(x+3)+B(x+1)(x+3)+C(x+1)(x-1)}{(x+1)(x-1)(x+3)}$$

$$\Rightarrow \quad x^2 \equiv A(x-1)(x+3)+B(x+1)(x+3)+C(x+1)(x-1) \quad \text{...(2)}$$

On putting $x = -1$ in (2), we get $(x + 1 = 0 \Rightarrow x = -1)$

$$(-1)^2 = (-1-1)(-1+3) + (B \times 0) + (C \times 0) \Rightarrow 1 = -4A \Rightarrow A = -\frac{1}{4}$$

On putting $x = 1$, in (2), we get $(x - 1 = 0 \Rightarrow x = 1)$

$$(1)^2 = (A \times 0) + B(1+1)(1+3) + (C \times 0) \Rightarrow 1 = 8B \Rightarrow B = \frac{1}{8}$$

On putting $x = -3$, in (2), we get $(x + 3 = 0 \Rightarrow x = -3)$

$$(-3)^2 = (A \times 0) + (B \times 0) + C(-3+1)(-3-1) \Rightarrow 9 = 8C \Rightarrow C = \frac{9}{8}$$

On substituting the values of A, B and C in (1), we get

$$\frac{x^2}{(x+1)(x-1)(x+3)} = \frac{-\frac{1}{4}}{x+1} + \frac{\frac{1}{8}}{x-1} + \frac{\frac{9}{8}}{x+3}$$

Hence, the partial fractions are $\dfrac{-1}{4(x+1)} + \dfrac{1}{8(x-1)} + \dfrac{9}{8(x+3)}$ **Ans.**

Example 7. *Resolve* $\dfrac{x^4 - 10x^2 + 20x + 13}{x^3 + 3x^2 - x - 3}$ *into partial fractions.*

Solution. Here the degree of the numerator is higher than that of denominator.

$\therefore$ Dividing $x^4 - 10x^2 + 20x + 13$ by $x^3 + 3x^2 - x - 3$, we get

$$\frac{x^4 - 10x^2 + 20x + 13}{x^3 + 3x^2 - x - 3} = x - 3 + \frac{20x + 4}{x^3 + 3x^2 - x - 3} \qquad \text{... (1)}$$

Let $$\frac{20x + 4}{(x-1)(x+1)(x+3)} = \frac{A}{x-1} + \frac{B}{x+1} + \frac{C}{x+3} \qquad \text{... (2)}$$

$$\begin{array}{r|l} & x - 3 \\ \hline x^3 + 3x^2 - x - 3 & x^4 - 10x^2 + 20x + 13 \\ & x^4 \pm 3x^3 \mp x^2 \mp 3x \\ \hline & -3x^3 - 9x^2 + 23x + 13 \\ & \mp 3x^3 \mp 9x^2 \pm 3x \pm 9 \\ \hline & 20x + 4 \end{array}$$

$$\Rightarrow \quad 20x + 4 = A(x+1)(x+3) + B(x-1)(x+3) + C(x-1)(x+1) \qquad \text{... (3)}$$

Putting $x = 1$ in (3); we get $24 = A \times 8 \Rightarrow 8A = 24; \Rightarrow A = \dfrac{24}{8} = 3.$

Again putting $x = -1$ in (3); we get

$$16 = B(-2)(2) \Rightarrow -4B = -16 \Rightarrow B = \frac{16}{4} = 4.$$

Again putting $x = -3$ in (3); we get $-56 = C(-4)(-2)$

$$\Rightarrow \quad 8C = 56 \Rightarrow C = \frac{56}{8} = 7.$$

Putting the values of A, B and C in (2) and result in (1).

Hence, $x - 3 + \dfrac{3}{x-1} + \dfrac{4}{x+1} + \dfrac{7}{x+3}$ are required partial fractions. **Ans.**

EXERCISE 5.1

Resolve into partial fractions:

1. $\dfrac{x+1}{(x-2)(x-3)}$ **Ans.** $\dfrac{-3}{x-2}+\dfrac{4}{x-3}$

2. $\dfrac{5x+1}{x^2+x-2}$ **Ans.** $\dfrac{3}{x+2}+\dfrac{2}{x-1}$

3. $\dfrac{x^2+9x+7}{x^2+15x+56}$ **Ans.** $1-\dfrac{7}{x+7}+\dfrac{1}{x+8}$

4. $\dfrac{2x^2-61}{x^2-x-30}$ **Ans.** $2+\dfrac{1}{x+5}+\dfrac{1}{x-6}$

5. $\dfrac{5x^2-46x+102}{x^2-9x+20}$ **Ans.** $5+\dfrac{2}{x-4}-\dfrac{3}{x-5}$

6. $\dfrac{15x^2-58x+51}{(x-1)(x-2)(x-3)}$ **Ans.** $\dfrac{4}{x-1}+\dfrac{5}{x-2}+\dfrac{6}{x-3}$

7. $\dfrac{5x+7}{(x-2)(x-4)(x-6)}$ **Ans.** $\dfrac{17}{8(x-2)}-\dfrac{27}{4(x-4)}+\dfrac{37}{8(x-6)}$

8. $\dfrac{x^2-3x+4}{(x-2)(x+2)(x+4)}$ **Ans.** $\dfrac{1}{12(x-2)}-\dfrac{7}{4(x+2)}+\dfrac{8}{3(x+4)}$

9. $\dfrac{9x^2+34x+29}{x^3+6x^2+11x+6}$ **Ans.** $\dfrac{2}{x+1}+\dfrac{3}{x+2}+\dfrac{4}{x+3}$

10. $\dfrac{36x^2-37x+9}{6x^3-11x^2+6x-1}$ **Ans.** $\dfrac{4}{x-1}+\dfrac{2}{2x-1}+\dfrac{3}{3x-1}$

11. $\dfrac{x^3}{x^2-3x+2}$ **[B.T.E. Delhi Jan. 2009]** **Ans.** $x+3-\dfrac{1}{x-1}+\dfrac{8}{x-2}$

12. $\dfrac{(x-1)(x-2)(x-3)}{(x-4)(x-5)(x-6)}$ **Ans.** $1+\dfrac{3}{x-4}-\dfrac{24}{x-5}-\dfrac{30}{x-6}$

13. $\dfrac{6x^3+5x^2-7}{3x^2-2x-1}$ **Ans.** $2x+3+\dfrac{5}{3x+1}+\dfrac{1}{x-1}$

(ii) Type II

The denominator contains repeated linear factors

In case a factor is repeated three times in the denominator then the corresponding three partial fractions will have three denominators.

Example 8. *Resolve* $\dfrac{3x^2+x-2}{(x-1)^2(1-2x)}$ *into partial fractions.*

Solution. Let $\dfrac{3x^2+x-2}{(x-1)^2(1-2x)} = \dfrac{A}{1-2x}+\dfrac{B}{x-1}+\dfrac{C}{(x-1)^2}$ ...(1)

$$= \frac{A(x-1)^2+B(x-1)(1-2x)+C(1-2x)}{(1-2x)(x-1)^2}$$

$\therefore \quad 3x^2+x-2 = A(x-1)^2+B(x-1)(1-2x)+C(1-2x)$ ... (2)

On putting $x = \dfrac{1}{2}$ in (2), we get

$\therefore \quad \dfrac{3}{4}+\dfrac{1}{4}-2 = A\left(\dfrac{1}{2}-1\right)^2+(B\times 0)+(C\times 0) \Rightarrow -\dfrac{3}{4}=\dfrac{A}{4} \Rightarrow A=-3$

Again putting $x = 1$ in (2), we get

$\therefore \quad 3+1-2 = (A\times 0)+(B\times 0)+C(1-2) = -C \Rightarrow C=-2$

To find B, equating the coefficient of x^2 from both sides of (2), we get

$$3 = A-2B \Rightarrow 3 = -3-2B \Rightarrow 2B = -6 \Rightarrow B = -3. \quad [\because A=-3]$$

On putting the values of A, B and C in (1), we get

$$\frac{3x^2+x-2}{(x-1)^2(1-2x)} = \frac{-3}{1-2x}+\frac{-3}{x-1}+\frac{-2}{(x-1)^2}$$

Hence, the partial fractions are $-\dfrac{3}{1-2x}-\dfrac{3}{x-1}-\dfrac{2}{(x-1)^2}$ **Ans.**

Example 9. *Resolve into partial fractions:* $\dfrac{4x}{(x+1)^2(x-1)^2}$

Solution. Let $\dfrac{4x}{(x+1)^2(x-1)^2} \equiv \dfrac{A}{x+1}+\dfrac{B}{(x+1)^2}+\dfrac{C}{x-1}+\dfrac{D}{(x-1)^2}$ ... (1)

$4x \equiv A(x+1)(x-1)^2+B(x-1)^2+C(x-1)(x+1)^2+D(x+1)^2$ ... (2)

On putting $x = 1$ in (2), we get

$4 = D(1+1)^2 \Rightarrow D = 1$

On putting $x = -1$ in (2), we get

$-4 = B(-1-1)^2 \Rightarrow B = -1$

Comparing the coefficients of x^3 of on both sides of (2), we have

$0 = A + C$ ... (3)

Comparing the constant terms on both sides of (2), we have

$0 = A + B - C + D$

$0 = A-1-C+1 \Rightarrow A-C=0 \Rightarrow A=C$ ... (4)

From (3) and (4), we have $A = C = 0$

On putting the values of A, B, C and D in (1), we get

$$\frac{4x}{(x+1)^2(x-1)^2} = \frac{-1}{(x+1)^2} + \frac{1}{(x-1)^2}$$

Hence, these are the required partial fractions. **Ans.**

Example 10. *Resolve into partial fractions* $\frac{2x+3}{(x-1)^3(x-2)}$.

Solution. Let $$\frac{2x+3}{(x-1)^3(x-2)} \equiv \frac{A}{(x-1)} + \frac{B}{(x-1)^2} + \frac{C}{(x-1)^3} + \frac{D}{(x-2)} \quad \text{... (1)}$$

Multiplying both sides of (1) by L.C.M., we get

$$(2x-3) \equiv A(x-1)^2(x-2) + B(x-1)(x-2) + C(x-2) + D(x-1)^3 \quad \text{... (2)}$$

Putting $x = 1$ in (2), we get

$$2(1) + 3 = C(1-2) \quad \therefore \quad C = -5$$

Putting $x = 2$ in (2), we get

$$2(2) + 3 = D(2-1)^3 \quad \therefore \quad D = 7$$

Comparing the coefficients of x^3 on both sides of (2), we get

$$0 = A + D \quad \therefore \quad A = -D = -7$$

Comparing the constant terms on both sides of (2), we get

$$3 = -2A + 2B - 2C - D$$
$$2B = 2A + 2C + D + 3$$
$$= -14 - 10 + 7 + 3 = -14 \quad \therefore \quad B = -7$$

Putting the values of A, B, C and D in (1), the required partial fractions are

$$\frac{2x+3}{(x-1)^3(x-2)} = \frac{-7}{(x-1)} - \frac{7}{(x-1)^2} - \frac{5}{(x-1)^3} + \frac{7}{x-2}$$

Hence, these are the partial fractions. **Ans.**

EXERCISE 5.2

Resolve into partial fractions:

1. $\frac{x^2+1}{(x-2)^2(x+3)}$ **Ans.** $\frac{3}{5(x-2)} + \frac{1}{(x-2)^2} + \frac{2}{5(x+3)}$

2. $\frac{1}{(x^2+x)(x^2-1)}$ **Ans.** $-\frac{1}{x} + \frac{1}{4(x-1)} + \frac{3}{4(x+1)} + \frac{1}{2(x+1)^2}$

3. $\frac{3x^2+x-2}{(x-2)^2(1-2x)}$ **Ans.** $\frac{-5}{3(x-2)} - \frac{4}{(x-2)^2} - \frac{1}{3(1-2x)}$

4. $\frac{2x+1}{(x+2)^2(x-3)^2}$ **Ans.** $\frac{-3}{25(x+2)} + \frac{3}{25(x-3)} + \frac{7}{5(x-3)^2}$

5. $\frac{x}{(x+1)^2(x^2-1)}$ **Ans.** $\frac{1}{8(x-1)} - \frac{1}{8(x+1)} - \frac{1}{4(x+1)^2} + \frac{1}{2(x+1)^3}$

6. $\dfrac{x}{(x^2-1)^2}$ **Ans.** $\dfrac{1}{4}\left[\dfrac{1}{(x-1)^2}-\dfrac{1}{(x+1)^2}\right]$

7. $\dfrac{x+4}{(x-2)^3(x+1)}$ **Ans.** $\dfrac{1}{9(x-2)}-\dfrac{1}{3(x-2)^2}+\dfrac{2}{(x-2)^3}-\dfrac{1}{9(x+1)}$

(iii) Type III

When the denominator contains nonrepeated irreducible quadratic factor

The following steps should be remembered:

Given fraction is $\dfrac{1}{(x+a)(x+b)(x^2+cx+d)}$

Step 1. Let $\dfrac{1}{(x+a)(x+b)(x^2+cx+d)}=\dfrac{A}{x+a}+\dfrac{B}{x+b}+\dfrac{Cx+D}{x^2+cx+d}$ ... (1)

Step 2. Multiply by L.C.M.

Step 3. Find out A, B, C and D.

Step 4. Substitute the values of A, B, C and D in (1).

Example 11. *Resolve* $\dfrac{2x+1}{(x-1)(x^2+1)}$ *into partial fractions.*

Solution. Let $\dfrac{2x+1}{(x-1)(x^2+1)}=\dfrac{A}{x-1}+\dfrac{Bx+C}{x^2+1}$ ... (1)

$$=\frac{A(x^2+1)+(Bx+C)(x-1)}{(x-1)(x^2+1)}$$

$\Rightarrow$ $2x+1=A(x^2+1)+(Bx+C)(x-1)$ ... (2)

On putting $x = 1$ in (2), we get

$$2+1=A(1+1) \Rightarrow 2A=3 \Rightarrow A=\frac{3}{2}$$ ... (3)

Equating coefficient of x^2 and x from both sides of (3), we get

$$A+B=0 \Rightarrow B=-A \quad \therefore \quad B=-\frac{3}{2} \qquad \left[\because A=\frac{3}{2}\right]$$

Equating the coefficient of x on both sides of (3), we get

$$2=C-B \text{ or } C=2+B=2-\frac{3}{2}=\frac{1}{2} \qquad \left[\because C=\frac{1}{2}\right]$$

Putting the values of A, B and C in (1), we have

$$\frac{2x+1}{(x-1)(x^2+1)}=\frac{3}{2(x-1)}+\frac{-\frac{3}{2}x+\frac{1}{2}}{x^2+1}=\frac{3}{2(x-1)}-\frac{3x-1}{3(x^2+1)}$$

Hence, partial fractions are $\dfrac{3}{2(x-1)}-\dfrac{3x-1}{2(x^2+1)}$ **Ans.**

Example 12. *Resolve* $\dfrac{x^3+2x^2}{(x+1)(x^2+2x+2)}$ *into partial fractions.*

Solution. Divide the numerator by denominator as degree of both is same.

$$\frac{x^3+2x^2}{(x+1)(x^2+2x+2)} = 1-\frac{x^2+4x+2}{(x+1)(x^2+2x+2)}$$

Let $\dfrac{x^2+4x+2}{(x+1)(x^2+2x+2)} = \dfrac{A}{x+1}+\dfrac{Bx+C}{x^2+2x+2}$ ... (1)

$\therefore$ $x^2 + 4x + 2 = A(x^2 + 2x + 2) + (Bx + C)(x + 1)$ ... (2)

On putting $x = -1$ in (1), we get

$$(-1)^2 + 4(-1) + 2 = A(1 - 2 + 2) \Rightarrow A = -1$$

Equating the coefficients of x^2 on both sides of (2), we get $1 = A + B$

But $A = -1$ $\therefore$ $B = 2$

Equating constant terms on both sides of (2), we get $2 = 2A + C$

But $A = -1$ $\therefore$ $C = 4$

Thus $\dfrac{x^3+2x^2}{(x+1)(x^2+2x+2)} = 1+\dfrac{1}{x+1}-\dfrac{2x+4}{x^2+2x+2}$

Hence, these are the required partial fractions. **Ans.**

Example 13. *Resolve into partial fractions* $\dfrac{3x-2}{(x+2)(x^2+2)}$ **[S.B.T.E. Dec. 2015]**

Solution. Let $\dfrac{3x-2}{(x+2)(x^2+2)} \equiv \dfrac{A}{x+2}+\dfrac{Bx+C}{x^2+2}$ ... (1)

Multiplying both sides of (1) by L.C.M., we get

$$(3x - 2) \equiv A(x^2 + 2) + (Bx + C)(x + 2) \quad \text{... (2)}$$

Putting $x = -2$ in (2), we get

$$3(-2) - 2 = A(4 + 2)$$

$\Rightarrow$ $-8 = 6A \Rightarrow A = -\dfrac{4}{3}$

Comparing the coefficient of x^2 on both sides of (2), we get

$$0 = A+B \Rightarrow B = -A = -\left(-\frac{4}{3}\right) = \frac{4}{3}$$

Comparing constant terms on both sides of (2), we get

$$-2 = 2A + 2C$$

$\Rightarrow$ $2C = -2 - 2A$

$\Rightarrow$ $2C = -2-2\left(-\dfrac{4}{3}\right) = -2+\dfrac{8}{3} = \dfrac{2}{3} \Rightarrow C = \dfrac{1}{3}$

Putting the values of A, B, C in (1), we get

$$\frac{3x-2}{(x+2)(x^2+2)} = \frac{-4}{3(x+2)} + \frac{\frac{4}{3}x+\frac{1}{3}}{x^2+2}$$

$$= \frac{-4}{3(x+2)} + \frac{4x+1}{3(x^2+2)}$$

Hence, these are the required partial fractions. **Ans.**

Example 14. *Resolve into partial fractions* $\dfrac{x}{(x^2+1)(x^2+x-2)}$

Solution. $\dfrac{x}{(x^2+1)(x^2+x-2)} = \dfrac{x}{(x^2+1)(x+2)(x-1)}$

Let $\dfrac{x}{(x^2+1)(x^2+x-2)} \equiv \dfrac{Ax+B}{x^2+1} + \dfrac{C}{x+2} + \dfrac{D}{x-1}$ ... (1)

Multiplying both sides of (1) by L.C.M., we get

$$x \equiv (Ax+B)(x+2)(x-1) + C(x^2+1)(x-1) + D(x+2)(x^2+1) \quad ...\ (2)$$

Putting $x = -2$ in (2), we get

$$-2 = C(4+1)(-2-1) \Rightarrow C = \frac{2}{15}$$

Putting $x = 1$ in (2), we get

$$1 = D(1+1)(1+2) \Rightarrow D = \frac{1}{6}$$

Comparing coefficient of x^3 on both sides of (2), we get

$$0 = A + C + D$$

$$\therefore \quad A = -C - D = -\frac{2}{15} - \frac{1}{6} = -\frac{3}{10}$$

Comparing constant terms on both sides of (2), we get

$$0 = -2B - C + 2D \Rightarrow 2B = -C + 2D$$

$$\Rightarrow \quad 2B = -\frac{2}{15} + \frac{1}{3} = \frac{1}{5} \Rightarrow B = \frac{1}{10}$$

Putting the values of A, B, C and D in (1), the required partial fractions are

$$\frac{x}{(x^2+1)(x^2+x-2)} = \frac{-\frac{3}{10}x+\frac{1}{10}}{x^2+1} + \frac{2}{15(x+2)} + \frac{1}{6(x-1)}$$

$$= -\frac{3x-1}{10(x^2+1)} + \frac{2}{15(x+2)} + \frac{1}{6(x+1)}$$ **Ans.**

Example 15. *Resolve into partial fractions* $\dfrac{1}{x^3+1}$.

Solution. We have, $\dfrac{1}{x^3+1} = \dfrac{1}{(x+1)(x^2-x+1)}$ $\quad [\because a^3 + b^3 = (a+b)(a^2 - ab + b^2)]$

Let $\dfrac{1}{(x+1)(x^2-x+1)} \equiv \dfrac{A}{x+1} + \dfrac{Bx+C}{x^2-x-1}$ ... (1)

Multiplying both sides of (1) by L.C.M., we get

$$1 = A(x^2 - x + 1) + (Bx + C)(x + 1) \quad \text{... (2)}$$

Putting $x = -1$ in (2), we get

$$1 = A(1 + 1 + 1) \Rightarrow A = \frac{1}{3}$$

Comparing the coefficient of x^2 on both sides of (2), we get

$$0 = A + B \Rightarrow B = -A = -\frac{1}{3}$$

Comparing constant terms on both sides of (2), we get

$$1 = A + C \Rightarrow C = 1 - A = 1 - \frac{1}{3} = \frac{2}{3}$$

Putting the values of A, B and C in (1) the required partial fractions are

$$\frac{1}{x^3+1} = \frac{1}{3(x+1)} - \frac{x-2}{3(x^2-x+1)}$$ **Ans.**

EXERCISE 5.3

Resolve into partial fractions:

1. $\dfrac{42-19x}{(x-4)(x^2+1)}$ **Ans.** $\dfrac{-2}{x-4} + \dfrac{2x-11}{x^2+1}$

2. $\dfrac{8x^2-9}{(3x+2)(x^2+5)}$ **Ans.** $-\dfrac{1}{3x+2} + \dfrac{3x-2}{x^2+5}$

3. $\dfrac{2x+3}{x^3+1}$ **Ans.** $\dfrac{1}{3(x+1)} + \dfrac{-x+8}{3(x^2-x+1)}$

4. $\dfrac{x}{(x^3-1)(x+2)}$ **Ans.** $\dfrac{1}{9(x-1)} + \dfrac{2}{9(x+2)} - \dfrac{x}{3(x^2+x+1)}$

5. $\dfrac{x+2}{(x^2+x-6)(x^2+x+6)}$ **Ans.** $\dfrac{1}{60(x+3)} + \dfrac{1}{15(x-2)} - \dfrac{x+2}{12(x^2+x+6)}$

6. $\dfrac{x}{(1-x)(1+x)(x^2+4)}$ **Ans.** $\dfrac{1}{10(1-x)} - \dfrac{1}{10(1+x)} + \dfrac{x}{5(x^2+4)}$

7. $\dfrac{x^5}{x^4-1}$ **Ans.** $x + \dfrac{1}{4(x-1)} + \dfrac{1}{4(x+1)} - \dfrac{x}{2(x^2+1)}$

8. $\dfrac{19-17x-2x^2}{(x+3)(x^2-4x+5)}$ **Ans.** $\dfrac{2}{x+3} - \dfrac{4x-3}{x^2-4x+5}$

9. $\dfrac{1}{(x^2+x+1)(x^2+1)}$ **Ans.** $\dfrac{x+1}{x^2+x+1} - \dfrac{x}{x^2+1}$

(iv) Type IV

When the denominator contains repeated irreducible quadratic factors

In fact this is the combination of types II and III.

Example 16. *Resolve into partial fractions* $\dfrac{x}{(x^2+1)^2(x-1)}$.

Solution. Let $\dfrac{x}{(x^2+1)^2(x-1)} \equiv \dfrac{Ax+B}{x^2+1}+\dfrac{Cx+D}{(x^2+1)^2}+\dfrac{E}{x-1}$ ... (1)

Multiplying both sides of (1) by L.C.M., we get

$$x = (Ax+B)(x^2+1)(x-1)+(Cx+D)(x-1)+E(x^2+1)^2 \quad \text{... (2)}$$

Putting $x = 1$ in (2), we get

$$1 = E(1+1)^2 \Rightarrow E = \frac{1}{4}$$

Comparing the coefficient of x^4 on both sides of (2), we get

$$0 = A + E \Rightarrow A = -E = -\frac{1}{4}$$

Comparing the coefficient of x^3 on both sides of (2), we get

$$0 = -A + B \Rightarrow B = A = -\frac{1}{4}$$

Comparing the coefficient of x^2 on both sides of (2), we get

$$0 = -A - B + C + 2E$$

$$\Rightarrow \quad C = -A + B - 2E = \frac{1}{4}-\frac{1}{4}-\frac{2}{4}=-\frac{1}{2}$$

Comparing constant terms on both sides of (2), we get

$$0 = -B - D + E$$

$$\Rightarrow \quad D = -B + E = \frac{1}{4}+\frac{1}{4}=\frac{1}{2}$$

Putting the values of A, B, C, D and E in (1), the required partial fractions are

$$\frac{-\frac{1}{4}x-\frac{1}{4}}{x^2+1}+\frac{-\frac{1}{2}x-\frac{1}{2}}{(x^2+1)^2}+\frac{\frac{1}{4}}{x-1}$$

i.e. $$\frac{x+1}{4(x^2+1)}-\frac{x-1}{2(x^2+1)^2}+\frac{1}{4(x-1)}$$ **Ans.**

Example 17. *Break up into partial fractions:* $\dfrac{x^3+x^2-x+1}{(x^2+1)(x-1)^2}$

Solution. Let $\dfrac{x^3+x^2-x+1}{(x^2+1)(x-1)^2} = \dfrac{A}{x-1}+\dfrac{B}{(x-1)^2}+\dfrac{Cx+D}{x^2+1}$ ... (1)

$$\text{L.C.M.} = (x^2+1)(x-1)^2$$

Multiplying both sides by $(x^2 + 1)(x - 1)^2$, we get

$$x^3 + x^2 - x + 1 = A(x-1) + B(x^2+1) + (Cx+D)(x-1)^2 \quad \text{... (2)}$$

Putting $x = 1$ in (2), we get

$$1 + 1 - 1 + 1 = A(1 - 1) + B(1 + 1) + (Cx + D)(1 - 1)^2$$

$$\Rightarrow \quad 2 = 2B \Rightarrow B = 1 \quad \text{... (3)}$$

Comparing the coefficients of x^3, x^2 and the constant terms on both sides of (2), we get

$$1 = A + C \quad \text{... (4)}$$

$$1 = -A + B - 2C + D \quad \text{... (5)}$$

$$1 = -A + B + D \quad \text{... (6)}$$

Solving (4), (5), (6), simultaneously, we get

$$A = 1, B = 1, C = 0, D = 1$$

Putting these values in (1), we get the partial fractions as

$$\frac{1}{x-1} + \frac{1}{(x-1)^2} + \frac{1}{x^2+1} \quad \textbf{Ans.}$$

EXERCISE 5.4

1. $\dfrac{x}{(1-x)(1+x^2)^2}$ **Ans.** $\dfrac{1}{4(1-x)} + \dfrac{x+1}{4(1+x^2)} + \dfrac{x-1}{2(1+x^2)^2}$

2. $\dfrac{2x^4 + 2x^2 + x + 1}{x(x^2+1)^2}$ **Ans.** $\dfrac{1}{x} + \dfrac{1-x}{(1+x^2)^2} + \dfrac{x}{1+x^2}$

3. $\dfrac{x^2}{(1+x)(1+x^2)^2}$ **Ans.** $\dfrac{1}{4(1+x)} - \dfrac{x-1}{4(1+x^2)} + \dfrac{x-1}{2(1+x^2)^2}$

4. Resolve $\dfrac{4x}{(1+x)^2(x-1)^2}$ into partial fractions. **[S.B.T.E. 2007]**

   Ans. $\dfrac{1}{(x-1)^2} - \dfrac{1}{(1+x)^2}$

5. $\dfrac{3x^2 + x - 2}{(x-2)^2(1-2x)}$ into partial fractions. **[S.B.T.E. 2008]**

   Ans. $\dfrac{-5}{3(x-2)} - \dfrac{4}{(x-2)^2} + \dfrac{1}{3(2x-1)}$

6. $\dfrac{x}{(x^2+x-2)(x^2-x+2)}$ **Ans.** $\dfrac{x+1}{4(x^2+x-2)} + \dfrac{x-1}{4(x^2-x+2)}$

7. $\dfrac{1}{(x^2+x-2)(x^2-x+2)}$ **Ans.** $\dfrac{1}{6(x-1)} - \dfrac{1}{24(x+2)} - \dfrac{x-1}{8(x^2-x+2)}$

8. $\dfrac{x}{(x-1)(x-2)}$ **Ans.** $\dfrac{1}{3(x+1)} + \dfrac{2}{3(x-2)}$

(v) Type V

When only even powers of *x* occurs in both numerator and denominator of the given fraction

Substitution: Put $x^2 = y$

Example 18. *Resolve into partial fractions* $\dfrac{x^2}{(x^2+4)(x^2+1)}$.

Solution. Put $x^2 = y$

$$\frac{x^2}{(x^2+4)(x^2+1)} = \frac{y}{(y+4)(y+1)}$$

Let $$\frac{y}{(y+4)(y+1)} = \frac{A}{y+4} + \frac{B}{y+1} \quad \ldots (1)$$

Multiplying both sides of (1) by L.C.M., we get

$$y \equiv A(y+1) + B(y+4) \quad \ldots (2)$$

Putting $y = -4$ in (2), we get

$$-4 = A(-4+1) \Rightarrow A = \frac{4}{3}$$

Putting $y = -1$ in (2), we get

$$-1 = B(3) \Rightarrow B = -\frac{1}{3}$$

Putting the values of A and B in (1), we get

$$\frac{y}{(y+4)(y+1)} = \frac{4}{3(y+4)} - \frac{1}{3(y+1)} \quad \ldots (3)$$

Again putting $y = x^2$ in (3), we get

$$\frac{x^2}{(x^2+4)(x^2+1)} = \frac{4}{3(x^2+4)} - \frac{1}{3(x^2+1)}$$ **Ans.**

Example 19. *Resolve into partial fractions* $\dfrac{x^2}{(x^2+2)(x^2-9)}$.

Solution. Since all the powers of x are even, put $x^2 = y$, so that the expression becomes $\dfrac{y}{(y+2)(y-9)}$

Let $$\frac{y}{(y+2)(y-9)} = \frac{A}{y+2} + \frac{B}{y-9} \quad \ldots (1)$$

$$\text{L.C.M.} = (y+2)(y-9)$$

Multiplying both sides by L.C.M., we get

$$y = A(y-9) + B(y+2) \quad \ldots (2)$$

Putting $y = -2$ in (2), we get

$$-2 = -11A \Rightarrow A = \frac{2}{11}$$

Putting $y = 9$ in (2), we get

$$9 = 11B \Rightarrow B = \frac{9}{11}$$

$$\therefore \quad \frac{y}{(y+2)(x-9)} = \frac{2}{11}\cdot\frac{1}{y+2} + \frac{9}{11}\cdot\frac{1}{y-9}$$

Putting these values in (1), we get

$$\frac{x^2}{(x^2+2)(x^2-9)} = \frac{2}{11}\times\frac{1}{x^2+2}\times\frac{9}{11}\times\frac{1}{x^2-9} \quad \text{... (3)}$$

Also, $\frac{1}{x^2-9}$ can be resolved into partial fractions as,

$$\frac{1}{x^2-9} = \frac{1}{6}\left[\frac{1}{x-3} - \frac{1}{x+3}\right]$$

Putting the partial fractions of $\frac{1}{x^2-9}$ in (3), we get the partial fractions as,

$$\frac{2}{11(x^2+2)} + \frac{3}{22(x-3)} - \frac{3}{22(x+3)} \quad \textbf{Ans.}$$

Example 20. *Resolve* $\frac{1}{x^4+x^2+1}$ *into partial fractions.* **[S.B.T.E. Dec. 2016, 2015]**

Solution. Here we have

$$\frac{1}{x^4+x^2+1} = \frac{1}{(x^2+x+1)(x^2-x+1)} = \frac{Ax+B}{x^2+x+1} + \frac{Cx+D}{x^2-x+1} \quad \text{... (1)}$$

$$\Rightarrow \quad 1 = (Ax+B)(x^2-x+1) + (Cx+D)(x^2+x+1)$$

$$\Rightarrow \quad 1 = (A+C)x^3 + (-A+B+C+D)x^2 + (A-B+C+D)x + (B+D) \quad \text{... (2)}$$

Equating the coefficient of x^3, x^2, x and the constant term both sides of (2), we get

$$A + C = 0 \quad \text{... (3)}$$

$$-A + B + C + D = 0 \quad \text{... (4)}$$

$$A - B + C + D = 0 \quad \text{... (5)}$$

$$B + D = 0 \quad \text{... (6)}$$

On solving (3), (4), (5) and (6), we get

$$A = \frac{1}{2}, B = \frac{1}{2}, C = -\frac{1}{2}, D = \frac{1}{2}$$

putting the values of A, B, C and D in (1), we get

$$\frac{1}{x^4+x^2+1} = \frac{x+1}{2(x^2+x+1)} + \frac{(-x+1)}{2(x^2-x+1)} \quad \textbf{Ans.}$$

EXERCISE 5.5

1. $\dfrac{x^2}{(x^2+a^2)(x^2+b^2)}$ **Ans.** $\dfrac{1}{(a^2-b^2)}\left[\dfrac{a^2}{x^2+a^2}-\dfrac{b^2}{x^2+b^2}\right]$

2. $\dfrac{4x^4+3}{(x^2+4)(x^2+1)(x^2+3)}$ **Ans.** $\dfrac{67}{3(x^2+4)}+\dfrac{7}{6(x^2+1)}-\dfrac{39}{2(x^2+3)}$

3. $\dfrac{x^2+3}{(x^2+5)(x^2-7)(x^2+9)}$ **Ans.** $\dfrac{1}{24(x^2+5)}+\dfrac{5}{96(x^2-7)}-\dfrac{3}{32(x^2+9)}$

4. $\dfrac{8x^3+46x^2-24x+120}{(x^2-4)(x^2+4)}$ **Ans.** $\dfrac{10}{x-2}-\dfrac{9}{x+2}+\dfrac{7x+8}{x^2+4}$

5. $\dfrac{3x+7}{(x+3)(x^2+1)}$ **Ans.** $-\dfrac{1}{5(x+3)}+\dfrac{x+12}{5(x^2+1)}$

6. $\dfrac{1}{x^3+1}$ **Ans.** $\dfrac{1}{3(x+1)}+\dfrac{(-x+2)}{3(x^2-x+1)}$

Permutations

6.1 INTRODUCTION

In this chapter, we shall discuss the problems of arranging of certain things, taking a particular number of things at a time. For example, 12 and 21 are different arrangements of 1 and 2 but represents the same group of 1 and 2. The order of things plays an important role in arrangements.

Suppose, we have four digits 1, 2, 3, 4. Now we want to know how many numbers can be formed by these four digits. We start listing all possible arrangements of four digits 1, 2, 3 and 4. But this method is very tedious because the number of arrangements may be large. Here in this chapter, we will learn the basic counting techniques. With the help of these techniques we can find the number of arrangements without listing them.

6.2 FUNDAMENTAL PRINCIPLE OF COUNTING (FPC)

(1) **Addition law:** If there are two operations such that they can be performed independently in m and n ways respectively, then either of the two operations can be performed in $m + n$ ways.

Example 1. *Suppose there are three flags of different countries. Keeping two of them in a particular order represents a particular signal. The following 6 signals (arrangements) can be formed by these 3 different flags.*

Solution: Total number of signals = 6 = 3 × 2 **Ans.**

Example 2. *Now, suppose we have 3 shirts and 4 pairs of pants. In how many possible ways can we dress up by wearing a shirt and a pair of pants. In this case, we can wear any of three shirts. After wearing one shirt we can wear any one of these pairs of pants with it. If we label the shirts as S_1, S_2, S_3 and the pant as P_1, P_2, P_3, and P_4 then the different ways of dressing up can be as under.*

Pants / Shirts	P_1	P_2	P_3	P_4
S_1	S_1P_1	S_1P_2	S_1P_3	S_1P_4
S_2	S_2P_1	S_2P_2	S_2P_3	S_2P_4
S_3	S_3P_1	S_3P_2	S_3P_3	S_3P_4

Solution:

Total number of ways = 4 + 4 + 4 = 12.

In this example, we add the number of ways in which we can wear a shirt and the number of ways in which we can wear a pair of pants.

(2) **Multiplication Law:** If one operation can be performed in m ways, and if corresponding to each of the m ways of performing this operation, there are n ways of performing a second operation, then the number of ways of performing the two operations together is $m \times n$.

Example 3. In a class there are 20 boys and 16 girls. The teacher wants to select a boy and a girl to represent the class in a function. In how many ways can the teacher make this selection?

Solution: Here, the teacher has to perform two operations:

(*i*) Selecting a boy from among 20 boys, and

(*ii*) Selecting a girl from among 16 girls.

The first of these can be performed in 20 ways and the second in 16 ways. Therefore, by the fundamental principle of multiplication, the required number of ways is $20 \times 16 = 320$.

Note: If we want to assemble a computer, there are m types of monitors, n types of keyboards, r types of C.P.U., etc. Then the number of ways of assembling a computer is $m \times n \times r \times ...$

Example 4. *How many three digit odd numbers can be formed from the digits 1, 2, 3, 4, 5. If the digits can be repeated.*

Solution: There will be as many ways as ways of filling three vacant squares in succession by the given digits.

First square can be filled by any five given digits in succession. □□□

Second square can be filled by any five digits and third square can be filled by three odd numbers 1, 3, 5. Thus, the numbers of ways in which three places can be filled by $5 \times 5 \times 3 = 75$ ways.

Hence, the required number of three digits odd number is 75. **Ans.**

Example 5. *Find the number of four letter words with or without meaning which can be formed out of the letter RULE, where the repetition of the letters is not allowed.*

Solution: There are as many words as there are ways of filling in four vacant squares by the four letters.

□ □ □ □

We have to remember that the repetition is not allowed.

First square can be filled in four different ways any one of the four letters R, U, L, E.

Second square can be filled by any of the remaining three letters in three different ways.

Third square can be filled by the remaining two letters in two ways.

Fourth square can be filled by the remaining one letter in one way.

Thus the number of ways in which four squares can be filled is $4 \times 3 \times 2 \times 1 = 24$

Hence, the required number of words = 24. **Ans.**

Example 6. *In how many ways can 3 persons be seated in a row containing 7 seats.*

Solution: First person can be seated in 7 ways.

When the first person has taken his seat, the number of seats for the second person $= 7 - 1 = 6$

$\therefore$ Second person can be seated in 6 ways and the third can be seated in 5 ways.

By the principal of F.P.C., total number of ways in which three persons can be seated in seven seats in a row = $(7 \times 6 \times 5)$ ways = 210 ways **Ans.**

Example 7. *How many 3-digit number can be formed from the digits 1, 2, 3, 4 and 5 assuming.*

(i) repetition of digits allowed

(ii) repetition of digits not allowed ?

Solution: (*i*) There are five digits 1, 2, 3, 4 and 5.

Every digit can be selected any number of times.

Hence, we can select first digit 5 times; the second digit 5 times and the third digit 5 times.

Hence, the number of ways in which the selection of three digits can be made

$$= 5 \times 5 \times 5 = 125 \text{ ways.}$$

(*ii*) Under the restriction, first digit can be selected in 5 ways.

After the selection of first digit, four digits are left.

Second digit can be selected in 4 ways and the third digit can be selected in 3 ways.

According to F.P.C. total number of ways = $5 \times 4 \times 3 = 60$ ways **Ans.**

Example 8. *How many 4 letter codes can be formed using the first 10 letters of the English alphabets, if no letter can be repeated.*

Solution: The first place can be filled by 10 different letters of the English alphabet in 10 different ways. Following which, the second place can be filled in by any one of the remaining 9 letters in 9 different ways.

Following which the third place can be filled in 8 different ways and the 4th place can be filled in 7 different ways.

Thus, the number of ways in which the 4-letters code can be formed by (multiplication principle)

$$= 10 \times 9 \times 8 \times 7 = 5040 \text{ ways}$$ **Ans.**

EXERCISE 6.1

1. A coin is tossed twice and the outcomes are recorded. Find the number of possible outcomes? **Ans.** 4
2. the digit from 0 to 9, are written on a slip of paper and placed in a box. Three of the slips of paper are drawn and placed in order. How many different outcomes are possible. **Ans.** 720
3. Given 4 flags of different colours, how many different signals can be generated if the signal requires the use of two flags, one below the other. **Ans.** 12
4. How many words (with or without meaning) of three distinct letters of the English alphabet are there? **Ans.** 15600
5. How many three-digit numbers more than 600 can be formed by using the digits 2, 3, 4, 6 and 7? **Ans.** 50
6. How many 2-digit even numbers can be formed from the digits 1, 2, 3, 4, and 5, if the digits can be repeated. **Ans.** 10
7. How many numbers are there between 100 and 1000 such that every digit is either 2 or 9. **Ans.** 8
8. How many numbers are there between 100 and 1000 such that 7 is in the unit place. **Ans.** 90
9. How many odd numbers less than 1000 can be formed using the digits 0, 1, 8, 9 (repetitions of digits is allowed). **Ans.** 32
10. Raj proposes to go to his friend Santosh's house in a town which is connected with his village by three different routes. From there he will go to the city where his uncle lives. The city is connected with the town by two different routes. List out the various possible routes which Raj can choose from his village to go to the city. **Ans.** 6
11. Let there be three jobs *a*, *b*, *c* and three persons *A*, *B* and *C*. Find the possible number of assignments of jobs so that one person is assigned only one job. **Ans.** 6
12. A coin is tossed twice and the outcomes are recorded. Find the number of possible outcomes. **Ans.** 4
13. Numbers 1, 2 and 3 are written on three cards. How many two-digits numbers can be formed by placing two cards side by side? **Ans.** 6
14. How many 3-letter code words are possible using the first 10 letters of English alphabet if

    (*i*) no letter can be repeated? (*ii*) letters are repeated? **Ans.** (*i*) 720, (*ii*) 1000
15. How many 2-digit even numbers can be formed from the digits 1, 2, 3, 4, and 5 if the digits can be repeated? **Ans.** 10
16. In how many ways can 5 persons sit in a car, when 2 including the driver sit in the front seat and 3 in the back seat, if 2 particular persons out of the 5 are to avoid the driver's seat. **Ans.** 72
17. Find the number of different signals that can be generated by arranging at least two flags in order (one below the other) on a vertical shaft, if five different flags are available. **Ans.** 320
18. How many 4 letter code words are possible using the first 10 letters of the English alphabet if no letter can be repeated? **Ans.** 5040
19. The digit from 0 to 9, are written on slips of paper and placed in a box. Three of the slips of paper are drawn and places in order. How many different outcomes are possible? **Ans.** 720

20. A sample of 3 bulbs is tested. A bulb is labelled as 'G' if it is good and 'D' if it is defective. Find the number of all the possible outcomes. **Ans.** 8
[**Hint:** GDD is one such outcome]

21. Given 4 flags of different colours, how many different signals can be generated if the signal requires the use of two flags, one below the other? **Ans.** 12

22. How many numbers can be formed the digits 1, 2, 3, 9 if repetition of digits is not allowed? **Ans.** 64

23. One team consists of 6 boys and 4 girls and the other team has 5 boys and 3 girls. How many single matches can be arranged between the two teams when a boy plays against a boy and a girl plays against a girl? **Ans.** 42

24. How many words (with or without meaning) of three distinct letters of English alphabets are there? **Ans.** 25

25. Find the total number of ways of answering 5 objective type questions, each question having 4 choices. **Ans.** 45

26. How many numbers are there between 100 and 1000 in which all the digits are distinct. **Ans.** 648

27. How many numbers are there between 100 and 1000 such that every digit is either 2 or 9. **Ans.** 8

28. How many numbers are there between 100 and 1000 such that 7 is in the units place. **Ans.** 90

29. How many three-digit numbers more than 600 can be formed by using the digits 2, 3, 4, 6 and 7? **Ans.** 50

30. How many odd numbers less than 1000 can be formed using the digits 0, 1, 8, 9 (repetitions of digits is allowed). **Ans.** 32

31. How many different numbers below 1000 can be formed from the digits 3, 7, 6, 4 and 8 if

(*i*) no digit is repeated (*ii*) digits can be repeated **Ans.** (*i*) 85, (*ii*) 155

32. It has been decided that the flag of a newly formed forum will be in three blocks, each coloured differently. If there are seven different colours on the whole to choose from, how many such designs are possible? **Ans.** $7 \times 6 \times 5 = 210$

33. From the digits 1, 2, 3, 4, 5, 6, how many three digit odd numbers can be formed when:

(*i*) the repetition of digits is allowed.

(*ii*) the repetition of digits is not allowed.

Ans. (*i*) $6 \times 6 \times 3 = 108$ (ii) $3(4 \times 5 \times 1) = 60$

Hints to Selected Questions

1. Number of times a coin is tossed = 2.

∴ If we consider a coin, it has two different faces, *i.e.* head (H) and tail (T). So in two tosses the possible outcomes could be as follows:

(1) Either it could be H in first toss and H in second = HH

(2) Either H in first or T in second = HT

(3) Either T in first or H in Second = TH

(4) Either T in first or T in second = TT

So all outcomes are HH, HT, TH and TT = 4 outcomes

5. As shown, we can use any of the five digits for tens and unit places, *i.e.*

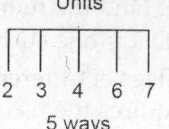

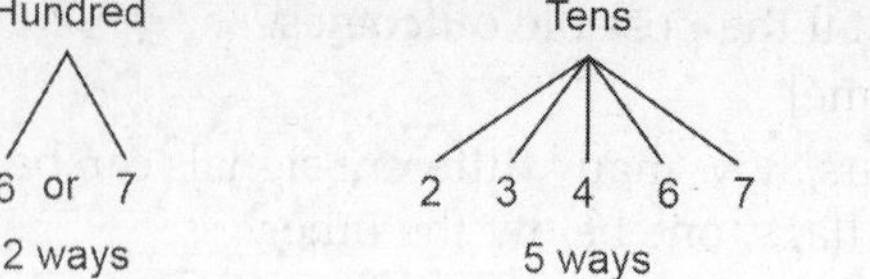

So, Total number of numbers = 2 × 5 × 5 = 50 numbers

6. As we have to consider only even numbers, So, the digit at the ones place can either be 2 or 4 only in this case.

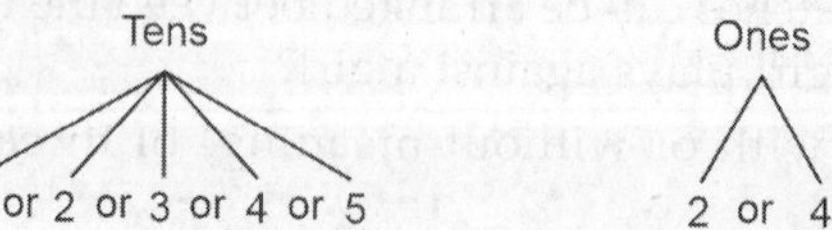

(a) So we can say digit in ones place can be selected in 2 ways.

(b) Digit in tens place can be selected any of the 5 digit, so number of ways for tens digit = 5 ways.

So, By FPC, total number of ways = 5 × 2 = 10 ways.

7. As shown all the three places can be filled in 2 ways each. Hence

Total number of numbers = 2 × 2 × 2 = 8 numbers

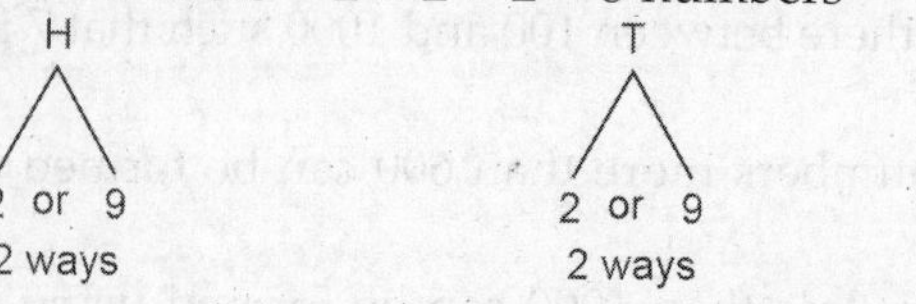

8.

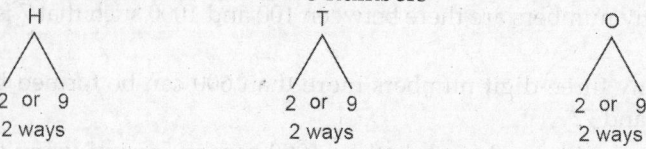

H Hundred	T Tens	O Unit
Any of the 10 digit except '0'	Any of the ten digits	7 (fixed) only 1 digit

Since the number has to be a three digit number so at hundred place we cannot use '0', *i.e.* 9 digits only.

We can use any of the ten digits at tens place as we can repeat the digits in this case. And we can have only one digit *i.e.* '7' as is required by the question.

Hence, Total number of digits = 9 × 10 × 1 = 90 numbers

9. Now we can make single digit, two digit and three digit odd numbers using these four digits.

[∴ the number have to be less than 1000]

So single digit odd numbers = 2

[∴ either 1 or 9 can be used as odd numbers in this case]

Two digit odd numbers can be made by choosing 1 or 9 any ones place and 1, 8, 9 at tens place as we cannot use '0' at tens place.

So,

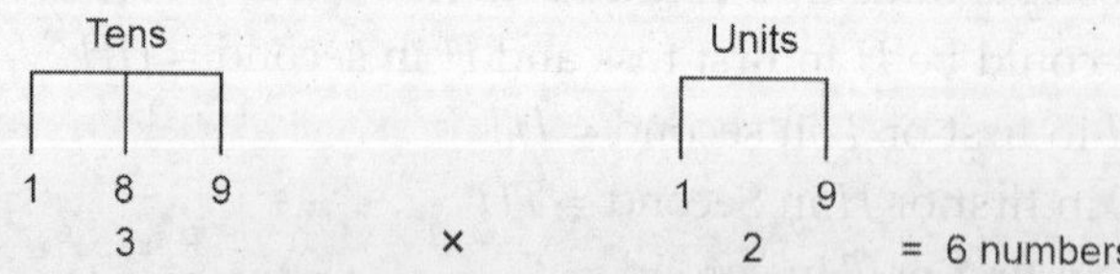

In three digit numbers unit place can be filled by 1 or 9 [∴ number has to be odd]

Tens place by any of the four digit and hundred place by only 1, 8, 9 as '0' if used at hundred place will make it 3 digit number. So,

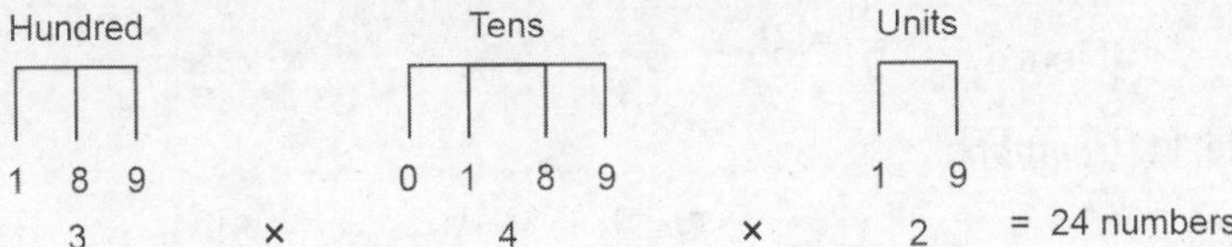

So, by FPC, total number of numbers = 24 + 6 + 2 = 32 numbers.

6.3 FACTORIAL NOTATION

Many times we come across products of the form 1×2, $1 \times 2 \times 3$, $1 \times 2 \times 3 \times 4$, ...

For our convenience, we use a special notation instead of writing all the factors of such a product. We write,

$$1! = 1$$
$$2! = 1 \times 2$$
$$3! = 1 \times 2 \times 3$$
$$n! = 1 \times 2 \times 3 \times ... \times n$$

Definition. *The continued product of first n natural numbers is called the n factorial and is denoted by n! or* $\underline{|n}$

For example ; $4! = 4 \times 3 \times 2 \times 1 = 24$, $5! = 5 \times 4 \times 3 \times 2 \times 1 = 120$

Note: $n!$ is defined for positive integer only.

6.4 DEDUCTION

$$n! = n(n-1)(n-2)(n-3) ... 3.2.1$$
$$= n[(n-1)(n-2)(n-3)... 3.2.1] = n[(n-1)!]$$

Thus
$$5! = 5 \times 4!$$
$$1! = 1 \times (0!) \Rightarrow 0! = 1$$

Example 9. *Evaluate:*

(*i*) 8! (*ii*) 4! – 3! (*iii*) $2 \times 6! - 3 \times 5!$

Solution: (*i*) We have,

$$8! = 8 \times 7 \times 6 \times 5 \times 4 \times 3 \times 2 \times 1 = 40320$$

(*ii*) We have,

$$4! - 3! = 4 \times 3! - 3! = 3!(4-1) = 3! \times 3 = 1 \times 2 \times 3 \times 3 = 18$$

(*iii*) $2 \times 6! - 3 \times 5! = 2 \times 6 \times 5! - 3 \times 5! = 5!(2 \times 6 - 3) = 5!(12 - 3)$

$$= (5 \times 4 \times 3 \times 2 \times 1) \times 9 = 120 \times 9 = 1080$$ **Ans.**

Example 10. (*i*) *Compute:* $\frac{8!}{4!}$. (*ii*) $\frac{8!}{4!} = 2!$?

Solution: We have,

(*i*) $$\frac{8!}{4!} = \frac{8 \times 7 \times 6 \times 5 \times 4!}{4!} = 8 \times 7 \times 6 \times 5 = 1680$$

(*ii*) Again $2! = 2 \times 1 = 2 \neq 1680$

Hence $\frac{8!}{4!} \neq 2!$

Example 11. *Compute:*

(*i*) $\frac{20!}{18!(20-18)!}$ (*ii*) $\frac{7!}{4! \times 3!}$ (*iii*) $\frac{8!}{6! \times 2!}$ **Ans.**

Solution: (*i*) We have, $\frac{20!}{18!(20-18)!} = \frac{20!}{18!(2)!} = \frac{20 \times 19 \times 18!}{18! \times 2 \times 1} = \frac{20 \times 19}{2 \times 1} = 190$

(*ii*) We have, $\frac{7!}{4! \times 3!} = \frac{7 \times 6 \times 5 \times 4!}{4! \times 3 \times 2 \times 1} = \frac{7 \times 6 \times 5}{3 \times 2 \times 1} = 35$

(*iii*) We have, $\frac{8!}{6! \times 2!} = \frac{8 \times 7 \times 6!}{6! \times 2 \times 1} = 28$ **Ans.**

Example 12. *If* $\frac{1}{6!} + \frac{1}{7!} = \frac{x}{8!}$; *find x.*

Solution: We have,

$$\frac{1}{6!} + \frac{1}{7!} = \frac{x}{8!} \Rightarrow \frac{8 \times 7}{8 \times 7 \times 6!} + \frac{8}{8 \times 7!} = \frac{x}{8!} \Rightarrow \frac{56}{8!} + \frac{8}{8!} = \frac{x}{8!}$$

$$\Rightarrow 56 + 8 = x \Rightarrow x = 64$$

Ans.

Example 13. *Evaluate:*

(*i*) $(n-r)!$, When $n = 6, r = 2$

(*ii*) $\frac{n!}{(n-r)!}$, when $n = 9, r = 5$

Solution: (*i*) Here, $n = 6$ and $r = 2$. Then

$$(n-r)! = (6-2)! = 4! = 4 \times 3 \times 2 \times 1 = 24$$

(*ii*) Here, $n = 9$, $r = 5$. Then

$$\frac{n!}{(n-r)!} = \frac{9!}{(9-5)!} = \frac{9!}{4!} = \frac{9 \times 8 \times 7 \times 6 \times 5 \times 4!}{4!} = 9 \times 8 \times 7 \times 6 \times 5 = 15120$$ **Ans.**

Example 14. *Evaluate:*

(*i*) $\frac{n!}{(n-r)!}$, when $r = 3$ (*ii*) $\frac{n!}{r!(n-r)!}$, when $n = 15, r = 12$

Solution: (*i*) Here, $r = 3$. Then

$$\frac{n!}{(n-r)!} = \frac{n!}{(n-3)!} = \frac{n(n-1)(n-2)(n-3)!}{(n-3)!}$$

$$= n(n-1)(n-2) = n(n^2 - 3n + 2) = n^3 - 3n^2 + 2n$$ **Ans.**

(*ii*) Here, $n = 15, r = 12$; Then

$$\frac{n!}{r!(n-r)!} = \frac{15!}{12!(15-12)!} = \frac{15!}{12! \times 3!} = \frac{12 \times 14 \times 13 \times 12!}{12! \times 3!} = \frac{15 \times 14 \times 13}{3 \times 2 \times 1}$$

$$= 5 \times 7 \times 13 = 455$$ **Ans.**

Example 15. *Convert the following product into factorials:*

(*i*) $6 \cdot 7 \cdot 8 \cdot 9 \cdot 10$ (*ii*) $2 \cdot 4 \cdot 6 \cdot 8 \cdot 10$

Solution: (*i*) We have,

$$6 \cdot 7 \cdot 8 \cdot 9 \cdot 10 = \frac{1 \cdot 2 \cdot 3 \cdot 4 \cdot 5 \cdot 6 \cdot 7 \cdot 8 \cdot 9 \cdot 10}{1 \cdot 2 \cdot 3 \cdot 4 \cdot 5} = \frac{10!}{5!}$$

(*ii*) We have,

$$2 \cdot 4 \cdot 6 \cdot 8 \cdot 10 = (2 \times 1) \cdot (2 \times 2) \cdot (2 \times 3) \cdot (2 \times 4) \cdot (2 \times 5)$$

$$= 2^5 \times (1 \cdot 2 \cdot 3 \cdot 4 \cdot 5) = 2^5 \times 5!$$ **Ans.**

EXERCISE 6.2

Evaluate:

1. $4!$ **Ans.** 24

2. $6!$ **Ans.** 720

3. $6! - 5!$ **Ans.** 600

4. $6! - 8!$ **Ans.** – 39600

5. $3 \times 6! - 4 \times 5!$ **Ans.** 1680

6. $3 \times 4! + 7 \times 4!$ **Ans.** 240

7. $(n - r)!$ when $n = 9, r = 5$ **Ans.** 24

8. Prove that $n!(n + 2) = n! + (n + 1)!$

9. Compute : $\dfrac{52!}{(47!)(5!)}$ **Ans.** 2598960

10. Evaluate: $\dfrac{n!}{r!(n-r)!}$, when (*i*) $n = 6, r = 2$, (*ii*) $n = 7, r = 4$ **Ans.** (*i*) 15, (*ii*) 35

11. Find x, if: (*i*) $\dfrac{1}{4!} + \dfrac{1}{5!} = \dfrac{1}{6!} + \dfrac{x}{7!}$ (*ii*) $\dfrac{1}{10!} + \dfrac{x}{2(11!)} = \dfrac{252}{12!}$ **Ans.** (*i*) 245, (*ii*) 20

12. Find x, if $\dfrac{(2x+1)!}{4!(x-1)} = \dfrac{11!}{8^2\, 9}$. **Ans.** 5

13. If $(n + 2)! = 2550\,(n!)$, find n **Ans.** 49

Prove that :

14. $\dfrac{n!}{(n-r)!} = n\,(n-1)\,(n-2) \ldots [n - (r-1)]$

15. $\dfrac{n!}{(n-r)!\,r!} + \dfrac{n!}{(n-r+1)!\,(r-1)!} = \dfrac{(n+1)!}{r!\,(n-r+1)!}$

16. $\dfrac{(2n+1)!}{n!} = 2^n [1 \cdot 3 \cdot 5 \ldots (2n-1)\,(2n+1)]$

Hints to Selected Questions

12. $\dfrac{(2x+1)!}{4!(x-1)!} = \dfrac{11!}{64 \times 9} \Rightarrow \dfrac{(2x+1)!}{(x-1)!} = \dfrac{11!}{8 \times 3} \Rightarrow \dfrac{(2x+1)!}{(x-1)!} = \dfrac{11!}{4!}$

On putting $x = 5$, we get L.H.S. = R.H.S. Hence $x = 5$.

15. L.H.S. $= \dfrac{n!}{(n-r)!\,r!} + \dfrac{n!}{(n-r+1)!\,(r-1)!} = n!\left[\dfrac{r+(n-r+1)}{(n-r+1)!\,r!}\right]$

$$= n!\left[\frac{n+1}{(n-r+1)!\,r!}\right] = \frac{(n+1)!}{(n+r+1)!\,r!} = \text{R.H.S.} \qquad [\because (n+1)! = (n+1)\,n!]$$

16. L.H.S. $= \dfrac{(2n+1)!}{n!} = \dfrac{1\cdot 2\cdot 3\cdot 4\cdot 5\cdot 6 \ldots (2n-1)\,(2n)\,(2n+1)}{n!}$

$$= \frac{[1\cdot 3\cdot 5\cdot 7\,(2n-1)\,(2n+1)]\,[2\cdot 4\cdot 6\cdot 8 \ldots 2n]}{n!}$$

$$= \frac{[1\cdot 3\cdot 5\cdot 7 \ldots (2n-1)\,(2n+1)]\ \cdot 2^n\ n!}{n!}$$

$$= 2^n\ [1\cdot 3\cdot 5\cdot 7 \ldots (2n-1)\,(2n+1)] = \text{R.H.S.}$$

6.5 PERMUTATIONS

Each of the different arrangements which can be made by taking some or all of a number of things at a time is called permutation.

Notation. The number of permutations of n things taken r at a time is denoted by ^nP_r or $P(n, r)$. The letter P is an abbreviation of the word permutations.

Thus, $^6P_4$ denotes the number of permutations or arrangements of 6 things taken 4 at a time.

A few examples of permutations are:

1. Arrangement of books on a self.
2. Formation of numbers with the given digits.
3. Formation of words with the given letters.

6.6 THE VALUE OF ^nP_r

To find the number of permutations of n different things, taken r at a time or to determine ^nP_r. The number of permutations of n things taken r at a time will be the same as the number of ways in which r blank place can be filled up with n given things.

As the first place can be filled in by any one of the n things so there are n ways of filling up the first place.

After having filled in the first place by any one of the n things, there are $(n - 1)$ things left. Hence, the second place can be filled in $(n - 1)$ ways. Now, as for every one way of filling up the first place, there are $(n - 1)$ ways of filling up the second place, so the first two places can by filled in $n\,(n - 1)$ ways.

Position of the object	1st	2nd	...	$(n-1)$th	nth
Number of ways	n	$n-1$	...	2	1

After having filled in the first two places in any one of the above ways, there are $(n-2)$ things left and so the third place can be filled in $(n-2)$ ways. Now for every one way of filling up the first two places there are $(n-2)$ ways of filling up the third place and so the first three places can be filled up in $n(n-1)(n-2)$ ways.

It may be observed that

(a) At every stage the number of factors is equal to the number of places filled up.

(b) Every factor is less than its preceding factor.

Hence, the number of ways of filling up all the r places, *i.e.* the number of permutations of n different things taken r at a time is $n(n-1)(n-2) \ldots r$ factors

$$= n(n-1)(n-2) \ldots (n-(r-1))$$

Hence, $\boxed{{}^nP_r = n(n-1)(n-2) \ldots (n-r-1)}$

Theorem 1. Prove that $P(n, r) = {}^nP_r = \dfrac{n!}{(n-r)!}$

Proof: We have, $P(n, r) = n(n-1)(n-2)(n-3) \ldots (n-r+1)$

$$= \frac{n(n-1)(n-2)(n-3) \ldots (n-(r-1))(n-r)(n-(r+1)) \ldots 3 \cdot 2 \cdot 1}{(n-r)(n-(r+1)) \ldots 3 \cdot 2 \cdot 1}$$

$$\boxed{P(n, r) = \frac{n!}{(n-r)!}}$$

Proved.

Theorem 2. Prove that: $P(n, n) = n!$

Proof: The number of all permutations of n distinct things, taken all at a time is same as the number of ways of filling n places when we have n distinct things at our disposal.

We know that, $P(n, r) = n(n-1)(n-2) \ldots [n-(r-1)]$

$$\Rightarrow \quad P(n, n) = n(n-1)(n-2) \ldots [n-(n-1)]$$

$$= n(n-1)(n-2) \ldots 3 \cdot 2 \cdot 1 = n!$$

$$\Rightarrow \quad \boxed{P(n, n) = n!}$$

Proved.

Theorem 3. Prove that: $0! = 1$

Proof: We have, $P(n, r) = \dfrac{n!}{(n-r)!} \Rightarrow P(n, n) = \dfrac{n!}{0!}$ [Putting $r = n$]

$$\Rightarrow \quad n! = \frac{n!}{0!} \qquad [\because P(n, n) = n!]$$

$$\Rightarrow \quad 0! = \frac{n!}{n!} = 1 \Rightarrow \boxed{0! = 1}$$

Proved.

Type I. To find the value of ${}^nP_r$ *or* $P(n, r)$:

Example 16. Evaluate the followings:

(*i*) ${}^4P_2$ (*ii*) $P(5, 3)$ (*iii*) $P(6, 6)$

Solution: We know that,

$$P(n, r) = {}^nP_r = \frac{n!}{(n-r)!}$$

(*i*) We have, $\quad {}^4P_2 = \dfrac{4!}{(4-2)!} = \dfrac{4\times3\times2\times1}{2\times1} = 12$

(*ii*) We have, $\quad P(5,3) = \dfrac{5!}{(5-3)!} = \dfrac{5!}{2!} = \dfrac{5\times4\times3\times2!}{2!} = 60$

(*iii*) We have, $\quad P(6,6) = \dfrac{6!}{(6-6)!} = \dfrac{6!}{0!} = 6! \qquad [\because 0! = 1]$

$$= 6\times5\times4\times3\times2\times1 = 720$$

Ans.

Type II. To find the unknown when a relation connecting *P* (*n*, *r*) is given.

Example 17. *Find r if:*

(*i*) ${}^5P_r = 2\cdot{}^6P_{r-1}$ $\qquad$ (*ii*) $5\times{}^4P_r = 6\times{}^5P_{r-1}$

Solution: (*i*) We have,

$$P(5,r) = 2P(6,r-1) \Rightarrow \frac{5!}{(5-r)!} = \frac{2(6!)}{[6-(r-1)]!}$$

$$\Rightarrow \frac{5!}{(5-r)!} = \frac{2(6!)}{(7-r)!} \Rightarrow \frac{5!}{(5-r)!} = \frac{2\times6(5!)}{(7-r)(6-r)(5-r)!}$$

$$\Rightarrow \frac{5!}{(5-r)!} = \frac{12(5!)}{(7-r)(6-r)(5-r)!} \Rightarrow (7-r)(6-r) = 12$$

$$\Rightarrow 42-13r+r^2 = 12 \Rightarrow r^2-13r+30 = 0 \Rightarrow r^2-3r-10r+30 = 0$$

$$\Rightarrow r(r-3)-10(r-3) = 0 \Rightarrow (r-3)(r-10) = 0$$

$$\Rightarrow r = 3 \text{ or } r = 10, \text{ but } r \neq 10$$

Hence, $r = 3$ **Ans.**

(*ii*) We have, $5\cdot P(4,r) = 6\cdot P(5,r-1)$

$$\Rightarrow \quad 5\frac{4!}{(4-r)!} = 6\frac{5!}{[5-(r-1)]!} \Rightarrow 5\frac{4!}{(4-r)!} = \frac{6\times5(4!)}{(6-r)!}$$

$$\Rightarrow \quad \frac{1}{(4-r)!} = \frac{6}{(6-r)!} \Rightarrow (6-r)! = 6(4-r)!$$

$$\Rightarrow (6-r)(5-r)(4-r)! = 6(4-r)! \Rightarrow (6-r)(5-r) = 6$$

$$\Rightarrow \quad 30-11r+r^2 = 6 \Rightarrow r^2-11r+24 = 0 \Rightarrow r^2-8r-3r+24 = 0$$

$$\Rightarrow \quad r(r-8)-3(r-8) = 0 \Rightarrow (r-3)(r-8) = 0$$

$$\Rightarrow \quad r = 3 \text{ or } r = 8 \text{ but } r \neq 8$$

Hence, $\quad r = 3$ **Ans.**

Example 18. *Find n if* ${}^{n-1}P_3 : {}^nP_4 = 1:9$.

Solution: We have,

$$P(n-1,3) : P(n,4) = 1:9$$

$$\Rightarrow \frac{(n-1)!}{(n-1-3)!} : \frac{n!}{(n-4)!} = 1:9 \Rightarrow \frac{(n-1)!}{(n-4)!}\times\frac{(n-4)!}{n!} = \frac{1}{9}$$

$$\Rightarrow \quad \frac{(n-1)!}{n\times(n-1)!}=\frac{1}{9} \quad \Rightarrow \quad \frac{1}{n}=\frac{1}{9} \quad \Rightarrow \quad n=9$$

Example 19. *If* ${}^{22}P_{r+1} : {}^{20}P_{r+2} = 11:52$*, find r.*

Solution: We have, ${}^{22}P_{r+1} : {}^{20}P_{r+2} = 11:52$

$$\Rightarrow \quad \frac{22!}{(21-r)!}:\frac{20!}{(18-r)!}=11:52 \quad \Rightarrow \quad \frac{22!}{(21-r)!}\times\frac{(18-r)!}{20!}=\frac{11}{52}$$

$$\Rightarrow \quad \frac{22\times 21\times 20!}{(21-r)(20-r)(19-r)\cdot(18-r)!}\times\frac{(18-r)!}{20!}=\frac{11}{52}$$

$$\Rightarrow \quad \frac{22\times 21}{(21-r)(20-r)(19-r)}=\frac{11}{52}$$

$$\Rightarrow \quad (21-r)(20-r)(19-r)=2\times 21\times 52$$

$$\Rightarrow \quad (21-r)(20-r)(19-r)=2\times 3\times 7\times 4\times 13$$

$$\Rightarrow \quad (21-r)(20-r)(19-r)=14\times 13\times 12$$

$$\Rightarrow \quad (21-r)(20-r)(19-r)=(21-7)(20-7)(19-7)$$

$$\Rightarrow \quad r=7$$

Ans.

Type III. To prove results related to ${}^nP_r$ *or* $P(n, r)$

Example 20. *Prove that:*

(*i*) ${}^nP_n = 2\cdot{}^nP_{n-2}$ (*ii*) ${}^{10}P_3 = {}^9P_3 + 3\,{}^9P_2$

Solution: (*i*) $\text{L.H.S} = P(n,n) = \dfrac{n!}{(n-n)!}$ $\left[\because P(n,r)=\dfrac{n!}{(n-r)!}\right]$

$= \dfrac{n!}{0!} = n!$ $[\because 0! = 1]$

$$\text{R.H.S.} = 2.P(n, n-2) = \frac{2(n!)}{[n-(n-2)]!} = \frac{2(n!)}{(n-n+2)!} = \frac{2(n!)}{2!} = n!$$

Here, L.H.S. = R.H.S. = $n!$

Hence, ${}^nP_n = 2\cdot{}^nP_{n-2}$ **Proved.**

(*ii*) $$\text{L.H.S.} = P(10,3) = \frac{10!}{(10-3)!} = \frac{10!}{7!} = \frac{10\times 9\times 8\times 7!}{7!} = 10\times 9\times 8 = 720$$

$R.H.S. = P(9,3) + 3P(9,2) = \dfrac{9!}{(9-3)!} + 3.\dfrac{9!}{(9-2)!}$ $\left[\because P(n,r)=\dfrac{n!}{(n-r)!}\right]$

$$= \frac{9\times 8\times 7\times 6!}{6!} + 3\times\frac{9\times 8\times 7!}{7!}$$

$$= (9\times 8\times 7) + (3\times 9\times 8) = 72\,(7+3) = 720$$

Here, L.H.S. = R.H.S. = 720

Hence, ${}^{10}P_3 = {}^{9}P_3 + 3 \cdot {}^{9}P_2$ **Proved.**

EXERCISE 6.3

Evaluate the following:

1. ${}^{5}P_2$ **Ans.** 20
2. ${}^{7}P_3$ **Ans.** 210
3. ${}^{10}P_4$ **Ans.** 5040

Find *n* if:

4. ${}^{n}P_2 = 30$ **Ans.** 6
5. ${}^{n}P_4 : {}^{n-1}P_3 = 9:1$ **Ans.** 9
6. $2P(\text{n}, 3) = P(n+1, 3)$ **Ans.** 5

Find *r* if:

7. $4 \cdot {}^{6}P_r = {}^{6}P_{r+1}$ **Ans.** 2
8. ${}^{11}P_r = {}^{12}P_{r-1}$ **Ans.** 9
9. $5P(4, r) = 6P(5, r-1)$ **Ans.** 3
10. Find all r and n such that $1 \le r \le n \le 12$ and such that $P(n, r)$ is a prime number. **Ans.** $r = 1$, $n = 2, 3, 5, 7, 11$

Prove that:

11. $P(n, r) = (n - r + 1)\, P(n, r - 1)$
12. $P(n, r) = n \cdot P(n - 1, r - 1)$
13. $P(12, 7)$ is divisible by $P(12, 4)$

Find '*n*' if

14. ${}^{2n}P_{n+1} : {}^{2n-2}P_n = 56:3$ **Ans.** 4
15. ${}^{2n}P_3 = 100 \cdot {}^{n}P_2$ **Ans.** 13
16. $P(10, r+1) : P(11, r) = 30:11$ **Ans.** 5
17. If $P(m + n, 2) = 56$ and $P(m - n, 2) = 12$, find m and n. **Ans.** $m = 6$, $n = 2$
18. If $P(n + 5, n + 1) = \dfrac{11(n-1)}{2} P(n + 3, n)$, find n. **Ans.** 6, 7
19. $1 \cdot P(1, 1) + 2 \cdot P(2, 2) + 3 \cdot P(3, 3) + \ldots {}^{n}P(n, n) = P(n + 1, n + 1) - 1$
20. $P(12, 7)$ is divisible by $P(12, 4)$
21. Show that $\sum_{r=1}^{10} r.{}^{r}P_r = {}^{11}P_{11} - 1$

Hints to Selected Questions

16. $P(10, r+1) : P(11, r) = 30:11$

$$ {}^{10}P_{r+1} : {}^{4}P_r = \frac{30}{11} \Rightarrow \frac{10!}{[10-(r+1)]!} : \frac{11!}{(11-r)!} = \frac{30}{11}$$

$$\Rightarrow \frac{10!}{(9-r)!} \times \frac{(11-r)!}{11!} = \frac{30}{11} \Rightarrow 10! \frac{(11-\text{r})\,(10-r)\,(9-r)!}{(9-r)!\,11.10!} = \frac{30}{11}$$

$$\Rightarrow (11-r)(10-r) = 30 \Rightarrow 110 - 10r - 11r + r^2 = 30$$

$$\Rightarrow \quad 110 - 21r + r^2 = 30 \quad \Rightarrow \quad r^2 - 21r + 80 = 0$$

$$\Rightarrow \quad r^2 - 16r - 5r + 80 = 0 \quad \Rightarrow \quad r(r-16) - 5(r-16) = 0$$

$$\Rightarrow \quad (r-5)(r-16) = 0$$

So $r = 5$ or $r = 16$

Since $r \ngtr n$ [*i.e.* r cannot be greater than n]

Therefore $r = 5$

6.7 APPLICATIONS OF PERMUTATIONS WHEN ALL THE OBJECTS ARE DISTINCT

Example 21. *How many different signals can be generated from 6 flags of different colours if each signal makes use of all the flags at a time, placed one below the other.*

Solution: We have,

Number of flags, $n = 6$

Number of flags used, $r = 6$

If P be the number of signals generated, then

$$P = P(6, 6) = \frac{6!}{(6-6)!} = \frac{6!}{0!} = 6! = 6 \times 5 \times 4 \times 3 \times 2 \times 1 = 720$$ **Ans.**

Example 22. *Four books, one each in chemistry, physics, biology and mathematics, are to be arranged in a shelf. In how many ways can this be done?*

Solution: Here, $n = r = 4$. If P is the number of arrangements, then

$$P = P(4, 4) = \frac{4!}{(4-4)!} = \frac{4!}{0!} = 4! \qquad [\because 0! = 1]$$

$$= 4 \times 3 \times 2 \times 1 = 24$$ **Ans.**

Example 23. *Find the number of 4-digit numbers that can be formed using the digits 1, 2, 3, 4, 5 if no digit is used more than once in a number. How many of these numbers will be even?*

Solution: According to the problem, $n = 5$ and $r = 4$.

Part a:

If P is the number of four digit numbers from the given digits. Then,

$$P = {}^5P_4 = \frac{5!}{(5-4)!} = \frac{5!}{1!} = 5! \qquad [\because 1! = 1]$$

$$= 5 \times 4 \times 3 \times 2 \times 1 = 120$$

Part b:

If the formed number is an even number, then the unit's place should be filled by 2 or 4. The number of ways in which unit's place can be filled

$${}^2P_1 = \frac{2!}{(2-1)!} = 2! = 2$$

Remaining three places can be filled in

$${}^4P_3 = \frac{4!}{(4-3)!} = \frac{4!}{1!} = 4! = 4 \times 3 \times 2 \times 1 = 24$$

If p is total number of even numbers, then

$p = {}^2P_1 \times {}^4P_3 = 2 \times 24 = 48$ **Ans.**

Example 24. *From a committee of 8 persons, in how many ways can we choose a chairman and a vice chairman assuming one person cannot hold more than one position?*

Solution: Chairman can be chosen from a committee of 8 persons in 8P, different ways.

Vice chairman can be chosen from the remaining 7 committee members in 7P, ways.

Thus, the number of ways in which the chairman, vice chairman can be chosen, by multiplication principle, is $8P_1 \times 7P_1$

$= 8 \times 7 = 56$ ways **Ans.**

Example 25. *How many four digit numbers can be formed by using the digits 0 to 9 with no digit repeated ?*

Solution: We have ten digits, viz., 0, 1, 2, 3, 4, 5, 6, 7, 8 and 9. 1000th place can be filled by any of the digits except '0'.

Hence, the place is filled in $P(9,1) = \dfrac{9!}{(9-1)!} = \dfrac{9(8!)}{8!} = 9$ ways

In the 100th place and at the tenth or units place zero can be used. Hence the remaining three places can be filled in

$$^9P_3 = \frac{9!}{(9-3)!} = \frac{9!}{6!} = \frac{9 \times 8 \times 7\,(6!)}{6!} = 9 \times 8 \times 7 = 504 \text{ ways}$$

Hence, total number of four digit numbers thus formed

$= 9 \times 504 = 4536.$ **Ans.**

Example 26. *How many words can be formed using all letters of the word EQUATION, using each letter exactly once?*

Solution: We have,

Number of letters in equation, $n = 8$

Number of letters to be taken at a time, $r = 8$

The number of words thus formed

$$^8P_8 = \frac{8!}{(8-8)!} = \frac{8!}{0!} = 8! = 8 \times 7 \times 6 \times 5 \times 4 \times 3 \times 2 \times 1 = 40320$$ **Ans.**

Example 27. *How many 3 letter words can be made using the letters of the word ORIENTAL?*

Solution: Total number of letters in the given words, $n = 8$

Number of letters to be used in forming words, $r = 3$

If the total number of words thus formed is $^8P_3$, then

$$^8P_3 = \frac{8!}{(8-3)!} = \frac{8!}{5!} = \frac{8 \times 7 \times 6 \times 5!}{5!} = 8 \times 7 \times 6 = 336$$ **Ans.**

EXERCISE 6.4

1. How, many numbers can be formed from the digits 1, 2, 3, 9 if repetition of digits is not allowed? **Ans.** 64
2. In how many ways three different rings can be worn in four fingers with at most one in each finger ? **Ans.** 24

3. Seven songs are to be rendered in a programme. In how many different orders could they be rendered? **Ans.** 5040
4. There are 6 items in column A and 6 items in column B. A student is asked to match each item in column A with an item in column B. How many possible answers (correct or incorrect) are there? **Ans.** 720
5. How many even numbers of three digits each can be made with the digits 1, 2, 3, 4, 6, 7 if no digit is repeated ? **Ans.** 60
6. Ten horses are running a race. In how many ways can these horses come in the first, second and third place, assuming no ties? **Ans.** 720
7. From a pool of 12 candidates, in how many ways can we select president, vice president, secretary and a treasurer if each of the 12 candidates can hold any office? **Ans.** 11880
8. How many different signals can be made by 5 flags from 8 flags of different colours? **Ans.** 6720
9. How many numbers lying between 100 and 1000 can be formed with the digits 1, 2, 3, 4, 5 if the repetition of digits is not allowed? **Ans.** 60
10. How many 4-letter words, with or without meaning, can be formed out of the letters of the word, "LOGARITHMS", if repetition of letters is not allowed ? **Ans.** 5040
11. How many words, with or without meaning, can be formed by using the letters of the word "TRIANGLE"? **Ans.** 8!
12. There are six periods on each working day of a school. In how many ways can one arrange 5 subjects such that each subject is allowed at least one period ? What value do you expect? **Ans.** 3600 **Value:** Each subject is important.
13. Amit wants to arrange 3 Economics, 2 History and 4 English books on a shelf. If the books on same subject are different, determine the number of possible arrangements. What value do you expect?
    Ans. 9! = 362880 **Value:** A well considered and careful arrangement liked by all.
14. In how many ways can the letters of the word 'DELHI' be arranged so that the letters, E and H occupy only even places? **Ans.** 12
15. In how many ways can the letters of the word 'TOWER' be arranged so that the letters O and E occupy only even places? **Ans.** $2 \times 6 = 12$
16. How many different arrangement, each consisting of 4 different letters can be formed from the letters of the words 'PERSONAL', if each arrangement is to begin with and end with a vowel? **Ans.** ${}^3P_2 \times {}^6P_2 = 180$
17. In how many ways can the letters of the word 'FRACTION' be arranged so that no two vowels are together? **Ans.** 14400
18. How many different words can be formed of the letters of the word 'COMBINE' so that
    (*i*) vowels always remain together. (*ii*) no two vowels are together.
    (*iii*) vowels may occupy odd places.
    Ans. (i) $5! \cdot 3! = 720$ (*ii*) $4! \cdot {}^5P_3 = 1440$ (iii) ${}^4P_3 \cdot 4! = 576$
19. Sachin wants to arrange 3 economics, 2 history and 4 english books on a shelf. If the books on the same subject are different, determine the number of possible arrangement, if all the books on a subject are to be together?
    Ans. $3! \cdot 3! \cdot 2! \cdot 4! = 1728$
20. In how many ways can 5 boys and 3 girls be arranged so that no two girls may sit together? **Ans.** 14400

21. Six men and five women are to sit in a row so that the women occupy the even places. Find the number of all possible arrangements? **Ans.** $5! \cdot 6! = 86400$

22. There are 8 students appearing in an examination of which 3 have to appear in a mathematics paper and the remaining 5 in different subjects. In how many ways can they be made to sit in a row if no two candidates in mathematics sit next to each other? **Ans.** $6! \cdot (6 \times 5 \times 4) = 14400$

23. It is required to seat 5 men and 4 women in a row so that women occupy the even places. How many such arrangements are possible? **Ans.** 2880

24. In how many ways can 9 examination papers be arranged so that the best and the worst papers are never together? [**Hint:** Required ways = $9! - 2! \times 8!$] **Ans.** 282240

25. When a group photograph is taken, all the seven teachers should be in the first row and all the twenty students should be in the second row. If the two corners of the second row are reserved for the two tallest students, interchangeable only between them, and if the middle seat of the front row is reserved for the principal, how many arrangements are possible? [**Hint:** Required arrangements = $6! \times (18! \times 2!)$]
Ans. $18! \times 1440$

26. In a class of 10 students there are 3 girls A, B, C. In how many different ways can they be arranged in a row such that no two of the three girls are consecutive.
[**Hint:** Number of arrangements = $7! \times {}^8P_3$] **Ans.** $7! \times 336$

27. In how many ways can a lawn tennis mixed double be made up from seven married couples if no husband and wife play in the same set? **Ans.** 840

28. M men and N women are to be seated in a row so that no two women sit together. If $M > N$ then show that the number of ways in which they can be seated as

$$\frac{M!(M+1)!}{(M-N+1)!}$$

29. How many words (with or without dictionary meaning) can be made from the letters in the word 'MONDAY', assuming that no letter is repeated, if

(*i*) 4 letters are used at a time

(*ii*) all letters are used at a time

(*iii*) all letters are used but the first is a vowel. **Ans.** (*i*) 360 (*ii*) 720 (*iii*) 240

Hints to Selected Questions

1. Now, we have 4 digits mainly 1, 2, 3 and 9 but we cannot repeat them.

Total number of four digit words $= 4! = 4 \times 3 \times 2 \times 1 = 24$

Total number of three digit words $= {}^4P_3 = \dfrac{4!}{(4-3)!} = \dfrac{4 \times 3 \times 2 \times 1}{1} = 24 \left[\because {}^nP_r = \dfrac{n!}{(n-r)!}\right]$

Total number of two letter words $= {}^4P_2 = \dfrac{4!}{(4-2)!} = \dfrac{4 \times 3 \times 2 \times 1}{2 \times 1} = 4 \times 3 = 12$

Total number of one letter words $= {}^4P_1 = \dfrac{4!}{(4-1)!} = \dfrac{4 \times 3 \times 2 \times 1}{3 \times 2 \times 1} = 4$

So, total number of ways $= 4 + 12 + 24 + 24 = 64$ ways.

4. Items in column $A = 6$, Items in column $B = 6$

Total possible answers after matching both the columns

$$= {}^6P_6 = \frac{6!}{(6-6)!} = \frac{6!}{0!} = \frac{6!}{1} = 6 \times 5 \times 4 \times 3 \times 2 \times 1 = 720 \qquad [\because 0! = 1]$$

5. Total digits available are 1, 2, 3, 4, 6, 7.

So to construct a three digit even number unit place can be filled in only 3 ways

[∵ only 2, 4 and 6 if used in unit place will make the number even]

The tens place digit can be filled in 5 ways as we cannot repeat the digits and one digit would be occupied by the unit place.

Similarly, hundred place can be filled by 4 digits only.

So Total number of number = 4 × 5 ×3 = 20 × 3 = 60 numbers,

7. Total number of candidates, $n = 12$

Total number of posts, $r = 4$

Total number of ways 12 candidates can be selected in these four posts = ${}^{12}P_4$

$$= \frac{12!}{(12-4)!} = \frac{12!}{8!} = \frac{12 \times 11 \times 10 \times 9 \times 8!}{8!} = 11880$$

9. Total number of digits available, $n = 5$ (*i.e.* 1, 2, 3, 4 & 5)

We have to make numbers between 100 & 1000 *i.e.* we have to make three digit numbers using above 5 digits. Therefore, $r = 3$

Now total number of three digit numbers available = ${}^5P_3$

$$= \frac{5!}{(5-3)!} = \frac{5!}{2!} = \frac{5 \times 4 \times 3 \times 2!}{2!} = 60 \text{ ways}$$

12. Total number of arrangements $= {}^6P_5 \times 5$

Combinations

7.1 INTRODUCTION

In the previous section, during the study about different permutations of the objects, we found that the order of occurence of the objects was important. We now consider problems involving combinations, where the order of occurrence does not matter.

7.2 COMBINATIONS

Each of the different groups or selections which can be made by taking some or all of a number of a things at a time (irrespective of the order) is called a combination.

For example the different combinations formed of three letters A, B, C are:

$$AB, AC, BC.$$

7.3 DIFFERENCE BETWEEN PERMUTATIONS AND COMBINATIONS

The process of selecting things is called combination and that of arranging things is called permutations.

If we have 4 objects A, B, C and D the possible selection (or combinations) and arrangements (or permutations) of 3 objects out of 4 are given below. This will help you understand clearly the difference between permutations and combinations.

	Selection ↓ Combination	Arrangement ↓ Permutation
	ABC	ABC, ACB, BAC, BCA, CAB, CBA
	ABD	ABD, ADB, BAD, BDA, DAB, DBA
	ACD	ACD, ADC, CAD, CDA, DAC, DCA
	BCD	BCD, BDC, CBD, CDB, DBC, DCB
Total	4 combinations	24 permutations

A few examples related to combinations are:

(*i*) Formation of a team from a number of players.

(*ii*) Formation of a particular committee from a number of members.

No.	Combinations	Permutations
1.	Concerns only the selection	Concerns selections as well as arrangement.
2.	Ordering of the selected item is immaterial	Ordering is essential

7.4 SIGNIFICANCE OF ^nC_r

The number of combinations of n things taken r at a time defines the number of groups of r things which can be formed from n things, and denoted by symbol ^nC_r or $C(n, r)$.

7.5 COMBINATIONS OF n DIFFERENT THINGS TAKEN r AT A TIME

The number of all combinations of n distinct objects, taken r at a time is given by

$$^nC_r = \frac{n!}{r!(n-r)!}$$

Proof: Let $\quad ^nC_r = x$

Each one of these x-combinations contain r things and can be arranged among themselves in $r!$ ways. Hence, one combination gives $r!$ permutations.

Therefore, x combinations will give rise to $(x.\ r!)$ permutations. But the number of permutations of n things taking r at a time is $= \dfrac{n!}{(n-r)!}$

$$\therefore \quad x \cdot r! = \frac{n!}{(n-r)!} \Rightarrow x = \frac{n!}{r!(n-r)!} \Rightarrow {}^nC_r = \frac{n!}{r!(n-r)!}$$

$$\Rightarrow \quad \boxed{^nC_r = \frac{n!}{r!(n-r)!}}, 1 \le r \le n$$

Corollary 1.
$$^nC_r = \frac{n!}{r!(n-r)!} = \frac{n(n-1)\dots(n-r+1)(n-r)\dots 3\cdot 2\cdot 1}{r!(n-r)!}$$

$$= \frac{n(n-1)(n-2)\dots(n-r+1)}{r!}$$

$$\Rightarrow \quad \boxed{^nC_r = \frac{n(n-1)\dots(n-r+1)}{1\cdot 2\dots r}}$$

This form of ^nC_r is generally used in practical problems, because it does not involve factorials. From the above form of ^nC_r we have

$$^nC_r = \frac{n(n-1)(n-2)\dots r \text{ factors}}{1\cdot 2\cdot 3\dots r}$$

For example:
$$^8C_4 = \frac{8\times 7\times 6\times 5}{1\times 2\times 3\times 4} = 70$$

Corollary 2.
$$^nC_0 = \frac{n!}{0!(n-0)!} = \frac{n!}{1\times n!} = 1$$

and
$$^nC_n = \frac{n!}{n!(n-n)!} = \frac{1}{0!} = 1$$

$$\Rightarrow \qquad \boxed{{}^nC_0 = {}^nC_n = 1}$$

Corollary 3. $\boxed{{}^nC_r = \dfrac{n!}{r!(n-r)!} = \dfrac{1}{r!}\left(\dfrac{n!}{(n-r)!}\right) = \dfrac{{}^nP_r}{r!}}$

SOME IMPORTANT THEOREMS

Theorem 1. *Prove that ${}^nC_r = {}^nC_{n-r}$, i.e. the number of combinations of n different things taken r at a time is equal to the number of combinations of n different things taken (n – r) at a time.*

Proof: Method (*i*). Every time we select a group of *r* things we leave behind another group of $(n - r)$ things. Thus for every combination of $(n - r)$ things, there corresponds a combination of *r* things.

$$\therefore \qquad {}^nC_r = {}^nC_{n-r}$$

Method (*ii*).
$${}^nC_r = \frac{n!}{r!(n-r)!} \qquad \text{...(1)}$$

and
$${}^nC_{n-r} = \frac{n!}{(n-r)![n-(n-r)]!} = \frac{n!}{(n-r)!\times r!} \qquad \text{...(2)}$$

From (1) and (2), we get

$$\boxed{{}^nC_r = {}^nC_{n-r}}$$

Theorem 2. *Prove that (Pascal's Rule):* ${}^nC_r + {}^nC_{r-1} = {}^{n+1}C_r \ (1 \le r \le n)$.

Proof: L.H.S. $= {}^nC_r + {}^nC_{r-1}$

$$= \frac{n!}{r!(n-r)!} + \frac{n!}{(r-1)!\,(n-r+1)!}$$

$$= \frac{n!\,(n-r+1)}{r!\,(n-r+1)(n-r)!} + \frac{n!\cdot r}{r\,(r-1)!\,(n-r+1)!}$$

$$= \frac{n!\,(n-r+1)}{r!\,(n-r+1)!} + \frac{n!\,r}{r!\,(n-r+1)!}$$

$$= n!\left\{\frac{n-r+1+r\,)}{r!\,(n-r+1)!}\right\}$$

$$= \frac{(n+1)\,n!}{r!\,(n+1-r)!} = {}^{n+1}C_r = \text{R.H.S.} \qquad \textbf{Proved.}$$

Note. ${}^nC_r$ is greatest if (*i*) $r = \dfrac{n}{2}$, when *n* is even

(*ii*) $r = \dfrac{n-1}{2}$, or $\dfrac{n+1}{2}$, when *n* is odd.

Theorem 3. *Prove that:* ${}^nC_r = \dfrac{n}{r}\cdot {}^{n-1}C_{r-1}, 1 \le r \le n.$

Proof: We have,

$$\text{L.H.S.} = {}^nC_r = \frac{n!}{r!(n-r)!} = \frac{n(n-1)!}{r(r-1)!\,\{(n-1)-(r-1)\}!}$$

$$= \frac{n}{r} \frac{(n-1)!}{(r-1)!\{(n-1)-(r-1)\}!}$$

$$= \frac{\boldsymbol{n}}{\boldsymbol{r}} \cdot {}^{n-1}C_{r-1} = \text{R.H.S.}$$

$\Rightarrow$ $$\boxed{{}^nC_r = \frac{n}{r} \cdot {}^{n-1}C_{r-1}}$$ **Proved.**

Remark: By using the above property, we obtain that

$${}^nC_r = \frac{n}{r} \cdot \frac{n-1}{r-1} \cdot \frac{n-2}{r-2} \cdots \frac{n(r-1)}{2} \cdot \frac{n-(r-1)}{1}$$

For example: $${}^9C_4 = \frac{9}{4} \times \frac{8}{3} \times \frac{7}{2} \times \frac{6}{1} = 126$$

Theorem 4. *Prove that:* $n \cdot {}^{n-1}C_{r-1} = (n - r + 1) \cdot {}^nC_{r-1}$, where $1 \leq r \leq n$.

Proof: We have,

$$\text{L.H.S.} = n \cdot {}^{n-1}C_{r-1} = n \cdot \frac{(n-1)!}{(r-1)!\{(n-1)-(r-1)\}!}$$

$$= \frac{n!}{(r-1)!\,(n-r)!}$$

$$= \frac{(n-r+1) \cdot n!}{(r-1)!\,(n-r+1)(n-r)!}$$

$$= (n-r+1)\left[\frac{n!}{(r-1)!\,(n-r+1)!}\right]$$

$$= (n-r+1)\left[\frac{n!}{(r-1)!\,\{n-(r-1)\}!}\right]$$

$\Rightarrow$ $$\boxed{n \cdot {}^{n-1}C_{r-1} = (n-r+1)\,{}^nC_{r-1}}$$

$\Rightarrow$ L.H.S. = R.H.S. **Proved.**

Theorem 5. *If* ${}^nC_x = {}^nC_y$, *then either* $x = y$ *or* $x + y = n$.

Proof: We have

${}^nC_x = {}^nC_y$

$\Rightarrow$ ${}^nC_x = {}^nC_y = {}^nC_{n-y}$ $[\because {}^nC_r = {}^nC_{n-r}]$

$\Rightarrow$ $x = y$ or $x = n - y$

$\Rightarrow$ $x = y$ or $x + y = n$

Remark. If ${}^nC_x = {}^nC_y$ and $x \neq y$, then $x + y = n$

For example: If ${}^{20}C_x = {}^{20}C_{x+8}$, find the value of x,

Solution. We have ${}^{20}C_x = {}^{20}C_{x+8}$ $\Rightarrow x + x + 8 = 20 \Rightarrow 2x = 12$

$\Rightarrow x = 6$

Example 1. *Evaluate:*

(*i*) $^{13}C_6 + {}^{13}C_5$ (*ii*) $^{25}C_{22} - {}^{24}C_{21}$ (*iii*) $^{31}C_{26} - {}^{30}C_{26}$

Solution. (*i*) We have,

(*i*) $^{13}C_6 + {}^{13}C_5 = {}^{14}C_6$ $[\because {}^nC_r + {}^nC_{r-1} = {}^{n+1}C_r]$

$$= \frac{14!}{6!(14-6)!} = \frac{14!}{6!8!} \qquad \left[{}^nC_r = \frac{n!}{r!(n-r)!}\right]$$

$$= \frac{14 \times 13 \times 12 \times 11 \times 10 \times 9 \times 8!}{6 \times 5 \times 4 \times 3 \times 2 \times 1 \times 8!}$$

$= 7 \times 13 \times 11 \times 3 = 3003$ **Ans.**

(*ii*) $$^{25}C_{22} - {}^{24}C_{21} = \frac{25!}{22!(25-22)!} - \frac{24!}{21!(24-21)!}$$

$$= \frac{25!}{22!3!} - \frac{24!}{21!3!}$$

$$= \frac{25 \times 24 \times 23 \times 22!}{22! \times 3 \times 2 \times 1} - \frac{24 \times 23 \times 22 \times 21!}{21! \times 3 \times 2 \times 1}$$

$= (25 \times 4 \times 23) - (4 \times 23 \times 22)$

$= 23\ (100 - 88) = 23 \times 12 = 276$ **Ans.**

(*iii*) $$^{31}C_{26} - {}^{30}C_{26} = \frac{31!}{26!(31-26)!} - \frac{30!}{26!(30-26)!}$$

$$= \frac{31!}{26!5!} - \frac{30!}{26!4!}$$

$$= \frac{31 \times 30 \times 29 \times 28 \times 27 \times 26!}{26! \times 5 \times 4 \times 3 \times 2 \times 1} - \frac{30 \times 29 \times 28 \times 27 \times 26!}{26! \times 4 \times 3 \times 2 \times 1}$$

$= (31 \times 29 \times 7 \times 27) - (5 \times 29 \times 7 \times 27)$

$= 29 \times 7 \times 27\ (31 - 5) = 29 \times 7 \times 27 \times 26 = 142506$ **Ans.**

Example 2. *If* $^nC_{10} = {}^nC_{12}$*, determine n and hence* $^nC_5$*.*

Solution. We have $^nC_{10} = {}^nC_{12} \Rightarrow 10 + 12 = n$ $[\because {}^nC_x = {}^nC_y \Rightarrow n = x + y]$

$\Rightarrow$ $n = 22$

Now, $$^nC_5 = {}^{22}C_5 = \frac{22!}{5!(22-5)!} \qquad \text{...(1)}$$

$$= \frac{22!}{5!\,17!} = \frac{22 \times 21 \times 20 \times 19 \times 18 \times 17!}{5 \times 4 \times 3 \times 2 \times 1 \times 17!} \qquad [\because n = 22]$$

$= 11 \times 21 \times 19 \times 6 = 26334$ **Ans.**

Example 3. *If* $^{2n}C_3 : {}^nC_2 = 12 : 1$

Solution. We have,

$^{2n}C_3 : {}^nC_2 = 12 : 1$

$$\Rightarrow \qquad \frac{(2n)!}{3!(2n-3)!} : \frac{n!}{2!(n-2)!} = \frac{12}{1}$$

$$\Rightarrow \quad \frac{(2n)!}{3!(2n-3)!} \times \frac{2!(n-2)!}{n!} = \frac{12}{1}$$

$$\Rightarrow \quad \frac{2n(2n-1)(2n-2)(2n-3)!}{3 \times 2.(2n-3)!} \times \frac{2!(n-2)!}{n(n-1)(n-2)!} = \frac{12}{1}$$

$$\Rightarrow \quad \frac{2n(2n-1)(2n-2)}{3} \times \frac{1}{n(n-1)} = \frac{12}{1}$$

$$\Rightarrow \quad \frac{2(n)(2n-1)\,2(n-1)}{3} \times \frac{1}{n(n-1)} = \frac{12}{1}$$

$$\Rightarrow \quad \frac{4(2n-1)}{3} = \frac{12}{1}$$

$$\Rightarrow \quad 2n - 1 = 12 \times \frac{3}{4} \Rightarrow 2n - 1 = 9 \Rightarrow 2n = 9 + 1$$

$$\Rightarrow \quad 2n = 10 \Rightarrow n = 5$$

Ans.

Example 4. *Verify:* $2 \times {}^7C_4 = {}^8C_4$.

Solution. L.H.S. $= 2 \times {}^7C_4 = 2 \times \frac{7!}{4!(7-4)!} = 2 \times \frac{7!}{4!(3)!}$

$$= 2 \times \frac{7 \times 6 \times 5 \times 4!}{4! \times 3 \times 2 \times 1} = 2 \times 7 \times 5 = 70$$

R.H.S. $= {}^8C_4 = \frac{8!}{4!\,(8-4)!} = \frac{8!}{4!\,4!}$

$$= \frac{8 \times 7 \times 6 \times 5 \times 4!}{4! \times 4!} = \frac{8 \times 7 \times 6 \times 5 \times 4!}{4 \times 3 \times 2 \times 1 \times 4!}$$

$$= 2 \times 7 \times 5 = 70$$

L.H.S. = R.H.S **Proved.**

Example 5. *Prove that:*

(*i*) ${}^2C_1 + {}^3C_1 + {}^4C_1 = {}^3C_2 + {}^4C_2$ (*ii*) $1 + {}^3C_1 + {}^4C_2 = {}^5C_3$

Solution. L.H.S $= {}^2C_1 + {}^3C_1 + {}^4C_1 = \frac{2!}{1!(2-1)!} + \frac{3!}{1!(3-1)!} + \frac{4!}{1!(4-1)!}$

$$= 2 + 3 + 4 = 9$$

R.H.S $= {}^3C_2 + {}^4C_2 = \frac{3!}{2!(3-2)!} + \frac{4!}{2!(4-2)!}$

$$= \frac{3!}{2!1!} + \frac{4!}{2!2!} = \frac{3 \times 2!}{2!} + \frac{4 \times 3 \times 2!}{2! \times 2!}$$

$$= 3 + 6 = 9$$

L.H.S. = R.H.S **Proved.**

(*ii*) L.H.S. $= 1 + {}^3C_1 + {}^4C_2 = 1 + \frac{3!}{1!(3-1)!} + \frac{4!}{2!(4-2)!}$

$$= 1 + \frac{3!}{2!1!} + \frac{4!}{2!.2!} = 1 + \frac{3 \times 2!}{2!} + \frac{4 \times 3 \times 2!}{2 \times 1 \times 2!}$$

$$= 1 + 3 + 6 = 10$$

$$\text{R.H.S.} = {}^5C_3 = \frac{5!}{3!\,(5-3)!} = \frac{5 \times 4 \times 3!}{3! \times 2 \times 1} = 10$$

$$\text{L.H.S.} = \text{R.H.S}$$

Proved.

Example 6. *Prove that:* $\sum_{r=1}^{5} {}^5C_r = 31$

Solution. $\text{L.H.S.} = \sum_{r=1}^{5} {}^5C_r$

$$= {}^5C_1 + {}^5C_2 + {}^5C_3 + {}^5C_4 + {}^5C_5$$

$$\text{R.H.S.} = \frac{5!}{1!\,(5-1)!} + \frac{5!}{2!\,(5-2)!} + \frac{5!}{3!\,(5-3)!} + \frac{5!}{4!\,(5-4)!} + \frac{5!}{5!\,(5-5)!}$$

$$= \frac{5!}{4!} + \frac{5!}{2!3!} + \frac{5!}{3!2!} + \frac{5!}{4!\,1!} + \frac{5!}{5!\,0!}$$

$$= \frac{5 \times 4!}{4!} + \frac{5 \times 4 \times 3!}{2 \times 1 \times 3!} + \frac{5 \times 4 \times 3!}{3! \times 2 \times 1} + \frac{5 \times 4!}{4!} + \frac{5!}{5!}$$

$$= 5 + 10 + 10 + 5 + 1$$

$$= 31$$

Proved.

Example 7. *Prove that:*

(*i*) $r.\,{}^nC_r = n.\,{}^{n-1}C_{r-1}$

(*ii*) ${}^nC_r \times {}^rC_s = {}^nC_s \times {}^{n-s}C_{r-s}$

Solution. (*i*)

$$\text{L.H.S} = r.\,{}^nC_r = r \cdot \frac{n!}{r!(n-r)} = r \cdot \frac{n!}{r\,(r-1)!\,(n-r)!}$$

$$= \frac{n!}{(r-1)!\,.(n-r)!} \qquad \text{...(1)}$$

$$\text{R.H.S} = n \cdot {}^{n-1}C_{r-1} = n \cdot \frac{(n-1)!}{(r-1)!\{(n-1)-(r-1)\}!}$$

$$= \frac{n!}{(r-1)!\,(n-r)!} \qquad \text{...(2)}$$

From (1) and (2), we obtain

$$\text{L.H.S.} = \text{R.H.S}$$

Proved.

(*ii*)

$$\text{L.H.S.} = {}^nC_r \times {}^rC_s = \frac{n!}{r!(n-r)!} \times \frac{r!}{s!(r-s)!}$$

$$= \frac{n!}{s!(n-r)!\,(r-s)!} \qquad \text{...(1)}$$

$$\text{R.H.S.} = {}^nC_s \times {}^{n-s}C_{r-s} = \frac{n!}{s!(n-s)!} \times \frac{(n-s)!}{(r-s)!\{(n-s)-(r-s)\}!}$$

$$= \frac{n!}{s!(n-s)!} \times \frac{(n-s)!}{(r-s)!\,(n-r)!}$$

$$= \frac{n!}{s!(n-r)!\,(r-s)!} \qquad ...(2)$$

L.H.S. = R.H.S **Proved.**

Example 8. *If* $^{n-1}C_r : {}^nC_r : {}^{n+1}C_r = 6 : 9 : 13$, *find n and r.*

Solution. We have,

$$^{n-1}C_r : {}^nC_r : {}^{n+1}C_r = 6 : 9 : 13$$

$$\Rightarrow \quad \frac{^{n-1}C_r}{6} = \frac{^nC_r}{9} = \frac{^{n+1}C_r}{13}$$

If $$\frac{^{n-1}C_r}{6} = \frac{^nC_r}{9}$$

$$\Rightarrow \quad 3(^{n-1}C_r) = 2(^nC_r) \Rightarrow 3\cdot\frac{(n-1)!}{r!(n-1-r)!} = 2\cdot\frac{n!}{r!\,(n-r)!}$$

$$\Rightarrow \quad 3\cdot\frac{(n-1)!}{r!(n-r-1)!} = 2\frac{n(n-1)!}{r!\,(n-r)(n-r-1)!}$$

$$\Rightarrow \quad 3(n-r) = 2n \Rightarrow 3n - 2n = 3r \Rightarrow n = 3r \qquad ...(1)$$

Again if $$\frac{^nC_r}{9} = \frac{^{n+1}C_r}{13}$$

$$\Rightarrow \quad 13\,^nC_r = 9\,^{n+1}C_r$$

$$\Rightarrow \quad 13.\frac{n!}{r!(n-r)!} = 9.\frac{(n+1)!}{r!\,(n+1-r)!}$$

$$\Rightarrow \quad 13.\frac{n!}{r!(n-r)!} = 9.\frac{(n+1)(n)!}{r!\,(n+1-r)(n-r)!}$$

$$\Rightarrow \quad 13(n+1-r) = 9(n+1)$$

$$\Rightarrow \quad 13n - 9n = 13r - 13 + 9 \qquad ...(2)$$

$$\Rightarrow \quad 4n = 13r - 4$$

Putting $n = 3r$ from (1) in (2), we get

$$4(3r) = 13r - 4 \Rightarrow 12r - 13r = -4 \Rightarrow r = 4$$

Putting $r = 4$ in (1), we get

$$n = 3(4) = 12$$

Hence, $n = 12$ and $r = 4$ **Ans.**

Example 9. *If* $^nP_r = {}^nP_{r+1}$ *and* $^nC_r = {}^nC_{r-1}$, *find the value of n and r.*

Solution. We have,

$$^nP_r = {}^nP_{r+1}$$

$$\Rightarrow \quad \frac{n!}{(n-r)!} = \frac{n!}{(n-r-1)!}$$

$$\Rightarrow \quad \frac{1}{(n-r)(n-r-1)!} = \frac{1}{(n-r-1)!}$$

$$\Rightarrow \quad n - r = 1 \qquad ...(1)$$

and $$^nC_r = {}^nC_{r-1}$$

$$\Rightarrow \qquad \frac{n!}{r!(n-r)!} = \frac{n!}{(r-1)!(n-r+1)!}$$

$$\Rightarrow \qquad \frac{n!}{r(r-1)!(n-r)!} = \frac{n!}{(r-1)!(n-r+1)(n-r)!}$$

$$\Rightarrow \qquad \frac{1}{r} = \frac{1}{n-r+1}$$

$$\Rightarrow \qquad n - r + 1 = r \Rightarrow n - 2r = -1 \qquad ...(2)$$

Solving (1) and (2), we get $\quad n = 3$ and $r = 2$ Ans.

EXERCISE 7.1

Evaluate each of the following:

1. ${}^{10}C_4 + {}^{10}C_5$ **Ans.** 462
2. ${}^{61}C_{57} - {}^{60}C_{56}$ **Ans.** 34220
3. If ${}^{n}C_8 = {}^{n}C_6$, find ${}^{n}C_2$. **Ans.** 91
4. Determine n if ${}^{2n}C_3 : {}^{n}C_3 = 11 : 1$ **Ans.** 6
5. If ${}^{4}P_2 = n.\ {}^{4}C_2$, find n. **Ans.** 2
6. If ${}^{n}C_4 = {}^{n}C_6$, find n. **Ans.** 10
7. If ${}^{2n}C_3 : {}^{n}C_2 = 12 : 1$, find n. **Ans.** 5
8. If ${}^{n}C_r : {}^{n}C_{r+1} = 1 : 2$ find ${}^{n}C_{r+1} : {}^{n}C_{r+2} = 2 : 3$, determine the values of n and r. **Ans.** $n = 14, r = 4$
9. If ${}^{2n}C_r = {}^{2n}C_{r+2}$, find r in term of n. **Ans.** $r = n - 1$
10. If ${}^{n}C_{n-4} = 15$, find the value of n. **Ans.** 6
11. For all positive integers n, show that ${}^{2n}C_n + {}^{2n}C_{n-1} = \frac{1}{2}\left({}^{2n+2}C_{n+1}\right)$

REVISION EXERCISE

Short Answer Questions (SAQs)

1. Eight chairs are numbered 1 to 8. Two women and three men wish to occupy one chair each. First the women choose the chairs from amongst the chairs 1 to 4 and then men select from the remaining chairs. Find the total number of possible arrangements.[**Hint:** 2 women occupy the chairs, from 1 to 4 in ${}^{4}P_2$ ways and 3 men occupy the remaining chairs in ${}^{6}P_3$ ways.] **Ans.** 1440
2. If the letters of the word RACHIT are arranged in all possible ways as listed in dictionary. Then what is the rank of the word RACHIT? **Ans.** 481
3. A candidate is required to answer 7 questions out of 12 questions, which are divided into two groups, each containing 6 questions. He is not permitted to attempt more than 5 questions from either group. Find the number of different ways of doing questions. **Ans.** 780
4. Out of 18 points in a plane, no three are in the same line except five points which are collinear. Find the number of lines that can be formed joining the point. [**Hint:** Number of straight lines = ${}^{18}C_2 - {}^{5}C_2 + 1$] **Ans.** 144
5. We wish to select 6 persons from 8, but if the person A is chosen, then B must be chosen. In how man ways can selections be made? **Ans.** 22

6. How many committee of five persons with a chairperson can be selected from 12 persons. **Ans.** 3960

7. How many automobiles license plates can be made if each plate contains two different digits? **Ans.** 468000

8. A bag contains 5 black and 6 red balls. Determine the number of ways in which 2 black and 3 red balls can be selected from the lot. **Ans.** 200

9. Find the number of permutations of n distinct things taken r together, in which 3 particular things must occur together. **Ans.** ${}^{n-3}P_{r-3}(r-2)!\,3!$

10. Find the number of different words that can be formed from the letters of the word 'TRIANGLE' so that no vowels are together. **Ans.** 14400

11. Find the number of positive integers greater than 6000 and less than 7000 which are divisible by 5, provided that no digit is to be repeated. **Ans.** 112

12. There are 10 persons named $P_1, P_2, P_3, ..., P_{10}$. Out of 10 persons, 5 persons are to be arranged in a line such that in each arrangement P_1 must occur whereas P_4 and P_5 do not occur. Find the number of such possible arrangements.

13. There are 10 lamps in a hall. Each one of them can be switched on independently. Find the number of ways in which the hall can be illuminated.

[**Hint:** Required number = $2^{10} - 1$]

14. A box contains two white, three black and four red balls. In how many ways can three balls be drawn from the box, if atleast one black ball is to be included in the draw. [**Hint:** Required number of ways = ${}^3C_1 \times {}^6C_2 + {}^3C_2 \times {}^6C_2 + {}^3C_3$]

15. If ${}^nC_{r-1} = 36$, ${}^nC_r = 84$ and ${}^nC_{r+1} = 126$, then find ${}^rC_2$.

[**Hint:** Form equation using $\frac{{}^nC_r}{{}^nC_{r+1}}$ and $\frac{{}^nC_r}{{}^nC_{r-1}}$ to find the value of r.] **Ans.** $r = 3$

16. Find the number of integers greater than 7000 that can be formed with the digits 3, 5, 7, 8 and 9 where no digits are repeated.

[**Hint:** Besides 4 digit integers greater than 7000, five digit integers are always greater than 7000] **Ans.** 192

17. If 20 lines are drawn in a plane such that no two of them are parallel and no three are concurrent, in how many points will they intersect each other? **Ans.** 190

18. In a certain city, all telephone number have six digits, the first two digits always being 41 or 42 or 46 or 62 or 64. How many telephone numbers have all six digits distinct? **Ans.** 8400

19. In an examination, a student has to answer 4 questions out of 5 questions; questions 1 and 2 are however compulsory. Determine the number of ways in which the student can make the choice. **Ans.** 3

20. A convex polygon has 44 diagonals. Find the number of its sides.

[**Hint:** Polygon of n sides has (${}^nC_2 - n$) number of diagonals.] **Ans.** 11

Long Answer Questions (LAQs)

21. If 18 mice were placed in two experimental groups and one control group, with all groups equally large. In how many ways can the mice be placed into three groups? **Ans.** $\frac{18!}{(6!)^3}$

22. A bag contains six white marbles and five red marbles. Find the number of ways in which four marbles can be drawn from the bag if (*a*) they can be of any colour (*b*) two must be white and two red (*c*) they must all be of the same colour.

Ans. (*a*) $^{11}C_4$ (*b*) $^6C_2 \times {}^5C_2$ (*c*) $^6C_4 + {}^5C_4$

23. In how many ways can a football team of 11 players be selected from 16 players? How many teams will

(*a*) include 2 particular players? (*b*) exclude 2 particular players?

Ans. (*a*) $^{14}C_9$ (*b*) $^{14}C_{11}$

24. A sports team of 11 students is to be constituted, selecting at least 5 from class XI and atleast 5 from class XII. If there are 20 students in each of these classes, in how many ways can the team be constituted? **Ans.** $2(^{20}C_5 \times {}^{20}C_6)$

25. A group consists of 4 girls and 7 boys. In how many ways can a team of 5 members be selected if the team has

(*a*) no girls (*b*) at least one boy and one girl (*c*) at least three girls.

Ans. (*a*) 21 (*b*) 441 (*c*) 91

Multiple Choice Questions (MCQs)

Choose the correct answer out of the given four options against each of the Questions from 26 to 40 (MCQ)

26. If $^nC_{12} = {}^nC_8$, then n is equal to

(*a*) 20 (*b*) 12 (*c*) 6 (*d*) 30 **Ans.** (*a*)

27. The number of possible outcomes when a coin is tossed 6 times is

(*a*) 36 (*b*) 64 (*c*) 12 (*d*) 32 **Ans.** (*b*)

28. The number of different four digit numbers that can be formed with the digits 2, 3, 4, 7 and using each digit only once is

(*a*) 120 (*b*) 96 (*c*) 24 (*d*) 100 **Ans.** (*c*)

29. The sum of the digits in unit place of all the numbers formed with the help of 3, 4, 5 and 6 taken all at a time is

(*a*) 432 (*b*) 108 (*c*) 36 (*d*) 18 **Ans.** (*b*)

30. Total number of words formed by 2 vowels and 3 consonants taken from 4 vowels and 5 consonants will be

(*a*) 60 (*b*) 120 (*c*) 7200 (*d*) 720 **Ans.** (*c*)

31. A five digit number divisible by 3 is to be formed using the numbers, 0, 1, 2, 3, 4 and 5 without repetition. The total number of ways this can be done is

(*a*) 216 (*b*) 600 (*c*) 240 (*d*) 3125 **Ans.** (*a*)

[**Hint:** 5 digit numbers can be formed using digits 0, 1, 2, 4, 5 or by using digits 1, 2, 3, 4, 5 only since sum of digits in these cases is divisible by 3]

32. If everybody in a room shakes hands with everyone. Then total number of persons in the room is

(*a*) 11 (*b*) 12 (*c*) 13 (*d*) 14 **Ans.** (*b*)

33. The number of triangles that are formed by choosing the vertices from a set of 12 points, seven of which lie on the same lines is

(*a*) 105 (*b*) 15 (*c*) 175 (*d*) 185 **Ans.** (*d*)

34. The number of parallelograms that can be formed from a set of four parallel lines intersecting another set of three parallel lines is

(*a*) 6 (*b*) 18 (*c*) 12 (*d*) 9 **Ans.** (*b*)

35. The number of ways in which a team of eleven players can be selected from 22 players always including 2 of them and excluding 4 of them is

(*a*) $^{16}C_{11}$ (*b*) $^{16}C_5$ (*c*) $^{16}C_9$ (*d*) $^{20}C_9$ **Ans.** (*c*)

36. The number of 5-digit telephone numbers having atleast one of their digits repeated is

(*a*) 90000 (*b*) 10000 (*c*) 30240 (*d*) 69760 **Ans.** (*d*)

37. The number of ways in which we can choose a committee from four men and six women so that the committee includes at least two men exactly twice as many women as men is

(*a*) 94 (*b*) 126 (*c*) 128 (*d*) None **Ans.** (*a*)

38. The total number of 9 digit numbers which have all different digits is

(*a*) 10! (*b*) 9! (*c*) $9 \times 9!$ (*d*) $10 \times 10!$ **Ans.** (*c*)

39. The number of words which can be formed out of the letters of the word ARTICLE, so that vowels occupy the even place is

(*a*) 1440 (*b*) 144 (*c*) 7! (*d*) $^4C_4 \times {}^3C_3$ **Ans.** (*b*)

40. Given 5 different green dyes, four different blue dyes and three different red dyes, the number of combinations of dyes which can be chosen taking at least one green and one blue dye is

(*a*) 3600 (*b*) 3720 (*c*) 3800 (*d*) 3600 **Ans.** (*b*)

[**Hint:** Possible numbers of choose is not choosing 5 green dyes, 4 blue dyes and 3 red dyes are 2^5, 2^4 and 2^3, respectively]

Fill in the blanks (Questions 41 to 50)

41. If $^nP_r = 840$, $^nC_r = 35$, then $r =$ ________. **Ans.** $n = 7$

42. $^{15}C_8 + {}^{15}C_9 - {}^{15}C_6 - {}^{15}C_7 =$ ________. **Ans.** 0

43. The number of permutations of n different objects, taken r at a time, when repetitions are allowed, is ________. **Ans.** n^r

44. The number of different words that can be formed from the letters of the word INTERMEDIATE such that two vowels never come together is ________.

[**Hint:** Number of ways of arranging 6 consonants of which two are alike is $\frac{6!}{2!}$ and number of ways of arranging vowels $= {}^7P_6 \times \frac{1}{3!} \times \frac{1}{2!}$.] **Ans.** 151200

45. Three balls are drawn from a bag containing 5 red, 4 white and 3 black balls. The number of ways in which this can be done if at least 2 are red is ________. **Ans.** 80

46. The number of six-digit numbers, all digits of which are odd is ________. **Ans.** 5^6

47. In a football championship, 153 matches were played. Every two teams played one match with each other. The number of teams, participating in the championship is ________. **Ans.** 18

48. The total number of ways in which six '+' and four '–' sign can be arranged in a line such that no two signs occur together is ________. **Ans.** 35

49. A committee of 6 is to be chosen from 10 men and 7 women so as to contain atleast 3 men and 2 women. In how many different ways can this be done if two particular women refuse to serve on the same committee.

[**Hint:** At least 3 men and 2 women: The number of ways = ${}^{10}C_3 \times {}^{7}C_3 + {}^{10}C_4 \times {}^{10}C_2$. For 2 particular women to be always there: the number of ways = ${}^{10}C_4 + {}^{10}C_3 \times {}^{5}C_1$. The total number of committees when two particular women are never together = Total – together] **Ans.** 7800

50. A box contains 2 white balls, 3 black balls and 4 red balls. The number of ways three balls be drawn from the box if at least one black ball is to be included in the draw is ________. **Ans.** 64

State whether the statements in Questions 51 to 59 are True or False? Also give justification.

51. There are 12 points in a plane of which 5 points are collinear, then the number of lines obtained by joining these points in pairs is ${}^{12}C_2 - {}^{5}C_2$. **Ans.** False

52. There letters can be posted in five letterboxes in 3^5 ways. **Ans.** False

53. In the permutations of n things, r taken together, the number of permutations in which m particular things occur together is ${}^{n-m}P_{r-m} \times {}^{r}P_{m}$. **Ans.** False

54. In a ship there are stalls for 12 animals, out of them horses, cows and calves (not less than 12 each) ready to be shipped. They can be loaded in 3^{12} ways. **Ans.** True

55. If some or all of n objects are taken at a time, the number of combinations is $2^n - 1$. **Ans.** True

56. There will be only 24 selections containing at least one red ball out of a bag containing 4 red and 5 black balls. It is being given that the balls of the same colour are identical. **Ans.** True

57. Eighteen guests are to be seated, half on each side of a long table. Four particular guests desire to sit on one particular side and three others on other side of the table. The number of ways in which the seating arrangements can be mad is $\frac{11!}{5!6!}(9!)(9!)$.

[**Hint:** After sending 4 on ones side and 3 on the other side, we have to select out of 11; 5 on one side and 6 n two other. Now there are 9 on each side of the long table and each can be arranged in 9 ways.] **Ans.** True

58. A candidate tis required to answer 7 questions out of 12 which are divided into two groups, each containing 6 questions. He is not permitted to attempt more than 5 questions from either group. He can choose seven questions in 650 ways.

Ans. False

59. To fill 12 vacancies there are 25 candidates of which 5 are from scheduled castes. If 3 of the vacancies are reserved for scheduled caste candidates while the rest are open to all, the number of ways in which the selection can be made is ${}^{5}C_3 \times {}^{20}C_9$. **Ans.** False

In each of the Questions from 60 to 64; match each item given under the column C_1 to its correct answer given under the column C_2.

60. There are 3 books on Mathematics, 4 on Physics and 5 on English. How many different collections can be made such that each collection consists of:

C_1	C_2
(*a*) One book of each subject	(*i*) 3968
(*b*) At least one book of each subject	(*ii*) 60
(*c*) One book of each subject	(*iii*) 3255

Ans. (*a*) ↔ (*ii*) (*b*) ↔ (*iii*) (*c*) ↔ (*i*)

61. Five boys and five girls form a line. Find the number of ways of making the seating arrangement under the following condition:

C_1	C_2
(*a*) Boys and girls alternate	(*i*) $5! \times 6!$
(*b*) No two girls sit together	(*ii*) $10! - 5!\, 6!$
(*c*) All the girls sit together	(*iii*) $(5!)^2 + (5!)^2$
(*d*) All the girls are never together	(*iii*) $2!\, 5!\, 5!$

Ans. (*a*) ↔ (*iii*) (*b*) ↔ (*i*) (*c*) ↔ (*iv*) (*d*) ↔ (*ii*)

62. There are 10 professors and 20 lecturers out of whom a committee of 2 professors and 3 lecturers is to be formed. Find:

C_1	C_2
(*a*) in how many ways committee: can be formed	(*i*) ${}^{10}C_2 \times {}^{19}C_3$
(*b*) in how many ways a particular: professor is included	(*ii*) ${}^{10}C_2 \times {}^{19}C_2$
(*c*) in how many ways a particular: lecturer is included	(*iii*) ${}^{9}C_1 \times {}^{20}C_3$
(*d*) in how many ways a particular: lecturer is excluded	(*iii*) ${}^{10}C_2 \times {}^{20}C_3$

Ans. (*a*) ↔ (*iv*) (*b*) ↔ (*iii*) (*c*) ↔ (*ii*) (*d*) ↔ (*i*)

63. Using the digits 1, 2, 3, 4, 5, 6, 7, a number of 4 different digit is formed. Find

C_1	C_2
(*a*) how many numbers are formed?	(*i*) 840
(*b*) how many numbers are exactly divisible by 2?	(*ii*) 200
(*c*) how many numbers are exactly divisible by 25?	(*iii*) 360
(*d*) how many of these are exactly divisible by 4?	(*iii*) 40

Ans. (*a*) ↔ (*i*) (*b*) ↔ (*iii*) (*c*) ↔ (*iv*) (*d*) ↔ (*ii*)

64. How many words (with or without dictionary meaning) can be made from the letters of the word MONDAY, assuming that no letter is repeated, if

C_1	C_2
(*a*) 4 letters are used at a time	(*i*) 720
(*b*) All letters are used at a time	(*ii*) 240
(*c*) All letters are used but the first is a vowel	(*iii*) 360

Ans. (*a*) ↔ (*iii*) (*b*) ↔ (*i*) (*c*) ↔ (*ii*)

Binomial Theorem

8.1 INTRODUCTION

The students already know the multiplication of algebraic expressions. They can find $(a + b)^2$ by multiplying $(a + b)$ by $(a + b)$, i.e.

$$(a + b)^2 = (a + b) \times (a + b) = a^2 + 2ab + b^2$$

Similarly,
$$(a + b)^3 = (a + b)^2 (a + b) = (a^2 + 2ab + b^2)(a + b)$$
$$= a^3 + 3a^2 b + 3ab^2 + b^3$$
$$(a + b)^4 = (a + b)^2 (a + b)^2$$
$$= (a^2 + 2ab + b^2) \times (a^2 + 2ab + b^2)$$
$$= a^4 + 4a^3b + 6a^2b^2 + 4ab^3 + b^4$$
$$= a^4 + 4a^3b + 6a^2b^2 + 4ab^3 + b^4$$

However, it becomes difficult to find the value of $(a + b)^{10}$ or $(a + b)^{16}$ as the index become bigger. Therefore, we look for a general formula which will help us in finding $(a + b)^n$, where n is a positive integer.

8.2 BINOMIAL EXPRESSION

An expression consisting of two terms is called a Binomial expression.

For example: $a + b$, $2a + 3b$, $a^2 + b^2$, $a^3 - b^3$ are all Binomial expressions.

8.3 BINOMIAL THEOREM

The general form of the Binomial expression is $(a + b)$ and the expansion of $(a + b)^n$, $n \in N$ is called the Binomial theorem. This theorem was first given by Sir Issac Newton, valid for any rational exponent.

8.4 DEVELOPMENT OF BINOMIAL EXPANSION

By actual multiplication, we may obtain the following expansions:

$$(a + b)^1 = a + b$$
$$(a + b)^2 = a^2 + 2ab + b^2 = {}^2C_0a^2 + {}^2C_1ab + {}^2C_2b^2$$
$$(a + b)^3 = a^3 + 3a^2b + 3ab^2 + b^3 = {}^3C_0a^3 + {}^3C_1a^2b + {}^3C_2ab^2 + {}^3C_3b^3$$
$$(a + b)^4 = a^4 + 4a^3b + 6a^2b^2 + 4ab^3 + b^4$$
$$= {}^4C_0a^4 + {}^4C_1a^3b + {}^4C_2a^2b^2 + {}^4C_3ab^3 + {}^4C_4b^4 \ldots$$
$$(a + b)^5 = a^5 + 5a^4b + 10a^3b^2 + 10a^2b^3 + 5ab^4 + b^5$$
$$= {}^5C_0a^5 + {}^5C_1a^4b + {}^5C_2a^3b^2 + {}^5C_3a^2b^3 + {}^5C_4ab^4 + {}^5C_5b^5 \ldots$$

We have observed that the coefficients in the above expansions follow a particular pattern as elaborated below.

8.5 PASCAL'S TRIANGLE

A keen observation will show that each term in the table is derived by adding together the two terms in the line above, which lie on either side of it. Thus, in the line for $n = 5$, the term 10 is found by adding together the terms 4 and 6 in the line $n = 4$

Exponent	Coefficients in the expansion of $(x + a)^n$
0	1
1	1 1
2	1 2 1
3	1 3 3 1
4	1 4 6 4 1
5	1 5 10 10 5 1

These coefficients in combinatorial form, called Binomial coefficients and may be rewritten as

$$
\begin{array}{ccccccccc}
 & & & & {}^0C_0 & & & & \\
 & & & {}^1C_0 & & {}^1C_1 & & & \\
 & & {}^2C_0 & & {}^2C_1 & & {}^2C_2 & & \\
 & {}^3C_0 & & {}^3C_1 & & {}^3C_2 & & {}^3C_3 & \\
{}^4C_0 & & {}^4C_1 & & {}^4C_2 & & {}^4C_3 & & {}^4C_4
\end{array}
$$

We observe that :

(*i*) The first term in the expansion (first term of the Binomial)^{index} is a^n or ${}^nC_0a^n$.

(*ii*) The second term is $n\,a^{n-1}\,b$ or ${}^nC_1a^{n-1}b$.

(*iii*) As the expansion progresses, the exponent of a decreases by one and the exponent of b increases by one.

(*iv*) *The total number of terms* in the expansion is one more than the index, i.e. $n + 1$.

(*v*) The last term is the $(n + 1)$th term, i.e. b^n or ${}^nC_nb^n$.

8.6 BINOMIAL THEOREM FOR POSITIVE INTEGERS

Theorem: If a and b are real numbers, then for all $n \in N$.

$$(a + b)^n = {}^nC_0a^nb^0 + {}^nC_1a^{n-1}b^1 + {}^nC_2a^{n-2}b^2 + \ldots + {}^nC_ra^{n-r}b^r + \ldots + {}^nC_{n-1}a^1b^{n-1} + {}^nC_na^0b^n$$

i.e., $$(a + b)^n = \sum_{r=0}^{n} {}^nC_ra^{n-r}b^r$$

Method I: By principle of mathematical induction

Proof: We shall prove the Binomial theorem by using the principle of mathematical induction. Let $P(n)$ be the statement:

$$(a + b)^n = {}^nC_0a^nb^0 + {}^nC_1a^{n-1}b + {}^nC_2a^{n-2}b^2 + \ldots + {}^nC_ra^{n-r}b^r + \ldots + {}^nC_{n-1}a^1b^{n-1} + {}^nC_na^0b^n$$

Step 1. For $n = 1$, we have

$$(a + b)^1 = a + b = {}^1C_0a^1b^0 + {}^1C_1a^0b^1$$

Thus, $P(1)$ is true.

Step 2. Let $P(m)$ be true. Then,

$$(a + b)^m = {}^mC_0a^mb^0 + {}^mC_1a^{m-1}b^1 + {}^mC_2a^{m-2}b^2 + ... + {}^mC_{m-1}a^1b^{m-1} + {}^mC_ma^0b^m$$

Thus, $P(m)$ is true.

Step 3. We shall show that $p(m + 1)$ is true for this we have to show that

$$(a + b)^{m+1} = {}^{m+1}C_0a^{m+1}b^0 + {}^{m+1}C_1a^mb^1 + {}^{m+1}C_2a^{m-1}b^2 + ... + {}^{m+1}C_ma^1b^m + {}^{m+1}C_{m+1}a^0b^{m+1}$$

Now,

$$\begin{aligned}(a + b)^{m+1} &= (a+b)\,(a + b)^m \\ &= (a+b)\,[{}^mC_0a^mb^0 + {}^mC_1a^{m-1}b^1 + ... + {}^mC_ra^{m-r}b^r + ... + {}^mC_{m-1}a^1b^{m-1} + {}^mC_ma^0b^m] \\ &= {}^mC_0a^{m+1}\,b^0 + ({}^mC_1 + {}^mC_0)\,a^mb^1 + ({}^mC_2 + {}^mC_1)\,a^{m-1}\,b^2 + ... \\ &\quad + ({}^mC_r + {}^mC_{r-1})a^{m-1+1}\,b^r + ... + ({}^mC_{m-1} + {}^mC_m)\,a^1.b^m + {}^mC_ma^{m+1} \\ &= {}^{m+1}C_0a^{m+1}\,b^0 + {}^{m+1}C_1a^mb^1 + {}^{m+1}C_2a^{m-1}\,b^2 + ... \\ &\quad + {}^{m+1}C_ra^{(m+1)-r}b^r + ... + {}^{m+1}C_ma^1b^m + {}^{m+1}C_{m+1}b^{m+1}\end{aligned}$$

$$[\because {}^mC_{r-1} + {}^mC_r = {}^{m+1}C_r]$$

Thus $P(m + 1)$ is true

Thus, $P(m)$ is true $\Rightarrow P(m+1)$ is true.

Hence, by the principle of mathematical induction the theorem is true for all $n \in N$.

Method II: By combinatorial method

Proof : We have,

$(a + b)^n = (a + b)\,(a + b)\,(a + b)\,(a + b)\,(a + b)\,(a + b)$ ... n times.

R.H.S. is the sum of the products that can be obtained by multiplying together one term from each bracket.

(*i*) a is multiplied from each bracket (n brackets), product is a^n.

(*ii*) We may take b from one bracket and a from each of the remaining $(n - 1)$ brackets and multiply them, product is $a^{n-1}b$ and this can be done in n ways. Thus, the final product is $na^{n-1}b$.

(*iii*) On choosing r brackets, b taken is out of each of them and a out of the remaining $(n - r)$ brackets, on multiplying them, we get $a^{n-r}\,b^r$. But r brackets containing b can be selected from the total number n brackets in ${}^nC_r$. Hence, sum of all the products $a^{n-r}\,b^r$ is ${}^nC_ra^{n-r}b^r$.

(*iv*) On taking b from each of the n brackets and multiplying them, we get b^n.

Hence, the R.H.S. is the sum of all the terms obtained on multiplying the n brackets is $a^n + {}^nC_1a^{n-1}b + {}^nC_2a^{n-2}b^2 + ... + {}^nC_ra^{n-r}b^r + ... + b^n$

$\therefore$ $\mathbf{(a + b)^n = a^n + {}^nC_1a^{n-1}b + {}^nC_2a^{n-2}b^2 + ... + {}^nC_ra^{n-r}b^r + ... + b^n}$ **Proved.**

8.7 SOME IMPORTANT RESULTS FROM THE BINOMIAL THEOREM

We have;

$$(a + b)^n = \sum_{r=0}^{n} {}^nC_ra^{n-r}b^r$$

$\Rightarrow$ $(a+b)^n = {}^nC_0a^nb^0 + {}^nC_1a^{n-1}b^1 + {}^nC_2a^{n-2}b^2 + \ldots + {}^nC_ra^{n-r}b^r + \ldots + {}^nC_na^0b^n$ ...(1)

1. Total number of terms in the expansion is $(n + 1)$, because r can have values from 0 to n.
2. The sum of the indices of a and b in each term is n.
3. Since, ${}^nC_r = {}^nC_{n-r}$ for $r = 0, 1, 2, \ldots, n$

$\Rightarrow$ ${}^nC_0 = {}^nC_n, {}^nC_1 = {}^nC_{n-1}, {}^nC_2 = {}^nC_{n-2}, \ldots$

So, the coefficients of terms equidistant from the beginning and end are equal. These coefficients are known as the binomial coefficients.

8.8 PARTICULAR CASES OF BINOMIAL EXPANSION

1. **Expansion of $(a-b)^n$:** Changing b into $-b$ in the theorem, we have

$$(a-b)^n = {}^nC_0a^n - {}^nC_1a^{n-1}b + {}^nC_2a^{n-2}b^2 - \ldots + (-1)^r\,{}^nC_ra^{n-r}b^r + \ldots + (-1)^n\,C_nb^n \quad \ldots(2)$$

2. **Expansion of $(a+b)^n + (a-b)^n$:** Adding (1) and (2), we get

$$(a+b)^n + (a-b)^n = 2[{}^nC_0a^nb^0 + {}^nC_2a^{n-2}b^2 + {}^nC_4a^{n-4}b^4 + \ldots]$$

3. **Expansion of $(a+b)^n - (a-b)^n$:** Subtracting (2) from (1), we get

$$(a+b)^n - (a-b)^n = 2[{}^nC_1a^{n-1}b^1 + {}^nC_3a^{n-3}b^3 + \ldots]$$

Note: If n is odd, then $[(a+b)^n + (a-b)^n]$ and $[(a+b)^n + (a-b)^n]$ both have the same number of terms equal to $\left(\frac{n+1}{2}\right)$ whereas if n is even, then $[(a+b)^n + (a-b)^n]$ has $\left(\frac{n}{2}+1\right)$ terms and $[(a+b)^n - (a-b)^n]$ has $\left(\frac{n}{2}\right)$ terms.

4. **Expansion of $(1+x)^n$:** Replacing a by 1 and b by x in the theorem, we have

$(1+x)^n = {}^nC_01^n + {}^nC_1x + {}^nC_2x^2 + \ldots + {}^nC_rx^r + \ldots + {}^nC_nx^n$

$$(1+x)^n = 1 + nx + \frac{n(n-1)}{2!}x^2 + \ldots + x^n$$

Note: **(1)** The coefficient of $(r + 1)$th term in the expansion of $(1 + x)^n$ is ${}^nC_r$

(2) The coefficient of x^r in the expansion of $(1 + x)^n$ is ${}^nC_r$.

5. **Expansion of $(1-x)^n$:** Replacing a by 1 and b by $-x$ in the theorem, we get

$(1-x)^n = {}^nC_01^n - {}^nC_1x + {}^nC_2x^2 + \ldots + (-1)^r\,{}^nC_rx^r + \ldots + (-1)^n\,{}^nC_nx^n$

$$(1-x)^n = 1 - nx + \frac{n(n-1)}{2!}x^2 - \ldots + (-1)x^n$$

Type I. To find the number of terms

Example 1. *Find the number of terms in the folllowing expansions:*

[S.B.T.E. 2016, 2015]

(*i*) $(3x + 2y)^{19}$ (*ii*) $\left(3x - \frac{1}{x^3}\right)^{10}$

(*iii*) $(1 + 2x + x^2)^{20}$ (*iv*) $(\sqrt{x} + \sqrt{y})^{10} + (\sqrt{x} - \sqrt{y})^{10}$

(*v*) $[(3x + y)^8 - (3x - y)^8]$ (*vi*) $(1 + 2\sqrt{3}x)^9 + (1 - 2\sqrt{3}x)^9$

Solution. We know that the number of terms in the expansion of $(a + b)^n$ is $(n + 1)$. Therefore,

(*i*) The number of terms in the expansion of $(3x + 2y)^{19} = 19 + 1 = 20$

(*ii*) The number of terms in the expansion of $\left(3x - \frac{1}{x^3}\right)^{10} = 10 + 1 = 11$

(*iii*) We have, $(1 + 2x + x^2)^{20} = \{(1 + x)^2\}^{20} = (1 + x)^{40}$

$\therefore$ the number of terms in the given expression $= 40 + 1 = 41$

(*iv*) If n is even, then the expansion of $(a + b)^n + (a - b)^n$ has $\left(\frac{n}{2} + 1\right)$ terms, so $(\sqrt{x} + \sqrt{y})^{10} + (\sqrt{x} - \sqrt{y})^{10}$ has $\left(\frac{10}{2} + 1\right)$, i.e. 6 terms.

(*v*) If n is even, then $(a + b)^n - (a - b)^n$ has $\frac{n}{2}$ terms. So $(3x + y)^8 - (3x - y)^8$ has 4 terms.

(*vi*) If n is odd, then the expansion of $(a + b)^n + (a - b)^n$ contains $\left(\frac{n + 1}{2}\right)$ terms. So, the expansion of $(1 + 2\sqrt{3}x)^9 + (1 - 2\sqrt{3}x)^9$ has $\left(\frac{9 + 1}{2}\right) = 5$ terms. **Ans.**

Type II. To expand the given expressions

Example 2. *Expand the following expressions:*

(*i*) $(1 - x)^6$ (*ii*) $\left(\frac{2}{x} - \frac{x}{2}\right)^5, (x \neq 0)$

Solution. We know that,

$$(a + b)^n = {}^nC_0a^n + {}^nC_1a^{n-1}b + {}^nC_2a^{n-2}b^2 + \ldots + {}^nC_ra^{n-r}b^r + \ldots + {}^nC_nb^n$$

Therefore,

(*i*)
$$(1 - x)^6 = {}^6C_0(1)^6 + {}^6C_1 1^5(-x) + {}^6C_2 1^4(-x)^2 + {}^6C_3 1^3(-x)^3 + {}^6C_4 1^2(-x)^4 + {}^6C_5(1)^1(-x)^5 + {}^6C_6(-x)^6$$

$$= \frac{6!}{0!6!}(1) + \frac{6!}{1!(6-1)!}(-x) + \frac{6!}{2!(6-2)!}(-x)^2 + \frac{6!(-x)^3}{3!(6-3)!} + \frac{6!}{4!(6-4)!}(-x)^4 + \frac{6!}{5!(6-5)!}(-x)^5 + \frac{6!}{6!(6-6)!}(-x)^6$$

$$= \frac{6!}{0!6!} + \frac{6!}{1!5!}(-x) + \frac{6!}{2!4!}(-x)^2 + \frac{6!}{3!3!}(-x)^3 + \frac{6!}{4!2!}(-x)^4 + \frac{6!}{5!1!}(-x)^5 + \frac{6!}{6!0!}(-x)^6$$

$$= 1 - 6x + 15x^2 - 20x^3 + 15x^4 - 6x^5 + x^6$$

(*ii*)
$$\left(\frac{2}{x} - \frac{x}{2}\right)^5 = {}^5C_0\left(\frac{2}{x}\right)^5 + {}^5C_1\left(\frac{2}{x}\right)^4\left(-\frac{x}{2}\right)^1 + {}^5C_2\left(\frac{2}{x}\right)^3\left(-\frac{x}{2}\right)^2 + {}^5C_3\left(\frac{2}{x}\right)^2\left(-\frac{x}{2}\right)^3 + {}^5C_4\left(\frac{2}{x}\right)^1\left(-\frac{x}{2}\right)^4 + {}^5C_5\left(-\frac{x}{2}\right)^5$$

$$= 1\left(\frac{2}{x}\right)^5 + 5\left(\frac{2}{x}\right)^4\left(-\frac{x}{2}\right) + 10\left(\frac{2}{x}\right)^3\left(-\frac{x}{2}\right)^2 + 10\left(\frac{2}{x}\right)^2\left(-\frac{x}{2}\right)^3 + 5\left(\frac{2}{x}\right)\left(-\frac{x}{2}\right)^4 + \left(-\frac{x}{2}\right)^5$$

$$= 32x^{-5} - 40x^{-3} + 20x^{-1} - 5x + \frac{5}{8}x^3 - \frac{1}{32}x^5$$ **Ans.**

Example 3. *Prove that:* $\sum_{r=0}^{n} 3^r.{}^nC_r = 4^n$

Solution. L.H.S. $= \sum_{r=0}^{n} 3^r.{}^nC_r = 3^0\,{}^nC_0 + 3^1\,{}^nC_1 + 3^2\,{}^nC_2 + \ldots + 3^n\,{}^nC_n$

$$= 1 + {}^nC_1(3)^1 + {}^nC_2(3)^2 + {}^nC_3(3)^3 + \ldots + {}^nC_n(3)^n$$

$$= (1 + 3)^n$$

$$= 4^n = \text{R.H.S.}$$ **Proved.**

Example 4. *Expand the following expressions :*

(*i*) $(1 + x + x^2)^3$ (*ii*) $\left(x - \frac{1}{y}\right)^{11}, y \neq 0$

Solution. (*i*) Let $y = x + x^2$. Then,

$$(1 + x + x^2)^3 = (1 + y)^3 = {}^3C_0 + {}^3C_1y^1 + {}^3C_2y^2 + {}^3C_3y^3$$

$$= 1 + 3y + 3y^2 + y^3 = 1 + 3(x + x^2) + 3(x + x^2)^2 + (x + x^2)^3$$

$$= 1 + 3(x + x^2) + 3(x^2 + 2x^3 + x^4) + {}^3C_0x^3(x^2)^0 + {}^3C_1x^{3-1}(x^2)^1 + {}^3C_2x^{3-2}(x^2)^2 + {}^3C_3x^0(x^2)^3\}$$

$$= 1 + 3(x + x)^2 + 3(x^2 + 2x^3 + x^4) + (x^3 + 3x^4 + 3x^5 + x^6)$$

$$= x^6 + 3x^5 + 6x^4 + 7x^3 + 6x^2 + 3x + 1$$ **Ans.**

(*ii*) We have,

$$\left(x - \frac{1}{y}\right)^{11} = {}^{11}C_0x^{11}\left(-\frac{1}{y}\right)^0 + {}^{11}C_1x^{10}\left(-\frac{1}{y}\right)^1 + {}^{11}C_2x^9\left(-\frac{1}{y}\right)^2 + {}^{11}C_3x^8\left(-\frac{1}{y}\right)^3 +$$

$${}^{11}C_4x^7\left(-\frac{1}{y}\right)^4 + {}^{11}C_5x^6\left(-\frac{1}{y}\right)^5 + {}^{11}C_6x^5\left(-\frac{1}{y}\right)^6 + {}^{11}C_7x^4\left(-\frac{1}{y}\right)^7$$

$$+ {}^{11}C_8x^3\left(-\frac{1}{y}\right)^8 + {}^{11}C_9x^2\left(-\frac{1}{y}\right)^9 + {}^{11}C_{10}x^1\left(-\frac{1}{y}\right)^{10} + {}^{11}C_{11}x^0\left(-\frac{1}{y}\right)^{11}$$

$$= x^{11} - 11\frac{x^{10}}{y} + 55\frac{x^9}{y^2} - 165\frac{x^8}{y^3} + 330\frac{x^7}{y^4} - 462\frac{x^6}{y^5}$$

$$+ 462\frac{x^5}{y^6} - \frac{330x^4}{y^7} + 165\frac{x^3}{y^8} - 55\frac{x^2}{y^9} + 11\frac{x}{y^{10}} - \frac{1}{y^{11}}$$

$$= x^{11} - 11x^{10}y^{-1} + 55x^9y^{-2} - 165x^8y^{-3} + 330x^7y^{-4} - 462x^6y^{-5}$$

$$+ 462x^5y^{-6} - 330x^4y^{-7} + 165x^3y^{-8} - 55x^2y^{-9} + 11xy^{-10} - y^{-11}$$ **Ans.**

Example 5. *Find the expansion of* $(3x^2 - 2ax + 3a^2)^3$

Solution. We have,

$$(3x^2 - 2ax + 3a^2)^3 = \{3x^2 - a(2x - 3a)\}^3$$
$$= {}^3C_0(3x^2)^3 - {}^3C_1(3x^2)^2\{a(2x - 3a)\} + {}^3C_2(3x^2)\, a^2(2x - 3a)^2 - {}^3C_3a^3(2x - 3a)^3$$
$$= (3x^2)^3 - 3(3x^2)^2 a(2x - 3a) + 3(3x^2)a^2(2x - 3a)^2 - a^3(2x - 3a)^3$$
$$= 27x^6 - 9x^4 . 3a(2x - 3a) + 9x^2a^2(4x^2 - 12ax + 9a^2y) - a^3.(8x^3 - 36x^2a + 54xa^2 - 27a^3)$$
$$= 27x^6 - 54ax^5 + 117a^2x^4 - 116a^3x^3 + 117a^4x^2 - 54a^5x + 27a^6 \quad \textbf{Ans.}$$

Example 6. *Expand using binomial theorem* $\left(1+\frac{x}{2}-\frac{2}{x}\right)^4, x \neq 0$

Solution. We have,

$$\left(1+\frac{x}{2}-\frac{2}{x}\right)^4 = \left\{1+\left(\frac{x}{2}-\frac{2}{x}\right)\right\}^4$$

$$= {}^4C_0 1^4 + {}^4C_1\left(\frac{x}{2}-\frac{2}{x}\right) + {}^4C_2\left(\frac{x}{2}-\frac{2}{x}\right)^2 + {}^4C_3\left(\frac{x}{2}-\frac{2}{x}\right)^3 + {}^4C_4\left(\frac{x}{2}-\frac{2}{x}\right)^4$$

$$= 1 + 4\left(\frac{x}{2}-\frac{2}{x}\right) + \frac{4(4-1)}{2!}\left(\frac{x}{2}-\frac{2}{x}\right)^2 + \frac{4(4-1)(4-2)}{3!}\left(\frac{x}{2}-\frac{2}{x}\right)^3 + \frac{4(4-1)(4-2)(4-3)}{4!}\left(\frac{x}{2}-\frac{2}{x}\right)^4$$

$$= 1+4\left(\frac{x}{2}-\frac{2}{x}\right)+6\left(\frac{x^2}{4}+\frac{4}{x^2}-2\right) +4\left\{{}^3C_0\left(\frac{x}{2}\right)^3 + {}^3C_1\left(\frac{x}{2}\right)^2\left(-\frac{2}{x}\right)^1 + {}^3C_2\left(\frac{x}{2}\right)^1\left(-\frac{2}{x}\right)^2 + {}^3C_3\left(-\frac{2}{x}\right)^3\right\}$$
$$+\left\{{}^4C_0\left(\frac{x}{2}\right)^4 + {}^4C_1\left(\frac{x}{2}\right)^3\left(-\frac{2}{x}\right)^1 + {}^4C_2\left(\frac{x}{2}\right)^2\left(-\frac{2}{x}\right)^2 + {}^4C_3\left(\frac{x}{2}\right)^1\left(-\frac{2}{x}\right)^3 + {}^4C_4\left(-\frac{2}{x}\right)^4\right\}.$$

$$= 1 + 4\left(\frac{x}{2}-\frac{2}{x}\right) + 6\left(\frac{x^2}{4}+\frac{4}{x^2}-2\right) +4\left\{\left(\frac{x}{2}\right)^3 + \left(\frac{x}{2}\right)^2\left(-\frac{2}{x}\right) + \frac{3(3-1)}{2!}\left(\frac{x}{2}\right)^1\left(-\frac{2}{x}\right)^2 + \frac{3(3-1)(3-2)}{3!}\left(-\frac{2}{x}\right)^3\right\}$$
$$+\left\{\left(\frac{x}{2}\right)^4 + 4\left(\frac{x}{2}\right)^3\left(-\frac{2}{x}\right) + \frac{4(4-1)}{2!}\left(\frac{x}{2}\right)^2\left(-\frac{2}{x}\right)^2 + \frac{4(4-1)(4-2)}{3!}\left(\frac{x}{2}\right)\left(-\frac{2}{x}\right)^3 + \frac{4(4-1)(4-2)(4-3)}{4!}\left(-\frac{2}{x}\right)^4\right\}$$

$$= 1+4\left(\frac{x}{2}-\frac{2}{x}\right)+6\left(\frac{x^2}{4}+\frac{4}{x^2}-2\right)+4\left\{\frac{x^3}{8}-\frac{3x}{2}+\frac{6}{x}-\frac{8}{x^3}\right\}+\left\{\frac{x^4}{16}-x^2+6-\frac{16}{x^2}+\frac{16}{x^4}\right\}$$

$$= 1+x(2-6)+x^2\left(\frac{3}{2}-1\right)+x^3\left(\frac{1}{2}\right)+x^4\left(\frac{1}{16}\right)+\frac{1}{x}(-8+24)+\frac{1}{x^2}(24-16)$$

$$+\frac{1}{x^3}(-32)+\frac{1}{x^4}(16)-12+6$$

$$= -5-4x+\frac{x^2}{2}+\frac{x^3}{2}+\frac{x^4}{2}+\frac{16}{x}+\frac{8}{x^2}-\frac{32}{x^3}+\frac{16}{x^4}$$ **Ans.**

Example 7. Using Binomial theorem, expand $\{(x + y)^5 + (x - y)^5\}$ and hence find the value of $\{(\sqrt{3}+1)^5 - (\sqrt{3}-1)^5\}$

Solution. We have,

$$(x + y\}^5 - (x - y)^5 = 2[{}^5C_1x^{5-1}y + {}^5C_3x^{5-3}y^3 + {}^5C_5x^{5-5}y^5]$$

$$= 2\left[\frac{5!}{1!4!}x^4y + \frac{5!}{3!2!}x^2y^3 + \frac{5!}{5!0!}x^0y^5\right]$$

$\Rightarrow$ $$(x + y)^5 - (x - y)^5 = 2\,[5x^4y + 10x^2y^3 + y^5] \quad ...(1)$$

Putting $x = \sqrt{3}$ and $y = 1$ in (1), we get

$$(\sqrt{3}+1)^5 - (\sqrt{3}-1)^5 = 2[5(\sqrt{3})^4(1) + 10(\sqrt{3})^2(1)^3 + (1)^5]$$

$$= 2\,[45 + 30 + 1] = 2 \times 76 = 152$$ **Ans.**

Example 8. *Evaluate* : $(\sqrt{3}+\sqrt{2})^6 - (\sqrt{3}-\sqrt{2})^6$

Solution. We know that:

$$(a + b)^n - (a - b)^n = 2[{}^nC_1a^{n-1}\,b + {}^nC_3a^{n-3}\,b^3 + ...]$$

Putting $a = \sqrt{3}, b = \sqrt{2}$ and $n = 6$ in the above, we get

$$(\sqrt{3}+\sqrt{2})^6 - (\sqrt{3}-\sqrt{2})^6 = 2\left[{}^6C_1(\sqrt{3})^5(\sqrt{2})^1 + {}^6C_3(\sqrt{3})^{6-3}(\sqrt{2})^3 + {}^6C_5{}^{6-5}(\sqrt{2})^5\right]$$

$$= 2[6(9\sqrt{3})(\sqrt{2}) + 20(3\sqrt{3})(2\sqrt{2}) + 6(\sqrt{3})(4\sqrt{2})]$$

$$= 2[54\sqrt{6} + 120\sqrt{6} + 24\sqrt{6}] = 396\sqrt{6}$$ **Ans.**

Example 9. *Let O denote the sum of odd terms and E denote the sum of even terms in the binomial expansion of $(x + a)^n$. Then, prove that*

(i) $O^2 - E^2 = (x^2 - a^2)^n$ *(ii)* $4OE = (x + a)^{2n} - (x - a)^{2n}$

Solution. We have,

$$(x + a)^n = {}^nC_0x^na^0 + {}^nC_1x^{n-1}a^1 + {}^nC_2x^{n-2}a_2 + ... + {}^nC_{n-1}xa^{n-1} + {}^nC_na^n$$

$\Rightarrow$ $$(x + a)^n = ({}^nC_0x^na^0 + {}^nC_2x^{n-2}a^2 + ...) + ({}^nC_1x^{n-1}a^1 + {}^nC_3x^{n-3}a^3 + ...)$$

$\Rightarrow$ $$(x + a)^n = O + E \quad ...(1)$$

and $$(x - a)^n = {}^nC_0x^n - {}^nC_1x^{n-1}a^1 + {}^nC_2x^{n-2}a^2 - {}^nC_3x^{n-3}a^3 + ...$$
$$+ {}^nC_{n-1}x(-1)^{n-1}a^{n-1} + {}^nC_n(-1)^na^n$$

$\Rightarrow$ $$(x + a)^n = ({}^nC_0x^n + {}^nC_2x^{n-2}a^2 + ...) - ({}^nC_1x^{n-1}a^1 + {}^nC_3x^{n-3}a^3 + ...)$$

$\Rightarrow$ $$(x - a)^n = O - E \quad ...(2)$$

(i) Multiplying (1) and (2), we get

$$(x + a)^n (x - a)^n = (O + E)(O - E)$$

$\Rightarrow \quad (x^2 - a^2)^n = O^2 - E^2$ **Proved.**

(ii) We have, $\quad 4OE = (O + E)^2 - (O - E)^2$

$\Rightarrow \quad 4OE = \{(x + a)^n\}^2 - \{(x - a)^n\}^2$ [Using (1) and (2)]

$\Rightarrow \quad 4OE = (x + a)^{2n} - (x - a)^{2n}$ **Proved.**

EXERCISE 8.1

Find the number of terms in the following expansions:

1. $(2x - 3y)^9$ **Ans.** 10

2. $(1 + 5\sqrt{2}x)^9 + (1 - 5\sqrt{2}x)^9$ **Ans.** 5

3. $(\sqrt{a} + \sqrt{b})^{20} + (\sqrt{a} - \sqrt{b})^{20}$ **Ans.** 11

4. $(1 + 2a + a^2)^{10}$ **Ans.** 21

5. $\left(2x - \dfrac{3}{x^3}\right)^{10}$ **Ans.** 11

Using Binomial theorem write down the expansions of the following:

6. $(3x - y)^4$ **Ans.** $81x^4 - 108x^3y + 54x^2y^2 - 12xy^3 + y^4$

7. $(3 + 2x^2)^4$ **Ans.** $81 + 216x^2 + 94x^6 + 16x^8$

8. $\left(x - \dfrac{y}{2}\right)^4$ **Ans.** $x^4 - 2x^3y + \dfrac{3}{2}x^2y^2 - \dfrac{1}{2}xy^3 + \dfrac{1}{16}y^4$

9. $\left(2x - \dfrac{1}{x}\right)^4, x \neq 0$ **Ans.** $16x^4 - 32x^2 + 24 - \dfrac{8}{x^2} + \dfrac{1}{x^4}$

10. $\left(\dfrac{2}{y} - \dfrac{y}{2}\right)^8, y \neq 0$ **Ans.** $\dfrac{256}{y^8} + \dfrac{512}{y^6} + \dfrac{448}{y^4} + \dfrac{224}{y^2} + 70 + 14y^2 + \dfrac{7}{4}y^4 + \dfrac{y^6}{8} + \dfrac{y^8}{256}$

11. $\left(\dfrac{x}{3} - \dfrac{4}{3x^2}\right)^5, x \neq 0$ **Ans.** $\dfrac{x^5}{243} - \dfrac{20}{243}x^2 + \dfrac{160}{243}\cdot\dfrac{1}{x} - \dfrac{640}{243}\cdot\dfrac{1}{x^4} + \dfrac{1280}{243}\cdot\dfrac{1}{x^7} - \dfrac{1024}{243}\cdot\dfrac{1}{x^{10}}$

12. $\left(\sqrt{\dfrac{x}{a}} - \sqrt{\dfrac{a}{x}}\right)^6$ **Ans.** $\left(\dfrac{x}{a}\right)^3 - 6\left(\dfrac{x}{a}\right)^2 + 15\left(\dfrac{x}{a}\right) - 20 + 15\left(\dfrac{a}{x}\right) - 6\left(\dfrac{a}{x}\right)^2 + \left(\dfrac{a}{x}\right)^3$

13. $(\sqrt[3]{x} - \sqrt[3]{a})^6$ **Ans.** $x^2 - 6x^{5/3}a^{1/3} + 15x^{4/3}a^{2/3} - 20ax + 15x^{2/3}a^{4/3} - 6x^{1/3}a^{5/3} + a^2$

14. $(1 + 2x - 3x^2)^5$

Ans. $1 + 10x + 25x^2 - 40x^3 - 190x^4 + 92x^5 + 570x^6 - 360x^7 - 675x^8 + 810x^9 - 243x^{10}$

15. $\left(x + 1 - \dfrac{1}{x}\right)^3$ **Ans.** $x^3 + 3x^2 - 5 + \dfrac{3}{x^2} - \dfrac{1}{x^3}$

Evaluate the following :

16. $\left(\sqrt{2} + 1\right)^6 - \left(\sqrt{2} - 1\right)^6$ **Ans.** $140\sqrt{2}$

17. $\left(2+\sqrt{5}\right)^2+\left(2-\sqrt{5}\right)^5$ **Ans.** 1364

18. $\left(3+\sqrt{2}\right)^5-\left(3-\sqrt{2}\right)^5$ **Ans.** $1178\sqrt{2}$

19. $\left(2+\sqrt{3}\right)^7+\left(2-\sqrt{3}\right)^7$ **Ans.** 1084

20. $\left(\sqrt{x+1}+\sqrt{x-1}\right)^6+\left(\sqrt{x+1}-\sqrt{x-1}\right)^6$ **Ans.** $16x\ (4x^2-3)$

8.9 GENERAL TERM IN A BINOMIAL EXPANSION

$$(a+b)^n = {}^nC_0a^nb^0+{}^nC_1a^{n-1}b^1+{}^nC_2a^{n-2}b^2+\ldots+{}^nC_ra^{n-r}b^r+\ldots+{}^nC_na^nb^n$$

We observe that:

$$\text{First term} = {}^nC_0a^nb^0$$
$$\text{Second term} = {}^nC_1a^{n-1}b^1$$
$$\text{Third term} = {}^nC_2a^{n-2}b^2$$
$$\text{Fourth term} = {}^nC_3a^{n-3}b^3$$

Thus, we find that the suffix of C in any term is one less than the number of term of the index of a is n minus the suffix of C and the index of b is same as the suffix of C.

Therefore, the $(r+1)$th term is given by

$${}^nC_ra^{n-r}b^r$$

Thus, if T_{r+1} denotes the $(r+1)$th term, then

$$\boxed{T_{r+1} = {}^nC_ra^{n-r}b^r}$$

This is called the general term.

Binomial expansion	General term
$(a+b)^n$	${}^nC_ra^{n-r}b^r$
$(a-b)^n$	$(-1)^r\,{}^nC_ra^{n-r}b^r$
$(1+x)^n$	${}^nC_rx^r$
$(1-x)^n$	$(-1)^r\,{}^nC_rx^r$

Note. In the binomial expansion $(a+b)^n$, the rth term from the end is $[(n+1)-(r-1)] = (n-r+2)$th term from the beginning.

Type I. To find the general term

Example 10. *Write the general term in the expansion of*

(*i*) $(x^2-y)^6$ (*ii*) $(1-x^2)^{12}$ (*iii*) $\left(x^2-\dfrac{1}{x}\right)^{12}, x \neq 0$

Solution. (*i*) We have

$$(x^2-y)^6 = [x^2+(-y)]^6$$

The general term in the expansion of the above Binomial is given by

$$T_{r+1} = {}^6C_r(x^2)^{6-r}\,(-y)^r \qquad \left[\because T_{r+1} = {}^nC_ra^{n-r}b^r\right]$$

$\Rightarrow$ $$T_{r+1} = (-1)^r\,{}^6C_rx^{12-2r}\,y^r$$ **Ans.**

(*ii*) We have,

$$(1 - x^2)^{12} = [1 + (-x^2)]^{12}$$

The general term in the expansion of the above Binomial is given by

$$T_{r+1} = {}^{12}C_r(1)^{12-r}\,(-x^2)^r$$
$$= (-1)^r\, {}^{12}C_r x^{2r}$$ **Ans.**

(*iii*) We have,

$$\left(x^2 - \frac{1}{x}\right)^{12} = \left[x^2 + \left(-\frac{1}{x}\right)\right]^{12}$$

The general term in the expansion of the above Binomial is given by

$$T_{r+1} = {}^{12}C_r(x^2)^{12-r}\left(-\frac{1}{x}\right)^r$$
$$= (-1)^r\, {}^{12}C_r x^{24-2r}\, x^{-r} = (-1)^r\, {}^{12}C_r x^{24-3r}$$ **Ans.**

Example 11. *Find the 4th term in the expansion of* $(x - 2y)^{12}$

Solution. We know that the $(r + 1)$th term in the expansion of $(a + b)^n$ is given by

$$T_{r+1} = {}^nC_r a^{n-r}\, b^r$$

$\therefore$ In the expansion of $(x - 2y)^{12}$, we have

$$T_4 = T_{3+1} = {}^{12}C_3(x)^{12-3}\,(-2y)^3$$

[here $n = 12$, $r = 3$, $a = x$, $b = -2y$]

$$= \frac{12!}{3!(12-3)!}x^9(-8y^3)$$

$$= -\frac{12 \times 11 \times 10 \times 9!}{3 \times 2 \times 1 \times 9!}x^9 \times 8y^3$$

$$= -2 \times 11 \times 10 \times 8 \times x^9y^3$$

$$= -1760x^9y^3$$ **Ans.**

Example 12. *Find the 13th term in the expansion of* $\left(9x - \frac{1}{3\sqrt{x}}\right)^{18}$, $x \neq 0$

Solution. We know that the $(r + 1)^{th}$ term in the expansion of $(a + b)^n$ is given by

$$T_{r+1} = {}^nC_r a^{n-r}\, b^r$$

$\therefore$ In the expansion of $\left(9x - \frac{1}{3\sqrt{x}}\right)^{18}$, we have

$$T_{13} = T_{12+1} = {}^{18}C_{12}(9x)^{18-12}\left(-\frac{1}{3\sqrt{x}}\right)^{12}$$

$$= \frac{18!}{12!(18-12)!}(9x)^6\left(\frac{1}{3\sqrt{x}}\right)^{12} = \frac{18!}{12! \times 6!}(9x)^6\,\frac{1}{3^{12}.x^6}$$

$$= \frac{18 \times 17 \times 16 \times 15 \times 14 \times 13 \times 12!}{12! \times 6 \times 5 \times 4 \times 3 \times 2 \times 1} \times \frac{9^6}{3^{12}} = 18564$$ **Ans.**

Example 13. *Find the rth term from the end in $(x + a)^n$.*

Solution. There are $(n + 1)$ terms in the expansion.

rth term from the end = $\{n + 1 - (r - 1)\}$th term from the beginning = $(n + 2 - r)$th term from the beginning.

Hence,

$$T_{n+2-r} = {}^nC_{n+1-r}\, x^{n-(n+1-r)}\, a^{n+1-r}$$
$$= {}^nC_{n+1-r}\, x^{n-n+r-1}\, a^{n+1-r}$$
$$= {}^nC_{n+1-r}\, x^{r-1}\, a^{n+1-r} \quad \textbf{Ans.}$$

Example 14. *Find the 4th term from the end in the expansion of* $\left(\frac{3}{x^2} - \frac{x^3}{6}\right)^7$

Solution. Clearly, the given expansion contains 8 terms.

So, 4th term from the end = (8 – 4 + 1)th = 5th term from the beginning.

$$\therefore \quad \text{Required term} = T_5 = T_{4+1} = {}^7C_4\left(\frac{3}{x^2}\right)^{7-4}\left(-\frac{x^3}{6}\right)^4$$

$$= {}^7C_3\left(\frac{3}{x^2}\right)^3\left(\frac{x^3}{6}\right)^4$$

$$= \frac{7\times 6\times 5}{3\times 2\times 1}\left(\frac{3^3}{x^6}\right)\left(\frac{x^{12}}{6^4}\right) = \frac{35}{48}x^6 \quad \textbf{Ans.}$$

EXERCISE 8.2

Find general terms in the following expansion:

1. $\left(2x + \frac{1}{x}\right)^5$ **Ans.** ${}^5C_r\,(2)^{5-r}\,x^{5-2r}, 0 \le r \le n$

2. $\left(\frac{4x}{5} - \frac{5}{2x}\right)^9$ **Ans.** ${}^9C_r(-1)^r\,\frac{2^{18-3r}}{5^{9-2r}}\cdot x^{9-2r}, 0 \le r \le n$

Find the specified term of the expansion in each of the following Binomials:

3. Tenth term of $(2x - y)^{11}$ **Ans.** $-220x^2y^9$

4. Fifth term of $(2a + 3b)^{12}$. Evalute it when $a = \frac{1}{3}$ and $b = \frac{1}{4}$ **Ans.** $\frac{55}{9}$

5. Sixth term of $\left(2x - \frac{1}{x^2}\right)^7$ **Ans.** $\frac{-84}{x^8}$

6. Tenth term of $\left(2x^2 + \frac{1}{x}\right)^{12}$ **Ans.** $\frac{1760}{x^3}$

7. Seventh term of $\left(\sqrt{x} - \frac{3}{x^2}\right)^{10}$ **Ans.** ${}^{10}C_4\,\frac{3^6}{x^{10}}$

EXERCISE 8.3

Write the middle terms of the following expansions:

1. $\left(x^2 - \frac{1}{x}\right)^6$ **Ans.** $-20x^3$
2. $\left(3a - \frac{a^3}{6}\right)^9$ **Ans.** $\frac{189}{8}a^{17}, -\frac{21}{16}a^{19}$
3. $\left(2x - \frac{1}{y}\right)^8$ **Ans.** $\frac{1120x^4}{y^4}$
4. $\left(\frac{a}{x} + \frac{x}{a}\right)^{10}$ **Ans.** 252
5. $\left(x^4 - \frac{1}{x^3}\right)^{11}$ **Ans.** $-462x^9, 462x^2$
6. $\left(x - \frac{2}{x}\right)^{11}$ **Ans.** $T_6 = -{}^{11}C_5.2^5.x = -14784x,\ T_7 = {}^{11}C_6.2^6.\frac{1}{x} = \frac{29568}{x}$
7. $(1 - 2x + x^2)^n$ **Ans.** $\frac{(2n)!}{(n!)^2}(-1)^n x^n$
8. $(1 + 3x + 3x^2 + x^3)^{2n}$ **Ans.** $\frac{(6n)!}{(3n!)^2}x^{3n}$
9. $\left(\frac{a}{x} + bx\right)^{12}$ **Ans.** $924a^6b^6$
10. $\left(x - \frac{1}{x}\right)^{2n+1}$ **Ans.** $(-1)^n\,{}^{2n+1}C_n x, (-1)^{n+1}\,{}^{2n+1}C_n.\frac{1}{x}$
11. $\left(3x - \frac{2}{x^2}\right)^{15}$ **Ans.** $\frac{-6435 \times 3^8 \times 2^7}{x^6}, \frac{6435 \times 3^7 \times 2^8}{x^9}$
12. Prove that the middle term in the expansion of $\left(2x + \frac{3}{x}\right)^{20}$ is $19 \times 17 \times 13 \times 11 \times 3^{10} \times 2^{12}$.

Type III. To find the coefficient:

Example 18. *Find the coefficient of x^5 in the binomial expansion of $(x + 3)^8$*

Solution. General term in the expansion $(x + 3)^8$ is $T_{r+1} = {}^8C_r x^{8-r}(3)^r$

Putting $8 - r = 5$, *i.e.* $r = 3$ in above, we get

$$T_{3+1} = T_4 = {}^8C_3 x^{8-3}(3)^3$$
$$= {}^8C_3 x^5 (27)$$

Hence, the required coefficient is

$$= \frac{8!}{3!(8-3)!}(27) = \frac{8!}{3! \times 5!} \times (27)$$
$$= \frac{8 \times 7 \times 6 \times 5!}{1 \times 2 \times 3 \times 5!}(27) = 8 \times 7 \times 27$$
$$= 1512$$ **Ans.**

Example 19. *Find the coefficient of $a^5 b^7$ in $(a - 2b)^{12}$*

Solution. We have

$$(a - 2b)^{12} = [a + (-2b)]^{12}$$

The general term in the expansion of $[a + (-2b)]^{12}$ is

$$T_{r+1} = {}^{12}C_7 a^{12-7}(-2b)^r \quad ...(1)$$

Putting $12 - r = 5$, i.e. $r = 7$ in (1) we, get

$$\begin{aligned} T_{7+1} &= {}^{12}C_7 a^{12-7}(-2b)^7 \quad ...(2) \\ &= {}^{12}C_7 a^5(-2)^7\, b^7 \\ &= {}^{12}C_7(-2)^7\, a^5\, b^7 \end{aligned}$$

Hence, the required coefficient is

$$\begin{aligned} {}^{12}C_7(-2)^7 &= -\frac{12!}{7!(12-7)!}.2^7 \\ &= -\frac{12\times11\times10\times9\times8\times7}{5\times4\times3\times2\times1\times7!}\cdot 2^7 \\ &= -11 \times 9 \times 8 \times 128 = -101376 \end{aligned}$$

Ans.

Example 20. *Find the coefficient of x^{16} in the binomial expansion of $(2x^2 - x)^{10}$*

(B.T.E. Delhi Mathematics 1, Jan. 2009)

Solution. We have,

$$(2x^2 - x)^{10} = [2x^2 + (-x)]^{10}$$

The general term in the expansion of $[2x^2 + (-x)]^{10}$ is

$$\begin{aligned} T_{r+1} &= {}^{10}C_r(2x^2)^{10-r}(-x)^r = {}^{10}C_r 2^{10-r}.x^{20-2r}.x^r.\,(-1)^r \\ &= (-1)^r.\, 2^{10-r}.\,{}^{10}C_r.x^{20-r} \quad ...(1) \end{aligned}$$

Putting $20 - r = 16$ i.e.; $r = 4$ in (1), we get

$$T_{4+1} = (-1)^4\, 2^{10-4}.\ {}^{10}C_4.x^{16}$$

Thus, coefficient of

$$\begin{aligned} x^{16} &= (-1)^4\,.2^6.\ {}^{10}C_4 \\ &= 64.\frac{10!}{4!(10-4)!} = 64\times\frac{10\times9\times8\times7\times6!}{4\times3\times2\times1\times6!} \\ &= 64 \times 10 \times 3 \times 7 = 13440 \end{aligned}$$

Ans.

Example 21. *The second, third and fourth terms in the expansion of $(x + a)^n$ are 240, 720 and 1080 respectively. Find n, x and a.*

Solution. We have, $T_2 = 240$, $T_3 = 720$, $T_4 = 1080$

Now,

$$T_2 = 240 \Rightarrow T_2 = {}^nC_1\, x^{n-1}\, a = 240 \quad ...(1)$$

$$T_3 = 720 \Rightarrow T_3 = {}^nC_2\, x^{n-2}\, a^2 = 720 \quad ...(2)$$

and

$$T_4 = 1080 \Rightarrow T_4 = {}^nC_3\, x^{n-3}\, a^3 = 1080 \quad ...(3)$$

Dividing (2) by (1), we get

$$\frac{{}^nC_2 x^{n-2} a^2}{{}^nC_1 x^{n-1} a} = \frac{720}{240}$$

$$\Rightarrow \quad \frac{{}^nC_2}{{}^nC_1}\left(\frac{a}{x}\right) = \frac{3}{1}$$

$$\Rightarrow \quad {}^nC_2\left(\frac{a}{x}\right) = 3\,{}^nC_1$$

$$\Rightarrow \quad \frac{n!}{2!(n-2)!}\left(\frac{a}{x}\right) = 3\frac{n!}{1!(n-1)!}$$

$$\Rightarrow \quad \frac{a}{x(2!)(n-2)!} = \frac{3}{(n-1)(n-2)!}$$

$$\Rightarrow \quad \frac{a}{2x} = \frac{3}{n-1} \qquad \text{...(4)}$$

Dividing (3) by (2), we get

$$\frac{{}^nC_3 x^{n-3} a^3}{{}^nC_2 x^{n-2} a^2} = \frac{1080}{720}$$

$$\Rightarrow \quad \frac{{}^nC_3}{{}^nC_2}\left(\frac{a}{x}\right) = \frac{3}{2}$$

$$\Rightarrow \quad {}^nC_3\left(\frac{a}{x}\right) = \frac{3}{2}\,{}^nC_2$$

$$\Rightarrow \quad \frac{n!}{(n-3)!\,3!}\left(\frac{a}{x}\right) = \frac{3}{2}\frac{n!}{2!(n-2)!}$$

$$\Rightarrow \quad \frac{1}{3(2!)(n-3)!}\left(\frac{a}{x}\right) = \frac{3}{2}\cdot\frac{1}{2!(n-2)(n-3)!}$$

$$\Rightarrow \quad \frac{a}{3x} = \frac{3}{2(n-2)} \qquad \text{...(5)}$$

From (4) and (5), we have

$$\frac{a}{x} = \frac{6}{n-1} = \frac{9}{2(n-2)}$$

$$\Rightarrow \quad \frac{2}{n-1} = \frac{3}{2(n-2)}$$

$$\Rightarrow \quad 4(n-2) = 3(n-1)$$

$$\Rightarrow \quad 4n - 8 = 3n - 3 \Rightarrow n = 5$$

Putting $n = 5$ in (4), we get

$$\frac{a}{2x} = \frac{3}{5-1} \Rightarrow a = \frac{3x}{2}$$

On putting the values of $n = 5$ and $a = \dfrac{3x}{2}$ in (1), we get

$${}^5C_1 . x^{5-1} . \frac{3x}{2} = 240 \Rightarrow 5 \times \frac{3}{2}x^5 = 240 \Rightarrow x^5 = \frac{240 \times 2}{15}$$

$$\Rightarrow \quad x^5 = 32 \Rightarrow x = 2$$

But

$$a = \frac{3x}{2} = \frac{3 \times 2}{2} = 3$$

Here, $n = 5$, $x = 2$ and $a = 3$ **Ans.**

Example 22. *Show that the coefficient of the middle term of $(1 + x)^{2n}$ is equal to the sum of the coefficients of the two middle terms of $(1 + x)^{2n-1}$*

Solution. The middle term in the expansion of $(1 + x)^{2n}$ is given by

$$T_{n+1} = {}^{2n}C_n x^n$$

So, the coefficient of the middle term in the expansion of $(1 + x)^{2n}$ is ${}^{2n}C_n$

Now, consider the expansion $(1 + x)^{2n-1}$

Here, the index $(2n - 1)$ is odd. So, $\left(\frac{(2n-1)+1}{2}\right)$th and $\left(\frac{(2n-1)+1}{2}+1\right)$th + 1, i.e. nth and $(n + 1)$th terms are middle terms.

$$\text{Now,} \quad T_n = T_{(n-1)+1} = {}^{2n-1}C_{n-1}(1)^{(2n-1)-(n-1)}x^{n-1}$$

$$\Rightarrow \quad T_n = {}^{2n-1}C_{n-1}\,x^{n-1}$$

$$\text{and} \quad T_{n+1} = {}^{2n-1}C_n(1)^{(2n-1)-n}x^n = {}^{2n-1}C_n x^n$$

So, the coefficients of the two middle terms in the expansion of $(1 + x)^{2n-1}$ are ${}^{2n-1}C_{n-1}$ and ${}^{2n-1}C_n$

$$\text{Therefore, sum of these coefficients} = {}^{2n-1}C_{n-1} + {}^{2n-1}C_n$$

$$= {}^{(2n-1)+1}C_n \qquad \left[\because {}^nC_{r-1} + {}^nC_r = {}^{n+1}C_r\right]$$

$$= {}^{2n}C_n$$

= Coefficient of middle term in the expansion of $(1 + x)^{2n}$ **Proved.**

Example 23. *Prove that there is no term involving x^6 in the expansion of $\left(2x^2 - \frac{3}{x}\right)^{11}$. where $r \neq 0$.*

Solution. Suppose x^6 occurs in $(r + 1)$th term in the expansion of $\left(2x^2 - \frac{3}{x}\right)^{11}$.

Now,

$$T_{r+1} = {}^{11}C_r(2x^2)^{11-r}\left(-\frac{3}{x}\right)^r$$

$$= {}^{11}C_r(-1)^r\,2^{11-r}\,3^r\,x^{22-3r} \qquad \text{...(1)}$$

For this term to contain x^6, we must have

$$22 - 3r = 6 \Rightarrow r = \frac{16}{3}, \text{ which is a fraction}$$

But, r is a natural number. Hence, there is no term containing x^6 **Ans.**

Type IV. Problems related to coefficients in binomial expansion

Binomial expansion	**Term**	**Coefficient**
$(1 + x)^n$	$(r + 1)^{\text{th}}$	${}^nC_r$
$(1 + x)^n$	x^r	${}^nC_r$
$(1 - x)^n$	$(r + 1)^{\text{th}}$	$(-1)^r\,{}^nC_r$
$(1 - x)^n$	x^r	$(-1)^r\,{}^nC_r$

In the above expansion, let us denote the coefficients ${}^nC_0, {}^nC_1, {}^nC_2, \ldots {}^nC_r, \ldots, {}^nC_n$, by $C_0, C_1, C_2, \ldots C_r, \ldots C_n$ respectively. Some properties concerning these coefficients a_x given below:

Property I. The coefficients of terms equidistant from the beginning and end in the expansion of $(1 + x)^n$ are equal.

Proof. We have,

$$(1 + x)^n = C_0 + C_1x + C_2x^2 + \ldots. + C_rx^r + \ldots + C_{n-1}x^{n-1} + C_nx^n.$$

The coefficients of $(r + 1)$th term from the beginning $= {}^nC_r$ ...(1)

The $(r + 1)$th term from the end is $(n - r + 1)$th term from the beginning.

$\therefore$ Coefficient of $(r + 1)$th term from the end $= {}^nC_{n-r}$ ...(2)

But $\qquad {}^nC_r = {}^nC_{n-r}$ ...(3)

From (1), (2) and (3), we conclude that the coefficient of terms are equidistant from the beginning and the end are equal.

Property II. In the expansion of $(1 + x)^n$, the sum of the binomial coefficients is 2^n.

i.e. $\qquad C_0 + C_1 + C_2 + \ldots + C_n = 2^n$

Proof: We have,

$$(1 + x)^n = C_0 + C_1x + C_2x^2 + C_3x^3 + \ldots C_{n-1}x^{n-1} + C_nx^n$$

Putting $x = 1$ in this expansion, we get

$$(1 + 1)^n = C_0 + C_1(1) + C_2(1)^2 + C_3(1)^3 + \ldots + C_{n-1}(1)^{n-1} + C_n(1)^n$$

$\Rightarrow C_0 + C_1 + C_2 + C_3 + \ldots C_{n-1} + C_n = 2^n$

Remark : Since ${}^nC_n = C_0 = 1$, therefore,

$$C_1 + C_2 + C_3 + \ldots + C_{n-1} + C_n = 2^n - 1$$

Property III. In the expansion of $(1 + x)^n$, the sum of the coefficients of odd terms is equal to the sum of the coefficients of even terms and each is equal to 2^{n-1}.

Proof. We have

$$(1 + x)^n = C_0 + C_1x + C_2x^2 + C_3x^3 + \ldots + C_{n-1}x^{n-1} + C_nx^n$$

Putting $x = 1$ and $x = -1$ in above expansion, we get

$$2^n = C_0 + C_1 + C_2 + C_3 + \ldots + C_{n-1} + C_n \quad \ldots(1)$$

and $\qquad 0 = C_0 - C_1 + C_2 - C_3 + \ldots + (-1)^{n-1}C_{n-1} + (-1)^n C_n$ ...(2)

Adding and subtracting (1) and (2), we get

$$2^n = 2(C_0 + C_2 + C_4 + \ldots) \quad \ldots(3)$$

and $\qquad 2^n = 2(C_1 + C_3 + C_5 + \ldots)$ ...(4)

From (3) and (4), we get

$$C_0 + C_2 + C_4 + \ldots = 2^{n-1} = C_1 + C_3 + C_5 + \ldots$$

Things to remember :

1.	${}^nC_0 + {}^nC_1 + {}^nC_2 + \ldots + {}^nC_n = 2^n$
2.	${}^nC_0 + {}^nC_2 + {}^nC_4 + \ldots + {}^nC_1 + {}^nC_3 + {}^nC_5 + \ldots = 2^{n-1}$
3.	${}^nC_1 + {}^nC_2 + {}^nC_3 + \ldots + {}^nC_n = 2^n - 1$

Property IV. The sum of the squares of the coefficients in the expansion of $(1 + x)^n$ is $\dfrac{2n!}{(n!)^2}$.

Proof: We have,

$$(1 + x)^n = C_0 + C_1x + C_2x^2 + C_3x^3 + ... + C_nx^n \quad ...(1)$$

$$(x + 1)^n = C_0x^n + C_1x^{n-1} + C_2x^{n-2} + ... + C_n \quad ...(2)$$

Multiplying (1) and (2), we get

$$(1 + x)^{2n} = [C_0 + C_1x + C_2x^2 + C_3x^3 + ... + C_nx^n] \times [C_0x^n + C_1x^{n-1} + C_2x^{n-2} + ...C_n] ...(3)$$

Equating coefficients of x^n on both sides of (3), we get

$$^{2n}C_n = C_0^{\ 2} + C_1^{\ 2} + C_2^{\ 2} + ... + C_n^{\ 2}$$

Hence, sum of the squares of coefficients

$$^{2n}C_n = \frac{2n!}{(n!)^2}$$

Example 34. *If $C_0, C_1, C_2, ..., C_n$ are the binomial coefficients in the expansion of $(1 + x)^n$, prove that*

(i) $C_1 + 2C_2 + 3C_3 + ... + nC_n = n{\cdot}2^{n-1}$

(ii) $C_0 + 2C_1 + 3C_2 + ... + (n+1)\, C_n = 2^n + n{\cdot}2^{n-1}$

Solution. *We know that,*

(i) L.H.S. $= C_1 + 2C_2 + 3C_3 + ... + nC_n$

$$= n + 2.\frac{n(n-1)}{2!} + 3.\frac{n(n-1)(n-2)}{3!} + ... + n$$

$$= n\left[1 + (n-1) + \frac{(n-1)(n-2)}{2!} + ... + 1\right]$$

$$= n\left[{}^{n-1}C_0 + {}^{n-1}C_1 + {}^{n-1}C_2 + ... + {}^{n-1}C_{n-1}\right]$$

$= n\, 2^{n-1} =$ R.H.S. **Proved.**

(ii) L.H.S. $= C_0 + 2C_1 + 3C_2 + ... + (n+1)C_n$

$= [C_0 + C_1 + C_2 + ... + C_n] + [C_1 + 2.C_2 + 3C_3 + ... + nC_n]$

$= 2^n + n.2^{n-1} =$ R.H.S. **Proved.**

$$\left[\because C_0 + C_1 + C_2 + ... + C_n = 2^n\right]$$

Example 35. *If C_r denotes ^nC_r, prove that*

(i) $\dfrac{C_1}{C_0} + 2\dfrac{C_2}{C_1} + 3\dfrac{C_3}{C_2} + ... + n\dfrac{C_n}{C_{n-1}} = \dfrac{n(n+1)}{2}$

(ii) $(C_0 + C_1)(C_1 + C_2)\,(C_2 + C_3)...\,(C_{n-1} + C_n) = \dfrac{C_0C_1C_2...C_{n-1}(n+1)^n}{n!}$

Solution. *(i)* We have, $C_0 = 1,\ C_1 = n,\ C_2 = \dfrac{n(n-1)}{2\cdot 1}, C_3 = \dfrac{n(n-1)(n-2)}{3\cdot 2\cdot 1}$

$$C_{n-1} = {}^nC_{n-1},\ {}^nC_1 = n,\ C_n = 1$$

$\therefore$ L.H.S. $= \dfrac{C_1}{C_0} + 2\dfrac{C_2}{C_1} + 3\dfrac{C_3}{C_2} + ... + n\dfrac{C_n}{C_{n-1}}$

$$= \frac{n}{1} + 2\frac{\frac{n(n-1)}{2\cdot1}}{n} + 3\cdot\frac{\frac{n(n-1)(n-2)}{3\cdot2\cdot1}}{\frac{n(n-1)}{2.1}} + \ldots + n\cdot\frac{1}{n}$$

$= n + (n-1) + (n-2) + \ldots + 1$

$= 1 + 2 + 3 \ldots + n$ (writing in reverse order)

$=$ Sum of first n natural numbers

$= \frac{n(n+1)}{2}$ = R.H.S. **Proved.**

(*ii*) We have, $C_0 + C_1 = C_0\left(1+\frac{C_1}{C_0}\right) = C_0\left(1+\frac{n}{1}\right)$

$\Rightarrow \quad (C_0 + C_1) = C_0(1+n)$ ...(1)

$$C_1 + C_2 = C_1\left(1+\frac{C_2}{C_1}\right) = C_1\left(1+\frac{\frac{n(n-1)}{2}}{n}\right)$$

$$= C_1\left[1+\frac{n-1}{2}\right]$$

$(C_1 + C_2) = C_1\left(\frac{1+n}{2}\right)$ ...(2)

Similarly $(C_2 + C_3) = C_2\left(\frac{1+n}{3}\right)$ ...(3)

and $(C_{n-1} + C_n) = C_{n-1}\left(\frac{n+1}{n}\right)$

On multiplying (1), (2), (3), we get

$$(C_0 + C_1)(C_1 + C_2)(C_2 + C_3)\ldots(C_{n-1} + C_n) = C_0(1+n)C_1\left(\frac{1+n}{2}\right)C_2\left(\frac{1+n}{3}\right)\ldots C_{n-1}\left(\frac{n+1}{n}\right)$$

$$= \frac{C_0\,C_1\,C_2\ldots C_{n-1}(n+1)^n}{n!}$$ **Proved**

Example 36. *Find the coefficient of x^4 in the expansion of $(1+x)^n(1-x)^n$. Deduce that*

$$C_2 = C_0C_4 - C_1C_3 + C_2C_2 - C_3C_1 + C_4C_0$$

Solution. We have,

$$(1+x)^n(1-x)^n = [(1+x)(1-x)]^n = (1-x^2)^n$$
$$= C_0 - C_1x^2 + C_2(x^2)^2 - \ldots$$
$$= C_0 - C_1x^2 + C_2x^4 - \ldots$$

$\therefore$ Coefficient of x^4 in the expansion of $(1+x)^n(1-x)^n$ is C_2

Now, $(1+x)^n = C_0 + C_1x + C_2x^2 + C_3x^3 + C_4x^4 + \ldots + C_nx^n$

Also, $(1-x)^n = C_0 - C_1x + C_2x^2 - C_3x^3 + C_4x^4 + \ldots + (-1)^nC_nx^n$

Multiplying these, we get

$$(1+x)^n(1-x)^n = (C_0 + C_1x + C_2x^2 + C_3x^3 + C_4x^4 + \ldots + C_nx^n)$$
$$\times (C_0 - C_1x + C_2x^2 - C_3x^3 + C_4x^4 + \ldots + (-1)^nC_nx^n)$$

Equating coefficient of x^4 on both sides, we get

$$C_2 = C_0C_4 - C_1C_3 + C_2C_2 - C_3C_1 + C_4C_0. \quad \textbf{Proved}$$

Example 37. *If C_r denotes ${}^nC_r$, then prove that*

$$(i)\ C_0 - \frac{C_1}{2} + \frac{C_2}{3} - ... = \frac{1}{n+1}$$

$$(ii)\ 2C_0 + 2^2\frac{C_1}{2} + 2^3\frac{C_2}{3} + ... + 2^{n+1}\frac{C_n}{n+1} = \frac{3^{n+1}-1}{n+1}$$

Solution. *We have,*

$$(i)\ \text{L.H.S.} = C_0 - \frac{C_1}{2} + \frac{C_2}{3} - \frac{C_3}{4} + ... + (-1)^n\frac{C_n}{n+1}$$

$$= 1 - \frac{n}{2} + \frac{n(n-1)}{3(2!)} - \frac{n(n-1)(n-2)}{4(3!)} + ... + \frac{(-1)^n 1}{n+1}$$

$$= \frac{1}{(n+1)}\left[(1+n) - \frac{(n+1)n}{2!} + \frac{(n+1)n(n-1)}{3!} - \frac{(n+1)n(n-1)(n-2)}{4!} + ... + (-1)^n\right]$$

$$= \frac{1}{n+1}\left[{}^{n+1}C_1 - {}^{n+1}C_2 + {}^{n+1}C_3 - {}^{n+1}C_4 + ... + (-1)^n\ {}^{n+1}C_{n+1}\right]$$

$$= \frac{1}{n+1}\left[1-(1-{}^{n+1}C_1 + {}^{n+1}C_2 - {}^{n+1}C_3 - {}^{n+1}C_4 - ... + (-1)^{n+1}\ {}^{n+1}C_1)\right]$$

$$= \frac{1}{n+1}\left[1-(1-1)^{n+1}\right] = \frac{1}{n+1} = \text{R.H.S.} \quad \textbf{Proved}$$

$$(ii)\ \text{L.H.S.} = 2C_0 + 2^2\frac{C_1}{2} + 2^3\frac{C_2}{3} + 2^4\frac{C_3}{4} + ... + 2^{n+1}\frac{C_n}{n+1}$$

$$= 2.1 + 2^2.\frac{n}{2} + \frac{2^3 n(n-1)}{3(2!)} + \frac{2^4 n(n-1)(n-2)}{4(3!)} + ... + \frac{2^{n+1}}{n+1}.1$$

$$= \frac{1}{n+1}\left[2.1(n+1) + \frac{2^2(n+1)4}{2!} + \frac{2^3(n+1)n(n-1)}{3!} + \frac{2^4(n+1)n(n-1)(n-2)}{4!} + ... + 2^{n+1}\right]$$

Adding and subracting 1, we have

$$= \frac{1}{n+1}\left[1 + (n+1).2 + \frac{(n+1)}{2!}2^2 + \frac{(n+1)n(n-1)}{3!}2^3 + ... + 2^{n+1} - 1 + ...\right]$$

$$= \frac{1}{n+1}\left[(1+2)^{n+1} - 1\right]$$

[$\because$ in the expansion of $(1+x)^n$, n is equal to $(n+1)$ and $x = 2$]

$$= \frac{3^{n+1}-1}{n+1} = \text{R.H.S.} \quad \textbf{Proved}$$

Example 38. *If C_r, denotes ${}^nC_r$, then prove that*

$$(i)\ C_0 + \frac{C_2}{3} + \frac{C_4}{5} + ... = \frac{2^n}{n+1} \quad (ii)\ 2C_0 + 5C_1 + 8C_2 + ... + (3n+2)C_n = (3n+4)2^{n-1}$$

$$= a\left(\sum_{r=0}^{n} {}^nC_r\right) + bn\left(\sum_{r=1}^{n} {}^{n-1}C_{r-1}\right)$$

$$\left[\because \sum_{r=0}^{n} {}^nC_r = 2^n, \sum_{r=1}^{n} {}^{n-1}C_{r-1} = 2^{n-1}\right]$$

$$= a.2^n + bn.2^{n-1}$$

$$= (2a + bn)2^{n-1}$$

Proved

(*ii*) We have, $C_3 + 2C_4 + 3C_5 + ... + (n-2)\,C_n$

$$= \sum_{r=3}^{n} (r-2)C_r = \sum_{r=3}^{n} (r-2)\,{}^nC_r = \sum_{r=3}^{n} r.{}^nC_r - \sum_{r=3}^{n} 2\,{}^nC_r$$

$$= \sum_{r=3}^{n} r.\frac{n}{r}.{}^{n-1}C_{r-1} - 2\sum_{r=3}^{n} {}^nC_r = n\left[\sum_{r=3}^{n} {}^{n-1}C_{r-1}\right] - 2\left[\sum_{r=3}^{n} {}^nC_r\right]$$

$$= n[({}^{n-1}C_2 + {}^{n-1}C_3 + ... + {}^{n-1}C_{n-1}) - 2\,({}^nC_3 + {}^nC_4 + ... + {}^nC_n)]$$

$$= n[({}^{n-1}C_0 + {}^{n-1}C_1 + {}^{n-1}C_2 + ... + {}^{n-1}C_{n-1}) - ({}^{n-1}C_0 + {}^{n-1}C_1)]$$

$$= -2[({}^nC_0 + {}^nC_1 + {}^nC_2 + {}^nC_3 + ... + {}^nC_n) - ({}^nC_0 + {}^nC_1 + {}^nC_2)]$$

$$= n[2^{n-1} - (1 + (n-1))] - 2\left[2n - \left\{1 + n + \frac{n(n-1)}{2}\right\}\right]$$

$$= n[2^{n-1} - n] - 2\left[2^n - \left(\frac{n^2 + n + 2}{2}\right)\right]$$

$$= n.2^{n-1} - n^2 - 2.\,2^n + n^2 + n + 2$$

$$= n.2^{n-1} - 2^{n+1} + n + 2 = (n-4)\,2^{n-1} + n + 2$$

Proved

EXERCISE 8.6

If $C_0, C_1, C_2, C_3, ..., C_n$ denote the binomial coefficients in the expansion of $(1 + x)^n$, prove the following:

1. $C_0 + 3C_1 + 5C_2 + ... + (2n+1)C_n = (n+1).2^n$
2. $C_0 - C_1 + C_2 - C_3 + ... + (-1)^n C_n = 0$
3. $C_1 - 2C_2 + 3C_3 - 4C_4 + ... + n(-1)^{n-1} C_n = 0$
4. $a - (a-1)C_1 + (a-2)C_2 - (a-3)C_3 + ... + (-1)^n (a-n)C_n = 0$
5. $C_0^2 + C_1^2 + C_2^2 + ... + C_n^2 = \dfrac{(2n)!}{(n!)^2}$
6. $C_0C_2 + C_1C_3 + ...C_{n-2}C_n = \dfrac{(2n)!}{(n-2)!\,(n+2)!}, (n \geq 2)$
7. $C_0 + \dfrac{C_1}{2} + \dfrac{C_2}{3} + ... + \dfrac{C_n}{n+1} = \dfrac{2^{n+1} - 1}{n+1}$

8. $C_0C_1 + C_1C_2 + C_2C_3 + ... + C_{n-1}C_n = \dfrac{(2n)!}{(n+1)!(n-1)!}$

9. $C_0 - C_1 + C_2 - C_3 + ... + (-1)^r C_r = \dfrac{(-1)^r (n-1)!}{r!(n-r-1)!}$

10. $C_0^2 - C_1^2 + C_2^2 - C_3^2 + ... + (-1)^n C_n^2 = 0$ or $(-1)^{n/2}\,{}^{n}C_{n/2}$ for n is odd or even.

11. $C_0 + 5C_1 + 5^2 C_2 + 5^3 C_3 + ... + 5^n C_n = 6^n$

12. $3C_0 - 8C_1 + 13C_2 - 18C_3 + ...$ upto $(n + 1)$ terms = 0

13. $\sum_{i=1}^{n} 3^r\, {}^nC_r = 4^n$

8.12 APPLICATIONS OF BINOMIAL THEOREM

In this section, we shall discuss some problems based on Binomial theorem.

Example 41. *Using binomial theorem, evaluate:*

(*i*) $(999)^5$ (*ii*) $(102)^6$ (*iii*) $(10.1)^5$

Solution. (*i*) We have,

$$(999)^5 = (1000 - 1)^5 = (10^3 - 1)^5$$

Expanding by binomial theorem, we get

$$(999)^5 = (10^3 - 1)^5$$

$$= {}^5C_0(10^3)^5 - {}^5C_1(10^3)^4(1) + {}^5C_2(10^3)^3(1)^2 - {}^5C_3(10^3)^2(1)^3 + {}^5C_4(10^3)(1)^4 - {}^5C_5(1)^5$$

$$= (10)^{15} - 5 \times 10^{12} + (10)^9 - 10(10)^6 + 5(10)^3 - 1$$

$$= [(10)^{15} - 5 \times 10^{12} + (10)^{10}] - (10)^7 + 5(10)^3 - 1$$

$$= 10^{10}[10^5 - 5 \times 100 + 1] - 10^3[10^4 - 5] - 1$$

$$= 10^{10}[100000 - 500 + 1] - 10^3[10000 - 5] - 1$$

$$= 99501 \times 10^{10} - 9995 \times 10^3 = 975009990004$$ **Ans**

(*ii*)

$$(102)^6 = (100 + 2)^6$$

$$= {}^6C_0 \times (100)^6 + {}^6C_1 \times (100)^5 \times 2 + {}^6C_2 \times (100)^4 \times 2^2 + {}^6C_3 \times (100)^3 \times 2^3 + {}^6C_4 \times (100)^2 \times 2^4 + {}^6C_5 \times (100)^1 \times 2^5 + {}^6C_6 \times (100)^0 \times 2^6$$

$$= (100)^6 + 6 \times (100)^5 \times 2 + 15 \times (100)^4 \times 2^2 + 20 \times (100)^3 \times 2^3 + 15 \times (100)^2 \times 2^4 + 6 \times (100)^1 \times 2^5 + 2^6$$

$$= 10^{12} + 12 \times 10^{10} + 6 \times 10^9 + 16 \times 10^7 + 24 \times 10^5 + 192 \times 10^2 + 64$$

$$= 1126162419264$$ **Ans**

(*iii*)

$$(10.1)^5 = (10 + 0.1)^5$$

$$= {}^5C_0 \times (10)^5 \times (0.1)^0 + {}^5C_1(10)^4(0.1)^1 + {}^5C_2(10)^3(0.1)^2 + {}^5C_3(10)^2(0.1)^3 + {}^5C_4(10)^1(0.1)^4 + {}^5C_5(10)^0(0.1)^5$$

$$= (10)^5 + 5 \times 10^4 \times 0.1 + 10 \times 10^3 \times (0.1)^2 + 10 \times (10)^2 \times (0.1)^3 + 5 \times 10 \times (0.1)^4 + (0.1)^5$$

$$= 10^5 + 5 \times 10^3 + 10^2 + 1 + 5 \times 0.001 + 0.00001$$

$$= 100000 + 5000 + 100 + 1 + 0.005 + 0.00001$$

$$= 105101.00501$$ **Ans**

(iii) $(1-x)^n = [1+(-x)]^n = 1 + n(-x) + \frac{n(n-1)}{1.2}(-x)^2 + \frac{n(n-1)(n-2)}{1.2.3}(-x)^3 + ...$

$$= 1 - nx + \frac{n(n-1)}{1.2}x^2 - \frac{n(n-1)(n-2)}{1.2.3}x^3 + ..., \text{where}, |-x|<1 |, \text{i.e.} |x|<1$$

(iv) $(1-x)^{-n} = [1+(-x)]^{-n} = 1 + (-n)(-x) + \frac{(-n)(-n-1)}{1.2}(-x)^2$

$$+ \frac{(-n)(-n-1)(-n-2)}{1.2.3}(-x)^3 + ...$$

$$= 1 + nx + \frac{n(n+1)}{1.2}(x)^2 + \frac{(n)(n+1)(n+2)}{1.2.3}x^3 + ...$$

where, $|-x| < 1$, *i.e.* $|x| < 1$

Type I: To find expansion:

Example 45. *Expand the following :*

(i) $(1 + x^4)^{-3}, |x| < 1$ (ii) $(1-2x)^{-1}, |x| < \frac{1}{2}$

Solution. (i) We know that,

$$(1 + x)^n = 1 + nx + \frac{n(n-1)}{2!}x^2 + \frac{n(n-1)(n-2)}{3!}x^3 + ...$$

$$\therefore \quad (1 + x^4)^{-3} = 1 + (-3)(x^4) + \frac{(-3)(-3-1)}{2!}(x^4)^2 + \frac{(-3)(-3-1)(-3-2)}{3!}(x^4)^3 + ...$$

$$\Rightarrow \quad (1 + x^4)^{-3} = 1 - 3x^4 + 6x^8 - 10x^{12} + ...$$ **Ans.**

(ii) We know that,

$$(1 + x)^n = 1 + nx + \frac{n(n-1)}{2!}x^2 + ...$$

$$\therefore \quad (1 - 2x)^{-1} = [1 + (-2x)]^{-1}$$

$$= 1 + (-1)(-2x) + \frac{(-1)(-2)}{2!}(-2x)^2 + \frac{(-1)(-2)(-3)}{3!}(-2x)^3 + ...$$

$$\Rightarrow \quad (1 - 2x)^{-1} = 1 + 2x + 4x^2 + 8x^3 + ...$$ **Ans.**

Example 46. *Expand* $\frac{1}{(4-3x^2)^{\frac{1}{3}}}$ *to four terms. For what values of x is the expansion valid?*

Solution. We have,

$$\frac{1}{(4-3x^2)^{\frac{1}{3}}} = (4-3x^2)^{-\frac{1}{3}} = \left\{4\left(1 - \frac{3}{4}x^2\right)\right\}^{-\frac{1}{3}}$$

$$= 4^{-\frac{1}{3}}\left(1 - \frac{3}{4}x^2\right)^{-\frac{1}{3}} = 4^{-\frac{1}{3}}\left(1 + \left(-\frac{3}{4}x^2\right)\right)^{-\frac{1}{3}}$$

$$= 4^{-\frac{1}{3}}\left[1+\left(-\frac{1}{3}\right)\left(-\frac{3x^2}{4}\right)+\frac{\left(-\frac{1}{3}\right)\left(-\frac{1}{3}-1\right)}{2!}\left(-\frac{3x^2}{4}\right)^2\right.$$

$$\left.+\frac{\left(-\frac{1}{3}\right)\left(-\frac{1}{3}-1\right)\left(-\frac{1}{3}-2\right)}{3!}\left(-\frac{3x^2}{4}\right)^3+\ldots\right]$$

$$= 4^{-\frac{1}{3}}\left[1+\frac{x^2}{4}+\frac{x^4}{8}+\frac{7}{96}x^6+\ldots\right]$$

The above expansion is valid if

$$\left|\frac{3x^2}{4}\right|<1 \Rightarrow |x^2|<\frac{4}{3} \Rightarrow |x|<\frac{2}{\sqrt{3}} \Rightarrow -\frac{2}{\sqrt{3}}<x<\frac{2}{\sqrt{3}}$$ **Ans.**

Type II: To find the general term:

Example 47. *Find the general term in the expansion of* $(1-x)^{-4}$.

Solution. Let T_{r+1} be the $(r+1)$th term in the expansion of $(1-x)^{-4}$.

$$T_{r+1} = \frac{(-4)(-4-1)(-4-2)\ldots(-4-r+1)}{r!}(-x)^r$$

$$= \frac{(-1)^r\cdot 4\cdot 5\cdot 6\ldots(r+3)}{r!}(-1)^r(x)^r$$

$$= \frac{(-1)^{2r}\cdot 4\cdot 5\cdot 6\ldots(r+3)}{r!}x^r$$

$$= \frac{4\cdot 5\cdot 6\ldots r(r+1)(r+2)(r+3)}{1.2.3.4\ldots(r-1)^r}$$

$$= \frac{(r+1)(r+2)(r+3)}{1.2.3}x^r$$ **Ans**

Example 48. *Find the general term of the expansion* $(2-3x)^{\frac{3}{2}}, |x|<\frac{2}{3}$

Solution. We have

$$(2-3x)^{\frac{3}{2}} = 2^{\frac{3}{2}}\left[1+\left(\frac{-3x}{2}\right)\right]^{\frac{3}{2}}$$

Let T_{r+1} be the $(r+1)$th term in the expansion of $(2-3x)^{\frac{3}{2}}$.

$$\Rightarrow \quad T_{r+1} = 2^{\frac{3}{2}}\left[\frac{\left(\frac{3}{2}\right)\left(\frac{3}{2}-1\right)\left(\frac{3}{2}-2\right)\left(\frac{3}{2}-3\right)\ldots\left(\frac{3}{2}-r+1\right)}{r!}\right]\left(\frac{-3x}{2}\right)^r$$

$$= \frac{2^{\frac{3}{2}} \times \frac{3}{2} \times \left(\frac{1}{2}\right)\left(-\frac{1}{2}\right)\left(-\frac{3}{2}\right)...\left(\frac{3-2r+2}{2}\right)}{r\,!}\left(-\frac{3x}{2}\right)^r$$

$$= \frac{2^{\frac{3}{2}}\left(\frac{3}{2}\right)\left(\frac{1}{2}\right)\left(-\frac{1}{2}\right)...\left(\frac{5-2r}{2}\right)}{r\,!}\left(-\frac{3x}{2}\right)^r$$

$$= \frac{\left(2^{\frac{3}{2}}\right)(3)(1)(-1)(-3)...(5-2r)}{2^r\,r!}(-r)^r \cdot 3^r \cdot \frac{1}{2^r} \cdot x^r$$

$$= \frac{\left(2^{\frac{3}{2}}\right)(3)(-1)^{r-2}...(1)(3)(5)...(2r-5)}{r!}(-r)^r \cdot 3^r \cdot \frac{1}{2^r} \cdot x^2$$

$$= \frac{(-1)^{r-2}.1.3.5...(2r-5)(-1)^r}{r!}.2.^{\frac{3-4r}{2}}.3^{r+1}.x^r$$

$$= \frac{1.3.5...(2r-5)}{r!}2.^{\frac{3-4r}{2}}.3^{r+1}.x^r$$

Example 49. *Find the 8th term of the expansion* $\left(1-3x^2\right)^{\frac{16}{3}}, |x| < \frac{1}{\sqrt{3}}$.

Solution. General term of the expansion $(1 + x)^n$ is

$$T_{r+1} = \frac{n(n-1)(n-2)(n-3)...(n+1-r)}{r\,!}x^r$$

Now putting $r + 1 = 8 \Rightarrow r = 7$ and $n = \frac{16}{3}$ in the above, we get

$$T_8 = T_{7+1} = \frac{\frac{16}{3}\left(\frac{16}{3}-1\right)\left(\frac{16}{3}-2\right)\left(\frac{16}{3}-3\right)\left(\frac{16}{3}-4\right)\left(\frac{16}{3}-5\right)\left(\frac{16}{3}-6\right)}{7!}(-3x^2)^7$$

$$= \frac{\left(\frac{16}{3}\right)\left(\frac{13}{3}\right)\left(\frac{10}{3}\right)\left(\frac{7}{3}\right)\left(\frac{4}{3}\right)\left(\frac{1}{3}\right)\left(\frac{-2}{3}\right)}{7!} \times (-1)^7 \times 3^7 \times x^{14}$$

$$= \frac{16\times13\times10\times7\times4\times1\times(-2)\times3^{-7}}{7!}\times(-1)^7.3^7\times x^{14}$$

$$= \frac{16\times13\times10\times7\times4\times1\times2x^{14}}{7!} = \frac{208}{9}x^{14}$$ **Ans**

Example 50. *Find the first negative term in the expansion of* $\left(1+\frac{3}{4}x\right)^{\frac{13}{3}}, 0 < x < 1.$

Solution. Let $(r + 1)$th term in the expansion of $\left(1+\frac{3}{4}x\right)^{\frac{13}{3}}$ be the negative term

Then

$$T_{r+1} = \frac{\frac{13}{3}\left(\frac{13}{3}-1\right)\left(\frac{13}{3}-2\right)...\left(\frac{13}{3}-(r-1)\right)}{r!}\left(\frac{3}{4}x\right)^r \quad ...(1)$$

This will be the first negative term if

$$\frac{13}{3}-(r-1) < 0 \Rightarrow r > 5\frac{1}{3} \Rightarrow r \geq 6$$

Hence, seventh term is the first negative term.

Putting $r = 6$ in (1), we get

$$T_7 = T_{6+1}$$

$$= \frac{\frac{13}{3}\cdot\frac{10}{3}\cdot\frac{7}{3}\cdot\frac{4}{3}\cdot\frac{1}{3}\left(-\frac{2}{3}\right)}{6!}\left(\frac{3}{4}x\right)^6$$

$$= \frac{13\times10\times7\times4\times(-2)\times3^{-6}\times3^6\times x^6}{6!\times4^6} = -\frac{91}{36864}x^6 \quad \textbf{Ans}$$

Type III: To find the coefficient of x^n.

Example 51. *Find the coefficient of* x^6 *in the expansion of* $(1-2x)^{-\frac{5}{2}}, |x| < \frac{1}{2}.$

Solution. Suppose $(r + 1)$th term in the expansion of $(1-2x)^{-\frac{5}{2}}$ contains x^6.

Now
$$T_{r+1} = \frac{\left(-\frac{5}{2}\right)\left(-\frac{5}{2}-1\right)\left(-\frac{5}{2}-2\right)...\left(-\frac{5}{2}-(r-1)\right)}{r!}(-2x)^r$$

$$= \frac{(-1)^r\left(\frac{5}{2}\right)\left(\frac{7}{2}\right)\left(\frac{9}{2}\right)...\left(\frac{2r+3}{2}\right)}{r!}(-1)^r 2^r x^r$$

$$= (-1)^{2r}\frac{5\cdot7\cdot9\cdot11...(2r+3)}{2^r\cdot r!}2^r.x^r$$

$$= \frac{5\cdot7\cdot9\cdot11...(2\ \ +3)}{}$$

For this term to contain x^6, we must have $r = 6$

$$\therefore \quad \text{Coefficient of } x^6 = \frac{5\cdot7\cdot9\cdot11\cdot13\cdot15}{6!} = \frac{15015}{16} \quad \textbf{Ans}$$

$\therefore$ Coefficient of $x^n = C_0 \times 1 - 2C_1 + 3 \times C_2 - 4 \times C_3 + ... + (-1)^n (n+1)C_n$

$= C_0 - 2C_1 + 3C_2 - 4C_3 + ... + (-1)^n (n+1)C_n$ ...(1)

Now,

$(x+1)^n (1+x)^{-2} = (x+1)^{n-2}$

$\Rightarrow$ Coefficient of x^n in $(x+1)^n (1+x)^{-2}$ = Coefficient of x^n in $(1+x)^{n-2} = 0$ ...(2)

From (1) and (2), we get

$C_0 - 2C_1 + 3C_2 - 4.C_3 + ... + (-1)^n (n+1)C_n = 0$ **Ans**

Example 57. *Find the coefficient of x^6 in the expression $(1 + 2x + 3x^2 + 4x^3 + ...)^{1/2}$*

Solution. We have,

$$(1 + 2x + 3x^2 + 4x^3 + ...)^{\frac{1}{2}} = [(1-x)^{-2}]^{\frac{1}{2}}$$

$$= (1-x)^{-1} = 1 + x + x^2 + ... + x^n + ...$$

Hence, the coefficient of x^6 in the given expression is equal to 1. **Ans**

Type IV: To find unknown values

Example 58. *Find a negative value of m if the coefficient of x^2 in the expansion of $(1+x)^m$ is 6.*

Solution. $$(1+x)^m = 1 + mx + \frac{m(m-1)}{2!}x^2 + ...$$

Coefficient of x^2 in the expansion $(1+x)^m$ is

$$\frac{m(m-1)}{2!} = 6 \quad \text{(given)}$$

$\Rightarrow$ $m^2 - m = 6 \times 2$

$\Rightarrow$ $m^2 - m - 12 = 0$

$\Rightarrow$ $m^2 - 4m + 3m - 12 = 0$

$\Rightarrow$ $m(m-4) + 3(m-4) = 0 \Rightarrow (m+3)(m-4) = 0$

$\Rightarrow$ $m = -3$ or $m = 4$

Hence, negative value of $m = -3$. **Ans**

Example 59. *If the binomial expansion of $(a + bx)^{-2}$ is $\frac{1}{4} - 3x + ...$, then find the value of a and b.*

Solution. We have,

$$(a+bx)^{-2} = \frac{1}{4} - 3x$$

$$\Rightarrow \quad a^{-2}\left[1 + \left(\frac{b}{a}\right)x\right]^{-2} = \frac{1}{4} - 3x$$

$$\Rightarrow \quad a^{-2}\left[1 - 2\left(\frac{b}{a}\right)x + ...\right] = \frac{1}{4} - 3x$$

$$\Rightarrow \quad \frac{1}{a^2} - \left(\frac{2b}{a^3}\right)x + ... = \frac{1}{4} - 3x$$

Equating the like terms, we get

$$\frac{1}{a^2} = \frac{1}{4} \Rightarrow a = \pm 2$$

and $$-\frac{2b}{a^3}x = -3x \Rightarrow 2b = 3a^3 = 3(\pm 2)^3$$

$$\Rightarrow \quad b = \pm 12$$

Hence, $a = \pm 2$ and $b = \pm 12$ **Ans**

EXERCISE 8.8

Find the condition of validity of the expansion of the following :

1. $\left(11+\frac{1}{c}x\right)^{-\frac{3}{4}}$ **Ans.** $\left|\frac{x}{11c}\right|<1$

2. $\left(a^2-\frac{1}{a}y\right)^{-\frac{1}{4}}$ **Ans.** $\left|\frac{y}{a^3}\right|<1$

3. $\frac{1}{(4-3x^2)^{\frac{1}{3}}}$ **Ans.** $|x|<\frac{2}{\sqrt{3}}$

4. $\frac{1}{(3-2x^2)^{\frac{2}{3}}}$ **Ans.** $|x|<\sqrt{\frac{3}{2}}$

Expand upto 3 terms in the expansion, in ascending powers of x, each of the following :

5. $\sqrt{2-3x}$ **Ans.** $\sqrt{2}-\frac{3\sqrt{2}}{4}+\frac{9\sqrt{2}}{32}x^2+...$

6. $(27-6x)^{-\frac{2}{3}}$ **Ans.** $\frac{1}{9}+\frac{4x}{243}+\frac{20x^2}{6591}+...$

7. $\left(x^2+\frac{1}{x}\right)^{\frac{4}{3}}$ **Ans.** $x^{\frac{4}{3}}-\frac{4}{3}x^{\frac{13}{3}}+\frac{14}{9}x^{\frac{22}{3}}+...$

8. $\left(1-\frac{x}{2}\right)^{-\frac{1}{2}}$ **Ans.** $1+\frac{x}{4}+\frac{3x^2}{32}+...$

9. $\frac{1}{\sqrt{5+4x}}$ **Ans .** $5^{-\frac{1}{2}}\left(1-\frac{2}{5}x+\frac{6}{25}x^2+...\right)$

10. $\frac{1}{(4-3x)^{\frac{1}{3}}}$ **Ans.** $4^{-\frac{1}{3}}\left(1+\frac{x}{4}+\frac{x^2}{8}+...\right)$

Write the general term in each of the following :

11. $(1-2x)^{\frac{3}{4}}$ **Ans.** $-\frac{3}{2^r(r!)}x^r$

12. $(1-x^2)^{-4}$ **Ans.** $\frac{(r+1)(r+2)(r+3)}{1\cdot 2\cdot 3}x^{2r}$

13. $\left(1-\frac{2x}{3}\right)^{-\frac{1}{2}}$ **Ans.** $\frac{(2r)!}{6^r(r!)}x^r$

14. $(1+x)^{\frac{5}{3}}$ **Ans.** $(-1)^{r-2}\left(\frac{10}{3^r}\right)\frac{1\cdot 4\cdot 7...(3r-8)}{r!}x^r$

Find the coefficient of :

15. x^3 in the expansion of $\frac{(1+3x)^2}{1-2x}$ **Ans.** 50

16. x^{10} in the expansion of $\dfrac{1+3x^2}{(1+x)^3}$ **Ans.** 201

17. x^5 in the expansion of $\dfrac{x^2}{\left(1-\dfrac{x}{2}\right)^5}+\dfrac{x}{1-3x}$ **Ans.** $85\dfrac{3}{8}$

18. x^7 in the expansion of $\left(\dfrac{1}{2}\sqrt{x}-\dfrac{2}{\sqrt{x}}\right)^{-14}$, if $|x| < 4$ **Ans.** 2^{-14}

19. Write first four terms in the expansion of $\dfrac{1}{(3-4x^2)^{\frac{1}{2}}}$

Ans. $\dfrac{1}{\sqrt{3}}, \dfrac{2}{3\sqrt{3}}x^2, \dfrac{2}{3\sqrt{3}}x^4, \dfrac{20}{27\sqrt{3}}x^6$

20. Find two values of m such that the coefficient of x^2 in the expansion of $(1-x)^m$ is 3.

Ans. $m = 3, -2$

21. Find the rational exponent m for which the third term in the expansion of $(1+x)^m$ is $-\dfrac{1}{8}x^2$. **Ans.** $m = \dfrac{1}{2}$

22. Find the first negative term in the expansion of $(1+x)^{\frac{7}{2}}$. **Ans.** $-\dfrac{7}{256}x^5$

23. If all the coefficients of $(a+bx)^{-2}$ are positive, prove that a and b are of the opposite signs.

24. Find 5th term in the expansion of $(1-2x^3)^{\frac{11}{2}}$ **Ans.** $\dfrac{1155}{8}x^{12}$

25. Find the term involving x^2 in the expansion of $\left(1-2x^{\frac{1}{3}}\right)^{-1}$ **Ans.** $64x^2$

8.17 APPROXIMATIONS

We have,

$$(1+x)^n = 1+nx+\frac{n(n-1)}{1\times 2}+\frac{n(n-1)(n-2)}{1\times 2\times 3}x^3+...$$

As $x < 1$, so the terms of the above expansion go on decreasing and if x be small, a stage may reach when we may neglect the terms containing higher powers of x of the expansion.

1. If x is numerically so small that its square and higher powers may be neglected, then

$$(1+x)^n = 1+nx \text{ (approximate)}$$

2. If x is numerically so small that its cube and higher powers may be neglected, then

$$(1+x)^n = 1+nx+\frac{n(n-1)}{1.2}x^2 \text{ (approximate)}$$

And so on.

Example 60. *If x is numerically so small that x^2 and the higher powers of x may be neglected, then prove that*

$$\frac{\left(1+\frac{3}{4}x\right)^{-4}(16-3x)^{\frac{1}{2}}}{(8+x)^{\frac{2}{3}}} = 1-\frac{305}{96}x$$

Solution. We have,

$$\Rightarrow \text{ L.H.S.} = \frac{\left(1+\frac{3}{4}x\right)^{-4}(16-3x)^{\frac{1}{2}}}{(8+x)^{\frac{2}{3}}} = \frac{\left(1+\frac{3}{4}x\right)^{-4}(16)^{\frac{1}{2}}\left(1-\frac{3x}{16}\right)^{\frac{1}{2}}}{(8)^{\frac{2}{3}}\left(1+\frac{x}{8}\right)^{\frac{2}{3}}}$$

$$= \frac{4}{4}\left(1+\frac{3}{4}x\right)^{-4}\left(1-\frac{3x}{16}\right)^{\frac{1}{2}}\left(1+\frac{x}{8}\right)^{-\frac{2}{3}}$$

$$= \left\{1+(-4)\left(\frac{3}{4}x\right)\right\}\left\{1+\frac{1}{2}\left(-\frac{3x}{16}\right)\right\}\left\{1+\left(-\frac{2}{3}\right)\left(\frac{x}{8}\right)\right\} \qquad \left[\because (1+x)^n \simeq 1+nx\right]$$

$$= (1-3x)\left(1-\frac{3}{32}x\right)\left(1-\frac{x}{12}\right) \qquad \text{[neglecting } x^2\text{]}$$

$$= \left(1-3x-\frac{3}{32}x\right)\left(1-\frac{1}{12}x\right)$$

$$= \left(1-\frac{99}{32}x\right)\left(1-\frac{x}{12}\right) = 1-\frac{99}{32}x-\frac{x}{12} \qquad \text{[neglecting } x^2\text{]}$$

$$= 1-\frac{305}{96}x = \text{R.H.S.}$$

Proved

Example 61. *If x = 0.001, then prove that* $\frac{(1-2x)^{\frac{2}{3}}(4+5x)^{\frac{3}{2}}}{\sqrt{1-x}} = 8.01$ *up to two decimal places.* ***[S.B.T.E 2012, 2011]***

Solution. We have,

$$\frac{(1-2x)^{\frac{2}{3}}(4+5x)^{\frac{3}{2}}}{\sqrt{1-x}} = \frac{(1-2x)^{\frac{2}{3}}4^{\frac{3}{2}}\left(1+\frac{5}{4}x\right)^{\frac{3}{2}}}{\sqrt{1-x}}$$

$$= \frac{\left(1 - \frac{4}{3}x + ...\right) 8\left(1 + \frac{15}{8}x + ...\right)}{(1-x)^{\frac{1}{2}}} \quad \text{[neglecting higher powers of } x\text{]}$$

$$= 8\left[1 - \frac{4x}{3} + \frac{15}{8}x + ...\right](1-x)^{-\frac{1}{2}} = 8\left[1 + \frac{13}{24}x\right](1-x)^{-\frac{1}{2}}$$

$$= 8\left[1 + \frac{13}{24}x\right]\left[1 + \frac{x}{2} + ...\right] \quad \text{[neglecting higher powers of } x\text{]}$$

$$= 8\left[1 + \frac{x}{2} + \frac{13}{24}x\right] = 8\left[1 + \frac{25}{24}x\right]$$

$$= 8 + \frac{25}{3}x = 8 + \frac{25}{3}(0.001) \quad [\because x = 0.001]$$

$$= 8.01 \text{ (approximate)}$$

Ans

Example 62. *If* $\frac{(1-3x)^{\frac{1}{2}} + (1-x)^{\frac{5}{3}}}{\sqrt{4-x}}$ *is approximately equal to $a + bx$ for all small values of x, then find a and b.*

Solution. We have,

$$\frac{(1-3x)^{\frac{1}{2}} + (1-x)^{\frac{5}{3}}}{\sqrt{4-x}} = \frac{(1-3x)^{\frac{1}{2}} + (1-x)^{\frac{5}{3}}}{2\left(1-\frac{x}{4}\right)^{\frac{1}{2}}}$$

$$= \frac{\left\{1 + \frac{1}{2}(-3x)\right\} + \left\{1 + \frac{5}{3}(-x)\right\}}{2\left\{1 + \frac{1}{2}\left(-\frac{x}{4}\right)\right\}} \quad \left[\begin{array}{l}\text{neglecting } x^2 \text{ and higher} \\ \text{power of } x, \text{ we have,} \\ (1+x)^n = 1 + nx\end{array}\right]$$

$$= \frac{1}{2}\frac{\left(1-\frac{3}{2}x\right)+\left(1-\frac{5}{3}x\right)}{\left(1-\frac{x}{8}\right)} = \frac{1}{2}\left[\left(1-\frac{3}{2}x\right)+\left(1-\frac{5}{3}x\right)\right]\left(1-\frac{x}{8}\right)^{-1}$$

$$= \frac{1}{2}\left(2 - \frac{3}{2}x - \frac{5}{3}x\right)\left(1 + (-1)\left(-\frac{x}{8}\right)\right) = \frac{1}{2}\left(2 - \frac{19}{6}x\right)\left(1 + \frac{x}{8}\right)$$

$$= \left(1 - \frac{19}{12}x\right)\left(1 + \frac{x}{8}\right) = 1 - \frac{19}{12}x + \frac{x}{8} = 1 - \frac{35}{24}x$$

$$\therefore \quad \frac{(1-3x)^{\frac{1}{2}} + (1-x)^{\frac{5}{3}}}{\sqrt{4-x}} = a + bx$$

$$\Rightarrow \quad 1-\frac{35}{24}x = a+bx \Rightarrow a=1, b=-\frac{35}{24} \quad \textbf{Ans}$$

Example 63. *If x is very small in magnitude when compared with a, show that*

$$\left(\frac{a}{a+x}\right)^{\frac{1}{2}}+\left(\frac{a}{a-x}\right)^{\frac{1}{2}}=2+\frac{3}{4a^2}x^2, \text{ nearly}$$

Solution. L.H.S. $= \left(\frac{a}{a+x}\right)^{\frac{1}{2}}+\left(\frac{a}{a-x}\right)^{\frac{1}{2}}$

$$= \left(\frac{1}{1+\frac{x}{a}}\right)^{\frac{1}{2}}+\left(\frac{1}{1-\frac{x}{a}}\right)^{\frac{1}{2}} = \left(1+\frac{x}{a}\right)^{-\frac{1}{2}}+\left(1-\frac{x}{a}\right)^{-\frac{1}{2}}$$

$$= \left[1+\left(-\frac{1}{2}\right)\left(\frac{x}{a}\right)+\frac{\left(-\frac{1}{2}\right)\left(-\frac{3}{2}\right)}{1.2}\left(\frac{x}{a}\right)^2+\ldots\right]$$

$$+\left[1+\left(-\frac{1}{2}\right)\left(-\frac{x}{a}\right)+\frac{\left(-\frac{1}{2}\right)\left(-\frac{3}{2}\right)}{1.2}\left(-\frac{x}{a}\right)^2+\ldots\right]$$

$$= \left(1-\frac{x}{2a}+\frac{3}{8a^2}x^2\right)+\left(1+\frac{x}{2a}+\frac{3}{8a^2}x^2\right) \quad \left[\text{neglecting } \left(\frac{x}{a}\right)^3.\left(\frac{x}{a}\right)^4 \ldots\right]$$

$$= 2+\frac{3}{4a^2}x^2, \text{nearly}$$

= R.H.S. **Proved**

Example 64. *If p is nearly equal to q and n > 1, show that*

$$\frac{(n+1)p+(n-1)q}{(n-1)p+(n+1)q} = \left(\frac{p}{q}\right)^{\frac{1}{n}}$$

Hence, find the approximate value of $\left(\frac{99}{101}\right)^{\frac{1}{6}}$.

Solution. Let $p = q + h$, where h is so small that its square and higher powers may be neglected. Then,

$$\frac{(n+1)p+(n-1)q}{(n-1)p+(n+1)q} = \frac{(n+1)(q+h)+(n-1)q}{(n-1)(q+h)+(n+1)q}$$

$$= \frac{2nq + (n+1)h}{2nq + (n-1)h} = \frac{1 + \left(\frac{n+1}{2nq}\right)h}{1 + \left(\frac{n-1}{2nq}\right)h} \qquad \left[\begin{array}{l}\text{on dividing numerator} \\ \text{and denominator by } 2nq\end{array}\right]$$

$$= \left\{1 + \left(\frac{n+1}{2nq}\right)h\right\}\left\{1 + \left(\frac{n-1}{2nq}\right)h\right\}^{-1} = \left\{1 + \left(\frac{n+1}{2nq}\right)h\right\}\left\{1 - \left(\frac{n-1}{2nq}\right)h\right\}$$

$$= 1 + \left(\frac{n+1}{2nq}\right)h - \left(\frac{n-1}{2nq}\right)h \qquad \left[\begin{array}{l}\text{expanding the 2nd} \\ \text{expression and neglecting } h^2 \\ \text{and higher powers of } h\end{array}\right]$$

$$= \left(1 + \frac{h}{nq}\right) \qquad \text{...(1)}$$

Also, $\left(\frac{p}{q}\right)^{\frac{1}{n}} = \left(\frac{q+h}{q}\right)^{\frac{1}{n}} = \left(1 + \frac{h}{q}\right)^{\frac{1}{n}}$

$$\Rightarrow \quad \left(\frac{p}{q}\right)^{\frac{1}{n}} = \left(1 + \frac{h}{nq}\right) \qquad \text{...(2)}$$

$$\left[\begin{array}{l}\text{Expanding and neglecting} \\ h^2 \text{ and higher powers of } h\end{array}\right]$$

From (1) and (2), we get

$$\frac{(n+1)p + (n-1)q}{(n-1)p + (n+1)q} = \left(\frac{p}{q}\right)^{\frac{1}{n}} \qquad \text{...(3) Proved.}$$

Deduction: Putting $p = 99$, $q = 101$ and $n = 6$ in (3), we get

$$\left(\frac{99}{100}\right)^{\frac{1}{6}} = \frac{(6+1)\times 99 + (6-1)\times 101}{(6-1)\times 99 + (6+1)\times 10} = \frac{(7\times 99) + (5\times 101)}{(5\times 99) + (7\times 101)} = \frac{599}{601} \qquad \textbf{Ans}$$

Example 65. *If x is very nearly equal to 1. Prove that :*

(i) $\frac{mx^m - nx^n}{m - n} = x^{m+n}$ *(ii)* $\frac{ax^b - bx^a}{x^b - x^a} = \frac{1}{1-x}$ *(nearly)*

Proof: Since x is nearly equal to 1, so let $x = 1 + h$, where h is so small that its square and higher powers may be neglected.

(i) L.H.S. $= \frac{mx^m - nx^n}{m - n} = \frac{m(1+h)^m - n(1+h)^n}{m - n}$

$$= \frac{m(1+mh) - n(1+nh)}{m-n} \qquad [\text{neglecting } h^2 \text{ and higher powers of } h]$$

$$= \frac{(m-n) + (m^2 - n^2)h}{m-n}$$

$$= \frac{(m-n)\{1+(m+n)h\}}{m-n} = 1+(m+n)h$$

$$\text{R.H.S.} = x^{m+n} = (1+h)^{m+n} = 1+(m+n)h$$

Here, L.H.S. = R.H.S. **Proved**

(*ii*) $$\text{L.H.S.} = \frac{ax^b - bx^a}{x^b - x^a} = \frac{a(1+h)^b - b(1+h)^a}{(1+h)^b - (1+h)^a}$$

$$= \frac{a(1+bh) - b(1+ah)}{(1+bh)-(1+ah)} \qquad \left[\because (1+x)^n = 1+nx\right]$$

$$= \frac{(a-b)+abh-abh}{(b-a)h} = -\frac{(a-b)}{(a-b)h} = -\frac{1}{h}$$

$$\text{R.H.S.} = \frac{1}{1-x} = \frac{1}{1-(1+h)} = -\frac{1}{h}$$

Here, L.H.S. = R.H.S. **Proved**

EXERCISE 8.9

1. If x is numerically so small that x^2 and the higher powers of x may be neglected, then prove that

$$\frac{(1-3x)^{\frac{1}{2}} + (4+5x)^{\frac{5}{3}}}{\sqrt{1-x}} = 8 + \frac{25}{3}x$$

2. When x is numerically small, show that $\sqrt{x^2+4} - \sqrt{x^2+1} = 1 - \frac{x^2}{4} + \frac{7x^2}{64}$

3. If for small values of x, $\frac{\sqrt{1+2x} + \sqrt[4]{16+3x}}{\sqrt[4]{1-2x}}$ is very nearly equal to $a + bx$, find the values of a and b. **Ans.** $a = 3, b = \frac{83}{32}$

4. If the binomial expansion of $(a+bx)^{-2}$ is $\frac{1}{4} - 3x + ...,$ find the values of a and b.
Ans. $a = 2$, $b = 12$

5. If x be a quantity, so small that x^3 may be neglected in comparison to a^3, prove that
$\sqrt{\frac{a}{a+x}} + \sqrt{\frac{a}{a-x}} = 2 + \frac{3x^2}{4a^2}.$

6. If p is nearly equal to q, prove that $\frac{5p+4q}{4p+5q)} = \left(\frac{p}{q}\right)^{\frac{1}{9}}$

7. If the square and higher powers of the difference between m and M can be neglected, show that $\sqrt{\frac{M}{m}} = \frac{M}{m+M} + \frac{m+M}{4m}$

8. If x be so small that its squares and higher powers may be neglected, find the approximate value of $\dfrac{\sqrt[3]{1-\frac{3}{7}x}+\left(1-\frac{3}{5}x\right)^{-5}}{\sqrt[4]{1+\frac{1}{2}x}+\sqrt[7]{1+\frac{7}{3}x}}$ **Ans.** $1+\dfrac{515}{336}x$

9. If c be a small quantity in comparison to l, show that $\sqrt{\dfrac{l}{l+c}}+\sqrt{\dfrac{l}{l-c}}=2+\dfrac{3}{4}\dfrac{c^2}{l^2}$ nearly.

10. If x is so small that its second and higher powers may be neglected find the coefficient of x in the expansion of $(1+4x)^{-\frac{5}{4}}(1+2x)^{\frac{1}{2}}$ **Ans.** –4

8.18 EVALUATION OF A ROOT

Suppose, we have to find the nth root of any number N. We express N in the form $a^n + b$, where a^n is nearest to N. Then, b which is either positive or negative is very small in comparison to a^n.

$$\therefore \quad N^{\frac{1}{n}} = (a^n+b)^{\frac{1}{n}} = (a^n)^{\frac{1}{n}}\left(1+\frac{b}{a^n}\right)^{\frac{1}{n}}$$

Now, on expanding $\left(1+\dfrac{b}{a^n}\right)^{\frac{1}{n}}$, an approximate value of the root can be obtained.

Example 66. *Using binomial theorem, evaluate* $\dfrac{1}{\sqrt{0.9}}$ *to four places of the decimal.*

Solution. We have,

$$\frac{1}{\sqrt{0.9}} = (0.9)^{-\frac{1}{2}} = (1-0.1)^{-\frac{1}{2}}$$

$$= 1+\left(-\frac{1}{2}\right)(-0.1)+\frac{\left(-\frac{1}{2}\right)\left(-\frac{3}{2}\right)}{2\times 1}(-0.1)^2+\frac{\left(-\frac{1}{2}\right)\left(-\frac{3}{2}\right)\left(-\frac{5}{2}\right)}{3\times 2\times 1}(-0.1)^3$$

$$= 1 + 0.05 + 0.00375 + 0.0003125 + \ldots$$

$$= 1.\,0540625 = 1.0541 \text{ (approximately)}$$

Hence, value of $\dfrac{1}{\sqrt{0.9}}$ correct to four places of decimal is 1.0541. **Ans**

Example 67. *Find the cube root of 1003 correct to five decimal places.*

Solution. We have,

$$(1003)^{\frac{1}{3}} = (1000+3)^{\frac{1}{3}} = (10^3+3)^{\frac{1}{3}} = 10\left[1+\frac{3}{10^3}\right]^{\frac{1}{3}}$$

$$= 10[1+0.003]^{\frac{1}{3}}$$

$$= 10\left[1+\frac{1}{3}(0.003)+\frac{\frac{1}{3}\left(\frac{1}{3}-1\right)}{2\times 1}(0.003)^2+\frac{\frac{1}{3}\left(\frac{1}{3}-1\right)\left(\frac{1}{3}-2\right)}{3\times 2\times 1}(0.003)^3+\ldots\right]$$

$$= 10\left[1+0.001+\frac{-\frac{2}{9}}{2\times 1}(0.000009)+\frac{\frac{10}{27}}{3\times 2\times 1}(0.000000027)+\ldots\right]$$

$$= 10\left[1+0.001-\frac{1}{9}(0.000009)+\frac{5}{81}(0.000000027)+\ldots\right]$$

$$= 10\ [1 + 0.001 - 0.000001 + 0.00000000\ 167\ \ldots]$$

$$= 10\ [1.00100000167 - 0.000001] = 10\ [1.000999]$$

$$= 10.00999$$ **Ans**

Example 68. *Evaluate* $(244)^{\frac{1}{5}}$, *correct upto 3 decimal places.*

Solution. We have,

$$(244)^{\frac{1}{5}} = (243+1)^{\frac{1}{5}} = (3^5+1)^{\frac{1}{5}} = 3\left(1+\frac{1}{3^5}\right)^{\frac{1}{5}}$$

$$= 3\left[1+\frac{1}{5}\times\frac{1}{3^5}+\frac{\left(\frac{1}{5}\right)\left(-\frac{4}{5}\right)}{2\times 1}\left(\frac{1}{3^5}\right)^2+\ldots\right]$$

$$= 3\left[1+\frac{1}{5}.\frac{1}{3^5}-\frac{2}{25}.\frac{1}{3^{10}}+\ldots\right]$$

$$= 3+\frac{1}{5}\times\frac{1}{3^4}-\frac{2}{25}\times\frac{1}{3^9}+\ldots$$

$$= 3 + 0.00246914 - 0.00000824 = 3.00247738$$

$$= 3.002 \text{ (approx.)}$$ **Ans**

Example 69. *Evaluate* $\frac{1}{\sqrt{47}}$ *to four decimal places.*

Solution. We have,

$$\frac{1}{\sqrt{47}} = \frac{1}{\sqrt{49-2}} = \frac{1}{7}.\frac{1}{\sqrt{1-\frac{2}{49}}}$$

$$= \frac{1}{7}\left(1-\frac{2}{49}\right)^{-\frac{1}{2}} = \frac{1}{7}\left(1-\frac{2}{7^2}\right)^{-\frac{1}{2}}$$

$$= \frac{1}{7}\left[1+\frac{1}{2}\left(\frac{2}{7^2}\right)+\frac{-\frac{1}{2}\left(-\frac{1}{2}-1\right)}{2\times 1}\left(-\frac{2}{7^2}\right)^2+\ldots\right]$$

$$= \frac{1}{7}+\frac{1}{7^3}+\frac{3}{2}.\frac{1}{7^5}+\ldots$$

$$= 0.14286+0.00292+\frac{3}{2}(0.000059)$$

$$= 0.14286 + 0.00292 + 0.0000885$$

$$= 0.1458685 = 0.1459$$

Here, value of $\frac{1}{\sqrt{47}}$ correct to four decimal places is 0.1459. **Ans**

EXERCISE 8.10

Use Binomial theorem to evaluate the following correct to 3 decimal places :

1. $\sqrt{53}$ **Ans.** 7.280
2. $(1.02)^{\frac{1}{2}}$ **Ans.** 1.01
3. $\sqrt[3]{999}$ **Ans.** 9.997
4. $(1.04)^{-3}$ **Ans.** 0.889
5. $(1.025)^{-\frac{1}{3}}$ **Ans.** 0.992
6. $(1008)^{\frac{1}{3}}$ **Ans.** 10.027
7. $(1010)^{\frac{1}{3}}$ **Ans.** 10.033
8. $(217)^{\frac{1}{3}}$ **Ans.** 6.009
9. $(626)^{\frac{1}{4}}$ **Ans.** 5.002
10. $(126)^{\frac{1}{3}}$ **Ans.** 5.013
11. $(0.98)^{-1}$ **Ans.** 1.063
12. $(128)^{\frac{1}{3}}$ **Ans.** 5.040

Determinant

9.1 INTRODUCTION

In engineering mathematics, solution of simultaneous equations is very important. In this chapter, we shall discuss the system of linear equations with emphasis on their solution by means of determinants.

9.2 DETERMINANT

The notation of determinant arises from the process of elimination of the unknown of simultaneous linear equations.

Consider the two linear equations in x,

$$a_1x + b_1 = 0 \quad \text{... (1)}$$

$$a_2x + b_2 = 0 \quad \text{... (2)}$$

From (1) $\qquad x = -\dfrac{b_1}{a_1}$

Substituting the value of x in (2), we get the eliminant.

$$a_2\left(-\frac{b_1}{a_1}\right) + b_2 = 0$$

$$\Rightarrow \qquad a_1b_2 - a_2b_1 = 0 \quad \text{... (3)}$$

From (1) and (2) by suppressing x, the eliminant is written as

$$\begin{vmatrix} a_1 & b_1 \\ a_2 & b_2 \end{vmatrix} = 0 \quad \text{... (4)}$$

The two rows of a_1, b_1, and a_2, b_2 enclosed by two vertical bars is called a determinant of second order.

$\begin{vmatrix} a_1 \\ a_2 \end{vmatrix}$ and $\begin{vmatrix} b_1 \\ b_2 \end{vmatrix}$ are two columns.

Column 1 $\qquad$ Column 2

Row 1 → $\begin{vmatrix} a_1 & b_1 \\ a_2 & b_2 \end{vmatrix}$
Row 2 →

Each quantity a_1, b_1, a_2, b_2 in a column or in a row is called an **element** or a **constituent.**
From (3) and (4), we conclude that

$$\begin{vmatrix} a_1 & b_1 \\ a_2 & b_2 \end{vmatrix} = a_1b_2 - a_2b_1$$

$a_1b_2 - a_2b_1$ is called the expansion of the determinant of $\begin{vmatrix} a_1 & b_1 \\ a_2 & b_2 \end{vmatrix}$

Example 1. *Expand the determinant:*

$$\begin{vmatrix} 3 & 2 \\ 6 & 7 \end{vmatrix}$$

Solution. $\begin{vmatrix} \overset{+}{3} & \overset{-}{2} \\ 6 & 7 \end{vmatrix} = (3) \times (7) - (2) \times (6) = 21 - 12 = 9.$ **Ans.**

EXERCISE 9.1

Expand the following determinants:

1. $\begin{vmatrix} 4 & 6 \\ 2 & 5 \end{vmatrix}$ **Ans.** 8

2. $\begin{vmatrix} -3 & 7 \\ 2 & 4 \end{vmatrix}$ **Ans.** – 26

3. $\begin{vmatrix} 8 & 5 \\ 3 & 1 \end{vmatrix}$ **Ans.** – 7

4. $\begin{vmatrix} 5 & -2 \\ 4 & 3 \end{vmatrix}$ **Ans.** 23

Choose the correct alternative:

5. If $f(x) = \begin{vmatrix} \frac{1}{\sqrt{2}} & \sin x & 1 \\ \frac{1}{\sqrt{2}} & \cos x & x \\ 1 & 1 & x^2 \end{vmatrix}$, then $f\left(\frac{\pi}{4}\right)$ is

(*a*) 0 (*b*) 1 (*c*) 2 (*d*) 3 **Ans.** (*a*)

[Diploma IETE, June 2005]

Example 2. *Evaluate* $\begin{vmatrix} x & 7 \\ x & 5x+1 \end{vmatrix}$ **[S.B.T.E. 2016]**

Solution. Here, we have

$$\begin{vmatrix} x & 7 \\ x & 5x+1 \end{vmatrix}$$

$$x(5x + 1) - 7x = 5x^2 - 6x$$ **Ans.**

Example 3. *Determine x if* $\begin{vmatrix} 5x & 2 \\ 1 & 3 \end{vmatrix} = 13$ **[S.B.T.E. 2014, 2013]**

Solution. Here, we have $\begin{vmatrix} 5x & 2 \\ 1 & 3 \end{vmatrix} = 13$

$$15x - 2 = 13 \Rightarrow x = \frac{15}{15} = 1$$ **Ans.**

Example 4. *If* $\begin{vmatrix} x+1 & x-1 \\ x-3 & x+2 \end{vmatrix} = \begin{vmatrix} 4 & -1 \\ 1 & 3 \end{vmatrix}$, *find x.*

Solution. Here, we have **[S.B.T.E. 2015]**

$$\begin{vmatrix} x+1 & x-1 \\ x-3 & x+2 \end{vmatrix} = \begin{vmatrix} 4 & -1 \\ 1 & 3 \end{vmatrix}$$

$$[(x + 1)((x + 2) - (x - 1)(x - 3)] = 12 + 1$$

$$\Rightarrow \quad \cancel{x^2} + 2x + x + 2 - \cancel{x^2} + 3x + x - 3 = 13$$

$$7x - 1 = 13$$

$$\therefore \quad x = 2$$ **Ans.**

Example 5. *Evaluate:* $\begin{vmatrix} \sin 10^\circ & -\cos 10^\circ \\ \sin 80^\circ & \cos 80^\circ \end{vmatrix}$ **[S.B.T.E. 2014, 2013]**

Solution. Here, we have

$$\begin{vmatrix} \sin 10^\circ & -\cos 10^\circ \\ \sin 80^\circ & \cos 80^\circ \end{vmatrix}$$

$$\Rightarrow \quad \sin 10^\circ \cdot \cos 80^\circ + \sin 80^\circ \cdot \cos 10^\circ$$

$$\therefore \quad \sin(10^\circ + 80^\circ) = \sin 90^\circ = 1$$ **Ans.**

Example 6. *Evaluate* $\begin{vmatrix} \cos 15^\circ & \sin 75^\circ \\ \sin 15^\circ & \cos 75^\circ \end{vmatrix}$ **[S.B.T.E. 2016, 2013]**

Solution. Here, we have

$$\begin{vmatrix} \cos 15^\circ & \sin 75^\circ \\ \sin 15^\circ & \cos 75^\circ \end{vmatrix}$$

$$\Rightarrow \quad \cos 15^\circ \cdot \cos 75^\circ - \sin 15^\circ \cdot \sin 75^\circ = \cos(15^\circ + 75^\circ)$$

$$= \cos 90^\circ = 0$$ **Ans.**

9.3 DETERMINANT AS AN ELIMINANT

Consider the following three equations having three unknowns, x, y and z.

$$a_1x + b_1y + c_1z = 0 \quad \text{... (1)}$$

$$a_2x + b_2y + c_2z = 0 \quad \text{... (2)}$$

$$a_3x + b_3y + c_3z = 0 \quad \text{... (3)}$$

From (2) and (3), by cross-multiplication, we get

$$\frac{x}{b_2c_3 - b_3c_2} = \frac{y}{a_3c_2 - a_2c_3} = \frac{z}{a_2b_3 - a_3b_2} = k \text{ (say)}$$

$$x = (b_2c_3 - b_3c_2)\,k$$

$$y = (a_3c_2 - a_2c_3)\,k$$

and $$z = (a_2b_3 - a_3b_2)\,k$$

Substituting the values of x, y and z in (1), we get the eliminant

$$a_1(b_2c_3 - b_3c_2)k + b_1(a_3c_2 - a_2c_3)k + c_1(a_2b_3 - a_3b_2)k = 0$$

$$\Rightarrow \quad a_1(b_2c_3 - b_3c_2) - b_1(a_2c_3 - a_3c_2) + c_1(a_2b_3 - a_3b_2) = 0 \quad \text{... (4)}$$

From (1), (2) and (3) by suppressing x, y, z, the remaining can be written in the determinant as

$$\begin{vmatrix} a_1 & b_1 & c_1 \\ a_2 & b_2 & c_2 \\ a_3 & b_3 & c_3 \end{vmatrix} = 0 \quad \text{... (5)}$$

This is determinant of third order.

As (4) and (5) both are the eliminant of the same equations.

$$\begin{vmatrix} a_1 & b_1 & c_1 \\ a_2 & b_2 & c_2 \\ a_3 & b_3 & c_3 \end{vmatrix} = a_1(b_2c_3 - b_3c_2) - b_1(a_2c_3 - a_3c_2) + c_1(a_2b_3 - a_3b_2) = 0$$

$$\begin{vmatrix} a_1 & b_1 & c_1 \\ a_2 & b_2 & c_2 \\ a_3 & b_3 & c_3 \end{vmatrix} = a_1\begin{vmatrix} b_2 & c_2 \\ b_3 & c_3 \end{vmatrix} - b_1\begin{vmatrix} a_2 & c_2 \\ a_3 & c_3 \end{vmatrix} + c_1\begin{vmatrix} a_2 & b_2 \\ a_3 & b_3 \end{vmatrix}$$

9.4 MINOR

The minor of an element is defined as a determinant obtained by deleting the row and column containing the element.

Thus, the minors of a_1, b_1 and c_1 are

$$\begin{vmatrix} b_2 & c_2 \\ b_3 & c_3 \end{vmatrix}, \begin{vmatrix} a_2 & c_2 \\ a_3 & c_3 \end{vmatrix} \text{ and } \begin{vmatrix} a_2 & b_2 \\ a_3 & b_3 \end{vmatrix} \text{ respectively.}$$

Thus

$$\begin{vmatrix} a_1 & b_1 & c_1 \\ a_2 & b_2 & c_2 \\ a_3 & b_3 & c_3 \end{vmatrix} = a_1 \text{ (minor of } a_1) - b_1 \text{ (minor of } b_1) + c_1 \text{ (minor of } c_1)$$

9.5 COFACTOR (LAPLACE EXPANSION)

The cofactor of any element of mth row and n^{th} column is $(-1)^{m+n}$ minor.

Thus the cofactor of $a_1 = (-1)^{1+1}(b_2c_3 - b_3c_2) = +(b_2c_3 - b_3c_2)$

$b_1 = (-1)^{1+2}(a_2c_3 - a_3c_2) = -(a_2c_3 - a_3c_2)$

$c_1 = (-1)^{1+3}(a_2b_3 - a_3b_2) = +(a_2b_3 - a_3b_2)$

The determinant $= a_1$ (cofactor of a_1) $- a_2$ (cofactor of a_2) $+ a_3$ (cofactor of a_3)

Example 7. *Find:*

(i) Minors

(ii) Cofactors of the elements of the first row of the determinant.

$$\begin{vmatrix} 2 & 3 & 5 \\ 4 & 1 & 0 \\ 6 & 2 & 7 \end{vmatrix}$$

[S.B.T.E. 2016]

Solution. (*i*) The minor of the element 2 is

$$\begin{vmatrix} 2 & 3 & 5 \\ 4 & 1 & 0 \\ 6 & 2 & 7 \end{vmatrix} = \begin{vmatrix} 1 & 0 \\ 2 & 7 \end{vmatrix} = (1) \times (7) - (0) \times (2) = 7 - 0 = 7$$

The minor of the element 3 is

$$\begin{vmatrix} 2 & 3 & 5 \\ 4 & 1 & 0 \\ 6 & 2 & 7 \end{vmatrix} = \begin{vmatrix} 4 & 0 \\ 6 & 7 \end{vmatrix} = (4) \times (7) - (0) \times (6) = 28 - 0 = 28$$

The minor of the element 5 is

$$\begin{vmatrix} 2 & 3 & 5 \\ 4 & 1 & 0 \\ 6 & 2 & 7 \end{vmatrix} = \begin{vmatrix} 4 & 1 \\ 6 & 2 \end{vmatrix} = (4) \times (2) - (1) \times (6) = 8 - 6 = 2$$

(*ii*) The cofactor of $2 = (-1)^{1+1}\ (7) = +7$

The cofactor of $3 = (-1)^{1+2}\ (28) = -28$

The cofactor of $5 = (-1)^{1+3}\ (2) = +2.$ **Ans.**

Example 8. *Expand the determinant:* $\begin{vmatrix} 6 & 2 & 3 \\ 2 & 3 & 5 \\ 4 & 2 & 1 \end{vmatrix}$

Solution. $\begin{vmatrix} 6 & 2 & 3 \\ 2 & 3 & 5 \\ 4 & 2 & 1 \end{vmatrix} = 6$ (cofactor of 6) – 2 (cofactor of 2) + 3 (cofactor of 3).

$$= 6\,(3 \times 1 - 5 \times 2) - 2\,(2 \times 1 - 4 \times 5) + 3\,(2 \times 2 - 3 \times 4)$$

$$= 6\,(3 - 10) - 2\,(2 - 20) + 3\,(4 - 12)$$

$$= 6\,(-7) - 2\,(-18) + 3\,(-8)$$

$$= -42 + 36 - 24$$

$$= -30$$ **Ans.**

Example 9. *Expand the fourth order determinant*

$$\begin{vmatrix} 0 & 1 & 2 & 3 \\ 1 & 0 & 2 & 0 \\ 2 & 0 & 1 & 3 \\ 1 & 2 & 1 & 0 \end{vmatrix}$$

Solution. Given determinant

$$= 0\,(-1)^{1+1}\begin{vmatrix} 0 & 2 & 0 \\ 0 & 1 & 3 \\ 2 & 1 & 0 \end{vmatrix} + 1(-1)^{1+2}\begin{vmatrix} 1 & 2 & 0 \\ 2 & 1 & 3 \\ 1 & 1 & 0 \end{vmatrix} + 2(-1)^{1+3}\begin{vmatrix} 1 & 0 & 0 \\ 2 & 0 & 3 \\ 1 & 2 & 0 \end{vmatrix} + 3(-1)^{1+4}\begin{vmatrix} 1 & 0 & 2 \\ 2 & 0 & 1 \\ 1 & 2 & 1 \end{vmatrix}$$

$$= 0 - \begin{vmatrix} 1 & 2 & 0 \\ 2 & 1 & 3 \\ 1 & 1 & 0 \end{vmatrix} + 2\begin{vmatrix} 1 & 0 & 0 \\ 2 & 0 & 3 \\ 1 & 2 & 0 \end{vmatrix} - 3\begin{vmatrix} 1 & 0 & 2 \\ 2 & 0 & 1 \\ 1 & 2 & 1 \end{vmatrix}$$

Now, $\begin{vmatrix} 1 & 2 & 0 \\ 2 & 1 & 3 \\ 1 & 1 & 0 \end{vmatrix} = 1(1 \times 0 - 3 \times 1) - 2(2 \times 0 - 3 \times 1) + 0(2 \times 1 - 1 \times 1) = -3 + 6 + 0 = 3$

$$\begin{vmatrix} 1 & 0 & 0 \\ 2 & 0 & 3 \\ 1 & 2 & 0 \end{vmatrix} = 1(0 \times 0 - 3 \times 2) - 0(2 \times 0 - 3 \times 1) + 0(2 \times 2 - 0 \times 1) = -6$$

$$\begin{vmatrix} 1 & 0 & 2 \\ 2 & 0 & 1 \\ 1 & 2 & 1 \end{vmatrix} = 1(0 \times 1 - 1 \times 2) - 0(2 \times 1 - 1 \times 1) + 2(2 \times 2 - 0 \times 1) = -2 - 0 + 8 = 6$$

$$\begin{vmatrix} 0 & 1 & 2 & 3 \\ 1 & 0 & 2 & 0 \\ 2 & 0 & 1 & 3 \\ 1 & 2 & 1 & 0 \end{vmatrix} = -3 + 2(-6) - 3(6) = -3 - 12 - 18 = -33$$

Ans.

EXERCISE 9.2

Expand the following determinants

1. $\begin{vmatrix} 2 & -3 & 4 \\ 5 & 1 & -6 \\ -7 & 8 & -9 \end{vmatrix}$ **Ans.** 5

2. $\begin{vmatrix} 5 & 0 & 7 \\ 8 & -6 & -4 \\ 2 & 3 & 9 \end{vmatrix}$ **Ans.** 42

3. $\begin{vmatrix} a & h & g \\ h & b & f \\ g & f & c \end{vmatrix}$ **Ans.** $abc + 2fgh + af^2 - bg^2 - ch^2$

4. $\begin{vmatrix} 3 & 2 & 5 & 7 \\ -1 & -4 & -3 & 0 \\ 6 & 4 & 2 & -1 \\ 2 & -1 & 0 & 3 \end{vmatrix}$ **Ans.** 96

9.6 RULES OF SARRUS

For third order determinants only.

After writing the determinant, repeat the first two columns as below:

$$\begin{vmatrix} a_1 & b_1 & c_1 \\ a_2 & b_2 & c_2 \\ a_3 & b_3 & c_3 \end{vmatrix} = \begin{vmatrix} a_1 & b_1 & c_1 \\ a_2 & b_2 & c_2 \\ a_3 & b_3 & c_3 \end{vmatrix}\begin{matrix} a_1 & b_1 \\ a_2 & b_2 \\ a_3 & b_3 \end{matrix}$$

(diagonals marked $+ \; + \; +$ and $- \; - \; -$)

$$= (a_1b_2c_3 + b_1c_2a_3 + c_1a_2b_3) + (-c_1b_2a_3 - a_1c_2b_3 - b_1a_2c_3)$$

Example 10. *Expand the determinant*

$$\Delta = \begin{vmatrix} 2 & 3 & 4 \\ 1 & 5 & 3 \\ 3 & 0 & 5 \end{vmatrix} \text{ by rule of Sarrus.}$$

Solution.

$$\Delta = \begin{vmatrix} 2 & 3 & 4 \\ 1 & 5 & 3 \\ 3 & 0 & 5 \end{vmatrix}\begin{matrix} 2 & 3 \\ 1 & 5 \\ 3 & 0 \end{matrix}$$

(diagonals marked $+ \; + \; +$ and $- \; - \; -$)

$$= (2) \times (5) \times (5) + (3) \times (3) \times (3) + (4) \times (1) \times (0)$$
$$- (4) \times (5) \times (3) - (2) \times (3) \times (0) - (3) \times (1) \times (5)$$
$$= 50 + 27 + 0 - 60 - 0 - 15 = 2$$

Ans.

EXERCISE 9.3

Expand the following determinants by rule of Sarrus.

1. $\begin{vmatrix} 3 & 2 & -4 \\ 5 & 1 & -1 \\ -2 & 6 & 7 \end{vmatrix}$ **Ans.** -155

2. $\begin{vmatrix} 1 & 4 & 2 \\ 2 & 5 & 3 \\ 3 & 6 & 4 \end{vmatrix}$ **Ans.** 0

3. $\begin{vmatrix} 6 & 3 & 7 \\ 32 & 13 & 37 \\ 10 & 4 & 11 \end{vmatrix}$ **Ans.** 10

4. $\begin{vmatrix} 9 & 25 & 6 \\ 7 & 13 & 5 \\ 9 & 23 & 6 \end{vmatrix}$ **Ans.** 6

5. If $a + b + c = 0$, solve the equation

$$\begin{vmatrix} a-x & c & b \\ c & b-x & a \\ b & a & c-x \end{vmatrix} = 0$$

Ans. $x = a + b + c,\ x = \pm\sqrt{a^2 + b^2 + c^2 - ab - bc - ca}$

6. Show that $x = 2$ is one root of the determinant $\begin{vmatrix} x & -6 & -1 \\ 2 & -3x & x-3 \\ 3 & 2x & x+2 \end{vmatrix} = 0$, and find other two roots.

[Diploma I.E.T.E. June 2005] **Ans.** $-1 \pm \sqrt{\dfrac{170}{5}}\, i$

Choose the correct alternative: **[Diploma I.E.T.E. June 2006]**

7. The value of the determinant $\begin{vmatrix} -1 & 1 & 1 \\ 1 & -1 & 1 \\ 1 & 1 & -1 \end{vmatrix}$ is equal to

(*a*) – 4 (*b*) 0 (*c*) 1 (*d*) 4 **Ans.** (*d*)

9.7 PROPERTIES OF DETERMINANTS

(*i*) The value of a determinant remains unaltered if the rows are changed into columns (or vice versa).

Consider the determinant:

$$\Delta = \begin{vmatrix} a_1 & b_1 & c_1 \\ a_2 & b_2 & c_2 \\ a_3 & b_3 & c_3 \end{vmatrix}$$

$$= a_1(b_2c_3 - b_3c_2) - b_1(a_2c_3 - a_3c_2) + c_1(a_2b_3 - a_3b_2)$$

$$= a_1b_2c_3 - a_1b_3c_2 - a_2b_1c_3 + a_3b_1c_2 + a_2b_3c_1 - a_3b_2c_1$$

$$= (a_1b_2c_4 - a_1b_3c_2) - (a_2b_1c_3 - a_2b_3c_1) + (a_3b_1c_2 - a_3b_2c_1)$$

$$= a_1(b_2c_3 - b_3c_2) - a_2(b_1c_3 - b_3c_1) + a_3(b_1c_2 - b_2c_1)$$

$$= \begin{vmatrix} a_1 & a_2 & a_3 \\ b_1 & b_2 & b_3 \\ c_1 & c_2 & c_3 \end{vmatrix}$$

Proved

(*ii*) If two rows (or two columns) of a determinant are interchanged, the sign of the value of the determinant changes.

Interchanging the first two rows of Δ, we get

$$\Delta' = \begin{vmatrix} a_2 & b_2 & c_2 \\ a_1 & b_1 & c_1 \\ a_3 & b_3 & c_3 \end{vmatrix}$$

$$= a_2(b_1c_3 - b_3c_1) - b_2(a_1c_3 - a_3c_1) + c_2(a_1b_3 - a_3b_1)$$

$$= a_2b_1c_3 - a_2b_3c_1 - a_1b_2c_3 + a_3b_2c_1 + a_1b_3c_2 - a_3b_1c_2$$
$$= -[(a_1b_2c_3 - a_1b_3c_2) - (a_2b_1c_3 - a_3b_1c_2) + (a_2b_2c_1 - a_3b_2c_1)]$$
$$= -[a_1(b_2c_3 - b_3c_2) - b_1(a_2c_3 - a_3c_2) + c_1(a_2b_3 - a_3b_2)]$$
$$= -\begin{vmatrix} a_1 & b_1 & c_1 \\ a_2 & b_2 & c_2 \\ a_3 & b_3 & c_3 \end{vmatrix} = -\Delta$$ **Proved**

(*iii*) If two rows (or columns) of a determinant are identical, the value of the determinant is zero.

Let $$\Delta = \begin{vmatrix} a_1 & b_1 & c_1 \\ a_1 & b_1 & c_1 \\ a_3 & b_3 & c_3 \end{vmatrix},$$ so that the first two rows are identical.

By interchanging the first two rows, we get the same determinant Δ.

By property (*ii*), on interchanging the rows, the sign of the determinant changes.

$\Rightarrow \quad \Delta = -\Delta$

$\Rightarrow \quad 2\,\Delta = 0$

$\Rightarrow \quad \Delta = 0$ **Proved**

(*iv*) If the elements of any row (or column) of a determinant be each multiplied by the same number, the determinant is multiplied by that number.

Let $$\Delta' = \begin{vmatrix} ka_1 & kb_1 & kc_1 \\ a_2 & b_2 & c_2 \\ a_3 & b_3 & c_3 \end{vmatrix}$$

$$= ka_1(b_2c_3 - b_3c_2) - kb_1(a_2c_3 - a_3c_2) + kc_1(a_2b_3 - a_3b_2)$$
$$= k[a_1(b_2c_3 - b_3c_2) - b_1(a_2c_3 - a_3c_2) + c_1(a_2b_3 - a_3b_2)$$
$$= k\begin{vmatrix} a_1 & b_1 & c_1 \\ a_2 & b_2 & c_2 \\ a_3 & b_3 & c_3 \end{vmatrix} = k\Delta.$$ **Proved**

Example 11. *Prove that* $\begin{vmatrix} a^2 & a & bc \\ b^2 & b & ca \\ c^2 & c & ab \end{vmatrix} = -\begin{vmatrix} 1 & 1 & 1 \\ a^2 & b^2 & c^2 \\ a^3 & b^3 & c^3 \end{vmatrix}$

Solution. $\begin{vmatrix} a^2 & a & bc \\ b^2 & b & ca \\ c^2 & c & ab \end{vmatrix}$

By multiplying R_1, R_2, R_3 by a, b and c respectively, we get

$$= \frac{1}{abc}\begin{vmatrix} a^3 & a^2 & abc \\ b^3 & b^2 & abc \\ c^3 & c^2 & abc \end{vmatrix} = \frac{abc}{abc}\begin{vmatrix} a^3 & a^2 & 1 \\ b^3 & b^2 & 1 \\ c^3 & c^2 & 1 \end{vmatrix} = \begin{vmatrix} a^3 & a^2 & 1 \\ b^3 & b^2 & 1 \\ c^3 & c^2 & 1 \end{vmatrix}$$

$$\text{Interchanging } C_1 \text{ and } C_3 = -\begin{vmatrix} 1 & a^2 & a^3 \\ 1 & b^2 & b^3 \\ 1 & c^2 & c^3 \end{vmatrix}$$

$$= -\begin{vmatrix} 1 & 1 & 1 \\ a^2 & b^2 & c^2 \\ a^3 & b^3 & c^3 \end{vmatrix}$$ By changing rows into columns. **Proved**

Example 12. *Without expanding and/or evaluating, show that*

$$\begin{vmatrix} a^2 & a & 1 & bcd \\ b^2 & b & 1 & cba \\ c^2 & c & 1 & dab \\ d^2 & d & 1 & abc \end{vmatrix} = \begin{vmatrix} a^3 & a^2 & a & 1 \\ b^3 & b^2 & b & 1 \\ c^3 & c^2 & c & 1 \\ d^3 & d^2 & d & 1 \end{vmatrix}$$

Solution.

$$\begin{vmatrix} a^2 & a & 1 & bcd \\ b^2 & b & 1 & cba \\ c^2 & c & 1 & dab \\ d^2 & d & 1 & abc \end{vmatrix} = \frac{1}{abcd}\begin{vmatrix} a^3 & a^2 & a & abcd \\ b^3 & b^2 & b & abcd \\ c^3 & c^2 & c & abcd \\ d^3 & d^2 & d & abcd \end{vmatrix}$$ By multplying R_1 by a, R_2 by b, R_3 by c, R_4 by d,

$$= \frac{abcd}{abcd}\begin{vmatrix} a^3 & a^2 & a & 1 \\ a^3 & b^2 & b & 1 \\ c^3 & c^2 & c & 1 \\ d^3 & d^2 & d & 1 \end{vmatrix}$$ Taking out abcd common from the 4th column

$$= \begin{vmatrix} a^3 & a^2 & a & 1 \\ a^3 & b^2 & b & 1 \\ c^3 & c^2 & c & 1 \\ d^3 & d^2 & d & 1 \end{vmatrix}$$ **Proved**

Example 13. *Prove that*

$$\begin{vmatrix} 1 & a & a^2 \\ 1 & b & b^2 \\ 1 & c & c^2 \end{vmatrix} = \begin{vmatrix} 1 & a & bc \\ 1 & b & ca \\ 1 & c & ab \end{vmatrix}$$

Solution. Try yourself (**Hint:** Multiply the first column by *abc*)

(*v*) If each element of a row (or column) of a determinant consists of the algebraic sum of *n* terms, the determinant can be expressed as the sum of *n* determinants.

Let $$\Delta' = \begin{vmatrix} a_1 + p_1 + q_1 & b_1 & c_1 \\ a_2 + p_2 + q_2 & b_2 & c_2 \\ a_3 + p_3 + q_3 & b_3 & c_3 \end{vmatrix}$$

$$= (a_1 + p_1 + q_1)(b_2c_3 - b_3c_2) - (a_2 + p_2 + q_2)(b_1c_3 - b_3c_1) + (a_3 + p_3 + q_3)(b_1c_2 - b_2c_1)$$

$$= a_1(b_2c_3 - b_3c_2) - a_2(b_1c_3 - b_3c_1) + a_3(b_1c_2 - b_2c_1) + p_1(b_2c_3 - b_3c_2) - p_2(b_1c_3 - b_3c_1) + p_3(b_1c_2 - b_2c_1) + q_1(b_2c_3 - b_3c_2) - q_2(b_1c_3 - b_3c_1) + q_3(b_1c_2 - b_2c_1)$$

$$= \begin{vmatrix} a_1 & b_1 & c_1 \\ a_2 & b_2 & c_2 \\ a_3 & b_3 & c_3 \end{vmatrix} + \begin{vmatrix} p_1 & b_1 & c_1 \\ p_2 & b_2 & c_2 \\ p_3 & b_3 & c_3 \end{vmatrix} + \begin{vmatrix} q_1 & b_1 & c_1 \\ q_2 & b_2 & c_2 \\ q_3 & b_3 & c_3 \end{vmatrix}$$ **Proved**

Example 14. *If* $\begin{vmatrix} a & a^2 & a^3 - 1 \\ b & b^2 & b^3 - 1 \\ c & c^2 & c^3 - 1 \end{vmatrix} = 0$, *prove that* $abc = 1$.

Solution. $\begin{vmatrix} a & a^2 & a^3 - 1 \\ b & b^2 & b^3 - 1 \\ c & c^2 & c^3 - 1 \end{vmatrix} = 0$ or $\begin{vmatrix} a & a^2 & a^3 \\ b & b^2 & b^3 \\ c & c^2 & c^3 \end{vmatrix} + \begin{vmatrix} a & a^2 & -1 \\ b & b^2 & -1 \\ c & c^2 & -1 \end{vmatrix} = 0$

$$\Rightarrow \quad abc\begin{vmatrix} 1 & a & a^2 \\ 1 & b & b^2 \\ 1 & c & c^2 \end{vmatrix} - \begin{vmatrix} a & a^2 & 1 \\ b & b^2 & 1 \\ c & c^2 & 1 \end{vmatrix} = 0$$

(Taking out common a, b, c from R_1, R_2 and R_3 from 1st determinant)

$$\Rightarrow \quad abc\begin{vmatrix} 1 & a & a^2 \\ 1 & b & b^2 \\ 1 & c & c^2 \end{vmatrix} + \begin{vmatrix} a & 1 & a^2 \\ b & 1 & b^2 \\ c & 1 & c^2 \end{vmatrix} = 0$$

(Interchanging C_2 and C_3)

$$\Rightarrow \quad abc\begin{vmatrix} 1 & a & a^2 \\ 1 & b & b^2 \\ 1 & c & c^2 \end{vmatrix} - \begin{vmatrix} 1 & a & a^2 \\ 1 & b & b^2 \\ 1 & c & c^2 \end{vmatrix} = 0$$

(Interchanging C_1 and C_2 of the second determinant)

$$\Rightarrow \quad (abc - 1)\begin{vmatrix} 1 & a & a^2 \\ 1 & b & b^2 \\ 1 & c & c^2 \end{vmatrix} = 0$$

$$\Rightarrow \quad abc - 1 = 0$$

$$\Rightarrow \quad abc = 1$$ **Proved**

Example 15. *Prove that* $\begin{vmatrix} 2\alpha & \alpha+\beta & \alpha+\gamma & \alpha+\delta \\ \beta+\alpha & 2\beta & \beta+\gamma & \beta+\delta \\ \gamma+\alpha & \gamma+\beta & 2\gamma & \gamma+\delta \\ \delta+\alpha & \delta+\beta & \delta+\gamma & 2\delta \end{vmatrix} = 0$

Solution. Given determinant

$$= \begin{vmatrix} \alpha+\alpha & \alpha+\beta & \alpha+\gamma & \alpha+\delta \\ \beta+\alpha & \beta+\beta & \beta+\gamma & \beta+\delta \\ \gamma+\alpha & \gamma+\beta & \gamma+\gamma & \gamma+\delta \\ \delta+\alpha & \delta+\beta & \delta+\gamma & \delta+\delta \end{vmatrix}$$

The above determinant can be expressed as the sum of 16 determinants.

Each of the 16 determinants has either two of its columns identical or two of its columns become identical after any of the common factors α, β, γ, δ is/are taken out.

$\therefore$ Each of the resulting determinant is zero. Hence the result.

(*vi*) The value of the determinant remains unaltered if to the elements of one row (or column) be added any constant multiple of the corresponding elements of any other row (or column) respectively.

Let $$\Delta = \begin{vmatrix} a_1 & b_1 & c_1 \\ a_2 & b_2 & c_2 \\ a_3 & b_3 & c_3 \end{vmatrix}$$

On multiplying the second column by l and the third column by m and adding to the first column, we get

$$\Delta' = \begin{vmatrix} a_1 + lb_1 + mc_1 & b_1 & c_1 \\ a_2 + lb_2 + mc_2 & b_2 & c_2 \\ a_3 + lb_3 + mc_3 & b_3 & c_3 \end{vmatrix}$$

$$= \begin{vmatrix} a_1 & b_1 & c_1 \\ a_2 & b_2 & c_2 \\ a_3 & b_3 & c_3 \end{vmatrix} + l\begin{vmatrix} b_1 & b_1 & c_1 \\ b_2 & b_2 & c_2 \\ b_3 & b_3 & c_3 \end{vmatrix} + m\begin{vmatrix} c_1 & b_1 & c_1 \\ c_2 & b_2 & c_2 \\ c_3 & b_3 & c_3 \end{vmatrix}$$

$= \Delta + 0 + 0$ (Since columns are identical)

$= \Delta$ **Proved**

Example 16. *Evaluate*

$$\Delta = \begin{vmatrix} 3 & 2 & 1 & 4 \\ 15 & 29 & 2 & 14 \\ 16 & 19 & 3 & 17 \\ 33 & 39 & 8 & 38 \end{vmatrix}$$

Solution. $$\Delta = \begin{vmatrix} 0 & 0 & 1 & 0 \\ 9 & 25 & 2 & 6 \\ 7 & 13 & 3 & 5 \\ 9 & 23 & 8 & 6 \end{vmatrix} \begin{matrix} \\ C_1 \to C_1 - 3C_3 \\ C_2 \to C_2 - 2C_3 \\ C_4 \to C_4 - 4C_3 \end{matrix}$$

On expanding by first row

$$= 0 + 0 + 1 \times \begin{vmatrix} 9 & 25 & 6 \\ 7 & 13 & 5 \\ 9 & 23 & 6 \end{vmatrix} + 0 = \begin{vmatrix} 0 & 2 & 0 \\ 7 & 13 & 5 \\ 9 & 23 & 6 \end{vmatrix} R_1 \to R_1 - R_3$$

On expanding by first row

$$= -2\begin{vmatrix} 7 & 5 \\ 9 & 6 \end{vmatrix} = -2(42 - 45)$$

$$= -2 \times (-3) = 6$$

Ans.

Example 17. *Prove that*

$$\begin{vmatrix} x & l & m & 1 \\ \alpha & x & n & 1 \\ \alpha & \beta & x & 1 \\ \alpha & \beta & \gamma & 1 \end{vmatrix} = (x - \alpha)(x - \beta)(x - \gamma)$$

Solution. Given determinant $= \begin{vmatrix} x - \alpha & l & m & 1 \\ 0 & x & n & 1 \\ 0 & \beta & x & 1 \\ 0 & \beta & \gamma & 1 \end{vmatrix} C_1 \to C_1 - \alpha C_4$

On expanding by first column $= (x - \alpha)\begin{vmatrix} x & m & 1 \\ \beta & x & 1 \\ \beta & \gamma & 1 \end{vmatrix}$

$$= (x - \alpha)\begin{vmatrix} x - \beta & n & 1 \\ 0 & x & 1 \\ 0 & \gamma & 1 \end{vmatrix} C_1 \to C_1 - \beta C_3$$

On expanding by first column $= (x - \alpha)(x - \beta)(x - \gamma)$ **Proved**

Example 18. *Show that $x = -(a + b + c)$ is one root of the equation.*

$$\begin{vmatrix} x + a & b & c \\ b & x + c & a \\ c & a & x + b \end{vmatrix} = 0 \text{ and solve the equation completely.}$$

Solution. By $C_1 \to C_1 + C_2 + C_3$, we get

$$\begin{vmatrix} x + a + b + c & b & c \\ x + a + b + c & x + c & a \\ x + a + b + c & a & x + b \end{vmatrix} = 0$$

$$\Rightarrow (x + a + b + c)\begin{vmatrix} 1 & b & c \\ 1 & x + c & a \\ 1 & a & x + b \end{vmatrix} = 0, (x + a + b + c)\begin{vmatrix} 1 & b & c \\ 0 & x - b + c & a - c \\ 0 & a - b & x + b - c \end{vmatrix} \begin{matrix} R_2 \to R_2 - R_1 \\ R_3 \to R_3 - R_1 \end{matrix}$$

On expanding by first column

$$(x + a + b + c)[(x - b + c)(x + b - c) - (a - b)(a - c) = 0$$

$$(x + a + b + c)[(x^2 - (b - c)^2 (a^2 - ac - ab + bc)] = 0$$

$$(x+a+b+c)(x^2-b^2-c^2+2bc-a^2+ac+ab-bc]=0$$

$$(x+a+b+c)(x^2-b^2-c^2+ab+bc+ca)=0$$

Either $\qquad x + a + b + c = 0$

$\Rightarrow \qquad x = -(a + b + c)$

$\Rightarrow \quad x^2 - a^2 - b^2 - c^2 + ab + bc + ca = 0$

$\Rightarrow \qquad x = \pm\sqrt{a^2+b^2+c^2-ab-bc-ca}$ **Ans.**

Example 19. *Solve the determinant equation*

$$\begin{vmatrix} 2x-1 & x+7 & x+4 \\ x & 6 & 2 \\ x-1 & x+1 & 3 \end{vmatrix} = 0$$

Solution. By $R_1 \to R_1 - (R_2 + R_3)$, we get

$$\begin{vmatrix} 0 & 0 & x-1 \\ x & 6 & 2 \\ x-1 & x+1 & 3 \end{vmatrix} = 0$$

On expanding by first row, we get

$$(x-1)(x^2+x-6x+6)=0$$

$\Rightarrow \qquad (x-1)(x-2)(x-3)=0$

$\Rightarrow \qquad x = 1, 2, 3.$ **Ans.**

Example 20. *Prove that* **[S.B.T.E. 2015, 2013]**

$$\begin{vmatrix} 1+a_1 & 1 & 1 \\ 1 & 1+a_2 & 1 \\ 1 & 1 & 1+a_3 \end{vmatrix} = a_1a_2a_3\left(1+\frac{1}{a_1}+\frac{1}{a_2}+\frac{1}{a_3}\right)$$

Proof. Taking out a_1, a_2, a_3 common from C_1, C_2, C_3 respectively, we get

$$\Delta = a_1\,a_2\,a_3\begin{vmatrix} \frac{1}{a_1}+1 & \frac{1}{a_2} & \frac{1}{a_3} \\ \frac{1}{a_1} & \frac{1}{a_2}+1 & \frac{1}{a_3} \\ \frac{1}{a_1} & \frac{1}{a_2} & \frac{1}{a_3}+1 \end{vmatrix}$$

Adding C_2 and C_3 to C_1 *i.e.* $C_1 \to C_1 + C_2 + C_3$

$$= a_1\,a_2\,a_3\begin{vmatrix} 1+\frac{1}{a_1}+\frac{1}{a_2}+\frac{1}{a_3} & \frac{1}{a_2} & \frac{1}{a_3} \\ 1+\frac{1}{a_1}+\frac{1}{a_2}+\frac{1}{a_3} & \frac{1}{a_2}+1 & \frac{1}{a_3} \\ 1+\frac{1}{a_1}+\frac{1}{a_2}+\frac{1}{a_3} & \frac{1}{a_2} & \frac{1}{a_3}+1 \end{vmatrix}$$

Taking out $\left(1+\frac{1}{a_1}+\frac{1}{a_2}+\frac{1}{a_3}\right)$ common from first column

$$= a_1 a_2 a_3 \left(1+\frac{1}{a_1}+\frac{1}{a_2}+\frac{1}{a_3}\right)\begin{vmatrix} 1 & \frac{1}{a_2} & \frac{1}{a_3} \\ 1 & \frac{1}{a_2}+1 & \frac{1}{a_3} \\ 1 & \frac{1}{a_2} & \frac{1}{a_3}+1 \end{vmatrix}$$

Subtracting 1st row from 2nd and 3rd row

$$= a_1 a_2 a_3 \left(1+\frac{1}{a_1}+\frac{1}{a_2}+\frac{1}{a_3}\right)\begin{vmatrix} 1 & \frac{1}{a_2} & \frac{1}{a_3} \\ 0 & 1 & 0 \\ 0 & 0 & 1 \end{vmatrix} \begin{matrix} \\ R_2 \to R_2 - R_1 \\ R_3 \to R_3 - R_1 \end{matrix}$$

Expanding from 1st column, we get

$$= a_1 a_2 a_3 \left(1+\frac{1}{a_1}+\frac{1}{a_2}+\frac{1}{a_3}\right)\times 1$$

$$= a_1 a_2 a_3 \left(1+\frac{1}{a_1}+\frac{1}{a_2}+\frac{1}{a_3}\right)$$ **Proved**

Example 21. *Evaluate:*

$$\begin{vmatrix} a-b-c & 2a & 2a \\ 2b & b-c-a & 2b \\ 2c & 2c & c-a-b \end{vmatrix}$$

Solution. By $R_1 \to R_1 + R_2 + R_3$, we get

$$\begin{vmatrix} a+b+c & a+b+c & a+b+c \\ 2b & b-c-a & 2b \\ 2c & 2c & c-a-b \end{vmatrix} = (a+b+c)\begin{vmatrix} 1 & 1 & 1 \\ 2b & b-c-a & 2b \\ 2c & 2c & c-a-b \end{vmatrix}$$

$$= (a+b+c)\begin{vmatrix} 1 & 0 & 0 \\ 2b & -(a+b+c) & 0 \\ 2c & 0 & -(a+b+c) \end{vmatrix} \begin{matrix} \\ C_2 \to C_2 - C_1 \\ C_3 \to C_3 - C_1 \end{matrix}$$

On expanding by first row = $(a + b + c)\,(a + b + c)^2 = (a + b + c)^3$. **Ans.**

Example 22. *Show without expanding*

$$\begin{vmatrix} 1 & 1 & 1 \\ x & y & z \\ x^2 & y^2 & z^2 \end{vmatrix} = (x-y)(y-z)(z-x)$$

Solution. By $C_1 \to C_1 - C_2$, $C_2 \to C_2 - C_3$, we get

$$= \begin{vmatrix} 0 & 0 & 1 \\ x-y & y-z & z \\ x^2-y^2 & y^2-z^2 & z^2 \end{vmatrix}$$

On expanding by first row, we get

$$= \begin{vmatrix} x-y & y-z \\ x^2-y^2 & y^2-z^2 \end{vmatrix}$$

$$= (x-y)\,(y-z) \begin{vmatrix} 1 & 1 \\ x+y & y+z \end{vmatrix}$$

$$= (x-y)\,(y-z)\,(y+z-x-y)$$

$$= (x-y)\,(y-z)\,(z-x)$$

Proved

Example 23. *Find the value of*

$$\begin{vmatrix} (b+c)^2 & a^2 & a^2 \\ b^2 & (c+a)^2 & b^2 \\ c^2 & c^2 & (a+b)^2 \end{vmatrix}$$

[S.B.T.E. 2014, 2013]

Solution. By $C_1 \to C_1 - C_3$, $C_2 \to C_2 - C_3$, we get

$$\begin{vmatrix} (b+c)^2-a^2 & a^2-a^2 & a^2 \\ b^2-b^2 & (c+a)^2-b^2 & b^2 \\ c^2-(a+b)^2 & c^2-(a+b)^2 & (a+b)^2 \end{vmatrix}$$

$$= \begin{vmatrix} (a+b+c)(b+c-a) & 0 & a^2 \\ 0 & (a+b+c)(c+a-b) & b^2 \\ (a+b+c)(c-a-b) & (a+b+c)(c+a-b) & (a+b)^2 \end{vmatrix}$$

On taking out $(a + b + c)$ as common from 1st and 2nd column, we get

$$= (a+b+c)^2 \begin{vmatrix} b+c-a & 0 & a^2 \\ 0 & c+a-b & b^2 \\ c-a-b & c-a-b & (a+b)^2 \end{vmatrix}$$

$$= (a+b+c)^2 \begin{vmatrix} -a+b+c & 0 & a^2 \\ 0 & a-b+c & b^2 \\ -2b & -2a & 2ab \end{vmatrix} R_3 \to R_3 - (R_1 + R_2)$$

$$= -2(a+b+c)^2 \begin{vmatrix} -a+b+c & 0 & a^2 \\ 0 & a-b+c & b^2 \\ b & a & -ab \end{vmatrix}$$

On expanding by first row

$$= -2(a+b+c)^2\,[(-a+b+c)\{-ab(a-b+c)-ab^2\}+a^2\{0-b(a-b+c\}]$$

$$= -2(a+b+c)^2\,[(-a+b+c)\,(-a^2b-abc)-a^2b(a-b+c)]$$

$$= -2ab(a+b+c)^2\,[(-a+b+c)(-a-c)-a(a-b+c)]$$

$$= -2ab(a+b+c)^2\,[(a^2+ac-ab-bc-ac-c^2-a^2+ab-ac)]$$

$$= -2ab(a+b+c)^2\,(-bc-ac-c^2)$$
$$= 2abc(a+b+c)^2\,(b+a+c)$$
$$= 2abc(a+b+c)^3$$

Ans.

EXERCISE 9.4

Expand the following determinants, using properties of the determinants.

1. $\begin{vmatrix} 1 & 3 & 7 \\ 4 & 9 & 1 \\ 2 & 7 & 6 \end{vmatrix}$ **Ans.** 51

2. Prove that $\begin{vmatrix} x & a & a \\ a & x & a \\ a & a & x \end{vmatrix} = (x+2a)(x-a)^2$

3. $\begin{vmatrix} 1 & 2 & 1 & 3 \\ 3 & 4 & 2 & 5 \\ 6 & 1 & 7 & 1 \\ 4 & 3 & 9 & 2 \end{vmatrix}$ **Ans.** 75

4. Solve the equation $\begin{vmatrix} x^3-a^3 & x^2 & x \\ b^3-a^3 & b^2 & b \\ c^2-a^3 & c^2 & c \end{vmatrix} = 0,$ $b \neq c, bc \neq 0$ **Ans.** $x = \dfrac{a^3}{bc}, x = b, x = 0$

5. Show that zero is one of the roots of the equation

$$\begin{vmatrix} 0 & x-a & x-b \\ x+a & 0 & x-c \\ x+b & x-c & 0 \end{vmatrix} = 0\cdot$$

6. Show without evaluating that determinant $\begin{vmatrix} 1 & x & y+z \\ 1 & y & x+z \\ 1 & z & x+y \end{vmatrix} = 0.$

7. Prove that $\begin{vmatrix} 1 & a^2 & a^3 \\ 1 & b^2 & b^3 \\ 1 & c^2 & c^3 \end{vmatrix} = -(a-b)(b-c)(c-a)(ab+bc+ca).$ **[Diploma I.E.T.E. Dec. 2005]**

8. Evaluate $\begin{vmatrix} 1 & \omega & \omega^2 \\ \omega^2 & 1 & \omega \\ \omega & \omega^2 & 1 \end{vmatrix}$ where ω is a complex cube root of unity.

9. Prove that $\begin{vmatrix} 1+a & 1 & 1 \\ 1 & 1+a & 1 \\ 1 & 1 & 1+a \end{vmatrix} = a^3 + 3a^2$ **[Mathematics I, Jan. 2009]**

10. Prove that $\begin{vmatrix} a+b & a & b \\ a & a+c & c \\ b & c & b+c \end{vmatrix} = 4abc$ **[S.B.T.E. 2017]**

Choose the correct alternative:

11. The value of the determinant $\begin{vmatrix} 1 & \omega^3 & \omega^5 \\ \omega^3 & 1 & \omega^4 \\ \omega^5 & \omega^4 & 1 \end{vmatrix}$, where ω is an imaginary cube root of unity is

(*a*) $(1-\omega)^2$ (*b*) 3 (*c*) − 3 (*d*) 4 **Ans.** (*b*)

12. The value of $\begin{vmatrix} x+a & x & x \\ x & x+a & x \\ x & x & x+a \end{vmatrix}$ is equal to

(*a*) $3a^2x$ (*b*) $a^2(3x-a)$ (*c*) $a^2(3x+a)$ (*d*) $3ax^2$ **Ans.** (*c*)

9.8 FACTOR THEOREM

If the elements of a determinant are polynomials in a variable x and if the substitution $x = a$ makes two rows (or columns) identical, then $(x - a)$ is a factor of the determinant.

When two rows are identical, the value of the determinant is zero. The expansion of the determinant being a polynomial in x vanish on putting $x = a$, then $x - a$ is its factor by the remainder theorem.

Example 24. *Show that* $\begin{vmatrix} 1 & 1 & 1 \\ x & y & z \\ x^2 & y^2 & z^2 \end{vmatrix} = (x-y)(y-z)(z-x)$ **[S.B.T.E. 2017, 2012]**

Solution. If we put $x = y$; $y = z$; $z = x$ then in each case two columns become identical and the determinant vanishes.

∴ $(x - y)$; $(y - z)$; $(z - x)$ are the factors.

Since the determinant is of third degree, the other factors can be numerical only say k.

$$\begin{vmatrix} 1 & 1 & 1 \\ x & y & z \\ x^2 & y^2 & z^2 \end{vmatrix} = k(x-y)(y-z)(z-x)$$

This leading term (product of the elements of the diagonal elements) in the given determinant is yz^2 and in the expansion

$k(x-y)(y-z)(z-x)$, we get kyz^2

Equating the coefficient of yz^2, we have

$$k = 1$$

Hence the expansion = $(x-y)(y-z)(z-x)$. **Proved.**

Example 25. *Show that* $\begin{vmatrix} x & x^2 & x^3 \\ y & y^2 & y^3 \\ z & z^2 & z^3 \end{vmatrix} = xyz(x-y)(y-z)(z-x)$ **[S.B.T.E. 2016, 2014, 2013]**

Solution. $\begin{vmatrix} x & x^2 & x^3 \\ y & y^2 & y^3 \\ z & z^2 & z^3 \end{vmatrix} = xyz\begin{vmatrix} 1 & x & x^2 \\ 1 & y & y^2 \\ 1 & z & z^2 \end{vmatrix}$

$= xyz(x-y)(y-z)(z-x)$ (see example 24) **Proved.**

Example 26. *Show that*

$$\begin{vmatrix} x^3 & x^2 & x & 1 \\ \alpha^3 & \alpha^2 & \alpha & 1 \\ \beta^3 & \beta^2 & \beta & 1 \\ \gamma^3 & \gamma^2 & \gamma & 1 \end{vmatrix} = (x-\alpha)(x-\beta)(x-\gamma)(\alpha-\beta)(\beta-\gamma)(\alpha-\gamma)$$

Proof. If we put $x = \alpha$; $x = \beta$; $x = \gamma$; $\alpha = \beta$, $\beta = \gamma$; $\gamma = \alpha$ then two rows become identical and the determinant vanishes.

$\therefore$ $(x - \alpha)$; $(x - \beta)$; $(x - \gamma)$; $(\alpha - \beta)$; $(\beta - \gamma)$; $(\alpha - \gamma)$ are the factors.

Since the determinant is six degree the other factor can be numerical only say k.

$$\begin{vmatrix} x^3 & x^2 & x & 1 \\ \alpha^3 & \alpha^2 & \alpha & 1 \\ \beta^3 & \beta^2 & \beta & 1 \\ \gamma^3 & \gamma^2 & \gamma & 1 \end{vmatrix} = (x-\alpha)(x-\beta)(x-\gamma)(\alpha-\beta)(\beta-\gamma)(\alpha-\gamma)$$

The leading term is $x^3\alpha^2\beta$.

And in the expansion it is $kx^3(-\alpha^2)\beta$.

$\therefore$ $k = -1$

Hence the expansion $= -(x-\alpha)(x-\beta)(x-\gamma)(\alpha-\beta)(\beta-\gamma)(\gamma-\alpha)$ **Ans.**

Example 27. *Factorize* $\Delta = \begin{vmatrix} 1 & 1 & 1 \\ a^2 & b^2 & c^2 \\ a^3 & b^3 & c^3 \end{vmatrix}$

Solution. Putting $a = b$, $C_1 = C_2$ and hence $\Delta = 0$.

$\therefore$ $a - b$ is a factor of Δ

Similarly $b - c$, $c - a$ are also factors of Δ

$\therefore$ $(a-b)(b-c)(c-a)$ is a third degree factor of Δ which itself of the fifth degree as is judged from the leading term b^2c^3

$\therefore$ The remaining factor must be of the second degree. As Δ is symmetrical in a, b, c the remaining factor must, therefore, be of the form

$$k(a^2 + b^2 + c^2) + l(ab + bc + ca)$$

$$\therefore \quad \Delta = (a-b)(b-c)(c-a)\{k(a^2+b^2+c^2) + l(ab+bc+ca)\}$$

If $k \neq 0$, we shall get terms like a^4b, b^4c, etc. which do not occur in Δ. Hence k must be zero.

$$\therefore \quad \Delta = (a-b)(b-c)(c-a)\{0 + l(ab+bc+ca)\}$$

$$\text{or} \quad \Delta = l(a-b)(b-c)(c-a)(ab+bc+ca)$$

The leading term in $\Delta = b^2c^3$

The corresponding term on R.H.S. $= lb^2c^3$

$\therefore$ $l = 1$

Hence $\Delta = (a-b)(b-c)(c-a)(ab+bc+ca)$. **Ans.**

Example 28. *Evaluate* $\begin{vmatrix} b-c & c-a & a-b \\ c-a & a-b & b-c \\ a-b & b-c & c-a \end{vmatrix}$ **[S.B.T.E. 2015]**

Solution. Here, we have

Let $$\Delta = \begin{vmatrix} b-c & c-a & a-b \\ c-a & a-b & b-c \\ a-b & b-c & c-a \end{vmatrix} R_1 \to R_1 + R_2 + R_3$$

$$\Delta = \begin{vmatrix} 0 & 0 & 0 \\ c-a & a-b & b-c \\ a-b & b-c & c-a \end{vmatrix}$$

First row elements are zero

So, $\Delta = 0$ **Ans.**

Example 29. *Without expanding show that*

$$\begin{vmatrix} 1/a & a^2 & bc \\ 1/b & b^2 & ca \\ 1/c & c^2 & ab \end{vmatrix} = 0$$ **[S.B.T.E. 2015]**

Solution. Here, we have

Let $\Delta = \begin{vmatrix} 1/a & a^2 & bc \\ 1/b & b^2 & ca \\ 1/c & c^2 & ab \end{vmatrix} \begin{matrix} R_1 \to aR_1 \\ R_2 \to bR_2 \\ R_3 \to cR_3 \end{matrix} = \begin{vmatrix} 1 & a^3 & abc \\ 1 & b^3 & abc \\ 1 & c^3 & abc \end{vmatrix} = \frac{abc}{abc} \begin{vmatrix} 1 & a^3 & 1 \\ 1 & b^3 & 1 \\ 1 & c^3 & 1 \end{vmatrix} = 0$ [using property (*iii*)].

Ans.

Example 30. *Show that:* $\begin{vmatrix} 1 & x & x^2 \\ x^2 & 1 & x \\ x & x^2 & 1 \end{vmatrix} = (1 - x^3)^2$ **[S.B.T.E. 2015]**

Solution. Here, we have

$$\text{L.H.S.} = \begin{vmatrix} 1 & x & x^2 \\ x^2 & 1 & x \\ x & x^2 & 1 \end{vmatrix} R_1 \to R_1 + R_2 + R_3 = \begin{vmatrix} 1+x^2+x & x+1+x^2 & x^2+x+1 \\ x^2 & 1 & x \\ x & x^2 & 1 \end{vmatrix}$$

$$= (1+x+x^2) \begin{vmatrix} 1 & 1 & 1 \\ x^2 & 1 & x \\ x & x^2 & 1 \end{vmatrix} \begin{matrix} R_1 \to R_1 - R_2 \\ R_2 \to R_2 - R_3 \end{matrix}$$

$$= (1+x+x^2) \begin{vmatrix} 1-x^2 & 0 & 1-x \\ x^2-x & 1-x^2 & x-1 \\ x & x^2 & 1 \end{vmatrix}$$

$$= (1+x+x^2)(1-x)(1-x)\begin{vmatrix} 1+x & 0 & 1 \\ -x & 1+x & -1 \\ x & x^2 & 1 \end{vmatrix}$$

$\Rightarrow \quad (1 + x + x^2)\,(1 - x)\,(1 - x)\,[(1 + x)\,\{(1 + x) + x^2\} + \{- x^3 - x\,(1 + x)\}]$

$\Rightarrow \quad (1 - x^3)\,(1 - x^3) = (1 - x^3)^2$

Proved

Example 31. *Show that* $\begin{vmatrix} a & b & c \\ a^2 & b^2 & c^2 \\ b+c & c+a & a+b \end{vmatrix} = (b-c)(c-a)(a-b)(a+b+c)$

[DIP.IETE. June, 2018]

Solution. Here we have

$$\text{L.H.S.} = \begin{vmatrix} a & b & c \\ a^2 & b^2 & c^2 \\ b+c & c+a & a+b \end{vmatrix} R_3 \to R_1 + R_3 = \begin{vmatrix} a & b & c \\ a^2 & b^2 & c^2 \\ a+b+c & a+b+c & a+b+c \end{vmatrix}$$

$$\Rightarrow \quad (a+b+c)\begin{vmatrix} a & b & c \\ a^2 & b^2 & c^2 \\ 1 & 1 & 1 \end{vmatrix} \begin{matrix} C_2 \to C_2 - C_1 \\ C_3 \to C_3 - C_1 \end{matrix}$$

$$\Rightarrow \quad (a+b+c)\begin{vmatrix} a & b-a & c-a \\ a^2 & b^2-a^2 & c^2-a^2 \\ 1 & 0 & 0 \end{vmatrix}$$

$$\Rightarrow \quad (a+b+c)(b-a)(c-a)\begin{vmatrix} a & 1 & 1 \\ a^2 & b+a & c+a \\ 1 & 0 & 0 \end{vmatrix} \qquad \text{[Expanding along } R_3]$$

$$= (a - b)\,(b - c)\,(c - a)\,(a + b + c)$$

Proved

Example 32. *Show that* $\begin{vmatrix} 1 & b+c & b^2+c^2 \\ 1 & c+a & c^2+a^2 \\ 1 & a+b & a^2+b^2 \end{vmatrix} = (a-b)(b-c)(c-a)$ **[S.B.T.E. 2017]**

Solution. Here we have

$$\begin{vmatrix} 1 & b+c & b^2+c^2 \\ 1 & c+a & c^2+a^2 \\ 1 & a+b & a^2+b^2 \end{vmatrix} \begin{matrix} R_2 \to R_2 - R_1 \\ R_3 \to R_3 - R_2 \end{matrix} = \begin{vmatrix} 1 & b+c & b^2+c^2 \\ 0 & a-b & a^2-b^2 \\ 0 & b-c & b^2-c^2 \end{vmatrix}$$

$$(a-b)(b-c)\begin{vmatrix} 1 & b+c & b^2+c^2 \\ 0 & 1 & a+b \\ 0 & 1 & b+c \end{vmatrix} = (a - b)\,(b - c)\,[1\,(b + c - a - b)]$$

$$\therefore \qquad (a - b)\,(b - c)\,(c - a)$$

Proved

EXERCISE 9.5

1. Evaluate, without expanding $\begin{vmatrix} a & a^2 & 1+a^3 \\ b & b^2 & 1+b^3 \\ c & c^2 & 1+c^3 \end{vmatrix}$ **Ans.** $(a-b)(b-c)(c-a)(1+abc)$

2. Show that $\begin{vmatrix} a^2+1 & ab & ac \\ ab & b^2+1 & bc \\ ac & bc & c^2+1 \end{vmatrix} = 1+a^2+b^2+c^2.$ **[Diploma IETE Dec. 2006]**

3. Without expanding $\Delta = \begin{vmatrix} (a-x)^2 & (a-y)^2 & (a-z)^2 \\ (b-x)^2 & (b-y)^2 & (b-z)^2 \\ (c-x)^2 & (c-y)^2 & (c-z)^2 \end{vmatrix}$ show that

$$\Delta = 2(a-b)(b-c)(c-a)(x-y)(y-z)(z-x)$$

4. Show (without expanding) that

$$\begin{vmatrix} bc & a^2 & a^2 \\ b^2 & ca & b^2 \\ c^2 & c^2 & ab \end{vmatrix} = \begin{vmatrix} bc & ab & ca \\ ab & ca & bc \\ ca & bc & ab \end{vmatrix} = -\frac{1}{2}(ab+bc+ca)[(ab-bc)^2+(bc-ca)^2+(ca-ab)^2].$$

9.9 CONJUGATE ELEMENTS

Two equidistant elements lying on a line perpendicular to the leading diagonal are said to be conjugate.

In the determinant $\begin{vmatrix} a_1 & b_1 & c_1 \\ a_2 & b_2 & c_2 \\ a_3 & b_3 & c_3 \end{vmatrix}$ $a_2, b_1; a_3, c_1; b_3, c_2$ are pairs of conjugate elements.

9.10 SPECIAL TYPES OF DETERMINANTS

(*i*) Orthosymmetric Determinant

If every element of the leading diagonal is the same and the conjugate elements are equal, then the determinant is said to be orthosymmetric determinant.

$$\begin{vmatrix} a & h & g \\ h & a & f \\ g & f & a \end{vmatrix}$$

(*ii*) Skew symmetric Determinant

If the elements of the leading diagonal are all zero and every other element is equal to its cojugate with sign changed, the determinant is said to be skew symmetric.

$$\begin{vmatrix} 0 & -a & -b \\ a & 0 & -c \\ b & c & 0 \end{vmatrix}$$

Property 1. A skew symmetric determinant of odd order vanishes.

Example 33. *Prove that*

$$\Delta = \begin{vmatrix} 0 & -a & -b \\ a & 0 & -c \\ b & c & 0 \end{vmatrix} = 0$$

Solution. Taking out (–1) common from each of the three columns

$$\Delta = (-1)^3 \begin{vmatrix} 0 & a & b \\ -a & 0 & c \\ -b & -c & 0 \end{vmatrix}$$

Changing rows into column

$$\Delta = (-1)^3 \begin{vmatrix} 0 & -a & -b \\ a & 0 & -c \\ b & c & 0 \end{vmatrix} = (-1)^3 \Delta = -\Delta$$

$\Rightarrow$ $\quad 2\Delta = 0 \;\Rightarrow\; \Delta = 0$ **Proved**

Property 2. A skew symmetric determinant of even order is a perfect square.

Example 34. *Prove that*

$$\begin{vmatrix} 0 & x & y & z \\ -x & 0 & c & b \\ -y & -c & 0 & a \\ -z & -b & -a & 0 \end{vmatrix} = (ax - by + cz)^2$$

Solution. By multiplying column 2nd by a, the given determinant is

$$= \frac{1}{a} \begin{vmatrix} 0 & ax & y & z \\ -x & 0 & c & b \\ -y & -ac & 0 & a \\ -z & -ab & -a & 0 \end{vmatrix}$$

By $C_2 \to C_2 - bC_3 + cC_4$, we get

$$= \frac{1}{a} \begin{vmatrix} 0 & ax - by + cz & y & z \\ -x & 0 & c & b \\ -y & 0 & 0 & a \\ -z & 0 & -a & 0 \end{vmatrix}$$

On expanding by column 2, we get

$$= \frac{-(ax - by + cz)}{a} \begin{vmatrix} -x & c & b \\ -y & 0 & a \\ -z & -a & 0 \end{vmatrix} = \frac{(ax - by + cz)}{a} \begin{vmatrix} x & c & b \\ y & 0 & a \\ z & -a & 0 \end{vmatrix}$$

$$= \frac{(ax - by + c)}{a \times a} \begin{vmatrix} ax & ac & ab \\ y & 0 & a \\ z & -a & 0 \end{vmatrix}$$

By $R_1 \to R_1 - bR_2 + cR_3$

$$= \frac{ax - by + cz}{a^2}\begin{vmatrix} ax - by + cz & ac - ac & ab - ab \\ y & 0 & a \\ z & -a & 0 \end{vmatrix}$$

$$= \frac{(ax - by + cz)}{a^2}\begin{vmatrix} ax - by + cz & 0 & 0 \\ y & 0 & a \\ z & -a & 0 \end{vmatrix} = \frac{(ax - by + cz)}{a^2}(ax - by + cz)(a^2)$$

$= (ax - by + cz)^2$ **Proved**

9.11 SOLUTION OF SIMULTANEOUS LINEAR EQUATIONS BY DETERMINANTS (CRAMER'S RULE)

Let us solve the following equations.

$$a_1x + b_1y + c_1z = d_1$$
$$a_2x + b_2y + c_2z = d_2$$
$$a_3x + b_3y + c_3z = d_3$$

Let
$$\Delta = \begin{vmatrix} a_1 & b_1 & c_1 \\ a_2 & b_2 & c_2 \\ a_3 & b_3 & c_3 \end{vmatrix} \Rightarrow x\,\Delta = \begin{vmatrix} a_1x & b_1 & c_1 \\ a_2x & b_2 & c_2 \\ a_3x & b_3 & c_3 \end{vmatrix}$$

Multiplying 2nd column by y and 3rd column by z and adding to the 1st column, we get

$$x\Delta = \begin{vmatrix} a_1x + b_1y + c_1z & b_1 & c_1 \\ a_2x + b_2y + c_2z & b_2 & c_2 \\ a_3x + b_3y + c_3z & b_3 & c_3 \end{vmatrix}$$

$$x\Delta = \begin{vmatrix} d_1 & b_1 & c_1 \\ d_2 & b_2 & c_2 \\ d_3 & b_3 & c_3 \end{vmatrix}$$

$$\Rightarrow \qquad x = \frac{\begin{vmatrix} d_1 & b_1 & c_1 \\ d_2 & b_2 & c_2 \\ d_3 & b_3 & c_3 \end{vmatrix}}{\begin{vmatrix} a_1 & b_1 & c_1 \\ a_2 & b_2 & c_2 \\ a_3 & b_3 & c_3 \end{vmatrix}} = \frac{\Delta_1}{\Delta}$$

$$y = \frac{\begin{vmatrix} a_1 & d_1 & c_1 \\ a_2 & d_2 & c_2 \\ a_3 & d_3 & c_3 \end{vmatrix}}{\begin{vmatrix} a_1 & b_1 & c_1 \\ a_2 & b_2 & c_2 \\ a_3 & b_3 & c_3 \end{vmatrix}} = \frac{\Delta_2}{\Delta}$$

$$z = \frac{\begin{vmatrix} a_1 & b_1 & d_1 \\ a_2 & b_2 & d_2 \\ a_3 & b_3 & d_3 \end{vmatrix}}{\begin{vmatrix} a_1 & b_1 & c_1 \\ a_2 & b_2 & c_2 \\ a_3 & b_3 & c_3 \end{vmatrix}} = \frac{\Delta_3}{\Delta}$$

$$x = \frac{\Delta_1}{\Delta}, y = \frac{\Delta_2}{\Delta}, z = \frac{\Delta_3}{\Delta}$$ **Ans.**

Example 35. *Solve, using Cramer's rule*

$$3x - 2y + 4z = 5$$
$$x + y + 3z = 2$$
$$-x + 2y - z = 1$$

[S.B.T.E. 2015]

Solution.

$$\Delta = \begin{vmatrix} 3 & -2 & 4 \\ 1 & 1 & 3 \\ -1 & 2 & -1 \end{vmatrix} = -5$$

$$\Delta_1 = \begin{vmatrix} 5 & -2 & 4 \\ 2 & 1 & 3 \\ 1 & 2 & -1 \end{vmatrix} = -33$$

$$\Delta_2 = \begin{vmatrix} 3 & 5 & 4 \\ 1 & 2 & 3 \\ -1 & 1 & -1 \end{vmatrix} = -13$$

$$\Delta_3 = \begin{vmatrix} 3 & -2 & 5 \\ 1 & 1 & 2 \\ -1 & 2 & -1 \end{vmatrix} = 12$$

$$x = \frac{\Delta_1}{\Delta} = \frac{-33}{-5} = \frac{33}{5}$$

$$y = \frac{\Delta_2}{\Delta} = \frac{-13}{-5} = \frac{13}{5}$$

$$z = \frac{\Delta_3}{\Delta} = \frac{12}{-5} = \frac{-12}{5}$$ **Ans.**

Example 36. *Solve, by determinants, the following set of simultaneous equations:*

$$5x - 6y + 4z = 15$$
$$7x + 4y - 3z = 19$$
$$2x + y + 6z = 46$$

Solution.

$$\Delta = \begin{vmatrix} 5 & -6 & 4 \\ 7 & 4 & -3 \\ 2 & 1 & 6 \end{vmatrix} = 419$$

$$\Delta_1 = \begin{vmatrix} 15 & -6 & 4 \\ 19 & 4 & -3 \\ 46 & 1 & 6 \end{vmatrix} = 1257$$

$$\Delta_2 = \begin{vmatrix} 5 & 15 & 4 \\ 7 & 19 & -3 \\ 2 & 46 & 6 \end{vmatrix} = 1676$$

$$\Delta_3 = \begin{vmatrix} 5 & -6 & 15 \\ 7 & 4 & 19 \\ 2 & 1 & 46 \end{vmatrix} = 2514$$

$$x = \frac{\Delta_1}{\Delta} = \frac{1257}{419} = 3$$

$$y = \frac{\Delta_2}{\Delta} = \frac{1676}{419} = 4$$

$$z = \frac{\Delta_3}{\Delta} = \frac{2514}{419} = 6.$$

Ans.

Example 37. *Solve the following set of equations by using Cramer's rule.*

$$2x - y + 3z = 9$$
$$x + y + z = 6$$
$$x - y + z = 2$$

[S.B.T.E. 2015, 2001]

Solution. Let $\Delta = \begin{vmatrix} 2 & -1 & 3 \\ 1 & 1 & 1 \\ 1 & -1 & 1 \end{vmatrix} = 2\,(1 + 1) + 1\,(1 - 1) + 3\,(-1 - 1) = 4 - 6 = -2$

$$\Delta_x = \begin{vmatrix} 9 & -1 & 3 \\ 6 & 1 & 1 \\ 2 & -1 & 1 \end{vmatrix} = 9\,(1 + 1) + 1\,(6 - 2) + 3\,(-6 - 2) = 18 + 4 - 24 = -2$$

$$\Delta_y = \begin{vmatrix} 2 & 9 & 3 \\ 1 & 6 & 1 \\ 1 & 2 & 1 \end{vmatrix} = 2\,(6 - 2) - 9\,(1 - 1) + 3\,(2 - 6) = 8 - 12 = -4$$

$$\Delta_z = \begin{vmatrix} 2 & -1 & 9 \\ 1 & 1 & 6 \\ 1 & -1 & 2 \end{vmatrix} = 2\,(2 + 6) + (2 - 6) + 9\,(-1 - 1) = 16 - 4 - 18 = -6$$

Now, $x = \dfrac{\Delta_x}{\Delta} = \dfrac{-2}{-2} = 1$

$$y = \frac{\Delta_y}{\Delta} = \frac{-4}{-2} = 2$$

$$z = \frac{\Delta_z}{\Delta} = \frac{-6}{-2} = 3$$

Hence, $x = 1$, $y = 2$ and $z = 3$. **Ans.**

EXERCISE 9.6

1. Solve the following system of equations using determinants only:

$$x + 2y + 5z = 23$$
$$3x + y + 4z = 26$$
$$6x + y + 7z = 47$$

Ans. $x = 4, y = 2, z = 3$

2. Solve the equations (by Cramer's rule):

$$x + y + z = 1$$
$$3x + 5y + 6z = 4$$
$$9x + 2y - 36z = 17$$

Ans. $x = \frac{1}{3}, y = 1, z = -\frac{1}{3}$

3. Using determinants, solve the following system of equations:

$$2y - z = 0$$
$$x + 3y = -4$$
$$3x + 4y = 3$$

Ans. $x = 5, y = -3, z = -6$

4. Apply Cramer's rule to solve the following equations:

$$x + y + z = -1$$
$$x + 2y + 3z = -4$$
$$x + 3y + 4z = -6$$

Ans. $x = 1, y = -1, z = -1$

5. Solve the equations (by Cramer's rule): **[S.B.T.E. 2014]**

$$x + y + z = 1$$
$$ax + by + cz = k$$
$$a^2x + b^2y - c^2z = k^2$$

provided that $a \neq b, b \neq c, c \neq a$.

Ans. $x = \frac{(b-k)(c-k)}{(b-a)(c-a)}, y = \frac{(a-k)(c-k)}{(a-b)(c-b)}, z = \frac{(a-k)(b-k)}{(a-c)(b-c)}$

6. Show that there are three real values of λ for which the equations:

$$(a - \lambda)\, x + by + cz = 0$$
$$bx + (c - \lambda)\, y + az = 0$$
$$cz + ay + (b - \lambda)\, z = 0$$

are simultaneously true, and that the product of these values of λ is

$$D = \begin{vmatrix} a & b & c \\ b & c & a \\ c & a & b \end{vmatrix}$$

9.12 RULE FOR MULTIPLICATION OF TWO DETERMINANTS

Multiply the elements of the first row of Δ_1 with the corresponding elements of the first, the second and the third row of Δ_2 respectively, to form first row of $\Delta_1\Delta_2$.

Their respective sums form the elements of the first row of $\Delta_1\Delta_2$. Similarly multiply the elements of the second row of Δ_1 with the corresponding elements of first, second and third row of Δ_2 to form the second row of $\Delta_1\Delta_2$ and so on.

Example 38. *Find the product*

$$\begin{vmatrix} a_1 & b_1 & c_1 \\ a_2 & b_2 & c_2 \\ a_3 & b_3 & c_3 \end{vmatrix} \times \begin{vmatrix} \alpha_1 & \beta_1 & \gamma_1 \\ \alpha_2 & \beta_2 & \gamma_2 \\ \alpha_3 & \beta_3 & \gamma_3 \end{vmatrix}$$

Solution. Product of the given determinants

$$\begin{vmatrix} a_1\alpha_1 + b_1\beta_1 + c_1\gamma_1 & a_1\alpha_2 + b_1\beta_2 + c_1\gamma_2 & a_1\alpha_3 + b_1\beta_3 + c_1\gamma_3 \\ a_2\alpha_1 + b_2\beta_1 + c_2\gamma_1 & a_2\alpha_2 + b_2\beta_2 + c_2\gamma_2 & a_2\alpha_3 + b_2\beta_3 + c_2\gamma_3 \\ a_3\alpha_1 + b_3\beta_1 + c_3\gamma_1 & a_3\alpha_1 + b_3\beta_1 + c_3\gamma_2 & a_3\alpha_3 + b_3\beta_3 + c_3\gamma_3 \end{vmatrix}$$ **Ans.**

Example 39. *Find* $\begin{vmatrix} a & b & c \\ b & c & a \\ c & a & b \end{vmatrix} \times \begin{vmatrix} -a & c & b \\ -b & a & c \\ -c & b & a \end{vmatrix}$

and hence show that

$$= \begin{vmatrix} 2bc - a^2 & c^2 & b^2 \\ c^2 & 2ca - b^2 & a^2 \\ b^2 & a^2 & 2ab - c^2 \end{vmatrix} = (a^3 + b^3 + c^3 - 3abc)^2$$

Solution. Product of the given determinants

$$= \begin{vmatrix} -a^2 + bc + bc & -ab + ab + c^2 & -ac + b^2 + ac \\ -ab + c^2 + ab & -b^2 + ac + ac & -bc + bc + a^2 \\ -ca + ca + b^2 & -bc + a^2 + bc & -c^2 + ab + ab \end{vmatrix}$$

$$= \begin{vmatrix} 2bc - a^2 & c^2 & b^2 \\ c^2 & 2ca - b^2 & a^2 \\ b^2 & a^2 & 2ab - c^2 \end{vmatrix}$$

Now $\begin{vmatrix} -a & c & b \\ -b & a & c \\ -c & b & a \end{vmatrix} = (-1)^2 \begin{vmatrix} a & b & c \\ b & c & a \\ c & a & b \end{vmatrix}$

$$= a(bc - a^2) - b(b^2 - ac) + c(ab - c^2)$$

$$= (a^3 + b^3 + c^3 - 3abc)$$

Product $= (a^3 + b^3 + c^3 - 3abc)^2$ **Proved**

Example 40. *Prove that the determinant*

$\begin{vmatrix} 2b_1 + c_1 & c_1 + 3a_1 & 2a_1 + 3b_1 \\ 2b_2 + c_2 & c_2 + 3a_2 & 2a_2 + 3b_2 \\ 2b_3 + c_3 & c_3 + 3a_3 & 2a_3 + 3b_3 \end{vmatrix}$ *is a multiple of the determinant*

$\begin{vmatrix} a_1 & b_1 & c_1 \\ a_2 & b_2 & c_2 \\ a_3 & b_3 & c_3 \end{vmatrix}$ *and find the other factor.*

Solution. $\begin{vmatrix} 2b_1+c_1 & c_1+3a_1 & 2a_1+3b_1 \\ 2b_2+c_2 & c_2+3a_2 & 2a_2+3b_2 \\ 2b_3+c_3 & c_3+3a_3 & 2a_3+3b_3 \end{vmatrix} = \begin{vmatrix} a_1 & b_1 & c_1 \\ a_2 & b_2 & c_2 \\ a_3 & b_3 & c_3 \end{vmatrix} \times \begin{vmatrix} 0 & 2 & 1 \\ 3 & 0 & 1 \\ 2 & 3 & 0 \end{vmatrix}$ **Ans.**

Example 41. *Prove that*

$$\begin{vmatrix} 1 & \cos(\beta-\alpha) & \cos(\gamma-\alpha) \\ \cos(\alpha-\beta) & 1 & \cos(\gamma-\beta) \\ \cos(\alpha-\gamma) & \cos(\beta-\gamma) & 1 \end{vmatrix} = 0$$

Solution. Let $\begin{vmatrix} \cos\alpha & \sin\alpha & 0 \\ \cos\beta & \sin\beta & 0 \\ \cos\gamma & \sin\gamma & 0 \end{vmatrix} \times \begin{vmatrix} \cos\alpha & \sin\alpha & 0 \\ \cos\beta & \sin\beta & 0 \\ \cos\gamma & \sin\gamma & 0 \end{vmatrix} = 0$

$$\Rightarrow \begin{vmatrix} \cos^2\alpha+\sin^2\alpha & \cos\alpha\cos\beta+\sin\alpha\sin\beta & \cos\alpha\cos\gamma+\sin\alpha\sin\gamma \\ \cos\beta\cos\alpha+\sin\beta\sin\alpha & \cos^2\beta+\sin^2\beta & \cos\beta\cos\gamma+\sin\beta\sin\gamma \\ \cos\gamma\cos\alpha+\sin\gamma\sin\alpha & \cos\gamma\cos\beta+\sin\gamma\sin\beta & \cos^2\gamma+\sin^2\gamma \end{vmatrix} = 0$$

$\Rightarrow \begin{vmatrix} 1 & \cos(\beta-\alpha) & \cos(\gamma-\alpha) \\ \cos(\alpha-\beta) & 1 & \cos(\gamma-\beta) \\ \cos(\alpha-\gamma) & \cos(\beta-\gamma) & 1 \end{vmatrix} = 0$ **Proved**

EXERCISE 9.7

1. For elements A, B, C, P, Q and R find the value of determinant

$\begin{vmatrix} \cos(A-P) & \cos(A-Q) & \cos(A-R) \\ \cos(B-P) & \cos(B-Q) & \cos(B-R) \\ \cos(C-P) & \cos(C-Q) & \cos(C-R) \end{vmatrix}$ **[Diploma IETE June 2006]**

2. *If A_1, A_2, A_3, B_1, B_2, B_3, C_1, C_2, C_3 are cofactors of the elements a_1, a_2, a_3, b_1, b_2, b_3, c_1, c_2, c_3 respectively of the determinant $(a_1\ b_2\ c_3)$, show that*

$$\begin{vmatrix} A_1 & B_1 & C_1 \\ A_2 & B_2 & C_2 \\ A_3 & B_3 & C_3 \end{vmatrix} = \begin{vmatrix} a_1 & b_1 & c_1 \\ a_2 & b_2 & c_2 \\ a_3 & b_3 & c_3 \end{vmatrix}^2$$

Matrices

10.1 INTRODUCTION

It is defined as a rectangular array of numbers, which acts as an information store. Matrices were discovered by Cayley, a French mathematician in 1860. The horizontal lines are known as **rows** and the vertical lines as **columns**.

Notation. The general element of *i*th row and *j*th column of a matrix is denoted by the letter a_{ij}, while matrix is denoted by the capital letter A or $[a_{ij}]$.

Example 1. *Here are the airline connections between the airports of two countries A and B.*

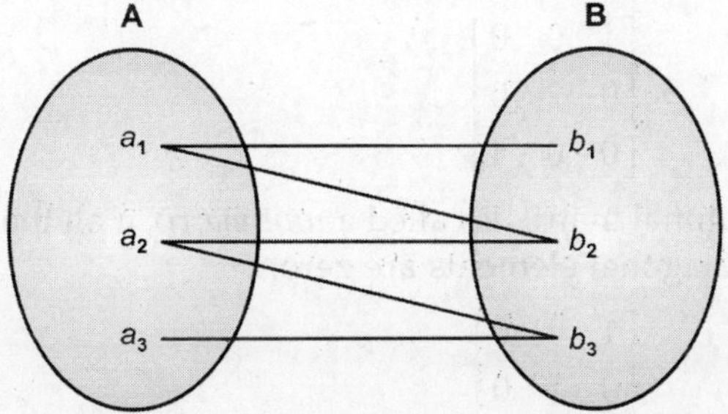

Solution. This information can be stored in the following manner:

$$\begin{array}{c} \\ a_1 \\ a_2 \\ a_3 \end{array}\begin{array}{c} \begin{array}{ccc} b_1 & b_2 & b_3 \end{array} \\ \begin{bmatrix} 1 & 1 & 0 \\ 0 & 1 & 1 \\ 0 & 0 & 1 \end{bmatrix} \end{array}$$

This system of numbers arranged in a rectangular array in rows and columns and bounded by the brackets [ ] is called a matrix.

Example 2. *Consider a system of equations:*

$$3x + 4y - 5z = 0$$
$$2x + 6y + 7z = 0$$

Solution. The coefficient of x, y, z can be written in following manner:

$$\begin{bmatrix} 3 & 4 & -5 \\ 2 & 6 & 7 \end{bmatrix}$$

which is a matrix of order 2×3 having 2 rows and 3 columns. (read as '2 by 3' matrix).

10.2 VARIOUS TYPES OF MATRICES

1. **Row Matrix:** If a matrix has only one row and any number of columns, it is called a *row matrix, i.e.*

$$[2 \quad 7 \quad 3 \quad 9]$$

2. **Column Matrix:** A matrix having one column and any number of rows is called a *column matrix, i.e.*

$$\begin{bmatrix} 1 \\ 3 \\ 2 \end{bmatrix}$$

3. **Square Matrix:** A matrix in which the number of rows is equal to the number of columns is called a *square matrix.*

$$\begin{bmatrix} 2 & 4 & 5 \\ 4 & 1 & 7 \\ 4 & 8 & 6 \end{bmatrix} \text{ which is a square matrix of order } 3 \times 3.$$

4. **Diagonal Matrix:** A square matrix is called a *diagonal matrix* if all its nondiagonal elements are zero, *i.e.*

$$\begin{bmatrix} 1 & 0 & 0 \\ 0 & 2 & 0 \\ 0 & 0 & 4 \end{bmatrix}$$

5. **Unit Matrix:** A diagonal matrix is called a *unit matrix* if all the diagonal elements are unity and nondiagonal elements are zero.

$$\begin{bmatrix} 1 & 0 & 0 \\ 0 & 1 & 0 \\ 0 & 0 & 1 \end{bmatrix}$$

6. **Zero Matrix or Null Matrix:** Any matrix in which all the elements are zero is called a *zero matrix or null matrix.*

$$\begin{bmatrix} 0 & 0 & 0 & 0 \\ 0 & 0 & 0 & 0 \end{bmatrix}$$

7. **Transpose of a Matrix:** A transpose of a matrix is obtained by interchanging the rows and the corresponding columns of a given matrix. The transpose of the matrix A is denoted by A', if

$$A = \begin{bmatrix} 2 & 3 & 4 \\ 5 & 6 & 7 \end{bmatrix} \text{ then } A' = \begin{bmatrix} 2 & 5 \\ 3 & 6 \\ 4 & 7 \end{bmatrix}$$

8. **Symmetric Matrix:** A square matrix is called symmetric, if $a_{ij} = a_{ji}$ or $A' = A$ for all values of i and j.

$$\begin{bmatrix} a & h & g \\ h & b & f \\ g & f & c \end{bmatrix}$$

9. **Skewsymmetric Matrix:** A square matrix is called skewsymmetric matrix if

(*i*) $a_{ij} = -a_{ji}$ for all values of i and j.

(*ii*) All diagonal elements are zero.

$$A' = -A$$

$$\begin{bmatrix} 0 & -h & -g \\ h & 0 & -f \\ g & f & 0 \end{bmatrix}$$

10. **Triangular Matrix:** A square matrix, all of whose elements below the leading diagonal are zero, is called an *upper triangular matrix*. A square matrix, all of whose elements above the leading diagonal are zero, is called a lower triangular matrix.

$$\begin{bmatrix} 1 & 2 & 3 \\ 0 & 4 & 1 \\ 0 & 0 & 2 \end{bmatrix}, \qquad \begin{bmatrix} 2 & 0 & 0 \\ 3 & 4 & 0 \\ 1 & 5 & 2 \end{bmatrix}$$

Upper triangular matrix   Lower triangular matrix

11. **Orthogonal Matrix:** A square matrix A is called an *orthogonal matrix* if the product of the matrix A and the transpose matrix A' is an identity matrix, *i.e.*

$$AA' = I$$

If $|A| = I$, matrix A is proper.

12. **Conjugate of a Matrix:**

$$\text{Let } A = \begin{bmatrix} 1+i & 2-3i & 4 \\ 7+2i & -i & 3-2i \end{bmatrix}$$

Elements	Conjugate element
$1+i$	$1-i$
$2-3i$	$2+3i$
4	4
$7+2i$	$7-2i$
$-i$	$+i$
$3-2i$	$3+2i$

Conjugate of matrix $A = \overline{A}$

$$\therefore \quad \overline{A} = \begin{bmatrix} 1-i & 2+3i & 4 \\ 7-2i & i & 3+2i \end{bmatrix}$$

13. **Matrix A^{θ}:** Transpose of the conjugate of a matrix A is denoted by A^{θ}.

$$\text{Let} \quad A = \begin{bmatrix} 1+i & 2-3i & 4 \\ 7+2i & -i & 3-2i \end{bmatrix}$$

$$\overline{A} = \begin{bmatrix} 1-i & 2+3i & 4 \\ 7-2i & +i & 3+2i \end{bmatrix}$$

$$(\bar{A})' = \begin{bmatrix} 1-i & 7-2i \\ 2+3i & i \\ 4 & 3+2i \end{bmatrix}$$

$\Rightarrow$ $$A^{\theta} = \begin{bmatrix} 1-i & 7-2i \\ 2+3i & i \\ 4 & 3+2i \end{bmatrix}$$

14. Unitary Matrix: A square matrix A is said to be *unitary* if

$$A^{\theta} A = I$$

e.g. $$A = \begin{bmatrix} \frac{1+i}{2} & \frac{-1+i}{2} \\ \frac{1+i}{2} & \frac{1-i}{2} \end{bmatrix}$$

15. Hermitian Matrix: A square matrix $A = (a_{ij})$ is called *Hermitian matrix* if every i-jth element of A is equal to conjugate complex of j-ith element of A.

In other words $a_{ij} = \bar{a}_{ji}$

e.g. $$\begin{bmatrix} 1 & 2+3i & 3+i \\ 2-3i & 2 & 1-2i \\ 3-i & 1+2i & 5 \end{bmatrix}$$

Necessary and sufficient condition for a matrix A to be Hermitian is that $A = A^{\theta}$ *i.e.* conjugate transpose of A $\Rightarrow A = (\bar{A})'$.

16. Skew Hermitian Matrix: A square matrix $A = (a_{ij})$ is called a *skew Hermitian matrix* if every i-jth element of A is equal to negative conjugate complex of j-ith element of A.

In other words $a_{ij} = -\bar{a}_{ji}$

All the elements in the principal diagonal are of the form:

$$a_{ii} = -\bar{a}_{ii}$$

$\Rightarrow$ $$a_{ii} + \bar{a}_{ii} = 0$$

If $a_{ii} = a + ib$ then $\bar{a}_{ii} = a - ib$

$$(a + ib) + (a - ib) = 0 \Rightarrow 2a = 0 \Rightarrow a = 0$$

so a_{ii} is pure imaginary $\Rightarrow a_{ii} = 0$

Hence all the diagonal elements of a skew Hermitian matrix are either zeros or pure imaginary.

e.g. $$\begin{bmatrix} i & 2-3i & 4+5i \\ -(2+3i) & 0 & 2i \\ -(4+5i) & 2i & -3i \end{bmatrix}$$

The necessary and sufficient condition for a matrix A to be skew Hermitian is that $A^{\theta} = -A$

$\Rightarrow$ $(\bar{A})' = -A$

17. **Idempotent Matrix:** A matrix such that $A^2 = A$ is called idempotent matrix.

e.g.
$$\begin{bmatrix} 2 & -2 & -4 \\ -1 & 3 & 4 \\ 1 & -2 & -3 \end{bmatrix}$$

18. **Periodic Matrix:** A matrix A will be called a periodic matrix if
$$A^{k+1} = A$$
where k is a + ve integer. If k is the least + ve integer for which $A^{k+1} = A$ then k is said to be the period of A. If we choose $k = 1$ we get $A^2 = A$ and is called to be idempotent matrix.

19. **Nilpotent Matrix:** A matrix will be called a nilpotent matrix if $A^k = 0$ (null matrix) where k is a + ve integer; however if k is the least + ve integer for which $A^k = 0$, then k is the *index* of the Nilpotent matrix.

e.g.
$$A = \begin{bmatrix} ab & b^2 \\ -a^2 & -ab \end{bmatrix}$$

A is a nilpotent matrix whose index is 2.

20. **Involuntary Matrix:** A matrix A will be called an involuntary matrix if $A^2 = I$ (unit matrix). Since $I^2 = I$ always

$\therefore$ Unit matrix is involuntary.

10.3 ADDITION OF MATRICES

If A, B, be two matrices of the same order, then their sum $A + B$ is defined as the matrix, each element of which is the sum of the corresponding elements of A and B.

Thus
$$\begin{bmatrix} 1 & 2 \\ 3 & 5 \\ 2 & 4 \end{bmatrix} + \begin{bmatrix} 3 & 4 \\ 4 & 3 \\ 5 & 6 \end{bmatrix} = \begin{bmatrix} 1+3 & 2+4 \\ 3+4 & 5+3 \\ 2+5 & 4+6 \end{bmatrix}$$
$$= \begin{bmatrix} 4 & 6 \\ 7 & 8 \\ 7 & 10 \end{bmatrix}$$

If
$$A = [a_{ij}],\ B = [b_{ij}]$$
then
$$A + B = [a_{ij} + b_{ij}]$$

Note: Only the matrices of the same order can be added.

Properties of Matrix Addition

1. **Commutative Law:** Let $A = [a_{ij}]$ and $B = [b_{ij}]$ be two matrices of the same order say $m \times n$.

   Then $A + B = [a_{ij} + b_{ij}]$ and $B + A = [b_{ij} + a_{ij}]$

   But $a_{ij} + b_{ij} = b_{ij} + a_{ij}$ as commutative law of addition holds good in real numbers.

   $A + B = B + A.$

   Hence matrix addition is commutative.

2. **Associative Law:** $A + (B + C) = (A + B) + C$

$$A + (B + C) = a_{ij} + (b_{ij} + c_{ij})$$

$$(A + B) + C = (a_{ij} + b_{ij}) + c_{ij}$$

But $a_{ij} + (b_{ij} + c_{ij}) = (a_{ij} + b_{ij}) + c_{ij}$

As associative law of addition holds good in real numbers.

$\therefore \quad A + (B + C) = (A + B) + C$

Hence matrix addition is associative.

10.4 SCALAR MULTIPLE OF A MATRIX

If a matrix is multiplied by a scalar quantity k, then each element is multiplied by k, *i.e.*

$$A = \begin{bmatrix} 2 & 3 & 4 \\ 4 & 5 & 6 \\ 6 & 7 & 9 \end{bmatrix}$$

$$3A = 3\begin{bmatrix} 2 & 3 & 4 \\ 4 & 5 & 6 \\ 6 & 7 & 9 \end{bmatrix} = \begin{bmatrix} 3\times2 & 3\times3 & 3\times4 \\ 3\times4 & 3\times5 & 3\times6 \\ 3\times6 & 3\times7 & 3\times9 \end{bmatrix} = \begin{bmatrix} 6 & 9 & 12 \\ 12 & 15 & 18 \\ 18 & 21 & 27 \end{bmatrix}$$

Example 3. *Find (i) $A + B$ (ii) $A - B$ (iii) $3A + 2B$.*

If $A = \begin{bmatrix} 5 & 4 & 5 \\ 1 & 2 & 1 \\ 2 & 2 & 3 \\ 1 & 1 & 1 \end{bmatrix}$ *and* $B = \begin{bmatrix} 3 & 2 & 3 \\ 2 & 1 & 1 \\ 3 & 3 & 4 \\ 1 & 0 & 0 \end{bmatrix}$

Solution.

$$(i)\ A + B = \begin{bmatrix} 5 & 4 & 5 \\ 1 & 2 & 1 \\ 2 & 2 & 3 \\ 1 & 1 & 1 \end{bmatrix} + \begin{bmatrix} 3 & 2 & 3 \\ 2 & 1 & 1 \\ 3 & 3 & 4 \\ 1 & 0 & 0 \end{bmatrix} = \begin{bmatrix} 8 & 6 & 8 \\ 3 & 3 & 2 \\ 5 & 5 & 7 \\ 2 & 1 & 1 \end{bmatrix}$$

$$(ii)\ A - B = \begin{bmatrix} 5 & 4 & 5 \\ 1 & 2 & 1 \\ 2 & 2 & 3 \\ 1 & 1 & 1 \end{bmatrix} - \begin{bmatrix} 3 & 2 & 3 \\ 2 & 1 & 1 \\ 3 & 3 & 4 \\ 1 & 0 & 0 \end{bmatrix} = \begin{bmatrix} 2 & 2 & 2 \\ -1 & 1 & 0 \\ -1 & -1 & -1 \\ 0 & 1 & 1 \end{bmatrix}$$

$$(iii)\ 3A + 2B = 3\begin{bmatrix} 5 & 4 & 5 \\ 1 & 2 & 1 \\ 2 & 2 & 3 \\ 1 & 1 & 1 \end{bmatrix} + 2\begin{bmatrix} 3 & 2 & 3 \\ 2 & 1 & 1 \\ 3 & 3 & 4 \\ 1 & 0 & 0 \end{bmatrix}$$

$$= \begin{bmatrix} 15 & 12 & 15 \\ 3 & 6 & 3 \\ 6 & 6 & 9 \\ 3 & 3 & 3 \end{bmatrix} + \begin{bmatrix} 6 & 4 & 6 \\ 4 & 2 & 2 \\ 6 & 6 & 8 \\ 2 & 0 & 0 \end{bmatrix} = \begin{bmatrix} 21 & 16 & 21 \\ 7 & 8 & 5 \\ 12 & 12 & 17 \\ 5 & 3 & 3 \end{bmatrix}$$

Example 4. *Prove that (A + B)′ = A′ + B′.*

$$A = \begin{bmatrix} 1 & 1 & 3 \\ 3 & 2 & 1 \\ 4 & 3 & 5 \end{bmatrix} \text{ and } B = \begin{bmatrix} 1 & 2 & 0 \\ 0 & 1 & 2 \\ 2 & 0 & 1 \end{bmatrix}$$

Solution. $$A = \begin{bmatrix} 1 & 1 & 3 \\ 3 & 2 & 1 \\ 4 & 3 & 5 \end{bmatrix} \therefore A' = \begin{bmatrix} 1 & 3 & 4 \\ 1 & 2 & 3 \\ 3 & 1 & 5 \end{bmatrix}$$

$$B = \begin{bmatrix} 1 & 2 & 0 \\ 0 & 1 & 2 \\ 2 & 0 & 1 \end{bmatrix} \therefore B' = \begin{bmatrix} 1 & 0 & 2 \\ 2 & 1 & 0 \\ 0 & 2 & 1 \end{bmatrix}$$

$$A + B = \begin{bmatrix} 1 & 1 & 3 \\ 3 & 2 & 1 \\ 4 & 3 & 5 \end{bmatrix} + \begin{bmatrix} 1 & 2 & 0 \\ 0 & 1 & 2 \\ 2 & 0 & 1 \end{bmatrix} = \begin{bmatrix} 2 & 3 & 3 \\ 3 & 3 & 3 \\ 6 & 3 & 6 \end{bmatrix}$$

$$[A + B]' = \begin{bmatrix} 2 & 3 & 6 \\ 3 & 3 & 3 \\ 3 & 3 & 6 \end{bmatrix} \quad \text{... (1)}$$

$$A' + B' = \begin{bmatrix} 1 & 3 & 4 \\ 1 & 2 & 3 \\ 3 & 1 & 5 \end{bmatrix} + \begin{bmatrix} 1 & 0 & 2 \\ 2 & 1 & 0 \\ 0 & 2 & 1 \end{bmatrix} = \begin{bmatrix} 2 & 3 & 6 \\ 3 & 3 & 3 \\ 3 & 3 & 6 \end{bmatrix} \quad \text{... (2)}$$

From (1) and (2), we get $(A + B)' = A' + B'$ **Proved.**

Example 5. *Define equal matrices. Find the values, x, y, z and a which satisfy the matrix equation.*

$$\begin{bmatrix} x+3 & 2y+x \\ z-1 & 4a-6 \end{bmatrix} = \begin{bmatrix} 0 & -7 \\ 3 & 2a \end{bmatrix}$$

Solution. Two matrices are equal if the corresponding elements of the matrices are equal.

$x + 3 = 0$, *i.e.* $x = -3$

$z - 1 = 3$, *i.e.* $z = 4$

$2y + x = -7$, *i.e.* $y = -2$

$4a - 6 = 2a$, *i.e.* $a = 3$ **Ans.**

Example 6. *Find the values of x, y and z if*

$$\begin{bmatrix} x & 3 \\ 4 & 2 \end{bmatrix} + \begin{bmatrix} 2 & y \\ 1 & -2 \end{bmatrix} - \begin{bmatrix} 1 & 1 \\ -2 & z \end{bmatrix} = \begin{bmatrix} 5 & 0 \\ 7 & 3 \end{bmatrix}$$

Solution. We have,

$$\begin{bmatrix} x & 3 \\ 4 & 2 \end{bmatrix} + \begin{bmatrix} 2 & y \\ 1 & -2 \end{bmatrix} - \begin{bmatrix} 1 & 1 \\ -2 & z \end{bmatrix} = \begin{bmatrix} 5 & 0 \\ 7 & 3 \end{bmatrix}$$ (by interchanging row and columns)

$$\Rightarrow \quad \begin{bmatrix} x+2-1 & 3+y-1 \\ 4+1+2 & 2-2-z \end{bmatrix} = \begin{bmatrix} 5 & 0 \\ 7 & 3 \end{bmatrix}$$

$$\Rightarrow \quad \begin{bmatrix} x+1 & y+2 \\ 7 & -z \end{bmatrix} = \begin{bmatrix} 5 & 0 \\ 7 & 3 \end{bmatrix}$$

We know that two matrices are equal if the corresponding elements of the matrices are equal

$$\Rightarrow \quad x+1 = 5 \quad \Rightarrow \quad x = 4$$

$$y + 2 = 0 \quad \Rightarrow \quad y = -2$$

$$-z = 3 \quad \Rightarrow \quad z = -3$$

Ans.

Example 7. *Show that any square matrix can be expressed as the sum of two matrices, one symmetric and the other skew symmetric.*

(Diploma IETE, Dec, 2005, June 2005)

Solution. Let A be a given square matrix

Then $\quad A = \frac{1}{2}(A + A') + \frac{1}{2}(A - A')$

Now $\quad (A + A')' = A' + A = A + A'$

$\therefore \quad A + A'$ is a symmetric matrix.

$\Rightarrow \quad \frac{1}{2}(A + A')$ is a symmetric matrix.

Also, $\quad (A - A') = A' - A$

$\quad = -(A - A')$

$\therefore \quad A - A$ or $\frac{1}{2}(A - A')$ is a skew symmetric matrix.

$\therefore \quad A = \frac{1}{2}(A + A') + \frac{1}{2}(A - A')$

= Symmetric matrix + skew symmetric matrix. **Proved.**

Symmetric matrix = $\frac{1}{2}(A + A')$

Skew symmetric matrix = $\frac{1}{2}(A - A')$

Example 8. *Express the following matrix as a sum of a symmetric and skew symmetric.*

$$\begin{bmatrix} -1 & 7 & 1 \\ 2 & 3 & 4 \\ 5 & 0 & 5 \end{bmatrix}$$

Solution. Let $\quad A = \begin{bmatrix} -1 & 7 & 1 \\ 2 & 3 & 4 \\ 5 & 0 & 5 \end{bmatrix}$, then $A' = \begin{bmatrix} -1 & 2 & 5 \\ 7 & 3 & 0 \\ 1 & 4 & 5 \end{bmatrix}$

$$A + A' = \begin{bmatrix} -1 & 7 & 1 \\ 2 & 3 & 4 \\ 5 & 0 & 5 \end{bmatrix} + \begin{bmatrix} -1 & 2 & 5 \\ 7 & 3 & 0 \\ 1 & 4 & 5 \end{bmatrix} = \begin{bmatrix} -2 & 9 & 6 \\ 9 & 6 & 4 \\ 6 & 4 & 10 \end{bmatrix}$$

$$A - A' = \begin{bmatrix} -1 & 7 & 1 \\ 2 & 3 & 4 \\ 5 & 0 & 5 \end{bmatrix} - \begin{bmatrix} -1 & 2 & 5 \\ 7 & 3 & 0 \\ 1 & 4 & 5 \end{bmatrix} = \begin{bmatrix} 0 & 5 & -4 \\ -5 & 0 & 4 \\ 4 & -4 & 0 \end{bmatrix}$$

We know that

$$A = \frac{1}{2}(A + A') + \frac{1}{2}(A - A')$$

$$\begin{bmatrix} -1 & 7 & 1 \\ 2 & 3 & 4 \\ 5 & 0 & 5 \end{bmatrix} = \frac{1}{2}\begin{bmatrix} -2 & 9 & 6 \\ 9 & 6 & 4 \\ 6 & 4 & 10 \end{bmatrix} + \frac{1}{2}\begin{bmatrix} 0 & 5 & -4 \\ -5 & 0 & 4 \\ 4 & -4 & 0 \end{bmatrix}$$ **Proved.**

Symmetric matrix Skew symmetric matrix

Example 9. *Express the matrix* $A = \begin{bmatrix} 3 & 2 & 3 \\ 4 & 5 & 3 \\ 2 & 4 & 5 \end{bmatrix}$ *as the sum of symmetric and skew symmetric matrix.* ***(S.B.T.E. 2016, 2015, 2014)***

Solution. Here we have

$$A = \begin{bmatrix} 3 & 2 & 3 \\ 4 & 5 & 3 \\ 2 & 4 & 5 \end{bmatrix}, A' = \begin{bmatrix} 3 & 4 & 2 \\ 2 & 5 & 4 \\ 3 & 3 & 5 \end{bmatrix}$$

$$\text{Symmetric} = \frac{A + A'}{2} = \frac{1}{2}\left\{\begin{bmatrix} 3 & 2 & 3 \\ 4 & 5 & 3 \\ 2 & 4 & 5 \end{bmatrix} + \begin{bmatrix} 3 & 4 & 2 \\ 2 & 5 & 4 \\ 3 & 3 & 5 \end{bmatrix}\right\} = \frac{1}{2}\begin{bmatrix} 6 & 6 & 5 \\ 6 & 10 & 7 \\ 5 & 7 & 10 \end{bmatrix}$$

$$\text{Skew symmetric} = \frac{A - A'}{2} = \frac{1}{2}\left\{\begin{bmatrix} 3 & 2 & 3 \\ 4 & 5 & 3 \\ 2 & 4 & 5 \end{bmatrix} - \begin{bmatrix} 3 & 4 & 2 \\ 2 & 5 & 4 \\ 3 & 3 & 5 \end{bmatrix}\right\}$$

$$= \frac{1}{2}\begin{bmatrix} 0 & -2 & 1 \\ 2 & 0 & 1 \\ -1 & 1 & 0 \end{bmatrix}$$ **Ans.**

EXERCISE 10.1

1. If $A = \begin{bmatrix} 2 & 1 & 2 \\ 1 & 0 & 1 \\ 3 & 1 & 1 \end{bmatrix}$. Find (*i*) $2A$ (*ii*) A' **Ans.** (*i*) $\begin{bmatrix} 4 & 2 & 4 \\ 2 & 0 & 2 \\ 6 & 2 & 2 \end{bmatrix}$ (*ii*) $\begin{bmatrix} 2 & 1 & 3 \\ 1 & 0 & 1 \\ 2 & 1 & 1 \end{bmatrix}$

2. Express the given matrix as a sum of symmetric and skew-symmetric matrices:

$$\begin{bmatrix} 5 & 3 & 2 \\ 1 & 6 & 7 \\ 4 & 0 & 9 \end{bmatrix}$$

Ans. $\begin{bmatrix} 5 & 3 & 2 \\ 2 & 6 & \frac{7}{2} \\ 3 & \frac{7}{2} & 9 \end{bmatrix} + \begin{bmatrix} 0 & 1 & -1 \\ -1 & 0 & \frac{7}{2} \\ 1 & \frac{-7}{2} & 0 \end{bmatrix}$

3. If $A = \begin{bmatrix} 1 & 4 & 7 \\ 2 & 5 & 8 \\ 3 & 6 & 9 \end{bmatrix}, B = \begin{bmatrix} 1 & 2 & 3 \\ 4 & 5 & 6 \\ 7 & 8 & 0 \end{bmatrix}, C = \begin{bmatrix} 0 & 1 & 2 \\ 3 & 4 & 5 \\ 6 & 7 & 8 \end{bmatrix}$

Prove that (*i*) $A + (B + C) = (A + B) + C$ (*ii*) $(A + B)' = A' + B'$

4. Find x, y, z, w if

$$3\begin{bmatrix} x & y \\ z & w \end{bmatrix} = \begin{bmatrix} x & 6 \\ -1 & 2w \end{bmatrix} + \begin{bmatrix} 4 & x+y \\ z+w & 3 \end{bmatrix}$$

Ans. $x = 2, y = 4, z = 1, w = 3$

5. Find x, y, z, w if

$$\begin{bmatrix} x-y & 2x+z \\ 2x-y & 3z+w \end{bmatrix} = \begin{bmatrix} -1 & 5 \\ 0 & 13 \end{bmatrix}$$

(*B.T.E. December 2006*) **Ans.** $x = 1, y = 2, z = 3, w = 4$

6. If $A = \begin{bmatrix} 1 & 5 & 6 \\ -6 & 7 & 0 \end{bmatrix}$ and $B = \begin{bmatrix} 1 & -5 & 7 \\ 8 & -7 & 7 \end{bmatrix}$ find $A + B$. **(*S.B.T.E. 2015*)**

Ans. $A + B = \begin{bmatrix} 1 & 0 & 13 \\ 2 & 0 & 7 \end{bmatrix}$

7. If $A + B = \begin{bmatrix} 1 & 5 & 6 \\ -6 & 7 & 0 \end{bmatrix}$ and $B = \begin{bmatrix} 1 & -5 & 7 \\ 8 & -7 & 7 \end{bmatrix}$ find $A - B$. **(*S.B.T.E. 2015*)**

Ans. $A - B = \begin{bmatrix} 0 & 10 & -1 \\ -14 & 14 & -7 \end{bmatrix}$

8. Find a, b, c and d if $\begin{bmatrix} 2a+b & a-2b \\ 5c-d & 4c+3d \end{bmatrix} = \begin{bmatrix} 4 & -3 \\ 11 & 24 \end{bmatrix}$ **(*S.B.T.E. 2015*)**

Ans. $a = 1, b = 1, c = 3, d = 4$

9. If $x\begin{bmatrix} 2 \\ 3 \end{bmatrix} + y\begin{bmatrix} -1 \\ 1 \end{bmatrix} = \begin{bmatrix} 10 \\ 5 \end{bmatrix}$, find $x + y$. **(*S.B.T.E. 2015*)**

Ans. $x + y = 3 - 4 = -1$

10. Compute AB, if $A = \begin{bmatrix} 1 & 2 & 3 \\ 4 & 5 & 6 \end{bmatrix}, B = \begin{bmatrix} 2 & 5 & 3 \\ 3 & 6 & 4 \\ 4 & 7 & 5 \end{bmatrix}$ **(*S.B.T.E. 2015*)**

Ans. $AB = \begin{bmatrix} 20 & 38 & 26 \\ 47 & 92 & 62 \end{bmatrix}$

11. Show that $A = \begin{bmatrix} 0 & 5 & 7 \\ -5 & 0 & 8 \\ -7 & -8 & 0 \end{bmatrix}$ is skew symmetric matrix. **(S.B.T.E. 2016)**

12. Is it possible to define the matrix $A + B$, If
    (*i*) A has 2 columns and B has 1 column. **Ans.** No
    (*ii*) A has 2 rows and B has 3 rows. **Ans.** No
    (*iii*) Both A and B are square matrices of the same order. **Ans.** Always

Choose the correct alternative:

13. If A is any real square matrix then $A + A'$ is
    (*a*) Hermitian (*b*) Skew-Hermitian (*c*) Symmetric (*d*) Skew-symmetric
    (Diploma IETE, Dec. 2005) **Ans.** (c)

10.5 MULTIPLICATION OF MATRIX

The product of two matrices A and B is only possible if the number of columns in A is equal to the number of rows in B, then the matrices A and B are said to be *conformable*.

The method of multiplication will be clear from the following examples.

Example 10. *If* $A = \begin{bmatrix} 1 & 0 & 2 \\ 1 & 2 & 1 \\ 2 & 3 & 1 \end{bmatrix}$ *and* $B = \begin{bmatrix} 1 & 2 \\ 3 & 0 \\ 4 & 1 \end{bmatrix}$

Find AB and BA.

Solution. $AB = \begin{bmatrix} [1 \; 0 \; 2] & R_1 \\ [1 \; 2 \; 1] & R_2 \\ [2 \; 3 \; 1] & R_3 \end{bmatrix} \times \begin{bmatrix} \overset{C_1}{\begin{bmatrix}1\\3\\4\end{bmatrix}} & \overset{C_2}{\begin{bmatrix}2\\0\\1\end{bmatrix}} \end{bmatrix}$

R_1, R_2, R_3 are rows of A and C_1, C_2 are columns of B.

$$\therefore \quad AB = \begin{bmatrix} R_1C_1 & R_1C_2 \\ R_2C_1 & R_2C_2 \\ R_3C_1 & R_3C_2 \end{bmatrix}$$

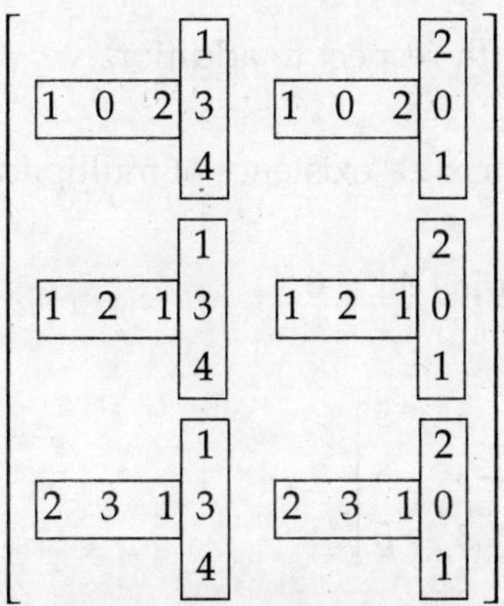

For convenience of multiplication, we write the columns in horizontal rectangles.

$$= \begin{bmatrix} \boxed{\begin{matrix}1 & 0 & 2\\ 1 & 3 & 4\end{matrix}} & \boxed{\begin{matrix}1 & 0 & 2\\ 2 & 0 & 1\end{matrix}} \\ \boxed{\begin{matrix}1 & 2 & 1\\ 1 & 3 & 4\end{matrix}} & \boxed{\begin{matrix}1 & 2 & 1\\ 2 & 0 & 1\end{matrix}} \\ \boxed{\begin{matrix}2 & 3 & 1\\ 1 & 3 & 4\end{matrix}} & \boxed{\begin{matrix}2 & 3 & 1\\ 2 & 0 & 1\end{matrix}} \end{bmatrix}$$

$$= \begin{bmatrix} 1\times1+0\times3+2\times4 & 1\times2+0\times0+2\times1 \\ 1\times1+2\times3+1\times4 & 1\times2+2\times0+1\times1 \\ 2\times1+3\times3+1\times4 & 2\times2+3\times0+1\times1 \end{bmatrix} = \begin{bmatrix} 1+0+8 & 2+0+2 \\ 1+6+4 & 2+0+1 \\ 2+9+4 & 4+0+1 \end{bmatrix} = \begin{bmatrix} 9 & 4 \\ 11 & 3 \\ 15 & 5 \end{bmatrix} \textbf{ Ans.}$$

$$BA = \begin{bmatrix} 1 & 2 \\ 3 & 0 \\ 4 & 1 \end{bmatrix} \times \begin{bmatrix} 1 & 0 & 2 \\ 1 & 2 & 1 \\ 2 & 3 & 1 \end{bmatrix}$$

But the number of columns in B is not equal to the number of rows in A, *i.e.* B and A are not conformable so multiplication of B and A is not possible.

Hence BA does not exist.

In general, if

A be $m \times n$ matrix and B be $n \times p$ matrix, and $AB = C$,

$$A = [a_{ij}],\ B = [b_{ij}],\ C = [c_{ij}]$$

where
$$c_{ij} = a_{i1}\, b_{1j} + a_{i2}\, b_{2j} + a_{i3}\, b_{3j} + \dots\ a_{in}\, b_{nj}$$

$$= \sum_{k=1}^{n} a_{ik}\, b_{kj}$$

Properties of Matrix Multiplication

1. Multiplication of matrices is not commutative.
$$AB \neq BA$$
2. Matrix multiplication is associative if conformability is assured.
$$A\,(BC) = (AB)\,C$$
3. Matrix multiplication is distributive with respect to addition.
$$A\,(B + C) = AB + AC$$
4. Multiplication of matrix A by a unit matrix, *i.e.* existence of multiplicative identity.
$$AI = IA = A$$
5. Multiplicative inverse of a matrix exist if $|A| \neq 0$.
$$A\,A^{-1} = A^{-1}\,A = I$$

Example 11. *If* $A = \begin{bmatrix} 2 & -3 & -5 \\ -1 & 4 & 5 \\ 1 & -3 & -4 \end{bmatrix}, B = \begin{bmatrix} -1 & 3 & 5 \\ 1 & -3 & -5 \\ -1 & 3 & 5 \end{bmatrix}$

find the product matrix AB.

Solution. Here, we have

$$A=\begin{bmatrix}2 & -3 & -5\\ -1 & 4 & 5\\ 1 & -3 & -4\end{bmatrix}, B=\begin{bmatrix}-1 & 3 & 5\\ 1 & -3 & -5\\ -1 & 3 & 5\end{bmatrix}$$

$$AB=\begin{bmatrix}-2-3+5 & 6+9-15 & 10+15-25\\ 1+4-5 & -3-12+15 & -5-20+25\\ -1-3+4 & 3+9-12 & 5+15-20\end{bmatrix}=\begin{bmatrix}0 & 0 & 0\\ 0 & 0 & 0\\ 0 & 0 & 0\end{bmatrix}=0$$ **Ans.**

Example 12. *If* $A=\begin{bmatrix}1 & 2 & 0\\ 2 & 0 & 6\\ 3 & 4 & 7\end{bmatrix}, B=\begin{bmatrix}3 & 2 & 1\\ 4 & 1 & 2\\ 5 & 0 & 1\end{bmatrix}$

compute AB and BA and prove that $AB \neq BA$.

Proof. $AB=\begin{bmatrix}1 & 2 & 0\\ 2 & 0 & 6\\ 3 & 4 & 7\end{bmatrix}\begin{bmatrix}3 & 2 & 1\\ 4 & 1 & 2\\ 5 & 0 & 1\end{bmatrix}$

$$=\begin{bmatrix}1\times3+2\times4+0\times5 & 1\times2+2\times1+0\times0 & 1\times1+2\times2+0\times1\\ 2\times3+0\times4+6\times5 & 2\times2+0\times1+6\times0 & 2\times1+0\times2+6\times1\\ 3\times3+4\times4+7\times5 & 3\times2+4\times1+7\times0 & 3\times1+4\times2+7\times1\end{bmatrix}$$

$$=\begin{bmatrix}3+8+0 & 2+2+0 & 1+4+0\\ 6+0+30 & 4+0+0 & 2+0+6\\ 9+16+35 & 6+4+0 & 3+8+7\end{bmatrix}=\begin{bmatrix}11 & 4 & 5\\ 36 & 4 & 8\\ 60 & 10 & 18\end{bmatrix}$$

$$BA=\begin{bmatrix}3 & 2 & 1\\ 4 & 1 & 2\\ 5 & 0 & 1\end{bmatrix}\begin{bmatrix}1 & 2 & 0\\ 2 & 0 & 6\\ 3 & 4 & 7\end{bmatrix}$$

$$=\begin{bmatrix}3\times1+2\times2+1\times3 & 3\times2+2\times0+1\times4 & 3\times0+2\times6+1\times7\\ 4\times1+1\times2+2\times3 & 4\times2+1\times0+2\times4 & 4\times0+1\times6+2\times7\\ 5\times1+0\times2+1\times3 & 5\times2+0\times0+1\times4 & 5\times0+0\times6+1\times7\end{bmatrix}$$

$$=\begin{bmatrix}3+4+3 & 6+0+4 & 0+12+7\\ 4+2+6 & 8+0+8 & 0+6+14\\ 5+0+3 & 10+0+4 & 0+0+7\end{bmatrix}=\begin{bmatrix}10 & 10 & 19\\ 12 & 16 & 20\\ 8 & 14 & 7\end{bmatrix}$$

$\therefore$ $AB \neq BA$

Hence multiplication of matrices is not commutative. **Proved.**

Example 13. *If* $A=\begin{bmatrix}1 & -1 & 1\\ -3 & 2 & -1\\ -2 & 1 & 0\end{bmatrix}, B=\begin{bmatrix}1 & 2 & 3\\ 2 & 4 & 6\\ 1 & 2 & 3\end{bmatrix}$

Compute AB and prove that AB = 0 does not imply A = 0 or B = 0.

(*Diploma IETE, June 2005*)

Solution. $AB = \begin{bmatrix} 1 & -1 & 1 \\ -3 & 2 & -1 \\ -2 & 1 & 0 \end{bmatrix} \begin{bmatrix} 1 & 2 & 3 \\ 2 & 4 & 6 \\ 1 & 2 & 3 \end{bmatrix}$

$$= \begin{bmatrix} 1\times1+(-1)\times2+1\times1 & 1\times2+(-1)\times4+1\times2 & 1\times3+(-1)\times6+1\times3 \\ -3\times1+2\times2+(-1)\times1 & (-3)\times2+2\times4+(-1)\times2 & (-3)\times3+2\times6+(-1)\times3 \\ (-2)\times1+1\times2+0\times1 & (-2)\times2+1\times4+0\times2 & (-2)\times3+1\times6+0\times3 \end{bmatrix}$$

$$= \begin{bmatrix} 1-2+1 & 2-4+2 & 3-6+3 \\ -3+4-1 & -6+8-2 & -9+12-3 \\ -2+2+0 & -4+4+0 & -6+6+0 \end{bmatrix} = \begin{bmatrix} 0 & 0 & 0 \\ 0 & 0 & 0 \\ 0 & 0 & 0 \end{bmatrix} = 0$$ **Proved.**

$AB = 0$, *i.e.* AB is a null matrix, does not necessarily imply that either A or $B = 0$ as shown above because neither A nor B is a null matrix whereas AB is a null matrix.

Example 14. *Prove that*

$$(AB)' = B' \cdot A'$$ **(Diploma IETE, Dec. 2005)**

Proof. Let $A = [a_{ij}]_{m \times n}$

$B = [b_{ij}]_{n \times p}$

Hence $C = AB = [c_{ij}]_{m \times p}$

$c_{ij} = i - j$th element of AB

$$= \sum_{k=I}^{n} a_{ik}\, b_{kj}$$

$\therefore$ $j - i$th element of $(AB)'_{p \times m}$

$= i - j$th element of $(AB)_{m \times p}$

$$= \sum_{k=I}^{n} a_{ik}\, b_{kj} \qquad \text{... (1)}$$

Now B' is $p \times n$ matrix.

A' is $n \times m$ matrix.

$j - i$th element of $B' \cdot A'$ = (jth row of B') $\times$ (ith column of A')

= (jth column of B) $\times$ (ith row of A)

= (ith row of A) $\times$ (jth column of B)

$$= \sum_{k=I}^{k=n} a_{ik}\, b_{kj} \qquad \text{... (2)}$$

Hence from (1) and (2) we conclude

$j - i$th element of $(AB)' = j - i$th element of $B' \cdot A'$.

$$(AB)' = B' \cdot A'$$ **Proved**

Example 15. *Prove that*

$$(AB)^n = A^n \cdot B^n \quad \text{if } AB = BA$$

Proof. $(AB)^1 = (AB) = (A) \cdot (B)$

$(AB)^2 = (AB) \cdot (AB)$

$= (ABA) \cdot B$ $\qquad (\because BA = AB)$

$$= (AAB) \cdot B$$
$$= (A^2B) \cdot B$$
$$= A^2 (B \cdot B)$$
$$= A^2 \cdot B^2$$

Suppose

$$(AB)^n = A^n \cdot B^n$$
$$(AB)^{n+1} = (AB^n) \cdot (AB)$$
$$= (A^n \cdot B^n) \cdot (AB)$$
$$= A^n \cdot (B^n \cdot A) \cdot B$$
$$= A^n \cdot (B^{n-1} \cdot BA) \cdot B$$
$$= A^n \cdot (B^{n-1} \cdot AB) \cdot \mathrm{B}$$
$$= A^n \cdot (B^{n-2} \cdot BA \cdot B) \cdot B$$
$$= A^n \cdot (B^{n-2}\, AB \cdot B) \cdot B$$
$$= A^n \cdot (B^{n-2} \cdot AB^2) \cdot B$$

Continuing the process n times

$$(AB)^{n+1} = A^n\,(A \cdot B^n)\,B$$
$$= A^n \cdot (A \cdot B^{n+1})$$
$$= A^{n+1} \cdot B^{n+1}.$$

Hence taking the above to be true for $n = n$, we have shown that it is true for $n = n + 1$ and also it was true for $n = 1, 2$ and hence it is universally true. **Proved.**

Example 16. *Compute the elements a_{43} and a_{22} of the matrix:*

$$A = \begin{bmatrix} 0 & 1 & 0 \\ 2 & 0 & 2 \\ 0 & 3 & 0 \\ 4 & 0 & 4 \end{bmatrix} \begin{bmatrix} 2 & -1 \\ -3 & 2 \\ 4 & 3 \end{bmatrix} \begin{bmatrix} 0 & 1 & -1 & 2 & -2 \\ 3 & -3 & 4 & -4 & 0 \end{bmatrix}$$

Solution.

$$A = \begin{bmatrix} 0 & 1 & 0 \\ 2 & 0 & 2 \\ 0 & 3 & 0 \\ 4 & 0 & 4 \end{bmatrix} \begin{bmatrix} -3 & 5 & -6 & 8 & -4 \\ 6 & -9 & 11 & -14 & 6 \\ 9 & -5 & 8 & -4 & -8 \end{bmatrix}$$

$$a_{43} = \text{(4th row of I matrix)} \times \text{(3rd column of II matrix)}$$
$$= (4)(-6) + (11)(0) + (4)(8) = -24 + 32 = 8$$
$$a_{22} = \text{(2nd row of I matrix)} \times \text{(2nd column of II matrix)}$$
$$= (2)(5) + (0)(-9) + (2)(-5) = 10 - 10 = 0.$$

Ans.

Example 17. *Show that the matrix*

$$A = \begin{bmatrix} 1 & 2 & 2 \\ 2 & 1 & 2 \\ 2 & 2 & 1 \end{bmatrix} \text{ satisfies the equation}$$

$A^2 - 4A - 5I = 0$ and hence find A^{-1}, where I is a unit matrix of 3×3.

Solution. We have

$$A = \begin{bmatrix} 1 & 2 & 2 \\ 2 & 1 & 2 \\ 2 & 2 & 1 \end{bmatrix}$$

$$A^2 = \begin{bmatrix} 1 & 2 & 2 \\ 2 & 1 & 2 \\ 2 & 2 & 1 \end{bmatrix}\begin{bmatrix} 1 & 2 & 2 \\ 2 & 1 & 2 \\ 2 & 2 & 1 \end{bmatrix} = \begin{bmatrix} 1+4+4 & 2+2+4 & 2+4+2 \\ 2+2+4 & 4+1+4 & 4+2+2 \\ 2+4+2 & 4+2+2 & 4+4+1 \end{bmatrix}$$

$$= \begin{bmatrix} 9 & 8 & 8 \\ 8 & 9 & 8 \\ 8 & 8 & 9 \end{bmatrix}$$

$$-4A = (-4)\begin{bmatrix} 1 & 2 & 2 \\ 2 & 1 & 2 \\ 2 & 2 & 1 \end{bmatrix} = \begin{bmatrix} -4 & -8 & -8 \\ -8 & -4 & -8 \\ -8 & -8 & -4 \end{bmatrix}$$

and $$-5I = (-5)\begin{bmatrix} 1 & 0 & 0 \\ 0 & 1 & 0 \\ 0 & 0 & 1 \end{bmatrix} = \begin{bmatrix} -5 & 0 & 0 \\ 0 & -5 & 0 \\ 0 & 0 & -5 \end{bmatrix}$$

Now, $$A^2 - 4A - 5I = \begin{bmatrix} 9 & 8 & 8 \\ 8 & 9 & 8 \\ 8 & 8 & 9 \end{bmatrix} + \begin{bmatrix} -4 & -8 & -8 \\ -8 & -4 & -8 \\ -8 & -8 & -4 \end{bmatrix} + \begin{bmatrix} -5 & 0 & 0 \\ 0 & -5 & 0 \\ 0 & 0 & -5 \end{bmatrix}$$

$$= \begin{bmatrix} 9-4-5 & 8-8+0 & 8-8+0 \\ 8-8+0 & 9-4-5 & 8-8+0 \\ 8-8+0 & 8-8+0 & 9-4-5 \end{bmatrix} = \begin{bmatrix} 0 & 0 & 0 \\ 0 & 0 & 0 \\ 0 & 0 & 0 \end{bmatrix} = 0$$

Hence, matrix A satisfies the equation $A^2 - 4A - 5I = 0$.

Now, $A^2 - 4A - 5I = 0 \quad \Rightarrow \quad A^2 - 4A = 5I$

$\Rightarrow \quad A^{-1}(A^2 - 4A) = 5A^{-1} I \quad \Rightarrow \quad A^{-1} A^2 - 4A^{-1} A = 5(A^{-1} I)$

$\Rightarrow \quad A - 4I = 5A^{-1}$

$$\Rightarrow \quad A^{-1} = \frac{1}{5}\left\{\begin{bmatrix} 1 & 2 & 2 \\ 2 & 1 & 2 \\ 2 & 2 & 1 \end{bmatrix} - \begin{bmatrix} 4 & 0 & 0 \\ 0 & 4 & 0 \\ 0 & 0 & 4 \end{bmatrix}\right\} = \frac{1}{5}\begin{bmatrix} -3 & 2 & 2 \\ 2 & -3 & 2 \\ 2 & 2 & -3 \end{bmatrix}$$ **Ans.**

Example 18. *Show that*

$$A = \begin{bmatrix} \cos\theta & 0 & \sin\theta \\ 0 & 1 & 0 \\ -\sin\theta & 0 & \cos\theta \end{bmatrix}$$ *is orthogonal. Find the value of* $|A|$

Proof. $$A \cdot A' = \begin{bmatrix} \cos\theta & 0 & \sin\theta \\ 0 & 1 & 0 \\ -\sin\theta & 0 & \cos\theta \end{bmatrix}\begin{bmatrix} \cos\theta & 0 & -\sin\theta \\ 0 & 1 & 0 \\ \sin\theta & 0 & \cos\theta \end{bmatrix}$$

$$= \begin{bmatrix} \cos^2\theta+\sin^2\theta & 0+0+0 & -\cos\theta\cdot\sin\theta+0+\sin\theta\cdot\cos\theta \\ 0+0+0 & 0+1+0 & 0+0+0 \\ -\sin\theta\cdot\cos\theta+\sin\theta\cdot\cos\theta & 0+0+0 & \sin^2\theta+0+\cos^2\theta \end{bmatrix}$$

$$= \begin{bmatrix} 1 & 0 & 0 \\ 0 & 1 & 0 \\ 0 & 0 & 1 \end{bmatrix} = I$$

Hence A is an orthogonal matrix.

$$|A| = \cos\theta(\cos\theta - 0) - 0(0-0) + \sin\theta(0+\sin\theta)$$
$$= \cos^2\theta + \sin^2\theta = 1.$$ **Proved.**

Example 19. *Find the value of $A + 3B$ if*

$$A = \begin{bmatrix} 2 & -1 \\ -3 & -4 \end{bmatrix} \text{ and } B = \begin{bmatrix} -2 & 0 \\ -1 & 3 \end{bmatrix}$$ **(S.B.T.E. 2015)**

Solution. Here, we have

$$A = \begin{bmatrix} 2 & -1 \\ -3 & -4 \end{bmatrix}, B = \begin{bmatrix} -2 & 0 \\ -1 & 3 \end{bmatrix}$$

$$A + 3B = \begin{bmatrix} 2 & -1 \\ -3 & -4 \end{bmatrix} + 3\begin{bmatrix} -2 & 0 \\ -1 & 3 \end{bmatrix} = \begin{bmatrix} -4 & -1 \\ -6 & 5 \end{bmatrix}$$ **Ans.**

Example 20. *Show that the matrix*

$$A = \begin{bmatrix} 2 & 3 & 4 \\ 3 & 4 & 7 \\ 4 & 7 & 8 \end{bmatrix} \text{ is symmetric.}$$ **(S.B.T.E. 2017, 2016)**

Solution. Here, we have

$$A = \begin{bmatrix} 2 & 3 & 4 \\ 3 & 4 & 7 \\ 4 & 7 & 8 \end{bmatrix}, A' = \begin{bmatrix} 2 & 3 & 4 \\ 3 & 4 & 7 \\ 4 & 7 & 8 \end{bmatrix}$$

$A' = A$, thus the given matrix is symmetric. **Ans.**

Example 21. *Find x, y, z and w if,*

$$\begin{bmatrix} 3x & 3y \\ 3z & 3w \end{bmatrix} = \begin{bmatrix} x & 6 \\ -1 & 2w \end{bmatrix} + \begin{bmatrix} 4 & x+y \\ z+w & 3 \end{bmatrix}$$ **(S.B.T.E. 2015)**

Solution. Here, we have

$$\begin{bmatrix} 3x & 3y \\ 3z & 3w \end{bmatrix} = \begin{bmatrix} x & 6 \\ -1 & 2w \end{bmatrix} + \begin{bmatrix} 4 & x+y \\ z+w & 3 \end{bmatrix}$$

By equality of matrices, we get

$\Rightarrow \quad 3x = x + 4 \quad \Rightarrow \quad 2x = 4 \therefore x = 2$

$\Rightarrow \quad 3y = 6 + x + y \quad \Rightarrow \quad 3y = 6 + 2 + y$

$\Rightarrow \quad 2y = 8 \quad \therefore y = 4$

$3w = 2w + 3 \quad \Rightarrow \quad w = 3$

$$3z = -1 + z + w = -1 + z + 3$$
$$2z = 2 \quad \therefore z = 1$$

Hence $x = 2$, $y = 4$, $z = 1$ and $w = 3$ **Ans.**

Example 22. *If* $A = \begin{bmatrix} 1 & 1 \\ -3 & 2 \end{bmatrix}$ *and* $B = \begin{bmatrix} 3 & 1 \\ 4 & 4 \end{bmatrix}$, *find AB.* **(S.B.T.E. 2015, 2014)**

Solution. Here, we have

$$A = \begin{bmatrix} 1 & 1 \\ -3 & 2 \end{bmatrix} \text{ and } B = \begin{bmatrix} 3 & 1 \\ 4 & 4 \end{bmatrix}$$

$$AB = \begin{bmatrix} 1 & 1 \\ -3 & 2 \end{bmatrix}\begin{bmatrix} 3 & 1 \\ 4 & 4 \end{bmatrix} = \begin{bmatrix} 3+4 & 1+4 \\ -9+8 & -3+6 \end{bmatrix} = \begin{bmatrix} 7 & 5 \\ -1 & 3 \end{bmatrix}$$ **Ans.**

Example 23. *Find* $\begin{bmatrix} a & b \\ -b & a \end{bmatrix} \times \begin{bmatrix} a & -b \\ b & a \end{bmatrix}$ **(S.B.T.E. 2015, 2014)**

Solution. Here, we have

$$\begin{bmatrix} a & b \\ -b & a \end{bmatrix} \times \begin{bmatrix} a & -b \\ b & a \end{bmatrix}$$

$$\Rightarrow \begin{bmatrix} a^2+b^2 & -ab+ab \\ -ab+ab & a^2-b^2 \end{bmatrix} = \begin{bmatrix} a^2+b^2 & 0 \\ 0 & a^2-b^2 \end{bmatrix}$$ **Ans.**

Example 24. *If* $x + y = \begin{bmatrix} 2 & 5 \\ 2 & 1 \end{bmatrix}$ *and* $x - y = \begin{bmatrix} 4 & 3 \\ 2 & 5 \end{bmatrix}$, *then find x and y.* **(S.B.T.E. 2014)**

Solution. Here, we have

$$x + y = \begin{bmatrix} 2 & 5 \\ 2 & 1 \end{bmatrix} \quad \text{... (1)}$$

$$x - y = \begin{bmatrix} 4 & 3 \\ 2 & 5 \end{bmatrix} \quad \text{... (2)}$$

(1) + (2), we get

$$2x = \begin{bmatrix} 6 & 8 \\ 4 & 6 \end{bmatrix} \Rightarrow x = \begin{bmatrix} 3 & 4 \\ 2 & 3 \end{bmatrix}$$

Putting the value x in (1), we get

$$\begin{bmatrix} 3 & 4 \\ 2 & 3 \end{bmatrix} + y = \begin{bmatrix} 2 & 5 \\ 2 & 1 \end{bmatrix} \Rightarrow y = \begin{bmatrix} -1 & 1 \\ 0 & -2 \end{bmatrix}$$

$$\therefore \quad x = \begin{bmatrix} 3 & 4 \\ 2 & 3 \end{bmatrix} \text{ and } y = \begin{bmatrix} -1 & 1 \\ 0 & -2 \end{bmatrix}$$ **Ans.**

Example 25. *If* $\begin{bmatrix} x & 3 \\ 4 & 2 \end{bmatrix} + \begin{bmatrix} 2 & y \\ 1 & -2 \end{bmatrix} - \begin{bmatrix} 3 & 1 \\ -2 & 3z \end{bmatrix} = \begin{bmatrix} 5 & 0 \\ 7 & 3 \end{bmatrix}$ *find x, y, z.* **(S.B.T.E. 2014)**

Solution. Here, we have

$$\begin{bmatrix} x & 3 \\ 4 & 2 \end{bmatrix} + \begin{bmatrix} 2 & y \\ 1 & -2 \end{bmatrix} - \begin{bmatrix} 3 & 1 \\ -2 & 3z \end{bmatrix} = \begin{bmatrix} 5 & 0 \\ 7 & 3 \end{bmatrix}$$

$$x + 2 - 3 = 5 \Rightarrow x = 6$$

$$3 + y - 1 = 0 \Rightarrow y = -2$$

$$2 - 2 + 3z = 3 \Rightarrow z = 1$$

$\therefore$ $x = 6, y = -2, z = 1$ **Ans.**

Example 26. *If* $A = \begin{bmatrix} 1 & -1 \\ 2 & -1 \end{bmatrix}, B = \begin{bmatrix} a & 1 \\ b & -1 \end{bmatrix}$ *and* $(A + B)^2 = A^2 + B^2$*, find a and b.* **(S.B.T.E. 2015)**

Solution. Here, we have

$$A = \begin{bmatrix} 1 & -1 \\ 2 & -1 \end{bmatrix}, B = \begin{bmatrix} a & 1 \\ b & -1 \end{bmatrix}$$

$$A + B = \begin{bmatrix} 1 & -1 \\ 2 & -1 \end{bmatrix} + \begin{bmatrix} a & 1 \\ b & -1 \end{bmatrix} = \begin{bmatrix} a+1 & 0 \\ b+2 & -2 \end{bmatrix}$$

$$(A + B)^2 = (A + B) \cdot (A + B) = \begin{bmatrix} a+1 & 0 \\ b+2 & -2 \end{bmatrix}\begin{bmatrix} a+1 & 0 \\ b+2 & -2 \end{bmatrix}$$

$$= \begin{bmatrix} a^2 + 2a + 1 & 0 \\ ab + 2a - b - 2 & 4 \end{bmatrix}$$

$$A^2 = A \cdot A = \begin{bmatrix} 1 & -1 \\ 2 & -1 \end{bmatrix}\begin{bmatrix} 1 & -1 \\ 2 & -1 \end{bmatrix} = \begin{bmatrix} -1 & 0 \\ 0 & -1 \end{bmatrix}$$

$$B^2 = B \cdot B = \begin{bmatrix} a & 1 \\ b & -1 \end{bmatrix}\begin{bmatrix} a & 1 \\ b & -1 \end{bmatrix} = \begin{bmatrix} a^2 + b^2 & a-1 \\ ab - b & b+1 \end{bmatrix}$$

$$A^2 + B^2 = \begin{bmatrix} -1 & 0 \\ 0 & -1 \end{bmatrix} + \begin{bmatrix} a^2 + b^2 & a-1 \\ ab - b & b+1 \end{bmatrix} = \begin{bmatrix} a^2 + b - 1 & a-1 \\ ab - b & b \end{bmatrix}$$

Given $(A + B)^2 = A^2 + B^2$

$$\begin{bmatrix} a^2 + 2a + 1 & 0 \\ ab + 2a - b - 2 & 4 \end{bmatrix} = \begin{bmatrix} a^2 + b - 1 & a-1 \\ ab - b & b \end{bmatrix}$$

By equality of matrices

$a - 1 = 0 \Rightarrow a = 1, b = 4$ **Ans.**

Example 27. *For which value of k, the matrix* $\begin{bmatrix} 0 & k & 2 \\ -3 & 0 & -1 \\ -2 & 1 & 0 \end{bmatrix}$ *is skew-symmetric.* **(S.B.T.E. 2016)**

Solution. Here, we have

Let $$A = \begin{bmatrix} 0 & k & 2 \\ -3 & 0 & -1 \\ -2 & 1 & 0 \end{bmatrix}$$

if the matrix is a skew symmetric then $A^T = -A$
if $k = 3$ the given matrix is skew-symmetric. **Ans.**

Example 28. *Find the value of x for which the matrix* $A = \begin{bmatrix} 1+x & 7 \\ 3-x & 8 \end{bmatrix}$ *singular.*

(S.B.T.E. 2015, 2013)

Solution. Here, we have

$$A = \begin{bmatrix} 1+x & 7 \\ 3-x & 8 \end{bmatrix}$$

The matrix A is singular then

$$|A| = 0$$
$$|A| = 8(x+1) - 7(3-x) = 15x - 13 = 0$$
$$\Rightarrow \quad 15x = 13, x = 13/15$$ **Ans.**

Example 29. *If* $A = \begin{bmatrix} -2 & 3 & -1 \\ -1 & 2 & -1 \\ -6 & 9 & -4 \end{bmatrix}, B = \begin{bmatrix} 1 & 3 & -1 \\ 2 & 2 & -1 \\ 3 & 0 & -1 \end{bmatrix}$*, verify that* $AB = BA = I_3$

(S.B.T.E. 2017)

Solution. Here, we have

$$A = \begin{bmatrix} -2 & 3 & -1 \\ -1 & 2 & -1 \\ -6 & 9 & -4 \end{bmatrix}, B = \begin{bmatrix} 1 & 3 & -1 \\ 2 & 2 & -1 \\ 3 & 0 & -1 \end{bmatrix}$$

$$AB = \begin{bmatrix} -2 & 3 & -1 \\ -1 & 2 & -1 \\ -6 & 9 & -4 \end{bmatrix} \begin{bmatrix} 1 & 3 & -1 \\ 2 & 2 & -1 \\ 3 & 0 & -1 \end{bmatrix} = \begin{bmatrix} 1 & 0 & 0 \\ 0 & 1 & 0 \\ 0 & 0 & 1 \end{bmatrix}$$

$$BA = \begin{bmatrix} 1 & 3 & -1 \\ 2 & 2 & -1 \\ 3 & 0 & -1 \end{bmatrix} \begin{bmatrix} -2 & 3 & -1 \\ -1 & 2 & -1 \\ -6 & 9 & -4 \end{bmatrix} = \begin{bmatrix} 1 & 0 & 0 \\ 0 & 1 & 0 \\ 0 & 0 & 1 \end{bmatrix}$$

$$\therefore \quad AB = BA = I_3$$ **Proved**

10.6 TRANSPOSE OF A MATRIX

If $A = [a_{ij}]$ be an $m \times n$ matrix, then the matrix obtained by interchanging the rows and columns of A is called the transpose of A. It is denoted by A' or (A^T).

In symbolic form, if $A = [a_{ij}]_{m \times n}$, then $A' = [a_{ij}]_{n \times m}$

For example: If $A = \begin{bmatrix} 1 & 4 \\ 2 & -5 \\ 3 & 6 \end{bmatrix}_{3\times 2}$ then $A' = \begin{bmatrix} 1 & 2 & 3 \\ 4 & -5 & 6 \end{bmatrix}_{2\times 3}$

Properties of Transpose of Matrices

(*i*) $(A')' = A$ (ii) $(kA)' = kA'$ (where k is any constant)

(*iii*) $(A + B)' = A' + B'$ (iv) $(AB)' = B'A'$

Example 30. *Find the transpose of each of the following matrices:*

$$(i)\ \begin{bmatrix} 5 \\ \frac{1}{2} \\ -1 \end{bmatrix} \quad (ii)\ \begin{bmatrix} 1 & -1 \\ 2 & 3 \end{bmatrix} \quad (iii)\ \begin{bmatrix} -1 & 5 & 6 \\ \sqrt{3} & 5 & 6 \\ 2 & 3 & -1 \end{bmatrix}$$

Solution.

(*i*) Let $A = \begin{bmatrix} 5 \\ \frac{1}{2} \\ -1 \end{bmatrix}$, then $A' = \begin{bmatrix} 5 & \frac{1}{2} & -1 \end{bmatrix}$

(*ii*) Let $A = \begin{bmatrix} 1 & -1 \\ 2 & 3 \end{bmatrix}$, then $A' = \begin{bmatrix} 1 & 2 \\ -1 & 3 \end{bmatrix}$

(*iii*) Let $A = \begin{bmatrix} -1 & 5 & 6 \\ \sqrt{3} & 5 & 6 \\ 2 & 3 & -1 \end{bmatrix}$, then $A' = \begin{bmatrix} -1 & \sqrt{3} & 2 \\ 5 & 5 & 3 \\ 6 & 6 & -1 \end{bmatrix}$ **Ans.**

Example 31. *If* $A = \begin{bmatrix} -1 & 2 & 3 \\ 5 & 7 & 9 \\ -2 & 1 & 1 \end{bmatrix}$ and $B = \begin{bmatrix} -4 & 1 & -5 \\ 1 & 2 & 0 \\ 1 & 3 & 1 \end{bmatrix}$, *then verify that:*

(*i*) $(A + B)' = A' + B'$ (*ii*) $(A - B)' = A' - B'$

Solution. We have,

$$A = \begin{bmatrix} -1 & 2 & 3 \\ 5 & 7 & 9 \\ -2 & 1 & 1 \end{bmatrix} \text{ and } B = \begin{bmatrix} -4 & 1 & -5 \\ 1 & 2 & 0 \\ 1 & 3 & 1 \end{bmatrix}$$

$$\Rightarrow \quad A' = \begin{bmatrix} -1 & 5 & -2 \\ 2 & 7 & 1 \\ 3 & 9 & 1 \end{bmatrix} \text{ and } B' = \begin{bmatrix} -4 & 1 & 1 \\ 1 & 2 & 3 \\ -5 & 0 & 1 \end{bmatrix}$$

Now, (*i*) $(A+B) = \begin{bmatrix} -1 & 2 & 3 \\ 5 & 7 & 9 \\ -2 & 1 & 1 \end{bmatrix} + \begin{bmatrix} -4 & 1 & -5 \\ 1 & 2 & 0 \\ 1 & 3 & 1 \end{bmatrix} = \begin{bmatrix} -5 & 3 & -2 \\ 6 & 9 & 9 \\ -1 & 4 & 2 \end{bmatrix}$

$$\Rightarrow \quad (A+B)' = \begin{bmatrix} -5 & 6 & -1 \\ 3 & 9 & 4 \\ -2 & 9 & 2 \end{bmatrix} \quad \text{... (1)}$$

$$A' + B' = \begin{bmatrix} -1 & 5 & -2 \\ 2 & 7 & 1 \\ 3 & 9 & 1 \end{bmatrix} + \begin{bmatrix} -4 & 1 & 1 \\ 1 & 2 & 3 \\ -5 & 0 & 1 \end{bmatrix} = \begin{bmatrix} -5 & 6 & -1 \\ 3 & 9 & 4 \\ -2 & 9 & 2 \end{bmatrix} \quad \text{... (2)}$$

From (1) and (2), we have

$(A + B)' = A' + B'$ **Proved.**

(*ii*) $$A - B = \begin{bmatrix} -1 & 2 & 3 \\ 5 & 7 & 9 \\ -2 & 1 & 1 \end{bmatrix} - \begin{bmatrix} -4 & 1 & -5 \\ 1 & 2 & 0 \\ 1 & 3 & 1 \end{bmatrix} = \begin{bmatrix} 3 & 1 & 8 \\ 4 & 5 & 9 \\ -3 & -2 & 0 \end{bmatrix}$$

$$\Rightarrow \quad (A - B)' = \begin{bmatrix} 3 & 4 & -3 \\ 1 & 5 & -2 \\ 8 & 9 & 0 \end{bmatrix} \qquad \ldots (3)$$

$$A' - B' = \begin{bmatrix} -1 & 5 & -2 \\ 2 & 7 & 1 \\ 3 & 9 & 1 \end{bmatrix} - \begin{bmatrix} -4 & 1 & 1 \\ 1 & 2 & 3 \\ -5 & 0 & 1 \end{bmatrix} = \begin{bmatrix} 3 & 4 & -3 \\ 1 & 5 & -2 \\ 8 & 9 & 0 \end{bmatrix} \qquad \ldots (4)$$

From (3) and (4), we have

$(A - B)' = A' - B'$ **Verified**

Example 32. *If* $A' = \begin{bmatrix} 3 & 4 \\ -1 & 2 \\ 0 & 1 \end{bmatrix}$ *and* $B = \begin{bmatrix} -1 & 2 & 1 \\ 1 & 2 & 3 \end{bmatrix}$, *then verify that*

(*i*) $(A + B)' = A' + B'$ (*ii*) $(A - B)' = A' - B'$

Proof. We have,

$$A' = \begin{bmatrix} 3 & 4 \\ -1 & 2 \\ 0 & 1 \end{bmatrix} \quad \Rightarrow \quad A = (A')' = \begin{bmatrix} 3 & -1 & 0 \\ 4 & 2 & 1 \end{bmatrix}$$

and $$B = \begin{bmatrix} -1 & 2 & 1 \\ 1 & 2 & 3 \end{bmatrix} \quad \Rightarrow \quad B' = \begin{bmatrix} -1 & 1 \\ 2 & 2 \\ 1 & 3 \end{bmatrix}$$

(*i*) $$A + B = \begin{bmatrix} 3 & -1 & 0 \\ 4 & 2 & 1 \end{bmatrix} + \begin{bmatrix} -1 & 2 & 1 \\ 1 & 2 & 3 \end{bmatrix} = \begin{bmatrix} 2 & 1 & 1 \\ 5 & 4 & 4 \end{bmatrix}$$ $\Rightarrow A = (A')' = \ldots$

$$\Rightarrow \quad (A + B)' = \begin{bmatrix} 2 & 5 \\ 1 & 4 \\ 1 & 4 \end{bmatrix} \qquad \ldots (1)$$

$$A' + B' = \begin{bmatrix} 3 & 4 \\ -1 & 2 \\ 0 & 1 \end{bmatrix} + \begin{bmatrix} -1 & 1 \\ 2 & 2 \\ 1 & 3 \end{bmatrix} = \begin{bmatrix} 2 & 5 \\ 1 & 4 \\ 1 & 4 \end{bmatrix} \qquad \ldots (2)$$

From (1) and (2), we have

$(A + B)' = A' + B'$ **Proved.**

(*ii*) $$A - B = \begin{bmatrix} 3 & -1 & 0 \\ 4 & 2 & 1 \end{bmatrix} - \begin{bmatrix} -1 & 2 & 1 \\ 1 & 2 & 3 \end{bmatrix} = \begin{bmatrix} 4 & -3 & -1 \\ 3 & 0 & -2 \end{bmatrix}$$

$$\Rightarrow \qquad (A-B)' = \begin{bmatrix} 4 & 3 \\ -3 & 0 \\ -1 & -2 \end{bmatrix} \qquad \text{... (3)}$$

$$A' - B' = \begin{bmatrix} 3 & 4 \\ -1 & 2 \\ 0 & 1 \end{bmatrix} - \begin{bmatrix} -1 & 1 \\ 2 & 2 \\ 1 & 3 \end{bmatrix} = \begin{bmatrix} 4 & 3 \\ -3 & 0 \\ -1 & -2 \end{bmatrix} \qquad \text{... (4)}$$

From (3) and (4), we have

$$(A-B)' = A' - B'$$ **Verified**

Example 33. *If* $A = \begin{bmatrix} \cos\alpha & \sin\alpha \\ -\sin\alpha & \cos\alpha \end{bmatrix}$, *verify that* $AA' = I_3 = A'A$.

Solution. Here $A' = \begin{bmatrix} \cos\alpha & -\sin\alpha \\ \sin\alpha & \cos\alpha \end{bmatrix}$,

$$\therefore \quad AA' = \begin{bmatrix} \cos\alpha & \sin\alpha \\ -\sin\alpha & \cos\alpha \end{bmatrix} \times \begin{bmatrix} \cos\alpha & -\sin\alpha \\ \sin\alpha & \cos\alpha \end{bmatrix}$$

$$= \begin{bmatrix} \cos^2\alpha + \sin^2\alpha & -\cos\alpha\sin\alpha + \sin\alpha\cos\alpha \\ -\sin\alpha\cos\alpha + \cos\alpha\sin\alpha & \sin^2\alpha + \cos^2\alpha \end{bmatrix}$$

$$= \begin{bmatrix} 1 & 0 \\ 0 & 1 \end{bmatrix} = I_3$$

Similarly, we can prove that

$$A'A = \begin{bmatrix} \cos\alpha & -\sin\alpha \\ \sin\alpha & \cos\alpha \end{bmatrix} \times \begin{bmatrix} \cos\alpha & \sin\alpha \\ -\sin\alpha & \cos\alpha \end{bmatrix}$$

$$= \begin{bmatrix} \cos^2\alpha + \sin^2\alpha & \cos\alpha\sin\alpha - \sin\alpha\cos\alpha \\ \sin\alpha\cos\alpha - \cos\alpha\sin\alpha & \sin^2\alpha + \cos^2\alpha \end{bmatrix}$$

$$= \begin{bmatrix} 1 & 0 \\ 0 & 1 \end{bmatrix} = I_2$$

Hence $AA' = I_2 = A'A$. **Verified.**

EXERCISE 10.2

1. If $A = \begin{bmatrix} 1 & 0 \\ 0 & 1 \end{bmatrix}$ find A^{10} *(S.B.T.E. 2015)* **Ans.** $A^{10} = \begin{bmatrix} 1 & 0 \\ 0 & 1 \end{bmatrix}$

2. If $A = \begin{bmatrix} 5x & 2 \\ -10 & 1 \end{bmatrix}$ is singular find x. *(S.B.T.E. 2015)* **Ans.** $x = -4$

3. Compute AB, if $A = \begin{bmatrix} 1 & 2 & 3 \\ 4 & 5 & 6 \end{bmatrix}$ and $B = \begin{bmatrix} 2 & 5 & 3 \\ 3 & 6 & 4 \\ 4 & 7 & 5 \end{bmatrix}$ is BA defined? **(B.T.E. Delhi June 2015)**

Ans. $\begin{bmatrix} 20 & 38 & 26 \\ 47 & 92 & 62 \end{bmatrix}$, No

4. If $A = \begin{bmatrix} 1 & -2 & 3 \\ 2 & 3 & -1 \\ -3 & 1 & 2 \end{bmatrix}$, $B = \begin{bmatrix} 1 & 0 & 2 \\ 0 & 1 & 2 \\ 1 & 2 & 0 \end{bmatrix}$ then show that $AB \neq BA$.

5. If $A = \begin{bmatrix} 0 & 1 & 2 \\ 1 & 2 & 3 \\ 2 & 3 & 4 \end{bmatrix}$ and $B = \begin{bmatrix} 1 & -2 \\ -1 & 0 \\ 2 & -1 \end{bmatrix}$ obtain the product AB and explain why BA is not defined.

Ans. $\begin{bmatrix} 3 & -2 \\ 5 & -5 \\ 7 & -8 \end{bmatrix}$

BA is not defined because number of columns in B is not equal to number of rows in A.

6. Find the value of A^2, if $A = \begin{bmatrix} 4 & -1 & -4 \\ 3 & 0 & -4 \\ 3 & -1 & -3 \end{bmatrix}$ **Ans.** $\begin{bmatrix} 1 & 0 & 0 \\ 0 & 1 & 0 \\ 0 & 0 & 1 \end{bmatrix}$

7. For what value of x is the matrix $\begin{bmatrix} 3-x & 2 & 2 \\ 1 & 4-x & 1 \\ -2 & -4 & -1-x \end{bmatrix}$ singular?

(B.T.E. Delhi, Jan. 2009, May, 2008)

[**Hint.** Matrix A is singular if $|A| = 0$] **Ans.** 1, 2, 3

8. Verify that $A = \begin{bmatrix} 5 & 3 \\ -1 & -2 \end{bmatrix}$ satisfies its characteristic equation $x^2 - 3x - 7 = 0$.

(Diploma IETE, June 2005)

9. Find $A^2 - 3A + 6I$, I is the unit matrix of order 3 and $A = \begin{bmatrix} 1 & -2 & 3 \\ 2 & 3 & -1 \\ -3 & 1 & 2 \end{bmatrix}$

Ans. $\begin{bmatrix} -9 & 1 & 2 \\ 5 & 1 & 4 \\ 2 & 8 & -6 \end{bmatrix}$

10. Let $f(x) = x^2 - 5x + 6$, find $f(A)$ if $A = \begin{bmatrix} 2 & 0 & 1 \\ 2 & 1 & 3 \\ 1 & -1 & 0 \end{bmatrix}$ **Ans.** $\begin{bmatrix} 1 & -1 & -3 \\ -1 & -1 & -10 \\ -5 & 4 & 4 \end{bmatrix}$

11. If $A = \begin{bmatrix} 0 & 1 & 0 \\ 0 & 0 & 1 \\ p & q & r \end{bmatrix}$ and I is the unit matrix of order 3, show that $A^3 + pI + qA + rA^2 = 0$

12. Given $A = \begin{bmatrix} 2 & -1 & 3 \\ 1 & 2 & 4 \\ 3 & 1 & 1 \end{bmatrix}$ Show that $A^3 - 5A^2 - 4A + 30I = 0$.

13. If $A = \begin{bmatrix} 3 & 2 & 1 \\ -1 & 2 & 0 \\ 1 & 2 & 5 \end{bmatrix}$, and $B = \begin{bmatrix} 2 & -1 \\ 0 & 1 \\ 1 & 0 \end{bmatrix}, C = \begin{bmatrix} -3 & 2 \\ 2 & 1 \\ -1 & 3 \end{bmatrix}$

show that $A(B + C) = AB + AC$.

14. Let $A = \begin{bmatrix} 1 & -1 & 1 \\ 2 & 0 & 1 \\ 3 & 0 & 1 \end{bmatrix}, B = \begin{bmatrix} 2 & -2 \\ 1 & 3 \\ 4 & 4 \end{bmatrix}$ verify that $A(AB) = A^2B$.

15. For the following matrices A, B, C, verify the result $A(BC) = (AB)C$.

$$A = \begin{bmatrix} 2 & 3 & -1 \\ 3 & 0 & 2 \end{bmatrix}, B = \begin{bmatrix} 1 \\ -1 \\ 2 \end{bmatrix}, C = [1, -2]$$

16. Verify that $(AB)^t = B^t \cdot A^t$ for the following matrices A and B.

$$A = \begin{bmatrix} 0 & 1 & 1 \\ 1 & 0 & 2 \\ 1 & 2 & 0 \end{bmatrix}, B = \begin{bmatrix} 3 & -1 \\ 7 & 2 \\ -5 & 9 \end{bmatrix}$$

17. If $A = \begin{bmatrix} 2 & 3 & -1 \\ 3 & 2 & 0 \\ 1 & -2 & -3 \end{bmatrix}, B = \begin{bmatrix} -1 & 2 \\ 2 & 5 \\ 0 & 1 \end{bmatrix}$ verify that $(AB)' = B' \cdot A'$.

18. If α and β differ by an odd multiple of $\pi/2$, prove that the product of two matrices given below is a null matrix.

$$A = \begin{bmatrix} \cos^2\alpha & \cos\alpha\sin\alpha \\ \cos\alpha\sin\alpha & \sin^2\alpha \end{bmatrix}, B = \begin{bmatrix} \cos^2\beta & \cos\beta\sin\beta \\ \cos\beta\sin\beta & \sin^2\beta \end{bmatrix}$$ **(B.T.E. Delhi. May 2005)**

19. Show that $\begin{bmatrix} \cos\theta & -\sin\theta \\ \sin\theta & \cos\theta \end{bmatrix} = \begin{bmatrix} 1 & -\tan\theta/2 \\ \tan\theta/2 & 1 \end{bmatrix} \begin{bmatrix} 1 & -\tan\theta/2 \\ -\tan\theta/2 & 1 \end{bmatrix}^{-1}$

20. If $A = \begin{bmatrix} \cos\theta & \sin\theta \\ -\sin\theta & \cos\theta \end{bmatrix}$, show that $A^n = \begin{bmatrix} \cos n\theta & \sin n\theta \\ -\sin n\theta & \cos n\theta \end{bmatrix}$ n being a positive integer.

21. If $A=\begin{bmatrix} 3 & -4 \\ 1 & -1 \end{bmatrix}$, then show that $A^n=\begin{bmatrix} 1+2n & -4n \\ n & 1-2n \end{bmatrix}$

22. Show that $\begin{bmatrix} \cos\phi & 0 & \sin\phi \\ \sin\theta\sin\phi & \cos\theta & -\sin\theta\cos\phi \\ -\cos\theta\sin\phi & \sin\theta & \cos\theta\cos\phi \end{bmatrix}$ is an orthogonal matrix. Is it proper or improper?

[**Hint.** Matrix is proper, if $|A| = I$]

23. If A and B are matrices, state the conditions under which the rule.
$A^2 - B^2 = (A + B)(A - B)$ holds **Ans.** $A = B$ or $AB = BA$

24. If A and B be square matrices of the same order, explain why in general $A = B$ as $AB = BA$
   (*i*) $(A + B)^2 \neq A^2 + 2AB + B^2$
   (*ii*) $(A - B)^2 \neq A^2 - 2AB + B^2$
   (*iii*) $(A + B)(A - B) \neq A^2 - B^2$. **Ans.** $AB \neq BA$

25. Three shopkeepers A, B and C go to a store to buy stationary. Shopkeeper A purchases 12 dozen notebooks, 5 dozen pens and 6 dozen pencils. Shopkeeper B purchases 10 dozen notebooks, 6 dozen pens and 7 dozen pencils. Shopkeeper C purchases 11 dozen notebooks, 13 dozen pens and 8 dozen pencils. A notebook costs 40 paise, a pen costs Rs. 1.25 and a pencil 35 paise each.
   (*i*) Represent the purchase of each of these individuals by means of a row vector and the prices by means of a column vector.
   (*ii*) Use matrix multiplication to calculate each individual's bill.

Ans. $\begin{bmatrix} 144 & 60 & 72 \\ 120 & 72 & 84 \\ 132 & 156 & 96 \end{bmatrix}\begin{bmatrix} 0.40 \\ 1.25 \\ 0.35 \end{bmatrix}=\begin{bmatrix} 157.80 \\ 167.40 \\ 281.40 \end{bmatrix}$

26. Determine the value of α, β, γ when $\begin{bmatrix} 0 & 2\beta & \gamma \\ \alpha & \beta & -\gamma \\ \alpha & -\beta & \gamma \end{bmatrix}$ is orthogonal.

Ans. $\alpha=\pm\frac{1}{\sqrt{2}}, \beta=\pm\frac{1}{\sqrt{6}}, \gamma=\pm\frac{1}{\sqrt{3}}$

27. A matrix x has $a + b$ rows and $a + 2$ columns while the matrix y has $b + 1$ rows and $a + 3$ columns. Both matrices xy and yx exist. Find a and b. Can you say xy and yx are of same type? Are they equal? **Ans.** $a = 2$, $b = 3$, No., No.

Choose the correct alternative (Questions 28 to 30).

28. If ω is complex cube root of unity, and $A=\begin{bmatrix} \omega & 0 \\ 0 & \omega \end{bmatrix}$, then A^{100} is equal to
   (*a*) 0 (*b*) $-A$ (*c*) A (*d*) none of these
   (Diploma, IETE, June 2006) **Ans.** (*c*)

29. If A and B are symmetric matrices, then $AB + BA$ is a

(*a*) diagonal matrix (*b*) null matrix

(*c*) symmetric matrix (*d*) skew-symmetric matrix.

(*Diploma IETE, June 2006*) Ans. (*c*)

30. If $A = \begin{pmatrix} 1 & 0 \\ 1 & 0 \end{pmatrix}$ and $B = \begin{pmatrix} 0 & 0 \\ 1 & 1 \end{pmatrix}$ then AB is equal to

(*a*) $\begin{pmatrix} 0 & 0 \\ 0 & 0 \end{pmatrix}$ (*b*) $\begin{pmatrix} 0 & 1 \\ 1 & 0 \end{pmatrix}$ (*c*) $\begin{pmatrix} 0 & 1 \\ 1 & 0 \end{pmatrix}$ (*d*) $\begin{pmatrix} 1 & 1 \\ 1 & 1 \end{pmatrix}$

Ans. (*a*)

31. If $A = \begin{bmatrix} 1 & 2 & 3 \\ 2 & 3 & 1 \\ 3 & 1 & 2 \end{bmatrix}, B = \begin{bmatrix} 0 & 1 & 2 \\ 1 & 2 & 3 \\ 2 & 3 & 0 \end{bmatrix}$, verify that $(A - B)' = A' - B'$

32. If $A' = \begin{bmatrix} -2 & 3 \\ 1 & 2 \end{bmatrix}$ and $B = \begin{bmatrix} -1 & 0 \\ 1 & 2 \end{bmatrix}$, then find $(A + 2B)'$

33. If $A = \begin{bmatrix} 3 & \sqrt{3} & 2 \\ 4 & 2 & 0 \end{bmatrix}$ and $B = \begin{bmatrix} 2 & -1 & 2 \\ 1 & 2 & 4 \end{bmatrix}$, then verify that

(*i*) $(A')' = A$ (*ii*) $(A + B)' = A' + B'$ (*iii*) $(kB)' = kB'$, where k is any constant

34. If $A = \begin{bmatrix} 2 & 3 \\ 1 & 2 \end{bmatrix}, B = \begin{bmatrix} 3 & 0 \\ 1 & 2 \end{bmatrix}$, verify that $(AB)' = B'A'$, where A', B' are transposes of A and B.

35. If $A = \begin{bmatrix} 1 & -1 & 1 \\ 2 & 1 & 3 \\ 4 & 1 & 8 \end{bmatrix}$ and $B = \begin{bmatrix} 4 & 1 & 0 \\ 2 & -3 & 1 \\ 1 & 1 & -1 \end{bmatrix}$ then verify that $(AB)' = B'A'$.

36. If $A = \begin{bmatrix} 2 & 4 & -1 \\ -1 & 0 & 2 \end{bmatrix}$ and $B = \begin{bmatrix} 3 & 4 & 5 \\ -1 & 2 & 7 \\ 2 & 1 & 0 \end{bmatrix}$ prove that $(AB)'$ and $B'A'$ are equal.

37. If $A = \begin{bmatrix} 2 & 3 \\ 0 & 1 \end{bmatrix}, B = \begin{bmatrix} 3 & 4 \\ 2 & 1 \end{bmatrix}$, then verify that $[AB]^t = B^t A^t$

38. If $A = \begin{bmatrix} 1 & -1 & 0 \\ 2 & 1 & 3 \\ 4 & 1 & 8 \end{bmatrix}$ and $B = \begin{bmatrix} 4 & 1 & 0 \\ 2 & -3 & 1 \\ 1 & 1 & -1 \end{bmatrix}$ then verify that $(AB)^t = B^t A^t$.

39. Find $A^2 + 2A + 3I$, if I is the unit matrix of order 3.

$$A = \begin{bmatrix} 2 & 1 & -3 \\ 1 & 0 & 5 \\ -2 & 4 & 1 \end{bmatrix}$$

(*S.B.T.E. 20015, 2013*)

40. If $A=\begin{bmatrix}1 & 2 & 3\\ 2 & -1 & 4\\ 3 & 1 & 1\end{bmatrix}$, show that $A^3 - A^2 - 18A - 30I = 0$ **(*S.B.T.E. 2017, 2014*)**

10.7 ELEMENTARY TRANSFORMATIONS

Any one of the following operations on a matrix is called an elementary transformation.

1. Interchanging any two rows (or columns). This transformation is indicated by R_{ij}, if the ith and jth rows are interchanged.
2. Multiplication of the elements of any row R_i (or column) by a non-zero scalar quantity k is denoted by $(k.R_i)$.
3. Addition of constant multiplication of the elements of any row R_j to the corresponding elements of any other row R_i is denoted by $(R_i + kR_j)$.

If a matrix B is obtained from a matrix A by one or more E-operations, then B is said to be equivalent to A. They symbol ~ is used for equivalence; *i.e.* $A \sim B$.

Example 34. *Reduce the following matrix to upper triangular form (Echelon form):*

$$\begin{bmatrix}1 & 2 & 3\\ 2 & 5 & 7\\ 3 & 1 & 2\end{bmatrix}$$

Solution. *Upper triangular matrix.* If in a square matrix, all the elements below the principal diagonal are zero, the matrix is called an upper triangular matrix.

$$\begin{bmatrix}1 & 2 & 3\\ 2 & 5 & 7\\ 3 & 1 & 2\end{bmatrix}\begin{matrix}\\ R_2 \to R_2 - 2R_1\\ R_3 \to R_3 - 3R_1\end{matrix} \sim \begin{bmatrix}1 & 2 & 3\\ 0 & 1 & 1\\ 0 & -5 & -7\end{bmatrix}\begin{matrix}\\ \\ R_3 \to R_3 + 5R_2\end{matrix} \sim \begin{bmatrix}1 & 2 & 3\\ 0 & 1 & 1\\ 0 & 0 & -2\end{bmatrix} \quad \textbf{Ans.}$$

Example 35. *Transform* $\begin{bmatrix}1 & 3 & 3\\ 2 & 4 & 10\\ 3 & 8 & 4\end{bmatrix}$ *into a unit matrix.*

Solution.
$$\begin{bmatrix}1 & 3 & 3\\ 2 & 4 & 10\\ 3 & 8 & 4\end{bmatrix}\begin{matrix}\\ R_2 \to R_2 - 2R_1\\ R_3 \to R_3 - 3R_1\end{matrix} \sim \begin{bmatrix}1 & 3 & 3\\ 0 & -2 & 4\\ 0 & -1 & -5\end{bmatrix} R_2 \to -\frac{1}{2}R_2$$

$$\sim \begin{bmatrix}1 & 3 & 3\\ 0 & 1 & -2\\ 0 & -1 & -5\end{bmatrix}\begin{matrix}R_1 \to R_1 - 3R_2\\ \\ R_3 \to R_3 + R_2\end{matrix} \sim \begin{bmatrix}1 & 0 & 9\\ 0 & 1 & -2\\ 0 & 0 & -7\end{bmatrix} R_3 \to -\frac{1}{7}R_3$$

$$\sim \begin{bmatrix}1 & 0 & 9\\ 0 & 1 & -2\\ 0 & 0 & 1\end{bmatrix}\begin{matrix}R_1 \to R_1 - 9R_3\\ R_2 \to R_2 + 2R_3\\ \\ \end{matrix} \sim \begin{bmatrix}1 & 0 & 0\\ 0 & 1 & 0\\ 0 & 0 & 1\end{bmatrix} \quad \textbf{Ans.}$$

10.8 ELEMENTARY MATRICES

A matrix obtained from a unit matrix by a single elementary transformation is called elementary matrix.

$$I = \begin{bmatrix} 1 & 0 & 0 \\ 0 & 1 & 0 \\ 0 & 0 & 1 \end{bmatrix}$$

Consider the matrix by $R_2 \to R_2 + 3R_1$

$\begin{bmatrix} 1 & 0 & 0 \\ 3 & 1 & 0 \\ 0 & 0 & 1 \end{bmatrix}$ is called the elementary matrix.

Theorem

Every elementary row transformation of a matrix can be affected by pre-multiplication with the corresponding elementary matrix.

Consider the matrix $\quad A = \begin{bmatrix} 2 & 3 & 4 \\ 5 & 6 & 7 \\ 3 & 5 & 9 \end{bmatrix}$

Let us apply row transformation $R_3 \to R_3 + 4R_1$ and we get a matrix B.

$$B = \begin{bmatrix} 2 & 3 & 4 \\ 5 & 6 & 7 \\ 11 & 17 & 25 \end{bmatrix}$$

Now we shall show that pre-multiplication of A by corresponding elementary matrix $R_3 \to R_3 + 4R_1$ will give us B.

Now, if $I = \begin{bmatrix} 1 & 0 & 0 \\ 0 & 1 & 0 \\ 0 & 0 & 1 \end{bmatrix}$ then, Elementary matrix $= \begin{bmatrix} 1 & 0 & 0 \\ 0 & 1 & 0 \\ 4 & 0 & 1 \end{bmatrix} (R_3 \to R_3 + 4R_1)$

$$\therefore \text{ Elementary matrix} \times A = \begin{bmatrix} 1 & 0 & 0 \\ 0 & 1 & 0 \\ 4 & 0 & 1 \end{bmatrix} \times \begin{bmatrix} 2 & 3 & 4 \\ 5 & 6 & 7 \\ 3 & 5 & 9 \end{bmatrix} = \begin{bmatrix} 2 & 3 & 4 \\ 5 & 6 & 7 \\ 11 & 17 & 25 \end{bmatrix} = B$$

Similarly, we can show that every elementary column transformation of a matrix can be affected by post-multiplication with the corresponding elementary matrix.

10.9 TO COMPUTE THE INVERSE OF A MATRIX FROM ELEMENTARY MATRICES (GAUSS–JORDAN METHOD)

If A is reduced to I by elementary transformation then

$$PA = I \quad \text{where} \quad P = P_n P_{n-1} \cdots P_2 P_1$$

$$\therefore \quad P = A^{-1} \quad = \text{Elementary matrix.}$$

Working rule. Write $A = IA$. Perform elementary row transformation on A of the left side and on I of the right hand side so that A is reduced to I and I of right hand side is reduced to P getting $I = PA$.

Then P is inverse of A.

10.10 THE INVERSE OF A SYMMETRIC MATRIX

The elementary transformation are to be transformed so that the property of being symmetric is preserved. This requires that the transformations occur in pairs, a row transformation must be followed immediately by the same column transformation.

Example 36. *Find the inverse of the following matrix employing elementary transformations:*

$$\begin{bmatrix} 3 & -3 & 4 \\ 2 & -3 & 4 \\ 0 & -1 & 1 \end{bmatrix}$$

Solution. The given matrix is $A = \begin{bmatrix} 3 & -3 & 4 \\ 2 & -3 & 4 \\ 0 & -1 & 1 \end{bmatrix}$.

$$\begin{bmatrix} 3 & -3 & 4 \\ 2 & -3 & 4 \\ 0 & -1 & 1 \end{bmatrix} = \begin{bmatrix} 1 & 0 & 0 \\ 0 & 1 & 0 \\ 0 & 0 & 1 \end{bmatrix} A$$

$$\Rightarrow \begin{bmatrix} 1 & -1 & \frac{4}{3} \\ 2 & -3 & 4 \\ 0 & -1 & 1 \end{bmatrix} = \begin{bmatrix} \frac{1}{3} & 0 & 0 \\ 0 & 1 & 0 \\ 0 & 0 & 1 \end{bmatrix} A \qquad R_1 \to \frac{R_1}{3}$$

$$\Rightarrow \begin{bmatrix} 1 & -1 & \frac{4}{3} \\ 0 & -1 & \frac{4}{3} \\ 0 & -1 & 1 \end{bmatrix} = \begin{bmatrix} \frac{1}{3} & 0 & 0 \\ -\frac{2}{3} & 1 & 0 \\ 0 & 0 & 1 \end{bmatrix} A \qquad R_2 \to R_2 - 2R_1$$

$$\Rightarrow \begin{bmatrix} 1 & -1 & \frac{4}{3} \\ 0 & 1 & -\frac{4}{3} \\ 0 & -1 & 1 \end{bmatrix} = \begin{bmatrix} \frac{1}{3} & 0 & 0 \\ \frac{2}{3} & -1 & 0 \\ 0 & 0 & 1 \end{bmatrix} A \qquad R_2 \to -R_2$$

$$\Rightarrow \begin{bmatrix} 1 & -1 & \frac{4}{3} \\ 0 & 1 & -\frac{4}{3} \\ 0 & 0 & -\frac{1}{3} \end{bmatrix} = \begin{bmatrix} \frac{1}{3} & 0 & 0 \\ \frac{2}{3} & -1 & 0 \\ \frac{2}{3} & -1 & 1 \end{bmatrix} A \qquad R_3 \to R_3 + R_2$$

$$\Rightarrow \begin{bmatrix} 1 & -1 & \frac{4}{3} \\ 0 & 1 & -\frac{4}{3} \\ 0 & 0 & 1 \end{bmatrix} = \begin{bmatrix} \frac{1}{3} & 0 & 0 \\ \frac{2}{3} & -1 & 0 \\ -2 & 3 & -3 \end{bmatrix} A \qquad R_3 \to -3R_3$$

$$\Rightarrow \begin{bmatrix} 1 & -1 & 0 \\ 0 & 1 & 0 \\ 0 & 0 & 1 \end{bmatrix} = \begin{bmatrix} 3 & -4 & 4 \\ -2 & 3 & -4 \\ -2 & 3 & -3 \end{bmatrix} A \qquad \begin{matrix} R_1 \to R_1 - 4/3R_3 \\ R_2 \to R_2 + 4/3R_3 \end{matrix}$$

$$\Rightarrow \begin{bmatrix} 1 & 0 & 0 \\ 0 & 1 & 0 \\ 0 & 0 & 1 \end{bmatrix} = \begin{bmatrix} 1 & -1 & 0 \\ -2 & 3 & -4 \\ -2 & 3 & -3 \end{bmatrix} A \qquad R_1 \to R_1 + R_2 \qquad \text{Hence, } A^{-1} = \begin{bmatrix} 1 & -1 & 0 \\ -2 & 3 & -4 \\ -2 & 3 & -3 \end{bmatrix} \textbf{ Ans.}$$

Example 37. *Find the inverse of the matrix M by applying elementary transformations*

$$\begin{bmatrix} 0 & 2 & 1 & 3 \\ 1 & 1 & -1 & -2 \\ 1 & 2 & 0 & 1 \\ -1 & 1 & 2 & 6 \end{bmatrix}.$$

Solution. Here, we have $A = \begin{bmatrix} 0 & 2 & 1 & 3 \\ 1 & 1 & -1 & -2 \\ 1 & 2 & 0 & 1 \\ -1 & 1 & 2 & 6 \end{bmatrix}$

Let
$$\begin{bmatrix} 0 & 2 & 1 & 3 \\ 1 & 1 & -1 & -2 \\ 1 & 2 & 0 & 1 \\ -1 & 1 & 2 & 6 \end{bmatrix} = \begin{bmatrix} 1 & 0 & 0 & 0 \\ 0 & 1 & 0 & 0 \\ 0 & 0 & 1 & 0 \\ 0 & 0 & 0 & 1 \end{bmatrix} A$$

$$\Rightarrow \begin{bmatrix} 1 & 1 & -1 & -2 \\ 0 & 2 & 1 & 3 \\ 1 & 2 & 0 & 1 \\ -1 & 1 & 2 & 6 \end{bmatrix} = \begin{bmatrix} 0 & 1 & 0 & 0 \\ 1 & 0 & 0 & 0 \\ 0 & 0 & 1 & 0 \\ 0 & 0 & 0 & 1 \end{bmatrix} A \qquad R_1 \leftrightarrow R_2$$

$$\Rightarrow \begin{bmatrix} 1 & 1 & -1 & -2 \\ 0 & 2 & 1 & 3 \\ 0 & 1 & 1 & 3 \\ 0 & 2 & 1 & 4 \end{bmatrix} = \begin{bmatrix} 0 & 1 & 0 & 0 \\ 1 & 0 & 0 & 0 \\ 0 & -1 & 1 & 0 \\ 0 & 1 & 0 & 1 \end{bmatrix} A \qquad \begin{matrix} R_3 \to R_3 - R_1 \\ R_4 \to R_4 + R_1 \end{matrix}$$

$$\Rightarrow \begin{bmatrix} 1 & 1 & -1 & -2 \\ 0 & 1 & 1 & 3 \\ 0 & 2 & 1 & 3 \\ 0 & 2 & 1 & 4 \end{bmatrix} = \begin{bmatrix} 0 & 1 & 0 & 0 \\ 0 & -1 & 1 & 0 \\ 1 & 0 & 0 & 0 \\ 0 & 1 & 0 & 1 \end{bmatrix} A \qquad R_3 \leftrightarrow R_2$$

$$\Rightarrow \begin{bmatrix} 1 & 1 & -1 & -2 \\ 0 & 1 & 1 & 3 \\ 0 & 0 & -1 & -3 \\ 0 & 0 & -1 & -2 \end{bmatrix} = \begin{bmatrix} 0 & 1 & 0 & 0 \\ 0 & -1 & 1 & 0 \\ 1 & 2 & -2 & 0 \\ 0 & 3 & -2 & 1 \end{bmatrix} A \qquad \begin{matrix} R_3 \to R_3 - 2R_2 \\ R_4 \to R_4 - 2R_2 \end{matrix}$$

$$\Rightarrow \begin{bmatrix} 1 & 1 & -1 & -2 \\ 0 & 1 & 1 & 3 \\ 0 & 0 & -1 & -3 \\ 0 & 0 & 0 & 1 \end{bmatrix} = \begin{bmatrix} 0 & 1 & 0 & 0 \\ 0 & -1 & 1 & 0 \\ 1 & 2 & -2 & 0 \\ -1 & 1 & 0 & 1 \end{bmatrix} A \qquad R_4 \to R_4 - R_3$$

$$\Rightarrow \begin{bmatrix} 1 & 1 & -1 & -2 \\ 0 & 1 & 1 & 3 \\ 0 & 0 & 1 & 3 \\ 0 & 0 & 0 & 1 \end{bmatrix} = \begin{bmatrix} 0 & 1 & 0 & 0 \\ 0 & -1 & 1 & 0 \\ -1 & -2 & 2 & 0 \\ -1 & 1 & 0 & 1 \end{bmatrix} A \qquad R_3 \to -R_3$$

$$\Rightarrow \begin{bmatrix} 1 & 1 & -1 & 0 \\ 0 & 1 & 1 & 0 \\ 0 & 0 & 1 & 0 \\ 0 & 0 & 0 & 1 \end{bmatrix} = \begin{bmatrix} -2 & 3 & 0 & 2 \\ 3 & -4 & 1 & -3 \\ 2 & -5 & 2 & -3 \\ -1 & 1 & 0 & 1 \end{bmatrix} A \qquad \begin{matrix} R_1 \to R_1 + 2R_4 \\ R_2 \to R_2 - 3R_4 \\ R_3 \to R_3 - 3R_4 \end{matrix}$$

$$\Rightarrow \begin{bmatrix} 1 & 1 & 0 & 0 \\ 0 & 1 & 0 & 0 \\ 0 & 0 & 1 & 0 \\ 0 & 0 & 0 & 1 \end{bmatrix} = \begin{bmatrix} 0 & -2 & 2 & -1 \\ 1 & 1 & -1 & 0 \\ 2 & -5 & 2 & -3 \\ -1 & 1 & 0 & 1 \end{bmatrix} A \qquad \begin{matrix} R_1 \to R_1 + R_3 \\ R_2 \to R_2 - R_3 \end{matrix}$$

$$\Rightarrow \begin{bmatrix} 1 & 0 & 0 & 0 \\ 0 & 1 & 0 & 0 \\ 0 & 0 & 1 & 0 \\ 0 & 0 & 0 & 1 \end{bmatrix} = \begin{bmatrix} -1 & -3 & 3 & -1 \\ 1 & 1 & -1 & 0 \\ 2 & -5 & 2 & -3 \\ -1 & 1 & 0 & 1 \end{bmatrix} A \qquad R_1 \to R_1 - R_2$$

$$I = A^{-1}A$$

Hence, $$A^{-1} = \begin{bmatrix} -1 & -3 & 3 & -1 \\ 1 & 1 & -1 & 0 \\ 2 & -5 & 2 & -3 \\ -1 & 1 & 0 & 1 \end{bmatrix}$$ **Ans.**

EXERCISE 10.3

Reduce the matrices to triangular form:

1. $A = \begin{bmatrix} 1 & 2 & 3 \\ 2 & 5 & 7 \\ 3 & 1 & 2 \end{bmatrix}$ **Ans.** $\begin{bmatrix} 1 & 2 & 3 \\ 0 & 1 & 1 \\ 0 & 0 & -2 \end{bmatrix}$ **2.** $\begin{bmatrix} 3 & 1 & 4 \\ 1 & 2 & -5 \\ 0 & 1 & 5 \end{bmatrix}$ **Ans.** $\begin{bmatrix} 0 & 1 & 4 \\ 0 & 5 & -19 \\ 0 & 0 & 22 \end{bmatrix}$

Find the inverse of the following matrices:

3. $\begin{bmatrix} 1 & 3 & 3 \\ 1 & 4 & 3 \\ 1 & 3 & 4 \end{bmatrix}$ **Ans.** $\begin{bmatrix} 7 & -3 & -3 \\ -1 & 1 & 0 \\ -1 & 0 & 1 \end{bmatrix}$ **4.** $\begin{bmatrix} 1 & -1 & 1 \\ 4 & 1 & 0 \\ 8 & 1 & 1 \end{bmatrix}$ **Ans.** $\begin{bmatrix} 1 & 2 & -1 \\ -4 & -7 & 4 \\ -4 & -9 & 5 \end{bmatrix}$

5. Use elementary row operations to find inverse of $A = \begin{bmatrix} 1 & 1 & 3 \\ 1 & 3 & -3 \\ -2 & -4 & -4 \end{bmatrix}$

Ans. $\dfrac{1}{4}\begin{bmatrix} 12 & 4 & 6 \\ -5 & -1 & -3 \\ -1 & -1 & -1 \end{bmatrix}$ *(AMIETE, June 2010)*

6. $\begin{bmatrix} 2 & 1 & -1 & 2 \\ 1 & 3 & 2 & -3 \\ -1 & 2 & 1 & -1 \\ 2 & -3 & -1 & 4 \end{bmatrix}$ **Ans.** $\dfrac{1}{18}\begin{bmatrix} 2 & 5 & -7 & 1 \\ 5 & -1 & 5 & -2 \\ -7 & 5 & 11 & 10 \\ 1 & -2 & 10 & 5 \end{bmatrix}$

7. $\begin{bmatrix} 1 & 2 & 3 & 1 \\ 1 & 3 & 3 & 2 \\ 2 & 4 & 3 & 3 \\ 1 & 1 & 1 & 1 \end{bmatrix}$ **Ans.** $\begin{bmatrix} 1 & -2 & 1 & 0 \\ 1 & -2 & 2 & -3 \\ 0 & 1 & -1 & 1 \\ -2 & 3 & -2 & 3 \end{bmatrix}$

8. $\begin{bmatrix} 2 & -6 & -2 & -3 \\ 5 & -13 & -4 & -7 \\ -1 & 4 & 1 & 2 \\ 0 & 1 & 0 & 1 \end{bmatrix}$ **Ans.** $\begin{bmatrix} -2 & 1 & 0 & 1 \\ 1 & 0 & 2 & -1 \\ -4 & 1 & -3 & 1 \\ -1 & 0 & -2 & 2 \end{bmatrix}$

10.11 ADJOINT OF A SQUARE MATRIX

The determinant $|A|$ of the square matrix A.

If $A = \begin{bmatrix} a_1 & a_2 & a_3 \\ b_1 & b_2 & b_3 \\ c_1 & c_2 & c_3 \end{bmatrix}$ Then $|A| = \begin{vmatrix} a_1 & a_2 & a_3 \\ b_1 & b_2 & b_3 \\ c_1 & c_2 & c_3 \end{vmatrix}$

The matrix formed by the cofactors of the elements in $|A|$ is

$$\begin{bmatrix} A_1 & A_2 & A_3 \\ B_1 & B_2 & B_3 \\ C_1 & C_2 & C_3 \end{bmatrix}$$

Cofactor $_{ij}$ = $(-1)^{i+j}$ Minor $_{ij}$

where $A_1 = \begin{vmatrix} b_2 & b_3 \\ c_2 & c_3 \end{vmatrix} = b_2c_3 - b_3c_2$, $\quad A_2 = -\begin{vmatrix} b_1 & b_3 \\ c_1 & c_3 \end{vmatrix} = -b_1c_3 + b_3c_1$

$$A_3 = \begin{vmatrix} b_1 & b_2 \\ c_1 & c_2 \end{vmatrix} = b_1c_2 - b_2c_1, \quad B_1 = -\begin{vmatrix} a_2 & a_3 \\ c_2 & c_3 \end{vmatrix} = -a_2c_3 + a_3c_2$$

$$B_2 = \begin{vmatrix} a_1 & a_3 \\ c_1 & c_3 \end{vmatrix} = a_1c_3 - a_3c_1, \quad B_3 = -\begin{vmatrix} a_1 & a_2 \\ c_1 & c_2 \end{vmatrix} = -a_1c_2 + a_2c_1$$

$$C_1 = \begin{vmatrix} a_2 & a_3 \\ b_2 & b_3 \end{vmatrix} = a_2b_3 - a_3b_2, \quad C_2 = -\begin{vmatrix} a_1 & a_3 \\ b_1 & b_3 \end{vmatrix} = -a_1b_3 + a_3b_1$$

$$C_3 = \begin{vmatrix} a_1 & a_2 \\ b_1 & b_2 \end{vmatrix} = a_1b_2 - a_2b_1$$

Then the transpose of the matrix of cofactors

$$\begin{bmatrix} A_1 & B_1 & C_1 \\ A_2 & B_2 & C_2 \\ A_3 & B_3 & C_3 \end{bmatrix}$$

is called the adjoint of the matrix A and is written as adj. A.

Property of adjoint A. The product of a matrix A and its adjoint is equal to unit matrix multiplied by the determinant A.

(adjoint A) $\cdot A = A \cdot$ (Adjoint A) $= |A| \cdot I$

Let $A = \begin{bmatrix} a_1 & a_2 & a_3 \\ b_1 & b_2 & b_3 \\ c_1 & c_2 & c_3 \end{bmatrix}$ and adj. $A = \begin{bmatrix} A_1 & B_1 & C_1 \\ A_2 & B_2 & C_2 \\ A_3 & B_3 & C_3 \end{bmatrix}$

$$A \cdot (\text{Adj} \cdot A) = \begin{bmatrix} a_1 & a_2 & a_3 \\ b_1 & b_2 & b_3 \\ c_1 & c_2 & c_3 \end{bmatrix} \times \begin{bmatrix} A_1 & B_1 & C_1 \\ A_2 & B_2 & C_2 \\ A_3 & B_3 & C_3 \end{bmatrix}$$

$$= \begin{bmatrix} a_1A_1 + a_2A_2 + a_3A_3 & a_1B_1 + a_2B_2 + a_3B_3 & a_1C_1 + a_2C_2 + a_3C_3 \\ b_1A_1 + b_2A_2 + b_3A_3 & b_1B_1 + b_2B_2 + b_3B_3 & b_1C_1 + b_2C_2 + b_3C_3 \\ c_1A_1 + c_2A_2 + c_3A_3 & c_1B_1 + c_2B_2 + c_3B_3 & c_1C_1 + c_2C_2 + c_3C_3 \end{bmatrix}$$

$$= \begin{bmatrix} |A| & 0 & 0 \\ 0 & |A| & 0 \\ 0 & 0 & |A| \end{bmatrix} = |A| \begin{bmatrix} 1 & 0 & 0 \\ 0 & 1 & 0 \\ 0 & 0 & 1 \end{bmatrix} = |A| \cdot I$$

10.12 INVERSE OF A MATRIX

If A and B are two square matrices of the same order such that

$$AB = BA = I \qquad (I = \text{unit matrix})$$

then B is called the inverse of A, *i.e.* $B = A^{-1}$ and A is the inverse of B.

Properties. *The inverse of a matrix is unique*

We suppose that B and C are the two inverse matrices of a given matrix A

then $\quad AB = BA = I \quad$ [$\because$ B is inverse of A.]

and $\quad AC = CA = I \quad$ [$\because$ C is inverse of A.]

But $\quad C \cdot (AB) = (CA) \cdot B \quad$ [Associative law]

$\Rightarrow \quad C \cdot I = I \cdot B$

$\because \quad C = B$

Hence, the inverse of matrix A is unique.

Condition for a square matrix A to posses an inverse is that matrix A is nonsingular.

i.e. $\quad |A| \neq 0$

If A is a square matrix and B be its inverse, then

$$AB = I$$

Taking determinant of both sides, we get

$$|AB| = |I|$$

i.e. the matrix A is a nonsingular.

To Find the Inverse Matrix by the Help of Adjoint Matrix

We know that

$$A \cdot (\text{Adj} \cdot A) = |A| I$$

$\Rightarrow \quad A \cdot \dfrac{(\text{Adj} \cdot A)}{|A|} = I \quad$ Provided $|A| \neq 0 \quad$ ... (1)

and $\quad A \cdot A^{-1} = I \quad$ ... (2)

From (1) and (2), we have $A \cdot A^{-1} = A \cdot \dfrac{(\text{Adj} \cdot A)}{|A|}$

$$\boxed{\mathbf{A^{-1} = \frac{1}{|A|}(Adj \cdot A)}}$$

Working Rule to Find the Inverse of a Matrix

Step 1. Find out minors and then cofactors of the given matrix.

Step 2. Replace the elements of the matrix by the corresponding cofactors.

Step 3. Write down the adjoint matrix by transposing the matrix of cofactors.

Step 4. Find out the value of the determinant of the matrix, *i.e.* $|A|$

Step 5. Inverse matrix = $\dfrac{1}{|A|}$ [adjoint A.]

Example 38. *Find the inverse of the matrix* $\begin{bmatrix} 5 & 2 \\ 3 & 1 \end{bmatrix}$.

Solution. Let $A = \begin{bmatrix} 5 & 2 \\ 3 & 1 \end{bmatrix}$

Matrix of cofactors = $\begin{bmatrix} 1 & -3 \\ -2 & 5 \end{bmatrix}$

$$\text{Adjoint } A = \begin{bmatrix} 1 & -2 \\ -3 & 5 \end{bmatrix}$$

$$|A| = 5 \times 1 + 2 \times (-3) = 5 - 6 = -1$$

$$A^{-1} = \frac{1}{|A|} \text{Adj} \cdot A$$

$$= -\begin{bmatrix} 1 & -2 \\ -3 & 5 \end{bmatrix}$$

$$= \begin{bmatrix} -1 & 2 \\ 3 & 5 \end{bmatrix}$$ **Ans.**

Example 39. *Find the adjoint of the matrix:*

$$\begin{bmatrix} 3 & 1 & 2 \\ 2 & -3 & -1 \\ 1 & 2 & 1 \end{bmatrix}$$ **(S.B.T.E. 2013)**

Solution. Let $A = \begin{bmatrix} 3 & 1 & 2 \\ 2 & -3 & -1 \\ 1 & 2 & 1 \end{bmatrix}$

The cofactors of 3, 1, 2, the elements of the first row are

$$\begin{vmatrix} -3 & -1 \\ 2 & 1 \end{vmatrix}, \quad -\begin{vmatrix} 2 & -1 \\ 1 & 1 \end{vmatrix}, \quad \begin{vmatrix} 2 & -3 \\ 1 & 2 \end{vmatrix}$$

$$= (-3 + 2), \quad -(2 + 1), \quad (4 + 3)$$

$$= -1, -3, 7$$

The cofactors of the elements 2, – 3, – 1 of the second row are

$$-\begin{vmatrix} 1 & 2 \\ 2 & 1 \end{vmatrix}, \quad \begin{vmatrix} 3 & 2 \\ 1 & 1 \end{vmatrix}, \quad -\begin{vmatrix} 3 & 1 \\ 1 & 2 \end{vmatrix}$$

$$= -(1 - 4), \quad (3 - 2), \quad -(6 - 1)$$

$$= 3, 1, -5$$

The cofactors of the elements 1, 2, 1 of the third row are

$$\begin{vmatrix} 1 & 2 \\ -3 & -1 \end{vmatrix}, \quad -\begin{vmatrix} 3 & 2 \\ 2 & -1 \end{vmatrix}, \quad \begin{vmatrix} 3 & 1 \\ 2 & -3 \end{vmatrix}$$

$$= (-1 + 6), \quad -(-3 - 4), \quad (-9 - 2)$$

$$= 5, 7, -11$$

$$\text{Matrix of cofactors} = \begin{bmatrix} -1 & -3 & 7 \\ 3 & 1 & -5 \\ 5 & 7 & -11 \end{bmatrix}$$

On transposing the matrix of cofactors, we get

Thus, $$\text{Adj. } A = \begin{bmatrix} -1 & 3 & 5 \\ -3 & 1 & 7 \\ 7 & -5 & -11 \end{bmatrix}$$ **Ans.**

Example 40. *Find the inverse of the matrix.*

$$A = \begin{bmatrix} 1 & 2 & 3 \\ 4 & 1 & 5 \\ 3 & 6 & 9 \end{bmatrix}$$

Solution. $|A| = 1(9-30) - 2(36-15) + 3(24-3)$

$$= -21 - 42 + 63 = 0$$

Since $|A| = 0$ hence matrix A is singular, and therefore it does not possess an inverse.

Ans.

Example 41. *Compute the adjoint of the matrix*

$$A = \begin{bmatrix} 1 & 4 & 5 \\ 3 & 2 & 6 \\ 0 & 1 & 0 \end{bmatrix}$$ *and verify that* $A\,(\text{adj } A) = (\text{adj } A)\,A = |A|\,I.$

(S.B.T.E. 2016, 2010, 2007)

Solution. $A = \begin{bmatrix} 1 & 4 & 5 \\ 3 & 2 & 6 \\ 0 & 1 & 0 \end{bmatrix}$

$$|A| = 1(-6) - 4(0) + 5(3) = 9$$

$$\text{Matrix of cofactors} = \begin{bmatrix} -6 & 0 & 3 \\ 5 & 0 & -1 \\ 14 & 9 & -10 \end{bmatrix}$$

$$\text{Adj } A = \begin{bmatrix} -6 & 5 & 14 \\ 0 & 0 & 9 \\ 3 & -1 & -10 \end{bmatrix}$$

$$A \cdot (\text{Adj } A) = \begin{bmatrix} 1 & 4 & 5 \\ 3 & 2 & 6 \\ 0 & 1 & 0 \end{bmatrix} \begin{bmatrix} -6 & 5 & 14 \\ 0 & 0 & 9 \\ 3 & -1 & -10 \end{bmatrix}$$

$$= \begin{bmatrix} -6+0+15 & 5+0-5 & 14+36-50 \\ -18+0+18 & 15+0-6 & 42+18-60 \\ 0+0+0 & 0+0-0 & 0+9-0 \end{bmatrix}$$

$$= \begin{bmatrix} 9 & 0 & 0 \\ 0 & 9 & 0 \\ 0 & 0 & 9 \end{bmatrix} = 9 \begin{bmatrix} 1 & 0 & 0 \\ 0 & 1 & 0 \\ 0 & 0 & 1 \end{bmatrix} = |A|\,I \qquad \dots (1)$$

$$(\text{Adj } A) \cdot A = \begin{bmatrix} -6 & 5 & 14 \\ 0 & 0 & 9 \\ 3 & -1 & -10 \end{bmatrix} \begin{bmatrix} 1 & 4 & 5 \\ 3 & 2 & 6 \\ 0 & 1 & 0 \end{bmatrix}$$

$$= \begin{bmatrix} -6+15+0 & -24+10+14 & -30+30+0 \\ 0+0+0 & 0+0+9 & 0+0+0 \\ 3-3-0 & 12-2-10 & 15-6-0 \end{bmatrix}$$

$$= \begin{bmatrix} 9 & 0 & 0 \\ 0 & 9 & 0 \\ 0 & 0 & 9 \end{bmatrix} = 9\begin{bmatrix} 1 & 0 & 0 \\ 0 & 1 & 0 \\ 0 & 0 & 1 \end{bmatrix} = |A|\,I \qquad \text{... (2)}$$

From (1) and (2), we get

$$A \cdot (\text{Adj } A) = (\text{Adj } A) \cdot A = |A|\,I$$ **Verified.**

Example 42. *An automobile company uses three types of steel S_1, S_2, S_3 to produce three types of cars C_1, C_2, C_3. Steel requirements (in tons) for each type of car are given below:*

Cars

		C_1	C_2	C_3
	S_1	2	3	4
Steel	S_2	1	1	2
	S_3	3	2	1

Using matrices, determine the numbers of cars of each type which can be produced using 29, 13 and 16 tons of three types of steel respectively.

Solution. Let the number of cars produced be C_1, C_2 and C_3.

$$\begin{bmatrix} 2 & 3 & 4 \\ 1 & 1 & 2 \\ 3 & 2 & 1 \end{bmatrix}\begin{bmatrix} C_1 \\ C_2 \\ C_3 \end{bmatrix} = \begin{bmatrix} 29 \\ 13 \\ 16 \end{bmatrix}$$

$$\begin{bmatrix} C_1 \\ C_2 \\ C_3 \end{bmatrix} = \begin{bmatrix} 2 & 3 & 4 \\ 1 & 1 & 2 \\ 3 & 2 & 1 \end{bmatrix}^{-1}\begin{bmatrix} 29 \\ 13 \\ 16 \end{bmatrix}$$

$$= \frac{1}{5}\begin{bmatrix} -3 & 5 & 2 \\ 5 & -10 & 0 \\ -1 & 5 & -1 \end{bmatrix}\begin{bmatrix} 29 \\ 13 \\ 16 \end{bmatrix} = \frac{1}{5}\begin{bmatrix} -87 & +65 & 32 \\ 145 & -130 & +0 \\ -29 & +65 & -16 \end{bmatrix}$$

$$= \begin{bmatrix} 2 \\ 3 \\ 4 \end{bmatrix}$$ **Ans.**

Example 43. *Prove that: $(AB)^{-1}) = B^{-1} \cdot A^{-1}$*

Proof. $(AB) \cdot (B^{-1} \cdot A^{-1}) = [(AB) \cdot B^{-1}] \cdot A^{-1}$

$$= [A \cdot (B \cdot B^{-1})] \cdot A^{-1}$$

$$= [A \cdot I] \cdot A^{-1}$$

$$= A \cdot A^{-1}$$

$$= I$$

$\therefore\quad B^{-1} \cdot A^{-1}$ is the inverse of AB.

$$B^{-1} \cdot A^{-1} = (AB)^{-1}.$$ **Proved.**

Example 44. *If a square matrix A satisfies*

$$I + A + A^2 + A^3 + ... + A^k = 0 \text{ then prove that } A^{-1} = A^k$$

Proof. $I + A + A^2 + A^3 + ... + A^k = 0$

$\Rightarrow\quad A^{-1} \cdot [I + A + A^2 + A^3 + ... + A^k] = A^{-1} \cdot 0$

$\Rightarrow\quad A^{-1} + I + A + A^2 + ... + A^{k-1} = 0$

Adding A^k on both sides, we get

$$A^{-1} + I + A + A^2 + ... + A^{k-1} + A^k = A^k$$

$$A^{-1} + 0 = A^k$$

$\Rightarrow\quad A^{-1} = A^k$ **Proved.**

Example 45. *Prove that: Adj (AB) = (Adj B) · (Adj A)*

Proof. We know that

$$A \cdot (\text{Adj } A) = |A|I = (\text{Adj } A) \cdot A$$

Now AB is also a n-squared matrix as

A and B are n-squared matrices.

$\therefore\quad AB\,(\text{Adj } AB) = |AB|I = (\text{Adj } AB) \cdot AB$

Now consider

$$\begin{aligned} AB\,(\text{Adj } B)\,(\text{Adj } A) &= A\,(B \text{ Adj } B) \text{ Adj } A \\ &= A\,[|B|I] \text{ Adj } A \\ &= |B|\,(AI) \text{ Adj } A \\ &= |B|\,(A \cdot \text{Adj } A) \\ &= |B|\,|A|I \\ &= |AB|I \\ &= AB \cdot (\text{Adj } AB) \end{aligned}$$

$$(\text{Adj } B)\,(\text{Adj } A) = (\text{Adj } AB) \qquad ...(1)$$

Similarly, $(\text{Adj } B)\,(\text{Adj } A) \cdot AB = \text{Adj } B\,(\text{Adj } A \cdot A) \cdot B$

$$\begin{aligned} &= \text{Adj } B\,(|A|I) \cdot B \\ &= |A|\,(\text{Adj } B) \cdot IB \\ &= |A|\,(\text{Adj } B \cdot B) \\ &= |A|\,|B|I \\ &= |AB|I \\ &= (\text{Adj } AB) \cdot AB \end{aligned}$$

$$(\text{Adj } B)\,(\text{Adj } A) = (\text{Adj } AB) \qquad ...(2)$$

From (1) and (2), we have

$$\text{Adj } (AB) = (\text{Adj } B) \cdot (\text{Adj } A).$$ **Proved.**

Example 46. *Prove that* $(A')^{-1} = (A^{-1})'$

Proof. Let A be non-singular square matrix *i.e.* $|A| \neq 0$.

Again we know that

$$|A| = |A'|, \qquad \therefore |A'| \neq 0$$

and hence A' is also non-singular and as such as an inverse.

Now $\qquad AA^{-1} = A^{-1} \cdot A = I$

$\therefore \qquad (A \cdot A^{-1})' = (A^{-1} \cdot A)' = I'$

But $\qquad (AB)' = B' \cdot A'$

$\therefore \qquad (A^{-1})' \cdot A' = A' \cdot (A')^{-1} = I \quad (\because I' = I)$

$(A^{-1})' = (A')^{-1}$ **Proved.**

Example 47. *If A is non-singular square matrix of order 3 and $|A| = 7$. Write the value of $|Adj \cdot A|$.* **(S.B.T.E. 2015)**

Proof. Here we have

$$|A| = 7, n = 3$$

We know that for a square matrix A of order n.

$$|\text{Adj} \cdot A| = |A|^{n-1} = (7)^{3-1} = 7^2 = 49$$ **Ans.**

Example 48. *If A is square matrix of order 3 such that $A\ (Adj\ A) = 4I_3$, find the value of $|A^{-1}|$.* **(S.B.T.E. 2014, 2013)**

Solution. Here we have

$$A \cdot (\text{Adj } A) = 4I_3 \qquad \text{... (1)}$$

We know that

$$A \cdot (\text{Adj } A) = |A| I_n \qquad \text{... (2)}$$

Comparing (1) and (2), we get

$$|A| = 4$$

$$|A^{-1}| = \frac{1}{|A|}$$

$$|A^{-1}| = \frac{1}{4}$$ **Ans.**

Example 49. *If* $A = \begin{bmatrix} 3 & 2 \\ 7 & 5 \end{bmatrix}$ *and* $B = \begin{bmatrix} 4 & 6 \\ 3 & 2 \end{bmatrix}$ *find $(AB)^{-1}$ and B^{-1}.* **(S.B.T.E. 2017, 2014)**

Solution. Here, we have

$$A = \begin{bmatrix} 3 & 2 \\ 7 & 5 \end{bmatrix}, B = \begin{bmatrix} 4 & 6 \\ 3 & 2 \end{bmatrix}$$

$$AB = \begin{bmatrix} 3 & 2 \\ 7 & 5 \end{bmatrix}\begin{bmatrix} 4 & 6 \\ 3 & 2 \end{bmatrix} = \begin{bmatrix} 12+6 & 18+4 \\ 28+15 & 42+10 \end{bmatrix} = \begin{bmatrix} 18 & 22 \\ 43 & 52 \end{bmatrix}$$

$$|AB| = 936 - 946 = -10$$

$$\text{Cofactor matrix} = \begin{bmatrix} 52 & -43 \\ -22 & 18 \end{bmatrix}$$

$$\text{Adj} \cdot AB = \begin{bmatrix} 52 & -22 \\ -43 & 18 \end{bmatrix}$$

$$(AB)^{-1} = \frac{1}{|AB|} \cdot \text{Adj} \cdot AB = \frac{1}{-10}\begin{bmatrix} 52 & -22 \\ -43 & 18 \end{bmatrix}$$

$$B = \begin{bmatrix} 4 & 6 \\ 3 & 2 \end{bmatrix}, \ |B| = 8 - 18 = -10$$

$$\text{Cofactor matrix} = \begin{bmatrix} 2 & -3 \\ -6 & 4 \end{bmatrix}, \text{Adj}\, B = \begin{bmatrix} 2 & -6 \\ -3 & 4 \end{bmatrix}$$

$$B^{-1} = \frac{1}{|B|}\text{Adj}\cdot B = \frac{1}{-10}\begin{bmatrix} 2 & -6 \\ -3 & 4 \end{bmatrix}$$

Example 50. *Find the inverse of the matrix*

$$A = \begin{bmatrix} 3 & 2 & 4 \\ 2 & 1 & 1 \\ 1 & 3 & 5 \end{bmatrix}$$ *by using adjoint of the matrix.* **(*DIPIETE June.* 2019)**

Solution. Here, we have

$$A = \begin{bmatrix} 3 & 2 & 4 \\ 2 & 1 & 1 \\ 1 & 3 & 5 \end{bmatrix}$$

$$|A| = 3(5-3) - 2\,(10-1) + 4\,(6-1)$$

or $$= 6 - 18 + 20 = 8$$

The cofactors of 3, 2, 4, the elements of the first row are:

$$\begin{vmatrix} 1 & 1 \\ 3 & 5 \end{vmatrix}, \quad -\begin{vmatrix} 2 & 1 \\ 1 & 5 \end{vmatrix}, \quad \begin{vmatrix} 2 & 1 \\ 1 & 3 \end{vmatrix},$$

$= \quad (5-3), \quad -(10-1), \quad (6-1)$

$= \quad 2, -9, 5$

The cofactors of 2, 1, 1 the elements of the second row are:

$$-\begin{vmatrix} 2 & 4 \\ 3 & 5 \end{vmatrix}, \quad \begin{vmatrix} 3 & 4 \\ 1 & 5 \end{vmatrix}, \quad -\begin{vmatrix} 3 & 2 \\ 1 & 3 \end{vmatrix}$$

$\Rightarrow\ -(10-12),\ (15-4),\ -(9-2),$ *i.e.* $2, 11, -7$

The cofactors of 1, 3, 5 the elements of the third row are:

$$\begin{vmatrix} 2 & 4 \\ 1 & 1 \end{vmatrix}, \quad -\begin{vmatrix} 3 & 4 \\ 2 & 1 \end{vmatrix}, \quad \begin{vmatrix} 3 & 2 \\ 2 & 1 \end{vmatrix}$$

$\Rightarrow\ (2-4), \quad -(3-8), \quad (3-4),$ *i.e.* $-2, 5, -1$

$$\text{Matrix of cofactors} = \begin{bmatrix} 2 & -9 & 5 \\ 2 & 11 & -7 \\ -2 & 5 & -1 \end{bmatrix}$$

$$\text{Adj}\cdot A = \begin{bmatrix} 2 & 2 & -2 \\ -9 & 11 & 5 \\ 5 & -7 & -1 \end{bmatrix}$$

$$A^{-1} = \frac{1}{|A|}(\text{Adj}\cdot A) = \frac{1}{8}\begin{bmatrix} 2 & 2 & -2 \\ -9 & 11 & 5 \\ 5 & -7 & -1 \end{bmatrix}$$ **Ans.**

Example 51. *Find the inverse of the matrix* $A = \begin{bmatrix} 1 & -2 & 3 \\ 2 & 3 & -1 \\ -3 & 1 & 2 \end{bmatrix}$ **(S.B.T.E. 2017, 2014)**

Solution. Here, we have

$$A = \begin{bmatrix} 1 & -2 & 3 \\ 2 & 3 & -1 \\ -3 & 1 & 2 \end{bmatrix}$$

$$|A| = 1\,(6+1) + 2\,(4-3) + 3\,(2+9) = 7 + 2 + 33 = 42 \neq 0,\ A^{-1} \text{ exist.}$$

The cofactors of 1, – 2, 3, the elements of the first row are:

$$\begin{vmatrix} 3 & -1 \\ 1 & 2 \end{vmatrix}, \quad -\begin{vmatrix} 2 & -1 \\ -3 & 2 \end{vmatrix}, \quad \begin{vmatrix} 2 & 3 \\ -3 & 1 \end{vmatrix}$$

$\Rightarrow \quad (6+1), \qquad -(4-3), \qquad (2+9),$ *i.e.* 7, –1, 11

The cofactors of 2, 3, –1, the elements of the second row are:

$$-\begin{vmatrix} -2 & 3 \\ 1 & 2 \end{vmatrix}, \quad \begin{vmatrix} 1 & 3 \\ -3 & 2 \end{vmatrix}, \quad -\begin{vmatrix} 1 & -2 \\ -3 & 1 \end{vmatrix}$$

$\Rightarrow \quad -(-4-3), \qquad (2+9), \qquad -(1-6),$ *i.e.* 7, 11, 5

The cofactors of – 3, 1, 2, the elements of the third row are:

$$\begin{vmatrix} -2 & 3 \\ 3 & -1 \end{vmatrix}, \quad -\begin{vmatrix} 1 & 3 \\ 2 & -1 \end{vmatrix}, \quad \begin{vmatrix} 1 & -2 \\ 2 & 3 \end{vmatrix}$$

$= \quad (2-9), \qquad -(1-6), \qquad (3+4)$

$= \quad -7, 7, 7$

$$\text{Cofactor matrix} = \begin{bmatrix} 7 & -1 & 11 \\ 7 & 11 & 5 \\ -7 & 7 & 7 \end{bmatrix}$$

$$\text{Adj} \cdot A = \begin{bmatrix} 7 & 7 & -7 \\ -1 & 11 & 7 \\ 11 & 5 & 7 \end{bmatrix}$$

$$A^{-1} = \frac{1}{|A|} \text{Adj} \cdot A = \frac{1}{42} \begin{bmatrix} 7 & 7 & -7 \\ -1 & 11 & 7 \\ 11 & 5 & 7 \end{bmatrix}$$ **Ans.**

Example 52. *Find the inverse of the matrix*

$$A = \begin{bmatrix} 3 & -3 & 4 \\ 2 & -3 & 4 \\ 0 & -1 & 1 \end{bmatrix}$$ **(S.B.T.E. 2015)**

Solution. Here we have

$$A = \begin{bmatrix} 3 & -3 & 4 \\ 2 & -3 & 4 \\ 0 & -1 & 1 \end{bmatrix}, \ |A| = 3(-3+4) + 3(2-0) + 4(-2-0) = 1 \neq 0$$

$$\text{Cofactor matrix } (A) = \begin{bmatrix} 1 & -2 & -2 \\ -1 & 3 & 3 \\ 0 & -4 & -3 \end{bmatrix}$$

$$\text{Adj} \cdot A = \begin{bmatrix} 1 & -1 & 0 \\ -2 & 3 & -4 \\ -2 & 3 & -3 \end{bmatrix}$$

$$A^{-1} = \frac{1}{|A|} \text{Adj} \cdot A = \frac{1}{1} \begin{bmatrix} 1 & -1 & 0 \\ -2 & 3 & -4 \\ -2 & 3 & -3 \end{bmatrix}$$

$$\therefore \quad A^{-1} = \begin{bmatrix} 1 & -1 & 0 \\ -2 & 3 & -4 \\ -2 & 3 & -3 \end{bmatrix} \quad \textbf{Ans.}$$

Example 53. *If* $A = \begin{bmatrix} 5 & 2 \\ 3 & 1 \end{bmatrix}$, *find* A^{-1}. **(*S.B.T.E. 2017, 2014*)**

Solution. Here, we have

$$A = \begin{bmatrix} 5 & 2 \\ 3 & 1 \end{bmatrix}, \ |A| = 5 - 6 = -1 \neq 0, \ A^{-1} \text{ exist.}$$

$$\text{Cofactor matrix } (A) = \begin{bmatrix} 1 & -3 \\ -2 & 5 \end{bmatrix}$$

$$\text{Adj} \cdot A = \begin{bmatrix} 1 & -2 \\ -3 & 5 \end{bmatrix}$$

$$A^{-1} = \frac{1}{|A|} \text{Adj} \cdot A = -1 \begin{bmatrix} 1 & -2 \\ -3 & 5 \end{bmatrix} = \begin{bmatrix} -1 & 2 \\ 3 & -5 \end{bmatrix} \quad \textbf{Ans.}$$

Example 54. *Find the matrix A satisfying the equation*

$$\begin{bmatrix} 2 & 1 \\ 3 & 3 \end{bmatrix} A \begin{bmatrix} 5 & 3 \\ 3 & 2 \end{bmatrix} = \begin{bmatrix} 1 & 0 \\ 0 & 1 \end{bmatrix}$$

Solution. Let $|B| = \begin{vmatrix} 2 & 1 \\ 3 & 3 \end{vmatrix}$ and $|C| = \begin{vmatrix} 5 & 3 \\ 3 & 2 \end{vmatrix}$. Then,

$$|B| = \begin{vmatrix} 2 & 1 \\ 3 & 3 \end{vmatrix} = 6 - 3 = 3 \neq 0 \text{ and } |C| = \begin{vmatrix} 5 & 3 \\ 3 & 2 \end{vmatrix} = 10 - 9 = 1 \neq 0$$

So, B and C are invertible matrices. The given matrix equation is

$$BAC = 1$$

On pre-multiplying by B^{-1} and post-multiplying by C^{-1}, we get

$$\Rightarrow \quad B^{-1}(BAC)\,C^{-1} = B^{-1} I\, C^{-1}$$

$$\Rightarrow \quad (B^{-1}B)\,A(CC^{-1}) = B^{-1} C^{-1}$$

$$\Rightarrow \quad I\,A\,I = B^{-1} C^{-1}$$

$$\Rightarrow \quad A = B^{-1} C^{-1} \qquad \text{... (1)}$$

$$\text{Matrix of cofactors of } B = \begin{bmatrix} 3 & -3 \\ -1 & 2 \end{bmatrix}$$

$$\text{Adjoint } B = \begin{bmatrix} 3 & -1 \\ -3 & 2 \end{bmatrix}$$

$$B^{-1} = \frac{1}{|B|}\text{Adj}\,B = \frac{1}{3}\begin{bmatrix} 3 & -1 \\ -3 & 2 \end{bmatrix} \qquad \text{... (2)}$$

$$\text{Matrix of cofactors of } C = \begin{bmatrix} 2 & -3 \\ -3 & 5 \end{bmatrix}$$

$$\text{Adjoint } C = \begin{bmatrix} 2 & -3 \\ -3 & 5 \end{bmatrix}$$

$$C^{-1} = \frac{1}{|C|}\text{Adj}\,C = \begin{bmatrix} 2 & -3 \\ -3 & 5 \end{bmatrix} \qquad \text{... (3)}$$

On putting the value of B^{-1} from (2) and C^{-1} from (3) in (1), we get

$$\therefore \quad A = B^{-1} C^{-1} = \begin{bmatrix} 3 & -1 \\ -3 & 2 \end{bmatrix}\begin{bmatrix} 2 & -3 \\ -3 & 5 \end{bmatrix} = \frac{1}{3}\begin{bmatrix} 6+3 & -9-5 \\ -6-6 & 9+10 \end{bmatrix}$$

$$= \frac{1}{3}\begin{bmatrix} 9 & -14 \\ -12 & 19 \end{bmatrix} = \begin{bmatrix} 3 & -\frac{14}{3} \\ -4 & \frac{19}{3} \end{bmatrix}$$ **Ans.**

EXERCISE 10.4

1. Find the Adjoint of $\begin{bmatrix} 1 & 2 \\ 3 & -5 \end{bmatrix}$. ***(S.B.T.E. 2015)***

Ans. $\begin{bmatrix} -5 & -2 \\ -3 & 1 \end{bmatrix}$

2. If $A^{-1} = \begin{bmatrix} 5 & 3 \\ -2 & -1 \end{bmatrix}$. Find $\left(A^T\right)^{-1}$. ***(S.B.T.E. 2016, 2015)***

Ans. $\begin{bmatrix} 5 & -2 \\ 3 & -1 \end{bmatrix}$

3. If $A = \begin{bmatrix} 2 & 3 \\ 5 & -2 \end{bmatrix}$, show that $A^{-1} = \frac{1}{19}A$. ***(S.B.T.E. 2014, 2013)***

4. If $A=\begin{bmatrix}3 & 1\\ 2 & 2\end{bmatrix}$, and $B=\begin{bmatrix}1 & 3\\ 2 & 4\end{bmatrix}$, show that $(AB)^{-1}=B^{-1}\cdot A^{-1}$. ***(S.B.T.E. 2016, 2013)***

Compute the inverse of the following matrices.

5. $\begin{bmatrix}8 & 4 & 2\\ 2 & 9 & 4\\ 1 & 2 & 8\end{bmatrix}$ **Ans.** $\dfrac{1}{454}\begin{bmatrix}64 & -28 & -2\\ -12 & 62 & -28\\ -5 & -12 & 64\end{bmatrix}$

6. $\begin{bmatrix}1 & 2 & 5\\ 2 & 3 & 1\\ -1 & 1 & 1\end{bmatrix}$ **Ans.** $\dfrac{1}{21}\begin{bmatrix}2 & 3 & -13\\ -3 & 6 & 9\\ 5 & -3 & -1\end{bmatrix}$

7. $\begin{bmatrix}3 & -4 & 2\\ 0 & 5 & 9\\ -4 & 8 & 1\end{bmatrix}$ **Ans.** $-\dfrac{1}{17}\begin{bmatrix}-67 & 20 & -46\\ -36 & 11 & -27\\ 20 & -8 & 15\end{bmatrix}$

8. $\begin{bmatrix}1 & 3 & -2\\ -3 & 0 & -5\\ 2 & 5 & 0\end{bmatrix}$ **Ans.** $\dfrac{1}{25}\begin{bmatrix}25 & -10 & -15\\ -10 & 4 & 11\\ -15 & 1 & 9\end{bmatrix}$

9. $\begin{bmatrix}-2 & 6 & 4\\ 1 & -3 & 2\\ 1 & 5 & 2\end{bmatrix}$ ***(S.B.T.E. 2015)*** **Ans.** $\dfrac{1}{8}\begin{bmatrix}-2 & 1 & 3\\ 0 & -1 & 1\\ 1 & 2 & 0\end{bmatrix}$

10. $\begin{bmatrix}0 & 1 & 2\\ 1 & 2 & 3\\ 3 & 1 & 1\end{bmatrix}$ **Ans.** $-\dfrac{1}{2}\begin{bmatrix}-1 & 1 & -1\\ 8 & -6 & 2\\ -5 & 3 & -1\end{bmatrix}$

11. Find A satisfying the matrix equation

$$\begin{bmatrix}2 & 1\\ 3 & 2\end{bmatrix}A\begin{bmatrix}-3 & 2\\ 5 & -3\end{bmatrix}=\begin{bmatrix}-2 & 4\\ 3 & -1\end{bmatrix}$$

(B.T.E. Delhi Jan. 2009) **Ans.** $A=\begin{bmatrix}24 & 13\\ -34 & -18\end{bmatrix}$

12. If $A=\begin{bmatrix}3 & 2\\ 7 & 5\end{bmatrix}$ and $B=\begin{bmatrix}6 & 7\\ 8 & 9\end{bmatrix}$, verify that : $(AB)^{-1}=B^{-1}A^{-1}$.

13. If $A=\begin{bmatrix}1 & 1 & 2\\ 1 & 9 & 3\\ 1 & 4 & 4\end{bmatrix}, B=\begin{bmatrix}1 & 2 & 0\\ 2 & 3 & -1\\ 1 & -1 & 3\end{bmatrix}$ verify that $(AB)^{-1}=B^{-1}\cdot A^{-1}$.

14. If $A=\begin{bmatrix}3 & -3 & 4\\ 2 & -3 & 4\\ 0 & -1 & 1\end{bmatrix}$ (*i*) Find A^{-1} (*ii*) Show that $A^3=A^{-1}$. **Ans.** (*i*) $\begin{bmatrix}1 & -1 & 0\\ -2 & 3 & -4\\ -2 & 3 & -3\end{bmatrix}$

15. Verify the theorem A (Adj A) = (Adj A) A = $|A|I$

for the matrix $A = \begin{bmatrix} 1 & 2 & 3 \\ 1 & 3 & 4 \\ 1 & 4 & 3 \end{bmatrix}$ **(S.B.T.E. 2017)**

16. Compute AB and BA where:

$A = \begin{bmatrix} 1 & -2 & 3 \\ 2 & 3 & -1 \\ -3 & 1 & -4 \end{bmatrix}$ and B is 3×3 matrix whose (i, j)th element is cofactor of (j, i)th element in A. **Ans.** $\begin{bmatrix} 0 & 0 & 0 \\ 0 & 0 & 0 \\ 0 & 0 & 0 \end{bmatrix}$

17. Define non-singular and invertible matrices. Prove that a square matrix is invertible if and only if it is non-singular.

18. If a square matrix A satisfies a relation $A^2 + A - I = 0$. Prove that A^{-1} exists and that $A^{-1} = I + A$, I being an identity matrix. **(Diplome IETE, June 2005)**

19. If $I + A + A^2 + ... + A^k = 0$, then A^{-1} is equal to **(Diploma IETE Dec. 2005)**

(*a*) A^k (*b*) A^{k-1} (*c*) A^{k+1} (*d*) $I + A$ **Ans.** (*a*)

20. The inverse of a diagonal matrix is **(Diploma IETE, Dec. 2006)**

(*a*) Not defined (*b*) a skew-symmetrix matrix

(c) a diagonal matrix (*d*) a unit matrix **Ans.** (*c*)

10.13 RANK OF A MATRIX

The rank of a matrix is said to be r if

(*a*) It has at least one non-zero minor of order r

(*b*) Every minor of A of order higher than r is zero.

Note: (*i*) Non-zero row is that row in which all the elements are not zero.

(*ii*) The rank of the product matrix AB of two matrices A and B is less than the rank of either of the matrices A and B.

(*iii*) Corresponding to every matrix A of rank r, there exist non-singular matrices P and Q such that $PAQ = \begin{bmatrix} I_r & 0 \\ 0 & 0 \end{bmatrix}$

Normal Form (Canonical Form)

By performing elementary transformation, any non-zero matrix A can be reduced to one of the following four forms, called the Normal form of A:

(*i*) I_r (*ii*) $[I_r, 0]$ (*iii*) $\begin{bmatrix} I_r \\ 0 \end{bmatrix}$ (*iv*) $\begin{bmatrix} I_r & 0 \\ 0 & 0 \end{bmatrix}$

The number r so obtained is called the rank of A and we write $\rho(A)$ r. The form $\begin{bmatrix} I_r & 0 \\ 0 & 0 \end{bmatrix}$ is called first canonical form of A. Since both row and column transformations may be used here, the element 1 of the first row obtained can be moved in the first column. Then both the first row and first column can be cleared of other non-zero elements. Similarly, the element 1 of the second row can be brought into the second column, and so on.

Rank of 2 × 2 matrix

Example 55: $A = \begin{bmatrix} 2 & 1 \\ 1 & 3 \end{bmatrix}$

Solution. $A = \begin{bmatrix} 2 & 1 \\ 1 & 3 \end{bmatrix} R_2 \to 2R_2 - R_1 = \begin{bmatrix} 2 & 1 \\ 0 & 5 \end{bmatrix}$

The number of non-zero row is 2

∴ Rank of matrix A is 2

Example 56. *Find the rank of the matrix.*

$$A = \begin{bmatrix} 1 & 3 & 5 & 7 \\ -3 & 2 & 4 & 5 \end{bmatrix}$$

Solution. Here, we have

$$A = \begin{bmatrix} 1 & 3 & 5 & 7 \\ -3 & 2 & 4 & 5 \end{bmatrix}$$

$$= \begin{bmatrix} 1 & 3 & 5 & 7 \\ 0 & 11 & 19 & 26 \end{bmatrix} R_2 \to R_2 + 3R_1$$

The number of non-zero row is 2, therefore Rank $(A) = 2$ **Ans.**

Example 57. *Find the rank of the matrix*

$$A = \begin{bmatrix} 1 & 2 & 3 & -4 \\ -2 & 3 & 7 & -1 \\ 1 & 9 & 16 & -13 \end{bmatrix}$$

Solution. Here, we have

$$A = \begin{bmatrix} 1 & 2 & 3 & -4 \\ -2 & 3 & 7 & -1 \\ 1 & 9 & 16 & -13 \end{bmatrix}$$

$$= \begin{bmatrix} 1 & 2 & 3 & -4 \\ 0 & 7 & 13 & -9 \\ 0 & 7 & 13 & -9 \end{bmatrix} \begin{matrix} \\ R_2 \to R_2 + 2R_1 \\ R_3 \to R_3 - R_1 \end{matrix}$$

$$= \begin{bmatrix} 1 & 2 & 3 & -4 \\ 0 & 7 & 13 & -9 \\ 0 & 0 & 0 & 0 \end{bmatrix} R_3 \to R_3 - R_2$$

Here the number of non-zero row is 2, therefore

Rank $(A) = 2$ **Ans.**

Example 58. *Reduce the matrix A to diagonal form* *(U.P.I semester Dec: 2005)*

$$A = \begin{bmatrix} -1 & 2 & -2 \\ 1 & 2 & 1 \\ -1 & -1 & 2 \end{bmatrix}$$

Solution. We have, $A = \begin{bmatrix} -1 & 2 & -2 \\ 1 & 2 & 1 \\ -1 & -1 & 2 \end{bmatrix}$

$$= \begin{bmatrix} -1 & 2 & -2 \\ 0 & 4 & -1 \\ 0 & -3 & 4 \end{bmatrix} \begin{matrix} \\ R_2 \to R_2 + R_1 \\ R_3 \to R_3 - R_1 \end{matrix}$$

$$= \begin{bmatrix} -1 & 2 & -2 \\ 0 & 4 & -1 \\ 0 & 0 & 13/4 \end{bmatrix} R_3 \to R_3 + \frac{3}{4} R_2$$

$$= \begin{bmatrix} -1 & 2 & -2 \\ 0 & 4 & -1 \\ 0 & 0 & 0 \end{bmatrix} R_3 \to \frac{4}{13} R_3$$

$$= \begin{bmatrix} -1 & 2 & 0 \\ 0 & 4 & 0 \\ 0 & 0 & 1 \end{bmatrix} \begin{matrix} R_1 \to R_1 + 2R_3 \\ R_2 \to R_2 + R_3 \\ \end{matrix}$$

$$= \begin{bmatrix} -1 & 2 & 0 \\ 0 & 1 & 0 \\ 0 & 0 & 1 \end{bmatrix} R_2 \to \frac{1}{4} R_2$$

$$= \begin{bmatrix} -1 & 0 & 0 \\ 0 & 1 & 0 \\ 0 & 0 & 1 \end{bmatrix} \begin{matrix} R_1 \to R_1 - 2R_2 \\ \\ \end{matrix}$$

$$= \begin{bmatrix} 1 & 0 & 0 \\ 0 & 1 & 0 \\ 0 & 0 & 1 \end{bmatrix} \begin{matrix} R_1 \to (-1)R_1 \\ \\ \end{matrix}$$

Ans.

EXERCISE 10.5

Find non singular matrices P and Q such that PAQ is normal form

1. $\begin{bmatrix} 1 & 2 & 3 \\ 3 & 1 & 2 \end{bmatrix}$ **Ans.** $p = \begin{bmatrix} 1 & 0 \\ -3 & 1 \end{bmatrix}, Q = \begin{bmatrix} 1 & \frac{2}{5} & \frac{-1}{5} \\ 0 & \frac{-1}{5} & \frac{-7}{5} \\ 0 & 0 & 1 \end{bmatrix}$

2. $\begin{bmatrix} 1 & 1 & 2 \\ 1 & 2 & 3 \\ 0 & -1 & -1 \end{bmatrix}$ **Ans.** $p = \begin{bmatrix} 1 & 0 & 0 \\ -1 & 1 & 0 \\ -1 & 1 & 1 \end{bmatrix}, Q = \begin{bmatrix} 1 & -1 & -1 \\ 0 & 1 & -1 \\ 0 & 0 & 1 \end{bmatrix}$

3. $\begin{bmatrix} 1 & 2 & 3 & -2 \\ 2 & -2 & 1 & 3 \\ 3 & 0 & 4 & 1 \end{bmatrix}$ **Ans.** $P = \begin{bmatrix} 1 & 0 & 0 \\ -2 & 1 & 0 \\ -1 & -1 & 1 \end{bmatrix}$, $Q = \begin{bmatrix} 1 & \frac{1}{3} & \frac{4}{15} & \frac{-1}{21} \\ 0 & -\frac{1}{6} & \frac{1}{6} & \frac{1}{6} \\ 0 & 0 & -\frac{1}{5} & 0 \\ 0 & 0 & 0 & \frac{1}{7} \end{bmatrix}$

Triangular Form

Rank = Number of non-zero row is upper triangular matrix.

Note. Non-zero row is that row which does not contain all the elements as zero.

Example 59. *Find the rank of the matrix*

$$\begin{bmatrix} 1 & 2 & 3 & 2 \\ 2 & 3 & 5 & 1 \\ 1 & 3 & 4 & 5 \end{bmatrix}$$

Solution. $\begin{bmatrix} 1 & 2 & 3 & 2 \\ 2 & 3 & 5 & 1 \\ 1 & 3 & 4 & 5 \end{bmatrix} \sim \begin{bmatrix} 1 & 2 & 3 & 2 \\ 0 & -1 & -1 & -3 \\ 0 & 1 & 1 & 3 \end{bmatrix} \begin{matrix} \\ R_2 \to R_2 - 2R_1 \\ R_3 \to R_3 - R_1 \end{matrix}$

$$\sim \begin{bmatrix} 1 & 2 & 3 & 2 \\ 0 & -1 & -1 & -3 \\ 0 & 0 & 0 & 0 \end{bmatrix} \begin{matrix} \\ \\ R_3 \to R_3 + R_2 \end{matrix}$$

Rank = Number of non zero rows = 2. **Ans.**

Example 60. *Find the rank of the matrix* $\begin{bmatrix} -1 & 2 & 3 & -2 \\ 2 & -5 & 1 & 2 \\ 3 & -8 & 5 & 2 \\ 5 & -12 & -1 & 6 \end{bmatrix}$

Solution. $\begin{bmatrix} -1 & 2 & 3 & -2 \\ 2 & -5 & 1 & 2 \\ 3 & -8 & 5 & 2 \\ 5 & -12 & -1 & 6 \end{bmatrix} \sim \begin{bmatrix} -1 & 2 & 3 & -2 \\ 0 & -1 & 7 & -2 \\ 0 & -2 & 14 & -4 \\ 0 & -2 & 14 & -4 \end{bmatrix} \begin{matrix} \\ R_2 \to R_2 + 2R_1 \\ R_3 \to R_3 + 3R_1 \\ R_4 \to R_4 + 5R_1 \end{matrix}$

$$\sim \begin{bmatrix} -1 & 2 & 3 & -2 \\ 0 & -1 & 7 & -2 \\ 0 & 0 & 0 & 0 \\ 0 & 0 & 0 & 0 \end{bmatrix} \begin{matrix} \\ \\ R_3 \to R_3 - 2R_2 \\ R_4 \to R_4 - 2R_2 \end{matrix}$$

Here the 4th order and 3rd order minors are zero. But a minor or second order

$$\begin{vmatrix} 3 & -2 \\ 7 & -2 \end{vmatrix} = -4 + 14 + 8 \neq 0$$

Rank = Number of non-zero rows = 2. $\therefore R(A) = 2$ **Ans.**

Example 61. *Use elementary transformation to reduce the following matrix A to triangular v form and hence find the rank of A.*

$$A = \begin{bmatrix} 2 & 3 & -1 & -1 \\ 1 & -1 & -2 & -4 \\ 3 & 1 & 3 & -2 \\ 6 & 3 & 0 & -7 \end{bmatrix}$$

(AMIETC, DEC, 2015)(R.G.P.V., Bhopal, June 2007, I Semester, Dec. 2005)

Solution. We have,

$$A = \begin{bmatrix} 2 & 3 & -1 & -1 \\ 1 & -1 & -2 & -4 \\ 3 & 1 & 3 & -2 \\ 6 & 3 & 0 & -7 \end{bmatrix} \sim \begin{bmatrix} 1 & -1 & -2 & -4 \\ 2 & 3 & -1 & -1 \\ 3 & 1 & 3 & -2 \\ 6 & 3 & 0 & -7 \end{bmatrix} R_1 \leftrightarrow R_2$$

$$\sim \begin{bmatrix} 1 & -1 & -2 & -4 \\ 0 & 5 & 3 & 7 \\ 0 & 4 & 9 & 10 \\ 0 & 9 & 12 & 17 \end{bmatrix} \begin{matrix} \\ R_2 \to R_2 - 2R_1 \\ R_3 \to R_3 - 3R_1 \\ R_4 \to R_4 - 6R_1 \end{matrix} \sim \begin{bmatrix} 1 & -1 & -2 & -4 \\ 0 & 5 & 3 & 7 \\ 0 & 0 & 33/5 & 22/5 \\ 0 & 0 & 33/5 & 22/5 \end{bmatrix} \begin{matrix} \\ \\ R_3 \to R_3 - 4/5R_2 \\ R_4 \to R_4 - 9/5R_2 \end{matrix}$$

$$\sim \begin{bmatrix} 1 & -1 & -2 & -4 \\ 0 & 5 & 3 & 7 \\ 0 & 0 & 33/5 & 22/5 \\ 0 & 0 & 0 & 0 \end{bmatrix} R_4 \to R_4 - R_3$$

$R(A)$ = Number of non-zero rows.

$\Rightarrow R(A) = 3$ **Ans.**

EXERCISE 10.6

Find the rank of the following matrices:

1. $\begin{bmatrix} 1 & 2 & 3 \\ 2 & 4 & 7 \\ 3 & 6 & 10 \end{bmatrix}$ **Ans.** 2

2. $\begin{bmatrix} 1 & 2 & 1 \\ -1 & 0 & 2 \\ 2 & 1 & -3 \end{bmatrix}$ **Ans.** 3

3. $\begin{bmatrix} 0 & 1 & 2 & -2 \\ 4 & 0 & 2 & 6 \\ 2 & 1 & 3 & 1 \end{bmatrix}$ **Ans.** 2

4. $\begin{bmatrix} 2 & 4 & 3 & -2 \\ -3 & -2 & -1 & 4 \\ 6 & -1 & 7 & 3 \end{bmatrix}$ **Ans.** 3

5. $\begin{bmatrix} 3 & 4 & 1 & 1 \\ 2 & 4 & 3 & 6 \\ -1 & -2 & 6 & 4 \\ 1 & -1 & 2 & -3 \end{bmatrix}$ **Ans.** 4

6. $\begin{bmatrix} 1 & 4 & 3 & -2 & 1 \\ -2 & -3 & -1 & 4 & 3 \\ -1 & 6 & 7 & 2 & 9 \\ -3 & 3 & 6 & 6 & 12 \end{bmatrix}$ **Ans.** 2

Reduce the following matrices to Echelon form and find out the rank:

7. $\begin{bmatrix} 1 & 1 & 2 \\ 1 & 2 & 2 \\ 2 & 2 & 3 \end{bmatrix}$ **Ans.** $\begin{bmatrix} 1 & 0 & 0 \\ 0 & 1 & 0 \\ 0 & 0 & 1 \end{bmatrix}$, Rank 3

8. $\begin{bmatrix} 1 & 2 & 3 & 0 \\ 2 & 4 & 3 & 2 \\ 3 & 2 & 1 & 3 \\ 6 & 8 & 7 & 5 \end{bmatrix}$ **Ans.** $\begin{bmatrix} I_3 & 0 \\ 0 & 0 \end{bmatrix}$, Rank = 3

9. $\begin{bmatrix} 2 & 3 & 5 & 7 & 12 \\ 1 & 1 & 2 & 3 & 5 \\ 3 & 3 & 6 & 9 & 15 \end{bmatrix}$ **Ans.** $\begin{bmatrix} I_2 & 0 \\ 0 & 0 \end{bmatrix}$, Rank = 2

10. $\begin{bmatrix} 2 & -4 & 3 & 1 & 0 \\ 1 & -2 & 1 & -4 & 2 \\ 0 & 1 & -1 & 3 & 1 \\ 4 & -7 & 4 & -4 & 5 \end{bmatrix}$ **Ans.** $\begin{bmatrix} I_3 & 0 \\ 0 & 0 \end{bmatrix}$, Rank = 3

Using elementary transformations, reduce the following matrices to the canonical form (or row-reduced Echelon form):

11. $A = \begin{bmatrix} 0 & 0 & 0 & 0 & 0 \\ 0 & 1 & 2 & 3 & 4 \\ 0 & 2 & 3 & 4 & 1 \\ 0 & 3 & 4 & 1 & 2 \end{bmatrix}$ **Ans.** $\begin{bmatrix} 1 & 0 & 0 & 0 \\ 0 & 1 & 0 & 0 \\ 0 & 0 & 1 & 0 \end{bmatrix}$

12. $A = \begin{bmatrix} 0 & 4 & -12 & 8 & 9 \\ 0 & 2 & -6 & 2 & 5 \\ 0 & 1 & -3 & 6 & 4 \\ 0 & -8 & 24 & 3 & 1 \end{bmatrix}$ **Ans.** $\begin{bmatrix} 1 & 0 & 0 & 0 \\ 0 & 1 & 0 & 0 \\ 0 & 0 & 1 & 0 \end{bmatrix}$

Using elementary transformations, reduce the following matrices to the normal form:

13. $A = \begin{bmatrix} 1 & 2 & 0 & -1 \\ 3 & 4 & 1 & 2 \\ -2 & 3 & 2 & 5 \end{bmatrix}$ **Ans.** $\begin{bmatrix} 1 & 0 & 0 & 0 \\ 0 & 1 & 0 & 0 \\ 0 & 0 & 1 & 0 \end{bmatrix}$

14. $A = \begin{bmatrix} 1 & 2 & 3 & 4 \\ 3 & 4 & 1 & 2 \\ 4 & 3 & 1 & 2 \end{bmatrix}$ **Ans.** $\begin{bmatrix} 1 & 0 & 0 & 0 \\ 0 & 1 & 0 & 0 \\ 0 & 0 & 1 & 0 \end{bmatrix}$

Obtain a matrix N in the normal form equivalent to

15. $A = \begin{bmatrix} 0 & 0 & 0 & 0 & 0 \\ 0 & 4 & 5 & 0 & 0 \\ 0 & 9 & 1 & -1 & 2 \\ 0 & 10 & 0 & 1 & 11 \end{bmatrix}$

Hence find non-singular matrices P and Q such that $PAQ = N$.

16. $\begin{bmatrix} 1 & -3 & 1 & 2 \\ 0 & 1 & 2 & 3 \\ 3 & 4 & 1 & -2 \end{bmatrix}$

Find the rank of the following matrix by reducing it into normal form:

17. $A = \begin{bmatrix} 1 & 3 & 2 & 5 & 1 \\ 2 & 2 & -1 & 6 & 3 \\ 1 & 1 & 2 & 3 & -1 \\ 0 & 2 & 5 & 2 & -3 \end{bmatrix}$ **Ans.** 3

18. $A = \begin{bmatrix} 1 & 2 & 3 & 1 \\ 1 & 3 & 3 & 2 \\ 2 & 4 & 3 & 3 \\ 1 & 1 & 1 & 1 \end{bmatrix}$ **Ans.** 4

19. Rank of matrix $A = \begin{bmatrix} 1 & 2 & 3 \\ 1 & 4 & 2 \\ 2 & 6 & 5 \end{bmatrix}$ is

(*a*) 0 (*b*) 1 (*c*) 3 (*d*) 2

(*AMIETE, June 2009*) **Ans.** (*d*)

20. For which value of '*b*' the rank of the matrix $A =$ is

(*a*) 1 (*b*) 2 (*c*) 3 (*d*) 0

(*AMIETE, Dec. 2009*) **Ans.** (*b*)

10.14 SIMULTANEOUS EQUATIONS WITH MORE THAN THREE UNKNOWNS

We have already solved simultaneous equations of two or three unknowns. When the number of unknowns in simultaneous equations is large, then it becomes tedious to solve them by the known methods. Simultaneous equations of large number of unknowns are very important in the field of science and engineering. We can use the following methods to solve such simultaneous equations.

(*a*) Gauss elimination method (*b*) Gauss–Jordan method

10.15 GAUSS ELIMINATION METHOD

In this method the unknowns of equations below are eliminated and the system is reduced to an upper triangular system. The unknowns are obtained by back substitution.

Let a system of simultaneous equations in n unknowns $x_1, x_2 \ldots\ldots\ldots\ldots x_n$ be

$$a_{11}x_1 + a_{12}x_2 + \ldots\ldots\ldots\ldots + a_{1n}x_n = b_1 \quad \ldots(1)$$

$$a_{21}x_1 + a_{22}x_2 + \ldots\ldots\ldots\ldots + a_{2n}x_n = b_2 \quad \ldots(2)$$

....

....

$$a_{n1}x_1 + a_{n2}x_2 + \ldots\ldots\ldots\ldots + a_{nn}x_n = b_n \quad \ldots(3)$$

Method to solve the above equations

Step 1. We eliminate x_1 from 2nd, 3rd .... nth equation with the help first equation

$$a_{11}x_1 + a_{12}x_2 + \ldots\ldots\ldots + a_{1n}x_n = b_1$$

$$a'_{22}x_2 + \ldots\ldots\ldots + a'_{2n}x_n = b_2'$$

....

....

$$a'_{n2}x_2 + \ldots\ldots\ldots + a'_{nn}x_n = b_n'$$

Step 2. We again eliminate x_2 from 3rd, 4th .... nth equation with the help of second equation.

$$a_{11}x_1 + a_{12}x_2 + a_{13}x_3 + \ldots\ldots\ldots + a_{1n}x_n = b_1$$

$$a'_{22}x_2 + a'_{23}x_3 + \ldots\ldots\ldots + a'_{2n}x_n = b_2'$$

$$a''_{33}x_3 + \ldots\ldots\ldots + a''_{3n}x_n = b_3''$$

....

....

$$a''_{n3}x_3 + \ldots\ldots\ldots + a''_{nn}x_n = b_n''$$

Step 3. We will eliminate x_3 and in fourth step x_4 and so on.

Finally the system of equations will be of the following form.

$$a_{11}x_1 + a_{12}x_2 + \ldots\ldots\ldots + a_{1n}x_n = b_1$$

$$a_{22}x_2 + \ldots\ldots\ldots + a'_{2n}x_n = b_2'$$

....

....

$$c_{nn}x_n = d_n$$

The given system is reduced to the above form *i.e.* triangular form.

Backward Substitution

We first find out the value of x_n from the last equation, then substitute the value of x_n in the $(n - 1)$th equation to get the value of x_{n-1}. Again substitute the value of x_{n-1} in $(n - 2)$th equation to get the value of x_{n-2}. By this backward substitution we can find the values of all the unknowns.

Example 62. *Solve the following equations by using Gauss-elimination method:*

$$2x_1 + 4x_2 + x_3 = 3$$

$$3x_1 + 2x_2 - 2x_3 = -2$$

$$x_1 - x_2 + x_3 = 6$$

(*R. G. P. V. Bhopal. III Semester, June 2006*)

Solution. When third equation is written as first equation, the system becomes

$$x_1 - x_2 + x_3 = 6 \quad ...(1)$$

$$2x_1 + 4x_2 + x_3 = 3 \quad ...(2)$$

$$3x_1 + 2x_2 - 2x_3 = -2 \quad ...(3)$$

Step 1. Subtracting 2 (1) from (2), and 3 (1) from (3), we get

$$x_1 - x_2 + x_3 = 6$$

$$6x_2 - x_3 = -9 \quad ...(4)$$

$$5x_2 - 5x_3 = -20 \quad ...(5)$$

Step 2. Operate $\frac{6}{5}$ (5) – (4)

$$x_1 - x_2 + x_3 = 6 \quad ...(1)$$

$$6x_2 - x_3 = -9 \quad ...(6)$$

$$-5x_3 = -15$$

Step 3. Backward substitution

From (7), $\quad x_3 = \frac{-15}{-5} = 3$

From (6), $\quad 6x_2 - 3 = -9 \Rightarrow 6x_2 = -6 \Rightarrow x_2 = -1$

From (1), $\quad x_1 - (-1) + 3 = 6 \Rightarrow x_1 = 6 - 3 - 1 = 2$

Hence, $\quad x_1 = 2, x_2 = -1, x_3 = 3$ **Ans.**

Example 63. *Solve the following equations by using Gauss elimination method:*

$$x + 2y + 3z + u = 3$$

$$4x - 6y - z - u = 27$$

$$3x - 2y - 3z + 2u = 13$$

$$x + y + z - u = 3$$

Solution. We have,

$$x + 2y + 3z + u = 3 \quad ...(1)$$

$$4x - 6y - z - u = 27 \quad ...(2)$$

$$3x - 2y - 3z + 2u = 13 \quad ...(3)$$

$$x + y + z - u = 3 \quad ...(4)$$

Step 1. Operate (2) – 4(1), (3) – 3(1) and (4) – 1(1),

$$x + 2y + 3z + u = 3 \quad ...(1)$$

$$-14y - 13z - 5u = 15 \quad ...(5)$$

$$-8y - 12z - u = 4 \quad ...(6)$$

$$-y - 2z - 2u = 0 \quad ...(7)$$

Step 2. Operate (6) $-\frac{4}{7}(5)$ and (7) $-\frac{1}{14}(5)$

$$x + 2y + 3z + u = 3 \quad \ldots(1)$$

$$-14y - 13z - 5u = 15 \quad \ldots(5)$$

$$-\frac{32}{7}z + \frac{32}{7}u = -\frac{32}{7} \Rightarrow -32z + 13u = -32 \quad \ldots(8)$$

$$\frac{-15}{14}z - \frac{23}{14}u = \frac{-15}{14} \Rightarrow 15z + 23u = 15 \quad \ldots(9)$$

Step 3. Operate (9) + $\frac{15}{32}$(8)

$$x + 2y + 3z + u = 3 \quad \ldots(1)$$

$$-14y - 13z - 5u = 15 \quad \ldots(5)$$

$$-32z + 13u = -32 \quad \ldots(8)$$

$$\frac{931}{32}u = 0 \quad \ldots(10)$$

Step 4. Backward substitution:

From (10), $\frac{931}{32}u = 0 \Rightarrow u = 0$

From (8), $-32z + 0 = -32 \Rightarrow z = 1$

From(5), $-14y - 13(1) - 0 = 15 \Rightarrow y = -2$

From(1), $x + 2(-2) + 3(1) + 0 = 3 \Rightarrow x = 4$

Hence, $x = 4, y = -2, z = 1, u = 0$ **Ans.**

EXERCISE 10.7

Solve the following system by Gauss elimination method.

1. $x - y + z = 1, -3x + 2y - 3z = -6, 2x - 5y + 4z = 5$ **Ans.** $x = -2, y = 3, z = 6$
2. $x + 10y + z = 12, x + y + 10z = 12, 10x + y + z = 12$ **Ans.** $x = 1, y = 1, z = 1$
3. $x - 2y + 9z = 8, 2x - 8y + z = -5, 3x + y - z = 3$ **Ans.** $x = 1, y = 1, z = 1$
4. $x + 3y + 10z = 23.89, 2x + 17y + 4z = 34.84, 28x + 4y - z = 31.88$
   Ans. $x = 0.99, y = 1.50, z = 1.84$
5. $2x + 6y - z = -11.98, 5x - y + z = 11.01, 4x - y + 3z = 10.01$
   Ans. $x = 1.64, y = -2.49, z = 0.32$
6. $x + y + z - u = 2, 7x + y + 3z + u = 12, 8x - y + z - 3u = -5, 10x + 5y + 3z + 2u = 20$
   Ans. $x = 1, y = 1, z = 1, u = 1$
7. $2z + 2y - z + u = 4, 4x + 3y - z + 2u = 6, 8x + 5y - 3z + 4u = 12, 3x + 3y - 2z + 2u = 6$
   Ans. $x = 1, y = 1, z = -1, u = -1$

8. $x_1 + 2x_2 - 12x_3 + 8x_4 = 27,\ 5x_1 + 4x_2 + 7x_3 - 2x_4 = 4,\ 6x_1 - 12x_2 - 8x_3 + 3x_4 = 49,\ 3x_1 - 7x_2 - 9x_3 - 5x_4 = -11$ **Ans.** $x_1 = 3,\ x_2 = -2,\ x_3 = 1,\ x_4 = 5$

9. $x_1 + 2x_2 + x_3 - x_4 = -2,\ 2x_1 + 3x_2 - x_3 + 2x_4 = 7,\ x_1 + x_2 + 3x_3 - 2x_4 = -6,\ x_1 + x_2 + x_3 + x_4 = 2$ **Ans.** $x_1 = 1,\ x_2 = 0,\ x_3 = -1,\ x_4 = 2$

10. $x_1 + 2x_2 + 3x_3 + 4x_4 = 32,\ 2x_1 - x_2 + 2x_3 - x_4 = 3,\ 3x_1 + 2x_2 + 4x_3 - x_4 = 17.7,\ 6x_1 + 7x_2 + 8x_3 - 5x_4 = 3$ **Ans.** $x_1 = 1.1,\ x_2 = 2.2,\ x_3 = 3.5,\ x_4 = 4$

10.16 GAUSS-JORDAN METHOD

(R.G.P.V. Bhopal, III Semester, Dec. 2007, AKTU. 2016 - 2017)

By this method we eliminate unknowns not only from the equations below but also from the equations above. In this way the system is reduced to a diagonal matrix.

Finally each equation consists of only one unknown and thus, we get the solution. Here, the labour of backward substitution for finding the unknowns is saved.

Gauss-Jordan method is modification of Gauss elimination method.

Example 64. *Apply Gauss-Jordan method to solve the equations*

$$x + y + z = 9$$

$$2x - 3y + 4z = 13$$

$$3x + 4y + 5z = 40$$

(R.G.P.V. Bhopal, III Semester. Dec. 2007)

Solution. The following system of linear equations can be written in matrix form:

$$AX = B$$

where $A = \begin{bmatrix} 1 & 1 & 1 \\ 2 & -3 & 4 \\ 3 & 4 & 5 \end{bmatrix},\ X = \begin{bmatrix} x \\ y \\ z \end{bmatrix},\ B = \begin{bmatrix} 9 \\ 13 \\ 40 \end{bmatrix}$

By using Gauss-Jordon method we have

$$R_2 \to R_2 - 2R_1$$
$$R_3 \to R_3 - 3R_1$$

$$\Rightarrow \begin{bmatrix} 1 & 1 & 1 \\ 2 & -3 & 4 \\ 3 & 4 & 5 \end{bmatrix}\begin{bmatrix} x \\ y \\ z \end{bmatrix} = \begin{bmatrix} 9 \\ 13 \\ 40 \end{bmatrix} \quad \Rightarrow \begin{bmatrix} 1 & 1 & 1 \\ 0 & -5 & 2 \\ 0 & 1 & 2 \end{bmatrix}\begin{bmatrix} x \\ y \\ z \end{bmatrix} = \begin{bmatrix} 9 \\ -5 \\ 13 \end{bmatrix}$$

$$R_3 \to R_3 + \frac{1}{5}R_2 \qquad\qquad R_1 \to R_1 + \frac{1}{5}R_2$$

$$\Rightarrow \begin{bmatrix} 1 & 1 & 1 \\ 0 & -5 & 2 \\ 0 & 0 & \frac{12}{5} \end{bmatrix}\begin{bmatrix} x \\ y \\ z \end{bmatrix} = \begin{bmatrix} 9 \\ -5 \\ 13 \end{bmatrix} \quad \Rightarrow \begin{bmatrix} 1 & 0 & \frac{7}{5} \\ 0 & -5 & 2 \\ 0 & 0 & \frac{12}{5} \end{bmatrix}\begin{bmatrix} x \\ y \\ z \end{bmatrix} = \begin{bmatrix} 8 \\ -5 \\ 12 \end{bmatrix}$$

$R_1 \to R_1 - \frac{7}{12} R_3$ $\qquad$ $R_2 \to -\frac{1}{5} R_2$

$R_2 \to R_2 - \frac{5}{6} R_3$ $\qquad$ $R_3 \to -\frac{1}{5} R_3$

$$\Rightarrow \begin{bmatrix} 1 & 0 & 0 \\ 0 & -5 & 0 \\ 0 & 0 & \frac{12}{5} \end{bmatrix} \begin{bmatrix} x \\ y \\ z \end{bmatrix} = \begin{bmatrix} 1 \\ -15 \\ 12 \end{bmatrix} \qquad \Rightarrow \begin{bmatrix} 1 & 0 & 0 \\ 0 & 1 & 0 \\ 0 & 0 & 1 \end{bmatrix} \begin{bmatrix} x \\ y \\ z \end{bmatrix} = \begin{bmatrix} 1 \\ 2 \\ 5 \end{bmatrix}$$

where $x = 1, y = 3, z = 5$ **Ans.**

Example 65. *Solve the system of equations by Gauss-Jordan method:*

$$x + y + z + u = 2$$
$$2x - y + 2z - u = -5$$
$$3x + 2y + 3z + 4u = 7$$
$$x - 2y - 3z + 2u = 5$$

Solution. The following of equations can be written in matrix form as

$$AX = B$$

where
$$A = \begin{bmatrix} 1 & 1 & 1 & 1 \\ 2 & -1 & 2 & -1 \\ 3 & 2 & 3 & 4 \\ 1 & -2 & -3 & 2 \end{bmatrix} \begin{bmatrix} x \\ y \\ z \\ u \end{bmatrix} = \begin{bmatrix} 2 \\ -5 \\ 7 \\ 5 \end{bmatrix} \begin{matrix} \\ R_2 \to R_2 - 2R_1 \\ R_3 \to R_3 - 3R_1 \\ R_4 \to R_4 - R_1 \end{matrix}$$

Using Gauss-Jordan method, we have

$$\begin{bmatrix} 1 & 1 & 1 & 1 \\ 0 & -3 & 0 & -3 \\ 0 & -1 & 0 & 1 \\ 0 & -3 & -4 & 1 \end{bmatrix} \begin{bmatrix} x \\ y \\ z \\ u \end{bmatrix} = \begin{bmatrix} 2 \\ -9 \\ 1 \\ 3 \end{bmatrix} R_2 \leftrightarrow R_3$$

$$\Rightarrow \begin{bmatrix} 1 & 1 & 1 & 1 \\ 0 & -1 & 0 & 1 \\ 0 & -3 & 0 & -3 \\ 0 & -3 & -4 & 1 \end{bmatrix} \begin{bmatrix} x \\ y \\ z \\ u \end{bmatrix} = \begin{bmatrix} 2 \\ 1 \\ -9 \\ 3 \end{bmatrix} \begin{matrix} \\ \\ R_3 \to R_3 - 3R_2 \\ R_4 \to R_4 - 3R_2 \end{matrix}$$

$$\Rightarrow \begin{bmatrix} 1 & 1 & 1 & 1 \\ 0 & -1 & 0 & 1 \\ 0 & 0 & 0 & -6 \\ 0 & 0 & -4 & -2 \end{bmatrix} \begin{bmatrix} x \\ y \\ z \\ u \end{bmatrix} = \begin{bmatrix} 2 \\ 1 \\ -12 \\ 0 \end{bmatrix} R_3 \leftrightarrow R_4$$

$$\Rightarrow \begin{bmatrix} 1 & 1 & 1 & 1 \\ 0 & -1 & 0 & -1 \\ 0 & 0 & -4 & -2 \\ 0 & 0 & 0 & -6 \end{bmatrix} \begin{bmatrix} x \\ y \\ z \\ u \end{bmatrix} = \begin{bmatrix} 2 \\ 1 \\ 0 \\ -12 \end{bmatrix} \begin{matrix} R_2 \to -R_2 \\ \\ R_3 \to -\frac{1}{4} R_3 \\ \\ R_4 \to -\frac{1}{6} R_4 \end{matrix}$$

$$\Rightarrow \begin{bmatrix} 1 & 1 & 1 & 1 \\ 0 & 1 & 0 & -1 \\ 0 & 0 & 1 & \frac{1}{2} \\ 0 & 0 & 0 & 1 \end{bmatrix} \begin{bmatrix} x \\ y \\ z \\ u \end{bmatrix} = \begin{bmatrix} 2 \\ -1 \\ 0 \\ 2 \end{bmatrix} \begin{matrix} R_1 \to R_1 - R_4 \\ R_2 \to R_2 + R_4 \\ R_3 \to R_3 - \frac{1}{2} R_4 \end{matrix}$$

$$\Rightarrow \begin{bmatrix} 1 & 1 & 1 & 0 \\ 0 & 1 & 0 & 0 \\ 0 & 0 & 1 & 0 \\ 0 & 0 & 0 & 1 \end{bmatrix} \begin{bmatrix} x \\ y \\ z \\ u \end{bmatrix} = \begin{bmatrix} 0 \\ 1 \\ -1 \\ 2 \end{bmatrix} R_1 \to R_1 - R_3$$

$$\Rightarrow \begin{bmatrix} 1 & 1 & 0 & 0 \\ 0 & 1 & 0 & 0 \\ 0 & 0 & 1 & 0 \\ 0 & 0 & 0 & 1 \end{bmatrix} \begin{bmatrix} x \\ y \\ z \\ u \end{bmatrix} = \begin{bmatrix} 1 \\ 1 \\ -1 \\ 2 \end{bmatrix} R_1 \to R_1 - R_2$$

$$\Rightarrow \begin{bmatrix} 1 & 0 & 0 & 0 \\ 0 & 1 & 0 & 0 \\ 0 & 0 & 1 & 0 \\ 0 & 0 & 0 & 1 \end{bmatrix} \begin{bmatrix} x \\ y \\ z \\ u \end{bmatrix} = \begin{bmatrix} 0 \\ 1 \\ -1 \\ 2 \end{bmatrix}$$

$x = 0, y = 1, z = -1, u = 2$ Ans.

EXERCISE 10.8

Solve the following system by Gauss–Jordan method.

1. $2x - 6y + 8z = 24, 5x + 4y - 3z = 2, 3x + y + 2z = 16$ **Ans.** $x = 1, y = 3, z = 5$

2. $x + 2y + z = 8, 2x + 3y + 4z = 20, 4x + 3y + 2z = 16$ **Ans.** $x = 1, y = 2, z = 3$

3. $3x + 4y + 5z = 18, 2x - y + 8z = 13, 5x - 2y + 7z = 20$ **Ans.** $x = 3, y = 1, z = 1$

4. $2x - y + 3z = 9, x + y + z = 6, x - y + z = 2$ **Ans.** $x = 1, y = 2, z = 3$

5. $10x + y + 2z = 13, 3x + 10y + z = 14, 2x + 3y + 10z = 15$ **Ans.** $x = 1, y = 1, z = 1$

6. $2x_1 + 2x_2 + x_3 = 6, 4x_1 + 2x_2 + 3x_3 = 4, x_1 + x_2 + x_3 = 2$ **Ans.** $x_1 = 5, x_2 = 1, x_3 = -6$

7. $x + 3y + 6z = 2, 3x - y + 4z = 9, x - 4y + 2z = 7$ **Ans.** $x = 2, y = -1, z = \frac{1}{2}$

8. $x + y + z = 6.6, x - y + z = 2.2, x + 2y + 3z = 15.2$ **Ans.** $x = 1.2, y = 2.2, z = 3.2$

9. $9x - 2y + z = 49.78, x + 5y - 3z = 17.99, -2x + 2y + 7z = 18.97$
Ans. $x = 6.13, y = 4.31, z = 3.23$

10. $2x_1 - x_2 + x_3 = -3, 2x_2 - x_3 + x_4 = 1, x_1 + 2x_3 - x_4 = -1, x_1 + x_2 + 3x_4 = 5$
Ans. $x_1 = -1, x_2 = 0, x_3 = 1, x_4 = 2$

10.17 SOLUTION OF SIMULTANEOUS EQUATIONS

Let the equations be

$$a_1x + a_2y + a_3z = d_1$$
$$b_1x + b_2y + b_3z = d_2$$
$$c_1x + c_2y + c_3z = d_3$$

We write the above equations in the matrix form

$$\begin{bmatrix} a_1 & a_2 & a_3 \\ b_1 & b_2 & b_3 \\ c_1 & c_2 & c_3 \end{bmatrix} \times \begin{bmatrix} x \\ y \\ z \end{bmatrix} = \begin{bmatrix} d_1 \\ d_2 \\ d_3 \end{bmatrix}$$

$$AX = B$$

where $A = \begin{bmatrix} a_1 & a_2 & a_3 \\ b_1 & b_2 & b_3 \\ c_1 & c_2 & c_3 \end{bmatrix}, X = \begin{bmatrix} x \\ y \\ z \end{bmatrix}$ and $B = \begin{bmatrix} d_1 \\ d_2 \\ d_3 \end{bmatrix}$

$$X = A^{-1} B.$$

Example 66. *Solve the equation using matrices*

$$x + y + z = 6$$
$$x - y + z = 2$$
$$2x + y - z = 1$$

Solution. Here, we have

$$x + y + z = 6$$
$$x - y + z = 2$$
$$2x + y - z = 1$$

The given equations are written in the matrix form:

$$\begin{bmatrix} 1 & 1 & 1 \\ 1 & -1 & 1 \\ 2 & 1 & -1 \end{bmatrix} \begin{bmatrix} x \\ y \\ z \end{bmatrix} = \begin{bmatrix} 6 \\ 2 \\ 1 \end{bmatrix}$$

$\Rightarrow \quad AX = B$

$\Rightarrow \quad X = A^{-1} B$

$$\text{Matrix of cofactors of } A = \begin{bmatrix} 0 & 3 & 3 \\ 2 & -3 & 1 \\ 2 & 0 & -2 \end{bmatrix}$$

$$|A| = (1 \times 0) + (1 \times 3)) + (1 \times 3) = 0 + 3 + 3 = 6$$

$$\text{Adjoint } A = \begin{bmatrix} 0 & 2 & 2 \\ 3 & -3 & 0 \\ 3 & 1 & -2 \end{bmatrix}$$

$$A^{-1} = \frac{1}{6} \begin{bmatrix} 0 & 2 & 2 \\ 3 & -3 & 0 \\ 3 & 1 & -2 \end{bmatrix}$$

Now, $\quad X = A^{-1} B$

$$\Rightarrow \quad \begin{bmatrix} x \\ y \\ z \end{bmatrix} = \frac{1}{6} \begin{bmatrix} 0 & 2 & 2 \\ 3 & -3 & 0 \\ 3 & 1 & -2 \end{bmatrix} \begin{bmatrix} 6 \\ 2 \\ 1 \end{bmatrix} = \frac{1}{6} \begin{bmatrix} 0+4+2 \\ 18-6+0 \\ 18+2-2 \end{bmatrix} = \frac{1}{6} \begin{bmatrix} 6 \\ 12 \\ 18 \end{bmatrix} = \begin{bmatrix} 1 \\ 2 \\ 3 \end{bmatrix}$$

$\therefore \quad x = 1, y = 2$ and $z = 3$ **Ans.**

Example 67. *Solve by matrix inversion method, the following system of equations.*

$$2x - y + z = 6$$
$$3x + 2y - z = 3$$
$$7x + 3y - 4z = 7$$

Solution. Here, we have

$$2x - y + z = 6$$
$$3x + 2y - z = 3$$
$$7x + 3y - 4z = 7$$

The given equations are written in the matrix form.

$$\begin{bmatrix} 2 & -1 & 1 \\ 3 & 2 & -1 \\ 7 & 3 & -4 \end{bmatrix}\begin{bmatrix} x \\ y \\ z \end{bmatrix} = \begin{bmatrix} 6 \\ 3 \\ 7 \end{bmatrix}$$

$$AX = B$$

$$X = A^{-1} B$$

$$\text{Matrix of cofactors of } A = \begin{bmatrix} -5 & 5 & -5 \\ -1 & -15 & -13 \\ -1 & 5 & 7 \end{bmatrix}$$

$$|A| = -10 - 5 - 5 = -20$$

$$\text{Adjoint } A = \begin{bmatrix} -5 & -1 & -1 \\ 5 & -15 & 5 \\ -5 & -13 & 7 \end{bmatrix}$$

$$A^{-1} = \frac{1}{-20}\begin{bmatrix} -5 & -1 & -1 \\ 5 & -15 & 5 \\ -5 & -13 & 7 \end{bmatrix}$$

$$X = A^{-1} B$$

$$\begin{bmatrix} x \\ y \\ z \end{bmatrix} = -\frac{1}{20} \cdot \begin{bmatrix} -5 & -1 & -1 \\ 5 & -15 & 5 \\ -5 & -13 & 7 \end{bmatrix}\begin{bmatrix} 6 \\ 3 \\ 7 \end{bmatrix}$$

$$= -\frac{1}{20}\begin{bmatrix} -30-3-7 \\ 30-45+35 \\ -30-39+49 \end{bmatrix} = -\frac{1}{20}\begin{bmatrix} -40 \\ 20 \\ -20 \end{bmatrix} = \begin{bmatrix} 2 \\ -1 \\ 1 \end{bmatrix}$$

$\therefore$ $x = 2,\ y = -1,\ z = 1$ **Ans.**

Example 68. *Solve the following equations:*

$$3x + y + 2z = 3$$
$$2x - 3y - z = -3$$
$$x + 2y + z = 4$$

Solution. The given equations in the matrix form are written as below:

$$\begin{bmatrix} 3 & 1 & 2 \\ 2 & -3 & -1 \\ 1 & 2 & 1 \end{bmatrix} \times \begin{bmatrix} x \\ y \\ z \end{bmatrix} = \begin{bmatrix} 3 \\ -3 \\ 4 \end{bmatrix}$$

$$\begin{bmatrix} x \\ y \\ z \end{bmatrix} = \begin{bmatrix} 3 & 1 & 2 \\ 2 & -3 & -1 \\ 1 & 2 & 1 \end{bmatrix}^{-1} \times \begin{bmatrix} 3 \\ -3 \\ 4 \end{bmatrix}$$

$$= \frac{1}{8}\begin{bmatrix} -1 & 3 & 5 \\ -3 & 1 & 7 \\ 7 & -5 & -11 \end{bmatrix} \times \begin{bmatrix} 3 \\ -3 \\ 4 \end{bmatrix}$$

$$= \frac{1}{8}\begin{bmatrix} -3 & -9 & 21 \\ -9 & -3 & 28 \\ 21 & 15 & -44 \end{bmatrix}$$

$$\begin{bmatrix} x \\ y \\ z \end{bmatrix} = \begin{bmatrix} 1 \\ 2 \\ -1 \end{bmatrix} \qquad \therefore x = 1, y = 2, z = -1$$ **Ans.**

Example 69. *Solve the following system by matrix method.*

$$x + y + z = 6$$
$$x + 2y + 3z = 10$$
$$x + 2y + 4z = 8$$

(S.B.T.E. 2017)

Solution. Here we have

$$x + y + z = 6$$
$$x + 2y + 3z = 10$$
$$x + 2y + 4z = 8$$

The given equations are written in the matrix form

$$\begin{bmatrix} 1 & 1 & 1 \\ 1 & 2 & 3 \\ 1 & 2 & 4 \end{bmatrix}\begin{bmatrix} x \\ y \\ z \end{bmatrix} = \begin{bmatrix} 6 \\ 10 \\ 8 \end{bmatrix}, \quad |A| = 1(8-6) - 1(4-3) + 1(2-2) = 2 - 1 = 1 \neq 0$$

$$AX = B$$
$$X = A^{-1} B \qquad \dots (1)$$

$$\text{Cofactor matrix } (A) = \begin{bmatrix} 2 & -1 & 0 \\ -2 & 3 & -1 \\ 1 & -2 & 1 \end{bmatrix},$$

$$\text{Adj} \cdot A = \begin{bmatrix} 2 & -2 & 1 \\ -1 & 3 & -2 \\ 0 & -1 & 1 \end{bmatrix}$$

$$A^{-1} = \frac{1}{|A|} \cdot \text{Adj} \cdot A = \frac{1}{1}\begin{bmatrix} 2 & -2 & 1 \\ -1 & 3 & -2 \\ 0 & -1 & 1 \end{bmatrix}$$

Putting the value of A^{-1} in (1), we get

$$\begin{bmatrix} x \\ y \\ z \end{bmatrix} = \begin{bmatrix} 2 & -2 & 1 \\ -1 & 3 & -2 \\ 0 & -1 & 1 \end{bmatrix}\begin{bmatrix} 6 \\ 10 \\ 8 \end{bmatrix} = \begin{bmatrix} 12-20+8 \\ -6+30-16 \\ 0-10+8 \end{bmatrix} = \begin{bmatrix} 0 \\ 8 \\ -2 \end{bmatrix}$$

Hence $x = 0, y = 8, z = -2$ **Ans.**

Example 70. *Determine the product*

$$\begin{bmatrix} -4 & 4 & 4 \\ -7 & 1 & 3 \\ 5 & -3 & -1 \end{bmatrix} \times \begin{bmatrix} 1 & -1 & 1 \\ 1 & -2 & -2 \\ 2 & 1 & 3 \end{bmatrix}$$

and use it to solve the equations:

$$x - y + z = 4$$
$$x - 2y - 2z = 9$$
$$2x + y + 3z = 1.$$

Solution. $\begin{bmatrix} -4 & 4 & 4 \\ -7 & 1 & 3 \\ 5 & -3 & -1 \end{bmatrix} \times \begin{bmatrix} 1 & -1 & 1 \\ 1 & -2 & -2 \\ 2 & 1 & 3 \end{bmatrix}$

$$= \begin{bmatrix} -4+4+8 & 4-8+4 & -4-8+12 \\ -7+1+6 & 7-2+3 & -7-2+9 \\ 5-3-2 & -5+6-1 & 5+6-3 \end{bmatrix}$$

$$= \begin{bmatrix} 8 & 0 & 0 \\ 0 & 8 & 0 \\ 0 & 0 & 8 \end{bmatrix} = 8\begin{bmatrix} 1 & 0 & 0 \\ 0 & 1 & 0 \\ 0 & 0 & 1 \end{bmatrix} = 8I$$

Hence inverse of $\begin{bmatrix} 1 & -1 & 1 \\ 1 & -2 & -2 \\ 2 & 1 & 3 \end{bmatrix} = \frac{1}{8}\begin{bmatrix} -4 & 4 & 4 \\ -7 & 1 & 3 \\ 5 & -3 & -1 \end{bmatrix}$

Let us solve the equations

$$x - y + z = 4$$
$$x - 2y - 2z = 9$$
$$2x + y + 3z = 1$$

$$\Rightarrow \begin{bmatrix} 1 & -1 & 1 \\ 1 & -2 & -2 \\ 2 & 1 & 3 \end{bmatrix}\begin{bmatrix} x \\ y \\ z \end{bmatrix} = \begin{bmatrix} 4 \\ 9 \\ 1 \end{bmatrix}$$

$$\begin{bmatrix} x \\ y \\ z \end{bmatrix} = \begin{bmatrix} 1 & -1 & 1 \\ 1 & -2 & -2 \\ 2 & 1 & 3 \end{bmatrix}^{-1} \begin{bmatrix} 4 \\ 9 \\ 1 \end{bmatrix}$$

$$= \frac{1}{8}\begin{bmatrix} -4 & 4 & 4 \\ -7 & 1 & 3 \\ 5 & -3 & -1 \end{bmatrix} \begin{bmatrix} 4 \\ 9 \\ 1 \end{bmatrix}$$

$$= \frac{1}{8}\begin{bmatrix} -16+36+4 \\ -28+9+3 \\ 20-27-1 \end{bmatrix}$$

$$= \frac{1}{8}\begin{bmatrix} 24 \\ -16 \\ -8 \end{bmatrix} = \begin{bmatrix} 3 \\ -2 \\ -1 \end{bmatrix}$$

$x = 3, y = -2, z = -1.$ **Ans.**

EXERCISE 10.9

Solve the following equations using matrices:

1. $2x - 3y = 7$
$4x - 3y = 10$ **Ans.** $x = 3/2, y = -4/3$

2. $2x + y = 1$
$x - 2y = 8$ **Ans.** $x = 2, y = -3$

3. $2x + 3y = 10$
$x + 6y = 4$ **Ans.** $x = 16/3, y = -2/9$

4. $5x + 2y = 3$
$3x + 2y = 5$ **Ans.** $x = -1, y = 4$

5. $7x - 2y = -7$
$2x - y = 1$ **Ans.** $x = -3, y = -7$

6. $x - 2y = 4$
$-3x + 5y = -7$ **Ans.** $x = -6, y = -5$

7. $2x - 3y + 2z = 5$
$4x + 2y - 3z = 2$
$x - 2y + 5z = 3$ **Ans.** $x = 1, y = -1, z = 0$

8. $2x + 8y + 5z = 5$
$x + y + z = -2$
$x + 2y - z = 2$ **Ans.** $x = -3, y = 2, z = -1$

9. $x + 4y + 3z = 1$
$2x + 5y + 4z = 4$
$x - 3y - 2z = 5$ **Ans.** $x = 3, y = -2, z = 2$

10.
$$x + y + z = 3$$
$$x + 2y + 3z = 4$$
$$x + 4y + 9z = 6$$
(S.B.T.E. 2017, 2016) **Ans.** $x = 2, y = 1, z = 0$

11.
$$2x_1 + 3x_2 + x_3 = 9$$
$$x_1 + 2x_2 + 3x_3 = 6$$
$$3x_1 + x_2 + 2x_3 = 8$$
Ans. $x_1 = \frac{35}{18}, x_2 = \frac{29}{18}, x_3 = \frac{5}{18}$

12.
$$2x - 2y + z = 2$$
$$3x + y - z = 0$$
$$x + 3y + 2z = 2$$
Ans. $x = \frac{3}{8}, y = -\frac{1}{8}, z = 1$

13.
$$x + y + 2z = 4$$
$$2x - y + 3z = 9$$
$$3x - y - z = 2$$
Ans. $x = 1, y = -1, z = 2$

14. Write the following system of equations in the matrix form $AX = B$ and solve this for X by finding A^{-1}. ***(Diploma IETE, Dec. 2006)***
$$2x_1 - x_2 + x_3 = 4$$
$$x_1 + x_2 + x_3 = 1$$
$$x_1 - 3x_2 - 2x_3 = 2$$
Ans. $x = 1, y = -1, z = 1$

15. Solve the system of equations ***(Diploma IETE, June 2006)***
$$x + y + z = 6$$
$$x - y + 2z = 5$$
$$3x + y + z = 8$$
by using inverse of a suitable matrix. **Ans.** $x = 1, y = 2, z = 3$

Eigen Values and Eigen Vectors

11.1 EIGEN VALUES

$$\text{Let} \begin{bmatrix} a_{11} & a_{12} & a_{13} & \cdots & a_{1n} \\ a_{21} & a_{22} & a_{23} & \cdots & a_{2n} \\ a_{31} & a_{32} & a_{33} & \cdots & a_{3n} \\ \cdots & \cdots & \cdots & \cdots & \cdots \\ a_{n1} & a_{n2} & a_{n3} & \cdots & a_{nn} \end{bmatrix} \begin{bmatrix} x_1 \\ x_2 \\ x_3 \\ \vdots \\ x_n \end{bmatrix} = \begin{bmatrix} y_1 \\ y_2 \\ y_3 \\ \vdots \\ y_n \end{bmatrix}$$

$$AX = y \qquad \ldots(1)$$

Where A is the matrix, X is the column vector and y is also column vector.

Here column vector X is transformed into the column vector y by means of the square matrix A.

Let X be a such vector which transforms into λX by means of the transformation (1). Suppose the linear transformation $y = AX$ transforms X into a scalar multiple of itself, i.e. λX.

$$AX = y = \lambda X$$

$$AX - \lambda IX = 0$$

$$(A - \lambda I)X = 0 \qquad \ldots(2)$$

Thus the unknown scalar λ is known as an eigen value of the matrix A and the corresponding non zero vector X as **eigen vector**.

The eigen values are also called characteristic values or proper values or latent values.

$$\text{Let } A = \begin{bmatrix} 2 & 2 & 1 \\ 1 & 3 & 1 \\ 1 & 2 & 2 \end{bmatrix}$$

$$A - \lambda I = \begin{bmatrix} 2 & 2 & 1 \\ 1 & 3 & 1 \\ 1 & 2 & 2 \end{bmatrix} - \lambda \begin{bmatrix} 1 & 0 & 0 \\ 0 & 1 & 0 \\ 0 & 0 & 1 \end{bmatrix} = \begin{bmatrix} 2-\lambda & 2 & 1 \\ 1 & 3-\lambda & 1 \\ 1 & 2 & 2-\lambda \end{bmatrix} \text{ is characteristic matrix}$$

Characteristic Polynomial: The determinant $|A - \lambda I|$ when expanded will give a polynomial, which we call as characteristic polynomial of matrix A.

For example; $\begin{bmatrix} 2-\lambda & 2 & 1 \\ 1 & 3-\lambda & 1 \\ 1 & 2 & 2-\lambda \end{bmatrix}$

$$= (2-\lambda)(6-5\lambda+\lambda^2-2) - 2(2-\lambda-1) + 1(2-3+\lambda)$$

$$= -\lambda^3 + 7\lambda^2 - 11\lambda + 5$$

Characteristic Equation: The equation $|A-\lambda I| = 0$ is called the characteristic equation of the matrix, A e.g.

$$\lambda^3 - 7\lambda^2 + 11\lambda - 5 = 0$$

Characteristic Roots or Eigen Values: The roots of characteristic equation $|A-\lambda I| = 0$ are called characteristic roots of matrix A. e.g.

$$\lambda^3 - 7\lambda^2 + 11\lambda - 5 = 0$$

$$\Rightarrow \quad (\lambda-1)(\lambda-1)(\lambda-5) = 0 \quad \therefore \lambda = 1, 1, 5$$

Characteristic roots are 1, 1, 5.

Some Important Properties of Eigen Values

(AMIETE, Dec. 2009)

(1) Any square matrix A and its transpose A' have the same eigen values.

Note. The sum of the elements on the principal diagonal of a matrix is called the **trace** of the matrix.

(2) The sum of the eigen values of a matrix is equal to the **trace** of the matrix.

(3) The product of the eigen values of a matrix A is equal to the **determinant** of A.

(4) If $\lambda_1, \lambda_2, \ldots \lambda_n$, are the eigen values of A, then the eigen values of

(i) kA are $k\lambda_1, k\lambda_2, \ldots, k\lambda_n$ *(ii)* A^m are $\lambda_1{}^m, \lambda_2{}^m, \ldots\ldots, \lambda_n{}^m$

(iii) A^{-1} are $\frac{1}{\lambda_1}, \frac{1}{\lambda_2}, \ldots, \frac{1}{\lambda_n}$.

Example 1. *Find the characteristic roots of the matrix* $\begin{bmatrix} 6 & -2 & 2 \\ -2 & 3 & -1 \\ 2 & -1 & 3 \end{bmatrix}$

Solution. The characteristic equation of the given matrix is

$$\begin{vmatrix} 6-\lambda & -2 & 2 \\ -2 & 3-\lambda & -1 \\ 2 & -1 & 3-\lambda \end{vmatrix} = 0$$

$$\Rightarrow \quad (6-\lambda)(9-6\lambda+\lambda^2-1) + 2(-6+2\lambda+2) + 2(2-6+2\lambda) = 0$$

$$\Rightarrow \quad -\lambda^3 + 12\lambda^2 - 36\lambda + 32 = 0$$

By trial, $\lambda = 2$ is a root of this equation.

$$\Rightarrow \quad (\lambda-2)(\lambda^2-10\lambda+16) = 0 \Rightarrow (\lambda-2)(\lambda-2)(\lambda-8) = 0$$

$\Rightarrow \lambda = 2, 2, 8$ are the characteristic roots or Eigen values. **Ans.**

Example 2. *The matrix A is defined as* $A = \begin{bmatrix} 1 & 2 & -3 \\ 0 & 3 & 2 \\ 0 & 0 & -2 \end{bmatrix}$

Find the eigen values of $3A^3 + 5A^2 - 6A + 2I$.

Solution. $|A - \lambda I| = 0$

$$\begin{bmatrix} 1-\lambda & 2 & -3 \\ 0 & 3-\lambda & 2 \\ 0 & 0 & -2-\lambda \end{bmatrix} = 0$$

$\Rightarrow \quad (1 - \lambda)(3 - \lambda)(-2 - \lambda) = 0$ or $\lambda = 1, 3, -2$

Eigen values of $A^3 = 1, 27, -8$; Eigen values of $A^2 = 1, 9, 4$

Eigen values of $A = 1, 3, -2$; Eigen values of $I = 1, 1, 1$

$\therefore$ Eigen values of $3A^3 + 5A^2 - 6A + 2I$

First eigen value $= 3(1)^3 + 5(1)^2 - 6(1) + 2(1) = 4$

Second eigen value $= 3(27) + 5(9) - 6(3) + 2(1) = 110$

Third eigen value $= 3(-8) + 5(4) - 6(-2) + 2(1) = 10$

Required eigen values are 4, 110, 10 **Ans.**

Example 3. *If* λ *is an eigen value of an orthogonal matrix, then* $\frac{1}{\lambda}$ *is also eigen value.*

[**Hint:** $AA' = I$ if λ is the eigen value of A, then $\lambda^2 = 1, \lambda = \frac{1}{\lambda}$]

Example 4. *Find the eigen values of the orthogonal matrix.*

$$B = \frac{1}{3}\begin{bmatrix} 1 & 2 & 2 \\ 2 & 1 & -2 \\ 2 & -2 & 1 \end{bmatrix}$$

Solution. The characteristic equation of $A = \begin{bmatrix} 1 & 2 & 2 \\ 2 & 1 & -2 \\ 2 & -2 & 1 \end{bmatrix}$ is $\begin{bmatrix} 1-\lambda & 2 & 2 \\ 2 & 1-\lambda & -2 \\ 2 & -2 & 1-\lambda \end{bmatrix} = 0$

$\Rightarrow \quad (1 - \lambda)[(1 - \lambda)(1- \lambda) - 4] - 2[2(1 - \lambda) + 4] + 2[-4 - 2(1 - \lambda)] = 0$

$\Rightarrow \quad (1 - \lambda)(1 - 2\lambda + \lambda^2 - 4) - 2(2 - 2\lambda + 4) + 2(- 4 - 2 + 2\lambda) = 0$

$\Rightarrow \quad \lambda^3 - 3\lambda^2 - 91 + 27 = 0$

$\Rightarrow \quad (\lambda - 3)^2(\lambda + 3) = 0$

The eigen values of A are 3, 3, –3, so the eigen values of $B = \frac{1}{3}A$ are 1, 1, –1.

Note. If $\lambda = 1$ is an eigen value of B then its reciprocal $\frac{1}{\lambda} = \frac{1}{1} = 1$ is also an eigen value of B. **Ans.**

EXERCISE 11.1

Show that, for any square matrix A:

1. If λ be an eigen value of a non singular matrix A, show then $\frac{|A|}{\lambda}$ is an eigen value of the matrix adj A.
2. There are infinitely many eigen vectors corresponding to a single eigen value.
3. Find the product of the eigen values of the matrix $\begin{bmatrix} 2 & -3 & 3 \\ 2 & 1 & 1 \\ 1 & 5 & 6 \end{bmatrix}$ **Ans.** 18
4. Find the sum of the eigen values of the matrix $\begin{bmatrix} 3 & 2 & 1 \\ 1 & 3 & 2 \\ 4 & 1 & 5 \end{bmatrix}$ **Ans.** 11
5. Find the eigen value of the inverse of the matrix $\begin{bmatrix} 4 & 6 & 6 \\ 1 & 3 & 2 \\ -1 & -4 & -3 \end{bmatrix}$ **Ans.** $-1, 1, \frac{1}{4}$
6. Find the eigen values of the square of the matrix $\begin{bmatrix} 1 & 0 & -1 \\ 1 & 2 & 1 \\ 2 & 2 & 3 \end{bmatrix}$ **Ans.** 1, 4, 9
7. Find the eigen values of the matrix $\begin{bmatrix} 3 & 1 & 4 \\ 0 & 2 & 6 \\ 0 & 0 & 5 \end{bmatrix}^3$ **Ans.** 8, 27, 125
8. The sum and product of the eigen values of the matrix $A = \begin{bmatrix} 2 & 2 & 1 \\ 1 & 3 & 1 \\ 1 & 2 & 2 \end{bmatrix}$ are respectively

   (*a*) 7 and 7 (*b*) 7 and 5 (*c*) 7 and 6 (*d*) 7 and 8

   (*AMIETE, June 2010*) **Ans.** (*b*)

11.2 CAYLEY-HAMILTON THEOREM

Satement. Every square matrix satisfies its own characteristic equation.

If $|A - \lambda I| = (-1)^n(\lambda^n + a_1\lambda^{n-1} + a_2\lambda^{n-2} + \ldots + a_n)$ be the characteristic polynomial of $n \times n$ matrix $A = (a_{ij})$ then the matrix equation $X^n + a_1X^{n-1} + a_2X^{n-2} + \ldots + a_nI = 0$ is satisfied by $X = A$, *i.e.* $A^n + a_1A^{n-1} + a_2A^{n-2} + \ldots + a_nI = 0$

Proof. Since the elements of the matrix $A - \lambda I$ are at most of the first degree in λ, the elements of adj. $(A - \lambda I)$ are at most degree $(n - 1)$ in λ. Thus, adj. $(A - \lambda I)$ may be written as a matrix polynomial in λ, given by

$$Adj(A - \lambda I) = B_0\lambda^{n-1} + B_1\lambda^{n-2} + \ldots + B_{n-1}$$

when $B_0, B_1, \ldots B_{n-1}$ are $n \times n$ matrices, their elements being polynomial in λ.

We know that

$$(A - \lambda I)Adj(A - \lambda I) = |A - \lambda I|I$$

$$(A - \lambda I)(B_0\lambda^{n-1} + B_1\lambda^{n-2} + \ldots + B_{n-1}) = (-1)^n(\lambda^n + a_1\lambda^{n-1} + \ldots + a_n)I$$

Equating coefficients of like powers of λ on both sides, we get

$$-IB_0 = (-1)^n I$$

$$AB_0 - IB_1 = (-1)^n a_1 I$$

$$AB_1 - IB_2 = (-1)^n a_2 I$$

....

$$AB_{n-1} = (-1)^n a_n I$$

On multiplying the equation by $A^n, A^{n-1}, \ldots, I$ respectively and adding, we obtain

$$0 = (-1)^n[A^n + a_1A^{n-1} + \ldots + a_nI]$$

Thus $\quad A^n + a_1A^{n-1} + \ldots + a_nI = 0$

for example, Let A be square matrix and if

$$\lambda^3 - 2\lambda^2 + 3\lambda - 4 = 0 \qquad \ldots(1)$$

be its characteristic equation, then according to Cayley-Hamilton theorem (1) is satisfied by A.

$$A^3 - 2A^2 + 3A - 4I = 0 \qquad \ldots(2)$$

We can find out A^{-1} from equation (2). On premultiplying equation (2) by A^{-1}, we get

$$A^2 - 2A + 3A - 4I = 0$$

$$A^{-1} = \frac{1}{4}[A^2 - 2A + 3I]$$

Example 5. *Find the characteristic equation of the symmetric matrix*

$$A = \begin{bmatrix} 2 & -1 & 1 \\ -1 & 2 & -1 \\ 1 & -1 & 2 \end{bmatrix}$$ *and verify that it is satisfied by A and hence obtain* A^{-1}.

Express $A^6 - 6A^5 + 9A^4 - 2A^3 - 12A^2 + 23A - 9I$ *in linear polynomial in A.*

Solution. Characteristic equation is $|A - \lambda I| = 0$

$$\begin{vmatrix} 2-\lambda & -1 & 1 \\ -1 & 2-\lambda & -1 \\ 1 & -1 & 2-\lambda \end{vmatrix} = 0$$

$$(2 - \lambda)[(2 - \lambda)^2 - 1] + 1[-2 + \lambda + 1] + 1[1 - 2 + \lambda] = 0$$

or $\quad (2 - \lambda)^3 - (2 - \lambda) + \lambda - 1 + \lambda - 1 = 0$

or $\quad (2 - \lambda)^3 - 2 + \lambda + \lambda - 1 + \lambda - 1 = 0$ or $(2 - \lambda)^3 + 3\lambda - 4 = 0$

or $\quad 8 - \lambda^3 - 12\lambda + 6\lambda^2 + 3\lambda - 4 = 0$

or $\quad -\lambda^3 + 6\lambda^2 - 9\lambda + 4 = 0$ or $\lambda^3 - 6\lambda^2 + 9\lambda - 4 = 0$

By Cayley-Hamilton theorem $\quad A^3 - 6A^2 + 9A - 4I = 0 \qquad \ldots(1)$

Verification:

$$A^2 = \begin{bmatrix} 2 & -1 & 1 \\ -1 & 2 & -1 \\ 1 & -1 & 2 \end{bmatrix}\begin{bmatrix} 2 & -1 & 1 \\ -1 & 2 & -1 \\ 1 & -1 & 2 \end{bmatrix}$$

$$= \begin{pmatrix} 4+1+1 & -2-2-1 & 2+1+2 \\ -2-2-1 & 1+4+1 & -1-2-2 \\ 2+1+2 & -1-2-2 & 1+1+4 \end{pmatrix}$$

$$= \begin{pmatrix} 6 & -5 & 5 \\ -5 & 6 & -5 \\ 5 & -5 & 6 \end{pmatrix}$$

$$A^3 = A^2.A = \begin{pmatrix} 6 & -5 & 5 \\ -5 & 6 & -5 \\ 5 & -5 & 6 \end{pmatrix}\begin{pmatrix} 2 & -1 & 1 \\ -1 & 2 & -1 \\ 1 & -1 & 2 \end{pmatrix}$$

$$= \begin{pmatrix} 12+5+5 & -6-10-5 & 6+5+10 \\ -10-6-5 & 5+12+5 & -5-6-10 \\ 10+5+6 & -5-10-6 & 5+5+12 \end{pmatrix}$$

$$= \begin{pmatrix} 22 & -21 & 21 \\ -21 & 22 & -21 \\ 21 & -21 & 22 \end{pmatrix}$$

$A^3 - 6A^2 + 9A - 4I$

$$= \begin{pmatrix} 22 & -21 & 21 \\ -21 & 22 & -21 \\ 21 & -21 & 22 \end{pmatrix} - 6\begin{pmatrix} 6 & -5 & 5 \\ -5 & 6 & -5 \\ 5 & -5 & 6 \end{pmatrix} + 9\begin{pmatrix} 2 & -1 & 1 \\ -1 & 2 & -1 \\ 1 & -1 & 2 \end{pmatrix} - 4\begin{pmatrix} 1 & 0 & 0 \\ 0 & 1 & 0 \\ 0 & 0 & 1 \end{pmatrix}$$

$$= \begin{pmatrix} 22-36+18-4 & -21+30-9-0 & 21-30+9-0 \\ -21+30-9-0 & 22-36+18-4 & -21+30-9-0 \\ 21-30+9+0 & -21+30-9-0 & 22-36+18-4 \end{pmatrix}$$

$$= \begin{pmatrix} 0 & 0 & 0 \\ 0 & 0 & 0 \\ 0 & 0 & 0 \end{pmatrix} = 0$$

So it is verified that the characteristic equation (1) is satisfied by A.

Inverse of Matrix A,

$$A^3 - 6A^2 + 9A - 4I = 0$$

On multiplying by A^{-1}, we get

$$A^2 - 6A + 9I - 4A^{-1} = 0 \quad \text{or} \quad 4A^{-1} = A^2 - 6A + 9I$$

or $$4A^{-1} = \begin{pmatrix} 6 & -5 & 5 \\ -5 & 6 & -5 \\ 5 & -5 & 6 \end{pmatrix} - 6\begin{pmatrix} 2 & -1 & 1 \\ -1 & 2 & -1 \\ 1 & -1 & 2 \end{pmatrix} + 9\begin{pmatrix} 1 & 0 & 0 \\ 0 & 1 & 0 \\ 0 & 0 & 1 \end{pmatrix}$$

$$= \begin{pmatrix} 6-12+9 & -5+6+0 & 5-6+0 \\ -5+6+0 & 6-12+9 & -5+6+0 \\ 5-6+0 & -5+6+0 & 6-12+9 \end{pmatrix} \Rightarrow A^{-1} = \frac{1}{4}\begin{pmatrix} 3 & 1 & -1 \\ 1 & 3 & 1 \\ -1 & 1 & 3 \end{pmatrix}$$ **Ans.**

Now $A^6 - 6A^5 + 9A^4 - 2A^3 - 12A^2 + 23A - 9I$

$= A^3(A^3 - 6A^2 + 9A - 4I) + 2(A^3 - 6A^2 + 9A - 4I) + 5A - I = 5A - I$ **Ans.**

Example 6. *Find the characteristic equation of the matrix* $A = \begin{bmatrix} 2 & 1 & 1 \\ 0 & 1 & 0 \\ 1 & 1 & 2 \end{bmatrix}$. *Verify Cayley-Hamilton theorem and hence prove that:*

$$A^8 - 5A^7 + 7A^6 - 3A^5 + A^4 - 5A^3 + 8A^2 - 2A + I = \begin{bmatrix} 8 & 5 & 5 \\ 0 & 3 & 0 \\ 5 & 5 & 8 \end{bmatrix}$$

(*A.M.I.E.T.E. Dec. 2016; Gujarat, II Semester, June 2009*)

Solution. Characteristic equation of the matrix A is

$$\begin{vmatrix} 2-\lambda & 1 & 1 \\ 0 & 1-\lambda & 0 \\ 1 & 1 & 2-\lambda \end{vmatrix} = 0$$

$\Rightarrow\ (2 - \lambda)[(1 - \lambda)(2 - \lambda)] - (0) + 1(0 - 1 + \lambda) \ \Rightarrow\ \lambda^3 - 5\lambda^2 + 7\lambda - 3 = 0$

According to Cayley-Hamilton theorem

$$A^3 - 5A^2 + 7A - 3I = 0 \qquad \ldots(1)$$

We have to verify the equation (1).

$$A^2 = \begin{bmatrix} 2 & 1 & 1 \\ 0 & 1 & 0 \\ 1 & 1 & 2 \end{bmatrix}\begin{bmatrix} 2 & 1 & 1 \\ 0 & 1 & 0 \\ 1 & 1 & 2 \end{bmatrix} = \begin{bmatrix} 5 & 4 & 4 \\ 0 & 1 & 0 \\ 4 & 4 & 5 \end{bmatrix}$$

$$A^3 = A.A^2 = \begin{bmatrix} 2 & 1 & 1 \\ 0 & 1 & 0 \\ 1 & 1 & 2 \end{bmatrix}\begin{bmatrix} 5 & 4 & 4 \\ 0 & 1 & 0 \\ 4 & 4 & 5 \end{bmatrix} = \begin{bmatrix} 14 & 13 & 13 \\ 0 & 1 & 0 \\ 13 & 13 & 14 \end{bmatrix}$$

$$A^3 - 5A^2 + 7A - 3I = \begin{bmatrix} 14 & 13 & 13 \\ 0 & 1 & 0 \\ 13 & 13 & 14 \end{bmatrix} - 5\begin{bmatrix} 5 & 4 & 4 \\ 0 & 1 & 0 \\ 4 & 4 & 5 \end{bmatrix} + 7\begin{bmatrix} 2 & 1 & 1 \\ 0 & 1 & 0 \\ 1 & 1 & 2 \end{bmatrix} - 3\begin{bmatrix} 1 & 0 & 0 \\ 0 & 1 & 0 \\ 0 & 0 & 1 \end{bmatrix}$$

$$= \begin{bmatrix} 14-25+14-3 & 13-20+7+0 & 13-20+7+0 \\ 0+0+0+0 & 1-5+7-3 & 0-0+0-0 \\ 13-20+7+0 & 13-20+7-0 & 14-25+14-3 \end{bmatrix} = \begin{bmatrix} 0 & 0 & 0 \\ 0 & 0 & 0 \\ 0 & 0 & 0 \end{bmatrix} = 0$$

Hence Cayley Hamilton Theorem is verified.

Now $A^8 - 5A^7 + 7A^6 - 3A^5 + A^4 - 5A^3 + 8A^2 - 2A + I$

$$= A^5(A^3 - 5A^2 + 7A - 3I) + A(A^3 - 5A^2 + 7A - 3I) + A^2 + A + I$$

$$= A^5 \times O + A \times O + A^2 + A + I = A^2 + A + I$$

$$= \begin{bmatrix} 5 & 4 & 4 \\ 0 & 1 & 0 \\ 4 & 4 & 5 \end{bmatrix} + \begin{bmatrix} 2 & 1 & 1 \\ 0 & 1 & 0 \\ 1 & 1 & 2 \end{bmatrix} + \begin{bmatrix} 1 & 0 & 0 \\ 0 & 1 & 0 \\ 0 & 0 & 1 \end{bmatrix}$$

$$= \begin{bmatrix} 5+2+1 & 4+1+0 & 4+1+0 \\ 0+0+0 & 1+1+1 & 0+0+0 \\ 4+1+0 & 4+1+0 & 5+2+1 \end{bmatrix} = \begin{bmatrix} 8 & 5 & 5 \\ 0 & 3 & 0 \\ 5 & 5 & 8 \end{bmatrix}$$ **Proved.**

EXERCISE 11.2

1. Find the characteristic polynomial of the matrix

$$A = \begin{bmatrix} 3 & 1 & 1 \\ -1 & 5 & -1 \\ 1 & -1 & 3 \end{bmatrix}$$

Verify Cayley-Hamilton Theorem for this matrix. Hence find A^{-1}.

Ans. $A^{-1} = \frac{1}{20}\begin{bmatrix} 7 & -2 & -3 \\ 1 & 4 & 1 \\ -2 & 2 & 8 \end{bmatrix}$

2. Use Cayley-Hamilton Theorem to find the inverse of the matrix

$$\begin{bmatrix} \cos\theta & \sin\theta \\ -\sin\theta & \cos\theta \end{bmatrix}$$

Ans. $\begin{bmatrix} \cos\theta & -\sin\theta \\ \sin\theta & \cos\theta \end{bmatrix}$

3. Using Cayley-Hamilton Theorem, find A^{-1}, given that

$$A = \begin{bmatrix} 2 & -1 & 3 \\ 1 & 0 & 2 \\ 4 & -2 & 1 \end{bmatrix}$$

Ans. $-\frac{1}{5}\begin{bmatrix} 4 & -5 & -2 \\ 7 & -10 & -1 \\ -2 & 0 & 1 \end{bmatrix}$

4. Using Cayley-Hamilton Theorem, find the inverse of the matrix

$$\begin{bmatrix} 5 & -1 & 5 \\ 0 & 2 & 0 \\ -5 & 3 & -15 \end{bmatrix}$$

Ans. $\frac{1}{10}\begin{bmatrix} 3 & 0 & 1 \\ 0 & 5 & 0 \\ -1 & 1 & -1 \end{bmatrix}$

5. Find the characteristic equation of the matrix

$$A = \begin{bmatrix} 1 & 3 & 7 \\ 4 & 2 & 3 \\ 1 & 2 & 1 \end{bmatrix}$$

and show that the equation is also satisfied by A. **Ans.** $\lambda^3 - 4\lambda^2 - 20\lambda - 35 = 0$

6. Find the eigenvalues of the matrix

$$\begin{bmatrix} 2 & -3 & 1 \\ 3 & 1 & 3 \\ -5 & 2 & -4 \end{bmatrix}$$

Ans. Eigen values are 0, 1, –2

7. Using Cayley-Hamilton Theorem obtain the inverse of the matrix

$$\begin{bmatrix} 1 & 1 & 3 \\ 1 & 3 & -3 \\ -2 & -4 & -4 \end{bmatrix}$$

Ans. $\frac{1}{8}\begin{bmatrix} 24 & 8 & 12 \\ -10 & -2 & -6 \\ -2 & -2 & -2 \end{bmatrix}$

8. Show that the matrix $A = \begin{bmatrix} 1 & -2 & 2 \\ 1 & 2 & 3 \\ 0 & -1 & 2 \end{bmatrix}$ satisfies its characteristic equation. Hence find A^{-1}.

Ans. $\frac{1}{9}\begin{bmatrix} 7 & 2 & -10 \\ -2 & 2 & -1 \\ -1 & 1 & 4 \end{bmatrix}$

9. Use Cayley Hamilton Theorem to find the inverse of

$$A = \begin{bmatrix} 1 & 2 & 4 \\ -1 & 0 & 3 \\ 3 & 1 & -2 \end{bmatrix}$$

Ans. $\frac{1}{7}\begin{bmatrix} -3 & 8 & 6 \\ 7 & -14 & -7 \\ -1 & 5 & 2 \end{bmatrix}$

10. Verify Cayley-Hamilton Theorem for the matrix

$$A = \begin{bmatrix} 1 & 1 & 2 \\ 3 & 1 & 1 \\ 2 & 3 & 1 \end{bmatrix}$$ Hence evaluate A^{-1}

Ans. $\frac{1}{11}\begin{bmatrix} -2 & 5 & -1 \\ -1 & -3 & 5 \\ 7 & -1 & -2 \end{bmatrix}$

11. $A = \begin{bmatrix} 1 & 4 \\ 2 & 3 \end{bmatrix}$, then express $A^5 - 4A^4 - 7A^3 + 11A^2 - A - 10I$ in terms of A.

Ans. $A + 5I$

12. If λ_1, λ_2 and λ_3 are the eigenvalues of the matrix

$\begin{bmatrix} -2 & -9 & 5 \\ -5 & -10 & 7 \\ -9 & -21 & 14 \end{bmatrix}$ then $\lambda_1 + \lambda_2 + \lambda_3$ is equal to

(*i*) –16 (*ii*) 2 (*iii*) –6 (*iv*) –14 **Ans.** (*ii*)

13. The matrix $A = \begin{bmatrix} 1 & 0 \\ 2 & 4 \end{bmatrix}$ is given. The eigenvalues of $4A^{-1} + 3A + 2I$ are

(A) 6, 15 (B) 9, 12 (C) 9, 15 (D) 7, 15 **Ans.** (C)

14. $A(3 \times 3)$ real matrix has an eigenvalue i, then its other two eigenvalues can be

(A) 0, 1 (B) –1, i (C) $2i$, $-2i$ (D) 0, $-i$

15. Verify Cayley-Hamilton theorem for the matrix

$$A = \begin{bmatrix} 1 & -2 & 3 \\ 2 & 4 & -2 \\ -1 & 1 & 2 \end{bmatrix}$$

16. Find adj. A by using Cayley-Hamilton theorem where A is given by

$$A = \begin{bmatrix} 1 & 2 & 1 \\ 0 & 1 & -1 \\ 3 & -1 & 1 \end{bmatrix}$$ (*R.G.P.V., Bhopal, April 2010*) **Ans.** $\begin{bmatrix} 0 & -3 & -3 \\ -3 & -2 & 1 \\ -3 & 7 & 1 \end{bmatrix}$

17. If a matrix $A = \begin{bmatrix} 1 & 0 & 0 \\ 0 & -1 & 0 \\ 1 & 0 & 1 \end{bmatrix}$, find the matrix A^{32}, using Cayley Hamilton Theorem.

Ans. $\begin{bmatrix} 1 & 0 & 0 \\ 0 & 1 & 0 \\ 32 & 0 & 1 \end{bmatrix}$

11.3 CHARACTERISTIC VECTORS OR EIGEN VECTORS

A column vector X is transformed into column vector y by means of a square matrix A. Now we want to multiply the column vector X by a scalar quantity λ so that we can find the same transformed column vector y.

i.e., $AX = \lambda X$

X is known as eigen vector.

Example 7. *Show that the vector (1, 1, 2) is an eigen vector of the matrix*

$$A = \begin{bmatrix} 3 & 1 & -1 \\ 2 & 2 & -1 \\ 2 & 2 & 0 \end{bmatrix}$$ corresponding to the eigen value 2.

Solution. Let X = (1, 1, 2).

Now,
$$AX = \begin{bmatrix} 3 & 1 & -1 \\ 2 & 2 & -1 \\ 2 & 2 & 0 \end{bmatrix}\begin{bmatrix} 1 \\ 1 \\ 2 \end{bmatrix} = \begin{bmatrix} 3+1-2 \\ 2+2-2 \\ 2+2+0 \end{bmatrix} = \begin{bmatrix} 2 \\ 2 \\ 4 \end{bmatrix} = 2\begin{bmatrix} 1 \\ 1 \\ 2 \end{bmatrix} = 2X$$

Corresponding to each characteristic root λ, we have a corresponding non-zero vector X which satisfies the equation $[A - \lambda I]X = 0$. The non-zero vector X is called characteristic vector or Eigen vector.

11.4 PROPERTIES OF EIGEN VECTORS

1. The eigen vector X of a matrix A is not unique.
2. $\lambda_1, \lambda_2, \ldots, \lambda_n$ be distinct eigen values of an $n \times n$ matrix then corresponding eigen vectors $X_1, X_2, \ldots, X_n$ form a linearly independent set.
3. If two or more eigen values are equal it may or may not be possible to get linearly independent eigen vectors corresponding to the equal roots.
4. Two eigen vectors X_1 and X_2 are called orthogonal vectors $X_1'X_2 = 0$.
5. Eigen vectors of a symmetric matrix corresponding to different eigen values are orthogonal.

Normalised form of vectors. To find normalised form of $\begin{bmatrix} a \\ b \\ c \end{bmatrix}$, we divide each element by $\sqrt{a^2+b^2+c^2}$.

For example, normalised form of $\begin{bmatrix} 1 \\ 2 \\ 2 \end{bmatrix}$ is $\begin{bmatrix} 1/3 \\ 2/3 \\ 3/3 \end{bmatrix}$ $\left[\sqrt{1^2+2^2+2^2}=3\right]$

11.5 NON-SYMMETRIC MATRICES WITH NON-REPEATED EIGEN VALUES

Example 8. *Find the eigen values and eigen vectors of matrix* $A = \begin{bmatrix} 3 & 1 & 4 \\ 0 & 2 & 6 \\ 0 & 0 & 5 \end{bmatrix}$

Solution. The characteristic equation of matrix A is given by

$$|A - \lambda I| = \begin{vmatrix} 3-\lambda & 1 & 4 \\ 0 & 2-\lambda & 6 \\ 0 & 0 & 5-\lambda \end{vmatrix} = (3-\lambda)(2-\lambda)(5-\lambda) = 0$$

$$\Rightarrow \qquad \lambda = 2, 3, 5$$

Thus the eigen values of matrix A are 2, 3, 5.

The eigen vectors of the matrix A corresponding to the eigen value λ is given by the non-zero solution of the equation $[A - \lambda I]X = 0$

$$\text{or} \quad \begin{bmatrix} 3-\lambda & 1 & 4 \\ 0 & 2-\lambda & 6 \\ 0 & 0 & 5-\lambda \end{bmatrix} \begin{bmatrix} x_1 \\ x_2 \\ x_3 \end{bmatrix} = \begin{bmatrix} 0 \\ 0 \\ 0 \end{bmatrix} \qquad \ldots(1)$$

When $\lambda = 2$ the corresponding eigen vector is given by

$$\begin{bmatrix} 3-2 & 1 & 4 \\ 0 & 2-2 & 6 \\ 0 & 0 & 5-2 \end{bmatrix} \begin{bmatrix} x_1 \\ x_2 \\ x_3 \end{bmatrix} = \begin{bmatrix} 0 \\ 0 \\ 0 \end{bmatrix}$$

$$\Rightarrow \qquad \begin{bmatrix} 1 & 1 & 4 \\ 0 & 0 & 6 \\ 0 & 0 & 3 \end{bmatrix} \begin{bmatrix} x_1 \\ x_2 \\ x_3 \end{bmatrix} = \begin{bmatrix} 0 \\ 0 \\ 0 \end{bmatrix}$$

$$\Rightarrow \qquad x_1 + x_2 + 4x_3 = 0,$$

$$0x_1 + 0x_2 + 6x_3 = 0$$

$$\frac{x_1}{6-0} = \frac{x_2}{0-6} = \frac{x_3}{0-0} = k \;\Rightarrow\; \frac{x_1}{1} = \frac{x_2}{-1} = \frac{x_3}{0} = k \;\Rightarrow\; x_1 = k,\ x_2 = -k,\ x_3 = 0$$

Hence $X_1 = \begin{bmatrix} k \\ -k \\ 0 \end{bmatrix} = k\begin{bmatrix} 1 \\ -1 \\ 0 \end{bmatrix}$ can be taken as an eigen vector of A corresponding to the eigen value $\lambda = 2$

When $\lambda = 3$, substituting in (1), the corresponding eigen vector is given by

$$\begin{bmatrix} 0 & 1 & 4 \\ 0 & -1 & 6 \\ 0 & 0 & 2 \end{bmatrix}\begin{bmatrix} x_1 \\ x_2 \\ x_3 \end{bmatrix} = \begin{bmatrix} 0 \\ 0 \\ 0 \end{bmatrix}$$

$$0x_1 + x_2 + 4x_3 = 0,$$

$$0x_1 - x_2 + 6x_3 = 0$$

$$\frac{x_1}{6+4} = \frac{x_2}{0-0} = \frac{x_3}{0-0} \Rightarrow \frac{x_1}{10} = \frac{x_2}{0} = \frac{x_3}{0} = \frac{k}{10}$$

$$x_1 = k, x_2 = 0, x_3 = 0$$

Hence $X_2 = \begin{bmatrix} k \\ 0 \\ 0 \end{bmatrix} = k\begin{bmatrix} 1 \\ 0 \\ 0 \end{bmatrix}$ can be taken as an eigen vector of A corresponding to the eigen value $\lambda = 3$.

When $\lambda = 5$.

Again, when $\lambda = 5$ substituting in (1), the corresponding eigen vector is given by

$$\begin{bmatrix} -2 & 1 & 4 \\ 0 & -3 & 6 \\ 0 & 0 & 0 \end{bmatrix}\begin{bmatrix} x_1 \\ x_2 \\ x_3 \end{bmatrix} = \begin{bmatrix} 0 \\ 0 \\ 0 \end{bmatrix}$$

$$-2x_1 + x_2 + 4x_3 = 0,$$

$$-3x_2 + 6x_3 = 0$$

By cross-multiplication method, we have

$$\frac{x_1}{6+12} = \frac{x_2}{0+12} = \frac{x_3}{6-0} \Rightarrow \frac{x_1}{18} = \frac{x_2}{12} = \frac{x_3}{6} \Rightarrow \frac{x_1}{3} = \frac{x_2}{2} = \frac{x_3}{1}$$

$$x_1 = 3k, x_2 = 2k, x_3 = k$$

Hence $X_3 = \begin{bmatrix} 3k \\ 2k \\ k \end{bmatrix} = k\begin{bmatrix} 3 \\ 2 \\ 1 \end{bmatrix}$, can be taken as an eigen vector of A corresponding to the eigen value $\lambda = 5$ **Ans.**

EXERCISE 11.3

Non-symmetric matrix with difference eigen values:

Find the eigen values and the corresponding eigen vectors for the following matrices:

1. $\begin{bmatrix} 1 & 1 & -2 \\ -1 & 2 & 1 \\ 0 & 1 & -1 \end{bmatrix}$ *(A.M.I.E.T.E., June 2006)* **Ans.** $-1, 1, 2; \begin{bmatrix} 1 \\ 0 \\ 1 \end{bmatrix}, \begin{bmatrix} 3 \\ 2 \\ 1 \end{bmatrix}, \begin{bmatrix} 1 \\ 3 \\ 1 \end{bmatrix}$

2. $\begin{bmatrix} 4 & 2 & -2 \\ -5 & 3 & 2 \\ -2 & 4 & 1 \end{bmatrix}$ **Ans.** 1, 2, 5; $\begin{bmatrix} 2 \\ 1 \\ 4 \end{bmatrix}, \begin{bmatrix} 1 \\ 1 \\ 2 \end{bmatrix}, \begin{bmatrix} 0 \\ 1 \\ 0 \end{bmatrix}$

3. $\begin{bmatrix} 2 & -2 & 3 \\ 1 & 1 & 1 \\ 1 & 3 & -1 \end{bmatrix}$ **Ans.** –2, 1, 3; $\begin{bmatrix} 11 \\ 1 \\ 14 \end{bmatrix}, \begin{bmatrix} -1 \\ 1 \\ 1 \end{bmatrix}, \begin{bmatrix} 1 \\ 1 \\ 1 \end{bmatrix}$

4. $\begin{bmatrix} -9 & 2 & 6 \\ 5 & 0 & -3 \\ -16 & 4 & 11 \end{bmatrix}$ **Ans.** –1, 1, 2; $\begin{bmatrix} 2 \\ -1 \\ 3 \end{bmatrix}, \begin{bmatrix} 1 \\ -1 \\ 2 \end{bmatrix}, \begin{bmatrix} 2 \\ -1 \\ 4 \end{bmatrix}$

5. $\begin{bmatrix} 4 & 6 & 6 \\ 1 & 3 & 2 \\ -1 & -4 & -3 \end{bmatrix}$ **Ans.** –1, 1, 4; $\begin{bmatrix} -6 \\ -2 \\ 7 \end{bmatrix}, \begin{bmatrix} 0 \\ 1 \\ -1 \end{bmatrix}, \begin{bmatrix} 3 \\ 1 \\ -1 \end{bmatrix}$

6. $\begin{bmatrix} 1 & 1 & 1 \\ 1 & 2 & 1 \\ 3 & 2 & 3 \end{bmatrix}$ **Ans.** 0, 1, 5; $\begin{bmatrix} 0 \\ -1 \\ 1 \end{bmatrix}, \begin{bmatrix} -1 \\ 0 \\ 1 \end{bmatrix}, \begin{bmatrix} 4 \\ 5 \\ 11 \end{bmatrix}$

7. $\begin{bmatrix} -2 & 1 & 1 \\ -11 & 4 & 5 \\ -1 & 1 & 0 \end{bmatrix}$ **Ans.** –1, 1, 2; $\begin{bmatrix} 0 \\ -1 \\ 1 \end{bmatrix}, \begin{bmatrix} 1 \\ 2 \\ 1 \end{bmatrix}, \begin{bmatrix} 2 \\ 3 \\ 1 \end{bmatrix}$

11.6 NON-SYMMETRIC MATRIX WITH REPEATED EIGEN VALUES

Example 9. *Find all the Eigen values and Eigen vectors of the matrix*

$$A = \begin{bmatrix} -2 & 2 & -3 \\ 2 & 1 & -6 \\ -1 & -2 & 0 \end{bmatrix}$$ *(AMIETE, Dec. 2009)*

Solution. Characteristic equation of A is $|A - \lambda I| = 0$

$$\begin{vmatrix} -2-\lambda & 2 & -3 \\ 2 & 1-\lambda & -6 \\ -1 & -2 & 0-\lambda \end{vmatrix} = 0$$

$\Rightarrow (-2-\lambda)[-\lambda + \lambda^2 - 12] - 2(-2\lambda - 6) - 3(-4 + 1 - \lambda) = 0$

$$\Rightarrow \quad \lambda^3 + \lambda^2 - 21\lambda - 45 = 0 \quad \ldots(1)$$

By trial: If $\lambda = -3$, then $-27 + 9 + 63 - 45 = 0$, so $(\lambda + 3)$ is one factor of equation (1).

The remaining factors are obtained on dividing equation (1) by $(\lambda + 3)$

–3	1	1	–21	–45	
		–3	6	45	(By synthetic division)
	1	–2	–15	0	

$$\Rightarrow \qquad \lambda^2 - 2\lambda - 15 = 0 \ \Rightarrow \ (\lambda - 5)(\lambda + 3) = 0$$

$$\Rightarrow \qquad (\lambda + 3)(\lambda + 3)(\lambda - 5) = 0 \ \Rightarrow \ \lambda = 5, -3, -3$$

To find the eigen vectors for corresponding eigen values, we will consider the matrix equation

$$[A - \lambda I]X = 0 \quad i.e., \quad \begin{bmatrix} -2-\lambda & 2 & -3 \\ 2 & 1-\lambda & -6 \\ -1 & -2 & 0-\lambda \end{bmatrix} \begin{bmatrix} x \\ y \\ z \end{bmatrix} = \begin{bmatrix} 0 \\ 0 \\ 0 \end{bmatrix} \qquad \text{...(2)}$$

On putting $\lambda = 5$ in eq. (2), we get $\begin{bmatrix} -7 & 2 & -3 \\ 2 & -4 & -6 \\ -1 & -2 & -5 \end{bmatrix} \begin{bmatrix} x \\ y \\ z \end{bmatrix} = \begin{bmatrix} 0 \\ 0 \\ 0 \end{bmatrix}$

$$\Rightarrow \qquad -7x + 2y - 3z = 0,$$

$$2x - 4y - 6z = 0$$

$$\frac{x}{-12-12} = \frac{y}{-6-42} = \frac{z}{28-4} \quad \text{or} \quad \frac{x}{-24} = \frac{y}{-48} = \frac{z}{24} \quad \text{or} \quad \frac{x}{1} = \frac{y}{2} = \frac{z}{-1} = k$$

$$x = k,\ y = 2k,\ z = -k$$

Hence, the eigen vector $X_1 = \begin{bmatrix} k \\ 2k \\ -k \end{bmatrix} = k \begin{bmatrix} 1 \\ 2 \\ -1 \end{bmatrix}$

Put $\lambda = -3$ in eq. (2), we get $\begin{bmatrix} 1 & 2 & -3 \\ 2 & 4 & -6 \\ -1 & -2 & 3 \end{bmatrix} \begin{bmatrix} x \\ y \\ x \end{bmatrix} = \begin{bmatrix} 0 \\ 0 \\ 0 \end{bmatrix}$

$$\Rightarrow \qquad x + 2y - 3z = 0,$$

$$2x + 4y - 6z = 0,$$

$$-x - 2y + 3z = 0$$

Here first, second and third equations are the same.

Let $x = k_1$, $y = k_2$ then $z = \frac{1}{3}(k_1 + 2k_2)$

Hence, the eigen vector is $\begin{bmatrix} k_1 \\ k_2 \\ \frac{1}{3}(k_1 + 2k_2) \end{bmatrix}$

Let $k_1 = 0$, $k_2 = 3$ Hence $X_2 = \begin{bmatrix} 0 \\ 3 \\ 2 \end{bmatrix}$

Since the matrix is non-symmetric, the corresponding eigen vectors X_2 and X_3 must be linearly independent. This can be done by choosing

$k_1 = 3, k_2 = 0$, and Hence $X_3 = \begin{bmatrix} 3 \\ 0 \\ 1 \end{bmatrix}$

Hence, $X_1 = \begin{bmatrix} 1 \\ 2 \\ -1 \end{bmatrix}, X_2 = \begin{bmatrix} 0 \\ 3 \\ 2 \end{bmatrix}, X_3 = \begin{bmatrix} 3 \\ 0 \\ 1 \end{bmatrix}.$ **Ans.**

EXERCISE 11.4

Non-symmetric matrices with repeated eigen values

Find the eigen values and eigen vectors of the following matrices:

1. $\begin{bmatrix} 2 & -2 & 2 \\ 1 & 1 & 1 \\ 1 & 3 & -1 \end{bmatrix}$ **Ans.** $-2, 2;$ $\begin{bmatrix} -4 \\ -1 \\ 7 \end{bmatrix}, \begin{bmatrix} 0 \\ 1 \\ 1 \end{bmatrix}$

2. $\begin{bmatrix} 2 & 2 & 1 \\ 1 & 3 & 1 \\ 1 & 2 & 2 \end{bmatrix}$ **Ans.** $1, 1;$ $\begin{bmatrix} 0 \\ 1 \\ -2 \end{bmatrix}, \begin{bmatrix} 1 \\ 0 \\ -1 \end{bmatrix}, \begin{bmatrix} 1 \\ 1 \\ 1 \end{bmatrix}$

3. $\begin{bmatrix} 2 & 1 & 1 \\ 2 & 3 & 2 \\ 3 & 3 & 4 \end{bmatrix}$ **Ans.** $1, 1, 7;$ $\begin{bmatrix} 0 \\ 1 \\ -1 \end{bmatrix}, \begin{bmatrix} 1 \\ 0 \\ -1 \end{bmatrix}, \begin{bmatrix} 1 \\ 2 \\ 3 \end{bmatrix}$

4. $\begin{bmatrix} -9 & 4 & 4 \\ -8 & 3 & 4 \\ -16 & 8 & 7 \end{bmatrix}$ **Ans.** $-1, -1, 3;$ $\begin{bmatrix} 0 \\ 1 \\ -1 \end{bmatrix}, \begin{bmatrix} 1 \\ 1 \\ 1 \end{bmatrix}, \begin{bmatrix} 1 \\ 1 \\ 2 \end{bmatrix}$

5. $\begin{bmatrix} 1 & 1 & 0 \\ 0 & 1 & 0 \\ 0 & 0 & 1 \end{bmatrix}$ *(AMIETE, Dec. 2010)* **Ans.** $1, 1, 1,$ $\begin{bmatrix} 1 \\ 0 \\ 1 \end{bmatrix}$

11.7 SYMMETRIC MATRICES WITH NON REPEATED EIGEN VALUES

Example 10. *Find the eigen values and the corresponding eigen vectors of the matrix*

$$\begin{bmatrix} -2 & 5 & 4 \\ 5 & 7 & 5 \\ 4 & 5 & -2 \end{bmatrix}$$

Solution. Characteristic equation of given matrix is $|A - \lambda I| = 0$.

$$\Rightarrow \quad \begin{vmatrix} -2-\lambda & 5 & 4 \\ 5 & 7-\lambda & 5 \\ 4 & 5 & -2-\lambda \end{vmatrix} = 0 \Rightarrow \quad \lambda^3 - 3\lambda^2 - 90\lambda - 216 = 0$$

By trial: Take $\lambda = -3$ then $-27 - 27 + 270 - 216 = 0$

By synthetic division

$$\begin{array}{r|rrrr} -3 & 1 & -3 & -90 & -216 \\ & & -3 & 18 & 216 \\ \hline & 1 & -6 & -72 & 0 \end{array}$$

$\Rightarrow \lambda^2 - 6\lambda - 72 = 0 \;\Rightarrow\; (\lambda - 12)(\lambda + 6) = 0 \;\Rightarrow\; \lambda = -3, -6, 12$

Matrix equation for eigen vectors $[A - \lambda I]X = 0$

$$\begin{bmatrix} -2-\lambda & 5 & 4 \\ 5 & 7-\lambda & 5 \\ 4 & 5 & -2-\lambda \end{bmatrix}\begin{bmatrix} x \\ y \\ z \end{bmatrix} = \begin{bmatrix} 0 \\ 0 \\ 0 \end{bmatrix} \quad \ldots(1)$$

On putting $\lambda = -3$ in (1), we get

$$\begin{bmatrix} 1 & 5 & 4 \\ 5 & 10 & 5 \\ 4 & 5 & 1 \end{bmatrix}\begin{bmatrix} x \\ y \\ z \end{bmatrix} = \begin{bmatrix} 0 \\ 0 \\ 0 \end{bmatrix} \Rightarrow \begin{matrix} x + 5y + 4z = 0 \\ 5x + 10y + 5z = 0 \end{matrix}$$

$$\frac{x}{25-40} = \frac{y}{20-5} = \frac{z}{10-25} \text{ or } \frac{x}{1} = \frac{y}{-1} = \frac{z}{1}$$

Hence the eigen vector $X_1 = \begin{bmatrix} 1 \\ -1 \\ 1 \end{bmatrix}$

On putting $\lambda = -6$ in (1), we get

$$\begin{bmatrix} 4 & 5 & 4 \\ 5 & 13 & 5 \\ 4 & 5 & 4 \end{bmatrix}\begin{bmatrix} x \\ y \\ z \end{bmatrix} = \begin{bmatrix} 0 \\ 0 \\ 0 \end{bmatrix} \Rightarrow \begin{matrix} 4x + 5y + 4z = 0 \\ 5x + 13y + 5z = 0 \end{matrix}$$

$$\frac{x}{25-52} = \frac{y}{20-20} = \frac{z}{52-25} \text{ or } \frac{x}{1} = \frac{y}{0} = \frac{z}{-1}$$

Hence the eigen vector $X_2 = \begin{bmatrix} 1 \\ 0 \\ -1 \end{bmatrix}$

On putting $\lambda = 12$ in (1), we get

$$\begin{bmatrix} -14 & 5 & 4 \\ 5 & -5 & 5 \\ 4 & 5 & -14 \end{bmatrix}\begin{bmatrix} x \\ y \\ z \end{bmatrix} = \begin{bmatrix} 0 \\ 0 \\ 0 \end{bmatrix} \Rightarrow \begin{matrix} -14x + 5y + 4z = 0 \\ 5x - 5y + 5z = 0 \end{matrix}$$

$$\frac{x}{25+20} = \frac{y}{20+70} = \frac{z}{70-25} \text{ or } \frac{x}{1} = \frac{y}{2} = \frac{z}{1}$$

Hence the eigen vector $X_3 = \begin{bmatrix} 1 \\ 2 \\ 1 \end{bmatrix}$ **Ans.**

EXERCISE 11.5

Symmetric matrices with non-repeated eigen values

Find the eigen values and eigen vectors of the following matrices:

1. $\begin{bmatrix} 5 & 0 & 1 \\ 0 & -2 & 0 \\ 1 & 0 & 5 \end{bmatrix}$ **Ans.** $-2, 4, 6;$ $\begin{bmatrix} 0 \\ 1 \\ 0 \end{bmatrix}, \begin{bmatrix} 1 \\ 0 \\ -1 \end{bmatrix}, \begin{bmatrix} 1 \\ 0 \\ 1 \end{bmatrix}$

2. $\begin{bmatrix} 3 & -1 & 1 \\ -1 & 5 & -1 \\ 1 & -1 & 3 \end{bmatrix}$ *(A.M.I.E.T.E., June 2017, 2016)* **Ans.** $2, 3, 6;$ $\begin{bmatrix} -1 \\ 0 \\ 1 \end{bmatrix}, \begin{bmatrix} 1 \\ 1 \\ 1 \end{bmatrix}, \begin{bmatrix} 1 \\ -2 \\ 1 \end{bmatrix}$

3. $\begin{bmatrix} 8 & -6 & 2 \\ -6 & 7 & -4 \\ 2 & -4 & 3 \end{bmatrix}$ *(U.P., I Semester, Jan 2011)* **Ans.** $0, 3, 15;$ $\begin{bmatrix} 1 \\ 2 \\ 2 \end{bmatrix}, \begin{bmatrix} 2 \\ 1 \\ -2 \end{bmatrix}, \begin{bmatrix} 2 \\ -2 \\ 1 \end{bmatrix}$

4. $\begin{bmatrix} 2 & 4 & -6 \\ 4 & 2 & -6 \\ -6 & -6 & -15 \end{bmatrix}$ *(A.M.I.E.T.E., Dec. 2017)* **Ans.** $-2, 9, -18;$ $\begin{bmatrix} 1 \\ -1 \\ 0 \end{bmatrix}, \begin{bmatrix} 2 \\ 2 \\ -1 \end{bmatrix}, \begin{bmatrix} 1 \\ 1 \\ 4 \end{bmatrix}$

5. $\begin{bmatrix} 1 & 1 & 3 \\ 1 & 5 & 1 \\ 3 & 1 & 1 \end{bmatrix}$ **Ans.** $-2, 3, 6;$ $\begin{bmatrix} -1 \\ 0 \\ 1 \end{bmatrix}, \begin{bmatrix} 1 \\ -1 \\ 1 \end{bmatrix}, \begin{bmatrix} 1 \\ 2 \\ 1 \end{bmatrix}$

11.8 SYMMETRIC MATRICES WITH REPEATED EIGEN VALUES

Example 11. *Find all the eigen values and eigen vectors of the matrix*

$$\begin{bmatrix} 2 & -1 & 1 \\ -1 & 2 & -1 \\ 1 & -1 & 2 \end{bmatrix}$$

Solution. The characteristic equation is $\begin{vmatrix} 2-\lambda & -1 & 1 \\ -1 & 2-\lambda & -1 \\ 1 & -1 & 2-\lambda \end{vmatrix} = 0$

$$\Rightarrow \quad (2-\lambda)[(2-\lambda)^2 - 1] + 1[-2 + \lambda + 1] + 1[1 - 2 + \lambda] = 0$$

$$\Rightarrow \quad (2-\lambda)(4 - 4\lambda + \lambda^2 - 1) + (\lambda - 1) + \lambda - 1 = 0$$

$$\Rightarrow \quad 8 - 8\lambda + 2\lambda^2 - 2 - 4\lambda + 4\lambda^2 - \lambda^3 + \lambda + 2\lambda - 2 = 0$$

$$\Rightarrow \quad -\lambda^3 + 6\lambda^2 - 9\lambda + 4 = 0$$

$$\Rightarrow \quad \lambda^3 - 6\lambda^2 + 9\lambda - 4 = 0 \qquad \ldots(1)$$

On putting $\lambda = 1$ in (1), equation (1) is satisfied. Therefore $(\lambda - 1)$ is one factor of equation (1). The other factor $(\lambda^2 - 5\lambda + 4)$ is obtained on dividing (1) by $(\lambda - 1)$.

$$\Rightarrow \quad (\lambda - 1)(\lambda^2 - 5\lambda + 4) = 0 \quad \Rightarrow \quad (\lambda - 1)(\lambda - 1)(\lambda - 4) = 0$$

$$\Rightarrow \quad \lambda = 1, 1, 4$$

Then eigen values are 1, 1, 4.

Matrix equation for eigen vectors $[A - \lambda I]X = 0$

$$\text{i.e.} \quad \begin{bmatrix} 2-\lambda & -1 & 1 \\ -1 & 2-\lambda & -1 \\ 1 & -1 & 2-\lambda \end{bmatrix} \begin{bmatrix} x_1 \\ x_2 \\ x_3 \end{bmatrix} = \begin{bmatrix} 0 \\ 0 \\ 0 \end{bmatrix} \quad \ldots(2)$$

When $\lambda = 4$, equation (2) becomes

$$\begin{bmatrix} 2-4 & -1 & 1 \\ -1 & 2-4 & -1 \\ 1 & -1 & 2-4 \end{bmatrix} \begin{bmatrix} x_1 \\ x_2 \\ x_3 \end{bmatrix} = \begin{bmatrix} 0 \\ 0 \\ 0 \end{bmatrix} \Rightarrow \begin{bmatrix} -2 & -1 & 1 \\ -1 & -2 & -1 \\ 1 & -1 & -2 \end{bmatrix} \begin{bmatrix} x_1 \\ x_2 \\ x_3 \end{bmatrix} = \begin{bmatrix} 0 \\ 0 \\ 0 \end{bmatrix}$$

$$\Rightarrow \quad -2x_1 - x_2 + x_3 = 0,$$

$$x_1 - x_2 - 2x_3 = 0$$

$$\frac{x_1}{2+1} = \frac{x_2}{1-4} = \frac{x_3}{2+1} \Rightarrow \frac{x_1}{1} = \frac{x_2}{-1} = \frac{x_3}{1} = k \Rightarrow x_1 = k, x_2 = -k, x_3 = k$$

$$\text{The eigen vector } X_1 = \begin{bmatrix} k \\ -k \\ k \end{bmatrix} = k \begin{bmatrix} 1 \\ -1 \\ 1 \end{bmatrix} \quad \text{or} \quad X_1 = \begin{bmatrix} 1 \\ -1 \\ 1 \end{bmatrix}$$

When $\lambda = 1$, equation (2) becomes

$$\begin{bmatrix} 2-1 & -1 & 1 \\ -1 & 2-1 & -1 \\ 1 & -1 & 2-1 \end{bmatrix} \begin{bmatrix} x_1 \\ x_2 \\ x_3 \end{bmatrix} = 0$$

$$\Rightarrow \begin{bmatrix} 1 & -1 & 1 \\ -1 & 1 & -1 \\ 1 & -1 & 1 \end{bmatrix} \begin{bmatrix} x_1 \\ x_2 \\ x_3 \end{bmatrix} = 0 \Rightarrow \begin{pmatrix} 1 & -1 & 1 \\ 0 & 0 & 0 \\ 0 & 0 & 0 \end{pmatrix} \begin{pmatrix} x_1 \\ x_2 \\ x_3 \end{pmatrix} = 0, \; R_2 \to R_2 + R_1, \; R_3 \to R_3 - R_1$$

$$\Rightarrow \quad x_1 - x_2 + x_3 = 0$$

Let $x_1 = k_1$ and $x_2 = k_2$

$k_1 - k_2 + x_3 = 0$ or $x_3 = k_2 - k_1$

$$\text{The eigen vector } X_2 = \begin{bmatrix} k_1 \\ k_2 \\ k_2 - k_1 \end{bmatrix} \Rightarrow X_2 = \begin{bmatrix} 1 \\ 1 \\ 0 \end{bmatrix} \qquad \begin{bmatrix} k_1 = 1 \\ k_2 = 1 \end{bmatrix}$$

$$\text{Let} \quad X_3 = \begin{bmatrix} l \\ m \\ n \end{bmatrix}$$

As X_3 is orthogonal to X_1 since the given matrix is symmetric

$$[1, -1, 1]\begin{bmatrix} l \\ m \\ n \end{bmatrix} = 0 \quad \text{or} \quad l - m + n = 0 \qquad \ldots(3)$$

As X_3 is orthogonal to X_2 since the given matrix is symmetric

$$[1, 1, 0]\begin{bmatrix} l \\ m \\ n \end{bmatrix} = 0 \quad \text{or} \quad l + m + 0 = 0 \qquad \ldots(4)$$

Solving (3) and (4), we get $\frac{l}{0-1} = \frac{m}{1-0} = \frac{n}{1+1} \Rightarrow \frac{l}{-1} = \frac{m}{1} = \frac{n}{2}$

The eigen vector $X_3 = \begin{bmatrix} -1 \\ 1 \\ 2 \end{bmatrix}$ **Ans.**

EXERCISE 11.6

Symmetric matrices with repeated eigen values

Find the eigen values and the corresponding eigen vectors of the following matrices:

1. $\begin{bmatrix} 1 & 2 & 3 \\ 2 & 4 & 6 \\ 3 & 6 & 9 \end{bmatrix}$ **Ans.** 0, 0, 14; $\begin{bmatrix} -2 \\ 1 \\ 0 \end{bmatrix}, \begin{bmatrix} 3 \\ 6 \\ -5 \end{bmatrix}, \begin{bmatrix} 1 \\ 2 \\ 3 \end{bmatrix}$

2. $\begin{bmatrix} 2 & 0 & 1 \\ 0 & 3 & 0 \\ 1 & 0 & 2 \end{bmatrix}$ **Ans.** 1, 3, 3; $\begin{bmatrix} 1 \\ 0 \\ -1 \end{bmatrix}, \begin{bmatrix} 1 \\ 1 \\ 1 \end{bmatrix}, \begin{bmatrix} 1 \\ -2 \\ 1 \end{bmatrix}$

3. $\begin{bmatrix} 6 & -2 & 2 \\ -2 & 3 & -1 \\ 2 & -1 & 3 \end{bmatrix}$ **Ans.** 8, 2, 2; $\begin{bmatrix} 2 \\ -1 \\ 1 \end{bmatrix}, \begin{bmatrix} 1 \\ 0 \\ -2 \end{bmatrix}, \begin{bmatrix} 1 \\ 2 \\ 0 \end{bmatrix}$

4. $\begin{bmatrix} 6 & -3 & 3 \\ -3 & 6 & -3 \\ 3 & -3 & 6 \end{bmatrix}$ **Ans.** 3, 3, 12

5. Choose the correct or the best of the answers given in the following Parts;

(*i*) Two of the eigen values of a 3 × 3 matrix, whose determinant equals 4, are –1 and 2. The third eigen value of the matrix is equal to

(*a*) –2 (*b*) –1 (*c*) 1 (*d*) 2

(*ii*) If a square matrix A has an eigen value λ, then an eigen value of the matrix $(kA)^T$ where, $k \neq 0$, is a scalar is

(*a*) λ/k (*b*) k/λ (*c*) $k\,\lambda$ (*d*) None of these (*iii*) An eigen value of a square matrix A is $\lambda = 0$. Then

(*a*) $|A| \neq 0$ (*b*) A is symmetric

(*c*) A is singular (*d*) A s skew-symmetric

(*iv*) The matrix A is defined as $\begin{bmatrix} -1 & 0 & 0 \\ 2 & -3 & 0 \\ 1 & 4 & 2 \end{bmatrix}$. The eigen values of A^2 are

(*a*) –1, –9, –4 (*b*) 1, 9, 4 (*c*) –1, –3, 2 (*d*) 1, 3, –2

(*v*) If the matrix is $A = \begin{bmatrix} -1 & 2 & 3 \\ 0 & 3 & 5 \\ 0 & 0 & -2 \end{bmatrix}$ then the eigen values of $A^3 + 5A + 8I$, are

(*a*) –1, 27, –8 (*b*) –1, 3, –2 (*c*) 2, 50, –10 (*d*) 2, 50, 10

(*vi*) The matrix A has eigen values $\lambda_i \neq 0$. Then $A^{-1} - 2I + A$ has eigen values

(*a*) $1 + 2\lambda_i + \lambda_i^2$ (*b*) $\dfrac{1}{\lambda_i} - 2 + \lambda_i$ (*c*) $1 - 2\lambda_i + \lambda_i^2$ (*d*) $1 - \dfrac{2}{\lambda_i} + \dfrac{1}{\lambda_i^2}$

(*vii*) The eigen values of a matrix A are 1, –2, 3. The eigen values of $3I - 2A + A^2$ are

(*a*) 2, 11, 6 (*b*) 3, 11,18 (*c*) 2, 3, 6 (*d*) 6, 3, 11

Ans. (*i*) (*b*), (*ii*) (*c*), (*iii*) (*c*), (*iv*) (*b*), (*v*) (*c*), (*vi*) (*b*), (*vii*) (*a*)

Complex Numbers

12.1 COMPLEX NUMBERS

A number of the form $a + ib$ is called a complex number when a and b are real numbers and $i = \sqrt{-1}$. We call 'a' the real part and 'b' the imaginary part of the complex number $a + ib$. If $a = 0$ the number ib is said to be purely imaginary, if $b = 0$, the number a is real.

A pair of complex numbers $a + ib$ and $a - ib$ are said to be conjugate of each other and the sum and product of a complex number and its conjugate complex are both real.

Let $x + iy$ be a complex number and $x - iy$ its conjugate complex.

Sum = $(x + iy) + (x - iy) = 2x$ (Real)

Product = $(x + iy) . (x - iy) = x^2 + y^2$ (Real)

Note: Let a complex number be z. Then the conjugate complex number is denoted by $\bar{z}$

Let $a + ib$ and $c + id$ be two complex numbers. Then

Addition. $(a + ib) + (c + id) = (a + c) + i\,(b + d)$

Subtracton. $(a + ib) - (c + id) = a - c + i\,(b - d)$

Multiplication. $(a + ib) \times (c + id) = ac - bd + i\,(ad + bc)$

Division. $\dfrac{a+ib}{c+id} = \dfrac{a+ib}{c+id}\cdot\dfrac{c-id}{c-id}$

$$= \frac{ac+bd}{c^2+d^2} + i\frac{bc-ad}{c^2+d^2}$$

For example, if two complex numbers $a + ib$ and $c + id$ are equal. Prove that

$$a = c \quad \text{and} \quad b = d$$

Proof. $a + ib = c + id$

$\Rightarrow \quad a - c = i\,(d - b)$

$$(a - c)^2 = -\,(d - b)^2$$

$$(a - c)^2 + (d - b)^2 = 0$$

Here sum of two positive numbers is zero. This is only possible if each number is zero.

$$(a - c)^2 = 0 \Rightarrow a = c$$

and $\quad (d - b)^2 = 0 \Rightarrow b = d$ **Proved.**

Example 1. *Express* $\dfrac{(1+i)(2+i)}{3+i}$ *in the form a + ib* **[S.B.T.E. 2015]**

Solution. $\dfrac{(1+i)(2+i)}{3+i} = \dfrac{2+i+2i-1}{3+i} = \dfrac{1+3i}{3+i}$

$$= \frac{(1+3i)(3-i)}{(3+i)(3-i)} = \frac{3-i+9i+3}{9+1}$$

$$= \frac{6+8i}{10}$$

$$= \frac{3}{5} + \frac{4}{5}i$$ **Ans.**

12.2 ARGAND DIAGRAM

Mathematician Argand represented a complex number in a diagram known as Argand diagram. A complex number $x + iy$ can be represented by a point P whose coordinates are (x, y). The axis of x is called the real axis and the axis of y the imaginary axis. The distance OP is the modulus and the angle OP makes with the x-axis is the argument of $x + iy$.

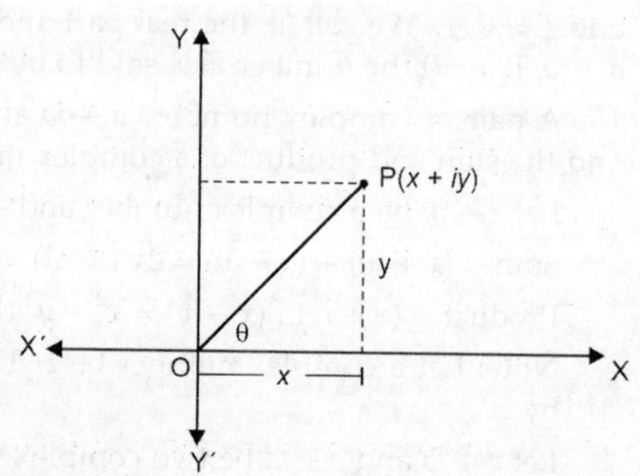

Example 2. *Express* $\dfrac{(6+i)\cdot(2-i)}{(4+3i)\cdot(1-2i)}$ *in the form a + ib.*

Solution. $\dfrac{(6+i)\cdot(2-i)}{(4+3i)\cdot(1-2i)} = \dfrac{12+1+i(2-6)}{4+6+i(3-8)} = \dfrac{13-4i}{10-5i}$

$$= \frac{(13-4i)(10+5i)}{(10-5i)(10+5i)} = \frac{150+25i}{100+25}$$

$$= \frac{6+i}{5} = \frac{6}{5} + \frac{1}{5}i.$$ **Ans.**

EXERCISE 12.1

Express the following in the form a + ib, where a and b are real:

1. If $z = 1 + i$, find (*i*) z^2 (ii) $\dfrac{1}{z}$ and plot them on the Argand diagram. **Ans.** (*i*) $2i$ (*ii*) $\dfrac{1}{2} - \dfrac{i}{2}$

2. $\dfrac{2-3i}{4-i}$ **Ans.** $\dfrac{11}{17} - \dfrac{10}{17}i$

3. $\dfrac{(3+4i)(2+i)}{1+i}$ **Ans.** $\dfrac{13}{2} + \dfrac{9}{2}i$

4. $\dfrac{(1+2i)^3}{(1+i)(2-i)}$ **Ans.** $-\dfrac{7}{2} + \dfrac{1}{2}i$

5. $\dfrac{3+2i}{(4-5i)(2+i)}$ **Ans.** $\dfrac{27}{205} + \dfrac{44}{205}i$

6. The points A, B, C represent the complex numbers z_1, z_2, z_3 respectively, and G is the centroid of the triangle ABC. If $4z_1 + z_2 + z_3 = 0$, show that the origin is the mid point of AG.

7. If $(x + iy)^{1/3} = a + ib$, then show that $4(a^2 - b^2) = \frac{x}{a} + \frac{y}{b}$.

8. If $(x + iy)^3 = u + iv$, then show that $\frac{u}{x} + \frac{v}{y} = 4(x^2 - y^2)$.

9. Find the values of x and y, if $\frac{(1+i)x-2i}{3+i} + \frac{(2-3i)y+i}{3-i} = i$. **Ans.** $x = 3$ and $y = -1$

10. If $a + ib = \frac{(x+i)^2}{2x^2+1}$, prove that $a^2 + b^2 - \frac{(x^2+1)^2}{(2x^2+1)^2}$.

11. $ABCD$ is a parallelogram on the Argand plane. The affixes of A, B, C are $8 + 5i$, $-7 - 5i$, $-5 + 5i$, respectively. Find the affix of D. **Ans.** $10 + 15i$

Choose the correct or best alternative in the following:

12. Let $z_1 = 2 - 5i;$

$z_2 = -1 + 4i;$

$z_3 = 6 + i;$

and $z_4 = 3 - 7i.$

13. Express $\frac{(z_1 + z_2)z_3}{z_4}$ in the form $a + bi$, $a, b \in R$.

(*a*) $\frac{208}{29} + \frac{27}{29}i$ (*b*) $\frac{208}{29} - \frac{27}{29}i$ (*c*) $\frac{28}{29} - \frac{27}{29}i$ (*d*) $\frac{28}{29} + \frac{17}{29}i$

(*Diploma I.E.T.E. June 2006*) Ans. (*d*)

12.3 MODULUS AND ARGUMENT

Let $x + iy$ be a complex number.

Putting $x = r \cos \theta$ and $y = r \sin \theta$ so that $r = \sqrt{x^2 + y^2}$

$$\cos \theta = \frac{x}{\sqrt{x^2 + y^2}} \quad \text{and} \quad \sin \theta = \frac{y}{\sqrt{x^2 + y^2}}$$

the positive value of the root being taken.

Then r is called the *modulus* or absolute value of the complex number $x + iy$ and is denoted by $|x + iy|$. The angle θ is called the **argument** or **amplitude** of the complex number $x + iy$ and is denoted by Arg. $(x + iy)$.

It is clear that θ will have infinite number of values differing by multiples of 2π. The values of θ lying in the range $-\pi < \theta \le \pi$ is called the *principal value* of the argument. A complex number $x + iy$ is denoted by a single letter z. The number $x - iy$ (conjugate) is denoted by $\bar{z}$. The complex number in polar is $r(\cos \theta + i \sin \theta)$.

Example 3. *Find the modulus and principal argument of the complex number*

$$\frac{1+2i}{1-(1-i)^2}$$

Solution. $\dfrac{1+2i}{1-(1-i)^2} = \dfrac{1+2i}{1-(1-1-2i)} = \dfrac{1+2i}{1+2i} = 1 = 1 + 0i$

$$\left|\frac{1+2i}{1-(1-i)^2}\right| = 1$$

Principal argument of $\dfrac{1+2i}{1-(1-i)^2} = \tan^{-1}\left(\dfrac{0}{1}\right) = \tan^{-1}(0) = 0°$ **Ans.**

12.4 TYPES OF COMPLEX NUMBERS

1. Cartesion form: $x + iy$
2. Polar form: $r(\cos\theta + i\sin\theta)$
3. Exponential or Eulerian form: $re^{i\theta}$

Example 4. *Express the following complex numbers in Eulerian form.*

(*i*) $-2 + 2i$ (*ii*) $1 + i$ **[S.B.T.E. 2016, 2013, 2010]**

Solution. Here, we have

(*i*) $-2 + 2i$

Let
$$z = -2 + 2i$$
$$r = |z| = \sqrt{x^2 + y^2} = \sqrt{(-2)^2 + (2)^2} = 2\sqrt{2}$$
$$\tan\theta = \frac{y}{x} = \frac{2}{-2} = -1 = \tan\left(\frac{3\pi}{4}\right)$$
$$\text{Arg}(z) = \theta = \frac{3\pi}{4}$$
$$\text{Eulerian form} = (z) = re^{i\theta} = 2\sqrt{2}e^{i3\pi/4}$$ **Ans.**

(*ii*) Let
$$z = 1 + i$$
$$r = |z| = \sqrt{1^2 + 1^2} = \sqrt{2}$$
$$\tan\theta = 1 = \tan\frac{\pi}{4}$$
$$\text{Arg}(z) = \theta = \frac{\pi}{4}$$
$$\text{Eulerian form}(z) = re^{i\theta} = \sqrt{2}e^{i\pi/4}$$ **Ans.**

Example 5. *If* $(x + iy) = \sqrt{\dfrac{a+ib}{c+id}}$*, prove that,* $(x^2 + y^2)^2 = \dfrac{a^2 + b^2}{c^2 + d^2}$

[S.B.T.E. 2016, 2010, 2008]

Solution. Here, we have

$$z = \sqrt{\frac{a+ib}{c+id}} \quad \text{... (1)} \quad [\because z = x + iy]$$

$$\overline{z} = \sqrt{\frac{a-ib}{c-id}} \quad \text{... (2)} \quad [\because \overline{z} = x - iy]$$

$$z \cdot \overline{z} = \left(\sqrt{\frac{a+ib}{c+id}}\right) \times \left(\sqrt{\frac{a-ib}{c-id}}\right)$$

$$(x+iy)\cdot(x-iy) = \sqrt{\frac{a^2+b^2}{c^2+d^2}}$$

$$\therefore \qquad (x^2+y^2)^2 = \frac{a^2+b^2}{c^2+d^2}$$ **Proved.**

Example 6. *Using Euler's exponential form prove that* $\sin^2 x + \cos^2 x = 1$

[S.B.T.E. 2014, 2012]

Solution. Here, we have

$$\text{L.H.S.} = \sin^2 x + \cos^2 x$$

$$\left(\frac{e^{ix}-e^{-ix}}{2}\right)^2 + \left(\frac{e^{ix}+e^{-ix}}{2}\right)^2 = \frac{e^{2ix}+e^{-2ix}-2}{4i^2} + \frac{e^{2ix}+e^{-2ix}+2}{4}$$

$$= \frac{1}{4}\left[-\cancel{e^{2ix}} - \cancel{e^{-2ix}} + 2 + \cancel{e^{2ix}} + \cancel{e^{-2ix}} + 2\right]$$

$$= \frac{1}{4}\times 4 = 1$$ **Proved.**

Example 7. *Express* $(1 + \sin\theta + i\cos\theta)$ *in modulus and argument form.*

[B.T.E. Delhi. May 2008]

Solution. Let $1 + \sin\theta + i\cos\theta = r(\cos\alpha + i\sin\alpha)$

Then equating real and imaginary parts, we get

$$r\cos\alpha = 1 + \sin\theta \qquad \ldots (1)$$

and $$r\sin\alpha = \cos\theta \qquad \ldots (2)$$

Squaring and adding (1) and (2), we get

$$r^2\cos^2\alpha + r^2\sin^2\alpha = (1+\sin\theta)^2 + \cos^2\theta$$

$$\Rightarrow \quad r^2(\cos^2\alpha + \sin^2\alpha) = 1 + \sin^2\theta + 2\sin\theta + \cos^2\theta$$

$$\Rightarrow \quad r^2 = 2(1+\sin\theta)$$

$$\Rightarrow \quad r = \sqrt{2(1+\sin\theta)} = \sqrt{2\left[1+\cos\left(\frac{\pi}{2}-\theta\right)\right]}$$

$$= \sqrt{2\left[1+2\cos^2\left(\frac{\pi}{4}-\frac{\theta}{2}\right)-1\right]} = \sqrt{2\left[2\cos^2\left(\frac{\pi}{4}-\frac{\theta}{2}\right)\right]}$$

$$= 2\cos\left(\frac{\pi}{4}-\frac{\theta}{2}\right)$$ **Ans.**

Dividing (2) by (1), we get

$$\frac{r\sin\alpha}{r\cos\alpha} = \frac{\cos\theta}{1+\sin\theta} \Rightarrow \tan\alpha = \frac{\cos\theta(1-\sin\theta)}{(1+\sin\theta)(1-\sin\theta)}$$

$$\tan\alpha = \frac{\cos\theta(1-\sin\theta)}{1-\sin^2\theta} = \frac{\cos\theta(1-\sin\theta)}{\cos^2\theta}$$

$$= \frac{1}{\cos\theta}(1-\sin\theta) = \sec\theta - \tan\theta$$ **Ans.**

Example 8. *Put the complex number* $\left(\frac{2+i}{3-i}\right)^2$ *into polar form.*

Solution. $\left(\frac{2+i}{3-i}\right)^2 = \left(\frac{3+4i}{8-6i}\right)\cdot\left(\frac{8+6i}{8+6i}\right) = \frac{1}{2}i$

Let $\frac{1}{2}i = r(\cos\theta + i\sin\theta)$

$$r\cos\theta = 0, \quad \text{or} \quad r\sin\theta = \frac{1}{2}$$

Squaring and adding $r^2(\cos^2\theta + \sin^2\theta) = \frac{1}{4} \Rightarrow r = \frac{1}{2}$

$$\cos\theta = 0 \text{ and } \sin\theta = 1 \quad \therefore \quad \theta = \frac{\pi}{2}$$

$$\frac{1}{2}i = \frac{1}{2}\left(\cos\frac{\pi}{2} + i\sin\frac{\pi}{2}\right)$$ **Ans.**

Example 9. *Convert* 12 ∠ – 60° *to the rectangular form.*

Solution. Let 12 ∠ – 60° = $x + iy$ ... (1)

$$r = 12, \quad \theta = -60°$$

$$x = r\cos\theta = 12\cos(-60°) = 12\left(\frac{1}{2}\right) = 6$$

$$y = r\sin\theta = 12\sin(-60°) = 12\left(-\frac{\sqrt{3}}{2}\right) = -6\sqrt{3}$$

Putting the value of x and y in (1), we have

$$2\angle -60° = 6 - i\,6\sqrt{3}$$ **Ans.**

Example 10. *Express the following complex numbers in the polar form:*

(*i*) $\frac{1+i}{1-i}$ (*ii*) $\frac{2+6\sqrt{3}i}{5+\sqrt{3}i}$ **[SBTE. 2016, 2015, 2012, 2010]**

Solution. Here, we have

(*i*) Let $z = \frac{1+i}{1-i}$

$\Rightarrow$ $z = \left(\frac{1+i}{1-i}\right) \times \left(\frac{1+i}{1+i}\right) = 0 + i$

$\Rightarrow$ $r = |z| = \sqrt{x^2+y^2} = \sqrt{(0)^2+(1)^2} = 1$

$\Rightarrow$ $\tan\theta = \frac{y}{x} = \frac{1}{0} = \infty = \tan\frac{\pi}{2}$

$\Rightarrow$ $\arg(z) = \theta = \frac{\pi}{2}$

polar form $z = r(\cos\theta + i\sin\theta)$

$$z = 1\left(\cos\frac{\pi}{2} + i\sin\frac{\pi}{2}\right)$$

$\therefore$
$$= \cos\frac{\pi}{2} + i\sin\frac{\pi}{2}$$ **Ans.**

(ii) Let
$$z = \left(\frac{2+6\sqrt{3}i}{5+\sqrt{3}i}\right) \times \left(\frac{5-\sqrt{3}i}{5-\sqrt{3}i}\right) = 1+\sqrt{3}i$$

$$r = \sqrt{x^2+y^2} = \sqrt{1^2+(\sqrt{3})^2} = 2$$

$$\tan\theta = \frac{y}{x} = \frac{\sqrt{3}}{1}$$

$$\tan\theta = \tan\left(\frac{\pi}{3}\right) \quad \text{or} \quad \theta = \left(\frac{\pi}{3}\right)$$

So, polar form $z = r(\cos\theta + i\sin\theta)$

$$z = 2\left(\cos\frac{\pi}{3} + i\sin\frac{\pi}{3}\right)$$ **Ans.**

EXERCISE 12.2

Find the modulus and principal argument of

1. $-\sqrt{3} - i$ **Ans.** $2, \frac{-5\pi}{6}$

2. $\frac{1+2i}{1-3i}$ **Ans.** $\frac{1}{\sqrt{2}}, \frac{3\pi}{4}$

3. $\frac{(1+i)^2}{1-i}$ **Ans.** $\sqrt{2}, \frac{3\pi}{4}$

4. $\sqrt{\left(\frac{1+i}{1-i}\right)}$ **Ans.** $1, \frac{\pi}{4}$

5. $\tan(\alpha - i)$ **Ans.** $\sec\alpha, \left(\frac{\pi}{2} - \alpha\right)$

6. $1 - \cos\alpha + i\sin\alpha$ ***(Diploma IETE, June 2005)*** **Ans.** $2\sin\frac{\alpha}{2}, \frac{\pi-\alpha}{2}$

7. Show that vertices represented by the complex numbers $(1+i)$, $(2+i)$, $(2+3i)$, $(1+3i)$ form a rectangle.

8. Prove that: $|z_1+z_2|^2 + |z_1-z_2|^2 = 2[|z_1|^2 + |z_2|^2]$

9. If z is any complex number and $\bar{z}$ is its complex conjugate then show that $z.\bar{z} = |z|^2$.

Convert the following to rectangular form and draw them on the Argand diagram:

10. $10 \angle -30°$ **Ans.** $5\sqrt{3} - 5i$

11. $5 \angle 90°$ **Ans.** $5i$

12. $20 \angle 45°$ **Ans.** $10\sqrt{2} + 10\sqrt{2}i$

13. $4 \angle 300°$ **Ans.** $2 - 2\sqrt{3}i$

Convert the following complex numbers into polar form:

14. $\frac{1+i}{1-i}$ **Ans.** $\cos\frac{\pi}{2} + i\sin\frac{\pi}{2}$

15. $\frac{-35+5i}{4\sqrt{2}+3\sqrt{2}i}$ **Ans.** $5\left(\cos\frac{3\pi}{4} + i\sin\frac{3\pi}{4}\right)$

16. $\dfrac{3(-4-\sqrt{3}+4i-i)}{8+2i}$ **Ans.** $r = 2.35$; $\theta = \tan^{-1}\{-(0.889)\}$

17. $\dfrac{2+6\sqrt{3}i}{5+\sqrt{3}i}$ **Ans.** $2\left(\cos\dfrac{\pi}{3}+i\sin\dfrac{\pi}{3}\right)$

18. $\dfrac{2+3i}{3-7i}$ **Ans.** $r = \dfrac{\sqrt{754}}{58}, \theta = \tan^{-1}\left(-\dfrac{23}{15}\right)$

19. $\left(\dfrac{4-5i}{2+3i}\right)\cdot\left(\dfrac{3+2i}{7+i}\right)$ **Ans.** 0.905, $\theta = \tan^{-1}(-7.2)$

20. $\dfrac{(2+5i)(-3+i)}{(1-2i)^2}$ **Ans.** $\dfrac{\sqrt{290}}{5}, \tan^{-1}\left(-\dfrac{1}{17}\right)$

Example 11. *Find the smallest positive integer n for which*

$$\left(\frac{1+i}{1-i}\right)^n = 1$$

[Diploma IETE, Dec. 2006]

Solution.
$$\left[\frac{1+i}{1-i}\right]^n = 1$$

$$\left[\frac{1+i}{1-i}\times\frac{1+i}{1+i}\right]^n = 1$$

$$\left(\frac{1-1+2i}{1+1}\right)^n = 1$$

$$(i)^n = 1 = (i)^4$$

$$n = 4$$ **Ans.**

Example 12. *If* $(x^2y-2)+i(x+2xy-5)=0$*, find the values of x and y.*

Solution. $(x^2y-2)+i(x+2xy-5) = 0$

$$x^2y - 2 = 0 \Rightarrow y = \frac{2}{x^2} \quad \text{... (1)}$$

and $$x + 2xy - 5 = 0 \quad \text{... (2)}$$

Putting the value of y in (2), we get

$$x + 2x\left(\frac{2}{x^2}\right) - 5 = 0$$

$$\Rightarrow \quad x^2 + 4 - 5x = 0$$

$$\Rightarrow \quad x^2 - 5x + 4 = 0$$

$$\Rightarrow \quad (x-1)(x-4) = 0$$

$$\Rightarrow \quad x = 1, x = 4$$

when $$x = 1, y = \frac{2}{(1)^2} = 2,$$

when $$x = 4,\ y = \frac{2}{(4)^2} = \frac{1}{8}$$ **Ans.**

Example 13. *Solve for θ such that the expression* $\frac{3+2i\sin\theta}{1-2i\sin\theta}$ *is imaginary.*

Solution. $$\frac{3+2i\sin\theta}{1-2i\sin\theta} = \frac{(3+2i\sin\theta)(1+2i\sin\theta)}{(1-2i\sin\theta)(1+2i\sin\theta)}$$

$$= \frac{3-4\sin^2\theta+8i\sin\theta}{1+4\sin^2\theta}, \text{ if } 3 - 4\sin^2\theta = 0 \text{ then}$$

$$= \frac{8i\sin\theta}{1+4\sin^2\theta} = \text{purely imaginary}$$

$$\sin^2\theta = \frac{3}{4} \Rightarrow \sin\theta = \frac{\sqrt{3}}{2}$$

$$\theta = \frac{\pi}{3}$$ **Ans.**

Example 14. *If* $a^2+b^2+c^2=1,\ b+ic=(1+a)\,z,$ *prove that*

$$\frac{a+ib}{1+c} = \frac{1+iz}{1-iz}.$$ **[Diploma IETE, June 2006]**

Solution. $$b + ic = (1 + a)\,z$$

$$\Rightarrow \quad z = \frac{b+ic}{1+a}$$

$$\frac{1+iz}{1-iz} = \frac{1+i\dfrac{b+ic}{1+a}}{1-i\dfrac{b+ic}{1+a}} = \frac{1+a+ib-c}{1+a-ib+c}$$

$$= \frac{[(1+a+ib)-c]}{(1+a+c-ib)} \times \frac{(1+a+c+ib)}{(1+a+c+ib)} = \frac{(1+a+ib)^2-c^2}{(1+a+c)^2+b^2}$$

$$= \frac{1+a^2-b^2+2a+2ib+2iab-c^2}{1+a^2+c^2+2a+2c+2ac+b^2}$$

$$= \frac{1+a^2-b^2-c^2+2a+2ib+2iab}{1+(a^2+b^2+c^2)+2a+2c+2ac}$$

Putting the value of $a^2 + b^2 + c^2 = 1$

$$= \frac{1+a^2-(1-a^2)+2a+2ib+2iab}{1+1+2a+2c+2ac}$$

$$\therefore \quad \frac{1+iz}{1-iz} = \frac{2(a^2+a+ib+iab)}{2(1+a+c+ac)} = \frac{2(1+a)(a+ib)}{2(1+a)(1+c)} = \frac{a+ib}{1+c}$$ **Ans.**

Similar Question

If $a = \cos\theta + i \sin\theta$, prove that $1 + a + a^2 = (1 + 2\cos\theta)(\cos\theta + i\sin\theta)$.

12.5 ALGEBRA OF COMPLEX NUMBER

(i) ADDITION

Let $z_1 = x_1 + iy_1$ and $z_2 = x_2 + iy_2$ be two complex numbers represented by the points P and Q on the Argand diagram.

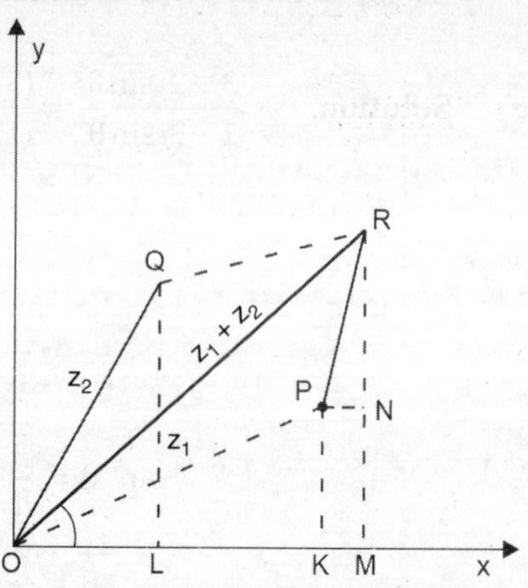

Complete the parallelogram OPRQ

Draw PK, RM, QL, $\perp$ s on OX.

Also draw $PN \perp$ to RM, *i.e.* $OM = OK + KM$

$$= OK + OL$$

$$= x_1 + x_2$$

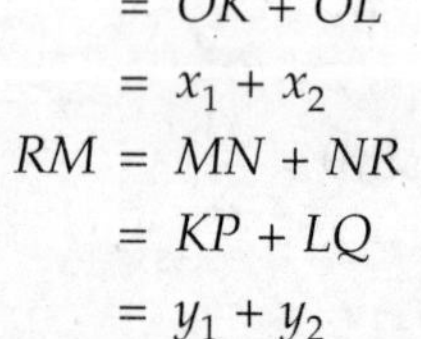

and $RM = MN + NR$

$$= KP + LQ$$

$$= y_1 + y_2$$

$\therefore$ The coordinates of R are $(x_1 + x_2, y_1 + y_2)$ and it represents the complex number.

$$(x_1 + x_2) + i(y_1 + y_2) = (x_1 + iy_1) + (x_2 + iy_2)$$

Thus the sum of two complex numbers is represented by the extremity of the diagonal of the parallelogram formed by OP (z_1) and OQ (z_2) as adjacent sides.

$$|z_1 + z_2| = OR \text{ and } \operatorname{Arg}(z_1 + z_2) = \angle ROM$$

(ii) SUBTRACTION

Let P and Q represent two complex numbers $z_1 = x_1 + iy_1$ and $z_2 = x_2 + iy_2$.

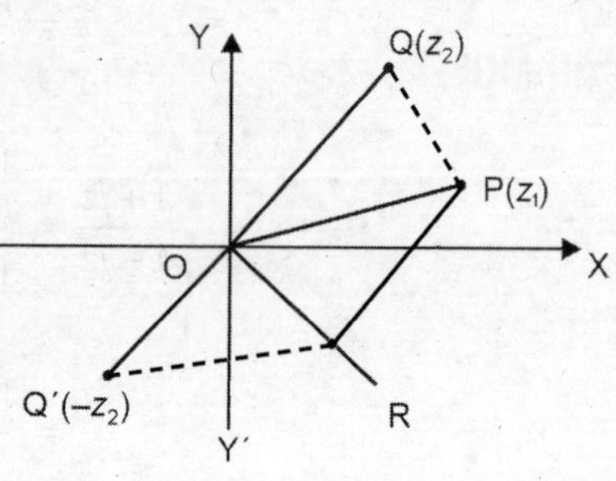

Then $z_1 - z_2 = z_1 + (-z_2)$ or $z_1 - z_2$ means the addition of z_1 and $-z_2$.

$-z_2$ is represented by OQ' formed by producing OQ to OQ' such that $OQ = OQ'$

Complete the parallelogram OPRQ′, then the sum of z_1 and $-z_2$ is represented by OR.

Example 15. *Prove that*

$$\textbf{(i)}\ |\mathbf{z}_1 + \mathbf{z}_2| \le |\mathbf{z}_1| + \mathbf{z}_2| \qquad \textbf{(ii)}\ |\mathbf{z}_1 - \mathbf{z}_2| \ge |\mathbf{z}_1| - \mathbf{z}_2|$$

Solution. Let $z_1 = x_1 + iy_1$, and $z_2 = x_2 + iy_2$ be the two complex numbers by geometry see Section 12.5 to figure.

$$|z_1| = OP, \quad |z_2| = OQ$$

(*i*) Since in a triangle any side is less than the sum of the other two.

In $\Delta\ OPR$, $OR < OP + PR$

$$OR < OP + OQ$$

$\Rightarrow$ $|z_1 + z_2| < |z_1| + |z_2|$

$OR = OP + PR$ if O, P, R are collinear.

$\Rightarrow$ $|z_1 + z_2| = |z_1| + |z_2|$

(*ii*) Again any side of a triangle is greater than the difference between the other two, we have

In $\Delta\, OPR$, $\quad OR > OP - PR$

$$OR > OP - OQ$$

$\Rightarrow \quad |z_1 - z_2| > |z_1| - |z_2|$ **Proved.**

By Algebra

(*i*)

$$|z_1 + z_2|^2 = (x_1 + x_2)^2 + (y_1 + y_2)^2$$

$$= (x_1^2 + y_1^2) + (x_2^2 + y_2^2) + 2(x_1x_2 + y_1y_2)$$

$$= (x_1^2 + y_1^2) + (x_2^2 + y_2^2) + 2\sqrt{(x_1x_2 + y_1y_2)^2}$$

$$= |z_1|^2 + |z_2|^2 + 2\sqrt{x_1^2x_2^2 + y_1^2y_2^2 + 2x_1x_2y_1y_2}$$

$$\le |z_1|^2 + |z_2|^2 + 2\sqrt{x_1^2x_2^2 + y_1^2y_2^2 + x_1^2y_2^2 + x_2^2y_1^2}$$

$$[\because (x_1y_2 - x_2y_1)^2 \ge 0 \;\Rightarrow\; x_1^2y_2^2 + x_2^2y_1^2 \ge 2x_1x_2\, y_1y_2]$$

$$\le |z_1|^2 + |z_2|^2 + 2\sqrt{(x_1^2 + y_1^2)\,(x_2^2 + y_2^2)}$$

$$\le |z_1|^2 + |z_2|^2 + 2|z_1|\,|z_2|$$

$$\le \{|z_1| + |z_2|\}^2$$

$$|z_1 + z_2| \ge |z_1| + |z_2|$$

(*ii*)

$$|z_1| = |(z_1 - z_2) + z_2| \le |z_1 - z_2| + |z_2|$$

$$|z_1| - |z_2| \le |z_1 - z_2|$$

$$|z_1 - z_2| \ge |z_1| - |z_2|$$

Proved.

(ii) MULTIPLICATION (By Algebra)

Let $x_1 = r_1 \cos\theta_1,\; y_1 = r_1 \sin\theta_1$

$$x_2 = r_2 \cos\theta_2,\; y_2 = r_2 \sin\theta_2$$

$$z_1 = x_1 + iy_1 = r_1 (\cos\theta_1 + i\sin\theta_1),\; |z_1| = r_1$$

$$z_2 = x_2 + iy_2 = r_2 (\cos\theta_2 + i\sin\theta_2),\; |z_2| = r_2$$

$$z_1z_2 = r_1r_2 (\cos\theta_1 + i\sin\theta_1)(\cos\theta_2 + i\sin\theta_2)$$

$$= r_1r_2[\cos(\theta_1 + \theta_2) + i\sin(\theta_1 + \theta_2)],\quad |z_1z_2| = r_1r_2$$

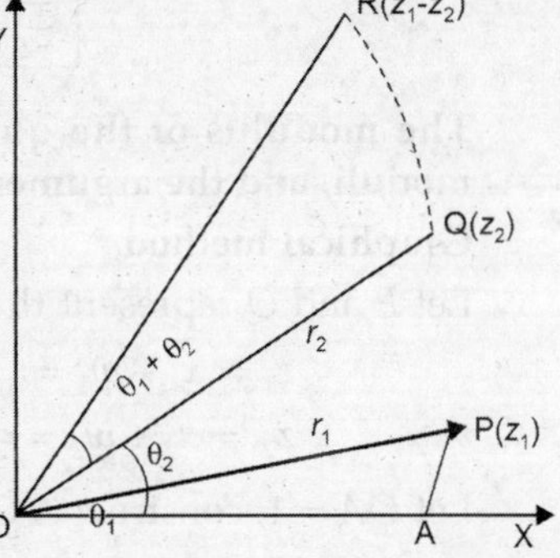

The modulus of the product of two complex numbers is the product of their moduli, and the argument of the product is the sum of the their arguments.

Graphical method

Let *P*, *Q* represent the complex numbers

$$z_1 = x_1 + iy_1$$

$$= r_1 (\cos\theta_1 + i\cos\theta_1)$$

$$z_2 = x_2 + iy_2$$

$$= r_1 (\cos\theta_2 + i\cos\theta_2)$$

Cut of $OA = 1$ along x-axis. Construct $\Delta\, ORQ$ on OQ similar to $\Delta\, OAP$

So that $\quad \dfrac{OR}{OP} = \dfrac{OQ}{OA}$

$$OR = OP \cdot OQ = r_1 r_2$$
$$\angle XOR = \angle AOQ + \angle OQR$$
$$= \theta_2 + \theta_1$$

Hence the product of two complex numbers z_1, z_2 is represented by the point R, such that

(*i*) $$|z_1 \cdot z_2| = |z_1| \cdot |z_2|$$

(*ii*) $$\text{Arg.}(z_1 \cdot z_2) = \text{Arg.}(z_1) + \text{Arg.}(z_2)$$

Cor. Multiplication of a complex number by i.

Let $$z = x + iy = r(\cos\theta + i\sin\theta)$$

$$i = 0 + i \cdot 1 = \left[\cos\frac{\pi}{2} + i\sin\frac{\pi}{2}\right]$$

$$i \cdot z = r(\cos\theta + i\sin\theta) \cdot \left[\cos\frac{\pi}{2} + i\sin\frac{\pi}{2}\right]$$

$$= r\left[\cos\left(\theta + \frac{\pi}{2}\right) + i\sin\left(\theta + \frac{\pi}{2}\right)\right]$$

Hence a complex number multiplied by i results in the rotation of the complex number $\pi/2$ in anticlockwise direction without change in magnitude.

(iv) DIVISION (By Algebra) [Diploma IETE, 2004]

$$\frac{z_1}{z_2} = \frac{r_1(\cos\theta_1 + i\sin\theta_1)}{r_1(\cos\theta_2 + i\sin\theta_2)}$$

$$= \frac{r_1(\cos\theta_1 + i\sin\theta_1)(\cos\theta_2 - i\sin\theta_2)}{r_1(\cos\theta_2 + i\sin\theta_2)(\cos\theta_2 - i\sin\theta_2)}$$

$$\boxed{\frac{z_1}{z_2} = \frac{r_1}{r_2}[\cos(\theta_1 - \theta_2) + i\sin(\theta_1 - \theta_2)]}$$

The modulus of the quotient of two complex numbers is the quotient of their moduli, and the argument is the difference of their arguments.

Graphical method

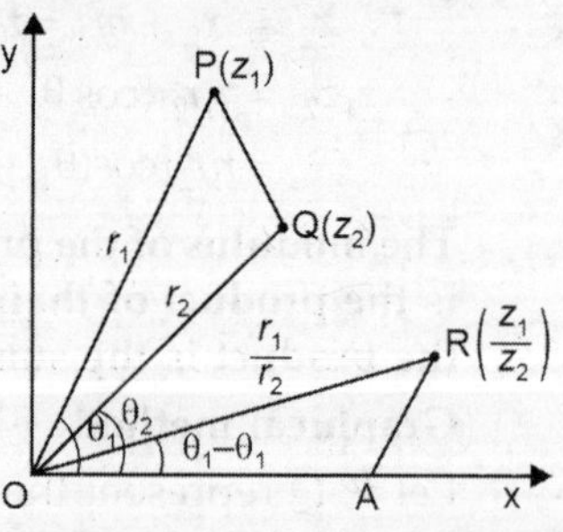

Let P and Q represent the complex numbers.

$$z_1 = x_1 + iy_1 = r_1(\cos\theta_1 + i\sin\theta_1)$$
$$z_2 = x_2 + iy_2 = r_2(\cos\theta_2 + i\sin\theta_2)$$

Let $OA = 1$, construct $\Delta\, OAR$ on OA similar to $\Delta\, OPQ$,

so that $$\frac{OR}{OA} = \frac{OP}{OQ}$$

$$\Rightarrow \quad \frac{OR}{1} = \frac{PO}{OQ} \quad \text{or} \quad OR = \frac{OP}{OQ} = \frac{r_1}{r_2}$$

$$\angle AOR = \angle QOP = \angle AOP - \angle AOQ = \theta_1 - \theta_2$$

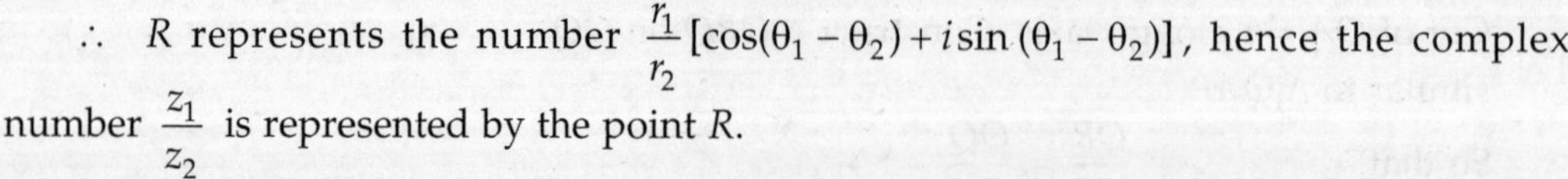

$\therefore$ R represents the number $\frac{r_1}{r_2}[\cos(\theta_1 - \theta_2) + i\sin(\theta_1 - \theta_2)]$, hence the complex number $\frac{z_1}{z_2}$ is represented by the point R.

(*i*) $\left|\dfrac{z_1}{z_2}\right| = \dfrac{|z_1|}{|z_2|}$ (*ii*) $\text{Arg.}\left(\dfrac{z_1}{z_2}\right) = \text{Arg.}(z_1) - \text{Arg.}(z_2)$

12.6 EXPONENTIAL AND CIRCULAR FUNCTIONS OF COMPLEX VARIABLES

If $z = x + iy$ then we define:

$$e^z = 1 + z + \frac{z^2}{2!} + \frac{z^3}{3!} + \frac{z^4}{4!} + \ldots \quad \text{... (1)}$$

$$\sin z = z - \frac{z^3}{3!} + \frac{z^5}{5!} - \frac{z^7}{7!} + \ldots \quad \text{... (2)}$$

$$\cos z = 1 - \frac{z^2}{2!} + \frac{z^4}{4!} - \frac{z^6}{6!} + \ldots \quad \text{... (3)}$$

From (2) and (3), we have

$$\cos z + i \sin z = \left(1 - \frac{z^2}{2!} + \frac{z^4}{4!} - \frac{z^6}{6!} + \ldots\ldots\right) + i\left(z - \frac{z^3}{3!} + \frac{z^5}{5!} - \ldots\right)$$

$$= 1 + \frac{(iz)^1}{1!} + \frac{(iz)^2}{2!} + \frac{(iz)^3}{3!} + \ldots$$

$$= e^{iz}$$

$$\Rightarrow \quad \cos z + i \sin z = e^{iz} \quad \text{... (4)}$$

Similarly, $\cos z - i \sin z = e^{-iz}$ ... (5)

From (4) and (5), we have

$$\cos z = \frac{e^{iz} + e^{-iz}}{2} \quad \text{... (6)}$$

$$\sin z = \frac{e^{iz} - e^{-iz}}{2i} \quad \text{... (7)}$$

Hyperbolic Functions

(*i*) $\sinh x = \dfrac{e^x - e^{-x}}{2}$ (*ii*) $\cosh x = \dfrac{e^x + e^{-x}}{2}$

(*iii*) $\tanh x = \dfrac{e^x - e^{-x}}{e^x + e^{-x}}$ (*iv*) $\coth x = \dfrac{e^x + e^{-x}}{e^x - e^{-x}}$

(*v*) $\text{sech}\, x = \dfrac{2}{e^x + e^{-x}}$ (*vi*) $\text{cosech}\, x = \dfrac{2}{e^x - e^{-x}}$

(*vii*) $\cosh x + \sinh x = \dfrac{e^x + e^{-x}}{2} + \dfrac{e^x - e^{-x}}{2} = e^x$

(*viii*) $\cosh x - \sinh x = \dfrac{e^x + e^{-x}}{2} - \dfrac{e^x - e^{-x}}{2} = e^{-x}$

(*ix*) $(\cosh x + \sinh x)^n = \cosh nx + \sinh nx$

12.7 SQUARE ROOTS OF A COMPLEX NUMBER

Let $a + ib$ be a complex number such that $\sqrt{a + ib} = x + iy$, where x and y are real numbers.

Then, $\sqrt{a + ib} = x + iy$

$$\Rightarrow \quad (a + ib) = (x + iy)^2$$

$$\Rightarrow \quad a + ib = (x^2 - y^2) + 2i\,xy$$

On equating real and imaginary parts, we get

$$x^2 - y^2 = a \qquad \text{... (1)}$$

and $$2\,xy = b \qquad \text{... (2)}$$

Now, $$(x^2 + y^2)^2 = (x^2 - y^2)^2 + 4x^2y^2$$

$$\Rightarrow \quad (x^2 + y^2)^2 = a^2 + b^2$$

$$\Rightarrow \quad (x^2 + y^2) = \sqrt{a^2 + b^2} \qquad [\because x^2 + y^2 \geq 0] \qquad \text{... (3)}$$

Solving the equations (1) and (3), we get

$$x^2 = \left(\frac{1}{2}\right)\left\{\sqrt{a^2 + b^2} + a\right\} \text{ and } y^2 = \pm\left(\frac{1}{2}\right)\left\{\sqrt{a^2 + b^2} - a\right\}$$

$$\Rightarrow \quad x = \pm\sqrt{\left(\frac{1}{2}\right)\left\{\sqrt{a^2 + b^2} + a\right\}} \text{ and } y = \pm\sqrt{\left(\frac{1}{2}\right)\left\{\sqrt{a^2 + b^2} - a\right\}}$$

If b is positive, then by equation (*ii*), x and y are of the same sign.

Hence, $$\sqrt{a + ib} = \pm\left[\sqrt{\frac{1}{2}\left\{\sqrt{a^2 + b^2} + a\right\}} + i\sqrt{\frac{1}{2}\left\{\sqrt{a^2 + b^2} - a\right\}}\right]$$

If b is negative, then by equation (*ii*), x and y are of different sign.

Hence, $$\sqrt{a + ib} = \pm\left[\sqrt{\frac{1}{2}\left\{\sqrt{a^2 + b^2} + a\right\}} - i\sqrt{\frac{1}{2}\left\{\sqrt{a^2 + b^2} - a\right\}}\right]$$

Remark: *It is evident from above that for any complex number z, we have*

(*i*) $$\sqrt{z} = \pm\left\{\sqrt{\frac{|z| + \text{Re}(z)}{2}} + i\sqrt{\frac{|z| - \text{Re}(z)}{2}}\right\}, \text{ if Im}(z) > 0$$

(*ii*) $$\sqrt{z} = \pm\left\{\sqrt{\frac{|z| + \text{Re}(z)}{2}} - i\sqrt{\frac{|z| - \text{Re}(z)}{2}}\right\}, \text{ if Im}(z) < 0$$

Example 16. *Find the square roots of the following:*

(*i*) $7 - 24i$ (*ii*) $5 + 12i$

Solution. Let $\sqrt{7 - 24i} = x + iy$. Then,

(*i*) $$\sqrt{7 - 24i} = x + iy$$

$$\Rightarrow \quad 7 - 24i = (x + iy)^2$$

$$\Rightarrow \quad 7 - 24i = (x^2 - y^2) + 2i\,xy$$

$$\Rightarrow \quad x^2 - y^2 = 7 \qquad \text{... (1)}$$

and $$2xy = -24 \qquad \text{... (2)}$$

now, $$(x^2 + y^2)^2 = (x^2 - y^2)^2 + 4\,x^2y^2$$

$$\Rightarrow \quad (x^2 + y^2)^2 = 49 + 576 = 625$$

$$\Rightarrow \quad x^2 + y^2 = 25 \qquad [\because x^2 + y^2 > 0] \qquad \text{... (3)}$$

On solving (1) and (3), we get

$$x^2 = 16 \text{ and } y^2 = 9 \Rightarrow x = \pm 4 \text{ and } y = \pm 3$$

From (2), $2xy$ is negative. So, x and y are of opposite signs.

$\therefore \quad (x = 4 \text{ and } y = -3) \text{ or } (x = -4 \text{ and } y = 3)$

Hence, $\sqrt{7-24i} = \pm(4-3i)$

Aliter Let $z = 7 - 24i$. Then, Re $(z) = 7$ and $|z| = \sqrt{49+576} = 25$

$$\therefore \quad \sqrt{7-24i} = \pm\left\{\sqrt{\frac{|z|+\text{Re}(z)}{2}} - i\sqrt{\frac{|z|-\text{Re}(z)}{2}}\right\} \qquad [\because \text{Im}(z) < 0]$$

$$\Rightarrow \quad \sqrt{7-24i} = \pm\left\{\sqrt{\frac{25+7}{2}} - i\sqrt{\frac{25-7}{2}}\right\} = \pm(4-3i)$$

(*ii*) Let $\sqrt{5+12i} = x + iy$.

$\Rightarrow \quad 5 + 12i = (x+iy)^2$

$\Rightarrow \quad 5 + 12i = (x^2 - y^2) + 2ixy$

$\Rightarrow \quad x^2 - y^2 = 5 \qquad \text{... (1)}$

and $\quad 2xy = 12 \qquad \text{... (2)}$

Now, $\quad (x^2+y^2)^2 = (x^2-y^2)^2 + 4x^2y^2$

$\Rightarrow \quad (x^2+y^2)^2 = 5^2 + 12^2 = 169$

$\Rightarrow \quad x^2 + y^2 = 13 \qquad [\because x^2 + y^2 > 0] \quad \text{... (3)}$

On solving (1) and (3), we get

$x^2 = 9 \text{ and } y^2 = 4 \Rightarrow x = \pm 3 \text{ and } y = \pm 2$

From (2), $2xy$ is positive. So, x and y are of the same signs.

$\therefore \quad (x = 3 \text{ and } y = 2) \text{ or } (x = -3 \text{ and } y = -2)$

Hence, $\sqrt{5+12i} = \pm(3 \pm 2i)$

Aliter Let $z = 5 + 12i$. Then, Re $(z) = 5$, and $|z| = \sqrt{25+144} = 13$

$$\therefore \quad \sqrt{5+12i} = \pm\left\{\sqrt{\frac{|z|+\text{Re}(z)}{2}} + i\sqrt{\frac{|z|-\text{Re}(z)}{2}}\right\} \qquad [\because \text{Im}(z) > 0]$$

$$\Rightarrow \quad \sqrt{5+12i} = \pm\left\{\sqrt{\frac{13+5}{2}} + i\sqrt{\frac{13-5}{2}}\right\}$$

$$\therefore \quad \sqrt{5+12i} = \pm(3+2i)$$

Example 17. *Find the square roots of* $-15-8i$.

Solution. Let $\sqrt{-15-8i} = x + iy$.

$\Rightarrow$ Then, $\quad -15 - 8i = (x+iy)^2$

$\Rightarrow \quad -15 - 8i = (x^2-y^2) + 2ixy$

$\Rightarrow \quad -15 = x^2 - y^2 \qquad \text{... (1)}$

and $\quad 2xy = -8 \qquad \text{... (2)}$

Now, $\quad (x^2+y^2)^2 = (x^2-y^2)^2 + 4x^2y^2$

$\Rightarrow \quad (x^2+y^2)^2 = (-15)^2 + 64 = 289$

$\Rightarrow \qquad x^2 + y^2 = 17 \qquad ...(3)$

On solving (1) and (3), we get

$$x^2 = 1 \text{ and } y^2 = 16 \Rightarrow x = \pm 1 \text{ and } y = \pm 4$$

From (2), $2xy$ is negative. So, x and y are of the opposite signs.

$\therefore \qquad (x = 1 \text{ and } y = -4) \text{ or } (x = -1 \text{ and } y = 4)$

Hence, $\sqrt{-15-8i} = \pm (1 - 4i)$

Example 18. *Find the square root of i.*

Solution. Let $\sqrt{i} = x + iy.$

$\Rightarrow$ Then, $i = (x + iy)^2$

$\Rightarrow \qquad (x^2 - y^2) + 2ixy = 0 + i$

$\Rightarrow \qquad x^2 - y^2 = 0 \qquad ...(1)$

and $\qquad 2xy = 1 \qquad ...(2)$

Now, $\qquad (x^2 + y^2)^2 = (x^2 - y^2)^2 + 4x^2y^2$

$\Rightarrow \qquad (x^2 + y^2)^2 = 0 + 1 = 1$

$\Rightarrow \qquad x^2 + y^2 = 1 \qquad [\because x^2 + y^2 > 0] \quad ...(3)$

Solving (1) and (3), we get

$$x^2 = \frac{1}{2} \text{ and } y^2 = \frac{1}{2} \Rightarrow x = \pm\frac{1}{\sqrt{2}} \text{ and } y = \pm\frac{1}{\sqrt{2}}$$

From (2), $2xy$ is positive. So, x and y are of the same signs.

$$\therefore \quad (x = \frac{1}{\sqrt{2}} \text{ and } y = \frac{1}{\sqrt{2}}) \text{ or } (x = -\frac{1}{\sqrt{2}} \text{ and } y = -\frac{1}{\sqrt{2}})$$

$$\therefore \quad \sqrt{i} = \pm\left\{\sqrt{\frac{|z| + \text{Re}(z)}{2}} + i\sqrt{\frac{|z| - \text{Re}(z)}{2}}\right\} \qquad [\because \text{Im}(z) > 0]$$

$$\Rightarrow \quad \sqrt{i} = \pm\left\{\sqrt{\frac{1+0}{2}} + i\sqrt{\frac{1-0}{2}}\right\} = \pm\left(\frac{1}{\sqrt{2}} + i\frac{1}{\sqrt{2}}\right) = \pm\frac{1}{\sqrt{2}}(1+i) \qquad \textbf{Ans.}$$

EXERCISE 12.3

Determine square roots of the following complex numbers:

1. $-5 + 12i$ **Ans.** $\pm (2 + 3i)$
2. $-7 - 24i$ **Ans.** $\pm (3 - 4i)$
3. $1 - i$ **Ans.** $\pm\left\{\left(\sqrt{\frac{\sqrt{2}+1}{2}}\right) - \left(\sqrt{\frac{\sqrt{2}-1}{2}}\right)i\right\}$
4. $-8 - 6i$ **Ans.** $\pm (1 - 3i)$
5. $8 - 15i$ **Ans.** $\pm\frac{1}{\sqrt{2}}(5 - 3i)$
6. $-11 - 60i$ **Ans.** $\pm (5 - 6i)$
7. $1 + 4\sqrt{-3}$ **Ans.** $\pm(2 + \sqrt{3}i)$
8. $4i$ **Ans.** $\pm\sqrt{2}(1+i)$
9. $-i$ **Ans.** $\pm\frac{1}{\sqrt{2}}(1-i)$

12.8 CUBE ROOTS OF UNITY

Let z be the cube root of unity

Then $z^3 = 1$

$\Rightarrow$ $z^3 - 1 = 0$

$\Rightarrow$ $(z - 1)(z^2 + z + 1) = 0$

$\Rightarrow$ $z - 1 = 0$ or $z^2 + z + 1 = 0$

$\Rightarrow$ $z = 1$ or $z^2 + z + 1 = 0$

Now solving; $z^2 + z + 1 = 0$

$$\Rightarrow \quad z = \frac{-1 \pm \sqrt{1 - 4(1)(1)}}{2}$$

$$\Rightarrow \quad z = \frac{-1 \pm \sqrt{-3}}{2}$$

$$\Rightarrow \quad z = \frac{-1 \pm \sqrt{3}\,i}{2}$$

Thus cube roots of unity are 1, $\frac{-1 + \sqrt{3}\,i}{2}$ and $\frac{-1 - \sqrt{3}\,i}{2}$

Note: $\omega = \frac{-1 + \sqrt{3}\,i}{2}$ **and** $\omega^2 = \frac{-1 - \sqrt{3}\,i}{2}$

Properties of Cube Roots of Unity

(*i*) Sum of all cube roots of unity is zero.

(*ii*) Product of all cube roots of unity is 1.

(*iii*) Each complex cube root of unity is square of the other.

Proof: (*i*) Sum of all cube roots of unity is zero.

Since 1, ω and ω^2 are cube roots of unity

$$\text{Now;} \quad 1 + \omega + \omega^2 = 1 + \left(\frac{-1 + \sqrt{3}\,i}{2}\right) + \left(\frac{-1 - \sqrt{3}\,i}{2}\right)$$

$$= \frac{2 - 1 + \sqrt{3}\,i - 1 - \sqrt{3}\,i}{2} = \frac{0}{2} = 0$$

(*ii*) Product of all cube roots of unity is 1.

Since 1, ω and ω^2 are cube roots of unity

$$1 \cdot \omega \cdot \omega^2 = 1 \cdot \left(\frac{-1 + \sqrt{3}\,i}{2}\right) \cdot \left(\frac{-1 - \sqrt{3}\,i}{2}\right)$$

$$1 \cdot \omega \cdot \omega^2 = \left(\frac{-1 + \sqrt{3}\,i}{2}\right) \cdot \left(\frac{-1 - \sqrt{3}\,i}{2}\right)$$

$$1\cdot\omega\cdot\omega^2 = \left(\frac{1-3i^2}{4}\right)$$

$$1\cdot\omega\cdot\omega^2 = \left(\frac{1+3}{4}\right) \quad \because i^2 = -1$$

$$1\cdot\omega\cdot\omega^2 = \frac{4}{4}$$

$$\therefore \quad \mathbf{1\cdot\omega\cdot\omega^2 = 1} \qquad [\because \omega^3 = 1]$$

Example 19. *Find cube roots of* 8, – 8, 27, – 27, 64

Solution. Let z be cube root of 8

Then $\quad z^3 = 8$

$\Rightarrow \quad z^3 - 8 = 0$

$\Rightarrow \quad z^3 - 2^3 = 0$

$\Rightarrow \quad (z-2)(z^2 + 2z + 4) = 0$

$\Rightarrow \quad z - 2 = 0 \text{ or } z^2 + 2z + 4 = 0$

$\Rightarrow \quad z = 2 \text{ or } z^2 + 2z + 4 = 0$

Now solving; $\quad z^2 + 2z + 4 = 0$

$\Rightarrow \quad z = \dfrac{-2 \pm \sqrt{4 - 4(1)(4)}}{2} \Rightarrow = \dfrac{-2 \pm 2\sqrt{3}\,i}{2}$

$\Rightarrow \quad z = 2\left(\dfrac{-1 \pm \sqrt{3}\,i}{2}\right)$

$\Rightarrow$ 2, 2ω and $2\omega^2$ are cube roots of 8. **Ans.**

Let z be cube root of – 8

Then $\quad z^3 = -8 \Rightarrow z^3 + 8 = 0 \Rightarrow z^3 + 2^3 = 0$

$\Rightarrow \quad (z+2)(z^2 - 2z + 4) = 0$

$\Rightarrow \quad z + 2 = 0 \text{ or } z^2 - 2z + 4 = 0$

$\Rightarrow \quad z = -2 \text{ or } z^2 - 2z + 4 = 0$

Now solving; $\quad z^2 - 2z + 4 = 0$

$\Rightarrow \quad z = \dfrac{2 \pm \sqrt{4 - 4(1)(4)}}{2} \Rightarrow z = \dfrac{2 \pm \sqrt{12}i}{2} \Rightarrow z = \dfrac{2 \pm 2\sqrt{3}\,i}{2}$

$\Rightarrow \quad z = -2\left(\dfrac{-1 \pm \sqrt{3}\,i}{2}\right)$

$\Rightarrow$ -2, -2ω and $-2\omega^2$ are cube roots of 8. **Ans.**

Let z be cube root of 27

Then $\quad z^3 = 27$

$\Rightarrow \quad z^3 - 27 = 0$

$\Rightarrow \quad z^3 - 3^3 = 0$

$\Rightarrow \quad (z-3)(z^2 + 3z + 9) = 0$

$\Rightarrow \quad z - 3 = 0 \text{ or } z^2 + 3z + 9 = 0$

$\Rightarrow \quad z = 3 \text{ or } z^2 + 3z + 9 = 0$

Now solving; $\quad z^2 + 3z + 9 = 0$

$$\Rightarrow \qquad z = \frac{-3 \pm \sqrt{9-4(1)(9)}}{2}$$

$$\Rightarrow \qquad z = \frac{-3 \pm 3\sqrt{3}\,i}{2}$$

$$\Rightarrow \qquad z = 3\left(\frac{-1 \pm \sqrt{3}\,i}{2}\right)$$

$\Rightarrow$ 3, 3ω and $3\omega^2$ are cube roots of 27. **Ans.**

Let z be cube root of – 27

Then $z^3 = -27$

$\Rightarrow \quad z^3 + 27 = 0$

$\Rightarrow \quad z^3 + 3^3 = 0$

$\Rightarrow \quad (z+3)(z^2 - 3z + 9) = 0$

$\Rightarrow \quad z + 3 = 0 \quad \text{or} \quad z^2 - 3z + 9 = 0$

$\Rightarrow \quad z = -3 \quad \text{or} \quad z^2 - 3z + 9 = 0$

Now solving; $z^2 - 3z + 9 = 0$

$$\Rightarrow \qquad z = \frac{3 \pm \sqrt{9-4(1)(9)}}{2}$$

$$\Rightarrow \qquad z = \frac{3 \pm \sqrt{-27}}{2} \quad \Rightarrow \quad z = \frac{3 \pm 3\sqrt{3}\,i}{2}$$

$$\Rightarrow \qquad z = -3\left(\frac{-1 \pm \sqrt{3}\,i}{2}\right)$$

$\Rightarrow$ – 3, – 3ω and – $3\omega^2$ are cube roots of – 27. **Ans.**

Let z be cube root of 64

Then $z^3 = 64 \Rightarrow z^3 - 64 = 0 \Rightarrow z^3 - 4^3 = 0$

$\Rightarrow \quad (z-4)(z^2 + 4z + 16) = 0$

$\Rightarrow \quad z - 4 = 0 \quad \text{or} \quad z^2 + 4z + 16 = 0$

$\Rightarrow \quad z = 4 \quad \text{or} \quad z^2 + 4z + 16 = 0$

Now solving; $z^2 + 4z + 16 = 0$

$$\Rightarrow \qquad z = \frac{-4 \pm \sqrt{16-4(1)(16)}}{2}$$

$$\Rightarrow \qquad z = \frac{-4 \pm \sqrt{-48}}{2}$$

$$\Rightarrow \qquad z = \frac{-4 \pm 4\sqrt{3}\,i}{2}$$

$$\Rightarrow \qquad z = 4\left(\frac{-1 \pm \sqrt{3}\,i}{2}\right)$$

$\Rightarrow$ 4, 4ω and $4\omega^2$ are cube roots of 64. **Ans.**

Example 20. *Evaluate*

(*i*) $(1 + \omega - \omega^2)^8$

Solution. $(1 + \omega - \omega^2)^8 = (-\omega^2 - \omega^2)^8$

$= (-2\omega^2)^8$

$= (-2\omega^2)^8$

$= 2^8\omega^{16}$

$= 256\omega \cdot \omega^{15}$

$= 256\omega \cdot (\omega^3)^5$

$= 256\omega$ $[\because \omega^3 = 1]$ **Ans.**

(*ii*) $\omega^{28} + \omega^{29} + 1$

Solution. $\omega^{28} + \omega^{29} + 1 = \omega \cdot \omega^{27} + \omega^2 \cdot \omega^{27} + 1$

$= \omega \cdot (\omega^3)^9 + \omega^2 \cdot (\omega^3)^9 + 1$

$= \omega \cdot (1)^9 + \omega^2 \cdot (1)^9 + 1$

$= \omega + \omega^2 + 1$

$= 1 + \omega + \omega^2$

$= 0$ **Ans.**

(*iii*) $(1 + \omega - \omega^2)(1 - \omega + \omega^2)$

Solution. $(1 + \omega - \omega^2)(1 - \omega + \omega^2)$

$= (-\omega^2 - \omega^2)(1 + \omega^2 - \omega)$

$= (-\omega^2 - \omega^2)(-\omega - \omega)$

$= (-2\omega^2)(-2\omega)$

$= 4\omega^3 = 4$ $[\because \omega^3 = 1]$ **Ans.**

(*iv*) $\left(\frac{-1+\sqrt{-3}}{2}\right)^7 + \left(\frac{-1-\sqrt{-3}}{2}\right)^7$

Solution. $\left(\frac{-1+\sqrt{-3}}{2}\right)^7 + \left(\frac{-1-\sqrt{-3}}{2}\right)^7$

$= (\omega)^7 + (\omega^2)^7$

$= \omega^7 + \omega^{14}$

$= \omega \cdot \omega^6 + \omega^2 \cdot \omega^{12}$

$= \omega \cdot (\omega^3)^2 + \omega^2 : (\omega^3)^4$

$= \omega \cdot (1)^2 + \omega^2 \cdot (1)^4$

$= \omega + \omega^2$

$= -1$ **Ans.**

(*v*) $(-1+\sqrt{-3})^5 + (-1-\sqrt{-3})^5 = (-1+\sqrt{-3})^5 + (-1-\sqrt{-3})^5$

$= \left(2 \cdot \frac{-1+\sqrt{-3}}{2}\right)^5 + \left(2 \cdot \frac{-1-\sqrt{-3}}{2}\right)^5$

$= 32(\omega)^5 + 32(\omega^2)^5$

$= 32\omega^5 + 32\omega^{10}$

$= 32\omega^5(1 + \omega^5)$

$= 32\omega^2 \cdot \omega^3 (1 + \omega^2 \cdot \omega^3)$

$$= 32\omega^2(1 + \omega^2)$$
$$= 32\omega^2(-\omega)$$
$$= -32\omega^3$$
$$= -32$$ **Ans.**

Example 21. *Show that*

(i) $x^3 - y^3 = (x - y)(x - \omega y)(x - \omega^2 y)$

Solution. R.H.S. $= (x - y)(x - \omega y)(x - \omega^2 y)$
$$= (x - y)(x^2 - \omega xy - \omega^2 xy + \omega^2 y^2)$$
$$= (x - y)(x^2 - (\omega + \omega^2)xy + y^2)$$
$$= (x - y)(x^2 + xy + y^2)$$
$$= x^3 - y^3$$
= L.H.S. **Proved.**

(ii) $x^3 + y^3 + z^3 - 3xyz = (x + y + z)(x + \omega y + \omega^2 z)(x + \omega^2 y + \omega z)$

Solution. R.H.S. $= (x + y + z)(x + \omega y + \omega^2 z)(x + \omega^2 y + \omega z)$
$$= (x + y + z)(x^2 + \omega^2 xy + \omega xz + \omega xy + \omega^3 y^2 + \omega^2 yz + \omega^2 zx + \omega^4 yz + \omega^3 z^2)$$
$$= (x + y + z)[x^2 + y^2 + z^2 + (\omega^2 + \omega)xy + (\omega^2 + \omega)zx + (\omega^2 + \omega)yz]$$
$$= (x + y + z)[x^2 + y^2 + z^2 + (-1)xy + (-1)zx + (-1)yz]$$
$$= (x + y + z)(x^2 + y^2 + z^2 - xy - yz - zx)$$
$$= x^3 + y^3 + z^3 - 3xyz$$
= R.H.S. **Proved.**

(iii) $(1 + \omega)(1 + \omega^2)(1 + \omega^4)(1 + \omega^8)$ ... 2*n factors* = 1

Solution. $(1 + \omega)(1 + \omega^2)(1 + \omega^4)(1 + \omega^8)$ ... $2n$ factors = 1

L.H.S. $= (1 + \omega)(1 + \omega^2)(1 + \omega^4)(1 + \omega^8)$ ... $2n$ factors
$$= (1 + \omega)(1 + \omega^2)(1 + \omega\cdot\omega^3)(1 + \omega^2\cdot\omega^6) \text{ ... } 2n \text{ factors}$$
$$= (1 + \omega)(1 + \omega^2)(1 + \omega)(1 + \omega^2) \text{ ... } 2n \text{ factors}$$
$$= (-\omega^2)(-\omega^2)(-\omega^2)(-\omega) \text{ ... } 2n \text{ factors}$$
$$= \omega^3\cdot\omega^3 \text{ ... } n \text{ factors}$$
$$= 1.1.1 \text{ ... } n \text{ factors} = 1$$
= R.H.S. **Proved.**

Example 22. *If* ω *is a root of* $x^2 + x + 1 = 0$, *show that its other root is* ω^2 *and prove that* $\omega^3 = 1$.

Solution. Since ω is a root of $x^2 + x + 1 = 0$

$\Rightarrow$ $\omega^2 + \omega + 1 = 0$ ... (1)

Now to show that ω^2 is root of $x^2 + x + 1 = 0$

we have to show that $\omega^4 + \omega^2 + 1 = 0$

$$\omega^4 + \omega^2 + 1 = (\omega^2)^2 + 1 + \omega^2$$
$$= (\omega^2)^2 + 1 + 2\omega^2 + \omega^2 - 2\omega^2$$
$$= (\omega^2 + 1)^2 - \omega^2$$
$$= (\omega^2 + 1 + \omega)(\omega^2 + 1 - \omega)$$
$$= (1 + \omega + \omega^2)(\omega^2 - \omega + 1)$$
$$= (0)(\omega^2 - \omega + 1)$$ [from (1)]

Hence $\omega^4 + \omega^2 + 1 = 0$ ... (2)

Now subtracting equations (1) from (2)

$$(\omega^4 + \omega^2 + 1) - (\omega^2 + \omega + 1) = 0$$

$$\Rightarrow \quad \omega^4 - \omega = 0$$

$$\Rightarrow \quad \omega(\omega^3 - 1) = 0$$

$$\omega \neq 0 \;\Rightarrow\; (\omega^3 - 1) = 0$$

$$\Rightarrow \quad \omega^3 = 1$$

Proved.

Example 23. *Prove that complex cube roots of* -1 *are* $\left(\frac{1+\sqrt{3}i}{2}\right)$ *and* $\left(\frac{1-\sqrt{3}i}{2}\right)$*; and hence prove that* $\left(\frac{1+\sqrt{-3}}{2}\right)^9 + \left(\frac{1-\sqrt{-3}}{2}\right)^9 = -2$

Solution. Let z be cube root of -1

$$\Rightarrow \quad z^3 = -1$$

$$\Rightarrow \quad z^3 + 1 = 0$$

$$\Rightarrow \quad (z+1)(z^2 - z + 1) = 0$$

$$\Rightarrow \quad z + 1 = 0 \quad \text{or} \quad z^2 - z + 1 = 0$$

$$\Rightarrow \quad z = -1 \quad \text{or} \quad z^2 - z + 1 = 0$$

Now solving; $z^2 - z + 1 = 0$

$$\Rightarrow \quad z = \frac{1 \pm \sqrt{1 - 4(1)(1)}}{2}$$

$$\Rightarrow \quad z = \frac{1 \pm \sqrt{-3}}{2}$$

$$\Rightarrow \quad z = \frac{1 \pm \sqrt{3}i}{2}$$

Thus cube roots of -1 are -1, $\frac{1+\sqrt{3}i}{2}$ and $\frac{1-\sqrt{3}i}{2}$

Hence complex cube roots of -1 are $\frac{1+\sqrt{3}i}{2}$ and $\frac{1-\sqrt{3}i}{2}$

Now we have to prove $\left(\frac{1+\sqrt{-3}}{2}\right)^9 + \left(\frac{1-\sqrt{-3}}{2}\right)^9 = -2$

$$\text{L.H.S.} = \left(\frac{1+\sqrt{-3}}{2}\right)^9 + \left(\frac{1-\sqrt{-3}}{2}\right)^9$$

$$= \left(-1\cdot\frac{-1-\sqrt{-3}}{2}\right)^9 + \left(-1\cdot\frac{-1+\sqrt{-3}}{2}\right)^9$$

$$= (-\omega^2)^9 + (-\omega)^9$$

$$= -\omega^{18} - \omega^9$$

$$= -(\omega^3)^6 - (\omega^3)^3$$

$$= -(1)^6 - (1)^3$$

$$= -1 - 1$$

$$= -2 = \text{R.H.S.}$$

Proved.

Example 24. *If ω is a cube root of unity, form an equation whose roots are 2ω and 2ω²*

Solution. An equation whose roots are 2ω and $2\omega^2$ is

$$z^2 - Sz + P = 0 \quad \text{... (1)}$$

Now
$$S = 2\,\omega + 2\omega^2$$
$$= 2\,(\omega + \omega^2)$$
$$= 2\,(-1)$$
$$= -2$$

$\Rightarrow$
$$P = 2\omega \cdot 2\omega^2$$
$$P = 4\omega^3$$
$$P = 4$$

Putting values in (1), we get

$\therefore$ $z^2 + 2z + 4 = 0$ is required equation. **Ans.**

EXERCISE 12.4

1. If z_1, z_2, z_3 are three complex numbers and

$$a_1 = z_1 + z_2 + z_3$$
$$b_1 = z_1 + \omega z_2 + \omega^2 z_3$$
$$c_1 = z_1 + \omega^2 z_2 + \omega z_3$$

show that

$$|a_1|^2 + |b_1|^2 + |c_1|^2 = 3\{|z_1|^2 + |z_2|^2 + |z_3|^2\}$$

where ω, ω^2 are cube roots of unity.

2. Given that $z_1 + z_2 + z_1 = A, z_1 + z_2\omega + z_3\omega^2 = B$ and $z_1 + z_2\omega^2 + z_3\omega = C$, where ω is a cube root of unity. Express z_1, z_2, z_3 in terms of A, B, C and ω.

(*Diploma IETE, June 2006*)

[**Hint:** $z_1 + z_2 + z_3 = A$ ... (1)

$z_1 + z_2\omega + z_3\omega^2 = B$ ... (2)

$z_1 + z_2\omega^2 + z_3\omega = C$ ... (3)

Adding (1), (2) and (3), we get

$$3z_1 + z_2(1 + w + w^2) + z_3(1 + w + w^2) = A + B + C$$

$$3z_1 + 0 + 0 = A + B + C$$

$$z_1 = \frac{A + B + C}{3} \quad \text{... (4)}$$

Putting the value of z_1 in (2) and (3) and then solve them for z_2 and z_3.]

3. Show that the roots of $(x + 1)^6 + (x - 1)^6 = 0$ are given by

$i\cot\frac{(2n+1)}{12}$, $n = 0, 1, 2, 3, 4, 5$. Deduce $\tan^2\left(\frac{\pi}{12}\right) + \tan^2\left(\frac{3\pi}{12}\right) + \tan^2\left(\frac{5\pi}{12}\right) = 15$

4. Show that all the roots of $(x + 1)^7 = (x - 1)^7$ are given by $\pm\, i \cot\left(\frac{n\pi}{7}\right)$, where $n = 1, 2, 3$, why $n \neq 0$.

12.9 RELATION BETWEEN CIRCULAR AND HYPERBOLIC FUNCTIONS

Circular f^{ns}	*Hyperbolic* f^{ns}	*Hyperbolic* f^{ns}	*Circular* f^{ns}
$\sin x$	$= -i \sinh (ix)$	$\sinh x$	$= -i \sin (ix)$
$\cos x$	$= \cosh (ix)$	$\cosh x$	$= \cos (ix)$
$\tan x$	$= -i \tanh (ix)$	$\tanh x$	$= -i \tan (ix)$
$\cot x$	$= i \coth (ix)$	$\coth x$	$= i \cot (ix)$
$\sec x$	$= \text{sech} (ix)$	$\text{sech}\, x$	$= \sec (ix)$
$\text{cosec}\, x$	$= i\, \text{cosech} (ix)$	$\text{cosech}\, x$	$= i\, \text{cosec} (ix)$

Example 25. *If cosh u = sec θ, show that*

$$u = \log \tan\left(\frac{\pi}{4}+\frac{\theta}{2}\right)$$ **[B.T.E. Delhi. May 2007]**

Solution. Here, we have

$$\cosh u = \sec \theta$$

$$\Rightarrow \quad \frac{e^u + e^{-u}}{2} = \frac{1}{\cos\theta} \Rightarrow \cos\theta\,(e^u + e^{-u}) = 2$$

$$\Rightarrow \quad \cos\theta\, e^u + \cos\theta\, e^{-u} = 2$$

$$\Rightarrow \quad \cos\theta\, e^{2u} + \cos\theta = 2e^u \qquad \text{(Multiplying by } e^u\text{)}$$

$$\Rightarrow \quad \cos\theta\, e^{2u} - 2e^u + \cos\theta = 0 \qquad \text{(Quadratic equation in } e^u\text{)}$$

$$e^u = \frac{2 \pm \sqrt{4 - 4\cos^2\theta}}{2\cos\theta} \Rightarrow e = \frac{1 \pm \sqrt{1-\cos^2\theta}}{\cos\theta} = \frac{1 \pm \sin\theta}{\cos\theta}$$

$$\Rightarrow \quad u = \log\left(\frac{\sin^2\frac{\theta}{2} + \cos^2\frac{\theta}{2} + 2\sin\frac{\theta}{2}\cos\frac{\theta}{2}}{\cos\theta}\right) = \log\frac{\left(\sin\frac{\theta}{2}+\cos\frac{\theta}{2}\right)^2}{\cos^2\frac{\theta}{2} - \sin^2\frac{\theta}{2}}$$

$$= \log\left(\frac{\sin\frac{\theta}{2}+\cos\frac{\theta}{2}}{\sin\frac{\theta}{2}-\cos\frac{\theta}{2}}\right) = \log\left(\frac{1+\tan\frac{\theta}{2}}{1-\tan\frac{\theta}{2}}\right) = \log\left(\frac{\tan\frac{\pi}{4}+\tan\frac{\theta}{2}}{1-\tan\frac{\pi}{4}\tan\frac{\theta}{2}}\right)$$

$$= \log\tan\left(\frac{\pi}{4}+\frac{\theta}{2}\right)$$ **Proved.**

12.10 DE MOVIER'S THEOREM

$$(\cos\theta + i\sin\theta)^n = \cos n\theta + i\sin n\theta$$

Proof: We know that $e^{i\theta} = \cos\theta + i\sin\theta$

$$(e^{i\theta})^n = (\cos\theta + i\sin\theta)^n$$

$$e^{in\theta} = (\cos\theta + i\sin\theta)^n$$

$$(\cos n\theta + i\sin n\theta) = (\cos\theta + i\sin\theta)^n$$ **Proved.**

If n is a fraction then $\cos n\theta + i\sin n\theta$ is one of the values of $(\cos\theta + i\sin\theta)^n$.

Example 26. *Express:* $\dfrac{(\cos\theta + i\sin\theta)^8}{(\sin\theta + i\cos\theta)^4}$ *in the form* $(x + iy)$. **[Diploma IETE Dec. 2005]**

Solution. $\dfrac{(\cos\theta + i\sin\theta)^8}{(\sin\theta + i\cos\theta)^4} = \dfrac{(\cos\theta + i\sin\theta)^8}{(i)^4\left(\cos\theta + \dfrac{1}{i}\sin\theta\right)^4} = \dfrac{(\cos\theta + i\sin\theta)^8}{(\cos\theta - i\sin\theta)^4}$

$$= \frac{(\cos\theta + i\sin\theta)^8}{[(\cos\theta + i\sin\theta)^{-1}]^4} = \frac{(\cos\theta + i\sin\theta)^8}{(\cos\theta + i\sin\theta)^{-4}}$$

$$= (\cos\theta + i\sin\theta)^{12}$$

$$= \cos 12\theta + i\sin 12\theta$$ **Ans.**

Example 27. *Simplify* $\left[\cos\dfrac{2\pi}{3} + i\sin\dfrac{2\pi}{3}\right]^{1/4}$

and express the result in a form free from trigonometrical expressions.

Solution. $\left[\cos\dfrac{2\pi}{3} + i\sin\dfrac{2\pi}{3}\right]^{1/4} = \left[\cos\dfrac{1}{4}\dfrac{2\pi}{3} + i\sin\dfrac{1}{4}\dfrac{2\pi}{3}\right] = \cos\dfrac{\pi}{6} + i\sin\dfrac{\pi}{6}$

$$= \frac{\sqrt{3}}{2} + \frac{i}{2}$$ **Ans.**

Example 28. *Prove that*

$$(1 + \cos\theta + i\sin\theta)^n + (1 + \cos\theta - i\sin\theta)^n = 2^{n+1}\cos^n\frac{\theta}{2}\cos\frac{n\theta}{2}$$

where n is an integer. **[Diploma IETE, June 2005]**

Proof. L.H.S.$= (1 + \cos\theta + i\sin\theta)^n + (1 + \cos\theta - i\sin\theta)^n$

$$= \left[1 + 2\cos^2\frac{\theta}{2} - 1 + 2i\sin\frac{\theta}{2}\cos\frac{\theta}{2}\right]^n + \left[1 + 2\cos^2\frac{\theta}{2} - 1 - 2i\sin\frac{\theta}{2}\cos\frac{\theta}{2}\right]^n$$

$$= \left[2\cos^2\frac{\theta}{2} + 2i\sin\frac{\theta}{2}\cos\frac{\theta}{2}\right]^n + \left[2\cos^2\frac{\theta}{2} - 2i\sin\frac{\theta}{2}\cos\frac{\theta}{2}\right]^n$$

$$= \left(2\cos\frac{\theta}{2}\right)^n\left[\cos\frac{\theta}{2} + i\sin\frac{\theta}{2}\right]^n + \left(2\cos\frac{\theta}{2}\right)^n\left[\cos\frac{\theta}{2} - i\sin\frac{\theta}{2}\right]^n$$

$$= 2^n\cos^n\frac{\theta}{2}\left[\cos\frac{n\theta}{2} + i\sin\frac{n\theta}{2}\right] + 2^n\cos^n\frac{\theta}{2}\left[\cos\frac{n\theta}{2} - i\sin\frac{n\theta}{2}\right]$$

$$= 2^n\cos^n\frac{\theta}{2}\left[\cos\frac{n\theta}{2} + i\sin\frac{n\theta}{2} + \cos\frac{n\theta}{2} - i\sin\frac{n\theta}{2}\right]$$

$$= 2^n\cos^n\frac{\theta}{2}\left(2\cos\frac{n\theta}{2}\right) = 2^{n+1}\cos^n\frac{\theta}{2}\cos\frac{n\theta}{2} = \text{R.H.S.}$$ **Proved.**

Example 29. *If n is a positive integer, prove that*

$$(\sqrt{3} + i)^n + (\sqrt{3} - i)^n = 2^{n+1}\cos\frac{n\pi}{6}, \quad (i = \sqrt{-1})$$

Proof. Let $\sqrt{3}+i = r(\cos\alpha + i\sin\alpha)$

$$r = \sqrt{3+1} = 2, \alpha = \tan^{-1}\left(\frac{1}{\sqrt{3}}\right) = \frac{\pi}{6}$$

$$\sqrt{3}+i = 2\left[\cos\frac{\pi}{6} + i\sin\frac{\pi}{6}\right]$$

$$(\sqrt{3}+i)^n + (\sqrt{3}-i)^n = \left[2\left(\cos\frac{\pi}{6} + i\sin\frac{\pi}{6}\right)\right]^n + \left[2\left(\cos\frac{\pi}{6} - i\sin\frac{\pi}{6}\right)\right]^n$$

$$= 2^n\left(\cos\frac{n\pi}{6} + i\sin\frac{n\pi}{6}\right) + 2^n\left(\cos\frac{n\pi}{6} - i\sin\frac{n\pi}{6}\right)$$

$$= 2(2)^n\cos\frac{n\pi}{6}$$

$$= 2^{n+1}\cos\frac{n\pi}{6}$$

Proved.

Example 30. *Let* (r, θ) *denote the point* $r(\cos\theta + i\sin\theta)$ *in the Argand plane. If* $a \equiv (1, \alpha)$, $b \equiv (1, \beta)$, $c \equiv (1, \gamma)$ *and* $a + b + c = 0$, *show that* $a^{-1} + b^{-1} + c^{-1} = 0$.

Solution. $(r, \theta) = r(\cos\theta + i\sin\theta)$

$$a \equiv (1, \alpha) = 1\cdot(\cos\alpha + i\sin\alpha)$$

$$b \equiv (1, \beta) = 1\cdot(\cos\beta + i\sin\beta)$$

$$c \equiv (1, \gamma) = 1\cdot(\cos\gamma + i\sin\gamma)$$

$$a + b + c = 0$$

$\Rightarrow$ $(\cos\alpha + i\sin\alpha) + (\cos\beta + i\sin\beta) + (\cos\gamma + i\sin\gamma) = 0$

$\Rightarrow$ $(\cos\alpha + \cos\beta + \cos\gamma) + i(\sin\alpha + \sin\beta + \sin\gamma) = 0$

Real part of L.H.S. = Real part of R.H.S.

$\cos\alpha + \cos\beta + \cos\gamma = 0$ ... (1)

Imaginary part of L.H.S. = Imaginary part of R.H.S.

$\sin\alpha + \sin\beta + \sin\gamma = 0$ ... (2)

Now $a^{-1} + b^{-1} + c^{-1} = \frac{1}{a} + \frac{1}{b} + \frac{1}{c}$

$$= \frac{1}{(\cos\alpha + i\sin\alpha)} + \frac{1}{(\cos\beta + i\sin\beta)} + \frac{1}{(\cos\gamma + i\sin\gamma)}$$

$$= \frac{(\cos\alpha - i\sin\alpha)}{(\cos\alpha + i\sin\alpha)(\cos\alpha - i\sin\alpha)} + \frac{\cos\beta - i\sin\beta}{(\cos\beta + i\sin\beta)(\cos\beta - i\sin\beta)} + \frac{\cos\gamma - i\sin\gamma}{(\cos\gamma + i\sin\gamma)(\cos\gamma - i\sin\gamma)}$$

$$= \frac{\cos\alpha - i\sin\alpha}{\cos^2\alpha + \sin^2\alpha} + \frac{\cos\beta - i\sin\beta}{\cos^2\beta + \sin^2\beta} + \frac{\cos\gamma - i\sin\gamma}{\cos^2\gamma + \sin^2\gamma}$$

$$= (\cos\alpha - i\sin\alpha) + (\cos\beta - i\sin\beta) + (\cos\gamma - i\sin\gamma)$$

$$= (\cos\alpha + \cos\beta + \cos\gamma) - i(\sin\alpha + \sin\beta + \sin\gamma)$$

Putting the values of $(\cos\alpha + \cos\beta + \cos\gamma)$ and $i(\sin\alpha + \sin\beta + \sin\gamma)$ from (1) and (2), we get

$$a^{-1} + b^{-1} + c^{-1} = 0 - i\,(0)$$
$$= 0$$ **Proved.**

Example 31. *Prove that the general value of θ which satisfies the equation.*

$$(\cos\theta + i\sin\theta)\cdot(\cos 2\theta + i\sin 2\theta)\ldots(\cos n\theta + i\sin n\theta) = 1 \text{ is}$$

$\dfrac{4m\pi}{n(n+1)}$, *where m is any integer.*

Solution. $(\cos\theta + i\sin\theta)(\cos 2\theta + i\sin 2\theta)\ldots(\cos n\theta + i\sin n\theta) = 1$

$$(\cos\theta + i\sin\theta)(\cos\theta + i\sin\theta)^2 \ldots (\cos\theta + i\sin\theta)^n = 1$$

$\Rightarrow$ $(\cos\theta + i\sin\theta)^{1+2} + \ldots + n = 1$

$\Rightarrow$ $(\cos\theta + i\sin\theta)^{\frac{n(n+1)}{2}} = (\cos 2m\pi + i\sin 2m\pi)$

$\Rightarrow$ $\cos\dfrac{n(n+1)}{2}\theta + i\sin\dfrac{n(n+1)}{2}\theta = \cos 2m\pi + i\sin 2m\pi$

$\Rightarrow$ $\dfrac{n(n+1)}{2}\theta = 2\,m\pi$

$\therefore$ $\theta = \dfrac{4m\pi}{n(n+1)}$ **Proved.**

Example 32. *If* $(a_1 + ib_1)\cdot(a_2 + ib_2)\ldots(a_n + ib_n) = A + iB$

Prove that

(*i*) $\tan^{-1}\left(\dfrac{b_1}{a_1}\right) + \tan^{-1}\left(\dfrac{b_2}{a_2}\right) + \ldots + \tan^{-1}\left(\dfrac{b_n}{a_n}\right) = \tan^{-1}\left(\dfrac{B}{A}\right)$

(*ii*) $(a_1^2 + b_1^2)(a_2^2 + b_2^2)\ldots(a_n^2 + b_n^2) = A^2 + B^2$

Proof. Let
$$a_1 = r_1\cos\alpha_1,\quad b_1 = r_1\sin\alpha_1$$
$$a_2 = r_2\cos\alpha_2,\quad b_2 = r_2\sin\alpha_2$$
$$a_3 = r_3\cos\alpha_3,\quad b_3 = r_3\sin\alpha_3$$
$$A = R\cos\theta,\quad B = R\sin\theta$$

$$(a_1 + ib_1)\cdot(a_2 + ib_2)\ldots(a_n + ib_n) = A + iB$$

$$r_1(\cos\alpha_1 + i\sin\alpha_1)\,r_2(\cos\alpha_2 + i\sin\alpha_2)\ldots\ldots r_n(\cos\alpha_n + i\sin\alpha_n) = R(\cos\theta + i\sin\theta)$$

$\Rightarrow$ $r_1r_2\ldots\ldots r_n[\cos(\alpha_1 + \alpha_2 + \ldots\alpha_n) + i\sin(\alpha_1 + \alpha_2 + \ldots\alpha_n)] = R(\cos\theta + i\sin\theta)$

$\therefore$ $r_1r_2\ldots\ldots r_n = R$

$\Rightarrow$ $(a_1^2 + b_1^2)(a_2^2 + b_2^2)\ldots(a_n^2 + b_n^2) = A^2 + B^2$

and $\alpha_1 + \alpha_2 + \ldots + \alpha_n = \theta$

$\therefore$ $\tan^{-1}\left(\dfrac{b_1}{a_1}\right) + \tan^{-1}\left(\dfrac{b_2}{a_2}\right) + \ldots\ldots \tan^{-1}\left(\dfrac{b_n}{a_n}\right) = \tan^{-1}\left(\dfrac{A}{B}\right)$ **Proved.**

Example 33. *Find the different values of* $(1 + i)^{1/3}$

Proof. Let $\quad 1 + i = r(\cos\theta + i\sin\theta)$

$$1 = r\cos\theta, \quad 1 = r\sin\theta$$

$$1^2 + 1^2 = r^2\cos^2\theta + r^2\sin^2\theta$$

$$\Rightarrow \quad 2 = r^2(\cos^2\theta + \sin^2\theta)$$

$$\Rightarrow \quad 2 = r^2, \quad r = \sqrt{2}$$

$$\Rightarrow \quad \frac{1}{1} = \frac{r\sin\theta}{r\cos\theta} \Rightarrow 1 = \tan\theta \Rightarrow \tan\left(\frac{\pi}{4}\right) = \tan\theta$$

$$\Rightarrow \quad \theta = \frac{\pi}{4}$$

$$\Rightarrow \quad 1 + i = \sqrt{2}\left(\cos\frac{\pi}{4} + i\sin\frac{\pi}{4}\right)$$

$$\Rightarrow \quad (1+i)^{1/3} = \left[\sqrt{2}\left(\cos\frac{\pi}{4} + i\sin\cos\frac{\pi}{4}\right)\right]^{1/3}$$

$$= \left[\sqrt{2}\cos\left(2n\pi + \frac{\pi}{4}\right) + i\sin\left(2n\pi + \frac{\pi}{4}\right)\right]^{1/3}$$

$$= (2)^{1/6}\left[\cos\frac{1}{3}\left(2n\pi + \frac{\pi}{4}\right) + i\sin\frac{1}{3}\left(2n\pi + \frac{\pi}{4}\right)\right]$$

Putting $\quad n = 0, 1, 2$ we get three values

$$(2)^{1/6}\left[\cos\frac{\pi}{12} + i\sin\frac{\pi}{12}\right], (2)^{1/6}\left[\cos\frac{9\pi}{12} + i\sin\frac{9\pi}{12}\right], (2)^{1/6}\left[\cos\frac{17\pi}{12} + i\sin\frac{17\pi}{12}\right]$$ **Ans.**

EXERCISE 12.5

1. Find the value of $(1 + i)^{1/5}$

   Ans. $(2)^{1/10}\left[\cos\frac{1}{5}\left(2n\pi + \frac{\pi}{4}\right) + i\sin\frac{1}{5}\left(2n\pi + \frac{\pi}{4}\right)\right]$, where $n = 0, 1, 2, 3, 4$

2. Find the value of $(1 + \sqrt{-3})^{3/4}$

   Ans. $\left[\cos\frac{3}{4}\left(2n\pi + \frac{\pi}{3}\right) + i\sin\frac{3}{4}\left(2n\pi + \frac{\pi}{3}\right)\right]$, where $n = 0, 1, 2, 3$

3. Find value of $(1 + i)^{2/3}$

   Ans. $2^{1/3}\left[\cos\left(\frac{4n\pi}{3} + \frac{\pi}{6}\right) + i\sin\left(\frac{4n\pi}{3} + \frac{\pi}{6}\right)\right]$, where $n = 0, 1, 2$

4. Find the value of $\left(\frac{1}{2} + \frac{\sqrt{-3}}{2}\right)^{3/4}$ **[Diploma IETE, Dec. 2006]**

   Ans. $2^{3/4}\left[\cos\left(\frac{3n\pi}{2} + \frac{\pi}{4}\right) + i\sin\left(\frac{3n\pi}{2} + \frac{\pi}{4}\right)\right]$ where $n = 0, 1, 2, 3$

5. Solve the equation with the help of De Moivre's theorem $x^7 - 1 = 0$

   Ans. $\cos\frac{2n\pi}{7} + i\sin\frac{2n\pi}{7}$ where $n = 0, 1, 2, 3, 4, 5, 6$

6. Find the roots of the equation $x^3 + 8 = 0$

Ans. $2\left[\cos\left(\frac{2n\pi+\pi}{3}\right)+i\sin\left(\frac{2n\pi+\pi}{3}\right)\right]$, where $n = 0, 1, 2$

7. If 'n' is a positive integer, then show that:

$$\left[\frac{1+\cos\theta+i\sin\theta}{1+\cos\theta-i\sin\theta}\right]^n = \cos n\theta + i\sin n\theta$$ **[DIP IETE, 2018]**

8. Find the square roots of the complex number $3 + 4i$. **Ans.** $\pm (2 + i)$

9. Write down all the values of $(1 + i)^{1/4}$ **[Diploma IETE, Dec. 2005]**

Ans. $2^{1/8}\left[\cos\frac{1}{4}\left(2n\pi+\frac{\pi}{4}\right)+i\sin\frac{1}{4}\left(2n\pi+\frac{\pi}{4}\right)\right]$, where $n = 0, 1, 2, 3$

10. If $z = \cos\theta + i\sin\theta$ then find $z^n + \frac{1}{z^n}$ **Ans.** $2\cos n\theta$, $n = 0, 1, 2, 3$

11. If $a_r = \cos\left(\frac{\pi}{2^r}\right)+i\sin\left(\frac{\pi}{2^r}\right)$, $r = 1, 2, 3,\ldots$ then show that $a_1 a_2 a_3 \ldots$ up to infinite form $= -1$.

Choose the correct alternative

12. A square root of $3 + 4i$ is

(*a*) $\sqrt{3}+i$ (*b*) $2 - i$ (*c*) $2 + i$ (*d*) None of these **Ans.** (*c*)

13. If z_1 and z_2 are two complex numbers then $|z_1 + z_2|$ is

(*a*) $|z_1|+|z_2|$ (*b*) $\le |z_1|+|z_2|$ (*c*) $\le |z_1|-|z_2|$ (*d*) $\ge |z_1|+|z_2|$ **Ans.** (*b*)

[Diploma IETE, Dec. 2005]

12.11 SEPARATION OF REAL AND IMAGINARY PART

$x + iy$ is the form of complex number, where x is real part and y is the imaginary part of complex number.

Example 34. *Separate the following into real and imaginary parts:*

(*i*) $\sin(x + iy)$ (*ii*) $\cos(x + iy)$ (*iii*) $\tan(x + iy)$

(*iv*) $\cot(x + iy)$ (*v*) $\sec(x + iy)$ (*vi*) $\text{cosec}(x + iy)$

Solution. (*i*) $\sin(x + iy) = \sin x \cos iy + \cos x \sin(iy)$

$$= \sin x \cosh y + i\cos x \sinh y$$

(*ii*) $\cos(x + iy) = \cos x \cos(iy) - \sin x \sin(iy)$

$$= \cos x \cosh y - i\sin x \sinh y$$

(*iii*)
$$\tan(x+iy) = \frac{\sin(x+iy)}{\cos(x+iy)} = \frac{2\sin(x+iy)\cos(x-iy)}{2\cos(x+iy)\cos(x-iy)}$$

$$= \frac{\sin 2x + \sin(2iy)}{\cos 2x + \cos(2iy)} \quad \left[\begin{array}{l}\because 2\sin A.\cos B = \sin(A+B)+\sin(A-B)\\ 2\cos A.\cos B = \cos(A+B)+\cos(A-B)\end{array}\right]$$

(*iv*)
$$\cot(x+iy) = \frac{\cos(x+iy)}{\sin(x+iy)} = \frac{2\cos(x+iy)\sin(x-iy)}{2\sin(x+iy)\cos(x-iy)}$$

$$= \frac{\sin 2x - \sin(2iy)}{\cos(2iy) - \cos 2x} = \frac{\sin 2x - i\sinh 2y}{\cosh 2y - \cos 2x}$$

(*v*) $$\sec(x+iy) = \frac{1}{\cos(x+iy)} = \frac{2\cos(x-iy)}{2\cos(x+iy)\cos(x-iy)}$$

$$= \frac{2[\cos x\cos(iy) + \sin x\cdot\sin(iy)]}{\cos 2x + \cos(2iy)}$$

$$= \frac{2[\cos x\cdot\cosh y + i\sin x\sinh y]}{\cos 2x + \cosh 2y}$$

(*vi*) $$\operatorname{cosec}(x+iy) = \frac{1}{\sin(x+iy)}$$

$$= \frac{2\sin(x-iy)}{2\sin(x+iy)\sin(x-iy)} = \frac{2[\sin x\cos(iy) - \cos x\sin(iy)]}{\cos(2iy) - \cos 2x}$$

$$= \frac{2[\sin x\cdot\cosh y - i\cos x\cdot\sinh y]}{\cosh 2y - \cos 2x}$$ **Ans.**

Example 35. *Separate the following into real and imaginary parts:*

(*a*) $\sinh(x+iy)$ (*b*) $\cosh(x+iy)$ (*c*) $\tanh(x+iy)$

Solution. (*a*) $\sinh(x+iy) = \sinh x\cosh(iy) + \cosh x\sinh(iy)$

$= \sinh x\cos y - \sinh x\sin y.$ **Ans.**

(*b*) $\cosh(x+iy) = \cosh x\cosh iy - \sinh x.\sinh iy$

$= \cosh x.\cosh y - i\sinh x.\sin y$

(*c*) $$\tanh(x+iy) = \frac{\sinh(x+iy)}{\cosh(x+iy)} = \frac{-i\sin i(x+iy)}{\cos i(x+iy)} = \frac{-i\sin(ix-y)}{\cos(ix-y)}$$

$$= \frac{-i\,2\sin(ix-y)\cdot\cos(ix+y)}{2\cos(ix-y)\cos(ix+y)}$$ (Note this step)

$$= -i\frac{\sin 2ix - \sin 2y}{\cos 2ix + \cos 2y} = -i\frac{i\sinh 2x - \sin 2y}{\cosh 2x + \cos 2y} = \frac{\sinh 2x + i\sin 2y}{\cosh 2x + \cos 2y}$$

$$= \frac{\sinh 2x}{\cosh 2x + \cos 2y} + i\frac{\sin 2y}{\cosh 2x + \cos 2y}$$ **Ans.**

Example 36. *Separate* $\tan^{-1}(a+ib)$ *into real and imaginary parts.*

[B.T.E. DELHI May/June 2007]

Solution. Let $\tan^{-1}(a+ib) = x+iy$ ... (1)

$\therefore$ $\tan(x+iy) = a+ib$

On both sides for i write $-i$

$\therefore$ $\tan(x-iy) = a-ib$

Now $\tan 2x = \tan[(x+iy)+(x-iy)]$

$$= \frac{\tan(x+iy)+\tan(x-iy)}{1-\tan(x+iy)\tan(x-iy)} = \frac{a+ib+a-ib}{1-(a+ib)(a-ib)} = \frac{2a}{1-a^2-b^2}$$

$$2x = \tan^{-1}\left[\frac{2a}{1-a^2-b^2}\right]$$

$$x = \frac{1}{2}\tan^{-1}\left[\frac{2a}{1-a^2-b^2}\right] \quad \text{... (2)}$$

and $\tan(2yi) = \tan[(x+iy)-(x-iy)]$

$$= \frac{\tan(x+iy)-\tan(x-iy)}{1+\tan(x+iy)\tan(x-iy)} = \frac{a+bi-a+bi}{1+(a+bi)(a-bi)}$$

$$\Rightarrow \quad i\tanh 2y = \frac{2bi}{1+a^2+b^2}$$

$$\Rightarrow \quad \tanh 2y = \frac{2b}{1+a^2+b^2}$$

$$\Rightarrow \quad 2y = \tanh^{-1}\left[\frac{2b}{1+a^2+b^2}\right]$$

$$\Rightarrow \quad y = \frac{1}{2}\tanh^{-1}\left[\frac{2b}{1+a^2+b^2}\right] \quad \text{... (3)}$$

Putting the values of x and y from (2) and (3) in (1), we get

$$\therefore \quad \tan^{-1}(a+ib) = \frac{1}{2}\tan^{-1}\left[\frac{2a}{1-a^2-b^2}\right] + \frac{1}{2}\tanh^{-1}\left[\frac{2b}{1+a^2+b^2}\right]$$ **Ans.**

Example 37. *Prove that* $\sin 7\theta = 7\sin\theta - 56\sin^3\theta + 112\sin^5\theta - 64\sin^7\theta$

Proof. $\cos 7\theta + i\sin 7\theta = (\cos\theta + i\sin\theta)^7$

$$= \cos^7\theta + {}^7C_1\cos^6\theta\,(i\sin\theta) + {}^7C_2\cos^5\theta\,(i\sin\theta)^2 + {}^7C_3\cos^4\theta\,(i\sin\theta)^3 +$$
$${}^7C_4\cos^3\theta\,(i\sin\theta)^4 + {}^7C_5\cos^2\theta\,(i\sin\theta)^5 + {}^7C_6\cos\theta\cdot(i\sin\theta)^6 + {}^7C_7\,(i\sin\theta)^7$$

Equating imaginary parts, we get

$$\sin 7\theta = 7\cos^6\theta\sin\theta - 35\cos^4\theta\sin^3\theta + 21\cos^2\theta\sin^5\theta - \sin^7\theta$$
$$= 7(1-\sin^2\theta)^3\sin\theta - 35(1-\sin^2\theta)^2\sin^3\theta + 21(1-\sin^2\theta)\sin^5\theta - \sin^7\theta$$
$$= 7(1-3\sin^2\theta+3\sin^4\theta-\sin^6\theta)\sin\theta - 35(1-2\sin^2\theta+\sin^4\theta)\sin^3\theta +$$
$$21\sin^5\theta - 21\sin^7\theta - \sin^7\theta$$
$$= 7\sin\theta - 21\sin^3\theta + 21\sin^5\theta - 7\sin^7\theta - 35\sin^3\theta$$
$$+ 70\sin^5\theta - 35\sin^7\theta + 21\sin^5\theta - 22\sin^7\theta$$

$$\therefore \sin 7\theta = 7\sin\theta - 56\sin^3\theta + 112\sin^5\theta - 64\sin^7\theta$$ **Proved.**

EXERCISE 12.6

Separate into real and imaginary parts.

1. $\operatorname{sech}(x+iy)$ **Ans.** $\dfrac{2\cosh x\cos y - 2i\sinh x\sin y}{\cosh 2x + \cos 2y}$

2. $\coth i(x+iy)$ **Ans.** $\dfrac{-\sinh 2y - i\sin 2x}{\cosh 2x - \cos 2y}$

3. coth $(x + iy)$ **Ans.** $\dfrac{\sinh 2x - i\sin 2y}{\cosh 2x - \cos 2y}$

4. If $\sin(\theta + i\phi) = p(\cos\alpha + i\sin\alpha)$, prove that

$p^2 = \dfrac{1}{2}[\cosh 2\phi - \cos 2\theta]$, $\tan\alpha = \tanh\phi\cot\theta$

5. If $\sin(\alpha + i\beta) = x + iy$, prove that $x^2 \operatorname{sech}^2\beta + y^2 \operatorname{cosech}^2\beta = 1$ and $x^2 \operatorname{cosec}^2\alpha - y^2 \sec^2\alpha = 1$

6. If $\cos(\theta + i\phi) = r(\cos\alpha + i\sin\alpha)$, prove that $\theta = \dfrac{1}{2}\log\left[\dfrac{\sin(\theta - \alpha)}{\sin(\theta + \alpha)}\right]$

7. If $\tan\left(\dfrac{\pi}{6} + i\alpha\right) = x + iy$, prove that $x^2 + y^2 + \dfrac{2x}{\sqrt{3}} = 1$

8. If $\tan(A + B) = \alpha + i\beta$, show that $\dfrac{1-(\alpha^2+\beta^2)}{1+(\alpha^2+\beta^2)} = \dfrac{\cos 2A}{\cosh 2B}$

9. If $\dfrac{x+iy-c}{x+iy+c} = e^{u+iv}$, prove that

$x = -\dfrac{c\sinh u}{\cosh u - \cos v}$, $\quad y = \dfrac{c\sinh v}{\cosh u - \cos v}$

Further, if $v = (2n+1)\dfrac{\pi}{2}$, prove that $x^2 + y^2 = c^2$, where n is an integer.

10. If $\dfrac{u-1}{u+1} = \sin(x + iy)$, where $u = \alpha + i\beta$ show that the argument of u is $\theta + \phi$ where

$\tan\theta = \dfrac{\cos x\sinh y}{1+\sin x\cosh y}$ and, $\tan\phi = \dfrac{\cos x\sinh y}{1-\sin x\cosh y}$

11. If $A + iB = C\tan(x + iy)$, prove that $\tan 2x = \dfrac{2CA}{C^2 - A^2 - B^2}$

12. If $\cosh(\alpha + i\beta) = x + iy$, prove that

(*a*) $\dfrac{x^2}{\cosh^2\alpha} + \dfrac{y^2}{\sinh^2\alpha} = 1$ (*b*) $\dfrac{x^2}{\cos^2\beta} - \dfrac{y^2}{\sin^2\beta} = 1$

13. If $\cos(\theta + i\phi) = R(\cos\alpha + i\sin\alpha)$, prove that $\phi = \dfrac{1}{2}\log_e\left[\dfrac{\sin(\theta-\alpha)}{\sin(\theta+\alpha)}\right]$

14. If $\cos(\alpha + i\beta)\cos(\gamma + i\delta) = 1$, prove that $\tanh^2\delta\cosh^2\beta = \sin^2\alpha$

15. If $\dfrac{u-1}{u+1} = \sin(x + iy)$, find u. **Ans.** $\tan^{-1}\left(\dfrac{2\cos x\sinh y}{\cos^2 x - \sinh^2 y}\right)$

12.12 LOGARITHM OF A COMPLEX NUMBER

If z and w are two complex numbers and $z = e^w$ then $w = \log z$ and if $w = \log z$; then $z = e^w$ Here, log z is a many valued function. General value of log z is defined by log z, where $\log z = \log z + 2n\pi i$.

Example 38. *Separate log $(x + iy)$ into its real and imaginary parts.*

Solution. Let $\quad x = r\cos\theta \quad$ ... (1)

and $\quad y = r\sin\theta \quad$ ... (2)

Squaring and adding (1) and (2), we have $x^2 + y^2 = r^2$

$$\therefore \quad r = \sqrt{x^2 + y^2},$$

Dividing (2) by (1), we have $\tan\theta = \dfrac{y}{x}$

$$\Rightarrow \quad \theta = \tan^{-1}\left(\frac{y}{x}\right)$$

$$\therefore \quad \log(x + iy) = \log[r(\cos\theta + i\sin\theta)]$$

$$= [\log r + \log(\cos\theta + i\sin\theta)]$$

$$= \log r + \log[\cos(2n\pi + \theta) + i\sin(2n\pi + \theta)]$$

$$= \log r + \log e^{i(2n\pi + \theta)} = \log r + i(2n\pi + \theta)$$

$$= \log\left(\sqrt{x^2 + y^2}\right) + i\left[2n\pi + \tan^{-1}\frac{y}{x}\right]$$

$$\log(x + iy) = \log\left(\sqrt{x^2 + y^2}\right) + i\left[2n\pi + \tan^{-1}\frac{y}{x}\right]$$

and $$\log(x + iy) = \log\left(\sqrt{x^2 + y^2}\right) + i\tan^{-1}\left(\frac{y}{x}\right) \quad \textbf{Ans.}$$

Example 39. *Write $i^{(1-i)}$ in the form $r(\cos\theta + i\sin\theta)$.*

Solution. $$i^{(1-i)} = e^{\log i^{1-i}} = e^{(1-i)\log i} = e^{(1-i)\log\left(\cos\frac{\pi}{2} + i\sin\frac{\pi}{2}\right)}$$

$$= e^{(1-i)\log\left\{\cos\left(2n\pi + \frac{\pi}{2}\right) + i\sin\left(2n\pi + \frac{\pi}{2}\right)\right\}}$$

$$= e^{(1-i)\log e^{i(2n\pi + \pi/2)}} = e^{(1-i)i\left(2n\pi + \frac{\pi}{2}\right)} = e^{(i+1)\left[2n\pi + \frac{\pi}{2}\right]}$$

$$= e^{i\left(2n\pi + \frac{\pi}{2}\right) + \left(2n\pi + \frac{\pi}{2}\right)} = e^{i\left(2n\pi + \frac{\pi}{2}\right)} \times e^{2n\pi + \frac{\pi}{2}} = e^{2n\pi + \frac{\pi}{2}} \times e^{i\left(2n\pi + \frac{\pi}{2}\right)}$$

$$= e^{2n\pi + \frac{\pi}{2}}\left[\cos\left(2n\pi + \frac{\pi}{2}\right) + i\sin\left(2n\pi + \frac{\pi}{2}\right)\right]$$

$$= e^{2n\pi + \frac{\pi}{2}}\left[\cos\left(\frac{\pi}{2}\right) + i\sin\left(\frac{\pi}{2}\right)\right]$$

$$= e^{2n\pi + \frac{\pi}{2}} \times i = ie^{2n\pi + \frac{\pi}{2}} \quad \textbf{Ans.}$$

EXERCISE 12.7

1. Find the general value of $\log i$. **Ans.** $(4n + 1)\dfrac{\pi i}{2}$

2. Express $\log(-5)$ in terms of $a + ib$. **Ans.** $\log 5 + i(2n + 1)\pi$

3. Find the value of z if

(*a*) $\cos z = 2$. **Ans.** $z = 2n\pi \pm i\log(2 + \sqrt{3})$

(*b*) $\cosh z = -1$. **Ans.** $z = (2n + 1)\pi i$

4. Find the general and principal values of i^i **Ans.** $e^{-\left(2n\pi+\frac{\pi}{2}\right)}, e^{-\frac{\pi}{2}}$

5. If $i^{(\alpha + i\beta)} = x + iy$, prove that $x^2 + y^2 = e^{-(4m + 1)\pi\theta}$

6. Prove that $\log\left(\frac{1}{l-e^{i\theta}}\right) = \log\left(\frac{1}{2}\operatorname{cosec}\theta\right) + i\left(\frac{\pi}{2} - \frac{\theta}{2}\right)$

7. Show that $\log \sin (x + iy) = \frac{1}{2}\log\left(\frac{\cosh 2y - \cos 2x}{2}\right) + i\tan^{-1}(\cot x \tanh y)$.

8. Prove that $\tan\left[i\log\frac{a-ib}{a+ib}\right] = \frac{2ab}{a^2-b^2}$. 9. $\log\left[\frac{\cos(x-iy)}{\cos(x+iy)}\right] = 2\,i\tan^{-1}(\tan x \tanh y)$.

10. Separate $i^{(1+i)}$ into real and imaginary parts. **Ans.** $ie^{-\frac{\pi}{2}}$

12.13 EXPANSION OF *cos nθ*, *sin nθ* IN TERMS OF *cos θ* AND *sin θ*

Method as an Example:

For Example. *Expand cos 6θ and sin 6θ in terms of cos θ and sin θ.*

Solution. $\cos 6\theta + i \sin 6\theta = (\cos\theta + i\sin\theta)^6$

Expansion by Binomial theorem

$$\cos 6\theta + i\sin 6\theta = \cos^6\theta + {}^6C_1\cos^5\theta\,(i\sin\theta) + {}^6C_2\cos^4\theta\,(i\sin\theta)^2 + {}^6C_3\cos^3\theta\,(i\sin\theta)^3 + {}^6C_4\cos^2\theta\,(i\sin\theta)^4 + {}^6C_5\cos\theta\,(i\sin\theta)^5 + {}^6C_6\,(i\sin\theta)^6$$

$$= \cos^6\theta + i\cdot 6\cos^5\theta\sin\theta - 15\cos^4\sin^2\theta - i.20\cos^3\theta\sin^3\theta + 15\cos^2\sin^4\theta + i\,6\cos\theta\sin^5\theta - \sin^6\theta$$

Equating real and imaginary parts, we have

$$\cos 6\theta = \cos^6\theta - 15\cos^4\theta\sin^2\theta + 15\cos^2\theta\sin^4\theta - \sin^6\theta$$

$$\sin 6\theta = 6\cos^5\theta\sin\theta - 20\cos^3\theta\sin^3\theta + 6\cos\theta\sin^5\theta$$ **Ans.**

Example 40. *Show that for all real* μ,

$$\cos(6\mu) = 32\cos^6(\mu) - 48\cos^4(\mu) + 18\cos^2(\mu) - 1$$

[Diploma I.E.T.E. June 2006]

Solution. From the above example, we know that

$$\cos 6\mu = \cos^6\mu - 15\cos^4\mu\sin^2\mu + 15\cos^2\mu\sin^4\mu - \sin^6\mu$$

$$\Rightarrow \cos 6\mu = \cos^6\mu - 15\cos^4\mu\,(1-\cos^2\mu) + 15\cos^2\mu\,(1-\cos^2\mu)^2 - (1-\cos^2\mu)^3$$

$$= \cos^6\mu - 15\cos^4\mu + 15\cos^6\mu + 15\cos^2\mu\,(1+\cos^4\mu - 2\cos^2\mu) - (1 - 3\cos^2\mu + 3\cos^4\mu - \cos^6\mu)$$

$$= \cos^6\mu - 15\cos^4\mu + 15\cos^6\mu + 15\cos^2\mu + 15\cos^6\mu - 30\cos^4\mu - 1 + 3\cos^2\mu - 3\cos^4\mu + \cos^6\mu$$

$$= 32\cos^6\mu - 48\cos^4\mu + 18\cos^2\mu - 1$$ **Proved.**

Example 41. *Expand tan 9θ in powers of tan θ.*

Solution. We know that,

$$\tan n\theta = \frac{{}^nC_1\tan\theta - {}^nC_3\tan^3\theta + {}^nC_5\tan^5\theta.....}{1 - {}^nC_2\tan^2\theta + {}^nC_4\tan^4\theta - {}^nC_6\tan^6\theta + ...}$$

$$\tan 9\theta = \frac{{}^9C_1 \tan\theta - {}^9C_3 \tan^3\theta + {}^9C_5 \tan^5\theta - {}^9C_7 \tan^7\theta + \tan^9\theta}{1 - {}^9C_2 \tan^2\theta + {}^9C_4 \tan^4\theta - {}^9C_6 \tan^6\theta + {}^9C_8 \tan^8\theta}$$

$$= \frac{9\tan\theta - 84\tan^3\theta + 126\tan^5\theta - 36\tan^7\theta + \tan^9\theta}{1 - 36\tan^2\theta + 126\tan^4\theta - 84\tan^6\theta + 9\tan^8\theta} \quad \textbf{Ans.}$$

EXERCISE 12.8

Prove that

1. $\cos 6\theta = \cos^6\theta - 15\cos^4\theta\sin^2\theta + 15\cos^2\theta\sin^4\theta - \sin^6\theta$.

2. $\dfrac{\sin 6\theta}{\cos 6\theta} = 32\sin^5\theta - 32\sin^3\theta + 6\sin\theta$.

3. $\sin 7\theta = 7\cos^6\theta\sin\theta - 35\cos^4\theta\sin^3\theta + 21\cos^2\theta\sin^5\theta - \sin^7\theta$.

4. $1 + \cos 10\theta = 2(16\cos^5\theta - 20\cos^3\theta + 5\cos\theta)^2$.

5. $1 - \cos 10\theta = 2(16\sin^5\theta - 20\sin^3\theta + 5\sin\theta)^2$.

6. $\tan 5\theta = \dfrac{5\tan\theta - 10\tan^3\theta + \tan^5\theta}{1 - 10\tan^2\theta + 5\tan^4\theta}$

12.14 EXPANSION OF $\cos^n\theta$, $\sin^n\theta$ IN TERMS OF SINES AND COSINES AS MULTIPLES OF θ

Method:

Let $\quad x = \cos\theta + i\sin\theta$, then $\dfrac{1}{x} = \cos\theta - i\sin\theta$

$$x + \frac{1}{x} = 2\cos\theta \quad \text{and} \quad x - \frac{1}{x} = 2i\sin\theta$$

Again $\quad x^n = \cos n\theta + i\sin n\theta, \ \dfrac{1}{x^n} = \cos n\theta - i\sin n\theta$

$$x^n + \frac{1}{x^n} = 2\cos n\theta \quad \text{and} \quad x^n - \frac{1}{x^n} = 2i\sin n\theta$$

To expand $\cos^n\theta$: Start from $(2\cos\theta)^n = \left(x + \dfrac{1}{x}\right)^n$

To expand $\sin^n\theta$: Start from $(2i\sin\theta)^n = \left(x - \dfrac{1}{x}\right)^n$

Example 42. *Express $\sin^5\theta$ in terms of sines of multiples of θ.*

Solution. $(2i\sin\theta)^5 = \left(x - \dfrac{1}{x}\right)^5 \qquad \left[\begin{array}{l} x = \cos\theta + i\sin\theta \\ \dfrac{1}{x} = \cos\theta - i\sin\theta \end{array}\right]$

$$i\,32\sin^5\theta = x^5 + 5x^4\left(-\frac{1}{x}\right) + 10x^3\left(-\frac{1}{x}\right)^2 + 10x^2\left(-\frac{1}{x}\right)^3 + 5x\left(-\frac{1}{x}\right)^4 + \left(-\frac{1}{x}\right)^5$$

(By binomial theorem)

$$= \left(x^5 - \frac{1}{x^5}\right) - 5\left(x^3 - \frac{1}{x^3}\right) + 10\left(x - \frac{1}{x}\right)$$

$$= 2i \sin 5\theta - 5\,(2i \sin 3\theta) + 10\,(2i \sin \theta)$$

$$16 \sin^5 \theta = \sin 5\theta - 5 \sin 3\theta + 10 \sin \theta$$

$$\sin^5 \theta = \frac{1}{16}(\sin 5\theta - 5 \sin 3\theta + 10 \sin \theta)$$ **Ans.**

Example 43. *Prove that*

$$-2^{12} \cos^6 \theta \sin^7 \theta = \sin 13\theta - \sin 11\theta - 6 \sin 9\theta + 6 \sin 7\theta + 15 \sin 5\theta - 15 \sin 3\theta - 20 \sin \theta$$

Solution. $x^n = (\cos\theta + i\sin\theta)^n = \cos n\theta + i \sin n\theta$ $\left[x = \cos\theta + i\sin\theta,\ \frac{1}{x} = \cos\theta - i\sin\theta\right]$

$\Rightarrow \quad \frac{1}{x^n} = (\cos\theta + i\sin\theta)^{-n} = \cos n\theta - i\sin n\theta$

$\Rightarrow \quad x^n + \frac{1}{x^n} = 2\cos n\theta$ and $x^n - \frac{1}{x^n} = 2i\sin n\theta$ $\left[x + \frac{1}{x} = 2\cos\theta,\ x - \frac{1}{x} = 2i\sin\theta\right]$

$$\Rightarrow \quad (2\cos\theta)^6 (2i\sin\theta)^7 = \left(x + \frac{1}{x}\right)^6 \left(x - \frac{1}{x}\right)^7 = \left(x^2 - \frac{1}{x^2}\right)^6 \left(x - \frac{1}{x}\right)$$

$$= \left[x^{12} + 6x^{10}\left(-\frac{1}{x^2}\right) + 15x^8\left(-\frac{1}{x^2}\right)^2 + 20x^6\left(-\frac{1}{x^2}\right)^3 + 15x^4\left(-\frac{1}{x^2}\right)^4 + 6x^2\left(-\frac{1}{x^2}\right)^5 + \left(-\frac{1}{x^2}\right)^6\right]\left(x - \frac{1}{x}\right)$$

$$= \left[x^{12} - 6x^8 + 15x^4 - 20 + \frac{15}{x^4} - \frac{6}{x^8} + \frac{1}{x^{12}}\right]\left[x - \frac{1}{x}\right]$$

$$= x^{13} - 6x^9 + 15x^5 - 20x + \frac{15}{x^3} - \frac{6}{x^7} + \frac{1}{x^{11}} - x^{11} + 6x^7 - 15x^3 + \frac{20}{x} - \frac{15}{x^5} + \frac{6}{x^9} - \frac{1}{x^{13}}$$

$$= \left(x^{13} - \frac{1}{x^{13}}\right) - \left(x^{11} - \frac{1}{x^{11}}\right) - 6\left(x^9 - \frac{1}{x^9}\right) + 6\left(x^7 - \frac{1}{x^7}\right) + 15\left(x^5 - \frac{1}{x^5}\right) - 15\left(x^3 - \frac{1}{x^3}\right) - 20\left(x - \frac{1}{x}\right)$$

$$= 2i\sin 13\theta - 2i\sin 11\theta - 6(2i\sin 9\theta) + 6(2i\sin 7\theta) + 15(2i\sin 5\theta) - 15(2i\sin 3\theta) - 20(2i\sin\theta)$$

$$\therefore \quad -2^{12}\cos^6\theta\sin^7\theta = \sin 13\theta - \sin 11\theta - 6\sin 9\theta + 6\sin 7\theta + 15\sin 5\theta - 15\sin 3\theta - 20\sin\theta$$ **Proved.**

EXERCISE 12.9

1. Express $\sin^7 \theta$ as a sum of sines of multiples of θ.

Ans. $\dfrac{-1}{64}[\sin 7\theta - 7\sin 5\theta + 21\sin 3\theta - 35\sin\theta]$

2. Express $\cos^8 \theta$ as a sum of cosines of multiples of θ.

Ans. $\dfrac{-1}{128}[\cos 8\theta + 8\cos 6\theta + 28\sin 6\theta + 56\cos 2\theta + 35]$

3. Prove that $2^7 \cos^3 \theta \sin^5 \theta = \sin 8\theta - 2\sin 6\theta - 2\sin 4\theta + 6\sin 2\theta$.

4. Prove that $\cos^4 \theta \sin^3 \theta = -\dfrac{1}{64}[\sin 7\theta + \sin 5\theta - 3\sin 3\theta - \sin\theta]$.

5. Prove that $-256 \sin^7 \theta \cos^2 \theta = \sin 9\theta - 5\sin 7\theta + 8\sin 5\theta - 14\sin\theta$.

Applications:

Example 44. *The centre of a regular hexagon is at the origin and one vertex is given by $\sqrt{3} + i$ on the Argand diagram. Find the complex number represented by the other vertices.*

Solution. Let the vertex A represents $\sqrt{3} + i$.

$$\sqrt{3} + i = 2\left(\cos\frac{\pi}{6} + i\sin\frac{\pi}{6}\right)$$

On turning OA anticlockwise through $\dfrac{\pi}{3}$, we get OB.

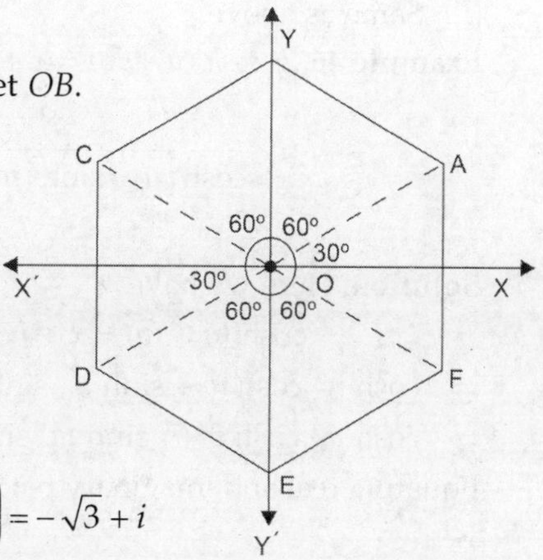

B represents $2\left[\cos\left(\dfrac{\pi}{6}+\dfrac{\pi}{3}\right) + i\sin\left(\dfrac{\pi}{6}+\dfrac{\pi}{3}\right)\right]$

$\Rightarrow$ $2\left(\cos\dfrac{\pi}{2} + i\sin\dfrac{\pi}{2}\right) = 0 + 2i$

Similarly,

C represents $2\left[\cos\left(\dfrac{\pi}{2}+\dfrac{\pi}{3}\right) + i\sin\left(\dfrac{\pi}{2}+\dfrac{\pi}{3}\right)\right]$

$\Rightarrow$ $2\left[\cos\dfrac{5\pi}{6} + i\sin\dfrac{5\pi}{6}\right] = 2\left(-\dfrac{\sqrt{3}}{2} + \dfrac{i}{2}\right) = -\sqrt{3} + i$

D represents $2\left[\cos\left(\dfrac{5\pi}{6}+\dfrac{\pi}{3}\right) + i\sin\left(\dfrac{5\pi}{6}+\dfrac{\pi}{3}\right)\right]$

$\Rightarrow$ $2\left[\cos\dfrac{7\pi}{6} + i\sin\dfrac{7\pi}{6}\right] = 2\left(\dfrac{-\sqrt{3}}{2} - \dfrac{i}{2}\right) = -\sqrt{3} - i$

E represents $2\left[\cos\left(\dfrac{7\pi}{6}+\dfrac{\pi}{3}\right) + i\sin\left(\dfrac{7\pi}{6}+\dfrac{\pi}{3}\right)\right]$

$\Rightarrow$ $2\left[\cos\dfrac{3\pi}{2} + i\sin\dfrac{3\pi}{2}\right] = 2(0 - i) = -2i$

F represents $2\left[\cos\left(\dfrac{3\pi}{2}+\dfrac{\pi}{3}\right) + i\sin\left(\dfrac{3\pi}{2}+\dfrac{\pi}{3}\right)\right]$

$\Rightarrow 2\left[\cos\dfrac{11\pi}{6} + i\sin\dfrac{11\pi}{6}\right] = 2\left(\dfrac{\sqrt{3}}{2} - \dfrac{i}{2}\right) = \sqrt{3} - i$ **Ans.**

Example 45. *If $x + iy = \sin(A + iB)$ prove that*

$$(i)\ \frac{x^2}{\cosh^2 B} + \frac{y^2}{\sinh^2 B} = 1 \qquad (ii)\ \frac{x^2}{\sin^2 A} - \frac{y^2}{\cos^2 A} = 1$$

[SBTE. 2017, 2015, 2013, 2010]

Solution. Here, we have

$$x + iy = \sin(A + iB)$$
$$x + iy = \sin A \cdot \cos iB + \cos A \cdot \sin iB$$
$$x + iy = \sin A \cdot \cosh B + i \cos A \cdot \sinh B$$
$$x = \sin A \cdot \cosh B,\ y = \cos A \cdot \sinh B$$
$$x^2 = \sin^2 A \cdot \cosh^2 B,\ y^2 = \cos^2 A \cdot \sinh^2 B$$

(*i*)
$$\text{L.H.S.} = \frac{x^2}{\cosh^2 B} + \frac{y^2}{\sinh^2 B}$$
$$= \frac{\sin^2 A \cdot \cosh^2 B}{\cosh^2 B} + \frac{\cos^2 A \cdot \sinh^2 B}{\sinh^2 B}$$
$$= \sin^2 A + \cos^2 A = 1$$
Proved.

(*ii*) Same as above.

Example 46. *If $\cosh(u + iv) = x + iy$, prove that*

$$(i)\ \frac{x^2}{\cosh^2 u} + \frac{y^2}{\sinh^2 u} = 1 \qquad (ii)\ \frac{x^2}{\cos^2 v} - \frac{y^2}{\sin^2 v} = 1$$

[SBTE. 2013, 2012, 2010]

Solution. Here we have

$$\cosh(u + iv) = x + iy$$
$$\Rightarrow \cosh u \cdot \cosh iv + \sinh u \cdot \sinh iv = x + iy$$
$$\Rightarrow \cosh u \cdot \cosh v + i \sinh u \cdot \sin v = x + iy$$

$$\left[\begin{array}{l} \because \cos i(u + iv) = x + iy \\ \cos(-u + iv) = x + iy \\ \Rightarrow \cos u \cdot \cos iv + \sin v \cdot \sin iu = x + iy \end{array}\right]$$

Equating real and imaginary part both sides, we get

$$x = \cosh u \cos v, \quad y = \sinh u \cdot \sin v$$
$$x^2 = \cosh^2 u \cdot \cos^2 v, \quad y^2 = \sinh^2 u \cdot \sin^2 v$$

(*i*)
$$\text{L.H.S.} = \frac{x^2}{\cosh^2 u} + \frac{y^2}{\sinh^2 u}$$
$$= \frac{\cosh^2 u \cdot \cos^2 v}{\cosh^2 u} + \frac{\sinh^2 u \cdot \sin^2 v}{\sinh^2 u}$$
$$= \cos^2 v + \sin^2 v = 1$$
Proved.

(*ii*)
$$\frac{\cosh^2 u \cdot \cos^2 v}{\cos^2 v} - \frac{\sinh^2 u \cdot \sin^2 v}{\sin^2 v}$$
$$= \cosh^2 u - \sinh^2 u = 1$$
Proved.

Example 47. *In an Argand diagram, one vertex of an equilateral triangle is given by $(1+i\sqrt{3})$. Find the complex numbers represented by other vertices, the origin being the circumcentre of the triangle.*

Solution. One vertex A is given by $(1+i\sqrt{3})$

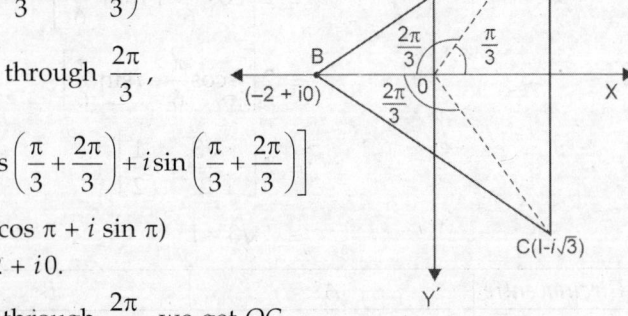

$$\Rightarrow \qquad 2\left(\cos\frac{\pi}{3}+i\sin\frac{\pi}{3}\right)$$

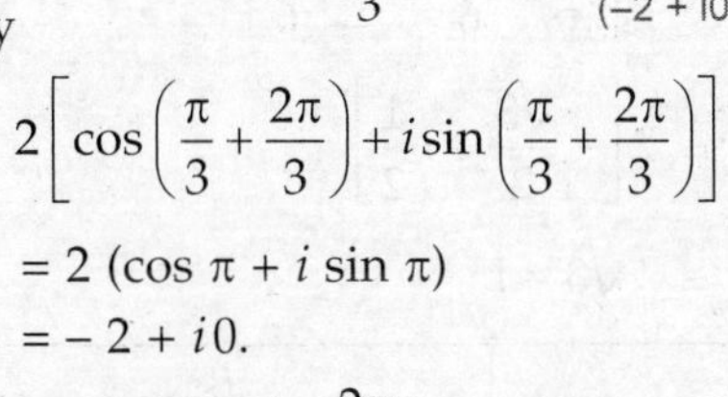

On turning OA anticlockwise through $\frac{2\pi}{3}$, we get OB. B is given by

$$2\left[\cos\left(\frac{\pi}{3}+\frac{2\pi}{3}\right)+i\sin\left(\frac{\pi}{3}+\frac{2\pi}{3}\right)\right]$$

$$= 2\,(\cos\pi + i\sin\pi)$$

$$= -2 + i0.$$

On turning OB anticlockwise through $\frac{2\pi}{3}$, we get OC.

C is given by
$$= 2\left[\cos\left(\pi+\frac{2\pi}{3}\right)+i\sin\left(\pi+\frac{2\pi}{3}\right)\right]$$

$$= 2\left[\cos\frac{5\pi}{3}+i\sin\frac{5\pi}{3}\right]$$

$$= 2\left[\cos\frac{\pi}{3}-i\sin\frac{\pi}{3}\right]$$

$$= 2\left[\frac{1}{2}-i\frac{\sqrt{3}}{2}\right]$$

$$= (1-i\sqrt{3})$$ **Ans.**

Example 48. *One vertex of an equilateral triangle is given by 4 + 5i and the circumcentre by $4-\sqrt{3}+6i$, in an Argand diagram. Find the complex numbers represented by other vertices.*

Solution. On shifting origin to the circumcentre, one vertex A is given by

$$(4+5i)-(4-\sqrt{3}+6i) = \sqrt{3}-i$$

$$\Rightarrow \qquad 2\left(\cos\frac{\pi}{6}-i\sin\frac{\pi}{6}\right) \Rightarrow 2\left[\cos\left(-\frac{\pi}{6}\right)+i\sin\left(-\frac{\pi}{6}\right)\right]$$

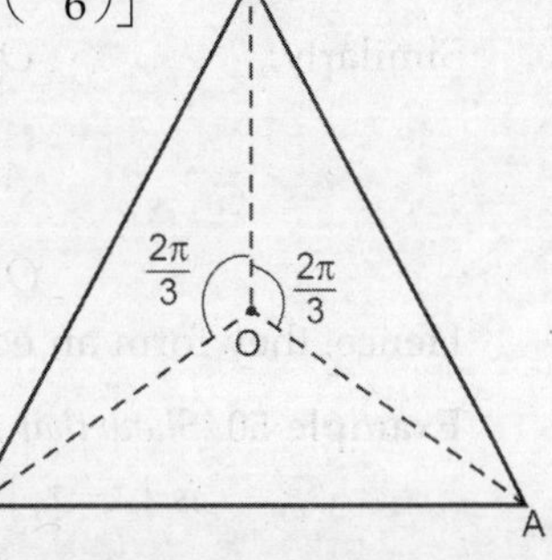

On turning OA anticlockwise through $\frac{2\pi}{3}$, we get OB.

B is given by
$$2\left[\cos\left(-\frac{\pi}{6}+\frac{2\pi}{3}\right)+i\sin\left(\frac{-\pi}{6}+\frac{2\pi}{3}\right)\right]$$

$$\Rightarrow \qquad 2\left[\cos\frac{\pi}{2}+i\sin\frac{\pi}{2}\right] = 2\,[0+i] = 0+2i$$

On turning OB anticlockwise through $\frac{2\pi}{3}$, we get OC.

C is given by $\quad 2\left[\cos\left(\frac{\pi}{2}+\frac{2\pi}{3}\right)+i\sin\left(\frac{\pi}{2}+\frac{2\pi}{3}\right)\right]$

$$= 2\left[\cos\frac{7\pi}{6}+i\sin\frac{7\pi}{6}\right]$$

$$= 2\left[-\cos\frac{\pi}{6}-i\sin\frac{\pi}{6}\right]$$

$$= 2\left[-\frac{\sqrt{3}}{2}-i\frac{1}{2}\right]$$

$$= -\sqrt{3}-i$$

Circumcentre	*A*	*B*	C
Origin	$\sqrt{3}-i$	$0+2i$	$-\sqrt{3}-i$
$4-\sqrt{3}+6i$	$(\sqrt{3}-i)+4-\sqrt{3}+6i$ $=4+5i$	$(0+2i)+(4-\sqrt{3}+6i)$ $=4-\sqrt{3}+8i$	$(-\sqrt{3}-i)+(4-\sqrt{3}+6i)$ $=4-2\sqrt{3}+5i$

Example 49. *Show that the origin and the complex numbers represented by the roots of the equation $z^2+az+b=0$ form an equilateral triangle if $a^2=3b$.*

Solution. We have, $\quad z^2+az+b=0$

$$\Rightarrow \qquad z=\frac{-a\pm\sqrt{a^2-4b}}{2} \qquad (\because\ a^2=3b)$$

$$=\frac{-a\pm\sqrt{3b-4b}}{2}=\frac{-a\pm i\sqrt{b}}{2}$$

Let us find out the distances among the origin, and $\frac{-a+i\sqrt{b}}{2}$ and $\frac{-a-i\sqrt{b}}{2}$.

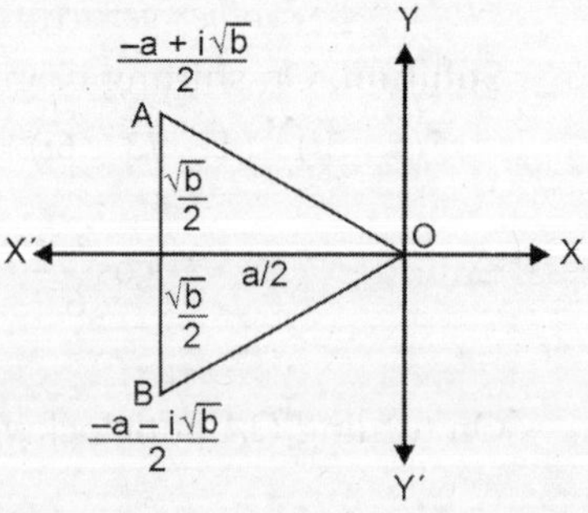

$$OA=\sqrt{\left(\frac{a}{2}\right)^2+\left(\frac{\sqrt{b}}{2}\right)^2}$$

$$=\sqrt{\frac{a^2+b}{4}}=\sqrt{\frac{3b+b}{4}}=\sqrt{b}$$

Similarly, $\quad OB=\sqrt{b}$

$$AB=\frac{\sqrt{b}}{2}+\frac{\sqrt{b}}{2}=\sqrt{b}$$

$$OA=OB=AB$$

Hence, they form an equilateral triangle. **Proved.**

Example 50. *Show that the equation of a circle on a line segment joining z_1 and z_2 as diameter is $|z-z_1|^2+|z-z_2|^2=|z_1-z_2|^2$*

Solution. Let AB be a diameter of the circle. Points A and B denote z_1 and z_2. Let P(z) be any point on the circumference of the circle.

APB is a rt. angled triangle $(AP)^2 + (BP)^2 = (AB)^2$

$$|z - z_1|^2 + |z - z_2|^2 = |z_1 - z_2|^2$$ **Proved.**

Example 51. *Show that the general equation of a real circle in a Argand Plane is of the form* $z\bar{z} + az + \bar{a}\bar{z} + b = 0$

Solution. $z\bar{z} + az + \bar{a}\bar{z} + b = 0$

$$\Rightarrow \quad (x + iy)(x - iy) + (a_1 + ia_2)(x + iy) + (a_1 - ia_2)(x - iy) + b = 0$$

$$\Rightarrow \quad x^2 + y^2 + a_1x + ia_1y + ia_2x - a_2y + a_1x - ia_1y - ia_2x - a_2y + b = 0$$

$$\Rightarrow \quad x^2 + y^2 + 2a_1x - 2a_2y + b = 0$$

This is the equation of circle. **Proved.**

Example 52. *What locus is represented by* $|z - 5 - 6i| = 4$?

Solution.
$$|z - 5 - 6i| = 4$$
$$|x + iy - 5 - 6i| = 4$$
$$\sqrt{(x-5)^2 + (y-6)^2} = 4$$
$$(x-5)^2 + (y-6)^2 = 4^2$$

It is a circle with centre (5, 6) and radius 4. **Ans.**

Example 53. *What region in z plane is represented by* $2 < |z - 4 - 5i| < 3$.

Solution. $|z - 4 - 5i| = 2$ represents a circle of radius 2 with centre at (4, 5).

$2 < |z - 4 - 5i|$ represents the exterior of the circle ... (1)

Similarly, $|z - 4 - 5i| = 3$ represents a circle of radius 3 with the same centre.

$|z - 4 - 5i| < 3$ represents R interior of the circle ... (2)

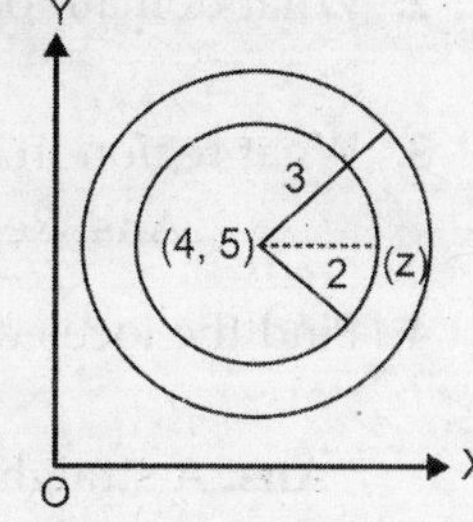

Hence, from (1) and (2).

$2 < |z - 4 - 5i| < 3$ represents the region lying between circles.

Example 54. *If* $|z_1 + z_2| = |z_1 - z_2|$. *Prove that the difference of the amplitudes of z and* z_2 *is* $\frac{\pi}{2}$.

Solution. Let $\quad z_1 = x_1 + iy_1, z_2 = x_2 + iy_2$

$$\Rightarrow \quad z_1 + z_2 = (x_1 + iy_1) + (x_2 + iy_2) = (x_1 + x_2) + i(y_1 + y_2)$$

$$\Rightarrow \quad z_1 - z_2 = (x_1 + iy_1) - (x_2 + iy_2) = (x_1 - x_2) + i(y_1 - y_2)$$

$$|z_1 + z_2|^2 = (x_1 + x_2)^2 + (y_1 + y_2)^2$$

$$|z_1 - z_2|^2 = (x_1 - x_2)^2 + (y_1 - y_2)^2$$

It is given that

$$|z_1 + z_2| = |z_1 - z_2|$$

$$(x_1 + x_2)^2 + (y_1 + y_2)^2 = (x_1 - x_2)^2 + (y_1 - y_2)^2$$

$$x_1^2 + x_2^2 + 2x_1x_2 + y_1^2 + y_2^2 + 2y_1y_2 = x_1^2 + x_2^2 - 2x_1x_2 + y_1^2 + y_2^2 - 2y_1y_2$$

$$\Rightarrow \quad 4x_1x_2 = -4y_1y_2$$

$$\Rightarrow \quad \left(\frac{y_1}{x_1}\right)\left(\frac{y_2}{x_2}\right) = -1 \qquad \text{... (1)}$$

Let α and β be the amplitudes of z_1 and z_2 and θ be the difference of their amplitudes.

$$\theta = \alpha - \beta$$

$$\tan\theta = \tan(\alpha - \beta)$$

$$= \frac{\tan\alpha - \tan\beta}{1 + \tan\alpha \cdot \tan\beta}$$

$$= \frac{\dfrac{y_1}{x_1} - \dfrac{y_2}{x_2}}{1 + \left(\dfrac{y_1}{x_1}\right)\left(\dfrac{y_2}{x_2}\right)} = \frac{\dfrac{y_1}{x_1} - \dfrac{y_2}{x_2}}{1 - 1} = \infty \qquad \text{[from (1)]}$$

$$\therefore \quad \theta = \frac{\pi}{2}$$

Proved.

EXERCISE 12.10

1. Find the locus of the point z if $|z - 2| = 3$

   Ans. Circle of radius 3 with centre (2, 0)

2. What domain of z plane is represented by $|z - 3| < 4$.

   Ans. Interior of the circle of radius 4 with centre (3, 0)

3. What region in Argand diagram is represented, by $1 < |z - 2 - i| < 2$.

   Ans. Region between two circles of radii 1 and 2 with the same centre (2, 1)

4. Find the locus of point z if arg. $(z - 4 - i) = \dfrac{\pi}{6}$

   Ans. A straight line passing through (4, 1) and making an angle of $\dfrac{\pi}{6}$ with x-axis.

5. What domain of z-plane is represented by $|z + 1| + |z - 1| < 3$.

   Ans. Interior of the ellipse having foci ± 1 and major axis 3 units.

6. Indicate graphically the set of points $z = x + iy$ which satisfy the following condition $|z + 1| = |z - 1|$.

   Ans. Imaginary axis.

7. Find the locus of a point z moving on Argand diagram so that $|z - z_1| = |z - z_2|$.

   Ans. The right bisector of the line joining the points z_1 and z_2

8. Find the locus of the point z when $\arg\left(\dfrac{z - a}{z - b}\right) = \alpha$.

   Ans. Circle passing through the points a and b making $\angle azb = \alpha$.

9. If the roots of $z^3 + iz^2 + 2i = 0$ represent vertices of a triangle in the Argand plane, then find area of the triangle. **Ans.** 2 units. **[Diploma I.E.T.E. Dec. 2006, June 2005]**

10. If the complex numbers z_1, z_2, z_3 be the vertices of an equilateral triangle, prove that
$z_1^2 + z_2^2 + z_3^2 = z_1 z_2 + z_2 z_3 + z_3 z_1$ **[Diploma IETE, June 2005]**

Choose the correct alternative

11. The complex numbers z_1, z_2 and z_3 satisfying $\dfrac{z_1 - z_2}{z_2 - z_3} = \dfrac{1 - i\sqrt{3}}{2}$ are vertices of the triangle which is

(*a*) acute-angled and isosceles (*b*) right-angled and isosceles

(*c*) obtuse-angled and isosceles (*d*) equilateral **Ans.** (*d*)

[Diploma IETE, June 2006]

12.15 PHASOR

Phasor is a uniformly rotating arrow which generates a sinusoidal function *A* sin ω*t*. The length of the arrow is taken equal to amplitude *A* of the sinusoidal wave. The regular velocity of the rotation of the arrow is taken equal to ω radians per sec. The time taken by the arrow for a complete rotation is $\dfrac{2\pi}{w}$ which is equal to the period of the wave.

The arrow starts from the horizontal axis. It makes an angle ω*t* with the horizontal axis in *t* seconds. The projection on the vertical axis at this moment is *A* sin ω*t*. As the time increases the vertical projection (*A* sin ω*t*) is the function as shown in the figure below:

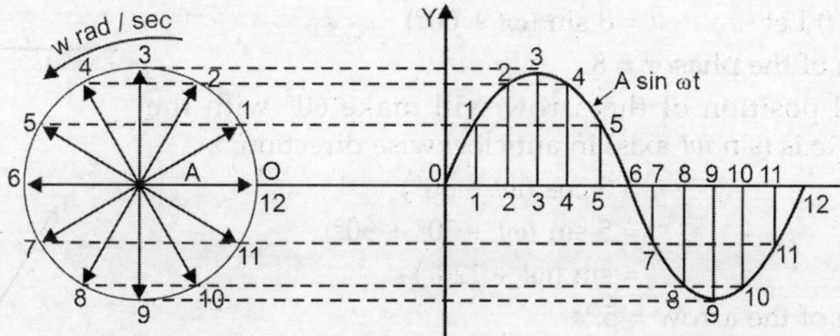

Thus the arrow (phasor) is a generator of the sinusoidal function *A* sin *wt*.

Example 55. *Construct the phasor diagram for the sinusoidal functions*

(*i*) *A* sin (ω*t* + α) (*ii*) 2 cos (ω*t* + 50°)

Solution. (*i*) $y = A \sin(\omega t + \alpha)$

ω rad /sec
Initial position
Y
S
O
t = 0
α
P

The phase angle α is the initial position of the phasor with respect to *t* = 0. The phasor will start from the initial position *PS*. The projection of *S* on *y*-axis is *O*.

So the corresponding point of *S* at *t* = 0 is *O* on the *y*-axis. The curve of *A* sin (ω*t* + α) will start from *O* in the first quadrant.

The length of the arrow is equal to *A*.

(*ii*) It will rotate with the angular velocity ω.

$$y = 2\cos(\omega t + 50°) = 2\sin(\omega t + 50° + 90°)$$
$$= 2\sin(\omega t + 140°)$$

The length of the arrow = 2, draw a circle of radius = 2

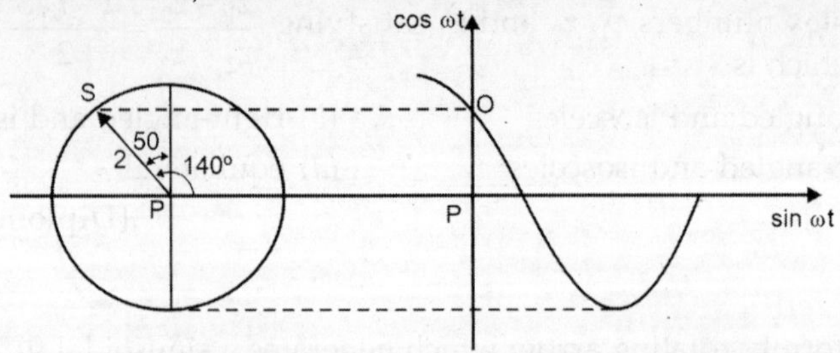

The phasor *PS* will make an angle of 140° with horizontal axis (sin ω*t* axis) or 50° with the vertical axis (cos ω*t* axis) in the initial position. The projection of the point *S* (tip of arrow) on *y*-axis is *O*. So the curve of 2 cos (ω*t* + 50°) will start from *O*.

Example 56. *Construct the phasors corresponding to the following functions.*

[SBTE. 2017, 2015]

(*i*) $8\sin(\omega t + 60°)$ (*ii*) $5\cos(\omega t + 30°)$

(*iii*) $-3\sin(\omega t + 45°)$ (*iv*) $-6\cos(\omega t - 40°)$

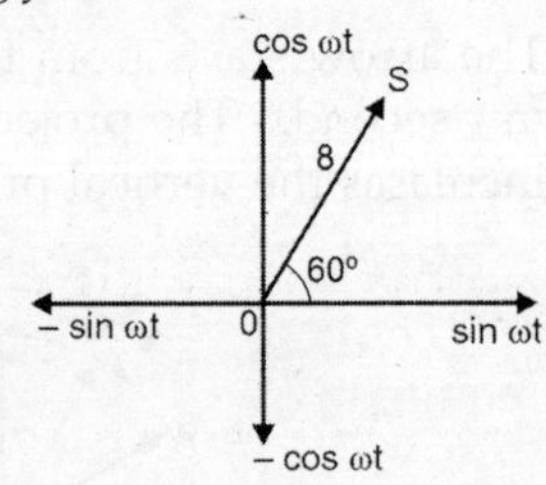

Solution. (*i*) Let $y = 8\sin(\omega t + 60°)$

The length of the phasor = 8.

The initial position of the arrow will make 60° with the horizontal axis (sin *wt* axis) in anticlockwise direction.

(*ii*) Let
$$y = 5\cos(\omega t + 30°)$$
$$= 5\sin(\omega t + 30° + 90°)$$
$$= \sin(\omega t + 120°)$$

The length of the arrow = 5.

The initial position of the arrow will make an angle of 120° with the sin ω*t* axis or 30° with the cos ω*t* axis in the anticlockwise direction.

(*iii*) $y = -3\sin(\omega t + 45°)$

The length of the arrow = 3

The phasor leads – sin ω*t* axis by 45°. Draw an arrow *OS* making an angle 45° with – sin ω*t* axis in positive direction.

Then *OS* is the required phasor.

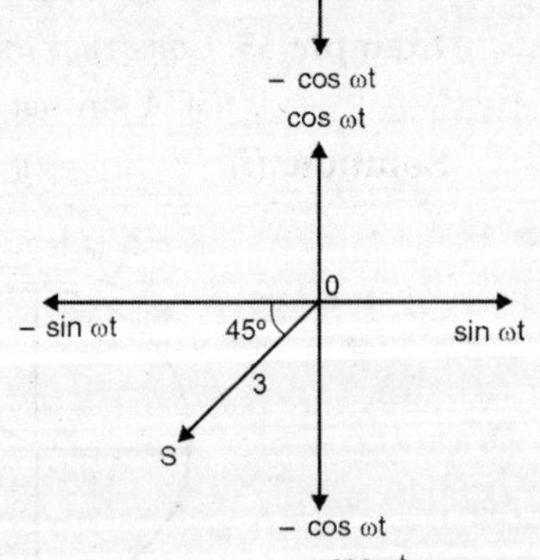

(*iv*) $y = -6\cos(\omega t - 40°)$

The length of the arrow = 6.

The phasor leads – cos ω*t* axis by 40°. Draw the line *OS* making an angle of 40° with – cos ω*t* axis clockwise.

Then *OS* is required phasor.

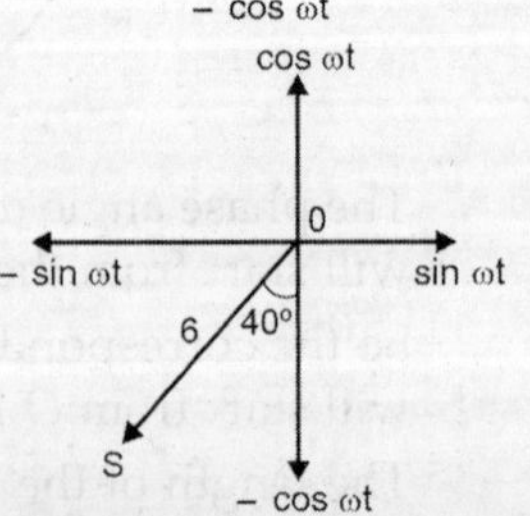

Example 57. *Construct the phasor diagrams for the following voltage and current:*

(*a*) $i = 3\sin(5000t - 15°)$

(*b*) $v = 150\sin(5000t + 45°)$

Solution. (*a*) $i = 3 \sin (5000t - 15°)$

The length of the phasor = 3

The phasor lags the sin ωt axis by 15°.

Draw a line *OS* below the sin ωt axis making an angle of 15° with it.

The phasor will rotate with 5000 radians per sec.

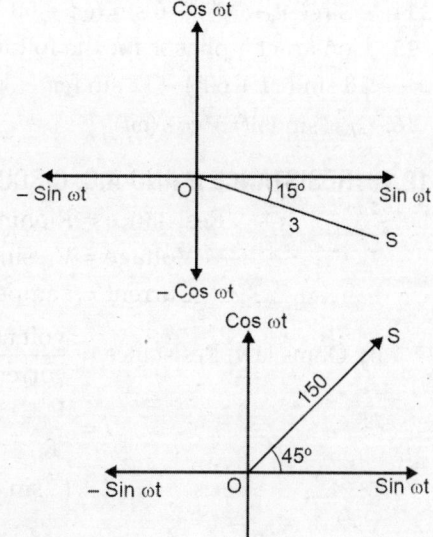

(*b*) $v = 150 \sin (5000t + 45°)$

The length of the phasor *OS* = 150.

The phasor *OS* leads the sin ωt axis by 45°.

Draw a line *OS* making 45° with the sin ωt axis in the positive direction. The phasor will rotate with 5000 radians per second.

Example 58. *Construct a phasor for the function*

$$4 \sin (\omega t + 20°) + 2 \sin (\omega t + 70°)$$

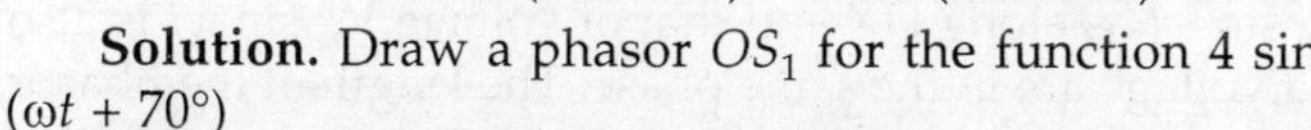

Solution. Draw a phasor OS_1 for the function 4 sin $(\omega t + 70°)$

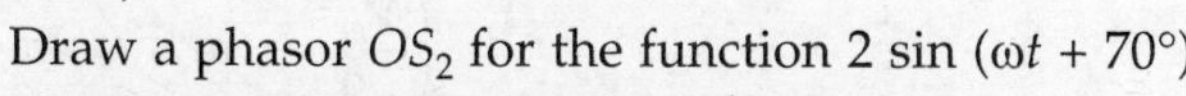

Draw a phasor OS_2 for the function 2 sin $(\omega t + 70°)$

Complete the parallelogram $OS_1\ RS_2$.

The diagonal of the parallelogram, *OR* is the required phasor. Horizontal component:

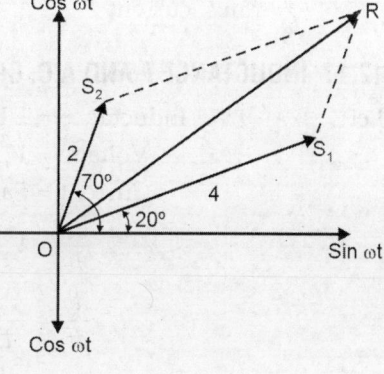

$$= 4 \cos 20° + 2 \cos 70°$$
$$= (4 \times 0.9397) + (2 \times 0.3420)$$
$$= 4.4428$$

Vertical component $= 4 \sin 20° + 2 \sin 70°$

$$= (4 \times 0.3420) + (2 \times 0.9397)$$
$$= 3.2474$$

$$OR = \sqrt{(4.4428)^2 + (3.2474)^2} = 5.503$$

The angle of the lead from sin ωt axis

$$= \tan^{-1}\left(\frac{3.2474}{4.4428}\right) = 36°10'$$

$\therefore\quad 4 \sin (\omega t + 20°) + 2 \sin (\omega t + 70°) = 5.503 \sin (\omega t + 36°10')$ **Ans.**

EXERCISE 12.11

Draw the phasor diagrams corresponding to the following functions:

1. $10 \sin (\omega t + 45°)$ **2.** $-5 \cos (\omega t - 30°)$ **3.** $10 \cos (\omega t + 60°)$

4. $-5 \sin (\omega t - 20°)$ **5.** $3 \sin (\omega t - 60°)$ **6.** $-3 \sin (\omega t - 40°)$

7. $5 \cos (\omega t - 60°)$ **8.** $-4 \cos (\omega t + 50°)$ **9.** $-7 \cos (\omega t - 45°)$

10. $4 \sin (\omega t + 10°)$ **[S.B.T.E. 2014, 2015]**

Determine a phasor for the follwing:

11. $10 \sin \omega t + 5 \cos \omega t$ **[B.T.E., May/June 2007]** **Ans.** $5\sqrt{5} \sin(\omega t + 26°34')$

12. $10 \sin (\omega t + 60°) + 8 \sin (\omega t + 30°)$ **Ans.** $18.8\ (\omega t + 41°47')$

13. $10 \cos (\omega t + 40°) - 15 \cos (\omega t - 60°)$ **Ans.** $19.42 \cos (\omega t + 89°30')$

14. $4 \sin(\omega t + 30°) + 6 \sin(\omega t + 60°)$ **Ans.** $9.673 \sin(\omega t + 48°4')$

15. Construct a phasor for the following functions:
$10 \sin(\omega t + 60°) - 12 \sin(\omega t - 30°)$ **Ans.** $17.115 \sin(\omega t + 83°16')$

16. $\sqrt{3} \sin(\omega t) + \cos(\omega t)$ **Ans.** $2 \sin(\omega t + 30°)$

12.16 RESISTANCE *R* AND A.C. CIRCUIT

Let Resistance = R ohm

Voltage = $V_m \sin \omega t$ volt

Current = i ampere

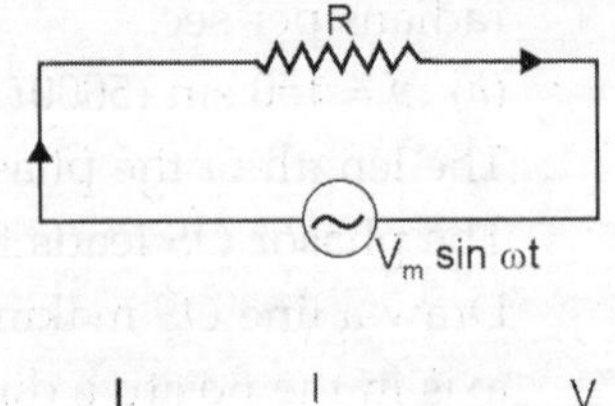

By Ohms law, Resistance = $\dfrac{\text{voltage}}{\text{current}}$

$$\Rightarrow \quad i = \frac{V_m}{R}$$

$$i = I_m \sin wt, \text{ where } I_m = \frac{V_m}{R}$$

Phasor diagram of current $I_m \sin wt$ as along OX and that of voltage $V_m \sin wt$ is also along OX. Hence, current and voltage are in the same phase. The length of the phasor

$= \dfrac{I_m}{\sqrt{2}}$ r.m.s. current $\left(\dfrac{V_m}{\sqrt{2}} = \text{r.m.s. voltage}\right)$

12.17 INDUCTANCE *L* AND A.C. CIRCUIT

Let Inductance = L henry

Voltage = $V_m \sin \omega t$ voltage

Current = i ampere

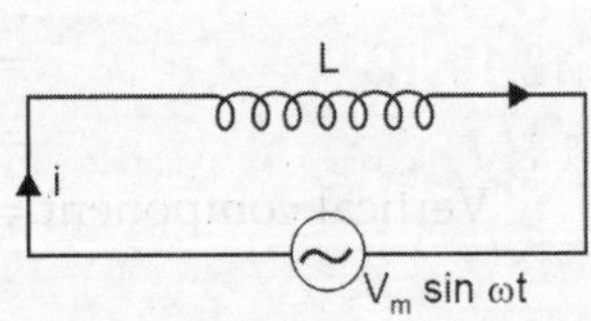

$$L \frac{di}{dt} = V_m \sin \omega t$$

$$\Rightarrow \quad \frac{di}{dt} = \frac{V_m}{L} \sin \omega t$$

Integrating $i = -\dfrac{V_m}{L\omega} \cos \omega t$ (No constant of integration in steady state)

$$\Rightarrow \quad i = -L_m \cos \omega t, \text{ where } I_m = \frac{V_m}{L\omega}.$$

The phasor of voltage is along $\sin \omega t$ axis, and the phasor of current i is along $-\cos \omega t$ axis.

Thus the current lags the voltage by $\dfrac{\pi}{2}$.

The position of Lw in the formula $I_m = \dfrac{V_m}{L\omega}$ is the same as that of resistance in the formula $I = \dfrac{V}{R}$

Lw is called *inductive reactance* of the coil. Its unit is ohm.

12.18 CAPACITORS *C* AND A.C. CIRCUIT

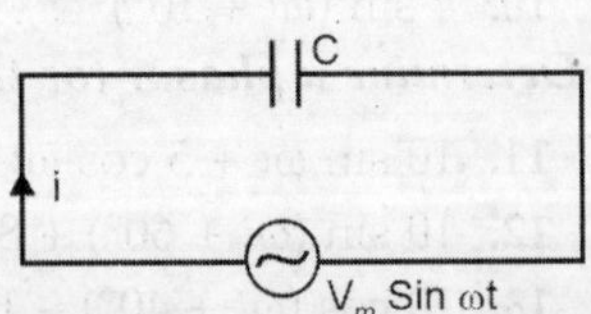

Let Capacitance = C farad

Charge on capacity = q coulomb

$$\text{Voltage} = V_m \sin \omega t \text{ volt}$$

$$\therefore \quad \frac{q}{C} = V_m \sin \omega t$$

$$\Rightarrow \quad q = c\, V_m \sin \omega t$$

Differentiating w.r.t. 't', we get

$$\frac{dq}{dt} = \omega c\, V_m \cos \omega t$$

$$\Rightarrow \quad i = \frac{V_m}{\frac{1}{\omega c}} \cdot \cos \omega t$$

$$i = I_m \cos \omega t \quad \text{where } I_m = \left(\frac{V_m}{\frac{1}{\omega c}}\right)$$

The phasor of $V_m \sin \omega t$ is along $\sin \omega t$ axis.

The phasor of i ($I_m \cos \omega t$) is along $\cos \omega t$ axis.

Thus the current leads the voltage by $\frac{\pi}{2}$.

The position of $\frac{1}{\omega c}$ in the formula $I_m = \frac{V_m}{\frac{1}{wc}}$ is the same as that of R (resistance) in the formula $I = \frac{V}{R}$.

$\frac{1}{\omega c}$ is called capacitive reactance X_c. Its unit is ohm.

12.19 R – C AND A.C. CIRCUIT

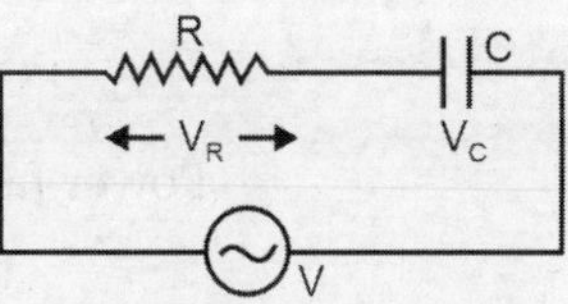

Let I be the current that flows through R and C.

Let voltage drop across $R = V_R$

$$V_R = IR$$

Voltage drop across $C = V_C$

$$\therefore \quad V_C = I \cdot X_C \qquad \text{[where } X_c = \text{capacitive reactance]}$$

Let us draw the phasor diagram

Since I is common, so I is taken as the reference phasor.

V_C has I by $\frac{\pi}{2}$ (By article 12.18)

V_R is in phase with I. (By article 12.15)

Let V be the voltage of the source, so voltage phasor V is represented by the result of the phasor V_C and V_R.

$$V^2 = V_R^2 + V_C^2$$

$$V = \sqrt{V_R^2 + V_C^2}$$

$$= \sqrt{(IR)^2 + (IX_C)^2}$$

$$= I\sqrt{R^2 + X_C^2}$$

$$\tan \theta = \frac{V_C}{V_R}$$

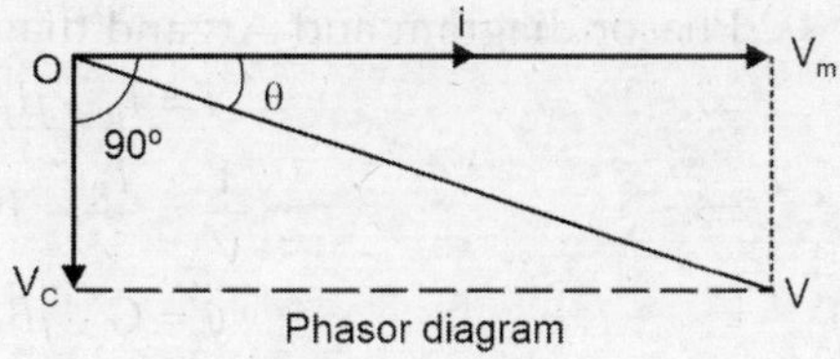

Phasor diagram

$$\therefore \qquad \theta = \tan^{-1}\left(\frac{V_C}{V_R}\right) = \tan^{-1}\left(\frac{X_C}{R}\right)$$

The right angled triangle formed by

$\frac{V_R}{I}, \frac{V_C}{I}, \frac{V}{I}$ as shown in the figure is known as impedance triangle.

Impedance z is given $|z| = \sqrt{R^2 + X_C^2}$ and $\tan\theta = \frac{X_C}{R}$.

12.20 IMPEDANCE

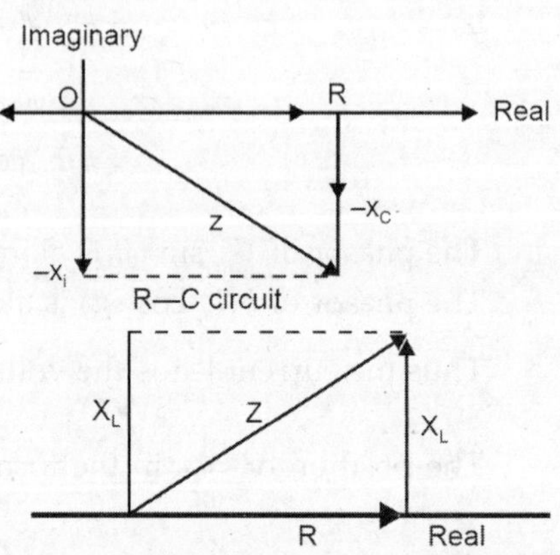

As we have seen in the last article impedance is a phasor. Phasor diagram and Argand diagram are similar.

Horizontal axis is called real axis in Argand diagram and V_R or R-axis in phasor diagram. Vertical axis is imaginary axis in Argand diagram and V_C or X_C axis in phasor diagram.

Let us consider the figure (*a*)

$$z = R - j\,XC$$

where $$j = \sqrt{-1}$$

Hence, impedance is a complex number.

In R and A.C. circuit, figure (*b*) $z = R + j\,X_L$

Similarly $R - L - C$ and A.C. circuit, figure (*c*)

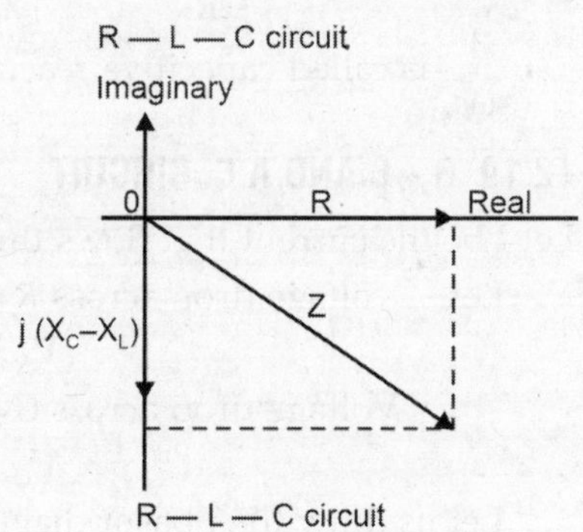

$$z = R + j\,X_L - j\,X_C$$

$$\text{Power} = I^2 \cdot R \text{ watts}$$

$$\text{Power factor} = \frac{\text{Resistance}}{\text{Impedance}}$$

$$\Rightarrow \qquad = \frac{R}{|Z|}$$

12.21 R – L IN PARALLEL CIRCUIT

Let resistance R and inductance L be connected in parallel. The voltage V across resistance R and inductance L is the same. So the phasor V is the reference phasor. Current phasor I_R is in phase with voltage V. But current phasor I_L phasor lags voltage V by $\frac{\pi}{2}$.

Let the total current phasor I be the resultant of phasors I_R and I_L.

Phasor diagram and Argand diagrams are similar

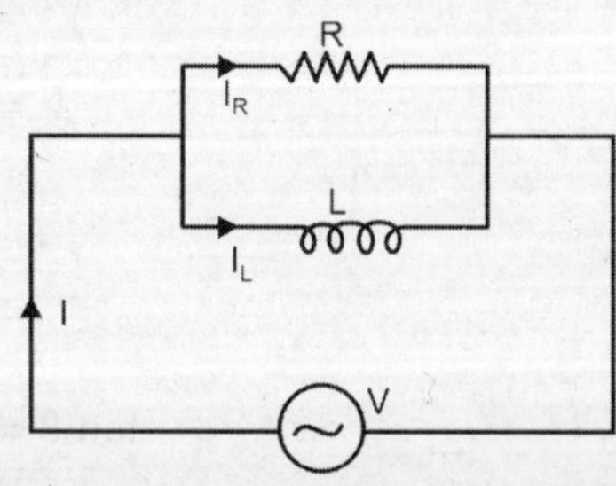

$$\Rightarrow \qquad I = I_R - jI_L$$

$$\Rightarrow \qquad \frac{1}{V} = \frac{I_R}{V} - j\frac{I_L}{V}$$

$$\Rightarrow \qquad y = G - jB_L$$

y is a complex number and is called admittance of the circuit. As admittance is reciprocal of impedance, its unit is mho.

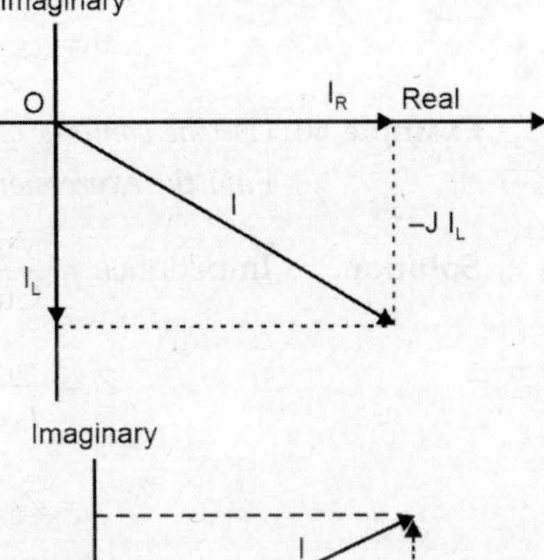

G is the reciprocal of the resistance and is called the *conductance.* Its unit is also mho. B_L is the reciprocal of inductive reactance and is called *inductive susceptance.* Its unit is also mho.

If we consider a circuit containing R and C in parallel. Phasor I_R is along phasor V and phasor I_C leads phasor V by $\frac{\pi}{2}$.

$$I = I_R + jI_C$$

$$\Rightarrow \quad \frac{I}{V} = \frac{I_R}{V} + j\frac{I_C}{V}$$

$$\Rightarrow \quad y = G + jB_C$$

Let us consider a parallel circuit containing R, L and C.

$$y = G - j\,(B_L - B_C)$$

If $\quad B_L > B_C$.

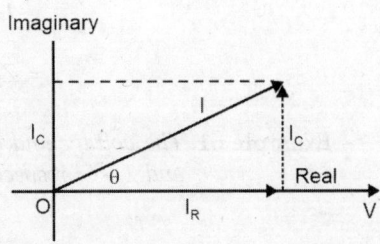

Example 59. *State in symbolic notation, the impedance of each of the following circuits at frequency of 50 cycles:*

(i) Resistance of 20 Ω in series with an inductance of 0.1 H.

(ii) A resistance of 50 Ω in series with a capacitance of 40 μF.

If the terminal voltage is 230 volts, find the value of current in each case and phase of each current relative to the applied voltage.

Solution. (*i*)

$$z = R + jLw$$
$$= 20 + j\,(0.1)\,(100\pi)$$
$$= 20 + j\,10\pi$$
$$w = 2\pi f = 2\pi \times 50 = 100\pi$$
$$i = \frac{V}{Z}$$
$$= \frac{230}{20 + j10\pi} = \frac{23}{2 + j\pi} = \frac{23(2 - j\pi)}{(2 + j\pi)\,(2 - j\pi)}$$
$$= \frac{23(2 - j\pi)}{4 + \pi^2} = \frac{23}{13.87}(2 - j\pi) = 1.66\,(2 - j\pi) = 3.32 - j\,5.22$$
$$\theta = \tan^{-1}\left(\frac{-5.22}{3.32}\right) = \tan^{-1}(-1.57) = -57.5° \qquad \textbf{Ans.}$$

(*ii*)

$$Z = R - j\frac{1}{cw}$$
$$= 50 - j\frac{1}{40 \times 10^{-6} \times 100\pi} = 50 - j\frac{250}{\pi} = 50 - j\,79.58$$
$$i = \frac{230}{50 - j79.58} = \frac{230(50 + j79.58)}{(50 - j79.58)\,(50 + j79.58)}$$
$$= \frac{230(50 + j79.58)}{8833} = 0.026\,(50 + j79.58) = 1.30 + j\,2.07$$

$$\theta = \tan^{-1}\left(\frac{2.07}{1.3}\right) = \tan^{-1}(1.592) = 57.9° \qquad \textbf{Ans.}$$

Example 60. *Use the complex current $I = I_0\, e^{i(wt+\alpha)}$ and voltage $V = V_0\, e^{i(wt+\beta)}$ Find the expression for impedance in a R – L circuit.*

Solution. Impedance $= \dfrac{\text{Voltage}}{\text{Current}}$

$$Z = \frac{V_0\, e^{i(wt+\beta)}}{I_0\, e^{i(wt+\alpha)}} = \frac{V_0}{I_0}\, e^{i(\beta-\alpha)}$$

$$= Z_0\, e^{i(\beta-\alpha)} \qquad \left(\text{where } Z_0 = \frac{V_0}{I_0}\right) \qquad \textbf{Ans.}$$

Example 61. *The voltage and current of a circuit are given by the complex number 3 + 4j and 2 – 5j respectively. Find the complex number of the impedance the circuit.* **[SBTE 2015, 2014]**

Solution. Impedance $= \dfrac{\text{Voltage}}{\text{Current}} = \dfrac{3+4j}{2-5j} = \dfrac{3+4j}{2-5j} \times \dfrac{2+5j}{2+5j}$

$$= \frac{6+15j+8j-20}{4+25} = \frac{-14+23j}{29}$$

$$= \frac{-14}{29} + \frac{23}{29}j \qquad \textbf{Ans.}$$

Example 62. *The admittance and current of a circuit are given by the complex numbers 7 + 5j and 17 – 6j respectively. Find the voltage of the circuit.*

Solution. Admittance $= 7 + j\,5$

Current $= 17 - 6\,j$

$$\text{Voltage} = \frac{\text{Current}}{\text{Admittance}} = \frac{17-j6}{7+j5}$$

$$= \frac{(17-j6)(7-j5)}{(7+j6)(7-j5)}$$

$$= \frac{119-j85-j42-30}{49+25} = \frac{89-j127}{74}$$

$$= \frac{89}{74} - j\frac{127}{74} \qquad \textbf{Ans.}$$

Example 63. *The current in a circuit is 10 – j2 amperes when the voltage across the circuit is 60 + j 20 volts. Find the magnitude of*

(i) admittance; and (ii) the current of the circuit

Solution. (*i*) Admittance $= \dfrac{\text{Current}}{\text{Voltage}} = \dfrac{10-j2}{60+j20}$

$$= \frac{5-j}{30+j10} \times \frac{30-j10}{30-j10} = \frac{140-j80}{900+100}$$

$$= 0.14 - j\,0.08$$

(*ii*) $$\text{Current} = 10 - j\,2 = \sqrt{104}\,\angle \tan^{-1}\left(-\frac{1}{5}\right)$$

Magnitude of current = $\sqrt{104}$ amperes. **Ans.**

Example 64: *Two impedance $z_1 = 8 + j\,6$ ohms and $z_2 = 6 - j\,8$ ohms are connected in parallel across 200 volts, 50 cycles per sec. A.C. mains. Calculate the magnitude of current in each branch and the total currernt in the circuit by symbolic method.*

[SBTE 2008, 2007]

Solution.

$$z_1 = 8 + j\,6$$

$$z_2 = 6 - j\,8$$

$$i_1 = \frac{V}{z_1} = \frac{200}{8 + j6} = 4(4 - j3)$$

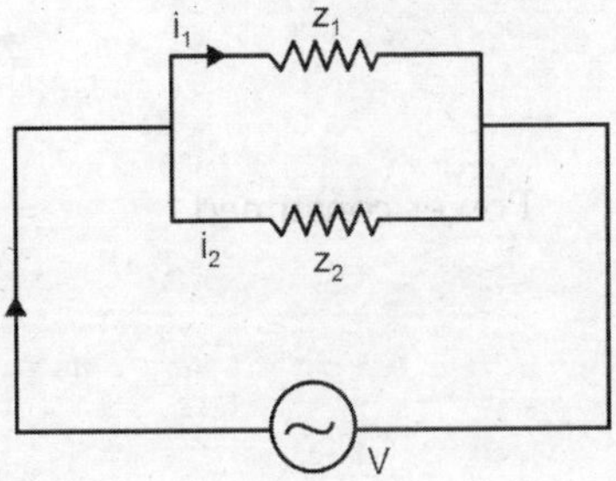

Magnitude of $i_1 = 4\sqrt{16 + 9} = 20$ amperes

$$i_2 = \frac{V}{z_2} = \frac{200}{6 - j8} = 4(3 + j4)$$

Magnitude of $i_2 = 4\sqrt{9 + 16} = 20$ amperes

Total current $= i_1 + i_2 = (16 - j\,12) + (12 + j\,16) = (28 + j\,4)$

Magnitude of the total $= \sqrt{28^2 + 4^2} = 20\sqrt{2}$ amperes **Ans.**

Example 65. *Two circuits of impedances $2 + j\,4$ ohms and $3 + j\,4$ ohms are connected in parallel and a.c. voltage of 100 volts is applied across the parallel combination. Calculate the magnitude of the currents as well as power factor for each circuit and the magnitude of the total current for the parallel combination and its power factor.*

Solution. $z_1 = 2 + j\,4,\ \ z_2 = 3 + j\,4$

$$i_1 = \frac{V}{z_1} = \frac{100}{2 + j4} = \frac{50}{1 + j2} = \frac{50(1 - j2)}{(1 + j2)(1 - j2)}$$

$$= \frac{50(1 - j2)}{1 + 4} = 10(1 - j2) = 10 - j20$$

Magnitude of $i_1 = \sqrt{100 + 400} = 10\sqrt{5}$ ampere

Power factor $= \dfrac{R_1}{|z_1|} = \dfrac{2}{\sqrt{20}} = \dfrac{1}{\sqrt{5}} = 0.447$

and $$i_2 = \frac{V}{z_2} = \frac{100}{3 + j4} = \frac{100(3 - j4)}{(3 + j4)(3 - j4)} = \frac{100(3 - j4)}{9 + 16}$$

$$= 4\,(3 - j\,4) = 12 - 16\,j$$

Magnitude of $i_2 = \sqrt{144 + 256} = 20$ amperes

Power factor $= \dfrac{R_2}{|z_2|} = \dfrac{3}{5} = 0.6$

$$i = i_1 + i_2 = (10 - j\,20) + (12 - 16\,j) = 22 - 36\,j$$

Magnitude $= \sqrt{484 + 1296} = 2\sqrt{445}$ **Ans.**

Example 66. *The impedance $z_1 = 10 - j\,60$ and $z_2 = 10 + j\,20$ are connected in parallel across a 200 volts a.c. supply. Calculate:*

(i) Currents in each branch and the total current… along with

(ii) power consumed in each branch

Solution. $z_1 = 10 - j\,60$ and $z_2 = 10 + j\,20$

$$i_1 = \frac{V}{z_1} = \frac{200}{10 - j60} = \frac{20}{1 - j6} = \frac{20(1 + j6)}{(1 - j6)(1 + j6)}$$

$$= \frac{20(1 + 6j)}{1 + 36} = \frac{20}{37} + j\frac{120}{37}$$

Power consumed $= i_1^2\,R_1 = \dfrac{1}{(37)^2}(400 + 14400)\,10 = \dfrac{148000}{1369} = 108.1$ watts

$$i_2 = \frac{V}{z_2} = \frac{200}{10 + j20} = \frac{20}{1 + j2} = \frac{20(1 - j2)}{(1 + j2)(1 - j2)}$$

$$= \frac{20(1 - j2)}{1 + 4} = 4(1 - j2) = 4 - j8$$

Power consumed $= i_2^{\,2}\,R_2$

$= (16 + 64)\,10 = 800$ watts

$$i = i_1 + i_2 = \left(\frac{20}{37} + j\frac{120}{37}\right) + (4 - j8)$$

$$= \frac{168}{37} - j\frac{176}{37}$$ **Ans.**

Example 67. *A resistance of 20 ohms, and inductance of 0.2 H and a capacitance of 100 μF are connected in series across 220 V and 50 cycles/sec. mains. Determine*

(a) impedance *(b) current* *(c) voltage across L, R and C*

(d) power in watt *(e) power factor*

Solution. Resistance (R) = 20 ohms   Voltage = 200 V

Inductance (L) = 0.2 H   Frequency = 50 cycles/sec

Capacitance (C) = 100 μF

(a) Impedance (Z) $= R - jX_C + jZ_L$

$$= 20 - j\frac{1}{\omega c} + j\,L_\omega$$

$$= 20 - j\frac{1}{50 \times 2\pi \times 100 \times 10^{-6}} + j(0.2) \times 2\pi \times 50$$

$$= 20 - j\frac{100}{\pi} + j\,20\pi$$

$= 20 - j\,31.831 + j\,62.8319$

$= 20 + j\,31$

$$|Z| = \sqrt{400 + 961} = \sqrt{1361} = 36.89 \text{ ohms}$$

(*b*) $$i = \frac{V}{|Z|} = \frac{200}{36.89} = 5.42 \text{ ampere}$$

(*c*) $$V_L = i\,X_L = 5.42 \times 2\pi f \times 0.2 = 5.42 \times 2 \times \pi \times 50 \times 0.2$$
$$= 340.55 \text{ volts}$$
$$V_R = iR = 5.42 \times 20 = 108.4 \text{ volts}$$
$$V_C = i \times \frac{1}{2\pi f\, X_C}$$
$$= \frac{5.42 \times 1}{2 \times \pi \times 50 \times 100 \times 10^{-6}} = 172.52 \text{ volts}$$

(*d*) $$\text{Power} = i^2 R = (5.42)^2 \times 20 = 57.828 \text{ watts}$$

(*e*) $$\text{Power factor} = \frac{R}{|Z|} = \frac{20}{36.89} = 0.542$$ **Ans.**

Example 68. *A sinusoidal voltage of r.m.s. value 100 volts and frequency 50 cycles per second (Hz) is applied to the circuit which has two resistances of 3 ohms and 4 ohms, two inductances of 0.0159 H and 0.0477 H and a capacitor of 318 μF. Calculate the current in the circuit and find its phase relative of that of the applied voltage.*

Solution. Voltage (v) = 100 volts, Freuency (f) = 50 hertz

Resistance (R_1) = 3 ohms, Resistance (R_2) = 4 ohms

Inductance (L_1) = 0.0159 H, Inductance (L_2) = 0.0477 H

Capacitance (C) = 318 μF

$$Z = R - jX_C + jX_L$$
$$Z = (R_1 + R_2) - j\frac{1}{wc} + j\,Lw$$
$$= (3+4) - j\frac{1}{50 \times 2\pi \times 318 \times 10^{-6}} + j\,(0.0159 + 0.0477)\, 50 \times 2\pi$$
$$= 7 - j\frac{1000}{318\pi} + j0.0636 \times 100\pi = 7 - j\,10.0097 + j\,19.981$$
$$= 7 + j\,9.9713$$
$$i = \frac{V}{Z} = \frac{100}{7 + j9.9713}$$
$$= \frac{100\,(7 - j9.9713)}{(7 + j9.9713)\,(7 - j9.9713)}$$
$$= \frac{100(7 - j9.9713)}{49 + 99.4268}$$
$$= \frac{700}{148.4268} - j\frac{997.13}{148.4268}$$
$$= 4.716 - j\,6.718$$

$$i = \sqrt{(4.716)^2 + (6.718)^2}$$

$$= \sqrt{22.241 + 45.132}$$

$$= \sqrt{67.373}$$

$$= 8.21$$

$$\theta = \tan^{-1}\left(\frac{-6.718}{4.716}\right) = \tan^{-1}(-1.4245)$$

$$= -54.93^\circ$$ **Ans.**

Example 69. *A coil of resistance 4 ohms and inductive reactance 4 ohms is connected in parallel with a resistance 15 ohms and capacitive reactance of 18 ohms. This parallel circuit is connected across 220 V mains. Find (a) the current taken by each circuit (b) and the total current (c) the power factor.*

Solution. $R_1 = 4$ ohms, $L_1 = 42$ ohms

$R_2 = 15$ ohms, $C_1 = 18$ ohms

$Z_1 = 4 + j\,42,$ $Z_2 = 15 - j\,18$

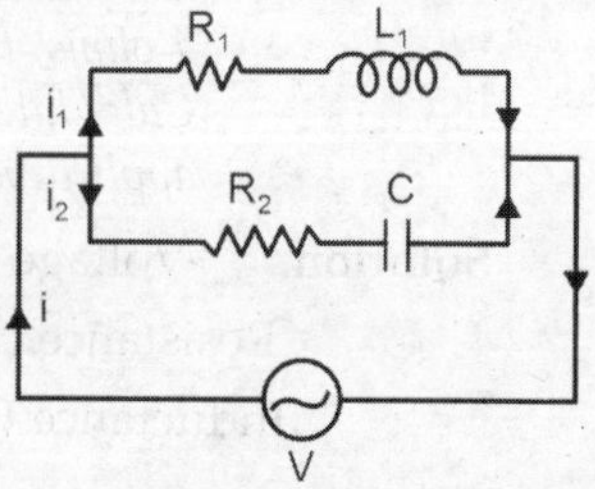

(a)

$$i_1 = \frac{V}{Z_1}$$

$$= \frac{220}{4 + j42} = \frac{110}{2 + j21}$$

$$= \frac{110(2 - j21)}{(2 + j21)(2 - j21)}$$

$$= \frac{110}{445}(2 - j21) = \frac{22}{89}(2 - j21)$$

$$= \frac{84}{89} - j\frac{462}{89} = 0.5 - j5.19$$

$$|i_1| = \sqrt{0.25 + 26.9361} = \sqrt{27.1861} = 5.21$$

$$i_2 = \frac{V}{Z_2}$$

$$= \frac{220}{15 - j18} = \frac{220(15 + j18)}{(15 - j18)(15 + j18)} = \frac{220}{549}(15 + j18)$$

$$= \frac{220}{183}(5 + j6) = \frac{1100}{183} + j\frac{1320}{183}$$

$$= 6 + j\,7.21$$

$$|i_2| = \sqrt{36 + 51.9841} = \sqrt{87.9841} = 9.38$$

(b)

$$i = i_1 + i_2$$

$$= (0.5 - j\,5.19) + (6 + j\,7.21)$$

$$= 6.5 + j\,2.02$$

$$|i| = \sqrt{42.25 + 4.0804} = \sqrt{46.3304} = 6.81$$

$$Z = \frac{Z_1 Z_2}{Z_1 + Z_2}$$

$$= \frac{(4 + j42)(15 - j18)}{(4 + j42) + (15 - j18)}$$

$$= \frac{816 + j558}{19 + i24} = \frac{(816 + j55)(19 - j24)}{(19 + j24)(19 - j24)}$$

$$= \frac{28896}{937} - j\frac{8982}{937} = 30.84 - j9.59$$

$$|z| = \sqrt{951.11 + 91.97} = \sqrt{1043.08} = 32.30$$

(*c*) $$\text{Power factor} = \frac{R}{|z|} = \frac{30.84}{32.30} = 0.95$$ **Ans.**

Example 70. *A resistance of 400 ohms, inductance of 2 H and a capacitance of 10 μF are connected in parallel. Find an equivalent series circuit containing pure resistance, and capacitance, at a frequency of 50 Hz.*

Solution. $R \simeq 400$ ohms, $L = 2\text{H}$, $C = 10\mu F$.

Let Z be the resultant impedance on connecting R, L and C in parallel.

$$\frac{1}{Z} = \frac{1}{R} + \frac{1}{jLw} + \frac{1}{-\frac{j}{cw}}$$

$$= \frac{1}{400} + \frac{1}{j2 \times 100\pi} - \frac{1}{\dfrac{j}{10 \times 10^6 \times 100\pi}}$$

$$= \frac{1}{400} - \frac{j}{200\pi} + \frac{j\pi}{1000}$$

$$= \frac{1}{400} - j\left(\frac{1}{200\pi} - \frac{\pi}{1000}\right)$$

$$= \frac{1}{400} - j(0.0015915 - 0.0031416)$$

$$= 0.0025 + j\,0.0015501$$

$$Z = \frac{1}{0.0025 + j0.0015501}$$

$$= \frac{1}{0.0025 + j0.0015501} \times \frac{0.0025 - j0.0015501}{0.0025 - j0.0015501}$$

$$= \frac{0.0025 - j0.0015501}{0.00000625 - j0.00000240}$$

$$= \frac{0.0025}{0.00000865} - j\frac{0.0015501}{0.00000865}$$

$$= 289.017 - j\,179.202$$

Equivalent series circuit containing pure resistance R and capacitance C.

$$R - \frac{j}{Cw} = 289.017 - j\,179.202$$

i.e.

$$R = 289.017$$

$$\frac{-j}{C \times 100\pi} = -j\,179.202$$

$$C = \frac{1}{100\pi \times 179.202} = 17.76 \times 10^{-6}$$

$$= 17.76\mu F$$

$$\left.\begin{aligned} \text{Pure resistance} &= 289.017 \text{ ohms} \\ \text{Capacitance} &= 17.76\ \mu F \end{aligned}\right\}$$ **Ans.**

EXERCISE 12.12

1. The voltage and current of a circuit are given by the complex numbers $70 + j\,20$ and $20 - j6$ respectively.
   Find the complex numbers representing
   (*i*) Admittance (*ii*) Impedance of the circuit.
   Ans. (*i*) $0.2415 - j\,0.1547$ (*ii*) $2.9358 + j\,1.8807$
2. The voltage and current of a circuit are given by the complex numbers $(3 + j\,2)$ and $(4 - j\,5)$ respectively. Find the complex numbers representing.
   (*i*) The impedance (*ii*) Admittance of the circuit
   Ans. (*i*) $0.04878 + j\,0.56098$ (*ii*) $0.15385 - j\,1.76923$
3. The current in a circuit is $10 - j\,3.5$ A. If the e.m.f. is $100 \angle 0°\ V$, find the admittance.
   Ans. $0.1 - j\,0.35$ mho
4. The e.m.f. in a series circuit is $100 + j\,50$ V. If the current is $10 \angle 0°\ A$, find the impedance in rectangular form, in polar form. **Ans.** $10 + j\,5$, $5\sqrt{5} \angle 26°\,34'$
5. The current in a circuit $20 - j\,6$ amperes, when voltage across the circuit is $70 + j\,20$ volts. Find the magnitude of
   (*i*) Admittance (*ii*) The current of the circuit.
   Ans. (*i*) 0.2858 mho (*ii*) 20.8806 amp.
6. The voltage and current of a circuit are given as $60 - j\,20$ volts and $12 - j\,3$ amperes respectively. Find
   (*i*) Magnitude of current (*ii*) Admittance (*iii*) Impedance.
   Ans. (*i*) 12.3693 amp. (*ii*) $0.195 + j\,0.015$ mho (*iii*) $5.098 - j\,0.392$
7. If two impedance $z_1 = 50 \angle -40°$ and $z_2 = 70 \angle 30°$ are connected in series, find the total impedance in polar form. **Ans.** $98.96 \angle 1.656°$ ohm
8. Two impedances $20 - j\,34.4$ and $89.6 + j\,50$ are connected in parallel. Find the resultant impedance. **Ans.** $29.04 - j\,22.76$
9. Two impedances $z_1 = 10 + 6\,j$ and $z_2 = 8 - 12\,j$ are connected in parallel across 200 volts, 50 cycles per sec. A.C. mains. Calculate the magnitude of the current in each branch and total current in the circuit.
   Ans. $17.132 \angle 149.02°$, $13.8675 \angle 56.26°$, $22.5531 \angle 6.9°$

Unit II: TRIGONOMETRY

Introduction to Trigonometry

13.1 INTRODUCTION

Trigonometry is a word derived from three Greek words.

Tri – Three
gonia – angle   } *i.e.* triangle measure.
metron – measure

Thus, trigonometry literally means measurements of triangles. It is considered to be that branch of mathematics which deals with the measurement of angles and the problems related with angles. In earlier times, this branch of mathematics was studied by sea captains for navigation, surveyors to map the lands and by engineers, etc. Now-a-days, trigonometry is used in science and technology.

In this chapter, we shall first review trigonometric ratios for acute angles, which are already known to us. Then we shall learn the trigonometric ratios to the angles more than 90° and study them as trigonometric functions.

13.2 TRIGONOMETRIC RATIOS OR FUNCTIONS (CIRCULAR FUNCTIONS)

Let a rotating ray starts from initial position *OA* and move about *O* and trace out an angle $\angle AOB = \theta$. Consider the right angled triangle *OAB*.

$$\sin\theta = \frac{\text{opposite side}}{\text{hypotenuse}} \quad \text{or} \quad \sin\theta = \frac{\text{perpendicular}}{\text{hypotenuse}}$$

$$\cos\theta = \frac{\text{adjacent side}}{\text{hypotenuse}} \quad \text{or} \quad \cos\theta = \frac{\text{base}}{\text{hypotenuse}}$$

$$\tan\theta = \frac{\text{opposite side}}{\text{adjacent side}} \quad \text{or} \quad \tan\theta = \frac{\text{perpendicular}}{\text{base}}$$

$$\cot\theta = \frac{\text{adjacent side}}{\text{opposite side}} \quad \text{or} \quad \cot\theta = \frac{\text{base}}{\text{perpendicular}}$$

$$\sec\theta = \frac{\text{hypotenuse}}{\text{adjacent side}} \quad \text{or} \quad \sec\theta = \frac{\text{hypotenuse}}{\text{base}}$$

$$\text{cosec}\,\theta = \frac{\text{hypotenuse}}{\text{opposite side}} \quad \text{or} \quad \text{cosec}\,\theta = \frac{\text{hypotenuse}}{\text{perpendicular}}$$

Note 1. sin A is one symbol and sin A does not mean sin × A is the same case for cos A and tan A.

2. In short, trigonometric ratios are written as t-ratios
3. Every t-ratio is a real number.
4. Trigonometrical ratio depends only upon the magnitude of the angle and the length of sides.
5. $(\sin\theta)^2$ means $(\sin\theta)\cdot(\sin\theta)$. It is also written as $\sin^2\theta$. Similarly, we write other trigonometric ratios also.

Reciprocal Relation

We have six trigonometric ratios and the reciprocal relations between them are as below:

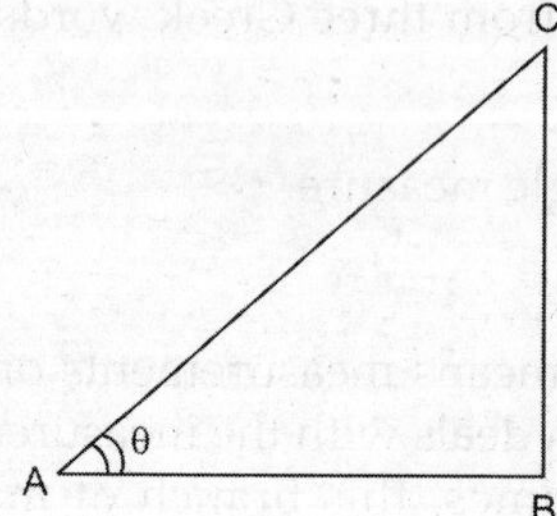

(i) $\operatorname{cosec} A = \dfrac{1}{\sin A}$; $\sin A = \dfrac{1}{\operatorname{cosec} A}$ $\left(\text{on } \sin A \times \operatorname{cosec} A = \dfrac{BC}{AC} \times \dfrac{AC}{BC} = 1\right)$

(ii) $\sec A = \dfrac{1}{\cos A}$; $\cos A = \dfrac{1}{\sec A}$ $\left(\text{on } \cos A \times \sec A = \dfrac{AB}{AC} \times \dfrac{AC}{AB} = 1\right)$

(iii) $\cot A = \dfrac{1}{\tan A}$; $\tan A = \dfrac{1}{\cot A}$ $\left(\tan A \times \cot A = \dfrac{BC}{AB} \times \dfrac{AB}{BC} = 1\right)$

Quotient Relation of *t* ratios

(i) $\tan A = \dfrac{\sin A}{\cos A}$ $\left[\tan A = \dfrac{BC}{AB} = \dfrac{\frac{BC}{AC}}{\frac{AB}{AC}} = \dfrac{\sin A}{\cos A}\right]$

(ii) $\cot A = \dfrac{\cos A}{\sin A}$ $\left[\cot A = \dfrac{AB}{BC} = \dfrac{\frac{AB}{AC}}{\frac{BC}{AC}} = \dfrac{\cos A}{\sin A}\right]$

Powers of Trigonometric Ratios

$\sin^2 A = (\sin A)^2$, $\sin^3 A = (\sin A)^3$, $\cos^3 A = (\cos A)^3$, etc.

13.3 TRIGONOMETRIC RATIOS IN TERMS OF COORDINATES

Let $B(x, y)$ be any point on its terminal side OA, such that $OB = r = \sqrt{x^2 + y^2}$ $(r > 0)$.

Let the distance from the origin to the point $B(x, y)$ be r. The distance r is positive at all times; x, the abscissa, and y, the ordinate are positive in the first quadrant.

If AOB is a right angled triangle, in which angle AOB is θ degrees, we define that

if $r = 1$, $\quad$ sine θ = sin θ $= \dfrac{y}{r} = y$

cosine θ = cos θ $= \dfrac{x}{r} = x$

tangent θ = tan θ $= \dfrac{y}{x}$

cotangent θ = cot θ $= \dfrac{x}{y}$

secant θ = sec θ $= \dfrac{r}{x} = \dfrac{1}{x}$

cosecant θ = cosec θ $= \dfrac{r}{y} = \dfrac{1}{y}$

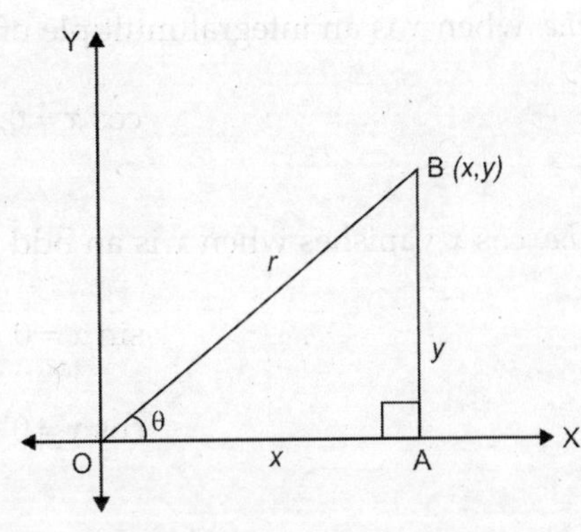

Draw a circle of radius 1 unit with centre at the origin. Let $P(a, b)$ be any point on the circle with angle $AOP = x$ radian

$$\sin x = \frac{AP}{OP} = \frac{b}{1} = y \text{ coordinate}$$

$$\cos x = \frac{OA}{OP} = \frac{a}{1} = x \text{ coordinate}$$

Y

r = 1

P (a,b)

x

X′ X

Y′

One complete rotation subtends an angle of 2π at the centre. Quarter rotation subtends an angle AOB equal to $\frac{\pi}{2}$. Half rotation subtends an angle $\angle AOC = \pi$ and three quarter rotation subtends $\angle AOD = \frac{3\pi}{2}$. The coordinates of the points A, B, C and D are respectively $(1, 0)$, $(0, 1)$, $(-1, 0)$ and $(0, -1)$.

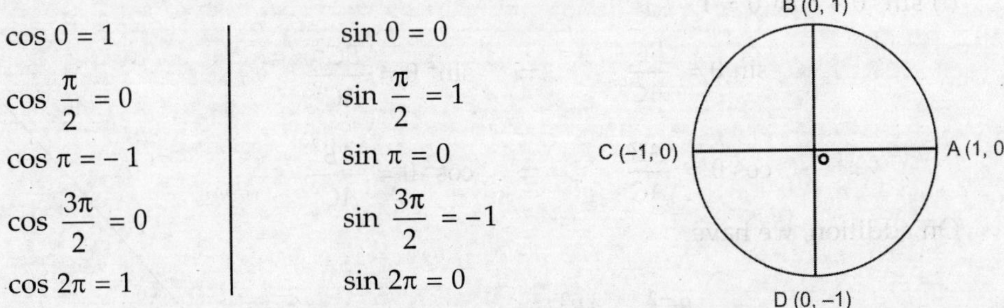

$\cos 0 = 1$	$\sin 0 = 0$
$\cos \dfrac{\pi}{2} = 0$	$\sin \dfrac{\pi}{2} = 1$
$\cos \pi = -1$	$\sin \pi = 0$
$\cos \dfrac{3\pi}{2} = 0$	$\sin \dfrac{3\pi}{2} = -1$
$\cos 2\pi = 1$	$\sin 2\pi = 0$

If we revolve the point P through one complete revolution, this point will come back to the same starting point P. So, the coordinates of the point P remains the same. Thus, we observe if angle x increases (or decreases) by an integral multiple of 2π, then values of sine and cosine do not change as being the coordinates of the point P.

$$\sin(2n\pi + x) = \sin x, n \in Z$$

$$\cos(2n\pi + x) = \cos x, n \in Z$$

If we observe the coordinates on the circle, then

$$\sin x = 0, \text{ if } x = 0, \pm\pi, \pm 2\pi, \pm 3\pi, \ldots$$

i.e. when x is an integral multiple of π.

$$\cos x = 0, \text{ if } x = \pm\frac{\pi}{2}, \pm\frac{3\pi}{2}, \pm\frac{5\pi}{2}, \ldots$$

i.e. $\cos x$ vanishes when x is an odd multiple of $\frac{\pi}{2}$. Thus,

$$\sin x = 0 \quad \Rightarrow \quad x = n\pi, n \in Z$$

$$\cos x = 0 \quad \Rightarrow \quad x = (2n+1)\frac{\pi}{2}, n \in Z$$

13.4 TRIGONOMETRIC IDENTITIES

An equation involving trigonometric ratios of an angle θ (say) is said to be a trigonometric identity, if it is satisfied for all the values of θ for which the given trigonometric ratios are defined.

For Example:

(a) $2\sin^2\theta - \sin\theta = \sin\theta\,(2\sin\theta - 1)$ is an identity.
But, $2\sin^2\theta - \sin\theta = 0$ is not an identity, but an equation, which is satisfied only for some particular values of θ.

(b) $3\cos^2\theta - 4\cos\theta = \cos\theta\,(3\cos\theta - 4)$ is an identity, whereas $3\cos^2\theta - 4\cos\theta = 0$ is an equation, but not an identity, since it is satisfied only for some particular values of θ and not for all values of θ.

Fundamental Identities

(*i*) $\sin^2\theta + \cos^2\theta = 1$ (*ii*) $1 + \tan^2\theta = \sec^2\theta$ (*iii*) $1 + \cot^2\theta = \text{cosec}^2\theta$

Proof. In right angled (ΔABC)

(*i*) $\sin^2\theta + \cos^2\theta = 1$

$$\sin\theta = \frac{BC}{AC} \quad \Rightarrow \quad \sin^2\theta = \frac{BC^2}{AC^2}$$

$$\cos\theta = \frac{AB}{AC} \quad \Rightarrow \quad \cos^2\theta = \frac{AB^2}{AC^2}$$

On addition, we have

$$\sin^2\theta + \cos^2\theta = \frac{BC^2}{AC^2} + \frac{AB^2}{AC^2}$$

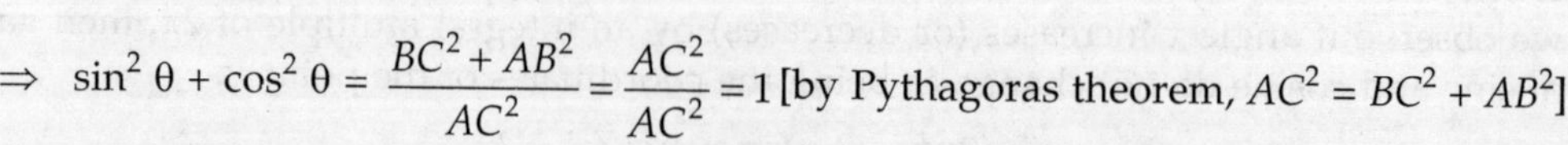

$$\Rightarrow \sin^2\theta + \cos^2\theta = \frac{BC^2 + AB^2}{AC^2} = \frac{AC^2}{AC^2} = 1 \text{ [by Pythagoras theorem, } AC^2 = BC^2 + AB^2]$$

$$\Rightarrow \quad \mathbf{\sin^2\theta + \cos^2\theta = 1}$$ **Proved.**

First Method

(*ii*) $\quad 1 + \tan^2\theta = \sec^2\theta$

$$\sec\theta = \frac{AC}{AB} \quad\Rightarrow\quad \sec^2\theta = \frac{AC^2}{AB^2}$$

$$\tan\theta = \frac{BC}{AB} \quad\Rightarrow\quad \tan^2\theta = \frac{BC^2}{AB^2}$$

$$1 + \tan^2\theta = 1 + \frac{BC^2}{AB^2} = \frac{AB^2 + BC^2}{AB^2}$$

$$\Rightarrow \qquad = \frac{AC^2}{AB^2} \qquad (\because AB^2 + BC^2 = AC^2, \text{ by Pythagoras theorem})$$

$\Rightarrow \quad$ **$1 + \tan^2\theta = \sec^2\theta$** **Proved.**

(*iii*) $\quad 1 + \cot^2\theta = \text{cosec}^2\theta$

$$\text{cosec}\,\theta = \frac{AC}{BC} \quad\Rightarrow\quad \text{cosec}^2\theta = \frac{AC^2}{BC^2}$$

$$\cot\theta = \frac{AB}{BC} \quad\Rightarrow\quad \cot^2\theta = \frac{AB^2}{BC^2}$$

$$1 + \cot^2\theta = 1 + \frac{AB^2}{BC^2} = \frac{BC^2 + AB^2}{BC^2}$$

$$\Rightarrow \qquad = \frac{AC^2}{BC^2} \qquad (\because BC^2 + AB^2 = AC^2, \text{ by Pythagoras theorem})$$

$\Rightarrow \quad$ **$1 + \cot^2\theta = \text{cosec}^2\theta$** **Proved.**

Second Method

(*ii*) $\quad 1 + \tan^2\theta = \sec^2\theta$

We have proved that $\sin^2\theta + \cos^2\theta = 1$

Dividing by $\cos^2\theta$, we get

$$\frac{\sin^2\theta}{\cos^2\theta} + \frac{\cos^2\theta}{\cos^2\theta} = \frac{1}{\cos^2\theta} \quad\Rightarrow\quad \tan^2\theta + 1 = \sec^2\theta$$ **Proved.**

(*iii*) $1 + \cot^2\theta = \text{cosec}^2\theta$

We know that $\sin^2\theta + \cos^2\theta = 1$

Dividing by $\sin^2\theta$, we get

$$\frac{\sin^2\theta}{\sin^2\theta} + \frac{\cos^2\theta}{\sin^2\theta} = \frac{1}{\sin^2\theta} \quad\Rightarrow\quad 1 + \cot^2\theta = \text{cosec}^2\theta$$ **Proved.**

In Brief:

1. $\sin^2\theta + \cos^2\theta = 1$
2. $\sin^2\theta = 1 - \cos^2\theta$
3. $\cos^2\theta = 1 - \sin^2\theta$
4. $1 + \tan^2\theta = \sec^2\theta$
5. $\tan^2\theta = \sec^2\theta - 1$, $\sec^2\theta - \tan^2\theta = 1$
6. $1 + \cot^2\theta = \text{cosec}^2\theta$
7. $\cot^2\theta = \text{cosec}^2\theta - 1$, $\text{cosec}^2\theta - \cot^2\theta = 1$

Relation among Trigonometric Ratios

t ratio	sin θ	cos θ	tan θ	cot θ	sec θ	cosec θ
sin θ	sin θ	$\sqrt{1-\cos^2\theta}$	$\frac{\tan\theta}{\sqrt{1+\tan^2\theta}}$	$\frac{1}{\sqrt{1+\cot^2\theta}}$	$\frac{\sqrt{\sec^2\theta-1}}{\sec\theta}$	$\frac{1}{\text{cosec}\,\theta}$
cos θ	$\sqrt{1-\sin^2\theta}$	cos θ	$\frac{1}{\sqrt{1+\tan^2\theta}}$	$\frac{\cot\theta}{\sqrt{1+\cot^2\theta}}$	$\frac{1}{\sec\theta}$	$\frac{\sqrt{\text{cosec}^2\theta-1}}{\text{cosec}\,\theta}$
tan θ	$\frac{\sin\theta}{\sqrt{1-\sin^2\theta}}$	$\frac{\sqrt{1-\cos^2\theta}}{\cos\theta}$	tan θ	$\frac{1}{\cot\theta}$	$\sqrt{\sec^2\theta-1}$	$\frac{1}{\sqrt{\text{cosec}^2\theta-1}}$
cot θ	$\frac{\sqrt{1-\sin^2\theta}}{\sin\theta}$	$\frac{\cos\theta}{\sqrt{1-\cos^2\theta}}$	$\frac{1}{\tan\theta}$	cot θ	$\frac{1}{\sqrt{\sec^2\theta-1}}$	$\sqrt{\text{cosec}^2\theta-1}$
sec θ	$\frac{1}{\sqrt{1-\sin^2\theta}}$	$\frac{1}{\cos\theta}$	$\sqrt{1+\tan^2\theta}$	$\frac{\sqrt{1+\cot^2\theta}}{\cot\theta}$	sec θ	$\frac{\text{cosec}\,\theta}{\sqrt{\text{cosec}^2\theta-1}}$
cosec θ	$\frac{1}{\sin\theta}$	$\frac{1}{\sqrt{1-\cos^2\theta}}$	$\frac{\sqrt{1+\tan^2\theta}}{\tan\theta}$	$\sqrt{1+\cot^2\theta}$	$\frac{\sec\theta}{\sqrt{\sec^2\theta-1}}$	cosec θ

13.5 COORDINATES IN DIFFERENT QUADRANTS

In the first quadrant, x and y both coordinates are positive. In the second quadrant, the abscissa x is negative and the ordinate y is positive. In the third quadrant, both coordinates x and y are negative. In the fourth quadrant, the abscissa x is positive and the ordinate y is negative.

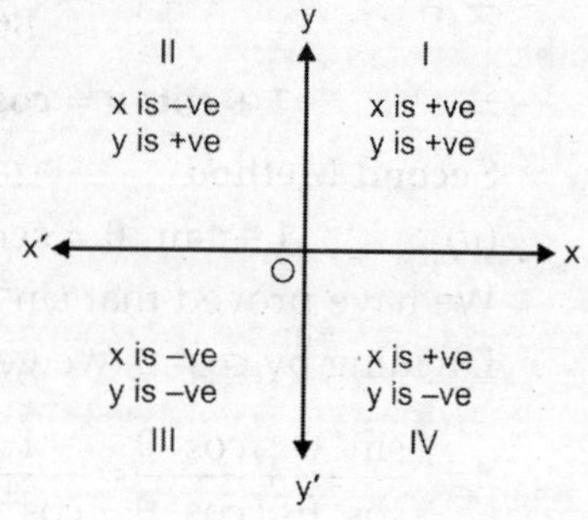

13.6 TRIGONOMETRIC RATIOS IN DIFFERENT QUADRANTS

Trigonometric ratios for the angles are the ratio of the sides of a right angled triangle. The sides are denoted by x and y. The coordinates x and y have different signs in different quadrants. So, the sign of trigonometric ratios depends upon the quadrant in which the angle lies.

In the first quadrant all the trigonometric ratios are positive as both x and y coordinates are positive and r is positive in all the quadrants.

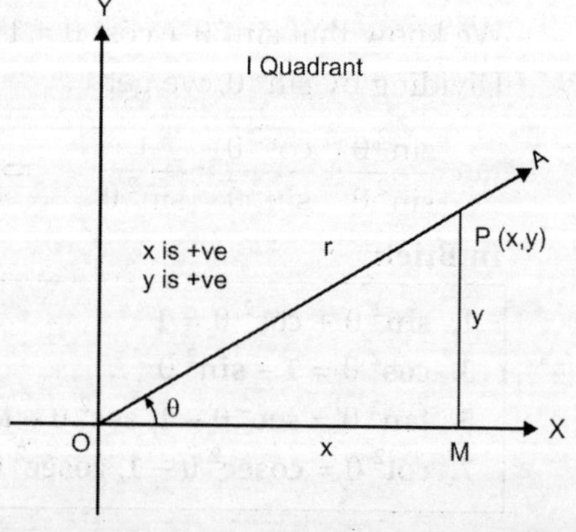

In the second quadrant

$\sin\theta = \frac{y}{r} = +ve$ as y and r are both positive.

$\cos\theta = \frac{-x}{r} = -ve$ as abscissa, x is negative and r is positive.

$\tan\theta = \frac{y}{-x} = -ve$ as abscissa x is negative and ordinate y is positive.

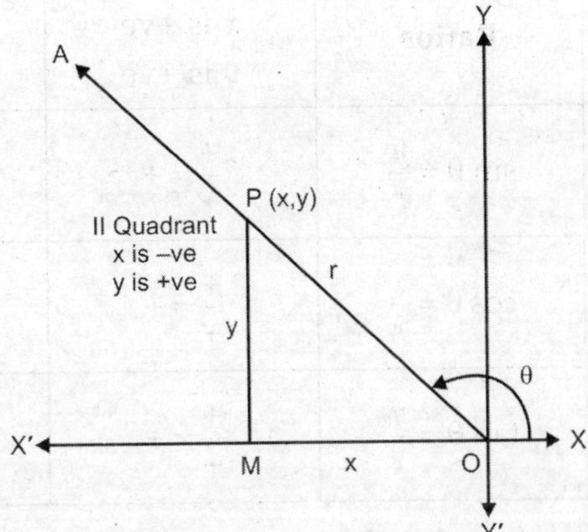

In the third quadrant

$\sin\theta = \frac{-y}{r} = -ve$, as ordinate y is negative and r is positive.

$\cos\theta = \frac{-x}{r} = -ve$, as abscissa x is negative and r is positive.

$\tan\theta = \frac{-y}{-x} = +ve$, as the ordinate and abscissa are both negative and the ratio of both the negative coordinates is positive.

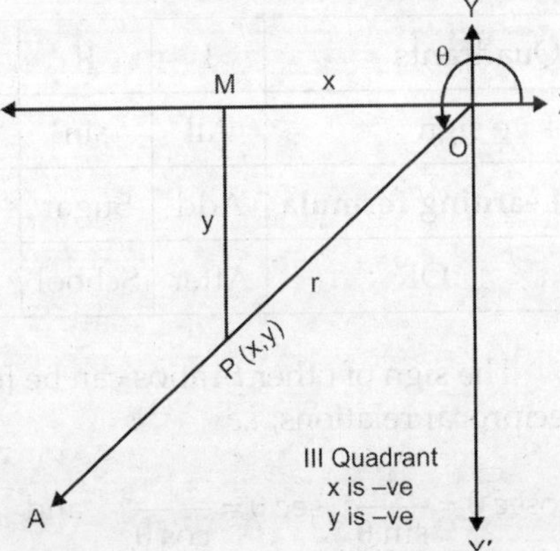

In the fourth quadrant

$\sin\theta = \frac{-y}{r} = -ve$, as ordinate y is negative and r is positive.

$\cos\theta = \frac{x}{r} = +ve$, as abscissa x and r are both positive.

$\tan\theta = \frac{-y}{x} = -ve$, as ordinate y is negative and abscissa x is positive.

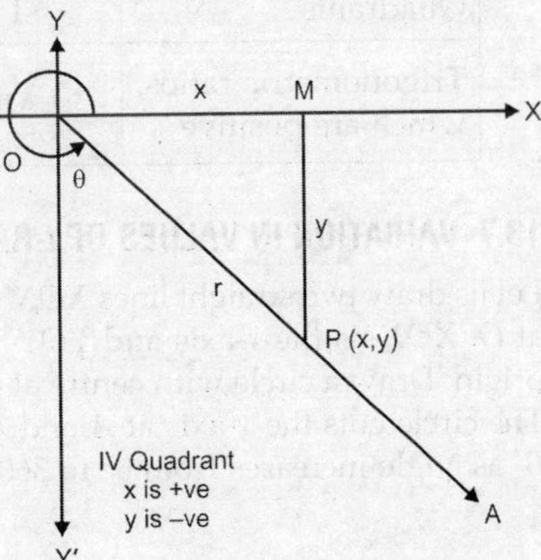

Trigonometric Ratios	Ist Quadrant	II Quadrant	III Quadrant	IV Quadrant
	x is +ve y is +ve	x is –ve y is +ve	x is –ve y is –ve	x is +ve y is –ve
$\sin\theta = \frac{y}{r}$	$\frac{+}{+} = +$	$\frac{+}{+} = +$	$\frac{-}{+} = -$	$\frac{-}{+} = -$
$\cos\theta = \frac{x}{r}$	$\frac{+}{+} = +$	$\frac{-}{+} = -$	$\frac{-}{+} = -$	$\frac{+}{+} = +$
$\tan\theta = \frac{y}{x}$	$\frac{+}{+} = +$	$\frac{+}{-} = -$	$\frac{-}{-} = +$	$\frac{-}{+} = -$

Simple rules to remember:

Quadrants	I	II	III	IV
+ ve sign	All	sin	tan	cos
Learning formula	Add	Sugar	To	Coffee
OR	After	School	To	College

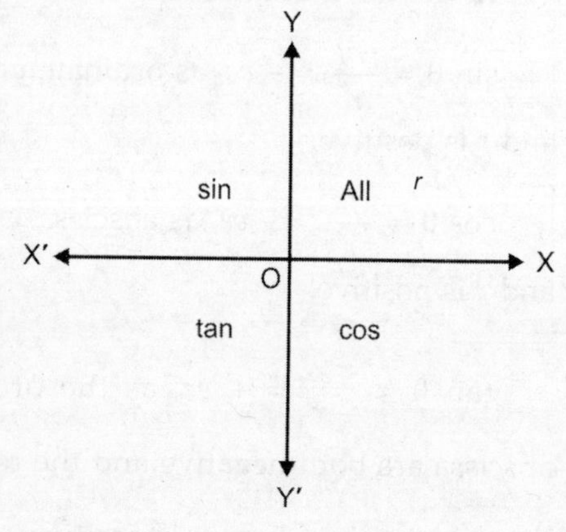

The sign of other t-ratios can be found by using reciprocal relations, *i.e.*

$$\text{cosec}\,\theta = \frac{1}{\sin\theta},\quad \sec\theta = \frac{1}{\cos\theta} \quad \text{and} \quad \cot\theta = \frac{1}{\tan\theta}$$

so, we have

Quadrants	I	II	III	IV
Trigonometric ratios, which are positive	All	$\sin\theta$	$\tan\theta$	$\cos\theta$
		cosec θ	$\cot\theta$	$\sec\theta$

13.7 VARIATION IN VALUES OF *t*-RATIOS

Let us draw two straight lines XOX' and YOY' intersecting at O. XOX' is the x-axis and YOY' is y-axis and O is the origin. Draw a circle with centre at origin O and radius 1. The circle cuts the x-axis at A and A' and y-axis at B and B' as angle increases from 0° to 360°.

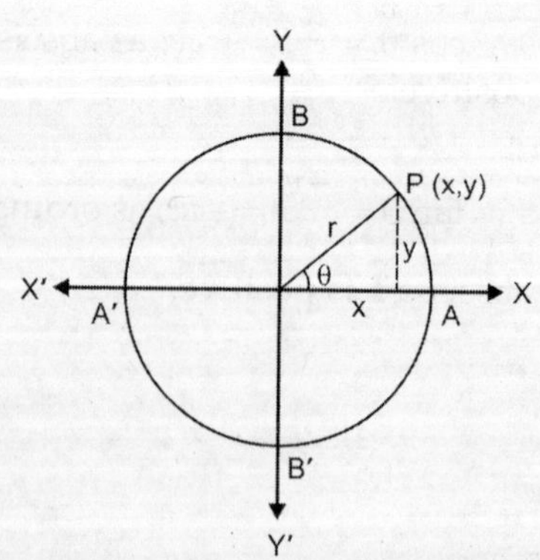

Quadrants / *t*-ratios	I Quadrant *x*-decreases from 1 to 0 *y*-increases from 0 to 1	II Quadrant *x*-decreases from 0 to –1 *y*-decreases from 1 to 0	III Quadrant *x*-increases from –1 to 0 *y*-decreases from 0 to –1	IV Quadrant *x*-increases from 0 to 1 *y*-increases from –1 to 0
$\sin\theta = \frac{y}{r} = \frac{y}{1} = y$	increases from 0 to 1	decreases from 1 to 0	decreases from 0 to –1	increases from –1 to 0
$\cos\theta = \frac{x}{r} = \frac{x}{1} = x$	decreases from 1 to 0	decreases from 0 to –1	increases from –1 to 0	increases from 0 to 1
$\tan\theta = \frac{y}{x}$	increases from 0 to ∞	increases from – ∞ to 0	increases from 0 to ∞	increases from – ∞ to 0
$\cot\theta = \frac{x}{y}$	decreases from ∞ to 0	decreases from 0 to – ∞	decreases from ∞ to 0	decreases from 0 to – ∞
$\sec\theta = \frac{r}{x} = \frac{1}{x}$	increases from 1 to ∞	increases from – ∞ to –1	decreases from –1 to – ∞	decreases from ∞ to 1
$\operatorname{cosec}\theta = \frac{r}{y} = \frac{1}{y}$	decreases from ∞ to 1	increases from 1 to ∞	increases from – ∞ to –1	decreases from –1 to – ∞

Example 1. *If* $\cos\theta = -\frac{1}{2}$, *θ lies in third quadrant, then find the other five trigonometric functions.*

Solution. We know that $\sin^2\theta + \cos^2\theta = 1 \Rightarrow \sin\theta = \pm\sqrt{1-\cos^2\theta}$

In third quadrant, sin θ is negative, therefore

$$\sin\theta = -\sqrt{1-\cos^2\theta} = -\sqrt{1-\left(-\frac{1}{2}\right)^2} = \sqrt{\frac{4-1}{4}} = -\frac{\sqrt{3}}{2} \Rightarrow \sin\theta = -\frac{\sqrt{3}}{2} \quad \text{... (1)}$$

$$\left[\because \cos\theta = -\frac{1}{2}\right]$$

Again
$$\tan\theta = \frac{\sin\theta}{\cos\theta} = \frac{-\frac{\sqrt{3}}{2}}{-\frac{1}{2}} = \sqrt{3}$$

The reciprocal relations are:

$$\operatorname{cosec}\theta = \frac{1}{\sin\theta} = \frac{1}{-\frac{\sqrt{3}}{2}} = -\frac{2}{\sqrt{3}} \qquad \left[\because \sin\theta = -\frac{\sqrt{3}}{2}\right]$$

$$\sec\theta = \frac{1}{\cos\theta} = \frac{1}{-\frac{1}{2}} = -2 \qquad \left[\because \cos\theta = -\frac{1}{2}\right]$$

and $$\cot\theta = \frac{1}{\tan\theta} = \frac{1}{\sqrt{3}}$$

Hence, other five trigonometric functions are:

$$\therefore \quad \sin\theta = -\frac{\sqrt{3}}{2},\ \operatorname{cosec}\theta = -\frac{2}{\sqrt{3}},\ \sec\theta = -2,\ \tan\theta = \sqrt{3},\ \cot\theta = \frac{1}{\sqrt{3}} \qquad \textbf{Ans.}$$

Example 2. *If* $\sin\theta = \frac{3}{5}$, *θ lies in second quadrant, then find the other five trigonometric functions.*

Solution. We have $\sin\theta = \frac{3}{5}$, θ lies in second quadrant.

We know that $\sin^2\theta + \cos^2\theta = 1$

$$\Rightarrow \qquad \cos\theta = \pm\sqrt{1-\sin^2\theta}$$

In second quadrant, cos θ is negative, therefore

$$\cos\theta = -\sqrt{1-\sin^2\theta}$$

$$= -\sqrt{1-\left(\frac{3}{5}\right)^2} = -\sqrt{\frac{25-9}{25}} = -\frac{4}{5} \qquad \left[\because \sin\theta = \frac{3}{5}\right]$$

$$\tan\theta = \frac{\sin\theta}{\cos\theta} = \frac{\frac{3}{5}}{-\frac{4}{5}} = -\frac{3}{4}$$

The reciprocal relations are:

$$\operatorname{cosec}\theta = \frac{1}{\sin\theta} = \frac{1}{\frac{3}{5}} = \frac{5}{3}$$

$$\sec\theta = \frac{1}{\cos\theta} = \frac{1}{-\frac{4}{5}} = -\frac{5}{3}$$

$$\cot\theta = \frac{1}{\tan\theta} = \frac{1}{-\frac{3}{4}} = -\frac{4}{3}$$

Hence, the other five trigonometric functions are:

$$\therefore \quad \operatorname{cosec}\theta = \frac{5}{3},\ \cos\theta = -\frac{4}{5},\ \sec\theta = -\frac{5}{4},\ \tan\theta = -\frac{3}{4},\ \cot\theta = -\frac{4}{3} \qquad \textbf{Ans.}$$

Example 3. *Find all trigonometric ratios, if* $\sin\theta = -\frac{2\sqrt{6}}{5}$ *and θ lies in IV quadrant.*

Solution. We have, $\sin\theta = -\frac{2\sqrt{6}}{5}$

We know that $\sin^2\theta + \cos^2\theta = 1$

$$\Rightarrow \qquad \cos\theta = \pm\sqrt{1-\sin^2\theta} = \pm\sqrt{1-\left(\frac{-2\sqrt{6}}{5}\right)^2} = \pm\sqrt{\frac{25-24}{25}} = \pm\sqrt{\frac{1}{25}} = \pm\frac{1}{5}$$

In the fourth quadrant, cos θ is positive, therefore $\cos\theta = \frac{1}{5}$

$$\tan\theta = \frac{\sin\theta}{\cos\theta} = \frac{-\frac{2\sqrt{6}}{5}}{\frac{1}{5}} = -2\sqrt{6}$$

The other trigonometric ratios are as follows:

$$\text{cosec}\,\theta = \frac{1}{\sin\theta} = \frac{1}{-\frac{2\sqrt{6}}{5}} = -\frac{5}{2\sqrt{6}}$$

$$\cot\theta = \frac{1}{\tan\theta} = \frac{1}{-2\sqrt{6}} = -\frac{1}{2\sqrt{6}}$$

$$\sec\theta = \frac{1}{\cos\theta} = \frac{1}{\frac{1}{5}} = 5$$

Ans.

Example 4. *If* $\sec\theta = \sqrt{2}$ *and* $\frac{3\pi}{2} < \theta < 2\pi$, *find the value of* $\frac{1+\tan\theta+\text{cosec}\,\theta}{1+\cot\theta-\text{cosec}\,\theta}$.

Solution. We have, $\sec\theta = \sqrt{2} \quad \Rightarrow \quad \cos\theta = \frac{1}{\sqrt{2}}$

$$\therefore \quad \sin\theta = \pm\sqrt{1-\cos^2\theta} = \pm\sqrt{1-\left(\frac{1}{\sqrt{2}}\right)^2} = \sqrt{1-\frac{1}{2}} = \pm\frac{1}{\sqrt{2}}$$

But, θ lies in the fourth quadrant in which sin θ is negative.

$$\therefore \qquad \sin\theta = -\frac{1}{\sqrt{2}} \quad \Rightarrow \quad \text{cosec}\,\theta = -\sqrt{2}$$

$$\text{And} \qquad \tan\theta = \frac{\sin\theta}{\cos\theta} \quad \Rightarrow \quad \tan\theta = \left(\frac{-1}{\sqrt{2}} \times \frac{\sqrt{2}}{1}\right) = -1$$

$$\Rightarrow \qquad \tan\theta = -1 \quad \Rightarrow \quad \cot\theta = -1$$

$$\text{Now,} \qquad \frac{1+\tan\theta+\text{cosec}\,\theta}{1+\cot\theta-\text{cosec}\,\theta} = \frac{1-1-\sqrt{2}}{1-1-(-\sqrt{2})} = \frac{-\sqrt{2}}{\sqrt{2}} = -1$$

Ans.

Example 5. *If* $\sin\theta = \frac{4}{5}$, *find the value of* $\frac{5\cos\theta+4\,\text{cosec}\,\theta+3\tan\theta}{4\cot\theta+3\sec\theta+5\sin\theta}$.

Solution. We have, $\quad \sin\theta = \frac{4}{5}$

We know that $\sin^2\theta + \cos^2\theta = 1$

$$\Rightarrow \qquad \cos\theta = \pm\sqrt{1-\sin^2\theta} = \pm\sqrt{1-\left(\frac{4}{5}\right)^2} = \pm\sqrt{\frac{25-16}{25}} = \pm\sqrt{\frac{9}{25}} = \pm\frac{3}{5}$$

since sin θ is positive, θ lies either in I quadrant or in II quadrant.

Case I: θ lies in I quadrant.

$$\therefore \qquad \cos\theta = \frac{3}{5}$$

$$\tan\theta = \frac{\sin\theta}{\cos\theta} = \frac{\frac{4}{5}}{\frac{3}{5}} = \frac{4}{3}, \quad \cot\theta = \frac{1}{\tan\theta} = \frac{3}{4}$$

$$\sec\theta = \frac{1}{\cos\theta} = \frac{5}{3}, \quad \text{cosec}\,\theta = \frac{1}{\sin\theta} = \frac{5}{4}$$

$$\therefore \qquad \frac{5\cos\theta + 4\,\text{cosec}\,\theta + 3\tan\theta}{4\cot\theta + 3\sec\theta + 5\sin\theta} = \frac{5\left(\frac{3}{5}\right)+4\left(\frac{5}{4}\right)+3\left(\frac{4}{3}\right)}{4\left(\frac{3}{4}\right)+3\left(\frac{5}{3}\right)+5\left(\frac{4}{5}\right)} = \frac{3+5+4}{3+5+4} = \frac{12}{12} = 1$$

Case II: θ lies in II quadrant.

$$\therefore \qquad \cos\theta = -\frac{3}{5}, \quad \tan\theta = \frac{\sin\theta}{\cos\theta} = \frac{\frac{4}{5}}{-\frac{3}{5}} = -\frac{4}{3}$$

$$\cot\theta = \frac{1}{\tan\theta} = -\frac{3}{4}, \quad \sec\theta = \frac{1}{\cos\theta} = -\frac{5}{3}$$

$$\text{cosec}\,\theta = \frac{1}{\sin\theta} = \frac{5}{4}$$

$$\frac{5\cos\theta + 4\,\text{cosec}\,\theta + 3\tan\theta}{4\cot\theta + 3\sec\theta + 5\sin\theta} = \frac{5\left(-\frac{3}{5}\right)+4\left(\frac{5}{4}\right)+3\left(-\frac{4}{3}\right)}{4\left(-\frac{3}{4}\right)+3\left(-\frac{5}{3}\right)+5\left(\frac{4}{5}\right)}$$

$$= \frac{-3+5-4}{-3-5+4} = \frac{-2}{-4} = \frac{1}{2}$$

Ans.

EXERCISE 13.1

Find the values of other five trigonometric functions in each of the following problems:

1. $\sin\theta = \frac{3}{5}$, θ lies in the first quadrant.

Ans. $\cos\theta = \frac{4}{5}, \tan\theta = \frac{3}{4}, \sec\theta = \frac{5}{4}, \cot\theta = \frac{4}{3}, \text{cosec}\,\theta = \frac{5}{3}$

2. $\sec\theta = \frac{13}{5}$, θ lies in fourth quadrant.

Ans. $\sin\theta = -\frac{12}{13}$, $\text{cosec}\,\theta = -\frac{13}{12}$, $\cos\theta = \frac{5}{13}$, $\tan\theta = -\frac{12}{5}$, $\cot\theta = \frac{-5}{12}$

3. $\tan\theta = \frac{4}{3}$, θ lies in third quadrant.

Ans. $\sin\theta = -\frac{4}{5}$, $\text{cosec}\,\theta = -\frac{5}{4}$, $\cos\theta = -\frac{3}{5}$, $\sec\theta = -\frac{5}{3}$, $\cot\theta = \frac{3}{4}$

4. $\cot\theta = -\frac{12}{5}$, θ lies in second quadrant.

Ans. $\sin\theta = \frac{5}{13}$, $\cos\theta = -\frac{12}{13}$, $\tan\theta = -\frac{5}{12}$, $\sec\theta = -\frac{13}{12}$, $\text{cosec}\,\theta = \frac{13}{5}$

5. If $\cos\theta = -\frac{1}{2}$, and $\pi < \theta < \frac{3\pi}{2}$, find the value of $4\tan^2\theta - 3\,\text{cosec}^2\,\theta$. **Ans.** 8

6. If $\sin\theta = \frac{21}{29}$, and $0 < \theta < \frac{\pi}{2}$, show that $\sec\theta + \tan\theta = 2\frac{1}{2}$.

7. If $\cos\theta = -\frac{3}{5}$ and $\pi < \theta < \frac{3\pi}{2}$, find the values of other five trigonometric functions and hence evaluate $\frac{\text{cosec}\,\theta + \cot\theta}{\sec\theta - \tan\theta}$. **Ans.** $\frac{1}{6}$

8. If $\sin\theta + \cos\theta = 0$ and θ lies in the fourth quadrant, find $\sin\theta$ and $\cos\theta$.

Ans. $-\frac{1}{\sqrt{2}}, \frac{1}{\sqrt{2}}$

9. If $\sin\theta = \frac{3}{5}$, $\tan\phi = \frac{1}{2}$ and $\frac{\pi}{2} < \theta < \pi < \phi < \frac{3\pi}{2}$, find the value of $8\tan\theta - \sqrt{5}\sec\phi$.

Ans. $-\frac{7}{2}$

10. The sine of an angle is to its cosine as 8 : 15. If θ lies in III quadrant, find the value of $\sin\theta$ and $\cos\theta$. **Ans.** $-\frac{8}{17}, -\frac{15}{17}$

11. If $\tan\theta = \frac{1}{\sqrt{7}}$, find the value of $\frac{\text{cosec}^2\,\theta - \sec^2\theta}{\text{cosec}^2\,\theta + \sec^2\theta}$. **Ans.** $\frac{3}{4}$

12. If $4\sin^2\theta = 1$, find the value of $\frac{2 + 3\cos^2\theta}{1 - 2\cos^2\theta}$. **Ans.** $-\frac{17}{2}$

13.8 TRIGONOMETRIC RATIOS OF SOME SPECIFIC ANGLES

Let ΔABC be a right angled triangle, right angled at B.

$$\sin\theta = \frac{BC}{AC}, \cos\theta = \frac{AB}{AC}, \tan\theta = \frac{BC}{AB}$$

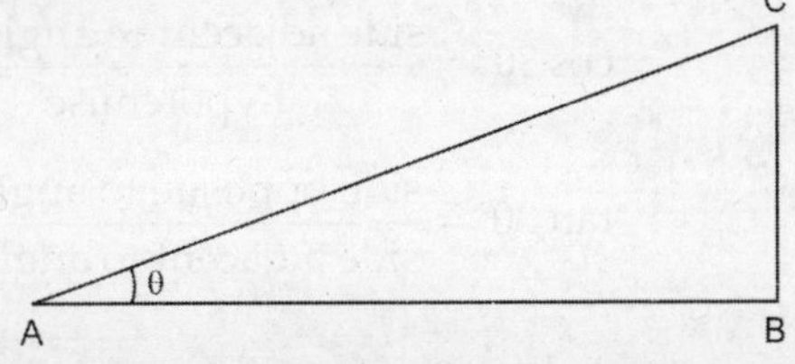

(*i*) Trigonometric ratios of 0° and 90°

As θ becomes smaller and smaller, line segment *BC* also becomes smaller and smaller. Finally, when θ becomes 0°, the point *C* coincides with *B*, *i.e.* *BC* = 0 and *AB* = *AC*.

$$\sin 0° = \frac{BC}{AC} = \frac{0}{AC} = 0 \qquad (BC = 0)$$

$$\cos 0° = \frac{AB}{AC} = \frac{AC}{AC} = 1 \qquad (AC = AB)$$

$$\tan 0° = \frac{BC}{AB} = \frac{0}{AB} = 0 \qquad (BC = 0)$$

Then, $\sin 0° = 0, \cos 0° = 1, \tan 0° = 0$

(*ii*) Trigonometric ratios of 90°

From Δ*ABC*, it is clear that as θ increases line segment *AB* becomes smaller and smaller. Finally, when θ is 90° the point *A* will coincide with *B*, *i.e.* *AB* = 0 and *AC* = *BC*.

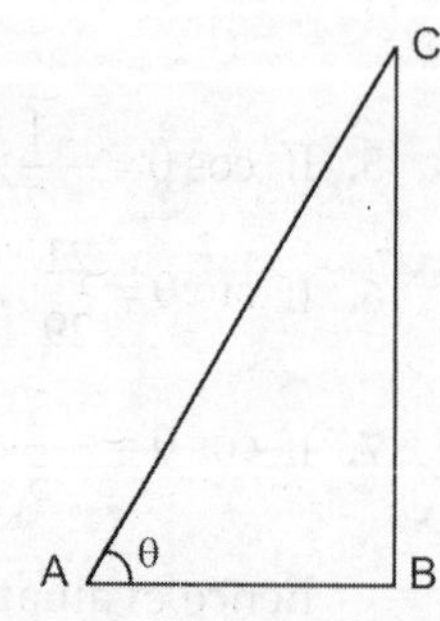

$$\sin 90° = \frac{BC}{AC} = \frac{BC}{BC} = 1$$

$$\cos 90° = \frac{AB}{AC} = \frac{0}{AC} = 0$$

$$\tan 90° = \frac{BC}{AB} = \frac{BC}{0} \text{ is not defined.}$$

(*iii*) Trigonometric ratios of 30° and 60°

Consider an equilateral triangle *ABC* with each side of length 2*a*.

Each angle of Δ*ABC* is of 60°. Let *AD* be the perpendicular from *A* on *BC*.

∴ *AD* is the bisector of ∠*A* and *D* is the mid-point of *BC*.

∴ *BD* = *DC* = *a* and ∠*BAD* = 30°

In Δ*ABD*, ∠*D* is a right angle, *AB* = 2*a* and *BD* = *a*

∴ By Pythagoras theorem,

$$AB^2 = AD^2 + BD^2 \quad \Rightarrow \quad (2a)^2 = AD^2 + (a)^2$$

$$\Rightarrow AD^2 = 4a^2 - a^2 = 3a^2 \Rightarrow AD = \sqrt{3}a.$$

(*iv*) Trigonometric ratios of 30°

In ΔABD, ∠*BAD* = 30°

$$\therefore \quad \sin 30° = \frac{\text{side opposite to angle } 30°}{\text{hypotenuse}} = \frac{BD}{AB} = \frac{a}{2a} = \frac{1}{2}$$

$$\cos 30° = \frac{\text{side adjacent to angle } 30°}{\text{hypotenuse}} = \frac{AD}{AB} = \frac{\sqrt{3}a}{2a} = \frac{\sqrt{3}}{2}$$

$$\tan 30° = \frac{\text{side opposite to angle } 30°}{\text{side adjacent to angle } 30°} = \frac{BD}{AD} = \frac{a}{\sqrt{3}a} = \frac{1}{\sqrt{3}}$$

$$\text{cosec } 30^\circ = \frac{1}{\sin 30^\circ} = 2, \ \sec 30^\circ = \frac{1}{\cos 30^\circ} = \frac{2}{\sqrt{3}}$$

$$\cot 30^\circ = \frac{1}{\tan 30^\circ} = \sqrt{3}$$

$$\Rightarrow \quad \sin 30^\circ = \frac{1}{2}, \ \cos 30^\circ = \frac{\sqrt{3}}{2}, \ \tan 30^\circ = \frac{1}{\sqrt{3}}.$$

(*v*) Trigonometric ratios of 60°

In ΔABD, $\angle B = 60^\circ$

$$\therefore \quad \sin 60^\circ = \frac{\text{side opposite to angle } 60^\circ}{\text{hypotenuse}} = \frac{AD}{AB} = \frac{\sqrt{3}a}{2a} = \frac{\sqrt{3}}{2}$$

$$\cos 60^\circ = \frac{\text{side adjacent to angle } 60^\circ}{\text{hypotenuse}} = \frac{BD}{AB} = \frac{a}{2a} = \frac{1}{2}$$

$$\tan 60^\circ = \frac{\text{side opposite to angle } 60^\circ}{\text{side adjacent to angle } 60^\circ} = \frac{AD}{BD} = \frac{\sqrt{3}a}{a} = \sqrt{3}$$

$$\text{cosec } 60^\circ = \frac{1}{\sin 60^\circ} = \frac{2}{\sqrt{3}}, \ \sec 60^\circ = \frac{1}{\cos 60^\circ} = 2, \ \cot 60^\circ = \frac{1}{\tan 60^\circ} = \frac{1}{\sqrt{3}}$$

$$\sin 60^\circ = \frac{\sqrt{3}}{2}, \ \cos 60^\circ = \frac{1}{2}, \ \tan 60^\circ = \sqrt{3}$$

(*vi*) Trigonometric ratios of 45°

Let ABC be a triangle, right angled at B, in which $\angle A = \angle C = 45^\circ$

$\therefore \quad BC = AB$

Let $\quad AB = BC = a$

Then, by Pythagoras Theorem, $AC^2 = AB^2 + BC^2 = a^2 + a^2 = 2a^2$

$\Rightarrow \quad AC = \sqrt{2}a$

In ΔABC, $\quad \angle C = 45^\circ$

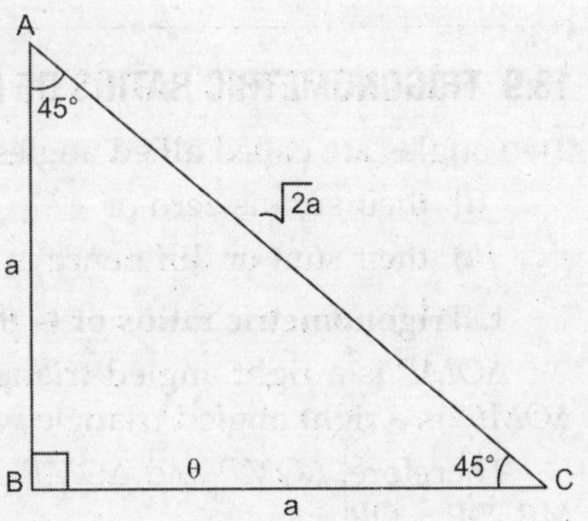

$$\therefore \quad \sin 45^\circ = \frac{\text{side opposite to angle } 45^\circ}{\text{hypotenuse}}$$

$$= \frac{AB}{AC} = \frac{a}{\sqrt{2}a} = \frac{1}{\sqrt{2}}$$

$$\cos 45^\circ = \frac{\text{side adjacent to angle } 45^\circ}{\text{hypotenuse}}$$

$$= \frac{BC}{AC} = \frac{a}{\sqrt{2}a} = \frac{1}{\sqrt{2}}$$

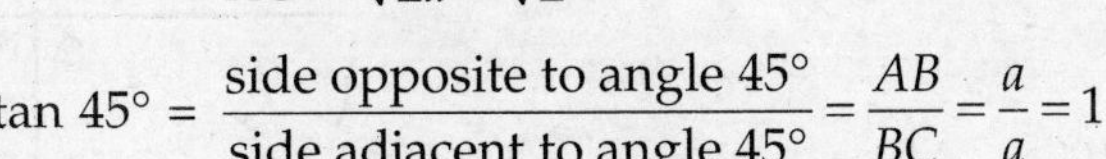

$$\tan 45^\circ = \frac{\text{side opposite to angle } 45^\circ}{\text{side adjacent to angle } 45^\circ} = \frac{AB}{BC} = \frac{a}{a} = 1$$

$$\text{cosec } 45^\circ = \frac{1}{\sin 45^\circ} = \sqrt{2}, \ \sec 45^\circ = \frac{1}{\cos 45^\circ} = \sqrt{2}, \ \cot 45^\circ = \frac{1}{\tan 45^\circ} = 1.$$

Values of trigonometric ratios, when $0 \leq \theta \leq 90°$

To remember the values of all trigonometric ratios for different values of θ, the following table will be useful.

Trigonometric table

θ / *t* ratios	0°	30°	45°	60°	90°
sin θ	0	$1/2$	$1/\sqrt{2}$	$\sqrt{3}/2$	1
cos θ	1	$\sqrt{3}/2$	$1/\sqrt{2}$	$1/2$	0
tan θ	0	$1/\sqrt{3}$	1	$\sqrt{3}$	Not defined
cosec θ	Not defined	2	$\sqrt{2}$	$2/\sqrt{3}$	1
sec θ	1	$2/\sqrt{3}$	$\sqrt{2}$	2	Not defined
cot θ	Not defined	$\sqrt{3}$	1	$1/\sqrt{3}$	0

Simple rule to remember: The following scheme will enable us to memorise the values of sine, cosine and tangent very easily.

Angle	0°	30°	45°	60°	90°
sin	$\sqrt{\frac{0}{4}}$	$\sqrt{\frac{1}{4}}$	$\sqrt{\frac{2}{4}}$	$\sqrt{\frac{3}{4}}$	$\sqrt{\frac{4}{4}}$
cos	$\sqrt{\frac{4}{4}}$	$\sqrt{\frac{3}{4}}$	$\sqrt{\frac{2}{4}}$	$\sqrt{\frac{1}{4}}$	$\sqrt{\frac{0}{4}}$
tan	$\sqrt{\frac{0}{4-0}}$	$\sqrt{\frac{1}{4-1}}$	$\sqrt{\frac{2}{4-2}}$	$\sqrt{\frac{3}{4-3}}$	Not defined

13.9 TRIGONOMETRIC RATIOS OF ALLIED ANGLES

Two angles are called allied angles, if either

(*i*) their sum is zero or

(*ii*) their sum or difference is a multiple of right angle.

1. Trigonometric ratios of (– θ) in terms of θ

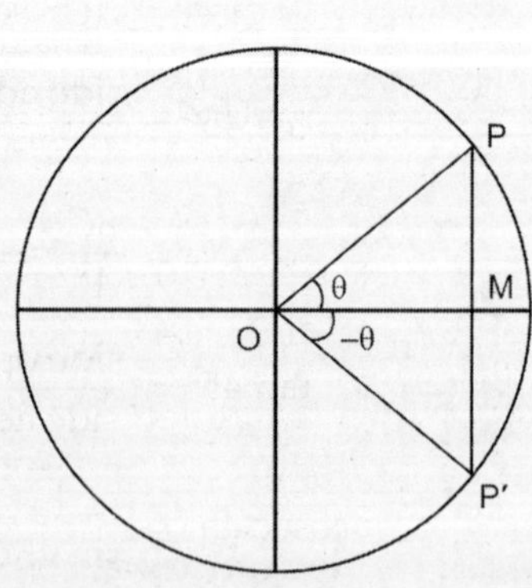

ΔOMP is a right angled triangle of acute angle θ and $\Delta OMP'$ is a right angled triangle with angle – θ.

Therefore, ΔOMP and $\Delta OMP'$ are similar. But $MP' = -MP$, $OP = OP'$

∴ In $\Delta OMP'$

$$\sin(-\theta) = \frac{MP'}{OP'} = -\frac{MP}{OP} = -\sin\theta$$

$$\cos(-\theta) = \frac{OM}{OP'} = \frac{OM}{OP} = \cos\theta$$

$$\tan(-\theta) = \frac{MP'}{OM} = -\frac{MP}{OM} = -\tan\theta$$

Similarly, $\operatorname{cosec}(-\theta) = -\operatorname{cosec}\theta$

$\sec(-\theta) = \sec\theta$

$\cot(-\theta) = -\cot\theta$

2. Trigonometric ratios of (90° + θ) in terms of θ

ΔOMP and $\Delta OP'M'$ are similar right angled triangle with $\angle MOP = \theta$.

$M'P' = OM$

But $OM' = -MP$ and $OP' = OP$

$\therefore$ In $\Delta M'OP'$

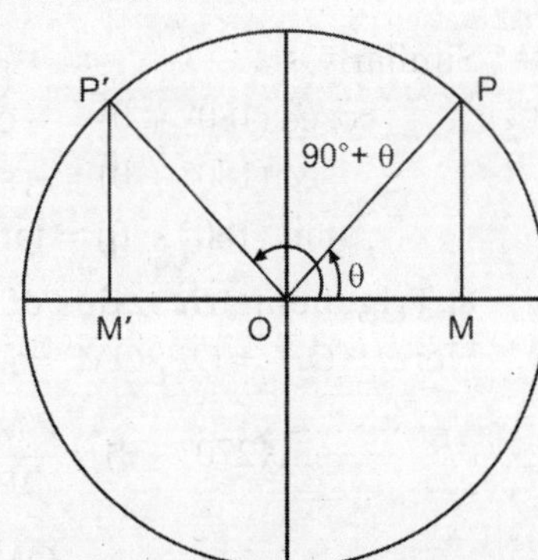

$$\sin(90° + \theta) = \frac{M'P'}{OP'} = \frac{OM}{OP} = \cos\theta$$

$$\cos(90° + \theta) = \frac{OM'}{OP'} = -\frac{MP}{OP} = -\sin\theta$$

$$\tan(90° + \theta) = \frac{M'P'}{OM'} = -\frac{OM}{MP} = -\cot\theta$$

Similarly,

$\operatorname{cosec}(90° + \theta) = \sec\theta$

$\sec(90° + \theta) = -\operatorname{cosec}\theta$

$\cot(90° + \theta) = -\tan\theta$

3. Trigonometric ratios of (90° – θ) in terms of θ

Changing θ into – θ in (90° + θ), we get

$\sin(90° - \theta) = \sin[90° + (-\theta)] = \cos(-\theta) = \cos\theta$ [using result (1) and (2)]

And $\cos(90° - \theta) = \cos[90° + (-\theta)] = -\sin(-\theta) = \sin\theta$

$\tan(90° - \theta) = \tan[90° + (-\theta)] = -\cot(-\theta) = \cot\theta$

Similarly, we can find out other trigonometric ratios.

4. Trigonometric ratios of (180° – θ) in terms of θ

Here $OM' = -OM$, $OP' = OP$, $M'P' = PM$

$$\sin(180° - \theta) = \frac{P'M'}{OP'} = \frac{PM}{OP} = \sin\theta$$

$$\cos(180° - \theta) = \frac{OM'}{OP'} = -\frac{OM}{OP} = -\cos\theta$$

$$\tan(180° - \theta) = \frac{P'M'}{OM'} = \frac{PM}{-OM} = -\tan\theta$$

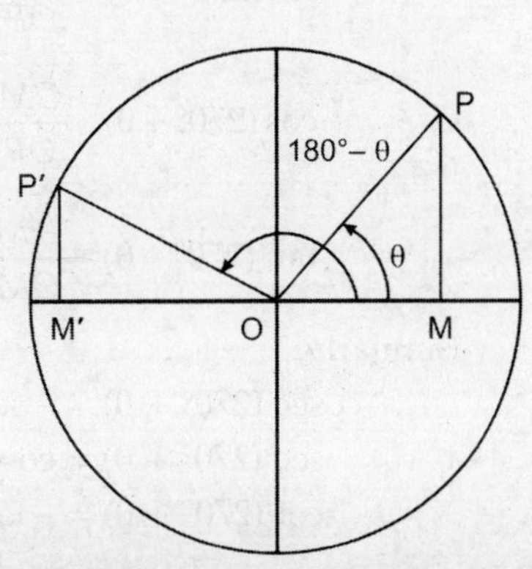

Similarly,

$\operatorname{cosec}(180° - \theta) = \operatorname{cosec}\theta$

$\sec(180° - \theta) = -\sec\theta$

$\tan(180° - \theta) = -\tan\theta$

5. Trigonometric ratios of (180° + θ) in terms of θ

ΔOMP and $OM'P'$ are similar right angled triangle; but

$$OM' = -OM \text{ and } M'P' = -MP,\ OP' = OP$$

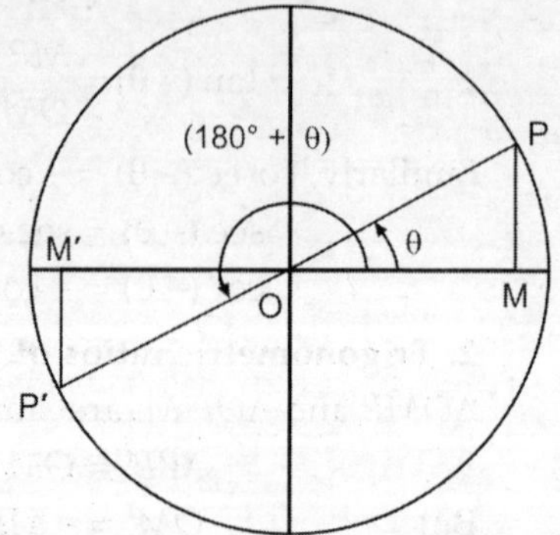

$$\sin(180° + \theta) = \frac{P'M'}{OP'} = \frac{-MP}{OP} = -\sin\theta$$

$$\cos(180° + \theta) = \frac{OM'}{OP'} = \frac{-OM}{OP} = -\cos\theta$$

$$\tan(180° + \theta) = \frac{P'M'}{OM'} = \frac{-PM}{-OM} = \frac{PM}{OM} = \tan\theta$$

Similarly,

$$\text{cosec}(180° + \theta) = -\text{cosec}\,\theta$$

$$\sec(180° + \theta) = -\sec\theta$$

$$\tan(180° + \theta) = \tan\theta$$

6. Trigonometric ratios of (270° – θ) in terms of θ

Here, $OP' = OP,\ OM' = -MP,\ M'P' = -OM$

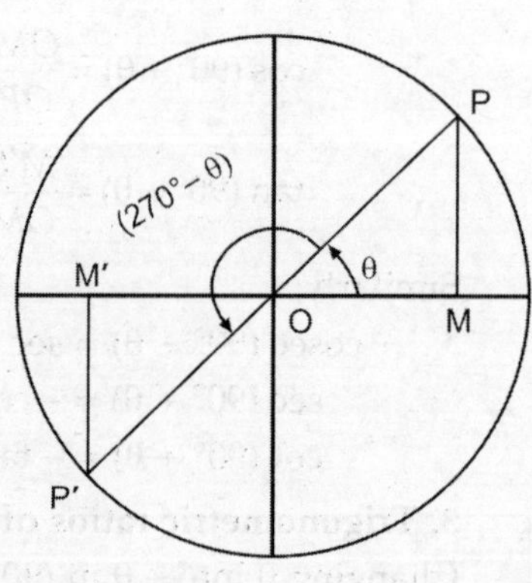

$$\therefore \quad \sin(270° - \theta) = \frac{P'M'}{OP'} = \frac{-OM}{OP} = -\cos\theta$$

$$\cos(270° - \theta) = \frac{OM'}{OP'} = \frac{-MP}{OP} = -\sin\theta$$

$$\tan(270° - \theta) = \frac{PM'}{OM'} = \frac{-OM}{-MP} = \frac{OM}{MP} = \cot\theta$$

Similarly,

$$\text{cosec}(270° - \theta) = -\sec\theta$$

$$\sec(270° - \theta) = -\text{cosec}\,\theta$$

$$\cot(270° - \theta) = \tan\theta$$

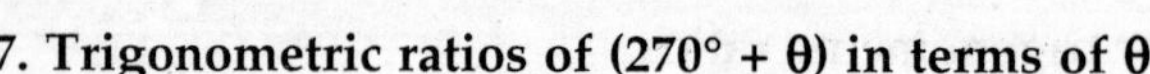

7. Trigonometric ratios of (270° + θ) in terms of θ

Here, $OM' = MP,\ M'P' = -OM,\ OP' = OP$

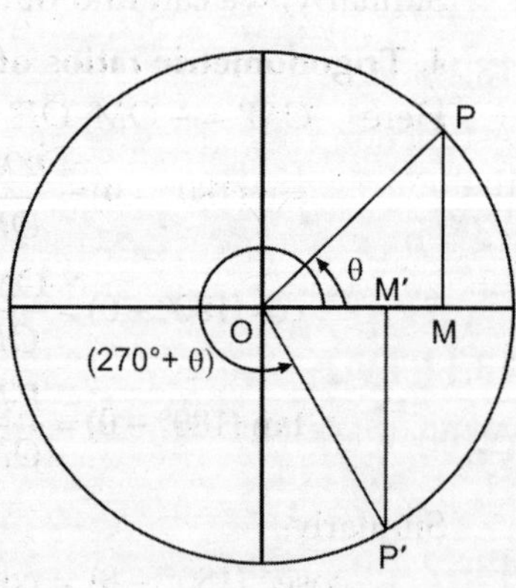

$$\sin(270° + \theta) = \frac{M'P'}{OP'} = \frac{-OM}{OP} = -\cos\theta$$

$$\cos(270° + \theta) = \frac{OM'}{OP'} = \frac{MP}{OP} = \sin\theta$$

$$\tan(270° + \theta) = \frac{P'M'}{OM'} = \frac{-OM}{MP} = -\cot\theta$$

Similarly,

$$\text{cosec}(270° + \theta) = -\sec\theta$$

$$\sec(270° + \theta) = \text{cosec}\,\theta$$

$$\cot(270° + \theta) = -\tan\theta$$

8. Trigonometric ratios of (360° – θ) in terms of θ

$\sin (360° - \theta) = \sin (-\theta)$ [∵ (360° – θ) and (– θ) are coterminal angles]

$= -\sin \theta$ [by (1)]

Similarly

$\cos (360° - \theta) = \cos (-\theta) = \cos \theta$

$\tan (360° - \theta) = \tan (-\theta) = -\tan \theta$

$\cot (360° - \theta) = \cot (-\theta) = -\cot \theta$

$\sec (360° - \theta) = \sec (-\theta) = \sec \theta$

$\text{cosec} (360° - \theta) = \text{cosec} (-\theta) = -\text{cosec}\, \theta$

9. Trigonometric ratios of (360° + θ) in terms of θ

After one complete rotation, we have the same Δ*OMP*, *i.e.* Δ's for θ and (360° + θ) is the same (*OMP*), and hence we get the same ratios as for θ, *i.e.*

$\sin (360° + \theta) = \sin \theta$ [(360° + θ) and θ are coterminal angles]

$\cos (360° + \theta) = \cos \theta$

$\tan (360° + \theta) = \tan \theta$

Similarly, we may get other *t* ratios.

Note: In fact, for any positive integer *n*, (360° × *n* + θ) is coterminal to θ. Therefore, for any positive integer *n*, we have,

$\sin (360° \times n + \theta) = \sin \theta, \cos (360° \times n + \theta) = \cos \theta$

$\tan (360° \times n + \theta) = \tan \theta, \text{cosec} (360° \times n + \theta) = \text{cosec}\, \theta$

$\sec (360° \times n + \theta) = \sec \theta, \cot (360° \times n + \theta) = \cot \theta$

Table of allied angles

t ratio	– θ	90° – θ	90° + θ	180° – θ	180° + θ	270° – θ	270° + θ	360° – θ	360° + θ	2*n*π – θ	2*n*π + θ
sin	– sin θ	cos θ	cos θ	sin θ	– sin θ	– cos θ	– cos θ	– sin θ	sin θ	– sin θ	sin θ
cos	cos θ	sin θ	– sin θ	– cos θ	– cos θ	– sin θ	sin θ	cos θ	cos θ	cos θ	cos θ
tan	– tan θ	cot θ	– cot θ	– tan θ	tan θ	cot θ	– cot θ	– tan θ	tan θ	– tan θ	tan θ

13.10 PERIODIC FUNCTION

A function *f* (*x*) is said to be a periodic function, if there exists a real number *a* > 0 such that *f* (*a* + *x*) = *f* (*x*) for all *x*, where *a* is called the fundamental period of *f* (*x*).

For example, sine and cosine functions are periodic with period 2π, because sin (2π + θ) = sin θ and cos (2π + θ) = cos θ.

13.11 EVEN FUNCTIONS

A function *f* (*x*) is said to be an even function, if *f* (– *x*) = *f* (*x*), for all *x* in its domain.

For example : cos θ and sec θ are even function, as cos (– θ) = cos θ and sec (– θ) = sec θ.

13.12 ODD FUNCTIONS

A function *f* (*x*) is said to be an odd function, if *f* (– *x*) = – *f* (*x*), for all *x* in its domain.

For example: sin θ, tan θ, cosec θ and cot θ are odd functions as sin (– θ) = – sin θ

13.13 WORKING RULE TO EXPRESS TRIGONOMETRIC RATIOS OF ANY ANGLE IN TERMS OF THOSE OF POSITIVE ACUTE ANGLE

Step 1. If the angle of the form $(-\theta)$, express it in terms of θ.

Step 2. If the angle is more than 360°, divide by 360° to separate the multiples of 360° from the given angle. Now, use the formula for $(n \times 360° + \theta)$.

Step 3. If the angle is now more than 180°, either use the formula for $(180 + \theta)°$ or for $(360 - \theta)°$ whichever may leave a smaller angle.

Step 4. If the angle is less than 180°, use the formula for $(180 - \theta)°$.

Example 6. *Find the values of the following trigonometric functions:*

(*i*) $\sin 765°$ (*ii*) $\text{cosec}\,(-1410°)$ (*iii*) $\tan\left(\frac{19\pi}{3}\right)$ (*iv*) $\cot\left(-\frac{15\pi}{4}\right)$

Solution. (*i*) We have, $\sin(765°) = \sin(2 \times 360° + 45°) = \sin 45° = \frac{1}{\sqrt{2}}$

(*ii*) We have,
$$\cos(-1410°) = -\text{cosec}\,1410° \quad [\because \text{cosec}\,(-\theta) = -\text{cosec}\,\theta]$$
$$= -\text{cosec}\,(360° \times 4 - 30°)$$
$$= -(-\text{cosec}\,30°) \quad [\text{cosec}\,(360° \times n - \theta) = -\text{cosec}\,\theta]$$
$$= \text{cosec}\,30° = 2$$
Ans.

(*iii*) We have,
$$\tan\left(\frac{19\pi}{3}\right) = \tan\left(6\pi + \frac{\pi}{3}\right) = \tan\left(3 \times 2\pi + \frac{\pi}{3}\right)$$
$$= \tan\left(\frac{\pi}{3}\right) = \sqrt{3} \quad [\because \tan(2\pi \times n + \theta) = \tan\theta]$$
Ans.

(*iv*) We have,
$$\cot\left(-\frac{15\pi}{4}\right) = -\cot\left(\frac{15\pi}{4}\right) \quad [\because \cot(-\theta) = -\cot\theta]$$
$$= -\cot\left(4\pi - \frac{\pi}{4}\right) = -\cot\left(2\pi \times 2 - \frac{\pi}{4}\right)$$
$$= -\left(-\cot\frac{\pi}{4}\right) = \cot\frac{\pi}{4} = 1 \quad [\because \cot(2n\pi - \theta) = -\cot\theta]$$
Ans.

Example 7. *If* $0 \le \theta \le \pi$, *find* θ *for which*

(*i*) $\sin\theta = \cos\theta$ (*ii*) $\sin\theta = \sin(-\theta)$

Solution. (*i*) We know that, $\sin\theta$ and $\cos\theta$ have the same sign, if θ lies in I quadrant or III quadrant. Since, $0 \le \theta \le \pi$, they will be equal in I quadrant.

$$\sin\theta = \cos\theta \Rightarrow \frac{\sin\theta}{\cos\theta} = 1 \Rightarrow \tan\theta = 1$$
$$\Rightarrow \tan\theta = \tan 45° \Rightarrow \theta = 45° \quad [\tan 45° = 1]$$

(*ii*) We have, $\sin\theta = \sin(-\theta)$
$$\Rightarrow \sin\theta = -\sin\theta \Rightarrow 2\sin\theta = 0 \Rightarrow \sin\theta = 0$$
$$\Rightarrow \theta = 0$$
$$\therefore \theta = 0, \pi \quad (\text{since}, 0 \le \theta \le \pi).$$
Ans.

Example 8. *Find all the angles between 0° and 360°, which satisfy the equation* $\sin^2\theta = \frac{3}{4}$.

Solution. The given equation is $\sin^2\theta = \frac{3}{4} \quad \Rightarrow \quad \sin\theta = \pm\frac{\sqrt{3}}{2}$

$\therefore$ Either $\quad \sin\theta = \frac{\sqrt{3}}{2}$ or $\sin\theta = -\frac{\sqrt{3}}{2}$

$\sin\theta = \frac{\sqrt{3}}{2}$ implies θ lies in I or II quadrant.

We have, $\quad \sin 60° = \frac{\sqrt{3}}{2}$ and $\sin(180 - 60°) = \sin 60° = \frac{\sqrt{3}}{2}$

$\therefore \quad \theta = 60°, 180° - 60° = 120°$

$\sin\theta = -\frac{\sqrt{3}}{2}$ implies θ lies in III or IV quadrant.

We have, $\sin(180° + 60°) = -\sin 60° = -\frac{\sqrt{3}}{2} \Rightarrow \theta = 180° + 60° = 240°$

$\sin(360° - 60°) = -\sin 60° = -\frac{\sqrt{3}}{2} \Rightarrow \theta = 360° - 60° = 300°$

Hence, the required angles are 60°, 120°, 240° and 300°. **Ans.**

Example 9. *Prove that:*

$$\tan 225°\cdot\cot 405° + \tan 765°\cdot\cot 675° = 0$$

Proof. L.H.S. $= \tan 225°\cdot\cot 405° + \tan 765°\cdot\cot 675°$

$= \tan(180° + 45°)\cot(360° + 45°) + \tan(2 \times 360° + 45°)\cot(2 \times 360° - 45°)$

$= \tan 45° \cot 45° + \tan 45° (-\cot 45°)$

$= \tan 45° \cot 45° - \tan 45° \cot 45°$

$= 0 =$ R.H.S. **Proved.**

Example 10. *Prove that:*

$$\tan 10° \cos 20° \tan 40° \tan 50° \operatorname{cosec} 70° \sec 80° \sin 80° = 1$$

Proof. L.H.S. $= \tan 10° \cos 20° \tan 40° \tan 50° \operatorname{cosec} 70° \sec 80° \sin 80°$

$= \tan 10° \cos 20° \tan 40° \tan(90° - 40°) \operatorname{cosec}(90° - 20°) \sec(90° - 10°) \sin(90° - 10°)$

$= \tan 10° \cos 20° \tan 40° \cot 40° \sec 20° \operatorname{cosec} 10° \cos 10°$

$= \frac{\sin 10°}{\cos 10°} \cos 20° \cdot \frac{\sin 40°}{\cos 40°} \cdot \frac{\cos 40°}{\sin 40°} \cdot \frac{1}{\cos 20°} \cdot \frac{\cos 10°}{\sin 10°}$

$= 1 =$ R.H.S. **Proved.**

Example 11. *In any quadrilateral ABCD, prove that:*

(*i*) $\sin(A + B) + \sin(C + D) = 0$ (*ii*) $\cos(A + B) = \cos(C + D)$

Proof. (*i*) We have, $A + B + C + D = 2\pi$ ($\because$ Sum of angles of quadrilateral is 2π)

$\Rightarrow \quad A + B = 2\pi - (C + D)$

$\Rightarrow \quad \sin(A + B) = \sin[2\pi - (C + D)]$

$\Rightarrow \quad \sin(A + B) = -\sin(C + D) \qquad [\because \sin(2\pi - \theta) = -\sin\theta]$

$\Rightarrow \quad \sin(A + B) + \sin(C + D) = 0$ **Proved.**

(*ii*) We have, $A + B + C + D = 2\pi$

$\Rightarrow \quad A + B = 2\pi - (C + D)$

$\Rightarrow \quad \cos(A + B) = \cos[2\pi - (C + D)]$

$\Rightarrow \quad \cos(A + B) = \cos(C + D) \qquad [\because \cos(2\pi - \theta) = \cos\theta]$ **Proved.**

Example 12. *Given angle C of a triangle ABC to be obtuse, find all angles, when sin (A + B)* $= \dfrac{\sqrt{3}}{2}$ *and cos (A – B)* $= \dfrac{1}{\sqrt{2}}$*, where A > B.*

Solution. We have $\sin(A + B) = \dfrac{\sqrt{3}}{2}$ But, $\dfrac{\sqrt{3}}{2} = \sin 60°$

$\Rightarrow \quad A + B = 60° \qquad \ldots (1)$

And $\quad \cos(A - B) = \dfrac{1}{\sqrt{2}}$ But, $\dfrac{1}{\sqrt{2}} = \cos 45°$

$\Rightarrow \quad A - B = 45° \qquad \ldots (2)$

Adding (1) and (2), we get

$$2A = 60° + 45° = 105°$$

$$\Rightarrow \quad A = \frac{105°}{2} = 52°30'$$

Subtracting (2) from (1), we get

$$2B = 60° - 45° = 15° \Rightarrow B = \frac{15°}{2} = 7°30'$$

Also, $\quad A + B + C = 180°$

$\Rightarrow \quad C = 180° - (A + B)$

$\quad = 180° - (52°30' + 7°30') = 180° - 60° = 120°$

Hence, the angles are 52°30′, 7°30′, 120°. **Ans.**

EXERCISE 13.2

Find the values of the following trigonometric functions:

1. $\cos 135°$ **Ans.** $-\dfrac{1}{\sqrt{2}}$
2. $\sec 120°$ **Ans.** -2
3. $\text{cosec } 150°$ **Ans.** 2
4. $\sin 315°$ **Ans.** $-\dfrac{1}{\sqrt{2}}$
5. $\sin(-330°)$ **Ans.** $\dfrac{1}{2}$
6. $\cos 405°$ **Ans.** $\dfrac{1}{\sqrt{2}}$
7. $\tan\left(\dfrac{11\pi}{6}\right)$ **Ans.** $-\dfrac{1}{\sqrt{3}}$
8. $\text{cosec}\left(-\dfrac{19\pi}{3}\right)$ **Ans.** $-\dfrac{2}{\sqrt{3}}$
9. $\sin 1845°$ **Ans.** $\dfrac{1}{\sqrt{2}}$
10. $\text{cosec}(-1200°)$ **Ans.** $-\dfrac{2}{\sqrt{3}}$
11. $\sin 4530°$ **Ans.** $-\dfrac{1}{2}$

12. If $\theta = \frac{\pi}{3}$, find the value of $\cos\theta + \cos 2\theta + \cos 3\theta + \cos 4\theta + \cos 5\theta + \cos 6\theta$.

Ans. 0

13. *Prove that:* $\sec\frac{\pi}{6}\tan\frac{\pi}{3} + \sin\frac{\pi}{4}\operatorname{cosec}\frac{\pi}{4} + \cos\frac{\pi}{6}\cot\frac{\pi}{3} = \frac{7}{2}$

13.14 FORMATION OF ANGLES

Let $\overrightarrow{OA}$ be a fixed ray. Let there be another ray coincident with $\overrightarrow{OA}$. This ray rotates about O and takes the position $\overrightarrow{OB}$. We say that $\overrightarrow{OB}$ has described an angle AOB.

The original ray is called the *initial side* and the final position of the ray after rotation is called the *terminal side* of the angle. The point of rotation is called the *vertex*.

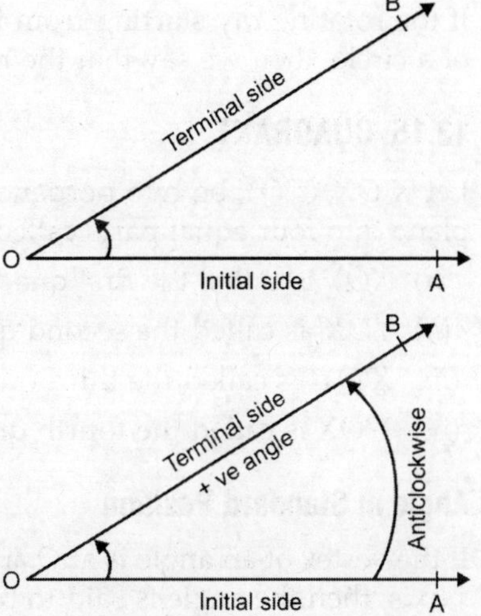

Positive angle: If the ray $\overline{OB}$ rotates about O from the initial position OA in the *anticlockwise* direction, we say that it describes a positive angle. In the figure $\angle AOB$ is a positive angle.

For example: (*i*) $\angle AOD + 60°$ (*ii*) $\angle XOY = +135°$

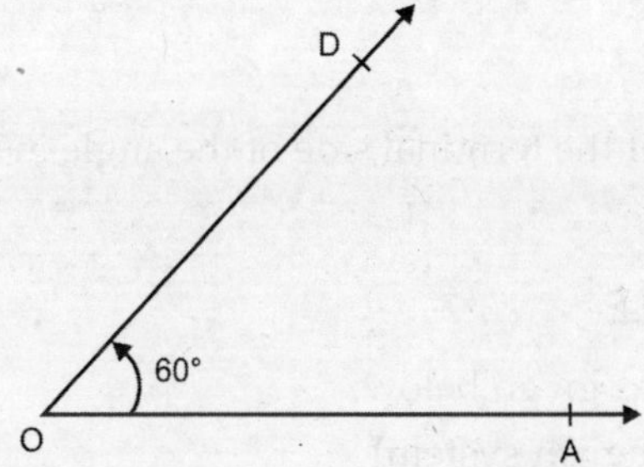

Negative angle: If the ray rotates about O from the position OA in the *clockwise* direction, we say that it describes a *negative angle*. In the figure, $\angle AOC$ is a negative angle.

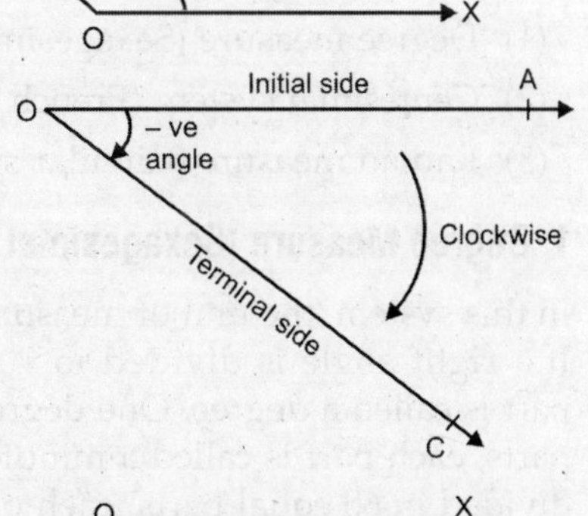

For example: (*i*) $\angle AOB = -30°$ (*ii*) $\angle XOY = -120°$

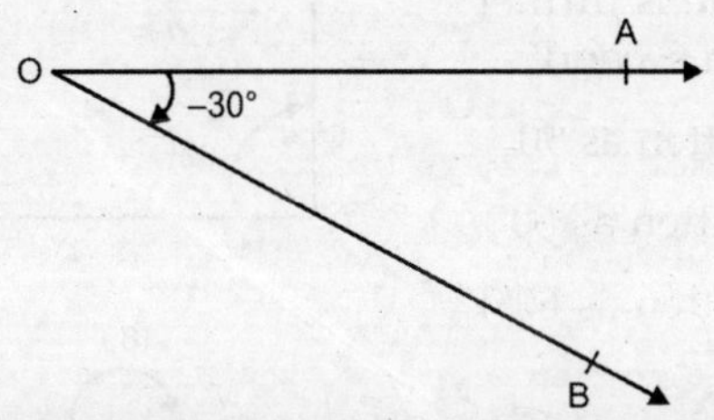

The point O of rotation is called the vertex

13.15 MEASURE OF AN ANGLE

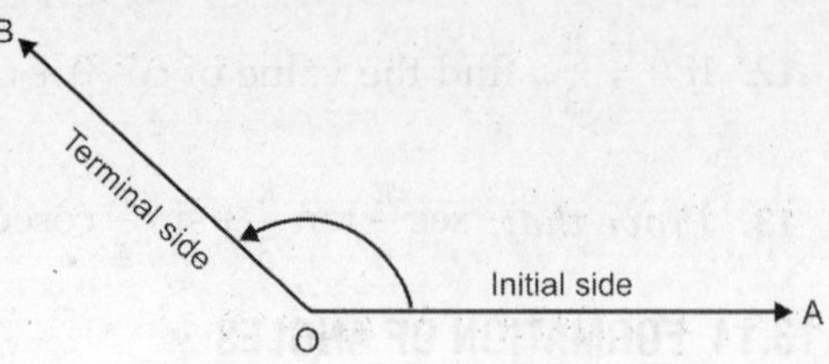

Angle is a measure of rotation of a given ray about its initial point. The measurement of an angle is the amount of rotation performed from initial side to the terminal side.

Right Angle

If the rotating ray starting from its initial position to final position describes one quarter of a circle, then we say that the measure of the angle formed is a right angle.

13.16 QUADRANT

Let $X'OX$, $Y'OY$ be two perpendicular lines intersecting at a point O. These divide the plane into four equal parts called quadrants.

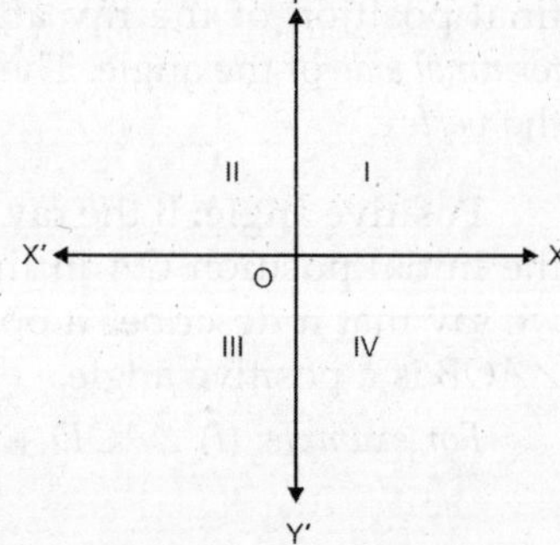

(*i*) XOY is called the first quadrant.

(*ii*) YOX' is called the second quadrant.

(*iii*) $X'OY'$ is called the third quadrant.

(iv) $Y'OX$ is called the fourth quadrant.

Angle in Standard Position

If the vertex of an angle is at O and its initial side lies along x-axis, then the angle is said to be in standard position.

Angle in a Quadrant

An angle is said to be in a particular quadrant, if the terminal side of the angle in standard position lies in that quadrant.

13.17 SYSTEM OF MEASUREMENT OF AN ANGLE

There are three systems for measuring angles as given below:

(1) Degree measure (Sexagesimal system) (English system)

(2) Centesimal system (French system)

(3) Radian measure (Circular system)

1. Degree Measure (Sexagesimal System):

In this system, the unit of measuring an angle is a degree. If a right angle is divided in 90 equal parts, then each part is called a degree. One degree is divided in 60 equal parts, each part is called a minute. One minute is further divided in 60 equal parts, each part is called a second.

$$\therefore \quad 1 \text{ right angle} = 90 \text{ degrees (written as } 90°)$$

$$1 \text{ degree} = 60 \text{ minutes (written as } 60')$$

$$1 \text{ minute} = 60 \text{ seconds (written as } 60'')$$

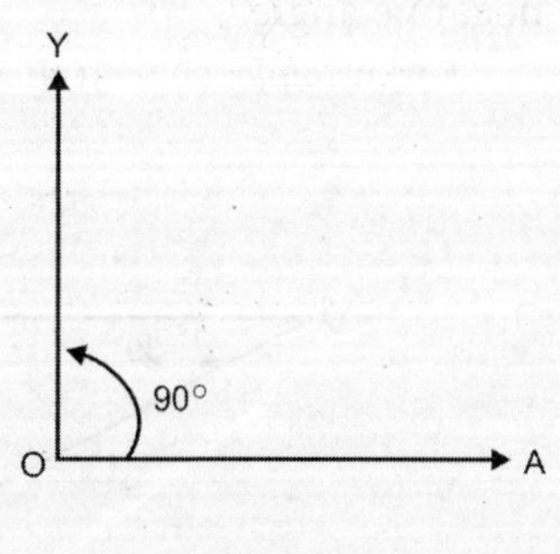

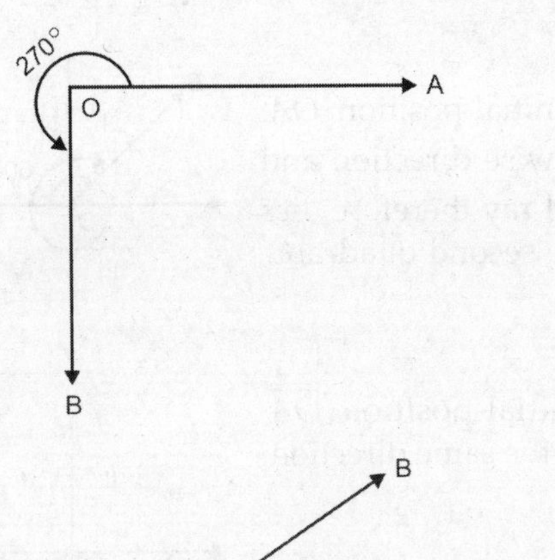

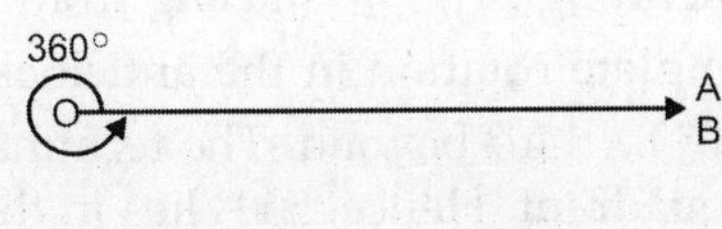

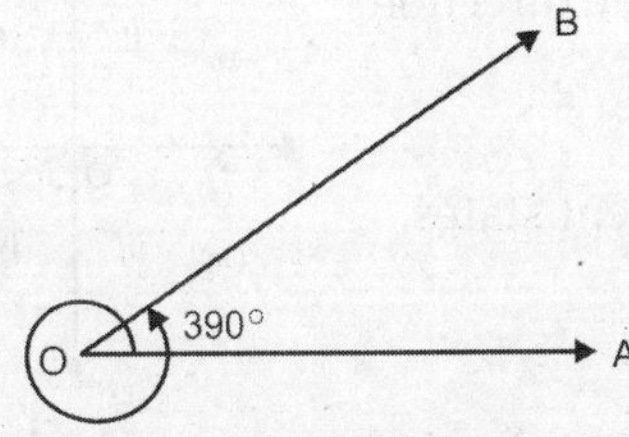

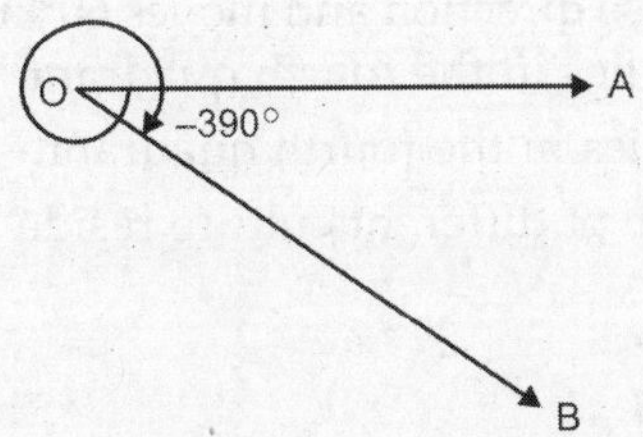

We denote the degrees, minutes, seconds by the symbols °, ′, ″ respectively.

Example 13. *Find the quadrants in which the following angles lie*

(*i*) 45° (*ii*) 120° (*iii*) 200° (*iv*) 300° (*v*) 500° (*vi*) – 60°

What value do you expect?

Solution.

(*i*) 45° lies in the I quadrant.

(*ii*) 120° lies in the II quadrant.

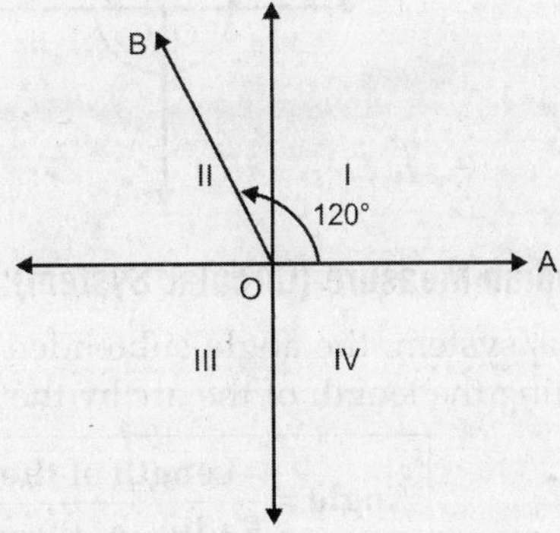

(*iii*) 200° lies in the III quadrant.

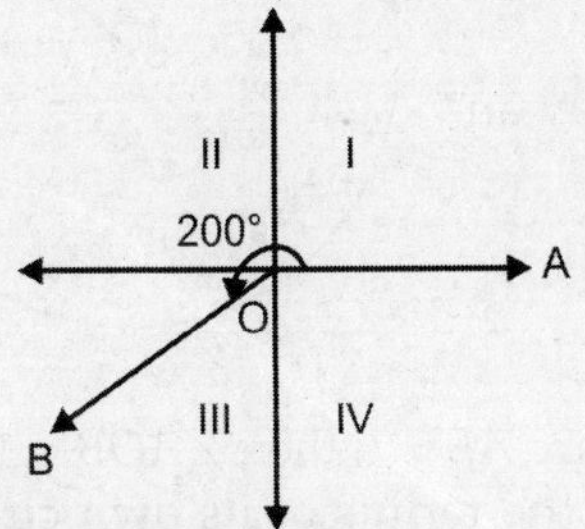

(*iv*) 300° lies in the IV quadrant.

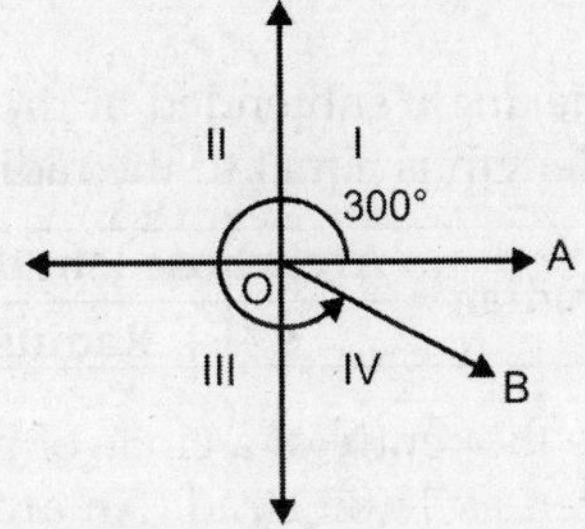

(v) $500° = (360° \times 1) + (90° + 50°)$

So, the generating ray $\overrightarrow{OB}$ starting from initial position $\overrightarrow{OA}$ makes one complete rotation in the anticlockwise direction and moves $90° + 50°$ *i.e.* $140°$ beyond. The terminal ray therefore, lies in the second quadrant. Hence, $500°$ lies in the second quadrant.

(vi) $-60°$

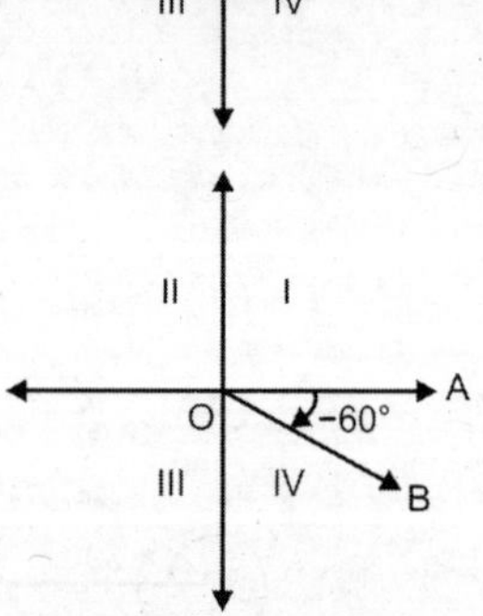

The generating ray $\overrightarrow{OB}$ starting from initial position $\overrightarrow{OA}$ rotates in clockwise direction and moves $60°$ in the same direction. The terminal ray lies in the fourth quadrant.

Hence $-60°$ lies in the fourth quadrant.

Value: People of different culture reside in different states.

Coterminal Angles

Two angles with different measures, but having the same initial sides and the same terminal sides are known as coterminal angles.

e.g. (i) The angles $60°$ and $-300°$ are coterminal.

(ii) The angles, $240°$ and $-120°$ are coterminal.

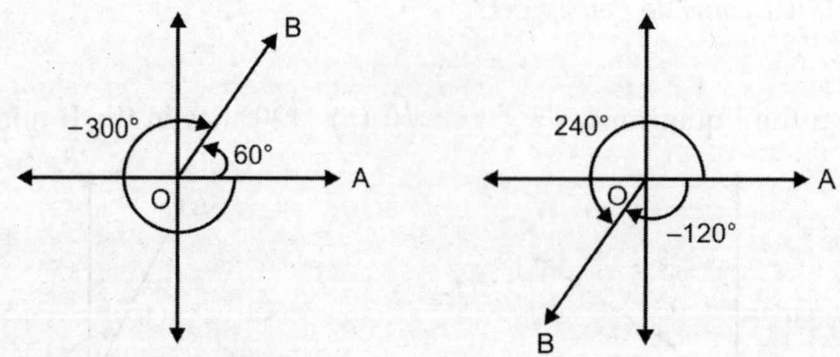

2. Radian Measure (Circular System):

In this system, the angle subtended by the arc at the centre of a circle is measured by dividing the length of the arc by the length of the radius.

$$\textbf{Angle} = \frac{\textbf{Length of the arc } \widehat{AB}}{\textbf{Radius of the circle, OA}}$$

In this system, the angle is measured in radians.

Radian

Radian is the angle subtended at the centre of a circle by an arc, whose length is equal to the radius.

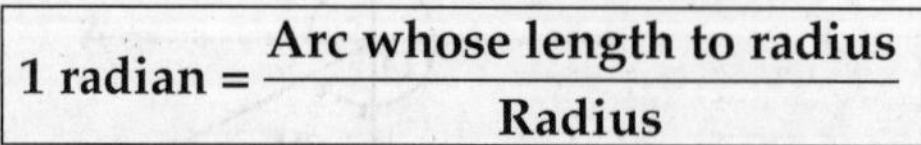

$$\textbf{1 radian} = \frac{\textbf{Arc whose length to radius}}{\textbf{Radius}}$$

Let O be the centre of a circle of radius r, cut off an arc $AB = r$, then $\angle AOB = 1$ radian and is written as $1°$ or 1 rad. An arc, which is 4 times the radius of its own circle will subtend an angle of 4 radians at the centre and so on.

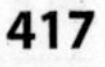

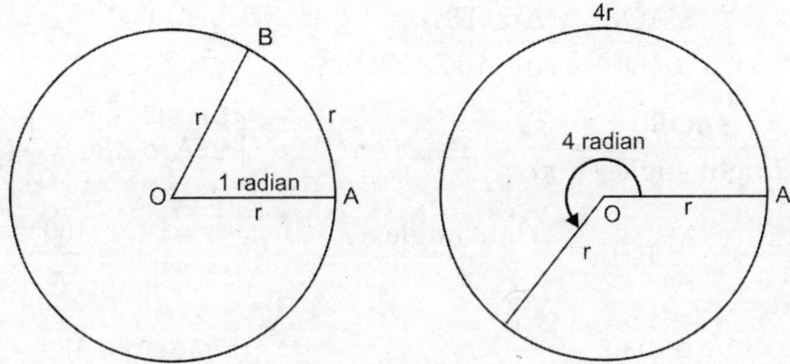

One complete rotation of the initial side subtends an angle of 2π radian at the centre.

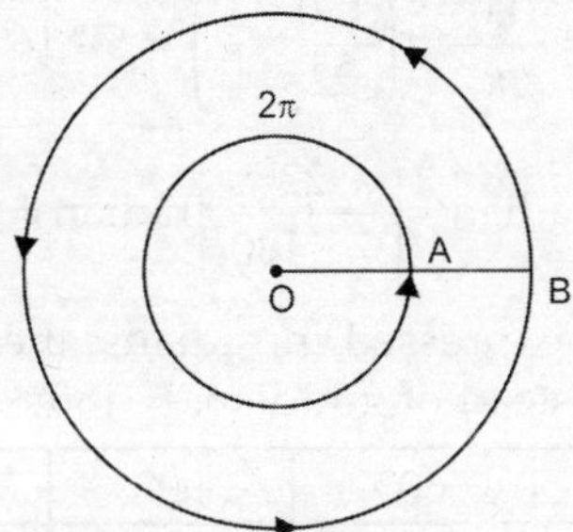

In one complete rotation, the length of the arc is equal to the circumference.

$$\text{Angle} = \frac{\text{Circumference}}{\text{Radius}} = \frac{2\pi r}{r} = 2\pi$$

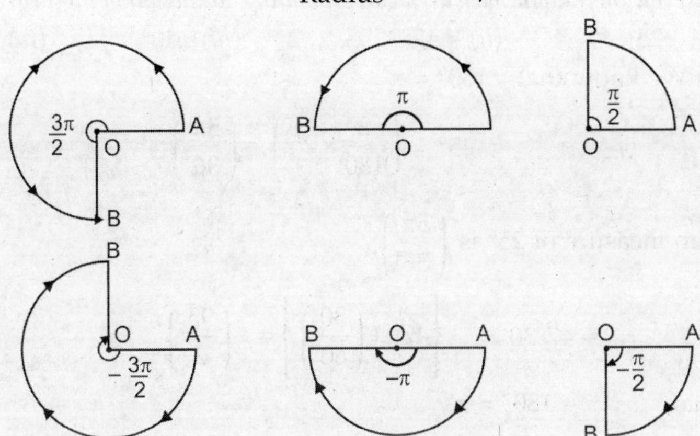

Note: **1.** '*c*' used in the notation of radian is the first letter of the word *circular system*.

2. When no unit is mentioned with an angle, it is always understood to be in radians.

13.18 RELATIONSHIP BETWEEN DEGREES AND RADIANS

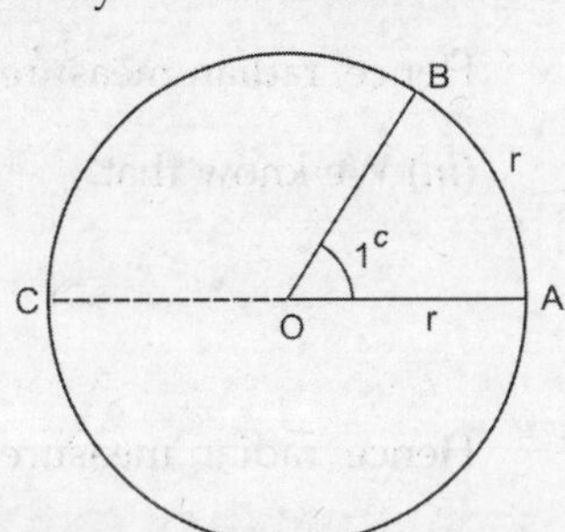

Draw any circle. Let O be its centre and r its radius. Let arc AB be equal to the radius. Then, $\angle AOB$ = 1 radian. Extend AO to meet the circle at C.

$\therefore$ $\angle AOC$ = a straight angle = 2 right angles

We know that the angles at the centre of a circle are proportional to the arcs subtending them.

$$\therefore \quad \frac{\angle AOB}{\angle AOC} = \frac{\text{Arc } AB}{\text{Arc } ABC}$$

$$\Rightarrow \quad \frac{\angle AOB}{2 \text{ right angles}} = \frac{r}{\pi r} \qquad \left[\because \text{Arc } ABC = \frac{1}{2}(\text{circumference})\right]$$

$$\Rightarrow \quad \angle AOB = \frac{2 \text{ right angles}}{\pi} \quad \Rightarrow \quad 1^c = \frac{180^\circ}{\pi}$$

Hence $\quad$ one radian $= \dfrac{180^\circ}{\pi} \quad \Rightarrow \quad \pi$ radian $= 180^\circ$

Corollary **1.** $\quad 1 \text{ radian} = \dfrac{180^\circ}{\pi} = \left(\dfrac{180}{22} \times 7\right)^\circ = 57^\circ 16'$ (approx.)

2. $\quad 1^\circ = \dfrac{\pi}{180}$ radian $= \left(\dfrac{22}{7 \times 180}\right)$ radian $= 0.01746$ radian.

Remark: When an angle is expressed in radians, the word radian is often omitted. Thus, we may write $\pi = 180^\circ$ instead of π radians $= 180^\circ$.

Degree	30°	45°	60°	90°	180°	270°	360°
Radians	$\frac{\pi}{6}$	$\frac{\pi}{4}$	$\frac{\pi}{3}$	$\frac{\pi}{2}$	π	$\frac{3\pi}{2}$	2π

Example 14. *Find the radian measures corresponding to the following degree measures*

(*i*) 25° $\quad$ (*ii*) $-47^\circ 30'$ $\quad$ (*iii*) 240° $\quad$ (*iv*) 520°

Solution. (*i*) We know that, $180^\circ = \pi^C$

$$\therefore \quad 25^\circ = \left(\frac{\pi}{180} \times 25\right)^c = \left(\frac{5\pi}{36}\right)^c$$

Hence, radian measure of 25° is $\left(\dfrac{5\pi}{36}\right)^c$. **Ans.**

(*ii*) We have, $\quad -47^\circ 30' = -\left\{47^\circ + \left(\dfrac{30}{60}\right)^\circ\right\} = -\left(\dfrac{95}{2}\right)^\circ \qquad [\because 1^\circ = 60']$

We know that, $\quad 180^\circ = \pi^c$

$$\therefore \quad -\left(\frac{95}{2}\right)^\circ = -\left(\frac{\pi}{180} \times \frac{95}{2}\right)^c = \left(\frac{19\pi}{72}\right)^c$$

Hence, radian measure of $-47^\circ 30'$ is $-\left(\dfrac{19\pi}{72}\right)^c$. **Ans.**

(*iii*) We know that, $\quad 180^\circ = \pi^c$

$$\therefore \quad 240^\circ = \left(\frac{\pi}{180} \times 240\right)^c = \left(\frac{4\pi}{3}\right)^c$$

Hence, radian measure of 240° is $\left(\dfrac{4\pi}{3}\right)^c$. **Ans.**

(*iv*) We know that, $180° = \pi^c$

$$\therefore \quad 520° = \left(\frac{\pi}{180} \times 520\right)^c = \left(\frac{26}{9}\pi\right)^c$$

Hence, radian measure of 520° is $\left(\frac{26}{9}\pi\right)^c$. **Ans.**

Example 15. *Find the degree measures corresponding to the following radian measures:*

(*i*) $\frac{5\pi}{3}$ (*ii*) $\frac{11}{16}$ (*iii*) -4

Solution. (*i*) We know that, $\pi^c = 180°$

$$\therefore \quad \left(\frac{5\pi}{3}\right)^c = \left(\frac{180}{\pi} \times \frac{5\pi}{3}\right)^\circ = 300°$$

Hence, the degree measure of $\frac{5\pi}{3}$ radian is 300°. **Ans.**

(*ii*) We know that, $\pi^c = 180°$

$$\therefore \quad \left(\frac{11}{16}\right)^c = \left(\frac{180}{\pi} \times \frac{11}{16}\right)^\circ = \left(\frac{180}{22} \times 7 \times \frac{11}{16}\right)^\circ = \left(\frac{315}{8}\right)^\circ = 39° + \left(\frac{3}{8}\right)^\circ$$

$$= 39° + \left(\frac{3 \times 60}{8}\right)' = 39° + 22' + \left(\frac{4}{8}\right)'$$

$$= 39° + 22' + \left(\frac{4 \times 60}{8}\right)'' = 39° + 22' + \left(\frac{240}{8}\right)'' = 39° + 22' + 30'' = 39°22'30''$$

Hence, degree measure of $\frac{11}{16}$ radian is 39°22′30″. **Ans.**

(*iii*) We know that, $\pi^c = 180°$ $\left(\pi = \frac{22}{7}\right)$

$$\therefore \quad -4^c = -\left(\frac{180}{\pi} \times 4\right) = -\frac{180 \times 7 \times 4}{22} = -\left(\frac{2520}{11}\right)^\circ = -\left(229\frac{1}{11}\right)^\circ$$

$$= -\left[229° + \left(\frac{60}{11}\right)'\right] = -\left[229° + \left(5\frac{5}{11}\right)'\right]$$

$$= -\left[229° + 5' + \left(\frac{5 \times 60}{11}\right)''\right] = -\left[229° + 5' + \left(27\frac{3}{11}\right)''\right] = -229°5'27'' \text{ (approx).}$$

Hence, degree measure of -4^c radian is $-229°5'27''$. **Ans.**

Example 16. *Express 5°37′30″ into radians.*

Solution. We know that, $30'' = \left(\frac{30}{60}\right)' = \left(\frac{1}{2}\right)'$

$$\therefore \quad 37'30'' = \left(37\frac{1}{2}\right)' = \left(\frac{75}{2 \times 60}\right)^\circ = \left(\frac{5}{8}\right)^\circ$$

So, $\quad 5°37'30'' = \left(5\frac{5}{8}\right)^\circ = \left(\frac{45}{8}\right)^\circ$

We know that, $\quad 180° = \pi^c$

$\therefore \quad \left(\frac{45}{8}\right)^\circ = \left(\frac{\pi}{180} \times \frac{45}{8}\right)^c = \left(\frac{\pi}{32}\right)^c$ **Ans.**

Example 17. *Find the angle between the minute hand and the hour hand of a clock at*

(i) 4:30 P.M. *(ii) 10:40 P.M.*

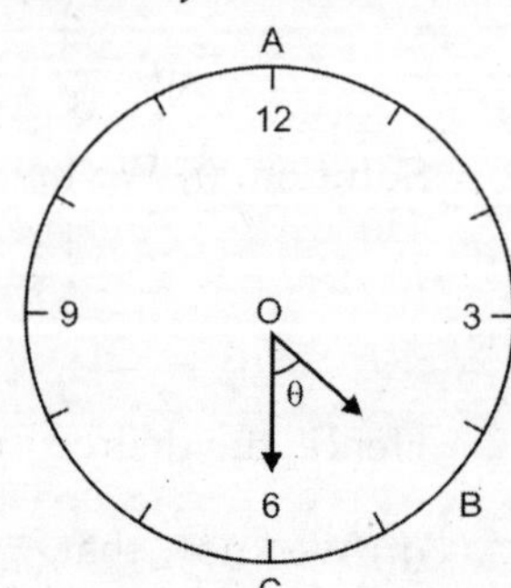

Solution. We know that the hour hand completes one rotation in 12 hours whereas the minute hand completes one rotation in one hour.

(*i*) At 4:30 P.M.: Let the hour hand and the minute hand be along *OB* and *OC* respectively.

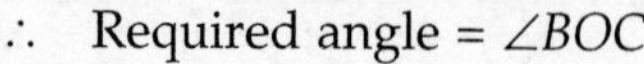

$\therefore$ Required angle = $\angle BOC$

= (Angle subtended by the minute hand in 30 minutes) – (angle subtended by the hour-hand in 4 hours 30 minutes, *i.e.* $\frac{9}{2}$ hours)

$$= \left(\frac{360}{60} \times 30\right)^\circ - \left(\frac{360}{12} \times \frac{9}{2}\right)^\circ = 180° - 135° = 45°$$ **Ans.**

(*ii*) At 10:40 P.M.: Let the hour hand and the minute hand be along *OC* and *OB* respectively.

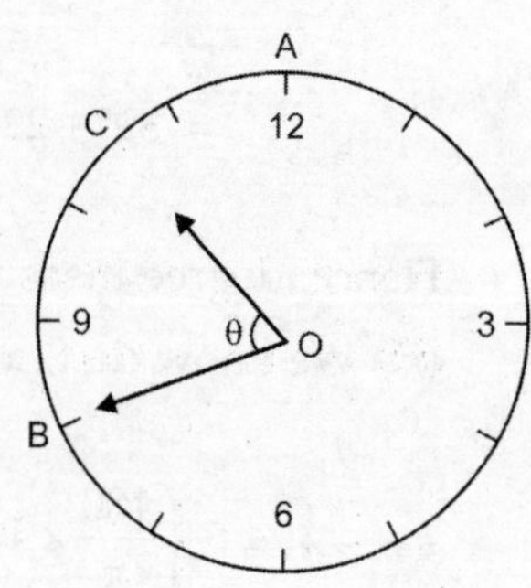

Required angle = $\angle BOC$

= (angle subtended by the hour hand in 10 hours 40 minutes *i.e.* $10\frac{2}{3}$ hours) – (angle subtended by the minute hand in 40 minutes)

$$= \left(\frac{360}{12} \times 10\frac{2}{3}\right)^\circ - \left(\frac{360}{60} \times 40\right)^\circ = \left(30 \times \frac{32}{3}\right)^\circ - (240)$$

$= 320° - 240° = 80°$ **Ans.**

Example 18. *Express in circular measures and also in degrees the angle of*

(i) a regular octagon *(ii) a regular polygon of 40 sides*

Solution. (*i*) A regular octagon has 8 equal sides.

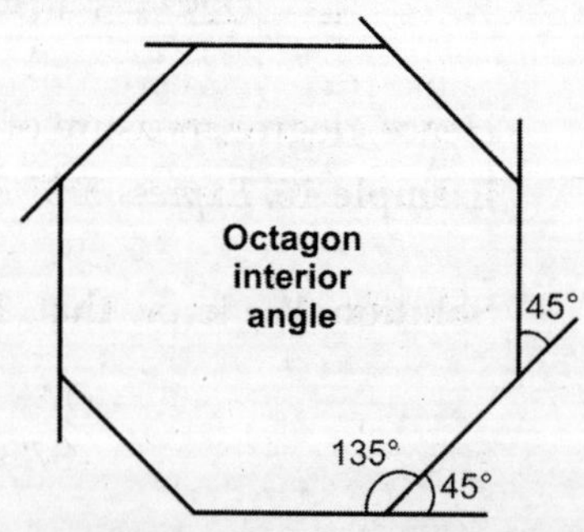

The sum of the 8 exterior angles = 360°

$\therefore$ Each exterior angle = $\frac{360°}{8} = 45°$

$\therefore$ Each interior angle = $180° - 45° = 135°$

Now, we know that $180° = \pi$ radian

$\therefore \quad 135° = \left(\frac{\pi}{180} \times 135\right)$ radians = $\frac{3\pi}{4}$ radians

Hence, the angle of regular octagon is $135° = \frac{3\pi}{4}$ radians.

(*ii*) Each exterior angle of a regular polygon of 40 sides $= \frac{360°}{40} = 9°$

$\therefore$ Each interior angle $= 180° - 9° = 171°$

Now, we know that, $180° = \pi$ radians

$$171° = \left(\frac{\pi}{180} \times 171\right) \text{radians} = \frac{19\pi}{20} \text{ radians}$$

Hence, the angle of a regular polygon of 40 sides is $\frac{19\pi}{20}$. **Ans.**

EXERCISE 13.3

1. In which quadrant do the following angles lie:

(*i*) 60° (*ii*) 135° (*iii*) – 45° (*iv*) 570°

Ans. (*i*) I (*ii*) II (*iii*) IV (*iv*) III

2. Find the radian measures corresponding to the following degree measures:

(*i*) 110° (*ii*) 490° (*iii*) – 270° (*iv*) 625°

Ans. (*i*) $\frac{11\pi}{18}$ (*ii*) $\frac{49\pi}{18}$ (*iii*) $-\frac{3\pi}{2}$ (*iv*) $\frac{125\pi}{36}$

3. Find the degree measures corresponding to the following radian measures:

(*i*) $\frac{\pi}{6}$ (*ii*) $-\frac{5\pi}{12}$ (*iii*) $\frac{18\pi}{5}$ (*iv*) $-\frac{7\pi}{18}$ (*v*) $\frac{1}{4}$

(*vi*) – 2 (*vii*) $\frac{1}{3}$

Ans. (*i*) 30° (*ii*) – 75° (*iii*) 648° (*iv*) – 70° (*v*) $14° \, 19' \, 5\frac{5''}{11}$ (*vi*) –114° 32′ 43.64″
(*vii*) 19° 5′ 27″

4. Find the radian measures corresponding to the following degree measures:

(*i*) 50° 37′ 30″ (*ii*) – 22° 30′ **Ans.** (*i*) $\frac{9\pi}{32}$ (*ii*) $-\frac{\pi}{8}$

5. Find in radians the angle of a regular:

(*i*) Pentagon (*ii*) Hexagon (*iii*) Decagon **Ans.** (*i*) $\frac{3\pi}{5}$ (*ii*) $\frac{2\pi}{3}$ (*iii*) $\frac{4\pi}{5}$

6. Find the angle between the minute hand of clock and the hour hand, when the time is 7 : 20 A.M. **Ans.** 100°

7. The difference of two angles is 20° and their sum is 2 radians. Find each angle in degrees. **Ans.** 67° 16′ 22″ and 47° 16′ 22″

Hints to Selected Questions

1. (*ii*) 135° = 90° + 45° (*iii*) The terminal ray will rotate in clockwise direction.

2. (*i*) $180° = \pi^c \Rightarrow 110° = \left(\frac{\pi}{180} \times 110\right) = \frac{11\pi}{18}$ radians

3. (*i*) $\pi^c = 180° \Rightarrow \left(\frac{\pi}{6}\right)^c = \left(\frac{180}{\pi} \times \frac{\pi}{6}\right)^\circ = 30°$

(*v*) $\pi^c = 180° \Rightarrow \frac{1}{4} = \left(\frac{180}{\pi} \times \frac{1}{4}\right)^\circ = \left(\frac{180 \times 7}{22} \times \frac{1}{4}\right)^\circ = \frac{315°}{22} = 14°\,19'\,5\frac{5''}{11}$

4. (*i*) $50°\,37'\,30'' = 50° + \left(\frac{37}{60}\right)^\circ + \left(\frac{30}{60 \times 60}\right)^\circ = 50° + \left(\frac{37}{60}\right)^\circ + \left(\frac{1}{120}\right)^\circ$

$$= \left(\frac{6000 + 74 + 1)}{120}\right)^\circ = \left(\frac{6075}{120}\right)$$

$\therefore$ Now, $\left(\frac{6075}{120}\right)^\circ = \left(\frac{\pi}{18} \times \frac{6075}{120}\right)^c = \frac{9\pi}{32}$ radians

(*ii*) $-22°\,30' = -\left(22° + \frac{30}{60}\right)^\circ = -\left(22 + \frac{1}{2}\right)^\circ = -\left(\frac{45}{2}\right)^\circ$

$\therefore$ Now $-\left(\frac{45}{2}\right)^\circ = -\left(\frac{\pi}{180} \times \frac{45}{2}\right)^c = -\frac{\pi}{8}$

5. (*i*) A regular pentagon has 5 equal sides.

The sum of the 5 equal exterior angles = 360°

$\therefore$ Each exterior angle $= \frac{360°}{2} = 72°$

$\therefore$ Each interior angle = 180° – 72° = 108°

Now, we know that 180° = π radians

(*ii*) A regular hexagon has 6 equal sides.

(*iii*) A regular decagon has 10 equal sides.

6. When the time in a clock is 7:20, the minute hand is at mark 4 and the hour hand has crossed $\frac{20}{60} = \frac{1}{3}$rd of the angle between 7 and 8.

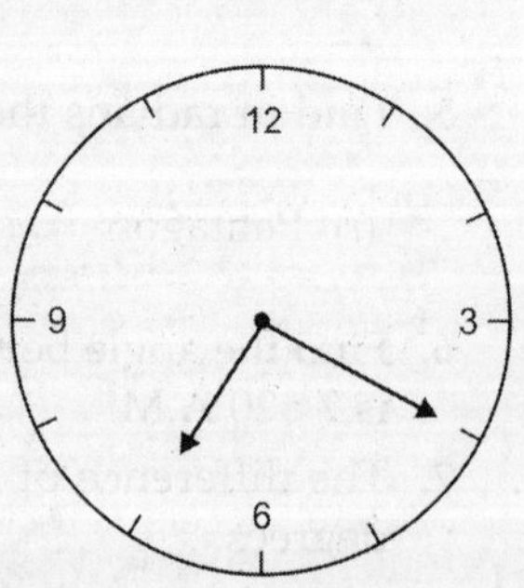

Now, the angle between two consecutive marks in a clock

$= \left(\frac{360}{12}\right)^\circ = 30°$

$\therefore$ The required angle between the two hands at the time 7 : 20

$= 3 \times 30° + \frac{1}{3}(30°) = 90° + 10° = 100°$

7. Let the angle be θ_1 and θ_2.

Given $\theta_1 - \theta_2 = 20°$ ... (1) $\qquad \pi^c = 180°$

$$\theta_1 + \theta_2 = 2 \text{ radians} = \left(\frac{1260}{11}\right)^\circ \quad \text{... (2)} \qquad 2^c = \left(\frac{180}{\pi}\times 2\right)^\circ = \left(\frac{180\times 7}{22}\times 2\right)^\circ$$

Adding (1) and (2), we get $\qquad = \left(\frac{1260}{11}\right)^\circ$

$$2\theta_1 = 20° + \left(\frac{1260}{11}\right)^\circ = \frac{1480}{11}$$

$$\Rightarrow \quad \theta_1 = \frac{1480}{2\times 11} = \frac{740}{11} = 67°16'22''$$

Putting $\theta_1 = 67°16'22''$ in (1), we get $67°16'22'' - \theta_2 = 20°$

$\Rightarrow \theta_2 = 67°16'22'' - 20° = 47°16'22''$

13.19 RELATIONSHIP AMONG ARC, RADIUS AND ANGLE

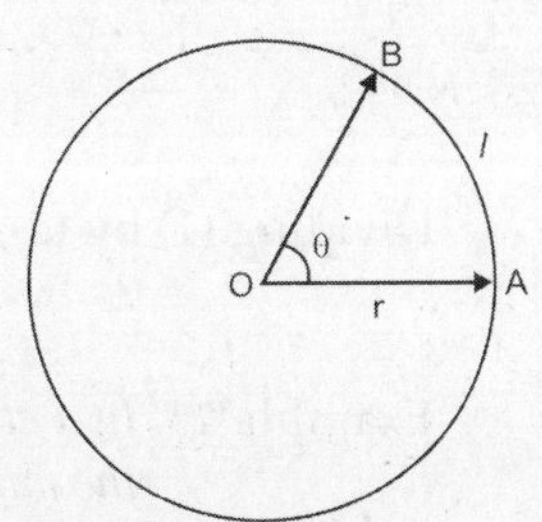

$$\text{Angle} = \frac{\text{Length of the arc}}{\text{Radius}}$$

$$\boxed{\theta = \frac{l}{r}}$$

Example 19. *Find the length of an arc of a circle of radius 3 cm, if the angle subtended at the centre is 30° (π = 3.14).*

Solution. Let l be the length of the arc. We know that, Angle $\theta = \frac{l}{r}$

Now, $\qquad r = 3,\ \theta = 30° = \left(\frac{30}{180}\pi\right)^c = \left(\frac{\pi}{6}\right)^c$

$$\Rightarrow \quad \frac{\pi}{6} = \frac{l}{3} \quad \Rightarrow \quad l = \frac{3.14\times 3}{6} = 1.57 \text{ cm}$$

Hence, the arc length is 1.57 cm. **Ans.**

Example 20. *If in two circles, arcs of the same length subtend angles of 60° and 75° at the centre, find the ratio of their radii.*

Solution. Let the radii of the two circles be r_1 and r_2 respectively. Also, let the length of arc in each case be l.

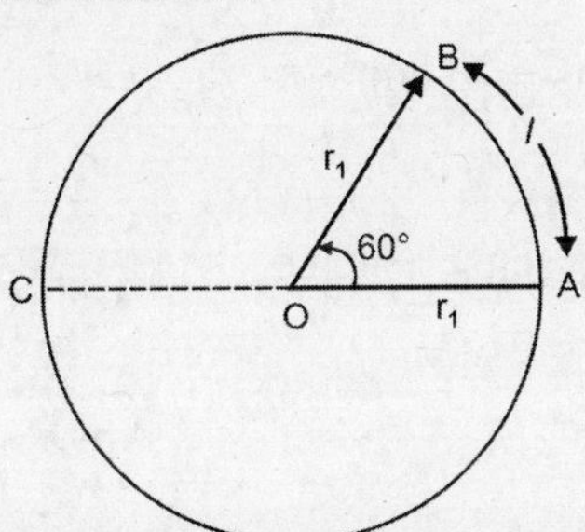

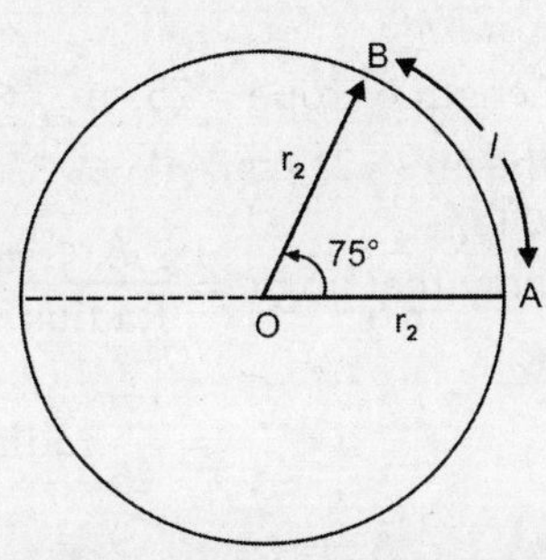

For first circle:

$$\theta = 60^\circ = 60 \times \frac{\pi}{180} = \frac{\pi}{3} \text{ radians.}$$

We know that, $\theta = \frac{l}{r} \Rightarrow r = \frac{l}{\theta} \Rightarrow r_1 = \frac{l}{\frac{\pi}{3}} \Rightarrow r_1 = \frac{3l}{\pi}$ ... (1)

***For second circle*:**

$$\theta = 75^\circ = 75 \times \frac{\pi}{180} = \frac{5\pi}{12} \text{ radians.}$$

We know that, $\theta = \frac{l}{r} \Rightarrow r = \frac{l}{\theta} \Rightarrow r_2 = \frac{l}{\frac{5\pi}{12}}$

$\Rightarrow$ $r_2 = \frac{12l}{5\pi}$ ... (2)

Dividing (1) by (2), we get $\frac{r_1}{r_2} = \frac{\frac{3l}{\pi}}{\frac{12l}{5\pi}} \Rightarrow \frac{r_1}{r_2} = \frac{15}{12} = \frac{5}{4}$ **Ans.**

Example 21. *In a circle of diameter 40 cm the length of a chord is 20 cm. Find the length of the minor arc of the chord.*

Solution. We have, radius $(r) = \frac{40}{2} = 20$ cm

And length of chord = 20 cm $\Rightarrow$ ΔOAB is an equilateral triangle

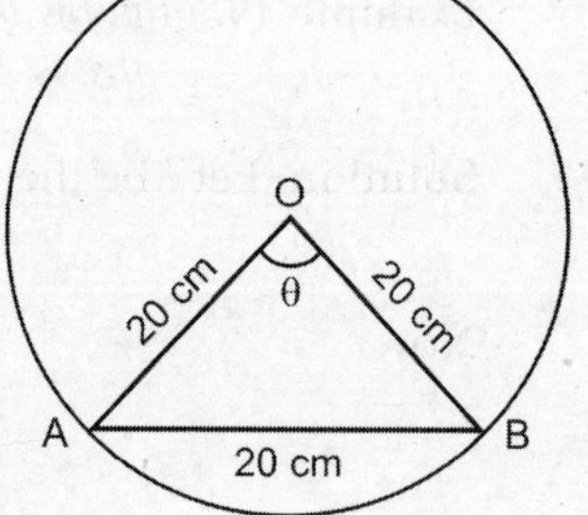

$\therefore$ $\theta = 60^\circ = \left(60 \times \frac{\pi}{180}\right) = \frac{\pi}{3}$

Also, we know that $\theta = \frac{\text{Arc } AB}{r}$

$\Rightarrow$ Arc $AB = r \times \theta = 20 \times \frac{\pi}{3} = \frac{20}{3}\pi$ cm. **Ans.**

Example 22. *Find the angle in radian through which a pendulum swings and its length is 75 cm and the tip describes an arc of length 21 m.*

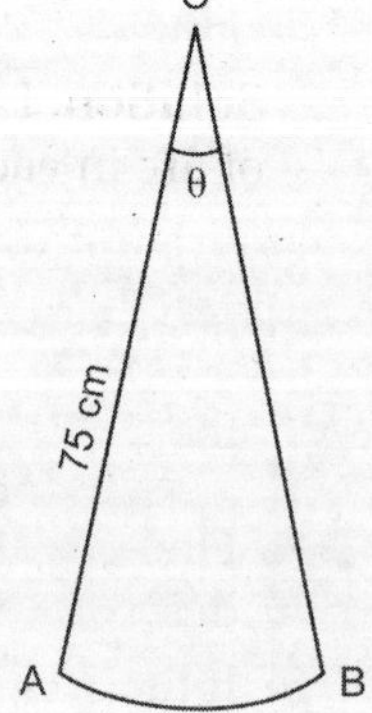

Solution. Length of rope = 75 cm. $\Rightarrow$ radius = 75 cm.

Length of the arc = 21 cm.

And, we know that angle $= \frac{\text{Arc}}{\text{Radius}} = \frac{21}{75}$ radians

$= \frac{7}{25}$ radians **Ans.**

Example 23. *A horse is tied to a post by a rope. If the horse moves along a circular path always keeping the rope tight and describes 88 metres when it has traced out 72° at the centre, find the length of the rope. Write the value found in it.*

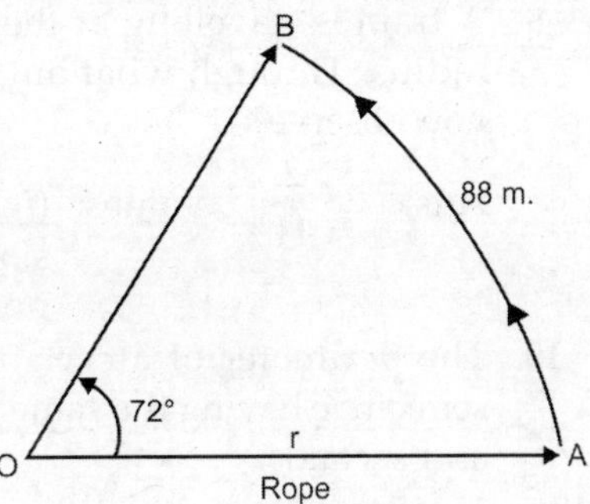

Solution. Let us denote the position of the post by the point O and let OA be the length of the rope in tight position. Suppose that the horse moves along the arc AB, so that $\angle AOB = 72°$ and arc $AB = 88$ m.

Let $$OA = r \text{ metres}$$

We know that $$\angle AOB = \frac{\text{Arc } AB}{\text{Radius}}$$

$$\Rightarrow \quad \text{Length of the rope } (r) = \frac{\text{Arc } AB}{\angle AOB} = \frac{88}{72 \times \frac{\pi}{180}} = \left(\frac{88 \times 5}{2\pi}\right) = 88 \times \frac{5}{2} \times \frac{7}{22} = 70 \text{ m}$$

Hence, the length of rope is 70 m. **Ans.**

Values: (*i*) We should be kind to animals (*ii*) Horses are stronger. (*iii*) A disciplined person never crosses his boundary.

EXERCISE 13.4

1. Find the length of arc intercepted by the given central angle in a circle, with radius as given below:

   (*i*) $\frac{\pi}{4}, r = 3$ (*ii*) $60°, r = 5$ (*iii*) $-120°, r = 2$ **Ans.** (*i*) $\frac{3\pi}{4}$ (*ii*) $\frac{5\pi}{3}$ (*iii*) $\frac{4\pi}{3}$

2. Find the degree measure of a central angle intercepting an arc l with length as given, in a circle with radius r as given below:

   (*i*) $l = \pi, r = 9$ (*ii*) $l = \frac{10\pi}{3}, r = 5$ **Ans.** (*i*) 20° (*ii*) 120°

3. The minute hand of a watch is 1.5 cm long. How far does its tip move in 50 minutes? (Use $\pi = 3.14$) **Ans.** 7.85 cm

4. A wheel makes 180 revolutions in one minutes. Through how many radians does it turn in one second? **Ans.** 6π radians

5. Find the degree measure of the angle subtended at the centre of a circle of diameter 200 cm. by an arc of length 22 cm. $\left(\text{Use } \pi = \frac{22}{7}\right)$ **Ans.** 12°36′

6. A circular wire of radius 12 cm is cut and bent, so as to lie along the circumference of a hoop whose radius is 96 cm. Find in radians the angle, which is subtended at the centre of the loop. **Ans.** $\frac{\pi}{4}$

7. Find the angle through which a pendulum swings if its length is 50 cm and the tip describes an arc of 16 cm. **Ans.** 18°19′38″

8. If the arcs of the same length in two circles subtend angles 75° and 120° at the centre, find the ratio of their radii. **Ans.** 8 : 5

9. A train is travelling at the rate of 20 km/h on a circular curve of a half kilometre radius. Through what angle in degrees has it turned in a minute? What value do you observe?

Ans. $\left(38\frac{2}{11}\right)^\circ$ Values: (*i*) Punctuality (*ii*) Safety comes first (*iii*) We should drive slow on circular curve to avoid accidents. (*iv*) Every Life is precious.

10. The perimeter of a certain sector of a circle is equal to the length of the arc of semicircle having the same radius. Express the angle of the sector in degrees, minutes and seconds. **Ans.** 65°27′16″

Hints to the Selected Questions

1. (*i*) Use $\theta = \frac{l}{r}$

3. $r = 1.5$ cm, $\theta = 50 \times 6° = 300° = \frac{5\pi}{3}$

4. Number of rotation made by the wheel in 1 sec. $= \frac{180}{60} = 3$

Radians turned by wheel in 3 rotation = $(3 \times 2\pi = 6\pi)$ radians

5. $r = \frac{200}{2} = 100$ cm, $l = 22$ cm, $\theta = \frac{l}{r} = \frac{22}{100}$ radians

6. $l = 12 \times 2\pi = 24\pi$, $r = 96$ cm, use $\theta = \frac{l}{r}$

8. For first circle; $r = r_1$ and $\theta = 75° = \frac{5}{12}\pi$ radians

For second circle; $r = r_2$, $\theta = 120° = \frac{2\pi}{3}$ radians use $\theta = \frac{l}{r}$

9. Here $r = \frac{1}{2}$ km = 500 m, Speed of train = 20 km/h

Distance moved in 1 hour (3600 seconds) = 20 km = 20,000 m.

l = Distance moved in 60 sec. (1 min) = $\left(\frac{20,000}{3600} \times 60\right)$ m $= \frac{2000}{6}$ m

$$\theta = \frac{l}{r} = \frac{2000}{6 \times 5000} = \frac{2}{3} \text{ radians}$$

$$= \left(\frac{2}{3} \times \frac{180°}{\pi}\right) = \left(\frac{2}{3} \times 180 \times \frac{7}{22}\right)^\circ = \frac{420}{11} = \left(38\frac{2}{11}\right)^\circ$$

10. We have perimeter of the sector = $r + r + s$.

But $\theta = \frac{s}{r} \Rightarrow s = \theta r$

Perimeter of the sector = $r + r + r\,\theta$

$\Rightarrow r(2 + \theta) = \pi\, r$ (given)

$\Rightarrow 2 + \theta = \pi$

$\Rightarrow \theta = (\pi - 2)$ radians

$$= \left(\frac{22}{7} - 2\right) \text{radians} = \left(\frac{8}{7} \times \frac{180}{\pi}\right)^\circ = \left(\frac{8}{7} \times \frac{180}{22} \times 7\right)^\circ$$

$= (65.455)° = 65° + 0.455° = 65° + (0.455 \times 60)'$

$= 65° + (27.3)' = 65° + 27' + (0.3)' = 65°27'18''.$

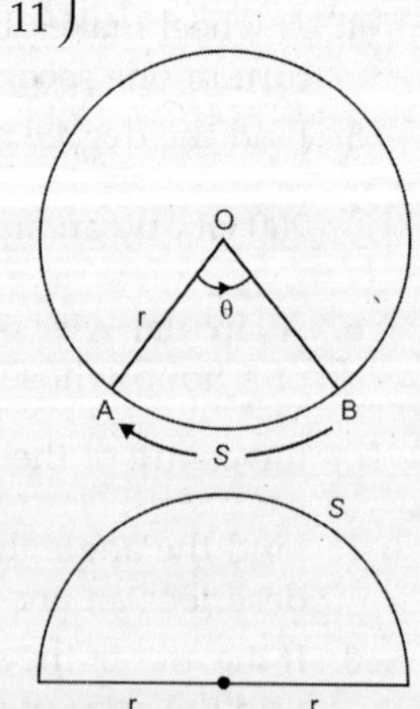

Factorization and Defactorization Formulae

14.1 INTRODUCTION

Compound angles are algebraic sum of two or more angles known as the constituent angles.

For example: If θ_1, θ_2, θ_3 are three constituent angles, then

$$\theta_1 \pm \theta_2,\ \theta_2 \pm \theta_3,\ \theta_1 \pm \theta_2 \pm \theta_3 \text{ are compound angles.}$$

In this chapter, we shall derive formulae which will express the trigonometric ratios of compound angles in terms of trigonometric ratios of constituent angles.

14.2 FACTORIZATION FORMULAE

Trigonometric Ratios of Sum and Difference of Angles

Summation Formulae

We shall prove that for all values of A and B

(*i*) $\sin(A + B) = \sin A \cdot \cos B + \cos A \cdot \sin B$

(*ii*) $\cos(A + B) = \cos A \cdot \cos B - \sin A \cdot \sin B$

(*iii*) $\tan(A + B) = \dfrac{\tan A + \tan B}{1 - \tan \cdot A \tan B}$

Proof: Let the rotating line start from initial line OX and trace out $\angle XOG = \angle A$ in the anticlockwise direction and let the rotating line further rotates to trace out

$$\angle GOP = \angle B, \text{ so that } \angle XOP = \angle (A + B).$$

From the point P draw $PM \perp OX$ and $PQ \perp OG$. From Q, draw $QL \perp PM$ and $QN \perp OX$. Then,

$$\angle QPL = 180° - 90° - (\angle PQL) = 90° - (90° - A) = A$$

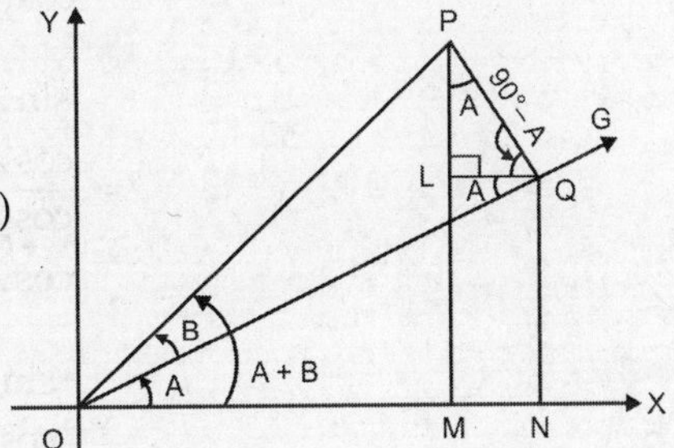

(*i*) $\sin(A + B) = \sin(\angle MOP) = \dfrac{PM}{OP} = \dfrac{PL + LM}{OP}$

$$= \frac{PL + QN}{OP} \qquad (\because LM = QN)$$

$$= \frac{PL}{OP} + \frac{QN}{OP} = \frac{PL}{PQ}\cdot\frac{PQ}{OP} + \frac{ON}{OQ}\cdot\frac{OQ}{OP}$$

$$= \cos A \sin B + \sin A \cos B.$$

$$\Rightarrow \quad \boxed{\sin(A+B) = \sin A \cdot \cos B + \cos A \cdot \sin B}$$

(*ii*) $$\cos(A+B) = \frac{OM}{OP} = \frac{ON - MN}{OP} = \frac{ON}{OP} - \frac{MN}{OP} = \frac{ON}{OP} - \frac{LQ}{OP} \quad (\because MN = LQ)$$

$$= \frac{ON}{OQ} \times \frac{OQ}{OP} - \frac{LQ}{PQ} \times \frac{PQ}{OP}$$

$$= \cos A \cos B - \sin A \sin B$$

$$\Rightarrow \quad \boxed{\cos(A+B) = \cos A \cdot \cos B - \sin A \cdot \sin B}$$

(*iii*) $$\tan(A+B) = \frac{PM}{OM} = \frac{PL + LM}{OM} = \frac{QN + PL}{ON - MN}$$

$$= \frac{QN + PL}{ON - LQ} \quad [\because MN = LQ]$$

$$= \frac{\frac{QN}{ON} + \frac{PL}{ON}}{1 - \frac{LQ}{ON}} = \frac{\frac{QN}{ON} + \frac{PL}{ON}}{1 - \frac{LQ}{PL} \cdot \frac{PL}{ON}} = \frac{\tan A + \tan B}{1 - \tan A \cdot \tan B}$$

$$\left(\because \frac{PL}{ON} = \frac{PL}{PQ} \cdot \frac{PQ}{OQ} \cdot \frac{OQ}{ON} = \cos A \cdot \tan B \cdot \sec A = \tan B\right)$$

$$\Rightarrow \quad \boxed{\tan(A+B) = \frac{\tan A + \tan B}{1 - \tan A \cdot \tan B}}$$

Note: (*i*) $\sin(A + B)$ is not equal to $\sin A + \sin B$ etc.

(*ii*) The formula for $\tan(A + B)$ is valid only when none of A, B and $A + B$ is a multiple of $\frac{\pi}{2}$.

Corollary: Derivation of formula for $\tan(A + B)$ from those of $\sin(A + B)$ and $\cos(A + B)$.

Proof: $$\tan(A+B) = \frac{\sin(A+B)}{\cos(A+B)} = \frac{\sin A + \cos B + \cos A \cdot \sin B}{\sin A + \cos B - \cos A \cdot \sin B}$$

$$= \frac{\frac{\sin A \cdot \cos B}{\cos A \cdot \cos B} + \frac{\cos A \cdot \sin B}{\cos A \cdot \cos B}}{\frac{\cos A \cdot \cos B}{\cos A \cdot \cos B} - \frac{\sin A \cdot \sin B}{\cos A \cdot \cos B}} \quad \left(\text{Dividing both numerator and denominator by } \cos A \cdot \cos B\right)$$

$$= \frac{\frac{\sin A}{\cos A} + \frac{\sin B}{\cos B}}{1 - \frac{\sin A}{\cos A} \cdot \frac{\sin B}{\cos B}} = \frac{\tan A + \tan B}{1 - \tan A \cdot \tan B}$$

Difference Formulae

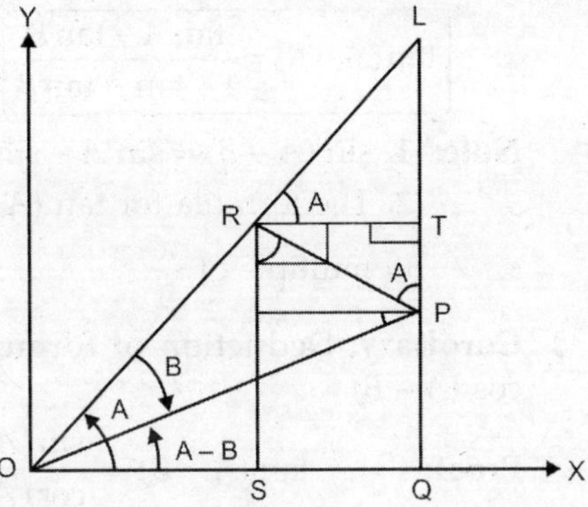

We shall prove that for all values of A and B

(*i*) $\sin(A-B) = \sin A\cdot\cos B - \cos A\cdot\sin B$

(*ii*) $\cos(A-B) = \cos A\cdot\cos B + \sin A\cdot\sin B$

(*iii*) $\tan(A-B) = \dfrac{\tan A - \tan B}{1+\tan A\tan B}$,

Proof: (*i*) Let the rotating line start from its initial position OX. Trace $\angle XOL = \angle A$ in anticlockwise direction and rotating back through angle B and occupy the final position OP, so that $\angle XOP = \angle(A-B)$.

From the point P, draw $PQ \perp OX$ and $PR \perp OL$, from R draw $RS \perp OX$ and RT perpendicular on QP (produced).

Now, $\angle TRL = \angle A$ (corresponding angles)

In both right angled ΔPRL and ΔRTP

$$\left.\begin{aligned}\angle LRT + \angle PRT = 90^\circ\\ \angle PRT + \angle RPT = 90^\circ\end{aligned}\right\} \Rightarrow \angle RPT = \angle LRT = \angle A$$

$$\therefore \quad \sin(A-B) = \sin\angle QOP = \frac{PQ}{OP} = \frac{TQ-TP}{OP}$$

$$= \frac{RS-TP}{OP} \qquad [\because TQ = RS]$$

$$= \frac{RS}{OP} - \frac{TP}{OP}$$

$$= \frac{RS}{OR}\times\frac{OR}{OP} - \frac{TP}{RP}\times\frac{RP}{OP} = \sin A\cos B - \cos A\sin B \qquad \textbf{Proved.}$$

$$\Rightarrow \quad \boxed{\sin(A-B) = \sin A\cdot\cos B - \cos A\cdot\sin B}$$

(*ii*)
$$\cos(A-B) = \frac{OQ}{OP} = \frac{OS+SQ}{OP} = \frac{OS+RT}{OP} \qquad [\because SQ = RT]$$

$$= \frac{OS}{OP} + \frac{RT}{OP} = \frac{OS}{OR}\times\frac{OR}{OP} + \frac{RT}{RP}\times\frac{RP}{OP}$$

$$= \cos A\cos B + \sin A\sin B \qquad \textbf{Proved.}$$

$$\Rightarrow \quad \boxed{\cos(A-B) = \cos A\cos B + \sin A\sin B}$$

(*iii*)
$$\tan(A-B) = \frac{PQ}{OQ} = \frac{TQ-TP}{OQ} = \frac{TQ-TP}{OS+SQ} = \frac{RS-TP}{OS+RT}$$

$$= \frac{\dfrac{RS}{OS} - \dfrac{TP}{OS}}{1+\dfrac{RT}{OS}} = \frac{\dfrac{RS}{OS} - \dfrac{TP}{OS}}{1+\dfrac{RT}{TP}\times\dfrac{TP}{OS}} = \frac{\tan A - \tan B}{1+\tan A\tan B} \qquad \textbf{Proved.}$$

$$\left(\because \frac{TP}{OS} = \frac{TP}{PR}\times\frac{PR}{OR}\times\frac{OR}{OS} = \cos A\cdot\tan B\sec A = \tan B\right)$$

$$\Rightarrow \quad \boxed{\tan(A-B)=\frac{\tan A-\tan B}{1+\tan A\tan B}}$$

Note: **1.** $\sin(A-B) \neq \sin A - \sin B$ etc.

2. The formula for $\tan(A-B)$ is valid only when none of A, B and $A-B$ is a multiple of $\frac{\pi}{2}$.

Corollary: Deduction of formula for $\tan(A-B)$ from those of $\sin(A-B)$ and $\cos(A-B)$.

Proof:

$$\tan(A-B)=\frac{\sin(A-B)}{\cos(A-B)}=\frac{\sin A\cdot\cos B-\cos A\cdot\sin B}{\cos A\cdot\cos B+\sin A\cdot\sin B}$$

$$=\frac{\dfrac{\sin A\cdot\cos B}{\cos A\cdot\cos B}-\dfrac{\cos A\cdot\sin B}{\cos A\cdot\cos B}}{\dfrac{\cos A\cdot\cos B}{\cos A\cdot\cos B}+\dfrac{\sin A\cdot\sin B}{\cos A\cdot\cos B}}\quad\left(\begin{array}{l}\text{Dividing both the}\\\text{numerator and}\\\text{denominator by}\\\cos A\cdot\cos B.\end{array}\right)$$

$$=\frac{\dfrac{\sin A}{\cos A}-\dfrac{\sin B}{\cos B}}{1+\dfrac{\sin A}{\cos A}\cdot\dfrac{\sin B}{\cos B}}=\frac{\tan A-\tan B}{1+\tan A\tan B}.$$

Proved.

Theorem: If none of angles, A, B and $(A \pm B)$ is a multiple of π, then prove that

(*i*) $\cot(A+B)=\dfrac{\cot A\cot B-1}{\cot B+\cot A}$ (*ii*) $\cot(A-B)=\dfrac{\cot A\cot B+1}{\cot B-\cot A}$

Proof: (*i*)

$$\cot(A+B)=\frac{\cos(A+B)}{\sin(A+B)}=\frac{\cos A\cdot\cos B-\sin A\cdot\sin B}{\sin A\cdot\cos B+\cos A\cdot\sin B}$$

$$=\frac{\dfrac{\cos A\cdot\cos B}{\sin A\cdot\sin B}-\dfrac{\sin A\cdot\sin B}{\sin A\cdot\sin B}}{\dfrac{\sin A\cdot\cos B}{\sin A\cdot\sin B}+\dfrac{\cos A\cdot\sin B}{\sin A\cdot\sin B}}=\frac{\dfrac{\cos A}{\sin A}\cdot\dfrac{\cos B}{\sin B}-1}{\dfrac{\cos B}{\sin B}+\dfrac{\cos A}{\sin A}}\quad\left(\begin{array}{l}\text{Dividing the}\\\text{numerator and}\\\text{denominator by}\\\sin A\sin B.\end{array}\right)$$

$$=\frac{\cot A\cot B-1}{\cot B+\cot A}.$$

Proved.

(*ii*)

$$\cot(A-B)=\frac{\cos(A-B)}{\sin(A-B)}=\frac{\cos A\cos B+\sin A\sin B}{\sin A\cos B-\cos A\sin B}$$

$$=\frac{\dfrac{\cos A\cos B}{\sin A\sin B}+\dfrac{\sin A\sin B}{\sin A\sin B}}{\dfrac{\sin A\cos B}{\sin A\sin B}-\dfrac{\cos A\sin B}{\sin A\sin B}}\quad\left(\begin{array}{l}\text{Dividing the numerator}\\\text{and denominator by}\\\sin A\sin B.\end{array}\right)$$

$$=\frac{\dfrac{\cos A}{\sin A}\cdot\dfrac{\cos B}{\sin B}+1}{\dfrac{\cos B}{\sin B}-\dfrac{\cos A}{\sin A}}=\frac{\cot A\cot B+1}{\cot B-\cot A}.$$

Proved.

14.3 SOME USEFUL FORMULAE

(*i*) $\sin(A+B)\cdot\sin(A-B) = \sin^2 A - \sin^2 B = \cos^2 B - \cos^2 A$

(*ii*) $\cos(A+B)\cdot\cos(A-B) = \cos^2 A - \sin^2 B = \cos^2 B - \sin^2 A$

(*iii*) $\tan(A+B+C) = \dfrac{\tan A + \tan B + \tan C - \tan A\cdot\tan B\cdot\tan C}{1-\tan A\cdot\tan B - \tan B\cdot\tan C - \tan C\cdot\tan A}$

Proof: (*i*) $\sin(A+B)\sin(A-B)$

$$= (\sin A\cdot\cos B + \cos A\cdot\sin B)(\sin A\cdot\cos B - \cos A\cdot\sin B)$$
$$= \sin^2 A\cos^2 B - \cos^2 A\sin^2 B$$
$$= \sin^2 A(1-\sin^2 B) - (1-\sin^2 A)(\sin^2 B)$$
$$= \sin^2 A - \sin^2 A\sin^2 B - \sin^2 B + \sin^2 A\sin^2 B = \sin^2 A - \sin^2 B$$
$$= (1-\cos^2 A) - (1-\cos^2 B) = \cos^2 B - \cos^2 A.$$

Proved.

(*ii*) $\cos(A+B)\cos(A-B)$

$$= (\cos A\cos B - \sin A\sin B)(\cos A\cos B + \sin A\sin B)$$
$$= \cos^2 A\cos^2 B - \sin^2 A\sin^2 B$$
$$= \cos^2 A(1-\sin^2 B) - (1-\cos^2 A)\sin^2 B$$
$$= \cos^2 A - \sin^2 B = (1-\sin^2 A) - (1-\cos^2 B)$$
$$= \cos^2 B - \sin^2 A$$

Proved.

(*iii*) $\tan(A+B+C) = \tan[(A+B)+C]$

$$= \frac{\tan(A+B)+\tan C}{1-\tan(A+B)\tan C}$$

$$= \frac{\dfrac{\tan A+\tan B}{1-\tan A+\tan B}+\tan C}{1-\left(\dfrac{\tan A+\tan B}{1-\tan A\tan B}\right)\tan C}$$

$$\Rightarrow \boxed{\mathbf{\tan(A+B+C) = \dfrac{\tan A+\tan B+\tan C-\tan A\tan B\tan C}{1-\tan A\tan B-\tan B\tan C-\tan C\tan A}}}$$

Proved.

Example 1. *Evaluate:*

(*i*) sin 75° (*ii*) cos 105° (*iii*) tan 75° + cot 75°

Solution. We have,

(*i*) $\sin 75° = \sin(45° + 30°)$

$$= \sin 45°\cos 30° + \cos 45°\sin 30°$$
$$= \left(\frac{1}{\sqrt{2}}\cdot\frac{\sqrt{3}}{2}\right)+\left(\frac{1}{\sqrt{2}}\cdot\frac{1}{2}\right) = \frac{\sqrt{3}}{2\sqrt{2}}+\frac{1}{2\sqrt{2}} = \frac{\sqrt{3}+1}{2\sqrt{2}} = \frac{(\sqrt{3}+1)\sqrt{2}}{2\sqrt{2}\times\sqrt{2}}$$
$$= \frac{\sqrt{6}+\sqrt{2}}{4}.$$

Ans.

(*ii*) $\cos 105° = \cos(60° + 45°)$

$$= \cos 60°\cos 45° - \sin 60°\sin 45°$$

$$= \left(\frac{1}{2}\times\frac{1}{\sqrt{2}}\right) - \left(\frac{\sqrt{3}}{2}\times\frac{1}{\sqrt{2}}\right)$$

$$= \frac{1}{2\sqrt{2}} - \frac{\sqrt{3}}{2\sqrt{2}} = \frac{1-\sqrt{3}}{2\sqrt{2}} = \frac{\sqrt{2}-\sqrt{6}}{4}$$ **Ans.**

(*iii*) $$\tan 75^\circ = \tan(45^\circ + 30^\circ) = \frac{\tan 45^\circ + \tan 30^\circ}{1 - \tan 45^\circ \tan 30^\circ}$$

$$= \frac{1+\frac{1}{\sqrt{3}}}{1-\frac{1}{\sqrt{3}}} = \frac{\sqrt{3}+1}{\sqrt{3}-1} \quad \text{... (1)}$$

and $$\cot 75^\circ = \frac{1}{\tan 75^\circ} = \frac{1}{\frac{\sqrt{3}+1}{\sqrt{3}-1}} = \frac{\sqrt{3}-1}{\sqrt{3}+1} \quad \text{[Using (1)]}$$

$\therefore$ $$\tan 75^\circ + \cot 75^\circ = \left(\frac{\sqrt{3}+1}{\sqrt{3}-1}\right) + \left(\frac{\sqrt{3}-1}{\sqrt{3}+1}\right)$$

$$= \frac{(\sqrt{3}+1)^2 + (\sqrt{3}-1)^2}{(\sqrt{3}-1)(\sqrt{3}+1)} = \frac{3+1+2\sqrt{3}+3+1-2\sqrt{3}}{3-1}$$

$$= \frac{8}{2} = 4$$ **Ans.**

Example 2. *Show that:*

(*i*) $\cos 70^\circ \cos 10^\circ + \sin 70^\circ \sin 10^\circ = \frac{1}{2}$

(*ii*) $\cos 130^\circ \cos 40^\circ + \sin 130^\circ \sin 40^\circ = 0$

Solution. (*i*) We have,

$$\text{L.H.S.} = \cos 70^\circ \cos 10^\circ + \sin 70^\circ \sin 10^\circ$$

$$= \cos(70^\circ - 10^\circ), \; [\because \cos(A - B) = \cos A \cos B + \sin A \sin B]$$

$$= \cos 60^\circ = \frac{1}{2} = \text{R.H.S.}$$ **Proved.**

(*ii*) $$\text{L.H.S.} = \cos 130^\circ \cdot \cos 40^\circ + \sin 130^\circ \cdot \sin 40^\circ$$

$$= \cos(130^\circ - 40^\circ) = \cos 90^\circ = 0 = \text{R.H.S.}$$ **Proved.**

Example 3. *Show that:*

(*i*) $\sin(40^\circ + \theta)\cos(10^\circ + \theta) - \cos(40^\circ + \theta)\sin(10^\circ + \theta) = \frac{1}{2}$

(*ii*) $\cos\left(\frac{\pi}{4} - \theta\right)\cos\left(\frac{\pi}{4} - \phi\right) - \sin\left(\frac{\pi}{4} - \theta\right)\sin\left(\frac{\pi}{4} - \phi\right) = \sin(\theta + \phi)$

Solution. (*i*) $$\text{L.H.S.} = \sin(40^\circ + \theta)\cos(10^\circ + \theta) - \cos(40^\circ + \theta)\sin(10^\circ + \theta)$$

$$= \sin\{(40^\circ + \theta) - (10^\circ + \theta)\}$$

$$[\because \sin(A - B) = \sin A \cos B - \cos A \sin B]$$

$$= \sin 30° = \frac{1}{2} = \text{R.H.S.}$$ **Proved.**

(*ii*) $$\text{L.H.S.} = \cos\left(\frac{\pi}{4}-\theta\right)\cos\left(\frac{\pi}{4}-\phi\right)-\sin\left(\frac{\pi}{4}-\theta\right)\sin\left(\frac{\pi}{4}-\phi\right)$$

$$= \cos\left\{\left(\frac{\pi}{4}-\theta\right)+\left(\frac{\pi}{4}-\phi\right)\right\} \quad [\because \cos(A+B) = \cos A \cos B - \sin A \sin B]$$

$$= \cos\left\{\frac{\pi}{2}-(\theta+\phi)\right\}$$

$$= \sin(\theta+\phi) = \text{R.H.S.} \quad \left[\because \cos\left(\frac{\pi}{2}-\theta\right) = \sin\theta\right]$$ **Proved.**

Example 4. *Find the value of tan 135°.* **[S.B.T.E. Delhi, 2015, 2011]**

Solution. We know that

$$\tan(180° - \theta) = -\tan\theta$$

$$\tan 135° = \tan(180° - 45°) = -\tan 45° = -1$$ **Ans.**

Example 5. *Find the value of cos 315°* **[S.B.T.E. Delhi, 2017]**

Solution. We know that

$$\cos(360° - \theta) = \cos\theta$$

$$\cos 315° = \cos(360° - 45°) = \cos 45° = \frac{1}{\sqrt{2}}$$ **Ans.**

Example 6. *Find the value of sin 15°* **[S.B.T.E. Delhi, 2017, 2016, 2015, 2013]**

Solution. Here, we have

$$\sin 15° = \sin(45° - 30°) = \sin 45.\cos 30 - \cos 45.\sin 30$$

$$= \left(\frac{1}{\sqrt{2}}\times\frac{\sqrt{3}}{2}\right)-\left(\frac{1}{\sqrt{2}}\times\frac{1}{2}\right)$$

$$= \frac{\sqrt{3}-1}{2\sqrt{2}} = \frac{\sqrt{2}(\sqrt{3}-1)}{4}$$ **Ans.**

Example 7. *Find the value of tan 15° and tan 75°* **[S.B.T.E. Delhi, 2015]**

Solution. Here, we have

$$\tan 15° = \tan(45° - 30°) = \frac{\tan 45° - \tan 30°}{1 + \tan 45° \cdot \tan 30°} = \frac{1-\frac{1}{\sqrt{3}}}{1+\frac{1}{\sqrt{3}}} = \frac{\sqrt{3}-1}{\sqrt{3}+1} = 2-\sqrt{3}$$ **Ans.**

$$\tan 75° = \tan(45° + 30°) = \frac{\tan 45° + \tan 30°}{1 - \tan 45° \cdot \tan 30°}$$

$$= \frac{1+\frac{1}{\sqrt{3}}}{1-1\cdot\frac{1}{\sqrt{3}}} = \frac{\sqrt{3}+1}{\sqrt{3}-1} = 2+\sqrt{3}$$ **Ans.**

Example 8. *If α + β = π/2, find the value of tan α . tan β.* **[S.B.T.E. Delhi, 2017, 2015]**

Solution. We know that

$$\tan(\alpha+\beta) = \frac{\tan\alpha+\tan\beta}{1-\tan\alpha\cdot\tan\beta}$$

$$\tan\pi/2 = \frac{\tan\alpha+\tan\beta}{1-\tan\alpha\cdot\tan\beta}$$

$$\infty = \frac{\tan\alpha+\tan\beta}{1-\tan\alpha\cdot\tan\beta} = \frac{1}{0} \qquad \left[\infty = \frac{1}{0}\right]$$

$\therefore \quad \tan\alpha \,.\, \tan\beta = 1$ **Ans.**

Example 9. *If sec θ – tan θ = P, find the value of tan θ.* **[DIPIETE 2018, S.B.T.E, 2015]**

Solution. We know that

$$\sec^2\theta - \tan^2\theta = 1$$

$$(\sec\theta+\tan\theta)(\sec\theta-\tan\theta) = 1 \qquad [\because a^2-b^2 = (a+b)(a-b)]$$

$$(\sec\theta+\tan\theta)\,P = 1$$

$$\sec\theta+\tan\theta = \frac{1}{P} \qquad \text{... (1)}$$

$$\sec\theta-\tan\theta = P \qquad \text{... (2) [Given]}$$

from (1) and (2), we get

$$2\tan\theta = \frac{1-P^2}{P} \Rightarrow \tan\theta = \frac{1-P^2}{2P}$$ **Ans.**

Example 10. *Find the value of tan (α + β), given that cot α =* $\frac{1}{2}, \alpha \in \left(\pi, \frac{3\pi}{2}\right)$ *and* $\sec\beta = -\frac{5}{3}, \beta \in \left(\frac{\pi}{2}, \pi\right)$.

Solution. We know that: $\tan(\alpha+\beta) = \dfrac{\tan\alpha+\tan\beta}{1-\tan\alpha\cdot\tan\beta}$ ... (1)

Given, $\cot\alpha = \frac{1}{2} \Rightarrow \tan\alpha = 2$

Also, $\sec\beta = -\frac{5}{3}$. Then, $\tan\beta = \sqrt{\sec^2\beta - 1} = \pm\sqrt{\frac{25}{9}-1} = \pm\frac{4}{3}$

But, $\beta \in \left(\frac{\pi}{2}, \pi\right) \Rightarrow \tan\beta = -\frac{4}{3}$ [∵ tan β is –ve in II quadrant]

Putting tan α = 2 and $\tan\beta = -\frac{4}{3}$ in (1), we get

$$\tan(\alpha+\beta) = \frac{2+\left(-\frac{4}{3}\right)}{1-2\cdot\left(-\frac{4}{3}\right)} = \frac{\frac{2}{3}}{\frac{11}{3}} = \frac{2}{11}$$ **Ans.**

Example 11. *Prove that* $\sin(n + 1)x \sin(n + 2)x + \cos(n + 1)x \cos(n + 2)x = \cos x$

Solution. L.H.S. $= \sin(n + 1)x \sin(n + 2)x + \cos(n + 1)x \cos(n + 2)x$

$= \cos(n + 1)x \cos(n + 2)x + \sin(n + 1)x \sin(n + 2)x$

$= \cos\{(n + 1)x - (n + 2)x\}$ $[\because \cos(A + B) = \cos A \cos B - \sin A \sin B]$

$= \cos\{nx + x - nx - 2x\} = \cos(-x)$

$= \cos x =$ R.H.S. **Proved.**

Example 12. *If* $3 \tan\theta \tan\phi = 1$, *prove that* $2\cos(\theta + \phi) = \cos(\theta - \phi)$

Solution. $3\tan\theta\tan\phi = 1 \Rightarrow \cot\theta\cot\phi = 3$

$$\Rightarrow \frac{\cos\theta\cos\phi}{\sin\theta\sin\phi} = \frac{3}{1}$$

By componendo and dividendo, we have

$$\frac{\cos\theta\cos\phi + \sin\theta\sin\phi}{\cos\theta\cos\phi - \sin\theta\sin\phi} = \frac{3+1}{3-1} \Rightarrow \frac{\cos(\theta-\phi)}{\cos(\theta+\phi)} = 2$$

$\Rightarrow 2\cos(\theta + \phi) = \cos(\theta - \phi)$. **Proved.**

Example 13. *If* $\cot\alpha\cot\beta = 2$, *show that* $\frac{\cos(\alpha+\beta)}{\cos(\alpha-\beta)} = \frac{1}{3}$

Solution. We have, $\cot\alpha \cdot \cot\beta = 2$

$$\Rightarrow \frac{\cos\alpha}{\sin\alpha}\cdot\frac{\cos\beta}{\sin\beta} = 2 \Rightarrow \frac{\cos\alpha\cdot\cos\beta}{\sin\alpha\cdot\sin\beta} = \frac{2}{1}$$

By componendo and dividendo method, we get

$$\frac{\cos\alpha\cos\beta + \sin\alpha\sin\beta}{\cos\alpha\cos\beta - \sin\alpha\sin\beta} = \frac{2+1}{2-1}$$

$$\Rightarrow \frac{\cos(\alpha-\beta)}{\cos(\alpha+\beta)} = \frac{3}{1} \Rightarrow \frac{\cos(\alpha+\beta)}{\cos(\alpha-\beta)} = \frac{1}{3}$$ **Proved.**

Example 14. *Show that:* $\tan 70° = 2\tan 50° + \tan 20°$

Solution. We have, $\tan 70° = \tan(50° + 20°)$

$$= \frac{\tan 50° + \tan 20°}{1 - \tan 50°\cdot\tan 20°}$$

$\Rightarrow \tan 70° - \tan 70° \tan 50° \tan 20° = \tan 50° + \tan 20°$

$$\Rightarrow \tan 70° - \frac{1}{\tan 20°}\cdot\tan 50° \tan 20° = \tan 50° + \tan 20°$$

$$\left[\because \tan 70° = \tan(90° - 20°) = \cot 20° = \frac{1}{\tan 20°}\right]$$

$\Rightarrow \tan 70° - \tan 50° = \tan 50° + \tan 20°$

$\tan 70° = 2\tan 50° + \tan 20°$ **Proved.**

Example 15. *Prove that* $\frac{\cos 4x + \cos 3x + \cos 2x}{\sin 4x + \sin 3x + \sin 2x} = \cot 3x$

Solution. L.H.S. $= \frac{(\cos 4x + \cos 2x) + \cos 3x}{(\sin 4x + \sin 2x) + \sin 3x}$

$$= \frac{2\cos\left(\frac{4x+2x}{2}\right)\cos\left(\frac{4x-2x}{2}\right)+\cos 3x}{2\sin\left(\frac{4x+2x}{2}\right)\cos\left(\frac{4x-2x}{2}\right)+\sin 3x} = \frac{2\cos 3x\cos x+\cos 3x}{2\sin 3x\cos x+\sin 3x}$$

$$= \frac{\cos 3x\,(2\cos x+1)}{\sin 3x\,(2\cos x+1)} = \frac{\cos 3x}{\sin 3x}$$

$= \cot 3x = \text{R.H.S.}$ **Proved.**

Example 16. *Prove that*

$$\cot x \cot 2x - \cot 2x \cot 3x - \cot 3x \cot x = 1$$

Solution. L.H.S. $= \cot x \cot 2x - \cot 2x \cot 3x - \cot 3x \cot x$

$$= \cot 2x\,(\cot x - \cot 3x) - \cot 3x \cot x \quad \text{... (1)}$$

Putting the value of $(\cot x - \cot 3x)$ in (1), we get

$$\left[\cot(3x-x) = \frac{1+\cot x\cot 3x}{\cot x - \cot 3x} \Rightarrow \cot x - \cot 3x = \frac{1+\cot x\cot 3x}{\cot(3x-x)}\right]$$

$$= \cot 2x\left[\frac{1+\cot x\cot 3x}{\cot 2x}\right] - \cot 3x\cot x$$

$$= 1 + \cot x\cot 3x - \cot 3x\cot x$$

$= 1 = \text{R.H.S.}$ **Proved.**

Example 17. *Prove that: $\tan 3A \tan 2A \tan A = \tan 3A - \tan 2A - \tan A$*

Solution. We have: $3A = 2A + A$

$\Rightarrow \quad \tan 3A = \tan(2A + A)$

$\Rightarrow \quad \tan 3A = \dfrac{\tan 2A + \tan A}{1 - \tan 2A\tan A}$

$\Rightarrow \quad \tan 3A(1 - \tan 2A\tan A) = \tan 2A + \tan A$

$\Rightarrow \quad \tan 3A - \tan 3A\tan 2A\tan A = \tan 2A + \tan A$

$\Rightarrow \quad \tan 3A - \tan 2A - \tan A = \tan 3A\tan 2A\tan A$ **Proved.**

Example 18. *If* $\tan A = \dfrac{m}{m-1}$ *and* $\tan B = \dfrac{1}{2m-1}$, *prove that* $A - B = \dfrac{\pi}{4}$.

Solution. We have, $\tan A = \dfrac{m}{m-1}$ and $\tan B = \dfrac{1}{2m-1}$

Now, $\quad \tan(A-B) = \dfrac{\tan A - \tan B}{1+\tan A\cdot\tan B} \quad \text{... (1)}$

Putting the values of $\tan A$ and $\tan B$ in (1), we get

$$\tan(A-B) = \frac{\frac{m}{m-1} - \frac{1}{2m-1}}{1+\left(\frac{m}{m-1}\right)\left(\frac{1}{2m-1}\right)}$$

$$= \frac{2m^2 - m - m + 1}{(m-1)(2m-1)} \times \frac{(m-1)(2m-1)}{2m^2 - 3m + 1 + m} = \frac{2m^2 - 2m + 1}{2m^2 - 2m + 1} = 1$$

$$\Rightarrow \qquad \tan(A-B) = \tan\frac{\pi}{4} \qquad \left[\because \tan\frac{\pi}{4}=1\right]$$

$$\Rightarrow \qquad A - B = \frac{\pi}{4}$$

Proved.

Example 19. *If* $\tan\beta = \dfrac{n\sin\alpha\cos\alpha}{1-n\sin^2\alpha}$ *; prove that* $\tan(\alpha-\beta) = (1-n)\tan\alpha$

Solution. L.H.S. $= \tan(\alpha-\beta) = \dfrac{\tan\alpha - \tan\beta}{1+\tan\alpha\tan\beta}$ ... (1)

Putting $\tan\beta = \dfrac{n\sin\alpha\cos\alpha}{1-n\sin^2\alpha}$ in (1), we get

$$\text{L.H.S.} = \frac{\dfrac{\sin\alpha}{\cos\alpha} - \dfrac{n\sin\alpha\cos\alpha}{1-n\sin^2\alpha}}{1+\dfrac{\sin\alpha}{\cos\alpha}\cdot\dfrac{n\sin\alpha\cos\alpha}{1-n\sin^2\alpha}} \qquad \left[\because \tan\alpha = \frac{\sin\alpha}{\cos\alpha}\right]$$

$$= \frac{\sin\alpha(1-n\sin^2\alpha) - n\sin\alpha\cos^2\alpha}{\cos\alpha(1-n\sin^2\alpha)+n\sin^2\alpha\cos\alpha}$$

$$= \frac{\sin\alpha - n\sin^3\alpha - n\sin\alpha\cos^2\alpha}{\cos\alpha - n\sin^2\alpha\cos\alpha + n\sin^2\alpha\cos\alpha}$$

$$= \frac{\sin\alpha - n\sin\alpha(\sin^2\alpha+\cos^2\alpha)}{\cos\alpha} = \frac{\sin\alpha - n\sin\alpha}{\cos\alpha}$$

$$[\because \sin^2\alpha + \cos^2\alpha = 1]$$

$$= \frac{\sin\alpha(1-n)}{\cos\alpha} = (1-n)\tan\alpha = \text{R.H.S.}$$

Proved.

Example 20. *Show that:*

(*i*) $\cos^2\left(\dfrac{\theta-\phi}{2}\right) - \sin^2\left(\dfrac{\theta+\phi}{2}\right) = \cos\theta\cdot\cos\phi$

(*ii*) $\cos^2 A + \cos^2 B - 2\cos A\cos B\cos(A+B) = \sin^2(A+B)$

Solution. (*i*) We know that, $\cos^2 A - \sin^2 B = \cos(A+B)\cdot\cos(A-B)$

$$\therefore \quad \text{L.H.S.} = \cos^2\left(\frac{\theta-\phi}{2}\right) - \sin^2\left(\frac{\theta+\phi}{2}\right) = \cos\left(\frac{\theta-\phi}{2}+\frac{\theta+\phi}{2}\right)\cos\left(\frac{\theta-\phi}{2}-\frac{\theta+\phi}{2}\right)$$

$$= \cos\theta\cdot\cos(-\phi) = \cos\theta\cos\phi = \text{R.H.S.}$$

Proved.

(*ii*)

$$\text{L.H.S.} = \cos^2 A + \cos^2 B - 2\cos A\cos B\cos(A+B)$$

$$= \cos^2 A + (1-\sin^2 B) - 2\cos A\cos B\cos(A+B)$$

$$= 1 + (\cos^2 A - \sin^2 B) - 2\cos A\cos B\cos(A+B)$$

$$= 1 + \cos(A+B)\cos(A-B) - 2\cos A\cos B\cos(A+B)$$

$$= 1 + \cos(A+B)[\cos(A-B) - 2\cos A\cos B]$$

$$= 1 + \cos(A+B)[\cos A\cos B + \sin A\sin B - 2\cos A\cos B]$$

$$= 1 + \cos(A+B)[\sin A\sin B - \cos A\cos B]$$

$$= 1 + \cos(A+B)[-\cos(A+B)]$$

$$= 1 - \cos^2(A+B) = \sin^2(A+B) = \text{R.H.S.}$$

Proved.

Example 21. *If $\theta + \phi = \alpha$ and tan $\theta = k$ tan ϕ, then prove that* $\sin(\theta - \phi) = \dfrac{k-1}{k+1} \sin\alpha$

Solution. We have, $\tan\theta = k \tan\phi$

$$\therefore \quad \frac{\tan\theta}{\tan\phi} = \frac{k}{1} \Rightarrow \frac{\tan\theta + \tan\phi}{\tan\theta - \tan\phi} = \frac{k+1}{k-1} \qquad \text{(By componendo and dividendo)}$$

$$\Rightarrow \frac{\dfrac{\sin\theta}{\cos\theta} + \dfrac{\sin\phi}{\cos\phi}}{\dfrac{\sin\theta}{\cos\theta} - \dfrac{\sin\phi}{\cos\phi}} = \frac{k+1}{k-1} \Rightarrow \frac{\sin\theta\cos\phi + \cos\theta\sin\phi}{\sin\theta\cos\phi - \cos\theta\sin\phi} = \frac{k+1}{k-1}$$

$$\Rightarrow \frac{\sin(\theta+\phi)}{\sin(\theta-\phi)} = \frac{k+1}{k-1} \Rightarrow \frac{\sin(\theta-\phi)}{\sin\alpha} = \frac{k-1}{k+1} \qquad [\because \alpha = \theta + \phi]$$

$$\Rightarrow \sin(\theta-\phi) = \frac{k-1}{k+1}\sin\alpha$$

Proved.

Example 22. *Prove that*

$$\frac{\sin(A-B)}{\sin A\cdot\sin B} + \frac{\sin(B-C)}{\sin B\cdot\sin C} + \frac{\sin(C-A)}{\sin C\cdot\sin A} = 0$$

[S.B.T.E. 2016]

Solution. L.H.S. $= \dfrac{\sin(A-B)}{\sin A\cdot\sin B} + \dfrac{\sin(B-C)}{\sin B\cdot\sin C} + \dfrac{\sin(C-A)}{\sin C\cdot\sin A}$

$$= \frac{\sin A\cdot\cos B - \cos A\cdot\sin B}{\sin A\cdot\sin B} + \frac{\sin B\cdot\cos C - \cos B\cdot\sin C}{\sin B\cdot\cos C} + \frac{\sin C\cdot\cos A - \cos C\cdot\sin A}{\sin C\cdot\sin A}$$

$$= \cot B - \cot A + \cot C - \cot B + \cot A - \cot C$$

$$= 0 = \text{R.H.S.}$$

Proved.

Example 23. *Prove that:*

$$\cos^2(A-B) + \cos^2B - 2\cos(A-B)\cos A.\cos B = \sin^2A$$

[S.B.T.E. 2015]

Solution. L.H.S. $= \cos^2(A-B) + \cos^2 B - 2\cos(A-B)\cos A.\cos B$

$$= \cos^2 B + \cos^2(A-B) - 2\cos(A-B)\cos A.\cos B$$

$$= \cos^2 B + \cos(A-B)\{\cos(A-B) - 2\cos A.\cos B\}$$

$$= \cos^2 B + \cos(A-B)\{\cos A.\cos B + \sin A.\sin B - 2\cos A.\cos B\}$$

$$= \cos^2 B + \cos(A-B)\{\sin A.\sin B - \cos A.\cos B\}$$

$$= \cos^2 B - \cos(A-B)\cos(A+B)$$

$$= \cos^2 B - (\cos^2 A - \sin^2 B)$$

$$= \sin^2 B + \cos^2 B - \cos^2 A$$

$$= 1 - \cos^2 A = \sin^2 A = \text{R.H.S.}$$

Proved.

Example 24. *If $\theta + \phi = \alpha$ and sin $\theta = k$ sin ϕ, prove that*

$$\tan\theta = \frac{k\sin\alpha}{1 + k\cos\alpha}, \tan\phi = \frac{\sin\alpha}{k + \cos\alpha}$$

Solution. We have,

$$\theta + \phi = \alpha \;\Rightarrow\; \phi = \alpha - \theta \qquad \text{... (1)}$$

and $\sin\theta = k\sin\phi$ ... (2)

$\Rightarrow$ $\sin\theta = k\sin(\alpha - \theta)$ [using (1)]

$= k[\sin\alpha\cos\theta - \cos\alpha\sin\theta]$

$\Rightarrow$ $\sin\theta = k\sin\alpha\cos\theta - k\cos\alpha\sin\theta$ ... (3)

Dividing both sides of (3) by $\cos\theta$, we get,

$\tan\theta = k\sin\alpha - k\cos\alpha.\tan\theta \;\Rightarrow\; \tan\theta + k\cos\alpha.\tan\theta = k\sin\alpha$

$\Rightarrow$ $\tan\theta(1 + k\cos\alpha) = k\sin\alpha$

$$\Rightarrow \tan\theta = \frac{k\sin\alpha}{1 + k\cos\alpha}$$ **Proved.**

Again, $\sin\theta = k\sin\phi \;\Rightarrow\; \sin(\alpha - \phi) = k\sin\phi$ $[\theta + \phi = \alpha \Rightarrow \theta = \alpha - \phi]$

$\Rightarrow$ $\sin\alpha\cos\phi - \cos\alpha.\sin\phi = k\sin\phi$ ... (4)

Dividing both sides of (4) by $\cos\phi$, we get

$\sin\alpha - \cos\alpha\tan\phi = k\tan\phi \;\Rightarrow\; (k + \cos\alpha)\tan\phi = \sin\alpha$

$$\Rightarrow \tan\phi = \frac{\sin\alpha}{k + \cos\alpha}$$ **Proved.**

Example 25. *If α and β are the solutions of the equation a tan θ + b sec θ = c, then show that*

$$\tan(\alpha + \beta) = \frac{2ac}{a^2 - c^2}$$

Solution. We have,

$a\tan\theta + b\sec\theta = c$ ... (1)

$\Rightarrow$ $c - a\tan\theta = b\sec\theta$

$\Rightarrow$ $(c - a\tan\theta)^2 = b^2\sec^2\theta$

$\Rightarrow$ $c^2 + a^2\tan^2\theta - 2ac\tan\theta = b^2(1 + \tan^2\theta)$

$\Rightarrow$ $(a^2 - b^2)\tan^2\theta - 2ac\tan\theta + (c^2 - b^2) = 0$ ... (2)

It is given that α and β are the solutions of (1). Therefore, tan α and tan β are roots of (2).

Hence, $\tan\alpha + \tan\beta = \dfrac{2ac}{a^2 - b^2}$ and $\tan\alpha.\tan\beta = \dfrac{c^2 - b^2}{a^2 - b^2}$

$$\Rightarrow \tan(\alpha + \beta) = \frac{\tan\alpha + \tan\beta}{1 - \tan\alpha.\tan\beta} = \frac{\dfrac{2ac}{a^2 - b^2}}{1 - \dfrac{c^2 - b^2}{a^2 - b^2}}$$

$$= \frac{2ac}{a^2 - b^2 - c^2 + b^2} = \frac{2ac}{a^2 - c^2}$$ **Proved.**

Example 26. (*i*) *If* $\sin x + \sin y = a, \cos x + \cos y = b$, *show that*

$$\cos(x - y) = \frac{1}{2}(a^2 + b^2 - 2)$$

(*ii*) *If* $\tan\theta + \tan\phi = a$ *and* $\cot\theta + \cot\phi = b$, *prove that* $\cot(\theta + \phi) = \frac{1}{a} - \frac{1}{b}$.

Solution. (*i*) We have, $a = \sin x + \sin y$ ... (1)

and $b = \cos x + \cos y$ ... (2)

Squaring (1) and (2) and adding, we get

$$a^2 + b^2 = (\sin^2 x + \sin^2 y) + 2\sin x \sin y + (\cos^2 x + \cos^2 y) + 2\cos x \cos y$$

$$\Rightarrow \quad a^2 + b^2 = 1 + 1 + 2(\cos x \cos y + \sin x \sin y)$$

$$\Rightarrow \quad a^2 + b^2 = 2 + 2\cos(x - y)$$

$$\Rightarrow \quad \cos(x - y) = \frac{1}{2}(a^2 + b^2 - 2)$$ **Proved.**

(*ii*) We have, $a = \tan\theta + \tan\phi$ ... (3)

and $b = \cot\theta + \cot\phi$ ... (4)

We know that,

$$\cot(\theta + \phi) = \frac{1}{\tan(\theta + \phi)} = \frac{1}{\frac{\tan\theta + \tan\phi}{1 - \tan\theta \cdot \tan\phi}} = \frac{1 - \tan\theta \tan\phi}{\tan\theta + \tan\phi}$$

$$= \frac{1}{\tan\theta + \tan\phi} - \frac{\tan\theta \cdot \tan\phi}{\tan\theta + \tan\phi} = \frac{1}{a} - \frac{\frac{1}{\cot\theta} \cdot \frac{1}{\cot\phi}}{\frac{1}{\cot\theta} + \frac{1}{\cot\phi}} = \frac{1}{a} - \frac{1}{\cot\theta + \cot\phi}$$

$$= \frac{1}{a} - \frac{1}{b}$$ [using (3) and (4)] **Proved.**

Example 27. *If* α, β *be two distinct real numbers satisfying the equation* $a\cos\theta + b\sin\theta = c$, *prove that:*

(*i*) $\sin(\alpha + \beta) = \frac{2ab}{a^2 + b^2}$ (*ii*) $\tan(\alpha + \beta) = \frac{2ab}{a^2 - b^2}$

(*iii*) $\cos(\alpha + \beta) = \frac{a^2 - b^2}{a^2 + b^2}$ (*iv*) $\cot(\alpha - \beta) = \frac{2c^2 - a^2 - b^2}{a^2 + b^2}$

Solution. We have,

$$a\cos\theta + b\sin\theta = c \quad \text{... (1)}$$

$$b\sin\theta = c - a\cos\theta$$

$$b^2\sin^2\theta = c^2 + a^2\cos^2\theta - 2ac\cos\theta$$

$$b^2(1 - \cos^2\theta) = c^2 + a^2\cos^2\theta - 2ac\cos\theta$$

$$\Rightarrow \quad (a^2 + b^2)\cos^2\theta - 2ac\cos\theta + c^2 - b^2 = 0 \quad \text{... (2)}$$

Since α, β are the roots of (1), so $\cos\alpha$, $\cos\beta$ are the roots of (2).

$$\therefore \quad \cos\alpha \cdot \cos\beta = \frac{c^2 - b^2}{a^2 + b^2} \quad \text{... (3)}$$

Squaring (1), we get

$$a^2\cos^2\theta + b^2\sin^2\theta + 2ab\sin\theta\cos\theta = c^2$$

$\Rightarrow \quad a^2 + b^2\tan^2\theta + 2ab\tan\theta = c^2\sec^2\theta$ [dividing by $\cos^2\theta$]

$\Rightarrow \quad a^2 + b^2\tan^2\theta + 2ab\tan\theta = c^2(1 + \tan^2\theta)$

$\Rightarrow \quad (b^2 - c^2)\tan^2\theta + 2ab\tan\theta + a^2 - c^2 = 0$ ... (4)

Since, $\tan\alpha$, $\tan\beta$ are the roots of (4).

$$\therefore \quad \tan\alpha + \tan\beta = -\frac{2ab}{b^2 - c^2} \text{ and } \tan\alpha \cdot \tan\beta = \frac{a^2 - c^2}{b^2 - c^2} \quad \text{... (5)}$$

$$\Rightarrow \quad \frac{\sin\alpha\sin\beta}{\cos\alpha\cos\beta} = \frac{a^2 - c^2}{b^2 - c^2}$$

$$\Rightarrow \quad \sin\alpha\sin\beta = \frac{a^2 - c^2}{b^2 - c^2}\cos\alpha\cos\beta = \left(\frac{a^2 - c^2}{b^2 - c^2}\right)\left(\frac{c^2 - b^2}{a^2 + b^2}\right) \quad \text{[using (3)]}$$

$$= \frac{c^2 - a^2}{a^2 + b^2} \quad \text{... (6)}$$

(*i*) We have, $\tan\alpha + \tan\beta = \dfrac{-2ab}{b^2 - c^2} \Rightarrow \dfrac{\sin\alpha}{\cos\alpha} + \dfrac{\sin\beta}{\cos\beta} = \dfrac{-2ab}{b^2 - c^2}$

$$\Rightarrow \quad \frac{\sin\alpha\cos\beta + \sin\beta\cos\alpha}{\cos\alpha\cos\beta} = \frac{-2ab}{b^2 - c^2}$$

$$\Rightarrow \sin(\alpha + \beta) = \frac{-2ab}{b^2 - c^2}(\cos\alpha\cos\beta) = \left(\frac{-2ab}{b^2 - c^2}\right)\left(\frac{c^2 - b^2}{a^2 + b^2}\right) = \frac{2ab}{a^2 + b^2} \quad \text{[using (3)]}$$

$$\Rightarrow \sin(\alpha + \beta) = \frac{2ab}{a^2 + b^2}$$ **Proved.**

(*ii*) $\tan(\alpha + \beta) = \dfrac{\tan\alpha + \tan\beta}{1 - \tan\alpha\tan\beta}$

$$= \frac{\frac{-2ab}{b^2 - c^2}}{1 - \frac{a^2 - c^2}{b^2 - c^2}} = \frac{\frac{-2ab}{b^2 - c^2}}{\frac{b^2 - c^2 - a^2 + c^2}{b^2 - c^2}} = \frac{\frac{-2ab}{b^2 - c^2}}{\frac{b^2 - a^2}{b^2 - c^2}} = \frac{2ab}{a^2 - b^2} \quad \text{[using (5)]}$$ **Proved.**

(*iii*) $\cos(\alpha + \beta) = \cos\alpha\cos\beta - \sin\alpha\sin\beta$

$$= \frac{c^2 - b^2}{a^2 + b^2} - \frac{c^2 - a^2}{a^2 + b^2} = \frac{a^2 - b^2}{a^2 + b^2} \quad \text{[using (3) and (6)]}$$ **Proved.**

(*iv*) $\cos(\alpha - \beta) = \cos\alpha\cos\beta + \sin\alpha\sin\beta$

$$= \frac{c^2 - b^2}{a^2 + b^2} + \frac{c^2 - a^2}{a^2 + b^2} = \frac{2c^2 - a^2 - b^2}{a^2 + b^2} \quad \text{[using (3) and (6)]}$$ **Proved.**

Example 28. *Prove that:* $(1 - \sin A)(1 + \sin A)\sec A = \cos A$ **[S.B.T.E. 2017]**

Solution. L.H.S. $= (1 - \sin A)(1 + \sin A)\sec A = [1 - \sin^2 A]\sec A$

$$= \cos^2 A \cdot \frac{1}{\cos A} = \cos A$$ **Proved.**

Example 29. *If $A + B + C = \pi$, show that*

$$\cot\frac{A}{2}+\cot\frac{B}{2}+\cot\frac{C}{2}=\cot\frac{A}{2}\cdot\cot\frac{B}{2}\cdot\cot\frac{C}{2}$$ **[DIPIETE 2018, 2014]**

Solution. We know that

$$\tan\frac{A}{2}\cdot\tan\frac{B}{2}+\tan\frac{B}{2}\cdot\tan\frac{C}{2}+\tan\frac{C}{2}\cdot\tan\frac{A}{2}=1$$

$$\Rightarrow \frac{1}{\cot\frac{A}{2}}\cdot\frac{1}{\cot\frac{B}{2}}+\frac{1}{\cot\frac{B}{2}}\cdot\frac{1}{\cot\frac{C}{2}}+\frac{1}{\cot\frac{C}{2}}\cdot\frac{1}{\cot\frac{A}{2}}=1$$

$$\Rightarrow \frac{\cot\frac{C}{2}+\cot\frac{A}{2}+\cot\frac{B}{2}}{\cot\frac{A}{2}\cdot\cot\frac{B}{2}\cdot\cot\frac{C}{2}}=1$$

$$\therefore \cot\frac{C}{2}+\cot\frac{A}{2}+\cot\frac{B}{2}=\cot\frac{A}{2}\cdot\cot\frac{B}{2}\cdot\cot\frac{C}{2}$$ **Proved.**

Example 30. *If $\cos(\alpha-\beta)+\cos(\beta-\gamma)+\cos(\gamma-\alpha)=-\frac{3}{2}$, prove that*

$$\cos\alpha+\cos\beta+\cos\gamma=\sin\alpha+\sin\beta+\sin\gamma=0$$

Solution. We have, $\cos(\alpha-\beta)+\cos(\beta-\gamma)+\cos(\gamma-\alpha)=-\frac{3}{2}$

$$\Rightarrow 2\cos\alpha\cos\beta+2\sin\alpha\sin\beta+2\cos\beta\cos\gamma+2\sin\beta\sin\gamma+2\cos\gamma\cos\alpha+2\sin\gamma\sin\alpha=-3$$

$$\Rightarrow (2\cos\alpha\cos\beta+2\cos\beta\cos\gamma+2\cos\gamma\cos\alpha)+(2\sin\alpha\sin\beta+2\sin\beta\sin\gamma+2\sin\gamma\sin\alpha)+3=0$$

$$\Rightarrow (2\cos\alpha\cos\beta+2\cos\beta\cos\gamma+2\cos\gamma\cos\alpha)+(2\sin\alpha\sin\beta+2\sin\beta\sin\gamma+2\sin\gamma\sin\alpha)+(\cos^2\alpha+\sin^2\alpha)+(\cos^2\beta+\sin^2\beta)+(\cos^2\gamma+\sin^2\gamma)=0$$

$$\Rightarrow (\cos^2\alpha+\cos^2\beta+\cos^2\gamma+2\cos\alpha\cos\beta+2\cos\beta\cos\gamma+2\cos\gamma\cos\alpha)+(\sin^2\alpha+\sin^2\beta+\sin^2\gamma+2\sin\alpha\sin\beta+2\sin\beta\sin\gamma+2\sin\gamma\sin\alpha)=0$$

$$\Rightarrow (\cos\alpha+\cos\beta+\cos\gamma)^2+(\sin\alpha+\sin\beta+\sin\gamma)^2=0$$

$$\Rightarrow \cos\alpha+\cos\beta+\cos\gamma=0 \text{ and } \sin\alpha+\sin\beta+\sin\gamma=0$$ **Proved.**

Example 31. *If $\tan(\pi\cos\theta)=\cot(\pi\sin\theta)$, prove that $\cos\left(\theta-\frac{\pi}{4}\right)=\pm\frac{1}{2\sqrt{2}}$.*

Solution. We have, $\tan(\pi\cos\theta)=\cot(\pi\sin\theta)$

$$\Rightarrow \frac{\sin(\pi\cos\theta)}{\cos(\pi\cos\theta)}=\frac{\cos(\pi\sin\theta)}{\sin(\pi\sin\theta)}$$

$$\Rightarrow \sin(\pi\cos\theta)\cdot\sin(\pi\sin\theta)=\cos(\pi\sin\theta)\cdot\cos(\pi\cos\theta)$$

$$\Rightarrow \sin(\pi\cos\theta)\cdot\sin(\pi\sin\theta)-\cos(\pi\sin\theta)\cdot\cos(\pi\cos\theta)=0$$

$$\Rightarrow \cos(\pi\sin\theta)\cos(\pi\cos\theta)-\sin(\pi\cos\theta).\sin(\pi\sin\theta)=0$$

$$\Rightarrow \cos(\pi\sin\theta+\pi\cos\theta)=0$$

$\Rightarrow \quad \pi\sin\theta + \pi\cos\theta = \pm\dfrac{\pi}{2}$ $\quad \left[\because \cos\left(\pm\dfrac{\pi}{2}\right) = 0\right]$

$\Rightarrow \quad \sin\theta + \cos\theta = \pm\dfrac{1}{2}$

$\Rightarrow \quad \dfrac{1}{\sqrt{2}}\cos\theta + \dfrac{1}{\sqrt{2}}\sin\theta = \pm\dfrac{1}{2\sqrt{2}}$ $\quad \left[\text{multiplying both sides by } \dfrac{1}{\sqrt{2}}\right]$

$\Rightarrow \quad \cos\theta\cdot\cos\dfrac{\pi}{4} + \sin\theta\cdot\sin\dfrac{\pi}{4} = \pm\dfrac{1}{2\sqrt{2}}$ $\quad \left[\cos\dfrac{\pi}{4} = \sin\dfrac{\pi}{4} = \dfrac{1}{\sqrt{2}}\right]$

$\Rightarrow \quad \cos\left(\theta - \dfrac{\pi}{4}\right) = \pm\dfrac{1}{2\sqrt{2}}$ **Proved.**

Example 32. *Prove that:* $\sin^2\left(\dfrac{\alpha+\beta}{2}\right) - \sin^2\left(\dfrac{\alpha-\beta}{2}\right) = \sin\alpha\cdot\sin\beta$ **[S.B.T.E. 2016]**

Solution. Here we have

$$\text{L.H.S.} = \sin^2\left(\frac{\alpha+\beta}{2}\right) - \sin^2\left(\frac{\alpha-\beta}{2}\right)$$

$$= \sin\left[\frac{(\alpha+\beta)}{2} + \frac{(\alpha-\beta)}{2}\right]\cdot\sin\left[\frac{(\alpha+\beta)}{2} - \frac{(\alpha-\beta)}{2}\right]$$

$$= \sin\left[\frac{\alpha}{2} + \frac{\beta}{2} + \frac{\alpha}{2} - \frac{\beta}{2}\right]\cdot\sin\left[\frac{\alpha}{2} + \frac{\beta}{2} - \frac{\alpha}{2} + \frac{\beta}{2}\right]$$

$$= \sin\left(\frac{2\alpha}{2}\right)\cdot\sin\left(\frac{2\beta}{2}\right) = \sin\alpha\cdot\sin\beta$$ **Proved.**

Example 33. *If A + B = 225°, prove that* $\dfrac{\cot A}{1+\cot A}\cdot\dfrac{\cot B}{1+\cot B} = \dfrac{1}{2}$ **[S.B.T.E. 2015, 2006]**

Solution. Here we have:

$$A + B = 225°$$

Taking tan on both sides, we get

$$\tan(A + B) = \tan 225°$$

$\Rightarrow \quad \dfrac{\tan A + \tan B}{1 - \tan A\cdot\tan B} = \tan(180° + 45°) = \tan 45° = 1$

$\Rightarrow \quad \tan A + \tan B = 1 - \tan A\cdot\tan B$

$\Rightarrow \quad \tan A + \tan B + \tan A\cdot\tan B = 1$

$\Rightarrow \quad \dfrac{1}{\cot A} + \dfrac{1}{\cot B} + \dfrac{1}{\cot A\cdot\cot B} = 1$

$\Rightarrow \quad \dfrac{\cot B + \cot A + 1}{\cot A\cdot\cot B} = 1$

$$\cot A + \cot B + 1 = \cot A\cdot\cot B \quad \text{... (1)}$$

$$\text{L.H.S} = \frac{\cot A}{1+\cot A}\cdot\frac{\cot B}{1+\cot B} = \frac{\cot A\cdot\cot B}{1+\cot B+\cot A+\cot A\cdot\cot B}$$

$$= \frac{\cot A \cdot \cot B}{(\cot A + \cot B + 1) + \cot A \cdot \cot B} = \frac{\cot A \cdot \cot B}{(\cot A \cdot \cot B) + \cot A \cdot \cot B}$$

$$= \frac{\cot A \cdot \cot B}{2 \cot A \cdot \cot B} = 1/2$$ **Proved.**

EXERCISE 14.1

Find the values of the following:

1. $\sin 120°$ **Ans.** $\frac{\sqrt{3}}{2}$

2. $\cos 540°$ **Ans.** -1

3. $\text{cosec } 4530°$ **Ans.** -2

4. $\sec\left(\frac{15\pi}{4}\right)$ **Ans.** $\sqrt{2}$

5. $\cos\left(\frac{83\pi}{6}\right)$ **Ans.** $\frac{\sqrt{3}}{2}$

6. $\cos\left(-\frac{15\pi}{2}\right)$ **Ans.** 0

Evaluate the following:

7. $\sin 22° \cos 38° + \cos 22° \cos 38°$ **Ans.** $\frac{\sqrt{3}}{2}$

8. Write the value of $\cos 80° \cdot \cos 70° + \sin 80° \cdot \sin 70°$ **[S.B.T.E 2016]** **Ans.** $\cos 10°$

9. $\sin 75° \cos 300° + \cos 1470° \sin(-1020°)$ **Ans.** 1

10. $\sin 780° \sin 480° + \cos 240° \cos 300°$ **Ans.** $\frac{1}{2}$

11. $\frac{\tan A - 1}{\tan A + 1}$ if $A = 75°$ **Ans.** $\frac{1}{\sqrt{3}}$

12. If $\text{cosec}\theta - \cot\theta = P$, find the value of $\sin\theta$ and $\cos\theta$ **[S.B.T.E 2015]**

Ans. $\sin\theta = \frac{2P}{1+P^2}$; $\cos\theta = \frac{1-P^2}{1+P^2}$

13. If $\sec\theta - \tan\theta = \frac{a+1}{a-1}$, find $\cos\theta$ **[S.B.T.E 2016, 2015]** **Ans.** $\cos\theta = \frac{1-a^2}{2a}$

14. If $\tan\theta = \frac{a}{b}$, find the value of $\frac{a\sin\theta - b\cos\theta}{a\sin\theta + b\cos\theta}$ **[S.B.T.E 2014]** **Ans.** $\frac{a^2 - b^2}{a^2 + b^2}$

15. What does 'X' denote in the relation $\cos^2 4A - \sin^2 3A = \cos X \cdot \cos A$. **[S.B.T.E 2016]**

Ans. $X = 7A$

Show that:

16. $\frac{\cos 11° + \sin 11°}{\cos 11° - \sin 11°} = \tan 56°$ **[S.B.T.E 2016]**

17. $\frac{\cos 8° - \sin 8°}{\cos 8° + \sin 8°} = \tan 37°$

18. If $\sin\alpha = \dfrac{3}{5}$, $\cos\beta = \dfrac{5}{13}$, α and β lie in I quadrant. Then, find:

(*i*) $\sin(\alpha+\beta)$ **Ans.** $\dfrac{63}{65}$ (*ii*) $\cos(\alpha+\beta)$ **Ans.** $\dfrac{-16}{65}$ (*iii*) $\tan(\alpha+\beta)$ **Ans.** $\dfrac{-63}{16}$

19. Find the values of (*i*) $\sin(\alpha-\beta)$, (*ii*) $\cos(\alpha-\beta)$ and (*iii*) $\tan(\alpha-\beta)$, given

(*a*) $\sin\alpha = \dfrac{8}{17}$, $\tan\beta = \dfrac{5}{12}$, α and β line in I quadrant.

Ans. (*i*) $\dfrac{21}{221}$ (*ii*) $\dfrac{220}{221}$ (*iii*) $\dfrac{21}{220}$

(*b*) $\cos\alpha = \dfrac{-12}{13}$, $\cot\beta = \dfrac{24}{7}$, α lies in II quadrant, β lies in I quadrant.

Ans. (*i*) $\dfrac{204}{325}$ (*ii*) $\dfrac{-253}{325}$ (*iii*) $\dfrac{-204}{253}$

20. If $\cos\alpha = \dfrac{-12}{13}$, $\cot\beta = \dfrac{24}{7}$, α lies in II quadrant. β lies in III quadrant. Find:

(*i*) $\sin(\alpha+\beta)$ **Ans.** $\dfrac{-36}{325}$ (*ii*) $\cos(\alpha+\beta)$ **Ans.** $\dfrac{323}{325}$ (*iii*) $\tan(\alpha+\beta)$ **Ans.** $\dfrac{-36}{323}$

Prove that:

21. $(\sin\alpha\cos\beta + \cos\alpha\sin\beta)^2 + (\cos\alpha\cos\beta - \cos\alpha\sin\beta)^2 = 1$

22. $\sin(A - 45°) = \dfrac{1}{\sqrt{2}}(\sin A - \cos A)$

23. $\sin^2(A+B) - \sin^2(A-B) = \sin 2A \sin 2B$

24. $\sin(n+1)\theta \sin(n-1)\theta + \cos(n+1)\theta\cos(n-1)\theta = \cos 2\theta$

25. $\dfrac{\tan(A-B)+\tan B}{1-\tan(A-B)\tan B} = \tan A$

26. $\dfrac{\sin(A-B)}{\cos A\cos B} + \dfrac{\sin(B-C)}{\cos B\cos C} + \dfrac{\sin(C-A)}{\cos C\cos A} = 0$

27. $\tan 75° - \tan 30° - \tan 75° \tan 30° = 1$

28. If $\tan x + \tan\left(x+\dfrac{\pi}{3}\right) + \tan\left(x+\dfrac{2\pi}{3}\right) = 3$. Prove that. $\dfrac{3\tan x - \tan^3 x}{1-3\tan^2 x} = 1$.

29. If $\cos\alpha + \sin\beta = a$ and $\sin\alpha + \cos\beta = b$, prove that $\sin(\alpha+\beta) = \dfrac{1}{2}(a^2+b^2-2)$

30. If $\sin(\alpha+\beta) = 1$ and $\sin(\alpha-\beta) = \dfrac{1}{2}$, where $0 \le \alpha, \beta \le \dfrac{\pi}{2}$, find the values of $\tan(\alpha+2\beta)$ and $\tan(2\alpha+\beta)$.

Ans. $-\sqrt{3}, -\dfrac{1}{\sqrt{3}}$

31. If $2\sin\alpha\cos\beta\sin\gamma = \sin\beta\sin(\alpha+\gamma)$, then show that tan α, tan β, tan γ are in H.P.

32. If $\tan\dfrac{\alpha}{2}$ and $\tan\dfrac{\beta}{2}$ are the roots or the equation $8x^2 - 26x + 15 = 0$, then and find the value of $\cos(\alpha+\beta)$.

Ans. $-\dfrac{627}{725}$

33. Prove that: $\dfrac{\tan^2 2\theta - \tan^2\theta}{1-\tan^2 2\theta\cdot\tan^2\theta} = \tan 3\theta\cdot\tan\theta$ **[S.B.T.E, 2016, 2014]**

34. If $A + B + C = \pi$, show that

$$\tan\frac{A}{2}\cdot\tan\frac{B}{2}+\tan\frac{C}{2}\cdot\tan\frac{B}{2}+\tan\frac{C}{2}\cdot\tan\frac{A}{2}=1$$ **[S.B.T.E 2017]**

35. Prove that:

$$\cos 2A\cos 2B+\sin^2(A-B)-\sin^2(A+B)=\cos(2A+2B)$$ **[S.B.T.E Delhi, 2016]**

36. $\dfrac{\sin(A-B)}{\sin(A+B)}=\dfrac{\tan A-\tan B}{\tan A+\tan B}$

37. $2\tan 70° = \tan 80° - \tan 10°$

14.4 DEFACTORIZATION FORMULAE FOR PRODUCTS INTO SUMS OR DIFFERENCES

1. $2\sin A\cos B = \sin(A + B) + \sin(A - B)$

2. $2\cos A\sin B = \sin(A + B) - \sin(A - B)$

Proof: We know that,

$$\sin A\cos B + \cos A\sin B = \sin(A + B) \quad \text{... (1)}$$

and $$\sin A\cos B - \cos A\sin B = \sin(A - B) \quad \text{... (2)}$$

1. Adding (1) and (2), we get

$$2\sin A\cos B = \sin(A + B) + \sin(A - B)$$

2. Subtracting (2) from (1), we get

$$2\cos A\sin B = \sin(A + B) - \sin(A - B)$$

Also, we know that

$$\cos A\cdot\cos B - \sin A\cdot\sin B = \cos(A + B) \quad \text{... (3)}$$

$$\cos A\cos B + \sin A\sin B = \cos(A - B) \quad \text{... (4)}$$

3. Adding (3) and (4), we get

$$2\cos A\cos B = \cos(A + B) + \cos(A - B)$$

4. Subtracting (3) from (4), we get

$$2\sin A\sin B = \cos(A - B) - \cos(A + B)$$

We have to remember the following important formulae.

1.	$2\sin A\cos B = \cos(A + B) + \sin(A - B)$
2.	$2\cos A\sin B = \sin(A + B) - \sin(A - B)$
3.	$2\cos A\cos B = \cos(A + B) + \cos(A - B)$
4.	$2\sin A\sin B = \cos(A - B) - \cos(A + B)$

Note: Formulae (1) to (4) are called product formulae on the basis of L.H.S. These formulae are also known as "*A*, *B*" formulae.

Caution: In the fourth formula:

R.H.S. = $\cos(A - B) - \cos(A + B)$ and not $\cos(A + B) - \cos(A - B)$.

Aid to memory

1. L.H.S. of these formulae starts with 2. *For example*, $2\sin A\cos B$.

2. R.H.S. contains sine and cosine of sum and difference of the angles with + and –.

1.	2sin × cos = sin(sum) + sin(difference)
2.	2cos × sin = sin(sum) – sin(difference)
3.	2cos × cos = cos(sum) + cos(difference)
4.	2sin × sin = cos(difference) – cos(sum)

Here, sum stands for sum of the angles and difference stands for difference of the angles.

Example 34. *Find the value of:*

(*i*) $2\sin 15°\cdot\cos 75°$ (*ii*) $2\cos 45°\cdot\sin 15°$ (*iii*) $2\cos 22\frac{1}{2}°\cdot\cos 67\frac{1}{2}°$

Solution. (*i*) We have,

$$2\sin 15°\cdot\cos 75° = \sin(15° + 75°) + \sin(15° - 75°)$$

$[\because 2\sin A\cos B = \sin(A + B) + \sin(A - B)$

$$= \sin 90° + \sin(-60°)$$

$$= \sin 90° - \sin 60° \qquad [\because \sin(-\theta) = -\sin\theta]$$

$$= 1 - \frac{\sqrt{3}}{2} = \frac{2-\sqrt{3}}{2}$$ **Ans.**

(*ii*) We know that,

$$2\cos A\sin B = \cos(A + B) - \sin(A - B)$$

$$\therefore \quad 2\cos 45°\sin 15° = \sin(45° + 15°) - \sin(45° - 15°)$$

$$= \sin 60° - \sin 30°$$

$$= \frac{\sqrt{3}}{2} - \frac{1}{2} = \frac{\sqrt{3}-1}{2}$$ **Ans.**

(*iii*) We know that,

$$2\cos A\cos B = \cos(A + B) + \cos(A - B)$$

$$\therefore \quad 2\cos 22\frac{1}{2}°\cdot\cos 67\frac{1}{2}° = \cos\left(22\frac{1}{2} + 67\frac{1}{2}\right)^{\circ} + \cos\left(22\frac{1}{2} - 67\frac{1}{2}\right)^{\circ}$$

$$= \cos 90° + \cos(-45°) = \cos 90° + \cos 45° [\because \cos(-\theta) = \cos\theta]$$

$$= 0 + \frac{1}{\sqrt{2}} = \frac{1}{\sqrt{2}}$$ **Ans.**

Example 35. *Find the value of:* (*i*) $2\sin 75°\sin 15°$ (*ii*) $\sin\frac{5\pi}{12}\sin\frac{\pi}{12}$

Solution. (*i*) We have,

$$2\sin 75°\sin 15° = \cos(75° - 15°) - \cos(75° + 15°)$$

$[\because 2\sin x\cdot\sin y = \cos(x - y) - \cos(x + y)]$

$$= \cos 60° - \cos 90° = \frac{1}{2} - 0 = \frac{1}{2}$$ **Ans.**

(*ii*) We have,

$$\sin\frac{5\pi}{12}\sin\frac{\pi}{12} = \frac{1}{2}\left[2\sin\frac{5\pi}{12}\sin\frac{\pi}{12}\right] = \frac{1}{2}\left[\cos\left(\frac{5\pi}{12} - \frac{\pi}{12}\right) - \cos\left(\frac{5\pi}{12} + \frac{\pi}{12}\right)\right]$$

$$= \frac{1}{2}\left(\cos\frac{\pi}{3} - \cos\frac{\pi}{2}\right) = \frac{1}{2}\left[\frac{1}{2} - 0\right] = \frac{1}{4}$$ **Ans.**

Example 36. *Prove that:*

(*i*) $\sin 20°\sin 40°\sin 80° = \frac{\sqrt{3}}{8}$ (*ii*) $\sin 10°\sin 30°\sin 50°\sin 70° = \frac{1}{16}$

Solution. (*i*) We have,

$$\text{L.H.S.} = \sin 20° \sin 40° \sin 80° = \frac{1}{2}\ [2\sin 20° \sin 40°]\cdot \sin 80°$$

$$= \frac{1}{2}\ [\cos (20° - 40°) - \cos (20° + 40°)] \sin 80°$$

$$[\because 2\cos A \cos B = \cos(A - B) - \cos(A + B)]$$

$$= \frac{1}{2}\ [\cos (-20°) - \cos 60°] \sin 80° = \frac{1}{2}\left(\cos 20° - \frac{1}{2}\right) \sin 80°$$

$$= \frac{1}{4}(2\cos 20° \sin 80° - \sin 80°) = \frac{1}{4}[\sin 100° - \sin (-60°) - \sin 80°]$$

$$= \frac{1}{4}(\sin (180° - 80°) + \sin 60° - \sin 80°] = \frac{1}{4}[\sin 80° + \sin 60° - \sin 80°]$$

$$= \frac{1}{4} \times \sin 60° = \frac{1}{4} \times \frac{\sqrt{3}}{2} = \frac{\sqrt{3}}{8} = \text{R.H.S.}$$ **Proved.**

(*ii*) $\text{L.H.S.} = \sin 10° \sin 30° \sin 50° \sin 70°$ **[S.B.T.E. 2015, 2014]**

$$= \sin 30° (\sin 10° \sin 50°) \sin 70° = \frac{1}{2}\cdot\frac{1}{2}\ [2\sin 10° \sin 50°] \sin 70°$$

$$= \frac{1}{4}\ [\cos (50° - 10°) - \cos (50° + 10°)] \sin 70°$$

$$[\because 2\sin A \sin B = \cos(A - B) - \cos(A + B)]$$

$$= \frac{1}{4}(\cos 40° - \cos 60°) \sin 70° = \frac{1}{4}[\sin 70° \cos 40° - \sin 70° \cos 60°]$$

$$= \frac{1}{4}\left[\sin 70° \cos 40° - \frac{1}{2}\sin 70°\right] = \frac{1}{8}[2\sin 70° \cos 40° - \sin 70°]$$

$$= \frac{1}{8}[\sin (70° + 40°) + \sin (70° - 40°) - \sin 70°]$$

$$[\because 2\sin A \cos B = \sin(A + B) + \sin(A - B)]$$

$$= \frac{1}{8}[\sin 110° + \sin 30° - \sin 70°] = \frac{1}{8}[\sin(180° - 70°) + \sin 30° - \sin 70°]$$

$$= \frac{1}{8}\left[\sin 70° + \frac{1}{2} - \sin 70°\right] = \frac{1}{8} \times \frac{1}{2} = \frac{1}{16} = \text{R.H.S.}$$ **Proved.**

Example 37. *Prove that:*

(*i*) $2\cos\frac{\pi}{13}\cos\frac{9\pi}{13} + \cos\frac{3\pi}{13} + \cos\frac{5\pi}{13} = 0$ (*ii*) $\tan 20° \tan 40° \tan 80° = \tan 60°$

Solution. (*i*) $\text{L.H.S.} = 2\cos\frac{\pi}{13}\cos\frac{9\pi}{13} + \cos\frac{3\pi}{13} + \cos\frac{5\pi}{13}$

$$= \cos\left(\frac{9\pi}{13} + \frac{\pi}{13}\right) + \cos\left(\frac{9\pi}{13} - \frac{\pi}{13}\right) + \cos\frac{3\pi}{13} + \cos\frac{5\pi}{13}$$

$$= \cos\frac{10\pi}{13} + \cos\frac{8\pi}{13} + \cos\frac{3\pi}{13} + \cos\frac{5\pi}{13}$$

$$= \cos\left(\pi - \frac{3\pi}{13}\right) + \cos\left(\pi - \frac{5\pi}{13}\right) + \cos\frac{3\pi}{13} + \cos\frac{5\pi}{13}$$

$$= -\cos\frac{3\pi}{13} - \cos\frac{5\pi}{13} + \cos\frac{3\pi}{13} + \cos\frac{5\pi}{13} \qquad [\because \cos(\pi - \theta) = -\cos\theta]$$

$= 0 =$ R.H.S. **Proved.**

(*ii*) L.H.S. $= \tan 20° \tan 40° \tan 80° = \dfrac{\sin 20° \sin 40° \sin 80°}{\cos 20° \cos 40° \cos 80°}$

$$= \frac{(2\sin 20° \sin 40°)\sin 80°}{(2\cos 20° \cos 40°)\cos 80°} = \frac{(\cos 20° - \cos 60°)\sin 80°}{(\cos 60° + \cos 20°)\cos 80°}$$

$$= \frac{\sin 80° \cos 20° - \frac{1}{2}\sin 80°}{\frac{1}{2}\cos 80° + \cos 80° \cos 20°} = \frac{2\sin 80° \cos 20° - \sin 80°}{\cos 80° + 2\cos 80° \cos 20°}$$

$$= \frac{\sin 100° + \sin 60° - \sin 80°}{\cos 80° + \cos 100° + \cos 60°} = \frac{\sin(180° - 80°) + \sin 60° - \sin 80°}{\cos 80° + \cos(180° - 80°) + \cos 60°}$$

$$= \frac{\sin 80° + \sin 60° - \sin 80°}{\cos 80° - \cos 80° + \cos 60°} = \frac{\sin 60°}{\cos 60°} = \tan 60° = \text{R.H.S.}$$ **Proved.**

EXERCISE 14.2

Express each of the following as sum or difference:

1. $2\cos 5\theta \cos 3\theta$ **Ans.** $\cos 8\theta + \cos 2\theta$
2. $2\sin 4\theta \sin 7\theta$ **Ans.** $\cos 3\theta - \cos 11\theta$
3. $\sin 5\theta \cos\theta$ **Ans.** $\frac{1}{2}(\sin 6\theta + \sin 4\theta)$
4. $\cos\theta \sin 8\theta$ **Ans.** $\frac{1}{2}(\sin 9\theta + \sin 7\theta)$
5. $2\cos(\alpha + \beta)\cos(\alpha - \beta)$ **Ans.** $\cos 2\alpha + \cos 2\beta$
6. $\sin\left(\dfrac{\alpha+\beta}{2}\right)\cos\left(\dfrac{\alpha-\beta}{2}\right)$ **Ans.** $\frac{1}{2}[\sin\alpha + \sin\beta]$

Prove that:

7. $2\sin\dfrac{5\pi}{12}\sin\dfrac{\pi}{12} = \dfrac{1}{2}$
8. $\cos(120° + \alpha)\cos(120° - \alpha) = \dfrac{2\cos 2\alpha - 1}{4}$
9. $\sin(60° + \alpha)\sin(420° - \alpha) = \dfrac{1 + 2\cos 2\alpha}{2}$
10. $\sin(60° - \alpha)\sin(780° + \alpha) = \dfrac{\sqrt{3} + 2\sin 2\alpha}{4}$
11. $\sec\left(\dfrac{\pi}{4} + \theta\right)\cdot\sec\left(\dfrac{\pi}{4} - \theta\right) = 2\sec 2\theta$
12. $2\sin(2\theta + \phi)\cos(\theta - 2\phi) = \sin(3\theta - \phi) + \sin(\theta + 3\phi)$
13. $\dfrac{2\sin(\alpha - \gamma)\cos\gamma - \sin(\alpha - 2\gamma)}{2\sin(\beta - \gamma)\cos\gamma - \sin(\beta - 2\gamma)} = \dfrac{\sin\alpha}{\sin\beta}$

14. $\cos 20° \cos 40° \cos 80° = \frac{1}{8}$ **15.** $\sin 10° \sin 30° \sin 50° \sin 70° = \frac{1}{16}$

16. $4 \cos 20° \cos 40° \cos 80° = \frac{1}{4}$ **17.** $\tan 20° \tan 40° \tan 80° = \tan 60°$

18. $\sin 20° \sin 40° \sin 80° = \frac{\sqrt{3}}{8}$ **19.** $\cos 10° \cos 50° \cos 70° = \frac{\sqrt{3}}{8}$

20. $\tan 20° \tan 40° \tan 60° \tan 80° = 3$

Show that:

21. $\tan(A+30°)+\cot(A-30°)=\dfrac{1}{\sin 2A-\sin 60°}$

22. $\tan(45° + \theta) - \tan(45° - \theta) = 2 \tan 2\theta$

23. $4 \cos A \cos(60° - A) \cos (60° + A) = \cos 3A$

24. $\sin A \sin (B - C) + \sin B \sin (C - A) + \sin C \sin (A - B) = 0$

25. $\sin^2 A + \sin^2(A - B) - 2\sin A \cos B \sin(A - B) = \sin^2 B$

14.5 TRANSFORMATION OF A SUM OR DIFFERENCE INTO A PRODUCT

In this section we shall prove:

1. $\sin C+\sin D=2\sin\left(\dfrac{C+D}{2}\right)\cos\left(\dfrac{C-D}{2}\right)$

2. $\sin C-\sin D=2\cos\left(\dfrac{C+D}{2}\right)\sin\left(\dfrac{C-D}{2}\right)$

3. $\cos C+\cos D=2\cos\left(\dfrac{C+D}{2}\right)\cos\left(\dfrac{C-D}{2}\right)$

4. $\cos C-\cos D=2\sin\left(\dfrac{C+D}{2}\right)\sin\left(\dfrac{D-C}{2}\right)$

Proof: Let $A + B = C$ and $A - B = D$. ... (1)

Then by addition and subtraction, we get

$$\left.\begin{aligned} &2A=C+D \Rightarrow A=\frac{C+D}{2} \\ \text{and} \quad &2B=C-D \Rightarrow B=\frac{C-D}{2} \end{aligned}\right\} \quad \text{... (2)}$$

On substituting $A+B=C, A-B=D, A=\dfrac{C+D}{2}$ and $B=\dfrac{C-D}{2}$ in the following formulae (proved in section 15.5)

1. $\sin (A + B) + \sin (A - B) = 2 \sin A \cos B$
2. $\sin (A + B) - \sin (A - B) = 2 \cos A \sin B$
3. $\cos (A + B) + \cos (A - B) = 2 \cos A \cos B$
4. $\cos (A - B) - \cos (A + B) = 2 \sin A \sin B$, we get

1. $\sin C + \sin D = 2\sin\left(\frac{C+D}{2}\right)\cos\left(\frac{C-D}{2}\right)$
2. $\sin C - \sin D = 2\cos\left(\frac{C+D}{2}\right)\sin\left(\frac{C-D}{2}\right)$
3. $\cos C + \cos D = 2\cos\left(\frac{C+D}{2}\right)\cos\left(\frac{C-D}{2}\right)$
4. $\cos C - \cos D = 2\sin\left(\frac{C+D}{2}\right)\sin\left(\frac{D-C}{2}\right)$

Caution. In the 4th formula, the R.H.S. contains $\sin\left(\frac{D-C}{2}\right)$ and not $\sin\left(\frac{C-D}{2}\right)$.

Note: **1.** These formulae are called '$C - D$' formulae.

2. They are also known as sum and difference formulae on the basis of L.H.S.

Aid to memory

1. (*a*) L.H.S. of each formula contains same *t*-ratios, *i.e.* $\sin C \pm \sin D$ or $\cos C \pm \cos D$

(*b*) If L.H.S. contains $\sin C \pm \cos D$, then either 'sin' is changed into 'cos' or 'cos' into 'sin' as given below:

$$\sin C \pm \cos D = \sin C \pm \sin\left(\frac{\pi}{2} - D\right) \text{ or } \sin C \pm \cos D = \cos\left(\frac{\pi}{2} - C\right) + \cos D.$$

2. R.H.S. of these formulae starts with 2, whereas sum and difference of the angles are divided by 2.

1. $\sin + \sin = 2\sin\left(\frac{\text{sum}}{2}\right)\cos\left(\frac{\text{difference}}{2}\right)$
2. $\sin - \sin = 2\cos\left(\frac{\text{sum}}{2}\right)\sin\left(\frac{\text{difference}}{2}\right)$
3. $\cos + \cos = 2\cos\left(\frac{\text{sum}}{2}\right)\cos\left(\frac{\text{difference}}{2}\right)$
4. $\cos - \cos = -2\sin\left(\frac{\text{sum}}{2}\right)\sin\left(\frac{\text{difference}}{2}\right)$

Here, sum stands for sum of angles and difference stands for difference of angles.

Example 38. *Express as product of t-ratios:*

(*i*) $\sin 75° + \sin 15°$ (*ii*) $\sin 50° - \cos 80°$ (*iii*) $\cos\frac{4\pi}{5} + \cos\frac{\pi}{5}$
(*iv*) $\cos 3\theta - \cos 7\theta$

Solution. (*i*) We know that,

$$\sin C + \sin D = 2\sin\left(\frac{C+D}{2}\right)\cos\left(\frac{C-D}{2}\right)$$

$$\therefore \quad \sin 75° + \sin 15° = 2\sin\left(\frac{75°+15°}{2}\right)\cos\left(\frac{75°-15°}{2}\right) = 2\sin 45° \cos 30°$$

$$= 2 \times \frac{1}{\sqrt{2}} \times \frac{\sqrt{3}}{2} = \sqrt{\frac{3}{2}}$$ **Ans.**

(*ii*) We know that,

$$\sin C - \sin D = 2\cos\left(\frac{C+D}{2}\right)\cdot\sin\left(\frac{C-D}{2}\right) \quad \text{... (1)}$$

$$\therefore \quad \sin 50° - \cos 80° = \sin 50° - \sin 10° \quad [\because \cos 80° = \cos(90° - 10°) = \sin 10°]$$

$$= 2\cos\left(\frac{50° + 10°}{2}\right)\sin\left(\frac{50° - 10°}{2}\right) \quad \text{[using (1)]}$$

$$= 2\cos 30° \sin 20° = \sqrt{3}\sin 20° \quad \textbf{Ans.}$$

(*iii*) We know that,

$$\cos C + \cos D = 2\cos\left(\frac{C+D}{2}\right)\cdot\cos\left(\frac{C-D}{2}\right)$$

$$\therefore \quad \cos\frac{4\pi}{5} + \cos\frac{\pi}{5} = 2\cos\left(\frac{\frac{4\pi}{5}+\frac{\pi}{5}}{2}\right)\cos\left(\frac{\frac{4\pi}{5}-\frac{\pi}{5}}{2}\right)$$

$$= 2\cos\frac{\pi}{2}\cdot\cos\frac{3\pi}{10} = 2\times 0\times\cos\frac{3\pi}{10} = 0 \quad \left[\because \cos\frac{\pi}{2} = 0\right] \textbf{Ans.}$$

(*iv*) We know that $\cos C - \cos D = 2\sin\left(\frac{C+D}{2}\right)\sin\left(\frac{D-C}{2}\right)$

$$\therefore \quad \cos 3\theta - \cos 7\theta = 2\sin\left(\frac{3\theta+7\theta}{2}\right)\sin\left(\frac{7\theta-3\theta}{2}\right) = 2\sin 5\theta\cdot\sin 2\theta \quad \textbf{Ans.}$$

Example 39. *Find the value of* $\sin\frac{3\pi}{2} + \cos\frac{2\pi}{3}$ **[S.B.T.E. 2015]**

Solution. Here we have

$$\sin\frac{3\pi}{2} + \cos\frac{2\pi}{3} = \sin\left(\pi + \frac{\pi}{2}\right) + \cos\left(\pi - \frac{\pi}{3}\right)$$

$$= -\sin\frac{\pi}{2} - \cos\frac{\pi}{3}$$

$$= -1 - \frac{1}{2} = -\frac{3}{2} \quad \textbf{Ans.}$$

Example 40. *Prove that:* sin 51° + cos 81° = cos 21° **[S.B.T.E. 2014]**

Solution. Here we have

$$\text{L.H.S.} = \sin 51° + \cos 81° = \sin 51° + \cos(90° - 9°)$$

$$= \sin 51° + \sin 9°$$

$$= 2\sin\frac{(51° + 9°)}{2}\cdot\cos\frac{(51° - 9°)}{2}$$

$$= 2\sin 30° \,.\, \cos 21°$$

$$= 2\times\frac{1}{2}\cos 21°$$

$$\therefore \quad \sin 51° + \cos 81° = \cos 21° \quad \textbf{Proved}$$

Example 41. *Express as sum or difference*

(*i*) $\sin\frac{5\theta}{2}\cdot\sin\frac{3\theta}{2}$ (*ii*) $\cos^2 75° - \cos^2 15°$ **[S.B.T.E. 2015, 2014]**

Solution. Here we have

(*i*) $$\sin\frac{5\theta}{2}\cdot\sin\frac{3\theta}{2} = \frac{1}{2}\left[2\sin\frac{5\theta}{2}\cdot\sin\frac{3\theta}{2}\right]$$

$$= \frac{1}{2}\left[\cos\left(\frac{5\theta}{2}-\frac{3\theta}{2}\right)-\cos\left(\frac{5\theta}{2}+\frac{3\theta}{2}\right)\right]$$

$$= \frac{1}{2}\left[\cos\frac{2\theta}{2}-\cos\frac{8\theta}{2}\right]$$

$\therefore$ $$\sin\frac{5\theta}{2}\cdot\sin\frac{3\theta}{2} = \frac{1}{2}[\cos\theta-\cos 4\theta]$$ **Ans.**

(*ii*) $$\cos^2 75° - \cos^2 15° = (1 - \sin^2 75°) - (1 - \sin^2 15°)$$

$$= \sin^2 15° - \sin^2 75°$$

$$= \sin (15° + 75°)\cdot\sin (15° - 75°)$$

$$= \sin 90°\cdot\sin(-60°)$$

$$= -\sin 90°\cdot\sin 60° = (-1)\cdot\frac{\sqrt{3}}{2}$$

$\therefore$ $$\cos^2 75° - \cos^2 15° = -\frac{\sqrt{3}}{2}$$ **Ans.**

Example 42. *Prove the following identities:*

(*i*) $\sin (150° + x) + \sin (150° - x) = \cos x$

(*ii*) $\cos\left(\frac{3\pi}{4}+x\right)-\cos\left(\frac{3\pi}{4}-x\right) = -\sqrt{2}\sin x$

Solution. (*i*) We have,

$$\text{L.H.S.} = \sin (150° + x) + \sin (150° - x)$$

$$= 2\sin\left(\frac{150°+x+150°-x}{2}\right)\cos\left(\frac{150°+x-150°+x}{2}\right)$$

$$= 2\sin 150° \cos x \qquad \left[\because \sin C + \sin D = 2\sin\frac{C+D}{2}\cos\frac{C-D}{2}\right]$$

$$= 2\sin(180° - 30°)\cdot\cos x = 2\sin 30°\cdot\cos x \qquad [\because \sin(180° - \theta) = \sin\theta]$$

$$= 2\times\frac{1}{2}\cdot\cos x = \cos x = \text{R.H.S.} \qquad \left[\because \sin 30° = \frac{1}{2}\right]$$ **Proved.**

(*ii*) $$\text{L.H.S.} = \cos\left(\frac{3\pi}{4}+x\right)-\cos\left(\frac{3\pi}{4}-x\right) = 2\sin\left(\frac{\frac{3\pi}{4}+x+\frac{3\pi}{4}-x}{2}\right)\sin\left(\frac{\frac{3\pi}{4}-x-\frac{3\pi}{4}-x}{2}\right)$$

$$\left[\because \cos C - \cos D = 2\sin\frac{C+D}{2}\sin\frac{D-C}{2}\right]$$

$$= 2\sin\left(\frac{3\pi}{4}\right)\cdot\sin(-x) = -2\sin\left(\pi-\frac{\pi}{4}\right)\cdot\sin x \qquad [\because \sin(-\theta) = -\sin\theta]$$

$$= -2\sin\frac{\pi}{4}\cdot\sin x \qquad [\because \sin(\pi - \theta) = \sin\theta]$$

$$= -2\times\frac{1}{\sqrt{2}}\sin x = -\sqrt{2}\sin x = \text{R.H.S.}$$ **Proved.**

Example 43. *Prove that:*

(*i*) $\sin 2x + 2\sin 4x + \sin 6x = 4\cos^2 x \sin 4x$

(*ii*) $\sin^2 6x - \sin^2 4x = \sin 2x \sin 10x$

Solution. (*i*) L.H.S. $= \sin 2x + 2\sin 4x + \sin 6x$

$$= \sin 2x + \sin 6x + 2\sin 4x$$

$$= 2\sin\left(\frac{2x+6x}{2}\right)\cos\left(\frac{2x-6x}{2}\right) + 2\sin 4x$$

$$\left[\because \sin C + \sin D = 2\sin\frac{C+D}{2}\cos\frac{C-D}{2}\right]$$

$$= 2\sin 4x \cos(-2x) + 2\sin 4x = 2\sin 4x\cos 2x + 2\sin 4x$$

$$= 2\sin 4x\,[\cos 2x + 1] \qquad [\because \cos(-\theta) = \cos\theta]$$

$$= 2\sin 4x\,[2\cos^2 x - 1 + 1] \qquad [\because \cos 2x = 2\cos^2 x - 1]$$

$$= 4\sin 4x.\cos^2 x = \text{R.H.S.}$$ **Proved.**

(*ii*) L.H.S $= \sin^2 6x - \sin^2 4x = (\sin 6x)^2 - (\sin 4x)^2$

$$= (\sin 6x + \sin 4x)(\sin 6x - \sin 4x)$$

$$= \left(2\sin\frac{6x+4x}{2}\cos\frac{6x-4x}{2}\right)\left(2\cos\frac{6x+4x}{2}\sin\frac{6x-4x}{2}\right)$$

$$= (2\sin 5x\cos x)(2\cos 5x.\sin x)$$

$$= (2\sin x\cos x)(2\sin 5x.\cos 5x)$$

$$= \sin 2x.\sin 10x = \text{R.H.S.} \qquad [\because \sin 2x = 2\sin x\cos x]$$ **Proved.**

Example 44. *Prove that:*

(*i*) $\dfrac{\sin 5\theta + \sin 3\theta}{\cos 5\theta + \cos 3\theta} = \tan 4\theta$ (*ii*) $\dfrac{\tan 5\theta + \tan 3\theta}{\tan 5\theta - \tan 3\theta} = 4\cos 2\theta\cos 4\theta$ **[S.B.T.E. 2015, 2014]**

Solution. (*i*) L.H.S. $= \dfrac{\sin 5\theta + \sin 3\theta}{\cos 5\theta + \cos 3\theta} = \dfrac{2\sin\left(\frac{5\theta+3\theta}{2}\right)\cos\left(\frac{5\theta-3\theta}{2}\right)}{2\cos\left(\frac{5\theta+3\theta}{2}\right)\cos\left(\frac{5\theta-3\theta}{2}\right)}$

$$= \frac{\sin 4\theta\cdot\cos\theta}{\cos 4\theta\cdot\cos\theta} = \frac{\sin 4\theta}{\cos 4\theta} \qquad \left[\begin{array}{l}\because \sin C + \sin D = 2\sin\left(\frac{C+D}{2}\right)\cos\left(\frac{C-D}{2}\right)\\ \text{and } \cos C + \cos D = 2\cos\left(\frac{C+D}{2}\right)\cos\left(\frac{C-D}{2}\right)\end{array}\right]$$

$$= \tan 4\theta = \text{R.H.S.}$$ **Proved.**

(*ii*) L.H.S. $= \dfrac{\tan 5\theta + \tan 3\theta}{\tan 5\theta - \tan 3\theta} = \dfrac{\frac{\sin 5\theta}{\cos 5\theta} + \frac{\sin 3\theta}{\cos 3\theta}}{\frac{\sin 5\theta}{\cos 5\theta} - \frac{\sin 3\theta}{\cos 3\theta}} = \dfrac{\sin 5\theta\cos 3\theta + \cos 5\theta\cdot\sin 3\theta}{\sin 5\theta\cdot\cos 3\theta - \cos 5\theta\sin 3\theta}$

$$= \frac{\sin(5\theta + 3\theta)}{\sin(5\theta - 3\theta)} \qquad \left[\begin{array}{l}\because \sin(A+B) = \sin A\cos B + \cos A\sin B \\ \text{and } \sin(A-B) = \sin A\cos B - \cos A\sin B\end{array}\right]$$

$$= \frac{\sin 8\theta}{\sin 2\theta} = \frac{2\sin 4\theta\cdot\cos 4\theta}{\sin 2\theta} = \frac{4\sin 2\theta\cdot\cos 2\theta\cdot\cos 4\theta}{\sin 2\theta}$$

$= 4\cos 2\theta\cdot\cos 4\theta =$ R.H.S. **Proved.**

Example 45. *Prove that:* $\cos 130° + \cos 110° + \cos 10° = 0$ **[S.B.T.E. 2017, 2015]**

Solution. Here we have

L.H.S. $= \cos 130° + \cos 110° + \cos 10°$

$$= 2\cos\left(\frac{130+110}{2}\right)\cdot\cos\left(\frac{130-110}{2}\right) + \cos 10°$$

$$= 2\cos 120°\cdot\cos 10° + \cos 10°$$

$$= 2\times\left(-\frac{1}{2}\right)\cos 10 + \cos 10$$

$= -\cos 10 + \cos 10 = 0$ **Proved.**

Example 46. *Prove that:* $\cos 220° + \cos 100° + \cos 20° = 0$ **[S.B.T.E 2015]**

Solution. Here we have

L.H.S. $= \cos 220° + \cos 100° + \cos 20°$

$$= 2\cos\left(\frac{220+100}{2}\right)\cos\left(\frac{220-100}{2}\right) + \cos 20°$$

$$= 2\cos 160°\cdot\cos 60° + \cos 20°$$

$$= 2\cos 160°\cdot\left(\frac{1}{2}\right) + \cos 20°$$

$$= \cos 160° + \cos 20° = 2\cos\left(\frac{160+20}{2}\right)\cos\left(\frac{160-20}{2}\right)$$

$$= 2\cos 90\cdot\cos 70°$$

$= 0$ **Proved.**

Example 47. *Prove that:*

(*i*) $\dfrac{\cos 3\theta + 2\cos 5\theta + \cos 7\theta}{\cos\theta + 2\cos 3\theta + \cos 5\theta} = \cos 2\theta - \sin 2\theta\tan 3\theta$

(*ii*) $\dfrac{\cos\theta + \cos 2\theta + \cos 3\theta + \cos 4\theta}{\sin\theta + \sin 2\theta + \sin 3\theta + \sin 4\theta} = \cot\dfrac{5}{2}\theta$

Solution. (*i*) L.H.S. $= \dfrac{(\cos 3\theta + \cos 7\theta) + 2\cos 5\theta}{(\cos\theta + \cos 5\theta) + 2\cos 3\theta}$

$$= \frac{2\cos 5\theta\cos 2\theta + 2\cos 5\theta}{2\cos 3\theta\cos 2\theta + 2\cos 3\theta} = \frac{2\cos 5\theta(\cos 2\theta + 1)}{2\cos 3\theta(\cos 2\theta + 1)} = \frac{\cos 5\theta}{\cos 3\theta}$$

$$= \frac{\cos(2\theta + 3\theta)}{\cos 3\theta} = \frac{\cos 2\theta\cos 3\theta - \sin 2\theta\sin 3\theta}{\cos 3\theta} = \frac{\cos 2\theta\cos 3\theta}{\cos 3\theta} - \frac{\sin 2\theta\cdot\sin 3\theta}{\cos 3\theta}$$

$= \cos 2\theta - \sin 2\theta\cdot\tan 3\theta =$ R.H.S. **Proved.**

(*ii*) L.H.S. $= \dfrac{(\cos\theta + \cos 4\theta) + \cos 2\theta + \cos 3\theta)}{(\sin\theta + \sin 4\theta) + (\sin 2\theta + \sin 3\theta)}$

$$= \frac{2\cos\frac{5\theta}{2}\cos\left(-\frac{3\theta}{2}\right) + 2\cos\frac{5\theta}{2}\cos\left(-\frac{\theta}{2}\right)}{2\sin\frac{5\theta}{2}\cos\left(-\frac{3\theta}{2}\right) + 2\sin\frac{5\theta}{2}\cos\left(-\frac{\theta}{2}\right)}$$

$$= \frac{2\cos\frac{5\theta}{2}\left(\cos\frac{3\theta}{2} + \cos\frac{\theta}{2}\right)}{2\sin\frac{5\theta}{2}\left(\cos\frac{3\theta}{2} + \cos\frac{\theta}{2}\right)} = \frac{\cos\frac{5\theta}{2}}{\sin\frac{5\theta}{2}} = \cot\frac{5\theta}{2} = \text{R.H.S.}$$ **Proved.**

Example 48. *Prove that:*

$$(\cos\alpha + \cos\beta)^2 + (\sin\alpha + \sin\beta)^2 = 4\cos^2\left(\frac{\alpha-\beta}{2}\right)$$

Solution. L.H.S. $= (\cos\alpha + \cos\beta)^2 + (\sin\alpha + \sin\beta)^2$

$$= \left(2\cos\frac{\alpha+\beta}{2}\cos\frac{\alpha-\beta}{2}\right)^2 + \left(2\sin\frac{\alpha+\beta}{2}\cos\frac{\alpha-\beta}{2}\right)^2$$

$$= 4\cos^2\left(\frac{\alpha+\beta}{2}\right)\cos^2\left(\frac{\alpha-\beta}{2}\right) + 4\sin^2\left(\frac{\alpha+\beta}{2}\right)\cos^2\left(\frac{\alpha-\beta}{2}\right)$$

$$= 4\cos^2\left(\frac{\alpha-\beta}{2}\right)\left(\cos^2\frac{\alpha+\beta}{2} + \sin^2\frac{\alpha+\beta}{2}\right) = 4\cos^2\left(\frac{\alpha-\beta}{2}\right)(1)$$

$$= 4\cos^2\left(\frac{\alpha-\beta}{2}\right) = \text{R.H.S.}$$ $[\because \sin^2\theta + \cos^2\theta = 1]$ **Proved.**

Example 49. *Prove that:*

$$\cos\alpha + \cos\beta + \cos\gamma + \cos(\alpha+\beta+\gamma) = 4\cos\left(\frac{\alpha+\beta}{2}\right)\cos\left(\frac{\beta+\gamma}{2}\right)\cos\left(\frac{\gamma+\alpha}{2}\right)$$

Solution. L.H.S. $= \cos\alpha + \cos\beta + \cos\gamma + \cos(\alpha+\beta+\gamma)$

$$= (\cos\alpha + \cos\beta) + [\cos\gamma + \cos(\alpha+\beta+\gamma)]$$

$$= 2\cos\left(\frac{\alpha+\beta}{2}\right)\cos\left(\frac{\alpha-\beta}{2}\right) + \cos\left(\frac{\alpha+\beta+\gamma+\gamma}{2}\right)\cdot\cos\left(\frac{\alpha+\beta+\gamma-\gamma}{2}\right)$$

$$= 2\cos\left(\frac{\alpha+\beta}{2}\right)\cos\left(\frac{\alpha-\beta}{2}\right) + 2\cos\left(\frac{\alpha+\beta}{2}\right)\cos\left(\frac{\alpha+\beta+2\gamma}{2}\right)$$

$$= 2\cos\left(\frac{\alpha+\beta}{2}\right)\left\{\cos\left(\frac{\alpha-\beta}{2}\right) + \cos\left(\frac{\alpha+\beta+2\gamma}{2}\right)\right\}$$

$$= 2\cos\left(\frac{\alpha+\beta}{2}\right)\left\{2\cos\left(\frac{\frac{\alpha-\beta}{2} + \frac{\alpha+\beta+2\gamma}{2}}{2}\right)\cos\left(\frac{\frac{\alpha+\beta+2\gamma}{2} - \frac{\alpha-\beta}{2}}{2}\right)\right\}$$

$$= 2\cos\left(\frac{\alpha+\beta}{2}\right)\left\{2\cos\left(\frac{\alpha+\gamma}{2}\right)\cos\left(\frac{\beta+\gamma}{2}\right)\right\}$$

$$= 4\cos\left(\frac{\alpha+\beta}{2}\right)\cos\left(\frac{\beta+\gamma}{2}\right)\cos\left(\frac{\gamma+\alpha}{2}\right) = \text{R.H.S.}$$ **Proved.**

Example 50. *Prove that:* $\cos\frac{\pi}{15}\cdot\cos\frac{2\pi}{15}\cdot\cos\frac{3\pi}{15}\cdot\cos\frac{4\pi}{15}\cdot\cos\frac{5\pi}{15}\cdot\cos\frac{6\pi}{15}\cdot\cos\frac{7\pi}{15} = \frac{1}{128}$

[S.B.T.E, 2016, 2015]

Solution. Here we have

$$\text{L.H.S.} = \cos\frac{\pi}{15}\cdot\cos\frac{2\pi}{15}\cdot\cos\frac{3\pi}{15}\cdot\cos\frac{4\pi}{15}\cdot\cos\frac{5\pi}{15}\cdot\cos\frac{6\pi}{15}\cdot\cos\frac{7\pi}{15},\quad \left[\because \cos\frac{5\pi}{15} = \cos 60° = \frac{1}{2}\right]$$

$$= \frac{1}{2}\left[\cos\frac{\pi}{15}\cdot\cos\frac{2\pi}{15}\cdot\cos\frac{4\pi}{15}\cdot\cos\left(\pi - \frac{8\pi}{15}\right)\right]\left[\cos\frac{3\pi}{15}\cdot\cos\frac{6\pi}{15}\right]$$

$$= -\frac{1}{2}\left[\cos\frac{\pi}{15}\cos 2\cdot\left(\frac{\pi}{15}\right)\cdot\cos 2^2\left(\frac{\pi}{15}\right)\cdot\cos 2^3\left(\frac{\pi}{15}\right)\right]\left[\cos\frac{3\pi}{15}\cdot\cos 2\cdot\frac{3\pi}{15}\right]$$

$$= -\frac{1}{2}\left[\frac{\sin 2^4\left(\frac{\pi}{15}\right)}{2^4\sin\left(\frac{\pi}{15}\right)}\right]\left[\frac{\sin 2^2\left(\frac{3\pi}{15}\right)}{2^2\sin\left(\frac{3\pi}{15}\right)}\right]$$

$$\left[\because \cos\cdot\cos 2A\cdot\cos 2^2 A\cdot\cos 2^3 A\cdots\cos^{n-1} A = \frac{\sin 2^n A}{2^n \sin A}\right]$$

$$= -\frac{1}{2}\left[\frac{\sin\frac{16\pi}{15}}{16\sin\frac{\pi}{15}}\right]\left[\frac{\sin\frac{12\pi}{5}}{4\sin\frac{3\pi}{15}}\right]$$

$$= -\frac{1}{2}\left[\frac{\sin\left(\pi + \frac{\pi}{15}\right)}{16\sin\frac{\pi}{15}}\right]\left[\frac{\sin\left(\pi - \frac{3\pi}{15}\right)}{4\sin\frac{3\pi}{15}}\right]$$

$$= -\frac{1}{2}\left[\frac{-\sin\frac{\pi}{15}}{16\sin\frac{\pi}{15}}\right]\left[\frac{\sin\frac{3\pi}{15}}{4\sin\frac{3\pi}{15}}\right] = -\frac{1}{2}\left(-\frac{1}{16}\right)\left(\frac{1}{4}\right) = \frac{1}{128}$$ **Proved.**

Example 51. *Prove that:*

$$\sin\alpha + \sin\beta + \sin\gamma - \sin(\alpha+\beta+\gamma) = 4\sin\left(\frac{\alpha+\beta}{2}\right)\sin\left(\frac{\beta+\gamma}{2}\right)\sin\left(\frac{\gamma+\alpha}{2}\right)$$

Solution. L.H.S. $= \sin\alpha + \sin\beta + \sin\gamma - \sin(\alpha+\beta+\gamma)$

$$= (\sin\alpha + \sin\beta) + [\sin\gamma - \sin(\alpha+\beta+\gamma)]$$

$$= 2\sin\left(\frac{\alpha+\beta}{2}\right)\cos\left(\frac{\alpha-\beta}{2}\right) + 2\cos\left(\frac{\gamma+\alpha+\beta+\gamma}{2}\right)\sin\left(\frac{\gamma-\alpha-\beta-\gamma}{2}\right)$$

$$= 2\sin\left(\frac{\alpha+\beta}{2}\right)\cos\left(\frac{\alpha-\beta}{2}\right) + 2\cos\left(\frac{\alpha+\beta+2\gamma}{2}\right)\sin\left(\frac{-\alpha-\beta}{2}\right)$$

$$= 2\sin\left(\frac{\alpha+\beta}{2}\right)\cos\left(\frac{\alpha-\beta}{2}\right) - 2\cos\left(\frac{\alpha+\beta+2\gamma}{2}\right)\sin\left(\frac{\alpha+\beta}{2}\right)$$

$[\because \sin(-\theta) = -\sin\theta]$

$$= 2\sin\left(\frac{\alpha+\beta}{2}\right)\left\{\cos\frac{\alpha-\beta}{2} - \cos\frac{\alpha+\beta+2\gamma}{2}\right\}$$

$$= 2\sin\left(\frac{\alpha+\beta}{2}\right)\left\{2\sin\left(\frac{\frac{\alpha-\beta}{2}+\frac{\alpha+\beta+2\gamma}{2}}{2}\right)\sin\left(\frac{\frac{\alpha+\beta+2\gamma}{2}-\frac{\alpha-\beta}{2}}{2}\right)\right\}$$

$$= 2\sin\left(\frac{\alpha+\beta}{2}\right)\left\{2\sin\left(\frac{\alpha-\beta+\alpha+\beta+2\gamma}{4}\right)\cdot\sin\left(\frac{\alpha+\beta+2\gamma-\alpha+\beta}{4}\right)\right\}$$

$$= 2\sin\left(\frac{\alpha+\beta}{2}\right)\left(2\sin\frac{\alpha+\gamma}{2}\cdot\sin\frac{\beta+\gamma}{2}\right)$$

$$= 4\sin\left(\frac{\alpha+\beta}{2}\right)\sin\left(\frac{\alpha+\gamma}{2}\right)\sin\left(\frac{\beta+\gamma}{2}\right) = \text{R.H.S.}$$ **Proved.**

Example 52. *If* $x\cos\theta = y\cos\left(\theta+\frac{2\pi}{3}\right) = z\cos\left(\theta+\frac{4\pi}{3}\right)$, *then show that* $xy + yz + zx = 0$.

Solution. Let $x\cos\theta = y\cos\left(\theta+\frac{2\pi}{3}\right) = z\cos\left(\theta+\frac{4\pi}{3}\right) = k$

$$\Rightarrow \quad \frac{1}{x} = \frac{\cos\theta}{k},\ \frac{1}{y} = \frac{\cos\left(\theta+\frac{2\pi}{3}\right)}{k},\ \frac{1}{z} = \frac{\cos\left(\theta+\frac{4\pi}{3}\right)}{k} \quad \text{... (1)}$$

Now, L.H.S. $= xy + yz + zx = \frac{xyz}{z} + \frac{xyz}{x} + \frac{xyz}{y} = xyz\left(\frac{1}{z}+\frac{1}{x}+\frac{1}{y}\right)$

$$= xyz\left[\frac{\cos\left(\theta+\frac{4\pi}{3}\right)}{k} + \frac{\cos\theta}{k} + \frac{\cos\left(\theta+\frac{2\pi}{3}\right)}{k}\right] \quad \text{[using (1)]}$$

$$= \frac{xyz}{k}\left[\cos\left(\theta+\frac{4\pi}{3}\right) + \cos\left(\theta+\frac{2\pi}{3}\right) + \cos\theta\right]$$

$$= \frac{xyz}{k}\left[2\cos\frac{2\theta+2\pi}{2}\cos\frac{2\pi}{6} + \cos\theta\right]$$

$$= \frac{xyz}{k}\left[2\cos(\pi+\theta)\cos\frac{\pi}{3}+\cos\theta\right] = \frac{xyz}{k}\left[-2\cos\theta\cdot\left(\frac{1}{2}\right)+\cos\theta\right]$$

$$= \frac{xyz}{k}[-\cos\theta+\cos\theta] = \frac{xyz}{k}(0) = 0 = \text{R.H.S.}$$

$\Rightarrow \quad xy + yz + zx = 0.$ **Proved.**

Example 53. *Prove that:*

$$\left(\frac{\cos A+\cos B}{\sin A-\sin B}\right)^n + \left(\frac{\sin A+\sin B}{\cos A-\cos B}\right)^n = \begin{cases} 2\cot^n\left(\frac{A-B}{2}\right), & \text{if } n \text{ is even.} \\ 0, & \text{if } n \text{ is odd.} \end{cases}$$

Solution. L.H.S. $= \left(\frac{\cos A+\cos B}{\sin A-\sin B}\right)^n + \left(\frac{\sin A+\sin B}{\cos A-\cos B}\right)^n$

$$= \left(\frac{2\cos\frac{A+B}{2}\cos\frac{A-B}{2}}{2\sin\frac{A-B}{2}\cos\frac{A+B}{2}}\right)^n + \left(\frac{2\sin\frac{A+B}{2}\cos\frac{A-B}{2}}{-2\sin\frac{A+B}{2}\sin\frac{A-B}{2}}\right)^n$$

$$= \left[\cot\left(\frac{A-B}{2}\right)\right]^n + \left[-\cot\left(\frac{A-B}{2}\right)\right]^n$$

$$= \cot^n\left(\frac{A-B}{2}\right) + (-1)^n\cot^n\left(\frac{A-B}{2}\right)$$

$$= \cot^n\left(\frac{A-B}{2}\right)\{1+(-1)^n\} = \begin{cases} 2\cot^n\left(\frac{A-B}{2}\right), & \text{if } n \text{ is even.} \\ 0, & \text{if } n \text{ is odd.} \end{cases}$$ **Proved.**

Example 54. *If $A + B + C = \pi$, prove that*

$$\frac{\cos A}{\sin B\sin C} + \frac{\cos B}{\sin C\sin A} + \frac{\cos C}{\sin A\sin B} = 2$$

Solution. We have, $A + B + C = \pi \Rightarrow A = \pi - (B + C)$ ... (1)

Now, $\frac{\cos A}{\sin B\sin C} = \frac{\cos(\pi-(B+C))}{\sin B\sin C}$ [using (1)]

$$= \frac{-\cos(B+C)}{\sin B\sin C} \quad [\because \cos(\pi-\theta) = -\cos\theta]$$

$$= \frac{-[\cos B\cos C-\sin B\sin C]}{\sin B\sin C} = \frac{-\cos B\cos C}{\sin B\sin C} + \frac{\sin B\sin C}{\sin B\sin C}$$

$\Rightarrow \quad \frac{\cos B}{\sin C\sin A} = 1 - \cot B\cot C$ ... (2)

Similarly, $\frac{\cos B}{\sin C\sin A} = 1 - \cot A\cot C$ ... (3)

and $$\frac{\cos C}{\sin A \sin B} = 1 - \cot A \cot B \qquad \text{... (4)}$$

Adding (2), (3) and (4), we get

$$\frac{\cos A}{\sin B \sin C} + \frac{\cos B}{\sin C \sin A} + \frac{\cos C}{\sin A \sin B} = 3 - (\cot B \cot C + \cot C \cot A + \cot A \cot B) \qquad \text{... (5)}$$

But $$\cot (A + B) = \frac{\cot A \cot B - 1}{\cot B + \cot A}$$

$$\Rightarrow \quad \cot (\pi - C) = \frac{\cot A \cot B - 1}{\cot B + \cot A} \qquad [\because A + B + C = \pi]$$

$$\Rightarrow \quad - \cot C [\cot B + \cot A] = \cot A \cot B - 1$$

$$\Rightarrow \quad - (\cot B \cot C + \cot C \cot A) - \cot A \cot B = 1$$

$$\Rightarrow \quad \cot B \cot C + \cot C \cot A + \cot A \cot B = 1 \qquad \text{... (6)}$$

Putting the value of $\cot B \cdot \cot C + \cot C \cdot \cot A + \cot A \cdot \cot B = 1$ from (6) in (5), we get

$$\frac{\cos A}{\sin B \sin C} + \frac{\cos B}{\sin C \sin A} + \frac{\cos C}{\sin A \sin B} = 3 - 1 = 2$$ **Proved.**

EXERCISE 14.3

Express each of the following as the product of sines and cosines:

1. $\sin 2\theta + \sin 4\theta$ **Ans.** $2 \sin 3\theta \cos \theta$

2. $\cos 79° + \cos 11°$ **Ans.** $2 \cos 45° \cos 34°$

3. $\sin 12\theta - \sin 4\theta$ **Ans.** $2 \cos 8\theta \sin 4\theta$

4. $\cos\frac{\pi}{13} - \cos\frac{2\pi}{13}$ **Ans.** $2\sin\frac{3\pi}{26}\sin\frac{\pi}{26}$

Prove that:

5. $\sin 38° + \sin 22° = \sin 82°$

6. $\sin\left(\frac{\pi}{4} + \theta\right) + \sin\left(\frac{\pi}{4} - \theta\right) = \sqrt{2}\cos\theta$

7. $\cos\left(\frac{2\pi}{3} + \theta\right) + \cos\left(\frac{\pi}{3} + \theta\right) = -\sqrt{3}\sin\theta$

8. $\cos 15° - \sin 15° = \frac{1}{\sqrt{2}}$

9. $\sin 61° - \cos 39° = 2 \cos 56° \sin 5°$

10. $\sin 4\theta + \cos 2\theta = 2 \sin(45° + \theta) \cos(45° - 3\theta)$

11. $\frac{\sin 75° - \sin 15°}{\cos 15° - \cos 75°} = 1$

12. $\frac{\cos 20° - \cos 70°}{\sin 70° - \sin 20°} = 1$

13. $\frac{\sin 7\theta + \sin 3\theta}{\cos 7\theta + \cos 3\theta} = \tan 5\theta$

14. $\frac{\cos 2B - \cos 2A}{\sin 2A + \sin 2B} = \tan(A - B)$

15. $\frac{\cos\left(\frac{\pi}{4} + \theta\right) - \cos\left(\frac{\pi}{4} - \theta\right)}{\sin\left(\frac{2\pi}{3} + \theta\right) - \sin\left(\frac{2\pi}{3} - \theta\right)} = \sqrt{2}$

16. $\frac{\sin(4A - 2B) + \sin(4B - 2A)}{\cos(4A - 2B) + \cos(4B - 2A)} = \tan(A + B)$

17. $\tan (45° + \theta) + \tan (45° - \theta) = 2 \sec 2\theta$

18. $\tan (45° + \theta) - \tan (45° - \theta) = 2 \tan 2\theta$

19. $\tan(\theta + 30°) + \cot(\theta - 30°) = \frac{1}{\sin 2\theta - \sin 60°}$

20. $\dfrac{\cos 4x + \cos 3x + \cos 2x}{\sin 4x + \sin 3x + \sin 2x} = \cot 3x$ **[S.B.T.E. 2014]**

21. $\dfrac{\cos \alpha + 2\cos 3\alpha + \cos 5\alpha}{\cos 3\alpha + 2\cos 5\alpha + \cos 7\alpha} = \cos 3\alpha \sec 5\alpha$

22. $\dfrac{\sin x + \sin 3x + \sin 5x + \sin 7x}{\cos x + \cos 3x + \cos 5x + \cos 7x} = \tan 4x$ **[B.T.E. Delhi; Jan. 2009]**

23. $\dfrac{\cos 2A \cos 3A - \cos 2A \cos 7A + \cos A \cos 10A}{\sin 4A \sin 3A - \sin 2A \sin 5A + \sin 4A \sin 7A} = \cot 6A \cot 5A$

24. $\dfrac{\sin(A-C) + 2\sin A + \sin(A+C)}{\sin(B-C) + 2\sin B + \sin(B+C)} = \dfrac{\sin A}{\sin B}$

25. If $b \sin \beta = a \sin (2\alpha + \beta)$, prove that $(b + a) \cot(\alpha + \beta) = (b - a) \cot \alpha$.

[**Hint:** $b \sin \beta = a \sin(2\alpha + \beta) \Rightarrow \dfrac{\sin(2\alpha+\beta)}{\sin \beta} = \dfrac{b}{a}$, use componendo–dividendo method]

26. If $\sin \theta = n \sin(\theta + 2\alpha)$; show that $\tan(\theta + \alpha) = \left(\dfrac{1+n}{1-n}\right) \tan \alpha$

27. Show that:
$\sin (y + z - x) + \sin (z + x - y) + \sin (x + y - z) - \sin (x + y + z) = 4 \sin x \sin y \sin z$

28. If $\sin \alpha + \sin \beta = \alpha$ and $\cos \alpha + \cos \beta = b$, find the values of

(*i*) $\tan\left(\dfrac{\alpha+\beta}{2}\right)$ **Ans.** $\dfrac{a}{b}$ (*ii*) $\tan\left(\dfrac{\alpha-\beta}{2}\right)$ **Ans.** $\pm\sqrt{\dfrac{4-a^2-b^2}{a^2+b^2}}$

14.6 TRIGONOMETRIC RATIOS OF MULTIPLE ANGLES

Theorem 1. *In this section, we have to prove that:*

(1) $\sin 2A = 2\sin A \cos A$

(2) $\cos 2A = \cos^2 A - \sin^2 A = 1 - 2 \sin^2 A = 2\cos^2 A - 1$ (3) $\tan 2A = \dfrac{2\tan A}{1-\tan^2 A}$

Proof:

(1) We know that $\sin(A + B) = \sin A \cos B + \cos A \sin B$

Replacing B by A, we get

$$\sin 2A = \sin (A + A) = \sin A \cos A + \cos A \sin A$$

$$\Rightarrow \boxed{\sin 2A = 2\sin A \cos A}$$

(2) We know that $\cos(A + B) = \cos A \cdot \cos B - \sin A \cdot \sin B$

Replacing B by A, we get

$$\cos 2A = \cos(A + A) = \cos A \cdot \cos A - \sin A \cdot \sin A$$

$$\Rightarrow \boxed{\cos 2A = \cos^2 A - \sin^2 A}$$

$$\Rightarrow \cos 2A = (1 - \sin^2 A) - \sin^2 A \qquad [\because \cos^2 A = 1 - \sin^2 A]$$

$$= 1 - \sin^2 A - \sin^2 A$$

$$\Rightarrow \boxed{\cos 2A = 1 - 2\sin^2 A}$$

Again, $\cos 2A = \cos^2 A - \sin^2 A = \cos^2 A - (1 - \cos^2 A) = \cos^2 A - 1 + \cos^2 A$

$\Rightarrow$ $\boxed{\mathbf{\cos 2A = 2\cos^2 A - 1}}$

(3) We know that $\tan(A + B) = \dfrac{\tan A + \tan B}{1 - \tan A \tan B}$

Replacing B by A, we get

$$\tan 2A = \tan(A + A) = \frac{\tan A + \tan A}{1 - \tan A \tan A}$$

$\Rightarrow$ $\boxed{\mathbf{\tan 2A = \dfrac{2\tan A}{1 - \tan^2 A}}}$

Corollary: *Here, we will prove that:*

(1) $1 - \cos 2A = 2\sin^2 A$ (2) $1 + \cos 2A = 2\cos^2 A$

Proof: (1) We have,

$\cos 2A = 1 - 2\sin^2 A \Rightarrow 1 - \cos 2A = 2\sin^2 A$ **Proved.**

Also, $\sin^2 A = \dfrac{1 - \cos 2A}{2} \Rightarrow \boxed{\mathbf{\sin A = \sqrt{\dfrac{1 - \cos 2A}{2}}}}$

(2) We have,

$\cos 2A = 2\cos^2 A - 1 \Rightarrow \cos 2A + 1 = 2\cos^2 A$ **Proved.**

Also, $\cos^2 A = \dfrac{\cos 2A + 1}{2} \Rightarrow \boxed{\mathbf{\cos A = \sqrt{\dfrac{1 + \cos 2A}{2}}}}$

Aid to Memory

1. sin (double angle) = 2sin (angle)·cos(angle)
2. cos (double angle) = 1 – 2sin²(angle)
3. cos (double angle) = 2cos²(angle) – 1

$$\boxed{\begin{aligned} \sin 2A &= 2\sin A\cos A \\ \cos 2A &= 2\cos^2 A - 1 \\ \cos 2A &= 1 - 2\sin^2 A \\ \cos 2A &= \cos^2 A - \sin^2 A \\ \tan 2A &= \frac{2\tan A}{1 - \tan^2 A} \end{aligned}}$$

Theorem 2. Prove that: (1) $\sin 2A = \dfrac{2\tan A}{1 + \tan^2 A}$ (2) $\cos 2A = \dfrac{1 - \tan^2 A}{1 + \tan^2 A}$

Proof: (1) R.H.S. $= \dfrac{2\tan A}{1 + \tan^2 A} = \dfrac{2\dfrac{\sin A}{\cos A}}{1 + \dfrac{\sin^2 A}{\cos^2 A}} = \dfrac{2\sin A}{\cos A} \times \dfrac{\cos^2 A}{\cos^2 A + \sin^2 A}$

$= \dfrac{2\sin A \cdot \cos A}{1} = \sin 2A = \text{L.H.S.}$ **Proved.**

(2) R.H.S. $= \dfrac{1-\tan^2 A}{1+\tan^2 A} = \dfrac{1-\dfrac{\sin^2 A}{\cos^2 A}}{1+\dfrac{\sin^2 A}{\cos^2 A}} = \dfrac{\cos^2 A - \sin^2 A}{\cos^2 A} \times \dfrac{\cos^2 A}{\cos^2 A + \sin^2 A}$

$= \cos^2 A - \sin^2 A = \cos 2A =$ L.H.S. **Proved.**

$$\boxed{\sin 2A = \frac{2\tan A}{1+\tan^2 A}} \qquad \boxed{\cos 2A = \frac{1-\tan^2 A}{1+\tan^2 A}}$$

Theorem 3. Prove that: (1) $\sin 3A = 3\sin A - 4\sin^3 A$

(2) $\cos 3A = 4\cos^3 A - 3\cos A$ (3) $\tan 3A = \dfrac{3\tan A - \tan^3 A}{1-3\tan^2 A}$

Proof: (1) We know that

$$\sin(A+B) = \sin A\cos B + \cos A\sin B$$

Replacing B by $2A$ on both sides, we get

$$\begin{aligned}\sin 3A = \sin(A+2A) &= \sin A\cos 2A + \cos A\sin 2A\\ &= \sin A(1-2\sin^2 A) + \cos A\cdot 2\sin A\cos A\\ &= \sin A - 2\sin^3 A + 2\sin A\cos^2 A\\ &= \sin A - 2\sin^3 A + 2\sin A(1-\sin^2 A)\\ &= \sin A - 2\sin^3 A + 2\sin A - 2\sin^3 A\end{aligned}$$

$\Rightarrow$ $\sin 3A = 3\sin A - 4\sin^3 A$ **Proved.**

(2) We know that

$$\cos(A+B) = \cos A\cos B - \sin A\sin B$$

Replacing B by $2A$, we get

$$\begin{aligned}\cos(A+B) = \cos(A+2A) &= \cos A\cos 2A - \sin A\sin 2A\\ &= \cos A(2\cos^2 A - 1) - \sin A\cdot 2\sin A\cos A\\ &= 2\cos^3 A - \cos A - 2\sin^2 A\cos A\\ &= 2\cos^3 A - \cos A - 2(1-\cos^2 A)\cos A\\ &= 2\cos^3 A - \cos A - 2\cos A + 2\cos^3 A\end{aligned}$$

$\therefore$ $\cos 3A = 4\cos^3 A - 3\cos A$ **Proved.**

(3) We know that

$$\tan(A+B) = \frac{\tan A + \tan B}{1-\tan A\tan B}$$

Replacing B by $2A$, we get

$$\tan 3A = \tan(A+2A) = \frac{\tan A + \tan 2A}{1-\tan A\tan 2A} = \frac{\tan A + \dfrac{2\tan A}{1-\tan^2 A}}{1-\tan A\cdot\dfrac{2\tan A}{1-\tan^2 A}}$$

$$= \frac{\tan A(1-\tan^2 A) + 2\tan A}{1-\tan^2 A - 2\tan^2 A}$$

$\Rightarrow$ $\tan 3A = \dfrac{3\tan A - \tan^3 A}{1-3\tan^2 A}$ **Proved.**

$$\sin 3A = 3\sin A - 4\sin^3 A; \quad \cos 3A = 4\cos^3 A - 3\cos A; \quad \tan 3A = \frac{3\tan A - \tan^3 A}{1 - 3\tan^2 A}$$

14.7 SUBMULTIPLE ANLGES

The angles $\frac{A}{2}, \frac{A}{3}, \ldots$ are called submultiple angles of A.

14.8 TRIGONOMETRIC RATIOS OF SUBMULTIPLE ANGLES

In this section, we shall prove that:

(1) $\sin A = 2\sin\frac{A}{2}\cos\frac{A}{2} = \dfrac{2\tan\frac{A}{2}}{1+\tan^2\frac{A}{2}}$

(2) $\cos A = \cos^2\frac{A}{2} - \sin^2\frac{A}{2} = 1 - 2\sin^2\frac{A}{2} = 2\cos^2\frac{A}{2} - 1 = \dfrac{1-\tan^2\frac{A}{2}}{1+\tan^2\frac{A}{2}}$

(3) $\tan A = \dfrac{2\tan\frac{A}{2}}{1-\tan^2\frac{A}{2}}$

Proof: (1) We know that $\quad \sin 2A = 2\sin A\cos A = \dfrac{2\tan A}{1+\tan^2 A}$ ... (1)

Changing A to $\frac{A}{2}$ in (1), we get $\quad \sin A = 2\sin\frac{A}{2}\cos\frac{A}{2} = \dfrac{2\tan\frac{A}{2}}{1+\tan^2\frac{A}{2}}$

(2) We know that, $\cos 2A = \cos^2 A - \sin^2 A = 1 - 2\sin^2 A = 2\cos^2 A - 1 = \dfrac{1-\tan^2 A}{1+\tan^2 A}$

Replacing A by $\frac{A}{2}$, we get

$$\cos A = \cos^2\frac{A}{2} - \sin^2\frac{A}{2} = 1 - 2\sin^2\frac{A}{2} = 2\cos^2\frac{A}{2} - 1 = \frac{1-\tan^2\frac{A}{2}}{1+\tan^2\frac{A}{2}}$$

(3) We know that $\tan 2A = \dfrac{2\tan A}{1-\tan^2 A}$

Replacing A by $\frac{A}{2}$, we get $\tan A = \dfrac{2\tan\frac{A}{2}}{1-\tan^2\frac{A}{2}}$ **Proved.**

14.9 TRIGONOMETRIC RATIOS OF SUBMULTIPLE ANGLES IN TERMS OF COS *A*

Theorem 1. (1) $\cos\frac{A}{2} = \pm\sqrt{\dfrac{1+\cos A}{2}}$ (2) $\sin\frac{A}{2} = \pm\sqrt{\dfrac{1-\cos A}{2}}$

(3) $\tan\frac{A}{2} = \sqrt{\dfrac{1-\cos A}{1+\cos A}}$

Proof: (1) We know that, $\cos A = 2\cos^2\frac{A}{2} - 1 \Rightarrow 2\cos^2\frac{A}{2} = 1 + \cos A$

$$\Rightarrow \boxed{\cos\frac{A}{2} = \pm\sqrt{\frac{1+\cos A}{2}}} \qquad \text{... (1)}$$

(2) We know that, $\cos A = 1 - 2\sin^2\frac{A}{2} \Rightarrow 2\sin^2\frac{A}{2} = 1 - \cos A$

$$\Rightarrow \boxed{\sin\frac{A}{2} = \pm\sqrt{\frac{1-\cos A}{2}}} \qquad \text{... (2)}$$

(3) We know that, $\tan\frac{A}{2} = \frac{\sin\frac{A}{2}}{\cos\frac{A}{2}} = \frac{\sqrt{\frac{1-\cos A}{2}}}{\sqrt{\frac{1+\cos A}{2}}}$ [Using (1) and (2)]

$$\Rightarrow \boxed{\tan\frac{A}{2} = \sqrt{\frac{1-\cos A}{1+\cos A}}}$$

Proved.

Aid to Memory

(1) $\cos(\text{any Angle}) = \pm\sqrt{\frac{1+\cos(\text{double angle})}{2}}$

(2) $\sin(\text{any Angle}) = \pm\sqrt{\frac{1-\cos(\text{double angle})}{2}}$

(3) $\tan(\text{any Angle}) = \sqrt{\frac{1-\cos(\text{double angle})}{1+\cos(\text{double angle})}}$

Example 55. *If* $\sin A = \frac{3}{5}$ *and is in I quadrant, find sin 2A, cos 2A, and tan 2A.*

[S.B.T.E. 2015]

Solution. We have, $\sin A = \frac{3}{5}$

$$\Rightarrow \cos A = \sqrt{1-\sin^2 A} = \sqrt{1-\frac{9}{25}} = \sqrt{\frac{16}{25}} = \frac{4}{5} \qquad [\because \cos^2 A = 1 - \sin^2 A]$$

and
$$\tan A = \frac{\sin A}{\cos A} = \frac{\frac{3}{5}}{\frac{4}{5}} = \frac{3}{4}$$

$$\therefore \quad \sin 2A = 2\sin A\cos A = 2\times\frac{3}{5}\times\frac{4}{5} = \frac{24}{25}$$

$$\cos 2A = 1 - 2\sin^2 A = 1 - 2\times\frac{9}{25} = 1 - \frac{18}{25} = \frac{7}{25}$$

$$\tan 2A = \frac{2\tan A}{1-\tan^2 A} = \frac{2\times\frac{3}{4}}{1-\frac{9}{16}} = \frac{24}{7} = 3\frac{3}{7}$$

Ans.

Example 56. *Find* $\sin\frac{x}{2}, \cos\frac{x}{2}$ *and* $\tan\frac{x}{2}$ *in each of the following:*

(*i*) $\sin x = \frac{1}{4}$, *x in II quadrant.* (*ii*) $\cos x = -\frac{1}{3}$, *x in III quadrant*

(*iii*) $\tan x = -\frac{4}{3}$, *x in II quadrant*

Solution. (*i*) We have, $\sin x = \frac{1}{4}, \ \frac{\pi}{2} < x < \pi \Rightarrow \frac{\pi}{4} < \frac{x}{2} < \frac{\pi}{2}$

i.e. $\frac{x}{2}$ lies in the I quadrant, so that all *t*-ratios of $\frac{x}{2}$ are positive.

Also, $\cos^2 x = 1 - \sin^2 x = 1 - \frac{1}{16} = \frac{15}{16}$ and cos *x* is negative in the II quadrant.

$$\Rightarrow \qquad \cos x = -\frac{\sqrt{15}}{4}$$

$$\therefore \quad \sin\frac{x}{2} = \sqrt{\frac{1-\cos x}{2}} = \sqrt{\frac{1+\frac{\sqrt{15}}{4}}{2}} = \sqrt{\frac{4+\sqrt{15}}{8}} \qquad \left[\because \cos x = \frac{-\sqrt{15}}{4}\right]$$

$$= \sqrt{\frac{8+2\sqrt{15}}{16}} = \frac{\sqrt{5}+\sqrt{3}}{4} \qquad [\because (\sqrt{5}+\sqrt{3})^2 = 5+3+2\sqrt{5}\times\sqrt{3} = 8+2\sqrt{15}]$$

$$\cos\frac{x}{2} = \sqrt{\frac{1+\cos x}{2}} = \sqrt{\frac{1-\frac{\sqrt{15}}{4}}{2}} = \sqrt{\frac{4-\sqrt{15}}{8}} = \sqrt{\frac{8-2\sqrt{15}}{16}} = \frac{\sqrt{5}-\sqrt{3}}{4}$$

$$[\because (\sqrt{5}-\sqrt{3})^2 = 5+3-2\sqrt{5}\times\sqrt{3} = 8-2\sqrt{15}]$$

$$\tan\frac{x}{2} = \frac{\sin\frac{x}{2}}{\cos\frac{x}{2}} = \frac{\sqrt{5}+\sqrt{3}}{\sqrt{5}-\sqrt{3}} = \frac{\sqrt{5}+\sqrt{3}}{\sqrt{5}-\sqrt{3}} \times \frac{\sqrt{5}+\sqrt{3}}{\sqrt{5}+\sqrt{3}}$$

$$= \frac{(\sqrt{5}+\sqrt{3})^2}{5-3} = \frac{8+2\sqrt{15}}{2} = 4+\sqrt{15}$$ **Ans.**

(*ii*) We have, $\cos x = -\frac{1}{3}, \pi < x < \frac{3\pi}{2} \Rightarrow \frac{\pi}{2} < \frac{x}{2} < \frac{3\pi}{4}$, *i.e.* $\frac{x}{2}$ lies in the II quadrant, so that $\sin\frac{x}{2} > 0, \cos\frac{\pi}{2} < 0$ and $\tan\frac{x}{2} < 0$.

$$\text{Now,} \quad \sin\frac{x}{2} = \pm\sqrt{\frac{1-\cos x}{2}} = \sqrt{\frac{1-\left(-\frac{1}{3}\right)}{2}} = \sqrt{\frac{2}{3}} = \frac{\sqrt{6}}{3}$$

$$\cos\frac{x}{2} = -\sqrt{\frac{1+\cos x}{2}} = -\sqrt{\frac{1-\frac{1}{3}}{2}} = \frac{1}{\sqrt{3}} = -\frac{\sqrt{3}}{3}$$

$$\tan\frac{x}{2} = \frac{\sin\frac{x}{2}}{\cos\frac{x}{2}} = \frac{\sqrt{\frac{2}{3}}}{-\frac{1}{\sqrt{3}}} = -\sqrt{2}$$ **Ans.**

(*iii*) We have, $\tan x = -\frac{4}{3}, \frac{\pi}{2} < x < \pi \Rightarrow \frac{\pi}{4} < \frac{x}{2} < \frac{\pi}{2}$ and $\frac{x}{2}$ lies in the I quadrant, so that all *t*-ratios of $\frac{x}{2}$ are positive.

Also, $\cos^2 x = \frac{1}{\sec^2 x} = \frac{1}{1+\tan^2 x} = \frac{1}{1+\left(\frac{16}{9}\right)} = \frac{9}{25}$ and $\cos x$ is negative in II quadrant.

$\Rightarrow$ $$\cos x = -\frac{3}{5}$$

$\therefore$ $$\sin\frac{x}{2} = \sqrt{\frac{1-\cos x}{2}} = \sqrt{\frac{1-\left(-\frac{3}{5}\right)}{2}} = \sqrt{\frac{4}{5}} = \frac{2}{\sqrt{5}} = \frac{2\sqrt{5}}{5}$$

$$\cos\frac{x}{2} = \sqrt{\frac{1+\cos x}{2}} = \sqrt{\frac{1-\frac{3}{5}}{2}} = \frac{1}{\sqrt{5}} = \frac{\sqrt{5}}{5}$$

$$\tan\frac{x}{2} = \frac{\sin\frac{x}{2}}{\cos\frac{x}{2}} = \frac{\frac{2}{\sqrt{5}}}{\frac{1}{\sqrt{5}}} = 2$$ **Ans.**

Example 57. *Find the values of:*

(*i*) $\sin 22\frac{1^\circ}{2}$ (*ii*) $\cos 22\frac{1^\circ}{2}$ (*iii*) $\sin 7\frac{1^\circ}{2}$

Solution. (*i*) We have, $\sin\frac{A}{2} = \pm\sqrt{\frac{1-\cos A}{2}}$ ... (1)

Putting $A = 45^\circ$ in (1), we get

$$\sin 22\frac{1^\circ}{2} = \sqrt{\frac{1-\cos 45^\circ}{2}} \qquad \left[\because \sin 22\frac{1^\circ}{2} \text{ is +ve}\right]$$

$\Rightarrow$ $$\sin 22\frac{1^\circ}{2} = \sqrt{\frac{1-\frac{1}{\sqrt{2}}}{2}} = \sqrt{\frac{\sqrt{2}-1}{2\sqrt{2}}}$$ **Ans.**

(*ii*) We have, $\cos\frac{A}{2} = \pm\sqrt{\frac{1+\cos A}{2}}$ ... (2)

Putting $A = 45^\circ$ in (2), we get

$\therefore$ $$\cos 22\frac{1^\circ}{2} = \sqrt{\frac{1+\cos 45^\circ}{2}} \qquad \left[\because \cos 22\frac{1^\circ}{2} \text{ is +ve}\right]$$

$$\Rightarrow \quad \cos 22\frac{1^\circ}{2} = \sqrt{\frac{1+\frac{1}{\sqrt{2}}}{2}} = \sqrt{\frac{\sqrt{2}+1}{2\sqrt{2}}}$$ **Ans.**

(*iii*) We have, $\quad \cos 15^\circ = \cos(45^\circ - 30^\circ) = \cos 45^\circ \cos 30^\circ + \sin 45^\circ \sin 30^\circ$

$$\Rightarrow \quad \cos 15^\circ = \left(\frac{1}{\sqrt{2}}\cdot\frac{\sqrt{3}}{2}\right) + \left(\frac{1}{\sqrt{2}}\cdot\frac{1}{2}\right) = \frac{\sqrt{3}+1}{2\sqrt{2}} \qquad \text{... (3)}$$

Putting $\quad A = 15^\circ$ in $\sin\frac{A}{2} = \pm\sqrt{\frac{1-\cos A}{2}}$, we get

$$\sin 7\frac{1^\circ}{2} = \sqrt{\frac{1-\cos 15^\circ}{2}} = \sqrt{\frac{1-\frac{\sqrt{3}+1}{2\sqrt{2}}}{2}} \qquad \left[\because \sin 7\frac{1^\circ}{2} \text{ is} + \text{ve}\right]$$

$$\sqrt{\frac{2\sqrt{2}-\sqrt{3}-1}{4\sqrt{2}}} = \sqrt{\frac{4-\sqrt{6}-\sqrt{2}}{8}} \qquad [\text{Multiplying num. and deno. by } \sqrt{2}\,]$$

$$= \frac{\sqrt{4-\sqrt{6}-\sqrt{2}}}{2\sqrt{2}}$$ **Ans.**

Example 58. *Show that:*

(*i*) $\cot 7\frac{1^\circ}{2} = \sqrt{2}+\sqrt{3}+\sqrt{4}+\sqrt{6}$ (*ii*) $\tan 142\frac{1^\circ}{2} = \sqrt{2}-\sqrt{3}+\sqrt{4}-\sqrt{6}$

Solution. (*i*) $\quad$ L.H.S. $= \cot 7\frac{1^\circ}{2} = \frac{\cos 7\frac{1^\circ}{2}}{\sin 7\frac{1^\circ}{2}} = \frac{2\cos^2 7\frac{1^\circ}{2}}{2\sin 7\frac{1^\circ}{2}\cos 7\frac{1^\circ}{2}} = \frac{1+\cos 15^\circ}{\sin 15^\circ}$

$$= \frac{1+\cos(45^\circ-30^\circ)}{\sin(45^\circ-30^\circ)} = \frac{1+[\cos 45^\circ \cos 30^\circ + \sin 45^\circ \sin 30^\circ]}{\sin 45^\circ \cos 30^\circ - \cos 45^\circ \sin 30^\circ}$$

$$= \frac{1+\left[\frac{1}{\sqrt{2}}\times\frac{\sqrt{3}}{2}+\frac{1}{\sqrt{2}}\times\frac{1}{2}\right]}{\frac{1}{\sqrt{2}}\times\frac{\sqrt{3}}{2}-\frac{1}{\sqrt{2}}\times\frac{1}{2}} = \frac{1+\frac{\sqrt{3}+1}{2\sqrt{2}}}{\frac{\sqrt{3}-1}{2\sqrt{2}}} = \frac{2\sqrt{2}+\sqrt{3}+1}{\sqrt{3}-1}\times\frac{\sqrt{3}+1}{\sqrt{3}+1}$$

$$= \frac{2\sqrt{6}+2\sqrt{2}+3+\sqrt{3}+\sqrt{3}+1}{3-1} = \frac{2\sqrt{6}+2\sqrt{2}+2\sqrt{3}+4}{2}$$

$$= \sqrt{6}+\sqrt{2}+\sqrt{3}+2 = \sqrt{2}+\sqrt{3}+\sqrt{4}+\sqrt{6} = \text{R.H.S.}$$ **Proved.**

(*ii*) L.H.S $= \tan 142\frac{1^\circ}{2} = \tan\left(180^\circ - 37\frac{1^\circ}{2}\right) = -\tan 37\frac{1^\circ}{2}$

$$= -\tan\left(45^\circ - 7\frac{1^\circ}{2}\right) = -\frac{\tan 45^\circ - \tan 7\frac{1^\circ}{2}}{1+\tan 45^\circ \tan 7\frac{1^\circ}{2}} = -\frac{1-\tan 7\frac{1^\circ}{2}}{1+\tan 7\frac{1^\circ}{2}}$$

$$= -\frac{\cos 7\frac{1°}{2} - \sin 7\frac{1°}{2}}{\cos 7\frac{1°}{2} + \sin 7\frac{1°}{2}} = -\frac{\left(\cos 7\frac{1°}{2} - \sin 7\frac{1°}{2}\right)^2}{\cos^2 7\frac{1°}{2} - \sin^2 7\frac{1°}{2}} =$$

$$= -\frac{\left(\cos^2 7\frac{1°}{2} + \sin^2 7\frac{1°}{2}\right) - 2\sin 7\frac{1°}{2}.\cos 7\frac{1°}{2}}{\cos 2\left(7\frac{1°}{2}\right)}$$

$$= -\frac{1 - \sin 15°}{\cos 15°} = -\frac{1 - \frac{\sqrt{3}-1}{2\sqrt{2}}}{\frac{\sqrt{3}+1}{2\sqrt{2}}} \qquad \left[\because \sin 15° = \frac{\sqrt{3}-1}{2\sqrt{2}}, \cos 15° = \frac{\sqrt{3}+1}{2\sqrt{2}}\right]$$

$$= -\frac{2\sqrt{2} - \sqrt{3} + 1}{\sqrt{3}+1} \times \frac{\sqrt{3}-1}{\sqrt{3}-1}$$

$$= \frac{-2\sqrt{6} + 2\sqrt{2} + 3 - \sqrt{3} - \sqrt{3} + 1}{3-1} = \frac{-2\sqrt{6} - 2\sqrt{3} + 2\sqrt{2} + 4}{2}$$

$$= -\sqrt{6} - \sqrt{3} + \sqrt{2} + 2 = \sqrt{2} - \sqrt{3} + \sqrt{4} - \sqrt{6} = \text{R.H.S.}$$ **Proved.**

Example 59. *Write the values of*

(*i*) $3\sin\frac{\pi}{12} - 4\sin^3\frac{\pi}{12}$ **[S.B.T.E. 2015]**

(*ii*) $4\cos^3\frac{\pi}{9} - 3\cos\frac{\pi}{9}$ **[S.B.T.E. 2016, 2014]**

Solution. Here we have

(*i*) $$3\sin\frac{\pi}{12} - 4\sin^3\frac{\pi}{12} = \sin 3\left(\frac{\pi}{12}\right) \qquad [\because \sin 3\theta = 3\sin\theta - 4\sin^3\theta]$$

$$= \sin\frac{\pi}{4} = \frac{1}{\sqrt{2}}$$ **Ans.**

(*ii*) $$4\cos^3\frac{\pi}{9} - 3\cos\frac{\pi}{9} = \cos 3\left(\frac{\pi}{9}\right) \qquad [\because \cos 3\theta = 4\cos^3\theta - 3\cos\theta]$$

$$= \cos\frac{\pi}{3} = \frac{1}{2}$$ **Ans.**

Example 60. *Write the value of:* $\dfrac{6\tan\frac{\pi}{9} - 2\tan^3\frac{\pi}{9}}{1 - 3\tan^2\frac{\pi}{9}}$ **[S.B.T.E. 2014, 2013]**

Solution. Here we have

$$\frac{6\tan\frac{\pi}{9} - 2\tan^3\frac{\pi}{9}}{1 - 3\tan^2\frac{\pi}{9}} = \frac{6\tan\frac{\pi}{9} - 2\tan^3\frac{\pi}{9}}{1 - 3\tan^3\frac{\pi}{9}}$$

$$= 2\cdot\frac{3\tan\frac{\pi}{9}-\tan^3\frac{\pi}{9}}{1-3\tan^2\frac{\pi}{9}} \qquad \left[\because \tan 3\theta = \frac{3\tan\theta-\tan^3\theta}{1-3\tan^2\theta}\right]$$

$$= 2\times\tan\frac{\pi}{3} = 2\times\sqrt{3} = 2\sqrt{3}$$ **Ans.**

Example 61. *Find the value of* $\tan\frac{\pi}{8}$. **[S.B.T.E. 2015]**

Solution. Here we have

$$\tan\frac{\pi}{8} = \sqrt{\frac{1-\cos\frac{\pi}{4}}{1+\cos\frac{\pi}{4}}} = \sqrt{\frac{1-\frac{1}{\sqrt{2}}}{1+\frac{1}{\sqrt{2}}}}$$

$$= \sqrt{\frac{\sqrt{2}-1}{\sqrt{2}+1}} = \sqrt{\frac{(\sqrt{2}-1)^2}{1}}$$

$$\therefore \quad \tan\frac{\pi}{8} = \sqrt{2}-1$$ **Ans.**

Example 62. *Prove that:*

$$\frac{\sec 8\theta - 1}{\sec 4\theta - 1} = \frac{\tan 8\theta}{\tan 2\theta}$$ **[S.B.T.E. 2015, 2014]**

Solution. Here we have

$$\text{L.H.S.} = \frac{\sec 8\theta-1}{\sec 4\theta-1} = \frac{\frac{1}{\cos 8\theta}-1}{\frac{1}{\cos 4\theta}-1}$$

$$= \frac{2\sin^2 4\theta\times\cos 4\theta}{\cos 8\theta\times 2\sin^2 2\theta} \qquad \left[\begin{array}{l}\because 1-\cos 8\theta = 2\sin^2 4\theta\\ 1-\cos 4\theta = 2\sin^2 2\theta\end{array}\right]$$

$$= \frac{2\sin 4\theta\cdot\sin 4\theta\cdot\cos 4\theta}{2\sin 2\theta\cdot\sin 2\theta\cdot\cos 8\theta}$$

$$= \frac{\sin 4\theta\,(2\sin 4\theta\cdot\cos 4\theta)}{2\sin 2\theta\cdot\sin 2\theta\cdot\cos 8\theta} \qquad [\because \sin 4\theta = \sin(2\times 2\theta)]$$

$$= \frac{2\sin 2\theta\cdot\cos 2\theta\cdot\sin 8\theta}{2\sin 2\theta\cdot\sin 2\theta\cdot\cos 8\theta}$$

$$= \frac{\tan 8\theta}{\tan 2\theta}$$ **Proved.**

Example 63. *Prove that:* $16\sin^5\theta - 20\sin^3\theta + 5\sin\theta = \sin 5\theta$.

Solution. R.H.S. $= \sin 5\theta = \sin(\theta+4\theta) = \sin\theta\cos 4\theta + \cos\theta\sin 4\theta$

$= \sin\theta(1-2\sin^2 2\theta) + \cos\theta\, 2\sin 2\theta\cos 2\theta$

$= \sin\theta[1-2\sin^2 2\theta] + 2\cos\theta\, 2\sin\theta\cos\theta(1-2\sin^2\theta)$

$$= \sin\theta[1 - 8\sin^2\theta\cos^2\theta] + 4(1-\sin^2\theta)\sin\theta(1-2\sin^2\theta)$$

$$= \sin\theta\left[1 - 8\sin^2\theta(1-\sin^2\theta) + 4(1-\sin^2\theta)(1-2\sin^2\theta)\right]$$

$$= \sin\theta\left[1 - 8\sin^2\theta + 8\sin^4\theta + 4 - 8\sin^2\theta - 4\sin^2\theta + 8\sin^4\theta\right]$$

$$= \sin\theta\left[16\sin^4\theta - 20\sin^2\theta + 5\right]$$

$$= 16\sin^5\theta - 20\sin^3\theta + 5\sin\theta = \text{L.H.S.}$$

Proved.

Example 64. *Prove that:* $\cos 5A = 16\cos^5 A - 20\cos^3 A + 5\cos A$

Solution. We have,

L.H.S. $= \cos 5A = \cos(3A + 2A)$

$$= \cos 3A\cos 2A - \sin 3A\sin 2A$$

$$= (4\cos^3 A - 3\cos A)(2\cos^2 A - 1) - (3\sin A - 4\sin^3 A)\times(2\sin A\cos A)$$

$$= (4\cos^3 A - 3\cos A)(2\cos^2 A - 1) - (3 - 4\sin^2 A)(2\sin^2 A\cos A)$$

$$= (4\cos^3 A - 3\cos A)(2\cos^2 A - 1) - \{3 - 4(1-\cos^2 A)\}\times\{2(1-\cos^2 A)\cos A\}$$

$$= (8\cos^5 A - 10\cos^3 A + 3\cos A) - 2\cos A(1-\cos^2 A)\times(4\cos^2 A - 1)$$

$$= (8\cos^5 A - 10\cos^3 A + 3\cos A) - 2\cos A(5\cos^2 A - 4\cos^4 A - 1)$$

$$= 8\cos^5 A - 10\cos^3 A + 3\cos A - 10\cos^3 A + 8\cos^5 A + 2\cos A$$

$$= 16\cos^5 A - 20\cos^3 A + 5\cos A = \text{R.H.S.}$$

Proved.

Example 65. *Prove that:* $\cos 6x = 32\cos^6 x - 48\cos^4 x + 18\cos^2 x - 1$

Solution. We know that,

$$\cos 3x = 4\cos^3 x - 3\cos x$$

$$\Rightarrow \cos 6x = 4\cos^3 2x - 3\cos 2x$$

$$= 4(2\cos^2 x - 1)^3 - 3(2\cos^2 x - 1)$$

$$= 4[8\cos^6 x - 12\cos^4 x + 6\cos^2 x - 1] - 6\cos^2 x + 3$$

$$= 32\cos^6 x - 48\cos^4 x + 24\cos^2 x - 4 - 6\cos^2 x + 3$$

$$= 32\cos^6 x - 48\cos^4 x + 18\cos^2 x - 1$$

$$= \text{L.H.S.}$$

Proved.

Example 66. *Prove that:* $\dfrac{1 + \sin\theta - \cos\theta}{1 + \sin\theta\cos\theta} = \tan\dfrac{\theta}{2}$. **[S.B.T.E. 2016]**

Solution. We have,

$$\text{L.H.S.} = \frac{1 + \sin\theta - \cos\theta}{1 + \sin\theta\cos\theta} = \frac{(1 - \cos\theta) + \sin\theta}{(1 + \cos\theta) + \sin\theta}$$

$$= \frac{2\sin^2\frac{\theta}{2} + 2\sin\frac{\theta}{2}\cos\frac{\theta}{2}}{2\cos^2\frac{\theta}{2} + 2\sin\frac{\theta}{2}\cos\frac{\theta}{2}} = \frac{\sin^2\frac{\theta}{2} + \sin\frac{\theta}{2}\cos\frac{\theta}{2}}{\cos^2\frac{\theta}{2} + \sin\frac{\theta}{2}\cos\frac{\theta}{2}}$$

$$= \frac{\sin\frac{\theta}{2}\left[\sin\frac{\theta}{2} + \cos\frac{\theta}{2}\right]}{\cos\frac{\theta}{2}\left[\cos\frac{\theta}{2} + \sin\frac{\theta}{2}\right]} = \frac{\sin\frac{\theta}{2}}{\sin\frac{\theta}{2}} = \tan\frac{\theta}{2} = \text{R.H.S.}$$ **Proved.**

Example 67. *Prove that:*

$$4(\cos^3 20° + \cos^3 40°) = 3(\cos 20° + \cos 40°).$$

Solution. Let us consider,

$\cos 60° = \cos 3(20°)$

$\Rightarrow \quad \cos 60° = 4\cos^3 20° - 3\cos 20°$

$\Rightarrow \quad \frac{1}{2} = 4\cos^3 20° - 3\cos 20° \quad \Rightarrow \quad 4\cos^3 20° = \frac{1}{2} + 3\cos 20°$ ... (1)

Also, $\cos 120° = \cos(3 \times 40°)$

$\Rightarrow \quad \cos 120° = \cos^3 40° - 3\cos 40°$

$\Rightarrow \quad -\frac{1}{2} = 4\cos^3 40° - 3\cos 40$

$\Rightarrow \quad 4\cos^3 40° = 3\cos 40° - \frac{1}{2}$ ... (2)

Now, $\quad \text{L.H.S.} = 4(\cos^3 20° + \cos^3 40°)$ ... (3)

Putting values of $4\cos^3 20°$ and $4\cos^3 40°$ from (1) and (2) respectively in (3), we get

$$= \left(\frac{1}{2} + 3\cos 20° + 3\cos 40° - \frac{1}{2}\right)$$

$$= 3(\cos 20° + \cos 40°) = \text{R.H.S.}$$ **Proved.**

Example 68. *Prove that:*

$$\tan\theta + 2\tan 2\theta + 4\tan 4\theta + 8\cot 8\theta = \cot\theta$$

Solution. Let us consider,

$$\cot\theta - \tan\theta = \frac{\cos\theta}{\sin\theta} - \frac{\sin\theta}{\cos\theta} = \frac{\cos^2\theta - \sin^2\theta}{\sin\theta.\cos\theta}$$

$$= \frac{2.\cos 2\theta}{2.\sin\theta\cos\theta} = \frac{2.\cos 2\theta}{\sin 2\theta} = 2\cot 2\theta$$

$\cot\theta - \tan\theta = 2\cot 2\theta$ ... (1)

$\text{L.H.S.} = \tan\theta + 2\tan 2\theta + 4\tan 4\theta + 8\cot 8\theta$

$= \cot\theta - (\cot\theta - \tan\theta) + 2\tan 2\theta + 4\tan 4\theta + 8\cot 8\theta$

$= \cot\theta - (2\cot 2\theta) + 2\tan 2\theta + 4\tan 4\theta + 8\cot 8\theta$ [using (1)]

$= \cot\theta - 2(\cot 2\theta - \tan 2\theta) + 4\tan 4\theta + 8\cot 8\theta$

$= \cot\theta - 2(2\cot 4\theta) + 4\tan 4\theta + 8\cot 8\theta$

$= \cot\theta - 4(\cot 4\theta - \tan 4\theta) + 8\cot 8\theta$

$\left[\begin{array}{l} \because \cot\theta - \tan\theta = 2\cot 2\theta \\ \Rightarrow \cot 2\theta - \tan 2\theta = 2\cot 4\theta \\ \quad \text{and so on ....} \end{array}\right]$

$= \cot\theta - 4(2\cot 8\theta) + 8\cot 8\theta$

$= \cot\theta - 8\cot 8\theta + 8\cot 8\theta = \cot\theta = \text{R.H.S.}$ **Proved.**

Example 69. *Prove that:* **[S.B.T.E. 2010]**

(*i*) $\sin A \cdot \sin(60° - A)\cdot \sin(60° + A) = \frac{1}{4}\sin 3A$

(*ii*) $\cos A \cdot \cos(60° - A)\cdot \cos(60° + A) = \frac{1}{4}\cos 3A$

Solution. (*i*) $\text{L.H.S.} = \sin \text{A} \sin(60° - \text{A}).\sin(60° + A)$

$= \frac{1}{2}\sin A[\cos 2A - \cos 120°] = \frac{1}{2}\sin A\left[1 - 2\sin^2 A + \frac{1}{2}\right]$

$= \frac{1}{4}\sin A[3 - 4\sin^2 A] = \frac{1}{4}[3\sin A - 4\sin^3 A]$

$= \frac{1}{4}\sin 3A = \text{R.H.S.}$ **Proved.**

(*ii*) $\text{L.H.S.} = \cos A \cdot \cos(60° - A)\cos(60° + A)$

$= \frac{1}{2}\cos A(\cos 120° + \cos 2A) = \frac{1}{2}\cos A\left[-\frac{1}{2} + 2\cos^2 A - 1\right]$

$= \frac{1}{4}\cos A[4\cos^2 A - 3]$

$= \frac{1}{4}[4\cos^3 A - 3\cos A] = \frac{1}{4}\cos 3A = \text{R.H.S.}$ **Proved.**

Example 70. *Prove that:* $\cos^2 2x - \cos^2 6x = \sin 8x \sin 4x$

Solution. $\text{L.H.S.} = \cos^2 2x - \cos^2 6x$

$= (\cos 2x - \cos 6x)(\cos 2x + \cos 6x)$ $[\because a^2 - b^2 = (a - b)(a + b)]$

$= \left[-2\sin\left(\frac{2x+6x}{2}\right)\sin\left(\frac{2x-6x}{2}\right)\right]\left[2\cos\left(\frac{2x+6x}{2}\right)\cos\left(\frac{2x-6x}{2}\right)\right]$

$= [-2\sin 4x \cdot \sin(-2x)[2\cos 4x \cdot \cos(-2x)]$

$= -(2\sin 4x \cdot \cos 4x)[2\sin(-2x)\cdot\cos(-2x)]$

$= -\sin 8x \cdot (-2\sin 2x\cos 2x)$

$= \sin 8x \cdot \sin 4x = \text{R.H.S.}$ **Proved.**

Example 71. *Prove that:* **[S.B.T.E, 2017, 2015, 2014, 2013]**

(*i*) $\cos 4x = 1 - 8\sin^2 x \cdot \cos^2 x$ (*ii*) $\tan 4\theta = \dfrac{4\tan\theta\,(1-\tan^2\theta)}{1-6\tan^2\theta+\tan^4\theta}$

Solution. (*i*) L.H.S. = cos 4*x*

$$= 1 - 2\sin^2 2x \qquad [\because \cos 2A = 1 - 2\sin^2 A]$$

$$= 1 - 2(\sin 2x)^2$$

$$= 1 - 2(2\sin x\cos x)^2 \qquad [\because \sin 2A = 2\sin A\cos A]$$

$$= 1 - 8\sin^2 x\cos^2 x = \text{R.H.S.}$$ **Proved.**

(*ii*) L.H.S. = tan 4θ = tan (2θ + 2θ)

$$= \frac{2\tan 2\theta}{1-\tan^2 2\theta} \quad \text{... (1)} \quad \left[\because \tan(A+B) = \frac{\tan A + \tan B}{1-\tan A\tan B}\right]$$

But, $$\tan 2\theta = \frac{2\tan\theta}{1-\tan^2\theta} \quad \text{... (2)}$$

From (1) and (2), we get

$$\tan 4\theta = \frac{2\left(\dfrac{2\tan\theta}{1-\tan^2\theta}\right)}{1-\left(\dfrac{2\tan\theta}{1-\tan^2\theta}\right)^2} = \frac{4\tan\theta\,(1-\tan^2\theta)}{[(1-\tan^2\theta)^2 - 4\tan^2\theta]}$$

$$= \frac{4\tan\theta\,(1-\tan^2\theta)}{1+\tan^4\theta - 2\tan^2\theta - 4\tan^2\theta} = \frac{4\tan\theta\,(1-\tan^2\theta)}{1-6\tan^2\theta+\tan^4\theta} = \text{R.H.S.}$$ **Proved.**

Example 72. *Prove that:*

$$2\tan 2x = \frac{\cos x + \sin x}{\cos x - \sin x} - \frac{\cos x - \sin x}{\cos x + \sin x}$$

Solution. R.H.S. $= \dfrac{\cos x + \sin x}{\cos x - \sin x} - \dfrac{\cos x - \sin x}{\cos x + \sin x} = \dfrac{(\cos x + \sin x)^2 - (\cos x - \sin x)^2}{\cos^2 x - \sin^2 x}$

$$= \frac{(\cos^2 x + \sin^2 x) + 2\cos x\sin x - (\cos^2 x + \sin^2 x - 2\cos x\sin x)}{\cos 2x}$$

$$= \frac{1 + \sin 2x - 1 + \sin 2x}{\cos 2x} = \frac{2\sin 2x}{\cos 2x} = 2\tan 2x = \text{L.H.S.}$$ **Proved.**

Example 73. *Prove that:*

$$\cos\theta\cos 2\theta\cos 4\theta \ldots \cos 2^{n-1}\theta = \frac{\sin(2^n\theta)}{2^n(\sin\theta)}$$

Solution. Here, we observe that each angle in L.H.S. is double of the preceding angle.

L.H.S. = $\cos\theta\cos 2\theta\cos 4\theta \ldots \cos 2^{n-1}\theta$

$$= \frac{1}{2 \sin \theta} (2 \sin \theta \cdot \cos \theta) \cos 2\theta \cdot \cos 4\theta \ldots \cos 2^{n-1} \theta.$$

$$= \frac{1}{2^2 \sin \theta} (2 \sin 2\theta \cos 2\theta) (\cos 4\theta \ldots \cos 2^{n-1} \theta)$$

$$= \frac{1}{2^3 \sin \theta} (2 \sin 4\theta \cdot \cos 4\theta) [\cos 8\theta \cos 16\theta \ldots \cos 2^{n-1} \theta]$$

$$= \frac{1}{2^4 \sin \theta} (2 \sin 8\theta \cdot \cos 8\theta) [\cos 16\theta \ldots \cos 2^{n-1} \theta]$$

....

....

$$= \frac{1}{2^n \sin \theta} \cdot [2 \sin 2^{n-1} \theta \cos 2^{n-1} \theta] = \frac{\sin (2^n \theta)}{2^n (\sin \theta)} = \text{R.H.S.}$$ **Proved.**

Example 74. *Prove that:* $\frac{\cos A}{1 - \sin A} = \tan\left(45° + \frac{A}{2}\right)$

Solution. L.H.S. $= \frac{\cos A}{1 - \sin A} = \frac{\cos^2 \frac{A}{2} - \sin^2 \frac{A}{2}}{\cos^2 \frac{A}{2} + \sin^2 \frac{A}{2} - 2 \sin \frac{A}{2} \cos \frac{A}{2}}$

$$= \frac{\left(\cos \frac{A}{2} - \sin \frac{A}{2}\right)\left(\cos \frac{A}{2} + \sin \frac{A}{2}\right)}{\left(\cos \frac{A}{2} - \sin \frac{A}{2}\right)^2} \qquad [\because \cos^2 \theta + \sin^2 \theta = 1]$$

$$= \frac{\left(\cos \frac{A}{2} + \sin \frac{A}{2}\right)}{\left(\cos \frac{A}{2} - \sin \frac{A}{2}\right)} = \frac{1 + \tan \frac{A}{2}}{1 - \tan \frac{A}{2}} \qquad \left[\text{Dividing num. and den. by } \cos \frac{A}{2}\right]$$

$$= \frac{\tan 45° + \tan \frac{A}{2}}{1 - \tan 45° \tan \frac{A}{2}} = \tan\left(45° + \frac{A}{2}\right) = \text{R.H.S.}$$ **Proved.**

Example 75. *Prove that:* $\sqrt{2 + \sqrt{2 + \sqrt{2 + 2 \cos 8\theta}}} = 2 \cos \theta$ **[S.B.T.E. 2014]**

Solution. L.H.S. $= \sqrt{2 + \sqrt{2 + \sqrt{2 + 2 \cos 8\theta}}} = \sqrt{2 + \sqrt{2 + \sqrt{2 (1 + \cos 8\theta)}}}$

$$= \sqrt{2 + \sqrt{2 + \sqrt{2 (2 \cos^2 4\theta)}}} \qquad [\because 1 + \cos 2\theta = 2 \cos^2 \theta]$$

$$= \sqrt{2 + \sqrt{2 + 2 \cos 4\theta}} = \sqrt{2 + \sqrt{2 (1 + \cos 4\theta)}}$$

$$= \sqrt{2+\sqrt{2(2\cos^2 2\theta)}} = \sqrt{2+2\cos 2\theta}$$

$$= \sqrt{2(1+\cos 2\theta)} = \sqrt{2.2\cos^2\theta} = \sqrt{4\cos^2\theta}$$

$$= 2\cos\theta = \text{R.H.S.}$$ **Proved.**

Example 76. *If* $\tan\frac{\theta}{2} = \sqrt{\frac{1-e}{1+e}}\tan\frac{\phi}{2}$, *prove that* $\cos\phi = \frac{\cos\theta - e}{1-e\cos\theta}$

Solution. We have, $\cos\phi = \dfrac{1-\tan^2\frac{\phi}{2}}{1+\tan^2\frac{\phi}{2}} = \dfrac{1-\frac{1+e}{1-e}\tan^2\frac{\theta}{2}}{1+\frac{1+e}{1-e}\tan^2\frac{\theta}{2}}$ $\left[\because \tan\frac{\phi}{2} = \sqrt{\frac{1+e}{1-e}}\tan\frac{\theta}{2}\right]$

$$= \frac{1-e-\tan^2\frac{\theta}{2} - e\tan^2\frac{\theta}{2}}{1-e+\tan^2\frac{\theta}{2}+e\tan^2\frac{\theta}{2}}$$

$$\Rightarrow \quad \cos\phi = \frac{1-\tan^2\frac{\theta}{2} - e\left[1+\tan^2\frac{\theta}{2}\right]}{1+\tan^2\frac{\theta}{2} - e\left[1-\tan^2\frac{\theta}{2}\right]} = \frac{\cos^2\frac{\theta}{2} - \sin^2\frac{\theta}{2} - e\left[\cos^2\frac{\theta}{2}+\sin^2\frac{\theta}{2}\right]}{\cos^2\frac{\theta}{2}+\sin^2\frac{\theta}{2} - e\left[\cos^2\frac{\theta}{2} - \sin^2\frac{\theta}{2}\right]}$$

$$= \frac{\cos\theta - e}{1-e\cos\theta} \qquad \left[\begin{array}{l}\because \cos 2A = \cos^2 A - \sin^2 A \\ \Rightarrow \cos A = \cos^2\frac{A}{2} - \sin^2\frac{A}{2}\end{array}\right]$$

$$= \text{R.H.S.}$$ **Proved.**

Example 77. *If* $\cos\theta = \frac{a\cos\phi + b}{a+b\cos\phi}$, *prove that* $\tan\frac{\theta}{2} = \sqrt{\frac{a-b}{a+b}}\tan\frac{\phi}{2}$.

Solution. We know that

$$\cos\theta = \frac{1-\tan^2\frac{\theta}{2}}{1+\tan^2\frac{\theta}{2}} \quad \text{and} \quad \cos\phi = \frac{1-\tan^2\frac{\phi}{2}}{1+\tan^2\frac{\phi}{2}} \qquad \text{...(1)}$$

Now, $\cos\theta = \dfrac{a\cos\phi + b}{a+b\cos\phi} \Rightarrow \dfrac{1-\tan^2\frac{\theta}{2}}{1+\tan^2\frac{\theta}{2}} = \dfrac{a\cdot\left(\frac{1-\tan^2\frac{\phi}{2}}{1+\tan^2\frac{\phi}{2}}\right)+b}{a+b\left(\frac{1-\tan^2\frac{\phi}{2}}{1+\tan^2\frac{\phi}{2}}\right)}$ [using (1)]

$$\Rightarrow \quad \frac{1-\tan^2\frac{\theta}{2}}{1+\tan^2\frac{\theta}{2}} = \frac{a\left[1-\tan^2\frac{\phi}{2}\right]+b\left[1+\tan^2\frac{\phi}{2}\right]}{a\left[1+\tan^2\frac{\phi}{2}\right]+b\left[1-\tan^2\frac{\phi}{2}\right]} = \frac{a-a\tan^2\frac{\phi}{2}+b+b\tan^2\frac{\phi}{2}}{a+a\tan^2\frac{\phi}{2}+b-b\tan^2\frac{\phi}{2}}$$

Applying componendo and dividendo, we get

$$\frac{2\tan^2\frac{\theta}{2}}{2}=\frac{2a\tan^2\left(\frac{\phi}{2}\right)-2b\tan^2\left(\frac{\phi}{2}\right)}{2a+2b}=\frac{(a-b)\tan^2\frac{\phi}{2}}{a+b}$$

$$\Rightarrow \qquad \tan\frac{\theta}{2}=\sqrt{\frac{a-b}{a+b}}\tan\frac{\phi}{2}$$ **Proved.**

Example. 78. *If* $2\cos\theta = x+\frac{1}{x}$, *prove that* $2\cos 3\theta = x^3+\frac{1}{x^3}$

Solution. We have, $2\cos\theta = x+\frac{1}{x}$ ... (1)

$$\therefore \qquad 2\cos 3\theta = 2(4\cos^3\theta-3\cos\theta)=8\cos^3\theta-6\cos\theta=(2\cos\theta)^3-3(2\cos\theta)$$

$$=\left(x+\frac{1}{x}\right)^3-3\left(x+\frac{1}{x}\right) \qquad \text{[Using (1)]}$$

$$=x^3+\frac{1}{x^3}+3x\cdot\frac{1}{x}\left(x+\frac{1}{x}\right)-3\left(x+\frac{1}{x}\right)=x^3+\frac{1}{x^3}.$$ **Proved.**

Example 79. *Prove that:*

$$\cos^3 A+\cos^3(120°+A)\cos^3(240°+A)=\frac{3}{4}\cos 3A$$

Solution. We have, $\cos 3A = 4\cos^3 A - 3\cos A \quad\Rightarrow\quad \cos^3 A=\frac{3}{4}\cos A+\frac{1}{4}\cos 3A$

Using this, we get

$$\text{L.H.S.}=\left(\frac{3}{4}\cos A+\frac{1}{4}\cos 3A\right)+\left(\frac{3}{4}\cos(120°+A)+\frac{1}{4}\cos(360°+3A)\right)$$

$$+\left[\frac{3}{4}\cos(240°+A)+\frac{1}{4}\cos(720°+3A)\right]$$

$$=\frac{3}{4}[(\cos A+\cos(120°+A)+\cos(240°+A)]+\frac{3}{4}(\cos 3A)$$

$$=\frac{3}{4}\left[\cos A+2\cos\left(\frac{360°+2A}{2}\right)\cos\left(-\frac{120°}{2}\right)\right]+\frac{3}{4}\cos 3A$$

$$=\frac{3}{4}\left[\cos A+2(-\cos A)\left(\frac{1}{2}\right)\right]+\frac{3}{4}\cos 3A$$

$$=\frac{3}{4}(\cos A-\cos A)+\frac{3}{4}\cos 3A=\frac{3}{4}\cos 3A=\text{R.H.S.}$$ **Proved.**

Example 80. *Prove that* $\cos\theta+\cos(120°+\theta)+\cos(\theta-120°)=0$, and hence, deduce that $\cos^3\theta+\cos^3(120°+\theta)+\cos^3(\theta-120°)=\frac{3}{4}\cos 3\theta$

Solution. $\text{L.H.S.}=\cos\theta+\cos(120°+\theta)+\cos(\theta-120°)$

$$=\cos\theta+2\cos\theta\cdot\cos(120°)$$

$$= \cos\theta + 2\cos\theta \cdot \cos(180^\circ - 60^\circ) = \cos\theta + 2\cos\theta\left(\frac{-1}{2}\right)$$

$$= \cos\theta - \cos\theta = 0 = \text{R.H.S.}$$ **Proved.**

Now, let $a = \cos\theta,\ b = \cos(120^\circ + \theta),\ c = \cos(\theta - 120^\circ)$

Here, $a + b + c = 0$

$\Rightarrow$ $a^3 + b^3 + c^3 = 3abc$ [From Algebra]

$\Rightarrow$ $\cos^3\theta + \cos^3(120^\circ + \theta) + \cos^3(\theta - 120^\circ) = 3\cos\theta\cos(120^\circ + \theta)\cos(\theta - 120^\circ)$

$$= \frac{3}{2}\cos\theta\,[\cos 2\theta + \cos 240^\circ] = \frac{3}{2}\cos\theta\left[\cos 2\theta - \frac{1}{2}\right]$$

$$= \frac{3}{2}\cos\theta\left[2\cos^2\theta - 1 - \frac{1}{2}\right] = \frac{3}{4}\cos\theta\,(4\cos^2\theta - 3)$$

$$= \frac{3}{4}(4\cos^3\theta - 3\cos\theta) = \frac{3}{4}\cos 3\theta$$ **Proved.**

Example 81. *If* $\tan\theta = \dfrac{x}{y}$, *find the value of* $x\sin 2\theta + y\cos 2\theta$. **[S.B.T.E. 2014]**

Solution. Here we have $\tan\theta = x/y$

$$x\sin 2\theta + y\cos 2\theta = x\left(\frac{2\tan\theta}{1+\tan^2\theta}\right) + y\left(\frac{1-\tan^2\theta}{1+\tan^2\theta}\right)$$

$$= \frac{2x\tan\theta + y(1-\tan^2\theta)}{(1+\tan^2\theta)}$$

$$= \frac{2x \times \dfrac{x}{y} + y\left(1 - x^2/y^2\right)}{1 + x^2/y^2}$$

$$= \frac{\dfrac{1}{y}(2x^2 + y^2 - x^2)}{\dfrac{1}{y^2}(x^2+y^2)}$$

$$= \frac{\dfrac{1}{y}(x^2+y^2)}{\dfrac{1}{y^2}(x^2+y^2)} = y$$ **Ans.**

14.10 FIND THE VALUE OF sin 18°

Let θ stands for 18°, so that $2\theta = 36^\circ$

$\Rightarrow$ $5\theta = 90^\circ$ $\Rightarrow$ $(3\theta + 2\theta) = 90^\circ$

$\Rightarrow$ $2\theta = 90^\circ - 3\theta$

$\Rightarrow$ $\sin 2\theta = \sin(90^\circ - 3\theta)$

$\Rightarrow \quad \sin 2\theta = \cos 3\theta \quad \Rightarrow \quad 2 \sin \theta \cos \theta = 4 \cos^3 \theta - 3 \cos \theta$

$\Rightarrow \quad 2 \sin \theta = 4 \cos^2 \theta - 3 \quad \Rightarrow \quad 2 \sin \theta = 4 (1 - \sin^2 \theta) - 3$

$\Rightarrow \quad 2 \sin \theta = 4 - 4 \sin^2 \theta - 3 \quad \Rightarrow \quad 4 \sin^2 \theta + 2 \sin \theta - 1 = 0$

$$\therefore \quad \sin \theta = \frac{-2 \pm \sqrt{(2)^2 - 4 \times 4 \times (-1)}}{2 \times 4} = \frac{-2 \pm \sqrt{20}}{8} = \frac{\pm \sqrt{5} - 1}{4}$$

Now, 18° being an angle of the I quadrant, sin 18° is necessarily positive, therefore

$$\boxed{\sin 18^\circ = \frac{\sqrt{5} - 1}{4}}$$

14.11 TO FIND THE VALUE OF cos 36°

We know that $\quad \cos 2\theta = 1 - 2 \sin^2 \theta \quad \text{... (1)}$

Putting $\theta = 18^\circ$ in (1), we get

$$\cos 36^\circ = 1 - 2 \sin^2 18^\circ$$

$$= 1 - 2\left(\frac{\sqrt{5} - 1}{4}\right)^2 \qquad \left[\because \sin 18^\circ = \frac{\sqrt{5} - 1}{4}\right]$$

$$= 1 - 2\left(\frac{5 + 1 - 2\sqrt{5}}{16}\right) = 1 - \left(\frac{6 - 2\sqrt{5}}{8}\right)$$

$$= 1 - \left(\frac{3 - \sqrt{5}}{4}\right) = \frac{4 - 3 + \sqrt{5}}{4} = \frac{\sqrt{5} + 1}{4}$$

$$\therefore \quad \boxed{\cos 36^\circ = \frac{\sqrt{5} + 1}{4}}$$

Corollary :

1. $$\cos 18^\circ = \sqrt{1 - \sin^2 18^\circ} = \sqrt{1 - \frac{5 + 1 - 2\sqrt{5}}{16}} \qquad \left[\because \sin 18^\circ = \frac{\sqrt{5} - 1}{4}\right]$$

$$= \frac{1}{4}\sqrt{10 + 2\sqrt{5}}$$

2. $$\sin 36^\circ = \sqrt{1 - \cos^2 36^\circ} = \sqrt{1 - \frac{5 + 1 + 2\sqrt{5}}{16}} \qquad \left[\because \cos 36^\circ = \frac{\sqrt{5} + 1}{4}\right]$$

$$= \frac{1}{4}\sqrt{10 - 2\sqrt{5}}$$

$$\boxed{\cos 18^\circ = \frac{1}{4}\sqrt{10 + 2\sqrt{5}}; \qquad \sin 36^\circ = \frac{1}{4}\sqrt{10 - 2\sqrt{5}}}$$

Corollary : To evaluate sin 72° and cos 72°

1. $\sin 72^\circ = \sin (90^\circ - 18^\circ) = \cos 18^\circ = \frac{1}{4}\sqrt{10 + 2\sqrt{5}}$

2. $\cos 72^\circ = \cos (90^\circ - 18^\circ) = \sin 18^\circ = \frac{1}{4}\left(\sqrt{5} - 1\right)$

Example 82. *Show that:* cos 36° cos 72° cos 108° cos 144° = $\frac{1}{16}$. **[S.B.T.E. 2014, 2013]**

Solution. L.H.S. = cos 36° cos 72° cos 108° cos 144°

$$= \cos 36° \cos (90° - 18°) \cos (90° + 18°) \cos (180° - 36°)$$

$$= \cos 36° \sin 18° (-\sin 18°) (-\cos 36°)$$

$$= \sin^2 18° \cdot \cos^2 36°$$

$$= \left(\frac{\sqrt{5}-1}{4}\right)^2 \left(\frac{\sqrt{5}+1}{4}\right)^2 = \left(\frac{5-1}{16}\right)^2 = \left(\frac{1}{4}\right)^2 = \frac{1}{16} = \text{R.H.S.}$$ **Proved.**

Example 83. *Prove that:*

(*i*) $\sin^2 72° - \sin^2 60° = \frac{\sqrt{5}-1}{8}$ (*ii*) $\sin\frac{\pi}{5}\sin\frac{2\pi}{5}\sin\frac{3\pi}{5}\sin\frac{4\pi}{5} = \frac{5}{16}$.

Solution. (*i*) We have,

L.H.S. = $\sin^2 72° - \sin^2 60°$

$$= \{\sin (90° - 18°)\}^2 - (\sin 60°)^2 = (\cos 18°)^2 - (\sin 60°)^2$$

$$= \left(\frac{\sqrt{10+2\sqrt{5}}}{4}\right)^2 - \left(\frac{\sqrt{3}}{2}\right)^2 \quad \left[\because \cos 18° = \frac{\sqrt{10+2\sqrt{5}}}{4}\right]$$

$$= \frac{10+2\sqrt{5}}{16} - \frac{3}{4} = \frac{10+2\sqrt{5}-12}{16} = \frac{2\sqrt{5}-2}{16} = \frac{\sqrt{5}-1}{8} = \text{R.H.S.}$$ **Proved.**

(*ii*) L.H.S. $= \sin\frac{\pi}{2}\sin\frac{2\pi}{5}\sin\frac{3\pi}{5}\sin\frac{4\pi}{5} = \sin\frac{\pi}{5}\sin\frac{2\pi}{5}\cdot\sin\left(\pi - \frac{2\pi}{5}\right)\sin\left(\pi - \frac{\pi}{5}\right)$

$$= \sin\frac{\pi}{5}\sin\frac{2\pi}{5}\cdot\sin\frac{2\pi}{5}\cdot\sin\frac{\pi}{5} = \left(\sin\frac{\pi}{5}\cdot\sin\frac{2\pi}{5}\right)^2$$

$$= (\sin 36° \sin 72°)^2 \quad \left[\because \frac{\pi}{5} = 36° \text{ etc.}\right]$$

$$= (\sin 36° \cdot \cos 18°)^2 \quad [\because \sin 72° = \sin (90° - 18°) = \cos 18°]$$

$$= \left(\frac{\sqrt{10-2\sqrt{5}}}{4}\cdot\frac{\sqrt{10+2\sqrt{5}}}{4}\right)^2$$

$$= \frac{(10-2\sqrt{5})(10+2\sqrt{5})}{16\times16} = \frac{100-20}{16\times16} = \frac{80}{16\times16} = \frac{5}{16} = \text{R.H.S.}$$ **Proved.**

Example 84. *Prove that:* sin 6° sin 42° sin 66° sin 78° = $\frac{1}{16}$ **[S.B.T.E. 2014]**

Solution. L.H.S. = sin 6° sin 42° sin 66° sin 78°

$$= \frac{1}{4}(2\sin 66° \sin 6°)(2 \sin 78° \sin 42°)$$

$$= \frac{1}{4}[\cos 60° - \sin 72°][\cos 36° - \cos 120°]$$

$$[\because 2 \sin A \sin B = \cos (A - B) - \cos (A + B)]$$

$$= \frac{1}{4}\left[\frac{1}{2} - \cos(90° - 18°)\right][\cos 36° - \cos(180° - 60°)]$$

$$= \frac{1}{4}\left(\frac{1}{2} - \sin 18°\right)[\cos 36° - (-\cos 60°)]$$

$$= \frac{1}{4}\left[\left(\frac{1}{2} - \sin 18°\right)\left(\cos 36° + \frac{1}{2}\right)\right]$$

$$= \frac{1}{4}\left(\frac{1}{2} - \frac{\sqrt{5}-1}{4}\right)\left(\frac{\sqrt{5}+1}{4} + \frac{1}{2}\right) = \frac{1}{4}\left(\frac{2-\sqrt{5}+1}{4}\right)\left(\frac{\sqrt{5}+1+2}{4}\right)$$

$$= \frac{1}{4}\left(\frac{3-\sqrt{5}}{4}\right)\left(\frac{3+\sqrt{5}}{4}\right) = \frac{1}{4}\left(\frac{9-5}{16}\right) = \frac{1}{16} = \text{R.H.S.}$$ **Proved.**

Example 85. *Prove that:* tan 6° tan 42° tan 66° tan 78° = 1

Solution. L.H.S. = tan 6° tan 42° tan 66° tan 78°

$$= \frac{\sin 6° \sin 42° \sin 66° \sin 78°}{\cos 6° \cos 42° \cos 66° \cos 78°} = \frac{(2\sin 66° \sin 6°)(2\sin 78° \sin 42°)}{(2\cos 66° \cos 6°)(2\cos 78° \cos 42°)}$$

$$= \frac{(\cos 60° - \cos 72°)(\cos 36° - \cos 120°)}{(\cos 60° + \cos 72°)(\cos 36° + \cos 120°)}$$

$$= \frac{(\cos 60° - \sin 18°)(\cos 36° + \sin 30°)}{(\cos 60° + \sin 18°)(\cos 36° - \sin 30°)} = \frac{\left(\frac{1}{2} - \frac{\sqrt{5}-1}{4}\right)\left(\frac{\sqrt{5}+1}{4} + \frac{1}{2}\right)}{\left(\frac{1}{2} + \frac{\sqrt{5}-1}{4}\right)\left(\frac{\sqrt{5}+1}{4} - \frac{1}{2}\right)}$$

$$= \frac{\left(\frac{2-\sqrt{5}+1}{4}\right)\left(\frac{\sqrt{5}+1+2}{4}\right)}{\left(\frac{2+\sqrt{5}-1}{4}\right)\left(\frac{\sqrt{5}+1-2}{4}\right)} = \frac{(3-\sqrt{5})(3+\sqrt{5})}{(\sqrt{5}+1)(\sqrt{5}-1)} = \frac{9-5}{5-1} = \frac{4}{4} = 1 = \text{R.H.S.}$$ **Proved.**

Example 86. *Show that:*

$$\left(1 + \cos\frac{\pi}{10}\right)\left(1 + \cos\frac{3\pi}{10}\right)\left(1 + \cos\frac{7\pi}{10}\right)\left(1 + \cos\frac{9\pi}{10}\right) = \frac{1}{16}.$$

Solution. L.H.S. $= \left(1 + \cos\frac{\pi}{10}\right)\left(1 + \cos\frac{3\pi}{10}\right)\left(1 + \cos\frac{7\pi}{10}\right)\left(1 + \cos\frac{9\pi}{10}\right)$

= (1 + cos 18°) (1 + cos 54°) (1 + cos 126°) (1 + cos 162°)

= (1 + cos 18°) (1 + cos 54°) [1 + cos (180° – 54°)] [1 + cos (180° – 18°)]

= (1 + cos 18°) (1 + cos 54°) (1 – cos 54°) (1 – cos 18°)

= $(1 - \cos^2 18°)(1 - \cos^2 54°)$

= $\sin^2 18° \sin^2 54° = \sin^2 18° [\sin(90° - 36°)]^2$

$$= \left(\frac{\sqrt{5}-1}{4}\right)^2 \left[\frac{1}{4}\sqrt{10-2\sqrt{5}}\right]^2 = \frac{1}{16}$$ **Proved.**

Example 87. *Prove that:*

$$\cos^4 \frac{\pi}{8} + \cos^4 \frac{3\pi}{8} + \cos^4 \frac{5\pi}{8} + \cos^4 \frac{7\pi}{8} = \frac{3}{2}$$ **[S.B.T.E. 2016, 2015, 2014]**

Solution. Here we have

$$\text{L.H.S.} = \cos^4 \frac{\pi}{8} + \cos^4 \frac{3\pi}{8} + \cos^4 \frac{5\pi}{8} + \cos^4 \frac{7\pi}{8}$$

$$= \cos^4 \frac{\pi}{8} + \cos^4 \frac{3\pi}{8} + \left\{\cos\left(\pi - \frac{3\pi}{8}\right)\right\}^4 + \left\{\cos\left(\pi - \frac{\pi}{8}\right)\right\}^4$$

$$= \cos^4 \frac{\pi}{8} + \cos^4 \frac{3\pi}{8} + \cos^4 \frac{3\pi}{8} + \cos^4 \frac{\pi}{8}$$

$$= 2\cos^4 \frac{\pi}{8} + 2\cos^4 \frac{3\pi}{8} = 2\left[\left(\cos^2 \frac{\pi}{8}\right)^2 + \left(\cos^2 \frac{3\pi}{8}\right)^2\right]$$

$$= 2\left[\left(\frac{1+\cos\frac{\pi}{4}}{2}\right)^2 + \left(\frac{1+\cos\frac{3\pi}{4}}{2}\right)^2\right] \qquad \left[\because \frac{1+\cos 2\theta}{2} = \cos^2\theta\right]$$

$$= \frac{1}{2}\left[\left(1+\frac{1}{\sqrt{2}}\right)^2 + \left(1-\frac{1}{\sqrt{2}}\right)^2\right]$$

$$= \frac{1}{2}\left(\frac{3}{2}+\frac{3}{2}\right) = \frac{3}{2}$$ **Proved.**

Example 88. *If $A + B + C = 180°$, prove that*

$$\cos 2A + \cos 2B + \cos 2C = -1 - 4\cos A \cdot \cos B \cdot \cos C$$

[S.B.T.E. 2014, 2013]

Solution. Here we have

$$\text{L.H.S.} = \cos 2A + \cos 2B + \cos 2C$$

$$= 2\cos\frac{(2A+2B)}{2} \cdot \cos\frac{(2A-2B)}{2} + \cos 2C \qquad \left[\begin{array}{l}\because A+B+C = 180° \\ A+B = 180° - C \\ \cos(A+B) = -\cos C\end{array}\right]$$

$$= 2\cos(A+B)\cos(A-B) + 2\cos^2 C - 1$$

$$= -2\cos C \cdot \cos(A-B) + 2\cos^2 C - 1$$

$$= -1 - 2\cos C\,[\cos(A-B) + \cos(A+B)]$$

$$= -1 - 2\cos C\,[2\cos A \cdot \cos B]$$

$$= -1 - 4\cos A \cdot \cos B \cdot \cos C$$ **Proved.**

Example 89. *When $A + B + C = 180°$, show that*

$$\sin 2A + \sin 2B + \sin 2C = 4\sin A \cdot \sin B \cdot \sin C$$

[S.B.T.E. 2017, 2016, 2015]

Solution. Here we have

$$\text{L.H.S.} = \sin 2A + \sin 2B + \sin 2C$$

$$= 2\sin\frac{(2A+2B)}{2}\cos\frac{(2A-2B)}{2}+\sin 2C$$

$$\left[\begin{array}{l} \because A+B+C=180^\circ \\ \Rightarrow A+B=180-C \\ \Rightarrow \sin(A+B)=\sin C \\ \therefore \cos(A+B)=-\cos C \end{array}\right]$$

$= 2\sin(A+B) + \cos(A-B) + 2\sin C\cdot\cos C$

$= 2\sin C\,[\cos(A-B) + \cos C]$

$= 2\sin C\,[\cos(A-B) - \cos(A+B)]$

$$= 2\sin C\cdot 2\sin\frac{(A+B)+(A+B)}{2}\sin\frac{(A+B)-(A-B)}{2}$$

$= 4\sin C\cdot\sin C\cdot\sin B$

$= 4\sin A\cdot\sin B\cdot\sin C$

L.H.S. = R.H.S. **Proved.**

EXERCISE 14.4

1. *Evaluate:*

(*i*) $2\sin 15^\circ\cos 15^\circ$ **Ans.** $\frac{1}{2}$ (*ii*) $1-2\sin^2 22.5^\circ$ **Ans.** $\frac{1}{\sqrt{2}}$

(*iii*) $2\cos^2 157.5^\circ - 1$ **Ans.** $\frac{1}{\sqrt{2}}$ (*iv*) $\cos^2\frac{\pi}{12}-\sin^2\frac{\pi}{12}$ **Ans.** $\frac{\sqrt{3}}{2}$

(*v*) $\frac{1}{2}-\sin^2\frac{7\pi}{12}$ **Ans.** $\frac{-\sqrt{3}}{4}$ (*vi*) $\cos\frac{\pi}{8}\sin\frac{\pi}{8}$ **Ans.** $\frac{\sqrt{2}}{4}$

(*vii*) $\dfrac{2\tan 22\frac{1^\circ}{2}}{1-\tan^2 22\frac{1^\circ}{2}}$ **Ans.** 1 (*viii*) $\sqrt{\dfrac{1+\cos 120^\circ}{2}}$ **Ans.** $\frac{1}{2}$

(*ix*) $\sqrt{\dfrac{1-\cos 300^\circ}{2}}$ **Ans.** $\frac{1}{2}$ (*x*) $8\cos^3\frac{\pi}{9}-6\cos\frac{\pi}{9}$ **Ans.** 1

2. Find the values of sin 2θ, cos 2θ and tan 2θ, given that

(*i*) $\sin\theta=-\frac{1}{2}$, θ lies in *IV quadrant.* **Ans.** $\frac{-\sqrt{3}}{2}, \frac{1}{2}, -\sqrt{3}$

(*ii*) $\tan\theta=-\frac{1}{5}$, θ lies in *II quadrant.* **Ans.** $\frac{-5}{13}, \frac{12}{13}, \frac{-5}{12}$

3. Given that $\tan A=\frac{1}{5}$, find the values of tan 2*A*, tan 4*A* and tan (45° – 4*A*).

Ans. $\frac{5}{12}, \frac{120}{119}, \frac{-1}{239}$

4. If $\frac{\pi}{2}<\theta<\pi$ and $\sin\theta=\frac{3}{5}$, find the values of the following functions.

(*i*) $\cos\frac{\theta}{2}$ **Ans.** $\frac{1}{\sqrt{10}}$ (*ii*) $\sec\frac{\theta}{2}$ **Ans.** $\sqrt{10}$

(*iii*) $\sin\frac{\theta}{2}$ **Ans.** $\frac{3}{\sqrt{10}}$ (*iv*) $\cot\frac{\theta}{2}$ **Ans.** $\frac{1}{3}$

(*v*) $\tan\frac{\theta}{2}$ **Ans.** 3

5. If $\sin\alpha = \frac{3}{5}$, find the values of (*a*) $\sin 3\alpha$ **Ans.** $\frac{117}{125}$

(*b*) $\cos 3\alpha$ **Ans.** $\frac{-44}{125}$, if α is in I Quadrant; $\frac{44}{125}$ if α is in II Quadrant

(*c*) $\tan 3\alpha$ **Ans.** $\frac{-117}{44}$, if α is in I Quadrant; $\frac{117}{44}$ if α is in II Quadrant.

6. Find $\sin 7\frac{1^\circ}{2}$, $\cos 7\frac{1^\circ}{2}$, and $\tan 11\frac{1^\circ}{4}$.

Ans. $\sqrt{\frac{4-\sqrt{2}-\sqrt{6}}{2\sqrt{2}}}, \sqrt{\frac{4+\sqrt{2}+\sqrt{6}}{2\sqrt{2}}}, -(\sqrt{2}+1)+\sqrt{4+2\sqrt{2}}$

7. Use appropriate formulae to find each of the following:

(*i*) $\sin 22\frac{1^\circ}{2}$ **Ans.** $\frac{\sqrt{2-\sqrt{2}}}{2}$ (*ii*) $\cos 22\frac{1^\circ}{2}$ **Ans.** $\frac{\sqrt{2+\sqrt{2}}}{2}$

(*iii*) $\cos 67\frac{1^\circ}{2}$ **Ans.** $\frac{1}{2}\sqrt{2-\sqrt{2}}$ (*iv*) $\sin 112\frac{1^\circ}{2}$ **Ans.** $\frac{1}{2}\sqrt{2+\sqrt{2}}$

(*v*) $\cos 112\frac{1^\circ}{2}$ **Ans.** $-\frac{1}{2}\sqrt{2-\sqrt{2}}$ (*vi*) $\tan 22\frac{1^\circ}{2}$ **Ans.** $\sqrt{2}-1$

8. If $\tan = \frac{1}{7}$, $\tan y = \frac{1}{3}$, prove that $\cos 2x = \sin 4y$.

9. Show that $\cot A + \tan A = 2\operatorname{cosec} 2A$; $\cot A - \tan A = 2\cot 2A$.

Deduce that $\tan 7\frac{1^\circ}{2} = \sqrt{6}-\sqrt{4}-\sqrt{3}+\sqrt{2}$

[**Hint.** For the second part, we have $(\cot A + \tan A) - (\cot A - \tan A) = 2(\operatorname{cosec} 2A - \cot 2A)$

$\Rightarrow \tan A = \operatorname{cosec} 2A - \cot 2A$. Now put $A = 7\frac{1^\circ}{2}$ in the above identity.]

10. If $2\cos\theta = x + \frac{1}{x}$, prove that $2\cos 3\theta = x^3 + \frac{1}{x^3}$

Prove that:

11. $\frac{3\cos\theta + \cos 3\theta}{3\sin\theta - \sin 3\theta} = \cot^3\theta$ **12.** $\sin A = \frac{\sin 3A}{1 + 2\cos 2A}$

13. $\frac{\cos 3A - \sin 3A}{\cos A + \sin A} = 1 - 2\sin 2A.$

14. $\cot \alpha - \tan \alpha = 2 \cot 2\alpha$.

15. $\cos^4 \theta - \sin^4 \theta = \cos 2\theta$.

16. $\dfrac{\sin 2A}{1-\cos 2A} = \cot A$

17. $\dfrac{\sin 3\theta}{\sin \theta} - \dfrac{\cos 3\theta}{\cos \theta} = 2$

18. $\dfrac{\cos A - \sin A}{\cos A + \sin A} = \sec 2A - \tan 2A$

19. $\dfrac{\sec 8A - 1}{\sec 4A - 1} = \dfrac{\tan 8A}{\tan 2A}$

20. $\cot A = \dfrac{1}{2}\left(\cot \dfrac{A}{2} - \tan \dfrac{A}{2}\right)$

21. $\dfrac{\cos A + \sin A}{\cos A - \sin A} - \dfrac{\cos A - \sin A}{\cos A + \sin A} = 2 \tan 2A$.

22. $2 \cos A = \sqrt{\left[2 + \sqrt{2(1+\cos 4A)}\right]}$.

23. $\tan 2A = (\sec 2A + 1)\sqrt{(\sec^2 A - 1)}$.

24. $(\cos A + \cos B)^2 + (\sin A + \sin B)^2 = 4 \cos^2 \left(\dfrac{A-B}{2}\right)$

25. $(\cos A - \cos B)^2 + (\sin A - \sin B)^2 = 4 \sin^2 \left(\dfrac{A-B}{2}\right)$

26. $\cot A + \cot (60° + A) + \cot (120° + A) = 3 \cot 3A$.

27. $\cos^2 \dfrac{\pi}{10} + \cos^2 \dfrac{2\pi}{5} + \cos^2 \dfrac{3\pi}{5} + \cos^2 \dfrac{9\pi}{10} = 2$

28. (*i*) $\cos^2 \dfrac{\pi}{8} + \cos^2 \dfrac{3\pi}{8} + \cos^2 \dfrac{5\pi}{8} + \cos^2 \dfrac{7\pi}{8} = 2$

(*ii*) $\sin^4 \dfrac{\pi}{8} + \sin^4 \dfrac{3\pi}{8} + \sin^4 \dfrac{5\pi}{8} + \sin^4 \dfrac{7\pi}{8} = \dfrac{3}{2}$

29. $\cos^3\left(x - \dfrac{2\pi}{3}\right) + \cos^3 x + \cos^3\left(x + \dfrac{2\pi}{3}\right) = \dfrac{3}{4} \cos 3x$.

30. $\cos^3 x \sin^2 x = \dfrac{1}{16}(2 \cos x - \cos 3x - \cos 5x)$

31. If $\theta = \dfrac{\pi}{2^n + 1}$ prove that $2^n \cos \theta \cos 2\theta \cos 2^2 \theta \ldots\ldots \cos 2^{n-1} \theta = 1$

[**Hint:** L.H.S. $= \dfrac{1}{\sin \theta}(2^n \sin \theta \cdot \cos \theta \cdot \cos 2\theta \cdot \cos 2^2 \theta \ldots\ldots \cos 2^{n-1} \theta)$

$= \dfrac{1}{\sin \theta}(2^{n-2} \sin 2^2 \theta \cdot \cos 2^2 \theta \ldots\ldots \cos 2^{n-1} \theta)$

$= \dfrac{2 \sin 2^{n-1} \theta \cos 2^{n-1} \theta}{\sin \theta} = \dfrac{\sin 2^n \theta}{\sin \theta} = \sin\left(\dfrac{2^n \pi}{2^n + 1}\right) \div \sin\left(\dfrac{\pi}{2^n + 1}\right)$

$= \sin\left(\pi - \dfrac{\pi}{2^n + 1}\right) \div \sin \dfrac{\pi}{2^n + 1} = 1$]

32. If $m \tan(\theta - 30°) = n \tan(\theta + 120°)$, show that $\cos 2\theta = \dfrac{m+n}{2(m-n)}$.

$$\left[\textbf{Hint: } \frac{m}{n} = \frac{\tan(\theta + 120°)}{\tan(\theta - 30°)} = \frac{3 - \tan^2\theta}{1 - 3\tan^2\theta} \quad \therefore \tan^2\theta = \frac{m - 3n}{3m - n} \text{ etc.}\right]$$

33. If $\sin \alpha = \lambda \sin(\theta - \alpha)$, then prove that: $\tan\left(\alpha - \dfrac{\theta}{2}\right) = \dfrac{\lambda - 1}{\lambda + 1} \tan \dfrac{\theta}{2}$.

34. If $\sec(\phi + \alpha)$, $\sec \phi$ and $\sec(\phi - \alpha)$ are in A.P., prove that $\cos\phi = \pm\sqrt{\left(2\cos^2 \dfrac{1}{2}\alpha\right)}$.

Solution of Trigonometric Equations

IMPORTANT FACT

Solution of Trigonometric Equations

Principal solutions | General solution

1. **Principal solutions.** The solutions of a trigonometric equation, for which $0 \leq \theta < 2\pi$ are called the principal solutions.

   For example: The principal solutions of $\sin \theta = 0$ are 0 and π.

2. **General solution.** The solution, consisting of all possible solutions of a trigonometric equation is called its general solution.

General solution

Trigonometric Equation

$\sin \theta = 0$	$\Rightarrow$	$\theta = n\pi$
$\cos \theta = 0$	$\Rightarrow$	$\theta = (2n+1)\frac{\pi}{2}$
$\tan \theta = 0$	$\Rightarrow$	$\theta = n\pi$
$\cot \theta = 0$	$\Rightarrow$	$\theta = (2n+1)\frac{\pi}{2}$
$\sec \theta = 0$	$\Rightarrow$	No solution
$\operatorname{cosec} \theta = 0$	$\Rightarrow$	No solution
$\sin \theta = \sin \alpha$	$\Rightarrow$	$\theta = n\pi + (-1)^n \alpha$
$\cos \theta = \cos \alpha$	$\Rightarrow$	$\theta = 2n\pi \pm \alpha$
$\tan \theta = \tan \alpha$	$\Rightarrow$	$\theta = n\pi \pm \alpha$

15.1 INTRODUCTION

We have studied about trigonometric functions, their properties and graphs. In this chapter, we shall study about solutions of trigonometric equations.

15.2 TRIGONOMETRIC EQUATION

An equation involving one or more trigonometric ratios of unknown angles is known as trigonometric equation.

For example: $\sin\theta = \frac{1}{2}$, $\cos 2\theta + \cos\theta = 0$ etc. are trigonometric equations.

15.3 SOLUTION OF A TRIGONOMETRIC EQUATION

A solution of a trigonometric equation is a value of the unknown angle that satisfies the equation.

For example: Let $\sin\theta = \frac{1}{2}$ be a trigonometric equation. Clearly, $\theta = \frac{\pi}{6}$ and $\theta = \frac{5\pi}{6}$ satisfy this equation. Therefore, these are its solutions.

We know that the values of sin x and cos x repeat after an interval of 2π and the values of tan x repeat after an interval of π. The solutions of a trigonometric equation for which $0 \le x \le 2\pi$ are called principal solutions. The solution involving integer n which gives all solutions of trigonometric equation is called general solution.

A trigonometric equation may have an unlimited number of solutions.

For example: If $\sin\theta = 0$, then the solutions of the equation are $\theta = 0, \pi, 2\pi, 3\pi, \ldots$

Solutions: There are two types of solutions of a trigonometric equation

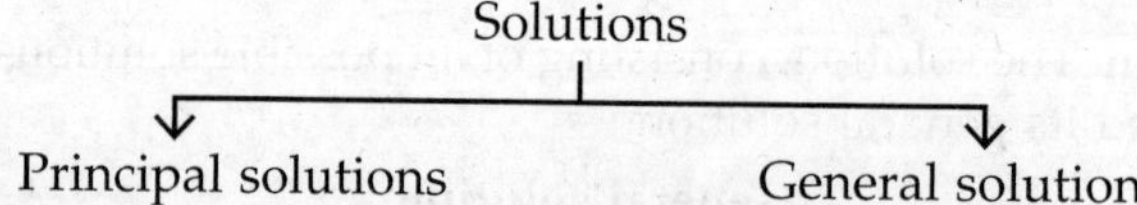

Principal solutions. The solutions of a trigonometric equation, for which $0 \le \theta < 2\pi$ are called the principal solutions.

For example: The principal solution of $\sin\theta = 0$ are 0 and π.

Example 1. *Find the principal solutions of the following equations.*

(*i*) $\cos\theta = \frac{\sqrt{3}}{2}$

(*ii*) $\sin\theta = \frac{-1}{\sqrt{2}}$

(*iii*) $\tan\theta = -\sqrt{3}$

Solution. (*i*) Since, cos θ is '+ve' therefore, θ lies in 1st or 4th quadrant.

Here, $\cos\theta = \frac{\sqrt{3}}{2}$

We know that, $\cos\frac{\pi}{6} = \frac{\sqrt{3}}{2}$ and $\cos = \left(2\pi - \frac{\pi}{6}\right) = \frac{\sqrt{3}}{2}$ $[\cos(2\pi - \theta) = \cos\theta]$

$\Rightarrow$ $\cos\left(\frac{11\pi}{6}\right) = \frac{\sqrt{3}}{2}$

$\therefore$ $\cos\frac{\pi}{6} = \cos\left(\frac{11\pi}{6}\right) = \frac{\sqrt{3}}{2}$

Therefore, principal solutions are $\frac{\pi}{6}$ and $\frac{11\pi}{6}$. **Ans.**

(*ii*) Since, sin θ is '–ve' therefore, θ lies in 3rd or 4th quadrant.

We know that, $\sin\dfrac{\pi}{4} = \dfrac{1}{\sqrt{2}}$

$\therefore\ \sin\left(\pi+\dfrac{\pi}{4}\right) = \dfrac{-1}{\sqrt{2}}$ and $\sin\left(2\pi-\dfrac{\pi}{4}\right) = \dfrac{-1}{\sqrt{2}}$ $\quad\left[\begin{array}{l}\sin(\pi+\theta) = -\sin\theta \\ \sin(2\pi-\theta) = -\sin\theta\end{array}\right]$

$\Rightarrow \sin\dfrac{5\pi}{4} = \sin\dfrac{7\pi}{4} = -\dfrac{1}{\sqrt{2}}$

Therefore, the principal solutions are $\dfrac{5\pi}{4}, \dfrac{7\pi}{4}$. **Ans.**

(*iii*) Since, tan θ is '–ve' therefore, θ lies in 2nd or 4th quadrant.

We know that, $\tan\dfrac{\pi}{3} = \sqrt{3}$

$\therefore\qquad \tan\left(\pi-\dfrac{\pi}{3}\right) = -\sqrt{3}$ and $\tan\left(2\pi-\dfrac{\pi}{3}\right) = -\sqrt{3}$

$\Rightarrow\qquad \tan\dfrac{2\pi}{3} = \tan\dfrac{5\pi}{3} = -\sqrt{3}$

Therefore principal solutions are $\dfrac{2\pi}{3}$ and $\dfrac{5\pi}{3}$. **Ans.**

Example 2. *Find the principal solutions of the following equations:*

(*i*) *tan* $x = \dfrac{1}{\sqrt{3}}$ (*ii*) *sec* $x = 2$

(*iii*) *cot* $x = -\sqrt{3}$ (*iv*) *cosec* $x = -2$

Solution. (*i*) We have, $\tan x = \dfrac{1}{\sqrt{3}}$

We know that, $\tan\dfrac{\pi}{6} = \dfrac{1}{\sqrt{3}}$

$x = \dfrac{\pi}{6}$ (In 1st quadrant)

and $\tan\left(\pi+\dfrac{\pi}{6}\right) = \tan\dfrac{\pi}{6} = \dfrac{1}{\sqrt{3}}$ (In third quadrant)

$\Rightarrow \tan\dfrac{\pi}{6} = \dfrac{1}{\sqrt{3}}$ and $\tan\dfrac{7\pi}{6} = \dfrac{1}{\sqrt{3}}$

So, $\dfrac{\pi}{6}$ and $\dfrac{7\pi}{6}$ are the required principal solutions. **Ans.**

(*ii*) $\sec x = 2 = \sec 60° \Rightarrow \cos x = \frac{1}{2} = \cos 60°$

$\therefore$ Principal value $= 60° = \frac{\pi}{3}$ radians.

For $\cos\theta = \cos\alpha,\ \theta = 2n\pi \pm \alpha.$

General value of $x = 2n\pi \pm \frac{\pi}{3}$ **Ans.**

(*iii*) $\cot x = -\sqrt{3} \Rightarrow \tan x = -\frac{1}{\sqrt{3}},\ \tan 30° = \frac{1}{\sqrt{3}};$

$\tan(180° - 30°) = -\tan 30° = -\frac{1}{\sqrt{3}}$ or $\tan 150° = -\frac{1}{\sqrt{3}}$

$\therefore$ Principal value of $x = 150° = \frac{5\pi}{6}$ radians

General value of $x = n\pi + \alpha = n\pi + \frac{5\pi}{6}$ **Ans.**

(*iv*) $\operatorname{cosec} x = -2$ or $\sin x = -\frac{1}{2}$

$\sin 30° = \frac{1}{2}$ or $\sin(-30°) = -\sin 30° = -\frac{1}{2}$

Principal value of $x = -30° = -\frac{\pi}{6}$ radians

General value of $x = n\pi + (-1)^n\alpha = n\pi + (-1)^n\left(-\frac{\pi}{6}\right) = n\pi - (-1)^n\left(\frac{\pi}{6}\right)$ **Ans.**

EXERCISE 15.1

Find the principal solutions of the following equations:

1. $\sin x = \frac{1}{2}$ **Ans.** $\frac{\pi}{6}, \frac{5\pi}{6}$

2. $\sin x \cos x = 0$ **Ans.** $0, \frac{\pi}{2}, \pi, \frac{3\pi}{2}$

3. $\sin\theta\cos\theta = \frac{1}{2}$ **Ans.** $\frac{\pi}{4}, \frac{5\pi}{4}$

4. $\cos 2x = \frac{-\sqrt{3}}{2}$ **Ans.** $\frac{5\pi}{12}, \frac{11\pi}{12}$

5. $\sin 3x = \frac{1}{\sqrt{2}}$ **Ans.** $\frac{\pi}{12}, \frac{\pi}{4}$

6. $\tan 4x = -\frac{1}{\sqrt{3}}$ **Ans.** $\frac{5\pi}{24}, \frac{11\pi}{24}$

7. $\cot 2x = \sqrt{3}$ **Ans.** $\frac{\pi}{12}, \frac{7\pi}{12}$

8. $\sec 3x = \sqrt{2}$ **Ans.** $\frac{\pi}{12}, \frac{7\pi}{12}$

9. $\operatorname{cosec} 5x = -2$ **Ans.** $\frac{7\pi}{30}, \frac{11\pi}{30}$

10. $2\sin^2\theta = 3\cos\theta$ **Ans.** $\frac{\pi}{3}, \frac{5\pi}{3}$

HINTS TO THE SELECTED QUESTIONS

2. $\sin x \cos x = 0 = \sin 0° \cos 0° = \sin\frac{\pi}{2} \cos\frac{\pi}{2} = \sin \pi \cos \pi = \sin\frac{3\pi}{2} \cos\frac{3\pi}{2}$

$\Rightarrow \quad x = 0, \frac{\pi}{2}, \pi, \frac{3\pi}{2}$ (sin x is +ve in I, II quadrant)

3. $\sin\theta \cos\theta = \frac{1}{2} = \cos\frac{\pi}{4} \sin\frac{\pi}{4}$

$\Rightarrow \quad x = \frac{\pi}{4}$ (sin x and cos x are +ve is 1st quadrant and $x = \pi + \frac{\pi}{4} = \frac{5\pi}{4}$)

4. $\cos 2x = \frac{-\sqrt{3}}{2} = -\cos\left(\frac{\pi}{6}\right) = \cos\left(\pi - \frac{\pi}{6}\right) = \cos\frac{5\pi}{6}$.

$\Rightarrow 2x = \frac{5\pi}{6} \Rightarrow x = \frac{5\pi}{12}$ (cos x is +ve in Ist quadrant)

and $2x = \pi + \frac{5\pi}{6} = \frac{11\pi}{6}$ (cos x is –ve in IIIrd quadrant) $\Rightarrow \quad x = \frac{11\pi}{12}$

6. $\tan 4x = -\frac{1}{\sqrt{3}} = -\tan\frac{\pi}{6} = \tan\left(\pi - \frac{\pi}{6}\right) = \tan\frac{5\pi}{6}$

$\Rightarrow 4x = \frac{5\pi}{6}$ (tan x is –ve in IInd quadrant) $\Rightarrow x = \frac{5\pi}{24}$

and $4x = 2\pi - \frac{\pi}{6}$ (tan x is –ve in IVth quadrant)

$\Rightarrow 4x = \frac{11\pi}{6} \Rightarrow x = \frac{11\pi}{24}$

9. $\text{cosec } 5x = -2 = -\text{cosec}\left(\frac{\pi}{6}\right) = \text{cosec}\left(\pi + \frac{\pi}{6}\right)$ (cosec $5x$ is –ve in IIIrd quadrant)

$\Rightarrow 5x = \frac{7\pi}{6} \Rightarrow x = \frac{7\pi}{30}$ and $5x = 2\pi - \frac{\pi}{6}$ [cosec $5x$ is –ve in IVth quadrant]

$\Rightarrow 5x = \frac{11\pi}{6} \Rightarrow x = \frac{11\pi}{30}$

10. $2\sin^2\theta = 3\cos\theta \quad \Rightarrow \quad 2(1 - \cos^2\theta) = 3\cos\theta$

$\Rightarrow 2 - 2\cos^2\theta = 3\cos\theta \quad \Rightarrow \quad 2\cos^2\theta + 3\cos\theta - 2 = 0$

$\Rightarrow 2\cos^2\theta + 4\cos\theta - \cos\theta - 2 = 0 \quad \Rightarrow \quad 2\cos\theta(\cos\theta + 2) - 1(\cos\theta + 2) = 0$

$\Rightarrow (2\cos\theta - 1)(\cos\theta + 2) = 0$ $[\cos\theta + 2 \neq 0]$

$\Rightarrow 2\cos\theta - 1 = 0$

$\Rightarrow \cos\theta = \frac{1}{2}$ (cos θ is +ve in Ist and IVth quadrants)

$\Rightarrow \theta = \frac{\pi}{3}$ and $\theta = 2\pi - \frac{\pi}{3} = \frac{5\pi}{3}$ (in IVth quadrant)

15.4 GENERAL SOLUTION

The values of sin x and cos x repeat after an interval of 2π and the values of tan x repeat after an interval of π. The solution involving integer n, which gives all solutions of trigonometric equation is called the general solution.

For example: General solution of the equation $\sin\theta = 0$ is $n\pi$, where n is 0 or any positive or negative integer.

(1) General Solution of sin θ = 0:

We have, $\sin\theta = 0$

In ΔPOM, By definition, we have

$$\sin\theta = \frac{PM}{OP} \qquad ...(1)$$

Putting $\sin\theta = 0$ in (1), we get

$$0 = \frac{PM}{OM}$$

$$\Rightarrow \qquad PM = 0$$

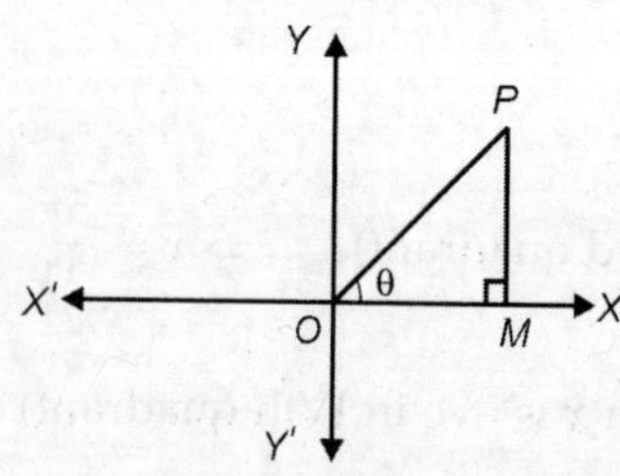

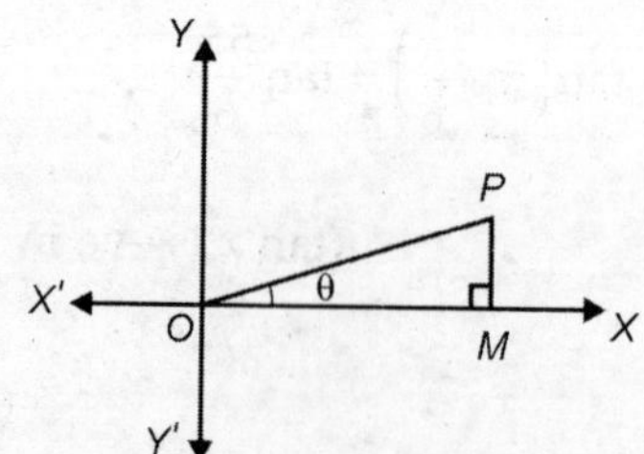

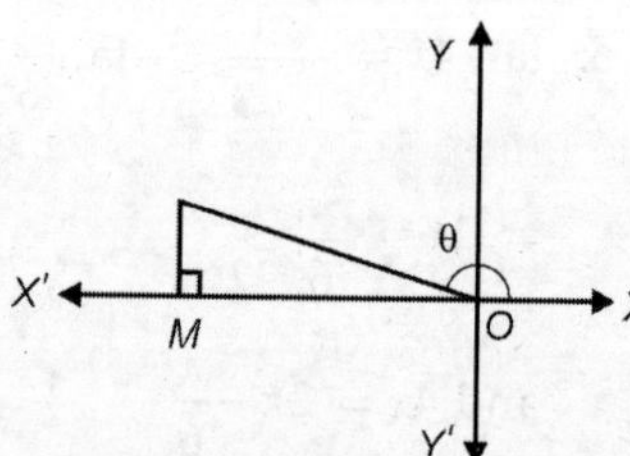

$\Rightarrow OP$ coincides with OX or OX' $\Rightarrow$ $\theta = 0, \pi, 2\pi, \ldots -\pi, -2\pi, -3\pi, \ldots$

$\sin\theta = 0 \Rightarrow \theta = n\pi$ where, $n = 0, \pm 1, \pm 2, \ldots$

(2) General Solutions of cos θ = 0:

We have $\cos\theta = 0$

in ΔPOM, By definition, we have

$$\cos\theta = \frac{OM}{OP} \qquad ...(1)$$

Putting $\cos\theta = 0$ in (1), we get

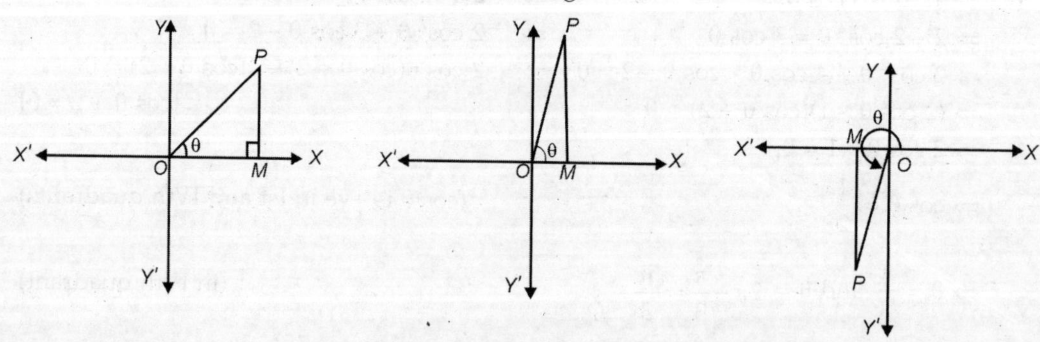

$$\frac{OM}{OP} = 0 \;\Rightarrow\; OM = 0$$

This is possible only when, *OP* coincides with *OY* or *OY′*

When, *OP* coincides with *OY*, $\theta = \frac{\pi}{2}, \frac{5\pi}{2}, \frac{9\pi}{2} \ldots$ or $\theta = -\frac{3\pi}{2}, -\frac{7\pi}{2}, \ldots$ ...(2)

When, *OP* coincides with *OY′*, $\theta = -\frac{\pi}{2}, -\frac{5\pi}{2}, -\frac{9\pi}{2} \ldots$ or $\theta = \frac{3\pi}{2}, \frac{7\pi}{2} \ldots$ ...(3)

From (2) and (3), we get

$\Rightarrow$ $\boxed{\cos\theta = 0 \Rightarrow \theta = (2n+1)\frac{\pi}{2}}$, where $n = 0, \pm 1, \pm 2$

(3) General Solution of tan θ = 0:

We have, $\tan\theta = 0$

In Δ*POM*, by definition, we have $\tan\theta = \frac{PM}{OM}$ ...(1)

Putting $\tan\theta = 0$ in (1), we get $\frac{PM}{OM} = 0$

$\Rightarrow$ $PM = 0$

$\Rightarrow$ *OP* coincides with *OX* or *OX′*

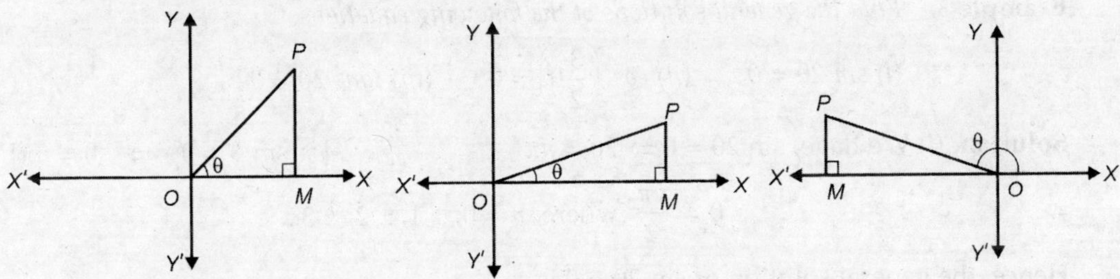

When, *OP* coincides with *OX*, $\theta = 0, 2\pi, 4\pi, \ldots$ or $-2\pi, -4\pi \ldots$ ...(2)

When, *OP* coincides with *OX′*, $\theta = 0, \pi, 3\pi, 5\pi\ldots$ or $-\pi, -3\pi, 5\pi, \ldots$ ...(3)

From (1) and (2), we get

$$\theta = 0, \pi, 2\pi, \ldots -\pi, -2\pi,$$

$\Rightarrow$ $\theta = n\pi$, where $n = 0, \pm 1, \pm 2, \ldots$

$$\boxed{\tan\theta = 0 \Rightarrow \theta = n\pi}$$

(4) General Solution of cot θ = 0:

We have, $\cot\theta = 0$

In Δ*POM*, by definition, we have, $\cot\theta = \frac{OM}{PM}$ ...(1)

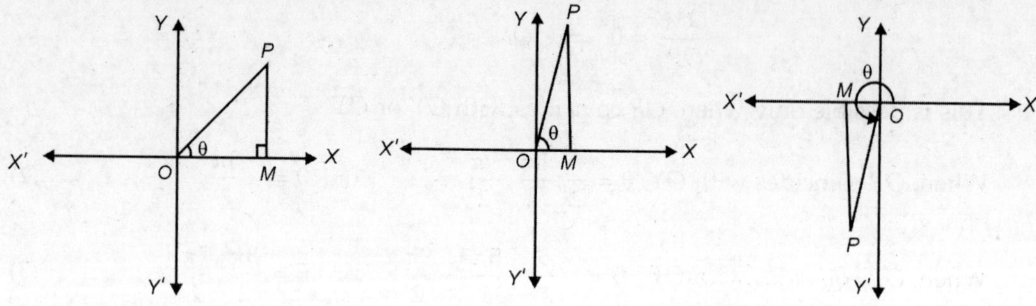

Putting cot θ = 0 in (1), we get $\frac{OM}{PM} = 0 \Rightarrow OM = 0$

⇒ OP coincides with OY and OY'

When OP coincides with OY, $\theta = \frac{\pi}{2}, \frac{5\pi}{2}, \frac{9\pi}{2}, \ldots$ or $\theta = \frac{-3\pi}{2}, \frac{-7\pi}{2}, \ldots$ ...(2)

When OP coincides with OY', $\theta = \frac{3\pi}{2}, \frac{7\pi}{2}, \ldots$ or $\theta = \frac{-\pi}{2}, \frac{-5\pi}{2}, \frac{-9\pi}{2}, \ldots$ ...(3)

From (2) and (3), we get

$$\boxed{\cot\theta = 0 \Rightarrow \theta = (2n+1)\frac{\pi}{2}}, \quad \text{where } n = 0, \pm 1, \pm 2, \pm 3, \ldots$$

Note. Since, sec θ ≥ 1 or sec θ ≤ – 1, therefore sec θ = 0 does not have any solution.

Similarly, cosec θ = 0 has no solution.

Example 3. *Find the general solutions of the following equations.*

(*i*) *sin* 2θ = 0 (*ii*) *cos* $\left(\frac{3}{2}\theta\right) = 0$ (*iii*) *tan*^{2} 2θ = 0

Solution. (*i*) We have, sin 2θ = 0 ⇒ 2θ = $n\pi$ [∵ sin θ = 0 ⇒ θ = $n\pi$]

$$\Rightarrow \qquad \theta = \frac{n\pi}{2} \text{ where, } n = 0, \pm 1, \pm 2, \pm 3\ldots$$

Hence, the general solution of sin 2θ = 0 is

$$\theta = \frac{n\pi}{2}, n \in Z$$

Ans.

(*ii*) We know that, the general solution of the equation cos θ = 0 is $\theta = (2n + 1)\frac{\pi}{2}$, $n \in Z$. Therefore,

$$\cos\left(\frac{3\theta}{2}\right) = 0 \Rightarrow \frac{3\theta}{2} = (2n+1)\frac{\pi}{2}$$

$$\Rightarrow \qquad \theta = (2n + 1)\frac{\pi}{3}, \text{ where } n = 0, \pm 1, \pm 2 \ldots..$$

which is the general solution of $\cos\left(\frac{3\theta}{2}\right) = 0$. **Ans.**

(*iii*) We know that the general solution of the equation tan θ = 0 is θ = $n\pi$, $n \in Z$.

Therefore, $\tan^2 2\theta = 0 \Rightarrow \tan 2\theta = 0 \Rightarrow 2\theta = n\pi$

$\Rightarrow \quad \theta = \dfrac{n\pi}{2}$, where $n = 0, \pm 1, \pm 2$ **Ans.**

which is the required solution.

EXERCISE 15.2

Find the general solutions of the following equations:

1. $\sin^2 7\theta = 0$ **Ans.** $\theta = \dfrac{n\pi}{7}, n \in Z$

2. $\cos 4\theta = 0$ **Ans.** $\theta = (2n+1)\dfrac{\pi}{8}, n \in Z$

3. $\cos^3 6\theta = 0$ **Ans.** $\theta = (2n+1)\dfrac{\pi}{12}, n \in Z$

4. $\tan 5\theta = 0$ **Ans.** $\theta = \dfrac{n\pi}{5}, n \in Z$

5. $\cot 3\theta = 0$ **Ans.** $\theta = (2n+1)\dfrac{\pi}{8}, n \in Z$

6. $\sec 4\theta = 0$ **Ans.** No solution

7. $\text{cosec}^2\,\theta = 0$ **Ans.** No solution

8. $\tan^2 6\theta = 0$ **Ans.** $\theta = \dfrac{n\pi}{6}$

9. $\sin^2 5\theta = 0$ **Ans.** $\theta = \dfrac{n\pi}{5}$

10. $\cos^2 3\theta = 0$ **Ans.** $\theta = (2n+1)\dfrac{\pi}{6}$

HINTS TO THE SELECTED QUESTIONS

3. $\cos^3 6\theta = 0 \Rightarrow \cos 6\theta = 0 \Rightarrow 6\theta = \cos^{-1}(0) = \dfrac{\pi}{2} \Rightarrow \theta = \dfrac{\pi}{12} \Rightarrow \theta = (2n+1)\dfrac{\pi}{12}, n \in Z$

6. $\sec 4\theta = 0 \Rightarrow \dfrac{1}{\cos 4\theta} = 0 \Rightarrow \cos 4\theta = \infty$

We know that general solution of $\cos 4\theta = \infty$ is not defined.

Hence, there is no solution.

10. $\cos^2 3\theta = 0 \Rightarrow \cos 3\theta = 0$

We know that, the general solution of the equation $\cos\theta = 0$ is $\theta = (2n+1)\dfrac{\pi}{2}, n \in Z$

Therefore, $\cos 3\theta = 0 \Rightarrow 3\theta = (2n+1)\dfrac{\pi}{2} \Rightarrow 0 = (2n+1), n \in Z.$

15.5 GENERAL SOLUTION OF *sin θ = sin α*

We have, $\sin\theta = \sin\alpha \Rightarrow \sin\theta - \sin\alpha = 0 \Rightarrow 2\cos\left(\dfrac{\theta+\alpha}{2}\right)\sin\left(\dfrac{\theta-\alpha}{2}\right) = 0$

$\Rightarrow$ Either $\cos\left(\dfrac{\theta+\alpha}{2}\right) = 0$ or $\sin\left(\dfrac{\theta-\alpha}{2}\right) = 0$

$\Rightarrow \quad \dfrac{\theta+\alpha}{2} = (2m+1)\dfrac{\pi}{2}, m \in Z$ $\quad\Rightarrow\quad \dfrac{\theta-\alpha}{2} = m\pi, m \in Z$

$\Rightarrow \quad \theta + \alpha = (2m+1)\pi$ $\quad\Rightarrow\quad \theta - \alpha = 2m\pi$

$\Rightarrow \quad \theta = (2m+1)\pi - \alpha \quad ...(1)$ $\quad\Rightarrow\quad \theta = 2m\pi + \alpha \quad ...(2)$

Thus, either $\theta = -\alpha$ + any odd multiple of π or $\theta = \alpha$ + any even multiple of π.

Combining these two results, the general value of θ such that sin θ = sin α is given by

$$\sin\theta = \sin\alpha \quad \Rightarrow \quad \theta = n\pi + (-1)^n\alpha, \text{ where } n \in Z \qquad ...(3)$$

Remark. The equation cosec θ = cosec α is equivalent to sin θ = sin α.

Thus, cosec θ = cosec α and sin θ = sin α have the same general solution.

15.6 GENERAL SOLUTION OF cos θ = cos α

We have, $\cos\theta = \cos\alpha$ or $\cos\theta - \cos\alpha = 0$

$$\Rightarrow -2\sin\left(\frac{\theta+\alpha}{2}\right)\sin\left(\frac{\theta-\alpha}{2}\right) = 0$$

$$\Rightarrow \text{Either } \sin\left(\frac{\theta+\alpha}{2}\right) = 0 \text{ or } \sin\left(\frac{\theta-\alpha}{2}\right) = 0$$

$$\Rightarrow \frac{\theta+\alpha}{2} = n\pi \qquad \Rightarrow \frac{\theta-\alpha}{2} = n\pi$$

$$\Rightarrow \theta + \alpha = 2n\pi \qquad \Rightarrow \theta - \alpha = 2n\pi$$

$$\Rightarrow \theta = 2n\pi - \alpha \qquad \Rightarrow \theta = 2n\pi + \alpha$$

Combining these two solutions, we get the general solution as

$$\cos\theta = \cos\alpha \quad \Rightarrow \quad \theta = 2n\pi \pm \alpha, \; n \in Z$$

Remark. Since, sec θ = sec α ⇔ cos θ = cos α. So, the general solutions of cos θ = cos α and sec θ = sec α are same.

15.7. GENERAL SOLUTION OF tan θ = tan α

We have, $\tan\theta = \tan\alpha$

$$\frac{\sin\theta}{\cos\theta} = \frac{\sin\alpha}{\cos\alpha}$$

$$\Leftrightarrow \quad \sin\theta\cos\alpha - \cos\theta\sin\alpha = 0$$

$$\sin(\theta - \alpha) = 0$$

$$\Leftrightarrow \quad \theta - \alpha = n\pi, \; n \in Z$$

$$\theta = n\pi + \alpha, \; n \in Z$$

Remark. Since, tan θ = tan α ⇔ cot θ = cot α.

So, the general solution of cot θ = cot α and tan θ = tan α are same.

15.8. WORKING RULE TO FIND THE GENERAL SOLUTIONS OF TRIGONOMETRIC EQUATIONS

Step 1: For a given equation, find a value of θ, which is between 0 and 2π or –π and π. This value of θ is called as α.

Step 2: General solution:

(*i*) sin θ = sin α; write $\theta = n\pi + (-1)^n\alpha, \; n \in Z$

(*ii*) cos θ = cos α; write $\theta = 2n\pi \pm \alpha, \; n \in Z$

(*iii*) tan θ = tan α; write $\theta = n\pi + \alpha, \; n \in Z$

Example 4. *Find the general solution for each of the following equations:*

(*i*) $\sec x = \sec (\pi + x)$ (*ii*) $\cos 4x = \cos 2x$

Solution. We have,

(*i*) $\sec x = \sec (\pi + x) \Rightarrow \dfrac{1}{\cos x} = \dfrac{1}{\cos(\pi + x)} \Rightarrow \cos x = \cos (\pi + x)$

The general solution is

$$x = 2n\pi \pm (\pi + x)$$

$\Rightarrow$	$x = 2n\pi + (\pi + x)$	or	$x = 2n\pi - (\pi + x)$
$\Rightarrow$	$x - x = \pi(2n + 1)$	or	$x + x = 2n\pi - \pi$
$\Rightarrow$	$0 = \pi(2n + 1)$	or	$2x = \pi(2n - 1)$
$\Rightarrow$			$x = \dfrac{\pi}{2}(2n - 1)$ **Ans.**

which is the general solution of the given equation.

(*ii*) We have $\cos 4x = \cos 2x$

The general solution is

$$4x = 2n\pi \pm 2x$$

$\Rightarrow$	$4x = 2n\pi + 2x$	or	$4x = 2n\pi - 2x$
$\Rightarrow$	$4x - 2x = 2n\pi$	or	$4x + 2x = 2n\pi$
$\Rightarrow$	$2x = 2n\pi$	or	$6x = 2n\pi$
$\Rightarrow$	$x = n\pi$	or	$x = \dfrac{1}{3} n\pi$, where $n \in Z$

The required general solution are $x = \dfrac{n\pi}{3}, n\pi, n \in Z$ **Ans.**

Example 5. *Solve the equation* $\sin \theta + \sin 3\theta + \sin 5\theta = 0$

Solution. We have,

	$\sin \theta + \sin 3\theta + \sin 5\theta = 0$	$\Rightarrow$	$(\sin 5\theta + \sin \theta) + \sin 3\theta = 0$
$\Rightarrow$	$2\sin 3\theta \cos 2\theta + \sin 3\theta = 0$	$\Rightarrow$	$\sin 3\theta (2\cos 2\theta + 1) = 0$
$\Rightarrow$	$\sin 3\theta = 0$ or $2 \cos 2\theta + 1 = 0$	$\Rightarrow$	$\sin 3\theta = 0$ or $\cos 2\theta = -\dfrac{1}{2}$
Now,	$\sin 3\theta = 0 \Rightarrow 3\theta = n\pi, n \in Z$	$\Rightarrow$	$\theta = \dfrac{n\pi}{3}, n \in Z$ ...(1)

And, $\cos 2\theta = -\dfrac{1}{2} \Rightarrow \cos 2\theta = \cos \dfrac{2\pi}{3}$

$\Rightarrow 2\theta = 2m\pi \pm \dfrac{2\pi}{3}, m \in Z \Rightarrow \theta = m\pi \pm \dfrac{\pi}{3}, m \in Z$ ...(2)

From (1) and (2), we have the general solution of the given equation as

$\theta = \dfrac{n\pi}{3}$ or $\theta = m\pi \pm \dfrac{\pi}{3}$, where $m, n \in Z$ **Ans.**

Example 6. *Solve: sin mθ + sin nθ = 0*

Solution. We have,

$$\sin m\theta + \sin n\theta = 0$$

$$\Rightarrow \quad \sin\left(\frac{m+n}{2}\right)\theta \cdot \cos\left(\frac{m-n}{2}\right)\theta = 0$$

$$\Rightarrow \quad \sin\left(\frac{m+n}{2}\right)\theta = 0 \quad \text{or} \quad \cos\left(\frac{m-n}{2}\right)\theta = 0$$

Now,

$$\cos\left(\frac{m+n}{2}\right)\theta = 0 \quad \Rightarrow \quad \left(\frac{m+n}{2}\right)\theta = r\pi,\ r \in Z \quad \ldots(1)$$

$$\theta = \frac{2r\pi}{m+n}$$

And

$$\cos\left(\frac{m-n}{2}\right)\theta = 0 \quad \Rightarrow \quad \cos\left(\frac{m-n}{2}\right)\theta = \cos\frac{\pi}{2}$$

$$\Rightarrow \quad \left(\frac{m-n}{2}\right)\theta = (2\pi - 1)\frac{\pi}{2}, \quad p \in Z$$

$$\Rightarrow \quad \theta = \left(\frac{2p+1}{m-n}\right)\pi, \quad p \in Z \quad \ldots(2)$$

From (1) and (2), we have

$\theta = \dfrac{2r\pi}{m+n}$ or $\theta = \left(\dfrac{2p+1}{m-n}\right)\pi$, where $m, n \in Z$ **Ans.**

Example 7. *Solve the following equations.*

(*i*) sin 2θ + cos θ = 0, (*ii*) sin 3θ + cos 2θ = 0

Solution. (*i*) We have, sin 2θ + cos θ = 0

$$\Rightarrow \quad \cos\theta = -\sin 2\theta \quad \Rightarrow \quad \cos\theta = \cos\left(\frac{\pi}{2} + 2\theta\right)$$

$$\Rightarrow \quad \theta = 2n\pi \pm \left(\frac{\pi}{2} + 2\theta\right),\ n \in Z$$

Taking positive sign, we have

$$\theta = 2n\pi + \left(\frac{\pi}{2} + 2\theta\right) \quad \Rightarrow \quad -\theta = 2n\pi + \frac{\pi}{2},\ n \in Z \quad \Rightarrow \quad \theta = -2n\pi - \frac{\pi}{2}$$

$$\Rightarrow \quad \theta = 2m\pi - \frac{\pi}{2}, \text{ where } m = -n \in Z \quad \ldots(1)$$

Taking negative sign, we have

$$\theta = 2n\pi - \left(\frac{\pi}{2} + 2\theta\right)$$

$\Rightarrow \quad 3\theta = 2n\pi - \dfrac{\pi}{2} \quad \Rightarrow \quad \theta = \dfrac{2n\pi}{3} - \dfrac{\pi}{6}, n \in Z$ ...(2)

From (1) and (2), we have

$\theta = 2n\pi - \dfrac{\pi}{2}$ or $\theta = \dfrac{2n\pi}{3} - \dfrac{\pi}{6}$ where $m, n \in Z$ **Ans.**

(*ii*) We have, $\sin 3\theta + \cos 2\theta = 0 \quad \Rightarrow \quad \cos 2\theta = -\sin 3\theta$

$\Rightarrow \quad \cos 2\theta = \cos\left(\dfrac{\pi}{2} + 3\theta\right) \quad \Rightarrow \quad 2\theta = 2n\pi \pm \left(\dfrac{\pi}{2} + 3\theta\right), n \in Z$

Taking positive sign, we have

$2\theta = 2n\pi + \dfrac{\pi}{2} + 3\theta \quad \Rightarrow \quad -\theta = 2n\pi + \dfrac{\pi}{2}$

$\Rightarrow \quad \theta = -2n\pi - \dfrac{\pi}{2} \quad \Rightarrow \quad \theta = 2m\pi - \dfrac{\pi}{2}$, where $-n = m \in Z$ ...(1)

Taking negative sign, we have

$2\theta = 2n\pi - \dfrac{\pi}{2} - 3\theta \quad \Rightarrow \quad 5\theta = 2n\pi - \dfrac{\pi}{2}$

$\Rightarrow \quad \theta = \dfrac{2n\pi}{5} - \dfrac{\pi}{10}, n \in Z$ ...(2)

From (1) and (2), we have

Hence, $\theta = \dfrac{2n\pi}{5} - \dfrac{\pi}{10}$ or $\theta = 2m\pi - \dfrac{\pi}{2}$, where $m, n \in Z$. Ans.

Example 8. *Solve the following equations.*

$\tan\theta + \tan 2\theta + \sqrt{3}\tan\theta\tan 2\theta = \sqrt{3}$

Solution. We have,

$\tan\theta + \tan 2\theta + \sqrt{3}\tan\theta\tan 2\theta = \sqrt{3}$

$\Rightarrow \quad \tan\theta + \tan 2\theta = \sqrt{3}(1 - \tan\theta\tan 2\theta)$

$\Rightarrow \quad \dfrac{\tan\theta + \tan 2\theta}{1 - \tan\theta\tan 2\theta} = \sqrt{3} \quad \Rightarrow \quad \tan(\theta + 2\theta) = \sqrt{3}$

$\Rightarrow \quad \tan 3\theta = \sqrt{3} \quad \Rightarrow \quad \tan 3\theta = \tan\dfrac{\pi}{3}$

$\Rightarrow \quad 3\theta = n\pi + \dfrac{\pi}{3}, n \in Z \quad \Rightarrow \quad \theta = \dfrac{n\pi}{3} + \dfrac{\pi}{9}, n \in Z$ **Ans.**

Example 9. *Solve the following:*

(*i*) $\tan\theta + \tan 2\theta + \tan\theta\tan 2\theta = 1$

(*ii*) $\tan\theta + \tan 2\theta + \tan 3\theta = \tan\theta\tan 2\theta\tan 3\theta$

Solution. (*i*) We have,

$\tan\theta + \tan 2\theta + \tan\theta\tan 2\theta = 1$

$$\tan\theta + \tan 2\theta = 1 - \tan\theta \tan 2\theta$$

$$\Rightarrow \quad \frac{\tan\theta + \tan 2\theta}{1 - \tan\theta \cdot \tan 2\theta} = 1$$

$$\Rightarrow \quad \tan 3\theta = 1 \quad \Rightarrow \quad \tan 3\theta = \tan\frac{\pi}{4}$$

$$\Rightarrow \quad 3\theta = n\pi + \frac{\pi}{4}, n \in Z \quad \Rightarrow \quad \theta = \frac{n\pi}{3} + \frac{\pi}{12}, \ n \in Z$$ **Ans.**

(*ii*) We have, $\tan\theta + \tan 2\theta + \tan 3\theta = \tan\theta \tan 2\theta \tan 3\theta$

$$\tan\theta + \tan 2\theta = -\tan 3\theta + \tan\theta \tan 2\theta \tan 3\theta$$

$$\Rightarrow \quad \tan\theta + \tan 2\theta = -\tan 3\theta(1 - \tan\theta \tan 2\theta)$$

$$\Rightarrow \quad \frac{\tan\theta + \tan 2\theta}{1 - \tan\theta \cdot \tan 2\theta} = -\tan 3\theta$$

$$\Rightarrow \quad \tan(\theta + 2\theta) = -\tan 3\theta$$

$$\Rightarrow \quad \tan 3\theta = -\tan 3\theta$$

$$\Rightarrow \quad 2\tan 3\theta = 0 \quad \Rightarrow \quad \tan 3\theta = 0$$

$$\Rightarrow \quad 3\theta = n\pi, n \in Z, \quad \Rightarrow \quad \theta = \frac{n\pi}{3}, n \in Z$$ **Ans.**

Example 10. *Solve:* $\tan^3 x - 3\tan x = 0$

Solution. We have, $\tan^3 x - 3\tan x = 0 \quad \Rightarrow \tan x(\tan^2 x - 3) = 0$

$\Rightarrow \quad \tan x = 0$ or $\tan^2 x - 3 = 0 \quad \Rightarrow \tan x = 0$ or $\tan x = \pm\sqrt{3}$

Now, $\quad \tan x = 0$

i.e., $\quad \tan x = \tan 0 \quad \Rightarrow x = n\pi, n \in I$ ...(1)

And $\quad \tan x = \sqrt{3} \Rightarrow \tan x = \tan\frac{\pi}{3} \Rightarrow x = \frac{\pi}{3} \Rightarrow x = n\pi + \frac{\pi}{3}, n \in I$ ...(2)

Also, $\quad \tan x = -\sqrt{3} \Rightarrow \tan x = \tan\left(\frac{-\pi}{3}\right) \Rightarrow x = -\frac{\pi}{3} \Rightarrow x = n\pi - \frac{\pi}{3}$ ...(3)

From (1), (2) and (3), we get the required general solution as

$$x = n\pi \quad \text{or} \quad x = n\pi \pm \frac{\pi}{3}$$ **Ans.**

Example 11. *Solve:* $4\sin x\cos x + 2\sin x + 2\cos x + 1 = 0$

Solution. We have,

$4\sin x\cos x + 2\sin x + 2\cos x + 1 = 0$

$\Rightarrow 2\sin x(2\cos x + 1) + 1(2\cos x + 1) = 0$

$\Rightarrow \quad (2\sin x + 1)(2\cos x + 1) = 0$

$$\Rightarrow \quad 2\sin x + 1 = 0 \quad \text{or} \quad 2\cos x + 1 = 0$$

$$\Rightarrow \quad \sin x = -\frac{1}{2} \quad \text{or} \quad \cos x = -\frac{1}{2}$$

Now, $\sin x = -\dfrac{1}{2} \Rightarrow \sin x = \sin\left(-\dfrac{\pi}{6}\right) \Rightarrow x = -\dfrac{\pi}{6}$

The general solution of this is

$$x = n\pi + (-1)^n\left(-\frac{\pi}{6}\right) = n\pi + (-1)^{n+1}\left(\frac{\pi}{6}\right) \Rightarrow x = \pi\left[n + \frac{(-1)^{n+1}}{6}\right] \quad \ldots(1)$$

And $\cos x = -\dfrac{1}{2} \Rightarrow \cos x = \cos\left(\pi - \dfrac{\pi}{3}\right) = \cos\dfrac{2\pi}{3} \Rightarrow x = \dfrac{2\pi}{3}$

The general solution of this is

$$x = 2n\pi \pm \frac{2\pi}{3} \quad \text{i.e.,} \quad x = 2\pi\left(n \pm \frac{1}{3}\right) \quad \ldots(2)$$

From (1) and (2), we have

$\pi\left[n + \dfrac{(-1)^{n+1}}{6}\right]$ and $2\pi\left(n \pm \dfrac{1}{3}\right)$ are the required solutions. **Ans.**

EXERCISE 15.3

Find the general solution of the following:

1. $\sin\theta = \dfrac{1}{2}$ **Ans.** $\theta = n\pi + (-1)^n\dfrac{\pi}{6}, n \in Z$

2. $\sin\theta = -\dfrac{\sqrt{3}}{2}$ **Ans.** $\theta = n\pi - (-1)^n\dfrac{\pi}{3}, n \in Z$

3. $\cos 2\theta = \dfrac{1}{2}$ **Ans.** $\theta = n\pi \pm \dfrac{\pi}{6}, n \in Z$

4. $\tan 2\theta = 1$ **Ans.** $\theta = \dfrac{n\pi}{2} + \dfrac{\pi}{8}, n \in Z$

5. $\tan 4\theta = -\sqrt{3}$ **Ans.** $\theta = \dfrac{n\pi}{4} - \dfrac{\pi}{12}, n \in Z$

6. $\sec 7\theta = \dfrac{2}{\sqrt{3}}$ **Ans.** $\theta = \dfrac{2n\pi}{7} \pm \dfrac{\pi}{42}, n \in Z$

7. $\operatorname{cosec} 4\theta = \dfrac{2}{\sqrt{3}}$ **Ans.** $\theta = \dfrac{n\pi}{4} + (-1)^n\dfrac{\pi}{12}, n \in Z$

8. $\sin 9\theta = \sin\theta$ **Ans.** $\theta = \dfrac{r\pi}{4}$ or $\theta = (2r+1)\dfrac{\pi}{10}$ where $r \in Z$

9. $\sin 2\theta = \cos 3\theta$ **Ans.** $\theta = (4n+1)\dfrac{\pi}{10}$ or $\theta = (4n-1)\dfrac{\pi}{2}$, where $n \in Z$

10. $\tan\theta = -\cot 2\theta$ **Ans.** $\theta = n\pi - \dfrac{\pi}{2}, n \in Z$

11. $\tan 3\theta = \cot \theta$ **Ans.** $\theta = \dfrac{n\pi}{4} + \dfrac{\pi}{8}, n \in Z$

12. $\tan 2\theta = \cot \theta$ **Ans.** $\theta = (2n+1)\dfrac{\pi}{6}, n \in Z$

13. $\tan m\theta = -\cot n\theta$ **Ans.** $\theta = \left(\dfrac{2r+1}{m+1}\right)\pi, r \in Z$

HINTS TO THE SELECTED QUESTIONS

6. $\sec 7\theta = \dfrac{2}{\sqrt{3}} \Rightarrow \cos 7\theta = \dfrac{\sqrt{3}}{2} = \cos\dfrac{\pi}{6}$

The general solution is $7\theta = 2n\pi \pm \left(\dfrac{\pi}{6}\right) \Rightarrow \theta = \dfrac{2n\pi}{7} \pm \dfrac{\pi}{42}, n \in Z.$

7. $\operatorname{cosec} 4\theta = \dfrac{2}{\sqrt{3}} \Rightarrow \sin 4\theta = \dfrac{\sqrt{3}}{2} = \sin\dfrac{\pi}{3}$

The general solution is $4\theta = n\pi + (-1)^n\left(\dfrac{\pi}{3}\right) \Rightarrow \theta = \dfrac{n\pi}{4} + (-1)^n \dfrac{\pi}{12}$

8. $\sin 9\theta = \sin\theta \Rightarrow \sin 9\theta - \sin\theta = 0$

$\Rightarrow 2\cos\left(\dfrac{9\theta+\theta}{2}\right)\sin\left(\dfrac{9\theta-\theta}{2}\right) = 0 \Rightarrow \cos 5\theta \sin 4\theta = 0$

$\Rightarrow$ Either, $\cos 5\theta = 0$ or $\sin 4\theta = 0$ $\Rightarrow$ The general solution is

Either, $5\theta = (2n+1)\dfrac{\pi}{2} \Rightarrow \theta = (2n+1)\dfrac{\pi}{10}, n \in Z$ or $4\theta = n\pi \Rightarrow \theta = \dfrac{n\pi}{4}, n \in Z$

9. $\sin 2\theta = \cos 3\theta \Rightarrow \cos\left(\dfrac{\pi}{2} - 2\theta\right) = \cos 3\theta \Rightarrow 3\theta = \dfrac{\pi}{2} - 2\theta$

Either (*i*)

$\Rightarrow 3\theta = 2n\pi \pm \left(\dfrac{\pi}{2} - 2\theta\right) \Rightarrow 3\theta + 2\theta = 2n\pi + \dfrac{\pi}{2} \Rightarrow 5\theta = \left(2n\pi + \dfrac{\pi}{2}\right)$

$\Rightarrow 5\theta = (4n+1)\dfrac{\pi}{2} \Rightarrow \theta = (4n+1)\dfrac{\pi}{10}$

or (*ii*) $3\theta = 2n\pi - \dfrac{\pi}{2} + 2\theta \Rightarrow 3\theta - 2\theta = 2n\pi - \dfrac{\pi}{2} \Rightarrow \theta = (4n-1)\dfrac{\pi}{2}$

10. $\tan\theta = -\cot 2\theta \Rightarrow \dfrac{\sin\theta}{\cos\theta} = -\dfrac{\cos 2\theta}{\sin 2\theta} \Rightarrow \cos 2\theta \cos\theta = -\sin 2\theta \sin\theta$

$\Rightarrow \cos 2\theta\cos\theta + \sin 2\theta \sin\theta = 0 \Rightarrow \cos(2\theta - \theta) = 0 \Rightarrow \cos\theta = 0 = \cos\dfrac{\pi}{2}$

$\Rightarrow \theta = \dfrac{\pi}{2} \Rightarrow \theta = (2n+1)\dfrac{\pi}{2}$

13. $\tan m\theta = -\cot n\theta \Rightarrow \dfrac{\sin m\theta}{\cos m\theta} = -\dfrac{\cos n\theta}{\sin n\theta}$

$\Rightarrow \cos m\theta \cos n\theta = -\sin m\theta \sin n\theta \quad \Rightarrow \quad \cos m\theta \cos n\theta + \sin m\theta \sin n\theta = 0$

$\Rightarrow \cos(m\theta - n\theta) = 0 = \cos\dfrac{\pi}{2} \quad \Rightarrow \quad m\theta - n\theta = \dfrac{\pi}{2} \quad \Rightarrow \quad (m-n)\theta = (2k+1)\dfrac{\pi}{2}$

$\Rightarrow \quad \theta = \dfrac{(2n+1)}{m-n}\dfrac{\pi}{2}$

Inverse Trigonometric Functions

16.1 INTRODUCTION

In this chapter, we shall use these concepts to define the inverse of all trigonometric functions and to study their properties.

16.2 INVERSE TRIGONOMETRIC FUNCTION

If $\sin\theta = x$, then $\theta = \sin^{-1}x$ (read as sine inverse x). Thus, we see that $\sin^{-1}x$ is a symbol which denotes an angle or a number, the value of whose sine is x. Similarly, $\cos^{-1}x$ denotes an angle whose cosine is x and so on. Thus,

$$\sin\frac{\pi}{6} = \frac{1}{2} \Rightarrow \frac{\pi}{6} = \sin^{-1}\left(\frac{1}{2}\right)$$

The expression $\sin^{-1}x$, $\cos^{-1}x$, $\tan^{-1}x$ are called *inverse trigonometric functions.*

Note. 1. The symbol arc sin x is also, used for $\sin^{-1}x$.

2. $\sin^{-1}x$ and $(\sin x)^{-1}$ have different meanings.

$$(\sin x)^{-1} = \frac{1}{\sin x}$$

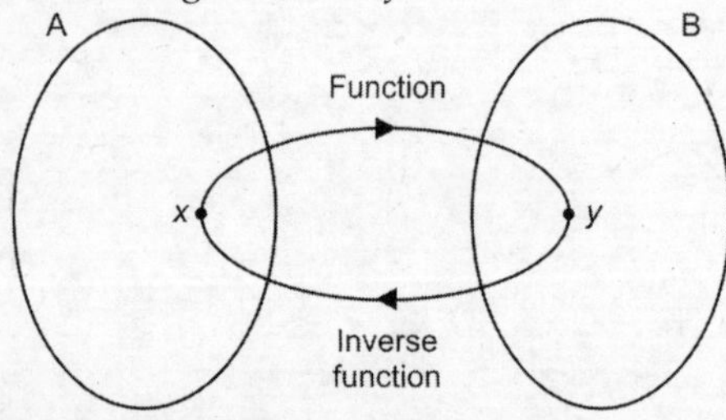

16.3 PRINCIPAL VALUES OF INVERSE TRIGONOMETRIC FUNCTIONS

Conditions for the principal value (angle) of inverse trigonometric functions:

1. The angle should be smallest.

2. If the angle lies in I or II quadrant, then the angle will be positive.

3. If the angle lies in III or IV quadrant, then the angle will be negative.

4. If the angle lies in II or III quadrant, then choose the angle of II quadrant.

For example

1. We know that

$$\sin 30^\circ = \frac{1}{2} \quad \Rightarrow \quad \sin^{-1}\left(\frac{1}{2}\right) = 30^\circ \qquad \left[\because \sin^{-1}\left(\frac{1}{2}\right) \text{ lies in the I quadrant}\right]$$

Also, $\sin 150^\circ = \sin(180^\circ - 30^\circ) = \sin 30^\circ = \frac{1}{2}$

$$\Rightarrow \quad \sin 150^\circ = \frac{1}{2} \quad \Rightarrow \quad \sin^{-1}\left(\frac{1}{2}\right) = 150^\circ$$

Again $\sin 390° = \sin(360° + 30°) = \sin 30° = \dfrac{1}{2}$

$\Rightarrow \quad \sin 390° = \dfrac{1}{2} \quad \Rightarrow \quad \sin^{-1}\left(\dfrac{1}{2}\right) = 390°$

It means $\sin^{-1}\left(\dfrac{1}{2}\right) = 30°, 150°, 390°, \ldots$, etc.

2. We know that

$\tan 120° = -\sqrt{3} \Rightarrow \tan^{-1}(-\sqrt{3}) = 120°$ ($\tan^{-1}(-\sqrt{3})$ lies in the II quadrant)

$\tan 240° = \tan(180° + 60°) = -\tan 60° = -\sqrt{3}$

$\tan 240° = -\sqrt{3} \Rightarrow \tan^{-1}(-\sqrt{3}) = 240°$

$\tan 300° = \tan(360° - 60°) = -\tan 60° = -\sqrt{3}$

thus, $\tan 300° = -\sqrt{3} \Rightarrow \tan^{-1}(-\sqrt{3}) = 300°$

So, $\tan^{-1}(-\sqrt{3}) = 120°, 240°, 300°, \ldots$, etc.

The smallest value of $\tan^{-1}(-\sqrt{3}) = 120°$

Thus principal value of $\tan^{-1}(-\sqrt{3})$, $= 120°$

3. Let us take the angle which lies in the III quadrant

$\cos 210° = \cos(360° - 150°) = \cos 150° = -\dfrac{\sqrt{3}}{2}$

$\cos 210° = -\dfrac{\sqrt{3}}{2} \quad \Rightarrow \quad \cos^{-1}\left(-\dfrac{\sqrt{3}}{2}\right) = 210°$

$\cos 150° = -\dfrac{\sqrt{3}}{2}$

$\cos(-150°) = -\dfrac{\sqrt{3}}{2} \Rightarrow \cos^{-1}\left(-\dfrac{\sqrt{3}}{2}\right) = -150°$

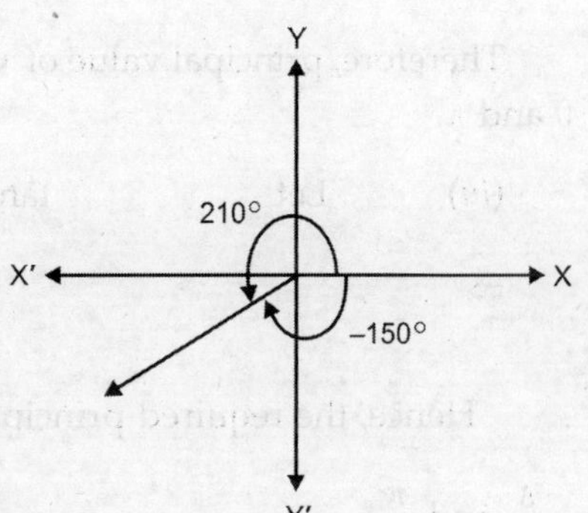

Thus, $\cos^{-1}\left(-\dfrac{\sqrt{3}}{2}\right) = 210°$ or $-150°$

As the angle lies in the III quadrant, the angle will be negative.

Thus, $\cos^{-1}\left(-\dfrac{\sqrt{3}}{2}\right) = -150°$

4. Now we take the angle in the IV quadrant.

$\text{cosec } 300° = \text{cosec}(360° - 60°) = -\text{cosec } 60° = \dfrac{-2}{\sqrt{3}}$

We know that, 300° means – 60°

Thus, $\text{cosec}^{-1}\left(-\dfrac{2}{\sqrt{3}}\right) = -60°$

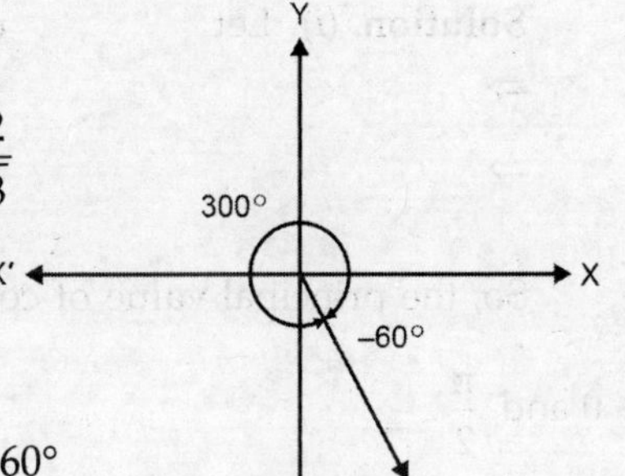

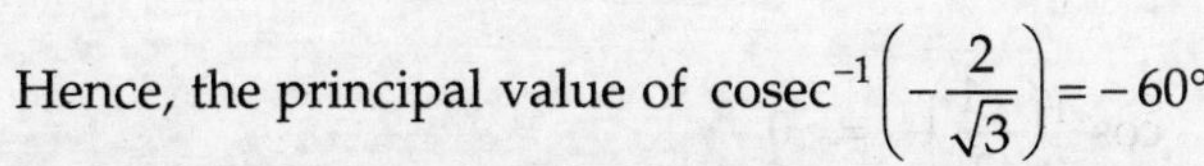

Hence, the principal value of $\text{cosec}^{-1}\left(-\dfrac{2}{\sqrt{3}}\right) = -60°$

Example 1. *Find the principal values of the following:*

(*i*) $\sin^{-1}\left(-\frac{1}{2}\right)$ (*ii*) $\cos^{-1}\left(\frac{\sqrt{3}}{2}\right)$ (*iii*) $\tan^{-1}(-\sqrt{3})$

Solution. (*i*) Let $\sin^{-1}\left(-\frac{1}{2}\right) = \theta$

$\Rightarrow \quad \sin\theta = -\frac{1}{2}$

$\Rightarrow \quad \sin\theta = -\sin\frac{\pi}{6} = \sin\left(-\frac{\pi}{6}\right)$

Therefore, principal value of $\sin^{-1}\left(-\frac{1}{2}\right)$ is $-\frac{\pi}{6}$ as principal value of $\sin^{-1}x$ lies between $-\frac{\pi}{2}$ and $\frac{\pi}{2}$. **Ans.**

(*ii*) Let $\cos^{-1}\left(\frac{\sqrt{3}}{2}\right) = \theta$

$\Rightarrow \quad \cos\theta = \frac{\sqrt{3}}{2} = \cos\frac{\pi}{6}$

Therefore, principal value of $\cos^{-1}\left(\frac{\sqrt{3}}{2}\right)$ is $\frac{\pi}{6}$ as principal value of $\cos^{-1}x$ lies between 0 and π. **Ans.**

(*iii*) Let $\tan^{-1}(-\sqrt{3}) = \theta$

$\Rightarrow \quad \tan\theta = -\sqrt{3} = \tan\left(\frac{-\pi}{3}\right)$

Hence, the required principal value is $\frac{-\pi}{3}$ as principal value of $\tan^{-1}x$ lies between $-\frac{\pi}{2}$ and $\frac{\pi}{2}$. **Ans.**

Example 2. *Find the principal values of the following:*

(*i*) $\operatorname{cosec}^{-1}(2)$ (*ii*) $\cos^{-1}\left(-\frac{1}{2}\right)$ (*iii*) $\tan^{-1}(-1)$

Solution. (*i*) Let $\operatorname{cosec}^{-1}(2) = \theta$

$\Rightarrow \quad \operatorname{cosec}\theta = 2$

$\Rightarrow \quad \operatorname{cosec}\theta = \operatorname{cosec}\left(\frac{\pi}{6}\right)$

So, the principal value of $\operatorname{cosec}^{-1}(2)$ is $\frac{\pi}{6}$ as principal value of $\operatorname{cosec}^{-1}x$ lies between 0 and $\frac{\pi}{2}$.

(*ii*) Let $\cos^{-1}\left(\frac{-1}{2}\right) = \theta$

$$\Rightarrow \qquad \cos\theta = -\frac{1}{2} = -\cos\left(\frac{\pi}{3}\right) = \cos\left(\pi - \frac{\pi}{3}\right)$$

$$\Rightarrow \qquad \cos\theta = \cos\frac{2\pi}{3}$$

So, the required principal value is $\frac{2\pi}{3}$ because the principal value of $\cos^{-1}\left(\frac{-1}{2}\right)$ lies between $\frac{\pi}{2}$ and π.

(*iii*) Let $\tan^{-1}(-1) = \theta$

$$\Rightarrow \qquad \tan\theta = -1 = \tan\left(-\frac{\pi}{4}\right)$$

Hence, the principal value of $\tan^{-1}(-1)$ is $-\frac{\pi}{4}$ because principal value of $\tan^{-1}x$ lies between $-\frac{\pi}{2}$ and $\frac{\pi}{2}$.

EXERCISE 16.1

Find the principal value of each of the following:

1. $\cos^{-1}\left(\frac{1}{2}\right)$ **Ans.** $\frac{\pi}{3}$

2. $\tan^{-1}(1)$ **Ans.** $\frac{\pi}{4}$

3. $\sin^{-1}(1)$ **Ans.** $\frac{\pi}{2}$

4. $\cot^{-1}(0)$ **Ans.** $\frac{\pi}{2}$

5. $\cot^{-1}(\sqrt{3})$ **Ans.** $\frac{\pi}{6}$

6. $\sin^{-1}\left(\frac{1}{\sqrt{2}}\right)$ **Ans.** $\frac{\pi}{4}$

7. $\tan^{-1}(0)$ **Ans.** 0

8. $\cot^{-1}\left(\frac{1}{\sqrt{3}}\right)$ **Ans.** $\frac{\pi}{3}$

9. $\sec^{-1}\left(\frac{2}{\sqrt{3}}\right)$ **Ans.** $\frac{\pi}{6}$

10. $\cos^{-1}\left(-\frac{1}{\sqrt{2}}\right)$ **Ans.** $\frac{3\pi}{4}$

11. $\sec^{-1}(-\sqrt{2})$ **Ans.** $\frac{3\pi}{4}$

12. $\sin^{-1}\left(-\frac{\sqrt{3}}{2}\right)$ **Ans.** $-\frac{\pi}{3}$

13. $\cot^{-1}(-\sqrt{3})$ **Ans.** $\frac{5\pi}{6}$

14. $\operatorname{cosec}^{-1}\left(\frac{-2}{\sqrt{3}}\right)$ **Ans.** $-\frac{\pi}{3}$

15. $\sec^{-1}\left(-\frac{2}{\sqrt{3}}\right)$ **Ans.** $\frac{5\pi}{6}$

16. $\operatorname{cosec}^{-1}(-\sqrt{2})$ **Ans.** $-\frac{\pi}{4}$

16.4 PROPERTIES OF INVERSE TRIGONOMETRIC FUNCTIONS

The following results are valid within the principal value branches of the corresponding inverse trigonometric functions and wherever they are defined.

Theorem 1

(*i*) $\sin^{-1}(\sin\theta) = \theta, \quad \theta \in \left[-\frac{\pi}{2}, \frac{\pi}{2}\right]$ (*ii*) $\cos^{-1}(\cos\theta) = \theta, \quad \theta \in [0, \pi]$

(*iii*) $\tan^{-1}(\tan\theta) = \theta \quad \theta \in \left(-\frac{\pi}{2}, \frac{\pi}{2}\right)$ (*iv*) $\cot^{-1}(\cot\theta) = \theta, \quad \theta \in (0, \pi)$

(*v*) $\sec^{-1}(\sec\theta) = \theta \quad \theta \in [0, \pi] - \left\{\frac{\pi}{2}\right\}$ (*vi*) $\text{cosec}^{-1}(\text{cosec}\,\theta) = \theta, \theta \in \left(-\frac{\pi}{2}, \frac{\pi}{2}\right) - \{0\}$

Theorem 2

(*i*) $\sin(\sin^{-1} x) = x, \quad x \in [-1, 1]$ (*ii*) $\cos(\cos^{-1} x) = x, \quad x \in [-1, 1]$

(*iii*) $\tan(\tan^{-1} x) = x, \quad x \in R$ (*iv*) $\cot(\cot^{-1} x) = x, \quad x \in R$

(*v*) $\sec(\sec^{-1} x) = x, \quad x \in R - [-1, 1]$ (*vi*) $\text{cosec}(\text{cosec}^{-1} x) = x, \quad x \in R - (-1, 1)$

Theorem 3

(*i*) $\sin^{-1}(-x) = -\sin^{-1} x, \quad x \in [-1, 1]$ (*ii*) $\cos^{-1}(-x) = \pi - \cos^{-1} x, \quad x \in [-1, 1]$

(*iii*) $\tan^{-1}(-x) = -\tan^{-1} x, \quad x \in R$ (*iv*) $\cot^{-1}(-x) = \pi - \cot^{-1} x, \quad x \in R$

(*v*) $\sec^{-1}(-x) = \pi - \sec^{-1} x, \quad x \in R - [-1, 1]$

(*vi*) $\text{cosec}^{-1}(-x) = -\text{cosec}^{-1} x, \quad x \in R - (-1, 1)$

Theorem 4

(*i*) $\text{cosec}^{-1} x = \sin^{-1}\left(\frac{1}{x}\right), \quad x \in R - (-1, 1)$

(*ii*) $\sec^{-1} x = \cos^{-1}\left(\frac{1}{x}\right), \quad x \in R - (-1, 1)$

(*iii*) $\cot^{-1} x = \begin{cases} \pi + \tan^{-1}\left(\frac{1}{x}\right), & x < 0 \\ \tan^{-1}\left(\frac{1}{x}\right), & x > 0 \end{cases}$

Theorem 5

(*i*) $\sin^{-1} x + \cos^{-1} x = \frac{\pi}{2}, \quad x \in [-1, 1]$

(*ii*) $\tan^{-1} x + \cot^{-1} x = \frac{\pi}{2}, \quad x \in R$

(*iii*) $\sec^{-1} x + \text{cosec}^{-1} x = \frac{\pi}{2}, \quad x \in R - (-1, 1)$

Proof: (*i*) Let $\sin^{-1} x = \theta$

Then, $\theta = \left[-\frac{\pi}{2}, \frac{\pi}{2}\right]$ ...(1)

$\Rightarrow \quad -\frac{\pi}{2} \le \theta \le \frac{\pi}{2} \Rightarrow -\frac{\pi}{2} \le -\theta \le \frac{\pi}{2}$ $(\because x \in [-1, 1])$

$\Rightarrow \quad 0 \le \frac{\pi}{2} - \theta \le \pi \Rightarrow \frac{\pi}{2} - \theta \in [0, \pi]$

Now, $\sin^{-1} x = \theta$

$\Rightarrow \quad x = \sin\theta \Rightarrow x = \cos\left(\frac{\pi}{2} - \theta\right)$

$\Rightarrow \quad \cos^{-1} x = \frac{\pi}{2}\theta \quad \left[\because x \in [-1, 1] \text{ and } \left(\frac{\pi}{2} - \theta\right) \in [0, \pi]\right]$

$$\Rightarrow \qquad \theta + \cos^{-1} x = \frac{\pi}{2} \qquad ...(2)$$

From (1) and (2), we get

$$\boxed{\sin^{-1} x + \cos^{-1} x = \frac{\pi}{2}}$$

(*ii*) Let $\tan^{-1} x = \theta$ ...(1)

Then, $\theta \in \left[-\frac{\pi}{2}, \frac{\pi}{2}\right]$ $[\because x \in R]$

$$\Rightarrow \quad -\frac{\pi}{2} < \theta < \frac{\pi}{2} \quad \Rightarrow \quad -\frac{\pi}{2} < -\theta < \frac{\pi}{2} \quad \Rightarrow \quad 0 < \frac{\pi}{2} - \theta < \pi$$

$$\Rightarrow \quad \left(\frac{\pi}{2} - \theta\right) \in (0, \pi)$$

Now, $\tan^{-1} x = \theta \;\Rightarrow\; x = \tan\theta \;\Rightarrow\; x = \cot\left(\frac{\pi}{2} - \theta\right)$

$$\Rightarrow \quad \cot^{-1} x = \frac{\pi}{2} - \theta \qquad \left[\because \frac{\pi}{2} - \theta \in (0, \pi)\right]$$

$$\Rightarrow \quad \theta + \cot^{-1} x = \frac{\pi}{2} \qquad ...(2)$$

From (1) and (2), we get

$$\boxed{\tan^{-1} x + \cot^{-1} x = \frac{\pi}{2}}$$

(*iii*) Let $\sec^{-1} x = \theta$ ...(1)

Then, $\theta \in [0, \pi] - \left\{\frac{\pi}{2}\right\}$ $[\because x \in R - (-1, 1)]$

$$\Rightarrow \quad 0 \le \theta\, \pi, \theta \ne \frac{\pi}{2} \quad \Rightarrow \quad -\pi \le -\theta \le 0, \theta \ne \frac{\pi}{2}$$

$$\Rightarrow \quad -\frac{\pi}{2} \le \frac{\pi}{2} - \theta \le \frac{\pi}{2}, \qquad \frac{\pi}{2} - \theta \ne 0$$

$$\Rightarrow \quad \left(\frac{\pi}{2} - \theta\right) \in \left[-\frac{\pi}{2}, \frac{\pi}{2}\right], \qquad \frac{\pi}{2} - \theta \ne 0$$

Now, $\sec^{-1} x = \theta$

$$\Rightarrow \quad x = \sec\theta$$

$$\Rightarrow \quad x = \operatorname{cosec}\left(\frac{\pi}{2} - \theta\right)$$

$$\Rightarrow \quad \operatorname{cosec}^{-1} x = \frac{\pi}{2} - \theta \qquad \left[\because \left(\frac{\pi}{2} - \theta\right) \in \left[-\frac{\pi}{2}, \frac{\pi}{2}\right], \frac{\pi}{2} - \theta \ne 0\right]$$

$$\Rightarrow \quad \theta + \operatorname{cosec}^{-1} x = \frac{\pi}{2} \qquad ...(2)$$

From (1) and (2), we get

$$\boxed{\sec^{-1} x + \operatorname{cosec}^{-1} x = \frac{\pi}{2}}$$

Theorem 6. Prove that:

$$\tan^{-1} x + \tan^{-1} y = \begin{cases} \tan^{-1}\left(\dfrac{x+y}{1-xy}\right), & \text{if } x > 0, y > 0 \text{ and } xy < 1 \\ \pi + \tan^{-1}\left(\dfrac{x+y}{1-xy}\right), & \text{if } x > 0, y > 0 \text{ and } xy > 1 \\ \tan^{-1}\left(\dfrac{x+y}{1-xy}\right) - \pi, & \text{if } x < 0, y < 0 \text{ and } xy > 1 \end{cases}$$

Proof. Let $\tan^{-1} x = A$ and $\tan^{-1} y = B$

$\Rightarrow \quad x = \tan A, \quad y = \tan B$ and $\quad A, B \in \left(-\dfrac{\pi}{2}, \dfrac{\pi}{2}\right)$

$$\therefore \quad \tan(A+B) = \frac{\tan A + \tan B}{1 - \tan A \tan B} = \frac{x+y}{1-xy} \qquad \text{...(1)}$$

Case I. $x > 0, y > 0, xy < 1$

$\Rightarrow \; x + y > 0, 1 - xy > 0 \;\Rightarrow\; \dfrac{x+y}{1-xy} > 0 \;\Rightarrow\; \tan(A+B) > 0$ [using (1)]

$\therefore \quad A + B$ lies in either I quadrant or III quadrant.

$$x > 0 \;\Rightarrow\; 0 < A < \frac{\pi}{2}, y > 0 \;\Rightarrow\; 0 < B < \frac{\pi}{2} \;\Rightarrow\; 0 < A + B < \pi$$

$$\Rightarrow \quad 0 < A + B < \frac{\pi}{2}$$

From (1), we have $A + B = \tan^{-1}\left(\dfrac{x+y}{1-xy}\right)$

$$\Rightarrow \quad \boxed{\tan^{-1} x + \tan^{-1} y = \tan^{-1}\left(\frac{x+y}{1-xy}\right)}$$

Case II. $x > 0, y > 0, xy > 1$

$\Rightarrow \; x + y > 0, 1 - xy < 0 \;\Rightarrow\; \dfrac{x+y}{1-xy} < 0 \;\Rightarrow\; \tan(A+B) < 0$

$\therefore \quad A + B$ lies in either II quadrant or IV quadrant.

$$x > 0 \;\Rightarrow\; 0 < A < \frac{\pi}{2},\; y > 0 \;\Rightarrow\; 0 < B < \frac{\pi}{2} \;\Rightarrow\; 0 < A + B < \pi$$

$$\Rightarrow \quad \frac{\pi}{2} < A + B < \pi \;\Rightarrow\; \frac{\pi}{2} - \pi < (A+B) - \pi < \pi - \pi$$

$$\Rightarrow \quad -\frac{\pi}{2} < A + B < -\pi < 0$$

$\therefore \quad \tan[(A+B) - \pi] = -\tan[\pi - (A+B)] = \tan(A+B) = \dfrac{x+y}{1-xy}$ [using (1)]

$$\Rightarrow \quad (A+B) - \pi = \tan^{-1}\left(\frac{x+y}{1-xy}\right) \;\Rightarrow\; (A+B) = \pi + \tan^{-1}\left(\frac{x+y}{1-xy}\right)$$

$$\Rightarrow \quad \boxed{\tan^{-1} x + \tan^{-1} y = \pi + \tan^{-1}\left(\frac{x+y}{1-xy}\right)}$$

Case III. $x < 0, y < 0, xy > 1$

$$\Rightarrow \quad x + y < 0, 1 - xy < 0 \quad \Rightarrow \quad \frac{x+y}{1-xy} > 0$$

$\Rightarrow$ $\tan(A + B) > 0$

$\therefore$ $A + B$ lies in either I quadrant or III quadrant.

$$x < 0 \Rightarrow -\frac{\pi}{2} < A < 0, \; y < 0 \Rightarrow -\frac{\pi}{2} < B < 0$$

$$\Rightarrow \quad -\pi < A + B < 0 \quad \Rightarrow \quad -\pi < A + B < -\frac{\pi}{2}$$

$$\Rightarrow \quad -\pi + \pi < (A + B) + \pi < -\frac{\pi}{2} + \pi$$

$$\Rightarrow \quad 0 < (A + B) + \pi < \frac{\pi}{2}$$

$$\Rightarrow \quad \tan[(A + B) + \pi] = \tan(A + B) = \frac{x+y}{1-xy} \qquad \text{[using (1)]}$$

$$\Rightarrow \quad (A + B) + \pi = \tan^{-1}\left(\frac{x+y}{1-xy}\right)$$

$$\Rightarrow \quad A + B = \tan^{-1}\left(\frac{x+y}{1-xy}\right) - \pi$$

$$\Rightarrow \quad \boxed{\tan^{-1} x + \tan^{-1} y = \tan^{-1}\left(\frac{x+y}{1-xy}\right) - \pi}$$

Theorem 7. Prove that:

$$2\tan^{-1} x = \tan^{-1}\left(\frac{2x}{1-x^2}\right)$$

Proof. Let $\tan^{-1} x = \theta$. Then, $x = \tan\theta$

Now, $$\tan 2\theta = \frac{2\tan\theta}{1-\tan^2\theta} \Rightarrow \tan 2\theta = \frac{2x}{1-x^2}$$

$$\Rightarrow \quad 2\theta = \tan^{-1}\left(\frac{2x}{1-x^2}\right)$$

$$\Rightarrow \quad \boxed{2\tan^{-1} x = \tan^{-1}\left(\frac{2x}{1-x^2}\right)}$$

Example 3. *Evaluate the following:*

(i) $\sin^{-1}\left(\sin\frac{2\pi}{3}\right)$ *(ii)* $\cos^{-1}\left(\cos\frac{7\pi}{6}\right)$ *(iii)* $\tan^{-1}\left(\tan\frac{3\pi}{4}\right)$

Solution. We know that, $\sin^{-1}(\sin\theta) = \theta$, if $-\frac{\pi}{2} < \theta < \frac{\pi}{2}$ $\cos^{-1}(\cos\theta) = \theta$, if $0 \le \theta \le \pi$

and $\tan^{-1}(\tan\theta) = \theta$, if $-\frac{\pi}{2} < \theta < \frac{\pi}{2}$. Therefore,

(*i*) $\sin^{-1}\left(\sin\frac{2\pi}{3}\right) \ne \frac{2\pi}{3}$, because $\frac{2\pi}{3}$ does not lie between $-\frac{\pi}{2}$ and $\frac{\pi}{2}$.

Now, $\sin^{-1}\left(\sin\frac{2\pi}{3}\right) = \sin^{-1}\left[\sin\left(\pi - \frac{\pi}{3}\right)\right] = \sin^{-1}\left[\sin\left(\frac{\pi}{3}\right)\right] = \frac{\pi}{3}$ $\quad \frac{\pi}{3} \in \left[-\frac{\pi}{2}, \frac{\pi}{2}\right]$

(*ii*) $\cos^{-1}\left(\cos\frac{7\pi}{6}\right) \ne \frac{7\pi}{6}$, because $\frac{7\pi}{6}$ does not lie between 0 and π

Now, $\cos^{-1}\left(\cos\frac{7\pi}{6}\right) = \cos^{-1}\left[\cos\left(2\pi - \frac{5\pi}{6}\right)\right]$ $\quad \left[\because \frac{7\pi}{6} = 2\pi - \frac{5\pi}{6}\right]$

$= \cos^{-1}\left(\cos\frac{5\pi}{6}\right)$ $\quad [\because \cos(2\pi - \theta) = \cos\theta]$

$= \frac{5\pi}{6}, \quad \frac{5\pi}{6} \in [0, \pi]$

(*iii*) $\tan^{-1}\left(\tan\frac{3\pi}{4}\right) \ne \frac{3\pi}{4}$, because $\frac{3\pi}{4}$ does not lie between $-\frac{\pi}{2}$ and $\frac{\pi}{2}$

Now, $\tan^{-1}\left(\tan\frac{3\pi}{4}\right) = \tan^{-1}\tan\left(\pi - \frac{\pi}{4}\right)$ $\quad \left[\because \frac{3\pi}{4} = \pi - \frac{\pi}{4}\right]$

$= \tan^{-1}\left(-\tan\frac{\pi}{4}\right)$ $\quad [\tan(\pi - \theta) = -\tan\theta]$

$= \tan^{-1}\left(\tan\left(-\frac{\pi}{4}\right)\right)$

$= -\frac{\pi}{4}$ because $-\frac{\pi}{4} \in \left(-\frac{\pi}{2}, \frac{\pi}{2}\right)$ **Ans.**

Example 4. *Evaluate:*

(*i*) $\sin\left(\cos^{-1}\frac{3}{5}\right)$ (*ii*) $\cos\left(\tan^{-1}\frac{3}{4}\right)$

(*iii*) $\sin\left(\frac{1}{2}\cos^{-1}\frac{4}{5}\right)$ (*iv*) $\sin\left[\frac{\pi}{3} - \sin^{-1}\left(-\frac{1}{2}\right)\right]$

(*v*) $\sin(\cot^{-1} x)$ (*vi*) $\tan\frac{1}{2}\left(\cos^{-1}\frac{\sqrt{5}}{3}\right)$

Solution. (*i*) Let $\cos^{-1}\left(\frac{3}{5}\right) = \theta$, then $\cos\theta = \frac{3}{5}$ ...(1)

$\Rightarrow \quad \sin\left(\cos^{-1}\frac{3}{5}\right) = \sin\theta \left(\sqrt{1 - \cos^2\theta}\right) = \sqrt{1 - \frac{9}{25}} = \frac{4}{5}$ [using (1)] (*ii*) Let $\tan^{-1}\left(\frac{3}{4}\right) = \theta$. Then, $\tan\theta = \frac{3}{4}$ ...(2)

$$\therefore \quad \cos\left(\tan^{-1}\frac{3}{4}\right) = \cos\theta = \frac{1}{\sec\theta} = \frac{1}{\sqrt{1+\tan^2\theta}}$$

$$= \frac{1}{\sqrt{1+\left(\frac{3}{4}\right)^2}} = \frac{1}{\sqrt{\frac{16+9}{16}}} = \frac{4}{5} \quad \text{[using (2)]}$$

(*iii*) Let $\cos^{-1}\left(\frac{4}{5}\right) = \theta$. Then, $\cos\theta = \frac{4}{5}$ ...(3)

$$\therefore \quad \sin\left(\frac{1}{2}\cos^{-1}\frac{4}{5}\right) = \sin\frac{1}{2}\theta = \frac{\sqrt{1-\cos\theta}}{2} = \sqrt{\frac{1-\frac{4}{5}}{2}} \quad \text{[using (3)]}$$

$$= \frac{\sqrt{5-4}}{10} = \frac{1}{\sqrt{10}}$$

(*iv*) $\sin\left[\frac{\pi}{3} - \sin^{-1}\left(-\frac{1}{2}\right)\right] = \sin\left[\frac{\pi}{3} + \sin^{-1}\left(\frac{1}{2}\right)\right]$ $[\because \sin^{-1}(-\theta) = -\sin^{-1}\theta]$

$$\Rightarrow \quad \sin\left(\frac{\pi}{3}+\frac{\pi}{6}\right) = \sin\frac{\pi}{2} = 1$$

(*v*) Let $\cot^{-1} x = \theta$ then, $\cot\theta = x$

$$\therefore \quad \sin\left(\cot^{-1} x\right) = \sin\theta = \frac{1}{\operatorname{cosec}\theta} = \frac{1}{\sqrt{1+\cot^2\theta}} = \frac{1}{\sqrt{1+x^2}}$$

(*vi*) Let $\cos^{-1}\left(\frac{\sqrt{5}}{3}\right) = \theta$ then, $\cos\theta = \frac{\sqrt{5}}{3}$

$$\therefore \quad \tan\frac{1}{2}\left(\cos^{-1}\frac{\sqrt{5}}{3}\right) = \tan\frac{\theta}{2} = \frac{\sin\frac{\theta}{2}}{\cos\frac{\theta}{2}} = \sqrt{\frac{1-\cos\theta}{1+\cos\theta}}$$

$$= \sqrt{\frac{1-(\sqrt{5}/3)}{1+(\sqrt{5}/3)}} = \sqrt{\frac{(3-\sqrt{5})}{(3+\sqrt{5})}\times\frac{(3-\sqrt{5})}{(3-\sqrt{5})}} = \sqrt{\frac{(3-\sqrt{5})^2}{9-5}} = \frac{3-\sqrt{5}}{2}. \quad \textbf{Ans.}$$

Example 5. *Prove that:* $\tan^{-1}(1) + \tan^{-1}(2) + \tan^{-1}(3) = \pi$

Solution. Here, we have

$$\text{L.H.S.} = \tan^{-1}(1) + \tan^{-1}(2) + \tan^{-1}(3)$$

$$= \tan^{-1}\frac{(1)+(2)}{1-(1)(2)} + \tan^{-1}(3)$$

$$= \tan^{-1}\left(\frac{3}{1-2}\right) + \tan^{-1}(3) = \tan^{-1}(-3) + \tan(3) = \tan^{-1}\left[\frac{(-3)+(3)}{1-(-3)(3)}\right]$$

$$= \tan^{-1}\left(\frac{0}{10}\right) = \tan^{-1}(0) = \pi = \text{R.H.S.}$$ **Proved.**

Example 6. *Prove that:*

(*i*) $2\tan^{-1}\left(\frac{1}{2}\right) + \tan^{-1}\left(\frac{1}{7}\right) = \tan^{-1}\left(\frac{31}{17}\right)$

(*ii*) $\tan^{-1}\left(\frac{3}{4}\right) + \tan^{-1}\left(\frac{3}{5}\right) - \tan^{-1}\left(\frac{8}{19}\right) = \frac{\pi}{4}$

Solution. (*i*) L.H.S. $= 2\tan^{-1}\left(\frac{1}{2}\right) + \tan^{-1}\left(\frac{1}{7}\right) = \tan^{-1}\left[\frac{2\times\frac{1}{2}}{1-\left(\frac{1}{2}\right)^2}\right] + \tan^{-1}\left(\frac{1}{7}\right)$

$$= \tan^{-1}\left(\frac{1}{\frac{3}{4}}\right) + \tan^{-1}\left(\frac{1}{7}\right) = \tan^{-1}\left(\frac{4}{3}\right) + \tan^{-1}\left(\frac{1}{7}\right)$$

$$= \tan^{-1}\left[\frac{\frac{4}{3}+\frac{1}{7}}{1-\frac{4}{3}.\frac{1}{7}}\right] = \tan^{-1}\left[\frac{\frac{28+3}{21}}{\frac{21-4}{21}}\right] = \tan^{-1}\left(\frac{31}{17}\right) = \text{R.H.S.}$$ **Proved.**

(*ii*) L.H.S. $= \tan^{-1}\left(\frac{3}{4}\right) + \tan^{-1}\left(\frac{3}{5}\right) - \tan^{-1}\left(\frac{8}{19}\right)$

$$= \left(\tan^{-1}\frac{3}{4} + \tan^{-1}\frac{3}{5}\right) - \tan^{-1}\left(\frac{8}{19}\right)$$

$$= \tan^{-1}\left[\frac{\frac{3}{4}+\frac{3}{5}}{1-\frac{3}{4}\times\frac{3}{5}}\right] - \tan^{-1}\left(\frac{8}{19}\right) = \tan^{-1}\left(\frac{27}{11}\right) - \tan^{-1}\left(\frac{8}{19}\right)$$

$$= \tan^{-1}\left[\frac{\frac{27}{11}-\frac{8}{19}}{1+\frac{27}{11}\times\frac{8}{19}}\right] = \tan^{-1}\left(\frac{425}{425}\right) = \tan^{-1}(1) = \frac{\pi}{4} = \text{R.H.S.}$$ **Proved.**

Example 7. *Prove the following:*

$$\cos^{-1}\left(\frac{12}{13}\right) + \sin^{-1}\left(\frac{3}{5}\right) = \sin^{-1}\left(\frac{56}{65}\right)$$

Solution. L.H.S. $= \cos^{-1}\left(\frac{12}{13}\right) + \sin^{-1}\left(\frac{3}{5}\right)$

$$= \sin^{-1}\left(\sqrt{1-\left(\frac{12}{13}\right)^2}\right) + \sin^{-1}\left(\frac{3}{5}\right) \quad \left[\because \cos^{-1}x = \sin^{-1}\left(\sqrt{1-x^2}\right)\right]$$

$$= \sin^{-1}\left(\sqrt{\frac{169-144}{169}}\right) + \sin^{-1}\left(\frac{3}{5}\right)$$

$$= \sin^{-1}\left(\frac{5}{13}\right) + \sin^{-1}\left(\frac{3}{5}\right)$$

$$= \sin^{-1}\left[\left(\frac{5}{13}\right)\sqrt{1-\left(\frac{3}{5}\right)^2} + \left(\frac{3}{5}\right)\sqrt{1-\left(\frac{5}{13}\right)^2}\right]$$

$$\left[\because \sin^{-1} x + \sin^{-1} y = \sin^{-1}\left(x\sqrt{1-y^2} + y\sqrt{1-x^2}\right)\right]$$

$$= \sin^{-1}\left[\left(\frac{5}{13}\right)\left(\frac{4}{5}\right) + \left(\frac{3}{5}\right)\left(\frac{12}{13}\right)\right]$$

$$= \sin^{-1}\left[\frac{20+36}{65}\right] = \sin^{-1}\left(\frac{56}{65}\right) = \text{R.H.S.}$$ **Proved.**

Example 8. *Find the value of each of the following:*

(*i*) $\cot(\tan^{-1} a + \cot^{-1} a)$ (*ii*) $\sin(\sin^{-1} x + \cos^{-1} x)$

Solution. (*i*) We have

$$\cot(\tan^{-1} a + \cot^{-1} a) = \cot\left(\tan^{-1} a + \tan^{1}\frac{1}{a}\right)$$

$$= \cot\left[\tan^{-1}\left(\frac{a+\frac{1}{a}}{1-a\frac{1}{a}}\right)\right]$$

$$= \cot[\tan^{-1}\infty] = \cot\frac{\pi}{2} = 0$$

(*ii*) We have,

$$\sin(\sin^{-1} x + \cos^{-1} x) = \sin\frac{\pi}{2} \qquad \left[\because \sin^{-1} x + \cos^{-1} x = \frac{\pi}{2}\right]$$

$$= 1$$ **Ans.**

Example 9. *Find the value of* $\tan\left\{\frac{1}{2}\sin^{-1}\left(\frac{2x}{1+x^2}\right) + \frac{1}{2}\cos^{-1}\left(\frac{1-y^2}{1+y^2}\right)\right\}$.

Solution. Let $x = \tan A$ and $y = \tan B$. Then,

$$\tan\left\{\frac{1}{2}\sin^{-1}\left(\frac{2x}{1+x^2}\right) + \frac{1}{2}\cos^{-1}\left(\frac{1-y^2}{1+y^2}\right)\right\} = \tan\left[\frac{1}{2}\sin^{-1}\left(\frac{2\tan A}{1+\tan^2 A}\right) + \frac{1}{2}\cos^{-1}\left(\frac{1-\tan^2 B}{1+\tan^2 B}\right)\right]$$

$$= \tan\left[\frac{1}{2}\sin^{-1}(\sin 2A) + \frac{1}{2}\cos^{-1}(\cos 2B)\right]$$

$$= \tan\left(\frac{2A}{2} + \frac{2B}{2}\right) = \tan(A+B)$$

$$= \frac{\tan A + \tan B}{1 - \tan A \tan B} = \frac{x+y}{1-xy}$$ **Ans.**

Example 10. *Solve the following equations:*

(*i*) $2\tan^{-1}(\cos x) = \tan^{-1}(2\,\text{cosec}\,x)$ (*ii*) $\tan^{-1}\left(\dfrac{1-x}{1+x}\right) = \dfrac{1}{2}\tan^{-1}x, (x > 0)$

Solution. (*i*) We have,

$$2\tan^{-1}(\cos x) = \tan^{-1}(2\,\text{cosec}\,x)$$

$$\Rightarrow \quad \tan^{-1}\left(\frac{2\cos x}{1-\cos^2 x}\right) = \tan^{-1}(2\,\text{cosec}\,x)$$

$$\Rightarrow \quad \frac{2\cos x}{\sin^2 x} = 2\,\text{cosec}\,x$$

$$\left| \begin{aligned} 2\tan^{-1}\theta &= \tan^{-1}\left(\frac{2\theta}{1-\theta^2}\right) \\ &= \tan^{-1}\left(\frac{2\cos x}{1-\cos^2 x}\right) \\ &\textit{if } \theta = \cos x \end{aligned} \right.$$

$$\Rightarrow \quad \cos x = \sin x \quad \Rightarrow \quad \tan x = 1 \quad \Rightarrow \quad x = \frac{\pi}{4}$$ **Ans.**

(*ii*) We have,

$$\tan^{-1}\left(\frac{1-x}{1+x}\right) = \frac{1}{2}\tan^{-1}(x), (x > 0)$$

$$\Rightarrow \quad \tan^{-1}1 - \tan^{-1}x = \frac{1}{2}\tan^{-1}x \qquad \left[\because \tan^{-1}\left(\frac{x-y}{1+xy}\right) = \tan^{-1}x - \tan^{-1}y\right]$$

$$\Rightarrow \quad \frac{\pi}{4} = \tan^{-1}x + \frac{1}{2}\tan^{-1}x = \frac{3}{2}\tan^{-1}x$$

$$\Rightarrow \quad \tan^{-1}x = \frac{2}{3}\times\frac{\pi}{4} = \frac{\pi}{6}$$

$$\Rightarrow \quad x = \tan\frac{\pi}{6} = \frac{1}{\sqrt{3}}$$ **Ans.**

Example 11. *Simplify each of the following expressions:*

(*i*) $\tan^{-1}\left(\dfrac{x}{\sqrt{a^2-x^2}}\right)$ (*ii*) $\tan^{1}\left(\dfrac{3a^2x - x^3}{a^3 - 3ax^2}\right)$ (*iii*) $\tan^{-1}\left(\dfrac{\cos x - \sin x}{\cos x + \sin x}\right)$

Solution. (*i*) Put $x = a\sin\theta \quad \Rightarrow \quad \sin\theta = \dfrac{x}{a}$

Now $$\frac{x}{\sqrt{a^2-x^2}} = \frac{a\sin\theta}{\sqrt{a^2 - a^2\sin^2\theta}} = \frac{a\sin\theta}{a\cos\theta} = \tan\theta \qquad ...(1)$$

$$\Rightarrow \quad \tan^{-1}\left(\frac{x}{a^2-x^2}\right) = \tan^{-1}(\tan\theta) = \theta = \sin^{-1}\left(\frac{x}{a}\right)$$ [using (1)] **Ans.**

(*ii*) We have, $$\tan^{-1}\left(\frac{3a^2x - x^3}{a^3 - 3ax^2}\right) = \tan^{-1}\left[\frac{3\left(\dfrac{x}{a}\right) - \left(\dfrac{x}{a}\right)^3}{1 - 3\left(\dfrac{x}{a}\right)^2}\right]$$

[Dividing the numerator and denominator by a^3]

$$= \tan^{-1}\left(\frac{3\tan\theta - \tan^3\theta}{1 - 3\tan^2\theta}\right)$$

$$\left[\text{Where, } \tan\theta = \frac{x}{a} \Rightarrow \theta = \tan^{-1}\left(\frac{x}{a}\right)\right]$$

$$= \tan^{-1}(\tan 3\theta) = 3\theta = 3\tan^{-1}\left(\frac{x}{a}\right)$$ **Ans.**

(*iii*) We have,

$$\tan^{-1}\left(\frac{\cos x - \sin x}{\cos x + \sin x}\right) = \tan^{-1}\left(\frac{1 - \tan x}{1 + \tan x}\right) = \tan^{-1}\left[\tan\left(\frac{\pi}{4} - x\right)\right]$$

$$= \frac{\pi}{4} - x$$ **Ans.**

Example 12. *If* $\cos^{-1}\left(\frac{x}{a}\right) + \cos^{-1}\left(\frac{y}{b}\right) = \alpha$, *prove that* $\frac{x^2}{a^2} - \frac{2xy}{ab}\cos\alpha + \frac{y^2}{b^2} = \sin^2\alpha$.

Solution. We have,

$$\cos^{-1}\left(\frac{x}{a}\right) + \cos^{-1}\left(\frac{y}{b}\right) = \alpha$$

$$\Rightarrow \quad \cos^{-1}\left[\frac{x}{a}.\frac{y}{b} - \sqrt{\left(1 - \frac{x^2}{a^2}\right)\left(1 - \frac{y^2}{b^2}\right)}\right] = \alpha$$

$$[\because \cos^{-1} x + \cos^{-1} y = \cos^{-1}(xy - \sqrt{(1 - x^2)(1 - y^2)})]$$

$$\Rightarrow \quad \frac{xy}{ab} - \sqrt{1 - \frac{x^2}{a^2} - \frac{y^2}{b^2} + \frac{x^2y^2}{a^2b^2}} = \cos\alpha$$

$$\Rightarrow \quad \frac{xy}{ab} - \cos\alpha = \sqrt{1 - \frac{x^2}{a^2} - \frac{y^2}{b^2} + \frac{x^2y^2}{a^2b^2}}$$

Squaring both sides, we have

$$\Rightarrow \quad \frac{x^2y^2}{a^2b^2} - \frac{2xy}{ab}\cos\alpha + \cos^2\alpha = 1 - \frac{x^2}{x^2} - \frac{y^2}{b^2} + \frac{x^2y^2}{a^2b^2}$$

$$\Rightarrow \quad \frac{x^2}{a^2} - \frac{2xy}{ab}\cos\alpha + \frac{y^2}{b^2} = 1 - \cos^2\alpha$$

$$\Rightarrow \quad \frac{x^2}{a^2} - \frac{2xy}{ab}\cos\alpha + \frac{y^2}{b^2} = \sin^2\alpha$$ **Ans.**

Example 13. *If* $\tan^{-1}(x) + \tan^{-1}(y) + \tan^{-1}(z) = \frac{\pi}{2}$, *prove that* $xy + yz + zx = 1$.

Solution. We have

$$\tan^{-1}(x) + \tan^{-1}(y) + \tan^{-1}(z) = \frac{\pi}{2} \Rightarrow \tan^{-1}(x) + \tan^{-1}(y) = \frac{\pi}{2} - \tan^{-1}(z)$$

$$\Rightarrow \quad \tan^{-1}\left(\frac{x+y}{1-xy}\right) = \cot^{-1}(z) = \tan^{-1}\frac{1}{z} \qquad [\because \tan^{-1}(z) + \cot^{-1}(z) = \frac{\pi}{2}]$$

$$\Rightarrow \quad \frac{x+y}{1-xy} = \frac{1}{z} \quad \Rightarrow \quad zx + yz = 1 - xy$$

$$\therefore \quad xy + yz + zx = 1$$

Proved.

Example 14. *Prove that:*

$$(i)\ \tan^{-1}\left(\sqrt{x}\right) = \frac{1}{2}\cos^{-1}\left(\frac{1-x}{1+x}\right) \qquad (ii)\ 2\left(\tan^{-1}\frac{1}{4} + \tan^{-1}\frac{2}{9}\right) = \tan^{-1}\left(\frac{4}{3}\right)$$

Solution. (*i*)

$$\text{R.H.S} = \frac{1}{2}\cos^{-1}\left(\frac{1-x}{1+x}\right)$$

$$= \frac{1}{2}\cos^{-1}\left(\frac{1-\tan^2\theta}{1+\tan^2\theta}\right) \qquad [\text{Put } x = \tan^2\theta]$$

$$= \frac{1}{2}\cos^{-1}(\cos 2\theta) = \frac{1}{2}(2\theta) = \theta$$

$$\text{L.H.S.} = \tan^{-1}\left(\sqrt{x}\right) = \tan^{-1}\left(\sqrt{\tan^2\theta}\right)$$

$$= \tan^{-1}(\tan\theta) \qquad [\text{Put } x = \tan^2\theta]$$

$$= \theta$$

Thus, $\quad$ L.H.S. = R.H.S. **Proved.**

(*ii*)

$$\text{L.H.S.} = 2\left(\tan^{-1}\frac{1}{4} + \tan^{-1}\frac{2}{9}\right)$$

$$= 2\tan^{-1}\left(\frac{1}{4}\right) + 2\tan^{-1}\left(\frac{2}{9}\right) = \tan^{-1}\left[\frac{2\times\frac{1}{4}}{1-\left(\frac{1}{4}\right)^2}\right] + \tan^{-1}\left[\frac{2\times\left(\frac{2}{9}\right)}{1-\left(\frac{2}{9}\right)^2}\right]$$

$$\left[\because 2\tan^{-1}x = \tan^{-1}\left(\frac{2x}{1-x^2}\right)\right]$$

$$= \tan^{-1}\left(\frac{\frac{1}{2}}{\frac{15}{16}}\right) + \tan^{-1}\left(\frac{\frac{4}{9}}{\frac{77}{81}}\right) = \tan^{-1}\left(\frac{8}{15}\right) + \tan^{-1}\left(\frac{36}{77}\right)$$

$$= \tan^{-1}\left(\frac{\frac{8}{15}+\frac{36}{77}}{1-\frac{8}{15}\times\frac{36}{77}}\right) \left[\because \tan^{-1}A + \tan^{-1}B = \tan^{-1}\left(\frac{A+B}{1-AB}\right)\right]$$

$$= \tan^{-1}\left(\frac{616+540}{15\times 77}\times\frac{15\times 77}{1155-288}\right) = \tan^{-1}\left(\frac{1156}{1}\times\frac{1}{867}\right)$$

$$= \tan^{-1}\left(\frac{1156}{867}\right) = \tan^{-1}\left(\frac{4}{3}\right)$$

Proved.

Example 15. *Prove that:*

$$\frac{9\pi}{8}-\frac{9}{4}\sin^{-1}\left(\frac{1}{3}\right)=\frac{9}{4}\sin^{-1}\left(\frac{2\sqrt{2}}{3}\right)$$

Solution. L.H.S $=\frac{9\pi}{8}-\frac{9}{4}\sin^{-1}\left(\frac{1}{3}\right)=\frac{9}{4}\left[\frac{\pi}{2}-\sin^{-1}\frac{1}{3}\right]$

$$=\frac{9}{4}\cos^{-1}\left(\frac{1}{3}\right)$$

$$\left[\because \sin^{-1}x+\cos^{-1}x=\frac{\pi}{2}\Rightarrow \cos^{-1}x=\frac{\pi}{2}-\sin^{-1}x \text{ or } \cos^{-1}\left(\frac{1}{3}\right)=\frac{\pi}{2}-\sin^{-1}\left(\frac{1}{3}\right)\right]$$

$$=\frac{9}{4}\sin^{-1}\left(\sqrt{1-\left(\frac{1}{3}\right)^2}\right)=\frac{9}{4}\sin^{-1}\left(\sqrt{1-\frac{1}{9}}\right)\left[\because \cos^{-1}x=\sin^{-1}\left(\sqrt{1-x^2}\right)\right]$$

$$=\frac{9}{4}\sin^{-1}\left(\frac{\sqrt{8}}{\sqrt{9}}\right)=\frac{9}{4}\sin^{-1}\left(\frac{2\sqrt{2}}{3}\right)$$

Proved.

Example 16. *Prove that:*

$$\tan^{-1}\left(\frac{1}{5}\right)+\tan^{-1}\left(\frac{1}{7}\right)+\tan^{-1}\left(\frac{1}{3}\right)+\tan^{-1}\left(\frac{1}{8}\right)=\frac{\pi}{4}$$

Solution. Let

$$\tan^{-1}\left(\frac{1}{5}\right)=\alpha \Rightarrow \tan\alpha=\frac{1}{5}$$

$$\tan^{-1}\left(\frac{1}{7}\right)=\beta \Rightarrow \tan\beta=\frac{1}{7}$$

$$\tan^{-1}\left(\frac{1}{3}\right)=\gamma \Rightarrow \tan\gamma=\frac{1}{3}$$

$$\tan^{-1}\left(\frac{1}{8}\right)=\delta \Rightarrow \tan\delta=\frac{1}{8}$$

$$\tan(\alpha+\beta)=\frac{\tan\alpha+\tan\beta}{1-\tan\alpha\,\tan\beta}=\frac{\frac{1}{5}+\frac{1}{7}}{1-\frac{1}{5}\times\frac{1}{7}}=\frac{\frac{12}{35}}{\frac{35-1}{35}}=\frac{12}{34}=\frac{6}{17}$$

$$\tan(\gamma+\delta)=\frac{\tan\gamma+\tan\delta}{1-\tan\gamma\,\tan\delta}=\frac{\frac{1}{3}+\frac{1}{8}}{1-\frac{1}{3}\times\frac{1}{8}}=\frac{\frac{11}{24}}{\frac{24-1}{24}}=\frac{11}{23}$$

$$\tan(\alpha+\beta+\gamma+\delta)=\frac{\tan(\alpha+\beta)+\tan(\gamma+\delta)}{1-\tan(\alpha+\beta)\tan(\gamma+\delta)}$$

$$=\frac{\frac{6}{17}+\frac{11}{23}}{1-\frac{6}{17}\times\frac{11}{23}}=\frac{\frac{138+187}{391}}{\frac{391-66}{391}}=\frac{\frac{325}{391}}{\frac{325}{391}}=\frac{325}{391}\times\frac{391}{325}=1=\tan\frac{\pi}{4}$$

$$= \alpha + \beta + \gamma + \delta = \frac{\pi}{4}$$

$$= \tan^{-1}\left(\frac{1}{5}\right) + \tan^{-1}\left(\frac{1}{7}\right) + \tan^{-1}\left(\frac{1}{3}\right) + \tan^{-1}\left(\frac{1}{8}\right) = \frac{\pi}{4}$$ **Proved.**

Example 17. *Prove that:* $2\tan^{-1}\left(\sqrt{\frac{a-b}{a+b}}\tan\frac{\theta}{2}\right) = \cos^{-1}\left(\frac{b + a\cos\theta}{a + b\cos\theta}\right)$.

Solution.

$\Rightarrow$ Let L.H.S. $= \frac{x}{2} = \tan^{-1}\left(\sqrt{\frac{a-b}{a+b}}\tan\frac{\theta}{2}\right) \Rightarrow \tan\frac{x}{2} = \sqrt{\frac{a-b}{a+b}}\tan\frac{\theta}{2}$ ...(1)

$\Rightarrow$ Now, $\cos x = \dfrac{1-\tan^2\frac{x}{2}}{1+\tan^2\frac{x}{2}} = \dfrac{1-\frac{a-b}{a+b}\tan^2\frac{\theta}{2}}{1+\frac{a-b}{a+b}\tan^2\frac{\theta}{2}}$ [using (1)]

$$= \frac{a+b-(a-b)\tan^2\frac{\theta}{2}}{a+b+(a-b)\tan^2\frac{\theta}{2}} = \frac{a\left(1-\tan^2\frac{\theta}{2}\right)+b\left(1+\tan^2\frac{\theta}{2}\right)}{a\left(1+\tan^2\frac{\theta}{2}\right)+b\left(1-\tan^2\frac{\theta}{2}\right)}$$

$$= \frac{a\left(\dfrac{1-\tan^2\frac{\theta}{2}}{1+\tan^2\frac{\theta}{2}}\right)+b}{a+b\left(\dfrac{1-\tan^2\frac{\theta}{2}}{1+\tan^2\frac{\theta}{2}}\right)} = \frac{a\cos\theta+b}{a+b\cos\theta}$$

$\Rightarrow$ $\cos x = \dfrac{a\cos\theta+b}{a+b\cos\theta}$

$\Rightarrow$ $x = \cos^{-1}\left(\dfrac{a\cos\theta+b}{a+b\cos\theta}\right) =$ R.H.S. **Proved.**

Example 18. *Solve for x :* $\tan^{-1}(2x) + \tan^{-1}(3x) = \frac{\pi}{4}$

Solution. We have, $\tan^{-1}(2x) + \tan^{-1}(3x) = \frac{\pi}{4}$

$\Rightarrow$ $\tan^{-1}\left(\dfrac{2x+3x}{1-(2x)(3x)}\right) = \dfrac{\pi}{4}$, $\left[\because \tan^{-1}x + \tan^{-1}y = \tan^{-1}\left(\dfrac{x+y}{1-xy}\right)\right]$

$\Rightarrow$ $\tan^{-1}\left(\dfrac{5x}{1-6x^2}\right) = \dfrac{\pi}{4} \Rightarrow \dfrac{5x}{1-6x^2} = \tan\dfrac{\pi}{4}$

$\Rightarrow$ $\dfrac{5x}{1-6x^2} = 1 \Rightarrow 5x = 1-6x^2$

$$\Rightarrow \quad 6x^2 + 5x - 1 = 0 \Rightarrow 6x^2 + 6x - x - 1 = 0$$

$$\Rightarrow \quad 6x(x + 1) - 1(x + 1) = 0 \Rightarrow (x + 1)(6x - 1) = 0$$

Either $x + 1 = 0 \Rightarrow x = -1$

or $6x - 1 = 0 \Rightarrow x = \frac{1}{6}$ **Ans.**

Example 19. *Solve the following equations*

(*i*) $\tan^{-1}\left(\frac{x-1}{x-2}\right) + \tan^{-1}\left(\frac{x+1}{x+2}\right) = \frac{\pi}{4}$ (*ii*) $\tan(\cos^{-1}x) = \sin(\tan^{-1} 2)$

Solution. (*i*) We have,

$$\tan^{-1}\left(\frac{x-1}{x-2}\right) + \tan^{-1}\left(\frac{x+1}{x+2}\right) = \frac{\pi}{4}$$

$$\Rightarrow \quad \tan^{-1}\left(\frac{\frac{x-1}{x-2} + \frac{x+1}{x+2}}{1 - \frac{x-1}{x-2}.\frac{x+1}{x+2}}\right) = \tan^{-1}(1)$$

$$\Rightarrow \quad \frac{\frac{x-1}{x-2} + \frac{x+1}{x+2}}{1 - \frac{x-1}{x-2}.\frac{x+1}{x+2}} = 1$$

$$\Rightarrow \quad \frac{(x-1)(x+2) + (x+1)(x-2)}{(x^2-4) - (x^2-1)} = 1$$

$$\Rightarrow \quad (x^2 + x - 2) + (x^2 - x - 2) = -3$$

$$\Rightarrow \quad 2x^2 - 4 = -3 \Rightarrow 2x^2 = 1 \Rightarrow x = \pm\frac{1}{\sqrt{2}}$$ **Ans.**

(*ii*) We have

$$\tan(\cos^{-1}x) = \sin(\tan^{-1}2)$$

Let $\theta = \cos^{-1}x \Rightarrow \cos\theta = x$

$$\therefore \text{ L.H.S.} = \tan(\cos^{-1}x) = \tan\theta = \frac{\sin\theta}{\cos\theta} = \frac{\sqrt{1-\cos^2\theta}}{\cos\theta} = \frac{\sqrt{1-x^2}}{x} \quad ...(1)$$

Let $\tan^{-1}(2) = \phi \Rightarrow \tan\phi = 2$

$$\Rightarrow \text{ R.H.S.} = \sin(\tan^{-1}2) = \sin\phi = \frac{1}{\text{cosec}\,\phi} = \frac{1}{\sqrt{1+\cot^2\phi}}$$

$$= \frac{1}{\sqrt{1+\left(\frac{1}{2}\right)^2}} = \frac{2}{\sqrt{5}} \quad ...(2)$$

From (1) and (2), we have $\frac{\sqrt{1-x^2}}{x} = \frac{2}{\sqrt{5}}$

$$\Rightarrow \quad \frac{1-x^2}{x^2} = \frac{4}{5} \Rightarrow 5 - 5x^2 = 4x^2 \Rightarrow 9x^2 = 5$$

$$\Rightarrow \quad x = \pm\frac{\sqrt{5}}{3}$$

$$x = -\frac{\sqrt{5}}{3} \Rightarrow \cos^{-1}x \text{ lies in II quadrant}$$

and thus $\tan(\cos^{-1}x)$ is –ve.

$\therefore$ L.H.S. is –ve, whereas R.H.S. is +ve.

$$\therefore \quad x = -\frac{\sqrt{5}}{3} \text{ is impossible.}$$

$$\Rightarrow \quad x = \frac{\sqrt{5}}{3}$$

Ans.

Example 20. *Prove that:*

$$\tan^{-1}\left(\frac{\sqrt{1+x^2}+\sqrt{1-x^2}}{\sqrt{1+x^2}-\sqrt{1-x^2}}\right) = \frac{\pi}{4} + \frac{1}{2}\cos^{-1}(x^2)$$

Solution. Let $x = \sqrt{\cos\theta}$

$$\text{L.H.S} = \tan^{-1}\left(\frac{\sqrt{1+x^2}+\sqrt{1-x^2}}{\sqrt{1+x^2}-\sqrt{1-x^2}}\right) = \tan^{-1}\left(\frac{\sqrt{1+\cos\theta}+\sqrt{1-\cos\theta}}{\sqrt{1+\cos\theta}-\sqrt{1-\cos\theta}}\right)$$

$$= \tan^{-1}\left(\frac{\sqrt{2\cos^2\frac{\theta}{2}}+\sqrt{2\sin^2\frac{\theta}{2}}}{\sqrt{2\cos^2\frac{\theta}{2}}-\sqrt{2\sin^2\frac{\theta}{2}}}\right)$$

$$= \tan^{-1}\left(\frac{\sqrt{2}\cos\frac{\theta}{2}+\sqrt{2}\sin\frac{\theta}{2}}{\sqrt{2}\cos\frac{\theta}{2}-\sqrt{2}\sin\frac{\theta}{2}}\right)$$

$$= \tan^{-1}\left(\frac{\cos\frac{\theta}{2}+\sin\frac{\theta}{2}}{\cos\frac{\theta}{2}-\sin\frac{\theta}{2}}\right) = \tan^{-1}\left(\frac{1+\tan\frac{\theta}{2}}{1-\tan\frac{\theta}{2}}\right)$$

$$= \tan^{-1}\left(\frac{\tan\frac{\pi}{4}+\tan\frac{\theta}{2}}{1-\tan\frac{\pi}{4}\tan\frac{\theta}{2}}\right)$$

$$= \tan^{-1}\left(\tan\left(\frac{\pi}{4}+\frac{\theta}{2}\right)\right)$$

$$= \frac{\pi}{4} + \frac{\theta}{2} = \frac{\pi}{4} + \frac{1}{2}\cos^{-1} x^2 \qquad \left[\because x = \sqrt{\cos\theta} \Rightarrow \cos^{-1} x^2 = \theta\right]$$

$$= \text{R.H.S.}$$

Proved.

Example 21. *Prove that:*

(i) $$\cot^{-1}\left(\frac{\sqrt{1+\sin x} + \sqrt{1-\sin x}}{\sqrt{1+\sin x} - \sqrt{1-\sin x}}\right) = \frac{x}{2}, \quad x \in \left(0, \frac{\pi}{4}\right)$$

(ii) $$\tan^{-1}\left(\frac{\sqrt{1+x} + \sqrt{1-x}}{\sqrt{1+x} - \sqrt{1-x}}\right) = \frac{\pi}{4} - \frac{1}{2}\cos^{-1} x, \quad \frac{-1}{\sqrt{2}} \le x \le 1.$$

Solution.

(i) L.H.S $$= \cot^{-1}\left(\frac{\sqrt{1+\sin x} + \sqrt{1-\sin x}}{\sqrt{1+\sin x} - \sqrt{1-\sin x}}\right)$$

$$= \cot^{-1}\left[\frac{\sqrt{\left(\cos\frac{x}{2} + \sin\frac{x}{2}\right)^2} + \sqrt{\left(\cos\frac{x}{2} - \sin\frac{x}{2}\right)^2}}{\sqrt{\left(\cos\frac{x}{2} + \sin\frac{x}{2}\right)^2} - \sqrt{\left(\cos\frac{x}{2} - \sin\frac{x}{2}\right)^2}}\right]$$

$$= \cot^{-1}\left[\frac{\left(\cos\frac{x}{2} + \sin\frac{x}{2}\right) + \left(\cos\frac{x}{2} - \sin\frac{x}{2}\right)}{\left(\cos\frac{x}{2} + \sin\frac{x}{2}\right) - \left(\cos\frac{x}{2} - \sin\frac{x}{2}\right)}\right]$$

$$= \cot^{-1}\left[\frac{2\cos\frac{x}{2}}{2\sin\frac{x}{2}}\right] = \cot^{-1}\left(\cot\frac{x}{2}\right) = \frac{x}{2} = \text{R.H.S.}$$

Proved.

(ii) L.H.S $$= \tan^{-1}\left(\frac{\sqrt{1+x} - \sqrt{1-x}}{\sqrt{1+x} + \sqrt{1-x}}\right) = \tan^{-1}\left(\frac{\sqrt{1+\cos 2\theta} - \sqrt{1-\cos 2\theta}}{\sqrt{1+\cos 2\theta} + \sqrt{1-\cos 2\theta}}\right)$$

[Put $x = \cos 2\theta$]

$$= \tan^{-1}\left(\frac{\sqrt{2\cos^2\theta} - \sqrt{2\sin^2\theta}}{\sqrt{2\cos^2\theta} + \sqrt{2\sin^2\theta}}\right)$$

$$= \tan^{-1}\left(\frac{\cos\theta - \sin\theta}{\cos\theta + \sin\theta}\right) = \tan^{-1}\left(\frac{1-\tan\theta}{1+\tan\theta}\right)$$

$$= \tan^{-1}\tan\left(\frac{\pi}{4} - \theta\right) = \frac{\pi}{4} - \theta \qquad \left[\because x = \cos 2\theta \Rightarrow \theta = \frac{1}{2}\cos^{-1} x\right]$$

$$= \frac{\pi}{4} - \frac{1}{2}\cos^{-1} x = \text{R.H.S.}$$

Proved.

EXERCISE 16.2

Find:

1. $\cos A$ if $\sec^{-1}\left(\frac{2}{\sqrt{3}}\right) = A$ **Ans.** $\frac{\sqrt{3}}{2}$

2. $\cos^{-1} x$ if $\sin^{-1} x = \frac{\pi}{3}$ **Ans.** $\frac{\pi}{6}$

3. x if $\sin^{-1}\left(\frac{1}{2}\right) = \tan^{-1} x$ **Ans.** $\frac{\pi}{6}$

Verify each of the following:

4. $\sin^{-1}\left(\frac{\sqrt{2}}{2}\right) - \sin^{-1}\left(\frac{1}{2}\right) = \frac{\pi}{12}$

5. $\cos^{-1}(0) + \tan^{-1}(-1) = \tan^{-1}(1)$

6. $\sin\left[\tan^{-1}\left(\sqrt{3}\right) + \cot^{-1}\left(\sqrt{3}\right)\right] = 1$

Show that:

7. $\tan\left(\sin^{-1}\frac{\sqrt{3}}{2} - \cos^{-1}\frac{\sqrt{3}}{2}\right) = \frac{\sqrt{3}}{2}$

8. $2\tan^{-1}\left(\frac{1}{2}\right) = \tan^{-1}\left(\frac{4}{3}\right)$

9. $\tan^{-1}(2) - \tan^{-1}(1) = \tan^{-1}\left(\frac{1}{3}\right)$

10. $2\tan^{-1}\left(\frac{1}{3}\right) = \tan^{-1}\left(\frac{3}{4}\right)$

11. $2\tan^{-1}\left(\frac{1}{2}\right) + \tan^{-1}\left(\frac{1}{3}\right) = \frac{\pi}{4}$

Write the following functions in the simplest form

12. $\tan^{-1}\left(\frac{\sin x}{1 + \cos x}\right)$ **Ans.** $\frac{x}{2}$

13. $\cos(\text{cosec}^{-1} x + \sec^{-1} x)$ **Ans.** 0

14. Find $\sin A$ if $\tan^{-1}\left(\frac{3}{4}\right) = A$ **Ans.** $\frac{3}{5}$

Show that

15. $\sin^{-1}\left(\sin^{-1}\frac{5}{13} + \sin^{-1}\frac{4}{5}\right) = \frac{63}{65}$

16. $\cos\left(\tan^{-1}\frac{15}{8} - \sin^{-1}\frac{7}{25}\right) = \frac{297}{425}$

17. $\sin\left(\sin^{-1}\frac{1}{2} + \cos^{-1}\frac{3}{5}\right) = \frac{3+4\sqrt{3}}{10}$

Prove that:

18. $\cos^{-1}\left(\frac{4}{5}\right) = \tan^{-1}\left(\frac{3}{4}\right)$

19. $2\tan^{-1}\left(\frac{1}{3}\right) + \tan^{-1}\left(\frac{1}{7}\right) = \frac{\pi}{4}$

20. $\tan^{-1}\left(\frac{m}{n}\right) - \tan^{-1}\left(\frac{m-n}{m+n}\right) = \frac{\pi}{4}$

21. $\cot^{-1}\left(\frac{ab+1}{a-b}\right) + \cot^{-1}\left(\frac{bc+1}{b-c}\right) + \cot^{-1}\left(\frac{ca+1}{c-a}\right) = 0$

Write the following in their simplest form:

22. $\tan^{-1}\left(\sqrt{\dfrac{1-\cos x}{1+\cos x}}\right)$ **Ans.** $\dfrac{x}{2}$

Show that

23. $\sin^{-1}\left(\dfrac{3}{5}\right)+\sin^{-1}\left(\dfrac{5}{13}\right)=\sin^{-1}\left(\dfrac{56}{65}\right)$ **24.** $\sin^{-1}\left(\dfrac{4}{5}\right)+2\tan^{-1}\left(\dfrac{1}{3}\right)=\dfrac{\pi}{2}$

25. $\cos^{-1}\left(\dfrac{4}{5}\right)+\cot^{-1}\left(\dfrac{5}{3}\right)=\tan^{-1}\left(\dfrac{27}{11}\right)$

26. $\tan^{-1}(1)+\tan^{-1}(2)+\tan^{-1}(3)=2\left[\tan^{-1}1+\tan^{-1}\dfrac{1}{2}+\tan^{-1}\dfrac{1}{3}\right]$

27. $\tan^{-1}x+\cot^{-1}(x+1)=\tan^{-1}(x^2+x+1)$

28. Simplify: $\cot^{-1}\left(\dfrac{\sqrt{1+\sin x}+\sqrt{1-\sin x}}{\sqrt{1+\sin x}-\sqrt{1-\sin x}}\right)$ **Ans.** $\dfrac{x}{2}$

Properties of a Triangle

17.1 INTRODUCTION

Let ABC be a triangle, the measures of the angle *BAC*, *CBA* and *ACB* are denoted by letters *A*, *B* and *C* respectively. The sides opposite to the angles *A*, *B*, *C* are denoted by *a*, *b*, *c* respectively. The angles and sides of a triangle are called its elements. A triangle which does not contain a right angle is called an oblique triangle.

17.2 LAW OF SINES (OR SINE FORMULA)

In any triangle, the sides are proportional to the sines of the opposite angles, i.e.

$$\frac{a}{\sin A} = \frac{b}{\sin B} = \frac{c}{\sin C}$$

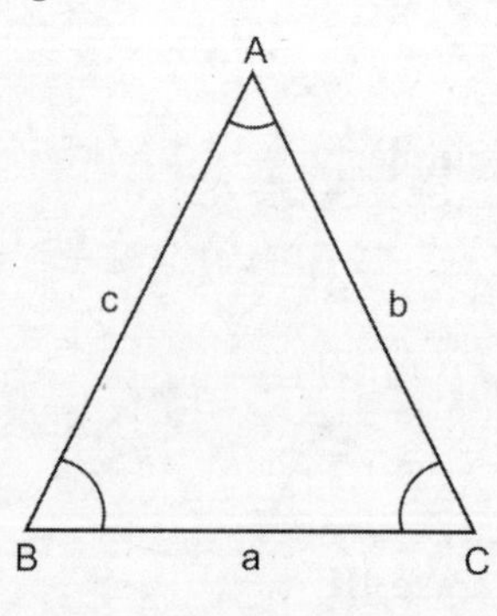

Proof: Let *ABC* be the triangle with $a = BC, b = CA$ and $c = AB$.
Then the following cases arise.

Case I: When $\Delta\ ABC$ is an acute angled triangle.

From A, draw $AD \perp BC$.

In right angled $\Delta\ ABD$, we have

$$\sin B = \frac{AD}{AB} \Rightarrow \sin B = \frac{AD}{c}$$

$$\Rightarrow \quad AD = c \sin B \qquad ...(1)$$

In right angled $\Delta\ ACD$, we have

$$\sin C = \frac{AD}{AC} \Rightarrow \sin C = \frac{AD}{b}$$

$$\Rightarrow \quad AD = b \sin C \qquad ...(2)$$

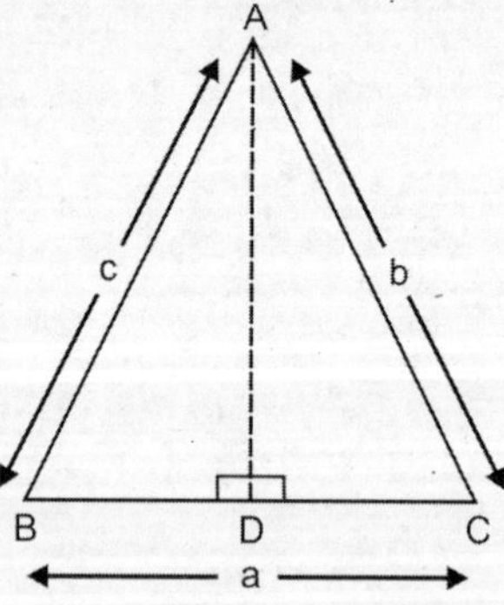

From (1) and (2), we get

$$c \sin B = b \sin C$$

$$\Rightarrow \quad \frac{b}{\sin B} = \frac{c}{\sin C} \qquad ...(3)$$

Similarly, by drawing perpendicular from C on *AB*, we have

$$\Rightarrow \quad \frac{a}{\sin A} = \frac{b}{\sin B} \qquad ...(4)$$

From (3) and (4), we have

$$\frac{a}{\sin A} = \frac{b}{\sin B} = \frac{c}{\sin C}$$

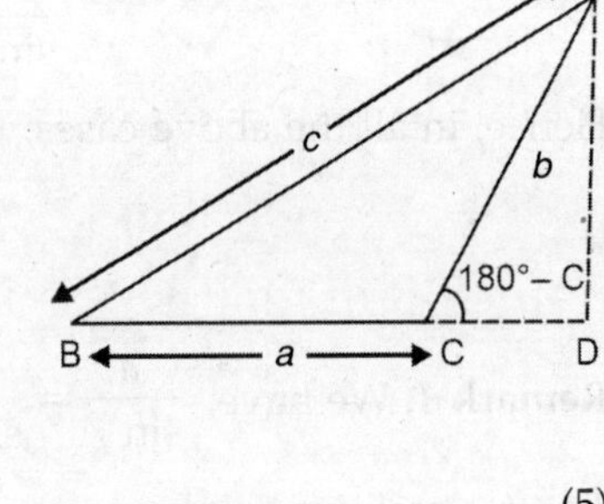

Case II. When ΔABC is an obtuse angled triangle.

From A, draw $AD \perp BC$ (produced)

In right angled ΔACD, we have

$$\sin(180° - C) = \frac{AD}{AC}$$

$$\Rightarrow \quad \sin C = \frac{AD}{b}$$

$$\Rightarrow \quad AD = b \sin C \qquad ...(5)$$

In right angled ΔABD, we have

$$\sin B = \frac{AD}{AB} \Rightarrow \sin B = \frac{AD}{c}$$

$$\Rightarrow \quad AD = c \sin B \qquad ...(6)$$

From (5) and (6), we have

$$b \sin C = c \sin B$$

$$\Rightarrow \quad \frac{b}{\sin B} = \frac{c}{\sin C} \qquad ...(7)$$

Similarly, by drawing perpendicular from C on AB, we have

$$\frac{a}{\sin A} = \frac{b}{\sin B} \qquad ...(8)$$

From (7) and (8), we have

$$\frac{a}{\sin A} = \frac{b}{\sin B} = \frac{c}{\sin C}$$

Case III. When ΔABC is an right angled triangle.

In right angled ΔABC, we have

$$\sin C = \sin\frac{\pi}{2} = 1 \qquad ...(9)$$

$$\sin A = \frac{BC}{AB} = \frac{a}{c}$$

$$\Rightarrow \quad c = \frac{a}{\sin A} \qquad ...(10)$$

$$\sin B = \frac{AC}{AB} = \frac{b}{c}$$

$$\Rightarrow \quad c = \frac{b}{\sin B} \qquad ...(11)$$

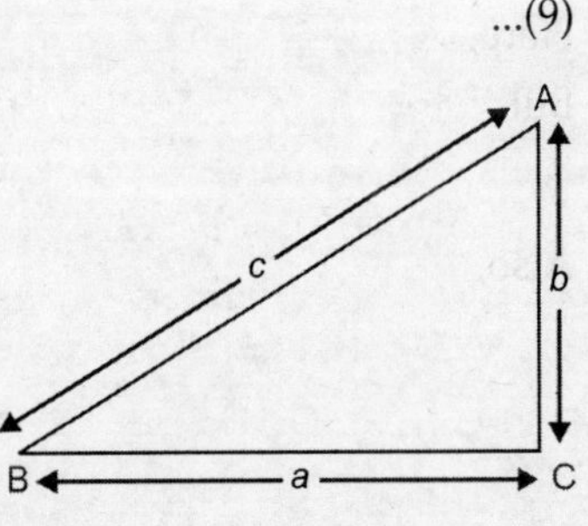

From (10) and (11), we have

$$\frac{a}{\sin A} = \frac{b}{\sin B} = \frac{c}{1}$$

$$\Rightarrow \quad \frac{a}{\sin A} = \frac{b}{\sin B} = \frac{c}{\sin C} \qquad \text{[using (9)]}$$

Hence, in all the above cases, we have

$$\boxed{\frac{\mathbf{a}}{\mathbf{sin\ A}} = \frac{\mathbf{b}}{\mathbf{sin\ B}} = \frac{\mathbf{c}}{\mathbf{sin\ C}}}$$

Remark 1. We have, $\frac{a}{\sin A} = \frac{b}{\sin B} = \frac{c}{\sin C} = k$ (say) $\Rightarrow a = k \sin A, b = k \sin B, c = k \sin C.$

These equalities help us to replace any side of a triangle by k times the sine of the corresponding angle, where k is a constant.

Example 1. *In* ΔABC, *if* $a = 2, b = 3$ *and* $\sin A = \frac{2}{3}$, *find* $\angle B$. **(S.B.T.E. 2015, 2013)**

Solution. We have, $a = 2, b = 3$ and $\sin A = \frac{2}{3}$,

We know that, $\frac{a}{\sin A} = \frac{b}{\sin B} = \frac{c}{\sin C}$ $\qquad \left[\therefore \frac{2}{\frac{2}{3}} = \frac{3}{\sin B} = \frac{c}{\sin C}\right]$

$$\Rightarrow \quad 3 = \frac{3}{\sin B} \Rightarrow \sin B = 1 \Rightarrow \angle B = 90° \qquad \textbf{Ans.}$$

Example 2. *In* ΔABC, *if* $a = 18, b = 24, c = 30$, *find* $\sin A, \sin B, \sin C$.

Solution. We have,

$$a = 18, b = 24, c = 30$$

We know that

$$\frac{\sin A}{a} = \frac{\sin B}{b} = \frac{\sin C}{c} = k \text{ (say)} \Rightarrow \frac{\sin A}{18} = \frac{\sin B}{24} = \frac{\sin C}{30} = k$$

$$\Rightarrow \quad \sin A = 18k, \quad \sin B = 24k, \quad \sin C = 30k \qquad ...(1)$$

Here, $(30)^2 = (24)^2 + (18)^2$. So it is a right angled triangle.

But, $\angle C = 90°$ as it is opposite to the biggest side c.

$\therefore$ $\sin C = \sin 90° = 1$

Also, $30k = 1 \Rightarrow k = \frac{1}{30}$ [using (1)]

$$\sin A = 18k = 18 \times \frac{1}{30} = \frac{18}{30} = \frac{3}{5}$$

$$\sin B = 24k = 24 \times \frac{1}{30} = \frac{24}{30} = \frac{4}{5}$$

and $\sin C = 1.$ **Ans.**

Example 3. *For any triangle ABC, prove that* **(S.B.T.E. 2013)**

$$\frac{\sin (B - C)}{\sin (B + C)} = \frac{b^2 - c^2}{a^2}$$

Solution. We know that,

$\Rightarrow \quad a = k \sin A, \quad b = k \sin B \quad$ and $\quad c = k \sin C$

On putting the values of a, b and c on R.H.S., we get

$$\text{R.H.S} = \frac{b^2 - c^2}{a^2} = \frac{k^2 \sin^2 B - k^2 \sin^2 C}{k^2 \sin^2 A}$$

$$= \frac{\sin^2 B - \sin^2 C}{\sin^2 A} = \frac{\sin(B+C)\sin(B-C)}{\sin^2 A}$$

$$= \frac{\sin(\pi - A)\sin(B-C)}{\sin^2 A}$$

$$= \frac{\sin A \sin(B-C)}{\sin^2 A}$$

$$= \frac{\sin(B-C)}{\sin A} = \frac{\sin(B-C)}{\sin(B+C)} = \text{L.H.S.}$$

Proved.

Example 4. *In a triangle ABC, if a cos A = b cos B, show that the triangle is either isosceles or right-angled.*

Solution. We know that,

$$\frac{a}{\sin A} = \frac{b}{\sin B} = \frac{c}{\sin C} = k$$

$\Rightarrow \quad a = k \sin A, \quad b = k \sin B \quad$ and $\quad c = k \sin C$ ...(1)

We have, $\quad a \cos A = b \cos B$ ...(2)

On putting the values of a and b from (1) in (2), we have

$\Rightarrow \quad k \sin A \cos A = k \sin B \cos B$

$\Rightarrow \quad 2 \sin A \cos A = 2 \sin B \cos B$

$\Rightarrow \quad \sin 2A = \sin 2B$ ...(3)

From (3), there are two possibilities

Either $2A = 2B \quad \Rightarrow \quad 2A = \pi - 2B$

$\therefore \quad A = B \quad \Rightarrow \quad 2A + 2B = \pi$

$\Rightarrow \quad A + B = \frac{\pi}{2}$

$\Rightarrow \quad C = \frac{\pi}{2}$

Hence, ΔABC is either isosceles or right angled. **Proved.**

Example 5. *In any triangle ABC, show that*

$$\frac{a-b}{a+b} = \frac{\tan\left(\frac{A-B}{2}\right)}{\tan\left(\frac{A+B}{2}\right)}$$

Solution. We know that,

$$\frac{a}{\sin A} = \frac{b}{\sin B} = \frac{c}{\sin C} = k$$

$$\Rightarrow \quad a = k \sin A, \qquad b = k \sin B \qquad \text{and} \quad c = k \sin C \qquad ...(1)$$

On putting the values of a and b from (1) on L.H.S., we get

$$\text{L.H.S.} = \frac{a-b}{a+b} = \frac{k \sin A - k \sin B}{k \sin A + k \sin B} = \frac{\sin A - \sin B}{\sin A + \sin B}$$

$$= \frac{2 \cos\left(\frac{A+B}{2}\right) \sin\left(\frac{A-B}{2}\right)}{2 \sin\left(\frac{A+B}{2}\right) \cos\left(\frac{A-B}{2}\right)}$$

$$= \cot\left(\frac{A+B}{2}\right) \tan\left(\frac{A-B}{2}\right)$$

$$= \frac{\tan\left(\frac{A-B}{2}\right)}{\tan\left(\frac{A+B}{2}\right)} = \text{R.H.S.}$$

Proved.

Example 6. *In any triangle ABC, show that*

$$a \cos\left(\frac{B-C}{2}\right) = (b+c) \sin \frac{A}{2}$$

Solution. We know that,

$$\frac{a}{\sin A} = \frac{b}{\sin B} = \frac{c}{\sin C} = k$$

$$\Rightarrow \quad a = k \sin A, \qquad b = k \sin B \qquad \text{and} \quad c = k \sin C \qquad ...(1)$$

On putting the values of a, b and c from (1) in $\frac{b+c}{a}$, we get

$$\frac{b+c}{a} = \frac{k \sin B + k \sin C}{k \sin A} \qquad \text{[Using (1)]}$$

$$= \frac{\sin B + \sin C}{\sin A} = \frac{2 \sin\left(\frac{B+C}{2}\right) \cos\left(\frac{B-C}{2}\right)}{\sin A}$$

$$= \frac{2 \sin\left(\frac{180° + A}{2}\right) \cos\left(\frac{B-C}{2}\right)}{\sin A} = \frac{2 \cos \frac{A}{2} \cos\left(\frac{B-C}{2}\right)}{2 \sin \frac{A}{2} \cos \frac{A}{2}}$$

$$= \frac{\cos\left(\frac{B-C}{2}\right)}{\sin \frac{A}{2}}$$

$$\therefore \quad a\cos\left(\frac{B-C}{2}\right) = (b+c)\sin\frac{A}{2}$$ **Proved.**

EXERCISE 17.1

For any triangle ABC, prove that:

1. $\dfrac{a+b}{c} = \dfrac{\cos\left(\dfrac{A-B}{2}\right)}{\sin\dfrac{C}{2}}$ **2.** $\dfrac{a-b}{c} = \dfrac{\sin\left(\dfrac{A-B}{2}\right)}{\cos\dfrac{C}{2}}$

3. $\sin\left(\dfrac{B-C}{2}\right) = \dfrac{b-c}{a}\cos\dfrac{A}{2}$

4. $a\sin(B-C) + b\sin(C-A) + c\sin(A-B) = 0$

5. $a(\sin B - \sin C) + b(\sin C - \sin A) + c(\sin A - \sin B) = 0$

6. $(b+c)\cos A + (c+a)\cos B + (a+b)\cos C = a+b+c$

7. $a\cos A + b\cos B + c\cos C = 2a\sin B\sin C$

8. $\dfrac{1+\cos(A-B)\cos C}{1+\cos(A-C)\cos B} = \dfrac{a^2+b^2}{a^2+c^2}$ **9.** $\dfrac{b-c}{b+c} = \dfrac{\tan\left(\dfrac{B-C}{2}\right)}{\tan\left(\dfrac{B+C}{2}\right)}$

10. $\dfrac{a\sin(B-C)}{b^2-c^2} = \dfrac{b\sin(C-A)}{c^2-a^2} = \dfrac{c\sin(A-B)}{a^2-b^2}$

11. $\dfrac{b^2-c^2}{\cos B + \cos C} + \dfrac{c^2-a^2}{\cos C + \cos A} + \dfrac{a^2-b^2}{\cos A + \cos B} = 0$

12. $\dfrac{a^2\sin(B-C)}{\sin B + \sin C} + \dfrac{b^2\sin(C-A)}{\sin C + \sin A} + \dfrac{c^2\sin(A-B)}{\sin A + \sin B} = 0$

13. $(b-c)\cot\dfrac{A}{2} + (c-a)\cot\dfrac{B}{2} + (a-b)\cot\dfrac{C}{2} = 0$

14. $a\cos A + b\cos B + c\cos C = 2b\sin A\sin C = 2c\sin A.\sin B$

15. $\dfrac{c}{a-b} = \dfrac{\tan\left(\dfrac{A}{2}\right) + \tan\left(\dfrac{B}{2}\right)}{\tan\left(\dfrac{A}{2}\right) - \tan\left(\dfrac{B}{2}\right)}$ **16.** $\dfrac{c}{a+b} = \dfrac{1-\tan\left(\dfrac{A}{2}\right)\tan\left(\dfrac{B}{2}\right)}{1+\tan\left(\dfrac{A}{2}\right)\tan\left(\dfrac{B}{2}\right)}$

17. If in any triangle the angles be to one another as 1 : 2 : 3, prove that the corresponding sides are $1:\sqrt{3}:2$

18. The angles of a triangle *ABC* are in A.P. and given that $b:c = \sqrt{3}:\sqrt{2}$. find $\angle A$.

Ans. 75°

19. In any $\Delta\, ABC, \angle A = 45°, \angle B = 60°$ *and* $\angle C = 75°$, find the ratio of its sides.

Ans. $2:\sqrt{6}:\sqrt{3}+1$

20. In any $\Delta\, ABC, \angle A = 45°, \angle B = 105°, a = 2,$ then find b. **Ans.** $2\sqrt{2}$

17.3 LAW OF COSINES (OR COSINE FORMULAE)

In any triangle ABC:

1. $a^2 = b^2 + c^2 - 2bc \cos A \quad \Rightarrow \quad \cos A = \dfrac{b^2 + c^2 - a^2}{2bc}$

2. $b^2 = c^2 + a^2 - 2ac \cos B \quad \Rightarrow \quad \cos B = \dfrac{a^2 + c^2 - b^2}{2ac}$

3. $c^2 = a^2 + b^2 - 2ab \cos C \quad \Rightarrow \quad \cos C = \dfrac{a^2 + b^2 - c^2}{2ab}$

Proof: Let ABC be a triangle. Then the following cases may arise:

Case I. When ΔABC is an acute angled triangle.

From A, draw $AD \perp BC$.

In right angled ΔABC, we have

$$\cos B = \frac{BD}{AB} \Rightarrow \cos B = \frac{BD}{c}$$

$$\Rightarrow \quad BD = c \cos B \quad \text{...(1)}$$

Now, in right angled ΔACD

Using pythagoras theorem, we have

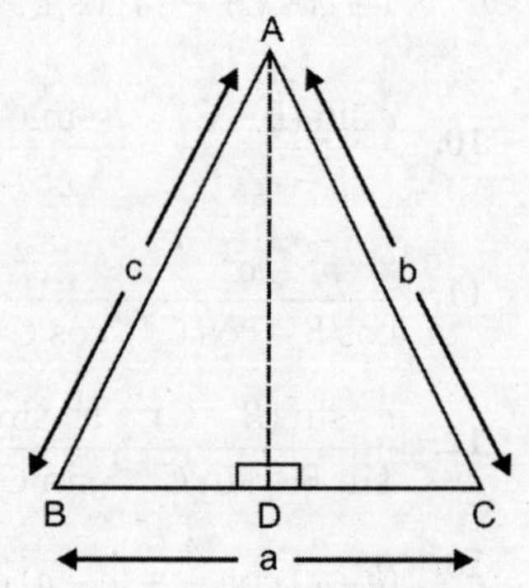

$$AC^2 = AD^2 + DC^2$$

$$AC^2 = AD^2 + (BC - BD)^2$$

$$AC^2 = AD^2 + BC^2 + BD^2 - 2BC.BD$$

$$= (AD^2 + BD^2) + BC^2 - 2BC.BD$$

$[\because \text{in } ABD, \angle D = 90° \Rightarrow AB^2 = AD^2 + BD^2]$

$$AC^2 = AB^2 + BC^2 - 2BC.BD$$

$$b^2 = c^2 + a^2 - 2ac \cos B \quad \text{[Using (1)]}$$

$$\cos B = \frac{c^2 + a^2 - b}{2ac}$$

Case II. When ΔABC is an obtuse angled triangle.

From A, draw $AD \perp BC$ (produced). In right angled ΔABD, we have

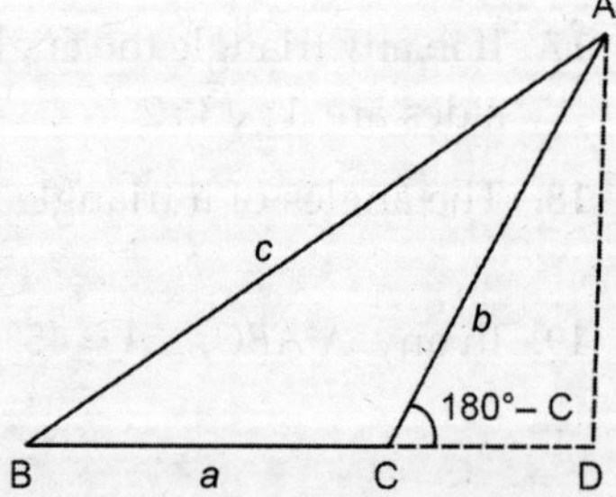

$$\cos (180° - B) = \frac{BD}{AB}$$

$$\Rightarrow \quad BD = -AB \cos B = -c \cos B \quad \text{...(1)}$$

By pythagoras theorem in right angled ΔACD, we have

$$AC^2 = AD^2 + CD^2$$

$$\Rightarrow \quad AC^2 = AD^2 + (BC + BD)^2$$

$$\Rightarrow \quad AC^2 = AD^2 + BC^2 + BD^2 + 2\,BC.BD$$

$$\Rightarrow \quad AC^2 = (AD^2 + BD^2) + BC^2 + 2\,BC.BD$$

$$\Rightarrow \quad AC^2 = AB^2 + BC^2 + 2BC.BD, [\text{In } \Delta\, ABD, AB^2 = AD^2 + BD^2] \quad \text{...(2)}$$

Putting the value of BD from (1) in (2), we get

$$\Rightarrow \quad b^2 = c^2 + a^2 + 2a(-c\cos B)$$

$$\Rightarrow \quad \cos B = \frac{c^2 + a^2 - b^2}{2ac}$$

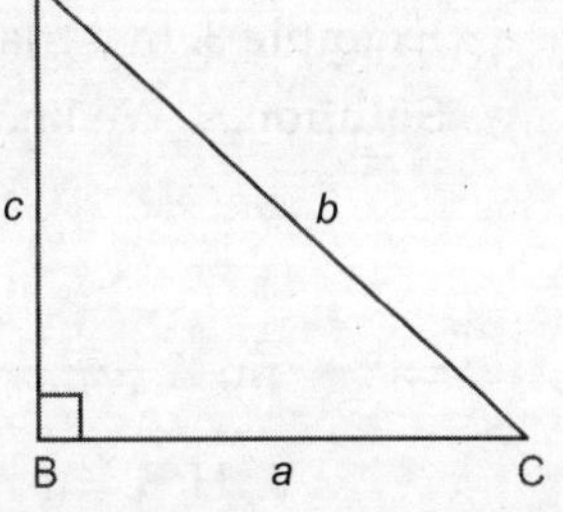

Case III. When ΔABC, is a right angled triangle.

Here, $\angle B = 90°$. Therefore using Pythagoras theorem, we have

$$AC^2 = AB^2 + BC^2$$

$$\Rightarrow \quad b^2 = c^2 + a^2$$

$$\Rightarrow \quad b^2 = c^2 + a^2 - 2ac.\cos B \quad [\because B = 90° \Rightarrow \cos B = 0]$$

$$\Rightarrow \quad \cos B = \frac{c^2 + a^2 - b^2}{2ac}$$

Hence, in all the above cases, we have

$$\cos B = \frac{c^2 + a^2 - b^2}{2ac}$$

Similarly, other result can be proved.

$$\boxed{\cos A = \frac{b^2 + c^2 - a^2}{2bc}} \qquad \boxed{\cos B = \frac{c^2 + a^2 - b^2}{2ca}} \qquad \boxed{\cos C = \frac{a^2 + b^2 - c^2}{2ab}}$$

Example 7. *In a triangle ABC, if a = 18, b = 24 and c = 30, find cos A, cos B and cos C.*

Solution. We have,

$a = 18$, $b = 24$ and $c = 30$ ...(1)

We know that,

$$\cos A = \frac{b^2 + c^2 - a^2}{2bc} \quad \text{...(2)}$$

Putting values of a, b and c from (1) in (2), we get

$$\cos A = \frac{(24)^2 + (30)^2 - (18)^2}{2(24)(30)}$$

$$= \frac{576 + 900 - 324}{1440} = \frac{1152}{1440} = \frac{4}{5}$$

Similarly, $\cos B = \dfrac{c^2 + a^2 - b^2}{2ca} = \dfrac{(30)^2 + (18)^2 - (24)^2}{2 \times 30 \times 18}$

$$= \frac{900 + 324 - 576}{1080} = \frac{648}{1080} = \frac{3}{5}$$

$$\cos C = \frac{a^2 + b^2 - c^2}{2ab} = \frac{(18)^2 + (24)^2 - (30)^2}{2 \times 18 \times 24}$$

$$= \frac{324 + 576 - 900}{864} = \frac{900 - 900}{864} = 0$$

Hence, $\cos A = \frac{4}{5}$, $\cos B = \frac{3}{5}$, $\cos C = 0$ **Ans.**

Example 8. In a triangle ABC, prove that $(b^2 - c^2) \cot A + (c^2 - a^2) \cot B + (a^2 - b^2) \cot C = 0$

Solution. We know that,

$$\frac{\sin A}{a} = \frac{\sin B}{b} = \frac{\sin C}{c} = k$$

$\Rightarrow \quad \sin A = ak, \quad \sin B = bk, \quad \sin C = ck$ ...(1)

L.H.S. $= (b^2 - c^2) \cot A + (c^2 - a^2) \cot B + (a^2 - b^2) \cot C$

$= (b^2 - c^2) \dfrac{\cos A}{\sin A} + (c^2 - a^2) \dfrac{\cos B}{\sin B} + (a^2 - b^2) \dfrac{\cos C}{\sin C}$ ...(2)

Putting the values of sin A, sin B, sin C from (1) and values of cos A, cos B and cos C in (2), we get

$$\text{L.H.S.} = \frac{(b^2 - c^2)}{ka}\left(\frac{b^2 + c^2 - a^2}{2bc}\right) + \frac{(c^2 - a^2)}{kb}\left(\frac{c^2 + a^2 - b^2}{2ac}\right) + \frac{(a^2 - b^2)}{kc}\left(\frac{a^2 + b^2 - c^2}{2ab}\right)$$

[using cosine formula]

$$= \frac{1}{2kabc}[(b^2 - c^2)(b^2 + c^2 - a^2)(c^2 - a^2)(a^2 + c^2 - b^2) + (a^2 - b^2)(a^2 + b^2 - c^2)]$$

$$= \frac{1}{2kabc}[(b^2 - c^2)(b^2 + c^2) - a^2(b^2 - c^2) + (c^2 - a^2)(c^2 + a^2) - b^2(c^2 - a^2)$$
$$+ (a^2 - b^2)(a^2 + b^2) - c^2(a^2 - b^2)]$$

$$= \frac{1}{2kabc}[(b^2 - c^2)(b^2 + c^2) + (c^2 - a^2)(c^2 + a^2) + (a^2 - b^2)(a^2 + b^2)$$
$$- a^2(b^2 - c^2) - b^2(c^2 - a^2) - c^2(a^2 - b^2)]$$

$$= \frac{1}{2kabc}[(b^4 - c^4) + (c^4 - a^4) + (a^4 - b^4) - (a^2b^2 - a^2c^2) - (b^2c^2 - b^2a^2) - (c^2a^2 - c^2b^2)]$$

$= \dfrac{1}{2kabc} \times 0 = 0 = \text{R.H.S.}$ **Proved.**

Example 9. *For any triangle ABC, prove that*

(*i*) $a(b\cos C - c\cos B) = b^2 - c^2$

(*ii*) $2(bc\cos A + ca\cos B + ab\cos C) = a^2 + b^2 + c^2.$

Solution. (*i*) L.H.S. $= a(b\cos C - c\cos B)$

$= ab\cos C - ac\cos B$

Putting the values of cos *B* and cos *C*, we get

$$\text{L.H.S.} = ab\left(\frac{a^2+b^2-c^2}{2ab}\right) - ac\left(\frac{c^2+a^2-b^2}{2ac}\right)$$

$$= \left(\frac{a^2+b^2-c^2}{2}\right) - \left(\frac{c^2+a^2-b^2}{2}\right)$$

$$= \frac{1}{2}[a^2+b^2-c^2-c^2-a^2+b^2]$$

$$= \frac{1}{2}[2b^2-2c^2] = b^2-c^2 = \text{R.H.S.}$$ **Proved.**

(ii) L.H.S. $= 2(bc\cos A + ca\cos B + ab\cos C)$ ...(1)

Putting the values of cos *A*, cos *B* and cos *C* in (1), we get

$$= 2bc\left(\frac{b^2+c^2-a^2}{2bc}\right) + 2ca\left(\frac{c^2+a^2-b^2}{2ca}\right) + 2ab\left(\frac{a^2+b^2-c^2}{2ab}\right)$$

$$= (b^2+c^2-a^2) + (c^2+a^2-b^2) + (a^2+b^2-c^2)$$

$$= a^2+b^2+c^2 = \text{R.H.S.}$$ **Proved.**

Example 10. *If in a triangle ABC,*

$\frac{2\cos A}{a} + \frac{\cos B}{b} + \frac{2\cos C}{c} = \frac{a}{bc} + \frac{b}{ca}$, *then prove that the triangle is right angled.*

Solution. We have,

$$\frac{2\cos A}{a} + \frac{\cos B}{b} + \frac{2\cos C}{c} = \frac{a}{bc} + \frac{b}{ca} \quad ...(1)$$

Putting the values of cos A, cos B and cos C in (1), we get

$$\Rightarrow \quad 2\left(\frac{b^2+c^2-a^2}{2abc}\right) + \left(\frac{c^2+a^2-b^2}{2abc}\right) + 2\left(\frac{a^2+b^2-c^2}{2abc}\right) = \frac{a}{bc} + \frac{b}{ca}$$

$$\Rightarrow \quad 2(b^2+c^2-a^2) + (c^2+a^2-b^2) + 2(a^2+b^2-c^2) = 2a^2+2b^2$$

$$\Rightarrow \quad b^2+c^2 = a^2$$

Hence, ΔBAC is a right angled triangle. **Proved.**

Example 11. *In a* ΔABC, *if* $\cos B = \frac{\sin A}{2\sin C}$, *show that the triangle is isosceles.*

Solution. We know that,

$$\cos B = \frac{c^2 + a^2 - b^2}{2ac} \quad ...(1)$$

Also; by sine rule, $\sin A = ak$, $\sin C = ck$ ...(2)

$$\cos B = \frac{\sin A}{2 \sin C}$$

Putting the values of cos B, sin A, sin C from (1) and (2) in (3), we get

$$\Rightarrow \quad \frac{c^2 + a^2 - b^2}{2ca} = \frac{ak}{2ck} \Rightarrow \frac{c^2 + a^2 - b^2}{2ca} = \frac{a}{2c} \Rightarrow c^2 + a^2 - b^2 = \frac{a}{2c} \times 2ca$$

$$\Rightarrow \quad c^2 + a^2 - b^2 = a^2 \Rightarrow c^2 = b^2 \Rightarrow c = b$$

Since, two sides of triangle are equal, therefore, triangle is isosceles. **Proved.**

Example 12. *With usual notations, if in a triangle ABC*

$$\frac{b+c}{11} = \frac{c+a}{12} = \frac{a+b}{13}, \textit{ then prove that } \frac{\cos A}{7} = \frac{\cos B}{19} = \frac{\cos C}{25}$$

Solution. Given that,

$$\frac{b+c}{11} = \frac{c+a}{12} = \frac{a+b}{13} = k \text{ (say)}$$

Then, $b + c = 11k$, $c + a = 12k$, $a + b = 13k$,

Adding these, we get

$$(a+c) + (c+a) + (a+b) = 11k + 12k + 13k$$

$$\Rightarrow \quad 2(a+b+c) = 36k \Rightarrow a+b+c = 18k$$

Now,

$$b + c = 11k \text{ and } a+b+c = 18k \Rightarrow a = 7k$$

$$c + a = 12k \text{ and } a+b+c = 18k \Rightarrow b = 6k$$

$$a + b = 13k \text{ and } a+b+c = 18k \Rightarrow c = 5k$$

$$\cos A = \frac{b^2 + c^2 - a^2}{2bc} = \frac{36k^2 + 25k^2 - 49k^2}{60k^2} = \frac{12}{60} = \frac{1}{5}$$

$$\cos B = \frac{c^2 + a^2 - b^2}{2ac} = \frac{25k^2 + 49k^2 - 36k^2}{70k^2} = \frac{38}{70} = \frac{19}{35}$$

and

$$\cos C = \frac{a^2 + b^2 - c^2}{2ab} = \frac{49k^2 + 36k^2 - 25k^2}{84k^2} = \frac{60}{84} = \frac{5}{7}$$

Now,

$$\cos A : \cos B : \cos C = \frac{1}{5} : \frac{19}{35} : \frac{5}{7} = 7 : 19 : 25$$

$$\Rightarrow \quad \frac{\cos A}{7} = \frac{\cos B}{19} = \frac{\cos C}{25}$$

Proved.

Example 13. *In any triangle ABC if* $\angle C = 60°$, *prove that*

$$\frac{1}{a+c} + \frac{1}{b+c} = \frac{3}{a+b+c}$$

Solution. We have, $\angle C = 60° \Rightarrow \cos C = \dfrac{1}{2}$

$$\Rightarrow \quad \frac{a^2+b^2-c^2}{2ab} = \frac{1}{2} \qquad \left[\cos C = \frac{a^2+b^2-c^2}{2ab}\right]$$

$$\Rightarrow \quad a^2+b^2-ab = c^2 \qquad ...(1)$$

Now, $\dfrac{1}{a+c}+\dfrac{1}{b+c} = \dfrac{3}{a+b+c}$

If $\dfrac{a+b+2c}{(a+c)(b+c)} = \dfrac{3}{a+b+c}$

i.e. If $(a+b+2c)(a+b+c) = 3(a+c)(b+c)$

i.e. If $(a+b)^2+2c^2+3c(a+b) = 3(ab+ac+bc+c^2)$

i.e. If $a^2+b^2+2ab+2c^2+3ac+3bc = 3ab+3ac+3bc+3c^2$

i.e. If $a^2+b^2-ab = c^2$, which is given. **Proved.**

Example 14. *In any triangle ABC, show that*

$$\left(\frac{b^2-c^2}{a^2}\right)\sin 2A + \left(\frac{c^2-a^2}{b^2}\right)\sin 2B + \left(\frac{a^2-b^2}{c^2}\right)\sin 2C = 0$$

Solution. We have,

$$\left(\frac{b^2-c^2}{a^2}\right)\sin 2A = \left(\frac{b^2-c^2}{a^2}\right)2\sin A\cos A = \frac{b^2-c^2}{a^2}.(2ka)\left(\frac{b^2+c^2-a^2}{2bc}\right)$$

$$= k\left(\frac{b^2-c^2}{a}\right)\left(\frac{b^2+c^2-a^2}{bc}\right)$$

$$= \frac{k}{abc}[(b^2-c^2)(b^2+c^2)-a^2(b^2-c^2)]$$

$$\Rightarrow \quad \left(\frac{b^2-c^2}{a^2}\right)\sin 2A = \frac{k}{abc}(b^4-c^4-a^2b^2+c^2a^2) \qquad ...(1)$$

Similarly, $$\left(\frac{c^2-a^2}{b^2}\right)\sin 2B = \frac{k}{abc}(c^4-a^4-b^2c^2+a^2b^2) \qquad ...(2)$$

and $$\left(\frac{a^2-b^2}{c^2}\right)\sin 2C = \frac{k}{abc}(a^4-b^4-c^2a^2+b^2c^2) \qquad ...(3)$$

Adding (1), (2) and (3), we get

$$\left(\frac{b^2-c^2}{a^2}\right)\sin 2A + \left(\frac{c^2-a^2}{b^2}\right)\sin 2B + \left(\frac{a^2-b^2}{c^2}\right)\sin 2C$$

$$= \frac{k}{abc}(b^4-c^4-a^2b^2+c^2a^2+c^4-a^4-b^2c^2$$

$$+a^2b^2+a^4-b^4-c^2a^2+b^2c^2)$$

$$= \frac{k}{abc}(0) = 0$$ **Proved.**

Example 15. *For any triangle ABC, prove that*

$$a(\cos C - \cos B) = 2(b - c)\cos^2 \frac{A}{2}$$

Solution. L.H.S. $= a(\cos C - \cos B)$

$$= a\left(\frac{a^2 + b^2 - c^2}{2ab} - \frac{a^2 + c^2 - b^2}{2ac}\right)$$

$$= a\left(\frac{a^2 c + b^2 c - c^3 - a^2 b - c^2 b + b^3}{2abc}\right)$$

$$= \frac{b^3 - c^3 + b^2 c - c^2 b + a^2 c - a^2 b}{2bc}$$

$$= \frac{(b^3 - c^3) + bc(b - c) - a^2(b - c)}{2bc}$$

$$= \frac{(b - c)(b^2 + c^2 + bc) + bc(b - c) - a^2(b - c)}{2bc}$$

$$= (b - c)\left(\frac{b^2 + c^2 + bc + bc - a^2}{2bc}\right) = (b - c)\left\{\frac{b^2 + c^2 - a^2}{2bc} + \frac{2bc}{2bc}\right\}$$

$$= (b - c)\{\cos A + 1\} = 2(b - c)\left\{\frac{\cos A + 1}{2}\right\}.$$

$$= 2(b - c)\cos^2 \frac{A}{2} = \text{R.H.S.} \left[\because \cos A = 2\cos^2 \frac{A}{2} - 1\right]$$ **Proved.**

Example 16. *Show that for any parallelogram, if a and b are the sides of two non-parallel sides, θ is the angle between these two sides and d is the length of the diagonal that has a common vertex with sides a and b, then* $d^2 = a^2 + b^2 + 2ab\cos\theta$.

Solution. Let $ABCD$ be the given parallelogram and $AB = a$, $AD = b$

$\angle BAD = \theta$ and $AC = d$.

Since, $ABCD$ is a parallelogram, so $BC = AD = b$

Also, $\angle BAD = \dfrac{2\pi - 2\theta}{2} = \pi - \theta$

[**Note:** In a parallelogram, opposite angles are equal and sum of angles in a quadrilateral is 360°]

$\therefore \quad \angle ABC + \angle ADC = 2\pi - 2\theta$, *i.e.* $2\angle ABC = 2\pi - 2\theta$].

Now, in ΔABC, $\cos(\angle ABC) = \cos(\pi - \theta) = -\cos\theta$

By cosine's law $\cos(\angle ABC) = -\cos\theta$

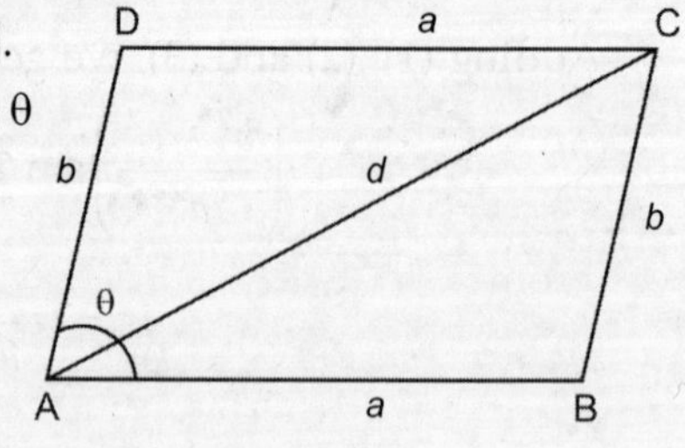

$$\Rightarrow \frac{a^2 + b^2 - d^2}{2ab} = -\cos\theta$$

$$\Rightarrow a^2 + b^2 - d^2 = -2ab\cos\theta$$

$$\Rightarrow a^2 + b^2 + 2ab\cos\theta = d^2$$

Proved.

EXERCISE 17.2

1. In a triangle ABC, if $a = 25$, $b = 52$, and $c = 63$, find $\cos A$. **Ans.** $\frac{12}{13}$
2. In a triangle ABC, if $a = 3$, $b = 5$, $c = 7$; find $\cos A$, $\cos B$ and $\cos C$. **Ans.** $\frac{13}{14}, \frac{11}{14}, -\frac{1}{2}$
3. If $b = 8$, $c = 5$ and $A = 60°$, find a. **Ans.** 7
4. If $a = 13$, $b = 8$, $c = 7$, find A. **Ans.** 120°
5. If the sides of a triangle ABC are $a = 9$, $b = 8$, $c = 4$, show that $6 \cos C = 4 + 3 \cos B$.
6. The sides of a triangle are $a = 4$, $b = 6$ and $c = 8$, show that $8 \cos A + 16 \cos B + 4 \cos C = 17$.
7. If the sides of a triangle are $a, b, \sqrt{a^2 + b^2 + ab}$, prove that the greatest angle of the triangle is 120°.

In any ΔABC, prove that:

8. $\frac{\cos A}{a} + \frac{\cos B}{b} + \frac{\cos C}{c} = \frac{a^2 + b^2 + c^2}{2abc}$
9. $a^2 = (b + c)^2 - 4bc \cos^2\left(\frac{A}{2}\right)$
10. $2\left(b \cos^2 \frac{C}{2} + c \cos^2 \frac{B}{2}\right) = a + b + c$
11. $(c^2 - a^2 + b^2) \tan A = (a^2 - b^2 + c^2) \tan B = (b^2 - c^2 + a^2) \tan C$
12. $(a - b)^2 \cos^2 \frac{C}{2} + (a + b)^2 \sin^2 \frac{C}{2} = c^2$
13. $\left(\frac{b^2 - c^2}{a^2}\right) \sin 2A + \left(\frac{c^2 - a^2}{b^2}\right) \sin 2B + \left(\frac{a^2 - b^2}{c^2}\right) \sin 2C = 0$
14. $(a - b)^2 \cos^2 \frac{C}{2} + (a + b)^2 \sin^2 \frac{C}{2} = c^2$
15. In a triangle ABC, if $\angle B = 60°$, prove that: $(a + b + c)(a - b + c) = 3ca$
16. In a triangle ABC, if $\frac{\cos A}{a} = \frac{\cos B}{b}$, show that the triangle is isosceles.

17.4 PROJECTION FORMULAE

Any side of a triangle is equal to the sum of the projections of other two sides on it, *i.e.*

(i) $a = b \cos C + c \cos B$

(ii) $b = c \cos A + a \cos C$

(iii) $c = a \cos B + b \cos A$

Case I. When ΔABC is an acute angled triangle.

From A, draw $AD \perp BC$.

In ΔABD, we have

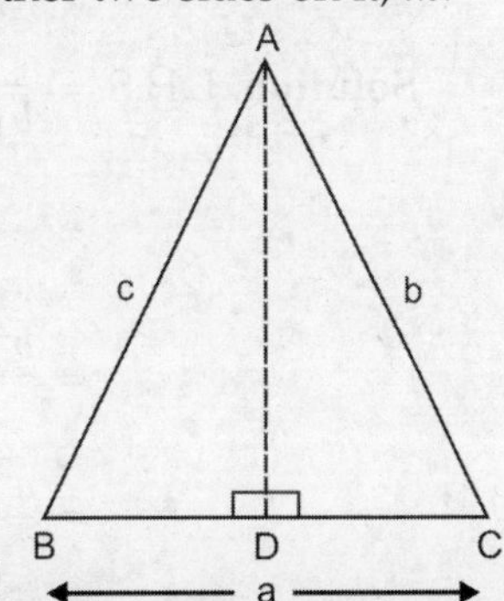

$$\cos B = \frac{BD}{AB} \Rightarrow \cos B = \frac{BD}{c}$$

$$\Rightarrow \quad BD = c \cos B \quad \text{... (1)}$$

In ΔACD, we have

$$\cos C = \frac{CD}{AC} \Rightarrow \cos C = \frac{CD}{b}$$

$$\Rightarrow \quad CD = b \cos C \quad \text{... (2)}$$

Also, $BC = BD + CD = c \cos B + b \cos C$ [Using (1) and (2)]

$$\Rightarrow \quad a = c \cos B + b \cos C$$

Case II. When ΔABC is an obtuse angled triangle.

From A, draw $AD \perp BC$ (produced)

In ΔABD, we have

$$\cos B = \frac{BD}{AB} \Rightarrow \cos B = \frac{BD}{c}$$

$$\Rightarrow \quad BD = c \cos B \quad \text{... (1)}$$

In ΔACD, we have

$$\cos (\pi - C) = \frac{CD}{AC} \Rightarrow -\cos C = \frac{CD}{b}$$

$$\Rightarrow \quad CD = -b \cos C \quad \text{... (2)}$$

Also, $BC = BD - CD = c \cos B - (-b \cos C)$ [Using (1) and (2)]

$$\Rightarrow \quad a = c \cos B + b \cos C$$

Case III. When ΔABC is a right angled triangle.

Here, $\angle C = 90° \Rightarrow \cos C = 0$ ... (1)

In ΔABC $\cos B = \frac{BC}{AB} \Rightarrow \cos B = \frac{a}{c}$

$$\Rightarrow \quad a = c \cos B$$

$$\Rightarrow \quad a = c \cos B + 0$$

$$\Rightarrow \quad a = c \cos B + b \cos C \quad [\text{From (1)}, \because \cos C = 0]$$

Hence, in all the above cases, we have

$$a = c \cos B + b \cos C$$

Similarly, other results can be proved.

$a = b \cos C + c \cos B$ | **$b = c \cos A + a \cos C$** | **$c = a \cos B + b \cos A$**

Example 17. *In any ΔABC, prove that*

$$\frac{\cos A}{b \cos C + c \cos B} + \frac{\cos B}{c \cos A + a \cos C} + \frac{\cos C}{A \cos B + b \cos A} = \frac{a^2 + b^2 + c^2}{2abc}$$

Solution. L.H.S. $= \frac{\cos A}{b \cos C + c \cos B} + \frac{\cos B}{c \cos A + a \cos C} + \frac{\cos C}{a \cos B + b \cos A}$

$$= \frac{\cos A}{a} + \frac{\cos B}{b} + \frac{\cos C}{c} \quad [\because a = b \cos C + c \cos B \text{ etc.}]$$

$$= \frac{b^2 + c^2 - a^2}{2abc} + \frac{a^2 + c^2 - b^2}{2abc} + \frac{a^2 + b^2 - c^2}{2abc} \quad \text{[Using cosine formulae]}$$

$$= \frac{a^2 + b^2 + c^2}{2abc} = \text{R.H.S.}$$

Proved.

Example 18. *In any ΔABC, prove that*

$$(i)\ 2\left(a\sin^2\frac{C}{2}+c\sin^2\frac{A}{2}\right)=a+c-b \quad (ii)\ 2\left(b\cos^2\frac{C}{2}+c\cos^2\frac{B}{2}\right)=a+b+c$$

Solution. (*i*) L.H.S. $= 2\left(a\sin^2\frac{C}{2}+c\sin^2\frac{A}{2}\right)=[a(1-\cos C)+c(1-\cos A)]$

$= a + c - (a \cos C + c \cos A) = a + c - b =$ R.H.S. **Proved.**

(*ii*) L.H.S. $= 2\left(b\cos^2\frac{C}{2}+c\cos^2\frac{B}{2}\right)=[b(1+\cos C)+c(1+\cos B)]$

$= [b + c + (b \cos C + c \cos B)]$

$= b + c + a = a + b + c =$ R.H.S. $[\because a = b\cos C + c\cos B]$ **Proved.**

Example 19. *In any triangle ABC, prove that*

$$a^3 \cos(B - C) + b^3 \cos(C - A) + c^3 \cos(A - B) = 3abc$$

Solution. We know that,

$$\frac{a}{\sin A}=\frac{b}{\sin B}=\frac{c}{\sin C}=k$$

Then, $a = k \sin A$, $b = k \sin B$, $c = k \sin C$

L.H.S. $= a^3 \cos(B - C) + b^3 \cos(C - A) + c^3 \cos(A - B)$

$= a^2 k \sin A \cos(B - C) + b^2 k \sin B \cos(C - A) + c^2 k \sin C \cos(A - B)$

$= \frac{k}{2}\,[a^2\{2\sin A\cos(B - C)\} + b^2\{2\sin B\cos(C - A)\} + c^2\{2\sin C\cos(A - B)\}]$

$= \frac{k}{2}\,[a^2\{2\sin(B + C)\cos(B - C)\} + b^2\{2\sin(C + A)\cos(C - A)\}$
$+ c^2\{2\sin(A + B)\cos(A - B)\}]$

$= \frac{k}{2}\,[a^2(\sin 2B + \sin 2C) + b^2(\sin 2C + \sin 2A) + c^2(\sin 2A + \sin 2B)]$

$= \frac{k}{2}\,[2a^2(\sin B\cos B + \sin C\cos C) + 2b^2(\sin C\cos C + \sin A\cos A)$
$+ 2c^2(\sin A\cos A + \sin B\cos B)]$

$= [a^2(k\sin B\cos B + k\sin C\cos C) + b^2(k\sin C\cos C + k\sin A\cos A)$
$+ c^2(k\sin A\cos A + k\sin B\cos B)]$

$= [a^2(b\cos B + c\cos C) + b^2(c\cos C + a\cos A) + c^2(a\cos A + b\cos B)]$

$= ab(a\cos B + b\cos A) + bc(b\cos C + c\cos B) + ca(a\cos C + c\cos A)$

$= abc + bca + cab = 3abc =$ R.H.S. [using projection formulae] **Proved.**

Example 20. *Dedeuce sine formulae by using projection formulae.*

Solution. By projection formulae, we have

$a = b\cos C + c\cos B$ ... (1)

and $b = c\cos A + a\cos C$ ... (2)

From (1), we have

$a - (\cos C)\,b - (\cos B)\,c = 0$ ... (3)

From (2), we have

$(\cos C)\,a - b + (\cos A)\,c = 0$ ... (4)

Solving (3) and (4) for a, b, c, we get

$$\frac{a}{-\cos A\cos C-\cos B}=\frac{b}{-\cos B\cos C-\cos A}=\frac{c}{-1+\cos^2 C}$$

$$\Rightarrow \quad \frac{a}{\cos A\cos C+\cos B}=\frac{b}{\cos B\cos C+\cos A}=\frac{c}{1-\cos^2 C}$$

$$\Rightarrow \quad \frac{a}{\cos A\cos C-\cos(A+C)}=\frac{b}{\cos B\cos C-\cos(B+C)}=\frac{c}{\sin^2 C}$$

$$[\because \cos B=\cos\{\pi-(A+C)\}=-\cos(A+C)\text{ etc.}]$$

$$\Rightarrow \quad \frac{a}{\cos A\cos C-(\cos A\cos C-\sin A\sin C)}$$

$$=\frac{b}{\cos B\cos C-(\cos B\cos C-\sin B\sin C)}=\frac{c}{\sin^2 C}$$

$$\Rightarrow \quad \frac{a}{\sin A\sin C}=\frac{b}{\sin B\sin C}=\frac{c}{\sin^2 C}$$

$$\Rightarrow \quad \frac{a}{\sin A}=\frac{b}{\sin B}=\frac{c}{\sin C}$$

This is the required sine formula. **Proved.**

EXERCISE 17.3

In a triangle ABC, prove the following:

1. $\dfrac{\sin B}{\sin C}=\dfrac{c-a\cos B}{b-a\cos C}$
2. $(b+c)\cos A+(c+a)\cos B+(a+b)\cos C=a+b+c$
3. $\sin^3 A\cos(B-C)+\sin^3 B\cos(C-A)+\sin^3 C\cos(A-B)=3\sin A\sin B\sin C$
4. $a(\cos B+\cos C-1)+b(\cos C+\cos A-1)+c(\cos A+\cos B-1)=0$
5. $\dfrac{c-b\cos A}{b-c\cos A}=\dfrac{\cos B}{\cos C}$
6. $2\left(a\sin^2\dfrac{C}{2}+c\sin^2\dfrac{A}{2}\right)=c+a-b$
7. If $c\cos^2\dfrac{A}{2}+a\cos^2\dfrac{C}{2}=\dfrac{3b}{2}$, then show that a, b, c are in A.P.
8. $\dfrac{a\sec A+b\sec B}{\tan A+\tan B}=\dfrac{b\sec B+c\sec C}{\tan B+\tan C}=\dfrac{c\sec C+a\sec A}{\tan C+\tan A}$
9. $2\left(b\cos^2\dfrac{C}{2}+c\cos^2\dfrac{B}{2}\right)=a+b+c$
10. $\dfrac{\cos C}{\cos A}=\dfrac{a-c\cos B}{c-a\cos B}$.

17.5 LAW OF TANGENTS (NAPIER'S ANALOGY)

In any triangle ABC,

(*i*) $\tan\left(\dfrac{B-C}{2}\right)=\left(\dfrac{b-c}{b+c}\right)\cot\dfrac{A}{2}$ (*ii*) $\tan\left(\dfrac{A-B}{2}\right)=\left(\dfrac{a-b}{a+b}\right)\cot\dfrac{C}{2}$

(*iii*) $\tan\left(\frac{C-A}{2}\right)=\left(\frac{c-a}{c+a}\right)\cot\frac{B}{2}$

Proof: We know that,

$$\frac{a}{\sin A}=\frac{b}{\sin B}=\frac{c}{\sin C}=k$$

$\Rightarrow$ $a = k\sin A,\ b = k\sin B,\ c = k\sin C$ ... (1)

Putting the values of b and c on R.H.S., we get

(*i*) R.H.S. $=\left(\frac{b-c}{b+c}\right)\cot\frac{A}{2}=\left(\frac{k\sin B-k\sin C}{k\sin B+k\sin C}\right)\cot\frac{A}{2}$ [using (1)]

$$=\left(\frac{\sin B-\sin C}{\sin B+\sin C}\right)\cot\frac{A}{2}=\left(\frac{2\sin\left(\frac{B-C}{2}\right)\cos\left(\frac{B+C}{2}\right)}{2\sin\left(\frac{B+C}{2}\right)\cos\left(\frac{B-C}{2}\right)}\right)\cot\frac{A}{2}$$

$$=\tan\left(\frac{B-C}{2}\right)\cot\left(\frac{B+C}{2}\right)\cot\frac{A}{2}$$

$$=\tan\left(\frac{B-C}{2}\right)\cot\left(\frac{\pi}{2}-\frac{A}{2}\right)\cot\frac{A}{2}$$

$$=\tan\left(\frac{B-C}{2}\right)\tan\frac{A}{2}\cot\frac{A}{2}=\tan\left(\frac{B-C}{2}\right)=\text{L.H.S.}$$

Similarly, (ii) and (iii) can be proved.

Example 21. *In any triangle ABC, if* $\angle A = 30°$, $b = 3$ *and* $c = 3\sqrt{3}$, *then find* $\angle B$ *and* $\angle C$.

Solution. Here $\angle A = 30°$

$\therefore$ $\frac{B+C}{2} = 90° - \frac{A}{2} = 90° - 15° = 75°$ ... (1)

Since, $c > b \Rightarrow \angle C > \angle B \Rightarrow B + C = 150°$ ... (2)

$$\tan\left(\frac{C-B}{2}\right)=\frac{c-b}{c+b}\cot\frac{A}{2}=\frac{c-b}{c+b}\tan\left(\frac{B+C}{2}\right)$$

$\Rightarrow$ $\tan\left(\frac{C-B}{2}\right)=\frac{3\sqrt{3}-3}{3(\sqrt{3}+1)}\tan 75°$ [using (1)]

$\Rightarrow$ $\tan\left(\frac{C-B}{2}\right)=\frac{3(\sqrt{3}-1)}{3(\sqrt{3}+1)}\tan(45°+30°)$

$$=\frac{(\sqrt{3}-1)}{(\sqrt{3}+1)}\left(\frac{\tan 45°+\tan 30°}{1-\tan 45°\tan 30°}\right)$$

$$=\frac{(\sqrt{3}-1)}{(\sqrt{3}+1)}\left(\frac{1+\frac{1}{\sqrt{3}}}{1-\frac{1}{\sqrt{3}}}\right)=\left(\frac{(\sqrt{3}-1)}{(\sqrt{3}+1)}\right)\left(\frac{(\sqrt{3}+1)}{(\sqrt{3}-1)}\right)=1$$

$$\Rightarrow \quad \frac{C-B}{2} = 45° \qquad [\because \tan 45° = 1]$$

$$\Rightarrow \quad C - B = 90° \qquad \text{... (3)}$$

Solving (2) and (3), we get

$$\angle B = 30° \text{ and } \angle C = 120°$$ **Ans.**

17.6 APPLICATION OF SINE AND COSINE FORMULA

Area of a Triangle

The area 'Δ' of a triangle *ABC* is given by

$$\Delta = \frac{1}{2} bc \sin A = \frac{1}{2} ca \sin B = \frac{1}{2} ab \sin C$$

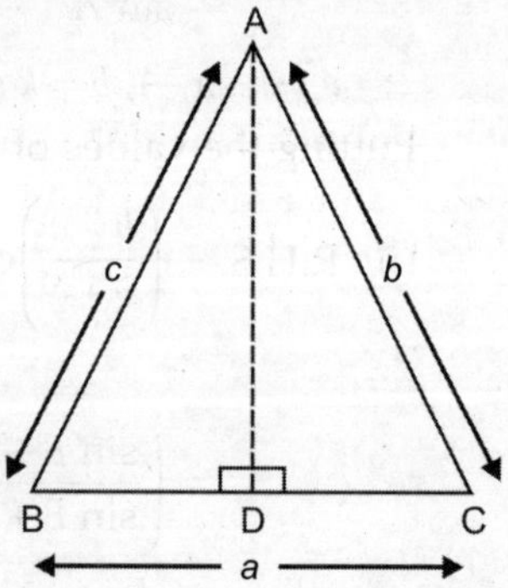

Proof: From *A*, draw $AD \perp BC$

In right angled ΔABD, we have

$$\sin B = \frac{AD}{AB} \Rightarrow \sin B = \frac{AD}{c}$$

$$\Rightarrow \quad AD = c \sin B \qquad \text{... (1)}$$

Now, Δ = Area of $\Delta ABC = \frac{1}{2} \times (\text{base}) \times (\text{height})$

$$\Rightarrow \quad \Delta = \frac{1}{2}(BC)(AD)$$

$$= \frac{1}{2}(a)(c \sin B) \qquad \text{[using (1)]}$$

$$\Rightarrow \quad \Delta = \frac{1}{2} ac \sin B$$

Similarly, it can be proved that

$$\Delta = \frac{1}{2} ab \sin C \text{ and } \Delta = \frac{1}{2} bc \sin A$$

Hence, $\boxed{\Delta = \frac{1}{2} bc \sin A}$ $\boxed{\Delta = \frac{1}{2} ca \sin B}$ $\boxed{\Delta = \frac{1}{2} ab \sin C}$ **Proved.**

Example 22. *Find the area of a triangle ABC in which $\angle C = 60°$, a = 5 cm and b = 6 cm.*

Solution. We have, $\angle C = 60°$, $a = 5$ cm, $b = 6$ cm

$$\text{Area of } \Delta ABC = \frac{1}{2} ab \sin C$$

$$= \frac{1}{2} \times 5 \times 6 \times \sin 60°$$

$$= 15 \times \frac{\sqrt{3}}{2} = \frac{15\sqrt{3}}{2} \text{ square cm.}$$ **Ans.**

Example 23. *In a triangle ABC, if a = 18, b = 24 and c = 30, find area of triangle ABC.*

Solution. Here, $a = 18$, $b = 24$ and $c = 30$.

We have,

$$\cos C = \frac{a^2 + b^2 - c^2}{2ab} = \frac{(18)^2 + (24)^2 - (30)^2}{2(18)(24)}$$

$$= \frac{324 + 576 - 900}{864} = \frac{900 - 900}{864} = 0$$

$$\Rightarrow \qquad \angle C = 90°$$

$$\therefore \qquad \Delta = \frac{1}{2} ab \sin C = \frac{1}{2} \times 18 \times 24 \times \sin 90° = 216 \text{ square units.}$$ **Ans.**

Example 24. *In any triangle ABC, prove that* $4\,\Delta \cot A = b^2 + c^2 - a^2$.

Solution.

$$\text{L.H.S.} = 4\Delta \cot A = 4 \cdot \frac{1}{2} bc \sin A \cdot \frac{\cos A}{\sin A}$$

$$= 2bc \cos A = 2bc \cdot \frac{b^2 + c^2 - a^2}{2bc} = b^2 + c^2 - a^2 = \text{R.H.S.}$$ **Proved.**

Example 25. *In any triangle ABC, prove that*

$$\frac{a^2 - b^2}{2} \cdot \frac{\sin A \sin B}{\sin (A - B)} = \Delta$$

Solution.

$$\text{L.H.S.} = \frac{a^2 - b^2}{2} \cdot \frac{\sin A \sin B}{\sin (A - B)} = \frac{(k^2 \sin^2 A - k^2 \sin^2 B) \sin A \sin B}{2 \sin (A - B)}$$

$$= \frac{k^2 (\sin^2 A - \sin^2 B) \sin A \sin B}{2 \sin (A - B)}$$

[using sine formula $a = k \sin A$ etc.]

$$= \frac{k^2 \sin (A + B) \sin (A - B) \cdot \sin A \sin B}{2 \sin (A - B)}$$

$$= \frac{k^2}{2} \cdot \sin (A + B) \sin A \sin B$$

$$= \frac{1}{2} (k \sin A)(k \sin B) \sin (\pi - C) \qquad [\because A + B = \pi - C]$$

$$= \frac{1}{2} ab \sin C = \Delta = \text{R.H.S.}$$ **Proved.**

Example 26. *In any triangle ABC, prove that*

$$a \cos A + b \cos B + c \cos C = 2a \sin B \sin C = \frac{8\Delta^2}{abc}.$$

Solution. Let

$$\frac{a}{\sin A} = \frac{b}{\sin B} = \frac{c}{\sin C} = k$$

$$\Rightarrow \qquad a = k \sin A,\ b = k \sin B,\ c = k \sin C \qquad \text{... (1)}$$

Now,

$$\text{L.H.S.} = a \cos A + b \cos B + c \cos C$$

$$= k \sin A \cos A + k \sin B \cos B + k \sin C \cos C \qquad \text{[using (1)]}$$

$$= \frac{k}{2} [\{\sin 2A + \sin 2B\} + \sin 2C]$$

$$= \frac{k}{2}[2\sin(A+B)\cos(A-B) + 2\sin C\cos C]$$

$$= \frac{k}{2}[2\sin(\pi - C)\cos(A-B) + 2\sin C\cos C]$$

$$= \frac{k}{2}[2\sin C\cos(A-B) + 2\sin C\cos C]$$

$$= \frac{k}{2}[2\sin C\{\cos(A-B) + \cos(\pi - \overline{A+B})\}]$$

$$= k\sin C[\cos(A-B) - \cos(A+B)]$$

$$= k\sin C[2\sin A\sin B]$$

$$= 2(k\sin A)\cdot\sin A\cdot\sin C$$

$$= 2a\sin B\sin C$$

$$= 2a\left(\frac{2\Delta}{ac}\right)\left(\frac{2\Delta}{ab}\right)$$

$$\left[\Delta = \frac{1}{2}ac\sin B \Rightarrow \sin B = \frac{2\Delta}{ac} \textit{ and } \sin C = \frac{2\Delta}{ab}\right]$$

$$= \frac{8\Delta^2}{abc} = \text{R.H.S.}$$ **Proved.**

EXERCISE 17.4

1. In a ΔABC, if $\angle A = 60°$, $b = 4$ and $c = \sqrt{3}$, show that its area is 3 sq. units.
2. Find the area of ΔABC, where $a = \sqrt{2}, b = \sqrt{3}$ and $c = \sqrt{5}$. **Ans.** $\frac{1}{2}\sqrt{6}$ square units
3. In a ΔABC, if $\angle A = 30°$ and $b : c = 2 : \sqrt{3}$, find $\angle B$. **Ans.** 90°

Prove that:

4. $b^2 \sin 2C + c^2 \sin 2B = 4\Delta$
5. $4\Delta(\cot A + \cot B + \cot C) = a^2 + b^2 + c^2$
6. If $a = 2b$ and $|A - B| = \frac{\pi}{3}$, find the measure of $\angle C$. **Ans.** $\frac{\pi}{3}$
7. If $a = \sqrt{3} + 1, b = \sqrt{3} - 1$ and $\angle C = 60°$, solve the triangle.
   Ans. $c = \sqrt{6}, \angle A = 105°, \angle B = 15°$
8. If $\angle B = 90°$, prove that $\tan\frac{A}{2} = \sqrt{\frac{b-c}{b+c}}$.
9. If $\angle C = 90°$ prove that $\tan\left(\frac{A-B}{2}\right) = \frac{a-b}{a+b}$.

17.7 FORMULAE for $\sin\frac{A}{2}, \sin\frac{B}{2}, \sin\frac{C}{2}$

In any triangle ABC, if $a + b + c = 2s$

then, $\sin\frac{A}{2} = \sqrt{\frac{(s-b)(s-c)}{bc}}, \sin\frac{B}{2} = \sqrt{\frac{(s-c)(s-a)}{ca}}, \sin\frac{C}{2} = \sqrt{\frac{(s-a)(s-b)}{ab}}$

Proof: We have,

$$\cos A = \frac{b^2 + c^2 - a^2}{2bc} \quad \text{... (1)}$$

We know that,

$$2\sin^2\frac{A}{2} = 1 - \cos A$$

$$= 1 - \frac{b^2 + c^2 - a^2}{2bc} \quad \text{[Using (1)]}$$

$$= \frac{2bc - b^2 - c^2 + a^2}{2bc} = \frac{a^2 - (b - c)^2}{2bc}$$

$$= \frac{(a + b - c)(a - b + c)}{2bc}$$

$$= \frac{(2s - 2c)(2s - 2b)}{2bc}$$

$$[\because a + b + c = 2s \Rightarrow a + b + c - 2c = 2s - 2c \Rightarrow a + b - c = 2s - 2c \text{ etc.}]$$

$$\Rightarrow \quad \sin^2\frac{A}{2} = \frac{(s - c)(s - b)}{bc}$$

Since $\frac{A}{2}$ is an acute angle, therefore, $\sin\frac{A}{2}$ will be positive.

$$\Rightarrow \quad \boxed{\sin\frac{A}{2} = \sqrt{\frac{(s - b)(s - c)}{bc}}}$$

Similarly, we can prove other results for $\sin\frac{B}{2}$ and $\sin\frac{C}{2}$.

17.8 FORMULAE for $\cos\frac{A}{2}, \cos\frac{B}{2}, \cos\frac{C}{2}$

In any triangle ABC, if $a + b + c = 2s$

then, $\cos\frac{A}{2} = \sqrt{\frac{s(s - a)}{bc}}, \cos\frac{B}{2} = \sqrt{\frac{s(s - b)}{ca}}, \cos\frac{C}{2} = \sqrt{\frac{s(s - c)}{ab}}$

Proof: We have, $\cos A = \frac{b^2 + c^2 - a^2}{2bc}$ ... (1)

We know that, $2\cos^2\frac{A}{2} = 1 + \cos A$

$$= 1 + \frac{b^2 + c^2 - a^2}{2bc} \quad \text{[Using (1)]}$$

$$= \frac{2bc + b^2 + c^2 - a^2}{2bc} = \frac{(b + c)^2 - a^2}{2bc}$$

$$= \frac{(b + c + a)(b + c - a)}{2bc}$$

$$[\because a + b + c = 2s \Rightarrow a + b + c - 2a = 2s - 2a \Rightarrow b + c - a = 2s - 2a]$$

$$\Rightarrow \qquad \cos^2 A = \frac{s(s-a)}{bc}$$

$$\Rightarrow \qquad \boxed{\cos\frac{A}{2} = \sqrt{\frac{s(s-a)}{bc}}} \qquad \left[\because A < 180° \Rightarrow \frac{A}{2} < 90° \Rightarrow \cos\frac{A}{2} > 0\right]$$

Similarly, we can prove that

$$\boxed{\cos\frac{B}{2} = \sqrt{\frac{s(s-b)}{ca}}} \text{ and } \boxed{\cos\frac{C}{2} = \sqrt{\frac{s(s-c)}{ab}}} \qquad \textbf{Ans.}$$

17.9 FORMULAE for $\tan\frac{A}{2}, \tan\frac{B}{2}, \tan\frac{C}{2}$

In any triangle ABC, if $a + b + c = 2s$

then

$$\tan\frac{A}{2} = \sqrt{\frac{(s-b)(s-c)}{s(s-a)}}, \tan\frac{B}{2} = \sqrt{\frac{(s-c)(s-a)}{s(s-b)}}$$

$$\tan\frac{C}{2} = \sqrt{\frac{(s-a)(s-b)}{s(s-c)}}$$

Proof: We know that,

$$\sin\frac{A}{2} = \sqrt{\frac{(s-b)(s-c)}{bc}} \qquad \text{... (1)}$$

and $$\cos\frac{A}{2} = \sqrt{\frac{s(s-a)}{bc}} \qquad \text{... (2)}$$

Also, $$\tan\frac{A}{2} = \frac{\sin\frac{A}{2}}{\cos\frac{A}{2}} = \frac{\sqrt{\frac{(s-b)(s-a)}{bc}}}{\sqrt{\frac{s(s-a)}{bc}}} \qquad \text{[using (1) and (2)]}$$

$$\Rightarrow \qquad \boxed{\tan\frac{A}{2} = \sqrt{\frac{(s-b)(s-a)}{s(s-a)}}}$$

Similarly, we can prove that

$$\tan\frac{B}{2} = \sqrt{\frac{(s-c)(s-a)}{s(s-b)}} \text{ and } \tan\frac{C}{2} = \sqrt{\frac{(s-a)(s-b)}{s(s-c)}} \qquad \textbf{Ans.}$$

Example 27. In a ΔABC, $a = 3$, $b = 5$, $c = 6$. *Calculate:*

(*i*) $\sin\frac{A}{2}$ (*ii*) $\cos\frac{A}{2}$ (*iii*) *area of triangle*

Solution. Here, $2s = a + b + c = 3 + 5 + 6 = 14 \Rightarrow s = 7$

(*i*) $$\sin\frac{A}{2} = \sqrt{\frac{(s-b)(s-c)}{bc}} = \sqrt{\frac{(7-5)(7-6)}{5\times 6}} = \sqrt{\frac{2\times 1}{5\times 6}} = \frac{1}{\sqrt{15}}$$

(*ii*) $$\cos\frac{A}{2} = \sqrt{\frac{s(s-a)}{bc}} = \sqrt{\frac{7(7-3)}{5\times 6}} = \sqrt{\frac{7\times 4}{5\times 6}} = \sqrt{\frac{14}{15}}$$

(*iii*) $$\text{Area of } \Delta = \sqrt{s(s-a)(s-b)(s-c)} = \sqrt{7(7-3)(7-5)(7-6)}$$

$$= \sqrt{7\times 4\times 2\times 1} = \sqrt{56} = 2\sqrt{14}$$ **Ans.**

Example 28. *In a triangle ABC, if a = 18, b = 24* and *c = 30, find* $\tan\frac{A}{2}, \tan\frac{B}{2}, \tan\frac{C}{2}$.

Solution. We have,

$$a = 18, b = 24, c = 30 \qquad \text{... (1)}$$

$$\Rightarrow \quad 2s = a + b + c = 18 + 24 + 30 = 72 \Rightarrow s = 36$$

$$\tan\frac{A}{2} = \sqrt{\frac{(s-c)(s-a)}{s(s-a)}}$$

$$= \sqrt{\frac{(36-24)(36-30)}{36(36-18)}} = \sqrt{\frac{12\times 6}{36\times 18}} = \sqrt{\frac{1}{9}}$$

$$\Rightarrow \quad \tan\frac{A}{2} = \frac{1}{3}$$

$$\tan\frac{B}{2} = \sqrt{\frac{(s-c)(s-a)}{s(s-b)}} = \sqrt{\frac{(36-30)(36-18)}{36(36-24)}} = \sqrt{\frac{6\times 18}{36\times 12}}$$

$$\Rightarrow \quad \tan\frac{B}{2} = \sqrt{\frac{1}{4}} = \frac{1}{2}$$

and $$\tan\frac{C}{2} = \sqrt{\frac{(s-a)(s-b)}{s(s-c)}} = \sqrt{\frac{(36-30)(36-24)}{36(36-30)}} = \sqrt{\frac{18\times 12}{36\times 6}}$$

$$\Rightarrow \quad \tan\frac{C}{2} = 1$$ **Ans.**

Example 29. *In a triangle ABC, prove that* $(a + b + c)\left(\tan\frac{A}{2} + \tan\frac{B}{2}\right) = 2c\cot\frac{C}{2}$.

Solution. $$\text{L.H.S.} = (a+b+c)\left(\tan\frac{A}{2} + \tan\frac{B}{2}\right)$$

$$= 2s\left[\sqrt{\frac{(s-b)(s-c)}{s(s-a)}} + \sqrt{\frac{(s-c)(s-a)}{s(s-b)}}\right]$$

$$= 2s\sqrt{\frac{s-c}{s}}\left[\sqrt{\frac{s-b}{s-a}} + \sqrt{\frac{s-a}{s-b}}\right]$$

$$= 2s\sqrt{\frac{s-c}{s}}\left[\frac{s-b+s-a}{\sqrt{s-a}\sqrt{s-b}}\right]$$

$$= \frac{2\sqrt{s}\sqrt{s-c}}{\sqrt{s-a}\sqrt{s-b}}(a+b+c-b-a)$$

$$= 2c\sqrt{\frac{s(s-c)}{(s-a)(s-b)}} = \frac{2c}{\tan\frac{C}{2}}$$

$$= 2c\cot\frac{C}{2} = \text{R.H.S.}$$ **Proved.**

Example 30. *In any triangle ABC, prove that* $(b+c-a)\left(\cot\frac{B}{2}+\cot\frac{C}{2}\right) = 2a\cot\frac{A}{2}$.

Solution. L.H.S. $= (b+c-a)\left(\cot\frac{B}{2}+\cot\frac{C}{2}\right)$

$[\because a+b+c=2s \Rightarrow b+c-a=2s-2a]$

$$= (2s-2a)\left[\sqrt{\frac{s(s-b)}{(s-c)(s-a)}}+\sqrt{\frac{s(s-c)}{(s-a)(s-b)}}\right]$$

$$= 2\sqrt{s(s-a)}\left\{\sqrt{\frac{s-b}{s-c}}+\sqrt{\frac{s-c}{s-b}}\right\}$$

$$= 2\sqrt{s(s-a)}\cdot\left\{\frac{(s-b)+(s-c)}{\sqrt{(s-c)(s-b)}}\right\}$$

$$= 2\sqrt{s(s-a)}\cdot\left\{\frac{2s-(b+c)}{\sqrt{(s-c)(s-b)}}\right\}$$

$$= 2\sqrt{s(s-a)}\cdot\left\{\frac{2s-(2s-a)}{\sqrt{(s-c)(s-b)}}\right\}$$

$$= 2\sqrt{s(s-a)}\cdot\frac{a}{\sqrt{(s-c)(s-b)}} = 2a\sqrt{\frac{s(s-a)}{(s-c)(s-b)}}$$

$$= 2a\cdot\cot\frac{A}{2} = \text{R.H.S.}$$ **Proved.**

Example 31. *In any triangle ABC, show that* $\cot\frac{A}{2}+\cot\frac{B}{2}+\cot\frac{C}{2} = \frac{a+b+c}{a+b-c}\cot\frac{C}{2}$.

Solution.

$$\text{L.H.S.} = \cot\frac{A}{2}+\cot\frac{B}{2}+\cot\frac{C}{2} = \sqrt{\frac{s(s-a)}{(s-b)(s-c)}}+\sqrt{\frac{s(s-b)}{(s-c)(s-a)}}+\sqrt{\frac{s(s-c)}{(s-a)(s-b)}}$$

$$= \frac{1}{\sqrt{s(s-a)(s-b)(s-c)}}\left(\sqrt{s^2(s-a)^2}+\sqrt{s^2(s-b)^2}+\sqrt{s^2(s-c)^2}\right)$$

$$= \frac{1}{\sqrt{s(s-a)(s-b)(s-c)}}[s(s-a+s-b+s-c)]$$

$$= \frac{1}{\sqrt{s(s-a)(s-b)(s-c)}}[s\{3s-(a+b+c)\}]$$

$$= \frac{1}{\sqrt{s(s-a)(s-b)(s-c)}}[s(3s-2s)]$$

$$\Rightarrow \quad \cot\frac{A}{2} + \cot\frac{B}{2} + \cot\frac{C}{2} = \frac{s^2}{\sqrt{s(s-a)(s-b)(s-c)}} \quad \text{... (1)}$$

Now, R.H.S. $= \dfrac{a+b+c}{a+b-c}\cot\dfrac{C}{2}$

$$= \frac{2s}{a+b+c-2c}\cot\frac{C}{2} = \frac{2s}{2s-2c}\cot\frac{C}{2} = \frac{s}{s-c}\cot\frac{C}{2}$$

$$= \frac{s}{s-c}\sqrt{\frac{s(s-c)}{(s-a)(s-b)}} = \frac{s^2}{s(s-a)(s-b)(s-c)}$$

$$\Rightarrow \quad \frac{a+b+c}{a+b-c}\cot\frac{C}{2} = \frac{s^2}{\sqrt{s(s-a)(s-b)(s-c)}} \quad \text{... (2)}$$

From (1) and (2), we have

$$\cot\frac{A}{2} + \cot\frac{B}{2} + \cot\frac{C}{2} = \frac{a+b+c}{a+b-c}\cot\frac{C}{2}$$ **Proved.**

Example 32. *In a triangle ABC, prove that*

$$\frac{(a+b+c)^2}{a^2+b^2+c^2} = \frac{\cot\frac{A}{2} + \cot\frac{B}{2} + \cot\frac{C}{2}}{\cot A + \cot B + \cot C}$$

Solution. R.H.S. $= \dfrac{\cot\frac{A}{2} + \cot\frac{B}{2} + \cot\frac{C}{2}}{\cot A + \cot B + \cot C}$ ... (1)

Now, Numerator $= \cot\dfrac{A}{2} + \cot\dfrac{B}{2} + \cot\dfrac{C}{2}$

$$= \sqrt{\frac{s(s-a)}{(s-b)(s-c)}} + \sqrt{\frac{s(s-b)}{(s-c)(s-a)}} + \sqrt{\frac{s(s-c)}{(s-a)(s-b)}}$$

$$= \frac{s(s-a) + s(s-b) + s(s-c)}{\sqrt{s(s-a)(s-b)(s-c)}}$$

$$= \frac{s(s-a+s-b+s-c)}{\Delta}$$

$$\left[\because \Delta = \sqrt{s(s-a)(s-b)(s-c)}\right]$$

$$= \frac{s\{3s-(a+b+c)\}}{\Delta} = \frac{s\{3s-2s\}}{\Delta} = \frac{s^2}{\Delta}$$

$$\Rightarrow \quad \cot\frac{A}{2} + \cot\frac{B}{2} + \cot\frac{C}{2} = \frac{s^2}{\Delta} \quad \text{... (2)}$$

Also denominator, $\cot A + \cot B + \cot C = \dfrac{\cos A}{\sin A} + \dfrac{\cos B}{\sin B} + \dfrac{\cos C}{\sin C}$

$$= \frac{b^2+c^2-a^2}{2bc}\cdot\frac{bc}{2\Delta} + \frac{c^2+a^2-b^2}{2ca}\cdot\frac{ca}{2\Delta} + \frac{a^2+b^2-c^2}{2ab}\cdot\frac{ab}{2\Delta}$$

$$\left[\because \Delta = \frac{1}{2}bc\sin A \Rightarrow \sin A = \frac{2\Delta}{bc}\right]$$

$$= \frac{1}{4\Delta}[(b^2+c^2-a^2)+(c^2+a^2-b^2)+(a^2+b^2-c^2)$$

$$\Rightarrow \cot A + \cot B + \cot C = \frac{b^2+c^2-a^2+c^2+a^2-b^2+a^2+b^2-c^2}{4\Delta} = \frac{a^2+b^2+c^2}{4\Delta} \quad \text{... (3)}$$

From (1) and (2) and (3), we get

$$\text{R.H.S.} = \frac{\text{Numerator}}{\text{Denominator}} = \frac{s^2}{\Delta}\times\frac{4\Delta}{a^2+b^2+c^2} = \frac{4s^2}{a^2+b^2+c^2} = \frac{(2s)^2}{a^2+b^2+c^2}$$

$$= \frac{(a+b+c)^2}{a^2+b^2+c^2} = \text{L.H.S.}$$

Proved.

EXERCISE 17.5

1. In a ΔABC, if $a = 24$, $b = 36$ and $c = 45$, find

(*i*) $\sin\dfrac{B}{2}$ **Ans.** 0.444 (*ii*) $\cos\dfrac{B}{2}$ **Ans.** 0.895

2. In a ΔABC, if $a = 13$, $b = 14$, $c = 15$, find :

(*i*) $\tan\dfrac{A}{2}$ **Ans.** $\dfrac{1}{2}$ (*ii*) area of triangle. **Ans.** 84 square units

3. In a ΔABC, if $a = 25$, $b = 52$, $c = 63$, find the value of $\tan\dfrac{A}{2}$, $\tan\dfrac{B}{2}$ and $\tan\dfrac{C}{2}$.

4. If $a = 2$ cm, $b = 3$ cm, $\angle c = 30°$ find area. **(S.B.T.E. 2017)** **Ans.** $\dfrac{4}{5}, \dfrac{1}{2}, \dfrac{9}{7}$

In any ΔABC, prove that:

5. $b\cos^2\dfrac{C}{2} + c\cos^2\dfrac{B}{2} = s$

6. $(b+c-a)\tan\dfrac{A}{2} = (c+a-b)\tan\dfrac{B}{2} = (a+b-c)\tan\dfrac{C}{2}$

7. $a\sin^2\dfrac{C}{2} + c\sin^2\dfrac{A}{2} = \dfrac{1}{2}(c+a-b)$.

8. $\tan\dfrac{A}{2}\tan\dfrac{B}{2} = \dfrac{a+b-c}{a+b+c}$

9. $(b-c)\cot\dfrac{A}{2} + (c-a)\cot\dfrac{B}{2} + (a-b)\cot\dfrac{C}{2} = 0$

10. $\left(\cot\dfrac{A}{2}+\cot\dfrac{B}{2}\right)\left(a\sin^2\dfrac{B}{2}+b\sin^2\dfrac{A}{2}\right)=c\cot\dfrac{C}{2}$

11. $2abc\cos\dfrac{A}{2}\cos\dfrac{B}{2}\cos\dfrac{C}{2}=(a+b+c)\,\Delta=2s\Delta$

12. $bc\cos^2\dfrac{A}{2}+ca\cos^2\dfrac{B}{2}+ab\cos^2\dfrac{C}{2}=s^2$

13. In any triangle ΔABC, if $(b + c) = 3a$, prove that $\cot\dfrac{B}{2}\cot\dfrac{C}{2}=2$

14. In a ΔABC, if a, b, c are in A.P., show that $2\sin\dfrac{A}{2}\sin\dfrac{C}{2}=\sin\dfrac{B}{2}$

15. In any ΔABC, if $3\tan\dfrac{A}{2}\tan\dfrac{C}{2}=1$, prove that a, b, c are in A.P.

16. In any ΔABC, if the sides a, b, c are in A.P. prove that

$\cos A\cot\dfrac{A}{2}, \cos B\cot\dfrac{B}{2}, \cos C\cot\dfrac{C}{2}$ are in A.P.

17.10 SOLUTION OF A TRIANGLE

When three elements out of six elements (three sides and three angles) of a triangle are given. Then the remaining three unknown elements of the triangle are calculated by using certain properties of a triangle. This is known as solution of a triangle.

There are four cases in the solution of a triangle, given below.

(*i*) Three sides are given.

(*ii*) Two sides and the included angle are given.

(*iii*) Two sides and one angle opposite to one of them are given.

(*iv*) One side and two angles are given.

Three Sides are Given

Procedure (*i*) We find A, B by using the following formulae.

$$\tan\frac{A}{2}=\sqrt{\frac{(s-b)(s-c)}{s(s-a)}} \qquad \left[\because\ s=\frac{a+b+c}{2}\right]$$

and
$$\tan\frac{B}{2}=\sqrt{\frac{(s-c)(s-a)}{s(s-b)}}$$

Angle C can be determined from the relation.

$$C = 180° - (A + B)$$

(*ii*) We can also find A, B by using the formulae.

$$\cos A=\frac{b^2+c^2-a^2}{2bc}$$

$$\cos B=\frac{c^2+a^2-b^2}{2ca}$$

This method is convenient if a, b, c are small otherwise this method is cumbersome as we cannot use log tables.

Example 33. *Find the greatest angle of the triangle, given that*

$$a = 24.2,\ b = 37.5,\ c = 28.9$$

Solution. Here $a = 24.2$, $b = 37.5$, $c = 28.9$

The greatest side is b, so angle B is the greatest.

$$s = \frac{1}{2}(a+b+c) = \frac{1}{2}(24.2+37.5+28.9) = 45.3$$

$$\tan\frac{B}{2} = \sqrt{\frac{(s-c)(s-a)}{s(s-b)}} = \sqrt{\frac{(45.3-28.9)(45.3-24.2)}{45.3(45.3-37.5)}} = \sqrt{\frac{16.4\times 21.1}{45.3\times 7.8}}$$

$$\log\left(\tan\frac{B}{2}\right) = \frac{1}{2}[\log 16.4 + \log 21.1 - \log 45.3 - \log 7.8]$$

$$= \frac{1}{2}[1.2148 + 1.3243 - 1.6561 - 0.8921]$$

$$= \frac{1}{2}[2.5391 - 2.5482] = \frac{1}{2}(-0.0091)$$

$$= -0.0046 = -0.0046 + 1 - 1 = \bar{1}.9954$$

$$\frac{B}{2} = 44°42' \text{ or } B = 89°24'.$$ **Ans.**

Example 34. *Solve the triangle, having given*

$$a = 45.73,\ b = 23.17,\ c = 40.52$$

Solution. Here $a = 45.73$, b 23.17, $c = 40.52$

$$s = \frac{1}{2}(a+b+c) = \frac{1}{2}(45.73+23.17+40.52)$$

$$= \frac{109.42}{2} = 54.71$$

$$\tan\frac{A}{2} = \sqrt{\frac{(s-b)(s-c)}{s(s-a)}}$$

$$= \sqrt{\frac{(54.71-23.17)(54.71-40.52)}{54.71(54.71-45.73)}}$$

$$= \sqrt{\frac{31.54\times 14.19}{54.71\times 8.98}}$$

$$\log\left(\tan\frac{A}{2}\right) = \frac{1}{2}[\log(31.54) + \log(14.19) - \log(54.71) - \log(8.98)]$$

$$= \frac{1}{2}[1.4989 + 1.1520 - 1.7381 - 0.9533]$$

$$= \frac{1}{2}[2.6509 - 2.6914]$$

$$= \frac{1}{2}(-0.0405) = -0.02025 = -0.0203 \text{ (approx)}$$

$$= -1 + 1 - 0.0203 = 1 + 0.9797 = \bar{1}.9797$$

$$\frac{A}{2} = 43°39'30'', A = 87°19'$$

Again $\tan\frac{B}{2} = \sqrt{\frac{(s-a)(s-c)}{s(s-b)}} = \sqrt{\frac{(54.71-45.73)(54.71-40.52)}{54.71(54.71-23.17)}} = \sqrt{\frac{8.98\times 14.19}{54.71\times 31.54}}$

$$\log\left(\tan\frac{B}{2}\right) = \frac{1}{2}[\log(8.98)+\log(14.19)-\log(54.71)-\log(31.54)]$$

$$= \frac{1}{2}[0.9533+1.1520-1.7381-1.4989]$$

$$= \frac{1}{2}[2.1053-3.2370] = \frac{1}{2}(-1.1317)$$

$$= -0.56585 = -1+1-0.56585 = \bar{1}.43415$$

$$\frac{B}{2} = 15°12' \Rightarrow B = 30°24'$$

$$C = 180° - (A+B) = 180° - (87°19' + 30°24') = 62°17'$$ **Ans.**

Example 35. *In triangle ABC, a = 5, b = 8 and c = 6. Solve the triangle, that is, find angles A, B and C.* ***(Math I, B.T.E. Delhi, Jan. 2009)***

Solution. We have, $a = 5$, $b = 8$ and $c = 6$

We know that,

A
c = 6
b = 8
B
a = 5
C

$$\cos A = \frac{b^2+c^2-a^2}{2bc}$$

$$\Rightarrow \quad \cos A = \frac{64+36-25}{2\times 8\times 6} = \frac{75}{96} = \frac{25}{32}$$

$$\Rightarrow \quad A = \cos^{-1}\left(\frac{25}{32}\right)$$

$$\Rightarrow \quad \cos B = \frac{c^2+a^2-b^2}{2ca} \Rightarrow \cos B = \frac{36+25-64}{2\times 6\times 5} = -\frac{3}{60} = -\frac{1}{20} \Rightarrow B = \cos^{-1}\left(-\frac{1}{20}\right)$$

$$\cos C = \frac{a^2+b^2-c^2}{2ab} = \frac{25+64-36}{2(5)(8)} = \frac{89-36}{80} = \frac{53}{80}$$

$$C = \cos^{-1}\left(\frac{53}{80}\right)$$ **Ans.**

EXERCISE 17.6

Solve the triangle, having given:

1. $a = 6, b = 7, c = 8.$ **Ans.** $\angle A = 46°34', \angle B = 57°54', \angle C = 75°32'$
2. $a = 15, b = 13, c = 14$, find B. **Ans.** $53°8'$
3. $a = 283, b = 317, c = 428.$ **Ans.** $\angle A = 41°23'30'', \angle B = 47°46'40'', \angle C = 90°49'50''$

4. $a = 1262, b = 1364, c = 1672$, find C. **Ans.** $\angle C = 79°$

5. $a = 24.4, b = 18.4, c = 26.4$. **Ans.** $\angle A = 63°8', \angle B = 42°32', \angle C = 74°20'$

6. Two forces $F_1 = 115$ kgf and $F_2 = 215$ kgf act on a particle and their resultant is 275 kgf. Find the angle between the forces. **Ans.** $109°6'$

Given Two Sides and the Included Angle

Given. b, c and angle A.

Procedure. $B + C = 180° - A$ ... (1)

(*i*) Using Napier's analogy.

$$\tan\left(\frac{B-C}{2}\right) = \frac{b-c}{b+c}\cot\frac{A}{2}$$

$$\tan\left(\frac{B-C}{2}\right) = \frac{b-c}{b+c}\tan\left(\frac{B+C}{2}\right) \qquad \left[\cot\frac{A}{2} = \cot\left(90° - \frac{B+C}{2}\right) = \tan\left(\frac{B+C}{2}\right)\right]$$

On putting the values of right hand side, we can find $B - C$. ... (2)

From (1) and (2) we can find B and C.

(*ii*) $\dfrac{a}{\sin A} = \dfrac{b}{\sin B}$

On substituting the values of b, A, B, we can find a.

Example 36. *Solve the triangle ABC, given that*

$b = 14, c = 11, A = 60°$.

Solution. $B + C = 180° - A = 180° - 60° = 120°$ ... (1)

$$\tan\left(\frac{B-C}{2}\right) = \frac{b-c}{b+c}\cot\frac{A}{2} = \frac{14-11}{14+11}\cot 30° = \frac{3}{25}(\sqrt{3})$$

$$\log\tan\left(\frac{B-C}{2}\right) = \log\left(3\sqrt{3}\right) - \log 25$$

$$= \log\left(3^{\frac{3}{2}}\right) - \log\left(5^2\right) = \frac{3}{2}\log 3 - 2\log 5$$

$$= \frac{3}{2}\,(0.4771) - 2(0.6990)$$

$$= 0.71565 - 1.3980 = -\,0.68235$$

$$= \bar{1}.31765 = \bar{1}.3177 \text{ (approx)}$$

$$\frac{B-C}{2} = 11°44' \text{ or } B - C = 23°28' \qquad \text{... (2)}$$

From (1) and (2), we have

$$B = 71°44', C = 48°16'$$

To find *a*

$$\frac{a}{\sin A} = \frac{b}{\sin B}$$

$$\frac{a}{\sin 60°} = \frac{14}{\sin 71°44'} \quad \Rightarrow \quad a = \frac{14 \sin 60°}{\sin 71°44'}$$

$$\log a = \log 14 + \log \sin 60° - \log \sin 71°44'$$

$$= 1.1461 + \bar{1}.9375 - \bar{1}.9776 = 1.1060$$

$$a = 12.76$$

$$a = 12.76,\ B = 71°44',\ C = 48°16'$$ **Ans.**

Example 37. *Solve the triangle ABC in which $b = 251$, $c = 147$ and $A = 47°$.*

Solution. Here, $b = 251$, $c = 147$ and $A = 47°$

$$B + C = 180° - A = 180° - 47° = 133° \quad \dots (1)$$

$$\tan\left(\frac{B-C}{2}\right) = \frac{b-c}{b+c} \cot \frac{A}{2} = \frac{251-147}{251+147} \cot \frac{47°}{2} = \frac{104}{398} \frac{1}{\tan 23°30'}$$

$$\log \tan\left(\frac{B-C}{2}\right) = \log 104 - \log 398 - \log \tan 23°30'$$

$$= 2.0170 - 2.5999 - (\bar{1}.6383)$$

$$= 2.0170 - 2.5999 + 1 - 0.6383$$

$$= -0.2212 = -1 + 1 - 0.2212 = \bar{1} + 0.7788$$

$$= \bar{1}.7788$$

$$\frac{B-C}{2} = 31° \text{ or } B - C = 62° \quad \dots (2)$$

Adding (1) and (2), we get

$$B = 97°30' \text{ and } C = 35°30'$$

To find *a*

$$\frac{a}{\sin A} = \frac{b}{\sin B}$$

$$\Rightarrow \quad \frac{a}{\sin 47°} = \frac{251}{\sin 97°30'} \quad \Rightarrow \quad a = \frac{\sin 47° \times 251}{\sin 97°30'}$$

$$\log a = \log \sin 47° + \log 251 - \log \sin 97°30'$$

$$= \bar{1}.8641 + 2.3997 - \bar{1}.9963$$

$$= 2.3997 - 1 + 0.8641 + 1 - 0.9963$$

$$= 2.2675$$

$$a = 185.1$$

$$a = 185.1,\ B = 97°30',\ C = 35°33'.$$ **Ans.**

EXERCISE 17.7

Solve the triangle ABC in which

1. $a = 21.35$, $b = 35.21$, $C = 50°48'$ **Ans.** $A = 37°18'$, $B = 91°54'$, $c = 27.3$

2. $a = 21$, $b = 11$, $C = 34°\ 42'\ 30''$, find A and B. **Ans.** $A = 117°38'45''$, $B = 27°38'45''$

3. $b = 5, c = 3, A = 120°$. **Ans.** $B = 38°12'27.4''$, $C = 21°47'\,32.6''$, $a = 7$
4. $b = 2.25, c = 1.75, A = 54°$, find B and C. **Ans.** $B = 76°47'2.2''$, $C = 49°12'57.8''$
5. $b = 130$, $c = 72$, and $A = 42°$ **Ans.** $B = 105°47'$, $C = 32°13'$, $a = 98.05$

Ambiguous Case

Given two sides and the angle opposite to one of them.

Let b, c be the given sides and B the given angle

(*i*) **To find C**

$$\frac{\sin C}{c} = \frac{\sin B}{b}$$

$$\sin C = \frac{c}{b} \sin B$$

$$\log \sin C = \log c + \log \sin B - \log b$$

On substituting the values of c, B and b we can find log sin C with the help of log tables. So C is known.

Note (*a*) If $\log \sin C > 0$

$\sin C > 1$ which is not possible, so there is no solution.

(*b*) If $\log \sin C = 0$

$\sin C = 1 \quad \therefore C = 90°$

(*c*) If $\log \sin C < 0$

$\sin C < 1$

Then there are two values of C, one acute say (C_1), other obtuse, $C_2 = 180° - C_1$.

Test if $B + C_2 > 180°$, C_2 is to be rejected.

If $B + C_2 < 180°$, then there are two solutions.

This is called ambiguous case.

(*ii*) **To find A**

If C has two values C_1 and C_2 then

A has also two values.

$$A_1 = 180° - (B + C_1)$$
$$A_2 = 180° - (B + C_2)$$

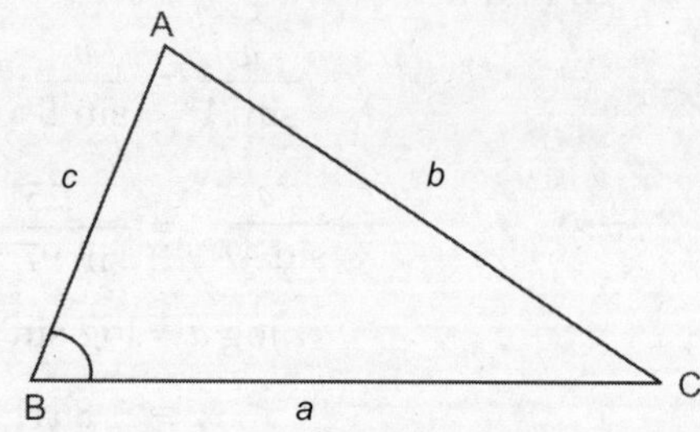

(*iii*) **To find *a***

$$\frac{a_1}{\sin A_1} = \frac{b}{\sin B} \quad \Rightarrow \quad a_1 = \frac{b \sin A_1}{\sin B}$$

$$\log a_1 = \log b + \log \sin A_1 - \log \sin B$$

Similarly, $\log a_2 = \log b + \log \sin A_2 - \log \sin B.$

We can find a_1 and a_2.

Example 38. *In a triangle ABC, b = 16 cm, c = 25 cm, and B = 33°15′. Find the angle C.*

Solution. Here, $b = 16$ cm, $c = 25$ cm, $B = 33°15'$.

We know that, $\frac{\sin C}{c} = \frac{\sin B}{b}$

$$\sin C = \frac{c}{b} \sin B$$

$$\log \sin C = \log c + \log \sin B - \log b = \log 25 + \log \sin 33°15' - \log 16$$

$$= 1.3979 + \bar{1}.7390 - 1.2041 = \bar{1}.9328$$

$$C_1 = 58°56'$$

$$C_2 = 180° - 58°56' = 121°4'$$

Ans.

Example 39. *Solve the triangle if*

$$b = 72.95,\ c = 82.31,\ B = 42°47'$$

Solution. (*i*) To find C

$$\frac{\sin C}{c} = \frac{\sin B}{b} \Rightarrow \quad \sin C = \frac{c \sin B}{b}$$

$$\log \sin C = \log c + \log \sin B - \log b$$

$$= \log (82.31) + \log \sin (42°47') - \log (72.95)$$

$$= 1.9155 + \bar{1}.8320 - 1.8630 = -0.1155 = \bar{1}.8845$$

$$C_1 = 50°2'.$$

$$C_2 = 180° - 50°2' = 129°58'$$

I Method

$$C = 50°2'$$

$$A = 180° - (B + C)$$

$$= 180° - (42°47' + 50°2')$$

$$= 180° - 92°49'$$

$$= 87°11'$$

To find *a*

$$\frac{a}{\sin A} = \frac{b}{\sin B}$$

$$a = \frac{b \sin A}{\sin B}$$

$$= \frac{72.95 \sin 87°11'}{\sin 42°27'}$$

$$\log a = \log (72.95) + \log \sin (87°11') - \log \sin (42°47')$$

$$= 1.8630 + \bar{1}.9995 - \bar{1}.8320$$

$$= 2.0305$$

$$a = 107.3$$

II Method

$$C = 129°58'$$

$$A = 180° - (B + C)$$

$$= 180° - (42°47' + 129°58')$$

$$= 180° - 172°45'$$

$$= 7°15'$$

To find *a*

$$\frac{a}{\sin A} = \frac{b}{\sin B}$$

$$a = \frac{b \sin A}{\sin B}$$

$$= \frac{72.95 \sin 7°15'}{\sin 42°27'}$$

$$\log a = \log (72.95) + \log \sin (7°15') - \log \sin (42°47')$$

$$= 1.8630 + \bar{1}.1011 - \bar{1}.8320$$

$$= 1.1321$$

$$a = 13.55$$

$\therefore$ Two solutions are

$C_1 = 50°2'$, $A_1 = 87°11'$, $a_1 = 107.3$

$C_2 = 129°58'$, $A_2 = 7°15'$, $a_2 = 13.55$

EXERCISE 17.8

Solve the triangle when

1. $a = 36.5, b = 31, A = 82°14'$ **Ans.** $B = 57°18', C = 40°28', c = 23.91$
2. $b = 15, c = 25, B = 32°15.$ **Ans.** $C_1 = 62°47', A_1 = 84°58', a_1 = 28$
   $C_2 = 117°13', A_2 = 30°32', a_2 = 14.29$
3. $b = 3, c = 3\sqrt{3}$ and $B = 30°$ **Ans.** $C_1 = 60°, A_1 = 90°, a_1 = 6$
   $C_2 = 120°, A_2 = 30°, a_2 = 3$
4. $a = 100, c = 100\sqrt{2}$ and $A = 30°.$ **Ans.** $C_1 = 45°, B_1 = 105°, b_1 = 50(\sqrt{6}+\sqrt{2})$
   $C_2 = 135°, B_2 = 15°, b_2 = 50(\sqrt{6}-\sqrt{2})$
5. In ΔABC, the sides a and b and angle A are given. If c_1 and c_2 be the two possible values of the third side, show that
   (*i*) $c_1 + c_2 = 2b \cos A$
   (*ii*) $c_1 \cdot c_2 = b^2 - a^2$
   (*iii*) $c_1 - c_2 = 2\sqrt{a^2 - b^2 \sin^2 A}$
6. *AC*, *AD* are two struts freely jointed at *A*, *AC* being 8.4 m long and $A = 28°$. A third strut *CB*, 5.8 m long is to be fastened at *C* at a point *B* on *AD*. Show that provided *AD* is sufficiently long, there are two possible locations for *B*. Calculate the distance of each position from *A*. **Ans.** 11.62 m, 3.16 m
7. An overhead conductor carrying electrical power runs straight from *A* to *B*, where $AB = 4$ km. *C* is the position of a factory some distance from *AB*. $\angle CAB = 46°45'$, $\angle CBA = 62°17'$.
   Find the distance of the factory from *A*. **Ans.** 3.745 km
8. Two forces 4 kgf , 5 kgf acting at a point have a resultant of 7 kgf. Find the angle between the two forces. **Ans.** 44°25′

Given One Side and Two Angles

Let a be the given side and B, C the given angles.

(*i*) To find A.

$$A = 180° - (B + C)$$

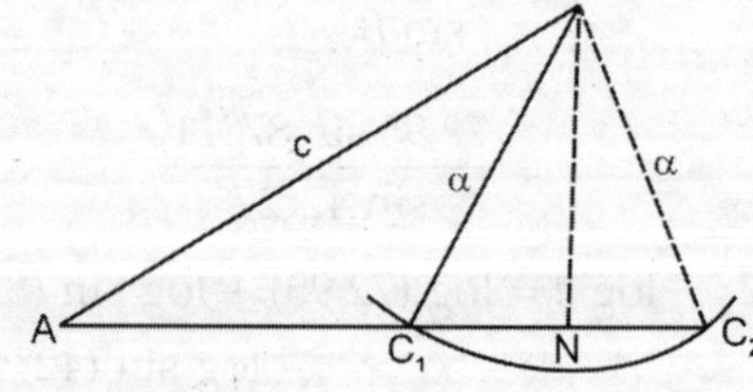

(*ii*) To find b.

$$\frac{b}{\sin B} = \frac{a}{\sin A} \quad \Rightarrow \quad b = \frac{a \sin B}{\sin A}$$

$$\log b = \log a + \log \sin B - \log \sin A.$$

On putting the values of a, B, A, we can find $\log b$ with the help of log table.

(*iii*) To find c.

$$\frac{c}{\sin C} = \frac{a}{\sin A} \quad \Rightarrow \quad c = a\frac{\sin C}{\sin A}$$

$$\log c = \log a + \log \sin C - \log \sin A$$

On substituting the values of a, C, A, we can find $\log c$ with the help of log tables.

Example 40. *In a triangle ABC, C = 38°20′, B = 45° and b = 64 cm. Find a.*

Solution. $A = 180° - (B + C)$

$$= 180° - (45° + 38°20') = 96°40'$$

$$\frac{a}{\sin A} = \frac{b}{\sin B} \Rightarrow a = \frac{b \sin A}{\sin B}$$

$$a = \frac{64 \sin (96°40')}{\sin 45°}$$

$$\log a = \log 64 + \log \sin (96°40') - \log \sin 45°$$

$$= \log 64 + \log \sin (180° - 83°20') - \log \sin 45°$$

$$= 6 \log 2 + \log \sin (83°20') - \log \left(\frac{1}{\sqrt{2}}\right)$$

$$= \log \sin (83°20') + 6 \log 2 + \frac{1}{2} \log 2$$

$$= \bar{1}.9971 + \frac{13}{2} \log 2 = \bar{1}.9971 + \frac{13}{2} \times (0.3010)$$

$$= \bar{1}.9971 + 1.9565 = 1.9536 \Rightarrow a = 89.86$$ **Ans.**

Example 41. *In a triangle ABC, B = 64°23′, C = 72°43′, a = 18.92, solve the triangle.*

Solution. $A = 180° - (B + C) = 180° - (64°23' + 72°43')$

$$= 42°54'$$

$$\frac{a}{\sin A} = \frac{b}{\sin B} \quad \text{or} \quad \frac{18.92}{\sin 42°54'} = \frac{b}{\sin 64°23'}$$

$$\Rightarrow \quad b = \frac{18.92 \sin 64°23'}{\sin 42°54'}$$

$$\Rightarrow \quad \log b = \log 18.92 + \log \sin 64°23' - \log \sin 42°54'$$

$$= 1.277 + \bar{1}.9551 - \bar{1}.8331 = 1.3990 \quad \text{or} \quad b = 25.06$$

$$\frac{a}{\sin A} = \frac{c}{\sin C} \quad \text{or} \quad \frac{18.92}{\sin 42°54'} = \frac{c}{\sin 72°43'}$$

$$\Rightarrow \quad c = \frac{18.92 \sin 72°43'}{\sin 42°54'}$$

$$\Rightarrow \quad \log c = \log 18.92 + \log \sin 72°43' - \log \sin 42°54'$$

$$= 1.2770 + \bar{1}.9799 - \bar{1}.8331 = 1.4238 \Rightarrow c = 26.54$$

Hence, $A = 42°54'$, $b = 25.06$, $c = 26.54$ **Ans.**

EXERCISE 17.9

Solve the triangle, if

1. $B = 60°, A = 30°, c = 13$ **Ans.** $C = 90°, b = 11.258, a = 6.5$
2. $A = 80°, B = 53°, a = 152$ **Ans.** $C = 47°, b = 123.2, c = 112.8$

3. $A = 49°11'$, $B = 21°15'$, $c = 5.23$ **Ans.** $C = 109°34'$, $a = 4.2$, $b = 2.012$

4. $B = 88°36'$, $C = 31°54'$, $a = 53$ **Ans.** $A = 59°30'$, $b = 61.51$, $c = 32.51$

5. $a = 36.5$, $b = 31$, $A = 82°14'$ **Ans.** $B = 57°18'$, $C = 40°28'$, $c = 23.908457$

6. $C = 38°20'$, $B = 45°$, $b = 64$, find a. **Ans.** 89.896

7. A mass of 98 gm is suspended by two strings, the ends of which are fixed at the same horizontal level. The triangle of forces is as shown below, where T_1 and T_2 are tensions in the strings. Calculate T_1 and T_2.

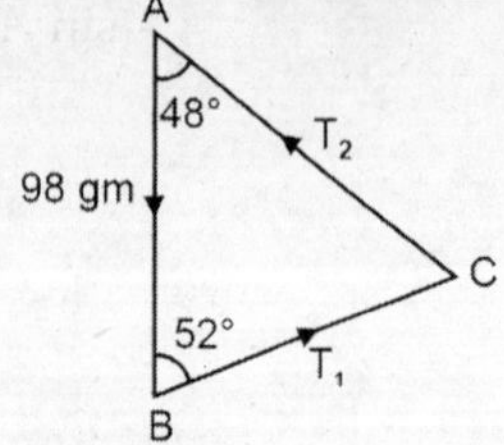

Ans. $T_1 = 73.94$ gm, $T_2 = 79.39$ gm

Height and Distance

18.1 INTRODUCTION

Ancient astronomers have used trigonometry to calculate the distance from the earth to the stars. Maps are constructed with the help of knowledge of trigonometry. Which is used in everyday life around us to find heights and distances of objects and their rotation.

In day to day life it was not possible to measure height and distance with a measuring tape like the height of a big tree, electric pole, apartment, tower, the width of a river and the distance between a ship and the lighthouse. We can determine these with the knowledge of trigonometry.

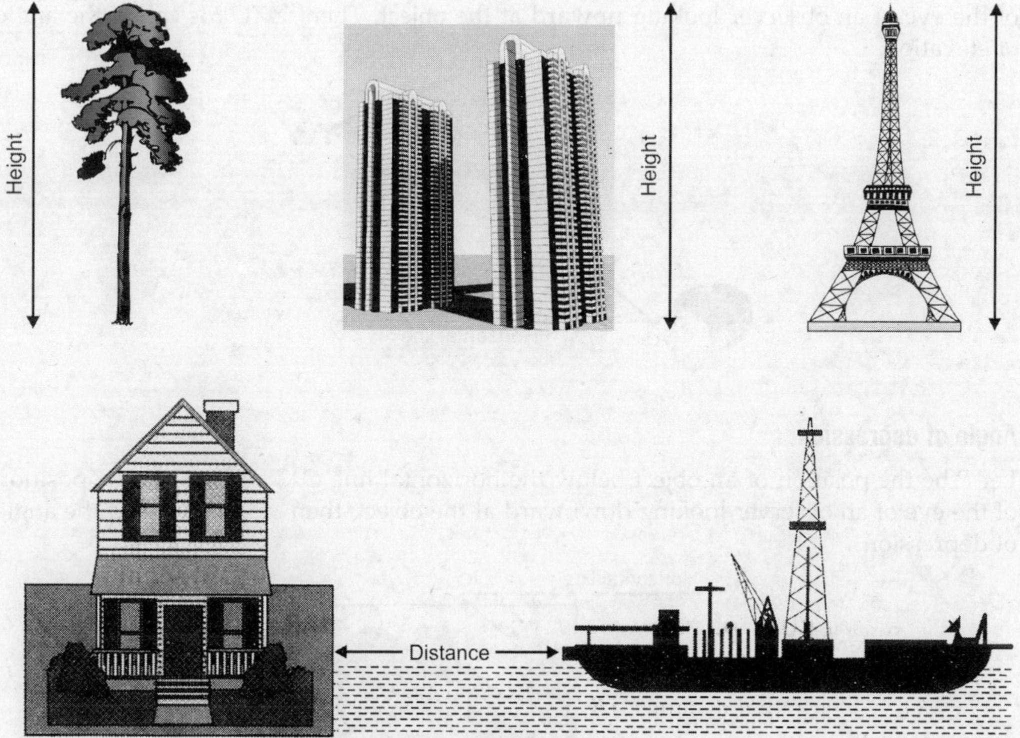

Let us define some basic terms.

Line of sight (line of vision)

The line of sight is the imaginary line drawn from the eye of the observer to the top of a minar, when a person is looking at the top of the minar.

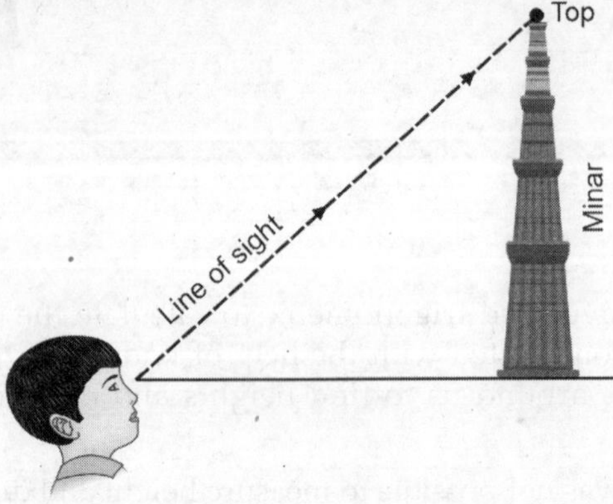

Angle of elevation

Let P be the position of an object above the horizontal line OX, where O is the position of the eye of an observer looking upward at the object. Then, $\angle XOP$ is called the angle of elevation.

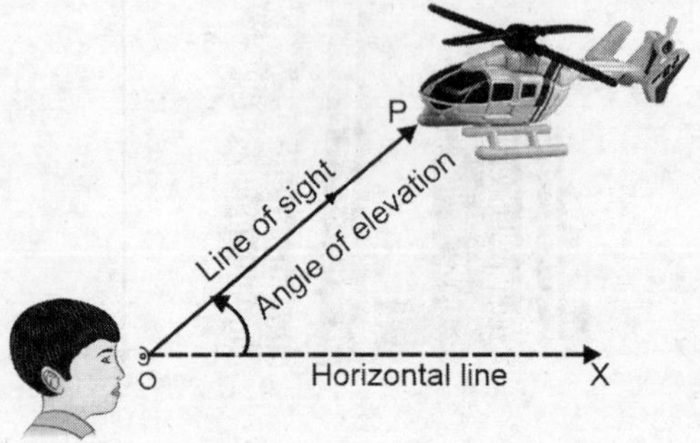

Angle of depression

Let P be the position of an object below the horizontal line OX, where O is the position of the eye of an observer looking downward at the object, then $\angle XOP$ is called the angle of depression.

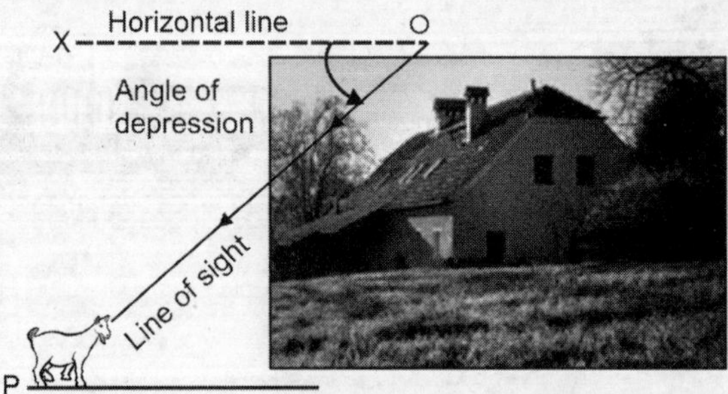

18.2 WORKING RULE TO FIND THE HEIGHT OF THE MINAR

We should know the following:

Step 1. Horizontal distance AB between the eye and the foot of the minar.

Step 2. Angle of elevation $\angle BAC$ of the top of the minar.

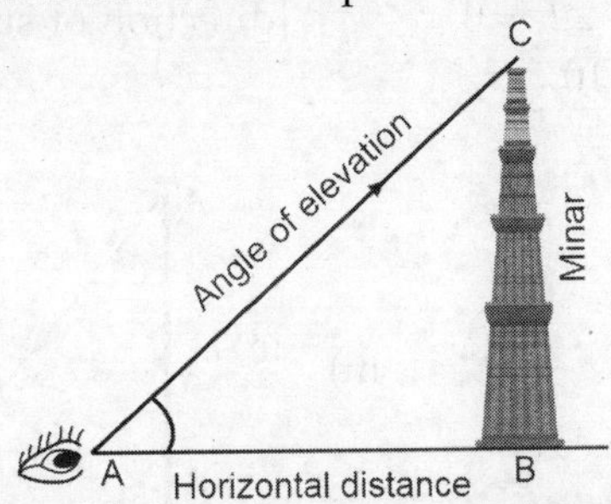

Example 1. *Find the height of a tower if the angle of elevation of top of the tower is 60° and the horizontal distance from eye to the foot of the tower is 100 m.*

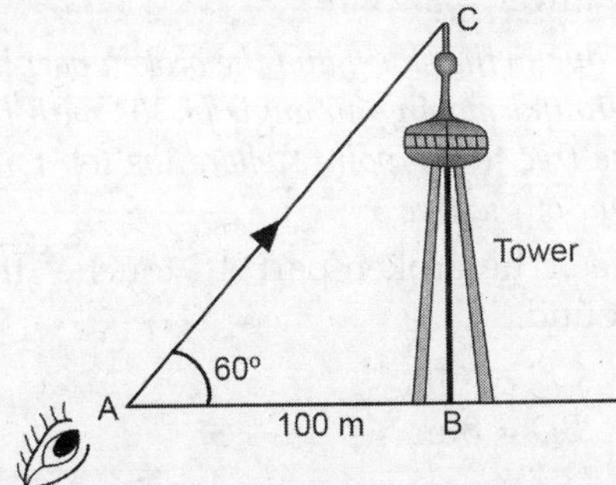

Solution. Let the height of the tower be BC. Horizontal distance $AB = 100$ m

In right ΔABC,

$$\tan\theta = \frac{BC}{AB} \quad \Rightarrow \quad \tan 60° = \frac{BC}{100} \quad \Rightarrow \quad \sqrt{3} = \frac{BC}{100} \quad \Rightarrow \quad BC = 100\sqrt{3} \text{ m}$$

Hence, the height of the tower is $100\sqrt{3}$ m. **Ans.**

Example 2. *A balloon is connected to a meteorological ground station by a cable of length 215 m inclined at 60° to the horizontal. Determine the height of the balloon from the ground. Assume that there is no slack in the cable.*

Solution. Length of cable $AC = 215$ m

Angle of elevation = 60°

Suppose height of balloon above the ground = x metre

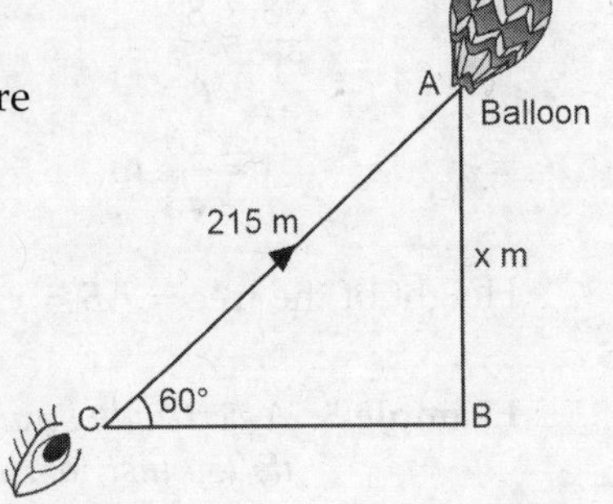

In right ΔABC $\quad \sin 60° = \dfrac{AB}{AC} \quad \Rightarrow \quad \dfrac{\sqrt{3}}{2} = \dfrac{x}{215}$

$\Rightarrow \quad 2x = 215\sqrt{3}$

$\therefore \quad x = \dfrac{215\sqrt{3}}{2} \quad \Rightarrow \quad x = 186.19$ m

Hence, the height of the balloon from the ground is 186.19 m. **Ans.**

Example 3. *A vertical stick 10 cm long casts a shadow 8 cm long. At the same time, a tower casts a shadow 30 m long Determine the height of the tower.*

Solution. Length of the stick $AB = 10$ cm

Length of the shadow $BC = 8$ cm

Let $\angle C = \theta \Rightarrow \angle F = \theta$ [direction of sun is the same in both cases.]

In ΔABC, $\tan\theta = \dfrac{AB}{BC} = \dfrac{10}{8}$

In ΔDEF, $\tan\theta = \dfrac{DE}{EF}$

$\Rightarrow \dfrac{10}{8} = \dfrac{x}{30}$ $(DE = x \text{ m})$

$\Rightarrow 8x = 300$

$\Rightarrow x = \dfrac{300}{8}\text{ m} = 37.5\text{ m}$

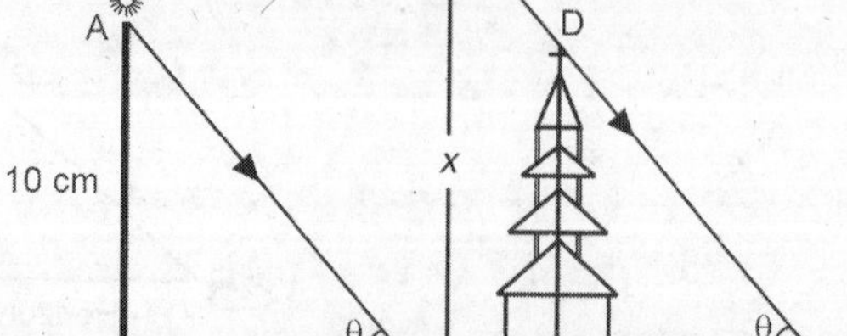

Hence, the height of the tower = 37.5 m **Ans.**

Example 4. *A tree breaks due to the storm and the broken part bends so that the top of the tree touches the ground making an angle of 30° with the ground. The distance from the foot of the tree to the point, where the top touches the ground is 8 metres. Find the height of the tree.*

Solution. Let AB be a tree. The broken part AC touches the ground at D, making an angle of 30° with the ground.

Let $AC = x$ m $\therefore CD = x$ m

and $BC = y$ m $BD = 8$ m (Given)

In right angled ΔDBC,

$$\tan 30° = \frac{BC}{BD}$$

$$\Rightarrow \frac{1}{\sqrt{3}} = \frac{y}{8}$$

$$\Rightarrow y = \frac{8}{\sqrt{3}}\text{ m}$$

Also $\cos 30° = \dfrac{BD}{CD}$

$$\Rightarrow \frac{\sqrt{3}}{2} = \frac{8}{x}$$

$$\Rightarrow x = \frac{16}{\sqrt{3}}\text{ m.}$$

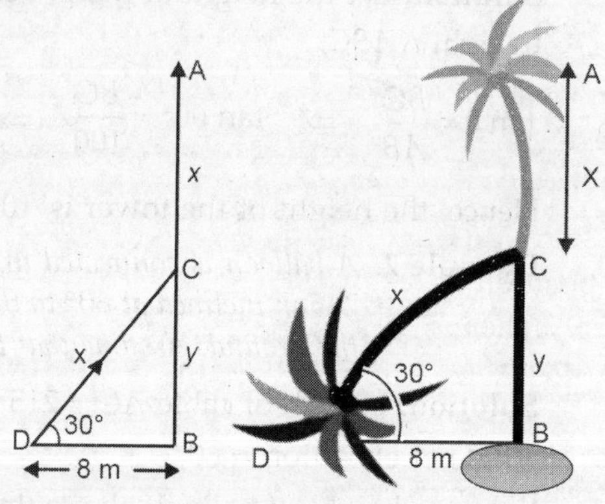

Height of the tree = $AB = x + y = \dfrac{16}{\sqrt{3}}\text{ m} + \dfrac{8}{\sqrt{3}}\text{ m} = \dfrac{24}{\sqrt{3}}\text{ m} = 8\sqrt{3}\text{ m}$ **Ans.**

Example 5. *A vertically straight tree, 15 m high is broken by the wind, in such a way that its top just touches the ground and makes an angle of 60° with the ground. At what height from the ground did the tree break? (Use $\sqrt{3} = 1.73$)*

Solution. Height of the tree $AB = 15$ m

In broke at C. Its top A touches the ground at D.

Now, $\quad AC = CD, \angle BDC = 60°$

Let $\quad BC = x$

$AC = AB - BC$

$\therefore \quad AC = 15 - x$

$\Rightarrow \quad CD = 15 - x \quad [\because AC = CD]$

In right ΔCBD,

$$\sin 60° = \frac{BC}{CD}$$

$$\Rightarrow \quad \frac{\sqrt{3}}{2} = \frac{x}{15 - x}$$

$$\Rightarrow \quad 2x = (15 - x)\sqrt{3}$$

$$\Rightarrow \quad 2x = 15\sqrt{3} - \sqrt{3}x$$

$$\Rightarrow \quad 2x + \sqrt{3}x = 15\sqrt{3}$$

$$\Rightarrow \quad x = \frac{15\sqrt{3}}{2 + \sqrt{3}} \qquad \Rightarrow \quad x = \frac{15\sqrt{3}}{2 + \sqrt{3}} \times \frac{2 - \sqrt{3}}{2 - \sqrt{3}}$$

$$\Rightarrow \quad x = \frac{30\sqrt{3} - 15 \times 3}{4 - 3} \qquad \Rightarrow \quad x = \frac{30 \times 1.73 - 45}{1}$$

$$\Rightarrow \quad x = \frac{51.9 - 45}{1} \qquad \Rightarrow \quad x = 6.9 \text{ m}$$

Hence, the tree broke at the height of 6.9 m. **Ans.**

Example 6. *A player sitting on the top of a tower of height 20 m observes the angle of depression of a ball lying on the ground as 60°. Find the distance between the foot of the tower and the ball.*

Solution. Let OQ be the tower and P be the position of the ball lying on the ground.

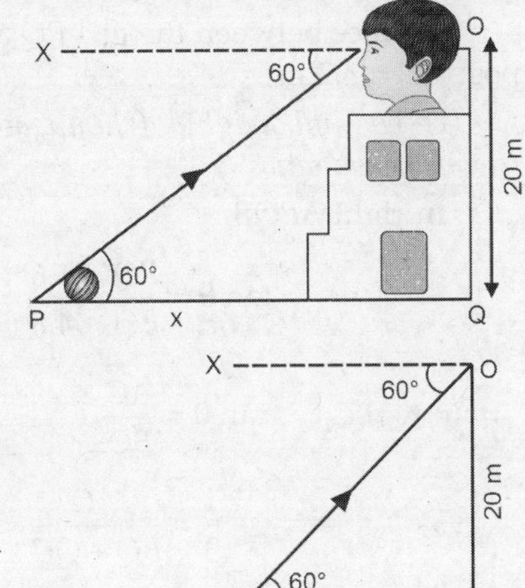

Then,

$OQ = 20$ m $\quad$ and $\quad \angle POX = 60° = \angle OPQ$.

Let $PQ = x$ metres.

$\therefore$ In right angled ΔPQO,

$$\tan 60° = \frac{OQ}{PQ}$$

$$\Rightarrow \quad \sqrt{3} = \frac{20}{x}$$

$$\Rightarrow \quad \sqrt{3}x = 20$$

$$\Rightarrow \quad x = \frac{20}{\sqrt{3}} = \frac{20}{1.732} = 11.55 \text{ m}$$

Hence, the distance between the foot of the tower and the ball = 11.55 m. **Ans.**

Example 7. *The angle of elevation of the top of a building from the foot of the tower is 30° and the angle of elevation of the top of the tower from the foot of the building is 60°. If the tower is 50 m high, find the height of the building.*

Solution. The height of the tower $AB = 50$ m

The height of the building $= CD$.

In right ΔABD,

$$\tan 60° = \frac{AB}{BD} \quad \Rightarrow \quad \sqrt{3} = \frac{50}{BD}$$

$$\Rightarrow \quad BD = \frac{50}{\sqrt{3}} \quad \text{... (1)}$$

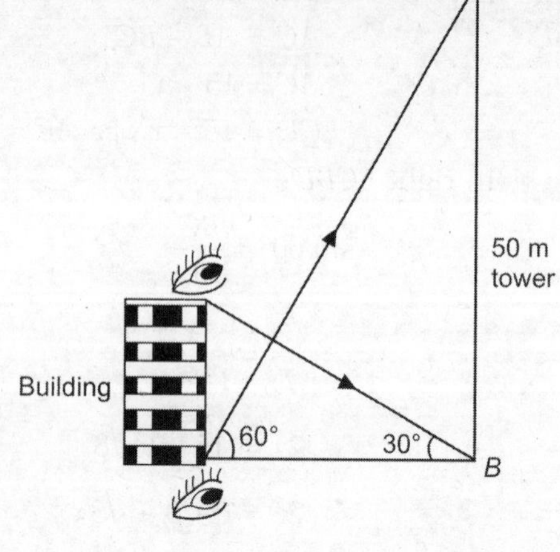

In right ΔBDC,

$$\tan 30° = \frac{CD}{BD}$$

$$\Rightarrow \quad \frac{1}{\sqrt{3}} = \frac{CD}{\frac{50}{\sqrt{3}}} \quad \text{[using (1)]}$$

$$\Rightarrow \quad \sqrt{3}CD = \frac{50}{\sqrt{3}}$$

$$\Rightarrow \quad CD = \frac{50}{3} = 16\frac{2}{3} = 16.67 \text{ m}$$

Hence the height of the building be 16.67 m. **Ans.**

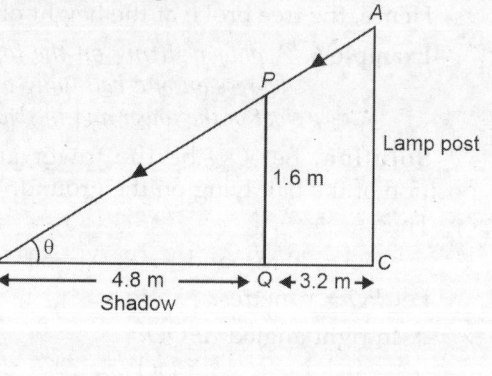

Example 8. *A 1.6 m tall girl stands at a distance of 3.2 m from a lamp post and casts a shadow of 4.8 m on the ground. Find the height of the lamp post by using (i) trigonometric ratios (ii) property of similar triangles.*

Solution. Height of the girl $PQ = 1.6$ m

Height of lamp post be AC.

Length of shadow $BQ = 4.8$ m

Distance between the girl PQ and lamp post $AC = 3.2$ m

(i) To find height of lamp post using trigonometric ratios

In right ΔPQB,

$$\tan\theta = \frac{PQ}{BQ} = \frac{1.6}{4.8} = \frac{1}{3} \quad \text{...(1)}$$

In ΔABC, $\quad \tan\theta = \dfrac{AC}{BC}$

$$\Rightarrow \quad \frac{1}{3} = \frac{AC}{BQ + QC} = \frac{AC}{4.8 + 3.2}$$

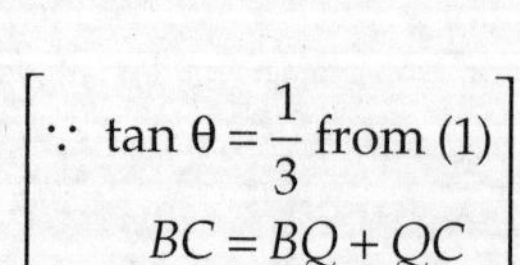
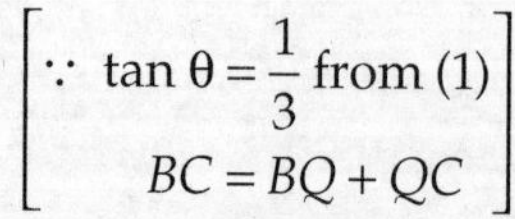

$$\left[\because \tan\theta = \frac{1}{3} \text{ from (1)} \quad BC = BQ + QC\right]$$

$$\Rightarrow \quad \frac{1}{3} = \frac{AC}{8.0} \quad \Rightarrow \quad AC = \frac{8}{3} = 2.66 \text{ m}$$

Ans.

(ii) To find the height of the lamp-post using property of similar triangle

In ΔPBQ and ΔABC

$$\angle PQB = \angle ACB \qquad \text{[each angle equal to } 90°]$$

$$\angle PBQ = \angle ABC = \theta \qquad \text{(common)}$$

$$\therefore \quad \Delta PBQ \sim \Delta ABC \qquad \text{(AA similarity)}$$

$$\therefore \quad \frac{AC}{PQ} = \frac{BC}{BQ} \quad \Rightarrow \quad \frac{AC}{1.6} = \frac{4.8 + 3.2}{4.8}$$

$$\Rightarrow \quad \frac{AC}{1.6} = \frac{8.0}{4.8} \quad \Rightarrow \quad AC = \frac{8 \times 1.6}{4.8} = \frac{8}{3} = 2.66 \text{ m}$$

Hence, the height of the lamp post is 2.66 m. **Ans.**

Example 9. *From a point on the ground the angles of elevation of the bottom and top of a transmission tower fixed at the top of a 20 m high building are 45° and 60° respectively. Find the height of the tower.* **[S.B.T.E. 2015]**

Solution. Height of the building $CD = 20$ m

Let height of tower $= AD$.

Angle of elevation of the bottom of the tower = 45°

Angle of elevation of the top of the tower = 60°

In right ΔBCD

$$\tan 45° = \frac{CD}{BC} \Rightarrow 1 = \frac{20}{BC} \Rightarrow BC = 20 \text{ m} \quad \text{... (1)}$$

Now, in right ΔABC,

$$\tan 60° = \frac{AC}{BC} \Rightarrow \sqrt{3} = \frac{AC}{20}$$

$$\Rightarrow \quad AC = 20\sqrt{3} \text{ m} \quad \text{... (2)}$$

Now, $AC = AD + CD$

$$\Rightarrow \quad AD = AC - CD$$

$$\Rightarrow \quad AD = 20\sqrt{3} - 20 = 20(\sqrt{3} - 1) \text{ m}$$

Hence, the height of tower

$$= 20(\sqrt{3} - 1) \text{ m}$$ **Ans.**

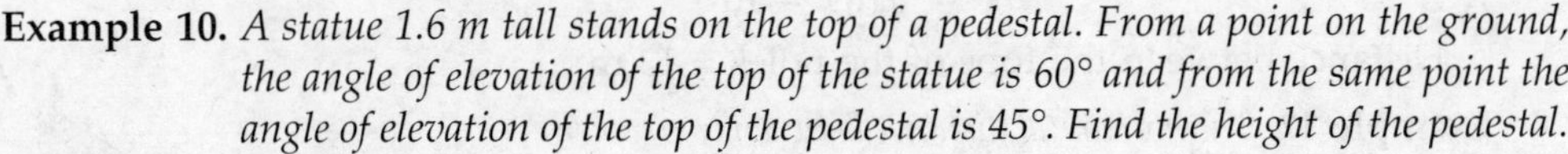

Example 10. *A statue 1.6 m tall stands on the top of a pedestal. From a point on the ground, the angle of elevation of the top of the statue is 60° and from the same point the angle of elevation of the top of the pedestal is 45°. Find the height of the pedestal.*

Solution. The height of the statue $AD = 1.6$ m

Let height of pedestal be CD.

Angle of elevation of the top of statue = 60°.

Angle of elevation of the top of pedestal = 45°.

In right ΔBCD,

$$\tan 45° = \frac{CD}{BC}$$

$$\Rightarrow \quad 1 = \frac{CD}{BC}$$

$$\Rightarrow \quad CD = BC \qquad \text{... (1)}$$

Now, in right ΔABC

$$\tan 60° = \frac{AC}{BC}$$

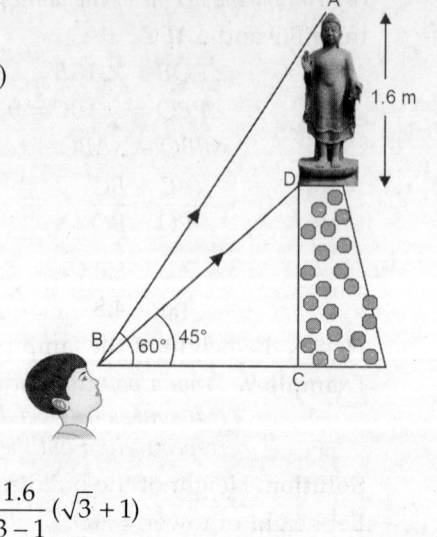

$$\Rightarrow \quad \sqrt{3} = \frac{AC}{BC} \quad \Rightarrow \quad AC = \sqrt{3}BC$$

$$\Rightarrow \quad AD + CD = \sqrt{3}BC$$

$$\Rightarrow \quad 1.6 + CD = \sqrt{3}CD$$

$$\Rightarrow \quad \sqrt{3}CD - CD = 1.6 \qquad [\because BC = CD]$$

$$\Rightarrow \quad (\sqrt{3}-1)\,CD = 1.6$$

$$\Rightarrow \quad CD = \frac{1.6}{\sqrt{3}-1} = \frac{1.6\,(\sqrt{3}+1)}{(\sqrt{3}-1)(\sqrt{3}+1)} = \frac{1.6}{3-1}(\sqrt{3}+1)$$

$$= \frac{1.6}{2}(\sqrt{3}+1)\text{ m} = 0.8\,(\sqrt{3}+1)\text{ m}$$

Hence, the height of pedestal is $0.8\,(\sqrt{3}+1)$ m. **Ans.**

Example 11. *Two poles of heights 6 m and 11 m stand vertically on the ground. If the distance between their base on the ground is 12 m, find the distance between their tops.*

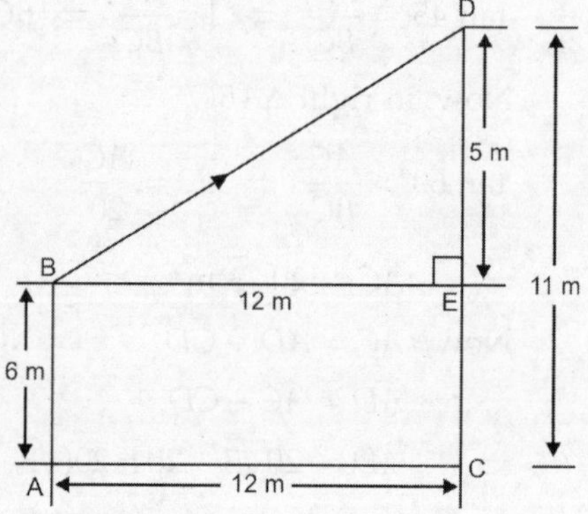

Solution. Let AB and CD represent the poles and AC is the distance between their feet.

Let $\quad BE \perp CD$

$\therefore \quad BE = AC = 12$ m

$DE = CD - EC = 11 - 6 = 5$ m.

In right ΔBED,

$$BD^2 = BE^2 + DE^2 \text{ (Pythagoras theorem)}$$

$$= 12^2 + 5^2 = 144 + 25 = 169$$

$$\therefore \quad BD = \sqrt{169} = 13$$

$\therefore$ Distance between the tops of the poles = 13 m **Ans.**

Example 12. *The horizontal distance between two towers is 140 m. The angle of depression of the top of the first tower, when seen from the top of the second tower is 30°. If the height of the first tower is 60 m, find the height of the second tower.*

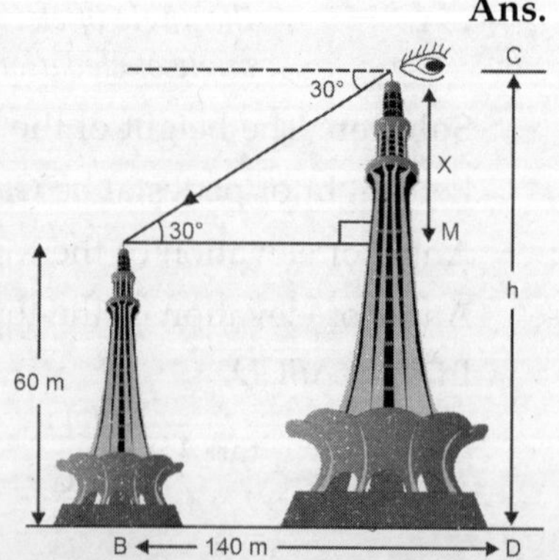

Solution. The height of the 1st tower $AB = 60$ m

Let the height of the second tower $CD = h$ m

Distance $BD = 140$ m

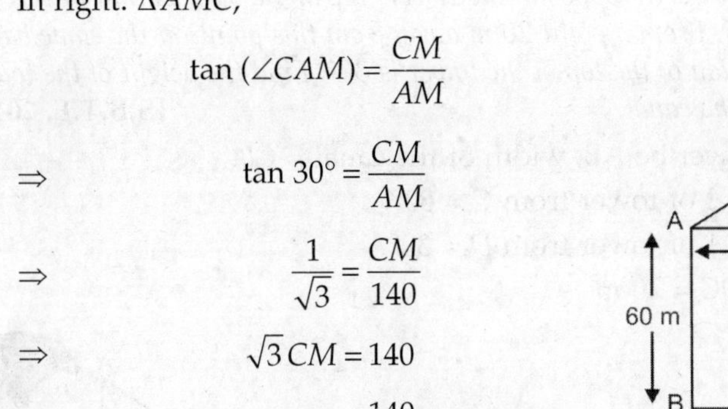

Now, $AM = BD = 140$ m, $\angle CAM = 30°$

In right. ΔAMC,

$$\tan(\angle CAM) = \frac{CM}{AM}$$

$$\Rightarrow \quad \tan 30° = \frac{CM}{AM}$$

$$\Rightarrow \quad \frac{1}{\sqrt{3}} = \frac{CM}{140}$$

$$\Rightarrow \quad \sqrt{3}\, CM = 140$$

$$\Rightarrow \quad CM = \frac{140}{\sqrt{3}} = 80.83 \text{ m}$$

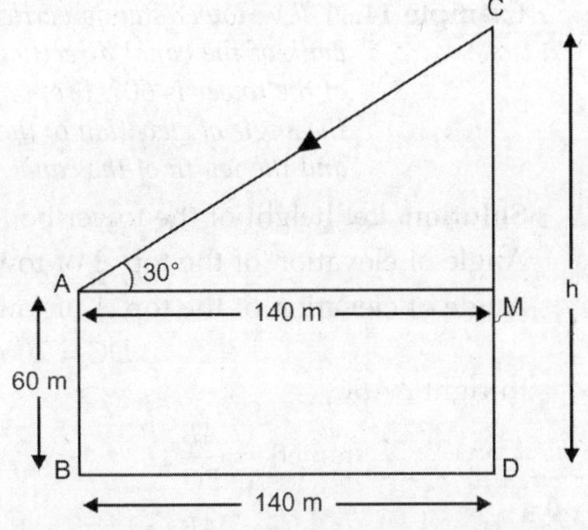

$\therefore$ Height of the second tower $CD = CM + MD = 80.83$ m $+ 60$ m $= 140.83$ m **Ans.**

Example 13. *A 1.5 m tall boy is standing at some distance from a 30 m tall building. The angle of elevation from his eyes to the top of the building increases from 30° to 60° as the walks towards the building. Find the distance he walked towards the building.*

Solution. The height of the building $AB = 30$ m

The height of boy $CD = 1.5$ m

$$BF = CD = 1.5 \text{ m}$$

$$AF = AB - BF = 30 - 1.5 = 28.5 \text{ m}$$

In right ΔAFE, $\tan 60° = \dfrac{AF}{EF}$

$$\Rightarrow \quad \sqrt{3} = \frac{28.5}{EF} \quad \Rightarrow \quad EF = \frac{28.}{\sqrt{3}}$$

Now, in right ΔAFC

$$\tan 30° = \frac{AF}{CF}$$

$$\Rightarrow \quad \frac{1}{\sqrt{3}} = \frac{28.5}{CF}$$

$$\Rightarrow \quad CE = 28.5\sqrt{3}$$

$$\Rightarrow \quad CE + EF = 28.5\sqrt{3}$$

$$\Rightarrow \quad CE + \frac{28.5}{\sqrt{3}} = 28.5\sqrt{3}$$

$$\Rightarrow \quad CE = 28.5\sqrt{3} - \frac{28.5}{\sqrt{3}} = \frac{28.5}{\sqrt{3}}(3-1) = \frac{28.5}{\sqrt{3}} \times 2 = \frac{28.5 \times 2\sqrt{3}}{3}$$

$$= 9.5 \times 2\sqrt{3} = 19\sqrt{3}$$

Hence, the distance he walked toward the building is $19\sqrt{3}$ m. **Ans.**

Example 14. *A T.V. tower stands vertically on a bank of a canal. From a point on the other bank of the canal directly opposite the tower, the angle of elevation of the top of the tower is 60°. From a point 20 m away from this point on the same bank, the angle of elevation of the top of the tower is 30°. Find the height of the tower and the width of the canal.* **[S.B.T.E. 2010]**

Solution. Let height of the tower be AB, width of the canal = CB.

Angle of elevation of the top A of tower from C = 60°.

Angle of elevation of the top A of tower from D = 30°.

$$DC = 20 \text{ m}$$

In right ΔABC,

$$\tan 60° = \frac{AB}{BC}$$

$$\Rightarrow \quad \sqrt{3} = \frac{AB}{BC}$$

$$\Rightarrow \quad AB = \sqrt{3}BC \quad \text{... (1)}$$

In right ΔABD,

$$\tan 30° = \frac{AB}{BD}$$

$$\Rightarrow \quad \frac{1}{\sqrt{3}} = \frac{AB}{BD}$$

$$\Rightarrow \quad BD = \sqrt{3}AB$$

$$BC + CD = \sqrt{3}AB \qquad [\because BD = BC + CD]$$

$$BC + 20 = \sqrt{3}(\sqrt{3}BC) \Rightarrow BC + 20 = 3BC \qquad [\text{From (1)}, AB = \sqrt{3}BC]$$

$$\Rightarrow \quad 2BC = 20 \Rightarrow BC = 10 \text{ m}$$

Putting $BC = 10$ in (1), we get

$$AB = 10\sqrt{3} \text{ m}$$

Hence, the height of the tower is $10\sqrt{3}$ m and width of the canal is 10 m. **Ans.**

Example 15. *The angle of elevation of the top of a tower, as seen from two points A and B situated in the same line and at distances p and q respectively, from the foot of the tower, are complementary. Prove that the height of the tower is* $\sqrt{pq}$. **[S.B.T.E. 2016]**

Solution. Let height of the tower $CD = h$ units

Let A and B be two points of observations.

And $\quad \angle CAD = \theta$

Then $\quad \angle CBD = 90° - \theta$

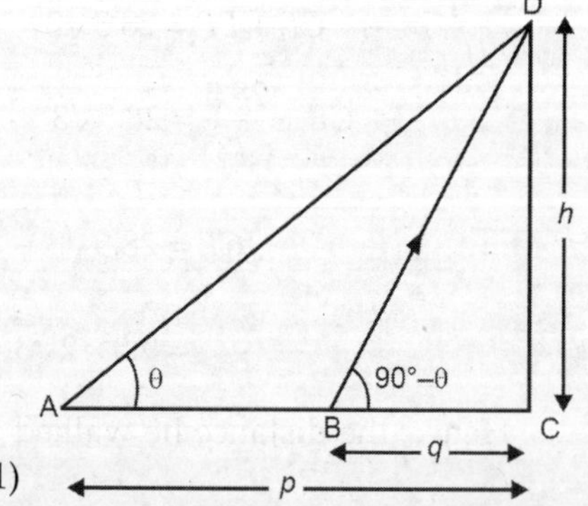

In right ΔACD, $\tan\theta = \dfrac{CD}{AC}$

$$\Rightarrow \quad \tan\theta = \frac{h}{p}$$

$$\Rightarrow \quad h = p \cdot \tan\theta \quad \text{... (1)}$$

In right ΔBCD,

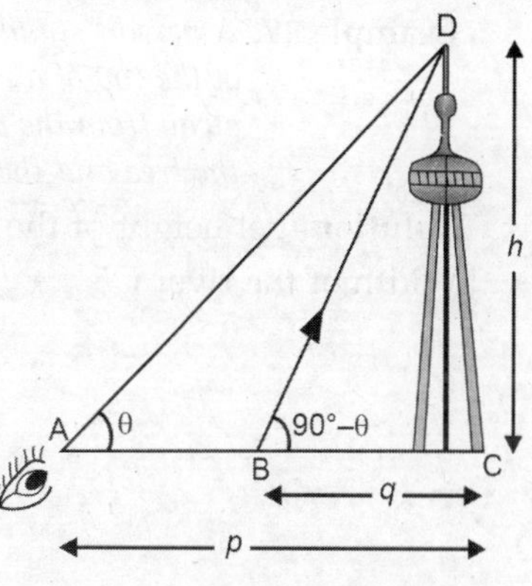

$$\tan(90° - \theta) = \frac{CD}{BC}$$

$$\Rightarrow \quad \cot\theta = \frac{h}{q} \quad [\because \tan(90° - \theta) = \cot\theta]$$

$$\Rightarrow \quad h = q\cot\theta \quad \ldots (2)$$

Multiplying (1) and (2), we get

$$h \cdot h = p\tan\theta \cdot q\cot\theta$$

$$\Rightarrow \quad h^2 = pq \quad [\because \tan\theta \cdot \cot\theta = 1]$$

$$\Rightarrow \quad h\text{ (height)} = \sqrt{pq} \text{ units.} \quad [\because \text{height can't be – ve}]$$

Proved.

Example 16. *At a point on level ground, the angle of elevation of a vertical tower is found to be such that its tangent is 5/12. On walking 192 metres towards the tower, the tangent of the angle is found to be 3/4. Find the height of the tower.*

Solution. Suppose height of the tower $CD = x$

Let A and B be the point of observations.

Let distance $BC = y$, $\tan A = \frac{5}{12}$, $\tan B = \frac{3}{4}$

Now in right ΔBCD,

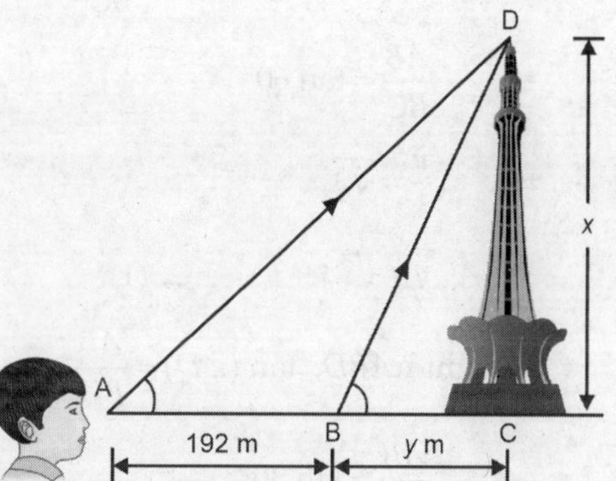

$$\tan B = \frac{CD}{BC} \quad \Rightarrow \quad \frac{x}{y} = \frac{3}{4} \quad \ldots (1)$$

Again in right ΔACD,

$$\tan A = \frac{CD}{AC} \quad \Rightarrow \quad \frac{CD}{AB + BC} = \tan A$$

$$\Rightarrow \quad \frac{x}{192 + y} = \frac{5}{12} \quad \ldots (2)$$

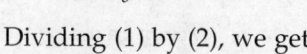

Dividing (1) by (2), we get

$$\frac{x}{y} \times \frac{192 + y}{x} = \frac{3}{4} \times \frac{12}{5} \quad \Rightarrow \quad \frac{192 + y}{y} = \frac{9}{5}$$

$$\Rightarrow \quad 9y = 5(192 + y) \quad \Rightarrow \quad 9y = 960 + 5y$$

$$\Rightarrow \quad 9y - 5y = 960 \quad \Rightarrow \quad 4y = 960$$

$$\Rightarrow \quad y = 240 \text{ m}$$

Putting the value of y in (1), we get

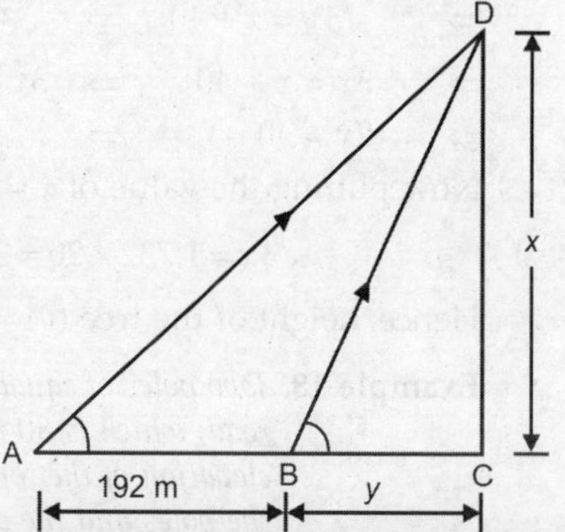

$$\frac{x}{240} = \frac{3}{4} \quad \Rightarrow \quad 4x = 720$$

$$\Rightarrow \quad x = 180$$

Hence, height of the tower = 180 m. **Ans.**

Example 17. *A person standing on the bank of a river, observes that the angle of elevation of the top of a tree, standing on the opposite bank is 60°. When he moves 40 m away from the bank, he finds the angle of elevation to be 30°. Find the height of the tree and the width of the river.* **[S.B.T.E. 2013]**

Solution. Let height of the tree $AB = y$

Width of the river $CB = x$

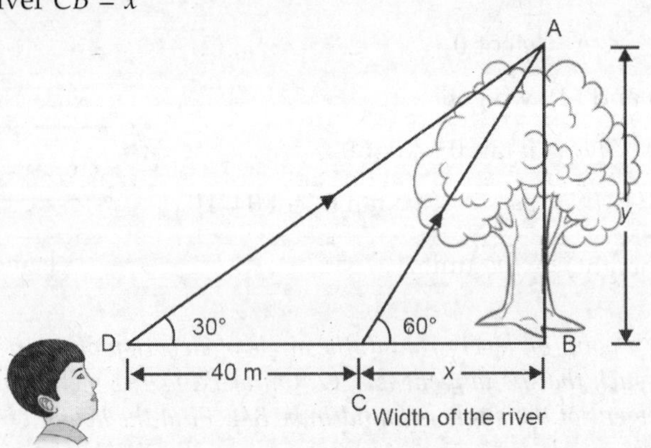

Let C be the point of observation and D be the other point of observation such

$$CD = 40 \text{ m}$$

In right ΔABC, $\tan(\angle C) = \dfrac{AB}{BC}$

$$\Rightarrow \quad \frac{AB}{BC} = \tan 60°$$

$$\Rightarrow \quad \frac{y}{x} = \sqrt{3}$$

$$\Rightarrow \quad y = \sqrt{3}x \qquad \ldots (1)$$

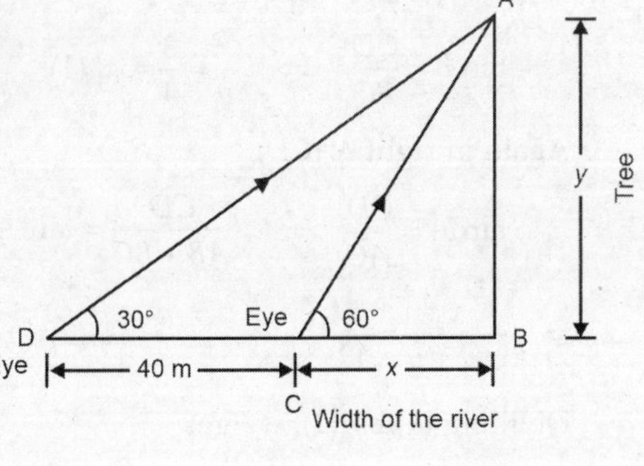

In right ΔABD, $\tan(\angle D) = \dfrac{AB}{BD}$

$$\Rightarrow \quad \frac{AB}{BD} = \tan 30° \quad \Rightarrow \quad \frac{y}{x+40} = \frac{1}{\sqrt{3}}$$

$$\Rightarrow \quad \sqrt{3}y = x + 40 \quad \Rightarrow \quad \sqrt{3} \cdot \sqrt{3}x = x + 40 \qquad \text{[From (1)]}$$

$$\Rightarrow \quad 3x = x + 40 \quad \Rightarrow \quad 3x - x = 40$$

$$\Rightarrow \quad 2x = 40 \quad \Rightarrow \quad x = 20$$

Now putting the value of $x = 20$ in (1), we get

$$y = \sqrt{3}x = 1.732 \times 20 = 34.64 \text{ m}$$

Hence, height of the tree $(y) = 36.64$ m and width of the river $(x) = 20$ m. **Ans.**

Example 18. *Two poles of equal heights are standing opposite each other on either side of the road, which is 80 m wide. From a point between them on the road the angles of elevation of the top of the poles are 60° and 30°, respectively. Find the height of the poles and the distances of the point from the poles.* **[S.B.T.E. 2016]**

Solution. Let AB and CD represent the poles.

Distance AC between their feet = 80 m

Let $AE = x$ then $EC = 80 - x$.

Given, $AB = CD$

In right ΔEAB,

$$\tan 30° = \frac{AB}{AE} \Rightarrow \frac{1}{\sqrt{3}} = \frac{AB}{x}$$

$$\Rightarrow \quad AB = \frac{x}{\sqrt{3}} \quad ...(1)$$

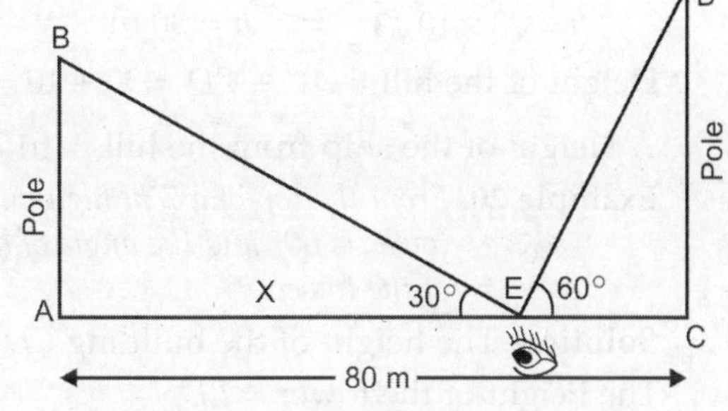

In right ΔDCE,

$$\tan 60° = \frac{DC}{EC} \Rightarrow \sqrt{3} = \frac{AB}{EC} \Rightarrow AB = \sqrt{3}EC \quad ...(2)$$

$[\because AB = DC]$

Putting the value of AB from (1) in (2), we get

$$\frac{x}{\sqrt{3}} = \sqrt{3}EC \Rightarrow x = 3EC \Rightarrow x = 3(80 - x)$$

$$\Rightarrow \quad x = 240 - 3x$$

$$\Rightarrow \quad 4x = 240 \Rightarrow x = 60 \text{ m}$$

Putting $x = 60$ in (1), we get

$$AB = \frac{60}{\sqrt{3}} = \frac{60\sqrt{3}}{3} = 20\sqrt{3} \text{ m}$$

Hence, the height of the poles are $20\sqrt{3}$ m.

Thus distance of point E from pole AB is 60 m and from pole DC is 80 – 60, *i.e.* 20 m.

Ans.

Example 19. *A man is standing on the deck of a ship, which is 10 m above water level. He observes the angle of elevation of the top of a hill as 60° and the angle of depression of the base of the hill as 30°. Calculate the distance of the hill from the ship and the height of the hill.*

Solution. Let B be position of the man, D be water level (or Base of the hill), AD be the hill.

Let $AC = h$

Let deck of ship $BC = x$, $CD = 10$ m

In right ΔACB, $\tan \angle ABC = \frac{AC}{BC} \Rightarrow \tan 60° = \frac{AC}{BC}$

$$\Rightarrow \quad \sqrt{3} = \frac{h}{x} \Rightarrow h = \sqrt{3}x \quad ...(1)$$

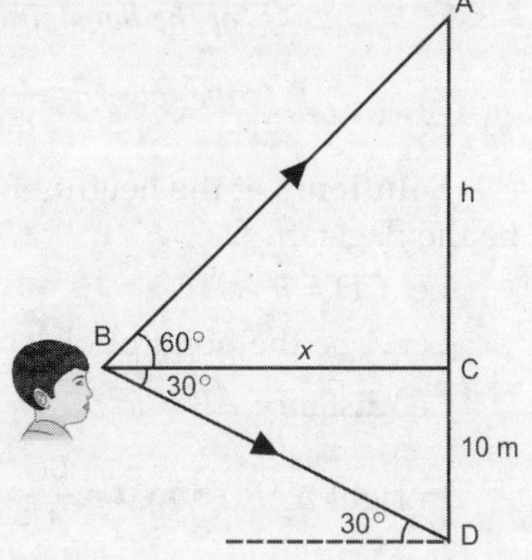

In right ΔBCD, $\tan 30° = \frac{CD}{BC}$

$$\Rightarrow \quad \frac{10}{x} = \frac{1}{\sqrt{3}} \Rightarrow x = 10\sqrt{3} \quad ...(2)$$

Putting the value of x from (2) in (1), we get

$$h=\sqrt{3}\times 10\sqrt{3} \quad \Rightarrow \quad h = 30 \text{ m}$$

Height of the hill = $AC + CD = 30 + 10 = 40$ cm

$\therefore$ Height of the ship from the hill = $10\sqrt{3}$ **Ans.**

Example 20. *From the top of a 7 m high building, the angle of elevation of the top of a cable tower is 60° and the angle of depression of its foot is 45°. Determine the height of the tower.* **[S.B.T.E. 2014, 2013]**

Solution. The height of the building $CD = 7$ m

The height of the tower = EF

Angle of elevation of the top F of the tower from $D = 60°$

Angle of depression of the foot of tower = 45°

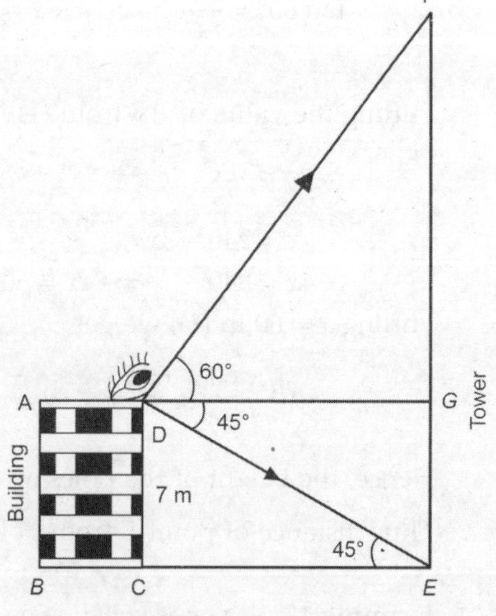

In right ΔDCE,

$$\tan 45° = \frac{DC}{CE}$$

$$\Rightarrow \quad 1 = \frac{7}{CE} \quad \Rightarrow \quad CE = 7 \text{ m}$$

In right ΔDGF,

$$\tan 60° = \frac{FG}{DG} \quad \Rightarrow \quad \sqrt{3} = \frac{FG}{7}$$

$$[\because CE = DG = 7 \text{ m}]$$

$$\Rightarrow \quad FG = 7\sqrt{3} \text{ m} \qquad (GE = DC = 7)$$

Hence, the height of the tower *i.e.* $FE = FG + GE$

$$= 7\sqrt{3} + 7 = 7(\sqrt{3}+1) \text{ m}$$ **Ans.**

Example 21. *A vertical tower stands on a horizontal plane and is surmounted by a vertical flagstaff of height h. At a point on the plane, the angle of elevation of the bottom of the flagstaff is α and that of the top of flagstaff is β. Prove that the height of the tower is* $\dfrac{h \tan\alpha}{\tan\beta - \tan\alpha}$.

Solution. Let the height of tower BC be x and CD be the flagstaff.

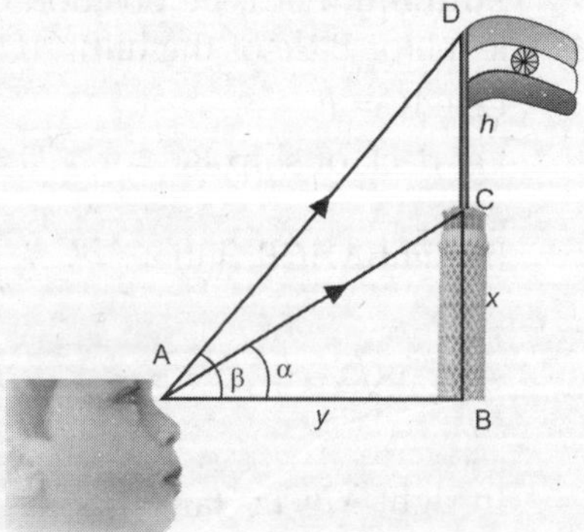

Let $CD = h$

Let A be the point of observation on the plane.

Let distance $AB = y$

In right ΔABC $\tan\alpha = \dfrac{BC}{AB} \Rightarrow \tan\alpha = \dfrac{x}{y}$ ... (1)

In right ΔABD,

$$\tan\beta = \frac{BD}{AB} \Rightarrow \tan\beta = \frac{BC+CD}{AB} \qquad [\because BD = BC + CD]$$

$$\tan \beta = \frac{x+h}{y} \qquad \text{... (2)}$$

Dividing (1) by (2), we get

$$\frac{\tan \alpha}{\tan \beta} = \left(\frac{x}{y}\right)\left(\frac{y}{x+h}\right)$$

$$\Rightarrow \quad \frac{\tan \alpha}{\tan \beta} = \frac{x}{x+h} \qquad \Rightarrow \quad x \tan \beta = x \tan \alpha + h \tan \alpha$$

$$\Rightarrow \quad x \tan \beta - x \tan \alpha = h \tan \alpha \qquad \Rightarrow \quad x (\tan \beta - \tan \alpha) = h \tan \alpha$$

$$\therefore \quad x \text{ (height of tower)} = \frac{h \tan \alpha}{\tan \beta - \tan \alpha}$$

Proved.

EXERCISE 18.1

1. The string of a kite is 100 m long and it makes an angle of 60° with the horizontal. Find the height of the kite, assuming that there is no slack in the string.

Ans. $50\sqrt{3}$ m

2. A circus artist is climbing from the ground along a rope stretched from the top of vertical pole and tied at the ground. The height of the pole is 12 m and the angle made by the rope with ground level is 30°. Calculate the distance covered by the artist in climbing to the top of the pole. **Ans.** 24 m

3. The upper part of a tree is broken by the wind and makes an angle of 30° with the ground. The distance from the root of the tree to the point, where the top touches the ground is 5 m. Find the height of the tree. **Ans.** 8.66 m

4. Two poles of heights 9 m and 12 m stand vertically on a level ground. The distance between their feet is 4 m. Determine the distance between their tops. **Ans.** 5 m

5. Two poles of height 7 m and 12 m stand on a plane ground. If the distance between their feet is 12 m, find the distance between their tops. **Ans.** 13 m

6. In the figure, *ABC* is a right-angled triangle. *D* is the mid-point of *BC*.

Show that $\dfrac{\tan \theta}{\tan \phi} = \dfrac{1}{2}$.

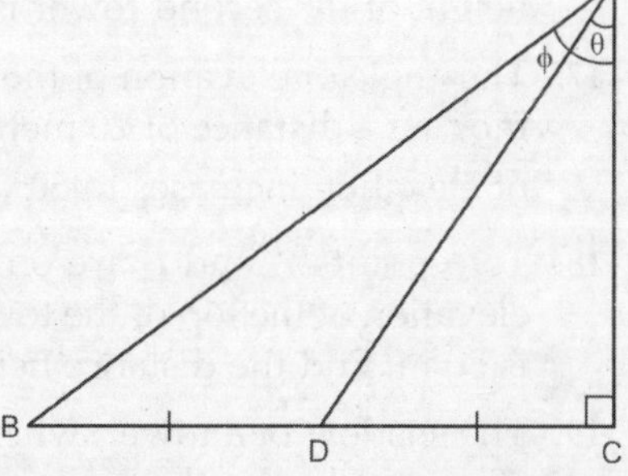

7. An electric pole is 10 m high. A steel wire tied to top of the pole is affixed at a point on the ground to keep the pole upright. If the wire makes an angle of 45° with the horizontal through the foot of the pole, find the length of the wire. **Ans.** 14.1 m

8. The angle of depression of two ships from the top of a light house are 45° and 30° towards east. If the ships are 200 metres apart, find the height of the light house.

Ans. 273.2 m

9. The angle of elevation of the top of a mountain at an unknown distance from the base is 30° and a distance 10 km further off from the mountain along the same line, the angle of elevation is 15°. Determine the height of the mountain.

(use tan 15° = 0.27) **Ans.** 5 km

10. A vertical tower stands on a horizontal plane and is surmounted by a flagstaff of height 12 m. At a point on the plane, the angle of elevation of the bottom of the flagstaff is 45° and of the top of the flagstaff is 60°. Determine the height of the tower and the horizontal distance (Write the answer correct to 2 decimal places).

Ans. 16.39 m, 16.39 m.

11. From the top of a building 12 m high, the angle of elevation of the top of a tower is found to be 45° and the angle of depression of the base of the tower as 30°. Find the height of the tower and its distance on the ground from the building.

Ans. $12(\sqrt{3}+1)\,\text{m}, 12\sqrt{3}\,\text{m}$

12. The angles of depression of the top and the bottom of a building 50 metres high as observed from the top of a tower are 30° and 60° respectively. Find the height of the tower and also the horizontal distance between the building and the tower.

Ans. $75\,\text{m}, 25\sqrt{3}\,\text{m}$

13. The angle of elevation of the top of a tower as observed from a point in a horizontal line through the foot of the tower is 30°. When the observer moves towards the tower a distance of 100 m, he finds the angle of elevation of the top to be 60°. Find the height of the tower and the distance of first position from the tower.

(B.T.E. *DELHI, Jan. 2009*) Ans. $50\sqrt{3}\,\text{m}, 50\,\text{m}$

14. A man on the roof of a house, which is 10 m high, observes the angle of elevation of the top of a building as 45° and angle of depression of the base of the building as 30°. Find the height of the building and its distance from the house.

Ans. $10(\sqrt{3}+1)\,\text{m}, 10\sqrt{3}\,\text{m}$

15. From the top of a building 15 m high, the angle of elevation of the top of a tower is found to be 30°. From the bottom of the same building, the angle of elevation of the top of the tower is found to be 60°. Find the height of the tower and the distance between the tower and the building. **Ans.** 22.5 m, 13 m

16. The horizontal distance between two towers is 70 m. The angles of depression of the top of the first tower, when seen from the top of the second tower is 30°. If the height of the second tower is 120 m, find the height of the first tower. **Ans.** 79.6 m

17. The angle of elevation of the top of a tower from a point A on the ground is 30°. On moving a distance of 20 metres towards the foot of the tower to a point B, the angle of elevation increases to 60°. Find the height of the tower. **Ans.** $10\sqrt{3}\,\text{m}$

18. Two points A and B are on the same side of a tower. They measure the angle of elevation of the top of the tower as 30° and 60° respectively. If the height of the tower is 80 m, find the distance between them. **Ans.** 92.38 m

19. The shadow of a tower, when the angle of elevation of the Sun is 45°, is found to be 10 metres longer than when it is 60°. Find the height of the tower. **Ans.** 23.65 m

20. The shadow of a vertical tower on level ground increases by 10 metres, when the altitude of the sun changes from angle of elevation 45° to 30°. Find the height of the tower, correct to one place of decimal. (Take $\sqrt{3}=1.73$) **Ans.** 13.7 m

21. A tower is 50 m high. Its shadow is x metres shorter when the Sun's altitude is 45° than when it is 30°. Find the value of x correct to nearest cm. **Ans.** 36.6 m

22. A 7 m long flagstaff is fixed on the top of a tower on the horizontal plane. From a point on the ground, the angles of elevation of the top and bottom of the flagstaff are 45° and 30° respectively. Find the height of the tower correct to one place of decimal.

Ans. 9.56 m

23. The height of a tower is half the height of the flagstaff on it. The angle of elevation of the top of the tower as seen from a point on the ground is 30°. Find the angle of elevation of the top of the flagstaff as seen from the same point. **Ans.** 60°

24. An aeroplane when 3500 m high passes vertically above another aeroplane at an instant when the angles of elevation of the two aeroplanes from the same point on the ground are 45° and 30° respectively. Find the vertical distance between the two aeroplanes. **Ans.** 1479.27 m

25. An aeroplane when 1200 m high passes vertically above another aeroplane at an instant when the angles of elevation of the two aeroplanes from the same point on the ground are 60° and 45° respectively. Find the vertical distance between the two aeroplanes. **Ans.** 507.18 m

26. The angle of elevation of an aeroplane from a point P on the ground is 60°. After a flight of 15 seconds, the angle of elevation changes to 30°. If the aeroplane is flying at a constant height of $1500\sqrt{3}$ m, find the speed of the aeroplane. **Ans.** 720 km/h

27. An aeroplane when 3000 metres high passes vertically above another aeroplane at an instance when their angles of elevation at the same observation point are 60° and 45° respectively. How many metres higher is the one than the other? **Ans.** 1268 m

28. The pilot of an aircraft flying horizontally at a speed of 1200 km/h observes that the angles of depression of a point on the ground changes from 30° to 45° in 15 seconds. Find the height at which the aircraft is flying. **Ans.** 6.83 km

29. The angle of elevation of the top Q of a vertical tower PQ from a point X on the ground is 60°. At a point Y, 40 m vertically above X, the angle of elevation is 45°. Find the height of the tower PQ and the distance XQ. **Ans.** 94.64 m, 54.64 m

30. From a point 100 m above the surface of a lake the angular elevation of the peak of a mountain is found to be 30° and the angle of depression of the image of the peak is 45°. Find the height of the peak. **Ans.** 373.2 m

31. A man on the deck of a ship is 16 m above water level. He observes that the angle of elevation of the top of a cliff is 45° and the angle of depression of the base is 30°. Calculate the distance of the cliff from the ship and the height of the cliff.

Ans. 27.713 m, 43.712 m

32. From the terrace of a house 8 m high, the angle of elevation of the top of a tower is 45° and the angle of depression of its reflection in a lake is found to be 60°. Determine the height of the tower from the ground. **Ans.** 12.62 m

33. The angle of elevation of the top of a tower from a point on the same level as the foot of the tower is 30°. On advancing 'p' metres towards the foot of the tower, the angle of elevation becomes 45°. Show that the height 'h' of the tower is given by

$$h = \frac{(\sqrt{3}+1)p}{2}.$$

Also, determine the height of the tower, if p = 150 metres, α = 30° and β = 60°

Ans. 129.9 m

34. *From a window* (60 metres high above the ground) of a house in a street the angles of elevation and depression of the top and the foot of another house on opposite side of street are 60° and 45° respectively. Show that the height of the opposite house is $60\,(1+\sqrt{3})$ metres.

35. The angle of elevation of an aeroplane as observed from a point 10 m above the ground is found to be 30° and angle of depression of its reflection in a lake is found to be 60°. Determine height of the aeroplane from the ground. **Ans.** 20 m

36. A round balloon of radius '*a*' subtends an angle θ at the eye of the observer while the angle of elevation of its centre is ϕ. Prove that the height of the centre of the balloon is $a \sin \phi \operatorname{cosec} \dfrac{\theta}{2}$.

37. A surveyor noted that the angle of elevation of a marker on the top of a hill was 30°. He walked 40 m towards the foot of the hill along level ground and found the angle of elevation of the marker as 45°. How far from surveyer's first position was the marker? **Ans.** $40(\sqrt{3}+1)$

38. A fire at a building *B* is reported on telephone to two fires stations F_1 and F_2, 10 km apart from each other on a straight road. F_1 observes that the fire is at angle of 60° to the road and F_2 observes that it is at angle of 30° from it. Which station should send his team and how much will it have to travel?

Ans. F_1, should send his team, 5 km

39. If a parachute is descending vertically and makes angles of depression of 45° and 60° at two observations points 100 m apart from each other on the left side of himself. Find, in metres, the approximate height from which the parachutist falls and also find, in metres the approximate distance of the point, where he falls on the ground from the first observation point. **Ans.** 236.6 m, 236.6 m

40. The line joining the top of a hill to the foot of the hill makes angle of 30° with the horizontal through the foot of the hill. There is one temple at the top of the hill and a guest house half way from the foot to the top. The tops of the temple and of the guest house both make an elevation of 45° at the foot of the hill. If the guest house is 100 m away from the foot of the hill along the hill, find the heights of the guest house and the temple. **Ans.** 36.6 m, 73.2 m

41. A boy is standing on the ground and flying a kite with 100 m string at an elevation of 30°. Another boy is standing on the roof of a 20 m high building and is flying his kite at an elevation of 30°. Both the boys are on opposite sides of both the kites. Find the length of the string such that the two kites meet. **Ans.** 60 m

42. Two stations due south of a leaning tower which leans towards north are at distances *a* and *b* from its foot. If α and β be the elevations of the top of the tower from these stations, prove that its inclination θ to the horizontal is given by $\cot\theta = \dfrac{b\cot\alpha - a\cot\beta}{b-a}$.

43. A man on the top of a vertical tower observes a car moving at a uniform speed coming directly towards it. If it takes 12 minutes for the angle of depression to change from 30° to 45° how soon after this, will the car reach the tower?

Ans. 16 minutes 23 secs

44. The shadow of a flagstaff is three times as long as the shadow of the flagstaff when the sunrays meet the ground at an angle of 60°. Find the angle between the sunrays and the ground at the time of longer shadow. **Ans.** 30°

45. The angle of elevation of a stationary cloud from a point 2500 m above a lake is 30° and the angle of depression of its reflection in the lake is 45°. What is the height of the cloud above the lake level? **Ans.** 9330 m

46. If the angle of elevation of a cloud from a point h metres above a lake is α and the angle of depression of its reflection in the lake is β, prove that the height of the cloud is $\dfrac{h(\tan\beta + \tan\alpha)}{\tan\beta - \tan\alpha}$. **[S.B.T.E. 2017, 2014]**

47. The horizontal distance between two trees of different heights is 60 m. The angle of depression of the top of the first tree, when seen from the top of the second tree is 45°. If the height of the second tree is 80 m, find the height of the first tree.

Ans. 20 m

48. From the top of a light house, the angles of depression of two ships on the opposite sides of it are observed to be α and β. If the height of the light house be h metres and the line joining the ships passes through the foot of the light house, show that the distance between the ships is $\dfrac{h(\tan\alpha + \tan\beta)}{\tan\alpha \tan\beta}$ metres.

49. A tower subtends an angle α at a point A in the plane of its base and the angle of depression of the foot of the tower at a point b metres just above A is β. Prove that the height of tower is $b \tan\alpha \cot\beta$.

50. PQ is a post of given height 'a' and AB is a tower at some distance, α and β are the angles of elevation of B, the top of the tower, at P and Q respectively. Find the height of the tower and its distance from the post.

Ans. Height $= \dfrac{a\tan\alpha}{\tan\alpha - \tan\beta}$, Distance $= \dfrac{a}{\tan\alpha - \tan\beta}$

51. Two boats approach a light house in mid sea from opposite directions. The angles of elevation of the top of the light house from two boats are 30° and 45° respectively. If the distance between two boats is 100 m, find the height of the light house.

Ans. $50(\sqrt{3} - 1)$

52. From an aeroplane vertically above a straight horizontal road, the angles of depression of two consecutive milestones on opposite sides of the aeroplane are observed to be α and β. Show that the height of the aeroplane above the road is $\dfrac{\tan\alpha \tan\beta}{\tan\alpha + \tan\beta}$.

53. A person observed the angle of elevation of the top of a tower to be 30°. He walked 50 m towards the foot of the tower along level ground and found the angle of elevation of the top of the tower to be 60°. Find the height of the tower.

Ans. 43.3 m

54. A path separates two walls. A ladder leaning against one wall rests at a point on the path. It reaches a height of 90 m on the wall and makes an angle of 60° with the ground. If while resting at the same point on the path, it were made to lean against the other wall, it would have made an angle of 45° with the ground. Find the height, it would have reached on the second wall. **Ans.** 73.48 m

55. A carpenter makes stools for electrician with a square top of side 0.5 m and at a height 1.5 m above the ground. Also each leg is inclined at an angle of 60° to the ground. Find the length of each leg and also the lengths of two steps to be put at equal distance in metres.

Ans. 1.732 m, length of steps = 1.654 m and 1.077 m

56. At the foot of a mountain the elevation of its summit is 45° after ascending 1000 m towards the mountain up a slope of 30° inclination, the elevation is found to be 60°. Find the height of the mountain. **Ans.** 1366 m

57. There is a small island in the middle of a 100 m wide river and a tall tree stands on the island. *P* and *Q* are points directly opposite to each other on the two banks and in line with the tree. If the angles of elevation of the top of the tree from *P* and *Q* are respectively 30° and 45°, find the height of the tree. **Ans.** 36.6 m

58. From a building 60 metres high, the angles of depression of the top and bottom of a lamp post are 30° and 60° respectively. Find the distance between the lamp post and the building. Also, find the difference of height between the building and the lamp post. **Ans.** 34.64 m, 40 m

59. A contractor planes to install two slides for the children to play in a park. For the children below the age of 5 years, she prefers to have a slide whose top is at a height of 1.5 m, and is inclined at an angle of 30° to the ground, whereas for elder children, she wants to have a steep slide at a height of 3 m, and inclined at an angle of 60° to the ground. What should be the length of the slide in each case.

Ans. $3\text{ m}, 2\sqrt{3}\text{ m}$

Plotting of Curves

19.1 INTRODUCTION

Sketching of simple curves is useful to know the limits which are required in finding the area of bounded regions by the curve $y = f(x)$. In the next chapter we have to find out the area bounded by the curve, x-axis and two ordinates. Directly no question will be asked in the examination from this chapter.

19.2 METHOD OF TRACING A CURVE

1. **Curve through origin:**

   The curve passes through the origin, if its equation does not contain constant term.

   For example, the curve $y^2 = 4ax$ passes through the origin.

2. **Symmetry:**

   (*i*) ***Symmetry about x-axis***

   A curve is symmetrical *about x-axis* if the equation remains the same by replacing y by $-y$. Here y should have even powers only. The shape of curve above x-axis is exactly identical to its shape below x-axis.

   For example, $y^2 = 4ax$ is symmetrical about x-axis.

   (*ii*) ***Symmetry about y-axis***

   It is symmetrical about y-axis if it contains only even powers of x.

   For example, $x^2 = 4ay$ is symmetrical about y-axis.

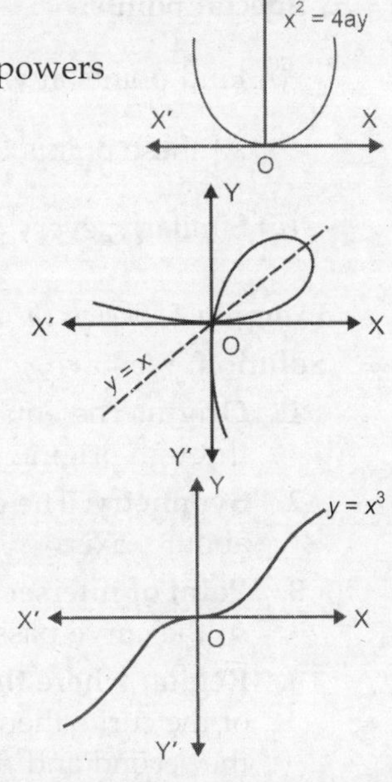

   (*iii*) ***Symmetry about the line y = x.***

   If on interchanging x and y, the equation remains the same then the curve is symmetrical about the line $y = x$.

   For example, $x^3 + y^3 = 3axy$ is symmetrical about the line $y = x$.

   (*iv*) ***Symmetry in opposite quadrants.***

   A curve is symmetrical in the opposite quadrants if its equation remains the same when x is replaced by $-x$ and y by $-y$.

   For example, $y = x^3$

3. The points of intersection with the axes:

(*i*) By putting $y = 0$ in the equation of the curve we get the coordinates of the points of intersection with the x-axis.

For example, $\frac{x^2}{a^2} + \frac{y^2}{b^2} = 1$ intersects the x-axis at the point whose abscissa is given by $\frac{x^2}{a^2} + \frac{0}{b^2} = 1$ or $\frac{x^2}{a^2} = 1$ or $x = \pm a$

$(a, 0)$ and $(-a, 0)$ are the coordinates of the required points.

(*ii*) By putting $x = 0$ in the equation of the curve, the ordinate of the point of intersection with the y-axis is obtained by solving the new equation.

4. Regions in which the curve does not lie:

If the value of y is imaginary for certain value of x then the curve does not exist for such values.

For example, 1. $y^2 = 4x$

For negative value of x, y is imaginary so there is no curve in second and third quadrant.

For example, 2. $a^2x^2 = y^3(2a - y)$

(*i*) *for $y > 2a$, x is imaginary. There is no curve beyond $y = 2a$.*

(*ii*) For negative values of y, x is imaginary. There is no curve in 3rd and 4th quadrant.

5. Table: To find some more points on the curve we can construct a table.

For example, $y^2 = 4x + 4$

x	-1	0	1	2	3
y	0	± 2	$\pm 2\sqrt{2}$	$\pm 2\sqrt{3}$	± 4

6. Special points:

(*i*) Find points at which $\frac{dy}{dx} = 0$.

At these points the tangent to the curve is parallel to x-axis.

(*ii*) Similarly, every point where $\frac{dx}{dy} = 0$ the tangent to the curve is parallel to y-axis.

Example 1. *Sketch the curve:* $y^2 = 4x$.

Solution. We have, $y^2 = 4x$

1. **Origin:** The equation of the curve does not contain constant term. So it passes through origin.
2. **Symmetry:** The equation contains even power of y so the curve is symmetrical about x-axis.
3. **Point of intersection with coordinate axis:** On putting $x = 0$, we have y also zero so the curve passes through origin.
4. **Region where the curve does not exist:** If we substitute $x = -1$ in the equation of the curve then $y^2 = -4 \Rightarrow y$ is imaginary. So, the curve does not lie or exist in the second and third quadrant.

5. **Table:** $y^2 = 4x$.

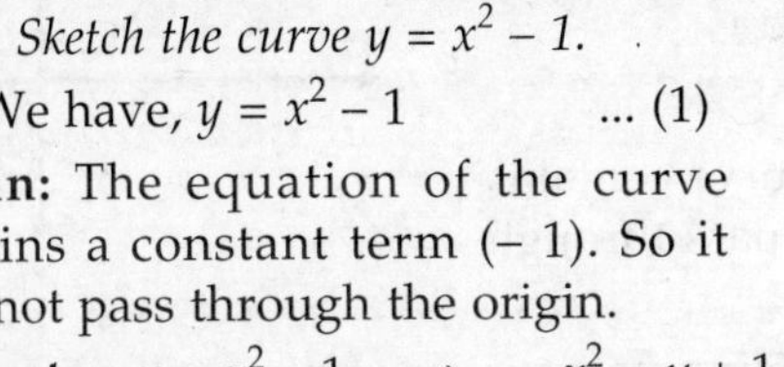

x	0	1	4	9
y	0	± 2	± 4	± 6

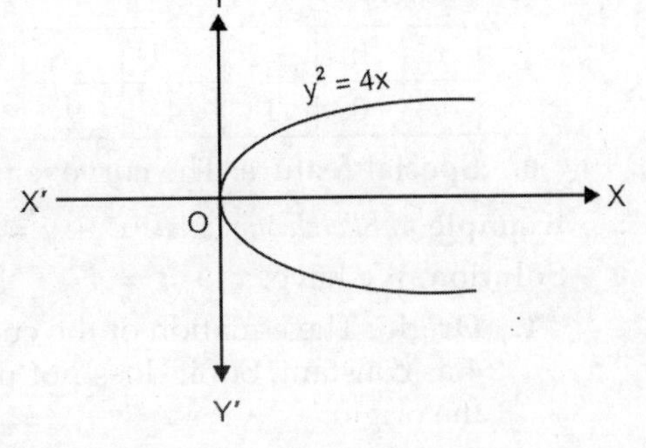

6. **Special feature:** Curve is monotonic.

Example 2. *Sketch the curve* $y = x^2 - 1$.

Solution. We have, $y = x^2 - 1$ ... (1)

1. **Origin:** The equation of the curve contains a constant term (– 1). So it does not pass through the origin.
2. **Symmetry:** $y = x^2 - 1 \Rightarrow x^2 = y + 1$
The equation contains even power of x, so the curve is symmetrical about y-axis.
3. **Point of intersection with coordinate axis:**
(*i*) On putting $x = 0$ in the equation of the curve, $y = -1$.
The curve crosses y-axis at the point $(0, -1)$.
(*ii*) On putting $y = 0$ in (1), we get
$x = \pm 1$. So the curve crosses x-axis at the points $(1, 0)$ and $(-1, 0)$.
4. **Region where the curve does not exist:**
If we substitute any value of y less then $-1(-2, \text{say})x$ becomes imaginary so the curve does not exist below $(0, -1)$.
5. **Table:**

x	0	1	– 1	2	– 2
y	– 1	0	0	3	3

The curve is as shown in the figure.

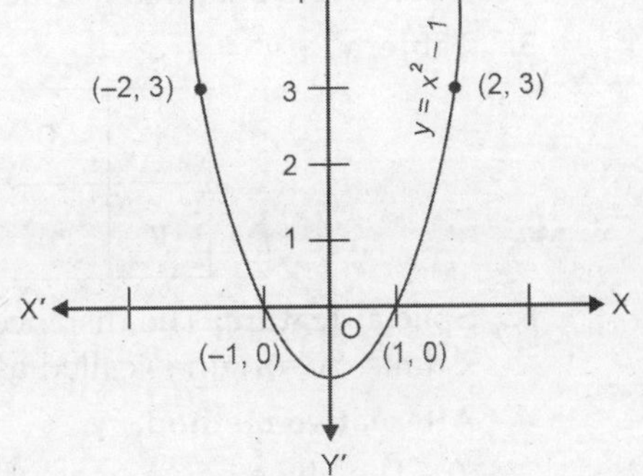

6. **Special points:** $y = x^2 - 1 \Rightarrow \dfrac{dy}{dx} = 2x$

If $\dfrac{dy}{dx} = 0 \Rightarrow 2x = 0 \Rightarrow x = 0$

At the point $(0, -1)$ on the curve the tangent is parallel to the x-axis. **Ans.**

Example 3. *Sketch the curve* $x^2 = 4y$.

Solution. We have, $x^2 = 4y$

1. **Origin:** The equation of the curve does not contain constant term. So, it passes through origin.
2. **Symmetry:** The equation contains even power of x. So, the curve is symmetrical about y-axis.
3. **Point of intersection with coordinate axis:** On putting $x = 0$ we have y also zero. So, the curve passed through the origin.
4. **Region where the curve does not exist:** If we substitute $y = -1$ in the equation of the curve then $x^2 = -4 \Rightarrow x$ is imaginary. So, the curve does not exist in third and fourth quadrant.

5. **Table:** $x^2 = 4y$

x	0	± 2	± 4	± 6
y	0	1	4	9

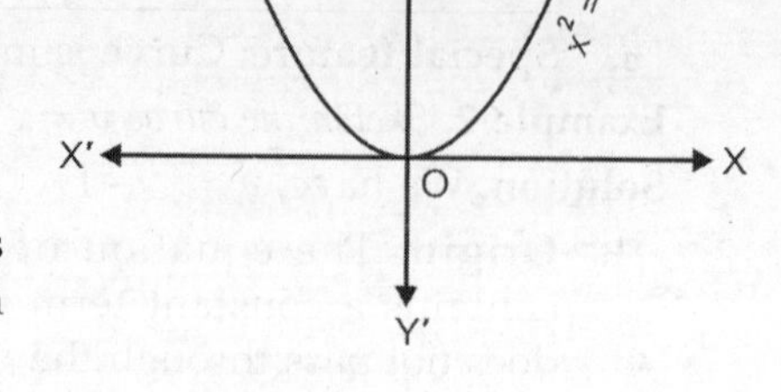

6. **Special feature:** The curve is monotonic.

Example 4. *Sketch the curve $x^2 + y^2 = 4$.*

Solution. We have, $x^2 + y^2 = 4$

1. **Origin:** The equation of the curve contains 4 as constant. So, it does not pass through the origin.
2. **Symmetry:** The equation contains even power of x as well as even powers of y. So, the curve is symmetrical about both the axes *i.e.* x-axis and y-axis.
3. **Point of intersection:** On putting $x = 0$, $y^2 = 4 \Rightarrow y = \pm 2$. So, the curve passes through (0, 2), (0, – 2).

   On putting $y = 0$, $x^2 = 4 \Rightarrow x = \pm 2$.

   So, the curve passes through the points (2, 0) and (– 2, 0)
4. **Region where the curve does not exist:**

   From step 3, it is clear that the curve that the curve passes in all the four quadrants.

   If $x > 2$, y becomes imaginary it means the curve will not pass beyond $x = 2$.

   If $y > 2$, x is imaginary it means the curve will not pass beyond $y = 2$.
5. **Table:** $x^2 + y^2 = 4$

x	0	1	± 2	$\pm\sqrt{3}$
y	± 2	$\pm\sqrt{3}$	0	± 1

6. **Special feature:** The distance from the origin to the curve remains the same *i.e.* 2 units. So, origin is called as centre of circle with radius 2 units.

   Alternative method.

   (*i*) The given equation is of second degree in x and y.

   (*ii*) Coefficient of x^2 = Coefficient of y^2.

   (*iii*) Coefficient of $x\,y = 0$.

   Thus, the given equation represents a circle.

   On comparing the given equation $x^2 + y^2 = 4$ with the standard equation of circle $x^2 + y^2 = a^2$, we get $a^2 = 4 \Rightarrow a = \pm 2$.

   The centre of the circle is (0, 0) and radius is 2.

   Hence, the given equation represents a circle with centre (0, 0) and radius 2.**Ans.**

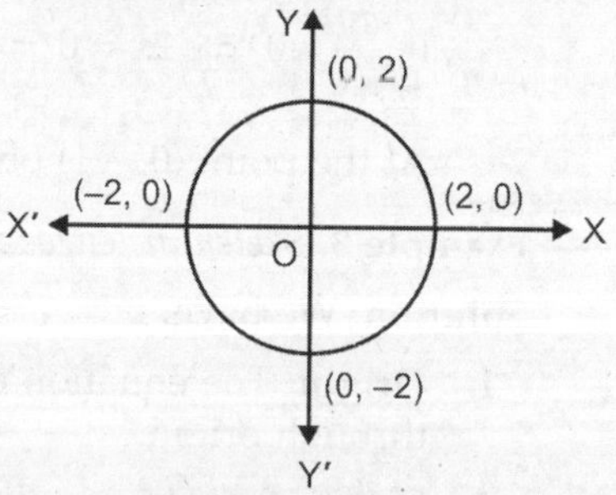

Example 5. *Sketch the graph of curve $(x - 1)^2 + y^2 = 1$.*

Solution. We have, $(x - 1)^2 + y^2 = 1$ *i.e.* $x^2 - 2x + y^2 = 0$

1. **Origin:** As the equation of the curve does not contain constant term. So, the curve passes through the origin.

2. **Symmetry:** The equation contains even power of y. So, it is symmetric about x-axis. Although the equation contains both even power (*i.e.*, 2) and odd power (*i.e.*, 1) of x. So, it is not symmetric about y-axis.
3. **Point of intersection with coordinate axes:**
   On putting $x = 0$ in the equation of the curve
   $(0, -1)^2 + y^2 = 1 \quad \Rightarrow \quad y = 0.$
   So, the curve will pass through origin.
   On putting $y = 0$ in the equation of the curve
   $(x - 1)^2 + 0 = 1 \quad \Rightarrow \quad x - 1 = \pm 1 \Rightarrow x = 0, 2$
   So, the curve passes through (2, 0) and origin.

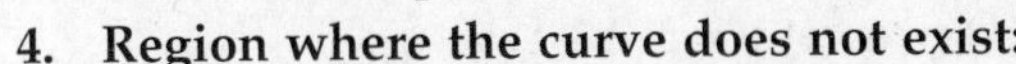

4. **Region where the curve does not exist:**
   If $x > 2$, or $x < 0$ becomes imaginary. So, the curve does not exist beyond $x > 2$ and $x < 0$.

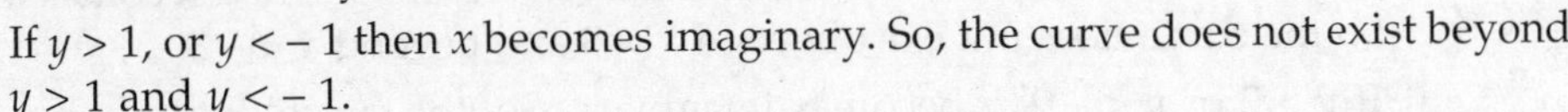

   If $y > 1$, or $y < -1$ then x becomes imaginary. So, the curve does not exist beyond $y > 1$ and $y < -1$.
5. **Table:** $x^2 - 2x + y^2 = 0$

x	0	1	2
y	0	± 1	0

6. **Special features:**
   $(x - 1)^2 + y^2 = 1$
   Differentiating both sides w.r.t. 'x', we get
   $$2(x-1) + 2y\frac{dy}{dx} = 0 \quad \Rightarrow \quad \frac{dy}{dx} = \frac{1-x}{y}$$
   If $\frac{dy}{dx} = 0$, then $x = 1$
   So, at points (1, 1) and (1, – 1) on the curve tangents are parallel to x-axis.

   Alternative method.
   (*i*) The given equation is of second degree in x and y.
   (*ii*) The coefficient of x^2 = The coefficient of y^2.
   (*iii*) Coefficient of $x\,y = 0$.
   Thus, the given equation represents a circle.
   $$x^2 + y^2 - 2x = 0$$
   Comparing this equation to the general equation of the circle *i.e.*
   $$x^2 + y^2 + 2gx + 2fy + c = 0, \text{ we get } g = -1, f = 0, c = 0$$
   $\therefore$ Centre of the circle is $(-g, -f) = (1, 0)$
   and radius is $\sqrt{g^2 + f^2 - c} = \sqrt{1 + 0 + 0} = 1$
   Hence, the equation represents a circle with centre (1, 0) and radius 1. **Ans.**

Example 6. *Draw a rough sketch of the curve:* $\frac{x^2}{9} + \frac{y^2}{4} = 1$

Solution. 1. Origin: $\frac{x^2}{9}+\frac{y^2}{4}=1$ ... (1)

Equation (1) contains constant term 1. Therefore it does not pass through the origin.

2. **Symmetry:** The equation (1) contains even power of x and y so it is symmetrical about both axes x and y. Since on replacing x by $-x$ and y by $-y$, the given equation remains unchanged, therefore the curve is symmetrical in opposite quadrants.
3. **Point of intersection of the curve with the coordinate axes:** On putting $y = 0$ in (1) we get, $x = \pm 3$.

   Similarly, on putting $x = 0$ in (1), we get $y = \pm 2$.

   So the curve passes through (3, 0), (– 3, 0) and (0, 2), (0, – 2).
4. **Region where the curve does not exist:** If $x > 3$ or $x < -3$, y becomes imaginary. So the curve does not exist beyond $x > 3$ and beyond $x < -3$.

   If $y > 2$ or $y < -2$, x becomes imaginary. So, the curve does not exist beyond $y > 2$ and beyond $y < -2$.
5. **Table:** $\frac{x^2}{9}+\frac{y^2}{4}=1$

x	0	3	– 3
y	± 2	0	0

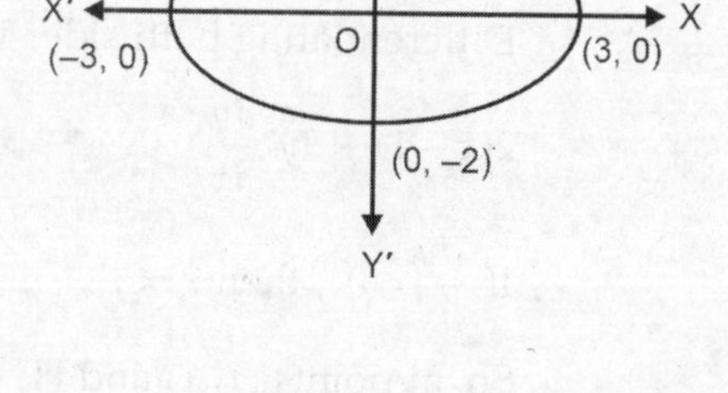

6. **Special points:** $\frac{x^2}{9}+\frac{y^2}{4}=1$

   Differentiating w.r.t. x, we get

   $$\frac{2x}{9}+\frac{2y}{4}\cdot\frac{dy}{dx}=0 \Rightarrow \frac{dy}{dx}=-\frac{4}{9}\frac{x}{y}$$

   At $x=0, \frac{dy}{dx}=0$

   At the points (0, 2) and (0, – 2) on the curve the tangents are parallel to the x-axis.

Ans.

Example 7. *Sketch the curve:* $\frac{x^2}{4}+\frac{y^2}{9}=1.$

Solution. We have $\frac{x^2}{4}+\frac{y^2}{9}=1$ ... (1)

1. **Origin:** The equation contains 1 as constant. Therefore it does not pass through the origin.
2. **Symmetry:** The equation of the curve contains even powers of x and y so it is symmetrical about both axes x and y.
3. **Point of intersection of curve with the coordinate axes:**

   On putting $x = 0$, we get $y = \pm 3$, the curve passes through (0, 3) and (0, – 3).

   On putting $y = 0$, we get $x = \pm 2$, the curve passes through (2, 0) and (– 2, 0).

4. Region where the curve does not exist:

If $x > 2$ or $x < -2$, y becomes imaginary. So the curve does not exist beyond $x > 2$ and $x < -2$.

If $y > 3$ or $y < -3$, x becomes imaginary.

So, the curve does not exist beyond $y > 3$ and $y < -3$.

5. Table: $\frac{x^2}{4}+\frac{y^2}{9}=1$

x	– 2	0	1	2
y	0	± 3	± 2.6	0

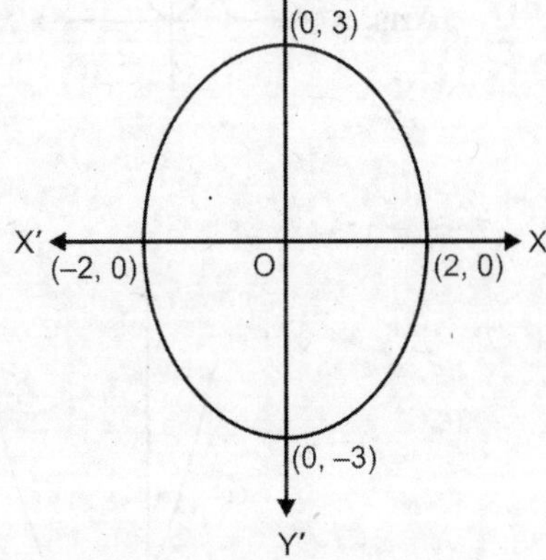

6. Special feature: $\frac{x^2}{4}+\frac{y^2}{9}=1$

Differentiating w.r.t. 'x', we get

$$\frac{2x}{4}+\frac{2y}{9}\frac{dy}{dx}=0 \quad\Rightarrow\quad \frac{dy}{dx}=-\frac{9}{4}\frac{x}{y}$$

At $x=0, \frac{dy}{dx}=0$

At the point (0, 3) and (0, – 3) on the curve the tangents are parallel to the x-axis.

Ans.

EXERCISE 19.1

Sketch the following curves:

1. $y^2 = x$

Ans.

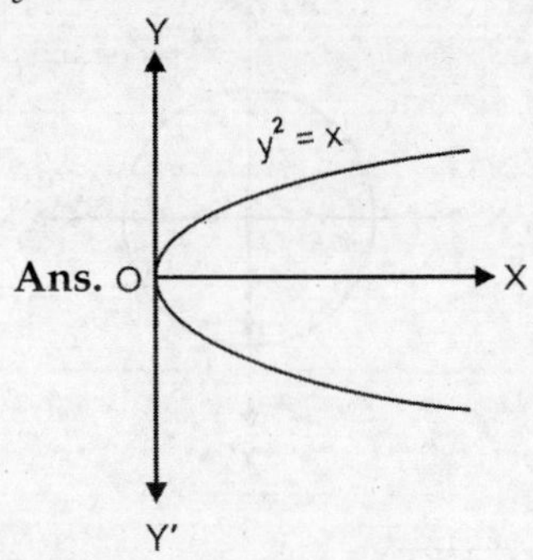

2. $2y^2 = 9x$

Ans.

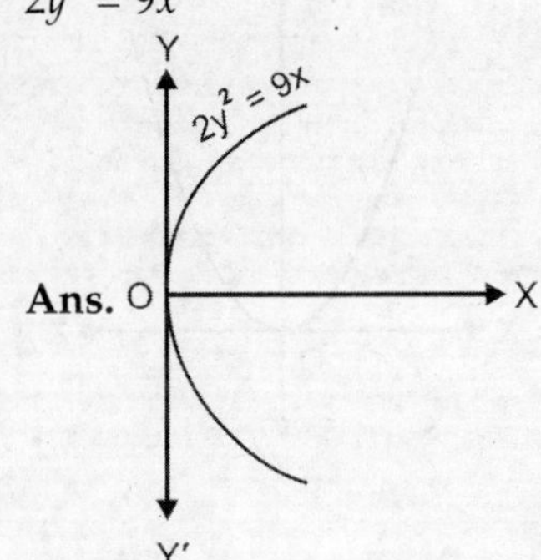

3. $y = x^2$

Ans.

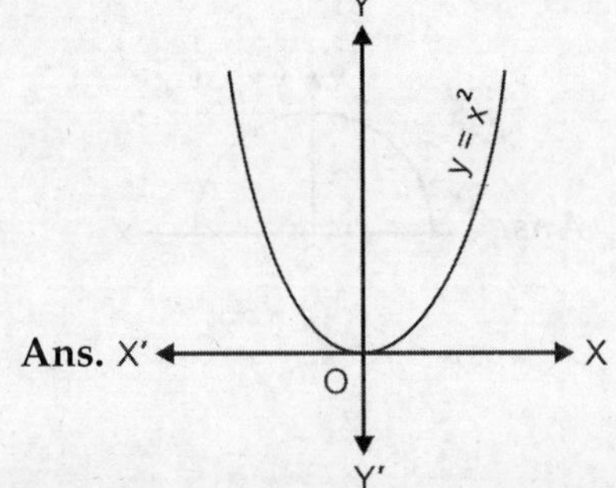

4. $y = x^2 + 2$

Ans.

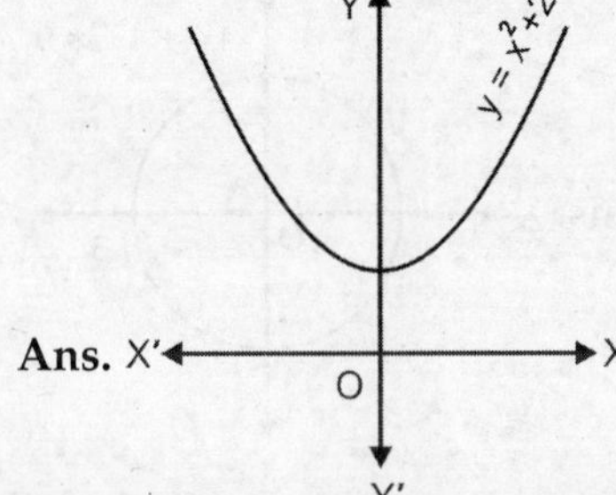

5. $4y = 3x^2$

Ans.

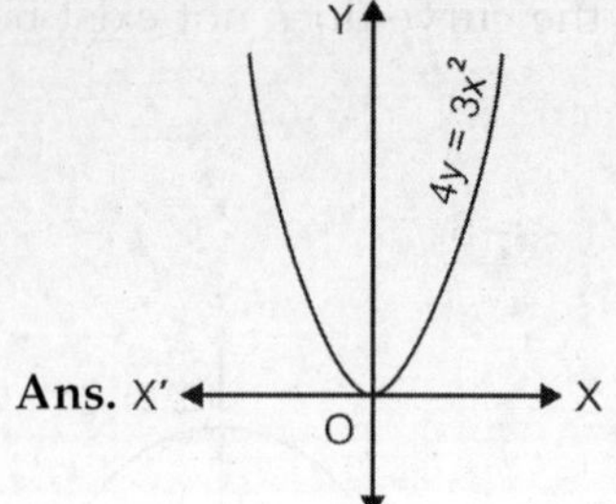

6. $y = x^2 - 2$

Ans.

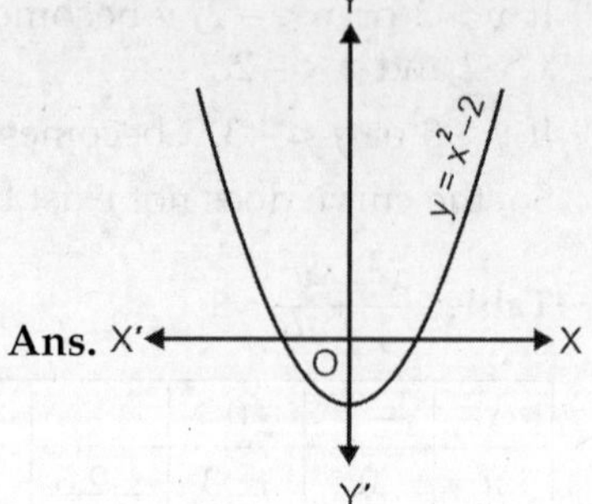

7. $y = x^2 - 2x$

Ans.

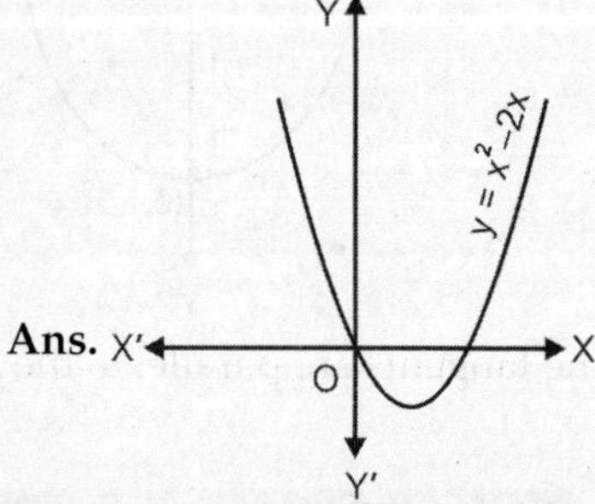

8. $y = 3x - x^2$

Ans.

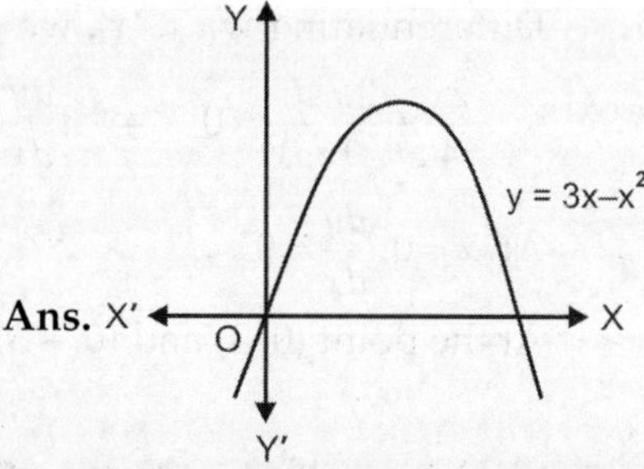

9. $y = x^4$

Ans.

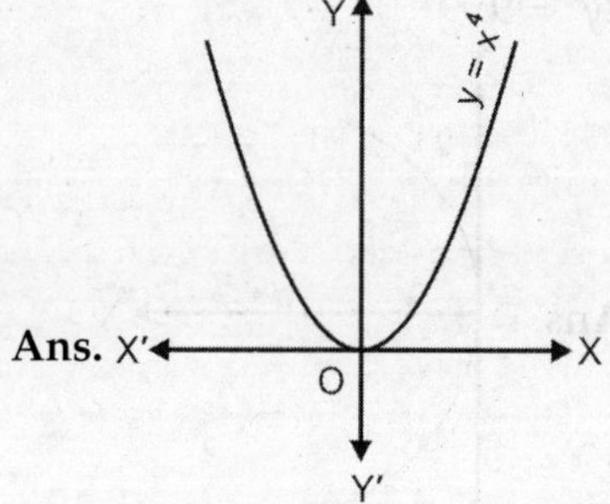

10. $x^2 + y^2 = 16$

Ans.

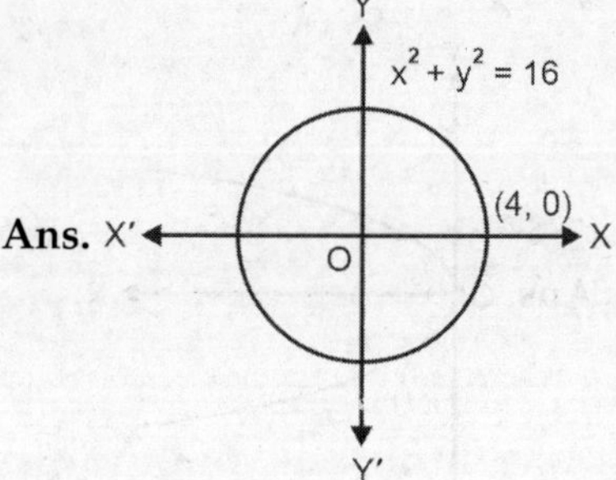

11. $4x^2 + 4y^2 = 9$

Ans.

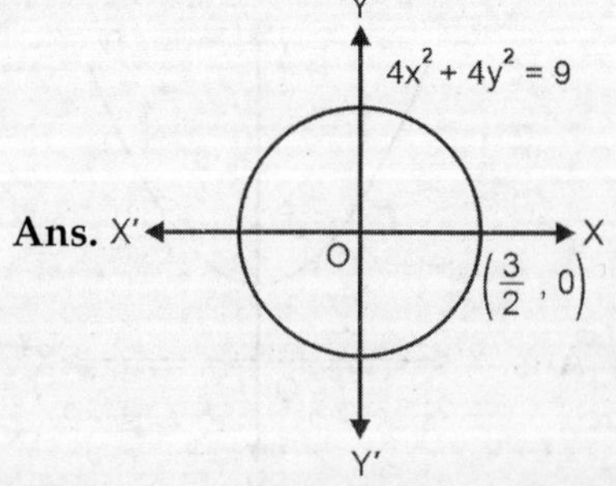

12. $y = \sqrt{9 - x^2}$

Ans.

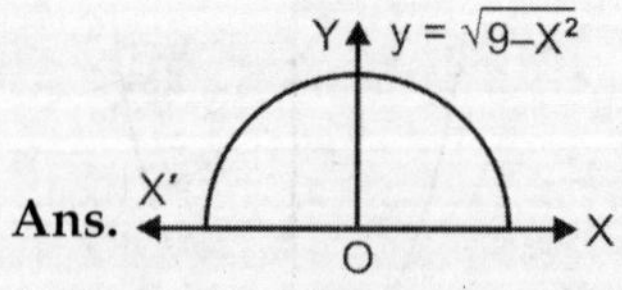

13. $\dfrac{x^2}{16}+\dfrac{y^2}{25}=1$

Ans.

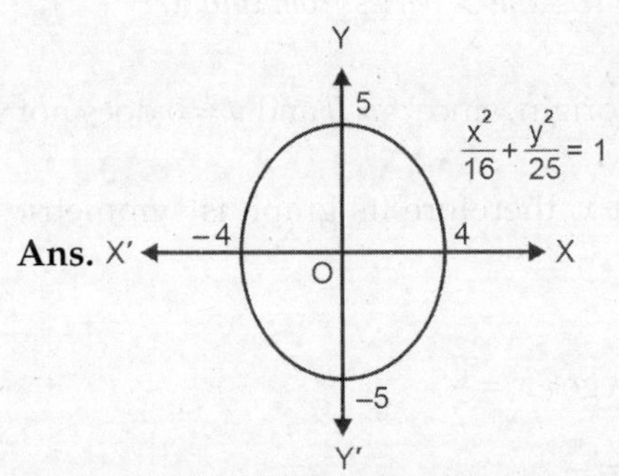

14. $\dfrac{x^2}{25}+\dfrac{y^2}{16}=1$

Ans.

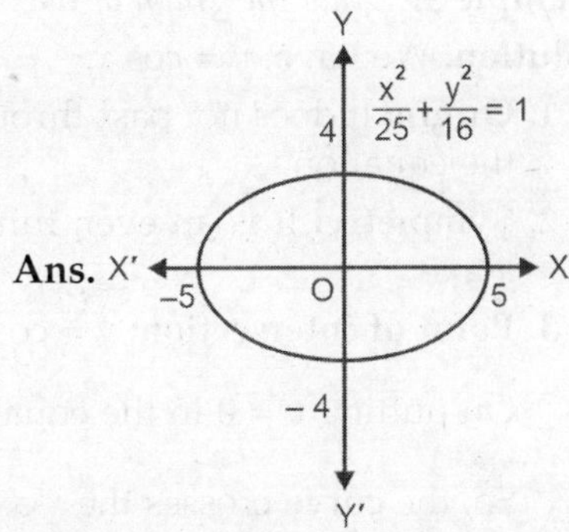

19.3 GRAPHS OF TRIGONOMETRIC FUNCTIONS

Approximations. The following approximations make the work of drawing graphs much easier.

$\sqrt{2}=1.41,$	$\pi=3.14,$	$\dfrac{1}{\sqrt{2}}=0.71,$	$\dfrac{\pi}{2}=1.57$
$\sqrt{3}=1.73,$	$\dfrac{\pi}{3}=1.05,$	$\dfrac{1}{\sqrt{3}}=0.58,$	$\dfrac{\sqrt{3}}{2}=0.87$

Example 8. *Draw the graph of the curve: $y = \sin x$*

Solution. We have, $y = \sin x$.

1. **Origin:** $x = 0$, $y = 0$ satisfy the equation, therefore curve passes through the origin.
2. **Symmetry:** It is an odd function of x, therefore its graph is symmetric about the origin.
3. **Point of intersection:** On putting $y = 0$ in the equation $\sin x = 0 \Rightarrow x = 0, \pi$; so the curve crosses x-axis at $x = 0$ and $x = \pi$.
4. **Region where the curve does not exist:**

   For all x, $-1 \le \sin x \le 1 \quad \Rightarrow \quad -1 \le y \le 1$

   Its graph lies in the horizontal strip determined by the lines $y = 1$ and $y = -1$.
5. **Table: $y = \sin x$**

x	0	$\dfrac{\pi}{6}$	$\dfrac{\pi}{4}$	$\dfrac{\pi}{3}$	$\dfrac{\pi}{2}$
$y = \sin x$	0	$\dfrac{1}{2}$	$\dfrac{1}{\sqrt{2}}$	$\dfrac{\sqrt{3}}{2}$	1

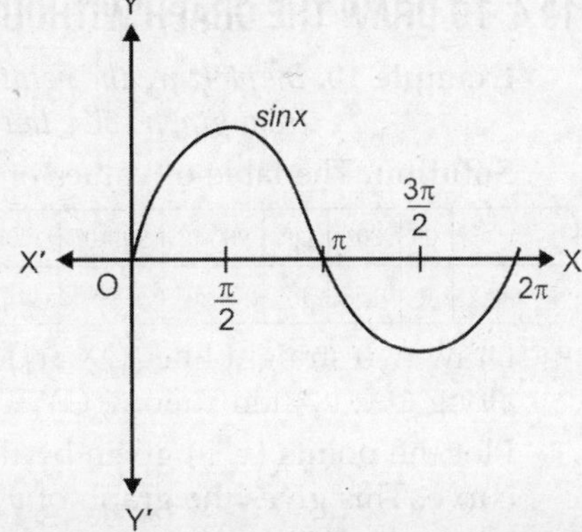

6. **Special points:**

   $y = \sin x$

   $$\frac{dy}{dx} = \cos x \quad \Rightarrow \quad \frac{dy}{dx} = 0, \text{ at } x = \frac{\pi}{2}$$

The tangent to the curve at $x = \frac{\pi}{2}$ is parallel to the x-axis.

Example 9. *Draw the graph of the curve $y = \cos x$ as x varies from 0 to π.*

Solution. We have, $y = \cos x$

1. **Origin:** It does not pass through the origin, since $x = 0$ and $y = 0$ does not satisfy the equation.
2. **Symmetry:** It is an even function of x, therefore its graph is symmetric about y-axis.
3. **Point of intersection:** $y = \cos x$

   On putting $y = 0$ in the equation, we get $x = \frac{\pi}{2}$.

   So, the curve crosses the x-axis at $x = \frac{\pi}{2}$.
4. **Region where the curve does not exist:**

   For all x, $-1 \le \cos x \le 1. \Rightarrow -1 \le y \le 1$

   ∴ The graph lies on horizontal strip determined by the lines $y = 1$ and $y = -1$.
5. **Table: $y = \cos x$**

x	0	$\frac{\pi}{6}$	$\frac{\pi}{4}$	$\frac{\pi}{3}$	$\frac{\pi}{2}$	$\frac{2\pi}{3}$	$\frac{3\pi}{4}$	$\frac{5\pi}{6}$	π
$y = \cos x$	1	$\frac{\sqrt{3}}{2}$	$\frac{1}{\sqrt{2}}$	$\frac{1}{2}$	0	$-\frac{1}{2}$	$-\frac{1}{\sqrt{2}}$	$\frac{-\sqrt{3}}{2}$	-1

6. **Special points:**

   $y = \cos x$

   $\frac{dy}{dx} = -\sin x$

   $\frac{dy}{dx} = 0 \quad \Rightarrow \quad -\sin x = 0 \quad \Rightarrow \quad x = 0$

   The tangent at $x = 0$ is parallel to the x-axis.

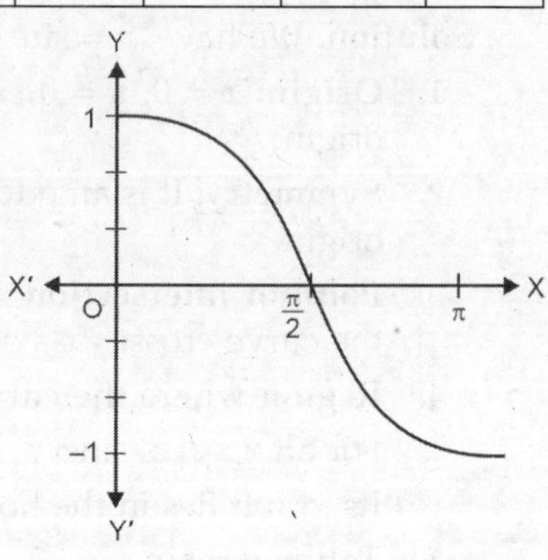

19.4 TO DRAW THE GRAPH WITHOUT USING STEPS

Example 10. *By plotting the points draw the graph of tan x as x varies from 0° to 360° and from graph solve tan x = 1.*

Solution. The table of values of x and $y = \tan x$ is

x	0°	30°	60°	90°–0°	90°+0°	120°	150°	180°	210°	240°	270°–0°	270°+0°	300°	330°	360°
$y = \tan x$	0	0.58	1.7	+∞	–∞	–1.7	–0.58	0	0.58	1.7	+∞	–∞	–1.7	–0.58	0

Draw two straight lines OX and OY mutually at rt. ∠s. Let the angle x be represented along OX, $y = \tan x$ along OY.

Plot the points (x, y) given by the above table and join them by a free hand smooth curve. This gives the graph of $y = \tan x$.

From the graph $x = 135°, 315°$ when $\tan x = -1$.

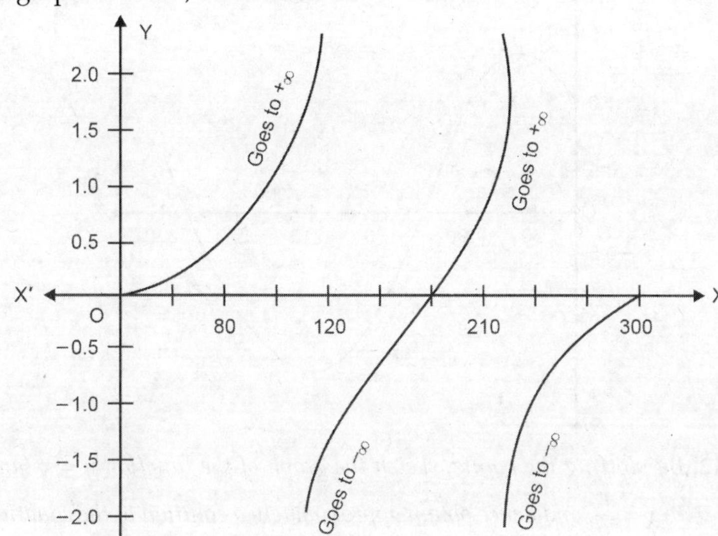

Note. While drawing the graph of $y = \tan x$, the following facts must be kept in mind.

1. **Variations of tan x in different quadrants.**
   In the first quadrant $\tan x$ increases from 0 to $+\infty$
   In the second quadrant $\tan x$ increases from $-\infty$ to 0
   In the third quadrant $\tan x$ increases from 0 to $+\infty$
   In the fourth quadrant $\tan x$ increases from $-\infty$ to 0
2. **Sign of tan x in different quadrants**
   $\tan x$ is + ve in the first and third quadrant and negative in second and fourth qudrants.
   $\therefore$ graph of $y = \tan x$ lies above OX in first and third quadrants and below OX in second and fourth quadrants.
3. **Similarity of shape in different quadrants**
   The shape of the graph of $y = \tan x$ is exactly same in first and third quadrants and also in second and fourth quadrants.

Example 11. By plotting the points draw the graph of $\sin x + \cos x$ as a varies from 0° to 360°.

Solution. Let $y = \sin x + \cos x$.

Prepare the following table for the corresponding values of x and y.

x	0°	30°	60°	90°	120°	150°	180°	210°	240°	270°	300°	330°	360°
$\sin x$	0	0.5	0.87	1	0.87	0.5	0	−0.5	−0.87	−1	−0.87	−0.5	0
$\cos x$	1	0.87	0.5	0	−0.5	−0.87	−1	−0.87	−0.5	0	0.5	0.87	1
$y = \sin x + \cos x$	1	1.37	1.37	1	0.37	−0.37	−1	−1.37	−1.37	−1	−0.37	0.37	1

Plot the points given by x and y. Join these points by a free hand curve and get the graph of $y = \sin x + \cos x$.

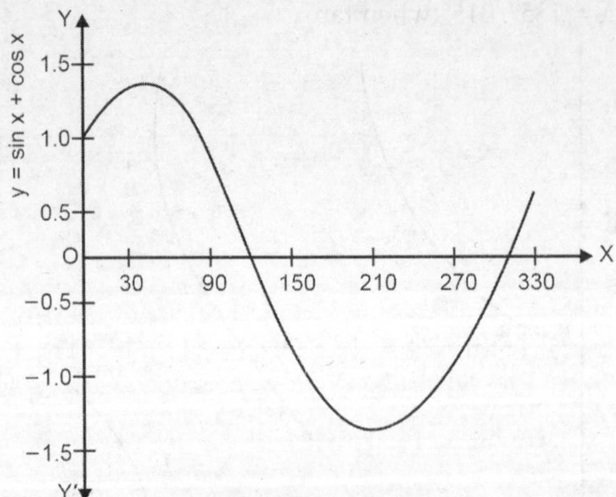

Example 12. *By plotting the points, sketch the graph of the function $y = x \tan x$ from $x = 0$ to $x = \frac{\pi}{2}$ and hence obtain approximately a solution of the equation $x \tan x = 1$.*

Solution. For the solution of $x \tan x = 1$, put each side equal to y, we get $x\tan x = y$ and $1 = y$. We have to plot two curves $y = x\tan x$ and $y = 1$, and their point of intersection will give the solution of the equation $x\tan x = 1$. For $x\tan x$ make the following table for x and $\tan x$ and multiply them. The values of x must be taken in radians. Plot these points on graph paper and join them in a smooth curves. The equation $y = 1$ represents a straight line parallel to x-axis XOX'. This line intercepts the curve at P. From P draw PA perpendicular to x-axis which will give the value of x and this comes out to be 49° (approximate). **Ans.**

Degree	0°	10°	20°	30°	40°	50°	60°	70°	80°	90°
Radian	0	0.174	0.349	0.524	0.698	0.873	1.047	1.222	1.396	1.571
$\tan x$	0	0.176	0.364	0.5774	0.8391	1.1918	1.732	2.7475	5.671	∞
$y = x\tan x$	0	0.0306	0.126	0.3025	0.5857	1.044	1.816	3.34	7.9167	∞

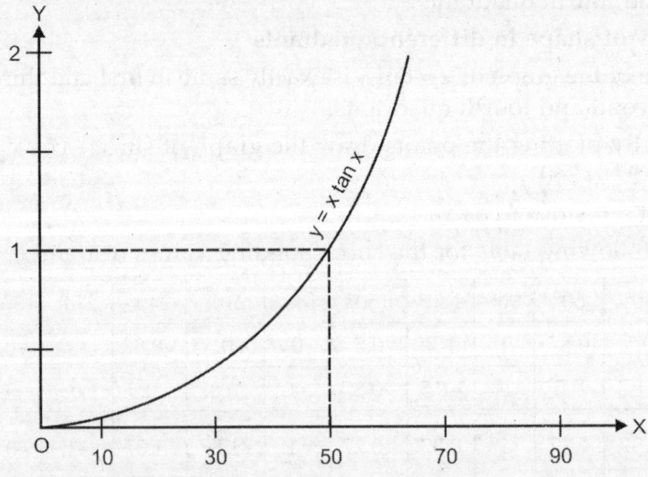

Example 13. *By plotting the points draw in the same figure the graphs of $y = \sin x$ and $y = 2 \cos x$ between 0° and 360°. Hence find out a value of x which is a solution of the equation $\tan x = 2$.*

Solution. We will draw two graphs $y = \sin x$ and $y = 2 \cos x$ independently with respect to the same axis and their point of intersection is the required solution.

For graph $y = \sin x$, give different values of x ranging between 0° and 360° and find the corresponding values of y as tabulated below. Join the graph in free hand smooth curve.

x	0°	30°	60°	90°	120°	150°	180°	210°	240°	270°	300°	330°	360°
$y = \sin x$	0	0.5	0.87	1	0.87	0.5	0	−0.5	−0.87	−1	−0.87	−0.5	0

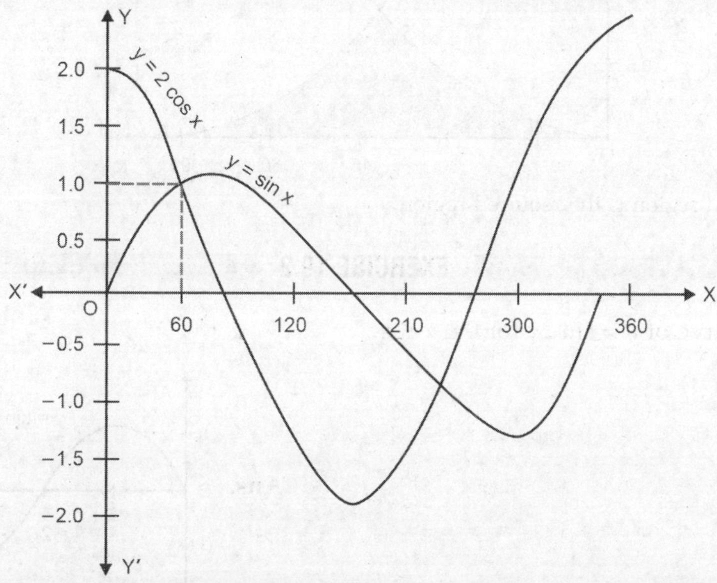

Also draw the graph $y = 2 \cos x$ from table given below on same axis of $y = \sin x$.

x	0°	30°	60°	90°	120°	150°	180°	210°	240°	270°	300°	330°	360°
$y = 1 \sin x$	1	0.87	0.5	0	−0.5	−0.87	−1	−0.87	−0.5	0	0.5	0.87	1
$y = 2 \cos x$	2	1.74	1	0	−1	−1.74	−2	−1.74	−1	0	1	1.74	2

The two curves $y = \sin x$ and $y = 2 \cos x$ intersect at P. From P draw a line PA parallel to OY which meets OX at A.

$x = 63° 26'$ **Ans.**

Example 14. *By plotting the points draw the graph of $y = \tan x$ and $y = \cot x$, $y = 0$ for $0 \le x \le \dfrac{\pi}{2}$ and shade the region enclosed by these curves.*

Solution. Prepare the following table, for $y = \tan x$ and $y = \cot x$ for $0 \le x \le \tan \dfrac{\pi}{2}$.

x	$0 + 0$	$\frac{\pi}{6}$	$\frac{\pi}{4}$	$\frac{\pi}{3}$	$\frac{\pi}{2} - 0$
$y = \tan x$	0	0.58	1	1.73	$+\infty$
$y = \cot x$	$+\infty$	1.73	1	0.58	0

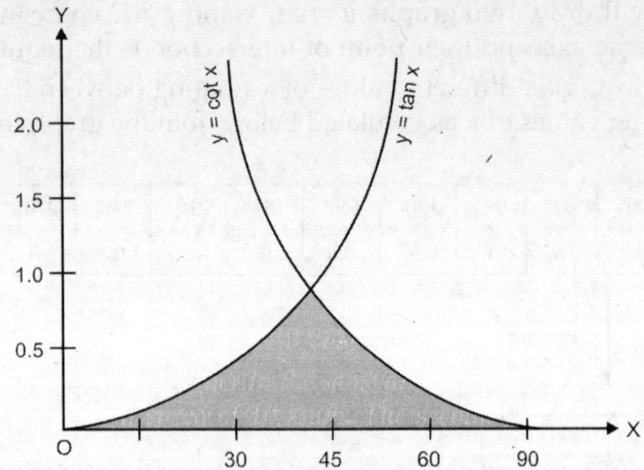

This shaded region is the required region. **Ans.**

EXERCISE 19.2

1. Plot a curve of $y = \sin 2x$ for $0 \le x \le \pi$.

Ans.

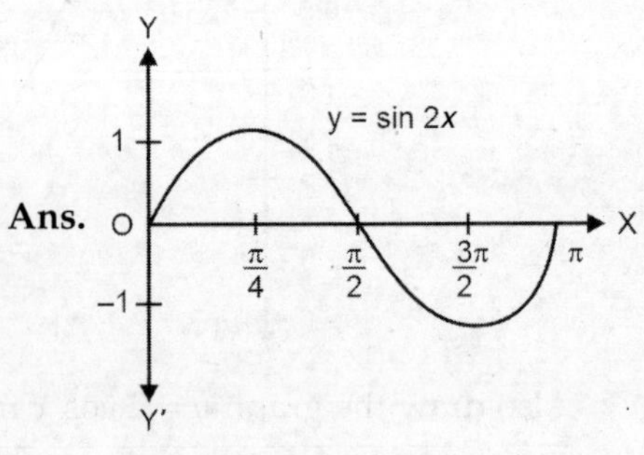

2. Plot a curve of $y = 2 \cos x$ for $0 \le x \le 2\pi$.

Ans.

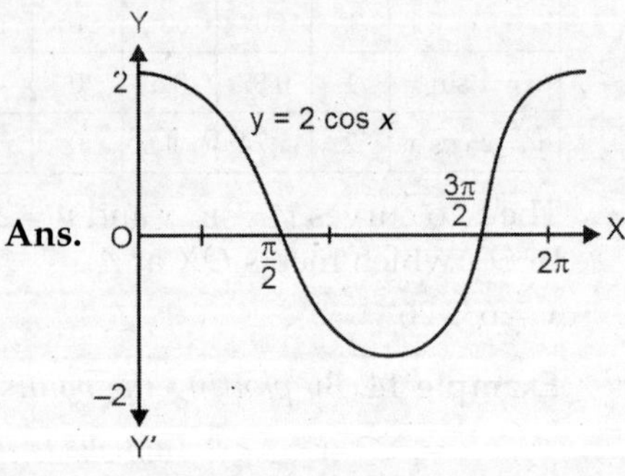

3. Sketch the graph of $y = | \sin x |$ for $0 \le x \le 2\pi$.

Ans.

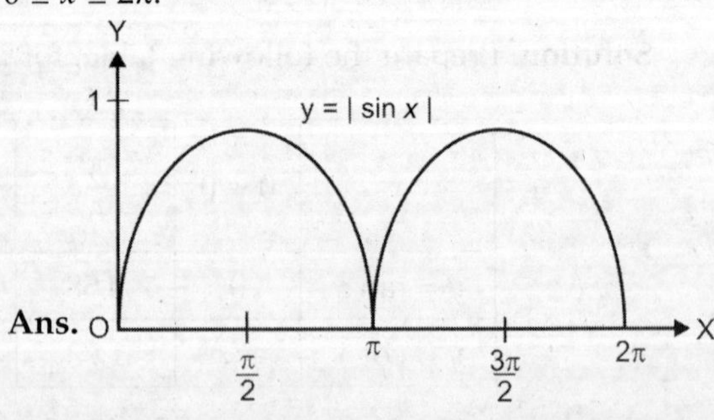

4. Sketch the graph of $y = \sin\left(3x + \frac{\pi}{4}\right)$ for $\frac{-\pi}{3} < x \le \frac{\pi}{3}$.

Ans.

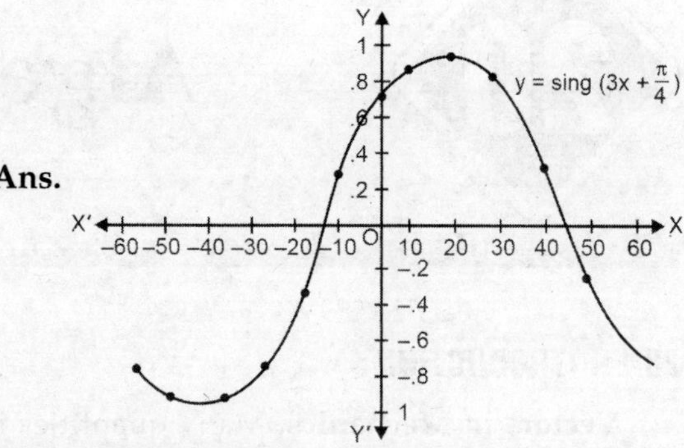

5. Draw a graph of tan x – cot x in the interval from 0 to 2π.

Ans.

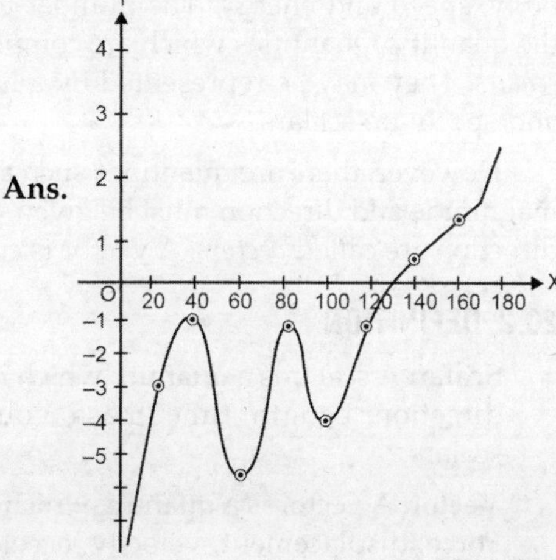

6. Solve graphically, the equation sin x = cot x. **Ans.** 52°

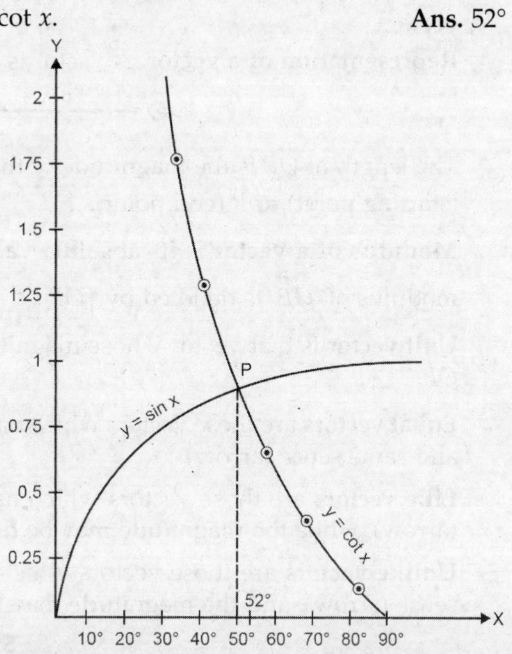

UNIT III: VECTORS

Algebra of Vectors

20.1 INTRODUCTION

Vectors in Mechanics. Many quantities in Physics and Mathematics are described completely by a number with a unit. *For example*, length of a line segment, mass of a body, speed and energy. The number together with the unit, is called the magnitude of the quantity. Quantities which are completely described by the magnitude alone are called *scalars*. They may be represented by a line segment drawn in any direction. The distance and speed are scalars.

However, there are quantities such as displacement, velocity and force, for which both magnitude and direction must be given, such quantities which posses both magnitude and direction are called *vectors*. A vector is completely represented by a directed line segment.

20.2 DEFINITION

Scalar: A scalar is a quantity which has magnitude only. It has no relation to a definite direction. Length, time, mass, volume, density, temperature, work, potential are scalars.

Vector: A vector is a quantity which has *magnitude* and is related to a definite *direction*. Force, displacement, velocity, acceleration, moment of the force, momentum are all vectors.

Representation of a vector. A vector is represented by a directed line segment $\overrightarrow{OP}$.

The length of OP is the magnitude of the vector $\overrightarrow{OP}$. The direction of $\overrightarrow{OP}$ is form O (starting point) to P (end point).

Modulus of a vector is its absolute value or its magnitude with positive sign. The modulus of $\overrightarrow{OP}$ is denoted by $|\overrightarrow{OP}|$.

Unit vector is that vector whose magnitude is unity. Unit vector of $\overrightarrow{OP}$ is written as OP.

Equal vectors are those vectors which have equal magnitude, same direction (parallel) and same sense (arrow).

Like vectors are those vectors which have the same direction (parallel), same sense (arrow), while the magnitude may be different.

Unlike vectors are those vectors which have the same direction (parallel), opposite sense (arrow), and the magnitude may be different.

Negative of vector is a vector whose magnitude is equal to that of the given vector, same direction (parallel) but opposite sense (arrow).

Zero vector or null vector is that vector whose magnitude is zero.

Collinear vectors: Two vectors are said to be collinear vectors, if they are parallel to the same line, irrespective to their magnitudes and directions.

Co-initial vectors: Vectors having the same initial point are called co-initial vectors.

Coterminous vectors: Vectors having the same terminal point are called coterminous vectors.

Coplanar vectors: Three (or more) vectors are said to be coplanar, if they lie in the same plane or are parallel to the same plane.

20.3 ADDITION OF VECTORS

A vector whose effect is the same as a set of two vectors, is called the sum or the resultant of the given vectors.

Let $\vec{a}$ and $\vec{b}$ be two given vectors. if $\overrightarrow{OA} = \vec{a}$ and $\overrightarrow{AB} = \vec{b}$ then the vector $\overrightarrow{OB}$ is called the sum of $\vec{a}$ and $\vec{b}$.

Symbolically

$$\overrightarrow{OA} + \overrightarrow{AB} = \overrightarrow{OB}$$

$$\Rightarrow \quad \vec{a} + \vec{b} = \overrightarrow{OB}$$

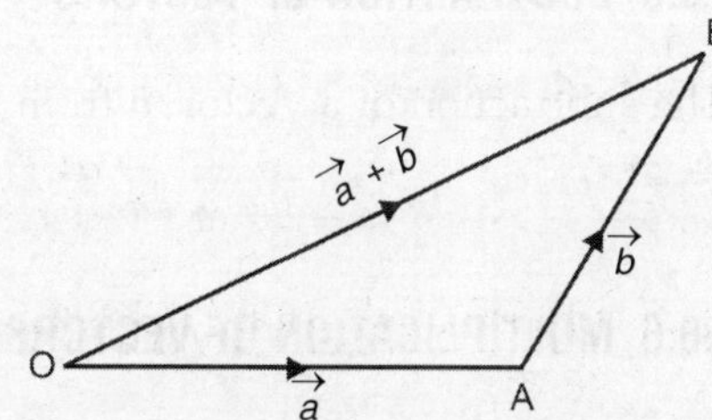

20.4 PROPERTIES OF ADDITION OF VECTORS

(*i*) Commutative law.
(*ii*) Associative law.

(*i*) Commutative law

$$\vec{a} + \vec{b} = \vec{b} + \vec{a}$$

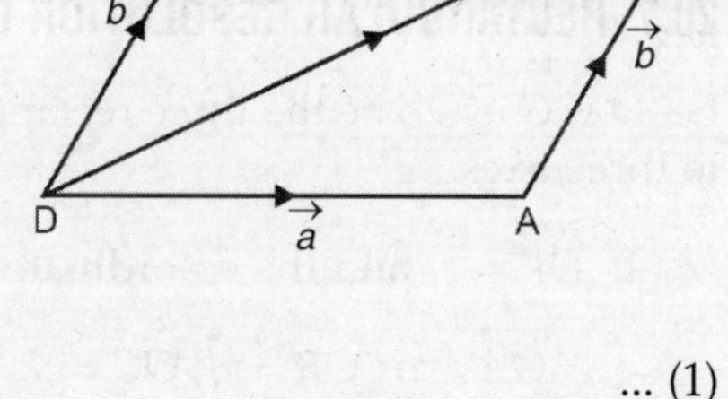

Let $\overrightarrow{OA} = \vec{a}$ and $\overrightarrow{AB} = \vec{b}$

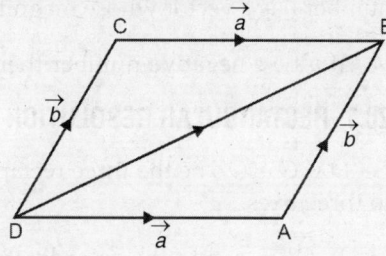

$$\vec{a} + \vec{b} = \overrightarrow{OA} + \overrightarrow{AB}$$

$$= \overrightarrow{OB} \quad \text{... (1)}$$

$$\vec{b} + \vec{a} = \overrightarrow{OC} + \overrightarrow{CB} = \overrightarrow{OB} \quad \text{... (2)}$$

From (1) and (2), we have

$$\vec{a} + \vec{b} = \vec{b} + \vec{a}$$

(*ii*) Associative law

$$\vec{a} + (\vec{b} + \vec{c}) = (\vec{a} + \vec{b}) + \vec{c}$$

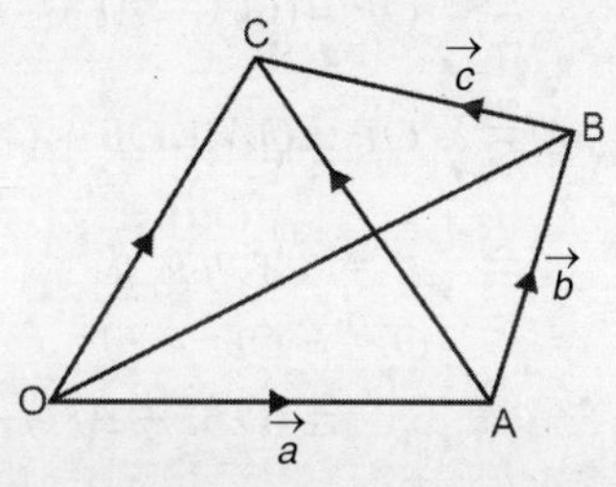

Let $\overrightarrow{OA} = \vec{a}$, $\overrightarrow{AB} = \vec{b}$ and $\overrightarrow{BC} = \vec{c}$

$$\vec{a} + (\vec{b} + \vec{c}) = \overrightarrow{OA} + (\overrightarrow{AB} + \overrightarrow{BC})$$

$$= \overrightarrow{OA} + \overrightarrow{AC}$$

$$= \overrightarrow{OC} \qquad \text{... (3)}$$

$$(\vec{a} + \vec{b}) + \vec{c} = (\overrightarrow{OA} + \overrightarrow{AB}) + \overrightarrow{BC}$$

$$= \overrightarrow{OB} + \overrightarrow{BC}$$

$$= \overrightarrow{OC} \qquad \text{... (4)}$$

From (3) and (4), we get

$$\boxed{\vec{a} + (\vec{b} + \vec{c}) = (\vec{a} + \vec{b}) + \vec{c}}$$

20.5 SUBTRACTION OF VECTORS

The subtraction of a vector $\vec{b}$ from $\vec{a}$ is the addition of $-\vec{b}$ (negative of $\vec{b}$) to $\vec{a}$

$$\vec{a} - \vec{b} = \vec{a} + (-\vec{b}).$$

20.6 MULTIPLICATION OF VECTORS BY A NUMBER

Let n be a real positive number and $\vec{a}$ be any vector. The product $n\,\vec{a}$, of a vector $\vec{a}$ and a number n is a vector whose magnitude is n times that of $\vec{a}$ and direction is the same to $\vec{a}$.

If n be a negative number then the direction of $n\,\vec{a}$ opposite to $\vec{a}$.

20.7 RECTANGULAR RESOLUTION OF A VECTOR

Let OX, OY, OZ be the three rectangular axes. Let $\hat{i}, \hat{j}, \hat{k}$ be three unit vectors and parallel to three axes.

If $\overrightarrow{OP} = \vec{r}$ and the co-ordinates of P be (x, y, z).

$$\overrightarrow{OA} = x\hat{i},\ \overrightarrow{OB} = y\hat{j},\ \overrightarrow{OC} = z\hat{k}$$

$$\overrightarrow{OP} = \overrightarrow{OF} + \overrightarrow{FP}$$

$$\Rightarrow \quad \overrightarrow{OP} = (\overrightarrow{OA} + \overrightarrow{AF}) + \overrightarrow{FP}$$

$$\Rightarrow \quad \overrightarrow{OP} = \overrightarrow{OA} + \overrightarrow{OB} + \overrightarrow{OC}$$

$$\Rightarrow \quad \vec{r} = x\hat{i} + y\hat{j} + z\hat{k}$$

$$OP^2 = OF^2 + FP^2$$

$$= (OA^2 + AF^2) + FP^2 = OA^2 + OB^2 + OC^2 = x^2 + y^2 + z^2$$

$$OP = \sqrt{x^2 + y^2 + z^2}$$

$$\boxed{|\overline{r}| = \sqrt{x^2 + y^2 + z^2}}$$

If OP makes α, β, γ angles with $\hat{i}, \hat{j}, \hat{k}$ direction, then its direction cosines are cos α, cos β, cos γ and are equal to

$$\boxed{\frac{x}{\sqrt{x^2+y^2+z^2}}, \frac{y}{\sqrt{x^2+y^2+z^2}}, \frac{z}{\sqrt{x^2+y^2+z^2}}}$$

Example 1. *In the adjoining fig. identify the following vectors.*

(i) Coinitial *(ii) Equal* *(iii) Collinear but not equal*

Solution. Here, we observe that

(i) $\vec{a}, \vec{b}$ are co-initial vectors

(ii) $\vec{d}$ and $\vec{b}$ are equal vectors

(iii) $\vec{a}$ and $\vec{c}$ are collinear but not equal vectors.

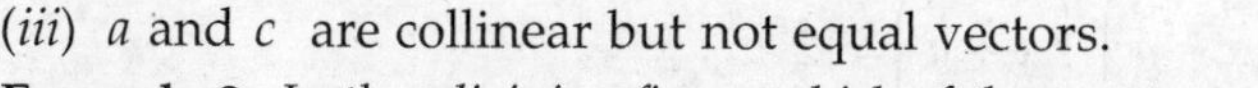

Example 2. *In the adjoining figure which of the vectors are:*

(i) Collinear *(ii) Equal* *(iii) Coinitial*

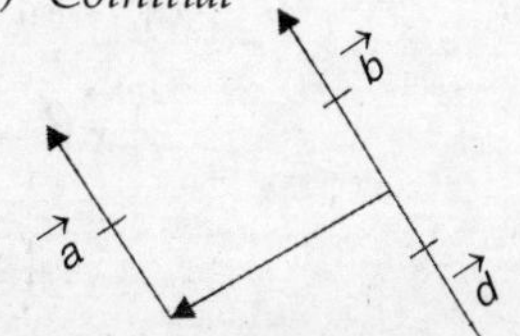

Solution.

(i) Collinear vectors are $\vec{a}$, $\vec{c}$ and $\vec{d}$.

(ii) Equal vectors are $\vec{a}$ and $\vec{c}$.

(iii) Coinitial vectors are $\vec{b}$, $\vec{c}$ and $\vec{d}$. **Ans.**

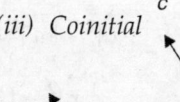

Example 3. *Calculate the modulus and the unit vector along the sum of vectors:*

$3\hat{i} + \hat{j} + 5\hat{k}, 2\hat{i} - \hat{j} + 6\hat{k}, 5\hat{i} + 2\hat{j} - 3\hat{k}$ **[B.T.E, Delhi Jan. 2009]**

Solution. Let $\vec{a} = 3\hat{i} + \hat{j} + 5\hat{k}$, $\vec{b} = 2\hat{i} - \hat{j} + 6\hat{k}$ and $\vec{c} = 5\hat{i} + 2\hat{j} - 3\hat{k}$

Then, $\vec{r} = \vec{a} + \vec{b} + \vec{c} = 10\hat{i} + 2\hat{j} + 8\hat{k}$

Now, $|\vec{r}| = \sqrt{10^2 + 2^2 + 8^2} = \sqrt{168}$

Unit vector along $\vec{r} = \dfrac{\vec{r}}{|\vec{r}|} = \dfrac{10\hat{i} + 2\hat{j} + 8\hat{k}}{\sqrt{168}} = \dfrac{5\hat{i} + \hat{j} + 4\hat{k}}{\sqrt{42}}$ **Ans.**

Example 4. *Find the value of p for which the vectors* $\vec{a} = 3\hat{i} + 2\hat{j} + 9\hat{k}$ and $\vec{b} = \hat{i} - p\hat{j} + 3\hat{k}$ *are parallel.* **[SBTE 2015]**

Solution. We know that, if two vectors are parallel then,

$$\vec{a} = p\vec{b}$$

$$(3\hat{i} + 2\hat{j} + 9\hat{k}) = p \cdot (\hat{i} - p\hat{j} + 3\hat{k})$$

$$\frac{3}{1} = \frac{2}{-p} = \frac{9}{3} \qquad \left[\vec{a} \text{ and } \vec{b} \text{ are parallel}\right]$$

$$p = \frac{-2}{3}$$ **Ans.**

Example 5. *If* $\vec{a} = \hat{i} + \hat{j}, \vec{b} = \hat{j} + \hat{k}, \vec{c} = \hat{k} + \hat{i}$ *, find the unit vector parallel to* $\vec{a} + 2\vec{b} + \vec{c}$ **[SBTE 2015]**

Solution. Here we have $\vec{a} + 2\vec{b} + \vec{c} = \hat{i} + \hat{j} + 2(\hat{j} + \hat{k}) + (\hat{k} + \hat{i}) = 2\hat{i} + 3\hat{j} + 3\hat{k}$

Unit vector parallel to $\left(\vec{a} + 2\vec{b} + \vec{c}\right) = \dfrac{2\hat{i} + 3\hat{j} + 3\hat{k}}{\sqrt{4+9+9}}$

$$= \frac{1}{\sqrt{22}}(2\hat{i} + 3\hat{j} + 3\hat{k})$$ **Ans.**

Example 6. *If* $\vec{a}$ *is a unit vector such that* $\left(\vec{x} - \vec{a}\right)\cdot\left(\vec{x} + \vec{a}\right) = 12$ find $|\vec{x}|$ **[SBTE 2015]**

Solution. Here we have $\left(\vec{x} - \vec{a}\right)\cdot\left(\vec{x} + \vec{a}\right) = 12$

$$\Rightarrow \quad |\vec{x}|^2 - |\vec{a}|^2 = 12$$

$$\Rightarrow \quad |\vec{x}|^2 - 1 = 12$$

$$\Rightarrow \quad |\vec{x}| = \sqrt{13}$$ **Ans.**

Example 7. *Show that the vectors* $3\hat{i} - 2\hat{j} + \hat{k}, \hat{i} - 3\hat{j} + 5\hat{k}$ and $2\hat{i} + \hat{j} - 4\hat{k}$ *form a right angled triangle.* **[SBTE 2016, 2015, 2010]**

Solution. Let $\vec{a} = 3\hat{i} - 2\hat{j} + \hat{k}, \vec{b} = \hat{i} - 3\hat{j} + 5\hat{k}, \vec{c} = 2\hat{i} + \hat{j} - 4\hat{k}$
if ΔABC is a rt. angled triangle

then $|\vec{b}|^2 = |\vec{a}|^2 + |\vec{c}|^2$ ... (1)

$$|\vec{a}| = \sqrt{9+4+1} = \sqrt{14}$$

$$|\vec{b}| = \sqrt{1+9+25} = \sqrt{35}$$

$$|\vec{c}| = \sqrt{4+1+16} = \sqrt{21}$$

from (1) 14 + 21 = 35

So, $\vec{a}, \vec{b}$ and $\vec{c}$ form a rt. angled triangle. **Proved.**

20.8 COLLINEARITY OF POINTS

Let A, B and C be the three collinear points with position vectors $\vec{a}, \vec{b}$ and $\vec{c}$ respectively. The necessary condition for these three points to be collinear is

$$x\vec{a} + y\vec{b} + z\vec{c} = 0$$

and $$x + y + z = 0$$

where three scalars x, y, z are not all zero simultaneously.

Example 8. *Show that the vectors* $2\hat{i} - 3\hat{j} + 4\hat{k}$ *and* $-4\hat{i} + 6\hat{j} - 8\hat{k}$ *are collinear.*

Solution. Let $\vec{a} = 2\hat{i} - 3\hat{j} + 4\hat{k}$

$$\vec{b} = -4\hat{i} + 6\hat{j} - 8\hat{k}$$

If $\vec{a}$ and $\vec{b}$ are collinear then

$$\vec{a} = \hat{\lambda}\,\vec{b}$$

$$\Rightarrow \quad 2\hat{i} - 3\hat{j} + 4\hat{k} = \lambda(-4\hat{i} + 6\hat{j} - 8\hat{k})$$

$$\Rightarrow \quad 2\hat{i} - 3\hat{j} + 4\hat{k} = -4\hat{\lambda}\hat{i} + 6\hat{\lambda}\hat{j} - 8\hat{\lambda}\hat{k})$$

$$2 = -4\lambda, \qquad -3 = 6\lambda, \qquad 4 = -8\lambda$$

$$\lambda = \frac{2}{-4} = \frac{1}{-2};\ \lambda = \frac{-3}{6} = \frac{4}{-8} = \lambda$$

Coefficients of component vectors are proportional.

Hence given vectors are collinear. **Ans.**

Example 9. *Show that the three points A(2, – 1, 3), B(4, 3, 1) and C(3, 1, 2) are collinear.*

[SBTE 2015, 2011]

Solution. Here we have

$$\overrightarrow{AB} = \overrightarrow{OB} - \overrightarrow{OA}$$

$$\overrightarrow{AB} = (4\hat{i} + 3\hat{j} + \hat{k}) - (2\hat{i} - \hat{j} + 3\hat{k}) = 2\hat{i} + 4\hat{j} - 2\hat{k}$$

$$\overrightarrow{BC} = \overrightarrow{OC} - \overrightarrow{OB}$$

$$= (3\hat{k} + \hat{j} + 2\hat{k}) - (4\hat{i} + 3\hat{j} + \hat{k}) = -\hat{i} - 2\hat{j} + \hat{k}$$

$$\overrightarrow{BC} = -\frac{1}{2}(2\hat{i} + 4\hat{j} - 2\hat{k})$$

$$\overrightarrow{BC} = -\frac{1}{2}\overrightarrow{AB}$$

$\overrightarrow{AB}$ and $\overrightarrow{BC}$ are parallel, B is common point

$\therefore A, B, C$ are collinear **Proved.**

Example 10. *Find a vector of magnitude 5 units which is parallel to the vector* $2\hat{i} - \hat{j}$.

[SBTE 2015]

Solution. Let $\vec{a} = 2\hat{i} - \hat{j}$

Unit vector parallel $a = \dfrac{\vec{a}}{|\vec{a}|} = \dfrac{1}{\sqrt{5}}(2\hat{i} - \hat{j})$

Magnitude 5 units parallel to

$$\vec{a} = 5 \times \frac{1}{\sqrt{5}}(2\hat{i} - \hat{j}) = \sqrt{5}\,(2\hat{i} - \hat{j})$$ **Ans.**

Example 11. *If $\vec{a}$ and $\vec{b}$ are the vectors determined by two adjacent sides of a regular shexagon, what are the vectors determined by the other sides taken in order.*

Solution. Let *ABCDEF* be a regular hexagon, such that

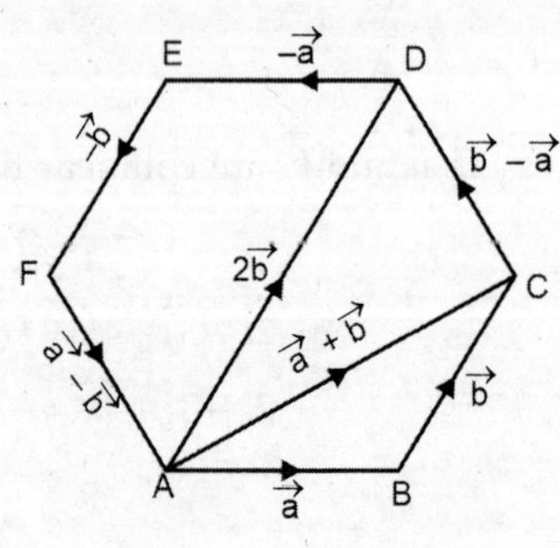

$\Rightarrow \quad \overrightarrow{AB} = \vec{a},\ \overrightarrow{BC} = \vec{b}$

$\Rightarrow \quad \overrightarrow{AC} = \overrightarrow{AB} + \overrightarrow{BC}$

$\quad = \vec{a} + \vec{b}$

$\Rightarrow \quad \overrightarrow{AD} = \overrightarrow{AC} + \overrightarrow{CD}$

$\Rightarrow \quad \overrightarrow{CD} = \overrightarrow{AD} - \overrightarrow{AC}$

$\quad = \overrightarrow{2b} - (\vec{a} + \vec{b})$

$\quad = \vec{b} - \vec{a}$

$\Rightarrow \quad \overrightarrow{DE} = -\overrightarrow{AB} = -\vec{a}$

$\Rightarrow \quad \overrightarrow{EF} = -\overrightarrow{BC} = -\vec{b}$

$\therefore \quad \overrightarrow{FA} = -\overrightarrow{CD} = -(\vec{b} - \vec{a}) = \vec{a} - \vec{b}$ **Ans.**

Example 12. *The velocity of a boat relative to water is represented by $3\hat{i} + 4\hat{j}$ and that of water relative to earth is $\hat{i} - 3\hat{j}$. What is the velocity of the boat relative to the earth if $\hat{i}$ and $\hat{j}$ represent one kilometer an hour East and North respectively.*

Solution. The velocity of boat relative to water = $3\hat{i} + 4\hat{j}$

$\Rightarrow$ Velocity of boat – Velocity of water = $3\hat{i} + 4\hat{j}$ ... (1)

The velocity of water relative to earth = $\hat{i} - 3\hat{j}$

$\Rightarrow$ Velocity of water – Velocity of earth = $\hat{i} - 3\hat{j}$ ... (2)

Adding (1) and (2), we have

Velocity of boat – Velocity of earth = $(3\hat{i} + 4\hat{j}) + (\hat{i} - 3\hat{j})$

$\Rightarrow$ Velocity of boat relative to earth = $4\hat{i} + \hat{j}$

It magnitude = $\sqrt{4^2 + 1^2} = \sqrt{17}$

Its direction is $\tan^{-1}\left(\frac{1}{4}\right)$ North of East. **Ans.**

Example 13. *Show that the vector $2\hat{i} - \hat{j} + \hat{k}, \hat{i} - 3\hat{j} - 5\hat{k}$ and $3\hat{i} - 4\hat{j} - 4\hat{k}$ form the sides of a right angled-triangle.*

Solution. Let $\overrightarrow{AB} = 2\hat{i} - \hat{j} + \hat{k}$

$\Rightarrow \quad \overrightarrow{BC} = \hat{i} - 3\hat{j} - 5\hat{k}$

$\Rightarrow \quad \overrightarrow{AC} = 3\hat{i} - 4\hat{j} - 4\hat{k}$

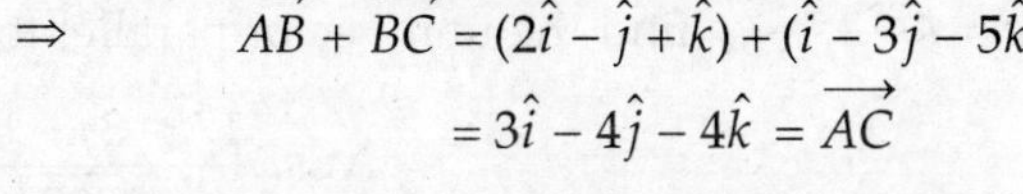

$$\Rightarrow \quad \overrightarrow{AB} + \overrightarrow{BC} = (2\hat{i} - \hat{j} + \hat{k}) + (\hat{i} - 3\hat{j} - 5\hat{k})$$

$$= 3\hat{i} - 4\hat{j} - 4\hat{k} = \overrightarrow{AC}$$

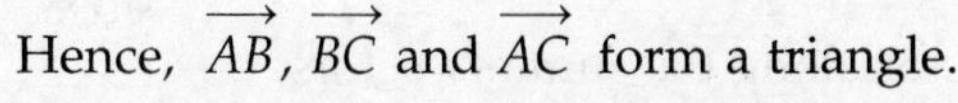

Hence, $\overrightarrow{AB}, \overrightarrow{BC}$ and $\overrightarrow{AC}$ form a triangle.

Now we have to prove that ΔABC is a rt. angled Δ.

$$\Rightarrow \quad |\overrightarrow{AB}| = \sqrt{(2)^2 + (-1)^2 + (1)^2} = \sqrt{6}$$

$$\Rightarrow \quad |\overrightarrow{BC}| = \sqrt{(1)^2 + (-3)^2 + (-5)^2} = \sqrt{35}$$

$$\Rightarrow \quad |\overrightarrow{AC}| = \sqrt{(3)^2 + (-4)^2 + (-4)^2} = \sqrt{41}$$

$$|\overrightarrow{AB}|^2 + |\overrightarrow{BC}|^2 = 6 + 35 = 41 = |\overrightarrow{AC}|^2$$

$\therefore$ ΔABC is a rt. angled triangle. **Proved.**

Example 14. *A particle if subjected to force 3 kg wt, 4 kg wt. 5 kg wt respectively acting in directions parallel to the edges AB, BC, CA of a equilateral ΔABC. Find the resulting force acting on the particle.*

Solution. Let $\overrightarrow{BC}$ direction be parallel to i and $j \perp$ to $\overrightarrow{BC}$ at B.

3 kg wt. force along $\overrightarrow{AB}$

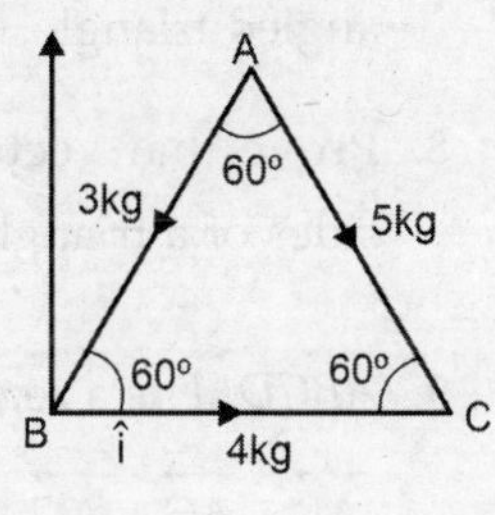

$$= 3\frac{\left(-\dfrac{\hat{i}}{2} - \dfrac{\sqrt{3}}{2}\hat{j}\right)}{\sqrt{\left(\dfrac{1}{2}\right)^2 + \left(\dfrac{\sqrt{3}}{2}\right)^2}}$$

$$= -\frac{3}{2}\hat{i} - \frac{3\sqrt{3}}{2}\hat{j}$$

4 kg wt. force along $BC = 4\hat{i}$

$$\text{5 kg wt. force along } CA = \frac{5\left(-\dfrac{\hat{i}}{2} + \dfrac{\sqrt{3}}{2}\hat{j}\right)}{\sqrt{\left(\dfrac{1}{2}\right)^2 + \left(\dfrac{\sqrt{3}}{2}\right)^2}} = -\frac{5}{2}\hat{i} + \frac{5\sqrt{3}}{2}\hat{j}$$

$$\text{Total force} = \left(-\frac{3\hat{i}}{2} - \frac{3\sqrt{3}}{2}\hat{j}\right) + 4\hat{i} + \left(-\frac{5\hat{i}}{2} + \frac{5\sqrt{3}}{2}\hat{j}\right)$$

$$= \sqrt{3}\,\hat{j} = \sqrt{3} \text{ kg wt.} \perp \text{ to } BC$$ **Ans.**

EXERCISE 20.1

1. If $\vec{a} = 2\hat{i} + \hat{j} - 8\hat{k}$ and $\vec{b} = \hat{i} + 3\hat{j} - 4\hat{k}$, find the magnitude and direction cosines of the vectors $\vec{a} + \vec{b}$ and $\vec{a} - 2\vec{b}$. **Ans.** Magnitude = 13, DCs 3/13, 4/13, – 12/13; Magnitude = 5, DCs. = 0, – 1, 0

2. If $\vec{a} = 3\hat{i} - \hat{j} - 4\hat{k}$, $\vec{b} = -2\hat{i} + 4\hat{j} - 3\hat{k}$, $\vec{c} = \hat{i} + 2\hat{j} - \hat{k}$, find the unit vector parallel to $3\vec{a} - 2\vec{b} + 4\vec{c}$. **Ans.** $\dfrac{17\hat{i} - 3\hat{j} - 10\hat{k}}{\sqrt{398}}$

3. Calculate the modulus and the unit vector along the sum of vectors.
$3\hat{i} + 2\hat{j} + 5\hat{k}, 2\hat{i} + \hat{j} + 6\hat{k}, 5\hat{i} + 2\hat{j} - 3\hat{k}$ **Ans.** $3\sqrt{21}, \dfrac{10\hat{i} + 5\hat{j} + 8\hat{k}}{3\sqrt{21}}$

4. Calculate the modulus and the unit vector along the sum of the vectors
$2\hat{i} + \hat{j} + 4\hat{k}, 3\hat{i} - 2\hat{j} + 7\hat{k}, 5\hat{i} + 2\hat{j} - 3\hat{k}$ **Ans.** $\dfrac{10\hat{i} + \hat{j} + 8\hat{k}}{\sqrt{165}}$

5. Calculate the modulus and the unit vector along the sum of vectors
$2\hat{i} + 2\hat{j} + 5\hat{k}, 3\hat{i} - \hat{j} + 6\hat{k}, 5\hat{i} + 2\hat{j} - 3\hat{k}$ **Ans.** $\dfrac{10\hat{i} + 3\hat{j} + 8\hat{k}}{\sqrt{173}}$

6. Calculate the unit vector along the sum of the vectors
$\hat{i} + \hat{j} + 3\hat{k}, 2\hat{i} - 2\hat{j} + 6\hat{k}, 4\hat{i} + 2\hat{j} - 4\hat{k}$ and find its modulus. **Ans.** $\dfrac{7\hat{i} + \hat{j} + 5\hat{k}}{\sqrt{75}}$

7. Show that the vectors $\vec{a} = 3\hat{i} - 2\hat{j} + \hat{k}$, $\vec{b} = \hat{i} - 3\hat{j} + 5\hat{k}$, $\vec{c} = 2\hat{i} + \hat{j} - 4\hat{k}$ form a right angled triangle. **[B.T.E, Delhi Jan. 2009]**

8. Prove that vectors $\vec{a} = 3\hat{i} + \hat{j} - 2\hat{k}$, $\vec{b} = -\hat{i} + 3\hat{j} + 4\hat{k}$, $\vec{c} = 4\hat{i} - 2\hat{j} - 6\hat{k}$ can form the sides of a triangle. Also find the length of the median bisecting the vector $\vec{c}$. **Ans.** $\sqrt{6}$

9. *ABCDEF* is a regular hexagon whose centroid is *O*. Show that
$\overrightarrow{AB} + \overrightarrow{AC} + \overrightarrow{AD} + \overrightarrow{AE} + \overrightarrow{AF} = 6\overrightarrow{AO}$.

10. Prove that the vectors:
$\vec{a} = 3\hat{i} + \hat{j} - 2\hat{k}$, $\vec{b} = -\hat{i} + 3\hat{j} + 4\hat{k}$ and $\vec{c} = 4\hat{i} - 2\hat{j} - 6\hat{k}$, can form the sides of a triangle. Also, find the length of the median to the largest side of the triangle. **Ans.** $\sqrt{6}$

11. Prove that the sum of all the vectors drawn from the centre of a regular octagon to its vertices is the zero vector. **[Diploma IETE Dec. 2006]**

Choose the correct alternative:

12. A unit vector parallel to $3\hat{i} + 4\hat{j} + 5\hat{k}$ is

(*a*) $-\dfrac{3}{5\sqrt{2}}\hat{i} - \dfrac{4}{5\sqrt{2}}\hat{j} + \dfrac{1}{\sqrt{2}}\hat{k}$ (*b*) $\dfrac{3}{5\sqrt{2}}\hat{i} - \dfrac{4}{5\sqrt{2}}\hat{j} - \dfrac{2}{\sqrt{2}}\hat{k}$

(c) $\frac{3}{5\sqrt{2}}\hat{i}+\frac{4}{5\sqrt{2}}\hat{j}+\frac{1}{\sqrt{2}}\hat{k}$ (d) $\frac{3}{5\sqrt{2}}\hat{i}-\frac{4}{5\sqrt{2}}\hat{j}+\frac{1}{\sqrt{2}}\hat{k}$ **Ans.** (*c*)

13. The points $2\hat{i}-\hat{j}+\hat{k}, \hat{i}-3\hat{j}-5\hat{k}, 3\hat{i}-4\hat{j}-4\hat{k}$ are the vertices of a triangle which is

(a) Equilateral (b) Isosceles.

(c) Right angled (d) None of these **Ans.** (*c*)

14. If *A* and *B* are the points (3, 4, 5) and (6, 8, 9) then the vector $\overrightarrow{AB}$ is

(*a*) $3\hat{i}+4\hat{j}+4\hat{k}$ (*b*) $3\hat{i}+4\hat{i}$

(*c*) $3\hat{i}-4\hat{i}-4\hat{k}$ (*d*) $3\hat{i}-4\hat{j}$ **Ans.** (*a*)

20.9 POSITION VECTOR OF A POINT

The position Vector of a point A with respect to origin *O* is the Vector $\overrightarrow{OA}$ which is used to specify the position of *A* w.r.t. *O*.

To find $\overrightarrow{AB}$ if the position Vectors of the point *A* and point *B* are given.

If the position vectors of *A* and *B* are $\vec{a}$ and $\vec{b}$. Let the origin be *O*.

Then $\overrightarrow{OA}=\vec{a}, \overrightarrow{OB}=\vec{b}$

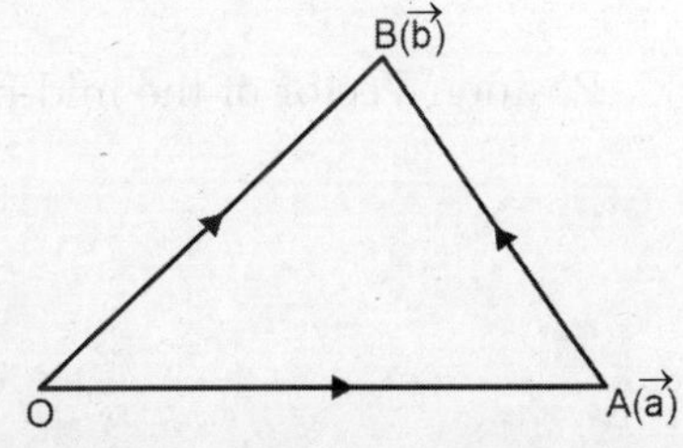

$$\overrightarrow{OA}+\overrightarrow{AB}=\overrightarrow{OB}$$

$$\overrightarrow{AB}=\overrightarrow{OB}-\overrightarrow{OA}$$

$$\Rightarrow \quad \overrightarrow{AB}=\vec{b}-\vec{a}$$

$\overrightarrow{AB}$ = **Position Vector of *B*** – ***Position vector of A***.

Example 15. *If A and B are (3, 4, 5) and (6, 8, 9) respectively. Find* $\overrightarrow{AB}$.

Solution. $\overrightarrow{AB}$ = Position Vector of *B* – Position Vector of *A*.

$$= (6\hat{i}+8\hat{j}+9\hat{k})-(3\hat{i}+4\hat{j}+5\hat{k})$$

$$= 3\hat{i}+4\hat{j}+4\hat{k}.$$ **Ans.**

20.10 RATIO FORMULA

To find the position vector of the point which divides the line joining two given points.

Let *A* and *B* be two points and *C* divides *AB* in the ratio of *m* : *n*.

Let *O* be the origin, then

$\overrightarrow{OA}=\vec{a}, \overrightarrow{OB}=\vec{b}, \overrightarrow{OC}=?$

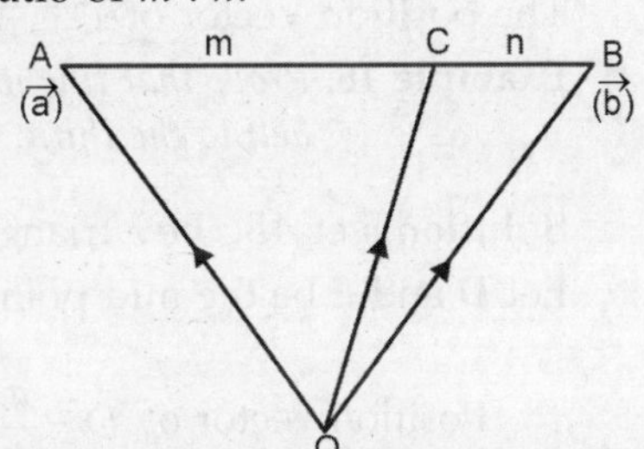

$$\overrightarrow{OC}=\overrightarrow{OA}+\overrightarrow{AC}$$

$$=\overrightarrow{OA}+\frac{m}{m+n}\overrightarrow{AB} \quad \left(\because AC=\frac{m}{m+n}AB\right)$$

$$= \vec{a} + \frac{m}{m+n}(\vec{b} - \vec{a}) \quad \left(\because \overrightarrow{AB} = \vec{b} - \vec{a}\right)$$

$$\boxed{\overrightarrow{OC} = \frac{m\vec{b} + n\vec{a}}{m+n}}$$

Cor. If $m = n = 1$, then C will be the mid-point, and

$$\boxed{\text{Middle point} = \frac{\vec{a} + \vec{b}}{2}}$$

Example 16. *If the position vectors of A and B are* $(\hat{i} + 2\hat{j} + 4\hat{k})$ *and* $(2\hat{i} + 3\hat{j} + 5\hat{k})$. *Find the position Vector of a point C that it divides AB in the ratio 2 : 3. Find also the mid-point of AB.* **[B.T.E, Delhi Jan. 2009]**

Solution. Position Vector of C

$$= \frac{3(\hat{i} + 2\hat{j} + 4\hat{k}) + 2(2\hat{i} + 3\hat{j} + 5\hat{k})}{2+3}$$

$$= \frac{1}{5}(7\hat{i} + 12\hat{j} + 22\hat{k})$$

Position Vector of the mid-point of *AB*

$$= \frac{(\hat{i} + 2\hat{j} + 4\hat{k}) + (2\hat{i} + 3\hat{j} + 5\hat{k})}{2}$$

$$= \frac{1}{2}(3\hat{i} + 5\hat{j} + 9\hat{k})$$ **Ans.**

Example 17. *If* $\vec{a}$ *and* $\vec{b}$ *are the position Vectors of the points A and B respectively then find the position Vector of a point C in AB produced such that* $\overrightarrow{AC} = 3\,\overrightarrow{AB}$.

Solution. Here we have

$$\overrightarrow{AC} = 3\,\overrightarrow{AB}$$

$\Rightarrow$ It means the point C divides AB in the ratio 3 : 2 externally

$$\vec{c} = \frac{3\vec{a} - 2\vec{b}}{3-2}$$

$$= 3\vec{a} - 2\vec{b}$$

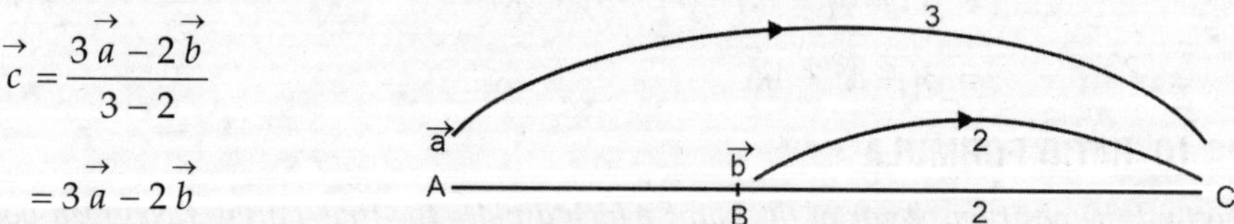

The position Vector of $\vec{C}$ is $3\vec{a} - 2\vec{b}$ **Ans.**

Example 18. *Prove that line joining the mid-points of two sides of a triangle is parallel and half to the third.* **[Diploma IETE, June 2006]**

Solution. Let *ABC* be a triangle and position vectors of *A*, *B*, *C* be $\vec{a}, \vec{b}, \vec{c}$ respectively. Let *D* and *E* be the mid points of *AB* and *AC* respectively.

$$\text{Position vector of } D = \frac{\vec{a} + \vec{b}}{2}$$

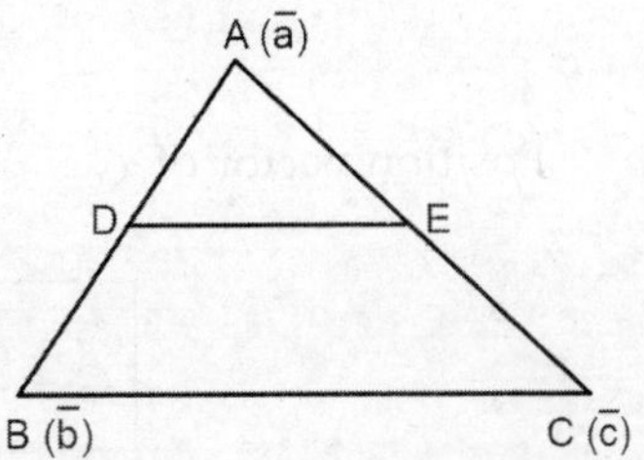

Position vector of $E = \dfrac{\vec{a} + \vec{c}}{2}$

$\overrightarrow{DE}$ = Position vector of E – Position vector of D

$$\overrightarrow{DE} = \frac{\vec{a} + \vec{c}}{2} - \frac{\vec{a} + \vec{b}}{2} = \frac{\vec{c} - \vec{b}}{2} = \frac{1}{2}\overrightarrow{BC}$$

i.e., DE is parallel to BC,

and $DE = \dfrac{1}{2} BC$. **Proved.**

Example 19. *If the mid-points of consecutive sides of any quadrilateral are connected by straight line. Prove that the resulting quadrilateral is a parallelogram.*

Solution. Let $ABCD$ be a quadrilateral, and P, Q, R and S be the mid-points of AB, BC, CD and DA. Let the position vectors of the vertices A, B, C and D be $\vec{A}, \vec{B}, \vec{C}$ and $\vec{D}$

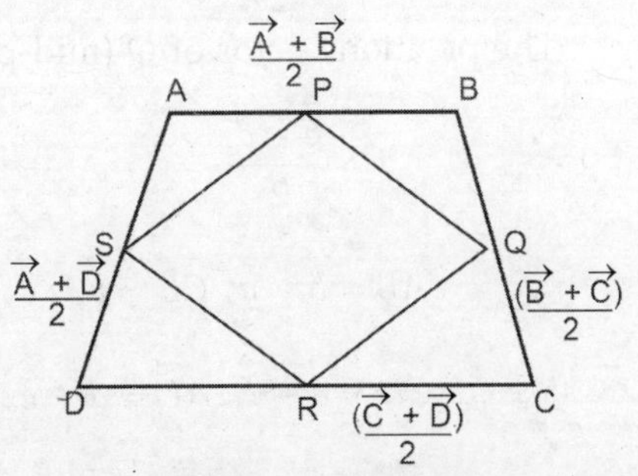

The position vector of $P = \dfrac{\vec{A} + \vec{B}}{2}$

The position vector of $Q = \dfrac{\vec{B} + \vec{C}}{2}$

The position vector of $R = \dfrac{\vec{C} + \vec{D}}{2}$

The position vector of $S = \dfrac{\vec{D} + \vec{A}}{2}$

$$\overrightarrow{PQ} = \vec{Q} - \vec{P} = \frac{\vec{B} + \vec{C}}{2} - \frac{\vec{A} + \vec{B}}{2} = \frac{\vec{C} - \vec{A}}{2} \quad \text{... (1)}$$

$$\overrightarrow{SR} = \vec{R} - \vec{S} = \frac{\vec{C} + \vec{D}}{2} - \frac{\vec{A} + \vec{D}}{2} = \frac{\vec{C} - \vec{A}}{2} \quad \text{... (2)}$$

From (1) and (2), we have

$\overrightarrow{PQ} = \overrightarrow{SR}$ Hence $PQ = SR$

and $PQ \parallel SR$.

[If one pair of opposite sides of a quadrilateral is equal and parallel then quadrilateral is a parallelogram].

Hence $PQRS$ is a parallelogram. **Proved.**

Example 20. *Find the position vector of the centroid of a triangle ABC, when position vectors of A, B and C are* $\vec{a}, \vec{b}$ *and* $\vec{c}$.

Solution. Let the centroid of the ΔABC be G and divides the median AD in the ratio of 2 : 1.

Position vector of $D = \dfrac{\vec{b} + \vec{c}}{2}$

Position vector of $G = \dfrac{1 \cdot \vec{a} + 2 \cdot \dfrac{\vec{b} + \vec{c}}{2}}{1+2}$

$$\boxed{\vec{G} = \frac{\vec{a} + \vec{b} + \vec{c}}{3}}$$

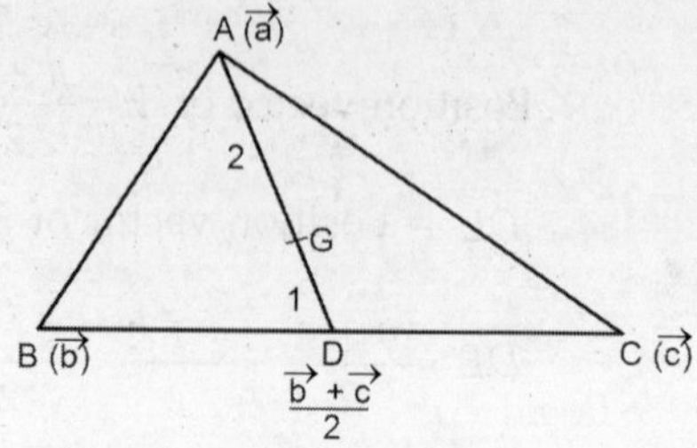

Example 21. *ABCD is a quadrilateral and P, Q are the mid-points of the diagonals AC and BD respectively. Show that*

$$\overrightarrow{AB} + \overrightarrow{CB} + \overrightarrow{CD} + \overrightarrow{AD} = 4\,\overrightarrow{PQ}$$

Solution. Let the position vectors of the points *A*, *B*, *C*, *D* be $\vec{a}, \vec{b}, \vec{c}, \vec{d}$ respectively.
The position vector of *P* (mid-point of *AC*)

$$= \frac{\vec{a} + \vec{c}}{2}$$

The position vector of *Q* (mid-point of *BD*)

$$= \frac{\vec{b} + \vec{d}}{2}$$

$$\overrightarrow{AB} = \vec{b} - \vec{a},\ \overrightarrow{CB} = \vec{b} - \vec{c},$$

$$\overrightarrow{CD} = \vec{d} - \vec{c},\ \overrightarrow{AD} = \vec{d} - \vec{a},$$

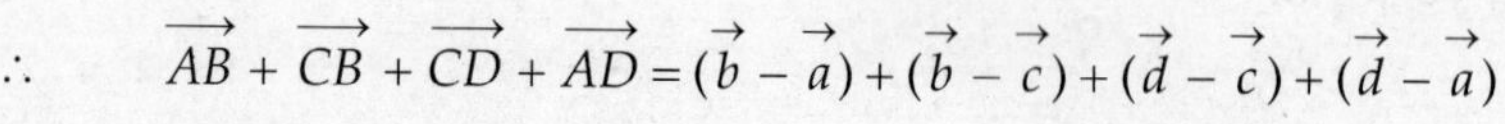

$$\therefore \quad \overrightarrow{AB} + \overrightarrow{CB} + \overrightarrow{CD} + \overrightarrow{AD} = (\vec{b} - \vec{a}) + (\vec{b} - \vec{c}) + (\vec{d} - \vec{c}) + (\vec{d} - \vec{a})$$

$$\Rightarrow \quad \overrightarrow{AB} + \overrightarrow{CB} + \overrightarrow{CD} + \overrightarrow{AD} = -2\vec{a} + 2\vec{b} - 2\vec{c} + 2\vec{d} \quad \text{... (1)}$$

Now $4\,\overrightarrow{PQ} = 4\left(\dfrac{\vec{b} + \vec{d}}{2} - \dfrac{\vec{a} + \vec{c}}{2}\right)$

$$= -2\vec{a} + 2\vec{b} - 2\vec{c} + 2\vec{d} \quad \text{... (2)}$$

From (1) and (2) we get required result. **Proved.**

Example 22. *Show that the line joining one vertex of a parallelogram to the mid-point of an opposite side trisects the diagonal and is trisected there at.*

Solution. Let *OACB* be a parallelogram, *D* is the mid-point of *OA*.

Let *O* be the origin, the position vector of *A* and *B* be $\vec{A}$ and $\vec{B}$.

$$\left[\begin{array}{l} \overrightarrow{OC} = \overrightarrow{OA} + \overrightarrow{AC} \\ \overrightarrow{OC} = \vec{A} + \vec{B} \end{array}\right]$$

The position vector of C is $\vec{A} + \vec{B}$.

Position vector of $D = \dfrac{\vec{O} + \vec{A}}{2} = \dfrac{\vec{A}}{2}$

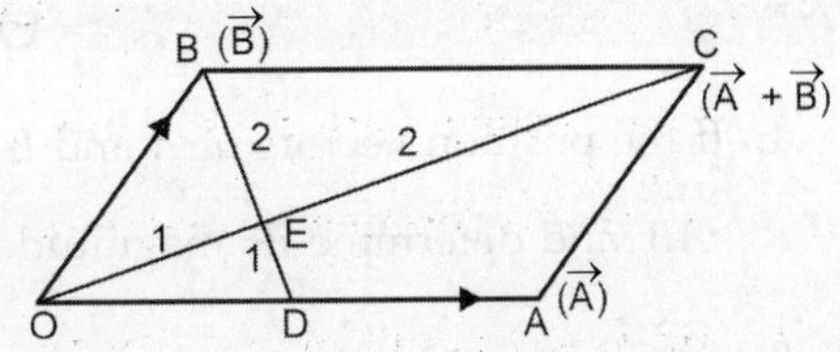

Position vector of point of trisection of $\overrightarrow{BD}$

$$= \frac{1 \cdot \vec{B} + 2 \cdot \frac{\vec{A}}{2}}{1+2} = \frac{\vec{A} + \vec{B}}{3}$$

Position vector of point of trisection of $\overrightarrow{OC}$

$$= \frac{1 \cdot (\vec{A} + \vec{B}) + 2 \cdot (\vec{O})}{1+2} = \frac{\vec{A} + \vec{B}}{3}$$

The position vector of the points of trisection of *BD* and *OC* is the same.

Hence *BD* trisects *OC* and *BD* is trisected at *E*. **Proved.**

Example 23. *In a trapezium, prove that the straight line joining the mid-points of the diagonals is parallel to the parallel sides and half their difference.*

Solution. Let *ABCD* be a trapezium with parallel sides *AB* and *DC*.

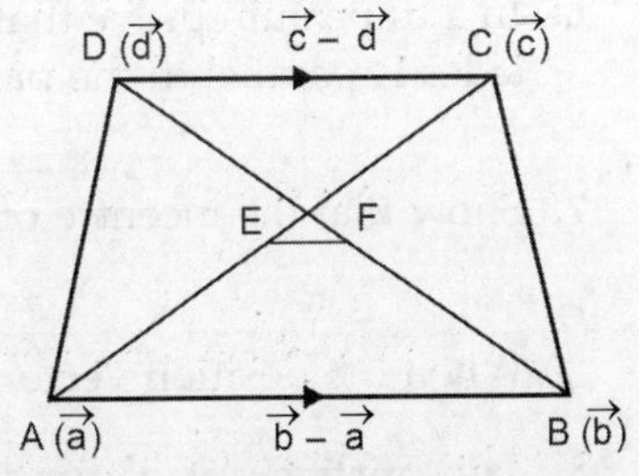

Let the position vectors of *A*, *B*, *C* and *D* be $\vec{a}, \vec{b}, \vec{c}$ and $\vec{d}$.

Since *DC* is parallel to *AB*.

$$\therefore \quad \overrightarrow{DC} = k\,\overrightarrow{AB}$$

$$\Rightarrow \quad (\vec{c} - \vec{d}) = k\,(\vec{b} - \vec{a})$$

The position vector of the mid-point *E* of $AC = \dfrac{\vec{a} + \vec{c}}{2}$

The position vector of the mid-point *F* of $BD = \dfrac{\vec{b} + \vec{d}}{2}$

$\overrightarrow{EF}$ = Position vector of *F* – Position vector of *E*

$$= \frac{\vec{b} + \vec{d}}{2} - \frac{\vec{a} + \vec{c}}{2}$$

$$= \frac{1}{2}[(\vec{b} - \vec{a}) - (\vec{c} - \vec{d})] \quad \text{... (1)}$$

$$= \frac{1}{2}[(\vec{b} - \vec{a}) - \hat{k}(\vec{b} - \vec{a})] \quad [\because (\vec{c} - \vec{d}) = \hat{k}(\vec{b} - \vec{a})]$$

$$= \frac{1}{2}(1 - \hat{k})(\vec{b} - \vec{a})$$

$$= \frac{1}{2}(1 - \hat{k})\,\overrightarrow{AB} \quad \text{... (2)}$$

Hence $\overrightarrow{EF}$ is parallel to $\overrightarrow{AB}$.

From (1), $\overrightarrow{EF} = \frac{1}{2}\left[\overrightarrow{AB} - \overrightarrow{DC}\right]$ **Proved.**

EXERCISE 20.2

1. If the position vectors of A and B are $2\hat{i} - 9\hat{j} - 4\hat{k}$ and $6\hat{i} - 3\hat{j} + 8\hat{k}$ respectively, find $\overrightarrow{AB}$ and determine its magnitude. **Ans.** $4\hat{i} + 6\hat{j} + 12\hat{k}$, Magnitude = 14

2. $ABCD$ is a parallelogram, P is the mid-point of BC and Q that of CD. If $\overrightarrow{AB} = \vec{a}$ and $\overrightarrow{AD} = \vec{b}$, find vectors $\overrightarrow{BD}$, $\overrightarrow{AP}$, $\overrightarrow{AQ}$, $\overrightarrow{PQ}$ and $\overrightarrow{PD}$ in terms of $\vec{a}$ and $\vec{b}$.

   Ans. $\overrightarrow{BD} = \vec{b} - \vec{a}$, $\overrightarrow{AP} = \vec{a} + \frac{\vec{b}}{2}$, $\overrightarrow{AQ} = \vec{b} + \frac{\vec{a}}{2}$, $\overrightarrow{PQ} = \frac{\vec{b}}{2} - \frac{\vec{a}}{2}$, $\overrightarrow{PD} = \frac{\vec{b}}{2} - \vec{a}$

3. Show that the three points whose position vectors are $2\hat{i} + 3\hat{j} - 4\hat{k}$, $\hat{i} - 2\hat{j} + 3\hat{k}$ and $-7\hat{i} + 10\hat{k}$ are collinear.

4. If the diagonals of a quadrilateral bisect each other, show by vector method, that the figure is a parallelogram.

5. Prove that the diagonals of a parallelogram bisect each other.

6. In a trapezium prove that the straight line joining the mid-points of the non-parallel sides is parallel to the parallel sides and half their sum.

7. Show that the incentre of a triangle ABC is $\dfrac{a\vec{\alpha} + b\vec{\beta} + c\vec{\gamma}}{a + b + c}$.

   When the position vectors of A, B, C are $\vec{\alpha}, \vec{\beta}, \vec{\gamma}$, $BC = a$, $CA = b$, $AB = c$.

8. The vertices of a quadrilateral are $A(\hat{i} + 2\hat{j} - \hat{k})$, $B(-4\hat{i} + 2\hat{j} - 2\hat{k})$, $C(4\hat{i} + \hat{j} - 5\hat{k})$, $D(2\hat{i} - \hat{j} + 3\hat{k})$. At the point A forces of magnitudes 2, 3, 2 gm wt. act along the lines AB, AC, AD respectively; find their resultant. **Ans.** $\frac{1}{\sqrt{26}}(\hat{i} - 9\hat{j} - 6\hat{k})$

9. Two forces act at a corner A of a quadrilateral $ABCD$ represented by $\overrightarrow{AB}$ and $\overrightarrow{AD}$ and two forces act at C represented by $\overrightarrow{CB}$ and $\overrightarrow{CD}$. Prove that their resultant is $4\overrightarrow{PQ}$ where P, Q are mid-points of AC and BD respectively.

10. Prove that the lines joining the mid-points of the opposite edges of a tetrahedron bisect each other.

20.11 PRODUCT OF TWO VECTORS

The product of two vectors results in two different ways, the one a number and other a vector. So there are two types of product of two vectors, namely scalar product and vector product. They are written as $\vec{a} \cdot \vec{b}$ and $\vec{a} \times \vec{b}$.

20.12 SCALAR OR DOT PRODUCT

The scalar, or dot product of two vectors $\vec{a}$ and $\vec{b}$ defined to be $|\vec{a}| \cdot |\vec{b}| \cos\theta$ (a scalar) where θ is the angle between $\vec{a}$ and $\vec{b}$.

Symbolically $\boxed{\vec{a} \cdot \vec{b} = |\vec{a}||\vec{b}| \cos \theta}$

Due to a dot between $\vec{a}$ and $\vec{b}$ this product is also called dot product.

The scalar product is commutative.

$$\vec{a} \cdot \vec{b} = \vec{b} \cdot \vec{a}$$

$$\therefore \qquad \vec{b} \cdot \vec{a} = |\vec{b}||\vec{a}| \cos(-\theta)$$

$$= |\vec{a}||\vec{b}| \cos \theta$$

$$\boxed{\vec{b} \cdot \vec{a} = \vec{a} \cdot \vec{b}} \qquad [\because \cos(-\theta) = \cos \theta]$$

2. Geometrical interpretation. The scalar product of two vectors is the product of one vector and the length of the projection of the other in the direction of the first.

Let $\qquad \overrightarrow{OA} = \vec{a}$ and $\overrightarrow{OB} = \vec{b}$

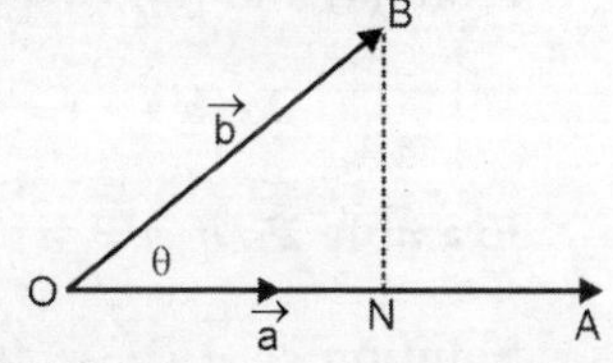

then $\qquad \vec{a} \cdot \vec{a} = OA \cdot OB \cdot \cos \theta$

$$= OA \cdot OB \cdot \frac{ON}{OB}$$

$$= OA \cdot ON$$

$$\boxed{\vec{a} \cdot \vec{b} = \textbf{Length of } \vec{a} \textbf{ and projection of } \vec{b} \textbf{ along } \vec{a}.}$$

Example 24. *Find the projection of the vector* $\hat{i} - 2\hat{j} + \hat{k}$ *on* $4\hat{i} - 4\hat{j} + 7\hat{k}$. **[SBTE 2015, 2013]**

Solution. Projection of the vector $\hat{i} - 2\hat{j} + \hat{k}$ on $4\hat{i} - 4\hat{j} + 7\hat{k}$

$$= (\hat{i} - 2\hat{j} + \hat{k}) \cdot \left(\frac{4\hat{i} - 4\hat{j} + 7\hat{k}}{\sqrt{16 + 16 + 49}} \right)$$

$$= \frac{4 + 8 + 7}{9} = \frac{19}{9} \qquad \textbf{Ans.}$$

20.13 USEFUL RESULTS

$\hat{i} \cdot \hat{i} = 1 \cdot 1 \cos 0 = 1$ $\qquad$ Similarly $\hat{j} \cdot \hat{j} = 1, \hat{k} \cdot \hat{k} = 1$

$\hat{i} \cdot \hat{j} = 1 \cdot 1 \cos 90° = 0$ $\qquad$ Similarly $\hat{j} \cdot \hat{k} = 0, \hat{k} \cdot \hat{i} = 0$

$\boxed{\hat{i} \cdot \hat{i} = \hat{j} \cdot \hat{j} = \hat{k} \cdot \hat{k} = 1}$ $\qquad$ $\boxed{\hat{i} \cdot \hat{j} = \hat{j} \cdot \hat{k} = \hat{k} \cdot \hat{i} = 0}$

Note. If the dot product of two vectors is zero then these vectors are perpendicular to each other.

20.14 DISTRIBUTIVE LAW

$$\vec{a} \cdot (\vec{b} + \vec{c}) = \vec{a} \cdot \vec{b} + \vec{a} \cdot \vec{c}$$

Let $\qquad \overrightarrow{OA} = \vec{a}, \overrightarrow{OB} = \vec{b}, \overrightarrow{BC} = \vec{c}$

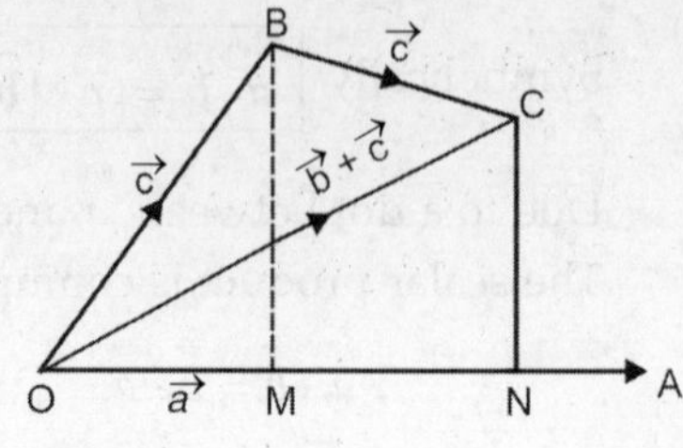

$$\text{R.H.S.} = \vec{a}\cdot\vec{b} + \vec{a}\cdot\vec{c}$$
$$= \overrightarrow{OA}\cdot\overrightarrow{OB} + \overrightarrow{OA}\cdot\overrightarrow{BC}$$
$$= OA\cdot OM + OA\cdot MN$$
$$= OA\,(OM + MN)$$
$$= OA\cdot ON \quad \text{... (1)}$$
$$\text{L.H.S.} = \vec{a}\cdot(\vec{b}+\vec{c})$$
$$= \overrightarrow{OA}\cdot(\overrightarrow{OB}+\overrightarrow{OC})$$
$$= \overrightarrow{OA}\cdot\overrightarrow{OC}$$
$$= OA\cdot ON \quad \text{... (2)}$$

From (1) and (2), we get

$$\vec{a}\cdot(\vec{b}+\vec{c}) = \vec{a}\cdot\vec{b} + \vec{a}\cdot\vec{c}$$

Proved.

Example 25. *If* $\vec{a} = 4\hat{i}+\hat{j}+\hat{k}$, $\vec{b} = 2\hat{i}+\hat{j}+2\hat{k}$ *and* $\vec{c} = 3\hat{i}+4\hat{j}+5\hat{k}$, *find* $(\vec{a}+\vec{b})\cdot(\vec{b}+\vec{c})$.

Solution. $\vec{a}+\vec{b} = (4\hat{i}+\hat{j}+\hat{k}) + (2\hat{i}+\hat{j}+2\hat{k}) = 6\hat{i}+2\hat{j}+3\hat{k}$

$$\vec{b}+\vec{c} = (2\hat{i}+\hat{j}+2\hat{k}) + (3\hat{i}+4\hat{j}+5\hat{k}) = 5\hat{i}+5\hat{j}+7\hat{k}$$

$$(\vec{a}+\vec{b})\cdot(\vec{b}+\vec{c}) = (6\hat{i}+2\hat{j}+3\hat{k})\cdot(5\hat{i}+5\hat{j}+7\hat{k})$$
$$= 30 + 10 + 21$$
$$= 61$$

Ans.

20.15 ANGLE BETWEEN TWO VECTORS $\vec{a}$ AND $\vec{b}$

Where $\vec{a} = a_1\hat{i}+a_2\hat{j}+a_3\hat{k}$ and $\vec{b} = b_1\hat{i}+b_2\hat{j}+b_3\hat{k}$

$$\Rightarrow \quad \vec{a}\cdot\vec{b} = (a_1\hat{i}+a_2\hat{j}+a_3\hat{k})\cdot(b_1\hat{i}+b_2\hat{j}+b_3\hat{k}) = \sqrt{a_1^2+a_2^2+a_3^2}\,\sqrt{b_1^2+b_2^2+b_3^2}\cos\theta$$

$$a_1b_1+a_2b_2+a_3b_3 = \sqrt{a_1^2+a_2^2+a_3^2}\,\sqrt{b_1^2+b_2^2+b_3^2}\cos\theta$$

$$\cos\theta = \frac{a_1b_1+a_2b_2+a_3b_3}{\sqrt{a_1^2+a_2^2+a_3^2}\,\sqrt{b_1^2+b_2^2+b_3^2}}$$

Example 26. *Find the angle between the vectors:* **[SBTE, 2015]**

$(2\hat{i}+6\hat{j}+3\hat{k})$ *and* $(12\hat{i}-4\hat{j}+3\hat{k})$.

Solution. Let the angle between the given vectors be θ then

$$(2\hat{i}+6\hat{j}+3\hat{k})\cdot(12\hat{i}+4\hat{j}-3\hat{k}) = \sqrt{2^2+6^2+3^2}\,\sqrt{12^2+(-4)^2+3^2}\cos\theta$$

$$\Rightarrow \quad 24 - 24 + 9 = 7.13\cos\theta$$

$$\Rightarrow \quad 9 = 91\cos\theta \;\Rightarrow\; \cos\theta = \frac{9}{91}$$

$$\Rightarrow \qquad \theta = \cos^{-1}\left(\frac{9}{91}\right).$$ **Ans.**

Example 27. *Show that* $2\hat{i} - \hat{j} + \hat{k}, \hat{i} - 3\hat{j} - 5\hat{k}, 3\hat{i} - 4\hat{j} - 4\hat{k}$ *form the sides of a right angled triangle.*

Solution. Let $\overrightarrow{AB} = 2\hat{i} - \hat{j} + \hat{k}$

$$\overrightarrow{BC} = \hat{i} - 3\hat{j} - 5\hat{k}$$

$$\overrightarrow{AB} + \overrightarrow{BC} = (2\hat{i} - \hat{j} + \hat{k}) + (\hat{i} - 3\hat{j} - 5\hat{k})$$

$$= 3\hat{i} - 4\hat{j} - 4\hat{k}$$

$$= \overrightarrow{AC}$$

Hence $\overrightarrow{AB}$, $\overrightarrow{BC}$ and $\overrightarrow{AC}$ from a triangle.

Now we have to prove that ΔABC is a rt. angled triangle.

$$\overrightarrow{AB} \cdot \overrightarrow{BC} = (2\hat{i} - \hat{j} + \hat{k}) \cdot (\hat{i} - 3\hat{j} - 5\hat{k})$$

$$= 2 + 3 - 5$$

$$= 0$$

$\therefore$ $\overrightarrow{AB}$ is perpendicular to $\overrightarrow{BC}$. **Proved.**

Example 28. *Prove that*

$$\cos(A + B) = \cos A \cos B - \sin A \sin B.$$

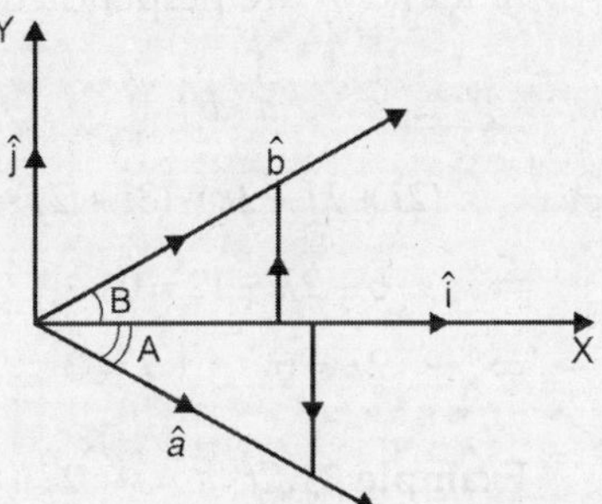

Solution. Let $\hat{a}$ and $\hat{b}$ be unit vectors. $\hat{a}$ and $\hat{b}$ make angles A and B with x-axis respectively.

$$\hat{a} = \cos A\hat{i} - \sin A\hat{j}$$

$$\hat{b} = \cos B\hat{i} + \sin B\hat{j}$$

$$\hat{a} \cdot \hat{b} = (\cos A\hat{i} - \sin A\hat{j}) \cdot (\cos B\hat{i} + \sin B\hat{j})$$

$$\Rightarrow \quad 1 \cdot 1 \cos(A + B) = \cos A \cos B - \sin A \sin B$$

$$\cos(A + B) = \cos A \cos B - \sin A \sin B.$$ **Proved.**

Example 29. *Prove that the altitudes of a triangle are concurrent.*

[Diploma IETE, Dec. 2005]

Solution. Let the altitudes AD and BE meet at O. Let $\vec{a}, \vec{b}, \vec{c}$ be the position vectors of A, B and C with respect to O as origin. Join CO.

Since $\overrightarrow{AO}$ is perpendicular to $\overrightarrow{BC}$.

$$\overrightarrow{AO} \cdot \overrightarrow{BC} = 0$$

$$\Rightarrow \quad -\vec{a} \cdot (\vec{c} - \vec{b}) = 0$$

$$-\vec{a}\cdot\vec{c}+\vec{a}\cdot\vec{b}=0 \quad \text{... (1)}$$

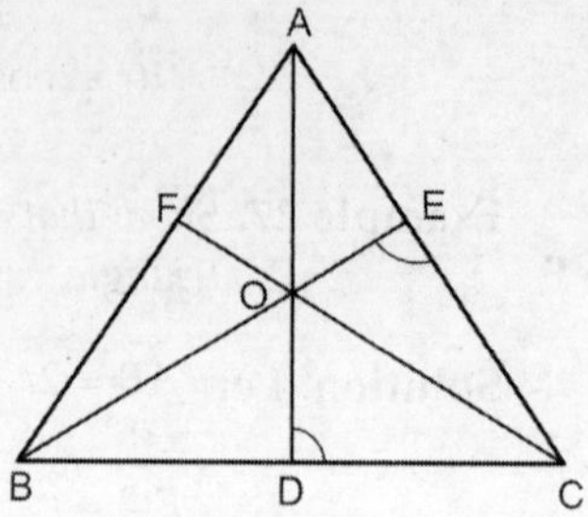

Since $\overrightarrow{BO}$ is perpendicular to $\overrightarrow{AC}$.

$$\overrightarrow{BO}\cdot\overrightarrow{AC}=0$$

$$\Rightarrow \quad -\vec{b}\cdot(\vec{c}-\vec{a})=0$$

$$\Rightarrow \quad -\vec{b}\cdot\vec{c}+\vec{b}\cdot\vec{a}=0 \quad \text{... (2)}$$

On subtracting (1) from (2), we have

$$\Rightarrow \quad -\vec{b}\cdot\vec{c}+\vec{a}\cdot\vec{c}=0$$

$$\Rightarrow \quad -\vec{c}\cdot(\vec{b}-\vec{a})=0$$

$$\overrightarrow{CO}\cdot\overrightarrow{AB}=0$$

$\therefore$ $\overrightarrow{CO}$ is perpendicular to $\overrightarrow{AB}$.

Hence, the three altitudes meet at a point. **Proved.**

Example 30. *Find the values of* λ, *if the vectors* $2\hat{i}+\lambda\hat{j}+3\hat{k}$ *and* $3\hat{i}+2\hat{j}-4\hat{k}$ *are perpendicular to each other.* **[SBTE 2015]**

Solution. Let $\vec{a}=2\hat{i}+\lambda\hat{j}+3\hat{k},\ \vec{b}=3\hat{i}+2\hat{j}-4\hat{k}$

If $\vec{a}$ and $\vec{b}$ are perpendicular, then

$$\vec{a}\cdot\vec{b}=0$$

$$(2\hat{i}+\lambda\hat{j}+3\hat{k})\cdot(3\hat{i}+2\hat{j}-4\hat{k})=0$$

$$\Rightarrow \quad 6+2\lambda-12=0$$

$$\Rightarrow \quad 2\lambda=6 \quad \therefore \lambda=3$$

Ans.

Example 31. *If* $\vec{a}=\hat{i}+2\hat{j}+3\hat{k},\ \vec{b}=-\hat{i}+2\hat{j}+\hat{k},\ \vec{c}=3\hat{i}+\hat{j}$, *find the value of t such that* $\vec{a}+t\vec{b}$ *is perpendicular to* $\vec{c}$. **[SBTE 2017, 2015]**

Solution. Here we have

$$\vec{a}=\hat{i}+2\hat{j}+3\hat{k},\ \vec{b}=-\hat{i}+2\hat{j}+\hat{k},\ \vec{c}=3\hat{i}+\hat{j}$$

$$\vec{a}+t\vec{b}=(\hat{i}+2\hat{j}+3\hat{k})+t\cdot(-\hat{i}+2\hat{j}+\hat{k})$$

$$\vec{a}+t\vec{b}=(1-t)\hat{i}+(2+2t)\hat{j}+(3+t)\hat{k}$$

$\vec{a}+t\vec{b}$ is perpendicular to $\vec{c}$

$$\left(\vec{a}+t\vec{b}\right)\cdot\vec{c}=0$$

$$\left[(1-t)\hat{i}+(2+2t)\hat{j}+(3+t)\hat{k}\right]\cdot\left[3\hat{i}+\hat{j}\right]=0$$

$$3 - 3t + 2 + 2t = 0$$

$$5 - t = 0$$

$$\therefore \quad t = 5$$

Ans.

Example 32. *Prove that in a triangle ABC* $\cos A = \dfrac{b^2 + c^2 - a^2}{2bc}$

Solution. $\overrightarrow{BC} = \overrightarrow{AC} - \overrightarrow{AB}$

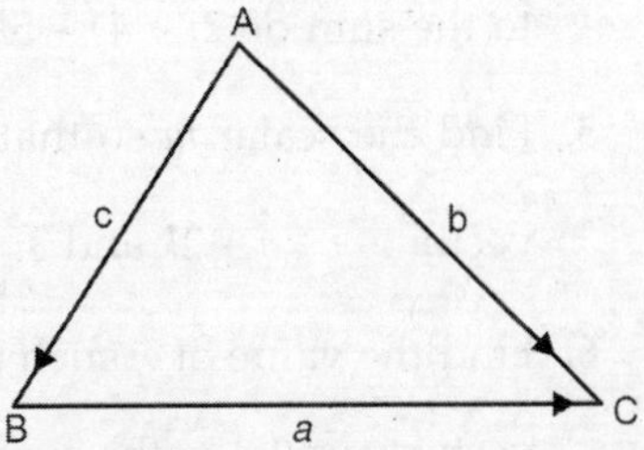

$$\overrightarrow{BC}\cdot\overrightarrow{BC} = (\overrightarrow{AC} - \overrightarrow{AB})\cdot(\overrightarrow{AC} - \overrightarrow{AB})$$

$$\Rightarrow \quad BC \cdot BC \cos\theta = \overrightarrow{AC}\cdot\overrightarrow{AC} + \overrightarrow{AB}\cdot\overrightarrow{AB} - 2\,\overrightarrow{AC}\cdot\overrightarrow{AB}$$

$$\Rightarrow \quad (BC)^2 = (AC)^2 + (AB)^2 - 2AC \cdot AB \cos A$$

$$\Rightarrow \quad a^2 = b^2 + c^2 - 2bc \cos A$$

$$\Rightarrow \quad 2bc \cos A = b^2 + c^2 - a^2$$

$$\cos A = \frac{b^2 + c^2 - a^2}{2bc}$$

Proved.

Example 33. *If* $|\vec{A}+\vec{B}| = 60, |\vec{A}-\vec{B}| = 40$ *and* $|\vec{B}| = 46$, *find* $|\vec{A}|$. **[SBTE 2014]**

Solution. $|\vec{A}+\vec{B}| = 60 \quad \Rightarrow \quad |\vec{A}+\vec{B}|^2 = 3600$

$$(\vec{A}+\vec{B})(\vec{A}+\vec{B}) = 3600$$

$$\Rightarrow \quad (\vec{A})^2 + (\vec{B})^2 + 2(\vec{A})\cdot(\vec{B}) = 3600$$

$$\Rightarrow \quad |\vec{A}|^2 + |\vec{B}|^2 + 2|\vec{A}||\vec{B}|\cos\theta = 3600 \qquad \text{... (1)}$$

And $|\vec{A}-\vec{B}| = 40$

Similarly, $|\vec{A}|^2 + |\vec{B}|^2 - 2|\vec{A}||\vec{B}|\cos\theta = 1600 \qquad \text{... (2)}$

Adding (1) and (2), we get

$$2|\vec{A}|^2 + 2|\vec{B}|^2 = 5200$$

$$\Rightarrow \quad |\vec{A}|^2 + |\vec{B}|^2 = 2600$$

$$|\vec{A}|^2 + 46^2 = 2600$$

$$\Rightarrow \quad |\vec{A}|^2 = 484$$

$$|\vec{A}| = 22$$

Ans.

EXERCISE 20.3

1. Find the value of $(2\hat{i}+3\hat{j}+4\hat{k})\cdot(\hat{i}+\hat{j}+\hat{k})$. **[SBTE 2017] Ans.** 9

2. Prove that $(\hat{i}+2\hat{j}+8\hat{k})$ and $(2\hat{i}+3\hat{j}-\hat{k})$ are perpendicular to each other.

3. Find the value of a which makes the vectors $a\hat{i}+2\hat{j}+3\hat{k}$ and $-\hat{i}+5\hat{j}+a\hat{k}$ perpendicular. **Ans.** – 5

4. Find the scalar m such that the scalar product of $\hat{i}+\hat{j}+\hat{k}$ with the unit vector parallel to the sum of $2\hat{i}+4\hat{j}-5\hat{k}$ and $m\hat{i}+2\hat{j}-3\hat{k}$ is equal to unity. **Ans.** $m = 1$

5. Find the scalar m so that the vector $2\hat{i}+\hat{j}-m\hat{k}$ is perpendicular to the sum of the vector $-\hat{i}+\hat{j}+2\hat{k}$ and $3\hat{i}+2\hat{j}+\hat{k}$. **Ans.** $m=\frac{7}{3}$

6. Find the value of λ such that the scalar product of the vector $\hat{i}+\hat{j}+\hat{k}$ with the unit vector parallel to the sum of the vectors $\lambda\hat{i}+2\hat{j}+3\hat{k}$ and $2\hat{i}-\lambda\hat{j}-5\hat{k}$ is equal to $\frac{1}{2}$. **[B.T.E, Delhi 2006] Ans.** $\lambda=\pm\sqrt{2}$

7. Find the projection of the vector $\hat{i}+\hat{j}+\hat{k}$ on $4\hat{i}+4\hat{j}+5\hat{k}$ **[SBTE, 2015] Ans.** $\frac{13}{57}(4\hat{i}+4\hat{j}+5\hat{k})$

8. If $|\vec{A}|=11, |\vec{B}|=23$ and $|\vec{A}-\vec{B}|=30$, find the value of $|\vec{A}+\vec{B}|$. **Ans.** 20

9. If $|\vec{A}+\vec{B}|=|\vec{A}-\vec{B}|$, find the angle between $\vec{A}$ and $\vec{B}$. What conclusion do you derive from the result? **Ans.** 90°

10. What is the meaning of $(\vec{a}+\vec{b})\cdot(\vec{a}-\vec{b})$ for the case where $|\vec{a}|^2=|\vec{b}|^2$. **Ans.** Perpendicular

11. Show that the acute angle between two diagonals of a cube has cosine $\frac{1}{3}$.

12. Show that the components of a vector $\vec{b}$ along and perpendicular to the vector $\vec{a}$ are $\frac{(\vec{a}\cdot\vec{b})\,\vec{a}}{\vec{a}\cdot\vec{a}}$ and $\frac{(\vec{a}\cdot\vec{a})\,\vec{b}-(\vec{a}\cdot\vec{b})\,\vec{a}}{\vec{a}\cdot\vec{a}}$ respectively.

13. Show that $3\hat{i}-2\hat{j}+\hat{k}, \hat{i}-3\hat{j}+5\hat{k}$ and $2\hat{i}+\hat{j}-4\hat{k}$ form a right angled triangle.

14. Prove that cos (α – β) = cos α cos β + sin α sin β.

15. Prove that $a = b\cos C + c\cos B$ in a triangle ABC.

16. Prove that the perpendicular bisectors of the sides of a triangle are concurrent.

17. Show that the mid-point of the hypotenuse of a right angled triangle is equidistant from its vertices.

18. Show that the sum of squares of the diagonals of a parallelogram is equal to the sum of squares of its sides.

19. Using vector method, prove that the angle in a semi-circle is a right angle.
20. In a tetrahedron, if two pairs of opposite edges are perpendicular, the third pair is also perpendicular to each other, and the sum of the squares of two opposite edges is the same for each pair.
21. In ΔOAB let $OA = \vec{a}, OB = \vec{b}$. Then find the vector representing AB and OM, where M is the mid-point of AB. **[Diploma IETE, June 2006]**

Choose the correct alternative in the following:

22. Any vector $\vec{a} = a_1\hat{i} + a_2\hat{j} + a_3\hat{k}$ is equal to

(a) $(a \cdot \hat{i})\,\hat{i} + (a \cdot \hat{j})\,\hat{j} + (a \cdot \hat{k})\,\hat{k}$ (b) $(a \cdot \hat{j})\,\hat{i} + (a \cdot \hat{k})\,\hat{j} + (a \cdot \hat{i})\,\hat{k}$

(c) $(a \cdot \hat{k})\,\hat{i} + (a \cdot \hat{i})\,\hat{j} + (a \cdot \hat{j})\,\hat{k}$ (d) $(a \cdot a)\,(\hat{i} + \hat{j} + \hat{k})$ **Ans.** (*a*)

[Diploma IETE, Dec. 2006]

23. If a and b are two unit vectors inclined at an angled θ and are such that $a + b$ is a unit vector, then θ is equal to

(a) $\frac{\pi}{4}$ (b) $\frac{\pi}{3}$ (c) $\frac{\pi}{2}$ (d) $\frac{2\pi}{3}$ **Ans.** (*d*)

[Diploma IETE, Dec. 2006]

24. If $\vec{a}$ and $\vec{b}$ be any two vectors, then show that

(a) $\left(\vec{a} + \vec{b}\right) \cdot \left(\vec{a} - \vec{b}\right) = \left|\vec{a}\right|^2 - \left|\vec{b}\right|^2$ (b) $\left|\vec{a} + \vec{b}\right|^2 = \left|\vec{a}\right|^2 + \left|\vec{b}\right|^2 + 2\vec{a} \cdot \vec{b}$

[Diploma IETE, June 2005]

20.16 WORK DONE AS A SCALAR PRODUCT

If a constant force F acting on a particle, displaces it from A to B, then,

$$\text{Work done} = (\text{component of } F \text{ along } AB) \cdot \text{Displacement}$$
$$= F \cos \theta \cdot AB$$
$$= \vec{F} \cdot \overrightarrow{AB}$$

Work done = Force · Displacement

Example 34. *Find the work done by a force $\vec{F} = -2\hat{i} + 3\hat{j} + 4\hat{k}$ when its point of application moves from the point $A(2, -1, -2)$ to the point $B(-1, 2, 3)$.* **[SBTE 2008]**

Solution. Here, we have $\vec{F} = -2\hat{i} + 3\hat{j} + 4\hat{k}$

The coordinates of the points A and B are (2, – 1, – 2) and (– 1, 2, 3) respectively.

$$\overrightarrow{AB} = \overrightarrow{OB} - \overrightarrow{OA} = (-\hat{i} + 2\hat{j} + 3\hat{k}) - (2\hat{i} - \hat{j} - 2\hat{k})$$

$$= -3\hat{i} + 3\hat{j} + 5\hat{k}$$

Work done $= \vec{F} \cdot \overrightarrow{AB}$

$= (-2\hat{i} + 3\hat{j} + 4\hat{k}) \cdot (-3\hat{i} + 3\hat{j} + 5\hat{k})$

$= 6 + 9 + 20$

$= 35$ Joule **Ans.**

Example 35. *The constant forces* $2\hat{i} - 5\hat{j} + 6\hat{k}, -\hat{i} + 2\hat{j} - \hat{k}$ *and* $2\hat{i} + 7\hat{j}$ *act on a particle which is displaced from position* $4\hat{i} - 3\hat{j} - 2\hat{k}$ *to position* $6\hat{i} + 7\hat{j} - 3\hat{k}$. *Find the total work done.* **[SBTE 2011]**

Solution. Total force $= (2\hat{i} - 5\hat{j} + 6\hat{k}) + (-\hat{i} + 2\hat{j} - \hat{k}) + (2\hat{i} + 7\hat{j})$

$= 3\hat{i} + 4\hat{j} + 5\hat{k}$

Displacement $= (6\hat{i} + 7\hat{j} - 3\hat{k}) - (4\hat{i} - 3\hat{j} - 2\hat{k})$

$= 2\hat{i} + 4\hat{j} - \hat{k}$

Work done = Force · Displacement

$= (3\hat{i} + 4\hat{j} + 5\hat{k}) \cdot (2\hat{i} + 4\hat{j} - \hat{k})$

$= 6 + 16 - 5 = 17$ Joule **Ans.**

Example 36. *A particle is acted upon by constant force* $4\hat{i} + \hat{j} - 3\hat{k}$ *and* $3\hat{i} + \hat{j} - \hat{k}$ *and is displaced from the point (1, 2, 3) to the point (5, 4, 1). Find the total work done.*

Solution. Total force $= (4\hat{i} + \hat{j} - 3\hat{k}) + (3\hat{i} + \hat{j} - \hat{k})$

$= 7\hat{i} + 2\hat{j} - 4\hat{k}$

Displacement from the point (1, 2, 3) to (5, 4, 1)

$= (5\hat{i} + 4\hat{j} + \hat{k}) - (\hat{i} + 2\hat{j} + 3\hat{k})$

$= 4\hat{i} + 2\hat{j} - 2\hat{k}$

Work done by the forces = Forces · displacement

$= (7\hat{i} + 2\hat{j} - 4\hat{k}) \cdot (4\hat{i} + 2\hat{j} - 2\hat{k})$

$= 28 + 4 + 8 = 40$ Joule **Ans.**

Example 37. *Forces of magnitudes 5 and 3 units acting in the directions* $6\hat{i} + 2\hat{j} + 3\hat{k}$ *and* $3\hat{i} - 2\hat{j} + 6\hat{k}$ *respectively act on a particle which is displaced from the point (2, 2, – 1) to (4, 3, 1). Find the work done by the forces.* **[SBTE 2017, 2013]**

Solution. First force of magnitude 5 units, acting in the direction $6\hat{i} + 2\hat{j} + 3\hat{k}$.

$$= 5\left(\frac{6\hat{i} + 2\hat{j} + 3\hat{k}}{\sqrt{6^2 + 2^2 + 3^2}}\right) = \frac{5}{7}(6\hat{i} + 2\hat{j} + 3\hat{k})$$

Second force of magnitude 3 units, acting in the direction $3\hat{i} - 2\hat{j} + 6\hat{k}$.

$$= 3\left(\frac{3\hat{i} - 2\hat{j} + 6\hat{k}}{\sqrt{3^2 + (-2)^2 + 6^2}}\right) = \frac{3}{7}(3\hat{i} - 2\hat{j} + 6\hat{k})$$

Resultant force

$$= \frac{5}{7}(6\hat{i} + 2\hat{j} + 3\hat{k}) + \frac{3}{7}(3\hat{i} - 2\hat{j} + 6\hat{k})$$

$$= \frac{1}{7}(39\hat{i} + 4\hat{j} + 33\hat{k})$$

Displacement from the point (2, 2, – 1) to (4, 3, 1)

$$= (4\hat{i} + 3\hat{j} + \hat{k}) - (2\hat{i} + 2\hat{j} - \hat{k})$$

$$= 2\hat{i} + \hat{j} + 2\hat{k}$$

Work done by the forces = Forced · Displacement

$$= \frac{1}{7}(39\hat{i} + 4\hat{j} + 33\hat{k}) \cdot (2\hat{i} + \hat{j} + 2\hat{k})$$

$$= \frac{1}{7}(78 + 4 + 66)$$

$$= \frac{148}{7} \text{ Joule}$$ **Ans.**

Example 38. *Find the projection of* $\hat{i} - 2\hat{j} + \hat{k}$ *on* $4\hat{i} - 4\hat{j} + 7\hat{k}$. **[SBTE 2015, 2008]**

Solution. Here we have

Let $\vec{a} = \hat{i} - 2\hat{j} + \hat{k},\ 4\hat{i} - 4\hat{j} + 7\hat{k}$

$$\vec{a} \cdot \vec{b} = 4 + 8 + 7 = 19$$

$$|\vec{b}| = \sqrt{16 + 16 + 49} = \sqrt{81} = 9$$

$$\text{Projection } \vec{a} \text{ on } \vec{b} = \frac{\vec{a} \cdot \vec{b}}{|\vec{b}|}$$

$$= \frac{19}{9}$$ **Ans.**

Example 39. *If with reference to right handed system of mutually perpendicular unit vectors* $\hat{i}, \hat{j}, \hat{k}$ *we have* $\vec{\alpha} = 3\hat{i} - \hat{j}$ *and* $\vec{\beta} = 2\hat{i} + \hat{j} - 3\hat{k}$ *express* $\vec{\beta}$ *in the form* $\vec{\beta} = \overrightarrow{\beta_1} + \overrightarrow{\beta_2}$ *where* $\overrightarrow{\beta_1}$ *is parallel to* $\vec{\alpha}$ *and* $\overrightarrow{\beta_2}$ *is* $\perp$ *to* $\vec{\alpha}$.

Solution. Here we have: $\vec{\alpha} = 3\hat{i} - \hat{j}$

$$\overrightarrow{\beta_1} \,||\, \vec{\alpha}$$

$$\therefore \quad \vec{\beta} = \lambda\vec{\alpha} \Rightarrow \overrightarrow{\beta_1} = \lambda(3\hat{i} - \hat{j}) \qquad \text{... (1)}$$

Also $\quad \vec{\beta} = \overrightarrow{\beta_1} + \overrightarrow{\beta_2} \Rightarrow \overrightarrow{\beta_2} = \vec{\beta} - \overrightarrow{\beta_1}$

$$\Rightarrow \quad \overrightarrow{\beta_2} = (2\hat{i} + \hat{j} - 3\hat{k}) - (3\lambda\hat{i} - \lambda\hat{j})$$

$$\vec{\beta_2} = (2 - 3\lambda)\hat{i} + (1 + \lambda)\hat{j} - 3\hat{k} \quad \text{... (2)}$$

Given β_2 is perpendicular to $\vec{\alpha}$

$$\overrightarrow{\beta_2} \cdot \vec{\alpha} = 0$$

$$\left[(2 - 3\lambda)\hat{i} + (1 + \lambda)\hat{j} - 3\hat{k}\right] \cdot (3\hat{i} - \hat{j}) = 0 \Rightarrow \lambda = \frac{1}{2}$$

Putting the value of λ in (1) and (2), we get

$$\overrightarrow{\beta_1} = \frac{1}{2}(3\hat{i} - \hat{j}),\ \overrightarrow{\beta_2} = \frac{1}{2}(\hat{i} + 3\hat{j} - 6\hat{k}) \quad \textbf{Ans.}$$

EXERCISE 20.4

1. A force of magnitude 6 units acting parallel to $2\hat{i} - 2\hat{j} + \hat{k}$ displaces the point of application from (1, 2, 3) to (5, 3, 7). Find the work done. **Ans.** 20

2. A particle acted on by two forces $4\hat{i} + \hat{j} - 3\hat{k}$ and $3\hat{i} + \hat{j} - \hat{k}$ is displaced from the point $\hat{i} + 2\hat{j} + \hat{k}$ to the point $5\hat{i} + 4\hat{j} + \hat{k}$. Find the work done by the force. **[SBTE 2012] Ans.** 32 Joules

3. A particle is acted upon by constant forces $4\hat{i} + \hat{j} - 3\hat{k}$ and $3\hat{i} + \hat{j} - \hat{k}$ and is displaced from the point (1, 2, 3) to the point (5, 4, 1). Find the total work spent by the forces. **Ans.** 40 Joules

4. Forces of magnitudes 5, 3, 1 N acting in the directions $6\hat{i} + 2\hat{j} + 3\hat{k}$, $3\hat{i} - 2\hat{j} + 6\hat{k}$, $2\hat{i} - 3\hat{j} - 6\hat{k}$ respectively act on a particle which is displaced from the point (2, – 1, – 3) to (5, – 1, 1) meter. Find the work done by the forces, the unit of length being metre. **[SBTE 2015] Ans.** 33 joules.

20.17 VECTOR PRODUCT OR CROSS PRODUCT

The vector or cross product of two vectors $\vec{a}$ and $\vec{b}$ is defined to be a vector such that

(*i*) its magnitude is $|\vec{a}|\,|\vec{b}| \sin\theta$, where θ is the angle between $\vec{a}$ and $\vec{b}$.

(*ii*) its direction is perpendicular to both vectors $\vec{a}$ and $\vec{b}$.

(*iii*) it forms with a right handed system.

Let $\hat{n}$ to be a unit vector perpendicular to both the vectors $\vec{a}$ and $\vec{b}$. $\vec{a}$, $\vec{b}$ and $\hat{n}$ forming a right handed system, then

$$\boxed{\vec{a} \times \vec{b} = |\vec{a}|\,|\vec{b}| \sin\theta \cdot \hat{n}}$$

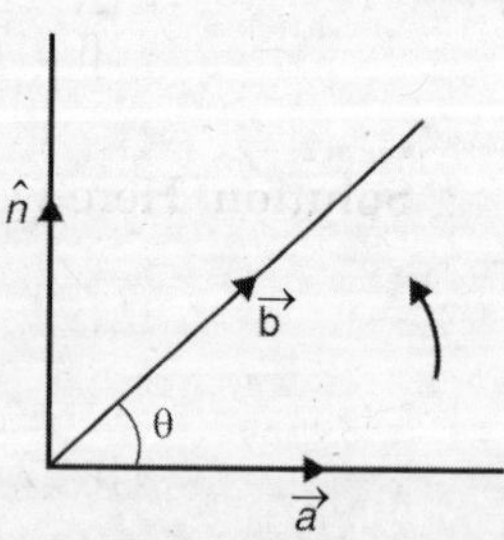

20.18 RIGHT HANDED SYSTEM

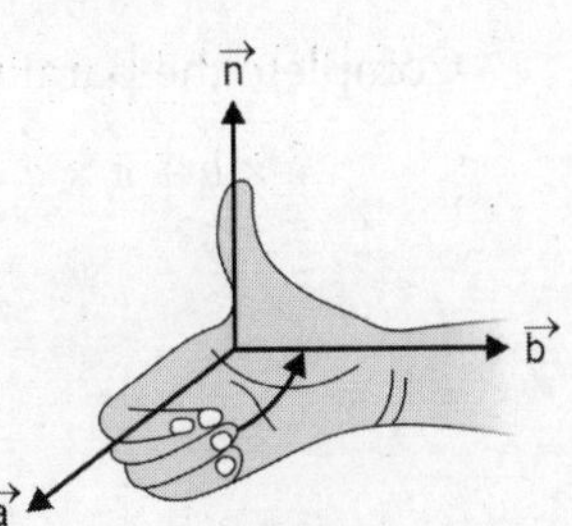

Stretch the fingers of right hand:

(*i*) The fingers are pointing in the direction of vector $\vec{a}$.

(*ii*) On curling the fingures, the curled fingures are pointing in the direction of vector $\vec{b}$.

(*iii*) The thumb will indicate the direction of vector $\vec{n}$.

20.19 RIGHT HANDED SCREW SYSTEM

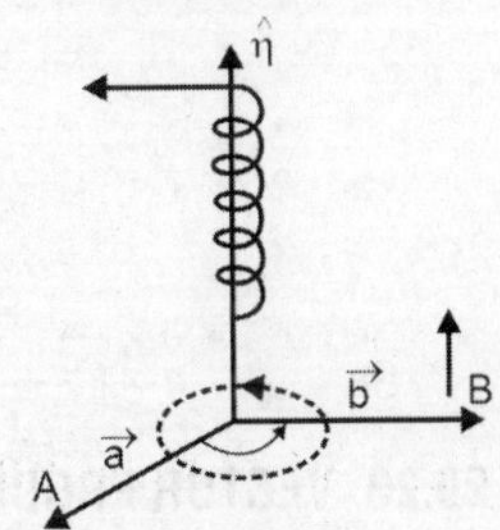

When the vector $\vec{a}$ is rotated counterclockwise towards the vector $\vec{b}$ the screw would advance in the direction of the positive vector $\vec{n}$ as shown in the Fig.

20.20 VECTOR PRODUCT IS NOT COMMUTATIVE

$$\boxed{\vec{a} \times \vec{b} \neq \vec{b} \times \vec{a}}$$

but $\quad \vec{a} \times \vec{b} = -\vec{b} \times \vec{a}$

$\because \quad \vec{b} \times \vec{a} = |\vec{b}||\vec{a}| \sin(-\theta) \cdot \hat{n}$ $\quad [\because \sin(-\theta) = -\sin\theta]$

$$= -|\vec{a}||\vec{b}| \sin\theta \cdot \hat{n} = -\vec{a} \times \vec{b}$$

Note. If two vectors $\vec{a}$ and $\vec{b}$ are parallel, then $\vec{a} \times \vec{b} = |\vec{a}||\vec{b}| \sin\theta° \cdot \hat{n}$

$\Rightarrow \quad \vec{a} \times \vec{b} = 0$

20.21 USEFUL RESULTS

Since, $\hat{i}, \hat{j}, \hat{k}$ are three mutually perpendicular unit vectors, then

$$\hat{i} \times \hat{i} = \hat{j} \times \hat{j} = \hat{k} \times \hat{k} = 0$$

$$\hat{i} \times \hat{j} = -\hat{j} \times \hat{i} = \hat{k}$$

$$\hat{j} \times \hat{k} = -\hat{k} \times \hat{j} = \hat{i}$$

$$\hat{k} \times \hat{i} = -\hat{i} \times \hat{k} = \hat{j}$$

20.22 DISTRIBUTIVE LAW

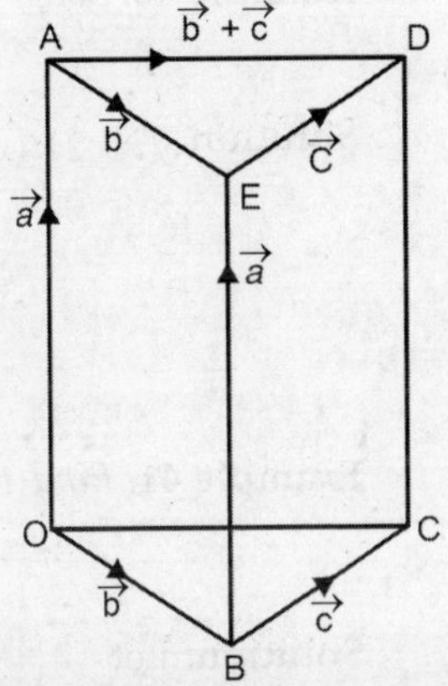

$$\vec{a} \times (\vec{b} + \vec{c}) = \vec{a} \times \vec{b} + \vec{a} \times \vec{c}$$

Let $\vec{a}, \vec{b}, \vec{c}$ be three vectors

$$\overrightarrow{OA} = \vec{a}, \quad \overrightarrow{OB} = \vec{b}, \quad \overrightarrow{BC} = \vec{c}$$

and $\quad \overrightarrow{BE} = \vec{a}, \quad \overrightarrow{AE} = \vec{b}, \quad \overrightarrow{ED} = \vec{c}$

Complete the parallelograms *OAEB*, *BCDE* and *OADC*.

$$\vec{a} \times \vec{b} + \vec{a} \times \vec{c} = \text{Vector Area of parallelogram } OAEB$$
$$+ \text{ Vector Area of parallelogram } BCDE$$
$$= \text{Vector area of } OAEDCB$$
$$= \text{Vector area of } OAEDCB - \text{Vector area of } \Delta OCB$$
$$+ \text{ Vector area of } \Delta ADE \qquad (\because \Delta OBC = \Delta ADE)$$
$$= \text{Vector area of } || \text{gm } OADC$$
$$= \overrightarrow{OA} \times \overrightarrow{AD}$$
$$= \vec{a} \times (\vec{b} + \vec{c})$$

So $\quad \vec{a} \times (\vec{b} + \vec{c}) = \vec{a} \times \vec{b} + \vec{a} \times \vec{c}$

20.23 VECTOR PRODUCT EXPRESSED AS A DETERMINANT

If $\quad \vec{a} = a_1\hat{i} + a_2\hat{j} + a_3\hat{k}$

$$\vec{b} = b_1\hat{i} + b_2\hat{j} + b_3\hat{k}$$

$$\vec{a} \times \vec{b} = (a_1\hat{i} + a_2\hat{j} + a_3\hat{k}) \times (b_1\hat{i} + b_2\hat{j} + b_3\hat{k})$$
$$= a_1b_1\hat{i} \times \hat{i} + a_1b_2\hat{i} \times \hat{j} + a_1b_3\hat{i} \times \hat{k} + a_2b_1\hat{j} \times \hat{i} + a_2b_2\hat{j} \times \hat{j}$$
$$+ a_2b_3\hat{j} \times \hat{k} + a_3b_1\hat{k} \times \hat{i} + a_3b_2\hat{k} \times \hat{j} + a_3b_3\hat{k} \times \hat{k}$$
$$= a_1b_2\hat{k} - a_1b_3\hat{j} - a_2b_1\hat{k} + a_2b_3\hat{i} + a_3b_1\hat{j} - a_3b_2\hat{i}$$
$$= (a_2b_3 - a_3b_2)\hat{i} - (a_1b_3 - a_3b_1)\hat{j} + (a_1b_2 - a_2b_1)\hat{k}$$
$$= \begin{vmatrix} \hat{i} & \hat{j} & \hat{k} \\ a_1 & a_2 & a_3 \\ b_1 & b_2 & b_3 \end{vmatrix}$$

Example 40. *Solve* $(\hat{i} + 2\hat{j} + 3\hat{k}) \times (2\hat{i} + \hat{j} - \hat{k})$. **[SBTE 2014]**

Solution. $(\hat{i} + 2\hat{j} + 3\hat{k}) \times (2\hat{i} + \hat{j} - \hat{k}) = \begin{vmatrix} \hat{i} & \hat{j} & \hat{k} \\ 1 & 2 & 3 \\ 2 & 1 & -1 \end{vmatrix}$

$$= (-2-3)\hat{i} - (-1-6)\hat{j} + (1-4)\hat{k}$$
$$= -5\hat{i} + 7\hat{j} - 3\hat{k}.$$ **Ans.**

Example 41. *Find the unit vector perpendicular to each of the vectors* $\hat{i} + 2\hat{j} + 3\hat{k}$ *and* $-3\hat{i} - 2\hat{j} + \hat{k}$ **[SBTE 2017]**

Solution. Let $\quad \vec{a} = \hat{i} + 2\hat{j} + 3\hat{k}$

$$\vec{b} = -3\hat{i} - 2\hat{j} + \hat{k}$$

$$\vec{a} \times \vec{b} = \begin{vmatrix} \hat{i} & \hat{j} & \hat{k} \\ 1 & 2 & 3 \\ -3 & -2 & 1 \end{vmatrix}$$

$$= (2+6)\hat{i} - (1+9)\hat{j} + (-2+6)\hat{k} = 8\hat{i} - 10\hat{j} + 4\hat{k}$$

$\vec{a} \times \vec{b}$ is a vector perpendicular to $\vec{a}$ and $\vec{b}$.

$$\text{Unit perpendicular vector} = \frac{8\hat{i} - 10\hat{j} + 4\hat{k}}{\sqrt{64+100+16}} = \frac{1}{\sqrt{180}}(8\hat{i} - 10\hat{j} + 4\hat{k})$$

$$= \frac{1}{\sqrt{45}}(4\hat{i} - 5\hat{j} + 2\hat{k})$$ **Ans.**

Example 42. *Find a unit vector perpendicular to both of the vectors* $\vec{A} = 2\hat{i} + \hat{j} - \hat{k}$ *and* $\vec{B} = \hat{i} - \hat{j} + 2\hat{k}.$ **[B.T.E Delhi Jan. 2009]**

Solution. $\vec{A} \times \vec{B} = \begin{vmatrix} \hat{i} & \hat{j} & \hat{k} \\ 2 & 1 & -1 \\ 1 & -1 & 2 \end{vmatrix} = (2-1)\hat{i} - (4+1)\hat{j} + (-2-1)\hat{k} = \hat{i} - 5\hat{j} - 3\hat{k}$

Unit vector $\vec{A} \times \vec{B}$ = Unit vector of $(\hat{i} - 5\hat{j} - 3\hat{k})$

$$= \frac{\hat{i} - 5\hat{j} - 3\hat{k}}{\sqrt{1+25+9}} = \frac{\hat{i} - 5\hat{j} - 3\hat{k}}{\sqrt{35}}$$ **Ans.**

EXERCISE 20.5

1. Find $\vec{a} \times \vec{b}$ if $\vec{a} = 2\hat{i} + 3\hat{j} + 4\hat{k}$ and $\vec{b} = \hat{i} + \hat{j} + \hat{k}.$ **Ans.** $-\hat{i} + 2\hat{j} - \hat{k}$
2. If $\vec{a} = 3\hat{i} - 2\hat{j} - 2\hat{k}$ and $\vec{b} = 2\hat{i} + 3\hat{j} + 4\hat{k}$, calculate $(\vec{a} + \vec{b}) \times (\vec{a} - \vec{b})$.
   Ans. $-8\hat{i} + 14\hat{j} + 26\hat{k}$
3. Show that $(\vec{a} - \vec{b}) \times (\vec{a} + \vec{b}) = 2\,\vec{a} \times \vec{b}$.
4. Show that $\vec{a} \times (\vec{b} + \vec{c}) + \vec{b} \times (\vec{c} + \vec{a}) + \vec{c} \times (\vec{a} + \vec{b}) = 0.$
5. Determine λ and μ by using vectors, such that the points $(-1, 3, 2)$, $(-4, 2, -2)$ and $(5, \lambda, \mu)$ lie on a straight line. **Ans.** $\lambda = 5, \mu = 10.$
6. What is the unit vector perpendicular to the plane of $\vec{a}$ and $\vec{b}$ if $\vec{a} = 4\hat{i} + 3\hat{j} + \hat{k}$ and $\vec{b} = 2\hat{i} - \hat{j} + 2\hat{k}.$ **Ans.** $\dfrac{7\hat{i} - 6\hat{j} - 10\hat{k}}{\sqrt{185}}$

7. Find a unit vector perpendicular to the plane of vector $\vec{a} = 2\hat{i} + \hat{j} - \hat{k}$ and $\vec{b} = \hat{i} - \hat{j} + 2\hat{k}$.

[Diploma IETE Dec. 2005] Ans. $\dfrac{\hat{i} - 5\hat{j} - 3\hat{k}}{\sqrt{35}}$

8. If $\vec{a} = \hat{i} - 2\hat{j} + 3\hat{k}$ and $\vec{b} = 3\hat{i} + \hat{j} + 2\hat{k}$, find a unit vector $\vec{c}$ which is perpendicular to $\vec{a}$ and $\vec{b}$. **Ans.** $\dfrac{-\hat{i} + \hat{j} + \hat{k}}{\sqrt{3}}$

9. Find a vector of magnitude 9 which is perpendicular to both the vectors $4\hat{i} - \hat{j} + 3\hat{k}$ and $-2\hat{i} + \hat{j} - 2\hat{k}$. **Ans.** $3(-\hat{i} + 2\hat{j} + 2\hat{k})$

10. Find a unit vector perpendicular to each of the vector $\vec{A} = 2\hat{i} - \hat{j} + \hat{k}$ and $\vec{B} = 3\hat{i} + 4\hat{j} - \hat{k}$ and obtain sine of the angle between vector $\vec{A}$ and $\vec{B}$.

Ans. $\dfrac{-3\hat{i} + 5\hat{j} + 11\hat{k}}{\sqrt{155}}, \sqrt{\dfrac{155}{156}}$

11. Find the angle between two vectors $\vec{a}$ and $\vec{b}$, if $|\vec{a} \times \vec{b}| = \vec{a} \cdot \vec{b}$.

[Diploma IETE Dec. 2005] Ans. $\dfrac{\pi}{4}$

20.24 GEOMETRICAL INTERPRETATION (AREA OF PARALLELOGRAM)

$\vec{a} \times \vec{b}$ represents the vector area of the parallelogram whose adjacent sides are $\vec{a}$ and $\vec{b}$.

$$\vec{a} \times \vec{a} = OA \cdot OB \sin\theta \cdot \hat{n}$$

$$= OA \cdot BM \cdot \hat{n}$$

$$= \text{Base} \cdot \text{height} \cdot \hat{n}$$

or $\boxed{\vec{a} \times \vec{b} = \textbf{vector area of parallelogram OALB}}$

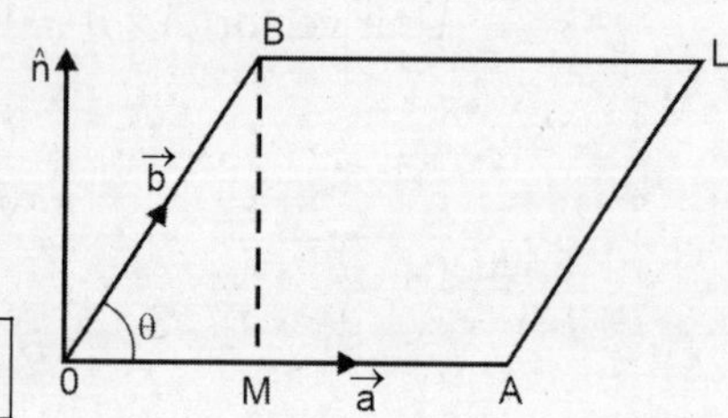

Example 43. *Find the area of parallelogram whose adjacent sides are* $\hat{i} - 2\hat{j} + 3\hat{k}$ *and* $2\hat{i} + \hat{j} - 4\hat{k}$.

Solution. Vector area of || gm = $\begin{vmatrix} \hat{i} & \hat{j} & \hat{k} \\ 1 & -2 & 3 \\ 2 & 1 & -4 \end{vmatrix}$

$$= (8-3)\hat{i} - (-4-6)\hat{j} + (1+4)\hat{k} = 5\hat{i} + 10\hat{j} + 5\hat{k}$$

Area of parallelogram = $\sqrt{(5)^2 + (10)^2 + (5)^2} = 5\sqrt{6}$ Sq. units. **Ans.**

Example 44. *Find the area of a triangle whose vertices are (1, 1, 1); (0, 1, 2) and (3, 2, 1).*

Solution. Let (1, 1, 1), (0, 1, 2) and (3, 2, 1) be the coordinates of A, B and C.

$\Rightarrow \quad \vec{A} = \hat{i} + \hat{j} + \hat{k}, \vec{B} = \hat{j} + 2\hat{k}$ and $\vec{C} = 3\hat{i} + 2\hat{j} + \hat{k}$

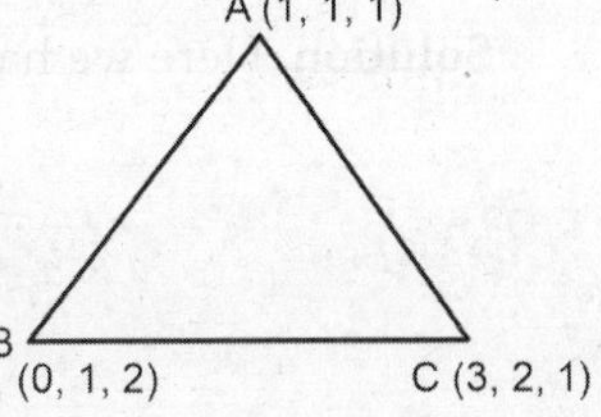

Area of $\Delta ABC = \frac{1}{2}|\overrightarrow{AB} \times \overrightarrow{AC}|$

Now, $\overrightarrow{AB}$ = Position vector of B – Position vector of A

$= (\hat{j} + 2\hat{k}) - (\hat{i} + \hat{j} + \hat{k}) = -\hat{i} + \hat{k}$

$\overrightarrow{AC}$ = Position vector of C – Position vector of A

$= \vec{C} - \vec{A} = (3\hat{i} + 2\hat{j} + \hat{k}) - (\hat{i} + \hat{j} + \hat{k}) = 2\hat{i} + \hat{j}$

$$\overrightarrow{AB} \times \overrightarrow{AC} = \begin{vmatrix} \hat{i} & \hat{j} & \hat{k} \\ -1 & 0 & 1 \\ 2 & 1 & 0 \end{vmatrix} = \hat{i}(0-1) - \hat{j}(0-2) + \hat{k}(-1-0)$$

$$\overrightarrow{AB} \times \overrightarrow{AC} = -\hat{i} + 2\hat{j} - \hat{k}$$

$$\Rightarrow \quad |\overrightarrow{AB} \times \overrightarrow{AC}| = \sqrt{(-1)^2 + (2)^2 + (-1)^2} = \sqrt{1+4+1} = \sqrt{6}$$

So, Area of $\Delta ABC = \frac{1}{2}|\overrightarrow{AB} \times \overrightarrow{AC}| = \frac{1}{2}\sqrt{6}$ Sq. units. **Ans.**

Example 45. *If* $|\vec{a}| = 10, |\vec{b}| = 2$ and $\vec{a} \cdot \vec{b} = 12$ find $|\vec{a} \times \vec{b}|$. **[SBTE 2014, 2011]**

Solution. Here we have

$$|\vec{a}| = 10, |\vec{b}| = 2, \vec{a} \cdot \vec{b} = 12.$$

$$|\vec{a} \times \vec{b}|^2 = |\vec{a}|^2 \cdot |\vec{b}|^2 - (\vec{a} \cdot \vec{b})^2$$

$$= 100 \times 4 - 144$$

$$|\vec{a} \times \vec{b}|^2 = 400 - 144 = 256$$

$$\therefore \quad |\vec{a} \times \vec{b}| = \sqrt{256} = 16$$ **Ans.**

Example 46. *If $\vec{a}$ and $\vec{b}$ two non-zero vectors such that $|\vec{a} \cdot \vec{b}| = |\vec{a} \times \vec{b}|$ then what is the angle between $\vec{a}$ and $\vec{b}$.* **[SBTE 2015]**

Solution. Here we have $|\vec{a} \cdot \vec{b}| = |\vec{a} \times \vec{b}|$

$$\Rightarrow \quad |\vec{a}||\vec{b}|\cos\theta = |\vec{a}||\vec{b}|\sin\theta$$

$$\Rightarrow \quad \cos\theta = \sin\theta$$

$$\Rightarrow \quad \tan\theta = 1 = \tan(\pi/4)$$

$$\therefore \quad \theta = \pi/4$$ **Ans.**

Example 47. *If $\vec{a} = \hat{i} - 2\hat{j} + 3\hat{k}$ and $\vec{b} = 2\hat{i} + 3\hat{j} - 5\hat{k}$, find $\vec{a} \times \vec{b}$. Also prove that $\vec{a}$ and $\vec{b}$ are perpendicular to $\vec{a} \times \vec{b}$. Find the area of parallelogram having $\vec{a}$ and $\vec{b}$ as diagonals.* **[SBTE 2014, 2013]**

Solution. Here we have $\vec{a} = \hat{i} - 2\hat{j} + 3\hat{k}, \vec{b} = 2\hat{i} + 3\hat{j} - 5\hat{k}$

$$\vec{a} \times \vec{b} = \begin{vmatrix} \hat{i} & \hat{j} & \hat{k} \\ 1 & -2 & 3 \\ 2 & 3 & -5 \end{vmatrix}$$

$$\vec{a} \times \vec{b} = \hat{i}(10-9) + \hat{j}(6+5) + \hat{k}(3+4) = \hat{i} + 11\hat{j} + 7\hat{k}$$

$$\vec{a} \cdot (\vec{a} \times \vec{b}) = (\hat{i} - 2\hat{j} + 3\hat{k}) \cdot (\hat{i} + 11\hat{j} + 7\hat{k}) = 1 - 22 + 21 = 0$$

$$\vec{b} \cdot (\vec{a} \times \vec{b}) = (2\hat{i} + 3\hat{j} - 5\hat{k}) \cdot (\hat{i} + 11\hat{j} + 7\hat{k}) = 2 + 33 - 35 = 0$$

$\therefore$ $\vec{a}$ and $\vec{b}$ are perpendicular *to* $\vec{a} \times \vec{b}$.

Area of || gm having $\vec{a}$ and $\vec{b}$ as diagonals

$$= \frac{1}{2}|\vec{a} \times \vec{b}| = \frac{1}{2}\left(\sqrt{1 + 121 + 49}\right)$$

$$= \frac{1}{2}\sqrt{171} \text{ sq. units.}$$ **Ans.**

EXERCISE 20.6

1. Find the area of a parallelogram whose sides are formed by the vectors $2\hat{i} - 3\hat{j} + \hat{k}$ and $\hat{i} + 4\hat{j} + 5\hat{k}$. **Ans.** $\sqrt{563}$ Sq. units.

2. Find the area of a parallelogram determined by $\vec{a} = \hat{i} + 2\hat{j} + 3\hat{k}$ and $\vec{b} = -3\hat{i} - 2\hat{j} + \hat{k}$. **Ans.** $6\sqrt{5}$ Sq. units.

3. Find the area of a parallelogram having diagonals $\vec{a} = 3\hat{i} + \hat{j} - 2\hat{k}$ and $\vec{b} = \hat{i} - 3\hat{j} + 4\hat{k}$. **[December 2006] Ans.** $5\sqrt{3}$ Sq. units.

4. Find the area of the triangle formed by the points whose position vectors are $3\hat{i} + \hat{j}$, $5\hat{i} + 2\hat{j} + \hat{k}, \hat{i} - 2\hat{j} + 3\hat{k}$. **Ans.** $\sqrt{29}$ Sq. units.

5. Show that perpendicular distance of the point $\vec{C}$ from the line joining $\vec{A}$ and $\vec{B}$ is $|\vec{B} \times \vec{C} + \vec{C} \times \vec{A} + \vec{A} \times \vec{B}| \div |\vec{B} - \vec{A}|$.

6. Find the area of the triangle whose vertices are (3, – 1, 2) (1, – 1, – 3), (4, – 3, 1). **[Diploma IETE, June 2005] Ans.** $\frac{\sqrt{197}}{2}$ Sq. units.

7. Show that area of the parallelogram with diagonals $\vec{a}$ and $\vec{b}$ is $\frac{1}{2}|\vec{a} \times \vec{b}|$.

[Diploma IETE June 2005]

20.25 MOMENT OF A FORCE

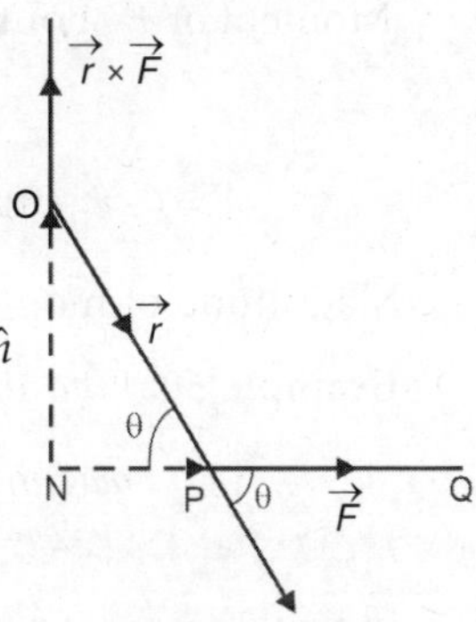

Let a force $F(\overrightarrow{PQ})$ act at a point P.

Moment of $\vec{F}$ about O

$= $ Product of force of F and perpendicular distance $ON \cdot \hat{n}$

$= (PQ)(ON)(\hat{n})$

$= (PQ)(OP)\sin\theta\,(\hat{n})$

$= \overrightarrow{OP} \times \overrightarrow{PQ} \Rightarrow \vec{M} = \vec{r} \times \vec{F}$

Example 48. *The force by* $5\hat{i} + \hat{j}$ *is acting through the point* $9\hat{i} - \hat{j} + 2\hat{k}$. *Find the moment about the point* $3\hat{i} + 2\hat{j} + \hat{k}$.

Solution. Let A be a given point about which the moment is to be obtained and let B be the point on the line of action of force. Then, the position vector of $B = 9\hat{i} - \hat{j} + 2\hat{k}$ and the position vector of $A = 3\hat{i} + 2\hat{j} + \hat{k}$.

$\therefore \quad \vec{r} = \overrightarrow{AB} = $ Position vector of B – Position vector of A

$= (9\hat{i} - \hat{j} + 2\hat{k}) - (3\hat{i} + 2\hat{j} + \hat{k}) = 6\hat{i} - 3\hat{j} + \hat{k}$

$\Rightarrow \quad \vec{r} = 6\hat{i} - 3\hat{j} + \hat{k}$ ... (1)

Let $\quad \vec{F} = 5\hat{i} + \hat{j}$ ... (2)

$\therefore$ Required moment vector $\vec{M} = \vec{r} \times \vec{F} = \begin{vmatrix} \hat{i} & \hat{j} & \hat{k} \\ 6 & -3 & 1 \\ 5 & 1 & 0 \end{vmatrix}$

$\Rightarrow \quad \vec{M} = \hat{i}(0-1) - \hat{j}(0-5) + \hat{k}(6+15)$

$= -\hat{i} + 5\hat{j} + 21\hat{k}$

Magnitude of moment $= |\vec{M}| = \sqrt{1 + 25 + 441} = \sqrt{467}$ units. **Ans.**

Example 49. *A force* $\vec{F} = 4\hat{i} + 4\hat{k}$ *acts through a point* $A(0, 2, 0)$. *Find the moment of* F *about a point* $B(4, 0, 4)$. **[S.B.T.E 2015]**

Solution. We have $\vec{F} = 4\hat{i} + 4\hat{k}$

Two points are $A(0, 2, 0)$ and $B(4, 0, 4)$

$\vec{r} = \overrightarrow{AB} = \overrightarrow{OB} - \overrightarrow{OA}$

$= (4\hat{i} + 4\hat{k}) - (2\hat{j})$

$$= 4\hat{i} - 2\hat{j} + 4\hat{k}$$

B (4, 0, 4)

$\vec{F} = 4\hat{i} + 4\hat{k}$

A (0, 2, 0)

Moment of $\vec{F}$ about the point $B = \vec{r} \times \vec{F} = \begin{vmatrix} \hat{i} & \hat{j} & \hat{k} \\ 4 & -2 & 4 \\ 4 & 0 & 4 \end{vmatrix}$

$$= (-8-0)\hat{i} - (16-16)\hat{j} + (0+8)\hat{k} = -8\hat{i} + 8\hat{k}$$

Magnitude of moment $= \sqrt{64+64} == \sqrt{128} = 8\sqrt{2}$ **Ans.**

Example 50. *Find the moment about the point M (– 2, 4, – 6) of the force represented in magnitude and position by $\overrightarrow{AB}$ where the point A and B have the coordinates (1, 2, – 3) and (3, – 4, 2) respectively.* **[Diploma IETE Dec. 2006]**

Solution. $\vec{F} = \overrightarrow{AB} = (3\hat{i} - 4\hat{j} + 2\hat{k}) - (\hat{i} + 2\hat{j} - 3\hat{k}) = 2\hat{i} - 6\hat{j} + 5\hat{k}$

$$\bar{r} = \overrightarrow{MA} = (\hat{i} + 2\hat{j} - 3\hat{k}) - (-2\hat{i} + 4\hat{j} - 6\hat{k}) = 3\hat{i} - 2\hat{j} + 3\hat{k}.$$

Moment, $\vec{M} = \bar{r} \times \vec{F}$

$\Rightarrow$ Moment of $\overrightarrow{AB}$ about $(-2, 4, -6) = \overrightarrow{MA} \times \overrightarrow{AB}$

$$= (3\hat{i} - 2\hat{j} + 3\hat{k}) \times (2\hat{i} - 6\hat{j} + 5\hat{k})$$

$$= \begin{vmatrix} \hat{i} & \hat{j} & \hat{k} \\ 3 & -2 & 3 \\ 2 & 6 & 5 \end{vmatrix} = (-10+18)\hat{i} - (15-6)\hat{j} + (-18+4)\hat{k}$$

$$= 8\hat{i} - 9\hat{j} - 14\hat{k}$$

Magnitude of the moment $= \sqrt{8^2 + (-9)^2 + (-14)^2}$

$$= \sqrt{64+81+196} = \sqrt{341}$$

$$\text{Dc's} = \frac{8}{\sqrt{341}}, \frac{-9}{\sqrt{341}}, \frac{-14}{\sqrt{341}}$$ **Ans.**

EXERCISE 20.7

1. Find the torque about the point $\hat{i} + 2\hat{j} - \hat{k}$, of a force $3\hat{i} + \hat{k}$ acting through the point $2\hat{i} - \hat{j} + 3\hat{k}$. **Ans.** $\sqrt{211}$
2. Find the torque about the point $2\hat{i} + \hat{j} - \hat{k}$ of a force represented by $4\hat{i} + \hat{k}$ acting through the point $\hat{i} - \hat{j} + 2\hat{k}$. **Ans.** $\sqrt{237}$
3. A force $\vec{F} = 4\hat{i} - 3\hat{k}$ passes through the point A whose position vector is $2\hat{i} - 2\hat{j} + 5\hat{k}$. Find the moment of $\vec{F}$ about the point B whose position vector is $\hat{i} - 3\hat{j} + \hat{k}$.

   [SBTE 2015] Ans. $-3\hat{i} + 19\hat{j} - 4\hat{k}$

4. A force is represented in magnitude and direction by the line joining the point $A(1, -2, 4)$ to the point $B(5, 2, 3)$. Find its moment about the point $(-2, 3, 5)$.

[SBTE 2015] Ans. $9\hat{i} - \hat{j} + 32\hat{k}$

5. Calculate the moment about the point (1, 1, 1) of a force of 5 kg. wt acting along the line $\overrightarrow{AB}$, where A and B are the point (2, 3, 4) and (3, 5, 6) respectively, the distances being measured in metres. **Ans.** $\frac{5}{3}(-2\hat{i} + \hat{j})$ joule

6. Force $2\hat{i} + 7\hat{j}, 2\hat{i} - 5\hat{j} + 6\hat{k}, -\hat{i} + 2\hat{j} - \hat{k}$ act on a point P whose position vector is $4\hat{i} - 3\hat{j} - 2\hat{k}$. Find the vector moment of resultant of three forces acting at P about the point Q, whose vector is $6\hat{i} + \hat{j} - 3\hat{k}$. **[SBTE 2015, 2014] Ans.** $-24\hat{i} + 13\hat{j} + 4\hat{k}$

7. Find the moment about a line through the origin having the direction of $2\hat{i} - 2\hat{j} + \hat{k}$ due to a 30 kgms, force acting at a point $(-4, 2, 5)$ in the direction of $12\hat{i} - 4\hat{j} - 3\hat{k}$.

Ans. $-\frac{760}{13}$

20.26 ANGULAR VELOCITY

Let a rigid body be rotating about the axis OA with the angular velocity ω which is a vector and its magnitude is ω radians per second and its direction is parallel to the axis of rotation OA.

Let P be any point on the body such that $\overrightarrow{OP} = \vec{r}$ and $\angle AOP = \theta$ and $AP \perp OA$. Let velocity of P be $\vec{V}$.

Let $\hat{n}$ be a unit vector perpendicular to $\vec{\omega}$ and $\vec{r}$

$$\vec{\omega} \times \vec{r} = (\omega r \sin \theta)\, \hat{n}$$

$$= (\omega\, AP)\, \hat{n}$$

$$= (\text{Speed of } P)\, \hat{n}$$

$$= \text{Velocity of } P \perp \text{ to } \vec{\omega} \text{ and } \vec{r} = \vec{V}$$

Hence $\vec{V} = \vec{\omega} \times \vec{r}$

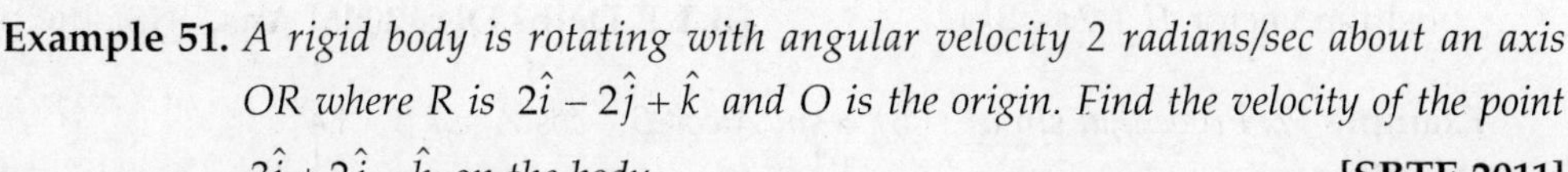

Example 51. *A rigid body is rotating with angular velocity 2 radians/sec about an axis OR where R is* $2\hat{i} - 2\hat{j} + \hat{k}$ *and O is the origin. Find the velocity of the point* $3\hat{i} + 2\hat{j} - \hat{k}$ *on the body.* **[SBTE 2011]**

Solution. Angular velocity = 2 rad/sec.

$$\overrightarrow{OR} = 2\hat{i} - 2\hat{j} + \hat{k}$$

$$\vec{\omega} = \frac{2(2\hat{i} - 2\hat{j} + \hat{k})}{\sqrt{4 + 4 + 1}}$$

$$= \frac{2}{3}(2\hat{i} - 2\hat{j} + \hat{k})$$

$$\vec{r} = \overrightarrow{OP} = 3\hat{i} + 2\hat{j} - \hat{k}$$

$$\vec{V} = \vec{\omega} \times \vec{r}$$

$$= \frac{2}{3}(2\hat{i} - 2\hat{j} - \hat{k}) \times (3\hat{i} + 2\hat{j} - \hat{k})$$

$$= \frac{2}{3}\begin{vmatrix} \hat{i} & \hat{j} & \hat{k} \\ 2 & -2 & 1 \\ 3 & 2 & -1 \end{vmatrix}$$

$$= \frac{2}{3}[(2-2)\hat{i} - (-2-3)\hat{j} + (4+6)\hat{k}]$$

$$= \frac{2}{3}[5\hat{j} + 10\hat{k}]$$

$$= \frac{10}{3}(\hat{j} + 2\hat{k}).$$ **Ans.**

EXERCISE 20.8

1. Angular velocity of a rotating rigid body about an axis of rotation is given by $\vec{\omega} = 4\hat{i} + \hat{j} - 2\hat{k}$. Find the linear velocity of a point P on the body whose position vector relative to a point on the axis of rotating is $2\hat{i} - 3\hat{j} + \hat{k}$. **Ans.** $-(5\hat{i} + 8\hat{j} + 14\hat{k})$
2. A rigid body is spinning with an angular velocity of 27 radians per second about an axis parallel to $2\hat{i} + \hat{j} - 2\hat{k}$ passing through the point $\hat{i} + 3\hat{j} - \hat{k}$. Find the velocity of the point whose position vector is $4\hat{i} + 8\hat{j} + \hat{k}$. **Ans.** $9(12\hat{i} - 10\hat{j} + 7\hat{k})$
3. A rigid body is spinning about the fixed point (3, – 1, – 2) with angular velocity of 5 radians per second, the axis of rotation being in the direction of (2, 1, – 2). Show that the velocities of the particles of the points (4, 1, 0) and (3, 2, 1) are 5 (2, – 2, 1) and 5 (3 – 2, 2) respectively.
4. A rigid body is spinning with angular velocity 20 radians per second about an axis parallel to $3\hat{i} - 4\hat{j}$ passing through the point (1, 0, 1). Find velocity of the point with position vector $(\hat{i} + 2\hat{j} - 3\hat{k})$. **[B.T.E Delhi Dec. 2006] Ans.** $8(8\hat{i} + 6\hat{j} + 3\hat{k})$

Example 52. *Prove that sin (A – B) = sin A cos B – cos A sin B*

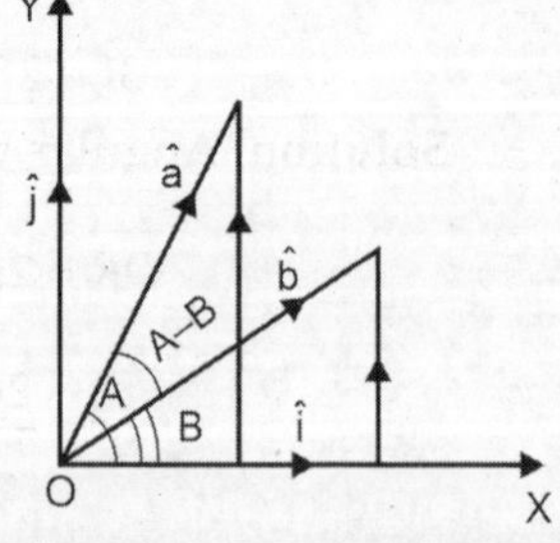

Solution. Let $\hat{a}$ and $\hat{b}$ be two unit vectors making angles A and B with x axis.

$$\hat{a} = (\cos A\hat{i} + \sin A\hat{j})$$

$$\hat{b} = (\cos B\hat{i} + \sin B\hat{j})$$

$$\hat{b} \times \hat{a} = (\cos B\hat{i} \times \sin B\hat{j}) \times (\cos A\hat{i} + \sin A\hat{j})$$

$$= \cos B \cos A(\hat{i} \times \hat{i}) + \cos B \sin A(\hat{i} \times \hat{j})$$

$$+ \sin B \cos A(\hat{j} \times \hat{i}) + \sin B \sin A(\hat{j} \times \hat{j})$$

$$= 1 \cdot 1 \cdot \sin (A - B)\hat{k} = \cos B \sin A\hat{k} - \sin B \cos A\hat{k}$$

$\therefore \quad \sin (A - B) = \sin A \cos B - \cos A \sin B.$ **Proved.**

EXERCISE 20.9

1. In a triangle ABC, prove that $\dfrac{a}{\sin A} = \dfrac{b}{\sin B} = \dfrac{c}{\sin C}$.
2. Prove that $\sin (A + B) = \sin A \cos B + \cos A \sin B$.
3. If $\vec{a} + \vec{b} + \vec{c} = 0$, prove that $\vec{a} \times \vec{b} = \vec{b} \times \vec{c} = \vec{c} \times \vec{a}$. Also interpret it geometrically.
4. If θ is the angle between the vectors $\vec{a}$ and $\vec{b}$, show that $\tan \theta = \dfrac{|\vec{a} \times \vec{b}|}{\vec{a} \cdot \vec{b}}$.
5. Show that the components of vector $\vec{b}$ parallel and perpendicular to $\vec{a}$ in the plane of $\vec{a}$ and $\vec{b}$ are (*i*) $\dfrac{(\vec{a} \cdot \vec{b})\,\vec{a}}{a^2}$ and (*ii*) $\dfrac{(\vec{a} \times \vec{b}) \times \vec{a}}{a^2}$.
6. Prove that $(\vec{a} \times \vec{b})^2 = (\vec{a} \cdot \vec{a})(\vec{b} \cdot \vec{b}) - (\vec{a} \cdot \vec{b})^2$.

20.27 SCALAR TRIPLE PRODUCT

Let $\vec{a}, \vec{b}, \vec{c}$ be three vectors then their dot product is written as $\vec{a}\,(\vec{b} \times \vec{c})$ or $[\vec{a}\ \vec{b}\ \vec{c}]$.

If $\quad \vec{a} = a_1\hat{i} + a_2\hat{j} + a_3\hat{k}$

$$\vec{b} = b_1\hat{i} + b_2\hat{j} + b_3\hat{k}$$

$$\vec{c} = c_1\hat{i} + c_2\hat{j} + c_3\hat{k}$$

$$\vec{a} \cdot (\vec{b} \times \vec{c}) = (a_1\hat{i} + a_2\hat{j} + a_3\hat{k}) \cdot [(b_1\hat{i} + b_2\hat{j} + b_3\hat{k}) \times (c_1\hat{i} + c_2\hat{j} + c_3\hat{k})]$$

$$= (a_1\hat{i} + a_2\hat{j} + a_3\hat{k}) \cdot [(b_2c_3 - b_3c_2)\hat{i} + (b_3c_1 - b_1c_3)\hat{j} + (b_1c_2 - b_2c_1)\hat{k}]$$

$$= a_1(b_2c_3 - b_3c_2) + a_2(b_3c_1 - b_1c_3) + a_3(b_1c_2 - b_2c_1).$$

$$= \begin{vmatrix} a_1 & a_2 & a_3 \\ b_1 & b_2 & b_3 \\ c_1 & c_2 & c_3 \end{vmatrix}$$

Similarly $\vec{b} \cdot (\vec{c} \times \vec{a})$ and $\vec{c} \cdot (\vec{a} \times \vec{b})$ have the same value.

$$\boxed{\vec{a} \cdot (\vec{b} \times \vec{c}) = \vec{b} \cdot (\vec{c} \times \vec{a}) = \vec{c} \cdot (\vec{a} \times \vec{b})}$$

The value of the product depends upon the cycle order of the vectors, but is independent of the position of the dot and cross. These may be interchanged.

The value of the product changes if the order is non-cycle.

Note. $\vec{a} \cdot (\vec{b} \cdot \vec{c})$, $\vec{a} \times (\vec{b} \cdot \vec{c})$ are meaningless.

20.28 GEOMETRICAL INTERPRETATION OF TRIPLE PRODUCT

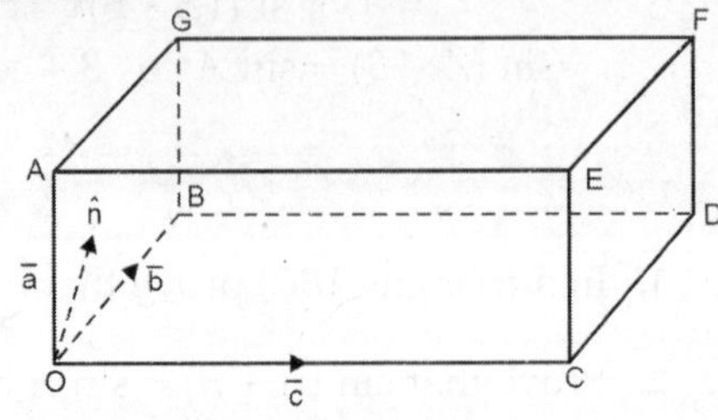

The scalar triple product $\vec{a} \cdot (\vec{b} \times \vec{c})$ represents the volume of the parallelepiped having $\vec{a}, \vec{b}, \vec{c}$ as its coterminus edges.

$$\vec{a} \cdot (\vec{b} \times \vec{c}) = a \cdot \text{Area of } \| \text{ gm } OBDC\ \hat{n}$$

= Area of ‖ gm $OBDC$ × perpendicular distance between the parallel faces $OBDC$ and $AEFG$.

$$\boxed{\vec{a} \cdot (\vec{b} \times \vec{c}) = \text{Volume of the parallelepiped}}$$

Note. (1) If $\vec{a} \cdot (\vec{b} \times \vec{c}) = 0$, then $\vec{a}, \vec{b}, \vec{c}$ are coplanar.

(2) ////////// $\frac{1}{6} \vec{a}\ \vec{b}\ \vec{c}$

Example 53. *Find the volume of parallelepiped*

If $\vec{a} = -3\hat{i} + 7\hat{j} + 5\hat{k}$, $\vec{b} = -3\hat{i} + 7\hat{j} - 3\hat{k}$ *and* $\vec{c} = 7\hat{i} - 5\hat{j} - 3\hat{k}$ *are the three coterminus edges of the parallelopiped.*

Solution. Volume = $\vec{a} \cdot (\vec{b} \times \vec{c}) = \begin{vmatrix} -3 & 7 & 5 \\ -3 & 7 & -3 \\ 7 & -5 & -3 \end{vmatrix}$

$= -3(-21 - 15) - 7(9 + 21) + 5(15 - 49)$

$= 108 - 210 - 170 = -272.$ cubic units. **Ans.**

EXERCISE 20.10

1. Find the volume of the parallelepiped with adjacent sides $\overrightarrow{OA} = 3\hat{i} - \hat{j}$, $\overrightarrow{OB} = \hat{j} + 2\hat{k}$, $\overrightarrow{OC} = \hat{i} + 5\hat{j} + 4\hat{k}$ extending from the origin of co-ordinates O. **Ans.** 20
2. Find the volume of the tetrahedron whose vertices are the points $A(2, -1, -3)$; $B(4, 1, 3)$; $C(3, 2, -1)$ and $D(1, 4, 2)$. **Ans.** $7\frac{1}{3}$
3. Find the volume of the tetrahedron whose vertices are the points $A(2, 1, 8)$; $B(3, 2, 9)$; $C(2, 1, 4)$ and $D(3, 3, 10)$. **Ans.** $\frac{2}{3}$
4. Prove that $[\vec{a} + \vec{b}, \vec{b} + \vec{c}, \vec{c} + \vec{a}] = 2[\vec{a}\ \vec{b}\ \vec{c}]$.

Choose the correct alternative:

5. Let $\vec{a} = (1, 2, 0), \vec{b} = (-3, 2, 0), \vec{c} = (2, 3, 4)$. Then $\vec{a} \cdot (\vec{b} \cdot \vec{c})$ equals

(*a*) 33 (*b*) 30 (*c*) 31 (*d*) 32

[Diploma IETE June 2006] Ans. (*d*)

20.29 CONDITION FOR COPLANARITY OF VECTORS

When scalar triple product of any three Vectors is zero, then the vectors are coplanar. Also if any three vectors are linearly dependent then they are coplanar vectors. Therefore multiple vectors are coplanar when only two vectors are lineary independent. Vectors among them.

$\therefore \quad \vec{a} \cdot \left(\vec{b} \times \vec{c}\right) = 0$, then $\vec{a}, \vec{b}, \vec{c}$ are coplanar.

Example 54. *Determine* λ *such that* $\vec{a} = \hat{i} + \hat{j} + \hat{k}, \vec{b} = 2\hat{i} - 4\hat{k}, \vec{c} = \hat{i} + \lambda\hat{j} + 3\hat{k}$ *are coplanar.*

[SBTE 2015]

Solution. Here we have $\vec{a} = \hat{i} + \hat{j} + \hat{k}, \vec{b} = 2\hat{i} - 4\hat{k}, \vec{c} = \hat{i} + \lambda\hat{j} + 3\hat{k}$

If $\vec{a}, \vec{b}$ and $\vec{c}$ are coplanar then $\vec{a} \cdot (\vec{b} \times \vec{c}) = 0$

$$(\hat{i} + \hat{j} + \hat{k}) \cdot (4\lambda\hat{i} - 10\hat{j} + 2\lambda\hat{k}) = 0$$

$$\Rightarrow \quad 4\lambda - 10 + 2\lambda = 0 \quad \Rightarrow \quad 6\lambda = 10$$

$$\therefore \quad \lambda = 5/3$$ **Ans.**

Example 55. *Prove that the four points*

$$4\hat{i} + 5\hat{j} + \hat{k}, -(\hat{j} + \hat{k}), 3\hat{i} + 9\hat{j} + 4\hat{k}, 4(-\hat{i} + \hat{j} + \hat{k})$$

are coplanar.

Solution. Let the points *A, B, C, D* be represented by the vectors

$$4\hat{i} + 5\hat{j} + \hat{k}, -(\hat{j} + \hat{k}), 3\hat{i} + 9\hat{j} + 4\hat{k}, 4(-\hat{i} + \hat{j} + \hat{k})$$

$$\overrightarrow{AB} = -(\hat{j} + \hat{k}) - (4\hat{i} + 5\hat{j} + \hat{k}) = -4\hat{i} - 6\hat{j} - 2\hat{k}$$

$$\overrightarrow{AC} = (3\hat{i} + 9\hat{j} + 4\hat{k}) - (4\hat{i} + 5\hat{j} + \hat{k}) = -\hat{i} + 4\hat{j} + 3\hat{k}$$

$$\overrightarrow{AD} = 4(-\hat{i} + \hat{j} + \hat{k}) - (4\hat{i} + 5\hat{j} + \hat{k}) = -8\hat{i} - \hat{j} + 3\hat{k}$$

$$\overrightarrow{AB} \cdot (\overrightarrow{AC} \times \overrightarrow{AD}) = \begin{vmatrix} -4 & -6 & -2 \\ -1 & 4 & 3 \\ -8 & -1 & 3 \end{vmatrix}$$

$$= -4(12 + 3) + 6(-3 + 24) - 2(1 + 32) = -60 + 126 - 66 = 0$$

Hence, the given four points are coplanar. **Proved.**

Example 56. *If four points whose position vectors are* $\vec{a}, \vec{b}, \vec{c}, \vec{d}$, *are coplanar, show that*

$$[\vec{a}\ \vec{b}\ \vec{c}] = [\vec{a}\ \vec{b}\ \vec{d}] + [\vec{a}\ \vec{d}\ \vec{c}] + [\vec{d}\ \vec{b}\ \vec{c}]$$

Solution. Let *A, B, C, D* be four points whose position vectors are $\vec{a}, \vec{b}, \vec{c}, \vec{d}$.

$$\overrightarrow{AD} = \vec{d} - \vec{a}$$

$$\overrightarrow{BD} = \vec{d} - \vec{b}$$

$$\overrightarrow{CD} = \vec{d} - \vec{c}$$

If $\overrightarrow{AD}, \overrightarrow{BD}, \overrightarrow{CD}$ are coplanar, then

$$\overrightarrow{AD} \cdot (\overrightarrow{BD} \times \overrightarrow{CD}) = 0$$

$$(\vec{d} - \vec{a})[(\vec{d} - \vec{b}) \times (\vec{d} - \vec{c})] = 0$$

$$(\vec{d} - \vec{a}) \cdot [\vec{d} \times \vec{d} - \vec{d} \times \vec{c} - \vec{b} \times \vec{d} + \vec{b} \times \vec{c}] = 0$$

$$(\vec{d} - \vec{a})[- \vec{d} \times \vec{c} - \vec{b} \times \vec{d} + \vec{b} \times \vec{c}] = 0$$

$$-\vec{d} \cdot (\vec{d} \times \vec{c}) - \vec{d} \cdot (\vec{b} \times \vec{d}) + \vec{d} \cdot (\vec{b} \times \vec{c}) + \vec{a} \cdot (\vec{d} \times \vec{c}) + \vec{a} \cdot (\vec{b} \times \vec{d}) - \vec{a} \cdot (\vec{b} \times \vec{c}) = 0$$

$$\Rightarrow \quad -0 + 0 + [\vec{d}\ \vec{b}\ \vec{c}] + [\vec{a}\ \vec{d}\ \vec{c}] + [\vec{a}\ \vec{b}\ \vec{d}] - [\vec{a}\ \vec{b}\ \vec{c}] = 0$$

$$\Rightarrow \quad [\vec{a}\ \vec{b}\ \vec{c}] = [\vec{a}\ \vec{b}\ \vec{d}] + [\vec{a}\ \vec{d}\ \vec{c}] + [\vec{d}\ \vec{b}\ \vec{c}]$$ **Proved.**

20.30 VECTOR PRODUCT OF THREE VECTORS

Let $\vec{a}, \vec{b}, \vec{c}$ be three vectors then their vector product is written as $\vec{a} \times (\vec{b} \times \vec{c})$

Let $\vec{a} = a_1\hat{i} + a_2\hat{j} + a_3\hat{k}$

$$\vec{b} = b_1\hat{i} + b_2\hat{j} + b_3\hat{k}$$

$$\vec{c} = c_1\hat{i} + c_2\hat{j} + c_3\hat{k}$$

$$\vec{a} \times (\vec{b} \times \vec{c}) = (a_1\hat{i} + a_2\hat{j} + a_3\hat{k}) \times [(b_1\hat{i} + b_2\hat{j} + b_3\hat{k}) \times (c_1\hat{i} + c_2\hat{j} + c_3\hat{k})]$$

$$= (a_1\hat{i} + a_2\hat{j} + a_3\hat{k}) \times [(b_2c_3 - b_3c_2)\hat{i} + (b_3c_1 - b_1c_3)\hat{j} + (b_1c_2 - b_2c_1)\hat{k}]$$

$$= [a_2(b_1c_2 - b_2c_1) - a_3(b_3c_1 - b_1c_3)]\hat{i} + [a_3(b_2c_3 - b_3c_2) - a_1(b_1c_2 - b_2c_1)]\hat{j}$$
$$+ [a_1(b_3c_1 - b_1c_3) - a_2(b_2c_3 - b_3c_2)]\hat{k}$$

$$= (a_1c_1 + a_2c_2 + a_3c_3)(b_1\hat{i} + b_2\hat{j} + b_3\hat{k}) - (a_1b_1 + a_2b_2 + a_3b_3)(c_1\hat{i} + c_2\hat{j} + c_3\hat{k})$$

$$= (\vec{a} \cdot \vec{c})\vec{b} - (\vec{a} \cdot \vec{b})\vec{c}.$$

$$\boxed{\therefore\ \vec{a} \times (\vec{b} \times \vec{c}) = (\vec{a} \cdot \vec{c})\vec{b} - (\vec{a} \cdot \vec{b})\vec{c}}$$

Example 57. *Let* $\vec{a} = \hat{i} + \hat{j} - \hat{k}$, $\vec{b} = \hat{i} - \hat{j} + \hat{k}$, $\vec{c} = \hat{i} - \hat{j} - \hat{k}$

Then find the vector $\vec{a} \times (\vec{b} \times \vec{c})$.

Solution. $\vec{a} \times (\vec{b} \times \vec{c}) = (\vec{a} \cdot \vec{c})\,\vec{b} - (\vec{a} \cdot \vec{b})\,\vec{c}$

$$= [(\hat{i} + \hat{j} - \hat{k}) \cdot (\hat{i} - \hat{j} - \hat{k})]\,(\hat{i} - \hat{j} + \hat{k}) - [(\hat{i} + \hat{j} - \hat{k}) \cdot (\hat{i} - \hat{j} + \hat{k})]\,(\hat{i} - \hat{j} - \hat{k})$$

$$= (1+1+1)\,(\hat{i} - \hat{j} + \hat{k}) - (1-1-1)\,(\hat{i} - \hat{j} - \hat{k})$$

$$= (\hat{i} - \hat{j} + \hat{k}) + (\hat{i} - \hat{j} - \hat{k})$$

$$= 2\hat{i} - 2\hat{j}$$ **Ans.**

EXERCISE 20.11

1. Determine λ such that $\vec{a} = \hat{i} + \hat{j} + \hat{k}$, $\vec{b} = 2\hat{i} - 4\hat{k}$, and $\vec{c} = \hat{i} + \lambda\hat{j} + 3\hat{k}$ are coplanar.
Ans. $\lambda = \frac{5}{3}$

2. Show that the four points $-6\hat{i} + 3\hat{j} + 2\hat{k}$, $3\hat{i} - 2\hat{j} + 4\hat{k}$, $5\hat{i} + 7\hat{j} + 3\hat{k}$ and $-13\hat{i} + 17\hat{j} - \hat{k}$ are coplanar.

3. Find the constant a such that the vectors $2\hat{i} - 3\hat{j} + \hat{k}$, $\hat{i} + 2\hat{j} - 3\hat{k}$, and $3\hat{i} + a\hat{j} + 5\hat{k}$, are coplanar. **Ans.** $a = -8$

4. Show that $\vec{a} \times (\vec{b} + \vec{c}) + \vec{b} \times (\vec{c} \times \vec{a}) + \vec{c} \times (\vec{a} \times \vec{b}) = 0$

5. Show that $\hat{i} \times (\vec{a} \times \vec{i}) + \hat{j} \times (\vec{a} \times j) + k \times (\vec{a} \times k) = 2\,\vec{a}$

6. Show that $\vec{a} \cdot (\vec{a} \times \vec{b}) = 0$

7. Find the value of $(\vec{a} \times \vec{b}) \times \vec{c}$ if
$\vec{a} = 3\hat{i} - \hat{j} + 2\hat{k}$, $\vec{b} = 2\hat{i} + \hat{j} - \hat{k}$, $\vec{c} = \hat{i} - 2\hat{j} + 2\hat{k}$ **[Diploma IETE Dec. 2006]**

Unit IV: COORDINATE GEOMETRY

Point and Distance

21.1 INTRODUCTION

The combination of algebra and geometry is known as analytic or coordinate geometry. It is the branch of mathematics, which treats geometry algebraically, *i.e.* in which geometric figures are studied by means of equations. Descartes is known as the founder of coordinate geometry.

Coordinate Geometry

It is a branch of geometry which sets up a definite correspondence between the position of a point in a plane and a pair of algebraic numbers, called coordinates.

Cartesian Coordinates (Rectangular Coordinates)

In cartesian coordinates the position of a point P is determined by knowing the distances from two perpendicular lines passing through fixed point. Let O be the fixed point called the orgin and XOX' and YOY', the two perpendicular lines through O, called cartesian or rectangular coordinates axes.

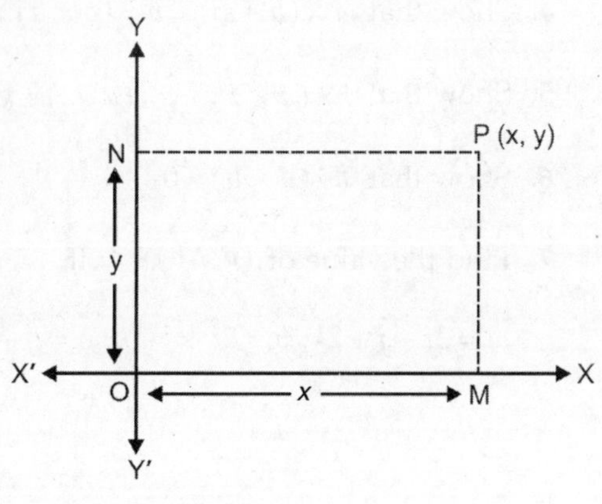

Draw PM and PN perpendiculars on OX and OY respectively. OM (or NP) is called the x-coordinate or abscissa of the point P, ON (or MP) is called the y-coordinate or the ordinate of the point P. if $OM = x$, $ON = y$ the coordinates of the point P are (x, y). These coordinates are called the cartesian coordinates of point P.

Notation: OM and ON are denoted by x and y respectively.

Axes of Coordinates

In the above figure, OX and OY are called as x-axis and y-axis respectively.

Orgin

It is the point O of intersection of the axes of coordinates.

Abscissa and Ordinate

OM is the Abscissa and ON is the ordinate.

Quadrant

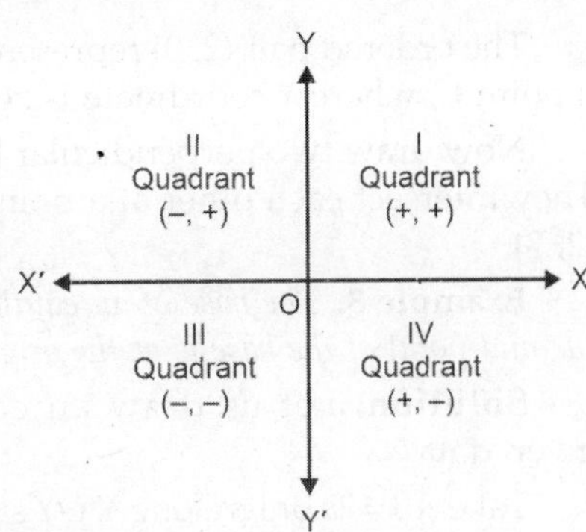

Let $X'OX$ and $Y'OY$ the coordinate axes divide the Euclidean plane, into four regions called the quadrants. The regions XOY, $X'OY$, $X'OY'$ and $Y'OX$ are known as the first, second, third and fourth quadrant respectively.

21.2 RULES FOR SIGNS OF COORDINATES

(*i*) In the first quadrant, both coordinates, *i.e.* abscissa and ordinate of a point are positive.

$\Rightarrow$ I quadrant : $x > 0, y > 0$

(*ii*) In the second quadrant, for a point, abscissa is negative and ordinate is positive

$\Rightarrow$ II quadrant : $x < 0, y > 0$

(*iii*) In the third quadrant, for a point, both abscissa and ordinate are negative.

$\Rightarrow$ III quadrant : $x < 0, y < 0$

(*iv*) In the fourth quadrant, for a point, the abscissa is positive and the ordinate is negative.

$\Rightarrow$ IV quadrant : $x > 0, y < 0$

Example 1. *In which quadrant do the following points lie ?*

(*i*) (6, 3) (*ii*) (–5, 4)
(*iii*) (–8, –5) (*iv*) (3, –6)

Also plot them.

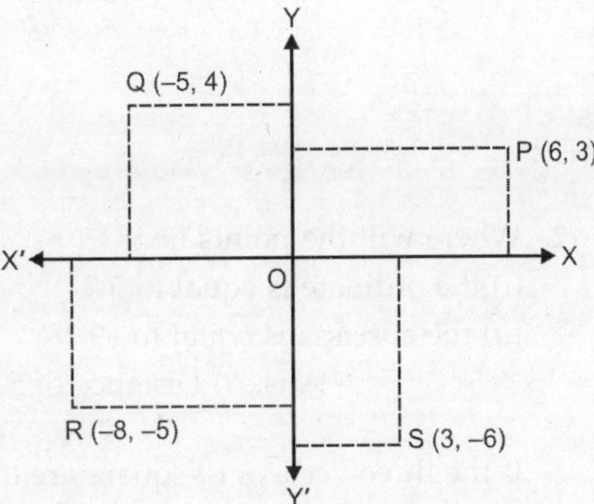

Solution.

(*i*) The abscissa and ordinate of the point (6, 3) both are positive. So it lies in the first quadrant.

(*ii*) The point (–5, 4) has negative abscissa and positive ordinate. So it lies in the second quadrant.

(*iii*) The point (–8, –5) has negative abscissa and negative ordinate. So it lies in the third quadrant.

(*iv*) The point (3, –6) has positive abscissa and negative ordinate. so it lies in the fourth quadrant.

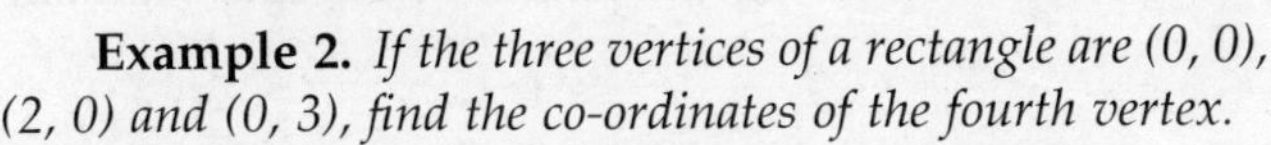

Example 2. *If the three vertices of a rectangle are (0, 0), (2, 0) and (0, 3), find the co-ordinates of the fourth vertex.*

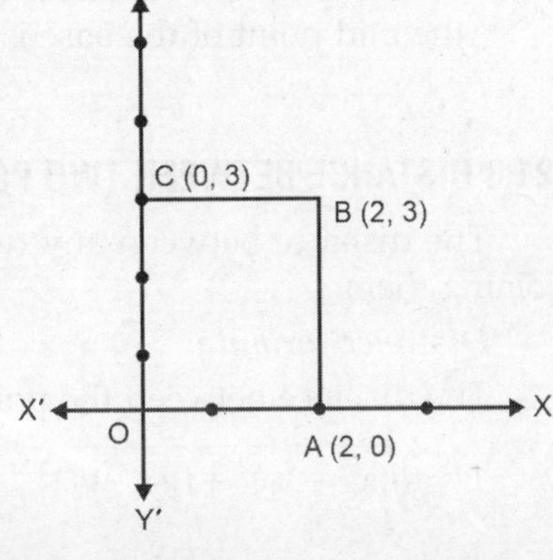

Solution. We have to plot the point according to the given co-ordinates. The ordered pair (0, 0) represents the origin O.

The ordered pair (2, 0) represents a point A on x-axis. The ordered pair (0, 3) represents a point C, where x-coordinate is zero and y coordinate = 3.

Now draw two perpendicular lines AB and CB passing through A and C respectively. They intersect each other at a point B and B is the required point whose coordinates are (2, 3). **Ans.**

Example 3. *The base of an equilateral triangle with side 2a lies along the y-axis, such that the mid-point of the base is at the origin. Find the vertices of the triangle.*

Solution. Let us draw an equilateral triangle with given data.

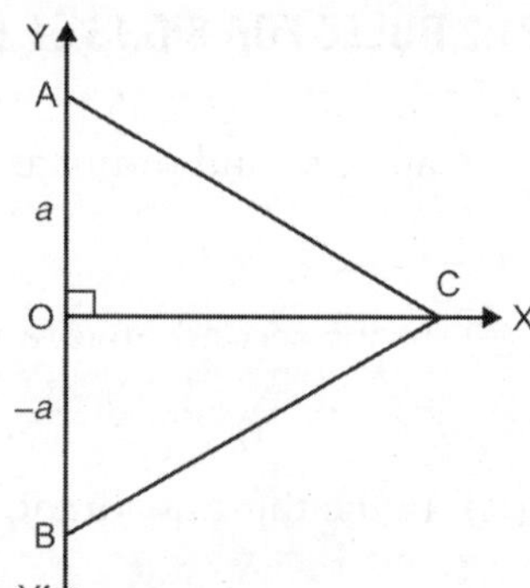

Take $AB = 2a$ units along $Y'OY$ such that $OA = OB = a$ units.

To find the height of ΔABC

$$CO^2 = AC^2 - OA^2$$

$$= (2a)^2 - a^2 = 4a^2 - a^2 = 3a^2$$

$$\therefore \quad OC = \sqrt{3}a$$

Hence, the vertices of the triangle ABC are

$A(0, a), B(0, -a)$ and $C(\sqrt{3}\,a, 0)$. **Ans.**

EXERCISE 21.1

1. Draw the quadrilateral whose vertices are (2, 3), (2, –3), (–2, –3) and (–2, 3).

Ans.

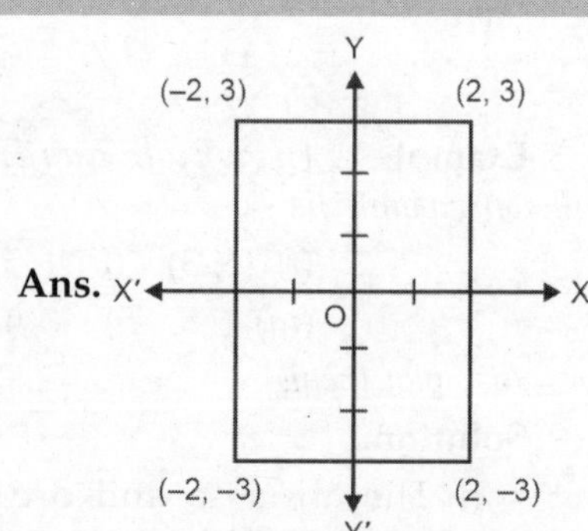

2. Where will the points lie if
   (*i*) the ordinate is equal to 5 ?
   (*ii*) the abscissa is equal to –9 ?

   Ans. (*i*) Distance of 5 units above the origin. (*ii*) Lies on a line parallel to the y-axis at a distance of 9 units to the left of the origin.

3. If the three vertices of square are (0, 0), (4, 0) and (0, 4), find the co-ordinates of the fourth vertex. **Ans.** (4, 4)

4. The base of an equilateral triangle with each side 4a lies along the x-axis such that the mid point of the base is at the origin. Find the vertices of the triangle.

   Ans. $(-2a, 0), (2a, 0), (0, 2\sqrt{3}\,a)$

21.3 DISTANCE BETWEEN TWO POINTS

The distance between any two points in the plane is the length of the line segment joining them.

***Distance formula*:**

The distance between the point $P(x_1, y_1)$ and $Q(x_2, y_2)$ is given by the formula

$$D = \sqrt{(x_2 - x_1)^2 + (y_2 - y_1)^2}$$

Example 4. *Find the distance between points A and B in each of the following:*

(*i*) $A(5, -12), B(9, -9)$ (*ii*) $A(6, -4), B(3, 0)$

(*iii*) $A(0, 0), B(-5, 12)$ (*iv*) $A(a \cos \alpha, a \sin \alpha)$, $B(a \cos \beta, a \sin \beta)$

Solution. (*i*) Here, $x_1 = 5, \; y_1 = -12$ *and* $x_2 = 9, \; y_2 = -9$

$$\therefore \quad AB = \sqrt{(9-5)^2 + (-9+12)^2} = \sqrt{4^2 + 3^2} = \sqrt{16+9} = \sqrt{25} = 5$$

(*ii*) Here, $x_1 = 6, y_1 = -4$ *and* $x_2 = 3, y_2 = 0$

$$\therefore \quad AB = \sqrt{(3-6)^2 + (0+4)^2} = \sqrt{9+16} = \sqrt{25} = 5$$

(*iii*) Here, $x_1 = 0, y_1 = 0$ *and* $x_2 = -5, y_2 = 12$

$$\therefore \quad AB = \sqrt{(-5-0)^2 + (12-0)^2} = \sqrt{(-5)^2 + (12)^2} = \sqrt{25+144} = \sqrt{169} = 13$$

(*iv*) Here, $x_1 = a \cos \alpha, y_1 = a \sin \alpha$ *and* $x_2 = a \cos \beta, y_2 = a \sin \beta$

$$\therefore \quad AB = \sqrt{(a \cos \beta - a \cos \alpha)^2 + (a \sin \beta - a \sin \alpha)^2}$$

$$= \sqrt{a^2 (\cos \beta - \cos \alpha)^2 + a^2 (\sin \beta - \sin \alpha)^2}$$

$$= a\sqrt{(\cos \beta - \cos \alpha)^2 + (\sin \beta - \sin \alpha)^2}$$

$$= a\sqrt{\cos^2 \beta + \cos^2 \alpha + \sin^2 \beta + \sin^2 \alpha - 2 \cos \alpha \cos \beta - 2 \sin \alpha \sin \beta}$$

$$= a\sqrt{(\cos^2 \beta + \sin^2 \beta) + (\cos^2 \alpha + \sin^2 \alpha) - 2 (\cos \alpha \cos \beta + \sin \alpha \sin \beta)}$$

$$= a\sqrt{1 + 1 - 2 \cos(\alpha - \beta)} = a\sqrt{2\{1 - \cos(\alpha - \beta)\}}$$

$$= a\sqrt{2.2 \sin^2\left(\frac{\alpha - \beta}{2}\right)} = 2a \sin\left(\frac{\alpha - \beta}{2}\right)$$ **Ans.**

Example 5. *Using distance formula show that the points (–3, –5), (1, –6) and (–7, –4) are collinear.*

Solution. Let the given points (–3, –5), (1, – 6) and (–7, – 4) be denoted by *A*, *B* and *C* respectively.

Now, $$AB = \sqrt{\{1 - (-3)\}^2 + \{(-6) - (-5)\}^2}$$

$$= \sqrt{(1+3)^2 + (-6+5)^2} = \sqrt{4^2 + (-1)^2} = \sqrt{16+1} = \sqrt{17}$$

$$BC = \sqrt{\{-7 - 1\}^2 + \{-4 - (-6)\}^2} = \sqrt{(-8)^2 + (2)^2} = \sqrt{64+4}$$

$$= \sqrt{68} = 2\sqrt{17}$$

$$CA = \sqrt{\{-7 - (-3)\}^2 + \{-4 - (-5)\}^2} = \sqrt{(-4)^2 + 1^2} = \sqrt{16+1} = \sqrt{17}$$

It is clear that $BC = CA + AB$

Therefore, the points *A*, *B* and *C* are collinear. **Proved.**

Example 6. *Prove that the points (–2, 2), (8, –2) and (–4, –3) are the vertices of a right angled triangle.*

Solution. Let the points (–2, 2), (8, –2) and (–4, –3) represent *A*, *B* and *C* respectively.

Now, $BC = \sqrt{(-4-8)^2 + (-3+2)^2} = \sqrt{144+1} = \sqrt{145}$

$AB = \sqrt{(8+2)^2 + (-2-2)^2} = \sqrt{100+16} = \sqrt{116}$

$CA = \sqrt{(-4+2)^2 + (-3-2)^2} = \sqrt{4+25} = \sqrt{29}$

Therefore, $BC^2 = 145$, $AB^2 = 116$, and $CA^2 = 29$

We observe that $BC^2 = CA^2 + AB^2$

Hence, by converse of Pythagoras theorem, ΔABC is a right angled triangle with right angle at A. **Proved.**

Example 7. *Show that the points (a, a), (–a, –a) and* $(-\sqrt{3}\,a, \sqrt{3}\,a)$ *are the vertices of an equilateral triangle.*

Solution. Let $A(a, a)$, $B(-a, -a)$ and $C(-\sqrt{3}\,a, \sqrt{3}\,a)$ be the given points. Then,

$$AB = \sqrt{(-a-a)^2 + (-a-a)^2} = \sqrt{4a^2 + 4a^2} = 2\sqrt{2}\,a$$

$$BC = \sqrt{(-\sqrt{3}\,a + a)^2 + (\sqrt{3}a + a)^2} = \sqrt{a^2\,(1-\sqrt{3})^2 + a^2\,(\sqrt{3}+1)^2}$$

$$= a\sqrt{(1-\sqrt{3})^2 + (1+\sqrt{3})^2} = a\sqrt{1+3-2\sqrt{3}+1+3+2\sqrt{3}} = a\sqrt{8} = 2\sqrt{2}a$$

$$AC = \sqrt{(-\sqrt{3}a - a)^2 + (\sqrt{3}a - a)^2} = \sqrt{a^2(\sqrt{3}+1)^2 + a^2\,(\sqrt{3}-1)^2}$$

$$= a\sqrt{(\sqrt{3}+1)^2 + (\sqrt{3}-1)^2} = a\sqrt{3+1+2\sqrt{3}+3+1-2\sqrt{3}} = a\sqrt{8} = 2\sqrt{2}a$$

Clearly, $AB = BC = CA$. Hence, the triangle ABC formed by the given points is an equilateral triangle. **Proved.**

Example 8. *Show that points A(1, 0), B(5, 3), C(2, 7) and D(–2, 4) are vertices of a square.*

Solution. We know that for a square, we have to prove that all the four sides are equal and two diagonals are also equal.

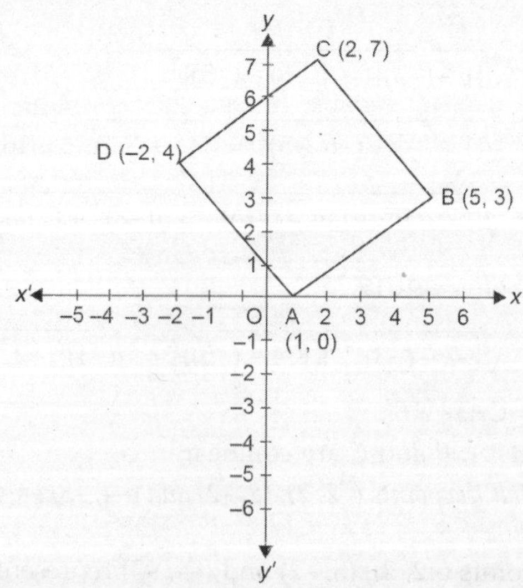

Now $AB = \sqrt{(5-1)^2 + (3-0)^2} = \sqrt{4^2 + 3^2} = \sqrt{16+9} = \sqrt{25} = 5$

$BC = \sqrt{(2-5)^2 + (7-3)^2} = \sqrt{3^2 + 4^2} = \sqrt{9+16} = \sqrt{25} = 5$

$CD = \sqrt{(-2-2)^2 + (4-7)^2} = \sqrt{(-4)^2 + (-3)^2} = \sqrt{16+9} = \sqrt{25} = 5$

$DA = \sqrt{\{1-(-2)\}^2 + (0-4)^2} = \sqrt{(3)^2 + 4^2} = \sqrt{9+16} = \sqrt{25} = 5$

Here, sides $AB = BC = CD = DA$ *i.e.* all are equal ...(1)

Again, $AC = \sqrt{(2-1)^2 + (7-0)^2} = \sqrt{1^2 + 7^2} = \sqrt{1+49} = \sqrt{50} = 5\sqrt{2}$ ...(2)

$BD = \sqrt{(-2-5)^2 + (4-3)^2} = \sqrt{(-7)^2 + (1)^2} = \sqrt{49+1} = \sqrt{50} = 5\sqrt{2}$ ...(3)

From (1), (2) and (3), all sides are equal and both diagonals are also equal. Hence, the given quadrilateral is a square. **Proved.**

Example 9. *Find the point on x-axis which is equidistant from (3, 2) and (–5, –2).*

Solution. Let the point $P(x, 0)$ on the x-axis be equidistant from $A(3, 2)$ and $B(-5, -2)$. Then, $PA = PB$

$$\Rightarrow \quad \sqrt{(x-3)^2 + (0-2)^2} = \sqrt{(x+5)^2 + (0+2)^2} \quad ...(1)$$

On squaring (1), we get

$$\Rightarrow \quad x^2 + 9 - 6x + 4 = x^2 + 25 + 10x + 4$$

$$\Rightarrow \quad -16x = 16 \quad \Rightarrow \quad x = -1$$

Hence, the required point on the x-axis is (–1, 0) **Ans.**

Example 10. *Find the coordinates of the circumcentre of the triangle, whose vertices are (8, 6) (8, –2) and (2, –2). Also, find its circumradius.*

Solution. Let $A(8, 6)$, $B(8, -2)$ and $C(2, -2)$ be the vertices of the given triangle and let $P(x, y)$ be the circumcentre of this triangle, then, $PA^2 = PB^2 = PC^2$

Now, $PA^2 = PB^2 \quad \Rightarrow \quad (x-8)^2 + (y-6)^2 = (x-8)^2 + (y+2)^2$

$\Rightarrow \quad (y-6)^2 = (y+2)^2$

$\Rightarrow \quad y^2 - 12y + 36 = y^2 + 4y + 4$

$\Rightarrow \quad 16y = 32 \quad \Rightarrow \quad y = 2$

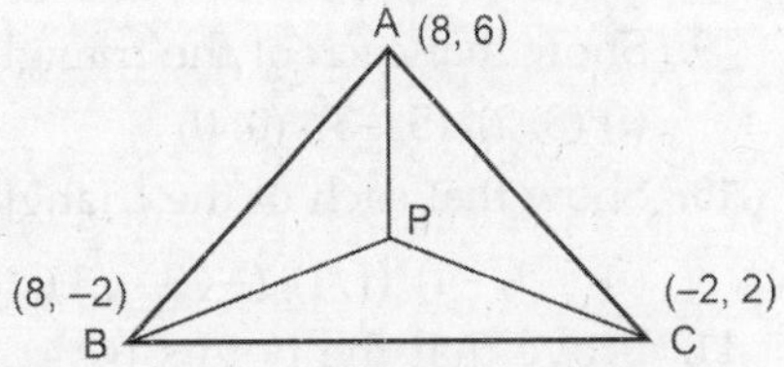

$\Rightarrow \quad PB^2 = PC^2$

$\Rightarrow \quad (x-8)^2 + (y+2)^2 = (x-2)^2 + (y+2)^2$

$\Rightarrow \quad (x-8)^2 = (x-2)^2 \quad \Rightarrow \quad x^2 - 16x + 64 = x^2 - 4x + 4$

$\Rightarrow \quad 12x = 60 \quad \Rightarrow \quad x = 5$

A (8, 6)
P
(8, –2) B
(–2, 2) C

So, the co-ordinates of the circumcentre P are (5, 2).

Also, circum-radius $= PA = PB = PC = \sqrt{(5-8)^2 + (2-6)^2}$

$= \sqrt{(-3)^2 + (-4)^2} = \sqrt{9+16} = \sqrt{25} = 5$ **Ans.**

Example 11. *If D is the mid-point of the side BC of a triangle ABC, prove that* $AB^2 + AC^2 = 2(AD^2 + DC^2)$

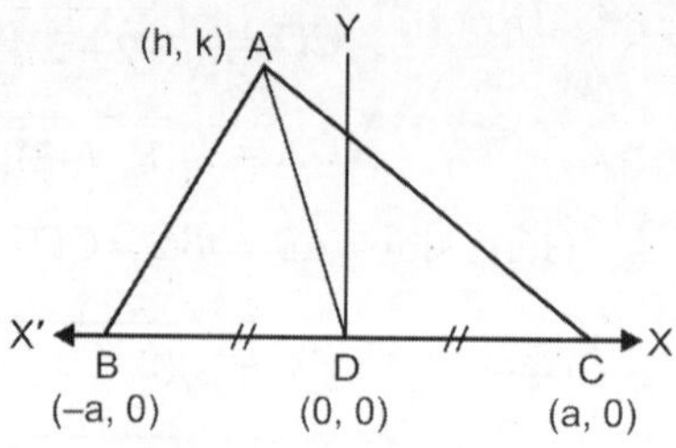

Solution. Let D be origin, DC is x-axis and $DY \perp DC$ is y-axis.

Again let $BC = 2a$, then the coordinates of B and C are $(-a, 0)$ and $(a, 0)$ respectively.

Let the coordinates of A be (h, k)

then $AB^2 + AC^2 = [(h+a)^2 + (k-0)^2] + [(h-a)^2 + (k-0)^2]$

$= 2h^2 + 2a^2 + 2k^2$

$= 2(h^2 + k^2 + a^2)$ ...(1)

Also, $2(AD^2 + DC^2) = 2([(h-0)^2 + (k-0)^2] + a^2) = 2(h^2 + k^2 + a^2)$ ...(2)

From (1) and (2), we get

$AB^2 + AC^2 = 2(AD^2 + DC^2)$ **Proved.**

EXERCISE 21.2

Find the distance between points A and B in the following.

1. $A(-3, 4), B(3, 0)$ **Ans.** $2\sqrt{13}$
2. $A(a+b, b+c), B(a-b, c-b)$ **Ans.** $2\sqrt{2}b$
3. $A(\cos\theta, \sin\theta), B(\sin\theta, \cos\theta)$ **Ans.** $\sqrt{2}\sqrt{1-\sin 2\theta}$
4. $A(a, 0), B(0, b)$ **Ans.** $\sqrt{a^2+b^2}$
5. A man walks from A (2, 4) to B (5, 4) to the east, then turning to the north he reaches C (5, 8). How far is he away from the point A? **Ans.** 5 units
6. Using distance formula, show that the points (3, 5), (1, 1) and (–2, –5) are collinear.
7. Check whether the points (–5, 7), (2, 5) and (1, –1) are all equidistant from the point (–2, 3). **Ans.** No
8. Show that the points (2, 3), (–4, –6) and $\left(1, \frac{3}{2}\right)$ do not form a triangle.
9. Show that each of the triangles whose vertices are given below is isosceles:
   (*i*) (8, 2), (5, –3), (0, 0) (*ii*) (0, 6), (–5, 3), (3, 1).
10. Show that each of the triangles whose vertices are given below is equilateral:
    (*i*) $(-1, -1), (1, 1), (-\sqrt{3}, \sqrt{3})$ (*ii*) $(2a, 4a), (2a, 6a), (2a+\sqrt{3}a, 5a)$.
11. Prove that the points (4, 4), (3, 5) and (–1, –1) are the vertices of a right angled triangle.
12. Find the radius of the circle that has its centre at (0, –4), and pass through $(\sqrt{13}, 2)$. **Ans.** 7
13. Using distance formula, show that (3, 3) is the centre of a circle passing through the points (6, 2), (0, 4) and (4, 6). Find the radius of the circle **Ans.** $\sqrt{10}$

14. Show that the quadrilaterals whose vertices are given below are parallelograms:

(*i*) (–2, –1), (1, 0), (4, 3), (1, 2) (*ii*) (–1, 0), (0, 3), (1, 3), (0, 0)

15. Show that the quadrilaterals whose vertices are given below are rectangles:

(*i*) (3, 2), (11, 8), (8, 12), (0, 6) (*ii*) (–2, –3), (6, 3), (3, 7), (–5, 1)

16. Show that the quadrilaterals whose vertices given below are rhombuses:

(*i*) (1, –2), (2, 3), (–3, 2), (–4, –3) (*ii*) (3, –4), (4, 2), (5, –4), (4, –10)

17. Show that the quadrilaterals whose vertices given below are squares:

(*i*) (2, 6), (5, 1), (0, –2), (–3, 3) (*ii*) (–2, –4), (1, –1), (–2, 2), (–5, –1)

21.4 SECTION FORMULAE

Formula for Internal Division

A point *C* divides the straight line joining two points $A(x_1, y_1)$ and $B(x_2, y_2)$ internally in the ratio $m_1 : m_2$

Thus, the coordinates of *C* are $\boxed{x = \dfrac{m_1x_2 + m_2x_1}{m_1 + m_2}, y = \dfrac{m_1y_2 + m_2y_1}{m_1 + m_2}}$

Corollary : Coordinates of the mid-point:

If *C* is the mid-point of *AB*, then it will divide *AB* in the ratio of 1 : 1, then the coordinates of *C* are

$$\frac{x_1 + x_2}{2}, \frac{y_1 + y_2}{2}$$

Aid to Memory:

The student should carefully note that in the ratio formula m_1 which corresponds to the segment *AC* is multiplied by the coordinates of *B* while m_2, which corresponds to *CB* is multiplied by the coordinates of *A*.

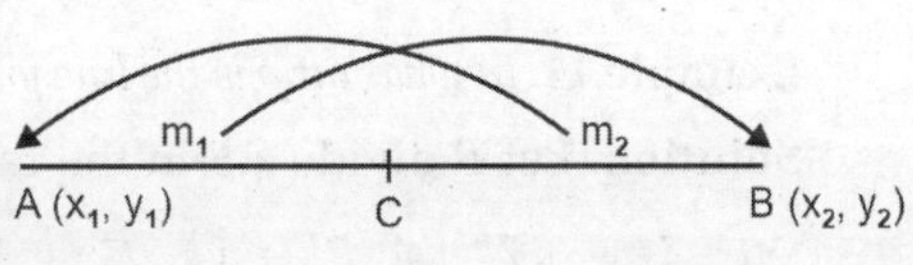

Formula for External Division

A point *C* divides the straight line joining two points $A(x_1, y_1)$ and $B(x_2, y_2)$ externally in the ratio $m_1 : m_2$; then the coordinates of *C* are:

$$x = \frac{m_1x_2 - m_2x_1}{m_1 - m_2}, \qquad y = \frac{m_1y_2 - m_2y_1}{m_1 - m_2}$$

Alternate method for obtaining coordinates for external division:

The coordinates of C in this case can be obtained by keeping in mind the fact that here m_1 and m_2 are measured in opposite directions and so if m_1 is reckoned positive, m_2 will be negative.

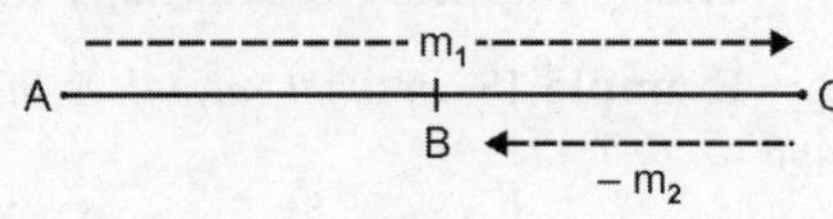

Note: If the point C is given and we are required to find the ratio in which C divides the line segment *AB*, it is convenient to take the ratio *k* : 1.

Then the coordinates of C are $\left(\dfrac{kx_2 + x_1}{k+1}, \dfrac{ky_2 + y_1}{k+1}\right)$

(*i*) If *k* is positive, then C divides *AB* internally.

(*ii*) If *k* is negative, then C divides *AB* externally.

Type I. When the ratio of division is given, then to find the point of division.

Example 12. *Find the co-ordinates of the point which divides: the join of (5, –4) and (–3, 2) internally in the ratio 1 : 2.* **[S.B.T.E. 2015]**

Solution: Let (x, y) be the coordinates of the point C of internal division.

Then $$x = \frac{1(-3)+2(5)}{1+2} = \frac{-3+10}{3} = \frac{7}{3}$$

$$y = \frac{1(2)+2(-4)}{1+2} = \frac{2+(-8)}{3} = \frac{-6}{3} = -2$$

Hence, the coordinates of the point of internal division are $\left(\frac{7}{3}, -2\right)$. **Ans.**

Example 13. *Find the coordinates of the point which divides: the join of (2, –6) and (4, 3) externally in the ratio 3 : 2.* **[S.B.T.E. 2008]**

Solution: Let (x, y) be the coordinates of the point C of external division.

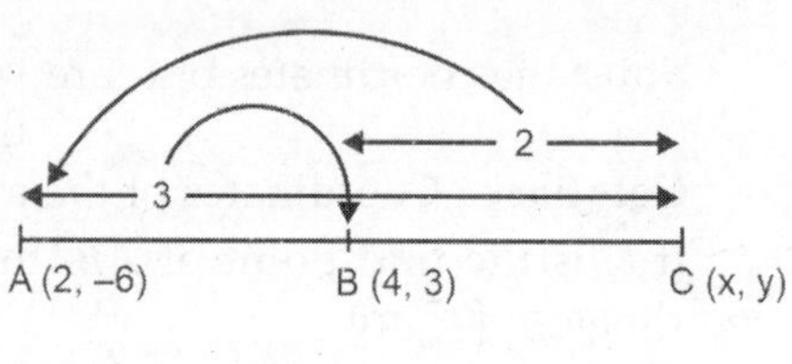

Then $$x = \frac{3(4)-2(2)}{3-2} = \frac{12-4}{1} = 8$$

$$y = \frac{3(3)-2(-6)}{3-2} = \frac{9+12}{1} = 21$$

Hence, the coordinates of the point of external division are (8, 21). **Ans.**

Type II. To find the ratio, when end points and point of division are given.

Example 14. *In what ratio is the line joining the points A(4, 4) and B(7, 7) divided by P(–1, –1)?*

Solution. Let P divide AB in the ratio of $k : 1$.

Then, P is $\left(\frac{7k+4}{k+1}, \frac{7k+4}{k+1}\right)$.

But, P is (–1, –1)

$$\therefore \quad \frac{7k+4}{k+1} = -1 \quad \Rightarrow \quad 7k+4 = -k-1$$

$$\Rightarrow \quad 8k = -5 \quad \Rightarrow \quad k = -\frac{5}{8}$$

Thus, P divides AB externally in the ratio 5 : 8. **Ans.**

Example 15. *In what ratio does the x-axis divide the line segment joining the points (2, –3) and (5, 6) ?* **[S.B.T.E. 2013]**

Solution. Let the required ratio be $k : 1$. Then coordinates of the point of division are $\left(\frac{5k+2}{k+1}, \frac{6k-3}{k+1}\right)$. But it is a point on x-axis on which y-coordinate of every point is zero.

$$\therefore \quad \frac{6k-3}{k+1} = 0 \Rightarrow 6k-3 = 0 \Rightarrow k = \frac{1}{2}$$

Thus, the required ratio is 1/2 : 1 or 1:2 **Ans.**

Example 16. *Determine the ratio in which the line $3x + y - 9 = 0$ divides the segment joining the points (1, 3) and (2, 7).*

Solution. Suppose the line $3x + y - 9 = 0$ divides the line segment joining $A(1, 3)$ and $B(2, 7)$ in the ratio $k : 1$ at point C. Then, the coordinates of C are $\left(\dfrac{2k+1}{k+1}, \dfrac{7k+3}{k+1}\right)$

But, C lies on $3x + y - 9 = 0$, therefore

$$3\left(\frac{2k+1}{k+1}\right) + \frac{7k+3}{k+1} - 9 = 0$$

$$\Rightarrow \qquad 6k + 3 + 7k + 3 - 9k - 9 = 0$$

$$\Rightarrow \qquad 4k - 3 = 0 \qquad \Rightarrow \qquad k = \frac{3}{4}$$

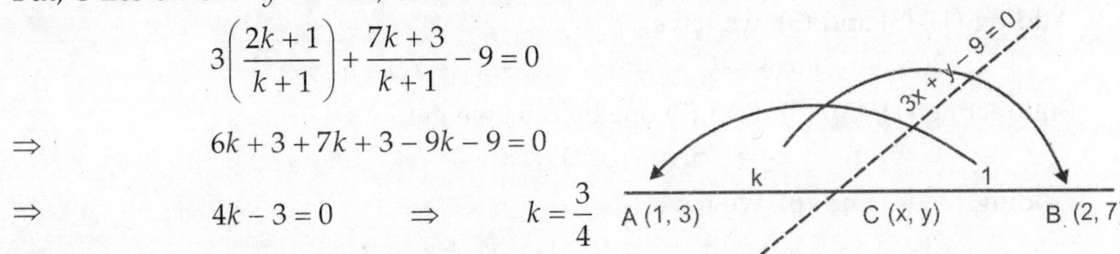

So, the required ratio is 3 : 4 internally. **Ans.**

TYPE III. To find out the unknown vertex.

Example 17. *Three consecutive vertices of a parallelogram ABCD are A(3, 0), B (5, 2), C(–2, 6). Find the fourth vertex D.* **[S.B.T.E. 2014]**

Solution. Let the fourth vertex be $D(x, y)$. Since, $ABCD$ is a parallelogram, the diagonals bisect each other.

$$\Rightarrow \qquad \text{Mid-point of } BD = \text{Mid-point of } AC.$$

Thus, $$\left(\frac{x+5}{2}, \frac{y+2}{2}\right) = \left(\frac{3+(-2)}{2}, \frac{0+6}{2}\right).$$

Equating abscissae and ordinates, we get

Thus, $$\frac{x+5}{2} = \frac{3-2}{2} \qquad \text{or} \qquad x + 5 = 1$$

Thus, $x = -4$.

and $$\frac{y+2}{2} = \frac{0+6}{2} \text{ or } y + 2 = 6 \text{ or } y = 4.$$

Hence, the fourth vertex $D(x, y)$ is $(-4, 4)$. **Ans.**

Example 18. *Find the vertices of a triangle, the mid-point of whose sides are (3, 1), (5, 6) and (–3, 2).*

Solution. Let $A(x_1, y_1), B(x_2, y_2), C(x_3, y_3)$ be the vertices of the ΔABC and D (3, 1), E (5, 6), F (–3, 2) the middle points of BC, CA, AB respectively.

$\because$ D is the mid-point of BC

$$\therefore \qquad \frac{x_2 + x_3}{2} = 3 \qquad \Rightarrow \qquad x_2 + x_3 = 6 \qquad ...(1)$$

and $$\frac{y_2 + y_3}{2} = 1 \qquad \Rightarrow \qquad y_2 + y_3 = 2 \qquad ...(2)$$

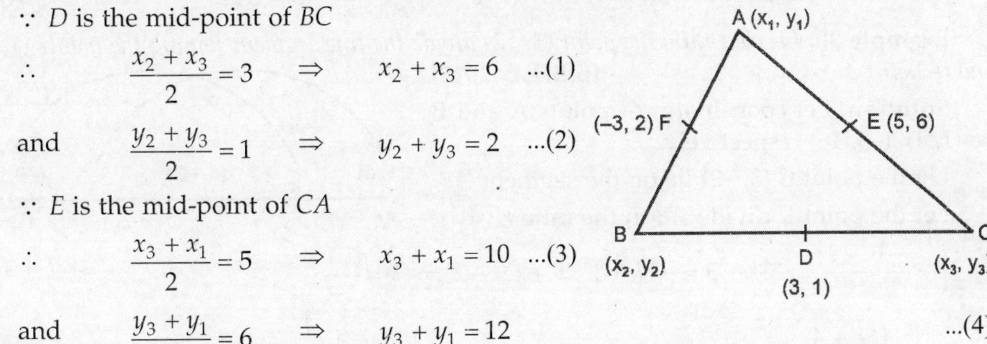

$\because$ E is the mid-point of CA

$$\therefore \qquad \frac{x_3 + x_1}{2} = 5 \qquad \Rightarrow \qquad x_3 + x_1 = 10 \qquad ...(3)$$

and $$\frac{y_3 + y_1}{2} = 6 \qquad \Rightarrow \qquad y_3 + y_1 = 12 \qquad ...(4)$$

$\because$ F is the mid-point of AB

$\therefore \quad \dfrac{x_1 + x_2}{2} = -3 \quad \Rightarrow \quad x_1 + x_2 = -6 \qquad ...(5)$

and $\quad \dfrac{y_1 + y_2}{2} = 2 \quad \Rightarrow \quad y_1 + y_2 = 4 \qquad ...(6)$

Adding (1), (3) and (5), we have

$$2(x_1 + x_2 + x_3) = 10 \quad \Rightarrow \quad (x_1 + x_2 + x_3) = 5 \qquad ...(7)$$

Subtracting (1), (3), (5) from (7) one by one, we get

$$x_1 = -1, \quad x_2 = -5, \quad x_3 = 11.$$

Adding (2), (4) and (6), we have

$$2(y_1 + y_2 + y_3) = 18 \quad \Rightarrow \quad y_1 + y_2 + y_3 = 9 \qquad ...(8)$$

Subtracting (2), (4), (6) from (8) one by one, we get

$$y_1 = 7, \quad y_2 = -3, \quad y_3 = 5.$$

Hence, the coordinates of the vertices are

$A(-1, 7), \quad B(-5, -3), \quad C(11, 5).$ **Ans.**

Example 19. *Find the length of a medians of a triangle whose vertices are (3, 5), (5, 3) and (7, 7).*

Solution. Median is a line joining the vertex of a triangle to the middle point of the opposite side. Let $A(3, 5)$, $B(5, 3)$, $C(7, 7)$ be the vertices of ΔABC.

Let D, E, F, be the mid-points of BC, CA, AB respectively, so that

Coordinates of D are $\left(\dfrac{5+7}{2}, \dfrac{3+7}{2}\right)$ *i.e.* (6, 5)

Length of median $AD = \sqrt{(3-6)^2 + (5-5)^2} = 3$

Coordinates of E are $\left(\dfrac{7+3}{2}, \dfrac{7+5}{2}\right)$ *i.e.* (5, 6)

Length of median $BE = \sqrt{(5-5)^2 + (3-6)^2} = 3$

Coordinates of F are $\left(\dfrac{3+5}{2}, \dfrac{5+3}{2}\right)$, *i.e.* (4, 4)

Length of median $CF = \sqrt{(7-4)^2 + (7-4)^2} = \sqrt{9+9} = \sqrt{18} = 3\sqrt{2}.$ **Ans.**

Example 20. *In what ratio the point (3, –2) divide the line segment joining the points (1, 4) and (–3, 16) ?* **[S.B.T.E. 2015]**

Solution. Let coordinates of points A and B be (1, 4), (–3, 16) respectively.

Let the point C (3, –2) lie on the segment AB.

Let the point C divide AB in the ratio $k : 1$

$$3 = \frac{-3k+1}{k+1} \Rightarrow 3k + 3 = -3k + 1 \Rightarrow 6k = -2 \Rightarrow k = -\frac{1}{3}$$

$$-2 = \frac{16k+4}{k+1} \Rightarrow -2k - 2 = 16k + 4 \Rightarrow 18k = -6 \Rightarrow k = -\frac{1}{3}.$$

Hence, the point (3, –2) divides the segment joining the points (1, 4) and (–3, 16) in the ratio 1:3 externally. **Ans.**

EXERCISE 21.3

Find the co-ordinates of the point, which internally divides the line joining the points.

1. A(2, –1) and B(–3, 4) in the ratio 2 : 3 **Ans.** (0, 1)
2. (p, q) and (q, p) in the ratio $p - q : p + q$ **Ans.** $\left(\frac{p^2 - q^2 + 2pq}{2p}, \frac{p^2 + q^2}{2p}\right)$

Find the co-ordinates of the point, which externally divides the line joining the points.

3. A (–3, –4) and B (2, 1) in the ratio 3 : 2. **Ans.** (12, 11)
4. A (–5, 9) and B (0, 8) in the ratio 3 : 1 **Ans.** $\left(\frac{5}{2}, \frac{15}{2}\right)$
5. Find the co-ordinates of the mid-point of the line joining the points (4, 7) and (6, 9). **Ans.** (5, 8)
6. In what ratio does the point (–5, 3) divide the joining line of (–3, –1) and (–8, 9) points? **Ans.** 2 : 3 Internally
7. Prove that the straight line $y - x + 2 = 0$ cuts the straight line joining (3, –1) and (8, 9) points in the ratio 2 : 3.
8. Determine the ratio in which $3x - 5y + 8 = 0$ divides the join of (4, 3) and (8, 7). **Ans.** 5 : 3 Internally
9. Find the ratio in which the line joining (–3, –10) and (3, 5) is divided by the
   (*i*) x-axis **Ans.** 2 : 1, (1, 0)
   (*ii*) y-axis. Also find the points of division **Ans.** 1 : 1, $\left(0, \frac{-5}{2}\right)$
10. In what ratio does y-axis divide the join of (7, 3) and (–5, –12)? **Ans.** 7 : 3, $\left(0, \frac{-23}{4}\right)$
11. Show that the points (3, –2), (4, 0), (6, –3) and (5, –5) are the vertices of a parallelogram.
12. Show that the quadrilateral with vertices (1, 4), (–2, 1), (0, –1) and (3, 2) is a parallelogram.
13. Prove that the points (–4, –1), (–2, –4) (4, 0) and (2, 3) are the vertices of a rectangle.
14. Prove that (4, 3), (6, 4) (5, 6) and (3, 5) are the angular points of a square.
15. The vertices of a triangle ABC are A(1, 2), B(4, 6) and C(6, 14). AD is the bisector of the angle A and meets BC at D.
    Find the co-ordinates of the point D. **Ans.** $\left(\frac{41}{9}, \frac{74}{9}\right)$
16. The points (3, –4) and (–6, 2) are the extremities of a diagonal of a parallelogram. If the third vertex is (–1, –3), find the co-ordinates of the fourth vertex. **Ans.** (–2, 1)
17. One end of diameter of a circle is (4, 1) and the centre is (3, 3).
    Find the co-ordinates of the other end of the diameter. **Ans.** (2, 5)
18. If the points (–2, –1), (1, 0), (x, 3) and (1, y) form a parallelogram.
    Find the values of x and y. **Ans.** $x = 4$, $y = 2$
19. If the co-ordinates of the mid-points of the sides of a triangle are (1, 1), (2, –3) and (3, 4), find the vertices of the triangle. **Ans.** (4, 0), (2, 8), (0, – 6)

21.5 CENTROID OF A TRIANGLE

The point of concurrency of three medians of a triangle is called its **centroid.**

Let $A(x_1, y_1), B(x_2, y_2), C(x_3, y_3)$ be the coordinates of the vertices of a triangle. Let D be the mid-point of BC.

$\therefore$ The co-ordinates of D are $\left(\dfrac{x_2 + x_3}{2}, \dfrac{y_2 + y_3}{2}\right)$

Let G (x, y) divide DA in the ratio 1 : 2 internally.

$$\therefore \quad x = \frac{1(x_1) + 2\left(\dfrac{x_2 + x_3}{2}\right)}{1 + 2} = \frac{x_1 + x_2 + x_3}{3}$$

and
$$y = \frac{1(y_1) + 2\left(\dfrac{y_2 + y_3}{2}\right)}{1 + 2} = \frac{y_1 + y_2 + y_3}{3}$$

$$\therefore \quad G = \left(\frac{x_1 + x_2 + x_3}{3}, \frac{y_1 + y_2 + y_3}{3}\right).$$

The symmetry of the co-ordinates of G shows that it also lies on the other two medians through B and C. Thus, all medians meet at G, *i.e.* it is the centroid.

$\therefore$ The centroid is $\left(\dfrac{\mathbf{x_1 + x_2 + x_3}}{\mathbf{3}}, \dfrac{\mathbf{y_1 + y_2 + y_3}}{\mathbf{3}}\right)$.

Example 21. *Find the centroid of the triangle formed by the straight lines.*

$x - 2y = 0, \qquad x + y = 3$ and $2x + y + 4 = 0$ **[B.T.E DELHI Jan 2009]**

Solution. We have,

$x - 2y = 0$ ...(1)

$x + y = 3$ ...(2)

$2x + y = -4$ ...(3)

Solving (1) and (2), we get $x = 2$ and $y = 1$

Let the coordinates of first vertex be A, then A (2, 1).

Solving (2) and (3), we get $x = -7$ and y = 10

Let the coordinates of second vertex be C, then C (–7, 10)

Solving (1) and (3), we get $x = -\dfrac{8}{5}$ and $y = -\dfrac{4}{5}$

Let the coordinates of third vertex be B,

then $B\left(\dfrac{-8}{5}, \dfrac{-4}{5}\right)$

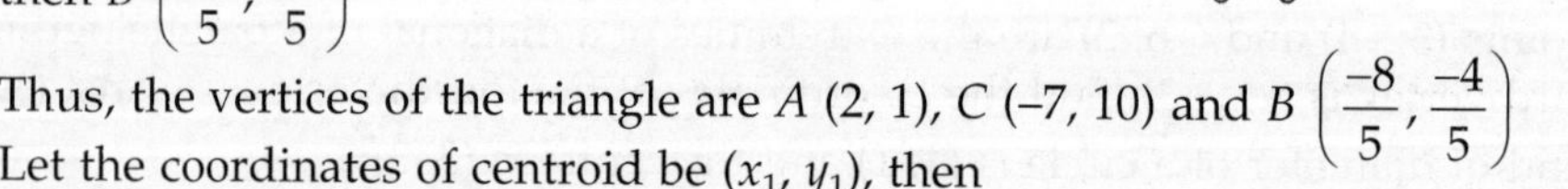

Thus, the vertices of the triangle are A (2, 1), C (–7, 10) and $B\left(\dfrac{-8}{5}, \dfrac{-4}{5}\right)$

Let the coordinates of centroid be (x_1, y_1), then

$$x_1 = \frac{1}{3}\left[2 - 7 - \frac{8}{5}\right] = \frac{1}{3}\left(\frac{-33}{5}\right) = -\frac{11}{5}$$

and
$$y_1 = \frac{1}{3}\left[1 + 10 - \frac{4}{5}\right] = \frac{1}{3}\left[\frac{51}{5}\right] = \frac{17}{5}$$

Hence, the centroid is $\left(\frac{-11}{5}, \frac{17}{5}\right)$ **Ans.**

Example 22. *The three vertices of a triangle ABC are A (1, 3), B(4, 7) and C(6, 15). Find the point of intersection of the bisector AD of ∠A and the side BC.*

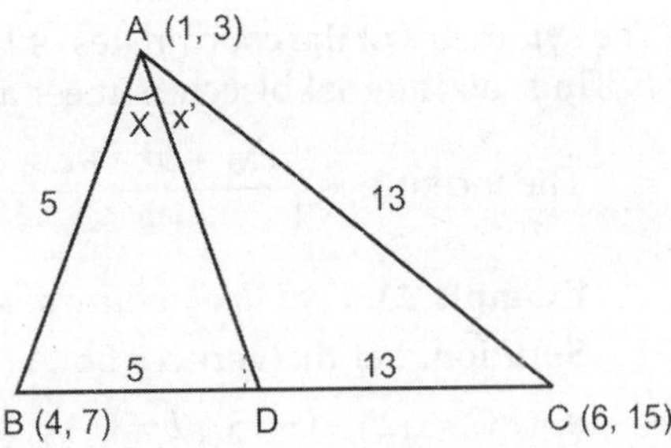

Solution. We have $A(1, 3)$, $B(4, 7)$ and $C(6, 15)$

$$AB = \sqrt{(4-1)^2 + (7-3)^2} = \sqrt{9+16} = 5.$$

$$AC = \sqrt{(6-1)^2 + (15-3)^2} = \sqrt{25+144} = 13$$

We know that AD bisector of $\angle A$ divides the opposite side BC at D in the ratio $AB : AC$ *i.e.* $5 : 13$.

$$\text{Co-ordinates of } D = \left(\frac{(5\times 6)+(13\times 4)}{5+13}, \frac{(5\times 15)+(13\times 7)}{5+13}\right) = \left(\frac{30+52}{18}, \frac{75+91}{18}\right)$$

$$= \left(\frac{82}{18}, \frac{166}{18}\right) = \left(\frac{41}{9}, \frac{83}{9}\right)$$ **Ans.**

21.6 INCENTRE OF A TRIANGLE

The point of concurrency of three internal bisectors of the angles of a triangle is called its **incentre.**

Let $A(x_1, y_1)$, $B(x_2, y_2)$, $C(x_3\ y_3)$ be the coordinates of the vertices of a triangle. Let AD, BE and CF be the internal bisectors of the angles A, B and C respectively.

Let the sides BC, CA and AB be denoted by a, b and c respectively.

By geometry, $$\frac{BD}{DC} = \frac{BA}{CA} = \frac{c}{b} \quad \text{...(1)}$$

Point D divides BC in the ratio $c : b$

∴ The coordinates of D are $\left(\frac{bx_2 + cx_3}{b+c}, \frac{by_2 + cy_3}{b+c}\right)$

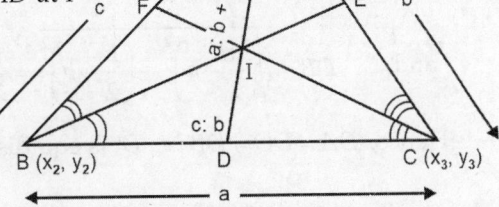

Now, CI is the bisector of $\angle C$ and intersects AD at I

Now, we have to find out $\frac{CD}{AC}$.

$$\therefore \quad \frac{CD}{AC} = \frac{ID}{AI}, \quad \frac{CD}{AC} = \frac{\frac{ab}{b+c}}{b} = \frac{a}{b+c}$$

∴ I divides AD internally in the ratio $a : b + c$.

∴ The coordinates of I are

$$\left(\frac{ax_1 + (b+c)\left(\frac{bx_2+cx_3}{b+c}\right)}{a+(b+c)}, \frac{ay_1 + (b+c)\left(\frac{by_2+cy_3}{b+c}\right)}{a+(b+c)}\right)$$

or $\left(\dfrac{ax_1 + bx_2 + cx_3}{a+b+c}, \dfrac{ay_1 + by_2 + cy_3}{a+b+c}\right)$

The symmetry of the coordinates of I shows that it also lies on the internal bisector through *B*. Thus, all internal bisectors meet at *I*, *i.e.* it is the incentre.

$\therefore$ The incentre is $\left(\dfrac{ax_1 + bx_2 + cx_3}{a+b+c}, \dfrac{ay_1 + by_2 + cy_3}{a+b+c}\right)$

Example 23. *Find the incentre of a triangle, whose vertices are (–1, 12), (–1, 0) and (4, 0).*

Solution. Let the vertices be $A(-1, 12)$, B $(-1, 0)$ and C $(4, 0)$. Then,

$$a = BC = \sqrt{25+0} = 5, \quad b = CA = \sqrt{25+144} = 13, \quad c = AB = \sqrt{0+144} = 12.$$

$\therefore$ The incentre of $\Delta\, ABC$ is $\left(\dfrac{ax_1 + bx_2 + cx_3}{a+b+c}, \dfrac{ay_1 + by_2 + cy_3}{a+b+c}\right)$

i.e. $\left(\dfrac{5\times(-1) + 13\times(-1) + 12\times 4}{5+13+12}, \dfrac{5\times 12 + 13\times 0 + 12\times 0}{5+13+12}\right)$ *or* (1, 2).

Example 24. *Use analytical geometry to prove that the mid-point of the hypotenuse of a right. angled triangle is equidistant from its vertices.*

Or

Prove that mid point of hypotenuse of a right angled triangle is equidistant from all three vertices.

Solution. Let *AOB* be a right angled triangle with base *OA* taken along x-axis and the perpendicular *OB* taken along y-axis. Let $OA = a$ and $OB = b$.

Let *D* be the mid-point of the hypotenuse *AB*. Then, the co-ordinates of *A*, *B* and *D* are respectively $(a, 0)$, $(0, b)$ and $(a/2, b/2)$.

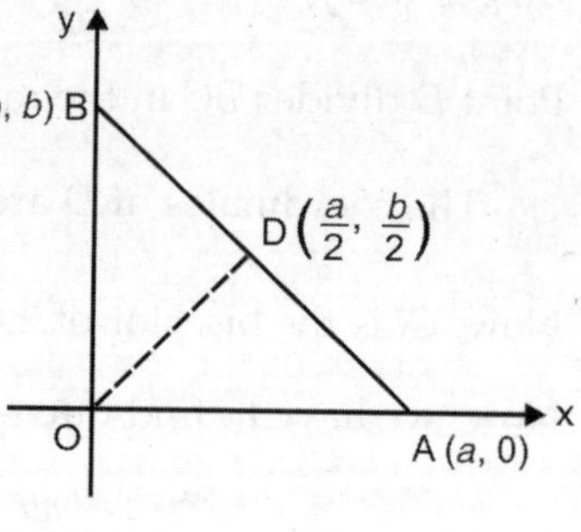

Now, $$DO = \sqrt{\left(\frac{a}{2} - 0\right)^2 + \left(\frac{b}{2} - 0\right)^2} = \frac{1}{2}\sqrt{a^2 + b^2},$$

$$DA = \sqrt{\left(\frac{a}{2} - a\right)^2 + \left(\frac{b}{2} - 0\right)^2} = \frac{1}{2}\sqrt{a^2 + b^2},$$

and, $$DB = \sqrt{\left(\frac{a}{2} - a\right)^2 + \left(\frac{b}{2} - b\right)^2} = \frac{1}{2}\sqrt{a^2 + b^2}$$

Hence, $DA = DO = DB$ *i.e.*, *D* is equidistant from the vertices of triangle *OAB*.

EXERCISE 21.4

1. Find the centroid of the triangle whose vertices are: (–4, 6), (2, –2) and (2, 5).

   Ans. (0, 3)

2. Find the incentre of the triangle given by the points (–36, 7), (20, 7) and (0, –8).

   Ans. (–1, 0)

3. Show that the middle point of the hypotenuse of a right-angled triangle is equidistant from the vertices.

4. Find the centroid of the triangle formed by the straight lines $x - 2y = 0, x + y = 3$ and $2x + y + 4 = 0.$ **Ans.** $\left(\frac{-11}{5}, \frac{17}{5}\right)$

5. If the coordinates of the mid-points of the sides of a triangle are (1, 1), (2, –3) and (3, 4). Find its
   (*i*) centroid **Ans.** (4, 0)
   (*ii*) incentre **Ans.** $\left[\frac{2\sqrt{13}+20\sqrt{2}}{\sqrt{13}+\sqrt{17}+4\sqrt{2}}, \frac{8\sqrt{13}-6\sqrt{17}}{\sqrt{13}+\sqrt{17}+5\sqrt{2}}\right]$

6. Two vertices of a triangle are (1, 2), (3, 5) and its centroid is at the origin. Find the coordinates of the third vertex. **Ans.** (–4, –7)
7. Prove analytically that the line segment joining the middle points of two sides of a triangle is equal to half of the third side.
8. Prove that the line joining middle points of the opposite sides of a quadrilateral and centre of the middle points of its diagonals meet at a point and bisect one another.
9. Find the coordinates of the incentre of the triangle whose vertices are (0, 6), (8, 12) and (8, 0). **Ans.** (5, 6)

Choose the correct alternative

10. The centroid of the triangle formed by the straight lines $y + x = 3, y - x = 3, y = 0$ is
    (a) (0, 0) (b) (1, 0) (c) (0, 1) (d) (1, 1) ***(Diploma IETE, Dec. 2006)*** **Ans.** (*c*)
11. The co-ordinates of the middle points of the sides of a triangle are (4, 2), (3, 3) and (2, 2). Then the co-ordinates of the centroid are
    (a) $\left(3, \frac{7}{3}\right)$ (b) (3, 3) (c) (4, 3) (d) (4, 7) **Ans.** (*a*)

21.7 AREA OF A TRIANGLE

$$\frac{1}{2}[x_1(y_2 - y_3) + x_2(y_3 - y_1) + x_3(y_1 - y_2)]$$

Where $A(x_1, y_1), B(x_2, y_2), C(x_3, y_3)$ are the vertices of the triangle.

The expression for the area can also be written in the determinant form as follows.

$$\Delta = \frac{1}{2}\begin{vmatrix} x_1 & y_1 & 1 \\ x_2 & y_2 & 1 \\ x_3 & y_3 & 1 \end{vmatrix}$$

Arrow Method

Step I. Write down the co-ordinates in two columns as shown in the figure and repeat the coordinates of the first vertex.

Step II. Match arrows diagonally. Each arrow head shows a product.

Step III. Downward arrows show plus sign and upward arrows show negative sign.

Step IV. Divide the sum of all products by 2.

For example

Area of triangle with vertices $(x_1, y_1), (x_2, y_2)$ and (x_3, y_3)

$= \frac{1}{2}$ [Sum of the product of downward arrows
–Sum of the products of the upward arrows]

$$= \frac{1}{2}[(x_1y_2 + x_2y_3 + x_3y_1) - (x_2y_1 + x_3y_2 + x_1y_3)]$$

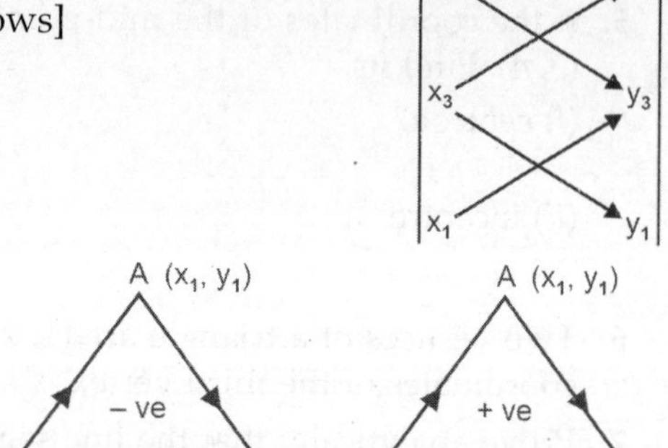

Sign of Area. If three points A, B, C are taken anticlockwise sense, then the calculated area of the triangle ABC will be positive, while if the points are taken in clockwise sense, then the calculated area will be negative.

In case the area calculated is negative we shall consider it positive.

21.8 CONDITION OF COLINEARITY OF THREE POINTS

$$x_1(y_2 - y_3) + x_2(y_3 - y_1) + x_3(y_1 - y_2) = 0$$

Three points $A(x_1, y_1), B(x_2, y_2)$ and $C(x_3, y_3)$ lie on a straight line, if the area of the triangle formed by them is zero.

i.e. $\quad \frac{1}{2}(x_1y_2 - x_2y_1 + x_2y_3 - x_3y_2 + x_3y_1 - x_1y_3) = 0$

$\Rightarrow \quad (x_1y_2 - x_2y_1 + x_2y_3 - x_3y_2 + x_3y_1 - x_1y_3) = 0$

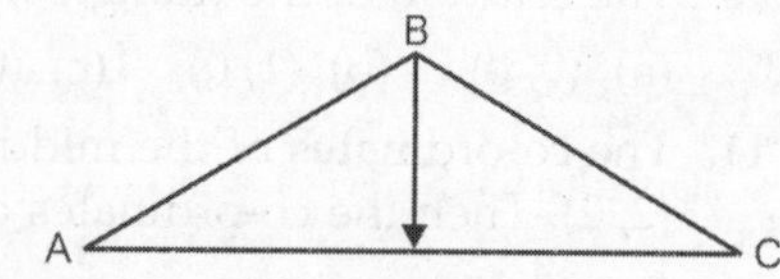

which can also be written in the from:

$$\boxed{x_1(y_2 - y_3) + x_2(y_3 - y_1) + x_3(y_1 - y_2) = 0}$$

TYPE I. On Area of Triangle

Example 25. *Find the area of a triangle whose vertices are (3, 8), (–4, 2) and (5, –1).* ***[S.B.T.E. 2015]***

Solution. Let $A(x_1, y_1) = (3, 8), \quad B(x_2, y_2) = (-4, 2) \quad$ and $\quad C(x_3, y_3) = (5, -1)$.

$$\text{ar}(\Delta ABC) = \frac{1}{2}\{x_1(y_2 - y_3) + x_2(y_3 - y_1) + x_3(y_1 - y_2)\}$$

$$\text{ar}(\Delta ABC) = \frac{1}{2}\{3(2+1) - 4(-1-8) + 5(8-2)\}$$

$$= \frac{1}{2}\{9 + 36 + 30\} = \frac{1}{2} \times 75 = 37.5 \text{ sq units}$$ **Ans.**

Alternate Method

$$\text{Area} = \frac{1}{2}[6 + 4 + 40) - (-32 + 10 - 3)]$$

$$= \frac{1}{2}[50 + 25] = 37.5 \text{ sq. units.}$$ **Ans.**

Area = $\frac{1}{2}$

3	8
–4	2
5	–1
3	8

Example 26. *Show that the three points (1, 1), (2, 1) and (4, 5) are collinear.* **[S.B.T.E. 2015]**

Solution. Here we have

$x_1 = 1, y_1 = -1, x_2 = 2, y_2 = 1, x_3 = 4, y_3 = 5$

if $A(x_1, y_1), B(x_2, y_2)$ and $C(x_3, y_3)$ are collinear then

$x_1(y_2 - y_3) + x_2(y_3 - y_1) + x_3(y_1 - y_2) = 0$

$\Rightarrow \quad 1(1-5) + 2(5+1) + 4(-1-1) = 0$

$\Rightarrow \quad -4 + 12 - 8 = 0$

$\therefore$ The given points are collinear **Ans.**

Example 27. *For what value of k are the points (k, 2–2k), (–k + 1, 2k) and (–4–k, 6–2k) are collinear.* **[S.B.T.E. 2015, 2008]**

Solution. Here we have

$x_1 = k, y_1 = 2 - 2k, x_2 = -k + 1, y_2 = 2k, x_3 = -4 - k, y_3 = 6 - 2k.$

If $A(x_1, y_1), B(x_2, y_2)$ and $C(x_3, y_3)$ are collinear then

$x_1(y_2 - y_3) + x_2(y_3 - y_1) + x_3(y_1 - y_2) = 0$ ...(1)

Putting the values of x_1, y_1, x_2, y_2, x_3 and y_3 in (1), we get

$k = -1, \frac{1}{2}$ **Ans.**

TYPE II. On Collinearity of Three Points:

Example 28. *Show that points (a, b + c), (b, c + a) and (c, a + b) are collinear.*

Solution. Let $A = (x_1, y_1) = (a, b + c), B = (x_2, y_2) = (b, c + a)$ and $C(x_3, y_3) = (c, a + b)$.

$$\Delta ABC = \frac{1}{2}[x_1(y_2 - y_3) + x_2(y_3 - y_1) + x_3(y_1 - y_2)]$$

$$= \frac{1}{2}\{a(c + a - a - b) + b(a + b - b - c) + c(b + c - c - a)\}$$

$$= \frac{1}{2}\{a(c - b) + b(a - c) + c(b - a)\}$$

$$= \frac{1}{2}\{ac - ab + ab - bc + bc - ac\} = 0$$

Since, the area is zero, the points must be collinear. **Proved.**

Example 29. *The area of a triangle is 5. Two of its vertices are (2, 1) and (3, –2). The third vertex is (x, y) where y = x + 3. Find the co-ordinates of the third vertex.*

Solution. Let the vertices be $A(x, y)$, $B(2, 1)$ and $C(3, -2)$.

We know that,

$$\text{ar}(\Delta ABC) = \frac{1}{2}\{x_1(y_2 - y_3) + x_2(y_3 - y_1) + x_3(y_1 - y_2)\}$$

$$\Rightarrow \quad \text{ar}(\Delta ABC) = \frac{1}{2}|\{x(1 + 2) + 2(-2 - y) + 3(y - 1)\}|$$

$$\Rightarrow \quad \text{ar}(\Delta ABC) = \frac{1}{2}|3x + y - 7|$$

$$\Rightarrow \quad \frac{1}{2}|3x + y - 7| = 5$$ [$\because$ given that $\Delta ABC = 5$ sq units]

$\Rightarrow \quad |3x + y - 7| = 10$

$\Rightarrow \quad |3x + y - 7| = 10 \quad$ and $\quad |-(3x + y - 7)| = 10$

$\Rightarrow \quad 3x + y - 7 = 10 \quad$ and $\quad 3x + y - 7 = -10$

$\Rightarrow \quad 3x + y = 10 + 7 = 17 \quad$ and $\quad 3x + y = -10 + 7 = -3$

Case I. When $3x + y = 17$ ...(1)

Also it is given that $y = x + 3$ ...(2)

Solving (1) and (2), we get $x = \frac{7}{2}, y = \frac{13}{2}$

Case II. When $3x + y = -3$

Solving (2) and (3), we get $x = -\frac{3}{2}, \quad y = \frac{3}{2}$

Hence, co-ordinates of third vertex are $\left(\frac{7}{2}, \frac{13}{2}\right)$ *or* $\left(-\frac{3}{2}, \frac{3}{2}\right)$ **Ans.**

EXERCISE 21.5

Find the area of triangles formed by the following points:

1. (3, 4), (2, –1), (4, –6) **Ans.** $\frac{15}{2}$ sq. units
2. $(a, c + a)$, (a, c) and $(-a, c - a)$ **Ans.** a^2 sq. units

Show that following points are collinear:

3. (1, –1), (2, 1) and (4, 5)
4. (–5, 1), (5, 5) and (10, 7)

For what values of x will the following lie on a line:

5. $(x, -1)$, (5, 7), (8, 11). **Ans.** $x = -1$
6. $(a, 0)$, $(0, b)$, $(3a, x)$ $(a \neq 0)$ **Ans.** $-2b$
7. For what value(s) of x, the area of the triangle formed by the point (5, –1), $(x, 4)$ and (6, 3) is 5.5 square units. **Ans.** $x = 9$ or $x = \frac{7}{2}$
8. Find the area of the triangle formed by the mid-points of sides of the triangle whose vertices are (2, 1), (–2, 3) and (4, –3). **Ans.** 1.5 sq. units
9. If three points $(x_1, y_1), (x_2, y_2)$ and (x_3, y_3) lie on the same line, prove that

$$\frac{y_2 - y_3}{x_2 x_3} + \frac{y_3 - y_1}{x_3 x_1} + \frac{y_1 - y_2}{x_1 x_2} = 0$$

Find the area of the quadrilateral whose vertices are:

10. (1, 2), (6, 2), (5, 3), and (3, 4) **Ans.** $\frac{11}{2}$ sq. units
11. (1, 1), (3, 4), (5, –2), and (4, –7) **Ans.** $\frac{41}{2}$ sq. units
12. Find the condition that the points (1, 1), (3, 5) and (a, b) are collinear. **Ans.** $2a - b - 1 = 0$
13. Find the area of the triangle formed by the straight lines $y = x$, $y = 2x$ and $y = 3x + 4$ **Ans.** 4

Choose the correct alternative:

14. Area of the triangle whose vertices are $(a, b)(a, a+b), (-a, -a+b)$ is

(a) a^2b^2 (b) a^2+b^2 (c) a^2 (d) b^2 **Ans.** (*c*)

Polar Coordinates

A polar coordinate system in a plane consists of a fixed point O, called the pole (or origin), and a ray emerging from the pole, called the polar axis. In such a coordinate system we can associate with each point P in the plane, a pair of polar coordinates (r, θ), where r is the distance from P to the pole and θ is an angle from the polar axis to the ray OP, 'r' is called the radial coordinate of P and 'θ' is called polar angle or angular coordinate of P.

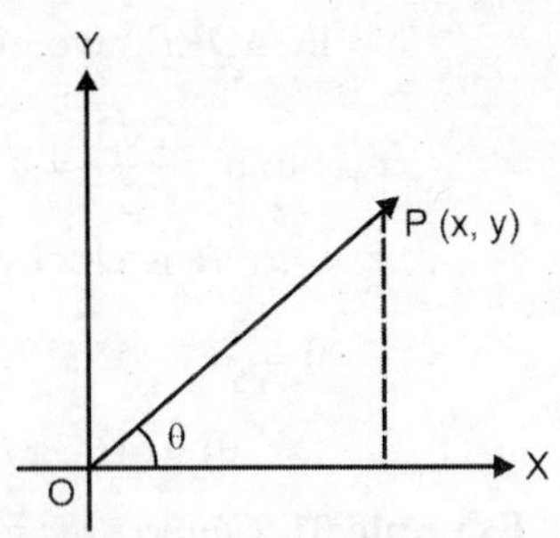

21.9 RELATION BETWEEN CARTESIAN AND POLAR COORDINATES OF A POINT

In ΔOPR

$$\cos\theta = \frac{OR}{OP} = \frac{x}{r}$$

$$x = r\cos\theta$$

$$\text{and } \sin\theta = \frac{PR}{OP} = \frac{y}{r}$$

$$y = r\sin\theta$$

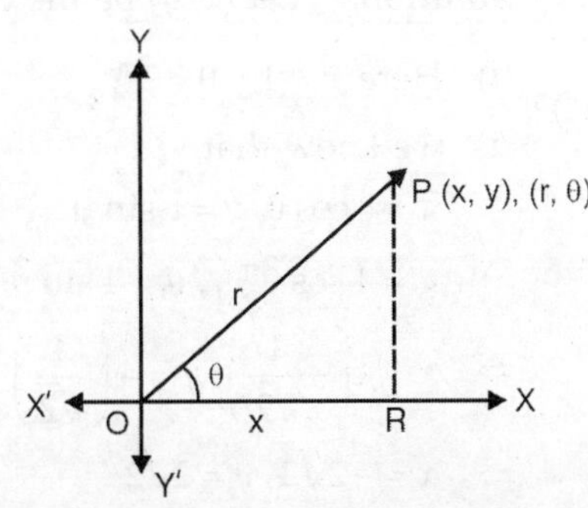

Therefore, $\boxed{x = r\cos\theta,\ y = r\sin\theta}$

where $\quad r = \sqrt{x^2+y^2}$

$$\tan\theta = y/x,\ \theta = \tan^{-1}\left(\frac{y}{x}\right)$$

$$\therefore \quad \boxed{r = \sqrt{x^2+y^2},\ \theta = \tan^{-1}\left(\frac{y}{x}\right)}$$

Example 30. *Convert the following Cartesian coordinates in to polar coordinates.*

(*i*) $(-1, \sqrt{3})$ (*ii*) $(2, -2\sqrt{3})$ **[S.B.T.E. 2015, 2007]**

Solution. (*i*) Here we have cartesian coordinates = $(-1, \sqrt{3})$

Let $P(r, \theta)$ be the polar coordinates

$$r = \sqrt{x^2+y^2} = \sqrt{(-1)^2+(\sqrt{3})^2} = 2$$

in ΔORP, we get

$$\tan\theta = \frac{\sqrt{3}}{1} = \sqrt{3} = \tan\frac{\pi}{3}$$

$$\theta = \pi - \frac{\pi}{3} = 2\pi/3$$

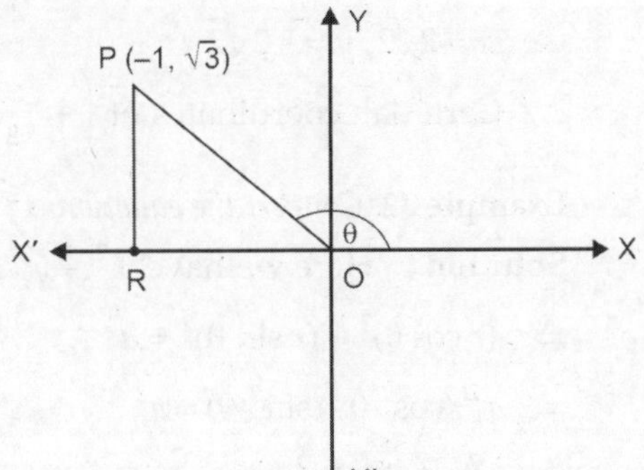

$\therefore$ polar co-ordinates of $(-1, \sqrt{3})$

$= (2, 2\pi/3)$ **Ans.**

(*ii*) Let $P(r, \theta)$ be the polar coordinates of $(2, -2\sqrt{3})$, then.

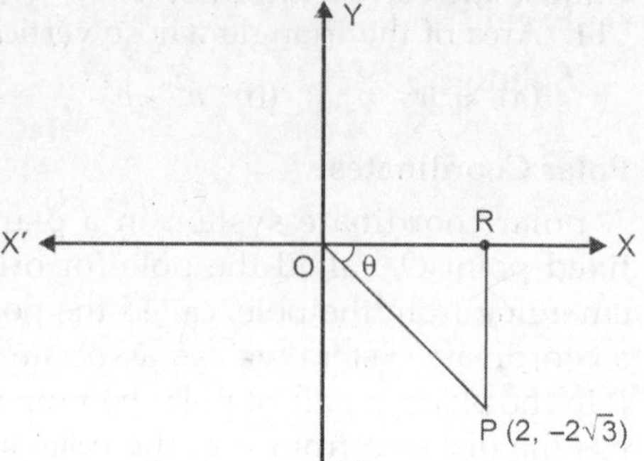

$$r = \sqrt{(2)^2 + (-2\sqrt{3})^2} = \sqrt{16} = 4$$

in ΔORP, we get

$$\tan\theta = \frac{2\sqrt{3}}{2} = \sqrt{3} = \tan\frac{\pi}{3}$$

$\therefore$ $\angle ROP$ is clockwise direction

$\therefore$ $\theta = \frac{\pi}{3}$

$P(r, \theta) = (4, -\pi/3)$ **Ans.**

Example 31. *Convert the following polar coordinates into cartesian coordinates.*

(*i*) $\left(4, \frac{3\pi}{4}\right)$ (*ii*) $\left(4, -\frac{\pi}{4}\right)$ **[S.B.T.E. 2015]**

Solution. Let (x, y) be the Cartesian coordinates of point $P\left(2, \frac{3\pi}{4}\right)$

(*i*) Here $r = 4$, $\theta = 3\pi/4$

we know that

$x = r\cos\theta,\ y = r\sin\theta$

$\Rightarrow$ $x = 4\cos 3\pi/4,\ y = 4\sin 3\pi/4$

$\Rightarrow$ $x = 4\left(-\frac{1}{\sqrt{2}}\right),\ y = 4\left(\frac{1}{\sqrt{2}}\right)$

$\Rightarrow$ $x = -2\sqrt{2},\ y = 2\sqrt{2}$

$\therefore$ Cartesian coordinates of $\left(4, \frac{3\pi}{4}\right) = (-2\sqrt{2}, 2\sqrt{2})$ **Ans.**

(*ii*) Here $r = 4$, $\theta = -\pi/4$

$x = r\cos\theta,\ y = r\sin\theta$

$\Rightarrow$ $x = 4\cos(-\pi/4),\ y = 4\sin(-\pi/4)$

$\Rightarrow$ $x = 4\left(\frac{1}{\sqrt{2}}\right),\ y = -4\left(\frac{1}{\sqrt{2}}\right)$ $[\because \cos(-\theta) = \cos\theta, \sin(-\theta) = -\sin\theta]$

$\Rightarrow$ $x = 2\sqrt{2},\ y = -2\sqrt{2}$

$\therefore$ Cartesian coordinates of $\left(4, \frac{-\pi}{4}\right) = (2\sqrt{2}, -2\sqrt{2})$ **Ans.**

Example 32. *Convert the equation* $x^2 - y^2 = a^2$ *in polar co-ordinates.* **[S.B.T.E. 2014, 2007]**

Solution. Here we have $x^2 - y^2 = a^2$

$\Rightarrow$ $(r\cos\theta)^2 - (r\sin\theta)^2 = a^2$

$\Rightarrow$ $r^2(\cos^2\theta - \sin^2\theta) = a^2$

$\Rightarrow$ $r^2\cos 2\theta = a^2$ $[\because \cos 2\theta = \cos^2\theta - \sin^2\theta]$

$\therefore$ polar equation of $x^2 - y^2 = a^2$ is $r^2\cos 2\theta = a^2$ **Ans.**

Example 33. *Change* $(x^2 + y^2)^2 = a^2 (x^2 - y^2)$ *in polar equation.* **[S.B.T.E. 2013, 2008]**

Solution. Here we have $(x^2 + y^2)^2 = a^2 (x^2 - y^2)$

$\Rightarrow$ $(r^2 \cos^2 \theta + r^2 \sin^2 \theta)^2 = a^2 (r^2 \cos^2 \theta - r^2 \sin^2 \theta)$

$\Rightarrow$ $r^2 \cos^2 \theta + r^2 \sin^2 \theta + 2 r \cos \theta . \sin \theta = a^2 r^2 (\cos^2 \theta - \sin^2 \theta)$

$\Rightarrow$ $(r^2)^2 = a^2 r^2 (\cos^2 \theta - \sin^2 \theta)$

$\Rightarrow$ $r^2 = a^2 \cos 2\theta$ **Ans.**

21.10 LOCUS

The locus of a variable point $P(x, y)$ is the path traced out by a moving point P under certain conditions and locus of a variable point $P(x, y)$ is called a cruve.

Working Rule to Find the Equation of the Locus

Step 1: Take general point $P(x, y)$.

Step 2: Write down the given geometrical conditions for a moving point.

Step 3: Express geometrical condition for a moving point $P(x, y)$ in terms of algebraic equation in x, y.

Step 4: The simplified equation so obtained is the required equation of the locus.

Example 34. *Find the equation of the set of points equidistant from (–1, –1) and (4, 2).*

Solution. Let the given point be A (–1, –1) and B (4, 2).

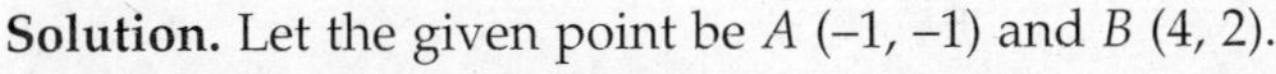

Let $P(x, y)$ be a general point on the locus.

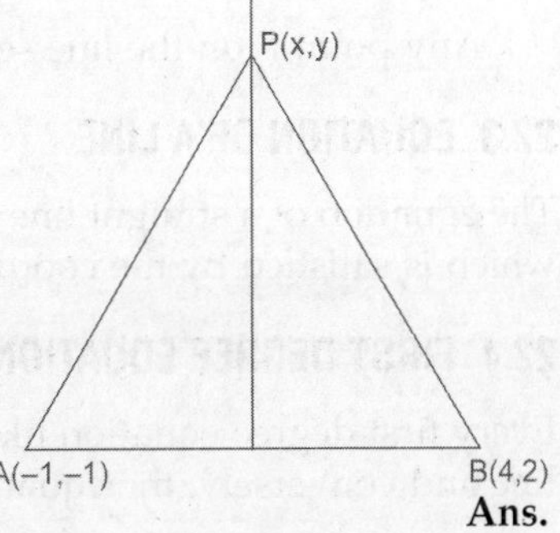

By the given condition, we have

$$PA = PB$$

$$\Rightarrow \sqrt{(x+1)^2 + (y+1)^2} = \sqrt{(x-4)^2 + (y-2)^2}$$

Squaring both the sides, we get

$$x^2 + 2x + 1 + y^2 + 2y + 1 = x^2 - 8x + 16 + y^2 - 4y + 4$$

$$\Rightarrow 2x + 8x + 2y + 4y + 2 - 20 = 0$$

$$\Rightarrow 10x + 6y - 18 = 0 \quad \Rightarrow \quad 5x + 3y - 9 = 0$$ **Ans.**

EXERCISE 21.6

1. Find the locus of a point which is equidistant from the points (1, 0) and (– 1, 0).
Ans. $x = 0$

2. Find the locus of a point which moves so that the sum of the squares of its distances from the points (2, 4) and (–3, –1) is 30. **Ans.** $x^2 + y^2 + x - 3y = 0$

3. $A(2, 0)$ and $B(4, 0)$ are two given points. A point P moves so that $PA^2 + PB^2 = 10$. Find the locus of P. **Ans.** $x^2 + y^2 - 6x + 5 = 0$

4. Find the locus of a point such that the line segments having end points (2, 0) and (–2, 0) subtend a right angle at that point. **Ans.** $x^2 + y^2 = 4$

5. A point moves so that the sum of its distances from $(ae, 0)$ and $(-ae, 0)$ is $2a$. Prove that the equation to its locus is $\frac{x^2}{a^2} + \frac{y^2}{b^2} = 1$, where $b^2 = a^2 (1 - e^2)$.

6. Two points A and B with co-ordinates (5, 3), (3, –2) are given. A point P moves so that the area of ΔPAB is constant and equal to 9 sq. units. Find the equation to the locus of the point P. **Ans.** $5x - 2y - 37 = 0, 5x - 2y - 1 = 0$

Straight Line

22.1 INTRODUCTION

In this chapter, we shall learn the methods of finding the equations of types of straight lines. The concept of 'slope' would be used quite often in this chapter.

22.2 STRAIGHT LINE

A straight line is a curve, such that all the points on the line segment joining any two points on it lies on it.

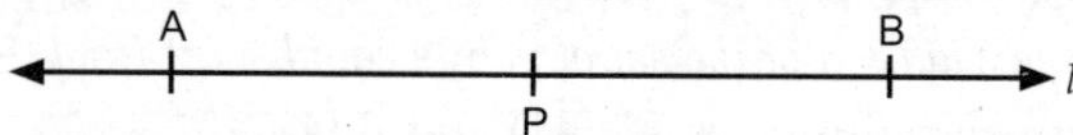

Any point P on the line segment joining any two points A and B, lies on the line l.

22.3 EQUATION OF A LINE

The equation of a straight line is the relation between x (the abscissa) and y (the ordinate), which is satisfied by the coordinates of each and every point on the line.

22.4 FIRST DEGREE EQUATION

Every first degree equation like $ax + by + c = 0$ would be the equation of a certain straight line and conversely, the equation of any straight line would always be of the type

$$ax + by + c = 0.$$

Note: A straight line is briefly written as a 'line'.

22.5 SLOPE OR GRADIENT OF A LINE

The horizontal line is parallel to the x-axis and vertical line is parallel to the y-axis in a plane. To know the direction of a line other than these two a number is associated with it and is called its *slope*.

Slope of a line is the measure of its rate of rise or fall.

For example: (*i*) The car is moving upward on a mountain with a slow speed.

(*ii*) The car is running down with a greater speed. It is due to rise and fall of mountain.

The slope of a line is the tangent of the angle made by the line in the anticlockwise direction from the x-axis.

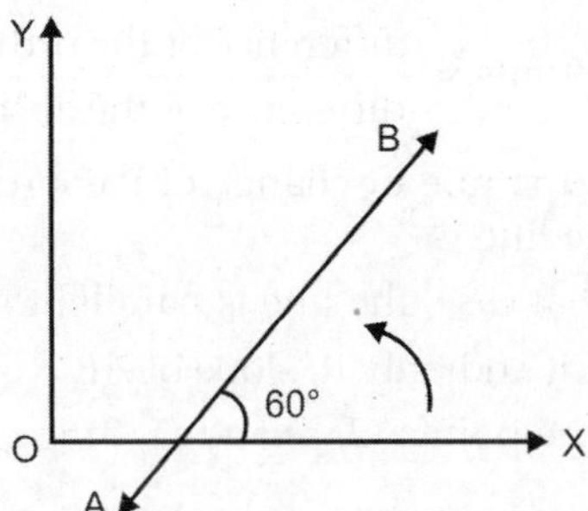

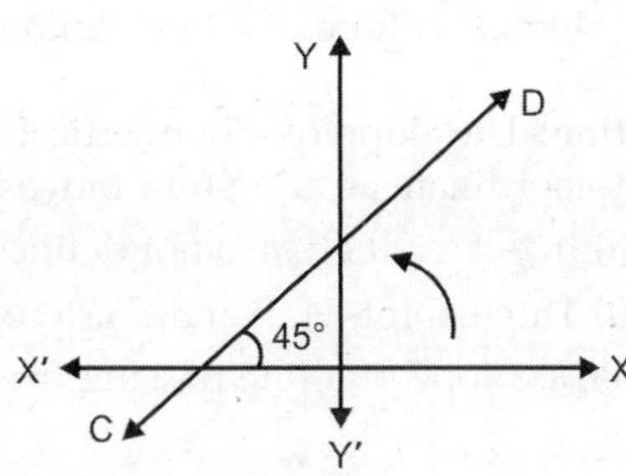

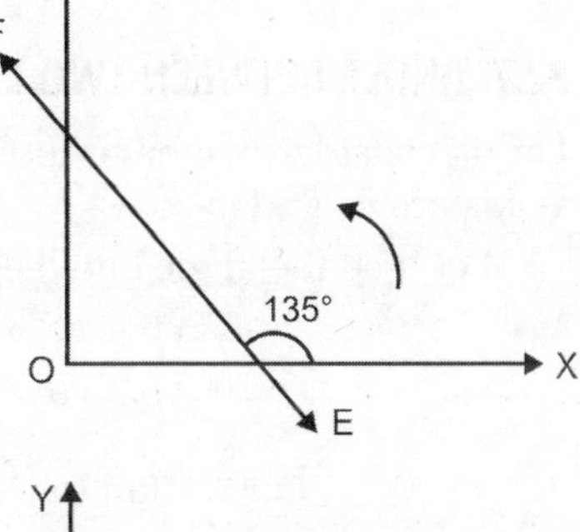

Thus, the slope of line AB is tan 60° *i.e.* $\sqrt{3}$, the slope of line CD is tan 45° = 1 and the slope of line EF is tan 135° *i.e.* – 1.

The slope of a line is usually denoted by m.

$\therefore$ $m = \tan\theta$, where θ is an angle made by a line in the anticlockwise direction from x-axis.

θ is also known as inclination of the line.

We observe that:

(*i*) If a line is parallel to x-axis, *i.e.* $\theta = 0$, then

$$m = \tan 0° = 0$$

(*ii*) If a line is parallel to y-axis, it is perpendicular to x-axis, *i.e.* $\theta = 90°$

$$\therefore m = \tan 90° = \infty, \text{ which is not defined.}$$

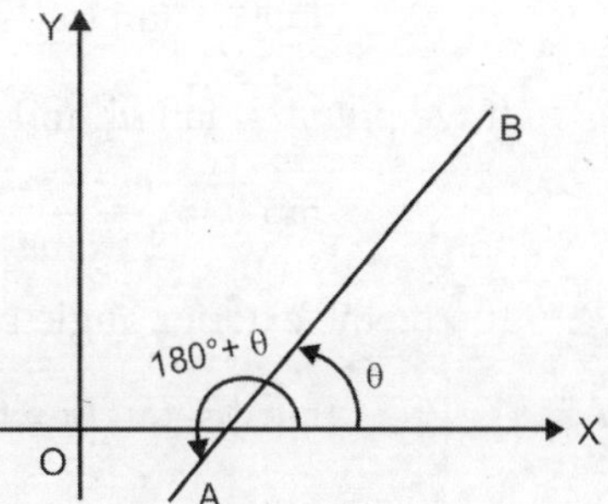

Thus, slope of x-axis is zero, and slope of y-axis is not defined.

Note: The slope of a line segment is independent of its sense as slope of $BA = \tan(180° + \theta) = \tan\theta$ = slope of AB.

22.6 SLOPE OF A LINE JOINING TWO POINTS

Let $A(x_1, y_1)$ and $B(x_2, y_2)$ be two given points. Let θ be the inclination of the line AB with x-axis, so that $m = \tan\theta$.

From A and B, draw AM and BN perpendiculars to the x-axis and draw $AC \perp BN$. Let BA meets OX in L.

$$\angle CAB = \angle XLB = \theta \quad \text{(corresponding angles)}$$

$$AC = MN = ON - OM = x_2 - x_1$$

$$BC = BN - CN = BN - AM = y_2 - y_1$$

From the right angle triangle ACB, we have

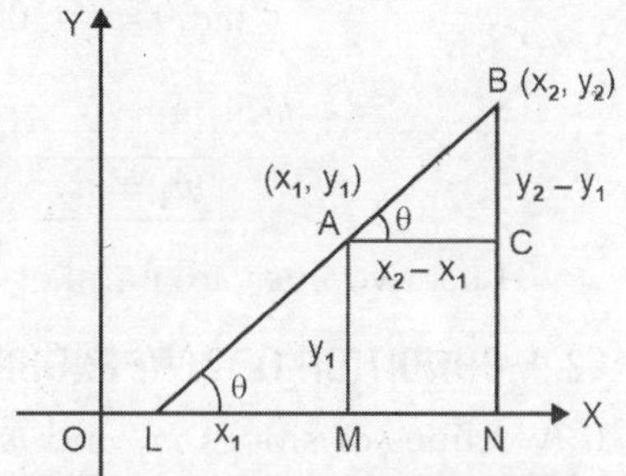

$$\tan\theta = \frac{BC}{AC} = \frac{y_2 - y_1}{x_2 - x_1}$$

$$\Rightarrow \quad \boxed{m = \frac{y_2 - y_1}{x_2 - x_1}}$$

Aid to Memory: Slope of a line joining two points $= \dfrac{\text{difference of the ordinates}}{\text{difference of the abscissae}}$

Definition: The slope of a nonvertical line is the rate of change of y coordinate with respect to x-coordinate as any point moves on the line.

Note: (*i*) If $x_2 = x_1$, then m is not defined. In that case, the line is parallel to the y-axis.

(*ii*) Three points A, B and C are collinear if and only if, slope of AB = slope of BC.

For example: Slope of a line passing through the points (2, 5) and (1, 3) is:

$$\frac{3-5}{1-2} = \frac{-2}{-1} = 2 \qquad \left(\text{Slope} = \frac{y_2 - y_1}{x_2 - x_1}\right)$$

22.7 ANGLE BETWEEN TWO LINES

Let us consider, we have to find the angle between two lines, whose inclinations to the x-axis are α_1 and α_2.

Let θ be the angle between the given lines, then

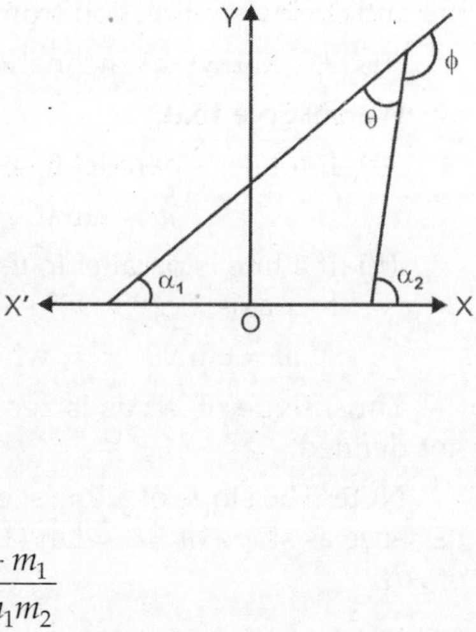

$$\alpha_2 = \theta + \alpha_1$$

$$\Rightarrow \quad \theta = \alpha_2 - \alpha_1$$

$$\Rightarrow \quad \tan\theta = \tan(\alpha_2 - \alpha_1) = \frac{\tan\alpha_2 - \tan\alpha_1}{1 + \tan\alpha_1 \tan\alpha_2}$$

If we put $m_1 = \tan\alpha_1$ and $m_2 = \tan\alpha_2$, we have

$$\tan\theta = \frac{m_2 - m_1}{1 + m_1 m_2}$$

If ϕ be the exterior angle between the lines then

$$\tan\phi = \tan(\pi - \theta) = -\tan\theta = -\frac{m_2 - m_1}{1 + m_2 m_1}$$

Hence, the complete angle formula is $\tan\theta = \pm\dfrac{m_2 - m_1}{1 + m_1 m_2}$

Aid to Memory: $\tan\theta = \pm\dfrac{\text{Difference of the slopes}}{1 + \text{product of the slopes}}$

22.8 CONDITION OF PARALLELISM OF LINES

If two lines of slopes m_1 and m_2 are parallel, then the angle θ between them is 0°.

$$\therefore \quad \tan\theta = \tan 0° \quad \Rightarrow \quad \frac{m_2 - m_1}{1 + m_1 m_2} = 0 \qquad \left[\text{Using } \tan\theta = \pm\frac{m_2 - m_1}{1 + m_1 m_2}\right]$$

$$\Rightarrow \quad m_2 - m_1 = 0$$

$$\Rightarrow \quad \boxed{m_1 = m_2}$$

Thus, two lines are parallel, if and only if their slopes are equal *i.e.* if $m_1 = m_2$.

22.9 CONDITION OF PERPENDICULARITY OF TWO LINES

If two lines of slopes m_1 and m_2 are perpendicular, then the angle θ between them is of 90°.

$\therefore \qquad \tan\theta = \tan 90°$

$$\Rightarrow \quad \frac{m_2 - m_1}{1 + m_1 m_2} = \frac{\sin 90°}{\cos 90°} \quad \Rightarrow \quad \frac{m_2 - m_1}{1 + m_1 m_2} = \frac{1}{0}$$

$$\Rightarrow \quad 1 + m_1 m_2 = 0 \quad \Rightarrow \quad m_1 m_2 = -1$$

Thus, two lines are perpendicular, if and only if their slopes m_1, m_2 satisfy the condition.

$$\boxed{m_1 m_2 = -1}$$

Example 1. *Find the slope of a line, whose inclination is:*
(*i*) 45° (*ii*) 150°

Solution. Let θ be the inclination of a line with *x*-axis, then its slope = tan θ.

(*i*) $\theta = 45° \Rightarrow$ slope $= \tan 45° = 1$

(*ii*) $\theta = 150° \Rightarrow$ slope $= \tan 150° = \tan(180° - 30°) = -\tan 30° = -\frac{1}{\sqrt{3}}$ **Ans.**

Example 2. *Find the slope of the line which makes an angle 30° with the positive direction of y-axis measured anticlockwise.*

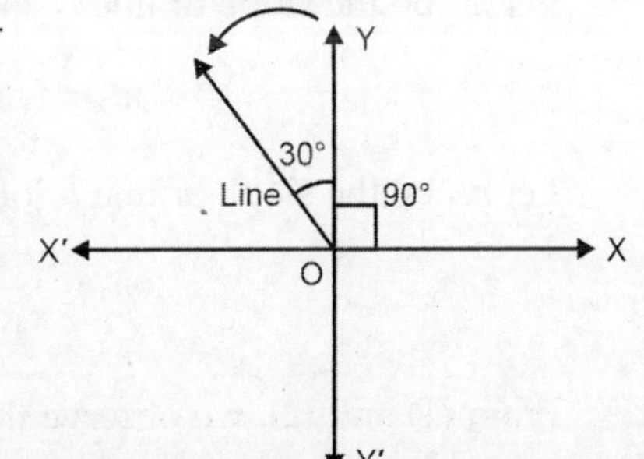

Solution. The line makes 30° with *y*-axis when measured anticlockwise.

The line makes (90° + 30°) *i.e.* 120° with the *x*-axis as shown in the figure.

Slope of the line = tan 120°

$= -\sqrt{3}$ **Ans.**

Example 3. *Find the slope of a line, which passes through the origin, and the mid-point of the line segment joining the points P(0, – 4) and B(8, 0).*

Solution. Co-ordinates of mid-point of the line segment joining the points *P*(0, – 4) and *B*(8, 0) are $\left(\frac{8+0}{2}, \frac{0-4}{2}\right)$ *i.e.*, (4, – 2).

Now, slope of the line passing through origin (0, 0) and (4, – 2)

$$= \frac{-2-0}{4-0} = \frac{-2}{4} = -\frac{1}{2}$$ **Ans.**

Example 4. *Find the slope of the line through the points:*
(*i*) (1, 2) and (4, 2) (*ii*) (0, – 4) and (– 6, 2).

Solution. (*i*) Here, $x_1 = 1, y_1 = 2, x_2 = 4, y_2 = 2$

We know that,

$$\text{Slope of line joining two points} = \frac{y_2 - y_1}{x_2 - x_1}$$

$$\therefore \qquad \text{Required slope} = \frac{2-2}{4-1} = \frac{0}{3} = 0$$

(*ii*) Here, $x_1 = 0, y_1 = -4, x_2 = -6$ and $y_2 = 2$

$\therefore$ Slope of the joining (0, – 4) and (– 6, 2) = $\dfrac{2+4}{-6-0} = \dfrac{6}{-6} = -1$ **Ans.**

Example 5. *Determine x, so that 2 is the slope of the line through points (2, 5) and (x, 3).*

Solution. Let $A = (2, 5)$ and $B = (x, 3)$ and slope of $AB = -2$

$$\Rightarrow \quad \frac{3-5}{x-2} = +2$$

$$\Rightarrow \quad 3 - 5 = 2(x-2) \quad \Rightarrow \quad -2 = 2x - 4$$

$$\Rightarrow \quad 2x = 2 \quad \Rightarrow \quad x = 1$$ **Ans.**

Example 6. *Show that the line joining the points (2, – 3) and (– 5, 1) is parallel to the line joining the points (7, – 1) and (0, 3) and perpendicular to the line joining (4, 5) and (0, – 2).*

Solution. Let m_1 be the slope of line l_1 joining points (2, – 3) and (– 5, 1).

$$\therefore \quad m_1 = \frac{y_2 - y_1}{x_2 - x_1} = \frac{1+3}{-5-2} = -\frac{4}{7} \quad \text{... (1)}$$

Let m_2 be the slope of line l_2 joining the points (7, – 1) and (0, 3).

$$\therefore \quad m_2 = \frac{y_2 - y_1}{x_2 - x_1} = \frac{3+1}{0-7} = -\frac{4}{7} \quad \text{... (2)}$$

Let m_3 be the slope of line l_3 joining the points (4, 5) and (0, – 2).

$$\therefore \quad m_3 = \frac{y_2 - y_1}{x_2 - x_1} = \frac{-2-5}{0-4} = \frac{7}{4} \quad \text{... (3)}$$

From (1) and (2), we observe that $m_1 = -\dfrac{4}{7} = m_2$

Thus, line l_1 and l_2 are parallel.

From (1) and (3), we observe that

$$m_1 \cdot m_3 = \left(-\frac{4}{7}\right)\left(\frac{7}{4}\right) = -1$$

Thus, line l_1 and l_3 are perpendicular. **Proved.**

Example 7. *Without using Pythagoras theorem, show that (4, 4), (3, 5) and (– 1, – 1) are the vertices of a right triangle.*

Solution. Let $A(4, 4)$, $B(3, 5)$ and $C(-1, -1)$ be the vertices of the ΔABC.

$$\therefore \quad \text{Slope of } AB = m_1 = \frac{y_2 - y_1}{x_2 - x_1} = \frac{5-4}{3-4} = -1$$

$$\text{Slope of } BC = m_2 = \frac{y_2 - y_1}{x_2 - x_1} = \frac{-1-5}{-1-3} = \frac{-6}{-4} = \frac{3}{2}$$

$$\text{Slope of } CA = m_3 = \frac{y_2 - y_1}{x_2 - x_1} = \frac{-1-4}{-1-4} = \frac{-5}{-5} = 1$$

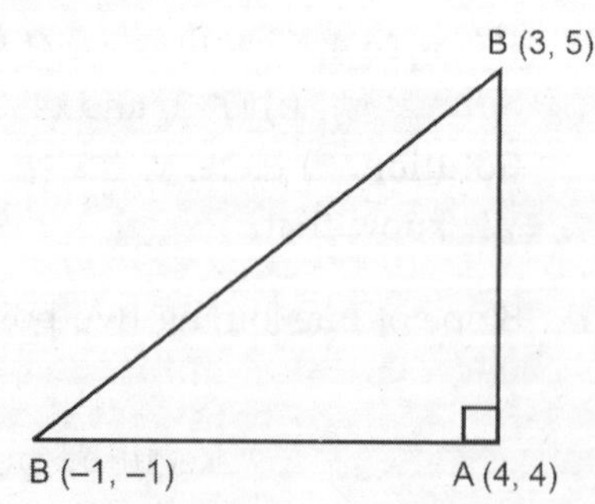

We observe that

$$m_1 \cdot m_3 = (-1)(1) = -1 \Rightarrow AB \perp CA$$

Hence, ΔABC is a right angled triangle. **Proved.**

Example 8. *A quadrilateral has the vertices at the points (– 4, 2), (2, 6), (8, 5) and (9, – 7). Show that the mid points of the sides of this quadrilateral are the vertices of a parallelogram.*

Solution. Let $A(-4, 2)$, $B(2, 6)$, $C(8, 5)$ and $D(9, -7)$ be the vertices of the quadrilateral $ABCD$.

Let E, F, G, H be the mid-points of AB, BC, CD and DA respectively.

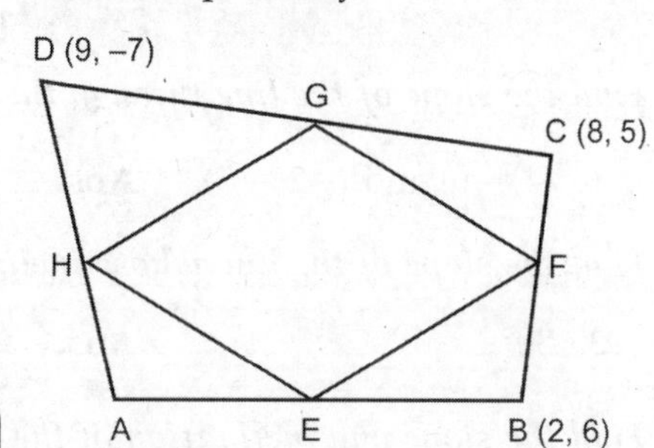

Coordinates of E are $\left(\dfrac{-4+2}{2}, \dfrac{2+6}{2}\right) = (-1, 4)$

Coordinates of F are $\left(\dfrac{2+8}{2}, \dfrac{6+5}{2}\right) = \left(5, \dfrac{11}{2}\right)$

Coordinates of G are $\left(\dfrac{8+9}{2}, \dfrac{5-7}{2}\right) = \left(\dfrac{17}{2}, -1\right)$

Coordinates of H are $\left(\dfrac{9-4}{2}, \dfrac{-7+2}{2}\right) = \left(\dfrac{5}{2}, -\dfrac{5}{2}\right)$

$$\text{Slope of } EF = \frac{y_2 - y_1}{x_2 - x_1} = \frac{\frac{11}{2} - 4}{5 - (-1)} = \frac{1}{4}, \quad \text{Slope of } GH = \frac{y_2 - y_1}{x_2 - x_1} = \frac{-\frac{5}{2} - (-1)}{\frac{5}{2} - \frac{17}{2}} = \frac{1}{4}$$

$\therefore$ Slope of EF = Slope of GH $\Rightarrow$ $EF \parallel GH$.

Similarly, slope of FG = Slope of EH $\Rightarrow$ $FG \parallel EH$

Since, $EF \parallel GH$ and $FG \parallel EH$, therefore $EFGH$ is a parallelogram. **Proved.**

Example 9. Three points $A(x_1, y_1)$, $B(x_2, y_2)$ and $C(x, y)$ are collinear. Prove that

$$(x - x_1)(y_2 - y_1) = (x_2 - x_1)(y - y_1)$$

A (x_1, y_1) ——— B (x_2, y_2) ——— C (x, y)

Solution. Since, $A(x_1, y_1)$, $B(x_2, y_2)$ and $C(x, y)$ are collinear. Therefore,

Slope of AB = Slope of AC

$$\Rightarrow \quad \frac{y_2 - y_1}{x_2 - x_1} = \frac{y - y_1}{x - x_1} \quad \Rightarrow \quad (x - x_1)(y_2 - y_1) = (x_2 - x_1)(y - y_1)$$ **Proved.**

Example 10. *The slope of a line is double of the slope of another line. If tangent of the angle between them is $\frac{1}{3}$, find the slope of the lines.*

Solution. If slope of one line is m, then the slope of the other line is $2m$.

Let angle between these two lines be θ.

Then, $\tan\theta = \dfrac{1}{3}$,

$$\text{But,} \quad \tan\theta = \frac{\tan\alpha_2 - \tan\alpha_1}{1 + \tan\alpha_1 \tan\alpha_2} \quad \Rightarrow \quad \frac{1}{3} = \frac{2m - m}{1 + m(2m)}$$

$$\Rightarrow \quad \frac{1}{3} = \frac{m}{1 + 2m^2} \quad \Rightarrow \quad 1 + 2m^2 = 3m$$

$$\Rightarrow \quad 2m^2 - 3m + 1 = 0 \quad \Rightarrow \quad (2m - 1)(m - 1) = 0$$

$\Rightarrow \quad 2m - 1 = 0 \qquad\qquad \Rightarrow \quad m - 1 = 0$

$\Rightarrow \quad m = \frac{1}{2} \qquad\qquad \Rightarrow \quad m = 1$

Thus the slope of these lines are $\frac{1}{2}$ and 1. **Ans.**

EXERCISE 22.1

Find the slope of the line through the points:

1. (4, – 6) and (– 2, – 5) **Ans.** $-\frac{1}{6}$
2. (– 2, 6) and (4, 8) **Ans.** $\frac{1}{3}$

Find the slope of the line whose inclination is:

3. 30° **Ans.** $\frac{1}{\sqrt{3}}$
4. 135° **Ans.** – 1

Find the slope and inclination of the line through each pair of the following points:

5. (1, 2) and (5, 6) **Ans.** 1, 45°
6. (10, 4) and (– 2, – 2) **Ans.** $\frac{1}{2}$, 26°34′
7. Find x, if the slope of the line joining (– 8, 11), (2, x) is $-\frac{4}{3}$. **Ans.** $-\frac{7}{3}$
8. If the line joining (– 5, 7) and (0, – 2) is perpendicular to the line joining (1, 3) and (4, x). Then find x. **Ans.** $\frac{14}{3}$
9. Find the slope of a line perpendicular to the line, which passes through each pair of the following points:
   (*i*) (0, 8) and (– 5, 2) **Ans.** $-\frac{5}{6}$ (*ii*) (1, – 11) and (5, 2) **Ans.** $-\frac{4}{13}$
10. What is the value of y so that the line through (3, y) and (2, 7) is parallel to the line through (– 1, 4) and (0, 6)? **Ans.** 9

State, whether the two lines AB and CD passing through the following points, are parallel, perpendicular or neither parallel nor perpendicular:

11. A(5, 6), B(2, 3) and C(9, – 2), D(6, – 5) **Ans.** Parallel
12. A(16, 6), B(3, 15) and C(8, 2), D(– 5, 3) **Ans.** Neither perpendicular nor parallel
13. A(2, – 5), B(– 2, 5) and C(6, 3), D(1, 1) **Ans.** Perpendicular
14. A(9, 5), B(– 1, 1) and C(8, – 3), D(3, – 5). **Ans.** Parallel
15. Without using Pythagoras Theorem, show that (12, 8), (– 2, 6) and (6, 0) are the vertices of a right angled triangle.
16. Without using distance formula, prove that the points A(1, 4), B(3, – 2) and C(– 3, 16) are collinear.

22.10 STRAIGHT LINES IN SIMPLEST FORMS

Equation of a line parallel to x-axis

Let AB be a straight line parallel to x-axis at a distance b from it. The straight line can be considered as the locus of a moving point $P(x, y)$ whose distance from the x-axis is equal to b for all position of P.

Draw PM perpendicular to the x-axis.

The, $OM = x$ and $MP = y$

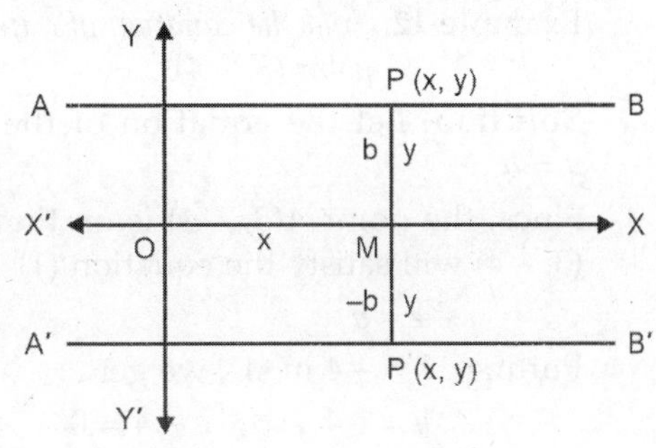

Case I. When line AB is above the x-axis, then putting $MP = b$, we get $y = b$ ... (1)

Case II. When line AB is below the x-axis, then putting $MP = -b$, we get $y = -b$ ... (2)

Combining (1) and (2), we get $y = \pm b$ which is the required equation of the straight line AB.

Corollary: The equation of the x-axis is $y = 0$, since $b = 0$ in this case.

Equation of a Line Parallel to y-axis

Let AB be a straight line parallel to y-axis at a distance 'a' from it. This straight line can be considered as the locus of a moving point $P(x, y)$ whose distance from y-axis is equal to a for all position of P.

Let PM be the perpendicular from P to the x-axis. Then, $OM = x$ and $MP = y$.

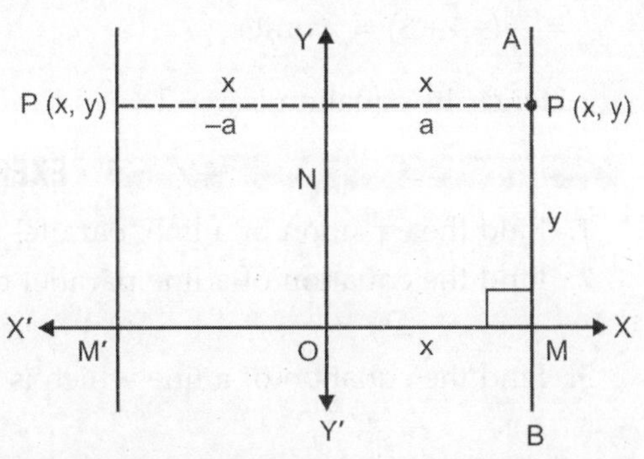

Case I. When line AB is right to the y-axis, then putting $OM' = a$, we get

$x = a$ ... (1)

Case II. When line AB is left to the y-axis, then putting $OM' = -a$, we get

$x = -a$ ... (2)

Combining (1) and (2), we get

$x = \pm a$

which is the required equation of the straight line AB.

Corollary: The equation of y-axis is $x = 0$, since $a = 0$ in this case.

Aid to Memory: The equation of a line parallel to x-axis does not contain x and the equation of a line parallel to y-axis does not contain y.

Example 11. *Write down the equations of the following lines:*

(i) A line parallel to x-axis and 2 units above it.

(ii) A line parallel to y-axis and 3 units to the left of it.

Solution. (*i*) The equation of a line AB parallel to x-axis is $y = b$. Now, the line is above x-axis and is at a distance of 2 units. Therefore, $b = 2$

Thus, the required equation is $y = 2$. **Ans.**

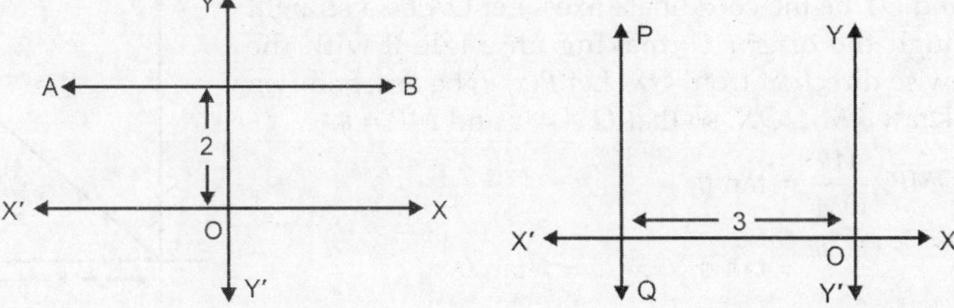

(*ii*) The equation of a line PQ parallel to y-axis is $x = a$. Now, the line is left to the y-axis and is at a distance of 3 units. Therefore, $a = -3$. Thus, the required equation is $x = -3$. **Ans.**

Example 12. *Find the equation of a line which is parallel to x-axis and passing through the point (3, – 4).*

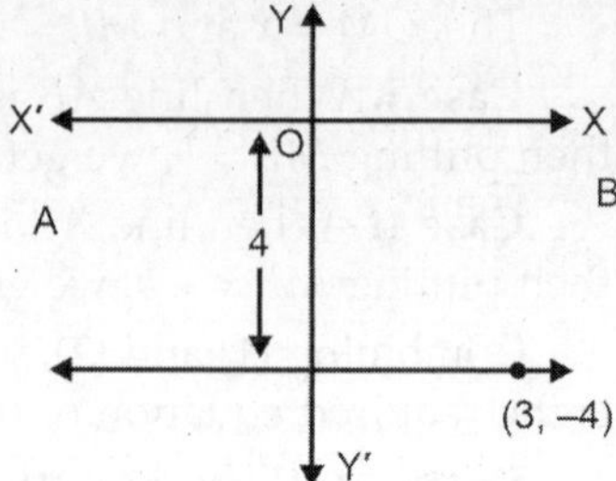

Solution. Let the equation of the required line be

$y = b$ ... (1)

Since, the point $A(3, -4)$ is on the line, so the point $(3, -4)$ will satisfy the equation (1).

$\therefore \quad -4 = b$

Putting $b = -4$ in (1), we get

$$y = -4 \Rightarrow y + 4 = 0$$

Hence, the required equation is $y + 4 = 0$. **Ans.**

Example 13. *Find the equation of a line which is equidistant from the lines x = – 4 and x = 8.*

Solution. Since, both the given lines are parallel to y-axis and the required line is equidistant from these lines, so it is also parallel to y-axis and its distance from y-axis

$= \frac{1}{2}(-4+8) = 2$ units.

Hence, its equation is $x = 2$. **Ans.**

EXERCISE 22.2

1. Find the equation of a line, parallel to x-axis and 5 units below it. **Ans.** $y = -5$
2. Find the equation of a line parallel to y-axis and 7 units to the right of it. **Ans.** $x = 7$
3. Find the equation of a line which is parallel to x-axis and passes through $(3, -9)$. **Ans.** $y = -9$
4. Find the equation of a line which is parallel to y-axis and passes through $(-4, 3)$. **Ans.** $x = -4$
5. Determine the equation of a line, parallel to x-axis and having intercept on y-axis as -2. **Ans.** $y + 2 = 0$
6. Find the equation of the line perpendicular to the x-axis and passing through the origin. **Ans.** $x = 0$

Find the equation of the straight line which is equidistant from the lines:

7. $x = -3$ and $x = 7$ **Ans.** $x = 2$
8. $y = -4$ and $y = 5$ **Ans.** $y = \frac{1}{2}$

22.11 EQUATION OF A LINE THROUGH THE ORIGIN

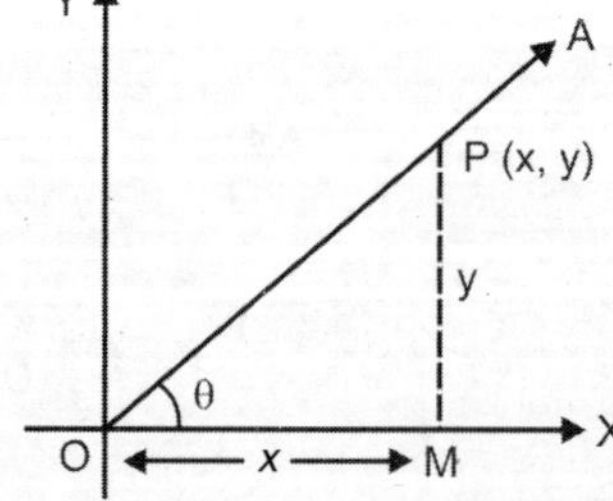

Let OX and OY be the coordinate axes. Let OA be a straight line through the origin O, making an angle θ with the anticlockwise direction from OX. Let $P(x, y)$ be any point on the line. Draw $PM \perp OX$, so that $OM = x$ and $MP = y$.

In ΔOMP, $\frac{MP}{OM} = \tan\theta$

$\Rightarrow \quad \frac{y}{x} = \tan\theta \Rightarrow y = x\tan\theta$

If we denote $\tan\theta$ by m, the equation takes the standard form,

$$\boxed{y = mx}$$

22.12 INTERCEPTS OF A LINE ON THE AXES

If a line AB meets x-axis in A and y-axis in B, then OA and OB are called the intercepts of the line AB on x-axis and y-axis respectively.

The intercept on x-axis, is positive, if A is to the right of the origin, as in Fig. (a) and negative, if A lies to the left of the origin as in Fig. (b).

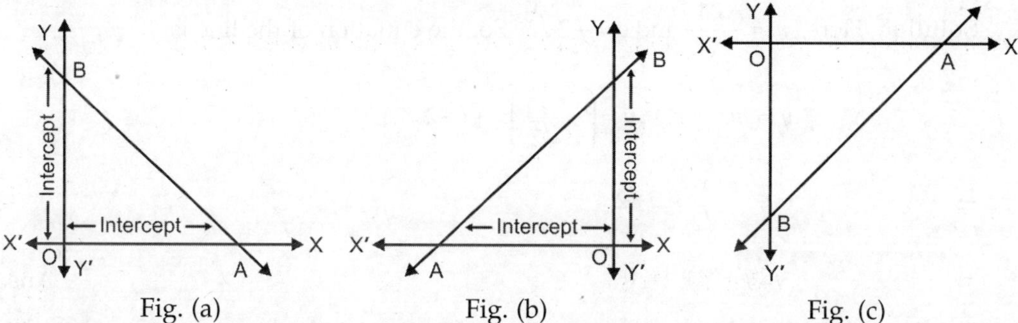

Fig. (a) Fig. (b) Fig. (c)

Also, intercept on y-axis is positive if B is above, origin and negative, if B lies below the origin.

Thus, OB is positive in Fig. (b) and negative in Fig. (c).

22.13 SLOPE-INTERCEPT FORM OF THE EQUATION OF A LINE

The equation of a line with slope m and making an intercept c on y-axis is $y = mx + c$.

Proof: Let OX and OY be the coordinate axes. Let the given line AB cut the axes in A and C.

So the $\angle XAB = \theta$ and its intercept on y-axis is equal to c, then $\tan\theta = m$.

Take any point $P(x, y)$ on the line and draw PM perpendicular to x-axis and CN parallel to x-axis to meet MP at N.

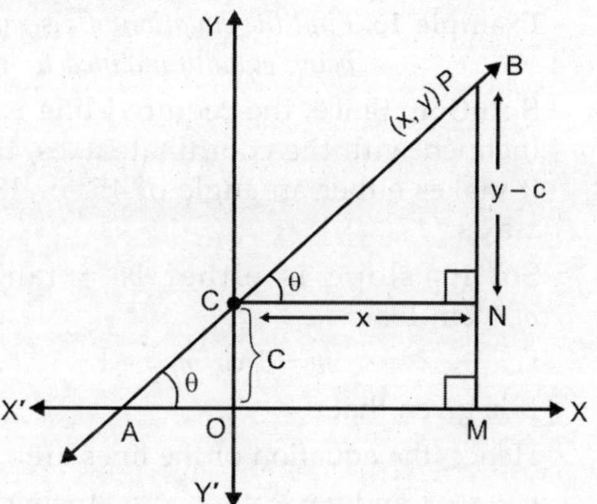

Then, $\tan\theta = \dfrac{NP}{CN} = \dfrac{MP - MN}{OM}$

$\Rightarrow \quad m = \dfrac{MP - OC}{OM}$ $\quad [\because \tan\theta = m \text{ and } MN = OC]$

$\Rightarrow \quad m = \dfrac{y - c}{x} \Rightarrow mx = y - c$

$$\Rightarrow \quad \boxed{y = mx + c}$$

which is the required equation and is called the slope intercept form.

Remark:

1. If the line passes through the origin, then $0 = m(0) + c \Rightarrow c = 0$. Therefore, the equation of the line passing through the origin is $y = mx$, where m is the slope of the line.

2. If the line is parallel to x-axis, then $m = 0$, therefore the equation of a line parallel to x-axis is $y = c$.

Example 14. *Find the equation of a line with slope* $-\dfrac{1}{\sqrt{2}}$ *and cutting off and intercept of* $2\sqrt{2}$ *units on negative direction of y-axis.*

Solution. Here, $m = -\dfrac{1}{\sqrt{2}}$ and $c = -2\sqrt{2}$. So, the equation of the line is

$$y = mx + c, \quad y = \left(-\frac{1}{\sqrt{2}}\right)x + (-2\sqrt{2})$$

$$\Rightarrow \quad y = -\frac{x}{\sqrt{2}} - 2\sqrt{2}$$

$$\Rightarrow \quad \sqrt{2}y + x + 4 = 0$$

Ans.

Example 15. *Write down the equation of the straight line, which makes an angle* $\tan^{-1}(2)$ *with x-axis and cuts off an intercept 5 from the positive direction of y-axis.*

Solution. Here $m = \tan(\tan^{-1} 2) = 2$, $c = 5$

Substituting these values in the slope intercept form $y = mx + c$, the required line is

$$y = 2x + 5.$$

Ans.

Example 16. *Find the equation of a straight line cutting off an intercept –1 from y-axis and being equally inclined to the axes.*

Solution. Since, the required line is equally inclined with the coordinate axes, therefore it makes either an angle of 45° or 135° with x-axis.

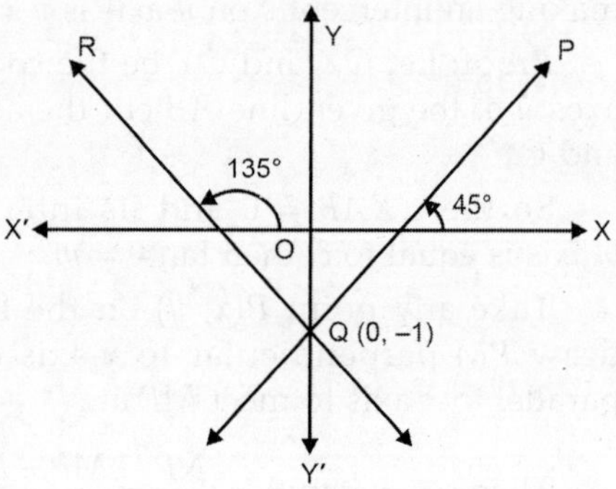

So, its slope is either $m = \tan 45°$ or $m = \tan 135°$

i.e. $\quad m = 1$ or $m = -1$

It is given that $c = -1$.

Hence, the equation of the lines are

$y = x - 1$ and $y = -x - 1$ $\quad$ [using $y = mx + c$] **Ans.**

Example 17. *Find the equation of a straight line, whose y-intercept is – 3 and which is parallel to the line joining the points (– 2, 3) and (4, – 5).*

Solution. Here c = y-intercept = – 3

Let m be the slope of the required line. Since, the required line is parallel to the line joining the points $A(-2, 3)$ and $B(4, -5)$.

$$\therefore \quad m = \text{slope of } AB = \frac{y_2 - y_1}{x_2 - x_1} = \frac{-5-3}{4-(-2)} = \frac{-8}{6} = -\frac{4}{3}$$

$\therefore$ The equation of the required line is $y = \left(-\dfrac{4}{3}\right)x + (-3)$

$$\Rightarrow \quad 3y = -4x - 9 \quad \Rightarrow \quad 4x + 3y + 9 = 0$$

Ans.

EXERCISE 22.3

Write the equation of a line which has the y-intercept:

1. 2 and slope 7 **Ans.** $7x - y + 2 = 0$
2. – 1 and is parallel to $y = 5x - 7$. **Ans.** $y = 5x - 1$
3. 2 and is inclined at 45° to the x-axis. **Ans.** $y = x + 2$
4. – 5 and is equally inclined to the axis. **Ans.** $y = x - 5$ or $y = -x - 5$
5. What will be the value of m and c, if the straight line $y = mx + c$ passes through the points (3, – 4) and (– 1, 2)? **Ans.** $m = -\frac{3}{2}, c = \frac{1}{2}$
6. Find the equations of the bisectors of the angles between the co-ordinates axes. **Ans.** $x \pm y = 0$
7. Find the equation of a line, which has y-intercept 2 and is perpendicular to $y = 2x + 1$. **Ans.** $x + 2y - 4 = 0$
8. Find the equation of a line which has y-intercept – 4 and perpendicular to $y + x \sin 45° + 2 = 0$. **Ans.** $y = \sqrt{2}x - 4$
9. Find the equation of line, which has y-intercept – 3 and is parallel to the line $x + y + 1 = 0$. **Ans.** $y = x + 3 = 0$
10. Find the equation of a straight line, which makes an angle $\tan^{-1}$ (2) with the x-axis and cuts off intercept 5 from the negative side of y-axis. **Ans.** $2x - y - 5 = 0$
11. Find the equation of a straight line, whose y intercept is – 4 and which is parallel to the line joining the points (– 3, 4) and (2, – 5). **Ans.** $5y + 9x + 4 = 0$
12. Find the equation of a straight line, whose y intercept is 5 and which is perpendicular to the line joining (1, – 2) and (– 2, – 3). **Ans.** $y + 3x - 5 = 0$

22.14 THE POINT SLOPE FORM OF A LINE

The equation of a line which passes through the point (x_1, y_1) and has the slope 'm' is

$$y - y_1 = m(x - x_1)$$

Proof: Let a line l makes an angle θ with OX such that

$$\tan \theta = m$$

Let $A(x_1, y_1)$ be a given point on the line and $P(x, y)$ be any point on the line.

Then, slope of $AP = \frac{y - y_1}{x - x_1}$

$$\Rightarrow \quad m = \tan \theta = \frac{y - y_1}{x - x_1}$$

$$\Rightarrow \quad \boxed{y - y_1 = m(x - x_1)}$$

which is required equation.

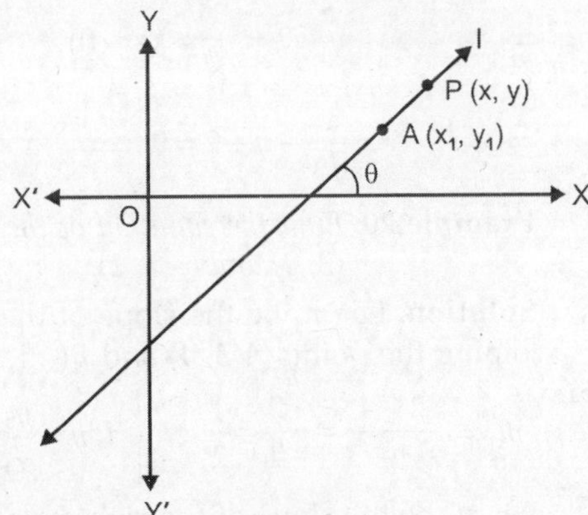

Example 18. *Find the equation of a line which satisfies the given conditions:*

(i) *Passing through (2, 2) and inclined to x-axis at 45°*

(ii) *Passing through (p cos α, p sin α) and making an angle (90° + α) with +ve direction of x-axis.*

Solution. (*i*) Here, $m = \tan 45° = 1, x_1 = 2, y_1 = 2$

We know that the equation of a line passing through (x_1, y_1) is $y - y_1 = m(x - x_1)$

$\therefore$ The equation of a line passing through (2, 2) is

$$y - 2 = 1(x - 2) \qquad [\because m = 1]$$

$$\Rightarrow \quad y - 2 = x - 2$$

$\Rightarrow \quad x - y = 0$ which is the required equation of line. **Ans.**

(*ii*) Here, $m = \tan(90° + \alpha) = -\cot\alpha, x_1 = p\cos\alpha, y_1 = p\sin\alpha$

So, the equation of a line passing through (x_1, y_1) is

$$y - y_1 = m(x - x_1)$$

$$\Rightarrow \quad y - p\sin\alpha = -\cot\alpha(x - p\cos\alpha)$$

$$\Rightarrow \quad y - p\sin\alpha = -\frac{\cos\alpha}{\sin\alpha}(x - p\cos\alpha)$$

$$\Rightarrow \quad y\sin\alpha - p\sin^2\alpha = -x\cos\alpha + p\cos^2\alpha$$

$$\Rightarrow \quad x\cos\alpha + y\sin\alpha - p(\sin^2\alpha + \cos^2\alpha) = 0$$

$\Rightarrow \quad x\cos\alpha + y\sin\alpha = p$, which is the required equation of the line. **Ans.**

Example 19. *Find the equation of line intersecting the y-axis at a distance of 2 units above the origin and making an angle of 30° with positive direction of the x-axis.*

Solution. The line intersect the y-axis at distance of 2 units above the origin. It implies that the line passing through (0, 2). It makes an angle 30° with the positive direction of x-axis. So, the slope of the line is

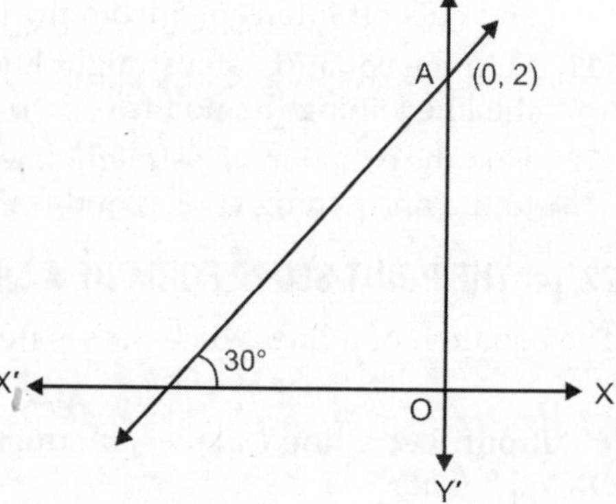

$$= \tan 30° = \frac{1}{\sqrt{3}}$$

Thus, the equation of the straight line

$$y - y_1 = m(x - x_1)$$

$$\Rightarrow \quad y - 2 = \frac{1}{\sqrt{3}}(x - 0)$$

$$\Rightarrow \quad \frac{x}{\sqrt{3}} - y + 2 = 0 \Rightarrow x - \sqrt{3}y + 2\sqrt{3} = 0$$ **Ans.**

Example 20. *Find the equation of the right bisector of the line segment joining the points (3, 4) and (– 1, 2).*

Solution. Let m_1 be the slope of the line segment joining the points $A(3, 4)$ and $B(-1, 2)$. Then,

$$m_1 = \frac{2-4}{-1-3} = \frac{-2}{-4} = \frac{1}{2} \qquad \left[\because m = \frac{y_2 - y_1}{x_2 - x_1}\right]$$

Let m_2 be the slope of the right bisector. Then.

$$m_2 = -2 \qquad [\because m_1 m_2 = -1]$$

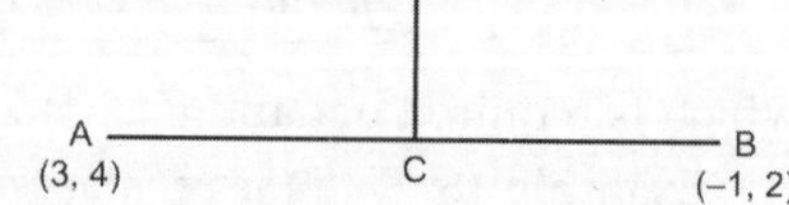

Co-ordinates of mid-point of the line segment joining the points (3, 4) and (– 1, 2) are

$$\left(\frac{3-1}{2}, \frac{4+2}{2}\right) \text{i.e.} (1, 3)$$

Now, the equation of the line passing through (1, 3) and having slope $m_2 = -2$ is

$$y - 3 = -2(x - 1)$$

$$\Rightarrow \quad 2x + y - 5 = 0$$

This is the required equation of the line. **Ans.**

Example 21. *Find the coordinates of the foot of perpendicular from the point (– 1, 3) to the line 3x – 4y – 16 = 0.*

Solution. We have,

$$3x - 4y - 16 = 0 \quad \text{... (1)}$$

$$\Rightarrow \quad 4y = 3x - 16$$

$$\Rightarrow \quad y = \frac{3}{4}x - 4$$

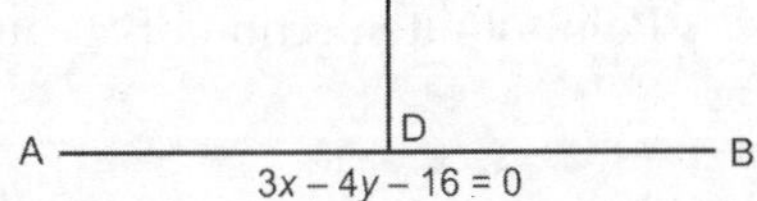

Let m_1 be the slope of this line. Then $m_1 = \frac{3}{4}$

If m_2 is the slope of any other line perpendicular to (1), then $m_2 = -\frac{4}{3}$ $[m_1 m_2 = -1]$

Equation of perpendicular line passing through (– 1, 3) and having slope $m_2 = -\frac{4}{3}$ is:

$$y - 3 = -\frac{4}{3}(x+1) \quad \Rightarrow \quad 3y - 9 = -4x - 4$$

$$\Rightarrow \quad 4x + 3y - 5 = 0 \quad \text{... (2)}$$

Solving (1) and (2), we get the co-ordinates of foot of perpendicular from the point (– 1, 3) to the line $3x - 4y - 16 = 0$: are

$$x = \frac{68}{25},\ y = -\frac{49}{25}$$

Hence, the coordinates of the foot of perpendicular are $\left(\frac{68}{25}, -\frac{49}{25}\right)$ **Ans.**

Example 22. *Find the coordinates of the foot of the perpendicular drawn from the point (2, 3) to the straight line 3x – y + 4 = 0.*

Solution. Here, we have the equation of a line *AB*

$$3x - y + 4 = 0 \quad \text{... (1)}$$

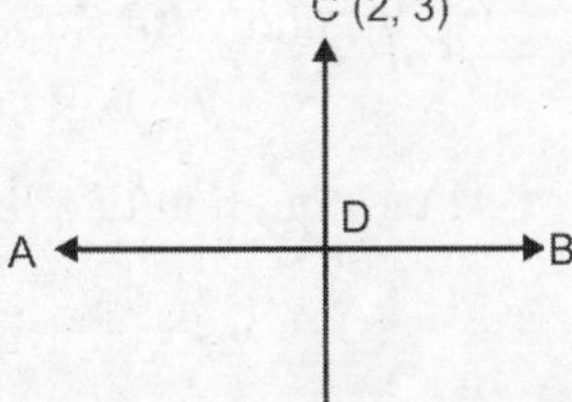

Its slope is 3.

Let the slope of a line perpendicular to (1) be m

$$m\,(3) = -1 \quad \Rightarrow \quad m = -\frac{1}{3}, \quad (m_1 m_2 = -1)$$

Equation of a line with slope $\left(-\frac{1}{3}\right)$ and passing through C(2, 3) is

$$y - 3 = -\frac{1}{3}(x - 2) \quad \Rightarrow \quad 3y - 9 = -x + 2$$

$$\Rightarrow \quad x + 3y = 11 \quad \text{... (2)}$$

On solving (1) and (2) we get the coordinates of a point of intersection of (1) and (2).

$$x = \frac{-1}{10}\ y = \frac{37}{10}$$

Hence the coordinates of the foot of perpendicular are $\left(\frac{-1}{10}, \frac{37}{10}\right)$. **Ans.**

Example 23. *The perpendicular from the origin to the line $y = mx + c$ meets it at the point $(-1, 2)$. Find the values of m and c.*

Solution. We have, $y = mx + c$ ... (1)

Equation of a line passing through (0, 0) and perpendicular to (1) is

$$y - 0 = -\frac{1}{m}(x - 0) \quad \Rightarrow \quad y = -\frac{x}{m} \qquad \text{... (2)}$$

Point of intersection of (1) and (2) is obtained by solving them

$$x = -\frac{mc}{m^2+1}, \qquad y = \frac{c}{m^2+1}$$

According to question, point of intersection of (1) and (2) is $(-1, 2)$.

Thus $\quad -1 = -\frac{mc}{m^2+1}$ and $2 = \frac{c}{m^2+1}$

$$\Rightarrow \qquad m = \frac{1}{2}, c = \frac{5}{2}$$ **Ans.**

Example 24. *The perpendicular from the origin to a line meets it at the point $(-2, 9)$, find the equation of the line.*

Solution. Let the perpendicular ON to the line AB meet it at $(-2, 9)$.

Slope of ON passing through (0, 0) and $(-2, 9)$

$$= \frac{9-0}{-2-0} = -\frac{9}{2}$$

Let slope of AB be *m*. Then,

$$-\frac{9}{2}m = -1 \ (\because m_1 m_2 = -1)$$

$$\Rightarrow \qquad m = \frac{2}{9}$$

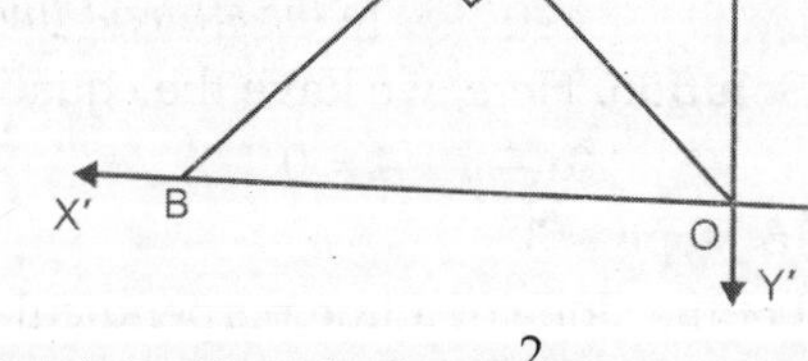

Equation of a line passing through a point (x_1, y_1) and with slope *m* is

$$y - y_1 = m(x - x_1)$$

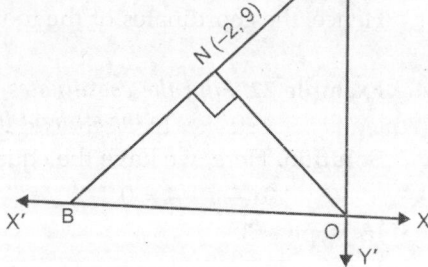

Thus, equation of a line passing through a point $(-2, 9)$ with slope $\frac{2}{9}$ is

$$y - 9 = \frac{2}{9}(x + 2)$$

$$9y - 81 = 2x + 4$$

$$2x - 9y + 85 = 0$$ Ans.

Example 25. *In triangle ABC with vertices A(2, 3), B(4, – 1), C(1, 2). Find the equation and length of altitude from the vertex A.*

Solution. There is a triangle *ABC* with vertices *A*(2, 3), *B*(4, – 1) and C(1, 2).

Let *AD* be the altitude drawn from the vertex *A* to *BC*.

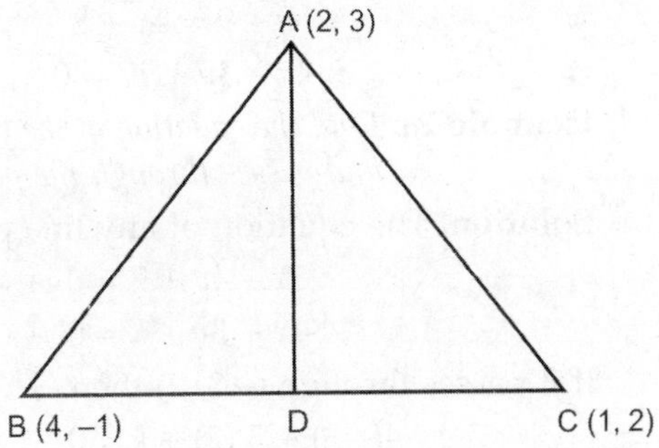

Slope of $BC = \frac{2+1}{1-4} = -\frac{3}{3} = -1$

Let the slope of $AD \perp BC = m$

$\Rightarrow$ $(-1)\, m = -1,\ m = 1\ (m_1 m_2 = -1)$

Equation of AD passing through $A(2, 3)$

with a slope $\frac{5}{3}$ is

A (2, 3)

B (4, –1) D C (1, 2)

$$y - 3 = (1)(x - 2) \quad \Rightarrow \quad x - y + 1 = 0 \quad \text{... (1)}$$

$$[y - y_1 = m(x - x_1)]$$

Equation of BC passing through $B(4, -1)$ and $C(1, 2)$ is

$$y + 1 = \frac{2+1}{1-4}(x-4) \quad \Rightarrow \quad y + 1 = -\frac{3}{3}(x-4)$$

$$\Rightarrow \quad x + y = 3 \qquad \text{... (2)}$$

Solving (1) and (2), we get the coordinates of D, *i.e.* $x = 1$, $y = 2$ *i.e.* (1, 2)

Length of $AD = \sqrt{(2-1)^2 + (3-2)^2} = \sqrt{1+1} = \sqrt{2}$ **Ans.**

Example 26. *Find the equation of a line passing through (– 3, 5) and perpendicular to the line through the points (2, 5) and (– 3, 6).*

Solution. Let m_1 be the slope of line AB joining the points $A(2, 5)$ and $B(-3, 6)$.

$$\therefore \quad m_1 = \frac{6-5}{-3-2} = -\frac{1}{5}$$

Let m_2 be the slope of the required line, which is perpendicular to the line AB.

$$\therefore \quad m_1 \cdot m_2 = -1$$

$$\Rightarrow \quad \left(-\frac{1}{5}\right) m_2 = -1 \quad \Rightarrow \quad m_2 = 5$$

Also, the equation of a line passing (– 3, 5) and having slope 5 is

$$y - 5 = 5(x + 3) \qquad [\because\ y - y_1 = m(x - x_1)]$$

$$\Rightarrow \quad y - 5 = 5x + 15 \quad \Rightarrow \quad 5x - y + 20 = 0$$

which is the required equation of the line. **Ans.**

Example 27. *Find the equation of the line passing through the point (– 2, 3) and perpendicular to the line 3x + 4y – 11 = 0.*

Solution. We have

$$3x + 4y - 11 = 0 \qquad \text{... (1)}$$

$$\text{Slope of (1)} = \frac{-3}{4}$$

Let the slope of the required line perpendicular to (1) is m.

$$m\left(\frac{-3}{4}\right) = -1,\ m = \frac{4}{3} \qquad (m_1 m_2 = -1)$$

Equation of a line passing through a point (– 2, 3) and with slope $\frac{4}{3}$ is

$$y - 3 = \frac{4}{3}(x + 2) \qquad [y - y_1 = m(x - x_1)]$$

$\Rightarrow$ $3y - 9 = 4x + 8$

$\Rightarrow$ $4x - 3y + 17 = 0$ **Ans.**

Example 28. *Find the equation of the line which is perpendicular to the line $3x - 4y + 7 = 0$ and passes through the point $(-3, 2)$.*

Solution. The equation of any line perpendicular to the line

$3x - 4y + 7 = 0$ is

$$4x + 3y + k = 0 \quad \text{... (1)}$$

If it passes through $(-3, 2)$ then

$$4(-3) + 3(2) + k = 0 \quad \Rightarrow \quad k = 6$$

Substituting this value of k in (1), we get

$$4x + 3y + 5 = 0$$

which is the required equation of the line. **Ans.**

Example 29. *Show that the perpendicular drawn from the point (4, 1) on the line joining (6, 5) and (2, – 1) divides it in the ratio 8 : 5.*

Solution. Let the points be $A(4, 1)$, $B(6, 5)$ and $C(2, -1)$.

Draw $AM \perp BC$

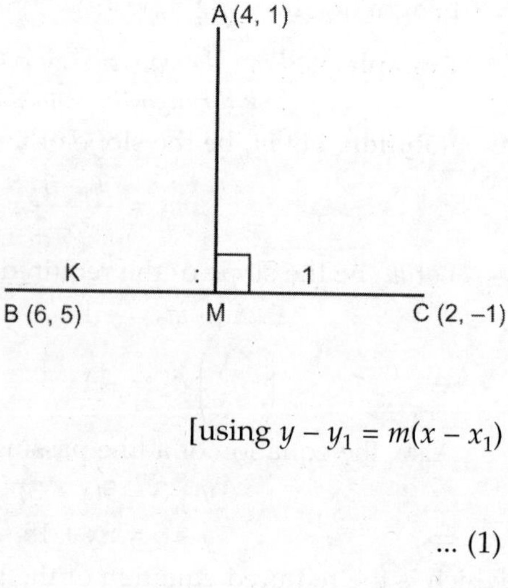

$$\text{Slope of } BC = \frac{-1-5}{2-6} = \frac{-6}{-4} = \frac{3}{2}$$

Now, (Slope of AM) × (Slope of BC) = – 1

$$(\text{Slope of } AM) \times \frac{3}{2} = 1$$

$$\Rightarrow \quad \text{Slope of } AM = -\frac{1}{3/2} = -\frac{2}{3}$$

Equation of AM is $y - 1 = -\frac{2}{3}(x - 4)$ [using $y - y_1 = m(x - x_1)$

$\Rightarrow$ $3y - 3 = -2x + 8$

$\Rightarrow$ $2x + 3y - 11 = 0$ ... (1)

Let M divides BC in the ratio $k : 1$ internally

$\therefore$ Coordinates of M are $\left(\frac{2k+6}{k+1}, \frac{-k+5}{k+1}\right)$

Since, M is on AM, therefore the coordinates of M will satisfy the equation (1).

$$\therefore \quad 2\left(\frac{2k+6}{k+1}\right) + 3\left(\frac{-k+5}{k+1}\right) - 11 = 0$$

$$\Rightarrow \quad 4k + 12 - 3k + 15 - 11k - 11 = 0$$

$$\Rightarrow \quad -10k + 16 = 0 \quad \Rightarrow \quad k = \frac{8}{5}$$

Hence, M divides BC in the ratio $\frac{8}{5} : 1$ internally, *i.e.* in the ratio of 8 : 5. Ans.

22.15 EQUATION OF TWO POINT FORM OF A LINE

The equation of a line passing through two points (x_1, y_1) and (x_2, y_2) is

$$y - y_1 = \frac{y_2 - y_1}{x_2 - x_1}(x - x_1)$$

Proof: Since the line passes through (x_1, y_1) and (x_2, y_2) therefore,

slope of the line $\qquad m = \dfrac{y_2 - y_1}{x_2 - x_1}$

Substituting this value of m in the point slope from *i.e.* $y - y_1 = m(x - x_1)$, we get

$$\boxed{y - y_1 = \frac{y_2 - y_1}{x_2 - x_1}(x - x_1)}$$ which is the required equation.

Example 30. *Find the equation of the line, which passing through the points (– 1, 1) and (2, – 4).* **[S.B.T.E. 2015]**

Solution. Here, $(x_1, y_1) \equiv (-1, 1)$ and $(x_2, y_2) \equiv (2, -4)$

We know that the equation of a line passing through two points (x_1, y_1) and (x_2, y_2) is

$$y - y_1 = \frac{y_2 - y_1}{x_2 - x_1}(x - x_1)$$

$$\therefore \quad y - 1 = \frac{-4-1}{2+1}(x+1) \quad \Rightarrow \quad y - 1 = \frac{-5}{3}(x+1)$$

$$\Rightarrow \quad 3y - 3 = -5x - 5 \quad \Rightarrow \quad 5x + 3y + 2 = 0$$ **Ans.**

Example 31. *The length L (in centimeters) of a copper rod is a linear function of its celsius temperature C. In an experiment, if L = 124.942 when C = 20 and L = 125.134 when C = 110, express L is term of C.*

Solution. Assuming L along x-axis and C along y-axis we have two points $A(124.942, 20)$ and $B(125.134, 110)$ in xy plane or LC plane.

By two-point form, the point (L, C) satisfies the equation

$$\left[y - y_1 = \frac{y_2 - y_1}{x_2 - x_1}(x - x_1)\right]$$

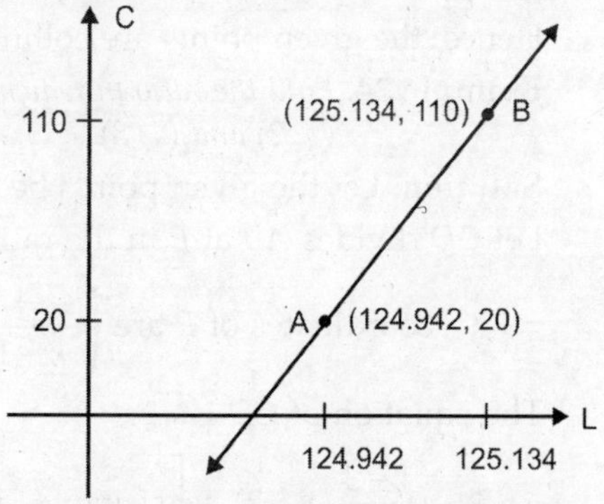

$$C - 20 = \frac{110 - 20}{125.134 - 124.942}(L - 124.942)$$

$$\Rightarrow \quad C - 20 = \frac{90}{0.192}(L - 124.942)$$

$$\Rightarrow \quad (C - 20)\frac{0.192}{90} = L - 124.942$$

$$\Rightarrow \quad L = \frac{0.192}{90}(C - 20) + 124.942$$ **Ans.**

C
110
20
(125.134, 110) B
A (124.942, 20)
L
124.942
125.134

Example 32. *The vertices of ΔPQR are P(2, 1), Q(– 2, 3) and R(4, 5). Find equation of the median through the vertex R.*

Solution. Here, we have ΔPQR whose vertices are $P(2, 1)$, $Q(-2, 3)$ and $R(4, 5)$.

Let S be the mid-point of the side PQ.

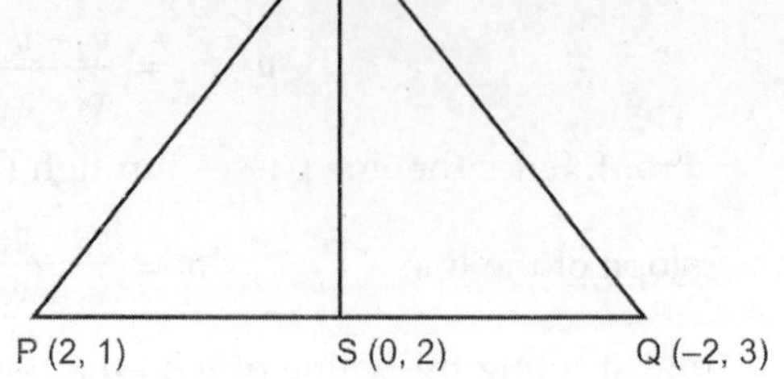

The coordinates of S are $\left(\frac{2-2}{2}, \frac{1+3}{2}\right)$

i.e. (0, 2).

RS is the median passing through $R(4, 5)$ and $S(0, 2)$.

The equation of RS is $y - 5 = \frac{2-5}{0-4}(x-4)$

$\Rightarrow \quad y - 5 = \frac{3}{4}(x-4)$

$\Rightarrow \quad 4y - 20 = 3x - 12$

$\Rightarrow \quad 3x - 4y + 8 = 0$

Hence, the equation of the median through R is $3x - 4y + 8 = 0$. **Ans.**

Example 33. *Show that the points $(at_1^2, 2at_1)$, $(at_2^2, 2at_2)$ and $(a, 0)$ are collinear if $t_1t_2 = -1$.*

Solution. Given point are $(at_1^2, 2at_1)$, $(at_2^2, 2at_2)$ and $(a, 0)$

The equation of the straight line passing through $(at_1^2, 2at_1)$ and $(at_2^2, 2at_2)$ is

$$y - 2at_1 = \frac{2at_2 - 2at_1}{at_2^2 - at_1^2}(x - at_1^2)$$

$$\Rightarrow \quad y - 2at_1 = \frac{2a(t_2 - t_1)}{a(t_2^2 - t_1^2)}(x - at_1^2) \qquad \left[\text{using } y - y_1 = \frac{y_2 - y_1}{x_2 - x_1}(x - x_1)\right]$$

$$\Rightarrow \quad y - 2at_1 = \frac{2}{t_2 + t_1}(x - at_1^2)$$

$$\Rightarrow \quad (t_1 + t_2)(y - 2at_1) = 2(x - at_1^2) \qquad \text{... (1)}$$

Given three points are collinear if $(a, 0)$ lies on (1).

$\Rightarrow$ if $(t_1 + t_2)(0 - 2at_1) = 2(a - at_1^2)$ $\quad \Rightarrow$ if $-2at_1^2 - 2at_1t_2 = 2a - 2at_1^2$

$\Rightarrow$ if $-2at_1t_2 = 2a$ $\quad \Rightarrow$ if $t_1t_2 = -1$

Hence, the given points are collinear if $t_1t_2 = -1$ **Ans.**

Example 34. *Find the ratio in which the line joining (2, 3) and (4, 1) divides the line joining (1, 2) and (4, 3).*

Solution. Let the given points be $A(1, 2)$, $B(4, 3)$, $C(2, 3)$ and $D(4, 1)$.

Let CD divides AB at P in the ratio $k : 1$ internally.

$\Rightarrow$ coordinates of P are $\left(\frac{4k+1}{k+1}, \frac{3k+2}{k+1}\right)$

The equation of CD is

$$y - 3 = \frac{1-3}{4-2}(x-2) \qquad \left[\text{Using } y - y_1 = \frac{y_2 - y_1}{x_2 - x_1}(x - x_1)\right]$$

$\Rightarrow \quad y - 3 = \frac{-2}{2}(x-2) \quad \Rightarrow \quad y - 3 = -x + 2$

$\Rightarrow \quad x + y - 5 = 0 \qquad \text{... (1)}$

Since, P lies on CD, the coordinates of P will satisfy the equation (1).

$$\therefore \quad \left(\frac{4k+1}{k+1}\right)+\left(\frac{3k+2}{k+1}\right)-5=0 \quad \Rightarrow \quad 4k+1+3k+2-5k-5=0$$

$$\Rightarrow \quad 2k-2=0 \quad \Rightarrow \quad k=1$$

Hence, the required ratio is 1 : 1. **Ans.**

EXERCISE 22.4

1. Find the equation of the line through the point (1, – 2) making an angle of 135° with the x-axis. **Ans.** $x + y + 1 = 0$
2. Find the equation of the straight line passing through the point (4, 3) and parallel to $3x + 4y = 12$. **Ans.** $3x + 4y - 24 = 0$
3. Find the equation of the line through the origin and perpendicular to $x + 2y = 4$. **Ans.** $2x - y = 0$
4. Find the equation of the line which is perpendicular to the line $\frac{x}{a}-\frac{y}{b}=1$ at the point, where it meets the x-axis. **Ans.** $ax + by = a^2$

Find the equation of the line joining the points:

5. $A(1, 1)$ and $B(2, 3)$ **Ans.** $2x - y - 1 = 0$
6. $A(3, 3)$ and $B(7, 6)$ **Ans.** $3x - 4y + 3 = 0$
7. ***The vertices of a triangle are* A(10, 4), B(– 4, 9) *and* C(– 2, – 1), *find:***
   (*a*) the equation of the side AB. **Ans.** $5x + 14y = 106$
   (*b*) the equation of the median through A. **Ans.** $y = 4$
   (*c*) the equation of the altitude through B. **Ans.** $12x + 5y + 3 = 0$
   (*d*) the equation of the perpendicular bisector of the side AB. **Ans.** $28x - 10y = 19$
8. Show that the points (1, 4), (3, – 2) and (– 3, 16) are collinear and find the equation of the line passing through these points. **Ans.** $3x + y - 7 = 0$
9. Find the equation of the straight line which bisects the joint of (5, 4) and (– 7, 0) and also bisects the join of (6, – 5) and (0, – 3). **Ans.** $3x + 2y - 1 = 0$
10. Mid points of the sides of a triangle are (2, 2), (2, 3), (4, 6). Find the equations of the sides. **Ans.** $3x - 2y - 2 = 0$, $2x - y - 1 = 0$, $x = 4$
11. Find the equation of the straight line which divides the join of the points (2, 3) and (– 5, 8) in the ratio 3 : 4 and is also perpendicular to it. **Ans.** $49x - 35y + 229 = 0$
12. The vertices of a quadrilateral are $A(-2, 6)$, $B(1, 2)$, $C(10, 4)$ and $D(7, 8)$. Find the equations of its diagonal. **Ans.** $x + 6y - 34 = 0$, $x - y + 1 = 0$

22.16 THE INTERCEPT FORM OF A LINE

The equation of a line which cuts off intercepts a and b respectively from the x and y-axis is $\frac{x}{a}+\frac{y}{b}=1.$

Proof: Let the straight line l meets the x-axis at A and the y-axis at B such that $OA = a$ and $OB = b$. Then the co-ordinates of A and B are $(a, 0)$ and $(0, b)$ respectively.

Take $P(x, y)$ on AB. Then,

By the point formula

$$y - 0 = \frac{b-0}{0-a}(x-a)$$

$\Rightarrow$ $$y = -\frac{b}{a}(x-a) \quad \text{... (1)}$$

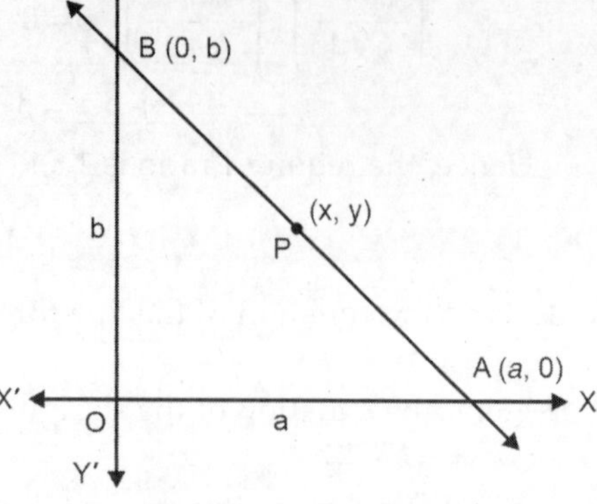

Dividing both sides of (1) by b, we get

$$\frac{y}{b} = -\frac{x}{a} + 1$$

$\Rightarrow$ $$\boxed{\frac{x}{a} + \frac{y}{b} = 1}$$

which is the required equation.

Aid to Memory: $\dfrac{x}{x\text{-intercept}} + \dfrac{y}{y\text{-intercept}} = 1$

Example 35. *Find the equation of the line which cuts off an intercept 5 on the negative direction of x-axis and an intercept 4 on positive direction of y-axis.*

Solution. Here, $a = -5$, $b = 4$.

So, the equation of the line is

$$\frac{x}{-5} + \frac{y}{4} = 1 \qquad \left[\text{Using } \frac{x}{a} + \frac{y}{b} = 1\right]$$

$\Rightarrow$ $4x - 5y = -20 \Rightarrow 4x - 5y + 20 = 0$

which is the required equation of the line. **Ans.**

Example 36. *Find the equation of the straight line passing through (2, 3) and cutting off equal intercepts along the positive directions of both the axes.*

Solution. Here, let $a = k$, $b = k$

So, the equation of the line is

$$\frac{x}{k} + \frac{y}{k} = 1 \qquad \left[\text{Using } \frac{x}{a} + \frac{y}{b} = 1\right]$$

$\Rightarrow$ $x + y = k$ ... (1)

Since, the line (1) passes through (2, 3)

$\therefore$ $2 + 3 = k \Rightarrow k = 5$

Substituting $k = 5$ in (1), we get $x + y = 5$

which is required equation of the line. **Ans.**

Example 37. *A straight line passes through the point (2, 3) and the portion of the line intercepted between the axes is bisected at this point. Find the equation of the line.* **[S.B.T.E. 2016, 2015]**

Solution. Let the intercepts of the line be a and b and the line intersects the x-axis and y-axis at A and B.

$\therefore$ $A(a, 0)$ and $B(0, b)$ are on the line and the mid point of AB is (2, 3).

$\therefore$ $$2 = \frac{a+0}{2} \Rightarrow a = 4$$

and $$3 = \frac{0+b}{2} \Rightarrow b = 6$$

Thus, the equation of the line in the intercept form is

$$\frac{x}{4}+\frac{y}{6}=1 \qquad \left[\text{Using } \frac{x}{a}+\frac{y}{b}=1\right]$$

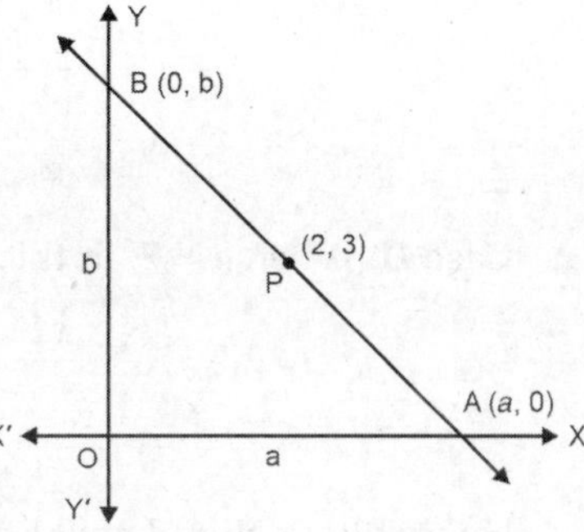

$\Rightarrow$ $3x + 2y - 12 = 0$ **Ans.**

Example 38. *Point R(h, k) divides a line segment between the axis in the ratio 1 : 2. Find the equation of the line.*

Solution. Let a line AB intersect x-axis at $(a, 0)$ and y-axis at $(0, b)$

By section formula

$$h = \frac{1\times 0+2a}{1+2}=\frac{2a}{3} \quad \Rightarrow \quad a = \frac{3h}{2}$$

$$k = \frac{1\times b+2\times 0}{3}=\frac{b}{3} \quad \Rightarrow \quad b = 3k$$

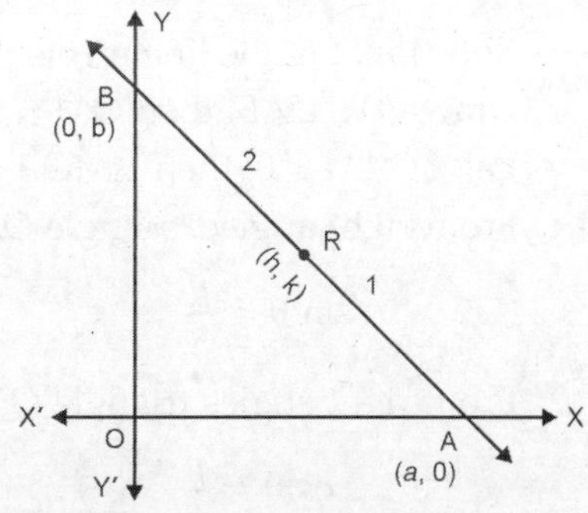

Equation of the line with intercepts a and b is

$$\Rightarrow \quad \frac{x}{a}+\frac{y}{b}=1$$

$$\Rightarrow \quad \frac{x}{\frac{3h}{2}}+\frac{y}{3k}=1$$

$$\Rightarrow \quad \frac{2x}{3h}+\frac{y}{3k}=1$$

$$\Rightarrow \quad \frac{2x}{h}+\frac{y}{k}=3 \quad \Rightarrow \quad 2kx + hy = 3kh$$

Ans.

Example 39. *Find the equation of the straight line which passes through (3, 4) and the sum of whose intercepts on the coordinates axes is 14.* **[S.B.T.E. 2015, 2013]**

Solution. Here, the sum of the intercepts = $(a + b) = 14$

$$\Rightarrow \quad b = 14 - a$$

The equation of the line in the intercept form is $\frac{x}{a}+\frac{y}{b}=1$

$$\therefore \quad \frac{x}{a}+\frac{y}{14-a}=1 \qquad \text{... (1)}$$

This line (1) passes through (3, 4).

$$\therefore \quad \frac{3}{a}+\frac{4}{14-a}=1$$

$$\Rightarrow \quad 42 - 3a + 4a = 14a - a^2 \quad \Rightarrow \quad a^2 - 13a + 42 = 0$$

$$\Rightarrow \quad (a - 6)(a - 7) = 0 \quad \Rightarrow \quad a = 6, 7$$

Case I. When $a = 6$, in this case (1) becomes

$$\frac{x}{6} + \frac{y}{14-6} = 1 \Rightarrow \frac{x}{6} + \frac{y}{8} = 1$$

$\Rightarrow$ $$4x + 3y - 24 = 0$$

Case II. When $a = 7$, in this case (1), becomes

$$\frac{x}{7} + \frac{y}{14-7} = 1 \Rightarrow \frac{x}{7} + \frac{y}{7} = 1$$

$\Rightarrow$ $$x + y - 7 = 0$$

Hence, the required equations of the lines are $4x + 3y - 24 = 0$ and $x + y - 7 = 0$. **Ans.**

Example 40. *If p be the measure of the perpendicular segment from the origin on the line whose intercepts on the axis are a and b, show that:*

$$\frac{1}{a^2} + \frac{1}{b^2} = \frac{1}{p^2}.$$

Solution. Let the line meets the x-axis in 'a' and y-axis in 'b', then $OA = a$ and $OB = b$. Draw $OM \perp AB$, then $OM = p$.

Let $\angle OAM = \theta$, then $\angle MOA = 90° - \theta$, $\angle BOM = \theta$.

From right angled triangle OMA,

$$\sin\theta = \frac{p}{a} \quad ...(1)$$

From right angles triangle OMB,

$$\cos\theta = \frac{p}{b} \quad ...(2)$$

Squaring (1) and (2) and adding, we get

$$\sin^2\theta + \cos^2\theta = \frac{p^2}{a^2} + \frac{p^2}{b^2} \Rightarrow 1 = p^2\left(\frac{1}{a^2} + \frac{1}{b^2}\right)$$

$\Rightarrow$ $$\frac{1}{a^2} + \frac{1}{b^2} = \frac{1}{p^2}$$ **Proved.**

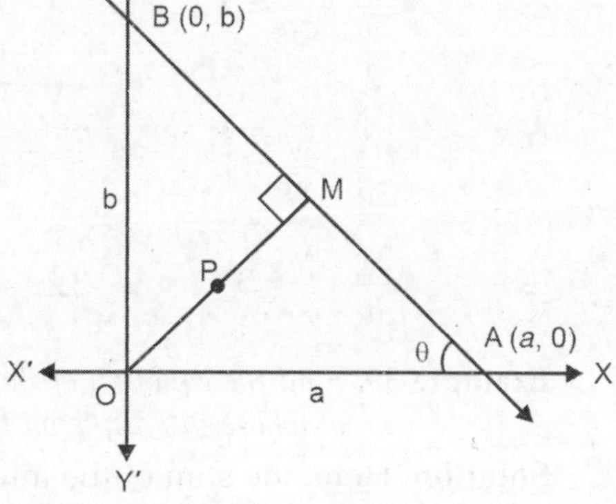

Example 41. *Find the equation of the straight line which pass through the origin and trisect the intercept of line $3x + 4y = 12$ between the axes.*

Solution. The given line is

$$3x + 4y = 12$$

$\Rightarrow$ $$\frac{x}{4} + \frac{y}{3} = 1 \quad ... (1)$$

Let the line (1) cuts x and y axes at A and B respectively, then

$$A \equiv (4, 0) \text{ and } B \equiv (0, 3)$$

Let the line AB be trisected at P and Q, then

$$AP : PB = 1 : 2$$

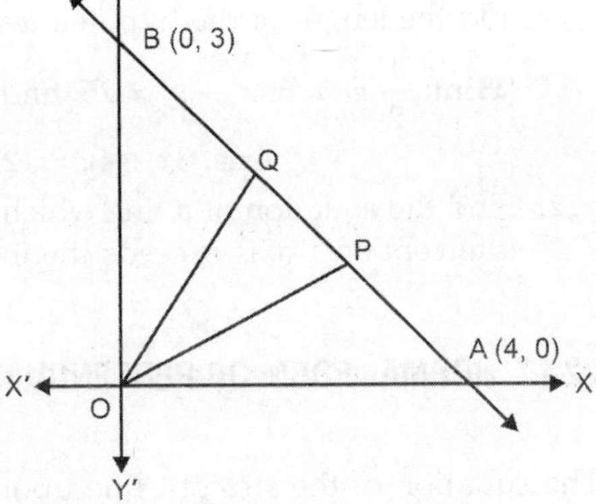

$$P \equiv \left(\frac{1\cdot 0+2\cdot 4}{1+2}, \frac{1\cdot 3+2\cdot 0}{1+2}\right)$$

$\Rightarrow \quad P \equiv \left(\frac{8}{3}, 1\right)$ and $AQ : QB = 2 : 1$

Also, $Q \equiv \left(\frac{2\cdot 0+1\cdot 4}{1+2}, \frac{2\cdot 3+1\cdot 0}{2+1}\right)$

$\Rightarrow \quad Q \equiv \left(\frac{4}{3}, 2\right)$

Now, equation of line OP passing through $(0, 0)$ is $y - 0 = \frac{1-0}{\frac{8}{3}-0}(x - 0)$

$\Rightarrow \quad y = \frac{3}{8}x \quad \Rightarrow \quad 3x - 8y = 0$

and equation of the line OQ passing through $(0, 0)$ is $y - 0 = \frac{2-0}{\frac{4}{3}-0}(x - 0)$

$\Rightarrow \quad 2y = 3x \quad \Rightarrow \quad 3x - 2y = 0$ **Ans.**

EXERCISE 22.5

Write down the equation of the straight line cutting of intercepts a and b from the axes where:

1. $a = -2, b = 3$ **Ans.** $3x - 2y + 6 = 0$
2. $a = 5, b = -6$ **Ans.** $6x - 5y - 30 = 0$
3. $a = -\frac{k}{m}, b = k$ **Ans.** $mx - y + k = 0$

Determine the x-intercept 'a' and the y-intercept 'b' of the following lines:

4. $3x + 5y - 15 = 0$ **Ans.** $a = 5, b = 3$
5. $x - y - 7 = 0$ **Ans.** $a = 7, b = -7$
6. Write down the equation of the line which makes an intercept of $2a$ on the x-axis and $3a$ on the y-axis. Given that the line passes through the point $(14, -9)$, find the numerical value of a. **Ans.** $3x + 2y = 6a, a = 4$
7. Find the equation of the straight line which passes through the point (5, 6) and has intercepts on the axes equal in magnitude but opposite in sign. **Ans.** $x - y + 1 = 0$
8. Find the equation of the straight line which passes through the point $(3, -2)$ and cuts off positive intercepts on the x-axis and y-axis which are in the ratio 4 : 3. **Ans.** $4y + 3x = 1$
9. Find the equations of the lines passing through the point (2, 2), such that the sum of their intercepts on the axes is 9. **Ans.** $x + 2y - 6 = 0, 2x + y - 6 = 0$
10. Find the equation of a line passing through the point $(-5, 4)$ and which is such that the portion of it between the axes is divided by the point in the ratio 1 : 2. **Ans.** $8x - 5y + 60 = 0, 2x - 5y + 30 = 0$

11. The area of the triangle formed by the coordinate axes and a line is 6 square units and the length of the hypotenues is 5 units. Find the equation of the line.

[**Hint:** $\frac{1}{2}ab = 6, a^2 + b^2 = 25$, find out the values of a and b]

Ans. $3x + 4y = 12$ or $3x + 4y = -12$ or $4x = 3y = 12$ or $4x + 3y = -12$

12. Find the equation of a line which passes through the point (22, – 6) and is such that intercept on x-axis exceeds the intercepts on y-axis by 5.

Ans. $6x + 11y - 66 = 0$ or $x + 2y - 10 = 0$

22.17 NORMAL FORM OR PERPENDICULAR FORM OF A LINE

[Diploma IETE, Dec. 2005]

The equation of the straight line upon which the length of the perpendicular from the origin is p and this perpendicular makes on angle α with x-axis is

$$x \cos \alpha + y \sin \alpha = p$$

Proof: Let the line AB such that the length of the perpendicular ON from the origin O to the line be p and $\angle XON = \alpha$.

Let $P(x, y)$ be any point on the line. Draw $PQ \perp OX$, $OM \perp ON$ and $PR \perp QM$. Then $OQ = x$ and $PQ = y$.

In ΔOQM,

$$\cos \alpha = \frac{OM}{OQ}$$

$$\Rightarrow \quad OM = OQ \cos \alpha = x \cos \alpha$$

In ΔPQR, $\sin \alpha = \frac{PR}{PQ} \Rightarrow PR = PQ \sin \alpha = y \sin \alpha$

$$\Rightarrow \quad MN = PR = y \sin \alpha$$

Now, $\quad p = ON = OM + MN = x \cos \alpha + y \sin \alpha$

Hence, the equation of the required line is

$$\boxed{x \cos \alpha + y \sin \alpha = p}$$

Example 42. *Write down the equation of the line for which : $p = 3$ and $\alpha = 120°$.*

Solution. Here, $p = 3$ and $\alpha = 120°$

The equation of the required line is

$$x \cos 120° + y \sin 120° = 3$$

$$\Rightarrow \quad x\left(-\frac{1}{2}\right) + y\left(\frac{\sqrt{3}}{2}\right) = 3 \Rightarrow x - y\sqrt{3} + 6 = 0$$ **Ans.**

Example 43. *The perpendicular distance of a line from the origin is 7 cm and its slope is – 1. Find the equation of the line.*

Solution. Here, $m = \tan \alpha = -1 \Rightarrow \alpha = \tan^{-1}(-1) = 135°$ and $p = 7$

Thus, the required line is

$$x \cos 135° + y \sin 135° = 7 \qquad \text{[Using } x \cos \alpha + y \sin \alpha = p\text{]}$$

$$\Rightarrow \quad x \cos (180° - 45°) + y \sin (180° - 45°) = 7$$

$\Rightarrow \qquad -x\cos 45° + y\sin 45° = 7$

$\Rightarrow \qquad -x\cdot\frac{1}{\sqrt{2}} + y\cdot\frac{1}{\sqrt{2}} = 7 \qquad \Rightarrow \qquad x - y + 7\sqrt{2} = 0$ **Ans.**

Example 44. *Find the equations of two straight lines which are at a distance $\frac{1}{2}$ from the origin and pass through the point (0, 1).*

Solution. The equation of any line which is at a distance $\frac{1}{2}$ from the origin is

$$x\cos\alpha + y\sin\alpha = \frac{1}{2} \qquad \text{... (1)}$$

It passes through the point (0, 1).

$\therefore \qquad (0)\cos\alpha + (1)\sin\alpha = \frac{1}{2} \qquad \Rightarrow \qquad \sin\alpha = \frac{1}{2}$

$\therefore \qquad \cos\alpha = \pm\sqrt{1-\sin^2\alpha} = \pm\sqrt{1-\left(\frac{1}{2}\right)^2} = \pm\sqrt{1-\frac{1}{4}} = \pm\frac{\sqrt{3}}{2}$

Putting these values of sin α and cos α in (1), we get

$$\pm\frac{\sqrt{3}}{2}x + \frac{1}{2}y = \frac{1}{2} \qquad \Rightarrow \qquad \pm\sqrt{3}x + y = 1$$

$\Rightarrow \qquad \sqrt{3}x + y - 1 = 0 \qquad \text{and} \qquad \sqrt{3}x - y + 1 = 0$

which are the required equations of the lines. **Ans.**

Example 45. *A line forms a triangle in the first quadrant with coordinate axes. If the area of the triangle is $54\sqrt{3}$ sq. unit and the perpendicular drawn from the origin to the line makes an angle 60° with x-axis, find the equation of the line.*

[S.B.T.E. 2014, 2011]

Solution. Here, $\angle MOB = 30°$

$\therefore \qquad \angle MOA = 60°$

Let $\qquad OM = p,\ OA = a,\ OB = b$

In ΔOMA, $\quad \frac{p}{a} = \cos 60° = \frac{1}{2}$

$\Rightarrow \qquad a = 2p$

In ΔOMB, $\quad \frac{p}{b} = \cos 30° = \frac{\sqrt{3}}{2}$

$\Rightarrow \qquad b = \frac{2p}{\sqrt{3}}$

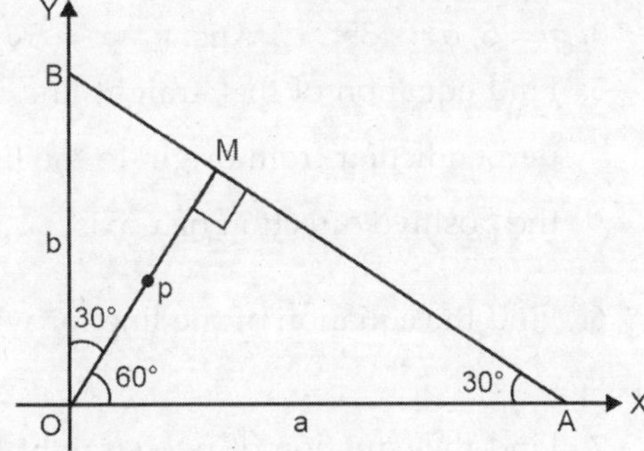

$\therefore$ Area of $\Delta OAB = \frac{1}{2}ab = \frac{1}{2}(2p)\left(\frac{2p}{\sqrt{3}}\right) = \frac{2p^2}{\sqrt{3}}$

$\Rightarrow \qquad \frac{2}{\sqrt{3}}p^2 = 54\sqrt{3} \quad \Rightarrow \quad p^2 = 81 \quad \Rightarrow \quad p = 9$ $\qquad [\because \Delta AOB = 54\sqrt{3}$ sq. units]

$\therefore$ $x\cos\alpha + y\sin\alpha = p$, the equation of the line AB is

$$x\cos 60° + y\sin 60° = 9 \qquad \Rightarrow \qquad x\left(\frac{1}{2}\right) + y\left(\frac{\sqrt{3}}{2}\right) = 9$$

$\Rightarrow \qquad x+\sqrt{3}y-18=0$ **Ans.**

Example 46. *A straight canal is $4\frac{1}{2}$ miles from a place and the shortest route from this place to the canal is exactly north-east. A village is 3 miles north and four miles east from the place. Does it lie by the nearest edge of the canal?*

Solution. Let the given place be O. Take this as the origin and the east and north directions through O as the x and y axes respectively.

Let AB be the nearest edge of the canal. From the question OL is perpendicular to AB such that $OL = 4\frac{1}{2}$ miles and $\angle LOA = 45°$.

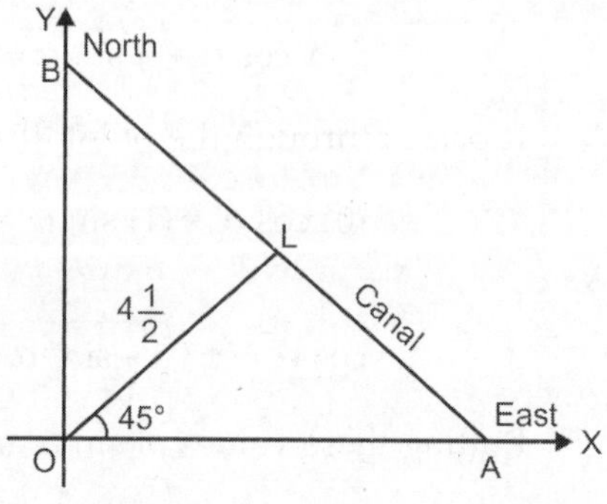

So, the equation of the canal is

$$x \cos 45° + y \sin 45° = 4\frac{1}{2}$$

$$\Rightarrow \qquad \frac{x}{\sqrt{2}}+\frac{y}{\sqrt{2}}=\frac{9}{2}$$

$$\Rightarrow \qquad \sqrt{2}\,(x+y)=9 \qquad \dots (1)$$

The position of the village is (4, 3). The village will lie on the edge of the canal if (4, 3) satisfies the equation (1). Clearly (4, 3) does not satisfy (1). Hence the village does not lie by the nearest edge of the canal. **Ans.**

EXERCISE 22.6

Find the equation of the straight line for which:

1. $p = 3, \alpha = 45°$ **Ans.** $x + y = 3\sqrt{2}$
2. $p = 5, \alpha = 30°$ **Ans.** $\sqrt{3}x+y=10$
3. $p = 5, \alpha = 135°$ **Ans.** $y - x = 5\sqrt{2}$
4. $p = 1, \alpha = 90°$ **Ans.** $y = 1$
5. Find equation of the straight line at a distance of 3 units from origin such that the perpendicular from origin to the line makes an angle α, given by $\tan \alpha = \frac{5}{12}$, with the positive direction of x-axis. **Ans.** $12x + 5y - 39 = 0, 12x + 5y + 39 = 0$
6. Find the equation of the line for which $p = 2$, $\sin \alpha = \frac{4}{5}$.
   Ans. $3x + 4y - 10 = 0, 3x - 4y + 10 = 0$
7. Find the equation of two straight lines which are at a distance of 3 units from the origin and passing through the point (6, 0). **Ans.** $x+\sqrt{3}y-6=0, x-\sqrt{3}y-6=0$
8. Find the equation of the straight line on which the perpendicular from origin makes an angle of 30° with x-axis and which forms a triangle of area $50/\sqrt{3}$ with the coordinate axes. **Ans.** $\sqrt{3}x+y=10$
9. (*i*) Show on a diagram the position of the straight line $x \cos 30° + y \sin 30° = 2$ in relation to the coordinate axes, indicating clearly which angle is 30° and which length is 2 units.

Find

(*ii*) The equation of the straight line parallel to the given line and passing through the points (4, 3).

(*iii*) The length of the perpendicular from the origin.

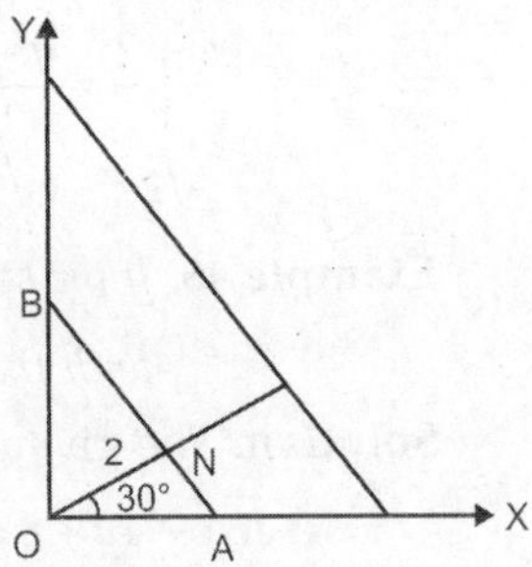

Ans. (*i*) $ON = 2$, $\angle XON = 30°$

(*ii*) $x \cos 30° + y \sin 30° = 5$

(*iii*) Length of perpendicular = 5

22.18 SYMMETRIC FORM AND PARAMETRIC EQUATION OF A LINE

The equation of the straight line passing through (x_1, y_1) and making an angle θ with the positive direction of x-axis is

$$\frac{x - x_1}{\cos\theta} = \frac{y - y_1}{\sin\theta} = r$$

where r is the distance of the point (x, y) on the line from the point (x_1, y_1).

Proof: Let a non-vertical line pass through the point $A(x_1, y_1)$ and having inclination θ *i.e.* making an angle θ with the positive direction of x-axis.

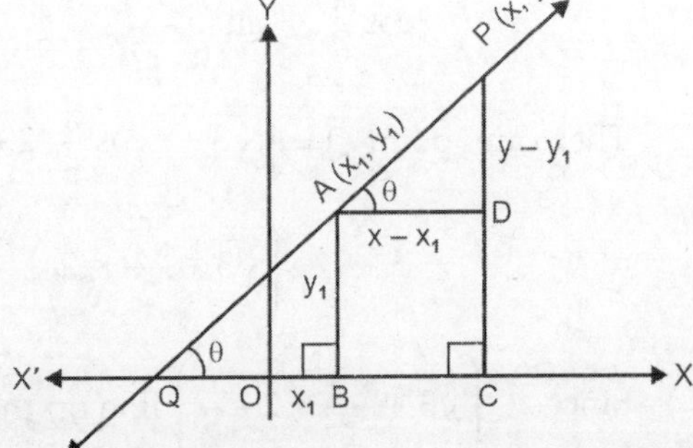

Let $P(x, y)$ be a general point on the line. Draw PC and AB perpendiculars to x-axis and also draw $AD \perp PC$.

Let $AP = r$

$\therefore \quad \angle PAD = \angle PQB = \theta$

From ΔAPD, we have

$$\cos\theta = \frac{AD}{AP} = \frac{BC}{AP} = \frac{OC - OB}{AP} = \frac{x - x_1}{r} \quad \text{... (1)}$$

and $$\sin\theta = \frac{PD}{AP} = \frac{PC - CD}{AP} = \frac{PC - AB}{AP} = \frac{y - y_1}{r} \quad \text{... (2)}$$

From (1) and (2), we have

$$\therefore \quad \frac{x - x_1}{\cos\theta} = r \text{ and } \frac{y - y_1}{\sin\theta} = r \quad \Rightarrow \quad \frac{x - x_1}{\cos\theta} = \frac{y - y_1}{\sin\theta} = r$$

This is the required equation of the line. **Proved.**

Remark: If $P(x, y)$ be a point at a distance of r units from a given point $Q(x_1, y_1)$, then $x = x_1 + r\cos\theta$ and $y = y_1 + r\sin\theta$.

Example 47. *Find the equation of the line through (– 2, 1) in symmetrical form when the angle made by the line with positive direction of x-axis is 45°.*

Solution. Here, $\theta = 45°$. The given point on the line is (– 2, 1).

Using the symmetric form, the equation of the line is

$$\frac{x + 2}{\cos 45°} = \frac{y - 1}{\sin 45°} \qquad \left[\frac{x - x_1}{\cos\theta} = \frac{y - y_1}{\sin\theta} = r\right]$$

$$\Rightarrow \quad \frac{x+2}{\frac{1}{\sqrt{2}}} = \frac{y-1}{\frac{1}{\sqrt{2}}} \quad \Rightarrow \quad x+2 = y-1 \quad \Rightarrow \quad x - y + 3 = 0$$ **Ans.**

Example 48. *If the straight line drawn through the point $P(\sqrt{3}, 2)$ and making an angle $\frac{\pi}{6}$ with the x-axis meets the line $\sqrt{3}x - 4y + 8 = 0$ at Q, find the length of PQ.*

Solution. The given line is

$$\sqrt{3}x - 4y + 8 = 0 \quad \text{... (1)}$$

and $\quad P \equiv (\sqrt{3}, 2)$

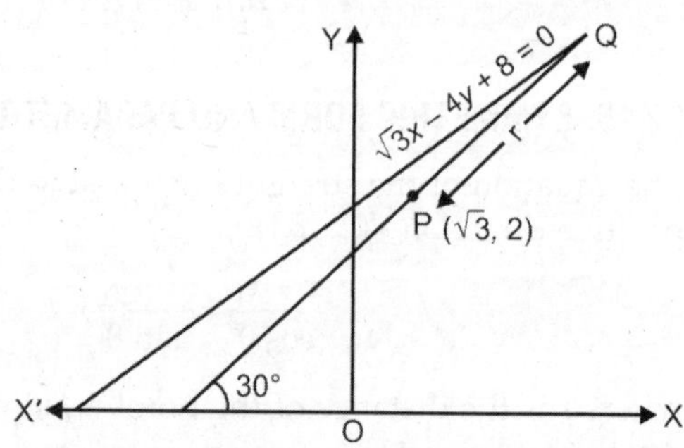

Let the line through $P(\sqrt{3}, 2)$ making an angle $\frac{\pi}{6}$ with the x-axis meet the line (1) at Q.

Let $PQ = r$, $\frac{x - \sqrt{3}}{\cos \frac{\pi}{6}} = \frac{y-2}{\sin \frac{\pi}{6}} = r$

Then, any point $Q = \left(\sqrt{3} + r\cos\frac{\pi}{6}, 2 + r\sin\frac{\pi}{6}\right)$

$$\Rightarrow \quad \left(\sqrt{3} + \frac{\sqrt{3}}{2}r, 2 + \frac{r}{2}\right) \quad \text{(Using } x = x_1 + r\cos\theta \text{ and } y = y_1 + r\sin\theta)$$

Since $Q\left(\sqrt{3} + \frac{\sqrt{3}}{2}r, 2 + \frac{r}{2}\right)$ line on the line (1).

$$\therefore \quad \sqrt{3}\left(\sqrt{3} + \frac{\sqrt{3}}{2}r\right) - 4\left(2 + \frac{r}{2}\right) + 8 = 0$$

$$\Rightarrow \quad 6 + 3r - 16 - 4r + 16 = 0 \quad \Rightarrow \quad r = 6$$

Hence, the length of PQ is 6. **Ans.**

Example 49. *Find the equation of the line through the point A(2, 3) and making an angle of 45° with the x-axis. Also, determine the length of intercept on it between A and the line x + y + 1 = 0.* **[S.B.T.E. 2015]**

Solution. The equation of a line through $A(2, 3)$ and making an angle of 45° with the x-axis is

$$\frac{x-2}{\cos 45^\circ} = \frac{y-3}{\sin 45^\circ} \quad \text{... (1)}$$

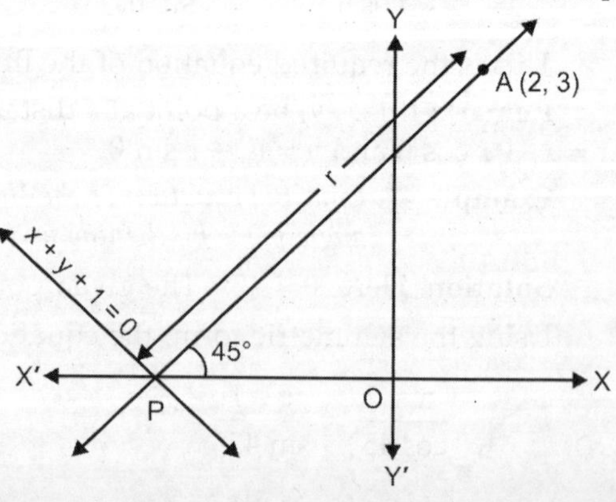

Suppose this line (1) meets the line $x + y + 1 = 0$ at P such that $AP = r$. Then the coordinates of P are given by

$$\frac{x-2}{\cos 45^\circ} = \frac{y-3}{\sin 45^\circ} = r$$

$\Rightarrow \quad x = 2 + r \cos 45°, y = 3 + 4 \sin 45°$

$\Rightarrow \quad x = 2+\frac{r}{\sqrt{2}}, y = 3+\frac{r}{\sqrt{2}}$

Thus, the coordinates of P are $\left(2+\frac{r}{\sqrt{3}}, 3+\frac{r}{\sqrt{2}}\right)$

Since, P lies on $x + y + 1 = 0$. Therefore $2+\frac{r}{\sqrt{2}}+3+\frac{r}{\sqrt{2}}+1=0$

$\Rightarrow \quad \sqrt{2}r=-6 \quad \Rightarrow \quad r=-3\sqrt{2}$

$\Rightarrow \quad$ Length $(AP) = |r| = 3\sqrt{2}$

Hence, the length of the intercept $= 3\sqrt{2}$. **Ans.**

Example 50. *The line joining two points A(2, 0), B(3, 1) is rotated about A in anticlockwise direction through an angle of 15°. Find the equation of the line in the new position. If B goes to C in the new position, what will be the coordinates of C?*

Solution. The slope m of the line AB is given by $m = \frac{1-0}{3-2} = 1$

So, AB makes an angle of 45° with x-axis. Now AB is rotated by 15° in anticlockwise direction and it makes an angle of 60° with x-axis in its new position AC.

Clearly AC passes through $A(2, 0)$ and makes an angle of 60° with x-axis, therefore the equation of AC is

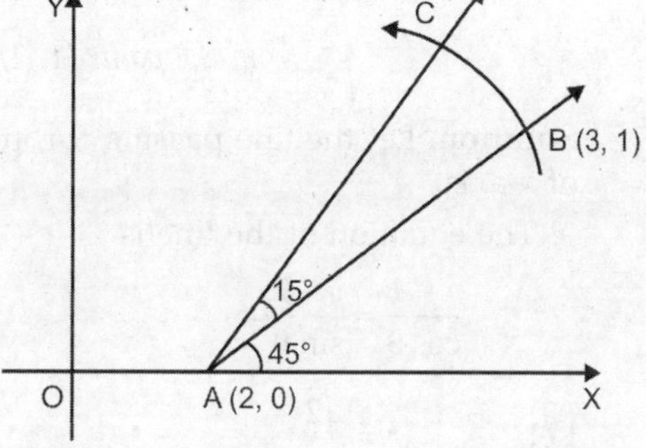

$$\frac{x-2}{\cos 60°}=\frac{y-0}{\sin 60°} \quad \Rightarrow \quad \frac{x-2}{\frac{1}{2}}=\frac{y-0}{\frac{\sqrt{3}}{2}}$$

We have, $AB = \sqrt{(3-2)^2+(1-0)^2}=\sqrt{2}$

So, the coordinates of C are given by

$$\frac{x-2}{\frac{1}{2}}=\frac{y-0}{\frac{\sqrt{3}}{2}}=\sqrt{2}$$

$\Rightarrow \quad x=2+\frac{1}{2}\sqrt{2}=2+\frac{1}{\sqrt{2}}$ and $y=\frac{\sqrt{3}}{2}\sqrt{2}=\frac{\sqrt{6}}{2}$

Hence, the coordinates of C are $\left(2+\frac{1}{\sqrt{2}}, \frac{\sqrt{6}}{2}\right)$ **Ans.**

Example 51. *Find the distance of the point (2, 5) from the line 3x + y + 4 = 0 measured parallel to a line having slope* $\frac{3}{4}$. **[S.B.T.E. 2017]**

Solution. Let AB be the distance of given point $A(2, 5)$ from line $3x + y + 4 = 0$ measured parallel to a line having slope $\frac{3}{4}$.

$\therefore$ Let $\tan \theta = \frac{3}{4}$

$\therefore$ θ lies in the first quadrant.

From the right angled triangle $\sin\theta = \dfrac{3}{5}$ and

$\cos\theta = \dfrac{4}{5}$

The equation of AB is

$$\frac{x-2}{\frac{4}{5}} = \frac{y-5}{\frac{3}{5}} = r \qquad \left[\text{Using } \frac{x-x_1}{\cos\theta} = \frac{y-y_1}{\sin\theta} = r\right]$$

$$\therefore \quad x = 2 + \frac{4}{5}r,\ y = 5 + \frac{3}{5}r$$

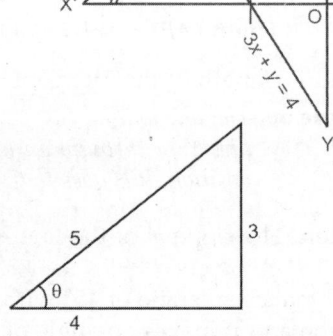

Let the coordinates of B be $\left(2 + \dfrac{4}{5}r, 5 + \dfrac{3}{5}r\right)$.

This point lies on $3x + y + 4 = 0$.

$$\Rightarrow \quad 3\left(2 + \frac{4}{5}r\right) + \left(5 + \frac{3}{5}r\right) + 4 = 0$$

$$\Rightarrow \quad 3r = -15 \ \Rightarrow\ r = -5$$

Hence, required distance AB = 5 units. **Ans.**

Example 52. *Find the direction in which a straight line must be drawn through the point (1, 2) so that its point of intersection with the line $x + y = 4$ may be at a distance $\sqrt{\frac{2}{3}}$ from the point (1, 2).*

Solution. Let the line passing through $A(1, 2)$ an angle θ with the positive direction of x-axis.

$\therefore$ The equation of the line is

$$\frac{x-1}{\cos\theta} = \frac{y-2}{\sin\theta} = r$$

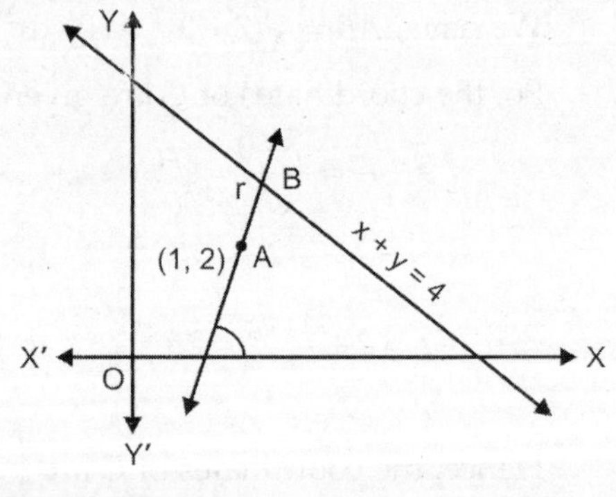

Let $\qquad r = \sqrt{\dfrac{2}{3}}$ (Given)

$$\therefore \quad \frac{x-1}{\cos\theta} = \frac{y-2}{\sin\theta} = \sqrt{\frac{2}{3}}$$

$$\Rightarrow \quad x = 1 + \sqrt{\frac{2}{3}}\cos\theta,\ y = 2 + \sqrt{\frac{2}{3}}\sin\theta$$

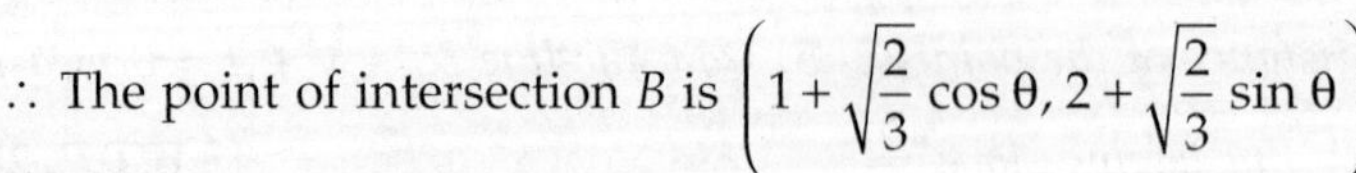

$\therefore$ The point of intersection B is $\left(1 + \sqrt{\dfrac{2}{3}}\cos\theta, 2 + \sqrt{\dfrac{2}{3}}\sin\theta\right)$

The point B also lies on the line $x + y = 4$.

$$\therefore \quad \left(1 + \sqrt{\frac{2}{3}}\cos\theta\right) + \left(2 + \sqrt{\frac{2}{3}}\sin\theta\right) = 4$$

$$\Rightarrow \quad \sqrt{\frac{2}{3}}(\cos\theta + \sin\theta) = 1 \ \Rightarrow\ \sin\theta + \cos\theta = \sqrt{\frac{3}{2}}$$

$$\Rightarrow \quad \frac{1}{\sqrt{2}}\cos\theta + \frac{1}{\sqrt{2}}\sin\theta = \frac{\sqrt{3}}{\sqrt{4}} = \frac{\sqrt{3}}{2}$$

$$\sin 45^\circ \cos\theta + \cos 45^\circ \sin\theta = \sin 60^\circ$$

$$\Rightarrow \quad \sin(45^\circ + \theta) = \sin 60^\circ \text{ or } \sin 120^\circ \quad \left(\because \sin 120^\circ = \sin(180^\circ - 60^\circ) = \sin 60^\circ = \frac{\sqrt{3}}{2}\right)$$

$$\Rightarrow \quad 45^\circ + \theta = 60^\circ \text{ or } 120^\circ$$

$$\Rightarrow \quad \theta = 60^\circ - 45^\circ = 15^\circ \text{ or } \theta = 120^\circ - 45^\circ = 75^\circ$$

Hence, the inclination of the line is either 15° or 75°. **Ans.**

EXERCISE 22.7

1. Find the equation of a line, which passes through the point (– 2, 3) and makes an angle of 30° with the positive direction of x-axis. **Ans.** $x + 2 = \sqrt{3}(y - 3)$

2. A straight line passes through the point $(2, \sqrt{3})$ and makes an angle of 60° with x-axis. Find the equation of the line. **Ans.** $\sqrt{3}x - y - \sqrt{3} = 0$

3. A line passes through a point $A(1, 2)$ and makes an angle of 60° with x-axis and intersects the line $x + y = 6$ at the point P. Find AP. **Ans.** $3(\sqrt{3} - 1)$

4. A line through (2, 3) makes an angle $\frac{3\pi}{4}$ with the negative direction of x-axis. Find the length of the line segment cut off between (2, 3) and the line $x + y - 7 = 0$. **Ans.** $AB = \sqrt{2}$

5. Find the distance of the point (2, 3) from the line $2x - 3y + 9 = 0$ measured along a line making an angle of 45° with x-axis. **Ans.** $4\sqrt{2}$

6. Find the distance of the point (3, 5) from the line $2x + 3y = 14$ measured parallel to a line having slope $\frac{1}{2}$. **Ans.** $\sqrt{5}$

7. A straight line drawn through the point $A(2, 1)$ making an angle $\frac{\pi}{4}$ with positive x-axis, intersects another line $x + 2y + 1 = 0$ at the point B. Find length AB. **Ans.** $\frac{5\sqrt{2}}{3}$

8. A line passes through a point (– 3, 2) and has a slope $-\sqrt{3}$, find the equation of this line in symmetric form. Also, find the coordinates of a point on the line at a distance of 2 units from the given point. **Ans.** $2y + 2\sqrt{3}x + (6\sqrt{3} - 4) = 0; (-4, 2 + \sqrt{3})$

9. Find the equation of a straight line, which passes through the point (2, 9) and makes an angle of 45° with x-axis. Also, find the points on the line which are at a distance of 5 units from (2, 9). **Ans.** $x - y + 7 = 0, \left(2 + \frac{5\sqrt{2}}{2}, 9 + \frac{5\sqrt{2}}{2}\right)$

10. Find the equation of the straight line passing through the point (3, 4) and inclined to the positive direction of x-axis at an angle of $\frac{3\pi}{4}$. Find, also the co-ordinates of two points on it on opposite sides of (3, 4) and at a distance of $\sqrt{2}$ from it.

Ans. $\dfrac{x-3}{-\frac{1}{\sqrt{2}}} = \dfrac{y-4}{\frac{1}{\sqrt{2}}} = r$ *i.e.* $x + y - 7 = 0$, (2, 5), (4, 3)

22.19 INTERSECTION OF STRAIGHT LINES

Let the equations of two given lines be

$$a_1x + b_1y + c_1 = 0 \quad \text{... (1)}$$

and

$$a_2x + b_2y + c_2 = 0 \quad \text{... (2)}$$

Since, the point of intersection of above two lines lies on both the lines, its coordinates satisfy both the equations. Solving (1) and (2) by cross-multiplication method, we get

$$\frac{x}{b_1c_2 - b_2c_1} = \frac{y}{c_1a_2 - c_2a_1} = \frac{1}{a_1b_2 - a_2b_1}$$

$$\Rightarrow \quad x = \frac{b_1c_2 - b_2c_1}{a_1b_2 - a_2b_1} \quad \text{and} \quad y = \frac{c_1a_2 - c_2a_1}{a_1b_2 - a_2b_1}$$

Provided $a_1b_2 - a_2b_2 \neq 0$

$\therefore$ The lines intersect at the point $\left(\dfrac{b_1c_2 - b_2c_1}{a_2b_2 - a_2b_1}, \dfrac{c_1a_2 - c_2a_1}{a_1b_2 - a_2b_1}\right)$ provided $a_1b_2 - a_2b_1 \neq 0$

i.e. $\dfrac{a_1}{a_2} \neq \dfrac{b_1}{b_2}$,

In case $\dfrac{a_1}{a_2} = \dfrac{b_1}{b_2}$, two possibilities arise:

(*i*) If $\dfrac{a_1}{a_2} = \dfrac{b_1}{b_2} \neq \dfrac{c_1}{c_2}$, then the lines are parallel and there is no point, which lies on both lines.

(*ii*) If $\dfrac{a_1}{a_2} = \dfrac{b_1}{b_2} = \dfrac{c_1}{c_2}$, then the lines are coincident and there are infinitely many points which lie on both lines.

Example 53. *Find the points of intersection of the following lines:*

(*i*) $2x + 3y - 6 = 0$ and $3x - 2y - 6 = 0$ (*ii*) $\dfrac{x}{3} - \dfrac{y}{4} = 0$ and $\dfrac{x}{2} + \dfrac{y}{3} = 1$

Solution. (*i*) We have,

$$2x + 3y - 6 = 0 \quad \text{... (1)}$$

$$3x - 2y - 6 = 0 \quad \text{... (2)}$$

Solving (1) and (2) simultaneously, we get

$$\frac{x}{-18-12} = \frac{y}{-18+12} = \frac{1}{-4-9}$$

$$\frac{x}{-30}=\frac{y}{-6}=\frac{1}{-13}$$

$\Rightarrow$ $$x=\frac{-30}{-13}=\frac{30}{13} \qquad \left[\because \frac{x}{-30}=\frac{1}{-13}\right]$$

And $$y=\frac{-6}{-13}=\frac{6}{13} \qquad \left[\because \frac{y}{-6}=\frac{1}{-13}\right]$$

Hence, the point of intersection is $\left(\frac{30}{13},\frac{6}{13}\right)$. **Ans.**

(*ii*) We have $\frac{x}{3}-\frac{y}{4}=0$

$\Rightarrow$ $4x - 3y + 0 = 0$

And $\frac{x}{2}+\frac{y}{3}=1 \Rightarrow 3x + 2y = 6 \Rightarrow 3x + 2y - 6 = 0$

Solving (1) and (2) simultaneously, we get

$$\frac{x}{18+0}=\frac{y}{0+24}=\frac{1}{8+9} \quad \Rightarrow \quad \frac{x}{18}=\frac{y}{24}=\frac{1}{17}$$

$\Rightarrow$ $$x=\frac{18}{17} \text{ and } y=\frac{24}{17}.$$

Hence, the point of intersection is $\left(\frac{18}{17},\frac{24}{17}\right)$. **Ans.**

Example 54. *Show that the line $3x + 3y + k = 0$ passes through the point of intersection of $3x + 4y + 6 = 0$ and $6x + 5y - 9 = 0$, if $k = -1$.*

Solution. We have, $3x + 4y + 6 = 0$ ... (1)

$6x + 5y - 9 = 0$ ... (2)

Solving (1) and (2) simultaneously, we get

$$\frac{x}{-36-30}=\frac{y}{36+27}=\frac{1}{15-24}$$

$\Rightarrow$ $$\frac{x}{-66}=\frac{y}{63}=\frac{1}{-9} \quad \Rightarrow \quad x=\frac{-66}{-9} \text{ and } y=\frac{63}{-9}$$

$\Rightarrow$ $$x=\frac{22}{3} \quad \text{and} \quad y=-7$$

Thus, the point of intersection of (1) and (2) is $\left(\frac{22}{3},-7\right)$.

If the line $3x + 3y + k = 0$ passes through $\left(\frac{22}{3},-7\right)$.

Then, $$3\left(\frac{22}{3}\right)+3(-7)+k=0$$

$\Rightarrow$ $22 - 21 + k = 0 \Rightarrow k = -1$ **Proved.**

Example 55. *The coordinates of points A, B and C are (1, 2), (– 2, 1) and (0, 6). Verify that the medians of the triangle ABC are concurrent. Also, find the coordinates of the point of concurrency (centroid).*

Solution. We have the coordinates of A, B and C are (1, 2), (– 2, 1) and (0, 6) respectively. Let D and E be the mid-points of BC and AC respectively.

$$\therefore \quad D=\left(\frac{-2+0}{2},\frac{1+6}{2}\right) \text{ and } E=\left(\frac{1+0}{2},\frac{2+6}{2}\right)$$

$$D=\left(-1,\frac{7}{2}\right),\ E=\left(\frac{1}{2},4\right)$$

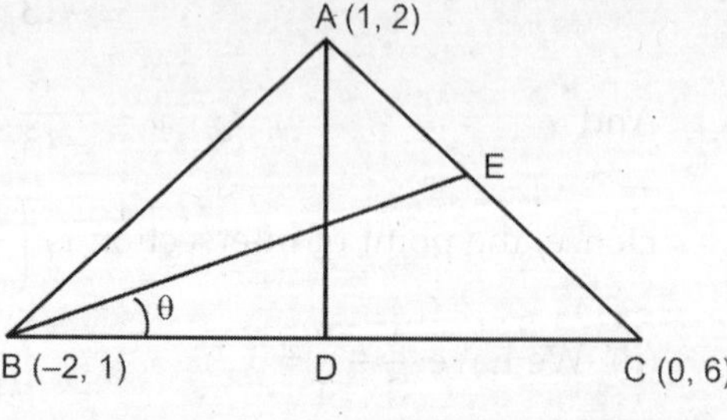

Equation of median AD passing through $A(1, 2)$ and $D\left(-1,\frac{7}{2}\right)$ is

$$y-2=\frac{\frac{7}{2}-2}{-1-1}(x-1)$$

$$\Rightarrow \quad y-2=-\frac{3}{4}(x-1) \quad \Rightarrow \quad 3x+4y-11=0 \qquad \text{... (1)}$$

Equation of median BE passing through $B(-2, 1)$ and $E\left(\frac{1}{2},4\right)$ is

$$y-1=\frac{4-1}{\frac{1}{2}+2}(x+2) \quad \Rightarrow \quad y-1=\frac{6}{5}(x+2) \quad \Rightarrow \quad 6x-5y+17=0 \qquad \text{... (2)}$$

On solving (1) and (2), we get the coordinates of the point of intersection $\left(-\frac{1}{3},3\right)$.

Ans.

Example 56. *Two consecutive sides of a parallelogram are $4x + 5y = 0$ and $7x + 2y = 0$. If the equation to one diagonal is $11x + 7y = 9$, find the equation of the other diagonal.*

Solution. Let the equation of sides AB and AD of a parallelogram $ABCD$ be

$$4x + 5y = 0 \qquad \text{... (1)}$$

And

$$7x + 2y = 0 \qquad \text{... (2)}$$

Solving (1) and (2) simultaneously, we get

$$x = 0 \text{ and } y = 0$$

⇒ The coordinates of A are (0, 0).

Equation of one diagonal of the parallelogram is

$$11x + 7y = 9 \qquad \text{... (3)}$$

Clearly $A(0, 0)$ does not lie on the diagonal (3), so (3) is the equation of diagonal BD.

We know that the diagonal of a parallelogram bisect each other.

On solving (1) and (3), we get

$$\frac{x}{-45-0}=\frac{y}{0+36}=\frac{1}{28-55} \quad \text{i.e.} \quad \frac{x}{-45}=\frac{y}{36}=\frac{1}{-27}$$

$$\Rightarrow \quad x=\frac{-45}{-27}=\frac{5}{3}, \qquad y=\frac{36}{-27}=-\frac{4}{3}$$

The coordinates of $B=\left(\frac{5}{3},-\frac{4}{3}\right)$

On solving (2) and (3), we get $x = -\frac{2}{3}, y = \frac{7}{3}$

$\Rightarrow$ Coordinates of $D = \left(-\frac{2}{3}, \frac{7}{3}\right)$.

Since, E is the mid-point of BD

$$\Rightarrow \qquad E = \left(\frac{1}{2}, \frac{1}{2}\right)$$

Now, the equation of diagonal AC which passes through $A(0, 0)$ and $E\left(\frac{1}{2}, \frac{1}{2}\right)$ is

$$y - 0 = \frac{\frac{1}{2} - 0}{\frac{1}{2} - 0}(x - 0) \Rightarrow y = x$$ **Ans.**

Example 57. *Find the area of the triangle formed by the lines $y - x = 0$, $x + y = 0$ and $x - k = 0$.*

Solution. Let ABC be the triangle whose sides are

$AB : x - k = 0$ ...(1)

$BC : y - x = 0$ ...(2)

$AC : x + y = 0$ ...(3)

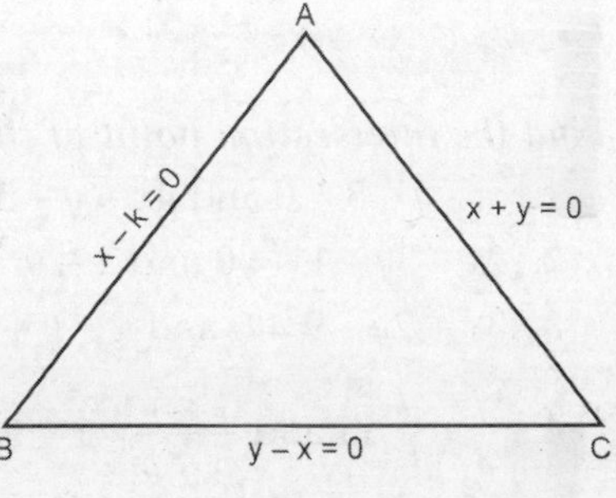

Solving (1) and (3), we get the coordinates of $A : (k, -k)$

Solving (1) and (2), we get the coordinates of $B : (k, k)$

Solving (2) and (3), we get the coordinates of $C : (0, 0)$.

$\therefore$ Area of triangle ABC

$$= \frac{1}{2}\{x_1(y_2 - y_3) + x_2(y_3 - y_1) + x_3(y_1 - y_2)\}$$

$$= \frac{1}{2}\{k(k - 0) + k(0 + k) + 0(-k - k)\}$$

$$= \frac{1}{2}\{k^2 + k^2\} = \frac{1}{2} \times 2k^2 = k^2$$ **Ans.**

Example 58. *Find the coordinates of the orthocentre of the triangle whose vertices are $(-1, 3)$, $(2, -1)$ and $(0, 0)$.*

Solution. Let the vertices of triangle ABC be $A(-1, 3)$, $B(2, -1)$ and $C(0, 0)$. Then orthocentre is the point of intersection of altitudes from the vertices to opposite sides. Let AD, BE and CF be the altitudes and $O(h, k)$ be the orthocentre of triangle ABC.

Since, $AO \perp BC$, therefore

(Slope of AO) $\times$ (Slope of BC) $= -1$

$$\Rightarrow \frac{k - 3}{h + 1} \times \frac{-1 - 0}{2 - 0} = -1$$

$$\Rightarrow \quad \frac{k-3}{h+1}\times\left(-\frac{1}{2}\right)=-1 \quad \Rightarrow \quad 2h+2=k-3$$

$$\Rightarrow \quad 2h-k+5=0 \qquad \ldots (1)$$

Also, $BO \perp AC$, therefore

(Slope of BO) × (Slope of AC) = – 1

$$\frac{k+1}{h-2}\times\frac{3-0}{-1-0}=-1 \quad \Rightarrow \quad \frac{k+1}{h-2}\times(-3)=-1$$

$$\Rightarrow \quad h-2=3k+3 \quad \Rightarrow \quad h-3k-5=0 \qquad \ldots (2)$$

On solving (1) and (2), we get

$$\frac{h}{5+15}=\frac{k}{5+10}=\frac{1}{-6+1}$$

$$\frac{h}{20}=\frac{k}{15}=\frac{1}{-5} \quad \Rightarrow \quad \frac{h}{20}=-\frac{1}{5} \text{ and } \frac{k}{15}=-\frac{1}{5}$$

$$h=-4,\ k=-3$$

Hence, the orthocentre is (– 4, – 3). **Ans.**

EXERCISE 22.8

Find the intersection point of the following lines:

1. $x+y-5=0$ and $4x-y-5=0$ **Ans.** (2, 3)
2. $2x-3y-11=0$ and $x-y-6=0$ **Ans.** (7, 1)
3. $3x+2y-9=0$ and $x-y+2=0$ **Ans.** (1, 3)
4. $\frac{x}{a}+\frac{y}{b}=1$ and $\frac{x}{b}+\frac{y}{a}=1$ **Ans.** $\left(\frac{ab}{a+b}, \frac{ab}{a+b}\right)$

Find which of the following pairs of lines are intersecting, parallel, coincident:

5. $2x-y+6=0$ and $x-2y+6=0$ **Ans.** Intersecting lines
6. $6x-y+7=0$ and $y-6x=8$ **Ans.** Parallel lines
7. $4x+y+9=0$ and $8x+2y=-18$ **Ans.** Coincident lines
8. $\frac{x}{a}+\frac{y}{b}=1$ and $\frac{x}{b}+\frac{y}{a}=1, (a \neq b)$ **Ans.** Intersecting lines
9. Find the area of triangle formed by the lines $x+4y=9$, $9x+10y+23=0$ and $7x+2y=11$. **Ans.** 26 units
10. For what value of k the lines $kx+2y+5=0$ will pass through the point of intersection of the lines $x-2y=3$ and $x+2y=9$? **Ans.** $k=-4/3$
11. Find the coordinates of the foot of perpendicular from a point (– 1, 3) to the line $3x-4y-16=0$. **Ans.** $\left(\frac{68}{25}, -\frac{49}{25}\right)$
12. If perpendiculars are drawn from origin to the straight lines $x+3y=3$ and $2x+3y=5$ then find the equation of the line joining the feet of these perpendiculars. **Ans.** $33x-61y+45=0$

13. Find the equations of the medians of a triangle, the equation of whose sides are: $3x + 2y + 6 = 0$; $2x - 5y + 4 = 0$ and $x - 3y - 6 = 0$

Ans. $41x - 112y - 70 = 0$, $16x - 59y - 120 = 0$, $25x - 53y + 50 = 0$

14. Show that the diagonals of the parallelogram formed by the four lines $3x + y = 0$, $3y + x = 0$, $3x + y = 4$ and $3y + x = 4$ are perpendicular.

22.20 CONCURRENT LINES

Three or more straight lines are said to be concurrent if they pass through a common point, i.e. they meet at a point. Thus, if three lines are concurrent the point of intersection of two lines, lies on the third line.

Condition of concurrency of three lines:

To show that the three lines $a_1x + b_1y + c_1 = 0$, $a_2x + b_2y + c_2 = 0$ and $a_3x + b_3y + c_3 = 0$ will be concurrent if

$$a_1(b_2c_3 - b_3c_2) + b_1(c_2a_3 - c_3a_2) + c_1(a_2b_3 - a_3b_2) = 0 \qquad \textbf{[S.B.T.E. 2016, 2014]}$$

Proof: Let $a_1x + b_1y + c_1 = 0$ ... (1)

$a_2x + b_2y + c_2 = 0$ ... (2)

$a_3x + b_3y + c_3 = 0$ ... (3)

be three concurrent lines. Then the point of intersection of (1) and (2) must lie on the third.

The coordinates of the point of intersection of (1) and (2) are $\left(\dfrac{b_1c_2 - b_2c_1}{a_1b_2 - a_2b_1}, \dfrac{c_1a_2 - c_2a_1}{a_1b_2 - a_2b_1}\right)$

[See Art. 22.19]

Since, the lines are concurrent, this point lies on (3).

$$\therefore \qquad a_3\left(\frac{b_1c_2 - b_2c_1}{a_1b_2 - a_2b_1}\right) + b_3\left(\frac{c_1a_2 - c_2a_1}{a_1b_2 - a_2b_1}\right) + c_3 = 0$$

$$\Rightarrow \qquad a_3(b_1c_2 - b_2c_1) + b_3(c_1a_2 - c_2a_1) + c_3(a_1b_2 - a_2b_1) = 0$$

Also, in the determinant form $\begin{vmatrix} a_1 & b_1 & c_1 \\ a_2 & b_2 & c_2 \\ a_3 & b_3 & c_3 \end{vmatrix}$

This is the required condition of concurrency of three lines.

Example 59. *Prove that the following lines are concurrent:*

$$5x - 3y = 1,\ 2x + 3y = 23,\ 42x + 21y = 257$$

Solution. Given lines are:

$5x - 3y - 1 = 0$ ... (1)

$2x + 3y - 23 = 0$ ... (2)

$42x + 21y - 257 = 0$ ... (3)

Solving (1) and (2), we get

$$\frac{x}{69+3} = \frac{y}{-2+115} = \frac{1}{15+6}$$

$$\Rightarrow \qquad \frac{x}{72} = \frac{y}{113} = \frac{1}{21} \qquad \Rightarrow \qquad x = \frac{72}{21} \text{ and } y = \frac{113}{21}$$

Thus, the lines (1) and (2) intersects at the point $\left(\frac{72}{21}, \frac{113}{21}\right)$.

$\left(\frac{72}{21}, \frac{113}{21}\right)$ lies on (3) if $42\left(\frac{72}{21}\right) + 21\left(\frac{113}{21}\right) - 257 = 0$

$\Rightarrow$ if $2 \times 72 + 113 - 257 = 0$ $\Rightarrow$ if $144 + 113 - 257 = 0$

$\Rightarrow$ if $257 - 257 = 0$, which is true.

Thus, $\left(\frac{72}{21}, \frac{113}{21}\right)$ is on all the lines. Hence, given lines are concurrent. **Proved.**

Example 60. *For what values of k are the three lines $4x + 7y - 9 = 0$, $5x + ky + 15 = 0$ and $9x - y + 6 = 0$ are concurrent.*

Solution. The given lines are

$$4x + 7y - 9 = 0 \quad \text{... (1)}$$

$$5x + ky + 15 = 0 \quad \text{... (2)}$$

$$9x - y + 6 = 0 \quad \text{... (3)}$$

Solving (1) and (3) simultaneously, we get

$$\frac{x}{42 - 9} = \frac{y}{-81 - 24} = \frac{1}{-4 - 63}$$

$$x = -\frac{33}{67}, \quad y = \frac{105}{67}$$

Thus, the point of intersection of (1) and (3) is $\left(-\frac{33}{67}, \frac{105}{67}\right)$.

Since, three lines are concurrent, this point of intersection lies on (2), *i.e.* $5x + ky + 15 = 0$.

$\Rightarrow$ $5\left(-\frac{33}{67}\right) + k\left(\frac{105}{67}\right) + 15 = 0$

$\Rightarrow$ $-165 + 105k + 1005 = 0$

$\Rightarrow$ $105k = -840$ $\Rightarrow$ $k = \frac{-840}{105} = -8$

Hence, for $k = -8$, the given lines are concurrent. **Ans.**

Example 61. *If lines whose equations are $y = m_1x + c_1$, $y = m_2x + c_2$ and $y = m_3x + c_3$ meet in a point then prove that*

$$m_1(c_2 - c_3) + m_2(c_3 - c_1) + m_3(c_1 - c_2) = 0$$

Solution. The equation of the given lines are

$$m_1x - y + c_1 = 0 \quad \text{... (1)}$$

$$m_2x - y + c_2 = 0 \quad \text{... (2)}$$

$$m_3x - y + c_3 = 0 \quad \text{... (3)}$$

Solving (1) and (2), we get

$$\frac{x}{-c_2 + c_1} = \frac{y}{m_2c_1 - m_1c_2} = \frac{1}{-m_1 + m_2}$$

Thus, the point of intersection of (1) and (2) is

$$\left(\frac{c_1 - c_2}{m_2 - m_1}, \frac{m_2c_1 - m_1c_2}{m_2 - m_1}\right)$$

The three lines will be concurrent if the point of intersection of (1) and (2) lies on (3).

$$\Rightarrow \quad m_3\left(\frac{c_1 - c_2}{m_2 - m_1}\right) - \left(\frac{m_2c_1 - m_1c_2}{m_2 - m_1}\right) + c_3 = 0$$

$$\Rightarrow \quad m_3(c_1 - c_2) - (m_2c_1 - m_1c_2) + c_3(m_2 - m_1) = 0$$

$$\Rightarrow \quad m_1(c_2 - c_3) + m_2(c_3 - c_1) + m_3(c_1 - c_2) = 0$$

Proved.

EXERCISE 22.9

Show that the following sets of three lines are concurrent.

1. $15x - 18y + 1 = 0$, $12x + 10y - 3 = 0$ and $6x + 66y - 11 = 0$
2. $3x - 5y - 11 = 0$, $5x + 3y - 7 = 0$ and $x + 2y = 0$
3. $2x + 3y - 4 = 0$, $x - 5y + 7 = 0$ and $6x - 17y + 24 = 0$
4. $\frac{x}{a} + \frac{y}{b} = 1$, $\frac{x}{b} + \frac{y}{a} = 1$ and $x = y$.
5. For what value of k are the three lines $2x - 5y + 3 = 0$, $5x - 9y + k = 0$ and $x - 2y + 1 = 0$ concurrent. **Ans.** $k = 4$
6. Find the orthocentre of the triangle whose vertices are (1, 2), (2, 3) and (4, 3). **Ans.** (1, 6)
7. Find circumcentre of the triangle whose vertices are (3, 1), (– 1, 5), (– 4, – 1). **Ans.** $\left(-\frac{5}{6}, \frac{7}{6}\right)$

22.21 EQUATIONS OF FAMILY OF LINES THROUGH THE INTERSECTION OF TWO LINES

$$A_1x + B_1y + C_1 = 0 \quad \text{and} \quad A_2x + B_2y + C_2 = 0$$

Proof: Here we have,

$$A_1x + B_1y + C_1 = 0 \qquad \text{... (1)}$$

And

$$A_2x + B_2y + C_2 = 0 \qquad \text{... (2)}$$

By combining (1) and (2), we have

$$A_1x + B_1y + C_1 + k(A_2x + B_2y + C_2) = 0 \qquad \text{... (3)}$$

where, k is constant and also called parameter.

(*i*) Equation (3) is of first degree in x and y. Hence it represents a family of lines.

(*ii*) The co-ordinates of a point which satisfy (1) and (2) simultaneously will also satisfy equation (3). What about value of k may be.

It means any point lying on (1) and (2) (point on intersection) will also lie on (3).

Hence, (3) passes through the point of intersection of (1) and (2).

(*iii*) A particular member of this family can be obtained on putting the value of k.

Example 62. *Find the equation of the line passing through the point of intersection of $x + 2y = 5$ and $x - 3y = 7$ and passing through the point (2, – 3).*

Solution. We have,

$$x + 2y = 5 \qquad \text{... (1)}$$

and

$$x - 3y = 7 \qquad \text{... (2)}$$

The equation of any line through the point of intersection of given lines (1) and (2) is

$$(x + 2y - 5) + k(x - 3y - 7) = 0 \quad \text{... (3)}$$

If the point (2, - 3) lies on (3), then

$$[2 + 2(-3) - 5] + k\,[2 - 3(-3) - 7] = 0$$

$$\Rightarrow \quad 2 - 6 - 5 + k(2 + 9 - 7) = 0 \quad \Rightarrow \quad -9 + 4k = 0$$

$$\Rightarrow \quad k = \frac{9}{4}$$

Substituting this value of k in equation (3), we get

$$(x + 2y - 5) + \frac{9}{4}(x - 3y - 7) = 0 \quad \Rightarrow \quad 4(x + 2y - 5) + 9(x - 3y - 7) = 0$$

$$\Rightarrow \quad 4x + 8y - 20 + 9x - 27y - 63 = 0 \quad \Rightarrow \quad 13x - 19y - 83 = 0$$

Ans.

Example 63. *Find the equation of the straight line which passes through the intersection of the lines $x - y - 1 = 0$ and $2x - 3y + 1 = 0$ and is parallel to the line $3x - 5y + 6 = 0$.*

Solution. We have the equation of three lines

$$x - y - 1 = 0 \quad \text{... (1)}$$

$$2x - 3y + 1 = 0 \quad \text{... (2)}$$

$$3x - 5y + 6 = 0 \quad \text{... (3)}$$

Any line passing through the point of intersection of (1) and (2) is

$$x - y - 1 + k\,(2x - 3y + 1) = 0$$

$$\Rightarrow \quad (2k + 1)\,x + (-1 - 3k)\,y + k - 1 = 0 \quad \text{... (4)}$$

Slope of the line (4) is $\dfrac{2k+1}{3k+1}$, Slope of the line (3) is $\dfrac{3}{5}$.

If line (3) is parallel to line (4), then

$$\frac{2k+1}{3k+1} = \frac{3}{5} \quad \Rightarrow \quad 10k + 5 = 9k + 3 \quad \Rightarrow \quad k = -2 \qquad (\because m_1 = m_2)$$

Putting the value of k in (4), we get $-3x + 5y - 3 = 0$

$$\Rightarrow \quad 3x - 5y + 3 = 0$$

Ans.

Example 64. *Find the equation of the line through the intersection of line $x + 2y - 3 = 0$ and $4x - y + 7 = 0$ and which is parallel to $5x + 4y - 20 = 0$.*

Solution. We have,

$$x + 2y - 3 = 0 \quad \text{... (1)}$$

and

$$4x - y + 7 = 0 \quad \text{... (2)}$$

The equation of the family of lines passing through the point of intersection of (1) and (2) is

$$(x + 2y - 3) + k\,(4x - y + 7) = 0 \quad \text{... (3)}$$

$$(1 + 4k)\,x + (2 - k)\,y + (-3 + 7k) = 0$$

$$\Rightarrow \quad y = -\frac{(1+4k)}{2-k}\,x - \frac{(-3+7k)}{2-k} \quad \text{... (4)}$$

The slope of each member of this family is given by

$$m_1 = -\frac{1+4k}{2-k}$$

Again, $5x + 4y - 20 = 0$

$$\Rightarrow \quad y = -\frac{5}{4}x + 5 \quad \text{... (5)}$$

Slope of (5), $m_2 = -\dfrac{5}{4}$

As (4) and (5) are parallel, therefore $m_1 = m_2$

$$\Rightarrow \quad -\frac{1+4k}{2-k} = -\frac{5}{4} \Rightarrow 4 + 16k = 10 - 5k \Rightarrow 21k = 6 \Rightarrow k = \frac{6}{21} = \frac{2}{7}$$

Substituting the value of k in (3), we get

$$(x + 2y - 3) + \frac{2}{7}(4x - y + 7) = 0 \Rightarrow 7x + 14y - 21 + 8x - 2y + 14 = 0$$

$$\Rightarrow \quad 15x + 12y - 7 = 0$$ **Ans.**

Example 65. *Find the equation of the straight line which passes through the intersection of the straight line $2x - 3y + 4 = 0$ and $3x + 4y + 5 = 0$ and is perpendicular to the straight line $6x - 7y + 8 = 0$.* **[Diplome IETE, Dec. 2005]**

Solution. The equation of lines lines are:

$$2x - 3y + 4 = 0 \quad \text{... (1)}$$

$$3x + 4y + 5 = 0 \quad \text{... (2)}$$

$$6x - 7y + 8 = 0 \quad \text{... (3)}$$

Equation of a line passing through the point of intersection of (1) and (2) is

$$2x - 3y + 4 + k(3x + 4y + 5) = 0$$

$$\Rightarrow \quad (2 + 3k)x + (-3 + 4k)y + 4 + 5k = 0 \quad \text{... (4)}$$

Slope of (4) $= \dfrac{3k+2}{3-4k}$; Slope of (3) $= \dfrac{6}{7}$

Line (4) is perpendicular to line (3), therefore

slope of line (4) × slope of line (3) = – 1

$$\frac{3k+2}{3-4k} \times \frac{6}{7} = -1 \Rightarrow \frac{18k+12}{21-28k} = -1$$

$$\Rightarrow \quad 18k + 12 = -21 + 28k$$

$$10k = 33 \Rightarrow k = \frac{33}{10}$$

Putting the value of k in (4), we get

$$2x - 3y + 4 + \frac{33}{10}(3x + 4y + 5) = 0$$

$$20x - 30y + 40 + 99x + 132y + 165 = 0$$

$$119x + 102y + 205 = 0$$ **Ans.**

Example 66. *Find the equation of the line passing through the point of intersection of the lines: $x - 2y + 3 = 0$ and $2x - 3y + 4 = 0$ and perpendicular to $x + y = 1$.*

[B.T.E. Delhi, Jan. 2009]

Solution. The equation of lines are

$$x - 2y + 3 = 0 \quad \text{... (1)}$$

$$2x - 3y + 4 = 0 \quad \text{... (2)}$$

$$x + y = 1 \quad \text{... (3)}$$

Equation of a line passing through the point of intersection of (1) and (2) is

$$x - 2y + 3 + k(2x - 3y + 4) = 0$$

$$\Rightarrow \quad (1 + 2k)x + (-2 - 3k)y + 3 + 4k = 0 \quad \text{... (4)}$$

Slope of (4) $= -\dfrac{1+2k}{(-2-3k)} = \dfrac{1+2k}{2+3k}$

Slope of (3) $= \dfrac{1}{-1} = -1$

Line (4) is perpendicular to line (3), therefore slope of line (4) × slope of line (3) = – 1

$$\Rightarrow \quad \left(\frac{1+2k}{2+3k}\right)(-1) = -1$$

$$\Rightarrow \quad \frac{1+2k}{2+3k} = 1 \quad \Rightarrow \quad 1 + 2k = 2 + 3k$$

$$\Rightarrow \quad 2k - 3k = 2 - 1 \Rightarrow \quad -k = 1 \quad \Rightarrow \quad k = -1$$

Putting the value of k in (4), we get

$$[1 + 2(-1)]x + [-2 - 3(-1)]y + 3 + 4(-1) = 0$$

$$\left.\begin{aligned} \Rightarrow \quad -x + y - 1 &= 0 \\ \Rightarrow \quad x - y + 1 &= 0 \end{aligned}\right\} \text{ is the required line.}$$ **Ans.**

Example 67. *Find the equation of the line passing through the intersection of the lines $4x + 7y - 3 = 0$ and $2x - 3y + 1 = 0$ that has equal intercepts on the axes.*

Solution. We have

$$4x + 7y - 3 = 0 \quad \text{... (1)}$$

and $$2x - 3y + 1 = 0 \quad \text{... (2)}$$

The equation of the family of lines passing through the point of intersection of (1) and (2) is

$$(4x + 7y - 3) + k(2x - 3y + 1) = 0 \quad \text{... (3)}$$

$$\Rightarrow \quad (4 + 2k)x + (7 - 3k)y + k - 3 = 0 \quad \text{... (4)}$$

Since (4) cuts off equal intercepts

$$\Rightarrow \quad (2k + 4)x + (7 - 3k)y = 3 - k \quad \Rightarrow \quad \frac{(2k+4)x}{3-k} + \frac{(7-3k)y}{3-k} = 1$$

$$\Rightarrow \quad \frac{x}{\frac{3-k}{2k+4}} + \frac{y}{\frac{3-k}{7-3k}} = 1 \text{ which is in the intercept form.}$$

But it is given that intercepts are equal.

So, $$\frac{3-k}{2k+4} = \frac{3-k}{7-3x} \quad \Rightarrow \quad \frac{1}{2k+4} = \frac{1}{7-3k}$$

$$\Rightarrow \quad 2k + 4 = 7 - 3k \quad \Rightarrow \quad 5k = 3 \Rightarrow \quad k = \frac{3}{5}$$

Substituting $k = \frac{3}{5}$ in (3), we get $(4x + 7y - 3) + \frac{3}{5}(2x - 3y + 1) = 0$

$\Rightarrow \quad 20x + 35y - 15 + 6x - 9y + 3 = 0 \quad \Rightarrow \quad 26x + 26y - 12 = 0$

$\Rightarrow \quad 13(x + y) - 6 = 0$ **Ans.**

EXERCISE 22.10

1. Find the equation of the line through the point of intersection $x + 2y = 5$ and $x - 3y = 7$ and passing through the point

(*i*) (0, 0) **Ans.** $2x + 29y = 0$ (*ii*) (1, 0) **Ans.** $x + 12y - 1 = 0$

(*iii*) (0, – 1) **Ans.** $3x - 29y - 29 = 0$

2. Find the equation of the line joining the point of intersection of $x - 2y + 3 = 0$ and $2x - 3y + 4 = 0$ to the point (4, – 5). **Ans.** $7x + 3y - 13 = 0$

3. Find the equation of the straight line passing through the intersection of two lines $x + 2y + 3 = 0$ and $3x + 4y + 7 = 0$ and parallel to the straight line $y - x = 3$.

Ans. $x - y = 0$.

4. Find the equation of the line which passes through the point of intersection of the lines $2x + 3y + 5 = 0$ and $3x + 4y - 18 = 0$ and parallel to the line $5x + 2y + 9 = 0$.

Ans. $5x + 2y = 268$

5. Find the equation of the line through the intersection of $y + x = 9$ and $2x - 3y + 7 = 0$ and perpendicular to the line $2y - 3x - 5 = 0$. **Ans.** $2x + 3y - 23 = 0$

6. Find the equation of line which passes through the point of intersection of $y - x - 1 = 0$ and $2x - y + 1 = 0$ and is perpendicular to the line $3x + 2y = 10$.

Ans. $2x - 3y + 3 = 0$.

7. A straight line passes through the point of intersection of the lines $x + 2y + 11 = 0$ and $3x - y + 5 = 0$ and makes an angle of 45° with the positive direction of the x-axis. Find its equation. **Ans.** $x - y - 1 = 0$

8. Find the equation of the line which passes through the point of intersection of the lines $3x - 4y + 6 = 0$ and $4x - y - 5 = 0$ and cuts off equal intercepts from the axes.

Ans. $x + y - 5 = 0$

9. Find the equation of line which passes through the intersection of the lines $x - y - 1 = 0$ and $2x - 3y + 1 = 0$ and is parallel to y-axis. **Ans.** $x = 4$

10. Find the equation of the straight line through the point of intersection of the lines $ax + by + c = 0$ and $a'x + b'y + c' = 0$ drawn parallel to x-axis.

Ans. $(a'b - ab')y + (a'c - ac') = 0$

11. Find the equation of the straight line which passes through the intersection of the straight lines $3x + 2y + 4 = 0$ and $x - y = 2$ and forms the triangle with the axes whose area is 8 square units. **Ans.** $x + 4y + 8 = 0$

12. Find the equations of the straight line passing through the intersection of lines $4x - 3y - 1 = 0$ and $2x - 5y + 3 = 0$ and equally inclined to the axes.

Ans. $x + y - 2 = 0$ and $x = y$

13. Find equation of the line joining the point (3, 5) to the point of intersection of lines $4x + y - 1 = 0$ and $7x - 3y - 35 = 0$ and prove that the line is equidistant from origin and point (8, 34). **[Diploma IETE Dec 2006]** **Ans.** $12x - y - 31 = 0$

14. Find the equation of tangent and normal to the curve $y = x^2 - 9$ at the point where it intersects the positive x-axis. **Ans.** $y = 6x - 18,\ x + 6y - 3 = 0$

15. Show that the equation of the straight line through the origin making angle ϕ with the line $y = mx + b$ is

$$\frac{y}{x} = \frac{m + \tan\phi}{1 - m\tan\phi}.$$

16. Find equations of lines which pass through the point (4, 5) and make an angle 45° with the line $2x + y + 1 = 0$. **Ans.** $x + 3y - 20 = 0$

Choose the correct alternative:

17. The equation of the straight line which makes equal intercepts on the axes and passes through the point (1, 2) is **[Diploma IETE Dec. 2005]**

(*a*) $x + y = 3$ (*b*) $x + 2y = 5$ (c) $x - y = 1$ (*d*) $2x + y + 4$ **Ans.** (*a*)

18. The distance between the parallel lines $3x + 4y + 5 = 0$ and $3x + 4y + 15 = 0$ is **[Diploma IETE Dec. 2005]**

(*a*) 1 (*b*) 2 (c) 3 (*d*) 5 **Ans.** (*b*)

22.22 ANGLE BETWEEN TWO LINES

We know that angle between two lines is given by

$$\tan\theta = \frac{m_2 - m_1}{1 + m_1 m_2}$$

where, m_1 and m_2 are slopes of the given lines and θ is angle between them.

Note: If $\tan\theta$ is negative, then angle between the lines is obtuse, while its supplement should be acute angle between the lines.

Example 68. *Find the angle between the lines* $y - \sqrt{3}x - 5 = 0$ and $\sqrt{3}y - x + 6 = 0$

Solution. Here, we have

$$y - \sqrt{3}x - 5 = 0 \quad \Rightarrow \quad y = \sqrt{3}x + 5 \quad \text{... (1)}$$

and $$\sqrt{3}y - x + 6 = 0 \quad \Rightarrow \quad y = \frac{1}{\sqrt{3}}x - 6 \quad \text{... (2)}$$

(2)
(1)
30°
150°

Let m_1 and m_2 be the slopes of lines (1) and (2) respectively.

Here, $m_1 = \sqrt{3}$ and $m_2 = \frac{1}{\sqrt{3}}$

We know that the angle θ between the two lines is given by

$$\tan\theta = \frac{m_2 - m_1}{1 + m_1 m_2} \Rightarrow \tan\theta = \frac{\sqrt{3} - \frac{1}{\sqrt{3}}}{1 + \sqrt{3}\cdot\frac{1}{\sqrt{3}}} = \frac{3 - 1}{2\sqrt{3}} = \frac{1}{\sqrt{3}}$$

$\Rightarrow \quad \theta = 30° \quad \Rightarrow$ Supplement angle = 180° – 30° = 150° **Ans.**

Example 69. *Find the tangent of the angle between the lines, whose intercepts on the axes are respectively p, – q and q, – p.*

Solution. Equation of a line whose intercepts are $p, -q$ is

$$\frac{x}{p} + \frac{y}{-q} = 1 \quad \Rightarrow \quad y = \frac{q}{p}x - q \quad \text{... (1)}$$

Equation of a line whose intercepts are $q, -p$ is

$$\frac{x}{q}+\frac{y}{-p}=1 \quad \Rightarrow \quad y=\frac{p}{q}x-p \qquad \text{... (2)}$$

Slope of (1) = $m_1=\frac{q}{p}$, Slope of (2) = $m_2=\frac{p}{q}$

$$\tan\theta=\frac{m_2-m_1}{1+m_1m_2} \quad \Rightarrow \quad \tan\theta=\frac{\frac{p}{q}-\frac{q}{p}}{1+\frac{p}{q}\cdot\frac{q}{p}} \quad \Rightarrow \quad \tan\theta=\frac{p^2-q^2}{2pq}$$

and the slope of the supplement angle = $\frac{p^2-q^2}{2pq}$ **Ans.**

Example 70. *Is the triangle, whose vertices are (5, – 6), (1, 2) and (– 7, – 2) a right angled, an acute triangle or an obtuse triangle?*

Solution. Let the vertices of triangles ABC be $A(5,-6)$, $B(1, 2)$ and $C(-7, -2)$.

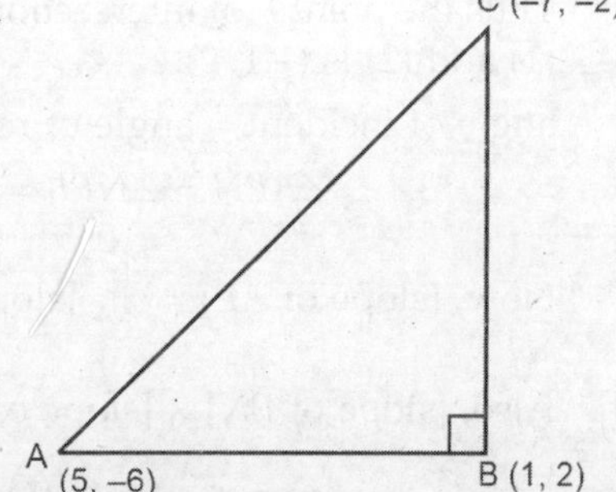

$$\text{Slope of } AC = m_1 = \frac{-6+2}{5+7}=\frac{-4}{12}=-\frac{1}{3}$$

$$\text{Slope of } AB = m_2 = \frac{2+6}{1-5}=\frac{8}{-4}=-2$$

$$\text{Slope of } BC = m_3 = \frac{-2-2}{-7-1}=\frac{-4}{-8}=\frac{1}{2}$$

Here, (slope of AB) × (slope of BC) = $(-2)\left(\frac{1}{2}\right)=-1=m_1m_2$

Therefore, AB is $\perp$ to BC $\Rightarrow$ $\angle B = 90°$

$\Rightarrow$ ΔABC is a right angled triangle. **Ans.**

Example 71. *Two lines passing through the point (2, 3) make an angle of 45°. If the slope of one of the lines is 2. Find the slope of the other.*

Solution. We have,

Slope of given line (m_1) = 2

Let slope of other line (m_2) = m

Angle between these two lines is 45°.

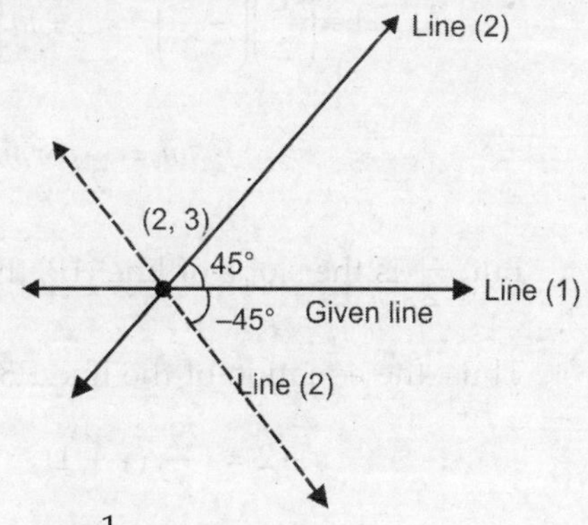

We know that $\tan\theta=\frac{m_1-m_2}{1+m_1m_2}$

$$\Rightarrow \tan 45° = \left|\frac{2-m}{1+2m}\right| \quad \Rightarrow \quad \pm 1=\frac{2-m}{1+2m}$$

$$\Rightarrow \quad 1+2m=2-m \quad \Rightarrow \quad 3m=1 \quad \Rightarrow \quad m=\frac{1}{3}$$

Also, for $\theta = -45°$, we have

$$\tan(-45°) = \frac{2-m}{1+2m} \quad \Rightarrow \quad -1 = \frac{2-m}{1+2m}$$

$$\Rightarrow \quad -1-2m = 2-m \quad \Rightarrow \quad -3 = m$$

Hence, the slope of second line is $\frac{1}{3}$ or -3. **Ans.**

Example 72. *A ray of light is sent along the line $x - 2y + 5 = 0$; upon reaching the line $3x - 2y + 7 = 0$ the ray is reflected back from it. Find the equation of the line containing the reflected ray.*

Solution. Equation of the incident ray AP is

$$x - 2y + 5 = 0 \quad \text{... (1)}$$

The equation of the mirror line is

$$3x - 2y + 7 = 0 \quad \text{... (2)}$$

Solving (1) and (2), we get $x = -1$ and $y = 2$. Thus the point P of intersection of these lines (1) and (2) is $(-1, 2)$.

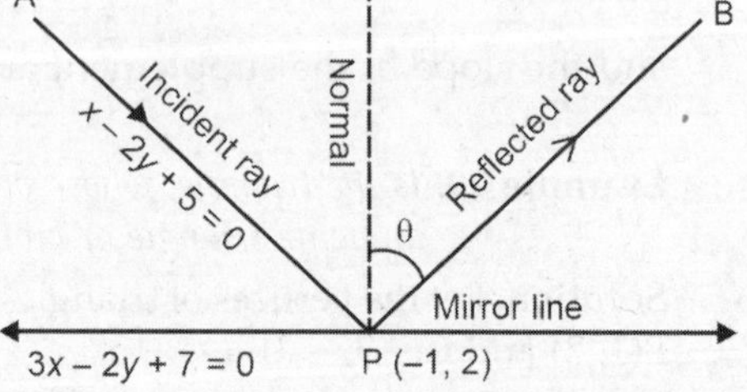

angle of incident = angle of reflection

$\Rightarrow \quad \angle APN = \angle NPB = \theta$ (say) [PN is normal]

Now, (slope of AP) $= \frac{1}{2}$, [slope of mirror line (2)] $= \frac{3}{2}$

Also, [slope of PN] × [slope of mirror line (2)] $= -1$

$$\Rightarrow \quad \text{Slope of } PN \times \left(\frac{3}{2}\right) = -1 \quad \Rightarrow \quad \text{Slope of } PN = -\frac{2}{3}$$

Let m be the slope of PB. Then,

$$\tan(\angle APN) = \tan(\angle BPN)$$

$$\Rightarrow \quad \frac{\frac{1}{2} - \left(-\frac{2}{3}\right)}{1 + \left(\frac{1}{2}\right)\left(-\frac{2}{3}\right)} = \pm \frac{m - \left(-\frac{2}{3}\right)}{1 + m\left(-\frac{2}{3}\right)} \quad \Rightarrow \quad 7(3-2m) = \pm 4(3m+2)$$

$$\Rightarrow \quad m = \frac{1}{2} \text{ or } m = \frac{29}{2}$$

But, $\frac{1}{2}$ is the slope of line AP, therefore slope of reflected ray $(PB) = \frac{29}{2}$.

Thus, the equation of the line PB (reflected ray), which passes through $P(-1, 2)$ is

$$y - 2 = \frac{29}{2}(x+1)$$

$$\Rightarrow \quad 29x - 2y + 33 = 0$$ **Ans.**

EXERCISE 22.11

Find the angles between each pairs of the following straight lines:

1. $\sqrt{3}x - y + 2 = 0$ and $-x + \sqrt{3}y + 4 = 0$ **Ans.** 30°

2. $3x - y + 5 = 0$ and $x - 3y + 1 = 0$ **Ans.** $\tan^{-1}\left(\frac{4}{3}\right)$

3. $3x + y + 12 = 0$ and $x + 2y - 1 = 0$ **Ans.** 45°

4. The line through (4, 3) and (– 6, 0) intersects the line $5x + y = 0$. Find the angle of intersection. **Ans.** $\tan^{-1}\left(\frac{53}{5}\right)$

5. Find the tangent of the angle between the lines, which have intercepts 3, 4 and 1, 8 on the axes respectively. **Ans.** $\frac{4}{5}$

6. $(m^2 - mn)y = (mn + n^2)x + n^3$ and $(mn + m^2)y = (mn - n^2)x + m^3$.

Ans. $\tan^{-1}\left(\frac{4m^2n^2}{m^4 - n^4}\right)$

7. Find the acute angle between the lines $2x - y + 3 = 0$ and $x + y + 2 = 0$.

Ans. $\tan^{-1}(3)$

8. Prove that the straight lines $(a + b)\,x + (a - b)\,y = 2ab$, $(a - b)\,x + (a + b)\,y = 2ab$ and $x + y = 0$ form an isosceles triangle, whose vertical angle is $2\tan^{-1}\left(\frac{a}{b}\right)$.

9. If the angle between two lines is $\frac{\pi}{4}$ and the slope of one of the lines is $\frac{1}{2}$, find the slope of the other line. **Ans.** Slope = 3

10. Prove that the points (2, – 1), (0, 2), (3, 3) and (5, 0) are the vertices of a parallelogram. Also, find the angle between its diagonals. **Ans.** $\tan^{-1}\left(\frac{22}{3}\right)$

11. Prove that the diagonals of the parallelogram formed by the four straight lines: $\sqrt{3}x + y = 0$, $\sqrt{3}y + x = 0$, $\sqrt{3}x + y = 1$ and $\sqrt{3}y + x = 1$ are at right angles to one another.

Ans. $7x - 3y - 10 = 0$, $3x + 7y + 4 = 0$

12. Find the equations of two straight lines passing through the point (1, – 1) and inclined at an angle of 45° to the line $2x - 5y + 7 = 0$.

Ans. $7x - 3y - 10 = 0$, $3x + y + 4 = 0$

22.23 EQUATIONS OF STRAIGHT LINES PASSING THROUGH A GIVEN POINT AND MAKING A GIVEN ANGLE WITH A GIVEN LINE

To prove that the equation of the straight lines, when pass through a given point (x_1, y_1) and make a given angle α with the given straight line $y = mx + c$ are

$$y - y_1 = \frac{m \pm \tan\alpha}{1 \mp m\tan\alpha}(x - x_1).$$

Proof: Let the given line be LMN, making an angle θ with the x-axis and let $P(x_1, y_1)$ be the given point. Then, $y = mx + c$ is the equation of the given line LMN.

Let PQ and PR be the required lines that make angle α with the given line. Let these lines meet the x-axis at Q and R and also make angles θ_1 and θ_2 respectively, with the positive direction of x-axis.

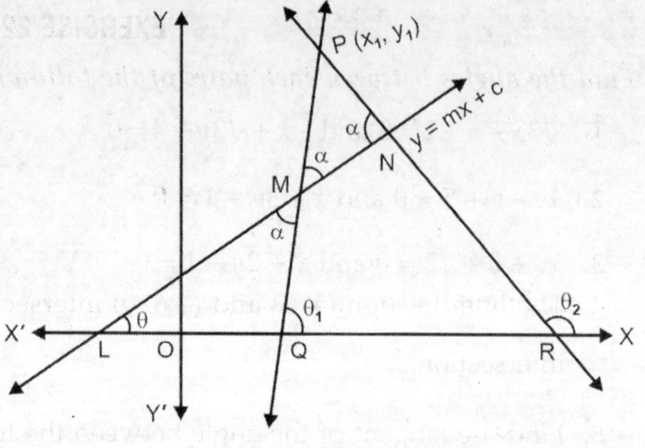

Then, the equation of required lines are

$$y - y_1 = \tan\theta_1 (x - x_1) \;...\; (1)$$

And $\quad y - y_2 = \tan\theta_2 (x - x_2) \;...\; (2)$

From ΔLMQ, we have

$$\theta_1 = \theta + \alpha$$

Also from ΔLNR, we have

$$\theta_2 = \theta + 180° - \alpha$$

Now, $\quad \theta_1 = \theta + \alpha$

$$\Rightarrow \quad \tan\theta = \tan(\theta + \alpha) = \frac{\tan\theta + \tan\alpha}{1 - \tan\theta\tan\alpha} = \frac{m + \tan\alpha}{1 - m\tan\alpha}$$

Again, $\quad \theta_2 = \theta + 180° - \alpha$

$$\Rightarrow \quad \tan\theta_2 = \tan(\theta + 180° - \alpha) = \tan(180° + \theta - \alpha) = \tan(\theta - \alpha)$$

$$= \frac{\tan\theta - \tan\alpha}{1 + \tan\theta\tan\alpha} = \frac{m - \tan\alpha}{1 + m\tan\alpha}$$

On substituting the values of $\tan\theta_1$ and $\tan\theta_2$ in (1) and (2), we get

$$y - y_1 = \frac{m + \tan\alpha}{1 - m\tan\alpha}(x - x_1) \text{ and } y - y_2 = \frac{m - \tan\alpha}{1 + m\tan\alpha}(x - x_2)$$

These are the equations of the two required lines. **Proved.**

Example 73. *Find the equations of the two lines passing through the point (1, – 1) and inclined at an angle of 45° with the given straight line 2x - 5y + 7 = 0.* **[SBTE 2014]**

Solution. We know that the equations of two straight lines, which pass through a point (x_1, y_1) and make a given angle a with the given straight line $y = mx + c$ are

P (1, –1)

$$y - y_1 = \frac{m \pm \tan\alpha}{1 \mp m\tan\alpha}(x - x_1)$$

Here, $x_1 = 1, y_1 = -1, \alpha = 45°$,

m = slope of the line $2x - 5y + 7 = 0$.

So, $m = \dfrac{2}{5}\left[\text{Using } m = -\dfrac{\text{Coefficient of } x}{\text{Coefficient of } y}\right]$

45° 45°
A Q 2x – 5y + 7 = 0 R B

So, the equations of the required lines are

$$y - (-1) = \frac{\frac{2}{5} + \tan 45°}{1 - \frac{2}{5}\tan 45°}(x - 1) \quad \text{and} \quad y - (-1) = \frac{\frac{2}{5} - \tan 45°}{1 + \frac{2}{5}\tan 45°}(x - 1)$$

$$\Rightarrow \quad y+1=\frac{\frac{2}{5}+1}{1-\frac{2}{5}(1)}(x-1) \quad \text{and} \quad y+1=\frac{\frac{2}{5}-1}{1+\frac{2}{5}(1)}(x-1)$$

$$\Rightarrow \quad y+1=\frac{7}{3}(x-1) \quad \text{and} \quad y+1=\frac{-3}{7}(x-1)$$

$$\Rightarrow \quad 3y+3=7x-7 \quad \text{and} \quad 7y+7=-3x+3$$

$$\Rightarrow \quad 7x-3y-10=0 \quad \text{and} \quad 3x+7y+4=0$$

Hence, the equations of the required lines are

$$7x-3y-10=0 \quad \text{and} \quad 3x+7y+4=0$$ **Ans.**

Example 74. *Find the equations of the straight lines through (3, 2) which make an acute angle of 45° with the line $x - 2y - 3 = 0$.*

Solution. Here, $x_1 = 3, y_1 = 2, \alpha = 45°$

And m = (slope of the line $x - 2y - 3 = 0$) $= \frac{1}{2}$, $\left[\because m = -\frac{\text{Coefficient of } x}{\text{Coefficient of } y}\right]$

So, the equations of the required lines are

$$y-2=\frac{\frac{1}{2}-\tan 45°}{1+\frac{1}{2}\tan 45°}(x-3) \quad \text{and} \quad y-2=\frac{\frac{1}{2}+\tan 45°}{1-\frac{1}{2}\tan 45°}(x-3)$$

$$\Rightarrow \quad y-2=-\frac{1}{3}(x-3) \quad \text{and} \quad y-2=3(x-3)$$

$$\Rightarrow \quad x+3y=9 \quad \text{and} \quad 3x-y=7$$

These are the required equations. **Ans.**

Example 75. *Two lines passing through the point (2, 3) intersects each other at an angle of 60°. If slope of one line is 2. Find equation of other line.*

Solution.

Angle between two lines is 60°. (*i.e.* $\theta = 60°$)

The slope of one line, *i.e.* $m_1 = 2$

Let the slope of other line be m_2. Then,

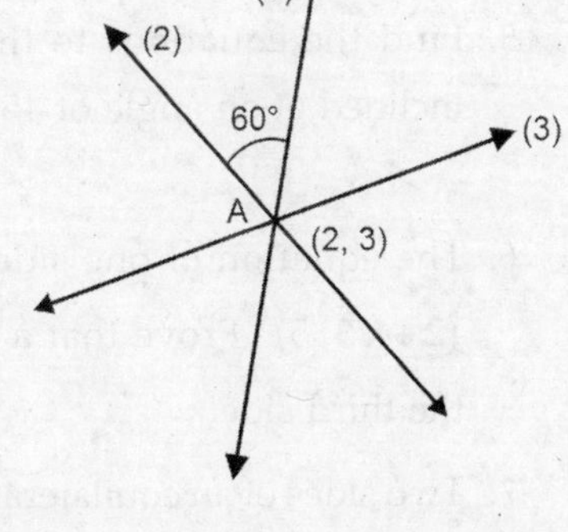

$$\tan\theta = \frac{m_2 - m_1}{1 + m_1 m_2} \quad \text{... (1)}$$

$$\Rightarrow \tan 60° = \frac{m_2 - 2}{1+2m_2} \quad \Rightarrow \quad \sqrt{3} = \frac{m_2-2}{1+2m_2}$$

$$\Rightarrow \sqrt{3}(1+2m_2) = m_2 - 2 \quad \Rightarrow \quad m_2(2\sqrt{3}-1) = -2-\sqrt{3}$$

$$\Rightarrow \quad m_2 = \frac{-(\sqrt{3}+2)}{(2\sqrt{3}-1)}$$

Thus, equation of a line passing through the point $A(2, 3)$ and having slope $\frac{-(\sqrt{3}+2)}{(2\sqrt{3}-1)}$ is

$$y-3=\frac{-(\sqrt{3}+2)}{(2\sqrt{3}-1)}(x-2)$$

$\Rightarrow \quad (2\sqrt{3}-1)(y-3) = -(\sqrt{3}+2)(x-2)$

$\Rightarrow \quad (\sqrt{3}+2)x + (2\sqrt{3}-1)y = 2\sqrt{3}+4+6\sqrt{3}-3$

$\Rightarrow \quad (\sqrt{3}+2)x + (2\sqrt{3}-1)y = 8\sqrt{3}+1$

If in (1) we take $m_1 - m_2$ in place of $m_2 - m_1$, then $m_2 = \dfrac{2-\sqrt{3}}{2\sqrt{3}+1}$

Thus, the equation of the other line passing through the point (2, 3) and having slope $\dfrac{2-\sqrt{3}}{2\sqrt{3}+1}$ is $y - 3 = \dfrac{2-\sqrt{3}}{2\sqrt{3}+1}(x-2)$

$\Rightarrow \quad (2\sqrt{3}+1)(y-3) = (2-\sqrt{3})(x-2)$

$\Rightarrow \quad (\sqrt{3}-2)x + (2\sqrt{3}+1)y = -4+2\sqrt{3}+6\sqrt{3}+3$

$\Rightarrow \quad (\sqrt{3}-2)x + (2\sqrt{3}+1)y = 8\sqrt{3}-1$ **Ans.**

EXERCISE 22.12

1. Find the equations of the two straight lines through (1, 1) and making an angle of 30° with the line $x - y - 2 = 0$.

   Ans. $(2+\sqrt{3})x - y - \sqrt{3} - 1 = 0,\ (2-\sqrt{3})x - y + \sqrt{3} - 1 = 0$

2. Find equations of the lines through the point (3, 2) which make an angle of 45° with the line $x - 2y = 3$. **Ans.** $3x - y - 7 = 0$ and $x + 3y - 9 = 0$

3. A diagonal of a square lies along the line $8x - 15y = 0$ and one vertex of the square is (1, 2). Find the equations of the sides of the square passing through this vertex.

   Ans. $23x - 7y - 9 = 0$ and $7x + 23y - 53 = 0$

4. Find the equations to the straight lines passing through the point (3, – 2) and inclined at 60° with the line $\sqrt{3}x + y = 1$. **Ans.** $y + 2 = 0$ and $\sqrt{3}x - y = 2 + 3\sqrt{3}$

5. Find the equations to the straight lines, which pass through the origin and are inclined at an angle of 45°to the straight line $2x + y + \sqrt{3}(y - x) = a$.

   Ans. $3y = (2\sqrt{3}-1)x, 3x + (2\sqrt{3}-1)y = 0$

6. The equation of one side of an equilateral triangle is $x - y = 0$ and one vertex is $(2+\sqrt{3}, 5)$. Prove that a second side is $y + (2-\sqrt{3})x = 6$ and find the equation of the third side. **Ans.** $y + (2+\sqrt{3})x = 12 + 4\sqrt{3}$

7. Two sides of an equilateral triangle are given by the equations $(2+\sqrt{3})x - y = 1 + 2\sqrt{3}$ and $(2-\sqrt{3})x - y = 1 - 2\sqrt{3}$ and its third side passes through the point (1, 1). Determine the equation of the third side. **Ans.** $x + y = 2$

8. Find the equations of two straight lines passing through (4, – 2) and making an angle of 60° with the line $2x + y = 0$.

   Ans. $(5\sqrt{3}-8)x + 11y + 54 - 20\sqrt{3} = 0,\ (5\sqrt{3}+8)x - 11y - 54 - 20\sqrt{3} = 0$

22.24 DISTANCE OF A POINT FROM A LINE

The distance of a point (x_1, y_1) from the line $x \cos \alpha + y \sin \alpha = p$.

Solution. Case I: We have given a line AB whose equation is

$$x \cos \alpha + y \sin \alpha = p \qquad \text{... (1)}$$

where α is the angle between the perpendicular and x-axis in anticlockwise direction. P is the length of $\perp$ from the origin to the line.

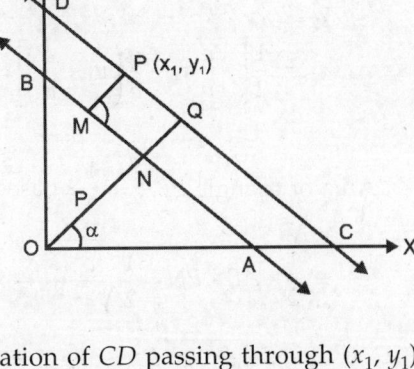

Any line CD parallel to (1) is

$$x \cos \alpha + y \sin \alpha = k \qquad \text{... (2)}$$

If a point $P(x_1, y_1)$ lies on (2) on the opposite side of line AB from O, then

$$x_1 \cos \alpha + y_1 \sin \alpha = k$$

On putting the value of k in (1), we get the equation of CD passing through (x_1, y_1) and parallel to (1) i.e.

$$x \cos \alpha + y \sin \alpha = x_1 \cos \alpha + y_1 \sin \alpha$$

Here, $x_1 \cos \alpha + y_1 \sin \alpha$ is the perpendicular distance from origin to the line CD.

Distance of (x_1, y_1) to line AB = Distance between the parallel lines AB and CD.

$$= PM = QN = OQ - ON$$

$$= (x_1 \cos \alpha + y_1 \sin \alpha) - p = |x_1 \cos \alpha + y_1 \sin \alpha - p|$$

Case II. If the equation of a line is in the general form

$$Ax + By + C = 0$$

Reducing the general equation to the normal form

$$\frac{A}{\sqrt{A^2+B^2}}x + \frac{B}{\sqrt{A^2+B^2}}y = \pm\frac{C}{\sqrt{A^2+B^2}}$$

From the result of **Case I**, distance between point (x_1, y_1) and the line CD

$$= \frac{Ax'}{\sqrt{A^2+B^2}} + \frac{By'}{\sqrt{A^2+B^2}} - \left(\frac{-C}{\sqrt{A^2+B^2}}\right) = \frac{Ax' + By' + C}{\sqrt{A^2+B^2}}$$

Second Method: The distance between a point (x_1, y_1) from the line

$$ax + by + c = 0 \text{ is } \frac{ax_1 + by_1 + c}{\sqrt{a^2+b^2}}$$

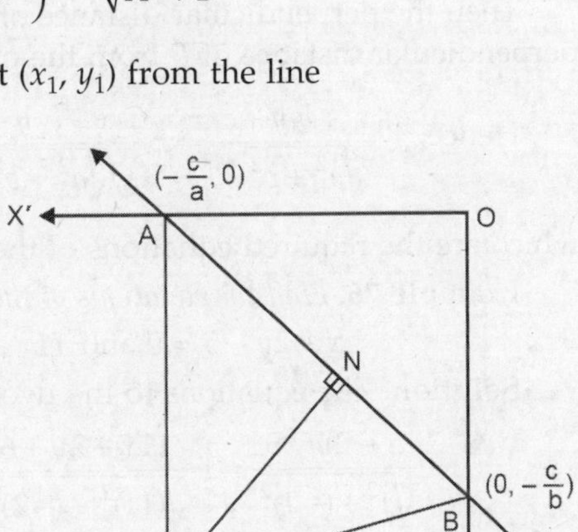

Solution. Given a point $P(x_1, y_1)$ and a line AB is $ax + by + c = 0$ ... (1)

Let us draw a perpendicular PN from P to AB.

Let the line (1) cut the coordinate axes at

$$A\left(-\frac{c}{a}, 0\right) \text{ and } B\left(0, -\frac{c}{b}\right).$$

Join PA and PB forming a triangle PAB.

Now, we have to find out the area of the triangle PBA.

$$\text{Area} = \frac{1}{2}\left[x_1\left(-\frac{c}{b}-0\right)+0\,(0-y_1)-\frac{c}{a}\left(y_1+\frac{c}{b}\right)\right]$$

$$= \frac{1}{2}\left[-\frac{c}{b}x_1-\frac{c}{a}\left(y_1+\frac{c}{b}\right)\right]=\frac{1}{2}\left[-\frac{c}{b}x_1-\frac{c}{a}y_1-\frac{c^2}{ab}\right] \quad \left[\because \Delta=\frac{1}{2}\begin{vmatrix} x_1 & y_1 & 1 \\ x_2 & y_2 & 1 \\ x_3 & y_3 & 1 \end{vmatrix}\right] \quad \text{... (2)}$$

$$\text{Area of triangle } PBA = \frac{1}{2}\times \text{base}\times\text{height}$$

$$= \frac{1}{2}\times AB\times PN=\frac{1}{2}\sqrt{\frac{c^2}{a^2}+\frac{c^2}{b^2}}\times PN \quad \text{... (3)}$$

$$= \frac{c}{2ab}\sqrt{a^2+b^2}\times PN$$

From (2) and (3), we have

$$\frac{c}{2ab}\sqrt{a^2+b^2}\times PN=\frac{1}{2}\left[-\frac{c}{b}x_1-\frac{c}{a}y_1-\frac{c^2}{ab}\right]$$

(x₁, y₁)

$= \dfrac{ax_1 + by_1 + c}{\sqrt{a^2 + b^2}}$

ax + by + c = 0

$$\Rightarrow \sqrt{a^2+b^2}\times PN=-[ax_1+by_1+c]$$

$$\Rightarrow \qquad PN=\frac{|\,ax_1+by_1+c\,|}{\sqrt{a^2+b^2}}$$

22.25 BISECTORS OF ANGLES BETWEEN TWO LINES

To find the equation of the bisectors of angles between the lines.

$$a_1x+b_1y+c_1=0 \quad \text{... (1)}$$

and $$a_2x+b_2y+c_2=0 \quad \text{... (2)}$$

Let $P(x, y)$ be any point on one of the bisectors.

Then the perpendicular distance of $P(x, y)$ from one line is equal in magnitude to the perpendicular distance of P from the other line

Hence, $$\frac{a_1x+b_1y+c_1}{\sqrt{a_1^2+b_1^2}}=\pm\frac{a_2x+b_2y+c_2}{\sqrt{a_2^2+b_2^2}}$$

which are the required equations of the bisectors.

Example 76. *Find the equations of the bisectors of the angles between the lines*

$$x-2y-5=0 \text{ and } 11x-2y+6=0$$

Solution. The equations to the two bisectors are

$$\frac{x-2y-5}{\sqrt{1^2+(-2)^2}}=\pm\frac{11x-2y+6}{\sqrt{(11)^2+(-2)^2}}$$

$$\Rightarrow \quad \frac{x-2y-5}{\sqrt{5}}=\pm\frac{11x-2y+6}{5\sqrt{5}}$$

Taking the positive sign, the equation of one bisectors is

$$6x + 8y + 31 = 0$$

Taking the negative sign, the equation of the other bisector is

$$16x - 12y - 19 = 0$$ **Ans.**

Example 77. *Find the coordinates of the incentre of the triangle formed by the lines whose equations are: $3x - 4y = 0$, $4x + 3y - 8 = 0$ and $24x - 7y - 12 = 0$.*

Solution. Let $AB : 3x - 4y = 0$ ... (1)

$BC : 4x + 3y - 8 = 0$ ... (2)

and $CA : 24x - 7y - 12 = 0$ ... (3)

be the sides of the triangle.

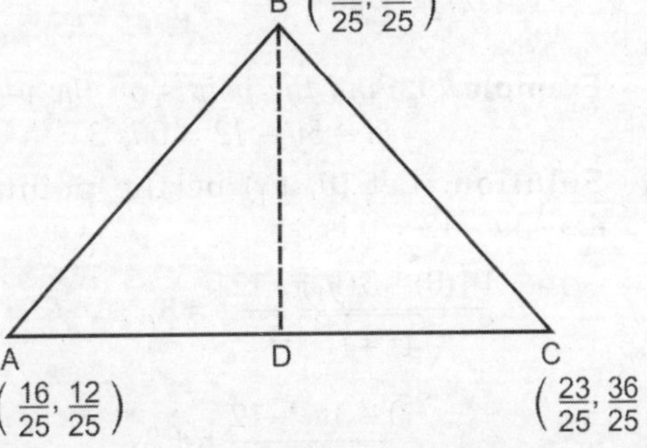

Solving (1) and (2), we have

$$\frac{x}{32} = \frac{y}{24} = \frac{1}{9+16}$$

i.e., $x = \frac{32}{25}, y = \frac{24}{25}$

$\therefore$ Co-ordinates of B are $\left(\frac{32}{25}, \frac{24}{25}\right)$

Solving (2) and (3) coordinates of C are $\left(\frac{23}{25}, \frac{36}{25}\right)$

Similarly, coordinates of A are $\left(\frac{16}{25}, \frac{12}{25}\right)$.

Equations of the bisectors of angle between AB and BC, *i.e.*

$3x - 4y = 0$ and $4x + 3y - 8 = 0$ are

$$\frac{3x - 4y}{\sqrt{3^2 + (-4)^2}} = \pm \frac{4x + 3y - 8}{\sqrt{4^2 + 3^2}}$$

$\Rightarrow$ $3x - 4y = \pm (4x + 3y - 8)$

$\Rightarrow$ $x + 7y - 8 = 0$ ... (4)

and $7x - y - 8 = 0$ ... (5)

Substituting coordinates of $A\left(\frac{16}{25}, \frac{12}{25}\right)$ and $C\left(\frac{23}{25}, \frac{36}{25}\right)$ in the left hand side of (4) we have

$$\frac{16}{25} + \frac{84}{25} - 8 = -4$$

and $$\frac{23}{25} + \frac{252}{25} - 8 = 3$$

which are of opposite signs.

Thus A and C lie on opposite sides of bisector (4).

Therefore $x + 7y - 8 = 0$ is the internal bisector of $\angle ABC$.
Similarly, the internal bisector of $\angle ACB$ is

$$11x + 2y - 13 = 0$$

Solving $x + 7y - 8 = 0$ and $11x + 2y - 13 = 0$ we get (1, 1) as the incentre. **Ans.**

Example 78. *Find the distance of the point (– 3, 4) from the line $3x + 4y - 5 = 0$.*

Solution. We have $3x + 4y - 5 = 0$ and point is (– 3, 4)

The required distance d is given by

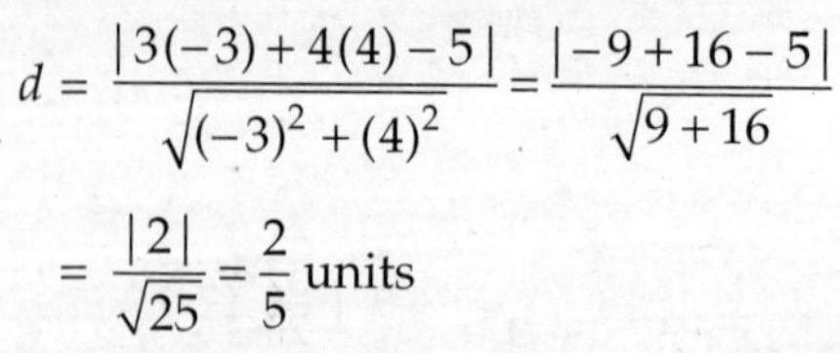

$$d = \frac{|3(-3)+4(4)-5|}{\sqrt{(-3)^2+(4)^2}} = \frac{|-9+16-5|}{\sqrt{9+16}}$$

$$= \frac{|2|}{\sqrt{25}} = \frac{2}{5} \text{ units}$$ **Ans.**

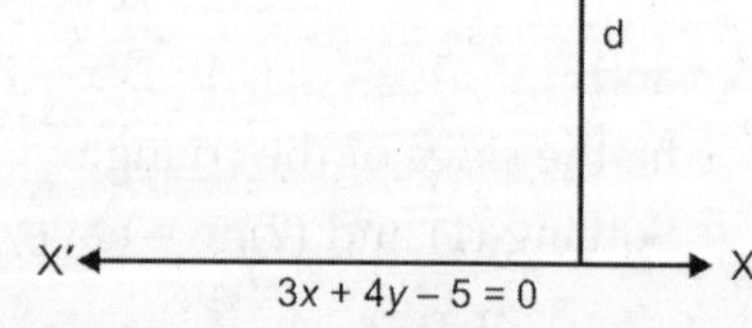

Example 79. *Find the points on the y-axis, whose perpendicular distance from the line $4x - 3y - 12 = 0$ is 3.*

Solution. Let $(0, y_1)$ be the point, whose perpendicular distance from line $4x - 3y - 12 = 0$ is 3.

$$\therefore \quad \frac{|4(0)-3(y_1)-12|}{\sqrt{4^2+(-3)^2}} = 3$$

$$\Rightarrow \quad \frac{0-3y_1-12}{\sqrt{16+9}} = \pm 3$$

$$\Rightarrow \quad \frac{-3y_1-12}{\sqrt{25}} = \pm 3$$

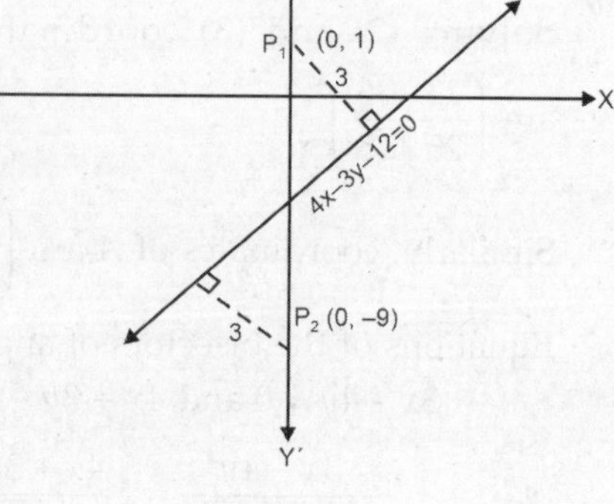

$$\Rightarrow \quad \frac{-3y_1-12}{5} = 3 \qquad\qquad \frac{-3y_1-12}{5} = -3$$

$$\Rightarrow \quad -3y_1 - 12 = 15 \qquad\qquad -3y_1 - 12 = -15$$

$$\Rightarrow \quad -3y_1 = +27 \qquad\qquad -3y_1 = -3$$

$$\Rightarrow \quad y_1 = -9 \qquad\qquad y_1 = 1$$

$\therefore$ Point is (0, – 9) $\qquad \therefore$ Point is (0, 1)

Thus the required points are (0, 1) and (0, – 9). **Ans.**

Example 80. *In a triangle with vertices A(2, 3), B(4, – 1) and C(– 1, 2), find the length of the altitude from the vertex A.*

Solution. We have vertices of the triangle ABC are $A(2, 3)$, $B(4, -1)$ and $C(-1, 2)$.

Now, equation of the line passing through $B(4, -1)$ and $C(-1, 2)$ is

$$y - (-1) = \frac{2-(-1)}{-1-4}(x-4)$$

$$\Rightarrow \quad y + 1 = -\frac{3}{5}(x-4)$$

$$\Rightarrow \quad 5y + 5 = -3x + 12$$

$$\Rightarrow \quad 3x + 5y - 7 = 0$$

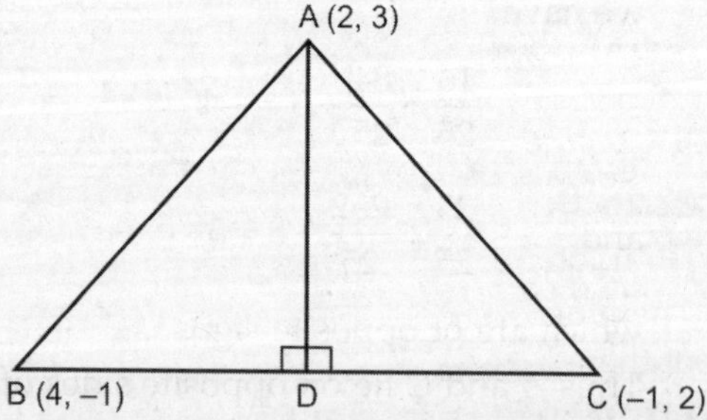

The length of altitude from the vertex $A(2, 3)$ to the line $3x + 5y - 7 = 0$ is

$$= \frac{|3(2)+5(3)-7|}{\sqrt{(3)^2+(5)^2}} = \frac{|6+15-7|}{\sqrt{9+25}} = \frac{14}{\sqrt{34}}$$ **Ans.**

Example 81. *Find the distance of the point $(-1, 1)$ from the line $12(x + 6) = 5(y - 2)$.*

Solution. We have, $12(x + 6) = 5(y - 2) \Rightarrow 12x - 5y + 82 = 0$ ... (1)

Comparing (1) with general equation of the line $Ax + By + C = 0$, we get

$A = 12, B = -5$ and $C = 82$.

Given point is $(x_1, y_1) \equiv (-1, 1)$. The distance of the given point from given line is

$$d = \frac{|Ax_1 + By_1 + C|}{\sqrt{A^2+B^2}} = \frac{|12(-1)+(-5)(1)+82|}{\sqrt{144+25}}$$

$$= \frac{|-12-5+82|}{\sqrt{169}} = \frac{65}{13} = 5 \text{ units.}$$ **Ans.**

Example 82. *If p and q are the lengths of perpendiculars from the origin to the lines $x \cos\theta - y \sin\theta = k \cos 2\theta$ and $x \sec\theta + y \operatorname{cosec}\theta = k$ respectively. Prove that $p^2 + 4q^2 = k^2$.* **[S.B.T.E. 2017, 2015, 2014]**

Solution. $x \cos\theta - y \sin\theta = k \cos 2\theta$ ... (1)

$x \sec\theta + y \operatorname{cosec}\theta = k$ ... (2)

Length of perpendicular from origin (0, 0) to (1) is

$$p = \frac{k \cos 2\theta}{\sqrt{\cos^2\theta + \sin^2\theta}} = k \cos 2\theta$$

$$p^2 = k^2 \cos^2 2\theta \quad \text{... (3)}$$

Length of perpendicular from (0, 0) to (2) is

$$q = \frac{k}{\sqrt{\sec^2\theta + \operatorname{cosec}^2\theta}} = \frac{k}{\sqrt{\frac{1}{\cos^2\theta} + \frac{1}{\sin^2\theta}}}$$

$$= \frac{k}{\sqrt{\frac{\sin^2\theta + \cos^2\theta}{\sin^2\theta\cos^2\theta}}} = \frac{k}{\frac{1}{\sin\theta\cos\theta}} = k \sin\theta\cos\theta$$

$$q = k^2 \sin^2\theta \cos^2\theta$$

$$4q^2 = 4k^2 \sin^2\theta\cos^2\theta = k^2(2\sin\theta\cos\theta)^2 = k^2\sin^2 2\theta \quad \text{... (4)}$$

On adding (3) and (4), we get

$$p^2 + 4q^2 = k^2\cos^2 2\theta + k^2 \sin^2 2\theta$$

$$= k^2(\cos^2 2\theta + \sin^2 2\theta) = k^2$$ **Proved.**

22.26 DISTANCE BETWEEN TWO PARALLEL LINES

If two lines are parallel, then they have the same distance between them throughout. Therefore to find the distance between two parallel lines choose an arbitrary point on one of them and find the length of the perpendicular on the other. To choose a point on a line give an arbitrary value to x or y and find the value of the other variable.

Working Rule to Find the the Distance between Two Parallel Lines:

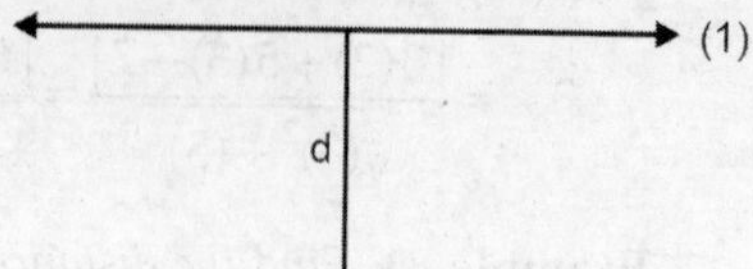

Step 1. Find the coordinates of any point on one of the given line, preferably by putting $x = 0$ and $y = 0$.

Step 2. The perpendicular distance of this point from the other line is the required distance between the lines.

Example 83. *Find the perpendicular distance between the lines:*

$$3x + 4y - 5 = 0 \text{ and } 6x + 8y - 45 = 0.$$

Solution. We have the parallel lines:

$$3x + 4y - 5 = 0 \quad \text{... (1)}$$

and $$6x + 8y - 45 = 0 \quad \text{... (2)}$$

Putting $x = 0$ in the equation (1), we get

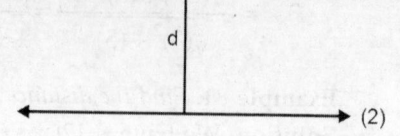

$$3(0) + 4y - 5 = 0 \quad \Rightarrow \quad y = \frac{5}{4}$$

Hence, $\left(0, \frac{5}{4}\right)$ is a point on the first line.

Now, perpendicular distance of this point from the line

$$6x + 8y - 45 = 0 \text{ is } \frac{\left|6 \times 0 + 8 \times \frac{5}{4} - 45\right|}{\sqrt{36 + 64}} = \frac{|-35|}{10} = \frac{7}{2}$$

which is the required distance. **Ans.**

Example 84. *Find the distance between the parallel lines $2x - 3y + 9 = 0$ and $4x - 6y + 1 = 0$.*

Solution. We have two parallel lines

$$2x - 3y + 9 = 0 \quad \text{... (1)}$$

$$4x - 6y + 1 = 0 \quad \text{... (2)}$$

The equations in the normal form of (1) and (2) are

$$\frac{2}{\sqrt{13}} x - \frac{3}{\sqrt{13}} y + \frac{9}{\sqrt{13}} = 0 \quad \text{... (3)}$$

and $$\frac{4x}{\sqrt{52}} - \frac{6y}{\sqrt{52}} + \frac{1}{\sqrt{52}} = 0 \quad \Rightarrow \quad \frac{2x}{\sqrt{13}} - \frac{3y}{\sqrt{13}} + \frac{1}{\sqrt{52}} = 0 \quad \text{... (4)}$$

The distance between (3) and (4) is

$$\frac{9}{\sqrt{13}} - \frac{1}{\sqrt{52}} = \frac{18}{\sqrt{52}} - \frac{1}{\sqrt{52}} = \frac{17}{\sqrt{52}}$$ **Ans.**

Example 85. *Find the distance between parallel lines:*

$$15x + 8y - 34 = 0 \text{ and } 15x + 8y + 31 = 0$$

Solution. We have two parallel lines

$$15x + 8y - 34 = 0 \quad \text{... (1)}$$

and $$15x + 8y + 31 = 0 \quad \text{... (2)}$$

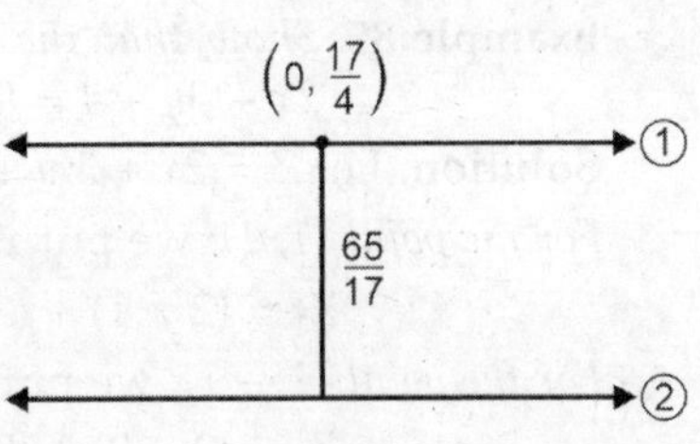

Putting $x = 0$ in the equation (1), we have

$$15(0) + 8y - 34 = 0 \Rightarrow y = \frac{17}{4}$$

So $\left(0, \frac{17}{4}\right)$ is a point on line (1).

Hence, the perpendicular distance from $\left(0, \frac{17}{4}\right)$ to the line (2) is

$$\frac{15 \times 0 + 8\left(\frac{17}{4}\right) + 31}{\sqrt{(15)^2 + (8)^2}} = \frac{34 + 31}{\sqrt{225 + 64}} = \frac{65}{\sqrt{289}} = \frac{65}{17}$$ **Ans.**

Example 86. *Find the distance between parallel lines:*

$$l(x + y) + p = 0 \text{ and } lx + ly - r = 0$$

Solution. We have two parallel lines:

$$lx + ly + p = 0 \quad \text{... (1)}$$

and

$$lx + ly - r = 0 \quad \text{... (2)}$$

Putting $x = 0$ in (1), we get

$$l(0) + ly + p = 0 \Rightarrow y = -\frac{p}{l}$$

Thus, $\left(0, -\frac{p}{l}\right)$ is a point on line (1).

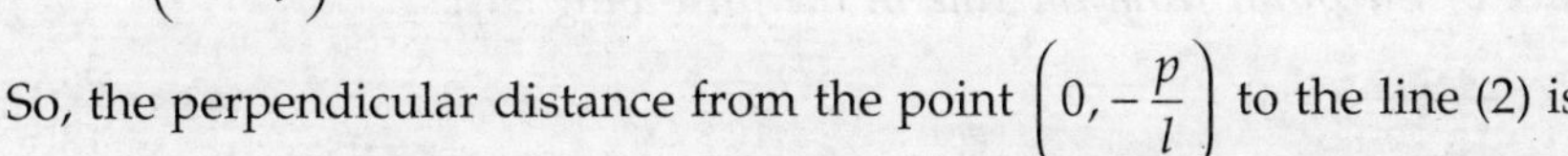

So, the perpendicular distance from the point $\left(0, -\frac{p}{l}\right)$ to the line (2) is

$$\left|\frac{l(0) + l\left(-\frac{p}{l}\right) - r}{\sqrt{l^2 + l^2}}\right| = \left|\frac{-p-r}{\sqrt{2}l}\right| = \frac{1}{\sqrt{2}}\left|\frac{p+r}{l}\right| \text{ units.}$$

Hence, the distance between lines (1) and (2) is $\frac{p+r}{\sqrt{2}l}$ units. **Ans.**

Remarks:

***(a)* One line and origin:**

If origin makes $ax + by + c$ as +ve, then this side is known as positive side of the line $ax + by + c = 0$.

If origin makes $ax + by + c$ as negative, then this side is known as negative side of the line.

Origin ⟶ +ve, then positive side
Origin ⟵ –ve, then negative side

***(b)* One line and two points:**

If two points (x_1, y_1) and (x_2, y_2) make L.H.S. of (1) *i.e.* $ax + by + c$ of the same sign, then both points lie on the same side.

If two points (x_1, y_1) and (x_2, y_2) make it of opposite signs, then the 2 points lie on the opposite sides of the line.

Example 87. *Show that the points (1, 1) and (2, – 1) lie on the same side of the line* $2x + 3y + 4 = 0$

Solution. Let $Z = 2x + 3y + 4 = 0$ ... (1)

For the point (1, 1); we put $x = 1, y = 1$ in (1) to get the value of Z_1 (say)

$$Z_1 = (2 \times 1) + (3 \times 1) + 4 = +9$$

For the point (2, – 1); we put $x = 2, y = -1$ in (1), to get the value of Z_2 (say)

$$Z_2 = (2 \times 2) + [3 \times (-1)] + 4 = 4 - 3 + 4 = +5$$

Here, the sign of both Z_1 and Z_2 are the same (+ ve).

Hence, both points lie on the same side of the given line **Proved.**

Example 88. *Show that the points (1, 4) and (0, – 3) lies on the opposite sides of the line* $x + 3y + 7 = 0$.

Solution. Let $Z = x + 3y + 7 = 0$ ... (1)

For point (1, 4)

We put $x = 1$ and $y = 4$ in (1), to get value of Z_1 (say).

$$\therefore \quad Z_1 = 1 + 3(4) + 7 = 1 + 12 + 7 = 20$$

For point (0, – 3)

We put $x = 0$ and $y = -3$ in (1), to get the value of Z_2 (say)

$$\therefore \quad Z^2 = 0 + 3(-3) + 7 = -9 + 7 = -2$$

Here the signs of Z_1 and Z_2 are + ve and – ve respectively *i.e.* opposite sign. Hence, the two points are on the opposite sides of the given line. **Ans.**

EXERCISE 22.13

Find the distance of the point from the line in the following cases:

1. (3, – 1); $12x - 5y - 7 = 0$ **Ans.** $\frac{34}{13}$ units

2. (– 3, – 4); $12(x + 6) = 5(y - 2)$ **Ans.** $\frac{66}{13}$ units

3. (b, a); $\frac{x}{a} - \frac{y}{b} = 1$ **Ans.** $\frac{a^2 - b^2 + ab}{\sqrt{a^2 + b^2}}$ units

4. (2, 3); $y = 4$ **Ans.** $d = 1$

5. Find the distance of the point (4, 2) from the line joining the points (4, 1) and (2, 3). **Ans.** $\frac{\sqrt{2}}{2}$ units

6. The vertices of a triangle are $A(-2, 1)$, $B(6, -2)$ and $C(4, 3)$. Find the length of the altitudes of the triangle. **Ans.** $\frac{34}{\sqrt{29}}, \frac{17}{\sqrt{10}}, \frac{34}{\sqrt{73}}$ units

7. Find the distance of the point of intersection of the lines $2x + 3y = 21$ and $3x - 4y + 11 = 0$ from the line $8x + 6y + 5 = 0$. **Ans.** $\frac{59}{10}$ units

8. Find the distance between the parallel straight lines $y = mx + c$ and $y = mx + d$. **Ans.** $\frac{|c - d|}{\sqrt{1 + m^2}}$

9. Find the distance between two parallel straight lines $4x - 3y - 9 = 0$ and $4x - 3y - 24 = 0$.
Ans. 3 units

10. Prove that the lines $2x + 3y = 19$ and $2x + 3y + 7 = 0$ are equidistant from the line $2x + 3y = 6$.

11. Show that the points $(3, -4)$ and $(2, 6)$ are opposite sides of the line $3x - 4y = 8$.

12. Show that the points $(-2, 5)$ and $(3, -1)$ lie on the opposite sides of the line $5x - 3y + 7 = 0$.

13. Show that the points $(-3, -2)$ and $(-1, 1)$ lie on the same side of the line $7x - y + 2 = 0$.

14. Show that the point $(1, 5)$ and the origin lies on the same side of the line $2x + y + 2 = 0$.

15. Which of the lines $3x + 4y - 7 = 0$ and $12x + 13y + 10 = 0$ is farther from the origin.
Ans. $3x + 4y - 7 = 0$

16. Which of the lines $x + 6y - 9 = 0$ and $2x - 5y + 8 = 0$ is farther from the point $(1, 5)$.
Ans. $x + 6y - 9 = 0$

17. A straight line is parallel to the line $3x - y - 3 = 0$ and $3x - y + 5 = 0$ and lies between them. Find the equation of the line if its distance from these lines are in the ratio 3 : 5.
Ans. $3x - y = 0$

18. Find the length of the perpendicular from the point $(4, -7)$ to the line joining the origin and the point of intersection of the lines $2x - 3y + 14 = 0$ and $5x + 4y - 7 = 0$.
Ans. 1 unit

19. A point moves so that the sum of squares of its distance from the two axes of coordinates is equal to the square of its distance from the line $x - y = 1$. Find the locus of the point.
Ans. $x^2 + y^2 + 2xy + 2x - 2y - 1 = 0$

20. Show that the perpendiculars let fall from any point on the straight line $7x + 4y = 16$ upon the two lines $3x - 4y = 4$ and $5x - 12y = 4$ are equal.

21. Find the points on the line $y = x$ which are at a distance of 5 units from $4x + 3y - 1 = 0$.
Ans. $\left(\frac{26}{7}, \frac{26}{7}\right)$ and $\left(-\frac{24}{7}, -\frac{24}{7}\right)$

22. What points on the x-axis are at a distance of 4 units from the line $3x - 4y - 5 = 0$.
Ans. $\left(\frac{25}{3}, 0\right), (-5, 0)$

23. What points on the line $x + y + 3 = 0$ are at a distance $\sqrt{5}$ from the line $x + 2y + 2 = 0$.
Ans. $(1, -4)$

Circles

23.1 DEFINITION

A circle is the locus of a point which moves so that its distance from a fixed point, called the centre is equal to a given distance. The given distance is called radius of the circle. The distance of the centre of a circle from any point on the circumference is always equal and fixed.

23.2 A CIRCLE WHOSE CENTRE IS THE ORIGIN

Standard Form

To find equation of a circle with origin as centre.

Let O be the centre of the circle and a its radius while P be any point on the circumference of the circle, and let its coordinates be x and y.

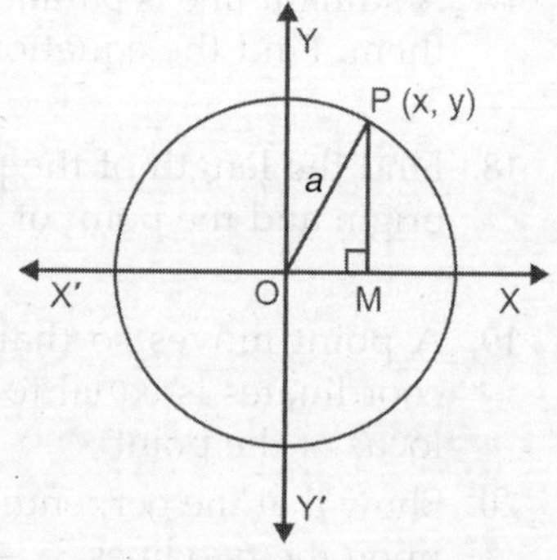

Then from the rt. angled $\Delta\ OMP$,

$$OM^2 + MP^2 = OP^2$$

$\boxed{x^2 + y^2 = a^2}$ is the required equation.

Central Form

To find equation of a circle whose centre and radius are given.

Solution. Let (h, k) be the centre C of the circle, and a the radius CP, where P is any point (x, y) on the circle.

Then $\quad CP = a$

$\Rightarrow \quad \sqrt{(x-h)^2 + (y-k)^2} = a$

On squaring both sides,

$\boxed{(x-h)^2 + (y-k)^2 = a^2}$ is the required equation.

Cor. If $h = 0$, $k = 0$, we get $x^2 + y^2 = a^2$, the form of the equation of section 22.2.

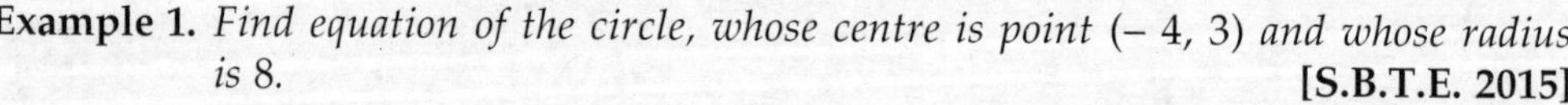

Example 1. *Find equation of the circle, whose centre is point* $(-4, 3)$ *and whose radius is* 8. **[S.B.T.E. 2015]**

Solution. The equation is $(x+4)^2 + (y-3)^2 = 8^2$. [using $(x-h)^2 + (y+k)^2 = a^2$]

$\Rightarrow \quad x^2 + y^2 + 8x - 6y - 39 = 0.$ **Ans.**

Example 2. *Obtain the equation of a circle which passes through the intersection of the lines:* $3x - 2y - 1 = 0$ *and* $4x + y - 27 = 0$ *and whose centre is the point* $(2, -3)$.

Solution. $3x - 2y - 1 = 0$ ... (1)

$4x + y - 27 = 0$ ... (2)

On solving (1) and (2), we get $x = 5, \quad y = 7$

The point of intersection of (1) and (2) is A (5, 7), C (2, – 3) is the centre of the circle.

A (5, 7)

C (2, –3)

$$\text{Radius} = \text{AC}$$

$$= \sqrt{(5-2)^2 + (7+3)^2} = \sqrt{9+100} = \sqrt{109}$$

Equation of circle with centre (2, – 3) and radius $\sqrt{109}$ is

$$(x - 2)^2 + (y + 3)^2 = 109$$ **Ans.**

Example 3. *Find the equation of the circle with its centre at (2, 3) and which passes through the point (5, 7).* **[S.B.T.E. 2015, 2014]**

Solution. $x_1 = 2, \quad y_1 = 3, \quad x_2 = 5, \quad y_2 = 7$

Distance between the point (x_1, y_1) and (x_2, y_2)

O(2, –3)

r

P (5, 7)

$$OP = \sqrt{(x_2 - x_1)^2 + (y_2 - y_1)^2}$$

$$OP = \sqrt{(5-2)^2 + (7-3)^2}$$

$$r = OP = \sqrt{25} = 5$$

Equation of the circle

$$(x - h)^2 + (y - k)^2 = r^2 \qquad [\because h = 2, k = 3, r = 5]$$

$$(x - 2)^2 + (y - 3)^2 = 5^2$$

$$\Rightarrow \quad x^2 + y^2 - 4x - 6y - 12 = 0$$ **Ans.**

Example 4. *Find the area of the circle $9x^2 + 9y^2 = 25$* **[S.B.T.E. 2015]**

Solution. We have $9x^2 + 9y^2 = 25 \quad \Rightarrow \quad x^2 + y^2 = \left(\frac{5}{3}\right)^2$

$$\text{Centre} = (0, 0), \text{ radius} = \frac{5}{3}$$

$$\text{Area of circle} = \pi r^2 = \pi\left(\frac{5}{3}\right)^2 = \frac{25}{9}\pi \text{ sq. units}$$ **Ans.**

Example 5. *Find equation of the circle which passes through the point (– 2, 4) and through the points in which the circle $x^2 + y^2 - 2x - 6y + 6 = 0$ is cut by the line $3x + 2y - 5 = 0$* **[DIPIETE, 2018]**

Solution. Let any point on the parabola be P (x, y)

Distance of P (x, y) from (– 1, 1) $= \sqrt{(x+1)^2 + (y-1)^2}$ ... (1)

Distance of P (x, y) from the line $x + y + 1 = 0$ is

$$\frac{x+y+1}{\sqrt{1^2+1^2}} = \frac{x+y+1}{\sqrt{2}} \quad \text{... (2)}$$

Equating (1) and (2), we get

$$\sqrt{(x+1)^2 + (y-1)^2} = \frac{x+y+1}{\sqrt{2}}$$

On squaring both sides, we get $(x + 1)^2 + (y - 1)^2 = \dfrac{(x + y + 1)^2}{2}$

$\Rightarrow \quad 2[x^2 + 2x + 1 + y^2 - 2y + 1] = x^2 + y^2 + 1 + 2xy + 2y + 2x$

$\Rightarrow \quad 2x^2 + 4x + 2y^2 - 4y + 4 = x^2 + y^2 + 2xy + 2x + 2y + 1$

$\Rightarrow \quad x^2 + y^2 - 2xy + 2x - 6y + 3 = 0$ **Ans.**

23.3 TO OBTAIN THE EQUATION OF A CIRCLE IN THE FORM

$$\boxed{x^2 + y^2 + 2gx + 2fy + c = 0}$$ **[Important]**

The equation of a circle whose centre is (h, k) and radius a is

$$(x - h)^2 + (y - k)^2 = a^2$$

$\Rightarrow \quad x^2 - 2hx + h^2 + y^2 - 2ky + k^2 = a^2$

$\Rightarrow \quad x^2 + y^2 - 2hx - 2ky + (h^2 + k^2 - a^2) = 0$

$\Rightarrow \quad x^2 + y^2 + 2gx + 2fy + c = 0$ where $g = -h, f = -k$ and $c = h^2 + k^2 - a^2$

$\boxed{\textbf{Centre} = (-g, -f)}$ and $\boxed{\textbf{Radius} = \sqrt{g^2 + f^2 - c}}$ **[Important]**

General Form

To prove that the equation $x^2 + y^2 + 2gx + 2fy + c = 0$ represents a circle and to find its centre and radius. **[Diploma IETE Dec., 2005]**

Proof. The equation may be written as

$$(x^2 + 2gx + g^2) + (y^2 + 2fy + f^2) = g^2 + f^2 - c$$

[Adding $g^2 + f^2$ to both sides and transposing c to the right hand]

$\Rightarrow \quad (x + g)^2 + (y + f)^2 = g^2 + f^2 - c$

$\Rightarrow \quad [x - (-g)]^2 + [y - (-f)]^2 = \left(\sqrt{g^2 + f^2 - c}\right)^2$

Comparing this with the equation $(x - h)^2 + (y - k)^2 = a^2$,

We find that the equation $x^2 + y^2 + 2gx + 2fy + c = 0$ represents a circle, whose centre is $(-g, -f)$ and radius is $\sqrt{g^2 + f^2 - c}$.

Example 6. *Find the centre and radius of circles:*

(*i*) $x^2 + y^2 - 6x + 8y + 9 = 0$

(*ii*) $4x^2 + 4y^2 - 10x + 5y + 5 = 0$

Solution. (*i*) Here $\quad g = -3, f = 4, c = 9$

$\therefore$ the co-ordinates of the centre are $(-g, -f)$ or $(3, -4)$ **Ans.**

And radius $= \sqrt{g^2 + f^2 - c} = \sqrt{9 + 16 - 9} = 4$ **Ans.**

(*ii*) The equation is $x^2 + y^2 - \dfrac{10}{4}x + \dfrac{5}{4}y + \dfrac{5}{4} = 0$

[**Note.** To make coefficients of x^2 and y^2 unity, we divide throughout by 4]

$$\therefore \quad g = -\frac{10}{8} = \frac{-5}{4} \text{ and } f = \frac{5}{8}, c = \frac{5}{4}$$

$\therefore$ coordinates of the centre are $\left(\frac{5}{4}, -\frac{5}{8}\right)$

and the radius = $\sqrt{\frac{25}{16} + \frac{25}{64} - \frac{5}{4}} = \sqrt{\frac{45}{64}} = \frac{3\sqrt{5}}{8}$ **Ans.**

Example 7. *Find the coordinates of the centre and radius of each of the following circles*

(*i*) $x^2 + y^2 - 8x + 6y - 24 = 0$ **[SBTE, 2017, 2014]**

(*ii*) $x^2 + y^2 + 8x - 6y = 0$ **[SBTE, 2014]**

(*iii*) $x^2 + y^2 - 10x + 8y - 3 = 0$ **[SBTE, 2017, 2014]**

(*iv*) $x^2 + y^2 - 6x - 4y - 24 = 0$ **[SBTE, 2016, 2015]**

Solution. Here we have

(*i*) $$x^2 + y^2 - 8x + 6y - 24 = 0 \quad \text{... (1)}$$

General equation of circle is:

$$x^2 + y^2 + 2gx + 2fy + c = 0 \quad \text{... (2)}$$

Comparing (2) with (1), we get

$$2g = -8, \quad g = -4, \quad 2f = 6, \quad f = 3, \quad c = -24$$

$$\text{Centre} = (-g, -f) = (4, -3)$$

$$\text{Radius} = \sqrt{g^2 + f^2 - c} = \sqrt{16 + 9 + 24} = 7$$ **Ans.**

(*ii*) Here we have

$$x^2 + y^2 + 8x - 6y = 0 \quad \text{... (1)}$$

General equation of circle is:

$$x^2 + y^2 + 2gx + 2fy + c = 0 \quad \text{... (2)}$$

Comparing (2) with (1), we get

$$g = 4, \quad f = -3, \quad c = 0$$

$$\text{Centre} = (-g, -f) = (-4, 3), \quad r = \sqrt{16 + 9} = 5$$ **Ans.**

Similarly (*iii*) and (*iv*) can be solved as above.

23.4 CONDITIONS FOR THE EQUATION OF A CIRCLE

The equation of a circle in the general form is

$$x^2 + y^2 + 2gx + 2fy + c = 0$$

Multiplying throughout by *a*, (To make it a particular case of the general equation),

we get $ax^2 + ay^2 + 2gax + 2fay + ac = 0$

$\therefore$ the conditions for an equation to represent a circle are:

(*i*) It should be an equation of the second degree in x and y.

(*ii*) The co-efficients of x^2 and y^2 should be equal.

(*iii*) There should be no term involving the product xy.

Example 8. *A point moves so that the sum of the squares of its distance from two fixed points is constant* ($= 2c^2$). *Prove that the locus is a circle.*

Solution. Let A, B be the points. Take AB as the x-axis and its middle point O as the origin. Draw $OY \perp AB$.

If $AB = 2a$, the co-ordinates of A and B are $(-a, 0)$ and $(a, 0)$ respectively.

If (x, y) be the moving point, then according to the question,

$$(x+a)^2 + (y-0)^2 + (x-a)^2 + (y-0)^2 = 2c^2$$

$$\Rightarrow \quad 2x^2 + 2a^2 + 2y^2 = 2c^2$$

$$\Rightarrow \quad x^2 + y^2 = c^2 - a^2$$

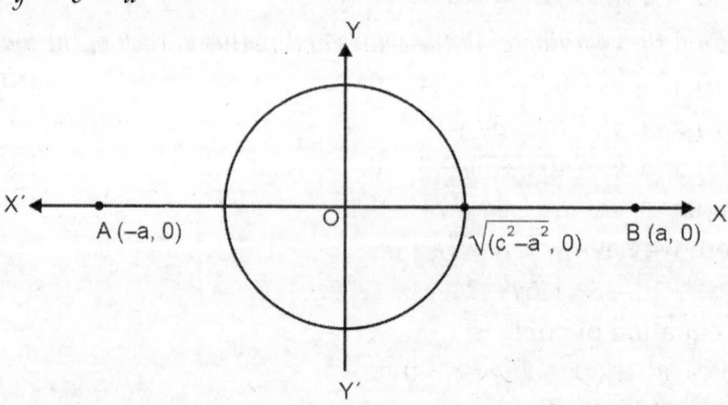

which is equation of a circle **Proved.**

EXERCISE 23.1

1. Find equation of a circle whose centre is at point (a, b) and whose radius is k.
Ans. $(x-a)^2 + (y-b)^2 = k^2$

2. Find the equation of a circle

(*a*) whose centre is (2, –3) and radius is 5. **Ans.** $x^2 + y^2 - 4x + 6y - 12 = 0$

(*b*) whose centre is (–3, –4) and radius is 7. **Ans.** $x^2 + y^2 + 6x + 8y - 24 = 0$

(*c*) whose centre is (a, b) and radius is $\sqrt{a^2 + b^2}$ **Ans.** $x^2 + y^2 - 2ax - 2by = 0$

3. Find the equation of the circle whose centre is (3, – 4) and

(*i*) passes through the point (4, 6); **Ans.** $x^2 + y^2 - 6x + 8y - 76 = 0$

(*ii*) passes through the intersection of the straight lines
$3x + 4y = 0$ and $4x + 3y = 0$; **Ans.** $x^2 + y^2 - 6x + 8y = 0$

(*iii*) passes through the intersection of the straight lines
$3x - 2y - 1 = 0$ and $4x + y - 27 = 0$ **Ans.** $x^2 + y^2 - 6x + 8y = 100$

4. Find the co-ordinates of the centre and radius of the circles:

(*i*) $x^2 + y^2 - 8x + 10y = 0$ **Ans.** $(4, -5), \sqrt{41}$

(*ii*) $x^2 + y^2 - 8x - 16y + 75 = 0$ **Ans.** $(4, 8), \sqrt{5}$

(*iii*) $4(x^2 + y^2) + 12ax - 6ay - a^2 = 0$ **Ans.** $\left(\frac{-3a}{2}, \frac{3a}{4}\right), \frac{7a}{4}$

5. Find equation of the circle whose centre lies on the intersection of $x - 2y + 3 = 0$ and $2x - 3y + 4 = 0$ and whose radius is 6 units. **[B.T.E. December 2006]**
Ans. $x^2 + y^2 - 2x - 4y = 31$

6. Find the equation of the circle which is concentric $x^2 + y^2 - 8x + 12y + 43 = 0$ and

(*i*) which passes through (6, 2); **Ans.** $x^2 + y^2 - 8x + 12y - 16 = 0$

(*ii*) has its radius equal to 7. **Ans.** $x^2 + y^2 - 8x + 12y + 3 = 0$

7. Find the equation of the circle concentric with the circle
$2x^2 + 2y^2 + 8x + 10y - 39 = 0$ and having its area equal to 16π.

Ans. $4x^2 + 4y^2 + 16x + 20y - 23 = 0$

8. Find the equation of the circle which passes through the centre of the circle
$x^2 + y^2 + 8x + 10y - 7 = 0$ and is concentric with the circle $2x^2 + 2y^2 - 8x - 12y - 9 = 0$

[Diploma IETE Dec. 2006] **Ans.** $x^2 + y^2 - 4x - 6y - 87 = 0$

9. Show that the points (1, – 6), (5, 2) (7, 0) and (– 1, – 4) are the vertices of a cyclic quadrilateral.

23.5 CIRCLE THROUGH THREE GIVEN POINTS

Example 9. *Find the equation of the circle passing through the points* (5, 7), (8, 1) *and* (1, 3). *Also find its centre and radius.*

Solution. Let the equation of the circle be

$$x^2 + y^2 + 2gx + 2fy + c = 0 \qquad \text{... (1)}$$

Since the points (5, 7) (8, 1) and (1, 3) lies on the circle, their co-ordinates must satisfy the equation (1).

$$\therefore \quad 25 + 49 + 10g + 14f + c = 0$$

$$\Rightarrow \quad 74 + 10g + 14f + c = 0 \qquad \text{... (2)}$$

and
$$64 + 1 + 16g + 2f + c = 0$$

$$\Rightarrow \quad 65 + 16g + 2f + c = 0 \qquad \text{... (3)}$$

and
$$1 + 9 + 2g + 6f + c = 0$$

$$\Rightarrow \quad 10 + 2g + 6f + c = 0 \qquad \text{... (4)}$$

Now we have to find the values of g, f and c from (2), (3) and (4).

Let us eliminate c.

Subtracting (3) from (2), we get

$$9 - 6g + 12f = 0 \;\Rightarrow\; 4f - 2g + 3 = 0 \qquad \text{... (5)}$$

Subtracting (4) from (3), we get

$$55 + 14g - 4f = 0 \;\Rightarrow\; 4f - 14g - 55 = 0 \qquad \text{... (6)}$$

Subtracting (6) from (5), we get

$$12g + 58 = 0$$

$$\therefore \quad g = -\frac{29}{6}$$

Substituting the value of g in (5), we get

$$4f = -\frac{29}{6} \times 2 - 3 = -\frac{76}{6}$$

$$\therefore \quad f = -\frac{19}{6}$$

$$\therefore \text{ from (4),} \quad c = -10 + \frac{29}{3} + 19$$

$$= \frac{-30 + 29 + 57}{3} = \frac{56}{3}$$

Now substituting these values of g, f and c in (1), we get

$$x^2 + y^2 + 2\left(-\frac{29}{6}\right)x + 2\left(-\frac{19}{6}\right)y + \frac{56}{3} = 0$$

$$\Rightarrow \quad 3(x^2 + y^2) - 29x - 19y + 56 = 0$$

The co-ordinates of the centre are $(-g, -f)$,

i.e., $\left(\frac{29}{6}, \frac{19}{6}\right)$ **Ans.**

$$\text{Radius} = \sqrt{\left(\frac{29}{6}\right)^2 + \left(\frac{19}{6}\right)^2 - \frac{56}{3}} = \frac{\sqrt{530}}{6}$$ **Ans.**

Example 10. *Find the circumcircle of the triangle formed by the lines*

$$x + y + 1 = 0$$
$$3x + y - 5 = 0$$
and $$2x + y - 5 = 0$$

Solution. Solving the given equations in pairs, the co-ordinates of the vertices of the triangle are $(3, -4)$, $(0, 5)$ and $(6, -7)$

Let the equation of the circumcircle be

$$x^2 + y^2 + 2gx + 2fy + c = 0$$

The coordinates of the vertices will satisfy it,

$$\therefore \quad 25 + 6g - 8f + c = 0 \quad \text{... (1)}$$
$$25 + 10f + c = 0 \quad \text{... (2)}$$
$$85 + 12g - 14f + c = 0 \quad \text{... (3)}$$

Subtracting (2) from (1), we get

$$6g - 18f = 0 \Rightarrow g = 3f \quad \text{... (4)}$$

Subtracting (2) from (3), we get

$$60 + 12f - 24f = 0$$
$$\Rightarrow \quad g - 2f + 5 = 0 \quad \text{... (5)}$$

Solving (4) and (5), we get $f = -5$, $g = -15$

$\therefore$ from (2), $c = -25 - 10(-5) = 25$

$\therefore$ the required equation is:

$$x^2 + y^2 - 30x - 10y + 25 = 0$$ **Ans.**

Note: The equation of the circle contains three constants; g, f and c; and therefore g, f and c can be found by using three conditions given in any problem.

Example 11. *Find the equation of the circle which passes through the points* (4, 1) *and* (6, 5) *and has its centre on the line* $4x + y = 16$.

Solution. Let the equation of the circle be

$$x^2 + y^2 + 2gx + 2fy + c = 0 \quad \text{... (1)}$$

Since (4, 1) and (6, 5) lie on (1), we have

$$17 + 8g + 2f + c = 0 \quad \text{... (2)}$$
and $$61 + 12g + 10f + c = 0 \quad \text{... (3)}$$

Also the centre $(-g, -f)$ of the circle lies on

$$4x + y - 16 = 0$$

$\therefore \qquad -4g - f - 16 = 0$

$\Rightarrow \qquad 4g + f + 16 = 0 \qquad \text{... (4)}$

Now we have to solve equations (2), (3) and (4) for g, f and c.

Subtracting (2) from (3), we get

$$4g + 8f + 44 = 0$$

$$g + 2f + 11 = 0 \qquad \text{... (5)}$$

Solving (4) and (5), we get

$$\frac{g}{11-32} = \frac{f}{16-44} = \frac{1}{8-1}$$

$$\therefore \qquad g = -\frac{21}{7} = -3,\ f = \frac{-28}{7} = -4$$

$\therefore$ from (2), $c = -17 + 24 + 8 = 15$

The equation of the circle is

$$x^2 + y^2 - 6x - 8y + 15 = 0 \qquad \textbf{Ans.}$$

Example 12. *Find the equation of the circle which passes through* (3, – 2), (– 2, 0) *and has its centre on the line* $2x - y = 3$. **[S.B.T.E. 2014, 2008]**

Solution. Let the equation of the circle be

$$x^2 + y^2 + 2gx + 2fy + c = 0 \qquad \text{... (1)}$$

As (1) passes through (3, – 2),

$$9 + 4 + 6g - 4f + c = 0$$

$$6g - 4f + c = -13 \qquad \text{... (2)}$$

As (1) also passes through (– 2, 0),

$$4 + 0 - 4g + c = 0$$

$$-4g + c = -4 \qquad \text{... (3)}$$

The centre $(-g, -f)$ of (1) lies on

$$2x - y = 3$$

$$-2g + f = 3 \qquad \text{... (4)}$$

On subtracting (3) from (2), we get

$$10g - 4f = -9 \qquad \text{... (5)}$$

From (4) and (5) $\qquad f = 6,\ g = \frac{3}{2}$

From (3) $\qquad c = 2$

Putting these values of g, f and c in (1), we get

$$x^2 + y^2 + 3x + 12y + 2 = 0 \qquad \textbf{Ans.}$$

Example 13. *Find the equation to a circle passing through the points* (1, 2) *and* (3, 0) *and cutting an intercept of 7 on the x-axis.*

Solution. Let the equation of the circle be

$$x^2 + y^2 + 2gx + 2fy + c = 0 \qquad \text{... (1)}$$

Circle (1) is passing through the points (1, 2) and (3, 0), so

$$1 + 4 + 2g + 4f + c = 0 \Rightarrow 2g + 4f + c = -5 \qquad \text{... (2)}$$

$$9 + 6g + c = 0 \Rightarrow 6g + c = -9 \qquad \text{... (3)}$$

Let the circle (1) cut an intercept AB equal to 7 on the x-axis. Let a perpendicular CD from the centre C fall on AB at D.

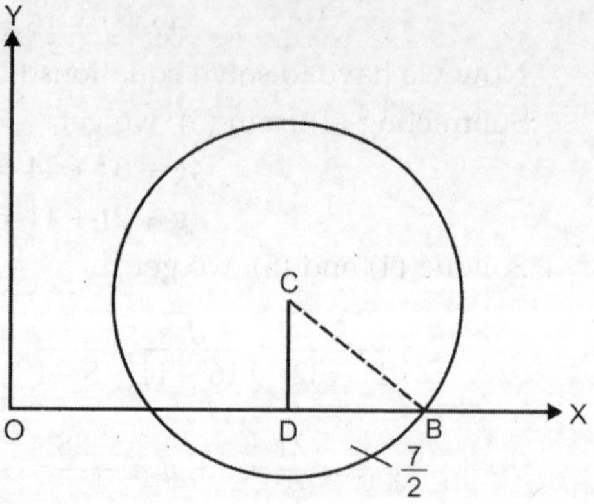

$$BD = \frac{1}{2}AB = \frac{7}{2}$$

$$BC^2 = CD^2 + BD^2 \quad [CD = f]$$

$$\Rightarrow \quad g^2 + f^2 - c = f^2 + \left(\frac{7}{2}\right)^2$$

$$\Rightarrow \quad g^2 - c = \frac{49}{4} \qquad \text{... (4)}$$

On adding (3) and (4), we get

$$g^2 + 6g = \frac{49}{4} - 9 \Rightarrow 4g^2 + 24g = 13$$

$$\Rightarrow \quad 4g^2 + 24g - 13 = 0 \Rightarrow 4g^2 + 26g - 2g - 13 = 0$$

$$\Rightarrow \quad 2g(2g + 13) - 1(2g + 13) = 0$$

$$\Rightarrow \quad (2g + 13)(2g - 1) = 0 \Rightarrow g = -\frac{13}{2}, \frac{1}{2}$$

Again, $\quad g^2 - c = \dfrac{49}{4}$

If $\quad g = -\dfrac{13}{2}$

$$\Rightarrow \quad \frac{169}{4} - c = \frac{49}{4}$$

$$\Rightarrow \quad c = 30$$

Now, $\quad 2g + 4f + c = -5$

$$\Rightarrow \quad 2\left(\frac{-13}{2}\right) + 4f + 30 = -5$$

$$\Rightarrow \quad 4f = -5 - 30 + 13$$

$$\Rightarrow \quad 4f = -22$$

$$\Rightarrow \quad f = -\frac{11}{2}$$

If $\quad g = \dfrac{1}{2}$

$$\frac{1}{4} - c = \frac{49}{4}$$

$$\Rightarrow \quad c = -12$$

$$2\left(\frac{1}{2}\right) + 4f - 12 = -5$$

$$\Rightarrow \quad 4f = -5 + 12 - 1$$

$$4f = 6$$

$$\Rightarrow \quad f = \frac{3}{2}$$

On putting the values of g, f and c in (1), we get

$$x^2 + y^2 - 13x - 11y + 30 = 0 \quad \text{or} \quad x^2 + y^2 + x + 3y - 12 = 0$$ **Ans.**

Example 14. *Find the equation of the circle which passes through the intersection of two circles.*

$x^2 + y^2 - 8x - 24y + 7 = 0$ and $x^2 + y^2 - 4x + 10y + 8 = 0$ *and has its centre on the x-axis.*

Solution. We have two circles

$$x^2 + y^2 - 8x - 24y + 7 = 0 \qquad \text{... (1)}$$

$$x^2 + y^2 - 4x + 10y + 8 = 0 \qquad \text{... (2)}$$

Let the equation of a circle passing through the points of intersection of (1) and (2) be

$$x^2 + y^2 - 8x - 24y + 7 + k(x^2 + y^2 - 4x + 10y + 8) = 0$$

$$(1 + k)\, x^2 + (1 + k)\, y^2 + (-8 - 4k)\, x + (-24 + 10k)\, y + 7 + 8k = 0$$

$$\Rightarrow \quad x^2 + y^2 + \left(\frac{-8-4k}{1+k}\right)x + \left(\frac{-24+10k}{1+k}\right)y + \frac{7+8k}{1+k} = 0 \quad \text{... (3)}$$

The centre of the circle (3) is

$$\left(\frac{4+2k}{1+k}, \frac{12-5k}{1+k}\right).$$

The centre lies on x-axis, so [$y = 0$]

$$\frac{12-5k}{1+k} = 0 \Rightarrow 12 - 5k = 0 \Rightarrow k = \frac{12}{5}$$

Putting the value of $k = \dfrac{12}{5}$ in (3), we have

$$x^2 + y^2 + \left(\frac{-8-\frac{48}{5}}{1+\frac{12}{5}}\right)x + \left(\frac{-24+24}{1+\frac{12}{5}}\right)y + \frac{7+\frac{96}{5}}{1+\frac{12}{5}} = 0$$

$$\therefore \qquad x^2 + y^2 - \frac{88}{17}x + \frac{131}{17} = 0 \qquad \textbf{Ans.}$$

23.6 DIAMETER FORM

Find equation of the circle when the join of points (x_1, y_1) and (x_2, y_2) is *a diameter*.

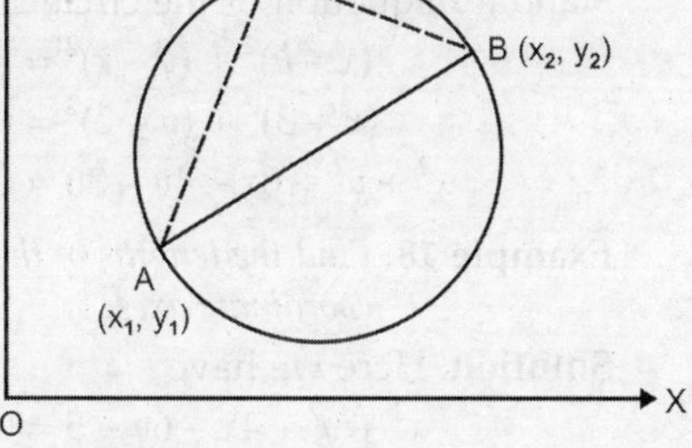

Let $A\,(x_1, y_1)$, $B\,(x_2, y_2)$ be the points.

Let (x, y) be any point on the circle.

Join AP, BP.

Then $\angle APB$, being in a semi-circle, is equal to one rt. angle

i.e., $\qquad AP \perp BP \qquad$ ...(1)

Now the slope of $AP = \dfrac{y-y_1}{x-x_1}$ and the slope of $BP = \dfrac{y-y_2}{x-x_2}$ from (1)

$$1 + \frac{y-y_1}{x-x_1} \cdot \frac{y-y_2}{x-x_2} = 0 \qquad (\because 1 + m_1 m_2 = 0)$$

$$\Rightarrow \quad \boxed{(x-x_1)(x-x_2) + (y-y_1)(y-y_2) = 0}$$

which is the required equation in diameter form.

Example 15. *Find the equation of the circle drawn on the line joining* (–1, 2) *and* (3, – 4) *as diameter.*

Solution. The required equation is

$$\{x - (-1)\}\,(x - 3) + (y - 2)\,\{y - (-4)\} = 0$$

$$\Rightarrow \quad (x + 1)(x - 3) + (y - 2)(y + 4) = 0$$

$$\Rightarrow \quad x^2 - 2x - 3 + y^2 + 2y - 8 = 0$$

$$\Rightarrow \quad x^2 + y^2 - 2x + 2y - 11 = 0 \qquad \textbf{Ans.}$$

Example 16. *Find the equation of a circle passing through the origin and making intercepts (4, 5) on the axes of co-ordinates.*

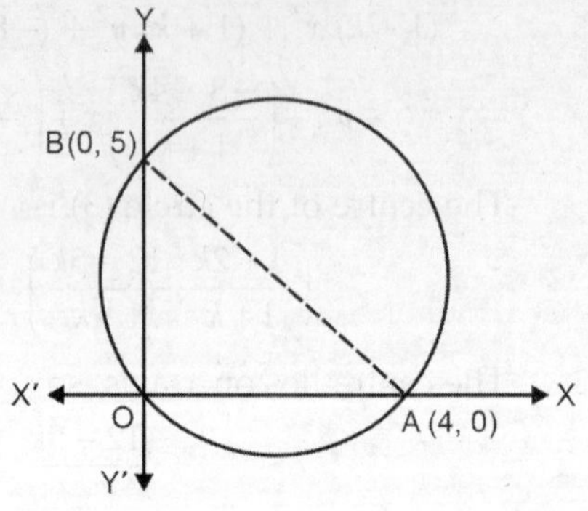

Solution. Let the intercepts be $OA = 4$, $OB = 5$.
the co-ordinates of A and B are (4, 0) and (0, 5) respectively.
Since $\angle AOB = 90°$, AB is a diameter.

$\therefore$ the required equation of the circle is

$$(x - 4)(x - 0) + (y - 0)(y - 5) = 0$$

$$\Rightarrow \quad x^2 + y^2 - 4x - 5y = 0$$ **Ans.**

Example 17. *Find the equation of the circle whose two diameters are $2x - 3y + 12 = 0$ and $x + 4y - 5 = 0$ and area is 154 sq. units.* **[SBTE. 2015, 2014, 2013, 2011]**

Solution. Here we have

$$2x - 3y + 12 = 0 \quad \text{... (1)}$$

and

$$2x + 4y - 5 = 0 \quad \text{... (2)}$$

Solving (1) and (2), we get

$$x = -3, y = 2$$

$$\pi r^2 = 154 \Rightarrow r^2 = \frac{154}{\pi} = \frac{154}{1} \times \frac{7}{22} = 49$$

$$\therefore \quad r = 7$$

$$\text{Centre} = (-3, 2), r = 7$$

Standard equation of the circle is

$$(x - h)^2 + (y - k)^2 = r^2$$

$$(x + 3)^2 + (y - 2)^2 = (7)^2$$

$$\therefore \quad x^2 + y^2 + 6x - 4y - 36 = 0$$ **Ans.**

Example 18. *Find the lengths of the intercepts made by circle $x^2 + y^2 - 4x - 6y - 5 = 0$ with coordinate axes.* **[SBTE. 2016, 2015]**

Solution. Here we have

$$x^2 + y^2 - 4x - 6y - 5 = 0$$

Here $g = -2, f = -3, c = -5$

Intercepts on X-axis = $2\sqrt{g^2 - c} = 2\sqrt{4 + 5} = 6$

Intercepts on Y-axis = $2\sqrt{f^2 - c} = 2\sqrt{14}$ **Ans.**

Example 19. *Find the equation of circle which passes through the centre of circle $x^2 + y^2 + 12x + 6y - 7 = 0$* and is concentric with the circle $2x^2 + 2y^2 + 6x - 8y - 8 = 0$. **[SBTE. 2017, 2016, 2015]**

Solution. Here we have

$$x^2 + y^2 + 12x + 6y - 7 = 0 \quad \text{... (1)}$$

$$2x^2 + 2y^2 + 6x - 8y - 8 = 0$$

or

$$x^2 + y^2 + 3x + 4y - 4 = 0 \quad \text{... (2)}$$

Centre of the circles (1) and (2) are (– 6, – 3) and $\left(\frac{-3}{2}, 2\right)$ respectively

Circles passes through (– 6, – 3) and concentric with (2)

$$\therefore \qquad r = \sqrt{\left(\frac{-3}{2}+6\right)^2 + (2+3)^2} = \frac{\sqrt{181}}{2}$$

Equation of the circle is

$$\left(x+\frac{3}{2}\right)^2 + (y-2)^2 = \left(\frac{\sqrt{181}}{2}\right)^2$$

$$\Rightarrow \qquad x^2 + \frac{9}{4} + 3x + y^2 + 4 - 4y = \frac{181}{4}$$

$$\Rightarrow \qquad x^2 + y^2 + 3x - 4y = \frac{181}{4} - \frac{9}{4} - 4 = 39$$

$$\therefore \qquad x^2 + y^2 + 3x - 4y - 39 = 0$$ **Ans.**

EXERCISE 23.2

1. Find the equation of circles which pass through the points: **[B.T.E Delhi, Jan. 2009]**

(*i*) (1, 2), (3, – 4) and (5, – 6) **Ans.** $x^2 + y^2 - 22x - 4y + 25 = 0$

(*ii*) (0, 0), (2, 1) and (– 3, 2) **Ans.** $x^2 + y^2 - 3x - 11y = 0$

(*iii*) (5, – 8), (– 2, 9) and (2, 1). Find the co-ordinates of its centre and radius also.

Ans. $x^2 + y^2 + 116x + 48y - 285 = 0$; (–58, –24); 65

2. Find the equation of the circle on the line joining the points (–1, 2) and (3, – 4) as diameter. Find the radius and the coordinates of the centre of this circle.

Ans. $x^2 + y^2 - 2x + 2y = 11$; (1, – 1); $\sqrt{13}$

3. Find the equation of the circle drawn on the joining the pairs of points given below as diameter.

(*i*) $(a - b, a + b)$ and $(a + b, a - b)$; **Ans.** $x^2 + y^2 - 2a(x + y) + 2(a^2 - b^2) = 0$

(*ii*) $(a \sin\theta, b\cos\theta)$ and $(a\cos\theta, - b\sin\theta)$.

Ans. $x^2 + y^2 - a(\sin\theta + \cos\theta)\, x - b(\cos\theta - \sin\theta)\, y + (a^2 - b^2)\sin\theta\cos\theta = 0$

4. Find the equation of the circles circumscribing the triangle formed by the lines:

(*i*) $x + y = 6$, $2x + y = 4$ and $x + 2y = 5$; **Ans.** $x^2 + y^2 - 17x - 19y + 50 = 0$

(*ii*) $2x + y - 3 = 0$, $x + y - 1 = 0$ and $3x + 2y - 5 = 0$ **Ans.** $x^2 + y^2 - 13x - 5y + 16 = 0$

5. Find the equation of the circle drawn on the line joining the centres of the circles as diameter. **[B.T.E. Delhi Jan. 2009]**

(*i*) $x^2 + y^2 - 6x + 8y - 20 = 0$ and $x^2 + y^2 + 8x - 6y = 0$ **Ans.** $x^2 + y^2 + x + y - 24 = 0$

(*ii*) $x^2 + y^2 = 16$ and $x^2 + y^2 - 10x + 12y + 70 = 0$ **Ans.** $x^2 + y^2 - 5x + 6y = 0$

6. Find the equation to the circle which passes through the origin and cuts off intercepts from the axes equal to:

(*i*) 3 and 4 **Ans.** $x^2 + y^2 - 3x - 4y = 0$ (*ii*) – 5 and 7 **Ans.** $x^2 + y^2 + 5x - 7y = 0$

7. Find the equation of the circle passing through the points (1, –2) and (4, –3) and having its centre on the line whose equation is $3x + 4y - 7 = 0$

Ans. $x^2 + y^2 - \frac{94}{15}x + \frac{6}{5}y + \frac{11}{3} = 0$

8. Find the equation of the circle which passes through the points:

(*i*) (3, 2) and (5, 4) and has its centre on $3x + 2y = 12$ **Ans.** $x^2 + y^2 + 4x - 18y + 11 = 0$

(*ii*) (4, 5), and (6, – 4) and has its centre on the x. **Ans.** $x^2 + y^2 - \frac{11}{2}x = 9$

(*iii*) (1, –3) and (2, –4) and has its radius equal to 5. **Ans.** $x^2 + y^2 + 4x + 14y + 28 = 0$

9. Find the equation of the circle whose centre lies on the line $x - 4y = 1$ and which passes through the points (3, 7) and (5, 5). **[Diploma IETE, June 2005]**

Ans. $x^2 + y^2 + 6x + 2y - 90 = 0$

10. Find the coordinates of the centre and the radius of the circle circumscribing the Δ formed by the $3x + 4y = 24$ and the axes of x and y. **Ans.** $x^2 + y^2 - 8x - 6y = 0$

11. Show that the four points (3, –2), (1, 0), (–1, – 2) and (1, – 4) are concyclic.

[Hint: Find the equation of the circle passing through any of the 3 given points and show that the 4th point lies on it].

12. Find equation of the circle concentric with the circle $x^2 + y^2 - 4x - 6y - 9 = 0$ and which passes through (–4, 5). **Ans.** $x^2 + y^2 - 4x - 6y - 27 = 0$

Choose the correct alternative:

13. The angle made by any diameter of a circle at any point on the circumference is

(*a*) 90° (*b*) 180° (*c*) 45° (*d*) 60°

[Diploma IETE June 2005] Ans. (*a*)

Conic Section

24.1 CONIC (GEOMETRICAL DEFINITION)

A conic is the locus of a point which moves so that its distance from a fixed point is in a constant ratio to its distance from a fixed straight line. It is a figure formed by intersection of a plane and a circular cone.

Focus. The fixed point is called the focus and is denoted by **S**.

Eccentricity. the constant ratio is called the eccentricity and is denoted by ***e***.

Directrix. the fixed straight line is called the directrix.

Type of conic. there are three types of conic.

(1) **Parabola**	If eccentricity	$e = 1$
(2) **Ellipse**	If "	$e < 1$
(3) **Hyperbola**	If "	$e > 1$
(4) **Circle**	If eccentricity	$e = 0$

Physical Interpretation

The section of a double cone with common vertex cut by plane is known as conic section. The shape of the conic section depends upon the position of the plane.

(1) If the plane is parallel to the base, the section is the ***circle***.

(2) If the plane is inclined to the base, the section is ***ellipse***.

(3) If the plane is inclined to the base and cuts the base, the section is ***parabola***.

(4) If the plane is parallel to the axis of the cone, the section is ***hyperbola***.

(5) If the plane contains the axis of the cone, the section has a ***pair of lines***.

24.2 PARABOLA

Definition. A parabola is the locus of a point which moves so that its distance from a fixed point is equal to its distance from a fixed straight line, directrix.

Standard equation of the parabola

$$\boxed{y^2 = 4ax}$$

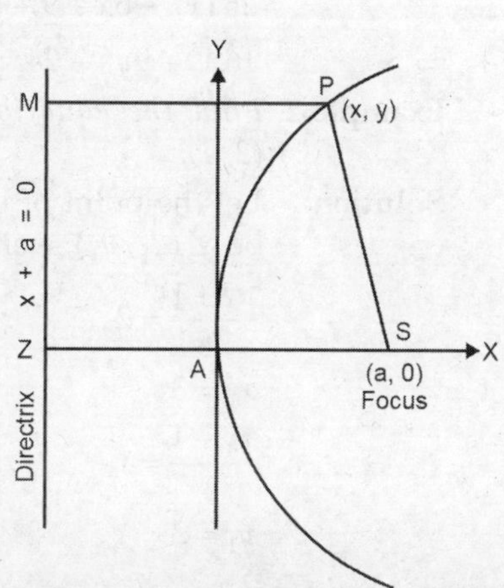

Let S be the focus and ZM the directrix of the parabola and P be a point on the parabola. Join S and P points and from P draw $PM \perp ZM$.

$SP = PM$ (to find the coordinates of focus and equation of directrix)

From S draw $SZ \perp$ on ZM.

Bisect SZ in A, i.e. $SA = AZ$. Then A lies on the parabola. Let $SZ = 2a$ then $AS = AZ = a$. Take A as the origin, AS as the x-axis, and $AY \perp$ to it as the y-axis.

Then the coordinates of S are $(a, 0)$ and the equation of the directrix is $x = -a$, or $x + a = 0$

Let the coordinates P be (x, y).

$$SP = PM$$

$$\sqrt{(x-a)^2 + (y-0)^2} = x + a$$

$$x^2 - 2ax + a^2 + y^2 = x^2 + 2ax + a^2$$

$$\Rightarrow \qquad y^2 = 4ax$$

which is the standard equation of a parabola.

Note:

Focus S is $(a, 0)$

Directrix is $x + a = 0$

Vertex A is $(0, 0)$

Axis AS of the parabola is $y = 0$.

Latus rectum LL' is $4a$ as calculated below:

Let the coordinate of L be (a, l).

The point L (a, l) lies on the parabola $y^2 = 4ax$.

$\therefore l^2 = 4a.a \quad \Rightarrow \quad l = 2a,$

Latus rectum $LSL' = 2l = 2(2a) = 4a$

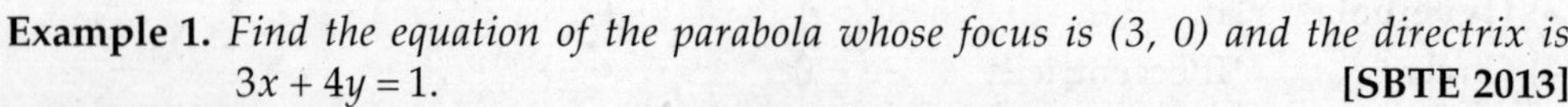

Example 1. *Find the equation of the parabola whose focus is (3, 0) and the directrix is* $3x + 4y = 1$. **[SBTE 2013]**

Solution. Let any point on the parabola be P (x, y).

Distance of P (x, y) from $(3, 0)$ = $\sqrt{(x-3)^2 + (y-0)^2}$

Distance of P (x, y) from tghe line $3x + 4y = 1$ is $\dfrac{3x + 4y - 1}{\sqrt{3^2 + 4^2}}$

Equating the two distances, we get

$$\sqrt{(x-3)^2 + y^2} = \frac{3x + 4y - 1}{5}$$

On squaring, we get

$$25(x^2 - 6x + 9 + y^2) = 9x^2 + 16y^2 + 1 + 24xy - 6x - 8y$$

$$\Rightarrow \qquad 16x^2 + 9y^2 - 24xy - 144x + 8y + 224 = 0$$ **Ans.**

Example 2. *Find the equation of parabola whose focus is (1, – 1) and vertex is (2, 1).* **[Diploma IETE June 2005]**

Solution. Let the point of intersection of the directrix and the axis of the parabola be $Z(x_1, y_1)$, vertex is the mid-point of ZS.

$$\therefore \quad \frac{x_1 + 1}{2} = 2 \qquad \therefore x_1 + 1 = 4$$

$$\Rightarrow \quad x_1 = 3$$

$$\therefore \quad \frac{y_1 - 1}{2} = 1 \qquad \therefore y_1 - 1 = 2$$

$$\Rightarrow \quad y_1 = 3$$

Directrix passes through (3, 3) and is ⊥ to axis of parabola.

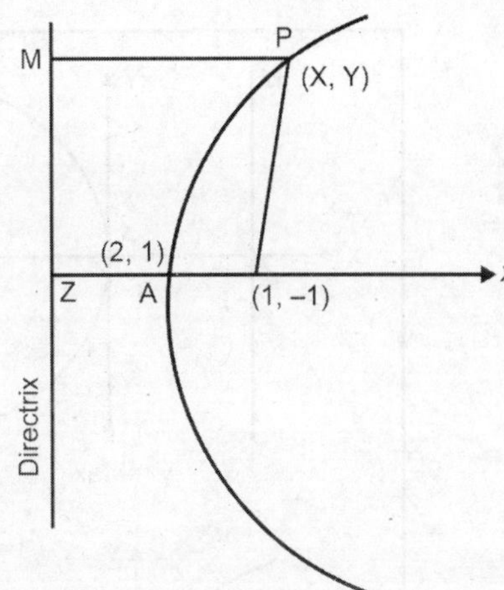

Slope of the axis of the parabola

$$= \frac{1-(-1)}{2-1} = \frac{2}{1}$$

∴ The slope of the directrix $= -\frac{1}{2}$

Equation of directrix is $y - 3 = -\frac{1}{2}(x - 3)$

or $x + 2y = 9$

Let (x, y) be any point on the parabola

$$SP = PM \text{ or } SP^2 = PM^2$$

$$(x-1)^2 + (y+1)^2 = \left(\frac{x+2y-9}{\sqrt{4+1}}\right)^2$$

$$x^2 - 2x + 1 + y^2 + 2y + 1 = \frac{1}{5}(x^2 + 4y^2 + 81 + 4xy - 18x - 36y)$$

$$\Rightarrow \quad 5(x^2 + y^2 - 2x + 2y + 2) = x^2 + 4y^2 + 4xy - 36y - 18x + 81$$

$$\Rightarrow \quad 4x^2 + y^2 - 4xy + 8x + 46y - 71 = 0$$ Ans.

EXERCISE 24.1

Find equation of the parabola whose:

1. Focus is (–3, 2) and the directrix is $x + y = 4$. **(B.T.E Delhi, Jan. 2009, Dec. 2006)**
   Ans. $x^2 + y^2 - 2xy + 20x + 10 = 0$
2. Focus is (1, 1) and the directrix is $x + y + 1 = 0$. **Ans.** $x^2 + 2xy + y^2 - 6x - 6y + 3 = 0$
3. Focus is (3, 0) and the directrix is 2x + y = 1. **Ans.** $x^2 - 4xy + 4y^2 - 26x + 2y + 44 = 0$

Find the equation of the parabola with:

4. Vertex (2, –3) and focus (0, 5). **Ans.** $16x^2 + y^2 + 8xy + 96x - 554y - 1879 = 0$
5. Focus (0, – 3) and the vertex is (0, 0). **Ans.** $x^2 = -12y$
6. Focus (0, –3) and the vertex is (–1, –3). **Ans.** $y^2 + 6y - 4x + 5 = 0$
7. Focus is $(a, 0)$ and the vertex is $(a', 0)$. **Ans.** $y^2 = -4(a' - a)(x - a')$

24.3 TYPES OF PARABOLAS: OTHER SHAPES AND SOME USEFUL RESULTS

Properties \ Equation of the parabola	$y^2 = 4ax$	$x^2 = 4ay$
Vertex (coordinates)	(0, 0)	(0, 0)
Focus (coordinates)	$(a, 0)$	$(0, a)$
Latus rectum (length)	$4a$	$4a$
Axis (equation)	$y = 0$	$x = 0$
Directrix (equation)	$x = -a$	$y = -a$

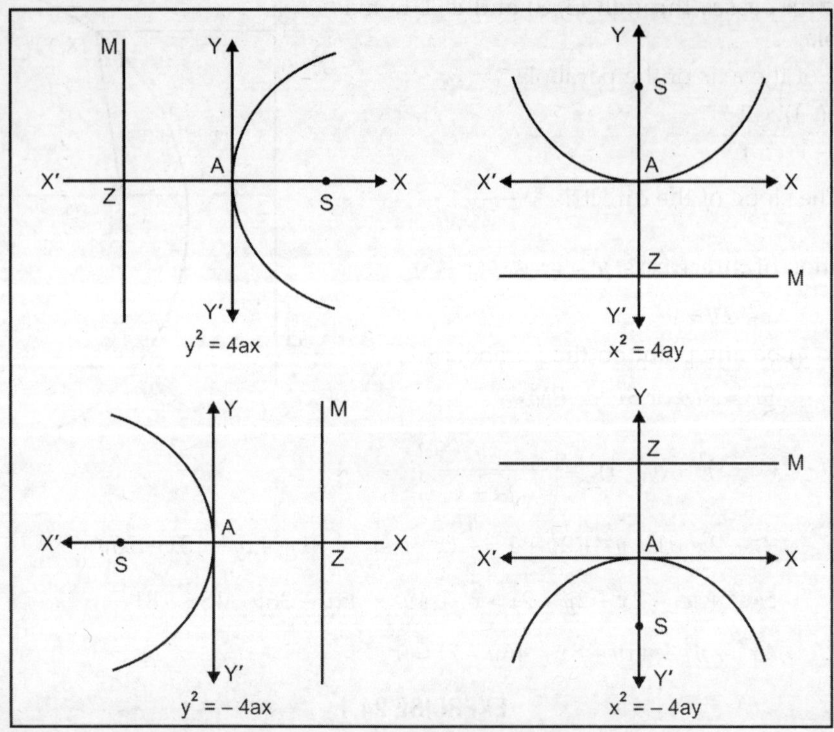

Example 3. *Find the vertex, focus, latus rectum, axis, and directrix of the parabola $y^2 = 10x$.*

Solution. Equation of the parabola is $y^2 = 10x = 4.\dfrac{5}{2}x$.

Comparing it with $y^2 = 4ax$, we have $a = \dfrac{5}{2}$

The coordinates of the vertex are (0, 0)

The coordinates of the focus $(a.\ 0)$ are $\left(\dfrac{5}{2}, 0\right)$

The length of latus rectum $(4a)$ is $4\dfrac{5}{2} = 10$

The equation of the axis $(y = 0)$ is $y = 0$.

The equation of the directrix $(x + a = 0)$ is $x + \dfrac{5}{2} = 0$.

$\mathbf{2x + 5 = 0}$ **Ans.**

Example 4. *Find the vertex, focus, axis directrix and latus rectum of the parabola $y^2 - 4y - 3x + 1 = 0$.* **[S.B.T.E. 2015]**

Solution. We have the equation of parabola

$$y^2 - 4y - 3x + 1 = 0 \quad ...(1)$$

$$\Rightarrow \quad y^2 - 4y + 4 - 3x + 1 - 4 = 0$$

$$\Rightarrow \quad (y - 2)^2 = 3x + 3$$

$$\Rightarrow \quad (y - 2)^2 = 3(x + 1)$$

$$\Rightarrow \quad Y^2 = 3X$$

where $Y = y - 2$ and $X = x + 1$

Equation of Parabola	Vertex	Focus	Directrix	Latus rectum
$y^2 = 3X$	(0, 0)	$\left(\frac{3}{4}, 0\right)$	$X = \frac{-3}{4}$	3
$y^2 - 4y - 3x + 1 = 0$	(–1, 2)	$\left(-\frac{1}{4}, 2\right)$	$x = -\frac{7}{4}$	3

Example 5. *Find the vertex, focus, latus rectum, axis and directrix of parabola* $y^2 - x - 2y + 2 = 0$. **[SBTE 2015]**

Solution. $y^2 - x - 2y + 2 = 0$

$$y^2 - 2y + 1 = x - 1$$

$$(y - 1)^2 = (x - 1)$$

Put $X = x - 1$ and $Y = y - 1$

$\therefore$ The transformed equation of the parabola is $Y^2 = X$

$$\Rightarrow \quad Y^2 = 4 \cdot \frac{1}{4} X$$

Equation	$Y^2 = 4 \cdot \frac{1}{4} X$	$(y - 1)^2 = (x - 1)$
Vertex	(0, 0)	$(1, 1), x = X + 1 = 0 + 1 = 1$ $y = Y + 1 = 0 + 1 = 1$
Focus	$\left(\frac{1}{4}, 0\right)$	$\left(\frac{5}{4}, 1\right), x = X + 1 = \frac{1}{4} + 1 = \frac{5}{4}$ $y = Y + 1 = 0 + 1 = 1$
Latus rectum axis	1 $Y = 0$	1 $y - 1 = 0, y = Y + 1 = 0 + 1 = 1$
Directrix	$X + \frac{1}{4} = 0$	$x - \frac{3}{4} = 0, x = X + 1 = -\frac{1}{4} + 1 = \frac{3}{4}$

Ans.

Example 6. *Find the vertex, focus, and directrix of the parabola* $4x^2 = 9y$. **[DIPIETE 2019]**

Solution. Here we have $4x^2 = 9y$ ...(1)

Form of parabola $x^2 = 4ay$ ...(2)

Comparing equations (2) and (1), we get.

$$4a = \frac{9}{4} \Rightarrow a = \frac{9}{16}$$

$$\text{Vertex} = (0, 0)$$

Directrix $y = -a \Rightarrow y = -9/16$

Focus $= (0, a) = \left(0, 9/16\right)$ **Ans.**

Example 7. *If the latus rectum of the parabola $x^2 = 2py$ is of length 16 units, find the value of p.* **[SBTE, 2016, 2015]**

Solution. Here we have $x^2 = 2py$

Length of latus rectum = $2p$

$\therefore$ $2p = 16 \Rightarrow p = 8$ **Ans.**

Example 8. *The focal distance of a point on the parabola $y^2 = 12x$ is 4. Find the coordinates of the point.* **[SBTE, 2017, 2016, 2014]**

Solution. Here we have.

$y^2 = 12x$ ...(1)

$y^2 = 4ax$ ...(2)

Comparing (1) with (2), we get

$a = 3$

Focal distance = $x + a$

$4 = x + a$ or $x = 4 - 3 = 1$

Putting the value of $x = 1$ in (1), we get,

$y^2 = 12 \Rightarrow y = \pm\sqrt{12} = \pm 2\sqrt{3}$

Coordinates of the points $\left(1, \pm 2\sqrt{3}\right)$ **Ans.**

Example 9. *Find the coordinates of the vertex and focus of the parabola $x^2 - 2x - 12y + 25 = 0$.* **[SBTE, 2017, 2016, 2015]**

Solution. Here we have $x^2 - 2x - 12y + 25 = 0$

$(x^2 - 2x + 1) - 12y + 24 = 0$

$(x - 1)^2 = 12(y - 2)$

$X^2 = 12Y$ ...(1)

where $X = x - 1, Y = y - 2$

$4a = 12$ [comparing with parabola]

$a = 3$

Vertex $x = 1, y = 2$ *i.e.* (1, 2)

Focus $x = 1, y = 5$ *i.e.* (1, 5) **Ans.**

Example 10. *Find the vertex, focus and directrix of the parabola.*

$4y^2 + 12x - 12y + 39 = 0$ **[SBTE, 2016, 2014, 2008]**

Solution. $4y^2 + 12x - 12y + 39 = 0$

i.e. $4y^2 - 12y = -12x - 39$

i.e. $y^2 - 3y = -3x - \dfrac{39}{4}$

i.e. $$y^2 - 3y + \frac{9}{4} = -3x - \frac{39}{4} + \frac{9}{4}$$

i.e. $$\left(y - \frac{3}{2}\right)^2 = -3x - \frac{15}{2}$$

i.e. $$\left(y - \frac{3}{2}\right)^2 = -3\left(x + \frac{5}{2}\right) \quad ...(1)$$

Put $Y = y - \frac{3}{2}$ and $X = x + \frac{5}{2}$ in (1), we have $Y^2 = -3X$

Equation	$Y^2 = 4\left(-\frac{3}{4}\right)X$	$\left(y - \frac{3}{2}\right)^2 = -3\left(x + \frac{5}{2}\right)$
Vertex	(0, 0)	$\left(-\frac{5}{2}, \frac{3}{2}\right)$
Focus	$\left(-\frac{3}{4}, 0\right)$	$\left(-\frac{13}{4}, \frac{3}{2}\right)$
Directrix	$X = \frac{3}{4}$	$x = -\frac{7}{4}$

Ans.

Example 11. *Determine the focus and the directrix of the parabola* $x^2 - 4x - 8y - 4 = 0$

Solution. We have the equation of parabola $x^2 - 4x - 8y - 4 = 0$

$$\Rightarrow \quad (x^2 - 4x + 4) - 8y - 4 - 4 = 0$$

$$(x - 2)^2 - 8y - 8 = 0$$

$$\Rightarrow \quad (x - 2)^2 = 8y + 8$$

$$\Rightarrow \quad (x - 2)^2 = 8(y + 1)$$

$$\Rightarrow \quad X^2 = 8Y$$

if $$X = x - 2, \quad Y = y + 1$$

Equation of parabola	Focus	Directrix
$X^2 = 8Y$	(0,. 2)	$Y = -2$
$x^2 - 4x - 8y - 4 = 0$	(–1, 4)	$y = -3$

Ans.

EXERCISE 24.2

1. Find the focus, vertex, length of latus rectum, equation of the axis and the equation of the directrix of the following:

(i) $y^2 = 8x$

Ans. Focus, (2, 0), Vertex (0, 0), latus rectum = 8, Axis $y = 0$, directrix, $x + 2 = 0$

(ii) $y^2 = -4x$

Ans. Focus, (–1, 0), Vertex (0, 0), latus rectum = 4, Axis y = 0, directrix, $x - 1 = 0$

(*iii*) $3x^2 = 8y$

Ans. Focus, $\left(0, \frac{2}{3}\right)$, Vertex (0, 0), latus rectum $= \frac{8}{3}$ Axis $x = 0$, directrix, $3y + 2 = 0$

(*iv*) $x^2 = -y$

Ans. Focus, $\left(0, -\frac{1}{4}\right)$, Vertex (0, 0), latus rectum = 1, Axis $x = 0$, directrix, $4y - 1 = 0$

2. Find the vertex, focus, latus rectum and directrix of the parabola.
$x^2 = 4x - y$ ***(Diploma IETE, Dec. 2005)*** **Ans.** (2, 4) $\left(2, \frac{15}{4}\right)$, 1, $4y - 17 = 0$

3. Find the focus, vertex, directrix and axis of the parabola
$y = -4x^2 + 3x.$ ***(Diploma IETE, Dec. 2006)***

Ans. $\left(\frac{3}{8}, \frac{1}{2}\right), \left(\frac{3}{8}, \frac{9}{16}\right), 8y - 5 = 0, 8x - 3 = 0$

4. Find the vertex, focus and directrix of the parabola
$(y + 3)^2 = 2(x + 2)$ **Ans.** $(-2, -3), \left(-\frac{3}{2}, -3\right), x = -\frac{5}{2}$

5. Find the focus, vertex, axis and directrix of parabola
$4(y - 1)^2 = -7(x - 3)$ **Ans.** $\left(\frac{41}{16}, 1\right)$, (3, 1), $y = 1, 16x = 55$

6. Find the vertex, focus and the directrix of the parabola
$y^2 = 4x + 4y$ **Ans.** (–1, 2), (0, 2), $x = -2$

7. Find the vertex, axis, focus and the directrix of the parabola
$y^2 = 5x - 4y - 9$ **Ans.** Vertex (1, –2); axis $y = -2$; focus $\left(\frac{9}{4}, -2\right)$; directrix $x = -\frac{1}{4}$

8. Find the vertex, the axis, the focus the directrix and the latus rectum of the parabola
$y^2 = 8x + 8y$ **Ans.** Vertex (–2, 4), axis: $y = 4$, (0, 4), $x + 4 = 0$, 8

9. Show that $y^2 - 8y - x + 19 = 0$ represents a parabola. Find its focus, vertex and directrix. **Ans.** $\left(\frac{13}{4}, 4\right)$, (3, 4), $4x = 11$

10. Find the vertex, axis, focus and directrix of the parabola
$x^2 + y = 9x - 14.$ **Ans.** $\left(\frac{9}{2}, \frac{25}{4}\right), 2x - 9 = 0 \quad \left(\frac{9}{2}, 6\right), 2y - 13 = 0$

11. Find the coordinates of the vertex, the focus and the equation of directrix of the parabola
$2x^2 + 5y - 6x = 4.$ **Ans.** Focus, $\left(\frac{3}{2}, \frac{43}{40}\right)$, Vertex $\left(\frac{3}{2}, \frac{17}{10}\right)$, directrix $40y - 93 = 0$

12. Find the vertex, axis, latus rectum and focus of the parabola $x^2 + 2y = 8x - 7$.

Ans. Focus (4, 4), Vertex $\left(4, \frac{9}{2}\right)$, latus rectum = 2, Axis = $x - 4 = 0$

13. Find the vertex, axis, focus, directrix of the parabola
$x^2 + 4x + 2y - 7 = 0$

Ans. Focus (–2, 5), Vertex (–2, 11/2), Axis $x + 2 = 0$, directrix $y - 6 = 0$

14. Find the vertex, focus and axis of the parabola

$3x^2 + 12x - 8y = 0$ **Ans.** (–2, –3/2), (–2, –5/6), $x + 2 = 0$

15. Obtain the co-ordinates of the focus and the equation of the directrix, if equation of the parabola is $x^2 - 8x + 2y - 10 = 0$. **Ans.** Focus $\left(4, \frac{25}{2}\right)$, directrix $2y - 27 = 0$

24.4 ELLIPSE

Ellipse is the locus of a point which moves so that its distance from a fixed point is in a constant ratio, less than one to its distance from a fixed straight line.

Focus. The fixed point is called the focus, and is denoted by S.

Eccentricity. The constant ratio is called eccentricity and is denoted by e, here $e < 1$.

Directrix. The fixed line is called the directrix.

Standard Equation. The standard equation of the ellipse is

$$\boxed{\frac{x^2}{a^2} + \frac{y^2}{b^2} = 1}$$

Let S be the focus and ZM be the directrix.

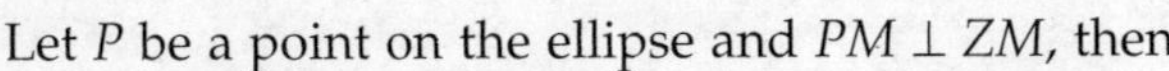

Let P be a point on the ellipse and $PM \perp ZM$, then

$$\frac{SP}{PM} = e \quad \text{...(1)}$$

Draw $SZ \perp$ on ZM. Divide SZ internally and externally at A and A' in the ratio $e : 1$, so that $\frac{SA}{AZ} = \frac{e}{1}, \frac{SA'}{A'Z} = \frac{e}{1}$

$$SA = e.\,AZ \quad \text{...(2)}$$

and $$SA' = e.\,A'Z \quad \text{...(3)}$$

Bisect AA' at C. Let $AA' = 2a$, then $AC = A'C = a$

To find the coordinates of S and Z.

Let C be the origin, CA as x-axis and $CY \perp$ to it as y-axis

On adding (2) and (3), we get

$$SA + SA' = e.\,(AZ + A'Z)$$

$$(CA - CS) + (CS' + CA') = e.\,[(CZ - CA) + (CA' + CZ)]$$

$$2\,CA = e.\,2CZ \quad \Rightarrow \quad CA = CA' = a$$

$$a = e.\,CZ \quad \Rightarrow \quad CZ = \frac{a}{e}$$

On substracting (3) from (2)

$$SA' - SA = e.\,(A'Z - AZ)$$

$$(CS + CA) - (CA - CS) = e.\,[(CA' + CZ) - (CZ - CA)]$$

$$2CS = e.\,2CA$$

$$CS = ae$$

$\therefore$ The coordinates of S and Z are

$(ae, 0)$ and $\left(\frac{a}{e}, 0\right)$ respectively.

Let the coordinates of P be (x, y).

From (1), $SP^2 = e^2\, PM^2$

$$(x - ae)^2 + (y - 0)^2 = e^2\left(\frac{a}{e} - x\right)^2$$

$$\Rightarrow \quad x^2 - 2aex + a^2e^2 + y^2 = e^2\left(x^2 + \frac{a^2}{e^2} - \frac{2xa}{e}\right)$$

$$\Rightarrow \quad x^2 - 2aex + a^2e^2 + y^2 = e^2\, x^2 + a^2 - 2aex$$

$$\Rightarrow \quad x^2\,(1 - e^2) + y^2 = a^2\,(1 - e^2)$$

$$\Rightarrow \quad \frac{x^2}{a^2} + \frac{y^2}{a^2\,(1 - e^2)} = 1$$

$$\Rightarrow \quad \frac{x^2}{a^2} + \frac{y^2}{b^2} = 1 \qquad [\text{Put } b^2 = a^2\,(1 - e^2)]$$

which is the standard equation of ellipse.

(*b*) To find the coordinates of B.

The equation of ellipse is $\boldsymbol{\frac{x^2}{a^2} + \frac{y^2}{b^2} = 1}$

B lies on y-axis $\quad \therefore \quad x$ coordinates of $B = 0$

$$\Rightarrow \quad \frac{0}{a^2} + \frac{y^2}{b^2} = 1$$

$$\Rightarrow \quad y = \pm b$$

$\therefore$ The coordinates of B and B' are $(0, b)$ and $(0, -b)$.

(*c*) A and A' are vertices.

AA' is called the *major axis.*

BB' is called the *minor axis.*

C is called the *centre.*

(*d*) Length of latus rectum.

Let the ends of the latus rectum be L and L'

Let the coordinate of L be (ae, l).

L lies on ellipse.

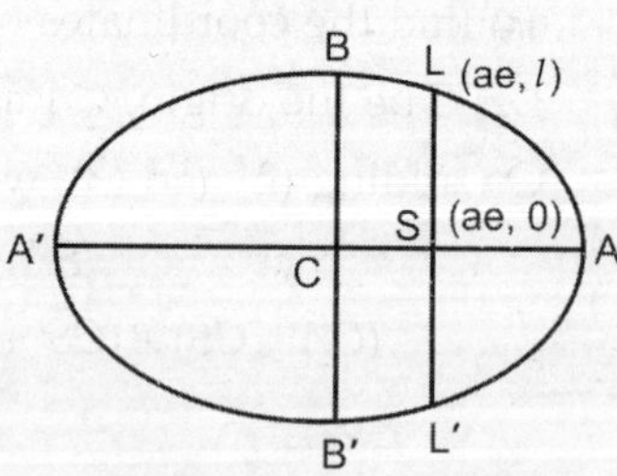

$$\therefore \quad \frac{a^2e^2}{a^2} + \frac{l^2}{b^2} = 1$$

$$\therefore \quad l^2 = b^2\,(1 - e^2) = b^2 \cdot \frac{b^2}{a^2}$$

$$l = \frac{b^2}{a}$$

Length of latus rectum $= LL' = \dfrac{2b^2}{a}$

24.5 FOCAL PROPERTY OF AN ELLIPSE

The sum of the focal distances of any point on an ellipse is constant and equal to the major axis, *i.e.* **SP + S′P = 2a**

Proof: Let S and S' be the foci and ZM, $Z'M'$ the corresponding directrics of the ellipse.

Let $P(x, y)$ be any point on the ellipse.

Join SP, $S'P$. from P draw $PM \perp$ on ZM, $PM' \perp$ on $Z'M'$ and $PN \perp$ on the x-axis. Then

(*i*) $SP = e.\,PM = e.ZN = e.\,(CZ - CN)$

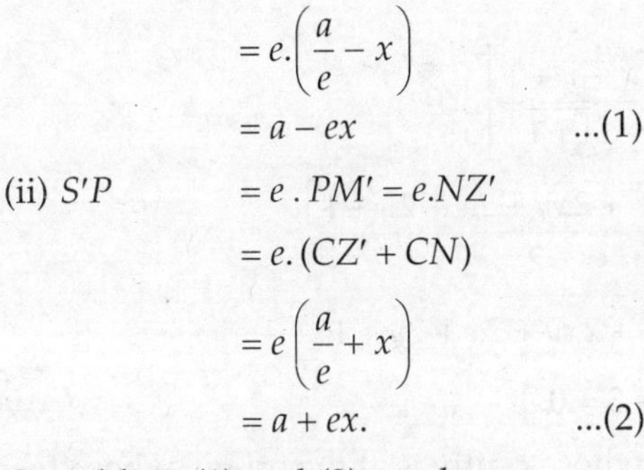

$$= e.\left(\frac{a}{e} - x\right)$$

$$= a - ex \qquad \ldots(1)$$

(ii) $S'P$ $= e\,.\,PM' = e.NZ'$

$$= e.\,(CZ' + CN)$$

$$= e\left(\frac{a}{e} + x\right)$$

$$= a + ex. \qquad \ldots(2)$$

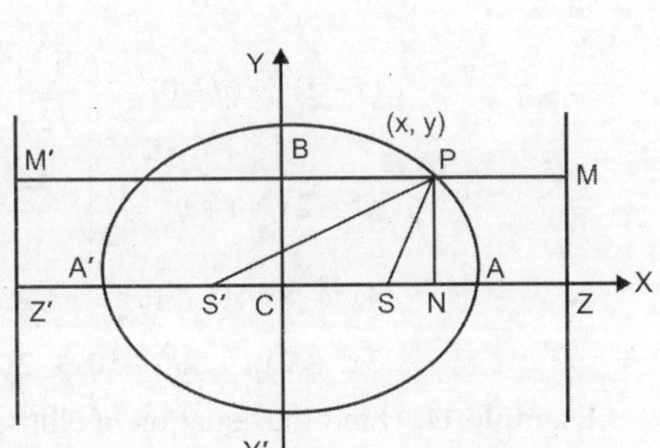

On adding (1) and (2), we have

$SP + S'P = (a - ex) + (a + ex)$

$SP + S'P = 2a$ (constant). **Proved.**

Example 12. *Obtain the equation of the ellipse whose two foci are (3, 2) and (1, –2) and whose major axis is of length 10 units.* **[S.B.T.E. 2014]**

Solution. We have, foci (3, 2) and (1, –2)

Major axis $= 2a = 10$

$P\,(x, y)$ be any point on the ellipse, then

$PS_1 + PS_2 =$ Major axis

$$\Rightarrow \sqrt{(x-1)^2 + (y+2)^2} + \sqrt{(x-3)^2 + (y-2)^2} = 10$$

$$\Rightarrow \sqrt{x^2 - 2x + 1 + y^2 + 4y + 4} = 10 - \sqrt{x^2 - 6x + 9 + y^2 - 4y + 4}$$

On squaring both the sides

$$x^2 - 2x + 1 + y^2 + 4y + 4 = 100 + x^2 - 6x + 9 + y^2 - 4y + 4$$

$$\Rightarrow \quad 4x + 8y - 108 = -20\sqrt{x^2 + y^2 - 6x - 4y + 13}$$

$$\Rightarrow \quad x + 2y - 27 = -5\sqrt{x^2 + y^2 - 6x - 4y + 13}$$

$$\Rightarrow \quad (x + 2y - 27)^2 = 25\,(x^2 + y^2 - 6x - 4y + 13)$$

$$\Rightarrow \quad x^2 + 4y^2 + 729 + 4xy - 54x - 108y = 25x^2 + 25y^2 - 150x - 100y + 325$$

$$\Rightarrow \quad 24x^2 + 21y^2 - 4xy - 96x + 8y - 404 = 0$$

Ans.

Example 13. *Find the equation of the ellipse whose focus is (I, 0), the directrix* $x + y + 1 = 0$ *and eccentricity is equal to* $\left(\frac{1}{\sqrt{2}}\right)$.

Solution. Let $S(1, 0)$ be the focus and P (x, y) any point on the ellipse and PM be the perpendicular from P to the directrix.

$x + y + 1 = 0$

For ellipse $SP = e \,.\, PM$

$\Rightarrow \quad SP^2 = e^2 \,(PM)^2$

$$\Rightarrow \quad (x-1)^2 + (y-0)^2 = \left(\frac{1}{\sqrt{2}}\right)^2 \left[\frac{x+y+1}{\sqrt{1+1}}\right]^2$$

$$\Rightarrow \quad x^2 - 2x + 1 + y^2 = \frac{1}{2}\left[\frac{x^2 + y^2 + 2xy + 2x + 2y + 1}{2}\right]$$

$$\Rightarrow \quad 4x^2 - 8x + 4 + 4y^2 = x^2 + y^2 + 2xy + 2x + 2y + 1$$

$$\Rightarrow \quad 3x^2 - 2xy + 3y^2 - 10x - 2y + 3 = 0.$$ **Ans.**

Example 14. Find the equation of ellipse whose centre is at the origin, whose foci are (1, 0) and (–1, 0) and eccentricity is $\frac{1}{2}$.

Solution. $SS' = 2ae$

$$\sqrt{(1+1)^2 + (0-0)^2} = 2a.\,\frac{1}{2}$$

$2 = a$

But $SP + S'P = 2a$

$$\Rightarrow \quad \sqrt{(x-1)^2 + (y-0)^2} + \sqrt{(x+1)^2 + (y-0)^2} = 2 \times 2$$

$$\Rightarrow \quad \sqrt{(x-1)^2 + y^2} = 4 - \sqrt{(x+1)^2 + y^2}$$

$$\Rightarrow \quad x^2 - 2x + 1 + y^2 = 16 + x^2 + 2x + 1 + y^2 - 8\sqrt{(x+1)^2 + y^2}$$

$$\Rightarrow \quad -4x - 16 = -8\sqrt{x^2 + 2x + 1 + y^2}$$

$$\Rightarrow \quad x + 4 = 2\sqrt{x^2 + y^2 + 2x + 1}$$

$$\Rightarrow \quad x^2 + 8x + 16 = 4(x^2 + y^2 + 2x + 1)$$

$$\therefore \quad 3x^2 + 4y^2 - 12 = 0$$ **Ans.**

Example 15. *Find the equation of the ellipse whose latus rectum is 5 and whose eccentricity is 2/3, referred to the principal axes as axes of coordinates.*

Solution. Latus rectum $= \frac{2b^2}{a} = 5$ ...(1)

Eccentricity $= e = \frac{2}{3}$ and $b^2 = a^2\,(1 - e^2)$

$$b^2 = a^2\left(1-\frac{4}{9}\right) \Rightarrow b^2 = \frac{5}{9}a^2 \qquad \left(e=\frac{2}{3}\right)$$

Putting the value of b^2 in (1), we get

$$2.\frac{5a^2}{9a} = 5 \Rightarrow \frac{2a}{9} = 1 \Rightarrow a = \frac{9}{2}$$

But $$b^2 = \frac{5}{9}.a^2 = \frac{5}{9}.\frac{81}{4} = \frac{45}{4}$$

Equation of the ellipse is

$$\frac{4x^2}{81} + \frac{4y^2}{45} = 1 \Rightarrow \frac{x^2}{\left(\frac{9}{2}\right)^2} + \frac{y^2}{\left(\frac{\sqrt{45}}{2}\right)^2} = 1$$ **Ans.**

Example 16. *Find the coordinates of the foci of the ellipse* $2x^2 + 3y^2 - 6 = 0$. *Also obtain the eccentricity, length of latus rectum.*

Solution. The given ellipse is

$$\frac{x^2}{3} + \frac{y^2}{2} = 1 \qquad ...(1)$$

Comparing (1) with $\frac{x^2}{a^2} + \frac{y^2}{b^2} = 1$, we get

$$a^2 = 3, b^2 = 2 \quad \text{or} \quad b^2 = a^2(1-e^2) \Rightarrow 2 = 3(1-e^2)$$

$$\Rightarrow \quad e^2 = \frac{1}{3} \Rightarrow e = \frac{1}{\sqrt{3}}$$

The foci of ellipse are $(ae, 0)$ and $(-ae, 0)$ *i.e.*

$$\left(\frac{1}{\sqrt{3}}.\sqrt{3}, 0\right) \text{ and } \left(-\frac{1}{\sqrt{3}}.\sqrt{3}, 0\right)$$

$\Rightarrow$ (1, 0) and (–1, 0)

The length of latus rectum $= \frac{2b^2}{a} = \frac{2\times 2}{\sqrt{3}} = \frac{4}{\sqrt{3}}$. **Ans.**

Example 17. *Find the eccentricity of the following:*

(*i*) $\frac{x^2}{16} + \frac{y^2}{4} = 1$ **[S.B.T.E. 2017]** (*ii*) $x^2 + y^2 = 1$ **[S.B.T.E. 2017]**

(*iii*) $3x^2 + 2y^2 = 6$ **[S.B.T.E. 2016]**

Solution. Here we have

(*i*) $\frac{x^2}{16} + \frac{y^2}{4} = 1 \Rightarrow a^2 = 16, b^2 = 4, a > b$

$$\text{Eccentricity } (e) = \sqrt{1-\frac{b^2}{a^2}} = \sqrt{1-\frac{4}{16}} = \sqrt{\frac{3}{4}} = \frac{\sqrt{3}}{2}$$ **Ans.**

(*ii*) Here we have

$$x^2 + y^2 = 1 \Rightarrow \frac{x^2}{1} + \frac{y^2}{1} = 1 \Rightarrow a^2 = 1, b^2 = 1, a = b$$

$$e = \sqrt{1 - \frac{1}{1}} = \sqrt{0} = 0$$

Ans.

(*iii*) Here we have

$$3x^2 + 2y^2 = 6 \Rightarrow \frac{x^2}{2} + \frac{y^2}{3} = 1 \Rightarrow a^2 = 2, b^2 = 3, a < b$$

$$e = \sqrt{1 - \frac{a^2}{b^2}} = \sqrt{1 - \frac{2}{3}} = \frac{1}{\sqrt{3}}$$

Ans.

Example 18. *Find the eccentricity of the ellipse.*

$$16(x+1)^2 + 9(y+1)^2 = 144$$

[SBTE 2016, 2014]

Solution. Here we have

$$\Rightarrow \quad 16(x+1)^2 + 9(y+1)^2 = 144$$

$$\Rightarrow \quad \frac{(x+1)^2}{9} + \frac{(y+1)^2}{16} = 1$$

$$\Rightarrow \quad \frac{X^2}{a^2} + \frac{Y^2}{b^2} = 1 \qquad [X = x+1, Y = y+1 \text{ shifting origin}]$$

$$a^2 = 9, b^2 = 16$$

$$b^2 > a^2 \text{ , i.e. } b > a$$

$$e = \sqrt{1 - \frac{a^2}{b^2}} = \sqrt{1 - \frac{9}{16}} = \frac{\sqrt{7}}{4}$$

Ans.

Example 19. *Find the equation of the ellipse whose axes are along the coordinate axes, vertices are (± 5,0) and foci at (± 4,0).* **[SBTE 2016, 2014]**

Solution. Here we have

vertices = (± 5, 0), Foci = (± 4, 0)

$$\therefore \quad a = 5 \quad \text{or} \quad ae = 4 \Rightarrow e = \frac{4}{5}$$

$$\text{or} \quad b^2 = a^2(1 - e^2) \Rightarrow b^2 = 25\left(1 - \frac{16}{25}\right) = 9$$

$$\therefore \quad \frac{x^2}{25} + \frac{y^2}{9} = 1$$

Ans.

Example 20. *Find the equation of the ellipse whose axes are along the co-ordinate axes, vertices are (0, ± 10) one eccentricity (e) =* $4/5$. **[DIPIETE June 2019]**

Solution. Here we given, vertices = (0, ± 10), $e = 4/5$.

Let the equation of the required ellipse be:

$$\frac{x^2}{a^2}+\frac{y^2}{b^2}=1 \qquad ...(1)$$

Coordinates of the vertices are $(0, \pm b)$

$b=10 \Rightarrow b^2=100$

$a^2=b^2(1-e^2) \Rightarrow a^2=100\left(1-\frac{16}{25}\right)=36$

Putting the values of a^2 and b^2 in (1), we get $\frac{x^2}{36}+\frac{y^2}{100}=1$ **Ans.**

Example 21. *Find the equation of the ellipse with focus at (1, 2) directrix 3x + 4y – 5 = 0 and e = 1/2.* **[SBTE 2016, 2014]**

Solution. Here we have:

Directrix = $3x+4y-5=0$

Let $P(x, y)$ be any point on ellipse

∴ $(\text{Focus})^2 = e^2 \times (\text{directrix})^2$

$$(x-1)^2+(y-2)^2=\frac{1}{4}\left[\frac{3x+4y-5}{\sqrt{25}}\right]^2$$

∴ $91x^2+84y^2-24xy-170x-325=0$ **Ans.**

Example 22. *Find the equation of the ellipse whose axes are parallel to the co-ordinate axes having centre at point (2, –3) one focus (3, –3) and one vertex (4, –3).* **[SBTE 2015, 2014]**

Solution. Here we have centre = (2, –3)

∴ $$\frac{(x-2)^2}{a^2}+\frac{(y+3)^2}{b^2}=1 \qquad ...(1)$$

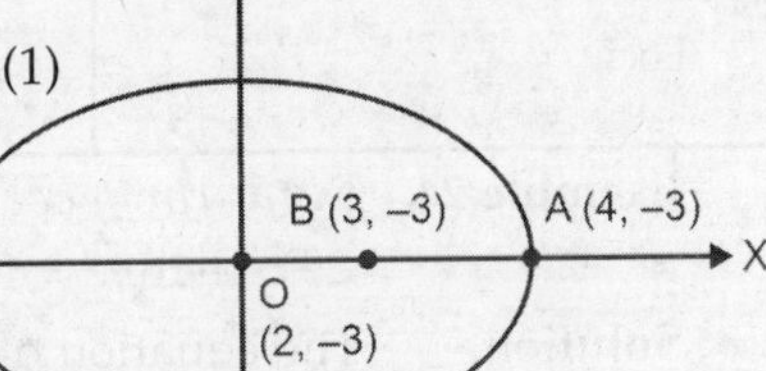

$OA=a$

$\sqrt{(4-2)^2+(-3+3)^2}=a$

or $a=2 \Rightarrow a^2=4$

and $\sqrt{(3-2)^2+(-3+3)^2}=ae \Rightarrow ae=1 \Rightarrow e=\frac{1}{2}$

$$b^2=a^2(1-e^2)=4\left(1-\frac{1}{4}\right)=3$$

∴ $$\frac{(x-2)^2}{4}+\frac{(y+3)^2}{3}=1 \qquad \text{[from (1)]}$$ **Ans.**

Example 23. *Find the centre, the length of the axes, eccentricity and foci of the ellipse* $x^2+2y^2-2x+12y+10=0$.

Solution. $x^2+2y^2-2x+12y+10=0$

$(x^2-2x+1)+2(y^2+6y+9)=-10+19$

$$\Rightarrow \quad (x-1)^2 + 2(y+3)^2 = 9$$

$$\Rightarrow \quad \frac{(x-1)^2}{9} + 2\frac{(y+3)^2}{9} = 1 \quad \Rightarrow \quad \frac{(x-1)^2}{(3)^2} + \frac{(y+3)^2}{\left(\frac{3}{\sqrt{2}}\right)^2} = 1$$

$$\Rightarrow \quad \frac{X^2}{3^2} + \frac{Y^2}{\left(\frac{3}{\sqrt{2}}\right)^2} = 1 \quad \left[\begin{array}{l} \textit{where } X = x - 1 \\ \quad\quad\;\; Y = y + 3 \end{array}\right]$$

$$\Rightarrow \quad b^2 = 9(1 - e^2)$$

$$\Rightarrow \quad \frac{9}{2} = 9(1 - e^2) \quad \Rightarrow \quad \frac{1}{2} = 1 - e^2$$

$$\Rightarrow \quad e^2 = \frac{1}{2} \quad \Rightarrow \quad e = \frac{1}{\sqrt{2}}$$

Equation	$\frac{X^2}{(3)^2} + \frac{Y^2}{\left(\frac{3}{\sqrt{2}}\right)^2} = 1$	$\frac{(x-1)^2}{(3)^2} + \frac{(y+3)^2}{\left(\frac{3}{\sqrt{2}}\right)^2} = 1$
Centre	(0, 0)	(1, – 3)
Length of major axis	6	6
Length of minor axis	$3\sqrt{2}$	$3\sqrt{2}$
Eccentricity	$\frac{1}{\sqrt{2}}$	$\frac{1}{\sqrt{2}}$
Foci	$\left(\pm\frac{3}{\sqrt{2}}, 0\right)$	$\left(\pm\frac{3}{\sqrt{2}} + 1, -3\right)$

Example 24. *Find the eccentricity, foci, length of latus rectum of ellipse* $25x^2 + 16y^2 = 400$. **[S.B.T.E. 2015]**

Solution. The equation of the ellipse is

$$25x^2 + 16y^2 = 400$$

$$\Rightarrow \quad \frac{x^2}{16} + \frac{y^2}{25} = 1$$

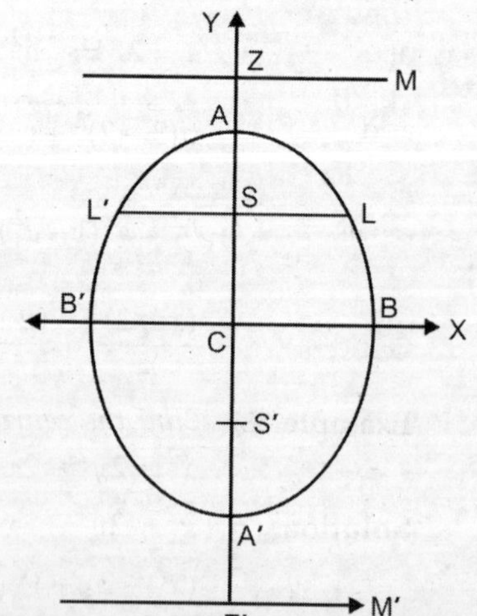

Here the denominator of $y^2 >$ the denominator of x^2.

$\therefore$ It is an ellipse with major axis vertical and it lies along the y-axis.

$$\therefore \quad a^2 = 25 \quad \text{and} \quad b^2 = 16$$

$$b^2 = a^2(1 - e^2) \Rightarrow 16 = 25(1 - e^2)$$

$$\Rightarrow \quad 25e^2 = 9$$

$$e = \frac{3}{5}.$$

Foci lie on the y-axis and are

$$\left(0, \pm 5 \times \frac{3}{5}\right) \Rightarrow (0, \pm 3)$$

The latus rectum $= 2.\frac{16}{5} = \frac{32}{5}$ **Ans.**

EXERCISE 24.3

Find the eccentricity, foci and the length of the latus rectum of the ellipse

1. $9x^2 + 16y^2 = 144$ ***(B.T.E, Delhi December 2006)*** **Ans.** $\frac{\sqrt{7}}{4}, (\pm\sqrt{7}, 0), \frac{9}{2}$

2. $x^2 + 4y^2 + 8y - 2x + 1 = 0$ **Ans.** $\frac{\sqrt{3}}{2}, (\pm\sqrt{3}, +1, -1), 1$

3. $3x^2 + 4y^2 - 12x - 8y + 4 = 0$ **Ans.** $\frac{1}{2}, (3, 1), (1, 1), 3$

4. Find the centre, the lengths of the axes, eccentricity and the foci of the ellipse

$x^2 + 4y^2 - 4x + 24y + 31 = 0.$ **Ans.** $(2, -3), 6, 3, \frac{\sqrt{3}}{2}, \left(2 \pm \frac{3\sqrt{3}}{2}, -3\right)$

5. $5x^2 + 4y^2 = 1$ **Ans.** $(0, 0), 1, \frac{2}{\sqrt{5}}, \frac{1}{\sqrt{5}}, \left(0, \pm \frac{1}{2\sqrt{5}}\right)$

6. $3x^2 + 2y^2 - 6 = 0$ **Ans.** $(0, 0), 2\sqrt{3}, 2\sqrt{2}, \frac{1}{\sqrt{3}}, (0, \pm 1)$

7. $25x^2 + 16y^2 = 1600$ **Ans.** $(0, 0), 16, 20, \frac{3}{5}, (0, \pm 6)$

8. $4x^2 + 3y^2 = 1$ **Ans.** $(0, 0), \frac{2}{\sqrt{3}}, 1, \frac{1}{2}, \left(0, \pm \frac{1}{2\sqrt{3}}\right)$

9. $4x^2 + y^2 - 8x + 2y + 1 = 0$ **Ans.** $(1, -1), 4, 2, \frac{\sqrt{3}}{2}, (1, \pm\sqrt{3} - 1)$

10. Find the eccentricity of an ellipse if its latus rectum is half its

(*a*) minor axis **Ans.** $\frac{\sqrt{3}}{2}$ (*b*) major axis **Ans.** $\frac{1}{\sqrt{2}}$

11. Find the equation of the ellipse whose focus is (1, 2), the directrix $3x + 4y = 5$ and eccentricity $= \frac{1}{2}$. **Ans.** $91x^2 + 84y^2 - 24xy - 170x - 360y + 475 = 0$

12. Find the equation of ellipse with focus at (1, 1) eccentricity $\frac{1}{2}$ and directrix $x - y + 3 = 0$ ***(B.T.E, Delhi, Jan. 2009)*** **Ans.** $7x^2 + 7y^2 + 2xy - 22x - 10y + 7 = 0$

13. Find the equation of ellipse whose focus is (2, 1), the directrix $x - 5 = 0$ and eccentricity $= \frac{1}{2}$. **Ans.** $3x^2 + 4y^2 - 6x - 8y - 5 = 0$

14. The foci of an ellipse are (± 2, 0) and its eccentricity is $\frac{1}{2}$, find its equation.

Ans. $\frac{x^2}{16} + \frac{y^2}{12} = 1$

15. Find the equation of the ellipse having its centre at the point (2, –3) one focus at (3, – 3) and one vertex at (4, – 3). **Ans.** $3x^2 + 4y^2 - 12x + 24y + 36 = 0$

16. Find the equation of the ellipse whose axes are the axes of co-ordinates and which passes through the points (–3, 1) and (2, –2). **Ans.** $3x^2 + 5y^2 = 32$

17. Find the equation of an ellipse with eccentricity equal to $\frac{1}{2}$, focus the point (1, – 2) and the equation of the corresponding directrix is $3x - 2y + 5 = 0$. Find the coordinates of its centre and the equations of its axes.

Ans. $43x^2 + 48y^2 + 12xy - 134x + 228y + 285 = 0$
$2x + 3y + 4 = 0$

24.6 HYPERBOLA

A hyperbola is the locus of a point which moves so that its distance from a fixed point is in a constant ratio, greater then one, of its distance from a fixed straight lline.

Focus. The fixed point is called the focus and denoted by S.

Eccentricity. The constant ratio is called the eccentricity and is denoted by e, here $e > 1$.

Directrix. the fixed line is called the directrix.

Standard equation. The standard equation of the hyperbola is

$$\boxed{\frac{x^2}{a^2} - \frac{y^2}{b^2} = 1}$$

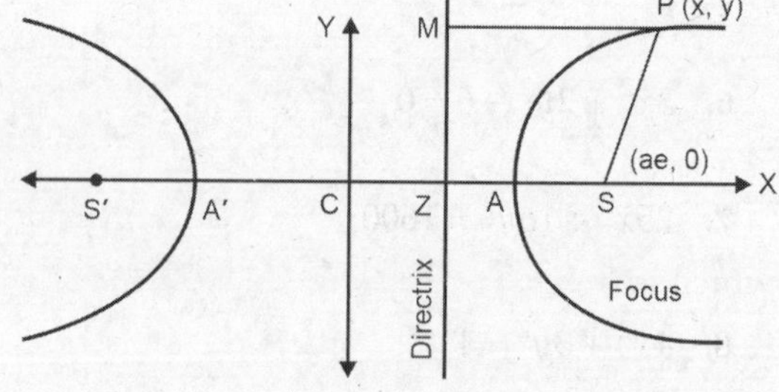

Let S be the focus and ZM be the directrix.

Let P be a point on the hyperbola and $PM \perp ZM$, then

$$\frac{SP}{PM} = \frac{e}{1} \qquad ...(1)$$

Draw $SZ \perp$ on ZM

Divide SZ internally and externally at A and A' in the ratio $e : 1$ so that

$$\frac{SA}{AZ} = \frac{e}{1} \qquad ...(2)$$

$$\frac{SA'}{A'Z} = \frac{e}{1} \qquad ...(3)$$

Bisect AA' at C. Let $AA' = 2\,a$, then

$$AC = CA' = a.$$

To find the co-ordinates of S and Z.

Let C be the orgin. CA as the x-axis and $CY \perp$ to it as y-axis.]

On adding (2) and (3), we get

$$SA + SA' = e\,(AZ + A'Z)$$

$$(CS - CA) + (CS + CA') = e\,[(CA - CZ) + (CZ + CA)]$$

$$2CS = 2e\,.\,CA \qquad [CA = CA' = a]$$

$$CS = ae$$

$\therefore$ The coordinates of S are $(ae, 0)$.

On subtracting (2) from (3), we get

$$SA' - SA = e\,(A'Z - AZ)$$

$$(CS + CA) - (CS - CA) = e\,[(CZ + CA) - (CA - CZ)]$$

$$2CA = e.\,2CZ$$

$$a = e\,.\,CZ$$

$$\frac{a}{e} = CZ$$

$\therefore$ The coordinates of Z are $\left(\frac{a}{e}, 0\right)$.

Let the coordinates of P be (x, y).

From (1) $$SP^2 = e^2\,.\,PM^2$$

$$(x - ae)^2 + (y - o)^2 = e^2\,.\left(x - \frac{a}{e}\right)^2$$

$\Rightarrow$ $$x^2 - 2aex + a^2e^2 + y^2 = e^2x^2 - 2aex + a^2$$

$\Rightarrow$ $$x^2\,(e^2 - 1) - y^2 = a^2\,(e^2 - 1)$$

$\Rightarrow$ $$\frac{x^2}{a^2} - \frac{y^2}{a^2\,(e^2 - 1)} = 1 \qquad [\text{Put } b^2 = a^2\,(e^2 - 1)]$$

$\Rightarrow$ $$\frac{x^2}{a^2} - \frac{y^2}{b^2} = 1$$

which is the standard equation of the hyperbola.

(b) A and A' are vertices.

AA' is called the transverse axis.

CY is called the conjugate axis.

Latus rectum is $\frac{2b^2}{a}$.

(c) If $a^2 = b^2$, the equation of hyperbola is

$$\frac{x^2}{a^2} - \frac{y^2}{a^2} = 1 \qquad \Rightarrow \quad x^2 - y^2 = a^2$$

This is called rectangular hyperbola.

$\therefore$ $$b^2 = a^2\,(e^2 - 1)$$

But $b^2 = a^2$ $\quad \therefore\ a^2 = a^2\,(e^2 - 1) \quad \Rightarrow \quad 1 = e^2 - 1$

$\Rightarrow$ $e^2 = 2 \quad \Rightarrow \quad e = \sqrt{2}.$

24.7 FOCAL PROPERTY OF A HYPERBOLA

The difference of the focal distances of any point on a hyperbola is constant, and equal to the transverse axis.

$$\boxed{S'P - SP = 2a}$$

Let S, S' be the foci and ZM and $Z'M'$ be the corresponding directrix of the hyperbola. Let $P(x, y)$ be any point on the hyperbola. Join SP and $S'P$.

Draw PM' and PM perpendiculars to ZM and $Z'M'$ respectively.

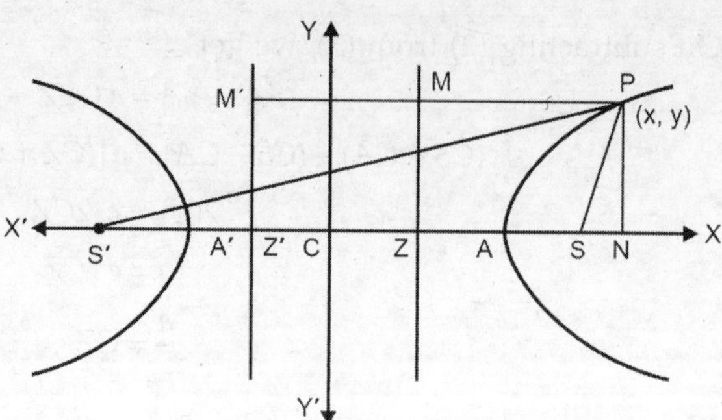

$$SP = e \cdot PM = e \cdot ZN = e \cdot (x - a/e) = ex - a \quad \text{...(1)}$$

$$S'P = e \cdot PM' = e \cdot Z'N = e \cdot (x + a/e) = ex + a \quad \text{...(2)}$$

On subtracting (1) from (2), we have

$$S'P - SP = (ex + a) - (ex - a)$$

$$\Rightarrow \quad S'P - SP = 2a$$

Example 25. *Find the equation of a hyperbola whose focus is (–1, 1), eccentricity is 3 and directrix is $x - y + 3 = 0$.*

Solution. Let P (x, y) be any point on the hyperbola whose focus is (–1, 1), directrix $x - y + 3 = 0$ and eccentricity $e = 3$.

Now, PM the length of perpendicular from P to ZM is

$$\frac{x - y + 3}{\sqrt{1^2 + 1^2}} \quad \text{and} \quad PS \text{ is } \sqrt{(x+1)^2 + (y-1)^2}$$

M P (x, y) $x - y + 3 = 0$ Z S (–1, 1)

By definition, we have

$$SP = e \cdot PM$$

$$SP^2 = e^2 \cdot PM^2$$

$$\Rightarrow \quad (x+1)^2 + (y-1)^2 = 9\left(\frac{x - y + 3}{\sqrt{2}}\right)^2$$

$$\Rightarrow \quad 2[x^2 + 2x + 1 + y^2 - 2y + 1] = 9[x^2 + y^2 + 9 - 2xy - 6y + 6x]$$

$$\Rightarrow \quad -7x^2 - 7y^2 + 18xy - 50x + 50y - 77 = 0$$

$$\therefore \quad 7x^2 + 7y^2 - 18xy + 50x - 50y + 77 = 0$$

This is the required equation of the hyperbola. **Ans.**

Example 26. *Find the equation of the hyperbola, referred to its axes of coordinates, the distance between whose foci is 16 and whose eccentricity is $\sqrt{2}$.*

Solution. Let the equation of the hyperbola be $\dfrac{x^2}{a^2} - \dfrac{y^2}{b^2} = 1$

Since the co-ordinates of the foci are $(ae, 0)$ and $(-ae, 0)$.

$$\therefore \quad (ae - 0) - (-ae - 0) = 16 \quad \text{or} \quad 2ae = 16$$

$$\Rightarrow \quad 2a \,.\, \sqrt{2} = 16 \;\Rightarrow\; a = 4\sqrt{2}, \quad (\because\ e = \sqrt{2})$$

Also, $$b^2 = a^2(e^2 - 1)$$

$$\Rightarrow \qquad b^2 = a^2 (2-1) \qquad (\because e = \sqrt{2})$$
$$b^2 = a^2 \quad \Rightarrow \quad b = a = 4\sqrt{2}$$

The equation is

$$\frac{x^2}{(4\sqrt{2})^2} - \frac{y^2}{(4\sqrt{2})^2} = 1 \quad \Rightarrow \quad \frac{x^2}{32} - \frac{y^2}{32} = 1$$

$$\therefore \qquad x^2 - y^2 = 32.$$ **Ans.**

Example 27. *Find the equation of the hyperbola whose vertices are (0, 0) (10, 0) and one of whose foci is (18, 0).*

Solution. Let the vertices be A' (0, 0) and A (10, 0).

Let $AA' = 2a \quad \therefore 2a = 10 \quad \Rightarrow \quad a = 5$

Let other focus be $(x, 0)$

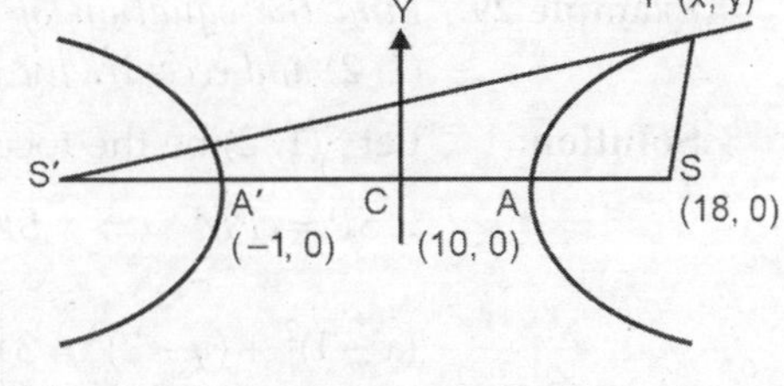

Mid-point of foci SS' = Mid-point of vertices AA'.

x-coordinate of mid-point SS' = x-coordinate of mid-point of AA'.

$$\frac{x+18}{2} = \frac{0+10}{2}$$

$$x + 18 = 10 \quad \Rightarrow \quad x = -8$$

$\therefore$ The coordinates of S' are $(-8, 0)$.

Let $P(x, y)$ be any point on the hyperbola.

By the property of hyperbola

$$S'P - SP = 2a$$

$$\Rightarrow \qquad \sqrt{(x+8)^2 + (y-0)^2} - \sqrt{(x-18)^2 + (y-0)^2} = 10$$

$$\Rightarrow \qquad \sqrt{(x+8)^2 + y^2} = 10 + \sqrt{(x-18)^2 + y^2}$$

Squaring both sides, we have

$$(x+8)^2 + y^2 = (10)^2 + (x-18)^2 + y^2 + 20\sqrt{(x-18)^2 + y^2}$$

$$\Rightarrow \qquad x^2 + 16x + 64 + y^2 = 100 + x^2 - 36x + 324 + y^2 + 20\sqrt{(x-18)^2 + y^2}$$

$$\Rightarrow \qquad 52x - 360 = 20\sqrt{(x-18)^2 + y^2}$$

$$\Rightarrow \qquad 13x - 90 = 5\sqrt{(x-18)^2 + y^2}$$

$$\Rightarrow \qquad (13x - 90)^2 = 25[(x-18)^2 + y^2]$$

$$\Rightarrow \qquad 169x^2 - 2340x + 8100 = 25(x^2 - 36x + 324 + y^2)$$

$$\Rightarrow \qquad 169x^2 - 2340x + 8100 = 25x^2 - 900x + 25y^2 + 8100$$

$$\Rightarrow \qquad 144x^2 - 1440x - 25y^2 = 0$$

$$\Rightarrow \qquad \frac{(x-5)^2}{25} - \frac{y^2}{144} = 1$$ **Ans.**

which is the required equation.

Example 28. *Find the equation of the hyperbola whose $e = \sqrt{5}$ and the sum of whose semi axes is 9.* **[S.B.T.E. 2015, 2014]**

Solution. Here we have $e = \sqrt{5}$

$$a + b = 9 \Rightarrow b = 9 - a \qquad ...(1)$$

We know $b^2 = a^2(e^2 - 1)$

$$(9 - a)^2 = a^2(5 - 1)$$

$$a^2 + 6a - 27 = 0 \Rightarrow a = 3, a = -9 \qquad [a < 0 \text{ Rejected}]$$

$$\therefore \quad \frac{x^2}{9} - \frac{y^2}{36} = 1 \qquad [\text{from (1)}, b = 6] \qquad \textbf{Ans.}$$

Example 29. *Find the equation of hyperbola whose directrix is $2x + y - 1 = 0$, focus (1, 2) and eccentricity $\sqrt{3}$.* **[S.B.T.E. 2016, 2015, 2014]**

Solution. Let $S(1, 2)$ be the focus and $P(x, y)$ by any point on the hyperbola.

$$\therefore SP = ePM \Rightarrow SP^2 = e^2 \cdot PM^2$$

$$(x-1)^2 + (y-2)^2 = 3\left[\frac{2x + y - 1}{\sqrt{4+1}}\right]^2$$

$$x^2 + y^2 - 2x - 4y + 5 = \frac{3}{5}(4x^2 + y^2 + 1 + 4xy - 4x - 2y)$$

$$\therefore \quad 7x^2 - 2y^2 - 2x + 14y + 12xy - 22 = 0 \qquad \textbf{Ans.}$$

Example 30. *Find the centre, the lengths of the axes, the eccentricity and the foci of the hyperbola $4x^2 - 9y^2 = 36$.*

Solution. The equation of the hyperbola can be wr:tten as

$$\frac{x^2}{9} - \frac{y^2}{4} = 1$$

$\therefore$ $a^2 = 9$ and $b^2 = 4 \Rightarrow a = 3$ and $b = 2$

$\therefore$ Transverse axis = 6 and conjugate axis = 4

$\therefore$ Centre = (0, 0)

We know that $b^2 = a^2(e^2 - 1)$

$$4 = 9(e^2 - 1) \Rightarrow e = \frac{\sqrt{13}}{3}$$

The coordinates of foci are $(\pm ae, 0)$, *i.e.* $(\pm\sqrt{13}, 0)$ **Ans.**

Example 31. *Find the centre, length of the axes, eccentricity directrices, foci and the length of the latus rectum of the hyperbola $9x^2 - 16y^2 = 144$.* **[S.B.T.E. 2015]**

Solution. The equation of the hyperbola can be written as

$$\frac{x^2}{16} - \frac{y^2}{9} = 1$$

$a^2 = 16$ and $b^2 = 9$

$a = 4$ and $b = 3$

$\therefore$ Transverse axis = $2a = 2 \times 4 = 8$

$\therefore$ Conjugate axis = $2b = 2 \times 3 = 6$

$\therefore$ Centre = (0, 0)

We know that $b^2 = a^2 (e^2 - 1)$

$\Rightarrow$ $9 = 16(e^2 - 1)$ $\Rightarrow$ $e = \sqrt{\frac{25}{16}} = \frac{5}{4}$

Directrices are

$$x = \frac{a}{e} \Rightarrow x = \frac{4}{\frac{5}{4}} \Rightarrow x = \frac{16}{5}$$

$$x = -\frac{a}{e} \Rightarrow x = \frac{-4}{\frac{5}{4}} \Rightarrow x = -\frac{16}{5}$$

The coordinates of foci are $(\pm ae, 0)$ *i.e.*; $\left(\pm 4 \times \frac{5}{4}, 0\right)$; *i.e.* $(\pm 5, 0)$

Length of latus rectum $= \frac{2b^2}{a} = \frac{2 \times 9}{4} = \frac{9}{2}$ **Ans.**

EXERCISE 24.4

1. Find the equation of hyperbola whose directrix is $x + y - 1 = 0$, and eccentricity 2.
   Ans. $x^2 + 4xy + y^2 - 4x + 2y - 7 = 0$
2. Find the equation of the hyperbola whose directrix $3x + 4y + 8 = 0$ and whose focus is (1, 1) and eccentricity is 2. **Ans.** $11x^2 + 96xy + 39y^2 + 242x + 306y + 206 = 0$
3. Find the equation of the hyperbola whose directrix is $2x + y = 1$ and whose focus is (1, 1) and eccentricity is $\sqrt{3}$. **Ans.** $7x^2 + 12xy - 2y^2 - 2x + 4y - 7 = 0$
4. Find the equation of hyperbola whose directrix $2x - y + a = 0$, focus $(a, 0)$ eccentricity $\frac{4}{3}$. **Ans.** $19x^2 - 64xy - 29y^2 + 154ax - 32ay - 29a^2 = 0$
5. Find the equation of the hyperbola whose focus is at the point (2, –1), eccentricity is 2 and the directrix is the straight line 2x + 3y = 1.
   Ans. $3x^2 + 23y^2 + 48xy + 36x - 50y - 61 = 0$
6. Find the eccentricity and the coordinates of the foci of the hyperbola $3x^2 - y^2 = 4$.
   Ans. 2, $\left(\pm \frac{4}{\sqrt{3}}, 0\right)$
7. find the eccentricity and the coordinates of the foci of the hyperbola $2x^2 - 3y^2 = 5$
   Ans. $\sqrt{\frac{5}{3}}, \left(\pm \frac{5}{\sqrt{6}}, 0\right)$

8. Find the centre, the lengths of axes, the eccentricity and foci of the hyperbola $x^2 - y^2 + 4x = 0$ **Ans.** $(-2, 0); 4, 4; \sqrt{2}; (-2 \pm 2\sqrt{2}, 0)$

9. Show that the ellipse $\frac{x^2}{16} + \frac{y^2}{7} = 1$ and hyperbola $\frac{x^2}{144} - \frac{y^2}{81} = \frac{1}{25}$ have the same foci.

10. Prove that the difference of focal distances of any point on a hyperbola is constant.

11. find the equation of hyperbola whose directrix is $2x + y = 1$, focus (1, 2) and eccentricity $\sqrt{3}$. **Ans.** $7x^2 + 12xy - 2y^2 - 2x + 14y - 22 = 0$

Functions

25.1 DEFINITION OF A FUNCTION

A function is a relation from a nonempty set A into a nonempty set B such that:

(i) All elements of set A are associated with the elements of set B.

(ii) An element of set A is associated with one and only one element of the set B.

$$f : A \to B \text{ such that } \{x, f(x) : x \in A \text{ and } f(x) \in B\}$$

A function from A to B is denoted by f and it is written as $f : A \to B$ or $A \xrightarrow{f} B$.
As the weight of a person is unique so a person $x \in A$ is associated with unique element (weight) of B. Thus an element of A cannot be associated with more than one element of B.

Example 1. *Let* $A = \{1, 2\}, B = \{3, 4, 5\}$, write down the function $f : A \to B$ such that $f = \{(a, b) ; a \in A, b \in B \text{ and } a + b \text{ is even}\}$

Solution. We have, $A = \{1, 2\}$ and $B = \{3, 4, 5\}$

$f : A \to B$

$f_1 = \{(1, 3), (2, 4)\}$ [Here, 1 should be associated with only

$f_2 = \{(1, 5), (2, 4)\}$ one element and not two elements 3 and 5]

Example 2. *Let* $A = \{3, 7\}$ and $B = \{5, 6\}$

Write down the function $f : A \to B$ *such that*

(i) $f = \{(a, b) : a \in A, b \in B \text{ and } a \times b \text{ is odd}\}$

(ii) $f = \{(a, b) : a \in A, b \in B \text{ and } a \times b \text{ is even}\}$

Solution. Let $A = \{3, 7\}$ and $B = \{5, 6\}$

(i) $f : A \to B$

$f = \{(3, 5), (7, 5)\}$

(ii) $f : A \to B$

$f = \{(3, 6), (7, 6)\}$ **Ans.**

25.2 SOME SPECIAL FUNCTIONS

1. Constant function:

A function $f(x)$ is constant, for example $f(x) = k$, where k is constant, $f(x) = 4$

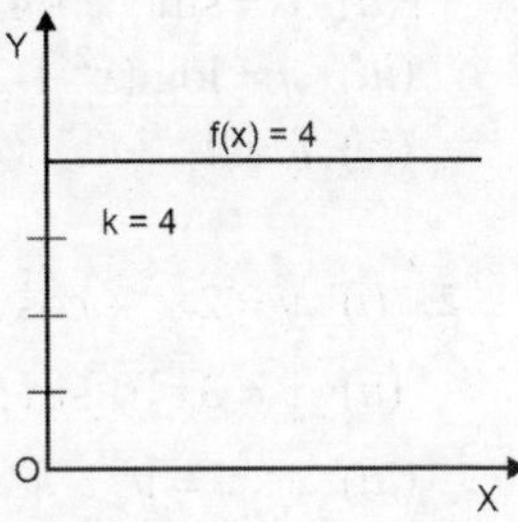

2. Absolute value function:

A function which is absolute value is known as absolute value function.

$$f(x) = |x| = \begin{cases} x, & \text{for } x \geq 0 \\ -x, & \text{for } x < 0 \end{cases}$$

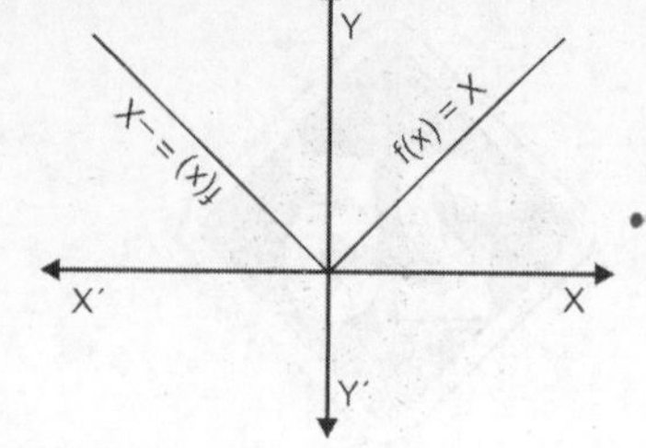

3. **Trigonometric function:**

   A function which involves sin x, cos x etc.

   For example $f(x) = 2\sin^2 x + 32\cos x + 1$

4. **Inverse Trigonometric function:**

   A function involving $\sin^{-1} x, \cos^{-1} x, \tan^{-1} x$ is called inverse trigonometric function.

5. **Logarithmic function:**

   A function involving $\log_a x, \log_2 x$, is called a logarithmic function.

   $f(x) = \log\left(\dfrac{x^2 + x + 1}{x^2 + 3x + 4}\right)$

6. **Exponential function:**

   A function whose base is constant and power is variable is called exponential function. *For example* $f(x) = a^{2x}$, $f(x) = e^{2x+3}$

7. **Parametric equations:**

   If a function involves two variables for example x and y and these x and y can further be expressed in terms of thrid variable θ. *For example* $x = a\cos\theta, y = a\sin\theta$.

8. **Explicit function:**

   If y can be expressed directly in terms of x then the function is called explicit function and *viceversa*. e.g. (*i*) $y = x^2 + 2x + 3$ (*ii*) $x = y^2 + \sqrt{y} + 4$

9. **Implicit function:**

   When neither y nor x can be expressed directly in terms of the other.

   For example $x^3 + x^2 y + y^2 x + x^2 + y^2 + 4 = 0$

10. **Even function:**

    If in a function of x, on replacing x by $-x$ the value of the function remains the same.

    For example $f(-x) = f(x)$ e.g., $y = x^2 + 1, y = 2\cos x$

11. **Odd function:**

    On replacing x in a function of x, the sign of the value of the function $f(x)$ changes; *i.e.*, $f(-x) = -f(x)$ *For example* $f(x) = x^3 + x, f(x) = 2\sin x + \sin 2x$

EXERCISE 25.1

Match the following with the correct type of function.

1. (*i*) $y = e^{x^2}$ — (a) Constant function

   (*ii*) $y = \sin^2 x + 4\cos x + 9$ — (b) Logarithmic function

   (*iii*) $y = \log(x^2 + 2x + 7)$ — (c) Trigonometric function

   (*iv*) $y = 4$ — (d) Exponential function

   Ans. (*i*) → (*d*), (*ii*) → (*c*), (*iii*) → (*b*), (*iv*) → (*a*)

2. (*i*) $y = 2x^2 + \cos x$ — (a) Implicit function

   (*ii*) $y = 3x^5 + \sin x$ — (b) Odd function

   (*iii*) $x^2 y = y^2 x + x^3 + y^2$ — (c) Inverse trigonometric function

(iv) $y = \tan^{-1}(x) + \sin^{-1}\left(\frac{1}{x}\right)$ (d) Even function

Ans. (*i*) → (*d*), (*ii*) → (*b*), (*iii*) → (*a*) (*iv*) → (*c*)

3. (*i*) $x = a \sin 2\theta, y = a \cos 2\theta$ (a) Absolute value function

(*ii*) $y = x^2 + 5x + 3$ (b) Explicit function

(*iii*) $y = |3x + 4|$ (c) Parametric function

(*iv*) $4x^2 + 4x^2 y + y^3 = 0$ (d) Implicit function

Ans. (*i*) → (*c*), (*ii*) → (*b*), (*iii*) → (*a*), (*iv*) → (*d*)

25.3 DOMAIN AND CO-DOMAIN ($f : A \to B$)

The set *A* is called the domain of function *f* and the set *B* is called the co-domain of *f*.

Open Interval

An open interval is an interval that does not include its end points. The open interval $\{x : a < x < b\}$ is denoted (a, b). Although the nonstandard notation $]a, b[$ is also used sometimes.

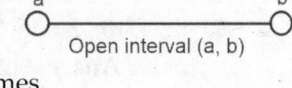

a b
Open interval (a, b)

Closed Interval

A closed interval is an interval that includes all of its limit points. If the endpoints of the interval are finite numbers *a* and *b*, then the interval $\{x : a \le x \le b\}$ is denoted $[a, b]$. If one of the endpoints is $\pm\infty$, then the interval still contains all of its limit points (although not all of its *endpoints*). Hence $[a, \infty)$ and $(-\infty, b]$ are also closed intervals. as is the interval $(-\infty, \infty)$.

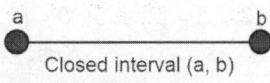

a b
Closed interval (a, b)

25.4 IMAGE

$$f : A \to B$$

If the element $x \in A$ corresponds to $y \in B$ under the function *f*, then we say that *y* is the image of *x* under *f* and we write $f(x) = y$.

Pre-image

If $f(x) = y$, then *x* is a pre-image of *y*.

25.5 RANGE

If *f* is a function from *A* to *B*, then each element of *A* corresponds to one and only one element of *B*. Whereas every element in *B* need not be the image of some *x* in *A*.

Example 3. *Let $f : A \to B$ be a function such that*

$$f = \{(a, b) : a \in A, b \in B \text{ and } b = a^2\}.$$

where $A = \{1, 2, 3\}$ *and* $B = \{1, 2, 3, 4, 5, 6, 7, 8, 9\}$

Find the domain, range and co-domain of the function.

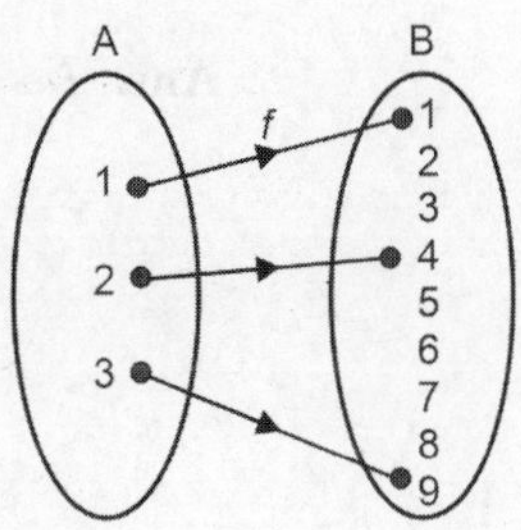

Solution. Here, we have $A = \{1, 2, 3\}$ and

$B = \{1, 2, 3, 4, 5, 6, 7, 8, 9\}$

$f = \{(1, 1), (2, 4), (3, 9)\}$

Domain = A = {1, 2, 3}

Range = {1, 4, 9}

Co-domain = B = {1, 2, 3, 4, 5, 6, 7, 8, 9} **Ans.**

Example 4. *Let* $f : A \to B$ *be a function such that* $f = \{(a, b) : a \in A, b \in B \text{ and } a + b = 4\}$, *where* $A = \{2, 3, 4\}, B = \{0, 1, 2, 3, 4\}$ *Find, domain, range and co-domain.*

Solution. Here, we have $A = \{2, 3, 4\}$ and $B = \{0, 1, 2, 3, 4\}$

We observe that $2 + 2 = 4$, $3 + 1 = 4$ and $4 + 0 = 4$

Thus, $f = \{(2, 2), (3, 1), (4, 0)\}$

Domain = {2, 3, 4}

Range = {2, 1, 0}

Co-domain = {0, 1, 2, 3, 4} **Ans.**

EXERCISE 25.2

Let f : A → B be a function, then find domain, range and co-domain in the following cases:

1. $A = \{2, 4, 10\}, B = \{6, 7, 18, 80, 102, 110\}$ and $f = \{(a, b) : a \in A \text{ and } b \in B \text{ and } b = a^2 + 2\}$

   Ans. $f = \{2, 6), (4, 18), (10, 102)\}$; Domain = {2, 4, 10}; Range = {6, 18, 102}; Co-domain = {6, 7, 18, 80, 102, 110}

2. $A = \{2, 4, 5\}, B = \{4, 8, 16, 64, 100, 125\}$ and $f = \{(a, b) : a \in A, b \in B \text{ and } b = a^3\}$

   Ans. $f = \{(2, 8), (4, 64), (5, 125)\}$; Domain = {2, 4, 5}; Range = {8, 64, 125}; Co-domain = {4, 8, 16, 64, 100, 125}

3. $A = \{0, 3, 5, 7\}, B = \{0, 1, 4, 13, 20, 31, 50, 57, 60\}$ *and*

   $f = \{(a, b) : a \in A, b \in B \text{ and } b = a^2 + a + 1\}$

   Ans. $f = \{(0, 1), (3, 13), (5, 31), (7, 57)\}$; Domain = {0, 3, 5, 7}; Range = {1, 13, 31, 57}; Co-domain = {0, 1, 4, 13, 20, 31, 50, 57, 60}

4. $A = \{2, 8, 9\}, B = \{1, 6, 100, 132, 150, 167, 176\}$ *and*

   $f = \{(a, b) : a \in A, b \in B \text{ and } b = 2a^2 + a - 4\}$

   Ans. $f = \{(2, 6), (8, 132), (9, 167)\}$; Domain = {2, 8, 9}; Range = {6, 132, 167}; Co-domain = {1, 6, 100, 132, 150, 167, 176}

5. $A = \{30°, 45°, 60°, 90°, 240°\}$,

   $$B = \left\{\frac{3}{2}, \frac{\sqrt{2}+1}{\sqrt{2}}, \frac{\sqrt{2}-1}{2}, \frac{\sqrt{3}+2}{2}, \frac{\sqrt{3}-2}{2}, 2, \frac{-\sqrt{3}+2}{2}, \frac{\sqrt{3}+2}{2}\right\}$$

   $f = \{(x, y) : x \in A, y \in B \text{ and } y = \sin x + 1\}$

   Ans. $$f = \left\{\left(30°, \frac{3}{2}\right), \left(45°, \frac{\sqrt{2}+1}{\sqrt{2}}\right), \left(60°, \frac{\sqrt{3}+2}{2}\right), (90°, 2), \left(240°, \frac{-\sqrt{3}+2}{2}\right)\right\};$$

   Domain = {30°, 45°, 60°, 90°, 240°};

   $$\text{Range} = \left\{\frac{3}{2}, \frac{\sqrt{2}+1}{\sqrt{2}}, \frac{\sqrt{3}+2}{2}, 2, \frac{-\sqrt{3}+2}{2}\right\};$$

   $$\text{Co-domain} = \left\{\frac{3}{2}, \frac{\sqrt{2}+1}{\sqrt{2}}, \frac{\sqrt{2}-1}{2}, \frac{\sqrt{3}+2}{2}, \frac{\sqrt{3}-2}{2}, 2, \frac{-\sqrt{3}+2}{2}, \frac{\sqrt{3}+2}{2}\right\}$$

6. $A = \{45°, 135°, 315°, 405°\}$, $B = \{1, 2, 3, 4\}$ $f = \{(x, y) : x \in A \text{ and } y \in B, \text{ and } y = \tan x + 2\}$

Ans. $f = \{(45°, 3), (135°, 1), (315°, 1), (405°, 3)\}$; Domain = $\{45°, 135°, 315°, 405°\}$; Range = $\{1, 3\}$ Co-domain = $\{1, 2, 3, 4\}$

25.6 IDENTIFICATION OF A FUNCTION

A function *f* is from a nonempty set *A* into a nonempty set *B* such that

(*i*) All elements of set *A* are associated with the elements of set *B*.

(*ii*) An elements of set *A* is associated with one and only one element of the set *B*.

$f : A \to B$ such that $\{x, f(x) : x \in A \text{ and } f(x) \in B\}$.

Note: In a function two or more elements of A can be associated with one element of B.

Illustrations

(*i*) Let f represent weight $y \in B$ of person $x \in A$. f represents a function since each element of *A* corresponds to exactly one element of *B*. Hence, *f* is a function from *A* to *B*.

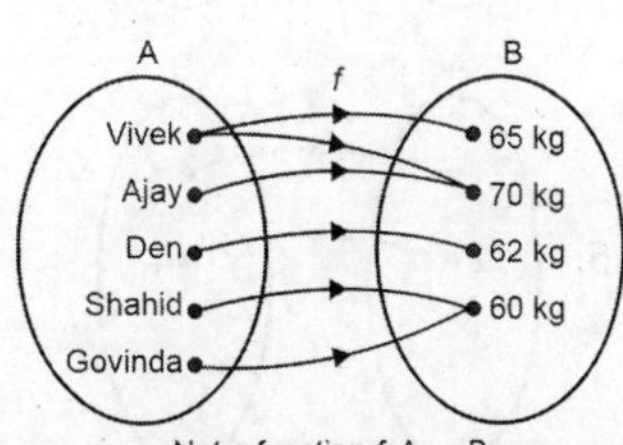

Function $f: A \to B$

(*ii*) Here, Vivek is associated with 65 kg and 70 kg. All person can have only unique weight, so the weight of Vivek cannot be 65 kg and 70 kg both. Hence, $A \to B$ is not a function.

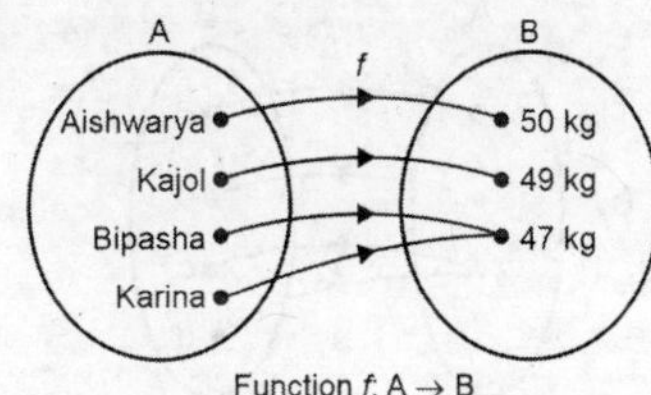

Not a function $f: A \to B$

(*iii*) Let, *f* assigns to each person of *A* to the elements (weight) of *B*. Here, Bipasha and Karina are associated with 47 kg. It is possible that the weight of two persons may be equal. Here, *f* is a function from *A* to *B*.

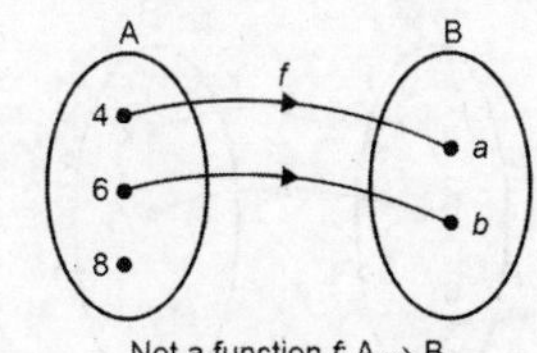

Function $f: A \to B$

(*iv*) Elements 4 and 6 of set *A* are associated with the elements *a* and *b* of set *B*. Element 8 of *A* is not associated to any element in *B*. Hence according to the definition of a function, all the elements of *A* are to be associated with the elements of *B*. Hence, this is not a function from *A* to *B*.

Not a function $f: A \to B$

(*v*) In this diagram, no arrow ends the element Meerut of *B*. It means there is no country in *A* whose capital is Meerut and under the definition of a function it is not necessary that all the elements of *B* should be associated with some element of *A*. Here, *f* is a function.

Here, Co-domain is {Delhi, London, Tokyo, Beijing, Meerut} but Range is {Delhi, London, Tokyo, Beijing}.

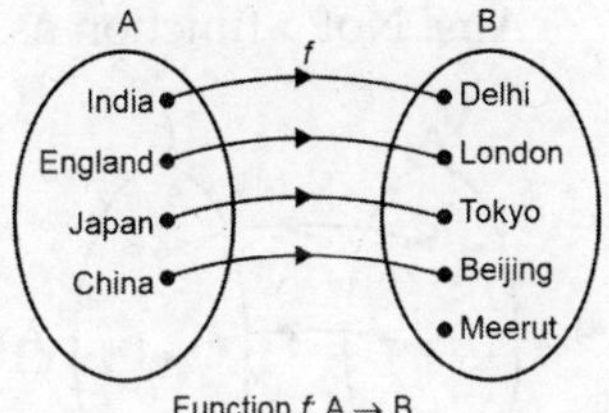

Function $f: A \to B$

EXERCISE 25.3

Determine whether *f* is a function or not in the following cases.

1. 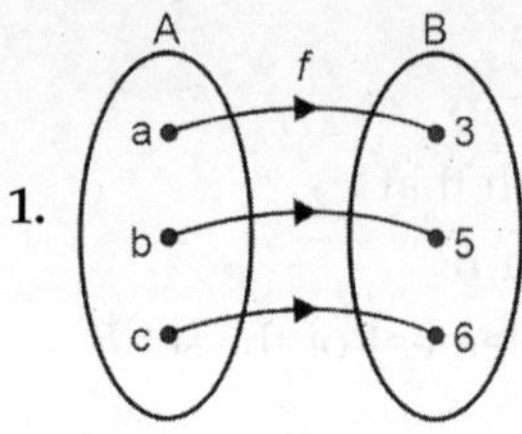

Ans. Function

2. 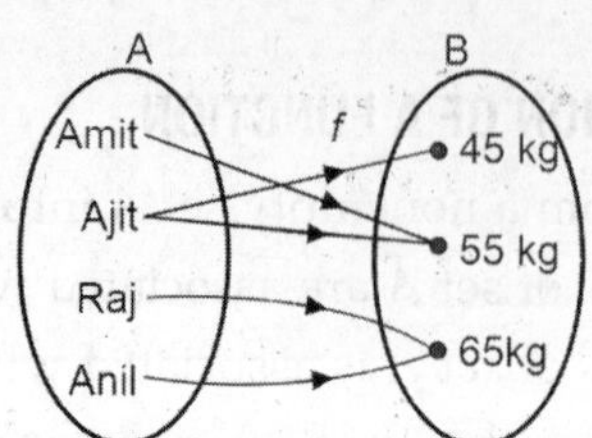

Ans. Not a function, as Ajit cannot have two weights 45 kg and 55 kg.

3. 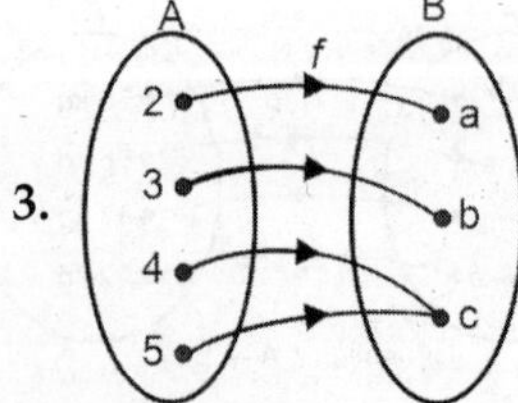

Ans. Function

4. 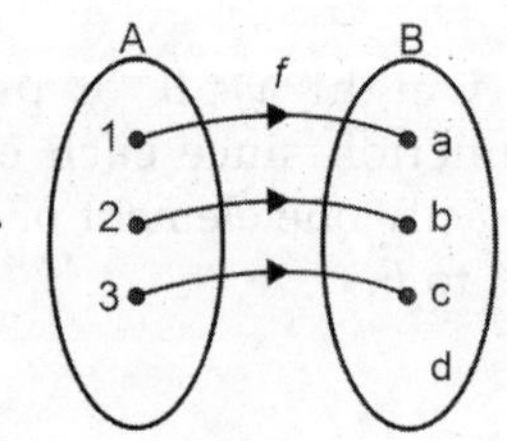

Ans. Function

5. 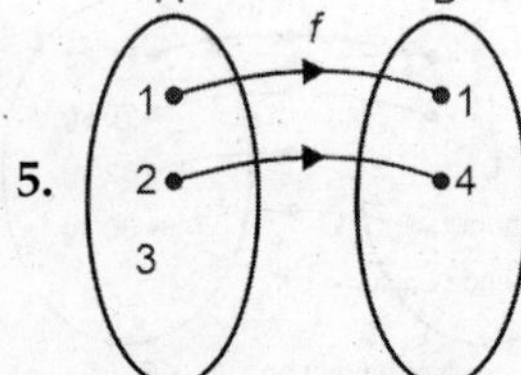

Ans. Not function, since there is no image of 3.

6. 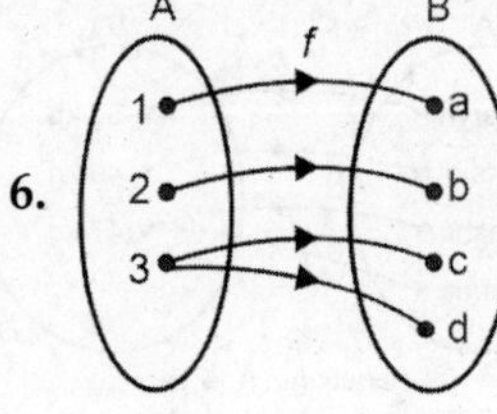

Ans. Not a function, since there cannot be two images *c* and *d* of 3.

7. 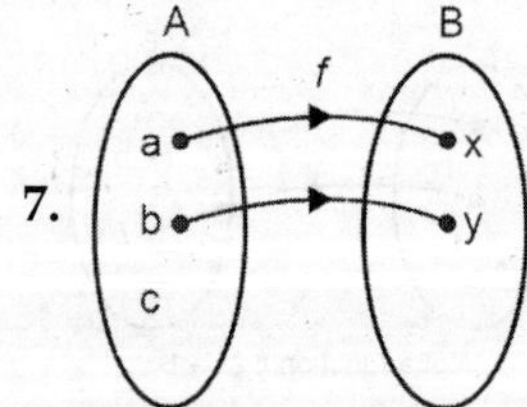

Ans. Not a function as there is no image of *c*.

8. 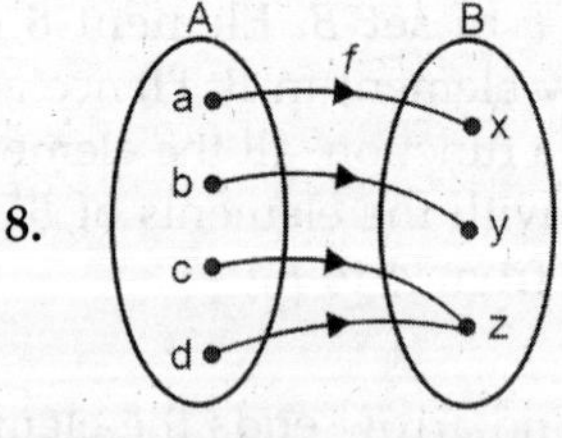

Ans. Function

9. 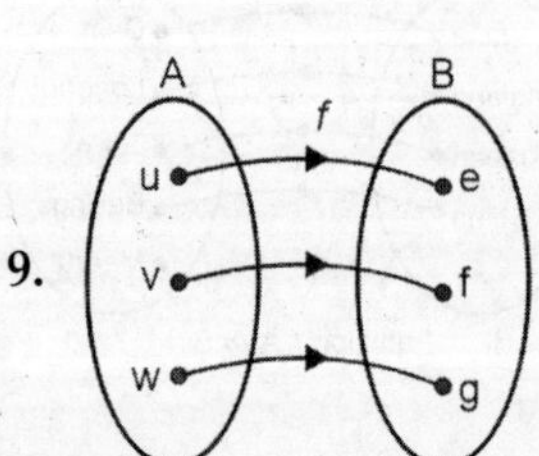

Ans. Function

10. 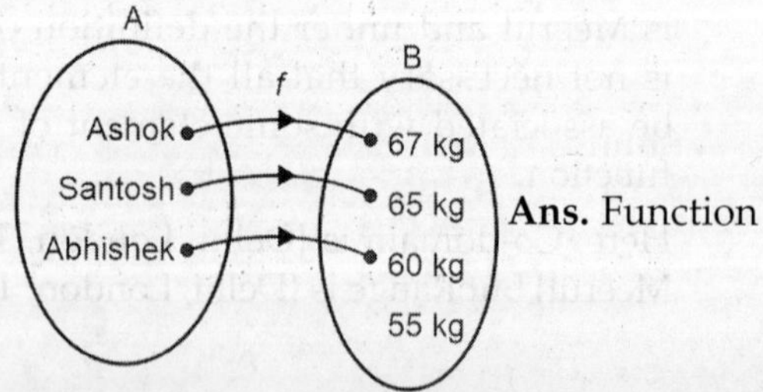

Ans. Function

25.7 VALUE OF A FUNCTION

In a function $f(x)$, if x is replaced by a then $f(a)$ is known as the value of function for $x = a$.

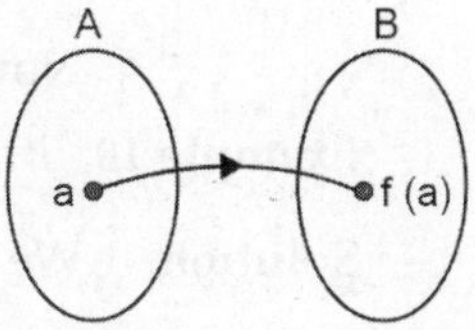

For example $f(x) = x^2 + x + 3$

$f(3) = 3^2 + 3 + 3 = 9 + 3 + 3 = 15$

Thus, value of the function $f(x)$ at $x = 3$ is 15.

Example 5. *If* $f(x) = 2x + 5$. *Find* $f(0) + f(1)$.

Solution. We have, $f(x) = 2x + 5$

$\therefore$ $f(0) = 2(0) + 5 = 0 + 5 = 5$

and $f(1) = 2(1) + 5 = 2 + 5 = 7$

$\therefore$ $f(0) + f(1) = 5 + 7 = 12$ **Ans.**

Example 6. *If* $f(x) = x^2 + 2x + 3$, *then find* $f(x + 2)$. **[S.B.T.E. 2017]**

Solution. We have, $f(x) = x^2 + 2x + 3$

$\Rightarrow$ $f(x + 2) = (x + 2)^2 + 2(x + 2) + 3$

$= x^2 + 4x + 4 + 2x + 4 + 3$

$= x^2 + 6x + 11$ **Ans.**

Example 7. *If* $f(x) = x^2 - 4x + 10$ *and* $z = y - 5$, *then find* $f(z)$.

Solution. We have, $f(x) = x^2 - 4x + 10$

$\Rightarrow$ $f(z) = z^2 - 4z + 10 = (y - 5)^2 - 4(y - 5) + 10$

$= y^2 - 10y + 25 - 4y + 20 + 10$

$= y^2 - 14y + 55$ **Ans.**

Example 8. *If* $f(x) = 3x^2 - 5x + 7$, *then show that* $f(-1) = 3f(1)$.

Solution. We have, $f(x) = 3x^2 - 5x + 7$

$\therefore$ $f(1) = 3(1)^2 - 5(1) + 7$

$\Rightarrow$ $f(1) = 3 - 5 + 7 = 10 - 5 = 5$

$\Rightarrow$ $f(1) = 5$ ...(1)

and $f(-1) = 5(-1)^2 - 5(-1) + 7$

$\Rightarrow$ $f(-1) = 3 + 5 + 7$

$\Rightarrow$ $f(-1) = 15$ ...(2)

From (1) and (2), we have

$f(-1) = 15 = 3 \times 5 = 3f(1)$

$\Rightarrow$ $f(-1) = 3f(1)$ **Proved.**

Example 9. *If* $f(x) = \dfrac{x^3 + 2x + 1}{x^2 + 3x}$. *Find* $f(1) + f(-1)$.

Solution. We have, $f(x) = \dfrac{x^3 + 2x + 1}{x^2 + 3x}$

$\therefore$ $f(1) = \dfrac{(1)^3 + 2(1) + 1}{(1)^2 + 3(1)} = \dfrac{1 + 2 + 1}{1 + 3} = \dfrac{4}{4} = 1$

and $$f(-1) = \frac{(-1)^3 + 2(-1) + 1}{(-1)^2 + 3(-1)} = \frac{-1-2+1}{1-3} = \frac{-2}{-2} = 1$$

Now $f(1) + f(-1) = 1 + 1 = 2$ **Ans.**

Example 10. *If $f(x) = x^2 + 1$, then find the value of x, for which $f(x) = f(3x+2)$.*

Solution. We have, $$f(x) = x^2 + 1$$

$$f(3x+2) = (3x+2)^2 + 1 = 9x^2 + 12x + 4 + 1$$

$$= 9x^2 + 12x + 5$$

Given that, $$f(x) = f(3x+2)$$

$$\therefore \quad x^2 + 1 = 9x^2 + 12x + 5$$

$$\Rightarrow \quad 9x^2 + 12x + 5 - x^2 - 1 = 0$$

$$\Rightarrow \quad 8x^2 + 12x + 4 = 0$$

$$\Rightarrow \quad 2x^2 + 3x + 1 = 0$$

$$\Rightarrow \quad 2x^2 + 2x + x + 1 = 0$$

$$\Rightarrow \quad 2x(x+1) + 1(x+1) = 0$$

$\Rightarrow$ Either $x + 1 = 0$ or $2x + 1 = 0$

$$\Rightarrow \quad x = -1 \text{ or } x = -\frac{1}{2}$$ **Ans.**

Example 11. *If $f(x) = \dfrac{1}{1+x}$, then find $f(x) + f(-x)$.*

Solution. We have, $$f(x) = \frac{1}{1+x} \qquad ...(1)$$

$$f(-x) = \frac{1}{1+(-x)} = \frac{1}{1-x} \qquad ...(2)$$

Adding (1) and (2), we get

$$f(x) + f(-x) = \frac{1}{1+x} + \frac{1}{1-x} = \frac{1-x+1+x}{(1+x)(1-x)} = \frac{2}{1-x^2}$$ **Ans.**

Example 12. *If $f(x) = \dfrac{1}{2x}$, then show that $f(4x^2 + 2x) = f(2x) - f(2x+1)$.*

Solution. We have, $$f(x) = \frac{1}{2x} \qquad ...(1)$$

$$f(2x) = \frac{1}{2(2x)} = \frac{1}{4x} \qquad ...(2)$$

$$f(2x+1) = \frac{1}{2(2x+1)} = \frac{1}{4x+2} \qquad ...(3)$$

From (2) and (3), we have

$$f(2x) - f(2x+1) = \frac{1}{4x} - \frac{1}{4x+2} = \frac{4x+2-4x}{4x(4x+2)}$$

$$= \frac{1}{2x(4x+2)} = \frac{1}{2(4x^2+2x)} \qquad ...(4)$$

Also, $$f(4x^2+2x) = \frac{1}{2(4x^2+2x)} \quad ...(5)$$

From (4) and (5), we have

$$f(4x^2+2x) = f(2x) - f(2x+1).$$ **Proved.**

Example 13. *If* $f(x) = 256^x - \log_2 x$, *then find* $f(1)$ *and* $f\left(\frac{1}{8}\right)$.

Solution. We have, $f(x) = 256^x - \log_2^x$

$\therefore$ $$f(1) = 256^1 - \log_2 \cdot 1$$
$$= 256 - 0 = 256 \qquad [\because \log_a 1 = 0]$$

and $$f\left(\frac{1}{8}\right) = (256)^{\frac{1}{8}} - \log_2\left(\frac{1}{8}\right) = (2^8)^{\frac{1}{8}} - \log_2(2)^{-3}$$
$$= 2^{8\times\frac{1}{8}} - (-3)\log_2(2) = 2^1 + 3\log_2(2)$$
$$= 2 + 3(1) \qquad [\because \log_a(a) = 1]$$
$$= 2 + 3 = 5$$ **Ans.**

Example 14. *If* $f(x) = \log\left(\frac{1+x}{1-x}\right)$ *then prove that* $f\left(\frac{2a}{1+a^2}\right) = 2f(a)$

Solution. We have $$f(x) = \log\left(\frac{1+x}{1-x}\right)$$

$\Rightarrow$ $$f(a) = \log\left(\frac{1+a}{1-a}\right) \quad ...(1)$$

Now, $$f\left(\frac{2a}{1+a^2}\right) = \log\left(\frac{1+\frac{2a}{1+a^2}}{1-\frac{2a}{1+a^2}}\right)$$
$$= \log\left(\frac{1+a^2+2a}{1+a^2-2a}\right) = \log\left(\frac{(1+a)^2}{(1-a)^2}\right)$$
$$= \log\left(\frac{1+a}{1-a}\right)^2 = 2\log\left(\frac{1+a}{1-a}\right)$$
$$= 2f(a)$$ [using (1)] **Proved.**

Example 15. If $f(x) = \log\left(\frac{x}{x-1}\right)$, then show that $f(m) + f(m+1) = \log\left(\frac{m+1}{m-1}\right)$

Solution. We have, $$f(x) = \log\left(\frac{x}{x-1}\right)$$

$\therefore$ $$f(m) = \log\left(\frac{m}{m-1}\right) \quad ...(1)$$

and $\quad f(m+1) = \log\left(\frac{m+1}{m+1-1}\right) = \log\left(\frac{m+1}{m}\right)$ ...(2)

Adding (1) and (2), we get

$$f(m) + f(m+1) = \log\left(\frac{m}{m-1}\right) + \log\left(\frac{m+1}{m}\right)$$

$$= \log\left(\frac{m}{m-1}\right)\left(\frac{m+1}{m}\right) = \log\left(\frac{m+1}{m-1}\right)$$ **Proved.**

Example 16. *If* $f(x) = \log_a x$, *then prove that*

(i) $f(mn) = f(m) + f(n)$ (ii) $f\left(\frac{m}{n}\right) = f(m) - f(n)$

(iii) $nf(m) = f(m^n)$

Solution. We have, $f(x) = \log_a(x)$

$\Rightarrow \quad f(m) = \log_a(m)$ ... (1)

and $\quad f(n) = \log_a(n)$ ... (2)

(i) $\quad f(mn) = \log_a(mn)$

$\Rightarrow \quad f(mn) = \log_a(m) + \log_a(n)$

$\Rightarrow \quad f(mn) = f(m) + f(n)$ [using (1) and (2)]

(ii) $\quad f\left(\frac{m}{n}\right) = \log_a\left(\frac{m}{n}\right)$

$\Rightarrow \quad f\left(\frac{m}{n}\right) = \log_a(m) - \log_a(n)$

$\Rightarrow \quad f\left(\frac{m}{n}\right) = f(m) - f(n)$ [using (1) and (2)]

(iii) $\quad n\,f(m) = n\log_a(m) = \log_a(m^n) = f(m^n)$ **Proved.**

Example 17. Write the domain of the function $f(x) = \frac{x^2-4}{x-2}$. **[S.B.T.E. 2017, 2015]**

Solution. Here we have,

$$f(x) = \frac{x^2-4}{x-2} = \frac{(x+2)(x-2)}{(x-2)}$$

$$f(x) = x + 2$$

$f(x)$ assumes real values for all real values for all x except $x = 2$

Hence domain $(f) = R - \{2\}$ **Ans.**

Example 18. *If* $f(x) = \sin 2x - \cos x$, *find* $f'(-\pi/3)$. **[S.B.T.E. 2017]**

Solution. Here we have,

$$f(x) = \sin 2x - \cos x$$

$$f'(x) = 2\cos 2x + \sin x$$

$$f'\left(\frac{-\pi}{3}\right) = 2\cos\left(\frac{-2\pi}{3}\right) + \sin\left(\frac{-\pi}{3}\right)$$

$$f'\left(\frac{-\pi}{3}\right) = 2\left(\frac{-1}{2}\right) - \frac{\sqrt{3}}{2} = -\frac{\left(2+\sqrt{3}\right)}{2}$$ **Ans.**

Example 19. *If* $f(x) = \log\left(\frac{1+x}{1-x}\right)$, *then show that* $f(m) + f(n) = f\left(\frac{m+n}{1+mn}\right)$.

Solution. We have, $$f(x) = \log\left(\frac{1+x}{1-x}\right)$$

$\therefore$ $$f(m) = \log\left(\frac{1+m}{1-m}\right) \quad ...(1)$$

$$f(n) = \log\left(\frac{1+n}{1-n}\right) \quad ...(2)$$

Adding (1) and (2), we get

$$L.H.S. = f(m) + f(n) = \log\left(\frac{1+m}{1-m}\right) + \log\left(\frac{1+n}{1-n}\right) = \log\left(\frac{1+m}{1-m}\right)\left(\frac{1+n}{1-n}\right)$$

$$= \log\left(\frac{1+m+n+mn}{1-m-n+mn}\right)$$

$$= \log\left(\frac{1+mn+(m+n)}{(1+mn)-(m+n)}\right)$$

$$= \log\left(\frac{1+\frac{m+n}{1+mn}}{1-\frac{m+n}{1+mn}}\right) \quad \left[\begin{array}{l}\text{Dividing both num.}\\ \text{and den. by } (1+mn)\end{array}\right]$$

$$= f\left(\frac{m+n}{1+mn}\right) = \text{R.H.S.}$$ **Proved.**

Example 20. If $f(x) = \log(1+\tan x)$, then show that

$$f\left(\frac{\pi}{4} - x\right) = \log 2 - f(x) = f\left(\frac{\pi}{4}\right) - f(x)$$

Solution. We have, $$f(x) = \log(1+\tan x) \quad ...(1)$$

$\therefore$ $$f\left(\frac{\pi}{4} - x\right) = \log\left[1 + \tan\left(\frac{\pi}{4} - x\right)\right]$$

$$= \log\left[1 + \frac{\tan\frac{\pi}{4} - \tan x}{1 + \tan\frac{\pi}{4}\tan x}\right] = \log\left[1 + \frac{1-\tan x}{1+\tan x}\right]$$

$$= \log\left[\frac{1+\tan x + 1 - \tan x}{1+\tan x}\right] = \log\left[\frac{2}{1+\tan x}\right]$$

$$= \log 2 - \log(1 + \tan x)$$

$$\Rightarrow \quad f\left(\frac{\pi}{4} - x\right) = \log 2 - f(x) \qquad \text{...(2) [using (1)]}$$

Also $\quad f\left(\frac{\pi}{4}\right) = \log\left(1 + \tan\frac{\pi}{4}\right)$

$$= \log(1+1) = \log 2 \qquad \text{...(3)}$$

Putting $\quad f\left(\frac{\pi}{4}\right) = \log 2$ from (3) in (2), we get

$$f\left(\frac{\pi}{4} - x\right) = f\left(\frac{\pi}{4}\right) - f(x) \qquad \text{...(4)}$$

Hence from (2) and (4), we get

$$f\left(\frac{\pi}{4} - x\right) = \log 2 - f(x) = f\left(\frac{\pi}{4}\right) - f(x)$$ **Proved.**

Example 21. *If* $f(x) = x - \frac{1}{x}$, *then show that* $[f(x)]^3 = f(x^3) + 3f\left(\frac{1}{x}\right)$

Solution. We have, $\quad f(x) = x - \frac{1}{x}$

$$\therefore \quad [f(x)]^3 = \left(x - \frac{1}{x}\right)^3$$

$$= x^3 - 3x^2\left(\frac{1}{x}\right) + 3x\left(\frac{1}{x}\right)^2 - \left(\frac{1}{x}\right)^3$$

$$= x^3 - 3x + 3x.\frac{1}{x^2} - \frac{1}{x^3}$$

$$= x^3 - 3x + \frac{3}{x} - \frac{1}{x^3}$$

$$= x^3 - \frac{1}{x^3} + 3\left(\frac{1}{x} - x\right) \qquad \text{...(1)}$$

Now, $\quad f(x^3) = x^3 - \frac{1}{x^3} \qquad \text{...(2)}$

$$f\left(\frac{1}{x}\right) = \frac{1}{x} - \frac{1}{\left(\frac{1}{x}\right)} = \frac{1}{x} - x$$

$$3f\left(\frac{1}{x}\right) = 3\left(\frac{1}{x} - x\right) \qquad \text{...(3)}$$

Adding (2) and (3)

$$\Rightarrow \quad f(x^3) + 3f\left(\frac{1}{x}\right) = x^3 - \frac{1}{x^3} + 3\left(\frac{1}{x} - x\right) \qquad \text{...(4)}$$

from (1) and (4)

$$\Rightarrow \quad [f(x)]^3 = f(x^3) + 3f\left(\frac{1}{x}\right)$$ **Proved.**

Example 22. *If* $f(x) = x^5 - x + x\cos x + \sin x$. Show that $f(x)$ is an odd function.

Solution. We have $f(x) = x^5 - x + x\cos x + \sin x$ ...(1)

$$f(-x) = (-x)^5 - (-x) + (-x)\cos(-x) + \sin(-x)$$

$$= -x^5 + x - x\cos x - \sin x \quad ...(2)$$

$$= -(x^5 - x + x\cos x + \sin x) = -f(x) \quad \text{[Using (1)]}$$

Since, $f(-x) = -f(x)$, therefore $f(x)$ is an odd function. **Proved.**

Example 23. *Test the function for even or odd:* $f(x) = 4x^6 - 3x^2 + \cos x$.

Solution. We have, $f(x) = 4x^6 - 3x^2 + \cos x$ ...(2)

$$f(-x) = 4(-x)^6 - 3(-x)^2 + \cos(-x)$$

$$= 4x^6 - 3x^2 + \cos x$$

$$= f(x) \quad \text{[Using (1)]}$$

Since, $f(-x) = f(x)$, therefore $f(x)$ is an even function. **Ans.**

Example 24. *State whether the function* $f(x) = \dfrac{e^x + e^{-x}}{5}$ *is even or odd.*

Solution. We have,

$$f(x) = \frac{e^x + e^{-x}}{5}$$

$$\Rightarrow \quad f(-x) = \frac{e^{-x} + e^{-(-x)}}{5} = \frac{e^{-x} + e^x}{5} = f(x)$$

Since, $f(-x) = f(x)$, therefore $f(x)$ is an even function. **Ans.**

Example 25. *A function f is defined on the set of integers as follows:*

$$f(x) = \begin{cases} 1 + x, & 1 \le x < 2 \\ 2x - 1, & 2 \le x < 4 \\ 3x - 10, & 4 \le x < 6 \end{cases}$$

find the value of $f(1)$, $f(1.5)$, $f(3)$ *and* $f(5)$.

Solution. Since 1 and 1.5 lies between 1 and 2.

$\therefore \quad f(x) = 1 + x, \quad 1 \le x < 2$

$\Rightarrow \quad f(1) = 1 + 1 = 2$

and $\quad f(1.5) = 1 + 1.5 = 2.5$

Since 3 lies between 2 and 4

$\therefore \quad f(x) = 2x - 1, 2 \le x < 4$

$\Rightarrow \quad f(3) = 2(3) - 1 = 6 - 1 = 5$

Since 5 lies between 4 and 6.

$\therefore \quad f(x) = 3x - 10, \quad 4 \le x < 6$

$\Rightarrow \quad f(5) = 3(5) - 10 = 15 - 10 = 5$ **Ans.**

Example 26. $f(x) = \frac{1}{1+2x}$ *then find* $f[f(x)]$.

Solution. We have $$f(x) = \frac{1}{1+2x}$$

$$f[f(x)] = \frac{1}{1+2[f(x)]} = \frac{1}{1+2\left(\frac{1}{1+2x}\right)} = \frac{1+2x}{1+2x+2} = \frac{1+2x}{3+2x}$$ **Ans.**

EXERCISE 25.4

1. If $f(x) = 3x^2 + 7$, then find $f(0)$ and $f(-1)$. **Ans.** 7, 10
2. If $f(x) = 4x^3 + 3x + 6$, then find $f(1)$, $f(-1)$. **Ans.** 13, –1.
3. If $f(x) = 2x^2 + x + 1$, then find $f(x+3)$. **Ans.** $2x^2 + 13x + 22$
4. If $f(x) = x^2 + 5x + 7$, then find $f(x+2)$. **Ans.** $x^2 + 9x + 21$
5. If $f(x) = x^2 + 6x + 3$, and $z = y + 1$, then find $f(z)$. **Ans.** $y^2 + 8y + 10$
6. If $f(x) = x^2 + 6x + 10$, then show that $f(2) = 28 - f(-2)$.
7. If $f(x) = \frac{x^2 + 3x + 2}{2x^3 + 5}$, find $f(0) + f(1)$. **Ans.** $\frac{44}{35}$
8. If $f(x) = x^2 + 4$, then find the value of x, for which $f(x) = f(2x+1)$. **Ans.** $-1, -\frac{1}{3}$
9. If $f(x) = \frac{1}{2(1-x)}$, find $f(x) + f(-x)$. **Ans.** $\frac{1}{1-x^2}$
10. If $f(x) = \frac{1}{x}$, show that $f(x+1) - f(x+2) = f(x^2 + 3x + 2)$.
11. If $f(x) = x^3 - \frac{1}{x^3}$, $x \neq 0$, show that $f(x) + f\left(\frac{1}{x}\right) = 0$
12. If f and g are real function defined by $f(x) = x^2 + 7$ and $g(x) = 3x + 5$, find value of:

    (*a*) $f(3) + g(-5)$ **Ans.** 6 (*b*) $f\left(\frac{1}{2}\right) \times g(14)$ **Ans.** $\frac{1364}{4}$

    (*c*) $f(-2) + g(-1)$ **Ans.** 13 (*d*) $f(t) - f(-2)$ **Ans.** $t^2 - 4$

    (*e*) $\frac{f(t) - f(5)}{t-5}$, it $t \neq 5$ **Ans.** $t + 5$
13. If $f(x) = \frac{x-1}{x+1}$, then show that (*i*) $f\left(\frac{1}{x}\right) = -f(x)$ (*ii*) $f\left(-\frac{1}{x}\right) = \frac{-1}{f(x)}$
14. If $f(x) = \frac{1}{1-x}$, then show that $f[f\{f(x)\}] = x$
15. If $f(x) = x^2$, find $\frac{f(1.1) - f(1)}{1.1 - 1}$ **Ans.** 2.1.
16. If $f : R \to R$ is defined by $f(x) = \frac{3x+1}{5x-3}$, $\left(x \neq \frac{3}{5}\right)$. Find $f\left(\frac{3x+1}{5x-3}\right)$ **Ans.** x

17. A function f is defined on the set of integers as follows:

$$f(x) = \begin{cases} 1+2x, & 0 \le x < 2 \\ 3x-2, & 2 \le x < 5 \\ 4x+5, & 5 \le x < 7 \end{cases}$$

Find the value of $f(0)$, $f(3)$ and $f(6)$. **Ans.** 1, 7, 29

18. The function $f: R \to R$ is defined by

$$f(x) = \begin{cases} 12x+5, & x > 1 \\ x-4, & x \le 1 \end{cases}$$

Find $f(0)$, $f(1)$, and $f(2)$. **Ans.** −4, −3, 29

19. If $f(x) = \log\left(\frac{1+x}{1-x}\right)$, then prove that $f\left(\frac{4a}{1+4a^2}\right) = 2f(2a)$

20. If $f(x) = 729^x - \log_3(2x)$, then find $f\left(\frac{1}{6}\right)$. **Ans.** −4

Limits

26.1 INTRODUCTION

In mathematics, limit of a function at a point is the value that the function approaches near to a given point, it is a fundamental concept in calculus and analysis. In this chapter we will discuss limit of a function when $x \to a$.

Important Limits

(1) $\lim\limits_{x \to a} \dfrac{x^n - a^n}{x - a} = n a^{n-1}$

(2) $\lim\limits_{x \to 0} \dfrac{\sin x}{x} = 1$

(3) $\lim\limits_{x \to 0} \dfrac{\tan x}{x} = 1$

(4) $\lim\limits_{x \to 0} \dfrac{e^x - 1}{x} = 1$

(5) $\lim\limits_{x \to 0} \dfrac{\log_e(1+x)}{x} = 1$

(6) $\lim\limits_{x \to \infty} \left(1 + \dfrac{1}{x}\right)^x = e$

(7) $\lim\limits_{x \to 0} \dfrac{a^x - 1}{x} = \log a$

26.2 INTUITIVE IDEA OF LIMITS

(*a*) Suppose we are travelling from kashmiri gate to connaught place by metro, which will reach connaught place at 10 a.m. As the time gets closer and closer to 10 a.m., the distance of the train from connaught place gets closer and closer to zero. If we consider time as independent variable denoted by t and the distance as function of time say $f(t)$, then we say that $f(t)$ approaches zero as t approaches zero. We can say that the limit of $f(t)$ is zero as t tends to zero.

(*b*) Let a regular polygon be described in a circle of given radius. We notice the following points from the geometry.

(*i*) The area of the polygon cannot be greater than the area of the circumscribed circle, no matter how, large the number of sides may be.

(*ii*) As the number of sides of the polygon increases indefinitely, the area of the polygon continuously approaches the area of the circle.

(*iii*) Ultimately the difference between the area of the circle and the area of the polygon can be made as small as we please by sufficiently increasing the number of sides of the polygon.

We can say that the limit of the area of the polygon inscribed in a circle is the area of the circle as the number of sides increases indefinitely.

26.3 MEANING OF $x \to a$

Let x be a variable and a be a constant. Since x is a variable, we can change its value at pleasure. It can be changed so that its value comes nearer and nearer to a. Then we say that x approaches a or x tends to a and it is denoted by $x \to a$:

We know that $|x - a|$ is the distance between x and a on the real number line and $0 < |x - a|$ if $|x \neq a$, "x tends to a" means

(*i*) $x \neq a$, *i.e.* $0 < |x - a|$, and

(*ii*) x takes up values nearer and nearer to a, *i.e.* the distance $|x - a|$ between x and a becomes smaller and smaller. One may ask "how much smaller"? The answer is, as much as we please. It may be less than 0.1, 0.00001, 0.0000001 and so on. In fact, we may choose any positive number δ. How so ever small it may be, $x - a|$ will always be less than δ. The above discussion leads up to the following definition of $x \to a$.

x → a^- a^+ ← x

a

Let x be a variable and a be a constant.

> Definition: Given a number $\delta > 0$ however small, if x takes up values, such that $0 < |x - a| < \delta$. Then x is said to tend to a, and is symbolically written as $x \to a$.

In simple language, we can say that if x can assume values such that the positive difference between x and a remains very small, then we say that x tends to a.

Note: If x approaches a from values less than a, *i.e.* from left side of a, we write $x \to a^-$.

If x approaches a from values greater than a, *i.e.* from right side of a, we write $x \to a^+$.

But $x \to a$ means both $x \to a^-$ and $x \to a^+$. So x approaches or tends to a means x approaches a from both sides right and left.

Neighbourhood of a point a

The set of all real numbers lying between $a - \delta$ and $a + \delta$ is called the neighbourhood of a.

Neighbourhood $a = (a - \delta, a + \delta)$

$x \in (a - \delta, a + \delta)$.

26.4 LIMIT OF A FUNCTION

Let us take some examples to find the limit of various functions.

Example 1. *Consider the function* $f(x) = \dfrac{x^2 - 4}{x - 2}$. *We investigate the behaviour of f(x) at the point x = 2 and near the point x = 2.*

Solution. $f(2) = \dfrac{4 - 4}{2 - 2} = \dfrac{0}{0}$, which is meaningless. Thus $f(x)$ is not defined at $x = 2$.

Now we try to evaluate the value of $f(x)$ when x is very near to 2. Some values of $f(x)$ for x less than 2 and then for x greater than 2 are discussed here.

$$f(1.9) = \frac{(1.9)^2 - 4}{1.9 - 2} = \frac{-0.39}{-0.1} = 3.9 \qquad f(2.1) = \frac{(2.1)^2 - 4}{2.1 - 2} = \frac{0.41}{0.1} = 4.1$$

$$f(1.99) = \frac{(1.99)^2 - 4}{1.99 - 2} = \frac{-0.0399}{-0.01} = 3.99 \qquad f(2.01) = \frac{(2.01)^2 - 4}{2.01 - 2} = \frac{0.0401}{0.01} = 4.01$$

$$f(1.999) = \frac{(1.999)^2 - 4}{1.999 - 2} = \frac{-0.003999}{-0.001} = 3.999 \qquad f(2.001) = \frac{(2.001)^2 - 4}{2.001 - 2} = \frac{0.004001}{0.001} = 4.001$$

$$f(x) = \frac{x^2 - 4}{x - 2}$$

3 3.9 3.99 3.999 → (4) ← 4.001 4.01 4.1

x = 1 1.9 1.99 1.999 (2) 2.001 2.01 2.1

It is clear from the above that as x gets nearer and nearer to 2 from either side, $f(x)$ gets closer and closer to 4 from either side.

When x approaches 2 from left hand side the function $f(x)$ tends to a definite number 4. Thus we say that as x tends to 2 the left hand limit of the function f exist and equal to *definite number* 4. Similarly, as x approaches 2 from right hand side, the function $f(x)$ tends to a *definite number* 4.

Again we say that as x approaches 2 from right hand side of 2, the right hand limit of f exists and equal to 4.

Example 2. *Discuss the limit of the function* $f(x) = \frac{1}{x}$ *at* $x = 0$ *and its graph.*

Solution. We have, $f(x) = \frac{1}{x}$

Let us draw the graph of the given function $f(x) = \frac{1}{x}$.

x	–1	–0.1	–0.01	–0.001	–0.0001	–0.00001	1	0.0001	0.001	0.01	0.1
$f(x)$	–1	–10	–100	–1000	–10,000	–10,00,000	1	10,000	1000	100	10

(*i*) As x approaches zero from left of zero the graph never approaches a finite number so we say that left hand limit of f at $x = 0$ does not exist *i.e.*

$$\lim_{x \to 0^-} f(x) \text{ does not exist.}$$

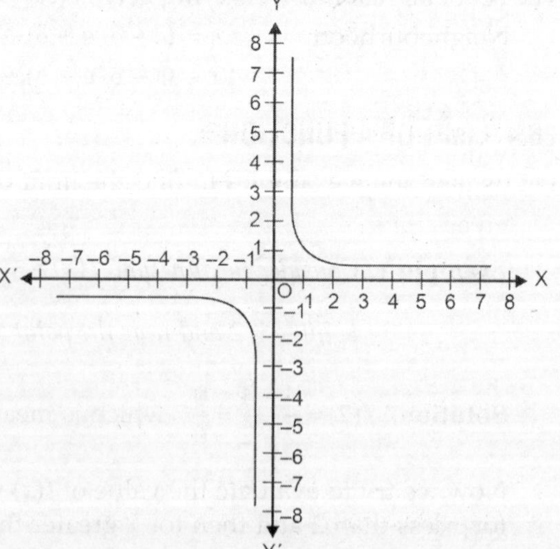

(*ii*) As x approaches zero from right of zero, the graph again does not approach *a* finite number.

Again we say that the right hand limit of f at $x = 0$ does not exist.

$$\Rightarrow \lim_{x \to 0^+} f(x) \text{ does not exist}$$

(*iii*) Left hand limit of f(at $x \to 0^-$) $\neq$ right hand limit of f(at $x \to 0^+$)

Example 3. *Evaluate:* $\lim_{x \to 3} (x+3)$.

Solution. We try to find out the $\lim_{x \to 3} (x+3)$ below:

Let us compute the value of function $f(x)$ for x very near to 3. Some of the points near and to the left of 3 are 2.9, 2.99, 2.999.

Values of the function are given in the table below. Similarly, some of the numbers near and right of 3 are 3.001, 3.01, 3.1. Value of the function at these points are also given in the table.

Increasing x → ← Decreasing x

x	2.9	2.99	2.999	③	3.001	3.01	3.1
$f(x)$	5.9	5.99	5.999	⑥	6.001	6.01	6.1

Increasing $f(x)$ → Limit ← Decreasing $f(x)$

From the table we deduce that the value of $f(x)$ at $x = 3$ should be greater than 5.999 and less than 6.001.

It is reasonable to assume that the value of function $f(x)$ at $x = 3$ from the left of 3 is 5.999.

$$\lim_{x \to 3^-} f(x) = 5.999 \qquad \text{... (1)}$$

Similarly, when x-approaches $x = 3$ from the right, $f(x)$ should be 6.001

$$\lim_{x \to 3^+} f(x) = 6.001 \qquad \text{... (2)}$$

From (1) and (2), we conclude that the limit is equal to 6.

$$\lim_{x \to 3^-} f(x) = \lim_{x \to 3^+} f(x) = \lim_{x \to 3} f(x) = 6$$

Ans.

Example 4. *Evaluate:* $\lim_{x \to 2} \frac{3x^2 - x - 10}{x^2 - 4}$

Solution. Let us try to the find out the limit of $\frac{3x^2 - x - 10}{x^2 - 4}$ below.

Let us compute the value of function $f(x)$ for x very near to 2. Some of the points near and to the left of 2 are 1.9, 1.99, 1.999.

Values of the function are given in the table below. Similarly, some of the numbers near and right of 2 are 2.001, 2.01, 2.1. Values of the function at these points are also given in the table.

Increasing x → ← Decreasing x

x	1.9	1.99	1.999	②	2.001	2.01	2.1
$f(x)$	2.743	2.749	2.7499	(2.75)	2.750	2.7506	2.756

Increasing $f(x)$ → Limit ← Decreasing $f(x)$

From the table we deduce that the value of $f(x)$ at $x = 2$ should be greater than 2.7499 and less than 2.750.

It is reasonable to assume that the value of function $f(x)$ at $x = 2$ from the left of 2 is 2.7499.

$$\therefore \qquad \lim_{x \to 2^-} f(x) = 2.7499 \qquad \text{... (1)}$$

Similarly, when x-approaches $x = 2$ from the right $f(x)$ should be 2.750.

$$\therefore \qquad \lim_{x \to 2^+} f(x) = 2.750 \qquad \text{... (2)}$$

From (1) and (2), we conclude that the limit is equal to 2.75

$$\lim_{x \to 2^-} f(x) = \lim_{x \to 2^+} f(x) = \lim_{x \to 2} f(x) = 2.75 \qquad \textbf{Ans.}$$

Example 5. *Evaluate:* $\lim_{x \to 3} \dfrac{x^4 - 81}{2x^2 - 5x - 3}$

Solution. Let us try to find $\lim_{x \to 3} \dfrac{x^4 - 81}{2x^2 - 5x - 3}$ below.

Let us compute the value of the function $f(x)$ for x very near to 3. Some of the points near and to the left 3 are 2.9, 2.99, 2.999.

Values of the function are given below in the table. Similarly, some numbers near and right of 3 are 3.001, 3.01, 3.1. Values of the function at these points are also tabulated below.

Increasing x → ← Decreasing x

x	2.9	2.99	2.999	(3)	3.001	3.01	3.1
$f(x)$	15.106	15.396	15.429	15.429	15.432	15.462	15.767

Increasing $f(x)$ → Limit ← Decreasing $f(x)$

From the table we deduce that the value of $f(x)$ at $x = 3$ should be greater than 15.426 and less than 15.432

It is resonable to assume that the value of function $f(x)$ at $x = 3$ from the left of 3 is

$$\lim_{x \to 3^-} f(x) = 15.426 \qquad \text{... (1)}$$

Similarly, when x approaches $x = 3$ from the right, $f(x)$ should be

$$\lim_{x \to 3^+} f(x) = 15.432 \qquad \text{... (2)}$$

From (1) and (2) $\lim_{x \to 3^-} f(x) = \lim_{x \to 3^+} f(x) = \lim_{x \to 3} f(x) = \dfrac{15.426 + 15.432}{2} = 15.429$ **Ans.**

Left hand limit of f at $x = a$: When x approaches a from left hand side of a, the function $f(x)$ tends to l "a definite number". This definite number l is said to be the left hand limit of f at $x = a$.

Right hand limit of f at $x = a$: When x approaches a from right hand side of a, the function $f(x)$ tends to l "a definite number". This definite number l is said be the right hand limit of f at $x = a$.

Existence of limit: Left hand limit of $f(x)$, (at $x = a$) = Right and limit of $f(x)$, (at $x = a$). Then the limit of $f(x)$ at $x = a$ exists.

26.5 WORKING RULE FOR EVALUATION OF RIGHT HAND LIMIT

Right hand limit of $f(x)$, when $x \to a = \lim_{x \to a^+} f(x)$

Step 1: Put $x = a + h$ and replace a^+ by $a + h$.

Step 2: Simplify $\lim_{h \to 0} f(a+h)$.

Step 3: Put $h = 0$ in the simplified form of $f(a + h)$

Step 4: The value obtained in step 3 is the right hand limit of $f(x)$ at $x = a$.

Illustration. *Evaluate the right hand limit of the function:*

$$f(x) = \begin{cases} \dfrac{|x-6|}{x-6}, & x \neq 6 \\ 0, & x = 6 \end{cases} \quad \text{at } x = 6$$

Solution. We have,

Right hand limit of $f(x)$ at $x = 6$

$$= \lim_{x \to 6^+} f(x) = \lim_{h \to 0} f(6+h) = \lim_{h \to 0} \frac{|6+h-6|}{6+h-6} = \lim_{h \to 0} \frac{|h|}{h} = \lim_{h \to 0} \frac{h}{h} = \lim_{h \to 0} 1 = 1 \qquad \textbf{Ans.}$$

26.6 WORKING RULE FOR EVALUATION OF LEFT HAND LIMIT

Left hand limit of $f(x)$, when $x \to a = \lim_{x \to a^-} f(x)$

Step 1: Put $x = a - h$ and replace a^- by $a - h$.

Step 2: Simplify $\lim_{h \to 0} f(a-h)$

Step 3: Put $h = 0$ in the simplified value of $f(a - h)$.

Step 4: The value obtained in step 3 is the left hand limit of $f(x)$ at $x = a$.

For Example: *Evaluate the left hand limit of the function:*

$$f(x) = \begin{cases} \dfrac{|x-4|}{x-4}, & x \neq 4 \\ 0, & x = 4 \end{cases} \quad \text{at } x = 4$$

Solution.

Left hand limit of $f(x)$ at $x = 4$

$$= \lim_{x \to 4^-} f(x) = \lim_{h \to 0} f(4-h) = \lim_{h \to 0} \frac{|4-h-4|}{4-h-4} = \lim_{h \to 0} \frac{|-h|}{-h} = \lim_{h \to 0} \frac{h}{-h} = \lim_{h \to 0} (-1) = -1 \qquad \textbf{Ans.}$$

26.7 VARIOUS CASES OF LIMITS

The following possibilities may arise:

(*i*) Left hand limit and right hand limit both exist and are equal.

(*ii*) Left hand limit and right hand limit both exist and are unequal.

(*iii*) The left hand limit exists but the right hand limit does not exist.

(*iv*) The right hand limit exists but the left hand limit does not exist.

(*v*) Neither the left hand limit nor the right hand limit exists.

In other words

A function $f(x)$ is said to possess a right hand limit as x approaches a from values higher than a and is expressed as

$$\lim_{x\to a^+} f(x) \text{ or } \lim_{x\to a^{+0}} f(x)$$

A function $f(x)$ is said to possess a left hand limit as x approaches a from values lower than a, and is expressed as:

$$\lim_{x\to a^-} f(x) \text{ or } \lim_{x\to a^{-0}} f(x).$$

Example 6. *Evaluate:* (*i*) $\lim_{x\to 0} f(x)$ (*ii*) $\lim_{x\to 1} f(x)$, *where* $f(x) = \begin{cases} 2x+3, & x \le 0 \\ 3(x+1), & x > 0 \end{cases}$

Solution. $f(x) = \begin{cases} 2x+3, & x \le 0 \\ 3(x+1), & x > 0 \end{cases}$

(*i*) We have to find $\lim_{x\to 0} f(x)$

$$\text{Left hand limit} = \lim_{x\to 0^-} f(x) = \lim_{x\to 0-h} 2(x)+3 = \lim_{h\to 0}[2(0-h)+3] = 3$$

$$\text{Right hand limit} = \lim_{x\to 0^+} f(x) = \lim_{x\to 0+h} 3(x+1) = \lim_{h\to 0}[3\{(0+h)+1\}] = 3$$

Here, $\lim_{x\to 0^-} f(x) = \lim_{x\to 0^+} f(x) = \lim_{x\to 0} f(x) = 3$ **Ans.**

(*ii*) $\lim_{x\to 1^-} f(x) = \lim_{x\to 1} 3(x+1) = 3(1+1) = 6$ **Ans.**

Example 7. *Evaluate:* $\lim_{x\to 0} f(x)$, *where* $f(x) = \begin{cases} \dfrac{|x|}{x}, & x \ne 0 \\ 0, & x = 0 \end{cases}$

Solution. We have to find $\lim_{x\to 0} f(x)$, where $f(x) = \begin{cases} \dfrac{|x|}{x}, & x \ne 0 \\ 0, & x = 0 \end{cases}$

$$\text{Left hand limit} = \lim_{x\to 0^-} f(x) = \lim_{x\to 0-h} \frac{|x|}{x} = \lim_{h\to 0} \frac{|-h|}{-h} = -1$$

$$\text{Right hand limit} = \lim_{x\to 0^+} f(x) = \lim_{x\to 0+h} \frac{|x|}{x} = \lim_{h\to 0} \frac{|h|}{h} = 1 \Rightarrow \lim_{x\to 0} f(x) \ne \lim_{x\to 0^+} f(x)$$

Limit does not exist at $x = 0$. **Ans.**

Example 8. *Suppose* $f(x) = \begin{cases} a+bx, & x<1 \\ 4, & x=1 \\ b-ax, & x>1 \end{cases}$ *and if* $\lim_{x\to 1} f(x) = f(1)$ *what are possible values of a and b?*

Solution. We have, $f(x) = \begin{cases} a+bx, & x<1 \\ 4, & x=1 \\ b-ax, & x>1 \end{cases}$

Left hand limit = $\lim\limits_{x\to 1^-} f(x) = \lim\limits_{x\to 1-h} (a+bx) = \lim\limits_{h\to 0} [a+b(1-h)] = a+b$

Right hand limit = $\lim\limits_{x\to 1^+} f(x) = \lim\limits_{x\to 1+h} (b-ax) = \lim\limits_{h\to 0} [b-a(1+h)] = b-a$

$\lim\limits_{x\to 1^-} f(x) = \lim\limits_{x\to 1^+} f(x) = \lim\limits_{x\to 1} f(x) = f(1)$

$$a+b = b-a = 4$$

$\Rightarrow$ $\left.\begin{array}{l} a+b=4 \\ b-a=4 \end{array}\right] \Rightarrow a=0, b=4$ **Ans.**

Example 9. *If the function f(x) satisfies* $\lim\limits_{x\to 1}\dfrac{f(x)-2}{x^2-1} = \pi$, *evaluate* $\lim\limits_{x\to 1} f(x)$

Solution. $\lim\limits_{x\to 1}\dfrac{f(x)-2}{x^2-1} = \pi \Rightarrow \dfrac{f(1)-2}{1-1} = \pi \Rightarrow f(1)-2=0 \Rightarrow f(1)=2$ **Ans.**

Example 10. *For what integers m and n, does* $\lim\limits_{x\to 0} f(x)$ *and* $\lim\limits_{x\to 1} f(x)$ *exist.*

$$\textit{exist if } f(x) = \begin{cases} mx^2+n, & x<0 \\ nx+m, & 0\le x\le 1 \\ nx^2+m, & x>1 \end{cases}$$

Solution. We have, $f(x) = \begin{cases} mx^2+n, & x<0 \\ nx+m, & 0\le x\le 1 \\ nx^2+m, & x>1 \end{cases}$

Limit at $x = 0$

$$\lim_{x\to 0^-} f(x) = \lim_{x\to 0-h} (mx^2+n) = \lim_{h\to 0} m(0-h)^2+n = n$$

$$\lim_{x\to 0^+} f(x) = \lim_{x\to (0+h)} (nx+m) = \lim_{h\to 0} [n(0+h)+m] = m$$

$\lim\limits_{x\to 0^-} f(x) = \lim\limits_{x\to 0^+} f(x) \Rightarrow n=m$, $\lim\limits_{x\to 0} f(x)$ exists if $m=n$ **Ans.**

Limit at $x = 1$

$$\lim_{x\to 1^-} f(x) = \lim_{x\to 1-h} (nx+m) = \lim_{h\to 0} n(1-h)+m = n+m$$

$$\lim_{x\to 1^+} f(x) = \lim_{x\to 1+h} (nx^2+m) = \lim_{h\to 0} [n(1+h)^2+m] = n+m$$

Hence, $\lim\limits_{x\to 1^-} f(x) = \lim\limits_{x\to 1^+} f(x) = \lim\limits_{x\to 1} f(x)$ exists for all $m, n \in Z$. **Ans.**

EXERCISE 26.1

Do the limits of the following functions exist? If the limit exists, write down its value.

1. $\lim\limits_{x\to 0}\dfrac{x}{|x|}$ **Ans.** No

2. $\lim\limits_{x\to 2}\dfrac{|x-2|}{x-2}$ **Ans.** No

3. $\lim_{x\to 0} \dfrac{x}{|x|+x^2}$ **Ans.** No

4. $\lim_{x\to 1} \dfrac{1}{|1-x|}$ **Ans.** Yes, ∞

5. $\lim_{x\to 0} \cos\dfrac{1}{x}$ **Ans.** No

6. If $f(x) = \begin{cases} x^2+4, \\ x^3, \end{cases}$ Find $\lim_{x\to 2} f(x) \begin{cases} \text{for } x<2, \\ \text{for } x>2 \end{cases}$ **Ans.** 8

7. If $f(x) = \begin{cases} \cos x, & x \geq 0 \\ x+k, & x<0, \end{cases}$

Find the value of k, given that $\lim_{x\to 0} f(x)$ exists. **Ans.** $k = 1$

HINTS TO THE SELECTED QUESTIONS

2. $f(x) = \dfrac{|x-2|}{x-2}$

$$\text{L.H.L.} = \lim_{x\to 2^-} f(x) = \lim_{h\to 0} \frac{|2-h-2|}{2-h-2} = \lim_{h\to 0} \frac{h}{-h} = -1$$

$$\text{R.H.L.} = \lim_{x\to 2^+} f(x) = \lim_{h\to 0} \frac{|2+h-2|}{2+h-2} = \lim_{h\to 0} \frac{h}{h} = 1$$

As L.H.L ≠ R.H.L ∴ Limit does not exist.

3. $\text{L.H.L.} = \lim_{h\to 0} \dfrac{(0-h)}{|0-h|+(0-h)^2} = \lim_{h\to 0} \dfrac{-h}{h+h^2} = \lim_{h\to 0} \dfrac{-1}{1+h} = \dfrac{-1}{1+0} = -1$

Yes the limit exists and is equal to –1.

$$\text{R.H.L.} = \lim_{h\to 0} \frac{(0+h)}{|0+h|+(0+h)^2} = \lim_{h\to 0} \frac{h}{h+h^2} = \lim_{h\to 0} \frac{1}{1+h} = 1$$

L.H.L ≠ R.H.L., Hence, $\lim_{x\to 0} \dfrac{x}{|x|+x^2}$ does not exist.

5. $\text{L.H.L.} = \lim_{x\to 0^-} \cos\dfrac{1}{x} = \lim_{h\to 0} \cos\left(\dfrac{1}{0-h}\right) = \lim_{x\to 0} \cos\dfrac{1}{h}$ = Oscillates between –1 and 1.

$\text{R.H.L.} = \lim_{x\to 0^+} \cos\dfrac{1}{x} = \lim_{h\to 0} \cos\left(\dfrac{1}{0+h}\right) = \lim_{h\to 0} \cos\dfrac{1}{h}$ = Oscillates between –1 and 1.

∴ $\lim_{x\to 0} \cos\dfrac{1}{x}$ does not exist.

7. $\text{L.H.L.} = \lim_{x\to 0^-} f(x) = \lim_{x\to 0^-} x+k = \lim_{h\to 0} f(0-h)$

$$= \lim_{h\to 0} 0-h+k = k$$

$$\text{R.H.L.} = \lim_{x\to 0^+} f(x) = \lim_{x\to 0^+} \cos x = \lim_{h\to 0} f(0+h) = \lim_{h\to 0} \cos(0+h)$$

$$= \lim_{h\to 0} \cos 0 \cos h - \sin 0 \sin h = \lim_{h\to 0} \cos h = 1$$

As limit exist, L.H.L. must be equal to R.H.L. So, $k = 1$.

Difference between the Value of a Function at a Point and Limit at a Point

Case 1: $\lim_{x\to a}(x)$ *and f(a) both exist but are not equal.*

For example

$$f(x) = \begin{cases} \dfrac{x^2-1}{x-1}, & x \neq 1 \\ 0, & x = 1 \end{cases}$$

$$\lim_{x\to 1} f(x) = \lim_{x\to 1} \frac{x^2-1}{x-1} = \lim_{x\to 1}(x+1) = 2, \ f(x) \text{ exists at } x = 1.$$

$f(1) = 0$, value of f also exists at $x = 1$.

But $\lim_{x\to 1} f(x) \neq f(1)$.

Case 2. $\lim_{x\to a} f(x)$ and $f(a)$ both exist and are equal.

For example $f(x) = x^2$

$\lim_{x\to 1} f(x) = \lim_{x\to 1}(x^2) = 1$, limit exists, and $f(1) = (1)^2 = 1$, value of f also exists.

$$\lim_{x\to 1} f(x) = f(1)$$

26.8 THEOREMS ON LIMITS

Let f and g be two real functions with common domain D, then

(*i*) $\lim_{x\to a}(f+g)(x) = \lim_{x\to a} f(x) + \lim_{x\to a} g(x)$

(*ii*) $\lim_{x\to a}(f-g)(x) = \lim_{x\to a} f(x) - \lim_{x\to a} g(x)$

(*iii*) $\lim_{x\to a}(c\cdot f)(x) = c\lim_{x\to a} f(x)$ [where c is constant]

(*iv*) $\lim_{x\to a}(f\cdot g)x = \lim_{x\to a} f(x)\cdot \lim_{x\to a} g(x)$

(*v*) $\lim_{x\to a}\left(\dfrac{f}{g}\right)(x) = \dfrac{\lim_{x\to a} f(x)}{\lim_{x\to a} g(x)}$ $\left[\text{If } \lim_{x\to a} g(x) \neq 0, \text{ for } x \in D\right]$

26.9 CANCELLATION OF COMMON FACTOR

Factorization Method

Let $\lim_{x\to a} \dfrac{f(x)}{g(x)}$

If by substituting $x = a$, $\dfrac{f(x)}{g(x)}$ reduces to the form $\dfrac{0}{0}$, then $(x - a)$ is a common factor of $f(x)$ and $g(x)$. So we first factorize $f(x)$ and $g(x)$ and then cancel out the common factor to evaluate the limit.

Working Rule

Step 1: Find out $\lim\limits_{x \to a}$, where $\lim\limits_{x \to a} f(x) = 0$ and $\lim\limits_{x \to a} g(x) = 0$

Step 2: Factorize $f(x)$ and $g(x)$.

Step 3: Cancel the common factor(s).

Step 4: Use substitution method to obtain the limit.

Formulae for factorization

(i) $a^2 - b^2 = (a - b)(a + b)$ (ii) $a^3 - b^3 = (a - b)(a^2 + ab + b^2)$

(iii) $a^3 + b^3 = (a + b)(a^2 - ab + b^2)$ (iv) $a^4 - b^4 = (a^2 - b^2)(a^2 + b^2) = (a - b)(a + b)(a^2 + b^2)$

(v) If $f(\alpha) = 0$, then $x - \alpha$ is a factor of $f(x)$.

Example 11. *Evaluate:* $\lim\limits_{x \to 3\sqrt{2}} \dfrac{\sqrt{11 - 2x} - (3 - \sqrt{2})}{x^2 - 18}$ **[S.B.T.E. June, 2017]**

Solution. Here we have: $\lim\limits_{x \to 3\sqrt{2}} \dfrac{\sqrt{11 - 2x} - (3 - \sqrt{2})}{x^2 - 18}$

$$\Rightarrow \lim_{x \to 3\sqrt{2}} \left(\frac{\sqrt{11 - 2x} - (3 - \sqrt{2})}{x^2 - 18} \times \frac{\sqrt{11 - 2x} + (3 - \sqrt{2})}{\sqrt{11 - 2x} + (3 - \sqrt{2})} \right)$$

$$= \lim_{x \to 3\sqrt{2}} \left[\frac{11 - 2x - (9 + 2 - 6\sqrt{2})}{(x^2 - 18)\left[\sqrt{11 - 2x} + (3 - \sqrt{2})\right]} \right]$$

$$= \lim_{x \to 3\sqrt{2}} \left[\frac{-2x + 6\sqrt{2}}{(x^2 - 18)\left[\sqrt{11 - 2x} + (3 - \sqrt{2})\right]} \right]$$

$$= \lim_{x \to 3\sqrt{3}} \left[\frac{-2(x - 3\sqrt{2})}{(x + 3\sqrt{2})(x - 3\sqrt{2})\left[\sqrt{11 - 2x} + (3 - \sqrt{2})\right]} \right]$$

$$\therefore = \frac{-2}{6\sqrt{2}\left[(3 - \sqrt{2}) + (3 - \sqrt{2})\right]} = \frac{-1}{3\sqrt{2}\left[2(3 - \sqrt{2})\right]} = \frac{-1}{6\sqrt{2}(3 - \sqrt{2})} \times \frac{3 + \sqrt{2}}{3 + \sqrt{2}}$$

$$= \frac{-(3 + \sqrt{2})}{42\sqrt{2}} = -\frac{3 + \sqrt{2}}{42\sqrt{2}} \times \frac{\sqrt{2}}{\sqrt{2}}$$

$$= -\frac{(2 + 3\sqrt{2})}{84}$$ **Ans.**

Example 12. *Evaluate:* $\lim_{x\to 0}\dfrac{e^{-x}-1}{x}$ **[S.B.T.E. June, 2017]**

Solution. Here we have $\lim_{x\to 0}\dfrac{e^{-x}-1}{x}$

$$\Rightarrow \quad \lim_{x\to 0}\left[\frac{e^{-x}-1}{-x}\times\frac{-x}{x}\right] \qquad \left[\because \lim_{x\to 0}\frac{e^{x}-1}{x}=1\right]$$

$$\Rightarrow \quad \lim_{x\to 0}\frac{e^{-x}-1}{x}=1\times(-1)=-1$$ **Ans.**

Example 13. *Evaluate:* $\dfrac{x^3-1}{x-1}$

Solution. If we put $x = 1$, the expression $\dfrac{x^3-1}{x-1}$ assumes the indeterminate form $\dfrac{0}{0}$. Therefore $(x - 1)$ is a common factor of $(x^3 - 1)$ and $(x - 1)$. Factorizing the numerator and denominator, we have,

$$\lim_{x\to 1}\frac{x^3-1}{x-1}=\lim_{x\to 1}\frac{(x-1)(x^2+x+1)}{(x-1)}=\lim_{x\to 1}(x^2+x+1) \qquad \left[\frac{0}{0}\text{ form}\right]$$

$$= 1^2 + 1 + 1 = 1 + 1 + 1 = 3$$ **Ans.**

Example 14. *Evaluate:* $\lim_{x\to 5}\dfrac{x^2-9x+20}{x^2-6x+5}$

Solution. If we put $x = 5$, the expression $\dfrac{x^2-9x+20}{x^2-6x+5}$ assumes the indeterminate form $-$. Therefore $(x - 5)$ is a common factor of numerator and denominator both. Factorising the numerator and denominator, we have $\left[\dfrac{0}{0}\text{ form}\right]$

$$\lim_{x\to 5}\frac{x^2-9x+20}{x^2-6x+5}=\lim_{x\to 5}\frac{(x-4)(x-5)}{(x-1)(x-5)}=\lim_{x\to 5}\frac{(x-4)}{(x-1)} \qquad [\text{After cancelling } (x-5)]$$

$$=\frac{5-4}{5-1}=\frac{1}{4}$$ **Ans.**

EXERCISE 26.2

Evaluate the following:

1. $\lim_{x\to 1}\dfrac{x^2-1}{x-1}$ **Ans.** 2
2. $\lim_{x\to 3}\dfrac{x^4-81}{x^2-9}$ **Ans.** 18
3. $\lim_{x\to 4}\dfrac{x^2-16}{\sqrt{x}-2}$ **Ans.** 32
4. $\lim_{x\to -1}\dfrac{x^3+1}{x+1}$ **Ans.** 3

5. $\lim\limits_{x\to 9}\dfrac{x^2-81}{\sqrt{x}-3}$ **Ans.** 108 6. $\lim\limits_{x\to 5}\dfrac{x^2-x-20}{x-5}$ **Ans.** 9

7. $\lim\limits_{x\to 1}\dfrac{x^2-1}{x^2-3x+2}$ **Ans.** – 2 8. $\lim\limits_{x\to \frac{1}{4}}\dfrac{4x-1}{2\sqrt{x}-1}$ **Ans.** 2

9. $\lim\limits_{x\to 0}\dfrac{(1+x)^4-1}{x}$ **Ans.** 4 10. $\lim\limits_{x\to 2}\dfrac{x^2-5x+6}{x^2-4}$ **Ans.** $-\dfrac{1}{4}$

HINTS TO THE SELECTED QUESTIONS

3. $\lim\limits_{x\to 4}\dfrac{(x^2-4^2)}{\sqrt{x}-2}=\lim\limits_{x\to 4}\dfrac{(x-4)(x+4)}{\sqrt{x}-2}=\lim\limits_{x\to 4}\dfrac{(x+4)[(\sqrt{x})^2-2^2]}{\sqrt{x}-2}$

$$=\lim_{x\to 4}\frac{(x+4)(\sqrt{x}+2)(\sqrt{x}-2)}{(\sqrt{x}-2)}=\lim_{x\to 4}(x+4)(\sqrt{x}+2)$$

$$=(4+4)(\sqrt{4}+2)=(8)(2+2)=8(4)=32.$$

8. $\lim\limits_{x\to \frac{1}{4}}\dfrac{(2\sqrt{x})^2-1^2}{2\sqrt{x}-1}=\lim\limits_{x\to \frac{1}{4}}\dfrac{(2\sqrt{x}-1)(2\sqrt{x}+1)}{(2\sqrt{x}-1)}=\lim\limits_{x\to \frac{1}{4}}(2\sqrt{x}+1)$

$$=2\cdot\sqrt{\frac{1}{4}}+1=2\cdot\frac{1}{2}+1=1+1=2.$$

9. $\lim\limits_{x\to 0}\dfrac{(1+x)^4-(1)^4}{x}=\lim\limits_{x\to 0}\dfrac{[(1+x)^2-1)][(1+x)^2+1)]}{x}$

$$=\lim_{x\to 0}\frac{(1+x^2+2x-1)(1+x^2+2x+1)}{x}=\lim_{x\to 0}\frac{(x^2+2x)(2+x^2+2x)}{x}$$

$$=\lim_{x\to 0}\frac{x(x+2)(x^2+2x+2)}{x}=\lim_{x\to 0}(x+2)(x^2+2x+2)$$

$$=(0+2)(0+2.0+2)=(2)(0+2)=(2)(2)=4.$$

26.10 THEOREM: $\lim\limits_{x\to a}\dfrac{x^n-a^n}{x-a}=na^{n-1}$

Let n be any positive integer. Then, $\lim\limits_{x\to a}\dfrac{x^n-a^n}{x-a}=na^{n-1}$

Proof: Putting $x = a + h$, we get

$$\frac{x^n-a^n}{x-a}=\frac{(a+h)^n-a^n}{a+h-a}=\frac{1}{h}\left[(a+h)^n-a^n\right]$$

$$=\frac{1}{h}\left[(a^n+{}^nC_1\,a^{n-1}\,h+\ldots+h^n-a^n\right]\text{[using Binomial theorem]}$$

$$= \frac{1}{h}\left[{}^nC_1\, a^{n-1}\, h + {}^nC_2\, a^{n-2} h^2 + ... + h^n\right]$$

$$= {}^nC_1\, a^{n-1} + {}^nC_2\, a^{n-2}\, h + ... + h^{n-1}$$

$$\therefore \quad \lim_{x \to a} \frac{x^n - a^n}{x - a} = \lim_{h \to 0}\left[{}^nC_1\, a^{n-1} + {}^nC_2\, a^{n-2}\, h + ...\, h^{n-1}\right] = na^{n-1} \quad (\text{as } {}^nC_1 = n) \quad \textbf{Proved}$$

Example 15. *Evaluate:* $\lim_{x \to a} \frac{x^m - a^m}{x^n - a^n}$

Solution. We have, $\lim_{x \to a} \frac{x^m - a^m}{x^n - a^n}$ $\left[\frac{0}{0} \text{ form}\right]$

$$= \lim_{x \to a}\left\{\frac{x^m - a^m}{x - a} \cdot \frac{x - a}{x^n - a^n}\right\} = \lim_{x \to a}\left\{\frac{x^m - a^m}{x - a} \div \frac{x^n - a^n}{x - a}\right\}$$

$$= \lim_{x \to a}\left\{\frac{x^m - a^m}{x - a}\right\} \div \lim_{x \to a}\left\{\frac{x^n - a^n}{x - a}\right\}$$

$$= ma^{m-1} \div na^{n-1} = \frac{ma^{m-1}}{na^{n-1}} = \frac{m}{n} a^{m-1-n+1} = \frac{m}{n} a^{m-n}$$ **Ans.**

Example 16. *Find all possible values of n, if* $\lim_{x \to 2} \frac{x^n - 2^n}{x - 2} = 80,\ n \in N.$

Solution. We have, $\lim_{x \to 2} \frac{x^n - 2^n}{x - 2} = 80$

$\Rightarrow \quad n.2^{n-1} = 80 \qquad \left[\because \lim_{x \to a} \frac{x^n - a^n}{x - a} = na^{n-1}\right]$

$n.2^{n-1} = 5.2^{5-1} \Rightarrow n = 5$ **Ans.**

Example 17. *If* $\lim_{x \to (-a)} \frac{x^7 + a^7}{x + a} = 7$ *find the value of a.*

Solution. We have, $\lim_{x \to (-a)} \frac{x^7 + a^7}{x + a} = 7$

$\Rightarrow \quad \lim_{x \to -a} \frac{x^7 - (-a)^7}{x - (-a)} = 7 \qquad \left[\lim_{x \to a} \frac{x^n - a^n}{x - a} = na^{n-1}\right]$

$\Rightarrow \quad 7\,(-a)^{7-1} = 7\,(-a)^6 = 7 \Rightarrow 7a^6 = 7$

$\Rightarrow \quad a^6 = \frac{7}{7} = 1 \Rightarrow a = \pm 1$ **Ans.**

EXERCISE 26.3

Evaluate the following:

1. $\lim\limits_{x\to 2}\dfrac{x^3-8}{x^2-4}$ **Ans.** 3

2. $\lim\limits_{x\to 2}\dfrac{x^7-128}{x-2}$ **Ans.** 448

3. $\lim\limits_{x\to 1}\dfrac{x^3-1}{x-1}$ **Ans.** 3

4. $\lim\limits_{x\to -1}\dfrac{x^3+1}{x+1}$ **Ans.** 3

5. $\lim\limits_{x\to 2}\dfrac{x^{10}-1024}{x^5-32}$ **Ans.** 64

6. $\lim\limits_{x\to 0}\dfrac{(1-x)^n-1}{x}$ **Ans.** $-n$

7. $\lim\limits_{2x\to -1}\dfrac{8x^3+1}{2x+1}$ **Ans.** 3

8. If $\lim\limits_{x\to 3}\dfrac{x^n-3^n}{x-3}=108$, find the value of n. **Ans.** 4

HINTS TO THE SELECTED QUESTIONS

1. As $\lim\limits_{2x\to a}\dfrac{x^n-a^n}{x^m-a^m}=\dfrac{n}{m}a^{n-m} \Rightarrow \lim\limits_{x\to 2}\dfrac{x^3-8}{x^2-4}=\lim\limits_{x\to 2}\dfrac{x^3-2^3}{x^2-2^2}=\dfrac{3}{2}(2)^{3-2}=\dfrac{3}{2}\times 2=3$

2. As $\lim\limits_{x\to a}\dfrac{x^n-a^n}{x-a}=na^{n-1} \Rightarrow \lim\limits_{x\to 2}\dfrac{x^7-2^7}{x-2}=7(2)^{7-1}=7.2^6=7\times 64=448$

6. Put $1-x=y$, as $x\to 0 \Rightarrow y\to 1$

$$\therefore \lim_{x\to 0}\frac{(1-x)^n-1}{x}=\lim_{y\to 1}\frac{y^n-1}{1-y}=-\lim_{y\to 1}\frac{y^n-1}{1-y}=-n(1)^{n-1}=-n$$

7. $$\lim_{2x\to -1}\frac{(2x)^3+(1)^3}{2x+1}=\frac{2^3}{2}\lim_{x\to -\frac{1}{2}}\frac{x^3+\left(\frac{1}{2}\right)^3}{x+\frac{1}{2}}=2^2\lim_{x\to -\frac{1}{2}}\frac{x^3-\left(-\frac{1}{2}\right)^3}{x-\left(-\frac{1}{2}\right)}$$

$$=4.3\left(-\frac{1}{2}\right)^{3-1}=4.3\left(-\frac{1}{2}\right)^2=3$$

8. $\lim\limits_{x\to 3}\dfrac{x^n-3^n}{x-3}=108$

As $\lim\limits_{x\to a}\dfrac{x^n-a^n}{x-a}=na^{n-1}\Rightarrow \lim\limits_{x\to 3}\dfrac{x^n-3^n}{x-3}=n(3)^{n-1}=108\Rightarrow n(3)^{n-1}=4(3)^{4-1}\Rightarrow n=4.$

To Evaluate the Trigonometric Limit

Case 1. When variable tends to 0

Theorem. Prove that $\lim\limits_{x \to 0} \dfrac{\sin x}{x} = 1$

Proof 1: By geometry

Draw a circle of radius unity and centre at O.

Let $\qquad \angle AOB = x$ radians.

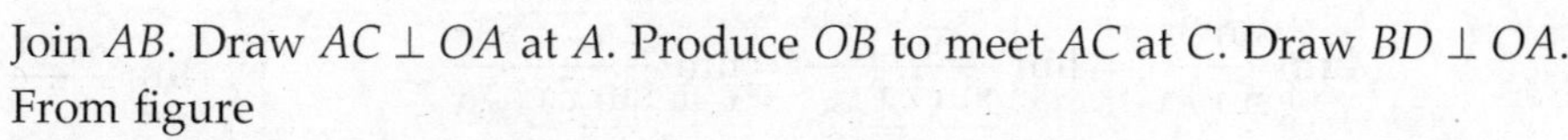

Join AB. Draw $AC \perp OA$ at A. Produce OB to meet AC at C. Draw $BD \perp OA$.

From figure

area of $\Delta OAB <$ area of sector $OBA <$ area of $\Delta\ OAC$.

$$\frac{1}{2}(OA)(BD) < \left(\frac{x}{2\pi}\right)\pi(OA)^2 < \frac{1}{2}(OA)\cdot(AC)$$

$$\Rightarrow \quad \frac{1}{2}[(OA)(OB)\sin x] < \frac{1}{2}(OA)^2 x < \frac{1}{2}(OA)\cdot(OA)\tan x \qquad \left[\because \frac{BD}{OB} = \sin x, \frac{AC}{OA} = \tan x\right]$$

$$\Rightarrow \quad \frac{1}{2}(1)(1)\sin x < \frac{1}{2}(1)^2\cdot x < \frac{1}{2}(1)\cdot\tan x$$

$$\Rightarrow \quad \sin x < x < \tan x$$

$$\Rightarrow \quad \sin x < x < \frac{\sin x}{\cos x}$$

$$\Rightarrow \quad 1 < \frac{x}{\sin x} < \frac{1}{\cos x} \text{ (dividing by } \sin x)$$

$$\Rightarrow \quad \frac{x}{\sin x} \text{ lies between 1 and } \frac{1}{\cos x}.$$

When $x \to 0$, $\cos x = 1$

When $x \to 0$, $\dfrac{x}{\sin x}$ lies between 1 and 1

$$\therefore \quad \lim_{x \to 0} \frac{x}{\sin x} = 1 \text{ or } \lim_{x \to 0} \frac{\sin x}{x} = 1$$

Proved.

Proof 2: By algebra

$$\lim_{x \to 0} \frac{\sin x}{x} = \lim_{x \to 0} \left(\frac{x - \frac{x^3}{3!} + \frac{x^5}{5!} - \dots}{x} \right) \qquad \text{[By sine series]}$$

$$= \lim_{x \to 0} \frac{\left(1 - \frac{x^2}{3!} + \frac{x^4}{5!} - \dots\right)}{x} = \lim_{x \to 0} \left(1 - \frac{x^2}{3!} + \frac{x^4}{5!} - \dots\right) = 1$$

Proved.

Example 18. *Prove that* $\lim\limits_{x \to 0} \dfrac{\tan x}{x} = 1.$

Solution. $\lim_{x\to 0}\frac{\tan x}{x} = \lim_{x\to 0}\frac{\sin x}{x}\cdot\frac{1}{\cos x} = \left(\lim_{x\to 0}\frac{\sin x}{x}\right)\cdot\left(\lim_{x\to 0}\frac{1}{\cos x}\right) = 1\times 1 = 1$ **Ans.**

Example 19. *Evaluate:* $\lim_{x\to 0}\frac{\sin 3x}{\sin 7x}$

Solution. We have,

$$\lim_{x\to 0}\frac{\sin 3x}{\sin 7x} = \lim_{x\to 0}\frac{\frac{\sin 3x}{3x}\cdot 3x}{\frac{\sin 7x}{7x}\cdot 7x} = \lim_{x\to 0}\frac{\frac{\sin 3x}{3x}}{\frac{\sin 7x}{7x}}\cdot\frac{3x}{7x} \qquad \left[\because \lim_{x\to 0}\frac{\sin x}{x} = 1\right]$$

$$= \frac{3}{7}\lim_{x\to 0}\frac{\lim_{x\to 0}\frac{\sin 3x}{3x}}{\lim_{x\to 0}\frac{\sin 7x}{7x}} = \frac{3}{7}\cdot\frac{1}{1} = \frac{3}{7}$$ **Ans.**

Example 20. *Evaluate* $\lim_{x\to 0}\frac{ax + x\cos x}{b\sin x}$

Solution. We have,

$$\lim_{x\to 0}\frac{ax + x\cos x}{b\sin x} = \lim_{x\to 0}\left(\frac{ax}{b\sin x} + \frac{x\cos x}{b\sin x}\right) = \frac{a}{b}\lim_{x\to 0}\frac{x}{\sin x} + \frac{1}{b}\lim_{x\to 0}\frac{x\cos x}{\sin x}$$

$$= \frac{a}{b}(1) + \frac{1}{b}\cdot\lim_{x\to 0}\frac{x}{\tan x} \qquad \left[\because \lim_{x\to 0}\frac{x}{\tan x} = 1\right]$$

$$= \left(\frac{a}{b}\right) + \frac{1}{b}(1) = \frac{a+1}{b}$$ **Ans.**

Example 21. *Evaluate:* $\lim_{x\to 0}\frac{\sin ax + bx}{ax + \sin bx}$; *a, b,* $\neq 0$

Solution. We have,

$$\Rightarrow \quad \lim_{x\to 0}\frac{\sin ax + bx}{ax + \sin bx} = \lim_{x\to 0}\frac{\frac{\sin ax}{x} + \frac{bx}{x}}{\frac{ax}{x} + \frac{\sin bx}{x}} = \frac{\lim_{x\to 0} a \times \frac{\sin ax}{ax} + \lim_{x\to 0} b}{\lim_{x\to 0} a + \lim_{x\to 0} b \times \frac{\sin bx}{bx}}$$

$$\Rightarrow \quad \frac{a\lim_{x\to 0}\frac{\sin ax}{ax} + \lim_{x\to 0} b}{\lim_{x\to 0} a + \lim_{x\to 0} b\frac{\sin bx}{bx}} = \frac{a(1) + b}{a + b(1)} = \frac{a+b}{a+b} = 1 \qquad \left[\because \lim_{x\to 0}\frac{\sin x}{x} = 1\right]$$ **Ans.**

Example 22. *Evaluate:* $\lim_{x\to 0}\frac{1-\cos x}{x^2}$.

Solution. We have, $\lim_{x\to 0}\frac{1-\cos x}{x^2} = \lim_{x\to 0}\frac{2\sin^2\frac{x}{2}}{x^2}$ $\qquad \left[\begin{array}{l}\cos x = 1 - 2\sin^2\frac{x}{2}\\ \Rightarrow 1 - \cos x = 2\sin^2\frac{x}{2}\end{array}\right]$

$$= 2\lim_{x\to 0}\frac{\sin\frac{x}{2}}{x}\cdot\frac{\sin\frac{x}{2}}{x} = 2\lim_{x\to 0}\frac{1}{2}\frac{\sin\frac{x}{2}}{\frac{x}{2}}\cdot\frac{1}{2}\lim_{x\to 0}\frac{\sin\frac{x}{2}}{\frac{x}{2}}$$

$$= \frac{2}{4}\lim_{\frac{x}{2}\to 0}\frac{\sin\frac{x}{2}}{\frac{x}{2}}\cdot\lim_{\frac{x}{2}\to 0}\frac{\sin\frac{x}{2}}{\frac{x}{2}} \qquad \left[\begin{array}{l}\text{As } x\to 0\\ \Rightarrow \frac{x}{2}\to 0\end{array}\right]$$

$$= \frac{1}{2}\cdot(1)\cdot(1) = \frac{1}{2} \qquad \textbf{Ans.} \quad \left[\because \lim_{x\to 0}\frac{\sin x}{x} = 1\right]$$

Example 23. *Evaluate:* $\lim_{x\to 0}\frac{1-\cos mx}{1-\cos nx}$.

Solution. We have,

$$\lim_{x\to 0}\frac{1-\cos mx}{1-\cos nx} = \lim_{x\to 0}\frac{2\sin^2\frac{mx}{2}}{2\sin^2\frac{nx}{2}} = \lim_{x\to 0}\left(\frac{\sin\frac{mx}{2}}{\sin\frac{nx}{2}}\right)^2 \qquad \left[\because \cos x = 1 - 2\sin^2\frac{x}{2}\right]$$

$$= \lim_{x\to 0}\left[\frac{\frac{\sin\frac{mx}{2}}{\frac{mx}{2}}\cdot\frac{mx}{2}}{\frac{\sin\frac{nx}{2}}{\frac{nx}{2}}\cdot\frac{nx}{2}}\right]^2 = \left[\frac{m}{n}\frac{\lim_{x\to 0}\frac{\sin\frac{mx}{2}}{\frac{mx}{2}}}{\lim_{x\to 0}\frac{\sin\frac{nx}{2}}{\frac{nx}{2}}}\right]^2 = \left[\frac{m}{n}\frac{\lim_{\frac{mx}{2}\to 0}\frac{\sin\frac{mx}{2}}{\frac{mx}{2}}}{\lim_{\frac{nx}{2}\to 0}\frac{\sin\frac{nx}{2}}{\frac{nx}{2}}}\right]^2$$

$$\left[\because x\to 0 \Rightarrow \frac{mx}{2}\to 0 \text{ and } \frac{nx}{2}\to 0\right]$$

$$= \left[\frac{m}{n}\cdot\frac{1}{1}\right]^2 = \frac{m^2}{n^2} \qquad \left[\because \lim_{x\to 0}\frac{\sin x}{x} = 1\right] \quad \textbf{Ans.}$$

EXERCISE 26.4

Evaluate the limits of the following functions:

1. $\lim_{x\to 0}\frac{\sin 4x}{\sin 6x}$ **Ans.** $\frac{2}{3}$

2. $\lim_{x\to 0}\frac{\sin px}{\sin qx}$ **Ans.** $\frac{p}{q}$

3. $\lim_{x\to 0}\frac{\tan 8x}{\sin 2x}$ **Ans.** 4

4. $\lim_{\to}\frac{\sin 5}{\tan 3}$ **Ans.** $\frac{5}{3}$

5. $\underset{n\to 0}{lt}\frac{\sin x^\circ}{x}$ **Ans.** $\frac{\pi}{180}$

6. $\lim_{\theta\to 0}\frac{1-\cos 4\theta}{1-\cos 6\theta}$ **Ans.** $\frac{4}{9}$

7. $\lim_{x\to 0}\frac{1-\cos 3x}{x^2}$ **Ans.** $\frac{9}{2}$

8. $\lim_{\theta\to 0}\frac{1-\cos\theta}{2\theta^2}$ **Ans.** –

9. $\lim_{x\to 0}\left\{\dfrac{1-\cos x}{2\sin^2 x}\right\}$ **Ans.** $\dfrac{1}{2}$ **10.** $\lim_{x\to 0}\dfrac{1-\cos 2x}{3\tan^2 x}$. **Ans.** $\dfrac{2}{3}$

HINTS TO THE SELECTED QUESTIONS

1. $$\lim_{x\to 0}\frac{\sin px}{\sin qx}=\lim_{x\to 0}\frac{\dfrac{\sin px}{px}\times px}{\dfrac{\sin qx}{qx}\times qx}=\frac{p}{q}\qquad\left[\text{As }\lim_{x\to 0}\frac{\sin x}{x}=1\right]$$

3. $$\lim_{x\to 0}\frac{\dfrac{\tan 8x}{8x}\times 8x}{\dfrac{\sin 2x}{2x}\times 2x}=\frac{1(8)}{1(2)}=\frac{8}{2}=4\qquad\left[\text{As }\lim_{x\to 0}\frac{\tan x}{x}=1\right]$$

5. As $180° = \pi$ radian, $x° = \dfrac{x\pi}{180}$ radian

So, $$\lim_{x\to 0}\frac{\sin x°}{x}=\lim_{x\to 0}\frac{\sin x\dfrac{\pi}{180}}{x\dfrac{\pi}{180}}\times\frac{\pi}{180}=(1)\frac{\pi}{180}=\frac{\pi}{180}$$

6. $$\lim_{\theta\to 0}\frac{1-\cos 4\theta}{1-\cos 6\theta}=\lim_{\theta\to 0}\frac{1-1+2\sin^2 2\theta}{1-1+2\sin^2 3\theta}=\frac{2\sin^2 2\theta}{2\sin^2 3\theta}=\frac{4}{9}\lim_{\theta\to 0}\left(\frac{\sin 2\theta}{2\theta}\right)^2\left(\frac{3\theta}{\sin 3\theta}\right)^2=\frac{4}{9}$$

9. $$\lim_{x\to 0}\left\{\frac{1-\cos x}{\sin^2 x}\right\}=\lim_{x\to 0}\frac{2\sin^2\dfrac{x}{2}}{\sin^2 x}=\lim_{x\to 0}\frac{\dfrac{2\sin^2\frac{x}{2}}{\frac{x^2}{4}}\times\dfrac{x^2}{4}}{\dfrac{\sin^2 x}{x^2}\times x^2}$$

$$=\lim_{x\to 0}\frac{2\left(\dfrac{\sin\frac{x}{2}}{\frac{x}{2}}\right)^2\times\dfrac{x^2}{4}}{\left(\dfrac{\sin x}{x}\right)^2\times x^2}=\frac{2(1)^2}{(1)^2}\times\frac{1}{4}=\frac{1}{2}$$

Case II. When the variable tends to a nonzero number (but not zero).

Now, we will learn to evaluate the trigonometric limit when the variable tends to nonzero number.

Working rule:

Step 1: Let the variable tends to a. $(x\to a)$

Step 2: Replace x by $a+h$, where $h\to 0$.

Step 3: Now the problem is tranformed in h where $h\to 0$. Use the method already discussed in previous exercise.

Example 24. *Evaluate* $\lim_{x \to 0} \frac{2\sin x - \sin 2x}{x^3}$ **[DIPIETE. June 2019]**

Solution. Here we have $\lim_{x \to 0} \frac{2\sin x - \sin 2x}{x^3} = \lim_{x \to 0} \frac{2\sin x - 2\sin x \cdot \cos x}{x^3}$

$$= \lim_{x \to 0} \frac{2\sin x\,(1 - \cos x)}{x^3} = \lim_{x \to 0} \frac{2\sin x \cdot 2\sin^2 \frac{x}{2}}{x^3}$$

$$= 4 \lim_{x \to 0} \frac{\sin x}{x} \cdot \lim_{x \to 0} \frac{\sin^2 \frac{x}{2}}{x^2}$$

$$= 4 \cdot 1 \lim_{x \to 0} \frac{1 \cdot \sin^2 \frac{x}{2}}{4 \cdot \left(\frac{x}{2}\right)^2} = \frac{4}{4} = 1$$ **Ans.**

Example 25. *Evaluate:* $\lim_{x \to 0} \frac{\sin 3x}{x}$ **[S.B.T.E. June 2017]**

Solution. Here we have

$$\lim_{x \to 0} \frac{\sin 3x}{x} = \lim_{x \to 0} \frac{\sin 3x}{3x} \times 3$$

$$= 3 \cdot (1) = 3$$ **Ans.** $\left[\because \lim_{x \to 0} \frac{\sin x}{x} = 1\right]$

Example 26. *Evaluate:* $\lim_{x \to 0} \left(1 + \frac{2}{x}\right)^{3/x}$ **[S.B.T.E. June 2017, 2016]**

Solution. Here we have $\lim_{x \to 0} \left(1 + \frac{2}{x}\right)^{\frac{3}{x}} = \lim_{x \to 0} \left[\left(1 + \frac{2}{x}\right)^{\frac{x}{2}}\right]^6$ $\left[\because \lim_{x \to 0} \left(1 + \frac{1}{x}\right)^x = e\right]$

$$= (e)^6 = e^6$$ **Ans.**

Example 27. *Evaluate:* $\lim_{x \to 3} \frac{\sqrt{10 - 3x} - \sqrt{2x - 5}}{\sqrt{x + 1} - 2}$ **[S.B.T.E. June 2017]**

Solution. Here we have $\lim_{x \to 3} \frac{\sqrt{10 - 3x} - \sqrt{2x - 5}}{\sqrt{x + 1} - 2}$

$$= \lim_{x \to 3} \left(\frac{\sqrt{10 - 3x} - \sqrt{2x - 5}}{\sqrt{x + 1} - 2} \times \frac{\sqrt{10 - 3x} + \sqrt{2x - 5}}{\sqrt{10 - 3x} + \sqrt{2x - 5}} \times \frac{\sqrt{x + 1} + 2}{\sqrt{x + 1} + 2}\right)$$

$$= \lim_{x \to 3} \frac{(10 - 3x) - (2x - 5)}{(x + 1) - 4} \times \frac{\sqrt{x + 1} + 2}{\sqrt{10 - 3x} + \sqrt{2x - 5}}$$

$$= \lim_{x \to 3} \left[\frac{-5(x - 3)}{x - 3} \times \frac{\sqrt{x + 1} + 2}{\sqrt{10 - 3x} + \sqrt{2x - 5}}\right]$$

$$= -5\times\frac{\sqrt{4}+2}{\sqrt{1}+\sqrt{1}} = -5\times 2 = -10$$ **Ans.**

Example 28. *Evaluate:* $\lim_{x\to 0}\frac{\tan x-\sin x}{x^3}$ **[SBTE. June 2017]**

Solution. Here we have: $\lim_{x\to 0}\frac{\tan x-\sin x}{x^3}$

$$= \lim_{x\to 0}\frac{1}{x^3}\left[\frac{\sin x}{\cos x}-\sin x\right]$$

$$= \lim_{x\to 0}\frac{1}{x^3}\left[\frac{\sin x-\sin x\cdot\cos x}{\cos x}\right]$$

$$= \lim_{x\to 0}\left[\frac{\sin x\,(1-\cos x)}{\cos x}\right]\cdot\frac{1}{x^3}$$

$$= \lim_{x\to 0}\frac{1}{x^3}\left[\frac{\sin x\times 2\sin^2\frac{x}{2}}{\cos x}\right]$$

$$= \lim_{x\to 0}\frac{1}{x^3}\left[\frac{x\times\frac{\sin x}{x}\times 2\left(\frac{\sin\frac{x}{2}}{\frac{x}{2}}\right)^2\times\left(\frac{x}{2}\right)^2}{\cos x}\right]$$

$$= \left(\frac{2}{4}\right)\cdot(1)\cdot(1)(1) = \frac{1}{2}$$ **Ans.**

Example 29. *Evaluate:* $\lim_{h\to 0}\frac{\sin(x+h)-\sin x}{h}$ **[SBTE. June 2016, 2015]**

Solution. Here we have: $\lim_{h\to 0}\frac{\sin(x+h)-\sin x}{h}$

$$= \lim_{h\to 0}\frac{2\cos\left(\frac{x+h+x}{2}\right)\cdot\sin\left(\frac{x+h-x}{2}\right)}{h}$$

$$= \lim_{h\to 0}\frac{2\cos\left(\frac{2x+h}{2}\right)\sin\left(\frac{h}{2}\right)}{2\times\frac{h}{2}}$$

$$= \frac{2\cdot\cos\left(\frac{2x+0}{2}\right)\times 1}{2}$$

$$\therefore\quad \lim_{h\to 0}\frac{\sin(x+h)-\sin x}{h} = \cos x$$ **Ans.**

Example 30. *Write the value of* $\lim\limits_{x\to 0}\dfrac{\sqrt{2+x}-\sqrt{2}}{x}$ **[SBTE. June 2016]**

Solution. Here we have

$$\lim_{x\to 0}\frac{\sqrt{2+x}-\sqrt{2}}{x} = \lim_{x\to 0}\left(\frac{\sqrt{2+x}-\sqrt{2}}{x}\times\frac{\sqrt{2+x}+\sqrt{2}}{\sqrt{2+x}+\sqrt{2}}\right)$$

$$= \lim_{x\to 0}\frac{(2+x)-2}{x(\sqrt{x+2}+2)} = \lim_{x\to 0}\frac{1}{\sqrt{2+x}+\sqrt{2}}$$

$$\therefore \quad \lim_{x\to 0}\frac{\sqrt{2+x}-\sqrt{2}}{x} = \frac{1}{\sqrt{2}+\sqrt{2}} = \frac{1}{4}\sqrt{2}$$ **Ans.**

Example 31. $\lim\limits_{x\to 0}\dfrac{x}{\sin 3x^\circ}$ **[SBTE. May, 2016]**

Solution. Here we have $\lim\limits_{x\to 0}\dfrac{x}{\sin 3x^\circ}$

$$= \lim_{x\to 0}\frac{x}{\sin\left(\frac{3\pi x}{180}\right)} = \lim_{x\to 0}\left(\frac{\frac{3\pi x}{180}}{\sin\frac{3\pi x}{180}}\right)\times\frac{180}{3\pi} = \frac{180}{3\pi}$$ **Ans.**

Example 32. *Evaluate:* $\lim\limits_{x\to 0}\dfrac{\tan x-\sin x}{\sin^3 x}$ **[SBTE. May 2016]**

Solution. Here we have $\lim\limits_{x\to 0}\dfrac{\tan x-\sin x}{\sin^3 x}$

$$= \lim_{x\to 0}\frac{1}{\sin^3 x}\left[\frac{\sin x}{\cos x}-\sin x\right]$$

$$= \lim_{x\to 0}\frac{1}{\sin^3 x}\left[\frac{\sin x(1-\cos x)}{\cos x}\right]$$

$$= \lim_{x\to 0}\frac{1}{\sin^2 x}\left[\frac{2\sin^2\frac{x}{2}}{\cos x}\right] \qquad \left[\because 1-\cos x = 2\sin^2\frac{x}{2}\right]$$

$$= \lim_{x\to 0}\frac{1}{4\sin^2\frac{x}{2}\cdot\cos^2\frac{x}{2}}\times\frac{2\sin^2\frac{x}{2}}{\cos x}$$

$$= \frac{2}{4}\lim_{x\to 0}\frac{1}{\cos^2\frac{x}{2}\cdot\cos x} = \frac{1}{2}\times\frac{1}{1\times 1} = \frac{1}{2}$$

$$\therefore \quad \lim_{x\to 0}\frac{\tan x-\sin x}{\sin^3 x} = \frac{1}{2}\times(1) = \frac{1}{2}$$ **Ans.**

Example 33. *If* $\lim\limits_{x\to 0}\dfrac{x}{\tan a^2 x}=\dfrac{1}{a}$, *find the value of a.* **[SBTE. May, 2016]**

Solution. Here we have $\lim\limits_{x\to 0}\dfrac{x}{\tan a^2 x}=\dfrac{1}{a}$

$$= \lim_{x\to 0}\frac{a^2 x}{a^2(\tan a^2 x)}=\frac{1}{a} \Rightarrow \frac{1}{a^2}=\frac{1}{a}$$

$$a^2 - a = 0$$

$$\therefore \quad a = 0, 1$$ **Ans.**

Example 34. *Evaluate:* $\lim\limits_{h\to 0}\dfrac{(a+h)\sin(a+h)-a\sin a}{h}$ **[SBTE. May 2016]**

Solution. Here we have: $\lim\limits_{h\to 0}\dfrac{(a+h)\sin(a+h)-a\sin a}{h}$

$$= \lim_{h\to 0}\frac{a\sin(a+h)+h\sin(a+h)-a\sin a}{h}$$

$$= \lim_{h\to 0}\left[\frac{a\{\sin(a+h)-\sin a}{h}+\sin(a+h)\right]$$

$$= \lim_{h\to 0}\left[\frac{a\cdot 2\cos\left(\frac{a+h+a}{2}\right)\cdot\sin\left(\frac{a+h-a}{2}\right)}{h}+\sin(a+h)\right]$$

$$= \lim_{h\to 0}\left[\frac{2a\cos\left(a+\frac{h}{2}\right)\sin\frac{h}{2}}{2\cdot\frac{h}{2}}+\sin(a+h)\right]$$

$$\therefore \quad \lim_{x\to 0}\frac{(a+h)\sin(a+h)-a\sin a}{h}=a\cos a+\sin a$$ **Ans.**

Example 35. *Evaluate:* $\lim\limits_{x\to 0}\dfrac{\sin 2x+\sin 6x}{\sin 5x-\sin 3x}$

Solution. Here we have: $\lim\limits_{x\to 0}\dfrac{\sin 2x+\sin 6x}{\sin 5x-\sin 3x}$

$$= \lim_{x\to 0}\frac{2\sin\left(\frac{6x+2x}{2}\right)\cdot\cos\left(\frac{6x-2x}{2}\right)}{2\cos\left(\frac{5x+3x}{2}\right)\cdot\sin\left(\frac{5x-3x}{2}\right)}$$

$$= \lim_{x\to 0}\frac{\sin 4x\cdot\cos 2x}{\cos 4x\cdot\sin x}=\lim_{x\to 0}\frac{\frac{\sin 4x}{4x}\cdot 4x\cdot\cos 2x}{\cos 4x\cdot\frac{\sin x}{x}\cdot x}$$

$$\therefore \quad \lim_{x\to 0}\frac{\sin 2x+\sin 6x}{\cos 5x-\sin 3x}=\frac{1\times 4\times 1}{1\times 1}=4$$ **Ans.**

Example 36. *Evaluate:* $\lim\limits_{x \to \pi} \dfrac{\sin(\pi - x)}{\pi(\pi - x)}$

Solution. We have, $\lim\limits_{x \to \pi} \dfrac{\sin(\pi - x)}{\pi(\pi - x)}$

$$= \frac{1}{\pi} \lim_{x \to \pi} \frac{\sin(\pi - x)}{(\pi - x)} = \frac{1}{\pi} \lim_{h \to 0} \frac{\sin h}{h} \qquad \left[\begin{array}{l} \because x \to \pi \\ \Rightarrow \quad \pi - x \to h \end{array}\right]$$

$$= \frac{1}{\pi} \cdot 1 = \frac{1}{\pi} \qquad \left[\because \lim_{x \to 0} \frac{\sin x}{x} = 1\right] \quad \textbf{Ans.}$$

Example 37. *Evaluate:* $\lim\limits_{x \to \frac{\pi}{4}} \dfrac{\sin x - \cos x}{x - \frac{\pi}{4}}$ **[SBTE. May 2016]**

Solution. We have, $\lim\limits_{x \to \frac{\pi}{4}} \dfrac{\sin x - \cos x}{x - \frac{\pi}{4}}$ $\left[\begin{array}{l} \text{Put } x = \frac{\pi}{4} + h \\ \text{As } x \to \frac{\pi}{4},\ h \to 0 \end{array}\right]$

$$= \lim_{h \to 0} \frac{\sin\left(\frac{\pi}{4} + h\right) - \cos\left(\frac{\pi}{4} + h\right)}{\frac{\pi}{4} + h - \frac{\pi}{4}} \qquad \left[\begin{array}{l} \cos\left(\frac{\pi}{4} + h\right) = \sin\left(\frac{\pi}{4} + h + \frac{\pi}{2}\right) \\ \qquad = \sin\left(+ \frac{}{}\right) \end{array}\right]$$

$$= \lim_{h \to 0} \frac{\sin\left(\frac{\pi}{4} + h\right) - \sin\left(h + \frac{3\pi}{4}\right)}{h} = \lim_{h \to 0} \frac{2\cos\left(h + \frac{\pi}{2}\right)\sin\left(-\frac{\pi}{4}\right)}{h}$$

$$\left[\begin{array}{l} \sin C - \sin D \\ = 2\cos\left(\frac{C + D}{2}\right)\sin\left(\frac{C - D}{2}\right) \end{array}\right]$$

$$= \lim_{h \to 0} \frac{2(-\sin h)\left[\sin\left(-\frac{\pi}{4}\right)\right]}{h} = \lim_{h \to 0} 2\sin\frac{\pi}{4}\left(\frac{\sin h}{h}\right)$$

$$= 2\sin\frac{\pi}{4}(1) = 2 \times \frac{1}{\sqrt{2}} = \sqrt{2} \qquad \textbf{Ans.}$$

EXERCISE 26.5

Evaluate the following limits:

1. $\lim\limits_{x \to \pi} \dfrac{\sin x}{x - \pi}$ **Ans.** – 1

2. $\lim\limits_{x \to \frac{\pi}{2}} \dfrac{2x - \pi}{\cos x}$ **Ans.** – 2

3. $\lim\limits_{x \to \frac{\pi}{2}} \left(\dfrac{\pi}{2} - x\right) \tan x$ **Ans.** 1

4. $\lim\limits_{x \to \pi} \dfrac{\sin 2x}{\sin x}$ **Ans.** – 2

5. $\lim\limits_{x \to \frac{\pi}{2}} \dfrac{\cos^2 x}{1 - \sin x}$ **Ans.** 2

6. *Evaluate*: $\lim\limits_{x \to 0} \dfrac{2^{x-1} - 1}{x - 1}$ **[SBTE. May 2016]** **Ans.** 2 log 2

$$\left[\textbf{Hint: } \lim_{x \to 0} \frac{2^x - 1}{\sqrt{1+x} - 1} \times \frac{\sqrt{1+x} - 1}{\sqrt{1+x} + 1}\right]$$

7. If $\lim\limits_{x \to 0} \dfrac{1 - \cos 2nx}{x^2} = 18$, find the values of n. **[SBTE. June 2015]** **Ans.** $n = \pm 3$

8. *Evaluate*: $\lim\limits_{x \to 0} \dfrac{\sin 2x}{\tan 3x}$ **[SBTE. June 2015]** **Ans.** $\dfrac{2}{3}$

9. *Evaluate*: $\lim\limits_{x \to 1} \dfrac{x^5 - 1}{x - 1}$ **[SBTE. June 2015, 2014]** **Ans.** 5

10. *Evaluate*: $\lim\limits_{h \to 0} \dfrac{\cos h - 1}{h}$ **[SBTE. June 2015, 2014]** **Ans.** 0

$$\left[\textbf{Hint: } \lim_{h \to 0} \frac{\cos h - 1}{h} = \lim_{h \to 0} -\frac{(1 - \cos h)}{h}\right]$$

11. If $\lim\limits_{x \to 0} \dfrac{x^3 - a^3}{x - a} = \lim\limits_{x \to 1} \dfrac{x^4 - 1}{x - 1}$, find all possible values of a. **[SBTE. May/June, 2015]**

$$\left[\textbf{Hint: } \lim_{x \to a} \frac{x^n - a^n}{x - a} = na^{n-1}\right]$$ **Ans.** $a = \pm \dfrac{2}{\sqrt{3}}$

12. *Evaluate*: $\lim\limits_{x \to 0} \dfrac{\sqrt{1 + \sin x} - \sqrt{1 - \sin x}}{x}$ **[SBTE. Dec. 2018] Ans.** $\dfrac{12}{5}$

HINTS TO THE SELECTED QUESTIONS

2. Put $x = \dfrac{\pi}{2} + h$ where $h \to 0$

$$\lim_{h \to 0} \frac{2\left(\frac{\pi}{2} + h\right) - \pi}{\cos\left(\frac{\pi}{2} + h\right)} = \lim_{h \to 0} \frac{\pi + 2h - \pi}{\cos\left(\frac{\pi}{2} + h\right)} = \lim_{h \to 0} \frac{2h}{\cos\left(\frac{\pi}{2} + h\right)} = \lim_{h \to 0} \frac{2h}{-\sin h}$$

$$= \lim_{h \to 0} -2 \frac{1}{\frac{\sin h}{h}} = -2 \cdot \frac{1}{\lim\limits_{h \to 0} \frac{\sin h}{h}} = \frac{-2}{1} = -2 \qquad \left[\because \lim_{h \to 0} \frac{\sin x}{x} = 1\right]$$

3. Put $x = \dfrac{\pi}{2} + h$ where $h \to 0$

$$\lim_{h\to 0}\left(\frac{\pi}{2}-\frac{\pi}{2}-h\right)\tan\left(\frac{\pi}{2}+h\right) = \lim_{h\to 0}(-h)\tan\left(\frac{\pi}{2}+h\right) = \lim_{h\to 0}(-h)(-\cot \text{h})$$

$$= \lim_{h\to 0} h\cot \text{h} = \lim_{h\to 0}\frac{h}{\tan h} = 1.$$

26.11 INFINITE FUNCTIONS

Now, we will discuss the evaluation of the limits of two functions

(*i*) Exponential function

(*ii*) Logarithmic function

1. Exponential functions

$$1+\frac{1}{1!}+\frac{1}{2!}+\frac{1}{3!}+\frac{1}{4!}+\ldots+\frac{1}{n!}+\ldots$$

This infinite series is denoted by e.

The domain of a function $f(x) = e^x$, $x \in R$ is R and the range is the set of positive real numbers.

The graph of $y = e^x$ is shown in the figure.

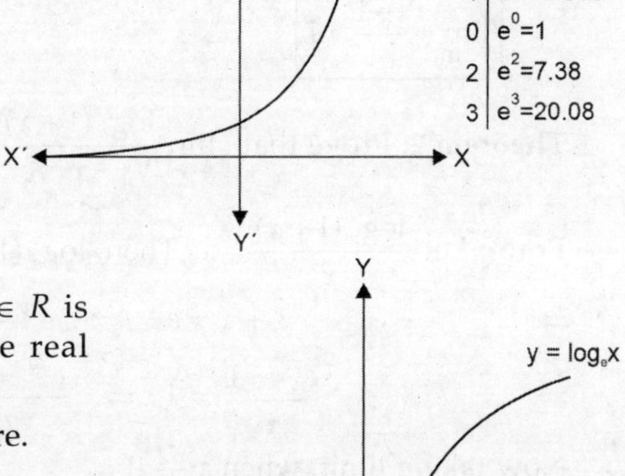

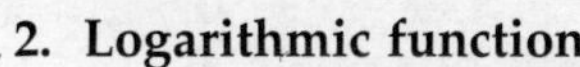

2. Logarithmic function

Let $e^y = x$ can be written as $\log_e x = y$

Its domain is $R^{\oplus}$ and the range is R.

The graph of the logarithmic of function is in the afiacent figure.

26.12 SOME IMPORTANT FUNCTIONS (SERIES)

1. $(1+x)^n = \left\{1 + nx + \dfrac{n(n-1)x^2}{2!} + \dfrac{n(n-1)(n-2)}{3!}x^3 + \ldots\right\}$ $[|x| < 1]$

2. $e^x = 1 + x + \dfrac{x^2}{2!} + \dfrac{x^3}{3!} + \ldots + \dfrac{x^n}{n!} + \ldots$

3. $\log(1+x) = x - \dfrac{x^2}{2} + \dfrac{x^3}{3} - \dfrac{x^4}{4} + \ldots$

4. $\log(1-x) = -x - \dfrac{x^2}{2} - \dfrac{x^3}{3} - \dfrac{x^4}{4} - \ldots$

Theorem 1. Prove that $\displaystyle\lim_{x\to 0}\frac{e^x - 1}{x} = 1$

Proof: We know that

$$e^x = 1 + \frac{x}{1!} + \frac{x^2}{2!} + \frac{x^3}{3!} + \frac{x^4}{4!} + \ldots \Rightarrow e^x - 1 = \frac{x}{1!} + \frac{x^2}{2!} + \frac{x^3}{3!} + \frac{x^4}{4!} + \ldots$$

$$\frac{e^x - 1}{x} = 1 + \frac{x}{2!} + \frac{x^2}{3!} + \frac{x^3}{4!} + \ldots$$

By the important inequality, we have

$$\frac{1}{1+|x|} \le \frac{e^x - 1}{x} \le 1 + (e-2)|x| \qquad x \in [-1, 1] - [0]$$

Also $\lim\limits_{x\to 0} \dfrac{1}{1+|x|} = \dfrac{1}{1+\lim\limits_{x\to 0}|x|} = \dfrac{1}{1+0} = 1$

and $\lim\limits_{x\to 0} [1+(e-2)|x|] = 1 + (e-2)\lim\limits_{x\to 0}|x| = 1+(e-2)0 = 1$

Therefore, by ***Sandwich Theorem***, we get

$$\boxed{\lim_{x\to 0} \frac{e^x - 1}{x} = 1}$$

Proved.

Theorem 2. Prove that $\lim\limits_{x\to 0} \dfrac{\log_e(1+x)}{x} = 1$

Proof: Let $\dfrac{\log_e(1+x)}{x} = y$. Then, $\log_e(1+x) = xy$

$$\Rightarrow \qquad 1 + x = e^{xy} \;\Rightarrow\; e^{xy} - 1 = x$$

$$\Rightarrow \qquad \frac{e^{xy} - 1}{x} = 1 \;\Rightarrow\; \frac{e^{xy} - 1}{xy} \cdot y = 1$$

Now taking limit, when $x \to 0$

$$\lim_{xy\to 0} \frac{e^{xy} - 1}{xy} \cdot \lim_{x\to 0} y = 1 \qquad [\text{Since } x \to 0 \Rightarrow xy \to 0]$$

$$\Rightarrow \qquad 1 \cdot \lim_{x\to 0} y = 1$$

$$\Rightarrow \qquad \lim_{x\to 0} y = 1 \;\Rightarrow\; \boxed{\lim_{x\to \infty} \frac{\log_e(1+x)}{x} = 1}$$

Proved.

Note: If no base of log is mentioned, then it is taken for granted that the base is *e*.
$\therefore$ log *a* is same as $\log_e a$.

Example 38. *Evaluate:* $\lim\limits_{x\to 0} \dfrac{e^x - \sin x - 1}{x}$

Solution. We have $\lim\limits_{x\to 0} \dfrac{e^x - \sin x - 1}{x} = \lim\limits_{x\to 0} \left[\dfrac{e^x - 1}{x} - \dfrac{\sin x}{x}\right]$

$$= \lim_{x\to 0} \frac{e^x - 1}{x} - \lim_{x\to 0} \frac{\sin x}{x} = 1 - 1 = 0$$

Ans.

Example 39. *Evaluate:* $\lim\limits_{x\to 1} \dfrac{\log_e x}{x-1}$

Solution. Put $x = 1 + h$, as $x \to 1 \;\Rightarrow\; h \to 0$. Therefore,

$$\lim_{x\to 1} \frac{\log_e x}{x-1} = \lim_{h\to 0} \frac{\log_e(1+h)}{h} = 1 \qquad \left(\because \lim_{x\to 0} \frac{\log_e(1+x)}{x} = 1\right)$$

Ans.

Example 40. *Evaluate* $\lim_{x\to 0}\frac{e^{4x}-1}{x}$

Solution. $\lim_{x\to 0}\frac{e^{4x}-1}{x} = \lim_{4x\to 0}\frac{e^{4x}-1}{4x}\cdot 4 = (1)\cdot 4 = 4$ **Ans.**

Example 41. *Evaluate* $\lim_{x\to 0}\frac{e^{\sin x}-1}{x}$

Solution. $\lim_{x\to 0}\frac{e^{\sin x}-1}{x} = \lim_{x\to 0}\left(\frac{e^{\sin x}-1}{\sin x}\times\frac{\sin x}{x}\right)$

$$= \lim_{x\to 0}\frac{e^{\sin x}-1}{\sin x}\times\lim_{x\to 0}\frac{\sin x}{x} = \lim_{\sin x\to 0}\frac{e^{\sin x}-1}{\sin x}\times 1 = 1\times 1 = 1$$ **Ans.**

Example 42. *Evalaute* $\lim_{x\to 0}\frac{a^x-b^x}{x}$ **[SBTE. June 2015]**

Solution. $\lim_{x\to 0}\frac{a^x-b^x}{x} = \lim_{x\to 0}\frac{(a^x-1)-(b^x-1)}{x}$

$$= \lim_{x\to 0}\left(\frac{a^x-1}{x}-\frac{b^x-1}{x}\right) = \lim_{x\to 0}\frac{a^x-1}{x} - \lim_{x\to 0}\frac{b^x-1}{x}$$

$$\log a - \log b = \log\left(\frac{a}{b}\right)$$ **Ans.**

Example 43. *Evaluate* $\lim_{x\to 0}\frac{e^x-e^{-x}}{x}$

Solution. $\lim_{x\to 0}\frac{e^x-e^{-x}}{x} = \lim_{x\to 0}\frac{e^x-\frac{1}{e^x}}{x} = \lim_{x\to 0}\frac{e^{2x}-1}{x}\lim_{x\to 0}\frac{1}{e^x} = 2\lim_{x\to 0}\frac{e^{2x}-1}{2x}\frac{1}{e^0}$

$$= 2\,(1)\left(\frac{1}{1}\right) = 2$$ **Ans.**

EXERCISE 26.6

Evaluate the following limits, if exist

1. $\lim_{x\to 0}\frac{e^{2+x}-e^2}{x}$ **Ans.** e^2
2. $\lim_{x\to 3}\frac{e^x-e^3}{x-3}$ **Ans.** e^3
3. $\lim_{x\to 0}\frac{x(e^x-1)}{1-\cos x}$ **Ans.** 2
4. $\lim_{x\to 0}\frac{\log_e(1+2x)}{x}$ **Ans.** 2
5. $\lim_{x\to 0}\frac{e^{ax}-e^{bx}}{x}$ **Ans.** $a-b$
6. $\lim_{x\to\pi/2}\frac{e^{\cos x}-1}{\cos x}$ **Ans.** 1

HINTS TO THE SELECTED QUESTIONS

1. $\lim\limits_{x\to 0}\dfrac{e^{2+x}-e^2}{x}=\lim\limits_{x\to 0}\dfrac{e^2(e^x-1)}{x}$

$$=\lim_{x\to 0}\frac{e^2\left(1+x+\frac{x^2}{2!}+\ldots-1\right)}{x}=\lim_{x\to 0}\frac{e^2\left(x+\frac{x^2}{2!}+\ldots\right)}{x}=e^2\lim_{x\to 0}\left(1+\frac{x}{2!}+\ldots\right)=e^2$$

4. $\lim\limits_{x\to 0}\dfrac{\log_e(1+2x)}{x}=\lim\limits_{x\to 0}\dfrac{2x-\frac{(2x)^2}{2}+\frac{(2x)^3}{3}-\frac{(2x)^4}{4}+\ldots}{x}$

$$=\lim_{x\to 0}\left(2-\frac{4x}{2}+\frac{8x^2}{3}-\frac{16x^3}{4}+\ldots\right)=2$$

Continuity and Differentiability

27.1 INTRODUCTION

Geometrically the definition of continuity of a function $x = a$ implies that the graph of the function has no break at $x = a$. The graph cannot jump immediately to a point above or below the line $y = f(a)$.

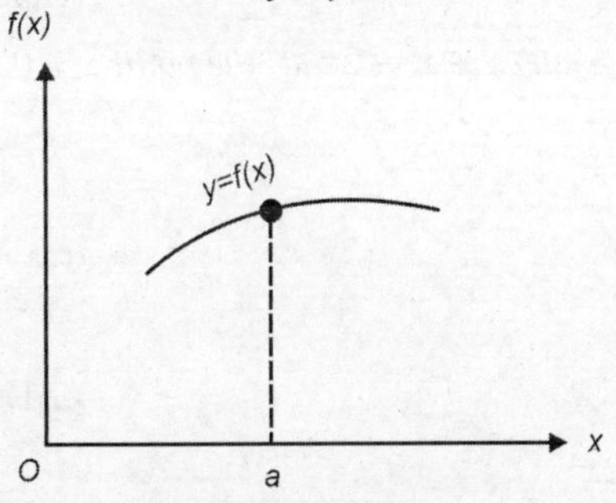

Continuous function at $x=a$

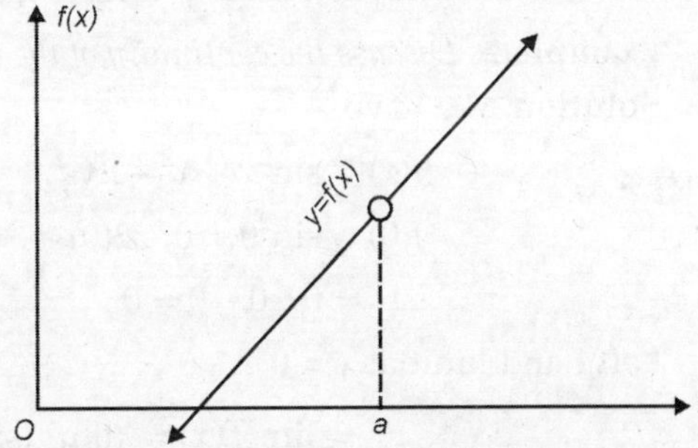

Discontinuous function at $x=a$

27.2 CONDITIONS FOR A FUNCTION TO BE CONTINUOUS

(*i*) Left hand limit of $f(x)$ at $x = a^-$ exists.

(*ii*) Right hand limit of $f(x)$ at $x = a^+$ exists.

(*iii*) Left hand limit of $f(x)$ = Right hand limit of $f(x)$ = Value of the function (at $x = a$).

Note: A function which is not continuous at $x \to a$ is known as discontinuous at $x = a$.

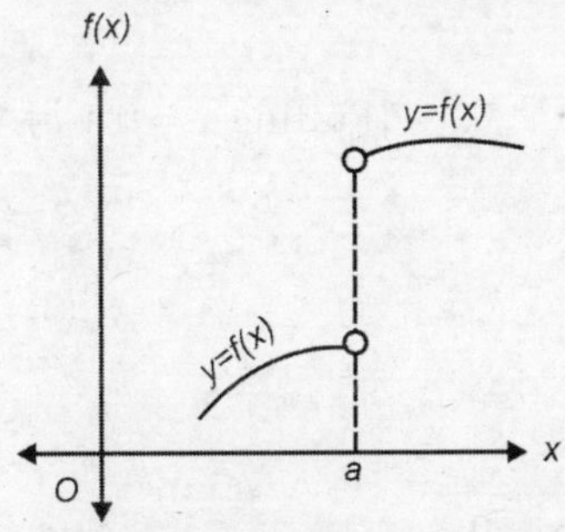

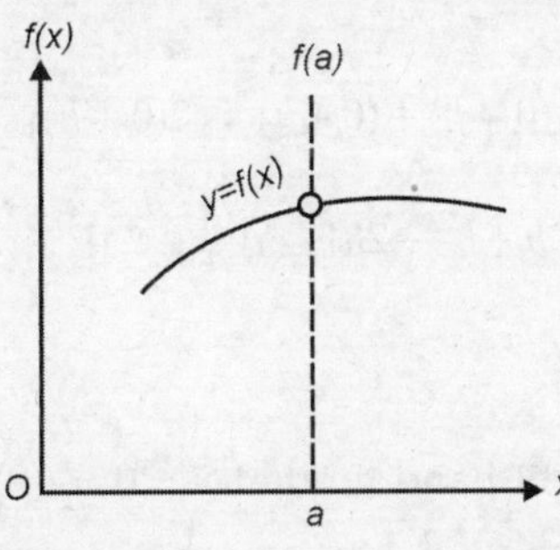

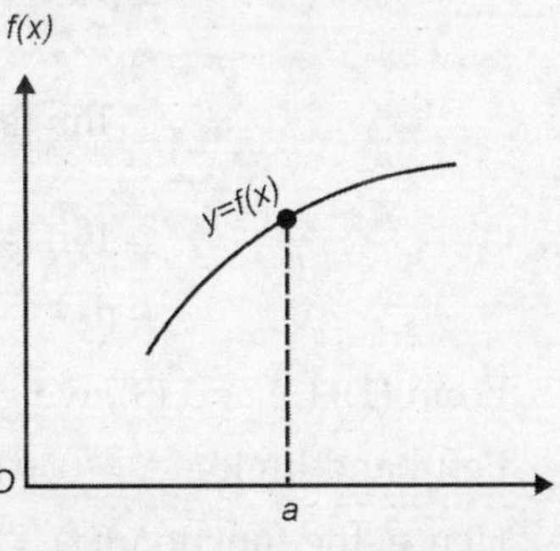

Left hand limit $f(x) \neq$ Right hand limit $f(x)$ Limit $f(x)$ does not exist $x \to a$ $f(x)$ is not continuous at $x = a$	Left hand limit $f(x)$ = Right hand limit $f(x) \neq f(a)$ Limit $f(x)$ exists at $x = a$ $f(x)$ is not continuous at $x = a$	Left hand limit $f(x)$ = Right hand limit $f(x) = f(a)$ The function $f(x)$ is continuous at $x = a$

Example 1. *Find the points at which the function* $f(x) = \dfrac{3x+7}{x^2-5x+6}$ *is continuous.*

Solution: The function $f(x) = \dfrac{3x+7}{x^2-5x+6} = \dfrac{3x+7}{x^2-3x-2x+6}$

$$= \frac{3x+7}{x(x-3)-2(x-3)} = \frac{3x+7}{(x-2)(x-3)}$$

The function is not defined if $x^2 - 5x + 6 = 0 \Rightarrow$ at $x = 2$ and $x = 3$.

The numerator of the function is continuous at every point. Also the function in the denominator is continuous at every point.

$\therefore$ $f(x)$ is continuous at every point $\in$R, except for $x = 2$ and $x = 3$ where it is not defined.

$\therefore$ $f(x)$ is continuous on the point R – {2, 3}. **Ans.**

Example 2. *Discuss the continuity of the function* $f(x) = \sin^2 x + x^2 - 2x$ *at, the point* $x = 0$.

Solution. We have,

$$f(x) = \sin^2 x + x^2 - 2x$$

$$\Rightarrow \quad f(0) = \sin^2 0 + 0 - 2(0)$$

$$= 0 + 0 - 0 = 0 \qquad \text{...(1)}$$

Left hand limit at $x = 0$

$$= \lim_{x\to 0} f(x) = \lim_{x\to(0-h)} (\sin^2 x + x^2 - 2x)$$

$$= \lim_{h\to 0} [\sin^2(0-h) + (0-h)^2 - 2(0-h)] \qquad \text{[Putting } x = 0 - h\text{)}$$

$$= \lim_{h\to 0} [\sin^2 h + h^2 + 2h] = 0 + 0 + 0 = 0 \qquad \text{...(2)}$$

Right hand limit at $x = 0$

$$= \lim_{x\to 0} f(x) = \lim_{x\to(0+h)} [\sin^2 x + x^2 - 2x]$$

$$= \lim_{h\to 0} [\sin^2(0+h) + (0+h)^2 - 2(0+h)] \qquad \text{[Putting } x = 0 + h\text{)}$$

$$= \lim_{h\to 0} [\sin^2 h + h^2 - 2h] = 0 + 0 - 0$$

$$= 0 \qquad \text{...(3)}$$

From (1), (2) and (3), we get

Left hand limit at $x = 0$ = Right hand limit at $x = 0 = f(0)$.

Hence, the function $f(x) = \sin^2 x + x^2 - 2x$ is continuous at $x = 0$. **Ans.**

Example 3. *A function f(x) is defined as*

$$f(x) = \begin{cases} 1+x & whenx < 2 \\ 5-x & whenx \geq 2 \end{cases}$$

Is the function continuous at $x = 2$?

Solution. Left hand limit = $\lim\limits_{x \to 2^-} f(x) = \lim\limits_{x \to 2-h} (1+x) = \lim\limits_{h \to 0}(1+2-h) = 3$ ...(1)

Right hand limit = $\lim\limits_{x \to 2^+} f(x) = \lim\limits_{x \to 2+h} (5-x) = \lim\limits_{h \to 0}(5-2-h) = 3$ ...(2)

$f(x) = 5 - x$

$\Rightarrow f(2) = 5 - 2 = 3$ ...(3)

From (1), (2) and (3), we get

$\lim\limits_{x \to 2^-} f(x) = \lim\limits_{x \to 2^+} f(x) = f(2)$

$\therefore$ The given function $f(x)$ is continuous at $x = 2$ **Ans.**

Example 4. *Show that the function defined by $g(x) = x - [x]$ is discontinuous at all integral points. Here, $[x]$ denotes the greatest integer less than or equal to x.*

Solution. Let $x = n, n \in I$. Then x is an integer.

Left hand limit = $\lim\limits_{\substack{x \to n^- \\ x<n}} \{x - [x]\} = \lim\limits_{x \to n}\{x - (n-1)\} = n - (n-1) = 1$

Right hand limit = $\lim\limits_{\substack{x \to n^+ \\ x>n}} \{x - [x]\} = \lim\limits_{x \to n}\{x - (n-1)\} = n - n = 0$

Since, left hand limit $\neq$ Right hand limit

$\therefore f(x)$ is discontinuous at all integers n.

When x is a real number.

Now, let $x = p$, where $n < p < n + 1$, n being an integer. Then

Left hand limit = $\lim\limits_{\substack{x \to p^- \\ x>p}} \{x - [x]\} = \lim\limits_{x \to p}\{x - n\} = p - n$

Right hand limit = $\lim\limits_{\substack{x \to p^+ \\ x>p}} \{x - [x]\} = \lim\limits_{x \to p}\{x - n\} = p - n$

$f(p) = p - [p] = p - n$

Hence, $f(x)$ is continuous at all non-integral points p. **Proved.**

Example 5. *Discuss whether the function*

$$f(x) = \begin{cases} x \tan^{-1} \dfrac{1}{x} & \text{when } x \neq 0 \\ 0 & \text{when } x = 0 \end{cases}$$

is continuous at $x = 0$

Solution. Left hand limit = $\lim\limits_{x \to 0^-} f(x) = \lim\limits_{x \to 0-h} x \tan^{-1}\left(\dfrac{1}{x}\right)$

$$= \lim_{x \to 0} (0-h)\tan^{-1}\left(\frac{1}{0-h}\right) = \lim_{h \to 0} h \tan^{-1}\frac{1}{h} \quad [\text{Putting } x = 0 - h]$$

$$= 0 \times \frac{\pi}{2} = 0$$

Right hand limit = $\lim_{x\to 0^+} f(x) = \lim_{x\to(0+h)} x\tan^{-1}\frac{1}{x}$

$= \lim_{h\to 0} h\tan^{-1}\frac{1}{h} = 0\times\frac{\pi}{2} = 0$ ...(2) [Putting $x = 0 + h$]

$f(0) = 0$ (given) ...(3)

From (1), (2) and (3), we get

$\lim_{h\to 0^-} f(x) = \lim_{h\to 0^+} f(x) = f(0)$

Hence, the given function is continuous at $x = 0$. **Ans.**

Example 6. *If* $f(x) = \begin{cases} \frac{|x|}{x}, & x \neq 0 \\ 0, & x = 0 \end{cases}$, *find whether f(x) is continuous at x = 0*

Solution. We have,

By definition, $f(0) = 0$ ...(1)

Also, $f(x) = \frac{|x|}{x}\ (x \neq 0)$

$$\Rightarrow \quad f(x) = \begin{cases} \frac{x}{x} = 1 & \text{when } x > 0 \\ \frac{x}{-x} = -1 & \text{when } x < 0 \end{cases}$$

$\therefore \quad \lim_{x\to 0^+} f(x) = 1$ ...(2)

and $\quad \lim_{x\to 0^-} f(x) = -1$ ...(3)

From (1), (2) and (3), we get

$\lim_{x\to 0^-} f(x) \neq \lim_{x\to 0^+} f(x) \neq f(0)$

Hence, f is not continuous at $x = 0$ **Proved.**

Example 7. *If* $f(x) = f(x) = \begin{cases} x\sin\left(\frac{1}{x}\right), & x \neq 0 \\ 0, & x = 0 \end{cases}$

Find whether f(x) is continuous at x = 0.

Solution. We have,

$$f(x) = \begin{cases} x\sin\left(\frac{1}{x}\right), & x \neq 0 \\ 0, & x = 0 \end{cases}$$

$\Rightarrow \quad f(0) = 0$

Left hand limit at $x = 0$ ...(1)

$$\lim_{x\to 0^-} f(x) = \lim_{x\to 0-h} x\sin\left(\frac{1}{x}\right)$$

$$= \lim_{h\to 0}(0-h)\sin\left(\frac{1}{0-h}\right) \qquad \text{[Putting } x = (0-h)\text{]}$$

$$= \lim_{h\to 0} h\sin\left(\frac{1}{h}\right) \qquad [\because \sin(-x) = -\sin x]$$

$$= 0 \qquad \ldots(2)\ [\because 0 \times \text{any number} = 0]$$

Right hand limit at $x = 0$

$$\lim_{h\to 0^+} f(x) = \lim_{x\to 0+h} x\sin\left(\frac{1}{x}\right)$$

$$= \lim_{h\to 0} h\sin\left(\frac{1}{h}\right) = 0 \qquad \ldots(3)\ \text{[Putting } x = (0+h)\text{]}$$

From (1), (2) and (3), we get

$$\lim_{h\to 0^-} f(x) = \lim_{x\to 0^+} f(x) = f(0)$$

Hence, f is continuous at $x = 0$. **Ans.**

Example 8. *A function f(x) is defined as*

$$f(x) = \begin{cases} \dfrac{1}{1+e^{\frac{1}{x}}} & \text{if } x \neq 0 \\ 0 & \text{if } x = 0 \end{cases}$$

Is the function continuous at $x = 0$?

Solution: Left hand limit $= \lim_{x\to 0^-} f(x) = \lim_{x\to 0-h} \dfrac{1}{1+e^{1/x}} = \lim_{h\to 0} \dfrac{1}{1+e^{1/0-h}}$

$$= \lim_{h\to 0} \frac{1}{1+\frac{1}{e^{1/h}}} = \frac{1}{1+\frac{1}{e^{\infty}}} = \frac{1}{1+0} = 1 \qquad \ldots(1)$$

Right hand limit $= \lim_{x\to 0^+} f(x) = \lim_{x\to 0+h} \dfrac{1}{1+e^{1/x}} = \lim_{h\to 0} \dfrac{1}{1+e^{1/0+h}}$

$$= \lim_{h\to 0} \frac{1}{1+e^{1/h}} = \frac{1}{1+e^{\infty}} = 0 \qquad \ldots(2)$$

From (1) and (2), we have

Left hand limit at $x = 0 \neq$ Right hand limit at $x = 0$ $\left[\because \lim_{x\to 0} f(x) \text{ does not exist}\right]$

Hence, the function f is not continuous at $x = 0$. **Ans.**

Example 9. *Show that the function given by*

$$f(x)=\begin{cases}\dfrac{e^{1/x}-1}{e^{1/x}+1} & \text{when } x\neq 0\\ 1 & \text{when } x=0\end{cases}$$

is discontinuous at x = 0.

Solution: Left hand limit = $\lim\limits_{x\to 0^-} f(x)=\lim\limits_{x\to 0-h}\dfrac{e^{\frac{1}{x}}-1}{e^{\frac{1}{x}}+1}=\lim\limits_{h\to 0}\dfrac{e^{1/0-h}-1}{e^{1/0-h}+1}$

$$=\lim_{h\to 0}\frac{e^{-1/h}-1}{e^{-1/h}+1}=\lim_{h\to 0}\frac{\frac{1}{e^{1/h}}-1}{\frac{1}{e^{1/h}}+1}=\frac{\frac{1}{e^{\infty}}-1}{\frac{1}{e^{\infty}}+1}=\frac{0-1}{0+1}=-1$$

Right hand limit = $\lim\limits_{x\to 0^+} f(x)=\lim\limits_{x\to 0+h}\dfrac{\frac{1}{e^{x}}-1}{\frac{1}{e^{x}}+1}=\lim\limits_{h\to 0}\dfrac{e^{1/0+h}-1}{e^{1/0+h}+1}$

$$=\lim_{h\to 0}\frac{1-e^{-\frac{1}{h}}}{1+e^{-\frac{1}{h}}}=\lim_{h\to 0}\frac{1-\frac{1}{e^{1/h}}}{1+\frac{1}{e^{1/h}}}=\frac{1-\frac{1}{e^{\infty}}}{1+\frac{1}{e^{\infty}}}=\frac{1-0}{1+0}=1$$

$\lim\limits_{x\to 0^-} f(x)\neq \lim\limits_{x\to 0^+} f(x)$

∴ The limit of the $f(x)$ at $x = 0$ does not exist.

Hence, the function is discontinuous at $x = 0$. **Proved.**

Example 10. *If the function*

$$f(x)=\begin{cases}3ax+b & \text{if } x>1\\ 11 & \text{if } x=1\\ 5ax-2b & \text{if } x<1.\end{cases}$$

is continuous at x = 1, find the values of a and b.

Solution. Given that the function f is continuous at $x = 1$,

⇒ Left hand limit = Right hand limit = $f(1)$

Now, $f(1) = 11$ ...(1) [∵$f(x)$=11, if $x = 1$]

Left hand limit at $x = 1$ = $\lim\limits_{x\to 1^-} f(x)=\lim\limits_{x\to 1-h}(5ax-2b)=\lim\limits_{h\to 0}[5a(1-h)-2b]$

[Put $x = 1 - h$]

$= 5a\,(1-0) - 2(b) = 5a - 2b$ ...(2)

Right hand limit at $x = 1 = \lim_{x \to 1^+} f(x) = \lim_{x \to 1+h} (3ax + b)$

$= \lim_{h \to 0} [3a(1+h) + b]$ [Put $x = 1 + h$]

$= 3a + b$ ...(3)

From (1), (2) and (3), we get

$$5a - 2b = 11 = 3a + b$$

$\Rightarrow$ $5a - 2b = 11$ ...(4)

and $3a + b = 11$ ...(5)

Solving (4) and (5), we get

$a = 3$ and $b = 2$ **Ans.**

Example 11. *For what value of k is the function.*

$$f(x) = \begin{cases} kx^2, & x \le 2 \\ 3, & x > 2 \end{cases} \quad \text{continuous at } x = 2.$$

Solution. First note that the function is defined at the given point $x = 2$ and its value is $4k$. Then find the limit of the function at $x = 2$.

Left hand limit $= \lim_{x \to 2^-} f(x) = \lim_{x \to 2-h} kx^2 = \lim_{h \to 0} k(2-h)^2$

$= k(2-0)^2 = 4k$ ...(1) [Put $x = 2 - h, h > 0$]

Right hand limit $= \lim_{x \to 2^+} f(x) = \lim_{x \to 2+h} 3 = 3$ ...(2)

Also, $f(2) = k(2)^2 = 4k$

The function $f(x)$ is continuous at $x = 2$.

So, left hand limit = Right hand limt = $f(2)$

$\Rightarrow$ $4k = 3 = 4k \quad \Rightarrow 4k = 3 \quad \Rightarrow k = \frac{3}{4}$

Hence, the function f is continuous for $k = \frac{3}{4}$

Example 12. *For what values of k, is the following function continuous at x = 0*

$$f(x) = \begin{cases} \frac{1 - \cos 4x}{8x^2} & , x \ne 0 \\ k & , x = 0 \end{cases}$$

Solution. Left hand limit $= \lim_{x \to 0^-} f(x) = \lim_{x \to 0^-} \frac{1 - \cos 4x}{8x^2}$

$$= \lim_{x \to 0-h} \frac{2\sin^2 2x}{8x^2} = \lim_{h \to 0} \frac{2\sin^2 2(0-h)}{8(0-h)^2} = \lim_{h \to 0} \left(\frac{\sin 2h}{2h}\right)^2 = 1^2 = 1$$

Right hand limit = $\lim_{x\to 0^+} f(x) = \lim_{x\to 0+h} \frac{1-\cos 4x}{8x^2}$

$$= \lim_{h\to 0}\frac{1-\cos 4(0+h)}{8(0+h)^2} = \lim_{h\to 0}\frac{1-[1-2\sin^2 2h]}{8h^2} = \lim_{h\to 0}\frac{2\sin^2 2h}{8h^2}$$

$$= \lim_{h\to 0}\left(\frac{\sin 2h}{2h}\right)^2 = 1$$

Also, $f(0) = k$

For continuity at $x = 0$, $\lim_{x\to 0} f(x) = f(0) \Rightarrow 1 = k \Rightarrow k = 1.$ **Ans.**

Hence, the function f is continuous, when $k = 1$

Example 13. *Find the value of k so that the function f is continuous at the indicated point:*

$$f(x) = \begin{cases} \dfrac{k\cos x}{\pi - 2x} & \text{if } x \neq \dfrac{\pi}{2} \\ 3 & \text{if } x = \dfrac{\pi}{2} \end{cases} \quad at\ x = \frac{\pi}{2}$$

Solution. We have,

$$f(x) = \begin{cases} \dfrac{k\cos x}{\pi - 2x} & \text{if } x \neq \dfrac{\pi}{2} \\ 3 & \text{if } x = \dfrac{\pi}{2} \end{cases}$$

Left hand limit = $\lim_{x\to \frac{\pi}{2}^-} f(x) = \lim_{x\to \frac{\pi}{2}-h} \frac{k\cos x}{\pi - 2x} = \lim_{h\to 0}\frac{k\cos\left(\frac{\pi}{2}-h\right)}{\pi - 2\left(\frac{\pi}{2}-h\right)}$

$$= \lim_{h\to 0}\frac{k\sin h}{2h} = \frac{k}{2}\lim_{h\to 0}\frac{\sin h}{h} = \frac{k}{2} \qquad \left[\text{Since, } \lim_{h\to 0}\frac{\sin h}{h} = 1\right]$$

Right hand limit = $\lim_{h\to \frac{\pi}{2}^+} f(x) = \lim_{x\to \frac{\pi}{2}+h} \frac{k\cos x}{\pi - 2x}$

$$= \lim_{h\to 0}\frac{k\cos\left(\frac{\pi}{2}+h\right)}{\pi - 2\left(\frac{\pi}{2}+h\right)} = \lim_{h\to 0}\frac{-k\sin h}{-2h}$$

$$= \frac{k}{2}\lim_{h\to 0}\frac{\sin h}{h} = \frac{k}{2} \qquad \left[\because \lim_{h\to 0}\frac{\sin h}{h} = 1\right]$$

Also, $f\left(\frac{\pi}{2}\right) = 3$

For continuity at $x = \frac{\pi}{2}$, $\lim_{x \to \frac{\pi}{2}} f(x) = f\left(\frac{\pi}{2}\right)$

$\Rightarrow$ $\frac{k}{2} = 3$

$\Rightarrow$ $k = 6$ **Ans.**

EXERCISE 27.1

Find all points of discontinuity of f(x), where f(x) is defined by:

1. $f(x) = \begin{cases} 2x+3, & x \le 2 \\ 2x-3, & x > 2 \end{cases}$ **Ans.** Discontinuous at $x = 2$

2. $f(x) = \begin{cases} \frac{x}{|x|} \text{ if } & x < 0 \\ -1 \text{ if } & x \ge 0 \end{cases}$ **Ans.** Continuous at $x = 0$

3. $f(x) = \begin{cases} |x|+3, & \text{if } x \le -3 \\ -2x & \text{if } -3 < x < 3 \\ 6x+2, & \text{if } x > 3 \end{cases}$ **Ans.** Discontinuous at $x = 3$.

4. $f(x) = \begin{cases} x+1, & \text{if } x \ge 1 \\ x^2+1, & \text{if } x < 1 \end{cases}$ **Ans.** Continuous at $x = 1$

5. $f(x) = \begin{cases} x^3 - 3, & \text{if } x \le 2 \\ x^2+1, & \text{if } x < 2, \end{cases}$ **Ans.** Continuous at $x = 2$

6. $f(x) = \begin{cases} x^{10} - 1, & \text{if } x \le 1 \\ x^2, & \text{if } x > 1 \end{cases}$ **Ans.** Not continuous at $x = 1$

7. Is the function defined by

$f(x) = \begin{cases} x+5 & \text{if } x \le 1 \\ x-5 & \text{if } x > 1 \end{cases}$ a continuous function? **Ans.** Not continuous at $x = 1$

Discuss the continuity of the function f(x), where f(x) is defined by:

8. $f(x) = \begin{cases} 3 & \text{if } 0 \le x \le 1 \\ 4 & \text{if } 1 < x < 3 \\ 5 & \text{if } 3 \le x \le 10 \end{cases}$ **Ans.** Not continuous at $x = 3$.

9. $f(x) = \begin{cases} 2x & \text{if } x \le 0 \\ 0 & \text{if } 0 \le x \le 1 \\ 4x & \text{if } x > 1 \end{cases}$ **Ans.** Continuous at $x = 0$ and discontinuous at $x = 1$.

10. $f(x) = \begin{cases} -2 & \text{if } x \le -1 \\ 2x & \text{if } -1 < x \le 1 \\ 2 & \text{if } x > 1 \end{cases}$ **Ans.** f is continuous function.

11. For what value of λ is the function

$$f(x) = \begin{cases} \lambda(x^2 - 2x) & \text{if } x \le 0 \\ 4x + 1 & \text{if } x > 0 \end{cases}$$

continuous at $x = 0$? What about continuity at $x = 1$?

Ans. f is not continuous at $x = 0$ for any value of λ but f is continuous at $x = 1$ for all values of λ.

12. Is the function defined by $x^2 - \sin x + 5$ continuous at $x = \pi$?

Ans. f is continuous at $x = \pi$

13. Discuss the continuity of the cosine, cosecant, secant and cotangent functions.

Ans. Cosine function is continuous for all values of $x \in R$

Secant function is not continuous at $x = \frac{\pi}{2}$ or at $x = (2n + 1)\frac{\pi}{2}$

Cosecant function is not continuous at $x = n\pi$, Cotangent function is continuous at all points $x \in R$, except $x = np$, $n \in Z$

14. Examine the continuity of f, where f is defined by

$$f(x) = \begin{cases} \sin x - \cos x & \text{if } x \ne 0 \\ -1 & \text{if } x = 0 \end{cases}$$

Ans. f is continuous at $x = 0$

15. The function is defined by

$f(x) = \begin{cases} kx + 1, & \text{if } x \le \pi \\ \cos x, & \text{if } x > \pi \end{cases}$ at $x = \pi$, is continuous. Find k. **Ans.** $k = \frac{-2}{\pi}$

16. The function is defined by

$f(x) = \begin{cases} kx + 1, & \text{if } x \le 5 \\ 3x - 5, & \text{if } x > 5 \end{cases}$ at $x = 5$, is continuous. Find k. **Ans.** $k = \frac{9}{5}$

17. Find the values of a and b such that the function defined by

$f(x) = \begin{cases} 5, & \text{if } x \le 2 \\ ax + b, & \text{if } 2 < x < 10 \\ 21, & \text{if } x \ge 10 \end{cases}$ is a continuous function. **Ans.** $a = 2$, $b = 1$.

Find all the points of discontinuity of f defined as follows:

18. $f(x) = |x| - |x + 1|$. **Ans.** Continuous for all $x \in R$.

19. $f(x) = \frac{1}{x}$ **Ans.** Not continuous at $x = 0$

20. $f(x) = \begin{cases} x, & \text{if } x < 0 \\ x^2, & \text{if } x \neq 0 \end{cases}$ at $x = 0$ **Ans.** Continuous at $x = 0$

21. $f(x) = \begin{cases} 2x - 1, & \text{if } x < 0 \\ 2x + 6, & \text{if } x \leq 0 \end{cases}$ at $x = 1$ **Ans.** Not continuous

22. $f(x) = \begin{cases} 5x - 4, & \text{if } x \leq 1 \\ 4x^2 + 3x, & \text{if } x > 1 \end{cases}$ at $x = 1$ **Ans.** Continuous

23. Test the function $f(x) = \dfrac{|2x-3|}{2x-3}$, $x \neq \dfrac{3}{2}$ and $f(3/2) = 0$ for continuity.

Ans. Discontinuous

24. If $f(x) = \begin{cases} \dfrac{x}{\sin 3x}, & \text{when } x \neq 0 \\ 3, & \text{when } x = 0 \end{cases}$ **Ans.** Discontinuous

Find whether function $f(x)$ is continuous at $x = 0$

25. Discuss the continuity of the function defined as

$f(x) = \begin{cases} 1 - x, & x < 0 \\ \sqrt{ax}, & x > 0 \\ 1, & x = 0 \end{cases}$ at $x = 0$ **Ans.** Discontinuous

26. Discuss the continuity of the function $f(x)$ at the point $x = 0$

$f(x) = \begin{cases} x, & x > 0 \\ 1, & x = 0 \\ -x, & x < 0 \end{cases}$ **Ans.** Discontinuous

27. If $f(x) = \begin{cases} \cos x, & 0 \leq x \leq \dfrac{\pi}{2} \\ x - \dfrac{\pi}{2}, & \dfrac{\pi}{2} < x \leq \pi. \\ 1, & x = \dfrac{\pi}{2} \end{cases}$ Discuss the continuity of $f(x)$ at $x = \dfrac{\pi}{2}$

Ans. Discontinuous

28. Examine the continuity of $f(x)$ at $x = 0$

$f(x) = \begin{cases} \dfrac{1}{1 - e^{1/x^2}}, & x \neq 0 \\ 1, & x = 0 \end{cases}$

Ans. Discontinuous at $x = 0$

29. Examine the continuity of $f(x)$ at $x = 0$

$f(x) = \begin{cases} \sin 1/x, & x \neq 0 \\ 0, & x = 0 \end{cases}$ **Ans.** Discontinuous.

30. Discuss the continuity of the following function.

$$f(x) = \begin{cases} \dfrac{x}{\sqrt{x^2}}, & x \neq 0 \\ 1, & x = 0 \end{cases}$$

Ans. Function is discontinuous

31. Prove that the function $f(x) = 5x - 3$ is continuous at $x = 0$, at $x = -3$ and at $x = 5$.

32. Show that $f(x) = [x]$ is not continuous at $x = n$, where n is an integer.

33. Show that the function $f(x) = \begin{cases} \dfrac{x}{|x|}, & x \neq 0 \\ 1, & x = 0 \end{cases}$ is discontinuous at $x = 0$.

Examine the continuity of the following functions:

34. $f(x) = \begin{cases} kx^2, & x \geq 1 \\ 4, & x < 1 \end{cases}$ at $x = 1$ **Ans.** $k = 4$

35. $f(x) = \begin{cases} k(x^2 - 2x), & \text{if } x < 0 \\ \cos x, & \text{if } x \geq 0 \end{cases}$ at $x = 0$ **Ans.** Not continuous for any value of k.

36. $f(x) = \begin{cases} kx + 5, & \text{if } x \leq 2 \\ x - 1, & \text{if } x > 2 \end{cases}$ at $x = 2$ **Ans.** $k = -2$

27.3 SOME CONTINUOUS FUNCTIONS

Theorem 1. *Every constant function is continuous.*

Proof. Let $f(x) = c$, where c is constant.

The domain of a constant function is R.

Let a be an arbitrary number in D, then

$$f(a) = c, \quad \ldots(1)$$

and $\lim_{x \to a} f(x) = \lim_{x \to a} c = c$ ...(2)

From (1) and (2), we have $\lim_{x \to a} f(x) = f(a)$

Thus, $f(x)$ is continuous at $x = a$, for all $a \in R$.

Since, a is an arbitrary number in R,

therefore $f(x)$ is continuous everywhere in R. **Proved**

Theorem 2. *The identity function is continuous.*

Proof. Let $f(x) = x$, for every $x \in R$

Let a be an arbitrary real number

Then, $f(a) = a$ ...(1)

and $\lim_{x \to a} f(x) = \lim_{x \to a} x = a$ ...(2)

From (1) and (2), we get

$$\lim_{x \to a} f(x) = f(a) = a$$

Thus, $f(x)$ is continuous at $x = a$, $a \in R$. Hence the identity function is continuous in R.

Theorem 3. *A polynomial function is everywhere continuous.*

Proof. Let $f(x) = a_0 + a_1x + a_2x^2 + \ldots a_nx^n$, $n \in I, n \geq 0, x \in R$ be a polynomial function. We shall prove the theorem by induction on n.

Step 1. When $n = 0$, we get $f(x) = a_0$, which is a constant function, therefore is continuous.

When $n = 1$, we get $\quad f(x) = a_0 + a_1x$

which is the sum of constant and a multiple of the identity function. It being the sum of two continuous functions is continuous everywhere.

Step 2. Let every polynomial function of degree at most n be everywhere continuous.

Consider a general polynomial function of degree $(n + 1)$, be

$$g(x) = a_0 + a_1x + a_2x^2 + \ldots a_nx^n + a_{n+1}x^{n+1}, \text{ where } a_{n+1} \neq 0$$

$$\Rightarrow \quad g(x) = a_0 + x(a_1 + a_2x + \ldots a_nx^{n-1} + a_{n+1}x^n)$$

It is the sum of a constant function a_0 (which is everywhere continuous) and the product of identity function x (which is everywhere continuous) and the polynomial function $a_1 + a_2x + \ldots a_{n+1}x^n$ of degree at most n (which is everywhere continuous by induction assumption). Therefore, $g(x)$ is everywhere continuous.

Hence, by the principle of mathematical induction, a polynomial function is everywhere continuous.

Theorem 4. *If f and g be continuous functions in D, then,*

(i) $f + g$ *is continuous*   *(ii)* $f - g$ *is continuous.*

(iii) $c f$ *is continuous*   *(iv)* fg is continuous.

(v) $\dfrac{f}{g}$ *is continuous in D except at those points where* $g(x) \neq 0$

(vi) $\dfrac{1}{f}$ *is continuous on* $D - \{x : f(x) \neq 0\}$

Proof. Let a be an arbitrary number in D. Since f and g are continuous on D, so they are continuous at a also.

$$\therefore \quad \lim_{x \to a} f(x) = f(a) \text{ and } \lim_{x \to a} g(x) = g(a)$$

(i) $$\lim_{x \to a}(f + g)(x) = \lim_{x \to a} f(x) + \lim_{x \to a} g(x)$$

$$= f(a) + g(a) = (f + g)(a) \qquad \ldots(1)$$

And $(f + g)(x) = (f + g)(a)$ $\qquad \ldots(2)$

From (1) and (2), we have

$(f + g)$ is continuous at $x = a$

Since a is an arbitrary number in D.

Hence, $(f + g)$ is continuous in D.

(*ii*) $\lim_{x\to a}(f-g)(x) = \lim_{x\to a}[f(x)-g(x)]$

$= \lim_{x\to a} f(x) - \lim_{x\to a} g(x)$

$= f(a) - g(a)$

$= (f-g)(a)$ ...(1)

And $(f-g)(x) = (f-g)(a)$ at $x = a$ ...(2)

From (1) and (2), we get

$\lim_{x\to a}(f-g)(x) = (f-g)(a)$

Thus, $(f-g)$ is continuous at $x = a$.

Since a is an arbitrary number in D.

Hence, $(f-g)$ is continuous in D.

(*iii*) $\lim_{x\to a}(c\,f)(x) = \lim_{x\to a}[c\,f(x)]$

$= c\lim_{x\to a} f(x)$

$= c\,f(a)$ ...(1)

And$(c\,f)(x) = c\,f(x)$

$= c\,f(a)$ at $x = a$

$= (c\,f)\,a$ ...(2)

From (1) and (2), we have

$\lim_{x\to a}(cf)(x) = (cf)(a)$

Thus, $(c\,f)$ is continuous at $x = a$

Since a is an arbitray number in D.

Therefore, $(c\,f)$;is continuous in D.

(*iv*) $\lim_{x\to a}(f.\,g)(x) = \lim_{x\to a} f(x)\cdot g(x)$

$= \lim_{x\to a} f(x)\cdot \lim_{x\to a} g(x)$

$= f(a)\cdot g(a)$

$= (f\cdot g)(a)$ ...(1)

At $x = a$,

$(f\cdot g)(x) = (f\cdot g)(a)$ ...(2)

From (1) and (2), we get

$$\lim_{x\to a}(f\cdot g)(x) = (f\cdot g)(a)$$

Thus, $(f\cdot g)$ is continuous at $x = a$.

Since a is an arbitrary number in D.

Hence, $(f\cdot g)$ is continuous in D.

(*v*) $\lim_{x\to a}\left(\dfrac{f}{g}\right)x = \lim_{x\to a}\dfrac{f(x)}{g(x)} = \dfrac{\lim_{x\to a} f(x)}{\lim_{x\to a} g(x)} = \dfrac{f(a)}{g(a)} = \left(\dfrac{f}{g}\right)(a)$ ...(1)

At $\quad x = a, \left(\frac{f}{g}\right)(x) = \left(\frac{f}{g}\right)(a)$ ...(2)

From (1) and (2), we have

$$\lim_{x \to a}\left(\frac{f}{g}\right)(x) = \left(\frac{f}{g}\right)(a)$$

Thus, $\left(\frac{f}{g}\right)$ is continuous at $x = a$.

Since a is an arbitrary number in D.

Therefore, $\left(\frac{f}{g}\right)$ is continuous in D.

(*vi*) Let $a \in D$ such that $f(a) \neq 0$.

We have:

$$\lim_{x \to a}\left(\frac{1}{f}\right)x = \lim_{x \to a}\frac{1}{f(x)} = \frac{1}{f(a)} = \frac{1}{(f)}(a) \qquad \text{...(1)}$$

$$\left(\frac{1}{f}\right)(x) = \frac{1}{(f)}(a) \qquad \text{at } x = a \qquad \text{...(2)}$$

From (1) and (2), we get

$$\lim_{x \to a}\left(\frac{1}{f}\right)x = \left(\frac{1}{f}\right)(a)$$

Thus $\frac{1}{f}$ is continuous at $x = a$.

Since a is arbitrary point in D such that $f(a) \neq 0$.

Hence, $\frac{1}{f}$ is continuous on $D - \{x : f(x) \neq 0\}$.

Theorem 5. *If f is continuous on its domain D, then $|f|$ is also continuous on D.*

Proof. Recall that $|f|$ (known as absolute function) is defined as:

$$|f|(x) = |f(x)|.$$

Let a be an arbitrary real number in D. Then, f is continuous at a.

$$f(x) = f(a) \text{ at } x = a \qquad \text{...(1)}$$

$$\therefore \quad \lim_{x \to a} f(x) = f(a)$$

Now, $\lim_{x \to a} |f|(x) = \lim_{x \to a} |f(x)|$ [By definition of $|f|$]

$$= \left|\lim_{x \to a} f(x)\right|$$

$$= |f(a)| = |f|(a) \qquad \text{...(2)}$$

From (1) and (2), we have

$$\lim_{x \to a}|f|(x) = f(a)$$

$\therefore$ $|f|$ is continuous at $x = a$.

Since a is an arbitrary point in D. Therefore, $|f|$ is continuous in D. **Proved**

Remark: The converse of the above theorem may not be true.

For example, consider the function

$$f(x) = \begin{cases} 1; & \text{if} \quad x \in I \\ -1, & \text{if} \quad x \in R - I \end{cases}$$

Let a be an arbitrary integer. Then,

Left hand limit: $\lim\limits_{x \to a^-} f(x) = \lim\limits_{h \to 0} f(a-h) = \lim\limits_{h \to 0}(-1) = -1$ ...(1)

[$\because h > 0, a - h \notin I$ as h is very small]

Right hand limit: $\lim\limits_{x \to a^+} f(x) = \lim\limits_{h \to 0} f(a+h) = \lim\limits_{h \to 0}(-1) = -1$ ...(2)

and $f(a) = 1$ ...(3)

From (1), (2) and (3), we have

$\therefore$ $\lim\limits_{x \to a^-} f(x) = \lim\limits_{x \to a^+} f(x) \neq f(a)$

So, f is discontinuous at $x = a$.

Now, $|f|(x) = |f(x)| = 1$ for all $x \in$ R

It is every where continuous. **Ans.**

Example 14. *Examine that sin $|x|$ is a continuous function.*

Solution. We know that, if $f(x)$ is continuous then $f|x|$ is also continuous.

Therefore, sin $|x|$ is continuous as sin x is a continuous function.

Example 15. *Show that the function defined by $f(x) = |\cos x|$ is a continuous function.*

Solution. We have, $f(x) = |\cos x|$

(*i*) We have proved that cos x is a continuous function.

(*ii*) We also have proved that if f is continuous then $|f|$ is also continuous. **(Theorem 5)**

From both the statements, we conclude that $|\cos x|$ is a continuous function. **Proved.**

Theorem 6. *The composition of two continuous functions is a continuous function.*

Proof. Let f and g be two real functions such that gof exists. Then,

Range (f) $\subseteq$ Domain (g)

Let a be an arbitrary point in the domain of f.

Then, $a \in$ Domain (f) $\Rightarrow f(a) \in$ Range (f)

$\Rightarrow$ $f(a) \in$ Domain (g) [$\because$ Range $f \subseteq$ Domain (g)]

Since f and g are continuous on their domains,

therefore $a \in$ domain f and $f(a) \in$ Domain (g)

$\Rightarrow$ f is continuous at $x = a$ and g is continuous at $f(a)$

$\Rightarrow \lim\limits_{x \to a} f(x) = f(a)$ and $\lim\limits_{y \to f(a)} g(y) = g(f(a))$

$\Rightarrow \lim\limits_{x \to a} f(x) = f(a)$ and $\lim\limits_{f(x) \to f(a)} g(f(x)) = g(f(a))$, where $y = f(x)$

$\Rightarrow \lim\limits_{x \to a} g(f(x)) = g(f(a))$ $[\because x \to a \Rightarrow f(x) \to f(a)]$

$\Rightarrow \lim\limits_{x \to a} gof(x) = gof(a)$

$\Rightarrow$ *gof* is continuous at $x = a$.

Since a is an arbitrary point in its domain. Hence, *gof* is continuous. **Proved.**

Theorem 7. The logarithmic function is continuous in its domain.

Proof. Let $f(x) = \log_c x$, where $c > 0$, be the logarithmic function. Domain of f is $(0, \infty)$

Let a be an arbitrary point in $(0, \infty)$, then,

Right hand limit at $x = a$.

$= \lim\limits_{x \to a^+} f(x) = \lim\limits_{h \to 0} f(a+h)$ [Putting $x = a + h$]

$$= \lim_{h \to 0} \log_c (a+h) = \lim_{h \to 0} \log_c a\left(1 + \frac{h}{a}\right)$$

$$= \lim_{h \to 0}\left[\log_c a + \log_c\left(1 + \frac{h}{a}\right)\right] = \lim_{h \to 0} \log_c a + \lim_{h \to 0}\left[\frac{\log_c\left(1 + \frac{h}{a}\right)}{\frac{h}{a}}\right] \cdot \frac{h}{a}$$

$$= \log_c a + \lim_{h \to 0} \frac{\log_c\left(1 + \frac{h}{a}\right)}{\frac{h}{a}} \times \lim_{h \to 0} \frac{h}{a} = \log_c a + 0$$

$= \log_c a = f(a)$ ...(1)

Similarly, $\lim\limits_{x \to a^-} f(x) = f(a)$ ...(2)

$f(x) = \log_c x \longrightarrow f(a) = \log_c a$ ...(3)

From (1), (2) and (3), we get

$$\lim_{x \to a^-} f(x) = \lim_{x \to a^+} f(x) = f(a)$$

So, $f(x)$ is continuous at $x = a$. But a is a real number in $(0, \infty)$

Hence, $\log x$ is continuous in $(0, \infty)$.

Example 16. Discuss the continuity of the following functions

(*a*) $f(x) = \sin x + \cos x$ (*b*) $f(x) = \sin x - \cos x$

(*c*) $f(x) = \sin x \,.\, \cos x$

Solution. We have proved that $\sin x$ and $\cos x$ are the continuous functions.

(*a*) $f(x) = \sin x + \cos x$ being the sum of two continuous functions is continuous wherever it is defined.

(*b*) $f(x) = \sin x - \cos x$ being the difference of the two continuous functions is continuous wherever it is defined.

(*c*) $f(x) = \sin x \,.\, \cos x$ being the product of the two continuous functions is continuous wherever it is defined. **Ans.**

Example 17. *Show that the function f defined by*

$$f(x) = |1 - x + |x||,\ x \in R \text{ is continuous.}$$

Solutions. Let us define two functions g and h

$$g(x) = 1 - x + |x|,\ x \in R$$

and $\quad h(x) = |x|,\ x \in R$

So, $\quad (hog)(x) = h[g(x)]$

$$= h[1 - x + |x|]$$

$$= |1 - x + |x||$$

$$= f(x),\ x \in R$$

As we know that $(1 - x)$ being a polynomial function is continuous. ...(1)

And $|x|$ being a modulus function is continuous. ...(2)

From (1) and (2), we have

$$g(x) = 1 - x + |x| \text{ is continuous}$$

$\therefore \quad |g(x)| = |1 - x + |x||$ is also continuous (by theorem 5) **Proved.**

Theorem 8. *The exponential function a^x, $a > 0$ is everywhere continuous.*

Proof. Let $\quad f(x) = a^x$, we have

$$\lim_{x\to 0} a^x = \lim_{x\to 0}\left[(a^x - 1) + 1\right] = \lim_{x\to 0}\left[\left(\frac{a^x - 1}{x}\right)x + 1\right]$$

$$= \lim_{x\to 0}\left(\frac{a^x - 1}{x}\right).\lim_{x\to 0} x + \lim_{x\to 0} 1 \qquad \left[\lim_{x\to 0}\left(\frac{a^x - 1}{x}\right) = \log_e a\right]$$

$$= \log_e a.\,(0) + 1 = 0 + 1 = 1$$

Let c be an arbitrary real number. Then,

Left hand limit at $x = c$

$$= \lim_{x\to c^-} f(x) = \lim_{h\to 0} f(c - h) \qquad \text{[Putting } x = c - h]$$

$$= \lim_{h\to 0} a^{(c-h)} = \lim_{h\to 0} a^c \,.\, a^{-h} = a^c \lim_{h\to 0}\frac{1}{a^h}$$

$$= a^c.\left(\frac{1}{a^0}\right) = a^c.\left(\frac{1}{1}\right) = a^c = f(c)$$

Right hand limit at $x = c$

$$= \lim_{x\to c^+} f(x) = \lim_{h\to 0} f(c + h) \qquad \text{[Putting } x = c + h]$$

$$= \lim_{h\to 0} a^{c+h} = \lim_{h\to 0} a^c \,.\, a^h = a^c \lim_{h\to 0} a^h$$

$$= a^c \,.\, a^0 = a^c(1) = a^c = f(c)$$

$\Rightarrow$ Left hand limit at $x = c$ = Right hand limit at $x = c = f(c)$

So, $f(x)$ is continuous at $x = c$. Since c is an arbitrary real number.

Hence, $f(x)$ is everywhere continuous. **Proved.**

Corollary: e^x is everywhere continuous.

Example 18. *If a function f is defined as*

$$f(x) = \begin{cases} \dfrac{|x-2|}{x-2}, & x \neq 2 \\ 0, & x = 2 \end{cases}$$

Show that f is everywhere continuous except $x = 2$.

Solution. We have,

$$f(x) = \begin{cases} \dfrac{|x-2|}{x-2}, & x \neq 2 \\ 0, & x = 2 \end{cases}$$

$$\Rightarrow \quad f(x) = \begin{cases} \dfrac{-(x-2)}{x-2} = -1, & x < 2 \\ \dfrac{x-2}{x-2} & x > 2 \\ 0; & x = 2 \end{cases} \qquad \left[\because |x-2| = \begin{cases} -(x-2), & x < 2 \\ x-2, & x > 2 \end{cases}\right]$$

Case 1. When $x < 2$, we have $f(x) = -1$, which , being a constant function, is continuous at each point $x < 2$.

Case 2. Also, when $x > 2$, we have $f(x) = 1$, which, being a constant function, is continuous at each point $x > 2$.

Case 3. Let us consider the point $x = 2$

We have, left hand limit at $x = 2 = \lim_{x\to 2^-} f(x) = \lim_{x\to 2^-} (-1) = -1$ ...(1)

and right hand limit at $x = 2 = \lim_{x\to 2^+} f(x) = \lim_{x\to 2^+} (1) = 1$ ...(2)

Also $f(2) = 0$ ...(3)

From (1), (2) and (3), we have

Left hand limit of $f(x) \neq$ Right hand limit of $f(x) \neq f(2)$

Thus, $f(x)$ is not continuous at $x = 2$

Hence, $f(x)$ is everywhere continuous, except at $x = 2$. **Ans.**

Example 19. *Find the values of a and b, so that the function.*

$$f(x)=\begin{cases} x+a\sqrt{2}\sin x, & 0\le x<\dfrac{\pi}{4} \\ 2x\cot x+b, & \dfrac{\pi}{4}\le x\le\dfrac{\pi}{2} \\ a\cos 2x-b\sin x, & \dfrac{\pi}{2}<x<\pi \end{cases}$$

is continuous for $0\le x\le\pi$.

Solution. Since f is continuous in $[0, \pi]$, therefore f must be continuous at $\dfrac{\pi}{4}$ and $\dfrac{\pi}{2}$.

$\therefore$ when $x=\dfrac{\pi}{4}$

Left hand limit = Right hand limit = $f\left(\dfrac{\pi}{4}\right)$

$$\Rightarrow \lim_{x\to\frac{\pi}{4}^-} f(x)=\lim_{x\to\frac{\pi}{4}^+} f(x)=f\left(\frac{\pi}{4}\right)$$

$$\Rightarrow \lim_{x\to\frac{\pi}{4}^-}\left[x+a\sqrt{2}\sin x\right]=\lim_{x\to\frac{\pi}{4}^+}[2x\cot x+b]=2\frac{\pi}{4}\cot\frac{\pi}{4}+b$$

$$\Rightarrow \lim_{h\to0}\left[\left(\frac{\pi}{4}-h\right)+a\sqrt{2}\sin\left(\frac{\pi}{4}-h\right)\right]=\lim_{h\to0}\left[2\left(\frac{\pi}{4}+h\right)\cot\left(\frac{\pi}{4}+h\right)+b\right]$$

$$=\frac{\pi}{2}\cot\frac{\pi}{4}+b$$

$$\Rightarrow \frac{\pi}{4}+a\sqrt{2}\sin\frac{\pi}{4}=2\left(\frac{\pi}{4}\right)\cot\left(\frac{\pi}{4}\right)+b=\frac{\pi}{2}\cot\frac{\pi}{4}+b$$

$$\Rightarrow \frac{\pi}{4}+a\sqrt{2}\left(\frac{1}{\sqrt{2}}\right)=\frac{\pi}{2}(1)+b$$

$$\Rightarrow \frac{\pi}{4}+a=\frac{\pi}{2}+b \Rightarrow a-b=\frac{\pi}{4} \qquad \ldots(1)$$

Also when $x=\dfrac{\pi}{2}$,

Left hand limit = Right hand limit = $f\left(\dfrac{\pi}{2}\right)$

$$\Rightarrow \lim_{x\to\frac{\pi}{2}^-} f(x)=\lim_{x\to\frac{\pi}{2}^+} f(x)=f\left(\frac{\pi}{2}\right)$$

$$\Rightarrow \lim_{x\to\frac{\pi}{2}^-}(2x\cot x+b)=\lim_{x\to\frac{\pi}{2}^+}(a\cos 2x-b\sin x)=\frac{2\pi}{2}\cot\frac{\pi}{2}+b$$

$$\Rightarrow \lim_{h\to 0}\left[2\left(\frac{\pi}{2}-h\right)\cot\left(\frac{\pi}{2}-h\right)+b\right]=\lim_{h\to 0}\left[a\cos 2\left(\frac{\pi}{2}+h\right)-b\sin\left(\frac{\pi}{2}+h\right)\right]=\pi\cot\frac{\pi}{2}+b$$

$$\Rightarrow \quad \pi\cot\frac{\pi}{2}+b=a\cos\pi-b\sin\frac{\pi}{2}=\pi\cot\frac{\pi}{2}+b$$

$$\Rightarrow \quad 0+b=a(-1)-b(1)=0+b \qquad \left[\because \cot\frac{\pi}{2}=0\right]$$

$$\Rightarrow \quad b=-a-b \Rightarrow a+2b=0 \qquad \ldots(2)$$

Subtracting (2) from (1), we get

$$-3b=\frac{\pi}{4} \Rightarrow b=\frac{-\pi}{12}$$

Putting the value of b in (2), we get

$$a-\frac{\pi}{6}=0 \Rightarrow a=\frac{\pi}{6}$$

Hence, $a=\frac{\pi}{6}, b=\frac{-\pi}{6}$ **Ans.**

EXERCISE 27.2

Discuss the continuity of the following functions:

1. $f(x) = \sin|x|$ **Ans.** Every where continuous

2. $f(x) = \begin{cases} x, & x \geq 0 \\ x^2, & x < 0 \end{cases}$ **Ans.** Continuous at $x = 0$.

3. Is the function f defined by $f(x) = \begin{cases} x, & \text{if } x \leq 1 \\ 5, & \text{if } x > 1 \end{cases}$ continuous at $x = 0$, at $x = 1$, $x = 2$? at $x = 2$
   Ans. Continuous at $x = 0$, at $x = 1$, at $x = 2$.

4. Examine the continuity of the function
   $f(x) = 2x^2 - 1$ at $x = 2$. **Ans.** Continuous at $x = 2$.

5. $f(x) = x - 5$ **Ans.** Continuous

6. $f(x) = \dfrac{1}{x-5}$ **Ans.** Continuous for $x \in R - \{5\}$.

7. $f(x) = \dfrac{x^2-25}{x+5}$ **Ans.** Continuous for all $x \in R - \{-5\}$.

8. $f(x) = |x - 5|$ **Ans.** Continuous for all $x \in R$

9. Prove that the function f(x) = x^n is continuous at $x = n$ when n is a positive integer.

10. Prove that the function $f(x) = 5x - 3$ is continuous at $x = 0$, at $x = -3$ and at $x = 5$.

11. Show that the function defined by $f(x) = \cos x^2$ is a continuous function.

12. Show that the function $f(x) = \begin{cases} \dfrac{x^n-1}{x-1} & \text{when } x \neq 0 \\ n & \text{when } x = 1 \end{cases}$ is continuous.

13. Show that the function $f(x) = \begin{cases} \dfrac{\sin x}{x}, & \text{when } x \neq 0 \\ 2, & \text{when } x = 0 \end{cases}$

is continuous at each point except $x = 0$.

14. Show that $f(x) = \begin{cases} \dfrac{|x-4|}{x-4}, & x \neq 4 \\ 0, & x = 4 \end{cases}$ is continuous at each point except $x = 4$.

15. Show that $f(x) = \begin{cases} \dfrac{x^3 - 64}{x^2 - 16}, & x \neq 4 \\ 1, & x = 4 \end{cases}$ is continuous at each point except $x = 4$.

16. Show that the function $f(x) = \begin{cases} \dfrac{x\left(e^{\frac{1}{x}} - e^{\frac{-1}{x}}\right)}{e^{\frac{1}{x}} + e^{-\frac{1}{x}}}, & x \neq 0 \\ 0, & x = 0 \end{cases}$ is continuous everywhere.

Find the value of constant so that the following functions are continuous:

17. $f(x) = \begin{cases} \dfrac{\sqrt{1+kx} - \sqrt{1-kx}}{x}, & \text{if } -1 \leq x < 0 \\ \dfrac{2x+1}{x-2}, & \text{if } 0 \leq x \leq 1 \end{cases}$ **Ans.** $k = -\dfrac{1}{2}$

18. $f(x) = \begin{cases} x^2 + \lambda, & \text{if } x \geq 0 \\ -x^2 - \lambda, & \text{if } x < 0 \end{cases}$ **Ans.** $\lambda = 0$

19. $f(x) = \begin{cases} x^2 + ax + b, & 0 \leq x < 2 \\ 3x + 2, & 2 \leq x \leq 4 \\ 2ax + 5b, & 4 < x \leq 8 \end{cases}$ is continuous on [0, 8] find the values of a and b.

Ans. $a = 3$, $b = -2$

20. If $f(x) = \begin{cases} \dfrac{x^2}{a}, & 0 \leq x < 1 \\ a, & 1 \leq x < \sqrt{2} \\ \dfrac{2b^2 - 4b}{x^2} & \sqrt{2} \leq x < \infty \end{cases}$

is continuous for $0 \leq x < \infty$, find a, b. **Ans.** $a = -1$, $b = 1$ and $a = 1$, $b = 1 \pm \sqrt{2}$

Multiple Choice Questions

Choose the correct alternative.

1. $\lim\limits_{x\to 0^-} \dfrac{1}{1+e^{\frac{1}{x}}} =$

(*i*) 0 (*ii*) 1 (*iii*) 2 (*iv*) limit does not exist **Ans.** (*ii*)

2. $\lim\limits_{x\to 1} \dfrac{x-1}{\sqrt{x^2+3}-2} =$

(*i*) 1 (*ii*) 2 (*iii*) 3 (*iv*) 4 **Ans.** (*ii*)

3. $\lim\limits_{x\to 0} \dfrac{1-\cos x}{x \sin x} =$

(*i*) 0 (*ii*) 1 (*iii*) $\dfrac{1}{2}$ (*iv*) $\dfrac{1}{4}$ **Ans.** (*iii*)

4. $\lim\limits_{n\to \infty} n\,\{\log(n+1) - \log n\} =$

(*i*) 1 (*ii*) 2 (*iii*) 3 (*iv*) 0 **Ans.** (*i*)

5. $\lim\limits_{x\to y} \dfrac{a^x - a^y}{x-y} =$

(*i*) $y^x \log_e a$ (*ii*) $x^y \log_e a$ (*iii*) $a^x \log_e y$ (*iv*) $a^x \log_e a$ **Ans.** (*iv*)

6. $\lim\limits_{x\to 0} \dfrac{1-\cos x}{\sin^2 x} =$

(*i*) 0 (*ii*) $\dfrac{1}{2}$ (*iii*) 1 (*iv*) 2 **Ans.** (*ii*)

7. $\lim\limits_{x\to \frac{\pi}{2}} (1+\cos x)^{2\sec x} =$

(*i*) e^4 (*ii*) e^{-3} (*iii*) $e^{\frac{1}{2}}$ (*iv*) e^2 **Ans.** (*iv*)

8. $\lim\limits_{n\to \infty} \left(\sqrt{n^2+n} - n\right) =$

(*i*) 3 (*ii*) $\dfrac{1}{2}$ (*iii*) 1 (*iv*) 0 **Ans.** (*ii*)

9. $\lim\limits_{x\to 0} \dfrac{(1+x)^{\frac{1}{x}} - e}{x} =$

(*i*) $-\dfrac{e}{2}$ (*ii*) $\dfrac{e}{2}$ (*iii*) e^2 (*iv*) e^3 **Ans.** (*i*)

10. $\lim\limits_{x\to 0} (1+\sin x)^{\cot x} =$

(*i*) e (*ii*) $e^{\frac{1}{2}}$ (*iii*) $\dfrac{e}{2}$ (*iv*) e^2 **Ans.** (*i*)

11. $\lim_{x\to 0}\left[\frac{1}{\sin x}-\frac{1}{\tan x}+\frac{\tan x}{x}\right]=$

(*i*) 3 (*ii*) 2 (*iii*) 1 (*iv*) 0 **Ans.** (*iii*)

12. $\lim_{x\to 0}\left[\frac{\log(1-x^2)}{\log\cos x}\right]=$

(*i*) 1 (*ii*) 2 (*iii*) 3 (*iv*) 4 **Ans.** (*ii*)

13. $\lim_{x\to 0}\frac{\sqrt{1+x}-1}{\sqrt[3]{1+x}-1}=$

(*i*) $\frac{3}{2}$ (*ii*) $\frac{2}{3}$ (*iii*) $\frac{3}{4}$ (*iv*) limit does not exist **Ans.** (*i*)

14. $\lim_{x\to 0}\frac{e^x-1}{x}=l,$ then l is equal to

(*i*) 0 (*ii*) 1 (*iii*) $\log l$ (*iv*) limit does not exist **Ans.** (*ii*)

15. $\lim_{x\to 0}\frac{\sqrt{1+x+x^2}-1}{\sin 4x}=$

(*i*) $\frac{1}{8}$ (*ii*) $\frac{1}{4}$ (*iii*) $\frac{1}{2}$ (*iv*) 1 **Ans.** (*i*)

16. $\lim_{x\to 0}([x]+1)=$

(*i*) 1 (*ii*) 0 (*iii*) –1 (*iv*) limit does not exist **Ans.** (*iv*)

17. $\lim_{x\to 0}\cos\frac{1}{x}=$

(*i*) 1 (*ii*) 0 (*iii*) 2 (*iv*) limit does not exist **Ans.** (*iv*)

18. $\lim_{x\to 1}\cos[x]=$

(*i*) 0 (*ii*) 1 (*iii*) 2 (*iv*) limit does not exist **Ans.** (*iv*)

19. $\lim_{x\to 0}\frac{x(1-x^2)^{\frac{1}{2x^2}}}{\sin x}=$

(*i*) $e^{-\frac{1}{2}}$ (*ii*) $e^{\frac{1}{2}}$ (*iii*) e^2 (*iv*) limit does not exist **Ans.** (*i*)

20. $\lim_{x\to \frac{\pi}{2}}\frac{1-\sin x}{x-\frac{\pi}{2}}=$

(*i*) –1 (*ii*) 0 (*iii*) 1 (*iv*) limit does not exist **Ans.** (*ii*)

21. $f(x)=\begin{cases}\dfrac{1-\sin x}{(\pi-2x)^2}, & x\neq\dfrac{\pi}{2}\\ k, & x=\dfrac{\pi}{2}\end{cases}$

If $f(x)$ is continuous at $x=\dfrac{\pi}{2}$. Then $k=$

(*i*) $\dfrac{1}{8}$ (*ii*) $\dfrac{1}{3}$ (*iii*) $\dfrac{2}{3}$ (*iv*) 8 **Ans.** (*i*)

22. The function $f(x)=\begin{cases}x^2\sin\dfrac{1}{x}, & \text{if } x\neq 0\\ 0, & \text{if } x=0\end{cases}$ is continuous for

(*i*) all x (*ii*) 0 (*iii*) all integer points (*iv*) None of these

Ans. (*ii*)

23. Let $f(x)=|x|+|x-1|$, then

(*i*) $f(x)$ is continuous at $x=0$ as well as at $x=1$

(*ii*) $f(x)$ is continuous at $x=0$ but not at $x=1$

(*iii*) $f(x)$ is continuous at $x=1$ but not at $x=0$

(*iv*) none of these **Ans.** (*i*)

24. The function $f(x)=[x]\cos\left(\dfrac{2x-1}{2}\right)\pi$, where [ ] denotes the greatest integer function is discontinuous at:

(*i*) all x (*ii*) all integer points

(*iii*) no x (*iv*) x which is not an integer. **Ans.** (*iii*)

25. The point of discontinuity of the function $f(x)=\begin{cases}\dfrac{|x^2-1|}{x-1}, & x\neq 1\\ 2, & x=1\end{cases}$

(*i*) $x=1$ (*ii*) all x (*iii*) all integer points (*iv*) None of these

Ans. (*i*)

26. If $f(x)=\begin{cases}\dfrac{\log(1+ax)-\log(1-bx)}{x}, & x\neq 0\\ k, & x=0\end{cases}$

and $f(x)$ is continuous at $x=0$, then the value of k is

(i) $a-b$ (ii) $a+b$ (*iii*) $\log a+\log b$ (*iv*) none of these **Ans.** (*ii*)

27. If the function $f(x)=\begin{cases}(\cos x)^{1/x}, & x\neq 0\\ k, & x=0\end{cases}$ is continuous at $x=0$, then the value of k is

(*i*) 0 (*ii*) 1 (*iii*) –1 (*iv*) e **Ans.** (*ii*)

28. To make $(x+1)^{\cot x}$ continuous at $x=0$, $f(0)$ must be defined as

(*i*) 0 (*ii*) e (*iii*) $1/e$ (*iv*) 1 **Ans.** (*ii*)

29. If the function $f(x) = \begin{cases} (1+|\sin x|)^{\frac{a}{|\sin x|}}, & -\frac{\pi}{6} < x < 0 \\ b & x = 0 \\ e^{\frac{\tan 2x}{\tan 3x}}, & 0 < x < \frac{\pi}{6} \end{cases}$ is continuous at $x = 0$, then

(*i*) $a = \log_e b,\ a = \frac{3}{2}$ (*ii*) $b = \log_e b,\ a = \frac{2}{3}$

(*iii*) $a = \log_e b,\ a = 2$ (*iv*) None of these **Ans.** (*i*)

30. The value of a, b, c for which the function $f(x) = \begin{cases} \frac{\sin(a+1)x + \sin x}{x}, & x < 0 \\ c, & x = 0 \\ \frac{\sqrt{x+bx^2} - \sqrt{x}}{bx^{\frac{3}{2}}}, & x > 0 \end{cases}$

is continuous at $x = 0$, are

(*i*) $a = \frac{-3}{2}$, $b = R - (0)$, $c = \frac{1}{2}$ (*ii*) $a = \frac{3}{2}$, $b = R$, $c = \frac{1}{2}$

(*iii*) $a = \frac{1}{2}$, $b = R - (0)$, $c = \frac{1}{2}$ (*iv*) $a = \frac{3}{2}$, $b = R - (0)$, $c = \frac{3}{2}$ **Ans.** (*i*)

31. If $f(x) = \begin{cases} k(x^2 - 2x), & \text{if } x \le 0 \\ 4x + 1, & \text{if } x > 0 \end{cases}$ is continuous at $x = 0$, then $k =$

(*i*) 1 only (*ii*) 2 only (*iii*) 3 only (*iv*) any value **Ans.** (*iv*)

32. If $f(x) = \begin{cases} \frac{1-\cos 4x}{x^2}, & \text{if } x < 0 \\ a, & \text{if } x = 0 \\ \frac{\sqrt{x}}{\sqrt{16+\sqrt{x}} - 4}, & \text{if } x > 0 \end{cases}$ is continuous at $x = 0$, then the value of a is

(*i*) 2 (*ii*) 4 (*iii*) 6 (*iv*) 8 **Ans.** (*iv*)

33. If the function $f(x) = \begin{cases} 3ax + b, & \text{for } x > 1 \\ 11, & \text{for } x = 1 \\ 5ax - 2b, & \text{for } x < 1 \end{cases}$

is continuous at $x = 1$, then the values of a and b are

(*i*) $a = 2, b = 3$ (*ii*) $a = -3, b = 2$

(*iii*) $a = 3, b = 2$ (*iv*) $a = 2, b = -3$ **Ans.** (*iii*)

34. If $f(x) = \begin{cases} a \sin \frac{\pi}{2}(x+1), & x \le 0 \\ \frac{\tan x - \sin x}{x^3} & x > 0 \end{cases}$

is continuous at $x = 0$, then a equals

(*i*) $\frac{1}{6}$ (*ii*) $\frac{1}{4}$ (*iii*) $\frac{1}{3}$ (*iv*) $\frac{1}{2}$ **Ans.** (*iv*)

35. If $f(x) = \begin{cases} ax^2 + b, & 0 \le x < 1 \\ 4, & x = 1 \\ x + 3, & 1 < x \le 2 \end{cases}$

then the value of (a, b) for which $f(x)$ cannot be continuous at $x = 1$, is

(*i*) (4, 0) (*ii*) (5, 2) (*iii*) (3, 1) (*iv*) (2, 2) **Ans.** (*ii*)

36. If $f(x) = \begin{cases} \frac{1 - \cos 10\,x}{x^2}, & x < 0 \\ a, & x = 0, \\ \frac{\sqrt{x}}{\sqrt{625 + \sqrt{x}} - 25} & x > 0 \end{cases}$

then the value of a so that $f(x)$ may be continuous at $x = 0$, is

(*i*) 50 (*ii*) –25 (*iii*) 25 (*iv*) None of these **Ans.** (*i*)

37. The point at which the function $\frac{x+1}{(x-2)(x-3)}$ is continuous is

(*i*) for all $x \in R - \{3, 4\}$ (*ii*) for all $x \in R$

(*iii*) for all $x \in R - \{2, 3\}$ (*iv*) None of these **Ans.** (*iii*)

38. The values of a and b for which the funciton $f(x) = \begin{cases} x + a\sqrt{2} \sin x, & 0 \le x < \frac{\pi}{4} \\ 2x \cot x + b, & \frac{\pi}{4} \le x \le \frac{\pi}{2} \\ a \cos 2x - b \sin x, & \frac{\pi}{2} < x \le \pi \end{cases}$

is continuous for $0 \le x \le \pi$ are

(*i*) $a = \frac{\pi}{3}, b = -\frac{\pi}{12}$ (*ii*) $a = \frac{\pi}{6}, b = -\frac{\pi}{12}$

(*iii*) $a = -\frac{\pi}{6}, b = \frac{\pi}{12}$ (*iv*) None of these **Ans.** (*ii*)

39. The function $f(x) = \dfrac{x^3 + x^2 - 16x + 20}{x - 2}$ is not defined for $x = 2$. In order to make $f(x)$ continuous at $x = 2$, $f(2)$ should be defined as

(*i*) 0 (*ii*) 2 (*iii*) 4 (*iv*) 6 **Ans.** (*i*)

40. If the function $f(x) = \begin{cases} kx^2, & x \le 2 \\ 3, & x > 2 \end{cases}$ is continuous at $x = 2$, then $k =$

(*i*) $\dfrac{4}{3}$ (*ii*) $\dfrac{3}{4}$ (*iii*) $\dfrac{2}{3}$ (*iv*) $\dfrac{5}{4}$ **Ans.** (*ii*)

41. If the funciton $f(x) = \begin{cases} 3x - 8, & x \le 5 \\ 2k, & x > 5 \end{cases}$ is continuous at $x = 5$, then $k =$

(*i*) $\dfrac{7}{2}$ (*ii*) $\dfrac{7}{3}$ (*iii*) $\dfrac{2}{7}$ (*iv*) $\dfrac{5}{2}$ **Ans.** (*i*)

42. If the funciton $f(x) = \begin{cases} x^2 + k, & x \ge 0 \\ -x^2 - k, & x < 0 \end{cases}$ is continuous at $x = 0$, then $k =$

(*i*) – 1 (*ii*) 1 (*iii*) 0 (*iv*) 2 **Ans.** (*iii*)

27.4 DIFFERENTIABILITY OF A FUNCTION

The function $f(x)$ is said to be differentiable at $x = a$, if

$$\lim_{h \to 0} \frac{f(a+h) - f(a)}{h} = \lim_{h \to 0} \frac{f(a-h) - f(a)}{-h}, \text{ or } Rf'(a) = Lf'(a)$$

The common value of $Rf'(a)$ and $Lf'(a)$ is denoted by $f'(a)$ and is known as derivative of $f(x)$ at $x = a$.

If however $Rf'(a) \ne Lf'(a)$ we say that $f(x)$ is not differentiable at $x = a$.

Note: (1) On both sides of the equation h is taken as positive and very small.

(2) Every differentiable function is continuous. But, every continuous function need not be differentiable.

We say $\dfrac{d}{dx}$ (sin x) = cos x.

Here, we derive the formula from the first principles

$$\frac{d}{dx}(\sin x) = \lim_{h \to 0} \frac{\sin(x+h) - \sin x}{h} = \lim_{h \to 0} \frac{\sin(x-h) - \sin x}{-h}$$

$$= \cos x$$

Whenever we define a derivative we put a condition "provided the limit exists". If the limit does not exist *i.e., for example,*

$$\frac{d|x|}{dx} = \lim_{h \to 0} \frac{|0+h| - |0|}{h} \ne \lim_{h \to 0} \frac{|0-h| - |0|}{-h}$$

$\Rightarrow$ $\dfrac{d|x|}{dx}$ is not differentiable at $x = 0$.

Derivative of $f(x)$ at the Point $x = a$

Derivative of function $f(x)$ with respect to x at the point $x = a$ is defined as

$$\lim_{h\to 0}\frac{f(a+h)-f(a)}{h},$$ provided the limit exists and denoted by $f'(a)$.

A function is said to be differentiable at $x = a$ if it has a derivative there. A function is said to be differentiable on an interval if it is differentiable at every point of the interval.

Existence of the Derivative

The limit must be the same whether h approaches 0 from the right or from the left. Then the derivative exists. If these limits are different, the derivative does not exist and the function will not be derivable.

The function is not derivable i.e. f' does not exist if either of the left handed or right handed limits
(*i*) does not exist, or (*ii*) both exist but they are not equal.

27.5 GEOMETRICAL MEANING OF DIFFERENTIABILITY AT A POINT $x = a$

Let us draw a curve $y = f(x)$. The co-ordinates of a point P on the curve are $[a, f(a)]$. Let a point $Q[a + h, f(a + h)]$ on the right side of P, and $R[a - h, f(a - h)]$ on the left side of P.

Draw perpendiculars PS and RT.

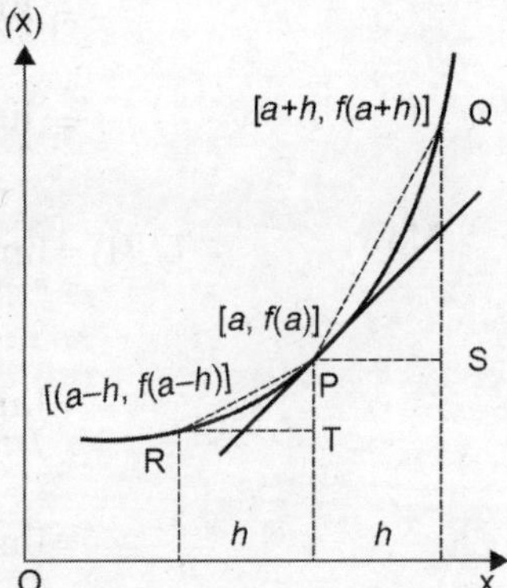

$$\text{Slope of chord } PQ = \frac{QS}{PS} = \frac{f(a+h)-f(a)}{h}$$

$$\text{Slope of chord } PR = \frac{PT}{RT} = \frac{f(a-h)-f(a)}{-h}$$

When Q tends to P, then the chord PQ becomes a tangent at P. As $h \to 0$, Points Q and R both tend to P from right hand side and from left hand side respectively. Slope of the tangent is given by

$$\lim_{Q\to P}[\text{Slope of chord } PQ] = \lim_{R\to P}[\text{Slope of chord} PR]$$

$$\lim_{h\to 0}\frac{f(a+h)-f(a)}{h} = \lim_{h\to 0}\frac{f(a-h)-f(a)}{-h}$$

Now $f(x)$ is differentiable at $x = a$.

$$\text{if } \lim_{h\to 0}\frac{f(a+h)-f(a)}{h} = \lim_{h\to 0}\frac{f(a-h)-f(a)}{-h}$$

Slope of the tangent at P, which is limiting position of the chords drawn on the left and right side of P is the same as the slope of the tangent at P.

Thus $f(x)$ is differentiable at the point P, iff there exists a unique tangent at P.

Example 20. *Let the function f(x) = |x|.*

or $f(x) = \begin{cases} x, & x > 0 \\ -x, & x < 0 \end{cases}$

We draw the curve of $y = |x|$.

The graph of $|x|$ is straight line AO and line OB.

On the right hand side of O, the slope of the tangent OA, $(y = x)$ is tan 45° *i.e.* 1

On the left hand side of O, the slope of the tangent OB $(y = -x)$ is tan 135° = –1

f(x)
f(x) = |–x|
f(x) = |x|
135°
45°
x¹
O
X

Here, the slope of OA (R. H. S.) ≠ the slope of OB (L.H.S.)

Hence, the function $f(x) = |x|$ is not differentiable at $x = 0$.

Example 21. *Show that $f(x) = x^2$ is differentiable at x = 1 and fund f'(1).*

Solution. We have $f(x) = x^2$

$$Rf'(1) = \lim_{h\to 0}\frac{f(1+h)-f(1)}{h}$$

$$= \lim_{h\to 0}\frac{(1+h)^2-1^2}{h} = \lim_{h\to 0}\frac{1+h^2+2h-1}{h}$$

$$= \lim_{h\to 0}\frac{h^2+2h}{h} = \lim_{h\to 0}\frac{h(h+2)}{h}$$

$$= \lim_{h\to 0}(h+2) = 2 \qquad \text{...(1) [Putting } h = 0]$$

Now, $$Lf'(1) = \lim_{h\to 0}\frac{f(1-h)-f(1)}{-h}$$

$$= \lim_{h\to 0}\frac{(1-h)^2-1^2}{-h} = \lim_{h\to 0}\frac{1+h^2-2h-1}{-h}$$

$$= \lim_{h\to 0}\frac{h(h-2)}{-h} = \lim_{h\to 0}-(h-2) = 2 \qquad \text{...(2) [Putting } h = 0]$$

From (1) and (2), we have

$$Lf'(1) = Rf'(1) = 2$$

This shows that $f(x)$ is differentiable at $x = 1$ and $f'(1) = 2$ **Proved**

Example 22. *If f is defined by*

$$f(x) = x^2 + 2x + 7, \text{ find } f'(3)$$

Solution: We have $f(x) = x^2 + 2x + 7$

$$Rf'(3) = \lim_{h\to 0}\frac{f(3+h)-f(3)}{h}$$

$$= \lim_{h\to 0}\frac{(3+h)^2+2(3+h)+7-[(3)^2+2(3)+7]}{h}$$

$$= \lim_{h\to 0}\frac{9+h^2+6h+6+2h+7-9-6-7}{h}$$

$$= \lim_{h\to 0}\frac{h(h+8)}{h} = \lim_{h\to 0}(h+8) = 8 \qquad \ldots(1)$$

$$Lf'(3) = \lim_{h\to 0}\frac{f(3-h)-f(3)}{-h}$$

$$= \lim_{h\to 0}\frac{(3-h)^2+2(3-h)+7-[(3)^2+2\times 3+7]}{-h}$$

$$= \lim_{h\to 0}\frac{9+h^2-6h+6-2h+7-9-6-7}{-h}$$

$$= \lim_{h\to 0}\frac{h^2-8h}{-h} = \lim_{h\to 0}\frac{h(h-8)}{-h}$$

$$= \lim_{h\to 0}(8-h) = 8 \qquad \ldots(2)$$

From (1) and (2), we have

$Lf'(3) = \mathrm{R}f'(3) = 8$ **Ans.**

Example 23. *If f is defined by*

$f(x) = x^3 + 7x^2 + 8x - 9$, find $f'(4)$

Solution: Given that,

$f(x) = x^3 + 7x^2 + 8x - 9,$

$$Rf'(4) = \lim_{h\to 0}\frac{f(4+h)-f(4)}{h}$$

$$= \lim_{h\to 0}\frac{(4+h)^3+7(4+h)^2+8(4+h)-9-[4^3+7(4)^2+8(4)-9]}{h}$$

$$= \lim_{h\to 0}\frac{[(64+48h+12h^2+h^3)+7(16+8h+h^2)+8(4+h)-9]-[64+112+32-9]}{h}$$

$$= \lim_{h\to 0}\frac{64+48h+12h^2+h^3+112+56h+7h^2+32+8h-9-64-112-32+9}{h}$$

$$= \lim_{h\to 0}\frac{h^3+19h^2+112h}{h} = \lim_{h\to 0}(h^2+19h+112) = 112 \qquad \ldots(1)$$

Similarly we can find,

$$Lf'(4) = \lim_{h\to 0}\frac{f(4-h)-f(4)}{-h} = 112 \qquad \ldots(2)$$

From (1) and (2), we get

$Rf'(4) = Lf'(4) = 112$ **Ans.**

Example 24. *If for the function f, given by $f(x) = kx^2 + 7x - 4$, $f'(5) = 97$, Find k.*

Solution. We have,

$$f(x) = kx^2 + 7x - 4$$

Then, $$f'(5) = \lim_{h\to 0} \frac{f(5+h)-f(5)}{h}$$

$$= \lim_{h\to 0} \frac{k(5+h)^2 + 7(5+h) - 4 - [k(5)^2 + 7(5) - 4]}{h}$$

$$= \lim_{h\to 0} \frac{k(25+h^2+10h)+35+7h-4-25k-35+4}{h}$$

$$= \lim_{h\to 0} \frac{25k+kh^2+10kh+7h-25k}{h}$$

$$= \lim_{h\to 0} \frac{kh^2+10kh+7h}{h} = \lim_{h\to 0} \frac{h(kh+10k+7)}{h}$$

$$= \lim_{h\to 0}(kh+10k+7h)$$

$\Rightarrow$ $f'(5) = 10\,k + 7$ ...(1)

But given that

$$f'(5) = 97$$

From (1) and (2), we get

$10k + 7 = 97$ ...(2)

$$10k = 97 - 7 = 90$$

$\Rightarrow$ $k = \frac{90}{10} = 9$ **Ans.**

Theorem 1. *Every differentiable function is continuous. But, every continuous function need not be differentiable.*

Proof. Let the function f be differentiable at $x = a$. Then by definition:

$$\lim_{h\to 0} \frac{f(a+h)-f(a)}{h} = f'(a) \quad ...(1)$$

Now, $$f(a+h) - f(a) = \frac{f(a+h)-f(a)}{h}.h$$

$$\Rightarrow \lim_{h\to 0}[f(a+h)-f(a)] = \lim_{h\to 0}\frac{f(a+h)-f(a)}{h}.\lim_{h\to 0}.h$$

$$\Rightarrow \lim_{h\to 0}[f(a+h)-f(a)] = f'(a).0 = 0 \quad \text{[From (1)]}$$

$$\Rightarrow \lim_{h\to 0} f(a+h) = f(a)$$

Therefore f is continuous at a. Since a is chosen arbitrary so f is continuous everywhere. Hence, every differentiable function is continuous. **Proved**

In order to show that a continuous function need not be differentiable, it is sufficient to give an example of a function which is continuous but not differentiable.

Consider $f(x) = |x|$ at $x = 0$.

For continuity:

Left hand limit at $x = 0$

$$\lim_{h\to 0} f(0-h) = \lim_{h\to 0} f(-h) = \lim_{h\to 0} |-h| = \lim_{h\to 0} (h) = 0$$

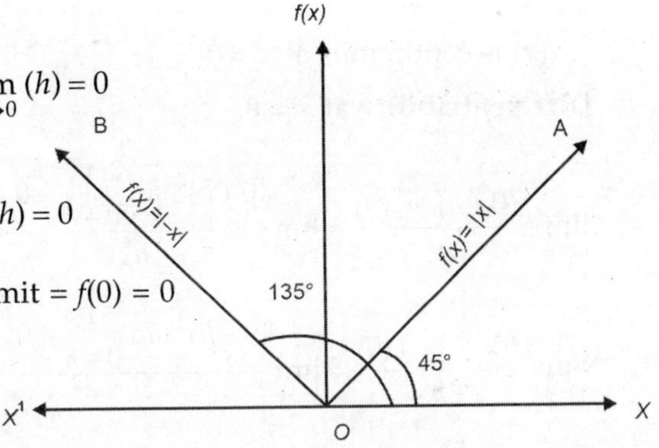

and Right hand limit at $x = 0$

$$\lim_{h\to 0} f(0+h) = \lim_{h\to 0} f(h) = \lim_{h\to 0} |h| = \lim_{h\to 0} (h) = 0$$

Here, Left hand limit = Right hand limit = $f(0) = 0$

f is continuous at $x = 0$.

Again for differentiability:

$$Rf'(0) = \lim_{h\to 0} \frac{f(0+h) - f(0)}{h}$$

$$= \lim_{h\to 0} \frac{|0+h| - |0|}{h} = \lim_{h\to 0} \frac{h}{h} = \lim_{h\to 0} (1) = 1 \qquad \ldots(1)$$

$$Lf'(0) = \lim_{h\to 0} \frac{f(0-h) - f(0)}{-h} = \lim_{h\to 0} \frac{f(-h) - f(0)}{-h}$$

$$= \lim_{h\to 0} \frac{|-h| - |0|}{-h} \lim_{h\to 0} \frac{h}{-h} = \lim_{h\to 0} (-1) = -1 \qquad \ldots(2)$$

From (1) and (2), we get

$$\mathrm{R}f'(0) \neq \mathrm{L}f'(0)$$

Therefore it is not differentiable at $x = 0$. Hence a continuous function need not be differentiable. **Proved**

Example 25. *Discuss the continuity and differentiability of*

$$f(x) = \begin{cases} x^2 \sin\dfrac{1}{x}, & \text{if } x \neq 0 \\ 0, & \text{if } x = 0 \end{cases}$$

Solution:. Let $\qquad f(x) = x^2 \sin \dfrac{1}{x}$

Continuity at $x = 0$.

$$\text{Left hand limit} = \lim_{h\to 0^-} f(x) = \lim_{x\to 0-h} \left(x^2 \sin\frac{1}{x}\right)$$

$$= \lim_{h\to 0} (0-h)^2 \sin\frac{1}{(0-h)} = 0 \qquad \ldots(1)$$

$$\text{Right hand limit} = \lim_{h\to 0^+} f(x) = \lim_{h\to 0+h} \left(x^2 \sin\frac{1}{x}\right)$$

$$= \lim_{h\to 0} (0+h)^2 \sin\frac{1}{(0+h)} = \lim_{h\to 0} h^2 \sin\frac{1}{h} = 0 \qquad \ldots(2)$$

From (1) and (2), we have

$$\lim_{h\to 0^-} f(x) = \lim_{h\to 0^+} f(x) = f(0)$$

$\therefore$ $f(x)$ is continuous at $x = 0$

Differentiability at $x = 0$. **Ans.**

$$\lim_{h\to 0}\frac{f(0+h)-f(0)}{h} = \lim_{h\to 0}\frac{(0+h)^2\sin\dfrac{1}{0+h}-0}{h} = \lim_{h\to 0} h\sin\frac{1}{h} = 0 \qquad \ldots(3)$$

$$\lim_{h\to 0}\frac{f(0-h)-f(0)}{-h} = \lim_{h\to 0}\frac{(0-h)^2\sin\dfrac{1}{0-h}-0}{-h}$$

$$= \lim_{h\to 0}(-h)\sin\frac{1}{-h} = \lim_{h\to 0} h\sin\frac{1}{h} = 0 \qquad \ldots(4)$$

From (3) and (4), we have

$$\therefore \lim_{h\to 0}\frac{f(0+h)-f(0)}{h} = \lim_{h\to 0}\frac{f(0-h)-f(0)}{-h}$$

$\therefore$ $f(x)$ is derivable at $x = 0$. **Ans.**

Example 26. *Show that the function*

$$f(x) = \begin{cases} 1+x, & \text{if } x \le 2 \\ 5-x, & \text{if } x > 2 \end{cases}$$

is not differentiable at $x = 2$.

Solution. We have

$$f(x) = \begin{cases} 1+x, & \text{if } x \le 2 \\ 5-x, & \text{if } x > 2 \end{cases}$$

$$R f'(2) = \lim_{h\to 0}\frac{f(2+h)-f(2)}{h} = \lim_{h\to 0}\frac{5-(2+h)-(1+2))}{h}$$

$$= \lim_{h\to 0}\frac{-h}{h} = \lim_{h\to 0}(-1) = -1 \qquad \ldots(1)$$

$$L f'(2) = \lim_{h\to 0}\frac{f(2-h)-f(2)}{-h} = \lim_{h\to 0}\frac{[1+(2-h)]-[1+2]}{-h}$$

$$= \lim_{h\to 0}\frac{3-h-3}{-h} = \lim_{h\to 0}\left(\frac{-h}{-h}\right) = \lim_{h\to 0}(1) = 1 \qquad \ldots(2)$$

From (1) and (2), we get

$$R f'(2) \ne L f'(2) = 2$$

Therefore, $f(x)$ is not differentiable at $x = 2$. **Proved.**

EXERCISE 27.3

1. If f is defined by $f(x) = x^2$, find $f'(2)$ **Ans.** 4

2. If $f(t) = 1 - 4t^2$, find $f'(1)$. **Ans.** –8

3. Find the derivative of the function f defined by $f(x) = mx + c$ at $x = 0$. **Ans.** m

4. Discuss the differentiability of the function

$$f(x) = \begin{cases} x^2, & x > 0 \\ \sin x, & x < 0 \end{cases}$$ at the point $x = 0$. **Ans.** Not differentiable

5. Find the right hand derivative and the left hand derivative of

$$f(x) = \begin{cases} x-1, & x < 2 \\ 2x-3, & x \geq 2 \end{cases}$$

at the point $x = 2$, and hence show that f is not differentiable at $x = 2$.
Ans. R $f'(2) = 2$, L $f'(2) = 1$

Discuss the differentiability of f(x) at the given point

6. $f(x) = \begin{cases} x^2, & \text{if } x \leq 1 \\ 2x-1, & \text{if } x > 1 \end{cases}$ at $x = 1$. **Ans.** Differentiable

7. $f(x) = \begin{cases} 3+2x, & -\frac{3}{2} \leq x < 0 \\ 3-2x, & 0 \leq x \leq \frac{3}{2} \end{cases}$ at $x = 0$ **Ans.** Not differentiable

8. $f(x) = \begin{cases} 2+x, & \text{if } x \geq 0 \\ 2-x, & \text{if } x < 0 \end{cases}$ at $x = 0$ **Ans.** Not differentiable

9. $f(x) = \begin{cases} 12x-13, & x \leq 3 \\ 2x^2+5, & x > 3 \end{cases}$ at $x = 3$ **Ans.** Differentiable

10. $f(x) = \begin{cases} -x, & x < 0 \\ x, & 0 \leq x \leq 1 \\ x^2-x+1, & x > 1 \end{cases}$ at $x = 1$ **Ans.** Not differentiable

11. $f(x) = \begin{cases} \dfrac{x-1}{2x^2-7x+5}, & x \neq 1 \\ \dfrac{-1}{3}, & x = 1 \end{cases}$ at $x = 1$ **Ans.** Not differentiable

12. Discuss the differentiability of $f(x) = x^{\frac{1}{2}}$ at $x = 0$. **Ans.** Not differentiable

13. Show that the function $f(x)$ defined as

$$f(x) = \begin{cases} x\cos\dfrac{1}{x}, & x \neq 0 \\ 0, & x = 0 \end{cases}$$

is continuous at a point $x = 0$ but not derivable at that point.

14. If for the function f, $f(x) = x^2 - 4x + 7$, show that $f'(5) = 2.\, f'\left(\dfrac{7}{2}\right)$.

15. Show that the derivative of the function f given by $f(x) = 2x^3 - 9x^2 + 12x + 9$, at $x = 1$ and at $x = 2$ are equal.

16. If $f(x) = x^4$, find $f'(0)$ and $f'\left(\dfrac{1}{2}\right)$ **Ans.** $0, -\dfrac{1}{2}$

17. For what values of a and b the function

$$f(x) = \begin{cases} x^2, & \text{if } x \leq 1 \\ 2ax + b, & \text{if } x > 1 \end{cases}$$ is differentiable at $x = 1$ **Ans.** $a = 1, b = -1$

18. A function $f(x)$ is defined as follows:

$$f(x) = \begin{cases} x, & x < 1 \\ 2 - x, & 1 \leq x \leq 2 \\ -2 + 3x - x^2 & x > 2 \end{cases}$$

Discuss its continuity and differentiability at $x = 1$ and $x = 2$.

Ans. Continuous at $x = 1$ and $x = 2$. Differentiable at $x = 2$ but not at $x = 1$.

19. Test the differentiability of the function,

$$f(x) = \begin{cases} x \cdot [x], & 0 \leq x < 2 \\ (x-1) \cdot [x], & 2 \leq x \leq 3 \end{cases}$$ at $x = 1$ and at $x = 2$

Ans. Not differentiable at $x = 1$ and $x = 2$.

20. If $f(x) = \dfrac{x}{1 + e^{1/x}}, x \neq 0$ and $f(0) = 0$, show that $f(x)$ is continuous at $x = 0$, but $f'(0)$ does not exist.

21. If $f(x) = \begin{cases} x, & \text{when } 0 \leq x < 1/2 \\ 1, & \text{when } x = 1/2 \\ 1 - x, & \text{when } 1/2 < x < 1 \end{cases}$

discuss the continuity of $f(x)$ at $x = \dfrac{1}{2}$. **Ans.** Not continuous at $x = \dfrac{1}{2}$

Differentiation

28.1 DIFFERENTIAL FROM THE FIRST PRINCIPLE

The limit of incremental ratio, *i.e.* lim $\frac{\delta y}{\delta x}$ as δx approaches zero is called the differential coefficient of y with respect to x and denoted by $\frac{dy}{dx}$.

$$\frac{dy}{dx} = \lim_{\delta x \to 0} \frac{\delta y}{\delta x}$$

$$\frac{d}{dx} f(x) = \lim_{\delta x \to 0} \frac{f(x+\delta x) - f(x)}{\delta x}$$

Example 1. *Find the differential coefficient of $y = x^2$ at $x = 2$.*

Solution. We have, $y = x^2$

$x + \delta x$	$y + \delta y$	δx	δy	$\frac{\delta y}{\delta x}$
2.0	4.00	0	0	$\frac{0}{0}$
2.1	4.41	0.1	0.41	$\frac{0.41}{0.1} = 4.1$
2.01	4.0401	0.01	0.0401	$\frac{0.0401}{0.01} = 4.01$
2.001	4.004001	0.001	0.004001	$\frac{0.004001}{0.001} = 4.001$
2.0001	4.00040001	0.0001	0.00040001	$\frac{0.00040001}{0.0001} = 4.0001$
		$\downarrow$ 0	$\downarrow$ 0	$\downarrow$ 4

As $\delta x \to 0$, hence $\delta y \to 0$. But the value of $\frac{dy}{dx}$ remains constant.

28.2 STANDARD RESULTS

A list of standard elementary formulae of differential calculus is given below:

1. $\frac{d}{dx}(x^n) = nx^{n-1}$

2. $\frac{d}{dx}(\log x) = \frac{1}{x}$

3. $\frac{d}{dx}(\sin x) = \cos x$

4. $\frac{d}{dx}(\cos x) = -\sin x$

5. $\frac{d}{dx}(\tan x) = \sec^2 x$

6. $\frac{d}{dx}(\cot x) = -\operatorname{cosec}^2 x$

7. $\frac{d}{dx}(\sec x) = \sec x \cdot \tan x$

8. $\frac{d}{dx}(\operatorname{cosec} x) = -\operatorname{cosec} x \cdot \cot x$

9. $\frac{d}{dx}(a^x) = a^x \log_e a$

10. $\frac{d}{dx}(e^x) = e^x$

11. $\frac{d}{dx}(\sin^{-1} x) = \frac{1}{\sqrt{1-x^2}}$

12. $\frac{d}{dx}(\cos^{-1} x) = -\frac{1}{\sqrt{1-x^2}}$

13. $\frac{d}{dx}(\sinh^{-1} x) = \frac{1}{\sqrt{1+x^2}}$

14. $\frac{d}{dx}(\cosh^{-1} x) = \frac{1}{\sqrt{x^2-1}}$

15. $\frac{d}{dx}(\tan^{-1} x) = \frac{1}{1+x^2}$

16. $\frac{d}{dx}(\cot^{-1} x) = -\frac{1}{1+x^2}$

17. $\frac{d}{dx}(\sec^{-1} x) = \frac{1}{x\sqrt{x^2-1}}$

18. $\frac{d}{dx}(\operatorname{cosec}^{-1} x) = -\frac{1}{x\sqrt{x^2-1}}$

19. $\frac{d}{dx}(uv) = u \cdot \frac{dv}{dx} + v \cdot \frac{du}{dx}$

20. $\frac{d}{dx}\left(\frac{u}{v}\right) = \frac{v \cdot \frac{du}{dx} - u \cdot \frac{dv}{dx}}{v^2}$

21. $\frac{dy}{dx} = \frac{dy}{du} \cdot \frac{du}{dx}$

22. $\frac{d}{dx}(\sinh x) = \cosh x$

23. $\frac{d}{dx}(\cosh x) = \sinh x$

24. $\frac{d}{dx}(\tanh x) = \operatorname{sech}^2 x$

25. $\frac{d}{dx}(\coth x) = -\operatorname{cosech}^2 x$

26. $\frac{d}{dx}(\operatorname{sech} x) = -\operatorname{sech} x \cdot \tanh x$

27. $\frac{d}{dx}(\operatorname{cosech} x) = -\operatorname{cosech} x \cdot \coth x$

28. $\frac{d}{dx}(\tanh^{-1} x) = \frac{1}{1-x^2}$

29. $\frac{d}{dx}(\coth^{-1} x) = \frac{1}{1-x^2}$

30. $\frac{d}{dx}(\operatorname{sech}^{-1} x) = -\frac{1}{x\sqrt{1-x^2}}$

31. $\frac{d}{dx}(\operatorname{cosech}^{-1} x) = -\frac{1}{x\sqrt{x^2+1}}$

Remember $\boxed{\mathbf{\sinh x = \frac{e^x - e^{-x}}{2}}}$

$$\boxed{\mathbf{\cosh x = \frac{e^x + e^{-x}}{2}}}$$

$$\cosh^2 x - \sinh^2 x = 1$$

$$\text{sech}^2 x = 1 - \tanh^2 x$$

$$\text{cosech}^2 x = -1 + \coth^2 x$$

$$\boxed{\mathbf{\sinh^{-1} x = \log [x + \sqrt{x^2 + 1}]}}$$

$$\boxed{\mathbf{\cosh^{-1} x = \log [x + \sqrt{x^2 - 1}]}}$$

Leibnitz's Theorem for nth derivative of the product of two functions.

$$\boxed{\frac{d^n}{dx^n}(u \cdot v) = {}^nC_0\, u_n v + {}^nC_1\, u_{n-1} v_1 + {}^nC_2\, u_{n-2} v_2 + \ldots + {}^nC_r\, u_{n-r} v_r + \ldots + {}^nC_n\, u v_n}$$

28.3 DIFFERENTIAL COEFFICIENT BY FIRST PRINCIPLE

DIFFERENTIAL COEFFICIENT OF x^n [S.B.T.E. MAY 2008]

Let $y = x^n$ ... (1)

Let δx be the increment in x and the corresponding increment δy in y.

$$y + \delta y = (x + \delta x)^n \quad \ldots (2)$$

Subtracting (1) from (2), we get

$$\delta y = (x + \delta x)^n - x^n$$

Dividing by δx, $\dfrac{\delta y}{\delta x} = \dfrac{(x + \delta x)^n - x^n}{\delta x} = \dfrac{x^n\left(1 + \dfrac{\delta x}{x}\right)^n - x^n}{\delta x}$

$$= \frac{x^n\left[1 + n \cdot \dfrac{\delta x}{x} + \dfrac{n(n-1)}{2!}\left(\dfrac{\delta x}{x}\right)^2 + \ldots - 1\right]}{\delta x}$$

$$= x^n\left[n \cdot \frac{1}{x} + \frac{n(n-1)}{2!}\frac{\delta x}{x^2} + \ldots\right]$$

$$\lim_{\delta x \to 0} \frac{\delta y}{\delta x} = \lim_{\delta x \to 0} x^n\left[\frac{n}{x} + \frac{n(n-1)}{2!}\frac{\delta x}{x^2} + \ldots\right]$$

$$\Rightarrow \quad \frac{dy}{dx} = x^n\left[\frac{n}{x}\right] = n\,x^{n-1}$$

$$\Rightarrow \quad \boxed{\mathbf{\frac{d(x^n)}{dx} = nx^{n-1}}}$$

Ans.

Example 2. Differentiate $[x]$ with respect to x, where $[x]$ represents the greatest integer less than or equal to x.

Solution. Here, we have

$$y = [x]$$

$$y = \text{greatest integer } a \text{ (say)} \qquad [x] = a = \text{constant}$$

$$\frac{dy}{dx} = 0$$ **Ans.**

Example 3. *Find* $\frac{dy}{dx}$ *if* (*a*) $y = (x - 2)^2 (2x - 3)$ (*b*) $y = \frac{1}{\sqrt{x + a}}$

Solution. (*a*) $y = (x - 2)^2 (2x - 3) = (x^2 - 4x + 4)(2x - 3)$

$$y = 2x^3 - 11x^2 + 20x^1 - 12$$

$$\frac{dy}{dx} = 2.3x^{3-1} - 11.2x^{2-1} + 20.1x^{1-1} - 0$$

$$= 6x^2 - 22x + 20.$$ **Ans.**

(*b*) $$y = \frac{1}{\sqrt{x + a}} = (x + a)^{-1/2}$$

$$\frac{dy}{dx} = -\frac{1}{2}(x + a)^{-1/2 - 1} = -\frac{1}{2}(x + a)^{-3/2}$$ **Ans.**

28.4 DIFFERENTIAL COEFFICIENT OF sin x [SBTE. June 2010]

Let $y = \sin x$

$\therefore$ $y + \delta y = \sin (x + \delta x)$

$$\delta y = \sin (x + \delta x) - \sin x$$

$$\frac{\delta y}{\delta x} = \frac{\sin (x + \delta x) - \sin x}{\delta x} = \frac{2 \cos \left(x + \frac{\delta x}{2}\right) \sin \frac{\delta x}{2}}{\delta x}$$

$$\lim_{\delta x \to 0} \frac{\delta y}{\delta x} = \lim_{\frac{\delta x}{2} \to 0} \cos \left(x + \frac{\delta x}{2}\right) \frac{\sin \frac{\delta x}{2}}{\frac{\delta x}{2}}$$

$$\frac{dy}{dx} = \cos x \qquad \left[\lim_{x \to 0} \frac{\sin x}{x} = 1\right]$$

$$\boxed{\frac{d}{dx}(\sin x) = \cos x.}$$ **Ans.**

28.5 DIFFERENTIAL COEFFICIENT OF cos x [SBTE. June 2009]

Let $y = \cos x$

$\Rightarrow$ $y + \delta y = \cos (x + \delta x)$

$\Rightarrow$ $\delta y = \cos (x + \delta x) - \cos x$

$$= 2 \sin \frac{(x - \overline{x + \delta x})}{2} \sin \frac{(x + \overline{x + \delta x})}{2} = -2 \sin \frac{\delta x}{2} \sin \left(x + \frac{\delta x}{2}\right)$$

$$\frac{\delta y}{\delta x} = -\frac{2\sin\frac{\delta x}{2}}{\delta x}\sin\left(x+\frac{\delta x}{2}\right) = -\frac{\sin\frac{\delta x}{2}}{\frac{\delta x}{2}}\sin\left(x+\frac{\delta x}{2}\right)$$

$$\frac{dy}{dx} = -\lim_{\delta x\to 0}\frac{\sin\frac{\delta x}{2}}{\frac{\delta x}{2}}\sin\left(x+\frac{\delta x}{2}\right) \qquad \left[\lim_{\frac{\delta x}{2}\to 0}\frac{\sin\frac{\delta x}{2}}{\frac{\delta x}{2}} = 1\right]$$

$$= -\sin x$$

$$\boxed{\frac{d}{dx}(\cos x) = -\sin x}$$

Ans.

28.6 DIFFERENTIAL COEFFICIENT OF tan x [SBTE. 2017, 2016, 2015, 2009]

Let $y = \tan x$

$$y + \delta y = \tan(x+\delta x) - \tan x$$

$$\delta y = \tan(x+\delta x) - \tan x = \frac{\sin(x+\delta x)}{\cos(x+\delta x)} - \frac{\sin x}{\cos x}$$

$$\delta y = \frac{\sin(x+\delta x)\cos x - \cos(x+\delta x)\sin x}{\cos(x+\delta x)\cos x}$$

$$\frac{\delta y}{\delta x} = \frac{\sin(x+\delta x - x)}{\delta x\cos(x+\delta x)\cos x} = \frac{1}{\cos(x+\delta x)\cos x}\cdot\frac{\sin\delta x}{\delta x}$$

$$\lim_{\delta x\to 0}\frac{\delta y}{\delta x} = \lim_{\delta x\to 0}\frac{1}{\cos(x+\delta x)\cos x}\cdot\frac{\sin\delta x}{\delta x} = \frac{1}{\cos^2 x}$$

$$\frac{dy}{dx} = \sec^2 x$$

$$\boxed{\frac{d}{dx}(\tan x) = \sec^2 x}$$

Ans.

28.7 DIFFERENTIAL COEFFICIENT OF cot x [SBTE. 2015, 2004]

Let $y = \cot x$

$$y + \delta y = \cot(x+\delta x)$$

$$\delta y = \cot(x+\delta x) - \cot x = \frac{\cos(x+\delta x)}{\sin(x+\delta x)} - \frac{\cos x}{\sin x}$$

$$= \frac{\cos(x+\delta x)\sin x - \sin(x+\delta x)\cos x}{\sin(x+\delta x)\sin x}$$

$$[\sin A\cos B - \cos A\sin B = \sin(A-B)]$$

$$= \frac{\sin(x - \overline{x+\delta x})}{\sin(x+\delta x)\sin x} = \frac{-\sin\delta x}{\sin(x+\delta x)\sin x}$$

$$\frac{\delta y}{\delta x} = -\frac{\sin \delta x}{\delta x} \frac{1}{\sin (x + \delta x) \sin x}$$

$$\frac{dy}{dx} = \lim_{\delta x \to 0} \left(\frac{-\sin \delta x}{\delta x} \right) \left(\frac{1}{\sin (x + \delta x) \sin x} \right) = -\frac{1}{\sin x \cdot \sin x} = -\operatorname{cosec}^2 x$$

$$\boxed{\frac{d}{dx} (\cot x) = -\operatorname{cosec}^2 x}$$ **Ans.**

28.8 DIFFERENTIAL COEFFICIENT OF sec x

Let $y = \sec x$

$$y + \delta y = \sec (x + \delta x)$$

$$\delta y = \sec (x + \delta x) - \sec x$$

$$= \frac{1}{\cos (x + \delta x)} - \frac{1}{\cos x} = \frac{\cos x - \cos (x + \delta x)}{\cos (x + \delta x) \cos x}$$

$$\left[\cos C - \cos D = 2 \sin \left(\frac{C + D}{2} \right) \sin \left(\frac{D - C}{2} \right) \right]$$

$$\frac{\delta y}{\delta x} = \frac{2 \sin \left(x + \frac{\delta x}{2} \right) \sin \frac{\delta x}{2}}{\delta x \cos (x + \delta x) \cos x}$$

$$\lim_{\delta x \to 0} \frac{\delta y}{\delta x} = \lim_{\delta x \to 0} \frac{\sin \left(x + \frac{\delta x}{2} \right) \sin \frac{\delta x}{2}}{\cos (x + \delta x) \cos x \cdot \frac{\delta x}{2}} = \frac{\sin x}{\cos^2 x}$$

$$\frac{dy}{dx} = \sec x \cdot \tan x \qquad \left[\lim_{x \to 0} \frac{\sin x}{x} = 1 \right]$$

$$\boxed{\frac{d}{dx} (\sec x) = \sec x \cdot \tan x}$$

28.9 DIFFERENTIAL COEFFICIENT OF cosec x

Similarly we can prove that

$$\boxed{\frac{d}{dx} (\operatorname{cosec} x) = -\operatorname{cosec} x \cdot \cot x}$$

28.10 DIFFERENTIAL COEFFICIENT OF a^x

[SBTE. 2009, 2007]

Let $y = a^x$

$$y + \delta y = a^{x + \delta x}$$

$$\frac{\delta y}{\delta x} = \frac{a^{x + \delta x} - a^x}{\delta x} = \frac{a^x (a^{\delta x} - 1)}{\delta x}$$

$$\lim_{\delta x \to 0} \frac{\delta y}{\delta x} = \lim_{\delta x \to 0} \frac{a^x (a^{\delta x} - 1)}{\delta x}$$

$$\frac{dy}{dx} = a^x \cdot \log_e a$$

$$\frac{d}{dx}(a^x) = a^x \cdot \log_e a$$

Cor. Let $\quad y = e^x$

Then $\quad \dfrac{dy}{dx} = e^x \log_e e$

$$\boxed{\frac{d}{dx}(e^x) = e^x}$$

Ans.

28.11 DIFFERENTIAL COEFFICIENT OF $log_e\, x$ [SBTE. 2017, 2009, 2008]

Let $\quad y = \log_e x$

$$y + \delta y = \log_e (x + \delta x)$$

$$\delta y = \log_e (x + \delta x) - \log_e x = \log_e \frac{(x + \delta x)}{x}$$

$$\frac{\delta y}{\delta x} = \frac{1}{\delta x} \log_e \left(1 + \frac{\delta x}{x}\right) = \frac{1}{x} \cdot \frac{x}{\delta x} \log_e \left(1 + \frac{\delta x}{x}\right) = \frac{1}{x} \log_e \left(1 + \frac{\delta x}{x}\right)^{\frac{x}{\delta x}}$$

$$\lim_{\delta x \to 0} \frac{\delta y}{\delta x} = \lim_{\delta x \to 0} \frac{1}{x} \log_e \left(1 + \frac{\delta x}{x}\right)^{\frac{x}{\delta x}}$$

$$\frac{dy}{dx} = \frac{1}{x} \log_e e = \frac{1}{x}$$

$$\boxed{\frac{d}{dx}(\log_e x) = \frac{1}{x}}$$

28.12 DIFFERENTIAL COEFFICIENT OF $\sin^{-1} x$

Let $\quad y = \sin^{-1} x$

$$\sin y = x$$

$\Rightarrow \quad \sin (y + \delta y) = x + \delta x$

$$\sin (y + \delta y) - \sin y = \delta x \qquad \left[\sin C - \sin D = 2 \cos \left(\frac{C + D}{2}\right) \sin \left(\frac{C - D}{2}\right)\right]$$

$$\frac{\delta y}{\delta x} = \frac{\delta y}{\sin (y + \delta y) - \sin y} = \frac{\delta y}{2 \cos \left(\dfrac{2y + \delta y}{2}\right) \sin \dfrac{\delta y}{2}}$$

$$\lim_{\delta x \to 0} \frac{\delta y}{\delta x} = \lim_{\delta y/2 \to 0} \frac{1}{\cos \left(y + \dfrac{\delta y}{2}\right)} \frac{\dfrac{\delta y}{2}}{\sin \dfrac{\delta y}{2}}$$

$$\frac{dy}{dx} = \frac{1}{\cos y} = \frac{1}{\sqrt{1 - \sin^2 y}} = \frac{1}{\sqrt{1 - x^2}}$$

$$\boxed{\frac{d}{dx}(\sin^{-1} x) = \frac{1}{\sqrt{1-x^2}}}$$

28.13 DIFFERENTIAL COEFFICIENT OF $\cos^{-1} x$

Similarly we can prove that

$$\frac{d}{dx}\cos^{-1} x = -\frac{1}{\sqrt{1-x^2}}$$

28.14 DIFFERENTIAL COEFFICIENT OF $\tan^{-1} x$

Let $\qquad y = \tan^{-1} x$

$$\tan y = x$$

$$\therefore \qquad \tan (y + \delta y) = x + \delta x$$

$$\delta x = \tan (y + \delta y) - \tan y$$

$$= \frac{\sin (y + \delta y)}{\cos (y + \delta y)} - \frac{\sin y}{\cos y}$$

$$\delta x = \frac{\sin (y + \delta y) \cos y - \cos (y + \delta y) \sin y}{\cos (y + \delta y) \cos y} = \frac{\sin (y + \delta y - y)}{\cos (y + \delta y) \cos y}$$

$$\frac{\delta y}{\delta x} = \frac{\delta y \cos (y + \delta y) \cos y}{\sin \delta y}$$

$$\lim_{\delta x \to 0} \frac{\delta y}{\delta x} = \lim_{\delta y \to 0} \cos (y + \delta y) \cos y \cdot \frac{\delta y}{\sin \delta y}$$

$$\frac{dy}{dx} = \cos^2 y = \frac{1}{\sec^2 y} = \frac{1}{1 + \tan^2 y} = \frac{1}{1 + x^2}$$

$$\boxed{\frac{d}{dx}(\tan^{-1} x) = \frac{1}{1 + x^2}}$$ **Ans.**

28.15 DIFFERENTIAL COEFFICIENT OF $\cot^{-1} x$

Similarly we can prove that

$$\boxed{\frac{d}{dx}(\cot^{-1} x) = -\frac{1}{1 + x^2}}$$ **Ans.**

28.16 DIFFERENTIAL COEFFICIENT OF $\sec^{-1} x$

Let $\qquad y = \sec^{-1} x$

$$\sec y = x$$

$$\sec (y + \delta y) = x + \delta x$$

$$\delta x = \sec (y + \delta y) - \sec y = \frac{1}{\cos (y + \delta y)} - \frac{1}{\cos y} = \frac{\cos y - \cos (y + \delta y)}{\cos (y + \delta y) \cos y}$$

$$\frac{\delta y}{\delta x} = \frac{\delta y \cos(y + \delta y) \cos y}{2 \sin\left(y + \frac{\delta y}{2}\right) \sin \frac{\delta y}{2}}$$

$$\lim_{\delta x \to 0} \frac{\delta y}{\delta x} = \lim_{\delta y \to 0} \frac{\cos(y + \delta y) \cos y}{\sin\left(y + \frac{\delta y}{2}\right)} \cdot \frac{\frac{\delta y}{2}}{\sin \frac{\delta y}{2}} \qquad \left(\begin{matrix} \text{As} & \delta x \to 0 \\ \text{hence} & \delta y \to 0 \end{matrix}\right)$$

$$\frac{dy}{dx} = \frac{\cos^2 y}{\sin y} = \cot y \cos y = \frac{1}{\sec y \tan y}$$

$$= \frac{1}{\sec y \sqrt{\sec^2 y - 1}} = \frac{1}{x\sqrt{x^2 - 1}}$$

$$\boxed{\frac{d}{dx}(\sec^{-1} x) = \frac{1}{x\sqrt{x^2 - 1}}}$$

Ans.

28.17 DIFFERENTIAL COEFFICIENT OF cosec^{-1} x

Similarly we can prove that

$$\boxed{\frac{d}{dx}(\text{cosec}^{-1} x) = -\frac{1}{x\sqrt{x^2 - 1}}}$$

Ans.

Example 4. *From the first principles find the differential coefficient of* $\sqrt{x}$.

Solution. Let $y = \sqrt{x}$

$$y + \delta y = \sqrt{x + \delta x}$$

$$\delta y = \sqrt{x + \delta x} - \sqrt{x}$$

$$\frac{\delta y}{\delta x} = \frac{\sqrt{x + \delta x} - \sqrt{x}}{\delta x}$$

$$\lim_{\delta x \to 0} \frac{\delta y}{\delta x} = \lim_{\delta x \to 0} \left(\frac{\sqrt{x + \delta x} - \sqrt{x}}{\delta x} \times \frac{\sqrt{x + \delta x} + \sqrt{x}}{\sqrt{x + \delta x} + \sqrt{x}} \right)$$

$$\frac{dy}{dx} = \lim_{\delta x \to 0} \frac{(x + \delta x) - x}{\delta x\,[\sqrt{x + \delta x} + \sqrt{x}]} = \lim_{\delta x \to 0} \left(\frac{1}{\sqrt{x + \delta x} + \sqrt{x}} \right)$$

$$= \frac{1}{2\sqrt{x}}$$

Ans.

Example 5. *Find the differential coefficient of sin* x^2 *from ab-initio.* **[SBTE. 2016, 2013]**

Solution. Let $y = \sin x^2$

$$y + \delta y = \sin (x + \delta x)^2$$

$$\frac{\delta y}{\delta x} = \frac{\sin (x + \delta x)^2 - \sin x^2}{\delta x}$$

$$= \frac{2\cos\left[x^2 + x\,\delta x + \frac{(\delta x)^2}{2}\right]\sin\left[x\cdot\delta x + \frac{(\delta x)^2}{2}\right]}{\delta x}$$

$$\lim_{\delta x\to 0}\frac{\delta y}{\delta x} = \lim_{\delta x\to 0} 2\cos\left[x^2 + x\cdot\delta x + \frac{(\delta x)^2}{2}\right]\times\frac{\sin\delta x\left(x+\frac{\delta x}{2}\right)}{\delta x\left(x+\frac{\delta x}{2}\right)}\times\left(x+\frac{\delta x}{2}\right)$$

$$\therefore \quad \frac{dy}{dx} = 2x\cos x^2 \qquad \textbf{Ans.}$$

Example 6. *Find the differential coefficient of* $\log_e x^2$ *from fundamentals.*

Solution. Let $y = \log_e x^2$

$$y + \delta y = \log_e (x+\delta x)^2$$

$$\frac{\delta y}{\delta x} = \frac{1}{\delta x}[\log(x+\delta x)^2 - \log x^2]$$

$$\frac{\delta y}{\delta x} = \frac{1}{\delta x}\log_e\frac{(x+\delta x)^2}{x^2}$$

$$\lim_{\delta x\to 0}\frac{\delta y}{\delta x} = \lim_{\delta x\to 0}\frac{1}{x}\frac{x}{\delta x}\log_e\left(1+\frac{\delta x}{x}\right)^2$$

$$\frac{dy}{dx} = \lim_{\delta x\to 0}\frac{2}{x}\log\left(1+\frac{\delta x}{x}\right)^{\frac{x}{\delta x}} = \frac{2}{x}\log_e e = \frac{2}{x} \qquad \text{Ans.}$$

Example 7. *Find the differential coefficient of* $\sqrt{\sin x}$ *from the first principles.*

[B.T.E. Delhi June 2005]

Solution. Let $y = \sqrt{\sin x}$

$$y + \delta y = \sqrt{\sin(x+\delta x)}$$

$$\frac{\delta y}{\delta x} = \frac{\sqrt{\sin(x+\delta x)} - \sqrt{\sin x}}{\delta x}$$

$$= \frac{\sqrt{\sin(x+\delta x)} - \sqrt{\sin x}}{\delta x}\times\frac{\sqrt{\sin(x+\delta x)} + \sqrt{\sin x}}{\sqrt{\sin(x+\delta x)} + \sqrt{\sin x}}$$

$$\frac{\delta y}{\delta x} = \frac{\sin(x+\delta x) - \sin x}{\delta x}\times\frac{1}{\sqrt{\sin(x+\delta x)} + \sqrt{\sin x}}$$

$$\lim_{\delta x\to 0}\frac{\delta y}{\delta x} = \lim_{\delta x\to 0}\cos\left(x+\frac{\delta x}{2}\right)\cdot\frac{\sin\frac{\delta x}{2}}{\frac{\delta x}{2}}\cdot\frac{1}{\sqrt{\sin(x+\delta x)} + \sqrt{\sin x}}$$

$$\frac{dy}{dx} = \cos x\cdot 1\,\frac{1}{\sqrt{\sin x} + \sqrt{\sin x}} = \frac{\cos x}{2\sqrt{\sin x}} \qquad \textbf{Ans.}$$

Example 8. *Differentiate the following function w.r.t x from first principles*

(*i*) $\sin^2 x$ **[SBTE. 2017, 2016, 2014]**

(*ii*) $\sin (x^2 + 1)$ **[SBTE. 2015]**

(*iii*) $\cos^2 x$ **[SBTE. 2015]**

Solution. (*i*) Let $y = f(x) = \sin^2 x$; then $f(x + h) = \sin^2 (x + h)$

By differentiation $\dfrac{dy}{dx} = \lim_{h \to 0} \dfrac{f(x+h) - f(x)}{h}$

$$\frac{d}{dx}(\sin^2 x) = \lim_{h \to 0} \frac{\sin^2 (x+h) - \sin^2 x}{h}$$

$$= \lim_{h \to 0} \frac{\sin (x+h+x) \cdot \sin (x+h-x)}{h}$$

$$= \lim_{h \to 0} \frac{\sin (2x+h) \cdot \sin (h)}{h}$$

$$= \lim_{h \to 0} \sin (2x+h) \times \lim_{h \to 0} \frac{\sinh}{h}$$

$$\therefore \quad \frac{d}{dx}(\sin^2 x) = \sin 2x = 2 \sin x \cdot \cos x.$$ **Ans.**

(*ii*) Let $y = f(x) = \sin (x^2 + 1)$

$$\frac{dy}{dx} = \lim_{h \to 0} \frac{f(x+h) - f(x)}{h}$$

$$\frac{d}{dx}\left[\sin (x^2+1)\right] = \lim_{h \to 0} \frac{\sin \{(x+h)^2 + 1\} - \sin (x^2+1)}{h}$$

$$= \lim_{h \to 0} \frac{2 \sin \frac{\{(x+h)^2 + 1 - (x^2+1)\}}{2} \cos \frac{\{(x+h)^2 + 1\} + (x^2+1)}{2}}{h}$$

$$= \lim_{h \to 0} \frac{2 \sin \left(\frac{2xh + h^2}{2}\right) \cos \left\{\frac{(x+h)^2 + 1 + (x^2+1)}{2}\right\}}{h}$$

$$= \lim_{h \to 0} \frac{2 \sin \frac{h(2x+h)}{2}}{\frac{h(2x+h)}{2}} \times \frac{2x+h}{2} \times \lim_{h \to 0} \cos \left\{\frac{(x+h)^2 + 1 + (x^2+1)}{2}\right\}$$

$$= 2 \times 1 \times \frac{2x+0}{2} \cos \left\{\frac{x^2 + 1 + (x^2+1)}{2}\right\}$$

$$\therefore \quad \frac{d}{dx}\left[\sin (x^2+1)\right] = 2x \cos (x^2+1)$$ **Ans.**

Example 9. *If $y = \sin^{-1} x$, then show that* $(1-x^2)\dfrac{d^2y}{dx^2} - x\dfrac{dy}{dx} = 0.$ **[SBTE. 2016]**

Solution. Here we have

$$y = \sin^{-1} x$$

$$\frac{dy}{dx} = \frac{1}{\sqrt{1-x^2}}$$

$$\left(\sqrt{1-x^2}\right)\frac{dy}{dx} = 1$$

$$\Rightarrow \quad \frac{d}{dx}\left[\frac{dy}{dx}\left(\sqrt{1-x^2}\right)\right] = 0$$

$$\Rightarrow \quad \left(\sqrt{1-x^2}\right)\frac{d^2y}{dx^2} + \frac{dy}{dx}\cdot\frac{1}{2\sqrt{1-x^2}}\cdot\frac{d}{dx}(1-x^2) = 0$$

$$\Rightarrow \quad \left(\sqrt{1-x^2}\right)\frac{d^2y}{dx^2} + \frac{dy}{dx}\left(\frac{-x}{\sqrt{1-x^2}}\right) = 0$$

$$\therefore \quad (1-x^2)\frac{d^2y}{dx^2} - x\frac{dy}{dx} = 0$$

Proved.

28.18 SOME GENERAL THEOREMS ON DIFFERENTIATION

Differential Coefficient of a Constant is Zero

Let $y = c$ [where c is a constant.]

$$y + \delta y = c$$

$$\therefore \quad \delta y = 0$$

$$\frac{\delta y}{\delta x} = \frac{0}{\delta x} \quad \Rightarrow \quad \frac{dy}{dx} = 0$$

$$\frac{d}{dx}(c) = 0$$

(let $c = 8$)

Thus $\dfrac{d}{dx}(8) = 0$

Differential Coefficient of the Product of a Constant and a Function

Let $y = cf(x)$ [where c is a constant.]

$$y + \delta y = cf(x + \delta x)$$

$$\frac{\delta y}{\delta x} = \frac{cf(x+\delta x) - cf(x)}{\delta x}$$

$$\lim_{\delta x \to 0}\frac{\delta y}{\delta x} = \lim_{\delta x \to 0}\frac{c[f(x+\delta x) - f(x)]}{\delta x}$$

$$\frac{dy}{dx} = c\cdot\frac{d}{dx}f(x) = c\cdot f'(x)$$

$$\frac{d}{dx}[cf(x)] = c \cdot f'(x)$$

Then $$\frac{d}{dx}(4x^3) = 4 \cdot \frac{d}{dx}(x^3) = 4 \times 3x^2 = 12x^2.$$

Differential Coefficient of the Sum of Functions

Let $y = u + v + ...$

and δx be the increment in x and corresponding increment in $u, v, ...$ and y be $\delta u, \delta v, ...$ and δy.

$$y + \delta y = (u + \delta u) + (v + \delta v) + ...$$

$$\delta y = \delta u + \delta v + ...$$

$$\frac{\delta y}{\delta x} = \frac{\delta u}{\delta x} + \frac{\delta v}{\delta x} + ...$$

On taking limits as $\delta x \to 0$, we have

$$\frac{dy}{dx} = \frac{du}{dx} + \frac{dv}{dx} + ...$$

Thus $$\frac{d}{dx}(6x^3 + 4x^2 + 7x + 8)$$

$$= \frac{d}{dx}(6x^3) + \frac{d}{dx}(4x^2) + \frac{d}{dx}(7x) + \frac{d}{dx}(8)$$

$$= 18x^2 + 8x + 7 + 0 = 18x^2 + 8x + 7.$$ **Ans.**

Differential Coefficient of a Function

Let y be a function of u and u be a function of x.

Let δx, δu and δy be the corresponding increments in x, u and y respectively.

$$\frac{\delta y}{\delta x} = \frac{\delta y}{\delta u} \cdot \frac{\delta u}{\delta x}$$

$$\lim_{\delta x \to 0} \frac{\delta y}{\delta x} = \lim_{\delta u \to 0} \frac{\delta y}{\delta u} \cdot \lim_{\delta x \to 0} \frac{\delta u}{\delta x} \qquad \left(\begin{array}{ll} \text{as} & \delta x \to 0 \\ \text{hence} & \delta u \to 0 \end{array}\right)$$

$$\boxed{\frac{dy}{dx} = \frac{dy}{du} \cdot \frac{du}{dx}}$$

Example 10. *If $y = \log(\sin x)$, find $\frac{dy}{dx}$.* **[SBTE. 2016, 2007]**

Solution. $y = \log u$ $\left[\begin{array}{l} \text{Put } u = \sin x \\ \frac{du}{dx} = \cos x \end{array}\right]$

$$\frac{dy}{dx} = \frac{1}{u}\frac{du}{dx},$$

$$\frac{dy}{dx} = \frac{1}{\sin x} \cdot \cos x = \cot x$$ **Ans.**

Example 11. *Differentiate the following with respect to x.*

$$(\log 5)^{\sqrt{x+1}} + \sec x^\circ$$

Solution. Let $y = (\log 5)^{\sqrt{x+1}} + \sec x°$

Converting $x°$ into radians

$$y = (\log 5)^{\sqrt{x+1}} + \sec\left(\frac{\pi x}{180}\right)$$

$$\frac{dy}{dx} = (\log 5)^{\sqrt{x+1}} \log\log 5 \left[\frac{1}{2}(x+1)^{-\frac{1}{2}}\right] + \frac{\pi}{180} \sec\frac{\pi x}{180} \tan\frac{\pi x}{180}$$

$$= \frac{1}{2\sqrt{x+1}} (\log 5)^{\sqrt{x+1}} \cdot \log\log 5 + \frac{\pi}{180} \sec\frac{\pi x}{180} \tan\frac{\pi x}{180}$$ **Ans.**

EXERCISE 28.1

Differentiate ab-initio the following:

1. $\frac{1}{3x+5}$ **Ans.** $-\frac{3}{(3x+5)^2}$

2. $\frac{ax+b}{cx+d}$ **Ans.** $\frac{(ad-bc)}{(cx+d)^2}$

3. $(2x^2+1)^{1/2}$ **Ans.** $2x(2x^2+1)^{-1/2}$

4. $\sin^2 x$ **Ans.** $\sin 2x$

5. $\cos^2 x$ **[Diploma IETE, June 2005]** **Ans.** $-\sin 2x$

6. $\frac{1}{\sqrt{x}}$ **Ans.** $-\frac{1}{2} x^{-3/2}$

7. $\sqrt{\sec x}$ **Ans.** $\frac{1}{2}\sqrt{\sec x} \tan x$

8. $\sqrt{ax+b}$ **Ans.** $\frac{a}{2\sqrt{ax+b}}$

9. $\sqrt{\log x}$ **Ans.** $\frac{1}{2x\sqrt{\log x}}$

Differentiate the following with respect to *x*:

10. $\sqrt[3]{x^2}$ **Ans.** $\frac{2}{3} x^{-1/3}$

11. $\frac{1}{\sqrt[7]{x^5}}$ **Ans.** $\frac{-5}{7} x^{-12/7}$

12. $\frac{x^4+8x+5}{x}$ **Ans.** $3x^2 - 5x^{-2}$

13. $2e^{3x} + \tan x - \cos 2x + 9\sin^{-1} x$ **Ans.** $6e^{3x} + \sec^2 x + 2\sin 2x + \frac{9}{\sqrt{1-x^2}}$

14. $(x+1)(x+2)$ **Ans.** $2x+3$

15. $(x+1)^{-5/2}$ **Ans.** $-\frac{5}{2}(x+1)^{-7/2}$

16. $(\sqrt{x}+1)\left(\frac{1}{x}+\frac{1}{\sqrt{x}}\right)$ **Ans.** $-(x^{-3/2}+x^{-2})$

17. $\sec^{-1}\left(\frac{a}{\sqrt{a^2-x^2}}\right)$ **Ans.** $\frac{1}{\sqrt{a^2-x^2}}$

18. If $y = \sqrt{1-x^2}$, calculate x, when $\frac{dy}{dx} = 1$. **Ans.** $x = \pm\frac{1}{\sqrt{2}}$

19. If $y = 1 + \frac{x}{1!} + \frac{x^2}{2!} + \frac{x^3}{3!} + \ldots + \frac{x^n}{n!}$, show that $\frac{dy}{dx} - y + \frac{x^n}{n!} = 0$.

20. If $f(x) = |x|$, then write $f'(2)$. **Ans.** 1

21. If $f(x) = x - [x]$, write the value of $f'\left(\frac{1}{2}\right)$. **Ans.** 1

22. If $y = \frac{2}{3}x^9 - \frac{5}{7}x^7 + 6x^3 - x$, find $\frac{dy}{dx}$ at $x = 1$. **Ans.** 18

23. If $\sqrt{y+x} + \sqrt{y-x} = c$, show that $\frac{dy}{dx} = \frac{y}{x} - \sqrt{\left(\frac{y^2}{x^2} - 1\right)}$

24. The volume V of a right cylinder of radius r and height h is given by $V = \pi r^2 h$, show that (*i*) $\frac{dV}{dr}$ is the area of the curved surface. (*ii*) $\frac{dV}{dh}$ is the area of the base.

25. Differentiate $\tan^{-1}\left(\sqrt{\frac{1-\cos x}{1+\cos x}}\right)$ with respect to x. **[Diploma, IETE, Dec. 2005] Ans.** $\frac{1}{2}$

26. $y = \sqrt{\cos^{-1}(e^x)}$ **Ans.** $-\frac{e^x}{2\sqrt{\cos^{-1}(e^x)}\sqrt{1-e^{2x}}}$.

27. The bending moment M of a beam supported at both ends is given by $M = \frac{W}{2}\left(\frac{l}{2} - x\right)$, where W, l are constants.

Find the shearing force S, assuming $S = \frac{dM}{dx}$. **Ans.** $-\frac{W}{2}$

Choose the correct alternative:

28. If $y = \sin^2 2x$, then $\frac{dy}{dx}$ is equal to **[Diploma IETE Dec. 2006]**

(*a*) 2 sin 4*x* (*b*) 4 sin 2*x* (*c*) sin 4*x* (*d*) 2 sin 2*x* **Ans.** (*a*)

Differential Coefficient of the Product of Two functions

Let $y = u \cdot v$, where u and v are functions of x.

$$y + \delta y = (u + \delta u) \cdot (v + \delta v)$$

$$\frac{\delta y}{\delta x} = \frac{(u+\delta u)\cdot(v+\delta v) - u\cdot v}{\delta x} = \frac{u\cdot\delta v + v\cdot\delta u + \delta u\cdot\delta v}{\delta x}$$

$$\lim_{\delta x \to 0}\frac{\delta y}{\delta x} = \lim_{\delta x \to 0}\left(u\cdot\frac{\delta v}{\delta x}\right) + \lim_{\delta x \to 0}\left(v\cdot\frac{\delta u}{\delta x}\right) + \lim_{\delta x \to 0}\left(\frac{\delta u}{\delta x}\cdot\delta v\right)$$

$$\frac{dy}{dx} = u\cdot\frac{dv}{dx} + v\cdot\frac{du}{dx} + \frac{du}{dx}\times 0 = u\cdot\frac{dv}{dx} + v\frac{du}{dx}$$

$$\boxed{\frac{d}{dx}(\text{I}\cdot\text{II}) = \text{I}\cdot\text{I}' + \text{I}'\cdot\text{II}}$$

Example 12. *If* $y = x^3 \cdot \sin x$, *find* $\frac{dy}{dx}$.

Solution. $\dfrac{dy}{dx} = x^3 \cdot \dfrac{d}{dx}(\sin x) + \sin x \dfrac{d}{dx}(x^3)$

$= x^3 \cdot \cos x + 3x^2 \cdot \sin x.$ **Ans.**

Example 13. *Find the derivative of absolute value function.*

Solution. Let $y = |x| = \sqrt{x^2} = (x^2)^{\frac{1}{2}}$ *(by definition)*

Differentiating w.r.t. 'x' we get

$$\frac{dy}{dx} = \frac{1}{2}(x^2)^{-\frac{1}{2}} \cdot \frac{d}{dx}(x^2) = \frac{1}{2\sqrt{x^2}} \cdot 2x$$

$$= \frac{x}{\sqrt{x^2}} = \frac{x}{|x|}$$

$\therefore \quad \dfrac{d}{dx}(|x|) = \dfrac{x}{|x|},\ x \neq 0.$ **Ans.**

Differential Coefficient of the Quotient of Two functions

Let $y = \dfrac{u}{v}$, where u, v are the functions of x.

$$y + \delta y = \frac{u + \delta u}{v + \delta v}$$

$$\delta y = \frac{u + \delta u}{v + \delta v} - \frac{u}{v} = \frac{u \cdot v + v \cdot \delta u - u \cdot v - u \cdot \delta v}{(v + \delta v) \cdot v}$$

$$= \frac{v \cdot \delta u - u \cdot \delta v}{v \cdot (v + \delta v)}$$

$$\lim_{\delta x \to 0} \frac{\delta y}{\delta x} = \lim_{\delta x \to 0} \frac{v \cdot \dfrac{\delta u}{\delta x} - u \cdot \dfrac{\delta v}{\delta x}}{v\,(v + \delta v)} \Rightarrow \frac{dy}{dx} = \frac{v \cdot \dfrac{du}{dx} - u \cdot \dfrac{dv}{dx}}{v^2}$$

$$\boxed{\frac{d}{dx}\left(\frac{I}{II}\right) = \frac{II \cdot I' - I \cdot II'}{(II)^2}}$$

Example 14. *If* $y = \dfrac{3x^2 + 5}{x^3 + 4}$, *find* $\dfrac{dy}{dx}$.

Solution. $\dfrac{dy}{dx} = \dfrac{(x^3 + 4) \cdot \dfrac{d}{dx}(3x^2 + 5) - (3x^2 + 5) \cdot \dfrac{d}{dx}(x^3 + 4)}{(x^3 + 4)^2}$

$$\frac{dy}{dx} = \frac{(x^3 + 4)(6x) - (3x^2 + 5)(3x^2)}{(x^3 + 4)^2}$$

$= \dfrac{-3x^4 - 15x^2 + 24x}{(x^3 + 4)^2}$ **Ans.**

Example 15. *Find* $\frac{dy}{dx}$, *if* $y = u^3 + 1, u = \frac{1}{\sqrt{v+2}}, v = 9x^3 + 8x^2 - 7x + 2.$

Solution. $y = u^3 + 1, \frac{dy}{du} = 3u^2$

$$u = \frac{1}{\sqrt{v+2}}, \frac{du}{dv} = -\frac{1}{2}(v+2)^{-3/2}$$

$$v = 9x^3 + 8x^2 - 7x + 2, \frac{dv}{dx} = 27x^2 + 16x - 7$$

$$\frac{dy}{dx} = \frac{dy}{du} \times \frac{du}{dv} \times \frac{dv}{dx}$$

$$= (3u^2)\left[-\frac{1}{2}(v+2)^{-3/2}\right](27x^2 + 16x - 7)$$

$$= -\frac{3}{2}\frac{u^2(27x^2 + 16x - 7)}{(v+2)^{3/2}}.$$ **Ans.**

Example 16. *If* $f(x) = |\cos x|$, *find* $f'\left(\frac{3\pi}{4}\right)$ **[SBTE 2015]**

Solution. Here we have $f(x) = |\cos x|$

$$f(x) = -\cos x$$

$$f'(x) = \sin x \qquad \left[\because \cos\left(\frac{3\pi}{4}\right) = -ve\right]$$

$$f'\left(\frac{3\pi}{4}\right) = \sin\left(\frac{3\pi}{4}\right) = \frac{1}{\sqrt{2}}$$ **Ans.**

Example 17. *If* $y = \sqrt{x^2 - 4x + 4}$, *find* $\frac{dy}{dx}$ *at* $x = 1$ **[SBTE 2014]**

Solution: Here we have

$$y = \sqrt{x^2 - 4x + 4}$$

or $$y = \sqrt{(x-2)^2}$$

$$= x - 2$$

$$dy/dx = 1 - 0 = 1.$$ **Ans.**

Example 18. *Find* $\frac{dy}{dx}$ *if* $x^3 + y^3 = 3axy$ **[SBTE 2016, 2013]**

Solution. Here we have

$$x^3 + y^3 = 3axy$$

$$3x^2 + 3y^2\frac{dy}{dx} = 3a\frac{d}{dx}(xy)$$

$\Rightarrow$ $$x^2 + y^2\frac{dy}{dx} = a\left\{x\frac{dy}{dx} + y \cdot 1\right\}$$

$$x^2 + y^2 \frac{dy}{dx} = ax\frac{dy}{dx} + ay$$

$$\frac{dy}{dx}(y^2 - ax) = ay - x^2$$

$$\therefore \quad \frac{dy}{dx} = \frac{ay - x^2}{y^2 - ax}$$ **Ans.**

Example 19. *Find* $\frac{dy}{dx}$ *of the following functions.*

(i) $y = x\,|x|, x < 0$ **[SBTE. 2016, 2015]**

(ii) $y = |x - 5|$ at $x = 0$ **[SBTE. 2018]**

(iii) $y = x^3 - 3|x| + \frac{2}{|x|}, x < 0$ **[SBTE. 2015]**

Solution.

(i) $y = x\,|x|, x < 0$

$\Rightarrow$ $y = x(-x)$ $\quad [\because |x| = -x, x < 0]$

$\Rightarrow$ $y = -x^2$

$\Rightarrow$ $\frac{dy}{dx} = -2x$ **Ans.**

(ii) $y = |x - 5|$ at $x = 0$

$y = -(x - 5)$ $\quad [\because |x - 5| = -(x - 5) \text{ at } x = 0]$

$\frac{dy}{dx} = -1 + 0 = -1$ **Ans.**

(iii) $y = x^3 - 3|x| + \frac{2}{|x|}, x < 0$

$\Rightarrow$ $y = x^3 - 3(-x) + \frac{2}{(-x)}$ $\quad [\because |x| = -x, x < 0]$

$\Rightarrow$ $y = x^3 + 3x - \frac{2}{x}$

$\Rightarrow$ $\frac{dy}{dx} = 3x^2 + 3 + \frac{2}{x^2}$ **Ans.**

Example 20. *Find the differential coefficient of cos (sin x)*

Solution. Let $y = \cos(\sin x)$

$$\frac{dy}{dx} = -\sin(\sin x) \cdot \frac{d}{dx}(\sin x)$$

$$= -\sin(\sin x) \cdot \cos x.$$ **Ans.**

Example 21. *Find* $\frac{dy}{dx}$, *if* $y = \sin[\cos(\tan x)]$.

Solution. $y = \sin[\cos(\tan x)]$

$$\frac{dy}{dx} = \cos[\cos(\tan x)] \cdot \frac{d}{dx}[\cos(\tan x)]$$

$$= \cos[\cos(\tan x)][-\sin(\tan x)] \cdot \frac{d}{dx}(\tan x)$$

$$= -\cos[\cos(\tan x)][\sin(\tan x)] \cdot \sec^2 x.$$ **Ans.**

Example 22. *Find* $\frac{dy}{dx}$ *if* $y = \log(\sin x)^{\cos x}$ **[SBTE. 2010]**

Solution. $y = \log(\sin x)^{\cos x} = \cos x \log \sin x$

$$\Rightarrow \quad \frac{dy}{dx} = \cos x \cdot \frac{d}{dx}\log(\sin x) + \log(\sin x) \cdot \frac{d}{dx}(\cos x)$$

$$= \cos x \cdot \left[\frac{1}{\sin x} \cdot \frac{d}{dx}(\sin x)\right] + \log(\sin x) \cdot (-\sin x)$$

$$= \cos x \left[\frac{\cos x}{\sin x}\right] - \sin x \log(\sin x)$$

$$= \frac{\cos^2 x}{\sin x} - \sin x \cdot \log(\sin x).$$ **Ans.**

Example 23. *Find the differential coefficient of $x^3 \sin^4 x (\log x)^5$.*

Solution. Let $y = x^3 \cdot \sin^4 x (\log x)^5$

$$\frac{dy}{dx} = x^3 \cdot \frac{d}{dx}[\sin^4 x (\log x)^5] + \sin^4 x (\log x)^5 \frac{d}{dx}(x^3)$$

$$= x^3 \left[\sin^4 x \cdot \frac{d}{dx}(\log x)^5 + (\log x)^5 \cdot \frac{d}{dx}\sin^4 x\right] + \sin^4 x (\log x)^5 \cdot 3x^2$$

$$= x^3 \left[\sin^4 x \cdot 5(\log x)^4 \frac{d}{dx}(\log x) + (\log x)^5 \cdot 4\sin^3 x \cdot \frac{d}{dx}\sin x\right] + 3x^2 \cdot \sin^4 x (\log x)^5$$

$$= 5x^3 \sin^4 x (\log x)^4 \frac{1}{x} + 4x^3 (\log x)^5 \sin^3 x \cos x + 3x^2 \sin^4 x (\log x)^5$$

$$= 5x^2 \sin^4 x (\log x)^4 + 4x^3 (\log x)^5 \sin^3 x \cos x + 3x^2 \sin^4 x (\log x)^5$$

$$= x^2 \sin^3 x \cdot (\log x)^4 \cdot [5\sin x + 4x \cos x \log x + 3\sin x \log x]$$ **Ans.**

Example 24. *Find* $\frac{dy}{dx}$ *if* $y = \log\{x - 3 + \sqrt{x^2 - 6x + 1}\}$

Solution. $y = \log\{x - 3 + \sqrt{x^2 - 6x + 1}\}$

$$\frac{dy}{dx} = \frac{1}{x - 3 + \sqrt{x^2 - 6x + 1}} \cdot \frac{d}{dx}\left\{x - 3 + \sqrt{x^2 - 6x + 1}\right\}$$

$$= \frac{1}{x-3+\sqrt{x^2-6x+1}}\left\{1-0+\frac{1}{2}(x^2-6x+1)^{-1/2}\frac{d}{dx}(x^2-6x+1)\right\}$$

$$= \frac{1}{x-3+\sqrt{x^2-6x+1}}\left\{1+\frac{1}{2\sqrt{x^2-6x+1}}\cdot(2x-6)\right\}$$

$$= \frac{1}{x-3+\sqrt{x^2-6x+1}}\cdot\frac{\sqrt{x^2-6x+1}+x-3}{\sqrt{x^2-6x+1}} = \frac{1}{\sqrt{x^2-6x+1}}.$$ **Ans.**

Example 25. *Find* $\frac{dy}{dx}$, *if* $y = \sin^{-1}\left(\frac{x}{\sqrt{x^2+a^2}}\right)$ **[SBTE 2012]**

Solution. $y = \sin^{-1}\left(\frac{x}{\sqrt{x^2+a^2}}\right)$

$$\frac{dy}{dx} = \frac{1}{\sqrt{1-\frac{x^2}{x^2+a^2}}}\cdot\frac{d}{dx}\left(\frac{x}{\sqrt{x^2+a^2}}\right)$$

$$= \frac{1}{\sqrt{\frac{x^2+a^2-x^2}{x^2+a^2}}}\cdot\frac{\sqrt{x^2+a^2}\cdot 1-\frac{1}{2}x(a^2+x^2)^{-1/2}\cdot(2x)}{x^2+a^2}$$

$$= \frac{\sqrt{x^2+a^2}}{a}\cdot\frac{\sqrt{x^2+a^2}-\frac{x^2}{\sqrt{x^2+a^2}}}{x^2+a^2} = \frac{\sqrt{x^2+a^2}}{a}\cdot\frac{x^2+a^2-x^2}{(x^2+a^2)\sqrt{x^2+a^2}}$$

$$= \frac{a}{x^2+a^2}.$$ **Ans.**

Example 26. *Differentiate the function w.r.t. x.* $\tan^{-1}\left(\frac{a\cos x-b\sin x}{b\cos x+a\sin x}\right)$ **[SBTE 2016]**

Solution: Here we have

Let $y = \tan^{-1}\left(\frac{a\cos x-b\sin x}{b\cos x+a\sin x}\right)$

$$y = \tan^{-1}\left(\frac{a-b\tan x}{b+a\tan x}\right)$$

$$y = \tan^{-1}\left(\frac{\frac{a}{b}-\tan x}{1+\frac{a}{b}\tan x}\right)$$

$$y = \tan^{-1}\left(\frac{a}{b}\right) - \tan^{-1}(\tan x)$$

$$y = \tan^{-1}\left(\frac{a}{b}\right) - x$$

$$\therefore \quad \frac{dy}{dx} = 0 - 1 = -1$$ **Ans.**

Example 27. *If* $y = \dfrac{x \sin^{-1} x}{\sqrt{1-x^2}} + \log\left(\sqrt{1-x^2}\right)$, prove that $\dfrac{dy}{dx} = \dfrac{\sin^{-1} x}{(1-x^2)^{3/2}}$ **[SBTE 2016, 2006]**

Solution. Here we have

$$\Rightarrow \quad y = \frac{x \sin^{-1} x}{\sqrt{1-x^2}} + \log\left(\sqrt{1-x^2}\right) = \frac{x \sin^{-1} x}{\sqrt{1-x^2}} + \frac{1}{2}\log(1-x^2)$$

$$\Rightarrow \quad \frac{dy}{dx} = x\left\{\frac{x \sin^{-1} x}{(1-x^2)\sqrt{1-x^2}} + \frac{1}{1-x^2}\right\} + \frac{\sin^{-1} x}{\sqrt{1-x^2}} - \frac{x}{1-x^2}$$

$$= \frac{x^2 \sin^{-1} x}{(1-x^2)\sqrt{1-x^2}} + \frac{x}{1-x^2} + \frac{\sin^{-1} x}{\sqrt{1-x^2}} - \frac{x}{1-x^2}$$

$$= \frac{\sin^{-1} x}{\sqrt{1-x^2}}\left(\frac{1}{1-x^2}\right)$$

$$\therefore \quad \frac{dy}{dx} = \frac{\sin^{-1} x}{(1-x^2)^{3/2}}$$ **Proved**

Example 28. Differentiate with respect to $x \cdot \tan^{-1}\left(\dfrac{\sqrt{1+\sin x} - \sqrt{1-\sin x}}{\sqrt{1+\sin x} + \sqrt{1-\sin x}}\right)$ **[SBTE 2017]**

Solution. Here we have

$$\Rightarrow \quad y = \tan^{-1}\left(\frac{\sqrt{1+\sin x} - \sqrt{1-\sin x}}{\sqrt{1+\sin x} + \sqrt{1-\sin x}}\right)$$

$$\Rightarrow \quad y = \tan^{-1}\left(\frac{\sqrt{(\cos x/2 + \sin x/2)^2} - \sqrt{(\cos x/2 - \sin x/2)^2}}{\sqrt{(\cos x/2 + \sin x/2)^2} + \sqrt{(\cos x/2 - \sin x/2)^2}}\right)$$

$$\Rightarrow \quad y = \tan^{-1}\left(\frac{2\sin x/2}{2\cos x/2}\right) = \tan^{-1} \cdot \tan(x/2)$$

$$y = x/2$$

$$\therefore \quad \frac{dy}{dx} = \frac{1}{2}$$ **Ans.**

Example 29. *Find* $\frac{dy}{dx}$, *if* $y = \sin x^\circ + \sin x$ **[SBTE. 2014]**

Solution. Here we have

$$y = \sin x^\circ + \sin x$$

$$y = \sin\left(\frac{\pi x}{180}\right) + \sin x$$

$$\frac{dy}{dx} = \cos\left(\frac{\pi x}{180}\right)\frac{d}{dx}\left(\frac{\pi x}{180}\right) + \cos x$$

$$= \cos\left(\frac{\pi x}{180}\right)\cdot\frac{\pi}{180} + \cos x$$

$$\left[\because\ 180^\circ = \pi,\quad 1^\circ = \frac{\pi}{180},\quad x^\circ = \frac{\pi x}{180}\right]$$

$$\therefore\quad \frac{dy}{dx} = \frac{\pi}{180}\cos\left(\frac{\pi x}{180}\right) + \cos x$$ **Ans.**

Example 30. *Find* $\frac{dy}{dx}$, *if* $y = \cos(x - y)$. **[SBTE. 2016, 2007]**

Solution. Here we have

$$y = \cos(x - y)$$

$$\Rightarrow\quad \frac{dy}{dx} = -\sin(x - y)\frac{d}{dx}(x - y)$$

$$= -\sin(x - y)\left\{1 - \frac{dy}{dx}\right\}$$

$$\Rightarrow\quad \frac{dy}{dx} = -\sin(x - y) + \sin(x - y)\frac{dy}{dx}$$

$$\Rightarrow\quad \frac{dy}{dx}\{1 - \sin(x - y)\} = -\sin(x - y)$$

$$\therefore\quad \frac{dy}{dx} = \frac{-\sin(x - y)}{1 - \sin(x - y)}$$ **Ans.**

Example 31. *Find* $\frac{dy}{dx}$ *if* $y = \sin x + a^x$ **[SBTE. 2015]**

Solution. Here we have $y = \sin x + a^x$

$$\frac{dy}{dx} = \cos x + a^x \log a$$ **Ans.**

Example 32. *Find* $\frac{dy}{dx}$ *if* $y = x^3 e^x$. **[SBTE. 2016]**

Solution. Here we have

$$y = x^3 e^x$$

$$\frac{dy}{dx} = x^3\frac{d}{dx}(e^x) + e^x\frac{d}{dx}(x^3) = x^3 e^x + e^x\cdot 3x^2$$

$$\therefore\quad \frac{dy}{dx} = x^2 e^x(x + 3)$$ **Ans.**

Example 33. *Find* $\frac{dy}{dx}$, *if* $y = \frac{e^x}{\sin x}$ **[SBTE. 2008]**

Solution. $\frac{dy}{dx} = \frac{\sin x\,(e^x) - e^x\,(\cos x)}{\sin^2 x} = \frac{e^x \cdot (\sin x - \cos x)}{\sin^2 x}$ **Ans.**

Example 34. *Find* $\frac{dy}{dx}$ *for* $\sin(xy) + \frac{x}{y} = x^2 - y^2$

Solution. We have, $\sin(xy) + \frac{x}{y} = x^2 - y^2$ ... (1)

Differentiating (1) w.r.t. 'x', we get

$$\cos(xy)\left[y + x\frac{dy}{dx}\right] + \frac{1}{y} - \frac{x}{y^2}\frac{dy}{dx} = 2x - 2y\frac{dy}{dx}$$

$$\Rightarrow \quad y\cos(xy) + x\cos(xy)\frac{dy}{dx} + \frac{1}{y} - \frac{x}{y^2}\frac{dy}{dx} = 2x - 2y\frac{dy}{dx}$$

$$\Rightarrow \quad \left[x\cos(xy) - \frac{x}{y^2} + 2y\right]\frac{dy}{dx} = -y\cos(xy) - \frac{1}{y} + 2x$$

$$\therefore \quad \frac{dy}{dx} = \frac{-y\cos(xy) - \frac{1}{y} + 2x}{x\cos(xy) - \frac{x}{y^2} + 2y}$$ **Ans.**

Example 35. *If* $f(x) = \sin\left(\frac{\pi}{2}[x] - x^5\right), 1 \le x < 2$ *and* $[x]$ *denotes the greatest integer less than or equal to x, find* $f'\left(\sqrt[5]{\frac{\pi}{2}}\right)$.

Solution. We have, $f(x) = \left(\frac{\pi}{2}[x] - x^5\right)$

since, $1 \le x < 2 \quad \therefore \quad [x] = 1 \quad \Rightarrow \quad f(x) = \sin\left(\frac{\pi}{2}[1] - x^5\right)$

$$\therefore \quad f(x) = \sin\left(\frac{\pi}{2} - x^5\right) = \cos x^5 \Rightarrow \quad f'(x) = -\sin x^5 \cdot 5x^4$$

$$\Rightarrow \quad f'\left(\sqrt[5]{\frac{\pi}{2}}\right) = -5\left(\sqrt[5]{\frac{\pi}{2}}\right)^{\frac{4}{5}} \cdot \sin\frac{\pi}{2} = -5\left(\frac{\pi}{2}\right)^4$$ **Ans.**

Example 36. *Find* $\frac{dy}{dx}$ *when* $y = \frac{1}{2}x\sqrt{a^2 - x^2} + \frac{1}{2}a^2\sin^{-1}\left(\frac{x}{a}\right)$ **[B.T.E. Delhi, May 2008]**

Solution. We have, $y = \frac{1}{2}x\sqrt{a^2 - x^2} + \frac{1}{2}a^2\sin^{-1}\left(\frac{x}{a}\right)$ ... (1)

Differentiating (1) w.r.t. 'x', we get

$$\frac{dy}{dx} = \frac{1}{2}\left[x\,\frac{1}{2}(a^2 - x^2)^{\frac{1}{2}-1}(-2x) + \sqrt{a^2 - x^2}\right] + \frac{1}{2}a^2\left[\frac{1}{\sqrt{1-\left(\frac{x}{a}\right)^2}}\cdot\frac{1}{a}\right]$$

$$= \frac{1}{2}\left[\frac{-x^2}{\sqrt{a^2 - x^2}} + \sqrt{a^2 - x^2}\right] + \frac{1}{2}a^2\left[\frac{a}{\sqrt{a^2 - x^2}}\cdot\frac{1}{a}\right]$$

$$= \frac{1}{2}\left[\frac{-x^2 + a^2 - x^2}{\sqrt{a^2 - x^2}}\right] + \frac{1}{2}a^2\left[\frac{1}{\sqrt{a^2 - x^2}}\right] = \frac{1}{2}\left[\frac{-2x^2 + a^2}{\sqrt{a^2 - x^2}} + \frac{a^2}{\sqrt{a^2 - x^2}}\right]$$

$$= \frac{1}{2}\,\frac{2a^2 - 2x^2}{\sqrt{a^2 - x^2}} = \sqrt{a^2 - x^2}$$

Ans.

EXERCISE 28.2

Find the differential coefficient of

1. $x \cos x$ **Ans.** $\cos x - x \sin x$ **2.** $e^x (a + b \cos x)$ **Ans.** $e^x (a + b \cos x - b \sin x)$

3. If $y = e^x \sin (3x + 2) \sin (2x - 1) \cos (2x + 1)$ find $\frac{dy}{dx}$. **[B.T.E. Delhi, May 2008]**

$$\left[\textbf{Hint. } y = \frac{e^x}{2}[\cos(x+3) - \cos(5x+1)]\cos(2x+1)\right.$$

$$= \frac{e^x}{2}[\cos(x+3)\cos(2x+1) - \cos(5x+1)\cos(2x+1)]$$

$$= \frac{e^x}{4}[\cos(3x+4) + \cos(x-2) - \cos(7x+2) - \cos(3x)]$$

$$\frac{dy}{dx} = \frac{e^x}{4}[\cos(3x+4) + \cos(x-2) - \cos(7x+2) - \cos(3x)$$

$$\left. - 3\sin(3x+4) - \sin(x-2) + 7\sin(7x+2) + 3\sin 3x]\right]$$

4. $\frac{1+x^2}{1-x^2}$ **Ans.** $\frac{4x}{(1-x^2)^2}$ **5.** $\frac{\log x}{\sin x}$ **Ans.** $\frac{\frac{\sin x}{x} - \log x \cdot \cos x}{\sin^2 x}$

6. $\frac{x \sin^{-1} x}{\sqrt{1-x^2}}$ **Ans.** $\frac{x(1-x^2)^{1/2} + \sin^{-1} x}{(1-x^2)^{3/2}}$

7. $(a + bx^2)^{3/2}$ **Ans.** $3bx\,(a + bx^2)^{1/2}$ **8.** e^{ax^2+b} **Ans.** $2axe^{ax^2+b}$

9. $(\sin^{-1} x^4)^4$ **Ans.** $\frac{16x^3(\sin^{-1} x^4)^3}{\sqrt{1-x^8}}$ **10.** $\log\left[x + \sqrt{x^2 + a^2}\right]$ **Ans.** $\frac{1}{\sqrt{x^2 + a^2}}$

11. $\log\left[x+2+\sqrt{x^2+4x+1}\right]$ **Ans.** $\dfrac{1}{\sqrt{x^2+4x+1}}$

12. If $y = x^4 \tan x$, find $\dfrac{dy}{dx}$. **Ans.** $x^4 \sec^2 x + 4x^3 \tan x$.

13. If $x\sqrt{(1+y)} + y\sqrt{(1+x)} = 0$, show that $\dfrac{dy}{dx} = -\dfrac{1}{(1+x)^2}$

28.19 LOGARITHMIC DIFFERENTIATION

1. Take the logarithm of the given function. **2.** Then differentiate

This method is useful for those functions in which

1. The base and index both are variables.

2. A number of functions are multiplied or divided.

Example 37. *Differentiate* $(\tan x)^{\sec x}$.

Solution. Let $y = (\tan x)^{\sec x}$.

Taking log of both sides

$$\log y = \log (\tan x)^{\sec x} = \sec x \cdot \log (\tan x)$$

On differentiation, we have

$$\frac{1}{y}\frac{dy}{dx} = \sec x \cdot \frac{\sec^2 x}{\sin x} + \sec x \tan x \log (\tan x)$$

$$\frac{dy}{dx} = y\left[\frac{1}{\cos x}\frac{\cos x}{\sin x}\sec^2 x + \sec x \cdot \tan x \log (\tan x)\right]$$

$$= (\tan x)^{\sec x}\left[\operatorname{cosec} x \cdot \sec^2 x + \sec x \cdot \tan x \cdot \log (\tan x)\right]$$ **Ans.**

Example 38. *Differentiate* $(\sin x)^{\log x}$

Solution. Let $y = (\sin x)^{\log x}$

$\Rightarrow$ $\log y = \log (\sin x)^{\log x} = \log x \cdot \log (\sin x)$

On differentiation, we have

$$\frac{1}{y}\frac{dy}{dx} = \log x\,\frac{\cos x}{\sin x} + \frac{1}{x}\cdot \log (\sin x)$$

$$\frac{dy}{dx} = y\left[\cot x \cdot \log x + \frac{1}{x}\log (\sin x)\right]$$

$$= (\sin x)^{\log x}\left[\cot x \cdot \log x + \frac{1}{x}\log (\sin x)\right]$$ **Ans.**

Example 39. *Find dy/dx given* $x^y + y^x = 10$. **[SBTE. 2013, 2007, 2005]**

Solution. $x^y + y^x = 10$

$u + v = 10$ [where $u = x^y$, $v = y^x$]

$$\frac{du}{dx} + \frac{dv}{dx} = 0 \qquad \dots (1)$$

$u = x^y$

$\Rightarrow$ $\log u = \log x^y = y \log x$

Differentiating w.r.t. "x", we get

$$\frac{1}{u}\frac{du}{dx} = \frac{y}{x} + \frac{dy}{dx}(\log x)$$

$$\frac{du}{dx} = u\left[\frac{y}{x} + \frac{dy}{dx}\log x\right] = x^y\left[\frac{y}{x} + \frac{dy}{dx}(\log x)\right]$$

$$= y\,x^{y-1} + x^y\frac{dy}{dx}(\log x) \quad \text{... (2)}$$

Again $\quad v = y^x \quad$ or $\quad \log v = \log y^x$

$\Rightarrow \quad \log v = x \log y$

Differentiating w.r.t 'x' we get

$$\frac{1}{v}\frac{dv}{dx} = x\left(\frac{1}{y}\frac{dy}{dx}\right) + 1\cdot\log y$$

$$\Rightarrow \quad \frac{dv}{dx} = v\left[\frac{x}{y}\frac{dy}{dx} + \log y\right]$$

$$\Rightarrow \quad \frac{dv}{dx} = y^x\left[\frac{x}{y}\frac{dy}{dx} + \log y\right]$$

$$= x\,y^{x-1}\frac{dy}{dx} + y^x\log y \quad \text{... (3)}$$

Putting the values of $\frac{du}{dx}$ and $\frac{dv}{dx}$ from (2) and (3) in (1), we get

$$y\,x^{y-1} + x^y\log x\frac{dy}{dx} + x^{x-1}\frac{dy}{dx} + y^x\log y = 0$$

$$(x^y\log x + x\,y^{x-1})\frac{dy}{dx} = -(y\,x^{y-1} + y^x\log y)$$

$$\frac{dy}{dx} = -\frac{y\,x^{y-1} + y^x\log y}{x\,y^{x-1} + x^y\log x} \quad \textbf{Ans.}$$

Example 40. *If* $y = (\tan x)^{\cot x} + (\cot x)^{\tan x}$, *find* $\frac{dy}{dx}$ **[SBTE. 2015, 2007]**

Solution. We have,

$$y = (\tan x)^{\cot x} + (\cot x)^{\tan x}$$

Let $\quad y = u + v \quad$ ... (1)

where, $\quad u = (\tan x)^{\cot x} \quad$ and $v = (\cot x)^{\tan x}$

$\Rightarrow \quad \log u = \cot x \log \tan x$

$$\Rightarrow \quad \frac{1}{u}\frac{du}{dx} = \cot x\frac{d}{dx}\log\tan x + \log\tan x\frac{d}{dx}\cot x$$

$$= \cot x\cdot\frac{1}{\tan x}\frac{d}{dx}\tan x + \log\tan x\,(-\operatorname{cosec}^2 x)$$

$$= \cot^2 x \cdot \sec^2 x + \log \tan x\,(-\operatorname{cosec}^2 x)$$

$$= \frac{\cos^2 x}{\sin^2 x} \cdot \frac{1}{\cos^2 x} - \operatorname{cosec}^2 x \log \tan x$$

$$= \operatorname{cosec}^2 x - \operatorname{cosec}^2 x \log \tan x = \operatorname{cosec}^2 x\,(1 - \log \tan x)$$

$$\Rightarrow \quad \frac{du}{dx} = u\,[\operatorname{cosec}^2 x\,(1 - \log \tan x)]$$

$$\Rightarrow \quad \frac{du}{dx} = (\tan x)^{\cot x}\,[\operatorname{cosec}^2 x\,(1 - \log \tan x)] \qquad \text{... (2)}$$

and $\quad v = (\cot x)^{\tan x}$

$$\Rightarrow \quad \log v = \tan x \log(\cot x)$$

$$\Rightarrow \quad \frac{1}{v}\frac{dv}{dx} = \tan x \frac{d}{dx}\log(\cot x) + \log(\cot x)\frac{d}{dx}(\tan x)$$

$$= \tan x \cdot \frac{1}{\cot x}(-\operatorname{cosec}^2 x) + \log(\cot x) \cdot \sec^2 x$$

$$= -\tan^2 x \cdot (\operatorname{cosec}^2 x) + \log(\cot x)\sec^2 x$$

$$= -\frac{\sin^2 x}{\cos^2 x} \cdot \frac{1}{\sin^2 x} + \log(\cot x)\sec^2 x$$

$$= -\sec^2 x + \sec^2 x \log(\cot x) = -\sec^2 x\,[1 - \log(\cot x)]$$

$$\Rightarrow \quad \frac{dv}{dx} = -v\,[\sec^2 x\,(1 - \log \cot x)]$$

$$\Rightarrow \quad \frac{dv}{dx} = -(\cot x)^{\tan x}\,[\sec^2 x\,(1 - \log \cot x)] \qquad \text{... (3)}$$

Now, from (1), $\quad \frac{dy}{dx} = \frac{du}{dx} + \frac{dv}{dx}$ ... (4)

Putting the values $\frac{du}{dx}$ and $\frac{dv}{dx}$ from (2) and (3) in (4), we get

$$\frac{dy}{dx} = (\tan x)^{\cot x}\,[\operatorname{cosec}^2 x\,(1 - \log \tan x)] - (\cot x)^{\tan x}\,[\sec^2 x\,(1 - \log \cot x)]$$ **Ans.**

Example 41. *Find* $\frac{dy}{dx}$ *if* $y = x^{x^x}$ **[SBTE. 2017]**

Solution. Here we have

$$y = x^{x^x} = e^{\log x^{x^x}}$$

$$\Rightarrow \quad y = e^{x^x \cdot \log x} \qquad \left[\because e^{\log x} = x\right]$$

$$\Rightarrow \quad \frac{dy}{dx} = e^{x^x \cdot \log x} \cdot \frac{d}{dx}(x^x \cdot \log x)$$

$$= x^{x^x}\frac{d}{dx}(e^{\log x^x} \times \log x)$$

$$= x^{x^x} \frac{d}{dx}(e^{x\log x} \times \log x)$$

$$= x^{x^x}\left[e^{x\log x} \cdot \frac{d}{dx}(\log x) + \log x \frac{d}{dx}(e^{x\log x})\right]$$

$$= x^{x^x}\left[e^{x\log x} \times \frac{1}{x} + \log x \times e^{x\log x} \cdot \frac{d}{dx}(x \cdot \log x)\right]$$

$$= x^{x^x}\left[e^{x\log x^x} \times \frac{1}{x} + \log x \times e^{\log x^x}\left\{x \times \frac{1}{x} + \log x\right\}\right]$$

$$\therefore \quad \frac{dy}{dx} = x^{x^x} \times x^x\left[\frac{1}{x} + \log x\,(1+\log x)\right]$$ **Ans.**

Example 42. *Find* $\frac{dy}{dx}$, *if* $y = x^x$ **[SBTE. 2016, 2008]**

Solution. Here we have

$$y = x^x$$

Taking log both sides, we get

$$\log y = x \cdot \log x$$

$$\frac{d}{dx}(\log y) = \frac{d}{dx}(x \cdot \log x)$$

$$\Rightarrow \quad \frac{1}{y}\frac{dy}{dx} = x \cdot \frac{1}{x} + \log x \cdot 1$$

$$\Rightarrow \quad \frac{1}{y}\frac{dy}{dx} = 1 + \log x$$

$$\Rightarrow \quad \frac{dy}{dx} = y(1+\log x) = (1+\log x)\cdot x^x$$ **Ans.**

Example 43. *Find* $\frac{dy}{dx}$ *if* $x^y = y^x$ **[SBTE. 2016, 2013, 2007]**

Solution. Here we have

$$x^y = y^x$$

Taking log on both sides, we get

$$\Rightarrow \quad y \cdot \log x = x \log y$$

Differentiating on both sides w.r.t. x, we get

$$\Rightarrow \quad y \cdot \frac{1}{x} + \log x \frac{dy}{dx} = x \cdot \frac{1}{y}\frac{dy}{dx} + \log y \cdot 1$$

$$\Rightarrow \quad \log x \frac{dy}{dx} - \frac{x}{y}\frac{dy}{dx} = \log y - \frac{y}{x} = \frac{x\log y - y}{x}$$

$$\Rightarrow \quad \frac{dy}{dx}\left(\frac{y\log x - x}{y}\right) = \frac{x\log y - y}{x}$$

$$\therefore \quad \frac{dy}{dx} = \frac{y\,(x\log y - y)}{x\,(y\log x - x)}$$ **Ans.**

Example 44. *If* $x^y = e^{x-y}$, *prove that* $\dfrac{dy}{dx} = \dfrac{\log x}{(1+\log x)^2}$ **[SBTE. 2016, 2015, 2014]**

Solution. Given $x^y = e^{x-y}$

Taking log both sides, we get

$$y\log x = \log(e^{x-y})$$

$$\Rightarrow \quad y\log x = (x-y)\log e \qquad [\because \log e = 1]$$

$$\Rightarrow \quad y\log x = x - y$$

$$\Rightarrow \quad y + y\log x = x$$

$$\Rightarrow \quad y = \frac{x}{1+\log x}$$

$$\therefore \quad \frac{dy}{dx} = \frac{\log x}{(1+\log x)^2}$$ **Proved.**

Example 45. *Differentiate* $\left(\dfrac{x^2-2ax}{b-cx^2}\right)^{1/2}$ *with respect to x.*

Solution. Let $y = \left(\dfrac{x^2-2ax}{b-cx^2}\right)^{1/2}$

Taking log of both sides, we obtain

$$\log y = \log\left(\frac{x^2-2ax}{b-cx^2}\right)^{1/2} = \frac{1}{2}\log\left(\frac{x^2-2ax}{b-cx^2}\right)$$

$$= \frac{1}{2}\log(x^2-2ax) - \frac{1}{2}\log(b-cx^2)$$

Differentiating, we get

$$\frac{1}{y}\frac{dy}{dx} = \frac{1}{2}\left(\frac{2x-2a}{x^2-2ax}\right) - \frac{1}{2}\left(\frac{-2cx}{b-cx^2}\right) = \frac{x-a}{x^2-2ax} + \frac{cx}{b-cx^2}$$

$$\frac{dy}{dx} = y\left[\frac{x-a}{x^2-2ax} + \frac{cx}{b-cx^2}\right]$$

$$= \left[\frac{x^2-2ax}{b-cx^2}\right]^{\frac{1}{2}}\left[\frac{x-a}{x^2-2ax} + \frac{cx}{b-cx^2}\right].$$ **Ans.**

Example 46. *Find* : $\frac{d}{dx}\log(\sec x + \tan x)$. **[B.T.E. Delhi Jan. 2009]**

Solution. $\frac{d}{dx}\log(\sec x + \tan x) = \frac{1}{\sec x + \tan x}\frac{d}{dx}(\sec x + \tan x)$

$$= \frac{1}{\sec x + \tan x}[\sec x \tan x + \sec^2 x]$$

$$= \frac{\sec x\,[\tan x + \sec x]}{[\sec x + \tan x]} = \sec x.$$ **Ans.**

Example 47. *If* $y = \left\{x + \sqrt{x^2 + a^2}\right\}^n$, *prove that* $\frac{dy}{dx} = \frac{ny}{\sqrt{x^2 + a^2}}$. **[SBTE. 2015, 2013, 2012]**

Solution. Here we have

$$y = \left\{x + \sqrt{x^2 + a^2}\right\}^n$$

Taking log both sides, we get

$$\log y = n\log\left\{x + \sqrt{x^2 + a^2}\right\}$$

$$\Rightarrow \quad \frac{1}{y}\frac{dy}{dx} = \frac{n}{x + \sqrt{x^2 + a^2}} \cdot \frac{d}{dx}\left\{x + \sqrt{x^2 + a^2}\right\}$$

$$\Rightarrow \quad \frac{dy}{dx} = \frac{ny}{x + \sqrt{x^2 + a^2}}\left\{1 + \frac{1}{2\sqrt{x^2 + a^2}}(2x + 0)\right\}$$

$$\Rightarrow \quad \frac{dy}{dx} = \frac{ny}{x + \sqrt{x^2 + a^2}}\left\{1 + \frac{x}{\sqrt{x^2 + a^2}}\right\}$$

$$\Rightarrow \quad \frac{dy}{dx} = \frac{ny}{x + \sqrt{x^2 + a^2}}\left\{\frac{\sqrt{x^2 + a^2} + x}{\sqrt{x^2 + a^2}}\right\}$$

$$\therefore \quad \frac{dy}{dx} = \frac{ny}{\sqrt{x^2 + a^2}}$$ **Proved**

Example 48. *For a positive value of 'a' find dy/dx*

where $y = a^{t + \frac{1}{t}}$ *and* $x = \left(t + \frac{1}{t}\right)^a$ **[SBTE. 2017]**

Solution. $y = a^{t + \frac{1}{t}}$

$$\frac{dy}{dt} = a^{\left(t + \frac{1}{t}\right)} \cdot \log a\left(1 - \frac{1}{t^2}\right)$$

and $\quad x = \left(1 + \frac{1}{t}\right)^a$

$$\frac{dx}{dt} = a\left(t+\frac{1}{t}\right)^{a-1} \cdot \left(1-\frac{1}{t^2}\right)$$

$$\frac{dy}{dx} = \frac{\frac{dy}{dt}}{\frac{dx}{dt}} = \frac{a^{\left(t+\frac{1}{t}\right)} \cdot \log a\left(1-\frac{1}{t^2}\right)}{a\left(t+\frac{1}{t}\right)^{a-1} \cdot \left(1-\frac{1}{t^2}\right)}$$

$$\frac{dy}{dx} = \frac{a^{\left(t+\frac{1}{t}\right)} \log a}{a\left(t+\frac{1}{t}\right)^{a-1}}$$ **Ans.**

EXERCISE 28.3

Differentiate the following:

1. $x^{\sin x}$ **Ans.** $x^{\sin x}\left(\frac{\sin x}{x} + \cos x \log x\right)$

2. $(\cos x)^x$ **[B.T.E. Delhi June 2005]** **Ans.** $(\cos x)^x\,(-x \tan x + \log \cos x)$

3. $(\sin x)^{\cos x}$ **Ans.** $(\sin x)^{\cos x}\,(\cot x \cos x - \sin x \log \sin x)$

4. $(\tan x)^{\log x}$ **Ans.** $(\tan x)^{\log x}\left(\frac{\log x}{\sin x \cos x} + \frac{\log \tan x}{x}\right)$

5. $(\sin x)^x$ **Ans.** $(\sin x)^x \cdot (\log \sin x + x \cot x)$

6. $(1 + \cos x)^x$ **Ans.** $(1+\cos x)^x\left[\log\left(2\cos^2\frac{x}{2}\right) - x\tan\frac{x}{2}\right]$

7. $x e^{\sqrt{\sin x}}$ **Ans.** $e^{\sqrt{\sin x}}\left[1+\frac{x \cos x}{2\sqrt{\sin x}}\right]$

8. If $(\cos x)^y = (\sin y)^x$, find $\frac{dy}{dx}$ **Ans.** $\frac{dy}{dx} = \frac{\log \sin y + y \tan x}{\log \cos x - x \cot y}$

9. If $y = x^{x^{x\ldots\infty}}$, find $\frac{dy}{dx}$. **[S.B.T.E. 2017]** **Ans.** $\frac{y^2}{x\,(1-y\log x)}$

10. If $y = (\log x)^x + (\sin x)^{\sin x}$, find $\frac{dy}{dx}$

Ans. $(\log x)^x\left(\log(\log x) + \frac{1}{\log x}\right) + (\sin x)^{\sin x}\,(\cos x + \cos x \log \sin x)$

11. If $x^x + x^{\sin x} = y$, find $\frac{dy}{dx}$. **[B.T.E. Delhi May 2008]**

Ans. $\frac{dy}{dx} = x^x\,(1+\log x) + x^{\sin x}\left[\frac{\sin x}{x} + \cos x \log x\right]$

12. If $x^y + y^x = (x+y)^{x+y}$ find $\frac{dy}{dx}$. **Ans.** $\frac{-x^{y-1} \cdot y - y^x \log y + (x+y)^{x+y} \cdot [1+\log(x+y)]}{x^y \log x + xy^{x-1} - (x+y)^{x+y}[1+\log(x+y)]}$

13. If $y = \dfrac{(x+1)^2 (x-2)^{1/2}}{(2x+1)^{1/3} (4x+5)^{2/5}}$, find $\dfrac{dy}{dx}$. **Ans.** $y\left[\dfrac{2}{x+1} + \dfrac{1}{2}\dfrac{1}{x-2} - \dfrac{2}{3}\dfrac{1}{2x+1} - \dfrac{8}{5}\dfrac{1}{4x+5}\right]$

14. If $y = x^{\sin x} + \sin^x x$ then find $\dfrac{dy}{dx}$.

Ans. $x^{\sin x}\left\{\cos x \cdot \log x + \dfrac{\sin x}{x}\right\} + (\sin x)^x \{\log \sin x + x \cot x\}$

28.20 DIFFERENTIATION OF IMPLICIT FUNCTIONS

If a function in x and y is given in such a way that x and y cannot be seperated easily, then the function is called an implicit function.

Remember. The derivative of x w.r.t. x is 1 but that of y w.r.t x is $\dfrac{dy}{dx}$. Do not forget to write $\dfrac{dy}{dx}$ when differentiating a function of y.

Example 49. *Find* $\dfrac{dy}{dx}$, *if* $ax^2 + 2h\,xy + by^2 + 2gx + 2fy + c = 0$

Solution. On differentiating $2ax + 2h\left(x\dfrac{dy}{dx} + y \cdot 1\right) + 2by\dfrac{dy}{dx} + 2g + 2f\dfrac{dy}{dx} = 0$

$$\Rightarrow \quad (2hx + 2by + 2f)\frac{dy}{dx} = -2ax - 2hy - 2g$$

$$\Rightarrow \quad (hx + by + f)\frac{dy}{dx} = -(ax + hy + g)$$

$$\Rightarrow \quad \frac{dy}{dx} = -\left(\frac{ax + hy + g}{hx + by + f}\right).$$

Ans.

Example 50. If $y = \cos(x - y)$ then find the value of $\dfrac{dy}{dx}$ **[SBTE. 2016]**

Solution. $y = \cos(x - y)$

$$\frac{dy}{dx} = -\sin(x-y)\frac{d}{dx}(x-y) = -\sin(x-y)\left(1 - \frac{dy}{dx}\right)$$

$$\Rightarrow \quad \frac{dy}{dx} = -\sin(x-y) + \sin(x-y)\frac{dy}{dx}$$

$$\Rightarrow \quad [\sin(x-y) - 1]\frac{dy}{dx} = \sin(x-y)$$

$$\therefore \quad \frac{dy}{dx} = \frac{\sin(x-y)}{\sin(x-y) - 1}$$

Ans.

EXERCISE 28.4

Differentiate the following w.r.t. "x".

1. $x^3 + y^3 - 3axy = 0$. **[B.T.E. Delhi, Jan. 2009; June 2005]** **Ans.** $\dfrac{x^2 - ay}{ax - y^2}$

2. $x^3 + 3x^2y + 6xy^2 + y^3 = 0$ **Ans.** $-\left(\dfrac{x^2+2xy+2y^2}{x^2+4xy+y^2}\right)$

3. $ax^2 + 2hxy + by^2 = 0$ **Ans.** $-\left(\dfrac{ax+hy}{hx+by}\right)$

4. $x^{3/2} + y^{3/2} = a^{3/2}$ **Ans.** $-\sqrt{\dfrac{x}{y}}$

5. $y = \log y^x$ **Ans.** $\dfrac{y \log y}{y - x}$

6. $x^y - y = 0$ **Ans.** $\dfrac{y^2}{x(1 - y\log x)}$

7. If $x^y = e^{x-y}$, prove that $\dfrac{dy}{dx} = \dfrac{\log x}{(1+\log x)^2}$ **[B.T.E. Delhi June 2005]**

8. If $\sin y = x \sin (a + y)$, prove that $\dfrac{dy}{dx} = \dfrac{\sin^2(a+y)}{\sin a}$.

9. If $\sqrt{(x+y)} + \sqrt{(y-x)} = c$, then show that $\dfrac{dy}{dx} = \dfrac{y}{x} - \sqrt{\left(\dfrac{y^2}{x^2}\right) - 1}$

28.21 PARAMETRIC EQUATIONS

If x and y are expressed in terms of a third variable (θ or t), then the third variable is called the parameter, equation containing parameter known as parametric equation.

Example 51. *If* $x = at^2$ *and* $y = 2at$, *find* $\dfrac{dy}{dx}$. **[SBTE. 2016, 2015]**

Solution. $x = at^2 \quad \therefore \quad \dfrac{dx}{dt} = 2at, \; y = 2at \quad \therefore \quad \dfrac{dy}{dt} = 2a$

$$\frac{dy}{dx} = \frac{\dfrac{dy}{dt}}{\dfrac{dx}{dt}} = \frac{2a}{2at} = \frac{1}{t}.$$ **Ans.**

Example 52. *If* $x = a(\theta + \sin\theta)$, $y = a\cos\theta$. *then show that* $\dfrac{dy}{dx} = -\tan\dfrac{\theta}{2}$

Solution. $x = a(\theta + \sin\theta)$ ... (1)

$y = a\cos\theta$ ... (2)

On differentiating (1) w.r.t. 'θ', we have $\dfrac{dx}{d\theta} = a(1 + \cos\theta)$

On differentiating (2) wr.t. 'θ', we get $\dfrac{dy}{d\theta} = -a\sin\theta$

We know that

$$\frac{dy}{dx}=\frac{\frac{dy}{d\theta}}{\frac{dx}{d\theta}}=\frac{-a\sin\theta}{a(1+\cos\theta)}=\frac{-\sin\theta}{1+\cos\theta}$$

$$=\frac{-2\sin\frac{\theta}{2}\cos\frac{\theta}{2}}{1+2\cos^2\frac{\theta}{2}-1}=\frac{-2\sin\frac{\theta}{2}\cos\frac{\theta}{2}}{2\cos^2\frac{\theta}{2}}=-\tan\frac{\theta}{2}$$ **Proved.**

Example 53. *If* $y=\frac{\sqrt{a^2+x^2}+\sqrt{a^2-x^2}}{\sqrt{a^2+x^2}-\sqrt{a^2-x^2}}$, *show that* $\frac{dy}{dx}=-\frac{2a^2}{x^3}\left\{1+\frac{a^2}{\sqrt{a^4-x^4}}\right\}$

[SBTE. 2016, 2014, 2012]

Solution. Given that: $y=\frac{\sqrt{a^2+x^2}+\sqrt{a^2-x^2}}{\sqrt{a^2+x^2}-\sqrt{a^2-x^2}}$

$$=\frac{\sqrt{a^2+x^2}+\sqrt{a^2-x^2}}{\sqrt{a^2+x^2}-\sqrt{a^2-x^2}}\times\frac{\sqrt{a^2+x^2}+\sqrt{a^2-x^2}}{\sqrt{a^2+x^2}+\sqrt{a^2-x^2}}$$

$$\therefore \quad y=\frac{2a^2+2\sqrt{a^4-x^4}}{2x^2}=\frac{a^2}{x^2}+\frac{\sqrt{a^4-x^4}}{x^2}$$

$$=a^2x^{-2}+\left(\sqrt{a^4-x^4}\right)x^{-2}$$

$$\frac{dy}{dx}=-2a^2x^{-3}+\sqrt{a^4-x^4}\times(-2x^{-3})+x^{-2}\times\frac{1}{2\sqrt{a^4-x^4}}\frac{d}{dx}(a^4-x^4)$$

$$=\frac{-2a^2}{x^3}-\frac{2\sqrt{a^4-x^4}}{x^3}-\frac{2x}{\sqrt{a^4-x^4}}$$

$$=\frac{-2a^2}{x^3}-\left\{\frac{2(a^4-x^4)+2x^4}{x^3\sqrt{a^4-x^4}}\right\}$$

$$=\frac{-2a^2}{x^3}-\left\{\frac{2a^4}{x^3\sqrt{a^4-x^4}}\right\}$$

$$\therefore \quad \frac{dy}{dx}=-\frac{2a^2}{x^3}\left\{1+\frac{a^2}{\sqrt{a^4-x^4}}\right\}$$ **Proved**

Example 54. *If* $y=\log\left\{x+\sqrt{x^2+a^2}\right\}$, *prove that* $(x^2+a^2)\left(\frac{dy}{dx}\right)^2=1$

[SBTE. 2016, 2015, BTE. Delhi. 2016, 2009, 2008, 2007]

Solution. Given: $y = \log\left\{x + \sqrt{x^2 + a^2}\right\}$

$$\frac{dy}{dx} = \frac{1}{x + \sqrt{x^2 + a^2}} \cdot \frac{d}{dx}\left(x + \sqrt{x^2 + a^2}\right)$$

$$\frac{dy}{dx} = \frac{1}{x + \sqrt{x^2 + a^2}}\left[1 + \frac{1}{2\sqrt{x^2 + a^2}} \cdot \frac{d}{dx}(x^2 + a^2)\right]$$

$$= \frac{1}{x + \sqrt{x^2 + a^2}}\left\{1 + \frac{x}{\sqrt{x^2 + a^2}}\right\}$$

$$= \frac{1}{x + \sqrt{x^2 + a^2}}\left\{\frac{\sqrt{x^2 + a^2} + x}{\sqrt{x^2 + a^2}}\right\}$$

$$\frac{dy}{dx} = \frac{1}{\sqrt{x^2 + a^2}} \quad \text{or} \quad \left(\frac{dy}{dx}\right)^2 = \frac{1}{x^2 + a^2}$$

$\therefore \qquad (x^2 + a^2)\left(\dfrac{dy}{dx}\right)^2 = 1$ **Proved**

Example 55. *If* $y = A\cos(\log x) + B\sin(\log x)$, *prove that* $x^2\dfrac{d^2y}{dx^2} + x\dfrac{dy}{dx} + y = 0$ **[SBTE. 2017]**

Solution. Here we have $y = A\cos(\log x) + B\sin(\log x)$ ... (1)

Differentiating (1) w.r.t. *x*, we get

$\Rightarrow \qquad \dfrac{dy}{dx} = -A\sin(\log x)\cdot\dfrac{1}{x} + B\cos(\log x)\cdot\dfrac{1}{x}$

$\Rightarrow \qquad x\dfrac{dy}{dx} = -A\sin(\log x) + B\cos(\log x)$ ...(2)

$\Rightarrow \quad x\dfrac{d^2y}{dx^2} + \dfrac{dy}{dx}\cdot 1 = -A\cos(\log x)\cdot\dfrac{1}{x} - B\sin(\log x)\cdot\dfrac{1}{x}$

$\Rightarrow \quad x^2\dfrac{d^2y}{dx^2} + x\dfrac{dy}{dx} = -A\cos(\log x) - B\sin(\log x)$

L.H.S. $= x^2\dfrac{d^2y}{dx^2} + x\dfrac{dy}{dx} + y$ [From (1)]

$= [-A\cos(\log x) - B\sin(\log x)] + [-A\sin(\log x) + B\cos(\log x)]$
$+ [A\cos x(\log x) + B\sin(\log x)]$

$= 0 =$ R.H.S. **Proved**

Example 56. *If* $y = a\cos nx + b\sin nx$, *such that* $\dfrac{d^2y}{dx^2} + ky = 0$, *find k.* **[SBTE. 2016, 2014]**

Solution. $\qquad y = a\cos nx + b\sin nx$

$$\Rightarrow \quad \frac{dy}{dx} = -an \sin nx + bn \cos nx$$

$$\Rightarrow \quad \frac{d^2y}{dx^2} = -an(n\cos nx) + bn(-n \sin nx)$$

$$\Rightarrow \quad = -n^2(a\cos nx + b \sin nx)$$

$$\Rightarrow \quad \frac{d^2y}{dx^2} = -n^2y \qquad \text{... (1)}$$

$$\Rightarrow \quad \frac{d^2y}{dx^2} + n^2y = 0$$

$$\Rightarrow \quad -n^2y + ky = 0 \qquad \text{[From (1)]}$$

$$\therefore \quad k = n^2$$

Ans.

Example 57. *If* $y = 1 + \frac{x}{1!} + \frac{x^2}{2!} + \frac{x^3}{3!} + \ldots$ *show that* $\frac{dy}{dx} = y.$ **[SBTE. 2016]**

Solution. $y = 1 + \frac{x}{1!} + \frac{x^2}{2!} + \frac{x^3}{3!} + \ldots\ldots$

$$\Rightarrow \quad \frac{dy}{dx} = 0 + 1 + \frac{2x}{2!} + \frac{3x^2}{3!} + \ldots\ldots$$

$$\Rightarrow \quad \frac{dy}{dx} = 1 + \frac{x}{1!} + \frac{x^2}{2!} + \ldots\ldots = y$$

Proved

Example 58. *If* $y = \log \tan\left(\frac{\pi}{4} + \frac{x}{2}\right)$, *prove that* $\frac{dy}{dx} = \sec x.$ **[SBTE. 2016]**

Solution. $y = \log \tan\left(\frac{\pi}{4} + \frac{x}{2}\right)$

$$\frac{dy}{dx} = \frac{1}{\tan\left(\frac{\pi}{4} + \frac{x}{2}\right)} \frac{d}{dx}\left[\tan\left(\frac{\pi}{4} - \frac{x}{2}\right)\right]$$

$$= \cot\left(\frac{\pi}{4} + \frac{x}{2}\right) \cdot \sec^2\left(\frac{\pi}{4} + \frac{x}{2}\right) \cdot \left[0 + \frac{1}{2}\right]$$

$$= \frac{\cos\left(\frac{\pi}{4} + \frac{x}{2}\right)}{\sin\left(\frac{\pi}{4} + \frac{x}{2}\right)} \cdot \frac{1}{\cos^2\left(\frac{\pi}{4} + \frac{x}{2}\right)} \cdot \left(\frac{1}{2}\right)$$

$$= \frac{1}{2\sin\left(\frac{\pi}{4} + \frac{x}{2}\right) \cdot \cos\left(\frac{\pi}{4} + \frac{x}{2}\right)} = \frac{1}{\sin 2\left(\frac{\pi}{4} + \frac{x}{2}\right)}$$

$$\therefore \quad \frac{dy}{dx} = \frac{1}{\sin\left(\frac{\pi}{2} + x\right)} = \frac{1}{\cos x} = \sec x$$

Proved

Example 59. *Find* $\frac{dy}{dx}$, *if* $x=\frac{1-t^2}{1+t^2}$ *and* $y=\frac{2t}{1+t^2}$ **[B.T.E. Delhi June 2005]**

Solution. We have,

$$x=\frac{1-t^2}{1+t^2} \qquad \text{Also,} \qquad y=\frac{2t}{1+t^2}$$

Put $t = \tan\theta$ Put $t = \tan\theta$

$$\Rightarrow \quad x=\frac{1-\tan^2\theta}{1+\tan^2\theta} \qquad \Rightarrow \quad y=\frac{2\tan\theta}{1+\tan^2\theta}$$

$$\Rightarrow \quad x = \cos 2\theta \qquad \Rightarrow \quad y = \sin 2\theta$$

$$\Rightarrow \quad \frac{dx}{d\theta} = -2\sin 2\theta \qquad \Rightarrow \quad \frac{dy}{d\theta} = 2\cos 2\theta$$

Now, $$\frac{dy}{dx}=\frac{dy/d\theta}{dx/d\theta}=\frac{2\cos 2\theta}{-2\sin 2\theta}=-\cot 2\theta=-\frac{1}{\tan 2\theta}$$

$$\Rightarrow \quad \frac{dy}{dx}=-\frac{1-\tan^2\theta}{2\tan\theta}=\frac{1-t^2}{2t}$$ **Ans.**

Example 60. *Differentiate a* $\cos^3 t$ *with respect to a* $\sin^3 t$.

Solution. Let $y = a\cos^3 t$ ... (1)

$z = a\sin^3 t$ ... (2)

Now we have to find $\frac{dy}{dz}$.

Differentiating (1) w.r.t., "t", we get

$$\frac{dy}{dt} = 3a\cos^2 t\,(-\sin t) = -3a\sin t\,(\cos^2 t)$$

On differentiating (2), we have

$$\frac{dz}{dt} = 3a\sin^2 t\,(\cos t)$$

$$\frac{dy}{dz}=\frac{\frac{dy}{dt}}{\frac{dz}{dt}}=\frac{-3a\sin t\cos^2 t}{3a\sin^2 t\cos t}$$

$$= -\cot t.$$ **Ans.**

Example 61. *If* $y=\tan^{-1}\left(\frac{\cos x+\sin x}{\cos x-\sin x}\right)$, *find* $\frac{dy}{dx}$. **[SBTE. 2006, 2005]**

Solution. $$y=\tan^{-1}\left(\frac{\cos x+\sin x}{\cos x-\sin x}\right)=\tan^{-1}\left(\frac{1+\tan x}{1-\tan x}\right)=\tan^{-1}\left(\frac{\tan\frac{\pi}{4}+\tan x}{1-\tan\frac{\pi}{4}\tan x}\right)$$

$$=\tan^{-1}\tan\left(\frac{\pi}{4}+x\right)=\frac{\pi}{4}+x$$ $[\because \tan^{-1}\tan(\theta)=\theta]$

$$\therefore \quad \frac{dy}{dx}=1$$ **Ans.**

Example 62. *If* $\sin y = x\sin(a+y)$*, prove that,* $\dfrac{dy}{dx} = \dfrac{\sin^2(a+y)}{\sin a}$.

[SBTE. 2017, 2015, 2014, 2013]

Solution. Here we have: $\sin y = x\sin(a+y)$

$$x = \frac{\sin y}{\sin(a+y)}$$

$$\frac{dx}{dy} = \frac{\sin(a+y)\cdot\cos y - \sin y\cdot\cos(a+y)\dfrac{d}{dy}(a+y)}{\sin^2(a+y)}$$

$$= \frac{\sin\{(a+y)-y\}}{\sin^2(a+y)} = \frac{\sin a}{\sin^2(a+y)}$$

$$\therefore \qquad \frac{dy}{dx} = \frac{\sin^2(a+y)}{\sin a}$$ **Proved.**

Example 63. *If* $x^2+y^2 = t - \dfrac{1}{t}$ *and* $x^4+y^4 = t^2+\dfrac{1}{t^2}$ *prove that* $\dfrac{dy}{dx} = \dfrac{1}{x^3y}$.

[SBTE. 2016, 2015]

Solution. $x^2+y^2 = t - \dfrac{1}{t}$ ... (1)

and $x^4+y^4 = t^2+\dfrac{1}{t^2}$ ... (2)

Squaring both sides equation (1), we get

$$x^4+y^4+2x^2y^2 = t^2+\frac{1}{t^2}-2 \Rightarrow x^4+y^4+2x^2y^2 = x^4+y^4-2 \quad \left[\because x^4+y^4 = t^2+\frac{1}{t^2}\right]$$

$$x^2y^2 = -1 \qquad \Rightarrow \qquad y^2 = -\frac{1}{x^2}$$

$$2y\frac{dy}{dx} = -\left(\frac{-2}{x^3}\right) = \frac{2}{x^3} \qquad \Rightarrow \qquad y\frac{dy}{dx} = \frac{1}{x^3}$$

$$\therefore \qquad \frac{dy}{dx} = \frac{1}{x^3y}$$ **Proved.**

Example 64. *If* $x\sqrt{1+y} + y\sqrt{1+x} = 0$*, prove that* $\dfrac{dy}{dx} = -\dfrac{1}{(x+1)^2}$.

[SBTE. 2017, 2015, 2014]

Solution. $x\sqrt{1+y} + y\sqrt{1+x} = 0$

$\Rightarrow \quad x^2(1+y) - y^2(1+x) = 0 \quad \Rightarrow \quad x^2 - y^2 + x^2y - xy^2 = 0$

$\Rightarrow \quad (x-y)(x+y+xy) = 0$

$\Rightarrow \quad x+y+xy = 0$ $\qquad [x - y \neq 0]$

$\Rightarrow \quad y(x+1) = -x \quad \Rightarrow \quad y = -\dfrac{x}{x+1}$

$$\frac{dy}{dx} = -\left[\frac{(x+1)\cdot 1 - x}{(x+1)^2}\right] = \frac{-1}{(x+1)^2}$$ **Proved.**

Example 65. *If* $x \cos(a+y) + \cos a \sin(a+y) = 0$, *show that*, $\frac{dy}{dx} = -\frac{\cos^2(a+y)}{\cos a}$.

[SBTE. 2017, 2013]

Solution. $x \cos(a+y) + \cos a \cdot \sin(a+y) = 0$

$$x = -\cos a \frac{\sin(a+y)}{\cos(a+y)} = -\cos a \cdot \tan(a+y)$$

$$\frac{dx}{dy} = -\cos a \cdot \sec^2(a+y) \frac{d}{dy}(a+y)$$

$$= -\cos a \cdot \frac{1}{\cos^2(a+y)}(0+1)$$

$$\frac{dx}{dy} = -\frac{\cos a}{\cos^2(a+y)}$$

$\therefore$ $$\frac{dy}{dx} = -\frac{\cos^2(a+y)}{\cos a}$$ **Proved**

Example 66. *If* $x = a(\theta - \sin\theta)$, $y = a(\sin\theta - \theta\cos\theta)$, *find* $\frac{dy}{dx}$. **[SBTE. 2017]**

Solution. $x = a(\theta - \sin\theta)$, $\frac{dx}{d\theta} = a(1 - \cos\theta)$

$$y = a(\sin\theta - \theta\cos\theta), \frac{dy}{d\theta} = a\theta\sin\theta$$

$$\frac{dy}{dx} = \frac{\frac{dy}{d\theta}}{\frac{dx}{d\theta}} = \frac{a\theta\sin\theta}{a(1-\cos\theta)} = \frac{\theta \cdot 2\sin\theta/2 \cdot \cos\theta/2}{2\sin^2\frac{\theta}{2}}$$

$$\frac{dy}{dx} = \theta\cot\frac{\theta}{2}$$ **Ans.**

Example 67. *If* $x = a\cos\theta$, $y = b\sin\theta$ *and* $\left(\frac{dy}{dx}\right)_{\theta=\pi/4} = 1$ *then find the relation between 'a' and 'b'.* **[DIP. IETE. 2018]**

Solution. $x = a\cos\theta$, $\frac{dx}{d\theta} = -a\sin\theta$

$\Rightarrow$ $y = b\sin\theta$, $\frac{dy}{d\theta} = b\cos\theta$

$\Rightarrow$ $$\frac{dy}{dx} = \frac{\frac{dy}{d\theta}}{\frac{dx}{d\theta}} = \frac{b\cos\theta}{-a\sin\theta} = -\frac{b}{a}\cot\theta$$

$$\Rightarrow \quad \left(\frac{dy}{dx}\right)_{\theta=\pi/4} = -\frac{b}{a}\cot\frac{\pi}{4} = -\frac{b}{a}$$

$$\Rightarrow \quad 1 = -\frac{b}{a} \qquad \left[\because \left(\frac{dy}{dx}\right)_{\theta=\pi/4} = 1\right]$$

$$\therefore \quad a + b = 0$$

Ans.

EXERCISE 28.5

Find $\frac{dy}{dx}$**, if**

1. $x = a(\theta - \sin\theta)$, $y = a(1 - \cos\theta)$ **[B.T.E. Delhi May 2007]** **Ans.** $\cot\frac{\theta}{2}$
2. $x = 3\cos\theta - \cos 3\theta$, $y = 3\sin\theta - \sin 3\theta$ **[B.T.E. Delhi June 2015]** **Ans.** $\tan 2\theta$
3. $x = a(\cos t + \log\tan t/2)$, $y = a\sin t$ **Ans.** $\tan t$
4. Differentiate $7x^5 - 11x^2$ with respect to $7x^2 - 15x$. **Ans.** $\frac{x^3(35x^3 - 22)}{14x - 15}$
5. Differentiate $\log x$ with respect to x^2. **Ans.** $\frac{1}{2x^2}$
6. Differentiate $\tan^{-1}\left[\frac{\sqrt{1+x^2}-1}{x}\right]$ w.r.t. $\tan^{-1} x$. **Ans.** $\frac{1}{2}$
7. Differentiate $\tan^{-1}\left(\frac{2x}{1-x^2}\right)$ w.r.t. $\cos^{-1}\left(\frac{1-x^2}{1+x^2}\right)$ **Ans.** 1
8. Differentiate $e^{\tan x}$ w.r.t. $\tan^{-1} x$. **Ans.** $\sec^2 x \tan x (1 + x^2)$

28.22 TRIGONOMETRICAL SUBSTITUTION AND TRIGONOMETRIC TRANSFORMATION

Sometimes trigonometrical substitution makes the differentiation easier.

Let $x = \sin^{-1} y$ ... (1)

$\sin x = y$ ... (2)

Putting the value of y from (2) into (1) we get

$\Rightarrow \quad x = \sin^{-1}(\sin x)$

Example 68. *Find the differential coefficient of* $\cos^{-1}(4x^3 - 3x)$ **[B.T.E. Delhi, Jan. 2009]**

Solution. Let $y = \cos^{-1}(4x^3 - 3x)$

$y = \cos^{-1}(4\cos^3\theta - 3\cos\theta)$ [Put $x = \cos\theta$]

$= \cos^{-1}\cos 3\theta = 3\theta = 3\cos^{-1} x$

$$\frac{dy}{dx} = -\frac{3}{\sqrt{1-x^2}}$$ **Ans.**

Example 69. *Differentiate* $\tan^{-1}\left(\frac{2x}{1-x^2}\right)$

Solution. Let $y = \tan^{-1}\left(\frac{2x}{1-x^2}\right)$ [Put $x = \tan\theta$]

$$y = \tan^{-1}\left(\frac{2\tan\theta}{1-\tan^2\theta}\right) = \tan^{-1}(\tan 2\theta) = 2\theta = 2\tan^{-1}(x)$$

$$\frac{dy}{dx} = \frac{2}{1+x^2}$$ **Ans.**

Example 70. *Differentiate* $\tan^{-1}\left(\frac{1-\cos x}{1+\cos x}\right)^{1/2}$ **[SBTE. 2010, 2005]**

Solution. Let $y = \tan^{-1}\left(\frac{1-\cos x}{1+\cos x}\right)^{1/2} = \tan^{-1}\left[\frac{1-\left(1-2\sin^2\frac{x}{2}\right)}{1+\left(2\cos^2\frac{x}{2}-1\right)}\right]^{1/2}$

$$= \tan^{-1}\left[\frac{2\sin^2\left(\frac{x}{2}\right)}{2\cos^2\left(\frac{x}{2}\right)}\right]^{1/2} = \tan^{-1}\left(\frac{\sin\frac{x}{2}}{\cos\frac{x}{2}}\right) = \tan^{-1}\left(\tan\frac{x}{2}\right)$$

$$= \frac{x}{2}$$

$\therefore$ $$\frac{dy}{dx} = \frac{1}{2}$$ **Ans.**

Example 71. *Find* $\frac{d}{dx}\cos^{-1}\left(\frac{1-x^2}{1+x^2}\right)$

Solution. $\frac{d}{dx}\cos^{-1}\left(\frac{1-x^2}{1+x^2}\right) = \frac{d}{dx}\cos^{-1}\left(\frac{1-\tan^2\theta}{1+\tan^2\theta}\right)$ [Put $x = \tan\theta \Rightarrow \theta = \tan^{-1}x$]

$$= \frac{d}{dx}\cos^{-1}(\cos 2\theta)$$ $\left[\because \cos 2\theta = \frac{1-\tan^2\theta}{1+\tan^2\theta}\right]$

$$= \frac{d}{dx}(2\theta)$$

$$= 2\frac{d}{dx}(\tan^{-1}x)$$ $[\because \theta = \tan^{-1}x]$

$$= \frac{2}{1+x^2}$$ **Ans.**

Example 72. *Differentiate* $\sin^2\left[\cot^{-1}\sqrt{\frac{1+x}{1-x}}\right]$.

Solution. Let $y = \sin^2\left[\cot^{-1}\sqrt{\frac{1+x}{1-x}}\right]$

Putting $x = \cos\theta$

$$y = \sin^2\left(\cot^{-1}\sqrt{\frac{1+\cos\theta}{1-\cos\theta}}\right) = \sin^2\cot^{-1}\left(\sqrt{\frac{1+2\cos^2\frac{\theta}{2}-1}{1-1+2\sin^2\frac{\theta}{2}}}\right)$$

$$= \sin^2\cot^{-1}\left(\sqrt{\frac{2\cos^2\frac{\theta}{2}}{2\sin^2\frac{\theta}{2}}}\right)$$

$$= \sin^2\left(\cot^{-1}\cot\frac{\theta}{2}\right) = \sin^2\frac{\theta}{2}$$

$$= \frac{1-\cos\theta}{2} = \frac{1}{2} - \frac{x}{2}$$

$$\frac{dy}{dx} = -\frac{1}{2}.$$ **Ans.**

EXERCISE 28.6

Differentiate following w.r.t. *x*.

1. $\sin^{-1}\left(\frac{2x}{1+x^2}\right)$ **Ans.** $\frac{2}{1+x^2}$ **2.** $\frac{1-\cos x}{1+\cos x}$ **Ans.** $\tan\frac{x}{2}\sec^2\frac{x}{2}$

3. $\sin\left(\frac{1+x^2}{1-x^2}\right)$ **[B.T.E. Delhi May 2007]** **Ans.** $\frac{4x}{(1-x^2)^2}\cos\left(\frac{1+x^2}{1-x^2}\right)$

4. $\tan^{-1}\left(\frac{\cos x}{1+\sin x}\right)$ **[B.T.E. Delhi May 2007]** **Ans.** $-\frac{1}{2}$

5. $\tan^{-1}\left(\frac{\sqrt{1+x^2}-1}{x}\right)$ (Put $x = \tan\theta$) **Ans.** $\frac{1}{2(1+x^2)}$

6. $\tan^{-1}\left(\frac{\sqrt{1+x}-\sqrt{1-x}}{\sqrt{1+x}+\sqrt{1-x}}\right)$ [**Hint.** Put $x = \cos 2\theta$] **Ans.** $\frac{1}{2\sqrt{1-x^2}}$

7. $\tan^{-1}\left(\frac{\sqrt{x}+\sqrt{a}}{1-\sqrt{ax}}\right)$ **[SBTE. 2012]** **Ans.** $\frac{1}{2\sqrt{x}\,(1+x)}$

8. $\sin^{-1}\left(\frac{a}{\sqrt{x^2+a^2}}\right)$. (Put $x = a\tan\theta$) **Ans.** $\frac{a}{a^2+x^2}$

9. Find $\frac{dy}{dx}$, if $y = \sin^{-1}\left(\frac{2\theta}{1+\theta^2}\right)$ and $x = \tan^{-1}\left(\frac{2\theta}{1-\theta^2}\right)$ **[IETE Dec. 2006]** **Ans.** 1

10. Differentiate $e^{\log\sin^{-1}(x^2)}$ **Ans.** $\frac{2x}{\sqrt{1-x^4}}$

11. If $y = \sin^{-1}(3x - 4x^3)$, find $\dfrac{dy}{dx}$. **Ans.** $\dfrac{3}{\sqrt{1-x^2}}$

12. If $y = \cos^{-1}\left(\sqrt{\dfrac{1+\cos x}{2}}\right)$, find $\dfrac{dy}{dx}$. **Ans.** $\dfrac{1}{2}$

13. If $y = \tan^{-1}\left(\dfrac{\cos x + \sin x}{\cos x - \sin x}\right)$, find $\dfrac{dy}{dx}$. **Ans.** 1

14. Differentiate $\tan^{-1}\left[\dfrac{1+x}{1-x}\right]$ w.r.t. x. **Ans.** $\dfrac{1}{1+x^2}$

15. If $y = \tan^{-1}\left(\dfrac{4+\sqrt{x}}{1-4\sqrt{x}}\right)$, find $\dfrac{dy}{dx}$. **Ans.** $\dfrac{1}{2\sqrt{x}\,(1+x)}$

16. If $y = \log_e \sqrt{\left(\dfrac{1+\sin x}{1-\sin x}\right)}$ find $\dfrac{dy}{dx}$. **Ans.** $\sec x$

28.23 HYPERBOLIC FUNCTIONS

Hyperbolic functions are defined as follows:

(*i*) $\sinh x = \dfrac{e^x - e^{-x}}{2}$, (*ii*) $\cosh x = \dfrac{e^x + e^{-x}}{2}$

(*iii*) $\tanh x = \dfrac{e^x - e^{-x}}{e^x + e^{-x}}$, (*iv*) $\coth x = \dfrac{e^x + e^{-x}}{e^x - e^{-x}}$

(*v*) $\operatorname{sech} x = \dfrac{2}{e^x + e^{-x}}$, (*vi*) $\operatorname{cosech} x = \dfrac{2}{e^x - e^{-x}}$

28.24 DERIVATIVES OF HYPERBOLIC FUNCTIONS

(*i*) $\dfrac{d}{dx}(\sinh x) = \dfrac{d}{dx}\left(\dfrac{e^x - e^{-x}}{2}\right) = \dfrac{e^x + e^{-x}}{2} = \cosh x$

(*ii*) $\dfrac{d}{dx}(\cosh x) = \dfrac{d}{dx}\left(\dfrac{e^x + e^{-x}}{2}\right) = \dfrac{e^x - e^{-x}}{2} = \sinh x$

(*iii*) $\dfrac{d}{dx}(\tanh x) = \dfrac{d}{dx}\left(\dfrac{\sinh x}{\cosh x}\right)$

$$\Rightarrow \quad \frac{d}{dx}(\tanh x) = \frac{\cosh x \dfrac{d}{dx}(\sinh x) - \sinh x \dfrac{d}{dx}(\cosh x)}{\cosh^2 x}$$

$$\Rightarrow \quad \frac{d}{dx}(\tanh x) = \frac{\cosh^2 x - \sinh^2 x}{\cosh^2 x} = \frac{1}{\cosh^2 x} = \operatorname{sech}^2 x \qquad (\because \cosh^2 x - \sinh^2 x = 1)$$

Similarly

(*iv*) $\frac{d}{dx}(\coth x) = -\operatorname{cosech}^2 x$ (*v*) $\frac{d}{dx}(\operatorname{sech} x) = -\operatorname{sech} x \cdot \tanh x$

(*vi*) $\frac{d}{dx}(\operatorname{cosech} x) = -\operatorname{cosech} x \cdot \coth x$

Note: Remember.

$$\cosh^2 x - \sinh^2 x = 1,\ \operatorname{sech}^2 x = 1 - \tanh^2 x,\ \operatorname{cosech}^2 x = -1 + \coth^2 x$$

$$\sinh^{-1} x = \log\left[x + \sqrt{x^2+1}\right],\ \cosh^{-1} x = \log\left[x + \sqrt{x^2-1}\right]$$

28.25 DERIVATIVES OF INVERSE HYPERBOLIC FUNCTIONS

(*i*) Let $y = \sinh^{-1} x$, then $\sinh y = x$

$$\Rightarrow \cosh y \frac{dy}{dx} = 1$$

$$\Rightarrow \frac{dy}{dx} = \frac{1}{\cosh y} = \frac{1}{\sqrt{1+\sinh^2 y}} = \frac{1}{\sqrt{x^2+1}}$$

$$\Rightarrow \frac{d}{dx}(\sinh^{-1} x) = \frac{1}{\sqrt{1+x^2}}$$

Similarly

(*ii*) $\frac{d}{dx}(\cosh^{-1} x) = \frac{1}{\sqrt{x^2-1}}$ (*iii*) $\frac{d}{dx}(\tanh^{-1} x) = \frac{1}{1-x^2}$, where $|x| < 1$,

(*iv*) $\frac{d}{dx}(\coth^{-1} x) = -\frac{1}{x^2-1}$, where $|x| < 1$.

(*v*) $\frac{d}{dx}(\operatorname{sech}^{-1} x) = -\frac{1}{x\sqrt{1-x^2}}$ (*vi*) $\frac{d}{dx}(\operatorname{cosech}^{-1} x) = \frac{-1}{x\sqrt{x^2+1}}$

28.26 SUCCESSIVE DIFFERENTIATION

If $y = f(x)$, its differential co-efficient $\frac{dy}{dx}$ is also a function of x. $\frac{dy}{dx}$ is further differentiated and the derivative of $\frac{dy}{dx}$ *i.e.* $\frac{d}{dx}\left(\frac{dy}{dx}\right)$ is called the second differential coefficient of y and is denoted by $\frac{d^2y}{dx^2}$.

Similarly third differential of y with respect to x is written as $\frac{d^3y}{dx^3}$.

Thus, $\frac{d^ny}{dx^n}$ is the nth derivative of y with respect to x.

Example 73. *Find the value of* $\frac{d^3y}{dx^3}$ *if* $y = \log(ax + b)$.

Solution. We have, $y = \log(ax + b)$

$$\Rightarrow \quad \frac{dy}{dx} = \frac{a}{ax + b}$$

Differentiating it again

$$\Rightarrow \quad \frac{d^2y}{dx^2} = \frac{-a^2}{(ax + b)^2}$$

Similarly, $\frac{d^3y}{dx^3} = \frac{2a^3}{(ax + b)^3}$. **Ans.**

Example 74. *Find the second derivative of* $y = e^{\sin x^2}$.

Solution. $y = e^{\sin x^2}$

On differentiating, w.r.t. 'x', we get

$$\Rightarrow \quad \frac{dy}{dx} = e^{\sin x^2} \cdot \cos x^2 \cdot 2x$$

$$\Rightarrow \quad \frac{dy}{dx} = 2xy \cos x^2$$

Again differentiating, we have

$$\Rightarrow \quad \frac{d^2y}{dx^2} = 2(1)\, y \cos x^2 + 2x \frac{dy}{dx} \cos x^2 + 2xy(-\sin x^2)(2x)$$

$$\Rightarrow \quad \frac{d^2y}{dx^2} = 2y \cos x^2 + 2x(2xy \cos x^2) \cos x^2 - 4x^2y \sin x^2$$

$$\Rightarrow \quad \frac{d^2y}{dx^2} = 2y \cos x^2 + 4x^2y \cdot \cos^2 x^2 - 4x^2y \sin x^2.$$ **Ans.**

Example 75. *If* $y = \log(x + \sqrt{1 + x^2})$, *prove that* $(1 + x^2)\frac{d^2y}{dx^2} + x\frac{dy}{dx} = 0$.

Solution. $y = \log(x + \sqrt{1 + x^2})$

$$\Rightarrow \quad \frac{dy}{dx} = \frac{1}{x + \sqrt{1 + x^2}} \left[1 + \frac{1}{2}(1 + x^2)^{-1/2} \cdot 2x\right]$$

$$= \frac{1}{x + \sqrt{1 + x^2}} \left[1 + \frac{x}{\sqrt{1 + x^2}}\right]$$

$$\Rightarrow \quad \frac{dy}{dx} = \frac{1}{x + \sqrt{1 + x^2}} \cdot \frac{\sqrt{1 + x^2} + x}{\sqrt{1 + x^2}} = \frac{1}{\sqrt{1 + x^2}}$$

$$\Rightarrow \qquad (1+x^2)\left(\frac{dy}{dx}\right)^2 = 1$$

Differentiating the equation again, we get

$$(1+x^2)\,2\frac{dy}{dx}\cdot\frac{d^2y}{dx^2} + 2x\cdot\left(\frac{dy}{dx}\right)^2 = 0$$

$$(1+x^2)\frac{d^2y}{dx^2} + x\frac{dy}{dx} = 0$$

Proved

Example 76. *If* $y = e^{a\sin^{-1}x}$, *show that* $(1-x^2)\,y_2 - xy_1 - a^2y = 0.$

Solution. We have, $y = e^{a\sin^{-1}x}$

Differentiating, we get $\dfrac{dy}{dx} = e^{a\sin^{-1}x}\dfrac{a}{\sqrt{1-x^2}}$

$$\left(\sqrt{1-x^2}\right)\frac{dy}{dx} = a\,e^{a\sin^{-1}x}$$

$$\left(\sqrt{1-x^2}\right)\frac{dy}{dx} = ay \qquad (\because\ y = e^{a\sin^{-1}x})$$

Squaring $\quad (1-x^2)\left(\dfrac{dy}{dx}\right)^2 = a^2y^2$

Again differentiating, we get

$$(1-x^2)\left(2\frac{dy}{dx}\frac{d^2y}{dx^2}\right) - 2x\left(\frac{dy}{dx}\right)^2 = a^2\left(2y\frac{dy}{dx}\right)$$

$$\Rightarrow \qquad (1-x^2)\frac{d^2y}{dx^2} - x\frac{dy}{dx} = a^2y$$

$$\Rightarrow \qquad (1-x^2)\,y_2 - xy_1 = a^2y.$$

Proved

Example 77. *If* $y = (x+\sqrt{1+x^2})^p$, *show that* $(1+x^2)\,y_2 + xy_1 - p^2y = 0$

Solution. $y = (x+\sqrt{1+x^2})^p$

$$y_1 = p(x+\sqrt{1+x^2})^{p-1}\left(1+\frac{1}{2}\frac{2x}{\sqrt{1+x^2}}\right)$$

$$= p(x+\sqrt{1+x^2})^{p-1}\frac{(\sqrt{1+x^2}+x)}{\sqrt{1+x^2}}$$

$$\sqrt{1+x^2}\,y_1 = p\,(x+\sqrt{1+x^2})^p$$

$$\sqrt{1+x^2}\,y_1 = py \qquad [\because\ y = (x+\sqrt{1+x^2})^p]$$

Squaring both the sides, we get

$$(1+x^2)\,y_1^2 = p^2 y^2$$

Differentiating again, we have

$$(1+x^2)\,2y_1y_2 + 2xy_1^2 = p^2\,(2yy_1)$$

$$(1+x^2)\,y_2 + xy_1 = p^2 y$$

$$(1+x^2)\,y_2 + xy_1 - p^2 y = 0$$

Proved

Example 78. *If* $y = \sin(m \sin^{-1} x)$, *prove that* $(1-x^2)\,y_2 - xy_1 + m^2 y = 0$.

Solution. $y = \sin(m \sin^{-1} x)$

Differentiating, we get

$$y_1 = \cos(m \sin^{-1} x) \cdot \frac{m}{\sqrt{1-x^2}}$$

$$\left(\sqrt{1-x^2}\right) y_1 = m \cos(m \sin^{-1} x)$$

On squaring, we get

$$(1-x^2)\,y_1^2 = m^2 \cos^2(m \sin^{-1} x)$$

$$= m^2\,[1 - \sin^2(m \sin^{-1} x)]$$

$$= m^2\,(1-y^2)$$

Differentiating again, we have

$$(1-x^2)\,2y_1y_2 - 2x\,y_1^2 = m^2\,(-2yy_1)$$

$$(1-x^2)\,y_2 - xy_1 + m^2 y = 0$$

Proved

Example 79. *If* $y = \sin(\sin x)$ *prove that* $\frac{d^2y}{dx^2} + \tan x \frac{dy}{dx} + y \cos^2 x = 0$

Solution. We have, $y = \sin(\sin x)$ ... (1)

Differentiating w.r.t. 'x', we get

$$\frac{dy}{dx} = \cos(\sin x) \frac{d}{dx}(\sin x) = \cos(\sin x) \cdot \cos x \qquad \text{... (2)}$$

Again differentiating w.r.t. 'x', we get

$$\frac{d^2y}{dx^2} = \cos(\sin x) \cdot (-\sin x) - \sin(\sin x) \cdot (\cos x) \cdot (\cos x)$$

$$= -\sin x \cdot \cos(\sin x) - \cos^2 x \cdot \sin(\sin x)$$

$$= -\sin x \cos(\sin x) - y \cos^2 x \qquad \text{... (3)} \quad [\because y = \sin(\sin x)]$$

Putting the values of $\frac{d^2y}{dx^2}, \frac{dy}{dx}$, and y from (3), (2) and (1) in $\frac{d^2y}{dx^2} + \tan x \frac{dy}{dx} + y \cos^2 x$, we get $\dfrac{d^2y}{dx^2} + \tan x \dfrac{dy}{dx} + y\cos^2 x = -\sin x \cdot \cos(\sin x) - y\cos^2 x$

$$+ \tan x \cdot \cos x \cdot \cos(\sin x) + y\cos^2 x$$

$$= -\sin x \cdot \cos(\sin x) + \frac{\sin x}{\cos x} \cdot \cos x \cdot \cos(\sin x)$$

$$= -\sin x \cos(\sin x) + \sin x \cdot \cos(\sin x)$$

$$= 0$$

Proved

Example 80. *Find A and B such that y = A sin 3x + B cos 3x satisfies the equation*

$$\frac{d^2y}{dx^2} + 4\frac{dy}{dx} + 3y = 10\cos 3x$$

Solution. $\dfrac{d^2y}{dx^2} + 4\dfrac{dy}{dx} + 3y = 10\cos 3x$ ... (1)

$\Rightarrow \qquad y = A\sin 3x + B\cos 3x$

$\Rightarrow \qquad \dfrac{dy}{dx} = 3A\cos 3x - 3B\sin 3x$

$\Rightarrow \qquad \dfrac{d^2y}{dx^2} = -9A\sin 3x - 9B\cos 3x$

Substituting the values of $y, \dfrac{dy}{dx}$ and $\dfrac{d^2y}{dx^2}$ in (1), we get

$-9A\sin 3x - 9B\cos 3x + 4(3A\cos 3x - 3B\sin 3x) + 3(A\sin 3x + B\cos 3x) = 10\cos 3x$

$\Rightarrow \quad (-9A - 12B + 3A)\sin 3x + (-9B + 12A + 3B - 10)\cos 3x = 0$

$\Rightarrow \quad (-6A - 12B)\sin 3x + (12A - 6B - 10)\cos 3x = 0.$

$\therefore \qquad -6A - 12B = 0$ ... (2)

$\qquad 12A - 6B - 10 = 0$ ... (3)

On solving (2) and (3), we get

$$A = \frac{2}{3} \text{ and } B = -\frac{1}{3}$$

Ans.

EXERCISE 28.7

1. If $y = x^5 + 7x^2 - 3x + 8$; find $\dfrac{d^4y}{dx^4}$. **Ans.** $120x$

2. If $x^2 + xy + 3y^2 = 1$, prove that $(x + 6y)^3 \dfrac{d^2y}{dx^2} + 22 = 0.$

3. If $y = e^{ax}\sin bx$, porve that $\dfrac{d^2y}{dx^2} - 2a\dfrac{dy}{dx} + (a^2 + b^2)y = 0.$

4. If $\cos^{-1}\left(\dfrac{y}{n}\right) = \log\left(\dfrac{x}{n}\right)^n$, prove that $x^2y_2 + xy_1 + n^2y = 0.$

5. If $y = a\cos(\log x) + b\sin(\log x)$, show that $x^2\dfrac{d^2y}{dx^2} + x\dfrac{dy}{dx} + y = 0$.

6. If $x = \sin t$, $y = \sin pt$, prove that $(1 - x^2)\dfrac{d^2y}{dx^2} - x\dfrac{dy}{dx} + p^2y = 0$.

7. If $x = a(\theta + \sin\theta)$ and $y = a(1 - \cos\theta)$, prove that $\dfrac{d^2y}{dx^2} = \dfrac{1}{4a}\sec^4\dfrac{\theta}{2}$.

8. If $x\sin(a + y) + \sin a\cos(a + y) = 0$, prove that $\dfrac{dy}{dx} = \dfrac{\sin^2(a + y)}{\sin a}$. [S.B.T.E. 2017]

28.27 DERIVATIVE OF ONE FUNCTION WITH RESPECT TO ANOTHER FUNCTION

Let $f(x)$ and $g(x)$ be two functions of x. To find the derivative of $f(x)$ with respect to $g(x)$, we put $u = f(x)$ and $v = g(x)$. Now, $\dfrac{du}{dv} = \left(\dfrac{du/dx}{dv/dx}\right)$ which is the required derivative.

Example 81. *Differentiate e^x with respect to x^2.*

Solution. Let $u = e^x$ and $v = x^2$

Differentiating with respect to x, we get

$$\frac{du}{dx} = e^x \text{ and } \frac{dv}{dx} = 2x$$

$$\therefore \quad \frac{du}{dx} = \frac{du/dx}{dv/dx} = \frac{e^x}{2x}$$ **Ans.**

Example 82. *Differentiate $\sin x^2$ with respect to x^2.*

Solution. Let $u = \sin x^2$ and $v = x^2$.

Differentiating with respect to x, we have

$$\frac{du}{dx} = \cos x^2 \frac{d}{dx}(x^2) = 2x\cos x^2 \quad \text{and} \quad \frac{dv}{dx} = 2x$$

$$\therefore \quad \frac{du}{dv} = \frac{du/dx}{dv/dx} = \frac{2x\cos x^2}{2x} = \cos x^2$$ **Ans.**

Example 83. *Differentiate:* (*i*) $\sin^{-1}(\theta)$ *with respect to* $\log(1 + \theta)$.

(*ii*) $a\sin\theta$ *with respect to* $a\left(\cos\theta + \log\tan\dfrac{\theta}{2}\right)$.

Solution. (*i*) Let $y = \sin^{-1}(\theta)$ and $x = \log(1 + \theta)$

Differentiating both w.r.t. θ, we get $\dfrac{dy}{d\theta} = \dfrac{1}{\sqrt{1-\theta^2}}$ and $\dfrac{dx}{d\theta} = \dfrac{1}{1+\theta}$.

$$\therefore \quad \frac{dy}{dx} = \frac{dy/d\theta}{dx/d\theta} = \frac{1}{\sqrt{1-\theta^2}} \cdot \frac{1+\theta}{1} = \sqrt{\frac{1+\theta}{1-\theta}}.$$ **Ans.**

(*ii*) Let $y = a\sin\theta$ and $x = a\left[\cos\theta + \log\tan\dfrac{\theta}{2}\right]$

Differentiating both w.r.t. 'θ', we get

$$\frac{dy}{d\theta} = a\cos\theta, \quad \frac{dx}{d\theta} = a\left[-\sin\theta + \frac{1}{2}\frac{\sec^2\frac{\theta}{2}}{\tan\frac{\theta}{2}}\right] = a\left[-\sin\theta + \frac{1}{2\sin\frac{\theta}{2}\cos\frac{\theta}{2}}\right]$$

$$= a\left[-\sin\theta + \frac{1}{\sin\theta}\right] = a\left(\frac{-\sin^2\theta + 1}{\sin\theta}\right) = a\frac{\cos^2\theta}{\sin\theta}$$

$$\therefore \quad \frac{dy}{dx} = \frac{\frac{dy}{d\theta}}{\frac{dx}{d\theta}} = \frac{a\cos\theta}{a\frac{\cos^2\theta}{\sin\theta}} = \tan\theta$$ **Ans.**

Example 84. *Differentiate* $\tan^{-1}x$ *with respect to* $\sin^{-1}x$.

Solution. Let $u = \tan^{-1}x$ and $v = \sin^{-1}x$

Differentiating both with respect to x, we get

$$\frac{du}{dx} = \frac{1}{1+x^2} \text{ and } \frac{dv}{dx} = \frac{1}{\sqrt{1-x^2}}$$

$$\therefore \quad \frac{du}{dv} = \frac{du/dx}{dv/dx} = \frac{\frac{1}{1+x^2}}{\frac{1}{\sqrt{1-x^2}}} = \frac{\sqrt{1-x^2}}{1+x^2}$$ **Ans.**

Example 85. *Differentiate* $\tan^{-1}\left(\frac{2x}{1-x^2}\right)$ *with respect to* $\sin^{-1}\left(\frac{2x}{1+x^2}\right)$

Solution. Let $u = \tan^{-1}\left(\frac{2x}{1-x^2}\right)$ and $v = \sin^{-1}\left(\frac{2x}{1+x^2}\right)$

Again, let $x = \tan\theta \Rightarrow \theta = \tan^{-1}(x)$

$$\Rightarrow \quad u = \tan^{-1}\left(\frac{2\tan\theta}{1-\tan^2\theta}\right) = \tan^{-1}(\tan 2\theta) = 2\theta = 2\tan^{-1}(x) \quad \text{... (1)}$$

$$\text{and} \quad v = \sin^{-1}\left(\frac{2\tan\theta}{1+\tan^2\theta}\right) = \sin^{-1}(\sin 2\theta) = 2\theta = 2\tan^{-1}(x) \quad \text{... (2)}$$

Differentiating (1) and (2) with respect to x, we get

$$\therefore \quad \frac{du}{dx} = 2\cdot\frac{d}{dx}(\tan^{-1}x) = \frac{2}{1+x^2}$$

$$\text{and} \quad \frac{dv}{dx} = 2\frac{d}{dx}(\tan^{-1}x) = \frac{2}{1+x^2}$$

$$\Rightarrow \quad \frac{du}{dv} = \frac{du/dx}{dv/dx} = \frac{\frac{2}{1+x^2}}{\frac{2}{1+x^2}} = 1.$$ **Ans.**

Example 86. *Differentiate* $\tan^{-1}\left(\dfrac{3x-x^3}{1-3x^2}\right)$ *with respect to* $\tan^{-1}\left(\dfrac{2x}{1-x^2}\right)$.

Solution. Let $u=\tan^{-1}\left(\dfrac{3x-x^3}{1-3x^2}\right)$ and $v=\tan^{-1}\left(\dfrac{2x}{1-x^2}\right)$

Putting $x=\tan\theta \;\Rightarrow\; \theta=\tan^{-1}x$

$$u=\tan^{-1}\left(\frac{3\tan\theta-\tan^3\theta}{1-3\tan^2\theta}\right)=\tan^{-1}(\tan 3\theta) \qquad \left[\because \tan 3\theta=\frac{3\tan\theta-\tan^3\theta}{1-3\tan^2\theta}\right]$$

$$u=3\theta=3\tan^{-1}x \qquad \text{... (1) } [\because \theta=\tan^{-1}x]$$

and $$v=\tan^{-1}\left(\frac{2x}{1-x^2}\right)=\tan^{-1}\left(\frac{2\tan\theta}{1-\tan^2\theta}\right)$$

$$=\tan^{-1}(\tan 2\theta)=2\theta \qquad \left[\because \tan 2\theta=\frac{2\tan\theta}{1-\tan^2\theta}\right]$$

$$\Rightarrow \quad v=2\tan^{-1}x \qquad \text{... (2) } [\because \theta=\tan^{-1}x]$$

Differentiating (1) and (2) with respect to x, we get

$$\frac{du}{dx}=3\cdot\frac{d}{dx}\tan^{-1}x=3\cdot\frac{1}{1+x^2}=\frac{3}{1+x^2} \qquad \text{... (3)}$$

and $$\frac{dv}{dx}=2\frac{d}{dx}\tan^{-1}x=2\cdot\frac{1}{1+x^2}=\frac{2}{1+x^2} \qquad \text{... (4)}$$

$$\Rightarrow \quad \frac{du}{dv}=\frac{du/dx}{dv/dx}=\frac{\dfrac{3}{1+x^2}}{\dfrac{2}{1+x^2}}=\frac{3}{2}$$ **Ans.**

Example 87. *Differentiate* $\tan^{-1}\left(\dfrac{\sqrt{1+x^2}-1}{x}\right)$ *with respect to* $\tan^{-1}x,\ x\neq 0$.

[SBTE. 2017, 2016]

Solution. Let $u=\tan^{-1}\left(\dfrac{\sqrt{1+x^2}-1}{x}\right)$ and $v=\tan^{-1}x$

Putting $x=\tan\theta \;\Rightarrow\; \theta=\tan^{-1}x$, we have

$$u=\tan^{-1}\left(\frac{\sqrt{1+\tan^2\theta}-1}{\tan\theta}\right)=\tan^{-1}\left(\frac{\sec\theta-1}{\tan\theta}\right)$$

$$=\tan^{-1}\left(\frac{1-\cos\theta}{\sin\theta}\right)=\tan^{-1}\left(\frac{2\sin^2\theta/2}{2\sin\theta/2\cdot\cos\theta/2}\right)=\tan^{-1}\left(\tan\frac{\theta}{2}\right)$$

$$=\frac{1}{2}\theta=\frac{1}{2}\tan^{-1}x$$

Now we have, $u = \frac{1}{2}\tan^{-1} x$ and $v = \tan^{-1} x$

$$\Rightarrow \quad \frac{du}{dx} = \frac{1}{2}\frac{d}{dx}(\tan^{-1} x) = \frac{1}{2}\frac{1}{1+x^2} \text{ and } \frac{dv}{dx} = \frac{d}{dx}(\tan^{-1} x) = \frac{1}{1+x^2}$$

$$\therefore \quad \frac{du}{dv} = \frac{du/dx}{dv/dx} = \frac{\frac{1}{2(1+x^2)}}{\frac{1}{1+x^2}} = \frac{1}{2(1+x^2)} \times \frac{(1+x^2)}{1} = \frac{1}{2}$$

Ans.

Example 88. *Differentiate* $\tan^{-1}\left\{\frac{\sqrt{1+x^2}-\sqrt{1-x^2}}{\sqrt{1+x^2}+\sqrt{1-x^2}}\right\}$ *with respect to* $\cos^{-1} x^2$.

[SBTE. 2013]

Solution. Let $u = \tan^{-1}\left\{\frac{\sqrt{1+x^2}-\sqrt{1-x^2}}{\sqrt{1+x^2}+\sqrt{1-x^2}}\right\}$ and $v = \cos^{-1} x^2$

Putting $x^2 = \cos\theta$, we have

$$u = \tan^{-1}\left\{\frac{\sqrt{1+\cos\theta}-\sqrt{1-\cos\theta}}{\sqrt{1+\cos\theta}+\sqrt{1-\cos\theta}}\right\}$$

$$= \tan^{-1}\left\{\frac{\sqrt{2\cos^2\frac{\theta}{2}}-\sqrt{2\sin^2\frac{\theta}{2}}}{\sqrt{2\cos^2\frac{\theta}{2}}+\sqrt{2\sin^2\frac{\theta}{2}}}\right\} \qquad \left[\begin{array}{l}\cos 2\theta = 2\cos^2\theta - 1\\ \cos 2\theta = 1 - 2\sin^2\theta\end{array}\right]$$

$$= \tan^{-1}\left\{\frac{\cos\frac{\theta}{2}-\sin\frac{\theta}{2}}{\cos\frac{\theta}{2}+\sin\frac{\theta}{2}}\right\} = \tan^{-1}\left\{\frac{1-\tan\frac{\theta}{2}}{1+\tan\frac{\theta}{2}}\right\} \qquad \left[\text{Dividing by } \cos\frac{\theta}{2}\right]$$

$$= \tan^{-1}\left\{\tan\left(\frac{\pi}{4}-\frac{\theta}{2}\right)\right\} = \left(\frac{\pi}{4}-\frac{\theta}{2}\right)$$

$$= \frac{\pi}{4} - \frac{1}{2}\cos^{-1} x^2 \qquad [\because x^2 = \cos\theta \Rightarrow \theta = \cos^{-1} x^2]$$

Differentiating w.r.t x, we have

$$\frac{du}{dx} = -\frac{1}{2}\frac{(-1)}{\sqrt{1-x^4}}\cdot\frac{d}{dx}(x^2) = \frac{1}{2}\cdot\frac{2x}{\sqrt{1-x^4}} = \frac{x}{\sqrt{1-x^4}}$$

and $\quad v = \cos^{-1} x^2 \Rightarrow \frac{dv}{dx} = \frac{-1}{\sqrt{1-x^4}}\cdot\frac{d}{dx}(x^2) = \frac{-2x}{\sqrt{1-x^4}}$

Therefore $\quad \frac{du}{dv} = \frac{du/dx}{dv/dx} = \frac{x}{\sqrt{1-x^4}} \times \frac{\sqrt{1-x^4}}{(-2x)} = -\frac{1}{2}$

Ans.

EXERCISE 28.8

Differentiate:

1. $\log(1 + x^2)$ with respect to $\tan^{-1} x$. **Ans.** $2x$

2. $e^{\sin x}$ with respect to $\cos x$. **Ans.** $-e^{\sin x} \cdot \cot x$

3. $\log \sec x$ with respect to $\tan x$. **Ans.** $\dfrac{\tan x}{\sec^2 x}$

4. $\sin^{-1}(2x\sqrt{1-x^2})$ with respect to $\sec^{-1}\left(\dfrac{1}{\sqrt{1-x^2}}\right)$, if $x \in \left(0, \dfrac{1}{\sqrt{2}}\right)$. **Ans.** -2

5. $\sin^{-1}\left(\dfrac{2x}{1+x^2}\right)$ with respect to $\cos^{-1}\left(\dfrac{1-x^2}{1+x^2}\right)$, if $0 < x < 1$. **Ans.** 1

6. $\tan^{-1}\left(\dfrac{x}{\sqrt{1-x^2}}\right)$ with respect to $\cos^{-1}(2x^2 - 1)$. **Ans.** $-\dfrac{1}{2}$

7. $\cos^{-1}\left(\dfrac{1-x^2}{1+x^2}\right)$ with respect to $\tan^{-1}\left(\dfrac{3x - x^3}{1-3x^2}\right)$. **Ans.** $\dfrac{2}{3}$

8. $\tan^{-1}\left(\dfrac{1+ax}{1-ax}\right)$ with respect to $\sqrt{1+a^2x^2}$. **Ans.** $\dfrac{1}{ax\sqrt{1+a^2x^2}}$

9. $\tan^{-1}\left(\dfrac{2\sqrt{x}}{1-x}\right)$ with respect to $\sin^{-1}\left(\dfrac{2\sqrt{x}}{1+x}\right)$. **Ans.** 1

10. $\tan^{-1}\left(\dfrac{x}{1+\sqrt{1-x^2}}\right)$ with respect to $\sin\left(2\cot^{-1}\sqrt{\dfrac{1+x}{1-x}}\right)$. **Ans.** $-\dfrac{1}{2x}$

11. $\sin^{-1}\left(2ax\sqrt{1-a^2x^2}\right)$ with respect to $\sqrt{1-a^2x^2}$. **Ans.** $-\dfrac{2}{ax}$

12. Prove that the derivative of $\tan^{-1}\left(\dfrac{\sqrt{1+x^2}-1}{x}\right)$ with respect to $\tan^{-1}\left(\dfrac{2x\sqrt{1-x^2}}{1-2x^2}\right)$ at $x = 0$ is $\dfrac{1}{4}$.

13. $\tan^{-1}\left(\dfrac{x-1}{x+1}\right)$ with respect to $\sin^{-1}(3x - 4x^3)$, if $\dfrac{-1}{2} < x < \dfrac{1}{2}$. **Ans.** $\dfrac{\sqrt{1-x^2}}{3(1+x^2)}$

14. $\tan^{-1}\left(\dfrac{\cos x}{1+\sin x}\right)$ with respect to $\sec^{-1} x$. **Ans.** $\dfrac{-x\sqrt{x^2-1}}{2}$

15. $\cos^{-1}(4x^3 - 3x)$ with respect to $\tan^{-1}\left(\dfrac{\sqrt{1-x^2}}{x}\right)$, if $\dfrac{1}{2} < x < 1$. **Ans.** 3

16. $\sin^{-1}\left(\sqrt{1-x^2}\right)$ with respect to $\cot^{-1}\left(\dfrac{x}{\sqrt{1-x^2}}\right)$, if $0 < x < 1$. **Ans.** 1

Hints to the Selected Questions

	The following formulae will be useful in solving the questions of this exercise.	
(i)	$\sin 2\theta = 2 \sin \theta \cos \theta$, $\sin 3\theta = 3 \sin \theta - 4 \sin^3 \theta$,	$\sin 2\theta = \dfrac{2 \tan \theta}{1+\tan^2\theta}$,
(ii)	$\cos 2\theta = 2 \cos^2 \theta - 1$, $\cos 3\theta = 4 \cos^3 \theta - 3 \cos \theta$	$\cos 2\theta = 1 - 2 \sin^2 \theta$; $\cos 2\theta = \dfrac{1-\tan^2\theta}{1+\tan^2\theta}$,
(iii)	$\tan 2\theta = \dfrac{2 \tan \theta}{1-\tan^2\theta}$, $1 + \tan^2 \theta = \sec^2 \theta$, $1 - \sin^2 \theta = \cos^2 \theta$	$\tan 3\theta = \dfrac{3 \tan \theta - \tan^3 \theta}{1 - 3 \tan^2 \theta}$,

1. Let $\quad y = \log (1 + x^2)$

$\therefore \quad \dfrac{dy}{dx} = \dfrac{2x}{1+x^2}$

and $\quad z = \tan^{-1} x$

$\dfrac{dz}{dx} = \dfrac{1}{1+x^2}$

Hence, $\quad \dfrac{dy}{dz} = \dfrac{dy/dx}{dz/dx} = \dfrac{2x/(1+x^2)}{1/(1+x^2)} = 2x.$

4. Let $\quad y = \sin^{-1}\left(2x\sqrt{1-x^2}\right)$

Assume $\quad x = \cos \theta$

Then $\quad y = \sin^{-1} (2 \sin \theta \cdot \cos \theta) = \sin^{-1} (\sin 2\theta) = 2\theta$

$\Rightarrow \quad y = 2 \cos^{-1} x$

$\therefore \quad \dfrac{dy}{dx} = \dfrac{-2}{\sqrt{1-x^2}}$

and $\quad z = \sec^{-1}\left(\dfrac{1}{\sqrt{1-x^2}}\right)$

Assume $\quad x = \sin \theta$

$z = \sec^{-1}\left(\dfrac{1}{\cos \theta}\right) = \sec^{-1} (\sec \theta) = \theta = \sin^{-1}(x)$

$$\frac{dz}{dx} = \frac{1}{\sqrt{1-x^2}}$$

$$\therefore \quad \frac{dy}{dz} = \frac{dy/dx}{dz/dx} = -2$$

5. Assume $x = \tan\theta$

6. Let $y = \tan^{-1}\left(\frac{x}{\sqrt{1-x^2}}\right)$

Assume $x = \sin\theta$

Then, $y = \tan^{-1}\left(\frac{\sin\theta}{\cos\theta}\right) = \tan^{-1}(\tan\theta) = \theta = \sin^{-1}(x).$

$$\therefore \quad \frac{dy}{dx} = \frac{1}{\sqrt{1-x^2}}$$

and $z = \cos^{-1}(2x^2 - 1)$

Assume $x = \cos\theta$

$$z = \cos^{-1}(2\cos^2\theta - 1) = \cos^{-1}(\cos 2\theta) = 2\theta = 2\cos^{-1} x$$

$$\frac{dz}{dx} = \frac{-2}{\sqrt{1-x^2}}$$

$$\frac{dy}{dz} = \frac{dy/dx}{dz/dx} = \frac{1/\sqrt{1-x^2}}{-2/\sqrt{1-x^2}} = \frac{-1}{2}.$$

7. Assume $x = \tan\theta$.

8. Assume $ax = \tan\theta$ and $1 = \tan\frac{\pi}{4}$

9. Assume $\sqrt{x} = \tan\theta$

10. Assume $x = \cos 2\theta$

11. Assume $ax = \sin\theta$

12. Assume $x = \tan\theta$, in second part put $x = \sin\theta$

13. Assume $x = \cos 2\phi$, $x = \sin\theta$ in second part.

14. Assume $x = \frac{\pi}{2} - \theta$

15. Assume $x = \cos\theta$

16. Assume $x = \cos\theta$

Multiple Choice Questions (MCQs)

Choose the correct alternative:

1. If $y = a\sin mx + b\cos mx$, then $\frac{dy}{dx}$ is equal to

(*i*) $m(a\cos mx + b\sin mx)$ (*ii*) $-my$

(*iii*) $a\cos mx - b\sin mx$ (*iv*) $m(a\cos mx - b\sin mx)$ **Ans.** (*iv*)

2. If $y = \sin^{-1}(\sqrt{x - ax} - \sqrt{a - ax})$, then $\frac{dy}{dx} =$

(*i*) $\frac{1}{2\sqrt{x}\sqrt{1-x}}$ (*ii*) $\frac{1}{\sin\sqrt{a-ax}}$ (*iii*) $\sin\sqrt{x}\sin\sqrt{a}$ (*iv*) None of these

Ans. (*i*)

3. If $y = \sqrt{x \log_e x}$, then $\frac{dy}{dx}$ at $x = e$ is

(*i*) $\sqrt{e}$ (*ii*) $\frac{1}{\sqrt{e}}$ (*iii*) $\frac{1}{e}$ (*iv*) None of these

Ans. (*ii*)

4. If $y = \cos^{-1}\left(\frac{2\cos x - 3\sin x}{\sqrt{13}}\right)$, then $\frac{dy}{dx}$ is equal to

(*i*) 0 (*ii*) -1 (*iii*) 1 (*iv*) None of these

Ans. (*iii*)

5. If $y = \cot(2x+1)$, then $\frac{dy}{dx} =$

(*i*) $2\,\text{cosec}^2(2x+1)$ (*ii*) $2\,\text{cosec}^2 2x$ (*iii*) $\text{cosec}^2(2x+1)$ (*iv*) $-2\,\text{cosec}^2(2x+1)$

Ans. (*iv*)

6. If $x = 3\cos\theta - \cos 3\theta$, $y = 3\sin\theta - \sin 3\theta$, then $\frac{dy}{dx} =$

(*i*) $\tan 2\theta$ (*ii*) $\tan 3\theta$ (*iii*) $\tan^2\theta$ (*iv*) None of these

Ans. (*i*)

7. If $x = a\cos t + \log\tan\frac{t}{2}$, $y = a\cos t$, then $\frac{dy}{dx} =$

(*i*) $\tan\frac{t}{2}$ (*ii*) $\tan^2 t$ (*iii*) $\tan t$ (*iv*) None of these

Ans. (*iii*)

8. If $y = \tan^{-1}\left(\frac{\sqrt{1+x^2}-1}{x}\right)$ and $t = \tan^{-1}x$, then $\frac{dy}{dx} =$

(*i*) $\frac{1}{2}$ (*ii*) 2 (*iii*) $-\frac{1}{2}$ (*iv*) N **Ans.** (*i*)

9. If $y = \tan^{-1}\left(\frac{2x}{1-x^2}\right)$ and $t = \cos^{-1}\left(\frac{1-x^2}{1+x^2}\right)$ then $\frac{dy}{dx} =$

(*i*) 0 (*ii*) 1 (*iii*) 3 (*iv*) 4 **Ans.** (*ii*)

10. If $x = a(\theta - \sin\theta)$ and $y = a(1 - \cos\theta)$, then $\frac{dy}{dx} =$

(*i*) $\cot\theta$ (*ii*) $\cot^2\theta$ (*iii*) $\cot\frac{\theta}{2}$ (*iv*) None of these

Ans. (*iii*)

11. If $x = \frac{1-t^2}{1+t^2}$ and $y = \frac{2t}{1+t^2}$, then $\frac{dy}{dx} =$

(*i*) $\frac{1+t^2}{t}$ (*ii*) $\frac{1+t^2}{2t}$ (*iii*) $\frac{1-t^2}{t}$ (*iv*) $\frac{1-t^2}{2t}$ **Ans.** (*iv*)

12. If $x = a\,\theta + \sin\theta$ and $y = a\cos\theta$, then $\frac{dy}{dx} =$

(*i*) $-\tan\frac{\theta}{2}$ (*ii*) $\tan\frac{\theta}{2}$ (*iii*) $\tan^2\theta$ (*iv*) $\tan^2\frac{\theta}{2}$ **Ans.** (*i*)

13. If $ax^2 + 2hxy + by^2 + 2gx + 2fy + c = 0$, then $\frac{dy}{dx} =$

(*i*) $\frac{ax+hy+g}{hx+by+f}$ (*ii*) $-\left(\frac{ax+hy+g}{hx+by+f}\right)$ (*iii*) $\frac{ax-hy-g}{hx-by-f}$ (*iv*) None of these

Ans. (*ii*)

14. If $ax^2 + 2hxy + by^2 = 0$, then $\frac{dy}{dx} =$

(*i*) $\frac{ax-hy}{hx-by}$ (*ii*) $\frac{ax+hy}{hx+by}$ (*iii*) $\frac{ax+hy}{hx-by}$ (*iv*) $-\frac{ax+hy}{hx+by}$ **Ans.** (*iv*)

15. If $y = x^{x^{x^{x^{\cdots \text{upto } \infty}}}}$, then $\frac{dy}{dx} =$

(*i*) $\frac{y}{x(1-y\log x)}$ (*ii*) $\frac{y^2}{x(1-y\log x)}$ (*iii*) $\frac{y^2}{x(1+y\log x)}$ (*iv*) None of these

Ans. (*ii*)

16. If $y = \log(\sec x + \tan x)$, then $\frac{dy}{dx} =$

(*i*) $-\sec x\tan x$ (*ii*) $\sec^2 x$ (*iii*) $\sec x$ (*iv*) None of these

Ans. (*iii*)

17. If $y = \log(x + 2 + \sqrt{x^2+4x+1})$, then $\frac{dy}{dx} =$

(*i*) $\frac{1}{\sqrt{x^2+4x+1}}$ (*ii*) $\sqrt{x^2+4x+1}$ (*iii*) $\frac{1}{x^2+4x+1}$ (*iv*) None of these

Ans. (*i*)

18. If $f(x) = \frac{5}{2}[x] - x^5$ and $1 \le x \le 2$, then $f'\left(\sqrt[5]{\frac{\pi}{2}}\right) =$

(*i*) $5\left(\frac{\pi}{2}\right)^{4/5}$ (*ii*) $5(\pi)^{\frac{4}{5}}$ (*iii*) $-5\left(\frac{\pi}{2}\right)^{\frac{4}{5}}$ (*iv*) None of these

Ans. (*iii*)

19. If $y = \sin^{-1}x + \cos^{-1}x$, then $\frac{dy}{dx} =$

(*i*) 0 (*ii*) 1 (*iii*) 2 (*iv*) 3 **Ans.** (*i*)

20. If $-\frac{\pi}{2} < x < 0$ and $y = \tan^{-1}\left(\sqrt{\frac{1-\cos 2x}{1+\cos 2x}}\right)$, then $\frac{dy}{dx} =$

(*i*) 0 (*ii*) 1 (*iii*) -1 (*iv*) None of these

Ans. (*iii*)

21. If $y = \sqrt{\sin x + y}$, then $\frac{dy}{dx} =$

(*i*) $\frac{\cos x}{2y-1}$ (*ii*) $\frac{\cos x}{2y+1}$ (*iii*) $\frac{\sin x}{2y-1}$ (*iv*) None of these

Ans. (*i*)

22. If $y = \tan^{-1}\left(\frac{\sin x + \cos x}{\cos x - \sin x}\right)$, then $\frac{dy}{dx} =$

(*i*) 1 (*ii*) 2 (*iii*) 3 (*iv*) 4 **Ans.** (*i*)

23. If $\sin^{-1}\left(\frac{x^2 - y^2}{x^2 + y^2}\right) = \log a$, then $\frac{dy}{dx} =$

(*i*) $\frac{x}{y}$ (*ii*) $\frac{y}{x}$ (*iii*) xy (*iv*) $x + y$ **Ans.** (*ii*)

24. If $y = \log\left(\sqrt{\tan x}\right)$, then $\frac{dy}{dx}$ at $x = \frac{\pi}{4}$ is

(*i*) 0 (*ii*) 2 (*iii*) 1 (*iv*) 4 **Ans.** (*iii*)

25. If $y = \frac{1}{1 + x^{a-b} + x^{c-b}} + \frac{1}{1 + x^{b-c} + x^{a-c}} + \frac{1}{1 + x^{b-a} + x^{c-a}}$, then $\frac{dy}{dx}$ is equal to

(*i*) 1 (*ii*) 0 (*iii*) 3 (*iv*) None of these

Ans. (*ii*)

26. If $y = \cos^{-1}(2t^2 - 1)$ and $x = \cos^{-1} t$, then $\frac{dy}{dx} =$

(*i*) $\frac{2}{x}$ (*ii*) $1 - x^2$ (*iii*) 0 (*iv*) 2 **Ans.** (*iv*)

27. If $3\sin(xy) + 4\cos(xy) = 5$, then $\frac{dy}{dx} =$

(*i*) $-xy$ (*ii*) $-\frac{y}{x}$ (*iii*) $-\frac{x}{y}$ (*iv*) $\frac{y}{x}$ **Ans.** (*ii*)

28. If $y = \sec(\tan^{-1} x)$, then $\frac{dy}{dx} =$

(*i*) $\frac{1}{\sqrt[x]{1+x^2}}$ (*ii*) $\frac{1}{\sqrt{1+x^2}}$ (*iii*) $\frac{x}{\sqrt{1+x^2}}$ (*iv*) $\frac{x}{1+x^2}$ **Ans.** (*iii*)

29. If $y = \sin\tan^{-1} x$, then $\frac{dy}{dx} =$

(*i*) $\frac{1}{(1+x^2)^{3/2}}$ (*ii*) $\frac{1}{\sqrt{1+x^2}}$ (*iii*) $\frac{1}{1+x^2}$ (*iv*) None of these

Ans. (*i*)

30. If $y = \tan^{-1}\left(\frac{1 - \cos x}{\sin x}\right)$, then $\frac{dy}{dx} =$

(*i*) 1 (*ii*) $-x$ (*iii*) -1 (*iv*) x **Ans.** (*iii*)

31. If $y = \log(\sin x)$, then $\dfrac{dy}{dx} =$

(*i*) $\tan x$ (*ii*) $\cot^2 x$ (*iii*) $\cot x$ (*iv*) $\tan^2 x$ **Ans.** (*iii*)

32. If $y = \sqrt{\operatorname{cosec} x}$, then $\dfrac{dy}{dx} =$

(*i*) $-\cot x \sqrt{\operatorname{cosec} x}$ (*ii*) $-\dfrac{1}{2}\cot x \sqrt{\operatorname{cosec} x}$

(*iii*) $\dfrac{1}{2}\cot x \sqrt{\operatorname{cosec} x}$ (*iv*) $-\dfrac{1}{2}\operatorname{cosec} x \sqrt{\cot x}$ **Ans.** (*ii*)

33. If $y = \cot^{-1}\left(\sqrt{\dfrac{1-\sin x}{1+\sin x}}\right)$, then $\dfrac{dy}{dx} =$

(*i*) $\dfrac{1}{2}$ (*ii*) $\dfrac{x}{2}$ (*iii*) $-\dfrac{1}{2}$ (*iv*) $-\dfrac{x}{2}$ **Ans.** (*i*)

34. If $y = \sec^{-1}\left(\dfrac{x^2+1}{x^2-1}\right)$, then $\dfrac{dy}{dx} =$

(*i*) $\dfrac{1}{\sqrt{1+x^2}}$ (*ii*) $-\dfrac{2}{1+x^2}$ (*iii*) $\dfrac{2}{\sqrt{1+x^2}}$ (*iv*) $-\dfrac{1}{\sqrt{1+x^2}}$ **Ans.** (*ii*)

35. If $y = \sin[x]$, then $\dfrac{dy}{dx} =$

(*i*) 0 (*ii*) $\cos[x]$ (*iii*) $\cos x$ (*iv*) None of these

Ans. (*i*)

Applications of Derivatives

29.1 INTRODUCTION

In fact applications of derivatives a part of own everyday life in the real mathematics used to calculate speed and distance travelled in motion, profit and loss in bussiness through graph, tangents and normal of curves or curve sketching, maxima and minima rate of change of quantity increasing and decreasing function and so on. Therefore, this chapter focus will be on applications of derivatives.

29.2 ERROR

Let y be a differentiable function of x, then

$$\lim_{\delta x \to 0} \frac{\delta y}{\delta x} = \frac{dy}{dx}$$

But the definition of a limit, when δx is very small, $\frac{\delta y}{\delta x}$ differs from its limit $\frac{dy}{dx}$ by a very small quantity which can be made as small as we please.

$$\therefore \quad \frac{\delta y}{\delta x} = \frac{dy}{dx} \quad \text{(approximately)}$$

or $$\delta y = \frac{dy}{dx} \times \delta x. \quad \text{(approximately)}$$

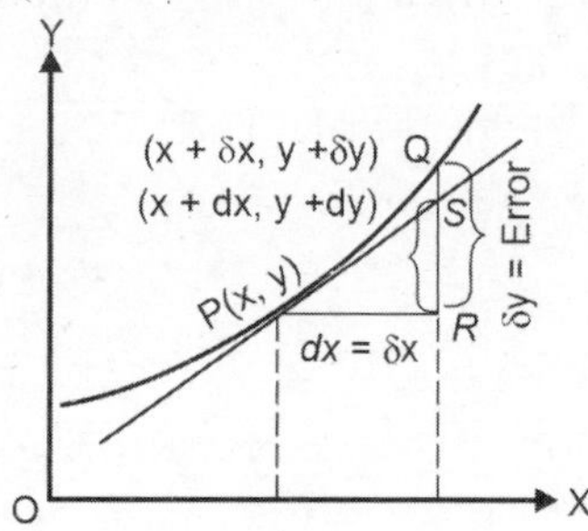

Small error in $y = \frac{dy}{dx} \cdot \delta x$

= Product of $\frac{dy}{dx}$ and the small error in x.

Absolute error in $x = \delta x$.

Relative error in $x = \frac{\delta x}{x}$

Percentage error in $x = \frac{\delta x}{x} \times 100$

Working Rule to Find Error

Step 1. First write down the relation between x and y in the form of equation.

Step 2. Differentiate both sides.

Step 3. Apply the formula, $\delta y = \left(\frac{dy}{dx}\right) \delta x$.

Example 1. *Find the percentage error in the area of a rectangle when an error of + 1 percent is made in measuring its length and breath.*

Solution. Let A be the area and l, b the length and breath

$$\Rightarrow \quad A = lb$$

$$\Rightarrow \quad \log A = \log l + \log b$$

On differentiating we get

$$\Rightarrow \quad \frac{\delta A}{A} = \frac{\delta l}{l} + \frac{\delta b}{b}$$

$$\Rightarrow \quad 100\frac{\delta A}{A} = 100\frac{\delta l}{l} + 100\frac{\delta b}{b}$$

$$= +1 + 1$$

Percentage error in area $= +2$ **Ans.**

Example 2. *If the radius of a sphere is measured as 7 m with an error of 0.02 m, then find the approximate error in calculating its volume.*

Solution. Let the radius of sphere be $r = 7\text{m}$, $\delta r = 0.02$ m and

volume of sphere $V = \dfrac{4\pi r^3}{3}$

$$\Rightarrow \quad \frac{dV}{dr} = 4\pi r^2$$

Therefore,

$$\delta V = \left(\frac{dV}{dr}\right)\delta r = (4\pi r^2)\delta r$$

$$= [4\pi(7)^2]\,(0.02) = (196\pi)\,(0.02) = 3.92\pi \text{ m}^3$$

Hence, the approximate error in calculating the volume is 3.92π m^3. **Ans.**

Example 3. *Find the percentage error in calculating the volume of cubical box if an error of 1% is made in measuring the length of edges of the cube.* **[SBTE. 2016, 2015, 2014]**

Solution. Let Δx be small error in 'x' and Δv be the small error in 'v'. then,

$$\frac{\Delta x}{x} \times 100 = 1 \qquad \text{[Given]}$$

$$\frac{dx}{x} \times 100 = 1 \qquad ...(1)$$

$$v = x^3$$

$$\Rightarrow \quad \frac{dv}{dx} = 3x^2$$

$$\Rightarrow \quad dv = 3x^2 dx$$

$$\Rightarrow \quad \frac{dv}{v} = \frac{3x^2}{v}dx$$

$$\Rightarrow \quad \frac{dv}{v} = \frac{3x^2}{x^3}dx \qquad [\because v = x^3]$$

$$\Rightarrow \quad \frac{dv}{v} \times 100 = 3\frac{dx}{x} \times 100 = 3 \times 1 \qquad \text{[]from (1)}$$

$$\Rightarrow \quad \frac{dv}{v} \times 100 = 3$$

% error of cube = 3% **Ans.**

Example 4. *The time period T of an oscillation of simple pendulum of length l is given by the equation* $T = 2\pi\sqrt{l/g}$, *where g is the constant. What is the percentage error in T. When l is increased by 1%* **[SBTE. 2015, 2013]**

Solution. Let Δl be a small error in l, while ΔT be the small error in T. then,

$$\frac{\Delta l}{l} \times 100 = 1 \qquad \text{[Given]}$$

$$\Rightarrow \quad \frac{dl}{l} \times 100 = 1 \qquad \ldots(1)\ [\because \Delta l = dl]$$

$$\Rightarrow \quad T = 2\pi\sqrt{l/g}$$

$$\Rightarrow \quad \log T = \log\left[2\pi\left(\frac{l}{g}\right)^{\frac{1}{2}}\right]$$

$$\Rightarrow \quad \log T = \log 2\pi + \log\left(\frac{l}{g}\right)^{\frac{1}{2}}$$

$$= \log 2\pi + \frac{1}{2}\log\left(\frac{l}{g}\right)$$

$$\Rightarrow \quad \log T = \log 2\pi + \frac{1}{2}[\log l - \log g]$$

$$\Rightarrow \quad \frac{1}{T}dT = 0 + \frac{1}{2}\frac{dl}{l}$$

$$= \frac{dT}{T} \times 100 = \frac{1}{2}\frac{dl}{l} \times 100 = \frac{1}{2} \times 1 \qquad \text{[From (1)]}$$

$$= \frac{dT}{T} \times 100 = \frac{1}{2}$$

$\therefore$ % error in T = $\frac{1}{2}$% **Ans.**

Example 5. *If the radius of a sphere is measured as 9 m with an error of 0.03 m. Then find the approximate error in calculating its surface area.*

Solution. Let r be the radius of the sphere and δr be the error in measuring the radius. Then, $r = 9$ m and $\delta r = 0.03$ m. Now, the surface area S of the sphere is given by

$$S = 4\pi r^2$$

$$\Rightarrow \quad \frac{dS}{dr} = 8\pi r$$

Therefore, $$\delta S = \left(\frac{dS}{dr}\right)\delta r$$

$$= (8\pi r)\,(\delta r)$$

$$= 8\pi(9)\,(0.03) = 2.16\pi \text{ m}^2$$

Hence, the approximate error in calculating the surface area is 2.16π m^2. **Ans.**

Example 6. *Find the approximate change in the volume V of a cube of side x metres caused by increasing the side by 1%.*

Solution. We know that, $V = x^3$

$$\Rightarrow \quad \frac{dV}{dx} = 3x^2$$

$$\Rightarrow \quad \delta V = \left(\frac{dV}{dx}\right)\delta x = 3x^2\,(\delta x)$$

$$= 3x^2\,\frac{x}{100} \qquad \left[\text{as } 1\% \text{ of } x \text{ is } \frac{x}{100}\right]$$

$$= 0.03\,x^3 \text{ m}^3$$

Thus, the approximate increase in volume is $0.03\,x^3$ m^3. **Ans.**

Example 7. *Find the approximate change in the surface area of a cube of side x meters caused by decreasing the side by 1%.*

Solution. Let the side of a cube be x meters.

Then surface area of area of cube be $S = 6x^2 \Rightarrow \dfrac{dS}{dx} = 12x$ ...(1)

$$\text{or} \quad \delta S = \left(\frac{dS}{dx}\right)\delta x = (12x)\,\delta x \qquad [\text{Using (1)}]$$

$$= (12x)\frac{x}{100} = \frac{12}{100}x^2 \qquad \left[\text{as } 1\% \text{ of } x \text{ is } \frac{x}{100}\right]$$

$$= 0.12\,x^2\text{m}^2$$

Thus, the approximate decrease in surface area is $0.12\,x^2$m^2. **Ans.**

Example 8. *If the radius of a spherical balloon increases by 0.2 percent. Find approximately the percentage increase in the volume.*

Solution. Let at any time x be the radius and y be the volume of the spherical balloon.

Then, $y = \dfrac{4}{3}\pi x^3$ ...(1)

Let δx be the change in the radius and let δy be the corresponding change in the volume of the spherical balloon. Then, given that

$$\frac{\delta x}{x} \times 100 = 0.2$$

Taking log of both sides of (1), we have

$$\log y = \log \frac{4}{3}\pi x^3$$

$$\Rightarrow \quad \log y = \log\frac{4}{3} + \log \pi + 3\log x \qquad ...(2)$$

Differentiating (2) *w.r.t.* 'x', we get

$$\frac{1}{y}\,\frac{dy}{dx} = \frac{3}{x}$$

$$\Rightarrow \quad \frac{1}{y}\,\frac{\delta y}{\delta x} = \frac{3}{x}$$

$$\Rightarrow \quad \frac{\delta y}{y} = \frac{3\delta x}{x} \Rightarrow \frac{100\delta y}{y} = \frac{3 \times 100 \times \delta x}{x} \qquad \left(\because \frac{dy}{dx} = \frac{\delta y}{\delta x} \text{ app.}\right)$$

$$\Rightarrow \quad \frac{100\delta y}{y} = 3 \times 0.2 = 0.6$$

Hence, percentage increase in volume = 0.6 **Ans.**

Example 9. *The quantity Q of water flowing over a V-notch is given by the formula $Q = CH^{5/2}$, where H is the height of water, C is the constant. Find the error in Q if the error in H is 1.5%.*

Solution. $Q = CH^{5/2}$

Taking log on both sides, we get

$$\log Q = \log CH^{5/2}$$

$$= \log C + \frac{5}{2}\log H$$

Differentiating both sides *w.r.t.* '*H*' we get

$$\frac{1}{Q}\frac{dQ}{dH} = 0 + \frac{5}{2}\frac{1}{H}$$

$$\Rightarrow \quad \frac{\delta Q}{Q} = 5\frac{\delta H}{2H}$$

$$\Rightarrow \quad \frac{\delta Q}{Q} \times 100 = \frac{5}{2}\frac{\delta H}{H} \times 100$$

$$= \frac{5}{2}(1.5)$$

$$= 3.75$$ **Ans.**

Example 10. *The deflection at the centre of a rod of length l and diameter d supported at its ends and loaded at the centre with a weight w varies as wl^3d^{-4}. What is the percentage increase in the deflection corresponding to the percentage increase in w, l and d of 3, 2 and 1 respectively?*

Solution. Let the deflection of the rod at the centre be *D*.

$$D = k\frac{wl^3}{d^4}$$

$$\log D = \log k + \log w + 3 \log l - 4 \log d$$

On differentiating $\quad \frac{\delta D}{D} = \frac{\delta w}{w} + 3\frac{\delta l}{l} - 4\frac{\delta d}{d}$

$$\Rightarrow \quad \frac{100\delta D}{D} = \frac{100\delta w}{w} + 3 \times \frac{100\delta l}{l} - 4 \times 100\frac{\delta d}{d}$$

$$= 3 + 3 \times 2 - 4 \times 1$$

$$= 3 + 6 - 4 = 5\%$$ **Ans.**

Example 11. *The diameter and altitude of a can in the shape of a right circular cylinder are measured as 40 cm and 64 cm respectively. If the possible error in each measurement is ± 5%. Find approximately the maximum possible error in the computed value for the volume and the lateral surface. Find the corresponding percentage error.*

Solution. Diameter of the can (*D*) = 40 cm

Altitude of the can (h) = 64 cm

$$\frac{100\delta D}{D} = \frac{100\delta h}{h} = \pm 5\%$$

$$V = \pi r^2 h$$

$$\log V = \log \pi + 2 \log r + \log h$$

$$\Rightarrow \quad \frac{\delta V}{V} = 0 + \frac{2\delta r}{r} + \frac{\delta h}{h}$$

$$\Rightarrow \quad \frac{\delta V}{V}100 = \frac{2\delta r}{r}100 + \frac{\delta h}{h}100$$

$$= 2(\pm 5) + (\pm 5) = \pm 15$$ **Ans.**

$$S = 2\pi r l$$

$$\log S = \log 2\pi + \log r + \log l$$

$$\Rightarrow \quad \frac{\delta S}{S} = 0 + \frac{\delta r}{r} + \frac{\delta l}{l}$$

$$\Rightarrow \quad \frac{\delta S}{S}100 = \frac{\delta r}{r}100 + \frac{\delta l}{l}100$$

$$= (\pm 5) + (\pm 5) = \pm 10$$ **Ans.**

Example 12. *In estimating the number of bricks in a pile which is measured to be $(5 \times 10 \times 5)$m, count of bricks is taken as 100 bricks per m^3. Find the error in the cost when the tape is stretched 2% beyond its standard length. The cost of bricks is Rs. 2,000 per thousand bricks.*

Solution. Volume $V = xyz$

$$\log V = \log x + \log y + \log z$$

Differentiating, we get

$$\frac{\delta V}{V} = \frac{\delta x}{x} + \frac{\delta y}{y} + \frac{\delta z}{z}$$

$$100\frac{\delta V}{V} = \frac{100\delta x}{x} + \frac{100\delta y}{y} + \frac{100\delta z}{z}$$

$$= 2 + 2 + 2 = 6$$

$$\delta V = \frac{6V}{100} = \frac{6(5 \times 10 \times 5)}{100} = 15 \text{ cubic metres.}$$

Number of bricks in $\delta V = 15 \times 100 = 1500$

Error in cost $= \dfrac{1500 \times 2000}{1000} = 3000$

This error in cost, a loss of the seller of bricks = ₹3000 **Ans.**

Example 13. *With the usual meaning for a, b, c and s if Δ be the area of a triangle, prove that the error in Δ resulting from a small error in the measurement of c, is given by*

$$\delta\Delta = \frac{\Delta}{4}\left[\frac{1}{s} + \frac{1}{s-a} + \frac{1}{s-b} - \frac{1}{s-c}\right]\delta c$$

Solution. We know that

$$\Delta = \sqrt{s(s-a)(s-b)(s-c)}$$

$$\log \Delta = \frac{1}{2}[\log s + \log(s-a) + \log(s-b) + \log(s-c)]$$

$$\frac{1}{\Delta}\frac{d\Delta}{dc} = \frac{1}{2}\left[\frac{1}{s}\frac{ds}{dc} + \frac{1}{s-a}\frac{d(s-a)}{dc} + \frac{1}{s-b}\frac{d(s-b).}{dc} + \frac{1}{s-c}\frac{d(s-c)}{dc}\right]$$

$$s = \frac{1}{2}(a+b+c), \frac{ds}{dc} = \frac{1}{2}$$

and $$s-c = \frac{1}{2}(a+b-c), \frac{d}{dc}(s-c) = \frac{ds}{dc} - 1 = \frac{1}{2} - 1 = -\frac{1}{2}$$

$\Rightarrow$ $$\frac{1}{\Delta}\frac{\delta\Delta}{\delta c} = \frac{1}{2}\left[\frac{1}{2}\frac{1}{s} + \frac{1}{2}\frac{1}{s-a} + \frac{1}{2}\frac{1}{s-b} - \frac{1}{2}\frac{1}{s-c}\right]$$

$\Rightarrow$ $$\delta\Delta = \frac{\Delta}{4}\left[\frac{1}{s} + \frac{1}{s-a} + \frac{1}{s-b} - \frac{1}{s-c}\right]\delta c$$

Proved.

Example 14. *Find the possible percentage error in computing the parallel resistance r of three resistances r_1, r_2, r_3 from the formula $\frac{1}{r} = \frac{1}{r_1} + \frac{1}{r_2} + \frac{1}{r_3}$ if r_1, r_2, r_3 are each in error by plus 1.2%.*

Solution. Here $$\frac{1}{r} = \frac{1}{r_1} + \frac{1}{r_2} + \frac{1}{r_3} \quad ...(1)$$

Differentiating, we get :

$$-\frac{1}{r^2}dr = -\frac{1}{r_1^2}dr_1 - \frac{1}{r_2^2}dr_2 - \frac{1}{r_3^2}dr_3$$

$\Rightarrow$ $$\frac{1}{r}\left(\frac{100dr}{r}\right) = \frac{1}{r_1}\left(\frac{100dr_1}{r_1}\right) + \frac{1}{r_2}\left(\frac{100dr_2}{r_2}\right) + \frac{1}{r_3}\left(\frac{100dr_3}{r_3}\right)$$

$$= \frac{1}{r_1}(1.2) + \frac{1}{r_2}(1.2) + \frac{1}{r_3}(1.2) = (1.2)\left[\frac{1}{r_1} + \frac{1}{r_2} + \frac{1}{r_3}\right]$$

$$= (1.2)\left(\frac{1}{r}\right) \quad \left[\because \text{From (1), } \frac{1}{r} = \frac{1}{r_1} + \frac{1}{r_2} + \frac{1}{r_3}\right]$$

$\Rightarrow$ $$\frac{100dr}{r} = 1.2\% = \text{the \% error in } r.$$

Ans.

EXERCISE 29.1

1. The radius of a sphere is found by measurement to be 10 cm with a possible error of 0.01 of a cm. Find the consequent error in the calculation of its surface
Ans. 2.51sq. cm
2. The radius and height of a cylinder are 5 cm and 10 cm respectively. Calculate approximate change in the volume with possible error of 0.01 cm in radius.
Ans. 3.14 cm^3
3. The radius of a cone is found by measurement to be 20 cm with a possible error of 0.1 cm in radius. Find the consequent error in its volume, if the height of the cone is 25 cm.
Ans. 104.76 cm^3

4. Find the approximate change in the volume of a cube of side x metres caused by increasing the side by 3%. **Ans.** $0.09x^3 m^3$

5. If there is an error of 0.1% in the measurement of the radius of a sphere, find approximately the percentage error in the calculation of the volume of the sphere. **Ans.** 0.3%

   [**Hint.** First take log and then differentiate]

6. Find the percentage error in calculating the volume of a cubical box if an error of 1% is made in measuring the lengths of edges of the cube. **Ans.** 3%

   [**Hint.** First take log and then differentiate]

7. If the density ρ of a body be inferred from its weights W, ω in air and water respectively, show that the relative error in ρ due to errors δW, $\delta\omega$ in W, ω is

$$\frac{\delta\rho}{\rho} = \frac{-\omega}{W-\omega} \cdot \frac{\delta W}{W} + \frac{\delta\omega}{W-\omega}$$

8. What error in the common logarithm of a number will be produced by an error of 1% in the number? **Ans.** 0.01

9. The indicated horse power I of an engine is calculated from the formula. $I = PLAN/33000$, where $A = \frac{\pi}{4}d^2$ assuming that errors of r per cent may have been made in measuring P, L, N and d. Find the greatest possible error in I. **Ans.** $5\,r\%$

10. The dimensions of a cone are, radius 4 cm, height 6 cm. What is the error in its volume if the scale used in taking the measurement is short by 0.01 cm per cm. **Ans.** $0.96\,\pi$ cm

11. The work that must be done to propel a ship of displacement D for a distance s in time t is proportional to $s^2D^{2/3}\,t^2$. Find approximately the percentage increase of work necessary when the displacement is increased by 1%, the time is diminished by 1% and the distance is increased by 3% **Ans.** $\frac{14}{3}\%$

12. The power P required to propel a ship of length l moving with a velocity V is given by $P = k\,V^3\,t^2$. Find the percentage increase in power if increase in velocity is 3% and increase in length is 4%. **Ans.** 17%

13. In estimating the cost of a pile of bricks measured as 2 m × 15 m × 1.2 m, the tape is stretched 1% beyond the standard length if the count is 450 bricks to 1 m^3 and bricks cost Rs. 530 per 1000, find the approximate error in the cost. **Ans.** Rs. 257.58

14. In estimating the cost of a pile of bricks measured as 6' × 50' × 4', the tape is stretched 1% beyond the standard length. If the count is 12 bricks to 1 ft^3, and bricks cost Rs. 100 per 1000, find the approximate error in the cost. **Ans.** 720 bricks, Rs. 25.20

15. Two sides of triangle are 70 cm and 40 cm in length, the angle between them being 64° 12'. But the angle is measured by mistake as 65°. What is the error in calculated length of the third side? **Ans.** 0.55 cm

16. In determining the specific gravity by the formula $S = \frac{A}{A-w}$, where A is the weight in air and w is the weight in water. A can be read within 0.01 gm and w within 0.02 gm. Find approximately the maximum error in S if the readings are A = 1.1 gm and w = 0.6 gm. **Ans.** – 0.112

17. The voltage V across a resistor is measured with error h, and the resistance R is measured with an error k. Show that the error in calculating the power $W(V,R) = \frac{V^2}{R}$ generated in the resistor is $\frac{V}{R^2}(2Rh - Vk)$. If V can be measured to an accuracy of 0.5 p.c. and R to an accuracy of 1 p.c., what is the approximate possible percentage error in W?

Ans. Zero percent

18. Find the possible percentage error in computing the parallel resistance r of two resistances r_1 and r_2 from the formula $\frac{1}{r} = \frac{1}{r_1} + \frac{1}{r_2}$, where r_1 and r_2 are both in error by + 2% each.

Ans. 2%

19. The time of swing t of a pendulum of length l under certain conditions is given by $t = 2\pi\sqrt{\frac{l}{g}}$, where $g' = g\left[\frac{r}{r+h}\right]^2$. Show that the % error in t due to error p% in h and q% in l is given by $\frac{1}{2}\left(q + \frac{2ph}{r}\right)$ where r is constant.

20. A balloon is in the form of right circular cylinder of radius 1.5 m and length 4 m and is surmounted by hemispherical ends. If the radius is increased by 0.01 m and the length by 0.05 m, find the percentage change in the volume of the balloon.

Ans. 2.389%

29.2 DERIVATIVE AS A RATE MEASURE

Derivative of one variable with respect to the other can be used to study the rate of change of variables connected by suitable relation.

$\frac{dy}{dx}$ is the rate of change of y with respect to x.

Example 15. *An inverted cone has a depth 10 cm and base of radius 5 cm. Water is poured into it at the rate of $\frac{3}{2}$ cm³/sec. Find the rate at which the level of the water in the cone is rising when the depth is 4 cm.* **(B.T.E. Delhi May 2008)**

Solution. Let θ be the semi-vertical angle of the cone CAB, whose depth is 10 cm. and radius OB = 5 cm.

We have, $$\tan\theta = \frac{OB}{OC} = \frac{5}{10} = \frac{1}{2}$$

Let V be the volume of the water in the cone means the volume of the cone $CA'B'$ after time t sec. and h be the depth of the water then

$$V = \frac{1}{3}\pi(O'B')^2\,(CO') = \frac{1}{3}\pi\,(h\tan\theta)^2\cdot h$$

$$= \frac{1}{3}\pi h^3 \tan^2\theta \quad \left[\because \tan\theta = \frac{O'B'}{O'C} = \frac{O'B'}{h} \Rightarrow O'B' = h\tan\theta\right]$$

$$= \frac{1}{3}\pi h^3\left(\frac{1}{4}\right) = \frac{1}{12}\pi h^3 \quad \left[\because \tan\theta = \frac{1}{2}\right]$$

$$\Rightarrow \quad \frac{dV}{dh} = \frac{1}{12} \times \pi \times 3h^2 = \frac{1}{4}\pi h^2 \quad ...(1)$$

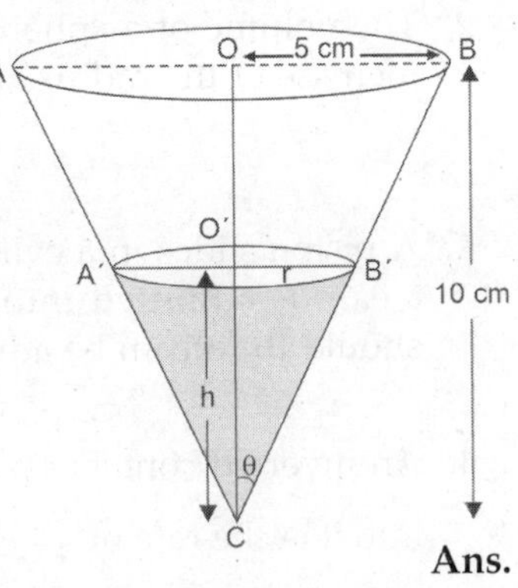

Differentiating *w.r.t.* 't' we get :

$$\frac{dV}{dt} = \frac{dV}{dh} \cdot \frac{dh}{dt} = \left(\frac{1}{4}\pi h^2\right)\frac{dh}{dt} \quad \text{[From (1)]}$$

$$\Rightarrow \quad \frac{3}{2} = \frac{\pi h^2}{4} \cdot \frac{dh}{dt} \quad \left[\because \frac{dV}{dt} = \frac{3}{2}\text{cm}^3/\text{sec.}\right]$$

$$\Rightarrow \quad \frac{dh}{dt} = \frac{6}{\pi h^2} \Rightarrow \left(\frac{dh}{dt}\right)_{h=4} = \frac{6}{\pi(4)^2} = \frac{3}{8\pi}\text{ cm/sec}$$

Ans.

Example 16. *If the rate of change of volume of a sphere is equal to the rate of change of its radius. Find its radius.* **[SBTE. 2016]**

Solution. Volume of sphere (V) = $\frac{4}{3}\pi r^3$

$$\frac{dv}{dt} = \frac{4}{3}\pi 3r^2 \frac{dr}{dt} = 4\pi r^2 \frac{dr}{dt}$$

$$\frac{dv}{dt} = \frac{dr}{dt} = 4\pi r^2 \frac{dr}{dt}$$

$$1 = 4\pi r^2$$

$$r = \frac{1}{2\sqrt{\pi}}$$

Ans.

Example 17. *Air is filled in a spherical balloon at the rate of 15 cm³/sec. At what rate is its surface area increasing when the radius is 5 cm.* **[SBTE. 2017]**

Solution. As $\quad V = \frac{4}{3}\pi r^3$

$$\Rightarrow \quad \frac{dv}{dt} = 4\pi r^2 \frac{dr}{dt}$$

$$\Rightarrow \quad 15 = 4\pi r^2 \frac{dr}{dt} \quad \left[\because \frac{dv}{dt} = 15\text{cm}^3/\text{sec}\right]$$

$$\Rightarrow \quad \frac{dr}{dt} = \frac{15}{4\pi r^2} \quad \text{or} \quad s = 4\pi r^2$$

$$\frac{ds}{dt} = 8\pi r\frac{dr}{dt} = 8\pi r \times \frac{15}{4\pi r^2} \quad [\because r = 5\text{ cm}]$$

$$= \frac{2 \times 15}{r} = \frac{30}{5} = 6\,\text{cm}^2/\text{sec}$$

Ans.

EXERCISE 29.2

1. A circular patch of oil spreads on water, and the area is growing at the rate of 16 sq. cm. per minute. How fast is the radius increasing when the radius is 6 cm (Take π = 22/7) **Ans.** $\frac{14}{33}$

2. The volume of a sphere is increasing at the rate of 20 cm per sec. Find the rate of increase of the radius and of the surface area when the volume is 36 π c. cm.

[B.T.E, Delhi May 2007] Ans. $\frac{1}{18\pi}$ cm / sec, $\frac{4}{3}$ sq.cm / sec.

3. A piston slides in a cylinder of 6 cms of diameter. At what rate does it move when steam is admitted into the cylinder at the rate of 19.008 cu. cm/s. At what rate should the steam be admitted to make piston move at the rate of 504 cm/s.

Ans. 672 cm/s., 14256 cu. cm/s.

4. An inverted cone has a depth of 10 cm and a base of radius 5 cm. Water is poured into it at the rate of $1\frac{1}{2}\text{cm}^3$ per minute. Find the rate at which the level of the water in the cone is rising when depth is 4 cm. **Ans.** $\frac{3}{8\pi}$ cm / s.

5. One end of a ladder 7 metres long is leaning against a wall. If the foot of the ladder be pulled away at the rate of 3 metres/min., how fast the top of the ladder is descending when the foot is 4 metres from the wall. **Ans.** $\frac{4}{11}\sqrt{33}$ metre / min.

[Hint. $x^2 + y^2 = 49$. Differentiate w.r.t."t"]

29.3 TANGENTS AND NORMALS

(*a*) Let P and Q be two neighbouring points on the curve $y = f(x)$.
Let the coordinates of P be (x, y) and Q $(x + \delta x, y + \delta y)$.
Join PQ and let it make an angle θ with x-axis at R.

$$\angle NPQ = \angle MRQ = \theta$$

$$\tan(\angle NPQ) = \frac{NQ}{NP} = \frac{MQ - MN}{LM} = \frac{MQ - MN}{OM - OL}$$

$$= \frac{y + \delta y - y}{x + \delta x - x}$$

$$\tan\theta = \frac{\delta y}{\delta x}$$

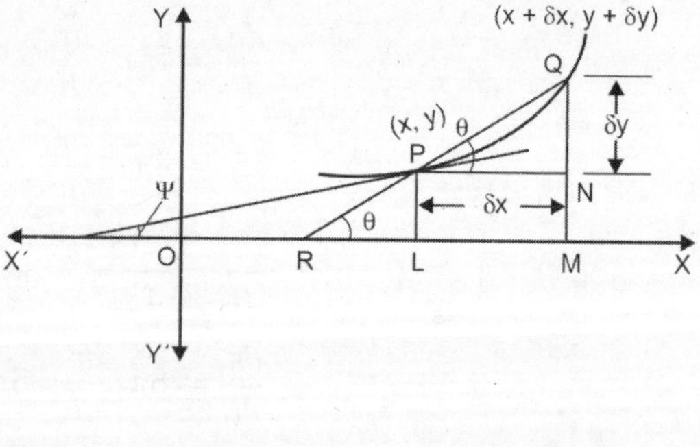

Now let $\delta x \to 0$, then $Q \to P$ and the secant PQ becomes tangent PT at P, and thus $\theta \to \psi$.

$$\lim_{P \to Q} \tan\theta = \lim_{\delta x \to 0} \frac{\delta y}{\delta x} \Rightarrow \tan\psi = \frac{dy}{dx}$$

$$\boxed{\frac{dy}{dx} = \text{Slope of the tangent at } P}$$

(b) **Equation of a tangent.** The equation of the tangent at (x_1, y_1) is

$$\boxed{y - y_1 = \frac{dy_1}{dx_1}(x - x_1)} \qquad \text{where } \frac{dy_1}{dx_1} = \frac{dy}{dx} \text{ at } (x_1, y_1)$$

Equation of the normal at (x_1, y_1) is

$$\boxed{y - y_1 = -\frac{dx_1}{dy_1}(x - x_1)}$$

Example 18. *Find the equation of the tangent and normal to the curve $y^2 = 3x^2 + 1$ at the point (1, 2).*

Solution.
$$y^2 = 3x^2 + 1$$

$$2y\frac{dy}{dx} = 6x \Rightarrow \frac{dy}{dx} = \frac{3x}{y}$$

$$\frac{dy}{dx} \text{ at the point}(1, 2) = \frac{3 \times 1}{2} = \frac{3}{2}$$

$\Rightarrow$ The slope of the tangent $= \frac{3}{2}$

Equation of tangent

$$y - 2 = \frac{3}{2}(x - 1)$$

$$\Rightarrow \quad 3x - 2y + 1 = 0.$$

The slope of the normal $= -\frac{2}{3}$

So the equation of the normal is

$$y - 2 = -\frac{2}{3}(x - 1)$$

$$2x + 3y - 8 = 0.$$

Ans.

Example 19. *Find the slope of normal to the curve $y^2 = 5x^2 - 2x + 3$ at (1, 6)* **[SBTE. 2014]**

Solution. Here we have

$$y^2 = 5x^2 - 2x + 3$$

$$2y\frac{dy}{dx} = 10x - 2$$

$$\Rightarrow \quad \frac{dy}{dx} = \frac{5x - 1}{y}$$

$$\text{Slope of normal at } (1, 6) = \frac{-1}{\left(\frac{dy}{dx}\right)_{at(1.6)}}$$

$$= \frac{-1}{\frac{5-1}{6}} = \frac{-1}{\frac{4}{6}} = \frac{-3}{2}$$

Ans.

Example 20. *Find the slope of tangent to the curve $y^2 = ax^3$ at (2, 3)* **[SBTE. 2015]**

Solution. Here we have

$$y^2 = ax^3$$

$$2y\frac{dy}{dx} = 3ax^2$$

$$\frac{dy}{dx} = \frac{3ax^2}{2y}$$

Slope of the tangent at (2, 3) $= \left(\frac{dy}{dx}\right)_{at(2,3)}$

$$= \left(\frac{3ax^2}{2y}\right)_{at(2,3)} = 2a$$ **Ans.**

Example 21. *Find the points on the circle $x^2 + y^2 + 2x - 4y = 0$ where the tangents are parallel to x-axis.*

Solution. $x^2 + y^2 + 2x - 4y = 0$

Defferentiating, we have

$$2x + 2y\frac{dy}{dx} + 2 - 4\frac{dy}{dx} = 0$$

$$(y-2)\frac{dy}{dx} = -x-1$$

$$\frac{dy}{dx} = -\frac{x+1}{y-2}$$

The tangent will be parallel to the x-axis if

$$\frac{dy}{dx} = 0 \Rightarrow -\frac{x+1}{y-2} = 0 \Rightarrow x = -1$$

Put $x = -1$ in the equation of circle

$$1 + y^2 - 2 - 4y = 0 \Rightarrow y^2 - 4y - 1 = 0$$

$$y = 2 \pm \sqrt{5}$$

The required points where the tangents will be parallel to x-axis are $(-1, 2 \pm \sqrt{5})$. **Ans.**

Example 22. *Find the equation of the tengent to the curve $x^2 + 2y = 8$ which is perpendicular to the line $x - 2y + 1 = 0$.*

Solution. The given curve is $x^2 + 2y = 8$ ... (1)

Differentiating w.r.t. 'x', we get

$$2x + 2\frac{dy}{dx} = 0 \Rightarrow \frac{dy}{dx} = -x$$

And the given line is

$$x - 2y + 1 = 0$$

Differentiating w.r.t. 'x', we get

$$1 - 2\frac{dy}{dx} = 0 \Rightarrow \frac{dy}{dx} = \frac{1}{2}$$

Since the tangent is perpendicular to the line therefore,

(Slope of tangent) (slope of line) = –1

$$\Rightarrow \qquad (-x)\left(\frac{1}{2}\right) = -1 \Rightarrow x = 2$$

Now, we have to find y-coordinate when $x = 2$.

On putting $x = 2$ in (1), we get

$$(2)^2 + 2y = 8 \;\Rightarrow\; 2y = 8 - 4 = 4 \;\Rightarrow\; y = 2$$

$\therefore$ Equation of tangent to (1) at the point (2, 2) is

$$y - 2 = -2\,(x - 2)$$

$$= -2x + 4$$

$\Rightarrow$ $$2x + y = 6$$ **Ans.**

Example 23. *The equation of the tangent at the point (2, 3) on the curve $y^2 = ax^3 + b$ is $y = 4x - 5$. Find the values of a and b.*

Solution. $$y^2 = ax^3 + b$$

$\therefore$ $$2y\frac{dy}{dx} = 3ax^2 \;\Rightarrow\; \frac{dy}{dx} = \frac{3ax^2}{2y}$$

$$\frac{dy}{dx} \text{ at the point } (2, 3) = \frac{3a(2)^2}{2 \times 3} = 2a$$

The slope of the tangent $y = 4x - 5$ is 4.

i.e. $$\frac{dy}{dx} = 4$$

or $$2a = 4 \quad \text{or} \quad a = 2$$

The point (2, 3) lies on the curve $y^2 = ax^3 + b$.

$\therefore$ $$(3)^2 = (2)^3\,a + b \quad \text{or} \quad 9 = 8a + b$$

But $$a = 2$$

$\therefore$ $$9 = 8 \times 2 + b \quad \text{or} \quad b = -7$$

Hence, $$a = 2; \quad b = -7$$ **Ans.**

Example 24. *Prove that the line $\frac{x}{a} + \frac{y}{b} = 1$ touches the curve $y = be^{-}$ at the point where the curve crosses the axis of y.* **[SBTE. 2016, 2015, 2014]**

Solution. Here $$y = be^{-\frac{x}{a}} \quad \text{... (1)}$$

If (1) crosses y-axis then x-coordinate is equal to zero. On putting $x = 0$ in (1), we get

$$y = be^{-\frac{0}{a}} = b$$

Now we have to find the equation of the tangent to the curve (1) at the point $(0, b)$.

Differentiating (1), we get $$\frac{dy}{dx} = \left(be^{-\frac{x}{a}}\right)\left(-\frac{1}{a}\right) = -\frac{b}{a}e^{-\frac{x}{a}} = -\frac{y}{a}$$

$$\frac{dy}{dx} \text{ at the point } (0, b) = -\frac{b}{a}$$

The equation of the line having the slope $\left(-\frac{b}{a}\right)$ and passing through $(0, b)$ is

$$y - y' = \text{slope}\,(x - x')$$

$$y - b = -\frac{b}{a}(x - 0)$$

$$\frac{y}{b} - 1 = -\frac{x}{a} \Rightarrow \frac{x}{a} + \frac{y}{b} = 1$$

Hence $\frac{x}{a} + \frac{y}{b} = 1$ touches the given curve at (0, *b*). **Proved.**

Example 25. *At what point on the curve $y = x^2$ does the tangent makes an angle of 45° with the X-axis.* **[SBTE. 2015]**

Solution. Here we have $y = x^2$ ... (1)

$$\frac{dy}{dx} = 2x \quad \text{... (2)}$$

$$\left(\frac{dy}{dx}\right)_{at\,(x_1,\,y_1)} = \tan 45° = 1 \quad \text{... (3)}$$

$$\left(\frac{dy}{dx}\right)_{at\,(x_1,\,y_1)} = 2x_1 \quad \text{[From (2)]}$$

$$1 = 2x_1 \quad \text{[From (3)]}$$

$$x_1 = \frac{1}{2}$$

$$y_1 = x_1^2 = \frac{1}{4} \quad \text{[From (1)]}$$

Thus, the required point is $\left(\frac{1}{2}, \frac{1}{4}\right)$ **Ans.**

Example 26. *Find the angle made by the tangent to the curve $xy = 1$ at (1, 1) with the positive X-axis.* **[SBTE. 2015]**

Solution. Here we have $xy = 1$

$$y = \frac{1}{x}$$

$$\frac{dy}{dx} = -\frac{1}{x^2}$$

$$\left(\frac{dy}{dx}\right)_{\text{at}\,(1,\,1)} = \frac{-1}{1} = -1$$

$$\tan\theta = \left(\frac{dy}{dx}\right)_{\text{at}\,(1,\,1)} = -1$$

$$\tan\theta = \tan\frac{3\pi}{4} \Rightarrow \theta = \frac{3\pi}{4}$$ **Ans.**

Example 27. *Prove that the curve $\left(\frac{x}{a}\right)^n + \left(\frac{y}{b}\right)^n = 2$ touches the straight line $\frac{x}{a} + \frac{y}{b} = 2$ at the point (a, b), whatever be the value of n.*

Solution. $$\frac{x^n}{a^n} + \frac{y^n}{a^n} = 2 \quad \text{... (1)}$$

On differentiating (1) w.r.t. 'x', we get

$$\frac{nx^{n-1}}{a^n} + \frac{ny^{n-1}}{b^n}\frac{dy}{dx} = 0$$

$$\Rightarrow \quad \frac{dy}{dx} = \left(-\frac{x^{n-1}}{a^n}\right)\left(\frac{b^n}{y^{n-1}}\right)$$

$$\left(\frac{dy}{dx}\right)_{\text{at }(a,b)} = \left(-\frac{a^{n-1}}{a^n}\right)\left(\frac{b^n}{b^{n-1}}\right)$$

$$= -\frac{b}{a}$$

Equation of a tangent which passes through (a, b) and has a slope $-\frac{b}{a}$.

$$y - b = -\frac{b}{a}(x - a)$$

$$\Rightarrow \quad \frac{y}{b} - \frac{b}{b} = -\frac{x}{a} + \frac{a}{a}$$

$$\Rightarrow \quad \frac{x}{a} + \frac{y}{b} = 2$$

Proved.

Example 28. *Show that the sum of the distances from the origin to the points where tangent to* $y = (\sqrt{a} - \sqrt{x})^2$ *meets the axis is constant.*

Solution. Equation of curve, $y = (\sqrt{a} - \sqrt{x})^2$ ... (1)

Differentiating (1), we have

$$\frac{dy}{dx} = 2(\sqrt{a} - \sqrt{x})\left(-\frac{1}{2\sqrt{x}}\right)$$

$$= -\frac{\sqrt{a} - \sqrt{x}}{\sqrt{x}} = -\frac{\sqrt{y}}{\sqrt{x}} \quad \text{... (2)}$$

Slope of the tangent at (α, β) is $-\sqrt{\frac{\beta}{\alpha}}$ ... (3)

Equation of the tangent at (α, β)

$$y - \beta = -\sqrt{\frac{\beta}{\alpha}}\,(x - \alpha) \quad \text{... (4)}$$

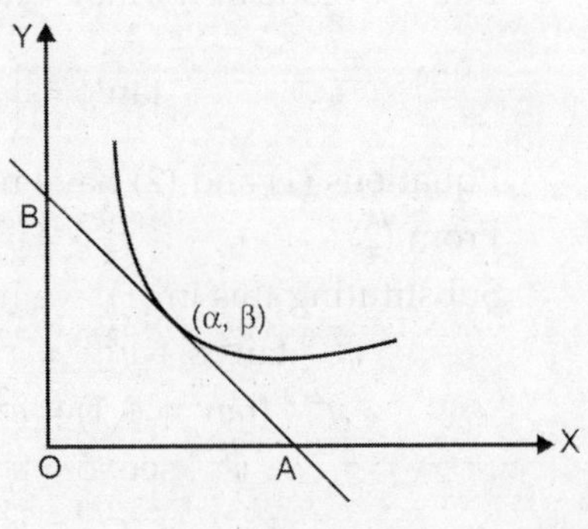

When the tangent (4) meets the y-axis then $x = 0$

$$y - \beta = -\frac{\sqrt{\beta}}{\sqrt{\alpha}}(0 - \alpha)$$

$$\Rightarrow \quad y - \beta = \sqrt{\alpha\beta} \quad \text{... (5)}$$

$$\Rightarrow \quad y = \beta + \sqrt{\alpha\beta}$$

$$\Rightarrow \quad OB = \beta + \sqrt{\alpha\beta}$$

When the tangent (4) meets the x-axis, then $y = 0$.

$$0 - \beta = -\frac{\sqrt{\beta}}{\sqrt{\alpha}}(x - \alpha)$$

$$\Rightarrow \qquad \beta = \frac{\sqrt{\beta}}{\sqrt{\alpha}}(x - \alpha)$$

$$\Rightarrow \qquad x = \alpha + \sqrt{\alpha\beta}$$

$$OA = \alpha + \sqrt{\alpha\beta}$$

$$OA + OB = \alpha + \sqrt{\alpha\beta} + \beta + \sqrt{\alpha\beta}$$

$$= \alpha + 2\sqrt{\alpha\beta} + \beta$$

$$= \left(\sqrt{\alpha^2} + \sqrt{\beta}\right)^2 = \left(\sqrt{a}\right)^2 = a$$

Hence, sum of the intercepts = constant. **Proved.**

Example 29. *If the normal to the curve $x^{2/3} + y^{2/3} = a^{2/3}$ makes an angle ϕ with the x-axis, show that its equation is $y \cos\phi - x \sin\phi = a \cos 2\phi$*

Solution. $x^{2/3} + y^{2/3} = a^{2/3}$ ... (1)

Differentiating (1) w.r.t. 'x', we get

$$\frac{2}{3}x^{-1/3} + \frac{2}{3}y^{-1/3}\frac{dy}{dx} = 0$$

$$\Rightarrow \qquad \frac{dy}{dx} = -\left(\frac{y}{x}\right)^{1/3}$$

$$\text{Slope of tangent} = -\left(\frac{y}{x}\right)^{1/3}$$

$$\text{Slope of normal} = \left(\frac{x}{y}\right)^{1/3}$$

But slope of normal = $\tan\phi$ (given)

$$\tan\phi = \left(\frac{x}{y}\right)^{1/3} \qquad \text{... (2)}$$

Equations (1) and (2) have now to be solved to find x and y.

From (2) $x^{1/3} = y^{1/3}\tan\phi$

Substituting this in (1), we have

$$y^{2/3}\tan^2\phi + y^{2/3} = a^{2/3}$$

$$y^{2/3}(\tan^2\phi + 1) = a^{2/3}$$

$$y^{2/3}\sec^2\phi = a^{2/3}$$

$$y^{2/3} = a^{2/3}\cos^2\phi$$

$$y = a\sec^3\phi$$

Substituting this value of y in (2), we get

$$x^{1/3} = a^{1/3}\cos\phi \cdot \frac{\sin\phi}{\cos\phi} = a^{1/3}\sin\phi$$

$$x = a \sin^3 \phi$$

Thus tan ϕ is the slope of the normal at $(a \sin^3 \phi, a \cos^3 \phi)$

The equation of the normal at the point is

$$y - a \cos^3 \phi = \frac{\sin \phi}{\cos \phi} (x - a \sin^3 \phi)$$

$$y \cos \phi - a \cos^4 \phi = x \sin \phi - a \sin^4 \phi$$

$$y \cos \phi - x \sin \phi = a \cos^4 \phi - a \sin^4 \phi$$

$$= a (\cos^4 \phi - \sin^4 \phi)$$

$$= a (\cos^2 \phi + \sin^2 \phi) (\cos^2 \phi - \sin^2 \phi)$$

$$= a (\cos^2 \phi - \sin^2 \phi)$$

$$= a \cos 2\phi$$ **Proved.**

Example 30. *The slope of the curve $2y^2 = ax^2 + b$ at $(1, -1)$ is -1 find 'a'.* **[SBTE. 2016, 2015, 2014]**

Solution. Here we have $2y^2 = ax^2 + b$

$$\Rightarrow \quad 4y \frac{dy}{dx} = 2ax$$

$$\Rightarrow \quad \frac{dy}{dx} = \frac{ax}{2y}$$

$$\Rightarrow \quad \left(\frac{dy}{dx}\right)_{\text{at }(1,-1)} = \left(\frac{ax}{2y}\right)_{\text{at }(1,-1)}$$

$$\Rightarrow \quad -1 = \frac{a}{-2} \qquad [\because \text{Slope} = -1]$$

$$\therefore \quad a = 2$$ **Ans.**

Example 31. *If the slope of the tangent of the curve $x = 1 - a \sin \theta$ and $y = b \cos^2 \theta$ at $\theta = \frac{\pi}{2}$ is 1, then find the value of $a - 2b$.* **[SBTE. 2017]**

Solution. Here we have

$$x = 1 - a \sin \theta, \frac{dx}{d\theta} = -a \cos \theta$$

$$y = b \cos^2 \theta, \frac{dy}{d\theta} = -2b \sin \theta \cdot \cos \theta$$

$$\frac{dy}{dx} = \frac{\frac{dy}{d\theta}}{\frac{dx}{d\theta}} = \frac{-2b \sin \theta \cdot \cos \theta}{-a \cos \theta} = \frac{2b}{a} \sin \theta$$

$$\left(\frac{dy}{dx}\right)_{\theta = \pi/2} = \frac{2b}{a} \sin \frac{\pi}{2} = \frac{2b}{a}$$

$$1 = \frac{2b}{a} \qquad \left[\because \left(\frac{dy}{dx}\right)_{\theta = \frac{\pi}{2}} = 1\right]$$

$$a = 2b$$

$$a - 2b = 0$$ **Ans.**

29.4 ANGLE OF INTERSECTION OF TWO CURVES

By the angle of intersection of two curves we mean the angle between the tangents at their common points of intersection.

Let the two curves CA and $C'B$ intersects each other at a point k.

If the equations of the tangent at k for two curves be

$$y = m_1x + c \quad \text{and} \quad y = m_2x + c$$

Then the angle between these lines is given by

$$\tan\theta = \frac{m_1 - m_2}{1 + m_1m_2}$$

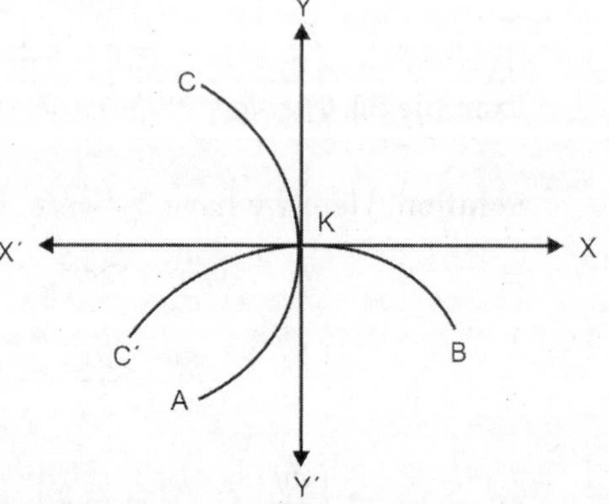

But m_1 and m_2 are the values of $\frac{dy}{dx}$ for the two curves at k. Therefore, if we replace m_1 and m_2 in the formula by the values of $\frac{dy}{dx}$ for the two curves at any of their point of intersection. We obtain the angle of intersection of the two curves at that point.

Note: (*i*) *The other angle of intersection is* $\pi - \theta$.

(*ii*) *The the two curves touch each other them* $m_1 = m_2$.

(*iii*) *If the two curves cut orthogonally then* $m_1 \cdot m_2 = -1$.

Example 32. *Find the angle of intersection of the parabolas* $y^2 = 32x$ *and* $x^2 = 32y$ *at their point of intersection other than the origin.*

Solution. The given equations of the curves are $y^2 = 32x$ ... (1)

$$x^2 = 32y \quad \text{... (2)}$$

From (2) $$y = \frac{x^2}{32}$$

Putting the value of y in (1), we get

$$\frac{x^4}{(32)^2} = 32x$$

$$\Rightarrow \quad x^4 = (32)^3 x$$

$$\Rightarrow \quad x^4 - (32)^3 x = 0$$

$$\Rightarrow \quad x[x^3 - (32)^3] = 0$$

$$\therefore \quad x = 0, 32$$ (rejecting imaginary values)

When $x = 0$ from (2), $y = 0$

When $x = 32$ from (2), $y = 32$

The points of intersection are (0, 0) and (32, 32)

For the curve (1) $2y \cdot \frac{dy}{dx} = 32$

$$\Rightarrow \qquad \frac{dy}{dx} = \frac{32}{2y}$$

For the curve (2), $2x = 32 \cdot \frac{dy}{dx}$

$$\therefore \qquad \frac{dy}{dx} = \frac{2x}{32}$$

At point (32, 32) : The slope of (1) is

$$\frac{dy}{dx} = \frac{32}{2y} = \frac{32}{2 \times 32} = \frac{1}{2}$$

At point (32, 32) : The slope of (2) is

$$\frac{dy}{dx} = \frac{2x}{32} = \frac{2 \times 32}{32} = 2$$

So at point (32, 32) slopes of the tangents are $\frac{1}{2}$ and 2

$$\tan\theta = \frac{2 - \frac{1}{2}}{1 + (2)\left(\frac{1}{2}\right)} = \frac{3}{4}$$

$$\theta = \tan^{-1}\left(\frac{3}{4}\right)$$ **Ans.**

Example 33. *If the curve $y = ae^x$ and $y = be^{-x}$ cut orthogonally, then find the value of ab.* **[SBTE. 2017]**

Solution. Here we have $y = ae^x, \frac{dy}{dx} = ae^x$

$$y = be^{-x}, \frac{dy}{dx} = -be^{-x}$$

Let the slope of the tangent to curve $y = ae^x$ and $y = be^{-x}$ at (x, y) are m_1 and m_2 respectively.

$$\left(\frac{dy}{dx}\right)_{\text{at }(x,y)} = m_1 = ae^x$$

and $$\left(\frac{dy}{dx}\right)_{\text{at }(x,y)} = m_2 = -be^{-x}$$

Both curves cut orthogonally, then

$$\tan\frac{\pi}{2} = \pm\left(\frac{m_2 - m_1}{1 + m_1 m_2}\right)$$

$$\infty = \frac{1}{0} = \pm\left(\frac{-be^{-x} - ae^x}{1 - abe^{-x} \cdot e^x}\right)$$

$$1 - ab = 0$$

$$\therefore \qquad ab = 1$$ **Ans.**

Example 34. *Find the equation of tangent of the curve given by*

$$x = a \sin^3 t,\ y = b \cos^3 t \text{ at } a \text{ point } t = \frac{\pi}{4}.$$ **[SBTE. 2016, 2015]**

Solution. Here we have $x = a\sin^3 t, \dfrac{dx}{dt} = 3a\sin^2 t \cdot \cos t$

$$y = b\cos^3 t, \frac{dy}{dt} = -3b\cos^2 t \cdot \sin t$$

$$\Rightarrow \quad \frac{dy}{dx} = \frac{\frac{dy}{dt}}{\frac{dx}{dt}} = \frac{-3b\cos^2 t \cdot \sin t}{3a\sin^2 t \cdot \cos t} = \frac{-b \cdot \cos t}{a \cdot \sin t}$$

$$\Rightarrow \quad \frac{dy}{dx} = \frac{-b}{a}\cot t$$

$$\Rightarrow \quad \left(\frac{dy}{dx}\right)_{t=\frac{\pi}{4}} = \frac{-b}{a}\cot\frac{\pi}{4} = \frac{-b}{a} \qquad \left[\because \cot\frac{\pi}{4} = 1\right]$$

Equation of the tangent at the point $(a \sin^3 t, b \cos^3 t)$

$$y - y_1 = \left(\frac{dy}{dx}\right)_{t=\frac{\pi}{4}}(x - x_1)$$

$$\Rightarrow \quad y - b\cos^3 t = \frac{-b}{a}(x - a\sin^3 t)$$

$$\Rightarrow \quad ay - ab\cos^3 t = -bx + ab\sin^3 t$$

$$\therefore \quad bx + ay - ab\cos^3 t - ab\sin^3 t = 0$$ **Ans.**

Example 35. *Prove that all normals to the curve $x = acost + atsint$, $y = asint - atcost$ are at a distance 'a' from the origin.* **[SBTE. 2016, 2015, 2013]**

Solution. Here we have $x = a \cos t + at \sin t$

$$\Rightarrow \quad \frac{dx}{dt} = -a\sin t + a(\sin t + t\cos t) = at\cos t$$

$$\Rightarrow \quad y = a\sin t - at\cos t$$

$$\Rightarrow \quad \frac{dy}{dt} = a\cos t - a(\cos t - t\sin t) = at\sin t$$

$$\Rightarrow \quad \frac{dy}{dx} = \frac{dy/dt}{dx/dt} = \frac{at\sin t}{at\cos t} = \tan t$$

Equation of normal at the point $(a \cos t + at \sin t, a \sin t - at \cos t)]$

$$y - y_1 = \frac{-1}{\left(\frac{dy}{dx}\right)}(x - x_1)$$

$$\Rightarrow \quad y - (a\sin t - at\cos t) = -\frac{1}{\tan t}[x - (a\cos t + at\sin t)]$$

$$\Rightarrow \quad y - a\sin t + at\cos t = \frac{-\cos t}{\sin t}(x - a\cos t - at\sin t)$$

$$\Rightarrow \quad y\sin t - a\sin^2 t + at\sin t\cos t = -x\cos t + a\cos^2 t + at\sin t\cdot\cos t$$

$$\Rightarrow \quad x\cos t + y\sin t - a = 0$$

Distance from origin to the normal x cos t + y sin $t - a = 0$

$$= \left|\frac{0\cdot\cos t + 0\sin t - a}{\sqrt{\cos^2 t + \sin^2 t}}\right| = a$$ **Proved.**

EXERCISE 29.3

1. Find the points on the curve $y = 12x - x^3$ at which the slope is zero. **Ans.** (2, 16), (–2, –16)
2. Find the points on the curve $y^2 = 2x^3$ at which the slope of the curve is 3. **Ans.** (2, 4), (0, 0), (2, –4)
3. Find the point on the curve $y = 2x^2 - 3x + 5$ at which the tangent makes an angle of 45° with the positive direction of x-axis. **Ans.** (1, 4)
4. Find where the tangent is perpendicular to the x-axis on the curve $y^2 = x^2(20 - x)$. **Ans.** (20, 0), (0, 0)
5. Find the point on the curve $y^2 = 4x$ at which the tangent to the curve is parallel to the line $y = x$. **Ans.** (1, 2)
6. Find the normals to the curve $xy + 2x - y = 0$ that are parallel to the line $2x + y = 0$. **Ans.** $2x + y + 3 = 0$, $2x + y - 3 = 0$
7. Find the values of a and b if the slope of the tangent to the curve $xy + ax + by = 0$, at (1, 1) is 2. Find also the slope of the normal and equation of the normal to the given curve at the given point. **Ans.** $a = 1$, $b = -2$, $-\frac{1}{2}$, $x + 2y = 3$
8. Find at what points on the curve $x^2 + y^2 = 13$, the tangent is parallel to the line $2x + 3y = 0$. **Ans.** (2, 3); (–2, –3)
9. Find the slope of the tangent at $\theta = \pi/2$ to the cycloid $x = a(\theta - \sin\theta)$, $y = a(1 - \cos\theta)$. **Ans.** 1
10. Find the equation of the tangent and the normal to the curve $y = 5x^2 - 2x + 3$ at (1, 6). **Ans.** $8x - y - 2 = 0$, $x + 8y - 49 = 0$
11. Find the equation of tangent to $16x^2 + 9y^2 = 144$ at (x_1, y_1), where $x_1 = 2$ and $y_1 > 0$. **[Diploma IETE June 2005]** **Ans.** $y + \frac{8}{3\sqrt{5}}x = \frac{12}{\sqrt{5}}$
12. Find the equation of the tangent to the curve $y^2 = 4x$ at a point $\left(\frac{1}{m^2}, \frac{2}{m}\right)$ **Ans.** $m^2x - my + 1 = 0$
13. Derive the equation of the tangent and the normal to the curve $y^2 = 4ax$ at the point $(at^2, 2at)$. **[Diploma IETE Dec. 2018]** **Ans.** $x - ty + at^2 = 0$, $y + tx = 2at + at^3$
14. Find the equation of the tangent and normal to the curve $y = x^3 - 3x^2 - x + 5$ at the point where $x = 3$. **Ans.** $8x - y - 22 = 0$, $x + 8y - 19 = 0$
15. Find the equation of the tangent to the curve $y = \frac{8a^2}{4a^2 + x^2}$ at the point where it cuts the y-axis. **Ans.** $y = 2$

16. For the plane curve $y = b \sin\left(\frac{\pi x}{a}\right)$, find the equations of the tangent and the normal at the point on the curve where $x = \frac{1}{4}a$.

Ans. $y - \frac{b}{\sqrt{2}} = \frac{b\pi}{\sqrt{2}a}\left(x - \frac{a}{4}\right), y - \frac{b}{\sqrt{2}} = -\frac{\sqrt{2}a}{b\pi}\left(x - \frac{a}{4}\right)$

17. Find the equations of the tangent and the normal to the curve $y(x-2)(x-3) - x + 7 = 0$ at the point where it cuts the axis of x. **[SBTE. 2016, 2014, 2010]**

Ans. $20y - x + 7 = 0,\ y + 20x - 140 = 0$

18. Prove that $\frac{x}{a} + \frac{y}{b} = 1$ touches the curve $y = be^{-x/a}$ at the point where the curve crosses the axis of y. **[SBTE. 2016]**

19. Find the equation of normal to the curve $x^{2/3} + y^{2/3}$ at the point $\left(\frac{a}{2\sqrt{2}}, \frac{a}{2\sqrt{2}}\right)$.

Ans. $y = x$

20. Show that the sum of the intercepts on the axes, of any tangent to the curve $x^{1/2} + y^{1/2} = a^{1/2}$ is constant.

21. In the curve $x^m y^n = a^{m+n}$ prove that the portion of the tangent intercepted between the axes is divided at the point of contact into segments which are in constant ratio.

22. Find the angle of intersection of the curves $x^2 + 4y^2 = 8$ and $x^2 - 2y^2 = 4$. **Ans.** $90°$

23. Show that the curves $\frac{x^2}{a^2 + \lambda_1} + \frac{y^2}{b^2 + \lambda_1} = 1$ and $\frac{x^2}{a^2 + \lambda_2} + \frac{y^2}{b^2 + \lambda_2}$ intersect at right angle.

29.5 INCREASING AND DECREASING FUNCTIONS

(*a*) If the value of a function $y = f(x)$ increases as x increases then the function $f(x)$ is said to be an increasing function of x. A function $\phi(x)$ is called to be a decreasing function if the value of $\phi(x)$ decreases as x increases, *i.e.*

$f(x) = x^2 + 5$ is an increasing function for all values of x.

$\phi(x) = \frac{1}{x}$ is a decreasing function for all values of x.

Example 36. *A circular plate expands by heat so that its radius is increasing at the rate of 0.2 cm per second. At what rate is the area increasing when the radius is 20 cm.*

Solution. $A = \pi r^2$

$$\frac{dA}{dt} = 2\pi r \frac{dr}{dt}$$

$$\frac{dr}{dt} = 0.2, \quad r = 20$$

$$\frac{dA}{dt} = 2 \times \frac{22}{7} \times 20 \times 0.2 = \frac{176}{7} \text{ sq. cm/sec}$$ **Ans.**

(*b*) **Condition for an increasing function**

Consider the increasing function $y = f(x)$.

Let P be a point (x, y) and Q $(x + \delta x, y + \delta y)$ on the curve, the value of the function at P and Q are $f(x)$ and $f(x + \delta x)$.

As $f(x)$ is increasing, $f(x + \delta x) > f(x)$

$$\Rightarrow f(x + dx) - f(x) > 0 \Rightarrow \frac{f(x+\delta x) - f(x)}{\delta x} > 0$$

$$f'(x) > 0 \Rightarrow \boxed{\frac{dy}{dx} > 0} \boxed{\frac{dy}{dx} = +ve}$$

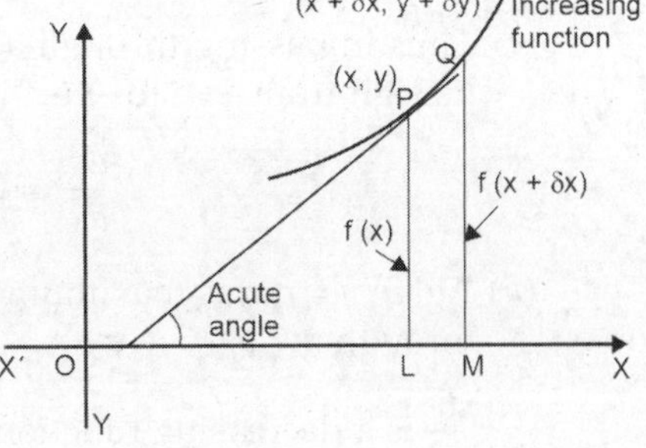

The tangent at any point of the curve makes an acute angle θ with the positive direction of x-axis.

$$\tan \theta = + \text{ve}.$$

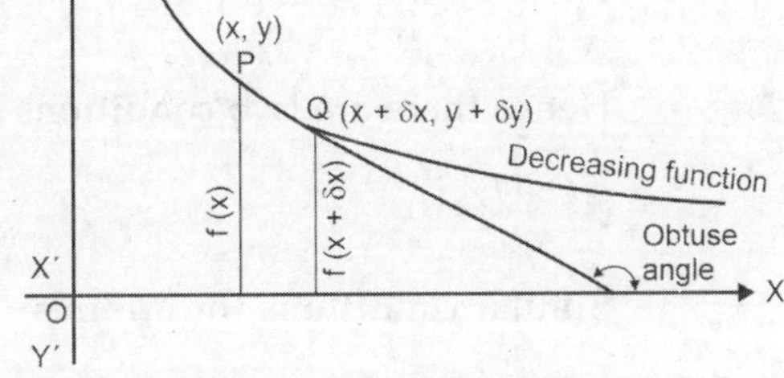

Similarly in case of decreasing function the tangent at any point of the curve makes an obtuse angle with the positive direction of x-axis.

$$\therefore \quad \boxed{\frac{dy}{dx} = - \text{ve}}$$

(*c*) **Turning point or stationary point**

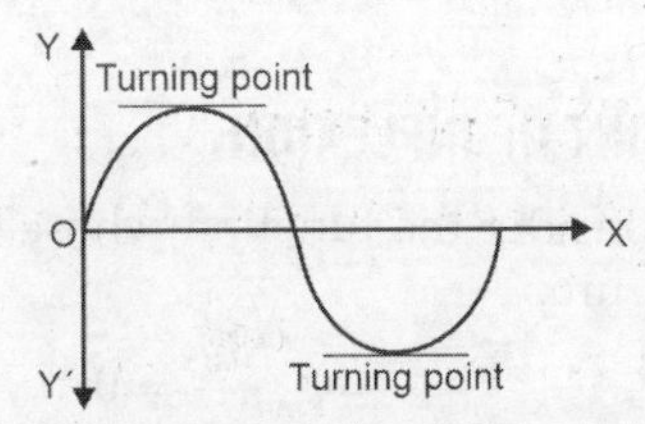

The point at which the function changes its nature is called the turning point. The increasing function may change to decreasing one and vice versa.

For example the function $\sin x$ is an increasing function between 0 and $\frac{\pi}{2}$ and decreasing between $\frac{\pi}{2}$ and $\frac{3\pi}{2}$. At $\frac{\pi}{2}$ the function is neither increasing nor decreasing but stationary. The tangent at the turning point is parallel to x-axis.

$$\Rightarrow \quad \frac{dy}{dx} = 0$$

29.6 MAXIMA AND MINIMA

(*a*) At a point where the function changes from an increasing function to a decreasing function, the function attains its maximum value. For example the function $\sin x$ has maximum value at turning point. Value of the function at the turning point is greater than all other values in the neighbourhood on either side of the turning point.

(*b*) **Conditions for maxima and minima**

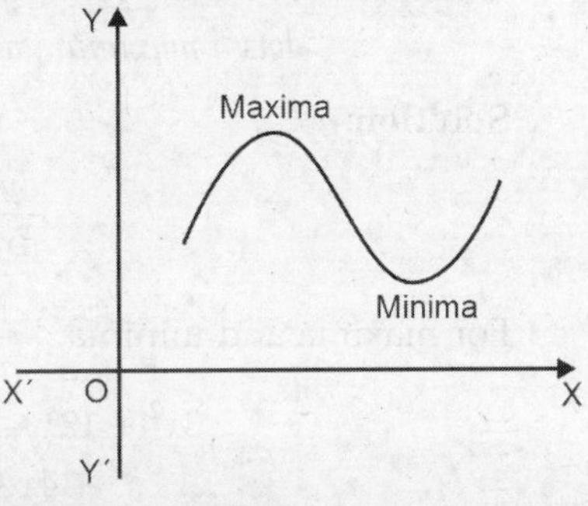

(*i*) Let $f(x)$ be maxima at $x = a$.

Now just before the maximum value, *i.e.* at $x = a$, the function is increasing.

$$\therefore \quad \frac{dy}{dx} = + \text{ve}$$

Just after the maximum value, *i.e.* at $x = a$, the function is decreasing.

$$\therefore \quad \frac{dy}{dx} = - \text{ve}.$$

Thus in passing through a maximum value at $x = a$ the derivative $\frac{dy}{dx}$ changes its sign from +ve to –ve.

$$\therefore \qquad \frac{dy}{dx} = 0 \text{ at } x = a.$$

(*ii*) Again $y = f(x)$ is maximum at $x = a$ here $\frac{dy}{dx}$ changes from +ve to –ve as it passes through $x = a$.

$\frac{dy}{dx}$ is a decreasing function so the derivative of $\frac{dy}{dx}$ should be negative at $x = 0$.

$$\because \qquad \frac{d^2y}{dx^2} = -\text{ve at } x = a.$$

Hence there are two conditions for maxima:

(1) $\frac{dy}{dx} = 0$ (2) $\frac{d^2y}{dx^2} = -\text{ ve}$

Similar conditions for minima:

(1) $\frac{dy}{dx} = 0$ (2) $\frac{d^2y}{dx^2} = +\text{ ve}$

29.7 POINT OF INFLEXION

Let us consider the function whose graph is shown in the figure.

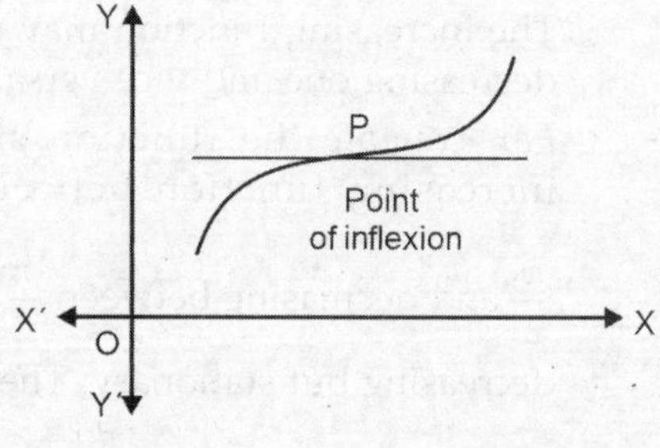

At P, $\frac{dy}{dx} = 0$

But before and after the point P the function is increasing.

$$\therefore \qquad \frac{dy}{dx} = +\text{ ve}$$

before and after the point P, $\frac{dy}{dx}$ does not change its sign while passing through P.

Condition for a point of inflexion

(*i*) $\frac{dy}{dx} = 0$ (*ii*) $\frac{d^2y}{dx^2} = 0$ (*iii*) $\frac{d^3y}{dx^3} \neq 0$

Example 37. *Find the points at which the function $y = x^3 + 6x^2 - 15x + 5$ has a maximum and minimum values.* **[SBTE. 2015]**

Solution. $$y = x^3 + 6x^2 - 15x + 5 \qquad \text{... (1)}$$

$$\Rightarrow \qquad \frac{dy}{dx} = 3x^2 + 12x - 15$$

For maxima and minima, $\frac{dy}{dx} = 0$

$$\Rightarrow \qquad 3x^2 + 12x - 15 = 0$$

$$\Rightarrow \qquad x^2 + 4x - 5 = 0 \Rightarrow x = 1, -5$$

Again differentiating (2), we have

$$\frac{d^2y}{dx^2} = 6x + 12$$

(*i*) When $x = 1, \frac{d^2y}{dx^2} = 6 + 12 = 18 = +\text{ve}$

$\therefore$ The given function is minimum at $x = 1$.

Minimum value of the function $= (1)^3 + 6\,(1)^2 - 15\,(1) + 5$

$= 1 + 6 - 15 + 5 = -3$ **Ans.**

(*ii*) When $x = -5, \frac{d^2y}{dx^2} = 6\,(-5) + 12 = -18 = -\text{ve}$

The given function is maximum at $x = -5$

Maximum value of the function $= (-5)^3 + 6\,(-5)^2 - 15\,(-5) + 5$

$= -125 + 150 + 75 + 5 = 105$ **Ans.**

Example 38. *Find the value of x for which the function* $(x-2)^3 \cdot (x-3)^2$ *is a maximum or minimum. Also find the point of inflexion.*

Solution. $y = (x-2)^3 \cdot (x-3)^2$

$$\Rightarrow \quad \frac{dy}{dx} = (x-2)^3 \cdot 2\,(x-3) + 3\,(x-2)^2 \cdot (x-3)^2$$

$$\Rightarrow \quad \frac{dy}{dx} = (x-2)^2\,(x-3)\,[2x - 4 + 3x - 9]$$

$$= (x-2)^2\,(x-3)\,(5x-13)$$

For maxima and minima, $\frac{dy}{dx} = 0$

$(x-2)^2(x-3)(5x-13) = 0 \quad \therefore \; x = 2, 3, 13/5$

Again differentiating

$$\frac{d^2y}{dx^2} = 2(x-2)\,(x-3)\,(5x-13) + (x-2)^2(5x-13) + (x-2)^2(x-3)\,5$$

$$= (x-2)\,[(10x^2 - 56x + 78) + (5x^2 - 23x + 26) + (5x^2 - 25x + 30)]$$

$$= (x-2)\,(20x^2 - 104x + 134)$$

(*i*) When $x = 2, \frac{d^2y}{dx^2} = 0$ and $\frac{d^3y}{dx^3} \neq 0$

The given function has a point of inflexion at $x = 2$. **Ans.**

(*ii*) When $x = 3, \frac{d^2y}{dx^2} = (3-2)\,(180 - 312 + 134) = +2 = +\text{ve}$

The given function is minimum at $x = 3$

The minimum value of the function $= (3-2)^3\,(3-3)^2 = 0$. **Ans.**

(*iii*) When $x = \frac{13}{5}$,

$$\Rightarrow \quad \frac{d^2y}{dx^2} = \left[\frac{13}{5} - 2\right] \times \left[20 \times \frac{169}{25} - 104 \times \frac{13}{5} + 134\right]$$

$$= \frac{3}{4} \times \frac{1}{5}(676 - 1352 + 670) = \frac{3}{25}(-6) = -\frac{18}{25} = -\text{ve}$$

The given function has maxima at $x = \dfrac{13}{5}$

The maximum value of the function $= \left[\dfrac{13}{5} - 2\right]^3 \times \left[\dfrac{13}{5} - 3\right]^2$

$$= \frac{27}{125} \times \frac{4}{25} = \frac{108}{3125}$$ **Ans.**

Example 39. *What is the maximum value of* $-4x^2 + 4x - 4$ on R ? **[SBTE. 2015]**

Solution. Here we have

$\Rightarrow$ $f(x) = -4x^2 + 4x - 4$

$\Rightarrow$ $f'(x) = -8x + 4$

for load maxima or minima

$\Rightarrow$ $f'(x) = 0$

$\Rightarrow$ $-8x + 4 = 0 \Rightarrow x = \dfrac{1}{2}$

Maximum value is

$\Rightarrow$ $f\left(\dfrac{1}{2}\right) = -4\left(\dfrac{1}{2}\right)^2 + 4\left(\dfrac{1}{2}\right) - 4$

$$= -4 \times \frac{1}{4} + 2 - 4 = -3$$ **Ans.**

Example 40. *If* $f(x) = a \sin x + \dfrac{1}{3} \sin 3x$ has extremum at $x = \dfrac{\pi}{3}$, find the value of *'a'*. **[SBTE. 2016]**

Solution. Here we have $f(x) = a \sin x + \dfrac{1}{3} \sin 3x$

$f'(x) = a \cos x + \cos 3x$

$\Rightarrow$ $f(x)$ has extremum at $x = \dfrac{\pi}{3}$

$\therefore$ $f'\left(\dfrac{\pi}{3}\right) = 0$

$\Rightarrow$ $f'\left(\dfrac{\pi}{3}\right) = a \cos \dfrac{\pi}{3} + \cos \pi = 0$

$\Rightarrow$ $\dfrac{a}{2} - 1 = 0 \Rightarrow a = 2$ **Ans.**

Example 41. *If* $f(x) = x^4 - 62x^2 + ax + 9$, attains a load extreme at $x = 1$ write the value of *'a'*. **[SBTE. 2017, 2016, 2015, 2013]**

Solution. Here we have $f(x) = x^4 - 62x^2 + ax + 9$

$f'(x) = 4x^3 - 124x + a$

$f(x)$ has extreme value $f'(1) = 0$

$f'(1) = 4 - 124 + a = 0 \Rightarrow a = 120$ **Ans.**

Example 42. *The function $f(x) = a \log_e x + bx^2 + x$ has extreme values at $x = 1$ and $x = 2$. Find a and b.*

Solution. Here we have $f(x) = a \log_e x + bx^2 + x$

$$\Rightarrow \quad f'(x) = \frac{a}{x} + 2bx + 1$$

$f(x)$ has extreme values at $x = 1$ and $x = 2$

then $f'(1) = 0$ and $f'(2) = 0$

$$f'(1) = a + 2b + 1 = 0$$

or $a + 2b = -1$ ... (1)

$$\Rightarrow \quad f'(2) = \frac{a}{2} + 4b + 1 = 0$$

or $a + 8b = -2$ ... (2)

On solving (1) and (2), we get

$$a = \frac{-2}{3}, b = -\frac{1}{6}$$ **Ans.**

EXERCISE 29.4

Find the points at which the following functions have a maximum and minimum values, also find the point of inflexion.

1. $y = 9x^3 - 45x^2 + 48x + 11.$ **Ans.** Minimum at $x = \frac{8}{3}$, minimum value $= \frac{31}{3}$

Maximum at $x = \frac{2}{3}$, maximum value $= \frac{77}{3}$

2. $y = 11 - 12x + 6x^2 + x^3.$ **Ans.** Point of inflexion at $x = 2$

3. $y = x^5 - 3x^4 + 5$ **[B.T.E. Delhi, May 2008]**

Ans. At $x = \frac{12}{5}$, Minimum

4. $y = x^5 - 5x^4.$ **Ans.** At $x = 4$, Minimum

5. $y = x^5 - 5x^4 + 5x^3 - 1.$ **Ans.** At $x = 0$, Pt. of inflexion

At $x = 1$, Maximum

At $x = 3$, Minimum

6. $y = 3 \sin^2 x + 4 \cos^2 x.$ **[Diploma IETE, Dec. 2005]**

Ans. Maximum at $x = 0$, maximum value = 4, minimum at $x = \pi/2$, minimum value = 3

7. Find the maximum value of $f(x) = (x - 1)(x - 2)(x - 3)$

Ans. Minimum at $x = \frac{6 \quad \sqrt{3}}{\quad}$ maximum at $x = \frac{6 - \sqrt{3}}{3}$

8. Find the minimum value of $f(x) = \sin x$ in the interval $\pi \le x \le 2\pi$. **Ans.** $x = \frac{3\pi}{2}$

9. Show that $\frac{\log x}{x}$ has a maximum value at $x = e$.

10. The strength S of a rectangular beam which can be cut from a circular log of radius r is given by $S = kx(4r^2 - x^2)$, where k is a constant and x is one side, find the value of x for which S is maximum. **Ans.** $x = \frac{2}{\sqrt{3}}r$

Choose the correct alternative:

11. The maximum value of $y = 2\cos 2x - \cos 4x$, $0 \le x \le \pi/2$ is

(*a*) -1 (*b*) $\frac{1}{2}$ (*c*) $\frac{3}{2}$ (d) 1 **Ans.** (*c*)

Example 43. *A figure consists of a semicircle with a rectangle on its diameter. Given that the perimeter of the figure is 20 metres, find its dimensions in order that its area may be maximum.*

Solution. Let l and b be the length and breadth of the rectangle.

Perimeter of the figure = 20 metres **[SBTE. 2017, 2013]**

$$\Rightarrow \quad l + 2b + \pi\frac{l}{2} = 20 \qquad \ldots (1)$$

$$\Rightarrow \quad 2l + 4b + \pi l = 40$$

$$\Rightarrow \quad 4b = 40 - 2l - \pi l$$

$$\text{Area of the figure} = lb + \frac{\pi}{2}\left\{\frac{l}{2}\right\}^2$$

$$\Rightarrow \quad A = \frac{l}{4}(40 - 2l - \pi l) + \frac{\pi l^2}{8}$$

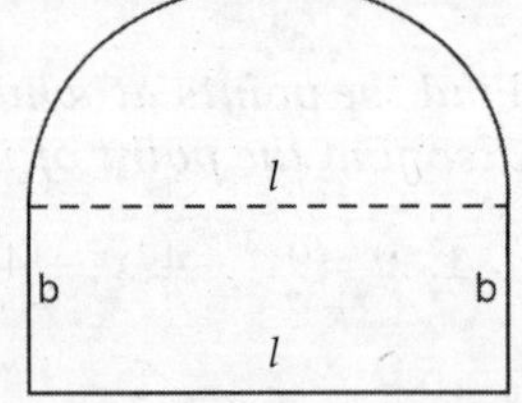

Differentiating with respect to 'l', we get

$$\Rightarrow \quad \frac{dA}{dl} = \frac{l}{4}(40 - 2l - \pi l) + \frac{l}{4}(-2 - \pi) + \frac{2\pi l}{8}$$

For maximum area, $\frac{dA}{dl} = 0$

$$\Rightarrow \quad 0 = \frac{1}{4}(40 - 2l - \pi l) + \frac{l}{4}(-2 - \pi) + \frac{\pi l}{4}$$

$$\Rightarrow \quad 0 = 40 - 2l - \pi l - 2l - \pi l + \pi l$$

$$\Rightarrow \quad 0 = 40 - 4l - \pi l$$

$$\Rightarrow \quad l = \frac{40}{4 + \pi}$$

Putting the value of l in (1), we get

$$\Rightarrow \quad b = 10 - \frac{1}{2}\frac{40}{4+\pi} - \frac{\pi}{4}\frac{40}{4+\pi} = 10 - \frac{20}{4+\pi} - \frac{10\pi}{4+\pi}$$

$$= \frac{40 + 10\pi - 20 - 10\pi}{4+\pi} = \frac{20}{4+\pi}$$

$$l = \frac{40}{\pi + 4}, b = \frac{20}{\pi + 4} \qquad \textbf{Ans.}$$

Example 44. *A rectangular sheet of metal of length 6 metres and width 2 metres is given. Four equal squares of side x metres are removed from the corners. The sides of this sheet is now turned up to form an open rectangular box. Find approximately, the height of the box, such that the volume of the box is maximum.*

[SBTE. 2017, 2015]

Solution. Length fo the sheet = 6 m

Width of the sheet = 2 m

Side of the square = x m

Height of the box = x m

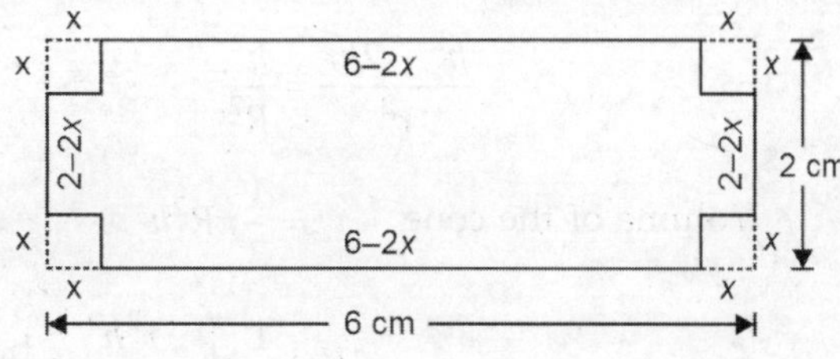

Volume of the rectangular box

$$= (6 - 2x)(2 - 2x)\,x$$

$$\Rightarrow \quad V = 4x(3 - x)(1 - x)$$

$$= 4(x^3 - 4x^2 + 3x)$$

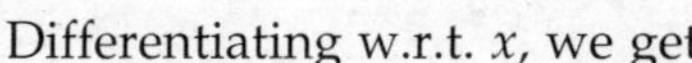

Differentiating w.r.t. x, we get

$$\frac{dV}{dx} = 4(3x^2 - 8x + 3)$$

For maximum, $\quad 0 = 4(3x^2 - 8x + 3)$

$$\Rightarrow \quad 0 = 3x^2 - 8x + 3$$

$$\Rightarrow \quad x = \frac{8 \pm \sqrt{64 - 36}}{6} = \frac{8 \pm \sqrt{28}}{6}$$

$$= \frac{4 - \sqrt{7}}{3}, \text{ and } \frac{4 + \sqrt{7}}{3} \text{ is not possible.}$$

$$\Rightarrow \quad \frac{d^2V}{dx^2} = 4(6x - 8)$$

$$\Rightarrow \quad \frac{d^2V}{dx^2}\left(\text{for } x = \frac{4 - \sqrt{7}}{3}\right) = 4(8 - 2\sqrt{7} - 8) = -8\sqrt{7} = -\text{ve}$$

So volume is maximum

$$\text{Height} = \frac{4 - \sqrt{7}}{3} \text{ metres.}$$ **Ans.**

Example 45. *A cone is circumscribed about a sphere of radius r. Show that when the volume of the cone is minimum, its altitude is 4r and its semivertical angle is* $\sin^{-1}\left(\frac{1}{3}\right)$. **[SBTE. 2017]**

Solution. Let R and h be the radius and height of the cone.

$$\Rightarrow \quad \sin\theta = \frac{OD}{AO} = \frac{BC}{AC}$$

$$\Rightarrow \quad \frac{r}{h - r} = \frac{R}{\sqrt{h^2 + R^2}}$$

$$\Rightarrow \quad \frac{r^2}{h^2 + r^2 - 2hr} = \frac{R^2}{h^2 + R^2}$$

$$\Rightarrow \quad \frac{h^2 + r^2 - 2hr}{r^2} = \frac{h^2 + R^2}{R^2}$$

$$\Rightarrow \quad \frac{h^2 - 2hr}{r^2} + \frac{r^2}{r^2} = \frac{h^2}{R^2} + \frac{R^2}{R^2}$$

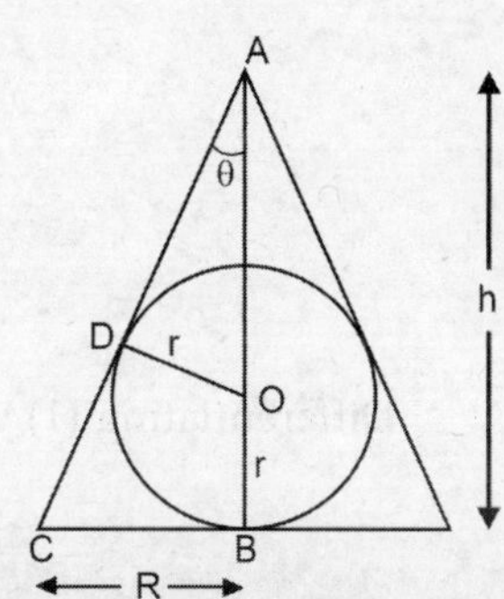

$$\Rightarrow \quad \frac{h^2 - 2hr}{r^2} = \frac{h^2}{R^2} \quad \Rightarrow \quad R^2 = \frac{r^2h^2}{h^2 - 2hr}$$

Volume of the cone $= \frac{1}{3}\pi R^2 h$

$$\Rightarrow \quad V = \frac{1}{3}\pi\left(\frac{r^2h^2}{h^2 - 2hr}\right)h$$

$$\Rightarrow \quad V = \frac{\pi r^2 h^2}{h - 2r} \times \frac{1}{3}$$

$$\Rightarrow \quad \frac{dV}{dh} = \frac{(h - 2r)\, 2\pi r^2 h - \pi\, r^2 h^2 (1)}{(h - 2r)^2}$$

For maxima and minima $0 = (h - 2r)\, 2\pi r^2 h - \pi\, r^2 h^2$

$$\Rightarrow \quad 0 = (h - 2r)\, 2 - h$$

$$\Rightarrow \quad 0 = 2h - 4r - h$$

$$\Rightarrow \quad h = 4r$$

Also at $h = 4r$, $\frac{d^2V}{dh^2} > 0$

$$\Rightarrow \quad \sin\theta = \frac{OD}{AO} = \frac{r}{h - r} = \frac{r}{4r - r} = \frac{1}{3}$$

$$\Rightarrow \quad \theta = \sin^{-1}\left(\frac{1}{3}\right)$$ **Proved.**

Example 46. *Show that the semivertical angle of the cone of maximum volume and given slant height is* $\tan^{-1}(\sqrt{2})$.

Solution. Let l be the given slant height of the cone and θ, its semivertical angle.

Height of cone (PO) $= l \cos\theta$

Radius of cone (OA) $= l \sin\theta$

Let V be the volume of the cone

$$\therefore \quad V = \frac{\pi}{3} r^2 h$$

$$\Rightarrow \quad V = \frac{\pi}{3}(l^2 - h^2)h \qquad (l^2 = r^2 + h^2)$$

$$\Rightarrow \quad V = \frac{\pi}{3}(l^2 h - h^3) \qquad \text{...(1)}$$

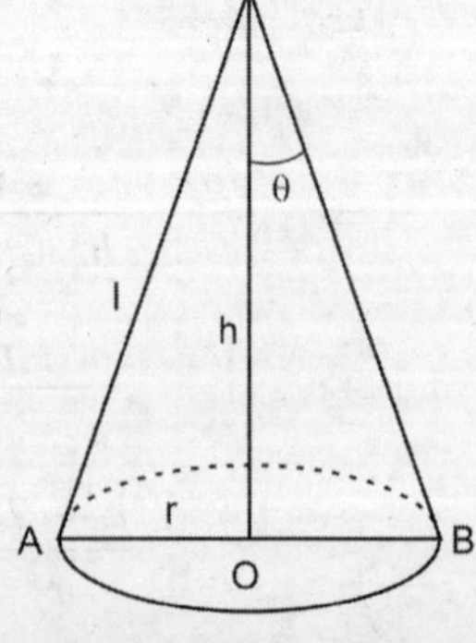

Differentiating (1) w.r.t. 'h', we obtain

$$\frac{dV}{dh} = \frac{\pi}{3}(l^2 - 3h^2)$$

For maximum and minimum value

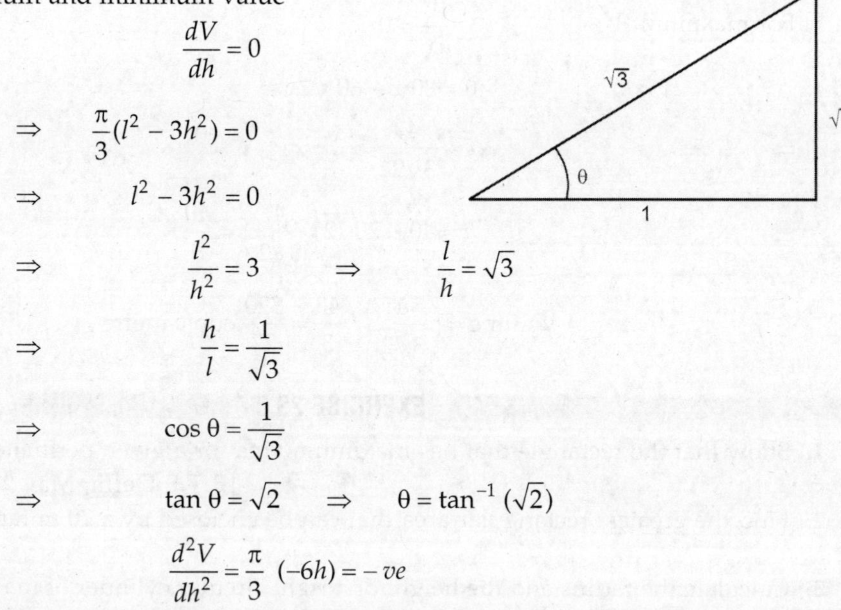

$$\frac{dV}{dh} = 0$$

$$\Rightarrow \quad \frac{\pi}{3}(l^2 - 3h^2) = 0$$

$$\Rightarrow \quad l^2 - 3h^2 = 0$$

$$\Rightarrow \quad \frac{l^2}{h^2} = 3 \quad \Rightarrow \quad \frac{l}{h} = \sqrt{3}$$

$$\Rightarrow \quad \frac{h}{l} = \frac{1}{\sqrt{3}}$$

$$\Rightarrow \quad \cos\theta = \frac{1}{\sqrt{3}}$$

$$\Rightarrow \quad \tan\theta = \sqrt{2} \quad \Rightarrow \quad \theta = \tan^{-1}(\sqrt{2})$$

$$\frac{d^2V}{dh^2} = \frac{\pi}{3}(-6h) = -ve$$

$\therefore$ V is maximum for $\theta = \tan^{-1}(\sqrt{2})$. **Proved.**

Example 47. *A log has the form of a frustum of a cone 20 metres long, the diameter of its ends being 2 m and 1 m. A beam of square section is to be cut from the log. Find its length, if the volume of the beam is maximum.*

Solution. Let x metre be one side of the beam and h metre be its height.

In the figure

$$\Rightarrow \quad \frac{AB}{BC} = \frac{CD}{DE}$$

$$\Rightarrow \quad \frac{20-h}{\frac{x}{\sqrt{2}} - \frac{1}{2}} = \frac{h}{1 - \frac{x}{\sqrt{2}}}$$

$$\Rightarrow \quad 20 - \frac{20x}{\sqrt{2}} - h + \frac{hx}{\sqrt{2}} = \frac{hx}{\sqrt{2}} - \frac{h}{2}$$

$$20 - \frac{20x}{\sqrt{2}} = \frac{h}{2}$$

$$\Rightarrow \quad h = 40 - 20\sqrt{2}x$$

Volume of the beam = Area of cross section × height

$$\Rightarrow \quad V = x^2 h$$

$$= x^2[40 - 20\sqrt{2}x]$$

$$\Rightarrow \quad V = 40x^2 - 20\sqrt{2}x^3$$

$$\Rightarrow \quad \frac{dV}{dx} = 80x - 60\sqrt{2}x^2$$

For maximum V, $\frac{dV}{dx} = 0$

$$0 = 80x - 60\sqrt{2}x^2$$

$$\Rightarrow \quad x = \frac{4}{3\sqrt{2}}$$

$$\Rightarrow \quad h = 40 - 20\sqrt{2}\left(\frac{4}{3\sqrt{2}}\right) = \frac{40}{3}$$

$$\therefore \quad \text{volume} = \left(\frac{4}{3\sqrt{2}}\right)^2 \frac{40}{3} = \frac{320}{27} \text{ cubic metre}$$ **Ans.**

EXERCISE 29.5

1. Show that the rectangle that has maximum area for a given perimeter is a square. **[B.T.E Delhi, May 2008; May 2007]**
2. Find the greatest rectangular area that can be enclosed by a 20 m fencing. **Ans.** 25 sq. m.
3. Calculate the radius and the height of a right circular cylinder of maximum volume which can be cut from a sphere of radius R. **Ans.** Height $= \frac{2R}{\sqrt{2}}$, Radius $= \sqrt{\frac{2}{3}}R$
4. Show that the right circular cylinder of given surface (including the ends), shall have maximum volume if its height is equal to the diameter of its base.
5. Show that the height of an open cylinder of given surface, that can contain maximum water, is equal to radius of its base.
6. An open cistern on a square base is to be made of tin to obtain 108 Cubic metres of water. Find the dimensions, so that minimum possible tin is used. **[B.T.E Delhi, Jan. 2009]**
   Ans. Height = 3 metres, side of base = 6 metres
7. An open rectangular tank with a square base and vertical sides is to be constructed of a sheet metal to hold a given quantity of water. Show that the cost of material will be least when the depth is half the width.
8. The sum of the perimeters of a circle and a square is k. Show that when the sum of their areas is least the side of the square is equal to the diameter of the circle.
9. The sum of surfaces of a cube and a sphere is given. Show that when the sum of their volumes is least, the diameter of the sphere is equal to the edge of the cube.
10. A log has the form of a frustum of a cone 30 m long the diameters of its ends being 2 m and 1 m. A beam of square section is to be cut from the log. Find its length, if the volume of the beam is maximum. **Ans.** Length = 20 m, Volume $= \frac{160}{9}$ cu. m
11. A wire 3 m long has to be bent into the form of a rectangle with an external circular loop at one corner, and the rectangle is to have one side double of the other. Find the dimensions of the circle and the rectangle so that the total area enclosed is minimum. **[SBTE. 2015]**
    Ans. Length $= \frac{9}{2\pi + 9}$ m, Breadth $= \frac{9}{4\pi + 18}$, Radius $= \frac{3}{2\pi + 9}$ m

12. A piece of the wire of length 10 m, is to be cut into two pieces, one of which is to be bent into the form of a square and the other into the form of a circle. Find Length and Radius when the sum of the areas of the circle and square is minimum.

Ans. Length $= \dfrac{10}{\pi + 4}$, Radius $= \dfrac{5}{\pi + 4}$

13. Show that the cone of the greatest volume which can be inscribed in a sphere has an altitude equal to $\dfrac{2}{3}$ of the diameter of the sphere.

14. Show that the volume of the greatest cylinder which can be inscribed in a cone of height h and semivertical angle α *is* $\dfrac{4}{27}\pi h^3 \tan^2 \alpha$. **[SBTE. 2016, 2013]**

15. Square pieces of equal size are cut from the corner of a sheet of metal measuring (80 × 50) cm and the edges folded to form an open tray. Find the side of each square which is cut form the corners for the volume to be maximum. **Ans.** 20 cm

16. A manufacturer intends to manufacture aluminium vessel of capacity $25\frac{1}{7}$ litres from a sheet of metal in the form of a right circular cylinder open at the top. Find the dimensions which use the least material. [Take $\pi = 22/7$]

Ans. Radius = Height = 2 cm

17. The efficiency e of a screw-jack is given by $e = \dfrac{\tan\theta}{\tan(\theta + \alpha)}$ where α is constant. Find θ for which the efficiency is maximum and find the maximum efficiency also.

$$\textbf{Ans. } \theta = \left(\frac{\pi}{4} - \frac{\alpha}{2}\right), e = \left(\frac{1 - \tan\frac{\alpha}{2}}{1 + \tan\frac{\alpha}{2}}\right)^2$$

Example 48. *Two towns are to get their water supply from a river. Both towns are on the same side of the river at a distance of 6 km and 18 km from the river bank. If the distance between the points on the river bank nearest to the towns be 10 km, find (i) where a single pumping station may be located to require the less amount of pipe, and (ii) how much pipe is needed.*

Solution. (*i*) Let A, B be two towns and CD be the bank of a river.

$$AC = 6 \text{ km}$$
$$BD = 18 \text{ km}$$
$$CD = 10 \text{ km}$$

Let a pumping station on the river be at P.

Suppose $PC = x$ km,

$$AP = \sqrt{AC^2 + PC^2} = \sqrt{36 + x^2}$$

$$BP = \sqrt{BD^2 + PD^2} = \sqrt{(18)^2 + (10 - x)^2}$$

Total length (l) of the pipe $= AP + BP$

$$l = \sqrt{36 + x^2} + \sqrt{424 - 20x + x^2}$$

$$\frac{dl}{dx} = \frac{x}{\sqrt{36 + x^2}} + \frac{-10 + x}{\sqrt{424 - 20x + x^2}}$$

For minimum l, $\quad \frac{dl}{dx} = 0$

$$\frac{x}{\sqrt{36 + x^2}} = \frac{10 - x}{\sqrt{424 - 20x + x^2}}$$

$$\Rightarrow \quad \frac{x^2}{36 + x^2} = \frac{x^2 - 20x + 100}{x^2 - 20x + 424}$$

$$\Rightarrow \quad \frac{36 + x^2}{x^2} = \frac{x^2 - 20x + 424}{x^2 - 20x + 100}$$

$$\Rightarrow \quad 1 + \frac{36}{x^2} = 1 + \frac{324}{x^2 - 20x + 100}$$

$$\Rightarrow \quad \frac{1}{x^2} = \frac{9}{x^2 - 20x + 100}$$

$$\Rightarrow \quad 9x^2 = x^2 - 20x + 100$$

$$\Rightarrow \quad 8x^2 + 20x - 100 = 0 \quad \text{or} \quad 2x^2 + 5x - 25 = 0$$

$$\Rightarrow \quad (2x - 5)(x + 5) = 0$$

$$\Rightarrow \quad x = 5/2, -5 \quad \text{and} \quad \frac{d^2 l}{dx^2} = +\text{ ve for } x = \frac{5}{2}$$

(*i*) The pumping station should be located on the bank at a distance of 2.5 km from C or on the other side of C, 5 km from it.

(*ii*) Total pipe line $= AP + BP$

$$= \sqrt{36 + \frac{25}{4}} + \sqrt{324 + \frac{225}{4}} = \frac{134}{2} + \frac{39}{2} = 26 \text{ km}$$ **Ans.**

Example 49. *In driving a motor boat assume that the petrol burnt (per hour) varies as the cube of its velocity. Show that the most economical speed going against a water current of C km per hour will be 3/2 C km. per hour.*

Solution. Petrol burnt per hour $\propto V^3$.

Petrol burnt in t hour $\propto V^3 . t$

Let the petrol burnt in t hour be Q

$$Q \propto V^3 . t$$

$$Q = kV^3 . t$$

Let the distance travelled in t hours be S

Actual speed $= V - C$

$$\text{Time} = \frac{\text{Distance}}{\text{velocity}}$$

$$t = \frac{S}{V - C}$$

Putting the value of t in (1), we get

$$Q = kV^3 \frac{S}{V - C}$$

Differentiating (2) w.r.t. 'V', we get

$$\frac{dQ}{dV} = kS \frac{(V - C)\,3V^2 - V^3.1}{(V - C)^2}$$

For Q to be minimum, $\frac{dQ}{dV} = 0$

$$0 = kS \frac{(V - C)\,3V^2 - V^3}{(V - C)^2}$$

$$0 = (V - C)\,3V^2 - V^3$$

$$0 = 3\,(V - C) - V$$

$$3V - V = 3C$$

$$2V = 3C$$

$V = \frac{3C}{2}$ and $\frac{d^2Q}{dV^2}$ at $V = \frac{3C}{2}$ is + ve **Proved.**

Example 50. *An electric bulb is hung up vertically from the ceiling at 1 meter height from the floor. Show that the point of maximum illumination on the floor lies at a distance* $\frac{a}{\sqrt{2}}$ *metres from the point on the floor vertically below the light. It is given that the illumination at a small area varies inversely as the square of the distance from the source of light and directly as the sine of the angle made by the rays of light with the normal to the area.*

Solution. Let an electric bulb hang at the point B in the ceiling.

A point C is on the floor, just vertically below the bulb

$$BC = a$$

P is a point on the floor at which the illumination I is maximum.

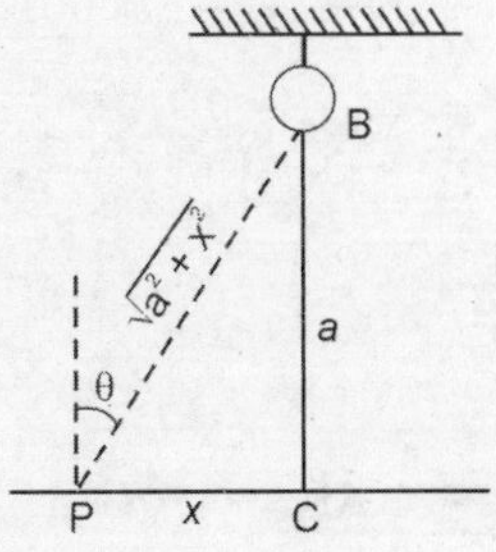

$$I \propto \frac{1}{BP^2}$$

$$\propto \sin\theta$$

$$I \propto \frac{\sin\theta}{BP^2} \qquad \text{(Let } PC = x\text{)}$$

$$\propto \frac{x}{\sqrt{a^2 + x^2}} \cdot \frac{1}{a^2 + x^2}$$

$$\propto \frac{x}{(a^2 + x^2)^{3/2}}$$

B
$\sqrt{a^2+x^2}$
a
θ
P x C

$$I = \frac{kx}{(a^2 + x^2)^{3/2}}$$

$$\frac{dI}{dx} = k\frac{(a^2 + x^2)^{3/2}.1 - \frac{3x}{2}(a^2 + x^2)^{1/2}.2x}{(a^2 + x^2)^3}$$

$$= k\frac{a^2 + x^2 - 3x^2}{(a^2 + x^2)^{5/2}} = k\frac{a^2 - 2x^2}{(a^2 + x^2)^{5/2}}$$

For maximum I, $\quad \frac{dI}{dx} = 0$

$$a^2 - 2x^2 = 0$$

$$x = \frac{a}{\sqrt{2}} \; also \; \frac{d^2I}{dx^2} < 0 \; for \; x = \frac{a}{\sqrt{2}}$$ **Proved.**

Example 51. *If in a submarine cable the range of signalling varies as* $x^2 \log_e\left(\frac{1}{x}\right)$, *where x is the ratio of the radius of the case to that of the cable. Find the value of x for which range of signalling is maximum.*

Solution. Range $\propto x^2 \log_e \frac{1}{x}$

$$\text{Range} = kx^2 \log_e\left(\frac{1}{x}\right)$$

$$R = -kx^2 \log_e x$$

Differentiating w.r.t. x, we get

$$\frac{dR}{dx} = -kx\left(\frac{}{x}\right) - k(2x)\log_e x = -kx - 2kx\log_e x$$

For maximum range, $\quad \frac{dR}{dx} = 0$

$\Rightarrow \quad 0 = -kx - 2kx \log_e x$

$\Rightarrow \quad 0 = -1 - 2\log_e x$

$\Rightarrow \quad \log_e x = -\frac{1}{2}$

$\Rightarrow \quad x = e^{-1/2} = \frac{1}{\sqrt{e}}$

$$\frac{d^2R}{dx^2} = -k - 2kx\left(\frac{1}{x}\right) - 2k(1)\log_e x = -3k - 2k\log_e x$$

At $\quad x = \frac{1}{\sqrt{e}}, \frac{d^2R}{dx^2} = -3k - 2k\left(-\frac{1}{2}\right) = -3k + k = -2k = -\text{ve}$

$\Rightarrow \quad x = \frac{1}{\sqrt{e}}$ **Ans.**

EXERCISE 29.6

1. The straight shore of a large lake runs east and west. If A and B are two points on this shore 12 km. apart. There is a town C, 9 km north of A and another town D, 15 km north of B. A single pumping station on the lake shore is to supply water to both towns. Where should it be located in order that the sum of its distance from C and D may be minimum. **Ans.** 4.5 km from A
2. An electric light is placed over the centre of a circular lawn, 100 m in diameter. Assuming that the intensity of light varies as the sine of the angle α which it makes with the ground is inversely proportional to the square of its distance. How high should the light be placed so that maximum intensity is obtained at the circumference of the plot. **Ans.** $25\sqrt{2}\ m$
3. A table of 7 m high is placed on a wall with its base 9 m, above the level of an observer's eye. How far from the wall, should an observer stand, so that the angle of vision subtended by the table may be maximum. **Ans.** 12 m
4. The horse power transmitted by a belt moving at v cm/s is proportional to $fv\dfrac{wv^3}{g}$, where f is maximum allowable stress in kg/sq cm, w = weight of belt per c.c. Find the velocity for the maximum horse power. **Ans.** $V = \sqrt{\dfrac{fg}{3w}}$ cm/s
5. A person being in a boat p km from nearest point of the beach, wishes to reach quickest possible point q km away from that point along the shore. The ratio of his rate of walking to his rate of rowing is $\sec\alpha$. Show that he should land at a distance $q - p\cot\alpha$ from the place to be reached.
6. The cost of fuel for running a train is proportional to the square of the speed generated in kilometers per hour and costs Rs. 48 per hour at 16 kilometers per hour. If the fixed charges are Rs. 300 per hour, show that the most economical speed is 40 kilometers per hour.

29.8 CURVATURE

Let PQ be a small arc of a curve of length δs, $\delta\psi$ is the angle between the tangents at P and Q called angle of contingence.

$\dfrac{\delta\psi}{\delta s}$ is called the average curvature of the arc PQ.

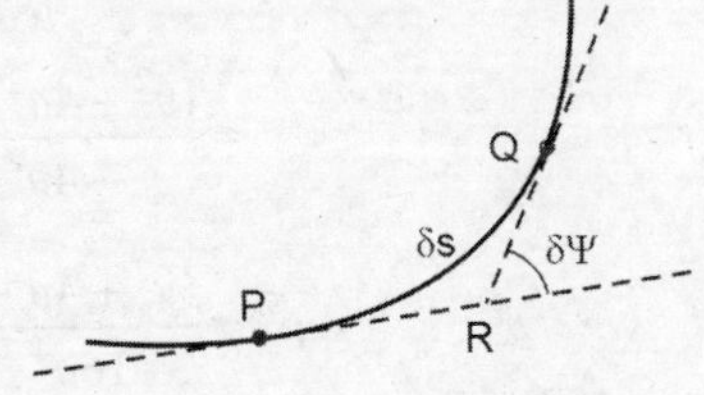

$$\textbf{Curvature of } P = \lim_{\delta s \to 0} \frac{\delta\psi}{\delta s}$$

$$= \frac{d\psi}{ds}$$

Reciprocal of curvature is called the radius of curvature.

Thus, **radius of curvature** $(\rho) = \dfrac{ds}{d\psi}$

29.9 RADIUS OF CURVATURE FOR CARTESIAN EQUATIONS

We know that $\tan\psi = \dfrac{dy}{dx}$

Differentiating w.r.t. *s*, we get

$$\sec^2\psi \cdot \frac{d\psi}{ds} = \frac{d^2y}{dx^2} \cdot \frac{dx}{ds}$$

or $$\frac{ds}{d\psi} = \frac{\sec^2\psi}{\frac{d^2y}{dx^2} \cdot \frac{dx}{ds}} = \frac{(1+\tan^2\psi)\frac{ds}{dx}}{\frac{d^2y}{dx^2}}$$

$$\rho = \frac{\left[1+\left(\frac{dy}{dx}\right)^2\right]\sqrt{1+\left(\frac{dy}{dx}\right)^2}}{\frac{d^2y}{dx^2}} \qquad \left[\because ds = \sqrt{1+\left(\frac{dy}{dx}\right)^2}\,dx\right]$$

$$\rho = \frac{\left[1+\left(\frac{dy}{dx}\right)^2\right]^{3/2}}{\frac{d^2y}{dx^2}}$$

Example 52. *If ρ is the radius of curvature at any point P on the parabola $y^2 = 4ax$ and S is the focus of the parabola, then show that ρ^2 varies as $(SP)^3$.*

Solution. $y^2 = 4ax$

$\Rightarrow$ $$2y\frac{dy}{dx} = 4a \text{ or } \frac{dy}{dx} = \frac{2a}{y}$$

$\Rightarrow$ $$\frac{d^2y}{dx^2} = -\frac{2a}{y^2}\frac{dy}{dx} = -\frac{2a}{y^2}\left(\frac{2a}{y}\right) = -\frac{4a^2}{y^3}$$

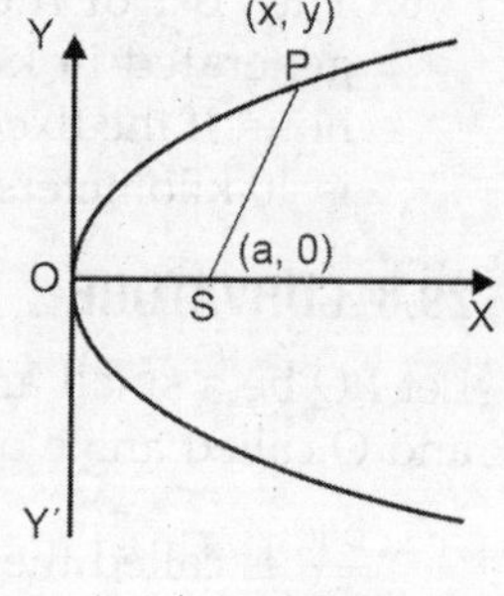

$\Rightarrow$ $$\rho = \frac{\left[1+\left(\frac{dy}{dx}\right)^2\right]^{3/2}}{\frac{d^2y}{dx^2}} = \frac{\left[1+\left(\frac{2a}{y}\right)^2\right]^{3/2}}{\frac{-4a^2}{y^3}}$$

$$= \frac{[y^2+4a^2]^{3/2}}{-4a^2}$$

$\Rightarrow$ $$\rho^2 = \frac{(y^2+4a^2)^3}{16a^4} = \frac{(4ax+4a^2)^3}{16a^4} = \frac{64a^3}{16a^4}(x+a)^3$$

$$= \frac{4}{a}(x+a)^3 \qquad \text{... (1)}$$

$\Rightarrow$ $$(SP)^2 = (x-a)^2 + (y-0)^2$$

$$= x^2 - 2ax + a^2 + y^2$$

$$= x^2 - 2ax + a^2 + 4ax$$

$$= (x^2 + 2ax + a^2) = (x+a)^2$$

$$(SP)^3 = (x+a)^3 \qquad \text{... (2)}$$

From (1) and (2), we have

$$\rho^2 = \frac{4}{a}(SP)^3$$

$$\Rightarrow \qquad \rho^2 \propto (SP)^3$$

Proved

Example 53. *Show that the radius of curvature of the hypocycloid $x^{2/3} + y^{2/3} = a^{2/3}$ at any point (b, c) is 3 $(abc)^{1/3}$.*

Solution. $x^{2/3} + y^{2/3} = a^{2/3}$

On differentiating, we get

$$\Rightarrow \quad \frac{2}{3}x^{-1/3} + \frac{2}{3}y^{-1/3}\frac{dy}{dx} = 0, \quad \frac{dy}{dx} = -\left(\frac{y}{x}\right)^{1/3}$$

$$\Rightarrow \quad \frac{d^2y}{dx^2} = -\frac{x^{1/3}\cdot 1/3\, y^{-2/3}\dfrac{dy}{dx} - y^{1/3}\cdot 1/3\, x^{-2/3}}{x^{2/3}}$$

$$= -\frac{x^{1/3}\cdot y^{-2/3}\cdot\left(-\dfrac{y^{1/3}}{x^{1/3}}\right) - x^{-2/3}\, y^{1/3}}{3x^{2/3}}$$

$$\Rightarrow \quad \frac{d^2y}{dx^2} = -\frac{-y^{-1/3} - x^{-2/3}\, y^{1/3}}{3x^{2/3}}$$

$$= \frac{x^{2/3} + y^{2/3}}{3x^{2/3}\,(x^{2/3}\, y^{1/3})} = \frac{a^{2/3}}{3x^{4/3}\, y^{1/3}}$$

$$\Rightarrow \quad \rho = \frac{\left[1 + \left(\dfrac{dy}{dx}\right)^2\right]^{3/2}}{\dfrac{d^2y}{dx^2}} = \frac{\left[1 + \dfrac{y^{2/3}}{x^{2/3}}\right]^{3/2}}{\dfrac{a^{2/3}}{3x^{4/3}\, y^{1/3}}}$$

$$= \frac{[x^{2/3} + y^{2/3}]^{3/2}\cdot 3x^{4/3}\, y^{1/3}}{x a^{2/3}}$$

$$= \frac{[a^{2/3}]^{3/2}\, 3x^{1/3}\, y^{1/3}}{a^{2/3}}$$

$$= 3a^{1/3}\, x^{1/3}\, y^{1/3}$$

$$= 3a^{1/3}\, b^{1/3}\, c^{1/3} \qquad (x = b, y = c)$$

$$= 3(abc)^{1/3}$$

Proved

Witch of Agnesi is a cubic plane curve defined from two diametrically opposite points of a circle figure with parameters $a = 1$, $a = 2$, $a = 4$ and $a = 8$.

Example 54. *Find the radius of curvature of the Witch of Agnesi*

$$y^2 = \frac{4a^2(2a - x)}{x} \quad \textit{where the curve meets x-axis.}$$

Solution. $y^2 = \dfrac{4a^2(2a-x)}{x}$ ... (1)

(1) meets the x-axis.

$$0 = \frac{4a^2(2a-x)}{x} \quad \text{or} \quad x = 2a$$

The point of intersection of (1) and x-axis is $(2a, 0)$.

Differentiating (1), we get

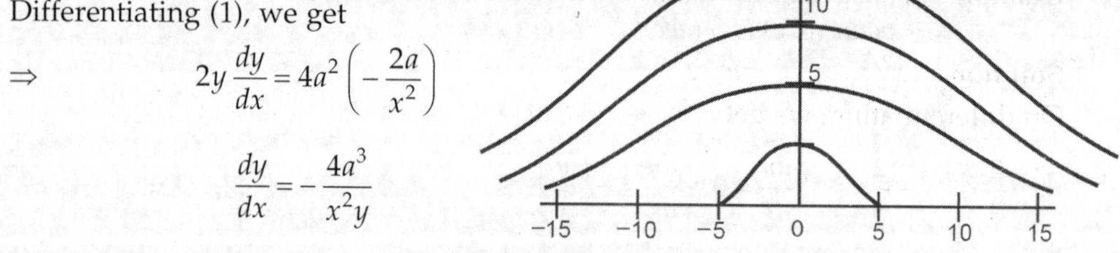

$$\Rightarrow \quad 2y\frac{dy}{dx} = 4a^2\left(-\frac{2a}{x^2}\right)$$

$$\frac{dy}{dx} = -\frac{4a^3}{x^2 y}$$

$$\Rightarrow \quad \left(\frac{dy}{dx}\right) \text{ at } (2a, 0) = -\infty \quad \text{or} \quad \frac{dx}{dy} = -\frac{x^2 y}{4a^3}$$

$$\Rightarrow \quad \left(\frac{dx}{dy}\right) \text{ at } (2a, 0) = 0$$

$$\Rightarrow \quad \frac{d^2x}{dy^2} = -\frac{1}{4a^3}\left[x^2 \cdot 1 + 2xy\frac{dx}{dy}\right]$$

$$\Rightarrow \quad \left(\frac{d^2x}{dy^2}\right) \text{ at } (2a, 0) = -\frac{1}{a}$$

$$\Rightarrow \quad \rho = \frac{\left[1+\left(\frac{dy}{dx}\right)^2\right]^{3/2}}{\frac{d^2y}{dx^2}} = \frac{\left[1+\left(\frac{dx}{dy}\right)^2\right]^{3/2}}{\frac{d^2x}{dy^2}} \qquad \left(\text{since } \frac{dy}{dx} = -\infty\right)$$

$$= \frac{(1+0)^{3/2}}{-\frac{1}{a}} = -a$$

Numerically, $\rho = -a$ **Ans.**

Example 55. *Find the points on the parabola ($x = at^2$, $y = 2at$) at which the radius of curvature is equal to its latus rectum.*

Solution. $x = at^2, \quad y = 2at$

$$\Rightarrow \quad \frac{dy}{dx} = \frac{dy}{dt}\frac{dt}{dx} = \frac{2a}{2at} = \frac{1}{t}$$

$$\Rightarrow \quad \frac{d^2y}{dx^2} = \frac{d}{dx}\left(\frac{dy}{dx}\right) = \frac{d}{dx}\left(\frac{1}{t}\right) = -\frac{1}{t^2}\frac{dt}{dx}$$

$$= -\frac{1}{t^2}\frac{1}{2at} = -\frac{1}{2at^3}$$

$$\Rightarrow \qquad \rho = \frac{\left[1+\left(\frac{dy}{dx}\right)^2\right]^{3/2}}{\frac{d^2y}{dx^2}} = \frac{\left[1+\frac{1}{t^2}\right]^{3/2}}{-\frac{1}{2at^3}}$$

$$= -2at^3\left(\frac{t^2+1}{t^2}\right)^{3/2}$$

$$\Rightarrow \qquad 4a = -2a\,(t^2+1)^{3/2} \qquad (\rho = 4a \text{ given})$$

$$\Rightarrow \qquad 2 = -(t^2+1)^{3/2}$$

$$\Rightarrow \qquad 4 = (t^2+1)^3$$

$$\Rightarrow \qquad (4)^{1/3} = t^2+1$$

$$\Rightarrow \qquad t^2 = (4)^{1/3} - 1 = 1.5875 - 1 = 0.5875$$

$$\Rightarrow \qquad t = \sqrt{0.5875} = 0.7665$$

$$x = at^2 = 0.5875a$$

$$y = 2at = 1.5330a$$

$$\therefore \qquad (x, y) = (0.5875a, 1.5330a)$$

Ans.

Example 56. *If ρ_1, ρ_2 be the radii of curvature at the extremities of two conjugate diameters of the ellipse $\frac{x^2}{a^2}+\frac{y^2}{b^2}=1$, then prove that $(\rho_1^{2/3}+\rho_2^{2/3})(ab)^{2/3} = a^2+b^2$.*

Solution. Let CP and CD be the two conjugate semi diameters. If the co-ordinates of P are $(a \cos \theta, b \sin \theta)$, then co-ordinates of D are

$$\Rightarrow \qquad \left[a\cos\left(\frac{\pi}{2}+\theta\right), b\sin\left(\frac{\pi}{2}+\theta\right)\right].$$

At P, $\qquad x = a\cos\theta, \frac{dx}{d\theta} = -a\sin\theta$

$$\Rightarrow \qquad y = b\sin\theta, \frac{dy}{d\theta} = b\cos\theta$$

$$\Rightarrow \qquad \frac{dy}{dx} = \frac{dy}{d\theta}\cdot\frac{d\theta}{dx} = \frac{b\cos\theta}{-a\sin\theta} = -\frac{b}{a}\cot\theta$$

$$\Rightarrow \qquad \frac{d^2y}{dx^2} = \frac{d}{dx}\left(\frac{dy}{dx}\right) = \frac{d}{dx}\left(-\frac{b}{a}\cot\theta\right) = \frac{b}{a}\operatorname{cosec}^2\theta\cdot\frac{d\theta}{dx}$$

$$= \frac{b}{a}\cdot\frac{\operatorname{cosec}^2\theta}{-a\sin\theta} = -\frac{b}{a^2}\operatorname{cosec}^3\theta$$

$$\Rightarrow \qquad \rho_1 \text{ at } P = \frac{\left[1+\left(\frac{dy}{dx}\right)^2\right]^{3/2}}{\frac{d^2y}{dx^2}} = \frac{\left(1+\frac{b^2}{a^2}\cot^2\theta\right)^{3/2}}{-\frac{b}{a^2}\operatorname{cosec}^3\theta}$$

Similarly, $$\rho_2 \text{ at } D = \frac{\left[1+\frac{b^2}{a^2}\cot^2\left(\frac{\pi}{2}+\theta\right)\right]^{3/2}}{-\frac{b}{a^2}\operatorname{cosec}^3\left(\frac{\pi}{2}+\theta\right)} = \frac{\left[1+\frac{b^2}{a^2}\tan^2\theta\right]^{3/2}}{-\frac{b}{a^2}\sec^3\theta}$$

$$\Rightarrow \quad \left(\rho_1^{2/3}+\rho_2^{2/3}\right) = \frac{\left[1+\frac{b^2}{a^2}\cot^2\theta\right]}{\left(-\frac{b}{a^2}\operatorname{cosec}^3\theta\right)^{2/3}} + \frac{1+\frac{b^2}{a^2}\tan^2\theta}{\left(-\frac{b}{a^2}\sec^3\theta\right)^{2/3}}$$

$$= \left(\frac{a^2+b^2\cot^2\theta}{a^2}\right)\left(\frac{a^{4/3}}{b^{2/3}}\sin^2\theta\right) + \left(\frac{a^2+b^2\tan^2\theta}{a^2}\right)\left(\frac{a^{4/3}}{b^{2/3}}\cos^2\theta\right)$$

$$= \frac{a^{4/3}}{b^{2/3}}\frac{1}{a^2}[a^2\sin^2\theta + b^2\cos^2\theta + a^2\cos^2\theta + b^2\sin^2\theta]$$

$$= \frac{1}{(ab)^{2/3}}(a^2+b^2)$$

$\therefore (\rho_1^{2/3}+\rho_2^{2/3})(ab)^{2/3} = a^2+b^2$ **Proved.**

Example 57. *Find the radius of curvature to the cycloid*

$$x = a(\theta + \sin\theta),\ y = a(1-\cos\theta)$$

at the point θ. *What will it be at* $\theta = \pi$?

Solution. $x = a(\theta+\sin\theta), \frac{dx}{d\theta} = a(1+\cos\theta)$

$$\Rightarrow \quad y = a(1-\cos\theta), \frac{dy}{d\theta} = a\sin\theta$$

$$\Rightarrow \quad \frac{dy}{dx} = \frac{dy}{d\theta}\frac{d\theta}{dx} = \frac{(a\sin\theta)}{a(1+\cos\theta)} = \frac{2\sin\frac{\theta}{2}\cos\frac{\theta}{2}}{1+2\cos^2\frac{\theta}{2}-1} = \tan\frac{\theta}{2}$$

$$\Rightarrow \quad \frac{d^2y}{dx^2} = \frac{d}{dx}\left(\tan\frac{\theta}{2}\right) = \left(\sec^2\frac{\theta}{2}\right)\left(\frac{1}{2}\frac{d\theta}{dx}\right)$$

$$= \frac{1}{2}\cdot\frac{\sec^2\frac{\theta}{2}}{a(1+\cos\theta)} = \frac{1}{2}\frac{\sec^2\frac{\theta}{2}}{2a\cos^2\frac{\theta}{2}} = \frac{\sec^4\frac{\theta}{2}}{4a}$$

$$\Rightarrow \quad \rho = \frac{\left[1+\left(\frac{dy}{dx}\right)^2\right]^{3/2}}{\frac{d^2y}{dx^2}}$$

$$= \frac{\left(1+\tan^2\frac{\theta}{2}\right)^{3/2}}{\frac{1}{4a}\cdot\sec^4\frac{\theta}{2}} = 4a\frac{\left(\sec^2\frac{\theta}{2}\right)^{3/2}}{\sec^4\frac{\theta}{2}}$$

$$= 4a\frac{\sec^3\frac{\theta}{2}}{\sec^4\frac{\theta}{2}} = 4a\cos\frac{\theta}{2}$$

At $\theta = \pi, \rho = 4a\cos\frac{\pi}{2} = 0$ **Ans.**

EXERCISE 29.7

Find the radius of curvature for the curves (1 – 4).

1. $\frac{1}{x^2}+\frac{1}{y^2} = 1$ at $\left(\frac{1}{4},\frac{1}{4}\right)$ **Ans.** $\frac{1}{\sqrt{2}}$

2. $\sqrt{x}+\sqrt{y} = \sqrt{a}$ at the points at which the curvature has an extremum. **Ans.** $\frac{\sqrt{2}}{a}$

3. *(i)* $\frac{x^2}{a^2}+\frac{y^2}{b^2} = 1$ at $(a, 0)$ **Ans.** $2a$ *(ii)* $x^3+y^3 = 3axy$ at $\left(\frac{3a}{2},\frac{3a}{2}\right)$ **Ans.** $\frac{3\sqrt{2}a}{16}$

(iii) If ρ_1 and ρ_2 be the radii of curvature at the ends of a focal chord of the parabola $y^2 = 4ax$, then prove that $\rho_1^{-2/3}+\rho_2^{-2/3} = (2a)^{-2/3}$.

(iv) $y^2 = 8x$ at its vertex. **Ans.** 4

4. $s = a\log\cot\left(\frac{\pi}{4}-\frac{\psi}{2}\right)+a\sin\psi\sec^2\psi$ **Ans.** $2a\sec^3\psi$

5. If CP and CD are conjugate semidiameters of the ellipse $\frac{x^2}{a^2}+\frac{y^2}{b^2} = 1$, prove that the radius of curvature at P is $\frac{CD^3}{ab}$.

6. Show that the radius of curvature for the curve $y = \frac{a}{2}\left(e^{\frac{x}{a}}+e^{-\frac{x}{a}}\right)$ is equal to portion of the normal intercepted between the curve and x-axis and it varies as the square of the ordinate.

7. If (x, y) be the coordinates of a point P on the curve $y = \log(\cos x)$, find the radius of curvature at P, and show that the projection of the radius of curvature on y-axis is constant.

8. Show that the radius of curvature at the point (S, ψ) of the curve $S = a\sec\psi\tan\psi + a\log(\sec\psi+\tan\psi)$ is $2a\sec^3\psi$.

9. Find the least value of radius of curvature for curve $y = \log x$, $x > 0$. **Ans.** $\frac{3\sqrt{3}}{2}$

29.10 RADIUS OF CURVATURE FOR POLAR EQUATIONS

Let $r = f(\theta)$ be the polar equation of a curve.

From the figure given, $\psi = \theta + \varphi$

$$\Rightarrow \quad \frac{d\psi}{ds} = \frac{d\theta}{ds} + \frac{d\varphi}{ds}$$

$$= \frac{d\theta}{ds} + \frac{d\varphi}{d\theta} \cdot \frac{d\theta}{ds}$$

$$= \frac{d\theta}{ds}\left(1 + \frac{d\varphi}{d\theta}\right) \qquad \text{... (1)}$$

We know that $\tan \varphi = r\frac{d\theta}{dr} = \frac{r}{\frac{dr}{d\theta}}$

Differentiating w.r.t. θ, we get

$$\Rightarrow \quad \sec^2 \varphi \frac{d\varphi}{d\theta} = \frac{\frac{dr}{d\theta} \cdot \frac{dr}{d\theta} - r \cdot \frac{d^2r}{d\theta^2}}{\left(\frac{dr}{d\theta}\right)^2}$$

or

$$\frac{d\varphi}{d\theta} = \frac{\left(\frac{dr}{d\theta}\right)^2 - r\frac{d^2r}{d\theta^2}}{\sec^2 \varphi \left(\frac{dr}{d\theta}\right)^2}$$

$$= \frac{\left(\frac{dr}{d\theta}\right)^2 - r \cdot \frac{d^2r}{d\theta^2}}{(1 + \tan^2 \varphi)\left(\frac{dr}{d\theta}\right)^2}$$

$$= \frac{\left(\frac{dr}{d\theta}\right)^2 - r\frac{d^2r}{d\theta^2}}{\left[1 + r^2\left(\frac{d\theta}{dr}\right)^2\right]\left(\frac{dr}{d\theta}\right)^2}$$

$$= \frac{\left(\frac{dr}{d\theta}\right)^2 - r\frac{d^2r}{d\theta^2}}{\left(\frac{dr}{d\theta}\right)^2 + r^2} \qquad \text{... (2)}$$

We also know that $\frac{ds}{d\theta} = \sqrt{r^2 + \left(\frac{dr}{d\theta}\right)^2}$ ... (3)

Putting the values of $\frac{d\theta}{ds}$ and $\frac{d\varphi}{d\theta}$ in (1) we get

$$\Rightarrow \quad \frac{d\psi}{ds} = \frac{1}{\sqrt{r^2 + \left(\frac{dr}{d\theta}\right)^2}} \left(1 + \frac{\left(\frac{dr}{d\theta}\right)^2 - r\frac{d^2r}{d\theta^2}}{\left(\frac{dr}{d\theta}\right)^2 + r^2}\right)$$

$$= \frac{\left(\frac{dr}{d\theta}\right)^2 + r^2 + \left(\frac{dr}{d\theta}\right)^2 - r\frac{d^2r}{d\theta^2}}{\left[r^2 + \left(\frac{dr}{d\theta}\right)^2\right]^{3/2}} = \frac{r^2 + 2\left(\frac{dr}{d\theta}\right)^2 - r\frac{d^2r}{d\theta^2}}{\left[r^2 + \left(\frac{dr}{d\theta}\right)^2\right]^{3/2}}$$

or
$$\frac{ds}{d\psi} = \frac{\left[r^2 + \left(\frac{dr}{d\theta}\right)^2\right]^{3/2}}{r^2 + 2\left(\frac{dr}{d\theta}\right)^2 - r\frac{d^2r}{d\theta^2}} \qquad \left(\rho = \frac{ds}{d\psi}\right)$$

or
$$\boxed{\rho = \frac{[r^2 + r_1^2]^{3/2}}{r^2 + 2r_1^2 - rr_2}}$$

Example 58. *Show that at any point on the equiangular spiral $r = ae^{\theta \cot \alpha}$. Show that the radius curvature is $\rho = r \operatorname{cosec} \alpha$.*

Solution. Here
$$\frac{dr}{d\theta} = a \cot \alpha\, e^{\theta \cot \alpha} = r \cot \alpha,$$

$$\Rightarrow \quad \frac{d^2r}{d\theta^2} = a \cot^2 \alpha\, e^{\theta \cot \alpha} = r \cot^2 \alpha$$

$$\Rightarrow \quad \left[r^2 + \left(\frac{dr}{d\theta}\right)^2\right]^{3/2} = [r^2 + r^2 \cot^2 \alpha]^{3/2} = r^3 \operatorname{cosec}^3 \alpha$$

and
$$r^2 + 2\left(\frac{dr}{d\theta}\right)^2 - r\frac{d^2r}{d\theta^2} = r^2 + 2r^2 \cot^2 \alpha - r^2 \cot^2 \alpha$$

$$= r^2 (1 + \cot^2 \alpha)$$

$$= r^2 \operatorname{cosec}^2 \alpha$$

$$\Rightarrow \quad \rho = \frac{\left[r^2 + \left(\frac{dr}{d\theta}\right)^2\right]^{3/2}}{r^2 + 2\left(\frac{dr}{d\theta}\right)^2 - r\frac{d^2r}{d\theta^2}}$$

$$\Rightarrow \quad \rho = \frac{r^3 \operatorname{cosec}^3 \alpha}{r^2 \operatorname{cosec}^2 \alpha} = r \operatorname{cosec} \alpha$$

Proved

Example 59. *For the curve* $r^n = a^n \cos n\,\theta$, *show that the radius of curvature is* $\dfrac{a^n r^{-n+1}}{n+1}$.

Solution. $r^n = a^n \cos n\,\theta$

or $\log r^n = \log a^n \cos n\,\theta$

or $n \log r = \log a^n + \log \cos n\,\theta$

By differentiation, we have $\dfrac{n}{r}\dfrac{dr}{d\theta} = -\dfrac{n \sin n\,\theta}{\cos n\,\theta}$

$$\Rightarrow \quad \frac{dr}{d\theta} = -r \tan n\,\theta$$

$$\Rightarrow \quad \frac{d^2r}{d\theta^2} = -\frac{dr}{d\theta}\tan n\,\theta - n\,r \sec^2 n\,\theta$$

$$= r \tan^2 n\,\theta - n\,r \sec^2 n\,\theta$$

$$\Rightarrow \quad \rho = \frac{\left\{r^2 + \left(\frac{dr}{d\theta}\right)^2\right\}^{3/2}}{r^2 + 2\left(\frac{dr}{d\theta}\right)^2 - r\frac{d^2r}{d\theta^2}}$$

$$= \frac{\{r^2 + r^2 \tan^2 n\,\theta\}^{3/2}}{r^2 + 2r^2 \tan^2 n\,\theta - r^2 \tan^2 n\,\theta + n\,r^2 \sec^2 n\,\theta}$$

$$= \frac{r^3 \sec^3 n\,\theta}{r^2 + r^2 \tan^2 n\,\theta + n r^2 \sec^2 n\,\theta}$$

$$= \frac{r \sec^3 n\,\theta}{\sec^2 n\,\theta + n \sec^2 n\,\theta} = \frac{r \sec n\,\theta}{1+n}$$

$$= \frac{r}{(1+n)\cos n\,\theta} = \frac{r\,a^n}{(1+n)\,r^n} = \frac{a^n r^{-n+1}}{(n+1)}$$ **Proved**

Example 60. *Show that the radius of curvature at any point of the curve* $r = a\,(1 - \cos\theta)$ *varies as* $\sqrt{r}$.

Solution. $r = a\,(1 - \cos\theta)$ ... (1)

Differentiating (1) w.r.t. 'θ', we get $\dfrac{dr}{d\theta} = a \sin\theta$

$$\Rightarrow \quad \frac{d^2r}{d\theta^2} = a \cos\theta$$

We know that $\rho = \dfrac{(r^2 + r_1^2)^{3/2}}{r^2 + 2r_1^2 - r\,r_2}$

$$= \frac{[a^2 (1-\cos\theta)^2 + a^2 \sin^2\theta]^{3/2}}{a^2 (1-\cos\theta)^2 + 2a^2 \sin^2\theta - a\,(1-\cos\theta)\cdot a\cos\theta}$$

$$= \frac{a^3 [1 - 2\cos\theta + \cos^2\theta + \sin^2\theta]^{3/2}}{a^2 [1 - 2\cos\theta + \cos^2\theta + 2\sin^2\theta - \cos\theta + \cos^2\theta]}$$

$$= \frac{a(2 - 2\cos\theta)^{3/2}}{(3 - 3\cos\theta)}$$

$$= \frac{a\, 2\sqrt{2}\,(1 - \cos\theta)^{3/2}}{3\,(1 - \cos\theta)}$$

$$= \frac{2\sqrt{2}}{3}\, a\, (1 - \cos\theta)^{\frac{1}{2}}$$

$$= \frac{2\sqrt{2}}{3}\, a \sqrt{\frac{r}{a}} \qquad \text{[from (1)]}$$

$$= \frac{2\sqrt{2}\,\sqrt{a}}{3}\sqrt{r}$$

Hence, the radius of curvature varies as $\sqrt{r}$. **Proved**

Example 61. *Show that at the points in which the curves $r = a\,\theta$ and $r\,\theta = a$ intersect, their curvatures are in the ratio $1 : 3$, $(0 < \theta < 2\pi)$.*

Solution. $r = a\,\theta$ ... (1)

$r\,\theta = a$ ... (2)

For intersection of the curve (1) and curve (2),

We put $r = a\,\theta$ in (2) and we get

$$(a\,\theta)\,\theta = a \text{ or } \theta^2 = 1,\ \theta = \pm 1$$

Let us find out radius of curvature of the curve (1) at $\theta = \pm 1$.

$$\Rightarrow \qquad \frac{dr}{d\theta} = a,\ \frac{d^2 r}{d\theta^2} = 0$$

$$\Rightarrow \qquad \rho_1 = \frac{[r^2 + r_1^2]^{3/2}}{r^2 + 2r_1^2 - r\,r_2}$$

$$= \frac{(a^2\theta^2 + a^2)^{3/2}}{a^2\theta^2 + 2a^2 - (a\theta)(0)}$$

$$\rho_1 \text{ at } (\theta = 1) = \frac{(a^2 + a^2)^{3/2}}{a^2 + 2a^2} = \frac{(2a^2)^{3/2}}{3a^2} = \frac{2\sqrt{2}}{3}a$$

Now we find radius of curvature of curve (2)

$$\Rightarrow \qquad \frac{dr}{d\theta}\theta + r = 0 \text{ or } \frac{dr}{d\theta} = -\frac{r}{\theta}$$

$$\Rightarrow \qquad \frac{d^2 r}{d\theta^2} = \frac{\theta\left(-\frac{dr}{d\theta}\right) - (-r)\cdot 1}{\theta^2} = \frac{\theta\left(+\frac{r}{\theta}\right) + r}{\theta^2} = \frac{2r}{\theta^2}$$

$$\Rightarrow \qquad \rho_2 = \frac{\left(\frac{a^2}{\theta^2} + \frac{r^2}{\theta^2}\right)^{3/2}}{\frac{a^2}{\theta^2} + 2\frac{r^2}{\theta^2} - \left(\frac{a}{\theta}\right)\left(\frac{2r}{\theta^2}\right)}$$

$$\Rightarrow \qquad \rho_2 \text{ at } (\theta = 1) = \frac{(a^2 + a^2)^{3/2}}{a^2 + 2a^2 - 2a^2} = \frac{(2a^2)^{3/2}}{a^2} = 2\sqrt{2}\,a$$

$$\therefore \qquad \rho_1 : \rho_2 = \frac{2\sqrt{2}}{3} a : 2\sqrt{2}\,a = 1:3$$

Proved

29.11 RADIUS OF CURVATURE FOR PEDAL EQUATIONS

Pedal equation for a plane curve is a relative between r and p

We know that $\quad p = r \sin \varphi$

Differentiating w.r.t. 'r',

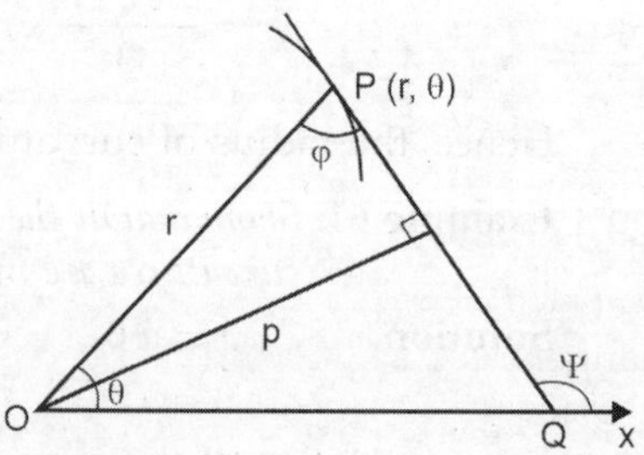

$$\Rightarrow \qquad \frac{dp}{dr} = r \cdot \cos \varphi \frac{d\varphi}{dr} + \sin \varphi$$

$$= r \cdot \frac{dr}{ds}\frac{d\varphi}{dr} + r\frac{d\theta}{ds}$$

$$= r \cdot \frac{d\varphi}{ds} + r \cdot \frac{d\theta}{ds}$$

$$\Rightarrow \qquad \frac{dp}{dr} = r\left(\frac{d\varphi}{ds} + \frac{d\theta}{ds}\right) \qquad \text{... (1)}$$

P (r, θ)
φ
r
p
θ
Ψ
O
Q
X

We also know that $\theta + \varphi = \psi$

$$\Rightarrow \qquad \frac{d\theta}{ds} + \frac{d\varphi}{ds} = \frac{d\psi}{ds} \qquad \text{... (2)}$$

Putting the value of $\frac{d\theta}{ds} + \frac{d\varphi}{ds}$ from (2) in (1), we get

$$\Rightarrow \qquad \frac{dp}{dr} = r\frac{d\psi}{ds} \quad \text{or} \quad \frac{ds}{d\psi} = r\frac{dr}{dp}$$

$$\Rightarrow \qquad \boldsymbol{\rho = r\frac{dr}{dp}}$$

Example 62. *Find the radius of curvature for a curve* $pa^m = r^{m+1}$.

Solution. $\quad pa^m = r^{m+1} \qquad \text{...(1)}$

Differentiating (1) w.r.t. 'r', we get

$$\Rightarrow \qquad \frac{dp}{dr}a^m = (m+1)\,r^m \qquad \Rightarrow \qquad \frac{dp}{dr} = (m+1)\frac{r^m}{a^m}$$

$$\Rightarrow \qquad \frac{dr}{dp} = \frac{a^m}{(m+1)\,r^m}$$

$$\Rightarrow \qquad r \cdot \frac{dr}{dp} = \frac{r \cdot a^m}{(m+1) \cdot r^m}$$

$$\therefore \qquad \rho = \frac{a^m}{(m+1)\, r^{m-1}} \qquad \textbf{Ans.}$$

EXERCISE 29.8

Find the radius of curvature

1. $r = 2^\theta$ at $\theta = 0$ **Ans.** $\sqrt{1+(\log 2)^2}$
2. $r^2 = a^2 \cos 2\theta$ **Ans.** $\frac{a^2}{3r}$
3. $\sqrt{r} \cos \frac{\theta}{2} = \sqrt{a}$ **Ans.** $2r\sqrt{\frac{r}{a}}$
4. $r^3 = 2ap^2$ **Ans.** $\frac{2\sqrt{2ar}}{3}$
5. $r^2 \cos 2\theta = 0$ **Ans.** $\frac{r^3}{a^2}$
6. $pr = 16$ **Ans.** $\frac{r^3}{16}$
7. $p^2 = ar$ **Ans.** $\frac{2p^3}{a^2}$
8. $\frac{1}{p^2} = \frac{1}{a^2} + \frac{1}{b^2} - \frac{r^2}{a^2 b^2}$ **Ans.** $\frac{a^2 b^2}{p^3}$

29.12 CIRCLE OF CURVATURE

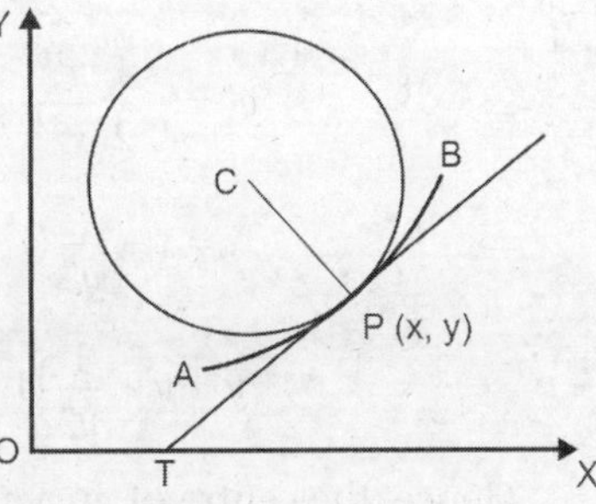

Let APB be a curve, P (x, y) any point on it, and PT be tangent at P.

Let a circle be drawn on the same side of the tangent as the curve, with its centre at C and having the same curvature as the curve at the point P.

This circle is known as the circle of curvature of the given curve at P. C is called the *centre of curvature.*

Any chord of the circle of curvature through the point of contact P is called the chord of curvature of the given curve at P.

29.13 CENTRE OF CURVATURE

Let $C(\bar{x}, \bar{y})$ be the centre of curvature of the curve at the point P (x, y). Draw CM and $PN \perp$ to x-axis. PT and CP are the tangent and normal at P.

$$\bar{x} = OM = ON - MN = x - LP \qquad \left[\begin{aligned} \frac{LP}{PC} &= \sin\psi \\ LP &= \rho \sin \psi \end{aligned}\right]$$

$$= x - \rho \sin \psi$$

$$= x - \frac{[1+y'^2]^{3/2}}{y''} \cdot \frac{y'}{\sqrt{1+y'^2}}$$

$$\tan \psi = y'$$

$$\sin \psi = \frac{y'}{\sqrt{1+y'^2}}$$

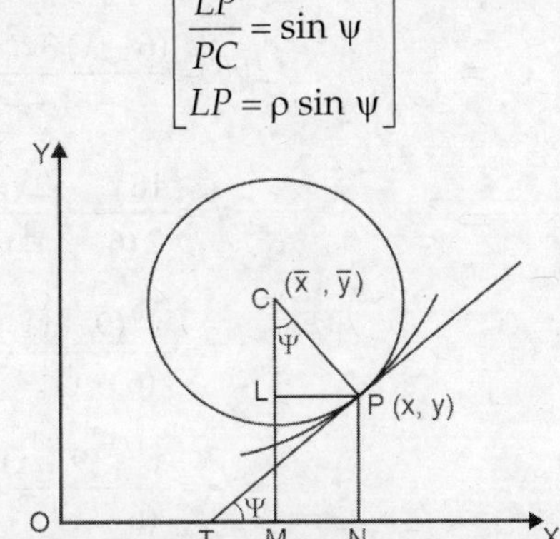

$$\boxed{\bar{x} = x - \frac{y'}{y''}(1 + y'^2)}$$

$$\bar{y} = CM = ML + LC = y + LC = y + \rho \cos\psi \qquad \left[\frac{LC}{PC} = \cos\psi,\ LC = \rho\cos\psi\right]$$

$$= y + \frac{(1 + y'^2)^{3/2}}{y''} \cdot \frac{1}{\sqrt{1 + y'^2}} \qquad \left[\tan\psi = y',\ \cos\psi = \frac{1}{\sqrt{1 + y'^2}}\right]$$

$$\boxed{\bar{y} = y + \frac{1}{y''}(1 + y'^2)}$$

Evolute is the locus at the centres of curvature of a curve is called its evolute and that curve is said to be an involute of evolute.

Example 63. *Find the centre of curvature of the parabola $y^2 = 4ax$ at the point $(a, 2a)$.*

Solution. $y^2 = 4ax$

$$\Rightarrow \quad y = 2a^{1/2}x^{1/2}$$

$$\Rightarrow \quad y' = \frac{2}{2}a^{1/2}x^{-1/2} = \frac{a^{1/2}}{x^{1/2}}, \qquad y' \text{ at } (a, 2a) = \frac{a^{1/2}}{a^{1/2}} = 1$$

$$\Rightarrow \quad y'' = -\frac{1}{2}\frac{a^{1/2}}{x^{3/2}}, \qquad y'' \text{ at } (a, 2a) = -\frac{1}{2}\frac{a^{1/2}}{a^{3/2}} = -\frac{1}{2a}$$

$$\Rightarrow \quad \bar{x} = x - \frac{y'}{y''}(1 + y'^2) = a - \frac{1}{-1/2a}(1 + 1) = a + 2a(1 + 1) = 5a$$

$$\Rightarrow \quad \bar{y} = y + \frac{1}{y''}(1 + y'^2) = 2a + \frac{1}{-1/2a}(1 + 1) = 2a - 2a(1 + 1) = -2a$$

Hence, the centre of curvature is $(5a, -2a)$ **Ans.**

Example 64. *Find the centre of curvature of the curve $x^3 + xy^2 - 6y^2 = 0$ at $(3, 3)$.*

Solution. $x^3 + xy^2 - 6y^2 = 0$ ... (1)

$$\Rightarrow \quad y^2 = \frac{x^3}{6 - x}$$

$$\Rightarrow \quad 2yy' = \frac{(6 - x)\,3x^2 - x^3(-1)}{(6 - x)^2}$$

$$\Rightarrow \quad y' = \frac{18x^2 - 2x^3}{2(6 - x)^2 y} = \frac{2x^2(9 - x)}{2(6 - x)^2 y}$$

$$= \frac{x^2(9 - x)}{(6 - x)^2} \cdot \frac{(6 - x)^{1/2}}{x^{3/2}}$$

$$= \frac{x^{1/2}(9 - x)}{(6 - x)^{3/2}}, \quad y' \text{ at } (3, 3) = 2$$

$$y'' = \frac{(6-x)^{3/2}\left(\frac{1}{2}\frac{9}{\sqrt{x}} - \frac{3}{2}\sqrt{x}\right) + x^{1/2}(9-x)\frac{3}{2}(6-x)^{1/2}}{(6-x)^3}$$

y'' at $(3, 3) = 1$

$$\overline{x} = x - \frac{y'}{y''}(1+y'^2) = 3 - \frac{2}{1}(1+4) = -7$$

$$\overline{y} = y + \frac{1}{y''}(1+y'^2) = 3 + \frac{1}{1}(1+4) = 8$$

Hence, the centre of the curvature = (– 7, 8) **Ans.**

Example 65. *Find the evolute of the parabola* $y^2 = 4ax$.

Solution. $y^2 = 4ax$

$$\Rightarrow \quad y = 2a^{1/2}x^{1/2},\; y' = \frac{a^{1/2}}{x^{1/2}}$$

$$\Rightarrow \quad y'' = -\frac{1}{2}\frac{a^{1/2}}{x^{3/2}}$$

If (X, Y) be the centre of curvature, we have

$$\Rightarrow \quad X = x - \frac{y'}{y''}(1+y'^2)$$

$$\Rightarrow \quad X = x - \frac{\frac{a^{1/2}}{x^{1/2}}}{-\frac{1}{2}\frac{a^{1/2}}{x^{3/2}}}\left(1+\frac{a}{x}\right)$$

$$\Rightarrow \quad X = x + 2x\left(1+\frac{a}{x}\right) = x + 2x + 2a = 3x + 2a \qquad \dots (1)$$

$$\Rightarrow \quad Y = y + \frac{1}{y''}(1+y'^2)$$

$$\Rightarrow \quad Y = y + \frac{1}{-\frac{1}{2}\frac{a^{1/2}}{x^{3/2}}}\left(1+\frac{a}{x}\right)$$

$$\Rightarrow \quad Y = y - \frac{2x^{3/2}}{a^{1/2}}\left(1+\frac{a}{x}\right) = y - \frac{2x^{3/2}}{a^{1/2}} - 2a^{1/2}x^{1/2}$$

$$\Rightarrow \quad Y = 2a^{1/2}x^{1/2} - \frac{2x^{3/2}}{a^{1/2}} - 2a^{1/2}x^{1/2}$$

$$\Rightarrow \quad Y = -2\frac{x^{3/2}}{a^{1/2}} \qquad \dots (2)$$

Eliminating x and y

$$Y = -2\frac{2}{a^{1/2}}\left(\frac{X-2a}{3}\right)^{3/2}$$

or $$Y^2 = \frac{4}{27a}(X - 2a)^3$$

or $$27aY^2 = 4\,(X - 2a)^3$$ **Ans.**

29.14 CHORD OF CURVATURE THROUGH ORIGIN

Let PQ be the Chord of curvature at P, through origin, PD be the diameter. Let the angle between the chord PQO and the tangent PT be φ.

$$\angle PDQ = \varphi$$

$$\angle PQD = 90°,\ PD = 2\rho$$

PQD is a rt. angled triangle.

$$\sin\varphi = \frac{PQ}{PD}$$

$$PQ = PD\sin\varphi = 2\rho\sin\varphi$$

Chord of curvature is $= 2\rho\sin\varphi$

Note. If chord of curvature is parallel to x-axis.

$$\varphi = \psi$$

$\therefore$ Chord of curvature $= 2\rho\sin\psi$

Similarly chord of curvature parallel to y-axis

$$= 2\rho\sin(90° - \psi)$$

$$= 2\rho\cos\psi$$

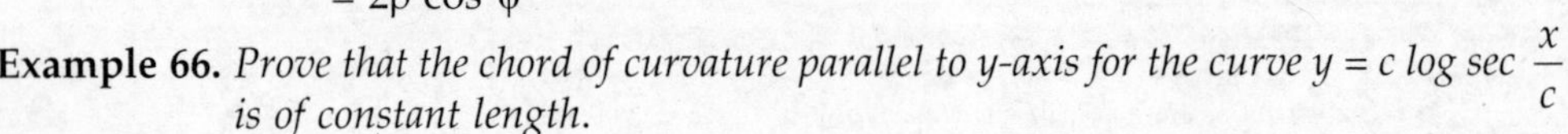

Example 66. *Prove that the chord of curvature parallel to y-axis for the curve* $y = c\log\sec\frac{x}{c}$ *is of constant length.*

Solution. Chord of curvature $= 2\rho\cos\psi$ ... (1)

Here $$y = c\log\sec\frac{x}{c}$$

$$y_1 = c\cdot\frac{1}{\sec\frac{x}{c}}\cdot\sec\frac{x}{c}\cdot\tan\frac{x}{c}\cdot\frac{1}{c}$$

$$\Rightarrow\quad y_1 = \tan\frac{x}{c}$$

$$\Rightarrow\quad y_2 = \frac{1}{c}\sec^2\frac{x}{c}$$

$$\Rightarrow\quad \rho = \frac{\left(1 + y'^2\right)^{3/2}}{y''} = \frac{\left(1 + \tan^2\frac{x}{c}\right)^{3/2}}{\frac{1}{c}\sec^2\frac{x}{c}} = \frac{\left(\sec^2\frac{x}{c}\right)^{3/2}}{\frac{1}{c}\sec^2\frac{x}{c}}$$

$$\Rightarrow\quad \rho = c\sec\frac{x}{c}$$

$$\Rightarrow\quad \tan\psi = y_1 = \tan\frac{x}{c} \qquad \left[\because\ \psi = \frac{x}{c}\right]$$

$$\therefore \qquad \cos\psi = \cos\frac{x}{c}$$

Putting the value of ρ and cos ψ in (1), we have

$$\text{Chord of curvature} = 2\left(c\sec\frac{x}{c}\right)\cos\frac{x}{c} = 2c = \text{Constant}$$ **Proved**

Example 67. *Prove that the chord of curvature through the pole of the equiangular spiral* $r = ae^{m\theta}$ *is* $2r$.

Solution. $r = ae^{m\theta}$

$$\Rightarrow \qquad r_1 = ame^{m\theta} = mr$$

$$\Rightarrow \qquad r_2 = mr_1 = m^2 r$$

$$\Rightarrow \qquad \rho = \frac{(r^2 + r_1^2)^{3/2}}{r^2 + 2r_1^2 - rr_2}$$

$$= \frac{(r^2 + m^2r^2)^{3/2}}{r^2 + 2m^2r^2 - m^2r^2}$$

$$= \frac{(1+m^2)^{3/2} r}{1 + 2m^2 - m^2}$$

$$= \frac{(1+m^2)^{3/2}\cdot r}{1+m^2}$$

$$= (1+m^2)^{1/2} r$$

$$\Rightarrow \qquad \tan\varphi = r\frac{d\theta}{dr} = \frac{r}{r_1} = \frac{r}{mr} = \frac{1}{m}$$

$$\Rightarrow \qquad \rho = (1+\cot^2\varphi)^{1/2}\, r = r\operatorname{cosec}\varphi$$

$$\text{Chord of curvature} = 2\rho\sin\varphi$$

$$= 2\,(r\operatorname{cosec}\varphi)\cdot\sin\varphi$$

$$= 2r$$ **Proved**

Example 68. *Find the chord of curvature through the pole of the cardioid* $r = a\,(1 + \cos\theta)$.

Solution. $r = a\,(1 + \cos\theta)$

We know that

$$\rho = \frac{2\sqrt{2ar}}{3}$$

$$\text{Chord of curvature} = 2\,\rho\sin\varphi$$

$$= 2\left(\frac{2\sqrt{2ar}}{3}\right)\left(r\frac{d\theta}{dr}\right)$$

$$= \frac{4\sqrt{2ar}}{3}\cdot\frac{r}{\sqrt{r^2 + \left(\frac{dr}{d\theta}\right)^2}}$$

$$= \frac{4\sqrt{2ar}}{3} \cdot \frac{r}{\sqrt{a^2 (1 + \cos\theta)^2 + a^2 \sin^2\theta}}$$

$$= \frac{4\sqrt{2ar}}{3} \cdot \frac{r}{\sqrt{2ar}} = \frac{4}{3} r$$ **Ans.**

Example 69. *Find the equation of the circle of curvature of the curve y = sin x at the point (π/2, 1). Also, find the lengths of the chords of curvature at that point.*

Solution. $y = \sin x$

$$\Rightarrow \quad \frac{dy}{dx} = \cos x, \frac{d^2y}{dx^2} = -\sin x$$

$$\Rightarrow \quad \rho = \frac{\left[1 + \left(\frac{dy}{dx}\right)^2\right]^{3/2}}{\frac{d^2y}{dx^2}} = \frac{[1 + \cos^2 x]^{3/2}}{-\sin x}$$

$$\Rightarrow \quad \rho \text{ at } \left(\frac{\pi}{2}, 1\right) = \frac{\left[1 + \cos^2 \frac{\pi}{2}\right]^{3/2}}{-\sin \frac{\pi}{2}} = -1$$

Radius = 1

$$\Rightarrow \quad \bar{x} = x - \frac{y'}{y''}(1 + y'^2)$$

$$\Rightarrow \quad \bar{x} = x - \frac{\cos x}{-\sin x}(1 + \cos^2 x)$$

$$\Rightarrow \quad \bar{x} \text{ at } \left(\frac{\pi}{2}, 1\right) = \frac{\pi}{2} - \frac{\cos \frac{\pi}{2}}{-\sin \frac{\pi}{2}}\left(1 + \cos^2 \frac{\pi}{2}\right) = \frac{\pi}{2}$$

$$\Rightarrow \quad \bar{y} = y + \frac{1}{y''}(1 + y'^2)$$

$$\Rightarrow \quad \bar{y} = y + \frac{1}{-\sin x}(1 + \cos^2 x)$$

$$\Rightarrow \quad \bar{y} \text{ at } \left(\frac{\pi}{2}, 1\right) = 1 + \frac{1}{-\sin \frac{\pi}{2}}\left(1 + \cos^2 \frac{\pi}{2}\right) = 0$$

Centre is at the point $\left(\frac{\pi}{2}, 0\right)$

Equation of the circle is $\left(x - \frac{\pi}{2}\right)^2 + (y - 0)^2 = 1^2$

or $\left(x-\frac{\pi}{2}\right)^2+y^2=1$ **Ans.**

Length of the chords of curvature = 2 ρ cos ψ

$$\tan\psi=\cos x=\cos\frac{\pi}{2}=0,\ \psi=0$$

Chord of curvature = 2 (1) cos 0 = 2. **Ans.**

EXERCISE 29.9

Find the coordinates of centre of curvature of the following curves.

1. $y^2 = x^3$ at the point (1, 1). **Ans.** $\left(-\frac{11}{2},\frac{16}{3}\right)$

2. $y = e^x$ at the point (0, 1). **Ans.** (– 2, 3)

3. $x^{2/3} + y^{2/3} = 1$ at the point (α, β). **Ans.** $\left(\alpha+3\alpha^{\frac{1}{3}}\beta^{\frac{2}{3}},\beta+3\alpha^{\frac{2}{3}}\beta^{\frac{1}{3}}\right)$

4. Show that $\left(x-\frac{3}{4}a\right)^2+\left(y-\frac{3}{4}a\right)^2=\frac{1}{2}a^2$ is the circle of curvature of the curve $\sqrt{x}+\sqrt{y}=\sqrt{a}$, at the point $\left(\frac{a}{4},\frac{a}{4}\right)$.

5. Show that the centre of curvature of any point (*x*, *y*) on the ellipse $\frac{x^2}{a^2}+\frac{y^2}{b^2}=1$ is

$$\left(\frac{a^2-b^2}{a^4}x^3,\frac{a^2-b^2}{b^4}y^3\right).$$

Hence show that the equation of evolute is $(ax)^{2/3} + (by)^{2/3} = (a^2 - b^2)^{2/3}$.

6. Prove that the evolute of the hyperbola $2xy = a^2$ is $(x + y)^{2/3} - (x - y)^{2/3} = 2a^{2/3}$

7. Show that evolute of the tractrix

$x=a\left(\cos t+\log\tan\frac{t}{2}\right),\ y=a\sin t$ is the catenary $y=a\cosh\frac{x}{a}$.

8. Find the evolute of astroid $x = a\cos^3\theta,\ y = a\sin^3\theta$. **Ans.** $(x+y)^{\frac{2}{3}}+(x-y)^{\frac{2}{3}}=2a^{2/3}$

9. If C_x and C_y be the chords of curvature parallel to *x*-axis and *y*-axis respectively at any point of the curve $y=ae^{\frac{x}{a}}$, show that $\frac{1}{(C_x)^2}+\frac{1}{(C_y)^2}=\frac{1}{(2aC_x)}$.

10. Show that the chord of curvature through the pole of the curve

$r^m = a^m\cos m\theta$ is $\frac{2r}{m+1}$.

11. Fill in the blanks:

(*a*) The radius of curvature is $\frac{ds}{d\psi}$, then the curvature is ....

(*b*) If the curvature of a curve increases then the radius of curvature ....

(*c*) In polar coordinates the formula for radius of curvature is ....

(*d*) For pedal equations the radius of curvature is ....

(*e*) Let $(\overline{x}, \overline{y})$ be the centre of curvature of the curve at the point (x, y), then

$\overline{y}$ = ... $\overline{x}$ = ...

(*f*) The locus of the centre of curvature of a curve is called ....

Ans. (*a*) $\dfrac{d\psi}{ds}$, (*b*) decreases, (*c*) $\dfrac{(r^2 + r_1^2)^{3/2}}{r^2 + 2r_1^2 - r r_2}$, (*d*) $r\dfrac{dr}{dp}$,

(*e*) $\overline{x} = x - \dfrac{y'}{y''}(1 + y'^2), \overline{y} = y + \dfrac{1}{y''}(1 + y'^2)$, (*f*) evolute

Partial Differentiation

30.1 INTRODUCTION

Area of a rectangle depends upon its length and breadth and hence we can say that area is the function of two variables, i.e. length and breadth. In a relation, z is called a function of two variables x and y if z has one definite value for every pair of values of x and y. Symbolically, it is written as

$$z = f(x, y)$$

Variables x and y are called independent variables while z is called the dependent variable. Similarly, we can define z as a function of more than two variables.

30.2 PARTIAL DERIVATIVES

Let $z = f(x, y)$ be a function of two independent variables x and y. If we keep y constant and x varies then z becomes a function of x only. The derivative of z with respect to x, keeping, y as constant is called partial derivative of "z" , w.r.t. "x" and is denoted by symbols

$$\boxed{\frac{\partial z}{\partial x}, \frac{\partial f}{\partial x}, f_x(x, y), \text{ etc.}}$$

Then $$\frac{\partial z}{\partial x} = \lim_{\delta x \to 0} \frac{f(x + \delta x, y) - f(x, y)}{\delta x}$$

The process of finding the partial differential coefficient of z w.r.t. "x" is that of ordinary differentiation, but with the difference only that we treat y as constant.

Similarly, the partial derivative of "z" w.r.t. "y" keeping x as constant is denoted by

$$\boxed{\frac{\partial z}{\partial y}, \frac{\partial f}{\partial y}, f_y, \text{ etc.}}$$

Then $$\frac{\partial z}{\partial y} = \lim_{\delta y \to 0} \frac{f(x, y + \delta y) - f(x, y)}{\delta y}$$

30.3 PARTIAL DERIVATIVES: GEOMETRICALLY

Geometrically, the partial derivative of z w.r.t. x gives the slope of the tangent drawn to the curve of intersection of the surface

$$z = f(x, y)$$

with a plane parallel to ZOX plane.

Example 1. *If* $f(x, y) = x^3y - xy^3$ *find* $\left[\dfrac{1}{\frac{\partial f}{\partial x}} \quad \dfrac{1}{\frac{\partial f}{\partial y}}\right]$

Solution. We have,

$$f(x, y) = x^3y - xy^3$$

$$\Rightarrow \quad \frac{\partial f}{\partial x} = 3x^2y - y^3$$

$$= 6 - 8 = -2 \text{ for } x = 1, y = 2$$

$$\Rightarrow \quad \frac{\partial f}{\partial y} = x^3 - 3xy^2$$

$$= 1 - 12 = -11 \text{ for } x = 1, y = 2$$

$$\left[\frac{1}{\frac{\partial f}{\partial x}} + \frac{1}{\frac{\partial f}{\partial y}}\right]_{\substack{x=1\\y=2}} = \left[\frac{1}{-2} + \frac{1}{-11}\right] = -\frac{13}{22}$$

Ans.

Example 2. *If* $u = \sin^{-1}\left(\frac{x}{y}\right) + \tan^{-1}\left(\frac{y}{x}\right)$, *then find the value of* $x\frac{\partial u}{\partial x} + y\frac{\partial u}{\partial y}$.

[S.B.T.E. 2017, 2013, 2010]

Solution.

$$u = \sin^{-1}\left(\frac{x}{y}\right) + \tan^{-1}\left(\frac{y}{x}\right)$$

$$\Rightarrow \quad \frac{\partial u}{\partial x} = \frac{1}{\sqrt{1-\left(\frac{x}{y}\right)^2}} \cdot \frac{1}{y} + \frac{1}{1+\left(\frac{y}{x}\right)^2} \cdot \left(-\frac{y}{x^2}\right)$$

$$= \frac{1}{\sqrt{y^2 - x^2}} - \frac{y}{x^2 + y^2}$$

$$\Rightarrow \quad x.\frac{\partial u}{\partial x} = \frac{x}{\sqrt{y^2 - x^2}} - \frac{xy}{x^2 + y^2} \qquad ...(1)$$

and

$$\frac{\partial u}{\partial y} = \frac{1}{\sqrt{1-\left(\frac{x}{y}\right)^2}}\left(-\frac{x}{y^2}\right) + \frac{1}{1+\left(\frac{y}{x}\right)^2}\left(\frac{1}{x}\right)$$

$$= -\frac{x}{y\sqrt{y^2 - x^2}} + \frac{x}{x^2 + y^2}$$

$$y.\frac{\partial u}{\partial y} = -\frac{x}{\sqrt{y^2 - x^2}} + \frac{xy}{x^2 + y^2} \quad ...(2)$$

On adding (1) and (2), we have

$$x.\frac{\partial u}{\partial x} + y.\frac{\partial u}{\partial y} = 0$$ **Ans.**

Example 3. *Find* $\frac{\partial u}{\partial r}$ *and* $\frac{\partial u}{\partial \theta}$ *if* $u = e^{r\cos\theta}\cos(r\sin\theta)$.

Solution. We have, $u = e^{r\cos\theta}. \cos(r \sin\theta)$

$\frac{\partial u}{\partial r} = e^{r\cos\theta}. [-\sin(r\sin\theta). \sin\theta] + [\cos\theta. e^{r\cos\theta}] \cos(r\sin\theta)$ [keeping θ as constant]

$= e^{r\cos\theta}. [-\sin(r\sin\theta). \sin\theta + \cos(r\sin\theta). \cos\theta]$

$= e^{r\cos\theta}. \cos(r\sin\theta + \theta)$ **Ans.**

$\frac{\partial u}{\partial \theta} = e^{r\cos\theta}. [-\sin(r\sin\theta). r\cos\theta] + (-r\sin\theta. e^{r\cos\theta}) \cos(r\sin\theta)$

[keeping r as constant]

$= -re^{r\cos\theta} [\sin(r\sin\theta)\cos\theta + \sin\theta\cos(r\sin\theta)]$

$= -re^{r\cos\theta} \sin[r\sin\theta + \theta]$ **Ans.**

Example 4. *If* $u = (1 - 2xy + y^2)^{-1/2}$, *prove that* $x\frac{\partial u}{\partial x} - y\frac{\partial u}{\partial y} = y^2u^3$ **[S.B.T.E. 2010]**

Solution. $u = (1 - 2xy + y^2)^{-1/2}$ ...(1)

Differentiating (1) partially w.r.t. "x", we get

$$\Rightarrow \frac{\partial u}{\partial x} = -\frac{1}{2}(1 - 2xy + y^2)^{-3/2}(-2y)$$

$$\Rightarrow x\frac{\partial u}{\partial x} = xy(1 - 2xy + y^2)^{-3/2} \quad ...(2)$$

Differentiating (1) partially w.r.t. "y", we get

$$\Rightarrow \frac{\partial u}{\partial y} = -\frac{1}{2}(1 - 2xy + y^2)^{-3/2}(-2x + 2y)$$

$$\Rightarrow y\frac{\partial u}{\partial y} = (xy - y^2)(1 - 2xy + y^2)^{-3/2} \quad ...(3)$$

Subtracting (3) from (2), we get

$$\Rightarrow x\frac{\partial u}{\partial x} - y\frac{\partial u}{\partial y} = xy(1-2xy+y^2)^{-3/2} - (xy-y^2)(1-2xy+y^2)^{-3/2}$$

$$= y^2(1 - 2xy + y^2)^{-3/2}$$

$$= y^2u^3$$ **Proved.**

Example 5. *If* $z = e^{ax+by}f(ax - by)$, *prove that* $b\frac{\partial z}{\partial x} + a\frac{\partial z}{\partial y} = 2abz$

Solution. $z = e^{ax+by}.f(ax - by)$

$\Rightarrow \quad \dfrac{\partial z}{\partial x} = ae^{ax+by}.f(ax-by)+e^{ax+by}.af'(ax-by)$

$\Rightarrow \quad b\dfrac{\partial z}{\partial x} = abe^{ax+by}.f(ax-by)+abe^{ax+by}.f'(ax-by)$ ...(1)

$\Rightarrow \quad \dfrac{\partial z}{\partial y} = be^{ax+by}.f(ax-by)+e^{ax+by}(-b)f'(ax-by)$

$\Rightarrow \quad a\dfrac{\partial z}{\partial y} = abe^{ax+by}.f(ax-by)-abe^{ax+by}f'(ax-by)$ ...(2)

On adding (1) and (2), we get

$\Rightarrow \quad b\dfrac{\partial z}{\partial x}+a\dfrac{\partial z}{\partial y} = 2abe^{ax+by} f(ax - by)$

$\Rightarrow \quad b\dfrac{\partial z}{\partial x}+a\dfrac{\partial z}{\partial y} = 2abz$ **Proved.**

30.4 PARTIAL DERIVATIVES OF HIGHER ORDER

Let $z = f(x, y)$ then $\dfrac{\partial z}{\partial x}$ and $\dfrac{\partial z}{\partial y}$ being the function of x and y can be further differentiated partially with respect to x and y.

Symbolically $\dfrac{\partial}{\partial x}\left(\dfrac{\partial z}{\partial x}\right)=\dfrac{\partial^2 z}{\partial x^2} \Rightarrow \dfrac{\partial^2 f}{\partial x^2}$ or f_{xx}

$\Rightarrow \quad \dfrac{\partial}{\partial y}\left(\dfrac{\partial z}{\partial x}\right)=\dfrac{\partial^2 z}{\partial y\partial x} \Rightarrow \dfrac{\partial^2 f}{\partial y\partial x}$ or f_{yx}

$\Rightarrow \quad \dfrac{\partial}{\partial x}\left(\dfrac{\partial z}{\partial y}\right)=\dfrac{\partial^2 z}{\partial x\partial y} \Rightarrow \dfrac{\partial^2 f}{\partial x\partial y}$ or f_{xy}

$\Rightarrow \quad \dfrac{\partial}{\partial y}\left(\dfrac{\partial z}{\partial y}\right)=\dfrac{\partial^2 z}{\partial y^2} \Rightarrow \dfrac{\partial^2 f}{\partial y^2}$ or f_{yy}

Note : $\dfrac{\partial^2 z}{\partial y\,\partial x} = \dfrac{\partial^2 z}{\partial x\,\partial y}$

Example 6. *If* $z = x^5+6x^4y + 10x^3y^2+2x^2y^3+5xy^4+y^5$, *then find* $\dfrac{\partial^2 z}{\partial x^2}, \dfrac{\partial^2 z}{\partial x\partial y}$ *and* $\dfrac{\partial^2 z}{\partial y^2}$.

Solution. $z = x^5 + 6x^4y + 10x^3y^2 + 2x^2y^3 + 5xy^4 + y^5$

$\Rightarrow \quad \dfrac{\partial z}{\partial x} = 5x^4 + 24x^3y + 30x^2y^2 + 4xy^3 + 5y^4$

$\Rightarrow \quad \dfrac{\partial z}{\partial y} = 6x^4 + 20x^3y + 6x^2y^2 + 20xy^3 + 5y^4$

$$\Rightarrow \quad \frac{\partial^2 z}{\partial x^2} = \frac{\partial}{\partial x}\left(\frac{\partial z}{\partial x}\right) = 20x^3 + 72x^2y + 60xy^2 + 4y^3$$

$$\Rightarrow \quad \frac{\partial^2 z}{\partial x \partial y} = \frac{\partial}{\partial x}\left(\frac{\partial z}{\partial y}\right) = 24x^3 + 60x^2y + 12xy^2 + 20y^3$$

$$\Rightarrow \quad \frac{\partial^2 z}{\partial y^2} = \frac{\partial}{\partial y}\left(\frac{\partial z}{\partial y}\right) = 20x^3 + 12x^2y + 60xy^2 + 20y^3$$ **Ans.**

Example 7. *Prove that:* $y = f(x + at) + g(x - at)$ *satisfies* $\frac{\partial^2 y}{\partial t^2} = a^2\left(\frac{\partial^2 y}{\partial x^2}\right)$ *here f and g are assumed to be at least twice differentiable and 'a' is any constant.*

Solution. We have,

$$y = f(x + at) + g(x - at) \qquad ...(1)$$

Differentiating (1) w.r.t. 'x' partially, we get

$$\Rightarrow \quad \frac{\partial y}{\partial x} = f'(x + at) + g'(x - at)$$

$$\Rightarrow \quad \frac{\partial^2 y}{\partial x^2} = f''(x + at) + g''(x - at)$$

Differentiating (1) w.r.t. 't' partially, we get

$$\Rightarrow \quad \frac{\partial y}{\partial t} = f'(x + at)\, a + g'(x - at)(-a)$$

$$\Rightarrow \quad \frac{\partial^2 y}{\partial t^2} = a^2 f''(x + at) + g''(x - at)\, a^2$$

$$= a^2 [f''(x + at) + g''(x - at)]$$

$$= a^2\left(\frac{\partial^2 y}{\partial x^2}\right)$$ **Proved.**

Example 8. *If* $z = \log(e^x + e^y)$, *show that* $rt - s^2 = 0$, *where* $r = \frac{\partial^2 z}{\partial x^2}, t = \frac{\partial^2 z}{\partial y^2}, s = \frac{\partial^2 z}{\partial x \partial y}$.

Solution. We have, $z = \log(e^x + e^y)$ ...(1)

Differentiating (1) partially w.r.t. 'x', we get

$$\Rightarrow \quad \frac{\partial z}{\partial x} = \frac{1}{e^x + e^y} . e^x \qquad ...(2)$$

$$\Rightarrow \quad r = \frac{\partial^2 z}{\partial x^2} = \frac{(e^x + e^y)e^x - e^x(e^x)}{(e^x + e^y)^2} = \frac{e^{x+y}}{(e^x + e^y)^2} \qquad ...(3)$$

Differentiating (1) w.r.t. "y", we get

$$\Rightarrow \quad \frac{\partial z}{\partial y} = \frac{1}{e^x + e^y} . e^y \qquad ...(4)$$

$$t = \frac{\partial^2 z}{\partial y^2} = \frac{(e^x + e^y)e^y - e^y(e^y)}{(e^x + e^y)^2} = \frac{e^{x+y}}{(e^x + e^y)^2} \quad ...(5)$$

Differentiating (2) w.r.t. "y", we get

$$s = \frac{\partial^2 z}{\partial y\, \partial x} = \frac{-e^x}{(e^x + e^y)^2} . e^y$$

$$= \frac{-e^{x+y}}{(e^x + e^y)^2}$$

Now, $$rt - s^2 = \frac{e^{2(x+y)}}{(e^x + e^y)^4} - \frac{e^{2(x+y)}}{(e^x + e^y)^4}$$

$$= 0$$ **Proved.**

Example 9. *If* $v = (x^2 + y^2 + z^2)^{-1/2}$, *show that* $\frac{\partial^2 v}{\partial x^2} + \frac{\partial^2 v}{\partial y^2} + \frac{\partial^2 v}{\partial z^2} = 0$.

Solution. $v = (x^2 + y^2 + z^2)^{-1/2}$

$$\Rightarrow \frac{\partial v}{\partial x} = -\frac{1}{2}(x^2 + y^2 + z^2)^{-3/2}.2x = -x(x^2 + y^2 + z^2)^{-3/2}$$

$$\Rightarrow \frac{\partial^2 v}{\partial x^2} = (-x)\left(-\frac{3}{2}\right)(x^2 + y^2 + z^2)^{-5/2}.(2x) + (-1).(x^2 + y^2 + z^2)^{-3/2}$$

$$= 3x^2 (x^2 + y^2 + z^2)^{-5/2} - (x^2 + y^2 + z^2)^{-3/2}$$

$$= (x^2 + y^2 + z^2)^{-5/2} [3x^2 - (x^2 + y^2 + z^2)]$$

$$= (x^2 + y^2 + z^2)^{-5/2} (2x^2 - y^2 - z^2) \quad ...(1)$$

Similarly, $$\frac{\partial^2 v}{\partial y^2} = (x^2 + y^2 + z^2)^{-5/2} (2y^2 - z^2 - x^2) \quad ...(2)$$

$$\Rightarrow \frac{\partial^2 v}{\partial z^2} = (x^2 + y^2 + z^2)^{-5/2} (2z^2 - x^2 - y^2) \quad ...(3)$$

On adding (1), (2) and (3), we get

$$\therefore \frac{\partial^2 v}{\partial x^2} + \frac{\partial^2 v}{\partial y^2} + \frac{\partial^2 v}{\partial z^2} = (x^2 + y^2 + z^2)^{-1/2} [2x^2 - y^2 - z^2 + 2y^2 - z^2 - x^2 + 2z^2 - x^2 - y^2$$

$$= (x^2 + y^2 + z^2)^{-5/2} (0) = 0$$ **Proved.**

Example 10. *If* $u = \log\left(\frac{x^2 + y^2}{xy}\right)$, *verify that* $\frac{\partial^2 u}{\partial x\, \partial y} = \frac{\partial^2 u}{\partial y\, \partial x}$ **[B.T.E. Delhi, May 2008]**

Solution. We have, $u = \log\left(\frac{x^2 + y^2}{xy}\right)$

$$\Rightarrow \quad \frac{\partial u}{\partial y} = \frac{1}{\frac{x^2+y^2}{xy}} \frac{\partial}{\partial y}\left(\frac{x^2+y^2}{xy}\right) = \frac{xy}{x^2+y^2}\left[\frac{xy(2y)-(x^2+y^2)(x)}{(xy)^2}\right]$$

$$= \frac{2xy^2 - x^3 - xy^2}{xy(x^2+y^2)} = \frac{xy^2 - x^3}{xy(x^2+y^2)} = \frac{y^2-x^2}{y(x^2+y^2)}$$

$$\Rightarrow \quad \frac{\partial^2 u}{\partial x \partial y} = \frac{\partial}{\partial x}\left[\frac{y^2-x^2}{y(x^2+y^2)}\right] = \frac{y(x^2+y^2)(-2x)-(y^2-x^2)(2xy)}{y^2(x^2+y^2)^2}$$

$$= \frac{-2xy[x^2+y^2+y^2-x^2]}{y^2(x^2+y^2)^2} = \frac{-4xy}{(x^2+y^2)^2} \qquad \text{...(1)}$$

Again, $$\frac{\partial}{\partial} = \frac{1}{\frac{x^2+y^2}{xy}}\left[\frac{xy(2x)-(x^2+y^2)(y)}{(xy)^2}\right]$$

$$= \frac{2x^2y - x^2y - y^3}{xy(x^2+y^2)} = \frac{x^2y - y^3}{xy(x^2+y^2)} = \frac{x^2-y^2}{x(x^2+y^2)}$$

$$\Rightarrow \quad \frac{\partial^2 u}{\partial y \partial x} = \frac{x(x^2+y^2)(-2y)-(x^2-y^2)(2xy)}{x^2(x^2+y^2)^2}$$

$$= \frac{-2xy[x^2+y^2+x^2-y^2]}{x^2(x^2+y^2)^2}$$

$$= \frac{-4x^3y}{x^2(x^2+y^2)} = \frac{-4xy}{(x^2+y^2)^2} \qquad \text{...(2)}$$

From (1) and (2), we have $\frac{\partial^2 u}{\partial x \partial y} = \frac{\partial^2 u}{\partial y \partial x}$

Example 11. *If* $z = \sin^{-1}\left(\frac{x}{y}\right)$, *write the value of* $\frac{\partial z}{\partial y}$ *at (1, 2)* **[S.B.T.E. 2017]**

Solution. Here, we have $$z = \sin^{-1}\left(\frac{x}{y}\right)$$

$$\Rightarrow \quad \frac{\partial z}{\partial y} = \frac{1}{\sqrt{1-\frac{x^2}{y^2}}} \cdot x\left(-\frac{1}{y^2}\right)$$

$$\Rightarrow \quad \left(\frac{\partial z}{\partial y}\right)_{at\,(1,2)} = \frac{1}{\sqrt{1-\frac{1}{4}}}\left(-\frac{1}{4}\right) = -\frac{1}{2\sqrt{3}}$$ **Ans.**

Example 12. *If* $f(x, y) = x^y$ *find* $\frac{\partial^2 f}{\partial y^2}$. **[S.B.T.E. 2015, 2014]**

Solution. Here we have

$$f(x, y) = x^y$$

$$\Rightarrow \quad \frac{\partial f}{\partial y} = x^y.\log x \qquad \left[\because \frac{d}{dx}(a^x) = a^x \log a\right]$$

$$\Rightarrow \quad \frac{\partial^2 f}{\partial y^2} = \log x \frac{\partial}{\partial y}(x^y)$$

$$= \log x.\ x^y.\log x$$

$$\therefore \quad \frac{\partial^2 f}{\partial y^2} = (\log x)^2.x^y \qquad \textbf{Ans.}$$

Example 13. *If* $u = x^y$, *show that* $\frac{\partial^3 u}{\partial x^2 \partial y} = \frac{\partial^3 u}{\partial x \partial y \partial x}$ **[S.B.T.E. 2017, 2016]**

Solution. $u = x^y$, $\log u = \log x^y = y \log x$

Differentiating partially, we get

$$\Rightarrow \quad \frac{1}{u}.\frac{\partial u}{\partial x} = \frac{y}{x} \qquad\qquad \Rightarrow \quad \frac{1}{u}.\frac{\partial u}{\partial y} = \log x$$

$$\Rightarrow \quad \frac{\partial u}{\partial x} = u\frac{y}{x} \qquad\qquad \Rightarrow \quad \frac{\partial u}{\partial y} = u \log x$$

$$\Rightarrow \quad \frac{1}{x}\cdot\left[u + y\cdot\frac{\partial u}{\partial y}\right] = \frac{v}{x} + \frac{y}{x}\cdot\frac{\partial u}{\partial y}$$

$$= \frac{v}{x} + \frac{y}{x}u\log x \qquad \left[\because \frac{\partial u}{\partial y} = u\log x\right]$$

$$\Rightarrow \quad \frac{\partial^3 u}{\partial x \partial y \partial x} = -\frac{u}{x^2} + \frac{1}{x}.\frac{\partial u}{\partial x} + y\left\{\frac{x.\left(\frac{\partial u}{\partial x}.\log x + \frac{u}{x}\right) - u \log x}{x^2}\right\}$$

$$= -\frac{u}{x^2} + \frac{1}{x}.\frac{\partial u}{\partial x} + \frac{y \log x}{x}\frac{\partial u}{\partial x} + \frac{uy}{x^2} - \frac{uy \log x}{x^2}$$

$$= -\frac{u}{x^2} + \frac{uy}{x^2} + \frac{uy^2 \log x}{x^2} + \frac{uy}{x^2} - \frac{uy \log x}{x^2}$$

$$= -\frac{u}{x^2} + \frac{2uy}{x^2} + \frac{uy^2 \log x}{x^2} - \frac{uy \log x}{x^2} \qquad ...(1)$$

$$\Rightarrow \quad \frac{\partial u}{\partial y} = u \log x$$

$$\Rightarrow \quad \frac{\partial^2 u}{\partial x \partial y} = \frac{u}{x} + \log x.\frac{\partial u}{\partial x} = \frac{u}{x} + \log x.\frac{uy}{x}$$

$$\Rightarrow \qquad \frac{\partial^3 y}{\partial x^2 \partial y} = -\frac{u}{x^2} + \frac{1}{x}.\frac{\partial u}{\partial x} + y.\frac{x.\left(\frac{u}{x} + \log x.\frac{\partial u}{\partial x}\right) - u \log x}{x^2}$$

$$= -\frac{u}{x^2} + \frac{uy}{x^2} + \frac{uy}{x^2} + \frac{y \log x}{x}\frac{\partial u}{\partial x} - \frac{uy \log x}{x^2}$$

$$= -\frac{u}{x^2} + \frac{2uy}{x^2} + \frac{y \log x}{x}\frac{uy}{x} - \frac{uy \log x}{x^2}$$

$$= -\frac{u}{x^2} + \frac{2uy}{x^2} + \frac{uy^2 \log x}{x^2} - \frac{uy \log x}{x^2} \qquad ...(2)$$

From (1) and (2), we get

$$\therefore \qquad \frac{\partial^3 u}{\partial x^2 \partial y} = \frac{\partial^3 u}{\partial x \, \partial y \, \partial x}$$ **Proved.**

Example 14. *If $u = \log (x^3 + y^3 + z^3 - 3xyz)$, show that*

$$\left(\frac{\partial}{\partial x} + \frac{\partial}{\partial y} + \frac{\partial}{\partial z}\right)^2 u = \frac{-9}{(x + y + z)^2}$$ **[S.B.T.E. 2017, 2016, 2014]**

Solution. We have, $u = \log (x^3 + y^3 + z^3 - 3xyz)$ ...(1)

Differentiating (1) partially w.r.t. "x", we get

$$\Rightarrow \qquad \frac{\partial u}{\partial x} = \frac{3x^2 - 3yz}{x^3 + y^3 + z^2 - 3xyz} \qquad ...(2)$$

$$\text{Similarly,} \qquad \frac{\partial u}{\partial y} = \frac{3y^2 - 3zx}{x^3 + y^3 + z^3 - 3xyz} \qquad ...(3)$$

$$\Rightarrow \qquad \frac{\partial u}{\partial z} = \frac{3z^2 - 3yx}{x^3 + y^3 + z^3 - 3xyz} \qquad ...(4)$$

On adding (2), 3) and (4), we get

$$\frac{\partial u}{\partial x} + \frac{\partial u}{\partial y} + \frac{\partial u}{\partial z} = \frac{3(x^2 + y^2 + z^2 - xy - yz - zx)}{x^3 + y^3 + z^3 - 3xyz}$$

$$= \frac{3(x^2 + y^2 + z^2 - xy - yz - zx)}{(x + y + z)(x^2 + y^2 + z^2 - xy - yz - zx)}$$

$$= \frac{3}{x + y + z}$$

$$\Rightarrow \qquad \left(\frac{\partial}{\partial x} + \frac{\partial}{\partial y} + \frac{\partial}{\partial z}\right) u = \frac{3}{x + y + z}$$

$$\Rightarrow \qquad \left(\frac{\partial}{\partial x} + \frac{\partial}{\partial y} + \frac{\partial}{\partial z}\right)^2 u = \left(\frac{\partial}{\partial x} + \frac{\partial}{\partial y} + \frac{\partial}{\partial z}\right)\left(\frac{\partial}{\partial x} + \frac{\partial}{\partial y} + \frac{\partial}{\partial z}\right) u$$

$$= \left(\frac{\partial}{\partial x}+\frac{\partial}{\partial y}+\frac{\partial}{\partial z}\right)\frac{3}{x+y+z}$$

$$= \frac{\partial}{\partial x}\frac{3}{x+y+z}+\frac{\partial}{\partial y}\frac{3}{x+y+z}+\frac{\partial}{\partial z}\frac{3}{x+y+z}$$

$$= -3(x+y+z)^{-2}-3(x+y+z)^{-2}-3(x+y+z)^{-2}$$

$$= \frac{-9}{(x+y+z)^2}$$ **Proved.**

30.5 WHICH VARIABLE IS TO BE TREATED AS CONSTANT

Let $x = r\cos\theta,\ y = r\sin\theta$

To find $\frac{\partial r}{\partial x}$, we need a relation between r and x.

$$r = x\sec\theta \quad \ldots(1)$$

and

$$r^2 = x^2+y^2 \quad \ldots(2)$$

Differentiating (1) w.r.t. 'x' keeping θ as constant

$$\frac{\partial r}{\partial x} = \sec\theta \quad \ldots(3)$$

Differentiating (2) w.r.t. "x" keeping y as constant

$$2r\frac{\partial r}{\partial x} = 2x$$

$$\Rightarrow \quad \frac{\partial r}{\partial x} = \frac{x}{r} = \cos\theta \quad \ldots(4)$$

From (3), $\frac{\partial r}{\partial x} = \sec\theta$ and from (4), $\frac{\partial r}{\partial x} = \cos\theta$. These two values of $\frac{\partial r}{\partial x}$ make confusion. To avoid any confusion we use the following notations.

Notation. (*i*) $\left(\frac{\partial r}{\partial x}\right)_\theta$ means partial derivative of r with respect to x, keeping θ as constant.

From (3) $\left(\frac{\partial r}{\partial x}\right)_\theta = \sec\theta$

(*ii*) $\left(\frac{\partial r}{\partial x}\right)_y$ means the partial derivative of r with respect to x keeping y as constant.

From (4) $\left(\frac{\partial r}{\partial x}\right)_y = \cos\theta$

(*iii*) When no indication is given regarding the variable to be treated constant, $\frac{\partial}{\partial x}$ means $\left(\frac{\partial}{\partial x}\right)_y$, $\frac{\partial}{\partial y}$ means $\left(\frac{\partial}{\partial y}\right)_x$, $\frac{\partial}{\partial r}$ means $\left(\frac{\partial}{\partial r}\right)_\theta$, $\frac{\partial}{\partial \theta}$ means $\left(\frac{\partial}{\partial \theta}\right)_r$

Example 15. *If $x = r\cos\theta$, $y = r\sin\theta$, then find*

(i) $\left(\dfrac{\partial x}{\partial r}\right)_\theta$ *(ii)* $\left(\dfrac{\partial y}{\partial \theta}\right)_r$ *(iii)* $\left(\dfrac{\partial r}{\partial x}\right)_y$ *(iv)* $\left(\dfrac{\partial \theta}{\partial y}\right)_x$

Solution. (i) $\left(\dfrac{\partial x}{\partial r}\right)_\theta$ means partial derivative of x with respect to r, taking θ as constant.

$$x = r\cos\theta$$

$$\Rightarrow \quad \left(\frac{\partial x}{\partial r}\right)_\theta = \cos\theta$$

(ii) $\left(\dfrac{\partial y}{\partial \theta}\right)_r$ means partial derivative of y with respect to θ, taking r as constant.

$$y = r\sin\theta$$

$$\Rightarrow \quad \left(\frac{\partial y}{\partial \theta}\right)_r = r\cos\theta$$

(iii) $\left(\dfrac{\partial y}{\partial x}\right)_y$ means partial derivative of r with respect to x, treating y as constant.

We have to express r as a function of x and y, free from θ.

$$r = \sqrt{x^2 + y^2} \qquad \text{(from the given equations)}$$

$$\Rightarrow \quad \left(\frac{\partial r}{\partial x}\right)_y = \frac{1}{2}\frac{1}{\sqrt{x^2+y^2}}.2x = \frac{x}{\sqrt{x^2+y^2}}$$

(iv) Before finding $\left(\dfrac{\partial \theta}{\partial y}\right)_x$ we have to express θ in terms x and y by eliminating r.

$$\theta = \tan^{-1}\left(\frac{y}{x}\right) \qquad \text{(from the given equatioins)}$$

$$\Rightarrow \quad \left(\frac{\partial \theta}{\partial y}\right)_x = \frac{1}{1+\dfrac{y^2}{x^2}}.\frac{1}{x} = \frac{x}{x^2+y^2} \qquad \textbf{Ans.}$$

EXERCISE 30.1

1. Find $\dfrac{\partial z}{\partial x}$, and $\dfrac{\partial z}{\partial y}$ if

(i) $z = x^4 + 6x^3y + 2x^2y^2 + 3xy^3 + 4y^4$

Ans. $\dfrac{\partial z}{\partial x} = 4x^3 + 18x^2y + 4xy^2 + 3y^3$; $\dfrac{\partial z}{\partial y} = 6x^3 + 4x^2y + 9xy^2 + 16y^3$

(ii) $z = e^{ax}\sin by$ **Ans.** $\dfrac{\partial z}{\partial x} = ae^{ax}\sin by, \dfrac{\partial z}{\partial y} = be^{ax}\cos by$

(iii) $z = \tan^{-1}(x + y)$ **Ans.** $\dfrac{\partial z}{\partial x} = \dfrac{1}{1+(x+y)^2}, \dfrac{\partial z}{\partial y} = \dfrac{1}{1+(x+y)^2}$

(iv) $z = \log(x^2 + y^2)$ **Ans.** $\dfrac{\partial z}{\partial x} = \dfrac{2x}{x^2 + y^2}, \dfrac{\partial z}{\partial y} = \dfrac{2y}{x^2 + y^2}$

(v) $z = \dfrac{(x - y)e^{xy}}{x + y}$ **Ans.** $\dfrac{\partial z}{\partial x} = \dfrac{(x^2 - y^2 + 2)ye^{xy}}{(x+y)^2}, \dfrac{\partial z}{\partial y} = \dfrac{(x^2 - y^2 + 2)xe^{xy}}{(x+y)^2}$

2. If $z = yf(x^2 - y^2)$, show that $y\dfrac{\partial z}{\partial x} + x\dfrac{\partial z}{\partial y} = \dfrac{xz}{y}$

3. Show that $x^2\dfrac{\partial^2 z}{\partial x^2} + 2xy\dfrac{\partial^2 z}{\partial x \partial y} + y^2\dfrac{\partial^2 z}{\partial y} = 0$ where $z = x\phi\left(\dfrac{y}{x}\right) + \psi\left(\dfrac{y}{x}\right)$

4. If $u = \log(x^2 + y^2) + \tan^{-1}\left(\dfrac{y}{x}\right)$. Show that $\dfrac{\partial^2 u}{\partial x^2} + \dfrac{\partial^2 u}{\partial y^2} = 0$

5. If $u(x, y, z) = \dfrac{1}{x^2 + y^2 + z^2}$, find value of $\dfrac{\partial^2 u}{\partial x^2} + \dfrac{\partial^2 u}{\partial y^2} + \dfrac{\partial^2 u}{\partial z^2}$.

Ans. $\dfrac{2}{(x^2 + y^2 + z^2)^2}$

6. If $u = x^2 \tan^{-1}\left(\dfrac{y}{x}\right) - y^2 \tan^{-1}\left(\dfrac{x}{y}\right)$, find the value of $\dfrac{\partial^2 u}{\partial x \partial y}$. **Ans.** $\dfrac{x^2 - y^2}{x^2 + y^2}$

7. If $x = e^{r\cos\theta}\cos(r\sin\theta)$ and $y = e^{r\cos\theta}\sin(r\sin\theta)$.

Prove that $\dfrac{\partial x}{\partial r} = \dfrac{1}{r}\dfrac{\partial y}{\partial \theta}, \dfrac{\partial y}{\partial r} = -\dfrac{1}{r}\dfrac{\partial x}{\partial \theta}$.

Hence deduce that $\dfrac{\partial^2 x}{\partial r^2} + \dfrac{1}{r}\dfrac{\partial y}{\partial \theta} + \dfrac{1}{r^2}\dfrac{\partial^2 x}{\partial \theta^2} = 0$

8. If $x = \cos\theta$, $y = r\sin\theta$, prove that $\dfrac{\partial r}{\partial x} = \dfrac{\partial x}{\partial r}, r\dfrac{\partial \theta}{\partial x} = \dfrac{1}{r}\cdot\dfrac{\partial x}{\partial \theta}$.

[Hint: $r = (x^2 + y^2)^{1/2}$, $\theta = \tan^{-1} y/x$]

9. If $x = r\cos\theta$, $y = r\sin\theta$, show that $\dfrac{\partial^2 r}{\partial x^2} + \dfrac{\partial^2 r}{\partial y^2} = \dfrac{1}{r}\left[\left(\dfrac{\partial r}{\partial x}\right)^2 + \left(\dfrac{\partial r}{\partial y}\right)^2\right]$.

30.6 HOMOGENEOUS FUNCTION

A function $f(x, y)$ is said to be homogeneous function in which the power of each term is the same.

A function $f(x, y)$ is a homogeneous function of order n, if the degree of each of its terms in x and y is equal to n. Thus

$$a_0x^n + a_1x^{n-1}y + a_2x^{n-2}y^2 + \dots + a_{n-1}xy^{n-1} + a_ny^n \quad \dots(1)$$

is a homogeneous function of order n.

The polynomial function (1) which can be written as

$$x^n\left[a_0 + a_1\left(\frac{y}{x}\right) + a_2\left(\frac{y}{x}\right)^2 + \ldots + a_{n-1}\left(\frac{y}{x}\right)^{n-1} + a_n\left(\frac{y}{x}\right)^n\right]$$

$$\boxed{f(x, y) = x^n \phi\left(\frac{y}{x}\right)} \qquad \ldots(2)$$

(*i*) The function $x^3\left[1 + \frac{y}{x} + 3\left(\frac{y}{x}\right)^2 + 5\left(\frac{y}{x}\right)^3\right]$ is a homogeneous function of order 3.

(*ii*) $\dfrac{\sqrt{x}+\sqrt{y}}{x^2+y^2} = \dfrac{\sqrt{x}\left[1+\sqrt{\frac{y}{x}}\right]}{x^2\left[1+\left(\frac{y}{x}\right)^2\right]} = x^{-3/2}\dfrac{1+\sqrt{\frac{y}{x}}}{1+\left(\frac{y}{x}\right)^2}$ is a homogeneous function of order –3/2.

(iii) $\sin^{-1}\left(\dfrac{\sqrt{x}+\sqrt{y}}{x^2+y^2}\right)$ is not a homogeneous function as it cannot be written in the form of (*ii*).

30.7 EULER'S THEOREM ON HOMOGENEOUS FUNCTIONS

Statement: If z is a homogeneous function of x, y of order n, then

$$\boxed{x\cdot\frac{\partial z}{\partial x} + y\cdot\frac{\partial z}{\partial y} = nz}$$

Proof. As z is a homogeneous function of x, y of order n.

$\therefore$ z can be written in the form $\qquad z = x^n.f\left(\frac{y}{x}\right) \qquad \ldots(1)$

Differentiating (1) partially w.r.t. x, we have

$$\frac{\partial z}{\partial x} = nx^{n-1}.f\left(\frac{y}{x}\right) + x^n.f'\left(\frac{y}{x}\right)\left(-\frac{y}{x^2}\right)$$

$$\Rightarrow \qquad \frac{\partial z}{\partial x} = nx^{n-1}.f\left(\frac{y}{x}\right) - x^{n-2}y.f'\left(\frac{y}{x}\right)$$

Multiplying both sides by x, we have

$$x\frac{\partial z}{\partial x} = nx^n.f\left(\frac{y}{x}\right) - x^{n-1}y.f'\left(\frac{y}{x}\right) \qquad \ldots(2)$$

Differentiating (1) partially w.r.t. y, we have $\dfrac{\partial z}{\partial y} = x^n f'\left(\dfrac{y}{x}\right)\dfrac{1}{x}$

Multiplying both sides by y, we get $y.\dfrac{\partial z}{\partial y} = x^{n-1}y\, f'\left(\dfrac{y}{x}\right) \qquad \ldots(3)$

Adding (2) and (3), we have $x.\dfrac{\partial z}{\partial x} + y.\dfrac{\partial z}{\partial y} = n\,x^n f\left(\dfrac{y}{x}\right)$

$$\Rightarrow \qquad x.\frac{\partial z}{\partial x} + y.\frac{\partial z}{\partial y} = nz$$ **Proved.**

Example 16. *Write the degree of homogeneous function* $f(x, y) = \dfrac{x^{\frac{1}{4}} + y^{\frac{1}{4}}}{x^{\frac{1}{5}} + y^{\frac{1}{5}}}$ **[S.B.T.E. 2014]**

Solution. Here we have $f(x, y) = \dfrac{x^{\frac{1}{4}} + y^{\frac{1}{4}}}{x^{\frac{1}{5}} + y^{\frac{1}{5}}}$

$$\Rightarrow \qquad f(x, y) = \frac{x^{\frac{1}{4}}\left[1+\left(\frac{y}{x}\right)^{\frac{1}{4}}\right]}{x^{\frac{1}{5}}\left[1+\left(\frac{y}{x}\right)^{\frac{1}{5}}\right]}$$

$$\Rightarrow \qquad f(x, y) = \frac{x^{\frac{1}{20}}\left[1+\left(\frac{y}{x}\right)^{\frac{1}{4}}\right]}{\left[1+\left(\frac{y}{x}\right)^{\frac{1}{5}}\right]} = x^{\frac{1}{20}}\phi\left(\frac{y}{x}\right)$$

$f(x, y)$ is homogeneous function degree $= \dfrac{1}{20}$ **Ans.**

Example 17. *If* $z = \dfrac{x^2y^2}{x + y}$, *find the value of* $x\dfrac{\partial^2 z}{\partial x^2} + y\dfrac{\partial^2 z}{\partial x \partial y} - 2\dfrac{\partial z}{\partial x}$

[S.B.T.E. 2016, 2015, 2014]

Solution. $z = \dfrac{x^2y^2}{x+y} = \dfrac{x^4\left(\frac{y}{x}\right)^2}{x\left[1+\left(\frac{y}{x}\right)\right]} = x^3\,\phi\left(\dfrac{y}{x}\right)$ ...(1)

z is a homogeneous function of degree 3.

By Euler's theorem $\qquad x\dfrac{\partial z}{\partial x} + y\dfrac{\partial z}{\partial y} = 3z$ ...(2)

Differentiating (2) w.r.t. 'x', we get

$$\Rightarrow \qquad \left(x\frac{\partial^2 z}{\partial x^2} + 1.\frac{\partial z}{\partial x}\right) + y\frac{\partial^2 z}{\partial x \partial y} = 3\frac{\partial z}{\partial x}$$

$$\Rightarrow \qquad x\frac{\partial^2 z}{\partial x^2} + y\frac{\partial^2 z}{\partial x \partial y} - 2\frac{\partial z}{\partial x} = 0$$

Example 18. *If* $u = \sin^{-1}\left(\dfrac{x^2 + y^2}{x + y}\right)$ *show that* $x.\dfrac{\partial u}{\partial x} + y.\dfrac{\partial u}{\partial y} = \tan u$

[S.B.T.E. 2016, 2015, 2014]

Solution. $u = \sin^{-1}\left(\dfrac{x^2 + y^2}{x + y}\right)$, here u is not a homogeneous function but if

$$z = \sin u = \frac{x^2 + y^2}{x + y}$$

then z is a homogeneous function of x and y of degree 1.

$\therefore$ **By Euler's theorem** $x.\dfrac{\partial z}{\partial x} + y.\dfrac{\partial z}{\partial y} = 1.z$

$$\Rightarrow \quad x\frac{dz}{du}\frac{\partial u}{\partial x} + y.\frac{dz}{du}\frac{\partial u}{\partial y} = z \quad ...(1)$$

Putting the values of $\dfrac{dz}{du}$ and z in (1), we get

$$\Rightarrow \quad x\cos u\frac{\partial u}{\partial x} + y\cos u\frac{\partial u}{\partial y} = \sin u$$

$$\Rightarrow \quad x\frac{\partial u}{\partial x} + y.\frac{\partial u}{\partial y} = \tan u.$$

Proved.

Example 19. *Given that* $F(u) = V(x, y, z)$ *where* V *is a homogeneous function in* x, y, z *of degree* n*, prove that*

$$\Rightarrow \quad x.\frac{\partial u}{\partial x} + y.\frac{\partial u}{\partial y} + z\frac{\partial u}{\partial z} = n\frac{F(u)}{F'(u)}$$

Solution. $F(u) = V(x, y, z)$ where V is a homogeneous function in x, y, z of degree n.

By Euler's theorem

$$\Rightarrow \quad x.\frac{\partial V}{\partial x} + y.\frac{\partial V}{\partial y} + z.\frac{\partial V}{\partial z} = nV \quad ...(1)$$

Now $\quad V(x, y, z) = F(u)$

$$\Rightarrow \quad \frac{\partial V}{\partial x} = F'(u).\frac{\partial u}{\partial x}$$

$$\Rightarrow \quad \frac{\partial V}{\partial y} = F'(u).\frac{\partial u}{\partial y}$$

$$\Rightarrow \quad \frac{\partial V}{\partial z} = F'(u)\frac{\partial u}{\partial z}$$

Substituting these values in (1), we get

$$F'(u)\frac{\partial u}{\partial x} + y.F'(u)\frac{\partial u}{\partial y} + z.F'(u)\frac{\partial u}{\partial z} = nF(u)$$

$$\therefore \quad \boxed{x.\frac{\partial u}{\partial x}+y.\frac{\partial u}{\partial y}+z.\frac{\partial u}{\partial z}=n\frac{F(u)}{F'(u)}}$$ **Proved.**

Example 20. *If z be a homogeneous function of degree n, then show that*

(i) $x.\dfrac{\partial^2 z}{\partial x^2}+y.\dfrac{\partial^2 z}{\partial x\partial y}=(n-1)\dfrac{\partial z}{\partial x}$; (ii) $x.\dfrac{\partial^2 z}{\partial x\partial y}+y.\dfrac{\partial^2 z}{\partial y^2}=(n-1)\dfrac{\partial z}{\partial y}$;

(iii) $x^2.\dfrac{\partial^2 z}{\partial x^2}+2xy.\dfrac{\partial^2 z}{\partial x\partial y}+y^2.\dfrac{\partial^2 z}{\partial y^2}=n(n-1)z.$

[Euler's theorem on homogeneous function for second derivatives]

Solution. First Method: By Euler's theorem $x.\dfrac{\partial z}{\partial x}+y\dfrac{\partial z}{\partial y}=nz$ ...(1)

Differentiating (1), partially w.r.t. x, we get

$$\Rightarrow \quad \frac{\partial z}{\partial x}+x\frac{\partial^2 z}{\partial x^2}+y\frac{\partial^2 z}{\partial x\partial y} = n\frac{\partial z}{\partial x}$$

(i) $$x.\frac{\partial^2 z}{\partial x^2}+y.\frac{\partial^2 z}{\partial x\partial y} = (n-1)\frac{\partial z}{\partial x}$$ ...(2) **Proved (1)**

Differentiating (1), partially w.r.t. y, we have

$$\Rightarrow \quad x.\frac{\partial^2 z}{\partial y\partial x}+\frac{\partial z}{\partial y}+y.\frac{\partial^2 z}{\partial y^2} = n\frac{\partial z}{\partial y}$$

(ii) $$x.\frac{\partial^2 z}{\partial y\partial x}+y.\frac{\partial^2 z}{\partial y^2} = (n-1)\frac{\partial z}{\partial y}$$ ...(3) **Proved (2)**

Multiplying (2) by x, we have

$$\Rightarrow \quad x^2.\frac{\partial^2 z}{\partial x^2}+xy.\frac{\partial^2 z}{\partial x\partial y} = (n-1)x\frac{\partial z}{\partial x}$$ ...(4)

Multiplying (3) by y, we have

$$\Rightarrow \quad xy.\frac{\partial^2 z}{\partial y\partial x}+y^2.\frac{\partial^2 z}{\partial y^2} = (n-1)y\frac{\partial z}{\partial y}$$ ...(5)

Adding (4) and (5), we get

(iii) $$x^2.\frac{\partial^2 z}{\partial x^2}+2xy.\frac{\partial^2 z}{\partial x\partial y}+y^2.\frac{\partial^2 z}{\partial y^2} = (n-1)\left(x\frac{\partial z}{\partial x}+y\frac{\partial z}{\partial y}\right)$$

$$= (n-1)nz$$ [from (1)]

$$\therefore \quad \boxed{x^2\frac{\partial^2 z}{\partial x^2}+2xy\frac{\partial^2 z}{\partial x\partial y}+y^2\frac{\partial^2 z}{\partial y^2}=n(n-1)z}$$ **Proved (3)**

Second Method

Second order derivatives of homogeneous functions : Since z is a homogeneous function of degree n, $\frac{\partial z}{\partial x}$ and $\frac{\partial z}{\partial y}$ are also homogeneous functions of degree $(n-1)$.

Applying Euler's theorem to the functions $\frac{\partial z}{\partial x}, \frac{\partial z}{\partial y}$, we have

$$\Rightarrow \quad x.\frac{\partial}{\partial x}\left(\frac{\partial z}{\partial x}\right) + y\frac{\partial}{\partial y}\left(\frac{\partial z}{\partial x}\right) = (n-1)\frac{\partial z}{\partial x}$$

$$\Rightarrow \quad x.\frac{\partial^2 z}{\partial x^2} + y.\frac{\partial^2 z}{\partial y\,\partial x} = (n-1)\frac{\partial z}{\partial x} \qquad ...(1)$$

and

$$x.\frac{\partial}{\partial x}\left(\frac{\partial z}{\partial y}\right) + y.\frac{\partial}{\partial y}\left(\frac{\partial z}{\partial y}\right) = (n-1)\frac{\partial z}{\partial y}$$

$$\Rightarrow \quad x.\frac{\partial^2 z}{\partial x\,\partial y} + y.\frac{\partial^2 z}{\partial y^2} = (n-1)\frac{\partial z}{\partial y} \qquad ...(2)$$

Multiply (1) by x and (2) by y and add, taking $\frac{\partial^2 z}{\partial x\,\partial y} = \frac{\partial^2 z}{\partial y\,\partial x}$

$$\Rightarrow \quad x^2.\frac{\partial^2 z}{\partial x^2} + 2xy.\frac{\partial^2 z}{\partial x\,\partial y} + y^2.\frac{\partial^2 z}{\partial y^2} = (n-1)\left[x.\frac{\partial z}{\partial x} + y\frac{\partial z}{\partial y}\right]$$

But by Euler's theorem $x.\frac{\partial z}{\partial x} + y.\frac{\partial z}{\partial y} = nz$

$$\therefore \quad x^2.\frac{\partial^2 z}{\partial^2 x} + 2xy.\frac{\partial^2 z}{\partial x\,\partial y} + y^2.\frac{\partial^2 z}{\partial y^2} = n(n-1)\,z \qquad \textbf{Proved.}$$

Example 21. *If* $u = \tan^{-1}\left(\frac{x^3+y^3}{x-y}\right)$

(i) Prove that $x.\frac{\partial u}{\partial x} + y.\frac{\partial u}{\partial y} = \sin 2u$ *(ii) Find* $x^2.\frac{\partial^2 u}{\partial x^2} + 2xy\frac{\partial^2 u}{\partial x\,\partial y} + y^2\frac{\partial^2 u}{\partial y^2}$.

Solution. Here u is not a homogeneous function. We however write

$$z = \tan u = \frac{x^3+y^3}{x-y} = \left[\frac{x^3\left[1+\left(\frac{y}{x}\right)^3\right]}{x\left[1-\left(\frac{y}{x}\right)\right]}\right] = x^2.\frac{1+\left(\frac{y}{x}\right)^3}{1-\left(\frac{y}{x}\right)}$$

so that z is a homogeneous function of x, y of order 2.

By Euler's Theorem

$$\therefore \quad x\frac{\partial z}{\partial x} + y.\frac{\partial z}{\partial y} = 2z \qquad ...(1)$$

$$\Rightarrow \quad x\frac{dz}{du}\frac{\partial u}{\partial x} + y\frac{dz}{du}\frac{\partial u}{\partial y} = 2z \quad ...(2)$$

Substituting the values of $\frac{dz}{du}$ and z in (2), we obtain

$$\Rightarrow \quad x.\sec^2 u\frac{\partial u}{\partial x} + y.\sec^2 u\frac{\partial u}{\partial y} = 2\tan u$$

Dividing by $\sec^2 u$, we get $x\frac{\partial u}{\partial x} + y.\frac{\partial u}{\partial y} = \frac{2\tan u}{\sec^2 u} = \frac{2\sin u}{\cos u}\frac{\cos^2 u}{1}$

$$(1) \quad x.\frac{\partial u}{\partial x} + y.\frac{\partial u}{\partial y} = \sin 2u \quad ...(3)$$ **Proved.**

Differentiating (3) w.r.t. x, we get

$$\Rightarrow \quad \left(x.\frac{\partial^2 u}{\partial x^2} + 1.\frac{\partial u}{\partial x}\right) + y\frac{\partial^2 u}{\partial x\partial y} = 2\cos 2u.\frac{\partial u}{\partial x} \quad ...(4)$$

Multiplying (4) by x, we have

$$\Rightarrow \quad x^2\frac{\partial^2 u}{\partial x^2} + x\frac{\partial u}{\partial x} + xy\frac{\partial^2 u}{\partial x\partial y} = 2\cos 2u.\left(x\frac{\partial u}{\partial x}\right) \quad ...(5)$$

Differentiating (3) w.r.t. y, we obtain

$$\Rightarrow \quad \left(x\frac{\partial^2 u}{\partial y\partial x}\right) + \left(y\frac{\partial^2 u}{\partial y^2} + 1.\frac{\partial u}{\partial y}\right) = 2\cos 2u.\frac{\partial u}{\partial y} \quad ...(6)$$

Multiplying (6) by y, we have

$$\Rightarrow \quad xy\frac{\partial^2 u}{\partial y\partial x} + y^2\frac{\partial^2 u}{\partial y^2} + y.\frac{\partial u}{\partial y} = 2\cos 2u.\left(y\frac{\partial u}{\partial y}\right) \quad ...(7)$$

Adding (5) and (7), we get

$$x^2\frac{\partial^2 u}{\partial x^2} + 2xy\frac{\partial^2 u}{\partial x\partial y} + y^2\frac{\partial^2 u}{\partial y^2} + \left(x\frac{\partial u}{\partial x} + y\frac{\partial u}{\partial y}\right) = 2\cos 2u\left(x\frac{\partial u}{\partial x}\right) + 2\cos 2u\left(y\frac{\partial u}{\partial y}\right)$$

$$\Rightarrow \quad x^2\frac{\partial^2 u}{\partial x^2} + 2xy\frac{\partial^2 u}{\partial x\partial y} + y^2\frac{\partial^2 u}{\partial y^2} + \sin 2u = 2\cos 2u\left(x\frac{\partial u}{\partial x} + \frac{\partial u}{\partial y}\right) \quad \text{[From (3)]}$$

$$\Rightarrow \quad x^2\frac{\partial^2 u}{\partial x^2} + 2xy\frac{\partial^2 u}{\partial x\partial y} + y^2\frac{\partial^2 u}{\partial y^2} + \sin 2u = 2\cos 2u\,(\sin 2u)$$

$$\Rightarrow \quad x^2\frac{\partial^2 u}{\partial x^2} + 2xy\frac{\partial^2 u}{\partial x\partial y} + y^2\frac{\partial^2 u}{\partial y^2} = 2\cos 2u.\sin 2u - \sin 2u$$

$$= \sin 2u\,(2\cos 2u - 1)$$ **Ans.**

Example 22. *If* $u = \tan^{-1}\left(\dfrac{x^3 + y^3}{x - y}\right)$ *prove that*

(i) $x\dfrac{\partial u}{\partial x} + y\dfrac{\partial u}{\partial y} = \sin 2u$ **[S.B.T.E. 2017, 2015, 2014]**

(ii) $x^2 \dfrac{\partial u^2}{\partial x^2} + 2xy \dfrac{\partial u^2}{\partial x \partial y} + y^2 \dfrac{\partial^2 u}{\partial y^2} = 2\cos 3u \sin u$

Solution. Here u is not a homogeneous function, we write

Let $$z = \tan u = \frac{x^3 + y^3}{x - y} = \frac{x^3\left[1 + \left(\dfrac{y}{x}\right)^3\right]}{x\left[1 - \left(\dfrac{y}{x}\right)\right]}$$

$$= x^2 \frac{\left[1 + \left(\dfrac{y}{x}\right)^3\right]}{\left[1 - \left(\dfrac{y}{x}\right)\right]} = x^2 \phi\left(\frac{y}{x}\right)$$

z is a homogeneous function, of order 2.

(i) By Euler's theorem,

$$\Rightarrow \quad x\frac{\partial u}{\partial x} + y.\frac{\partial u}{\partial y} = n.\frac{f(u)}{f'(u)} \qquad [f(u) = \tan u]$$

$$= \frac{2\tan u}{\sec^2 u} = \frac{2\sin u.\cos^2 u}{\cos u}$$

$$\therefore \quad x.\frac{\partial u}{\partial x} + y.\frac{\partial u}{\partial y} = 2\sin u \cos u = \sin 2u.$$ **Proved.**

(ii) We know that

$$\Rightarrow \quad x^2 \frac{\partial^2 u}{\partial x^2} + 2xy \frac{\partial^2 u}{\partial x \partial y} + y^2 \frac{\partial^2 u}{\partial y^2} = g(u)\,[g'(u) - 1]$$

$$= \sin 2u\,(2\cos 2u - 1)$$

$$= 2\sin 2u \,.\, \cos 2u - \sin 2u$$

$$= \sin 4u - \sin 2u$$

$$= 2\cos\left(\frac{4u + 2u}{2}\right)\sin\left(\frac{4u - 2u}{2}\right)$$

$$\therefore \quad x^2 \frac{\partial^2 u}{\partial x^2} + 2xy \frac{\partial^2 u}{\partial x \partial y} + y^2 \frac{\partial^2 u}{\partial y^2} = 2\cos 3u.\ \sin u$$ **Proved**

EXERCISE 30.2

1. **Verify Euler's theorem in the function relations** **[S.B.T.E. 2015]**
   (i) $f(x, y) = ax^2 + 2hxy + by^2$ (ii) $u = x^3 - 3x^2y + 5xy^2 + y^3$

2. If $v = \dfrac{x^3y^3}{x^3 + y^3}$, show that $x.\dfrac{\partial v}{\partial x} + y.\dfrac{\partial v}{\partial y} = 3v$.

3. If $u = \log\left(\dfrac{x^3 + y^3}{x^2 + y^2}\right)$, prove that $x.\dfrac{\partial u}{\partial x} + y.\dfrac{\partial u}{\partial y} = 1$.

4. If $u = \log\left(\dfrac{x^4 + y^4}{x + y}\right)$, show that $x\dfrac{\partial u}{\partial x} + y\dfrac{\partial u}{\partial y} = 3$. **[GATE. 2020]**

   $\left[\textbf{Hint. } \text{Let } z = e^u = \dfrac{x^4 + y^4}{x + y}\right]$

5. If $u = \sin\left(\dfrac{\sqrt{x} - \sqrt{y}}{\sqrt{x} + \sqrt{y}}\right)$, show that $x.\dfrac{\partial u}{\partial x} + y.\dfrac{\partial u}{\partial y} = 0$.

6. If $u = \sin^{-1}\left(\dfrac{x^3 + y^3}{x + y}\right)$, show that $x.\dfrac{\partial u}{\partial x} + y.\dfrac{\partial u}{\partial y} = 2\tan u$.

7. If $u = \sin^{-1}\left(\dfrac{x + y}{\sqrt{x} + \sqrt{y}}\right)$, prove that **[S.B.T.E. 2015, 2014]**

   (i) $x\dfrac{\partial u}{\partial x} + y\dfrac{\partial u}{\partial y} = \dfrac{1}{2}\tan u$ (ii) $x^2\dfrac{\partial^2 u}{\partial x^2} + 2xy\dfrac{\partial^2 u}{\partial x\partial y} + y^2\dfrac{\partial^2 u}{\partial y^2} = -\dfrac{\sin u \cos 2u}{4\cos^3 u}$.

8. Verify Euler's theorem for the function $u = x^n \sin\left(\dfrac{y}{x}\right)$.

TOTAL DIFFERENTIATION

30.8 TOTAL DIFFERENTIATION

In partial differentiation of a function of two or more variables, only one variable varies. But in total diferentiation, increments are given in all the variables.

30.9 TOTAL DIFFERENTIAL COEFFICIENT

Let
$$z = f(x, y) \quad ...(1)$$
If δx and δy be the icnrements in x and y respectively, let δz be the corresponding increment of z.

Then
$$z + \delta z = f(x + \delta x, y + \delta y) \quad ...(2)$$
Subtracting (1) from (2), we have
$$\delta z = f(x + \delta x, y + \delta y) - f(x, y) \quad ...(3)$$
Adding and subtracting $f(x, y + \delta y)$ on the R.H.S. of (3), we have
$$\delta z = f(x + \delta x, y + \delta y) - f(x, y + \delta y) + f(x, y + \delta y) - f(x, y)$$

$$\Rightarrow \qquad \delta z = \left[\frac{f(x+\delta x, y+\delta y) - f(x, y+\delta y)}{\delta x}\right]\times \delta x + \left[\frac{f(x, y+\delta y) - f(x,y)}{\delta y}\right]\times \delta y$$

On taking limit when $\delta x \to 0$ and $\delta y \to 0$

$$\delta z = \frac{\delta f}{\delta x}.\delta x + \frac{\delta f}{\delta y}.\delta y \qquad \textbf{[Remember]} \ ...(4)$$

dz is called as the total differential of z.

Corollary 1. Differentiation of composite function

If $z = f(x, y)$, where $x = \phi(t)$, $y = \psi(t)$

Here z is a composite function of t.

Dividing (4) by dt, we have $\boxed{\dfrac{dz}{dt} = \dfrac{\partial z}{\partial x}\dfrac{dx}{dt} + \dfrac{\partial z}{\partial y}\dfrac{dy}{dt}}$ **[Remember]** ...(5)

Then $\dfrac{dz}{dt}$ is called the total differential coefficient of z.

Corollary 2. Let $z = f(x, y)$

where $x = \phi(u, v)$

$y = \psi(u, v)$

Then from (5), we obtain

$$\Rightarrow \qquad \frac{\partial z}{\partial u} = \frac{\partial f}{\partial x}.\frac{\partial x}{\partial u} + \frac{\partial f}{\partial y}.\frac{\partial y}{\partial u} \qquad ...(6)$$

and

$$\frac{\partial z}{\partial v} = \frac{\partial f}{\partial x}.\frac{\partial x}{\partial v} + \frac{\partial f}{\partial y}.\frac{\partial y}{\partial v} \qquad ...(7)$$

Example 23. *If* $u = \sin\left(\frac{x}{y}\right), x = e^t, y = t^2$, *prove that* : $\dfrac{du}{dt} = \left(\dfrac{t-2}{t^3}\right)e^t \cos\left(\dfrac{e^t}{t^2}\right)$

(B.T.E Delhi June 2007)

Solution. Here, we have, $u = \sin\left(\frac{x}{y}\right), x = e^t, y = t^2$

$$\Rightarrow \qquad \frac{\partial u}{\partial x} = \cos\left(\frac{x}{y}\right).\frac{1}{y}, \frac{\partial u}{\partial y} = \cos\left(\frac{x}{y}\right)\left(-\frac{x}{y^2}\right); \frac{dx}{dt} = e^t, \frac{dy}{dt} = 2t$$

We know that,

$$\frac{du}{dt} = \frac{\partial u}{\partial x}\frac{dx}{dt} + \frac{\partial u}{\partial y}\frac{dy}{dt}$$

$$= \left[\cos\left(\frac{x}{y}\right).\frac{}{y}\right]e + \cos\left(\frac{x}{y}\right)\left(-\frac{x}{}\right)2t$$

$$= \left[\cos\left(\frac{e^t}{t^2}\right)\frac{1}{t^2}\right]e^t + \cos\left(\frac{e^t}{t^2}\right)\left(-\frac{e^t}{t^4}\right)2t,$$

$$= \cos\left(\frac{e^t}{t^2}\right)\frac{e^t}{t^2} + \cos\left(\frac{e^t}{t^2}\right)\left(-\frac{e^t}{t^3}\right)2$$

$$= \cos\left(\frac{e^t}{t^2}\right)\left[\frac{e^t}{t^2} - \frac{2e^t}{t^3}\right]$$

$$= \cos\left(\frac{e^t}{t^2}\right)e^t\left[\frac{1}{t^2} - \frac{2}{t^3}\right]$$

$$= \left(\frac{t-2}{t^3}\right)e^t\cos\left(\frac{e^t}{t^2}\right)$$ **Proved.**

Example 24. *If $u = x^3 + y^3$ where, $x = a\cos t$, $y = b\sin t$, find $\frac{du}{dt}$ and verify the result.* **(S.B.T.E. 2015)**

Solution. $u = x^3 + y^3$, $x = a\cos t$, $y = b\sin t$

$$\frac{du}{dt} = \frac{\partial u}{\partial x}\frac{dx}{dt} + \frac{\partial u}{\partial y}\frac{dy}{dt}$$

$$= (3x^2)(-a\sin t) + (3y^2)(b\cos t)$$

$$= -3a^3\cos^2 t\sin t + 3b^3\sin^2 t\cos t$$

Verification,

$$u = x^3 + y^3$$

$$= a^3\cos^3 t + b^3\sin^3 t$$

$$\frac{du}{dt} = -3a^3\cos^2 t\,.\,\sin t + 3b^3\sin^2 t\cos t.$$ **Proved.**

Example 25. *If $u = x^2 + y^2$ and $x = s + 3t$, $y = 2s - t$, find $\frac{\partial^2 u}{\partial s^2}, \frac{\partial^2 u}{\partial t^2}$.*

Solution.

$$u = x^2 + y^2$$

$$\frac{\partial u}{\partial x} = 2x, \quad \frac{\partial u}{\partial y} = 2y$$

We know that

$$\Rightarrow \quad \frac{\partial u}{\partial s} = \frac{\partial u}{\partial x}\frac{\partial x}{\partial s} + \frac{\partial u}{\partial y}\frac{\partial y}{\partial s} = (2x)(1) + (2y)(2) = 2x + 4y$$

$$\Rightarrow \quad \frac{\partial^2 u}{\partial s^2} = \frac{\partial}{\partial s}\left(\frac{\partial u}{\partial s}\right) = \frac{\partial}{\partial s}(2x + 4y) = 2\frac{\partial x}{\partial s} + 4\frac{\partial y}{\partial s} = (2\times 1) + (4\times 2) = 10$$

$$\Rightarrow \quad \frac{\partial u}{\partial t} = \frac{\partial u}{\partial x}\frac{\partial x}{\partial t} + \frac{\partial u}{\partial y}\frac{\partial y}{\partial t} = 2x\times 3 + 2y(-1) = 6x - 2y$$

$$\Rightarrow \quad \frac{\partial^2 u}{\partial t^2} = \frac{\partial}{\partial t}\left(\frac{\partial u}{\partial t}\right) = \frac{\partial}{\partial t}(6x - 2y) = 6\frac{\partial x}{\partial t} - 2\frac{\partial y}{\partial t}$$

$$= 6\times(3) + (-2)(-1) = 20$$ **Ans.**

Example 26. *If $w = f(x, y)$, $x = r \cos \theta$, $y = r \sin \theta$, show that*

$$\left(\frac{\partial w}{\partial r}\right)^2 + \frac{1}{r^2}\left(\frac{\partial w}{\partial \theta}\right)^2 = \left(\frac{\partial f}{\partial x}\right)^2 + \left(\frac{\partial f}{\partial y}\right)^2$$

Solution. Since, $x = r \cos \theta$, $y = r \sin \theta$

$$\Rightarrow \quad \frac{\partial x}{\partial r} = \cos\theta, \frac{\partial y}{\partial r} = \sin\theta$$

$$\Rightarrow \quad \frac{\partial x}{\partial \theta} = -r\sin\theta, \frac{\partial y}{\partial \theta} = r\cos\theta$$

Now
$$\frac{\partial w}{\partial r} = \frac{\partial f}{\partial x}\frac{\partial x}{\partial r} + \frac{\partial f}{\partial y}\frac{\partial y}{\partial r}$$

$$= \frac{\partial f}{\partial x}(\cos\theta) + \frac{\partial f}{\partial y}(\sin\theta) \qquad \text{...(1)}$$

$$\Rightarrow \quad \frac{\partial w}{\partial \theta} = \frac{\partial f}{\partial x}\frac{\partial x}{\partial \theta} + \frac{\partial f}{\partial y}\frac{\partial y}{\partial \theta}$$

$$= \frac{\partial f}{\partial x}(-r\sin\theta) + \frac{\partial f}{\partial y}(r\cos\theta)$$

$$\Rightarrow \quad \frac{1}{r}\frac{\partial w}{\partial \theta} = -\frac{\partial f}{\partial x}\sin\theta + \frac{\partial f}{\partial y}\cos\theta \qquad \text{...(2)}$$

Squaring (1) and (2) and adding, we obtain

$$\Rightarrow \quad \left(\frac{\partial w}{\partial r}\right)^2 + \frac{1}{r^2}\left(\frac{\partial w}{\partial \theta}\right)^2 = \left(\frac{\partial f}{\partial x}\right)^2 + \left(\frac{\partial f}{\partial y}\right)^2$$

Proved.

Example 27. *If $u = f(y - z, z - x, x - y)$, prove that:* $\frac{\partial u}{\partial x} + \frac{\partial u}{\partial y} + \frac{\partial u}{\partial z} = 0.$

Solution. Let $r = y - z$, $s = z - x$, $t = x - y$ so that

$$u = f(r, s, t)$$

$$\Rightarrow \quad \frac{\partial u}{\partial x} = \frac{\partial u}{\partial r}\frac{\partial r}{\partial x} + \frac{\partial u}{\partial s}\frac{\partial s}{\partial x} + \frac{\partial u}{\partial t}\frac{\partial t}{\partial x}$$

$$= \frac{\partial u}{\partial r}(0) + \frac{\partial u}{\partial s}(-1) + \frac{\partial u}{\partial t}(1) = -\frac{\partial u}{\partial s} + \frac{\partial u}{\partial t} \qquad \text{...(1)}$$

$$\Rightarrow \quad \frac{\partial u}{\partial y} = \frac{\partial u}{\partial r}\frac{\partial r}{\partial y} + \frac{\partial u}{\partial s}\frac{\partial s}{\partial y} + \frac{\partial u}{\partial t}\frac{\partial t}{\partial y} = \frac{\partial u}{\partial r}(1) + \frac{\partial u}{\partial s}(0) + \frac{\partial u}{\partial t}(-1)$$

$$= \frac{\partial u}{\partial r} - \frac{\partial u}{\partial t} \qquad \text{...(2)}$$

$$\Rightarrow \quad \frac{\partial u}{\partial z} = \frac{\partial u}{\partial r}\frac{\partial r}{\partial z} + \frac{\partial u}{\partial s}\frac{\partial s}{\partial z} + \frac{\partial u}{\partial t}\frac{\partial t}{\partial z} = \frac{\partial u}{\partial r}(-1) + \frac{\partial u}{\partial s}(1) + \frac{\partial u}{\partial t}(0)$$

$$= -\frac{\partial u}{\partial r} + \frac{\partial u}{\partial s} \qquad \text{...(3)}$$

Adding (1), (2) and (3), we get

$$\frac{\partial u}{\partial x}+\frac{\partial u}{\partial y}+\frac{\partial u}{\partial z} = 0$$ **Proved.**

Example 28. *If* $u = e^{xyz}$ *find* $\frac{\partial^3 u}{\partial x\,\partial y\,\partial z}$ **[S.B.T.E. 2015, 2014, 2010]**

Solution. Here we have $u = e^{xyz}$

$$\Rightarrow \quad \frac{\partial u}{\partial z} = xye^{xyz}, \frac{\partial^2 u}{\partial y\,\partial z} = x\left[e^{xyz} + xyze^{xyz}\right]$$

$$\Rightarrow \quad \frac{\partial^2 u}{\partial y\,\partial z} = x\left[e^{xyz} + xyze^{xyz}\right]$$

$$\Rightarrow \quad \frac{\partial^3 u}{\partial x\,\partial y\,\partial z} = \frac{\partial}{\partial x}\left[x.e^{xyz} + x^2yze^{xyz}\right]$$

$$= \frac{\partial}{\partial x}(x.e^{xyz}) + \frac{\partial}{\partial x}\left(x^2.yze^{xyz}\right)$$

$$= e^{xyz} + xyze^{xyz} + 2xyze^{xyz} + x^2y^2z^2e^{xyz}$$

$$\therefore \quad \frac{\partial^3 u}{\partial x\,\partial y\,\partial z} = e^{xyz}\left(1 + 3xyz + x^2y^2z^2\right)$$ **Ans.**

Example 29. *If* $u = u\left(\frac{y-x}{xy}, \frac{z-x}{xz}\right)$*, show that* $x^2\frac{\partial u}{\partial x} + y^2\frac{\partial u}{\partial y} + z^2\frac{\partial u}{\partial z} = 0$

Solution. We have, $u = u\left(\frac{y-x}{xy}, \frac{z-x}{zx}\right) = u(r, s)$

where $r = \frac{y-x}{xy}, s = \frac{z-x}{zx}$

$$= \frac{1}{x} - \frac{1}{y}, s = \frac{1}{x} - \frac{1}{z}$$

$$\Rightarrow \quad \frac{\partial r}{\partial x} = -\frac{1}{x^2}, \frac{\partial s}{\partial x} = -\frac{1}{x^2}$$

$$\Rightarrow \quad \frac{\partial r}{\partial y} = \frac{1}{y^2}, \frac{\partial s}{\partial z} = \frac{1}{z^2}$$

$$\Rightarrow \quad \frac{\partial r}{\partial z} = 0, \frac{\partial s}{\partial y} = 0$$

We know that $\frac{\partial u}{\partial x} = \frac{\partial u}{\partial r}\frac{\partial r}{\partial x} + \frac{\partial u}{\partial s}\frac{\partial s}{\partial x} = \frac{\partial u}{\partial r}\left(-\frac{1}{x^2}\right) + \frac{\partial u}{\partial s}\left(-\frac{1}{x^2}\right)$

$$= -\frac{1}{x^2}\frac{\partial u}{\partial r} - \frac{1}{x^2}\frac{\partial u}{\partial s}$$

$$\Rightarrow \qquad x^2 \frac{\partial u}{\partial x} = -\frac{\partial u}{\partial r} - \frac{\partial u}{\partial s} \qquad ...(1)$$

$$\frac{\partial u}{\partial y} = \frac{\partial u}{\partial r}\frac{\partial r}{\partial y} + \frac{\partial u}{\partial s}\frac{\partial s}{\partial y} = \frac{\partial u}{\partial r}\frac{1}{y^2} + \frac{\partial u}{\partial s} \times 0 = \frac{1}{y^2}\frac{\partial u}{\partial r}$$

$$\Rightarrow \qquad y^2 \frac{\partial u}{\partial y} = \frac{\partial u}{\partial r} \qquad ...(2)$$

$$\frac{\partial u}{\partial z} = \frac{\partial u}{\partial r}\frac{\partial r}{\partial z} + \frac{\partial u}{\partial s}\frac{\partial s}{\partial z} = \frac{\partial u}{\partial r} \times 0 + \frac{\partial u}{\partial s} \times \frac{1}{z^2} = \frac{1}{z^2}\frac{\partial u}{\partial s}$$

$$\Rightarrow \qquad z^2 \frac{\partial u}{\partial z} = \frac{\partial u}{\partial s} \qquad ...(3)$$

On adding (1), (2) and (3), we obtain

$$x^2 \frac{\partial u}{\partial x} + y^2 \frac{\partial u}{\partial y} + z^2 \frac{\partial u}{\partial z} = 0$$ **Proved.**

Example 30. *If* $\phi\,(cx - az,\, cy - bz) = 0$, *show that* $ap + bq = c$

where $p = \dfrac{\partial z}{\partial x}$ *and* $q = \dfrac{\partial z}{\partial y}$

Solution. $\phi\,(cx - az,\, cy - bz) = 0$

$$\phi\,(r, s) = 0$$

where $r = cx - az,\ s = cy - bz$

$$\frac{\partial r}{\partial x} = c - a\frac{\partial z}{\partial x},\ \frac{\partial r}{\partial y} = a\frac{\partial z}{\partial y}$$

$$\Rightarrow \qquad \frac{\partial s}{\partial x} = -b\,\frac{\partial z}{\partial x},\ \frac{\partial s}{\partial y} = c - b\,\frac{\partial z}{\partial y}$$

We know that

$$\frac{\partial \phi}{\partial x} = \frac{\partial \phi}{\partial r}\frac{\partial r}{\partial x} + \frac{\partial \phi}{\partial s}\frac{\partial s}{\partial x}$$

$$\Rightarrow \qquad 0 = \frac{\partial \phi}{\partial r}\left(c - a\,\frac{\partial z}{\partial x}\right) + \frac{\partial \phi}{\partial s}\left(-b\,\frac{\partial z}{\partial x}\right)$$

$$\Rightarrow \qquad 0 = c\,\frac{\partial \phi}{\partial r} + \frac{\partial z}{\partial x}\left(-a\,\frac{\partial \phi}{\partial r} - b\,\frac{\partial \phi}{\partial s}\right)$$

$$\Rightarrow \qquad a\,\frac{\partial z}{\partial x} = \frac{ac\,\dfrac{\partial \phi}{\partial r}}{a\,\dfrac{\partial \phi}{\partial r} + b\,\dfrac{\partial \phi}{\partial s}} \qquad ...(1)$$

Again

$$\frac{\partial \phi}{\partial y} = \frac{\partial \phi}{\partial r}\frac{\partial \phi}{\partial y} + \frac{\partial \phi}{\partial s}\frac{\partial s}{\partial y}$$

$$\Rightarrow \qquad 0 = \frac{\partial \phi}{\partial r}\left(-a\,\frac{\partial z}{\partial y}\right) + \frac{\partial \phi}{\partial s}\left(c - b\,\frac{\partial z}{\partial y}\right)$$

$$\Rightarrow \qquad 0 = c\frac{\partial\phi}{\partial s} - \frac{\partial z}{\partial y}\left(a\frac{\partial\phi}{\partial r} + b\frac{\partial\phi}{\partial s}\right)$$

$$\Rightarrow \qquad b\frac{\partial z}{\partial y} = \frac{bc\dfrac{\partial\phi}{\partial s}}{a\dfrac{\partial\phi}{\partial r} + b\dfrac{\partial\phi}{\partial s}} \qquad \text{...(2)}$$

Adding (1) and (2), we get

$$\Rightarrow \qquad a\frac{\partial z}{\partial x} + b\frac{\partial z}{\partial y} = \frac{ac\dfrac{\partial\phi}{\partial r} + bc\dfrac{\partial\phi}{\partial s}}{a\dfrac{\partial\phi}{\partial r} + b\dfrac{\partial\phi}{\partial s}}$$

$$\Rightarrow \qquad a\frac{\partial z}{\partial x} + b\frac{\partial z}{\partial y} = c \Rightarrow ap + bq = c$$ **Proved.**

Example 31. *A function $f(x, y)$ is rewritten in terms of new variables ξ and η given by $\xi = x\cos\alpha + y\sin\alpha$, $\eta = -x\sin\alpha + y\cos\alpha$*

Show that $\dfrac{\partial^2 f}{\partial x^2} + \dfrac{\partial^2 f}{\partial y^2} = \dfrac{\partial^2 f}{\partial \xi^2} + \dfrac{\partial^2 f}{\partial \eta^2}$.

Solution.

$$\xi = x\cos\alpha + y\sin\alpha$$

$$\eta = -x\sin\alpha + y\cos\alpha$$

$$\Rightarrow \qquad \frac{\partial f}{\partial x} = \frac{\partial f}{\partial \xi}\frac{\partial \xi}{\partial x} + \frac{\partial f}{\partial \eta}\frac{\partial \eta}{\partial x}$$

$$\Rightarrow \qquad \frac{\partial f}{\partial x} = \frac{\partial f}{\partial \xi}\cos\alpha + \frac{\partial f}{\partial \eta}(-\sin\alpha)$$

$$\Rightarrow \qquad \frac{\partial^2 f}{\partial x^2} = \frac{\partial}{\partial x}\left(\frac{\partial f}{\partial x}\right) = \frac{\partial}{\partial x}\left(\cos\alpha\frac{\partial f}{\partial \xi} - \sin\alpha\frac{\partial f}{\partial \eta}\right)$$

$$= \cos\alpha\frac{\partial^2 f}{\partial \xi^2}\frac{\partial \xi}{\partial x} - \sin\alpha\frac{\partial^2 f}{\partial \eta^2}\frac{\partial \eta}{\partial x}$$

$$= \cos\alpha\frac{\partial^2 f}{\partial \xi^2}(\cos\alpha) - \sin\alpha\frac{\partial^2 f}{\partial \eta^2}(-\sin\alpha)$$

$$= \cos^2\alpha\frac{\partial^2 f}{\partial \xi^2} + \sin^2\alpha\frac{\partial^2 f}{\partial \eta^2} \qquad \text{...(1)}$$

$$\Rightarrow \qquad \frac{\partial f}{\partial y} = \frac{\partial f}{\partial \xi}\frac{\partial \xi}{\partial y} + \frac{\partial f}{\partial \eta}\frac{\partial \eta}{\partial y} = \frac{\partial f}{\partial \xi}\sin\alpha + \frac{\partial f}{\partial \eta}\cos\alpha$$

$$\Rightarrow \qquad \frac{\partial^2 f}{\partial y^2} = \frac{\partial}{\partial y}\left(\frac{\partial f}{\partial y}\right) = \frac{\partial}{\partial y}\left(\sin\alpha\frac{\partial f}{\partial \xi} + \cos\alpha\frac{\partial f}{\partial \eta}\right)$$

$$= \sin\alpha\frac{\partial^2 f}{\partial \xi^2}\frac{\partial \xi}{\partial y} + \cos\alpha\frac{\partial^2 f}{\partial \eta^2}\frac{\partial \eta}{\partial y}$$

$$= \sin\alpha \frac{\partial^2 f}{\partial \xi^2}(\sin\alpha) + \cos\alpha \frac{\partial^2 f}{\partial \eta^2}(\cos\alpha)$$

$$= \sin^2\alpha \frac{\partial^2 f}{\partial \xi^2} + \cos^2\alpha \frac{\partial^2 f}{\partial \eta^2} \quad \text{...(2)}$$

Adding (1) and (2), we get

$$\Rightarrow \quad \frac{\partial^2 f}{\partial x^2} + \frac{\partial^2 f}{\partial y^2} = (\sin^2\alpha + \cos^2\alpha)\frac{\partial^2 f}{\partial \xi^2} + (\sin^2\alpha + \cos^2\alpha)\frac{\partial^2 f}{\partial \eta^2}$$

$$= \frac{\partial^2 f}{\partial \xi^2} + \frac{\partial^2 f}{\partial \eta^2}$$

Proved.

Example 32. *If $u = f(r)$ and $x = r\cos\theta$, $y = r\sin\theta$.*

Prove that: $\dfrac{\partial^2 u}{\partial x^2} + \dfrac{\partial^2 u}{\partial y^2} = f''(r) + \dfrac{1}{r}f'(r)$

Solution.

$$u = f(r)$$

$$x = r\cos\theta \quad \text{...(1)}$$

$$y = r\sin\theta \quad \text{...(2)}$$

Squaring and adding (1) and (2), we get

$$r^2 = x^2 + y^2 \text{ so that } \frac{\partial r}{\partial x} = \frac{x}{r}$$

$$\Rightarrow \quad \frac{\partial u}{\partial x} = \frac{df}{dr}.\frac{\partial r}{\partial x}$$

$$\Rightarrow \quad \frac{\partial u}{\partial x} = \frac{df}{dr}.\frac{x}{r}$$

Differentiating again w.r.t. x, we get

$$\Rightarrow \quad \frac{\partial^2 u}{\partial x^2} = \left(\frac{d^2 f}{dr^2}\frac{\partial r}{\partial x}\right).\frac{x}{r} + \frac{df}{dr}\left[\frac{r.1 - x\frac{\partial r}{\partial x}}{r^2}\right]$$

$$= \left(\frac{d^2 f}{dr^2}\frac{x}{r}\right).\frac{x}{r} + \frac{df}{dr}\left[\frac{r.1 - x.\frac{x}{r}}{r^2}\right]$$

$$= \frac{d^2 f}{dr^2}\frac{x^2}{r^2} + \frac{df}{dr}\frac{r^2 - x^2}{r^3} = \frac{d^2 f}{dr^2}\frac{x^2}{r^2} + \frac{df}{dr}\frac{y^2}{r^3} \quad \text{...(3)}$$

Similarly,

$$\frac{\partial^2 u}{\partial y^2} = \frac{d^2 f}{dr^2}\frac{y^2}{r^2} + \frac{df}{dr}\frac{x^2}{r^3} \quad \text{...(4)}$$

On adding (3) and (4), we get

$$\Rightarrow \quad \frac{\partial^2 u}{\partial x^2} + \frac{\partial^2 u}{\partial y^2} = \frac{d^2 f}{dr^2}\frac{x^2 + y^2}{r^2} + \frac{df}{dr}\frac{x^2 + y^2}{r^3} = \frac{d^2 f}{dr^2} + \frac{df}{dr}\frac{1}{r}$$

$$= f''(r) + \frac{1}{r} f'(r)$$ **Proved.**

Example 33. *If z be a function of x and y, and u and v be two other variables such that*

$$u = lx + my$$
$$v = ly - mx$$

show that $$\frac{\partial^2 z}{\partial x^2} + \frac{\partial^2 z}{\partial y^2} = (l^2 + m^2)\left(\frac{\partial^2 z}{\partial u^2} + \frac{\partial^2 z}{\partial v^2}\right)$$

Solution. $$z = f(u, v) \quad \text{... (1)}$$

$$u = lx + my, \quad \frac{\partial u}{\partial x} = l, \frac{\partial u}{\partial y} = m$$

$$v = ly - mx, \quad \frac{\partial v}{\partial x} = -m, \frac{\partial v}{\partial y} = l$$

Differentiating (1) partially w.r.t. x, we get

$$\Rightarrow \quad \frac{\partial z}{\partial x} = \frac{\partial z}{\partial u}\frac{\partial u}{\partial x} + \frac{\partial z}{\partial v}\frac{\partial v}{\partial x}$$

$$\Rightarrow \quad \frac{\partial z}{\partial x} = \frac{\partial z}{\partial u} l + \frac{\partial z}{\partial v}(-m) \Rightarrow \frac{\partial}{\partial x} = \left(l\frac{\partial}{\partial u} - m\frac{\partial}{\partial v}\right)$$

$$\Rightarrow \quad \frac{\partial^2 z}{\partial x^2} = \frac{\partial}{\partial x}\left(\frac{\partial z}{\partial x}\right) = \left(l\frac{\partial}{\partial u} - m\frac{\partial}{\partial v}\right)\left(l\frac{\partial z}{\partial u} - \frac{\partial z}{\partial v}\right)$$

$$= l\frac{\partial}{\partial u}\left(l\frac{\partial z}{\partial u} - m\frac{\partial z}{\partial v}\right) - m\frac{\partial}{\partial v}\left(l\frac{\partial z}{\partial u} - m\frac{\partial z}{\partial v}\right)$$

$$= l^2\frac{\partial^2 z}{\partial u^2} - lm\frac{\partial^2 z}{\partial u \partial v} - ml\frac{\partial^2 z}{\partial v \partial u} + m^2\frac{\partial^2 v}{\partial v^2}$$

$$= l^2\frac{\partial^2 z}{\partial u^2} - 2lm\frac{\partial^2 z}{\partial u \partial v} + m^2\frac{\partial^2 z}{\partial v^2} \quad \text{...(2)}$$

Now let us differentiate (1) partially w.r.t. 'y' and we get

$$\Rightarrow \quad \frac{\partial z}{\partial y} = \frac{\partial z}{\partial u}\frac{\partial u}{\partial y} + \frac{\partial z}{\partial v}\frac{\partial v}{\partial y}$$

$$\Rightarrow \quad \frac{\partial z}{\partial y} = \frac{\partial z}{\partial u} m + \frac{\partial z}{\partial v} l\,; \quad \frac{\partial}{\partial y} = m\frac{\partial}{\partial u} + l\frac{\partial}{\partial v}$$

$$\Rightarrow \quad \frac{\partial^2 z}{\partial y^2} = \frac{\partial}{\partial y}\left(\frac{\partial z}{\partial y}\right) = \left(m\frac{\partial}{\partial u} + l\frac{\partial}{\partial v}\right)\left(m\frac{\partial z}{\partial u} + l\frac{\partial z}{\partial v}\right)$$

$$= m\frac{\partial}{\partial u}\left(m\frac{\partial z}{\partial u} + l\frac{\partial z}{\partial v}\right) + l\frac{\partial}{\partial v}\left(m\frac{\partial z}{\partial u} + l\frac{\partial z}{\partial v}\right)$$

$$= m^2\frac{\partial^2 z}{\partial u^2} + ml\frac{\partial^2 z}{\partial u \partial v} + lm\frac{\partial^2 z}{\partial v \partial u} + l^2\frac{\partial^2 z}{\partial v^2}$$

$$= m^2 \frac{\partial^2 z}{\partial u^2} + 2ml \frac{\partial^2 z}{\partial u \partial v} + l^2 \frac{\partial^2 z}{\partial v^2} \qquad ...(3)$$

Adding (2) and (3), we obtain

$$\Rightarrow \quad \frac{\partial^2 z}{\partial x^2} + \frac{\partial^2 z}{\partial y^2} = (l^2 + m^2)\frac{\partial^2 z}{\partial u^2} + (l^2 + m^2)\frac{\partial^2 z}{\partial v^2}$$

$$= (l^2 + m^2)\left(\frac{\partial^2 z}{\partial u^2} + \frac{\partial^2 z}{\partial v^2}\right)$$ **Proved.**

EXERCISE 30.3

1. If $z = \sin^{-1}(x - y)$, $x = 3t$, $y = 4t^3$, show that $\frac{dz}{dt} = \frac{3}{\sqrt{1-t^2}}$

2. Find $\frac{dz}{dt}$ when $z = xy^2 + x^2y$, $x = at^2$, $y = 2at$
verify by direct substitution. **Ans.** $2a^3t^3(8 + 5t)$

3. If $u = xe^y z$ where $y = \sqrt{a^2 - x^2}$, $z = \sin^3 x$, find $\frac{du}{dx}$ **Ans.** $e^y z\left(1 - \frac{x^2}{y} + 3x\cot x\right)$

4. If $x^2 + y^2 + z^2 - 2xyz = 1$, show that

$$\frac{dx}{\sqrt{1-x^2}} + \frac{dy}{\sqrt{1-y^2}} + \frac{dz}{\sqrt{1-z^2}} = 0$$

[**Hint.** $x^2 + y^2 + z^2 - 2xyz = 1$, $y^2 - 2xyz = 1 - x^2 - z^2$
$y^2 - 2xyz + x^2z^2 = 1 - x^2 - z^2 + x^2z^2$
$(y - xz)^2 = (1 - x^2)(1 - z^2)$]

5. If $z = f(x, y)$, where $x = e^u \cos v$, $y = e^u \sin v$

show that $\left(\frac{\partial z}{\partial u}\right)^2 + \left(\frac{\partial z}{\partial v}\right)^2 = e^{2u}\left[\left(\frac{\partial z}{\partial x}\right)^2 + \left(\frac{\partial z}{\partial y}\right)^2\right]$ **[S.B.T.E. 2017] Ans.** 7

6. If $z = z(u, v)$, $u = x^2 - 2xy - y^2$ and $v = y$. Show that

$(x + y)\frac{\partial z}{\partial x} + (x - y)\frac{\partial z}{\partial y} = 0$ is equivalent to $\frac{\partial z}{\partial v} = 0$

7. If $u = f(x^2 + 2yz, y^2 + 2zx)$, prove that $(y^2 - zx)\frac{\partial u}{\partial x} + (x^2 - yz)\frac{\partial u}{\partial y} + (z^2 - xy)\frac{\partial u}{\partial z} = 0$

8. By changing the independent variables x and t to u and v by means of the relationships

$u = x - at$, $v = x + at$ show that $a^2\frac{\partial^2 y}{\partial x^2} - \frac{\partial^2 y}{\partial t^2} = 4a^2\frac{\partial^2 y}{\partial u \partial v}$.

9. If $x^2 = au + bv$, $y = au - bv$, prove that: $\left(\frac{\partial u}{\partial x}\right)_y \left(\frac{\partial x}{\partial u}\right)_v = \frac{1}{2} = \left(\frac{\partial v}{\partial y}\right)_x \left(\frac{\partial y}{\partial v}\right)_u$

10. If $z = f(x, y)$ where $x = uv$, $y = \frac{u+v}{u-v}$, show that $2x\frac{\partial z}{\partial x} = u\frac{\partial z}{\partial u} + \frac{\partial z}{\partial v}$

11. If $u = x\cos\frac{y}{z}, x = 3r^2 + 2s, y = 4r - 2x^3, z = 2r^2 - 3s^2$ find $\frac{\partial u}{\partial r}, \frac{\partial u}{\partial s}$

Ans. $\frac{\partial u}{\partial r} = 6r\cos\frac{y}{z} - \frac{4x}{z}\sin\frac{y}{z} + \frac{4xyr}{z^2}\sin\frac{y}{2}, \frac{\partial u}{\partial s} = 2\cos\frac{y}{z} + \frac{6xs^2}{z}\sin\frac{y}{z} - \frac{6xys}{z^2}\sin\frac{y}{z}$

12. If $z = f(x, y)$, where $x = e^u \cos v, y = e^u \sin v$.

Prove that: $\left(\frac{\partial f}{\partial x}\right)^2 + \left(\frac{\partial f}{\partial y}\right)^2 = e^{-2u}\left[\left(\frac{\partial f}{\partial u}\right)^2 + \left(\frac{\partial f}{\partial v}\right)^2\right]$

13. If $x = \frac{\cos\theta}{u}, y = \frac{\sin\theta}{u}$ and $z = f(x, y)$, then show that

$$\frac{\partial^2 z}{\partial x^2} + \frac{\partial^2 z}{\partial y^2} = u^2\frac{\partial^2 z}{\partial u^2} + u^4\frac{\partial^2 z}{\partial \theta^2}$$

14. If $u = z\sin^{-1}\left(\frac{y}{x}\right)$ where $x = 3r^2 + 2s, y = 4r - 2s^3, z = 2r^2 - 3s^2$ find $\frac{\partial u}{\partial r}, \frac{\partial u}{\partial s}$.

Ans. $\frac{\partial u}{\partial r} = -\frac{6ryz}{x\sqrt{x^2 - y^2}} + \frac{4z}{\sqrt{x^2 - y}} + 4r\sin^{-1}\left(\frac{y}{x}\right), \frac{\partial u}{\partial s} = \frac{-2yz}{x\sqrt{x^2 - y^2}} - \frac{6s^2 z}{\sqrt{x^2 - y^2}} - 6s\sin^{-1}\left(\frac{y}{x}\right)$

15. If $z = f(u, v)$ where $u = x\cos\theta - y\sin\theta, v = x\sin\theta + y\cos\theta$, show that

$$x\frac{\partial z}{\partial x} + y\frac{\partial z}{\partial y} = u\frac{\partial z}{\partial u} + v\frac{\partial z}{\partial v}, \theta \text{ being constant.}$$

16. Given $u = f(x, y), x = e^s \cos t, y = e^s \sin t$, show that $\frac{\partial^2 u}{\partial s^2} + \frac{\partial^2 u}{\partial t^2} = e^{2s}\left(\frac{\partial^2 u}{\partial x^2} + \frac{\partial^2 u}{\partial y^2}\right)$

17. If by the substitution $u = x^2 - y^2, v = 2xy, f(x, y) = \varphi(u, v)$

Show that $\frac{\partial^2 f}{\partial x^2} + \frac{\partial^2 f}{\partial y^2} = 4(x^2 + y^2)\left(\frac{\partial^2 \varphi}{\partial u^2} + \frac{\partial^2 \varphi}{\partial v^2}\right)$

18. Given the transformation $x = \cos h\,\xi \cos\eta, y = \sin\xi \sin\eta$
establish the following equation for the function u (function of x, y and also of ξ, η):

$$\frac{\partial^2 f}{\partial \xi^2} + \frac{\partial^2 u}{\partial \eta^2} = (\sinh^2\xi + \sin^2\eta)\left(\frac{\partial^2 u}{\partial x^2} + \frac{\partial^2 u}{\partial y^2}\right)$$

19. If $z = f(x, y)$, where $x = r\cos\theta, y = r\sin\theta$, prove that

$$\left(\frac{\partial z}{\partial x}\right)^2 + \left(\frac{\partial z}{\partial y}\right)^2 = \left(\frac{\partial z}{\partial r}\right)^2 + \frac{1}{r^2}\left(\frac{\partial z}{\partial \theta}\right)^2 \text{ and } \frac{\partial^2 z}{\partial x^2} + \frac{\partial^2 z}{\partial y^2} = \frac{\partial^2 z}{\partial r^2} + \frac{1}{r^2}\left(\frac{\partial^2 z}{\partial \theta^2}\right) + \frac{1}{r}\left(\frac{\partial z}{\partial r}\right)$$

20. If by substitution $u = x^2 - y^2, v = 2xy, f(x, y) = \varphi(u, v)$.

Show that $\frac{\partial^2 f}{\partial x^2} + \frac{\partial^2 f}{\partial y^2} = 4\sqrt{(u^2 + v^2)}\left(\frac{\partial^2 \varphi}{\partial u^2} + \frac{\partial^2 \varphi}{\partial v^2}\right)$

21. If $x = r \sin\theta \cos\varphi$, $y = r \sin\theta \sin\varphi$, $z = r\cos\theta$, $v = v\,(x, y, z)$,

prove that: $\left(\frac{\partial v}{\partial x}\right)^2 + \left(\frac{\partial v}{\partial y}\right)^2 + \left(\frac{\partial v}{\partial z}\right)^2 = \left(\frac{\partial v}{\partial r}\right)^2 + \left(\frac{1}{r}\frac{\partial v}{\partial \theta}\right)^2 + \left(\frac{1}{r\sin\theta}\frac{\partial v}{\partial \varphi}\right)^2$

22. If v be a potential function such that $v = v(r)$ and $r^2 = x^2 + y^2 + z^2$,

show that $\frac{\partial^2 v}{\partial x^2} + \frac{\partial^2 v}{\partial y^2} + \frac{\partial^2 v}{\partial z^2} = \frac{\partial^2 v}{\partial r^2} + \frac{2}{r}\frac{dv}{dr}$

23. Given that $w = x + 2y + z^2$, $x = r/s$, $y = r^2 + e^s$, and $z = 2r$, show that

$r\frac{\partial w}{\partial r} + s\frac{\partial w}{\partial s} = 12r^2 + 2se^s$.

24. Find $\frac{\partial w}{\partial v}$ when $u = 0$, $v = 0$,

if $w = (x^2 + y - 2)^4 + (x - y + 2)^3$, $x = u - 2v + 1$ and $y = 2u + v - 2$ **Ans.** 99

25. If $x = u + v + w$, $y = vw + wu + uv$, $z = uvw$ and F is a function of x, y, z, then show

that $u\frac{\partial F}{\partial u} + v\frac{\partial F}{\partial v} + w\frac{\partial F}{\partial w} = x\frac{\partial F}{\partial x} + 2y\frac{\partial F}{\partial y} + 3z\frac{\partial F}{\partial z}$.

26. If $u = x + ay$ and $v = x + by$ transforms the equation $2\frac{\partial^2 z}{\partial x^2} - 5\frac{\partial^2 z}{\partial x \partial y} + 3\frac{\partial^2 z}{\partial y^2} = 0$ into

the equation $\frac{\partial^2 z}{\partial u \partial v} = 0$, find the values of a and b. **Ans.** $a = 1, b = \frac{2}{3}; a = \frac{2}{3}, b = 1$

27. If $f(x, y) = e^{xy^2}$ the total differential of the function at the point (1, 2) is

(a) $e\,(dx + dv)$ (b) $e^4\,(dx + dy)$

(c) $e^4\,(4dx + dy)$ (d) $4e^4\,(dx + 4dy)$ **Ans.** (*d*)

Indefinite Integration

31.1 INTEGRATION AS INVERSE PROCESS OF DIFFERENTIATION

In the previous chapter, we found out the differential coefficient of a given function. In integration, if the differential coefficient is given, we have to find out the original function. Integration may be regarded as a process which is the inverse of differentiation.

In other words if $f(x)$ is given, we have to find out a function $\phi(x)$ whose derivative is $f(x)$. The function to be found is the *integral* of the given function. The given function is known as *integrand.* The process of finding out integral of a given function is known as *integration.*

31.2 DEFINITION

Let $f(x)$ be a given function of x. If we can find a function $\phi(x)$ such that

$$\frac{d}{dx}\phi(x) = f(x)$$

then we say that $\phi(x)$ is an integral of $f(x)$. We write

$$\int f(x)dx = \phi(x)$$

the sign $\int$ is called the integral sign. The symbol dx indicates that the integration is to be performed with respect to the variable x.

For example,

$$\frac{d}{dx}(\sin x) = \cos x$$

$$\int \cos x\, dx = \sin x + C$$

$$\frac{d}{dx}(x)^2 = 2x$$

$$\int 2x\, dx = x^2 + C$$

Note: (1) The symbol $\int$ is an elongated S which is the first letter of sum, as the process of integration originated from summation of infinite series.

(2) The symbol $\int$ and dx separated have no meaning. These two symbols may be regarded something like a pair of brackets in which the function to be integrated.

31.3 CONSTANT OF INTEGRATION

Let $$\frac{d}{dx}\,\phi(x) = f(x)$$

By definition $$\int f(x)dx = \phi(x) \qquad \text{... (1)}$$

We know that differential coefficient of constant C is zero.

$$\therefore \qquad \frac{d}{dx}[\phi(x) + C] = f(x)$$

By definition $$\int f(x)\,dx = \phi(x) + C \qquad \text{... (2)}$$

Here C is known as the constant of integration. $\phi(x) + C$ is called the indefinite integral of $f(x)$ due to indefiniteness of the constant C and $\phi(x)$ is known as the *principal primitive.* From (1) and (2) it follows that two integrals of the same function differ by a constant.

For example,

$$\frac{d}{dx}(x^4 + 2) = 4x^3 \qquad \therefore \qquad \int 4x^3\,dx = x^4 + 2$$

$$\frac{d}{dx}(x^4 + 5) = 4x^3 \qquad \therefore \qquad \int 4x^3\,dx = x^4 + 5$$

$x^4 + 2$ and $x^4 + 5$ are the two different integral of the given function and differ by a constant. For the sake of convenience we omit the constant of integration from the result but the existence of constant in the result is always to be understood.

31.4 STANDARD FORMULAE

Knowledge of differential calculus gives the following table of results, which may be taken as the standard formulae of integral calculus.

1. $\int x^n\,dx = \frac{x^{n+1}}{n+1} + C$ **2.** $\int \frac{1}{x}dx = \log x + C$

3. $\int e^x\,dx = e^x + C$ **4.** $\int a^x\,dx = \frac{a^x}{\log_e a} + C$

5. $\int \sin x\,dx = -\cos x + C$ **6.** $\int \cos x\,dx = \sin x + C$

7. $\int \tan x\,dx = \log \sec x + C$ **8.** $\int \cot x\,dx = \log \sin x + C$

9. $\int \sec x\,dx = \log(\sec x + \tan x) = \log \tan\left(\frac{x}{2} + \frac{\pi}{4}\right) + C$

10. $\int \operatorname{cosec} x\,dx = \log(\operatorname{cosec} x - \cot x) = \log \tan \frac{x}{2} + C$

11. $\int \sec^2 x\,dx = \tan x + C$ **12.** $\int \operatorname{cosec}^2 x\,dx = -\cot x + C$

13. $\int \sec x \tan x\,dx = \sec x + C$ **14.** $\int \operatorname{cosec} x \cot x\,dx = -\operatorname{cosec} x + C$

15. $\int \frac{dx}{\sqrt{a^2 - x^2}} = \sin^{-1}\left(\frac{x}{a}\right) + C$

16. $\int \frac{dx}{\sqrt{a^2 - x^2}} = -\cos^{-1}\left(\frac{x}{a}\right) + C$

17. $\int \frac{dx}{\sqrt{x^2 + a^2}} = \log\left(\frac{x + \sqrt{x^2 + a^2}}{a}\right) = \sinh^{-1}\left(\frac{x}{a}\right) + C$

18. $\int \frac{dx}{\sqrt{x^2 - a^2}} = \log\left(\frac{x + \sqrt{x^2 - a^2}}{a}\right) = \cosh^{-1}\left(\frac{x}{a}\right) + C$

19. $\int \frac{dx}{a^2 + x^2} = \frac{1}{a}\tan^{-1}\left(\frac{x}{a}\right) + C$

20. $\int \frac{dx}{a^2 + x^2} = \frac{-1}{a}\cot^{-1}\left(\frac{x}{a}\right) + C$

21. $\int \frac{dx}{a^2 - x^2} = \frac{1}{2a}\log\left(\frac{a + x}{a - x}\right) + C$

22. $\int \frac{dx}{x\sqrt{x^2 - a^2}} = \frac{1}{a}\sec^{-1}\left(\frac{x}{a}\right) + C$

23. $\int \frac{dx}{x\sqrt{x^2 - a^2}} = -\frac{1}{a}\operatorname{cosec}^{-1}\left(\frac{x}{a}\right) + C$

24. $\int \cosh x \, dx = \sinh x + C$

25. $\int \sinh x \, dx = \cosh x + C$

26. $\int \operatorname{sech}^2 x \, dx = \tanh x + C$

27. $\int \operatorname{cosech}^2 x \, dx = -\coth x + C$

Note: (*i*) Result (1) may be stated in words as follows:

To obtain the integral of x^n, increase the index of x by unity and divide by the new index.

(*ii*) If x is replaced by $(bx + c)$ in the above integrand, then the result is divided by b.

For example $\int \cos(bx + c)\, dx = \frac{1}{b}\cdot\sin(bx + c)$, and $\int e^{bx+c} dx = \frac{1}{b}e^{bx+c} + C$

31.5 INTEGRAL OF THE PRODUCT OF A CONSTANT AND A FUNCTION

Let C be a constant then $\int C\, f(x)\, dx = C\int f(x)\, dx$

The integral of the product of a constant and a function is equal to the product of the constant and the integral of the function,

i.e. $\int 8x^2\, dx = 8\int x^2\, dx$

31.6 INTEGRAL OF THE ALGEBRAIC SUM OF A NUMBER OF FUNCTIONS

$\int \{f_1(x) + f_2(x) + ...,\}\, dx = \int f_1(x)\, dx + \int f_2(x)\, dx + ...$

For example; $\int (\sin x + e^x)\, dx = \int \sin x\, dx + \int e^x\, dx$

The integral of the sum of the functions is equal to the sum of the integrals of the functions.

Example 1. *Write an antiderivative for each of the following functions using the method of inspection:*

(i) $\sin 2x$ *(ii)* $\cos 3x$ *(iii)* e^{2x} *(iv)* $(ax + b)^2$ *(v)* $\sin 2x - 4e^{2x}$

Solution. (*i*) We recall a function whose derivative is sin $2x$.

$$\frac{d}{dx}(\cos 2x) = -2\sin 2x$$

$$\Rightarrow \quad \therefore \quad \sin 2x = -\frac{1}{2}\frac{d}{dx}(\cos 2x) = \frac{d}{dx}\left(-\frac{1}{2}\cos 2x\right)$$

$\Rightarrow$ antiderivative of sin $2x$ is $-\frac{1}{2}\cos 2x$ **Ans.**

(*ii*) We recall a function whose derivative is cos $3x$.

$$(\sin 3x) = 3\cos 3x$$

$$\Rightarrow \quad \cos 3x = \frac{1}{3}\frac{d}{dx}(\sin 3x) = \frac{d}{dx}\left(\frac{1}{3}\sin 3x\right)$$

Therefore antiderivative of cos $3x$ is $\frac{1}{3}$ sin $3x$. **Ans.**

(*iii*) We recall a function whose derivative is e^{2x}.

$$\Rightarrow \quad \frac{d}{dx}(e^{2x}) = 2e^{2x}$$

$$\Rightarrow \quad e^{2x} = \frac{1}{2}\frac{d}{dx}(e^{2x}) = \frac{d}{dx}\left(\frac{1}{2}e^{2x}\right)$$

Therefore antiderivative of e^{2x} is $\frac{1}{2}e^{2x}$. Ans.

(*iv*) We recall a function whose derivative is $(ax + b)^2$.

$$\frac{d}{dx}(ax + b)^3 = 3(ax + b)^2(a)$$

$$\Rightarrow \quad (ax + b)^2 = \frac{1}{3a}\frac{d}{dx}(ax+b)^3 = \frac{d}{dx}\left[\frac{1}{3a}(ax+b)^3\right]$$

Therefore, antiderivative of $(ax + b)^2$ is $\frac{1}{3a}(ax + b)^3$. **Ans.**

(*v*) We recall a function whose derivative is sin $2x$.

$$\frac{d}{dx}(\cos 2x) = -2\sin 2x$$

$$\Rightarrow \quad \sin 2x = -\frac{1}{2}\frac{d}{dx}(\cos 2x) = \frac{d}{dx}\left(\frac{-1}{2}\cos 2x\right)$$

$\Rightarrow$ antiderivative of sin $2x$ is $-\frac{1}{2}$ cos $2x$.

$$\Rightarrow \quad \int \sin 2x\,dx = -\frac{1}{2}\cos 2x \qquad \ldots(1)$$

Again, we recall a function whose derivative is e^{2x}.

$$\frac{d}{dx}(e^{2x}) = 2e^{2x}$$

$$\Rightarrow \quad e^{2x} = \frac{1}{2}\frac{d}{dx}(e^{2x})$$

$$\Rightarrow \qquad 4e^{2x} = 2\frac{d}{dx}(e^{2x}) = \frac{d}{dx}(2e^{2x})$$

$\Rightarrow$ antiderivative of $4e^{2x}$ is $2e^{2x}$.

$$\Rightarrow \qquad \int 4e^{2x}dx = 2e^{2x} \qquad \ldots(2)$$

From (1) and (2), we have

$$\int(\sin 2x - 4e^{2x})\,dx = \int \sin 2x\,dx - \int 4e^{2x}\,dx$$

Hence, $\int(\sin 2x - 4e^{2x})\,dx = -\frac{1}{2}\cos 2x - 2e^{2x}$ **Ans.**

Example 2. *Evaluate the following integrals:*

(i) $\int x^2\,dx$ (ii) $\int \sqrt{x}\,dx$ (iii) $\int \frac{1}{\sqrt{x}}\,dx$

(iv) $\int 5x\,dx$ (v) $\int \frac{1}{3x^2}\,dx$ (vi) $\int \frac{1}{3e^{-x}}\,dx$

(vii) $\int \sqrt[7]{x^3}\,dx$ (viii) $\int \frac{2}{x}\,dx$ (ix) $\int 3.2^x\,dx$

Solution. We have,

(i) $\int x^2\,dx = \frac{x^{2+1}}{2+1} + C = \frac{x^3}{3} + C$

(ii) $\int \sqrt{x}\,dx = \int x^{\frac{1}{2}}\,dx = \frac{x^{\frac{1}{2}+1}}{\frac{1}{2}+1} + C = \frac{x^{3/2}}{\frac{3}{2}} + C = \frac{2}{3}x^{\frac{3}{2}} + C$ **Ans.**

Example 3. *Evaluate:* $\int(3x^2 - 14x + 5)\,dx$

Solution. $\int(3x^2 - 14x + 5)\,dx = 3\int x^2\,dx - 14\int x\,dx + 5\int dx$

$$= 3\cdot\frac{x^{2+1}}{2+1} - 14\cdot\frac{x^{1+1}}{1+1} + 5\cdot\frac{x^{0+1}}{0+1} + C$$

$$= x^3 - 7x^2 + 5x + C$$ **Ans.**

Example 4. *Evaluate* $\int\left(x^4 + 3x^3 + \frac{1}{x^4} + \frac{1}{x} + 4\right)dx$

Solution. $\int\left(x^4 + 3x^3 + \frac{1}{x^4} + \frac{1}{x} + 4\right)dx$

$$= \int x^4dx + 3\int x^3dx + \int x^{-4}dx + \int x^{-1}dx + 4\int dx$$

$$= \frac{x^{4+1}}{4+1} + 3\cdot\frac{x^{3+1}}{3+1} + \frac{x^{-4+1}}{-4+1} + \log x + 4x + C$$

$$= \frac{x^5}{5} + \frac{3}{4}x^4 - \frac{x^{-3}}{3} + \log x + 4x + C$$ **Ans.**

Example 5. *Evaluate* $\int \sin 4x \cos 5x \, dx$.

Solution. $\int \sin 4x \cos 5x \, dx = \frac{1}{2}\int 2\sin 4x \cdot \cos 5x \, dx$

$$= \frac{1}{2}\int (\sin 9x - \sin x) \, dx$$

$$= \frac{1}{2}\left(-\frac{\cos 9x}{9} + \cos x\right) = -\frac{1}{18}\cos 9x + \frac{1}{2}\cos x \qquad \textbf{Ans.}$$

Example 6. *Evaluate* $\int \frac{\sin^3 x - \cos^3 x}{\sin^2 x \cos^2 x} dx$

Solution. $\int \frac{\sin^3 x - \cos^3 x}{\sin^2 x \cos^2 x} dx = \int \left[\frac{\sin^3 x}{\sin^2 x \cos^2 x} - \frac{\cos^3 x}{\sin^2 x \cos^2 x}\right] dx$

$$= \int \left(\frac{\sin x}{\cos^2 x} - \frac{\cos x}{\sin^2 x}\right) dx = \int (\tan x \sec x - \cot x \operatorname{cosec} x) \, dx$$

$$= \sec x + \operatorname{cosec} x + C \qquad \textbf{Ans.}$$

Example 7. *If* $\int \sqrt{9 - x^2} \, dx = \frac{x}{2}\sqrt{9 - x^2} + K\sin^{-1}\left(\frac{x}{3}\right) + C$, *then find K.* **[SBTE. 2017]**

Solution. Here we have

$$\int \sqrt{9 - x^2} \, dx = \frac{x}{2}\sqrt{9 - x^2} + K\sin^{-1}\left(\frac{x}{3}\right) + C \qquad \text{... (1)}$$

We know that

$$\int \sqrt{a^2 - x^2} \, dx = \frac{x}{2}\sqrt{a^2 - x^2} + \frac{1}{2}a^2\sin^{-1}\left(\frac{x}{a}\right) + C \qquad \text{... (2)}$$

Comparing (1) with (2), we get $K = \frac{9}{2}$ **Ans.**

Example 8. *Evaluate:* $\int e^{-\log x} dx$ **[SBTE. 2017]**

Solution.

Let $I = \int e^{-\log x} dx = \int e^{\log \frac{1}{x}} dx = \int \frac{1}{x} dx$

$$= \log x + C \qquad \textbf{Ans.}$$

Example 9. *Evaluate:* $\int e^{3\log x} dx$ **[SBTE. 2016, 2015]**

Solution.

Let $I = \int e^{3\log x} dx = \int e^{\log x^3} dx \qquad \left[\because e^{\log x} = x\right]$

$$= \int x^3 dx = \frac{x^4}{4} + C \qquad \textbf{Ans.}$$

Example 10. *Evaluate the following integrals.*

(*i*) $\int \frac{x^2}{x+1}\,dx$ **[SBTE. 2016]** (*ii*) $\int \frac{2^x}{3^x}\,dx$ **[SBTE. 2015]**

Solution.

(*i*) Let $$I = \int \frac{x^2}{x+1}\,dx = \int \frac{x^2-1+1}{x+1}\,dx$$

$$= \int \left[x-1+\frac{1}{x+1} \right] dx$$

$\therefore$ $$I = \frac{x^2}{2} - x + \log(x+1) + C$$ **Ans.**

(*ii*) Let $$I = \int \frac{2^x}{3^x}\,dx = \int \left(\frac{2}{3}\right)^x dx$$

$$= \frac{\left(\frac{2}{3}\right)^x}{\log \frac{2}{3}} + C \qquad \left[\because \int a^x dx = \frac{a^x}{\log a} + C \right]$$

$\therefore$ $$I = \left(\frac{2}{3}\right)^x \left[\frac{1}{\log 2 - \log 3} \right] + C$$ **Ans.**

Example 11. *Evaluate:* $\int \frac{1}{\sqrt{3x+4} - \sqrt{3x+1}}\,dx$ **[SBTE. 2014]**

Solution.

Let $$I = \int \left(\frac{1}{\sqrt{3x+4} - \sqrt{3x+1}} \times \frac{\sqrt{3x+4} + \sqrt{3x+1}}{\sqrt{3x+4} + \sqrt{3x+1}} \right) dx$$

$$= \frac{1}{3} \int \left[(3x+4)^{1/2} + (3x+1)^{1/2} \right] dx$$

$$= \frac{1}{3} \left[\frac{2}{9} \left\{ (3x+4)^{3/2} + (3x+1)^{3/2} \right\} \right] + C$$

$\therefore$ $$I = \frac{2}{27} \left[(3x+4)^{3/2} + (3x+1)^{3/2} \right] + C$$ **Ans.**

Example 12. *Evaluate the following integrals.*

(*i*) $\int \sec x (\sec x - \tan x)\,dx$ **[SBTE. 2015, 2014, 2013]**

(*ii*) $\int \frac{5\cos^3 x + 6\sin^3 x}{2\sin^2 x \cdot \cos^2 x}\,dx$ **[SBTE. 2016]**

(*ii*) $\int \sec^2 (7-4x)$ **[SBTE. 2016]**

Solution.

(*i*) Let $I = \int \sec x (\sec x - \tan x)\, dx$

$$= \int \sec^2 x\, dx - \int \sec x \cdot \tan dx$$

$\therefore$ $I = \tan x - \sec x + C$ **Ans.**

(*ii*) Let $I = \int \frac{5\cos^3 x + 6\sin^3 x}{2\sin^2 x \cdot \cos^2 x} dx$

$$= \int \left[\frac{5}{2} \frac{\cos^3 x}{\sin^2 x \cdot \cos^2 x} + \frac{6}{2} \frac{\sin^3 x}{\sin^2 x \cdot \cos^2 x} \right] dx$$

$$= \frac{5}{2} \int \cot x \cdot \operatorname{cosec} x\, dx + 3 \int \tan x \cdot \sec x\, dx$$

$\therefore$ $= -\frac{5}{2} \operatorname{cosec} x + 3 \sec x + C$ **Ans.**

(*iii*) Let $I = \int \sec^2 (7 - 4x)\, dx$

$I = -\frac{1}{4} \tan (7 - 4x) + C$ **Ans.**

Example 13. *Evaluate* $\int \frac{1}{16 + y^2} dy$ **[SBTE. 2014]**

Solution. $\int \frac{dy}{4^2 + y^2} = \frac{1}{4} \tan^{-1} \left(\frac{y}{4} \right)$ **Ans.**

Example 14. *Evaluate* $\int (5x + 8)^{1/2}\, dx$.

Solution. $\int (5x + 8)^{1/2}\, dx = \frac{1}{5} \frac{(5x + 8)^{1/2+1}}{\frac{1}{2} + 1} = \frac{1}{5} \times \frac{2}{3} (5x + 8)^{3/2} = \frac{2}{15} (5x + 8)^{3/2}$ **Ans.**

Example 15. *Evaluate the integral of*

$$\frac{3\sin x}{2\cos^2 x} + \frac{4\cos x}{5\sin^2 x} + \frac{2}{\cos^2 x} + \frac{3}{\sin^2 x} + \frac{m}{1 + x^2} + \frac{1}{2x + 3}$$

Solution. $\int \left(\frac{3\sin x}{2\cos^2 x} + \frac{4\cos x}{5\sin^2 x} + \frac{2}{\cos^2 x} + \frac{3}{\sin^2 x} + \frac{m}{1 + x^2} + \frac{1}{2x + 3} \right) dx$

$$= \frac{3}{2} \int \tan x \sec x\, dx + \frac{4}{5} \int \cot x \operatorname{cosec} x\, dx + 2 \int \sec^2 x\, dx$$

$$+ 3 \int \operatorname{cosec}^2 x\, dx + m \int \frac{dx}{1 + x^2} + \int \frac{1}{2x + 3} dx$$

$$= \frac{3}{2} \sec x - \frac{4}{5} \operatorname{cosec} x + 2 \tan x - 3 \cot x + m \tan^{-1} x + \frac{1}{2} \log(2x + 3)$$ **Ans.**

EXERCISE 31.1

Write down the integral w.r.t. 'x' of the following:

1. $x^5 + \sin x$ **Ans.** $\frac{x^6}{6} - \cos x + C$

2. $(3x - 4)^2$ **Ans.** $\frac{1}{9}(3x - 4)^3 + C$

3. $\left(x + \frac{1}{x}\right)^3$ **Ans.** $\frac{x^4}{4} + \frac{3}{2}x^2 + 3\log x - \frac{1}{2x^2} + C$

4. $\frac{(1 - 3x)^2}{x^3}$ **Ans.** $-\frac{1}{2x^2} + \frac{6}{x} + 9\log x + C$

5. $\frac{(1 + x)^3}{x}$ **Ans.** $\log x + 3x + \frac{3x^2}{2} + \frac{x^3}{3} + C$

6. $\frac{(x + 3)^3}{x^2}$ **Ans.** $-\frac{27}{x} + 27\log x + 9x + \frac{x^2}{2} + C$

7. $\frac{(1 + 2x)^2}{x^3}$ **Ans.** $-\frac{1}{2x^2} - \frac{4}{x} + 4\log x + C$

8. $\frac{x^5 - 5x^4 + 6x^3 - 9}{x^4}$ **Ans.** $\frac{x^2}{2} - 5x + 6\log x + \frac{3}{x^3} + C$

9. $\frac{3 - 5x^2 + 7x^4 - 9x}{x^6}$ **Ans.** $-\frac{3}{5x^5} + \frac{5}{3x^3} - \frac{7}{x} + \frac{9}{4x^4} + C$

10. $\frac{1}{3x + 4}$ **Ans.** $\frac{1}{3}\log(3x + 4) + C$

11. $\frac{1}{25 + x^2}$ **Ans.** $\frac{1}{5}\tan^{-1}\left(\frac{x}{5}\right) + C$

12. $\frac{1}{\sqrt{25 - x^2}}$ **Ans.** $\sin^{-1}\left(\frac{x}{5}\right) + C$

13. $\frac{1}{\sqrt{x + 1} - \sqrt{x}}$ **Ans.** $\frac{2}{3}(x + 1)^{3/2} + \frac{2}{3}x^{3/2} + C$

14. $\frac{x - 1}{x + 1}$ [**Hint:** Divide numerator by denominator] **Ans.** $x - 2\log(x + 1) + C$

15. $\frac{x^2}{x + 1}$ **Ans.** $\frac{x^2}{2} - x + \log(x + 1) + C$

16. $\frac{x^2 - 4x + 3}{x - 2}$ **Ans.** $\frac{x^2}{2} - 2x - \log(x - 2) + C$

17. $\dfrac{4}{x\sqrt{x^2-1}}$ **Ans.** $4\sec^{-1}x + C$

18. $\sin 2x$ **Ans.** $-\dfrac{1}{2}\cos 2x + C$

19. $\dfrac{4-5\sin x}{\cos^2 x}$ **Ans.** $4\tan x - 5\sec x + C$

20. $\sin 2x \sin 3x$ **Ans.** $\dfrac{1}{2}\left[\sin x - \dfrac{1}{5}\sin 5x\right] + C$

21. $\dfrac{\sec x}{\sec 2x}$ **Ans.** $2\sin x - \log(\sec x + \tan x) + C$

22. $\tan^2 x$ **Ans.** $\tan x - x + C$

23. $\sin^2 x$ **Ans.** $\dfrac{x}{2} - \dfrac{\sin 2x}{4} + C$

24. $\cos^3 x$ **Ans.** $\dfrac{1}{4}\left[\dfrac{1}{3}\sin 3x + 3\sin x\right] + C$

25. $\dfrac{1}{1-\sin^2 x}$ **Ans.** $\tan x + C$

26. $\dfrac{1}{1+\sin x}$ **Ans.** $\tan x - \sec x + C$

27. $\dfrac{1}{1+\cos x}$ **Ans.** $-\cot x + \operatorname{cosec} x + C$

28. $\dfrac{1}{1-\cos x}$ **Ans.** $-\cot x - \operatorname{cosec} x + C$

29. $\sqrt{1+\sin 2x}$ **[SBTE. 2016]** **Ans.** $-\cos x + \sin x + C$

30. $\sqrt{1-\sin x}$ **Ans.** $-2\left(\cos\dfrac{x}{2} + \sin\dfrac{x}{2}\right) + C$

31. $\sqrt{1-\cos 2x}$ **Ans.** $-\sqrt{2}\cos x + C$

32. $\sqrt{1+\cos 2x}$ **[SBTE. 2015]** **Ans.** $\sqrt{2}\sin x + C$

33. Choose the correct option: $\int \dfrac{\sin^3 x - \cos^3 x}{\sin^2 x \cos^2 x}\,dx$ **[Diploma IETE. Dec., 2015]**

(*a*) $\tan x - \cot x$ (*b*) $\tan x + \cot x$ (*c*) $\sec x + \operatorname{cosec} x + C$ (*d*) $\sec x - \operatorname{cosec} x$ **Ans.** (*c*)

31.7 METHODS OF INTEGRATION

We use the following methods for integration:

(*i*) Substitution (*ii*) Integration by parts (*iii*) Partial fractions method

31.8 METHOD OF SUBSTITUTION

When the given functions cannot be integrated with the help of the formulae given in Section 32.4 we should try the method of substitution.

When a function and its derivative are given in the question then t is substituted for the given function so that the transformed function is easily integrable.

The following examples will make the method clear.

Example 16. *Evaluate* $\int x \cos x^2\, dx$. **[SBTE. 2015]**

Solution. Here $\frac{d}{dx}(x^2) = 2x$, the substitution is

$$x^2 = t \text{ so that } 2\,x\,dx = dt \;\Rightarrow\; x\,dx = \frac{dt}{2}$$

$$\therefore \quad \int x \cos x^2\, dx = \frac{1}{2}\int \cos t\, dt = \frac{1}{2}\sin t = \frac{1}{2}\sin x^2 + C$$ **Ans.**

Example 17. *Integrate* cot x.

Solution. Here $\frac{d}{dx}(\sin x) = \cos x$

$\therefore$ the substitution $\sin x = t$ so that $\cos x\, dx = dt$

$$\int \cot x\, dx = \int \frac{\cos x}{\sin x} dx = \int \frac{dt}{t} = \log t = \log \sin x + C.$$ **Ans.**

Example 18. *Evaluate the following:*

(*i*) $\int \frac{\cos x - \sin x}{1+\sin 2x} dx$ (*ii*) $\int \tan^3 x\, dx$ **[B.T.E. Delhi, May 2008]**

Solution. (*i*) Let $I = \int \frac{\cos x - \sin x}{1+\sin 2x} dx = \int \frac{\cos x - \sin x}{\sin^2 x + \cos^2 x + 2\sin x \cos x} dx$

$$= \int \frac{\cos x - \sin x}{(\sin x + \cos x)^2} dx \qquad \text{... (1)}$$

Putting $\sin x + \cos x = t$, so that $(\cos x - \sin x)\, dx = dt$ in (1), we get

$$I = \int \frac{dt}{t^2} = -\frac{1}{t} = \frac{-1}{(\sin x + \cos x)} + C$$ **Ans.**

(*ii*) Let $I = \int \tan^3 x\, dx \qquad = \int \tan x \cdot \tan^2 x\, dx$

$$= \int \tan x \cdot (\sec^2 x - 1)\, dx = \int \tan x \cdot \sec^2 x\, dx - \int \tan x\, dx$$

[Put $\tan x = t$ so that $\sec^2 x\, dx = dt$ in first integral we get]

$$= \int t\, dt - \int \tan x\, dx$$

$$= \frac{t^2}{2} - \log \sec x = \frac{\tan^2 x}{2} - \log \sec x + C$$ **Ans.**

Example 19. *Evaluate* $\int \frac{\sin(2+3\log x)}{x} dx$.

Solution. Here $\frac{d}{dx}(2 + 3\log x) = \frac{3}{x}$

∴ the substitution is

$$2 + 3 \log x = t \text{ so that } \frac{3}{x} dx = dt \Rightarrow \frac{dx}{x} = \frac{1}{3} dt$$

$$\int \frac{\sin (2 + 3\log x)}{x} dx = \frac{1}{3} \int \sin t = -\frac{1}{3} \cos (2 + 3 \log x) + C$$ **Ans.**

Example 20. *Evaluate* $\int \sqrt{a^2 - x^2}\, dx$ **[SBTE 2015]**

Solution. Here, we put $x = a \sin \theta$ so that

$$\sqrt{a^2 - x^2} = \sqrt{a^2 - a^2 \sin^2 \theta} = a \cos \theta$$

is a simple trigonometric ratio and can be easily integrated.

$$x = a \sin \theta \text{ so that } dx = a \cos \theta\, d\theta$$

$$\int \sqrt{a^2 - x^2}\, dx = \int \sqrt{a^2 - a^2 \sin^2 \theta}\, a \cos \theta\, d\theta = \int (a \cos \theta)\, a \cos \theta\, d\theta$$

$$= a^2 \int \cos^2 \theta\, d\theta = \frac{a^2}{2} \int (\cos 2\theta + 1)\, d\theta = \frac{a^2}{2} \left[\frac{\sin 2\theta}{2} + \theta \right] + C$$

$$= \frac{a^2}{2} [\theta + \sin \theta \cos \theta] + C = \frac{a^2}{2} \left[\sin^{-1} \frac{x}{a} + \frac{x}{a^2} \sqrt{a^2 - x^2} \right] + C$$

$$= \frac{a^2}{2} \sin^{-1} \left(\frac{x}{a} \right) + \frac{x}{2} \sqrt{a^2 - x^2} + C.$$ **Ans.**

Example 21. *Evaluate* $\int \frac{\sin(\tan^{-1} x)}{1 + x^2} dx$

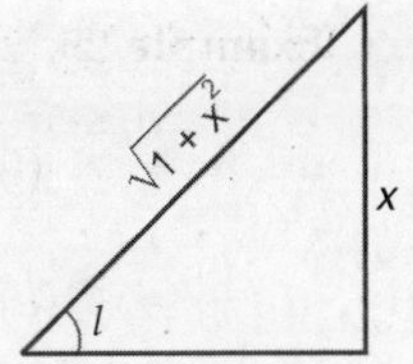

Solution. $\frac{d}{dx} (\tan^{-1} x) = \frac{1}{1 + x^2}$

∴ The substitution is $\tan^{-1} x = t$ so that $\frac{1}{1 + x^2} dx = dt$

$$\int \frac{\sin(\tan^{-1} x)}{1 + x^2} dx = \int \sin t\, dt = -\cos t + C = -\frac{1}{\sqrt{1 + x^2}} + C$$ **Ans.**

Example 22. *Evalaute* $\int \frac{e^x\, dx}{\sqrt{1 + e^{2x}}}$. **[SBTE. 2010]**

Solution. Put $e^x = t$, so that $e^x\, dx = dt$

$$\Rightarrow \quad \int \frac{e^x}{\sqrt{1 + e^{2x}}} dx = \int \frac{dt}{\sqrt{1 + t^2}} = \sinh^{-1} t = \sinh^{-1} (e^x) + C$$ **Ans.**

Example 23. *Evaluate* $\int \tan^{-1} \left(\frac{\sin 2x}{1 + \cos 2x} \right) dx$

Solution. $\int \tan^{-1} \left(\frac{\sin 2x}{1 + \cos 2x} \right) dx = \int \tan^{-1} \left(\frac{2 \sin x \cos x}{1 + 2\cos^2 x - 1} \right) dx$, $[\because \sin 2x = 2 \sin x \cdot \cos x]$

$$= \int \tan^{-1}\left(\frac{2\sin x\cos x}{2\cos^2 x}\right) dx$$

$$= \int \tan^{-1}\left(\frac{\sin x}{\cos x}\right) dx = \int \tan^{-1}(\tan x)\, dx$$

$$= \int x\, dx = \frac{x^2}{2} + C$$ **Ans.**

Example 24. *Evaluate:* $\int \frac{\sin 2x}{a^2 + b^2 \sin^2 x} dx$

Solution. Let $$I = \int \frac{\sin 2x}{a^2 + b^2 \sin^2 x} dx$$

$$= \frac{1}{b^2} \int \frac{\sin 2x}{\left(\frac{a}{b}\right)^2 + \sin^2 x}$$

Putting $\sin^2 x = t$ so that $2 \sin x \cos x\, dx = dt$ or $\sin 2x\, dx = dt$; we get

$$I = \frac{1}{b^2} \int \frac{dt}{\left(\frac{a}{b}\right)^2 + t} = \frac{1}{b^2} \log\left[\left(\frac{a}{b}\right)^2 + t\right] + C$$

$$= \frac{1}{b^2} \log\left[\left(\frac{a}{b}\right)^2 + \sin^2 x\right] + C$$ **Ans.**

Example 25. *Evaluate the following integrals.*

(*i*) $\int \tan^3 x \cdot \sec^2 x\, dx$ **[SBTE. 2016, 2015]**

(*ii*) $\int \sec^4 x \tan x\, dx$ **[SBTE. 2017]**

(*iii*) $\int x^4 d(x^2)$ **[SBTE. 2015]**

Solution.

(*i*) Let $$I = \int \tan^3 x \cdot \sec^2 x\, dx$$

put $\tan x = t$, $\sec^2 x\, dx = dt$

$$I = \int t^3 dt = \frac{t^4}{4} + C = \frac{\tan^4 x}{4} + C$$ **Ans.**

(*ii*) Let $$I = \int \sec^4 x \cdot \tan x\, dx = \int \sec^3 x \cdot \sec x \cdot \tan x\, dx$$

put $\sec x = t$, $\sec x \tan x\, dx = dt$

$$I = \int t^3 dt = \frac{t^4}{4} + C = \frac{\sec^4 x}{4} + C$$ **Ans.**

(*iii*) Let $$I = \int x^4 d(x^2)$$

put $x^2 = t$; $d(x^2) = dt$

$$I = \int t^2 dt = \frac{t^3}{3} + C$$

$\therefore$ $$I = \frac{x^6}{3} + C$$ **Ans.**

Example 26. *Evaluate the following integrals.*

(*i*) $\int \frac{1}{\sin(x-a)\cdot\sin(x-b)} dx$ **[SBTE. 2017, 2015, 2009]**

(*ii*) $\int \frac{1}{\cos(x-a)\cdot\cos(x-b)} dx$ **[SBTE. 2015, 2013]**

Solution.

(*i*) Let $I = \int \frac{1}{\sin(x-a)\cdot\cos(x-b)} dx$

$$I = \frac{1}{\sin(a-b)} \int \frac{\sin(a-b)}{\sin(x-a)\cdot\sin(x-b)} dx$$

$$I = \frac{1}{\sin(a-b)} \int \frac{\sin\{(x-b)-(x-a)\}}{\sin(x-a)\cdot\sin(x-b)} dx \; [\because \; \sin(A-B) = \sin A\cdot\cos B - \cos A\cdot\sin B]$$

$$= \frac{1}{\sin(a-b)} \int \{\cot(x-a) - \cot(x-b)\}\, dx$$

$$= \frac{1}{\sin(a-b)} [\log|\sin(x-a)| - \log|\sin(x-b)|] + C$$

$\therefore$ $$I = \frac{1}{\sin(a-b)} \left[\log\left|\frac{\sin(x-a)}{\sin(x-b)}\right|\right] + C$$ **Ans.**

(*ii*) Let $$I = \int \frac{1}{\cos(x-a)\cdot\cos(x-b)} dx$$

$$= \frac{1}{\sin(a-b)} \int \frac{\sin\{(x-b)-(x-a)\}}{\cos(x-a)\cdot\cos(x-b)} dx$$

$\therefore$ $$I = \frac{1}{\sin(a-b)} \log\left|\frac{\cos(x-a)}{\cos(x-b)}\right| + C$$ **Ans.**

EXERCISE 31.2

Evalutate the following w.r.t. x.

1. $\int \frac{2x}{1+x^2} dx$ **Ans.** **Ans.** $\log(1+x^2) + C$

2. $\int \frac{x\,dx}{\sqrt{1+x^2}}$ **Ans.** $\sqrt{1+x^2} + C$

3. $\int x(a^2 + x^2)^n\,dx$ **Ans.** $\frac{1}{2(n+1)}(a^2 + x^2)^{n+1} + C$

4. $\int \frac{x^3}{1+x^8}\,dx$ **Ans.** $\frac{1}{4}\tan^{-1}(x^4) + C$

5. $\int \left(\sqrt{\frac{2+x}{2-x}}\right) dx$ **Ans.** $2\sin^{-1}\left(\frac{x}{2}\right) - \sqrt{4-x^2} + C$

6. $\int x \sin x^2\,dx$ **Ans.** $-\frac{1}{2}\cos x^2 + C$

7. $\int e^x \sin e^x\,dx$ **Ans.** $-\cos e^x + C$

8. $\int \frac{-\text{cosec}^2 x}{a + b\cot x}\,dx$ **Ans.** $\frac{1}{b}\log(a + b\cot x) + C$

9. $\int \frac{\sqrt{\tan^{-1} x}}{1+x^2}\,dx$ **Ans.** $\frac{2}{3}(\tan^{-1} x)^{3/2} + C$

10. $\int \frac{dx}{1+\cos^2 x}$ **Ans.** $\frac{1}{\sqrt{2}}\tan^{-1}\left(\frac{\tan x}{\sqrt{2}}\right) + C$

11. $\int \frac{x\tan^{-1} x^2}{1+x^4}\,dx$ **Ans.** $\frac{1}{4}(\tan^{-1} x^2)^2 + C$

12. $\int \frac{\sqrt{(\tan x)}}{\sin x \cdot \cos x}\,dx$ **Ans.** $2\sqrt{(\tan x)} + C$

13. $\int \sin^3 x \cos x\,dx$ **Ans.** $\frac{(\sin x)^4}{4} + C$

14. $\int \frac{1}{x}\sin(\log x)\,dx$ **Ans.** $-\cos(\log x) + C$

15. $\int \sin x\, e^{\cos x}\,dx$ **Ans.** $-e^{\cos x} + C$

16. $\int \frac{\sin x}{a + b\cos x}\,dx$ **Ans.** $-\frac{1}{b}\log(a + b\cos x) + C$

17. $\int \frac{ax+b}{ax^2 + 2bx + c}\,dx$ **Ans.** $\frac{1}{2}\log(ax^2 + 2bx + c) + C$

18. $\int \frac{\log x}{x}\,dx$ **Ans.** $\frac{(\log x)^2}{2} + C$

19. $\int \tan x\,dx$ **Ans.** $\log(\sec x) + C$

20. $\int \sec x\, dx$ **Ans.** $\log(\sec x + \tan x)$ or $\log \tan\left(\frac{x}{2}+\frac{\pi}{4}\right) + C$

21. $\int \text{cosec}\, x\, dx$ **Ans.** $\log(\text{cosec}\, x - \cot x)$ or $\log \tan \frac{x}{2} + C$

22. $\int \tan x \sec^3 x\, dx$ **Ans.** $\frac{\sec^3 x}{3} + C$

23. $\int \frac{1}{x \log x}\, dx$ **Ans.** $\log \log x + C$

24. $\int \frac{1}{x \log x \log(\log x)}\, dx$ **Ans.** $\log\{\log(\log x)\} + C$

25. $\int \frac{\sin 2x\, dx}{a^2 + b^2 \sin^2 x}$ **Ans.** $\frac{1}{b^2} \log(a^2 + b^2 \sin^2 x) + C$

26. $\int \frac{\sin x \cos x}{\sqrt{1+\sin^2 x}}\, dx$ **Ans.** $\sqrt{1+\sin^2 x} + C$

27. $\int \frac{1-\tan x}{1+\tan x}\, dx$ **[B.T.E. Delhi May 2007]** **Ans.** $\log \cos\left(\frac{\pi}{4} - x\right) + C$

28. $\int \cos\left(2\cot^{-1}\sqrt{\frac{1-x}{1+x}}\right) dx$ **[Hint : Put $x = \cos\theta$]** **Ans.** $-\frac{x^2}{2} + C$

29. $\int \sqrt{\frac{2a-x}{x}}\, dx$ **[Hint: Put $x = 2a \sin^2\theta$]** **Ans.** $\sqrt{2ax - x^2} + 2a \sin^{-1}\left(\sqrt{\frac{x}{2a}}\right) + C$

30. $\int \frac{1}{\sin x + \cos x}\, dx$ **Ans.** $\frac{1}{\sqrt{2}} \log \tan\left(\frac{x}{2} + \frac{\pi}{8}\right) + C$

31. $\int x e^{x^2}\, dx$ **Ans.** $\frac{1}{2} e^{x^2} + C$

32. $\int \frac{1}{2\sqrt{x}(x+1)}\, dx$ **Ans.** $\tan^{-1}(\sqrt{x}) + C$

33. $\int \frac{x\, dx}{\sqrt{x^2+a^2} + \sqrt{x^2-a^2}}$ **Ans.** $\frac{1}{6a^2}[(x^2+a^2)^{3/2} - (x^2-a^2)^{3/2}] + C$

34. $\int \frac{e^x}{e^x - e^{-x}}\, dx$ **Ans.** $\frac{1}{2}\log(e^{2x} - 1) + C$

35. $\int \frac{1}{e^x - 1}\, dx$ **Ans.** $\log(1 - e^{-x}) + C$

36. $\int \frac{e^x(1+x)}{\sin^2(xe^x)}\, dx$ **Ans.** $-\cot(xe^x) + C$

[**Hint:** Put $xe^x = t$, so that $e^x (1 + x)\, dx = dt$]

37. $\int \frac{e^x}{\sqrt{1-e^{2x}}}\, dx$ **Ans.** $\sin^{-1}(e^x) + C$

38. $\int \frac{(\sin^{-1} x)^3}{\sqrt{1-x^2}}\, dx$ **Ans.** $\frac{1}{4}(\sin^{-1} x)^4 + C$

39. $\int \frac{1}{(\sec^{-1} x)\, x\sqrt{x^2-1}}\, dx$ **Ans.** $\log(\sec^{-1} x) + C$

40. $\int \frac{\sqrt{(a+b\sec^{-1} x)}}{x\sqrt{(x^2-1)}}\, dx$ **Ans.** $\frac{2}{3b}(a+b\sec^{-1} x)^{3/2} + C$

41. $\int \frac{dx}{x(1+\log x)^2}$ **[B.T.E. Delhi May 2008]** **Ans.** $-\frac{1}{1+\log x} + C$

42. $\int \frac{\sin(2+3\log x)}{x}\, dx$ **Ans.** $-\frac{1}{3}\cos(2+3\log x) + C$

43. $\int \frac{(x+1)}{x(x+\log x)}\, dx$ **Ans.** $\log(x+\log x) + C$

44. $\int \frac{x-\sin x}{1-\cos x}\, dx$ **[Diploma IETE. June 2005]** **Ans.** $-x\cot\frac{x}{2} + C$

45. $\int \frac{x^2\tan^{-1} x^3}{1+x^6}\, dx$ **[Diploma IETE. June 2005]** **Ans.** $\frac{1}{6}(\tan^{-1} x^3)^2 + C$

46. $\int e^x \sin e^x\, dx$ **[B.T.E. Delhi May 2008]** **Ans.** $-\cos e^x + C$

47. $\int \frac{e^x}{1+e^{2x}}\, dx$ is equal to **[Diploma IETE. June 2005]**

(*a*) $\tan^{-1}(e^x) + C$ (*b*) $\cot^{-1}(e^x) + C$ (*c*) $\sin^{-1}(e^x) + C$ (*d*) 0 **Ans.** (*a*)

48. $\int \frac{\cos^2 x - \sin^2 x}{\sin x \cos x}\, dx$ is equal to **[Diploma IETE. June 2005]**

(*a*) $\log(\sin 2x) + C$ (*b*) $\log(\cot 2x) + C$
(c) $\log(\cos 2x) + C$ (*d*) $\log(\tan 2x) + C$ **Ans.** (*a*)

31.9 SOME IMPORTANT FORMULAE

1. $\int \frac{dx}{a^2+x^2} = \frac{1}{a}\tan^{-1}\left(\frac{x}{a}\right) + C$

2. $\int \frac{dx}{a^2-x^2} = \frac{1}{2a}\log\left(\frac{a+x}{a-x}\right) + C$

3. $\displaystyle\int \frac{dx}{\sqrt{a^2 - x^2}} = \sin^{-1}\left(\frac{x}{a}\right) + C$ (Put $x = a \sin \theta$)

4. $\displaystyle\int \frac{dx}{\sqrt{a^2 + x^2}} = \sin^{-1}\left(\frac{x}{a}\right) = \log\left(\frac{x + \sqrt{x^2 + a^2}}{a}\right) + C$ (Put $x = a \sinh \theta$)

5. $\displaystyle\int \frac{dx}{\sqrt{x^2 - a^2}} = \cosh^{-1}\left(\frac{x}{a}\right) = \log\left(\frac{x + \sqrt{x^2 - a^2}}{a}\right) + C$ (Put $x = a \cosh \theta$)

6. $\displaystyle\int \sqrt{a^2 - x^2}\, dx = \frac{x}{2}\sqrt{a^2 - x^2} + \frac{a^2}{2}\sin^{-1}\left(\frac{x}{a}\right) + C$ (Put $x = a \sin \theta$)

7. $\displaystyle\int \sqrt{a^2 + x^2}\, dx = \frac{x}{2}\sqrt{a^2 + x^2} + \frac{a^2}{2}\sinh^{-1}\left(\frac{x}{a}\right) + C$ (Put $x = a \sinh \theta$)

8. $\displaystyle\int \sqrt{x^2 - a^2}\, dx = \frac{x}{2}\sqrt{x^2 - a^2} - \frac{a^2}{2}\cosh^{-1}\left(\frac{x}{a}\right) + C$ (Put $x = a \cosh \theta$)

31.10 INTEGRATION BY PARTS

This method is used for integration of the product of two functions.

Let u and v be two functions of x.

$$\frac{d}{dx}(u \cdot v) = u \cdot \frac{dv}{dx} + v \cdot \frac{du}{dx}$$

Integrating both sides, we have

$$u \cdot v = \int u \cdot \frac{dv}{dx} \cdot dx + \int v \cdot \frac{du}{dx} \cdot dx$$

$$\int u \frac{dv}{dx} \cdot dx = u \cdot v - \int v \cdot \frac{du}{dx} \cdot dx \quad \text{... (1)}$$

Now if $u = f(x)$ and $\dfrac{dv}{dx} = \phi(x)$ so that $v = \int \phi(x)\, dx$

Substituting these values in (1), we get

$$\int f(x) \cdot \phi(x) dx = f(x) \cdot \int \phi(x) dx - \int \left[f'(x) \cdot \left\{ \int \phi(x)\, dx \right\} \right] dx$$

$$\int I \cdot II = I \int II - \int I' \int II$$

In words, integral of the product of two functions

= First function × integral of the second – integral of

[diff. coeff. of first function × integral of the second]

Note 1: The success of this method depends upon choosing the first function in such a way that the second term on the right hand side may be easy to integrate.

Remember:

1. The formula of integration by parts can be repeated if necessary.

2. 1 is sometimes taken as one of the factors and chosen as the second function.
3. Sometimes the integral on the right hand side comes out to be the same as on the left hand side. This integral of the right hand sides to be transposed to the left hand side and the value of the integral is found.

Note 2: We can also choose the first function as the function which comes first in word **ILATE:** where

I = inverse trigonometric function
L = logorithmic functions
A = algebraic functions
T = trigonometric functions
E = exponential functions.

Example 27. *Evaluate* $\int x \cdot e^x \, dx$ **[BTE Delhi 2016]**

Solution. Let first function = x

Second function = e^x

$$\int I \cdot II = I \int II - \int I' \int II$$

$$\int x \cdot e^x \, dx = x \int e^x \, dx - \int \left\{ \frac{dx}{dx} \int e^x \, dx \right\} dx$$

$$= x \cdot e^x - \int 1 \cdot e^x \, dx = x \cdot e^x - e^x + C$$

Ans.

Example 28. *Evaluate:* $\int x^3 \log x \, dx$ **[B.T.E, Delhi May 2008]**

Solution. Let $I = \int x^3 \log x \, dx$

Integrating by parts, we get

$$I = \log x \cdot \left(\frac{x^4}{4} \right) - \int \frac{1}{x} \cdot \left(\frac{x^4}{4} \right) dx$$

$$= \frac{x^4}{4} \cdot \log x - \frac{1}{4} \int x^3 \, dx = \frac{x^4}{4} \log x - \frac{1}{4} \cdot \frac{x^4}{4} + C$$

$$= \frac{x^4}{4} \left[\log x - \frac{1}{4} \right] + C$$

Ans.

Example 29. *Integrate* log x, *w.r.t.* x. **[SBTE 2016, 2015, 2014]**

Solution. $\int \log x \, dx = \int \log x \cdot 1 \cdot dx$

Let first function = $\log x$; second function = 1

$$= \log x \int 1 \, dx - \int \left\{ \frac{d}{dx} (\log x) \int 1 \, dx \right\} dx$$

$$= \log x \cdot x - \int \frac{1}{x} \cdot x \, dx = x \log x - \int dx$$

$$= x \log x - x + C$$

Ans.

Example 30. *Evaluate:* $\int \cos^{-1} x \, dx$.

Solution. $\int \cos^{-1} x \, dx = \int \cos^{-1} x \cdot 1 \, dx = \cos^{-1} x \cdot x - \int \frac{-1}{\sqrt{1-x^2}} \cdot (x) \, dx$

$$= x \cos^{-1} x - \sqrt{1-x^2} + C$$ **Ans.**

Example 31. *Evaluate:* $\int \cos^{-1} (\sin x) \, dx$.

Solution. Let $I = \int \cos^{-1} (\sin x) \cdot 1 \, dx$

Integrating by parts, we get

$$\Rightarrow \quad I = \cos^{-1}(\sin x) \cdot x + \int \frac{\cos x}{\sqrt{1-\sin^2 x}} x \, dx$$

$$\Rightarrow \quad I = x \cos^{-1}(\sin x) + \int \frac{\cos x \cdot x}{\cos x} dx$$

$$= x \cos^{-1}(\sin x) + \int x \cdot dx$$

$$= x \cos^{-1}(\sin x) + \frac{x^2}{2} + C$$ **Ans.**

Example 32. *Evaluate* $\int x^2 \cos x \, dx$ **[SBTE. 2014]**

Solution. $\int x^2 \cos x \, dx = x^2 \int \cos x \, dx - \int \left\{ \frac{d}{dx} x^2 \int \cos x \, dx \right\} dx$

$$= x^2 \sin x - \int 2x \sin x \, dx$$

$$= x^2 \sin x - 2 \left[x \int \sin x \, dx - \int \left\{ \frac{d}{dx}(x) \int \sin x \, dx \right\} dx \right]$$

$$= x^2 \sin x + 2x \cos x - 2\int 1 \cdot \cos x \, dx$$

$$= x^2 \sin x + 2x \cos x - 2 \sin x + C$$ **Ans.**

Example 33. *Evaluate* $\int e^{ax} \cos bx \, dx$.

Solution. $\int e^{ax} \cos bx \, dx = e^{ax} \left(\frac{\sin bx}{b} \right) - \int (ae^{ax}) \left(\frac{\sin bx}{b} \right) dx$

$$= \frac{e^{ax} \sin bx}{b} - \frac{a}{b} \int e^{ax} \sin bx \, dx$$

$$= \frac{e^{ax} \sin bx}{b} - \frac{a}{b} \left[e^{ax} \left(-\frac{\cos bx}{b} \right) - \int (ae^{ax}) \left(-\frac{\cos bx}{b} \right) dx \right]$$

$$= \frac{e^{ax} \sin bx}{b} + \frac{a}{b^2} e^{ax} \cos bx - \frac{a^2}{b^2} \int e^{ax} \cos bx \, dx$$

$$\Rightarrow \left(1+\frac{a^2}{b^2}\right)\int e^{ax}\cos bx\,dx = \frac{e^{ax}\sin bx}{b} + \frac{a}{a^2}e^{ax}\cos bx$$

$$\Rightarrow \frac{a^2+b^2}{b^2}\int e^{ax}\cos bx\cdot dx = \frac{e^{ax}}{b^2}[b\sin bx + a\cos bx]$$

$$\Rightarrow \int e^{ax}\cos bx\cdot dx = \frac{e^{ax}}{a^2+b^2}[b\sin bx + a\cos bx]$$

Put $a = r\cos\alpha$, $b = r\sin\alpha$,

$a^2 + b^2 = r^2\cos^2\alpha + r^2\sin^2\alpha = r^2$

$$= \frac{e^{ax}}{r^2}[r\sin\alpha\sin bx + r\cos\alpha\cos bx]$$

$$= \frac{re^{ax}}{r^2}\cos(bx-\alpha)$$

$$= \frac{e^{ax}}{\sqrt{a^2+b^2}}\cos\left(bx - \tan^{-1}\frac{b}{a}\right) + C$$ **Ans.**

Example 34. *Evaluate:* $\int \frac{x\cdot\sin^{-1}x}{\sqrt{1-x^2}}dx$.

Solution. Here $\frac{d}{dx}(\sin^{-1}x) = \frac{1}{\sqrt{1-x^2}}$

$\therefore$ Put: $\sin^{-1}x = t$, we have $x = \sin t$

so that $\frac{1}{\sqrt{1-x^2}}dx = dt$

$$\therefore \int \frac{x\sin^{-1}x}{\sqrt{1-x^2}}dx = \int t\cdot\sin t\,dt$$

$$= t\cdot(-\cos t) - \int 1\cdot(-\cos t)\,dt$$

$$= -t.\cos t + \sin t$$

$$= -\sin^{-1}x\sqrt{1-x^2} + x$$

$$= x - \sqrt{1-x^2}\,\sin^{-1}x + C$$ **Ans.**

Example 35. *Evaluate:* $\int \frac{e^x}{x}(x\log x + 1)\,dx$

Solution. $\int \frac{e^x}{x}(x\log x+1)\,dx = \int\left(e^x\log x + \frac{e^x}{x}\right)dx$

$$= \int e^x\log x\,dx + \int\frac{e^x}{x}dx \quad \text{... (1)}$$

Integrating first integral by parts, we have

$$\int e^x \log x \cdot dx = \log x \cdot e^x - \int \frac{e^x}{x}\, dx \qquad \text{... (2)}$$

On putting the value of $\int e^x \log x \cdot dx$ in (1), we have

$$\int \frac{e^x}{x}(x \log x + 1)\, dx = \log x \cdot e^x - \int \frac{e^x}{x}\, dx + \int \frac{e^x}{x}\, dx = e^x \log x + C$$ **Ans.**

Example 36. *Evaluate:* $\int \frac{e^x (1 + \sin x)}{1 + \cos x}\, dx$ **[DIPIETE Dec. 2018]**

Solution. $\int \frac{e^x (1 + \sin x)}{1 + \cos x}\, dx$

$$= \int \frac{e^x + 2e^x \sin\frac{x}{2}\cos\frac{x}{2}}{2\cos^2\frac{x}{2}} = \frac{1}{2}\int e^x \sec^2\frac{x}{2}\, dx + \int e^x \tan\frac{x}{2}\, dx$$

$$= e^x \tan\frac{x}{2} - \int e^x \tan\frac{x}{2}\, dx + \int e^x \tan\frac{x}{2}\, dx$$

$$= e^x \tan\frac{x}{2} + C$$ **Ans.**

Example 37. *Evaluate integrals*

$\int x \cdot \tan^{-1} x\, dx$ **[SBTE. 2016, 2013, 2008]**

Solution. Let $I = \int \underset{II}{x} \cdot \underset{I}{\tan^{-1} x}\, dx$

$$= \tan^{-1} x \int x\, dx - \int \left[\frac{d}{dx}(\tan^{-1} x) \cdot \int x\, dx\right] dx$$

$$= \frac{x^2}{2}\tan^{-1} x - \frac{1}{2}\int \frac{x^2}{1 + x^2}\, dx + C$$

$$= \frac{x^2}{2}\tan^{-1} x - \frac{1}{2}\int \frac{(1 + x^2) - 1}{1 + x^2}\, dx + C$$

$\therefore$ $$I = \frac{x^2}{2}\tan^{-1} x - \frac{1}{2}\left[x - \tan^{-1} x\right] + C$$ **Ans.**

Example 38. *Evaluate:* $\int e^x \cdot \frac{x}{(x+1)^2}\, dx$ **[SBTE. 2012, 2010, 2009]**

Solution. Let $I = \int e^x \cdot \frac{x}{(x+1)^2}\, dx = \int e^x \left[\frac{(x+1) - 1}{(x+1)^2}\right] dx$

$$= \int e^x \cdot \frac{1}{x+1}\,dx - \int e^x \cdot \frac{1}{(x+1)^2}\,dx$$

$$= \frac{1}{x+1}\int e^x\,dx - \int\left[-\frac{1}{(x+1)^2}\cdot e^x\right]dx - \int e^x \cdot \frac{1}{(1+x)^2}\,dx + C$$

$$= e^x \cdot \frac{1}{(x+1)} + \int e^x \frac{1}{(x+1)^2}\,dx - \int e^x \cdot \frac{1}{(x+1)^2}\,dx + C$$

$\therefore$ $$I = e^x \cdot \frac{1}{x+1} + C$$ **Ans.**

Example 39. *Evaluate* $\int x^3 \log 2x\,dx$ **[SBTE. 2016]**

Solution. Let $$I = \int x^3 \cdot \log 2x\,dx$$

$$II \cdot I$$

$$= \log 2x \int x^3 dx - \int \frac{1}{2x}(2)\cdot\frac{x^4}{4}dx$$

$$= \frac{1}{4}x^4 \log 2x - \int \frac{1}{4}x^3\,dx + C$$

$\therefore$ $$= \frac{1}{4}x^4 \log 2x - \frac{x^4}{16} + C$$ **Ans.**

Example 40. *Find f(x) satisfying,* $\int e^x \sec x\,(1+\tan x)\,dx = e^x \cdot f(x) + C$ **[SBTE. 2016, 2015]**

Solution. Let $$I = \int e^x \sec x\,(1+\tan x)\,dx$$

$$= \int e^x \cdot \sec x\,dx + \int e^x \sec x \cdot \tan x\,dx$$

$$= \sec x \int e^x dx - \int\left[\sec x \cdot \tan e^x\right]dx + \int e^x \cdot \sec x \cdot \tan x\,dx$$

$\therefore$ $$I = e^x \cdot \sec x + C$$

$\therefore$ $$f(x) = \sec x$$ **Ans.**

Example 41. *If* $\int \frac{(\log x)^2}{x}dx = K(\log x)^3 + C$, *what is the value of K?* **[SBTE. 2015]**

Solution. Let $$I = \int \frac{(\log x)^2}{x}dx$$

put $\log x = t,\ \frac{1}{x}\,dx = dt$

$$I = \int t^2 dt = \frac{t^3}{3} + C$$

$$I = \frac{1}{3}(\log x)^3 + C$$

$\therefore \qquad K = \frac{1}{3}$ **Ans.**

Example 42. *Evaluate:* $\int e \frac{\quad}{(\quad 1)} dx$ **[SBTE. 2016, 2014]**

Solution. Let
$$I = \int e^x \frac{x^2+1}{(x+1)^2} dx$$
$$= \int e^x \frac{(x+1)^2 - 2x}{(x+1)^2} dx = \int e^x dx - 2\int e^x \frac{x}{(x+1)^2} dx$$
$$= e^x - 2\int e^x \left[\frac{(x+1)-1}{(x+1)^2}\right] dx$$
$$= e^x - 2\int e^x \cdot \frac{1}{x+1} dx + 2\int e^x \frac{1}{(x+1)^2} dx$$
$$= e^x - 2\cdot\frac{e^x}{x+1} - 2\int e^x \cdot \frac{1}{(x+1)^2} dx + 2\int e^x \cdot \frac{1}{(x+1)^2} dx$$
$\therefore$
$$I = e^x - 2\cdot\frac{e^x}{x+1} + C$$
Ans.

EXERCISE 31.3

Integrate the following functions w.r.t. x.

1. $x\,(x+9)^{3/2}$ **Ans.** $\frac{2}{5}x\,(x+9)^{5/2} - \frac{4}{35}(x+9)^{7/2} + C$
2. $x \sin x$ **[SBTE. 2016, 2015]** **Ans.** $-\,x \cos x + \sin x + C$
3. $x \cos 2x$ **Ans.** $\frac{x \sin 2x}{2} + \frac{\cos 2x}{4} + C$
4. $x \cos^2 x$ **Ans.** $\frac{x^2}{4} + \frac{x}{4}\sin 2x + \frac{1}{8}\cos 2x + C$
5. $\frac{x}{\cos^2 x}$ **Ans.** $x \tan x + \log(\cos x) + C$
6. $x \sin x \cos x$ **Ans.** $\frac{1}{8}(\sin 2x - 2x \cos 2x) + C$
7. $x^2 \sin x \cos x$ **Ans.** $\frac{-x^2 \cos 2x}{4} + \frac{x \sin 2x}{2} + \frac{\cos 2x}{4} + C$
8. $\sec^3 x$ **[SBTE. 2014]** **Ans.** $\frac{1}{2}[\sec x \tan x + \log(\sec x + \tan x)] + C$
9. $x \log(1+x)$ **Ans.** $\frac{1}{2}(x^2-1)\log(1+x) - \frac{x^2}{4} + \frac{x}{2} + C$

10. $x^2 \log x$ **[B.T.E. Delhi June, 2005]** **Ans.** $\frac{x^3}{9}\left(\log \frac{x^3}{e}\right) + C$

11. $x^2 e^x$ **[B.T.E. Delhi May 2008]** **Ans.** $e^x (x^2 - 2x + 2) + C$

12. $\log (x + \sqrt{x^2 + a^2})$ **[B.T.E. Delhi May 2008]** **Ans.** $x \log (x + \sqrt{x^2 + a^2}) - \sqrt{x^2 + a^2} + C$

13. $x^2 \tan^{-1} x$ **Ans.** $\frac{x^3}{3}\tan^{-1} x - \frac{x^2}{6} + \frac{1}{6}\log (x^2 + 1) + C$

14. $x^2 \sin^{-1} x$ **Ans.** $\frac{x^3}{3}\sin^{-1} x - \frac{1}{9}(1 - x^2)^{3/2} + \frac{1}{3}(1 - x^2)^{1/2} + C$

15. $x^n \log x$ **Ans.** $\frac{x^{n+1}}{(n+1)^2}[(n+1)\log x - 1] + C$

16. $x \sec^{-1} x$ **Ans.** $\frac{x^2}{2}\sec^{-1} x - \frac{1}{2}\sqrt{x^2 - 1} + C$

17. $\sin^{-1} x$ **[SBTE. 2010]** **Ans.** $x \sin^{-1} x + \sqrt{1 - x^2} + C$

18. $\tan^{-1} x$ **[SBTE. 2008]** **Ans.** $x \tan^{-1} x - \log \sqrt{1 + x^2} + C$

19. $\tan^{-1}\left(\frac{2x}{1 - x^2}\right)$ **Ans.** $2x \tan^{-1} x - \log (1 + x^2) + C$

20. $\sin^{-1}\left(\sqrt{\frac{x}{a + x}}\right)$ **[Hint:** Put $x = a \tan^2 \theta$**]** **Ans.** $x \tan^{-1}\left(\sqrt{\frac{x}{a}}\right) - \sqrt{ax} + a \tan^{-1}\left(\sqrt{\frac{x}{a}}\right) + C$

21. $\frac{x^2 \tan^{-1} x}{(1 + x^2)}$ **Ans.** $x \tan^{-1} x - \frac{1}{2}\log (1 + x^2) - \frac{1}{2}(\tan^{-1} x)^2 + C$

22. $\frac{x \cdot e^{m \sin^{-1} x}}{\sqrt{(1 - x^2)}}$ **Ans.** $\frac{e^{m \sin^{-1} x}}{m^2 + 1}\left[mx - \sqrt{1 - x^2}\right] + C$

23. $\frac{xe^x}{(1 + x)^2}$ **Ans.** $\frac{e^x}{x + 1} + C$

24. $e^x\left(\frac{1 - \sin x}{1 - \cos x}\right)$ **[SBTE. 2018]** **Ans.** $-e^x \cot \frac{x}{2} + C$

25. $\frac{x + \sin x}{1 + \cos x}$ **[Diploma IETE. Dec. 2006]** **Ans.** $x \tan \frac{x}{2} + C$

26. $(x + 1)\, e^x \log (xe^x)$ **Ans.** $xe^x [\log (xe^x) - 1] + C$

27. $\frac{x + \cos x}{1 - \sin x}$ **Ans.** $x \tan\left(\frac{x}{2} + \frac{\pi}{4}\right) + C$

28. $e^x (\cot x - \operatorname{cosec}^2 x)$ **Ans.** $e^x \cot x + C$

29. $e^{ax} \sin bx$ **Ans.** $\dfrac{e^{ax}}{\sqrt{a^2+b^2}} \sin\left(bx - \tan^{-1}\dfrac{b}{a}\right) + C$

30. $e^x \cos x$ **Ans.** $\dfrac{1}{2} e^x (\sin x + \cos x) + C$

31. $e^x (\sin x + \cos x)$ **Ans.** $\dfrac{e^x}{\sqrt{2}}\left[\sin\left(x - \dfrac{\pi}{4}\right) + \cos\left(x - \dfrac{\pi}{4}\right)\right] + C$

31.11 INTEGRATION BY PARTIAL FRACTIONS METHOD

The given rational algebraic fraction is split into a number of fractions and each fraction is integrated. The fractions obtained will be of the following type:

$$\frac{1}{x}, \frac{1}{bx+c}, \frac{1}{x^2+a^2}, \frac{px+q}{ax^2+bx+c}, \text{ etc.}$$

(i) The Denominator Containing Linear NonRepeated Factors Only

Example 43. *Evaluate:* $\displaystyle\int \frac{x^2}{(x^2-1)(x+3)} dx$

Solution. Let $\dfrac{x^2}{(x+1)(x-1)(x+3)} = \dfrac{A}{x+1} + \dfrac{B}{x-1} + \dfrac{C}{x+3}$

$\therefore$ $$x^2 = A(x-1)(x+3) + B(x+1)(x+3) + C(x+1)(x-1) \quad \ldots (1)$$

Putting $x = -1$, $x = 1$ and $x = -3$ in succession in (1), we get

$$(-1)^2 = A(-1-1)(-1+3) \quad \therefore \quad A = -\frac{1}{4}$$

$$(1)^2 = B(1+1)(1+3) \quad \therefore \quad B = \frac{1}{8}$$

$$(-3)^2 = C(-3+1)(-3-1) \quad \therefore \quad C = \frac{9}{8}$$

Hence, $\displaystyle\int \frac{x^2}{(x+1)(x-1)(x+3)} dx$

$$= -\frac{1}{4}\int \frac{1}{x+1} dx + \frac{1}{8}\int \frac{1}{x-1} dx + \frac{9}{8}\int \frac{1}{x+3} dx$$

$$= -\frac{1}{4}\log(x+1) + \frac{1}{8}\log(x-1) + \frac{9}{8}\log(x+3) + C$$ **Ans.**

Example 44. *Evaluate:* $\displaystyle\int \frac{x^2+1}{(2x+1)(x^2-1)} dx$

Solution. $\displaystyle\int \frac{x^2+1}{(2x+1)(x^2-1)} dx = \int \frac{x^2+1}{(2x+1)(x+1)(x-1)} dx$

Let $$\frac{x^2+1}{(2x+1)(x+1)(x-1)} \equiv \frac{A}{2x+1} + \frac{B}{x+1} + \frac{C}{x-1} \quad \ldots (1)$$

$$\Rightarrow \quad x^2+1 = A(x^2-1)+B(2x+1)(x-1)+C(2x+1)(x+1) \qquad \text{... (2)}$$

Putting $x = 1$ on both sides of (2), we get

$$1+1 = C\,(3)\,(2), \quad C = \frac{1}{3}$$

Putting $x = -\frac{1}{2}$ in (2), we get

$$\frac{1}{4}+1 = A\left(\frac{1}{4}-1\right) \Rightarrow \frac{5}{4} = \frac{-3}{4}A \Rightarrow A = \frac{-5}{3}$$

Equating the coefficients of x^2 on both sides of (2), we get

$$1 = A + 2B + 2C$$

$$1 = \frac{-5}{2} + 2B + \frac{2}{3} \Rightarrow 2B = 2 \Rightarrow B = 1$$

On putting the values of A, B and C in (1), we get

$$\int \frac{x^2+1}{(2x+1)\,(x+1)\,(x-1)}\,dx = \int \frac{\frac{-5}{3}}{2x+1}\,dx + \int \frac{1}{x+1}\,dx + \int \frac{\frac{1}{3}}{x-1}\,dx$$

$$= \frac{-5}{6}\log\,(2x+1) + \log\,(x+1) + \frac{1}{3}\log\,(x-1) + C \qquad \textbf{Ans.}$$

Example 45. *Evalaute:* $\int \frac{2x}{(x^2+1)\,(x^2+2)}\,dx$

Solution. Let $\quad I = \int \frac{2x}{(x^2+1)(x^2+2)}\,dx \qquad \text{... (1)}$

On putting $x^2 = t$ so that $2x\,dx = dt$ in (1), we have

$$I = \int \frac{dt}{(t+1)\,(t+2)}$$

Let $\quad \frac{1}{(t+1)\,(t+2)} = \frac{A}{t+1} + \frac{B}{t+2} \qquad \text{... (2)}$

$$\Rightarrow \quad 1 \equiv A(t+2) + B(t+1)$$

$$1 = A(-1+2) \Rightarrow A = 1 \qquad (t = -1)$$

$$1 = B(-2+1) \Rightarrow B = -1 \qquad (t = -2)$$

$$\frac{1}{(t+1)\,(t+2)} = \frac{1}{t+1} - \frac{1}{t+2}$$

$$\int \frac{1}{(t+1)\,(t+2)}\,dt = \int \frac{1}{t+1}\,dt - \int \frac{1}{t+2}\,dt$$

$$= \log\,(t+1) - \log\,(t+2) = \log\left(\frac{t+1}{t+2}\right)$$

$$= \log\left(\frac{x^2+1}{x^2+2}\right) + C \qquad \textbf{Ans.}$$

(ii) The Degree of Numerator is Equal or Higher to that of Denominator

Divide the numerator by the denominator or express the fraction as a quotient plus a proper fraction. The proper fraction is split into partial fractions. Then the partial fractions with quotient are integrated.

Example 46. $\int \frac{x^5}{x^3 - 2x^2 - 5x + 6} dx$

Solution. $\int \frac{x^5}{x^3 - 2x^2 - 5x + 6} = x^2 + 2x + 9 + \frac{22x^2 + 33x - 54}{x^3 - 2x^2 - 5x + 6}$

Now the factors of $x^3 - 2x^2 - 5x + 6$ are $(x - 1)$, $(x + 2)$, $(x - 3)$.

Let $\frac{22x^2 + 33x - 54}{(x-1)(x+2)(x-3)} = \frac{A}{x-1} + \frac{B}{x+2} + \frac{C}{x-3}$

$\therefore \quad 22x^2 + 33x - 54 = A(x+2)(x-3) + B(x-1)(x-3) + C(x-1)(x+2)$

Putting $x = 1$, $x = -2$, $x = 3$ in succession, we get

$\Rightarrow \quad 22 + 33 - 54 = A(1 + 2)(1 - 3) \quad \Rightarrow \quad A = -\frac{1}{6}$

$\Rightarrow \quad 88 - 66 - 54 = B(-2 - 1)(-2 - 3) \quad \Rightarrow \quad B = -\frac{32}{15}$

$\Rightarrow \quad 198 + 99 - 54 = C(3 - 1)(3 + 2) \quad \Rightarrow \quad C = \frac{243}{10}$

$$\therefore \int \frac{x^5}{x^3 - 2x^2 - 5x + 6} dx = \int \left[x^2 + 2x + 9 - \frac{1}{6(x-1)} - \frac{32}{15(x+2)} + \frac{243}{10(x-3)} \right] dx$$

$$= \frac{x^3}{3} + x^2 + 9x - \frac{1}{6} \log(x-1) - \frac{32}{15} \log(x+2) + \frac{243}{10} \log(x-3) + C$$

Ans.

(iii) The Denominator Contains Repeated Linear Factors

In case a factor is repeated three times in the denominator then the corresponding three partial fractions will have three denominators.

Example 47. *Evaluate* $\int \frac{4x\,dx}{(x+1)^2(x-1)^2}$

Solution. Let $\frac{4x}{(x+1)^2(x-1)^2} \equiv \frac{A}{x+1} + \frac{B}{(x+1)^2} + \frac{C}{x-1} + \frac{D}{(x-1)^2}$

$$4x \equiv A(x+1)(x-1)^2 + B(x-1)^2 + C(x-1)(x+1)^2 + D(x+1)^2$$

Putting $x = 1$, we get, $\quad 4 = D(1 + 1)^2 \Rightarrow D = 1$

Putting $x = -1$, we get, $\quad -4 = B(-1 - 1)^2 \Rightarrow B = -1$

Comparing the coefficients of x^3 and constants on both sides, we get

$0 = A + C \qquad \text{... (1)}$

$0 = A + B - C + D \Rightarrow 0 = A - C - 1 + 1 \Rightarrow A = C \qquad \text{... (2)}$

From (1) and (2) we have $A = C = 0$

$$\therefore \int \frac{4x\,dx}{(x+1)^2(x-1)^2} = \int \left[\frac{-1}{(x+1)^2} + \frac{1}{(x-1)^2}\right] dx = \frac{1}{x+1} - \frac{1}{x-1} = \frac{-2}{x^2-1} + C \qquad \textbf{Ans.}$$

Example 48. *Evaluate:* $\dfrac{\cos x\,dx}{(1+\sin x)^2(2+\sin x)}$

Solution. Let $I = \int \dfrac{\cos x\,dx}{(1+\sin x)^2\,(2+\sin x)}$

Putting $\sin x = t$ so that $\cos x\,dx = dt$, we get

$$I = \int \frac{dt}{(1+t)^2\,(2+t)} \qquad \text{... (1)}$$

Now, $\dfrac{1}{(1+t)^2(2+t)} = \dfrac{A}{(1+t)} + \dfrac{B}{(1+t)^2} + \dfrac{C}{(2+t)}$

$$\Rightarrow \quad 1 = A\,(1+t)\,(2+t) + B\,(2+t) + C\,(1+t)^2 \qquad \text{... (2)}$$

Putting $t = -1$ in (2), we get

$$1 = B\,(2-1) \;\Rightarrow\; B = 1$$

Again putting $t = -2$ in (2), we get

$$\Rightarrow \quad 1 = C\,(1-2)^2 \;\Rightarrow\; C = 1$$

Equating the coefficients of t^2 in (2), we get

$$0 = A + C \;\Rightarrow\; A + 1 = 0 \;\Rightarrow\; A = -1$$

$$\therefore \quad \frac{1}{(1+t)^2(2+t)} = \frac{-1}{(1+t)} + \frac{1}{(1+t)^2} + \frac{1}{(2+t)}$$

$$\Rightarrow \quad \int \frac{dt}{(1+t)^2(2+t)} = \int \frac{dt}{(1+t)} + \int \frac{dt}{(1+t)^2} + \int \frac{dt}{(2+t)}$$

$$\Rightarrow \quad I = -\log(1+t) - \frac{1}{1+t} + \log(2+t) = \log\left(\frac{2+t}{1+t}\right) - \frac{1}{1+t}$$

$$\therefore \quad \int \frac{\cos x\,dx}{(1+\sin x)^2(2+\sin x)} = \log\left(\frac{2+\sin x}{1+\sin x}\right) - \frac{1}{1+\sin x} + C \qquad \textbf{Ans.}$$

(iv) The Denominator Containing Quadratic Factors

Example 49. *Evaluate:* $\displaystyle\int \frac{x^3+2x^2}{(x+1)(x^2+2x+2)}\,dx$

Solution. Divide the numerator by denominator as the degree of both is the same.

$$\frac{x^3+2x^2}{(x+1)(x^2+2x+2)}\,dx = 1 - \frac{x^2+4x+2}{(x+1)(x^2+2x+2)}\,dx$$

Let $\dfrac{x^2+4x+2}{(x+1)(x^2+2x+2)} = \dfrac{A}{x+1} + \dfrac{Bx+C}{x^2+2x+2}$

$$\therefore \quad x^2 + 4x + 2 = A(x^2 + 2x + 2) + (Bx + C)\,(x + 1)$$

Putting $x = -1$, we get

$$(-1)^2 + 4(-1) + 2 = A(1 - 2 + 2) \Rightarrow A = -1$$

Equating the coefficients of x^2 and constant terms on both sides, we get

$$1 = A + B \quad \text{But } A = -1 \quad \therefore B = 2$$

$$2 = 2A + C, \quad \text{But } A = -1 \quad \therefore C = 4$$

Hence $\int \frac{x^3 + 2x^2}{(x+1)(x^2+2x+2)} dx = \int \left[\frac{2x+2}{x^2+2x+2} + \frac{2}{(x+1)^2+1} \right] dx$

$$= x + \log(x+1) - \int \left[\frac{2x+2}{x^2+2x+2} + \frac{2}{(x+1)^2+1} \right] dx$$

$$= x + \log(x+1) - \log(x^2+2x+2) - 2\tan^{-1}(x+1) + C \textbf{ Ans.}$$

EXERCISE 31.4

Integrate the following w.r.t. x.

1. $\frac{x}{(3-x)(3+2x)}$ **Ans.** $-\frac{1}{6}[2\log(3-x) + \log(3+2x)] + C$

2. $\frac{1}{x^2+5x+6}$ **[B.T.E. Delhi May 2008]** **Ans.** $\log\left(\frac{x+2}{x+3}\right) + C$

3. $\frac{\sec^2 x}{(1+\tan x)(2+\tan x)}$ **Ans.** $\log\left(\frac{\tan x + 1}{\tan x + 2}\right) + C$

4. $\frac{x^2-3x+4}{(x-2)(x+2)(x+4)}$ **Ans.** $\frac{1}{12}\log(x-2) - \frac{7}{4}\log(x+2) + \frac{8}{3}\log(x+4) + C$

5. $\frac{x^3+1}{x(x^2-1)}$ **Ans.** $x + \log\left(\frac{x-1}{x}\right) + C$

6. $\frac{3x+2}{(x+1)^2(x-2)}$ **Ans.** $\frac{8}{9}\log\left(\frac{x-2}{x+1}\right) - \frac{1}{3(x+1)} + C$

7. $\frac{1}{(x^2+a^2)(x^2+b^2)}$ **Ans.** $\frac{1}{a^2-b^2}\left[-\frac{1}{a}\tan^{-1}\frac{x}{a} + \frac{1}{b}\tan^{-1}\frac{x}{b}\right] + C$

8. $\frac{x^2}{(x^2+a^2)(x^2+b^2)}$ **Ans.** $\frac{1}{a^2-b^2}\left[a\tan^{-1}\frac{x}{a} - b\tan^{-1}\frac{x}{b}\right] + C$

9. $\frac{x^2+1}{(x-1)^2(x+1)^2}$ **Ans.** $\frac{x}{1-x^2} + C$

10. $\frac{8x+8}{(2x+1)^2(2x+3)^2}$ **Ans.** $\frac{-1}{(2x+1)(2x+3)} + C$

11. $\frac{2x+9}{(x+4)^2(x+5)^2}$ **Ans.** $\frac{-1}{(x+4)(x+5)} + C$

12. $\dfrac{x}{(x^2-2)(x^2+1)}$ **Ans.** $\dfrac{1}{6}\log\left(\dfrac{x^2-2}{x^2+1}\right)+C$

13. $\dfrac{1}{(x-1)(x^2+4)}$ **Ans.** $\dfrac{1}{5}\log(x-1)-\dfrac{1}{10}\log(x^2+4)-\dfrac{1}{10}\tan^{-1}\left(\dfrac{x}{2}\right)+C$

14. $\dfrac{1}{x(1+x^2)}$ **Ans.** $\log x-\dfrac{1}{2}\log(1+x^2)+C$

15. $\dfrac{x}{(1+x)(1+x^2)}$ **[B.T.E. Delhi May 2008]**

Ans. $\dfrac{1}{2}[-\log(1+x)+\dfrac{1}{2}\log(x^2+1)+\tan^{-1}x]+C$

16. $\dfrac{x^2}{(x^2+1)(3x^2+1)}$ **Ans.** $\dfrac{1}{2}\tan^{-1}(x)-\dfrac{1}{2\sqrt{3}}\tan^{-1}(\sqrt{3}x)+C$

17. $\dfrac{1}{x(x^n+1)}$ [**Hint:** Multiply numerator and denominator by x^{n-1}] **Ans.** $\dfrac{1}{n}\log\left(\dfrac{x^n}{x^n+1}\right)$

18. $\dfrac{x^2+3x+3}{x^3+x^2+x+1}$ **[B.T.E. Delhi May 2007]**

Ans. $\dfrac{1}{3}\log(x^3+x^2+x+1)+\dfrac{1}{6}\log(x+1)-\dfrac{1}{12}\log(x^2+1)+\dfrac{5}{2}\tan^{-1}(x)+C$

IMPORTANT FORMULAE

1. $\displaystyle\int\frac{1}{a^2+x^2}\,dx=\frac{1}{a}\tan^{-1}\left(\frac{x}{a}\right)+C$

2. $\displaystyle\int\frac{dx}{a^2-x^2}=\frac{1}{2a}\log\left(\frac{a+x}{a-x}\right)+C$

3. $\displaystyle\int\frac{dx}{x^2-a^2}=\frac{1}{2a}\log\left(\frac{x-a}{x+a}\right)+C$

4. $\displaystyle\int\frac{1}{\sqrt{a^2-x^2}}\,dx=\sin^{-1}\left(\frac{x}{a}\right)+C$

5. $\displaystyle\int\frac{1}{\sqrt{a^2+x^2}}\,dx=\log\left|x+\sqrt{a^2+x^2}\right|+C$

6. $\displaystyle\int\frac{1}{\sqrt{x^2-a^2}}\,dx=\log\left|x+\sqrt{x^2-a^2}\right|+C$

7. $\displaystyle\int\sqrt{a^2-x^2}\,dx=\frac{x}{2}\sqrt{a^2-x^2}+\frac{a^2}{2}\sin^{-1}\left(\frac{x}{a}\right)+C$

8. $\displaystyle\int\sqrt{a^2+x^2}\,dx=\frac{x}{2}\sqrt{a^2+x^2}+\frac{a^2}{2}\log\left(x+\sqrt{a^2+x^2}\right)+C$

9. $\displaystyle\int\sqrt{x^2-a^2}\,dx=\frac{x}{2}\sqrt{x^2-a^2}-\frac{a^2}{2}\log\left(x+\sqrt{x^2-a^2}\right)+C$

Now in all the cases of $\displaystyle\int\frac{px+q}{ax^2+bx+c}\,dx$ or $\displaystyle\int\frac{px+q}{\sqrt{ax^2+bx+c}}\,dx$

or $\displaystyle\int(px+q)\sqrt{ax^2+bx+c}\,dx$

Split $px + q$ so that $px + q = \frac{d}{dx}(ax^2 + bx + c) + k$

31.12 INTEGRATION OF RATIONAL AND IRRATIONAL FUNCTIONS

Integral of the following Forms

(*i*) $\int \frac{1}{ax^2 + bx + c}\,dx$ (*ii*) $\int \frac{1}{\sqrt{ax^2 + bx + c}}\,dx$ (*iii*) $\int \sqrt{ax^2 + bx + c}\,dx$

If the denominator cannot be factorised then we apply the following method.

Example 50. *Evaluate:* $\int \frac{dx}{3x^2 + 6x + 21}$

Solution. $\int \frac{dx}{3x^2 + 6x + 21} = \frac{1}{3}\int \frac{dx}{(x^2 + 2x + 1) + 6} = \frac{1}{3}\int \frac{dx}{(x+1)^2 + (\sqrt{6})^2}$

$$\left[\text{This is of the form } \int \frac{dx}{x^2 + a^2} = \frac{1}{a}\tan^{-1}\frac{x}{a}\right]$$

$$= \frac{1}{3}\frac{1}{\sqrt{6}}\tan^{-1}\left(\frac{x+1}{\sqrt{6}}\right) + C$$ **Ans.**

Example 51. *Evaluate:* $\int \frac{dx}{2x^2 + x + 3}$ **[SBTE. 2016, 2015]**

Solution. Let $I = \int \frac{dx}{2x^2 + x + 3} = \frac{1}{2}\int \frac{dx}{x^2 + \frac{1}{2}x + \frac{3}{2}}$

$$= \frac{1}{2}\int \frac{dx}{\left(x^2 + \frac{1}{2}x + \frac{1}{16}\right) + \frac{3}{2} - \frac{1}{16}} = \frac{1}{2}\int \frac{dx}{\left(x^2 + \frac{1}{4}\right)^2 + \frac{23}{16}}$$

$$= \frac{1}{2}\int \frac{dx}{\left(x + \frac{1}{4}\right)^2 + \left(\frac{\sqrt{23}}{4}\right)^2} = \frac{1}{2} \times \frac{4}{\sqrt{23}}\tan^{-1}\left(\frac{x + \frac{1}{4}}{\frac{\sqrt{23}}{4}}\right)$$

$$= \frac{2}{\sqrt{23}}\tan^{-1}\left(\frac{4x+1}{\sqrt{23}}\right) + C$$ **Ans.**

Example 52. *Evaluate:* $\int \frac{x^2 - 1}{x^4 + 1}\,dx$ **[SBTE. 2015, 2014]**

Solution. Let $I = \int \frac{x^2 - 1}{x^4 + 1}\,dx$

$$= \int \frac{1-\frac{1}{x^2}}{x^2+\frac{1}{x^2}}\,dx = \int \frac{\left(1-\frac{1}{x^2}\right)}{\left(x+\frac{1}{x}\right)^2-2}\,dx$$

Putting $x+\frac{1}{x}=t$ so that $\left(1-\frac{1}{x^2}\right)dx = dt,$ we get

$$I = \int \frac{dt}{t^2-2} = \int \frac{dt}{t^2-(\sqrt{2})^2}$$

$$= \frac{1}{2\sqrt{2}}\log\left(\frac{t-\sqrt{2}}{t+\sqrt{2}}\right)+C \qquad \left[\because \int \frac{1}{x^2-a^2}\,dx = \frac{1}{2a}\log\left(\frac{x-a}{x+a}\right)\right]$$

$$= \frac{1}{2\sqrt{2}}\log\left(\frac{x+\frac{1}{x}-\sqrt{2}}{x+\frac{1}{x}+\sqrt{2}}\right) = \frac{1}{2\sqrt{2}}\log\left(\frac{x^2+1-\sqrt{2}x}{x^2+1+\sqrt{2}x}\right)+C \qquad \textbf{Ans.}$$

Example 53. *Integrate:* $\int \frac{x^2+1}{x^4+1}\,dx$ **[SBTE. 2015]**

Solution. Here, we have $\int \frac{x^2+1}{x^4+1}\,dx$

$$= \int \frac{1+\frac{1}{x^2}}{x^2+\frac{1}{x^2}}\,dx = \int \frac{1+\frac{1}{x^2}}{x^2+\frac{1}{x^2}-2+2}\,dx = \int \frac{1+\frac{1}{x^2}}{\left(x-\frac{1}{x}\right)^2+2}\,dx \quad \ldots(1)$$

On putting $\left(x-\frac{1}{x}\right)=t$ so that $\left(1+\frac{1}{x^2}\right)dx = dt$ in (1), we have,

$$= \int \frac{dt}{t^2+2} = \int \frac{dt}{t^2+(\sqrt{2})^2} = \frac{1}{\sqrt{2}}\tan^{-1}\left(\frac{t}{\sqrt{2}}\right)$$

$$= \frac{1}{\sqrt{2}}\tan^{-1}\left(\frac{x-\frac{1}{x}}{\sqrt{2}}\right)+C \qquad \textbf{Ans.}$$

Example 54. *Evaluate:* $\int \frac{dx}{x^2-x-6}$

Solution. $\int \frac{dx}{x^2-x-6} = \int \frac{dx}{\left(x-\frac{1}{2}\right)^2-\left(\frac{5}{2}\right)^2}$

$\left[\text{This is of the form } \int \frac{dx}{x^2 - a^2} = \frac{1}{2a}\log\left(\frac{x-a}{x+a}\right)\right]$

$$\Rightarrow \quad \int \frac{dx}{x^2 - x - 6} = \frac{1}{2 \times 5/2} \log \frac{\left(x - \frac{1}{2}\right) - \left(\frac{5}{2}\right)}{\left(x - \frac{1}{2}\right) + \left(\frac{5}{2}\right)} = \frac{1}{5}\log\left(\frac{x-3}{x+2}\right) + C$$ **Ans.**

Example 55. *Evaluate:* $\int \frac{dx}{\sqrt{4 - 2x - x^2}}$

Solution. $\int \frac{dx}{\sqrt{4 - 2x - x^2}} = \int \frac{dx}{\sqrt{4 - (x^2 + 2x)}} = \int \frac{dx}{\sqrt{5 - (x+1)^2}}$

$$= \int \frac{dx}{\sqrt{(\sqrt{5})^2 - (x+1)^2}} \quad \left[\text{This is of the form } \int \frac{dx}{\sqrt{a^2 - x^2}} = \sin^{-1}\left(\frac{x}{a}\right)\right]$$

$$= \sin^{-1}\left(\frac{x+1}{\sqrt{5}}\right) + C$$ **Ans.**

Example 56. *Evaluate:* $\int \frac{dx}{\sqrt{2x^2 + 3x + 4}}$ **[SBTE. 2013]**

Solution. $\int \frac{dx}{\sqrt{2x^2 + 3x + 4}} = \frac{1}{\sqrt{2}} \int \frac{dx}{\sqrt{x^2 + \frac{3}{2}x + 2}}$

$$= \frac{1}{\sqrt{2}} \int \frac{dx}{\sqrt{\left(x + \frac{3}{4}\right)^2 + \frac{23}{16}}}$$

$$= \frac{1}{\sqrt{2}} \int \frac{dx}{\sqrt{\left(x + \frac{3}{4}\right)^2 + \left(\frac{\sqrt{23}}{4}\right)^2}}$$

$\left[\text{This is of the form } \int \frac{dx}{\sqrt{x^2 + a^2}} = \sinh^{-1}\left(\frac{x}{a}\right)\right]$

$$\int \frac{dx}{\sqrt{2x^2 + 3x + 4}} = \frac{1}{\sqrt{2}} \sinh^{-1}\left(\frac{x + \frac{3}{4}}{\frac{\sqrt{23}}{4}}\right) = \frac{1}{\sqrt{2}} \sinh^{-1}\left(\frac{4x+3}{\sqrt{23}}\right) + C$$ **Ans.**

Example 57. *Evaluate:* $\int \frac{dx}{\sqrt{x^2 + 3x + 1}}$

Solution. $\int \frac{dx}{\sqrt{x^2 + 3x + 1}} = \int \frac{dx}{\sqrt{\left(x + \frac{3}{2}\right)^2 + 1 - \frac{9}{4}}}$

$$= \int \frac{dx}{\sqrt{\left(x+\frac{3}{2}\right)^2 - \left(\frac{\sqrt{5}}{2}\right)^2}}$$

$$\left[\text{This is of the form} = \int \frac{dx}{\sqrt{x^2 - a^2}} = \cosh^{-1}\left(\frac{x}{a}\right)\right]$$

$$\therefore \quad \int \frac{dx}{\sqrt{x^2+3x+1}} = \cosh^{-1}\left(\frac{x+\frac{3}{2}}{\frac{\sqrt{5}}{2}}\right) = \cosh^{-1}\left(\frac{2x+3}{\sqrt{5}}\right) + C \qquad \textbf{Ans.}$$

Example 58. *Evaluate:* $\int \frac{x}{\sqrt{x^2+6x+10}}\,dx$

Solution. $\int \frac{x}{\sqrt{x^2+6x+10}}\,dx = \int \frac{\frac{1}{2}(2x+6)-3}{\sqrt{x^2+6x+10}}\,dx$

$$= \frac{1}{2}\int \frac{2x+6}{\sqrt{x^2+6x+10}}\,dx - 3\int \frac{1}{\sqrt{(x+3)^2+(1)^2}}\,dx$$

$$= \frac{1}{2}\int \frac{dt}{\sqrt{t}} - 3\sinh^{-1}(x+3), \text{ where } t = x^2+6x+10$$

$$= \sqrt{t} - 3\sinh^{-1}(x+3)$$

$$= \sqrt{x^2+6x+10} - 3\sinh^{-1}(x+3) + C \qquad \textbf{Ans.}$$

Example 59. *Evaluate:* $\int \frac{\cos x}{\sqrt{\sin^2 x - 2\sin x - 3}}\,dx$

Solution. Let $\quad I = \int \frac{\cos x}{\sqrt{\sin^2 x - 2\sin x - 3}}\,dx$

Putting $\sin x = t$, so that $\cos x\,dx = dt$

$$\therefore \quad I = \int \frac{dt}{\sqrt{t^2-2t-3}}$$

$$= \int \frac{dt}{\sqrt{(t^2-2t+1)-4}} = \int \frac{dt}{\sqrt{(t-1)^2-(2)^2}}$$

$$= \log\left(\frac{(t-1)+\sqrt{(t-1)^2-(2)^2}}{2}\right) + C$$

$$= \log\left[\frac{(\sin x - 1)+\sqrt{\sin^2 x - 2\sin x - 3}}{2}\right] + C \qquad \textbf{Ans.}$$

Example 60. *Evaluate* $\int \sqrt{1+x-2x^2}\,dx$

Solution. $\int \sqrt{1+x-2x^2}\,dx = \sqrt{2}\int \sqrt{\frac{}{2}+\frac{}{2}-x^2}\,dx$

$$= \sqrt{2}\int \sqrt{\frac{1}{2}-\left(x^2-\frac{x}{2}\right)}\,dx$$

$$= \sqrt{2}\int \sqrt{\frac{1}{2}-\left(x-\frac{1}{4}\right)^2+\frac{1}{16}}\,dx$$

$$= \sqrt{2}\int \sqrt{\left(\frac{3}{4}\right)^2-\left(x-\frac{1}{4}\right)^2}\,dx$$

$$\left[\text{This is the form} = \int \sqrt{a^2-x^2}\,dx = \frac{x}{2}\sqrt{a^2-x^2}+\frac{a^2}{2}\sin^{-1}\left(\frac{x}{a}\right)\right]$$

$$\Rightarrow \quad \int \sqrt{1+x-2x^2}\,dx = \sqrt{2}\left[\frac{1}{2}\left(x-\frac{1}{4}\right)\sqrt{\frac{9}{16}-\left(x-\frac{1}{4}\right)}+\frac{1}{2}\cdot\frac{9}{16}\sin{}^{1}\left(\frac{-}{-}\right)\right]$$

$$= \frac{1}{2}\left[x-\frac{1}{4}\right]\sqrt{1+x-2x^2}+\frac{9\sqrt{2}}{32}\sin^{-1}\left[\frac{4x-1}{3}\right]+C \qquad \textbf{Ans.}$$

Example 61. *Evaluate* $\int \frac{1}{\sqrt{1-x-x^2}}\,dx$ **[SBTE. 2015, 2014, 2010]**

Solution.

Let $I = \int \frac{1}{\sqrt{1-x-x^2}}\,dx$

$$= \int \frac{1}{\sqrt{-\left(x^2+x+\frac{1}{4}\right)+\frac{1}{4}+1}}\,dx$$

$$= \int \frac{1}{\sqrt{\frac{5}{4}-\left(x+\frac{1}{2}\right)^2}}\,dx = \int \frac{1}{\sqrt{\left(\frac{\sqrt{5}}{2}\right)^2-\left(x+\frac{1}{2}\right)^2}}\,dx$$

$$= \sin^{-1}\left(\frac{x+\frac{1}{2}}{\frac{\sqrt{5}}{2}}\right)+C$$

$\therefore \qquad I = \sin^{-1}\left(\frac{2x+1}{\sqrt{5}}\right)+C$ **Ans.**

Example 62. *Evaluate:* $\int \frac{1}{x\{6(\log x)^2+7\log x+2\}}dx$ **[SBTE. 2017]**

Solution. Let $I = \int \frac{1}{x\{6(\log x)^2+7\log x+2\}}dx$

put $\log x = t,\ \frac{1}{x}dx = dt$

$$I = \int \frac{1}{6t^2+7t+2}dt = \frac{1}{6}\int \frac{1}{t^2+\frac{7}{6}t+\frac{1}{3}}dt$$

$$= \frac{1}{6}\int \frac{1}{\left(t^2+\frac{7}{6}t+\frac{49}{144}\right)+\left(\frac{1}{3}-\frac{49}{144}\right)}dt$$

$$= \frac{1}{6}\int \frac{1}{\left(t+\frac{7}{12}\right)^2-\left(\frac{1}{12}\right)^2}dt$$

$$\therefore \qquad I = \log\left|\frac{\log x+\frac{1}{2}}{\log x+\frac{2}{3}}\right|+C$$ **Ans.**

Example 63. *Evaluate* $\int \frac{1}{\sqrt{x(1-2x)}}dx$ **[DIPIETE. June 2019]**

Solution. Let $I = \int \frac{1}{\sqrt{x(1-2x)}}dx$

$$= \int \frac{1}{\sqrt{x-2x^2}}dx = \frac{1}{\sqrt{2}}\int \frac{1}{\sqrt{-\left\{x^2-\frac{x}{2}+\left(\frac{1}{4}\right)^2-\left(\frac{1}{4}\right)^2\right\}}}dx$$

$$= \frac{1}{\sqrt{2}}\int \frac{1}{\sqrt{-\left\{\left(x-\frac{1}{4}\right)^2-\left(\frac{1}{4}\right)^2\right\}}}dx = \frac{1}{\sqrt{2}}\sin^{-1}\left(\frac{x-\frac{1}{4}}{\frac{1}{4}}\right)+C$$

$$I = \frac{1}{\sqrt{2}}\sin^{-1}(4x-1)+C$$ **Ans.**

EXERCISE 31.5

Evaluate the following:

1. $\int \frac{dx}{x^2+2x+5}$ **Ans.** $\frac{1}{2}\tan^{-1}\left(\frac{x+1}{2}\right)+C$

2. $\int \frac{dx}{2x^2 + x + 1}$ **Ans.** $\frac{2}{\sqrt{7}} \tan^{-1}\left(\frac{4x+1}{\sqrt{7}}\right) + C$

3. $\int \frac{dx}{3x^2 + 4x + 1}$ **Ans.** $\frac{1}{2} \log\left(\frac{3x+1}{x+1}\right) + C$

4. $\int \frac{dx}{2x^2 - 2x + 1}$ **Ans.** $\tan^{-1}(2x - 1) + C$

5. $\int \frac{dx}{2x^2 + x + 3}$ **Ans.** $\frac{2}{\sqrt{23}} \tan^{-1}\left(\frac{4x+1}{\sqrt{23}}\right) + C$

6. $\int \frac{dx}{\sqrt{(1 - x - x^2)}}$ **Ans.** $\sin^{-1}\left(\frac{2x+1}{\sqrt{5}}\right) + C$

7. $\int \frac{dx}{\sqrt{(x^2 + 2x + 3)}}$ **Ans.** $\sinh^{-1}\left(\frac{x+1}{\sqrt{2}}\right) + C$

8. $\int \frac{dx}{\sqrt{5x^2 + 8x - 4}}$ **Ans.** $\frac{1}{\sqrt{5}} \cosh^{-1}\left(\frac{5x+4}{6}\right) + C$

9. $\int \frac{(x+1)}{2x - x^2} dx$ **[B.T.E. Delhi May 2007]** **Ans.** $\frac{-1}{2} \log(2x - x^2) + \log\left(\frac{x}{2-x}\right) + C$

10. $\int \sqrt{(3 + 2x - x^2)}\, dx$ **Ans.** $\frac{(x-1)\sqrt{3 + 2x - x^2}}{2} + 2\sin^{-1}\left(\frac{x-1}{2}\right) + C$

11. $\int \sqrt{x^2 + x + 1}\, dx$ **Ans.** $\frac{(2x+1)\sqrt{x^2 + x + 1}}{4} + \frac{3}{8} \sinh^{-1}\left(\frac{2x+1}{\sqrt{3}}\right) + C$

12. $\int \sqrt{7x - 10 - x^2}\, dx$ **Ans.** $\frac{2x-7}{4}\sqrt{7x - 10 - x^2} + \frac{9}{8} \sin^{-1}\left(\frac{2x-7}{3}\right) + C$

13. $\int \sqrt{(4x^2 + 9)}\, dx$ **Ans.** $\frac{x}{2}\sqrt{4x^2 + 9} + \frac{9}{4} \sinh^{-1}\left(\frac{2x}{3}\right) + C$

14. $\int \sqrt{x - x^2}\, dx$ **Ans.** $\frac{(2x-1)\sqrt{x - x^2}}{4} + \frac{1}{8} \sin^{-1}(2x - 1) + C$

15. $\int \sqrt{4x^2 - 4x + 5}\, dx$ **Ans.** $\frac{(2x-1)\sqrt{4x^2 - 4x + 5}}{4} + \sinh^{-1}\left(\frac{2x-1}{2}\right) + C$

16. $\int 4x^3 \sqrt{5 - x^2}\, dx$ **Ans.** $\frac{2}{15}(5 - x^2)^{3/2}(5 - 6x^2) + C$

Type I: Integral of the form $\int \frac{dx}{a \sin x + b \cos x}$ **[Diploma IETE, Dec. 2005]**

Put $a = r\cos\alpha$, $b = r\sin\alpha$, so that $r^2 = a^2 + b^2$, $\tan\alpha = \frac{b}{a}$

$$\int \frac{dx}{a\sin x + b\cos x} = \int \frac{dx}{r\cos\alpha\,\sin x + r\sin\alpha\,\cos x}$$

$$= \frac{1}{r}\int \frac{dx}{\sin(x+\alpha)} = \frac{1}{r}\int \operatorname{cosec}(x+\alpha)\,dx$$

$$= \frac{1}{\sqrt{a^2+b^2}}\log\tan\left[\frac{x}{2} + \frac{1}{2}\tan^{-1}\frac{b}{a}\right] + C$$ **Ans.**

Example 64. *Evaluate:* $\int \frac{dx}{3\sin x + 4\cos x}$

Solution. Put $3 = r\cos\alpha$, $4 = r\sin\alpha$ so that $r = 5$, $\tan\alpha = \frac{4}{3}$

or $$\int \frac{dx}{3\sin x + 4\cos x} = \int \frac{dx}{r\cos\alpha\,\sin x + r\sin\alpha\,\cos x}$$

$$= \int \frac{dx}{r\sin(x+\alpha)} = \frac{1}{5}\int \operatorname{cosec}(x+\alpha)\,dx$$

$$= \frac{1}{5}\log\tan\left[\frac{x}{2} + \frac{\alpha}{2}\right] + C$$

$$= \frac{1}{5}\log\tan\left[\frac{x}{2} + \frac{1}{2}\tan^{-1}\frac{4}{3}\right] + C$$ **Ans.**

Type II: Integral of the form $\int \frac{dx}{a\sin x + b\cos x \pm \sqrt{a^2+b^2}}$

Example 65. *Evaluate:* $\int \frac{d\theta}{3\sin\theta - 4\cos\theta + 5}$

Solution. Put $3 = r\sin\alpha$, $4 = r\cos\alpha$ so that $r = 5$, $\tan\alpha = \frac{3}{4}$

$$\int \frac{d\theta}{3\sin\theta - 4\cos\theta + 5} = \int \frac{d\theta}{5\sin\alpha\sin\theta - 5\cos\alpha\cos\theta + 5}$$

$$= \frac{1}{5}\int \frac{d\theta}{-\cos(\theta+\alpha) + 1} = \frac{1}{10}\int \frac{d\theta}{\sin^2\left(\frac{\theta}{2} + \frac{\alpha}{2}\right)}$$

$$= \frac{1}{10}\int \operatorname{cosec}^2\left[\frac{\theta}{2} + \frac{\alpha}{2}\right] d\theta = -\frac{1}{10}\cdot\frac{2}{1}\cot\left[\frac{\theta}{2} + \frac{\alpha}{2}\right]$$

$$= -\frac{1}{5}\cot\left[\frac{\theta}{2} + \frac{\alpha}{2}\right] \quad \text{where } \tan\alpha = \frac{3}{4}$$ **Ans.**

Type III: Integral of the form $\int \frac{p\sin x + q\cos x}{a\sin x + b\cos x}\,dx$

Example 66. *Evaluate* $\int \frac{2\sin x + 3\cos x}{3\sin x + 4\cos x}\,dx.$

Solution. The given expression is arranged in the form of $A + B\frac{f'(x)}{f(x)}$

Let $\int \frac{2\sin x + 3\cos x}{3\sin x + 4\cos x} = A + \frac{B(3\cos x - 4\sin x)}{3\sin x + 4\cos x}$

$\therefore$ $2\sin x + 3\cos x = A(3\sin x + 4\cos x) + B(3\cos x - 4\sin x)$

$2\sin x + 3\cos x = (3A - 4B)\sin x + (4A + 3B)\cos x$

Equating the coefficients of sin x and cos x on both sides.

$$\left.\begin{aligned}2 &= 3A - 4B\\ 3 &= 4A + 3B\end{aligned}\right] \Rightarrow A = \frac{18}{25},\ B = \frac{1}{25}$$

$$\int \frac{2\sin x + 3\cos x}{3\sin x + 4\cos x}\,dx = \int \frac{18}{25}\,dx + \frac{1}{25}\int \frac{3\cos x - 4\sin x}{3\sin x + 4\cos x}\,dx$$

$$= \frac{18}{25}x + \frac{1}{25}\log(3\sin x + 4\cos x) + C$$ **Ans.**

Type IV: Integral of the form $\int \frac{dx}{a + b\sin x}$ or $\int \frac{dx}{a + b\cos x}$ **[B.T.E. Delhi, 2008]**

Put $\sin x = \frac{2\tan\frac{x}{2}}{1 + \tan^2\frac{x}{2}}$ and $\cos x = \frac{1 - \tan^2\frac{x}{2}}{1 + \tan^2\frac{x}{2}}$

After solving,

Put $\tan\frac{x}{2} = t$ so that $\frac{1}{2}\sec^2\frac{x}{2}\,dx = dt$

Example 67. *Evaluate* $\int \frac{dx}{2 - \sin x}$

Solution. $\int \frac{dx}{2 - \sin x} = \int \frac{dx}{2 - \frac{2\tan x/2}{1 + \tan^2 x/2}}$, $\left[\text{Put } \sin x = \frac{2\tan\frac{x}{2}}{1 + \tan^2\frac{x}{2}}\right]$

$$= \int \frac{(1 + \tan^2 x/2)\cdot dx}{2\left(1 + \tan^2\frac{x}{2}\right) - 2\tan\frac{x}{2}} = \frac{1}{2}\int \frac{\sec^2 x/2\cdot dx}{\tan^2\frac{x}{2} - \tan\frac{x}{2} + 1}$$

$$= \int \frac{dt}{t^2 - t + 1}$$ $\left[\begin{aligned}&\text{Put } \tan\frac{x}{2} = t\\ &\frac{1}{2}\sec^2\frac{x}{2}\,dx = dt\end{aligned}\right.$

$$= \int \frac{dt}{t^2 - t + \frac{1}{4} + 1 - \frac{1}{4}} = \int \frac{dt}{\left(t - \frac{1}{2}\right)^2 + \frac{3}{4}}$$

$$= \frac{2}{\sqrt{3}} \tan^{-1}\left(\frac{t - \frac{1}{2}}{\frac{\sqrt{3}}{2}}\right) = \frac{2}{\sqrt{3}} \tan^{-1}\left(\frac{2t-1}{\sqrt{3}}\right) + C$$

$$= \frac{2}{\sqrt{3}} \tan^{-1}\left(\frac{2\tan\frac{x}{2} - 1}{\sqrt{3}}\right) + C \qquad \textbf{Ans.}$$

EXERCISE 31.6

Evaluate the following:

1. $\int \frac{dx}{\sqrt{3}\sin x - \cos x}$ **Ans.** $\frac{1}{2}\log\tan\left(\frac{x}{2} - \frac{\pi}{12}\right) + C$

2. $\int \frac{\sec x}{\sqrt{3} + \tan x}\,dx$ **Ans.** $\frac{1}{2}\log\tan\left(\frac{x}{2} + \frac{\pi}{6}\right) + C$

3. $\int \frac{dx}{3\cos x - 4\sin x + 5}$ **Ans.** $\frac{1}{5}\tan\left(\frac{x}{2} + \frac{1}{2}\tan^{-1}\frac{4}{3}\right) + C$

4. $\int \frac{dx}{12\cos x - 5\sin x + 13}$ **Ans.** $\frac{1}{13}\tan\left(\frac{x}{2} + \frac{1}{2}\tan^{-1}\frac{5}{12}\right) + C$

5. $\int \frac{\sin x + 8\cos x}{2\sin x + 3\cos x}\,dx$ **Ans.** $2x + \log(2\sin x + 3\cos x) + C$

6. $\int \frac{dx}{2 + 3\tan x}$ **Ans.** $\frac{2x}{13} + \frac{3}{13}\log(2\cos x + 3\sin x) + C$

7. $\int \frac{dx}{3 + 4\sin x}$ **Ans.** $\frac{1}{\sqrt{7}}\log\left(\frac{3\tan\frac{x}{2} + 4 - \sqrt{7}}{3\tan\frac{x}{2} + 4 + \sqrt{7}}\right) + C$

8. $\int \frac{dx}{2\cos x + 1}\,dx$ **Ans.** $\frac{1}{\sqrt{3}}\log\left(\frac{\sqrt{3} + \tan\frac{x}{2}}{\sqrt{3} - \tan\frac{x}{2}}\right) + C$

9. $\int \frac{dx}{1 + \sin x}$ **Ans.** $-\tan\left(\frac{\pi}{4} - \frac{x}{2}\right) + C$

10. $\int \frac{dx}{3\cos x + 4\sin x + 6}$ **Ans.** $\frac{1}{\sqrt{11}}\tan^{-1}\left(\frac{3\tan x/2 + 4}{\sqrt{11}}\right) + C$

11. $\int \frac{1}{a^2 \sin^2 x + b^2 \cos^2 x} dx$ **Ans.** $\frac{1}{ab} \tan^{-1}\left[\frac{a \tan x}{b}\right] + C$

21.13 INTEGRALS OF THE FORM

(*i*) $\int \frac{dx}{a + b\cos^2 x}$ (*ii*) $\int \frac{dx}{a + b\sin^2 x}$ (*iii*) $\int \frac{1}{a\sin^2 x + b\cos^2 x} dx$

Method:

(1) Divide numerator and denominator by $\cos^2 x$

(2) Replace $\sec^2 x$ by $1 + \tan^2 x$.

(3) Put $\tan x = t$ so that $\sec^2 x dx = dt$

(4) Evaluate the integral obtained e.g. $\int \frac{1}{at^2 + bt + c} dt$, by using methods given in previous sections

Example 68: Evaluate: $\int \frac{1}{3 - 4\sin^2 x} dx$

Solution: Let $I = \int \frac{1}{3 - 4\sin^2 x} dx$

$= \int \frac{\sec^2 x}{3\sec^2 x - 4\tan^2 x} dx$ [Dividing num. and denom. by $\cos^2 x$]

Put $\tan x = t$

$\sec^2 x dx = dt$

$\therefore$ $I = \int \frac{dt}{3(1+t^2) - 4t^2} = \int \frac{dt}{3 - t^2} = \int \frac{1}{(\sqrt{3})^2 - t^2} dt$

$= \frac{1}{2\sqrt{3}} \log\left|\frac{\sqrt{3} + t}{\sqrt{3} - t}\right| + C$

$= \frac{1}{2\sqrt{3}} \log\left|\frac{\sqrt{3} + \tan x}{\sqrt{3} - \tan x}\right| + C$ **Ans.**

Example 69: Evaluate $\int \frac{dx}{a^2 \sin^2 x + b^2 \cos^2 x}$

Solution. We write

$$\int \frac{dx}{a^2 \sin^2 x + b^2 \cos^2 x} = \int \frac{\sec^2 x\, dx}{b^2 + a^2 \tan^2 x}$$

Putting $t = \tan x$, we see that the given integral

$$= \int \frac{dt}{b^2 + a^2 t^2} = \frac{1}{a^2} \int \frac{dt}{(t^2 + b^2 / a^2)}$$

$$= \frac{1}{a^2} \cdot \frac{a}{b} \tan^{-1}\left(\frac{at}{b}\right) = \frac{1}{ab} \tan^{-1}\left(\frac{a \tan x}{b}\right).$$

Example 70: *Evaluate* $\int \dfrac{\sin x \cos x}{a^2 \cos^2 x + b^2 \sin^2 x}\, dx.$

Solution. Dividing the numerator and denominator by $\cos^2 x$, we have

$$\int \frac{\sin x \cos x}{a^2 \cos^2 x + b^2 \sin^2 x}\, dx = \int \frac{\tan x\, dx}{a^2 + b^2 \tan^2 x}.$$

Putting $\tan^2 x = t$, we see that the integral

$$= \int \frac{dt}{2(1+t)(a^2+b^2 t)}$$

$$= \frac{1}{2}\int \left(\frac{1}{a^2+b^2} \cdot \frac{1}{1+t} + \frac{b^2}{b^2-a^2} + \frac{1}{a^2+b^2 t} \right) dt$$

$$= \frac{1}{2}\left[\frac{1}{a^2-b^2} \log(1+t) + \frac{b^2}{b^2-a^2} \cdot \frac{1}{b^2} \log(a^2 - b^2 t) \right]$$

$$= \frac{1}{2(a^2-b^2)} \log \left(\frac{\sec^2 x}{a^2 + b^2 \tan^2 x} \right)$$

$$= -\frac{1}{2(a^2-b^2)} \log(a^2 \cos^2 x + b^2 \sin^2 x).$$

Note. The integral may also be evaluated by putting

$$\sin^2 x = t, \text{ or } \cos^2 x = t.$$

Example 71: *Evaluate* $\int \dfrac{dx}{(2\sin x + \cos x)^2}$

Solution. $\int \dfrac{dx}{(2\sin x + \cos x)^2}$

Dividing numerator and denominator by $\cos^2 x$.

$$= \int \frac{\sec^2 x\, dx}{(2\tan x + 1)^2}$$

Put $\tan x = t$ so that $\sec^2 x\, dx = dt$

$$= \int \frac{dx}{(2t+1)^2} = \frac{-1}{2} \frac{1}{2t+1} = -\frac{1}{2} \cdot \frac{1}{2\tan x + 1} + C$$ **Ans.**

31.14 INTEGRALS OF THE FORM

$$\frac{a\sin x + b\cos x}{c\sin x + d\cos x} = A + B\frac{f'(x)}{f(x)}$$

Method

Put the numerator = *A*(Denominator) + *B*(Differential coefficient of denominator)

Method will be clear from the following solved examples.

Example 72. *Evaluate* $\int \frac{2\sin x + 3\cos x}{3\sin x + 4\cos x}\,dx$.

Solution. $\int \frac{2\sin x + 3\cos x}{3\sin x + 4\cos x}\,dx$

Put $\frac{2\sin x + 3\cos x}{3\sin x + 4\cos x} \equiv A + B\frac{(3\cos x - 4\sin x)}{3\cos x + 4\sin x}$

or $2\sin x + 3\cos x \equiv A(3\sin x + 4\cos x) + B(3\cos x - 4\sin x)$

Equating the coefficients of $\sin x$ and $\cos x$ on both sides

$$\left.\begin{array}{l} 2 = 3A - 4B \\ 3 = 4A + 3B \end{array}\right] \Rightarrow A = \frac{18}{25}, B = \frac{1}{25}$$

$$\therefore \quad \int \frac{2\sin x + 3\cos x}{3\sin x + 4\cos x}\,dx = \int\left(\frac{18}{25} + \frac{1}{25}\frac{3\cos x - 4\sin x}{3\sin x + 4\cos x}\right)dx$$

$$= \frac{18}{25}x + \frac{1}{25}\log(3\sin x + 4\cos x) + C$$ **Ans.**

Example 73. *Evaluate* $\int \frac{dx}{1 + \tan x}$.

Solution. $\int \frac{dx}{1+\tan x} = \int \frac{dx}{1 + \frac{\sin x}{\cos x}} = \int \frac{\cos x\,dx}{\cos x + \sin x}$

Put $\frac{\cos x}{\cos x + \sin x} = A + B\frac{f'(x)}{f(x)}$

$$= A + B\frac{-\sin x + \cos x}{\cos x + \sin x}$$

$$\cos x = A(\cos x + \sin x) + B(-\sin x + \cos x)$$

or $\cos x = (A + B)\cos x + (A - B)\sin x$

Equating the coefficients of $\cos x$ and $\sin x$ on both sides, we get

$$\left.\begin{array}{l} 1 = A + B \\ 0 = A - B \end{array}\right] \Rightarrow \begin{array}{l} A = \tfrac{1}{2} \\ B = \tfrac{1}{2} \end{array}$$

Hence $\int \frac{\cos x}{\cos x + \sin x}\,dx = \int\left[\frac{1}{2} + \frac{1}{2}\frac{-\sin x + \cos x}{\cos x + \sin x}\right]dx$

$$= \frac{x}{2} + \frac{1}{2}\log(\cos x + \sin x) + C$$ **Ans.**

EXERCISE 31.7

Evaluate the following integrals:

1. $\int \frac{\cos x}{1 + \sin^2 x}\,dx$, **Ans.** $\tan^{-1}(\sin x) + C$

2. $\int \frac{dx}{\sin x + \sin 2x}$ **Ans.** $\frac{1}{3}\log\left[\sin x\frac{(1+\cos x)}{(1+2\cos x)^2}\right] + C$

3. $\int \frac{2-\sin x}{\sin x(1-\cos x)}dx$ **Ans.** $\log \tan\frac{1}{2}x+\cot\frac{1}{2}x-\frac{1}{2}\text{coses}^2 x+\frac{1}{2}x+C$

4. $\int \frac{\sec x}{1+\operatorname{cosec} x}dx$ **Ans.** $\frac{1}{4}\log\left[\frac{1+\sin x}{1-\sin x}\right]+\frac{1}{2}(1+\sin x)+C$

5. $\int \frac{dx}{1+\cos^2 x}$ **Ans.** $\frac{1}{\sqrt{2}}\tan^{-1}\left(\tan\frac{x}{\sqrt{2}}\right)+C$

6. $\int \frac{dx}{1+3\sin^2 x}$ **Ans.** $\frac{1}{2}\tan^{-1}(2\tan x)+C$

7. $\int \frac{d\theta}{5+4\cos 2\theta}$ **Ans.** $\frac{1}{3}\tan^{-1}\left(\frac{1}{3}\tan\theta\right)+C$

8. $\int \frac{dx}{1-\cos^4 x}$ **Ans.** $-\frac{1}{2}\cot x+\frac{1}{2\sqrt{2}}\left(\tan^{-1}\frac{1}{\sqrt{2}}\tan x\right)+C$

9. $\int \frac{dx}{(2\sin x+\cos x)^2}$ **Ans.** $-\frac{\cos\frac{x}{2}}{2\sin x+\cos x}+C$

10. $\int \frac{26\sin x-13\cos x}{4\sin x-7\cos x}dx$ **Ans.** $3x + 2\log(4\sin x - 7\cos x)$

11. $\int \frac{\sin x+8\cos x}{2\sin x+3\cos x}dx$ **Ans.** $2x + \log(2\sin x + 3\cos x)$

12. $\int \frac{5\cos x+6}{2\cos x+\sin x+3}dx$ **Ans.** $2x + \log(2\cos x + \sin x + 3)$

13. $\int \frac{dx}{2+3\tan x}$ **Ans.** $\frac{2x}{13}+\frac{3}{13}\log(2\cos x + 3\sin x)$

14. $\int \frac{\sin x\, dx}{\sin x+\cos x}$ **Ans.** $\frac{x}{2}-\frac{1}{2}\log(\sin x+\cos x)$

15. Fill in the blanks

(*a*) $\int \frac{2\sin\theta+\cos\theta}{7\sin\theta-5\cos\theta}d\theta = \ldots\ldots$ **Ans.** $\frac{9\theta}{74}+\frac{17}{74}\log(7\sin\theta - 5\cos\theta)$

(*b*) $\int \frac{1}{\sqrt{3}\sin x-\cos x}dx = \ldots\ldots$ **Ans.** $\frac{1}{2}\log\tan\left(\frac{x}{2}-\frac{\pi}{12}\right)$

(*c*) $\int \frac{dx}{\sqrt{\sin^3 x\sin(x+\alpha)}} = \ldots\ldots$ **Ans.** $\frac{-2}{\sin\alpha}\left[\frac{\sin(x+\alpha)}{\sin x}\right]^{1/2}$

(*d*) $\int \frac{dx}{(1+2\cos x)^3} = \ldots\ldots$ **Ans.** $\frac{-2\sin x\cos x}{3(1+2\cos x)^2}-\frac{\sqrt{3}}{9}\log\frac{\cos\left(\frac{x}{2}+\frac{\pi}{6}\right)}{\cos\left(\frac{x}{2}-\frac{\pi}{6}\right)}$

(*e*) $\int\left\{\sqrt{(\sin x)}+\sqrt{(\cos x)}\right\}^{-4}dx = \ldots\ldots$ **Ans.** $\frac{-\left[3\sqrt{(\tan x)}+1\right]}{3\left[\sqrt{(\tan x)}+1\right]^3}$

31.15 REDUCTION FORMULA FOR $\int \sin^n x\,dx$ [Diploma, IETE, 2016]

Illustration. $\int \sin^n x\,dx = \int \sin^{n-1} x \cdot \sin x\,dx$

[Integrating by parts.]

$$= \sin^{n-1} x(-\cos x) - \int (n-1)\sin^{n-2} x \cos x\,(-\cos x)\,dx$$

$$= -\sin^{n-1} x \cos x + (n-1)\int \sin^{n-2} x \cos^2 x\,dx$$

$$= -\cos x \sin^{n-1} x + (n-1)\int \sin^{n-2} x(1-\sin^2 x)dx$$

$$= -\cos x \sin^{n-1} x + (n-1)\int \sin^{n-2} x\,dx - (n-1)\int \sin^n x\,dx$$

$\therefore \quad \int \sin^n x\,dx + (n-1)\int \sin^n x\,dx = -\cos x \sin^{n-1} x + (n-1)\int \sin^{n-2} x\,dx$

or $\quad n\int \sin^n x\,dx = -\cos x \sin^{n-1} x + (n-1)\int \sin^{n-2} x\,dx$

or $\quad \int \sin^n x\,dx = -\dfrac{1}{n}\cos x \sin^{n-1} x + \dfrac{n-1}{n}\int \sin^{n-2} x\,dx$

This is the required formula.

Note. Similarly we get

$$\boxed{\int \cos^n x\,dx = \frac{1}{n}\cos^{n-1} x \sin x + \frac{n-1}{n}\int \cos^{n-2} x\,dx}$$

Example 74: $\int \cos^4 x\,dx$

Solution: $\int \cos^4 x\,dx = \dfrac{\sin x \cos^3 x}{4} + \dfrac{3}{4}\int \cos^2 x\,dx$

$$= \frac{\sin x \cos^3 x}{4} + \frac{3}{4}\left[\frac{x}{2} + \frac{\sin 2x}{4}\right] + C \qquad \textbf{Ans.}$$

Example 75: *Integrate*

(*i*) $\sin^3 x$ (*ii*) $\sin^4 x$.

Solution: (*i*) The index, 3, being odd, we proceed by substitution.

We put $\cos x = t$, so that

$$\int \sin^3 x\,dx = -\int (1-t^2)\,dt$$

$$= -\left(t - \frac{t^2}{3}\right)$$

$$= -\left(\cos x - \frac{\cos^3 x}{3}\right)$$

$$= -\cos x + \frac{1}{3}\cos^3 x + C. \qquad \textbf{Ans.}$$

(*ii*) The index, 4, being even, we employ the reduction formula. To integrate $\sin^4 x$, we first obtain the reduction formula. Putting $n = 4, 2$ successively in the reduction formula, we obtain

$$\int \sin^4 x\,dx = -\frac{\cos x \sin^3 x}{4} + \frac{3}{4}\int \sin^2 x\,dx \qquad \ldots(i)$$

$$\int \sin^2 x\, dx = -\frac{\cos x \sin x}{2} + \frac{1}{2}\int \sin^0 x\, dx$$

$$= -\frac{\cos x \sin x}{2} + \frac{1}{2}x \qquad \ldots(ii)$$

From (*i*) and (*ii*), we obtain

$$\int \sin^4 x = -\frac{\cos x \sin^3 x}{4} - \frac{3}{8}\cos x \sin x + \frac{3}{8}x + C \qquad \textbf{Ans.}$$

EXERCISE 31.8

Evaluate:

1. (*i*) $\int \sin^2 x\, dx$ (*ii*) $\int \sin^5 x\, dx$ (*iii*) $\int \sin^6 x\, dx$.

Ans. (*i*) $\frac{1}{2}(x - \sin x \cos x) + C$ (*ii*) $-\left[\cos x - \frac{1}{3}\cos^3 x + \frac{1}{5}\cos^5 x\right] + C$

(*iii*) $\frac{1}{6}\cos x \sin^5 x - \frac{5}{24}\cos x \sin^3 x - \frac{5}{16}\cos x \sin x + \frac{5}{16}x + C$

2. (*i*) $\int \sin^7 x\, dx$ (*ii*) $\int \sin^8 x\, dx$ (*iii*) $\int \sin^9 x\, dx$.

Ans. (*i*) $\frac{\cos^7 x}{7} - \frac{3\cos^5 x}{5} + \cos^3 x - \cos x + C$

(*ii*) $-\frac{1}{8}\sin^7 x \cos x - \frac{7}{48}\sin^5 x \cos x + \frac{5}{64}\left(x + \frac{\sin 4x}{8} - \sin 2x\right) + C$

(*iii*) $-\cos x + \frac{4}{3}\cos^3 x - \frac{6}{5}\cos^5 x + \frac{4}{7}\cos^7 x - \frac{\cos^9 x}{9} + C$

3. (*i*) $\int \cos^7 x\, dx$ (*ii*) $\int \cos^6 2x\, dx$ **Ans.** (*i*) $\sin x - \sin^3 x + \frac{3}{5}\sin^5 x - \frac{1}{7}\sin^7 x + C$

(*ii*) $-\frac{1}{12}\sin^5 2x \cos 2x - \frac{5}{48}\sin^3 2x \cos 2x + \frac{15}{96}\sin 2x \cos 2x + \frac{15}{48}x + C$

Reduction formula for $\int \sin^m x \cos^n x\, dx$

The integral $\int \sin^m x \cos^n x\, dx$ is to be connected with

$\int \sin^{m-2} x \cos^n x\, dx$.

Step. (*i*) Here the smaller powers of sin x and cos x are $m - 2$ and n.

Take the new function whose powers are $(m - 2) + 1$, $n + 1$.

Let $P = \sin^{(m-2)+1} x \cdot \cos^{n+1} x$

(*ii*) Find $\frac{dP}{dx}$ and re-arrange the function.

(*iii*) Integrate both sides.

Solution. Let $\qquad P = \sin^{(m-2)+1} x \cdot \cos^{n+1} x$

or $\qquad P = \sin^{m-1} x \cos^{n+1} x$

$$\frac{dP}{dx} = \sin^{m-1} x.(n+1)\cos^n x(-\sin x) + (m-1)\sin^{m-2} x \cos x.\cos^{n+1} x$$

$$= -(n+1)\sin^m x \cos^n x + (m-1)\sin^{m-2} x \cos^{n+2} x$$

$$= -(n+1)\sin^m x \cos^n x + (m-1)\sin^{m-2} x \cos^n x(1-\sin^2 x)$$

$$= -(n+1)\sin^m x \cos^n x + (m-1)\sin^{m-2} x \cos^n x - (m-1)\sin^m x \cos^n x$$

$$= -(m+n)\sin^m x \cos^n x + (m-1)\sin^{m-2} x \cos^n x$$

Integrating both sides, we get

$$P = -(m+n)\int \sin^m x \cos^n x\, dx + (m-1)\int \sin^{m-2} x \cos^n x\, dx$$

or $(m+n)\int \sin^m x \cos^n x\, dx = (m-1)\int \sin^{m-2} x \cos^n x\, dx - P$

$$= (m-1)\int \sin^{m-2} x \cos^n x\, dx - \sin^{m-1} x \cos^{n+1} x$$

$$\int \sin^m x \cos^n x\, dx = \frac{m-1}{m+n}\int \sin^{m-2} x \cos^n x\, dx - \frac{1}{m+n}\sin^{m-1} x \cos^{n+1} x$$

$$= -\frac{\sin^{m-1} x \cos^{n+1} x}{m+n} + \frac{m-1}{m+n}\int \sin^{m-2} x \cos^n x\, dx$$

or $$I_{m,n} = -\frac{\sin^{m-1} x \cos^{n+1} x}{m+n} + \frac{m-1}{m+n} I_{m-2,n}$$

This is the required formula.

Working Rule for $\int \sin^m x \cos^n x\, dx$; $m, n \in N$.

Step 1. If the integral has power of sin x and odd positive integer, put cos $x = t$.

Step 2. If the integral has power of cos x an odd positive integer, put sin $x = t$.

Step 3. It the integral has power of sin x and cos x both odd positive integers, put sin $x = t$ or cos $x = t$.

Step 4. If the power of sin x and cos x are both even positive integers express $\sin^m x \cos^n x$ in terms of cosines of multiples of angles by using $2\sin^2 x = 1 - \cos 2x$ or $2\cos^2 x = 1 + \cos 2x$.

Step 5. If the sum of powers of sin x and cos x is an even negative integer, express the integrand as the product of the powers of tan x and sec x and put tan $x = t$.

Example 76: *Evaluate:*

$$(i)\ \int \sin^3 x \cdot \cos^2 x\, dx \quad (ii)\ \int \sin^2 x \cdot \cos^5 x\, dx \quad (iii)\ \int \cos^3 x \cdot \sin^9 x\, dx$$

Solution. (*i*) Let $I = \int \sin^3 x \cdot \cos^2 x\, dx = \int \sin^2 x \cdot \cos^2 x(\sin x\, dx)$

$$= \int (1-\cos^2 x)\cos^2 x(\sin x\, dx)$$

Here, power of sin x is odd, therefore we put

cos $x = t$ so that $-\sin x\, dx = dt$.

$\therefore \qquad I = -\int (1-t^2)\cdot t^2 \cdot dt = -\int (t^2 - t^4) dt = -\left[\frac{t^3}{3} - \frac{t^5}{5}\right] + C$

$= -\frac{1}{3}\cos^3 x + \frac{1}{5}\cos^5 x + C$ **Ans.**

(ii) Let $I = \int \sin^2 x \cdot \cos^5 x \, dx = \int \sin^2 x \cdot \cos^4 x (\cos x \, dx)$

$= \int \sin^2 x (1-\sin^2 x)^2 (\cos x)\, dx \qquad [\because \cos^2 x = 1 - \sin^2 x]$

Here, power of cos x is odd, therefore, we put

$\sin x = t$ so that $\cos x \, dx = dt.$

$\Rightarrow \qquad I = \int t^2 (1-t^2)^2 dt = \int t^2 (1 + t^4 - 2t^2)\, dt = \int (t^2 + t^6 - 2t^4)\, dt$

Hence $\qquad I = \frac{t^3}{3} + \frac{t^7}{7} - \frac{2t^5}{5} + C = \frac{\sin^3 x}{3} + \frac{\sin^7 x}{7} - 2\frac{\sin^5 x}{5} + C$ **Ans.**

(iii) Let $\qquad I = \int \cos^3 x \cdot \sin^9 x \, dx.$

Here, neither sin x nor cos x are even so we can put either cos $x = t$ or sin $x = t$. Since, power of sin x is higher So, we put sin $x = t$ so that cos x $dx = dt$.

$\Rightarrow \qquad I = \int \cos^3 x \cdot \sin^9 x \, dx = \int \cos^2 x \cdot \sin^9 x \cdot (\cos x \, dx)$

$= \int (1 - \sin^2 x)\sin^9 x \cdot (\cos x \, dx) \qquad [\because \cos^2 x = 1 - \sin^2 x]$

$= \int (1-t^2)\cdot t^9 \cdot dt = \int (t^9 - t^{11}) dt$

$= \frac{1}{10} t^{10} - \frac{1}{12} t^{12} + C = \frac{1}{10}\sin^{10} x - \frac{1}{12}\sin^{12} x + C$ **Ans.**

Example 77. *Evaluate:*

$$\int \cos^3 x \cdot e^{\log(\sin x)^2} \cdot dx$$

Solution. Let $\qquad I = \int \cos^3 x \cdot e^{\log(\sin x)^2} \cdot dx = \int \cos^3 x \cdot \sin^2 x \, dx$

Here, power of cos x is odd, therefore we put sin $x = t$ so that cos x $dx = dt$

$\Rightarrow \qquad I = \int \cos^2 x \cdot \sin^2 x \cdot \cos x \, dx = \int (1 - \sin^2 x)\cdot \sin^2 x (\cos x \, dx)$

$= \int (1-t^2) t^2 \cdot dt = \int (t^2 - t^4)\, dt = \frac{t^3}{3} - \frac{t^5}{5} + C = \frac{\sin^3 x}{3} - \frac{\sin^5 x}{5} + C$

EXERCISE 31.9

Evaluate the following integrals:

1. $\int \sin^3 x \cdot \cos^3 x \, dx$ **Ans.** $\frac{1}{6}\cos^6 x - \frac{1}{4}\cos^4 x + C$

2. $\int \sin^3 x \cdot \cos^4 x\, dx$ **Ans.** $-\dfrac{\cos^5 x}{5}+\dfrac{\cos^7 x}{7}+C$

3. $\int \sin^3 x \cdot \cos^5 x\, dx$ **Ans.** $-\dfrac{\cos^6 x}{6}+\dfrac{\cos^8 x}{8}+C$

4. $\int \sin^5 x \cdot \cos x\, dx$ **Ans.** $\dfrac{\sin^6 x}{6}+C$

5. $\int \sin^3 x \cdot \cos^6 x\, dx$ **Ans.** $-\left[\dfrac{\cos^7 x}{7}-\dfrac{\cos^9 x}{9}\right]+C$

6. $\int x\cos^3 x^2 \cdot \sin x^2\, dx$ **Ans.** $-\dfrac{1}{8}\cos^4 x^2+C$

7. $\int \sin x \cdot \cos^5 \dfrac{x}{2}\, dx$ **Ans.** $-\dfrac{4}{7}\cos^7 \dfrac{x}{2}+C$

8. $\int \cos^3(ax+b)\sin(ax+b)\, dx$ **Ans.** $-\dfrac{\cos^4(ax+b)}{4a}+C$

9. $\int \cos^2 x \cdot \sin^5 x\, dx$ **Ans.** $\dfrac{2}{5}\cos^5 x-\dfrac{1}{7}\cos^7 x-\dfrac{1}{3}\cos^3 x+C$

10. $\int x\sin^3 x^2 \cdot \cos x^2\, dx$ **Ans.** $\dfrac{1}{8}\sin^4 x^2+C$

Definite Integrals and Their Applications

32.1 DEFINITE INTEGRAL

If $\phi(x)$ be an integral of $f(x)$, then $\phi(b) - \phi(a)$ is called the definite integral of $f(x)$ between the limits a and b and is denoted by the symbol $\int_a^b f(x)\,dx$ where a is called the lower limit and b the upper limit.

The interval $[a, b]$ is called the range of integration.

$\phi(b) - \phi(a)$ is also denoted by $[\phi(x)]_a^b$

Thus we write $\int_a^b f(x)\,dx = [\phi(x)]_a^b = \phi(b) - \phi(a)$

Note. 1. The name is given as "definite integral" because the indefinite constant of integration C does not appear here. It is clear from the following:

$$\int_a^b f(x)\,dx = [\phi(x) + C]_a^b = [\phi(b) + C] - [\phi(a) + C]$$

$$\boxed{\int_a^b f(x)\,dx = \phi(b) - \phi(a)}$$

2. The meaning of the word "limit" is "end value". It is entirely different from the sense in the phase the limit of $f(x)$ as x tends to b is A.

3. $\int_a^b f(x)\,dx$ is read as the integral of $f(x)dx$ from a to b.

4. $\int_a^b f(x)\,dx$ geometrically represents an area bounded by the curve $y = f(x)$, the x-axis and the two ordinates $x = a$, $x = b$.

IMPORTANT PROPERTIES

$$\int_a^b f(x)dx = [\phi(x)]_a^b = \phi(b) - \phi(a), \text{ where } \phi(x) = \int f(x)dx$$

Theorems:

1. $\int_a^b f(x)dx = -\int_b^a f(x)dx$ (Interchanging the limits)

2. $\int_a^b f(x)dx = \int_a^c f(x)dx + \int_c^b f(x)dx$ (where $a < c < b$)

3. $\int_a^b f(x)dx = \int_a^b f(a+b-x)dx$

4. $\int_0^a f(x)dx = \int_0^a f(a-x)dx$ (Very important)

5. $\int_0^{2a} f(x)dx = \int_0^a f(x)dx + \int_0^a f(2a-x)dx$

6. $\int_0^{2a} f(x)dx = 2\int_0^a f(x)dx,$ if $f(2a-x) = f(x)$

$= 0,$ if $f(2a-x) = -f(x)$

7. $\int_{-a}^a f(x)dx = 2\int_0^a f(x)dx,$ if $f(x)$ is an even function.

$= 0,$ if $f(x)$ is an odd function.

8. $\int_a^b f(x)dx = \lim_{h\to 0} h\left[f(a) + f(a+h) + f(a+2h) + ... + f(a+\overline{n-1}h)\right]$

9. $\int_0^{\pi/2} \sin^n x\, dx = \dfrac{(n-1)(n-3)\ldots\ldots}{n(n-2)(n-4)\ldots\ldots} \times \dfrac{\pi}{2},$ if n is even.

32.2 EVALUATION OF DEFINITE INTEGRAL

The value of the definite integral is obtained by

(*a*) Integrating the given function, omiting the arbitrary constant.

(*b*) Substituting the first, upper limit for the variable x in the integral and then lower limit.

(*c*) Subtract second result from the first.

Example 1. *Evaluate* (*a*) $\int_2^3 x^3\, dx$ (*b*) $\int_{-\pi/2}^{+\pi/2} \cos x\, dx$ (*c*) $\int_1^3 \dfrac{x}{1+x^2}\, dx$

Solution. (*a*) $\int_2^3 x^3\, dx = \left[\dfrac{x^4}{4}\right]_2^3 = \left[\dfrac{(3)^4}{4} - \dfrac{(2)^4}{4}\right] = \dfrac{81}{4} - \dfrac{16}{4} = \dfrac{65}{4}$ **Ans.**

(*b*) $\int_{-\pi/2}^{\pi/2} \cos x\, dx = [\sin x]_{-\pi/2}^{\pi/2} = \left[\sin\dfrac{\pi}{2} - \sin\left(-\dfrac{\pi}{2}\right)\right]$

$= 1 - (-1) = 2$ **Ans.**

(*c*) Put $t = 1 + x^2$ so that $dt = 2x\, dx$

$$\int_1^3 \frac{x}{1+x^2}\, dx = \frac{1}{2}\cdot\int \frac{dt}{t} = \frac{1}{2}\log t$$

$$= \frac{1}{2}\left[\log(1+x^2)\right]_1^3 = \frac{1}{2}[\log 10 - \log 2] = \frac{1}{2}\log 5$$ **Ans.**

Example 2. *Evaluate:* $\int_0^{\pi/2} \dfrac{\sin x}{(1-\cos x)^2}\, dx.$

Solution. Put $1 - \cos x = t$, $\sin x\, dx = dt$

$t = 1 - \cos x = 1 - \cos \pi/2 = 1$ $\therefore$ Upper limit of $t = 1$

$t = 1 - \cos x = 1 - \cos 0 = 0$ $\therefore$ Lower limit of $t = 0$

$$\int_0^{\pi/2} \frac{\sin x}{(1-\cos x)^2}\, dx = \int_0^1 \frac{dt}{t^2} = \left[\frac{t^{-1}}{-1}\right]_0^1 = -\left[1 - \frac{1}{0}\right] = \infty$$ **Ans.**

Example 3. *Evaluate:* $\int_0^a \frac{dx}{(x^2+a^2)^2}$

Solution. Put $x = a \tan\theta$ so that $dx = a \sec^2\theta \, d\theta$

$\theta = \tan^{-1}(x/a) = \tan^{-1}(a/a) = \tan^{-1}(1) = \pi/4$ $\therefore$ Upper limit of $\theta = \pi/4$

$\theta = \tan^{-1}(x/a) = \tan^{-1}(0/a) = 0$ $\therefore$ Lower limit of $\theta = 0$

$$\int_0^a \frac{dx}{(x^2+a^2)^2} = \int_0^{\pi/4} \frac{a\sec^2\theta\, d\theta}{(a^2\tan^2\theta + a^2)^2} = \frac{1}{a^3}\int_0^{\pi/4} \cos^2\theta\, d\theta$$

$$= \frac{1}{2a^3}\int_0^{\pi/4} (\cos 2\theta + 1)d\theta$$

$$= \frac{1}{2a^3}\left[\frac{1}{2}\sin 2\theta + \theta\right]_0^{\pi/4} = \frac{1}{2a^3}\left[\frac{1}{2} + \frac{\pi}{4}\right]$$

$$= \frac{1}{8a^3}[\pi + 2]$$ **Ans.**

Example 4. *Evaluate:* $\int_0^1 x(\tan^{-1}x)^2 dx$

Solution. $\int_0^1 (\tan^{-1}x)^2 \cdot x\, dx$

$$= \left[(\tan^{-1}x)^2 \frac{x^2}{2}\right]_0^1 - \int_0^1 \frac{2\tan^{-1}x}{1+x^2}\cdot\frac{x^2}{2}\, dx$$

$$= \frac{\pi^2}{32} - \int_0^1 \frac{(x^2+1-1)\tan^{-1}(x)}{1+x^2}\, dx$$

$$= \frac{\pi^2}{32} - \int_0^1 \frac{x^2+1}{x^2+1}\tan^{-1}x\, dx + \int_0^1\left(\frac{\tan^{-1}x}{1+x^2}\right)dx$$

$$= \frac{\pi^2}{32} - \int_0^1 1\cdot\tan^{-1}x\, dx + \int_0^1\left(\frac{\tan^{-1}x}{1+x^2}\right)dx$$

$$= \frac{\pi^2}{32} - \left[\tan^{-1}x \cdot x - \int \frac{x\, dx}{1+x^2}\right]_0^1 + \left[\frac{(\tan^{-1}x)^2}{2}\right]_0^1$$

$$= \frac{\pi^2}{32} - \left[x\tan^{-1}x - \frac{1}{2}\log(1+x^2)\right]_0^1 + \frac{\pi^2}{32}$$

$$= \frac{\pi^2}{16} - \frac{\pi}{4} + \frac{1}{2}\log 2$$ **Ans.**

Example 5. *Evaluate the following integrals.*

(*i*) $\int_0^1 \frac{1}{x^2+1}\,dx$ **[SBTE. 2016, 2015, 2014]**

(*ii*) $\int_{\pi/6}^{\pi/3} \frac{1}{\sin 2x}\,dx$ **[SBTE. 2016]**

(*iii*) $\int_0^{\pi/4} \frac{1}{\cos^2 x}\,dx$ **[SBTE. 2015]**

(*iv*) $\int_0^{\pi/2} \sin^2 x\,dx$ **[SBTE. 2014]**

Solution. Here we have

(*i*) Let $I = \int_0^1 \frac{1}{x^2+1}\,dx = \left[\tan^{-1} x\right]_0^1 = \left[\tan^{-1}(1) - \tan^{-1}(0)\right]$

$\therefore\ I = \pi/4 - 0 = \pi/4$ **Ans.**

(*ii*) Let $I = \int_{\pi/6}^{\pi/3} \frac{1}{\sin 2x} = \int_{\pi/6}^{\pi/3} \operatorname{cosec} 2x\,dx = \frac{1}{2}\left[\log(\operatorname{cosec} 2x - \cot 2x\right]_{\pi/6}^{\pi/3}$

$$= \frac{1}{2}\left[\log\left(\operatorname{cosec}\frac{2\pi}{3} - \cot\frac{2\pi}{3}\right) - \log\left(\operatorname{cosec}\frac{\pi}{3} - \cot\frac{\pi}{3}\right)\right]$$

$$= \frac{1}{2}\left[\log\frac{3}{\sqrt{3}} - \log\frac{1}{\sqrt{3}}\right]$$

$\therefore\ I = \frac{1}{2}\log 3$ **Ans.**

(*iii*) Let $I = \int_0^{\pi/4} \frac{1}{\cos^2 x} = \int_0^{\pi/4} \sec^2 x\,dx = \left[\tan x\right]_0^{\pi/4} = 1$ **Ans.**

(*iv*) Let $I = \int_0^{\pi/2} \sin^2 x\,dx = \frac{1}{2}\int_0^{\pi/2}(1 - \cos 2x)\,dx = \frac{1}{2}\left[x + \frac{\sin 2x}{2}\right]_0^{\frac{\pi}{2}} = \frac{1}{2}\left[\frac{\pi}{2} + 0 - 0 - 0\right]$

$\therefore\ I = \pi/4$ **Ans.**

Example 6. *If* $\int_0^{36} \frac{1}{2x+9}\,dx = \log k$, *find the value of* k. **[SBTE. 2015]**

Solution. Here we have

$$\int_0^{36} \frac{1}{2x+9}\,dx = \log k$$

$$\Rightarrow \quad \frac{1}{2}\left[\log(2x+9)\right]_0^{36} = \log k$$

$$\Rightarrow \quad \frac{1}{2}\left[\log 81 - \log 9\right] = \log k$$

$$\Rightarrow \qquad \frac{1}{2}\log\frac{81}{9} = \log k$$

$$\Rightarrow \qquad \log(9)^{1/2} = \log k$$

$$\Rightarrow \qquad \log 3 = \log k$$

$$\therefore \qquad k = 3$$

Ans.

Example 7. *If* $\int_0^a 5x^4\,dx = 243$, *find the value of a.* **[SBTE. 2017, 2015]**

Solution. Here we have

$$5\int_0^a x^4\,dx = 243$$

$$\Rightarrow \qquad 5\left[\frac{x^5}{5}\right]_0^a = 243$$

$$\Rightarrow \qquad a^5 = 243$$

$$\Rightarrow \qquad a^5 = (3)^5$$

$$\therefore \qquad a = 3$$

Ans.

Example 8. *Evaluate the following integrals.*

(*i*) $\int_{-2}^{-1}\frac{|x|}{x}\,dx$ **[SBTE. 2015]**

(*ii*) $\int_{-2}^{0} x|x|\,dx$ **[SBTE. 2015]**

(*iii*) $\int_{\pi/2}^{\pi} |\cos x|\,dx$ **[SBTE. 2014]**

Solution.

(*i*) Let $I = \int_{-2}^{-1}\frac{|x|}{x}\,dx$

$$= \int_{-2}^{-1}\frac{-x}{x}\,dx \qquad \left[|x| = \begin{cases} x, & x \ge 0 \\ -x, & x < 0\end{cases}\right]$$

$$\therefore\ I = -[x]_{-2}^{-1} = -[(-1)-(-2)] = -1$$

Ans.

(*ii*) Let $I = \int_{-2}^{0} x|x| = \int_{-2}^{0} x(-x)\,dx$

$$\therefore\ I = -\left[\frac{x^3}{3}\right]_{-2}^{0} = \frac{-8}{3}$$

Ans.

(*iii*) Let $I = \int_{\pi/2}^{\pi} |\cos x|$

$$\left[\because\ |\cos x| = \begin{cases} \cos x, & 0 \le x \le \frac{\pi}{2} \\ -\cos x & \frac{\pi}{2} \le x \le \pi\end{cases}\right]$$

$$= \int_{\pi/2}^{\pi} (-\cos x)\,dx$$

$$= -[\sin x]_{\pi/2}^{\pi}$$

$$\therefore\ I = -[0-1] = 1$$ **Ans.**

Example 9. *Evaluate:* $\int_0^{\pi/4} \sqrt{1+\sin 2x}\,dx$ **[DIPIETE Dec. 2018]**

Solution. Here we have $\int_0^{\pi/4} \sqrt{1+\sin 2x}\,dx$

$$\Rightarrow \int_0^{\pi/4} \sqrt{\sin^2 x + \cos^2 x + 2\sin x \cdot \cos x\,dx} \qquad [\because \sin 2x = 2\sin x \cdot \cos x]$$

$$\Rightarrow \int_0^{\pi/4} \sqrt{(\sin x + \cos x)^2}\,dx$$

$$\Rightarrow \int_0^{\pi/4} (\sin x + \cos x)\,dx = -[\cos x]_0^{\pi/4} + [\sin x]_0^{\pi/4}$$

$$= -\left[\frac{1}{\sqrt{2}} - 1\right] + \left[\frac{1}{\sqrt{2}} - 0\right]$$

$$= \frac{\sqrt{2}-1}{\sqrt{2}} + \frac{1}{\sqrt{2}}$$

$$= \frac{\sqrt{2}-1+1}{\sqrt{2}}$$

$$\int_0^{\pi/4} \sqrt{1+\sin 2x}\,dx = \frac{\sqrt{2}}{\sqrt{2}} = 1$$ **Ans.**

Example 10. *Evaluate:* $\int_0^{\pi/4} \log(1+\tan x)\,dx$ **[SBTE. 2017, 2015]**

Solution. Let $I = \int_0^{\pi/4} \log(1+\tan x)\,dx$ $\left[\because \int_0^a f(x)\,dx = \int_0^a f(a-x)\,dx\right]$

$$= \int_0^{\pi/4} \log[1+\tan(\pi/4 - x)]\,dx$$

$$= \int_0^{\pi/4} \log\left[1+\frac{\tan \pi/4 - \tan x}{1+\tan \pi/4 \cdot \tan x}\right]dx$$

$$= \int_0^{\pi/4} \log\left[1+\frac{1-\tan x}{1+\tan x}\right]dx = \int_0^{\pi/4} \log\left[\frac{2}{1+\tan x}\right]dx$$

$$= \int_0^{\pi/4} [\log 2 - \log(1+\tan x)]\,dx$$

$$= \int_0^{\pi/4} \log 2\,dx - \int_0^{\pi/4} \log(1+\tan x)\,dx$$

$$I = \log 2 \int_0^{\pi/4} dx - I \quad \Rightarrow \quad 2I = \frac{\pi}{4}\log 2$$

$$\therefore\ I = \frac{\pi}{8}\log 2$$ **Ans.**

Example 11. *Evaluate the following integrals.*

(*i*) $\int_{-\pi/2}^{\pi/2} \sin^3 x\,dx$ **[SBTE. 2016]** (*ii*) $\int_{-\pi/2}^{\pi/2} x^{10}\sin^7 x\,dx$ **[SBTE. 2015]**

Solution. (*i*) Let $I = \int_{-\pi/2}^{\pi/2} \sin^3 x\,dx$ $\left[\because \int_{-a}^{a} f(x)dx = \begin{cases} 2\int_0^a f(x)dx, & f(x) \text{ is even} \\ 0, & f(x) \text{ is odd} \end{cases}\right]$

Let $f(x) = \sin^3 x$

$$f(-x) = \sin^3(-x) = [\sin(-x)]^3 = [-\sin x]^3 = -\sin^3 x$$

$$f(-x) = -f(x),\ f(x) \text{ is odd function}$$

$$I = \int_{-\pi/2}^{\pi/2} \sin^3 x\,dx = 0$$ **Ans.**

(*ii*) Let $I = \int_{-\pi/2}^{\pi/2} x^{10}\sin^7 x\,dx$

$$f(x) = x^{10}\sin^7 x$$

$$f(-x) = (-x)^{10}\cdot\sin^7(-x) = x^{10}[-\sin x]^7$$

$$f(-x) = -x^{10}\sin^7 x = -f(x)$$

$$\therefore\ I = 0$$ **Ans.**

32.3 REDUCTION FORMULA $\int_0^{\frac{\pi}{2}} \sin^n x\,dx$ **[D.I.P.I.E.T.E. 2015]**

Solution. We know that, $\int \sin^n x\,dx = -\frac{1}{n}\cos x \sin^{n-1} x + \frac{n-1}{n}\int \sin^{n-2} x\,dx$ (see Example 5)

$$\int_0^{\frac{\pi}{2}} \sin^n x\,dx = -\left[\frac{\cos x \sin^{n-1} x}{n}\right]_0^{\frac{\pi}{2}} + \frac{n-1}{n}\int_0^{\frac{\pi}{2}} \sin^{n-2} x\,dx$$

$$\int_0^{\frac{\pi}{2}} \sin^n x\,dx = \frac{n-1}{n}\int_0^{\frac{\pi}{2}} \sin^{n-2} x\,dx \quad \text{... (1)}$$

In (1) change n into $n-2$.

$$\int_0^{\frac{\pi}{2}} \sin^{n-2} x\,dx = \frac{n-3}{n-2}\int_0^{\frac{\pi}{2}} \sin^{n-4} x\,dx \quad \text{... (2)}$$

Putting the value of $\int_0^{\frac{\pi}{2}} \sin^{n-2} x\,dx$ from (2) in (1), we get

$$\int_0^{\frac{\pi}{2}} \sin^n x\,dx = \frac{n-1}{n} \times \frac{n-3}{n-2} \int_0^{\frac{\pi}{2}} \sin^{n-4} x\,dx$$

Similarly, $$\int_0^{\frac{\pi}{2}} \sin^n x\,dx = \frac{n-1}{n} \times \frac{n-3}{n-2} \times \frac{n-5}{n-4} \int_0^{\frac{\pi}{2}} \sin^{n-6} x\,dx$$

and so on.

Case I. If n is even, then

$$\int_0^{\frac{\pi}{2}} \sin^n x\,dx = \frac{n-1}{n} \times \frac{n-3}{n-2} \times \frac{n-5}{n-4} \times \ldots \frac{1}{2} \times \int_0^{\frac{\pi}{2}} (\sin x)^0\,dx$$

$$= \frac{(n-1)(n-3)(n-5)\ldots 1}{n(n-2)(n-4)\ldots 2} \int_0^{\frac{\pi}{2}} 1 \cdot dx$$

$$= \frac{(n-1)(n-3)(n-5)\ldots 1}{n(n-2)(n-4)\ldots 2} [x]_0^{\frac{\pi}{2}}$$

$$\boxed{\int_0^{\frac{\pi}{2}} \sin^n x\,dx = \frac{(n-1)(n-3)(n-5)\ldots 1}{n(n-2)(n-4)\ldots 2} \times \frac{\pi}{2}}$$ **Ans.**

Case II. If n is odd, then

$$\int_0^{\frac{\pi}{2}} \sin^n x\,dx = \frac{n-1}{n} \times \frac{n-3}{n-2} \times \frac{n-5}{n-4} \times \ldots \times \frac{2}{3} \times \int_0^{\frac{\pi}{2}} \sin x\,dx$$

$$= \frac{(n-1)(n-3)(n-5)\ldots 2}{n(n-2)(n-4)\ldots 3} [-\cos x]_0^{\frac{\pi}{2}}$$

$$\boxed{\int_0^{\frac{\pi}{2}} \sin^n x\,dx = \frac{(n-1)(n-3)(n-5)\ldots 2}{n(n-2)(n-4)\ldots 3}}$$ **Ans.**

Similarly, $$\int_0^{\frac{\pi}{2}} \cos^n x\,dx = \frac{(n-1)(n-3)(n-5)\ldots 1}{n(n-2)(n-4)\ldots 2} \times \frac{\pi}{2} \text{ when } n \text{ is even}$$

$$\int_0^{\frac{\pi}{2}} \cos^n x\,dx = \frac{(n-1)(n-3)(n-5)\cdots \times 2}{n(n-2)(n-4)\ldots \times 3} \text{ when } n \text{ is odd.}$$

Example 12. *Evaluate:* (*i*) $\int_0^{\frac{\pi}{2}} \sin^7 x\,dx$ (*ii*) $\int_0^{\frac{\pi}{2}} \sin^6 x\,dx$

Solution. (*i*) $\int_0^{\frac{\pi}{2}} \sin^7 x\,dx = \frac{6 \times 4 \times 2}{7 \times 5 \times 3} = \frac{16}{35}$ **Ans.**

(*ii*) $\int_0^{\frac{\pi}{2}} \sin^6 x\,dx = \frac{5 \times 3 \times 1}{6 \times 4 \times 1} \times \frac{\pi}{2} = \frac{5\pi}{32}$ **Ans.**

Example 13. *Evaluate:* (*i*) $\int_0^{\frac{\pi}{2}} \cos^9 x\,dx$ (*ii*) $\int_0^{\frac{\pi}{2}} \cos^{10} x\,dx$

Solution. (*i*) $\int_0^{\frac{\pi}{2}} \cos^9 x\,dx = \frac{8\times6\times4\times2}{9\times7\times5\times3} = \frac{128}{315}$ **Ans.**

(*ii*) $\int_0^{\frac{\pi}{2}} \cos^{10} x\,dx = \frac{9\times7\times5\times3\times1}{10\times8\times6\times4\times2}\times\frac{\pi}{2} = \frac{63\pi}{512}$ **Ans.**

32.4 REDUCTION FORMULA FOR $\int_0^{\frac{\pi}{2}} \sin^m x \cos^n x\,dx$

Case 1. If m and n both are not even

$$\int_0^{\frac{\pi}{2}} \sin^m x \cos^n x\,dx = \frac{[(m-1)\text{ go on subtracting } 2\ldots][(n-1)\text{ go on subtracting } 2\ldots]}{(m+n)\text{ go on subtracting } 2}$$

$$\boxed{\int_0^{\frac{\pi}{2}} \sin^m x \cos^n x\,dx = \frac{[(m-1)(m-3)(m-5)\ldots][(n-1)(n-3)(n-5)\ldots]}{(m+n)(m+n-2)(m+n-4)\ldots}}$$

Case 2. If m and n both are even

$$\int_0^{\frac{\pi}{2}} \sin^m x \cos^n x\,dx = \frac{[(m-1)\text{ go on subtracting } 2\ldots][(n-1)\text{ go on subtracting } 2\ldots]}{(m+n)\text{ go on subtracting } 2}\,\frac{\pi}{2}$$

$$\boxed{\int_0^{\frac{\pi}{2}} \sin^m x \cos^n x\,dx = \frac{[(m-1)(m-3)(m-5)\ldots][(n-1)(n-3)(n-5)\ldots]}{(m+n)(m+n-2)(m+n-4)\ldots}\,\frac{\pi}{2}}$$

Example 14. *Evaluate:* (*i*) $\int_0^{\frac{\pi}{2}} \sin^4 x \cos^3 x\,dx$ (*ii*) $\int_0^{\frac{\pi}{2}} \sin^6 x \cos^8 x\,dx$

Solution. (*i*) We have, $I_{4,3} = \int_0^{\frac{\pi}{2}} \sin^4 x \cos^3 x\,dx$

We know that when m is even and n is odd

$$\int_0^{\frac{\pi}{2}} \sin^m x \cos^n x\,dx$$

$$= \frac{(m-1)(m-3)(m-5)\ldots3.1}{(m+n)(m+n-2)\ldots(n+2)} \times \frac{(n-1)(n-3)\ldots2}{n(n-2)\ldots1}$$

$$\therefore\ I_{4,3} = \int_0^{\frac{\pi}{2}} \sin^4 x \cos^3 x\,dx = \frac{(4-1)(4-3)}{(4+3)(4+3-2)} \times \frac{(3-1)}{3.1} = \frac{3\times1}{7\times5}\times\frac{2}{3} = \frac{2}{35}$$ **Ans.**

(*ii*) $\int_0^{\frac{\pi}{2}} \sin^6 x \cos^8 x\,dx$

$$= \frac{(6-1)(6-3)(6-5)(8-1)(8-3)(8-5)(8-7)}{(6+8)(6+8-2)(6+8-4)(6+8-6)(6+8-8)(6+8-10)(6+8-12)} \times \frac{\pi}{2}$$

$$= \frac{5\times3\times1\times7\times5\times3\times1}{14\times12\times10\times8\times6\times4\times2} \times \frac{\pi}{2} = \frac{5\pi}{4096}$$ **Ans.**

32.5 ALTERNATIVE RULE TO EVALUATE $\int_0^{\frac{\pi}{2}} \sin^m x \cos^n x\,dx$

$$\int_0^{\frac{\pi}{2}} \sin^m x \cos^n x\,dx = \frac{\left|\frac{m+1}{2}\right.\left|\frac{n+1}{2}\right.}{2\left|\frac{m+n+2}{2}\right.}$$ where m and n are positive integers

It is necessary to remember that

(i) $\lceil n = (n-1)!$ or gamma n = factorial $(n-1)$ (ii) $\left|\frac{1}{2}\right. = \sqrt{\pi}$

(iii) $n! = n\,(n-1)!$ $\Rightarrow$ $\lceil(n+1) = n\lceil n$

e.g. $\left|\frac{5}{2}\right. = \frac{3}{2}\left|\frac{3}{2}\right. = \frac{3}{2}\cdot\frac{1}{2}\left|\frac{1}{2}\right. = \frac{3}{2}\cdot\frac{1}{2}\cdot\sqrt{\pi}$

Caution. The limits of integration must be 0 to $\frac{\pi}{2}$

Example 15. *Evaluate:* $\int_0^{\frac{\pi}{2}} \sin^5 x \cos^8 x\,dx.$

Solution. $$\int_0^{\frac{\pi}{2}} \sin^5 x \cos^8 x\,dx = \frac{\left|\frac{5+1}{2}\right.\left|\frac{1+8}{2}\right.}{2\left|\frac{5+1+8+1}{2}\right.}$$

$$= \frac{\lceil 3\left|\frac{9}{2}\right.}{2\left|\frac{15}{2}\right.} = \frac{(2.1)\cdot\left(\frac{7}{5}\times\frac{5}{2}\times\frac{3}{2}\times\frac{1}{2}\times\sqrt{\pi}\right)}{2\left(\frac{13}{2}\cdot\frac{11}{2}\cdot\frac{9}{2}\cdot\frac{7}{2}\cdot\frac{5}{2}\cdot\frac{3}{2}\cdot\frac{1}{2}\cdot\sqrt{\pi}\right)}$$

$$= \frac{1}{\frac{13}{2}\cdot\frac{11}{2}\cdot\frac{9}{2}} = \frac{8}{1287}$$ **Ans.**

Example 16. *Evaluate:* $\int_0^{\frac{\pi}{2}} \sin^4 x \cos^2 x\,dx$ **[B.T.E. Delhi, May 2008]**

Solution. $$\int_0^{\frac{\pi}{2}} \sin^4 x \cos^2 x\,dx = \frac{\left|\frac{4+1}{2}\right.\left|\frac{2+1}{2}\right.}{2\left|\frac{4+1+2+1}{2}\right.}$$

$$= \frac{\left|\frac{5}{2}\right.\left|\frac{3}{2}\right.}{2\lceil 4} = \frac{\left(\frac{3}{2}\cdot\frac{1}{2}\sqrt{\pi}\right)\cdot\left(\frac{1}{2}\cdot\sqrt{\pi}\right)}{2\cdot 3\cdot 2\cdot 1} = \frac{\pi}{32}$$ **Ans.**

Example 17. *Evaluate:* $\int_0^{\frac{\pi}{2}} \cos^7 x\,dx$ **[D.I.P.I.E.T.E. Dec. 2018]**

Solution. $\int_0^{\frac{\pi}{2}} \cos^7 x\,dx = \int_0^{\frac{\pi}{2}} \cos^7 x \sin^0 x\,dx$

$$= \frac{\Gamma\left(\frac{7+1}{2}\right)\Gamma\left(\frac{0+1}{2}\right)}{2\Gamma\left(\frac{7+1+0+1}{2}\right)} = \frac{\Gamma 4 \cdot \Gamma\frac{1}{2}}{2\Gamma\frac{9}{2}} = \frac{(3.2.1)\sqrt{\pi}}{2\cdot\left(\frac{7}{2}\cdot\frac{5}{2}\cdot\frac{3}{2}\cdot\frac{1}{2}\cdot\sqrt{\pi}\right)}$$

$$= \frac{16}{35}$$ **Ans.**

Example 18. *Evaluate:* $\int_0^{\frac{\pi}{2}} \sin^8 x\,dx$

Solution. $\int_0^{\frac{\pi}{2}} \sin^8 x\,dx = \int_0^{\frac{\pi}{2}} \sin^8 x \cos^0 x\,dx$

$$= \frac{\Gamma\left(\frac{8+1}{2}\right)\Gamma\left(\frac{0+1}{2}\right)}{2\Gamma\left(\frac{8+1+0+1}{2}\right)} = \frac{\Gamma\frac{9}{2}\Gamma\frac{1}{2}}{2\Gamma 5} = \frac{\left(\frac{7}{2}\cdot\frac{5}{2}\cdot\frac{3}{2}\cdot\frac{1}{2}\cdot\sqrt{\pi}\right)(\sqrt{\pi})}{2\cdot(4\cdot 3\cdot 2\cdot 1)}$$

$$= \frac{35\pi}{256}$$ **Ans.**

Example 19. *Evaluate the following integrals.*

(*i*) $\int_0^{\frac{\pi}{2}} \sin^3 x\,dx$ **[SBTE. 2016, 2014]** (*ii*) $\int_0^{\frac{\pi}{2}} \sin^2 x \cdot \cos^6 x\,dx$ **[SBTE. 2015, 2011]**

Solution.

(*i*) Let $I = \int_0^{\frac{\pi}{2}} \sin^3 x = \dfrac{\Gamma\left(\frac{3+1}{2}\right)\cdot\Gamma\left(\frac{0+1}{2}\right)}{2\cdot\Gamma\left(\frac{3+0+2}{2}\right)}$

$$I = \frac{\Gamma 2\cdot\Gamma\frac{1}{2}}{2\Gamma\frac{5}{2}} = \frac{1!\times\sqrt{\pi}}{\frac{3}{2}\sqrt{\pi}} = \frac{2}{3}$$ **Ans.**

(*ii*) Let $I = \int_0^{\frac{\pi}{2}} \sin^2 x \cdot \cos^6 x\,dx$

$$= \frac{\Gamma\left(\frac{2+1}{2}\right)\cdot\Gamma\left(\frac{6+1}{2}\right)}{2\cdot\Gamma\left(\frac{2+6+2}{2}\right)} = \frac{\Gamma\frac{3}{2}\cdot\Gamma\frac{7}{2}}{2\Gamma 5}$$

$$= \frac{\frac{1}{2}\Gamma\frac{1}{2}\cdot\frac{15}{8}\Gamma\frac{1}{2}}{2\times 4!} = \frac{5\pi}{256}$$ **Ans.**

Example 20. *Evaluate the following integrals.*

(*i*) $\int_0^\infty x^3 (4+x^2)^{-5/2}\, dx$ **[SBTE. 2016, 2015]** (*ii*) $\int_0^1 x^{3/2}(1-x)^{3/2}\, dx$ **[SBTE. 2014, 2012]**

Solution. Here we have

(*i*) $$I = \int_0^\infty x^3 (4+x^2)^{-5/2}\, dx$$

$$\begin{bmatrix} x = 0 \\ 2\tan\theta = 0, \theta = 0 \\ x = \infty \\ 2\tan\theta = \infty \\ \theta = \pi/2 \end{bmatrix}$$

Put $x = 2\tan\theta$

$$dx = 2\sec^2\theta\, d\theta$$

$$I = \int_0^{\frac{\pi}{2}} \left[(2\tan\theta)^3 \cdot (4+4\tan^2\theta)^{-5/2} \cdot 2\sec^2\theta\, d\theta\right]$$

$$= \frac{1}{2}\int_0^{\frac{\pi}{2}} \frac{\sin^3\theta}{\cos^3\theta} \cdot \cos^3\theta\, d\theta = \frac{1}{2}\int_0^{\frac{\pi}{2}} \sin^3\theta\, d\theta$$

$$= \frac{1}{2} \times \frac{\left|\overline{\frac{3+1}{2}} \cdot \right|\overline{\frac{0+1}{2}}}{2\left|\overline{\frac{3+0+2}{2}}\right.} = \frac{1}{4}\, \frac{\left|\overline{2} \cdot \right|\overline{\frac{1}{2}}}{\left|\overline{\frac{5}{2}}\right.}$$

$$= \frac{1}{4} \times \frac{\sqrt{\pi}}{\frac{3}{2} \times \frac{1}{2} \times \sqrt{\pi}} = \frac{1}{3}$$ **Ans.**

(*ii*) Here we have

$$I = \int_0^1 x^{3/2}(1-x)^{3/2}\, dx$$

Put $x = \sin^2\theta \Rightarrow dx = 2\sin\theta \cdot \cos\theta\, d\theta$

$$I = \int_0^{\frac{\pi}{2}} (\sin^2\theta)^{3/2}(1-\sin^2\theta)^{3/2}\, 2\sin\theta \cdot \cos\theta\, d\theta$$

$$\begin{bmatrix} x = 0 \\ \sin^2\theta = 0 \\ \theta = 0 \\ x = 1 \\ \theta = \pi/2 \end{bmatrix}$$

$$= 2\int_0^{\frac{\pi}{2}} \sin^4\theta \cdot \cos^4\theta\, d\theta$$

$$= 2 \times \frac{\left|\overline{\frac{4+1}{2}} \cdot \right|\overline{\frac{4+1}{2}}}{2\left|\overline{\frac{4+4+2}{2}}\right.} = \frac{\left|\overline{\frac{5}{2}} \cdot \right|\overline{\frac{5}{2}}}{\left|\overline{5}\right.}$$

$$= \frac{\frac{3}{2} \times \frac{1}{2} \cdot \left|\overline{\frac{1}{2}} \times \frac{3}{2} \times \frac{1}{2}\right|\overline{\frac{1}{2}}}{4!} = \frac{3\pi}{128}$$ **Ans.**

Example 21. *Evaluate:* $\int_0^a x^2\sqrt{a^2-x^2}\, dx$ **[S.B.T.E. 2017, 2016]**

Solution. Let $I = \int_0^a x^2\sqrt{a^2-x^2}\, dx$

Putting $x = a\sin\theta$, $dx = a\cos\theta\, d\theta$

$$I = \int_0^{\frac{\pi}{2}} a^2 \sin^2\theta \sqrt{a^2 - a^2\sin^2\theta} \cdot a\cos\theta\, d\theta$$

$$\left[\begin{array}{c} x = 0 \\ 0 = a\sin\theta \\ \theta = 0 \\ \text{and } x = a \\ a = a\sin\theta \\ \theta = \pi/2 \end{array}\right]$$

$$= \int_0^{\frac{\pi}{2}} a^3 \sin^2 \cdot \cos\theta \cdot a\cos\theta\, d\theta$$

$$= a^4 \int_0^{\frac{\pi}{2}} \sin^2\theta \cdot \cos^2\theta\, d\theta$$

$$= a^4 \cdot \frac{\left\lceil\frac{2+1}{2} \cdot \left\lceil\frac{2+1}{2}\right.\right.}{2 \cdot \left\lceil\frac{2+2+2}{2}\right.} = a^4 \cdot \frac{\left\lceil\frac{3}{2} \cdot \left\lceil\frac{3}{2}\right.\right.}{2 \cdot \lceil 3}$$

$$= a^4 \cdot \frac{\frac{1}{2} \cdot \left\lceil\frac{1}{2} \cdot \frac{1}{2}\left\lceil\frac{1}{2}\right.\right.}{2 \times 2!} = a^4 \cdot \frac{\frac{1}{2}\sqrt{\pi} \cdot \frac{1}{2}\sqrt{\pi}}{2 \times 2 \times 1} = a^4 \cdot \frac{\pi/4}{4}$$

$$\therefore\ I = \frac{\pi a^4}{16}$$ **Ans.**

Example 22. Evaluate: $\int_0^{\infty} \frac{1}{3+x^2}\, dx$ **[D.I.P.I.E.T.E. Dec. 2018]**

Solution. $\int_0^{\infty} \frac{1}{3+x^2}\, dx = \int_0^{\infty} \frac{1}{(\sqrt{3})^2 + x^2}\, dx$

$$= \left[\frac{1}{\sqrt{3}} \tan^{-1}\left(\frac{x}{\sqrt{3}}\right)\right]_0^{\infty}$$

$$= \frac{1}{\sqrt{3}}\left[\tan^{-1}(\infty) - 0\right]$$

$$= \frac{1}{\sqrt{3}} \times \frac{\pi}{2}$$

$$= \frac{\pi}{2\sqrt{3}}$$ **Ans.**

EXERCISE 32.1

Evaluate the following:

1. $\int_0^1 x^{10}\, dx$ **Ans.** $\frac{1}{11}$

2. $\int_0^{\frac{\pi}{2}} \sin\theta\cos\theta\, d\theta$ **Ans.** $\frac{1}{2}$

3. $\int_0^1 \frac{3x}{1+x^2}\, dx$ **Ans.** $\frac{3}{2}\log 2$

4. $\int_0^{\frac{\pi}{3}} \sec x \tan x\, dx$ **Ans.** 1

5. $\int_0^{\frac{\pi}{2}} \cos\left(x + \frac{\pi}{4}\right) dx$ **Ans.** $1 - \frac{1}{\sqrt{2}}$

6. $\int_0^{\frac{\pi}{4}} \tan^2 x\, dx$ **Ans.** $1 - \frac{\pi}{4}$

7. $\int_0^{\frac{\pi}{2}} \cos^2 t\,dt$ **Ans.** $\frac{\pi}{4}$ 8. $\int_1^{\sqrt{3}} \frac{dx}{1+x^2}$ **Ans.** $\frac{\pi}{12}$

9. $\int_{-1}^{3} \frac{dx}{2x+3}$ **Ans.** log 3 10. $\int_0^{a} x^2 \sin x^3\,dx$ **Ans.** $\frac{1}{3}(1-\cos a^3)$

11. $\int_0^1 \frac{(\tan^{-1} x)^2}{1+x^2}\,dx$ **Ans.** $\frac{1}{192}\pi^3$ 12. $\int_0^{\frac{\pi}{4}} \frac{\sec^2 x}{1+\tan x}\,dx$ **Ans.** log 2

13. $\int_0^1 \frac{5x^3}{(1+x^8)}\,dx$ **Ans.** $\frac{5}{16}\pi$ 14. $\int_0^{\frac{\pi}{2}} \frac{\cos x}{1+\sin^2 x}\,dx$ **Ans.** $\frac{\pi}{4}$

15. $\int_0^{\infty} \frac{e^x}{1+e^{2x}}\,dx$ **Ans.** $\frac{\pi}{2}$ 16. $\int_1^2 \sqrt{\left(\frac{x-1}{2-x}\right)}\,dx$ **Ans.** $\frac{\pi}{2}$

17. $\int_1^3 \frac{\cos(\log x)}{x}\,dx$ **Ans.** sin log 3 18. $\int_0^3 x\sqrt{x^2+16}\,dx$ **Ans.** $20\frac{1}{3}$

19. $\int_0^{\frac{\pi}{4}} x \sin 2x\,dx$ **Ans.** $\frac{1}{4}$

20. $\int_3^5 x \log x\,dx$ **Ans.** $\frac{25}{2}\log 5 - \frac{9}{2}\log 3 - 4$

21. $\int_0^{\frac{\pi}{2}} \frac{\tan x\,dx}{1+m^2\tan^2 x}$ **Ans.** $\frac{\log m^2 - 1}{2(m^2-1)}$ [**Hint.** convert tan x into sin x and cos x]

22. $\int_0^{\frac{\pi}{2}} \sin^5 x \cos^3 x\,dx$ **Ans.** $\frac{1}{24}$ 23. $\int_0^{\frac{\pi}{2}} \sin^2 x \cos^6 x\,dx$ **Ans.** $\frac{5\pi}{256}$

24. $\int_0^{\frac{\pi}{2}} \sin^5 x\,dx$ **Ans.** $\frac{8}{15}$ 25. $\int_0^{\frac{\pi}{2}} \cos^6 x\,dx$ **Ans.** $\frac{5\pi}{32}$

26. $\int_0^{\frac{\pi}{6}} \cos^4 3\phi \sin^2 6\phi\,d\phi$ **Ans.** $\frac{5\pi}{192}$

27. $\int_0^1 x^{3/2}(1-x)^{3/2}\,dx$ (Put $x = \sin^2\theta$) **[May 2008]** **Ans.** $\frac{3\pi}{128}$

28. $\int_0^1 x^4(1-x^2)^{3/2}\,dx$ **Ans.** $\frac{256}{15015}$

29. $\int_0^{2a} x^3\sqrt{2ax-x^2}\,dx$ **Ans.** $\frac{7\pi}{4}a^5$ 30. $\int_0^1 \frac{x^3}{(1+x^8)}\,dx$ **Ans.** $\frac{\pi}{16}$

31. $\int_0^{\infty} \frac{x^2}{(4+x^2)^{5/2}}\,dx$ **Ans.** $\frac{1}{12}$ 32. $\int_0^a x^4\sqrt{a^2-x^2}\,dx$ **Ans.** $\frac{\pi a^6}{32}$

33. $\int_0^2 \frac{(4-x)\,dx}{x(x^2-2x+2)}$ **Ans.** $2\log 2 - \frac{\pi}{2}$

34. $\int_0^{\frac{\pi}{2}} \frac{\cos x\, dx}{(1+\sin x)(2+\sin x)}$ **[B.T.E. Delhi May 2008]** **Ans.** log (4/3)

35. $\int_0^{\pi} \theta \sin^2\theta \cos\theta\, d\theta$ **Ans.** $-\frac{4}{9}$

32.6 THEOREMS ON DEFINITE INTEGRALS

In certain cases the value of definite integral can be found out easily with the help of theorems given below.

Let $\int f(x)dx = \phi(x)$, so that $\int_a^b f(x)dx = \phi(b) - \phi(a)$

Theorem 1. $\int_a^b f(x)dx = \int_a^b f(t)dt$

Proof. $\int_a^b f(t)dx = [\phi(x)]_a^b = \phi(b) - \phi(a)$

$\int_a^b f(t)dt = [\phi(t)]_a^b = \phi(b) - \phi(a)$ Hence the result **Proved.**

Verify $\int_1^2 x^2\, dx = \int_1^2 y^2\, dy$

Theorem 2. $\int_a^b f(x)dx = -\int_b^a f(x)dx$

Proof. $-\int_b^a f(x)dx = -[\phi(x)]_b^a = -[\phi(a) - \phi(b)]$

$= \phi(b) - \phi(a) = \int_a^b f(x)dx$

Interchanging the limits, changes the sign of the integral.
Hence the result. **Proved.**

Verify $\int_1^2 x^4\, dx = -\int_2^1 x^4\, dx$

Theorem 3. $\int_a^b f(x)dx = \int_a^c f(x)dx + \int_c^b f(x)dx$

Proof. $\int_a^c f(x)dx + \int_c^b f(x)dx = [\phi(x)]_a^c + [\phi(x)]_c^b$

$= [\phi(c) - \phi(a)] + [\phi(b) - \phi(c)]$

$= \phi(b) - \phi(a) = \int_a^b f(x)dx$ **Proved.**

In general $\int_a^b f(x)dx = \int_a^c f(x)dx + \int_c^d f(x)dx + \ldots + \int_k^b f(x)dx$

Verify $\int_1^3 x^2\, dx = \int_1^2 x^2\, dx + \int_2^3 x^2\, dx$

Theorem 4. $\int_0^a f(x)dx = \int_0^a f(a-x)dx$ (Very Important)

Proof. Put $x = a - t$ then $dx = -dt$

When $x = a,$ $t = a - x = a - a = 0$

When $\quad x = 0, \quad t = a - x = a - 0 = a$

$$\int_0^a f(x)\cdot dx = -\int_a^0 f(a-t)\,dt = \int_0^a f(a-t)\,dt \quad \text{(By Theorem 2)}$$

$$= \int_0^a f(a-x)\,dx \quad \text{(By Theorem 1)} \; \textbf{Proved.}$$

Example 23. *Evaluate:* $\int_0^{\frac{\pi}{2}} \frac{\sqrt{\sin x}}{\sqrt{\sin x}+\sqrt{\cos x}}\,dx$ **[SBTE. 2016, 2015]**

Solution. Let $I = \int_0^{\frac{\pi}{2}} \frac{\sqrt{\sin x}}{\sqrt{\sin x}+\sqrt{\cos x}}\,dx$ ... (1)

Then
$$I = \int_0^{\frac{\pi}{2}} \frac{\sqrt{\sin\left(\frac{\pi}{2}-x\right)}}{\sqrt{\sin\left(\frac{\pi}{2}-x\right)}+\sqrt{\cos\left(\frac{\pi}{2}-x\right)}}\,dx$$

$$I = \int_0^{\frac{\pi}{2}} \frac{\sqrt{\cos x}}{\sqrt{\cos x}+\sqrt{\sin x}}\,dx \quad \text{... (2)}$$

Adding (1) and (2), we get

$$2I = \int_0^{\frac{\pi}{2}} \frac{\sqrt{\sin x}}{\sqrt{\sin x}+\sqrt{\cos x}}\,dx + \int_0^{\frac{\pi}{2}} \frac{\sqrt{\cos x}}{\sqrt{\sin x}+\sqrt{\cos x}}\,dx$$

$$= \int_0^{\frac{\pi}{2}} \frac{\sqrt{\sin x}+\sqrt{\cos x}}{\sqrt{\sin x}+\sqrt{\cos x}}\,dx = \int_0^{\frac{\pi}{2}} dx = [x]_0^{\pi/2} = \frac{\pi}{2}$$

$\Rightarrow \quad I = \frac{\pi}{4}$ **Ans.**

Example 24. *Evaluate:* $\int_0^{\frac{\pi}{2}} \frac{\sin^2 x}{\sin x+\cos x}\,dx$ **[SBTE. 2016, 2015]**

Solution. Let $I = \int_0^{\frac{\pi}{2}} \frac{\sin^2 x}{\sin x+\cos x}\,dx$ ... (1)

$$\Rightarrow \quad I = \int_0^{\frac{\pi}{2}} \frac{\sin^2\left(\frac{\pi}{2}-x\right)}{\sin\left(\frac{\pi}{2}-x\right)+\cos\left(\frac{\pi}{2}-x\right)}\,dx \quad \textit{(By above theorem)}$$

$$\Rightarrow \quad I = \int_0^{\frac{\pi}{2}} \frac{\cos^2 x}{\cos x+\sin x}\,dx \quad \text{... (2)}$$

Adding (1) and (2), we get

$$2I = \int_0^{\frac{\pi}{2}} \frac{\sin^2 x}{\sin x+\cos x}\,dx + \int_0^{\frac{\pi}{2}} \frac{\cos^2 x}{\sin x+\cos x}\,dx$$

$$\Rightarrow \qquad 2I = \int_0^{\frac{\pi}{2}} \frac{\sin^2 x + \cos^2 x}{\sin x + \cos x}\, dx = \int_0^{\frac{\pi}{2}} \frac{1}{\sin x + \cos x}\, dx$$

$$\Rightarrow \qquad 2I = \frac{1}{\sqrt{2}} \int_0^{\frac{\pi}{2}} \frac{1}{\frac{1}{\sqrt{2}} \sin x + \frac{1}{\sqrt{2}} \cos x}\, dx$$

$$\Rightarrow \qquad 2I = \frac{1}{\sqrt{2}} \int_0^{\frac{\pi}{2}} \frac{1}{\cos \frac{\pi}{4} \sin x + \sin \frac{\pi}{4} \cos x}\, dx = \frac{1}{\sqrt{2}} \int_0^{\frac{\pi}{2}} \frac{1}{\sin\left(x + \frac{\pi}{4}\right)} dx$$

$$= \frac{1}{\sqrt{2}} \int_0^{\frac{\pi}{2}} \operatorname{cosec}\left(x + \frac{\pi}{4}\right) dx = \frac{1}{\sqrt{2}} \left[\log \tan\left(\frac{x}{2} + \frac{\pi}{8}\right)\right]_0^{\pi/2}$$

$$= \frac{1}{\sqrt{2}} \left[\log \tan \frac{3\pi}{8} - \log \tan \frac{\pi}{8}\right]$$

$$\Rightarrow \qquad I = \frac{1}{2\sqrt{2}} \log \left(\frac{\tan \frac{3\pi}{8}}{\tan \frac{\pi}{8}}\right) \qquad \textbf{Ans.}$$

Example 25. *Evaluate:* $\int_0^{\pi} \frac{x \sin x}{1 + \cos^2 x}\, dx$ **[SBTE. 2015, 2010]**

Solution. $$I = \int_0^{\pi} \frac{x \sin x}{1 + \cos^2 x}\, dx$$

$$= \int_0^{\pi} \frac{(\pi - x) \sin (\pi - x)}{1 + \cos^2 (\pi - x)}\, dx$$

$$= \int_0^{\pi} \frac{(\pi - x) \sin x}{1 + \cos^2 x}\, dx$$

$$= \int_0^{\pi} \frac{\pi \sin x}{1 + \cos^2 x}\, dx - \int_0^{\pi} \frac{x \sin x}{1 + \cos^2 x}\, dx$$

$$\therefore \qquad I = \pi \int_0^{\pi} \frac{\sin x\, dx}{1 + \cos^2 x} - I \quad \Rightarrow \quad 2I = \pi \int_0^{\pi} \frac{\sin x\, dx}{1 + \cos^2 x}$$

Put $\cos x = t,$ $\qquad -\sin x\, dx = dt$

When $x = \pi,$ $\qquad t = -1$

When $x = 0,$ $\qquad t = 1.$

$$2I = -\int_{+1}^{-1} \frac{\pi\, dt}{1 + t^2} = \pi \int_{-1}^{+1} \frac{dt}{1 + t^2} = \pi \left[\tan^{-1} t\right]_{-1}^{+1}$$

$$= \pi \left[\tan^{-1}(1) - \tan^{-1}(-1)\right]$$

$$= \pi\left[\frac{\pi}{4} - \left(-\frac{\pi}{4}\right)\right] = \frac{\pi^2}{2}$$

$$I = \frac{\pi^2}{4}$$ **Ans.**

Example 26. *Evaluate:* $\int_0^{\frac{\pi}{2}} \log \sin x \, dx$

Solution. Let $I = \int_0^{\frac{\pi}{2}} \log \sin x \, dx$ ... (1)

Then $I = \int_0^{\frac{\pi}{2}} \log \sin\left(\frac{\pi}{2} - x\right) dx$

$= \int_0^{\frac{\pi}{2}} \log \cos x \, dx$ ... (2)

Adding (1) and (2), we get

$$2I = \int_0^{\frac{\pi}{2}} \log \sin x \, dx + \int_0^{\frac{\pi}{2}} \log \cos x \, dx$$

$$= \int_0^{\frac{\pi}{2}} \log (\sin x \cdot \cos x) \, dx = \int_0^{\frac{\pi}{2}} \log\left(\frac{\sin 2x}{2}\right) dx \quad [\because \sin 2x = 2 \sin x \cdot \cos x]$$

$$= \int_0^{\frac{\pi}{2}} \log (\sin 2x) \, dx - \int_0^{\frac{\pi}{2}} \log 2 \cdot dx$$

$$= \int_0^{\frac{\pi}{2}} \log (\sin 2x) \, dx - \log 2 \int_0^{\frac{\pi}{2}} dx$$

$$= \int_0^{\frac{\pi}{2}} \log (\sin 2x) \, dx - \log 2 \, [x]_0^{\pi/2}$$

$$= \int_0^{\frac{\pi}{2}} \log (\sin 2x) \, dx - \frac{\pi}{2} \log 2$$

Now put, $2x = t$ so that $2dx = dt$

when $x = \frac{\pi}{2}$, $t = \pi$

when $x = 0$, $t = 0$

$$2I = \frac{1}{2} \int_0^{\pi} \log \sin t \, dt - \frac{\pi}{2} \log 2$$

$$= \frac{1}{2} \cdot 2 \int_0^{\frac{\pi}{2}} \log \sin t \, dt - \frac{\pi}{2} \log 2$$

$$= \int_0^{\frac{\pi}{2}} \log \sin t \, dt - \frac{\pi}{2} \log 2$$

$$= \int_0^{\frac{\pi}{2}} \log \sin x \, dx - \frac{\pi}{2} \log 2$$

$$2I = I - \frac{\pi}{2}\log 2$$

$$I = -\frac{\pi}{2}\log 2$$ **Ans.**

Cor. $\int_0^{\frac{\pi}{2}} \log\cos x\, dx = -\frac{\pi}{2}\log 2$

Theorem 5. $\int_0^{2a} f(x)dx = \int_0^a f(x)\,dx + \int_0^a f(2a-x)\,dx$

Proof. $\int_0^{2a} f(x)dx = \underset{(i)}{\int_0^a f(x)\,dx} + \underset{(ii)}{\int_a^{2a} f(x)\,dx}$ (By Theorem 3)

In (*ii*) we put $x = 2a - t$ so that $dx = -dt$

when $x = a,\quad t = 2a - a = a$

$x = 2a,\quad t = 2a - 2a = 0$

$$\int_a^{2a} f(x)dx = -\int_a^0 f(2a-t)\,dt = \int_0^a f(2a-t)\,dt$$

$$= \int_0^a f(2a-x)\,dx$$ (By Theorem 1)

Hence, $\int_0^{2a} f(x)\,dx = \int_0^a f(x)\,dx + \int_0^a f(2a-x)\,dx$ **Proved.**

Theorem 6. (*i*) $\int_0^{2a} f(x)\,dx = 2\int_0^a f(x)\,dx$ If $f(2a-x) = f(x)$

(*ii*) $\int_0^{2a} f(x)\,dx = 0$ If $f(2a-x) = -f(x)$

Proof. $\int_0^{2a} f(x)\,dx = \int_0^a f(x)\,dx + \int_0^a f(2a-x)\,dx$ ... (1)

(by Theorem 5)

(*i*) By putting $f(x) = f(2a-x)$ in (1), we get

$$\int_0^{2a} f(x)\,dx = \int_0^a f(x)\,dx + \int_0^a f(x)\,dx = 2\int_0^a f(x)\,dx$$

(*ii*) By putting $f(x) = -f(2a-x)$ in (1), we get

$$\int_0^{2a} f(x)\,dx = \int_0^a f(x)\,dx - \int_0^a f(x)\,dx = 0$$

This theorem is used to make the upper limit half. **Proved.**

In words. When the lower limit is zero.

(*i*) The integral is twice the integral of the same function between half the limit if $f(x) = f(2a-x)$.

(*ii*) The integral is equal to zero if $f(x) = -f(2a-x)$

Example 27. *Evaluate:* $\int_0^{\pi} \sin^8 x\, dx$

Solution. $\int_0^{\pi} \sin^8 x\, dx = 2\int_0^{\frac{\pi}{2}} \sin^8 x\, dx$ $\qquad \left[\because \sin^8(\pi - x) = \sin^8 x\right]$

$$= 2\cdot\frac{\left|\frac{9}{2}\right.\left|\frac{1}{2}\right.}{2\left|\frac{9+1}{2}\right.} = \frac{\left|\frac{9}{2}\right.\left|\frac{1}{2}\right.}{\left|5\right.} = \frac{\frac{7}{2}\cdot\frac{5}{2}\cdot\frac{3}{2}\cdot\frac{1}{2}\cdot\sqrt{\pi}\cdot\sqrt{\pi}}{4\cdot3\cdot2\cdot1} = \frac{35}{128}\pi \qquad \textbf{Ans.}$$

Example 28. *Evaluate:* $\int_0^{\pi} \theta \sin^6\theta \cos^4\theta\, d\theta$

Solution. Let $I = \int_0^{\pi} \theta \sin^6\theta \cos^4\theta\, d\theta$ (By Theorem 4)

$$I = \int_0^{\pi} (\pi - \theta)\sin^6(\pi - \theta)\cos^4(\pi - \theta)\, d\theta$$

$$= \int_0^{\pi} (\pi - \theta)\sin^6\theta \cos^4\theta\, d\theta$$

$$= \pi\int_0^{\pi} \sin^6\theta\cos^4\theta\, d\theta - \int_0^{\pi} \theta\sin^6\theta\cos^4\theta\, d\theta$$

$$= \pi\int_0^{\pi} \sin^6\theta\cos^4\theta\, d\theta - I$$

$\Rightarrow$
$$2I = \pi\int_0^{\pi} \sin^6\theta\cos^4\theta\, d\theta$$

$$= 2\pi\int_0^{\pi/2} \sin^6\theta\cos^4\theta\, d\theta \quad \left[\because \sin^6(\pi-\theta)\cos^4(\pi-\theta) = \sin^6\theta\cos^4\theta\right]$$

or
$$I = \pi\int_0^{\frac{\pi}{2}} \sin^6\theta\cos^4\theta\, d\theta = \frac{\left|\frac{7}{2}\right.\left|\frac{5}{2}\right.}{2\left|6\right.}$$

$$= \pi\frac{(5/2\cdot3/2\cdot\frac{1}{2}\cdot\sqrt{\pi})\cdot(3/2\cdot\frac{1}{2}\sqrt{\pi})}{2\cdot5\cdot4\cdot3\cdot2\cdot1} = \frac{3\pi^2}{512} \qquad \textbf{Ans.}$$

Theorem 7. (*i*) $\int_{-a}^{+a} f(x)dx = 2\int_0^a f(x)dx,$ if $f(-x) = +f(x)$ (even function)

(*ii*) $\int_{-a}^{+a} f(x)dx = 0,$ if $f(-x) = -f(x)$ (odd function)

Proof. $\int_{-a}^{a} f(x)dx = \int_{-a}^{0} f(x)dx + \int_0^a f(x)dx$ ... (1)

Now, $\int_{-a}^{0} f(x)dx = -\int_a^0 f(-t)dt$ when $x = -t$

$$= \int_0^a f(-t)dt = \int_0^a f(-x)dx$$

Putting the value of $\int_{-a}^{0} f(x)dx$ in (1), we get

$$\int_{-a}^{a} f(x)dx = \int_{0}^{a} f(x)dx + \int_{0}^{a} f(-x)dx$$

(*i*) $$\int_{-a}^{a} f(x)dx = \int_{0}^{a} f(x)dx + \int_{0}^{a} f(x)dx, \text{ if } f(-x) = f(x)$$

$$\int_{-a}^{a} f(x)dx = 2\int_{0}^{a} f(x)dx$$ **Proved.**

(*ii*) $$\int_{-a}^{a} f(x)dx = \int_{0}^{a} f(x)dx + \int_{0}^{a} f(-x)dx$$

$$= \int_{0}^{a} f(x)dx - \int_{0}^{a} f(x)dx \text{ if } f(-x) = -f(x)$$

$$= 0$$ **Proved.**

This theorem is used for changing the limits from $\int_{-a}^{+a}$ to $2\int_{0}^{+a}$

Example 29. *Evaluate:* $\int_{-\pi/2}^{\pi/2} \cos^5 x\, dx$ **[B.T.E. Delhi May 2007]**

Solution. $$\int_{-\pi/2}^{\pi/2} \cos^5 x\, dx = 2\int_{0}^{\pi/2} \cos^5 x\, dx \qquad [\because \cos(-x) = +\cos x]$$

$$= 2\cdot\frac{\left|\!\overline{\tfrac{6}{2}}\right.\left|\!\overline{\tfrac{1}{2}}\right.}{2\left|\!\overline{\tfrac{7}{2}}\right.} = \frac{\left|\!\overline{3}\right.\left|\!\overline{\tfrac{1}{2}}\right.}{\left|\!\overline{\tfrac{7}{2}}\right.} = \frac{2\cdot 1\cdot\sqrt{\pi}}{\frac{5}{2}\cdot\frac{3}{2}\cdot\frac{1}{2}\sqrt{\pi}} = \frac{16}{15}.$$ **Ans.**

Example 30. *Evaluate:* $\int_{-\pi/2}^{\pi/2} \sin^5 x\, dx$

Solution. $$\int_{-\pi/2}^{\pi/2} \sin^5 x\, dx = 0 \qquad \left[\because \sin^5(-x) = -\sin^5 x\right]$$ **Ans.**

EXERCISE 32.2

Evaluate the following:

1. $\int_{0}^{\frac{\pi}{2}} \frac{\sin x}{\sin x + \cos x} dx$ **Ans.** $\frac{\pi}{4}$

2. $\int_{0}^{\pi} \log(1+\cos x)\, dx$ **Ans.** $-\pi \log \frac{1}{2}$

3. $\int_{0}^{\pi/4} \log(1+\tan\theta)\, d\theta$ **Ans.** $\frac{\pi}{8}\log 2$

4. $\int_{0}^{\frac{\pi}{2}} \log \tan x\, dx$ **Ans.** 0

5. $\int_{0}^{\frac{\pi}{2}} \frac{dx}{1+\cot x}$ **Ans.** $\frac{\pi}{4}$

6. $\int_{0}^{\infty} \frac{\log(1+x^2)}{1+x^2} dx$ **Ans.** $\pi \log 2$

7. $\int_{0}^{\frac{\pi}{2}} \frac{dx}{a\sin^2 x + b\cos^2 x}$ **[B.T.E, Delhi May 2007]** **Ans.** $\sqrt{\frac{b}{a}}\cdot\frac{\pi}{2}$

8. $\int_{0}^{\pi} \frac{x\, dx}{a^2\cos^2 x + b^2\sin^2 x}$ **Ans.** $\frac{\pi^2}{2ab}$

9. $\int_{0}^{\frac{\pi}{2}} x\cot x\, dx$ **Ans.** $\frac{\pi}{2}\log 2$

10. $\int_0^{\frac{\pi}{2}} \frac{x}{\sin x + \cos x}\, dx$ **Ans.** $\frac{-\pi}{4\sqrt{2}} \log\left[\frac{\cot\left(\frac{3\pi}{8}\right)}{\cot\left(\frac{\pi}{8}\right)}\right]$

11. $\int_0^{\frac{\pi}{2}} \log \cos x\, dx$ **Ans.** $-\frac{\pi}{2}\log 2$ **12.** $\int_0^{\pi} x \sin^7 x\, dx$ **Ans.** $\frac{16\pi}{35}$

13. $\int_0^1 \frac{\sin^{-1} x}{x}\, dx$ **Ans.** $\frac{\pi}{2}\log 2$ **14.** $\int_0^{\pi} \sin^6 x \cos^5 x\, dx$ **Ans.** 0

15. $\int_0^{\pi} \sin^4 x \cos^6 x\, dx$ **Ans.** $\frac{3\pi}{256}$

16. $\int_0^{\frac{\pi}{2}} \sin^2 x \cos^4 x\, dx$ **[B.T.E, Delhi May 2007]** **Ans.** $\frac{\pi}{32}$

17. $\int_0^{\pi} \sin^9 x\, dx$ **Ans.** $\frac{256}{315}$ **18.** $\int_{-\pi}^{+\pi} \cos^6 x\, dx$ **Ans.** $\frac{5\pi}{8}$

19. Prove that $\int_0^1 x^{n-1}(1-x)^{m-1}\, dx = \int_0^1 x^{m-1}(1-x)^{n-1}\, dx$

32.7 AREA OF PLANE CURVES (QUADRATURE)

Let us consider the area bounded by the curve AB, perpendicular lines AD ($x = a$) and BC ($x = b$) and CD (a part of the x-axis).

Let AB be the curve $y = f(x)$.

Take two neighbouring points $P(x, y)$ and $Q(x + \delta x, y + \delta y)$ on the curve AB between A and B.

Draw PM and QN perpendiculars to x-axis.

Let the area $PQNM$ be δA. since δx is small, we can regard $PMNQ$ as a trapezium.

Area of trapezium $= \frac{1}{2}(PM + QN) \cdot MN$

$$\delta A = \frac{1}{2}(y + y + \delta y) \cdot \delta x$$

$$\frac{\delta A}{\delta x} = y + \frac{1}{2}\delta y$$

$$\lim_{\delta x \to 0} \frac{\delta A}{\delta x} = \lim_{\delta y \to 0}\left(y + \frac{1}{2}\delta y\right) \Rightarrow \frac{dA}{dx} = y = f(x)$$

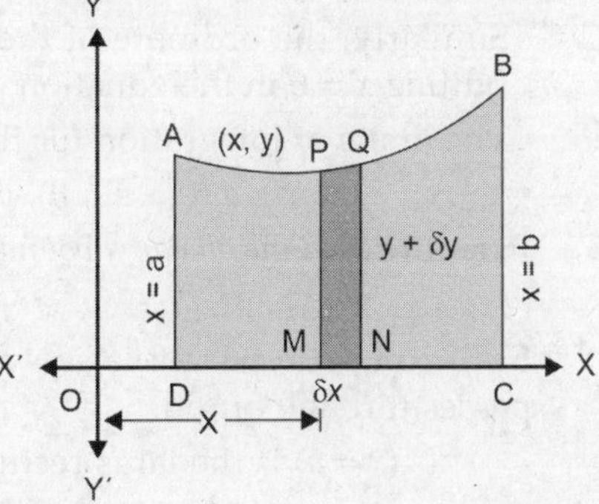

Integrating, we get $A = \int y\, dx = \int f(x) dx$

In case of the area $ABCD$, limits are a and b.

$$\text{Area } ABCD = A = \int_a^b y\, dx = \int_a^b f(x)\, dx \qquad A = \int_a^b y\, dx$$

Cor. Similarly the area bounded by the curve, y-axis and the straight lines $y = c$ and $y = d$ is given by

$$A = \int_c^d x\, dy$$

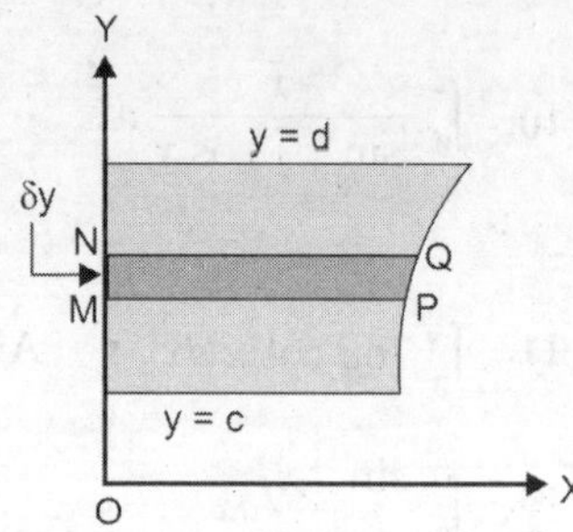

Area is always taken to be positive, even if its calculated value happens to be negative.

Limits of integration. In order to assign the limits of integration we have to form some rough idea of the shape of the curve. The following rules will help for tracing the curve.

32.8 PLOTTING OF A CURVE

1. **Symmetry**

   (*i*) A curve is symmetrical about x-axis if the equation does not change by replacing y by $-y$. Here y should have even powers only.
   For example $y^2 = 4ax$ is symmetrical about x-axis.

   (*ii*) It is symmetrical about y-axis, if it contains only even powers of x.
   For example $x^2 = 4ay$ is symmetrical about y-axis.

   (*iii*) If on interchanging x and y, the equation does not change then the curve is symmetrical about the line $y = x$.
   For example $x^2 + y^2 + xy + 1 = 0$ is symmetrical about the line $y = x$.

2. **Curve through origin.** If the equation of the curve does not contain constant term, the curve passes through the origin.
   For example, The curve $y^2 = 4x$ passes through origin

3. **The points of intersection with the axes.** By putting $y = 0$ in the equation of the curve we get the x co-ordinate of the point of intersection with the x-axis.

   For example, $\frac{x^2}{a^2}+\frac{y^2}{b^2}=1$ intersects the x-axis at the point whose abscissa is given by

$$\frac{x^2}{a^2}+\frac{y^2}{b^2}=1 \quad \Rightarrow \quad \frac{x^2}{a^2}=1 \quad \Rightarrow \quad x=\pm a$$

   $(a, 0)$ and $(-a, 0)$ are the required points.
   Similarly, the ordinate of the point of intersection with the y-axis is obtained on putting $x = 0$ in the equation of the curve.
   The limits of integration for the length of the cycloid

$$x = a\,(\theta + \sin\theta),\ y = a\,(1 - \cos\theta) \text{ are } -\pi \text{ to } \pi.$$

Example 31. *Find the area bounded by the parabola $y^2 = 4ax$ and its latus rectum.*

[SBTE. 2017, 2015, 2010]

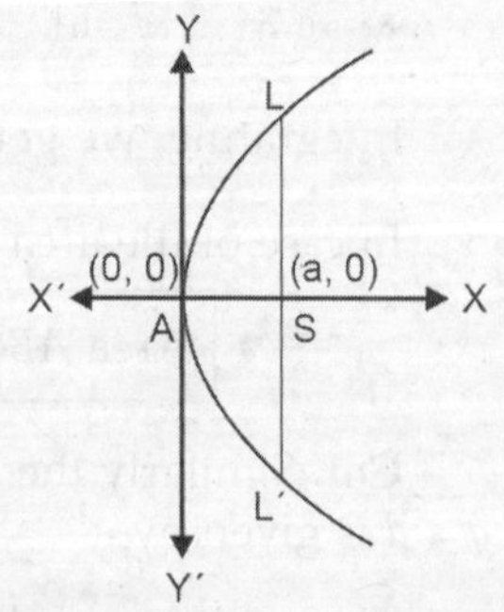

Solution. The parabola is symmetrical about x-axis and passes through origin.

LL' $(x = a)$ is the latus rectum

The area bounded by the parabola $y^2 = 4ax$ and latus rectum is $LAL'L$, and is double of the area $LASL$

The required area $LAL'L = 2$ area $LASL$

$$= 2\int_0^a y\,dx = 2\int_0^a 2\sqrt{ax}\,dx$$

$$= 4\sqrt{a}\int_0^a x^{1/2}\,dx = 4\sqrt{a}\left[\frac{x^{3/2}}{3/2}\right]_0^a$$

$$= 4\sqrt{a}\left[\frac{2}{3}a^{3/2} - 0\right] = \frac{8}{3}a^2 \text{ sq. units}$$ **Ans.**

Example 32. *Show that the curve $a^2y^2 = x^2 (a^2 - x^2)$ consists of two loops and find the area of each loop.*

Solution. $a^2y^2 = x^2 (a^2 - x^2) \Rightarrow y^2 = \frac{x^2}{a^2}(a^2 - x^2)$

(*i*) The equation does not include constant term. So it passes through origin.

(*ii*) The equation includes y^2, so it is symmetrical about x-axis. Similarly it is also symmetrical about y-axis.

(*iii*) If $x = \pm a$, then $y = 0$

Hence the curve consists of two loops.

Limits for upper half of one loop of the curve are $x = 0$ to $x = a$.

$$A = 2\int_0^a y \cdot dx$$

$$= 2\int_0^a \frac{x}{a}\sqrt{a^2 - x^2}\, dx$$

$$= \frac{2}{a}\int_0^a x\sqrt{a^2 - x^2}\, dx$$

$$= \frac{2}{a}\left(-\frac{1}{2}\right)\left[\frac{(a^2 - x^2)^{3/2}}{3/2}\right]_0^a$$

$$= -\frac{2}{3a}\left[(a^2 - x^2)^{3/2}\right]_0^a = -\frac{2}{3a}(0 - a^3)$$

$$= \frac{2a^2}{3}.$$ **Ans.**

Example 33. *Find the area included between the curve $xy^2 = a^2 (a - x)$ and its asymptote.*

Solution. $xy^2 = a^2 (a - x)$

$\Rightarrow$ $y^2 = \frac{a^2 (a - x)}{x}$

$y = 0$ when $x = a$

$y = \infty$ when $x = 0$

$\therefore$ $x = 0$ is its asymptote.

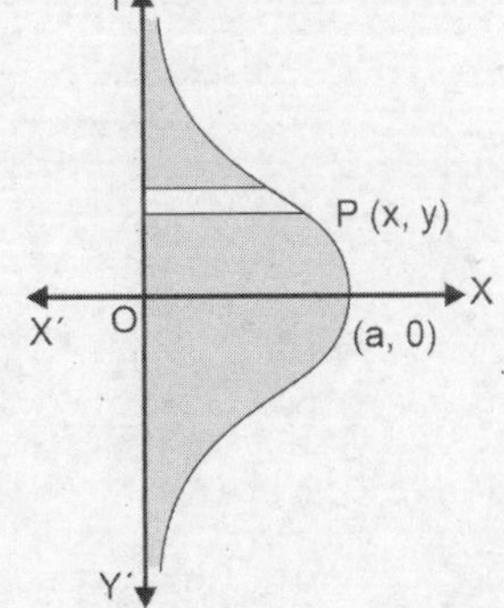

The curve is symmetrical about x-axis as it contains y^2.

Limits for upper half of the curve are $y = 0$ to $y = \infty$

$$xy^2 = a^2 (a - x)$$

$$xy^2 = a^3 - a^2 x$$

$$xy^2 + a^2x = a^3,\ x (y^2 + a^2) = a^3$$

$$x = \frac{a^3}{a^2 + y^2}$$

Required area $= 2\int_0^\infty x\, dy = 2\int_0^\infty \frac{a^3}{a^2 + y^2}\, dy$

$$= 2a^3 \int_0^{\infty} \frac{1}{a^2 + y^2}\, dy = 2a^3 \left[\frac{1}{a} \tan^{-1} \frac{y}{a}\right]_0^{\infty}$$

$$= 2a^2 \left[\tan^{-1} \infty - \tan^{-1} 0\right] = 2a^2 \left[\frac{\pi}{2} - 0\right]$$

$= a^2\pi$ sq. units **Ans.**

Example 34. *Find the area common to the parabola $y^2 = x$ and the circle $x^2 + y^2 = 2$.*

Solution. $y^2 = x$ ... (1)

$x^2 + y^2 = 2$ ... (2)

Let us solve (1) and (2).

Put the value of y^2 from (1) into (2), we get

$$x^2 + x = 2$$

$$\Rightarrow \quad x^2 + x - 2 = 0$$

$$\Rightarrow \quad (x + 2)(x - 1) = 0$$

$$x = -2, 1$$

$$\therefore \quad y = \pm\sqrt{2}\, i, \pm 1$$

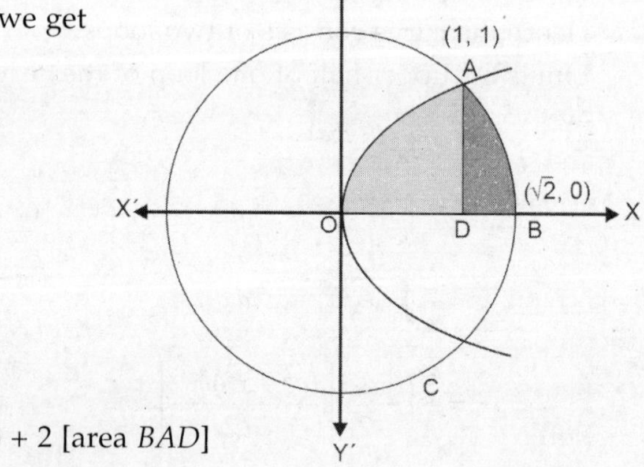

As (1) and (2) intersect at A (1, 1)

and (2) intersect the x-axis at $(\sqrt{2}, 0)$

Required area $OABC$ = 2 area (OAD) + 2 [area BAD]

$$= 2\int_0^1 y_1\, dx + 2\int_1^{\sqrt{2}} y_2\, dx$$

$$= 2\int_0^1 \sqrt{x}\, dx + 2\int_1^{\sqrt{2}} \sqrt{2 - x^2}\, dx$$

$$= 2\left[\frac{2}{3} x^{3/2}\right]_0^1 + 2\left[\frac{x}{2}\sqrt{2 - x^2} + \frac{2}{2}\sin^{-1}\left(\frac{x}{\sqrt{2}}\right)\right]_1^{\sqrt{2}}$$

$$= \frac{4}{3} + 2\left[\sin^{-1}(1) - \frac{1}{2}\sqrt{2 - 1} - \sin^{-1}\left(\frac{1}{\sqrt{2}}\right)\right]$$

$$= \frac{4}{3} + 2\left[\frac{\pi}{2} - \frac{1}{2} - \frac{\pi}{4}\right] = \frac{4}{3} + 2\left[\frac{\pi}{4} - \frac{1}{2}\right] = \frac{4}{3} + \frac{\pi}{2} - 1$$

$= \frac{\pi}{2} + \frac{1}{3}$ sq. units **Ans.**

Example 35. *Find the area bounded by $y^2 = 9x$ and $x^2 = 9y$* **[SBTE. 2017, 2015]**

Solution. $y^2 = 9x$ ... (1)

$x^2 = 9y$ ... (2)

On solving (1) and (2), we get the points of intersection of the curves, (0, 0) and (9, 9).

Limits for the required area are $x = 0$ to $x = 9$.

Required area = Area ($OCAB$) – Area ($ODAB$)

$$= \int y_1\,dx - \int y_2\,dx$$

$$= \int_0^9 3\sqrt{x}\,dx - \int_0^9 \frac{x^2}{9}\,dx$$

$$= 3\left[\frac{x^{3/2}}{3/2}\right]_0^9 - \frac{1}{9}\left[\frac{x^3}{3}\right]_0^9$$

$$= 2(27-0) - \frac{1}{27}(9\times 9\times 9 - 0)$$

$$= 54 - 27 = 27 \text{ sq. units.}$$

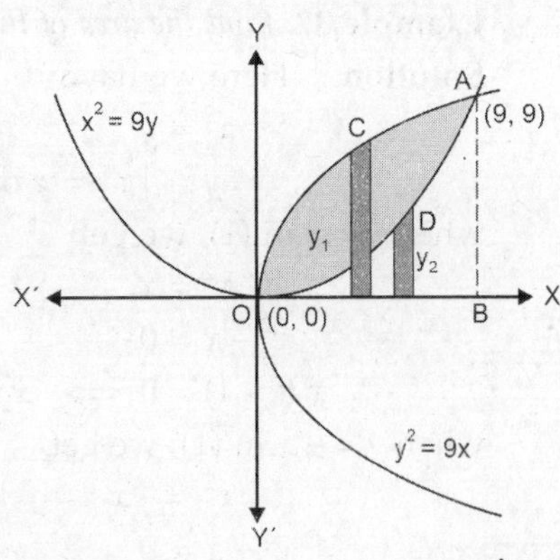

Ans.

Example 36. *Find the area bonded by the parabolas $y^2 = 5x + 6$ and $x^2 = y$.*

Solution. $y^2 = 5x + 6$ ... (1)

and $x^2 = y$ ... (2)

(1) and (2) are parabolas.

(1) is symmetrical about the x-axis

(2) is symmetrical about the y-axis.

(2) passes through origin as it does not include constant.

Let us solve the equation (1) and (2).

Putting $y = x^2$ in (1), we get

$x^4 = 5x + 6$

By trial $x = -1, 2$

Limits for the required area are $x = -1$ and $x = 2$.

Writing the equations as $y_1^2 = 5x + 6$, $y_2 = x^2$

The required area (shown shaded)

$$= \int_{-1}^{2} (y_1 - y_2)\,dx$$

$$= \int_{-1}^{2} (\sqrt{5x+6} - x^2)\,dx$$

$$= \left[\frac{(5x+6)^{3/2}}{5\times 3/2} - \frac{x^3}{3}\right]_{-1}^{2}$$

$$= \left[\frac{(16)^{3/2}\times 2}{15} - \frac{8}{3}\right] - \left[\frac{1\times 2}{15} - \frac{(-1)}{3}\right]$$

$$= \left(\frac{128}{15} - \frac{8}{3}\right) - \left(\frac{2}{15} + \frac{1}{3}\right)$$

$$= \frac{88}{15} - \frac{7}{15} = \frac{81}{15} = \frac{27}{5} \text{ sq. units}$$

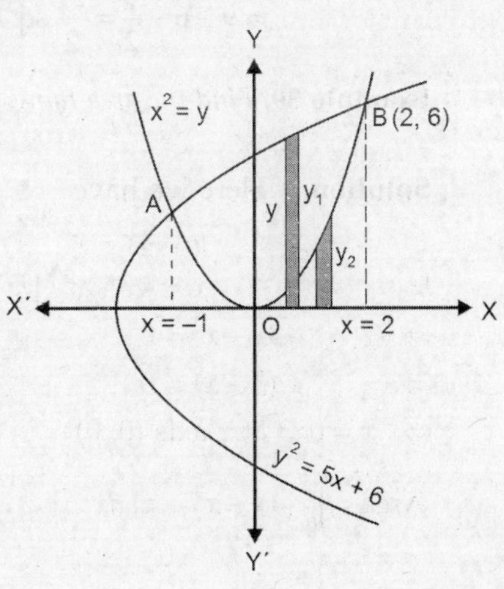

Ans.

Example 37. *Find the area of the region bounded by $x^2 = y$ and $y = |x|$.*

Solution. Here we have

$$x^2 = y \quad \text{... (1)}$$

$$y = |x| = x \text{ or } -x$$

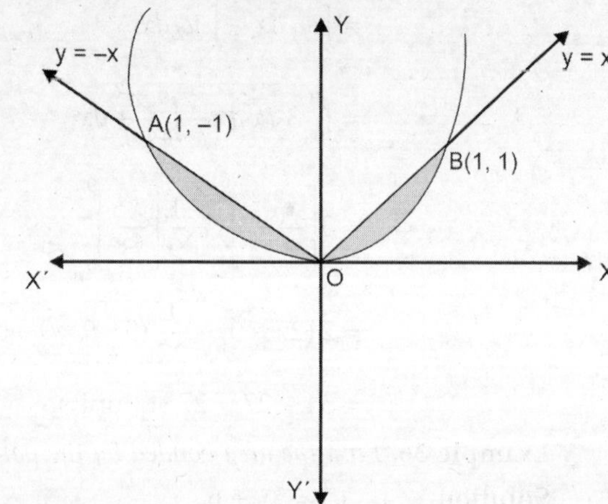

when $y = x$ in (1), we get

$$x^2 = x$$

$$x^2 - x = 0$$

$$x(x - 1) = 0 \Rightarrow x = 1, 0$$

when $y = -x$ in (1), we get

$$x^2 = -x$$

$$x^2 + x = 0$$

$$x(x + 1) = 0 \Rightarrow x = -1, 0$$

Required area $= 2\int_0^1 (x - x^2)\,dx$

$$= 2\left[\frac{x^2}{2} - \frac{x^3}{3}\right]_0^1 = 2\left[\frac{1}{2} - \frac{1}{3}\right] = \frac{1}{3} \text{ sq. units}$$ **Ans.**

Example 38. *Find the area of the region bounded by the curve $y = x^2 + 2$, $y = x$, $x = 0$ and $x = 3$.* **[SBTE. 2016, 2015]**

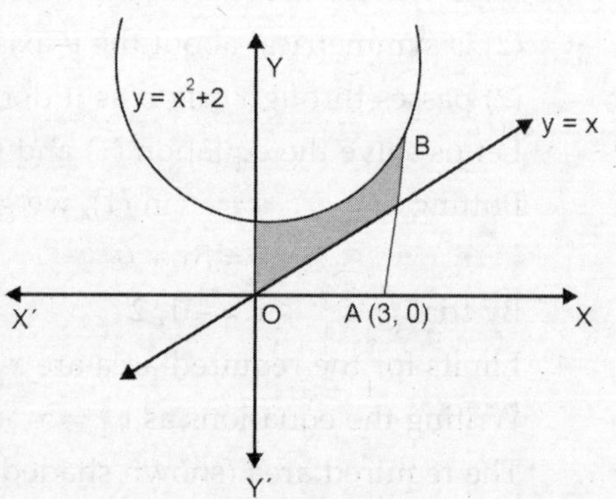

Solution. Required area $= \int_0^3 \left[(x^2 + 2) - x\right] dx$

$$= \left[\frac{x^3}{3} + 2x - \frac{x^2}{2}\right]_0^3$$

$$= 9 + 6 - \frac{9}{2} = \frac{21}{2} \text{ sq. units}$$ **Ans.**

Example 39. *Find the area lying between the parabola $y = 4x - x^2$ and the line $y = x$.*

[D.I.P.I.E.T.E. Dec. 2019]

Solution. Here we have

$$y = 4x - x^2$$

$$\Rightarrow \quad x = 4x - x^2 \quad [\because y = x]$$

$$\Rightarrow \quad 3x - x^2 = 0$$

$$\Rightarrow \quad x(3 - x) = 0$$

$\Rightarrow$ $x = 0, 3$ at x-axis $(0, 0)$

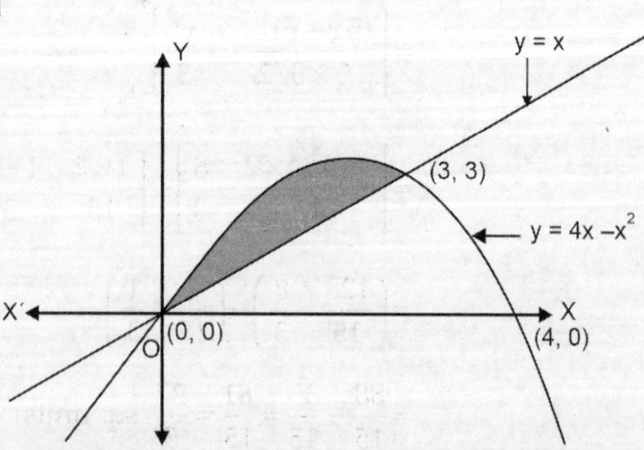

$$\text{Area} = \int_0^3 \{4x - x^2 - x\}\,dx$$

$$= \int_0^3 (3x - x^2)\,dx$$

$$= \frac{3}{2}\left[x^2\right]_0^3 - \frac{1}{3}\left[x^3\right]_0^3$$

$$= \left(\frac{3}{2} \times 9\right) - \left(\frac{1}{3} \times 27\right)$$

$$= \frac{27}{2} - \frac{27}{3} = \frac{81-54}{6} = \frac{27}{6} = \frac{9}{2}$$

Required area = $\frac{9}{2}$ sq. units **Ans.**

Example 40. *Find the area bounded by the parabola $y^2 = 4x$ and the line $y = 2x - 4$.*

Solution. $y^2 = 4x$ ... (1)

$y = 2x - 4$ ... (2)

Let us solve (1) and (2).

Put the value of y from (2) in (1), we get

$$(2x-4)^2 = 4x$$

$\Rightarrow$ $4x^2 - 16x + 16 = 4x$

$\Rightarrow$ $4x^2 - 20x + 16 = 0$

$\Rightarrow$ $x^2 - 5x + 4 = 0$

$\Rightarrow$ $(x-1)(x-4) = 0 \Rightarrow x = 1, 4$

If $x = 1, y = 2 - 4 = -2$

$x = 4, y = 8 - 4 = 4$

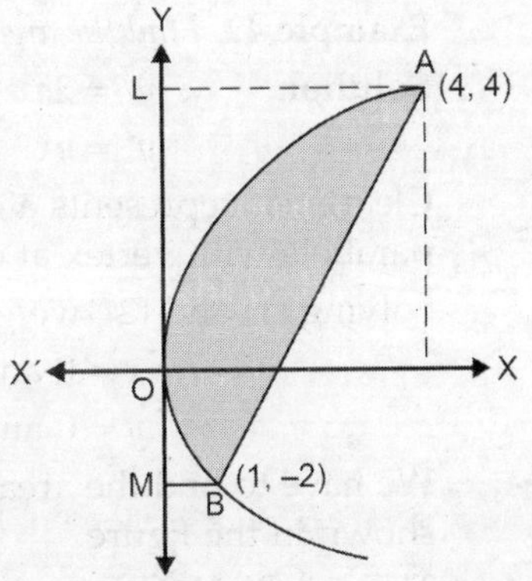

Let the parabola (1) intersect the line (2) at A (4, 4) and B (1, – 2).

Limits for the required area are $y = -2$, and $y = 4$.

Draw AL and $BM \perp$ to the y-axis.

Required area (AOB) = Area of trapezium $LABM$ – Area between the curve AB and the y-axis.

$$\text{Area } (AOB) = \frac{1}{2}(LA + MB)\, LM - \int_{-2}^{4} x\, dy$$

$$= \frac{1}{2}(4+1)\,6 - \int_{-2}^{4} \frac{y^2}{4}\, dy = 15 - \frac{1}{4}\left[\frac{y^3}{3}\right]_{-2}^{4}$$

$$= 15 - \frac{1}{12}[64 + 8] = 15 - 6$$

$= 9$ sq. units. **Ans.**

Example 41. *Find the area cut off from the parabola $4y = 3x^2$ by the line $2y = 3x + 12$.*

Solution. $4y = 3x^2$ ... (1)

$2y = 3x + 12$ ... (2)

On solving (1) and (2), we get $x = -2$ and $x = 4$.

The required area (ABO) = Area ($ABCD$) – Area ($AOBCD$)

$$= \int y_2\, dx - \int y_1\, dx$$

$$= \int_{-2}^{4} \frac{3x+12}{2}\,dx - \int_{-2}^{4} \frac{3x^2}{4}\,dx$$

$$= \left(\frac{3}{2}\frac{x^2}{2} + 6x\right)_{-2}^{4} - \left(\frac{3x^3}{12}\right)_{-2}^{4}$$

$$= \left(\frac{3}{4}\times 16 + 24\right) - \left(\frac{3}{4}\times 4 - 12\right) - \left(\frac{64}{4} + \frac{8}{4}\right)$$

$= 36 + 9 - 18$

$= 27$ Sq. units. **Ans.**

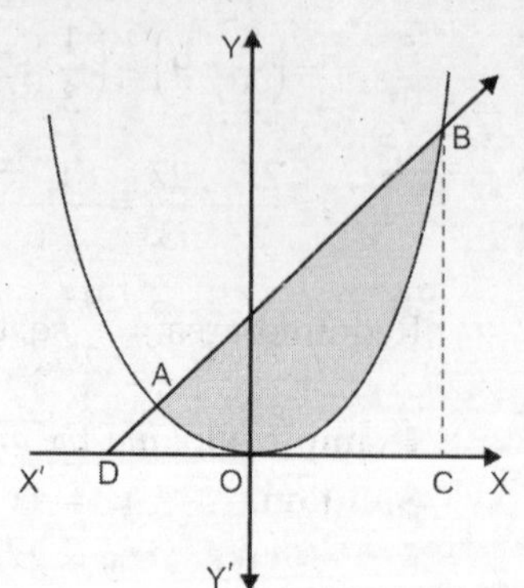

Example 42. *Find the area included between the parabola $y^2 = ax$ and the circle $y^2 = 2ax - x^2$.*

Solution. $y^2 = 2ax - x^2 \Rightarrow (x-a)^2 + y^2 = a^2$ ... (1)

$y^2 = ax$ ... (2)

Clearly (1) represents a circle whose centre is $(a, 0)$ and radius a and (2) represents a parabola with vertex at origin.

Solving (1) and (2), we get

$x = 0$ and $x = a$

$y = 0$ and $y = a$

We have to find the area of the shaded portion as shown in the figure.

The area $OLAMO$ = Area $(OLACO)$ – Area $(OMACO)$

$$= \int_0^a y_1\,dx - \int_0^a y_2\,dx \qquad (y_1 = LN,\ y_2 = MN)$$

$$= \int_0^a \sqrt{2ax - x^2}\,dx - \int_0^a \sqrt{ax}\,dx$$

$$= \int_0^a \sqrt{a^2 - (a-x)^2}\,dx - \sqrt{a}\int_0^a x^{1/2}\,dx$$

(I) (II)

Put $a - x = a\cos\theta$ in (I) so that $-dx = -a\sin\theta\,d\theta$

$$I = \int_0^{\pi/2} a\sin\theta\, a\sin\theta\, d\theta = \frac{a^2}{2}\int_0^{\pi/2} 2\sin^2\theta\,d\theta$$

$$= \frac{a^2}{2}\int_0^{\pi/2}(1-\cos 2\theta)\,d\theta = \frac{a^2}{2}\left[\theta - \frac{\sin 2\theta}{2}\right]_0^{\pi/2} = \frac{a^2\pi}{4}$$

$$II = \sqrt{a}\int_0^a x^{1/2}\,dx = \sqrt{a}\left[\frac{x^{3/2}}{3/2}\right]_0^a = \frac{2a^2}{3}$$

The area $(OLAMO) = \dfrac{\pi a^2}{4} - \dfrac{2}{3}a^2$

The required area $= 2\left[\frac{\pi a^2}{4} - \frac{2}{3}a^2\right]$ sq. units. **Ans.**

Example 43. *Find the area enclosed between the cycloid $x = a\ (t - \sin t),\ y = a\ (1 - \cos t)$. and its base.*

Solution. $x = a\ (t - \sin t)$ ... (1)

$y = a\ (1 - \cos t)$ ... (2)

For tracing the curve we prepare a table of x and for different values of t.

Table

t	0	$\frac{\pi}{2}$	π	$\frac{3\pi}{2}$	2π
x	0	$0.5a$	$3.1a$	$5.7a$	$6.3a$
y	0	a	$2a$	a	0

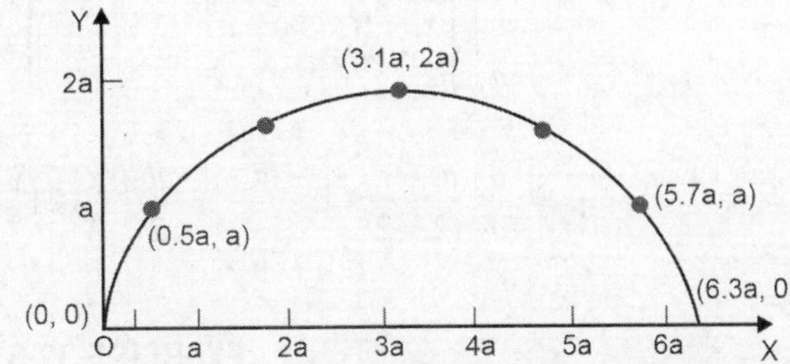

Limits for t are 0 to 2π.

Hence the required area $= \int y\, dx$

$$= \int_0^{2\pi} a\,(1 - \cos t) \cdot a\,(1 - \cos t)\, dt = a^2 \int_0^{2\pi} [1 - 2\cos t + \cos^2 t]\, dt$$

$$= a^2 \int_0^{2\pi} \left[1 - 2\cos t + \frac{1}{2}(\cos 2t + 1)\right] dt$$

$$= a^2 \left[t - 2\sin t + \frac{1}{4}\sin 2t + \frac{t}{2}\right]_0^{2\pi}$$

$$= a^2\,[2\pi + \pi] = 3\pi a^2 \text{ Sq. units.}$$ **Ans.**

Example 44. *Find the area of the smaller region bounded by the ellipse $\frac{x^2}{a^2} + \frac{y^2}{b^2} = 1$ and the line $\frac{x}{a} + \frac{y}{b} = 1$.* **[SBTE. 2016, 2015]**

Solution. Here we have

$$\frac{x^2}{a^2} + \frac{y^2}{b^2} = 1$$

$$\frac{y^2}{b^2} = 1 - \frac{x^2}{a^2}$$

$$y = \frac{b}{a}\sqrt{a^2 - x^2} \quad \text{... (1)}$$

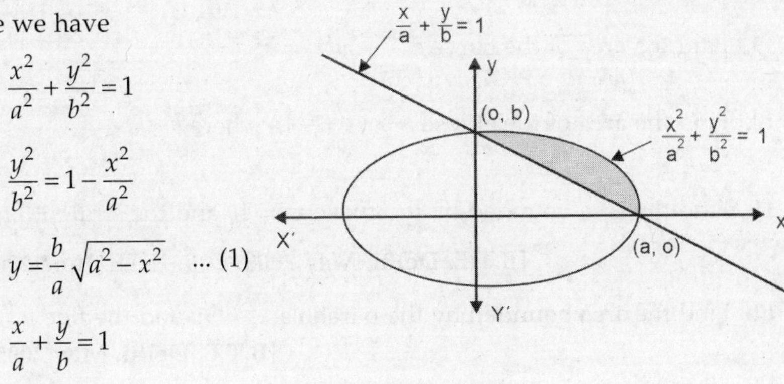

and $\frac{x}{a} + \frac{y}{b} = 1$

$$\frac{y}{b} = 1 - \frac{x}{a}$$

$$y = \frac{b}{a}\sqrt{a - x} \qquad \text{... (2)}$$

$$\text{Required area} = \int_0^a \left[\frac{b}{a}\sqrt{a^2 - x^2} - \frac{b}{a}(a - x)\right] dx$$

$$= \frac{b}{a}\left\{\int_0^a \sqrt{a^2 - x^2}\, dx - \int_0^a (a - x)\, dx\right\}$$

$$= \frac{b}{a}\left[\left\{\frac{x}{2}\sqrt{a^2 - x^2} + \frac{a^2}{2}\sin^{-1}\left(\frac{x}{a}\right)\right\}_0^a - \left\{ax - \frac{x^2}{2}\right\}_0^a\right]$$

$$= \frac{b}{a}\left[\frac{a^2}{2}\cdot\frac{\pi}{2} - 0 - \frac{a^2}{2}\right] = \frac{ab}{2}\left(\frac{\pi}{2} - 1\right) \text{ sq. units} \qquad \textbf{Ans.}$$

EXERCISE 32.3

1. Find the area bounded by the axis of x and the curve $y = (2 + x)^2$ between the ordinates $x = 1$ and $x = 5$. **Ans.** $\frac{316}{3}$ sq. units.

2. Find the area bounded by x axis and the curve $y = 1 - x^2$. **Ans.** $\frac{4}{3}$ sq. units.

3. Find the area of circle $x^2 + y^2 = a^2$. **Ans.** πa^2 sq. units.

4. Find the area bounded by the ellipse $\frac{x^2}{2} + \frac{y^2}{4} = 1$. **Ans.** $2\sqrt{2}\,\pi$ sq. units.

5. Find the area of the loop of the curve $ay^2 = x^2(a - x)$. **Ans.** $\frac{8a^2}{15}$ sq. units.

6. Find the area between Cissiod $y^2 = \frac{x^3}{2a - x}$ and its asymptote $x = 2a$. **Ans.** $3\pi a^2$ sq. units.

7. Find the area enclosed by the curve $xy^2 = 4(2 - x)$ and y-axis. **Ans.** 4π sq. units.

8. Trace the curve $a^2 x^2 = y^3(2a - y)$ and find the area enclosed by it. **Ans.** πa^2 sq. units.

9. Find the area of the curve $x^{2/3} + y^{2/3} = a^{2/3}$. **Ans.** $\frac{3}{8}\pi a^2$ sq. units.

10. Find the area of the ellipse $\frac{x^2}{a^2} + \frac{y^2}{b^2} = 1$, where $b < a$. **[SBTE. 2017 2016]** **Ans.** πab sq. units.

11. Find the area bounded by the curve $x^2 = 4y$ and the straight line $x = 4y - 2$. **[B.T.E. Delhi, May 2008; Dip. IETE, June 2005] Ans.** $\frac{9}{8}$ sq. units.

12. Find the area bounded by the parabola $x^2 = 9y$ and the line $y = 4x$. **[B.T.E. Delhi. May 2005] Ans.** 864 sq. units.

13. Find the area included between the parabola $y^2 = 4x$ and the line $x + y = 3$. **[SBTE. 2012, 2010] Ans.** $\frac{64}{3}$ sq. units.

14. Find the area bounded by the parabola $y^2 = 4x$ and the straight line $y = 2x - 4$.
[B.T.E. Delhi. May 2005)] Ans. 9 sq. units.

15. Find the area common to two curves $y^2 = 2x$ and $x^2 + y^2 = 4x$.

Ans. $\frac{2}{3}(3\pi - 8)$ sq. units.

16. Find the area between two parabolas $y^2 = 4ax$ and $x^2 = 4ay$.

[SBTE. 2017] Ans. $\frac{16a^2}{3}$ sq. units.

17. Find the area enclosed by circles $x^2 + y^2 - 2ax = 0$ and $x^2 + y^2 - 2ay = 0$.

[Diploma IETE, Dec. 2005] Ans. $\frac{a^2}{6}(4a - 3\pi - 6)$ sq. units.

18. Find the area bounded by the parabola, $y = x^2 - 7x + 6$, the x-axis and the lines $x = 2$ and $x = 6$.

Ans. $\frac{56}{3}$ sq. units.

19. Find the area of the segment cut off from the parabola $x^2 = 8y$ by the line $x - 2y + 8 = 0$.
[SBTE, 2016, 2015] Ans. 36 sq. units.

20. Find the area of the loop of the curve: $3ay^2 = x(x - a)^2$

Ans. $\frac{8a^2}{15\sqrt{3}}$ sq. units.

21. Find the area common to the two ellipse $\frac{x^2}{a^2} + \frac{y^2}{b^2} = 1$ *and* $\frac{x^2}{b^2} + \frac{y^2}{a^2} = 1$.

Ans. $4ab\tan^{-1}\left(\frac{b}{a}\right)$ sq. units.

32.9 VOLUME OF SOLIDS BY REVOLUTION OF DISC

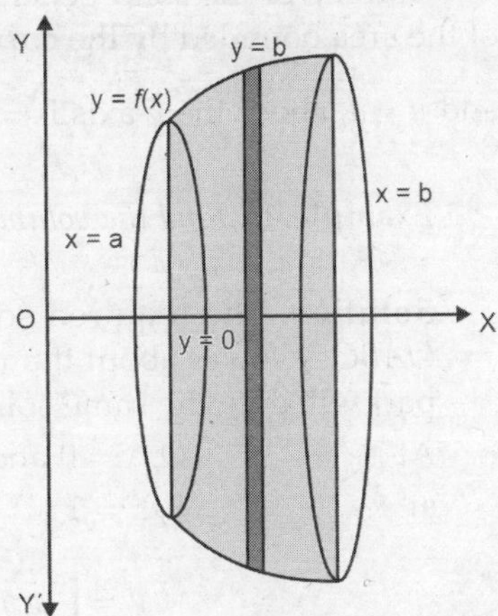

(1) Revolving the disc about x-axis

Area of the disc = ydx

(2) **Volume** of the solid generated by revolving the area of the disk bound by $x = a$, $x = b$ about the x-axis

$V = \pi \int_a^b y^2\, dx$

Proof. Let V be the volume of the solid generated by rotating about x-axis of the area bound by

(*i*) the curve AB, $y = f(x)$ (*ii*) the x-axis

(*iii*) coordinates AL $(x = a)$, BM $(x = b)$.

Let $P(x, y)$ be any point on the curve and $Q(x + \delta x, y + \delta y)$. Draw PR and $QS \perp$ s to the x-axis.

Let the volume of the solid generated by the rotation of the area $PQSR$ about the x-axis be δV.

Suppose the curve AB $[y = f(x)]$ is revolved about x-axis and we are required to find the volume V of the solid of revolution.

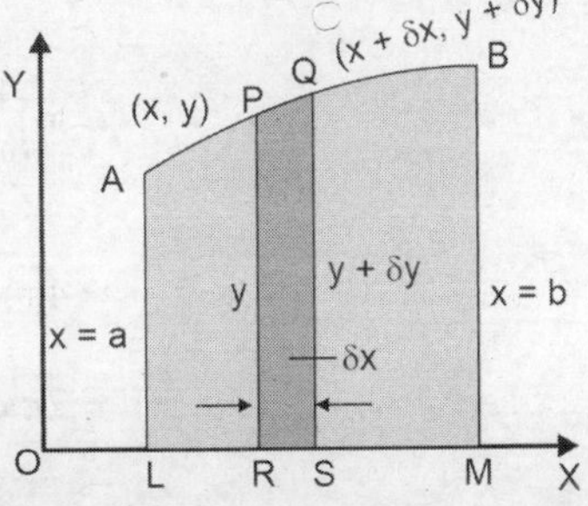

Let AD $(x = a)$ and BC $(x = b)$ be two perpendicular lines on x-axis. Take two neighbouring points $P(x, y)$ and $Q(x + \delta x, y + \delta y)$ on the curve AB. Draw PM $(= y)$ and QN $(= y + \delta y)$ perpendiculars on x-axis. The solid generated by revolution of the area *PMNQ* about *x*-axis may be regarded as a disc. The average radius of the disc is $\frac{y+(y+\delta y)}{2}$, *i.e.* $y+\frac{\delta y}{2}$. Let the volume of disc be δV.

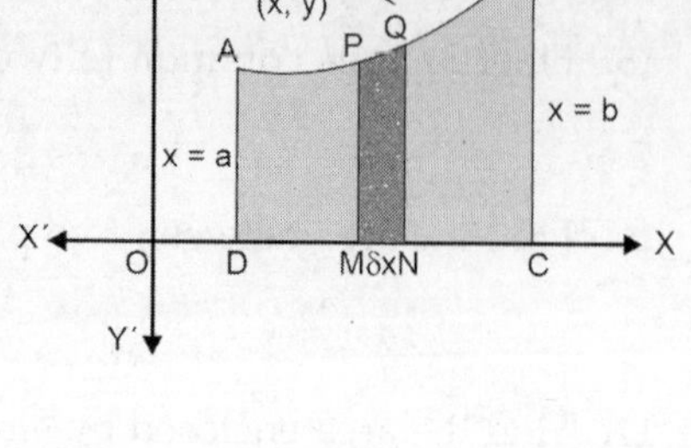

$$\left[V=\pi r^2 h, r=y+\frac{\delta y}{2}, h=MN=\delta x\right]$$

$$\Rightarrow \quad \delta V=\pi\left(y+\frac{\delta y}{2}\right)^2\cdot\delta x$$

$$\Rightarrow \quad \frac{\delta V}{\delta x}=\pi\left(y+\frac{\delta y}{2}\right)^2$$

$$\Rightarrow \quad \frac{dV}{dx}=\pi y^2 \qquad \text{(on taking limits as } \delta x\to 0 \therefore \delta y\to 0)$$

On integrating both sides w.r.t. *x* we get $V=\int_a^b \pi y^2\,dx$

$$\therefore \quad V=\pi\int_a^b y^2\,dx.$$

Cor. The volume of the solid generated by the rotation of the area about *y*-axis is

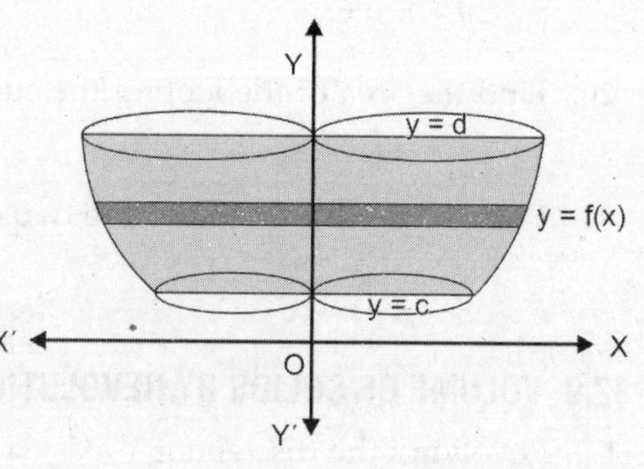

$$=\pi\int_a^b x^2\,dy$$

Area of the disc = xdy

Volume of the solid generated by rotating of the area bounded by the curve $x=f(y)$, $y=c$ and $y=d$, about the *y*-axis is $=\pi\int_c^d x^2\,dy=V$

Example 45. *Find the volume of revolution, about x-axis of the parabola $y^2=4x$ bounded by $x=1$.*

Solution. The required solid is generated when the area *OABO* revolves about the *x*-axis, the revolution of the other part will give the same solid.

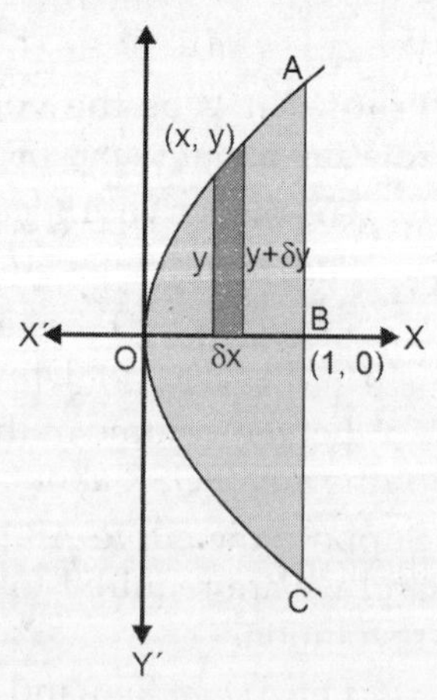

At   *O*, $x=0$ and

at   *B*, $x=1$

$$\therefore \quad V=\int_a^b \pi y^2\,dx$$

$$=\pi\int_0^1 4x\,dx=4\pi\left[\frac{x^2}{2}\right]_0^1$$

$$=4\pi\times\frac{1}{2}$$

$= 2\pi$ cubic units **Ans.**

32.10 VOLUME OF SOLIDS OF REVOLUTION BY WASHER METHOD

The area bounded by $y = f(x)$ and $y = a$ is revolved about x-axis. A solid is generated on the revolution of the bounded area. We have to find out the volume of the solid generated.

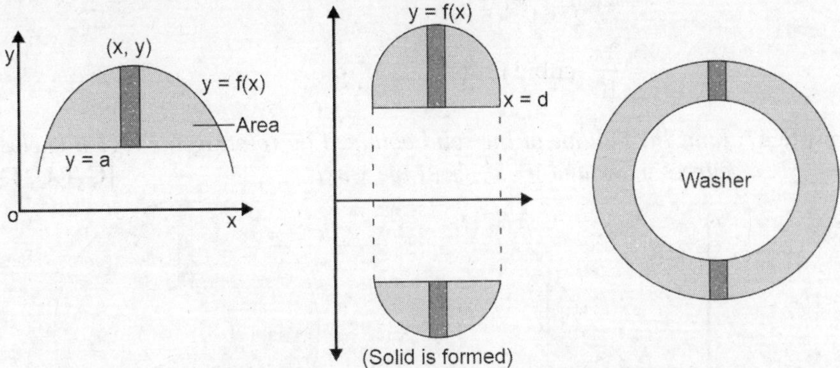

A strip $(y - a)\,\delta x$ is revolved about x-axis. A hollow disc (washer) is formed, its volume is $\pi(y - a)^2.\delta x$.

The volume of the solid generated is obtained by integrating $\pi(y - a)^2\,\delta x$ form c to d.

$$V = \int_c^d \pi(y - a)^2\,dx$$

Similarly the volume of the solid generated by revolving the bounded area about y-axis is

$$V = \int_e^f \pi(x - a)^2\,dy$$

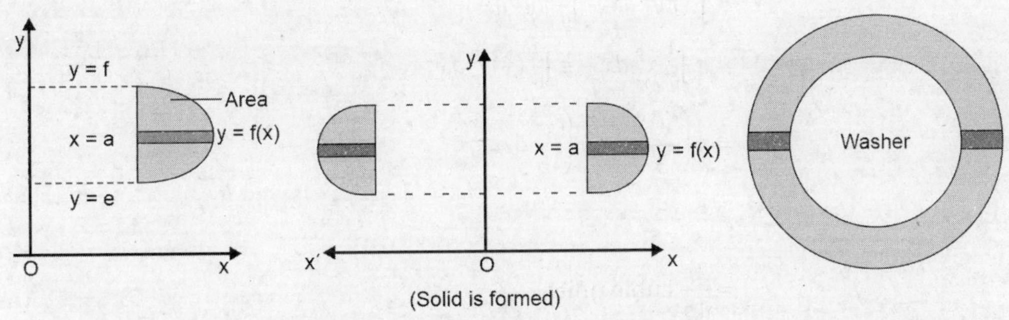

Example 46. *Find the volume of the solid that results when the region enclosed by the curves $y = x^2$ and $x = y^2$ is revolved about the y-axis.* **[G.T.U., Dec., 2015]**

Solution. Here, we have the curves

$$y = x^2,\ x = y^2$$

We will apply Washer method to solve the problem.

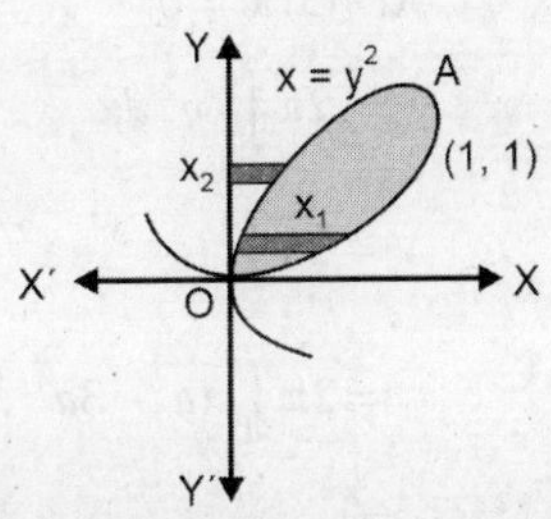

Areas $x_1 dy$ and $x_2 dy$ are revolved about y-axis. By revolving these areas, a solid is generated whose volume is

$$V = \pi \int_0^1 x_1^2\,dy - \pi \int_0^1 x_2^2\,dy$$

$$= \pi \int_0^1 y\,dy - \pi \int_0^1 (y^2)^2\,dy$$

$$= \pi\left(\frac{y^2}{2}\right)_0^1 - \pi\left(\frac{y^5}{5}\right)_0^1$$

$$= \pi\left(\frac{1}{2}\right) - \pi\left(\frac{1}{5}\right)$$

$$= \frac{3\pi}{10} \text{ cubic units}$$ **Ans.**

Example 47. *Find the volume of the solid obtained by rotating the region enclosed by the curves $y = x$ and $y = x^2$ about the x-axis.* **[G.T.U. Dec. 2013]**

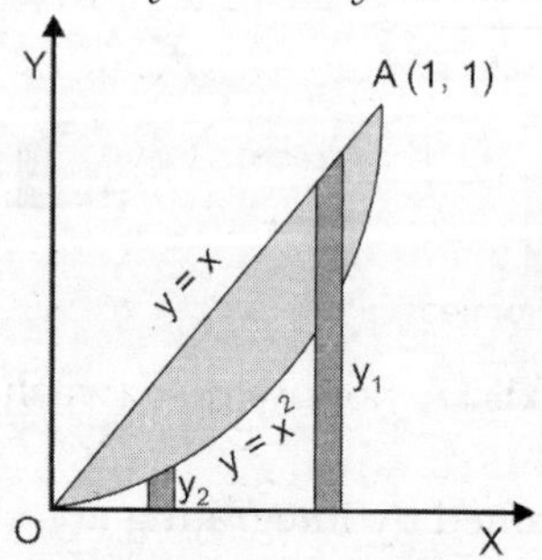

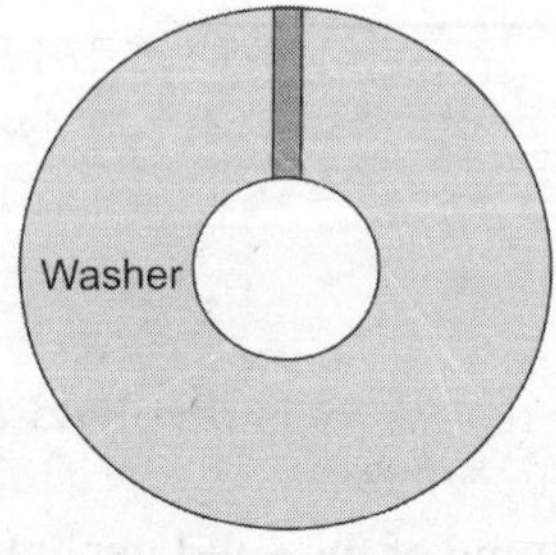

Solution. Here, we have the curves

$$y = x,\ y = x^2$$

Here we will apply the Washer method.

Area $y_1 dx$ and $y_2 dx$ are rotated about x-axis, solid is generated whose volume is

$$V = \int_0^1 \pi y_1^2\, dx - \int_0^1 \pi y_2^2\, dx$$

$$= \pi \int_0^1 x^2\, dx - \pi \int_0^1 (x^2)^2\, dx$$

$$= \pi\left(\frac{x^3}{3}\right)_0^1 - \pi\left(\frac{x^5}{5}\right)_0^1$$

$$= \frac{\pi}{3} - \frac{\pi}{5}$$

$$= \frac{2\pi}{15} \text{ cubic units}$$ **Ans.**

Example 48. *Find the volume of solid generated by the revolution of the hypocycloid $x^{2/3} + y^{2/3} = a^{2/3}$ about the x-axis.*

Solution. The volume required = 2 × the volume formed by the revolution of OAB about OA.

At $O, x = 0,$ at $A, x = a,$

$$V = 2\pi \int_0^a y^2\, dx$$

$$V = 2\pi \int_0^a (a^{2/3} - x^{2/3})^3\, dx.$$

$$= 2\pi \int_0^a (a^2 - 3a^{4/3}x^{2/3} + 3a^{2/3}x^{4/3} - x^2)\, dx$$

$$\left[\begin{array}{ll} \because & x^{2/3} + y^{2/3} = a^{2/3} \\ \Rightarrow & y^{2/3} = a^{2/3} - x^{2/3} \\ \Rightarrow & y^2 = (a^{2/3} - x^{2/3})^3 \end{array}\right]$$

$$= 2\pi\left[a^2 x - 3\frac{a^{4/3}x^{5/3}}{5/3} + 3a^{2/3}\frac{x^{7/3}}{7/3} - \frac{x^3}{3}\right]_0^a$$

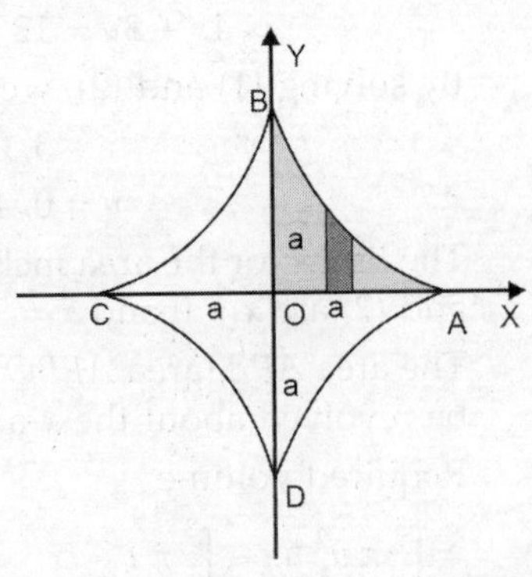

$$= 2\pi\left[a^2 - 3\times\frac{3}{5}a^{\frac{4}{3}}a^{\frac{5}{3}} + 3\times\frac{3}{7}a^{\frac{2}{3}}a^{\frac{7}{3}} - \frac{a^3}{3}\right]$$

$$= 2\pi\left[a^3 - \frac{9}{5}a^3 + \frac{9}{7}a^3 - \frac{a^3}{3}\right] = 2\pi a^3\left[1 - \frac{9}{5} + \frac{9}{7} - \frac{1}{3}\right]$$

$$= \frac{32\pi a^3}{105} \text{ cubic units.}$$ **Ans.**

Example 49. *Find the volume of the solid formed by the revolution of the curve* $xy^2 = 4(2 - x)$ *through four right angles about the y-axis.*

Solution. $xy^2 = 4(2 - x)$

$\Rightarrow$ $y^2 = \dfrac{8 - 4x}{x}$

(*i*) If $x = 0$, then $y = \pm\infty$

(*ii*) If $y = 0$ then $x = 2$

The limits are $y = -\infty$ to $y = \infty$.

Volume of the required solid

$$= \int_{-\infty}^{+\infty} \pi x^2\, dy$$

$$= \pi\int_{-\infty}^{+\infty} \frac{64}{(4 + y^2)^2}\, dy$$

$$\left[\begin{aligned} xy^2 &= 4(2 - x) \\ \Rightarrow\quad xy^2 &= 8 - 4x \\ \Rightarrow (y^2 + 4)x &= 8 \\ \Rightarrow\quad x &= \frac{8}{y^2 + 4} \end{aligned}\right]$$

Put $y = 2\tan\theta$

$$= 64\pi\int_{-\pi/2}^{+\pi/2} \frac{1}{(4 + 4\tan^2\theta)^2}(2\sec^2\theta\, d\theta)$$

$$= 64\pi\int_{-\pi/2}^{+\pi/2} \frac{2\sec^2\theta}{16\sec^4\theta}\, d\theta$$

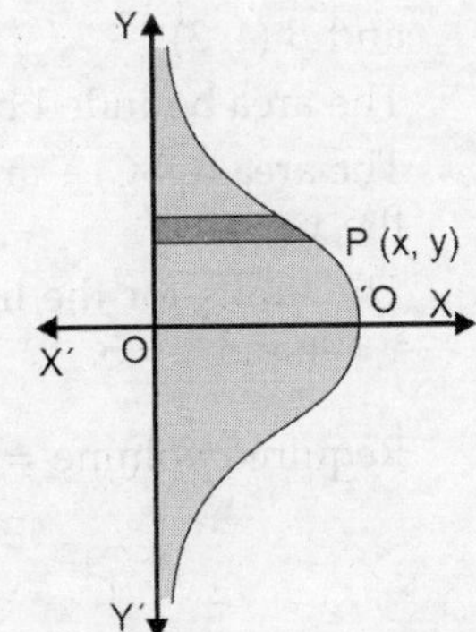

$$= 8\pi\int_{-\pi/2}^{+\pi/2} \cos^2\theta\, d\theta$$

$$= 4\pi\int_{-\pi/2}^{+\pi/2} (\cos 2\theta + 1)\, d\theta$$

$$= 4\pi\left[\frac{\sin 2\theta}{2} + \theta\right]_{-\pi/2}^{+\pi/2}$$

$$= 4\pi^2 \text{ cubic units}$$ **Ans.**

Example 50. *Find the volume generated by revolving about x-axis the area cut off from the parabola* $9y = 4(9 - x^2)$ *by the line* $4x + 3y = 12$.

Solution. $9y = 4(9 - x^2)$ ... (1)

$$4x + 3y = 12 \qquad \dots (2)$$

By solving (1) and (2), we get

$$x = 3, 0$$
$$y = 0, 4$$

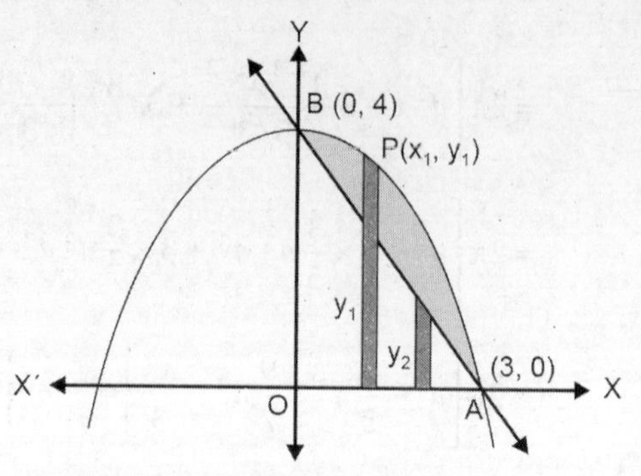

The limits for the area included between (1) and (2) are $x = 0$ and $x = 3$.

The area APB (area $APBO$ – area ABO) is to be revolved about the x-axis.

Required volume

$$= \int_0^3 \pi\, y_1^2\, dx - \int_0^3 \pi\, y_2^2\, dx$$

$$= \pi \int_0^3 \frac{16}{81}(9 - x^2)^2\, dx - \pi \int_0^3 \frac{1}{9}(12 - 4x)^2\, dx$$

$$= \frac{16\pi}{81}\int_0^3 (81 - 18x^2 + x^4)\, dx - \frac{16\pi}{9}\int_0^3 (9 - 6x + x^2)\, dx$$

$$= \frac{16\pi}{81}\left(81x - \frac{18x^3}{3} + \frac{x^5}{5}\right)_0^3 - \frac{16\pi}{9}\left(9x - \frac{6x^2}{2} + \frac{x^3}{3}\right)_0^3$$

$$= \frac{16\pi}{81}\left(243 - 162 + \frac{243}{5}\right) - \frac{16\pi}{9}(27 - 27 + 9)$$

$$= 16\pi\left(3 - 2 + \frac{3}{5}\right) - 16\pi = \frac{48\pi}{5} = 9.6\pi \text{ cubic units}$$ **Ans.**

Example 51. *Find the volume generated by revolving about OX and the area bounded by the following curves.* $x^2 + y^2 = 25$, $3x - 4y = 0$, $y = 0$ *lying in the first quadrant.*

Solution. $x^2 + y^2 = 25$ is a circle whose centre is at O and radius = 5, $3x - 4y = 0$ represents a straight line OB passing through O.

$y = 0$ is x-axis.

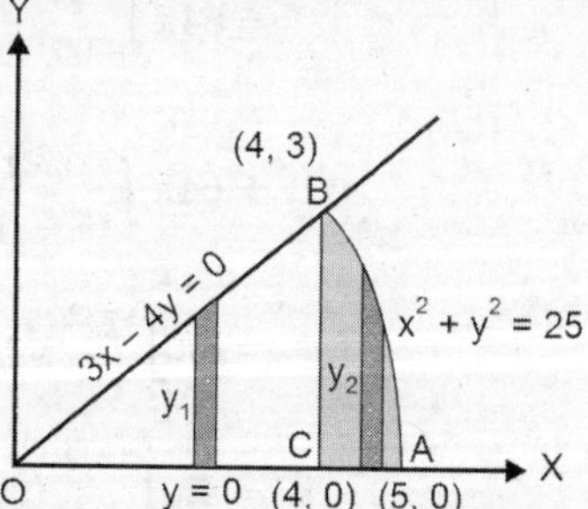

Points of intersection of the circle and line are A (5, 0) and B (4, 3).

The area bounded by these curves is OAB.

The area (OBC) + area (CBA) are to be revolved about the x-axis.

The limits for the line are $x = 0$, $x = 4$ and for circle, $x = 4$ and $x = 5$.

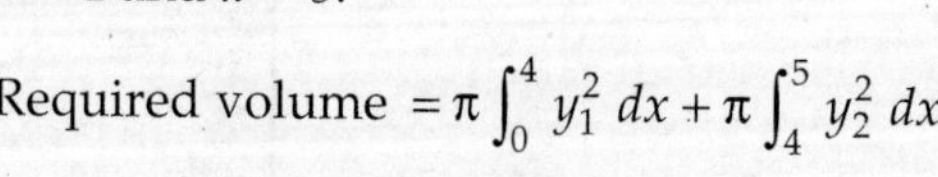

$$\text{Required volume} = \pi \int_0^4 y_1^2\, dx + \pi \int_4^5 y_2^2\, dx$$

$$= \pi \int_0^4 \left(\frac{3x}{4}\right)^2 dx + \pi \int_4^5 (25 - x^2)\, dx = \frac{9\pi}{16}\left[\frac{x^3}{3}\right]_0^4 + \pi\left[25x - \frac{x^3}{3}\right]_4^5$$

$$= \frac{3\pi}{16} \times 64 + \pi\left[125 - \frac{125}{3} - 100 + \frac{64}{3}\right] = \frac{50\pi}{3} \text{ cubic units}$$ **Ans.**

Example 52. *Show that the volume of the spherical cap of a sphere whose height is h and base has radius c is:* $\frac{\pi h}{6}(3c^2 + h^2)$.

Solution. Let a be the radius of the sphere. The spherical cap of a sphere is formed by revolving the area ACD, part of a circle $x^2 + y^2 = a^2$ about x-axis.

$DC = h \qquad \therefore \qquad OD = a - h.$

The limits of integration are $x = a - h$ to $x = a$

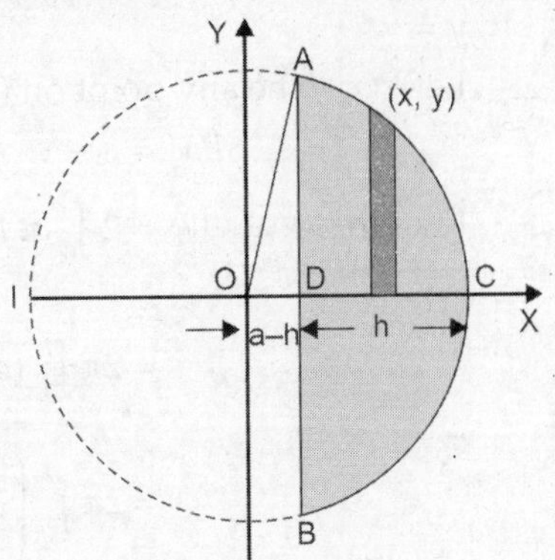

Required volume $= \pi \int_{a-h}^{a} y^2\, dx$

$$= \pi \int_{a-h}^{a} (a^2 - x^2)\,dx \qquad \left[\begin{array}{l} x^2 + y^2 = a^2 \\ \Rightarrow y^2 = a^2 - x^2 \end{array}\right]$$

$$= \pi\left[a^2 x - \frac{x^3}{3}\right]_{a-h}^{a} = \pi\left[a^3 - \frac{a^3}{3} - a^2(a-h) + \frac{(a-h)^3}{3}\right]$$

$$= \pi\left[a^3 - \frac{a^3}{3} - a^3 + a^2 h + \frac{a^3}{3} - a^2 h + ah^2 - \frac{h^3}{3}\right]$$

$$= \pi\left[ah^2 - \frac{h^3}{3}\right] \qquad \text{... (1)}$$

For the right angled triangle OAD.

$$OA^2 = AD^2 + OD^2$$

$$a^2 = c^2 + (a-h)^2$$

$$\Rightarrow \qquad a^2 = c^2 + a^2 + h^2 - 2ah$$

$$\Rightarrow \qquad a = \frac{c^2 + h^2}{2h}$$

Putting the value of a in (1), we get

$$\text{Required volume} = \pi\left[\frac{(c^2 + h^2)h^2}{2h} - \frac{h^3}{3}\right]$$

$$= \pi\left(\frac{3c^2 h + 3h^3 - 2h^3}{6}\right)$$

$$= \frac{\pi h}{6}(3c^2 + h^2)$$

Proved.

Example 53. *If the curve* $(a - x)\, y^2 = a^2 x$ *revolves about its asymptote. Find the volume so formed.*

Solution. $(a - x)\, y^2 = a^2 x \qquad \text{... (1)}$

$$y^2 = \frac{a^2 x}{a - x}$$

(*i*) The curve (1) passes through origin as it does not contain constant.

(*ii*) The curve (1) is symmetrical about x-axis as it contains y^2.

(*iii*) At $x = a$, $y = \infty$

The asymptote is $x = a$.

Hence the limits for upper half of the curve are $y = 0$, $y = \infty$.

Let $P(x, y)$ be any point on (1). PM is $\perp$ to the asymptote.

$$PM = a - x$$

$$\text{Required volume } = 2\int_0^{\infty} \pi\, PM^2\, dy$$

$$= 2\pi \int_0^{\infty} (a - x)^2\, dy \qquad \left(\text{From (1), } x = \frac{ay^2}{a^2 + y^2}\right)$$

$$= 2\pi \int_0^{\infty} \left(a - \frac{ay^2}{a^2 + y^2}\right)^2 dy$$

$$= 2\pi a^6 \int_0^{\infty} \frac{dy}{(a^2 + y^2)^2} \qquad \left[\begin{array}{ll}\text{Put} & y = a\tan\theta \\ \text{so that} & dy = a\sec^2\theta\, d\theta\end{array}\right]$$

$$= 2\pi a^6 \int_0^{\pi/2} \frac{a\sec^2\theta\, d\theta}{(a^2 + a^2\tan^2\theta)^2}$$

$$= \frac{2\pi a^7}{a^4} \int_0^{\pi/2} \frac{\sec^2\theta\, d\theta}{\sec^4\theta}$$

$$= 2\pi a^3 \int_0^{\pi/2} \cos^2\theta\, d\theta$$

$$= \pi a^3 \int_0^{\pi/2} (\cos 2\theta + 1)\, d\theta = \pi a^3 \left[\frac{\sin 2\theta}{2} + \theta\right]_0^{\pi/2}$$

$$= \pi a^3 \left(\frac{\pi}{2}\right) = \frac{\pi^2 a^3}{2} \text{ cubic units.}$$ **Ans.**

Example 54. *Find the volume of the solid generated by revolving the cardioid $r = a\,(1 + \cos\theta)$ about the initial line.*

Solution. $r = a\,(1 + \cos\theta)$

$$\text{Volume} = \pi \int_0^{2a} y^2\, dx$$

The curve $r = a\,(1 + \cos\theta)$ cuts the initial line at $x = 0$ and $x = 2a$.

Let us find out the values of y^2 and dx.

$$y = r\sin\theta = a\,(1 + \cos\theta)\sin\theta = a \cdot 2\cos^2\frac{\theta}{2}\, 2\sin\frac{\theta}{2}\cos\frac{\theta}{2} = 4a\cos^3\frac{\theta}{2}\sin\frac{\theta}{2}$$

$$y^2 = 16a^2\cos^6\frac{\theta}{2}\sin^2\frac{\theta}{2}$$

$x = r\cos\theta = a(1+\cos\theta)\cos\theta$

$x = a\cdot 2\cos^2\frac{\theta}{2}\left(2\cos^2\frac{\theta}{2}-1\right)$

$= 4a\cos^4\frac{\theta}{2} - 2a\cos^2\frac{\theta}{2}$

$\therefore \quad dx = 4a\cdot 4\cos^3\frac{\theta}{2}\left(-\sin\frac{\theta}{2}\right)\cdot\frac{1}{2}\,d\theta - 2a\cdot 2\cos\frac{\theta}{2}\left(-\sin\frac{\theta}{2}\right)\cdot\frac{1}{2}\,d\theta$

$= -8\cos^3\frac{\theta}{2}\sin\frac{\theta}{2}\,d\theta + 2a\cos\frac{\theta}{2}\sin\frac{\theta}{2}\,d\theta$

Volume $= \pi\int_0^{\pi}\left(16a^2\cos^6\frac{\theta}{2}\cdot\sin^2\frac{\theta}{2}\right)\cdot\left[-8a\cos^3\frac{\theta}{2}\sin\frac{\theta}{2}+2a\cos\frac{\theta}{2}\sin\frac{\theta}{2}\right]d\theta$

$= \pi a^3\int_0^{\pi}\left(-128\cos^9\frac{\theta}{2}\sin^3\frac{\theta}{2}+32\cos^7\frac{\theta}{2}\sin^3\frac{\theta}{2}\right)d\theta$

$= 2\pi a^3\int_0^{\pi/2}(-128\cos^9 t\sin^3 t + 32\cos^7 t\sin^3 t)\,dt$ Put $\frac{\theta}{2}=t$, so that $d\theta = 2dt$

$= 2\pi a^3\left[-128\times\frac{\lceil 5\lceil 2}{2\lceil 7} + 32\cdot\frac{\lceil 4\lceil 2}{2\lceil 6}\right]$

$= 2\pi a^3\left(-128\times\frac{4\times 3\times 2\times 1\times 1}{2\times 6\times 5\times 4\times 3\times 2\times 1}+32\times\frac{3\times 2\times 1\times 1}{2\times 5\times 4\times 3\times 2\times 1}\right)$

$= 2\pi a^3\left(-\frac{32}{15}+\frac{4}{5}\right) = -\frac{8\pi a^3}{3}$ cubic units. **Ans.**

EXERCISE 32.4

1. Find the volume of the solid generated by revolving the region bounded by $y=\sqrt{x}$ and the lines $x = 1$, $x = 4$ about the line $y = 0$. **[SBTE. 2017, 2010] Ans.** $\frac{15\pi}{2}$ cubic units
2. Find the volume of the solid generated by the revolution of the curve $y^2 = 25x$ between the ordinates $x = 1$ and $x = 2$, about x-axis. **Ans.** $75\pi/2$ cubic units
3. Find the volume of the solid generated by the revolution of the area between the parabola $y^2 = 4ax$ and its latus rectum about the x-axis. **[SBTE. 2015, 2013] Ans.** $2\pi a^3$ cubic units
4. Find the volume generated by revolving the ellipse $\frac{x^2}{49}+\frac{y^2}{25}=1$ about x-axis. **Ans.** $\frac{700}{3}\pi$ cubic units
5. Find the volume of the solid of revolution obtained by revolving the ellipse $\frac{x^2}{a^2}+\frac{y^2}{b^2}=1$ about (*a*) the minor axis; (*b*) major axis of the ellipse. **[SBTE. 2017, 2012]** **Ans.** (*a*) $\frac{4}{3}\pi a^2 b$ (*b*) $\frac{4}{3}\pi ab^2$ cubic units

6. Find the volume of the solid formed by rotating completely about the x-axis area enclosed between $y^2 = x^3 + 5x$ and the lines $x = 2$ and $x = 4$.

[SBTE. 2012, 2010] **Ans.** 90π cubic units

7. The curve $ay^2 = x^3$ revolves about the axis of y. Find the volume generated between the planes perpendicular to the axis of revolution, at the origin and through the point $27y = 8a$. **Ans.** $\frac{128}{5103}\pi a^3$ cubic units

8. Find the volume of the solid generated by revolution of the semi-circle of radius a, about its bounding diameter. **Ans.** $\frac{4\pi a^3}{3}$ cubic units

9. Find the volume of the solid generated by the revolution of the cycloid $x = a(\theta - \sin\theta)$, $y = a(1 - \cos\theta)$ about its base. **Ans.** $5\pi^2 a^3$ cubic units

10. Prove that the volume of the solid generated by the revolution, about the x-axis of the loop of the curve $x = t^2$ and $y = t - \frac{t^3}{3}$ is $\frac{3\pi}{4}$.

11. Find the volume of the solid generated by the revolution of the loop of the curve $x(x^2 + y^2) = a(x^2 - y^2)$ about the x-axis. **Ans.** $2\pi a^3\left(\log 2 - \frac{2}{3}\right)$ cubic units

12. Find the volume of the solid generated by the revolution of the area of the curve $x = a\cos^3\theta$, $y = a\sin^3\theta$ lying between $\theta = -\frac{\pi}{2}$ and $\theta = \frac{\pi}{2}$ about x-axis.

Ans. $\frac{16\pi a^3}{105}$ cubic units

13. A loop of the curve $y^2 = x^2(1 - x^2)$ is rotated about the y-axis. Find the volume generated. **Ans.** $\pi^2/4$ cubic units

Tick (✓) the correct answer

14. The curve $y = f(x)$ is between $y = c$ and $y = d$ rotated about a line $x = a$. The volume generated is

(i) $\pi\int_c^d (a - x)^2\, dy$ (ii) $\pi\int_c^d y^2\, dx$ (iii) $\pi\int_c^d x^2\, dy$ (iv) None of these **Ans.** (i)

15. The volume of the solid generated by revolving the area included between the parabola $y^2 = 4ax$ and its latus rectum about the x-axis is

(i) πa^3 (ii) $2\pi a^3$ (iii) $\frac{4}{3}\pi a^3$ (iv) $3\pi a^3$ **Ans.** (ii)

16. If a circle $x^2 + y^2 = a^2$ is rotated about x-axis, the volume generated is

(i) πa^3 (ii) $2\pi a^3$ (iii) $\frac{4}{3}\pi a^3$ (iv) $\frac{2}{3}\pi a^3$ **Ans.** (iii)

17. If a circle in positive quadrant is rotated about y-axis is

(i) $\frac{4}{3}\pi a^3$ (ii) $\frac{2}{3}\pi a^3$ (iii) $4\pi a^3$ (iv) πa^3 **Ans.** (ii)

18. If the area enclosed by $y = x$, $y = 0$ and $x = a$ is revolved about x-axis, the volume generated is

(*i*) πa^3 (*ii*) $2\pi a^3$ (*iii*) $\dfrac{2\pi a^3}{3}$ (*iv*) $\dfrac{\pi}{3}a^3$ **Ans.** (*iii*)

19. The volume of the solid generated by revolving the segment of $x + y = 2a$ between the axes about x-axis is

(*i*) $\dfrac{8}{3}\pi a^3$ (*ii*) $8\pi a^3$ (*iii*) $\dfrac{1}{3}\pi a^3$ (*iv*) $\dfrac{4}{3}\pi a^3$ **Ans.** (*i*)

32.11 MEAN VALUE OF A FUNCTION

$y = f(x)$ over the range (a, b) is

$$= \frac{\text{Area under the curve}}{\text{Length of the base}} = \frac{\int_a^b y\,dx}{b-a}$$

Example 55. *Find the mean value of the ordinate of $y^2 = 4x$ from (0, 0) to (4, 4) taking uniformly along the x-axis.* **[SBTE. 2016]**

Solution. $y^2 = 4x$

$$\text{Mean value} = \frac{\int_0^4 y\,dx}{4-0} = \frac{1}{4}\int_2^4 2\sqrt{x}\,dx = \frac{1}{2}\left[\frac{2}{3}x^{3/2}\right]_0^4 = \frac{8}{3}$$ **Ans.**

Example 56. *Find the mean value of $\sin^2 \omega t$ from $t = 0$ to $t = 2\pi/\omega$*

[B.T.E. Delhi Dec. 2014]

Solution. We have

$$y = \sin^2 \omega t$$

$$\text{Mean value} = \frac{\int_0^{\frac{2\pi}{\omega}} \sin^2 \omega t\,dt}{\frac{2\pi}{\omega}-0} = \frac{\omega}{2\pi}\int_0^{\frac{2\pi}{\omega}}\left(\frac{1-\cos 2\omega t}{2}\right)dt$$

$$= \frac{\omega}{2\pi}\int_0^{\frac{2\pi}{\omega}}\frac{1}{2}\,dt - \frac{\omega}{4\pi}\int_0^{\frac{2\pi}{\omega}}\cos 2\omega t\,dt$$

$$= \frac{\omega}{4\pi}[t]_0^{\frac{2\pi}{\omega}} - \frac{\omega}{4\pi}\cdot\frac{1}{2\omega}[\sin 2\omega t]_0^{\frac{2\pi}{\omega}}$$

$$= \frac{\omega}{4\pi}\left[\frac{2\pi}{\omega}-0\right] - \frac{1}{8\pi}[\sin 4\pi - \sin 0]$$

$$= \frac{1}{2} - \frac{1}{8\pi}(0)$$

$$= \frac{1}{2}$$ **Ans.**

Example 57. *Find the mean value of $\sin^4 pt$ over the range $t = 0$ to $t = \dfrac{\pi}{p}$.*

Solution. $$\text{Mean value} = \frac{\int_0^{\pi/p} \sin^4 pt\,dt}{\frac{\pi}{p}-0} = \frac{p}{\pi}\int_0^{\pi/p}(\sin^2 pt)^2\,dt$$

$$= \frac{p}{\pi} \int_0^{\pi/p} \left(\frac{1 - \cos 2pt}{2} \right)^2 dt$$

$$= \frac{p}{4\pi} \int_0^{\pi/p} (1 - 2\cos 2pt + \cos^2 2pt)\, dt$$

$$= \frac{p}{8\pi} \int_0^{\pi/p} (2 - 4\cos 2pt + \cos 4pt + 1)\, dt$$

$$= \frac{p}{8\pi} \int_0^{\pi/p} (3 - 4\cos 2pt + \cos 4pt)\, dt$$

$$= \frac{p}{8\pi} \left[3t - \frac{4\sin 2pt}{2p} + \frac{\sin 4pt}{4p} \right]_0^{\pi/p}$$

$$= \frac{p}{8\pi} \left[\frac{3\pi}{p} \right] = \frac{3}{8}$$ **Ans.**

Example 58. *A quantity of gas confined behind the piston expands adiabatically according to the law $pV^n = C$. Show that the average value of the pressure as volume increases from V_1 to V_2 is*

$$\frac{p_1V_1 - p_2V_2}{(n-1)(V_2 - V_1)}$$

Solution. $PV^n = C$

$$P = CV^{-n}$$

$$= \frac{\int_{v_1}^{v_2} P\, dV}{V_2 - V_1} = \frac{\int_{v_1}^{v_2} CV^{-n}\, dV}{V_2 - V_1}$$

$$= \left[\frac{CV^{-n+1}}{(-n+1)(V_2 - V_1)} \right]_{V_1}^{V_2} = \frac{CV_2^{-n+1} - CV_1^{-n+1}}{(-n+1)(V_2 - V_1)}$$

$$= \frac{CV_1^{-n} \cdot V_1 - CV_2^{-n} \cdot V_2}{(n-1)(V_2 - V_1)} \qquad (p = CV^{-n})$$

$$= \frac{p_1V_1 - p_2V_2}{(n-1)(V_2 - V_1)}$$ **Proved.**

Example 59. *Find the mean value of the following:*

(*i*) $f(x) = \sqrt{x}$ over the range $0 \le x \le 4$ **[SBTE. 2015]**

(*ii*) $f(x) = \sin x$ over the range $[0, 2\pi]$ **[SBTE. 2015, 2013]**

Solution. Here we have

(*i*) $f(x) = \sqrt{x},\ 0 \le x \le 4$

$$\text{Mean value (M.V.)} = \frac{\int_a^b f(x)\, dx}{b - a}$$

$$= \frac{\int_0^4 \sqrt{x}\, dx}{4-0} = \frac{1}{4}\left[\frac{x\sqrt{x}}{\frac{3}{2}}\right]_0^4$$

$$= \frac{1}{4} \times \frac{2}{3}\left[x\sqrt{x}\right]_0^4$$

$$\therefore \quad \text{M.V.} = \frac{1}{6}[8] = \frac{4}{3}$$ **Ans.**

(*ii*) Here we have

$f(x) = \sin x,\ 0 \le x \le 2\pi$

$$\text{M.V.} = \frac{\int_0^a f(x)\, dx}{b-a}$$

$$= \frac{\int_0^{2\pi} \sin x\, dx}{2\pi - 0} = \frac{1}{2\pi}[-\cos x]_0^{2\pi}$$

$$= -\frac{1}{2\pi}[\cos 2\pi - \cos 0]$$

$$\therefore \quad \text{M.V.} = -\frac{1}{2\pi}(1-1) = 0$$ **Ans.**

32.12 THE ROOT MEAN SQUARE VALUE

The root mean square value of a function $y = f(x)$ over the range (a, b) is

$$\sqrt{\frac{\int_a^b y^2\, dx}{b-a}}$$

Example 60. *Find R.M.S. of $\log_e x$ over the range $x = 1$ to $x = e$.* **[SBTE. 2012]**

Solution. $$\text{R.M.S.} = \sqrt{\frac{\int_1^e (\log x)^2\, dx}{e-1}} \qquad \text{... (1)}$$

Now, $$\int (\log x)^2 \cdot 1\, dx = (\log x)^2 \cdot x - 2\int \log x \frac{1}{x} \cdot x\, dx$$

$$= x(\log x)^2 - 2\int \log x\, dx$$

$$= x(\log x)^2 - 2\left[(\log x)\, x - \int \frac{1}{x} \cdot x\, dx\right]$$

$$= x(\log x)^2 - 2x\log x + 2x$$

$$\int_1^e (\log x)^2\, dx = \left[x(\log x)^2 - 2x\log x + 2x\right]_1^e$$

$$= [e(\log_e e)^2 - 2e\log_e e + 2e - 2]$$

$$= [e - 2e + 2e - 2] = e - 2$$

Putting the value of the integral in (1), we get

$$\text{R.M.S.} = \sqrt{\frac{e-2}{e-1}}$$

Example 61. *Show that the R.M.S. value of the expression a sin pt + b cos qt + c sin rt + d cos st + ... is*

$$\sqrt{\frac{1}{2}(a^2+b^2+c^2+d^2+...)}, \textit{ where } p, q, r, s \textit{ ...... are integers.}$$

Solution.

$$\text{R.M.S.} = \sqrt{\frac{\int_0^{2\pi} (a \sin pt + b\cos qt + c \sin rt + d \cos st + ...)^2\, dt}{2\pi - 0}}$$

$$= \frac{1}{\sqrt{2\pi}} \sqrt{\int_0^{2\pi} (a^2 \sin^2 pt + b^2 \cos^2 qt + c^2 \sin^2 rt + d^2 \cos^2 st + ... + 2ab \sin pt \cos qt + 2ac \sin pt \sin rt + ...)\, dt}$$

$$= \frac{1}{\sqrt{2\pi}} \sqrt{(a^2\pi + b^2\pi + c^2\pi^2 + d^2\pi^2 + ... + 0 + 0 + ...}$$

$$= \sqrt{\frac{1}{2}(a^2+b^2+c^2+d^2+...)}$$ **Proved.**

Example 62. *Find the R.M.S. value of the following:*

(*i*) $f(x) = \sin x$ over the interval $[0, 2\pi]$ **[SBTE. 2016, 2013]**

(*ii*) $f(t) = a \sin wt$ over its period **[SBTE. 2013]**

Solution. Here we have

(*i*) $f(x) = \sin x,\ 0 \le x \le 2\pi$

$$\text{R.M.S.V.} = \sqrt{\frac{1}{b-a}\int_a^b |f(x)|^2\, dx}$$

$$= \sqrt{\frac{1}{2\pi}\int_0^{2\pi} \sin^2 x\, dx}$$

$$= \sqrt{\frac{1}{2\pi}\cdot\frac{1}{2}\int_0^{2\pi} (1-\cos 2x)\, dx} \qquad [\because \sin 4\pi = 0]$$

$$= \sqrt{\frac{1}{4\pi}\left[x - \frac{\sin 2x}{2}\right]_0^{2\pi}} = \sqrt{\frac{1}{4\pi}\times 2\pi}$$

$$\therefore \quad = \sqrt{\frac{1}{2}} = \frac{1}{\sqrt{2}}$$ **Ans.**

(*ii*) Here we have

$$f(t) = a \sin wt,\ 0 \le t \le 2\pi/w$$

$$\text{R.M.S.V.} = \sqrt{\frac{1}{\frac{2\pi}{w}}\int_0^{2\pi/w} a^2 \sin^2 wt\, dt}$$

$$= \sqrt{\frac{a^2 w}{2\pi}\int_0^{2\pi/w}\left(\frac{1-\cos 2wt}{2}\right)dt}$$

$$= \sqrt{\frac{a^2 w}{4\pi}\left[x - \frac{\sin 2wt}{2w}\right]_0^{2\pi/w}} \qquad [\because \sin 4\pi = 0]$$

$$\therefore \qquad = \sqrt{\frac{a^2 w}{4\pi} \times \frac{2\pi}{w}} = \frac{a}{\sqrt{2}} \qquad \textbf{Ans.}$$

EXERCISE 32.5

1. Find the mean value of sin $(nx + \alpha)$ over the range $x = -\frac{\alpha}{n}$ to $x = -\frac{\alpha}{n} + \frac{\pi}{2n}$ where n and α are positive integers. **Ans.** $\frac{2}{\pi}$

2. A current of i amperes flowing through a resistance of R ohms produces heat at the rate of Ri^2 joules per second. Find the average rate at which heat is produced by an alternating current $i = A \sin wt$ during one cycle. **Ans.** $\frac{RA^2}{2}$

3. Find the mean value of

(*i*) x^2 from $x = 1$ to $x = 4$. **Ans.** 7

(*ii*) e^{-2x} from $x = 0$ to $x = a$. **Ans.** $\frac{1-e^{-2a}}{2a}$

(*iii*) $x^2 e^x$ from $x = 1$ to $x = 3$. **Ans.** 48.85

(*iv*) $x \log_e x$ from $x = 1$ to $x = a$. **Ans.** $\frac{a^2}{2(a-1)}\log_e a - \frac{a+1}{4}$

4. Find the mean values of the following

(*i*) $\cos(pt + \alpha)\cos(pt + \beta)$ **Ans.** $\frac{\cos(\alpha - \beta)}{2}$, period $= \frac{\pi}{p}$

(*ii*) $\sin 2pt \cos(3pt + \alpha)$ **Ans.** 0, period $= \frac{2\pi}{p}$

(*iii*) $(2 \sin 3pt + 5 \sin pt)^2$ **Ans.** 14.5, period $= \frac{\pi}{p}$

32.13 NUMERICAL INTEGRATION

Simpson's Rule

The rule is useful in finding the approximate area of a figure. The figure is divided into two parts by drawing a line as show in the figure. The area of the each part of the figure is calculated separated by Simpson's rule and the sum of the area of two parts is the area of the given figure.

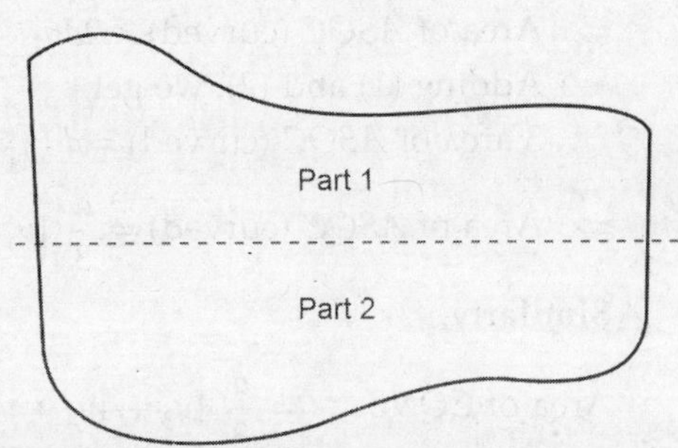

The area of part I is calculated as below:

The base line is divided into an even (here,

eight) number of equal parts. By drawing parallel ordinates $y_1, y_2, y_3 \dots y_9$ through the points of division the figure is divided into even number (here, eight) of section as shown in the figure.

Let the common distance between the consecutive ordinates be equal to d. Join SR and RQ. Through R draw a tangent TRW to the curve, meeting the produced ordinates AS and CQ at T and W.

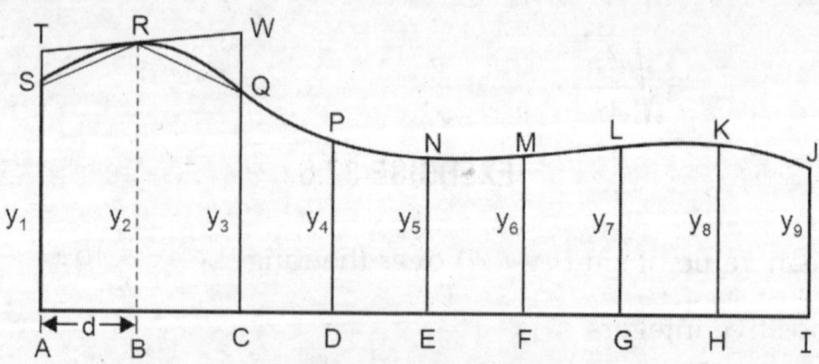

Area of the first section $ASRB$ (curved)

$$> \text{Area of trapezium } ASRB$$

$$> \frac{1}{2} AB\,(AS + BR)$$

$$> \frac{1}{2} d\,(y_1 + y_2)$$

$\therefore$ 2 area of the first section $ASRB$ (curved) $> d\,(y_1 + y_2)$

Similarly 2 area of the second $BRQC > d\,(y_2 + y_3)$

On adding, we have

2 (Area of the first section $ASRB$ + Area of the second section $BRQC$) $> d\,(y_1 + y_2) + d\,(y_2 + y_3)$

$\Rightarrow$ 2 area of $ASQC$ (curved) $> d\,(y_1 + 2y_2 + y_3)$ ...(1)

Again from the figure it is clear that

Area of $ASRB$ (curved) + Area of $BRQC$ (curved)

$$< \text{area of trapezium } ATWC$$

$$< \frac{1}{2} AC\,(AT + WC)$$

$$< \frac{1}{2} \cdot 2d\,(2BR) \qquad (\because AT + WC = 2BR)$$

$$< d\,(2y_2)$$

$\Rightarrow$ Area of $ASQC$ (curved) $< 2dy_2$ ... (2)

Adding (1) and (2), we get

3 area of $ASQC$ (curved) $= d\,(y_1 + 2y_2 + y_3) + 2dy_2$ (approximately)

$\Rightarrow$ Area of $ASQC$ (curved) $= \dfrac{d}{3}\,[y_1 + 4y_2 + y_3]$ ... (3)

Similarly,

Area of $CQNE$ $= \dfrac{d}{3}\,[y_3 + 4y_4 + y_5]$ ... (4)

Area of $ENLG$ $= \frac{d}{3}[y_5 + 4y_6 + y_7]$ ... (5)

Area of $GLJI$ $= \frac{d}{3}[y_7 + 4y_8 + y_9]$ ... (6)

On adding (3), (4), (5) and (6), we have

Total Area of $ASJI = \frac{d}{3}[(y_1 + y_9) + 2(y_3 + y_5 + y_7) + 4(y_2 + y_4 + y_6 + y_8)]$ approximately.

$$\textbf{Area} = \frac{d}{3}[(y_1 + y_l) + 2\Sigma y_o + 4\Sigma y_e]$$

where d = Common distance

y_1 = first ordinate

y_l = last ordinate

Σy_o = sum of remaining odd ordinates

Σy_e = sum of the even ordinates.

Proof By Integration

Assumption. Any curve can be built up from arcs of parabolas of vertical axis.

Proof. Let the given curve AG be

$$y = f(x) \quad \text{... (1)}$$

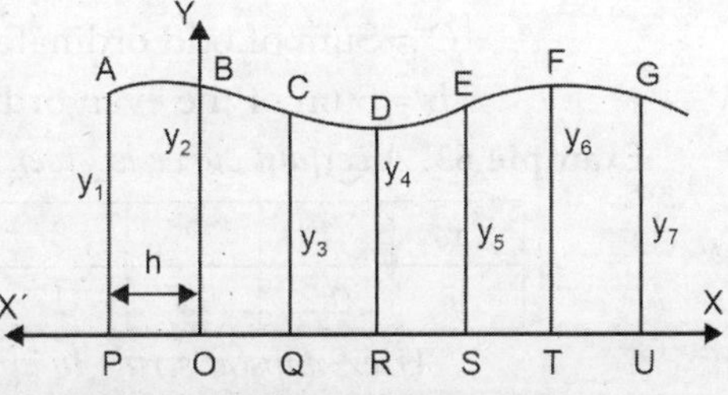

Draw ⊥s AP and BO to x-axis.

Divide PU into even number of equal parts say h and draw ordinates y_1, y_2, y_3 ......

Let O be the origin. The coordinates of P and Q are $(-h, 0)$, $(h, 0)$

$y = ax^2 + bx + c$ represents a parabola with its axis vertical.

$$\text{Area } (APQC) = \int_{-h}^{+h} y\, dx = \int_{-h}^{h} (ax^2 + bx + c)\, dx$$

$$= \left[\frac{ax^3}{3} + \frac{bx^2}{2} + cx\right]_{-h}^{h} = \frac{2h}{3}(ah^2 + 3c) \quad \text{... (2)}$$

$A(-h, y_1)$, $B(0, y_2)$, $C(h, y_3)$ lie on the parabola.

$\therefore$ $y_1 = ah^2 - bh + c$ ... (3)

$y_2 = c$ ... (4)

$y_3 = ah^2 + bh + c$ ... (5)

On adding (3) and (5), we get $y_1 + y_3 = 2ah^2 + 2c$

$\Rightarrow$ $y_1 + y_3 = 2ah^2 + 2y_2 \Rightarrow 2ah^2 = y_1 + y_3 - 2y_2$

$$ah^2 = \frac{1}{2}(y_1 + y_3 - 2y_2)$$

Put the value of ah^2 and c in (2), we get

$$\text{Area } (APQC) = \frac{2h}{3}\left[\frac{1}{2}(y_1 + y_3 - 2y_2) + 3y_2\right]$$

$$= \frac{h}{3}\ [y_1 + y_3 - 2y_2 + 6y_2]$$

$$= \frac{h}{3}\ [y_1 + 4y_2 + y_3] \quad \text{... (6)}$$

Similarly, the area of $CQSE = \frac{h}{3}\ [y_3 + 4y_4 + y_5]$ ... (7)

Similarly, the area of $ESUG = \frac{h}{3}\ [y_5 + 4y_6 + y_7]$ and so on ... (8)

Adding (6), (7) and (8), we get

$$\text{Whole area} = \frac{h}{3}\ [(y_1 + 4y_2 + y_3) + (y_3 + 4y_4 + y_5) + (y_5 + 4y_6 + y_7) + ...]$$

$$= \frac{h}{3}\ [(y_1 + y_7) + 2\ (y_3 + y_5 + ...) + 4\ (y_2 + y_4 + y_6 + ...)]$$

$$\boxed{\textbf{Area} = \frac{h}{3}[X + 2O + 4E]}$$ **Ans.**

where X = Sum of the first and the last ordinates.

O = Sum of odd ordinates.

E = Sum of the even ordinates.

Example 63. *A certain curve is given by the following points of rectangular ordinates.*

x	1	2	3	4	5	6	7	8	9
y	0.2	0.7	1	1.3	1.5	1.7	1.9	2.1	2.3

Use Simpson's rule to approximate the volume generated by revolving the area between the curve, x-axis and the ordinates x = 1 and x = 9.

Solution. The volume generated by revolving the area between the curve, the x-axis and the ordinates $x = 1$ and $x = 9$ is

$$\pi \int_1^9 y^2\, dx = \pi \frac{d}{3}\left[(y_1^2 + y_9^2) + 2(y_3^2 + y_5^2 + y_7^2) + 4(y_2^2 + y_4^2 + y_6^2 + y_8^2)\right]$$

$$= \pi \frac{1}{3}\left[\left\{(0.2)^2 + (2.3)^2\right\} + 2\left\{(1)^2 + (1.5)^2 + (1.9)^2\right\} + 4\left\{(0.7)^2 + (1.3)^2 + (1.7)^2 + (2.1)^2\right\}\right]$$

$$= \frac{\pi}{3}[(0.04 + 5.29) + 2(1 + 2.25 + 3.61) + 4(0.49 + 1.69 + 2.89 + 4.41)]$$

$$= \frac{\pi}{3}[5.33 + 13.72 + 37.92]$$

$$= \frac{\pi}{3} \times 56.97 = 18.99\pi = 59.6589 \text{ cubic units}$$ **Ans.**

Example 64. *A river is 80 meters wide. The depth d in meters at a distance x meters from one bank is given by the following table:*

x	0	10	20	30	40	50	60	70	80
d	0	4	7	9	12	15	14	8	3

Find approximately the area of cross-section of the river. **[SBTE. 2015, 2012]**

Solution.

x	0	10	20	30	40	50	60	70	80
d	0	4	7	9	12	15	14	8	3

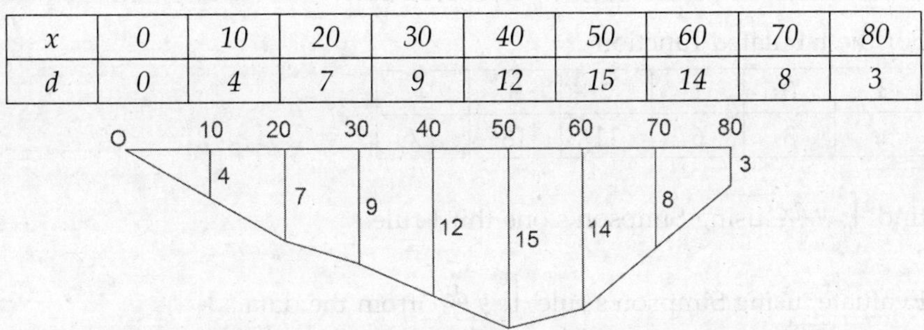

$$A=\frac{d}{3}[(y_1+y_9)+2(y_3+y_5+y_7)+4(y_2+y_4+y_6+y_8)]$$

$$=\frac{10}{3}[(0+3)+2(7+12+14)+4(4+9+15+8)]$$

$$=\frac{10}{3}[3+66+144]$$

$$=\frac{10}{3}\times 213=710 \text{ sq. metres.}$$ **Ans.**

Example 65. *The velocity v of particle at a distance from a point on its path is given by the table.*

S	0	10	20	30	40	50	60
v	47	58	64	65	61	52	38

Estimate the time taken to travel, 60 meters by using Simpson's one-third rule.

Solution. Here we have

S	0	10	20	30	40	50	60
v	47	58	64	65	61	52	38
$\frac{1}{v}$	$\frac{1}{47}$	$\frac{1}{58}$	$\frac{1}{64}$	$\frac{1}{65}$	$\frac{1}{61}$	$\frac{1}{52}$	$\frac{1}{38}$

$$\int y\,dx=\frac{d}{3}[(y_1+y_7)+2(y_3+y_5)+4(y_2+y_4+y_6)]$$

$$=\frac{10}{3}\left[\left(\frac{1}{47}+\frac{1}{38}\right)+2\left(\frac{1}{64}+\frac{1}{61}\right)+4\left(\frac{1}{58}+\frac{1}{65}+\frac{1}{52}\right)\right]$$

$$=\frac{10}{3}[0.0213+0.0263)+2(0.0156+0.0164)+4(0.0172+0.0154+0.0192)]$$

$$=\frac{10}{3}[0.0476+0.0640+0.2072]$$

$$=\frac{10}{3}\times 0.3188=1.06$$ **Ans.**

EXERCISE 32.6

1. For the tabulated function

x	0	1	2	3	4
y	3	6	11	18	27

find $\int_0^4 y\,dx$ using Simpson's one third rule. **Ans.** 49.33

2. Evaluate, using Simpson's rule $\int_0^6 y\,dx$, from the data.

x	0	1	2	3	4	5	6
y	0.146	0.161	0.176	0.190	0.204	0.217	0.230

Ans. 1.136

3. A curve is drawing to pass through the points given by the following table:

x	1	1.5	2	2.5	3	3.5	4
y	2	2.4	2.7	2.8	3	2.6	2.1

Estimate the area bounded by the curve, the x-axis and the lines.

$x = 1, x = 4.$ **Ans.** 7.783

4. The coordinates of curve $y = f(x)$ for given values of x are tabulated below:

x	1	2	3	4	5	6	7
y	2.105	2.808	3.614	4.604	5.851	7.451	9.467

Using Simpson's rule, evaluate the area bounded by the curve, the ordinates at $x = 1$ and $x = 7$ and the x-axis. **Ans.** 29.989

5. A plot of land lies between a straight fence and a stream. At distance x meters from one end of the fence the widths of the plot y are given below.

x	0	20	40	60	80	100	120
y	0	22	41	53	38	17	0

Use Simpson's parabolic rule to find approximately the area of the plot.

Ans. 3506.67 sq meters

6. The water under portion of a water tank is divided by horizontal planes one metre apart of the following area:

472, 398, 302, 198, 116, 60, 34, 12, 4 sq. m.

Find the volume in cubic m, between the two extreme areas. **Ans.** 1350.66 cu.m.

7. The cross-section of a tree is A sq. cm. at a distance x cm. from one end corresponding values A and x are

x	10	30	50	70	90	110	130	150	170
y	120	123	129	131	131	135	142	153	177

Find by Simpson's rule the volume of the tree in the cubic cm. between $x = 10$ and $x = 170$. **Ans.** 21873.3 c.c

8. The velocity of a train which starts from rest is given by the following table, the time being reckoned in minutes from the start and speed in kilometers per hour:

Time (in minutes)	2	4	6	8	10	12	14	16	18	20
Speed (km/h)	10	18	25	29	32	20	11	5	2	0

Estimate approximately by Simpson's rule, the total distance run in 20 minutes.

Ans. 5.156 km.

[**Hint:** (1) when $t = 0$, $x = 0$, (2) minutes are to be converted into hour]

9. Use trapezoidal rule to find the volume of solid of revolution formed by rotating about the x-axis the area bounded by the lines $x = 0$, $x = 1$, $y = 0$ and the curve passing through the points given below:

x	0	0.25	0.50	0.75	1
y	1	0.9896	0.9589	0.9089	0.8415

Ans. 5.464379

10. The following table gives the velocity v of a particle at time t:

(seconds):	0	2	4	6	8	10	12
(meters/sec):	4	6	16	34	60	94	136

Find (*i*) the distance moved by the particle in 12 seconds and (*ii*) the acceleration at $t = 2$ secs. **Ans.** (*i*) 252 m. (*ii*) 3 m/sec^2

11. A rocket is launched from the ground. Its acceleration is registered during the first 81 seconds and is given in the table below. Using Simpson's $\frac{1}{3}$rd rule, find the velocity of the rocket at $t = 80$ seconds.

t (s)	0	10	20	30	40	50	60	70	80
f (cm/s^2)	30	31.63	33.34	35.47	37.75	40.33	43.25	46.69	50.67

Ans. 3086.1 cm/sec^2

32.14 VALUE OF DEFINITE INTEGRAL BY SIMPSON'S RULE

Example 66. *Applying Simpson's rule, obtain an approximate value of* $\int_0^1 \frac{dx}{1+x^2}$ *taking four intervals and hence obtain an approximate value of p correct to four places of decimals.* **[SBTE. 2015, 2014, 2013, 2012]**

Solution. The points which divide the interval 0 to 1 in 4 equal parts are $0, \frac{1}{4}, \frac{2}{4}, \frac{3}{4}, 1$ and the length of each interval is $\frac{1}{4}$. The value of the ordinates are obtained by putting these values in the integral $\frac{1}{1+x^2}$.

x	0	$\frac{1}{4}$	$\frac{1}{2}$	$\frac{3}{4}$	1
$\frac{1}{1+x^2}$	1	$\frac{16}{17}=0.9411$	$\frac{4}{5}=0.8$	$\frac{16}{25}=0.64$	$\frac{1}{2}=0.5$
	y_1	y_2	y_3	y_4	y_5

Hence the value of the integral by Simpson's Rule,

$$\int_0^1 \frac{dx}{1+x^2} = \frac{h}{3}[X+2O+4E] = \frac{h}{3}[(y_1+y_5)+2(y_3)+4(y_2+y_4)]$$

$$= \frac{1}{3}\cdot\frac{1}{4}[(1+0.5)+2(0.8)+4(0.9411+0.64)]$$

$$= \frac{9.4244}{12} = 0.7854 \quad \text{... (1)}$$

Also, $$\int_0^1 \frac{dx}{1+x^2} = \left[\tan^{-1}x\right]_0^1 = \frac{\pi}{4} \quad \text{... (2)}$$

From (1) and (2), we have

$$\frac{\pi}{4} = 0.7854 \text{ or, } \pi = 3.1416$$ **Ans.**

Example 67. Using Simpson's rule to estimate the integral $\int_0^{\pi/2}\sqrt{\sin\theta}\,d\theta$, divide the interval $[0, \pi/2]$ into six equal parts. **[SBTE. 2017, 2015, 2007]**

Solution. Here we have

$$f(x) = \int_0^{\pi/2}\sqrt{\sin\theta}\,d\theta,$$

$$h = \frac{b-a}{n} = \frac{\pi/2-0}{6} = \frac{\pi}{12}$$

θ	$a=0$	$a+h$ $\pi/12=15°$	$a+2h$ $\pi/6=30°$	$a+3h$ $\pi/4$	$a+4h$ $\pi/3$	$a+5h$ $5\pi/12$	$b=90°$
$f(x)=\sqrt{\sin\theta}$	0	0.5087	0.7071	0.8409	0.9306	0.9828	1
	y_0	y_1	y_2	y_3	y_4	y_5	y_6

By Simpson's rule.

$$\int_0^{\pi/2}\sqrt{\sin\theta}\,d\theta = \frac{h}{3}\left[(y_0+y_6)+4(y_1+y_3+y_5)+2(y_2+y_4)\right]$$

$$= \frac{\pi}{36}[(0+1)+4(0.5087+0.8409+0.9828)+2(0.7071+0.9306)]$$

$$= \frac{\pi}{36}\times 13.605$$

$$\therefore \int_0^{\pi/2}\sqrt{\sin\theta}\,d\theta = 1.1877$$ **Ans.**

Example 68. *Evaluate the value of* $\log_e(2)$ *by finding* $\int_0^1 \frac{2x\,dx}{1+x^2}$, *using Simpson's rule, by dividing the interval into four equal parts.*

Solution. $$\int_0^1 \frac{2x}{1+x^2}\,dx = \left[\log_e(1+x^2)\right]_0^1 = \log_e(2)$$

$$f(x) = \frac{2x}{1+x^2}.$$

The following is the table of value of x and $f(x)$.

x	0	0.25	0.50	0.75	1.00
$f(x) = \frac{2x}{1+x^2}$	0	0.47	0.80	0.96	1.00

Here $h = 0.25$. Applying Simpson's rule,

$$\int_0^1 \frac{2x}{1+x^2}\,dx = \frac{h}{3}[(y_1+y_5)+2y_3+4(y_2+y_4)]$$

or $$\left[\log(1+x^2)\right]_0^1 = \frac{0.25}{3}[(0+1)+2\times 0.8+4(0.47+0.96)]$$

$\therefore$ $$\log_e(2) = 0.69$$ **Ans.**

Example 69. *Calculate by Simpson's rule an approximate value of* $\int_{-3}^{+3} x^4\,dx$ *by taking seven equidistant intervals.*

Solution. The seven equidistant parts are – 3, – 2, – 1, 0, 1, 2, 3, and the length of each interval is 1.

The values of the ordinates are obtained by putting these values in the integrand x^4.

x	–3	–2	–1	0	1	2	3
x^4	81	16	1	0	1	16	81

$$\int y\,dx = \frac{d}{3}[(y_1+y_7)+2(y_3+y_5)+4(y_2+y_4+y_6)]$$

$$\int_{-3}^{3} x^4\,dx = \frac{1}{3}[(81+81)+2(1+1)+4(16+0+16)]$$

$$= \frac{1}{3}[162+4+128]$$

$$= \frac{1}{3}\times 294 = 98$$

$$\text{Exact value} = \int_{-3}^{3} x^4\,dx = \left[\frac{x^5}{5}\right]_{-3}^{3} = \frac{1}{5}[243+243]$$

$$= \frac{1}{5}\times 486 = 97.2$$ **Ans.**

Example 70. *Find an approximate value of log 2 regarded as* $\int_0^1 \frac{dx}{1+x}$

[SBTE. 2017, 2016, 2015, 2013]

Solution. Let us have ten division

x	0	0.1	0.2	0.3	0.4	0.5	0.6	0.7	0.8	0.9	1.0
$\frac{1}{1+x}$	1	0.9090	0.8333	0.7692	0.7143	0.6667	0.6250	0.5882	0.5556	0.5263	0.5

By Simpson's rule

$$\int_0^1 \frac{1}{1+x}\,dx = \frac{h}{3}[(y_1 + y_l) + 2(y_3 + y_5 + \ldots) + 4(y_2 + y_4 + \ldots)]$$

$$= \frac{0.1}{3}[(1 + 0.5) + 2(0.8333 + 0.7143 + 0.6250 + 0.5556)$$

$$+ 4(0.9090 + 0.7692 + 0.6667 + 0.5882 + 0.5263)]$$

$$= \frac{0.1}{3}[1.5 + 5.4564 + 13.8376]$$

$$= \left(\frac{0.1}{3}\right) \times 20.794$$

Also $\int_0^1 \frac{1}{1+x}\,dx = [\log(1+x)]_0^1 = \log 2$

$\therefore$ $\log 2 = 0.6931$ **Ans.**

Example 71. Evaluate $\int_0^1 e^{x^2}\,dx$ *using Simpson's third rule by taking h = 0.1*

x	0	0.1	0.2	0.3	0.4	0.5	0.6	0.7	0.8	0.9	1
e^{x^2}	1	1.0101	1.0408	1.0942	1.1735	1.2840	1.4333	1.6323	1.8965	2.2479	2.7183

$$\int_0^1 e^{x^2}\,dx = \frac{h}{3}[(y_1 + y_l) + 2\Sigma y_0 + 4\Sigma y_e]$$

$$= \frac{0.1}{3}[(1 + 2.7183) + 2(1.0408 + 1.1735 + 1.4333 + 1.8965)$$

$$+ 4(1.0101 + 1.0942 + 1.2840 + 1.6323 + 2.2479)]$$

$$= \frac{0.1}{3}[3.7183 + 2(5.5441) + 4(7.2685)] = \frac{0.1}{3}[3.7183 + 11.0882 + 29.0740]$$

$$= \frac{0.1}{3} \times 43.8805 = 1.4627$$ **Ans.**

EXERCISE 32.7

1. Calculate $\int_0^{10} x^2\,dx$ by Simpson's one-third rule with eleven ordinates and compare the result obtained with the exact value. **Ans.** $333\frac{1}{3}$ and 333

2. Use Simpson's rule to evaluate $\int_1^2 \frac{dx}{x}$ by dividing the interval (1, 2) into four equal parts. Hence find the value of $\log_e 2$ **[SBTE. 2017, 2010] Ans.** $\log_e 2 = 0.693$

3. Evaluate $\int_{\frac{1}{2}}^1 \frac{dx}{x}$ by Simpson's one-third rule with five ordinates and from the result, find log 2 correct to three decimal places. **[SBTE. 2017, 2015] Ans.** 0.6931

4. Evaluate $\int_0^1 \frac{dx}{1+x}$ and find log 2. **Ans.** 0.6931

5. Evaluate $\int_{-1}^0 \frac{dx}{1-x}$, by Simpson's one-third rule taking 10 sub-intervals, correct to three decimal places. **Ans.** 0.6931

6. Evaluate $\int_0^2 e^{2x}\,dx$ by Simpson's $\frac{1}{3}$ rule by taking $h = 0.5$ given that $e = 2.718$, $e^2 = 7.389$, $e^3 = 20.086$, $e^4 = 54.598$ compare your result with exact value of the integral. **Ans.** 26.922, 26.799

7. Given: $e^0 = 1$, $e^1 = 2.72$, $e^2 = 7.39$, $e^3 = 20.09$ and $e^4 = 54.60$.

   Use Simpson's rule to find an approximate value of $\int_0^4 e^x\,dx$.

   Also compare your result with the exact value of the integral.

   [SBTE. 2016, 2015] Ans. 53.8733, 53.60

8. Using Simpson's one-third rule, find $\int_0^{\pi} \frac{\sin x}{x}\,dx$ by dividing $(0, \pi)$ in six equal parts. **Ans.** 1.8520

9. Compute the value of the following integral using Simpson's 1/3 rule with for sub-intervals, $\int_{1.2}^{1.6}\left(x+\frac{1}{x}\right)dx$ **Ans.** 0.8477

10. Simpson's rule for numerical integration requires the minimum number of ______ and as such the curve happens to be ______ **Ans.** 2, parabola

Choose the correct alternative:

11. The Simpson's one-third rule in numerical integration needs at least

    (*a*) 4 geometrical points (*b*) a parabolic curve

    (*c*) 3 geometrical points (*d*) a circular curve **Ans.** (*b*)

32.15 TRAPEZOIDAL RULE

$$\text{Area} = \frac{h}{2}[(y_1 + y_n) + 2(y_2 + y_3 + y_4 + \ldots\ldots y_{n-1})]$$

Proof: The Simple Trapezoidal Rule is based on approximating $f(x)$ by the straight line joining (x_1, y_1) and (x_2, y_2) ......

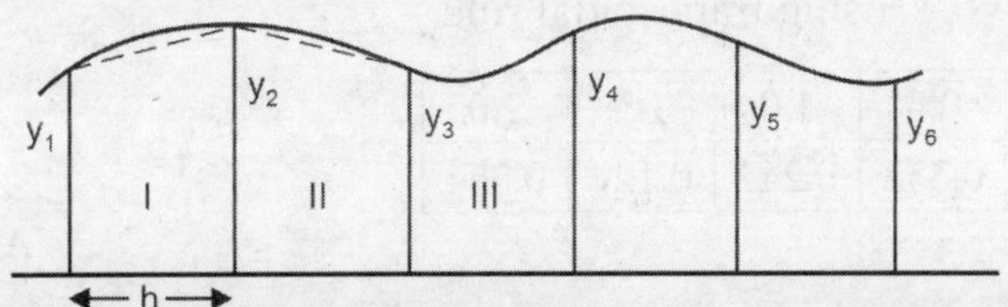

Area of trapezium = $\frac{1}{2}$ (sum of the parallel sides) ×

(perpendicular distance between them)

Area of I trapezium = $\frac{1}{2}(y_1 + y_2) \cdot h$ ... (1)

Area of II trapezium = $\frac{1}{2}(y_2 + y_3) \cdot h$ ... (2)

Area of III trapezium = $\frac{1}{2}(y_3 + y_4) \cdot h$ ... (3)

....

Area of $(n-1)$th trapezium = $\frac{1}{2}\{y_{n-1} + y_n\} \cdot h$

Adding (1), (2), (3) ...... and so on, we get

$$\textbf{Area} = \frac{h}{2}[(y_1 + y_n) + 2(y_2 + y_3 + y_4 + \ldots + y_{n-1})]$$ **Proved.**

Example 72. *By dividing the range (0, π) into ten equal parts, evaluate* $\int_0^{\pi} \sin x\, dx$ *by trapezoidal rule.*

Solution. Range = π – 0 = π, hence $h = \frac{\pi}{10}$

x	0	$\frac{\pi}{10}$	$\frac{2\pi}{10}$	$\frac{3\pi}{10}$	$\frac{4\pi}{10}$	$\frac{5\pi}{10}$	$\frac{6\pi}{10}$	$\frac{7\pi}{10}$	$\frac{8\pi}{10}$	$\frac{9\pi}{10}$	π
$y = \sin x$	0.0	0.3090	0.5878	0.8090	0.9511	1.0	0.9511	0.8090	0.5878	0.3090	0.0

By trapezoidal rule: $\int_0^{\pi} \sin x\, dx = \frac{h}{2}[(y_1 + y_n) + 2(y_2 + y_3 + y_4 + \ldots\ldots\ldots + y_{n-1})]$

$$\int_0^{\pi} \sin x\, dx = \frac{\pi}{20}[(0+0) + 2(0.3090 + 0.5878 + 0.8090$$

$$+ 0.9511 + 1 + 0.9511 + 0.8090 + 0.5878 + 0.3090)]$$

$$= 1.9843$$ **Ans.**

EXERCISE 32.8

1. Evaluate $\int_0^2 f(x)\, dx$ using trapezoidal rule.

x	0.0	0.5	1.0	1.5	2.0
$f(x)$	0.399	0.352	0.242	0.129	0.054

Ans. 0.4747 sq. units

2. A curve is drawn to pass through the points given by the following table.

x	1	1.5	2	2.5	3	3.5	4
y	2	2.4	2.7	2.8	3	2.6	2.1

Estimate the area bounded by the curve, the x-axis and the lines $x = 1$, $x = 4$.

Ans. 7.775 sq. units

3. A river is 50 metres wide. The depth of water d metres at a distance of x metres from one bank is given by the following table:

x	0	5	10	15	20	25	30	35	40	45	50
d	0	2	4	5	6	8	10	9	3	3	2

Find by Trapezoidal rule the area of the cross section of the river.

Ans. 225 sq. metres

Differential Equations of First Order and Their Applications

33.1 INTRODUCTION

An equation containing an independent variable, a dependent variable and the derivatives of the dependent variable is called a *differential equation*.

Examples each of the following are differential equations:

(*i*) $\frac{dy}{dx}+9y=e^{2x}$ (*ii*) $\frac{d^2y}{dx^2}+7\frac{dy}{dx}+5y=\cos x$ (*iii*) $\frac{dy}{dx}=\frac{x^3-y^3}{xy^2-x^2y}$

(*iv*) $x^2dx+y^2dy=0$ (*v*) $\frac{d^2y}{dx^2}=e^x$ (*vi*) $\frac{d^2y}{dx^2}+\frac{dy}{dx}+y=0$

33.2 ORDER OF A DIFFERENTIAL EQUATION

The order of the highest order derivative occurring in a differential equation is called the order of the differential equation.

For example $\frac{dy}{dx}+y\sin x=\cos x$ is a differential equation of first order.

Example 1. *Determine the order of each of the following differential equations.*

(*i*) $y'+5y=0$ (*ii*) $y'+y=e^x$ (*iii*) $y''=1+y+y^2$

(*iv*) $y'''+2y''+y'=0$ (*v*) $y''+(y')^2+2y=0$ (*vi*) $y^{iv}+y=\sin x$

Solution. (*i*) We have, $y'+5y=0 \Rightarrow \frac{dy}{dx}+5y=0$

In this equation, the highest order derivative is $\frac{dy}{dx}$ and its order is 1. So, it is a differential equation of order 1.

(*ii*) We have, $y'+y=e^x \Rightarrow \frac{dy}{dx}+y=e^x$

In this equation, the highest order derivative is $\frac{dy}{dx}$ and its order is 1. So it is a differential equation of order 1.

(*iii*) We have, $y''=1+y+y^2 \Rightarrow \frac{d^2y}{dx^2}=1+y+y^2$

In this equation, the highest order derivative is $\frac{d^2y}{dx^2}$ and its order is 2. So it is a differential equation of order 2.

(*iv*) We have, $y''' + 2y'' + y' = 0 \Rightarrow \frac{d^3y}{dx^3} + 2\frac{d^2y}{dx^2} + \frac{dy}{dx} = 0$

In this equation, the highest order derivative is $\frac{d^3y}{dx^3}$ and its order is 3. So it is a differential equation of order 3.

(*v*) We have, $y'' + (y')^2 + 2y = 0 \Rightarrow \frac{d^2y}{dx^2} + \left(\frac{dy}{dx}\right)^2 + 2y = 0$

In this equation, the highest order derivative is $\frac{d^2y}{dx^2}$ and its order is 2. So it is a differential equation of order 2.

(*vi*) We have, $y^{iv} + y = \sin x \Rightarrow \frac{d^4y}{dx^4} + y = \sin x$

In this equation, the highest order derivative is $\frac{d^4y}{dx^4}$ and its order is 4. So it is a differential equation of order 4.

33.3 DEGREE OF A DIFFERENTIAL EQUATION

The power of the highest order derivative occurring in the differential equation, after it is made free from radicals and fractions, is called the degree of the differential equation.

Example 2. *Find the degree of the following differential equations*

(*i*) $\left(\frac{dy}{dx}\right)^3 + 6y = \tan x$ (*ii*) $\frac{d^2y}{dx^2} + 3\left(\frac{dy}{dx}\right)^3 + 5y = 0$

Solution. (*i*) We have, $\left(\frac{dy}{dx}\right)^3 + 6y = \tan x$

In this equation, the highest order derivative is $\frac{dy}{dx}$. And the power of highest order derivative $\frac{dy}{dx}$ is 3.

Hence, the degree of the above equation is 3. **Ans.**

(*ii*) We have, $\frac{d^2y}{dx^2} + 3\left(\frac{dy}{dx}\right)^3 + 5y = 0$

In this equation, the highest order derivative is $\frac{d^2y}{dx^2}$. And the power of the highest order derivative is 1.

Hence, the degree of the above equation is 1. **Ans.**

Example 3. *Determine the order and degree of the following differential equations.*

(*i*) $y' + 6y^2 + y = 0$ (*ii*) $(y')^2 + y^3 + y = 0$

(*iv*) $y'' + 2y' + \sin y = 0$ (*iv*) $(y''')^2 + (y'')^3 + (y')^4 + y^5 = 0$

Solution. (*i*) We have, $y' + 6y^2 + y = 0 \Rightarrow \frac{dy}{dx} + 6y^2 + y = 0$

In this equation the highest order derivative is $\frac{dy}{dx}$ its power is 1. So, it is a differential equation of order 1 and degree 1. **Ans.**

(*ii*) We have, $(y')^2 + y^3 + y = 0 \Rightarrow \left(\frac{dy}{dx}\right)^2 + y^3 + y = 0$

In this equation, the highest order derivative is $\frac{dy}{dx}$ and its order is 1 and power is 2. So it is a differential equation of order 1 and degree 2. **Ans.**

(*iii*) We have, $y'' + 2y' + \sin y = 0 \Rightarrow \frac{d^2y}{dx^2} + 2\frac{dy}{dx} + \sin y = 0$

In this equation, the highest order derivative is $\frac{d^2y}{dx^2}$, its order is 2 and power is 1. So it is a differential equation of order 2 and degree 1.

(*iv*) We have, $(y''')^2 + (y'')^3 + (y')^4 + y^5 = 0$

$$\Rightarrow \left(\frac{d^3y}{dx^3}\right)^2 + \left(\frac{d^2y}{dx^2}\right)^3 + \left(\frac{dy}{dx}\right)^4 + y^5 = 0$$

In this equation, the highest order derivative is $\frac{d^3y}{dx^3}$ and its order is 3 and power is 2. So it is a differential equation of order 3 and degree 2. **Ans.**

Example 4. *Determine the order and degree of the differential equation:*

$$y = px + \sqrt{a^2p^2 + b^2} \text{ where } p = \frac{dy}{dx}$$

Solution. We have,

$$y = px + \sqrt{a^2p^2 + b^2}$$

$$\Rightarrow \quad y - px = \sqrt{a^2p^2 + b^2}$$

Squaring both sides, we get

$$\Rightarrow \quad (y - px)^2 = a^2p^2 + b^2$$

$$\Rightarrow \quad y^2 - 2xyp + p^2x^2 = a^2p^2 + b^2$$

$$\Rightarrow \quad (x^2 - a^2)p^2 - 2xyp + y^2 - b^2 = 0$$

$$\Rightarrow \quad (x^2 - a^2)\left(\frac{dy}{dx}\right)^2 - 2xy\left(\frac{dy}{dx}\right) + y^2 - b^2 = 0 \qquad \left[\because p = \frac{dy}{dx}\right]$$

Here, the highest order derivative is $\frac{dy}{dx}$ and its order is 1 and power is 2.

Hence, it is a differential equation of order 1 and degree 2. **Ans.**

EXERCISE 33.1

Find the order of each of the following differential equations:

1. $\left(\frac{dy}{dx}\right)^2 - \sin^2 y = 0$ **Ans.** 1

2. $\frac{d^2y}{dx^2} + \left(\frac{dy}{dx}\right)^2 + 2y = 0$ **Ans.** 2

3. $\left(\frac{d^2y}{dx^2}\right)^3 + \cos\frac{dy}{dx} = 0$ **Ans.** 2

4. $\frac{d^3y}{dx^3} + 2\frac{d^2y}{dx^2} - \frac{dy}{dx} + y^4 = 0$ **Ans.** 3

5. $\frac{dy}{dx}+3y^2+5=0$ **Ans.** 1

6. $\sqrt{\frac{dy}{dx}}+2y=\sin x$ **Ans.** 1

7. $\left(\frac{dy}{dx}\right)^4+y^2=y$ **Ans.** 1

8. $\frac{dy}{dx}=\sqrt{1+y+y^4}$ **Ans.** 1

9. $\frac{d^5y}{dx^5}+y^6=0$ **Ans.** 5

10. $\left(\frac{dy}{dx}\right)^4+2y\left(\frac{d^2y}{dx^2}\right)=0$ **Ans.** 2

Find the degree of each of the following differential equations:

11. $\frac{dy}{dx}+5y^2=0$ **Ans.** 1

12. $\left(\frac{dy}{dx}\right)^2+y^3-1=0$ **Ans.** 2

13. $\frac{dy}{dx}+\sin\frac{dy}{dx}=0$ **Ans.** Not defined

14. $\frac{d^2y}{dx^2}+y^2=0$ **Ans.** 1

15. $\frac{d^4y}{dx^4}+\frac{d^3y}{dx^3}+\frac{d^2y}{dx^2}+\frac{dy}{dx}+y^4=0$ **Ans.** 1

16. $\frac{d^5y}{dx^5}+y^2+e^{\frac{dy}{dx}}=0$ **Ans.** Not defined

Determine the order and degree of each of the following equations:

17. $x^4\left(\frac{dy}{dx}\right)-4xy-3x^3=0$ **Ans.** Order 1, Degree 1

18. $\frac{1}{x^2}\left(\frac{d^2y}{dx^2}\right)+9y=-4e^{-x}$ **Ans.** Order 2, Degree 1

19. $\left(\frac{d^2y}{dx^2}\right)^3+9\left(\frac{dy}{dx}\right)^4+10=\sin x$ **Ans.** Order 2, Degree 3

20. $x\cdot\left(\frac{dy}{dx}\right)+\frac{2}{\left(\frac{dy}{dx}\right)}=y^2$ **Ans.** Order 1, Degree 2

21. $(2x-2y+5)\,dx=(x-y+3)\,dy$ **Ans.** Order 1, Degree 1

22. $\frac{d^2y}{dx^2}=\left\{1+\left(\frac{dy}{dx}\right)^2\right\}^{\frac{3}{2}}$ **[SBTE, 2015]** **Ans.** Order 2, Degree 2

Determine the order and degree of each of the following differential equations. State also, whether they are linear or non linear.

		Order	Degree	Linear/Non-linear
23.	$\frac{d^2y}{dx^2}+4y=0$	**Ans.** 2	1	Linear
24.	$\left(\frac{dy}{dx}\right)^2+\frac{1}{\frac{dy}{dx}}=2$	**Ans.** 1	3	Nonlinear

25. $\sqrt[3]{\dfrac{d^2y}{dx^2}} = \sqrt{\dfrac{dy}{dx}}$ **Ans.** 2 2 Nonlinear

26. $y\dfrac{d^2x}{dy^2} = y^2 + 1$ **Ans.** 2 1 Linear

27. $\dfrac{d^3y}{dx^3} + \left(\dfrac{d^2y}{dx^2}\right)^3 + \dfrac{dy}{dx} + 4y = \sin x$ **[SBTE 2015]** **Ans.** 3 1 Nonlinear

28. $9\dfrac{d^2y}{dx^2} = \left[1 + \left(\dfrac{dy}{dx}\right)^2\right]^{\frac{3}{2}}$ **[SBTE, 2015]** **Ans.** 2 2 Nonlinear

29. $y = x\dfrac{dy}{dx} + a\sqrt{1 + \left(\dfrac{dy}{dx}\right)^2}$ **Ans.** 1 2 Nonlinear

30. $\left(\dfrac{d^2y}{dx^2}\right)^2 + \left(\dfrac{dy}{dx}\right)^2 = x \sin\left(\dfrac{d^2y}{dx^2}\right)$ **Ans.** 2 Undefined Nonlinear

33.4 FORMATION OF DIFFERENTIAL EQUATIONS

Consider $F(x, y, C_1, C_2, C_3, \ldots, C_n) = 0$ ... (1)

be the solution of a differential equation, where $C_1, C_2, \ldots C_n$ are n arbitrary constants. If we eliminate these n constants, we obtain the differential equation of nth order satisfied by (1). Equation (1) taken together with n relations obtained by differentiating equation (1), n times helps us to eliminate the n constants.

Working rule for the formation of differential equation

Step 1: If the given equation contains x, y and arbitrary constant 'a', then it is represented as $f_1(x, y, a) = 0$ ... (1)

Step 2: Differentiating equation (1), w.r.t. 'x', we get an equation involving y', y, x and a, *i.e.*

$$f_2(x, y, y', a) = 0 \qquad \ldots (2)$$

Step 3: On eleminating 'a' from (1) and (2), we get the required differential equation, *i.e.* $F(x, y, y') = 0$

Example 5. *Find the differential equation of the family of curves given by* $x^2 + y^2 = 2ax$.

Solution. The given equation is:

$$x^2 + y^2 = 2ax \qquad \ldots (1)$$

Here, a is the only arbitrary constant.

Differentiating (1) with respect to 'x', we get

$$2x + 2y\frac{dy}{dx} = 2a$$

Substituting the value of $2a$ in (1), we get

$$x^2 + y^2 = x\left(2x + 2y\frac{dy}{dx}\right)$$

$$\Rightarrow \qquad 2xy\frac{dy}{dx} + x^2 - y^2 = 0$$

This is the required differential equation. **Ans.**

Example 6. *Form the differential equation of the family of curves represented by the equation:*

$$(2x + a)^2 + y^2 = a^2$$

Solution. The given equation is : $(2x + a)^2 + y^2 = a^2$ ... (1)

Which contains only one constant, a on differentiating (1) w.r.t. x, we get

$$2(2x + a) \cdot 2 + 2y\frac{dy}{dx} = 0$$

$$\Rightarrow \qquad 2x + a + \frac{1}{2}y\frac{dy}{dx} = 0$$

$$\Rightarrow \qquad 2x + a = -\frac{1}{2}y\frac{dy}{dx}$$

$$\Rightarrow \qquad a = -2x - \frac{1}{2}y\frac{dy}{dx} \qquad \text{... (2)}$$

Eliminating a from (1) and (2), we get

$$\left(-\frac{1}{2}y\frac{dy}{dx}\right)^2 + y^2 = \left(-2x - \frac{1}{2}y\frac{dy}{dx}\right)^2$$

$$\Rightarrow \qquad y^2 - 4x^2 = 2xy\frac{dy}{dx}$$

This is the required differential equation. **Ans.**

Example 7. *Form the differential equation representing the family of curves given by* $(x - a)^2 + 2y^2 = a^2$, *where a is an arbitrary constant.*

Solution. The given equation is $(x - a)^2 + 2y^2 = a^2$ ... (1)

which contains only one constant, a.

Differentiating (1) w.r.t. x, we get

$$2(x - a) + 2.2y\frac{dy}{dx} = 0 \qquad \Rightarrow \quad (x - a) + 2y\frac{dy}{dx} = 0$$

$$\Rightarrow \qquad (x - a) = -2y\frac{dy}{dx} \quad \Rightarrow \quad -a = -2y\frac{dy}{dx} - x \qquad \text{... (2)}$$

Eliminating 'a' from (1) and (2), we get

$$\left(-2y\frac{dy}{dx}\right)^2 + 2y^2 = \left(-2y\frac{dy}{dx} - x\right)^2 \quad \Rightarrow \quad 4xy\frac{dy}{dx} = 2y^2 - x^2$$

This is the required differential equation. **Ans.**

Example 8. *Show that the differential equation of which* $y = 2(x^2 - 1) + ce^{-x^2}$ *is a solution is* $\frac{dy}{dx} + 2xy = 4x^3$

Solution. We have, $y = 2(x^2 - 1) + ce^{-x^2}$ ... (1)

Differentiating (1), with respect to x, we get

$$\frac{dy}{dx} = 2\cdot(2x) + ce^{-x^2}(-2x)$$

$$\Rightarrow \quad \frac{dy}{dx} - 4x = -2x\,ce^{-x^2}$$

$$= -2x\,[y - 2(x^2 - 1)] \qquad \left[\because y = 2(x^2-1) + ce^{-x^2} \Rightarrow ce^{-x^2} = y - 2(x^2-1)\right]$$

$$\Rightarrow \quad \frac{dy}{dx} - 4x = -2xy + 4x(x^2-1) \Rightarrow \frac{dy}{dx} - 4x = -2xy + 4x^3 - 4x$$

$$\Rightarrow \quad \frac{dy}{dx} + 2xy = 4x^3$$

This is the required differential equation. **Proved.**

Example 9. *Prove that $x^2 - y^2 = c(x^2 + y^2)^2$ is a general solution of the differential equation*

$$(x^3 - 3xy^2)\,dx = (y^3 - 3x^2y)\,dy$$

Solution. The solution of the differential equation is

$$x^2 - y^2 = c(x^2 + y^2)^2 \qquad \text{... (1)}$$

Differentiating (1), with respect to 'x', we get

$$2x - 2y\frac{dy}{dx} = c\cdot 2(x^2 + y^2)\left(2x + 2y\frac{dy}{dx}\right) \qquad \text{... (2)}$$

Substituting the value of c from (1) in (2), we get

$$\Rightarrow \quad (x - y)\frac{dy}{dx} = \frac{x^2 - y^2}{(x^2+y^2)^2}(x^2 + y^2)\left(2x + 2y\frac{dy}{dx}\right) \qquad \left[\because x^2 - y^2 = c(x^2+y^2)^2 \Rightarrow c = \frac{x^2 - y^2}{(x^2+y^2)^2}\right]$$

$$\Rightarrow \quad (x^2 + y^2)\left(x - y\frac{dy}{dx}\right) = (x^2 - y^2)\left(2x + 2y\frac{dy}{dx}\right)$$

$$\Rightarrow \quad [2y(x^2 - y^2) + y(x^2 + y^2)]\frac{dy}{dx} = x(x^2 + y^2) - 2x(x^2 - y^2)$$

$$\Rightarrow \quad (3x^2y - y^3)\frac{dy}{dx} = 3xy^2 - x^3$$

$$\Rightarrow \quad (x^3 - 3xy^2)dx = (y^3 - 3x^2y)\,dy$$

This is required differential equation. **Proved.**

Example 10. *Find the differential equation of which $y = Ae^x + Be^{3x} + Ce^{5x}$ is a solution.*

[Diploma IETE, June 2005]

Solution. We have, $\quad y = Ae^x + Be^{3x} + Ce^{5x}$ ... (1)

On differentiating (1) three times, w.r.t. x, we get

$$\Rightarrow \quad \frac{dy}{dx} = Ae^x + 3Be^{3x} + 5Ce^{5x} \qquad \text{... (2)}$$

$$\Rightarrow \quad \frac{d^2y}{dx^2} = Ae^x + 9Be^{3x} + 25Ce^{5x} \qquad \text{... (3)}$$

$$\Rightarrow \qquad \frac{d^3y}{dx^3} = Ae^x + 27Be^{3x} + 125Ce^{5x} \qquad \text{... (4)}$$

On subtracting equation (1) from equation (2), we get

(2) – (1), $\dfrac{dy}{dx} - y = 2Be^{2x} + 4Ce^{5x}$ ... (5)

Similarly,

(3) – (1) $\dfrac{d^2y}{dx^2} - y = 8Be^{2x} + 24Ce^{5x}$ ... (6)

(4) – (1) $\dfrac{d^3y}{dx^3} - y = 26Be^{2x} + 124Ce^{5x}$ ... (7)

(6) – 4 (5) $\dfrac{d^2y}{dx^2} - y - 4\dfrac{dy}{dx} + 4y = 8Ce^{5x}$ ... (8)

(7) – 13 (5) $\dfrac{d^3y}{dx^3} - y - 13\dfrac{dy}{dx} + 13y = 72Ce^{5x}$ ... (9)

(9) – 9 (8) $\dfrac{d^3y}{dx^3} - y - 13\dfrac{dy}{dx} + 13y - 9\dfrac{d^2y}{dx^2} + 9y + 36\dfrac{dy}{dx} - 36y = 0$

$$\Rightarrow \qquad \frac{d^3y}{dx^3} - 9\frac{d^2y}{dx^2} + 23\frac{dy}{dx} - 15y = 0 \qquad \textbf{Ans.}$$

Example 11. *Form the differential equation of the family of circles touching the y-axis at origin.*

Solution. Let C denotes the family of circles touching x-axis at origin. Let $(a, 0)$ be the coordinates of the centre of any member of the family.

Therefore equation of family C is

$$(x - a)^2 + y^2 = a^2 \;\Rightarrow\; x^2 + y^2 = 2ax \qquad \text{... (1)}$$

Where a is an arbitrary constant. Differentiating both sides of equation (1), with respect to x, we get

$$2x + 2y\frac{dy}{dx} = 2a \qquad \text{... (2)}$$

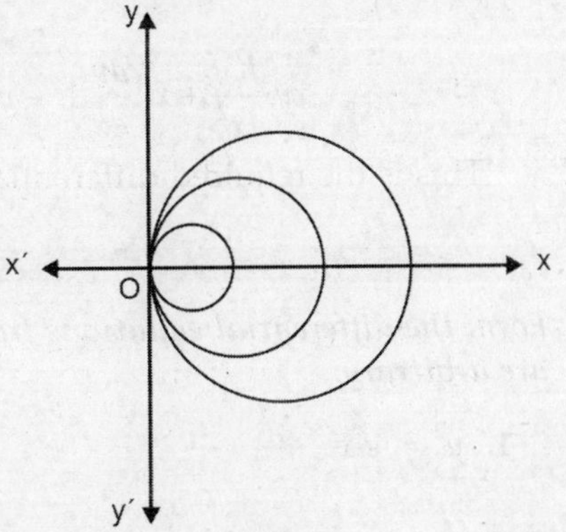

To eliminate a we substitute the value of $2a$ from equation (2), in equation (1), we get

$$x^2 + y^2 = \left(2x + 2y\frac{dy}{dx}\right)x$$

$$\Rightarrow \qquad \frac{dy}{dx} = \frac{x^2 + y^2 - 2x^2}{2yx}\;;\; \frac{dy}{dx} = \frac{y^2 - x^2}{2yx}$$

Y
X′
X
O
Y′

This is the required differential equation of the given family of circles. **Ans.**

Example 12. *Form the differential equation of the family of parabolas having vertex at origin and axis along positive y-axis.*

Solution. Let P denote the family of above said parabolas and let $(0, a)$ be the focus of a member of the given family where a is an arbitrary constant.

Therefore equation of family P is

$$x^2 = 4ay \qquad \text{... (1)}$$

Differentiating both sides of equation (1) with respect to y, we get

$$\Rightarrow \qquad 4a = 2x \cdot \frac{dx}{dy} \qquad \text{... (2)}$$

To eliminate a from equations (1) and (2). We substitute the value of $4a$ from (2) in equation (1), we get

$$\Rightarrow \qquad x^2 = \left(2x \cdot \frac{dx}{dy}\right) \cdot y$$

$$\Rightarrow \qquad x - 2xy\frac{dx}{dy} = 0$$

which is the differential equation of the given family of parabolas. **Ans.**

Example 13. *Form the differential equation of the family of ellipses having foci on y-axis and centre at origin.*

Solution. We know that the equation of family of ellipses is

$$\Rightarrow \qquad \frac{x^2}{a^2} + \frac{y^2}{b^2} = 1 \qquad \text{... (1)}$$

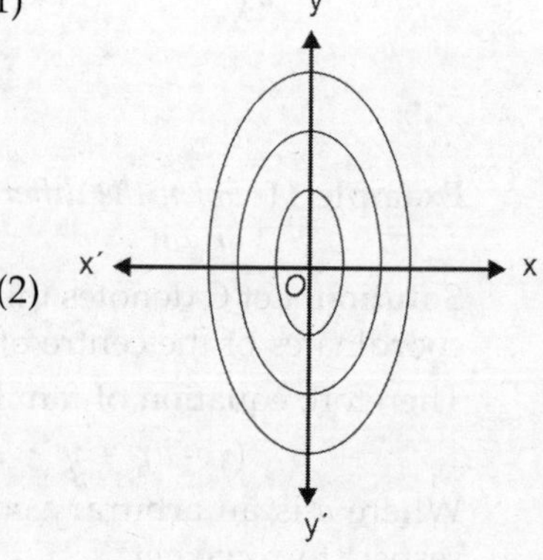

Differentiating (1), w.r.t., 'x', we get

$$\Rightarrow \qquad \frac{2x}{a^2} + \frac{2y}{b^2}\frac{dy}{dx} = 0$$

$$\Rightarrow \qquad \frac{y}{x}\left(\frac{dy}{dx}\right) = -\frac{b^2}{a^2} \qquad \text{... (2)}$$

Again differentiating (2), w.r.t., 'x', we get

$$\Rightarrow \qquad \left(\frac{y}{x}\right)\left(\frac{d^2y}{dx^2}\right) + \left(\frac{x\frac{dy}{dx} - y}{x^2}\right)\frac{dy}{dx} = 0$$

$$\Rightarrow \qquad xy\frac{d^2y}{dx^2} + x\left(\frac{dy}{dx}\right)^2 - y\frac{dy}{dx} = 0$$

This is the required differential equation. **Ans.**

EXERCISE 33.2

Form the differential equations from the following family of curves; where constants are arbitrary:

1. $y^2 = 4ax$ **Ans.** $2x\frac{dy}{dx} = y$

2. $y = cx + 2c^2 + c^3$ **Ans.** $y = x\frac{dy}{dx} + 2\left(\frac{dy}{dx}\right)^2 + \left(\frac{dy}{dx}\right)^3$

3. $xy = a^2$ **Ans.** $y + x\frac{dy}{dx} = 0$

4. $x^2 + (y - b)^2 = 1$ **Ans.** $x^2\left\{\left(\frac{dy}{dx}\right)^2 + 1\right\} = \left(\frac{dy}{dx}\right)^2$

5. $x^2 + y^2 = a^2$ **Ans.** $x + y\frac{dy}{dx} = 0$

6. $x^2 - y^2 = a^2$ **Ans.** $x - y\frac{dy}{dx} = 0$

7. $(x - a)^2 - y^2 = 1$ **Ans.** $y^2\left(\frac{dy}{dx}\right)^2 - y^2 = 0$

8. $y^2 = 4a(x - b)$ **Ans.** $y\left(\frac{d^2y}{dx^2}\right) + \left(\frac{dy}{dx}\right)^2 = 0$

9. $y = Ae^x + Be^{-x}$ **Ans.** $\frac{d^2y}{dx^2} - y = 0$

10. $y = a \sin x + b \cos x$ **Ans.** $\frac{d^2y}{dx^2} + y = 0$

11. $y = ax^2 + bx + c$ **Ans.** $\frac{d^3y}{dx^3} = 0$

12. $y = a \sin (x + b)$ **Ans.** $\frac{d^2y}{dx^2} + y = 0$

13. Find the differential equation of all the circles; which pass through the origin and whose centres lie on x-axis. **Ans.** $(x^2 - y^2) + 2xy\frac{dy}{dx} = 0$

14. Find the differential equation of all the parabolas with latus rectum '$4a$' and whose axis are parallel to x-axis. **Ans.** $2a \cdot \frac{d^2y}{dx^2} + \left(\frac{dy}{dx}\right)^3 = 0$

15. Find the differential equation of all non-vertical lines in a plane. **Ans.** $\frac{d^2y}{dx^2} = 0$

16. Find the differential equation of the family of all circles of radius r.

Ans. $\left\{1 + \left(\frac{dy}{dx}\right)^2\right\}^3 = r^2\left(\frac{d^2y}{dx^2}\right)^2$

17. Find the differential equation of the family of all circles in the first quadrant, which touches the coordinate axes. **Ans.** $(x - y)^2\left[\left(\frac{dy}{dx}\right)^2 + 1\right] = \left(x + y\frac{dy}{dx}\right)^2$

18. Form the differential equation corresponding to $y^2 = a(b - x)(b + x)$ by eliminating a and b. **Ans.** $x\left[y\frac{d^2y}{dx^2} + \left(\frac{dy}{dx}\right)^2\right] = y\frac{dy}{dx}$

19. Form the differential equation representing the family of curves $y = \tan^{-1} x + ce^{\tan^{-1} x}$ where c is an arbitrary constant. **Ans.** $(1+x^2)\frac{dy}{dx} - y + \tan^{-1} x - 1 = 0$

20. Form the differential equation of the family of curves $y = ae^x + be^{2x} + ce^{-3x}$, where a, b, c are arbitrary constants. **Ans.** $\frac{d^3y}{dx^3} - 7\frac{dy}{dx} + 6y = 0$

21. Form the differential equation from the equation $y = ax^3 + bx^2$, where a, and b are to be taken as parameters. **Ans.** $x^2\frac{d^2y}{dx^2} - 4x\frac{dy}{dx} + 6y = 0$

22. Form the differential equation from the equation $y = e^x (A \cos x + B \sin x)$, where A and B are to be taken as parameters. **Ans.** $\frac{d^2y}{dx^2} - 2\frac{dy}{dx} + 2y = 0$

23. Show that $y = e^x + e^{-2x}$ is a solution of $\frac{d^2y}{dx^2} + \frac{dy}{dx} - 2y = 0$ **[SBTE. 2007]**

24. Show that the differential equation that represents all parabolas having their axes or symmetry coincident with the axis of x is $y\frac{d^2y}{dx^2} + \left(\frac{dy}{dx}\right)^2 = 0$

Choose the correct alternative:

25. The differential equation of the curve $y = a \cos (x - b)$, where a and b are constants, is

(*a*) $\frac{d^2y}{dx^2} - y = 0$ (*b*) $\frac{d^2y}{dx^2} - \frac{dy}{dx} - y = 0$

(c) $\frac{d^2y}{dx^2} + \frac{dy}{dx} - y = 0$ (*d*) $\frac{d^2y}{dx^2} + y = 0$ **Ans.** (*d*)

33.5 SOLUTION OF DIFFERENTIAL EQUATION OF FIRST ORDER AND FIRST DEGREE

We will discuss the solution of differential equation of first order and first degree by separation of variables. The standard methods of solving the differential equations of the following type:

(*i*) Differential equations solvable by separation of the variables

(*ii*) Homogeneous differential equations

(*iii*) Linear differential equations of the first order

33.6 VARIABLE SEPARABLE

If a differential equation can be written in the form

$$f(y)\,dy = \phi(x)\,dx$$

We say that variables are separable, y on left hand side and x on right hand side.

We get the solution by integrating both sides.

For example, *Solve* $5y' = e^x y^4$.

We have $5\frac{dy}{dx} = e^x \cdot y^4$

On separating the variables, we get

$$\Rightarrow \qquad 5\frac{dy}{y^4} = e^x\,dx \;\Rightarrow\; \int e^x\,dx = 5\int y^{-4}\,dy$$

Integrating both sides.

$$\Rightarrow \qquad e^x + \frac{5}{3y^3} = c$$

is the required solution. **Ans.**

Example 14. *Solve:* $\dfrac{dy}{dx} = \sqrt{4-y^2}$

Solution. We have, $\dfrac{dy}{dx} = \sqrt{4-y^2}$

On separating the variables, we get

$$\frac{dy}{\sqrt{4-y^2}} = dx$$

Integrating both sides, we get

$$\int\frac{dy}{\sqrt{2^2-y^2}} = \int dx$$

Integrating both sides, we get.

$$\Rightarrow \qquad \sin^{-1}\left(\frac{y}{2}\right) = x + c$$

which is required solution. **Ans.**

Example 15. *Solve the following differential equations:*

(*i*) $(e^x + e^{-x})\,dy - (e^x - e^{-x})\,dx = 0$ (*ii*) $\dfrac{dy}{dx} = e^{x+y}$

Solution. (*i*) We have, $(e^x + e^{-x})\,dy - (e^x - e^{-x})\,dx = 0$

On separating the variables, we get

$$\Rightarrow \qquad dy = \left(\frac{e^x - e^{-x}}{e^x + e^{-x}}\right)dx$$

Integrating both sides, we get

$$\Rightarrow \qquad y = \log\,|e^x + e^{-x}| + c$$ **Ans.**

(*ii*) We have, $\dfrac{dy}{dx} = e^{x+y}$

$$\Rightarrow \qquad \frac{dy}{dx} = e^x \cdot e^y$$

On separating the variables, we get

$$e^{-y}\,dy = e^x\,dx$$

Integrating both sides, we get

$$\Rightarrow \qquad \int e^{-y}\,dy = \int e^x\,dx$$

$\Rightarrow \qquad -e^{-y} = e^{x} + c$

$\Rightarrow \qquad e^{x} + e^{-y} + c = 0$ **Ans.**

Example 16. *Solve:* $(x+1)\dfrac{dy}{dx} = x(y^2+1)$

Solution. $(x+1)\dfrac{dy}{dx} = x(y^2+1)$

$$\Rightarrow \quad \frac{dy}{y^2+1} = \frac{x\,dx}{x+1} \Rightarrow \frac{1}{1+y^2}dy = \left(1-\frac{1}{x+1}\right)dx$$

Integrating $\displaystyle\int \frac{1}{1+y^2}dy = \int dx - \int \frac{1}{x+1}dx$

$\Rightarrow \qquad \tan^{-1} y = x - \log(1+x) + C$ **Ans.**

Example 17. *Solve:* $(xy^2 + x)\,dx + (yx^2 + y)\,dy = 0$ **[SBTE 2016, 2015]**

Solution. $(xy^2 + x)\,dx + (yx^2 + y)\,dy = 0$

$$\Rightarrow \quad (y^2+1)\,x\,dx = -(x^2+1)\,y\,dy \Rightarrow \frac{y\,dy}{y^2+1} = -\frac{x}{x^2+1}dx$$

$$\Rightarrow \quad \frac{1}{2}\int \frac{2y}{y^2+1}dy = -\frac{1}{2}\int \frac{2x}{x^2+1}dx$$

$$\Rightarrow \quad \frac{1}{2}\log(y^2+1) = -\frac{1}{2}\log(x^2+1) + \frac{1}{2}\log C$$

$$\Rightarrow \quad \log(y^2+1) + \log(x^2+1) = \log C \Rightarrow \log(y^2+1)(x^2+1) = \log C$$

$$(y^2+1)(x^2+1) = C$$ **Ans.**

Example 18. *Solve:* $y^2(x^2-1)\dfrac{dy}{dx} - x^2(y^2-1) = 0$.

Solution. $y^2(x^2-1)\dfrac{dy}{dx} - x^2(y^2-1) \Rightarrow \dfrac{y^2dy}{y^2-1} = \dfrac{x^2}{x^2-1}dx$

$$\Rightarrow \quad \left(1+\frac{1}{y^2-1}\right)dy = \left(1+\frac{1}{x^2-1}\right)dx$$

$$\Rightarrow \quad \int\left(1+\frac{1}{y^2-1}\right)dy = \int\left(1+\frac{1}{x^2-1}\right)dx$$

$$\Rightarrow \quad y + \frac{1}{2}\log\left(\frac{y-1}{y+1}\right) = x + \frac{1}{2}\log\left(\frac{x-1}{x+1}\right) + C$$ **Ans.**

Example 19. *Solve the following differential equations.*

(*i*) $2y\,dx + x\,dy = 0$ **[SBTE 2017]**

(*ii*) $\dfrac{dy}{dx} = 1 + x + y + xy$ **[SBTE 2017, 2015]**

Solution. (*i*) $2\,y\,dx + x\,dy = 0$

$$2\frac{dx}{x} + \frac{dy}{y} = 0$$

Integrating both sides, we get

$\Rightarrow \qquad 2 \log x + \log y = \log C$

$\Rightarrow \qquad \log x^2y = \log C$

$\Rightarrow \qquad x^2y = C$ **Ans.**

(*ii*) Here we have $\dfrac{dy}{dx} = 1 + x + y + xy$

$$\frac{dy}{dx} = (1 + x)(1 + y)$$

$$\int \frac{dy}{1+y} = \int (1+x)\,dx$$

$$\log (1 + y) = x + \frac{x^2}{2} + C$$ **Ans.**

Example 20. *Solve:* $x (1 + y^2)\, dx - y (1 + x^2)\, dy = 0$ **[SBTE 2017]**

Solution. Here we have

$x(1 + y^2)dx - y(1 + x^2)dy = 0$

$\Rightarrow \qquad \dfrac{x}{1+x^2}dx = \dfrac{y}{1+y^2}dy$

Integrating both sides, we get

$$\int \frac{x}{1+x^2}dx = \int \frac{y}{1+y^2}dy$$

$\Rightarrow \qquad \dfrac{1}{2}\log (1+x^2) = \dfrac{1}{2}\log (1+y^2) + \log C$

$\Rightarrow \qquad \log (1 + x^2)^{1/2} = \log (1 + y^2)^{1/2}\cdot C$

$\Rightarrow \qquad (1 + x^2)^{1/2} = (1 + y^2)^{1/2}\cdot C$

$\Rightarrow \qquad 1 + x^2 = (1 + y^2)\, C^2$ **Ans.**

Example 21. *Solve* $\dfrac{dy}{dx} = ye^x$ **[SBTE 2015]**

Solution. $\qquad \dfrac{1}{y}dy = e^x dx$

Integrating both sides, we get $\displaystyle\int \frac{1}{y}dy = \int e^x dx$

$$\log y = e^x + C$$ **Ans.**

Example 22. *Solve:* $\sqrt{1+x^2+y^2+x^2y^2} + xy\dfrac{dy}{dx} = 0$ **[SBTE 2019, 2010, 2007]**

Solution. Here we have

$\sqrt{1+x^2+y^2+x^2y^2} + xy\dfrac{dy}{dx} = 0$

$\Rightarrow \qquad \sqrt{1+x^2+y^2+x^2y^2} = -\,xy\dfrac{dy}{dx}$

$$\Rightarrow \qquad \sqrt{(1+x^2)(1+y^2)} = -xy\frac{dy}{dx}$$

$$\Rightarrow \qquad \frac{\sqrt{1+x^2}}{x}dx = -\frac{y}{\sqrt{1+y^2}}dy$$

Integrating both sides, we get

$$\Rightarrow \qquad \underset{I_1}{\int \frac{\sqrt{1+x^2}}{x}dx} = \underset{I_2}{\int \frac{-y}{\sqrt{1+y^2}}dy} \qquad \text{... (1)}$$

put $\qquad 1 + x^2 = z^2 \Rightarrow x^2 = z^2 - 1$

$$\Rightarrow \qquad 2x\,dx = 2\,zdz,\; dx = \frac{z}{x}dz$$

$$\Rightarrow \qquad I_1 = \int \frac{z^2}{x^2}dz = \int \frac{z^2}{z^2-1}dz = \int \frac{(z^2-1)+1}{z^2-1}dz$$

$$= \int \left(1 + \frac{1}{z^2-1}\right)dz = z + \frac{1}{2}\log\left|\frac{z-1}{z+1}\right|$$

$$= \sqrt{1+x^2} + \frac{1}{2}\log\left|\frac{\sqrt{1+x^2}-1}{\sqrt{1+x^2}+1}\right|$$

$$\Rightarrow \qquad I_2 = -\int \frac{y}{\sqrt{1+y^2}}dy = -\sqrt{1+y^2}$$

Putting the value, I_1 and I_2 in (1), we get

$$\therefore \quad \sqrt{1+x^2} + \frac{1}{2}\log\left|\frac{\sqrt{1+x^2}-1}{\sqrt{1+x^2}+1}\right| = -\sqrt{1+y^2} + C \qquad \textbf{Ans.}$$

Example 23. *Solve:* $(x+y)^2\frac{dy}{dx} = a^2$ **[SBTE, 2009]**

Solution. Here we have $(x+y)^2\frac{dy}{dx} = a^2$ ... (1)

put $\qquad x + y = v,\; \frac{dy}{dx} = \frac{dv}{dx} - 1$

putting the value of $x + y$ and $\frac{dy}{dx}$ in (1), we get

$$\Rightarrow \qquad v^2\left(\frac{dv}{dx} - 1\right) = a^2$$

$$\Rightarrow \qquad v^2\frac{dv}{dx} = a^2 + v^2$$

$$\Rightarrow \qquad \frac{v^2}{a^2+v^2}\,dv = dx$$

$$\Rightarrow \qquad \frac{a^2+v^2-a^2}{a^2+v^2}\,dv = dx$$

$$\Rightarrow \qquad \left\{1-\frac{a^2}{a^2+v^2}\right\} dv = dx$$

Integrating both sides, we get

$$\Rightarrow \qquad v - a^2\cdot\frac{1}{a}\tan^{-1}\left(\frac{v}{a}\right) = x + C$$

$$\Rightarrow \quad (x+y) - a\tan^{-1}\left(\frac{x+y}{a}\right) = x + C$$ **Ans.**

Example 24. *Solve:* $\sin x\cos y\dfrac{dy}{dx} + \cos x\sin y = 0$.

Solution. $\sin x\cos y\dfrac{dy}{dx} + \cos x\sin y = 0$

$$\Rightarrow \qquad \sin x\cos y\,dy = -\cos x\sin y\,dx$$

$$\Rightarrow \qquad \frac{\cos y}{\sin y}\,dy = -\frac{\cos x}{\sin x}\,dx$$

Integrating both sides, we get

$\Rightarrow\ \log(\sin y) = -\log(\sin x) + \log C \Rightarrow \log(\sin y\sin x) = \log C \Rightarrow \sin y\sin x = C$ **Ans.**

Example 25. *Solve the following differential equation*

$$\frac{dy}{dx} = \frac{e^x(\sin^2 x + \sin 2x)}{y(2\log y + 1)}$$ **[SBTE 2010, 2009]**

Solution. Here, we have

$$\frac{dy}{dx} = \frac{e^x(\sin^2 x + \sin 2x)}{y(2\log y + 1)}$$

Separating the variables, we have

$$y\,(2\log y + 1)\,dy = e^x\,(\sin^2 x + \sin 2x)\,dx$$

$$\Rightarrow \qquad \int y(2\log y + 1)\,dy = \int e^x\sin^2 x\,dx + \int e^x\sin 2x\,dx$$

$$\Rightarrow \quad (2\log y + 1)\frac{y^2}{2} - \int\left(\frac{2}{y}\right)\left(\frac{y^2}{2}\right)dy = \sin^2 x\cdot e^x - \int 2\sin x\cos x e^x\,dx + \int e^x\sin 2x\,dx$$

$$\Rightarrow \qquad (2\log y + 1)\frac{y^2}{2} - \int y\,dy = e^x\cdot\sin^2 x - \int e^x\cdot\sin 2x\,dx + \int e^x\sin 2x\,dx$$

$$\Rightarrow \qquad \frac{y^2}{2}(2\log y + 1) - \frac{y^2}{2} = e^x\sin^2 x + C$$

$$y^2 \log y + \frac{y^2}{2} - \frac{y^2}{2} = e^x \sin^2 x + C$$

$$\therefore \quad y^2 \log y = e^x \sin^2 x + C \qquad \textbf{Ans.}$$

EXERCISE 33.3

Solve the following differential equations:

1. $(1+x^2)\dfrac{dy}{dx} = (1+y^2)$ **[SBTE 2017, 2015]** **Ans.** $y - x = a\,(1 + xy)$

2. $x\dfrac{dy}{dx} + \cot y = 0$, given that $y = \dfrac{\pi}{4}$ when $x = \sqrt{2}$ **Ans.** $\cos y = \dfrac{x}{2}$

3. $x \cos^2 y\, dx = y \cos^2 x\, dy$ **[SBTE 2016]**

Ans. $y \tan y + \log \cos y = x \tan x + \log \cos x + C$

4. $\dfrac{dy}{dx} = \dfrac{\sqrt{1-y^2}}{\sqrt{1-x^2}}$ **[SBTE 2013]** **Ans.** $\sin^{-1} y = \sin^{-1} x + \sin^{-1} C$

5. $\sec^2 x \tan y\, dx + \sec^2 y \tan x\, dy = 0$ **[SBTE 2017, 2015]** **Ans.** $\tan x + \tan y = C$

6. $(x^2 - yx^2)\, dy + (y^2 + xy^2)\, dx = 0$ **[SBTE 2016, 2015]** **Ans.** $\log\left(\dfrac{x}{y}\right) - \left(\dfrac{1}{x} + \dfrac{1}{y}\right) = C$

7. $y - x\dfrac{dy}{dx} = a\left(y^2 + \dfrac{dy}{dx}\right)$ **Ans.** $(ay - 1)(x + a) = Cy$

8. $3e^x \tan y\, dx + (1+e^x)\sec^2 y\, dy = 0$, given $y = \dfrac{\pi}{4}$, $x = \log 2$ **Ans.** $(1 + e^x)^3 \tan y = 27$

9. $(e^y + 1) \cos x\, dx + e^y \sin x\, dy = 0$ **[B.T.E, Delhi June 2005]** **Ans.** $(e^y + 1) \sin x = C$

10. $y \sec^2 x + (y+7) \tan x \dfrac{dy}{dx} = 0$ **Ans.** $y^7 \tan x = Ce^{-y}$

11. $\dfrac{dy}{dx} = e^{x-y} + x^2 e^{-y}$ **[SBTE 2016]** **Ans.** $e^y = e^x + \dfrac{x^3}{3} + C$

12. $\dfrac{dy}{dx} = e^{3x-2y} + x^2 e^{-2y}$ **[SBTE 2015]** **Ans.** $\dfrac{e^{2y}}{2} = \dfrac{e^{3x}}{3} + \dfrac{x^3}{3} + C$

13. $\dfrac{dy}{dx} = e^{3x-y} + x^2 e^{-y}$ **[Diploma IETE Dec. 2006]** **Ans.** $e^y = \dfrac{e^{3x}}{3} + \dfrac{x^3}{3} + C$

14. $(xy + x)\, dy - (xy + y)\, dx = 0$ **Ans.** $x = Cye^{y-x}$

15. $(1+y)\, xy \dfrac{dy}{dx} = (1-x^2)(1-y)$ **Ans.** $\log x + 2\log(1-y) = \dfrac{x^2}{2} - \dfrac{y^2}{2} - 2y + C$

16. $\dfrac{dy}{dx} = \left(\sin x + \dfrac{\log x}{x}\right) \Big/ (\cos y - \sec^2 y)$ **Ans.** $\sin y - \tan y = -\cos x + \dfrac{1}{2}(\log x)^2 + C$

17. $(2xy + 3y)\, dx + (x^2 + 3x)\, dy = 0$ **Ans.** $x^2 y + 3xy = C$

18. $\dfrac{dy}{dx} - x \tan(y - x) = 1$ **Ans.** $\log \sin(y - x) = \dfrac{x^2}{2} + C$

19. $y(1+x^2)^{1/2}dy + x\sqrt{1+y^2}\,dx = 0$ **Ans.** $\sqrt{1+y^2}+\sqrt{1+x^2} = C$

20. $\dfrac{dy}{dx} = \dfrac{x(2\log x+1)}{\sin y + y\cos y}$ **Ans.** $y\sin y = \dfrac{x^2}{2}(2\log x+1) - \dfrac{x^2}{2} + C$

21. $\dfrac{dy}{dx} = \cos(x+y) + \sin(x+y)$ **Ans.** $\log\left(\tan\dfrac{x+y}{2}+1\right) = x + C$

Choose the correct alternative:

22. The solution of the differential equation $\sqrt{y}\,dx + \sqrt{x}\,dy = 0$

[Diploma IETE, Dec. 2006]

(*a*) $\sqrt{x}+\sqrt{y}$ = constant (*b*) $\sqrt{xy}$ = constant

(*c*) $x\sqrt{y}+y\sqrt{x}$ = constant (*d*) $\dfrac{\sqrt{x}}{\sqrt{y}}$ = constant **Ans.** (*a*)

23. Solution of differential equation $\dfrac{dy}{dx} = e^{x-y}$ is

(*a*) $e^x + e^y$ = constant (*b*) $e^x - e^y$ = constant

(*c*) $e^x \cdot e^y$ = constant (*d*) e^x / e^y = constant **Ans.** (*b*)

33.7 EQUATION REDUCIBLE TO VARIABLES SEPARABLE FORM

The equation of the form:

$$\frac{dy}{dx} = f(ax+by+c) \qquad \text{... (1)}$$

can be reduced to variables separable form by putting

$$ax + by + c = t \;\Rightarrow\; a + b\frac{dy}{dx} = \frac{dt}{dx}$$

$$\Rightarrow \qquad \frac{dy}{dx} = \frac{1}{b}\left(\frac{dt}{dx} - a\right)$$

Thus (1), becomes

$$\frac{1}{b}\left(\frac{dt}{dx} - a\right) = f(t) \;\Rightarrow\; \frac{dt}{dx} = a + b\,f(t)$$

$$\Rightarrow \qquad \frac{dt}{a + b\,f(t)} = dx$$

The variables have been separated.

Integrating both sides, we get the required solution.

$$\int \frac{dt}{a + b\,f(t)} = x + C, \quad \text{where} \quad t = ax + by + c.$$

Example 26. *Find the solution of the differential equation* $\dfrac{dy}{dx} - x\tan(y-x) = 1$.

Solution. $\dfrac{dy}{dx} - x\tan(y-x) = 1$ ... (1)

On putting $y - x = t$ so that $\frac{dy}{dx} - 1 = \frac{dt}{dx}$ in (1), we get

$$\Rightarrow \quad \left(1 + \frac{dt}{dx}\right) - x \tan t = 1,$$

$$\Rightarrow \quad \frac{dt}{dx} = x \tan t, \quad \cot t \, dt = x \, dx$$

Integrating, we have $\log(\sin t) = \frac{x^2}{2} + C$

$$\Rightarrow \quad \log \sin(y - x) = \frac{x^2}{2} + C$$ **Ans.**

Example 27. *Solve:* $\cos(x + y)\, dy = dx$ **[SBTE 2007]**

Solution. We have $\cos(x + y)\, dy = dx \Rightarrow \frac{dy}{dx} = \sec(x + y)$ ... (1)

On putting $x + y = t$ so that $1 + \frac{dy}{dx} = \frac{dt}{dx} \Rightarrow \frac{dy}{dx} = \frac{dt}{dx} - 1$ in (1), we get

$$\Rightarrow \quad \frac{dt}{dx} - 1 = \sec t \Rightarrow \frac{dt}{dx} = 1 + \sec t$$

$$\Rightarrow \quad \frac{dt}{1 + \sec t} = dx \Rightarrow \frac{\cos t}{\cos t + 1} dt = dx$$

$$\Rightarrow \quad \int \frac{\cos t}{\cos t + 1} dt = \int dx \Rightarrow \int \left[1 - \frac{1}{\cos t + 1}\right] dt = x + C$$

$$\Rightarrow \quad \int \left[1 - \frac{1}{2\cos^2 \frac{t}{2} - 1 + 1}\right] dt = x + C$$

$$\Rightarrow \quad \int \left(1 - \frac{1}{2}\sec^2 \frac{t}{2}\right) dt = x + C \Rightarrow t - \tan\frac{t}{2} = x + C$$

$$\Rightarrow \quad x + y - \tan\left(\frac{x + y}{2}\right) = x + C$$

$$\Rightarrow \quad y - \tan\left(\frac{x + y}{2}\right) = C$$ **Ans.**

Example 28. *Solve:* $\frac{dy}{dx} = (4x + y + 1)^2$.

Solution. We have,

$$\Rightarrow \quad \frac{dy}{dx} = (4x + y + 1)^2 \quad \text{... (1)}$$

Putting $4x + y + 1 = t,$

so that $4 + \frac{dy}{dx} = \frac{dt}{dx} \Rightarrow \frac{dy}{dx} = \frac{dt}{dx} - 4$

Now, substituting $4x + y + 1 = t$ and $\frac{dy}{dx} = \frac{dt}{dx} - 4$ in (1), we get

$$\Rightarrow \quad \frac{dt}{dx} - 4 = t^2 \Rightarrow \frac{dt}{dx} = t^2 + 4$$

$$\Rightarrow \quad dt = (t^2 + 4)\, dx \Rightarrow \frac{dt}{t^2 + 4} = dx$$

Integrating both sides, we get

$$\Rightarrow \quad \int \frac{dt}{t^2 + 4} = \int dx$$

$$\Rightarrow \quad \frac{1}{2}\tan^{-1}\left(\frac{t}{2}\right) = x + C$$

$$\Rightarrow \quad \frac{1}{2}\tan^{-1}\left(\frac{4x + y + 1}{2}\right) = x + C$$

Ans.

Example 29. *Solve:* $\sin^{-1}\left(\frac{dy}{dx}\right) = x + y,$

Solution. We have, $\sin^{-1}\left(\frac{dy}{dx}\right) = x + y \Rightarrow \frac{dy}{dx} = \sin(x + y)$ ... (1)

Putting $x + y = t$, so that $1 + \frac{dy}{dx} = \frac{dt}{dx} \Rightarrow \frac{dy}{dx} = \frac{dt}{dx} - 1$

Now, substituting $x + y = t$ and $\frac{dy}{dx} = \frac{dt}{dx} - 1$ in (1), we get

$$\Rightarrow \quad \frac{dt}{dx} - 1 = \sin t \Rightarrow \frac{dt}{dx} = \sin t + 1$$

$$\Rightarrow \quad dx = \frac{dt}{1 + \sin t}$$

Integrating both sides, we get

$$\Rightarrow \quad \int dx = \int \frac{dt}{1 + \sin t}$$

$$\Rightarrow \quad \int dx = \int \frac{1 - \sin t}{1 - \sin^2 t}\, dt = \int \frac{1 - \sin t}{\cos^2 t}$$

$$\Rightarrow \quad \int dx = \int (\sec^2 t - \tan t \sec t)\, dt$$

$$\Rightarrow \quad x = \tan t - \sec t + C$$

$$\Rightarrow \quad x = \tan(x + y) - \sec(x + y) + C$$

Ans.

Example 30. *Solve:* $\frac{dy}{dx} = \sin(x + y) + \cos(x + y).$

Solution. We have, $\frac{dy}{dx} = \sin(x + y) + \cos(x + y)$ ... (1)

Putting $\qquad x + y = t \Rightarrow 1 + \frac{dy}{dx} = \frac{dt}{dx} \Rightarrow \frac{dy}{dx} = \frac{dt}{dx} - 1$

Now, substituting $x + y = t$ and $\frac{dy}{dx} = \frac{dt}{dx} - 1$ in (1), we get

$$\Rightarrow \qquad \frac{dt}{dx} - 1 = \sin t + \cos t$$

$$\Rightarrow \qquad \frac{dt}{dx} = \sin t + \cos t + 1$$

$$\Rightarrow \qquad \frac{dt}{1 + \sin t + \cos t} = dx \qquad \text{... (2)}$$

Again putting $\sin t = \frac{2\tan\frac{t}{2}}{1 + \tan^2\frac{t}{2}}$ and $\cos t = \frac{1 - \tan^2\frac{t}{2}}{1 + \tan^2\frac{t}{2}}$ in (2), we get

$$\Rightarrow \qquad \frac{dt}{1 + \frac{2\tan\frac{t}{2}}{1 + \tan^2\frac{t}{2}} + \frac{1 - \tan^2\frac{t}{2}}{1 + \tan^2\frac{t}{2}}} = dx$$

$$\Rightarrow \qquad \frac{dt}{1 + \frac{2\tan\frac{t}{2}}{\sec^2\frac{t}{2}} + \frac{1 - \tan^2\frac{t}{2}}{\sec^2\frac{t}{2}}} = dx$$

$$\Rightarrow \qquad \frac{\sec^2\frac{t}{2}}{\sec^2\frac{t}{2} + 2\tan\frac{t}{2} + 1 - \tan^2\frac{t}{2}}\, dt = dx$$

$$\Rightarrow \qquad \frac{\sec^2\frac{t}{2}}{2\left\{1 + \tan\frac{t}{2}\right\}}\, dt = dx$$

Integrating both sides, we get

$$\Rightarrow \qquad \frac{1}{2}\int \frac{\sec^2\frac{t}{2}}{1 + \tan\frac{t}{2}}\, dt = \int dx \Rightarrow \log\left|1 + \tan\frac{t}{2}\right| = x + C$$

$$\Rightarrow \qquad \log\left|1 + \tan\left(\frac{x + y}{2}\right)\right| = x + C$$ **Ans.**

Example 31. *Solve:* $\dfrac{dy}{dx}=\dfrac{x+2y-1}{x+2y+1}$.

Solution. We have, $\dfrac{dy}{dx}=\dfrac{x+2y-1}{x+2y+1}$ ... (1)

Putting $x + 2y = t$ so that $1+2\dfrac{dy}{dx}=\dfrac{dt}{dx}$

$$\Rightarrow \quad \frac{dy}{dx}=\frac{1}{2}\left(\frac{dt}{dx}-1\right)$$

Putting these values in (1), we get

$$\Rightarrow \quad \frac{1}{2}\left(\frac{dt}{dx}-1\right)=\frac{t-1}{t+1}$$

$$\Rightarrow \quad \left(\frac{dt}{dx}-1\right)=\frac{2(t-1)}{(t+1)}$$

$$\Rightarrow \quad \frac{dt}{dx}=\frac{2t-2}{t+1}+1=\frac{3t-1}{t+1}$$

$$\Rightarrow \quad \left(\frac{t+1}{3t-1}\right)dt=dx$$

Integrating both sides, we get

$$\Rightarrow \quad \int\left(\frac{t+1}{3t-1}\right)dt=\int dx$$

$$\Rightarrow \quad \frac{1}{3}\int\frac{(3t-1)+4}{3t-1}dt=\int dx$$

$$\Rightarrow \quad \int\left\{1+\frac{4}{3t-1}\right\}dt=3\int dx$$

$$\Rightarrow \quad t+\frac{4}{3}\log|3t-1|=3x+C$$

$$\Rightarrow \quad (x+2y)+\frac{4}{3}\log|3x+6y-1|=3x+C$$

$$\Rightarrow \quad 2(y-x)+\frac{4}{3}\log|3x+6y-1|=C$$

Ans.

EXERCISE 33.4

Solve each of the following differential equations:

1. $\dfrac{dy}{dx}=(x+y+1)^2$ **Ans.** $\tan^{-1}(x+y+1)=x+C$

2. $\dfrac{dy}{dx}=(x+y)^2$ **Ans.** $x+y=\tan(x+C)$

3. $\dfrac{dy}{dx}=\sec(x+y)$ **Ans.** $y=\tan\left(\dfrac{x+y}{2}\right)+C$

4. $\dfrac{dy}{dx} = \tan(x+y)$ **Ans.** $y - x + \log|\sin(x+y) + \cos(x+y)| = C$

5. $\dfrac{dy}{dx} = \tan^2(x+y)$ **Ans.** $2(y-x) + \sin 2(x+y) = C$

6. $\dfrac{dy}{dx} = \dfrac{x+y-1}{x+y+1}$ **Ans.** $x - y + C = \log|x+y|$

7. $\dfrac{dy}{dx} = \dfrac{x-y+3}{2(x-y)+5}$ **Ans.** $x - 2y + \log|x-y+2| = C$

8. $(x+y+1)\dfrac{dy}{dx} = 1$ **Ans.** $x = Ce^y - y - 2$

9. $\dfrac{dy}{dx} = \dfrac{2x+3y+4}{4x+6y+5}$ **Ans.** $9\log|14x+21y-22| = 42y - 21x + C$

10. $\dfrac{dy}{dx} + 1 = e^{x-y}$ **Ans.** $-\dfrac{1}{2}\log|2e^{y-x} - 1| = x + C$

11. $(x-y)^2\dfrac{dy}{dx} = 1$ **Ans.** $(x-y) + \dfrac{1}{2}\log\left|\dfrac{x-y-1}{x-y+1}\right| = x + C$

12. $\cos^2(x-2y) = 1 - 2\dfrac{dy}{dx}$ **Ans.** $x = \tan(x-2y) + C$

13. $(x+y+3)\,dy = (x+y-3)\,dx$ **Ans.** $-x + y + 3\log(x+y) = C$

14. $(2x-y+2)\,dx + (4x-2y-1)\,dy = 0$ **Ans.** $2(2x-y) - \log(2x-y) = 5x + C$

15. $(2x+y+1)\,dx + (4x+2y-1)\,dy = 0$ **Ans.** $2(2x+y) + \log|2x+y-1| = 3x + C$

16. $(x+2y)(dx-dy) = dx + dy$ **Ans.** $3x - 3y + C = 2\log|3x+6y-1|$

17. $(4x+y)^2\dfrac{dx}{dy} = 1$ **Ans.** $\tan^{-1}\left(\dfrac{4x+y}{2}\right) = 2x + C$

18. $\dfrac{dy}{dx} = 1 + \tan(y-x)$ **Ans.** $\sin(y-x) = e^{x+C}$

33.8 HOMOGENEOUS DIFFERENTIAL EQUATIONS

A differential equation of the form:

$$\boxed{\frac{dy}{dx} = \frac{f(x,y)}{\phi(x,y}}$$

is called a homogeneous differential equation, if each term of $f(x, y)$ and $\phi(x, y)$ is of the same degree, *e.g.* $\dfrac{dy}{dx} = \dfrac{3xy+y^2}{3x^2+xy}$

In such cases, we put $\boxed{y = vx}$ and $\boxed{\dfrac{dy}{dx} = v + x\dfrac{dv}{dx}}$

The reduced equation involves v and x only. This new differential equation can be solved by *variables separable method.*

Working Rule

Step 1: Write the differential equation in the form $\dfrac{dy}{dx} = \dfrac{f(x, y)}{\phi(x, y)}$

Step 2: Put $y = vx$ and $\dfrac{dy}{dx} = v + x\dfrac{dv}{dx}$ in the equation obtained in step 1, and cancel common factor (x) from the numerator as well as denominator of the right hand fraction.

Step 3: The equation formed in step 2 contains variables v and x. Separate the variables v and x.

Step 4: Integrate both sides to obtain the solution in terms of v and x.

Step 5: Replace v by $\dfrac{y}{x}$ in the solution obtained in step 4.

This is the required solution.

Example 32. *Solve:* $\dfrac{dy}{dx} = \dfrac{x - y}{x + y}$ **[Diploma IETE Dec. 2006]**

Solution.
$$\frac{dy}{dx} = \frac{x - y}{x + y} \quad \text{... (1)}$$

Put
$$y = vx$$

so that
$$\frac{dy}{dx} = v + x\frac{dv}{dx}$$

On substituting these values of y and $\dfrac{dy}{dx}$ in (1), we get

$$v + x\frac{dv}{dx} = \frac{x - vx}{x + vx}$$

$$\Rightarrow \quad v + x\frac{dv}{dx} = \frac{1 - v}{1 + v}$$

$$\Rightarrow \quad x\frac{dv}{dx} = \frac{1 - v}{1 + v} - v = \frac{1 - v - v - v^2}{1 + v} = \frac{1 - 2v - v^2}{1 + v}$$

$$\Rightarrow \quad \frac{v + 1}{v^2 + 2v - 1}dv = -\frac{dx}{x}$$

$$\Rightarrow \quad \frac{1}{2}\int \frac{2v + 2}{v^2 + 2v - 1}dv = -\int \frac{dx}{x}$$

$$\Rightarrow \quad \frac{1}{2}\log(v^2 + 2v - 1) = -\log x + \log c$$

$$\Rightarrow \quad \log(v^2 + 2v - 1) = \log\left(\frac{c^2}{x^2}\right)$$

$$\Rightarrow \quad v^2 + 2v - 1 = \frac{c^2}{x^2}$$

$$\Rightarrow \quad \frac{y^2}{x^2} + 2\frac{y}{x} - 1 = \frac{c^2}{x^2}$$

$$\therefore \quad y^2 + 2xy - x^2 = c^2$$

Ans.

Example 33. *Solve:* $x^2 dy + y\,(x + y)\,dx = 0$. **[SBTE 2015]**

Solution. $x^2 dy + y\,(x + y)\,dx = 0 \Rightarrow \dfrac{dy}{dx} = -\dfrac{xy + y^2}{x^2}$

Put $\quad y = vx$

so that $\quad \dfrac{dy}{dx} = v + x\dfrac{dv}{dx}$

The given equation becomes, on substitution,

$$\Rightarrow \quad v + x\frac{dv}{dx} = -\frac{vx^2 + v^2x^2}{x^2} = -\frac{v + v^2}{1}$$

$$\Rightarrow \quad x\frac{dv}{dx} = -2v - v^2$$

Separating the variables, we get

$$\Rightarrow \quad \frac{dv}{v^2 + 2v} = -\frac{dx}{x}$$

$$\Rightarrow \quad \frac{dv}{2v} - \frac{dv}{2(v+2)} = -\frac{dx}{x}$$

$$\Rightarrow \quad \frac{1}{2}\cdot\int\frac{dv}{v} - \frac{1}{2}\int\frac{dv}{v+2} = -\int\frac{dx}{x}$$

$$\Rightarrow \quad \frac{1}{2}\log v - \frac{1}{2}\log(v + 2) = -\log x + C$$

$$\log v - \log(v + 2) = -2\log x + 2C$$

$$\Rightarrow \quad \log\left(\frac{v}{v+2}\right) + \log x^2 = 2C$$

$$\Rightarrow \quad \log\left(\frac{vx^2}{v+2}\right) = \log a$$

$$\Rightarrow \quad \frac{vx^2}{v+2} = a$$

$$\Rightarrow \quad \frac{\frac{y}{x}\cdot x^2}{\frac{y}{x} + 2} = a, \qquad \left(\because v = \frac{y}{x}\right)$$

$$\Rightarrow \quad \frac{x^2 y}{2x + y} = a$$

$$\therefore \quad x^2 y = a\,(2x + y)$$

Ans.

Example 34. *Solve:* $(x^2 + y^2)\, dx - 2xy\, dy = 0.$ **[Diploma IETE, June 2019, 2018]**

Solution. $(x^2 + y^2)\, dx = 2\, xy\, dy$

$$\Rightarrow \quad \frac{dy}{dx} = \frac{x^2 + y^2}{2xy}$$

$$\left[\text{Put } y = vx \text{ so that } \frac{dy}{dx} = v + x\frac{dv}{dx}\right]$$

$$\Rightarrow \quad v + x\frac{dv}{dx} = \frac{x^2 + v^2x^2}{2x \cdot vx}$$

$$\Rightarrow \quad x\frac{dv}{dx} = \frac{1+v^2}{2v} - v = \frac{1+v^2-2v^2}{2v}$$

$$\Rightarrow \quad \frac{2v\,dv}{1-v^2} = \frac{dx}{x}$$

$$\Rightarrow \quad \int\frac{2v\,dv}{1-v^2} = \int\frac{dx}{x}$$

$$\Rightarrow \quad -\log(1 - v^2) = \log x + \log C$$

$$\Rightarrow \quad \log\left(\frac{1}{1-v^2}\right) = \log Cx$$

$$\Rightarrow \quad \frac{1}{1-v^2} = Cx$$

$$\Rightarrow \quad \frac{x^2}{x^2-y^2} = Cx$$

$$\Rightarrow \quad x = C(x^2 - y^2)$$ **Ans.**

Example 35. *Solve:* $\frac{dy}{dx} = \frac{y}{x}\left(\log\frac{y}{x} + 1\right).$

Solution. $\frac{dy}{dx} = \frac{y}{x}\left(\log\frac{y}{x} + 1\right)$ ... (1)

Put $y = vx$ so that $\frac{dy}{dx} = v + x\frac{dv}{dx}$

Equation (1) becomes

$$\Rightarrow \quad v + x\frac{dv}{dx} = \frac{vx}{x}\left(\log\frac{vx}{x} + 1\right)$$

$$= v(\log v + 1)$$

$$\Rightarrow \quad = v\log v + v$$

$$\Rightarrow \quad x\frac{dv}{dx} = v\log v \quad \text{or} \quad \int\frac{dv}{v\log v} = \int\frac{dx}{x}$$

Put $\log v = t, \quad \frac{dv}{v} = dt$

$$\Rightarrow \quad \int\frac{dt}{t} = \log x + \log C$$

$$\Rightarrow \quad \log t = \log Cx \quad \text{or} \quad t = Cx$$

$\Rightarrow \qquad \log v = Cx$

$\therefore \qquad \log\left(\frac{y}{x}\right) = Cx \;\Rightarrow\; \frac{y}{x} = e^{Cx} \;\Rightarrow\; y = xe^{Cx}$ **Ans.**

Example 36. *Solve the differential equation* $(x + y)\,dy - (x - y)\,dx = 0$ *given that* $y = 1, x = 1.$ **[DIPIETE 2018]**

Solution. Here we have $(x + y)\,dy - (x - y)\,dx = 0$

$\Rightarrow \qquad (x + y)\,dy = (x - y)\,dx$

$$\Rightarrow \qquad \frac{dy}{dx} = \frac{(x - y)}{x + y} \qquad \text{... (1)}$$

$$\text{or} \qquad \frac{dy}{dx} = v + x\frac{dv}{dx} \qquad \text{(Putting } y = vx\text{)}$$

putting the values of y and $\frac{dy}{dx}$ in (1), we get

$$\Rightarrow \qquad v + x\frac{dv}{dx} = \frac{(x - vx)}{x + vx}$$

$$\therefore \qquad x\frac{dv}{dx} = \frac{(1 - 2v - v^2)}{1 + v} = -\frac{v^2 + 2v - 1}{1 + v}$$

$$\Rightarrow \qquad -\int\frac{dx}{x} = \frac{1}{2}\int\frac{2v + 2}{v^2 + 2v - 1}$$

$$\Rightarrow \qquad \log\left(\frac{c^2}{x^2}\right) = \log(v^2 + 2v - 1)$$

$\therefore \qquad y^2 + 2xy - x^2 = c^2 \qquad \text{... (2)}$

putting $x = 1$, $y = 1$ in (2), we get

$\Rightarrow \qquad 1 + 2 - 1 = c^2$

$\Rightarrow \qquad c^2 = 2$

$\therefore \qquad y^2 + 2xy - x^2 = 2$ **Ans.**

Example 37. *Solve:* $(x^2 + xy)\,dy = (x^2 + y^2)\,dx$ **[SBTE 2016]**

Solution. Here we have: $(x^2 + xy)\,dy = (x^2 + y^2)\,dx$

$$\frac{dy}{dx} = \frac{x^2 + y^2}{x^2 + xy} \qquad \text{... (1)}$$

put $y = vx$, $\frac{dy}{dx} = v + x\frac{dv}{dx}$ in (1), we get

$$\Rightarrow \qquad v + x\frac{dv}{dx} = \frac{1 + v^2}{1 + v}$$

$$\Rightarrow \qquad x\frac{dv}{dx} = \frac{1 + v^2}{1 + v} - v = \frac{1 - v}{1 + v}$$

$$\Rightarrow \qquad \frac{1 + v}{1 - v}dv = \frac{1}{x}dx$$

Integrating both sides, we get

$$\Rightarrow \quad \int \frac{1+v}{1-v}\,dv = \int \frac{dx}{x}$$

$$\Rightarrow \quad \int \frac{2-(1-v)}{1-v}\,dv = \log x + \log C$$

$$\Rightarrow \quad -2\log(1-v) - v = \log xC$$

$$\Rightarrow \quad -v = \log xC + 2\log(1-v)$$

$$\Rightarrow \quad -v = \log xC + \log(1-v)^2$$

$$\Rightarrow \quad e^{-v} = xC\,(1-v)^2$$

$$\Rightarrow \quad e^{-y/x} = xC\,(1-y/x)^2$$

$$\Rightarrow \quad xe^{-y/x} = C\,(x-y)^2$$

Ans.

Example 38. *Solve:* $x\,dy - y\,dx = \sqrt{x^2+y^2}\,dx$ **[SBTE 2017, 2015]**

Solution. Here we have $x\,dy - y\,dx = \sqrt{x^2+y^2}\,dx$

$$\Rightarrow \quad \frac{dy}{dx} = \frac{y+\sqrt{x^2+y^2}}{x} \qquad \ldots (1)$$

Putting $y = vx$, $\frac{dy}{dx} = v + x\frac{dv}{dx}$ in (1), we have

$$\Rightarrow \quad v + x\frac{dv}{dx} = \frac{vx+\sqrt{x^2+v^2x^2}}{x}$$

$$\Rightarrow \quad v + x\frac{dv}{dx} = v + \sqrt{1+v^2}$$

$$\Rightarrow \quad x\frac{dv}{dx} = \sqrt{1+v^2}$$

$$\Rightarrow \quad \frac{dv}{\sqrt{1+v^2}} = \frac{dx}{x}$$

Integrating both sides, we get

$$\Rightarrow \quad \int \frac{dv}{\sqrt{1+v^2}} = \int \frac{dx}{x}$$

$$\Rightarrow \quad \log\left(v+\sqrt{1+v^2}\right) = \log x + \log C$$

$$\Rightarrow \quad \log\left(\frac{y}{x}+\sqrt{1+\frac{y^2}{x^2}}\right) = \log xC$$

$$\Rightarrow \quad \frac{y}{x} + \frac{\sqrt{x^2+y^2}}{x} = xC$$

$$\Rightarrow \quad y + \sqrt{x^2+y^2} = x^2C$$

Ans.

EXERCISE 33.5

Solve the following differential equations:

1. $(1 + x^2)\, dy - xy\, dx = 0$ **Ans.** $y^2 = C(1 + x^2)$
2. $\dfrac{dy}{dx} = \dfrac{x}{2y + x}$ **Ans.** $(x + y)(2y - x)^2 = c^3$
3. $x(x - y)\, dy + y^2\, dx = 0$ **Ans.** $y = x \log (yC)$
4. $(y^2 - xy)\, dx + x^2\, dy = 0$ **[B.T.E. Delhi June 2004]** **Ans.** $\dfrac{x}{y} = \log x + C$
5. $x(y - x)\dfrac{dy}{dx} = y(y + x)$ **[B.T.E. Delhi June 2005]** **Ans.** $\dfrac{y}{x} - \log (xy) = a$
6. $\dfrac{dy}{dx} + \dfrac{x - 2y}{2x - y} = 0$ **Ans.** $y - x = C(x + y)^3$
7. $\dfrac{dy}{dx} = \sin\left(\dfrac{y}{x}\right) + \dfrac{y}{x}$ **Ans.** $\tan\left(\dfrac{y}{2x}\right) = ax$
8. $\dfrac{dy}{dx} = \dfrac{3xy + y^2}{3x^2}$ **[SBTE 2015]** **Ans.** $3x + y \log x + Cy = 0$
9. $\dfrac{dy}{dx} = \dfrac{x^2 - 2y^2}{2xy}$ **[SBTE 2014, 2010]** **Ans.** $4y^2 - x^2 = C/x^2$
10. $(x^2 + y^2)\, dy = xy\, dx$ **Ans.** $\dfrac{-x^2}{2y^2} + \log y = C$
11. $(x^2 - y^2)\, dx + 2\, xy\, dy = 0$ **[SBTE 2017, 2016, 2015]** **Ans.** $x^2 + y^2 = ax$
12. $x^2y\, dx - (x^3 + y^3)\, dy = 0$ **Ans.** $\dfrac{-x^3}{3y^3} + \log y = C$
13. $(y^2 + 2xy)\, dx + (2x^2 + 3xy)\, dy = 0$ **Ans.** $xy^2(x + y) = C$
14. $(2xy^2 - x^3)\, dy + (y^3 - 2yx^2)\, dx = 0$ **Ans.** $x^2y^2(y^2 - x^2) = C$
15. $2xy^2\, dy - (x^3 + 2y^3)\, dx = 0$ **Ans.** $2y^3 = 3x^3 \log x + 3x^3C$
16. $x\sin\dfrac{y}{x}\, dy = \left(y\sin\dfrac{y}{x} - x\right) dx$ **Ans.** $\cos\left(\dfrac{y}{x}\right) = \log x + C$
17. $\left\{x\cos\dfrac{y}{x} + y\sin\dfrac{y}{x}\right\} y - \left\{y\sin\dfrac{y}{x} - x\cos\dfrac{y}{x}\right\} x\dfrac{dy}{dx} = 0$ **Ans.** $xy\cos\left(\dfrac{y}{x}\right) = c$
18. $\dfrac{dy}{dx} = \dfrac{y}{x} - \sqrt{\left(\dfrac{y^2}{x^2} - 1\right)}$ **Ans.** $y + \sqrt{y^2 - x^2} = c$
19. $y^2 + x^2\dfrac{dy}{dx} = xy\dfrac{dy}{dx}$ **[B.T.E. Delhi May 2008, 2005]**

    Ans. $y = x \log cy$
20. $3x^2\, dy = (3xy + y^2)\, dx$ **[B.T.E. Delhi May 2008, 2005]**

    Ans. $x = -y \log c\, x^{1/3}$

33.9 EQUATIONS REDUCIBLE TO HOMOGENEOUS FORM

The equations of the form

$$\frac{dy}{dx} = \frac{ax + by + c}{Ax + By + C}$$

can be reduced to the homogeneous form by the substitution

$$x = X + h,\ y = Y + k \qquad (h, k \text{ being constants})$$

$$\frac{dy}{dx} = \frac{dY}{dX}$$

The given differential equation reduces to

$$\frac{dY}{dX} = \frac{a(X+h) + b(Y+k) + c}{A(X+h) + B(Y+k) + C} = \frac{aX + bY + ah + bk + c}{AX + BY + Ah + Bk + C}$$

Choose h, k so that $ah + bk + c = 0$

$$Ah + Bk + C = 0$$

Then the given equation becomes homogeneous

$$\frac{dY}{dX} = \frac{aX + bY}{AX + BY}$$

Case of failure: If $\frac{a}{A} = \frac{b}{B}$ then the values of h, k will not be finite.

$$\frac{a}{A} = \frac{b}{B} = \frac{1}{m} \text{ (say)}$$

$$A = am, \quad B = bm$$

The given equation becomes $\frac{dy}{dx} = \frac{ax + by + c}{m(ax + by) + C}$

Now put $ax + by = z$ and apply the method of variables separable.

Example 39. *Solve:* $\frac{dy}{dx} = \frac{x + 2y - 3}{2x + y - 3}$

Solution. Put $x = X + h,\ y = Y + k$

The given equation reduces to

$$\frac{dY}{dX} = \frac{(X+h) + 2(Y+k) - 3}{2(X+h) + (Y+k) - 3} = \frac{X + 2Y + (h + 2k - 3)}{2X + Y + (2h + k - 3)} \quad \text{... (1)}$$

Now choose h and k so that $h + 2k - 3 = 0,\ 2h + k - 3 = 0$

Solving the equations, we get $h = k = 1$

$$\therefore \quad \frac{dY}{dX} = \frac{X + 2Y}{2X + Y} \quad \text{... (2)}$$

Put $\quad Y = vX$, so that $\frac{dY}{dX} = v + X\frac{dv}{dX}$

The equation (2) is transformed as

$$v + X\frac{dv}{dX} = \frac{X + 2vX}{2X + vX} = \frac{1 + 2v}{2 + v}$$

$$X\frac{dv}{dX}=\frac{1+2v}{2+v}-v=\frac{1-v^2}{2+v}$$

$$\Rightarrow \quad \left(\frac{2+v}{1-v^2}\right)dv=\frac{dX}{X}$$

$$\Rightarrow \quad \frac{1}{2}\frac{1}{(1+v)}dv+\frac{3}{2}\frac{1}{1-v}dv=\frac{dX}{X} \qquad \text{(Partial fractions)}$$

On integrating, we have

$$\frac{1}{2}\log(1+v)-\frac{3}{2}\log(1-v)=\log X+\log C$$

$$\Rightarrow \quad \log\frac{1+v}{(1-v)^3}=\log C^2X^2 \quad \Rightarrow \quad \frac{1+v}{(1-v)^3}=C^2X^2$$

$$\frac{1+\frac{Y}{X}}{\left(1-\frac{Y}{X}\right)^3}=C^2X^2 \quad \Rightarrow \quad \frac{X+Y}{(X-Y)^3}=C^2$$

$$\Rightarrow \quad X+Y=C^2(X-Y)^3$$

Put $\quad X=x-1 \quad \text{and} \quad Y=y-1$

$$x+y-2=a(x-y)^3$$ **Ans.**

EXERCISE 33.6

Solve the following differential equations:

1. $\dfrac{dy}{dx}=\dfrac{2x+9y-20}{6x+2y-10}$ **Ans.** $(2x-y)^2=C(x+2y-5)$

2. $(12x+21y-9)\,dx+(47x+40y+7)\,dy=0$ **Ans.** $(x+5y-4)^3(3x+2y+1)=C$

3. $(3y-7x+7)\,dx+(7y-3x+3)\,dy=0$ **Ans.** $(y-x+1)^2(y+x-1)^5=C$

4. $(x+y-10)\,dx+(x-y-2)\,dy=0$ **Ans.** $(y-4)^2-2(x-6)(y-4)-(x-6)^2=C^2$

5. $\dfrac{dy}{dx}=\dfrac{y-x+1}{y+x+5}$ **Ans.** $\log[(y+3)^2+(x+2)^2]+2\tan^{-1}\left(\dfrac{y+3}{x+2}\right)=a$

6. $(3x-4y-2)dx-(4x+3y-6)\,dy=0$

Ans. $3(5y-2)^2+(5x-6)(5y-2)-3(5x-6)^2=25C^2$

7. $2xy\dfrac{dy}{dx}=x^2+3y^2$ **Ans.** $y^2+x^2=Cx^3$

8. $x\,dy-y\,dx=\sqrt{x^2-y^2}\,dx$ **Ans.** $\sin^{-1}\left(\dfrac{y}{x}\right)=\log|x|+C$

9. $(3xy+y^2)\,dx=(x^2+xy)dy$ **Ans.** $\log\left|\dfrac{y}{x^3}\right|+\dfrac{y}{x}=C$

10. $(x^2-y^2)\,dx-2xy\,dy=0$ **Ans.** $x(x^2-3y^2)=C$

11. $(2x^2y+y^3)dx+(xy^2-3x^3)dy=0$ **Ans.** $xy=e^{-x^2y^2}$, $(xy\neq 0)$

12. $2xy\dfrac{dy}{dx} = x^2 + y^2$ **Ans.** $x = C(x^2 - y^2)$

13. $xy\dfrac{dy}{dx} = x^2 - y^2$ **Ans.** $(x^2 - 2y^2)\, x^2 = C$

14. $x^2\dfrac{dy}{dx} = x^2 + xy + y^2$ **Ans.** $\tan^{-1}\left(\dfrac{y}{x}\right) = C + \log|x|$

15. $ye^{\frac{x}{y}}\, dx = \left(xe^{\frac{x}{y}} + y^2\right) dy$ **Ans.** $e^{x/y} = y + C$

16. $x^2\dfrac{dy}{dx} = x^2 + 5xy + 4y^2$ **Ans.** $-\dfrac{1}{4}\left(\dfrac{y}{x} + \dfrac{1}{2}\right)^{-1} = \log|x| + C$

17. $(y^2 - 2xy)\, dx = (x^2 - 2xy)\, dy$ **Ans.** $x^2y - xy^2 = C$

18. $\left(1 + e^{\frac{x}{y}}\right) dx + e^{\frac{x}{y}}\left(1 + \dfrac{x}{y}\right) dy = 0$ **Ans.** $x + ye^{\frac{x}{y}} = C$

19. $3x^2\, dy = (3xy + y^2)\, dx$ **Ans.** $\dfrac{-3x}{y} = \log|x| + C$

20. $\dfrac{dy}{dx} = \dfrac{x}{2y + x}$ **Ans.** $(x + y)\,(2y - x)^2 = C$

33.10 LINEAR DIFFERENTIAL EQUATION

A differential equation of the form

$$\frac{dy}{dx} + Py = Q \quad \text{... (1)}$$

is called a linear differential equation in y, where P and Q, are functions of x (but not of y) or constants.

For example:

1. $\dfrac{dy}{dx} + x^2y = x^4$ 2. $\dfrac{dy}{dx} + \dfrac{y}{x} = x^2$ 3. $\dfrac{dy}{dx} + \dfrac{1}{x\log x}y = \dfrac{2}{x^2},\ x > 0$

In such case (1), multiply both sides of (1) by $^{\int P dx}$

$$\Rightarrow \quad e^{\int P\,dx}\left(\frac{dy}{dx} + Py\right) = Qe^{\int P\,dx}$$

The left hand side of (2) is $\dfrac{d}{dx}\left[y \cdot e^{\int P\,dx}\right]$

$$\Rightarrow \quad \frac{d}{dx}\left[y \cdot e^{\int P\,dx}\right] = Q \cdot e^{\int P\,dx}$$

Integrating both sides $y \cdot e^{\int P\,dx} = \int \left(Q \cdot e^{\int P\,dx}\right) dx$

This is the required solution.

Note: $e^{\int P dx}$ is called the integrating factor.

Solution is $\boxed{y \times [I.F.] = \int [Q \times I.F.]\, dx + C}$

Working rule for solving $\dfrac{dy}{dx} + Py = Q$

Step 1: Find the integrating factor $(I.F.) = e^{\int P dx}$

Step 2: Multiply both sides of the given equation by $I.F.$

Step 3: Integrate both sides of the equation obtained in step 2, to get the solution.

$$y(I.F.) = \int (Q \times I.F.)\, dx$$

This is the required solution.

Example 40. *Solve:* $x\dfrac{dy}{dx} - 3y = x^2$ **[SBTE 2016, 2015]**

Solution. $$\frac{dy}{dx} - \frac{3}{x}y = x$$

Here the integrating factor $= e^{\int -\frac{3}{x}dx} = e^{-3\log x} = e^{\log 1/x^3} = \dfrac{1}{x^3}$

The solution is

$$\Rightarrow \quad y\frac{1}{x^3} = \int \frac{1}{x^3}\, x\, dx$$

$$\Rightarrow \quad \frac{y}{x^3} = \int x^{-2} dx \;\Rightarrow\; \frac{y}{x^3} = -\frac{1}{x} + C$$

$$\therefore \quad y = -x^2 + Cx^3$$ **Ans.**

Example 41. *Solve:* $x^2(x^2-1)\dfrac{dy}{dx} + x(x^2+1)y = (x^2-1)$

Solution. $$\frac{dy}{dx} + \frac{x^2+1}{x(x^2-1)}y = \frac{1}{x^2}$$

Hence the integrating factor

$$= e^{\int \frac{x^2+1}{x(x^2-1)}dx} = e^{\int\left(-\frac{1}{x}+\frac{1}{x+1}+\frac{1}{x-1}\right)dx}$$

$$= e^{-\log x + \log(x+1) + \log(x-1)} = e^{\log\left(\frac{x^2-1}{x}\right)} = \frac{x^2-1}{x}$$

The solution is

$$\Rightarrow \quad y\left(\frac{x^2-1}{x}\right) = \int \frac{x^2-1}{x}\cdot\frac{1}{x^2}\, dx$$

$$= \int\left(\frac{1}{x} - \frac{1}{x^3}\right)dx$$

$$\Rightarrow \quad y\left(\frac{x^2-1}{x}\right) = \log x + \frac{1}{2x^2} + C$$

$$\Rightarrow \quad 2xy\,(x^2-1) = 2x^2 \log x + 2\,Cx^2 + 1$$ **Ans.**

Example 42. *Solve* $\sec x \dfrac{dy}{dx} = y + \sin x$ **[B.T.E. Delhi May 2008]**

Solution. We have,

$$\sec x \frac{dy}{dx} = y + \sin x$$

$$\Rightarrow \quad \frac{dy}{dx} - (\cos x)\, y = \sin x \cos x$$

This is the linear differential equation of first order.

$$\Rightarrow \qquad I.F. = e^{\int -\cos x\, dx} = e^{-\sin x}$$

Its solution is $\quad y\,(I.F.) = \int Q(I.F.) \cdot dx$

$$\Rightarrow \qquad y(e^{-\sin x}) = \int \sin x \cdot \cos x \cdot e^{-\sin x}\, dx$$

Putting $\sin x = t$ so that $\cos x\, dx = dt$, we get

$$\Rightarrow y(e^{-\sin x}) = \int t e^{-t}\, dt + C = t\left(\frac{e^{-t}}{-1}\right) - \int \frac{1 \cdot e^{-t}}{-1}\, dt$$

$$= -te^{-t} - e^{-t} + C = -e^{-t}(t+1) + C$$

$$\Rightarrow \qquad y(e^{-\sin x}) = -e^{-\sin x}(\sin x + 1) + C$$

$$\therefore \qquad y = -(\sin x + 1) + Ce^{\sin x}$$ **Ans.**

Example 43. *Solve the differential equation:* $\dfrac{dy}{dx} + 3y = e^{-2x}$

Solution. The given differential equation is linear in y.

Here: $\qquad P = 3$ and $Q = e^{-2x}$

$$\therefore \qquad I.F. = e^{\int P\, dx} = e^{\int 3\, dx} = e^{3x}$$

$\therefore$ The solution is

$$\Rightarrow \qquad y\,(I.F.) = \int (Q \times I.F.)\, dx$$

$$\Rightarrow \qquad y\, e^{3x} = \int e^{3x} \cdot e^{-2x} dx = \int e^{x} dx = e^{x} + C$$

$$\therefore \qquad y = e^{-2x} + Ce^{-3x}$$ **Ans.**

Example 44. *Solve the differential equation:* $x\dfrac{dy}{dx} + y = x^4$

Solution. $\qquad x\dfrac{dy}{dx} + y = x^4$

$$\Rightarrow \qquad \frac{dy}{dx} + \frac{1}{x} y = x^3 \qquad \text{... (1)}$$

This is a linear differential equation in y

Here, $\qquad P = \dfrac{1}{x}, \quad Q = x^3$

$$I.F. = e^{\int P\,dx} = e^{\int \frac{1}{x}dx} = e^{\log x} = x$$

The solution of (1) is

$$y\,(I.F.) = \int (Q \times I.F.)\,dx$$

$$\Rightarrow \quad y \cdot x = \int x^3 \cdot x\,dx$$

$$= \int x^4\,dx = \frac{1}{5}x^5 + C$$

$$\Rightarrow \quad y = \frac{1}{5}x^4 + \frac{C}{x}$$ **Ans.**

Example 45. *Solve:* $x\dfrac{dy}{dx} - y = x^2$

Solution. We have, $x\dfrac{dy}{dx} - y = x^2$

$$\Rightarrow \quad \frac{dy}{dx} - \frac{1}{x}y = x \qquad \text{... (1)}$$

This is a linear differential equation in y.

Here, $P = -\dfrac{1}{x}$ and $Q = x$

Now, $I.F. = e^{\int P dx} = e^{\int -\frac{1}{x}dx} = e^{-\log x} = e^{\log x^{-1}} = x^{-1} = \dfrac{1}{x}$

$\therefore$ The solution of (1) is $y(I.F.) = \int (Q \times I.F.)\,dx$

$$\Rightarrow \quad y \cdot \frac{1}{x} = \int \frac{1}{x} \cdot x\,dx = x + C$$

$$\Rightarrow \quad y = x^2 + Cx$$ **Ans.**

Example 46. *Solve:* $\dfrac{dp}{dr} = \dfrac{2a}{r^2} - \dfrac{2p}{r^2}$

Solution. We have, $\dfrac{dp}{dr} = \dfrac{2a}{r^2} - \dfrac{2p}{r^2}$

$$\Rightarrow \quad \frac{dp}{dr} + \frac{2p}{r} = \frac{2a}{r^2}$$

$$\Rightarrow \quad I.F. = e^{\int \frac{2}{r}dr} = e^{2\log r} = e^{\log r^2} = r^2$$

or $\quad p \cdot r^2 = \int \dfrac{2a}{r^2} \cdot r^2 dr$

$$= 2a\int dr$$

$$= 2ar + C$$ **Ans.**

Example 47. *Solve:* $x\log x\dfrac{dy}{dx}+y=2\log x$ **[SBTE. 2016]**

Solution. Here we have

$$\Rightarrow \quad x\log x\frac{dy}{dx}+y=2\log x$$

$$\Rightarrow \quad \frac{dy}{dx}+\frac{1}{x\log x}\cdot y=\frac{2}{x}$$

$$\Rightarrow \quad P=\frac{1}{x\log x},\quad Q=\frac{2}{x}$$

$$\Rightarrow \quad I.F.=e^{\int P dx}=e^{\int\frac{1}{x\log x}dx}=e^{\log(\log x)}=\log x$$

Required solution $y\cdot I.F.=\int(Q\cdot I.F.)dx+C$

$$\Rightarrow \quad y\times\log x=\int\left(\frac{2}{x}\cdot\log x\right)dx+C$$

$$\Rightarrow \quad y\log x=2\int\frac{1}{x}\log x\,dx+C$$

put $\log x=t \Rightarrow \dfrac{1}{x}\,dx=dt$

$$\Rightarrow \quad y\log x=2\int t\cdot dt+C$$

$$\Rightarrow \quad y\log x=\not{2}\frac{t^2}{\not{2}}+C$$

$$\therefore \quad y\log x=(\log x)^2+C$$ **Ans.**

Example 48. *Solve:* $\dfrac{dy}{dx}-2y=\cos 3x$ **[SBTE. 2016, 2015]**

Solution. Here we have

$$\Rightarrow \quad \frac{dy}{dx}-2y=\cos 3x$$

$P=-2,\quad Q=\cos 3x$

$$\Rightarrow \quad I.F.=e^{-\int 2dx}=e^{-2x}$$

Required solution

$$\Rightarrow \quad y\times I.F.=\int(Q\cdot I.F.)dx+C$$

$$\Rightarrow \quad ye^{-2x}=\int \underset{I}{e^{-2x}}\cdot\underset{II}{\cos 3x}\,dx$$

Let $$I=e^{-2x}\int\cos 3x+\frac{2}{3}\int e^{-2x}\cdot\sin 3x\,dx$$

$$= \frac{1}{3}e^{-3x}\sin 3x + \frac{2}{3}e^{-2x}\cdot\left(\frac{-\cos 3x}{3}\right) - \frac{2}{3}\int(-2)e^{-2x}\left(\frac{-\cos 3x}{3}\right)dx$$

$$= \frac{1}{3}e^{-2x}\sin 3x - \frac{2}{9}e^{-2x}\cos 3x - \frac{4}{9}\int e^{-2x}\cdot\cos 3x\,dx$$

$$= \frac{1}{3}e^{-2x}\sin 3x - \frac{2}{9}e^{-2x}\cos 3x - \frac{4}{9}\cdot I$$

$$\Rightarrow \quad I + \frac{4}{9}I = \frac{1}{3}e^{-2x}\sin 3x - \frac{2}{9}e^{-2x}\cos 3x$$

$$\Rightarrow \quad \frac{13}{9}I = \frac{1}{3}e^{-2x}\sin 3x - \frac{2}{9}e^{-2x}\cos 3x$$

$$\Rightarrow \quad I = \frac{1}{13}e^{-2x}(3\sin 3x - 2\cos 3x) + C$$

$$\therefore \quad ye^{-2x} = \frac{1}{13}e^{-2x}(3\sin 3x - 2\cos 3x) + C$$ **Ans.**

Example 49. *Solve:* $(1+x^2)\frac{dy}{dx} + 2xy - 4x^2 = 0$, *subject to the initial condition* $y(0) = 0$.

[SBTE. 2017, 2016, 2015, 2014]

Solution. Here we have: $(1+x^2)\frac{dy}{dx} + 2xy - 4x^2 = 0$

$$\Rightarrow \quad \frac{dy}{dx} + \left(\frac{2x}{1+x^2}\right)y = \frac{4x^2}{1+x^2}$$

$$\Rightarrow \quad P = \frac{2x}{1+x^2}, \quad Q = \frac{4x^2}{1+x^2}$$

$$\Rightarrow \quad I.F. = e^{2\int\frac{x}{1+x^2}dx} = e^{\log(1+x^2)} = 1+x^2$$

Required solution,

$$\Rightarrow \quad y \times I.F. = \int(Q\cdot I.F.)dx + C$$

$$\Rightarrow \quad y\,(1+x^2) = \int\left(\frac{4x^2}{1+x^2}\right)(1+x^2)\,dx + C$$

$$\Rightarrow \quad y\,(1+x^2) = 4\int x^2 dx + C$$

$$y\,(1+x^2) = \frac{4}{3}x^3 + C \qquad \text{... (1)}$$

$x = 0$, $y = 0$ in (1), we get

$$C = 0$$

$$\therefore \quad y\,(1+x^2) = \frac{4}{3}x^3$$ **Ans.**

EXERCISE 33.7

Solve the following differential equations:

1. $\dfrac{dy}{dx}+\dfrac{1}{x}y=x^2$ **Ans.** $4\,xy = x^4 + a$

2. $\dfrac{dy}{dx}+\dfrac{1}{x}y=x^3-3$ **Ans.** $xy=\dfrac{x^5}{5}-\dfrac{3x^2}{2}+C$

3. $\dfrac{dy}{dx}+y\cot x=\cos x$ **Ans.** $y\sin x=\dfrac{\sin^2 x}{2}+C$

4. $\dfrac{dy}{dx}+2xy=2e^{-x^2}$ **Ans.** $y = (2x + C)\, e^{-x^2}$

5. $\dfrac{dy}{dx}+3y=e^{2x}$ **Ans.** $y=\dfrac{1}{5}e^{2x}+Ce^{-3x}$

6. $\dfrac{dy}{dx}+y\sec x=\tan x$ **Ans.** $y=\dfrac{C-x}{\sec x+\tan x}+1$

7. $\cos^2 x\dfrac{dy}{dx}+y=\tan x$ **[Diploma IETE Dec. 2014]** **Ans.** $y = \tan x - 1 + Ce^{-\tan x}$

8. $x\,dy + (x + y)\,dx = 0$ **Ans.** $2xy + x^2 = a$

9. $(x+a)\dfrac{dy}{dx}-3y=(x+a)^5$ **Ans.** $2y = (x + a)^5 + 2C\,(x + a)^3$

10. $(x+1)\dfrac{dy}{dx}-y=e^{3x}(x+1)^2$ **Ans.** $3y = (1 + x)\,[e^{3x} + C]$

11. $x\cos x\dfrac{dy}{dx}+y\,(x\sin x+\cos x)=1$ **Ans.** $xy = \sin x + C\cos x$

12. $x\log x\dfrac{dy}{dx}+y=2\log x$ **Ans.** $y\log x = (\log x)^2 + C$

13. $t\,ds - (3t + 1)\,s\,dt = t^3 e^{3t}\,dt$ **Ans.** $s=\left(\dfrac{t^2}{2}+c\right)te^{3t}$

14. $\dfrac{dy}{dx}=-\dfrac{x+y\cos x}{1+\sin x}$ **Ans.** $y(1+\sin x)=\dfrac{-x^2}{2}+C$

15. $\dfrac{dy}{dx}+y\cot x=5\,e^{\cos x}$ **Ans.** $y\sin x = -5\,e^{\cos x} + C$

16. $(3e^{3x}\,y - 2x)\,dx + e^{3x}\,dy = 0$ **Ans.** $ye^{3x} = x^2 + C$

17. $x\dfrac{dy}{dx}+y=\log x$ **Ans.** $y\cdot x = x\log x - x + C$

18. $(1-x^2)\dfrac{dy}{dx}-xy=1$ **[B.T.E. Delhi May 2008]** **Ans.** $\sqrt{1-x^2}\,y=\sin^{-1}x+C$

19. $(x+1)\dfrac{dy}{dx}-2y=(x+1)^4$ **[B.T.E. 2008, 2005]** **Ans.** $y=\left(\dfrac{x^2}{2}+x+C\right)(x+1)^2$

20. $(1+x^2)\dfrac{dy}{dx}+2xy=\dfrac{1}{1+x^2}$ **[Diploma IETE, Dec. 2018]** **Ans.** $y(1+x^2)=\tan^{-1}x+C$

21. $x\dfrac{dy}{dx}+2y=x^2\log x$ **Ans.** $yx^2=\dfrac{x^4}{4}\log x-\dfrac{x^4}{16}+C$

22. $\dfrac{dy}{dx}+y\cos x=\dfrac{1}{2}\sin 2x$ **[SBTE, 2014, 2007]** **Ans.** $y=\sin x-1+Ce^{-\sin x}$

33.11 EQUATIONS REDUCIBLE TO THE LINEAR FORM

Bernoulli's equation of the form $\boxed{\dfrac{dy}{dx}+Py=Qy^n}$ ... (1)

where P and Q are constants or functions of x can be reduced to the linear form on dividing by y^n and substituting $\dfrac{1}{y^{n-1}}=z$.

On dividing both sides of (1) by y^n

$$\frac{1}{y^n}\frac{dy}{dx}+\frac{1}{y^{n-1}}P=Q \quad \text{... (2)}$$

Put $\dfrac{1}{y^{n-1}}=z$ so that $\dfrac{(1-n)}{y^n}\dfrac{dy}{dx}=\dfrac{dz}{dx}$

$\therefore$ (2) becomes $\dfrac{1}{1-n}\dfrac{dz}{dx}+Pz=Q$

$$\Rightarrow \quad \frac{dz}{dx}+P(1-n)z=Q(1-n)$$

which is a linear equation and can be solved easily by the previous method discussed in section 34.10.

Example 50. *Solve:* $\dfrac{dy}{dx}=y\tan x-y^2\sec x$

Solution. $\dfrac{dy}{dx}=y\tan x-y^2\sec x$

$$\Rightarrow \quad \frac{dy}{dx}-y\tan x=-y^2\sec x$$

$$\Rightarrow \quad -\frac{1}{y^2}\frac{dy}{dx}+\frac{1}{y}\tan x=\sec x \quad \text{... (1)}$$

Substituting $\dfrac{1}{y}=z$, so that $-\dfrac{1}{y^2}\cdot\dfrac{dy}{dx}=\dfrac{dz}{dx}$ in (1), we get

$$\Rightarrow \quad \frac{dz}{dx}+z\tan x=\sec x$$

$$\Rightarrow \quad I.F.=e^{\int \tan x\,dx}=e^{\log\sec x}=\sec x$$

Its solution is

$$z \cdot \sec x = \int (\sec x)(\sec x\, dx)$$

$$= \int \sec^2 x\, dx$$

$$= \tan x + c$$

$$\therefore \qquad \frac{\sec x}{y} = \tan x + c$$ **Ans.**

Example 51. *Solve* $x\dfrac{dy}{dx} = y\{\log y - \log x + 1\}$ **[B.T.E Delhi June 2005]**

Solution. $x\dfrac{dy}{dx} = y\{\log y - \log x + 1\}$

$$\Rightarrow \qquad \frac{1}{y}\frac{dy}{dx} = \frac{1}{x}\{\log y - \log x + 1\}$$

$$\Rightarrow \qquad \frac{1}{y}\frac{dy}{dx} - \frac{\log y}{x} = -\frac{\log x}{x} + \frac{1}{x} \qquad \text{... (1)}$$

Put $z = \log y$, so that $\dfrac{dz}{dx} = \dfrac{1}{y}\dfrac{dy}{dx}$

(1) becomes $\dfrac{dz}{dx} - \dfrac{z}{x} = -\dfrac{\log x}{x} + \dfrac{1}{x}$

$$I.F. = e^{-\int \frac{1}{x} dx} = e^{-\log x} = e^{\log\left(\frac{1}{x}\right)} = \frac{1}{x}$$

$\therefore$ The solution is $z \cdot \dfrac{1}{x} = \displaystyle\int \left(-\frac{\log x}{x} + \frac{1}{x}\right)\frac{1}{x}\, dx$

$$\Rightarrow \qquad \frac{1}{x}\log y = \int \left(-\frac{\log x}{x^2} + \frac{1}{x^2}\right) dx$$

$$= (\log x)\left(\frac{1}{x}\right) - \int \frac{1}{x} \cdot \frac{1}{x}\, dx + \int \frac{1}{x^2}\, dx$$

$$= \frac{1}{x}\log x + C$$

$$\Rightarrow \qquad \log y = \log x + Cx$$ **Ans.**

Example 52. *Solve:* $\sec^2 y \dfrac{dy}{dx} + x \tan y = x^3$.

Solution. $\sec^2 y \dfrac{dy}{dx} + x \tan y = x^3$ ... (1)

Putting $\tan y = z$, so that $\sec^2 y \dfrac{dy}{dx} = \dfrac{dz}{dx}$, (1) becomes

$$\Rightarrow \qquad \frac{dz}{dx} + xz = x^3$$

Here, the integrating factor is $e^{\int x\,dx} = e^{\frac{x^2}{2}}$

$\therefore$ The solution is $z \cdot e^{\frac{x^2}{2}} = \int \left(e^{\frac{x^2}{2}} \right) x^3 dx + C = \int x^2 \left(xe^{\frac{x^2}{2}} dx\right)$

$$\Rightarrow \qquad z \cdot e^{\frac{x^2}{2}} = x^2 e^{\frac{x^2}{2}} - 2\int x \cdot e^{\frac{x^2}{2}} dx$$

$$\Rightarrow \qquad z \cdot e^{\frac{x^2}{2}} = x^2 e^{\frac{x^2}{2}} - 2e^{\frac{x^2}{2}} + C \Rightarrow z = x^2 - 2 + ce^{-\frac{x^2}{2}}$$

$$\therefore \qquad \tan y = x^2 - 2 + ce^{-\frac{x^2}{2}}$$

Ans.

EXERCISE 33.8

Solve the following differential equations:

1. $\dfrac{1}{y^2}\dfrac{dy}{dx} - \dfrac{1}{y} = 2xe^{-x}$ **Ans.** $e^x + x^2 y + cy = 0$

2. $\dfrac{dy}{dx} + xy = xy^3$ **Ans.** $y^2 = \dfrac{1}{1 + ce^{x^2}}$

3. $\dfrac{dy}{dx} = y \tan x - y^2 \sec x$ **Ans.** $\sec x = (\tan x + c)\, y$

4. $\dfrac{dy}{dx} = 2y \tan x + y^2 \tan^2 x$. If $y = 1$ at $x = 0$. **Ans.** $\dfrac{1}{y}\sec^2 x = -\dfrac{\tan^3 x}{3} + 1$

5. $\sin y \dfrac{dy}{dx} = \cos x\,(2\cos y - \sin^2 x)$ **Ans.** $\cos y = \dfrac{1}{4}[2\sin^2 x - 2\sin x + 1] - ce^{-2\sin x}$

6. $\dfrac{dy}{dx} + \dfrac{y}{x} = x^2 y^6$ **[DIPIETE Dec. 2019]** **Ans.** $\dfrac{1}{x^5 y^5} = \dfrac{5}{2x^2} + c$

7. $x\dfrac{dy}{dx} + y \log y = xye^x$ **Ans.** $x \log y = e^x (x - 1) + c$

8. $(x - y^2)\, dx + 2xy\, dy = 0$ **Ans.** $\dfrac{y^2}{x} + \log x = c$

9. $e^y \left(\dfrac{dy}{dx} + 1\right) = e^x$ **Ans.** $e^{x+y} = \dfrac{e^{2x}}{2} + C$

10. $x^2 y - x^3 \dfrac{dy}{dx} = y^4 \cos x$ **Ans.** $x^3 = y^3 (3 \sin x + c)$

11. $y + 2\dfrac{dy}{dx} = y^3 (x - 1)$ **Ans.** $y^2 (x + ce^x) = 1$

12. $\dfrac{dy}{dx} + x \sin 2y = x^3 \cos^2 y$ **Ans.** $\tan y = \dfrac{1}{2}(x^2 - 1) + ce^{-x^2}$

13. $\dfrac{dy}{dx} - \dfrac{\tan y}{1 + x}(1 + x)e^x \sec y$ **Ans.** $\sin y = (e^x + c)\,(1 + x)$

14. $x\dfrac{dy}{dx}+\dfrac{y^2}{x}=y$ **Ans.** $\dfrac{x}{y}=\log x+c$

15. $\dfrac{1}{1+y^2}\dfrac{dy}{dx}+2x\tan^{-1}(y)=x^3$ **Ans.** $\tan^{-1}(y)=\dfrac{1}{2}(x^2-1)+ce^{-x^2}$

33.12 LINEAR DIFFERENTIAL EQUATION IN x

Example 53. *Solve:* $(1+y^2)\,dx=(\tan^{-1}y-x)\,dy$ **[SBTE 2017, 2016, 2015]**

Solution. $(1+y^2)\,dx=(\tan^{-1}y-x)\,dy$

$$\Rightarrow \quad \frac{dx}{dy}=\frac{\tan^{-1}y}{1+y^2}-\frac{x}{1+y^2}$$

$$\Rightarrow \quad \frac{dx}{dy}+\frac{x}{1+y^2}=\frac{\tan^{-1}y}{1+y^2}$$

This is a linear differential equation in x,

$$I.F.=e^{\int\frac{1}{1+y^2}dy}=e^{\tan^{-1}y}$$

$$x\cdot e^{\tan-1y}=\int e^{\tan^{-1}y}\frac{\tan^{-1}y}{1+y^2}dy,\quad \text{put } \tan^{-1}y=t$$

$$=\int e^t t\,dt$$

$$=t\,e^t-e^t+C=e^t(t-1)+C$$

$$=e^{\tan^{-1}y}(\tan^{-1}y-1)+C$$

$$x=(\tan^{-1}y-1)+Ce^{-\tan^{-1}y}$$ **Ans.**

Example 54. *Solve:* $(x+\tan y)\,dy=\sin 2y\,dx$

Solution. We have, $(x+\tan y)\,dy=\sin 2y\,dx$

$$\Rightarrow \quad \sin 2y\frac{dx}{dy}=x+\tan y$$

$$\Rightarrow \quad \frac{dx}{dy}=x\operatorname{cosec}2y+\frac{\tan y}{\sin 2y}$$

$$\Rightarrow \quad \frac{dx}{dy}-x\operatorname{cosec}2y=\frac{1}{2}\sec^2 y$$

$$\Rightarrow \quad I.F.=e^{-\int\operatorname{cosec}2y\,dy}$$

$$=e^{-\frac{1}{2}\log\tan y}=e^{\log\left(\frac{1}{\sqrt{\tan y}}\right)}=\frac{1}{\sqrt{\tan y}}$$

$$\Rightarrow \quad \frac{x}{\sqrt{\tan y}}=\int\frac{1}{2}\frac{\sec^2 y}{\sqrt{\tan y}}dy$$

$$\Rightarrow \qquad \frac{x}{\sqrt{\tan y}} = \sqrt{\tan y} + C$$

$$\Rightarrow \qquad x = \tan y + C\sqrt{\tan y}$$ **Ans.**

Example 55. *Solve:* $ye^y\,dx = (y^3 + 2\,xe^y)\,dy$

Solution. $\dfrac{dx}{dy} = \dfrac{y^3 + 2xe^y}{ye^y} = y^2e^{-y} + \dfrac{2x}{y}$

$$\Rightarrow \qquad \frac{dx}{dy} - \frac{2x}{y} = y^2e^{-y}$$

$$\Rightarrow \qquad I.F. = e^{\int -\frac{2}{y}dy} = e^{-2\log y} = e^{\log y^{-2}} = y^{-2}$$

Solution is $\qquad xy^{-2} = \int (y^2\,e^{-y})\,(y^{-2})\,dy$

$$= \int e^{-y}\,dy$$

$$= -e^{-y} + C$$ **Ans.**

Example 56. *Solve:* $y \log y \dfrac{dx}{dy} + x - \log y = 0.$ **[SBTE, 2016]**

Solution. $y \log y \cdot \dfrac{dx}{dy} + x - \log y = 0$

$$\Rightarrow \qquad y \log y \cdot \frac{dx}{dy} + x = \log y$$

$$\Rightarrow \qquad \frac{dx}{dy} + \frac{x}{y \log y} = \frac{1}{y}$$

$$I.F. = e^{\int \frac{dy}{y \log y}} = e^{\log(\log y)} = \log y$$

$\therefore$ The solution is $\; x \cdot \log y = \int \dfrac{1}{y}(\log y)\,dy$

$$\Rightarrow \qquad x \cdot \log y = \frac{1}{2}(\log y)^2 + C$$ **Ans.**

EXERCISE 33.9

Solve the following differential equations:

1. $(x + y + 1)\,\dfrac{dy}{dx} = 1$ **Ans.** $x + y + 2 = ce^y$

2. $(x + 2y^3)\,dy = y \cdot dx$ **[SBTE, 2015]** **Ans.** $x = y^3 + cy$

3. $(2x - 10y^3)\,\dfrac{dy}{dx} + y = 0$ **Ans.** $x = 2y^3 + cy^{-2}$

4. $e^{-y}\sec^2 y\,dy = dx + x\,dy$ **Ans.** $xe^y = \tan y + C$

5. $y \sin 2x\,dx - (y^2 + \cos^2 x)\,dy = 0$ **Ans.** $3y\cos^2 x + y^3 = C$

33.13 EXACT DIFFERENTIAL EQUATION

An exact differential equation is formed by directly differentiating its primitive (solution) without other process

$$Mdx + Ndy = 0$$

is said to be an exact differential equation if it satisfies the following condition

$$\boxed{\frac{\partial M}{\partial y} = \frac{\partial N}{\partial x}}$$

where $\frac{\partial M}{\partial y}$ denotes the differential coefficient of M with respect to y keeping x constant

and $\frac{\partial N}{\partial x}$, the differential coefficient of N with respect to x, keeping y constant.

Method for Solving Exact Differential Equations

Step 1: Integrate M w.r.t. x keeping y constant.

Step 2: Integrate w.r.t. y, only those terms of N which do not contain x.

Step 3: Result of 1 + result of 2 = Constant.

Example 57. *Solve:* $(5x^4 + 3x^2y^2 - 2xy^3)\,dx + (2x^3y - 3x^2y^2 - 5y^4)\,dy = 0$

Solution. Here, $\quad M = 5x^4 + 3x^2y^2 - 2xy^3,\ N = 2x^3y - 3x^2y^2 - 5y^4$

$$\Rightarrow \quad \frac{\partial M}{\partial y} = 6x^2y - 6xy^2,\ \frac{\partial N}{\partial x} = 6x^2y - 6xy^2$$

Since, $\frac{\partial M}{\partial y} = \frac{\partial N}{\partial x}$, the given equation is exact.

Now $\int M\,dx + \int$ (terms of N is not containing x) $dy = C$

$$\Rightarrow \quad \int (5x^4 + 3x^2y^2 - 2xy^3)\,dx + \int -5y^4 dy = C$$

$$\therefore \quad x^5 + x^3y^2 - x^2y^3 - y^5 = C$$ **Ans.**

Example 58. *Solve:* $\{2xy\cos x^2 - 2xy + 1\}\,dx + \{\sin x^2 - x^2 + 3\}\,dy = 0$

Solution. Here we have,

$$\{2xy\cos x^2 - 2xy + 1\}\,dx + \{\sin x^2 - x^2 + 3\}\,dy = 0 \quad \text{... (1)}$$

$$M\,dx + N\,dy = 0 \quad \text{... (2)}$$

Comparing (1) and (2), we get

$$\Rightarrow \quad M = 2xy\cos x^2 - 2xy + 1 \quad \Rightarrow \quad \frac{\partial M}{\partial y} = 2x\cos x^2 - 2x$$

$$\Rightarrow \quad N = \sin x^2 - x^2 + 3 \quad \Rightarrow \quad \frac{\partial N}{\partial x} = 2x\cos x^2 - 2x$$

$$\therefore \quad \frac{\partial M}{\partial y} = \frac{\partial N}{\partial x}$$

So the given differential equation is exact differential equation

Hence solution is $\displaystyle\int\limits_{y \text{ as const}} M\,dx + \int (\text{terms of } N \text{ not containing } x)\,dy = C$

$$\Rightarrow \quad \int (2xy\cos x^2 - 2xy + 1)\,dx + \int 3\,dy = C$$

$$\Rightarrow \quad \int [y(2x\cos x^2) - y(2x) + 1]\,dx + 3\int dy = C$$

$$\Rightarrow \quad y\int 2x\cos x^2 dx - y\int 2x\,dx + \int 1\,dx + 3\int dy = C$$

Put $x^2 = t$ so that $2x\,dx = dt$

$$\Rightarrow \quad y\int \cos t\,dt - 2y\frac{x^2}{2} + x + 3y = C$$

$$\Rightarrow \quad y\sin t - x^2 y + x + 3y = C$$

$$\therefore \quad y\sin x^2 - x^2 y + x + 3y = C$$ **Ans.**

Example 59. *Solve:* $(1 + e^{x/y}) + e^{x/y}\left(1 - \dfrac{x}{y}\right)\dfrac{dy}{dx} = 0$

[Nagpur University, Summer 2008, AMIETE, June 2018]

Solution. We have,

$$\Rightarrow \quad \left(1 + e^{\frac{x}{y}}\right) + e^{\frac{x}{y}}\left(1 - e^{\frac{x}{y}}\right)\frac{dy}{dx} = 0 \;\Rightarrow\; \left(1 + e^{\frac{x}{y}}\right)dx + \left(e^{\frac{x}{y}} - e^{\frac{x}{y}}\frac{x}{y}\right)dy = 0$$

$$\Rightarrow \quad M = 1 + e^{\frac{x}{y}} \;\Rightarrow\; \frac{\partial M}{\partial y} = -\frac{x}{y^2}e^{\frac{x}{y}}$$

$$\Rightarrow \quad N = e^{\frac{x}{y}} - e^{\frac{x}{y}}\frac{x}{y} \;\Rightarrow\; \frac{\partial N}{\partial x} = \frac{1}{y}e^{\frac{x}{y}} - \frac{1}{y}e^{\frac{x}{y}} - \frac{x}{y^2}e^{\frac{x}{y}} = -\frac{x}{y^2}e^{\frac{x}{y}}$$

$$\Rightarrow \quad \frac{\partial M}{\partial y} = \frac{\partial N}{\partial x}$$

$\therefore$ Given equation is exact.

Its solution is $\int\left(\text{[illegible]}\,e^{\frac{x}{y}}\right)dx + \int \quad N \quad x\,dy = C$

$$\Rightarrow \quad \int\left(1 + e^{\frac{x}{y}}\right)dx + \int 0\,dy = C \;\Rightarrow\; x + ye^{\frac{x}{y}} = C$$ **Ans.**

Example 60. *Solve:* $[1 + \log(xy)]dx + \left[1 + \dfrac{x}{y}\right]dy = 0$

Solution. $[1 + \log x + \log y]dx + \left[1 + \dfrac{x}{y}\right]dy = 0$

which is in the form $M\,dx + N\,dy = 0$

$$\Rightarrow \quad M = [1 + \log x + \log y] \quad \text{and} \quad N = 1 + \frac{x}{y}$$

$$\Rightarrow \quad \frac{\partial M}{\partial y} = \frac{1}{y} \text{ and } \frac{\partial N}{\partial x} = \frac{1}{y} \Rightarrow \frac{\partial M}{\partial y} = \frac{\partial N}{\partial x}$$

Hence the given differential equation is exact.

$$\Rightarrow \text{ Solution is } \int_{y \text{ constant}} M\,dx + \int N(\text{terms not containing } x)\,dy = C$$

$$\Rightarrow \quad \int (1 + \log x + \log y)\,dx + \int dy = C$$

$$\Rightarrow \quad x + \int \log x\,dx + \int \log y\,dx + y = C \qquad \text{... (1)}$$

Now, $\int \log x\,dx = \int \log x \cdot (1)\,dx = (\log x)x - \int \left[\frac{d}{dx}(\log x)x\right] dx = x\log x - \int \frac{1}{x} \cdot x\,dx$

$$= x\log x - \int dx = x\log x - x = x[\log x - 1]$$

$\therefore$ Equation (1) becomes $\Rightarrow x + x\log x - x + x\log y + y = C$

$x\,[\log x + \log y] + y = C \Rightarrow x\log xy + y = C$ **Ans.**

EXERCISE 33.10

Solve the following differential equations:

1. $(x + y - 10)\,dx + (x - y - 2)\,dy = 0$ **Ans.** $\frac{x^2}{2} + xy - 10x - \frac{y^2}{2} - 2y = C$

2. $(y^2 - x^2)\,dx + 2xy\,dy = 0$ **Ans.** $\frac{x^3}{3} = xy^2 + C$

3. $\left(1 + 3e^{x/y}\right)dx + 3e^{x/y}\left(1 - \frac{x}{y}\right)dy = 0$ **[RGPV, Bhopal, Winter 2010]**
Ans. $x + 3y\,e^{x/y} = C$

4. $(2x - y)\,dx = (x - y)\,dy$ **Ans.** $xy = x^2 + \frac{y^2}{2} + C$

5. $(y\sec^2 x + \sec x\tan x)\,dx + (\tan x + 2y)\,dy = 0$ **Ans.** $y\tan x + \sec x + y^2 = C$

6. $(ax + hy + g)\,dx + (hx + by + f)\,dy = 0$ **Ans.** $ax^2 + 2hxy + by^2 + 2gx + 2fy + C = 0$

7. $(x^4 - 2xy^2 + y^4)\,dx - (2x^2y - 4xy^3 + \sin y)\,dy = 0$ **Ans.** $\frac{x^5}{5} - x^2y^2 + xy^4 + \cos y = C$

8. $(2xy + e^y)\,dx + (x^2 + xe^y)\,dy = 0$ **Ans.** $x^2y + xe^y = C$

9. $(x^2 + 2ye^{2x})\,dy + (2xy + 2y^2e^{2x})\,dx = 0$ **Ans.** $x^2y + y^2e^{2x} = C$

10. $\left[y\left(1 + \frac{1}{x}\right) + \cos y\right]dx + (x + \log x - x\sin y)\,dy = 0$ **[MDU 2010]**

Ans. $y\,(x + \log x) + x\cos y = C$

11. $(x^3 - 3xy^2)\,dx + (y^3 - 3x^2y)\,dy = 0,\ y(0) = 1$ **Ans.** $x^4 - 6x^2y^2 + 4y^4 = 4C$

12. Differential equation $M(x, y)\,dx + N(x, y)\,dy = 0$ is an exact differential equation if

(*a*) $\frac{\partial M}{\partial y} + \frac{\partial N}{\partial x} = 0$ (*b*) $\frac{\partial M}{\partial y} = \frac{\partial N}{\partial x}$ (*c*) $\frac{\partial M}{\partial y} \times \frac{\partial N}{\partial x} = 1$ (*d*) None of the above

Ans. (*b*)

33.14 EQUATIONS REDUCIBLE TO THE EXACT EQUATIONS

Sometimes a differential equation which is not exact may become so, on multiplication by a suitable function known as the integrating factor.

Rule I: If $\dfrac{\dfrac{\partial M}{\partial y}-\dfrac{\partial N}{\partial x}}{N}$ is a function of x alone, say $f(x)$, then $I.F. = e^{\int f(x)dx}$

[AMIETE June 2011]

Example 61. *Solve:* $(2x \log x - xy)\, dy + 2y\, dx = 0$ ... (1)

Solution. $M = 2y, \quad N = 2x \log x - xy$

$$\Rightarrow \quad \frac{\partial M}{\partial y} = 2, \quad \frac{\partial N}{\partial x} = 2(1+\log x) - y$$

Here,
$$\frac{\dfrac{\partial M}{\partial y}-\dfrac{\partial N}{\partial x}}{N} = \frac{2-2-2\log x + y}{2x \log x - xy} = \frac{-(2\log x - y)}{x(2\log x - y)} = -\frac{1}{x} = f(x)$$

$$\Rightarrow \quad I.F. = e^{\int f(x)dx} = e^{\int -\frac{1}{x}dx} = e^{-\log x} = e^{\log x^{-1}} = x^{-1} = \frac{1}{x}$$

On multiplying the given differential equation (1) by $\frac{1}{x}$, we get

$$\Rightarrow \quad \frac{2y}{x}dx + (2\log x - y)dy = 0 \;\Rightarrow\; \int \frac{2y}{x}dx + \int -y\, dy = c$$

$$\therefore \quad 2y\log x - \frac{1}{2}y^2 = c$$ **Ans.**

EXERCISE 33.11

Solve the following differential equations:

1. $(y \log y)\, dx + (x - \log y)\, dy = 0$ **Ans.** $2x \log y = c + (\log y)^2$

2. $\left(y + \frac{1}{3}y^3 + \frac{1}{2}x^2\right)dx + \frac{1}{4}(1+y^2)x\,dy = 0$ **Ans.** $\frac{yx^4}{4} + \frac{y^3x^4}{12} + \frac{x^6}{12} = c$

3. $(y - 2x^3)\, dx - x\,(1 - xy)\, dy = 0$ **Ans.** $-\frac{y}{x} - x^2 + \frac{y^2}{2} = c$

4. $(x \sec^2 y - x^2 \cos y)\, dy = (\tan y - 3x^4)\, dx$ **Ans.** $-\frac{1}{x}\tan y - x^3 + \sin y = c$

5. $(x - y^2)\, dx + 2xy\, dy = 0$ **Ans.** $y^2 = cx - x \log x$

Rule II: If $\dfrac{\dfrac{\partial N}{\partial x}-\dfrac{\partial M}{\partial y}}{M}$ is a function of y alone, say $f(y)$, then

$$I.F. = e^{\int f(y)dy}$$

Example 62. *Solve:* $(y^4 + 2y)\, dx + (xy^3 + 2y^4 - 4x)\, dy = 0$

Solution. Here $M = y^4 + 2y; \quad N = xy^3 + 2y^4 - 4x$ ... (1)

$$\Rightarrow \qquad \frac{\partial M}{\partial y} = 4y^3 + 2; \quad \frac{\partial N}{\partial x} = y^3 - 4$$

$$\Rightarrow \qquad \frac{\frac{\partial N}{\partial x} - \frac{\partial M}{\partial y}}{M} = \frac{(y^3 - 4) - (4y^3 + 2)}{y^4 + 2y} = \frac{-3(y^3 + 2)}{y(y^3 + 2)} = -\frac{3}{y} = f(y)$$

$$\Rightarrow \qquad I.F. = e^{\int f(y)dy} = e^{\int -\frac{3}{y}dy} = e^{-3\log y} = e^{\log y^{-3}} = y^{-3} = \frac{1}{y^3}$$

On multiplying the given equation (1) by $\frac{1}{y^3}$ we get the exact differential equation.

$$\Rightarrow \left(y + \frac{2}{y^2}\right)dx + \left(x + 2y - \frac{4x}{y^3}\right)dy = 0$$

$$\therefore \int\left(y + \frac{2}{y^2}\right)dx + \int 2y\,dy = c \;\Rightarrow\; x\left(y + \frac{2}{y^2}\right) + y^2 = c$$ **Ans.**

EXERCISE 33.12

Solve the following differential equations:

1. $(3x^2y^4 + 2xy)\,dx + (2x^3y^3 - x^2)\,dy = 0$ **Ans.** $x^3y^2 + \frac{x^2}{y} = C$

2. $(xy^3 + y)\,dx + 2(x^2y^2 + x + y^4)\,dy = 0$ **Ans.** $\frac{x^2y^4}{2} + xy^2 + \frac{y^6}{3} = C$

3. $y(x^2y + e^x)\,dx - e^x dy = 0$ **Ans.** $\frac{x^3}{3} + \frac{e^x}{y} = C$

4. $(2x^4y^4e^y + 2xy^3 + y)\,dx + (x^2y^4e^y - x^2y^2 - 3x)\,dy = 0$ **Ans.** $x^2e^y + \frac{x^2}{y} + \frac{x}{y^3} = C$

Rule III: If M is of the form $M = yf_1(xy)$ and N is of the form $N = xf_2(xy)$ then

$$\text{I.F.} = \frac{1}{Mx + Ny}$$

Example 63. *Solve:* $y(xy + 2x^2y^2)dx + x(xy - x^2y^2)dy = 0$

Solution. $y(xy + 2x^2y^2)dx + x(xy - x^2y^2)dy = 0$ ... (1)

Dividing (1) by xy, we get

$$\Rightarrow \qquad y\,(1 + 2xy)\,dx + x\,(1 - xy)\,dy = 0 \qquad \text{... (2)}$$

$$\Rightarrow \qquad M = y\,f_1\,(xy), \quad N = x\,f_2\,(xy)$$

$$\Rightarrow \qquad I.F. = \frac{1}{Mx - Ny} = \frac{1}{xy(1 + 2xy) - xy(1 - xy)} = \frac{1}{3x^2y^2}$$

On multiplying (2) by $\frac{1}{3x^2y^2}$, we have an exact differential equation

$$\Rightarrow \left(\frac{1}{3x^2y}+\frac{2}{3x}\right)dx+\left(\frac{1}{3xy^2}-\frac{1}{3y}\right)dy=0 \Rightarrow \int\left(\frac{1}{3x^2y}-\frac{2}{3x}\right)dx+\int-\frac{1}{3y}dy=c$$

$$\Rightarrow \quad -\frac{1}{3xy}+\frac{2}{3}\log x-\frac{1}{3}\log y=c \Rightarrow -\frac{1}{xy}+2\log x-\log y=b$$ **Ans.**

EXERCISE 33.13

Solve the following differential equations:

1. $(y - xy^2)\,dx - (x + x^2y)\,dy = 0$ **Ans.** $\log\left(\frac{x}{y}\right) - xy = A$

2. $y\,(1 + xy)\,dx + x\,(1 - xy)\,dy = 0$ **Ans.** $xy\log\left(\frac{y}{x}\right) = cxy - 1$

3. $y\,(1 + xy)\,dx + x\,(1 + xy + x^2y^2)\,dy = 0$ **Ans.** $\frac{1}{2x^2y^2}+\frac{1}{xy}-\log y = c$

4. $(xy\sin xy + \cos xy)\,y\,dx + (xy\sin xy - \cos xy)\,x\,dy = 0$ **Ans.** $y\cos xy = cx$

Rule IV: For of this type of $x^m y^n(ay\,dx + bx\,dy) + x^{m'}y^{m'}(a'\,y\,dx + b'\,x\,dy) = 0,$ the integrating factor is $x^h \cdot y^k$.

where $\frac{m+h+1}{a}=\frac{n+k+1}{b}$ and $\frac{m'+h+1}{a'}=\frac{n'+k+1}{b'}$

Example 64. *Solve:* $(y^3 - 2x^2y)\,dx + (2xy^2 - x^3)\,dy = 0$

Solution. $(y^3 - 2x^2y)\,dx + (2xy^2 - x^3)\,dy = 0$

$y^2\,(y\,dx + 2x\,dy) + x^2\,(-2y\,dx - x\,dy) = 0$

Here $m = 0, n = 2, a = 1, b = 2, \quad m' = 2, n' = 0, a' = -2, b' = -1$

$$\Rightarrow \quad \frac{0+h+1}{1}=\frac{2+k+1}{2} \text{ and } \frac{2+h+1}{-2}=\frac{0+k+1}{-1}$$

$$\Rightarrow \quad 2h + 2 = 2 + k + 1 \text{ and } h + 3 = 2k + 2$$

$$\Rightarrow \quad 2h - k = 1 \text{ and } h - 2k = -1$$

On solving $h = k = 1$. Integrating Factor $= xy$

Multiplying the given equation by xy, we get

$$(xy^4 - 2x^3y^2)\,dx + (2x^2y^3 - x^4y)\,dy = 0$$

which is an exact differential equation.

$$\Rightarrow \quad \int(xy^4-2x^3y^2)\,dx = C \Rightarrow \frac{x^2y^4}{2}-\frac{2x^4y^2}{4}=C$$

$$\Rightarrow \quad x^2y^4 - x^4y^2 = C' \Rightarrow x^2y^2\,(y^2 - x^2) = C'$$ **Ans.**

Example 65. *Solve:* $(3y-2xy^3)\,dx+(4x-3x^2y^2)\,dy=0$. **[U.P. II Semester, June 2007]**

Solution. $(3y-2xy^3)\,dx+(4x-3x^2y^2)\,dy=0$

$$\Rightarrow \quad (3y\,dx+4x\,dy)+xy^2(-2y\,dx-3x\,dy)=0 \qquad \text{... (1)}$$

Comparing the coefficients of (1) with

$$x^m y^n (a\,y\,dx + b\,x\,dy) + x^{m'} y^{n'} (a'\,y\,dx + b'\,x\,dy) = 0, \text{ we get}$$

$$m = 0,\ n = 0,\ a = 3,\ b = 4$$

$$m' = 1,\ n' = 2,\ a' = -2,\ b' = -3$$

To find the integrating factor $x^h\, y^k$

$$\Rightarrow \qquad \frac{m+h+1}{a} = \frac{n+k+1}{b} \text{ and } \frac{m'+h+1}{a'} = \frac{n'+k+1}{b'}$$

$$\Rightarrow \qquad \frac{0+h+1}{3} = \frac{0+k+1}{4} \text{ and } \frac{1+h+1}{-2} = \frac{2+k+1}{-3}$$

$$\Rightarrow \qquad \frac{h+1}{3} = \frac{k+1}{4} \text{ and } \frac{h+2}{2} = \frac{k+3}{3} \Rightarrow 4h - 3k + 1 = 0 \qquad \text{... (2)}$$

and $\qquad 3h - 2k = 0 \Rightarrow h = \frac{2k}{3} \qquad \text{... (3)}$

Putting the value of h from (3) in (2), we get

$$\Rightarrow \qquad \frac{8k}{3} - 3k + 1 = 0 \Rightarrow -\frac{k}{3} + 1 = 0 \Rightarrow k = 3$$

Putting $k = 3$ in (2), we get $h = \frac{2k}{3} = \frac{2 \times 3}{3} = 2$

$$\Rightarrow \qquad I.F. = x^h\, y^k = x^2 y^3$$

On multiplying the given differential equation by x^2y^3, we get

$$\Rightarrow \quad x^2y^3(3y - 2xy^3)\,dx + x^2y^3(4x - 3x^2y^2)\,dy = 0$$

$$\Rightarrow \quad (3x^2y^4 - 2x^3y^6)\,dx + (4x^3y^3 - 3x^4y^5)\,dy = 0$$

This is the exact differential equation.

Its solution is $\int (3x^2y^4 - 2x^3y^6)\,dx = 0 \Rightarrow x^3y^4 - \frac{x^4}{2}y^6 = C$ **Ans.**

EXERCISE 33.14

Solve the following differential equations:

1. $(2y\,dx + 3x\,dy) + 2xy\,(3y\,dx + 4x\,dy) = 0$ **Ans.** $x^2y^3\,(1 + 2xy) = c$
2. $(y^2 + 2yx^2)\,dx + (2x^3 - xy)\,dy = 0$ **Ans.** $4(xy)^{1/2} - \frac{2}{3}\left(\frac{y}{x}\right)^{3/2} = c$
3. $(3x + 2y^2)\,y\,dx + 2x\,(2x + 3y^2)\,dy = 0$ **Ans.** $x^2y^4\,(x + y^2) = c$
4. $(2x^2y^2 + y)\,dx - (x^3y - 3x)\,dy = 0$ **Ans.** $\frac{7}{5}x^{10/7}y^{-5/7} - \frac{7}{4}x^{-4/7}y^{-12/7} = c$
5. $x\,(3y\,dx + 2x\,dy) + 8y^4\,(y\,dx + 3x\,dy) = 0$ **Ans.** $x^3y^2 + 4x^2y^6 = c$

Rule V: If the given equation $Mdx + Ndy = 0$ is homogeneous equation and $Mx + Ny \neq 0$, then $\frac{1}{Mx + Ny}$ is an integrating factor.

Example 66. *Solve:* $\dfrac{dy}{dx}=\dfrac{x^3+y^3}{xy^2}$

Solution. $(x^3 + y^3)\, dx - (xy^2)\, dy = 0$ ... (1)

Here $M = x^3 + y^3, \quad N = -xy^2$

$$I.F. = \frac{1}{Mx+Ny} = \frac{1}{x(x^3+y^3)-xy^2(y)} = \frac{1}{x^4}$$

Multiplying (1) by $\dfrac{1}{x^4}$ we get $\dfrac{1}{x^4}(x^3+y^3)dx + \dfrac{1}{x^4}(-xy^2)dy = 0$

$\Rightarrow \quad \left(\dfrac{1}{x}+\dfrac{y^3}{x^4}\right)dx - \dfrac{y^2}{x^3}dy = 0,$ which is an exact differential equation.

$\displaystyle\int\left(\frac{1}{x}+\frac{y^3}{x^4}\right)dx = c \;\Rightarrow\; \log x - \frac{y^3}{3x^3} - c$ **Ans.**

EXERCISE 33.15

Solve the following differential equations:

1. $x^2y\, dx - (x^3 + y^3)\, dy = 0$ **Ans.** $-\dfrac{x^3}{3y^3} + \log y = c$

2. $(y^3 - 3xy^2)\, dx + (2x^2y - xy^2)\, dy = 0$ **Ans.** $\dfrac{y}{x} + 3\log x - 2\log y = c$

3. $(x^2y - 2xy^2)\, dx - (x^3 - 3x^2y)\, dy = 0$ **Ans.** $\dfrac{x}{y} - 2\log x + 3\log y = c$

4. $(y^3 - 2yx^2)\, dx + (2xy^2 - x^3)\, dy = 0$ **Ans.** $x^2y^4 - x^4y^2 = c$

33.15 SIMPLE ELECTRIC CIRCUITS

We will consider circuits made up of

(*i*) Voltage source which may be a battery or a generator.

(*ii*) Resistance, inductance and capacitance.

(1) Table of elements, symbols and units

	Element	*Symbol*	*Unit*
1.	Charge	q	coulomb
2.	Current	i	ampere
3.	Resistance	R	ohm
4.	Inductance	L	henry
5.	Capacitance	C	farad
6.	Electromotive force or voltage (constant),	constant	volt
7.	Variable voltage	variable	volt

(2) Basic relations

(*i*) $i = \dfrac{dq}{dt}$

(*ii*) Voltage drop across resistance $R = Ri$

(*iii*) Voltage drop across inductance $L = L \cdot \dfrac{di}{dt}$

(*iv*) Voltage drop across capacitance $C = \dfrac{q}{C}$.

(3) Kirchhoff's law

I. Voltage Law: The algebraic sum of the voltage drop around a closed circuit is equal to the resultant electromotive force in the circuit.

II. Current Law: At a junction or node current coming (flowing into) is equal to current going (flowing out of).

33.16 APPLICATIONS OF DIFFERENTIAL EQUATION

The formation of differential equation for an electric circuit depends upon the above laws.

(*i*) **L – R series circuit:** Let i be the current flowing in the circuit containing resistance R and inductance L in series, with voltage source E, at any time.

By voltage law

$$\Rightarrow \quad Ri + L\frac{di}{dt} = E$$

$$\Rightarrow \quad \frac{di}{dt} + \frac{R}{L}i = \frac{E}{L} \qquad \text{... (1)}$$

This is the linear differential equation.

$$\Rightarrow \quad I.F. = e^{\int \frac{R}{L}dt} = e^{\frac{R}{L}t}$$

R
i
L
+ −
E

Its solution is, $\quad i \cdot e^{\frac{R}{L}t} = \int \dfrac{E}{L} e^{\frac{R}{L}t}\, dt + C$

$$\Rightarrow \quad i \cdot e^{\frac{R}{L}t} = \frac{E}{L}\frac{L}{R} e^{\frac{R}{L}t} + C$$

$$\Rightarrow \quad i = \frac{E}{R} + Ce^{-\frac{Rt}{L}} \qquad \text{... (2)}$$

At $t = 0$, $i = 0$, $\therefore$ $C = -\dfrac{E}{R}$

Thus (2) becomes $\quad i = \dfrac{E}{R}\left[1 - e^{-\frac{R}{L}t}\right]$

which shows that i increases with t and attains maximum value $\dfrac{E}{R}$.

(*ii*) **L – R – C series circuit.** Let i be current in the circuit containing resistance R, inductance L, and capacitance C in series with voltage source E, any time t.

By voltage law

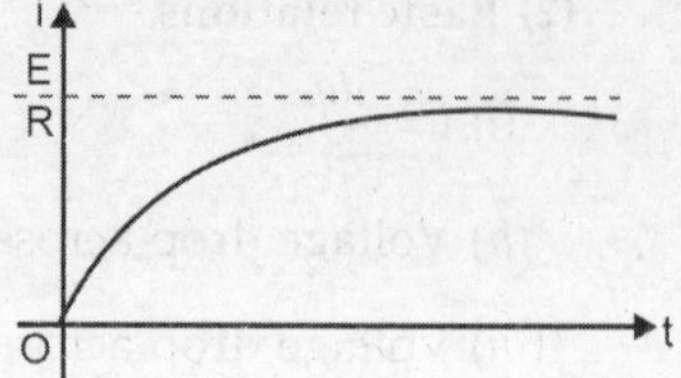

$$Ri + L\frac{di}{dt} + \frac{q}{c} = E \qquad \left(\because i = \frac{dq}{dt}\right)$$

$$\Rightarrow \qquad L\frac{d^2q}{dt^2} + R\frac{dq}{dt} + \frac{q}{c} = E$$

Example 67. *A resistance of 100 ohms, an inductance of 0.5 henry are connected in series with a battery of 20 volts. Find the current in the circuit as a function of time.*

Solution. By Kirchhoff's first law, we have

$$\Rightarrow \qquad Ri + L\frac{di}{dt} = E$$

$$\Rightarrow \qquad \frac{di}{dt} + \frac{R}{L}i = \frac{E}{L}$$

This is a linear differential equation and its solution as given section 34.10 is

$$\Rightarrow \qquad i = \frac{E}{R}\left[1 - e^{-\frac{R}{L}t}\right]$$

$$\Rightarrow \qquad R = 100 \text{ ohms}, \quad L = 0.5 \text{ henry}, \quad E = 20 \text{ volts}$$

$$\Rightarrow \qquad i = \frac{20}{100}\left[1 - e^{-\frac{100}{0.5}t}\right] = \frac{1}{5}\left[1 - e^{-200t}\right] \qquad \textbf{Ans.}$$

Example 68. *In a condenser discharging electricity the voltage V satisfies the equation $k\frac{dV}{dt} + V = 0$, where K is a constant and t is the time measured in seconds.*

Given k = 50, find the time t in which V decreases to one-tenth of its original value.

Solution. $$k\frac{dV}{dt} + V = 0 \qquad \text{[Given]}$$

$$\Rightarrow \qquad k\frac{dV}{dt} = -V$$

$$\Rightarrow \qquad \frac{dV}{V} = -\frac{1}{k}dt$$

Integrating, we get

$$\Rightarrow \qquad \int_v^{\frac{v}{10}} \frac{dV}{V} = -\frac{1}{k}\int_0^t dt$$

$$\Rightarrow \qquad [\log V]_v^{\frac{v}{10}} = -\frac{1}{k}[t]_0^t$$

$$\Rightarrow \qquad \log\left(\frac{V}{10}\right) - \log V = -\frac{t}{k}$$

$$\Rightarrow \qquad \log\left(\frac{V}{10}\right) \times \frac{1}{V} = -\frac{t}{k}$$

$$\Rightarrow \quad \log\left(\frac{1}{10}\right) = -\frac{t}{50} \qquad \text{(Given } k = 50\text{)}$$

$$\Rightarrow \quad -\log_e 10 = \frac{t}{50} \Rightarrow \frac{1}{\log_{10} e} = \frac{t}{50}$$

$$\Rightarrow \quad \frac{1}{\log_{10}(2.71828)} = \frac{t}{50} \Rightarrow \frac{1}{0.4343} = \frac{t}{50}$$

$$\Rightarrow \quad t = \frac{50}{0.4343}$$

$$= 115.13 \text{ seconds.}$$ **Ans.**

Example 69. *A condenser of capacity C farads with V_0 is discharged through a resistance R ohms. Show that if q coulomb is the charge on the condenser, i ampere the current and v the voltage at time t.* **[SBTE, 2005]**

$$q = CV,\ V = Ri \text{ and } i = \frac{dq}{dt}$$

hence show that $V = V_0\, e^{-\frac{t}{RC}}$

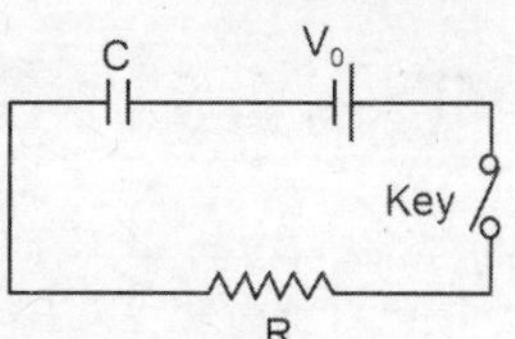

Solution. Voltage across $R = R\,i$

Voltage drop across capacitance $v = \dfrac{q}{C}$

$\therefore$ The equation of discharge of condenser can be written, when after release of key, the condenser gets discharged and at that time voltage across the battery gets zero so that

$$V_0 = 0$$

The differential equation of the above circuit is

$$\Rightarrow \quad Ri + \frac{q}{C} = 0$$

$$\Rightarrow \quad R\frac{dq}{dt} + \frac{q}{C} = 0 \qquad \left(\because \text{ as } i = \frac{dq}{dt}\right)$$

$$\Rightarrow \quad \frac{dq}{dt} + \frac{q}{RC} = 0 \Rightarrow \frac{dq}{dt} = -\frac{q}{RC}$$

$$\Rightarrow \quad \frac{dq}{q} = -\frac{1}{RC}\,dt$$

Integrating both sides, we get

$$\Rightarrow \quad \int \frac{dq}{q} = -\frac{1}{RC}\int dt$$

$$\Rightarrow \quad \log q = -\frac{1}{RC}t + a \qquad \text{... (1)}$$

But at $t = 0$, the charge at the condenser is q_0 such that

$$\Rightarrow \quad \log q_0 = -\frac{1}{RC}(0) + a$$

$\Rightarrow \quad a = \log q_0$ ... (2)

Putting the value of a from (2) in (1), we have

$$\Rightarrow \quad \log a = -\frac{1}{RC}t + \log q_0 \quad \Rightarrow \quad \log q - \log q_0 = -\frac{1}{RC}t$$

$$\Rightarrow \quad \log\left(\frac{q}{q_0}\right) = -\frac{1}{RC}t$$

$$\Rightarrow \quad \frac{q}{q_0} = e^{-\frac{t}{RC}}$$

$$\Rightarrow \quad q = q_0 e^{-\frac{t}{RC}} \quad \text{... (3)}$$

Dividing both sides of (3) by C, we get

$$\Rightarrow \quad \frac{q}{C} = \frac{q_0}{C} e^{-\frac{t}{RC}}$$

$$\therefore \quad V = V_0 e^{-\frac{t}{RC}} \quad \left[\text{as } \frac{q}{C} = V, \ \frac{q_0}{C} = V_0\right]$$ **Proved.**

Example 70. *An inductance of* 2 *henries and a resistance of* 20 *ohms are connected in series with an e.m.f. E volts. If the current is zero when t* = 0, *find the current at the end of* 0.01 sec *if (a)* E = 100 Volts *(b)* E = 100 sin 150 *t Volts.*

Solution. Differential equation of the above circuit is

Ist case

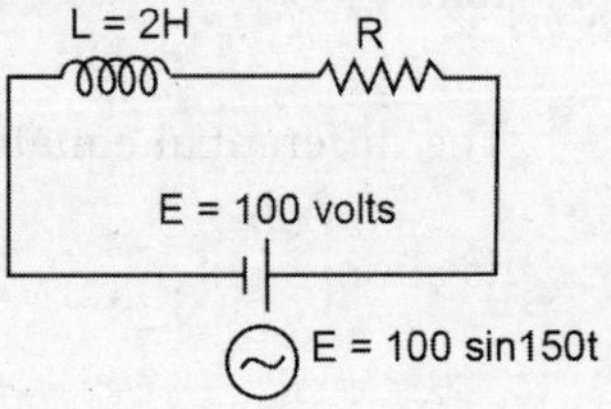

L = 2H

R

E = 100 volts

E = 100 sin150t

$$\Rightarrow \quad L\frac{dI}{dt} + RI = E$$

$$\Rightarrow \quad \frac{dI}{dt} + \frac{R}{L}I = \frac{E}{L}$$

$$\Rightarrow \quad \text{I.F.} = e^{\int \frac{R}{L}dt} = e^{\frac{Rt}{L}}$$

Its solution is $\quad I e^{\frac{Rt}{L}} = \frac{E}{L}\int e^{\frac{Rt}{L}}\, dt$

$$\Rightarrow \quad Le^{\frac{Rt}{L}} = \frac{E}{L}\frac{L}{R}e^{\frac{R}{L}t} + a$$

$$\Rightarrow \quad I e^{\frac{Rt}{L}} = \frac{E}{R}e^{\frac{R}{L}t} + a \quad \text{... (1)}$$

At $\quad t = 0, I = 0$

From (1), we have

$$\Rightarrow \quad 0 = \frac{E}{R} + a$$

$$\Rightarrow \quad a = -\frac{E}{R}$$

(1) becomes $\quad I e^{\frac{Rt}{L}} = \frac{E}{R} e^{\frac{Rt}{L}} - \frac{E}{R}$

$$\Rightarrow \quad I = \frac{E}{R} - \frac{E}{R} e^{-\frac{R}{L}t}$$

$$\Rightarrow \quad = \frac{E}{R}\left[1 - e^{-\frac{R}{L}t}\right]$$

On putting the values of E, R and L, we get

$$I = \frac{100}{20}\left[1 - e^{-\frac{20}{2}t}\right]$$

$$= 5\,[1 - e^{-10t}]$$

$$= 5\,[1 - e^{-10 \times 0.01}] \qquad \text{[at } t = 0.01 \text{ sec]}$$

$$= 5\,[1 - e^{-0.1}]$$

$$= 5\left[1 - \frac{1}{e^{0.1}}\right]$$

$$= 0.475 \text{ (Approx.)}$$ **Ans.**

2nd case

$$\Rightarrow \quad L\frac{dI}{dt} + RI = 100 \sin 150\, t$$

$$\Rightarrow \quad \frac{dI}{dt} + \frac{R}{L}I = \frac{100}{L} \sin 150\, t$$

$$\Rightarrow \quad I.F. = e^{\int \frac{R}{L} dt} = e^{\frac{R}{L}t}$$

Its solution is $\quad I e^{\frac{R}{L}t} = \frac{100}{L} \int \sin 150t \cdot e^{\frac{R}{L}t}\, dt$

$$\Rightarrow \quad I e^{\frac{R}{L}t} = \frac{100}{L}\left[\frac{e^{\frac{R}{L}t}}{(150)^2 + \frac{R^2}{L^2}}\left(\frac{R}{L}\sin 150t - 150 \cos 150t\right)\right] + a$$

$$\Rightarrow \quad I = \frac{100L}{(150L)^2 + R^2}\left(\frac{R}{L}\sin 150t - 150 \cos 150t\right) + ae^{-\frac{R}{L}t} \qquad \dots (2)$$

When $\quad I = 0$ at $t = 0$ ... (3)

Putting the values of (3) in (2), we get

$$\Rightarrow \quad 0 = a - \frac{100 \times 150L}{R^2 + (150)^2 L^2}$$

$$\Rightarrow \quad a = \frac{100 \times 150L}{R^2 + 150^2 L^2} \qquad \dots (4)$$

Putting the value of *a* from (4) in (2), we get

$$\Rightarrow \qquad I = \frac{100L}{R^2 + 150^2 L^2}\left[\frac{R}{L}\sin 150t - 150\cos 150t + 150e^{-\frac{R}{L}t}\right]$$

When $R = 20\ \Omega$, $L = 2$ henries, $t = 0.01$ sec

$$\Rightarrow \qquad I = \frac{100 \times 2}{400 + 90000}\left[\frac{20}{2}\sin(150 \times 0.01) - 150\cos(150 \times 0.01) + 150e^{-0.1}\right]$$

$$= \frac{200}{90400}[10\sin 1.5 - 150\cos 1.5 + 150\,e^{-0.1}]$$

Keeping the values of sine and cosine functions, we get

$$\Rightarrow \qquad I = \frac{1}{452}[135.3] \qquad \text{[Taking Approximate]}$$

$$= \frac{135.5}{452} = 0.299 \text{ Amp} \qquad \text{[Approximate]} \quad \textbf{Ans.}$$

Example 71. *A 20 ohms resistance is connected in series with a capacitor of 0.01 farad and e.m.f E Volts given by* $40\,e^{-3t} + 20\,e^{-6t}$. *If $q = 0$ at $t = 0$, show that the maximum charge on the capacitor is 0.25 coulomb.*

Solution. Equation of charge and discharge can be written as follows

$$\Rightarrow \qquad R\frac{dQ}{dt} + \frac{Q}{C} = 40e^{-3t} + 20\,e^{-6t}$$

$$\Rightarrow \qquad \frac{dQ}{dt} + \frac{Q}{RC} = \frac{40}{R}e^{-3t} + \frac{20}{R}e^{-6t}$$

$$\Rightarrow \qquad \frac{dQ}{dt} + \frac{Q}{RC} = \frac{40}{20}e^{-3t} + \frac{20}{20}e^{-6t} \qquad [\because R = 20\ \Omega, C = 0.01 \text{ given}]$$

$$\Rightarrow \qquad \frac{dQ}{dt} + \frac{Q}{20 + 0.01} = 2e^{-3t} + e^{-6t}$$

$$\Rightarrow \qquad \frac{dQ}{dt} + 5Q = 2e^{-3t} + e^{-6t}$$

$$I.F. = e^{5\int dt} = e^{5t}$$

Its solution is

$$Qe^{5t} = \int e^{5t}(2e^{-3t} + e^{-6t})\,dt + a = \int (2e^{2t} + e^{-t})\,dt + a$$

$$= e^{2t} - e^{-t} + a$$

$$\Rightarrow \qquad Q = e^{-3t} - e^{-6t} + ae^{-5t} \qquad \text{... (1)}$$

When $\qquad t = 0, Q = 0 \qquad$ ... (2)

Putting the values of (2) in (1), we get

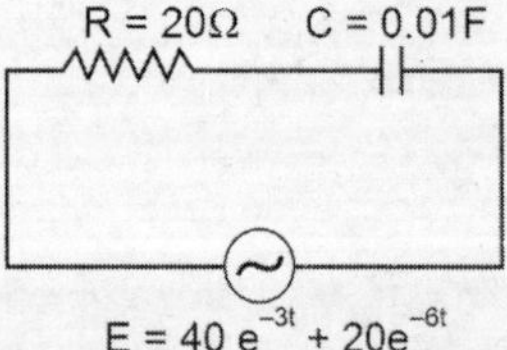

$$\Rightarrow \qquad 0 = ae^{-5\times 0} + e^{-3\times 0} - e^{-6\times 0}$$

$$0 = a$$

$$\Rightarrow \qquad Q = 0 + e^{-3t} - e^{-6t}$$

$$\Rightarrow \qquad Q = e^{-3t} - e^{-6t}$$

For maximum values, $\frac{dQ}{dt} = 0$

$$\Rightarrow \quad -3e^{-3t} - (-6)\,e^{-6t} = 0$$

$$\Rightarrow \quad 3e^{-3t} = 6e^{-6t}$$

$$\Rightarrow \quad e^{-3t} = 2e^{-6t} \Rightarrow \frac{1}{2} = e^{-3t}$$

$$\Rightarrow \quad 2 = e^{3t} \quad \Rightarrow \log 2 = 3t$$

$$\Rightarrow \quad t = \frac{1}{3}\log 2$$

$$\Rightarrow \quad t = \log (2^{1/3})$$

Maximum charge (By putting the value of t).

$$Q = e^{-3t} - e^{-6t}$$

$$= e^{-3(\log 2^{1/3})} - e^{-6(\log 2^{1/3})}$$

$$= e^{(\log 2^{1/3})^{-3}} - e^{\log (2^{1/3})^{-6}}$$

$$= e^{\log\frac{1}{2}} - e^{\log\frac{1}{4}}$$

$$= \frac{1}{2} - \frac{1}{4} = \frac{2-1}{4} = \frac{1}{4} = 0.25 \text{ amp}.$$

Example 72. *When a resistance R ohms and a capacitance C farads are connected in series an e.m.f. E volts, the current i amperes is given by*

$$R\frac{dI}{dt} + \frac{I}{C} = \frac{dE}{dt}$$

If R = 1000 ohms,*C* = 50 × 10^{-4} *farads, i* = 10 *amperes and t* = 0, *find the current for t* = 1 *and E* = 100 *sin* 120 π*t volts.*

Solution. $$R\frac{dI}{dt} + \frac{I}{C} = \frac{dE}{dt} \qquad \text{(given)}$$

$$\Rightarrow \quad R\frac{dI}{dt} + \frac{I}{C} = \frac{d}{dt}[100 \sin 120\,\pi t]$$

$$\Rightarrow \quad R\frac{dI}{dt} + \frac{I}{C} = 100 \times 120\pi \cos (120\,\pi t]$$

$$\Rightarrow \quad 1000\frac{dI}{dt} + \frac{I}{50 \times 10^{-4}} = 100 \times 120\,\pi \cos 120\,\pi t \qquad \left\{\begin{array}{l}\because R = 1000\Omega \\ C = 50 \times 10^{-4} F \text{ (given)}\end{array}\right\}$$

$$\Rightarrow \quad \frac{dI}{dt} + \frac{I}{5} = 12\,\pi \cos (120\,\pi t)$$

$$I.F. = e^{\frac{1}{5}\int dt} = e^{\frac{1}{5}t}$$

Its solution is $I e^{\frac{1}{5}t} = 12\pi \int e^{\frac{1}{5}t} \cos (120\pi t)\, dt$

$$= 12\pi \frac{e^{\frac{1}{5}t}}{\left(\frac{1}{5}\right)^2 + (120\pi)^2}\left[\frac{1}{5}\cos(120\pi t) + 120\pi \sin(120\pi t)\right] + a$$

$$I = \frac{12\pi}{\left(\frac{1}{5}\right)^2 + (120\pi)^2}\left[\frac{1}{5}\cos(120\pi t) + 120\pi \sin(120\pi t)\right] + ae^{-\frac{1}{5}t} \quad \dots (1)$$

If $\quad t = 0, \quad I = 10$ amps. $\quad \dots (2)$

Putting the values of t and I from (2) in (1), we get

$$\Rightarrow \quad 10 = \frac{\frac{12}{5}\pi}{\left(\frac{1}{5}\right)^2 + (120\pi)^2} + a \Rightarrow a = 10 - \frac{\frac{12}{5}\pi}{\left(\frac{1}{5}\right)^2 + (120\pi)^2}$$

$$\Rightarrow \quad a = 10 - \frac{60\pi}{1 + (600\pi)^2}$$

Putting the value of a in (1), we obtain

$$I = \frac{300\pi}{1 + (600\pi)^2}\left[\frac{1}{5}\cos(120\pi t) + 120\pi \sin(120\pi t)\right] + \left[10 - \frac{60\pi}{1 + (600\pi)^2}\right]e^{-1/5t} \quad \dots (3)$$

On putting $t = 1$ in (3), we have

$$\Rightarrow \quad I = \frac{300\pi}{1 + (600\pi)^2}\left[\frac{1}{5}\cos 120\pi + 120\pi \sin 120\pi\right] + \left[10 - \frac{60\pi}{1 + (600\pi)^2}\right]e^{-\frac{1}{5}}$$

$$= \frac{300\pi}{1 + (600\pi)^2}\left[\frac{1}{5}\right] + \left[10 - \frac{60\pi}{1 + (600\pi)^2}\right]e^{-1/5}$$

$$= \frac{60\pi}{1 + (600\pi)^2} + \left[10 - \frac{60\pi}{1 + (600\pi)^2}\right]e^{-1/5}$$

$= 12.21$ amp **Ans.**

Example 73. *A circuit consists of a resistance R ohms and an inductance of L henry, connected to a generator of E cos (wt + α) volts. Find the current in the circuit.*
($I = 0$, when $t = 0$).

Solution. Differential equation of the above circuit is given by

$$\Rightarrow \quad L\frac{dI}{dt} + RI = E\cos(wt + \alpha)$$

$$\Rightarrow \quad \frac{dI}{dt} + \frac{R}{L}I = \frac{E}{L}\cos(wt + \alpha)$$

$$\Rightarrow \quad I.F. = e^{\frac{R}{L}\int dt} = e^{\frac{R}{L}t}$$

L R
E cos (ωt + α)

Its solution is $$Ie^{\frac{R}{L}t} = \frac{E}{L}\int e^{\frac{R}{L}t} \cos(wt+\alpha)\,dt$$

$$= \frac{E}{L}\frac{e^{\frac{R}{L}t}}{\sqrt{\frac{R^2}{L^2}+w^2}}\cos\left(wt+\alpha-\tan^{-1}\frac{w}{\frac{R}{L}}\right)+a$$

$$\Rightarrow \qquad I = \frac{E}{\sqrt{R^2+L^2w^2}}\cos\left[wt+\alpha-\tan^{-1}\frac{Lw}{R}\right]+ae^{-\frac{R}{L}t} \qquad \text{... (1)}$$

If $$t = 0,\quad I = 0 \qquad \text{... (2)}$$

Putting the values of t and I from (2) in (1), we get

$$\Rightarrow \qquad 0 = a+\frac{E}{\sqrt{R^2+w^2L^2}}\cos\left(\alpha-\tan^{-1}\frac{wL}{R}\right)$$

$$a = -\frac{E}{\sqrt{R^2+w^2L^2}}\cos\left(\alpha-\tan^{-1}\frac{wL}{R}\right)$$

Putting the value of a in (1), the solution we get

$$\Rightarrow \qquad I = \frac{E}{\sqrt{R^2+w^2L^2}}\cos\left(wt+\alpha-\tan^{-1}\frac{wL}{R}\right)$$

$$-\frac{E}{\sqrt{R^2+w^2L^2}}e^{-\frac{R}{L}t}\cos\left(\alpha-\tan^{-1}\frac{wL}{R}\right) \qquad \textbf{Ans.}$$

Example 74. *An inductance of* 2 *henries and a resistance of* 20 *ohms are connected in series with an e.m.f.* 100 sin 150t *volts. Find the current in the circuit at any time t.*

Solution. The differential equation of the above circuit can be written as

$$\Rightarrow \qquad L\frac{dI}{dt}+RI = E$$

L = 2H R

E = 100 sin 150t

$$\Rightarrow \qquad L\frac{dI}{dt}+RI = 100\sin 150\,t$$

$$\Rightarrow \qquad \frac{dI}{dt}+\frac{R}{L}I = \frac{100}{L}\sin 150\,t$$

$$\Rightarrow \qquad \frac{dI}{dt}+10I = 50\sin 150\,t \qquad \left[\because \begin{array}{l} R = 20\,\Omega \\ L = 2\,\text{H}\end{array}\right]$$

$$I.F. = e^{10\int dt}\ a = e^{10t}$$

Its solution is $$I\cdot e^{10t} = 50\int e^{10t}\sin 150\,t\,dt$$

$$= 50\frac{e^{10t}}{(10)^2+(150)^2}[10\sin 150t-150\cos 150t]+a$$

$$\Rightarrow \qquad I = \frac{50}{22600}[10 \sin 150t - 150 \cos 150t] + ae^{-10t}$$

$$\Rightarrow \qquad I = \frac{5}{226}[\sin 150t - 15 \cos 150t] + ae^{-10t} \qquad \text{... (1)}$$

If $\qquad t = 0, \;\; I = 0 \qquad$ ... (2)

Putting the values of t and I in (1), we get

$$0 = \frac{5}{226}(-15) + a \;\Rightarrow\; a = \frac{75}{226}$$

On substituting the value of a in (1), we have

$$\Rightarrow \qquad I = \frac{5}{226}[\sin 150t - 15 \cos 150t] + \frac{75}{226}e^{-10t}$$

$$= \frac{5}{226}[\sin 150t - 15 \cos 150t + 15e^{-10t}] \qquad \textbf{Ans.}$$

Example 75. *A resistance R in series with inductance L is shunted by an equal resistance R in series with capacity C. An alternating e.m.f. E sin pt produces currents* i_1 *and* i_2 *in two branches. If* i_1 *and* i_2 *are zero when* $t = 0$*, determine* i_1 *and* i_2 *from the differential equations:*

$$L\frac{di_1}{dt} + Ri_1 = E \sin pt \quad ; \quad \frac{i_2}{C} + R\frac{di_2}{dt} = PE \cos pt$$

Solution. $\qquad L\frac{di_1}{dt} + Ri_1 = E \sin pt$

$$\Rightarrow \qquad \frac{di_1}{dt} + \frac{R}{L}i_1 = \frac{E}{L}\sin pt$$

$$\text{I.F.} = e^{\int \frac{R}{L}dt} = e^{\frac{Rt}{L}}$$

$\therefore$ Its solution is, $i_1 \cdot e^{\frac{Rt}{L}} = \int \frac{E}{L} \sin pt \, e^{\frac{Rt}{L}} dt + c_1$

$$= \frac{E}{L}\frac{\left[\frac{R}{L}e^{\frac{Rt}{L}} \sin pt - Pe^{\frac{Rt}{L}} \cos pt\right] + C_1}{p^2 + \frac{R^2}{L^2}}$$

$$\Rightarrow \qquad i_1 = \frac{E}{L}\frac{\frac{R}{L}\sin pt - P \cos pt}{p^2 + \frac{R^2}{L^2}} + C_1 e^{-\frac{Rt}{L}}$$

When $t = 0, \;\; i_1 = 0, \;\; C_1 = \frac{EP}{L} \div \left[P^2 + \frac{R^2}{L^2}\right]$

$$\Rightarrow \quad i_1 = \left[\frac{ER}{L^2}\sin pt - \frac{EP}{L}\cos pt + \frac{EP}{L}e^{-\frac{Rt}{L}}\right] \div \left[P + \frac{R^2}{L^2}\right]$$

Now $\quad R\frac{di_2}{dt} + \frac{i_2}{C} = PE\cos pt \quad \Rightarrow \quad \frac{di_2}{dt} + \frac{1}{RC}i_2 = \frac{PE}{R}\cos pt$

$$\Rightarrow \quad I.F. = e^{\int \frac{1}{RC}dt} = e^{\frac{t}{RC}}$$

Its solution is, $\quad i_2 \cdot e^{\frac{t}{RC}} = \int e^{\frac{t}{RC}} \frac{PE}{R}\cos pt\, dt + C_2$

$$= \frac{PE}{R}\left[\frac{1}{RC}e^{-\frac{t}{RC}}\cos pt + pe^{-\frac{t}{RC}}\sin pt\right] \div \left(P^2 + \frac{1}{R^2C^2}\right) + C_2$$

when $t = 0, \quad i_2 = 0 \quad \therefore \quad C_2 = -\left[\frac{PE}{RC^2} \div \left(P^2 + \frac{1}{R^2C^2}\right)\right]$

$$\Rightarrow \quad i_2 = \left[\frac{RC}{R^2C}\cos pt + \frac{P^2E}{R}\sin pt \cdot \frac{PE}{R^2C}e^{-\frac{t}{RC}}\right] \div \left[P^2 + \frac{1}{R^2C^2}\right] \quad \textbf{Ans.}$$

EXERCISE 33.16

1. A coil having a resistance of 15 ohms and an inductance of 10 henries is connected to 90 volts supply. Determine the value of current after 2 seconds.

   Ans. 5.95 amp.

2. A reistance of 70 ohms and an inductance of 0.80 henry are connected in series with a battery of 10 volts. Determine the expression for current as a function of time after $t = 0$.

   Ans. $i = \frac{1}{7}\left(1 - e^{-\frac{175}{2}t}\right)$

3. A circuit consits of resistance R ohms and a condenser of C farads connected to a constant e.m.f. E, if $\frac{q}{C}$ is the voltage of the condenser at time t after closing the circuit. Show that $\frac{q}{C} = E - Ri$ and hence show that the voltage at time t is $E\left(1 - e^{\frac{-t}{CR}}\right)$.

4. Show that the current $t = \frac{Q}{CR}e^{-\frac{t}{RC}}$ during the discharge of a condenser of charge Q coulomb through a resistance R ohms.

5. Solve $L\frac{di}{dt} + Ri = E\cos wt$. **Ans.** $i = \frac{E}{L^2w^2 + R^2}(R\cos wt + Lw\sin wt)\, Re^{-\frac{Rt}{L}}$

6. Show that the differential equation for the current i in an electric circuit containing inductance L and resistance R in series and acted on by an electromotive force E sin wt satisfies the equation $L\frac{di}{dt} + Ri = E\sin wt$.

   Find the value of current at any time t, if initially there is no current in the circuit.

$$\textbf{Ans. } i = \frac{EL}{R^2 + L^2w^2}\left[\frac{R}{L}\sin wt - w\cos wt + we^{-\frac{Rt}{L}}\right]$$

7. Find the current $I(t)$ in the RL – circuit under the following assumptions.

   $R = 1$ ohm, $L = 10$ henries, $E = 6$ volts,

   when $0 < t < 10$ sec, $E = 0$ when $t > 10$ sec. and $I(0) = 6$ amperes.

8. A coil of inductance of 1 henry and resistance 10 ohms is connected in series with an *e.m.f.* E_0 sin 10 t volts, where t (sec.) is time. When $t = 0$, the current I (amp) is zero. If $I = 5$ amp. When t = 0.1 sec., what must be the value of E_0?

Ans. 149.45 volts.

33.17 RATE OF COOLOING

Newton's law of cooling: The Newton's law of cooling states that the temperature of a body changes at a rate which is proportional to the difference in temperature between that of the surrounding medium and that of the body itself.

Example 76. *Water at temperature 100 °C cools in 10 minutes to 88°C in a room temperature of 25°C. Find the temperature of water after 20 minutes.*

Solution. Here we apply "Newton's law of cooling" which states that the rate of decrease of the temperature of a body is proportional to the difference between the temperature of the body and that of the medium;

i.e., $$\frac{dT}{dt} = -k(T - T_0)$$

where T is the temperature of the body at time t and T_0 the constant temperature of the medium.

Thus, $$\frac{dT}{dt} = -k(T - 25) \quad \text{or} \quad \frac{dT}{T-25} = -k\,dt$$

On integrating, $\log(T - 25) = -kt + c_1$ ...(1)

Substituting $T = 100$, $t = 0$ in (1), we get $\log 75 = c_1$

$\therefore$ (1) becomes, $\log(T - 25) = -kt + \log 75$ $\quad \log\left(\frac{T-25}{75}\right) = -kt$ ...(2)

Also $T = 88$ when $t = 10$ minutes,

$$\therefore \quad \log\left(\frac{88-25}{75}\right) = -10k, \text{ giving } k = \frac{1}{10}\log\left(\frac{75}{63}\right) = \frac{1}{10}\log\left(\frac{25}{21}\right)$$

Substituting in (2), we have $\log\left(\frac{T-25}{75}\right) = \left(-\frac{1}{10}\log\frac{25}{21}\right)t$ ...(3)

Putting $t = 20$ in (3), we have $\log\left(\frac{T-25}{75}\right) = -2\log\left(\frac{25}{21}\right) \quad \therefore \quad \frac{T-25}{75} = \left(\frac{25}{21}\right)^{-2}$

$$\therefore \qquad T = 25 + 75\left(\frac{25}{21}\right)^2 = 77.9°C \qquad \textbf{Ans.}$$

Example 77. *The rate at which a body cools is proportional to the difference between the temperature of the body and that of the surrounding air. If a body in air at 25°C will cool from 100°C to 75°C in one minute, find its temperature at the end of three minutes.*

Solution. Let temperature of the body be T°C.

$$\frac{dT}{dt} = k(T-25) \quad \text{or} \quad \frac{dT}{T-25} = k\,dt$$

$$\log(T-25) = kt + \log A \quad \text{or} \quad \log\left(\frac{T-25}{A}\right) = kt \quad \text{or} \quad T - 25 = Ae^{kt} \quad \ldots.(1)$$

When $\qquad t = 0$, then $T = 100$, from (1), $A = 75$

When $\qquad t = 1$, then $T = 75$ and $A = 75$, From (1),

$$75 - 25 = 75\,e^k \Rightarrow \frac{2}{3} = e^k$$

$\therefore$ (1) becomes $\qquad T = 25 + 75\,e^{kt}$

When $\qquad t = 3$, then $T = 25 + 75\,e^{3k} = 25 + 75 \times 8/27 = 47.22$ **Ans.**

Example 78. *The rate at which the ice melts is proportional to the amount of ice at the instant. Find the amount of ice left after 2 hours if half the quantity melts in 30 minutes.*

Solution. Let m be the amount of ice at any time t.

$$\therefore \qquad \frac{dm}{dt} = km \quad \text{or} \quad \frac{dm}{m} = k\,dt.$$

$$\int \frac{dm}{m} = k\int dt \quad \text{or} \quad \log m = kt + C \qquad \ldots(1)$$

Putting $\qquad t = 0$, $m = M$ in (1), we get

$$\log M = 0 + C \quad \text{or} \quad C = \log M$$

(1) becomes, on putting the value of C

$$\log m = kt + \log M \qquad \ldots(2)$$

On putting $\qquad m = M/2$ when $t = 1/2$ hour in (2), we get

$$\log\frac{M}{2} = \frac{k}{2} + \log M \Rightarrow \log\frac{M}{2M} = \frac{k}{2}$$

$$\Rightarrow \qquad \log\frac{1}{2} = \frac{k}{2} \quad \text{or} \quad k = 2\log\frac{1}{2}$$

On putting the value of k in (2), we have

$$\log m = \left(2\log\frac{1}{2}\right)t + \log M \qquad \ldots(3)$$

On putting $t = 2$ hours in (3), we have

$$\log m = 4\log \frac{1}{2} + \log M$$

$$\Rightarrow \quad \log \frac{m}{M} = \log \left(\frac{1}{2}\right)^4 \quad \Rightarrow \quad \frac{m}{M} = \frac{1}{16} \quad \Rightarrow \quad m = \frac{M}{16}$$

After 2 hours, amount of ice left = $\frac{1}{16}$ of the amount of ice at the beginning. **Ans.**

Example 79. *Water at temperature 100 °C cools in 10 minutes to 80°C in a room temperature of 25°C. Find:*

(i) the temperature of water after 20 minutes

(ii) the time when the temperature is 40°C. $\left[\log_e \frac{11}{15} = 0.3101, e^{-6.2} = .5379\right]$

Solution. Here we apply "Newton's law of cooling" which sates that the rate of decrease of the temperature of a body is proportional to the difference between the temperature of the body and that of the medium.

i.e., $$\frac{dT}{dt} = k(T - T_0)$$

where T is the temperature of the body at time t and T_0 the constant temperature of the medium.

Thus, $$\frac{dT}{dt} = k(T - 25) \Rightarrow \frac{dT}{T-25} = k\,dt$$

Integrating, we get $\log(T - 25) = kt + \log c$

$$T = 25 + ce^{kt} \quad \ldots(1)$$

Putting $t = 0$ and $T = 100$ in (1), we get

$$100 = 25 + ce^0 \Rightarrow c = 75$$

Substituting $c = 75$, in (1), we get

$$T = 25 + 75\,e^{kt} \quad \ldots(2)$$

Again putting $T = 80$, and $t = 10$ minutes in (2), we get

$$80 = 25 + 75\,e^{k(10)}$$

$$\Rightarrow \quad 55 = 75\,e^{10k}$$

$$\Rightarrow \quad \frac{55}{75} = e^{10k} \Rightarrow 10k = \log \frac{11}{15}$$

$$\Rightarrow \quad k = \frac{1}{10} \times \log\left(\frac{11}{15}\right) = \frac{1}{10} \times .3101 \qquad \left[\because \log\left(\frac{11}{15}\right) = .3101\right]$$

$$= 0.03101$$

(*i*) At $t = 20$ minutes

$$T = 25 + 75\,e^{(0.03101)20} \qquad \text{[From(2)]}$$

$$= 25 + 75\, e^{0.62}$$

$$= 25 + 75 \times 0.5379 \qquad [\because e^{0.62} = 0.5379]$$

$$= 25 + 40.3425 = 65.3425$$

Thus, the temperature of water after 20 minutes is 65.3425°C **Ans.**

(*ii*) Let the time t_1 has elasped to bring the temperature to 40°C.

$$\Rightarrow \quad 40 = 25 + 75e^{(0.03101)t_1} \Rightarrow 15 = 75\, e^{(0.03101)t_1} \Rightarrow 1 = 5\, e^{(0.03101)t_1}$$

$$\Rightarrow \quad (0.03101)t_1 = \log\frac{1}{5} = -\log 5 = -0.6990 \Rightarrow t_1 = -\frac{0.6990}{0.03101} = -22(\text{approx.})$$

$$= 22 \text{ (numerically)}$$

Thus, the time taken to have water at 40°C = (10 + 20 + 22) minutes. = 52 minutes. **Ans.**

Example 80. *A thermometer reading 80°F is taken outside. Five minutes later the thermometer reads 60°F. After another 5 minutes the thermometer reads 50°F, what is the temperature outside?*

Solution. Let at any time *t* the thermometer reading be T°F and the outside temperature be S°F. Then, by Newton's law of cooling:

$$\frac{dT}{dt} \propto (S - T) \quad \Rightarrow \quad \frac{dT}{dt} = -k(T - S) \quad \Rightarrow \quad \frac{dT}{T-S} = -k\, dt$$

Integrating both sides, we get

$$\log(T - S) = -kt + C \qquad \ldots(1)$$

It is given that $T = 80°\text{F}$ at $t = 0$

$$\therefore \quad \log(80 - S) = 0 + C \quad \Rightarrow \quad C = \log(80 - S)$$

Putting the value of C in (1), we get

$$\log(T - S) = -kt + \log(80 - S)$$

$$\Rightarrow \quad \log\left(\frac{T-S}{80-S}\right) = -kt \qquad \ldots(2)$$

It is given that $T = 60°\text{F}$ at $t = 5$ and $T = 50°\text{F}$ at $t = 10$

Substituting these values in (2), we get

$$\log\left(\frac{60-S}{80-S}\right) = -5k \quad \Rightarrow \quad 2\log\left(\frac{60-S}{80-S}\right) = -10k \qquad \ldots(3)$$

and

$$\log\left(\frac{50-S}{80-S}\right) = -10k \qquad \ldots(4)$$

From (3) and (4), we have

$$2\log\left(\frac{60-S}{80-S}\right) = \log\left(\frac{50-S}{80-S}\right)$$

$$\Rightarrow \quad \log\left(\frac{60-S}{80-S}\right)^2 = \log\left(\frac{50-S}{80-S}\right) \quad \Rightarrow \quad \left(\frac{60-S}{80-S}\right)^2 = \left(\frac{50-S}{80-S}\right)$$

$$\Rightarrow \qquad (60 - S)^2 = (50 - S)(80 - S)$$

$$\Rightarrow \qquad 3600 - 120S + S^2 = 4000 - 130S + S^2 \quad \Rightarrow \quad 10S = 400$$

$$\Rightarrow \qquad S = 40°\text{F}$$

Therefore, the outside temperature is 40°F **Ans.**

Example 81. *A wet porous substance in the open air loses its moisture at a rate proportional to the moisture content. If a sheet hung in the wind loses half of its moisture during the first hour, when will it have lost 95% moisture, weather conditions remaining the Same.*

Solution. By hypothesis,

$$\frac{dQ}{dt} \propto Q \Rightarrow \frac{dQ}{dt} = -kQ \ (k > 0)$$

where, $Q = Q(t)$ is the quantity of moisture present in the substance at any instant t (in hours)

$$\Rightarrow \qquad \frac{dQ}{dt} = -k\,dt$$

Integrating both sides, we get

$$\Rightarrow \qquad \log Q = -kt + \log C \quad \Rightarrow \quad \log \frac{Q}{C} = -kt \quad \Rightarrow \quad Q = Ce^{-kt} \qquad \ldots(1)$$

Let $\qquad Q = Q_0 \quad \text{when} \quad t = 0$

Putting these values in (1), we get

$$\therefore \qquad Q_0 = Ce^0 = C \quad \Rightarrow \quad Q = Q_0 e^{-kt} \qquad \ldots(2)$$

Again when $\qquad t = 1, Q = \dfrac{Q_0}{2}$

Putting these values in (2), we get

$$\Rightarrow \qquad \frac{Q_0}{2} = Q_0 e^{-k.1} \quad \Rightarrow \quad e^k = 2 \quad k = \log 2 \qquad \ldots(3)$$

We want to find t, when substance losses 95% moisture *i.e.*, when $\dfrac{Q}{Q_0} = \dfrac{5}{100} = \dfrac{1}{20}$

Putting $\dfrac{Q}{Q_0} = \dfrac{1}{20}$ in (2), we get

$$\Rightarrow \qquad \frac{1}{20} = e^{-kt} \quad \Rightarrow \quad e^{kt} = 20 \quad \Rightarrow \quad kt = \log 20$$

$$\Rightarrow \qquad t = \frac{\log 20}{k} = \frac{\log 20}{\log 2} \qquad [\because k = \log 2]$$

$$= 4.32 \text{ hours}$$ **Ans.**

EXERCISE 32.17

1. A body originally at 80°C cools down to 60°C in 20 minutes, the temperature of the air being 40°C. What will be the temperature of the body after 40 minutes from the original. **Ans.** 50°C

2. If the temperature of the air is 30°C and the substance cools from 100°C to 70°C in 15 minutes, find when the temperature will be 40°C. **Ans.** 52.16 minutes

3. A cup of coffee at temperature 100°C is placed in a room whose temperature is 15°C and it cools to 60°C in 5 minutes. Find its temperature after a further interval of 5 minutes. **Ans.** 38.9°C

4. If a thermometer is taken outdoors where the temperature is 0°C from a room in which the temperature is 21°C and the reading drops to 10°C in 1 minute, how long after its removal will the reading be 5°C. **Ans.** 1 minute, 56 seconds

5. The doctor took the temperature of a dead body at 11.30 pm. which was 94.6°F. He took the temperature of the body again after one hour, which was 93.4°F. If the temperature of the room was 70°F, estimate the time of death.

   Taking normal temperature of human body as 98.6°F.

   $\left[\text{Given: } \log\frac{143}{123} = 0.15066, \log\frac{123}{117} = 0.05\right]$ **Ans.** 8.30 pm. (Approx.)

6. If the air is maintained at 30°C and the temperature of the body cools from 80°C to 60°C in 12 minutes, find the temperature of the body after 24 minutes. **Ans.** 48°C

7. A wet porous substance in the open air loses its moisture at a rate proportional to the moisture content. If a sheet hung in the wind loses half its moisture during the fiat hour, when will it have lost 90% moisture, weather conditions remaining the same? **Ans.** 3.32 hours

8. The temperature T of a cooling object drops at a rate proportional to the difference $(T - S)$, where S is the constant temperature of the surrounding medium. Thus

   $$\frac{dT}{dt} = -k(T - S)$$

   where k (>0) is a constant and t is the time. Solve the differential equation if it is given that $T(0) = 150$. **Ans.** $\frac{T-S}{150-S} = e^{-kt}$

HINTS TO THE SELECTED QUESTIONS

1. Here we apply "Newton's law of cooling" which states that the rate of decrease of the temperature of a body is proportional to the difference between the temperature of the body and that of the medium;

   $$\frac{dT}{dt} = -k(T - T_0)$$

   Where T is the temperature of the body at time t and T_0 the constant temperature of the medium

   $$\frac{dT}{dt} = -k(T - 40) \text{ or } \frac{dT}{T-40} = -k\,dt$$

On integrating, $\log (T - 40) = -kt + C_1$ ...(1)

Substituting $T = 80$, $t = 0$ in (1), we get $\log 40 = C_1$

$\therefore$ (1) becomes, $\log (T - 40) = - kt + \log 40 \Rightarrow \log\left(\frac{T-40}{40}\right) = - kt$ ...(2)

Also, $T = 60, t = 20, \quad \therefore \quad \log\left(\frac{60-40}{40}\right) = -20k \Rightarrow k = \frac{1}{20}\log\frac{40}{20} = \frac{1}{20}\log 2$

Substituting in (2), we have $\log\left(\frac{T-40}{40}\right) = \frac{-t}{20}\log 2$ ...(3)

Putting $t = 40$ in (3) we have $\log\left(\frac{T-40}{40}\right) = -\frac{40}{20}\log 2$

$\Rightarrow \log\left(\frac{T-40}{40}\right) = \log 2^{-2} \quad \Rightarrow \quad \frac{T-40}{40} = \frac{1}{4} \quad \Rightarrow \quad T = 40 + 10 = 50°C$

2. Let at any time t the temperature of substance be T°C and the air temperature be T_0°C. Then, by Newton's law of cooling.

$$\frac{dT}{dt} \propto (T_0 - T) \Rightarrow \frac{dT}{dt} = - k(T - T_0) \Rightarrow \int\frac{dT}{T - T_0} = -k\int dt$$

$$\Rightarrow \quad \log(T - T_0) = -kt + c \quad ...(1)$$

Linear Differential Equations of Higher Order

34.1 INTRODUCTION

The general form of the linear differential equation of second order is

$$\frac{d^2y}{dx^2}+P\frac{dy}{dx}+Qy=R$$

where P and Q are constants and R is either a function of x or a contsant.

Differential operator. Symbol D stands for the operation of differentiation, *i.e.*

$$Dy=\frac{dy}{dx},\ D^2y=\frac{d^2y}{dx^2}$$

$\frac{1}{D}$ stands for the operation of integration and $\frac{1}{D^2}$ stands for the operation of integration twice.

$\frac{d^2y}{dx^2}+P\frac{dy}{dx}+Qy=R$ can be written in the operator form

$$D^2y+PDy+Qy=R$$

$$(D^2+PD+Q)y=R$$

34.2 COMPLETE SOLUTION = COMPLEMENTARY FUNCTION + PARTICULAR INTEGRAL

Let us consider a linear differential equation of the first order.

$$\frac{dy}{dx}+Py=Q \qquad ...(1)$$

Its solution is

$$ye^{\int Pdx}=\int Qe^{\int Pdx}dx+C$$

$$y=Ce^{-\int Pdx}+e^{-\int Pdx}\int(Qe^{\int Pdx})dx$$

$\Rightarrow$ $$y=Cu+v(\text{say}) \qquad ...(2)$$

where $$u=e^{-\int Pdx} \text{ and } v=e^{-\int Pdx}\int\left(Qe^{\int Pdx}\right)dx$$

(*a*) Now differentiating $u=e^{-\int Pdx}$ w.r.t. x.

$$\frac{du}{dx} = -Pe^{-\int Pdx} = -Pu \quad \Rightarrow \quad \frac{du}{dx} + Pu = 0$$

$$\Rightarrow \qquad \frac{d(cu)}{dx} + P(cu) = 0$$

which shows that $y = cu$ is the general solution of

$$\frac{dy}{dx} + Py = 0$$

(*b*) Differentiating $v = e^{-\int Pdx} \int \left(Qe^{\int Pdx}\right)dx$ with respect to x.

$$\frac{dv}{dx} = -Pe^{-\int Pdx} \int \left(Qe^{\int Pdx}\right)dx + e^{-\int Pdx} Qe^{\int Pdx}$$

$$\Rightarrow \qquad \frac{dv}{dx} = -Pv + Q$$

$\Rightarrow \qquad \dfrac{dy}{dx} + Py = Q$ which shows that $y = v$ is the solution of

$$\frac{dy}{dx} + Py = Q$$

Solution of the differential equation (1) is (2) consisting of two parts i.e. cu and v. cu is the solution of the differential equation whose R.H.S. is zero, cu is known as *complementary function*. Second part of (*b*) is v free from any arbitary constant and is known as *particular integral*.

Complete Solution = Complementary function + Particular Integral

$\Rightarrow \qquad$ **y = C.F. + P.I.**

34.3 METHOD FOR FINDING THE COMPLEMENTARY FUNCTION (C.F.)

In finding complementary function, R.H.S. of the given equation is replaced by zero.

Consider the differential equation

$$(D^n + a_1D^{n-1} + a_2D^{n-2} + \ldots + a_nD^0)y = 0 \qquad \ldots(1)$$

Where $a_1, a_2, \ldots a_n$ are all constants.

Let $\qquad y = C_1e^{mx}$ be its solution

$$Dy = D_1me^{mx}, \qquad D^2y = D_1m^2e^{mx}\ldots \qquad D^ny = D_1m^ne^{mx}\ldots$$

Putting these values in (1) we get

$$C_1(m^n + a_1m^{n-1} + a_2m^{n-2} + \ldots + a_n)e^{mx} = 0$$

Its auxiliary equation is $m^n + a_1m^{n-1} + a_2m^{n-2} + \ldots + a_n = 0 \qquad \ldots(2)$

This is Polynomial equation of degree n. So it has n roots i.e.

$$m_1, \quad m_2, \quad m_3, \quad \ldots \quad m_n$$

The complementary function of equation (1) depends upon the nature of the roots. The six nature of roots are as follows:

In brief Nature of roots and corresponding C.F.

Sl	*Nature of Roots of A.E.*	*Roots*	*C.F.*
1.	Real (ational) and Distinct roots	m_1, m_2, m_3	$C_1e^{m_1x} + C_2e^{m_2x} + C_3e^{m_3x}$
2.	Repeated roots	$m_1 = m_2$, $m_1 = m_2 = m_3$	$(C_1 + C_2x)\, e^{m_1x}$ $(C_1 + C_2x + C_3x^2)\, e^{m_1x}$
3.	Complex roots	$m_1 = \alpha + i\beta$ $m_2 = \alpha - i\beta$	$e^{\alpha x}[C_1 \cos \beta x + C_2 \sin \beta x]$
4.	Repeated Complex roots	$m_1 = m_2 = \alpha + i\beta$ $m_3 = m_4 = \alpha - i\beta$	$e^{\alpha x}[(C_1 + C_2x) \cos \beta x + (C_3 + C_4x) \sin \beta x]$
5.	Irrational roots	$m_1 = a + \sqrt{b}$ $m_2 = a - \sqrt{b}$	$e^{ax}\left[C_1 \cosh \sqrt{b}x + C_2 \sinh \sqrt{b}x\right]$
6.	Repeated irrational roots	$m_1 = m_2 = a + \sqrt{b}$ $m_3 = m_4 = a - \sqrt{b}$	$e^{ax}\left[(C_1 + C_2x)\cosh \sqrt{b}x + (C_3 + C_4x)\sinh \sqrt{b}x\right]$

Example 1. *Solve:* $\frac{d^2y}{dx^2} - 8\frac{dy}{dx} + 15y = 0$

Solution. Given equation can be written as:

$\Rightarrow$ $(D^2 - 8D + 15)\, y = 0$

Here auxiliary equation is $m^2 - 8m + 15 = 0$

$\Rightarrow$ $(m - 3)(m - 5) = 0$ $\Rightarrow$ $m = 3, 5$

Hence the required solution is

$y = C_1 e^{3x} + C_2e^{5x}$. **Ans.**

Example 2. *Solve:* $\frac{d^2y}{dx^2} - 6\frac{dy}{dx} + 9y = 0$ **[Diploma IETE Dec. 2005]**

Solution. Given equation can be written as:

$(D^2 - 6D + 9)\, y = 0$

A.E. is $m^2 - 6m + 9 = 0$

$\Rightarrow$ $(m - 3)^2 = 0$ $\Rightarrow$ $m = 3, 3$

Hence the required solution is

$y = (C_1 + C_2x)\, e^{3x}$ **Ans.**

Example 3. *Solve:* $\frac{d^2y}{dx^2} + 4\frac{dy}{dx} + 5y = 0$

Solution. Here the auxiliary equation is

$m^2 + 4m + 5 = 0$

Its roots are $-2 \pm i$.

The complementary function is

$$e^{-2x}(C_1 \cos x + C_2 \sin x)$$ **Ans.**

Example 4. *Solve:* $\dfrac{d^2y}{dx^2} - 3\dfrac{dy}{dx} + 2y = 0$ **[SBTE 2017, 2015]**

Subject to the conditions y (0) = 0 and y' (0) = 1.

Solution. $\dfrac{d^2y}{dx^2} - 3\dfrac{dy}{dx} + 2y = 0$

$\Rightarrow \quad D^2y - 3Dy + 2y = 0$

$\Rightarrow \quad (D^2 - 3D + 2)\, y = 0$

A.E. is $\quad m^2 - 3m + 2 = 0$

$(m - 1)(m - 2) = 0$

$m = 1, 2$

$$y = C_1e^x + C_2e^{2x} \qquad \text{...(1)}$$

Given $\quad y(0) = 0$, *i.e.*, if $x = 0$, then $y = 0$

Putting these values in (1), we have

$$0 = C_1 + C_2 \qquad \text{...(2)}$$

On differentiating (1), we get

$$\Rightarrow \quad \frac{dy}{dx} = C_1e^x + 2C_2e^{2x} \qquad \text{...(3)}$$

Given $\quad y'(0) = 1$, *i.e.* $\dfrac{dy}{dx} = 1$ if $x = 0$

Putting these values in (3), we get

$$1 = C_1 + 2C_2 \qquad \text{...(4)}$$

Solving (2) and (4), we get $C_1 = -1$, $C_2 = 1$

On putting these values of C_1 and C_2 in (1), we get

$$y = -e^x + e^{2x}$$ **Ans.**

EXERCISE 34.1

Solve the following equations:

1. $\dfrac{d^2y}{dx^2} + \dfrac{dy}{dx} - 30y = 0$ **Ans.** $y = C_1 e^{5x} + C_2 e^{-6x}$

2. $\dfrac{d^2y}{dx^2} + \dfrac{dy}{dx} + y = 0$ **Ans.** $y = e^{-x/2}\left[C_1 \cos\dfrac{\sqrt{3}}{2}x + C_2 \sin\dfrac{\sqrt{3}}{2}x\right]$

3. $\dfrac{d^3y}{dx^3} - 6\dfrac{d^2y}{dx^2} + 11\dfrac{dy}{dx} - 6y = 0$ **Ans.** $y = C_1 e^x + C_2 e^{2x} + C_3 e^{3x}$

4. $\dfrac{d^2y}{dx^2} - 2\dfrac{dy}{dx} + y = 0$ **Ans.** $y = (C_1 + C_2 x) e^x$

5. $\frac{d^2y}{dx^2} - 4\frac{dy}{dx} + 4y = 0$ **Ans.** $y = (c_1 + c_2 x)\, e^{2x}$

6. $\frac{d^2y}{dx^2} - 7\frac{dy}{dx} + 12y = 0$ **[Diploma IETE Dec. 2005]** **Ans.** $y = c_1 e^{3x} + c_2 e^{4x}$

7. $\frac{d^2y}{dx^2} - 8\frac{dy}{dx} + 16y = 0$ **Ans.** $y = (C_1 + C_2 x)\, e^{4x}$

8. $\frac{d^2y}{dx^2} + \mu^2 y = 0$ **Ans.** $y = C_1 \cos \mu x + C_2 \sin \mu x$

9. $\frac{d^3y}{dx^3} - 2\frac{d^2y}{dx^2} + 4\frac{dy}{dx} - 8y = 0$ **Ans.** $y = C_1 e^{2x} + C_2 \cos 2x + C_3 \sin 2x$

10. $\frac{d^4y}{dx^4} - 4\frac{d^3y}{dx^3} + 8\frac{d^2y}{dx^2} - 8\frac{dy}{dx} + 4y = 0$ **[Hint.** $(D^2 - 2D + 2)^2\, y = 0$**]**

Ans. $y = e^x\, [(C_1 + C_2 x) \cos x + (C_3 + C_4 x) \sin x]$

11. Solve: $\frac{d^2y}{dx^2} - 4\frac{dy}{dx} + 5y = 0$ given that $y = 1$ and $\frac{dy}{dx} = 2$ when $x = 0$

Ans. $y = e^{2x} \cos x$

12. If $\frac{d^2r}{dt^2} = w^2 r$, find the value of r in terms of t if $r = a$ and $\frac{dr}{dt} = 0$ when $t = 0$.

Ans. $r = \frac{a}{2}(e^{wt} + e^{-wt})$

13. If $\frac{d^2x}{dt^2} + \mu x = 0$, $\mu > 0$, $\frac{dx}{dt} = 0$, $x = a$, when $t = \frac{\pi}{\sqrt{\mu}}$. **Ans.** $x = -a\cos\sqrt{\mu}\, t$

14. The equation for the bending of a strut is $EI\frac{d^2y}{dx^2} + Py = 0$.

If $y = 0$ when $x = 0$ and $y = a$ when $x = \frac{l}{2}$, find y. **Ans.** $y = \dfrac{a \sin\sqrt{\frac{P}{EI}}\, x}{\sin\sqrt{\frac{P}{EI}\frac{l}{2}}}$

15. Solve: $(D^4 + 2D^2 + 1)\, y = 0$ **[Diploma IETE Dec. 2005]**

Ans. $y = (c_1 + c_2 x) \cos x + (c_3 + c_4 x) \sin x$

34.4 INVERSE OPERATOR $\frac{1}{f(D)}$

$\frac{1}{f(D)}\phi(x)$ is such a function which when operated on with $f(D)$ gives $\phi(x)$, i.e.

$$f(D)\left[\frac{1}{f(D)}\phi(x)\right] = \phi(x)$$

$\frac{1}{f(D)}\phi(x)$ satisfies the equation $f(D)\,y = \phi(x)$ and is, therefore, its particular integral.

$\therefore$ $f(D)$ and $\frac{1}{f(D)}$ are inverse operations.

$$\frac{1}{D}\phi(x) = \int \phi(x)\,dx$$

34.5 RULES TO FIND PARTICULAR INTEGRAL

$$\boxed{\frac{1}{f(D)}e^{ax} = \frac{1}{f(a)}e^{ax}}$$

$$\boxed{\text{If } f(a) = 0 \text{ then } \frac{1}{f(D)}.e^{ax} = x.\frac{1}{f'(a)}.e^{ax}}$$

$$\boxed{\text{If } f'(a) = 0 \text{ then } \frac{1}{f(D)}.e^{ax} = x^2.\frac{1}{f''(a)}.e^{ax}}$$

34.6 $\frac{1}{f(D)}e^{ax} = \frac{1}{f(a)}e^{ax}$

We know $De^{ax} = a\,e^{ax},\ D^2\,e^{ax} = a^2e^{ax},\ \ldots\ldots,\ D^n\,e^{ax} = a^n\,e^{ax}$

Let
$$\begin{aligned} f(D)\,e^{ax} &= (D^n + K_1\,D^{n-1} + \ldots\ K_n)\,e^{ax} \\ &= (a^n + K_1\,a^{n-1} + \ldots + K_n)\,e^{ax} \\ &= f(a)\,e^{ax} \end{aligned}$$

Operating both sides by $\frac{1}{f(D)}$

$$\Rightarrow \quad \frac{1}{f(D)}.f(D)e^{ax} = \frac{1}{f(D)}.f(a)e^{ax}$$

$$\Rightarrow \quad e^{ax} = f(a)\frac{1}{f(D)}.e^{ax}$$

$$\Rightarrow \quad \frac{1}{f(D)}e^{ax} = \frac{1}{f(a)}e^{ax}.$$

If $f(a) = 0$, the above rule fails.

Then $$\frac{1}{f(D)}e^{ax} = x.\frac{1}{f(D)}e^{ax} = x\frac{1}{f'(a)}e^{ax}.$$

If $$f'(a) = 0 \text{ then } \frac{1}{f(D)}e^{ax} = x^2\frac{1}{f''(a)}e^{ax}.$$

Example 5. *Solve* $\frac{d^2y}{dx^2} + 6\frac{dy}{dx} + 9y = 5e^{3x}$ **[SBTE. 2016, 2015]**

Solution. $(D^2 + 6D + 9)\,y = 5e^{3x}$

Auxiliary equation is $m^2 + 6m + 9 = 0 \quad \Rightarrow \quad (m + 3)^2 = 0$

$\Rightarrow \quad m = -3, -3, \quad \text{C.F.} = (C_1 + C_2x)e^{-3x}$

$$\text{P.I.} = \frac{1}{D^2 + 6D + 9}.5.e^{3x}$$

$$= 5\frac{e^{3x}}{(3)^2 + 6(3) + 9} = \frac{5e^{3x}}{36}$$

The complete solution is

$$y = (C_1 + C_2x)e^{-3x} + \frac{5e^{3x}}{36}$$ **Ans.**

Example 6. *Solve* $\frac{d^2y}{dx^2} - 2a\frac{dy}{dx} + a^2y = e^{ax}$

Solution. $(D^2 - 2aD + a^2)\, y = e^{ax}$

Auxiliary equation is $m^2 - 2am + a^2 = 0 \quad \Rightarrow \quad (m - a)^2 = 0$

$\Rightarrow \quad m = a, a, \quad \text{C.F} = (C_1 + C_2x)\, e^{ax}$

$$\text{P.I.} = \frac{1}{D^2 - 2aD + a^2}e^{ax} = \frac{1}{(D - a)^2}.e^{ax}$$

$$= x\frac{1}{2(D - a)}.\, e^{ax}$$ [Differentiating the denominator]

$$= \frac{x^2}{2}.\, e^{ax}$$ [Again differentiating]

The complete solution is

$$y = (C_1 + C_2x)e^{ax} + \frac{x^2}{2}e^{ax}.$$ **Ans.**

Example 7. *Solve* $(4D^2 + 4D - 3)y = e^{2x}$ **[SBTE 2015]**

Solution. Here we have

$\Rightarrow \quad (4D^2 + 4D - 3)y = e^{2x}$

A.E. is $\quad 4D^2 + 4D - 3 = 0$

$\Rightarrow \quad (2D - 1)(2D + 3) = 0$

$$D = \frac{1}{2}, -3$$

$\therefore \quad \text{C.F.} = c_1e^{\frac{1}{2}x} + c_2e^{-3x}$

$\Rightarrow \quad \text{P.I.} = \frac{1}{4D^2 + 4D - 3}e^{2x} = \frac{4}{4(2)^2 + 4(2) - 3}e^{2x} = \frac{1}{21}e^{2x}$

$$y = \text{C.S.} = \text{C.F} + \text{P.I.}$$

$\therefore \quad = C_1e^{\frac{1}{2}x} + C_2e^{-3x} + \frac{1}{21}e^{2x}$ **Ans.**

EXERCISE 34.2

Solve the following differential equations:

1. $\frac{d^2y}{dx^2}+5\frac{dy}{dx}+6y=e^x$ **Ans.** $y=C_1e^{-2x}+C_2e^{-3x}+\frac{e^x}{12}$

2. $\frac{d^2y}{dx^2}-3\frac{dy}{dx}+2y=e^{3x}$ **[SBTE, 2016]** **Ans.** $y=C_1e^x+C_2e^{2x}+\frac{e^{3x}}{2}$

3. $\frac{d^2y}{dx^2}+4y=e^{2x}$ **Ans.** $y=C_1\cos 2x+C_2\sin 2x+\frac{1}{8}e^{2x}$

4. $\frac{d^2y}{dx^2}-\frac{dy}{dx}+y=e^x+e^{-x}$ **Ans.** $y=e^{x/2}\left[c_1\cos\frac{\sqrt{3}}{2}x+c_2\sin\frac{\sqrt{3}}{2}x\right]+e^x+\frac{1}{3}e^{-x}$

5. $\frac{d^2y}{dx^2}+4\frac{dy}{dx}+5y=-2\cosh x$ **Ans.** $y=e^{-2x}(c_1\cos x+c_2\sin x)-\frac{1}{10}e^x-\frac{1}{2}e^{-x}$

6. $\frac{d^2x}{dt^2}+4\frac{dx}{dt}-5x=3e^{4t}$ **Ans.** $x=c_1e^t+c_2e^{-5t}+\frac{1}{9}e^{4t}$

7. $\frac{d^2y}{dx^2}-8\frac{dy}{dx}+16y=e^{4x}$ **Ans.** $y=(c_1+c_2x)e^{4x}+\frac{x^2}{2}e^{4x}$

8. $(D^3-2D^2-5D+6)\,y=e^{3x}$ **[SBTE 2016]** **Ans.** $y=c_1e^x+c_2e^{-2x}+c_3e^{3x}+\frac{xe^{3x}}{10}$

9. $\frac{d^2y}{dx^2}+4\frac{dy}{dx}+3y=e^{-3x}$ **Ans.** $y=c_1e^{-x}+c_2e^{-3x}-\frac{xe^{-3x}}{2}$

10. $\frac{d^3y}{dx^3}-\frac{d^2y}{dx^2}+4\frac{dy}{dx}-4y=e^x$ **Ans.** $y=c_1e^x+c_2\cos 2x+c_3\sin 2x+\frac{xe^x}{5}$

11. $\frac{d^2y}{dx^2}-6\frac{dy}{dx}+9y=e^{3x}$ **Ans.** $y=(c_1+c_2x)e^{3x}+\frac{x^2}{2}e^{3x}$

12. $\frac{d^2y}{dx^2}-4y=(1+e^x)^2$ **Ans.** $y=c_1e^{2x}+c_2e^{-2x}+\frac{x}{4}e^{2x}-\frac{2}{3}e^x-\frac{1}{4}$

13. $\frac{d^2y}{dx^2}+31\frac{dy}{dx}+240y=272e^{-x}$ **Ans.** $y=C_1e^{-15x}+C_2e^{-16x}+\frac{136}{105}e^{-x}$

14. $\frac{d^3y}{dx^3}+3\frac{d^2y}{dx^2}+3\frac{dy}{dx}+y=e^{-x}$ **Ans.** $y=(c_1+c_2x+c_3x^2)e^{-x}+\frac{x^3}{6}e^{-x}$

15. $\frac{d^2y}{dx^2}-4y=x\sinh x$ **Ans.** $y=C_1e^{2x}+C_2e^{-2x}-\frac{x}{3}\cosh x-\frac{2}{3}\sinh x$

$\left[\textbf{Hint. } \sinh x=\frac{e^x-e^{-x}}{2}\right]$

16. $\dfrac{d^2y}{dx^2}-\dfrac{dy}{dx}-6y=e^x\cosh 2x$ **Ans.** $y=c_1e^{3x}+c_2e^{-2x}+\dfrac{1}{10}xe^{3x}-\dfrac{1}{8}e^{-x}$

17. $(D-1)^2(D^2+1)^2\,y=e^x$ **Ans.** $y=(c_1+c_2x)e^x+(c_3+c_4x)(c_5\cos x+c_6\sin x)+\dfrac{x^2e^x}{8}$

Choose the correct alternative:

18. The solution of the differential equation $(D^2+4)\,y=e^x$ is **[DIP IETE Dec. 2006]**

(*a*) $c_2\cos 2x-c_2\sin 2x+\dfrac{e^x}{4}$ (*b*) $c_1\cos 2x+c_2\sin 2x+\dfrac{e^x}{4}$

(*c*) $c_1\cos 2x+c_2\sin 2x+\dfrac{e^x}{5}$ (*d*) $c_1\cos 4x-c_2\sin 4x+\dfrac{e^x}{5}$ **Ans.** (*c*)

34.7 $\dfrac{1}{f(D)}x^n=[f(D)]^{-1}x^n.$

Expand $[f(D)]^{-1}$ by the Binomial theorem in ascending powers of D as far as the result of operation on x^n is zero.

Example 8. *Solve:* $\dfrac{d^2y}{dx^2}-7\dfrac{dy}{dx}+12y=x$

Solution. $(D^2-7D+12)\,y=x$

Auxiliary equation is $m^2-7m+12=0 \quad\Rightarrow\quad (m-3)(m-4)=0$

$\Rightarrow \quad m=3,4 \qquad \therefore \quad \text{C.F}=c_1e^{3x}+c_2e^{4x}$

$$\text{P.I.}=\frac{1}{D^2-7D+12}.x=\frac{1}{12}\frac{1}{1-\frac{7}{12}D+\frac{D^2}{12}}.x$$

$$=\frac{1}{12}\left[1-\frac{7}{12}D+\frac{D^2}{12}\right]^{-1}x=\frac{1}{12}\left[1+\frac{7}{12}D-\frac{D^2}{12}+\ldots\ldots\right]x$$

$$=\frac{1}{12}\left[x+\frac{7}{12}\right]$$

The complete solution is $y=c_1e^{3x}+c_2e^{4x}+\dfrac{1}{12}\left[x+\dfrac{7}{12}\right]$ **Ans.**

Example 9. *Solve the following differential equation:*

$$\frac{d^2y}{dx^2}+5\frac{dy}{dx}+4y=3-2x$$ **[B.T.E. Delhi. June 2005]**

Solution. We have,

$$\frac{d^2y}{dx^2}+5\frac{dy}{dx}+4y=3-2x$$

$\Rightarrow \quad D^2y+5Dy+4y=3-2x$

$\Rightarrow \quad (D^2+5D+4)\,y=3-2x$

Auxiliary equation is $m^2 + 5m + 4 = 0 \quad \Rightarrow \quad (m + 1)(m + 4) = 0$

$\Rightarrow \quad m = -1, -4$

$$\text{C.F} = C_1 e^{-x} + C_2 e^{-4x}$$

$$\text{P.I.} = \frac{1}{D^2 + 5D + 4}(3 - 2x)$$

$$= \frac{1}{4}\left[\frac{1}{1 + \frac{5}{4}D + \frac{D^2}{4}}\right](3 - 2x)$$

$$= \frac{1}{4}\left[1 + \frac{5}{4}D + \frac{D^2}{4}\right]^{-1}(3 - 2x)$$

$$= \frac{1}{4}\left[1 - \frac{5D}{4} - \frac{D^2}{4} + ...\right](3 - 2x)$$

$$= \frac{1}{4}\left[(3 - 2x) - \frac{5}{4}D(3 - 2x) - \frac{1}{4}D^2(3 - 2x)....\right]$$

$$= \frac{1}{4}\left[3 - 2x - \frac{5}{4}(-2)\right]$$

$$= \frac{1}{4}\left[3 - 2x + \frac{5}{2}\right] = \frac{1}{4}\left[\frac{11}{2} - 2x\right]$$

Its complete solution is $\quad y = \text{C.F.} + \text{P.I}$

$$y = c_1 e^{-x} + c_2 e^{-4x} + \frac{1}{4}\left[\frac{11}{2} - 2x\right]$$ **Ans.**

Example 10. *Solve:* $(D^2 - 6D + 9)y = x^2 + 2e^{2x}$ **[SBTE 2015, 2006]**

Solution. Here we have,

$$(D^2 - 6D + 9)y = x^2 + 2e^{2x}$$

A.E. is $\quad D^2 - 6D + 9 = 0$

$$(D - 3)^2 = 0 \quad \Rightarrow \quad D = 3, 3$$

$\Rightarrow \quad C.F. = (C_1 + C_2 x)e^{3x}$

$\Rightarrow \quad P.I. = \frac{1}{D^2 - 6D + 9}x^2 + \frac{1}{D^2 - 6D + 9}2e^{2x}$

$$= \frac{1}{(D - 3)^2}x^2 + \frac{1}{4 - 12 + 9}2e^{2x}$$

$$= \frac{1}{9}\left[1 - \frac{D}{3}\right]^{-2} x^2 + 2e^{2x}$$

$$= \frac{1}{9}\left[1 + 2\left(\frac{D}{3}\right) + 3\left(\frac{D}{3}\right)^2 + ...\right].x^2 + 2e^{2x}$$

$$= \frac{1}{9}\left[x^2 + \frac{4x}{3} + \frac{2}{3}\right] + 2e^{2x}$$

$\therefore \qquad y = C.S. = (C_1 + C_2 x)e^{3x} + \frac{1}{9}\left[x^2 + \frac{4}{3}x + \frac{2}{9}\right] + 2e^{2x}$ **Ans.**

Example 11. *Solve:* $\frac{d^2y}{dx^2} - 4y = x^2$.

Solution. $\frac{d^2y}{dx^2} - 4y = x^2$

$\Rightarrow \qquad (D^2 - 4)\, y = x^2$

A.E. is $\qquad m^2 - 4 = 0$

$$m = \pm 2$$

$$\text{C.F.} = C_1 e^{2x} + C_2 e^{-2x}$$

$$\text{P.I.} = \frac{1}{D^2 - 4} x^2 = -\frac{1}{4}\frac{1}{1 - \frac{D^2}{4}} x^2$$

$$= -\frac{1}{4}\left(1 - \frac{D^2}{4}\right)^{-1} x^2 = -\frac{1}{4}\left[1 + \frac{D^2}{4} + \ldots\right] x^2$$

$$= -\frac{1}{4}\left[x^2 + \frac{1}{4}(2)\right] = -\frac{x^2}{4} - \frac{1}{8}$$

Complete solution is $y = C_1 e^{2x} + C_2 e^{-2x} - \frac{x^2}{4} - \frac{1}{8}$ **Ans.**

EXERCISE 34.3

Solve the following differential equations:

1. $\frac{d^2y}{dx^2} - y = 2 + 3x$ **Ans.** $y = c_1 e^x + c_2 e^{-x} - 3x - 2$

2. $\frac{d^2y}{dx^2} - 5\frac{dy}{dx} - 6y = 3 - 2x$ **Ans.** $y = c_1 e^{-x} + c_2 e^{6x} + \frac{x}{3} - \frac{7}{9}$

3. $(2D^2 + 3D + 4)\, y = x^2 - 2x$

Ans. $y = e^{-\frac{2}{4}x}\left[C_1 \cos\frac{\sqrt{23}}{4}x + C_2 \sin\frac{\sqrt{23}}{4}x\right] + \frac{1}{32}\left[8x^2 - 28x + 13\right]$

4. $\frac{d^2y}{dx^2} + 2\frac{dy}{dx} = 1 + x^2$. **Ans.** $y = c_1 + c_2 e^{-2x} + \frac{x}{12}\left(2x^2 - 3x + 9\right)$

5. $\frac{d^3y}{dx^3} - \frac{d^2y}{dx^2} - 6\frac{dy}{dx} = 1 + x^2$ **Ans.** $c_1 + c_2 e^{-2x} + c_3 e^{3x} - \frac{1}{108}\left(6x^3 - 3x^2 + 25x\right)$

6. $(D^2 - 4D + 3)\, y = x^3$ **Ans.** $y = c_1 e^x + c_2 e^{3x} + \frac{1}{27}\left(9x^3 + 36x^2 + 78x + 80\right)$

7. $\frac{d^4y}{dx^4} + 4y = x^4$ **Ans.** $y = e^x (c_1 \cos x + c_2 \sin x) + e^{-x}(c_3 \cos x + c_4 \sin x) + \frac{1}{4}(x^4 - 6)$

8. $\frac{d^3y}{dx^3}+3\frac{d^2y}{dx^2}+2\frac{dy}{dx}=x^2$ **Ans.** $y=c_1+c_2e^{-x}+c_3e^{-2x}+\frac{x^3}{6}-\frac{3x^2}{4}+\frac{7x}{4}$

9. $(D^3+2D^2+4D+8)\,y=x^2$ **Ans.** $y=c_1e^{-2x}+c_2\cos 2x+c_3\sin 2x+\frac{1}{8}(x^2-x)$

10. $\frac{d^2r}{d\theta^2}-8\frac{dr}{d\theta}+16r=2\theta^2$ **Ans.** $r=(c_1+c_2\theta)e^{4\theta}+\frac{1}{16}\left(2\theta^2+2\theta+\frac{3}{4}\right)$

11. $D^2(D^2+4)\,y=96x^2$ **Ans.** $y=c_1+c_2x+c_3\cos 2x+c_4\sin 2x+2x^2(x^2-3)$

34.8 $\frac{1}{f(D^2)}.\sin ax=\frac{\sin ax}{f(-a^2)}$, $\frac{1}{f(D^2)}.\cos ax=\frac{\cos ax}{f(-a^2)}$

$D(\sin ax)=a\,.\cos ax$, $D^2(\sin ax)=D(a\cos ax)=-a^2\,.\sin ax$

$D^4(\sin ax)=D^2\,.\,D^2(\sin ax)=D^2(-a^2\sin ax)=(-a^2)^2\sin ax$

$(D^2)^n\sin ax=(-a^2)^n\sin ax$

Hence $f(D^2)\sin ax=f(-a^2)\sin ax$

$$\Rightarrow \quad \frac{1}{f(D^2)}.f(D^2)\sin ax=\frac{1}{f(D^2)}.f(-a^2).\sin ax$$

$$\Rightarrow \quad \sin ax=f(-a^2).\frac{1}{f(D^2)}.\sin ax$$

$$\Rightarrow \quad \frac{1}{f(D^2)}.\sin ax=\frac{\sin ax}{f(-a^2)}$$

Similarly, $\frac{1}{f(D^2)}\cos ax=\frac{\cos ax}{f(-a^2)}$

If $f(-a^2)=0$ then above rule fails.

$$\Rightarrow \quad \frac{1}{f(D^2)}\sin ax=x\frac{\sin ax}{f'(-a^2)}$$

If $f'(-a^2)=0$ then, $\frac{1}{f(D^2)}\sin ax=x^2\frac{\sin ax}{f''(-a^2)}$

Example 12. *Solve:* $\frac{d^2y}{dx^2}+3\frac{dy}{dx}+2y=\sin 2x$ **[Diploma IETE Dec. 2005]**

Solution. $\frac{d^2y}{dx^2}+3\frac{dy}{dx}+2y=\sin 2x$

$$(D^2+3D+2)\,y=\sin 2x$$

A.E. is $m^2+3m+2=0$

$$m=-1,-2$$

$$\text{C.F}=C_1e^{-x}+C_2e^{-2x}$$

$$\text{P.I.} = \frac{1}{D^2+3D+2}\sin 2x$$

$$= \frac{1}{-4+3D+2}\sin 2x, \quad [D^2 = -(2)^2 = -4]$$

$$= \frac{1}{3D-2}\sin 2x$$

$$= \frac{3D+2}{9D^2-4}\sin 2x = \frac{3D+2}{9(-4)-4}\sin 2x$$

$$= -\frac{1}{40}(3D+2)\sin 2x$$

$$= -\frac{1}{40}[3D(\sin 2x)+2\sin 2x]$$

$$= -\frac{1}{40}[6\cos 2x+2\sin 2x]$$

$$= -\frac{1}{20}[3\cos 2x+\sin 2x]$$

Complete solution is $\quad y = C_1e^{-x}+C_2e^{-2x} - \frac{1}{20}(3\cos 2x+\sin 2x)$ **Ans.**

Example 13. *Solve:* $\frac{d^2y}{dx^2}+\frac{dy}{dx}+y = \cos 2x$

Solution. $\quad (D^2 + D + 1)\, y = \cos 2x$

A.E. is $\quad m^2 + m + 1 = 0$

$$m = \frac{-1\pm\sqrt{-3}}{2} = -\frac{1}{2}\pm\frac{\sqrt{3}}{2}i$$

$$\text{C.F.} = e^{-\frac{x}{2}}\left[C_1\cos\frac{\sqrt{3}}{2}x + C_2\sin\frac{\sqrt{3}}{2}x\right]$$

$$\text{P.I.} = \frac{1}{D^2+D+1}.\cos 2x$$

$$= \frac{1}{(-2^2)+D+1}.\cos 2x = \frac{1}{D-3}.\cos 2x$$

$$= \frac{D+3}{D^2-9}.\cos 2x = \frac{D+3}{(-2^2)-9}.\cos 2x$$

$$= -\frac{1}{13}(D+3)\cos 2x = -\frac{1}{13}(D\cos 2x+3\cos 2x)$$

$$= -\frac{1}{13}(-2\sin 2x+3\cos 2x)$$

Complete solution is $\quad$ = C.F. + P.I.

$$= e^{-\frac{x}{2}}\left[C_1\cos\frac{\sqrt{3}}{2}x + C_2\sin\frac{\sqrt{3}}{2}x\right] + \frac{1}{13}[2\sin 2x - 3\cos 2x]$$ **Ans.**

Example 14. *Solve:* $(D^2 + 4)\, y = \cos 2x$ **[DIPLOMA IETE 2019]**

Solution. $(D^2 + 4)\, y = \cos 2x$

Auxiliary equation is $m^2 + 4 = 0$

$\Rightarrow$ $m = \pm 2i$, C.F $= C_1 \cos 2x + C_2 \sin 2x$

$$\text{P.I.} = \frac{1}{D^2+4}\cos 2x = x.\frac{1}{2D}\cos 2x$$

$$= \frac{x}{2}\left(\frac{1}{2}\sin 2x\right) = \frac{x}{4}\sin 2x$$

Complete solution is $y = C_1 \cos 2x + C_2 \sin 2x + \frac{x}{4}\sin 2x.$ **Ans.**

Example 15. *Solve the differential equation:*

$$(D-1)^2 (D^2+1) y = \sin \frac{1}{2} x, \qquad D \equiv \frac{d}{dx}$$

Solution. $(D-1)^2 (D^2+1) y = \sin \frac{1}{2} x$

A.E. is $(m - 1)^2 (m^2 + 1) = 0$

$\Rightarrow$ Either $(m - 1)^2 = 0$ *i.e.*, $m = 1, 1,$

or $m^2 + 1 = 0$, *i.e.*, $m^2 = -1 \Rightarrow m = \pm i$

C.F $= (C_1 + C_2 x)\, e^x + C_3 \cos x + C_4 \sin x$

$$\text{P.I.} = \frac{1}{(D-1)^2 (D^2+1)} \sin \frac{1}{2} x$$

$$= \frac{1}{(D-1)^2\left(-\frac{1}{4}+1\right)} \sin \frac{1}{2} x \qquad \left(D^2 = -\frac{1}{2^2}\right)$$

$$= \frac{4}{3}\frac{1}{(D-1)^2} \sin \frac{x}{2} = \frac{4}{3}\frac{1}{D^2-2D+1}\sin\frac{x}{2}$$

$$= \frac{4}{3}\frac{1}{-\frac{1}{4}-2D+1}\sin\frac{x}{2} = \frac{4}{3}\frac{1}{-2D+\frac{3}{4}}\sin\frac{x}{2}$$

$$= \frac{4}{3}\frac{2D+\frac{3}{4}}{-4D^2+\frac{9}{16}}\sin\frac{x}{2} = \frac{4}{3}\frac{2D+\frac{3}{4}}{-4\left(-\frac{1}{4}\right)+\frac{9}{16}}\sin\frac{x}{2}$$

$$= \frac{4}{3}\cdot\frac{16}{25}\left[2D\sin\frac{x}{2}+\frac{3}{4}\sin\frac{x}{2}\right] = \frac{64}{75}\left(\cos\frac{x}{2}+\frac{3}{4}\sin\frac{x}{2}\right)$$

Complete solution is

$y = (C_1 + C_2 x)e^x + C_3 \cos x + C_4 \sin x + \frac{64}{75}\left(\cos\frac{x}{2}+\frac{3}{4}\sin\frac{x}{2}\right)$ **Ans.**

Example 16. *Solve:* $D^2y - 2Dy + 3y = x + \cos x.$ **[DIP IETE 2018]**

Solution. Here we have,

$D^2y - 2Dy + 3y = x + \cos x$

or $(D^2 - 2D + 3)y = x + \cos x$

A.E. is $D^2 - 2D + 3 = 0 \Rightarrow D = 1 \pm \sqrt{2}i$

C.F. $= e^x\left[C_1 \cos\sqrt{2}x + C_2 \sin\sqrt{2}x\right]$

$$\text{P.I.} = \frac{1}{D^2 - 2D + 3}x + \frac{1}{D^2 - 2D + 3}\cos x$$

$$= \left[\frac{1}{1 + \frac{D^2}{3} - \frac{2D}{3}}\right]x + \frac{1}{-1 - 2D + 3}\cos x$$

$$= \left[1 + \frac{D^2}{3} - \frac{2}{3}D\right]^{-1} x + \frac{1}{2 - 2D}\cos x$$

$$= \left[1 - \frac{D^2}{3} + \frac{2}{3}D - ....\right]x + \frac{2 + D}{4 - 4D^2}\cos x$$

$$= \left[x - 0 + \frac{2}{3}\right] + \frac{2 + D}{8}\cos x$$

$$= x + \frac{2}{3} + \frac{1}{8}(2\cos x - \sin x)$$

C.S. = C.F. + P.I.

$$= e^x\left[C_1 \cos\sqrt{2}x + C_2 \sin\sqrt{2}x\right] + x + \frac{2}{3} + \frac{1}{8}(2\cos x - \sin x)$$ **Ans.**

Example 17. *Solve the differential equation*

$$\frac{d^2y}{dx^2} + 4y = x^2 + \cos 2x$$ **[DIP IETE 2019]**

Solution. Here we have,

$$\Rightarrow \quad \frac{d^2y}{dx^2} + 4y = x^2 + \cos 2x$$

$$\Rightarrow \quad (D^2 + 4)y = x^2 + \cos 2x$$

A.E. is $D^2 + 4 = 0 \Rightarrow D^2 = -4 \quad \Rightarrow \quad D = \pm 2i$

C.F. $= C_1 \cos 2x + C_2 \sin 2x$

$$\text{P.I.} = \frac{1}{D^2 + 4}x^2 + \frac{1}{D^2 + 4}\cos 2x \qquad \left\{\because D^2 + 4\Big|_{at\,D^2 = -4} = 0\right\}$$

$$= \frac{1}{4\left(1 + \frac{D^2}{4}\right)}x^2 + \frac{x}{2D}\cos 2x$$

$$= \frac{1}{4}\left[1 + \frac{D^2}{4}\right]^{-1} x^2 + \frac{x}{2}\int \cos 2x$$

$$= \frac{1}{4}\left[1 - \frac{D^2}{4} - ....\right]x^2 + \frac{x}{2}\cdot\frac{1}{2}\sin 2x$$

$$= \frac{1}{4}\left[x^2 - \frac{1}{4}(2)\right] + \frac{x}{4}\sin 2x$$

$$\text{P.I.} = \frac{x^2}{4} - \frac{1}{8} + \frac{x}{4}\sin 2x$$

Complete solution is = C.F + P. I.

$$= C_1\cos 2x + C_2\sin 2x + \frac{x^2}{4} - \frac{1}{8} + \frac{x}{4}\sin 2x \qquad \textbf{Ans.}$$

Example 18. *Solve* $\frac{d^2y}{dx^2} + 2\frac{dy}{dx} + y = e^{2x} + \cos^2 x$ **[SBTE 2014, 2013, 2012]**

Solution. $\frac{d^2y}{dx^2} + 2\frac{dy}{dx} + y = e^{2x} + \cos^2 x$

or $(D^2 + 2D + 1)y = e^{2x} + \cos^2 x$

A.E. is $D^2 + 2D + 1 = 0$

$\Rightarrow$ $(D + 1)^2 = 0 \quad \Rightarrow D = -1, -1$

C.F. = $(C_1 + C_2x)e^{-x}$

$$\text{P.I.} = \frac{1}{D^2 + 2D + 1}e^{2x} + \frac{1}{D^2 + 2D + 1}\cos^2 x$$

$$= \frac{1}{(2)^2 + 2(2) + 1}e^{2x} + \frac{1}{D^2 + 2D + 1}\frac{1}{2}(1 + \cos 2x)$$

$$= \frac{1}{9}e^{2x} + \frac{1}{2}\cdot\frac{1}{D^2 + 2D + 1}e^{0x} + \frac{1}{2}\cdot\frac{1}{D^2 + 2D + 1}\cos 2x$$

$$= \frac{1}{9}e^{2x} + \frac{1}{2}\left[1 + \frac{1}{2D - 3} \times \frac{2D + 3}{2D + 3}\cos 2x\right]$$

$$= \frac{1}{9}e^{2x} + \frac{1}{2}\left[1 + \frac{2D(\cos 2x) + 3\cos 2x}{-25}\right]$$

$$= \frac{1}{9}e^{2x} + \frac{1}{2}\left[1 - \frac{1}{25}(3\cos 2x - 4\sin 2x)\right]$$

$$\therefore \quad y = (C_1 + C_2x)e^{-x} + \frac{1}{9}e^{2x} + \frac{1}{2}\left[1 - \frac{1}{25}(3\cos 2x - 4\sin 2x)\right] \qquad \textbf{Ans.}$$

Example 19. *Solve:* $\frac{d^2x}{dt^2} - 3\frac{dx}{dt} + 2x = 4t + e^{3t}$ *given* $x = 1$ *and* $\frac{dx}{dt} = -1$ *, when* $t = 0$

[SBTE 2017, 2011]

Solution. Here we have

$$\frac{d^2x}{dt^2} - 3\frac{dx}{dt} + 2x = 4t + e^{3t}$$

A.E. is $D^2 - 3D + 2 = 0$

$\Rightarrow$ $(D - 1)(D - 2) = 0 \quad \Rightarrow D = 1, 2$

C.F. = $C_1e^t + C_2e^{2t}$

$$\text{P.I.} = \frac{1}{D^2-3D+2}4t + \frac{1}{D^2-3D+2}e^{3t}$$

$$= \frac{4}{2\left\{1+\left(\frac{D^2-3D}{2}\right)\right\}}(t) + \frac{1}{9-9+2}(e^{3t})$$

$$= 2\left\{1+\left(\frac{D^2-3D}{2}\right)\right\}^{-1}(t) + \frac{1}{2}e^{3t}$$

$$= 2\left\{1-\left(\frac{D^2-3D}{2}\right)\right\}.t + \frac{1}{2}e^{3t}$$

$$= 2\left\{t - \frac{D^2}{2}(t) + \frac{3D}{2}(t)\right\} + \frac{1}{2}e^{3t} = 2t+3+\frac{1}{2}e^{3t}$$

x = C.S. = C.F. + P. I.

$$\therefore x = C_1e^t + C_2e^{2t} + 2t + 3 + \frac{1}{2}e^{3t} \quad ...(1)$$

$$\frac{dx}{dt} = C_1e^t + 2C_2e^{2t} + 2 + \frac{3}{2}e^{3t} \quad ...(2)$$

Given $t = 0$, $x = 1$ and $t = 0$, $\frac{dx}{dt} = -1$

From (1) and (2), we get

$$1 = C_1 + C_2 + 3 + \frac{1}{2} = C_1 + C_2 + \frac{7}{2} \quad ...(3)$$

$$-1 = C_1 + 2C_2 + 2 + \frac{3}{2} = C_1 + 2C_2 + \frac{7}{2} \quad ...(4)$$

On solving (3) and (4), we get

$$C_1 = -\frac{1}{2}, C_2 = -2$$

Putting the values of C_1 and C_2 in (1), we get

$$x = -\frac{1}{2}e^t - 2e^{2t} + 2t + 3 + \frac{1}{2}e^{3t}$$ **Ans.**

EXERCISE 34.4

Solve the following differential equations:

1. $(D^2 + 4)\,y = \sin 3x.$ **Ans.** $y = C_1 \cos 2x + C_2 \sin 2x - \frac{1}{5}\sin 3x$

2. $(D^2 + 9)\,y = \cos 3x.$ **(DIP IETE Dec. 2006) Ans.** $y = C_1 \cos 3x + C_2 \sin 3x + \frac{x \sin 3x}{6}$

3. $(D^2 + a^2)\,y = \sin ax.$ **Ans.** $y = C_1 \cos ax + C_2 \sin ax - \frac{x}{2a}\cos ax$

4. $\frac{d^2y}{dx^2} + 6y = \sin 4x$ **Ans.** $y = C_1 \cos \sqrt{6}x + C_2 \sin \sqrt{6}x - \frac{1}{10}\sin 4x$

5. $\dfrac{d^2y}{dx^2} + \dfrac{dy}{dx} + y = \sin 2x$

Ans. $y = e^{-x/2}\left(C_1 \cos \dfrac{\sqrt{3}x}{2} + C_2 \sin \dfrac{\sqrt{3}x}{2}\right) - \dfrac{1}{13}(2\cos 2x + 3\sin 2x)$

6. $\dfrac{d^2y}{dx^2} + 4\dfrac{dy}{dx} + 4y = 3\sin 2x - 4x^2$ **(B.T.E. Delhi May 2008)**

Ans. $y = (C_1 + C_2 x)e^{-2x} - \dfrac{3}{8}\cos 2x - x^2 + 2x - \dfrac{3}{2}$

7. $\dfrac{d^2y}{dx^2} + 3\dfrac{dy}{dx} + 2y = 3\sin x$ given that $y = 0.09$ and $\dfrac{dy}{dx} = -0.07$, when $x = 0$

(DIPIETE June. 2006) **Ans.** $y = 1.61e^{-x} - 0.62e^{-2x} - \dfrac{3}{10}(3\cos x - \sin x)$

8. $\dfrac{d^2y}{dx^2} - 7\dfrac{dy}{dx} + 6y = 2\sin 3x$ given that $y = 1$, $\dfrac{dy}{dx} = 0$ when $x = 0$

Ans. $y = -\dfrac{379}{375}e^{x} - \dfrac{13}{125}e^{6x} + \dfrac{1}{75}[7\cos 3x - \sin 3x]$

9. $\dfrac{d^4y}{dx^4} + 2\dfrac{d^2y}{dx^2} + y = \cos x$ **Ans.** $y = (C_1 + C_2 x)\cos x + (C_3 + C_4 x)\sin x - \dfrac{x^2}{8}\cos x$

10. $\dfrac{d^2y}{dx^2} + 2\dfrac{dy}{dx} + y = \cos^2 x$ **Ans.** $y = (C_1 + C_2 x)e^{-x} + \dfrac{1}{50}(4\sin 2x - 3\cos 2x) + \dfrac{1}{2}$

11. $(D^3 + 1)\,y = 2\cos^2 x$

Ans. $y = C_1 e^{-x} + e^{x/2}\left(C_2 \cos \dfrac{\sqrt{3}}{2}x + C_3 \sin \dfrac{\sqrt{3}}{2}x\right) + 1 + \dfrac{1}{65}(-8\sin 2x + \cos 2x)$

[**Hint.** $\cos 2x = 2\cos^2 x - 1 \Rightarrow 2\cos^2 x = \cos 2x + 1 = \cos 2x + e^{0x}$]

12. $\dfrac{d^2y}{dx^2} + y = \sin 3x \cos 2x$ **Ans.** $y = C_1 \cos x + C_2 \sin x + \dfrac{1}{48}[-\sin 5x - 12x\cos x]$

13. $\dfrac{d^2y}{dx^2} + y = \sin x \sin 2x$ **Ans.** $y = C_1 \cos x + C_2 \sin x + \dfrac{1}{16}(4x\sin x + \cos 3x)$

[**Hint.** $\sin x \sin 2x = \cos x - \cos 3x$]

14. $\dfrac{d^2y}{dx^2} + 4y = e^x + \sin 2x$ **Ans.** $y = C_1 \cos 2x + C_2 \sin 2x + \dfrac{1}{5}e^x - \dfrac{x}{4}\cos 2x$

15. $\dfrac{d^3y}{dx^3} - 2\dfrac{d^2y}{dx^2} - 19\dfrac{dy}{dx} + 20y = xe^x + 2e^{-4x}\sin x$ **Ans.** C.F. $= C_1 e^{5x} + C_2 e^{-4x} + C_3 e^{x}$

16. $\dfrac{d^2y}{dx^2} - 4\dfrac{dy}{dx} + 13y = 24e^{2x} + \sin 3x$

Ans. $y = e^{2x}[C_1 \cos 3x + C_2 \sin 3x] + \dfrac{8}{3}e^{2x} + \dfrac{1}{40}[3\cos 3x + \sin 3x]$

17. $\frac{d^2y}{dx^2} - 4\frac{dy}{dx} + 3y = e^{2x} + \cos x$ **Ans.** $y = C_1e^x + C_2e^{3x} + \frac{1}{10}(-2\sin x + \cos x) - e^{2x}$

18. $\frac{d^2y}{dx^2} - 2\frac{dy}{dx} + 3y = \cos x + x^2$

Ans. $y = e^x\left[C_1 \cos\sqrt{2}x + C_2 \sin\sqrt{2}x\right] + \frac{1}{4}(\cos x - \sin x) + \frac{1}{3}\left(x^2 + \frac{4}{3}x - \frac{2}{9}\right)$

19. $(D^3 - 3D^2 + 4D - 2)\, y = e^x + \cos x$

Ans. $y = (C_1 + C_2 \cos x + C_3 \sin x)\, e^x + xe^x + \frac{1}{10}(3\sin x + \cos x)$

20. $(D^3 - 4D^2 + 13D)\, y = 1 + \cos 2x$

Ans. $y = C_1 + e^{2x}(C_2 \cos 3x + C_3 \sin 3x) + \frac{1}{290}(9\sin 2x + 8\cos 2x) + \frac{x}{13}$

21. $(D^2 - 4D + 4)\, y = e^{2x} + x^3 + \cos 2x$

Ans. $y = (C_1 + C_2x)e^{2x} + \frac{1}{2}x^2e^{2x} + \frac{1}{8}(2x^3 + 6x^2 + 9x + 6) - \frac{1}{8}\sin 2x$

22. $\frac{d^2y}{dx^2} + n^2y = h \sin Px \; (P \neq n)$

where h, p and n are constants satisfying the conditions $y = a, \frac{dy}{dx} = b$ for $x = 0$.

Ans. $y = a\cos nx + \left(\frac{b}{n} - \frac{ph}{n(n^2 - p^2)}\right)\sin nx + \frac{h \sin px}{n^2 - p^2}$

23. $\frac{d^2y}{dx^2} + 3\frac{dy}{dx} = 4\sin^2 x$ **[DIPIETE. 2019] Ans.** $y = C_1 + C_2e^{-3x} - \frac{1}{5}(2\cos 2x + 3\sin 2x)$

34.9 ELECTRIC CIRCUITS (APPLICATIONS)

We have already discussed $R - L$ and $R - L - C$ electric circuits. Here we will do circuit-problems involving second order differential equations.

Example 20. *A condenser of capacity C is discharged through the inductance L and a resistance R in series and the charge q at the time t satisfies the equation.*

$$L\frac{d^2q}{dt^2} + R\frac{dq}{dt} + \frac{q}{C} = 0$$ **[SBTE 2012]**

Given that L = 0.25 H, R = 250 Ω, C = 2 × 10^{-6} *F and that when t = 0, the charge q is 0.02 coulombs, and the current* $\frac{dq}{dt} = 0$, *obtain the value of q in terms of t.* **(B.T.E. Delhi June, 2007)**

Solution. Substituting the given values, the equation becomes

$$\frac{d^2q}{dt^2} + \frac{250}{0.25}\frac{dq}{dt} + \frac{q}{2 \times 10^{-6} \times 0.25} = 0$$

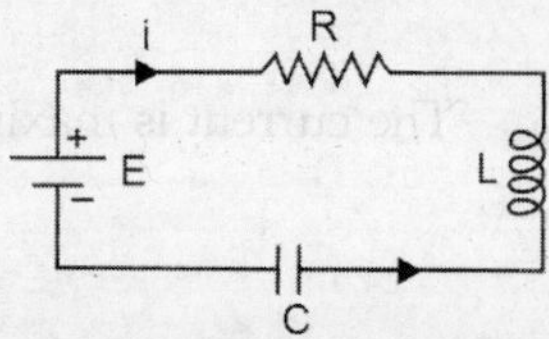

$$\Rightarrow \quad \frac{d^2q}{dt^2} + 1000\frac{dq}{dt} + 2 \times 10^6 q = 0$$

The auxiliary equation is $m^2 + 1000\,m + 2 \times 10^6 = 0 \Rightarrow m = -500 \pm 1323\,i$

$\therefore$ The solution is $q = e^{-500\,t}\,[A \cos 1323\,t + B \sin 1323\,t]$

when $t = 0, \quad q = 0.002 \quad \therefore A = 0.002$

$$\Rightarrow \quad \frac{dq}{dt} = -500\,e^{-500t}\,[A \cos 1323\,t + B \sin 1323\,t]$$

$$+ e^{-500\,t} \times 1323\,[-A \sin 1323\,t + B \cos 1323\,t]$$

when $t = 0, \quad \frac{dq}{dt} = 0 \quad \therefore B = 0.0008$

Hence, $q = e^{-500\,t}\,[0.002 \cos 1323\,t + 0.0008 \sin 1323\,t]$ **Ans.**

Example 21. *The differential equation for a circuit in which self-inductance neutralize each other is*

$$L\frac{d^2i}{dt^2} + \frac{i}{C} = 0$$

Find the current i as a function of t, given that I is the maximum current and i = 0 when t = 0.

Solution. $L\frac{d^2i}{dt^2} + \frac{i}{C} = 0$

$$\Rightarrow \quad \frac{d^2i}{dt^2} + \frac{i}{LC} = 0 \quad \Rightarrow \quad \left(D^2 + \frac{1}{LC}\right)i = 0 \qquad \left[\because \frac{d}{dt} = D, \quad \frac{d^2}{dt^2} = D^2\right]$$

A.E. is $m^2 + \frac{1}{LC} = 0$

$$\Rightarrow \quad m^2 - \frac{j^2}{LC} \qquad \left[\because j = \sqrt{-1}\right]$$

$$m = \pm\frac{j}{\sqrt{LC}}$$

Its solution can be written as

$$i = c_1 \sin\frac{1}{\sqrt{LC}}t + c_2 \cos\frac{1}{\sqrt{LC}}t \qquad \text{... (1)}$$

At $t = 0, i = 0$

$\Rightarrow \quad C_2 = 0$

From (1) $i = c_1 \sin\frac{t}{\sqrt{LC}}$

When $\sin\frac{t}{\sqrt{LC}} = 1$

The current is maximum, that is $I = c_1$

$$i = I \sin\frac{t}{\sqrt{LC}}$$ **Ans.**

Example 22. *An electric circuit consists of an inductance 0.1 henry, a resistance of 20 ohms and a condenser of capacitance 25 micro farads. Find the charge q and the current i at time t, given the initial conditions. q = 0.05 coloumbs. i = 0 when t = 0.*

Solution. The differential equation of the above given circuit can be written as

L = 0.1H R = 20Ω C = 25μF

$$\Rightarrow \quad L\frac{dI}{dt} + RI + \frac{q}{C} = 0$$

$$\Rightarrow \quad L\frac{d^2q}{dt^2} + R\frac{dq}{dt} + \frac{q}{C} = 0$$

$$\Rightarrow \quad \frac{d^2q}{dt^2} + \frac{R}{L}\frac{dq}{dt} + \frac{q}{LC} = 0$$

First we will solve the equation and then put the values of R, L and C.

For convenience we put

$$\Rightarrow \quad \frac{R}{L} = 2b \Rightarrow b = \frac{R}{2L} = \frac{20}{2 \times 0.1} = 100$$

$$\Rightarrow \quad \frac{1}{LC} = k^2$$

$$\Rightarrow \quad k = \sqrt{\frac{1}{LC}} = \sqrt{\frac{1}{0.1 \times 25 \times 10^{-6}}} = \sqrt{\frac{10^7}{25}} = 632.5 > 100$$

$$\Rightarrow \quad b < k$$

Our equation reduces to

$$\Rightarrow \quad \frac{d^2q}{dt^2} + 2b\frac{dq}{dt} + k^2q = 0$$

A.E. is $\quad m^2 + 2b\,m + k^2 = 0.$

So that
$$m = \frac{-2b \pm \sqrt{4b^2 - 4k^2}}{2}$$
$$= -b \pm \sqrt{b^2 - k^2} = -b \pm j\sqrt{k^2 - b^2}$$

C.F. is
$$q = e^{-bt}\left[A\cos\sqrt{k^2 - b^2}\,t + B\sin\sqrt{k^2 - b^2}\,t\right] \quad \text{...(1)}$$

On putting $\quad q = 0.05$ and $t = 0$ in (1),

we get $\quad 0.05 = A$

On differentiating (1), we get

$$\frac{dq}{dt} = -be^{-bt}\left[A\cos\sqrt{k^2 - b^2}\,t + B\sin\sqrt{k^2 - b^2}\,t\right]$$

$$+ e^{-bt}\left[-A\sqrt{k^2 - b^2}\sin\sqrt{k^2 - b^2}\,t + B\sqrt{k^2 - b^2}\cos\sqrt{k^2 - b^2}\,t\right] \quad \text{...(2)}$$

On putting $\frac{dq}{dt} = 0$ and $t = 0$ in (2), we get

$$0 = -b\,A + B\sqrt{k^2 - b^2} \quad \Rightarrow B = \frac{bA}{\sqrt{k^2 - b^2}} = \frac{0.05b}{\sqrt{k^2 - b^2}}$$

Substituting the values of A and B in (1), we have

$$q = e^{-bt}\left[0.05\cos\sqrt{k^2 - b^2}\,t + \frac{0.05b}{\sqrt{k^2 - b^2}}\sin\sqrt{k^2 - b^2}\,t\right] \quad \text{...(3)}$$

Now $\quad e^{-bt} = e^{-100\,t}$

and $\sqrt{k^2 - b^2} = \sqrt{\frac{10^7}{25} - (100)^2} = \sqrt{400000 - 100000} = \sqrt{390000} = 624.5$

On putting these values in (3), we have

$$q = e^{-100t}\left[0.05\cos 624.5t + \frac{0.05 \times 100}{624.5}\sin 624.5t\right]$$

$$\Rightarrow \quad q = e^{-100t}\left[0.05\cos 624.5t + 0.008\sin 624.5t\right] \quad \text{... (4)}$$

On differentiating (4), we have

$$\frac{dq}{dt} = -100\,e^{100t}\left[0.05\cos 624.5t + 0.008\sin 624.5t\right]$$

$$+ e^{-100t}\left[-0.05 \times 624.5\sin 624.5t + 0.008 \times 624.5\cos 624.5t\right]$$

$$\Rightarrow \quad i = e^{100t}\left[(-5 + 4.996)\cos 624.5t + (0.8 + 31.225)\sin 624.5t\right]$$

$$= e^{100t}\left[-0.004\cos 624.5t - 32.025\sin 624.5t\right]$$

$$= 32e^{-100t}\sin 624.5\,t \quad \text{(Approximately)}$$ **Ans.**

Example 23. *For an electric circuit with circuit constants, L, R, C the charge Q on a plate condenser is given by*

$$L\frac{d^2q}{dt^2} + R\frac{dq}{dt} + \frac{q}{C} = E \quad \textit{and the current by } i = \frac{dq}{dt}$$

Let $\quad L = 1\,H, \qquad C = 10^{-4}\,F,$

$\quad R = 100\,\Omega, \qquad E = 100\,V,$

Suppose that no charges are present and no current is flowing at time t = 0, when the e.m.f. is applied. Determine q and i at any time t.

Solution. The differential equation is

$$\Rightarrow \quad L\frac{d^2q}{dt^2} + R\frac{dq}{dt} + \frac{q}{C} = E$$

$$\Rightarrow \quad \frac{d^2q}{dt^2} + \frac{R}{L}\frac{dq}{dt} + \frac{q}{LC} = \frac{E}{L}$$

Put $\quad \frac{R}{L} = 2b$ and $\frac{1}{LC} = k^2$

The equation becomes

$$\Rightarrow \quad \frac{d^2q}{dt^2} + 2b\frac{dq}{dt} + k^2q = \frac{E}{L}$$

Rearranging the terms, we have

$$\frac{d^2q}{dt^2}+2b\frac{dq}{dt}+k^2\left[q-\frac{E}{k^2L}\right]=0$$

Put $$x=q-\frac{E}{k^2L}$$

so $$\frac{dx}{dt}=\frac{dq}{dt}$$

and $$\frac{d^2x}{dt^2}=\frac{d^2q}{dt^2}$$

Our equation reduces to

$$\frac{d^2x}{dt^2}+2b\frac{dx}{dt}+k^2x=0$$

This equation is exactly identical as we have solved the above equation in problem 2.

$$\Rightarrow \quad x=e^{-bt}\left[A\cos\sqrt{k^2-b^2}\,t+B\sin\sqrt{k^2-b^2}\,t\right]$$

$$\Rightarrow \quad q-\frac{E}{k^2L}=e^{-bt}\left[A\cos\sqrt{k^2-b^2}\,t+B\sin\sqrt{k^2-b^2}\,t\right]$$

$$\Rightarrow \quad q=\frac{E}{k^2L}+e^{-bt}\left[A\cos\sqrt{k^2-b^2}\,t+B\sin\sqrt{k^2-b^2}\,t\right] \quad ...(1)$$

On putting $q = 0$ and $t = 0$ in (1), we get

$$0=\frac{E}{k^2L}+A \Rightarrow A=-\frac{E}{k^2L}$$

Differentiating (1), we have

$$\Rightarrow \quad \frac{dq}{dt}=-be^{-bt}\left[A\cos\sqrt{k^2-b^2}\,t+B\sin\sqrt{k^2-b^2}\,t\right]$$

$$+e^{-bt}\left[-A\sqrt{k^2-b^2}\,t\sin\sqrt{k^2-b^2}\,t+B\sqrt{k^2-b^2}\cos\sqrt{k^2-b^2}\,t\right] \quad ...(2)$$

On putting $\frac{dq}{dt}=0$ and $t = 0$ in (2),

we have $$0=-bA+B\sqrt{k^2-b^2}$$

$$\Rightarrow \quad B=\frac{bA}{\sqrt{k^2-b^2}}=\frac{-\frac{bE}{k^2L}}{\sqrt{k^2-b^2}}$$

Substituting the values of A and B in (1), we have

$$\Rightarrow \quad q=\frac{E}{k^2L}+e^{-bt}\left[-\frac{E}{k^2L}\cos\sqrt{k^2-b^2}\,t-\frac{bE}{k^2L\sqrt{k^2-b^2}}\sin\sqrt{k^2-b^2}\,t\right]$$

$$\Rightarrow \qquad q = \frac{E}{k^2 L}\left[1 - e^{-bt}\left(\cos\sqrt{k^2 - b^2}\,t + \frac{b}{\sqrt{k^2 - b^2}}\sin\sqrt{k^2 - b^2}\,t\right)\right] \qquad ...(3)$$

Now
$$\frac{E}{k^2 L} = \frac{E}{\frac{1}{LC}.L} - EC = 100 \times 10^{-4} = \frac{1}{100}$$

$$b = \frac{R}{2L} = \frac{100}{2 \times 1} = 50$$

$$\Rightarrow \qquad \sqrt{k^2 - b^2} = \sqrt{\frac{1}{LC} - (50)^2} = \sqrt{\frac{1}{10^{-4}} - (50)^2}$$

$$\Rightarrow \qquad \sqrt{10000 - 2500} = \sqrt{7500} = 50\sqrt{3}$$

On putting these values in (3), we get

$$\therefore \qquad q = \frac{1}{100}\left[1 - e^{-50t}\,(\cos 50\sqrt{3}t + \frac{1}{\sqrt{3}}\sin 50\,\sqrt{3}t)\right] \qquad \textbf{Ans.}$$

EXERCISE 34.5

1. For an electric circuit with circuit constants L, R, C the charge q on the plate of the condenser is given by :

$L\frac{d^2q}{dt^2} + R\frac{dq}{dt} + \frac{q}{C} = 0$. Find q at any time t.

Discuss the case when R is negligible an show that q is oscillatory.

Calculate its period and frequency.

$$\textbf{Ans. } q = e^{\frac{R}{2L}t}\left[A\cos\frac{\sqrt{4CL - R^2C^2}}{2LC}t + B\sin\frac{\sqrt{4CL - R^2C^2}}{2LC}t\right]$$

$$q = A\cos\frac{1}{\sqrt{LC}}t + B\sin\frac{1}{\sqrt{LC}}t. \text{ Period} = 2\pi\sqrt{LC}, \text{ frequency} = \frac{1}{2\pi\sqrt{LC}}$$

2. A condenser of capacity C is discharged through an inductance L and a resistance R series and the charge q at any time is given by

$$L\frac{d^2q}{dt^2} + R\frac{dq}{dt} + \frac{q}{C} = 0$$

If L = 10 m, $R = 200\ \Omega$, C = 0.1 μF and also when $t = 0$, charge $q = 0.01$ C and current $\frac{dq}{dt} = 0$. Find the value of q at any time t and find the frequency of the circuit if the discharge is oscillatory.

$$\textbf{Ans. } q = e^{-10000t}\,[0.01\cos 3 \times 10^4\, t + 0.33 \times 10^{-2}\sin 3 \times 10^4\, t]$$

$$\text{Frequency} = \frac{3 \times 10^4}{2\pi}$$

3. A condenser of capacity C is discharged through L and a resistance R in series and the charge q at any time t given by the equation

$$L\frac{d^2q}{dt^2}+R\frac{dq}{dt}+\frac{q}{C}=0$$

If $L = 0.5$ H, $R = 300\ \Omega$, $C = 2 \times 10^{-6}$ F and also when $t = 0$, charge $q = 0.01$ C and current $\frac{dq}{dt}=0$, find the value of q in terms of t.

Ans. $q=e^{-300t}\left[0.01\cos 100\sqrt{91}\,t+0.0031\sin 100\sqrt{91}\,t\right]$

UNIT IX: LAPLACE TRANSFORMS

Laplace and Inverse Laplace Transform

35.1 INTRODUCTION

Laplace transforms help in solving the differential equations with boundary values without finding the general solution and the values of the arbitrary constants.

35.2 LAPLACE TRANSFORM

Let $f(t)$ be function defined for all positive values of t, then

$$F(s) = \int_0^\infty e^{-st} f(t)\, dt$$

provided the integral exists, is called the **Laplace transform** of $f(t)$. It is denoted as

$$\boxed{L[f(t)] = F(s) = \int_0^\infty e^{-st} f(t)\, dt}$$

35.3 IMPORTANT FORMULAE

1. $L(1) = \dfrac{1}{s}$

2. $L(t^n) = \dfrac{n!}{s^{n+1}}$, then $n = 0, 1, 2, 3...$

3. $L\left(e^{at}\right) = \dfrac{1}{s-a}$ $\qquad (s > a)$

4. $L(\cosh at) = \dfrac{s}{s^2 - a^2}$ $\qquad (s^2 > a^2)$

5. $L(\sinh at) = \dfrac{a}{s^2 - a^2}$ $\qquad (s^2 > a^2)$

6. $L(\sin at) = \dfrac{a}{s^2 + a^2}$ $\qquad (s > 0)$

7. $L(\cos at) = \dfrac{s}{s^2 + a^2}$ $\qquad (s > 0)$

Proof of the above important formulae

$$\boxed{(\,)\ -}$$

Proof. $L(1) = \int_0^\infty 1 \cdot e^{-st} dt = \left[\frac{e^{-st}}{-s}\right]_0^\infty = -\frac{1}{s}\left[\frac{1}{e^{st}}\right]_0^\infty = -\frac{1}{s}[0-1] = \frac{1}{s}$

Hence $L(1) = \frac{1}{s}$ **Proved.**

2. $\boxed{L\left(t^n\right) = \frac{n!}{s^{n+1}}}$ where n and s are positive.

Proof. $$L(t^n) = \int_0^\infty e^{-st} t^n dt$$

Putting $$st = x \text{ or } t = \frac{x}{s} \text{ or } dt = \frac{dx}{s}$$

Thus, we have $$L(t^n) = \int_0^\infty e^{-x}\left(\frac{x}{s}\right)^n \frac{dx}{s} \Rightarrow L(t^n) = \frac{1}{s^{n+1}}\int_0^\infty e^{-x}.x^n dx$$

$$\Rightarrow L(t^n) = \frac{\overline{|n+1}}{s^{n+1}} \Rightarrow L(t^n) = \frac{n!}{s^{n+1}} \qquad \left[\begin{array}{l}\overline{|n+1} = \int_0^\infty e^{-x}.x^n dx \\ \text{and} \quad \overline{|n+1} = n!\end{array}\right]$$ **Proved.**

3. $\boxed{L\left(e^{at}\right) = \frac{1}{s-a}}$ where $s > a$

Proof.
$$L(e^{at}) = \int_0^\infty e^{-st}.e^{at} dt = \int_0^\infty e^{-st+at} \cdot at$$
$$= \int_0^\infty e^{(-s+a)t} \cdot dt = \int_0^\infty e^{-(s-a)t} \cdot dt$$
$$= \left[\frac{e^{-(s-a)t}}{-(s-a)}\right]_0^\infty = -\frac{1}{s-a}\left[\frac{1}{e^{(s-a)t}}\right]_0^\infty$$
$$= \frac{-1}{(s-a)}(0-1) = \frac{1}{s-a}$$ **Proved.**

4. $\boxed{L(\cosh at) = \frac{s}{s^2-a^2}}$

Proof.
$$L(\cosh at) = L\left[\frac{e^{at}+e^{-at}}{2}\right] \qquad \left(\therefore \cosh at = \frac{e^{at}+e^{-at}}{2}\right)$$
$$= \frac{1}{2}L(e^{at}) + \frac{1}{2}L(e^{-at})$$
$$= \frac{1}{2}\left[\frac{1}{s-a} + \frac{1}{s+a}\right] \qquad \left[\because L\left(e^{at}\right) = \frac{1}{s-a}\right]$$
$$= \frac{1}{2}\left[\frac{s+a+s-a}{s^2-a^2}\right] = \frac{s}{s^2-a^2}$$ **Proved.**

5. $\boxed{L(\sinh at) = \dfrac{a}{s^2 - a^2}}$ $\left[\because \sinh at = \dfrac{e^{at} - e^{-at}}{2}\right]$

Proof.

$$L(\sinh at) = L\left[\frac{1}{2}\left(e^{at} - e^{-at}\right)\right]$$

$$= \frac{1}{2}\left[L\left(e^{at}\right) - L\left(e^{-at}\right)\right] = \frac{1}{2}\left[\frac{1}{s-a} - \frac{1}{s+a}\right]$$

$$= \frac{1}{2}\left[\frac{s+a-s+a}{s^2-a^2}\right]$$

$$= \frac{a}{s^2-a^2}$$

Proved.

6. $\boxed{L(\sin at) = \dfrac{a}{s^2 + a^2}}$

Proof.

$$L(\sin at) = L\left[\frac{e^{iat} - e^{-iat}}{2i}\right] \qquad \left[\because \sin at = \frac{e^{iat} - e^{-iat}}{2i}\right]$$

$$= \frac{1}{2i}\left[L\left(e^{iat} - e^{-iat}\right)\right]$$

$$= \frac{1}{2i}\left[L\left(e^{iat}\right) - L\left(e^{-iat}\right)\right]$$

$$= \frac{1}{2i}\left[\frac{1}{s-ia} - \frac{1}{s+ia}\right]$$

$$= \frac{1}{2i}\frac{s+ia-s+ia}{s^2+a^2}$$

$$= \frac{1}{2i}\frac{2ia}{s^2+a^2} = \frac{a}{s^2+a^2}$$

Proved.

7. $\boxed{L(\cos at) = \dfrac{s}{s^2 + a^2}}$

Proof.

$$L(\cos at) = L\left(\frac{e^{iat} + e^{-iat}}{2}\right) \left[\because \cos at = \frac{e^{iat} + e^{-iat}}{2}\right]$$

$$= \frac{1}{2}[L(e^{iat} + e^{-iat})] = \frac{1}{2}[L(e^{iat}) + L(e^{-iat})]$$

$$= \frac{1}{2}\left[\frac{1}{s-ia} + \frac{1}{s+ia}\right] = \frac{1}{2}\frac{s+ia+s-ia}{s^2+a^2}$$

$$= \frac{s}{s^2+a^2}$$

Proved.

Example 1. *Find the Laplace transform of f(t) defined as :*

$$f(t) = \begin{cases} \dfrac{t}{k}, \text{when} & 0<t<k \\ 1, \text{when} & t>k \end{cases}$$

Solution.

$$L[f(t)] = \int_0^k \frac{t}{k} e^{-st} dt + \int_k^\infty 1.e^{-st} dt$$

$$= \frac{1}{k}\left[\left(t\,\frac{e^{-st}}{-s}\right)_0^k - \int_0^k \frac{e^{-st}}{-s} dt\right] + \left[\frac{e^{-st}}{-s}\right]_k^\infty$$

$$= \frac{1}{k}\left[\frac{ke^{-ks}}{-s} - \left(\frac{e^{-st}}{s^2}\right)_0^k\right] + \frac{e^{-ks}}{s}$$

$$= \frac{1}{k}\left[\frac{ke^{-ks}}{-s} - \frac{e^{-sk}}{s^2} + \frac{1}{s^2}\right] + \frac{e^{-ks}}{s}$$

$$= -\frac{e^{-sk}}{s} - \frac{1}{k}\frac{e^{-ks}}{s^2} + \frac{1}{k}\frac{1}{s^2} + \frac{e^{-ks}}{s}$$

$$= \frac{1}{ks^2}\left[-e^{-ks} + 1\right]$$ **Ans.**

Example 2. *Find the Lapalce transform of*

$$f(t) = \begin{cases} t^2, & 0 < t < 2 \\ t-1, & 2 < t < 3 \\ 7, & t > 3 \end{cases}$$ **(U.P, II Semester, June 2007)**

Solution. $L[f(t)] = \int_0^\infty e^{-st} f(t)\, dt = \int_0^2 t^2 e^{-st} dt + \int_2^3 (t-1)e^{-st} dt + \int_3^\infty 7e^{-st} dt$

$$\left[\int I\cdot II = I\cdot II_1 - I'II_{11} + I''II_{111} + \ldots\right]$$

$$= \left[t^2\left(\frac{e^{-st}}{(-s)}\right) - 2t\,\frac{e^{-st}}{(-s)^2} + 2\frac{e^{-st}}{(-s)^3}\right]_0^2 + \left[(t-1)\left(\frac{e^{-st}}{(-s)}\right) - \frac{e^{-st}}{(-s)^2}\right]_2^3 + 7\left[\frac{e^{-st}}{-s}\right]_3^\infty$$

$$= \left[-4\left(\frac{e^{-2s}}{s}\right) - 4\left(\frac{e^{-2s}}{s^2}\right) - 2\left(\frac{e^{-2s}}{s^3}\right) + \frac{2}{s^3}\right] + \left[2\left(\frac{e^{-3s}}{s^2}\right) - \frac{e^{-3s}}{s^2} + \frac{e^{-2s}}{s} + \frac{e^{-2s}}{s^2}\right] + 7\left(0 + \frac{e^{-3s}}{s}\right)$$

$$= \frac{2}{s^3} + e^{-2s}\left[-\frac{4}{s} - \frac{4}{s^2} - \frac{2}{s^3}\right] + e^{-3s}\left[-\frac{2}{s} - \frac{1}{s^2}\right] + e^{-2s}\left[\frac{1}{s} + \frac{1}{s^2}\right] + e^{-3s}\left[\frac{7}{s}\right]$$

$$= \frac{2}{s^3} + e^{-2s}\left[-\frac{4}{s} - \frac{4}{s^2} - \frac{2}{s^3} + \frac{1}{s} + \frac{1}{s^2}\right] + e^{-3s}\left[-\frac{2}{s} - \frac{1}{s^2} + \frac{7}{s}\right]$$

$$= \frac{2}{s^3} + e^{-2s}\left[-\frac{3}{s} - \frac{3}{s^2} - \frac{2}{s^3}\right] + e^{-3s}\left[\frac{5}{s} - \frac{1}{s^2}\right]$$

$$= \frac{2}{s^3} - \frac{e^{-2s}}{s^3}(2 + 3s + 3s^2) + \frac{e^{-3s}}{2}(5s - 1)$$ **Ans.**

Example 3. *From the first principles find the Laplace transform of (1 + sin 2t).*

Solution. Laplace transform of (1 + sin 2t)

$$= \int_0^\infty e^{-st}(1 + \sin 2t)\, dt$$

$$= \int_0^\infty e^{-st}\left(1 + \frac{e^{2it} - e^{-2it}}{2i}\right) dt$$

$$= \frac{1}{2i}\int_0^\infty [2ie^{-st} + e^{(-s+2i)t} - e^{(-s-2i)t}\, dt$$

$$= \frac{1}{2i}\left[\frac{2ie^{-st}}{-s} + \frac{e^{(-s+2i)t}}{-s+2i} - \frac{e^{(-s-2i)t}}{-s-2i}\right]_0^\infty$$

$$= \frac{1}{2i}\left[\left(0 + \frac{2i}{s}\right) + \frac{1}{-s+2i}(0-1) - \frac{1}{-s-2i}(0-1)\right]$$

$$= \frac{1}{2i}\left[\frac{2i}{s} + \frac{1}{s-2i} - \frac{1}{s+2i}\right]$$

$$= \frac{1}{2}\left[\frac{2}{s} + \frac{4}{s^2+4}\right] = \frac{1}{s} + \frac{2}{s^2+4}$$ **Ans.**

Alternate method: $L(1 + \sin 2t) = L(1) + L(\sin 2t) = \frac{1}{s} + \frac{2}{s^2+4}$ **Ans.**

35.4 PROPERTIES OF LAPLACE TRANSFORM

(1) *Linearity* $L[af_1(t) + bf_2(t)] = a\,L\,[f_1(t)] + b\,L[f_2(t)]$

Proof. $L[af_1(t) + bf_2(t)] = \int_0^\infty e^{-st}[af_1(t) + bf_2(t)]\, dt$

$$= a\int_0^\infty e^{-st} f_1(t)\, dt + b\int_0^\infty e^{-st} f_2(t)\, dt$$

$$= a\,Lf_1(t) + b\,Lf_2(t)$$ **Proved.**

(2) *Change of scale property*

If $L\{f(t)\} = F(s)$ then $\boxed{L\{f(at)\} = \frac{1}{a}F\left(\frac{s}{a}\right)}$

Proof. $L\{f(at)\} = \int^\infty e^{\ st} f(at)\, dt$ $\left[\text{Put } at = u \Rightarrow dt = \frac{du}{a}\right]$

$$= \int_0^\infty e^{-\left(\frac{s}{a}\right)u} f(u)\frac{du}{a}$$

$$= \frac{1}{a}\int_0^{\infty} e^{-\left(\frac{s}{a}\right)u} f(u)\, du$$

$$= \frac{1}{a}\int_0^{\infty} e^{-\left(\frac{s}{a}\right)t} f(t)\, dt$$

$$= \frac{1}{a}\int_0^{\infty} e^{-St} f(t)dt = \frac{1}{a} Lf(t) = \frac{1}{a} F(S)$$

$$\left[\text{Put } S = \frac{s}{a}\right]$$

$$= \frac{1}{a} F\left(\frac{s}{a}\right)$$ **Proved.**

35.5 FIRST SHIFTING THEOREM

If $L\{f(t)\} = F(s)$, then $\boxed{L[e^{at} f(t)] = F(s-a)}$

Proof. $$L[e^{at} f(t)] = \int_0^{\infty} e^{-st} . e^{at} f(t)\, dt$$

$$= \int_0^{\infty} e^{-(s-a)t} f(t)\, dt$$

$$= \int_0^{\infty} e^{-rt} f(t) dt \qquad \text{where } r = s - a$$

$$= F(r) = F(s-a)$$ **Proved.**

With the help of this property, we can have the following important results:

1. $L(e^{at} t^n) = \dfrac{n!}{(s-a)^{n+1}}$
2. $L(e^{at} \cosh bt) = \dfrac{s-a}{(s-a)^2 - b^2}$
3. $L(e^{at} \sinh bt) = \dfrac{b}{(s-a)^2 - b^2}$
4. $L(e^{at} \sin bt) = \dfrac{b}{(s-a)^2 + b^2}$
5. $L(e^{at} \cos bt) = \dfrac{s-a}{(s-a)^2 + b^2}$

35.6 HEAVISIDE'S SHIFTING THEOREM (SECOND TRANSLATION PROPERTY)

(a) If $L\{f(t)\} = F(s)$ and $g(t) = \begin{cases} f(t-a), & t > a \\ 0, & 0 < t < a \end{cases}$ then prove that

$$\boxed{L\{g(t)\} = e^{-as} F(s)}$$ **(U.P. II Semester, Summer 2006)**

Proof. $$L\{g(t)\} = \int_0^{\infty} e^{-st} g(t) dt$$

$$= \int_0^a e^{-st} g(t)dt + \int_a^\infty e^{-st} g(t)dt$$

[$g(t) = 0$, when $0 < t < a$]

$$= 0 + \int_a^\infty e^{-st} . f(t-a)dt$$

$$= \int_a^\infty e^{-st} f(t-a)dt$$

[Put $t - a = u \Rightarrow dt = du$]

$$= \int_0^\infty e^{-s(u+a)} f(u)du$$

$$= e^{-sa} \int_0^\infty e^{-su} f(u)du = e^{-sa} \int_0^\infty e^{-st} f(t)dt \qquad \textbf{Proved.}$$

$$\boxed{L\{g(t)\} = e^{-as} F(s)}$$

Example 4. *Find the Laplace transform of* $\cos^2 t$. **[DIP IETE 2018]**

Solution. We know that

$$\cos 2t = 2\cos^2 t - 1$$

$$\cos^2 t = \frac{1}{2}[\cos 2t + 1]$$

$$L(\cos^2 t) = L\left[\frac{1}{2}(\cos 2t + 1)\right] = \frac{1}{2}[L(\cos 2t) + L(1)]$$

$$= \frac{1}{2}\left[\frac{s}{s^2 + (2)^2} + \frac{1}{s}\right] = \frac{1}{2}\left[\frac{s}{s^2 + 4} + \frac{1}{s}\right] \qquad \textbf{Ans.}$$

Example 5. *If* $L(\cos^2 t) = \dfrac{s^2 + 2}{s(s^2 + 4)}$, *find* $L(\cos^2 at)$ **(U.P. I Semester, Summer 2006)**

Solution. We have

$$L(\cos^2 t) = \frac{s^2 + 2}{s(s^2 + 4)}$$

By change of scale property, we have

$$L(\cos^2 at) = \frac{1}{a} . \frac{\left(\frac{s}{a}\right)^2 + 2}{\frac{s}{a}\left[\left(\frac{s}{a}\right)^2 + 4\right]} = \frac{1}{a}\left[\frac{s^2 + 2a^2}{\frac{s}{a}(s^2 + 4a^2)}\right]$$

$$= \frac{s^2 + 2a^2}{s(s^2 + 4a^2)} \qquad \textbf{Ans.}$$

Example 6. *Find the Laplace transform of* $t^{-\frac{1}{2}}$.

Solution. We know that

$$L(t^n) = \frac{\lceil n + 1}{s^{n+1}}$$

Put $n = -\frac{1}{2}, L\left(t^{-\frac{1}{2}}\right) = \frac{\left|\overline{-\frac{1}{2}+1}\right.}{s^{-1/2+1}} = \frac{\left|\overline{\frac{1}{2}}\right.}{\sqrt{s}} = \frac{\sqrt{\pi}}{\sqrt{s}}$ where $\left|\overline{\frac{1}{2}}\right. = \sqrt{\pi}$ **Ans.**

Example 7. *Find the Laplace transform of 2 sin 2t cos 4t.*

Solution. We have

$$\begin{aligned} f(t) &= 2 \sin 2t \cos 4t \\ &= \sin(2t + 4t) + \sin(2t - 4t) \\ &= \sin 6t - \sin 2t \\ \therefore \quad L[f(t)] &= L[\sin 6t] - L[\sin 2t] \\ &= \frac{6}{s^2 + 36} - \frac{2}{s^2 + 4} \end{aligned}$$ **Ans.**

Example 8. *Find the Laplace transform of 4 sin³ t.*

Solution. We have

$$\begin{aligned} f(t) &= 4 \sin^3 t \\ &= 3 \sin t - \sin 3t, \qquad [\sin 3t = 3 \sin t - 4 \sin^3 t] \\ Lf(t) &= 3L(\sin t) - L(\sin 3t) \\ &= \frac{3}{s^2 + 1} - \frac{3}{s^2 + 9} \end{aligned}$$ **Ans.**

Example 9. *Find the Laplace transform of 4 cosh 2t sin 4t*

Solution. We have

$$\begin{aligned} f(t) &= 4 \cosh 2t \sin 4t \\ &= 4\left(\frac{e^{2t} + e^{-2t}}{2}\right)\left(\frac{e^{4it} - e^{-4it}}{2i}\right) \\ &= -i\left[e^{(2+4i)t} - e^{(2-4i)t} + e^{(-2+4i)t} - e^{(-2-4i)t}\right] \\ Lf(t) &= -i\left[L\left(e^{(2+4i)t}\right) - L\left(e^{(2-4i)t}\right) + L\left(e^{(-2+4i)t}\right) - L\left(e^{(-2-4i)t}\right)\right] \\ &= -i\left[\frac{1}{s-2-4i} - \frac{1}{s-2+4i} + \frac{1}{s+2-4i} - \frac{1}{s+2+4i}\right] \\ &= -i\left[\left(\frac{1}{s-2-4i} - \frac{1}{s+2+4i}\right) - \left(\frac{1}{s-2+4i} - \frac{1}{s+2-4i}\right)\right] \\ &= -i\left[\frac{4+8i}{s^2 - (2+4i)^2} - \frac{4-8i}{s^2 - (2-4i)^2}\right] \end{aligned}$$ **Ans.**

EXERCISE 35.1

Find the Laplce transforms of the following:

1. $t + t^2 + t^3$ **Ans.** $\frac{1}{s^2} + \frac{2}{s^3} + \frac{6}{s^4}$ **2.** $\sin t \cos t$ **Ans.** $\frac{1}{s^2 + 4}$

3. t^3e^{-2t} **Ans.** $\dfrac{6}{(s+2)^4}$ 4. $\sin^3 2t$ **Ans.** $\dfrac{48}{(s^2+4)(s^2+36)}$

5. $e^{-t}\cos^2 t$ **Ans.** $\dfrac{1}{2s+2}+\dfrac{s+1}{2s^2+4s+10}$ 6. $\sin 2t \cos 3t$ **Ans.** $\dfrac{2(s^2-5)}{(s^2+1)(s^2+25)}$

7. $\sin 2t \sin 3t$ **Ans.** $\dfrac{12s}{(s^2+1)(s^2+25)}$

8. $\cos at \sin h\, at$ **Ans.** $\dfrac{1}{2}\left[\dfrac{s-a}{(s-a)^2+a^2}-\dfrac{s+a}{(s+a)^2+a^2}\right]$

9. $\sinh^3 t$ **Ans.** $\dfrac{6}{(s^2-1)(s^2-9)}$ 10. $\cos t \cos 2t$ **Ans.** $\dfrac{s(s^2+5)}{(s^2+1)(s^2+9)}$

11. $\cosh at \sin at$ **Ans.** $\dfrac{a(s^2+2a^2)}{s^4+4a^4}$

12. $f(t)=\begin{cases}\cos\left(t-\dfrac{2\pi}{3}\right), & t>\dfrac{2\pi}{3}\\ 0, & t<\dfrac{2\pi}{3}\end{cases}$ **Ans.** $e^{\frac{-2\pi r}{3}}\frac{r}{s^2+1}$

35.7 EXISTENCE THEOREM

According to this theorem $\int_0^{\infty} e^{-st} f(t)\,dt$ exists if $\int_0^{\lambda} e^{-st} f(t)\,dt$ can actually be evaluated and its limit as $\lambda \to \infty$ exists.

Otherwise, we may use the following theorem:

If $f(t)$ is continuous and $\lim\limits_{t\to\infty}\left[e^{-at} f(t)\right]$ is finite, the Laplace transform of $f(t)$, i.e. $\int\limits_0^{\infty} e^{-st} f(t)$ exists for $s > a$.

It should however, be kept in mind that the above aforesaid conditions are sufficient but not necessary.

For example $L\left(\dfrac{1}{\sqrt{t}}\right)$ exists though $\dfrac{1}{\sqrt{t}}$ is infinite at $t = 0$. Similarly a function $f(t)$ for which $\lim\limits_{t\to\infty}\left[e^{-at} f(t)\right]$ is finite and having a finite discontinuity will have a Laplace transform of $s > a$.

35.8 LAPLACE TRANSFORM OF THE DERIVATIVE OF f(t)

$$L[f'(t)] = sL[f(t)] - f(0), \text{ where } L[f(t)] = f(s)$$

Proof. $$L(f'(t)] = \int_0^{\infty} e^{-st} f'(t)dt$$

Integrating by parts, we get

$$L[f'(t)] = \left[e^{-st} f(t)\right]_0^{\infty} - \int_0^{\infty} (-se^{-st})\, f(t)\, dt$$

$$= -f(0) + s\int_0^{\infty} e^{-st} f(t)\, dt \quad [e^{-st}f(t) = 0, \text{ when } t = \infty]$$

$$= -f(0) - SL[f(t)]$$

$$\boxed{L[f'(t)] = sL[f(t)] - f(0).}$$ **Proved.**

Note. Roughly, Laplace transform of **derivative** of $f(t)$ corresponds to **multiplication** of the Laplace transform of $f(t)$ by s.

35.9 LAPLACE TRANSFORM OF DERIVATIVE OF ORDER n

$$L[f^{n}(t)] = s^n L[f(t)] - s^{n-1} f(0) - s^{n-2} f'(0) - s^{n-3} f''(0) - \ldots - f^{n-1}(0)$$

Proof. We have already proved in Section 37.8 that

$$L[f'(t)] = s\, L[f(t)] - f(0) \quad \ldots(1)$$

Replacing $f(t)$ by $f'(t)$ and $f'(t)$ by $f''(t)$ in (1), we get

$$L[f''(t)] = s\, L[f'(t)] - f'(0) \quad \ldots(2)$$

Putting the value of $L[f'(t)]$ from (1) in (2), we have

$$L[f''(t)] = s\,[s\, L[f(t)] - f(0)] - f'(0)$$

$$\Rightarrow \quad L[f''(t)] = s^2 L[f(t)] - s\, f(0) - f'(0)$$

Similarly,

$$L[f'''(t)] = s^3 L[f(t)] - s^2 f(0) - s\, f'(0) - f''(0)$$

$$L[f^{iv}(t)] = s^4 L[f(t)] - s^3 f(0) - s^2 f'(0) - sf''(0) - f'''(0)$$

$$\boxed{L[f^{n}(t)] = s^{n} L[f(t)] - s^{n-1} f(0) - s^{n-2} f'(0) - s^{n-3} f''(0) + \ldots - f^{n-1}_{(0)}}$$

35.10 LAPLACE TRANSFORM OF INTEGRAL OF $f(t)$

$$\boxed{L\left[\int_0^t f(t)dt\right] = \frac{1}{s} F(s)} \quad \text{where } L[f(t)] = F(s)$$

Proof. Let $\phi(t) = \int_0^t f(t)\,dt$ and $\phi(0) = 0$ then $\phi'(t) = f(t)$

We know that formula of Laplace transforms of $\phi'(t)$ i.e.

$$L[\phi'(t)] = s\, L[\phi(t)] - \phi(0)$$

$$\Rightarrow \quad L[\phi'(t)] = s\, L[\phi(t)] \quad [\phi(0) = 0]$$

$$\Rightarrow \quad L[\phi(t)] = \frac{1}{s} L[\phi'(t)]$$

Putting the values of $\phi(t)$ and $\phi'(t)$, we get

$$L\left[\int_0^1 f(t)\, dt\right] = \frac{1}{s} L[f(t)] \Rightarrow \boxed{L\left[\int_0^t f(t)\, dt\right] = \frac{1}{s} F(s)}$$ **Proved.**

Note. (1) Laplace transform of **integral** of $f(t)$ corresponds to the division of the Laplace transform of $f(t)$ by s.

(2) $$\int_0^t f(t)\, dt = L^{-1}\left[\frac{1}{s} F(s)\right]$$

35.11 LAPLACE TRANSFORM OF (t) (Multiplication by t) (U.P. I Semester Summer 2005)

If $\quad L[f(t)] = F(s)$, then

$$L[t^n f(t)] = (-1)^n \frac{d^n}{ds^n}[F(s)].$$

Proof.
$$L[f(t)] = F(s) = \int_0^{\infty} e^{-st} f(t)\, dt \qquad ...(1)$$

Differentiating (1) w.r.t. "s", we get

$$\Rightarrow \quad \frac{d}{ds}[F(s)] = \frac{d}{ds}\left[\int_0^{\infty} e^{-st} f(t)\, dt\right] = \int_0^{\infty} \frac{\partial}{\partial s}\left(e^{-st}\right) f(t)\, dt$$

$$= \int_0^{\infty}\left(-te^{-st}\right).f(t)\, dt = \int_0^{\infty} e^{-st}[-t.f(t)]\, dt$$

$$= L[-tf(t)] \text{ or } L[t\,f(t)] = (-1)^1 \frac{d}{ds}[F(s)]$$

Similarly,
$$L[t^2 f(t)] = (-1)^2 \frac{d^2}{ds^2}[F(s)]$$

$$\Rightarrow \quad L[t^3 f(t)] = (-1)^3 \frac{d^3}{ds^3}[F(s)]$$

$$\boxed{\boldsymbol{L\left[t^n f(t)\right] = (-1)^n \frac{d^n}{ds^n}[F(s)]}}$$ **Proved.**

35.12 INITIAL AND FINAL VALUE THEOREMS

(a) Initial Value Theorem $\quad L\{f(t)\} = F(s)$

$$\Rightarrow \quad \lim_{t \to 0} f(t) = \lim_{s \to \infty}[sF(s)],$$, provided the limit exists.

Proof.
$$L\{f'(t)\} = s\,L\{f(t)\} - f(0)$$

$$\Rightarrow \quad \int_0^{\infty} e^{-st} f'(t)\, dt = sF(s) - f(0)$$

$$\Rightarrow \quad \lim_{s \to \infty} \int_0^{\infty} e^{-st} f'(t)\, dt = \lim_{s \to \infty}[sF(s) - f(0)]$$

$$\Rightarrow \quad \lim_{s \to \infty}[sF(s)] = f(0) + \int_0^{\infty}\left(\lim_{s \to \infty} e^{-st}\right) f'(t)\, dt$$

$$= f(0) + \int_0^{\infty}(0) f'(t)\, dt \qquad \left(\because \lim_{s \to \infty} e^{-st} = 0\right)$$

$$= f(0) + 0 = f(0) = \lim_{t \to 0} f(t)$$

(b) Final Value Theorem $\quad L\{f(t)\} = F(s)$

$$\Rightarrow \quad \lim_{t \to \infty} f(t) = \lim_{s \to 0} f([sF(s)],$$ provided the limits exist.

Proof. $L\{f'(t)\} = sL\{f(t)\} - f(0) \quad \Rightarrow \int_0^\infty e^{-st} f'(t) dt = sF(s) - f(0)$

$$\Rightarrow \quad \lim_{s\to 0}\int_0^\infty e^{-st} f'(t)\, dt = \lim_{s\to 0}[sF(s) - f(0)]$$

$$\Rightarrow \quad \lim_{s\to 0}[sF(s)] - f(0) = \lim_{s\to 0}\int_0^\infty e^{-st} f'(t)\, dt$$

$$\Rightarrow \quad \lim_{s\to 0}[sF(s)] - f(0) = \int_0^\infty \lim_{s\to 0} e^{-st} f'(t)\, dt = \int_0^\infty (1) f'(t)\, dt$$

$$\left[\because \lim_{s\to 0} e^{-st} = 1\right]$$

$$\Rightarrow \quad \boxed{\lim_{s\to 0}[sF(s)] = \lim_{t\to\infty} f(t)}$$ **Proved**

35.13 EXPONENTIAL INTEGRAL FUNCTION $\int_t^\infty \left(\frac{e^{-x}}{x}\right) dx$

Let
$$f(t) = \int_t^\infty \frac{e^{-x}}{x} dx$$

$$\Rightarrow \quad f'(t) = -\frac{e^{-t}}{t} \Rightarrow tf'(t) = -e^{-t}$$

[Here –ve sign appears due to lower limit]

Taking Laplace transform of $tf'(t)$, we get

$$L\{tf'(t)\} = Lt\left(-\frac{e^{-t}}{t}\right) = L\{-e^{-t}\} = -L\{e^{-t}\}$$

$$\Rightarrow \quad -\frac{d}{ds}[sF(s) - f(0)] = -\frac{1}{s+1}$$

$$\Rightarrow \quad \frac{d}{ds}[s\,F(s)] = \frac{1}{s+1} \qquad \left[\because f(0) = \text{constant} \therefore \frac{d}{ds} f(0) = 0\right]$$

Integrating both sides, we get

$$s\,F(s) = \log(s+1) + C \qquad \text{...(1)}$$

Now, by final value theorem, we have

$$\lim_{s\to 0} sF(s) = \lim_{t\to\infty} f(t) \qquad \text{...(2)}$$

Hence,
$$\lim_{s\to 0}[sF(s)] = \lim_{s\to 0}[\log(s+1) + C] = 0 + C = C \qquad \text{...(3)}$$

Also,
$$\lim_{t\to\infty}[f(t)] = \lim_{t\to\infty}\int_t^\infty \left(\frac{e^{-x}}{x}\right) dx = 0 \qquad \text{...(4)}$$

Putting the values of $\lim_{s\to 0}[s\,F(s)]$ and $\lim_{t\to\infty}[f(t)]$ from (3) and (4) in (2), we get C = 0.

Hence from (1), $sF(s) = \log(s+1) \Rightarrow F(s) = \left\{\dfrac{\log(s+1)}{s}\right\}$

$$\Rightarrow \qquad \boxed{L\int_t^{\infty}\left(\frac{e^{-x}}{x}\right)dx = \left[\frac{\log(s+1)}{s}\right]}$$

Example 10. *Find the Laplace transform of t sin at*

Solution.

$$L(t\sin at) = L\left(t\frac{e^{iat}-e^{-iat}}{2i}\right) = \frac{1}{2i}[L(t\cdot e^{iat}) - L(te^{-iat})]$$

$$= \frac{1}{2i}\left[\frac{1}{(s-ia)^2} - \frac{1}{(s+ia)^2}\right] = \frac{1}{2i}\left[\frac{(s+ia)^2-(s-ia)^2}{(s-ia)^2(s+ia)^2}\right]$$

$$= \frac{1}{2i}\frac{(s^2+2ias-a^2)-(s^2-2ias-a^2)}{(s^2+a^2)^2}$$

$$= \frac{1}{2i}\frac{4ias}{(s^2+a^2)^2} = \frac{2as}{(s^2+a^2)^2} \qquad \textbf{Ans.}$$

Example 11. *Find the Laplace transform of t sinh at*

Solution.

$$L(\sinh at) = \frac{a}{s^2-a^2}$$

$$= -\frac{d}{ds}\left(\frac{a}{s^2-a^2}\right)$$

$$= \frac{2as}{(s^2-a^2)^2} \qquad \textbf{Ans.}$$

Example 12. *Find the Laplace transform of t^2 cos at*

Solution.

$$L(\cos at) = \frac{s}{s^2+a^2}$$

$$L(t^2\cos at) = (-1)^2\frac{d^2}{ds^2}\left[\frac{s}{s^2+a^2}\right] = \frac{d}{ds}\frac{(s^2+a^2)1-s(2s)}{(s^2+a^2)^2}$$

$$= \frac{d}{ds}\frac{a^2-s^2}{(s^2+a^2)^2}$$

$$= \frac{(s^2+a^2)^2(-2s)-(a^2-s^2)2(s^2+a^2)(2s)}{(s^2+a^2)^4}$$

$$= \frac{(s^2 + a^2)(-2s) - (a^2 - s^2)4s}{(s^2 + a^2)^3}$$

$$= \frac{-2s^3 - 2a^2s - 4a^2s + 4s^3}{(s^2 + a^2)^3} = \frac{2s(s^2 - 3a^2)}{(s^2 + a^2)^3}$$ **Ans.**

Example 13. *Obtain the Laplace transform of $t^2e^t.\sin 4t$*

Solution. $L(\sin 4t) = \dfrac{4}{s^2 + 16}$

$$\Rightarrow \quad L(e^t \sin 4t) = \frac{4}{(s-1)^2 + 16}$$

$$\Rightarrow \quad L(te^t \sin 4t) = -\frac{d}{ds}\left(\frac{4}{s^2 - 2s + 17}\right) = \frac{4(2s-2)}{(s^2 - 2s + 17)^2}$$

$$\Rightarrow \quad L(t^2e^t \sin 4t) = -\frac{d}{ds}\left(\frac{4(2s-2)}{(s^2 - 2s + 17)^2}\right)$$

$$= -4\frac{(s^2 - 2s + 17)^2 2 - (2s-2)2(s^2 - 2s + 17)(2s-2)}{(s^2 - 2s + 17)^4}$$

$$= -4\frac{(s^2 - 2s + 17)^2 2 - 2(2s-2)^2}{(s^2 - 2s + 17)^3}$$

$$= \frac{-4(2s^2 - 4s + 34 - 8s^2 + 16s - 8)}{(s^2 - 2s + 17)^3}$$

$$= \frac{-4(-6s^2 + 12s + 26)}{(s^2 - 2s + 17)^3} = \frac{8(3s^2 - 6s - 13)}{(s^2 - 2s + 17)^3}$$ **Ans.**

Example 14. *Find the Laplace transform of the function*

$$f(t) = te^{-t} \sin 2t$$

Solution. $L[\sin 2t] = \dfrac{2}{s^2 + 4}$

$$\Rightarrow \quad L[e^{-t} \sin 2t] = \frac{2}{(s+1)^2 + 4} = F(s) \text{ (say)}$$

$$\Rightarrow \quad L[te^{-t} \sin 2t] = -F'(s) = -\frac{d}{ds}\left[\frac{2}{(s+1)^2 + 4}\right] = \frac{2.2(s+1)}{\left[(s+1)^2 + 4\right]^2}$$

$$= \frac{4(s+1)}{\left[(s+1)^2 + 4\right]^2}$$ **Ans.**

EXERCISE 35.2

Find the Laplace transforms of the following.

1. $t \sin at$ **Ans.** $\dfrac{as}{(s^2 \quad a^2)^2}$

2. $t \cosh at$ **Ans.** $\dfrac{s^2 + a^2}{(s^2 - a^2)^2}$

3. $t \cos t$ **Ans.** $\dfrac{s^2 - 1}{(s^2 + 1)^2}$

4. $t \cosh t$ **Ans.** $\dfrac{s^2 + 1}{(s^2 - 1)^2}$

5. $t^2 \sin t$ **Ans.** $\dfrac{2(3s^2 - 1)}{(s^2 + 1)^3}$

6. $t^3 e^{-3t}$ **Ans.** $\dfrac{6}{(s + 3)^4}$

7. $t\sin^2 3t$ **[DIPIETE June 2019]** **Ans.** $\dfrac{1}{2}\left[\dfrac{1}{s^2} - \dfrac{s^2 - 36}{(s^2 + 36)^2}\right]$

8. $t\, e^{at} \sin at$ **Ans.** $\dfrac{2a(s - a)}{\left(s^2 - 2as + 2a^2\right)^2}$

9. $te^{-t} \cosh t$ **Ans.** $\dfrac{2(s^3 + 6s^2 + 9s + 2)}{(s^2 + 4s + 5)^3}$

10. $t^2 e^{-2t} \cos t$ **Ans.** $\dfrac{2(s^3 + 10s^2 + 25s + 22)}{(s^2 + 4s + 5)^3}$

11. $\int_0^t e^{-2t} t \sin^3 t \, dt$ **Ans.** $\dfrac{3(s + 2)}{2s}\left[\dfrac{1}{\left[(s + 2)^2 + 9\right]^2} - \dfrac{1}{\left[(s + 2)^2 + 1\right]^2}\right]$

12. If $f(t)$ is continuous, except for an ordinary discontinuity at $t = a$, $(a > 0)$ as given in the figure, then show that
$L[f'(t)] = s[f(t)] - f(0) - e^{-as}[f(a + 0) - f(a - 0)]$

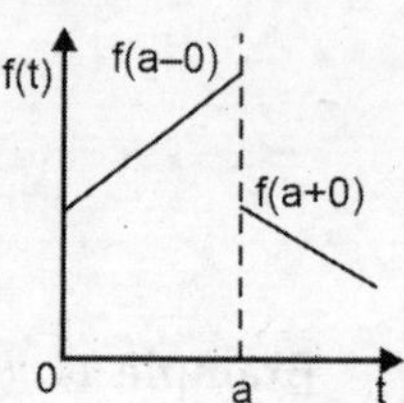

13. (a) Laplace transform of $t^n e^{-at}$ is

(i) $\dfrac{\overline{|n}}{(s + a)^n}$ (ii) $\dfrac{(n + 1)!}{(s + a)^{n+1}}$

(iii) $\dfrac{n!}{(s + a)^n}$ (iv) $\dfrac{\overline{|n + 1}}{(s + a)^{n+1}}$ **Ans.** (*iv*)

(b) Laplace transform of $f(t) = te^{at}.\ \sin(at),\ t > 0$

(i) $\dfrac{2a(s - a)}{\left[(s - a)^2 + a^2\right]^2}$ (ii) $\dfrac{a(s - a)}{(s - a)^2 + a^2}$

(iii) $\dfrac{s - a}{(s - a)^2 + a^2}$ (iv) $\dfrac{(s - a)^2}{(s - a)^2 + a^2}$ **Ans.** (*i*)

(c) If $f(x) = x^4 f(x)$, where $f(x)$ has derivatives of all orders, then $L\left[\dfrac{d^4 f(x)}{dx^4}\right]$ is given by

(i) $s^3L\,[f(x)]$ (ii) $s^4Lf(x)$

(iii) $s^4L\,[f^3(x)]$ (iv) none of these **Ans.** (ii)

(d) The Laplace transform of $te^{-t}\cos h\,2t$ is

(i) $\dfrac{s^2+2s+5}{(s^2+2s-3)^2}$ (ii) $\dfrac{s^2-2s+5}{(s^2+2s-3)^2}$

(iii) $\dfrac{4s+4}{(s^2+2s-3)^2}$ (iv) $\dfrac{4s-4}{(s^2+2s-3)^2}$ **Ans.** (i)

35.14 LAPLACE TRANSFORM OF $\frac{1}{t}f(t)$ (Division by t)

If $L[f(t)] = F(s)$, then $L\left[\dfrac{1}{t}f(t)\right] = \int_0^\infty F(s)ds$ **(U.P. II Semester Summer, 2007, 2005)**

Proof. We know that $L[f(t)] = F(s)$ or $F(s) = \int_0^\infty e^{-st} f(t)dt$ ... (1)

Integrating (1) w.r.t. 's', we have

$$\int_s^\infty F(s)ds = \int_s^\infty \left[\int_0^\infty e^{-st} f(t)dt\right] ds$$

$$= \int_0^\infty f(t)\left[\int_s^\infty e^{-st}ds\right]dt = \int_0^\infty f(t)\left[\frac{e^{-st}}{-t}\right]_s^\infty dt$$

$$= \int_0^\infty \frac{-f(t)}{t}\left[e^{-st}\right]_s^\infty dt = \int_0^\infty \frac{-f(t)}{t}\left[0-e^{-st}\right]dt$$

$$= \int_0^\infty e^{-st}\left\{\frac{1}{t}.f(t)\right\} dt = L\left[\frac{1}{t}f(t)\right]$$

$$\Rightarrow \quad \boxed{L\left[\frac{1}{t}f(t)\right] = \int_s^\infty F(s)ds}$$ **Proved.**

Cor. $$L^{-1}\int_s^\infty F(s)\,ds = \frac{1}{t}f(t)$$

Example 15. *Find the Laplace transform of* $\dfrac{\sin 2t}{t}$.

Solution. $$L(\sin 2t) = \frac{2}{s^2+4}$$

$$L\left(\frac{\sin 2t}{t}\right) = \int_s^\infty \frac{2}{s^2+4}ds = 2.\frac{1}{2}\left[\tan^{-1}\frac{s}{2}\right]_s^\infty = \left[\tan^{-1}\infty - \tan^{-1}\frac{s}{2}\right]$$

$$= \frac{\pi}{2} - \tan^{-1}\left(\frac{s}{2}\right) = \cot^{-1}\left(\frac{s}{2}\right)$$

Example 16. *Find the Laplace transform of* $f(t) = \int_0^t \frac{\sin at}{t} dt$

(U.P. II Semester Summer 2005)

Solution.
$$L(\sin at) = \frac{a}{s^2 + a^2}$$

$$\Rightarrow \quad L\left(\frac{\sin at}{t}\right) = \int_s^\infty \frac{a}{s^2 + a^2} ds = \left[\tan^{-1}\frac{s}{a}\right]_s^\infty$$

$$= \frac{\pi}{2} - \tan^{-1}\left(\frac{s}{a}\right) = \cot^{-1}\left(\frac{s}{a}\right)$$

Hence
$$L\left[\int_0^t \frac{\sin at}{t} dt\right] = \frac{1}{s}\cot^{-1}\left(\frac{s}{a}\right)$$
Ans.

Example 17. *Find the Laplace transform of* $\frac{\cos at - \cos bt}{t}$

(Uttarakhand II Semester June 2007)

Solution. Here,
$$f(t) = \frac{\cos at - \cos bt}{t}$$

We know that, $L(\cos at - \cos bt) = L(\cos at) - L(\cos bt) = \frac{s}{s^2 + a^2} - \frac{s}{s^2 + b^2}$

$$L\left(\frac{\cos at - \cos bt}{t}\right) = \int_s^\infty \left(\frac{s}{s^2 + a^2} - \frac{s}{s^2 + b^2}\right) ds$$

$$= \left[\frac{1}{2}\log\left(s^2 + a^2\right) - \frac{1}{2}\log\left(s^2 + b^2\right)\right]_s^\infty$$

$$= \frac{1}{2}\left[\log\left(\frac{s^2 + a^2}{s^2 + b^2}\right)\right]_s^\infty = \frac{1}{2}\left[\log\left(\frac{1 + \frac{a^2}{s^2}}{1 + \frac{b^2}{s^2}}\right)\right]_s^\infty$$

$$= \frac{1}{2}\log 1 - \frac{1}{2}\log\left(\frac{1 + \frac{a^2}{s^2}}{1 + \frac{b^2}{s^2}}\right)$$

$$= 0 - \frac{1}{2}\log\left(\frac{s^2 + a^2}{s^2 + b^2}\right) \qquad [\log 1 = 0]$$

$$= \frac{1}{2}\log\left(\frac{s^2 + b^2}{s^2 + a^2}\right)$$
Ans.

Example 18. *If* $f(t) = \dfrac{e^{at} - \cos bt}{t}$, *find the Laplace transform of* $f(t)$.

Solution.
$$f(t) = \frac{e^{at} - \cos bt}{t} = \frac{e^{at}}{t} - \frac{\cos bt}{t}$$

We know that,
$$L(e^{at} - \cos bt) = \left(\frac{1}{s-a} - \frac{s}{s^2+b^2}\right)$$

$$\Rightarrow \quad L\left(\frac{e^{at} - \cos bt}{t}\right) = \int_s^\infty \left(\frac{1}{s-a} - \frac{s}{s^2+b^2}\right) ds$$

$$= \left[\log(s-a) - \frac{1}{2}\log(s^2+b^2)\right]_s^\infty$$

$$= \left[\frac{2\log(s-a) - \log(s^2+b^2)}{2}\right]_s^\infty$$

$$= \frac{1}{2}\left[\log(s-a)^2 - \log(s^2+b^2)\right]_s^\infty$$

$$= \frac{1}{2}\left[\log\frac{(s-a)^2}{s^2+b^2}\right]_s^\infty = \frac{1}{2}\left[\log\left\{\frac{\left(1-\frac{a}{s}\right)^2}{1+\frac{b^2}{s^2}}\right\}\right]_s^\infty$$

$$= \frac{1}{2}\left[0 - \log\frac{\left(1-\frac{a}{s}\right)^2}{1+\frac{b^2}{s^2}}\right] = \frac{1}{2}\left[\log\frac{s^2+b^2}{(s-a)^2}\right]$$ **Ans.**

Example 19. *Find the Laplace transform of* $\dfrac{1-\cos t}{t^2}$. **[DIP IETE 2018, 2019]**

Solution.
$$L(1-\cos t) = L(1) - L(\cos t) = -$$

$$\Rightarrow \quad L\left[\frac{1-\cos t}{t}\right] = \int_s^\infty \left(\frac{1}{s} - \frac{s}{s^2+1}\right) ds = \left[\log s - \frac{1}{2}\log(s^2+1)\right]_s^\infty$$

$$= \frac{1}{2}\left[\log s^2 - \log\left(s^2+1\right)\right]_s^\infty = \frac{1}{2}\left[\log\left(\frac{s^2}{s^2+1}\right)\right]_s^\infty$$

$$= \frac{1}{2}\left[\log\frac{s^2}{s^2\left(1+\frac{1}{s^2}\right)}\right]_s^\infty = \frac{1}{2}\left[0 - \log\left(\frac{s^2}{s^2+1}\right)\right]$$

$$= -\frac{1}{2}\log\left(\frac{s^2}{s^2+1}\right)$$

Again, $$L\left[\frac{1-\cos t}{t^2}\right] = -\frac{1}{2}\int_s^\infty \log\left(\frac{s^2}{s^2+1}\right)ds$$

$$= -\frac{1}{2}\int_s^\infty \left(\log\frac{s^2}{s^2+1}.1\right)ds$$

Integrating by parts, we have

$$= -\frac{1}{2}\left[\log\left(\frac{s^2}{s^2+1}\right)s - \int\frac{s^2+1}{s^2}\frac{(s^2+1)2s-s^2(2s)}{(s^2+1)^2}.sds\right]_s^\infty$$

$$= -\frac{1}{2}\left[s\log\left(\frac{s^2}{s^2+1}\right) - 2\int\frac{1}{s^2+1}ds\right]_s^\infty$$

$$= -\frac{1}{2}\left[s\log\left(\frac{s^2}{s^2+1}\right) - 2\tan^{-1}s\right]_s^\infty$$

$$= -\frac{1}{2}\left[0-2\left(\frac{\pi}{2}\right) - s\log\left(\frac{s^2}{s^2+1}\right) + 2\tan^{-1}s\right]$$

$$= -\frac{1}{2}\left[-\pi - s\log\left(\frac{s^2}{s^2+1}\right) + 2\tan^{-1}s\right]$$

$$= \frac{\pi}{2} + \frac{s}{2}\log\left(\frac{s^2}{s^2+1}\right) - \tan^{-1}s$$

$$= \left(\frac{\pi}{2} - \tan^{-1}s\right) + \frac{s}{2}\log\left(\frac{s^2}{s^2+1}\right)$$

$$= \cot^{-1}s + \frac{s}{2}\log\left(\frac{s^2}{s^2+1}\right)$$ **Ans.**

Example 20. *Evaluate* $L\left[e^{-4t}\frac{\sin 3t}{t}\right]$.

Solution. $$L[\sin 3t] = \frac{3}{s^2+3^2}$$

$$\Rightarrow \quad L\left[\frac{\sin 3t}{t}\right] = \int_s^\infty \left(\frac{3}{s^2+9}\right)ds = \left[\frac{3}{3}\tan^{-1}\left(\frac{s}{3}\right)\right]_s^\infty = \left[\tan^{-1}\left(\frac{s}{3}\right)\right]_s^\infty$$

$$= \frac{\pi}{2} - \tan^{-1}\left(\frac{s}{3}\right) = \cot^{-1}\left(\frac{s}{3}\right)$$

$$L\left[e^{-4t}\frac{\sin 3t}{t}\right] = \cot^{-1}\left(\frac{s+4}{3}\right) = \tan^{-1}\left(\frac{3}{s+4}\right) \qquad \textbf{Ans.}$$

EXERCISE 35.3

Find the Laplace transform of the following.

1. $\frac{1}{t}(1-e^t)$ **Ans.** $\log\left(\frac{s-1}{s}\right)$

2. $\frac{1}{t}\left(e^{-at}-e^{-bt}\right)$ **Ans.** $\log\left(\frac{s+b}{s+a}\right)$

3. $\frac{1}{t}(1-\cos at)$ **Ans.** $-\frac{1}{2}\log\left(\frac{s^2}{s^2+a^2}\right)$

4. $\frac{1}{t}\sin^2 t$ **Ans.** $\frac{1}{4}\log\left(\frac{s^2+4}{s^2}\right)$

5. $\frac{1}{t}\sinh t$ **Ans.** $-\frac{1}{2}\log\left(\frac{s-1}{s+1}\right)$

6. $\frac{1}{t}\left(e^{-t}\sin t\right)$ **Ans.** $\cot^{-1}(s+1)$

7. $\frac{1}{t}(1-\cos t)$ **Ans.** $\frac{1}{2}\left[\log\left(s^2+1\right)-\log s\right]$

8. $\int_0^\infty \frac{1}{t}e^{-2t}\sin t\,dt$ **Ans.** $\frac{1}{s}\cot^{-1}(s+2)$

9. $\int_0^\infty \frac{e^{-t}-e^{-3t}}{t}dt$ **Ans.** $\log 3$

35.15 UNIT STEP FUNCTION

With the help of unit step functions, we can find the inverse transform of functions, which cannot be determined with previous methods.

The unit step function $u(t-a)$ is defined as follows :

$$u(t-a) = \frac{e^{-as}}{s}, \begin{cases} \textbf{0 when } t < a \\ \textbf{1 when } t \geq a \end{cases} \quad \text{where } a \geq 0$$

u (t–a)
1
A
0
a
t

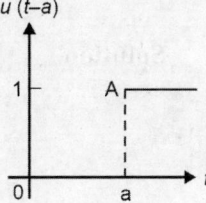

35.16 LAPLACE TRANSFORM OF UNIT FUNCTIION

$$L[u(t-a)] = \frac{e^{-as}}{s}$$

Proof. $$L[u(t-a)] = \int_0^\infty e^{-st} u(t-a)dt$$

$$= \int_0^a e^{-st}.0dt + \int_a^\infty e^{-st}.1dt = 0 + \left[\frac{e^{-st}}{-s}\right]_a^\infty$$

$$\boxed{L[u(t-a)] = \frac{e^{-as}}{s}}$$ **Proved.**

Example 21. *Express the following function in terms of unit step functions and find its Laplace transform :*

$$f(t) = \begin{cases} 8, & t < 2 \\ 6, & t \ge 2 \end{cases}$$

Solution. $$f(t) = \begin{cases} 8+0, & t < 2 \\ 8-2, & t \ge 2 \end{cases}$$

$$= 8 + \begin{bmatrix} 0, & t < 2 \\ -2, & t \ge 2 \end{bmatrix}$$

$$= 8 + (-2)\begin{cases} 0, & t < 2 \\ 1, & t \ge 2 \end{cases}$$

$$= 8 - 2u(t-2)$$

$$Lf(t) = 8L(1) - 2Lu(t-2) = \frac{8}{s} - 2\frac{e^{-2s}}{s}$$ **Ans.**

Example 22. *Draw the graph of $u(t-a) - u(t-b)$.*

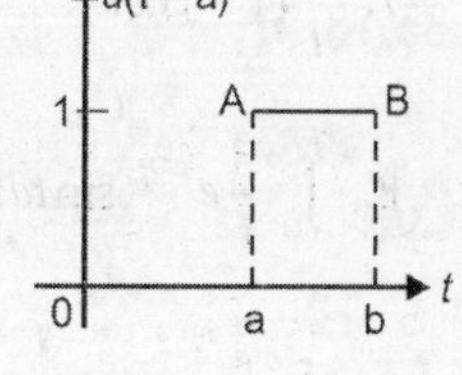

Solution. As in Section 35.15 the graph of $u(t-a)$ is a straight line parallel to t-axis from A to ∞.

Similarly, the graph of $u(t-b)$ is a straight line parallel to t-axis from B to ∞.

Hence, the graph of $u(t-a) - u(t-b)$ is AB. **Ans.**

Example 23. *Express the following function in terms of unit step function and find its Laplace transform :*

$$f(t) = \begin{cases} E, & a < t < b \\ 0, & t \ge b \end{cases}$$

Solution. $$f(t) = E\begin{cases} 1, & a < t < b \\ 0, & t \ge b \end{cases} \quad [L[f(t-a).u(t-a)] = e^{-as}F(s)]$$

$$Lf(t) = E\left[\frac{e^{-as}}{s} - \frac{e^{-bs}}{s}\right]$$ **Ans.**

35.17 SECOND SHIFTING THEOREM

If $L[f(t)] = F(s)$, then $\boxed{L[f(t-a).u(t-a)] = e^{-as}F(s)}$

Proof.

$$L[f(t-a).u(t-a))] = \int_0^\infty e^{-st}[f(t-a).u(t-a)]dt$$

$$= \int_0^a e^{-st} f(t-a).0dt + \int_a^\infty e^{-st} f(t-a)(1)dt$$

$$= \int_a^\infty e^{-st} f(t-a)dt$$

$$= \int_0^\infty e^{-s(u+a)} f(u)du, \qquad \text{where } u = t-a$$

$$= e^{-sa}\int_0^\infty e^{-su}.f(u)du = e^{-sa}F(s)$$ **Proved.**

Example 24. *Express the following function in terms of unit step function :*

$$f(t) = \begin{cases} t-1, & 1 < t < 2 \\ 3-t, & 2 < t < 3 \end{cases}$$

and find its Laplace transform.

Solution. $f(t) = \begin{cases} t-1, & 1 < t < 2 \\ 3-t, & 2 < t < 3 \end{cases}$

$= (t-1)\,[u\,(t-1) - u\,(t-2)] + (3-t)\,[u\,(t-2) - u(t-3)]$

$= (t-1)\,u(t-1) - (t-1)\,u(t-2) + (3-t)\,u(t-2) + (t-3)u(t-3)$

$= (t-1)\,u(t-1) - 2\,(t-2)\,u(t-2) + (t-3)\,u(t-3)$

$= e^{-s}\,L(t) - 2\,e^{-2s}\,L(t) - e^{-3s}\,L(t),\ [L[f(t-a)\,.\,u(t-a)] = e^{-as}F(s)]$

$$Lf(t) = \frac{e^{-s}}{s^2} - 2\frac{e^{-2s}}{s^2} + \frac{e^{-3s}}{s^2}$$ **Ans.**

35.18 THEOREM $Lf(t)\,u(t-a) = e^{-as}\,L[\,f(t+a)]$

Proof.

$$Lf(t).u(t-a) = \int_0^\infty e^{-st}[f(t).u(t-a)]\,dt$$

$$= \int_0^a e^{-st}[f(t).u(t-a)]\,dt + \int_a^\infty e^{-st}.[f(t).u(t-a)]\,dt$$

$$= 0 + \int_a^\infty e^{-st}.f(t)(1)\,dt$$

$$= \int_a^\infty e^{-s(y+a)}.f(y+a)\,dy$$

$$= e^{-as}\int_a^\infty e^{-sy}.f(y+a)dy \qquad (t-a=y)$$

$$= e^{-as}\int_a^\infty e^{-st} f(t+a)dt = e^{-as}Lf(t+a)$$ **Proved.**

Example 25. *Find the Laplace transform of $t^2u\,(t-3)$.*

Solution. $t^2.u(t-3) = [(t-3]^2 + 6\,(t-3) + 9]\;u\,(t-3)$

$= (t-3)^2.u(t-3) + 6(t-3).u(t-3) + 9u\,(t-3)$

$$Lt^2.u(t-3) = L(t-3)^2.u(t-3) + 6L(t-3).u(t-3) + 9Lu(t-3)$$

$$= e^{-3s}\left[\frac{2}{s^3}+\frac{6}{s^2}+\frac{9}{s}\right]$$

Aliter.

$$Lt^2u\,(t-3) = e^{-3s}L(t+3)^2 = e^{-3s}L(t^2+6t+9)$$

$$= e^{-3s}\left[\frac{2}{s^3}+\frac{6}{s^2}+\frac{9}{s}\right] \qquad \textbf{Ans.}$$

Example 26. *Find the Laplace transform of $e^{-2t}u_\pi(t)$ where*

$$u_\pi(t) = \begin{cases} 0; & t<\pi \\ 1; & t>\pi \end{cases}$$

Solution.

$$u_\pi(t) = \begin{cases} 0; & t<\pi \\ 1; & t>\pi \end{cases}$$

$$u_\pi(t) = u\,(t-\pi)$$

$$L[u_\pi(t)] = L[u\,(t-\pi)]$$

$$= \frac{e^{-\pi s}}{s}$$

$$\therefore \quad L[e^{-2t}\,u_\pi(t)] = \frac{e^{-\pi(s+2)}}{s+2} \qquad \textbf{Ans.}$$

Example 27. *Express the following function in terms of unit step function and find its Laplace transform :* $f(t) = \begin{cases} 0, & 0<t<1 \\ t-1, & 1<t<2 \\ 1, & 2<t \end{cases}$

Solution. The given function shown in the figure is expressed in algebraic form

$$f(t) = \begin{cases} 0, & 0<t<1 \\ t-1, & 1<t<2 \\ 1, & 2<t \end{cases}$$

$$f(t) = (t-1)\,[u(t-1) - u(t-2)] + u\,(t-2)$$

$$= (t-1)u(t-1) - u(t-2)(t-1-1)$$

$$= (t-1)u(t-1) - (t-2)u\,(t-2)$$

$$Lf(t) = L(t-1)u(t-1) - L(t-2)u\,(t-2)$$

$$= \frac{e^{-s}}{s^2} - \frac{e^{-2s}}{s^2} \qquad \textbf{Ans.}$$

Example 28. *Represent $f(t) = \sin 2t$, $2\pi < t < 4\pi$ and $f(t) = 0$ otherwise, in terms of unit step function and then find its Laplace transform.*

Solution.

$$f(t) = \begin{cases} \sin 2t, & 2\pi < t < 4\pi \\ 0, & \text{otherwise} \end{cases}$$

$$f(t) = \sin 2t\,[u(t-2\pi) - u(t-4\pi)]$$

$$Lf(t) = L\,[\sin 2t.u(t-2\pi)] - L\,[\sin 2t.u(t-4\pi)]$$

$$= L\,[\sin 2\,(t + 2\pi)] - e^{-4\pi s}\,L[\sin 2\,(t + 4\pi)]$$

$$= e^{-2\pi s}L\,[\sin 2t] - e^{-4\pi s}\,L\,[\sin\,(2t)]$$

$$= e^{-2\pi s}\left(\frac{2}{s^2+4}\right) - e^{-4\pi s}\left(\frac{2}{s^2+4}\right)$$

$$= \left(e^{-2\pi s} - e^{-4\pi s}\right)\frac{2}{s^2+4}$$ **Ans.**

Example 29. *A function $f(t)$ obeys the equation $f(t) + 2\int_0^t f(t)\,dt = \cosh 2t$.*

Find the Laplace transform of $f(t)$ **(U.P. II Semester Summer 2006)**

Solution. We have $f(t) + 2\int_0^t f(t)dt = \cosh 2t$

Taking Laplace transformation of both the sides, we get

$$L\{f(t)\} + 2L\int_0^t f(t)dt = L(\cosh 2t)$$

$$\Rightarrow \quad F(s) + 2.\frac{1}{s}F(s) = \frac{s}{s^2 - 4}$$

$$\Rightarrow \quad F(s)\left\{1 + \frac{2}{s}\right\} = \frac{s}{s^2 - 4}$$

$$\Rightarrow \quad F(s)\left\{\frac{s+2}{s}\right\} = \frac{s}{s^2 - 4}$$

$$\Rightarrow \quad F(s) = \left(\frac{s}{s^2 - 4}\right)\left(\frac{s}{s+2}\right)$$

$$\Rightarrow \quad F(s) = \frac{s^2}{(s^2-4)(s+2)}$$ **Ans.**

EXERCISE 35.4

Find the Laplace transform of the following.

1. $f(t) = \begin{cases} t-1, & 1 < t < 2 \\ 0, & \text{otherwise} \end{cases}$ **Ans.** $\dfrac{e^{-s} - e^{-2s}}{s^2} - \dfrac{e^{-2s}}{s}$

2. $e^t u(t-1)$ **Ans.** $\dfrac{e^{-(s-1)}}{s-1}$

3. $\dfrac{1-e^{2t}}{s} + tu(t) + \cosh t.\cos t$ **Ans.** $\log\left(\dfrac{s-2}{s}\right) + \dfrac{1}{s^2} + \dfrac{s^3}{s^4+4}$

4. $t^2 u(t-2)$ **Ans.** $\dfrac{e^{-2s}}{s^3}(4s^2 + 4s + 2)$

5. $\sin t\, u(t-4)$ **Ans.** $\dfrac{e^{-4s}}{s^2+1}[\cos 4 + s\sin 4]$

6. $f(t) = k(t-2)\,[u(t-2) - u(t-3)]$ **Ans.** $\dfrac{k}{s^2}\left[e^{-2s} - (s+1)e^{-3s}\right]$

7. $f(t) = \dfrac{k\sin \pi t}{T}[u(t-2T) - u(t-3T)]$ **Ans.** $\dfrac{k\pi T}{s^2T^2+\pi^2}\left(e^{-2sT} - e^{-3sT}\right)$

Express the following in terms of unit step functions and find Laplace transform.

8. $f(t) = \begin{cases} t, & 0 < t < 2 \\ 0, & 2 < t \end{cases}$ **Ans.** $u(t) - u(t-2), \dfrac{1-(2s+1)e^{-2s}}{s^2}$

9. $f(t) = \begin{cases} \sin t, & 0 < t < \pi \\ t, & t > \pi \end{cases}$ **Ans.** $\dfrac{1+e^{-\pi s}}{s^2+1} + \dfrac{e^{-\pi s}(\pi s+1)}{s^2}$

10. $f(t) = \begin{cases} 4, & 0 < t < 1 \\ -2, & 0 < t < 3 \\ 5, & t > 3 \end{cases}$ **Ans.** $\dfrac{4-6e^{-s}+7e^{-3s}}{s}$

11. The Laplace transform of $tu_2(t)$ is

(*i*) $\left(\dfrac{1}{s^2}+\dfrac{2}{s}\right)e^{-2s}$ (*ii*) $\dfrac{1}{s^2}e^{-2s}$

(*iii*) $\left(\dfrac{1}{s^2}-\dfrac{2}{s}\right)e^{-2s}$ (*iv*) $\dfrac{e^{-2s}}{s^2}$ **Ans.** (*i*)

35.19 PERIODIC FUNCTION

Let $f(t)$ be a periodic function with period T, then

$$\boxed{L[f(t)] = \frac{\int_0^T e^{-st} f(t)\, dt}{1-e^{-sT}}}$$

Proof. $L[f(t)] = \int_0^\infty e^{-st} f(t)\,dt = \int_0^T e^{-st} f(t)\,dt + \int_T^{2T} e^{-st} f(t)\,dt + \int_{2T}^{3T} e^{-st} f(t)\,dt + \ldots$

Substituting $t = u + T$ in second integral and $t = u + 2T$ in third integral, and so on.

$$L[f(t)] = \int_0^T e^{-st} f(t)\,dt + \int_0^T e^{-s(u+T)} f(u+T)\,du + \int_0^T e^{-s(u+2T)} f(u+2T)\,du + \ldots$$

$$= \int_0^T e^{-st} f(t)\,dt + e^{-sT}\int_0^T e^{-st} f(u)\,du + e^{-2sT}\int_0^T e^{-su} f(u)\,du + \ldots$$

$[f(u) = f(u+T) = f(u+2T) = f(u+3T) = \ldots]$

$$= \int_0^T e^{-st} f(t)\,dt + e^{-sT}\int_0^T e^{-sT} f(t) + e^{-2sT}\int_0^T e^{-st} f(t)\,dt + \ldots$$

$$= \left[1 + e^{-sT} + e^{-2sT} + e^{-3sT} + \ldots\right]$$

$$\int_0^T e^{-st} f(t)\, dt, \left[1 + a + a^2 + a^3 + \ldots = \frac{1}{1-a}\right]$$

$$= \frac{1}{1-e^{-sT}} \int_0^T e^{-st} f(t)\, dt.$$ **Proved.**

Example 30. *Find the Laplace transform of the waveform*

$$f(t) = \left(\frac{2t}{3}\right), 0 \le t \le 3.$$

Solution. $$L[f(t)] = \frac{1}{1-e^{-sT}} \int_0^T e^{-st} f(t)\, dt$$

$\Rightarrow$ $$L\left[\frac{2t}{3}\right] = \frac{1}{1-e^{-3s}} \int_0^3 e^{-st} \left(\frac{2}{3} t\right) dt$$

$$= \frac{1}{1-e^{-3s}} \frac{2}{3} \left[\frac{te^{-st}}{-s} - (1)\frac{e^{-st}}{s^2}\right]_0^3$$

$$= \frac{2}{3} \frac{1}{1-e^{-3s}} \left[\frac{3e^{-3s}}{-s} - \frac{e^{-3s}}{s^2} + \frac{1}{s^2}\right]$$

$$= \frac{2}{3} \cdot \frac{1}{1-e^{-3s}} \left[\frac{3e^{-3s}}{-s} + \frac{1-e^{-3s}}{s^2}\right]$$

$$= \frac{2e^{-3s}}{-s\left(1-e^{-3s}\right)} + \frac{2}{3s^2}$$ **Ans.**

Example 31. *Draw the graph and find the Laplace transform of the triangular wave function of period 2 C given by*

$$f(t) = \begin{cases} t, & 0 < t \le C \\ 2C - t, & C < t < 2C \end{cases}$$ **(Uttarakhand II Semester June 2007)**

Solution. Period = 2C = T

Laplace transform of periodic function $f(t)$

$$L\{f(t)\} = \frac{\int_0^T e^{-st} f(t)\, dt}{1-e^{-sT}}$$

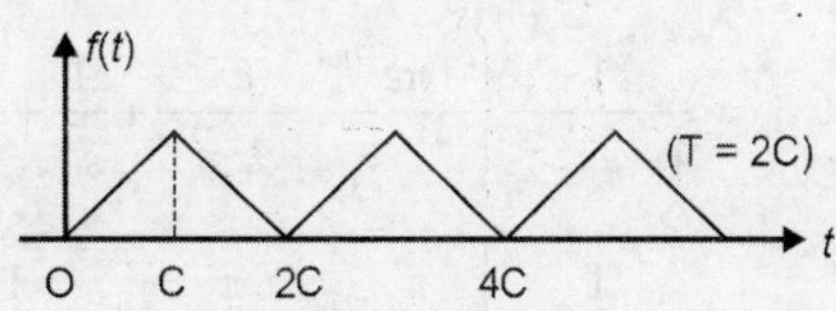

$$= \frac{1}{1-e^{-2Cs}} \int_0^{2C} e^{-st} f(t)\, dt$$

On putting the values of $f(t)$, we get

$$L[f(t)] = \frac{1}{1-e^{-2Cs}} \left[\int_0^C e^{-st}\, t dt + \int_C^{2C} e^{-st} (2C - t) dt\right]$$

$$= \frac{1}{1-e^{-2Cs}}\left[\left\{\frac{te^{-st}}{-s} - 1\cdot\frac{e^{-st}}{(-s)^2}\right\}_0^C + \left\{(2C-t)\frac{e^{-st}}{(-s)} - (-1)\frac{e^{-st}}{(-s)^2}\right\}_C^{2C}\right]$$

$$= \frac{1}{1-e^{-2Cs}}\left[\left\{\frac{C\cdot e^{-Cs}}{-s} - \frac{e^{-Cs}}{(-s)^2} - 0 + \frac{1}{s^2}\right\} + \left\{(2C-2C)\frac{e^{-2Cs}}{(-s)} + \frac{e^{-2Cs}}{s^2} - \left((2C-C)\frac{e^{-Cs}}{-s} + \frac{e^{-Cs}}{s^2}\right)\right\}\right]$$

$$= \frac{1}{1-e^{-2Cs}}\left\{-\frac{Ce^{-Cs}}{s} - \frac{e^{-Cs}}{s^2} + \frac{1}{s^2} + \frac{e^{-2Cs}}{s^2} + \frac{Ce^{-Cs}}{s} - \frac{e^{-Cs}}{s^2}\right\}$$

$$= \frac{1}{1-e^{-2Cs}}\left\{\frac{1}{s^2}(1 - 2e^{-Cs} + e^{-2Cs})\right\}$$

$$= \frac{\left(1-e^{-Cs}\right)^2}{s^2\left(1+e^{-Cs}\right)\left(1-e^{-Cs}\right)} = \frac{1-e^{-Cs}}{s^2\left(1+e^{-Cs}\right)}$$ **Ans.**

Example 32. *Draw the graph of the periodic function* $f(t) = \begin{cases} t, & 0 < t < \pi \\ \pi - t, & \pi < t < 2\pi \end{cases}$ *and find its Laplace transform.*

Solution. Period = 2π = T

Laplace transform of periodic function

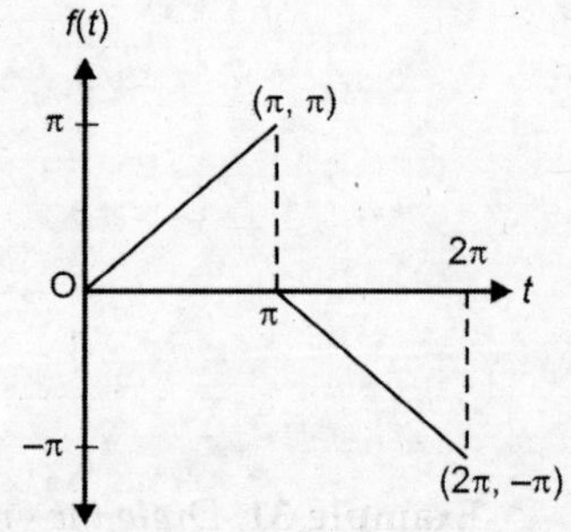

$$L\{f(t)\} = \frac{\int_0^T e^{-st} f(t)dt}{1-e^{-sT}}$$

$$= \frac{1}{1-e^{-2\pi s}}\int_0^{2\pi} e^{-st} f(t)dt$$

$$= \frac{1}{1-e^{-2\pi s}}\left[\int_0^{\pi} e^{-st}t\,dt + \int_{\pi}^{2\pi} e^{-st}(\pi - t)dt\right]$$

$$= \frac{1}{1-e^{-2\pi s}}\left\{\frac{te^{-st}}{-s} - 1\frac{e^{-st}}{(-s)^2}\right\}_0^{\pi} + \left\{(\pi - t)\frac{e^{-st}}{(-s)} - (-1)\frac{e^{-st}}{(-s)^2}\right\}_{\pi}^{2\pi}$$

$$= \frac{1}{1-e^{-2\pi s}}\left[\left\{\frac{\pi e^{-\pi s}}{-s} - \frac{e^{-\pi s}}{(-s)^2} - 0 + \frac{1}{s^2}\right\} + \left\{(\pi - 2\pi)\frac{e^{-2\pi s}}{-s} + \frac{e^{-2\pi s}}{s^2} - (\pi - \pi)\frac{e^{-\pi s}}{-s} + \frac{e^{-\pi s}}{s^2}\right\}\right]$$

$$= \frac{1}{1-e^{-2\pi s}}\left\{-\frac{\pi e^{-\pi s}}{s} - \frac{e^{-\pi s}}{s^2} + \frac{1}{s^2} + \pi\frac{e^{-2\pi s}}{s} + \frac{e^{-2\pi s}}{s^2} - 0 - \frac{e^{-\pi s}}{s^2}\right\}$$

$$= \frac{1}{1-e^{-2\pi s}}\left\{-\frac{\pi}{s}e^{-\pi s} + \frac{\pi}{s}e^{-2\pi s} + \frac{1}{s^2} - \frac{1}{s^2}e^{-\pi s} + \frac{1}{s^2}e^{-2\pi s} - \frac{e^{-\pi s}}{s^2}\right\}$$

$$= \frac{1}{1-e^{-2\pi s}}\left[\frac{\pi}{2}\left(e^{-2\pi s} - e^{-\pi s}\right) + \frac{1}{s^2}(1 + e^{-2\pi s} - 2e^{-\pi s})\right]$$

f(t)
π
(π, π)
O
π
2π
t
−π
(2π, −π)

$$= \frac{-\pi s e^{-\pi s}\left(1-e^{-\pi s}\right)+\left(1-e^{-\pi s}\right)^2}{s^2\left(1+e^{-\pi s}\right)\left(1-e^{-\pi s}\right)}$$

$$= \frac{-\pi s e^{-\pi s}+1-e^{-\pi s}}{s^2\left(1+e^{-\pi s}\right)}$$ **Ans.**

Example 33. *Find the Laplace transform of the function (half wave rectifier)*

$$f(t) = \begin{cases} \sin wt & \text{for} \quad 0 < t < \dfrac{\pi}{\omega} \\ 0 & \text{for} \quad \dfrac{\pi}{\omega} < t < \dfrac{2\pi}{\omega}. \end{cases}$$

Solution. $$L[f(t)] = \frac{1}{1-e^{-sT}}\int_0^T e^{-st} f(t)dt$$

$$= \frac{1}{1-e^{-\frac{2\pi s}{\omega}}}\int_0^{\frac{2\pi}{\omega}} e^{-st} f(t)dt \qquad \left[\begin{array}{l} f(t) \text{ is a periodic functiion} \\ T = \dfrac{2\pi}{\omega} \end{array}\right.$$

$$= \frac{1}{1-e^{-\frac{2\pi s}{\omega}}}\left[\int_0^{\frac{\pi}{\omega}} e^{-st} \sin \omega t dt + \int_{\frac{\pi}{\omega}}^{\frac{2\pi}{\omega}} e^{-st} \times 0 \times dt\right]$$

$$= \frac{1}{1-e^{-\frac{2\pi s}{\omega}}}\int_0^{\frac{\pi}{\omega}} e^{-st} \sin \omega t \, dt$$

$$\left[\int e^{ax} \sin bx \, dx = e^{ax}\frac{(a \sin bx - b \cos bx)}{a^2+b^2}\right]$$

$$L[f(t)] = \frac{1}{1-e^{-\frac{2\pi s}{\omega}}}\left[\frac{e^{-st}(-s\sin \omega t - \omega \cos \omega t)}{s^2+\omega^2}\right]_0^{\frac{\pi}{\omega}}$$

$$= \frac{1}{1-e^{-\frac{2\pi s}{\omega}}}\left[\frac{\omega e^{-\frac{\pi}{\omega}s}+\omega}{s^2+\omega^2}\right]$$

$$= \frac{\omega\left[1+e^{-\frac{\pi s}{\omega}}\right]}{(s^2+\omega^2)\left[1-e^{-\frac{2\pi s}{\omega}}\right]}$$

$$= \frac{\omega\left[1+e^{-\frac{\pi s}{\omega}}\right]}{(s^2+\omega^2)\left(1-e^{-\frac{\pi}{\omega}s}\right)\left(1+e^{-\frac{\pi}{\omega}s}\right)}$$

$$= \frac{\omega}{(s^2+\omega^2)\left[1-e^{-\frac{\pi s}{\omega}}\right]} \qquad \textbf{Ans.}$$

Example 34. *Find the Laplace transform of the periodic function (sawtooth wave).*

$$f(t) = \frac{kt}{T} \text{ for } 0 < t < T, \qquad f(t+T) = f(t)$$

Solution.

$$L[f(t)] = \frac{1}{1-e^{sT}}\int_0^T e^{-st} f(t)dt = \frac{1}{1-e^{sT}}\int_0^T e^{-st}\frac{kt}{T}dt$$

$$= \frac{1}{1-e^{sT}}\frac{k}{T}\int_0^T e^{-st}.tdt$$

$$= \frac{k}{T(1-e^{sT})}\left[t\frac{e^{-st}}{-s} - \int 1.\frac{e^{-st}}{-s}dt\right]_0^T \qquad \text{(Integration by parts)}$$

$$= \frac{k}{T(1-e^{sT})}\left[\frac{te^{-st}}{-s} - \frac{e^{-st}}{s^2}\right]_0^T$$

$$= \frac{k}{T(1-e^{sT})}\left[\frac{Te^{-sT}}{-s} - \frac{e^{-sT}}{s^2} + \frac{1}{s^2}\right]$$

$$= \frac{k}{T(1-e^{sT})}\left[\frac{Te^{-sT}}{-s} + \frac{1}{s^2}\left(1-e^{-sT}\right)\right]$$

$$= -\frac{ke^{-sT}}{T(1-e^{sT})} + \frac{k}{Ts^2} \qquad \textbf{Ans.}$$

Example 35. *Obtain Laplace transform of the rectangular wave given by*

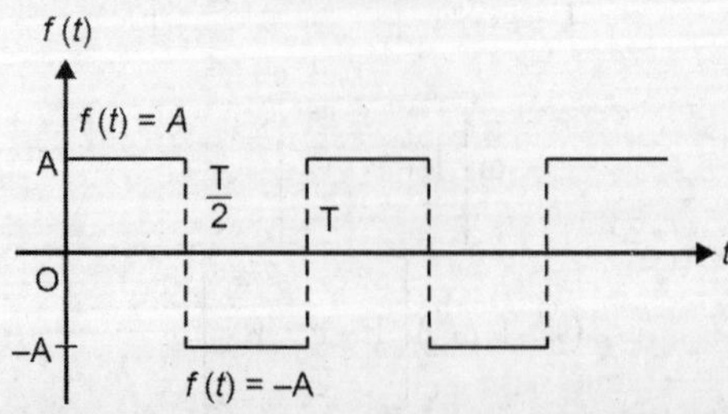

Solution. We know that Laplace transform of a periodic function, i.e.

$$L[f(t)] = \frac{\int_0^T e^{-st} f(t) dt}{1-e^{-sT}}$$

$$= \frac{\int_0^{\frac{T}{2}} e^{-st} A dt + \int_{\frac{T}{2}}^{T} e^{-st}(-A) dt}{1-e^{-sT}}$$

$$= A \frac{\left[\frac{e^{-st}}{-s}\right]_0^{\frac{T}{2}} - \left[\frac{e^{-st}}{-s}\right]_{\frac{T}{2}}^{T}}{1-e^{-sT}}$$

$$= \frac{A}{1-e^{-sT}}\left[-\frac{e^{-\frac{sT}{2}}}{s} + \frac{1}{s} + \frac{e^{-sT}}{s} - \frac{e^{-\frac{sT}{2}}}{s}\right]$$

$$= \frac{A}{s\left(1-e^{-sT}\right)}\left[1-2e^{-\frac{sT}{2}} + e^{-sT}\right]$$

$$= \frac{A}{s\left(1-e^{-sT}\right)}\left[1-e^{-\frac{sT}{2}}\right]^2$$

$$= \frac{A\left[1-e^{-\frac{sT}{2}}\right]^2}{s\left(1+e^{\frac{-sT}{2}}\right)\left(1-e^{\frac{-sT}{2}}\right)} = \frac{A}{s}\frac{\left(1-e^{\frac{-sT}{2}}\right)}{\left(1+e^{\frac{-sT}{2}}\right)}$$

$$= \frac{A\left(e^{\frac{sT}{4}} - e^{-\frac{sT}{4}}\right)}{s\left(e^{\frac{sT}{4}} + e^{-\frac{sT}{4}}\right)} = \frac{A}{s}\tanh\frac{sT}{4}$$ **Ans.**

Example 36. *A periodic square wave function f(t) in terms of unit step functions is written as*

$$f(t) = k\left[u_0(t) - 2u_a(t) + 2u_{2a}(t) - 2u_{3a}(t) + \ldots\right]$$

Show that the Laplace transform of f(t) is given by $L[f(t)] = \frac{k}{s}\tanh\left(\frac{as}{2}\right)$

Solution.
$$f(t) = k\left[u_0(t) - 2u_a(t) + 2u_{2a}(t) - 2u_{3a}(t) + \ldots\right]$$
$$f(t) = k[u(t-0) - 2u(t-a) + 2u(t-2a) - 2u(t-3a) + \ldots$$

$$L[f(t)] = k[Lu(t-0) - 2\,Lu(t-a) + 2Lu(t-2a) - 2\,Lu\,(t-3a) + ...$$

$$= k\left[\frac{1}{s} - 2\frac{e^{-as}}{s} + 2\frac{e^{-2as}}{s} - 2\frac{e^{-3as}}{s} + ...\right]$$

$$= \frac{k}{s}\left[1 - 2e^{-as} + 2e^{-2as} - 2e^{-3as} + ...\right]$$

$$= \frac{k}{s}\left[1 - 2(e^{-as} - e^{-2as} + e^{-3as} - ...\right]$$

$$= \frac{k}{s}\left[1 - 2\frac{e^{-as}}{1+e^{-as}}\right] = \frac{k}{s}\left[\frac{1+e^{-as} - 2e^{-as}}{1+e^{-as}}\right]$$

$$= \frac{k}{s}\left[\frac{1-e^{-as}}{1+e^{-as}}\right] = \frac{k}{s}\left[\frac{e^{\frac{as}{2}} - e^{\frac{-as}{2}}}{e^{\frac{as}{2}} + e^{-\frac{as}{2}}}\right]$$

$$= \frac{k}{s}\tanh\left(\frac{as}{2}\right)$$

Proved.

EXERCISE 35.5

1. Find the Laplace transform of the periodic functioin.

   $f(t) = e^t$ for $0 < t < 2\pi$ **Ans.** $\dfrac{e^{2(1-s)\pi} - 1}{(1-s)(1-e^{-2\pi s})}$

2. Obtain Laplace transform of full wave rectified sine wave given by

   $f(t) = \sin\omega t,\ 0 < t < \dfrac{\pi}{w}$ **Ans.** $\dfrac{\omega}{\left(s^2+\omega^2\right)}\coth\left(\dfrac{\pi s}{2\omega}\right)$

3. Find the Laplace transform of the staircase function

   $f(t) = kn,\ np < t < (n+1)\,p,\ n = 0, 1, 2, 3$ **Ans.** $\dfrac{ke^{ps}}{s(1-e^{-ps})}$

Find Laplace transform of the following :

4. $f(t) = t^2, 0 < t < 2, f(t+2) = f(t)$ **Ans.** $\dfrac{2 - e^{-2s} - 4se^{-2s} - 4s^2e^{-2s}}{s^3(1-e^{-2s})}$

5. $f(t) = \begin{cases} 1, & 0 \le t \le \dfrac{a}{2} \\ -1, & \dfrac{a}{2} \le t < a \end{cases}$ **Ans.** $\dfrac{1}{s}\tanh\left(\dfrac{as}{4}\right)$

6. $f(t) = \begin{cases} \cos\omega t, & 0 < t < \dfrac{\pi}{\omega} \\ 0, & \dfrac{\pi}{\omega} < t < \dfrac{2\pi}{\omega} \end{cases}$ **Ans.** $\dfrac{s}{\left(s^2+\omega^2\right)\left(1-e^{-\frac{\pi s}{\omega}}\right)}$

7. $f(t)=\begin{cases} t, & 0<t<1 \\ 0, & 1<t<2 \end{cases} \qquad f(t+2)=f(t)$ **Ans.** $\dfrac{1-e^{-s}(s+1)}{s^2(1-e^{-2s})}$

8. $f(t)=\begin{cases} \dfrac{2t}{T}, & 0\le t\le \dfrac{T}{2} \\ \dfrac{2}{T}(T-t), & \dfrac{T}{2}\le t\le T \end{cases} \qquad f(t+T)=f(t)$ **Ans.** $\dfrac{2}{Ts^2}\tanh\dfrac{sT}{4}-\dfrac{1}{s\left(e^{\frac{sT}{2}}+1\right)}$

35.20 IMPULSE FUNCTION

When a large force acts for a short time, then the product of the force and the time is called impulse in applied mechanics. The unit impulse function is the limiting function.

$$\delta(t-1)=\begin{cases} \dfrac{1}{\varepsilon}, & a<t<a+\varepsilon \\ 0, & \text{otherwise} \end{cases}$$

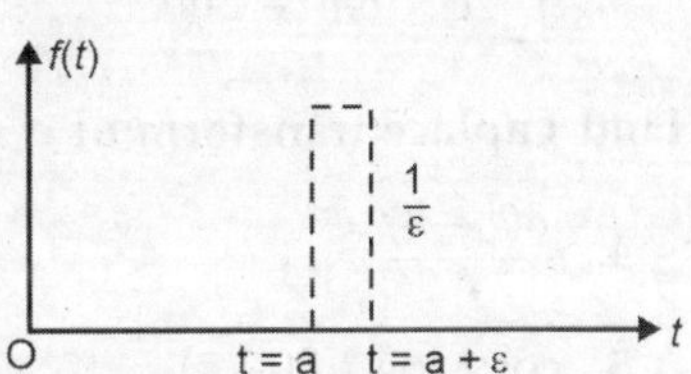

The value of the function (height of the strip in the figure) becomes infinite as $\varepsilon\to 0$ and the area of the rectangle is unity.

(1) The unit impulse function is defined as follows :

$$\delta(t-a) = \begin{cases} \infty & \text{for } t=a \\ 0 & \text{for } t\neq a \end{cases}$$

and $$\int_0^\infty \delta(t-a)\,dt = 1 \qquad \text{[Area of strip = 1]}$$

(2) Laplace transform of unit impulse function

$$\int_0^\infty f(t)\delta(t-a)\,dt = \int_a^{a+\varepsilon} f(t).\frac{1}{\varepsilon}dt \qquad \left\{\begin{array}{l}\text{Mean value theorem}\\ \int_a^b f(t)dt=(b-a)f(n)\end{array}\right.$$

$$= (a+\varepsilon-a)f(n).\frac{1}{\varepsilon} \qquad \text{where } a<n<a+\varepsilon$$

$$= f(n)$$

Property I. $\boxed{\int_0^\infty f(t)\delta(t-a)\,dt = f(a)}$ (as $\varepsilon\to 0$)

Note. If $f(t)=e^{-st}$ and $L[\delta(t-a)]=e^{-as}$

Example 37. *Evaluate* $\int_{-\infty}^{\infty} e^{-5t}\,\delta(t-2)\,dt$

Solution. $$\int_{-\infty}^{\infty} e^{-5t}\,\delta(t-2)\,dt = e^{-5\times 2}=e^{-10}$$ **Ans.**

Property II: $\boxed{\int_{-\infty}^{\infty} f(t)\delta'(t-a)\,dt = -f'(a)}$

Proof. $$\int_{-\infty}^{\infty} f(t)\delta'(t-a)\,dt = [f(t).\delta(t-a)]_{-\infty}^{\infty}-\int_{-\infty}^{\infty} f'(t)\delta(t-a)\,dt$$

$$= 0-0-f'(a)=-f'(a)$$

Example 38. *Find the Laplace transform of* $t^3\delta(t-4)$.

Solution. $$Lt^3\delta(t-4) = \int_0^\infty e^{-st}t^3\delta(t-4)\,dt = 4^3e^{-4s}$$ **Ans.**

EXERCISE 35.6

Evaluate the following:

1. $\int_0^\infty e^{-3t}\delta(t-4)\,dt$ **Ans.** e^{-12}

2. $\int_{-\infty}^{\infty} \sin 2t\,\delta\left(t-\frac{\pi}{4}\right) dt$ **Ans.** 1

3. $\int_{-\infty}^{\infty} e^{-3t}\,\delta'(t-2)\,dt$ **Ans.** $3e^{-6}$

Find Laplace transform of

4. $\frac{\delta(t-4)}{t}$ **Ans.** $\frac{e^{-4s}}{4}$

5. $\cos t \log t\ \delta(t-\pi)$ **Ans.** $-e^{-\pi s}\log \pi$

6. $e^{-4t}\ \delta\ (t-3)$ **Ans.** $e^{-3(s+4)}$

35.21 CONVOLUTION THEOREM

[U.P. II Semester Summer 2006]

If $$L[f_1(t)] = F_1(s) \text{ and } L[f_2(t)] = F_2(s)$$

then $$L\left\{\int_0^t f_1(x)f_2(t-x)dx\right\} = F_1(s).F_2(s)$$

or $$\boxed{L^{-1}(F_1(s)\cdot F_2(s)) = \int_0^t f_1(x)f_2(t-x)dx}$$

Proof. We have

$$L\left\{\int_0^t f_1(x)f_2(t-x)dx\right\} = \int_0^\infty e^{-st}\left[\int_0^t f_1(x)f_2(t-x)\,dx\right]dt \quad \text{(By definition)}$$

where the double integral is taken over the infinite region in the first quadrant lying between the lines $x = 0$ and $x = t$.

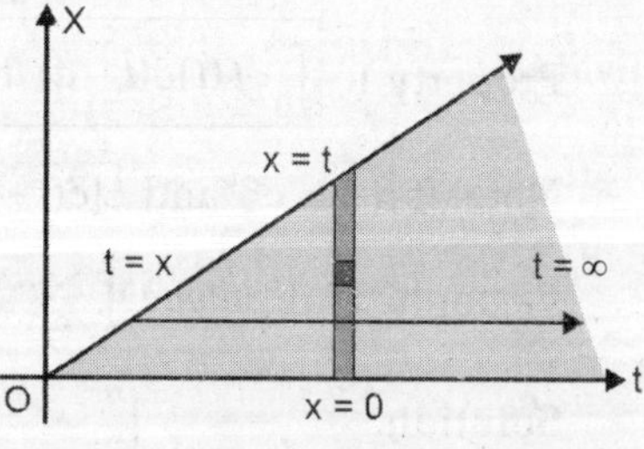

Here first we are integrating w.r.t. "x", within limits $x = 0$ and $x = t$, and then we will integrate w.r.t. "t" with limits $t = 0$ and $t = \infty$.

On chainging the order of integration first we integrate w.r.t. "t" with limits $t = x$ and $t = \infty$ and then w.r.t. "x" with limits $x = 0$ and $x = \infty$.

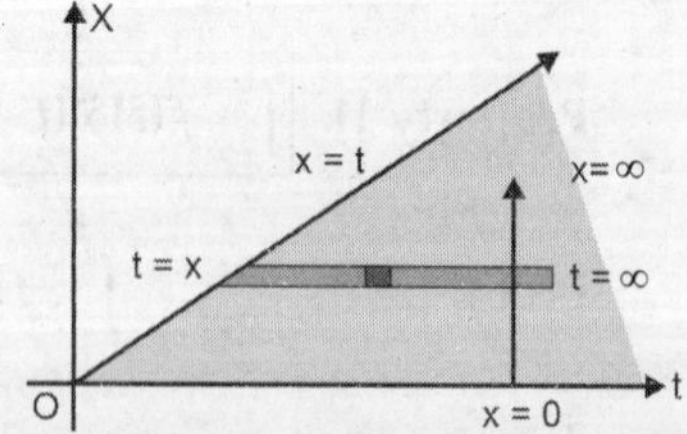

On changing the order of integration, the integral becomes

$$\int_0^\infty dx\left[\int_x^\infty e^{-st} f_1(x).f_2(t-x)\,dt\right]$$

$$= \int_0^\infty dx\left[\int_x^\infty e^{-s(t-x+x)} f_1.(x).f_2(t-x)\,dt\right]$$

$$= \int_0^\infty dx\left[\int_x^\infty e^{-s(t-x)}.e^{-sx} f_1(x).f_2(t-x)\,dt\right]$$

$$= \int_0^\infty e^{-sx} f_1(x)\,dx\left[\int_x^\infty e^{-s(t-x)} f_2(t-x)\,dt\right]$$

$$= \int_0^\infty e^{-sx} f_1(x)\,dx\left[\int_0^\infty e^{-sz} f_2(z)\,dz\right] \qquad [\text{Put } t-x=z \Rightarrow -dx=dz]$$

Lower limit $x - x = z \Rightarrow z = 0$

$$= \int_0^\infty e^{-sx} f_1(x)F_2(s)\,dx = \left[\int_0^\infty e^{-sx} f_1(x)\,dx\right]F_2(s)$$

$$= F_1(s)F_2(s)$$ **Proved.**

Example 39. *Find the Laplace transform of* $\int_0^t e^x.\sin(t-x)\,dx$

Solution. By Convolution theorem

$$L\int_0^t f_1(x)f_2(t-x)\,dx = F_1(s).F_2(s)$$

$$\Rightarrow \quad L\int_0^t e^x.\sin(t-x)\,dx = L(e^t)\,.\,L\,(\sin t)$$

$$= \frac{1}{s-1}\frac{1}{s^2+1} = \frac{1}{(s-1)(s^2+1)}$$ **Ans.**

EXERCISE 35.7

Evaluate the following by using Laplace transform:

1. $\int_0^\infty te^{-4t}\sin t\,dt$ **Ans.** $\frac{8}{289}$

2. $\int_0^\infty \frac{e^{-2t}\sinh t\,\sin t}{t}dt$ **Ans.** $\frac{1}{2}\tan^{-1}\left(\frac{1}{2}\right)$

3. $\int_0^\infty \frac{\sin^2 t}{t^2}dt$ **Ans.** $i\,\frac{5}{2}$

4. $\int_0^\infty \frac{e^{-t}-e^{-4t}}{t}dt$ **Ans.** $\log 4$

35.22 FORMULAE TABLE OF LAPLACE TRANSFORM

S. No.	*f*(*t*)	*F*(*s*)
1.	e^{at}	$\frac{1}{s-a}$
2.	t^n	$\frac{\lfloor n+1}{s^{n+1}}$ or $\frac{n!}{s^{n+1}}$
3.	$\sin at$	$\frac{a}{s^2+a^2}$
4.	$\cos at$	$\frac{s}{s^2+a^2}$
5.	$\sinh at$	$\frac{a}{s^2-a^2}$
6.	$\cosh at$	$\frac{s}{s^2-a^2}$
7.	$u(t-a)$	$\frac{e^{-as}}{s}$
8.	$\delta(t-a)$	e^{-as}
9.	$e^{bt}\sin at$	$\frac{a}{(s-b)^2+a^2}$
10.	$e^{bt}\cos at$	$\frac{s-b}{(s-b)^2+a^2}$
11.	$\frac{t}{2a}\sin at$	$\frac{s}{(s^2+a^2)^2}$
12.	$t\cos at$	$\frac{s^2-a^2}{(s^2+a^2)^2}$
13.	$\frac{1}{2a^3}(\sin at - at\cos at)$	$\frac{1}{(s^2+a^2)^2}$
14.	$\frac{1}{2a}(\sin at + at\cos at)$	$\frac{s^2}{(s^2+a^2)^2}$

35.23 TABLE OF PROPERTIES OF LAPLACE TRANSFORM

S. No.	Property	f(t)	F(s)
1.	Scaling	$f(at)$	$\frac{1}{a}F\left(\frac{s}{a}\right), a > 0$
2.	Derivative	$\frac{df(t)}{dt}$	$s\,F(s) - f(0), \quad s > 0$
		$\frac{d^2f(t)}{dt^2}$	$s^2\,F(s) - sf\,(0) - f'\,(0), s > 0$
		$\frac{d^3f(t)}{dt^3}$	$s^3\,F(s) - s^2f(0) - sf'(0) - f''\,(0), s > 0$
3.	Integral	$\int_0^t f(t)dt$	$\frac{1}{s}F(s), \quad s > 0$
4.	Initial value	$\lim_{t \to 0} f(t)$	$\lim_{s \to \infty} sF(s)$
5.	Final value	$\lim_{t \to \infty} f(t)$	$\lim_{s \to 0} sF(s)$
6.	First shifting	$e^{-at}f(t)$	$F(s + a)$
7.	Second shifting	$f(t)\,u\,(t - a)$	$e^{-a}\,Lf\,(t + a)$
8.	Multiplication by t	$t\,f(t)$	$-\frac{d}{ds}F(s)$
9.	Multiplication by t^n	$t^nf(t)$	$(-1)^n\frac{d^n}{ds^n}F(s)$
10.	Division by t	$\frac{1}{t}f(t)$	$\int_s^\infty F(s)\,ds$
11.	Periodic function	$f(t)$	$\frac{\int_0^T e^{-st}f(t)}{1-e^{-sT}}dt \; \{f(t+T) = f(t)\}$
12.	Convolution	$f(t)^*\,g(t)$	$F(s)\,G(s)$

35.24 INVERSE LAPLACE TRANSFORMS

If $F(s)$ is the Laplace transform of a function $f(t)$, then $f(t)$ is known as inverse laplace transform. Now we will discuss how to find f(t) when F(s) is given.

If $L\,[f(t)] = F(s)$, then $L^{-1}\,[F(s)] = f(t)$ where L^{-1} is called the Inverse Laplace transform operator.

From the application point of view, the Inverse Laplace transform is very useful.

Inverse Laplace transform is used in solving differential equations without finding the general solution and arbitary constants.

35.25 FORMULAE TABLE OF INVERSE LAPLACE TRANSFORM

No.	Formula	No.	Formula
1.	$L^{-1}\left(\frac{1}{s}\right) = 1$	2.	$L^{-1}\left(\frac{1}{s^n}\right) = \frac{t^{n-1}}{(n-1)!}$
3.	$L^{-1}\left(\frac{1}{s-a}\right) = e^{at}$	4.	$L^{-1}\left(\frac{s}{s^2-a^2}\right) = \cosh at$
5.	$L^{-1}\left(\frac{1}{s^2-a^2}\right) = \frac{1}{a}\sinh at$	6.	$L^{-1}\left(\frac{1}{s^2+a^2}\right) = \frac{1}{a}\sin at$
7.	$L^{-1}\left(\frac{s}{s^2+a^2}\right) = \cos at$	8.	$L^{-1}\{F(s-a)\} = e^{at} f(t)$
9.	$L\left[\frac{1}{(s-a)^2+b^2}\right] \frac{1}{\,}e^{at} \sin bt$	10.	$L^{-1}\left[\frac{s-a}{(s-a)^2+b^2}\right] = e^{at} \cos bt$
11.	$L^{-1}\left[\frac{1}{(s-a)^2-b^2}\right] = \frac{1}{b}e^{at} \sinh bt$	12.	$L^{-1}\left[\frac{s-a}{(s-a)^2+b^2}\right] = e^{at} \cosh bt$
13.	$L^{-1}\left[\frac{1}{(s^2+a)^2}\right] = \frac{1}{2a^3}(\sin at - at\cos at)$	14.	$L^{-1}\left[\frac{s}{(s^2+a)^2}\right] = \frac{1}{2a}t\sin at$
15.	$L^{-1}\left[\frac{s^2-a^2}{(s^2+a^2)^2}\right] = t\cos at$	16.	$\mathrm{L}^{-1}(1) = \delta(t)$
17.	$L^{-1}\left[\frac{s^2}{(s^2+a^2)^2}\right] = \frac{1}{2a}[\sin at + at\cos at]$	18.	$L^{-1}\left\{\frac{1}{s}F(s)\right\} = \int_0^t f(t)\,dt$

Example 40. *Show that:* $\frac{1}{s^{1/2}} = L\left[\frac{1}{\sqrt{\pi t}}\right]$ **(U.P. II Semester, Summer 2005)**

Solution. We have to show that $\frac{1}{s^{1/2}} = L\left[\frac{1}{\sqrt{\pi t}}\right]$

Now, $$L^{-1}\left\{\frac{1}{s^n}\right\} = \frac{t^{n-1}}{(n-1)!} = \frac{t^{n-1}}{\Gamma n}$$

So $$L^{-1}\left\{\frac{1}{s^{1/2}}\right\} = \frac{t^{\frac{1}{2}-1}}{\Gamma\frac{1}{2}} = \frac{t^{-\frac{1}{2}}}{\Gamma\frac{1}{2}} = \frac{t^{-\frac{1}{2}}}{\sqrt{\pi}}$$

$\Rightarrow$ $$L^{-1}\left\{\frac{1}{s^{1/2}}\right\} = \frac{1}{\sqrt{\pi t}} \Rightarrow \frac{1}{s^{1/2}} = L\left[\frac{1}{\sqrt{\pi t}}\right]$$ **Proved.**

Example 41. *Find the inverse Laplace transform of the following :*

(i) $\dfrac{1}{s-2}$ (ii) $\dfrac{1}{s^2-9}$ (iii) $\dfrac{1}{s^2-16}$

(iv) $\dfrac{1}{s^2+25}$ (v) $\dfrac{s}{s^2+9}$ (vi) $\dfrac{1}{(s-2)^2+1}$

(vii) $\dfrac{s-1}{(s-1)^2+4}$ (viii) $\dfrac{1}{(s+3)^2-4}$ (ix) $\dfrac{s+2}{(s+2)^2-25}$

(x) $\dfrac{}{2\quad 7}$

Solution.

(i) $L^{-1}\left(\dfrac{1}{s-2}\right) = e^{2t}$

(ii) $L^{-1}\left(\dfrac{1}{s^2-9}\right) = L^{-1}\left[\dfrac{1}{3}.\dfrac{3}{s^2-(3)^2}\right] = \dfrac{1}{3}\sinh 3t$

(iii) $L^{-1}\left(\dfrac{s}{s^2-16}\right) = L^{-1}\left[\dfrac{s}{s^2-(4)^2}\right] = \cosh 4t$

(iv) $L^{-1}\left(\dfrac{1}{s^2+25}\right) = \dfrac{1}{5}\dfrac{5}{s^2+(5)^2} = \dfrac{1}{5}\sin 5t$

(v) $L^{-1}\left(\dfrac{s}{s^2+9}\right) = \dfrac{s}{s^2+(3)^2} = \cos 3t$

(vi) $\left[\dfrac{}{(\ -2)\ +1}\right] = e^{2t}\sin t$

(vii) $L^{-1}\left[\dfrac{s-1}{(s-1)^2+4}\right] = e^{t}\cos 2t$

(viii) $L^{-1}\left[\dfrac{1}{(s+3)^2-4}\right] = \dfrac{1}{2}\dfrac{2}{(s+3)^2-(2)^2} = \dfrac{1}{2}e^{-3t}\sinh 2t$

(ix) $L^{-1}\left[\dfrac{s+2}{(s+2)^2-25}\right] = L^{-1}\left[\dfrac{(s+2)}{(s+2)^2-(5)^2}\right] = e^{-2t}\cosh 5t$

(x) $L^{-1}\left(\dfrac{1}{2s-7}\right) = \dfrac{1}{2}e^{\frac{7}{2}t}$ $\left[\because\ L^{-1}F(as) = \dfrac{1}{a}f\left(\dfrac{t}{a}\right)\right]$

Example 42. *Find Inverse Laplace transform of*

(a) $\left\{\dfrac{6}{2s-3} - \dfrac{3+4s}{9s^2-16} + \dfrac{8-6s}{16s^2+9}\right\}$ (b) $\dfrac{2s-5}{9s^2-25}$ (c) $\dfrac{s-2}{6s^2+20}$

Solution.

(a) $$L^{-1}\left\{\frac{6}{2s-3}-\frac{3}{9s^2-16}-\frac{4s}{9s^2-16}+\frac{8}{16s^2+9}-\frac{6s}{16s^2+9}\right\}$$

$$= L^{-1}\left\{\frac{3}{s-\frac{3}{2}}-\frac{\frac{1}{3}}{s^2-\left(\frac{4}{3}\right)^2}-\frac{\frac{4}{9}s}{s^2-\left(\frac{4}{3}\right)^2}+\frac{\frac{1}{2}}{s^2+\left(\frac{3}{4}\right)^2}-\frac{\frac{3}{8}s}{s^2+\left(\frac{3}{4}\right)^2}\right\}$$

$$= L^{-1}\left\{\frac{3}{s-\frac{3}{2}}-\frac{1}{4}\frac{\frac{4}{3}}{s^2-\left(\frac{4}{3}\right)^2}-\frac{4}{9}\frac{s}{s^2-\left(\frac{4}{3}\right)^2}+\frac{2}{3}\frac{\frac{3}{4}}{s^2+\left(\frac{3}{4}\right)^2}-\frac{3}{8}\frac{s}{s^2+\left(\frac{3}{4}\right)^2}\right\}$$

$$= 3.e^{\frac{3}{2}t}-\frac{1}{4}\sinh\frac{4}{3}t-\frac{4}{9}\cosh\frac{4}{3}t+\frac{2}{3}\sin\frac{3}{4}t-\frac{3}{8}\cos\frac{3}{4}t$$ **Ans.**

(b) $$L^{-1}\left(\frac{2s-5}{9s^2-25}\right) = L^{-1}\left[\frac{2s}{9s^2-25}-\frac{5}{9s^2-25}\right] = L^{-1}\left[\frac{2s}{9\left[s^2-\left(\frac{5}{3}\right)^2\right]}-\frac{5}{9\left[s^2-\left(\frac{5}{3}\right)^2\right]}\right]$$

$$= \frac{2}{9}\cosh\frac{5}{3}t-\frac{1}{3}L^{-1}\left(\frac{\frac{5}{3}}{s^2-\left(\frac{5}{3}\right)^2}\right) = \frac{2}{9}\cosh\frac{5t}{3}-\frac{1}{3}\sinh\left(\frac{5t}{3}\right)$$ **Ans.**

(c) $$L^{-1}\left(\frac{s-2}{6s^2+20}\right) = L^{-1}\left(\frac{s}{6s^2+20}\right)-L^{-1}\left(\frac{2}{6s^2+20}\right) = \frac{1}{6}L^{-1}\left(\frac{s}{s^2+\frac{10}{3}}\right)-\frac{1}{3}L^{-1}\left(\frac{1}{s^2+\frac{10}{3}}\right)$$

$$= \frac{1}{6}L^{-1}\left(\frac{s}{s^2+\frac{10}{3}}\right)-\frac{1}{3}\times\sqrt{\frac{3}{10}}L^{-1}\left(\frac{\sqrt{\frac{10}{3}}}{s^2+\frac{10}{3}}\right) = \frac{1}{6}\cos\sqrt{\frac{10}{3}}t-\frac{1}{\sqrt{30}}\sin\sqrt{\frac{10}{3}}t$$ **Ans.**

EXERCISE 35.8

Find the Inverse Laplace transform of the following :

1. $\dfrac{3s-8}{4s^2+25}$ **Ans.** $\dfrac{3}{4}\cos\dfrac{5t}{2}-\dfrac{4}{5}\sin\dfrac{5t}{2}$

2. $\dfrac{3\left(s^2-2\right)^2}{2s^5}$ **Ans.** $\dfrac{3}{2}-3t^2+\dfrac{1}{4}t^4$

3. $\dfrac{2s-5}{4s^2+25}+\dfrac{4s-18}{9-s^2}$ **Ans.** $\dfrac{1}{2}\left(\cos\dfrac{5t}{2}-\sin\dfrac{5t}{2}\right)-4\cosh 3t+6\sinh 3t$

4. $\dfrac{5s-10}{9s^2-16}$ **Ans.** $\dfrac{5}{9}\cosh\dfrac{4}{3}t-\dfrac{5}{6}\sinh\dfrac{4}{3}t$

5. $\dfrac{1}{4s}+\dfrac{16}{1-s^2}$ **Ans.** $\dfrac{1}{4}-16\sinh t$

35.26 MULTIPLICATION BY S

$$\boxed{L^{-1}[sF(s)]=\frac{d}{dt}f(t)+f(0)\delta(t)}$$

Example 43. *Find the Inverse Laplace transform of (i)* $\dfrac{s}{s^2+1}$ *(ii)* $\dfrac{s}{4s^2-25}$ *(iii)* $\dfrac{3s}{2s+9}$

Solution.

(*i*)
$$L^{-1}\left(\frac{1}{s^2+1}\right) = \sin t$$

$$L^{-1}\left(\frac{s}{s^2+1}\right) = \frac{d}{dt}(\sin t)+\sin(0)\delta(t)=\cos t \qquad \textbf{Ans.}$$

(*ii*)
$$L^{-1}\left(\frac{1}{4s^2-25}\right) = \frac{1}{4}L^{-1}\left(\frac{1}{s^2-\frac{25}{4}}\right)$$

$$= \frac{1}{4}\cdot\frac{2}{5}L^{-1}\left[\frac{\frac{5}{2}}{s^2-\left(\frac{5}{2}\right)^2}\right]=\frac{1}{10}\sinh\frac{5}{2}t \qquad \textbf{Ans.}$$

$$L^{-1}\left(\frac{s}{4s^2-25}\right) = \frac{1}{10}\frac{d}{dt}\sinh\frac{5}{2}t+\frac{1}{10}\sinh\frac{5}{2}(0)\delta(t)$$

$$= \frac{1}{10}\left(\frac{5}{2}\right)\cosh\frac{5}{2}t=\frac{1}{4}\cosh\frac{5}{2}t \qquad \textbf{Ans.}$$

(*iii*)
$$L^{-1}\left(\frac{3}{2s+9}\right) = \frac{3}{2}L^{-1}\left(\frac{1}{s+\frac{9}{2}}\right)=\frac{3}{2}e^{-\frac{9}{2}t}$$

$$L^{-1}\left(\frac{3s}{2s+9}\right) = \frac{3}{2}\frac{d}{dt}\left(e^{-\frac{9}{2}t}\right)+\frac{3}{2}e^{-\frac{9}{2}(0)\delta(t)} = \frac{3}{2}\left(-\frac{9}{2}\right)e^{-\frac{9}{2}t}+\frac{3}{2}$$

$$= -\frac{27}{4}e^{-\frac{9}{2}t}+\frac{3}{2} \qquad \textbf{Ans.}$$

EXERCISE 35.9

Find the inverse laplace transform of the following :

1. $\dfrac{s}{s+5}$ **Ans.** $-5\,e^{-5t}+1$

2. $\dfrac{2s}{3s+6}$ **Ans.** $\dfrac{2}{3}\left[-2e^{-2t}+1\right]$

3. $\dfrac{s}{2s^2-1}$ **Ans.** $\dfrac{1}{2}\cosh\dfrac{t}{\sqrt{2}}$

4. $\dfrac{s^2}{s^2+a^2}$ **Ans.** $-a\sin at+1$

5. $\dfrac{s^2+4}{s^2+9}$ **Ans.** $-\dfrac{5}{3}\sin 3t+1$

6. $L^{-1}\left[\dfrac{s^2}{\left(s^2+4\right)^2}\right]$ is

(i) $\sin 2t+\dfrac{t}{2}\cos 2t$ (ii) $\dfrac{1}{4}\sin 2t+\dfrac{t}{2}\cos 2t$

(iii) $\dfrac{1}{4}\sin 2t+t\cos 2t$ (iv) $\dfrac{1}{4}\sin 2t+\dfrac{t}{4}\cos 2t$ **Ans.** (ii)

35.27 DIVISION BY s (MULTIPLICATION BY $\frac{1}{s}$)

$$L^{-1}\left[\frac{F(s)}{s}\right]=\int_0^t\left[L^{-1}[F(s)]\right]dt=\int_0^t f(t)\,dt$$

Example 44. *Find the inverse Laplace transform of*

(i) $\dfrac{1}{s(s+a)}$ (ii) $\dfrac{1}{s\left(s^2+1\right)}$ (iii) $\dfrac{s^2+3}{s\left(s^2+9\right)}$

Solution.

(i)
$$L^{-1}\left(\frac{1}{s+a}\right)=e^{-at}$$

$$L^{-1}\left[\frac{1}{s(s+a)}\right]=\int_0^t L^{-1}\left(\frac{1}{s+a}\right)dt=\int_0^t e^{-at}\,dt=\left[\frac{e^{-at}}{-a}\right]_0^t$$

$$=\frac{e^{-at}}{-a}+\frac{1}{a}=\frac{1}{a}\left[1-e^{-at}\right]$$ **Ans.**

(ii)
$$L^{-1}\left(\frac{1}{s^2+1}\right)=\sin t$$

$$L^{-1}\left[\frac{1}{s}\left(\frac{1}{s^2+1}\right)\right]=\int_0^t L^{-1}\left(\frac{1}{s^2+1}\right)dt$$

$$\int_0^t \sin t dt = [-\cos t]_0^t = -\cos t + 1$$ **Ans.**

(*iii*)
$$L^{-1}\left[\frac{s^2+3}{s\left(s^2+9\right)}\right] = L^{-1}\left[\frac{s^2+9-6}{s(s^2+9)}\right]$$

$$= L^{-1}\left[\frac{1}{s} - \frac{6}{s(s^2+9)}\right]$$

$$= 1 - 2\int_0^t \sin 3t dt = 1 + 2 \times \frac{1}{3}[\cos 3t]_0^t$$

$$= 1 + \frac{2}{3}\cos 3t - \frac{2}{3}$$

$$= \frac{2}{3}\cos 3t + \frac{1}{3} = \frac{1}{3}[2\cos 3t + 1]$$ **Ans.**

EXERCISE 35.10

Find the inverse Laplace transform of the following :

1. $\frac{1}{2s(s-3)}$ **Ans.** $\frac{1}{2}\left[\frac{e^{3t}}{3} - \frac{1}{3}\right]$

2. $\frac{1}{s(s+2)}$ **Ans.** $\frac{1-e^{-2t}}{2}$

3. $\frac{1}{s(s^2-16)}$ **Ans.** $\frac{1}{16}[\cosh 4t - 1]$

4. $\frac{1}{s(s^2+a^2)}$ **Ans.** $\frac{1-\cos at}{a^2}$

5. $\frac{s^2+2}{s\left(s^2+4\right)}$ **Ans.** $\cos^2 t$

6. $\frac{1}{s^2(s+1)}$ **Ans.** $t - 1 + e^{-t}$

7. $\frac{1}{s^3(s^2+1)}$ **Ans.** $\frac{t^2}{2} + \cos t - 1$

8. $L^{-1}\left[\frac{1}{s\left(s^2+1\right)}\right]$ is

(*i*) $1 - \cos t$ (*ii*) $1 + \cos t$

(*iii*) $1 - \sin t$ (*iv*) $1 + \sin t$ **Ans.** (*i*)

35.28 FIRST SHIFTING PROPERTY

$L^{-1} F(s) = f(t)$, then $\boxed{L^{-1}F(s+a) = e^{-at}L^{-1}[F(s)]}$

Example 45. *Find the inverse Lapace transform of*

(i) $\dfrac{1}{(s+2)^5}$ (ii) $\dfrac{s}{s^2+4s+13}$ (iii) $\dfrac{1}{9s^2+6s+1}$

Solution.

(i)
$$L^{-1}\left(\frac{1}{s^5}\right) = \frac{t^4}{4!}$$

then
$$L^{-1}\left[\frac{1}{(s+2)^5}\right] = e^{-2t}.\frac{t^4}{4!}$$ **Ans.**

(ii)
$$\left(\frac{}{s \quad +4s+13}\right) = \left[\frac{+2-2}{(\quad+2)^2+(3)^2}\right]$$

$$= L^{-1}\left[\frac{s+2}{(s+2)^2+(3)^3}\right] - L^{-1}\left[\frac{2}{(s+2)^2+(3)^2}\right]$$

$$= e^{-2t}L^{-1}\left[\frac{s}{s^2+3^2}\right] - e^{-2t}L^{-1}\left[\frac{2}{3}\left(\frac{3}{s^2+3^2}\right)\right]$$

$$= e^{-2t}\cos 3t - \frac{2}{3}e^{-2t}\sin 3t$$ **Ans.**

(iii)
$$L^{-1}\left(\frac{1}{9s^2+6s+1}\right) = L^{-1}\left[\frac{1}{(3s+1)^2}\right] = \frac{1}{9}L^{-1}\left[\frac{1}{\left(s+\frac{1}{3}\right)^2}\right]$$

$$= \frac{1}{9}e^{-\frac{t}{3}}L^{-1}\left(\frac{1}{s^2}\right)$$

$$= \frac{1}{9}e^{-\frac{t}{3}}t = \frac{te^{-\frac{t}{3}}}{9}$$ **Ans.**

EXERCISE 35.11

Obtain the inverse Laplace transform of the following :

1. $\dfrac{s+8}{s^2+4s+5}$ **Ans.** $e^{-2t}(\cos t + 6\sin t)$

2. $\dfrac{s}{(s+3)^2+4}$ **Ans.** $e^{-3t}(\cos 2t - 1.5\sin 2t)$

3. $\dfrac{s}{(s+7)^4}$ **Ans.** $e^{-7t}\dfrac{t^2}{6}(3-7t)$

4. $\dfrac{s+2}{s^2-2s-8}$ **Ans.** $e^{t}(\cos h\,3t + \sin h\,3t)$

5. $\dfrac{s}{s^2+6s+25}$ **Ans.** $e^{-3t}\left[\cos 4t-\dfrac{3}{4}\sin 4t\right]$ **6.** $\dfrac{1}{2(s-1)^2+32}$ **Ans.** $\dfrac{e^t}{8}\sin 4t$

7. $\dfrac{s-4}{4(s-3)^2+16}$ **Ans.** $\dfrac{1}{4}e^{3t}\cos 2t-\dfrac{1}{8}e^{3t}\sin 2t$

35.29 SECOND SHIFTING PROPERTY

$$\boxed{L^{-1}\left[e^{-as}F(s)\right]f(-a)U(t-a)}$$

Example 46. *Obtain inverse Laplace transform of*

(i) $\dfrac{e^{-\pi s}}{(s+3)}$ *(ii)* $\dfrac{e^{-s}}{(s+1)^3}$

Solution.

(i)
$$L^{-1}\left(\frac{1}{s+3}\right) = e^{-3t}$$

$$L^{-1}\left(\frac{e^{-\pi s}}{s+3}\right) = e^{-3t(t-\pi)}\ u(t-\pi)$$ **Ans.**

(ii)
$$L^{-1}\left(s^{1/3}\right) = \frac{t^2}{2!} \Rightarrow L^{-1}\left[\frac{1}{(s+1)^3}\right]=e^{-t}\,\frac{t^2}{2!}$$

$$L^{-1}\left[\frac{e^{-s}}{(s+1)^3}\right] = e^{-(t-1)}\,\frac{(t-1)^2}{2!}.u(t-1)$$ **Ans.**

Example 47. *Evaluate* $L^{-1}\left[\dfrac{e^{-s}-3e^{-3s}}{s^2}\right]$

Solution.
$$L^{-1}\left[\frac{e^{-s}-3e^{-3s}}{s^2}\right] = L^{-1}\left[\frac{e^{-s}}{s^2}-\frac{3e^{-3s}}{s^2}\right] \quad ...(1)$$

We know that
$$L\,[u(t-a)] = \frac{e^{-as}}{s}$$

and
$$L[(t-a)\,u(t-a)] = \frac{e^{-as}}{s^2}$$

Using these results in (1), we get

$$\therefore \quad L^{-1}\left[\frac{e^{-s}-3e^{-3s}}{s^2}\right] = (t-1)\,u\,(t-1)-3\,(t-3)\,u\,(t-3)$$ **Ans.**

Example 48. *Find the inverse Laplace transform of* $\dfrac{se^{-\frac{s}{2}} + \pi e^{-s}}{s^2 + \pi^2}$ *in terms of unit step functions.*

Solution.

$$L^{-1}\left(\frac{\pi}{s^2+\pi^2}\right) = \sin \pi t$$

$$L^{-1}\left[e^{-s}\frac{\pi}{s^2+\pi^2}\right] = \sin\pi(t-1).u\,(t-1) = -\sin(\pi t).u(t-1) \quad ...(1)$$

and

$$L^{-1}\left(\frac{s}{s^2+\pi^2}\right) = \cos \pi t$$

$$L^{-1}\left[e^{-\frac{s}{2}}\frac{s}{s^2+\pi^2}\right] = \cos\pi\left(t-\frac{1}{2}\right)\cdot u\left(t-\frac{1}{2}\right)$$

$$= \sin \pi t\cdot u\left(t-\frac{1}{2}\right) \quad ...(2)$$

On adding (1) and (2), we get

$$L^{-1}\left[\frac{e^{-\frac{s}{2}}s + e^{-s}.\pi}{s^2+\pi^2}\right] = \sin(\pi t)\cdot u\left(t-\frac{1}{2}\right) - \sin(\pi t)\cdot u(t-1)$$

$$= \sin \pi t\left[u\left(t-\frac{1}{2}\right) - u(t-1)\right]$$ **Ans.**

Example 49. *Find the inverse Laplace transform of* $\dfrac{e^{-cs}}{s^2(s+a)}$, $c > 0$.

Solution. We have, $$L^{-1}\left[\frac{e^{-cs}}{s^2(s+a)}\right] = L^{-1}\left[-\frac{e^{-cs}}{a^2 s} + \frac{e^{-cs}}{as^2} + \frac{e^{-cs}}{a^2(s+a)}\right]$$

(By partial fractions)

$$= \left[\left(\frac{1}{a^2}\frac{e^{-cs}}{s}\right) + \left(\frac{1}{a}\right)\frac{e^{-cs}}{s^2} + \left(\frac{1}{a^2}\right)\frac{e^{-c(s+a)}}{e^{\;ca}(s\;\;a)}\right]$$

$$= -\frac{1}{a^2}lu(t-c) + \frac{1}{a}(t-c)u(t-c) + \frac{1}{a^2e^{-ca}}e^{at}u(t-c)$$

$$= u(t-c)\left[\frac{-1}{a^2} + \frac{1}{a}(t-c) + \frac{1}{a^2}e^{a(c+t)}\right],$$ where $u(t-c)$ is unit step function. **Ans.**

Example 50. *Find the value of* $L^{-1}\left\{\dfrac{1}{(s^2+a^2)^2}\right\}$. **[DIP IETE 2018]**

Solution.

$$\frac{1}{(s^2+a^2)^2} = \frac{1}{s}\cdot\frac{s}{(s^2+a^2)^2} = -\frac{1}{2s}\frac{d}{ds}\left(\frac{1}{s^2+a^2}\right)$$

$$\Rightarrow \quad L^{-1}\left\{\frac{1}{(s^2+a^2)}\right\} = L^{-1}\left\{-\frac{1}{2s}\frac{d}{ds}\left(\frac{1}{s^2+a^2}\right)\right\}$$

$$= -\frac{1}{2s}\left\{-t\frac{1}{a}\sin at\right\} = \frac{1}{2a}\frac{1}{s}(t\sin at)$$

$$= \frac{1}{2a}\int_0^t t\sin at\,dt$$

$$= \frac{1}{2a}\left[t\left(\frac{-\cos at}{a}\right) - \int\frac{-\cos at}{a}dt\right]_0^t$$

$$= \frac{1}{2a}\left[-\frac{t}{a}\cos at + \frac{\sin at}{a^2}\right]_0^t$$

$$= \frac{1}{2a^3}[-at\cos at + \sin at] \qquad \textbf{Ans.}$$

EXERCISE 35.12

Obtain inverse Laplace transform of the following :

1. $\dfrac{e^{-2s}}{(s+1)^3}$ **Ans.** $e^{-(t-2)}\dfrac{(t-2)^2}{2}U(t-2)$

2. $\dfrac{e^{-2s}}{(s+1)(s^2+2s+2)}$ **Ans.** $e^{-(t-2)}\,|\,1-\cos(t-2)\,|\,U(t-2)$

3. $\dfrac{e^{-s}}{\sqrt{s+1}}$ **Ans.** $\dfrac{e^{-(t-1)}}{\sqrt{\pi(t-1)}}.U(t-1)$

4. $\dfrac{e^{-\frac{\pi}{2}s}+e^{-\frac{3\pi}{2}s}}{s^2+1}$ **Ans.** $\cos t\left[U\left(t-\frac{3\pi}{2}\right)-U\left(t-\frac{\pi}{2}\right)\right]$

5. $\dfrac{e^{-4s}(s+2)}{s^2+4s+5}$ **Ans.** $e^{-2(t-4)}\cos(t-4)\,U(t-4)$

6. $\dfrac{e^{-as}}{s^2}$ **Ans.** $f(t)=\begin{cases}t-a \text{ when } t>a\\ 0 \text{ when } t<a\end{cases}$

7. $\dfrac{e^{-\pi s}}{s^2+1}$ **Ans.** $-\sin t.\,u(t-\pi)$

Tick (✓) the correct answers

8. (*a*) The inverse Laplace transform of $\frac{(e^{-3s})}{s^3}$ is

(*i*) $(t-3)\, u_3(t)$ (*ii*) $(t-3)^2\, u_3(t)$

(*iii*) $\frac{(t-3)^2}{2} u_3\,(t)$ (*iv*) $(t+3)\, u_3(t)$ **Ans.** (*iii*)

(*b*) If Laplace transform of a function $f(t)$ equals $\frac{(e^{-2s}-e^{-s})}{s}$.

(*i*) $f(t) = 1,\ t > 1$;

(*ii*) $f(t) = 1$, when $1 < t < 2$, and 0 otherwise;

(*iii*) $f(t) = -1$, when $1 < t < 2$, and 0 otherwise;

(*iv*) $f(t) = -1$, when $1 < t < 2$, and 0 otherwise; **Ans.** (*iv*)

(*c*) The Laplace inverse $L^{-1}\left[\frac{2}{s}\left(e^{-2s} - e^{-4s}\right)\right]$ equals

(*i*) 2, if $0 < t < 4$; 0 otherwise,

(*ii*) 2, if $t > 0$

(*iii*) 2, if $0 < t < 2$; 0 otherwise,

(*iv*) 2, if $2 < t < 4$; 0 otherwise, **Ans.** (*iv*)

(*d*) The Laplace transform of $t\, u_2\,(t)$ is

(*i*) $\left(\frac{1}{s^2}+\frac{2}{s}\right)e^{-2s}$ (*ii*) $\frac{1}{s^2}e^{-2s}$

(*iii*) $\left(\frac{1}{s^2}-\frac{2}{s}\right)e^{-2s}$ (*iv*) $\frac{1}{s^2}e^{-2s}$ **Ans.** (*i*)

(*e*) The inverse Laplace transform of $\frac{ke^{-as}}{s^2+k^2}$ is

(*i*) $\sin kt$ (*ii*) $\cos kt$

(*iii*) $u(t-a) \sin kt$ (*iv*) none of these. **Ans.** (*iv*)

(*f*) Inverse Laplace transform of 1 is :

(*i*) 1 (*ii*) $\delta(t)$

(*iii*) $\delta(t-1)$ (*iv*) $u(t)$ **Ans.** (*ii*)

35.30 INVERSE LAPLACE TRANSFORM OF DERIVATIVES

$$L^{-1}\left[\frac{d}{ds}F(s)\right] = -t\, L^{-1}\,[F(s)] = -t\, f(t)$$

$$\Rightarrow \qquad \boxed{L^{-1}\,[F(s)] = -\frac{1}{t}L^{-1}\left[\frac{d}{ds}F(s)\right]}$$

Example 51. *Find inverse Laplace transform of* $\tan^{-1}\left(\frac{1}{s}\right)$

Solution. $$L^{-1}\left(\tan^{-1}\frac{1}{s}\right) = -\frac{1}{t}L^{-1}\left[\frac{d}{ds}\tan^{-1}\left(\frac{1}{s}\right)\right]$$

$$= -\frac{1}{t}L^{-1}\left[\frac{1}{1+\frac{1}{s^2}}\left(-\frac{1}{s^2}\right)\right]$$

$$= \frac{1}{t}L^{-1}\left[\frac{1}{1+s^2}\right] = \frac{\sin t}{t}$$ **Ans.**

Example 52. *Find* $L^{-1}\left\{\log\frac{s+1}{s-1}\right\}$ **(Uttarakhand II Semester June 2007)**

Solution. $$L^{-1}\left\{\log\frac{s+1}{s-1}\right\} = -\frac{1}{t}L^{-1}\left[\frac{d}{ds}\log\left(\frac{s+1}{s-1}\right)\right]$$

$$= -\frac{1}{t}L^{-1}\left[\frac{d}{ds}\log(s+1) - \frac{d}{ds}\log(s-1)\right]$$

$$= -\frac{1}{t}L^{-1}\left[\frac{1}{s+1} - \frac{1}{s-1}\right]$$

$$= -\frac{1}{t}\left[e^{-t} - e^{t}\right] = \frac{1}{t}\left[e^{t} - e^{-t}\right]$$ **Ans.**

Example 53. *Find the inverse Laplace transform of* $F(s) = \log\left(\frac{s+a}{s+b}\right)$.

Solution. $$L^{-1}\log\left(\frac{s+a}{s+b}\right) = -\frac{1}{t}L^{-1}\left[\frac{d}{ds}\log\frac{s+a}{s+b}\right]$$

$$= -\frac{1}{t}L^{-1}\left[\frac{d}{ds}\log(s+a) - \frac{d}{ds}\log(s+b)\right]$$

$$= -\frac{1}{t}L^{-1}\left[\frac{1}{s+a} - \frac{1}{s+b}\right]$$

$$= -\frac{1}{t}\left[e^{-at} - e^{-bt}\right] = \frac{1}{t}\left(e^{-bt} - e^{-at}\right)$$ **Ans.**

Example 54. *Obtain the inverse Laplace transform of log* $\left(\frac{s^2-1}{s^2}\right)$.

Solution. $$L^{-1}\left[\log\left(\frac{s^2-1}{s^2}\right)\right] = -\frac{1}{t}L^{-1}\left[\frac{d}{ds}\log\left(\frac{s^2-1}{s^2}\right)\right]$$

$$= -\frac{1}{t}L^{-1}\left[\frac{d}{ds}\left\{\log\left(s^2-1\right) - 2\log s\right\}\right]$$

$$= -\frac{1}{t}L^{-1}\left[\frac{2s}{s^2-1} - \frac{2}{s}\right]$$

$$= -\frac{1}{t}[2\cosh t - 2] = \frac{2}{t}[1 - \cosh t]$$ **Ans.**

Example 55. *Find $L^{-1}[\cot^{-1}(1+s)]$*

Solution.
$$L^{-1}[\cot^{-1}(1+s)] = -\frac{1}{t}L^{-1}\left[\frac{d}{ds}\cot^{-1}(1+s)\right]$$
$$= -\frac{1}{t}L^{-1}\left[\frac{-1}{1+(s+1)^2}\right] = \frac{1}{t}L^{-1}\left[\frac{1}{(s+1)^2+1}\right]$$
$$= \frac{1}{t}e^{-t}\sin t \qquad \textbf{Ans.}$$

Example 56. *Obtain the inverse Laplace transform of* $\cot^{-1}\left(\frac{s+3}{2}\right)$.

Solution. We know that
$$L^{-1}[F(s)] = -\frac{1}{t}L^{-1}\left[\frac{d}{ds}F(s)\right]$$
$$\therefore \quad L^{-1}\left[\cot^{-1}\left(\frac{s+3}{2}\right)\right] = -\frac{1}{t}L^{-1}\left[\frac{d}{ds}\cot^{-1}\left(\frac{s+3}{2}\right)\right]$$
$$= -\frac{1}{t}L^{-1}\left\{\frac{-\frac{1}{2}}{1+\left(\frac{s+3}{2}\right)^2}\right\}$$
$$= \frac{1}{2t}L^{-1}\left\{\frac{4}{4+(s+3)^2}\right\}$$
$$= \frac{1}{t}L^{-1}\left\{\frac{2}{2^2+(s+3)^2}\right\} = \frac{1}{t}e^{-3t}L^{-1}\left(\frac{2}{2^2+s^2}\right)$$
$$= \frac{e^{-3t}}{t}\sin 2t \qquad \textbf{Ans.}$$

Example 57. *Find the inverse Laplace Transform of* $\frac{s+1}{(s^2+6s+13)^2}$.

Solution.
$$L^{-1}\left[\frac{s+1}{(s^2+6s+13)^2}\right] = L^{-1}\frac{s+1}{\left[(s+3)^2+2^2\right]^2}$$
$$= L^{-1}\left[\frac{s+3-2}{\left[(s+3)^2+2^2\right]^2}\right]$$
$$= L^{-1}\left[\frac{s+3}{\left[(s+3)^2+2^2\right]^2}\right] - L^{-1}\frac{2}{\left[(s+3)^2+2^2\right]^2}$$
$$= e^{-3t}L^{-1}\left[\frac{s}{(s^2+2^2)^2}\right] - e^{-3t}L^{-1}\left[\frac{2}{(s^2+2^2)^2}\right] \qquad ...(1)$$

First Part.

We know that

$$L^{-1}\left[\frac{-2s}{(s^2+2^2)^2}\right] = L^{-1}\frac{d}{ds}\left(\frac{1}{s^2+2^2}\right)$$

$$= -tL^{-1}\left(\frac{1}{s^2+2^2}\right)$$

$$= -\frac{t}{2}\sin 2t$$

$$\Rightarrow \quad L^{-1}\left[\frac{s}{(s^2+2^2)^2}\right] = \frac{t}{4}\sin 2t$$

$$\Rightarrow e^{-3t}L^{-1}\left[\frac{s}{(s^2+2^2)^2}\right] = \frac{1}{4}e^{-3t}t\sin 2t$$

$$\Rightarrow \quad L^{-1}\frac{s+3}{(s+3)^2+2^2} = \frac{1}{4}e^{-3t}\sin 2t$$

Second Part.

We know that

$$L^{-1}\frac{1}{(s^2+a^2)^2} = \frac{1}{2a^3}(\sin at - at\cos at)$$

$$L^{-1}\frac{2}{(s^2+2^2)^2} = \frac{2}{2(2)^3}[\sin 2t - 2t\cos 2t]$$

$$= \frac{1}{8}[\sin 2t - 2t\cos 2t]$$

$$e^{-3t}L^{-1}\frac{2}{(s^2+2^2)^2} = \frac{e^{-3t}}{8}[\sin 2t - 2t\cos 2t]$$

$$L^{-1}\left[\frac{-2}{[(s+3)^2+2^2]^2}\right] = -\frac{e^{-3t}}{8}[\sin 2t - 2t\cos 2t]$$

Putting the values of both parts in (1), we get

$$L^{-1}\left[\frac{s+1}{(s^2+6s+13)^2}\right] = \frac{t}{4}e^{-3t}\sin 2t - \frac{6^{-3t}}{8}(\sin 2t - 2t\cos 2t)$$

$$= \frac{e^{-3t}}{8}[2t.\sin 2t - \sin 2t + 2t\cos 2t] \quad \textbf{Ans.}$$

EXERCISE 35.13

Obtain inverse Laplace transform of the following:

1. $\log\left(1+\frac{\omega^2}{s^2}\right)$ **Ans.** $-\frac{2}{t}\cos\omega t + 2$

2. $\log\left(1+\frac{1}{s^2}\right)$ **Ans.** $\frac{2}{t}[1-\cos t]$

3. $\frac{s}{1+s^2+s^4}$ **Ans.** $\frac{2}{\sqrt{3}}\sin\frac{\sqrt{3}}{2}t\sinh\frac{t}{2}$

4. $\frac{s}{(s^2+a^2)^2}$ **Ans.** $\frac{t\sin at}{2a}$

5. $s\log\left(\frac{s}{\sqrt{s^2+1}}\right)+\cot^{-1}(s)$ **Ans.** $\frac{1-\cos t}{t^2}$

6. $\frac{1}{2}\log\left\{\frac{s^2+b^2}{(s-a)^2}\right\}$ **Ans.** $\frac{e^{-at}-\cos bt}{t}$

7. $\tan^{-1}(s+1)$ **Ans.** $-\frac{1}{t}e^{-t}\sin t$

35.31 INVERSE LAPLACE TRANSFORM OF INTEGRALS

$$L^{-1}\left[\int_s^\infty F(s)ds\right] = \frac{f(t)}{t} = \frac{1}{t}L^{-1}[F(s)]$$

or

$$\boxed{L^{-1}[F(s)] = t\,L^{-1}\int_s^\infty F(s)ds}$$

Example 58. *Obtain* $L^{-1}\left[\dfrac{2s}{\left(s^2+1\right)^2}\right]$

Solution.

$$L^{-1}\left[\frac{2s}{\left(s^2+1\right)^2}\right] = t\,L^{-1}\int_s^\infty \frac{2sds}{\left(s^2+1\right)^2} = tL^{-1}\left[-\frac{1}{s^2+1}\right]_s^\infty$$

$$= tL^{-1}\left[-0+\frac{1}{s^2+1}\right] = t\sin t \qquad \textbf{Ans.}$$

35.32 PARTIAL FRACTIONS METHOD

Example 59. *Find the inverse Laplace transform of* $\dfrac{1}{s^2-5s+6}$.

Solution. Let us convert the given function into partial fractions.

$$L^{-1}\left[\frac{1}{s^2-5s+6}\right] = L^{-1}\left[\frac{1}{s-3}-\frac{1}{s-2}\right]$$

$$= L^{-1}\left(\frac{1}{s-3}\right)-L^{-1}\left(\frac{1}{s-2}\right) = e^{3t}-e^{2t} \qquad \textbf{Ans.}$$

Example 60. *Find the inverse Laplace transform of* $\dfrac{s-1}{s^2-6s+25}$

Solution.

$$L^{-1}\left(\frac{s-1}{s^2-6s+25}\right) = L^{-1}\left[\frac{s-1}{(s-3)^2+(4)^2}\right] = L^{-1}\left[\frac{s-3+2}{(s-3)^2+(4)^2}\right]$$

$$= L^{-1}\left[\frac{s-3}{(s-3)^2+(4)^2}\right]+\frac{1}{2}L^{-1}\left[\frac{4}{(s-3)^2+(4)^2}\right]$$

$$= e^{3t}\cos 4t+\frac{1}{2}e^{3t}\sin 4t \qquad \textbf{Ans.}$$

Example 61. *Find the Inverse Laplace Transform of* $\dfrac{s+4}{s(s-1)(s^2+4)}$.

Solution. Let us first resolve $\dfrac{s+4}{s(s-1)(s^2+4)}$ into partial fractions.

$$\frac{s+4}{s(s-1)(s^2+4)} \equiv \frac{A}{s}+\frac{B}{s-1}+\frac{Cs+D}{s^2+4} \qquad \text{...(1)}$$

$s + 4 \equiv A(s-1)(s^2+4) + Bs(s^2+4) + (Cs+D)s(s-1)$

Putting $s = 0$, we get $4 = -4A \Rightarrow A = -1$

Putting $s = 1$, we get $5 = B \cdot 1 \cdot (1+4) \Rightarrow B = 1$

Equating the coefficients of s^3 on both sides of (1), we have

$$0 = A + B + C \Rightarrow 0 = -1 + 1 + C \Rightarrow C = 0$$

Equating the coefficients of s on both sides of (1), we get

$$1 = 4A + 4B - D \Rightarrow 1 = -4 + 4 - D \Rightarrow D = -1.$$

On putting the values of A, B, C, D in (1), we get

$$\Rightarrow \quad \frac{s+4}{s(s-1)(s^2+4)} = -\frac{1}{s} + \frac{1}{s-1} - \frac{1}{s^2+4}$$

$$\therefore \quad L^{-1}\left[\frac{s+4}{s(s-1)(s^2+4)}\right] = L^{-1}\left[-\frac{1}{s} + \frac{1}{s-1} - \frac{1}{s^2+4}\right]$$

$$= L^{-1}\left(\frac{1}{s}\right) + L^{-1}\left(\frac{1}{s-1}\right) - \frac{1}{2}L^{-1}\left(\frac{2}{s^2+2^2}\right)$$

$$= -1 + e^t - \frac{1}{2}\sin 2t$$ **Ans.**

Example 62. *Find the inverse Laplace transform of*

$$\frac{5s+3}{(s-1)(s^2+2s+5)}$$ **(U.P. II Semester Summer 2005)**

Solution. $L^{-1}\left\{\frac{5s+3}{(s-1)(s^2+2s+5)}\right\}$

Let $$\frac{5s+3}{(s-1)(s^2+2s+5)} = \frac{A}{s-1} + \frac{Bs+C}{s^2+2s+5}$$

$$\Rightarrow \quad 5s + 3 = A(s^2 + 2s + 5) + (Bs + C)(s-1)$$

$$\Rightarrow \quad 5s + 3 = s^2(A + B) + s(2A - B + C) + (5A - C)$$

Comparing the coefficients of s^2, s and constant, we get

$$A + B = 0 \quad \text{...(1)}$$

$$2A - B + C = 5 \quad \text{...(2)}$$

$$5A - C = 3 \quad \text{...(3)}$$

On adding equations (1) and (2), we have

$$3A + C = 5 \quad \text{...(4)}$$

Adding equations (3) and (4), we get

$$8A = 8 \Rightarrow A = 1$$

Putting $A = 1$ in (3), we get

$$C = 2$$

Putting $A = 1$, $C = 2$ in (2), we get

$$B = -1$$

Thus $\dfrac{5s+3}{(s-1)(s^2+2s+5)} = \dfrac{1}{s-1} + \dfrac{-s+2}{s^2+2s+5} = \dfrac{1}{s-1} - \dfrac{s-2}{(s+1)^2+2^2}$

$$= \frac{1}{s-1} - \frac{s+1}{(s+1)^2+2^2} + \frac{3}{(s+1)^2+2^2}$$

$$L^{-1}\left\{\frac{5s+3}{(s-1)(s^2+2s+5)}\right\} = L^{-1}\left\{\frac{1}{s-1}\right\} + L^{-1}\left\{\frac{3}{(s+1)^2+2^2}\right\} - L^{-1}\left\{\frac{s+1}{(s+1)^2+2^2}\right\}$$

$$= e^t + 3e^{-t}L^{-1}\left\{\frac{1}{s^2+2^2}\right\} - e^{-t}L^{-1}\left\{\frac{s}{s^2+2^2}\right\}$$

$$= e^t + 3e^{-t}.\frac{1}{2}\sin 2t - e^{-t}\cos 2t \quad \textbf{Ans.}$$

Example 63. *Find the inverse Laplace transform of* $\dfrac{s^2}{(s^2+a^2)(s^2+b^2)}$

Solution. Let us convert the given function into partial fractions.

$$L^{-1}\left[\frac{s^2}{(s^2+a^2)(s^2+b^2)}\right] = L^{-1}\left[\frac{a^2}{a^2-b^2}.\frac{1}{s^2+a^2} - \frac{b^2}{a^2-b^2}.\frac{1}{s^2+b^2}\right]$$

$$\frac{}{a^2-b^2}\left[\frac{a^2}{s^2+a^2} \quad \frac{b^2}{a^2+b^2}\right] = \frac{1}{a^2-b^2}\left[a^2\left(\frac{1}{a}\sin at\right) - b^2\left(\frac{1}{b}\sin bt\right)\right]$$

$$= \frac{1}{a^2-b^2}[a\sin at - b\sin bt] \quad \textbf{Ans.}$$

EXERCISE 35.14

Find the inverse Laplace transform by using partial fraction method :

1. $\dfrac{s^2+2s+6}{s^3}$ **Ans.** $1+2t+3t^2$

2. $\dfrac{1}{s^2-7s+12}$ **Ans.** $e^{4t}-e^{3t}$

3. $\dfrac{s+2}{s^2-4s+13}$ **Ans.** $e^{2t}\cos 3t + \dfrac{4}{3}e^{2t}\sin 3t$

4. $\dfrac{3s+1}{(s-1)(s^2+1)}$ **Ans.** $e^t - 2\cos t + \sin t$

5. $\dfrac{11s^2-2s+5}{2s^3-3s^2-3s+2}$ **Ans.** $2e^{-t}+5e^{2t}-\dfrac{3}{2}e^{\frac{t}{2}}$

6. $\dfrac{2s^2-6s+5}{(s-1)(s-2)(s-3)}$ **Ans.** $\dfrac{1}{2}e^t - e^{2t} + \dfrac{5}{2}e^{3t}$

7. $\dfrac{s-4}{(s-4)^2+9}$ **Ans.** $e^{4t}\cos 3t$

8. $\dfrac{16}{(s^2+2s+5)^2}$ **Ans.** $e^{-t}(\sin 2t - 2t\cos 2t)$

9. $\dfrac{1}{(s+1)(s^2+2s+2)}$ **Ans.** $e^{-t}(1-\cos t)$

10. $\dfrac{1}{(s-2)(s^2+1)}$ **Ans.** $\dfrac{1}{5}e^{2t} - \dfrac{1}{5}\cos t - \dfrac{2}{5}\sin t$

11. $\dfrac{s^2-6s+7}{(s^2-4s+5)^2}$ **Ans.** $te^{2t}\{\cos t - \sin t\}$

35.33 INVERSE LAPLACE TRANSFORM BY CONVOLUTION

$$L\left\{\int_0^t f_1(x) * f_2(t-x)dx\right\} = F_1(s).\ F_2(s)$$

or

$$\boxed{\int_0^t f_1(x).f_2(t-x)dx = L^{-1}F_1(s).F_2(s)}$$

Example 64. *State convolution theorem and hence find*

$$L^{-1}\left\{\frac{1}{(s+2)^2(s-2)}\right\}$$ **(Uttarakhand II Semester June 2007)**

Solution. Convolution theorem (see Section 37.21)

Let $L\{f_1(t)\} = F_1(s)$ and Let $L\{f_2(t)\} = F_2(s)$

$$\Rightarrow \quad F_1(s) = \frac{1}{(s+2)^2} \text{ and } F_2(s) = \frac{1}{s-2}$$

$$\Rightarrow \quad f_1(t) = L^{-1}\left[\frac{1}{(s+2)^2}\right] = te^{-2t}$$

$$\Rightarrow \quad f_2(t) = L^{-1}\left[\frac{1}{(s-2)}\right] = e^{2t}$$

According to convolution theorem

$$\Rightarrow \quad L^{-1}[F_1(s).F_2(s)] = \int_0^t f_1(x) f_2(t-x)\,dx$$

$$\left[\begin{aligned} L(e^{-2t}) &= \frac{1}{s+2} \\ L(te^{-2t}) &= -\frac{d}{ds}\frac{1}{(s+2)} = \frac{1}{(s+2)^2} \end{aligned}\right]$$

$$\Rightarrow \quad L^{-1}\left[\frac{1}{(s+2)^2(s-2)}\right] = \int_0^t x e^{-2x}.e^{2(t-x)}dx = \int_0^t xe^{2t-4x}dx$$

$$= \left[x\frac{e^{2t-4x}}{-4} - \int 1\cdot\frac{e^{(2t-4x)}}{-4}\,dx\right]_0^t$$

$$= \left[-\frac{x}{4} e^{2t-4x} + \frac{1}{4} \left\{ \frac{e^{(2t-4x)}}{-4} \right\} \right]_0^t$$

$$= \frac{-t}{4} e^{2t-4t} - \frac{1}{16} e^{2t-4t} + \frac{1}{16} e^{2t}$$

$$= \frac{-t}{4} e^{-2t} - \frac{1}{16} e^{-2t} + \frac{1}{16} e^{2t}$$

$$= \frac{e^{2t}}{16} - \frac{1}{16} e^{-2t} [4t+1]$$ **Ans.**

Example 65. *Using the convolution theorem find* $L^{-1} \left\{ \frac{s^2}{(s^2 + a^2)(s^2 + b^2)} \right\}, a \neq b$.

(U.P. II Semester Summer 2006)

Solution. We have, $L(\cos at) = \frac{s}{s^2 + a^2}$ and $L(\cos bt) = \frac{s}{s^2 + b^2}$

Hence, by the convolution theorem

$$L \left\{ \int_0^t \cos ax \cos b(t - x) dx \right\} = \frac{s^2}{(s^2 + a^2)(s^2 + b^2)}$$

Therefore,

$$L^{-1} \left\{ \frac{s^2}{(s^2 + a^2)(s^2 + b^2)} \right\} = \int_0^t \cos ax \cos b(t - x)\, dx$$

$$= \frac{1}{2} \int_0^t \{\cos (ax + bt - bx) + \cos (ax - bt + bx)\}\, dx$$

$$= \frac{1}{2} \int_0^t \cos[(a-b)x + bt]\, dx + \frac{1}{2} \int_0^t \cos[(a+b)x - bt]\, dx$$

$$= \left[\frac{\sin[(a-b)x + bt]}{2(a-b)} \right]_0^t + \left[\frac{\sin[(a+b)x - bt]}{2(a+b)} \right]_0^t$$

$$= \frac{\sin at - \sin bt}{2(a-b)} + \frac{\sin at + \sin bt}{2(a+b)}$$

$$= \frac{a \sin at - b \sin bt}{a^2 - b^2}$$ **Ans.**

Example 66. *Evaluate* $L^{-1} \left\{ \frac{s}{(s^2 + 1)(s^2 + 4)} \right\}$.

Solution. We know that $L^{-1} \left(\frac{s}{s^2 + 1} \right) = \cos t$ and $L^{-1} \left(\frac{2}{s^2 + 2^2} \right) = \sin 2t$

$$L^{-1}\left\{\frac{s}{(s^2+1)(s^2+4)}\right\} = \frac{1}{2}L^{-1}\left[\left(\frac{s}{s^2+1}\right)\left(\frac{2}{s^2+4}\right)\right]$$

$$= \frac{1}{2}\int_0^t \sin 2x\cos(t-x)\,dx \qquad \text{[By convolution theorem]}$$

$$= \int_0^t \sin x\cos x\{\cos t\cos x + \sin t\sin x\}dx$$

$$= \int_0^t \left[\sin x\cos^2 x\cos t + \sin^2 x\cos x\sin t\right]dx$$

$$= \left[-\frac{\cos^3 x}{3}\cos t + \frac{\sin^3 x}{3}\sin t\right]_0^t$$

$$= -\frac{\cos^4 t}{3} + \frac{\sin^4 t}{3} + \frac{\cos t}{3}$$

$$= \frac{1}{3}\left[\sin^4 t - \cos^4 t\right] + \frac{\cos t}{3}$$

$$= \frac{1}{3}(\sin^2 t + \cos^2 t)(\sin^2 t - \cos^2 t) + \frac{\cos t}{3}$$

$$= \frac{1}{3}(\sin^2 t - \cos^2 t) + \frac{\cos t}{3}$$

$$= -\frac{1}{3}\cos 2t + \frac{\cos t}{3}$$

$$= \frac{1}{3}(\cos t - \cos 2t) \qquad \textbf{Ans.}$$

Example 67. *Obtain* $L^{-1}\left[\frac{1}{s(s^2+a^2)}\right]$

Solution. $L^{-1}\left(\frac{1}{s}\right) = 1$ and $L^{-1}\left(\frac{1}{s^2+a^2}\right) = \frac{\sin at}{a}$

$$L^{-1}\{F_1(s).F_2(s)\} = \int_0^t f_1(x)f_2(t-x)dx \quad \text{(convolution theorem)}$$

Hence by the convolution theorem

$$L^{-1}\left[\frac{1}{s}\cdot\frac{1}{s^2+a^2}\right] = \int_0^t \frac{\sin a(t-x)}{a}dx = \left[\frac{-\cos(at-ax)}{-a^2}\right]_0^t$$

$$= \frac{1}{a^2}[1-\cos at] \qquad \textbf{Ans.}$$

Example 68. *Using convolution theorem, prove that*

$$L^{-1}\left[\frac{1}{s^3(s^2+1)}\right] = \frac{t^2}{2} + \cos t - 1$$ **(U.P. II Semester Summer 2005)**

Solution. We know that, $L^{-1}\left\{\frac{1}{s^3}\right\} = \frac{t^2}{2!}$

$$L^{-1}\left\{\frac{1}{s^2+1}\right\} = \sin t$$

Using convolution theorem,

$$L^{-1}\left\{\frac{1}{s^3(s^2+1)}\right\} = \int_0^t \frac{(t-x)^2}{2!}\sin x\,dx$$

$$= \frac{1}{2}\int_0^t \left(t^2+x^2-2tx\right)\sin x\,dx$$

$$= \frac{1}{2}\left[\left(t^2+x^2-2tx\right)(-\cos x) - \int(2x-2t)(-\cos x)dx\right]_0^t$$

$$= \frac{1}{2}\left[(t^2+x^2-2tx)(-\cos x) + 2\int(x-t)\cos x\,dx\right]_0^t$$

$$= \frac{1}{2}\left[(t^2+x^2-2tx)(-\cos x) + 2(x-t)\sin x + 2\cos x\right]_0^t$$

$$= \frac{1}{2}\left[(t^2+t^2-2t^2)(-\cos t) + 0 + 2\cos t + t^2\cos 0 - 2\cos 0\right]$$

$$= \frac{1}{2}[2\cos t + t^2 - 2] = \cos t + \frac{t^2}{2} - 1 = \frac{t^2}{2} + \cos t - 1$$ **Ans.**

EXERCISE 35.15

Obtain the inverse Laplace transform using convolution theorem:

1. $\frac{s^2}{\left(s^2+a^2\right)^2}$ **Ans.** $\frac{1}{2}t\cos at + \frac{1}{2a}\sin at$

2. $\frac{1}{\left(s^2+1\right)^3}$ **Ans.** $\frac{1}{8}\left[(3-t^2)\sin t - 3t\cos t\right]$

3. $\frac{s}{\left(s^2+a^2\right)^2}$ **Ans.** $\frac{t\sin at}{2a}$

4. $\frac{1}{s^2\left(s^2-a^2\right)}$ **Ans.** $\frac{1}{a^3}[-at + \sinh at]$

5. $\frac{1}{(s+1)(s^2+1)}$ **Ans.** $\frac{1}{2}\left(\cos t - \sin t - e^{-t}\right)$

35.34 HEAVISIDE INVERSE FORMULA OF $\frac{F(s)}{G(s)}$

If $F(s)$ and $G(s)$ be two polynomials in S. The degree of $F(s)$ is less than that of $G(s)$. Let $\alpha_1, \alpha_2, \alpha_3, \ldots \alpha_n$ be n roots of the equation $G(s) = 0$

Inverse Laplace formula of $\frac{F(s)}{G(s)}$ is given by

$$L^{-1}\left\{\frac{F(s)}{G(s)}\right\} = \sum_{i=1}^{n} \frac{F(\alpha_i)}{G'(\alpha_i)} e^{\alpha_i t}$$

Example 69. *Find* $L^{-1}\left\{\frac{2s^2+5s-4}{s^3+s^2-2s}\right\}$.

Solution. Let

$$
\begin{aligned}
F(s) &= 2s^2 + 5s - 4 \\
\Rightarrow \quad G(s) &= s^3 + s^2 - 2s = s(s^2 + s - 2) = s(s+2)(s-1) \\
\Rightarrow \quad G'(s) &= 3s^2 + 2s - 2 \\
\Rightarrow \quad G(s) &= 0 \text{ has three roots, } 0, 1, -2 \\
\Rightarrow \quad \alpha_1 &= 0, \alpha_2 = 1, \alpha_3 = -2
\end{aligned}
$$

By Heaviside inverse formula

$$
\begin{aligned}
\Rightarrow \quad L^{-1}\left\{\frac{F(s)}{G(s)}\right\} &= \sum_{i=1}^{n} \frac{F(\alpha_i)}{G'(\alpha_i)} e^{\alpha_i t} \\
&= \left\{\frac{F(\alpha_1)}{G'(\alpha_1)}\right\} e^{t\alpha_1} + \frac{F(\alpha_2)}{G'(\alpha_2)} e^{t\alpha_2} + \frac{F(\alpha_3)}{G'(\alpha_3)} e^{t\alpha_3} \\
&= \frac{F(0)}{G'(0)} e^0 + \frac{F(1)}{G'(1)} e^t + \frac{F(-2)}{G'(-2)} e^{-2t} \\
&= \frac{-4}{-2} e^0 + \frac{3}{3} e^t + \frac{(-6)}{(6)} e^{-2t} = 2 + e^t - e^{-2t} \quad \textbf{Ans.}
\end{aligned}
$$

Example 70. *Find* $L^{-1}\left[\frac{2s^2-6s+5}{s^3-6s^2+11s-6}\right]$.

Solution. Let

$$
\begin{aligned}
F(s) &= 2s^2 - 6s + 5 \\
\Rightarrow \quad G(s) &= s^3 - 6s^2 + 11s - 6 = (s-1)(s-2)(s-3) \\
\Rightarrow \quad G(s) &= 0 \text{ has three roots, } 1, 2, 3. \\
\Rightarrow \quad \alpha_1 &= 1, \alpha_2 = 2, \alpha_3 = 3 \\
\Rightarrow \quad G'(s) &= 3s^2 - 12s + 11
\end{aligned}
$$

By Heaviside inverse formula, we have $L^{-1}\left\{\frac{F(s)}{G(s)}\right\} = \sum_{i=1}^{n} \frac{F(\alpha_i)}{G'(\alpha_i)} e^{t\alpha_i}$

$$\Rightarrow \quad L^{-1}\left\{\frac{2s^2-6s+5}{s^3-6s^2+11s-6}\right\} = \frac{F(\alpha_1)}{G'(\alpha_1)} e^{t\alpha_1} + \frac{F(\alpha_2)}{G'(\alpha_2)} e^{t\alpha_2} + \frac{F(\alpha_3)}{G'(\alpha_3)} e^{t\alpha_3}$$

$$\frac{F(1)}{G'(1)}e^t + \frac{F(2)}{G'(2)}e^{2t} + \frac{F(3)}{G'(3)}e^{3t} = \frac{(1)}{(2)}e^t + \frac{(1)}{(-1)}e^{2t} + \frac{(5)}{(2)}e^{3t}$$

$$= \frac{1}{2}e^t - e^{2t} + \frac{5}{2}e^{3t} \qquad \textbf{Ans.}$$

EXERCISE 35.16

Using Heaviside expansion formula, find the inverse Laplace transform of the following :

1. $\dfrac{s-1}{s^2+3s+2}$ **Ans.** $-2e^{-t} + 3e^{-2t}$

2. $\dfrac{s}{(s-1)(s-2)(s-3)}$ **Ans.** $\frac{1}{2}e^t - 2e^{2t} + \frac{3}{2}e^{3t}$

3. $\dfrac{2s+3}{(s-2)(s-3)(s-4)}$ **Ans.** $\frac{7}{2}e^{2t} - 9e^{3t} + \frac{11}{2}e^{4t}$

4. $\dfrac{11s^2-2s+5}{2s^3-3s^2-3s+2}$ **Ans.** $2e^t + 5e^{2t} - \frac{3}{2}e^{\frac{t}{2}}$

35.35 SOLUTION OF DIFFERENTIAL EQUATIONS (BY LAPLACE TRANSFORMS)

Ordinary linear differential equations with constant coefficients can be easily solved by the Laplace Transform method, without finding the general solution and the arbitrary constants. The method will be clear from the following examples:

Example 71. *Using Laplace transforms, find the solution of the initial value prblem*

$$y'' - 4y' + 4y = 64 \sin 2t$$
$$y(0) = 0,\ y'(0) = 1.$$

Solution.

$$y'' - 4y' + 4y = 64 \sin 2t \qquad ...(1)$$
$$y(0) = 0,\ y^1(0) = 1.$$

Taking Laplace transform of both sides of (1), we have

$$[s^2\bar{y} - sy(0) - y'(0)] - 4(s\bar{y} - y(0)) + 4\bar{y} = \frac{64 \times 2}{s^2+4} \qquad ...(2)$$

On putting the values of $y(0)$ and $y'(0)$ in (2), we get

$$\Rightarrow \quad s^2\bar{y} - 1 - 4s\bar{y} + 4\bar{y} = \frac{128}{s^2+4}$$

$$\Rightarrow \quad (s^2 - 4s + 4)\bar{y} = 1 + \frac{128}{s^2+4}$$

$$\Rightarrow \quad (s-2)^2\bar{y} = 1 + \frac{128}{s^2+4}$$

$$\Rightarrow \quad \bar{y} = \frac{1}{(s-2)^2} + \frac{128}{(s-2)^2(s^2+4)}$$

$$= \frac{1}{(s-2)^2} - \frac{8}{s-2} + \frac{16}{(s-2)^2} + \frac{8s}{s^2+4}$$

$$y = L^{-1}\left[-\frac{8}{s-2} + \frac{17}{(s-2)^2} + \frac{8s}{s^2+4}\right]$$

$$y = -8e^{2t} + 17t\,e^{2t} + 8\cos 2t$$ **Ans.**

Example 72. *Using the Laplace transforms, find the solution of the initial value problem*

$$y'' + 25y = 10\cos 5t,\ y(0) = 2,\ y'(0) = 0.$$

Solution. Taking Laplace transforms of the given differential equation, we get

$$\Rightarrow \quad [s^2\bar{y} - sy(0) - y'(0)] + 25\bar{y} = 10\frac{s}{s^2+25} \qquad ...(1)$$

Putting the value of $y(0)$ and $y'(0)$ in (1), we get

$$s^2\bar{y} - 2s + 25\bar{y} = \frac{10s}{s^2+25}$$

$$\Rightarrow \quad (s^2+25)\bar{y} = 2s + \frac{10s}{s^2+25}$$

$$\Rightarrow \quad \bar{y} = \frac{2s}{s^2+25} + \frac{10s}{(s^2+25)^2}$$

$$\Rightarrow \quad y = L^{-1}\left[\frac{2s}{s^2+25} + \frac{10s}{(s^2+25)^2}\right]$$

$$= 2\cos 5t + L^{-1}\left[\frac{10s}{(s^2+25)^2}\right]$$

$$= 2\cos 5t + L^{-1}\frac{d}{ds}\left[\frac{-5}{(s^2+25)}\right]$$

$$\left[\text{On differentiating } \frac{-5}{s^2+25}, \text{ we get } \frac{10s}{(s^2+25)^2}.\right]$$

$$= 2\cos 5t - t\sin 5t.$$ **Ans.**

Example 73. *Applying convolution theorem, solve the following initial value problem*

$$y'' + y = \sin 3t,\ y(0) = 0,\ y'(0) = 0.$$

Solution. $$y'' + y = \sin 3t,$$

Taking Laplace transform of both sides, we have

$$\Rightarrow \quad \left[s^2\bar{y} - sy(0)y'(0)\right] + \bar{y} = \frac{3}{s^2+9} \qquad ...(1)$$

On putting the values of $y(0)$, $y'(0)$ in (1), we get

$$s^2\bar{y} + \bar{y} = \frac{3}{s^2+9}$$

$$(s^2+1)\bar{y} = \frac{3}{s^2+9}$$

$$\bar{y} = \frac{3}{(s^2+1)(s^2+9)}$$

$$= \frac{3}{8}\left[\frac{1}{s^2+1} - \frac{1}{s^2+9}\right]$$

Taking the inverse transform, we get

$$y = \frac{3}{8}L^{-1}\left(\frac{1}{s^2+1}\right) - \frac{3}{8}L^{-1}\left(\frac{1}{s^2+9}\right)$$

$$y = \frac{3}{8}\sin t - \frac{3}{8} \times \frac{1}{3}\sin 3t$$

$$= \frac{3}{8}\sin t - \frac{1}{8}\sin 3t$$ **Ans.**

Example 74. *Using Laplace transforms, find the solution of the initial value problem*

$y'' + 9y = 6y \cos 3t$, $y(0) = 2$, $y'(0) = 0$ **(U.P. II Semester Summer 2006)**

Solution.

$$y'' + 9y = 6 \cos 3t \quad ...(1)$$

$$y(0) = 2,\ y'(0) = 0$$

Taking Laplace transform of (1), we get

$$\left[s^2\bar{y} - sy(0) - y'(0)\right] + 9\bar{y} = 6\left(\frac{s}{s^2+9}\right) \quad ...(2)$$

Putting the values of $y(0)$ and $y'(0)$ in (2), we have

$$\Rightarrow \quad s^2\bar{y} - 2s + 9\bar{y} = \frac{6s}{s^2+9}$$

$$\Rightarrow \quad (s^2+9)\bar{y} = 2s + \frac{6s}{s^2+9}$$

$$\Rightarrow \quad \bar{y} = \frac{2s}{s^2+9} + \frac{6s}{(s^2+9)^2}$$

$$\Rightarrow \quad y = L^{-1}\left[\frac{2s}{s^2+9}\right] + L^{-1}\left[\frac{6s}{(s^2+9)^2}\right]$$

$$= 2\cos 3t + L^{-1}\frac{d}{ds}\left[\frac{-3}{(s^2+9)}\right]$$

$$= 2\cos 3t - t\sin 3t$$ **Ans.**

Example 75. *Using Laplace transformation solve the following differential equation:*

$$\frac{d^2x}{dt^2} + 9x = \cos 2t, \text{ if } x(0) = 1,\ x\left(\frac{\pi}{2}\right) = -1$$

Solution.
$$\frac{d^2x}{dt^2} + 9x = \cos 2t \quad ...(1)$$

Taking Laplace transform of both the sides of (1), we get

$$\Rightarrow \quad L\left(\frac{d^2x}{dt^2}\right) + 9L(x) = L(\cos 2t)$$

$$\Rightarrow \quad s^2\bar{x} - sx(0) - x'(0) + 9\bar{x} = \frac{s}{s^2+4} \quad ...(2)$$

On putting $x(0) = 1$ in (2), we get

$$\Rightarrow \quad s^2\bar{x} - s + 9\bar{x} - x'(0) = \frac{s}{s^2+4}$$

$$\Rightarrow \quad (s^2+9)\bar{x} = s + \frac{s}{s^2+4} + x'(0) = \frac{s\left(s^2+4\right)+s}{s^2+4} + x'(0)$$

$$= \frac{s^3+5s}{s^2+4} + x'(0)$$

$$\Rightarrow \quad \bar{x} = \frac{(s^3+5s)}{(s^2+4)(s^2+9)} + \frac{x'(0)}{s^2+9}$$

$$= \frac{1}{5}\frac{s}{s^2+4} + \frac{4}{5}\frac{s}{s^2+9} + \frac{x'(0)}{s^2+9}$$

Taking the inverse Laplace transform, we get

$$\Rightarrow \quad x(t) = \frac{1}{5}L^{-1}\left(\frac{s}{s^2+4}\right) + \frac{4}{5}L^{-1}\left(\frac{s}{s^2+9}\right) + L^{-1}\left(\frac{x'(0)}{s^2+9}\right)$$

$$\Rightarrow \quad x(t) = \frac{1}{5}\cos 2t + \frac{4}{5}\cos 3t + \frac{x'(0)\sin 3t}{3} \quad ...(3)$$

On putting $x\left(\frac{\pi}{2}\right) = -1$ in (3), we get

$$\Rightarrow \quad -1 = -\frac{1}{5} + 0 - \frac{x'(0)}{3}$$

$$\Rightarrow \quad x'(0) = \frac{12}{5}$$

On putting the value of $x'(0)$ in (3), we get

$$x = \frac{1}{5}\cos 2t + \frac{4}{5}\cos 3t + \frac{12}{5}\frac{\sin 3t}{3} = \frac{1}{5}[\cos 2t + 4\cos 3t + 4\sin 3t]$$ **Ans.**

Example 76. *Using Laplace transform technique solve the following initial value problem*

$$\frac{d^2y}{dt^2} + 2\frac{dy}{dt} + 2y = 5 \sin t, \; y(0) = y'(0) = 0$$

Solution.

$$y'' + 2y' + 2y = 5 \sin t$$
$$y(0) = y'(0) = 0$$

Taking the Laplace transform of both sides, we have

$$[s^2\bar{y} - sy(0) - y'(0)] + 2[s\bar{y} - y(0)] + 2\bar{y} = 5 \times \frac{1}{s^2+1} \quad ...(1)$$

On substituting the values of $y(0)$, and $y'(0)$ in (1), we get

$$\Rightarrow \quad s^2\bar{y} + 2s\bar{y} + 2\bar{y} = \frac{5}{s^2+1}$$

$$\Rightarrow \quad [s^2 + 2s + 2]\bar{y} = \frac{5}{s^2+1}$$

$$\Rightarrow \quad \bar{y} = \frac{5}{(s^2+2s+2)(s^2+1)}$$

Resolving into partial fractions, $\bar{y} = \dfrac{2s+3}{s^2+2s+2} + \dfrac{-2s+1}{s^2+1}$

Taking the inverse transform, we get

$$\Rightarrow \quad y = L^{-1}\left(\frac{2s+3}{s^2+2s+2}\right) + L^{-1}\left(\frac{-2s+1}{s^2+1}\right)$$

$$= L^{-1}\left[\frac{2(s+1)+1}{(s+1)^2+1}\right] + L^{-1}\left(\frac{-2s}{s^2+1}\right) + L^{-1}\left(\frac{1}{s^2+1}\right)$$

$$= L^{-1}\left[\frac{2(s+1)}{(s+1)^2+1}\right] + L^{-1}\left[\frac{1}{(s+1)^2+1}\right] - 2\cos t + \sin t$$

$$= 2e^{-t}\cos t + e^{-t}\sin t - 2\cos t + \sin t$$ **Ans.**

Example 77. *Solve the initial value problem* **[DIP IETE. 2019]**

$2y'' + 5y' + 2y = e^{-2t}, \quad y(0) = 1, y'(0) = 1$

using the Laplace transforms

Solution. $2y'' + 5y' + 2y = e^{-2t}, \quad y(0) = 1, y'(0) = 1$

Taking the Laplace transform of both sides, we get

$$2[s^2\bar{y} - sy(0) - y'(0)] + 5[s\bar{y} - y(0)] + 2\bar{y} = \frac{1}{s+2} \quad ...(1)$$

On substituting the values of $y(0)$ and $y'(0)$ in (1), we get

$$\Rightarrow \quad 2[s^2\bar{y} - s - 1] + 5[s\bar{y} - 1] + 2\bar{y} = \frac{1}{s+2}$$

$$\Rightarrow \quad [2s^2 + 5s + 2]\bar{y} - 2s - 2 - 5 = \frac{1}{s+2}$$

$$\Rightarrow \quad (2s^2 + 5s + 2)\bar{y} = 2s + 7 + \frac{1}{s+2}$$

$$\Rightarrow \quad \bar{y} = \frac{1}{(s+2)(2s^2+5s+2)} + \frac{2s+7}{2s^2+5s+2}$$

$$\Rightarrow \quad \frac{1 + 2s \;\; + 7s + 4s + 14}{(2s \;\; + 5s + 2)(s+2)} = \frac{2s^2 + 11s + 15}{(2s+1)(s+2)^2}$$

$$= \frac{\frac{40}{9}}{2s+1} - \frac{\frac{11}{9}}{s+2} - \frac{\frac{1}{3}}{(s+2)^2}$$

$$= \frac{40}{9}\frac{1}{2}\frac{1}{s+\frac{1}{2}} - \frac{11}{9}\frac{1}{s+2} - \frac{1}{3}\frac{1}{(s+2)^2}$$

$$\Rightarrow \quad y = \frac{20}{9}\frac{1}{s+\frac{1}{2}} - \frac{11}{9}\frac{1}{(s+2)} + \frac{1}{3}\frac{d}{ds}\frac{1}{(s+2)}$$

$$\Rightarrow \quad y = \frac{20}{9}e^{-\frac{1}{2}t} - \frac{11}{9}e^{-2t} - \frac{1}{3}te^{-2t}$$ **Ans.**

Example 78. *Solve* $\frac{d^2y}{dx^2} + 2\frac{dy}{dx} + 5y = e^{-x}\sin x$ *where* $y(0) = 0,\ y'(0) = 1.$

Solution. $$\frac{d^2y}{dx^2} + 2\frac{dy}{dx} + 5y = e^{-x}\sin x$$

Taking the Laplace transform of both the sides, we get

$$\Rightarrow [s^2\bar{y} - sy(0) - y'(0)] + 2[s\bar{y} - y(0)] + 5\bar{y} = L(e^{-x}\sin x)$$

$$\Rightarrow [s^2\bar{y} - sy(0) - y'(0)] + 2[s\bar{y} - y(0)] + 5\bar{y} = \frac{1}{(s+1)^2+1} \qquad ...(1)$$

On substituting the values of $y(0)$ and $y'(0)$ in (1), we get

$$\Rightarrow \quad (s^2\bar{y} - 1) + 2(s\,\bar{y}) + 5\bar{y} = \frac{1}{s^2+2s+2}$$

$$\Rightarrow \quad (s^2 + 2s + 5)\bar{y} = 1 + \frac{1}{s^2+2s+2} = \frac{s^2+2s+3}{s^2+2s+2}$$

$$\Rightarrow \quad \bar{y} = \frac{s^2+2s+3}{(s^2+2s+5)(s^2+2s+2)}$$

On resolving the R.H.S. into partial fractions, we get

$$\bar{y} = \frac{2}{3}\frac{1}{s^2 + 2s + 5} + \frac{1}{3}\frac{1}{s^2 + 2s + 2}$$

On inversion, we obtain

$$y = \frac{2}{3}L^{-1}\left(\frac{1}{s^2+2s+5}\right) + \frac{1}{3}L^{-1}\left(\frac{1}{s^2+2s+2}\right)$$

$$\Rightarrow \quad y = \frac{1}{3}L^{-1}\left[\frac{2}{(s+1)^2+(2)^2}\right] + \frac{1}{3}L^{-1}\left[\frac{1}{(s+1)^2+(1)^2}\right]$$

$$\Rightarrow \quad y = \frac{1}{3}e^{-x}\sin 2x + \frac{1}{3}e^{-x}\sin x$$

$$\Rightarrow \quad y = \frac{1}{3}e^{-x}(\sin x + \sin 2x)$$ **Ans.**

Example 79. *Solve the equation by the transform method:*

$$\frac{dy}{dt} + 2y + \int_0^t y\,dt = \sin t,\ y(0) = 1$$

Solution. Taking Laplace transform of the given equations, we get

$$\Rightarrow \quad [s\bar{y} - y(0)] + 2\bar{y} + \frac{\bar{y}}{s} = \frac{1}{s^2+1} \quad ...(1)$$

$$\left[\because \left\{\int_0^t y\,dt = \frac{\bar{y}}{s}\right\}\right]$$

Putting the value of $y(0) = 1$ in (1), we get

$$\Rightarrow \quad [s\,\bar{y} - 1] + 2\bar{y} + \frac{\bar{y}}{s} = \frac{1}{s^2+1}$$

$$\Rightarrow \quad \bar{y}\left(s + 2 + \frac{1}{s}\right) = 1 + \frac{1}{s^2+1} \quad [\because y(0) = 1]$$

$$\Rightarrow \quad \frac{1}{s}\bar{y}(s^2 + 2s + 1) = \frac{s^2+1+1}{s^2+1}$$

$$\Rightarrow \quad \frac{1}{s}\bar{y}(s+1)^2 = \frac{s^2+2}{s^2+1}$$

$$\Rightarrow \quad \bar{y}(s+1)^2 = \frac{s^3+2s}{s^2+1}$$

$$\Rightarrow \quad \bar{y} = \frac{s^3+2s}{(s+1)^2(s^2+1)}$$

$$= \frac{1}{s+1} - \frac{3}{2(s+1)^2} + \frac{1}{2(s^2+1)} \quad \text{[By partial fractions]}$$

Taking inverse Laplace transform, we have

$$y = e^{-t} - \frac{3}{2}t\,e^{-t} + \frac{1}{2}\sin t \qquad \textbf{Ans.}$$

Example 80. *Using Laplace transforms, find the solution of the initial value problem*

$$y'' + 9y = 9u\,(t-3), \quad y(0) = y'(0) = 0$$

where u (t – 3) is the unit step functions.

Solution.

$$y'' + 9y = 9u\,(t-3) \qquad \text{...(1)}$$

Taking Laplace transform of (1), we have

$$\Rightarrow \quad s^2\bar{y} - sy(0) - y'(0) + 9\bar{y} = 9\frac{e^{-3s}}{s} \qquad \text{...(2)}$$

Putting the values of $y(0) = 0$ and $y'(0) = 0$ in (2), we get

$$\Rightarrow \quad s^2\bar{y} + 9\bar{y} = 9\left(\frac{e^{-3s}}{s}\right)$$

$$\Rightarrow \quad (s^2+9)\bar{y} = 9\left(\frac{e^{-3s}}{s}\right)$$

$$\Rightarrow \quad \bar{y} = 9\left[\frac{e^{-3s}}{s(s^2+9)}\right]$$

$$\Rightarrow \quad y = L^{-1}\left[\frac{e^{-3s}}{s(s^2+9)}\right] \qquad \text{...(3)}$$

We know that

$$L^{-1}\left(\frac{3}{s^2+9}\right) = \sin 3t$$

$$\Rightarrow \quad 3L^{-1}\left[\frac{3}{s(s^2+9)}\right] = 3\int_0^t \sin 3t\,dt = -[\cos 3t]_0^t = 1 - \cos 3t \qquad \text{...(4)}$$

[Using second shifting theorem]

Using (4), we get the inverse of (3)

$$\therefore \quad y = [1 - \cos 3\,(t-3)]\,u(t-3)] \qquad \textbf{Ans.}$$

35.36 ELECTRIC CIRCUIT

Consider an electric circuit consisting of a resistance R, inductance L, a condensor of capacity C and electromotive power of voltage E in a series. A switch is also connected in the circuit. Here, $i = \frac{dq}{dt}$

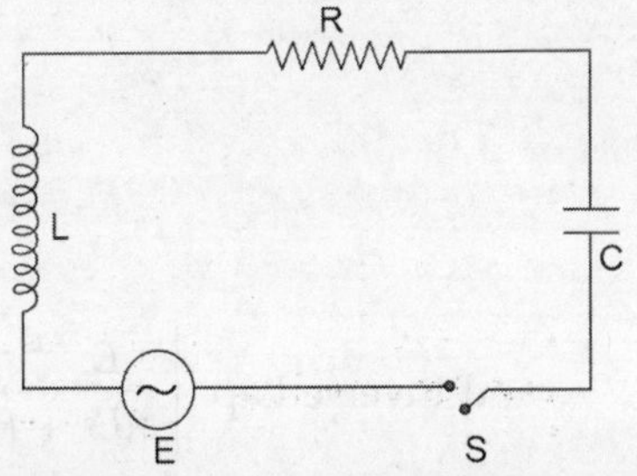

Voltage developed by $Ri, L\frac{di}{dt}$ and $\frac{q}{C}$

By Kirchhoff law $\boxed{L\frac{di}{dt} + Ri + \frac{q}{C} = E}$

Example 81. *A resistance R in series with inductance L is connected with e.m.f. E (t). The current is given by* $L\frac{di}{dt} + Ri = E$

If the switch is connected at t = 0 and disconnected at t = a, find the current i in terms of t.

Solution. Conditions under which current i flows are

$$E(t) = \begin{cases} E, & 0 < t < a \\ 0, & t > a \end{cases}$$

Given equation is
$$L\frac{di}{dt} + Ri = E \qquad ...(1)$$

Taking Laplace transform of (1), we get

$$\Rightarrow \quad L[s\bar{i} - i(0)] + R\bar{i} = \int_0^\infty e^{-st} E dt$$

$$\Rightarrow \quad L(s\bar{i}) + R\bar{i} = \int_0^\infty e^{-st} E dt \qquad [\because i(0) = 0]$$

$$\Rightarrow \quad (Ls + R)\bar{i} = \int_0^\infty e^{-st} .E dt = \int_0^a e^{-st} E\, dt + \int_a^\infty e^{-st} E\, dt$$

$$= E\left[\frac{e^{-st}}{-s}\right]_0^a + 0 = \frac{E}{s}[1 - e^{-as}] = \frac{E}{s} - \frac{E}{s}e^{-as}$$

$$\Rightarrow \quad \bar{i} = \frac{E}{s(Ls + R)} - \frac{Ee^{-as}}{s(Ls + R)}$$

Taking inverse Laplace transform, we obtain

$$i = \text{Inverse Lap.}\left[\frac{E}{s(Ls+R)}\right] - \text{Inverse Lap.}\left[\frac{E e^{-as}}{s(Ls+R)}\right] \qquad ...(2)$$

Now, we have to find the value of inverse Lap. $\left[\frac{E}{s(Ls + R)}\right]$

$$\text{Inverse Lap.}\left[\frac{E}{s(Ls + R)}\right] = \frac{E}{L} \text{ inverse Lap.}\left[\frac{1}{s\left(s + \frac{R}{L}\right)}\right]$$

$$= \frac{E}{L}\frac{L}{R} \text{ inverse Lap.}\left[\frac{1}{s} - \frac{1}{s + \frac{R}{L}}\right]$$

$$= \frac{E}{R}\left[1 - e^{-\frac{R}{L}t}\right] \qquad \text{(Resolving into partial fractions)}$$

and $\text{Inverse Lap.}\left[\frac{Ee^{-as}}{s(Ls + R)}\right] = \frac{E}{R}\left[1 - e^{-\frac{R}{L}(t-a)}\right]u(t - a)$ (By second shifting theorem)

On substituting the values of the inverse transforms in (2), we get

$$\Rightarrow \qquad i = \frac{E}{R}\left[1 - e^{-\frac{R}{L}t}\right] - \frac{E}{R}\left[1 - e^{-\frac{R}{L}(t-a)}\right]u(t-a)$$

Hence $$i = \frac{E}{R}\left[1 - e^{-\frac{R}{L}t}\right] \text{ for } 0 < t < a, \qquad [u(t-a) = 0]$$

$$= \frac{E}{R}\left[1 - e^{-\frac{R}{L}t}\right] - \frac{E}{R}\left[1 - e^{-\frac{R}{L}(t-a)}\right] \text{ for } t > a \qquad [u(t-a) = 1]$$

$$= \frac{E}{R}\left[e^{-\frac{R}{L}(t-a)} - e^{-\frac{R}{L}t}\right]$$

$$= \frac{E}{R}e^{-\frac{R}{L}t}\left[e^{\frac{Ra}{L}} - 1\right]$$ **Ans.**

Example 82. *Using the inverse Laplace transform, find the current i(t) in the LC - Circuit. Assuming L = 1 henry. C = 1 farad, zero initial current and charge on the capacitor, and*

v(t) = t when 0 < t < 1

= 0 otherwise.

Solution. The differential equation for L and C circuit is

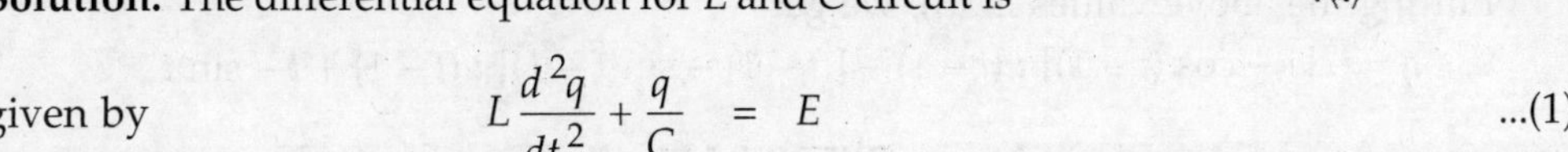

given by $$L\frac{d^2q}{dt^2} + \frac{q}{C} = E \qquad ...(1)$$

Putting $L = 1$, $C = 1$, $E = v(t)$ in (1), we get

$$\Rightarrow \qquad \frac{d^2q}{dt^2} + q = v(t) \qquad ...(2)$$

Taking Laplace transform of (2), we have

$$\Rightarrow \qquad s^2\bar{q} - sq(0) - q'(0) + \bar{q} = \int_0^\infty v(t)e^{-st}\,dt$$

Substituting $q(0) = 0$, and $q'(0) = 0$, we get

$$\Rightarrow \qquad s^2\bar{q} + \bar{q} = \int_0^1 te^{-st}\,dt + \int_0^\infty 0e^{-st}\,dt$$

$$\Rightarrow \qquad (s^2 + 1)\bar{q} = \left[t\frac{e^{-st}}{-s}\right]_0^1 - \int_0^1 \frac{e^{-st}}{-s}\,dt$$

$$= \frac{e^{-s}}{-s} - \left[\frac{e^{-st}}{s^2}\right]_0^1 = \frac{e^{-s}}{s} - \frac{e^{-s}}{s^2} + \frac{1}{s^2}$$

$$\Rightarrow \quad \bar{q} = \frac{1}{s^2+1}\left[-\frac{e^{-s}}{s} - \frac{e^{-s}}{s^2} + \frac{1}{s^2}\right]$$

$$\Rightarrow \quad \bar{q} = \frac{-e^{-s}}{s(s^2+1)} - \frac{e^{-s}}{s^2(s^2+1)} + \frac{1}{s^2(s^2+1)}$$

Taking inverse Laplace transform, we get

$$q = \text{Inverse Lap. } \frac{-e^{-s}}{s(s^2+1)} - \text{inverse Lap. } \frac{e^{-s}}{s^2(s^2+1)} + \text{inverse Lap. } \frac{1}{s^2(s^2+1)} \qquad ...(3)$$

We know that

$$\text{Inverse Lap. } [e^{-as} F(s)] = f(t-a)u(t-a)$$

$$\text{Inverse Lap. } \frac{1}{s(s^2+1)} = \int_0^1 \sin t dt = [-\cos t]_0^t = 1 - \cos t \qquad ... (4)$$

$$\text{Inverse Lap. } \frac{1}{s^2(s^2+1)} = \int_0^1 (1 - \cos t)\, dt = t - \sin t \qquad ...(5)$$

In view of this, we have

$$\text{Inverse Lap. } \left[\frac{-e^{-s}}{s(s+1)}\right] = [1 - \cos (t-1)]\, u\, (t-1) \qquad \text{[from (4)]}$$

$$\Rightarrow \quad \frac{e^{-s}}{s^2(s^2+1)} = [(t-1) - \sin (t-1)]\, u(t-1) \qquad \text{[from (5)]}$$

Putting the above values in (3), we get

$\therefore \quad q = -[1 - \cos (t-1)]\, u(t-1) - [(t-1) - \sin(t-1)]\, u(t-1) + t - \sin t$ **Ans.**

EXERCISE 35.17

Solve the following differential equations using Laplace transform:

1. $\frac{d^2y}{dx^2} + y = 0$ where $y = 1$ and $\frac{dy}{dx} = -1$ at $x = 0$. **Ans.** $y = \cos x - \sin x$

2. $\frac{d^2y}{dx^2} - 4y = 0$, where $y = 0$ and $\frac{dy}{dx} = -6$ at $x = 0$ **Ans.** $y = \frac{3}{2}e^{2x} + \frac{3}{2}e^{-2x}$

3. $\frac{d^2y}{dx^2} + y = 0$, where $y = 1$, $\frac{dy}{dx} = 1$ at $x = 0$ **Ans.** $y = \sin x + \cos x$

4. $\frac{d^2y}{dx^2} + 2\frac{dy}{dx} + 5y = 0$, where $y = 2$ $\frac{dy}{dx} = -4$ at $x = 0$. **Ans.** $y = e^{-x}(2 \cos 2x - \sin 2x)$

5. $\frac{d^3y}{dx^3} + 2\frac{d^2y}{dx^2} - \frac{dy}{dx} - 2y = 0$, given $y = \frac{dy}{dx} =, \frac{d^2y}{dx^2} = 6$ at $x = 0$. **Ans.** $y = e^x - 3e^x + 2e^{-2x}$

6. $\dfrac{d^2y}{dx^2} + y = 3\cos 2x$, where $y = \dfrac{dy}{dx} = 0$ at $x = 0$, **Ans.** $y = \cos x - \cos 2x$.

7. $\dfrac{d^2y}{dx^2} + \dfrac{dy}{dx} - 2y = 1 - 2x$, given $y = 0$, $\dfrac{dy}{dx} = 4$ at $x = 0$. **Ans.** $y = e^x - e^{-2x} + x$

8. $\dfrac{d^2y}{dx^2} - 3\dfrac{dy}{dx} + 2y = 4e^{2x}$, given $y = -3$, and $\dfrac{dy}{dx}$ -5 at $x = 0$
Ans. $y = -7e^x + 4e^{2x} + 4xe^{2x}$

9. $\dfrac{d^2y}{dx^2} - 3\dfrac{dy}{dx} + 2y = 4x + e^{2x}$, where $y = 1$, $\dfrac{dy}{dx} = -1$ at $x = 0$.

Ans. $y = 3 + 2x + \dfrac{1}{2}e^{3x} - 2e^{2x} - \dfrac{1}{2}e^x$

10. $\dfrac{d^3y}{dx^3} + 2\dfrac{d^2y}{dx^2} - \dfrac{dy}{dx} - 2y = 0$, where $y = 1$, $\dfrac{dy}{dx} = 2$, $\dfrac{d^2y}{dx^2} = 2$ at $x = 0$.

Ans. $y = \dfrac{5}{3}e^x - e^{-x} + \dfrac{1}{2}e^{-2x}$

Statistics

36.1 INTRODUCTION

Statistics is a branch of science dealing with the collection of data, organising, summarising, presenting and analysing data for drawing valid conclusions and thereafter making reasonable decisions on the basis of such analysis.

36.2 FREQUENCY DISTRIBUTION

Frequency distribution is the arranged data, summarised by distributing it into classes or categories with their frequencies.

Wages of 100 workers in thousands

Wages in ₹	0–10	10–20	20–30	30–40	40–50
Number of workers	12	23	35	20	10

36.3 GRAPHICAL REPRESENTATION

It is often useful to represent frequency distribution by means of a diagram. The different types of diagrams are:

1. Histogram
2. Frequency polygon
3. Frequency curve
4. Cummulative frequency curve or Ogive
5. Bar chart
6. Circles or Pie diagrams.

Histogram consists of a set of rectangles having their heights proportional to the class-frequencies, for equal class-intervals. For unequal class-intervals, the areas of rectangles are proportional to the frequencies.

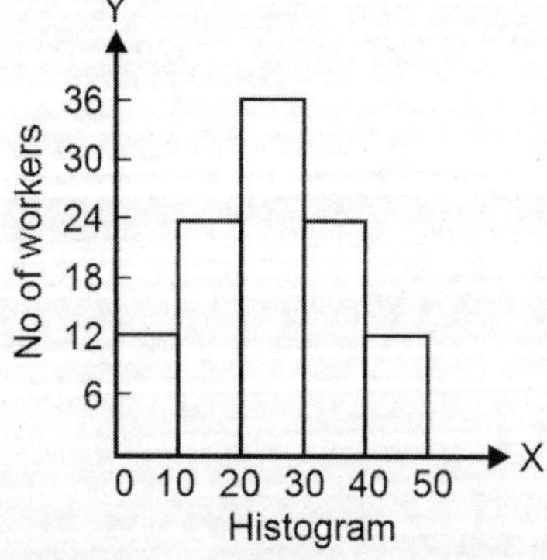

Histogram

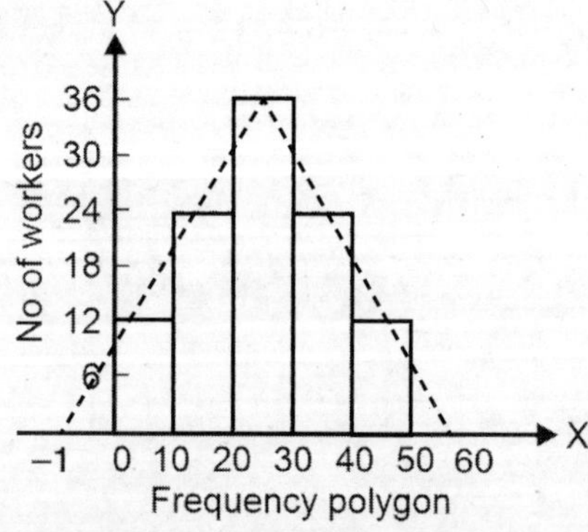

Frequency polygon

Frequency polygon is a line graph of class-frequency plotted class-mark. It can be obtained by connecting mid-points on the tops of the rectangles in the histogram.

Cumulative frequency curve or the Ogive If the various points are plotted according to the upper limit of the class as x-coordinate and the cumulative frequency as y-coordinate and these points are joined by a free hand smooth curve. The curve obtained is known as cummulative frequency curve or the Ogive.

36.4 MEASURES OF CENTRAL TENDENCY OR AVERAGES

An average is a value which is representative of a set of data. Average value may also be termed as measure of central tendency. There are five types of common measures of Central Tendency.

(*i*) Arithmetic average or Mean (*ii*) Median
(*iii*) Mode (*iv*) Geometric Mean
(*v*) Harmonic Mean

(i) Arithmetic Mean

If $x_1, x_2, x_3, \ldots. x_n$ are n numbers, then their arithmetic mean is defined by

$$A.M. = \frac{x_1 + x_2 + x_3 + \ldots\ldots + x_n}{n} = \frac{\sum x_n}{n} \qquad \ldots. (1)$$

If the numbers x_1 occurs f_1 times, x_2 occurs f_2 times and so on, then

$$A.M. = \frac{f_1x_1 + f_2x_2 + \ldots\ldots + f_nx_n}{f_1 + f_2 + \ldots\ldots + f_n} = \frac{\sum f\,x}{\sum f} \qquad \ldots (2)$$

(*a*) **Direct Method** uses above two formulae.

Example 1. *Find the Mean of 20, 22, 25, 28, 30.*

Solution. $A.M. = \frac{20 + 22 + 25 + 28 + 30}{5} = \frac{125}{5} = 25$ **Ans.**

Example 2. *Find the Mean of the following:*

Number	8	10	15	20
Frequency	5	8	8	4

Solution. $\Sigma fx = (8 \times 5) + (10 \times 8) + (15 \times 8) + (20 \times 4) = 40 + 80 + 120 + 80 = 320$

$\Sigma f = 5 + 8 + 8 + 4 = 25$

$$A.M. = \frac{\sum fx}{\sum f} = \frac{320}{25} = 12.8$$ **Ans.**

(*b*) **Short cut method**

Let a be the assumed mean, d the deviation of the variate x from a.

Then $$\frac{\sum fd}{\sum f} = \frac{\sum f(x-a)}{\sum f} = \frac{\sum fx}{\sum f} - \frac{\sum fa}{\sum f} = A.M. - \frac{a\sum f}{\sum f} = A.M. - a$$

$$\therefore \quad A.M. = a + \frac{\sum fd}{\sum f}$$

Example 3. *Find the arithmetic mean for the following distribution:*

Class	0–10	10–20	20–30	30–40	40–50
Frequency	7	8	20	10	5

Solution. Let assumed mean $(a) = 25$.

Class	*Mid-value* (x)	*Frequency* (f)	$x - 25 = d$	fd
0 – 10	5	7	– 20	– 140
10 – 20	15	8	– 10	– 80
20 – 30	25	20	0	0
30 – 40	35	10	+ 10	+ 100
40 – 50	45	5	+ 20	+ 100
Total		$\Sigma f = 50$		$\Sigma fd = -20$

$$\text{A.M.} = a + \frac{\sum fd}{\sum f} = 25 + \frac{-20}{50} = 24.6 \qquad \textbf{Ans.}$$

(*c*) **Step deviation method**

Let a be the assumed mean, i the width of the class interval and

$$d = \frac{x-a}{i}, \quad \text{A.M.} = a + \frac{\sum fd}{\sum f}(i)$$

Example 4. *Find arithmetic mean of the data given in example 3 by step deviation method.*

Solution. Let assumed mean $a = 25$

Class	*Mid-value* (x)	*Frequency* (f)	$d = \dfrac{x-a}{i}$	$f.d$
0 – 10	5	7	– 2	– 14
10 – 20	15	8	– 1	– 8
20 – 30	25	20	0	0
30 – 40	35	10	+ 1	+ 10
40 – 50	45	5	+ 2	+ 10
Total		$\Sigma f = 50$		$\Sigma fd = -2$

$$\text{A.M.} = a + \frac{\sum fd}{\sum f}(i) = 25 + \frac{-2}{50} \times 10 = 24.6 \qquad \textbf{Ans.}$$

(ii) Median

Median is defined as the measure of the central item when they are arranged in ascending or descending order of magnitude.

When the total number of the items is odd and equal to say n, then the value of $\frac{1}{2}(n+1)^{\text{th}}$ item gives the median.

When the total number of the frequencies is even, say n, then there are two middle items, and so the mean of the values of $\left(\frac{n}{2}\right)$ th and $\left(\frac{n}{2}+1\right)$ th items is the median.

Example 5. *Find the median of 6, 8, 9, 10, 11, 12, 13.*

Solution. Total number of items = 7

The middle item $= \frac{1}{2}(7+1)^{th} = 4^{th}$

Median = Value of the 4th item = 10 **Ans.**

For grouped data, $\boxed{\textbf{Median} = l + \left(\frac{\frac{N}{2} - C}{f}\right) \times i}$

where l is the lower limit of the median class, f is the frequency of the class, i is the width of the class-interval, C is the cumulative frequency of the class preceding the median-class and N is total frequency of the data.

Example 6. *Find the value of median from the following data:*

No. of days for which absent (less than)	5	10	15	20	25	30	35	40	45
No. of students	29	224	465	582	634	644	650	653	655

Solution. The given cumulative frequency distribution will first be converted into ordinary frequency as under.

Class-Interval	*Cumulative frequency*	*Ordinary frequency*
0 – 5	29	29
5 – 10	224	224 – 29 = 195
5 – 15	465	465 – 224 = 241
15 – 20	582	582 – 465 = 117
20 – 25	634	634 – 582 = 52
25 – 30	644	644 – 634 = 10
30 – 35	650	650 – 644 = 6
35 – 40	653	653 – 650 = 3
40 – 45	655	655 – 653 = 2

$$\text{Median} = \text{size of } \frac{655}{2} \text{ or } 327.5 \text{ th item}$$

327.5th item lies in 10 – 15 which is the median class.

$$\text{Median} = l + \left(\frac{\frac{N}{2} - C}{f}\right) \times i$$

where l *stands* for lower limit of median class, N stands for the total frequency, C stands for the cumulative frequency of the class preceding the median class, i stands for class interval f stands for frequency for the median class.

$$\text{Median} = 10 + \left(\frac{\frac{655}{2} - 224}{241}\right) \times 5$$

$$= 10 + \left(\frac{103.5 \times 5}{241}\right) = 10 + 2.15 = 12.15 \qquad \textbf{Ans.}$$

Example 7. *Determine the median wage graphically from the following data:*

Wages (in ₹)	20–40	40–60	60–80	80–100	100–120	120–140	140–160
No. of workers	4	6	10	16	12	7	3

Solution. Method 1. Draw two cumulative frequency curves one by 'less than' method and other by 'more than' method. From the point where both these curves meet, draw a perpendicular on the x-axis and the point where it meets the X-axis is the median.

Daily wages less than (₹)	*No. of workers (f)*	*Daily wages more than (₹)*	*No. of workers (f)*
40	4	20	58
60	10	40	54
80	20	60	48
100	36	80	38
120	48	100	22
140	55	120	10
160	58	140	3
		160	0

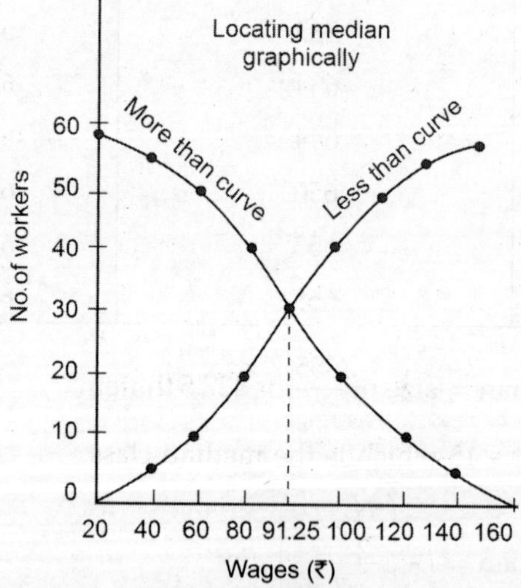

Method 2. If we draw only one Ogive, say, by 'less than' method, we can also determine the value of median from it. This is shown by the following graph:

$$\text{Median} = \text{Size of } \frac{58}{2} = 29\text{th item}$$

Take 29 on the y-axis and draw a perpendicular on the ogive. From the point where it meets the ogive, draw another perpendicular on the x-axis.

Locating median graphically

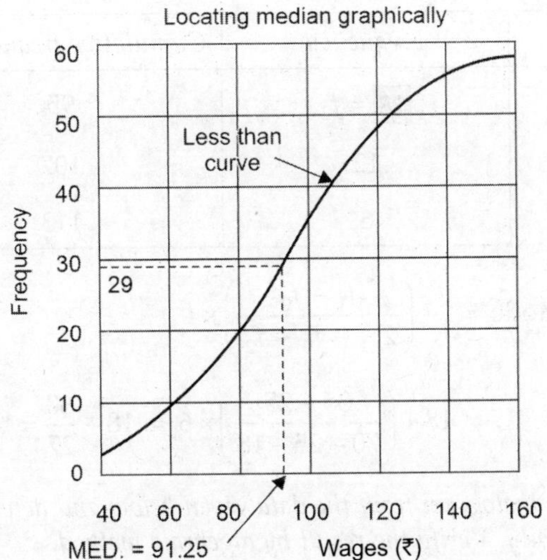

It is clear from the above graph that the daily median wage is ₹ 91.25. **Ans.**

(iii) Mode

Mode is defined to be the size of the variable which occurs most frequently.

Example 8. *Find the mode of the following items:*

0, 1, 6, 7, 2, 3, 7, 6, 6, 2, 6, 0, 5, 6, 0.

Solution. As 6 occurs 5 times and no other item occurs 5 or more than 5 times, hence the mode is 6. **Ans.**

For grouped data, $\textbf{Mode} = l + \left(\dfrac{f_1 - f_0}{2f_1 - f_0 - f_2}\right) \times i$

where l is the lower limit of the modal class, f_1 is the frequency of the modal class, f_0 is the frequency of the class preceding modal class, f_2 is the frequency of the class of succeeding modal class and i is the width of modal class.

Emperical formula

Mean – Mode = 3 [Mean – Median]

Example 9. *Find the mode from the following data:*

Age	*0–6*	*6–12*	*12–18*	*18–24*	*24–30*	*30–36*	*36–42*
Frequency	*6*	*11*	*25*	*35*	*18*	*12*	*6*

Solution.

Age	*Frequency*	*Cumulative frequency*
0–6	6	6
6–12	11	17
12–18	25 $= f_0$	42
18 – 24	35 $= f_1$	77

Age	Frequency	Cumulative frequency
24–30	18 $= f_2$	95
30–36	12	107
36–42	6	113

$$\text{Mode} = l + \left(\frac{f_1 - f_0}{2f_1 - f_0 - f_2}\right) \times i$$

$$= 18 + \left(\frac{35 - 25}{70 - 25 - 18}\right) \times 6 = 18 + \frac{60}{27} = 18 + 2.22 = 20.22 \quad \textbf{Ans.}$$

Example 10. *Draw a histogram from the data given below and determine the value of mode graphically. Verify the result by algebraic method.*

Size	*140–150*	*150–160*	*160–170*	*170–180*	*180–190*	*190–200*	*200–210*
Frequency	*40*	*60*	*100*	*180*	*40*	*30*	*20*

[Osmania Univ. 2007]

Solution.

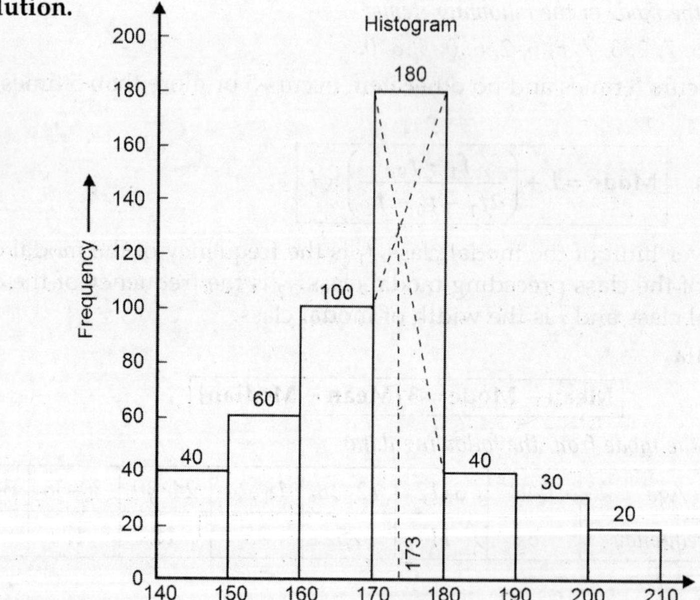

It is clear from the above histogram that the modal value is 173 approximate.

Direct calculation:

Mode lies in the class 170–180

$$\text{Mode} = l + \frac{f_1 - f_0}{2f_1 - f_0 - f_2} \times i$$

$f_1 = 180, f_0 = 100, f_2 = 40, l = 170, i = 10$

$$\therefore \quad \text{Mode} = 170 + \frac{180-100}{360-100-40} \times 10 = 170 + \frac{80}{220} \times 10 = 170 + 3.64 = 173.64 \quad \textbf{Ans.}$$

Example 11. *The following table gives the weight (in kg) of 60 students. Find the value of mode graphically.*

Weight (in kg)	*No. of Students*
29.5–34.5	*3*
34.5–39.5	*5*
39.5–44.5	*12*
44.5–49.5	*18*
49.5–54.5	*14*
54.5–59.5	*6*
59.5–64.5	*2*

Solution. The modal line touches the x-axis at 47.5. Hence, the value of mode is 47.5 kg. If we calculate mode directly by the formula, it would be:

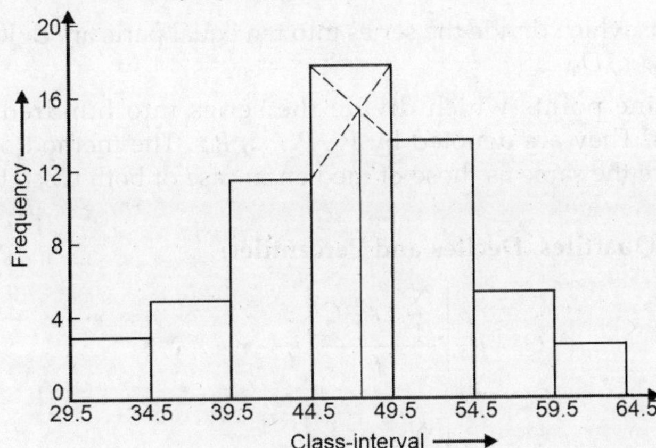

$$Mode = l + \left(\frac{f_1 - f_0}{2f_1 - f_0 - f_2}\right) \times i \text{; modal class is 44.5–49.5}$$

$$Mode = 44.5 + \frac{18-12}{36-12-14} \times (5) = 44.5 + \left(\frac{6}{10} \times 5\right) = 47.5 \quad \textbf{Ans.}$$

(iv) Geometric Mean

If $x_1, x_2, x_3, ..., x_n$ be n values of variate x, then the geometric mean

$$G = (x_1 \times x_2 \times x_3 \times x_4 \times ..... \times x_n)^{\frac{1}{n}}$$

Example 12. *Find the geometric mean of 4, 8, 16.*

Solution. $G.M. = (4 \times 8 \times 16)^{1/3} = 8.$ **Ans.**

(v) Harmonic Mean

Harmonic mean of a series of values is defined as the reciprocal of the arithmetic mean of their reciprocals. Thus if H be the harmonic mean, then

$$\frac{1}{H} = \frac{1}{n}\left[\frac{1}{x_1} + \frac{1}{x_2} + \ldots\ldots + \frac{1}{x_n}\right]$$

Example 13. *Calculate the harmonic mean of 4, 8, 16.*

Solution.
$$\frac{1}{H} = \frac{1}{3}\left[\frac{1}{4} + \frac{1}{8} + \frac{1}{16}\right] = \frac{7}{48}$$

$$H = \frac{48}{7} = 6.853$$ **Ans.**

36.5 PARTITION VALUES

These are the values which divide the series into a number of equal parts.

The three points which divide the series into four equal parts are called **quartiles.** They are denoted by Q_1, Q_2 and Q_3. The first quartile Q_1 exceeds 25% of the observations, the second quartile Q_2 coincides with the median and the third quartile Q_3 exceeds 75% of the observations.

The nine points which divide the series into ten equal parts are called **deciles** and are denoted by $D_1, D_2, \ldots, D_9$.

The ninety nine points which divide the series into hundred equal parts are called **percentiles.** They are denoted by $P_1, P_2, \ldots, P_{99}$. The method of computation of partition values are the same as those of median in case of both discrete and continuous distributions.

Formulae for Quartiles, Deciles and Percentiles:

$$Q_r = l + \left[\frac{\frac{N}{4} \times r - C}{f}\right] i, \quad r = 1, 2, 3$$

$$D_r = l + \left[\frac{\frac{N}{10} \times r - C}{f}\right] i, \quad r = 1, 2, \ldots\ldots 9$$

$$P_r = l + \left[\frac{\frac{N}{100} \times r - C}{f}\right] i, \quad r = 1, 2, 3, \ldots\ldots 99$$

36.6 MEASURES OF DISPERSION

Average of a date discussed earlier fail to reveal full details of the distribution. Two or three distributions may have the same average but still they may differ from each other in many ways. In such cases, further statistical analysis of the data is necessary so that these difference between various series can be studied and accounted for the calculation. Such analysis will make our result more accurate and we shall be more confident of our conclusions.

Suppose, there are three series of nine items each as follows:

	Series A	*Series B*	*Series C*
	50	48	5
	50	50	15
	50	46	20
	50	49	25
	50	47	35
	50	52	80
	50	53	85
	50	51	90
	50	54	95
Total	450	450	450
Mean	50	50	50

In the first series, the mean is 50 and the value of all the items is identical. The items are not at all scattered, and the mean is the representative of this distribution. However, in the second case, though the mean is 50 yet all the items of the series have different values. But the items are not very much scattered as the minimum value of the series is 46 and the maximum is 54 in the range. In this case also, mean is a good representative of the series because the difference between the mean and other items is not very significant. In the third series also, the mean is 50 and the values of different items are also different, but here the values are very widely scattered. Though the mean is the same in all the three series, yet the series differ widely from each other in their formation. Obviously, the average does not satisfactorily represent the individual items in this group and to know about the series completely, further analysis is essential. The scatter among the items in the first case nil, in the second case it varies within a small range, while in the third case the values range between a very big span and they are widely scattered. It is evident from the above, that a study of the extent of the scatter around average should also be made to throw more light on the composition of a series. The name given to this scatter *dispersion*.

Definition

Some important definitions of dispersion are given below:

(*i*) "Dispersion or spread is the degree of the scatter or variation of the variable about a central value." – *Brooks and Dick*

(*ii*) "Dispersion is the measure of the variations of the items" – *A. L. Bowley*

(*iii*) "The degree to which numerical data tend to spread about an average value is called the variation or dispersion of the data." – *Spiegel*

(*iv*) "Measures of variability are usually used to indicate how tightly bunched the sample values are around the mean." – *Dyckman and Thomas*

From the above definitions, it is clear that *in a general sense* the term dispersion refers to the variability in the size of items. If the variation is substantial, dispersion is said to be considerable and if the variation is very little, dispersion is insignificant.

Different Measures of Dispersion

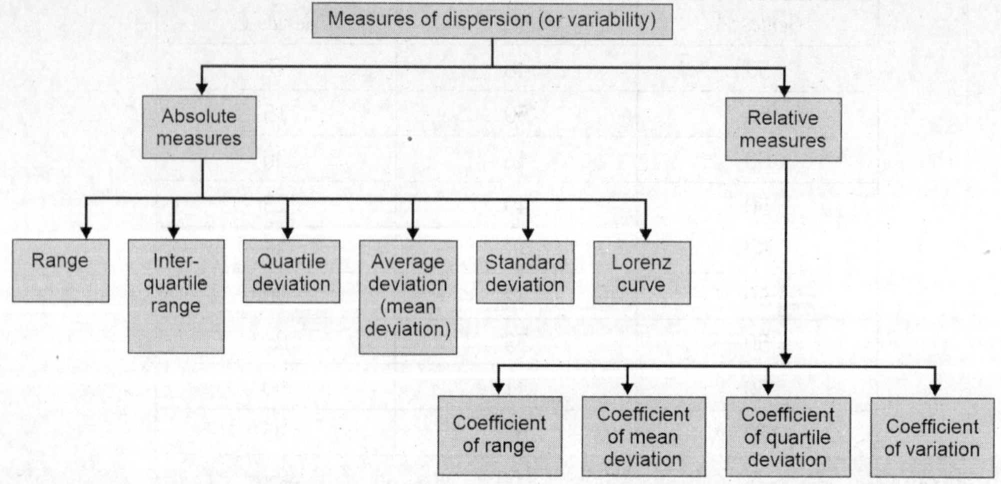

36.7 RANGE

Range is the simplest measure of dispersion. It is the difference between two extreme observations of the distribution.

$$\text{Range} = L - S$$

where L is the largest value and S is the smallest value in a series.

With all its limitations, range is commonly used in quality control, weather forecasting, variation in money, sales, share values, gold prices, etc.

36.8 QUARTILE DEVIATION

$$\text{Q.D.} = \frac{\text{Inter quartile range}}{2} = \frac{Q_3 - Q_1}{2}$$

where Q_1 and Q_3 are the first and third quartiles of the distribution. It is an absolute measure of dispersion. If it is divided by the average value of the two quartiles, a relative measure of dispersion is obtained.

$$\therefore \quad \text{Coefficient of Q.D.} = \frac{Q_3 - Q_1}{Q_3 + Q_1}$$

36.9 AVERAGE DEVIATION OR MEAN DEVIATION

It is the mean of the absolute values of the deviations of a given set of numbers from their arithmetic mean.

If $x_1, x_2, x_3, ..., x_n$ be a set of numbers with frequencies $f_1, f_2, ..., f_n$ respectively. Let $\overline{x}$ be the arithmetic mean of the numbers $x_1, x_2, ..., x_n$, then

$$\textbf{Mean deviation} = \frac{\sum f_i |x_i - \overline{x}|}{\sum f_i}$$

Example 14. *Find the mean deviation of the following frequency distribution.*

Class	0–6	6–12	12–18	18–24	24–30
Frequency	8	10	12	9	5

Solution. Let $a = 15$

Class	*Mid-value* (x)	*Frequency* (f)	$d = x - a$	fd	$\lvert x - \bar{x} \rvert$	$f\lvert x - \bar{x} \rvert$
0–6	3	8	– 12	– 96	11	88
6–12	9	10	– 6	– 60	5	50
12–18	15	12	0	0	1	12
18–24	21	9	+ 6	54	7	63
24–30	27	5	+ 12	60	13	65
Total		$\Sigma f = 44$		$\Sigma fd = -42$		$\Sigma f\lvert x - 14 \rvert = 278$

$$\text{Mean } (\bar{x}) = a + \frac{\sum fd}{\sum f} = 15 - \frac{42}{44} = 14 \quad \text{(nearly)}$$

$$\text{Average deviation} = \frac{\sum f\lvert x - \bar{x}\rvert}{\sum f} = \frac{278}{44} = 6.3$$ **Ans.**

36.10 STANDARD DEVIATION

Standard deviation is defined as the square root of the mean of the squares of the deviations from the arithmetic mean.

$$S.D. = \sigma = \sqrt{\frac{\sum f(x - \bar{x})^2}{\sum f}}$$

Note: 1. The square of the standard deviation σ^2 is called variance.

2. σ^2 is called the second moment about the mean and is denoted by μ_2.

36.11 SHORTEST METHOD FOR CALCULATING STANDARD DEVIATION

We know that $$\sigma^2 = \frac{1}{N}\sum f(x - \bar{x})^2 = \frac{1}{N}\sum f(x - a - \overline{x - a})^2$$

$$= \frac{1}{N}\sum f(d - \overline{x - a})^2 \quad \text{Where } x - a = d$$

$$= \frac{1}{N}\sum fd^2 - 2(\bar{x} - a)\frac{1}{N}\sum fd + (\bar{x} - a)^2 \frac{1}{N}\sum f$$

$$= \frac{1}{N}\sum fd^2 - 2(\bar{x} - a)\frac{1}{N}\sum fd + (\bar{x} - a)^2 \qquad \left[\because \sum f = N\right]$$

$$\bar{x} = a + \frac{\sum fd}{N} \quad \text{or} \quad \bar{x} - a = \frac{\sum fd}{N}$$

$$\sigma^2 = \frac{1}{N}\sum fd^2 - 2\left(\frac{\sum fd}{N}\right)\left(\frac{1}{N}\sum fd\right) + \left(\frac{\sum fd}{N}\right)^2$$

$$= \frac{1}{N}\sum fd^2 - \left(\frac{\sum fd}{N}\right)^2$$

$$S.D. = \sigma = \sqrt{\frac{\sum fd^2}{N} - \left(\frac{\sum fd}{N}\right)^2}$$

Example 15. *Calculate the mean and standard deviation for the following data:*

Size of item	6	7	8	9	10	11	12
Frequency	3	6	9	13	8	5	4

Solution. Assumed mean (a) = 9

x	f	$d = x - a$	$f.d.$	$f{\cdot}d^2$
6	3	− 3	− 9	27
7	6	− 2	− 12	24
8	9	− 1	− 9	9
9	13	0	0	0
10	8	+ 1	8	8
11	5	+ 2	10	20
12	4	+ 3	12	36
	$\sum f = 48$		$\sum fd = 0$	$\sum fd^2 = 124$

$$\text{Mean} = a + \frac{\sum fd}{\sum f} = 9 + 0 = 9$$

$$\text{S.D.} = \sqrt{\frac{\sum f(x-\bar{x})^2}{\sum f}} = \sqrt{\frac{\sum fd^2}{\sum f} - \left(\frac{\sum fd}{\sum f}\right)^2} = \sqrt{\frac{124}{48}} = 1.6 \qquad \textbf{Ans.}$$

Example 16. *From the following frequency distribution, compute the standard deviation of 100 students:*

Mass in kg	60–62	63–65	66–68	69–71	72–74
Number of students	5	18	42	27	8

Solution. Assumed mean (a) = 67

Mass in kg	*No. of students (f)*	x	$d = x - a$	$f{\cdot}d$	$f{\cdot}d^2$
60–62	5	61	− 6	− 30	180
63–65	18	64	− 3	− 54	162
66–68	42	67	0	0	0
69–71	27	70	3	81	243
72–74	8	73	6	48	288
	$\sum f = 100$			$\sum fd = 45$	$\sum fd^2 = 873$

$$\text{S.D.} = \sqrt{\frac{\sum fd^2}{\sum f} - \left(\frac{\sum fd}{\sum f}\right)^2} = \sqrt{\frac{873}{100} - \left(\frac{45}{100}\right)^2}$$

$$= \sqrt{8.73 - 0.2025} = \sqrt{8.5275} = 2.9202 \qquad \textbf{Ans.}$$

Example 17. *Compute the standard deviation for the following frequency distribution:*

Class interval	*0–4*	*4–8*	*8–12*	*12–16*
Frequency	*4*	*8*	*2*	*1*

Solution. Assumed mean = 6

Class interval	f	x	$d = x - 6$	fd	fd^2
0–4	4	2	– 4	– 16	64
4–8	8	6	0	0	0
8–12	2	10	+ 4	8	32
12–16	1	14	+ 8	8	64
	$\sum f = 15$			$\sum fd = 0$	$\sum fd^2 = 873$

$$\text{S.D.} = \sqrt{\frac{\sum fd^2}{\sum f} - \left(\frac{\sum fd}{\sum f}\right)^2} = \sqrt{\frac{160}{15} - 0} = 3.266 \qquad \textbf{Ans.}$$

36.12 COEFFICIENT OF VARIATION

The standard deviation discussed above is an absolute measure of dispersion. The corresponding relative measure is known as the coefficient of variation. This measure developed by Karl Pearson is the most commonly used measure of relative variation. It is used in such problems where we want to compare the variability of two or more than two series. That series (or group) for which the coefficient of variation is greater is said to be more variable or less consistent, less uniform, less stable or less homogeneous. On the other hand, the series for which coefficient of variation is less is said to be less variable or more consistent, more uniform, more stable or more homogeneous. Coefficient of variation is denoted by C.V. and is obtained as follows:

$$\textbf{Coefficient of variation or C.V.} = \frac{\sigma}{\overline{X}} \times 100$$

It may be pointed out that although any measure of dispersion can be used in conjunction with any average in computing relative dispersion. Statisticians, almost always, use the standard deviation as the measure of dispersion and the arithmetic mean as the average. When the relative dispersion is stated in terms of the arithmetic mean and the standard deviation, the resulting percentage is known as the coefficient of variation or coefficient of variability.

Example 18. *Calculate the coefficient of variation from the following data relating to wages of workers of a factory.*

Wages (in ₹)	*No. of workers*	*Wages (in ₹)*	*No. of workers*
40–60	*10*	*120–140*	*20*
60–80	*15*	*140–160*	*10*
80–100	*28*	*160–180*	*8*
100–120	*32*	*180–200*	*5*

[B.Com., Nagarjuna Univ., 2006]

Solution.

Calculation of Coefficient of Variation

Wages	*mid value* (x)	f	$d = (x - 110)/10$	fd	fd^2
40–60	50	10	-6	-60	360
60–80	70	15	-4	-60	240
80–100	90	28	-2	-56	112
100–120	110	32	0	0	0
120–140	130	20	$+2$	$+40$	80
140–160	150	10	$+4$	$+40$	160
160–180	170	8	$+6$	$+48$	288
180–200	190	5	$+8$	$+40$	320
		$N = 128$		$\Sigma fd = -8$	$\Sigma fd^2 = 1560$

$$\text{C.V.} = \frac{\sigma}{\overline{X}} \times 100$$

Mean: $$\overline{X} = a + \frac{\Sigma fd}{N} \times i = 110 - \frac{8}{128} \times 20 = 110 - 125 = 108.75$$

$$\sigma = \sqrt{\frac{\Sigma fd^2}{N} - \left(\frac{\Sigma fd}{N}\right)^2} \times i = \sqrt{\frac{1560}{128} - \left(\frac{-8}{128}\right)^2} \times 20$$

$$= \sqrt{12.1875 - 0.0039} \times 20 = \sqrt{12.1836} \times 20 = 69.81$$

$$\text{C.V.} = \frac{69.81}{108.75} \times 100 = 64.19\%.$$ **Ans.**

Example 19. *The following table gives the figures of profits of two companies. A and B for the last 10 years. Which of the two companies has greater consistency in profits:*

Year	*Profits A (in ₹)*	*Profits B (in ₹)*
1973	*700*	*550*
1974	*675*	*600*
1975	*725*	*575*
1976	*625*	*550*
1977	*650*	*650*
1978	*700*	*600*
1979	*650*	*550*
1980	*700*	*525*
1981	*600*	*625*
1982	*650*	*600*

Solution. **Calculation of mean and standard deviation**

Company A				*Company B*			
(X_1)	*Deviation from* $a = 700$ $(X_1 - 700)$	*Step Dev.* $\frac{X_1 - 700}{25} = d_1$	d_1^2	(X_2)	*Deviation from* $a = 625$ $(X_2 - 625)$	*Step Dev.* $\frac{X_2 - 625}{25} = d_2$	d_2^2
700	0	0	0	550	– 75	– 3	9
625	– 25	– 1	1	600	– 25	– 1	1
725	+ 25	+ 1	1	575	– 50	– 2	4
625	– 75	– 3	9	550	– 75	– 3	9
650	– 50	– 2	4	650	+ 25	+ 1	1
700	0	0	0	600	– 25	– 1	1
650	– 50	– 2	4	550	– 75	– 3	9
700	0	0	0	525	– 100	– 4	16
600	– 100	– 4	16	625	0	0	0
650	– 50	– 2	4	600	– 25	– 1	1
		– 13	39			– 17	51

Company A. Mean or $\bar{X}_1 = a + \left(\frac{\Sigma d_1}{N} \times i\right) = 700 + \left(\frac{-13}{10} \times 25\right) = 700 - 32.5 =$ ₹ 667.5

Standard Deviation or $\sigma_1 = \sqrt{\frac{\Sigma d_1^2}{N} - \left(\frac{\Sigma d_1}{N}\right)^2} \times i = \sqrt{\frac{39}{10} - \left(\frac{-13}{10}\right)^2} \times 25 = \sqrt{3.9 - 1.69} \times 25$

$= 37.15$

Coefficient of Variation or C.V. $= \frac{\sigma_1}{\bar{X}_1} \times 100 = \frac{37.15}{667.5} \times 100 = 5.56$

Company B. Mean or $\bar{X}_2 = a + \left(\frac{\Sigma d_2}{N} \times i\right) = 625 + \left(\frac{-17}{10} \times 25\right) = 625 - 42.5 =$ ₹ 582.5

Standard Deviation or $\sigma_2 = \sqrt{\frac{\Sigma d_2^2}{N} - \left(\frac{\Sigma d_2}{N}\right)^2} \times i = \sqrt{\frac{51}{10} - \left(\frac{-17}{10}\right)^2} \times 25 = \sqrt{5.1 - 2.89} \times 25$

$= 37.15$

Coefficient of Variation or C.V. $= \frac{\sigma_2}{\bar{X}_2} \times 100 = \frac{37.15}{582.5} \times 100 = 6.38$

Company *A* is more consistent in profits as the C.V. there is 5.56 as compared to 6.38 in case of Company *B*. Conversely, we can say that in Company *A* variability of profits is less than the variability in Company *B*. **Ans.**

In the above example, the standard deviation of the two series is **equal**. However, since Standard Deviation is an absolute measure, the comparison would be fallacious. We should calculate either coefficient of the standard deviation or Coefficient of variation.

Example 20. *The life of the bulbs produced by Companies A and B are given below:*

Length of life (in hours)	*Company A (No. of Lamps)*	*Company B (No. of Lamps)*
500–700	5	4
700–900	11	30
900–1100	26	12
1100–1300	10	8
1300–1500	8	6
	60	60

The bulbs of which Company are more consistent from the point of view of length of life?

Solution. **Calculation of mean and Standard Deviation**

Length of Life (Hour)	*Mid-values (X)*	*Step-Dev. from 1000 (i = 200)* $\frac{X-1000}{200}=d$	*Factory (A)*			*Factory (B)*		
			(f)	*(fd)*	*(fd²)*	*(f)*	*(fd)*	*(fd²)*
500–700	600	– 2	5	– 10	20	4	– 8	16
700–900	800	– 1	11	– 11	11	30	– 30	30
900–1100	1000	0	26	0	0	12	0	0
1100–1300	1200	+ 1	10	+ 10	10	8	+ 8	8
1300–1500	1400	+ 2	8	+ 16	32	6	+ 12	24
			60	+ 5	73	60	– 18	78

Company A

$$\text{Mean} = 1000+\left(\frac{5}{60}\times 200\right)=1{,}016.67 \text{ hours}$$

$$\text{S.D.} = \sqrt{\frac{73}{60}-\left(\frac{5}{60}\right)^2}\times 200 = 220 \text{ hours}$$

$$\text{C.V.} = \frac{220}{1016.67}\times 100 = 21.64$$

Company B

$$\text{Mean} = 1000+\left(\frac{-18}{60}\times 200\right)=940 \text{ hours}$$

$$\text{S.D.} = \sqrt{\frac{78}{60}-\left(\frac{-18}{60}\right)^2}\times 200 = 220 \text{ hours}$$

$$\text{C.V.} = \frac{220}{940}\times 100 = 23.41$$

Bulbs of Company A are more consistent than those of Company B. **Ans.**

Example 21. *Samples of polythene bags from two manufactures, A and B, are tested by a prospective buyer for bursting pressure and the results are as follows:*

Bursting Pressure (lb)	*Number of Bags*	
	A	*B*
5.0–9.9	2	9
10.0–14.9	9	11
15.0–19.9	29	18
20.0–24.9	54	32
25.0–29.9	11	27
30.0–34.9	5	13

Which set of bags has more uniform pressure? If prices are the same, which manufacturer's bags would be preferred by the buyer? Why?

Solution. For finding out which set of bags has more uniform pressure, we shall compare coefficient of variation by converting given data to exclusive method.

Calculation of coefficient of variation (Manufacturer A)

Bursting Pressure (lb)	*Mid value (m)*	(f)	*(m – 22.45)/5 (d)*	fd	fd^2
4.95–9.95	7.45	2	– 3	– 6	18
9.95–14.95	12.45	9	– 2	– 18	36
14.95–19.95	17.45	29	– 1	– 29	29
19.95–24.95	22.45	54	0	0	0
24.95–29.95	27.45	11	+ 1	+ 11	11
29.95–34.95	32.45	5	+ 2	+ 10	20
		$N = 110$		$\Sigma fd = -32$	$\Sigma fd^2 = 114$

$$\text{C.V.} = \frac{\sigma}{\overline{X}} \times 100$$

$$\overline{X} = a + \frac{\Sigma fd}{N} \times i = 22.45 - \frac{32}{110} \times 5 = 22.45 - 1.46 = 20.99$$

$$\sigma = \sqrt{\frac{\Sigma fd^2}{N} - \left(\frac{\Sigma fd}{N}\right)^2} \times i = \sqrt{\frac{114}{110} - \left(\frac{-32}{110}\right)^2} \times 5$$

$$= \sqrt{1.036 - 0.085} \times 5 = 4.88$$

$$\text{C.V.} = \frac{4.88}{20.99} \times 100 = 23.25\ \%.$$

Calculation of coefficient of variation (Manufacturer B)

Bursting Pressure (lb)	*Mid value (m)*	(f)	*(m – 22.45)/5 (d)*	fd	fd^2
4.95–9.95	7.45	9	– 3	– 27	81
9.95–14.95	12.45	11	– 2	– 22	44
14.95–19.95	17.45	18	– 1	– 18	18
19.95–24.95	22.45	32	0	0	0
24.95–29.95	27.45	27	+ 1	+ 27	27
29.95–34.95	32.45	13	+ 2	+ 26	52
		$N = 110$		$\Sigma fd = -14$	$\Sigma fd^2 = 222$

$$\bar{X} = a + \frac{\Sigma fd}{N} \times i = 22.45 - \frac{14}{110} \times 5 = 22.45 - 64 = 21.81$$

$$\sigma = \sqrt{\frac{\Sigma fd^2}{N} - \left(\frac{\Sigma fd}{N}\right)^2} \times i = \sqrt{\frac{222}{110} - \left(\frac{-14}{110}\right)^2} \times 5$$

$$= \sqrt{2.018 - 0.16} \times 5 = 1.415 \times 5 = 7.07$$

$$\text{C.V.} = \frac{7.07}{21.81} \times 100 = 32.42\%$$

Since coefficient of variation is less in case of bags of manufacturer *A* hence they have more uniform pressure. If prices are the same, the buyer should prefer the bags of manufcturer *A* because they have more uniform pressure. **Ans.**

EXERCISE 36.1

Arithmetic Mean

1. The monthly income of ten families (in rupees) in a certain locality are given below:

Family	*A*	*B*	*C*	*D*	*E*	*F*	*G*	*H*	*I*	*J*
Income	30	70	10	75	500	8	42	250	40	36

Calculate the arithmetic mean by (*a*) direct method, and (*b*) short cut method.

Ans. 106.1

2. Given the following information:

Year	National income	Per capita	Net domestic
2001–2002	34,412	636.1	34,696
2002–2003	34,871	629.4	37.019
2003–2004	34,323	606.4	40,693
2004–2005	36,183	626.0	50,821
2005–2006	36,455	617.9	59,696
2006–2007	39,626	659.3	60,851
2007–2008	40,164	655.2	64,534

Find the average national income, per capita income. **Ans.** 36, 576.285 ; 639.2

3. From the following data of the marks obtained by 60 students of a class, calculate arithmetic mean

Marks	No. of students	Marks	No. of students
20	8	50	10
30	12	60	6
40	20	70	4

Ans. 41

4. The following data related to the distance travelled by 520 villagers to buy their weekly requirements:

Miles travelled	2	4	6	8	10	12	14	16	18	20
Number of villagers	38	104	140	78	48	42	28	24	16	2

Calculate the arithmetic mean. **Ans.** 7.7769

5. Given the following frequency distribution, calculate the arithmetic mean:

Monthly wages in ₹	No. of workers	Monthly wages in ₹	No. of workers
12.5–17.5	2	37.5–42.5	4
17.5–22.5	22	42.5–47.5	6
22.5–27.5	10	47.5–52.5	1
27.5–32.5	14	52.5–57.5	1
32.5–37.5	3		Total 63

Ans. 28.25

6. Find mean from the following distribution:

Class interval	Frequency
15–25	4
25–35	11
35–45	19
45–55	14
55–65	0
65–75	2

Ans. 40.2

7. Calculate arithmetic mean from the following data:

Temp. °C	No. of days	Temp °C	No. of days
−40 to −30	10	0 to 10	65
−30 to −20	28	10 to 20	180
−20 to −10	30	20 to 30	10
−10 to 0	42	—	

Ans. 4.288

8. Calculate arithmetic mean from the following data:

Marks (less than)	80	70	60	50	40	30	20	10
No. of students	100	90	80	60	32	20	13	5

Ans. 45

9. The monthly profit in rupees of 100 shops are distributed as follows:

Profit per shop	0–100	0–200	0–300	0–400	0–500	0–600
No. of shops	12	30	57	77	94	100

Find the average profit per shop. **Ans.** 217

10. The monthly profit (in rupees) of 100 shops are distributed as follows:

Profit per shop	0–1000	1000–2000	2000–3000	3000–4000	4000–5000	5000–6000	6000–7000
No. of shops	10	15	27	20	18	7	3

On scrutiny, the following mistakes of tabulation were revealed:

Shop	A	B	C	D	E	F	G	H
Correct profit	6120	3621	5400	8600	3689	2150	4800	800
Entery profit	3360	2950	4850	6800	3780	2045	5200	1250

Revise the frequency table and calculate mean profit per shop. **Ans.** 3080

11. If the average is 16.82, find the missing frequency from the following data:

Class interval	Frequency
0–5	10
5–10	12
10–15	16
15–20	—
20–25	14
25–30	10
30–35	8

Ans. 18

12. The following table gives the weekly wages in rupees of workers in certain commercial organisation. The frequency of the class-interval 49-52 is missing.

Weekly wages (₹)	40–43	43–46	46–49	49–52	52–55
No. of workers	31	58	60	?	27

If mean of the above frequency distribution is ₹ 47.2 find the missing frequency. **Ans.** 44

13. Mean of 20 values is 45. If one of these values is to be taken 64 instead of 46, find the corrected mean. **Ans.** 44.1

14. There are 500 workers working in a factory. Their mean was calculated as ₹ 200. Later on, it was discovered that the wages of two workers were misread as 180 and 20 in place of 80 and 220. Find the correct average. **Ans.** 200.2

15. The mean salary paid to 1,000 employees of an establishment was found to be ₹ 180.40. Later on, after disbursement of salary, it was discovered that the salary of two employees was wrongly entered as ₹ 297 and ₹ 165. Their correct salaries were ₹ 197 and ₹ 185. Find the correct arithmetic mean. **Ans.** 180.32

16. The mean yearly salary of employees of a company was ₹ 20,000. The mean yearly salaries of male and female employees were ₹ 20,800, ₹ 16,800 respectively. Find out the percentage of males and females employed by the company. **Ans.** 20%

17. The mean age of a group of 100 children was 9.35 years. The mean age of 25 of them was 8.75 years and that of another 65 was 10.51 years. What was the mean age for the remaining children? **Ans.** 9.81

18. The mean wage of 150 labourers working in a factory running three shifts of 60, 40 and 50 labourers is ₹ 114.00. The mean wage of 60 labourers working in the first shift is ₹ 121.50 and that of 40 labourers working in the second shift is ₹ 107.75. Find the mean wage of the labourers working in the third shift. **Ans.** 110.00

19. Average monthly production of minerals in the eight months in a country was 407.5 thousand tonnes. For the next months, the average was 412.5 thousand tonnes. Is it correct to say that the average monthly production for the whole year was 410.0 thousand tonnes? **Ans.** 409.3

20. The mean wage of 300 labourers working in a factory running in three shifts of 120, 80, and 100 labourers is ₹ 230. The mean wage of 120 labourers working in the first shift is ₹ 242 and that of 80 labourers working in the second shift is ₹ 215. Find the mean wage of the labourers working in the third shift. **Ans.** 227.6

21. The mean age of a combined group of men and women is 30 years. If the mean age of the group of men is 32 and that of the group of women is 27. Find out the percentage of the men and women is the group. **Ans.** Men = 40%; Women = 60%

22. The average monthly sales for the first eleven months of the year in respect of a certain salesman were ₹ 12,000, but due to his illness during the last month, the average monthly sales for the whole year came down to ₹ 11,375. What was the value of his sales during the last month? **Ans.** ₹ 4,500

Mode and Median

23. Calculate mean, mode and median from the following data of the heights (in inches) of a group of students:

61, 62, 62, 63, 61, 63, 64, 64, 60, 65, 63, 64, 65, 66, 64

Now a group of students whose heights are 60, 66, 59, 68, 67 and 70 inches, is added to the original group.

Find mean, and median of the combined group.

Ans. *Mean* = 63.133, *Mode* = 64, *Median* = 63

24. Calculate mode and median from the following data:

33, 20, 35, 50, 37, 33, 35, 25, 35, 34, 35. **Ans.** *Mode* = 35, *Median* = 35

25. Calculate the mode from the following series.

Size	Frequency	Size	Frequency	Size	Frequency
4	40	10	57	16	63
5	48	11	55	17	52
6	52	12	50	18	48
7	57	13	52	19	40
8	60	14	41		
9	63	15	57		

Ans. *Mode* = 9

26. Calculate the mode from the following frequency distribution:

Central size of fields (in acres)	10	20	30	40	50	60	70
No. of fields	7	12	17	29	31	5	3

Ans. *Mode* = 41.46

27. Find the mode from the following table:

Class	Frequency	Class	Frequency
4–8	10	24–28	8
8–12	12	28–32	17
12–16	16	32–36	5
16–20	14	36–40	4
20–24	10		

Ans. *Mode* = 14.67

28. Calculate the Mode from the following data:

Size of the items	**Number of items**	**Size of the items**	**Number of items**
Below 50	97	Below 30	60
Below 45	95	Below 25	30
Below 40	90	Below 20	12
Below 35	80	Below 15	4

Ans. *Mode* = 27.72

29. Find the modal wage from the following:

Wages (in ₹)	**Number of employees**	**Wage (in ₹)**	**No. of employees**
50.00–59.99	8	90.00–99.99	10
60.00–69.99	10	100.00–109.99	5
70.00–79.99	16	110.00–119.99	2
80.00–89.99	14		

Ans. *Mode* = 77.50

30. Form an ordinary frequency table from the following cumulative distribution of marks obtained by 22 students and calculate (*i*) Mean (*ii*) Median (*iii*) Mode.

Marks	**No. of Students**
Below 10	3
Below 20	8
Below 30	17
Below 40	20
Below 50	22

Ans. $\overline{X}$ = 23.19, *Md* = 23.33, *Mode* = 24

31. From the following table, find the median and the mode:

Income (₹)	100–200	100–300	100–400	100–500	100–600
No. of persons	15	33	63	83	100

Ans. ₹ 356.67, ₹ 354.54

Standard Deviation and Coefficient of Variation

32. The following data give the annual yield in kilograms from 10 experimental farms:
2,020 2,100 2,040 2,030 2,070 2,060 2,080 2,050 2,110 2,090
Find the mean and standard deviation of the annual yield. **Ans.** $\overline{X}$ = 11, σ = 2.24

33. Ten students of B. Com. have obtained the following marks in statistics out of 100 marks. Calculate the standard deviation of marks obtained.

Serial No.	**Marks**	**Serial No.**	**Marks**
1	5	6	42
2	10	7	45
3	20	8	48
4	25	9	70
5	40	10	80

34. The following table gives the number of finished articles turned out per day by different number of workers in a factory. Find the mean value and the standard deviation of the daily output of finished articles:

No. of articles	No. of workers	No. of articles	No. of workers
18	3	23	17
19	7	24	13
20	11	25	8
21	14	26	5
22	18	27	4

Ans. $\overline{X} = 22.78, \sigma = 2.12$

35. Calculate standard deviation for the following distribution of net profits earned by a group of companies:

Profit (in ₹)	20–30	30–40	40–50	50–60	60–70	70–80	80–90	90–100
No. of Companies	30	58	62	85	112	70	57	26

Ans. 18.71

36. Daily sales (recorded in rupees) of a retail shop are given below:

Daily sales (in ₹)	102	106	110	114	118	122	126
No. of days	3	9	25	35	17	10	1

Calculate the mean and the standard deviation of the above data and explain what they indicate about the distribution of daily sales? **Ans.** Mean = 113.52, S.D. = 4.94

37. Find out the standard deviation of the following series:

Size of holdings (Hectares)	No. of farmers
0–2	1,000
2–4	2,300
4–6	3,600
6–8	2,400
8–10	1,700
10–12	3,000
12–14	500

What conclusion do you draw from your results? **Ans.** $\sigma = 3.31$

38. From the following data, calculate arithmetic mean, median, mode, standard deviation and the percentage of workers who get salary between ₹ 125 and ₹ 175.

Salary in (₹)	100–110	110–120	120–130	130–140	140–150	150–160	160–170	170–180	180–190
No. of employees	12	34	72	120	156	112	82	36	16

Ans. (*i*) $\overline{X} = 145.53$ (*ii*) $\sigma = 17.34$ (*iii*) *Median* = 144.5 (*iv*) *Mode* = 145.26 (*v*) 81.8752%

39. Calculate coefficient of variation from the following data:

Income (₹)	**No. of Families**	**Income (₹)**	**No. of Families**
Less than 500	8	Less than 800	50
Less than 600	20	Less than 900	58
Less than 700	38	Less than 1000	60

Ans. C.V. = 20.09 %

40. From the following table of marks obtained by *A* and *B* in 10 tests of 150 marks each, find out who is more intelligent and who is more consistent?

A	25	50	45	30	70	42	36	48	35	60
B	10	70	50	20	95	55	42	60	48	80

Ans. $A : \overline{X} = 44$; S.D. = 13.08 ; C.V. = 29.73 %

$B : \overline{Y} = 53$; S.D. = 24.35 ; C.V. = 35.94 %

B is more intelligent and A is more consistent

41. The number of runs scored by two batsmen *A* and *B* in different innings is as follows:

A	12	115	6	73	7	19	119	36	84	29
B	47	12	76	42	4	51	37	48	13	0

Who is the better run-getter? Which batsman is more consistent?

Ans. $A : \overline{X} = 50$; C.V. = 83.66 %

$B : \overline{Y} = 33$; C.V. = 70.9 %

A is better run-getter B is more consistent.

42. Goals scored by two teams *A* and *B* in a football season are as follows:

No. of goals scored in a match:		0	1	2	3	4
No. of matches:	A	27	9	8	5	4
	B	17	9	6	5	3

Find which team may be considered more consistent.

Ans. C.V. of *A* = 123.6%, C.V. of *B* = 109%;Team B is more consistent

43. From the following data, find the standard deviation and coefficient of variation:

Wages (Less than)	10	20	30	40	50	60	70	80
No. of Persons	12	30	65	107	157	202	222	230

Ans. S.D. = 17.26; C.V. = 42.7%

44. In the following table, distribution of students is shown according to their weights in kg. Find the coefficient of variation of each. Which series has greater variation?

Weights in kg	20–30	30–40	40–50	50–60	60–70	Total
Class *A*	7	10	20	18	7	62
Class *B*	5	9	21	15	6	56

Ans. C.V. of *A* = 25%, C.V. of *B* = 23.5%, Series A has greater variation

Probability and Distributions

37.1 PROBABILITY

Probability is a concept which numerically measure the degree of uncertainty and therefore, of certainty of the occurrence of events.

If an event A can happen in m ways, and fail in n ways, all these ways being equally likely to occur, then the probability of the **happening of** A is

$$= \frac{\text{Number of favourable cases}}{\text{Total number of mutually exclusive and equally likely cases}} = \frac{m}{m+n}$$

and that of its failing is defined as $\dfrac{n}{m+n}$ **[Bihar SBTE 2010]**

If the probability of the **happening** $= p$

and the probability of **not happening** $= q$

then $p+q = \dfrac{m}{m+n} + \dfrac{n}{m+n} = \dfrac{m+n}{m+n} = 1$ or $p+q=1$

For instance, on tossing a coin, the probability of getting tail or a head is $\frac{1}{2}$.

37.2 DEFINITIONS

1. **Die:** It is a small cube. Dots are . .. ... :: ::. ::: marked on its six faces. Plural of the die is dice. On throwing a die, the outcome is the number of dots on its upper face.
2. **Cards:** A pack of cards consists of four suits, *i.e.* spades, hearts, diamonds and clubs. Each suit consists of 13 cards, nine cards numbered 2, 3, 4, ..., 10, and ace, a king, a queen and a jack or knave. Colour of spades and clubs is black and that of hearts and diamonds is red. Aces, kings, queens, and jacks are known as *face* cards.
3. **Exhaustive Events or Sample Space:** The set of all possible outcomes of a single performance of an experiment is exhaustive events or sample space. Each outcome is called a sample point. In case of tossing a coin once, $S = (H, T)$ is the *sample space*. Two outcomes head and tail constitute an exhaustive event because no other outcome is possible.
4. **Random Experiment:** There are experiments, in which results may be altogether different, even though they are performed under identical conditions. They are known as random experiments. Tossing a coin or throwing a die is random experiment.

5. **Continuous Random Variables:** It is the one which can assume any value within a number, *i.e.* all values of continuous scale. For example (*i*) the weights (in kg) of a group of individuals (*ii*) the heights of a group of individuals.
6. **A discrete Random Variable** is one which can assume only isolated values. For example, The number of heads in 4 tosses of a coin is a discrete random variable as it cannot assume values other than 0, 1, 2, 3, 4.
7. **Trial and Event:** Performing a random experiment is called a trial and outcome is termed as event. Tossing of a coin is a trial and the turning up of head or tail is an event.
8. **Equally Likely Events:** Two events are said to be '*equally likely*', if one of them cannot be expected in preference to the other. For instance, if we draw a card from well-shuffled pack, we may get any card, then the 52 different cases are equally likely.
9. **Independent Event:** Two events may be *independent*, when the actual happening of one does not influence in any way the probability of the happening of the other. **Example.** The event of getting head on first coin and the event of getting tail on the second coin in a simultaneous throw of two coins are independent.
10. **Mutually Exclusive Events:** Two events are known as *mutually exclusive*, when the occurrence of one of them excludes the occurrence of the other. For example, on tossing of a coin, either we get head or tail, but not both.
11. **Compound Event:** When two or more events occur in composition with each other, the simultaneous occurrence is called a compound event. When a die is thrown, getting a 5 or 6 is a compound event.
12. **Favourable Events:** The events, which ensure the required happening, are said to be favourable events. For example, in throwing a die, to have the even numbers 2, 4 and 6 are favourable cases.
13. **Conditional Probability:** The probability of happening an event A, such that event B has already happened, is called the conditional probability of happening of A on the condition that B has already happened. It is usually denoted by $P(A/B)$.
14. **Odds In Favour Of an Event and Odd Against an Event:** If number of favourable ways = m, number of not favourable events = n (*i*) odds in favour of the event = $\frac{m}{n}$, (*ii*) Odd against the event = $\frac{n}{m}$
15. **Classical Definition of Probability:** If there are N equally likely, mutually exclusive and exhaustive events of an experiment and m of these are favourable, then the probability of the happening of the event is defined as $\frac{m}{N}$.
16. **Expected value:** If $p_1, p_2, p_3, ..., p_n$ are the probabilities of the events $x_1, x_2, x_3, ..., x_n$ respectively the expected value

$$E(x) = p_1 x_1 + p_2 x_2 + p_3 x_3 + ... + p_n x_n = \sum_{r=1}^{n} p_r x_r$$

Example 1. *If two dice are thrown. Find the probability that the sum of the numbers coming up on them is 9, if it is known that the number 5 always occurs on the first die.*

[Bihar SBTE 2011]

Solution. Sample space, *i.e.* $S = \{(5, 1), (5, 2), (5, 3), (5, 4), (5, 5), (5, 6) \Rightarrow n(S) = 6$

E = Favourable outcome = $(5, 4) \Rightarrow n(E) = 1$

$$P(E) = \frac{n(E)}{n(S)} = \frac{1}{6}$$ **Ans.**

Example 2. *With an ordinary six faced die, find the probability of throwing*
(a) 5 (b) an even number

Solution. (*a*) There are 6 possible ways in which the die can fall and there is only one way of throwing 5.

$$\text{Probability} = \frac{\text{Number of favourable ways}}{\text{Total number of equally likely ways}} = \frac{1}{6}$$

(*b*) Total number of ways of throwing a die = 6

Total number of ways of getting an even number, *i.e.* 2, 4, 6, = 3

The required probability $= \frac{3}{6} = \frac{1}{2}$ **Ans.**

Example 3. *Find the probability of throwing 9 with two dice.*

Solution. Total number of possible ways of throwing two dice

$= 6 \times 6 = 36$

Number of ways getting 9, *i.e.* $(3 + 6), (4 + 5), (5 + 4), (6 + 3) = 4.$

$\therefore$ The required probability $= \frac{4}{36} = \frac{1}{9}$ **Ans.**

Example 4. *From a pack of 52 cards, one is drawn at random. Find the probability of getting a king.*

Solution. A king can be chosen in 4 ways.

But a card can be drawn in 52 ways.

$\therefore$ The required probability $= \frac{4}{52} = \frac{1}{13}.$ **Ans.**

Example 5. *One card is drawn from a pack of cards. Find the probability that the card is either red or a king or both.* **[Bihar SBTE 2009]**

Solution. The probability of drawing a red card $= \frac{26}{52} = \frac{1}{2}$

The probability of drawing a king card $= \frac{4}{52} = \frac{1}{13}$

The probability of drawing a red king (both) $= \frac{1}{2} \times \frac{1}{13} = \frac{1}{26}$

Required probability $= \frac{1}{2} + \frac{1}{13} - \frac{1}{26} = \frac{7}{13}$ **Ans.**

EXERCISE 37.1

1. In a class of 12 students if 5 are boys and the rest are girls. Find the probability that a student selected will be a girl. **Ans.** $\frac{7}{12}$

2. A bag contains 7 red and 8 black balls. Find the probability of drawing a red ball.

Ans. $\frac{7}{15}$

3. Three of the six vertices of a regular hexagon are chosen at random. Find the probability that the triangle with three vertices is equilateral.

Ans. $\frac{1}{10}$

4. What is the probability that a leap year, selected at random, will contain 53 sundays.

Ans. $\frac{2}{7}$

Fill in the blanks with appropriate correct answer

5. Chance of throwing 6 at least once in four throws with single dice is ....

Ans. $\frac{671}{1296}$

6. A pair of fair dice is thrown and one die shows a four. The probability that the other die shows 5 is ....

Ans. $\frac{1}{36}$

Choose the correct answer:

7. In a given race, the odds in favour of horses A, B, C, D are $1 : 3$, $1 : 4$, $1 : 5$, $1 : 6$ respectively. The probability that horse C wins the race is

(*i*) $\frac{1}{4}$ (*ii*) $\frac{1}{5}$ (*iii*) $\frac{1}{6}$ (*iv*) $\frac{1}{7}$ **Ans.** (*iii*)

8. In tossing a fair die, the probability of getting an odd number or a number less than 4 is

(*i*) 2 (*ii*) 1/2 (*iii*) 2/3 (*iv*) 3/4 **Ans.** (*iii*)

37.3 ADDITION LAW OF PROBABILITY **[Bihar SBTE 2008, 2010]**

If $p_1, p_2, \dots p_n$ be separate probabilities of mutually exclusive events, then the probability P that any of these events will happen is given by $P = p_1 + p_2 + p_3 + \dots + p_n$

[Bihar SBTE 2011, 2010]

Proof. Let $A, B, C, \dots\dots$ be the events, where probabilities are respectively $p_1, p_2, \dots, p_n$.

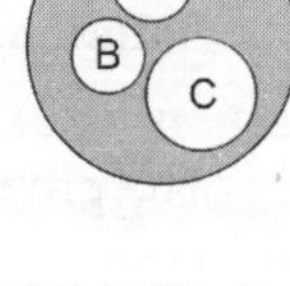

Let n be the total number of favourable cases to either A or B or C or .... $= m_1 + m_2 + m_3 + \dots\dots + m_n$

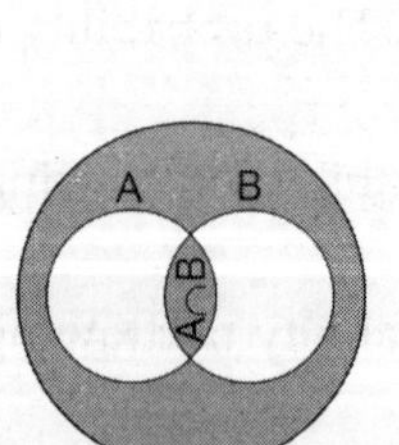

$$\text{Hence, } P(A + B + C + \dots) = \frac{m_1 + m_2 + m_3 + \dots + m_n}{n}$$

$$= \frac{m_1}{n} + \frac{m_2}{n} + \frac{m_3}{n} + \dots\dots + \frac{m_n}{n}$$

$$= P(A) + P(B) + P(C) + \dots\dots$$

$$P = p_1 + p_2 + p_3 + \dots + p_n$$

Proved.

Not Mutually Exclusive Events

Consider the case where two events A and B are not mutually exclusive. The probability of the event that either A or B or both occur is given as

$$P(A \cup B) = P(A) + P(B) - P(A \cap B)$$

Example 6. *A bag contains four white and two black balls and a second bag contains three of each colour. A bag is selected at random, and a ball is then drawn at random from the bag chosen. What is the probability that the ball drawn is white?*

Solution. There are two mutually exclusive cases,

(*i*) when the first bag is chosen, (*ii*) when the second bag is chosen.

Now the chance of choosing the first bag is $\frac{1}{2}$ and if this bag is chosen, the probability of drawing a white ball is 4/6. Hence the probability of drawing a white ball from first bag

$$= \frac{1}{2} \times \frac{4}{6} = \frac{1}{3}$$

Similarly the probability of drawing a white ball from second bag is

$$= \frac{1}{2} \times \frac{3}{6} = \frac{1}{4}$$

Since the events are mutually exclusive the required probability

$$= \frac{1}{3} \times \frac{1}{4} = \frac{7}{12}$$ **Ans.**

Example 7. *Three machines I, II and III manufacture 0.4, 0.5 and 0.1 of the total production respectively. The percentage of defective items produced by I, II and III is 2, 4 and 1 percent respectively. For an item chosen at random, what is the probability that it is defective?*

Solution. The defective item produced by machine I $= \frac{0.4 \times 2}{100} = \frac{0.8}{100}$

The defective item produced by machine II $= \frac{0.5 \times 4}{100} = \frac{2}{100}$

The defective item produced by machine III $= \frac{0.1 \times 1}{100} = \frac{0.1}{100}$

The total defective items produced by machines I, II, III

$$= \frac{0.8}{100} + \frac{2}{100} + \frac{0.1}{100} = \frac{2.9}{100} = 0.029$$

The required probability $= \frac{0.029}{1} = 0.029$ **Ans.**

Example 8. *An urn contains 10 black and 10 white balls. Find the probability of drawing two balls of the same colour.* **[Bihar SBTE 2009]**

Solution. Probability of drawing two black balls $= \frac{^{10}C_2}{^{20}C_2}$

$\therefore$ Probability of drawing two white balls $= \frac{^{10}C_2}{^{20}C_2}$

∴ Probability of drawing two balls of the same colour

$$= \frac{^{10}C_2}{^{20}C_2} + \frac{^{10}C_2}{^{20}C_2} = 2 \cdot \frac{^{10}C_2}{^{20}C_2} = 2 \cdot \frac{\frac{10 \times 9}{2 \times 1}}{\frac{20 \times 19}{2 \times 1}} = \frac{9}{19}$$ **Ans.**

37.4 MULTIPLICATION LAW OF PROBABILITY

If there are two independent events the respective probabilities of which are known, then the probability that both will happen is the product of the probabilities of their happening respectively.

$$P(AB) = P(A) \times P(B)$$ **[Bihar SBTE 2010, 2009, 2008]**

Proof. Suppose A and B are two independent events. Let A happen in m_1 ways and fail in n_1 ways

$$P(A) = \frac{m_1}{m_1 + n_1}$$

Also let B happen in m_2 ways and fail in n_2 ways

∴ $$P(B) = \frac{m_2}{m_2 + n_2}$$

Now there are four possibilities

A and B both may happen, then the number of ways $= m_1 \cdot m_2$.

A may happen and B may fail, then the number of ways $= m_1 \cdot n_2$.

A may fail and B may happen, then the number of ways $= n_1 \cdot m_2$

A and B both may fail, then the number of ways $= n_1 \cdot n_2$

Thus, the total number of ways $= m_1m_2 + m_1n_2 + n_1m_2 + n_1n_2 = (m_1 + n_1)(m_2 + n_2)$

Hence the probabilities of the happening of both A and B

$$P(AB) = \frac{m_1m_2}{(m_1 + n_1)(m_2 + n_2)} = \frac{m_1}{m_1 + n_1} \cdot \frac{m_2}{m_2 + n_2} = P(A) \cdot P(B)$$ **Proved.**

Example 9. *Find the probability of drawing an ace, a king and a queen in three successive draws from a pack of cards if the cards are not replaced after each draw.*

Solution. The probability of drawing an ace $= \frac{4}{52} = \frac{1}{13}$

If the card is not replaced, the pack will now have 51 cards.

Thus the probability of drawing a king $= \frac{4}{51}$

The probability of drawing a queen $= \frac{4}{50} = \frac{2}{25}$

∴ The probability of drawing three cards $= \frac{1}{13} \times \frac{4}{51} \times \frac{2}{25} = \frac{8}{16575}$ **Ans.**

Example 10. *A card is drawn from an ordinary pack and a gambler bets that it is a spade or a king. What is the probability of his Not winning the bet?*

[Bihar SBTE 2012]

Solution. P (king) $= \frac{4}{52} = \frac{1}{13}$

P (spade) $= \frac{13}{52} = \frac{1}{4}$, P (king of spade) $= \frac{1}{52}$

Required P, *i.e.* $= \frac{1}{13} + \frac{1}{4} - \frac{1}{52} = \frac{4}{13}$

P (of his Not winning the bet) $= 1 - \frac{4}{13} = \frac{9}{13}$ **Ans.**

Example 11. *An article manufactured by a company consists of two parts A and B. In the process of manufacture of part A, 9 out of 100 are likely to be defective. Similarly, 5 out of 100 are likely to be defective in the manufacture of part B. Calculate the probability that the assembled article will not be defective (assuming that the events of finding the part A non-defective and that of B are independent).*

Solution. Probability that part A will be defective $= \frac{9}{100}$

Probability that part A will not be defective $= \left(1 - \frac{9}{100}\right) = \frac{91}{100}$

Probability that part B will be defective $= \frac{5}{100}$

Probability that part B will not be defective $= \left(1 - \frac{5}{100}\right) = \frac{95}{100}$

Probability that the assembled article will not be defective = (Probability that part A will not be defective) × (Probability that part B will not be defective)

$$= \left(\frac{91}{100}\right) \times \left(\frac{95}{100}\right) = 0.8645$$ **Ans.**

Example 12. *The probability that machine A will be performing an usual function in 5 years' time is $\frac{1}{4}$, while the probability that machine B will still be operating usefully at the end of the same period is $\frac{1}{3}$.*

Find the probability in the following cases that in 5 years time:

(i) Both machines will be performing an usual function.

(ii) Neither will be operating. *(iii) Only machine B will be operating.*

(iv) At least one of the machines will be operating.

Solution. P (A operating usefully) $= p(A) = \frac{1}{4}$, so $q(A) = 1 - \frac{1}{4} = \frac{3}{4}$

P (B operating usefully) $= p(B) = \frac{1}{3}$, so $q(B) = 1 - \frac{1}{3} = \frac{2}{3}$

(i) P (both A and B will operate usefully) $= p(A) \cdot p(B) = \left(\frac{1}{4}\right) \times \left(\frac{1}{3}\right) = \frac{1}{12}$

(ii) P (neither will be operating) $= q(A) \cdot q(B) = \left(\frac{3}{4}\right) \times \left(\frac{2}{3}\right) = \frac{1}{2}$

(*iii*) P (only B will be operating) $= p(B) \times q(A) = \left(\frac{1}{3}\right) \times \left(\frac{3}{4}\right) = \frac{1}{4}$

(*iv*) P (at least one of the machines will be operating)

$$= 1 - P \text{ (none of them operates) } = 1 - \frac{1}{2} = \frac{1}{2}$$ **Ans.**

Example 13. *There are two groups of subjects one of which consists of 5 science and 3 engineering subjects and the other consists of 3 science and 5 engineering subjects. An unbiased die is cast. If number 3 or number 5 turns up, a subject is selected at random from the first group, otherwise the subject is selected at random from the second group. Find the probability that an engineering subject is selected ultimately.*

Solution. Probability of turning up 3 or 5 $= \frac{2}{6} = \frac{1}{3}$

Probability of selecting engineering subject from first group $= \frac{3}{8}$

Now the probability of selecting engineering subject from first group on turning up 3 or 5 $= \left(\frac{1}{3}\right) \times \left(\frac{3}{8}\right) = \frac{1}{8}$ ... (1)

Probability of not turning up 3 or 5 $= 1 - \frac{1}{3} = \frac{2}{3}$

Probability of selecting engineering subject from second group $= \frac{5}{8}$

Now probability of selecting engineering subject from second group on not turning up 3 or 5 $= \frac{2}{3} \times \frac{5}{8} = \frac{5}{12}$ ... (2)

Probability of the selection of engineering subject $= \frac{1}{8} + \frac{5}{12}$ [From (1) and (2)]

$$= \frac{13}{24}$$ **Ans.**

Example 14. *An urn A contains 2 white and 4 black balls. Another urn B contains 5 white and 7 black balls. A ball is transferred from the urn A to the urn B, then a ball is drawn from urn B. Find the probability that it is white.*

Solution. Urn A contains 2 white and 4 black balls.

Urn B contains 5 white and 7 black balls.

Now there are two cases of transferring a ball from A to B.

Case I. When a white ball is transferred from A to B

P (Transfer of a white ball) $= \frac{2}{2+4} = \frac{1}{3}$

After transfer of a white ball, urn B contains 6 white balls and 7 black balls.

P (Drawing a white ball from urn B after transfer)

$= P$ (Transfer of a white ball) $\times$ P (Drawing of a white ball)

$$= \left(\frac{1}{3}\right)\left(\frac{6}{6+7}\right) = \frac{1}{3} \times \frac{6}{13} = \frac{2}{13}$$

Case II. When a black ball is transferred from A to B.

P (Transfer of a black ball) $= \frac{4}{2+4} = \frac{2}{3}$

After transfer of a black ball, urn B contains 5 white and 8 black balls.

P (Drawing a white ball from urn B after transfer)

$= P$ (Transfer of a black ball) $\times$ P (Drawing of a white ball)

$$= \frac{2}{3}\left(\frac{5}{5+8}\right) = \frac{10}{39}$$

Required probability $= \frac{2}{13} + \frac{10}{39} = \frac{16}{39}$ **Ans.**

Example 15. *A bag contains 5 white and 7 black balls. Two balls are drawn in succession. What is the probability that first is white and second is black?* **[Bihar SBTE 2010]**

Solution. Let A be the event of a white ball and a black ball $n(A) = {}^5C_1 \times {}^7C_1$

Required probability $= \frac{{}^5C_1 \times {}^7C_1}{12_2} = \frac{5 \times 7}{66} = \frac{35}{66}$ **Ans.**

Example 16. *A bag contains 10 white and 15 black ball. Two balls are drawn in successively. What is the probability that both are of different colour?* **[Bihar SBTE 2009]**

Solution. Probability of drawing a white ball $= \frac{10}{25}$

After a white ball has been drawn there remains 15 black and 9 white balls.

Now, probability of drawing black ball $= \frac{15}{24}$

$\therefore$ The probability of drawing first a white and 2nd black $= \frac{10}{25} \times \frac{15}{24} = \frac{1}{4}$

Similarly the probability of drawing first a black and then a white ball $= \frac{15}{25} \times \frac{10}{24} = \frac{1}{4}$

Thus the required probability $= \frac{1}{4} + \frac{1}{4} = \frac{1}{2}$ **Ans.**

Example 17. *Three groups of children contains 3 girls and 1 boy; 2 girls and 2 boys; 1 girl and 3 boys respectively. One child is selected at random from each group. Find the chance of selecting 1 girl and 2 boys.*

Solution. There are three ways of selecting 1 girl and two boys.

I way: Girl is selected from first group, boy from second group and second boy from third group. Probability of the selection of (Girl + Boy + Boy)

$$= \frac{3}{4} \times \frac{2}{4} \times \frac{3}{4} = \frac{18}{64}$$

II way: Boy is selected from first group, girl from second group and second boy from third group.

Probability of the selection of (Boy + Girl + Boy)

$$= \frac{1}{4} \times \frac{2}{4} \times \frac{3}{4} = \frac{6}{64}$$

III way: Boy is selected from first group, second boy from second group and the girl from the third group. Probability of selection of (Boy + Boy + Girl)

$$=\frac{1}{4}\times\frac{2}{4}\times\frac{1}{4}=\frac{2}{64}$$

Total probability $=\frac{18}{64}+\frac{6}{64}+\frac{2}{64}=\frac{26}{64}=\frac{13}{32}$ **Ans.**

Example 18. *The number of children in a family in a region are either 0, 1 or 2 with probability 0.2, 0.3 and 0.5 respectively. The probability of each child being a boy or girl 0.5. Find the probability that a family has no boy.*

Solution. Here there are three types of families

(*i*) Probability of zero child (boys) = 0.2

(*ii*)

Boy	Girl
0	1
1	0

Probability of zero boy in case II

$= 0.3 \times 0.5 = 0.15$

(*iii*)

Boy	Girl
0	2
1	1
2	0

In this case probability of zero boy $=0.5\times\frac{1}{3}=0.167$

Considering all the three cases, the probability of zero boy

$= 0.2 + 0.15 + 0.167 = 0.517$ **Ans.**

Example 19. *A husband and wife appear in an interview for two vacancies for the same post. If the probability of husband's selection is $\frac{1}{7}$ while wife's selection is $\frac{1}{5}$. What is the probability that*

(*i*) *both of them will be selected* (*ii*) *only one of them will be selected, and*

(*iii*) *none of them will be selected?*

Solution. P (husband's selection) $=\frac{1}{7}$, P (wife's selection) $=\frac{1}{5}$

(*i*) P (both selected) $=\frac{1}{7}\times\frac{1}{5}=\frac{1}{35}$

(*ii*) P (only one selected) = P (only husband's selection) + P (only wife's selection)

$$=\left(\frac{1}{7}\times\frac{4}{5}\right)+\left(\frac{1}{5}\times\frac{6}{7}\right)=\frac{10}{35}=\frac{2}{7}$$

(*iii*) P (none of them will be selected) $=\frac{6}{7}\times\frac{4}{5}=\frac{24}{35}$ **Ans.**

Example 20. *A problem of statistics is given to three students A, B and C whose chances of solving it are $\frac{1}{2}, \frac{1}{3}$ and $\frac{1}{4}$ respectively. What is the probability that the problem will be solved?* **[Bihar SBTE 2011, 2010]**

Solution. The probability that A can solve the problem $=\frac{1}{2}$

The probability that A cannot solve the problem $=1-\frac{1}{2}$

Similarly the probability that B and C cannot solve the problem are $\left(1-\frac{1}{3}\right)$ and $\left(1-\frac{1}{4}\right)$

$\therefore$ The probability that A, B, C cannot solve the problem $=\left(1-\frac{1}{2}\right)\times\left(1-\frac{1}{3}\right)\times\left(1-\frac{1}{4}\right)$

$$=\frac{1}{2}\times\frac{2}{3}\times\frac{3}{4}=\frac{1}{4}$$

Hence, the probability that the problem can be solved $=1-\frac{1}{4}=\frac{3}{4}$ **Ans.**

Example 21. *A student takes his examination in four subjects α, β, γ, δ. He estimates his chances of passing in α as $\frac{4}{5}$, in β as $\frac{3}{4}$, in γ as $\frac{5}{6}$ and in δ as $\frac{2}{3}$. To qualify, he must pass in α and at least two other subjects. What is the probability that he qualifies?*

Solution. $P(\alpha)=\frac{4}{5}, P(\beta)=\frac{3}{4}, P(\gamma)=\frac{5}{6}, P(\delta)=\frac{2}{3}$

There are four possibilities of passing at least two subjects.

(*i*) Probability of passing β, γ and failing $\delta=\frac{3}{4}\times\frac{5}{6}\times\left(1-\frac{2}{3}\right)=\frac{3}{4}\times\frac{5}{6}\times\frac{1}{3}=\frac{5}{24}$

(*ii*) Probability of passing γ, δ and failing $\beta=\frac{5}{6}\times\frac{2}{3}\times\left(1-\frac{3}{4}\right)=\frac{5}{6}\times\frac{2}{3}\times\frac{1}{4}=\frac{5}{36}$

(*iii*) Probability of passing δ, β and failing $\gamma=\frac{2}{3}\times\frac{3}{4}\times\left(1-\frac{5}{6}\right)=\frac{2}{3}\times\frac{3}{4}\times\frac{1}{6}=\frac{1}{12}$

(*iv*) Probability of passing $\beta,\gamma,\delta=\frac{3}{4}\times\frac{5}{6}\times\frac{2}{3}=\frac{5}{12}$

Probability of passing at least two subjects $=\frac{5}{24}+\frac{5}{36}+\frac{1}{12}+\frac{5}{12}=\frac{61}{72}$

Probability of passing α and at least two subjects $=\frac{4}{5}\times\frac{61}{72}=\frac{61}{90}$ **Ans.**

Example 22. *A box contains 9 tickets numbered 1 to 9 inclusive. If 3 tickets are drawn from the box one at a time, find the probability that they are alternatively either odd, even, odd or even, odd, even.* **[MDU Dec. 2009]**

Solution. Total number of tickets = 9

Number of odd tickets = 5

Number of even tickets = 4

$P(\text{odd, even, odd}) = P(\text{odd})\cdot P(\text{even})\cdot P(\text{odd})=\frac{5}{9}\times\frac{4}{8}\times\frac{4}{7}=\frac{10}{63}$

$P(\text{even, odd, even}) = P(\text{even})\cdot P(\text{odd})\cdot P(\text{even})=\frac{4}{9}\times\frac{5}{8}\times\frac{3}{7}=\frac{5}{42}$

Required probability = P(odd, even, odd or even, odd, even)

$= P(\text{odd, even, odd}) + P(\text{even, odd, even}) = \frac{10}{63} + \frac{5}{42} = \frac{5}{8}$ **Ans.**

Example 23. *Bag A contains 10 red and 5 white balls. Bag B contains 8 red and 7 white balls. If any one ball (red or white) is transferred from A to bag B, find the probability of drawing one white ball from bag B.* **[Bihar SBTE 2012]**

Solution. Now there are two case of (transferring a ball from A to B).

Case 1. When a white ball is transferred from A to B

$$P\text{ (transfer of a white ball)} = \frac{5}{10+5} = \frac{1}{3}$$

P (drawing a white ball from bag. B after transferring

$$= P\text{ (transfer of white ball)} \times P\text{ (drawing of a white ball)} = \left(\frac{1}{3}\right)\left(\frac{8}{8+8}\right) = \frac{1}{3} \times \frac{1}{2} = \frac{1}{6}$$

Case 2. When a red ball is transfer from A to B

$$P\text{ (transfer of } R\text{ Red ball)} = \frac{10}{10+5} = \frac{10}{15} = \frac{2}{3}$$

P (drawing a white ball from bag B after transfer)

$= P$ (transfer of a red ball) × P (drawing of a white II)

$$= \frac{2}{3}\left(\frac{7}{7+9}\right) = \frac{2}{3} \times \frac{7}{16} = \frac{14}{48}$$

$\therefore$ Required Productivity $= \frac{1}{6} + \frac{14}{40} = \frac{31}{60}$ **Ans.**

Example 24. *A bag contains 10 black and 10 white balls. Find the probability of drawing two balls of the same colour.* **[Bihar SBTE 2009]**

Solution. Here number of ways in which 2 balls are selected = ${}^{20}C_2 = 190$

$\Rightarrow$ Number of ways in which 2 white balls are selected = ${}^{10}C_2 = 45$

$\Rightarrow$ $P(\text{drawing 2 white balls}) = \frac{45}{190} = P(\text{drawing 2 black balls})$

The probability of drawing two balls of the same colour $= \frac{45}{190} + \frac{45}{190}$

$= \frac{90}{190} = \frac{9}{19}$ **Ans.**

Example 25. *A committee is to be formed by choosing two boys and four girls out of a group of five boys and six girls. What is the probability that a particular boy named A and a particular girl named B are selected in the committee?*

Solution. Two boys are to be selected out of 5 boys. A particular boy A is to be included in the committee. It means that only 1 boy is to be selected out of 4 boys.

Number of ways of selection = ${}^4C_1$

Similarly a girl B is to be included in the committee.

Then only 3 girls are to be selected out of 5 girls.

Number of ways of selection = ${}^5C_3$

Required probability $= \dfrac{{}^4C_1 \times {}^5C_3}{{}^5C_2 \times {}^6C_4} = \dfrac{4\times 10}{10\times 15} = \dfrac{4}{15}$ **Ans.**

Example 26. *There are 6 positive and 8 negative numbers. Four numbers are chosen at random, without replacement and multiplied. What is the probability that the product is a positive number?*

Solution. To get from the product of four numbers, a positive number, the possible combinations are as follows:

S. No.	*Out of 6 positive numbers*	*Out of 8 negative numbers*	*Positive numbers*
1.	4	0	${}^6C_4 \times {}^8C_0 = \dfrac{6\times 5}{1\times 2}\times 1 = 15$
2.	2	2	${}^6C_2 \times {}^8C_2 = \dfrac{6\times 5}{1\times 2}\times\dfrac{8\times 7}{1\times 2} = 420$
3.	0	4	${}^6C_0 \times {}^8C_4 = 1\times\dfrac{8\times 7\times 6\times 5}{1\times 2\times 3\times 4} = 70$
			Total = 505

$$\text{Required probability} = \frac{{}^6C_4\times{}^8C_0 + {}^6C_2\times{}^8C_2 + {}^6C_0\times{}^8C_4}{{}^{14}C_4}$$

$$= \frac{15+420+70}{\dfrac{14\times 13\times 12\times 11}{1\times 2\times 3\times 4}} = \frac{505\times 4\times 3\times 2\times 1}{14\times 13\times 12\times 11} = \frac{505}{1001}$$ **Ans.**

Example 27. *A six-faced die is so biased that, when thrown, it is twice as likely to show an even number than an odd number. If it is thrown twice, what is the probability that the sum of two numbers thrown is odd.*

Solution. A biased die, when thrown, shows even number twice than an odd number.

Probability of showing even number $= \dfrac{2}{2+1} = \dfrac{2}{3}$

Probability of showing odd number $= \dfrac{1}{1+2} = \dfrac{1}{3}$

Sum of two numbers is odd if the first is even and the second is odd or vice versa.

Probability of sum to be odd = Probability of an even number × Probability of an odd number + Probability of an odd number × Probability of an even number.

$$= \left(\frac{2}{3}\times\frac{1}{3}\right) + \left(\frac{1}{3}\times\frac{2}{3}\right) = \frac{2}{9}+\frac{2}{9} = \frac{4}{9}$$ **Ans.**

Example 28. *A can hit a target 3 times in 5 shots, B can hit a target 2 times in 5 shots and C three times in 4 shots. All of them fire one shot each simultaneously at the target. What is the probability that*

(i) Exacly 2 shots hit *(ii) At least two shots hit?*

Solution. Probability of A hitting the target $= \frac{3}{5}$

Probability of B hitting the target $= \frac{2}{5}$

Probability of C hitting the target $= \frac{3}{4}$

(*i*) Probability that exactly 2 shots hit the target

$$= P(A)\,P(B)\,P(\overline{C}) + P(A)\,P(C)\,P(\overline{B}) + P(B)\,P(C)\,P(\overline{A})$$

$$= \frac{3}{5}\times\frac{2}{5}\times\left(1-\frac{3}{4}\right)+\frac{3}{5}\times\frac{3}{4}\times\left(1-\frac{2}{5}\right)+\frac{2}{5}\times\frac{3}{4}\times\left(1-\frac{3}{5}\right)$$

$$= \left(\frac{6}{25}\times\frac{1}{4}\right)+\left(\frac{9}{20}\times\frac{3}{5}\right)+\left(\frac{6}{20}\times\frac{2}{5}\right)$$

$$= \frac{6+27+12}{100} = \frac{45}{100} = \frac{9}{20}$$ **Ans.**

(*ii*) Probability of at least two shots hitting the target

= Probability of 2 shots + probability of 3 shots hitting the target

$$= \frac{9}{20} + P(A)\,P(B)\,P(C) = \frac{9}{20} + \frac{3}{5}\times\frac{2}{5}\times\frac{3}{4} = \frac{63}{100}$$ **Ans.**

Example 29. *A factory, manufacturing televisions has four units A, B, C, D. Units A, B, C, D manufacture 15%, 20%, 30% and 35% of the total output respectively. It was found that out of their outputs 1%, 2%, 2% and 3% are defective. A television is chosen at random from the total output and found to be defective. What is the probability that it came from unit D?* **[Bihar SBTE 2010, 2009]**

Solution. Let the factory manufacture 1000 television.

Unit A manufactures = 15 TV, Unit B Manufactures = 20 TV

Unit C manufactures = 30 TV, Unit D manufactures = 35 TV

Defective TV manufactured by unit $A = 15 \times \frac{1}{100} = 0.15$

Defective TV manufactured by unit $B = 20 \times \frac{2}{100} = 0.4$

Defective TV manufactured by unit $C = 30 \times \frac{2}{100} = 0.6$

Defective TV manufactured by unit $D = 35 \times \frac{3}{100} = 1.05$

Total defective TV = 0.15 + 0.4 + 0.6 + 1.05 = 2.20

Probability of defective TV from unit $D = \frac{1.05}{2.20} = \frac{21}{44}$ **Ans.**

Example 30. *A and B take turns in throwing two dice, the first to throw 10 being awarded the prize. Show that if A has the first throw, their chances of winning are in the ratio 12 : 11.*

Solution. The combinations of throwing 10 from two dice can be

(6 + 4), (4 + 6), (5 + 5).

The number of combination is 3.

Total combination from two dice = 6 × 6 = 36.

$\therefore$ The probability of throwing $10 = p = \frac{3}{36} = \frac{1}{12}$

The probability of not getting $10 = q = 1 - \left(\frac{1}{12}\right) = \frac{11}{12}$

If *A* has to win, he should throw 10 in either the first, the third, the fifth, ... throws.

Their respective probabilities are $= p, q^2p, q^4p, ... = \frac{1}{12}, \left(\frac{11}{12}\right)^2 \frac{1}{12}, \left(\frac{11}{12}\right)^4 \frac{1}{12} \cdots$

A's total probability of winning $= \frac{1}{12} + \left(\frac{11}{12}\right)^2 \cdot \frac{1}{12} + \left(\frac{11}{12}\right)^4 \cdot \frac{1}{12} + ...$

$$= \frac{\frac{1}{12}}{1 - \left(\frac{11}{12}\right)^2} = \frac{12}{23} \left[\text{This is infinite G.P. Its sum} = \frac{a}{1-r}\right]$$

B can win in either 2nd, 4th, 6th ... throws.

So *B*'s total chance of winning $= qp + q^3p + q^5p + ...$

$$= \left(\frac{11}{12}\right)\left(\frac{1}{12}\right) + \left(\frac{11}{12}\right)^3\left(\frac{1}{12}\right) + \left(\frac{11}{12}\right)^5\left(\frac{1}{12}\right) + ... = \frac{\left(\frac{11}{12}\right)\left(\frac{1}{12}\right)}{1 - \left(\frac{11}{12}\right)^2} = \frac{11}{23}$$

Hence A's chance to B's chance $= \frac{12}{23} : \frac{11}{23} = 12 : 11$ **Proved.**

Example 31. *A and B throw alternatively a pair of dice. A wins if he throws 6 before B throws 7 and B wins if he throws 7 before A throws 6. Find their respective chances of winning, if A begins.*

Solution. Number of ways of throwing 6

i.e. (1 + 5), (2 + 4), (3 + 3), (4 + 2), (5 + 1) = 5

Probability of throwing $6 = \frac{5}{36} = p_1, q_1 = \frac{31}{36}$

Number of ways of throwing 7

i.e. (1 + 6), (2 + 5), (3 + 4), (4 + 3), (5 + 2), (6 + 1) = 6

Probability of throwing $7 = \frac{6}{36} = \frac{1}{6} = p_2, q_2 = \frac{5}{6}$

$$\Rightarrow \quad P(A) = p_1 + q_1 q_2 \, p_1 + q_1^2 q_2^2 \, p_1 + ...$$

$$\Rightarrow \quad P(B) = q_1 p_2 + q_1^2 q_2 \, p_2 + q_1^3 q_2^2 \, p_2 + ...$$

Probability of A's winning $= p_1 + q_1 q_2 p_1 + q_1^2 q_2^2 p_1 + ...$

$$= \frac{p_1}{1 - q_1 q_2} = \frac{\frac{5}{36}}{1 - \frac{31}{36} \times \frac{5}{6}} = \frac{5}{36} \times \frac{36 \times 6}{61} = \frac{30}{61}$$

Probability of B's winning $= q_1 p_2 + q_1^2 q_2 p_2 + q_1^3 q_2^2 p_2 + ...$

$$= \frac{q_1 p_2}{1 - q_1 q_2} = \frac{\frac{31}{36} \times \frac{1}{6}}{1 - \left(\frac{31}{36}\right)\left(\frac{5}{6}\right)} = \frac{31}{36 \times 6} \times \frac{36 \times 6}{61} = \frac{31}{61}$$ **Ans.**

EXERCISE 37.2

1. The probability that Nirmal will solve a problem is $\frac{2}{3}$ and the probability that Satyajit will solve it is $\frac{3}{4}$. What is the probability that (*a*) the problem will be solved (*b*) neither can solve it. **Ans.** (*a*) $\frac{11}{12}$, (*b*) $\frac{1}{12}$

2. An urn contains 13 balls numbering 1 to 13. Find the probability that a ball selected at random is a ball with number that is a multiple of 3 or 4. **Ans.** $\frac{6}{13}$

3. Four persons are chosen at random from a group containing 3 men, 2 women, and 4 children. Show that the probability that exactly two of them will be children is $\frac{10}{21}$.

4. A five digit number is formed by using the digits 0, 1, 2, 3, 4 and 5 without repetition. Find the probability that the number is divisible by 6. **Ans.** $\frac{4}{25}$

5. The chances that doctor A will diagnose a disease X correctly is 60%. The chances that a patient will die by his treatment after correct diagnosis is 40% and the chances of death by wrong diagnosis is 70%. A patient of doctor A, who had disease X, died, what is the chance that his disease was diagnosed correctly. **Ans.** $\frac{6}{13}$

6. An anti-aircraft gun can take a maximum of four shots on enemy's plane moving from it. The probabilities of hitting the plane at first, second, third and fourth shots are 0.4, 0.3, 0.2 and 0.1 respectively. Find the probability that the gun hits the plane. **Ans.** 0.6976

7. An electronic component consists of three parts. Each part has probability 0.99 of performing satisfactorily. The component fails if two or more parts do not perform satisfactorily. Assuming that the parts perform independently, determine the probability that the component does not perform satisfactorily. **Ans.** 0.000298

8. If face cards are removed from a full pack, then out of the remaining 40 cards, 4 are drawn at random. What is the probability that they belong to different suits? **Ans.** $\frac{1000}{9139}$

9. Of the cigarette smoking population, 70% are men and 30% women, 10% of these men and 20% of these women smoke 'WILLS.' What is the probability that a person seen smoking a 'WILLS' will be a man?

Ans. $\frac{7}{13}$

10. A machine contains a component C that is vital to its operation. The reliability of component C is 80%. To improve the reliability of a machine, a similar component is used in parallel to form a system S. The machine will work provided that one of these components functions correctly. Calculate the reliability of the system S.

Ans. 96%

11. In a bolt factory, machines A, B and C manufacture 25%, 35% and 40% of the total output respectively. Of their outputs, 5%, 4% and 2% are defective bolts. A bolt is chosen at random and found to be defective. What is the probability that the bolt came from machine A ? B ? C ?

Ans. $\frac{25}{69}, \frac{28}{69}, \frac{16}{69}$

12. One bag contains four white and two black beads and another contains three of each colour. A bead is drawn from each bag. What is the probability that one is white and one is black?

Ans. $\frac{1}{2}$

13. The odds that a book will be favourably reviewed by three independent critics are 5 to 2, 4 to 3, 3 to 4 respectively. What is the probability that of the three reviews, a majority will be favourable?

Ans. $\frac{209}{343}$

14. Let E and F be independent events. The probability that both E and F happen is $\frac{1}{12}$ and the probability that neither E nor F happen is $\frac{1}{2}$. Then find $P(E)$ and $P(F)$.

Ans. $P(E) = \frac{1}{3}, P(F) = \frac{1}{4}$

15. Given a random variable whose range set is (1, 2) and whose probability is $f(1) = \frac{1}{4}$ and $f(2) = \frac{3}{4}$. Find the mean and variance of the distribution.

Ans. Mean $= \frac{7}{4}$, Var $= \frac{3}{16}$

16. A man takes a step forward with probability 0.4 and backward with probability 0.6. Find the probability that at the end of 11 steps, he is just one step away from the starting point. **Ans.** 0.5263

17. What would be the expectation of the number of failures preceding the first success is an infinite series of independent trials with the constant probability of success p?

Hint: The probabilities of success in 1st, 2nd, 3rd trials respectively are p, qp, $q^2 p$, $q^3 p$, ...

The expected number of failures preceding the first success

$$E(x) = (0 \cdot p) + (1 \cdot qp) + (2 \cdot q^2 p) + \text{....} \infty$$

$$= qp\,[1 + 2q + 3q^2 + \text{....} \infty] \text{ where } q < 1$$

$$= \frac{qp}{(1-q)^2} = \frac{qp}{p^2} = \frac{q}{p}$$

56

18. A candidate is selected for interview for three posts. For the first post there are three candidates, for the second there are 4, and for the third are 2. What is the chance of getting at least one post? **Ans.** $\frac{3}{4}$

19. The chance of hitting a target by a bomb is 50% when 4 bombs are dropped, what is the probability of destroying the target, if one bomb is just sufficient to destroy it. **Ans.** $\frac{15}{16}$

20. A pair of dice is tossed twice. Find the probability of scoring 7 points (*i*) once, (*ii*) at least once (*iii*) twice. **[K.U. Dec. 2009] Ans.** (*i*) $\frac{5}{8}$ (*ii*) $\frac{11}{36}$ (*iii*) $\frac{1}{36}$

21. Fill in the blanks:

(*a*) If the probabilities of n independent events are $p_1, p_2, p_3, ..., p_n$, then the probability that at least one of the event will happen is ....

(*b*) For a biased die, the probabilities for the different faces to turn up are:

Face	1	2	3	4	5	6
Prob.	0.1	0.32	0.21	0.15	0.05	0.17

The die is tossed and you are told that either face 1 or face 2 has turned up. Then the probability that it is face 1, is ....

(*c*) The probability of getting a ticket of number of multiple of 5 in a random draw from a bag containing tickets of even numbers from 1 to 100, is ....

(*d*) A town has two doctors X and Y operating independently. If the prob. the doctors X is available, is 0.9 and that for Y is 0.8, then the prob. that at least one doctor is available, when needed is ....

(*e*) From a pack of well shuffled cards, one card is drawn randomly. A gambler bets it as a diamond or a king. The odds in favour of his winning the bet are ....

(*f*) From a pack of cards, 2 cards are drawn, the first being replaced before the second is drawn. The probability that the first is a diamond and the second is a king will be ....

(*g*) From an urn containing 12 white and 8 black balls two balls are drawn at random. The probability that both the balls will turn to be black is ....

(*h*) A ball is taken out of a pot containing 6 white and 12 red balls. The probability that the ball is white is ....

(*i*) A speaks truth in 75% and B in 80% of the cases. The percentage of cases in which they are likely to contradict each other narrating the same incident is ....

Ans. (*a*) $1 - (1 - p_1)(1 - p_2) ...... (1 - p_n)$ (*b*) $\frac{5}{21}$ (*c*) $\frac{1}{5}$ (*d*) 0.98 (*e*) 4 : 9

(*f*) $\frac{1}{52}$ (*g*) $\frac{14}{95}$ (*h*) $\frac{1}{3}$ (*i*) 35%

22. Tick (✓) the correct answer:

(*i*) The probability that at least one of the events A and B occurs is 0.8 and the probability that both the events occur simultaneously is 0.25. Then probability $P(A) + P(B)$ is

(*a*) 0.65 (*b*) 0.75 (*c*) 0.85 (*d*) 0.95 **Ans.** (*b*)

(*ii*) A, B, C are independent events such that $P(A) = P(B)$ and probability that at least one of them happens is 1/2. The probability that A or B happens given that at least one of A, B, or C happens is $\frac{2}{9}$. Find $P(A)$ and $P(C)$.

(*a*) 1/11 (*b*) $P(A) = 1 - \frac{\sqrt{7}}{3},\ P(C) = \frac{5}{14}$.

(*c*) None of the above (*d*) Can't answer **Ans.** (*b*)

(*iii*) An unbiased coin is tossed five times. Given that heads were obtained in two of the tosses, the probability that these were obtained in the first two tosses is

(*a*) 1/10 (*b*) 1/4 (*c*) 1/32 (*d*) None of these **Ans.** (*a*)

(*iv*) Groups are formed of 4 persons out of 12 persons. The probability that one particular person is never included is

(*a*) 2/3 (*ii*) 1/3 (*iii*) 1/4 (*iv*) None of these **Ans.** (*a*)

(*v*) 50 tickets are serially numbered 1 to 50. One ticket is drawn from these at random. The probability of its being a multiple of 3 or 4 is

(*a*) 12/25 (*b*) 14/25 (*c*) 2/5 (*d*) None of these **Ans.** (*a*)

(*vi*) The probabilities of occurring of two events E, F are 0.25 and 0.5 respectively and of occurring both simultaneously is 0.14. Then the probability of the occurrence of the neither event is

(*a*) 0.61 (*b*) 0.39 (*c*) 0.89 (*d*) None of these **Ans.** (*b*)

(*vii*) A bag contains 5 black and 4 white balls. Two balls are drawn at random. The probability that they match, is

(*a*) 7/12 (*b*) 5/8 (*c*) 5/9 (*d*) 4/9 **Ans.** (*d*)

(*viii*) A, B, C in order toss a coin, the first to throw a head wins. Assuming the game continues indefinitely their respective chances of

(*a*) $\frac{4}{7}, \frac{2}{7}, \frac{1}{7}$ (*b*) $\frac{1}{7}, \frac{4}{7}, \frac{2}{7}$ (*c*) $\frac{2}{7}, \frac{4}{7}, \frac{1}{7}$ (*d*) None of these **Ans.** (*a*)

(*ix*) A purse contains 4 copper coins, 3 silver coins, the second purse contains 6 copper coins and 2 silver coins. A coin is taken out of any purse, the probability that it is a copper coin is

(*a*) 4/7 (*b*) 3/4 (*c*) 3/7 (*d*) 37/56 **Ans.** (*d*)

(*x*) In rolling two fair dice, the probability of getting equal numbers or numbers with an even product is

(*a*) 6/36 (*b*) 30/36 (*c*) 27/36 (*d*) 3/36 **Ans.** (*b*)

(*xi*) One of the two events must occur. If the chance of one is 2/3 of the other, then odds in favour of the other are

(*a*) 1 : 3 (*b*) 2 : 3 (*c*) 3 : 1 (*d*) None of these **Ans.** (*d*)

(*xii*) The probability that a certain beginner at golf gets a good shot if he uses the correct club is 1/3, and the probability of a good shot with an incorrect club is 1/4. In his bag are 5 different clubs, only one of which is correct for the shot in question. If he chooses a club at random and takes a stroke, the probability that he gets a good shot is

(*a*) $\frac{1}{3}$ (*b*) $\frac{1}{12}$ (*c*) $\frac{4}{15}$ (*d*) $\frac{7}{12}$ **Ans.** (*c*)

(*xiii*) India plays two matches each with West Indies and Australia. In any match, the probabilities of India getting points 0, 1 and 2, are 0.45, 0.05 and 0.50 respectively. Assuming that the outcomes are independent, the probability of India getting at least 7 points is

(*a*) 0.8750 (*b*) 0.0875 (*c*) 0.625 (*d*) 0.0250. **Ans.** (*b*)

(*xiv*) A bag contains 10 bolts, 3 of which are defective. Two bolts are drawn without replacement. The probability that both the bolts drawn are not defective is

(*a*) $\frac{49}{100}$ (*b*) $\frac{7}{15}$ (*c*) $\frac{4}{9}$ (*d*) $\frac{3}{10}$ **Ans.** (*b*)

(*xv*) The probability that a family has k children is $(0.5)^{k+1}$, $k = 0, 1, 2, ...$ If four families are chosen at random, the probability that each family has at least one child is

(*a*) 1/16 (*b*) 1/256 (*c*) 3/16 (*d*) 3/256 **Ans.** (*a*)

(*xvi*) The random variable X has $N(1, 4)$ distribution, then

(*a*) $P(x > 3) > P(x > 1)$ (*b*) $P(x > 3) < P(x < 1)$

(*c*) $P(x < 3) < P(x > 1)$ (*d*) $P(x < 3) < P(x < 1)$ **Ans.** (*b*)

(*xvii*) Two distinguishable dice are tossed simultaneously. The probability that multiple of 2 does not occur on the first die or multiple of 3 does not occur on the second die is

(*a*) $\frac{5}{36}$ (*b*) $\frac{10}{36}$ (*c*) $\frac{20}{36}$ (*d*) $\frac{30}{36}$ **Ans.** (*d*)

(*xviii*) An unbiased die with faces marked 1, 2, 3, 4, 5, 6 is rolled 4 times, out of four face values obtained, the probability that the minimum face value is not less than 2 and the maximum face value is not greater than 5 is then

(*a*) $\frac{16}{81}$ (*b*) $\frac{2}{9}$ (*c*) $\frac{80}{81}$ (*d*) $\frac{8}{9}$ **Ans.** (*a*)

(*xix*) There are q persons sitting in a row. Two of them are selected at random, the probability that the two selected persons are not together is

(*a*) $\frac{2}{q}$ (*b*) $1-\frac{2}{q}$

(*c*) $\frac{q(q-1)}{(q+1)(q+2)}$ (*d*) None of these **Ans.** (*b*)

(*xx*) Probability of any event can not be greater than .... and less than ....

(*a*) 1, 0 (*b*) 2, 1 (*c*) 3, 1 (*d*) 0, 1 **Ans.** (*a*)

37.5 BINOMIAL DISTRIBUTION $P(r) = {}^nC_r\, p^r \cdot q^{n-r}$

To find the probability of the happening of an event once, twice, thrice, ... r times ... exactly in n trials.

Let the probability of the happening of an event A in one trial be p while its probability of not happening be $1 - p = q$.

We assume that there are n trials and the happening of the event A is r times and its not happening is $n - r$ times.

$$\underbrace{A\,A \ldots\ldots A}_{r \text{ times}} \qquad \underbrace{\overline{A} \cdot \overline{A} \ldots\ldots \overline{A}}_{n-r \text{ times}} \qquad \ldots (1)$$

A indicates its happening, $\overline{A}$ its failure and $P(A) = p$ and $P(\overline{A}) = q$.

We see that (1) has the probability

$$\underbrace{pp \ldots p}_{r \text{ times}} \quad \underbrace{q \cdot q \ldots q}_{n-r \text{ times}} = p^r \cdot q^{n-r} \qquad \ldots (2)$$

Clearly (1) is merely one order of arranging rA's.

The probability of (1) $= p^r\, q^{n-r} \times$ Number of different arrangements of rA's and $(n - r)\, \overline{A}$'s.

The number of different arrangement of rA's and $(n - r)\, \overline{A}$'s $= {}^nC_r$

$\therefore$ Probability of the happening of an event r times is ${}^nC_r\, P^r \cdot q^{n-r}$.

$$\Rightarrow \quad P(r) = {}^nC_r\, p^r \cdot q^{n-r} \; (r = 0, 1, 2, \ldots, n).$$

$$= (r + 1)\text{th term of } (q + p)^n$$

If $r = 0$, probability of happening of an event 0 times

$$= {}^nC_0\, q^n\, p^0 = q^n$$

If $r = 1$, probability of happening of an event 1 times $= {}^nC_1\, q^{n-1}\, p$

If $r = 2$, probability of happening of event 2 times $= {}^nC_2\, q^{n-2}\, p^2$

If $r = 3$, probability of happening of an event 3 times $= {}^nC_3\, q^{n-3}\, p^3$ and so on.

These terms are clearly the successive terms in the Binomial expansion of $(q + p)^n$.

Hence it is called binomial sistribution.

Example 32. *Find the probability of getting 4 heads in 6 tosses of a fair coin.*

[Bihar SBTE 2011]

Solution. $p = \frac{1}{2}, q = \frac{1}{2}, n = 10, r = 4$

We know that $P(r) = {}^nC_r\, q^{n-r}\, p^r$

$$P(4) = {}^{10}C_4\, q^{10-4}\, p^4$$

$$= \frac{10 \times 9 \times 8 \times 7}{4 \times 3 \times 2} \cdot \left(\frac{1}{2}\right)^6 \left(\frac{1}{2}\right)^4 = 210 \left(\frac{1}{2}\right)^{10} = \frac{210}{1024} = \frac{105}{512} \qquad \textbf{Ans.}$$

Example 33. *If on an average one ship in every ten is wrecked, find the probability that out of 5 ships expected to arrive, at least 4 will arrive safely.*

Solution. Out of 10 ships, one ship is wrecked.

i.e., Nine ships out of ten ships are safe.

$$P\,(\text{safety}) = \frac{9}{10}$$

P(At least 4 ships out of 5 are safe) = P (4 or 5) = P (4) + P (5)

$$= {}^5C_4\, p^4\, q^{5-4} + {}^5C_5\, p^5\, q^0 = 5\left(\frac{9}{10}\right)^4\left(\frac{1}{10}\right)+\left(\frac{9}{10}\right)^5=\left(\frac{9}{10}\right)^4\left(\frac{5}{10}+\frac{9}{10}\right)=\frac{7}{5}\left(\frac{9}{10}\right)^4 \quad \textbf{Ans.}$$

Example 34. *The overall percentage of failures in a certain examination is 20. If six candidates appear in the examination, what is the probability that at least five pass the examination?*

Solution. Probability of failures = 20% = $\dfrac{20}{100}=\dfrac{1}{5}$

Probability of pass (P) = $1-\dfrac{1}{5}=\dfrac{4}{5}$

Probability of at least five pass = P (5 or 6)

$= P\,(5) + P\,(6) = {}^6C_5\, p^5\, q + {}^6C_6\, p^6\, q^0$

$$= 6\left(\frac{4}{5}\right)^5\left(\frac{1}{5}\right)+\left(\frac{4}{5}\right)^6=\left(\frac{4}{5}\right)^5\left[\frac{6}{5}+\frac{4}{5}\right]=2\left(\frac{4}{5}\right)^5=\frac{2048}{3125}=0.65536 \quad \textbf{Ans.}$$

Example 35. *Ten percent of screws produced in a certain factory turn out to be defective. Find the probability that in a sample of 10 screws chosen at random, exactly two will be defective.*

Solution. $p=\dfrac{1}{10}, q=\dfrac{9}{10}, n=10, r=2$

$$\Rightarrow \qquad P(r) = {}^nC_r\, p^r\, q^{n-r}$$

$$\Rightarrow \qquad P(2) = {}^{10}C_2\left(\frac{1}{10}\right)^2\left(\frac{9}{10}\right)^{10-2}=\frac{10\times 9}{1\times 2}\left(\frac{1}{10}\right)^2\left(\frac{9}{10}\right)^8$$

$$= \frac{1}{2}\cdot\left(\frac{9}{10}\right)^9 = 0.1937 \qquad \textbf{Ans.}$$

Example 36. *The probability that a man aged 60 will live to be 70 is 0.65. What is the probability that out of 10 men, now 60, at least 7 will live to be 70?*

Solution. The probability that a man aged 60 will live to be 70 = p = 0.65

$$\Rightarrow \qquad q = 1 - p = 1 - 0.65 = 0.35$$

Number of men = n = 10

Probability that at least 7 men (7 or 8 or 9 or 10) will live to 70

$= P\,(7) + P\,(8) + P\,(9) + P\,(10) = {}^{10}C_7\, q^3\, p^7 + {}^{10}C_8\, q^2\, p^8 + {}^{10}C_9\, q\, p^9 + p^{10}$

$$= \frac{10\times 9\times 8}{1\times 2\times 3}(0.35)^3\,(0.65)^7 + \frac{10\times 9}{1\times 2}(0.35)^2\,(0.65)^8 + 10\,(0.35)\,(0.65)^9 + (0.65)^{10}$$

$= (0.65)^7\,[120\,(0.35)^3 + 45\,(0.35)^2\,(0.65) + 10\,(0.35)\,(0.65)^2 + (0.65)^3]$

$= (0.65)^7 \times 125\,[120 \times (0.07)^3 + 45 \times (0.07)^2\,(0.13) + 10\,(0.07)\,(0.13)^2 + (0.13)^3]$

$= 0.04902 \times 125\,[0.04 + 0.028665 + 0.011830 + 0.002197]$

$= 6.1275 \times 0.082692 = 0.5067$ **Ans.**

Example 37. *If 10% of bolts produced by a machine are defective. Determine the probability that out of 10 bolts, chosen at random (i) 1 (ii) none (iii) at the most 2 bolts will be defective.*

Solution. Probability of defective bolts = p = 10% = 0.1

Probability of not defective bolts = $q = 1 - p = 1 - 0.1 = 0.9$

Total number of bolts = $n = 10$

(*i*) Probability of 1 defective bolt $= {}^{10}C_1 \, (0.1)^1 \, (0.9)^9 = 0.3874$

(*ii*) Probability that none is defective = Probability of 0 defective bolt

$= P\,(0) = {}^{10}C_0 \, (0.1)^0 \, (0.9)^{10} = 0.3487$

(*iii*) Probability of 2 defective $= {}^{10}C_2 \, (0.1)^2 \, (0.9)^8 = 0.1937$

Probability of at most 2 defective $= P\,(0 \text{ or } 1 \text{ or } 2)$

$= P\,(0) + P\,(1) + P\,(2) = 0.3487 + 0.3874 + 0.1937$

$= 0.9298$ **Ans.**

Example 38. *A die is thrown 8 times and it is required to find the probability that 3 will show (i) Exactly 2 times (ii) At least seven times (iii) At least once.*

Solution. The probability of throwing 3 in a single trial = $p = \dfrac{1}{6}$

The probability of not throwing 3 in a single trial = $q = \dfrac{5}{6}$

(*i*) P (getting 3, exactly 2 times)

$$= {}^8C_2 \, q^6 \, p^2 = 28\left(\frac{5}{6}\right)^6\left(\frac{1}{6}\right)^2 = \frac{28 \times 5^6}{6^8}$$

(*ii*) P (getting 3, at least seven times) = P (getting 3, at 7 or 8 times)

$$= P\,(7) + P\,(8) = {}^8C_7 \, q^1 \, p^7 + {}^8C_8 \, q^0 \, p^8$$

$$= 8\left(\frac{5}{6}\right)\left(\frac{1}{6}\right)^7 + \left(\frac{1}{6}\right)^8 = \frac{41}{6^8}$$

(*iii*) P (getting 3 at least once)

$= P$ (getting 3, at 1 or 2 or 3 or 4 or 5 or 6 or 7 or 8 times)

$= P\,(1) + P\,(2) + P\,(3) + P\,(4) + P\,(5) + P\,(6) + P\,(7) + P\,(8)$

$= P\,(0) + P\,(1) + P\,(2) + P\,(3) + P\,(4) + P\,(5) + P\,(6) + P\,(7) + P\,(8) - P\,(0)$

$= 1 - P\,(0) = 1 - {}^8C_0 \, q^8 \, p^0$

$$= 1 - \left(\frac{5}{6}\right)^8$$ **Ans.**

Example 39. *An underground mine has 5 pumps installed for pumping out storm water, the probability of any one of the pumps failing during the storm is $\frac{1}{8}$. What is the probability that (i) at least 2 pumps will be working; (ii) all the pumps will be working during a particular storm?* **[Bihar SBTE. 2010]**

Solution. (*i*) Probability of pump failing $= \dfrac{1}{8}$

Probability of pump working $= 1 - \dfrac{1}{8} = \dfrac{7}{8}, p = \dfrac{7}{8}, q = \dfrac{1}{8}, n = 5$

(*i*) P (At least 2 pumps working) = P (2 or 3 or 4 or 5 pumps working)

$= P\,(2) + P\,(3) + P\,(4) + P\,(5) = {}^5C_2 \, p^2 \, q^3 + {}^5C_3 \, p^3 \, q^2 + {}^5C_4 \, p^4 \, q + {}^5C_5 \, p^5 \, q^0$

$$= 10\left(\frac{7}{8}\right)^2\left(\frac{1}{8}\right)^3 + 10\left(\frac{7}{8}\right)^3\left(\frac{1}{8}\right)^2 + 5\left(\frac{7}{8}\right)^4\left(\frac{1}{8}\right) + \left(\frac{7}{8}\right)^5$$

$$= \frac{1}{8^5}[10 \times 49 + 10 \times 343 + 5 \times 2401 + 16807]$$

$$= \frac{1}{8^5}[490 + 3430 + 12005 + 16807] = \frac{32732}{8^5} = \frac{8183}{8192}$$

(*ii*) P (All the 5 pumps working) $= P(5) = {}^5C_5\, p^5\, q^0 = \left(\frac{7}{8}\right)^5 = \frac{16807}{32768}$ **Ans.**

Example 40. *Assuming that 20% of the population of a city are literate, so that the chance of an individual being literate is $\frac{1}{5}$ and assuming that 100 investigators each take 10 individuals to see whether they are literate, how many investigators would you expect to report 3 or less were literate.*

Solution. $p = \frac{1}{5}, n = 10$

P (3 or less) $= P$ (0 or 1 or 2 or 3) $= P(0) + P(1) + P(2) + P(3)$

$$= {}^{10}C_0\left(\frac{1}{5}\right)^0\left(\frac{4}{5}\right)^{10} + {}^{10}C_1\left(\frac{1}{5}\right)^1\left(\frac{4}{5}\right)^9 + {}^{10}C_2\left(\frac{1}{5}\right)^2\left(\frac{4}{5}\right)^8 + {}^{10}C_3\left(\frac{1}{5}\right)^3\left(\frac{4}{5}\right)^7$$

$$= \left(\frac{4}{5}\right)^{10} + \frac{10}{5}\left(\frac{4}{5}\right)^9 + \frac{45}{25}\left(\frac{4}{5}\right)^8 + \frac{120}{125}\left(\frac{4}{5}\right)^7$$

$$= \left(\frac{4}{5}\right)^7 [(0.8)^3 + 2\,(0.8)^2 + 1.8\,(0.8) + 0.96]$$

$$= 0.2097152\,[0.512 + 1.28 + 1.44 + 0.96]$$

$$= 0.2097152 \times 4.192 = 0.879126118$$

Required number of investigators $= 0.879126118 \times 100 = 87.91261$

$= 88$ (approximate) **Ans.**

Example 41. *Write two-three areas where Binomial distribution is applied. The probability of entering student in Chartered Accountant will be graduate is 0.5. Determine the probability that out of 10 students (i) none (ii) one or (iii) at least one will graduate.*

Solution. Binomial distribution applied in various areas are (*i*) Probability of Winning a Lottery or Gamble (*ii*) Numbers of students Passing or failing (*iii*) Electrons outcome (*iv*) Game of Playing Cards.

Given, the probability of an entering student in Chartered Accountant will graduate is $p = 0.5$

$\therefore$ The probability of an entering student in Chartered Accountant will not graduate is $q = 0.5$.

Therefore

(*i*) The probability of none will graduate out of 10 students

$P(0) = {}^{10}C_0\, p^0\, q^{10} = {}^{10}C_0\,(0.5)^0\,(0.5)^{10}$

$= 9.765625 \times 10^{-4}$ **Ans.**

(*ii*) The probability of exactly one student will be graduate out of 10 students.

$P(1) = {}^{10}C_1\,(0.5)^1\,(0.5)^9 = 10 \times 0.5 \times (0.5)^9$

$= 9.765625 \times 10^{-3}$ **Ans.**

(*iii*) The probability of at least one will be graduate out of 10 students

P (At least one) = 1 – (probability of none will graduate)

$= 1 - 9.765625 \times 10^{-4} = 0.99$ **Ans.**

Example 42. *Assuming half the population of a town consumes chocolates and that 100 investigators each take 10 individuals to see whether they are consumers, how many investigators would you expect to report that three people or less were consumers?*

Solution. The chance for an individual to be consumer is $p = \dfrac{1}{2}$

The chance of not being a consumer $= q = 1 - \dfrac{1}{2} = \dfrac{1}{2}$.

Here we have to find the probabilities of 0, 1, 2, and 3 successes.

$$P(r \le 3) = P(0) + P(1) + P(2) + P(3)$$

$$= {}^{10}C_0\, q^{10} p^0 + {}^{10}C_1\, q^9 p^1 + {}^{10}C_2\, q^8 p^2 + {}^{10}C_3\, q^7 p^3$$

$$= \left(\frac{1}{2}\right)^{10} + 10\left(\frac{1}{2}\right)^9\left(\frac{1}{2}\right) + 45\left(\frac{1}{2}\right)^8\left(\frac{1}{2}\right)^2 + 120\left(\frac{1}{2}\right)^7\left(\frac{1}{2}\right)^3$$

$$= \left(\frac{1}{2}\right)^{10}[1+10+45+120] = \frac{176}{1024}$$

The number of investigators to report that three or less people were consumers of chocolates is given by $\dfrac{176}{1024} \times 100 = 17.2$

Hence, 17 investigators would report that 3 or less people are consumers. **Ans.**

Example 43. *The probability that a bomb dropped from a plane will strike the target is* $\dfrac{1}{5}$. *If six bombs are dropped, find the probability that:*

(*i*) *Exactly two will strike the target.* **[MDU May 2007]**

(*ii*) *At least two will stroke the target.* **[RGPV Bhopal II Sem. Feb. 2006]**

Solution. Here, $p = \dfrac{1}{5}, q = 1 - \dfrac{1}{5} = \dfrac{4}{5}, n = 6$

We know that $P(r) = {}^nC_4\, p^r q^{n-r}$

$$\Rightarrow \quad P(2) = {}^6C_2\left(\frac{1}{5}\right)^2\left(\frac{4}{5}\right)^{6-2} = 15\left(\frac{256}{15625}\right) = \frac{768}{3125} = 0.24576$$

$$\Rightarrow \quad P(\text{at least } 2) = P(2, 3, 4, 5, 6)$$

$$= P(2) + P(3) + P(4) + P(5) + P(6)$$

$$= P(0) + P(1) + P(2) + P(3) + P(4) + P(5) + P(6) - P(0) - P(1)$$

$$= 1 - [P(0) + P(1)]$$

$$= 1 - \left[{}^6C_0\left(\frac{1}{5}\right)^0\left(\frac{4}{5}\right)^6 + {}^6C_1\left(\frac{1}{5}\right)\left(\frac{4}{5}\right)^5\right] = 1 - \left[\frac{4096}{15625} + 6\left(\frac{1024}{15625}\right)\right]$$

$$= 1 - \frac{10240}{15625} = \frac{5385}{15625} = \frac{1077}{3125} = 0.34464$$

Hence (*i*) $P = 0.24576$ (*ii*) $P = 0.34464$ **Ans.**

Example 44. *Three defective bulbs are mixed with 7 good ones. Find the probability distribution of the number of defective bulbs, if bulbs are drawn at random.*

Solution. Here p = Probability of defective bulb $= \frac{3}{10}$

$\Rightarrow$ q = Probability of non-defective bulb $= 1 - \frac{3}{10} = \frac{7}{10}$

$\therefore$ The probability distribution of number of defective bulb $= {}^3C_r \left(\frac{3}{10}\right)^r \left(\frac{7}{10}\right)^{3-r}$ **Ans.**

Example 45. *Two cards are drawn successively with replacement from a well-shuffled pack of 52 cards. Find the probability distribution of the number of kings.*

Solution. Here $P_1 = \frac{{}^4C_1}{{}^{52}C_1} = \frac{4}{52} = \frac{1}{13}$, $\quad P_2 = \frac{{}^4C_1}{{}^{52}C_1} = \frac{4}{52} = \frac{1}{13}$

Let P = probability that the two cards are drawn successively with replacement of the king.

$$= \frac{1}{13} \times \frac{1}{13} = \frac{1}{169}$$

$$q = 1 - p = 1 - \frac{1}{169} = \frac{168}{169}$$

$\therefore$ Probability distribution of the number of kings $= {}^2C_r \left(\frac{1}{169}\right)^r \left(\frac{168}{169}\right)^{2-r}$ **Ans.**

Example 46. *A bag contains 7 white and 3 black balls. A ball is drawn and replaced. What is the probability of 2 white and 3 black balls in five drawings?*

Solution. Let S is the sample space.

W is the event of drawing a white ball and B is the event of drawing a black ball.

According to question $n(S) = 10$, $n(W) = 7$, $n(B) = 3$

If P = probability of drawing a white ball in one trial.

Then, $p = \frac{n(W)}{n(S)} = \frac{7}{10}$

q = probability of drawing not a white ball in one trial.

i.e., the probability of drawing a black ball.

$$= 1 - \frac{7}{10} = \frac{3}{10}$$

$\therefore$ The probability of drawing 2 white 3 black balls in five drawing $= {}^5C_2 \cdot p^2 q^3$

$$= {}^5C_2 \cdot \left(\frac{7}{10}\right)^2 \cdot \left(\frac{3}{10}\right)^3$$ **Ans.**

Example 47. *The probability that a Television manufactured by a company will be defective is $\frac{1}{10}$. If 12 such Televisions are manufactured, find the probability that:*

(a) Exactly two will be defective (b) At least two will be defective (c) None will be defective **[Bihar SBTE 2011]**

Solution. Here, $p = \dfrac{1}{10}$

$\Rightarrow \quad q = 1 - \dfrac{1}{10} = \dfrac{9}{10}$

(*a*) $P(2) = {}^nC_2\, p^2 q^{n-2} = {}^{12}C_2 \left(\dfrac{1}{10}\right)^2 \left(\dfrac{9}{10}\right)^{10} = \dfrac{66 \times 9^{10}}{10^{12}}$ **Ans.**

(*b*) P (At least two defective) $= 1 - P(0) - P(1)$

$$= 1 - {}^{12}C_0 \left(\frac{1}{10}\right)^0 \left(\frac{9}{10}\right)^{12} - {}^{12}C_1 \left(\frac{1}{10}\right)^1 \left(\frac{9}{10}\right)^{11} = 1 - \left(\frac{9}{10}\right)^{12} - \frac{12}{10}\left(\frac{9}{10}\right)^{11}$$ **Ans.**

(*c*) P (None is defective) $= {}^nC_0\, p^0 q^{12}$

$$= {}^{12}C_0 \left(\frac{1}{10}\right)^0 \left(\frac{9}{10}\right)^{12} = \left(\frac{9}{10}\right)^{12}$$ **Ans.**

EXERCISE 37.3

1. If 20% of the bolts produced by a machine are defective, determine the probability that out of 4 bolts chosen at random (*a*) 1 (*b*) 0 (*c*) At most 2 bolts will be defective. **Ans.** (*a*) 0.4096 (*b*) 0.4096 (*c*) 0.9728.
2. Six dice are thrown 729 times. How many times do you expect at least three dice to show a five or a six? **Ans.** 233
3. Find the probability of getting a total of 7 at least once in 4 tosses of a pair of fair dice? **Ans.** $\dfrac{671}{1296}$
4. If the chance that any one of the 10 telephone lines is busy at any instant is 0.2, what is the chance that 5 of the lines are busy? What is the probability that all the lines are busy? **Ans.** ${}^{10}C_5\,(0.2)^5\,(0.8)^5,\ (0.2)^{10}$
5. An insurance salesman sells policies to 5 men, all of identical age in good health. According to the actuarial tables the probability that a man of this particular age will be alive 30 years hence is $\dfrac{3}{2}$. Find the probability that in 30 years.
   (*a*) All 5 men (*b*) At least 3 men (c) Only 2 men (*d*) At least 1 man will be alive.
   Ans. (*a*) $\dfrac{32}{243}$ (*b*) $\dfrac{192}{243}$ (*c*) $\dfrac{40}{243}$ (*d*) $\dfrac{242}{243}$
6. A box contains 10 screws, 3 of which are defective. Two screws are drawn at random without replacement. Find the probability that none of the two screws is defective. **Ans.** $\dfrac{7}{15}$
7. Out of 800 families with four children each, how many families would be expected to have:
   (*a*) 2 boys and 2 girls (*b*) At least one boy (c) no girl (*d*) at most two girls?
   Assume equal probabilities for boys and girls. **Ans.** (*i*) 300, (*ii*) 750, (*iii*) 50, (*iv*) 550

8. In a hurdle race, a player has to cross 10 hurdles. The probability that he will clear each hurdle is 5/6. What is the probability that he will knock down less than 2 hurdles? **Ans.** $\frac{8}{3}\left(\frac{5}{6}\right)^9$

9. In a lot of 500 solenoids 25 are defective. Find the probability of 0, 1, 2, 3, defective solenoids in a random sample of 20 solenoids.

[MDU May 2008] **Ans.** 0.3585, 0.3774, 0.1887, 0.0596.

10. An electronic component consists of three parts. Each part has probability 0.99 of performing satisfactorily. The component fails if 2 or more parts do not perform satisfactorily. Assuming that the parts perform independently, determine the probability that the component does not perform satisfactorily. **Ans.** 0.000298

11. The incidence of occupational disease in an industry is such that the workers have 20% chance of suffering from it. What is the probability that out of 6 workers 4 or more will catch the disease? **[MDU 2006, AMIE Winter 2005]** **Ans.** $\frac{53}{2125}$

12. Among 10,000 random digits, find the probability p that the digit 3 appears at most 950 times. **Ans.** $\sum_{x=0}^{950} 10{,}000\, C_r \left(\frac{1}{10}\right)^5 \left(\frac{9}{10}\right) 10{,}000 - r$

13. In a bombing action there is 50% chance that any bomb will strike the target. Two direct hits are needed to destroy the target completely. How many bombs are required to be dropped to give a 99% chance or better of completely destroying the target. **[RGPV Bhopal June 2008]** **Ans.** 11

14. Fill in the blanks:

(*a*) A coin is biased so that a head is twice as likely to occur as a tail. If the coin is tossed 3 times, the probability of getting exactly 2 tails, is .... **Ans.** $\frac{2}{9}$

(*b*) The probability of getting number 5 exactly two times in five throws of an unbiased die is .... **Ans.** $10 \cdot \frac{5^3}{6^5}$

(*c*) A die is thrown 6 times. The probability to get greater than 4 appears at least once is .... **Ans.** $\frac{665}{729}$

(*d*) For what, one should be?

(*i*) Obtaining 6 at least once in 4 throws of a die.

or (*ii*) Obtaining a double-six at least once in 24 throws with two dice. **Ans.** (*i*)

(*e*) The probability of producing a defective bolt is 0.1. The probability that out of 5 bolts one will be defective is .... **Ans.** $\frac{1}{2}\left(\frac{9}{10}\right)^4$

(*f*) If the probability of hitting a target is 5% and 5 shots are fired independently, the probability that the target will be hit at least once is .... **Ans.** $1 - (0.95)^5$

15. Tick (✓) the correct answer:

(*a*) If a coin is tossed 6 times in succession, the probability of getting at least one head is

(*i*) 1/64 (*ii*) 3/32 (*iii*) 63/64 (*iv*) 1/2 **Ans.** (*iii*) (*b*) A coin is tossed until a tail appears or at the most five times. Given that the tail does not appear on the first two tosses, the probability that the coin will be tossed 5 times, is

(*i*) 1/2 (*ii*) 3/5 (*iii*) 1/3 (*iv*) 1/4 **Ans.** (*iv*)

(*c*) In a certain manufacturing process it is known that on an average, 1 in every 100 items is defective. What is the probability that 5 items are inspected before a defective item is found?

(*i*) 0.0096 (*ii*) 0.96 (*iii*) 0.096 (*iv*) None of these **Ans.** (*i*)

(*d*) The probability that a marksman will hit a target is given as $\frac{1}{5}$. Then his probability of at least one hit in 10 shots is

(*i*) $1-\left(\frac{4}{5}\right)^{10}$ (*ii*) $\frac{1}{5^{10}}$ (*iii*) $1-\frac{1}{5^{10}}$ (*iv*) None of these. **Ans.** (*i*)

(*e*) The probability of having at least one tail in 4 throws with a coin is

(*i*) $\frac{15}{16}$ (*ii*) $\frac{1}{16}$ (*iii*) $\frac{1}{4}$ (*iv*) 1. **Ans.** (*i*)

(*f*) A coin is tossed 3 times. The probability of obtaining two heads will be

(*i*) $\frac{3}{8}$ (*ii*) $\frac{1}{2}$ (*iii*) 1 (*iv*) 2. **Ans.** (*i*)

(*g*) 8 coins are tossed simultaneously. The probability of getting at least 6 heads is

(*i*) $\frac{57}{64}$ (*ii*) $\frac{229}{256}$ (*iii*) $\frac{7}{64}$ (*iv*) $\frac{37}{256}$. **Ans.** (*iv*)

(*h*) Three unbiased coins are tossed simultaneously. This is repeated four times. The probability of getting at least one head each time is

(*i*) $\left(\frac{3}{4}\right)^4$ (*ii*) $\left(\frac{7}{8}\right)^4$ (*iii*) $\left(\frac{1}{8}\right)^4$ (*iv*) $\left(\frac{1}{4}\right)^4$ **Ans.** (*ii*)

(*i*) In rolling two fair dice, the probability of getting equal numbers or numbers with an even product

(*i*) $\frac{6}{36}$ (*ii*) $\frac{30}{36}$ (*iii*) $\frac{27}{36}$ (*iv*) $\frac{3}{36}$ **Ans.** (*ii*)

37.6 MEAN OF BINOMIAL DISTRIBUTION

$$(q+p)^n = q^n + {}^nC_1\, q^{n-1}\, p^1 + {}^nC_2\, q^{n-2}\, p^2 + {}^nC_3\, q^{n-3}\, p^3 + \ldots\ldots + {}^nC_r\, q^{n-r}\, p^r + \ldots\ldots + p^n$$

Successes (r)	*Frequency* (f)	*Product* ($r^2 f$)
0	q^n	0
1	$n\,q^{n-1}\,p$	$n\,q^{n-1}\,p$
2	$\frac{n(n-1)}{2}\,q^{n-2}\,p^2$	$n(n-1)\,q^{n-2}\,p^2$
3	$\frac{n(n-1)(n-2)}{6}\,q^{n-3}\,p^3$	$\frac{n(n-1)(n-2)}{2}\,q^{n-3}\,p^3$
....	......	......
n	p^n	np^n

[Bihar SBTE 2009]

$$\Sigma fr = nq^{n-1}\,p + n\,(n-1)\,q^{n-2}\,p^2 + \frac{n\,(n-1)\,(n-2)}{2}\,q^{n-3}\,p^3 + ... + n\,p^n$$

$$= np\,[q^{n-1} + \frac{(n-1)}{1!}\,q^{n-2}\,p + \frac{(n-1)\,(n-2)}{2}\,q^{n-3}\,p^2 + ... + p^{n-1}]$$

$$= np\,(q+p)^{n-1} = np$$

$$\Sigma f = q^n + nq^{n-1}p + \frac{n\,(n-1)}{2}\,q^{n-2}\,p^2 + ... + p^n$$

$$= (q+p)^n = 1 \quad \text{(since } q+p=1)$$

Hence, Mean $= \dfrac{\Sigma fr}{\Sigma f} = \dfrac{np}{1}$

37.7 VARIANCE AND STANDARD DEVIATION OF BINOMIAL DISTRIBUTION

Successes (*r*)	*Frequency* (*f*)	*Product* ($r^2 f$)
0	q^n	0
1	$n\,q^{n-1}\,p$	$n\,q^{n-1}\,p$
2	$\frac{n\,(n-1)}{2}\,q^{n-2}\,p^2$	$2n\,(n-1)\,q^{n-2}\,p^2$
3	$\frac{n\,(n-1)\,(n-2)}{6}\,q^{n-3}\,p^3$	$\frac{3n\,(n-1)\,(n-2)}{2}\,q^{n-3}\,p^3$
....	......	......
n	p^n	$n^2 p^n$

[Bihar SBTE 2009, 2010, 2011, 2014]

We know that $$\sigma^2 = \frac{\Sigma fr^2}{\Sigma f} - \left(\frac{\Sigma fr}{\Sigma f}\right)^2 \quad ... (1)$$

r is the deviation of items (successes) from 0.

$$\Sigma f = 1,\ \Sigma fr = np$$

$$\Sigma fr^2 = 0 + nq^{n-1}\,p + 2n\,(n-1)\,q^{n-2}\,p^2 + \frac{3n\,(n-1)\,(n-2)}{2}\,q^{n-3}\,p^3 + ... + n^2\,p^n$$

$$= np\left[q^{n-1} + \frac{2\,(n-1)}{1!}\,q^{n-2}\,p + \frac{3\,(n-1)\,(n-2)}{2!}\,q^{n-3}\,p^2 + ... + np^{n-1}\right]$$

$$= np\left[q^{n-1} + \frac{(n-1)\,q^{n-2}\,p}{1!} + \frac{(n-1)\,(n-2)}{2!}\,q^{n-3}\,p^2 + ... + p^{n-1}\right.$$

$$\left. + \frac{(n-1)\,q^{n-2}\,p}{1!} + \frac{2\,(n-1)\,(n-2)}{2!}\,q^{n-3}\,p^2 + ... + (n-1)\,p^{n-1}\right]$$

$$= np\left[q^{n-1} + (n-1)\,q^{n-2}\,p + \frac{(n-1)\,(n-2)}{2!}\,q^{n-3}\,p^2 + ... + p^{n-1}\right.$$

$$\left. + (n-1)\,p\left\{q^{n-2} + (n-2)q^{n-3}\;p + \frac{(n-2)\,(n-3)}{2!}\,q^{n-4}\;p^2 + ... + p^{n-2}\right\}\right]$$

$$= np\,[(q+p)^{n-1} + (n-1)\,p\,(q+p)^{n-2}]$$

$$= np\,[1 + (n-1)\,p]$$

$$= np\,[np + (1-p)] = np\,[np + q] = n^2\,p^2 + npq$$

Putting these values in (1), we have

$$\textbf{Variance} = \sigma^2 = \frac{n^2\,p^2 + n\,pq}{1} - \left(\frac{np}{1}\right)^2 = npq,$$

$$\Rightarrow \qquad S.D. = \sigma = \sqrt{n\,p\,q}$$

Hence for the binomial distribution,

Mean = np, $\mu_2 = \sigma^2 = n\;p\;q$

Example 48. *If the probability of a defective bolt is 0.1, find (a) the mean (b) the standard deviation for the distribution bolts in a total of 400.*

Solution. $n = 400$, $p = 0.1$, Mean $= np = 400 \times 0.1 = 40$

Standard deviation $= \sqrt{npq} = \sqrt{400 \times 0.1\,(1-0.1)} = \sqrt{400 \times 0.1 \times 0.9} = 20 \times 0.3 = 6$ **Ans.**

Example 49. *A die is tossed thrice. A success is getting 1 or 6 on a toss. Find the mean and variance of the number of successes.* **[AMIETE Dec. 2010]**

Solution. $n = 3, p = \dfrac{1}{3}, q = \dfrac{2}{3}$

Mean $= np = 3 \times \dfrac{1}{3} = 1$, variance $= npq = 3 \times \dfrac{1}{3} \times \dfrac{2}{3} = \dfrac{2}{3}$ **Ans.**

Example 50. *If mean and variance of a binomial distribution are 4 and 2 respectively, find the probability of (i) exactly 2 successes (ii) less than 2 successes (iii) at least 2 successes.* **[RGPV Bhopal II Sem. June 2005]**

Solution. Mean = 4 $\Rightarrow$ $np = 4$ ... (1)

Variance = 2 $\Rightarrow$ $npq = 2$ ... (2)

Dividing (2) by (1), we get

$$\Rightarrow \qquad \frac{npq}{np} = \frac{2}{4} \qquad \Rightarrow \qquad q = \frac{1}{2}$$

$$p = 1 - q = 1 - \frac{1}{2} = \frac{1}{2}$$

Putting the values of p in (1), we get $n\left(\dfrac{1}{2}\right) = 4 \quad \Rightarrow \quad n = 8$

(*i*) Probability of r successes $= {}^nC_r\,p^r\,q^{n-r}$

$$P(2) = {}^8C_2\left(\frac{1}{2}\right)^2\left(\frac{1}{2}\right)^{8-2} = {}^8C_2\left(\frac{1}{2}\right)^8 = \frac{8 \times 7}{2}\,\frac{1}{256} = \frac{7}{64}$$

(*ii*) P (less than 2 successes) = $P(0) + P(1) = {}^8C_0\, p^0 q^8 + {}^8C_1\, p^1 q^7 = \frac{1}{256} + 8\frac{1}{2}\left(\frac{1}{2}\right)^7 = \frac{9}{256}$

(*iii*) P (at least 2 successes) = $P(2) + P(3) + \ldots + P(8)$

$= P(0) + P(1) + P(2) + P(3) + \ldots + P(8) - P(0) - P(1)$

$= 1 - P(0) - P(1) = 1 - [P(0) + P(1)] = 1 - \frac{9}{256} = \frac{247}{256}$ **Ans.**

Example 51. *Fit a binomial distribution to the following data:*

x	*0*	*1*	*2*	*3*	*4*
f	*30*	*62*	*46*	*10*	*2*

[M.D.U. Dec. 2009]

Solution. We have

x	f	fx
0	30	0
1	62	62
2	46	92
3	10	30
4	2	8
	$\Sigma f = 150$	$\Sigma fx = 192$

Mean of given data $= \frac{\Sigma f x}{\Sigma f} = \frac{192}{150} = 1.28$

$\Rightarrow \quad np = 1.28 \qquad (\because n = 4)$

$\Rightarrow \quad 4p = 1.28 \quad \Rightarrow \quad p = 0.32$ and $q = 1 - p = 1 - 0.32 = 0.68$

Also, $N = 150 \qquad [N = \Sigma f]$

Hence, the binomial distribution is $N(q + p)^n = 150\,(0.68 + 0.32)4$ **Ans.**

Example 52. *Fit a Binomial distribution for the following data and compare the theoretical frequencies with actual ones:*

x	*0*	*1*	*2*	*3*	*4*	*5*
f	*2*	*14*	*20*	*34*	*22*	*8*

[RGPV Bhopal II Sem. June 2006]

Solution

x	f	fx	$P(r) = {}^5C_r\, p^r q^{5-r}$	*Theoretical Frequency*
0	2	0	${}^5C_0\,(0.568)^0\,(0.432)^5 = 0.015$	$100 \times 0.015 = 1.5$
1	14	14	${}^5C_1\,(0.568)^1\,(0.432)^4 = 0.099$	$100 \times 0.099 = 9.9$
2	20	40	${}^5C_2\,(0.568)^2\,(0.432)^3 = 0.260$	$100 \times 0.260 = 26.0$
3	34	102	${}^5C_3\,(0.568)^3\,(0.432)^2 = 0.342$	$100 \times 0.342 = 34.2$
4	22	88	${}^5C_4\,(0.568)^4\,(0.432)^1 = 0.225$	$100 \times 0.225 = 22.5$
5	8	40	${}^5C_5\,(0.568)^5\,(0.432)^0 = 0.0591$	$100 \times 0.0591 = 5.91$
	100	284		

$$\Sigma f = 100, \quad \Sigma f = 284$$

$$\text{Mean} = \frac{\Sigma fx}{\Sigma f} = \frac{284}{100} = 2.84$$

$$\text{Mean} = np = 2.84$$

$$\Rightarrow \quad 5p = 2.84 \Rightarrow p = \frac{2.84}{5} = 0.568$$

$$\Rightarrow \quad q = 1 - p = 1 - 0.568 = 0.432$$

Binomial distribution = $100\,(0.432 + 0.568)^5$ **Ans.**

37.8 RECURRENCE RELATION FOR THE BINOMIAL DISTRIBUTION

For Binomial distribution, $P\,(r) = {}^nC_r\, p^r\, q^{n-r}$ ... (1)

$$P\,(r+1) = {}^nC_{r+1}\, p^{r+1}\, q^{n-r-1} \quad \text{... (2)}$$

On dividing (2) by (1), we get

$$\frac{P\,(r+1)}{P\,(r)} = \frac{{}^nC_{r+1}}{{}^nC_r}\frac{p^{r+1}q^{n-r-1}}{p^r\,q^{n-r}} = \frac{n\,(n-1)(n-2)\ldots(n-r)}{(r+1)!}\,\frac{r!}{n\,(n-1)(n-2)\ldots(n-r+1)}\,\frac{P}{q}$$

$$\frac{P\,(r+1)}{P\,(r)} = \frac{n-r}{r+1}\frac{p}{q} \quad \text{or} \quad \boxed{P\,(r+1) = \frac{n-r}{r+1}\frac{p}{q}\,P\,(r)}$$ **Ans.**

EXERCISE 37.4

1. Fit a Binomial distribution to the following frequency data:

x	0	1	3	4
f	2.8	62	10	4

Ans. $P\,(r) = {}^{104}C_r\,(0.00999)^r\,(0.99111)^{104-r}$

2. Fit a Binomial distribution to the following frequency distribution:

x	0	1	2	3	4	5	6
f	13	25	52	58	32	16	4

[KU Dec. 2009]

Ans. $200\,(0.554 + 0.446)^6$

3. Fill in the blanks:

(*a*) If three persons selected at random are stopped on a street, then the probability that all of them were born on Sunday is .... **Ans.** $\frac{1}{343}$

(*b*) The mean, standard deviation and skewness of Binomial distribution are ...., ... and .... **Ans.** $np, \sqrt{npq}$

(*c*) If n and p are the parameters of a binomial distribution the standard deviation is .... **Ans.** $\sqrt{npq}$

(*d*) The Binomial distribution of mean 5 and variance $\frac{10}{3}$ is ....

Ans. ${}^{15}C_r\left(\frac{1}{3}\right)^r\left(\frac{2}{3}\right)^{15-r}$

4. Tick (✓) the correct answer:

(*a*) The variance for a Binomial distribution is:

(*i*) np (*ii*) $\sqrt{np}$ (*iii*) npq (*iv*) $\sqrt{npq}$

[RGPV Bhopal II Sem. June 2007] Ans. (*iii*)

(*b*) For the Binomial distribution $(p + q)^n$, the relation of mean and variance is:

(*i*) means = variance (*ii*) means < variance

(*iii*) mean > variance (*iv*) $(\text{mean})^2$ = variances

[RGPV Bhopal II Semester June 2006] Ans. (*iii*)

(*c*) In usual notation, for Binomial distribution, npq, is

(*i*) $< np$ (*ii*) $= np$ (*iii*) $> np$ (*iv*) None of the above

[AMIE Winter 2005] Ans. (*i*)

(*d*) The standard deviation of Binomial distribution ${}^nC_r\, p^r\, q^{n-r}$ $(r = 0, 1, 2, \ldots\ldots, n)$ is

(*i*) $\sqrt{np}$ (*ii*) $\sqrt{nq}$ (*iii*) $\sqrt{pq}$ (*iv*) $\sqrt{npq}$

[Bihar SBTE 2014, 2012, 2008] Ans. (*iv*)

37.9 POISSON DISTRIBUTION

Poisson distribution is a particular limiting form of the Binomial distribution when p (or q) is very small and n is large enough but np is finite.

Probabilities of Poisson distribution are given by $\boxed{P(r) = \dfrac{m^r e^{-m}}{r!}}$

where m is the mean of the distribution. **[Bihar SBTE, 2011, 2010]**

Proof. In Binomial distribution.

$$P(r) = {}^nC_r\, q^{n-r}\, p^r = {}^nC_r\,(1-p)^{n-r}\, p^r \qquad \left(\text{since mean} = m = np,\ p = \frac{m}{n}\right)$$

$$= {}^nC_r \left(1 - \frac{m}{n}\right)^{n-r} \left(\frac{m}{n}\right)^r \qquad (m \text{ is constant})$$

$$= \frac{n\,(n-1)\,(n-2)\ldots(n-\overline{r-1})}{r!} \left(\frac{m}{n}\right)^r \left(1 - \frac{m}{n}\right)^{n-r}$$

$$= \frac{\frac{n}{n}\left(\frac{n}{n} - \frac{1}{n}\right)\left(\frac{n}{n} - \frac{2}{n}\right)\ldots\left(\frac{n}{n} - \frac{r-1}{n}\right) m^r \left(1 - \frac{m}{n}\right)^n}{r!\left(1 - \frac{m}{n}\right)^r} = \frac{1\left(1 - \frac{1}{n}\right)\left(1 - \frac{2}{n}\right)\ldots\left(1 - \frac{r-1}{n}\right) m^r \left(1 - \frac{m}{n}\right)^n}{r!\left(1 - \frac{m}{n}\right)^r}$$

Taking limits, when n tends to infinity

$$\lim_{n \to \infty}\left(1 - \frac{m}{n}\right)^n = \lim_{n \to \infty}\left[\left(1 - \frac{m}{n}\right)^{-\frac{n}{m}}\right]^{-m} = e^{-m}$$

$$P(r) = \frac{m^r}{r!} e^{-m}$$

$$\boxed{P(r) = \frac{e^{-m} \cdot m^r}{r!}}$$

Example 53. *Prove that the variance of a Poisson distribution is equal to its mean. Also, if x is a Poisson variate so that P (0) = P (1) = k, then prove that* $k = \frac{1}{e}$.

Solution. Poisson distribution is

$$P(r) = \frac{e^{-m} m^r}{r!}$$

$$P(0) = \frac{e^{-m} m^0}{0!} = e^{-m}$$

$$P(1) = \frac{e^{-m} m^1}{1!} = me^{-m}$$

Given $P(0) = P(1)$

$\Rightarrow \quad e^{-m} = me^{-m} \quad \Rightarrow \quad m = 1$

$P(0) = P(1) = k$

$e^{-1} = 1 \cdot e^{-1} = k \quad \Rightarrow \quad k = \frac{1}{e}$ **Proved.**

37.10 MEAN OF POISSON DISTRIBUTION

$$P(r) = \frac{e^{-m} \cdot m^r}{r!}$$

Successes (r)	*Frequency (f)*	$f \cdot r$
0	$\frac{e^{-m} m^0}{0!}$	0
1	$\frac{e^{-m} m^1}{1!}$	$e^{-m} \cdot m$
2	$\frac{e^{-m} m^2}{2!}$	$e^{-m} \cdot m^2$
3	$\frac{e^{-m} m^3}{3!}$	$\frac{e^{-m} \cdot m^3}{2!}$
...	...	...
r	$\frac{e^{-m} m^r}{r!}$	$\frac{e^{-m} \cdot m^r}{(r-1)!}$
...	...	...

[Bihar SBTE 2012, 2010, 2008]

$$\Sigma fr = 0 + e^{-m} \cdot m + e^{-m} \cdot m^2 - e^{-m} \cdot \frac{m^3}{2!} + ... + e^{-m} \frac{m^r}{(r-1)!} + ...$$

$$= e^{-m} \cdot m \left[1 + \frac{m}{1!} + \frac{m^2}{2!} + ... + \frac{m^{r-1}}{(r-1)!} \right] = m \cdot e^{-m} \cdot [e^m] = m$$

$$\text{Mean} = \frac{\Sigma fr}{\Sigma f} = \frac{m}{1}$$

$$\boxed{\text{Mean} = m}$$ **Ans.**

37.11 STANDARD DEVIATION OF POISSON DISTRIBUTION

$$P(r) = \frac{e^{-m} m^r}{r!}$$

Successes (r)	*Frequency* (f)	rf	r^2f
0	$\frac{e^{-m} m^0}{0!}$	0	0
1	$\frac{e^{-m} m^1}{1!}$	$e^{-m} \cdot m$	$e^{-m} \cdot m$
2	$\frac{e^{-m} m^2}{2!}$	$e^{-m} \cdot m^2$	$2e^{-m} \cdot m^2$
3	$\frac{e^{-m} m^3}{3!}$	$\frac{e^{-m} \cdot m^3}{2!}$	$3e^{-m} \cdot \frac{m^3}{2!}$
r	$\frac{e^{-m} m^r}{r!}$	$\frac{e^{-m} \cdot m^r}{(r-1)!}$	$\frac{re^{-m} \cdot m^r}{(r-1)!}$
....	....	....	...
	$\Sigma f = 1$	$\Sigma fr = m$	

[Bihar SBTE 2012, 2010]

$$\Sigma fr^2 = 0 + e^{-m} \cdot m + 2e^{-m} \cdot m^2 + 3 \cdot e^{-m} \cdot \frac{m^3}{2} + ... + \frac{r\,e^{-m} \cdot m^r}{(r-1)!} + ...$$

$$= m \cdot e^{-m} \left[1 + 2m + \frac{3m^2}{2!} + \frac{4m^3}{3!} + ... \frac{r \cdot m^{r-1}}{(r-1)!} + ... \right]$$

$$= m \cdot e^{-m} \left[1 + m + \frac{m^2}{2!} + \frac{m^3}{3!} + ... \frac{m^{r-1}}{(r-1)!} + ... + m + 2\frac{m^2}{2!} + \frac{3m^3}{3!} + ... + \frac{(r-1)m^{r-1}}{(r-1)!} + ... \right]$$

$$= m \cdot e^{-m} \left[\left\{ 1 + m + \frac{m^2}{2!} + \frac{m^3}{3!} + \ldots \frac{m^{r-1}}{(r-1)!} + \ldots \right\} + m \left\{ 1 + \frac{m}{1!} + \frac{m^2}{2!} + \ldots + \frac{m^{r-2}}{(r-2)!} + \ldots \right\} \right]$$

$$= m \cdot e^{-m} [e^m + m e^m] = m + m^2$$

$$\sigma^2 = \frac{\Sigma fr^2}{\Sigma f} - \left(\frac{\Sigma fr}{\Sigma f} \right)^2 = \frac{m + m^2}{1} - (m)^2 = m \text{ or } \sigma = \sqrt{m}$$

$$\boxed{\sigma = \text{S.D.} = \sqrt{m}}$$

Hence mean and variance of a Poisson distribution are each equal to m. Similarly, we can obtain

$$\mu_3 = m, \quad \mu_4 = 3m^2 + m$$

$$\beta_1 = \frac{1}{m}, \quad \beta_2 = 3 + \frac{1}{m}$$

$$\gamma_1 = \frac{1}{\sqrt{m}}, \; \gamma_2 = \frac{1}{m}$$

37.12 MEAN DEVIATION

Measures of dispersion: It is the mean of the absolute values of the deviation of a given set of numbers from their arithmetic mean. If $x_1, x_2, x_3 \ldots\ldots x_n$ be a set of numbers with frequencies $f_1, f_2, \ldots f_n$ respectively. Let $\bar{x}$ be the arithmetic mean of the numbers $x_1, x_2, \ldots\ldots x_n$, then

$$\text{Mean deviation} = \frac{\Sigma f_i \,|\, x_i - \bar{x} \,|}{\Sigma f_i}$$ **[Bihar SBTE 2012]**

Show that in a Poisson distribution with unit mean, the mean deviation about the mean is $\left(\frac{2}{e}\right)$ times the standard deviation.

Explanation. $P(r) = \frac{m^r}{r!} e^{-m}$ But mean = 1, *i.e.* $m = 1$ and S.D. = $\sqrt{m} = 1$

Hence, $P(r) = \frac{e^{-1}}{r!} = \frac{1}{e} \cdot \frac{1}{r!}$

r	$P(r)$	$\|r-1\|$	$P(r)\,\|r-1\|$
0	$\frac{1}{e}$	1	$\frac{1}{e}$
1	$\frac{1}{e}$	0	0
2	$\frac{1}{e}\frac{1}{2!}$	1	$\frac{1}{e}\frac{1}{2!}$

r	$P(r)$	$\lvert r-1\rvert$	$P(r)\ \lvert r-1\rvert$
3	$\frac{1}{e}\frac{1}{3!}$	2	$\frac{1}{e}\frac{2}{3!}$
4	$\frac{1}{e}\frac{1}{4!}$	3	$\frac{1}{e}\frac{3}{4!}$
r	$\frac{1}{e}\frac{1}{r!}$	$r-1$	$\frac{1}{e}\frac{r-1}{r!}$

$$\Sigma P(r)\lvert r-1\rvert = \frac{1}{e}+0+\frac{1}{e}\frac{1}{2!}+\frac{1}{e}\frac{2}{3!}+\frac{1}{e}\frac{3}{4!}+\ldots+\frac{1}{e}\frac{r-1}{r!}+\ldots$$

$$=\frac{1}{e}\left[1+0+\frac{1}{2!}+\frac{2}{3!}+\frac{3}{4!}+\ldots+\frac{r-1}{r!}+\ldots\right]$$

$$=\frac{1}{e}\left[1+\left(\frac{1}{1!}-\frac{1}{1!}\right)+\left(\frac{2}{2!}-\frac{1}{2!}\right)+\left(\frac{3}{3!}-\frac{1}{3!}\right)+\left(\frac{4}{4!}-\frac{1}{4!}\right)+\ldots+\left(\frac{r}{r!}-\frac{1}{r!}\right)+\ldots\right]$$

$$=\frac{1}{e}\left[1+\frac{1}{1!}+\frac{2}{2!}+\frac{3}{3!}+\frac{4}{4!}+\ldots+\frac{r}{r!}+\ldots-\frac{1}{1!}-\frac{2}{2!}-\frac{1}{3!}-\frac{1}{4!}\ldots-\frac{1}{r!}-\ldots\right]$$

$$=\frac{1}{e}\left[1+\left\{1+\frac{1}{1!}+\frac{1}{2!}+\frac{1}{3!}+\ldots+\frac{1}{(r-1)!}+\ldots\right\}-\left\{1+\frac{1}{1!}+\frac{1}{2!}+\frac{1}{3!}+\frac{1}{4!}+\ldots+\frac{1}{r!}\ldots\right\}+1\right]$$

$$=\frac{1}{e}[1+e-e+1]=\frac{2}{e}=\frac{2}{e}(1)=\frac{2}{e}\text{ S.D.}$$ **Proved.**

37.13 RECURRENCE FORMULA FOR POISSON DISTRIBUTION

For Poisson distribution $P(r)=\dfrac{e^{-m}\cdot m^{r}}{r!}$ ... (1)

$\therefore \quad P(r+1)=\dfrac{e^{-m}\,m^{r+1}}{(r+1)!}$ ... (2)

dividing (2) by (1), we get

$$\frac{P(r+1)}{P(r)}=\frac{e^{-m}\,m^{r+1}}{(r+1)!}\,\frac{r!}{e^{-m}\cdot m^{r}}=\frac{m}{r+1}$$

$$\boxed{\boldsymbol{P(r+1)=\frac{m}{r+1}\,P(r)}}$$

Example 54. *If the variance of the Poisson distribution is 2, find the probabilities for r = 1, 2, 3, 4 from the recurrence relation of Poisson distribution. Also find P (r ≥ 4).*

Solution. Variance = m = 2;

Mean = 2

$$P(r+1) = \frac{m}{r+1} P(r) \qquad \text{[Recurrence relation]}$$

Now $$P(r+1) = \frac{2}{r+1} P(r) \qquad (m = 2)$$

If $r = 0, P(1) = \frac{2}{0+1} P(0) = \frac{2}{0+1}(0.1353) = 0.2706 \qquad P(0) = e^{-m} = e^{-2} = 0.1353$

If $r = 1, P(2) = \frac{2}{1+1} P(1) = \frac{2}{2}(0.2706) = 0.2706$

If $r = 2, P(3) = \frac{2}{2+1} P(2) = \frac{2}{3}(0.2706) = 0.1804$

If $r = 3, P(4) = \frac{2}{3+1} P(3) = \frac{1}{2}(0.1804) = 0.0902$

$P(r \geq 4) = P(4) + P(5) + P(6) + ... = 1 - [P(0) + P(1) + P(2) + P(3)]$

$= 1 - [0.1353 + 0.2706 + 0.2706 + 0.1804] = 1 - 0.8569 =$ **0.1431** **Ans.**

Example 55. *Assume that the probability of an individual coal miner being killed in a mine accident during a year is* $\frac{1}{2400}$. *Use appropriate statistical distribution to calculate the probability that in a mine employing 200 miners, there will be at least one fatal accident in a year.*

Solution. $p = \frac{1}{2400}, n = 200$

$$m = np = \frac{200}{2400} = \frac{1}{12}$$

$P(\text{At least one}) = P(1 \text{ or } 2 \text{ or } 3 \text{ or } .... \text{ or } 200) = P(1) + P(2) + P(3) + ... + P(200)$

$= 1 - P(0) = 1 - \frac{e^{-m} \cdot m^0}{0!} = 1 - e^{-\frac{1}{12}} = 1 - 0.92 = 0.08$ **Ans.**

Example 56. *Suppose 3% of bolts made by a machine are defective, the defects occurring at random during production. If bolts are packaged 50 per box, find (a) exact probability and (b) Poisson approximation to it, that a given box will contain 5 defective botts.*

Solution. $p = \frac{3}{100} = 0.03$

(*a*) $q = 1 - p = 1 - 0.03 = 0.97$

Hence the probability for 5 defective bolts in a lot of 50

$= {}^{50}C_5 (0.03)^5 (0.97)^{45} = 0.013074$ (Binomial distribution)

(*b*) To get Poisson approximation $m = np = 50 \times \frac{3}{100} = \frac{3}{2} = 1.5$

Required Poisson approximation $= \frac{m^r e^{-m}}{r!} = \frac{(1.5)^5 e^{-1.5}}{5!} = 0.01412$ **Ans.**

Example 57. *The number of arrivals of customers during any day follows Poisson distribution with a mean of 5. What is the probability that the total number of customers on two days selected at random is less than 2?*

Solution. $m = 5$

$$P(r) = \frac{e^{-m}\, m^r}{r!},\ P(r) = \frac{e^{-5}(5)^r}{r!}$$

If the number of customers on two days < 2 = 1 or 0

First day	*Second day*	*Total*
0	0	0
0	1	1
1	0	1

Required probability = $P(0)\, P(0) + P(0)\, P(1) + P(1)\, P(0)$

$$= \frac{e^{-5}(5)^0}{0!} \cdot \frac{e^{-5}(5)^0}{0!} + \frac{e^{-5}(5)^0}{0!} \cdot \frac{e^{-5}(5)^1}{1!} + \frac{e^{-5}(5)^1}{1!} \frac{e^{-5}(5)^0}{0!}$$

$$= e^{-5} \cdot e^{-5} + e^{-5} \cdot e^{-5} \cdot 5 + e^{-5} \cdot 5 \cdot e^{-5}$$

$$= e^{-10}\,[1 + 5 + 5] = 11e^{-10} = 11 \times 4.54 \times 10^{-5} = 4.994 \times 10^{-4}$$ **Ans.**

Example 58. *Using Poisson distribution, find the probability that the ace of spades will be drawn from a pack of well-shuffled cards at least once in 104 consecutive trials.*

Solution. Probability of the ace of spades = $p = \frac{1}{52}, n = 104$

$$m = np = 104 \times \frac{1}{52} = 2$$

$$P(r) = e^{-m} \cdot \frac{m^r}{r!} = e^{-2} \cdot \frac{2^r}{r!} = \frac{1}{e^2} \frac{2^r}{r!}$$

P (at least once) = $P(1) + P(2) + P(3) + \ldots + P(104) = 1 - P(0)$

$$= 1 - \frac{1}{e^2} \times \frac{2^\circ}{0!} = 1 - \frac{1}{e^2} = 1 - 0.135 = 0.865$$ **Ans.**

Example 59. *In a certain factory producing cycle tyres, there is a small chance of 1 in 500 tyres to be defective. The tyres are supplied in lots of 10. Using Poisson distribution, calculate the approximate number of lots containing no defective, one defective and two defective tyres, respectively, in a consignment of 10,000 lots.*

Solution. $p = \frac{1}{500}, n = 10$

$$m = np = 10 \times \frac{1}{500} = \frac{1}{50} = 0.02,\ P(r) = \frac{e^{-m} \cdot m^r}{r!}$$

S. No.	Probability of defective	Number of lots containing defective
1	$P(0) = \frac{e^{-0.02}(0.02)^0}{0!} = e^{-0.02} = 0.9802$	$10000 \times 0.9802 = 9802$ lots

S. No.	Probability of defective	Number of lots containing defective
2	$P(1) = \dfrac{e^{-0.02}(0.02)^1}{1!}$ $= 0.9802 \times 0.02 = 0.019604$	$10000 \times 0.019604 = 196$ lots
3	$P(2) = \dfrac{e^{-0.02}(0.02)^2}{2!}$ $= 0.9802 \times 0.0002 = 0.00019604$	$10000 \times 0.000196 = 2$ lots

Ans.

Example 60. *A car hire firm has two cars which it hires out day by day. The number of demands for a car on each day is distributed as Poisson distribution with mean 1.5. Calculate the number of days in a year on which (i) neither car is on demand (ii) a car demand is refused.* ($e^{-15} = 0.2231$)

[MDU Dec. 2010, June 2009]

Solution. $m = 1.5$

(*i*) If the car is not used, then demand (r) = 0

$$P(r) = \frac{e^{-m} \cdot m^r}{r!}, \; P(0) = \frac{e^{-1.5}(1.5)^0}{0!} = e^{-1.5} = 0.2231$$

Number of days in a year when the demand is zero = $365 \times 0.231 = 81.4315$

Ans. 81 days

(*ii*) Some demand is refused if the number of demands is more than two *i.e.* $r > 2$.

$P(r > 2) = P(3) + P(4) + \ldots = 1 - [P(0) + P(1) + P(2)]$

$$= 1 - \left[\frac{e^{-1.5}(1.5)^0}{0!} + \frac{e^{-1.5}(1.5)^1}{1!} + \frac{e^{-1.5}(1.5)^2}{2!}\right]$$

$= 1 - [e^{-1.5} + e^{-1.5} \times 1.5 + e^{-1.5} \times 1.125] = 1 - e^{-1.5}[1 + 1.5 + 1.125] = 1 - e^{-1.5} \times 3.625$

$= 1 - 0.2231 \times 3.625 = 1 - 0.8087375 = 0.1912625$ **Ans.**

Number of days in a year when some demand of car is refused

$= 365 \times 0.1912625 = 69.81 = 70$ days. **Ans.**

Example 61. *If the probability that an individual suffers a bad reaction from a certain injection is 0.001. Determine the probability that out of 2000 individuals*

(a) exactly 3 (b) more than 2 individuals (c) None (d) More than one individual will suffer a bad reaction.

Solution. $p = 0.001, n = 2000$

$m = np = 2000 \times 0.001 = 2$

$$\therefore \quad P(r) = \frac{e^{-m} m^r}{r!} = e^{-2}\frac{2^r}{r!} = \frac{1}{e^2} \times \frac{2^r}{r!}$$

$$P(3) = \frac{1}{e^2} \cdot \frac{2^3}{3!} = \frac{1}{(2.718)^2} \times \frac{8}{6} = (0.135) \times \frac{4}{3} = 0.18$$

(*b*) P (more than 2) = P (3) + P (4) + P (5) + ... + P (2000)

$$= 1 - [P(0) + P(1) + P(2) = 1 - \left[\frac{e^{-2}(2)^0}{0!} + \frac{e^{-2}(2)^1}{1!} + \frac{e^{-2}(2)^2}{2!}\right]$$

$$= 1 - e^{-2}[1 + 2 + 2] = 1 - \frac{5}{e^2} = 1 - 5 \times 0.135 = 1 - 0.675 = 0.325 \quad \textbf{Ans.}$$

(*c*) P (none) = $P(0) = \dfrac{e^{-2}(2)^0}{0!} = 0.135$

(*d*) P (more than 1) = P (2) + P (3) + P (4) + ... + P (2000) = 1 – [P (0) + P (1)]

$$= 1 - \left[\frac{e^{-2}(2)^0}{0!} + \frac{e^{-2}(2)^1}{1!}\right] = 1 - 3e^{-2} = 1 - 3 \times 0.135 = 1 - 0.405 = 0.595 \quad \textbf{Ans.}$$

Example 62. *A manufacturer knows that the razor blades he makes contain on an average 0.5% of defectives. He packs them in packets of 5. What is the probability that a packet picked at random will contain 3 or more faulty blades?*

Solution. $p = 0.5\% = 0.005, n = 5$

$m = np = 5 \times 0.005 = 0.025$

$$P(r) = \frac{e^{-m} \cdot m^r}{r!} = \frac{e^{-0.025}(.025)^r}{r!}$$

$$P(3 \text{ or more}) = P(3) + P(4) + P(5) = \frac{e^{-0.025}(0.025)^3}{3!} + \frac{e^{-0.025}(0.025)^4}{4!} + \frac{e^{-0.025}(0.025)^5}{5!}$$

$$= \frac{e^{-0.025}(0.025)^3}{5!}[20 + 5(0.025) + (0.025)^2] = \frac{0.975 \times 0.000015625 \times 20.125625}{120}$$

$= 0.000002555.$ **Ans.**

Example 63. *In a certain factory turning out razor blades, there is a small chance of 0.002 for any blade to be defective. The blades are supplied in packets of 10. Use appropriate and suitable distribution to calculate the approximate number of packets containing no defective, one defective and two defective blades respectively in a consignment of 5000 packets.*

[KU 2009, RGPV Bhopal II Sem. June 2006]

Solution. Here $p = 0.002, \quad n = 10$

$m = np \Rightarrow m = 10 \times 0.002 = 0.020$

$$P(r) = \frac{e^{-m} \cdot (m)^r}{r!}$$

r	$P(r) = \dfrac{e^{-0.02}(0.02)^r}{r!}$	Number of packets = 50000 $P(r)$
0	$P(0) = \dfrac{e^{-0.02}(0.02)^0}{r!} = 0.980$	50000 × (0.980) = 49000

1	$P(1) = \dfrac{e^{-0.02}\,(0.02)^1}{1!} = 0.0196$	$50000 \times (0.0196) = 980$
2	$P(2) = \dfrac{e^{-0.02}\,(0.02)^2}{2!} = 0.000196$	$50000 \times (0.000196) = 9.8$

Hence, number of packets containing no defective razor blades = 49000.

Number of packets containing one defective razor blade = 980

Number of packets containing two defective razor blades = 9.8 **Ans.**

Example 64. *Suppose that a book of 600 pages contains 40 printing mistakes. Assume that these errors are randomly distributed throughout the book and x the number of errors per page has a Poisson distribution. What is the probability that 10 pages selected at random will be free of errors?*

Solution. $p = \dfrac{40}{600} = \dfrac{1}{15},\ n = 10$

$$m = np = 10 \times \frac{1}{15} = \frac{2}{3} \quad \Rightarrow \quad P(r) = \frac{e^{-m} \cdot m^r}{r!} = \frac{e^{\frac{-2}{3}}\left(\frac{2}{3}\right)^r}{r!}$$

$$P(0) = \frac{e^{\frac{-2}{3}}\left(\frac{2}{3}\right)^0}{0!} = e^{\frac{-2}{3}} = 0.51$$ **Ans.**

Example 65. *If there are 3 misprints in a book of 1000 pages find the probability that a given page will contain (i) no misprint (ii) more than 2 misprints.*

[U.P. III Sem. Dec. 2009]

Solution. Total number of pages = 1000

No. of misprints = 3

$$p = \frac{3}{1000} = 0.003,\ n = 1,\ m = np = 1 \times 0.003 = 0.003$$

Poisson distribution

$$P(r) = \frac{e^{-m} \cdot m^r}{r!},\ P(0) = \frac{e^{-0.003}\,(0.003)^0}{0!} = e^{-0.003} = 0.997$$

$$P(r > 2) = P(3) = \frac{e^{-0.003}\,(0.003)^3}{3!} = 0.0000000045$$

Hence (*i*) the probability that a page will contain no error = 0.997

(*ii*) the probability that a page will contain more than two misprints = 0.0000000045

Ans.

Example 66. *A manufacturer knows that the condensers he makes contain on an average 1% of defectives. He packs them in boxes of 100. What is the probability that a box picked out at random will contain 4 or more faulty condensers?*

[MDU May 2007]

Solution. $p = 1\% = 0.01,\ n = 100,\ m = np = 100 \times 0.01 = 1$

$$P(r) = \frac{e^{-m} \cdot (m)^r}{r!} = \frac{e^{-1}(1)^r}{r!} = \frac{e^{-1}}{r!}$$

P (4 or more faulty condensers) $= P(4) + P(5) + \ldots + P(100)$

$$= 1 - [P(0) + P(1) + P(2) + P(3)]$$

$$= 1 - \left[\frac{e^{-1}}{0!} + \frac{e^{-1}}{1!} + \frac{e^{-1}}{2!} + \frac{e^{-1}}{3!}\right] = 1 - e^{-1}\left[1 + 1 + \frac{1}{2} + \frac{1}{6}\right] = 1 - \frac{8}{3e} = 1 - 0.981 = 0.019$$ **Ans.**

Example 67. *An insurance company found that only 0.01% of the population is involved in a certain type of accident each year. If its 1000 policy holders were randomly selected from the population, what is the probability that not more than two of its clients are involved in such an accident next year? (given that $e^{0.1} = 0.9048$)*

Solution. $p = 0.01\% = \frac{1}{100} \times \frac{1}{100} = \frac{1}{10000}$, $n = 1000$

$$m = np = (1000) \times \frac{1}{10000} = \frac{1}{10} = 0.1$$

$$P(r) = \frac{e^{-m} m^r}{r!}$$

P (not more than 2) $= P$ (0, 1 and 2) $= P(0) + P(1) + P(2)$

$$= \frac{e^{-0.1}(0.1)^0}{0!} + \frac{e^{-0.1}(0.1)^1}{1!} + \frac{e^{-0.1}(0.1)^2}{2!} = e^{-0.1}\left(1 + 0.1 + \frac{0.01}{2}\right)$$

$$= 0.9048 \times 1.105 = 0.9998$$ **Ans.**

Example 68. *Fit a Poisson distribution to the set of observations:*

x	0	1	2	3	4
f	122	60	15	2	1

[MDU 2006 Dec. 2007, May 2008, RGPV Bhopal II Sem. Dec. 2007, June 2007]

Solution. The mean number $= \frac{\Sigma f \cdot x}{\Sigma f}$.

x	f	fx
0	122	0
1	60	60
2	15	30
3	2	6
4	1	4
Total	200	100

$$\text{Mean} = \frac{\Sigma f x}{\Sigma f} = \frac{100}{200} = \frac{1}{2}$$

x	$P(x)=\dfrac{e^{-1/2}(1/2)^x}{x!}$	Theoretical frequency	Given frequency
0	$P(0)=\dfrac{e^{-\frac{1}{2}}\left(\frac{1}{2}\right)^0}{1!}=0.6065$	0.6065 × 200 = 121.3	122
1	$P(1)=\dfrac{e^{-\frac{1}{2}}\left(\frac{1}{2}\right)^1}{1!}=\dfrac{0.6065}{2}=0.3033$	0.3033 × 200 = 60.7	60
2	$P(2)=\dfrac{e^{-\frac{1}{2}}\left(\frac{1}{2}\right)^2}{2!}=\dfrac{0.6065}{8}=0.0758$	0.0758 × 200 = 15.2	15
3	$P(3)=\dfrac{e^{-\frac{1}{2}}\left(\frac{1}{2}\right)^3}{3!}=\dfrac{0.6065}{48}=0.0126$	0.0216 × 200 = 2.5	2
4	$P(4)=\dfrac{e^{-\frac{1}{2}}\left(\frac{1}{2}\right)^4}{4!}=\dfrac{0.6065}{348}=0.0016$	0.0016 × 200 = 0.32	1

Ans.

Example 69. *A skilled typist, on routine work, kept a record of mistakes made per day during 300 working days.*

Mistakes per day	*0*	*1*	*2*	*3*	*4*	*5*	*6*
No. of days	*143*	*90*	*42*	*12*	*9*	*3*	*1*

Fit a Poisson distribution to the above data and hence calculate the theoretical frequencies.

Solution. The mean number of mistakes

$$= \frac{1}{300}\ (143 \times 0 + 90 \times 1 + 42 \times 2 + 12 \times 3 + 9 \times 4 + 3 \times 5 + 1 \times 6)$$

$$= \frac{1}{300}\ (90 + 84 + 36 + 36 + 15 + 6) = \frac{267}{300} = 0.89$$

Number of mistakes	Probability $P(r)=\dfrac{e^{-0.89}\times(0.89)^r}{r!}$	Theoretical frequency	Given frequency
0	$\dfrac{e^{-0.89}\times(0.89)^0}{0!}=0.411$	$0.411 \times 300 = 123.3 \approx 123$	143
1	$\dfrac{e^{-0.89}\times(0.89)^1}{1!}=0.365$	$0.365 \times 300 = 109.5 \approx 110$	90

Number of mistakes	Probability $P(r)=\dfrac{e^{-0.89}\times(0.89)^r}{r!}$	Theoretical frequency	Given frequency
2	$\dfrac{e^{-0.89}\times(0.89)^2}{2!}=0.163$	$0.163 \times 300 = 48.9 \approx 49$	42
3	$\dfrac{e^{-0.89}\times(0.89)^3}{3!}=0.048$	$0.048 \times 300 = 14.4 \approx 14$	12
4	$\dfrac{e^{-0.89}\times(0.89)^4}{4!}=0.011$	$0.011 \times 300 = 3.3 \approx 3$	9
5	$\dfrac{e^{-0.89}\times(0.89)^5}{5!}=0.002$	$0.002 \times 300 = 0.6 \approx 1$	3
6	$\dfrac{e^{-0.89}\times(0.89)^6}{6!}=0.0003$	$0.0003 \times 300 = 0.09 \approx 0$	1

Example 70. *Fit a Poisson distribution to the following data which gives the number of yeast cells per square for 400 squares.*

No. of cells per square (x)	*0*	*1*	*2*	*3*	*4*	*5*	*6*	*7*	*8*	*9*	*10*	*Total*
No. of squares (f)	*103*	*143*	*98*	*42*	*8*	*4*	*2*	*0*	*0*	*0*	*0*	*400*

It is given that $e^{-1.32} = 0.2674$

Solution.

x	0	1	2	3	4	5	6	7	8	9	10	Total
f	103	143	98	42	8	4	2	0	0	0	0	400
$f \cdot x$	0	143	196	126	32	20	12	0	0	0	0	529

$$m = \text{Mean} = \frac{\Sigma f \cdot x}{\Sigma f} = \frac{529}{400} = 1.32 = 1.32$$

But Poisson distribution is $P(x)=\dfrac{e^{-m}\cdot m^x}{x!}=\dfrac{e^{-1.32}(1.32)^x}{x!}$ or $P(x)=\dfrac{0.2674\,(1.32)^x}{x!}$

No. of cells	Probability $P(x)=\dfrac{0.2674\,(1.32)^x}{x!}$	Theoretical frequency	Given frequency
0	$\dfrac{0.2674\,(1.32)^0}{0!}=0.2674$	$0.257 \times 400 = 107$	103
1	$\dfrac{0.2674\,(1.32)^1}{1!}=0.353$	$0.353 \times 400 = 141$	143

No. of cells	Probability $P(x) = \frac{0.2674\,(1.32)^x}{x!}$	Theoretical frequency	Given frequency
2	$\frac{0.2674\,(1.32)^2}{2!} = 0.233$	$0.233 \times 400 = 93.2 \approx 93$	98
3	$\frac{0.2674\,(1.32)^3}{3!} = 0.1025$	$0.1025 \times 400 = 41$	42
4	$\frac{0.2674\,(1.32)^4}{4!} = 0.0338$	$0.0338 \times 400 = 13.52 \approx 14$	8
5	$\frac{0.2674\,(1.32)^5}{5!} = 0.00893$	$0.00893 \times 400 = 3.57 \approx 4$	4
6	$\frac{0.2674\,(1.32)^6}{6!} = 0.00196$	$0.00196 \times 400 = 0.784 \approx 1$	2
7	$\frac{0.2674\,(1.32)^7}{7!} = 0.00037$	$0.00037 \times 400 = 0.148 \approx 0$	0
8	$\frac{0.2674\,(1.32)^8}{8!} = 0.00006$	$0.00006 \times 400 = 0.024 \approx 0$	0
9	$\frac{0.2674\,(1.32)^9}{9!} = 0.00000897$	$0.00000897 \times 400 = 0.003588 \approx 0$	0
10	$\frac{0.2674\,(1.32)^{10}}{10!} = 0.00000118$	$0.00000118 \times 400 = 0.000472 \approx 0$	0

Example 71. *Data was collected over a period of 10 years, showing number of deaths from horse kicks in each of the 200 army corps. The distribution of deaths was as follows:*

No. of deaths	*0*	*1*	*2*	*3*	*4*	*Total*
Frequency	*109*	*65*	*22*	*3*	*1*	*200*

Fit a Poisson distribution to the above data and hence calculate the theoretical frequencies. **[MDU May 2009]**

Solution. Mean of given distribution $= \frac{\Sigma f x}{\Sigma f} = \frac{65 + 44 + 9 + 4}{200} = \frac{122}{200} = 0.61$

This is the parameter (m) of the Poisson distribution.

$\therefore$ Required Poisson distribution is $N \cdot \frac{m^r e^{-m}}{r!}$, where $N = \Sigma f = 200$

$$= 200e^{-0.61}\frac{(0.61)^r}{r!} = 200 \times 0.5433\,\frac{(0.61)^r}{r!} = 108.7 \times \frac{(0.61)^r}{r!}$$

r	*P (r)*	*Theoretical Frequency*	*Given Frequency*
0	108.7	109	109
1	$108.7 \times 0.61 = 66.3$	66	65
2	$108.7 \times \frac{(0.61)^2}{2!} = 20.2$	20	22
3	$108.7 \times \frac{(0.61)^3}{3!} = 4.1$	4	3
4	$108.7 \times \frac{(0.61)^4}{4!} = 0.7$	1	1
	Total	200	200

Ans.

EXERCISE 37.5

1. Find the probability that at most 5 defective fuses will be found in a box of 200 fuses if experience shows that 2 per cent of such fuses are defective. **Ans.** 0.785
2. The number of accidents during a year in a factory has the Poisson distribution with mean 1.5. The accidents during different years are assumed independent. Find the probability that only 2 accidents take place during 2 years time. **Ans.** 0.224
3. A manufacturer of cotter pins knows that 5% of his product is defective. If he sells cotter pins in boxes of 100 and guarantee that not more than 10 pins will be defective, what is the approximate probability that a box will fail to meet the guaranteed quality. [$e^{-0.5} = 0.006738$] **Ans.** 0.0136875
4. Suppose the number of telephone calls on an operator received from 9.00 to 9.05 follow a Poisson distribution with mean 3. Find the probability that

   (*i*) the operator will receive no calls in that time interval tomorrow,

   (*ii*) in the next three days the operator will receive a total of 1 call in that time interval. [$e^{-3} = 0.04978$] **Ans.** (*i*) e^{-3} (*ii*) $3 \times (e^{-3})^2\,(e^{-3} \cdot 3)$.
5. On the basis of past record it has been found that there is 70% chance of power cut in a city on any particular day. What is the probability that from the first to the 10th date of the month, there are 5 or more days without power cut.

   Ans. $\left[\frac{3^5}{5!} + \frac{3^6}{6!} + \frac{3^7}{7!} + \frac{3^8}{8!} + \frac{3^9}{9!} + \frac{3^{10}}{10!}\right] e^{-3}$
6. The distribution of typing mistakes committed by a typist is given below. Assuming a Poisson model, find out the expected frequencies.

Mistakes per page	0	1	2	3	4	5
No. of pages	142	156	69	27	5	1

Ans. 147, 147, 74, 25, 6, 1 pages.

7. Let x be the number of cars per minute passing a certain crossing of roads between 5.00 P.M. and 7.00 P.M. on a holiday. Assume x has a Poisson distribution with mean 4. Find the probability of observing atmost 3 cars during any given minute between 5.00 P.M. and 7 P.M. (given $e^{-4} = 0.0183$) **Ans.** 0.4331

8. Number of customers arriving at a service counter during a day has a Poisson distribution with mean 100. Find the probability that at least one customer will arrive on each day during a period of five days. Also find the probability that exactly 3 customers will arrive during two days. **Ans.** $(1 - e^{-100})^5, e^{-200} \times \frac{4\,(100)^3}{3}$

9. In a normal summer, a truck driver gets on an average on puncture in 1000 km. Applying Poisson distribution, find the probability that he will have
(*i*) no puncture, (*ii*) two punctures in a journey of 3000 kms. **Ans.** (*i*) e^{-3} (*ii*) $4.5\, e^{-3}$

10. Wireless sets are manufactured with 25 soldered joints each. On an average, 1 joint in 500 is defective. How many sets can be expected to be free from defective joints in a consignment of 10000 sets? **Ans.** 9512

11. In a certain factory turning out razor blades, there is small chance $\frac{1}{500}$ for any blade to be defective. The blades are supplied in packets of 10. Using Poisson's distribution, calculate the approximate number of packets containing (*i*) no defective (*ii*) one defective and (*iii*) two defective blades respectively in a consignment of 10,000 packets. ($e^{-0.02} = 0.9802$). **Ans.** (*i*) 9802 (*ii*) 196 (*iii*) 2

12. If m and μ_r denoted by the mean and central rth moment of a Poisson distribution, then prove that

$$\mu_{r+1} = r\,m\,\mu_{r-1} + m\,\frac{d\mu_r}{dm}. \qquad \left[\textbf{Hint. } \mu_r = \sum_{r=0}^{\infty} (x-m)^r \frac{e^{-m} m^x}{x!}, \text{ find } \frac{d\mu_r}{dm}\right]$$

13. A certain screw-making machine produces an average 2 defective screws out of 100, and pack them in boxes of 500. Find the probability that a box contains 15 defective screws. **[MDU Dec. 2005, AMIE Winter 2005]** **Ans.** 0.0347

14. The distribution of the number of road accidents per day in a city is Poisson with mean 4. Find the number of days out of 100 days when there will be:

(*i*) no accident (*ii*) at least 2 accidents

(*iii*) at most 3 accidents (*iv*) between 2 and 5 accident.

Ans. (*i*) 2 days (*ii*) 91 days (*iii*) 43 days (*iv*) 39 days.

15. Fill in the blanks:

(*a*) If a random variable x follows Poisson distribution such that $P(x = 1) = P(x = 2)$ then the mean of the distribution is .... **Ans.** 2

(*b*) Mean and variance of a Poisson distribution are .... **Ans.** equal

(*c*) If the probability of a defective fuse is 0.05, the variance for the distribution of defective fuses in a total of 40 is .... **Ans.** 2

(*d*) The probability of the king of hearts drawn from a pack of cards once in 52 trials is .... **Ans.** $\frac{1}{e}$

(*e*) If the standard deviation of the Poisson distribution is $\sqrt{2}$, the probability for $r = 2$ is .... **Ans.** $\frac{2}{e^2}$

(*f*) If x has a modified Poisson distribution

$P_k = P(x = k) = \frac{(e^m - 1)^{-1} m^k}{k!}$, $(k = 1, 2, 3, ...)$, then the expectation of x is .... **Ans.** $\frac{me^m}{e^m - 1}$

(*g*) If x has a poisson distribution such that $P(x = k) = P(x = k + 1)$ for some positive integer k then mean of x is .... **Ans.** $k + 1$

Choose the correct answer:

16. In the Poisson distribution if $P(x = k) = P(x = k + 1)$, then the mean is:

(*a*) k (*b*) $2k$ (*c*) $k + 1$ (*d*) $k - 1$

[RGPV Bhopal II Sem. June 2007] Ans. (*c*)

17. The value of measure of skewness of Poisson distribution is:

(*a*) m (*b*) $\sqrt{m}$ (*c*) $\frac{1}{m}$ (*d*) $\frac{1}{\sqrt{m}}$

[RGPV Bhopal II Sem. June 2016] Ans. (*d*)

18. Poisson distribution with unit mean, mean-deviation about the mean is:

(*a*) $\frac{1}{e}$ (*b*) $\frac{\sigma}{e}$ (*c*) $\frac{2\sigma}{e}$ (*d*) $\frac{2}{e}$

[RGPV Bhopal II Sem. Feb. 2006] Ans. (*d*)

19. In the Poisson distribution if $2P(x = 1) = P(x = 2)$, then the variance is:

(*a*) 0 (*b*) –1 (*c*) 4 (*d*) 2

[RGPV Bhopal II Sem. June 2007] Ans. (*c*)

20. Let X be a Poisson random variable, such that $2P(X = 0) = P(X = 2)$. Then standard deviation of x is

(*a*) 4 (*b*) 2 (*c*) $-\sqrt{2}$ (*d*) $\sqrt{2}$ **Ans.** (*d*)

21. A card is drawn from a well shuffled pack of cards. A sequence of 156 consecutive trials are made. Using Poisson distribution, the probability that the Queen of clubs will be drawn at least once is obtained as

(*a*) e^{-3} (*b*) $1 - e^{-3}$ (*c*) $e^{-\frac{1}{3}}$ (*d*) $1 - e^{-\frac{1}{3}}$ **Ans.** (*b*)

22. The random variable X has a Poisson distribution. If $P(x = 3) = \frac{1}{6}$, $P(x = 2) = \frac{1}{3}$, then $P(x = 0)$ is

(*a*) exp (– 3/2) (*b*) exp (3/2) (*c*) exp (– 3) (*d*) exp (– 1/2)

Ans. (*a*)

23. For the Poisson distribution if $P(x = 2) = \frac{2}{3}\ P(x = 1)$, then mean of Poisson's distribution will be

(*a*) 0 (*b*) $\frac{4}{3}$ (*c*) $\frac{3}{4}$ (*d*) None of these

Ans. (*b*)

24. In Poisson distribution standard deviation is equal to ....

(*a*) Mean (*b*) Square of mean

(*c*) Square root of mean (*d*) Variance

[Bihar SBTE, 2012] **Ans.** (*c*)

25. Which result is true for Poisson distribution?

(*a*) Mean = Variance (*b*) Mean = $\sqrt{\text{Variance}}$

(*c*) Mean = Standard deviation (*d*) Mean = $\sqrt{\text{Standard deviation}}$

[Bihar SBTE. 2011] **Ans.** (*a*)

26. Standard deviation of Poisson Distribution is equal to ....

(*a*) $\sqrt{np}$ (*b*) np (*c*) npq (*d*) $\sqrt{npq}$

[Bihar SBTE. 2011] **Ans.** (*d*)

27. Mean and Variance of the Poisson Distribution are ....

(*a*) Unequal (*b*) Equal

(*c*) Both (a) and (b) (*d*) None of these

[Bihar SBTE. 2010] **Ans.** (*a*)

37.14 NORMAL DISTRIBUTION (GAUSSIAN DISTRIBUTION)

Normal distribution is a continuous distribution. It is derived as the limiting form of the Binomial distribution for large values of n where neither p nor q is very small.

The normal distribution is given by the equation

$$f(x) = \frac{1}{\sigma\sqrt{(2\pi)}}\, e^{-\frac{(x-\mu)^2}{2\sigma^2}} \qquad \text{... (1)}$$

where μ = mean, σ = standard deviation, π = 3.14159 ..., [e = 2.71828 ...]

$$P(x_1 < x < x_2) = \int_{x_1}^{x_2} \frac{1}{\sigma\sqrt{(2x)}}\, e^{-\frac{(x-\mu)^2}{2\sigma^2}}\, dx$$

On substitution $z = \frac{x-\mu}{\sigma}$ in (1), we get $f(z) = \frac{1}{\sqrt{2\pi}}\, e^{-\frac{1}{2}z^2}$ ... (2)

Here mean = 0, standard deviation = 1.

Equation (2) is known as standard form of normal distribution.

Theorem. To derive normal distribution as a limiting case of Binomial distribution where $p \neq q$ but $p \approx q$. **[U.P. III Sem. Dec. 2006]**

Statement. The limiting case of Binomial Distribution $(p + q)^n$, as $n \to \infty$ and neither p nor q are very small, generates the Normal Distribution.

Proof. The frequencies for r and $(r + 1)$ successes in Binomial distribution are

$$f(r) = N \cdot {}^nC_r\, p^r\, q^{n-r} \text{ and } f(r+1) = N \cdot {}^nC_{r+1}\, p^{r+1}\, q^{n-(r+1)}$$

The frequency of r successes > frequency of $(r + 1)$ successes if

$$f(r) > f(r+1) \quad \Rightarrow \quad \frac{f(r)}{f(r+1)} > 1$$

$$\Rightarrow \quad \frac{N \cdot {}^nC_r\, p^r\, q^{n-r}}{N \cdot {}^nC_{r+1}\, p^{r+1}\, q^{n-r-1}} > 1 \quad \Rightarrow \quad \frac{\dfrac{n!}{r!\,(n-r)!} \cdot p^r \cdot q^{n-r}}{\dfrac{n!}{(r+1)!\,(n-r-1)!}\, p^{r+1}\, q^{n-r-1}} > 1$$

$$\Rightarrow \quad \frac{n!\, p^r \cdot q^{n-r}\,(r+1)!\,(n-r-1)!}{r!\,(n-r)! \cdot n!\, p^{r+1} \cdot q^{n-r-1}} > 1$$

$$\Rightarrow \quad \frac{q \cdot (r+1)}{(n-r)\,p} > 1 \quad \Rightarrow \quad qr + q > np - pr$$

$$\Rightarrow \quad q > np - r\,(p+q)$$

$$\Rightarrow \quad r > np - q \qquad \text{... (1)}$$

Again, similarly the frequency of r successes > the frequency of $(r - 1)$ successes if

$$f(r) > f(r-1) \quad \Rightarrow \quad \frac{f(r)}{f(r-1)} > 1$$

$$\Rightarrow \quad \frac{N\, {}^nC_r\, p^r\, q^{n-r}}{N \cdot {}^nC_{r-1}\, p^{r-1}\, q^{n-(r-1)}} > 1 \quad \Rightarrow \quad \frac{\dfrac{n!}{r!\,(n-r)!} \cdot p^r \cdot q^{n-r}}{\dfrac{n!}{(r-1)!\,(n-r+1)!}\, p^{r-1}\, q^{n-r+1}} > 1$$

$$\Rightarrow \quad \frac{n!\, p^r\, q^{n-r}\,(r-1)!\,(n-r+1)!}{r!\,(n-r)!\, n!\, p^{r-1}\, q^{n-r+1}} > 1 \quad \Rightarrow \quad \frac{p(n-r+1)}{rq} > 1$$

$$\Rightarrow \quad pn - pr + p > rq \quad \Rightarrow \quad pn + p > pr + qr$$

$$\Rightarrow \quad pn + p > r\,(p+q) \quad \Rightarrow \quad pn + p > r \qquad \text{... (2)}$$

$$[\because p + q = 1]$$

from (1) and (2), we have

$$pn + p > r > np - q$$

$$pn + p + q > r > np$$

$$np + 1 > r > np$$

Since a possible value of r is np, therefore, without loss of generality we can assume that np is an integer as $n \to \infty$. Hence the frequency of np successes can be assumed to be maximum frequency. Let y_0 be the frequency of np successes and y_x be the frequency of $(np + x)$ successes.

Then

$$y_0 = f(np) = N \cdot {}^nC_{np}\, p^{np}\, q^{n-np} \qquad \text{[from (1), for r = } np\text{]}$$

$$= N \frac{n!}{(np)!\,(n-np)!}\, p^{np}\, q^{n-np}$$

$$= N \frac{n!}{(np)!\,(nq)!} p^{np} q^{n-np} \qquad \text{... (3)} \quad [\because q = 1-p]$$

and
$$y_x = N \cdot \frac{n!}{(np+x)!\,(nq-x)!} p^{np+x} q^{nq-x} \qquad \text{... (4)}$$

Dividing (4) by (3), we get

$$\frac{y_x}{y_0} = \frac{(np)!\,(nq)!}{(np+x)!\,(nq-x)!} p^x q^{-x} \qquad \text{... (5)}$$

For n being large, then according to James Stirling's approximation formula for factorials, we have

$$n! = e^{-n} h^{n+1/2} \sqrt{(2\pi)},$$

From (5)
$$\frac{y_x}{y_0} = \frac{e^{-np} (np)^{np+1/2} \sqrt{2\pi}\, e^{-nq} (nq)^{nq+1/2} \sqrt{2\pi}\, p^x q^{-x}}{e^{-(np+x)} (np+x)^{np+x+1/2} \sqrt{2\pi}\, e^{-(nq-x)} (nq-x)^{nq-x+1/2} \sqrt{2\pi}}$$

$$= \frac{(np)^{np+1/2} (nq)^{nq+1/2} (nq/nq)^x}{(np)^{np+x+1/2} \left\{1 + \frac{x}{np}\right\}^{np+x+1/2} (np)^{nq-x+1/2} \left\{1 - \frac{x}{nq}\right\}^{nq-x+1/2}}$$

$$= \frac{1}{\left\{1 + \frac{x}{np}\right\}^{np+x+1/2} \left\{1 - \frac{x}{nq}\right\}^{nq+x+1/2}}$$

$$\therefore \quad \log\left(\frac{y_x}{y_0}\right) = -\left(np + x + \frac{1}{2}\right) \log\left(1 + \frac{x}{np}\right) - \left(nq - x + \frac{1}{2}\right) \log\left(1 - \frac{x}{nq}\right)$$

$$= -\left(np + x + \frac{1}{2}\right)\left(\frac{x}{np} - \frac{x^2}{2n^2p^2} + \frac{x^3}{3n^3p^3} - \ldots\right)$$

$$+ \left(nq - x + \frac{1}{2}\right)\left(\frac{x}{nq} + \frac{x^2}{2n^2q^2} + \frac{x^3}{3n^3q^3} + \ldots\right)$$

$$= x\left(1 - \frac{1}{2np} + 1 + \frac{1}{2np}\right) + x^2\left(\frac{1}{2np} - \frac{1}{np} + \frac{1}{4n^2p^2} + \frac{1}{2nq} - \frac{1}{nq} + \frac{1}{4n^2q^2}\right)$$

$$+ x^3\left(\frac{1}{3n^2q^2} + \frac{1}{6n^3q^3} - \frac{1}{2n^2q^2} - \frac{1}{2n^2p^2} - \frac{1}{3n^2q^2} + \frac{1}{6n^3p^3}\right) - \ldots$$

$$= \frac{p-q}{2npq} x + \frac{p^2+q^2}{4n^2p^2q^2} x^2 - \frac{x^2}{2npq} + \ldots + \text{terms of higher orders.}$$

Neglecting terms containing $1/n^2$, we have

$$\therefore \quad \log\left(\frac{y_x}{y_0}\right) = -\frac{q-p}{2npq} x - \frac{x^2}{2npq}$$

Since $p < 1$, $q < 1$ and so $q - p$ is very small as compared to n. Therefore 1st term may be neglected. $(q - p = 0)$.

$$\therefore \quad \log\left(\frac{y_x}{y_0}\right) = -\frac{x^2}{2npq} = -\frac{x^2}{2\sigma^2} \qquad [\because \sigma^2 = npq, \text{ the variance of Binomial distribution}]$$

$$\Rightarrow \qquad y_x = y_0\, e^{-x^2/2\sigma^2}$$

Proved.

Example 72. *In a normal distribution, 31% of the items are under 45 and 8% are over 64. Find the mean and standard deviation of the distribution.*

[AMIETE Dec. 2010, AKTU 2014-15]

Solution. Let $\bar{x}$ be the mean and σ the S.D.

If $x = 45$, $\quad z = \dfrac{45 - \bar{x}}{\sigma}$

If $x = 64$, $\quad z = \dfrac{64 - \bar{x}}{\sigma}$

Area between 0 and $z = \dfrac{45 - \bar{x}}{\sigma} = 0.50 - 0.31 = 0.19$

[From the table, for the area 0.19, $z = 0.496$]

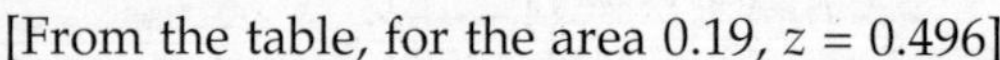

$$\frac{45 - \bar{x}}{\sigma} = -0.496 \qquad \text{... (1)}$$

Area between $z = 0$ and $z = \dfrac{64 - \bar{x}}{\sigma} = 0.5 - 0.08 = 0.42.$

(From the table, for area 0.42, $z = 1.4050$)

$$\frac{64 - \mu}{\sigma} = 1.405 \qquad \text{... (2)}$$

Solving (1) and (2) we get $\mu = 50$, $\sigma = 10$. **Ans.**

Example 73. *The income of a group of 10,000 persons was found to be normally distributed with mean ₹ 750 p.m. and standard deviation of ₹ 50. Show that, of this group, about 95% had income exceeding ₹ 668 and only 5% had income exceeding ₹ 832. Also find the lowest income among the richest 100.*

Solution. Mean = μ = 750

Standard deviation = σ = 50

and $\quad z = \dfrac{x - \mu}{\sigma}$

(*i*) If $x_1 = 668$, then $\quad z = \dfrac{668 - 750}{50} = -1.64$

$$\begin{aligned} P(x_1 > 668) &= P(z_1 > -1.64) \\ &= 0.5 + P(-1.64 \le z \le 0) \\ &= 0.5 + P(0 \le z < 1.64) \\ &= 0.5 + 0.4495 \\ &= 0.9495 \end{aligned}$$

$\therefore$ Percentage of persons having income exceeding ₹ 668 = 94.95% ≈ 95%

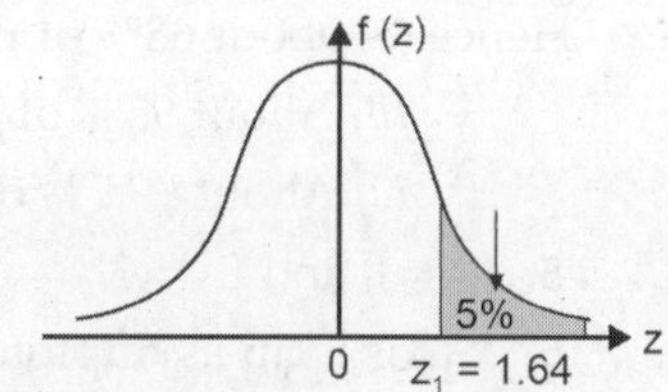

(*ii*) If $x = 832$, then $$z = \frac{832 - 750}{50} = 1.64$$

$$P(x_2 > 832) = P(z_2 > 1.64)$$
$$= 0.5 - 0.4495$$
$$= 0.0505$$

$\therefore$ Percentage of persons having income exceeding ₹ 832 = 5.05% ≈ 5%

(*iii*) Let x be the lowest income among the richest 100 persons.

100 persons = 1% of 10,000

100 persons represents 1% area under the curve on the right hand side.

Thus the area between 0 and z

$$= 0.5 - 0.01 = 0.49$$

From the table z for area 0.49 is 2.33

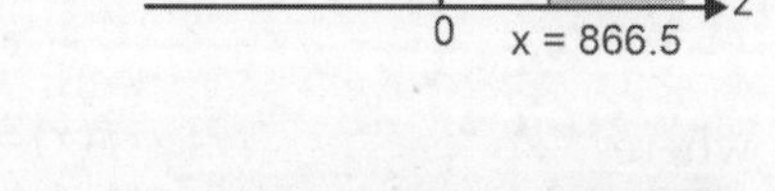

$$z = \frac{x - \mu}{\sigma}$$

$$\Rightarrow \quad 2.33 = \frac{x - 750}{50} \quad \Rightarrow \quad x - 750 = 50 \times 2.33$$

$$\Rightarrow \quad x - 750 = 116.5 \quad \Rightarrow \quad x = 866.5$$

Hence, the minimum income among the 100 richest persons is equal to ₹ 866.5. **Ans.**

37.15 NORMAL CURVE

A normal curve shows Binomial distribution graphically of a continuous random variable. The probabilities of heads in 10 tosses are ${}^{10}C_0\, q^{10}\, p^0,\ {}^{10}C_1\, q^9\, p^1,\ {}^{10}C_2\, q^8\, p^2,\ {}^{10}C_3\, q^7\, p^3,\ {}^{10}C_4\, q^6\, p^4,\ {}^{10}C_5\, q^5\, p^5,\ {}^{10}C_6\, q^4\, p^6,\ {}^{10}C_7\, q^3\, p^7,\ {}^{10}C_8\, q^2\, p^8,\ {}^{10}C_9\, q^1\, p^9,\ {}^{10}C_{10}\, q^0\, p^{10}$.

$p = \frac{1}{2}, q = \frac{1}{2}$. It is shown in the figure given below.

If the variates (heads here) are treated as if they were continuous, the required probability curve will be a *normal curve* as shown in the given figure by dotted lines.

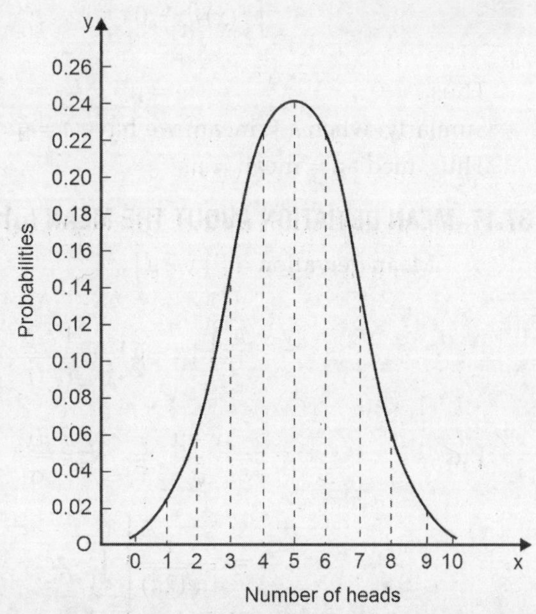

Properties of the normal curve

$$y = \frac{1}{\sigma\sqrt{2\pi}}\, e^{-\frac{(x-\mu)^2}{2\sigma^2}}$$

1. The curve is symmetrical about the line $x = \mu$.
2. The mean, median and mode coincide.
3. y decreases rapidly as x increases numerically. The curve extends to infinity on either side of the origin.
4. (*a*) $P(\mu - \sigma < x < \mu + \sigma) = 0.6826$

   (*b*) $P(\mu - 2\sigma < x < \mu + 2\sigma) = 0.9544$

   (c) $P(\mu - 3\sigma < x < \mu + 3\sigma) = 0.9973$

Hence (*a*) about 68% of the values lie between $(\mu - \sigma)$ and $(\mu + \sigma)$

(*b*) About 95% of the values lie between $(\mu - 2\sigma)$ and $(\mu + 2\sigma)$.

(*c*) About 99.7% of the values will be between $(\mu - 3\sigma)$ and $(\mu + 3\sigma)$.

5. $\beta_1 = 0$ and $\beta_2 = 3$.

6. *x*-axis is an asymptote to the curve. No portion of the curve lies below the *x*-axis.

7. The points of inflexion are $x = \mu \pm \sigma$.

8. Mean deviation about mean $\approx \frac{4}{5}\sigma$ and quartile deviation $\approx \frac{2}{3}\sigma$.

37.16 MEDIAN OF THE NORMAL DISTRIBUTION

If *a* is the median, then it divides the total area into two equal halves so that

$$\int_{-\infty}^{a} f(x)\,dx = \frac{1}{2} = \int_{a}^{\infty} f(x)\,dx$$

where $$f(x) = \frac{1}{\sigma\sqrt{(2\pi)}} e^{-\frac{(x-\mu)^2}{2\sigma^2}}$$

Suppose Median $a >$ mean μ then

$$\int_{-\infty}^{\mu} f(x)\,dx + \int_{\mu}^{a} f(x)\,dx = \frac{1}{2}$$

$$\frac{1}{2} + \int_{\mu}^{a} f(x)\,dx = \frac{1}{2} \qquad \left[\text{but } \int_{-\infty}^{\mu} f(x)\,dx = \frac{1}{2}\right]$$

$$\int_{\mu}^{a} f(x)\,dx = 0$$

Thus $$a = \mu$$

Similarly, when $a <$ mean, we have $a = \mu$.

Thus, median = mean = μ.

37.17 MEAN DEVIATION ABOUT THE MEAN (μ) **[U.P. III Sem. Dec. 2009]**

Mean deivation $= E\,|x - \mu|$

$$= \int_{-\infty}^{\infty} |x-\mu| \frac{1}{\sigma\sqrt{(2\pi)}} e^{-\frac{(x-\mu)^2}{2\sigma^2}}\,dx$$

Put $$z = \frac{x-\mu}{\sigma} \Rightarrow dz = \frac{dx}{\sigma}$$

$$= \sigma\frac{1}{\sqrt{(2\pi)}}\left[\int_{-\infty}^{0} -ze^{\frac{-z^2}{2}}\,dz + \int_{0}^{\infty} ze^{\frac{-z^2}{2}}\,dz\right] = \frac{2\sigma}{\sqrt{(2\pi)}}\int_0^{\infty} ze^{\frac{-z^2}{2}}\,dz$$

Put $$\frac{z^2}{2} = t \Rightarrow zdz = dt$$

$$= \sigma\sqrt{\frac{2}{\pi}} \int_0^\infty e^{-t}\, dt = \sigma\sqrt{\frac{2}{\pi}}\left[-e^{-t}\right]_0^\infty$$

$$= \sigma\sqrt{\frac{2}{\pi}}[-0+1]$$

$$\text{Mean deviation} = \sigma\sqrt{\frac{2}{\pi}} \simeq \frac{4}{5}\sigma$$

37.18 MODE OF THE NORMAL DISTRIBUTION

We know that mode is the value of the variate x for which $f(x)$ is maximum. Thus, by differential calculus $f(x)$ is maximum if $f'(x) = 0$ and $f''(x) < 0$

where $$f(x) = \frac{1}{\sigma\sqrt{(2\pi)}} e^{-\frac{(x-\mu)^2}{2\sigma^2}}$$

Clearly $f(x)$ will be maximum when the exponent will be maximum which will be the case

$$\frac{(x-\mu)}{2\sigma^2} = 0 \Rightarrow (x-\mu)^2 = 0 \Rightarrow x = \mu$$

Thus mode is μ, and modal ordinate $= \dfrac{1}{\sigma\sqrt{(2\pi)}}$

37.19 MOMENTS OF NORMAL DISTRIBUTION

$$\mu_{2n+1} = \int_{-\infty}^{\infty} (x-\mu)^{2n+1} f(x)\, dx$$

$$= \frac{1}{\sigma\sqrt{(2\pi)}} \int_{-\infty}^{\infty} (x-\mu)^{2n+1} e^{-\frac{(x-\mu)^2}{2\sigma^2}} dx$$

$$= \frac{1}{\sqrt{(2\pi)}} \int_{-\infty}^{\infty} (\sigma z)^{2n+1} e^{-\frac{z^2}{2}} dz \qquad \left[z = \frac{x-\mu}{\sigma}\right]$$

$$= \frac{\sigma^{2n+1}}{\sqrt{(2\pi)}} \int_{-\infty}^{\infty} z^{2n+1} e^{-\frac{z^2}{2}} dz = 0 \qquad \left(\text{since } z^{2n+1} e^{-\frac{z^2}{2}} \text{ is an odd function}\right)$$

$$\mu_{2n} = \int_{-\infty}^{\infty} (x-\mu)^{2n} f(x)\, dx$$

$$= \frac{1}{\sqrt{(2\pi)}} \int_{-\infty}^{\infty} (\sigma z)^{2n} e^{-\frac{z^2}{2}} dz = \frac{\sigma^{2n}}{\sqrt{2\pi}} \int_{-\infty}^{\infty} z^{2n} e^{-\frac{z^2}{2}} dz$$

$$= \frac{2\sigma^{2n}}{\sqrt{(2\pi)}} \int_0^\infty z^{2n} e^{-\frac{z^2}{2}} dz \qquad \left[z^{2n} \cdot e^{-\frac{z^2}{2}} \text{ is an even function}\right]$$

$$= \frac{2\sigma^{2n}}{\sqrt{(2\pi)}} \int_0^\infty (2t)^n e^{-t} \frac{1}{\sqrt{2}} t^{-\frac{1}{2}} dt \qquad \left[\frac{z^2}{2} = t \Rightarrow dz = \frac{1}{\sqrt{2}} t^{-\frac{1}{2}} dt\right]$$

$$= \frac{2^n \sigma^{2n}}{\sqrt{\pi}} \int_0^\infty t^{\left(n+\frac{1}{2}-1\right)} e^{-t} dt = \frac{2^n \sigma^{2n}}{\sqrt{\pi}} \int_0^\infty e^{-t} t^{\left(n-\frac{1}{2}\right)} dt$$

$$= \frac{2^n \sigma^{2n}}{\sqrt{\pi}} \overline{\left| n+\frac{1}{2} \right.}$$

Changing n to $(n-1)$, we get

$$\mu_{2n-2} = \frac{2^{n-1} \sigma^{2n-2}}{\sqrt{\pi}} \overline{\left| n-\frac{1}{2} \right.}$$

On dividing, we get

$$\frac{\mu_{2n}}{\mu_{2n-2}} = 2\sigma^2 \frac{\overline{\left| n+\frac{1}{2} \right.}}{\overline{\left| n-\frac{1}{2} \right.}} = \frac{2\sigma^2 \left(n-\frac{1}{2}\right) \overline{\left| n-\frac{1}{2} \right.}}{\overline{\left| n-\frac{1}{2} \right.}} = 2\sigma^2 \left(n-\frac{1}{2}\right)$$

$$\mu_{2n} = \sigma^2 (2n-1) \mu_{2n-2}$$

which gives the recurrence relation for the moments of normal distribution

$$\mu_{2n} = [(2n-1)\sigma^2][(2n-3)\sigma^2]\mu_{2n-4}$$

$$= [(2n-1)\sigma^2][(2n-3)\sigma^2][(2n-5)\sigma^2]\mu_{2n-6}$$

$$= [(2n-1)\sigma^2][(2n-3)\sigma^2][(2n-5)\sigma^2] \ldots\ldots (3\sigma^2)(1 \cdot \sigma^2)\mu_0$$

$$= (2n-1)(2n-3)(2n-5) - \ldots\ldots 1 \cdot \sigma^{2n} \qquad (\mu_0 = 1)$$

$$= 1 \cdot 3 \cdot 5 \cdot 7 \ldots\ldots (2n-5)(2n-3)(2n-1)\sigma^{2n}$$

37.20 MOMENT GENERATING FUNCTION OF NORMAL DISTRIBUTION

Normal distribution function is given by

$$f(x) = \frac{1}{\sigma\sqrt{2\pi}} e^{-\frac{(x-\mu)^2}{2\sigma^2}} \qquad [-\infty < x < \infty]$$

Moment generating function about the origin $= M_X(t)$

$$= \int_{-\infty}^{\infty} e^{tx} \left(\frac{1}{\sigma\sqrt{2\pi}}\right) e^{-\frac{1}{2}\left(\frac{x-\mu}{\sigma}\right)^2} dx \qquad \ldots (1)$$

On putting $\frac{x-\mu}{\sigma} = z$ so that $dx = \sigma\, dz$ in (1), we get

$$M_X(t) = \frac{1}{\sigma\sqrt{2\pi}} \int_{-\infty}^{\infty} e^{t(\sigma z+\mu)} e^{-\frac{z^2}{2}} \sigma\, dz$$

$$= \frac{\sigma e^{\mu t}}{\sigma\sqrt{2\pi}} \int_{-\infty}^{\infty} e^{t\sigma z} \cdot e^{-\frac{z^2}{2}} dz = \frac{e^{\mu t}}{\sqrt{2\pi}} \int_{-\infty}^{\infty} e^{-\frac{1}{2}(z^2 - 2t\sigma z)} dz$$

$$= \frac{e^{\mu t}}{\sqrt{2\pi}} \int_{-\infty}^{\infty} e^{-\frac{1}{2}[z^2 - 2t\sigma z + t^2\sigma^2] + \frac{1}{2}t^2\sigma^2} dz$$

$$= \frac{e^{\mu t + \frac{1}{2}t^2\sigma^2}}{\sqrt{2\pi}} \int_{-\infty}^{\infty} e^{-\frac{1}{2}(z - t\sigma)^2} dz \qquad \text{...(2)}$$

On putting $\frac{1}{2}(z - t\sigma)^2 = y^2$ in (2) so that

$$(z - t\sigma)\, dz = 2y\, dy$$

$$\Rightarrow \quad \sqrt{2}\, y dz = 2y\, dy$$

$$\Rightarrow \quad dz = \sqrt{2}\, dy$$

$$M_X(t) = \frac{e^{\mu t + \frac{1}{2}t^2\sigma^2}}{\sqrt{2\pi}} \int_{-\infty}^{\infty} e^{-y^2} \sqrt{2}\, dy = \frac{e^{\mu t + \frac{1}{2}t^2\sigma^2}}{\sqrt{\pi}} \int_{-\infty}^{\infty} e^{-y^2} dy$$

$$= e^{\mu t + \frac{1}{2}t^2\sigma^2} \frac{1}{\sqrt{\pi}}(\sqrt{\pi}) = e^{\mu t + \frac{1}{2}t^2\sigma^2} \qquad \left[\because \int_{-\infty}^{\infty} e^{-y^2} dy = \sqrt{\pi}\right]$$

37.21 AREA UNDER THE NORMAL CURVE

By taking $z = \frac{x - \bar{x}}{\sigma}$, the standard normal curve is formed.

The total area under this curve is 1. The area under the curve is divided into two equal parts by $z = 0$. Left hand side area and right hand side area to $z = 0$ is 0.5. The area between the ordinate $z = 0$ and any other ordinate can be noted from the table–1 on last page of the chapter.

Example 74. *In mathematics final examination, if the mean was 72, and the standard deviation was 15. Determine the standard scores of students receiving grades.*

(*a*) 60 (*b*) 93 (*c*) 72

Solution. Here, $\bar{x} = 72, \sigma = 15$

$$(a)\ z = \frac{x - \bar{x}}{\sigma} = \frac{60 - 72}{15} = -0.8 \quad (b)\ z = \frac{93 - 72}{15} = 1.4 \quad (c)\ z = \frac{72 - 72}{15} = 0$$ **Ans.**

Example 75. *Find the area under the normal curve in each of the cases*

(*a*) $z = 0$ and $z = 1.2$ (*b*) $z = -0.68$ and $z = 0$

(*c*) $z = -0.46$ and $z = 2.21$ (*d*) $z = 0.81$ and $z = 1.94$

(*e*) to the left of $z = -0.6$ (*f*) to the right of $z = -1.28$.

Solution. See Table 1, last page of the chapter.

(*a*) Area between $z = 0$ and $z = 1.2$
$= .3849$ **Ans.**

(*b*) Area between $z = 0$ and $z = -0.68$
$= 0.2518$ **Ans.**

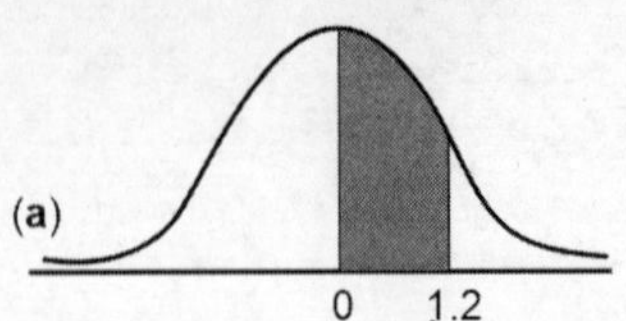

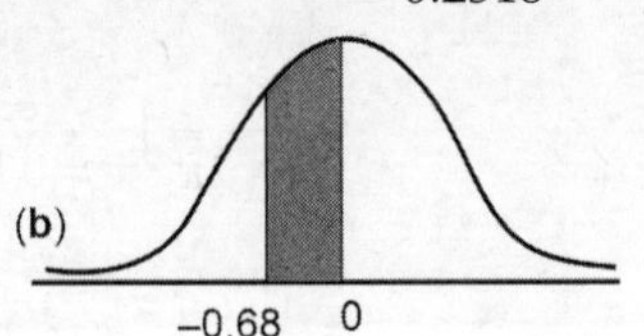

(*c*) Required area = (Area between $z = 0$ and $z = 2.21$)
+ (Area between $z = 0$ and $z = -0.46$)
= (Area between $z = 0$ and $z = 2.21$)
+ (Area between $z = 0$ and $z = 0.46$)
$= 0.4865 + 0.1772 = 0.6637.$ **Ans.**

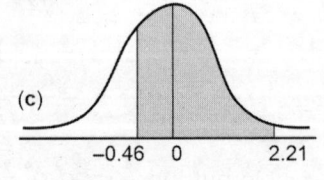

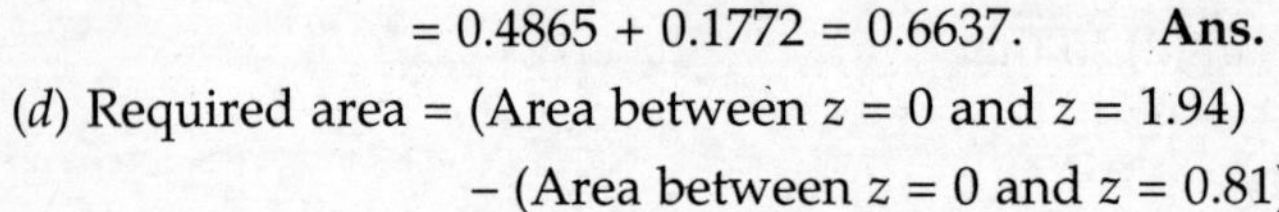

(*d*) Required area = (Area between $z = 0$ and $z = 1.94$)
− (Area between $z = 0$ and $z = 0.81$)
$= 0.4738 - 0.2910 = 0.1828$ **Ans.**

(*e*) Required area = 0.5 − (Area between $z = 0$ and $z = -0.6$)
$= 0.5 - 0.2257 = 0.2743$ **Ans.**

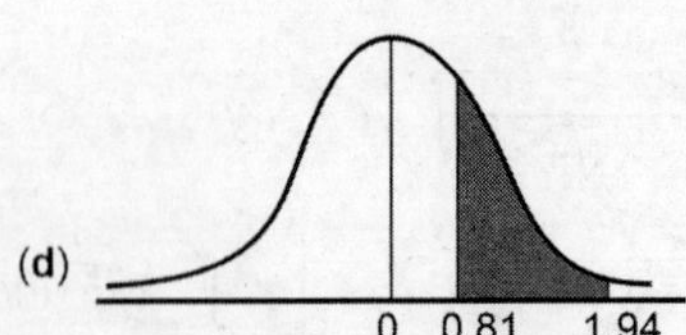

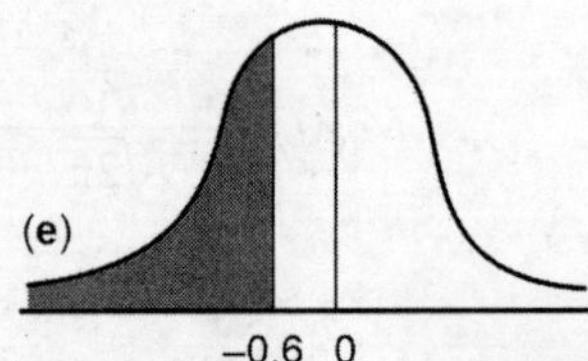

(*f*) Required area = (Area between $z = 0$ and $z = -1.28$) + 0.5
$= 0.3997 + 0.5$
$= 0.8997.$ **Ans.**

Example 76. *Find the value of z in each of the cases*
(*a*) *Area between 0 and z is 0.3770*
(*b*) *Area to the left of z is 0.8621*

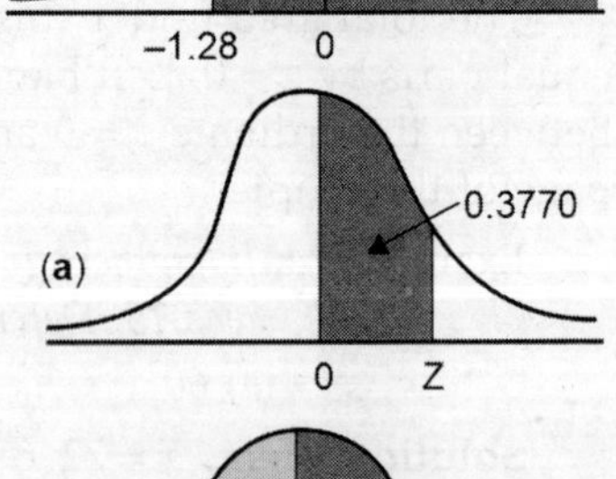

Solution.

(*a*) $z = \pm 1.16$

(*b*) Since the area is greater than 0.5.
Area between 0 and z.
$= 0.8621 - 0.5 = 0.3621$
from the Table $z = 1 + 0.09 = 1.09$ **Ans.**

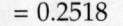

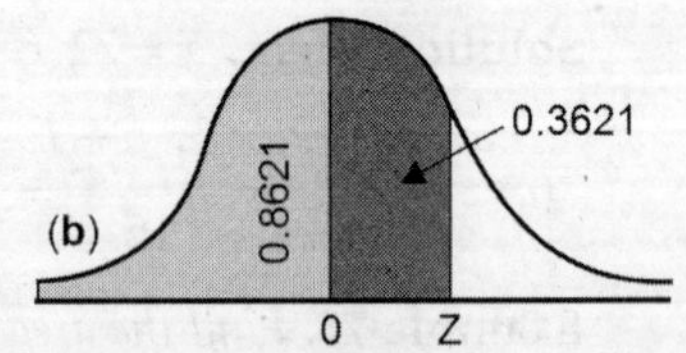

Example 77. *Students of a class were given an aptitude test their marks were found to be normally distributed with mean 60 and standard deviation 5. What percentage of students scored more than 60 marks?*

Solution. $x = 60, \bar{x} = 60, \sigma = 5$

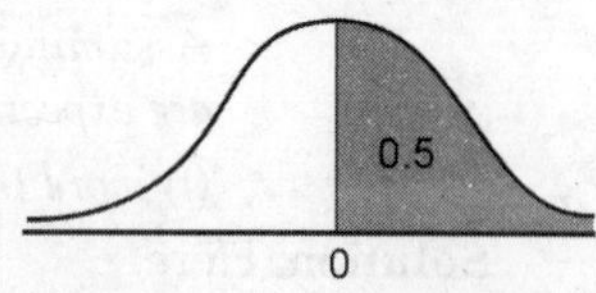

$$z = \frac{x - \bar{x}}{\sigma} = \frac{60 - 60}{5} = 0$$

if $x > 60$ then $z > 0$

Area lying to the right of $z = 0$ is 0.5.

The percentage of students getting more than 60 marks = 50% **Ans.**

Example 78. *Assume mean height of soldiers to be 68.22 inches with a variance of 10.8 inches square. How many soldiers in a regiment of 1,000 would you expect to be over 6 feet tall, given that the area under the standard normal curve between x = 0 and x = 0.35 is 0.1368 and between x = 0 and x = 1.15 is 0.3746.*

Solution. Mean = $\bar{x}$ = 68.22 inch

Variance = σ^2 = 10.8 inches squares

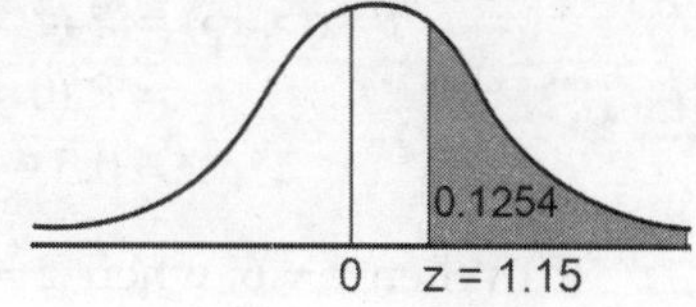

If $x = 72$ inches then $z = \frac{x - \bar{x}}{\sigma} = \frac{72 - 68.22}{\sqrt{10.8}} = 1.15$

$$P(x > 72) = P(z > 1.15)$$
$$= 0.5 - P(0 \le z \le 1.15)$$
$$= 0.5 - 0.3746 = 0.1254$$

Number of soldiers = 1000 × 0.1254 = 125.4 = 125 **Ans.**

Example 79. *If the height of 300 students are normally distributed with mean 64.5 inches and standard deviation 3.3 inches, find the height below which 99% of the student lie.*

Solution. Mean ($\bar{x}$) = 64.5 inches

S.D. = σ = 3.3 inches

$$z = \frac{x - \bar{x}}{\sigma}$$

$$\Rightarrow \quad z = \frac{x - 64.5}{3.3} \qquad \text{... (1)}$$

Area = 0.99 – 0.5 = 0.49

From the table, z for area 0.49 is 2.327.

Putting the value of z in (1), we get

$$\Rightarrow \quad \frac{x - 64.5}{3.3} = 2.327$$

$$\Rightarrow \quad x - 64.5 = 3.3 \times 2.327$$

$$x - 64.5 = 7.68$$

$$\Rightarrow \quad x = 7.68 + 64.5$$
$$= 72.18 \text{ inches}$$

Hence 99% students are of height less than 72.18 inches. **Ans.**

Example 80. *A sample of 100 dry battery cells tested to find the length of life produced the following results:*

$\bar{x}$ = 12 hours, σ = 3 hours

Assuming the data to be normally distributed, what percentage of battery cells are expected to have life

(i) more than 15 hours (ii) less than 6 hours (iii) between 10 and 14 hours?

Solution. Here, Mean = $\bar{x}$ = 12 hours

and Standard deviation = σ = 3 hours

x denotes the length of life of dry battery cells.

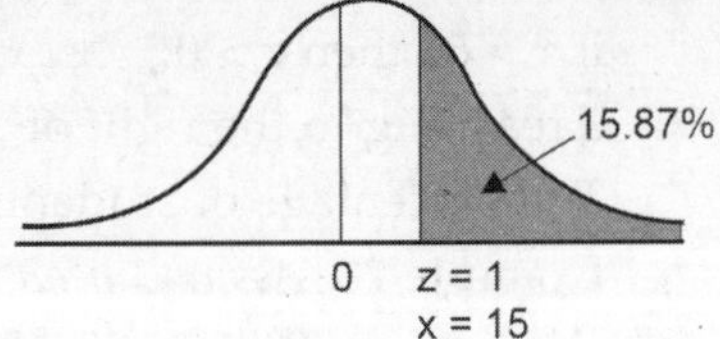

$$z = \frac{x - \bar{x}}{\sigma}$$

(*i*) When $x = 15$, then $z = \frac{15 - 12}{3} = 1$

$\therefore \quad P(x > 15) = P(z > 1)$

$= P(0 < z < \infty) - P(0 < z < 1)$

$= 0.5 - 0.3413 = 0.1587 = 15.87\%$

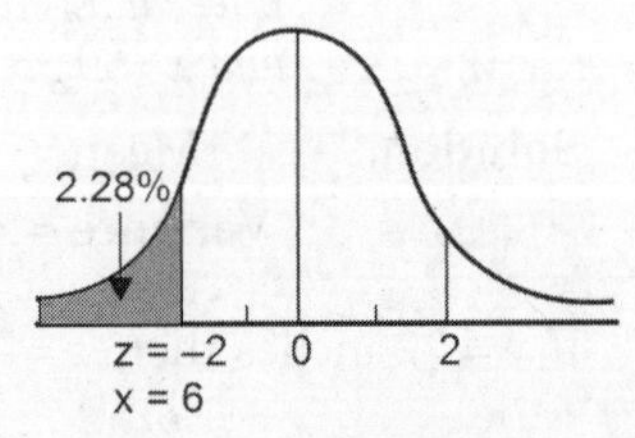

(*ii*) When $x = 6$, when $z = \frac{6 - 12}{3} = \frac{-6}{3} = -2$

$P(x < 6) = P(z < -2)$

$= P(z > 2) = 0.5 - P(0 < z < 2)$

$= 0.5 - 0.4772 = 0.0228 = 2.28\%$

(*iii*) When $x = 10$, then $z = \frac{10 - 12}{3} = \frac{-2}{3} = -0.67$

When $x = 14$, then $z = \frac{14 - 12}{3} = \frac{2}{3} = 0.67$

$P(10 < x < 14) = P(-0.67 < z < 0.67)$

$= 2P(0 < z < 0.67) = 2 \times 0.2485$

$= 0.4970 = 49.70\%$

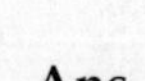

Ans.

Example 81. *The mean yield per plot of a crop is 17 kg and standard deviation is 3 kg. If distribution of yield per plot is normal, find the percentage of plots giving yields:*

(i) Between 15.5 kg and 20 kg; and (ii) More than 20 kg.

[U.P. MBA 2005]

Solution. Mean = μ = 17 kg

S.D. = σ = 3 kg

Standard Normal variable $z = \frac{x - \mu}{\sigma}$

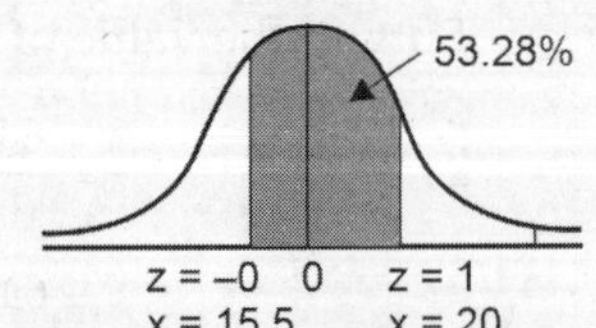

(*i*) When $x_1 = 15.5$, $z_1 = \frac{x_1 - \mu}{\sigma} = \frac{15.5 - 17}{3} = -0.5$

When $x_2 = 20$, $z_2 = \frac{x_2 - \mu}{\sigma} = \frac{20 - 17}{3} = 1$

$\therefore \quad P(15.5 < x < 20) = P(-0.5 < z < 1)$

$= P(0 < z < 0.5) + P(0 < z < 1)$

$= 0.1915 + 0.3413$

$= 0.5328$

$\therefore$ Required percentage of plots = 53.28% **Ans.**

(*ii*) When $x = 20,\ z = \dfrac{20-17}{3} = 1$

$$P\,(x > 20) = P\,(z > 1)$$
$$= 0.5 - P\,(0 < z < 1)$$
$$= 0.5 - 0.3413$$
$$= 0.1587$$

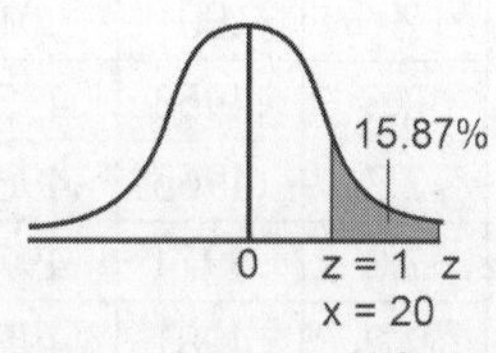

Ans.

TABLE 37.1 Area under standard normal curve from $Z = 0$ to $Z = \dfrac{x - \bar{x}}{\sigma}$

↓Z→	.00	.01	.02	.03	.04	0.5	.06	.07	.08	.09
0.0	.0000	.0040	.0080	.0120	.0160	.0199	.0239	.0279	.0319	.0359
0.1	.0398	.0438	.0478	.0517	.0557	.0596	.0636	.0675	.0714	.0753
0.2	.0793	.0832	.0871	.0910	.0948	.0987	.1026	.1064	.1103	.1141
0.3	.1179	.1217	.1255	.1293	.1331	.1368	.1406	.1443	.1480	.1517
0.4	.1554	1591	.1628	.1664	.1700	.1736	.1772	.1808	.1844	.1879
0.5	.1915	.1950	.1985	.2019	.2054	.2088	.2123	.2157	.2190	.2224
0.6	.2257	.2291	.2324	.2357	.2389	.2422	.2454	.2486	.2517	.2549
0.7	.2580	.2611	.2642	.2673	.2703	.2734	.2764	.2794	.2823	.2852
0.8	.2881	.2910	.2939	.2967	.2995	.3023	.3051	.3078	.3106	.3133
0.9	.3159	.3186	.3212	.3238	.3264	.3289	.3315	.3340	.3365	.3389
1.0	.3413	.3438	.3461	.3485	.3508	.3531	.3554	.3577	.3599	.3621
1.1	.3643	3665	.3686	.3708	.3729	.3749	.3770	.3790	.3810	.3830
1.2	.3849	.3869	.3888	.3907	.3925	.3944	.3962	.3980	.3997	.4015
1.3	.4032	.4049	.4066	.4082	.4099	.4115	.4131	.4147	.4162	.4177
1.4	.4192	.4207	.4222	.4236	.4251	.4265	.4279	.4292	.4306	.4319
1.5	.4332	.4345	.4357	.4370	.4382	.4394	.4406	.4418	.4429	.4441
1.6	.4452	.4463	.4474	.4484	.4495	.4505	.4515	.4525	.4535	.4545
1.7	.4554	.4564	.4573	.4582	.4591	.4599	.4608	.4616	.4625	.4633
1.8	.4641	.4649	.4656	.4664	.4671	.4678	.4686	.4693	.4699	.4706
1.9	.4713	.4719	.4726	.4732	.4738	.4744	.4750	.4756	.4761	.4767
2.0	.4772	.4778	.4783	.4788	.4793	.4798	.4803	.4808	.4812	.4817
2.1	.4821	.4826	.4830	.4834	.4838	.4842	.4846	.4850	.4854	.4857
2.2	.4861	.4864	.4868	.4871	.4875	.4878	.4881	.4884	.4887	.4890
2.3	.4893	.4896	.4898	.4901	.4904	.4906	.4909	.4911	.4913	.4916
2.4	.4918	.4920	.4922	.4925	.4927	.4929	.4931	.4932	.4934	.4936
2.5	.4938	.4940	.4941	.4943	.4945	.4946	.4948	.4949	.4951	.4952

↓Z→	.00	.01	.02	.03	.04	0.5	.06	.07	.08	.09
2.6	.4953	.4955	.4956	.4957	.4959	.4960	.4961	.4962	.4963	.4964
2.7	.4965	.4966	.4967	.4968	.4969	.4970	.4971	.4972	.4973	.4974
2.8	.4974	.4975	.4976	.4977	.4977	.4978	.4979	.4979	.4980	.4981
2.9	.4981	.4982	.4982	.4983	.4984	.4984	.4985	.4985	.4986	.4986
3.0	.4987	.4987	.4987	.4988	.4988	.4989	.4989	.4989	.4990	.4990

* An entry in the Table is the proportion under the entire curve which is between Z = 0 and a positive value of Z. Area for negative values of Z are obtained by symmetry. For different values of Z, table gives area (shown shaded in the figure) under normal curve.

EXERCISE 37.6

1. In a regiment of 1000, the mean height of the soldiers is 68.12 units and the standard deviation is 3.374 units. Assuming a normal distribution, how many soldiers could be expected to be more than 72 units? It is given that

   $P\ (z = 1.00) = 0.3413$, $P\ (z = 1.15) = 0.3749$ and

   $P\ (z = 1.25) = 0.3944$, where z is the standard normal variable. **Ans.** 125°

2. The lifetime of radio tubes manufactured in a factory is known to have an average value of 10 years. Find the probability that the lifetime of a tube taken randomly (*i*) exceeds 15 years, (*ii*) is less than 5 years, assuming that the exponential probability law is followed. **Ans.** (*i*) 0.2231, (*ii*) 0.3935

3. Analysis of past data show that hub thickness of a particular type of gear is normally distributed about a mean thickness of 2.00 cm with a standard deviation of 0.04 cm.

   (*i*) What is the probability that a gear chosen at random will have a thickness greater than 2.06 cm.?

   (*ii*) How many gears in a production run of 600 such gears will have a thickness between 1.89 and 1.95 cm?

   Given $\phi\ (1.5) = 0.4332$, $\phi\ (2.75) = 0.4970$, $\phi\ (1.25) = 0.3944$

   Ans. (*i*) 0.0668, (*ii*) 62 (61.56) app.

4. The breaking strength X of a cotton fabric is normally distributed with $E\ (X) = 16$ and $\sigma\ (X) = 1$. The fabric is said to be good if $X \geq 14$. What is the probability that a fabric chosen at random is good. Given that $\phi\ (2) = 0.9772$. **Ans.** 0.9772

5. A manufacturer knows from experience that resistance of resistors he produces is normal with mean $\mu = 140\ \Omega$ and standard deviation $\sigma = 5\ \Omega$. Find the percentage of resistors that will have resistance between 138 Ω and 142 Ω. (Given $\phi\ (0.4) = 0.6554$, where z is standard normal variate). **Ans.** 31.08%

6. A manufacturing company packs pencils in fancy plastic boxes. The length of the pencils is normally distributed with $\mu = 6''$ and $\sigma = 0.2''$. The internal length of the boxes is 6.4″. What is the probability that the box would be too small for the pencils? (Given that a value of the standardized normal distribution function is $\phi\ (2) = 0.9772$). **Ans.** 0.0228.

7. A manufacturer produces airmail envelopes, whose weight is normal with mean $\mu = 1.95$ gm and standard deviation $\sigma = 0.05$ gm. The envelopes are sold in lots of

1000. How many envelopes in a lot will be heavier than 2 gm? Use the fact that

$$\frac{1}{\sqrt{2\pi}}\int_2^1 \exp\left(\frac{-x^2}{2}\right)dx = 0.3413$$ **Ans.** 159

8. The mean height of 500 students is 151 cm and the standard deviation is 15 cm. Assuming that the heights are normally distributed, find height of how many students, lie between 120 cm and 155 cm. **Ans.** 294

9. A large number of measurements is normally distributed with a mean of 65.5″ and S.D. of 6.2″. Find the percentage of measurements that fall between 54.8″ and 68.8″. **Ans.** 66.01%

10. Find the mean and variance of the density function $f(x) = \lambda e^{-\lambda x}$ **Ans.** $\frac{1}{\lambda}, \frac{1}{\lambda^2}$

11. If x is normally distributed with mean 1 and variance 4,
(*i*) Find $P\,(-3 \le x \le 3)$; (*ii*) Obtain k if $P\,(x \le k) = 0.90$ **Ans.** (*i*) 0.8185, (*ii*) 3.56

12. A normal variable x has mean 1 and variance 4. Find the probability that $x \ge 3$. (Given: z is the standard normal variable and $\phi\,(0) = 0.5$, $\phi\,(0.5) = 0.6915$, $\phi\,(1) = 0.8413$, $\phi\,(1.5) = 0.9332$) **Ans.** 0.1587

13. (*a*) If x is normally distributed with mean 4 and variance 9; find
(*i*) $P\,(2.55 \le x \le 5.5)$ (*ii*) Obtain k if $P\,(x \le k) = 0.9$
Use $P\,(z \le .5) = 0.691$ and $P\,(z \le 1.3) = 0.90$. **Ans.** (*i*) 0.382, (*ii*) 7.9.

(*b*) If $\log_e x$ is normally distributed with mean 1 and variance 4, find $P\left(\frac{1}{2} < x < 2\right)$ given that $\log_e(2) = 0.693$. **Ans.** 0.24

(*c*) For a standard normal variate z, $P\,(-0.72 \le z \le 0) =$ .... **Ans.** 0.2642

14. The random variable x is normally distributed with $E\,(x) = 2$ and variance $V\,(x) = 4$. Find a number p (approximately), such that $P\,(x > p) = 2P\,(x \le p)$ [The values of the standard normal distribution are $\phi\,(-0.43) = 0.3336$, and $\phi\,(-0.44) = 0.3300$]. **Ans.** 1.13834

15. The continuous random variable x is normally distributed with $E\,(x) = \mu$ and $V\,(x) = \mu^2$. If $Y = cx + d$, then find $V\,(Y)$. **Ans.** $c^2\,\mu^2$

16. The pdf of X is given by $f(X) = \lambda e^{-\lambda x}$ $x \ge 0$, λ 0. (*i*) Calculate $P\,[X > E\,(X)]$.
(*ii*) If $X \sim N\,(75, 25)$, find $P\,[X > 80/X > 77]$

(*iii*) If $X \sim N\,(10, 4)$ find $P\,[\,|X| \ge 5]$ **Ans.** (*i*) $\frac{1}{e}$, (*ii*) $\frac{1}{5\sqrt{2\pi}}\,e^{-\frac{(x-75)^2}{2(0.5)}}$, (*iii*) 0.062

17. A random variable x has a standard normal distribution ϕ.
Prove that $P\,(1 > |X| > k) = 2\,[1 - \phi\,(k)]$.

18. The random variable x has the probability density function $f(x) = kx$ if $0 \le x \le 2$.
Find k. Find x such that (*i*) $P\,(X \le x) = 0.1$ (*ii*) $P\,(X \le x) = 0.95$

Ans. $k = \frac{1}{2}$, (*i*) $x = 0.632$ (*ii*) $x = 1.949$

Latest Examination Papers
Part I

June 2019

Engineering Mathematics I

DE51/DC51/DE101/DC101 **ET/CS (Current and New Scheme)**

Time: 3 Hours **Max. Marks: 100**

NOTE: There are 9 Questions in all

- **Qustion 1 is compulsory and carries 20 marks. Answer to Q.1 must be written in the space provided for it in the answer book supplied and nowhere else.**
- **The answer sheet for Q.1 will be collected by the invigilator after 45 minutes of the commencement of the examination.**
- **Out of the remaining EIGHT Questions answer any FIVE Questions. Each question carries 16 marks.**
- **Any required data not explicitly given, may be suitabley assumed and stated.**

Q.1 Choose the correct or the best alternative in the following: (2×10)

(a) The value of $\underset{x\to 0}{\text{Lt}}\ \frac{\sin^2 x/3}{x^2}$ is

(A) $\frac{2}{9}$ (B) $\frac{4}{9}$ (C) $\frac{1}{9}$ (D) 0

(b) If $y = \sin(x+a)\cos(x+a)$, then $\frac{dy}{dx}$ is

(A) $-\cos(x+a)$ (B) $\sin(x+a)$ (C) $\sin 2(x+a)$ (D) $\cos 2(x+a)$

(c) The value of $\int \sin(ax \quad b)\,dx$ is.

(A) $-\frac{1}{a}\cos(ax+b)+C$ (B) $\frac{1}{a}\cos(ax+b)+C$

(C) $\cos(ax+b)+C$ (D) $-\cos(ax+b)+C$

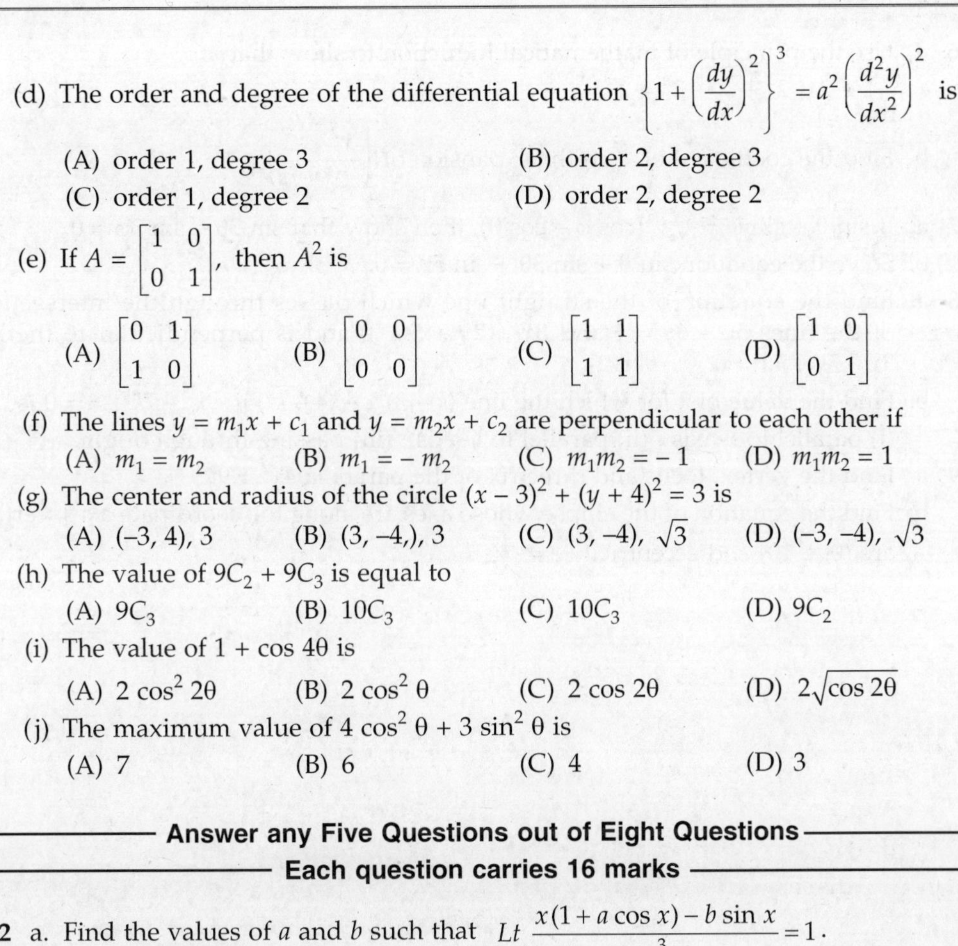

(d) The order and degree of the differential equation $\left\{1+\left(\frac{dy}{dx}\right)^2\right\}^3 = a^2\left(\frac{d^2y}{dx^2}\right)^2$ is

(A) order 1, degree 3 (B) order 2, degree 3
(C) order 1, degree 2 (D) order 2, degree 2

(e) If $A = \begin{bmatrix} 1 & 0 \\ 0 & 1 \end{bmatrix}$, then A^2 is

(A) $\begin{bmatrix} 0 & 1 \\ 1 & 0 \end{bmatrix}$ (B) $\begin{bmatrix} 0 & 0 \\ 0 & 0 \end{bmatrix}$ (C) $\begin{bmatrix} 1 & 1 \\ 1 & 1 \end{bmatrix}$ (D) $\begin{bmatrix} 1 & 0 \\ 0 & 1 \end{bmatrix}$

(f) The lines $y = m_1x + c_1$ and $y = m_2x + c_2$ are perpendicular to each other if
(A) $m_1 = m_2$ (B) $m_1 = -m_2$ (C) $m_1m_2 = -1$ (D) $m_1m_2 = 1$

(g) The center and radius of the circle $(x-3)^2 + (y+4)^2 = 3$ is
(A) (–3, 4), 3 (B) (3, –4,), 3 (C) (3, –4), $\sqrt{3}$ (D) (–3, –4), $\sqrt{3}$

(h) The value of $9C_2 + 9C_3$ is equal to
(A) $9C_3$ (B) $10C_3$ (C) $10C_3$ (D) $9C_2$

(i) The value of $1 + \cos 4\theta$ is
(A) $2\cos^2 2\theta$ (B) $2\cos^2\theta$ (C) $2\cos 2\theta$ (D) $2\sqrt{\cos 2\theta}$

(j) The maximum value of $4\cos^2\theta + 3\sin^2\theta$ is
(A) 7 (B) 6 (C) 4 (D) 3

Answer any Five Questions out of Eight Questions

Each question carries 16 marks

Q.2 a. Find the values of a and b such that $\underset{x\to 0}{Lt}\ \frac{x(1+a\cos x)-b\sin x}{x^3} = 1$. **(8)**

b. Find the nth derivative of $e^x(2x+3)^3$. **(8)**

Q.3 a. Integrate $\int e^x\left(\frac{1-\sin x}{1-\cos x}\right)dx$. **(8)**

b. Evaluate $\int \frac{x}{(x+2)^2(x-1)}dx$. **(8)**

Q.4 a. Discuss the consistency of the following system of equations:

$$2x + 3y + 4z = 11,\ x + 5y + 7z = 15,\ 3x + 11y + 13z = 25.$$

If found consistent, then solve them **(8)**

b. Find the Inverse of the matrix $A = \begin{bmatrix} 3 & 2 & 4 \\ 2 & 1 & 1 \\ 1 & 3 & 5 \end{bmatrix}$ by using adjoint of the matrix. **(8)**

Q.5 a. Solve the differential equation $\left(x\tan{}^y\!/_x - y\sec^2{}^y\!/_x\right)dx - x\sec^2{}^y\!/_x\,dy = 0$. **(8)**

b. Solve the differential equation $x\frac{dy}{dx} + y = x^3y^6$. **(8)**

Q.6 a. Use the principle of mathematical induction to show that
$1 + 2 + 2^2 + ... + 2^n = 2^{n-1}$ **(8)**

b. Find the coefficient of x^7 in the expansion of $\left(\frac{x^2}{2} - \frac{2}{x}\right)^8$. **(8)**

Q.7 a. If $\sin\theta + \sin\varphi = \sqrt{3}\,(\cos\varphi - \cos\theta)$, then show that $\sin 3\theta + \sin 3\varphi = 0$. **(8)**

b. Solve the equation $\sin\theta + \sin 3\theta + \sin 5\theta = 0$. **(8)**

Q.8 a. Find the equation of the straight line which passes through the intersections of the lines $5x - 6y = 1$ and $3x + 2y + 5 = 0$ and is perpendicular to the line $3x - 5y + 11 = 0$. **(8)**

b. Find the value of k for which the line $(k - 3)\,x - (4 - k^2)\,y + k^2 - 7k + 6 = 0$ is
(i) parallel to x-axis (ii) parallel to y-axis (iii) passing through origin. **(8)**

Q.9 a. Find the vertex, focus and directrix of the parabola $4x^2 = 9y$. **(8)**

b. Find the equation of the ellipse whose axes are along the coordinate axes, vertices are $(0, \pm 10)$ and eccentricity $e = 4/5$. **(8)**

June 2019

Applied Mathematics I

3K-BSN-02A **ET/CS (Current and New Scheme)**

Time: 3 Hours **Max. Marks: 100**

NOTE:

- **Part 'A' may be attempted in first 6 pages of Answer Sheet.**
- **Part 'B' in rest of the Sheets of Answer Sheet.**
- **Answers may be given in English or Hindi.**

PART A

Q.1 Attempt any 10 parts: **(10×2=20)**

(a) Write the order and degree of the differential equation: $\frac{d^4y}{dx^4} - \sin\left(\frac{d^3y}{dx^3}\right) = 0$.

(b) If $y = \sin^2 x$, differentiate it with respect to x.

(c) Write the modulus and argument of $-1 - \sqrt{3}\,i$.

(d) Draw the phasor diagram of the function $f(t) = -5\cos(\cot t - 30°)$.

(e) Evaluate $\lim_{x\to 1} \frac{x^3 - 1}{x - 1}$.

(f) If $f(x, y) = \sin x \cos y$, find $\frac{\partial f}{\partial x}$ and $\frac{\partial f}{\partial y}$.

(g) If $\int_0^a (x^2 + 1)\,dx = 72$, find the value of a.

(h) Find $\lim_{x\to 0} \frac{\sin ax}{\sin bx}$.

(i) Solve $\int x \cos 2x\, dx$.

(j) Write the general solution of the differential equation $(D^2 + 2D + 1)\,y = 0$.

(k) Find the integral: $\int \frac{dx}{x^2 - 16}$.

(l) Find the solution of the differential equation: $(2 - y)\,dy - (x + 1)\,dx = 0$.

(m) Find the slope of the tangent to the curve $y = x^3 - x$ at $x = 2$.

(n) Evaluate $\lim_{x\to 0} \frac{\sin 2x}{\sin 3x}$.

Q.2 Attempt any five parts: **(5×4=20)**

(a) $\lim_{x \to 0} \frac{x^n - 3^n}{x - 3} = 108, n \in \mathrm{N}$. Find n.

(b) Use first principle to find the derivative of $\sin x$.

(c) Evaluate $\int_0^{\pi/2} \sin^7 x \, dx$.

(d) Differentiate $\sin(\cos(x^2))$ with respect to x.

(e) If $x = a(\theta - \cos\theta)$ and $y = a(1 + \theta\sin\theta)$, find $\left(\frac{dy}{dx}\right)_{\theta=\frac{\pi}{2}}$.

(f) Find $\int \cos 6x\sqrt{1 + \sin 6x}\, dx$.

(g) Find the general solution of $\frac{d^2y}{dx^2} + 3\frac{dy}{dx} + 6y = 0$.

(h) Find all the points of local maxima of the function given by

$$f(x) = 2x^3 - 6x^2 + 6x + 5.$$

PART B

Attempt any 3 questions: **(3×20=60)**

Q.3 a. Find the first order partial derivatives of the given function:

$$f(x, y) = \tan^{-1}\left(\frac{x}{y}\right)$$

b. If $y = \frac{x}{2}\sqrt{a^2 - x^2} + \frac{a^2}{2}\sin^{-1}\left(\frac{x}{a}\right)$, find $\frac{dy}{dx}$.

Q.4 a. Evaluate $\int \frac{1}{(x-1)(x^2+1)} dx$.

b. Find the area of the region bounded by the curve $y = x^2$ and the line $y = 4$.

Q.5 a. Solve the following differential equation:

$$(x^2 + y^2)dx + xy\, dy = 0, \quad y(1) = 1$$

b. Evaluate $\int_0^a x^3\sqrt{a^2 - x^2}\, dx$.

Q.6 a. Evaluate $\lim_{x \to 0} \frac{\tan x - \sin x}{\sin^3 x}$.

b. Use Simpson's 1/3rd rule, to evaluate $\int_0^1 \frac{dx}{3 + 2x}$ taking 4 intervals.

Q.7 a. Solve the differential equation: $(1 + y^2) + (x - e^{\tan^{-1}y})\frac{dy}{dx} = 0$.

b. Show that of all rectangles that can be inscribed in a given circle, the square has the maximum area.

June 2019

Applied Mathematics I

3K-BSN-02 **ET/CS (Current and New Scheme)**

Time: 3 Hours **Max. Marks: 100**

NOTE:

- **Part 'A' may be attempted in first 6 pages of Answer Sheet.**
- **Part 'B' in rest of the Sheets of Answer Sheet.**
- **Answers may be given in English or Hindi.**

PART A

Q.1 This question consists of 14 parts each of 2 marks. Attempt any 10 parts: (10 × 2 = 20)

i. Express $\sqrt{3}+i$ in polar form.

ii. If $f(x) = 2\log e^x$, find $f(e)$.

iii. Evaluate $\lim_{x\to 0}\frac{e^{\sin 3x}-1}{x}$.

iv. Differentiate $\operatorname{cosec}^2 x$ with respect to x.

v. Write the integrating factor of the differential equation: $\frac{dy}{dx}+y\cot x=\sec x$.

vi. Find the slope of the normal to the curve $y = 2x^2 + 3\sin x$, at $x = 0$.

vii. Evaluate $\int \sin^2 x\,dx$.

viii. Evaluate $\int \frac{1+\tan x}{x+\log(\sec x)}\,dx$.

ix. Evaluate $\int_0^{\pi}\sin x\,dx$.

x. Evaluate $\int 2^3 \log_2 x\,dx$.

xi. Find the area bounded by the curve $y = 2e^x$, the ordinates at $x = 0$, $x = -1$ and the x-axis.

xii. Evaluate $\int_0^1\left(\frac{e^x}{1+e^{2x}}\right)dx$.

xiii. Find the degree and order of the differential equation $\left(y-\left(\frac{dy}{dx}\right)^2\right)^{1/5}=\left(\frac{d^2y}{dx^2}+\frac{dy}{dx}\right)^{1/3}$

xiv. If $z = x^y$, find $\frac{\partial z}{\partial x}$.

Q.2 This question consists of 8 parts each of 4 marks. Attempt any 5 parts: (5 × 4 = 20)

i. Find $\int x \sin^{-1} x \, dx$.

ii. Differentiate sin x from the first principles with respect to x.

iii. Find $\frac{dy}{dx}$, if $\cos y = x \cos (a + y)$.

iv. Evaluate $\int_0^a x^2 \sqrt{a^2 - x^2} \, dx$.

v. Show that the line $\frac{x}{a} + \frac{y}{b} = 1$ touches the curve $y = be^{x/a}$ at the point where it crosses the y-axis.

vi. Find the area of the region bounded by the line $y - 1 = x$, the x-axis and the ordinates $x = -2$ and $x = 3$.

vii. If $f(x, y) = x^3 - 2x^2y^2$, find $f_x(-1, 2)$ and $f_y(2, -1)$.

viii. Solve the differential equation: $\frac{dy}{dx} = \frac{\sqrt{x^2 + y^2} + y}{x}$.

PART-B

Part B consists of 5 questions each of 20 marks. Attempt any 3 questions: (3×20=60)

Q.3 a. Evaluate $\lim_{x \to 0} \frac{\tan x - \sin x}{x^3}$.

b. If the normal to the curve $x^{2/3} + y^{2/3} = a^{2/3}$ makes an angle ϕ with the x-axis, show that its equation is $y \cos \phi - x \sin \phi = a \cos 2\phi$.

Q.4 a. If $y = (x \cos x)^x + (x \sin x)^{1/x}$, find $\frac{dy}{dx}$.

b. Show that the height of the closed right circular cylinder of given surface and maximum volume is equal to the diameter of its base.

Q.5 a. Evaluate $\int \frac{dx}{3 + 4 \sin x}$.

b. Prove that $\int_0^1 \frac{\log (1 + x)}{1 + x^2} dx = \frac{\pi}{8} \log_e 2$.

Q.6 a. The area enclosed between the line $y = 5x$ and lines $y = 1$ and $y = 3$ is revolved about the y-axis. Find the volume of the solid thus generated.

b. Using the trapezoidal rule, obtain the approximate value of the integral $\int_0^1 \left(\frac{1}{1 + x^2} \right) dx$ taking 6 equal intervals.

Q.7 a. If $u = \sin^{-1} \left(\frac{x + y}{\sqrt{x} + \sqrt{y}} \right)$, show that $x \frac{\partial u}{\partial x} + y \frac{\partial u}{\partial y} = \frac{1}{2} \tan u$.

b. Solve the differential equation: $\frac{d^2y}{dx^2} - 2\frac{dy}{dx} + y = e^{2x}$.

June 2019

Engineering Mathematics I

DE51/DC51/DE101/DC101 **ET/CS (Current and New Scheme)**

Time: 3 Hours **Max. Marks: 100**

NOTE: There are 9 Questions in all

- **Qustion 1 is compulsory and carries 20 marks. Answer to Q.1 must be written in the space provided for it in the answer book supplied and nowhere else.**
- **The answer sheet for the Q.1 will be collected by the invigilator after 45 minutes of the commencement of the examination.**
- **Out of the remaining EIGHT Questions answer any FIVE Questions. Each question carries 16 marks.**
- **Any required data not explicitly given, may be suitabley assumed and stated.**

Q.1 Choose the correct or the best alternative in the following: **(2 × 10)**

a. If $n_{C_{10}} = n_{C_8}$ then $n_{C_{16}}$ is equal to

(A) 153 (B) 151 (C) 157 (D) 155

b. If the 7th and 13th term of a progression is 23 and 41 respectively, then the 21st term is

(A) 65 (B) 60 (C) 55 (D) 50

c. If $2 \sin^2 x = 3 \cos x$ then x is

(A) 60° (B) 50° (C) 40° (D) 30°

d. The value of $\sqrt{3}$ cosec 20° – sec 20° is

(A) 4 (B) 5 (C) 6 (D) 7

e. If the distance between the points $(x, -7)$, $(3, -3)$ is 5 unit, The value of x is

(A) 0 (B) 0 or 6 (C) 6 (D) none of these

f. The length of the perpendicular from the straight line $x - 2y - 5 = 0$ drawn from the point (–3, –5) is

(A) $\frac{1}{\sqrt{5}}$ (B) $\frac{2}{\sqrt{5}}$ (C) $\frac{3}{\sqrt{5}}$ (D) $\frac{4}{\sqrt{5}}$

g. The area of the triangle with vertices (–3, 5), (3, –6), (7, 2) is

(A) 41 units (B) 40 units (C) 46 units (D) 45 units

h. $\left(\frac{dx}{dy}\right)^2 + 5\sqrt[3]{y} = x$ is

(A) linear of degree 2 (B) nonlinear of order 1 and degree 2

(C) nonlinear of order 1 and degree 6 (D) None of these

i. The function $f(x) = \begin{cases} x\sin\frac{1}{x}, & \text{if } x \neq 0 \\ 0, & \text{if } x = 0 \end{cases}$ is

(A) discontinuous at $x = 0$ (B) $\underset{x \to 0}{\text{Lt}} f(x)$ does not exists

(C) continuous at $x = 0$ (D) none of these

j. The value of $\int_0^1 \left(\frac{1-x}{1+x}\right) dx$ is

(A) $2 \log 2 - 1$ (B) $\log 2$ (C) $\log 2 - 1$ (D) $2 \log 2$

Answer any Five Questions out of Eight Questions

Each question carries 16 marks

Q.2 a. Using the principle of mathematical induction, prove that $(2n + 7) < (n + 3)^2$ for all values of $n \in N$. **(8)**

b. If $(1 + x)^n = C_0 + C_1x + C_2x^2 + ... + C_nx^n$ then show that

$$C_0^2 + C_1^2 + C_2^2 + C_3^2 + C_4^2 + ... + C_n^2 = \frac{(2n)!}{n!\,n!}.$$ **(8)**

Q.3 a. If $A + B + C = \pi$, show that $\cos A + \cos B + \cos C = 1 + 4 \sin\frac{A}{2} \sin\frac{B}{2} \sin\frac{C}{2}$. **(8)**

b. Prove that $\left(\frac{\cos A + \cos B}{\sin A - \sin B}\right)^n + \left(\frac{\sin A + \sin B}{\cos A - \cos B}\right)^n = \begin{cases} 2\cot^n\left(\frac{A-B}{2}\right), & \text{if } n \text{ is even} \\ 0, & \text{if } n \text{ is odd} \end{cases}$

Q.4 a. If the area of a triangle formed by the straight line L and the coordinate axis is 5 unit where L is perpendicular to the straight line $5x - y = 1$, find the equation of the straight line L. **(8)**

b. The inclination of a straight line passing through the point (4, 5) is 30°. Find the coordinate of the point lying on that line whose distance from the given point is 3 unit. **(8)**

Q.5 a. Solve the following systems of linear equations by matrix method.

$5x - 7y + z = 11$, $6x - 8y - z = 5$, $3x + 2y - 6z = 7$. **(8)**

b. Given $A = \begin{pmatrix} 1 & 2 & 2 \\ 2 & 1 & 2 \\ 2 & 2 & 1 \end{pmatrix}$ show that $A^2 - 4A - 5I = 0$. Hence find A^{-1}. **(8)**

Q.6 a. Solve: $(x^2 - y^2)\, dx = 2xy dy$. **(8)**

b. Solve: $(x^2y - 2xy^2)\, dx = (x^3 - 3x^2y)\, dy$. **(8)**

Q.7 a. Evaluate $\int_\alpha^\beta \sqrt{(x-\alpha)(\beta-x)}\, dx$. **(8)**

b. Integrate $\int \tan^{-1}(1 + x + x^2)\, dx$. **(8)**

Q.8 a. Find the *n*th derivative of $\frac{1}{x^2 + a^2}$. **(8)**

b. Find the values of *a* and *b* such $\underset{x \to 0}{Lt} \frac{x(1 + a\cos x) - b\sin x}{x^3} = 1.$

Q.9 a. Find the equation of the parabola whose focus is at (–1, 1) and directrix is $x + y + 1 = 0$. **(8)**

b. If *e* and *e′* are the eccentricities of two conjugate hyperbola prove that

$$\frac{1}{e^2} + \frac{1}{e'^2} = 1.$$ **(8)**

June 2018

Engineering Mathematics I

DE51/DC51/DE101/DC101 **ET/CS (Current and New Scheme)**

Time: 3 Hours **Max. Marks: 100**

NOTE: There are 9 Questions in all

- **Qustion 1 is compulsory and carries 20 marks. Answer to Q.1 must be written in the space provided for it in the answer book supplied and nowhere else.**
- **The answer sheet for the Q.1 will be collected by the invigilator after 45 minutes of the commencement of the examination.**
- **Out of the remaining EIGHT Questions answer any FIVE Questions. Each question carries 16 marks.**
- **Any required data not explicitly given, may be suitably assumed and stated.**

Q.1 Choose the correct or the best alternative in the following: **(2 × 10 = 20)**

a. If $y = x^2 - \cos x - \frac{1}{x^2}$, then $\frac{dy}{dx}$ is

(A) $x - \cos x + \frac{2}{x^3}$ (B) $2x - \sin x + \frac{2}{x^3}$

(C) $2x + \sin x - \frac{2}{x^3}$ (D) $2x + \sin x + \frac{2}{x^3}$

b. Let A and B be two matrices, then the relation $(AB)^n = A^nB^n$, if

(A) $AB = BA$ (B) $AB \neq BA$ (C) $A = B$ (D) $A^{-1} = B$

c. The differential equation of the family of curves $y = e^x (A \cos x + B \sin x)$, where A and B are arbitrary constants, is

(A) $\frac{dy}{dx} + Ax + By = 0$ (B) $\frac{dy}{dx} - Ax - By = 0$

(C) $\frac{d^2y}{dx^2} - 2\frac{dy}{dx} + 2y = 0$ (D) none of these

d. If $\sin A = \frac{3}{5}$ and $\cos B = \frac{9}{41}$ where $0 < A < \frac{\pi}{2}, 0 < B < \frac{\pi}{2}$ then the value of $\cos (A - B)$ is

(A) $\frac{106}{107}$ (B) $-\frac{156}{205}$ (C) $\frac{156}{205}$ (D) none of these

e. If $\Delta = \begin{vmatrix} 2x-1 & x+7 & x+4 \\ x & 6 & 2 \\ x-1 & x+1 & 3 \end{vmatrix}$, then the value of Δ in the respect of x is equal to,

(A) 2, 3, 5 (B) 1, 3, 4 (C) 1, 2, 3 (D) 1, 2, 4

f. The order (O) and degree (D) of differential equation of $y\dfrac{d^2x}{dy^2} = y^2 + 1$ is

(A) O = 2, D = 1 (B) O = 0, D = 1 (C) O = 1, D = 1 (D) O = 1, D = 2

g. $\int \sin^2 x \cdot \cos^2 x \, dx$ is equal to

(A) $\dfrac{1}{8}\left(x - \dfrac{\sin 4x}{4}\right) + C$ (B) $\dfrac{1}{8}\left(x + \dfrac{\sin 4x}{4}\right) + C$

(C) $\dfrac{1}{8}\left(x - \dfrac{\cos 4x}{4}\right) + C$ (D) $\dfrac{1}{8}\left(x + \dfrac{\cos 4x}{4}\right) + C$

h. The solution of the differential equation $\dfrac{dy}{dx} = xy^2 - xy$ is equal to

(A) $\log \dfrac{y-1}{y} = \dfrac{x^2}{2} + C$ (B) $\log \dfrac{y-1}{y} = x + 2 + C$

(C) $\log \dfrac{y+1}{y} = \dfrac{x^2}{2} + C$ (D) $\log \dfrac{y+1}{y} = x - 2 + C$

i. If $y = \tan^{-1}\left(\dfrac{\cos x}{1 + \sin x}\right)$ then $\dfrac{dy}{dx}$ is

(A) $-\dfrac{1}{2}$ (B) $\dfrac{1}{2}$ (C) 1 (D) – 1

j. The equation of the line which makes intercepts – 4 and 5 on the axis is:

(A) $4x + 5y + 20 = 0$ (B) $4x - 5y + 20 = 0$

(C) $5x - 4y + 20 = 0$ (D) $5x + 4y + 20 = 0$

Answer any Five Questions out of Eight Questions

Each question carries 16 marks

Q.2 a. Differentiate the following functions: **(8)**

(*i*) $x^n e^x \log_e x$ (*ii*) $\text{cosec}^{-1}\left(\dfrac{x^2+1}{x^2-1}\right) + \cos^{-1}\left(\dfrac{x^2-1}{x^2+1}\right)$

b. Find the equation of the tangent to the curve $x^2 + 2y = 8$ which is perpendicular to the line $x - 2y + 1 = 0$. **(8)**

Q.3 a. Evaluate $\int x \cos^3 x \, dx$. **(8)**

b. Evaluate $\int \dfrac{1}{\sqrt{x(1-2x)}} dx$ **(8)**

Q.4 a. Apply Cramer's rule to solve the following system of linear equations:

$$3x - 2y + 4z = 5$$
$$x + y + 3z = 2$$
$$-x + 2y - z = 1$$

(8)

b. Show that $\begin{vmatrix} a & b & c \\ a^2 & b^2 & c^2 \\ b+c & c+a & a+b \end{vmatrix} = (b-c)(c-a)(a-b)(a+b+c)$ **(8)**

Q.5 a. Solve the differential equation $(x + y)\,dy + (x - y)\,dx = 0$ given that $y = 1$ when $x = 1$ **(8)**

b. Solve the differential equation $\dfrac{dy}{dx} + \dfrac{1+\cos 2y}{1-\cos 2x} = 0$. **(8)**

Q.6 a. Find the term independent of x in the expansion of $\left(2x^2 - \dfrac{1}{x}\right)^{12}$. **(8)**

b. Find three numbers in A.P. whose sum is 21 and their product is 315. **(8)**

Q.7 a. Prove that, $\cos 2A \cdot \cos 2B + \sin^2(A - B) - \sin^2(A + B) = \cos(2A + 2B)$. **(8)**

b. If $A + B + C = \pi$, prove that $\cot\dfrac{A}{2} + \cot\dfrac{B}{2} + \cot\dfrac{C}{2} = \cot\dfrac{A}{2}\cot\dfrac{B}{2}\cot\dfrac{C}{2}$. **(8)**

Q.8 a. Find the equation of the lines through the origin and making an angle of 60° with the line $x + \sqrt{3}y + 3\sqrt{3} = 0$. **(8)**

b. Find the area of the triangle formed by the lines $y = x$, $y = 2x$ and $y = 3x + 4$ **(8)**

Q.9 a. Find the equation to the circle, which passes through the point $(-2, 4)$ and through the points in which the circle $x^2 + y^2 - 2x - 6y + 6 = 0$ is cut by the line $3x + 2y - 5 = 0$.

b. Find the equation of the parabola with focus $(3, -4)$ and the directrix $6x - 7y + 5 = 0$ **(8)**

N3031 **1600301**

Bihar S.B.T.E. 2018 (Odd)

Time : 3 Hrs. **Sem.III (G) App.Maths-I**

Full Marks : 70 **Pass Marks : 28**

- *Answer all 20 questions from Group A, each question carries 1 marks.*
- *Answer all five questions from Group B, each question carries 4 marks.*
- *Answer all five questions from Group C, each question carries 6 marks.*
- *All parts of a question must be answered at one place in sequence, otherwise they may not be evaluated.*
- *The figure in right hand margin indicate marks.*

GROUP-A

Q.1 Choose the most suitable answer from the following options: **(1×20=20)**

i. The value of $\int e^{3x-1} \cdot dx$ is

(A) $e^{3x-1} + C$ (B) $\frac{1}{3} e^{3x-1} + C$ (C) $3 \cdot e^{3x-1} + C$ (D) none of these

ii. The value of $\int \log x \cdot dx$

(a) $x \log x + x + C$ (b) $\log x - x + C$ (c) $x \log x - x + C$ (d) none of these

iii. The value of $\int_0^{\pi/2} \cos x \cdot dx$ is

(a) 0 (b) 1 (c) 2 (d) none of these

iv. The value of $\int_{-2}^{2} \frac{|x|}{x} \cdot dx$ is

(a) 1 (b) 2 (c) 0 (d) none of these

v. The value of $\int_0^1 \frac{\tan^{-1} x}{1+x^2} \cdot dx$ is

(a) $\frac{\pi}{32}$ (b) $\frac{\pi^2}{32}$ (c) $\frac{32}{\pi^2}$ (d) none of these

vi. The orthogonal trajectory of the family of curves $y^2 = 4ax$ is

(a) $x^2 + y^2 = 2c$ (b) $x^2 + 2y^2 = 2c$ (c) $2x^2 + y^2 = 2c$ (d) None of these

vii. The order and degree of the differential equation $\left(\frac{d^2y}{dx^2}\right)^2 + x^3\left(\frac{dy}{dx}\right)^3 = x^4$ is

(a) 2 and 3 (b) 3 and 2 (c) 2 and 2 (d) None of these

viii. The differential equation of the family of curves $y = a \cos (nx + b)$ where a and b are parameters is ....

(a) $y_1 + n^2y = 0$ (b) $y_2 + n^2y = 0$ (c) $y_1 + n^2y_2 = 0$ (d) None of these

ix. Solution of differential equation $\sec^2 x \cdot \tan y \cdot dx + \sec^2 y \cdot \tan x dy = 0$ is

(a) $\tan x \cdot \tan y = k$ (b) $\tan x + \tan y = k$

(c) $\tan x - \tan y = k$ (d) None of these

x. The integrating factor of the differential equation $x\dfrac{dy}{dx} + y = x^3$ is

(a) $\log x$ (b) x (c) x^2 (d) none of these

xi. The differential equation $M(x \cdot y)dx + N(x, y)dy = 0$ is an exact differential equation of ....

(a) $\dfrac{\partial M}{\partial y} + \dfrac{\partial N}{\partial x} = 0$ (b) $\dfrac{\partial M}{\partial y} - \dfrac{\partial N}{\partial x} = 0$ (c) $\dfrac{\partial M}{\partial y} \times \dfrac{\partial N}{\partial x} = 1$ (d) none of these

xii. $L\{\cosh at\}$ is equal to: (when $s > 0$)

(a) $\dfrac{s}{s^2 - a^2}$ (b) $\dfrac{s}{s^2 + a^2}$ (c) $\dfrac{1}{s^2 - a^2}$ (d) none of these

xiii. If $L\{f(t)\} = F(s)$, then $L\{e^{at}.f(t)\}$ is equal to

(a) $F(s + a)$ (b) $F(s - a)$ (c) $F(s^2 - a^2)$ (d) none of these

xiv. The Laplace transform of $t \cdot \cos at$ that is $L[t \cdot \cos at]$ is ....

(a) $\dfrac{a^2 + s^2}{(a^2 + s)^2}$ (b) $\dfrac{a^2 - s^2}{a^2 + s^2}$ (c) $\dfrac{a^2 - s^2}{(a^2 + s^2)^2}$ (d) none of these

xv. Inverse Laplace transform of $\dfrac{s}{s^2 + 1}$ is

(a) $\sin t$ (b) $\cos t$ (c) $-\sin t$ (d) none of these

xvi. The inverse Laplace transform of $\dfrac{1}{(s+2)^3}$ is

(a) $e^{2t} \cdot \dfrac{t^2}{\underline{|2}}$ (b) $e^{2t} \cdot \dfrac{t^3}{\underline{|3}}$ (c) $e^{-2t} \cdot \dfrac{t^2}{\underline{|2}}$ (d) none of these

xvii. If $f(x)$ is an even function then, its Fourier expansion contains only

(a) sine terms (b) cosine terms

(c) sine and cosine terms both (d) none of these

xviii. Solution of simultaneous linear algebraic equation obtained by ......

(a) Newton - Raphson method (b) Gauss elimination method

(c) Bisection method (d) none of these

xix. The equation $x^3 - 5x - 11 = 0$ has at least one root lies between

(a) 2 and 3 (b) 0 and 1 (c) 1 and 2 (d) none of these

xx. If $f(x) = 0$ is an algebraic equation then Newton - Raphson method is given by

$x_{n+1} = x_n - \dfrac{f(x_n)}{?}$

(a) $f(x_{n-1})$ (b) $f'(x_{n-1})$ (c) $f'(x_n)$ (d) none of these

GROUP B

Answer all Five Question: **(5×4=20)**

Q.2 Evaluate: $\int (5x+3)\sqrt{2x-1} \cdot dx$ **(4)**

OR

Evaluate: $\int \frac{x \sin^{-1} x}{\sqrt{1-x^2}} \cdot dx$ **(4)**

Q.3 Integrate: $\int_0^1 \frac{1-x}{1+x} \cdot dx$

OR

Integrate: $\int_0^{\pi/2} \frac{\sqrt{\tan x}}{1+\sqrt{\tan x}} \cdot dx$

Q.4 Solve the following differential equation $(x+y)^2 \frac{dy}{dx} = a^2$ **(4)**

OR

Solve the following differential equation $\frac{dy}{dx} = \frac{y}{x} + \tan \frac{y}{x}$ **(4)**

Q.5 Find the Laplace transform of e^{-t} (3 sin 2t – 5 cosh 2t) **(4)**

OR

Find the Inverse Laplace transform of $\frac{s-2}{s^2+5s+6}$. **(4)**

Q.6 Find the real root of the equation $x^3 - 2x - 5 = 0$ by Newton Raphson method (three iteration only). **(4)**

OR

Find the real root of the equation $x^3 - 4x + 1 = 0$ by Regula Falsi method (three iteration only). **(4)**

GROUP C

Answer all Five Question: **(5×6=30)**

Q.7 Find the area of the portion of the parabola $y^2 = 16\,ax$ cut off by the line $y = 4mx$. **(6)**

OR

Find the area of the region bounded by the ellipse

$$\frac{x^2}{4} + \frac{y^2}{9} = 1$$ **(6)**

Q.8 Solve the following differential equation **(6)**

$$x\frac{dy}{dx} + y = y^2 \log x$$

OR

Solve the following differential equation **(6)**

$$\frac{dy}{dx} = \frac{6x - 2y - 7}{3x - y + 4}$$

Q.9 Evaluate $L^{-1}\left\{\frac{1}{(s-2)(s+2)^2}\right\}$ by using convolution theorem

OR

Solve the following equation by Laplace transform method $y'' - 3y' + 2y = 1 - e^{2t}$ if $y(0) = 1, y'(0) = 0$ **(6)**

Q.10 Find the half range cosine series $f(x) = \sin x$ in the interval $(0, \pi)$

$$\frac{\pi}{4} = 1 - \frac{1}{3} + \frac{1}{5} - \frac{1}{7} + \ldots\ldots\ldots\ldots$$ **(6)**

OR

Find the Fourier series of $f(x) = e^{ax}$ in the interval $(-\pi, \pi)$ **(6)**

Q.11 Solve the following equation by Gauss elimination method

$2x + 3y + z = 13, x - y - 2z = -1, 3x + y + 4z = 15$ **(6)**

OR

Solve the following equation by Jocobi's iteration method **(6)**

$10x + y + 2z = 13, 3x + 10y + z = 14, 2x + 3y + 10z = 15$

Applied Mathematics I

97-ASN-02

Time: 3 Hours **Max. Marks: 100**

NOTE:
- **Part 'A' may be attempted in first 6 pages of Answer Sheet.**
- **Part 'B' in rest of the Sheets of Answer Sheet.**
- **Answers may be given in English or Hindi.**

PART A

Q.1 Attempt any 10 questions: **(10×2=20)**

a. For what values of a are $\vec{r_1} = a\hat{i} - 2\hat{j} + \hat{k}$ and $\vec{r_2} = 2a\hat{i} + a\hat{j} - 4\hat{k}$ perpendicular?

b. If $\vec{r_1} = 2\hat{i} - \hat{j} + \hat{k}, \vec{r_2} = \hat{i} + 3\hat{j} - 2\hat{k}, \vec{r_3} = -2\hat{i} + \hat{j} - 3\hat{k}, \vec{r_4} = 3\hat{i} + 2\hat{j} + 5\hat{k}$, find scalars a, b, c such that $\vec{r_4} = a\vec{r_1} + b\vec{r_2} + c\vec{r_3}$.

c. If $\vec{r_1} = 3\hat{i} - \hat{j} + 2\hat{k}, \vec{r_2} = 2\hat{i} + \hat{j} - \hat{k}, \vec{r_3} = \hat{i} - 2\hat{j} + 2\hat{k}$, Find $\left|\vec{r}\right|$ where $\vec{r} = \vec{r_1} + 2\vec{r_2} - \vec{r_3}$.

d. Compute $AB - BA$, where $A = \begin{bmatrix} 3 & 4 \\ 5 & 7 \end{bmatrix}, B = \begin{bmatrix} 2 & 4 \\ 3 & 1 \end{bmatrix}$.

e. If $A = \begin{bmatrix} 3 & 1 \\ 1 & 4 \end{bmatrix}$, then show that $A^2 - 7A + 11\,I = O$ where I is a unit matrix of second order.

f. Show that the matrix $\begin{bmatrix} 1 & 2 & 3 \\ 2 & -4 & 5 \\ 3 & 5 & 7 \end{bmatrix}$ is symmetric.

g. Show that $E = 1 + \Delta$.

h. Find the equation of the circle whose centre is at (–1, 2) and radius is 4.

PART B

Attempt any 3 questions: **(3×20=60)**

Q.3 a. Construct a forward difference table from the following data

x	:	0	5	10	15	20	25
$f(x)$	:	7	11	14	18	24	32

b. The following table gives corresponding values of x and y:

x	:	0	1	2	3	4
$f(x)$	:	3	6	11	18	27

Q.4 a. Solve the system of equations

$$x + 2y + 3z = 1, \quad 2x + 3y + 2z = 2, \quad 3x + 3y + 4z = 1$$

b. Show that $A = \begin{bmatrix} 1 & 2 & 3 \\ 2 & -1 & 4 \\ 3 & 1 & 1 \end{bmatrix}$ satisfies the matrix equation $A^3 - A^2 - 18A - 30I = 0$

Q.5 a. Find the equation of the line which passes through the point (3, – 4) and the sum of its intercepts on the coordinate axes is – 5.

b. Find the equation of the line passing through the point of intersection of the lines $x - 2y - 4 = 0$ and $4x - y - 4 = 0$ and is perpendicular to the line $9x + 22y - 8 = 0$.

Q.6 a. Find the equation of the circle passing through the points (2, 3), (–1, 6) and whose centre lies on $2x + 5y + 1 = 0$.

b. For the parabola $y^2 + 4y + 20x + 4 = 0$ find the coordinates of the vertex and equation of the directrix.

Q.7 a. Find a unit vector parallel to the resultant of vectors $\vec{r_1} = 2\hat{i} + 4\hat{j} - 5\hat{k}, \vec{r_2} = \hat{i} + 2\hat{j} + 3\hat{k}$

b. If $\vec{r_1} = \hat{i} + \hat{j} + \hat{k}, \vec{r_2} = \hat{j} - \hat{k},$ then find a vector $\vec{r_3}$ satisfying the equations

$\vec{r_1} \times \vec{r_3} = \vec{r_2}, \ \vec{r_1} \cdot \vec{r_3} = 3.$

Applied Mathematics I

2K4-ASN-02

Time: 3 Hours **Max. Marks: 100**

NOTE:
- **Part 'A' may be attempted in first 6 pages of Answer Sheet.**
- **Part 'B' in rest of the Sheets of Answer Sheet.**
- **Answers may be given in English or Hindi.**

PART A

Q.1 Attempt any 10 questions: **(10×2=20)**

a. If $\vec{a} = 2\hat{i} + \hat{j} - \hat{k}$ and $\vec{b} = \hat{i} - 2\hat{j} + 3\hat{k}$ determine $\left|\vec{a} + 2\vec{b}\right|$.

b. Find $3\vec{a} - \vec{b}$, if $\vec{a} = \hat{i} + 2\hat{j} + \hat{k}$ and $\vec{b} = 2\hat{i} + \hat{j} - 3\hat{k}$.

c. Find $\vec{a} \cdot \vec{b}$ if $\vec{a} = -2\hat{i} + 2\hat{j} - \hat{k}$ and $\vec{b} = 2\hat{i} - 3\hat{j} + 4\hat{k}$.

d. Find the equation of the straight line with slope 3 and y intercept –2.

e. Find the equation of the circle centred at (1, –1) having radius 7.

f. Find the coordinates of the vertex and focus for the parabola $2x^2 - 14y + 16 = 0$.

g. Find the slope of a line perpendicular to the line $5x + 6y + 15 = 0$.

h. Find the polar coordinates of the point having Cartesian coordinates $(-1, \sqrt{3})$.

i. If $f(x) = 12x^5 + 7x^3 - 5x + 2$ then find $f''(x)$.

j. Evaluate $\lim_{x \to 1}(x^2 - 3x + 10)$.

k. If $f(x) = \dfrac{3x - 2}{2x^2 - 1}$, then find $f''(x)$.

l. If $f(x) = e^{\sin 2x}$ then find $f''(x)$.

m. If $f(x) = \log(\log x^2)$ then find $f''(x)$.

n. Evaluate $\lim_{x \to 0} \dfrac{\sin x}{x}$.

Q.2 Attempt any five questions: **(5×4=20)**

a. If If $f(x) = x^2 \sin 2x$ then find $f''(x)$.

b. $f(x) = 5^x e^{2+5x}$ then find $f'(x)$.

c. Evaluate $\lim_{x \to 0} \dfrac{a^x - b^x}{x}$.

d. Find the equation of the line which passes through the point (–1, –2) and perpendicular to the line through (–2, 3) and (–5, –6).

e. Find the equation of the parabola whose vertex is at origin, focus is on x-axis and passes through (–2, 6).

f. If the sum of two unit vectors is a unit vector, prove that the magnitude of their difference is $\sqrt{3}$.

g. Show that the vectors $\vec{a} = 3\hat{i} - 2\hat{j} + \hat{k}$, $\vec{b} = \hat{i} - 3\hat{j} + 5\hat{k}$, $\vec{c} = 2\hat{i} + \hat{j} - 4\hat{k}$ form a right angled triangle.

h. If $\vec{a} = \hat{i} + 2\hat{j} - 3\hat{k}$, $\vec{b} = 3\hat{i} - \hat{j} + 2\hat{k}$, find the angle between $(2\vec{a} + \vec{b})$ and $(\vec{a} + 2\vec{b})$.

PART B

Attempt any 3 questions: **(3×20=60)**

Q.3 a. If $y = e^{ax} \sin bx$, then show that $y_2 - 2ay_1 + (a^2 + b^2)\, y = 0$, where y_1 and y_2 are first order and second order differential of y.

b. If $\hat{a}, \hat{b}$ are unit vectors and θ is the angle between them, show that $\sin\frac{\theta}{2} = \frac{1}{2}\left|\hat{a} - \hat{b}\right|$.

Q.4 a. Evaluate $\lim\limits_{x\to 0} \dfrac{\sin(\pi \cos^2 x)}{x^2}$.

b. Find the equation of the circle passing through the points (5, 1) (4, 6) and (2, –2).

Q.5 a. Find the sides and angles of the triangle whose vertices are $\hat{i} - 2\hat{j} + 2\hat{k}$, $2\hat{i} + \hat{j} - \hat{k}$, $3\hat{i} - \hat{j} + 2\hat{k}$.

b. Find the coordinates of the centre, the vertices, the foci, the eccentricity and equation of the directrices for the hyperbola $x^2 - 4y^2 + 6x + 16y - 11 = 0$.

Q.6 a. Find the coordinates of the centre, the vertices, the foci, the eccentricity and the lengths of the major and minor axes of the ellipse $16x^2 + 9y^2 + 32x - 36y - 92 = 0$.

b. If $2^x + 2^y = 2^{x+y}$, then find $\dfrac{dy}{dx}$.

Q.7 a. If $y = \dfrac{x^2}{\log(1 - 4x^2)}$, then find $\dfrac{dy}{dx}$.

b. Show that the three points A, B, C with position vectors $-2\vec{a} + 3\vec{b} + 5\vec{c}$, $\vec{a} + 2\vec{b} + 3\vec{c}$, $7\vec{a} - \vec{c}$ are collinear.

Applied Mathematics I

2K5-AS-02

Time: 3 Hours **Max. Marks: 100**

NOTE:

- **Part 'A' may be attempted in first 6 pages of Answer Sheet.**
- **Part 'B' in rest of the Sheets of Answer Sheet.**
- **Answers may be given in English or Hindi.**

PART A

Q.1 Attempt any 10 questions: **(10×2=20)**

a. Find the inverse of the matrix $A = \begin{bmatrix} 5 & -2 \\ 4 & -3 \end{bmatrix}$.

b. Find the product AB of the matrices $A = \begin{bmatrix} 1 & -2 \\ 3 & 1 \end{bmatrix}$ and $B = \begin{bmatrix} 2 & -2 \\ 1 & -4 \end{bmatrix}$.

c. Determine the intercepts on x and y-axis for the straight line $3x + 4y - 12 = 0$.

d. Find the equation of the circle centred at (2, –2) having radius 6.

e. Find the coordinates of the vertex and focus for the parabola $x^2 + 12y = 0$.

f. Find the slope of a line perpendicular to the line $2x + 5y + 1 = 0$.

g. If A is a square matrics such that $|A^T| = 8$, write the value of $|A^{-1}|$.

h. If $\begin{bmatrix} a & 14 \\ 1 & b \end{bmatrix} + 2\begin{bmatrix} 2 & -4 \\ 3 & 1 \end{bmatrix} = \begin{bmatrix} 8 & 5 \\ 7 & 7 \end{bmatrix}$, then find the values of a and b.

i. If $\vec{a} = 2\hat{i} + \hat{j} - \hat{k}$ and $\vec{b} = 2\hat{i} - 2\hat{j} + 3\hat{k}$, determine $|\vec{a} + \vec{b}|$

j. Find $5\vec{a} - \vec{b}$, if $\vec{a} = \hat{i} - 2\hat{j} + \hat{k}$ and $\vec{b} = 2\hat{i} + \hat{j} - 3\hat{k}$.

k. Find $\frac{d}{dx}(e^{ax-b})$.

l. Find $\frac{}{dx}(\sin\quad)$.

m. Evaluate $\lim_{x\to 2}(2x^2 - 4)$.

n. Find $\frac{d}{dx}(5x^3)$.

Q.2 Attempt any 5 questions: **(5×4=20)**

a. Convert the Cartesian coordinates $(-3, -\sqrt{3})$ in to the corresponding polar coordinates.

b. Find the equation of the line which passes through the point (2, –4) and parallel to the line $8x + 2y + 3 = 0$.

c. Evaluate the determinant of the matrix $A = \begin{bmatrix} 3 & 1 & 3 \\ 3 & 1 & 2 \\ 1 & 2 & 3 \end{bmatrix}$.

d. If $\vec{a} = 2\hat{i} + \hat{j} + 3\hat{k}$, $\vec{b} = 3\hat{i} + 2\hat{j} - \hat{k}$, then find $\vec{a} . \vec{b}$.

e. If $y = \log(\text{cosec } x - \cot x)$, find $\frac{dy}{dx}$.

f. Find the derivative of cos x from the first principles.

g. Compute AB-BA, where $A = \begin{bmatrix} 1 & 8 \\ 2 & 5 \end{bmatrix}, B = \begin{bmatrix} 3 & 2 \\ 1 & -2 \end{bmatrix}$.

h. Obtain the equation of the circle having the points (1, –2) and (–1, 1) as the end points of a diameter.

PART B

Attempt any 3 questions: **(3×20=60)**

Q.3 a. Evaluate $\lim_{x \to 0} \frac{\tan x - \sin x}{x^3}$.

b. If $A = \begin{bmatrix} 2 & 1 & 3 \\ 3 & 1 & 2 \\ 1 & 2 & 3 \end{bmatrix}$, verify that $A\,(adjA) = (adjA)\ A = |A| I$, where I is the identity matrix of the third order.

Q.4 a. A triangle is formed by the lines $x + y = 3$, $2x - 3y = 6$ and $7x - 3y + 9 = 0$, find the coordinates of the centroid of the triangle.

b. Find the equation of the circle with centre (–1, 1) and touching the line $2x - y = 8$.

Q.5 a. Find a unit vector perpendicular to each of the vectors $3\hat{i} - 2\hat{j} + \hat{k}$ and $\hat{i} + \hat{j} - 3\hat{k}$.

b. Find the coordinates of the centre, vertices, foci and equations of the directrices for the hyperbola $x^2 - 3y^2 + 12x + 6y + 18 = 0$.

Q.6 a. If $y = \frac{x}{2}\sqrt{a^2 - x^2} + \frac{a^2}{2}\sin^{-1}\left(\frac{x}{a}\right)$, show that $\frac{dy}{dx} = \sqrt{a^2 - x^2}$.

b. If $\vec{a}$ and $\vec{b}$ be two – zero vectors such that $\left|\vec{a} + \vec{b}\right| = \left|\vec{a} - \vec{b}\right|$, show that $\vec{a}$ is perpendicular to $\vec{b}$.

Q.7 a. Solve the following system of equations using matrices:

$$x + y + z = 3, \quad x - 2y + 3z = 4, \quad x + 4y + 9z = 6.$$

b. Show that $\begin{vmatrix} 1 & 1 & 1 \\ a & b & c \\ a^3 & b^3 & c^3 \end{vmatrix} = (a-b)(b-c)(c-a)(a+b+c)$ applying the properties of determinants.

Applied Mathematics I

3K3-AAM-04

Time: 3 Hours **Max. Marks: 100**

NOTE:
- **Part 'A' may be attempted in first 6 pages of Answer Sheet.**
- **Part 'B' in rest of the Sheets of Answer Sheet.**
- **Answers may be given in English or Hindi.**

PART A

Q.1 Attempt any 10 questions: **(10×2=20)**

a. Evaluate cos 150°.

b. Write the value of $1 - 2\sin^2\left(\frac{\pi}{8}\right)$.

c. Express sin 3α in terms of sin α.

d. Find the equation of the circle centred at (0, 0) having radius 8.

e. Find the coordinates of the vertex and focus for the parabola $y^2 = 12x$.

f. Find the slope of a line perpendicular to the line $3x - 2y + 1 = 0$.

g. Prove that $(1 - \cos A)(1 + \sec A)\cot A = \sin A$.

h. If $A = \begin{bmatrix} 2 & -1 \\ 4 & 3 \end{bmatrix}, B = \begin{bmatrix} 3 & 1 \\ 1 & -2 \end{bmatrix}$, then find $2A + B$.

i. If $A = \begin{bmatrix} 2 & 3 \\ 1 & 5 \end{bmatrix}$, then find the transpose of A.

j. Find the inverse of the matrix $A = \begin{bmatrix} 4 & 2 \\ -1 & 3 \end{bmatrix}$.

k. Find the product of the matrices $A = \begin{bmatrix} 1 & 2 \\ -4 & 3 \end{bmatrix}$, and $B = \begin{bmatrix} 2 & 2 \\ 1 & -4 \end{bmatrix}$.

l. If $f(x) = x^4 - 3x^2 + 8x + 6$, then find $f'''(x)$.

m. Evaluate $\lim_{x\to 2} (4x^2 - 5x)$.

n. If $f(x) = \frac{2}{x^2 - 1}$, then find $f''(x)$.

Q.2 Attempt any 5 questions: **(5×4=20)**

a. $A + B + C = 180°$, show that $\sin 2A + \sin 2B + \sin 2C = 4 \sin A \sin B \sin C$.

b. Prove that $\cos 10° \cos 50° \cos 60° \cos 70° = \frac{\sqrt{3}}{16}$.

c. If $f(x) = x^5 (1 - x^2)^4$, then find $f''(x)$.

d. Find $\lim_{x \to 2} \frac{x^3 - 8}{x - 2}$.

e. Find the equation of the line which passes through the point (–2, –4) and parallel to the line $8x - 2y + 3 = 0$.

f. Evaluate the determinant of the matrix $A = \begin{bmatrix} 3 & 1 & 3 \\ 3 & 1 & 2 \\ 1 & 2 & 3 \end{bmatrix}$.

g. Compute AB-BA, where $A = \begin{bmatrix} 1 & 8 \\ 2 & 5 \end{bmatrix}$, $B = \begin{bmatrix} 3 & 2 \\ 1 & -2 \end{bmatrix}$.

h. Find the equation of the parabola whose vertex is at origin and directrix is $y + 3 = 0$.

PART B

Q.3 Attempt any 3 questions: (3×20=60)

a. If $y = \sqrt{\sin x + \sqrt{\sin x + \sqrt{\sin x \ldots\ldots}}}$ to ∞ show that $\frac{dy}{dx} = \frac{\cos x}{2y - 1}$.

b. Express the matrix $A = \begin{bmatrix} -1 & 7 & 1 \\ 2 & 3 & 4 \\ 5 & 0 & 5 \end{bmatrix}$ as sum of symmetric and skew-symmetric matrix.

Q.4 a. Find the inverse of the matrix $A = \begin{bmatrix} 2 & 1 & 3 \\ 3 & 1 & 2 \\ 1 & 2 & 3 \end{bmatrix}$.

b. Find the equation of the circle passing through the points (5, 1), (4, 6) and (2, –2).

Q.5 a. Prove that $\cos 130° + \cos 110° + \cos 10° = 0$.

b. Find the coordinates of the centre, the vertices, the foci, the eccentricity and equation of the directrices for the hyperbola $x^2 - 3y^2 + 6x + 6y + 18 = 0$.

Q.6 a. Find the coordinates of the centre, the vertices, the foci, the eccentricity and the lengths of the major and minor axes of the ellipse $9x^2 + 25y^2 - 36x - 150y + 36 = 0$.

b. Prove that $\cos^2 x \sin^3 x = \frac{1}{16}(2 \sin x + \sin 3x - \sin 5x)$.

Q.7 a. If $y = (\sin x)^{\cos x} + \alpha^x$, then find $\frac{dy}{dx}$.

b. Solve the following system of equation using matrix method.
$x + y + z = 3$, $x + 2y + 3z = 4$, $x + 4y + 9z = 6$.

Applied Mathematics I

96-AS-03

Time: 3 Hours **Max. Marks: 100**

NOTE:

- **Part 'A' may be attempted in first 6 pages of Answer Sheet.**
- **Part 'B' in rest of the Sheets of Answer Sheet.**
- **Answers may be given in English or Hindi.**

PART A

Q.1 Attempt any 10 parts: **(10×2=20)**

a. If $\begin{bmatrix} x & -2 \\ 2 & y \end{bmatrix} + \begin{bmatrix} 1 & 1 \\ 0 & -2 \end{bmatrix} = \begin{bmatrix} 1 & -1 \\ 2 & 3 \end{bmatrix}$, then find the values of x and y.

b. Find the inverse of the matrix $A = \begin{bmatrix} 1 & 2 \\ 3 & 1 \end{bmatrix}$.

c. Find the product of the matrices $A = \begin{bmatrix} 1 & 1 \\ -3 & -2 \end{bmatrix}$ and $B = \begin{bmatrix} 3 & 1 \\ 1 & 2 \end{bmatrix}$.

d. Determine the intercept on x and y-axis for the straight line $6x + 4y + 12 = 0$.

e. Find the equation of the circle centred at (2, –1) having radius 3.

f. Find the coordinates of the vertex and focus for the parabola $2y^2 - 8x + 2 = 0$.

g. Find the slope of a line perpendicular to the line $x - y + 1 = 0$.

h. Find the values of x and y so that the vectors $2\hat{i} + 3\hat{j}$ and $x\hat{i} + y\hat{j}$ are equal.

i. Find the value of μ if $\vec{a} = 3\hat{i} - 3\hat{j} + \mu\hat{k}$ and $\vec{b} = \hat{i} + 3\hat{j} + 2\hat{k}$ are orthogonal.

j. Find $2\vec{a} - 3\vec{b}$ if $\vec{a} = 2\hat{i} + 2\hat{j} + \hat{k}$, and $\vec{b} = -\hat{i} + \hat{j} - 3\hat{k}$.

k. Find $\frac{d}{dx}(e^{ax+b})$.

l. Find $\frac{d}{dx}(\cos^2 x)$.

m. Evaluate $\lim_{x \to 2} (x^2 - 2)$.

n. Find $\frac{d}{dx}(7x^{5/7})$.

Q.2 This question consists of 8 parts each of 4 marks. Attempt any 5 parts: **(5×4=20)**

a. If $A = \begin{bmatrix} 1 & -1 \\ 0 & 2 \end{bmatrix}, B = \begin{bmatrix} 1 & 0 \\ 1 & 2 \end{bmatrix}$, verify that $(A + B)^2 \neq A^2 + B^2 + 2AB$.

b. Find the equation of the line which passes through the point (–2, –4) and parallel to the line $6x + 2y + 3 = 0$.

c. Prove that $\begin{vmatrix} a+b & a & b \\ a & a+c & c \\ b & c & b+c \end{vmatrix} = 4abc.$

d. If $\vec{a} = \hat{i} + \hat{j} + 3\hat{k}$, $\vec{b} = \hat{i} - 3\hat{j} + \hat{k}$, then find the angle between $\vec{a} + \vec{b}$ and $\vec{a} - \vec{b}$.

e. Evaluate $\int (1 + \sin(2x))dx$.

f. Find the equation of the parabola whose focus is at (3, 2) and directrix is $y + 3 = 0$.

g. Find the vector parallel to the vector $\hat{i} - 2\hat{j}$ and has the magnitude 10 units.

h. Find y', if $y = \dfrac{(2x^2 + 3)^{1/2}}{x} \cos x$.

PART B

Attempt any 3 questions: **(3×20=60)**

Q.3 a. Find the equations of tangent and normal to the curve $y = x^2 - 5x + 7$ at the point (1, 3).

b. Show that $A = \begin{bmatrix} 1 & 2 & 3 \\ 2 & -1 & 4 \\ 3 & 1 & 1 \end{bmatrix}$ satisfies $A^3 - A^2 - 18A - 30I = 0$.

Q.4 a. Evaluate $\int \left\{ \dfrac{\cos 5x + \cos 4x}{1 - 2\cos 3x} \right\} dx$.

b. Find the equation of the circle which passes through the points (1, –6), (2, 1) and (5, 2).

Q.5 a. Find the coordinates of the centre, the vertices, the foci, the eccentricity and equation of the directrices for the hyperbola $x^2 - 3y^2 + 9x + 8y + 16 = 0$.

b. Forces of magnitude 1 and 3 newtons acting in the directions of $2\hat{i} - 3\hat{j} - 6\hat{k}$ and $3\hat{i} - 2\hat{j} + 6\hat{k}$ respectively act on a particle which is displaced from point (2, –1, –3) to (5, –1, 1). Find the work done by the forces, the unit of length being metre.

Q.6 a. Find the coordinates of the centre, the vertices, the foci, the eccentricity and the lengths of the major and minor axes of the ellipse $9x^2 + 25y^2 - 32x - 150y + 34 = 0$.

b. Show that the vectors $\vec{a} - 2\vec{b} + 3\vec{c}, -2\vec{a} + 3\vec{b} - 4\vec{c}, -\vec{b} + 2\vec{c}$ are coplanar.

Q.7 a. Solve the following differential equations:

$$\frac{dy}{dx} + x^2 y = \sin x.$$

b. Prove that $\begin{vmatrix} x & x^2 & yz \\ y & y^2 & zx \\ z & z^2 & xy \end{vmatrix} = (x - y)(y - z)(z - x)(xy + yz + zx).$

Applied Mathematics I

3K-ASN-02A

Time: 3 Hours **Max. Marks: 100**

NOTE:

- **Part 'A' may be attempted in first 6 pages of Answer Sheet.**
- **Part 'B' in rest of the Sheets of Answer Sheet.**
- **Answers may be given in English or Hindi.**

PART A

Q.1 Attempt any 10 parts: **(10×2=20)**

a. How many terms are there in the series 3, 6, 9, 12, ... 36.

b. Find the sixth term of G.P. 12, –4, 4/3, –4/9, ...

c. Write the ninth term of $(1 + 2x)^{12}$.

d. Find the equation of the straight line passing through the point (–1, –2) with slope $\frac{4}{5}$.

e. Find the equation of the circle centred at (1, –1) having radius 5.

f. Find the slope of a line perpendicular to the line $2x + y + 3 = 0$.

g. Find the value of $1 - 2 \sin^2 45°$.

h. Prove that $(\sec A - 1)(\sec A + 1) \cot A = \tan A$.

i. If $A = \begin{bmatrix} 2 & -4 \\ 3 & 1 \end{bmatrix}, B = \begin{bmatrix} 4 & -1 \\ 5 & 3 \end{bmatrix}$, then find $2A\text{-}3B$.

j. What is the value of $(\tan 45°)^2$

k. Find the inverse of the matrix $A = \begin{bmatrix} 1 & 9 \\ 7 & 6 \end{bmatrix}$.

l. Find the product of the matrix $A = \begin{bmatrix} 3 & -1 \\ 5 & 2 \end{bmatrix}$ and $B = \begin{bmatrix} 5 & 7 \\ -1 & -1 \end{bmatrix}$.

m. Find the unit vector in the direction of $\vec{a} = 2\hat{i} - 3\hat{j} + \hat{k}$.

n. Find $3\vec{a} + 4\vec{b}$, if $\vec{a} = \hat{i} - 2\hat{j} + \hat{k}$, and $\vec{b} = 3\hat{i} + \hat{j} + 3\hat{k}$.

Q.2 Attempt any five questions: **(5×4=20)**

a. The 5th term of a G.P. is 81 and second term is 24, find the series.

b. Find the value of n if the coefficients of the sixth and the sixteenth terms in the expansion of $(a + b)^n$ are equal.

c. Find the equation of the line which passes through the point (–2, –4) and parallel to the line $8x - 2y + 3 = 0$.

d. Resolve into partial fractions $\frac{2x-3}{x^2+5x+6}$.

e. If $\vec{a} = 2\hat{i} + \hat{j} + \hat{k}$, $\vec{b} = 3\hat{i} - 5\hat{j} + \hat{k}$, then find $\vec{a} \cdot \vec{b}$.

f. Prove that sin 47° + cos 77° = cos 17°.

g. Compute *AB-BA*, where $A = \begin{bmatrix} -6 & 2 \\ 5 & 4 \end{bmatrix}, B = \begin{bmatrix} 3 & 1 \\ -2 & 4 \end{bmatrix}$.

h. Find the equation of the parabola whose vertex is (1, 3) and focus is (–1, 3).

PART B

Attempt any three questions: **(3×20=60)**

Q.3 a. Find the sum of all integers between 200 and 1000 that are divisible by 4.

b. Express the matrix $A = \begin{bmatrix} 4 & 2 & -3 \\ 1 & 3 & -6 \\ -5 & 0 & -7 \end{bmatrix}$ as sum of symmetric and skew-symmetric matrix.

Q.4 a. Find the inverse of the matrix $A = \begin{bmatrix} 2 & 5 & 3 \\ 3 & 1 & 2 \\ 1 & 2 & 1 \end{bmatrix}$.

b. Find the equation of the circle having radius $\sqrt{13}$ and tangent to the line $2x - 3y + 1 = 0$ at (1, 1).

Q.5 a. Prove that cos 130° + cos 110° + cos 10° = 0.

b. Find the coordinates of the centre, the vertices, the foci, the eccentricity and equation of the directrices for the hyperbola $x^2 - 4y^2 + 6x + 16y - 11 = 0$.

Q.6 a. Find the coordinates of the centre, the vertices, the foci, the eccentricity and the lengths of the major and minor axes of the ellipse $x^2 + 2y^2 + 4x - 12y + 20 = 0$.

b. If $\vec{a} = \hat{i} - 2\hat{j} + 3\hat{k}$, $\vec{b} = -\hat{i} + 2\hat{j} + 2\hat{k}$ and $\vec{c} = 3\hat{i} + \hat{j}$, find t such that $\vec{a} + t\vec{b}$ is perpendicular to $\vec{c}$.

Q.7 a. Solve the following system of equations by cramer's Rule.

$$x + y + z = 3,\ x - 2y + 3z = 4,\ x + 4y + 9z = 6.$$

b. Show that $\begin{vmatrix} 1+a & 1 & 1 \\ 1 & 1+b & 1 \\ 1 & 1 & 1+c \end{vmatrix} = abc\left(1 + \frac{1}{a} + \frac{1}{b} + \frac{1}{c}\right)$.

Dec. 2017

Applied Mathematics I

3K4-MAN-02 **Mechanical/Automobile Branch**

Time: 3 Hours **Max. Marks: 100**

NOTE:

- **Part 'A' may be attempted in first 6 pages of Answer Sheet.**
- **Part 'B' in rest of the Sheets of Answer Sheet.**
- **Answers may be given in English or Hindi language.**

PART A

Q.1 Attempt any 10 parts: **(10×2=20)**

i. Find the 12th term of G.P. 2, 8, 32 ....

ii. Find the value of k such that the sequence $k - 1, k + 3$ and $3k - 1$ are in A.P.

iii. Determine the intercept on x and y-axes for the straight line $6x + 2y + 12 = 0$.

iv. Find the equation of the circle centred at (1, – 2) having radius 6.

v. Find the coordinates of the vertex and focus for the parabola $y^2 - 24x = 0$.

vi. Find the slope of a line perpendicular to the line $2x + 3y + 6 = 0$.

vii. Find the value of sin 30°.

viii. Prove that $(1 - \sin A)(1 + \sin A) \sec A = \cos A$.

ix. If $\begin{bmatrix} x & -2 \\ 2 & y \end{bmatrix} + \begin{bmatrix} 1 & 1 \\ 0 & -2 \end{bmatrix} = \begin{bmatrix} 1 & -1 \\ 2 & 4 \end{bmatrix}$, then find the values of x and y.

x. Resolve into partial fractions $\dfrac{x}{(x-1)(x-2)}$.

xi. Find the inverse of the matrix $A = \begin{bmatrix} 5 & 2 \\ 3 & 1 \end{bmatrix}$.

xii. Find the product of the matrices $A = \begin{bmatrix} 1 & 1 \\ -3 & 2 \end{bmatrix}$ and $B = \begin{bmatrix} 3 & 1 \\ 4 & 4 \end{bmatrix}$.

xiii. Find the value of μ if $\vec{a} = 2\hat{i} - 3\hat{j} + \mu\hat{k}$ and $\vec{b} = \hat{i} + 3\hat{j} + 2\hat{k}$ are orthogonal.

xiv. Find $2\vec{a} - 4\vec{b}$, if $\vec{a} = \hat{i} + 2\hat{j} + \hat{k}$, and $\vec{b} = 2\hat{i} + \hat{j} - 3\hat{k}$.

Q.2 This question consists of 8 parts each of 4 marks. Attempt any 5 questions: **(5×4=20)**

i. Find the equation of the line which passes through the point (– 2, – 4) and parallel to the line $6x + 2y + 3 = 0$.

ii. Find the sum of thirteen terms of the A.P. 3, 8, 13, 18 ....

iii. Prove that $\begin{vmatrix} a+b & a & b \\ a & a+c & c \\ b & c & b+c \end{vmatrix} = 4\,abc.$

iv. If $\vec{a} = 2\hat{i} + \hat{j} + 3\hat{k}$, $\vec{b} = \hat{i} - 3\hat{j} + \hat{k}$, then find $\vec{a} \cdot \vec{b}$.

v. When $A + B + C = 180°$ show that $\tan A + \tan B + \tan C = \tan A \tan B \tan C$.

vi. Find the value of sin 225°.

vii. Compute $AB - BA$, where $A = \begin{bmatrix} 5 & 4 \\ -1 & 3 \end{bmatrix}, B = \begin{bmatrix} 2 & 3 \\ -4 & 1 \end{bmatrix}$.

viii. Find the equation of the parabola whose vertex is at (3, 2) and directrix is $y + 4 = 0$.

PART B

Attempt any three questions: **(3×20=60)**

Q.3 a. Determine the middle term in the expansion of $\left(\frac{1}{x} - x^2\right)^{12}$.

b. If $A = \begin{bmatrix} -2 & 3 & -1 \\ -1 & 2 & -1 \\ -6 & 9 & -4 \end{bmatrix}, B = \begin{bmatrix} 1 & 3 & -1 \\ 2 & 2 & -1 \\ 3 & 0 & -1 \end{bmatrix}$, verify that $AB = BA = I_3$.

Q.4 a. Find the inverse of the matrix $A = \begin{bmatrix} 3 & -3 & 4 \\ 2 & -3 & 4 \\ 0 & -1 & 1 \end{bmatrix}$.

b. Find the equation of the circle passes through the points (1, –6), (2, 1) and (5, 2).

Q.5 a. Prove that cos 130° + cos 110° + cos 10° = 0.

b. Find the coordinates of the centre, the vertices, the foci, the eccentricity and equation of the directrices for the hyperbola $x^2 - 3y^2 + 6x + 6y + 18 = 0$.

Q.6 a. Find the coordinates of the centre, the vertices, the foci, the eccentricity and the lengths of the major and minor axes of the ellipse

$$9x^2 + 25y^2 - 36x - 150y + 36 = 0.$$

b. A force of magnitude 6 units acting parallel to $2\hat{i} - 2\hat{j} + \hat{k}$ displace the point of application from (1, 2, 3) to (5, 3, 7).

Q.7 a. Solve the following system of equations.

$$x + y + z = 6,\ x + 2y + 3z = 10,\quad x + 2y + 5z = 12.$$

b. Show that $A = \begin{bmatrix} 1 & 2 & 3 \\ 2 & -1 & 4 \\ 3 & 1 & 1 \end{bmatrix}$ satisfies $A^3 - A^2 - 18A - 30I = 0$.

Dec. 2017

Applied Mathematics I

3K4-CAM-02 **Civil Engineering**

Time: 3 Hours **Max. Marks: 100**

NOTE:
- **Part 'A' may be attempted in first 6 pages of Answer Sheet.**
- **Part 'B' in rest of the Sheets of Answer Sheet.**
- **Answers may be given in English or Hindi language.**

PART-'A'

Q.1 Attempt any ten parts: **(2×10=20)**

i. If $\frac{2x-1}{(x-2)(x+1)} = \frac{A}{x-2} + \frac{B}{x+1}$, Find A and B.

ii. The third term of a G.P. is 4. Find the product of its first five terms.

iii. Write the number of terms in the expansion of $(1 + 2x + x^2)^{25}$.

iv. If $\vec{a} = \hat{i} + 2\hat{j} + 3\hat{k}$, $\vec{b} = \hat{i} + 2\hat{j} + \hat{k}$, $\vec{c} = 3\hat{i} + \hat{j}$, find t such $\vec{a} + t\vec{b}$ is perpendicular to $\vec{c}$.

v. Write the value of cos 50° cos 40° + sin 50° sin 40°.

vi. Find the value of sin 15°.

vii. Find the coordinates of the centre and radius of circle $x^2 + y^2 - 8x - 4y = 29$.

viii. Find the eccentricity of the ellipse $x^2 + 5y^2 = 25$.

ix. Find the length of the latus rectum of the parabola $x^2 = 36y$.

x. Write the equation of the x-axis.

xi. Find out whether the matrix is symmetric or not $\begin{bmatrix} 2 & 3 & 4 \\ 3 & 4 & 7 \\ 4 & 7 & 8 \end{bmatrix}$.

xii. Evaluate the determinant of matrix $\begin{bmatrix} 2 & -4 \\ 3 & 2 \end{bmatrix}$.

xiii. If $\alpha + \beta = \frac{\pi}{2}$, find $\tan \alpha \tan \beta$.

xiv. If in a ΔABC, $\angle C = 30°$, $a = 2$ cm, $b = 3$ cm, find its area.

xv. Find the eccentricity of the hyperbola $9x^2 - 4y^2 = 169$.

Q.2 Attempt any five parts: **(4×5=20)**

i. Using the Binomial theorem, find the first four terms of the expression $\left(3a - \frac{1}{x}\right)^9$

ii. If the sum of first n terms of the A.P. 85 + 90 + 95 + ... is equal to the sum of the first $3n$ terms of the A.P. 9 + 11 + 13 + ... find the value of n.

iii. Prove that $1 - 2\cos^2 x = \dfrac{\tan \quad 1}{\tan \cdot \quad 1}$.

iv. If $A = \begin{bmatrix} 1 & 3 \\ 2 & 1 \\ 4 & 2 \end{bmatrix} B = \begin{bmatrix} 1 & 3 & 2 \\ 2 & 4 & 1 \end{bmatrix}$ find $A \times B$.

v. Without expanding, show that

$$\begin{vmatrix} 1 & b+c & b^2+c^2 \\ 1 & c+a & c^2+a^2 \\ 1 & a+b & a^2+b^2 \end{vmatrix} = (b-c)\ (c-a)\ (a-b).$$

vi. If $\vec{a} = 2\hat{i} - 3\hat{j} - \hat{k}$, $\vec{b} = \hat{i} + 4\hat{j} - 2\hat{k}$, find $\vec{a} \times \vec{b}$ and a unit vector perpendicular to both $\vec{a}$ and $\vec{b}$.

vii. If $A + B + C = 180°$, then show that $\sin 2A + \sin 2B + \sin 2C = 4 \sin A \sin B \sin C$.

viii. Find the coordinates of the vertex and focus of the parabola $y^2 + 4y + 20x + 4 = 0$.

PART B

Attempt any three questions: **(3×20=60)**

Q.3 a. Find the equation of the parabola whose vertex is (–4, 3) and focus is (–6, 3).

b. Find the inverse of the matrix

$$A = \begin{bmatrix} 1 & -2 & 3 \\ 2 & 3 & -1 \\ -3 & 1 & 2 \end{bmatrix}.$$

Q.4 a. Using matrix method, solve the system of linear equation.

$$x + y + z = 6$$
$$x + 2y + 3z = 10$$
$$x + 2y + 4z = 8$$

b. Find the equation of the circle which passes through the centre of the circle $x^2 + y^2 + 6x + 12y - 7 = 0$ and is concentric with the circle. $2x^2 + 2y^2 + 6x - 8y - 8 = 0$

Q.5 a. Show that the vectors $2\hat{i} - \hat{j} + \hat{k}$, $\hat{i} - 3\hat{j} - 5\hat{k}$ and $3\hat{i} - 4\hat{j} - 4\hat{k}$ from the sides of a right angle triangle.

b. Find the equation of the ellipse with focus at the point (1, 1), eccentricity $\frac{1}{2}$ and direction $x - y + 3 = 0$

Q.6 a. Show that the equation $16x^2 - 36y^2 - 96x + 144y = 720$ represents a hyperbola. Find the coordinate of the centre, the vertices, the foci, the lengths of the axes, the eccentricity and the equation of the directrix of the hyperbola.

b. Show that the middle term in the binomial expansion of $(1 + x)^{2n}$ is

$$\frac{1.3.5 ......(2x-1)}{n} 2^n \cdot x^n.$$

Q.7 a. If the angle of elevation of a cloud from a point λ metres above a lake is α and the angle of depression of its reflection in the lake is β, prove that the height of the cloud is $\dfrac{\lambda(\tan\beta + \tan\alpha)}{\tan\beta - \tan\alpha}$.

b. If p and q be the perpendiculars from the origin upon the straight lines $x \sec\theta + y \operatorname{cosec}\theta = a$ and $x \cos 2\theta - y \sin\theta = a \cos 2\theta$, prove that $4p^2 + q^2 = a^2$.

Dec. 2017

Applied Mathematics I

3K-ASN-02A **Electrical Engineering Branch**

Time: 3 Hours **Max. Marks: 100**

NOTE:

- **Part 'A' may be attempted in first 6 pages of Answer Sheet.**
- **Part 'B' in rest of the Sheets of Answer Sheet.**
- **Answers may be given in English or Hindi language.**

PART A

Q.1 Attempt any ten questions: **(10×2=20)**

i. How many terms are the series 3, 6, 9, 12, ... 36.

ii. Find the value of k so that sequence $2k - 5$, $k - 4$, $10 - 3k$ form a G.P.

iii. Resolve into partial fractions $\dfrac{1}{(x-2)(x+1)}$.

iv. Find the equation of the straight line passes through the point (–1, –2) with slope 3/4.

v. Find the equation of the circle centred at (1, –1) having radius 5.

vi. Find the coordinates of the vertex and focus for the parabola $y^2 = 8x$.

vii. Find the slope of a line perpendicular to the line $2x - 5y + 10 = 0$.

viii. Find the value of sin 60°.

ix. Prove that $(1 - \cos A)(1 + \sec A)\cot A = \sin A$.

x. If $A = \begin{bmatrix} 2 & -4 \\ -9 & 1 \end{bmatrix}, B = \begin{bmatrix} 4 & -1 \\ 5 & 3 \end{bmatrix}$, then find $A + 5B$.

xi. Prove that $\sin^2 A + \cos^2 A = 1$.

xii. Find the product of the matrices $A = \begin{bmatrix} 3 & 4 \\ 2 & 1 \end{bmatrix}, B = \begin{bmatrix} 5 & 7 \\ 2 & 1 \end{bmatrix}$.

xiii. Find the unit vector in the direction of $\vec{a} = 2\hat{i} - 3\hat{j} + \hat{k}$.

xiv. Find $3\vec{a} + 4\vec{b}$, if $\vec{a} = \hat{i} - 2\hat{j} + \hat{k}$, and $\vec{b} = 3\hat{i} + \hat{j} + 3\hat{k}$.

Q.2 Attempt any five parts: **(4×5=20)**

i. Find the sum of fifteen terms of the A.P. 3, 8, 13, 18 ....

ii. Find the seventh term in $\left(x^{\frac{1}{4}} - \dfrac{3}{x^{\frac{1}{2}}}\right)^{-1}, \left(|x| > 3^{\frac{4}{3}}\right)$.

iii. Find the equation of the line which passes through the point (–2, –4) and parallel to the line $4x + 2y + 3 = 0$.

iv. Evaluate the determinant of the matrix, $A = \begin{bmatrix} 1 & 2 & 5 \\ 3 & 1 & 4 \\ 1 & 5 & 2 \end{bmatrix}$.

v. If $\vec{a} = 2\hat{i} + \hat{j} + \hat{k}, \vec{b} = 3\hat{i} + 5\hat{j} + \hat{k}$, then find $\vec{a} \cdot \vec{b}$.

vi. Prove that $\dfrac{2\tan\frac{x}{2}}{1 + \tan^2\frac{x}{2}} = \sin x$.

vii. Find the value of cos 315°.

viii. Compute $AB - BA$, Where $A = \begin{bmatrix} 6 & 2 \\ 5 & 4 \end{bmatrix}$ and $B = \begin{bmatrix} 3 & 2 \\ -5 & 4 \end{bmatrix}$.

PART B

Attempt any three questions: **(3×20=60)**

Q.3 a. Find the sum of all integers between 200 and 1000 that are divisible by 3.

b. Express the matrix $A = \begin{bmatrix} -1 & 7 & 1 \\ 2 & 3 & 4 \\ 5 & 0 & 5 \end{bmatrix}$ as sum of symmetric and skew-symmetric matrix.

Q.4 a. Find the inverse of the matrix $A = \begin{bmatrix} 1 & 2 & 0 \\ 2 & 3 & -1 \\ 1 & -1 & 3 \end{bmatrix}$.

b. Find the equation of the circle passes through the points (4, 1) and tangent at (2, –1) is $2x - 3y - 7 = 0$.

Q.5 a. If $A + B + C = 180°$, prove that $\tan\frac{A}{2}\tan\frac{B}{2} + \tan\frac{B}{2}\tan\frac{C}{2} + \tan\frac{A}{2}\tan\frac{C}{2} = 1$.

b. Find the coordinates of the centre, the vertices, the foci, the eccentricity and equation of the directrices for the hyperbola $16x^2 - 25y^2 + 400 = 0$.

Q.6 a. Find the coordinates of the centre, the vertices,the foci, the eccentricity and the length of the major and minor axes of the ellipse $3x^2 + 2y^2 - 30x - 4y + 23 = 0$.

b. Forces of magnitude 5 and 3 units act on a particle in the directions $6\hat{i} + 2\hat{j} + \hat{k}3$ and $3\hat{i} + 2\hat{j} + 6\hat{k}$ respectively, act on a particle which is displaced from the point (2, 2, –1) to (4, 3, 1). Find the work done by the forces.

Q.7 a. Solve the following system of equations:

$$x + y + z = 3, \quad x + 2y + 3z = 4, x + 4y + 9z = 6$$

b. Show that $\begin{vmatrix} 1 & b+c & b^2+c^2 \\ 1 & c+a & c^2+a^2 \\ 1 & a+b & a^2+b^2 \end{vmatrix} = (a-b)(b-c)(c-a)$.

Dec. 2017

Applied Mathematics I

3K-ASN-02 **Elec Comm, Digital Electronics, Medical Electronics**

Time: 3 Hours **Max. Marks: 100**

NOTE:

- **Part 'A' may be attempted in first 6 pages of Answer Sheet.**
- **Part 'B' in rest of the Sheets of Answer Sheet.**
- **Answers may be given in English or Hindi language.**

PART A

Q.1 Attempt any ten parts: **(10×2=20)**

i. Find the value of K so that the sequence $2K - 5, K - 4, 10 - 3K$ forms a G.P.

ii. Find the Fifteenth term of the sequence 3, 8, 13, 18 ....

iii. If $A + B + C = 180°$, than show that $\sin\frac{1}{2}(B+C) = \cos\frac{1}{2}A$.

iv. Find the value of sin 15° using formulae.

v. Express $\cos 4\alpha$ in terms of $\cos \alpha$.

vi. Determine the x and y-intercepts of the equation of line $3x + 4y + 12 = 0$.

vii. Write the equations of the circle having centre at (0, 0) and radius equal to 5.

viii. Find the eccentricity of the ellipse $\frac{x^2}{16} + \frac{y^2}{4} = 1$.

ix. Find the centre and radius of the circle $x^2 + y^2 - 10x + 8y + 5 = 0$.

x. The focal distance of a point of the parabola $y^2 = 12x$ is 4. Find the abscissa of this point.

xi. Evaluate the determinant of the matrices $A = \begin{bmatrix} 2 & 3 \\ 4 & 5 \end{bmatrix}$.

xii. Find the sum of the two matrices $A = \begin{bmatrix} 1 & 2 \\ 3 & 4 \end{bmatrix}$ and $B = \begin{bmatrix} 2 & 4 \\ 3 & 7 \end{bmatrix}$.

xiii. Find the product of the given matrices $A = [1 \ 3 \ 2]$ and $B = \begin{bmatrix} 1 & 2 & 3 \\ 3 & 1 & 2 \\ 2 & 4 & 1 \end{bmatrix}$.

xiv. Find the dot product of the following vectors $\vec{A} = 2\hat{i} - 3\hat{j}$ and $\vec{B} = -4\hat{i} + 2\hat{j}$.

Q.2 Attempt any five parts: **(5×4=20)**

i. The fourth term of an A.P. is 14, and the ninth term is 34. Find the Thirteenth term.

ii. Find the sum of all even positive integers less than 200 which are not divisible by 6.

iii. If $A + B + C = 180°$, show that

$\sin 2A + \sin 2B + \sin 2C = 4 \sin A \sin B \sin C$.

iv. Find the equation of an ellipse whose vertices are $(\pm 4, 0)$ and major axis is 4.

v. Solve following systems of simultaneous by Cramer's Rule $\begin{cases} 2x + y = 4 \\ 3x + y = 1 \end{cases}$.

vi. Evaluate the determine of $A = \begin{bmatrix} 1 & 2 & 3 \\ 2 & -3 & 4 \\ -3 & 4 & 5 \end{bmatrix}$.

vii. Find the equation of the straight line which passes through $(-1, -2)$ and perpendicular to the line through $(-2, 3)$ and $(-5, -6)$.

viii. Find the equation of the circle which passes through the points $(5, 1)$, $(4, 6)$ and $(2, -2)$.

PART B

Attempt any three questions: **(3×20=60)**

Q.3 a. Find the term independent of y in the expansion of $(xy^{1/6} - y^{2/3})^5$.

b. If $A + B + C = 180°$, then prove that

$\sin^2 A + \sin^2 B + \sin^2 C = 2 \sin A \sin B \sin C$.

Q.4 a. Find the equation of the ellipse whose foci are $(\pm 10, 0)$ and eccentricity is 5/6.

b. Show that $\begin{bmatrix} 1 & 1 & 1 \\ a & b & c \\ a^2 & b^2 & c^2 \end{bmatrix} = (a - b)(b - c)(c - a)$ (by using properties of determinants).

Q.5 a. Find the coordinates of the centre, the vertices, the foci, the eccentricity and the equation of the directrix of the hyperbola $144x^2 - 25y^2 - 576x + 200y + 3776 = 0$.

b. Express the following matrix as the sum of the symmetric and skew symmetric matrix $\begin{bmatrix} 3 & -2 & -4 \\ 3 & -2 & -5 \\ 1 & 1 & 2 \end{bmatrix}$.

Q.6 a. If $\vec{A} = \hat{i} + 2\hat{j} + 3\hat{k}, \vec{B} = -\hat{i} + 2\hat{j} + \hat{k}, \vec{C} = 3\hat{i} + \hat{j}$, Find λ such that $\vec{A} + \lambda\vec{B}$ is perpendicular to $\vec{C}$.

b. Find the equation of the line through the point (1, 2, 3) and parallel to the planes

$$x + y + z = 6 \quad \text{and} \quad 2x - y - 3z = 0.$$

Q.7 a. Find the equation of the circle passes through two points (2, 3), (–1, 6) and centre lying on $2x + 5y + 1 = 0$.

b. Find the equation of the line which passes through the point (3, –4) and the sum of its intercepts on the coordinate axes is –5.

Part II

June 2019

Engineering Mathematics II

DE 105/DC 105 **ET/CS (New Scheme)**

Time: 3 Hours **Max. Marks: 100**

NOTE: There are 9 Questions in all

- **Question 1 is compulsory and carries 20 marks. Answer to Q.1 must be written in the space provided for it in the answer book supplied and nowhere else.**
- **The answer sheet for the Q.1 will be collected by the invigilator after 45 minutes of the commencement of the examination.**
- **Out of the remaining EIGHT Questions answer any FIVE Questions. Each question carries 16 marks.**
- **Any required data not explicitly given, may be suitably assumed and stated.**

Q.1 Choose the correct or best alternative in the following: **(2 ×10)**

a. The value of $\underset{x\to 0}{\text{Lt}} \dfrac{1-\cos x}{3x^2}$ is

(A) 1 (B) 0 (C) $\dfrac{1}{6}$ (D) $\dfrac{1}{3}$

b. The value of the definite integral $\int_{-a}^{a} |x|\, dx$ is equal to

(A) a (B) a^2 (C) 0 (D) $2a$

c. The particular Integral of the differential equation $(D-2)^2 y = e^{2x}$ is

(A) $\dfrac{x^2}{2}e^{2x}$ (B) $-\dfrac{x^2}{2}e^{2x}$ (C) $\dfrac{x^2}{2}e^{-2x}$ (D) $-\dfrac{x^2}{2}e^{-2x}$

d. The imaginary part of $(\sin x + i \cos x)^5$ is

(A) $-\cos 5x$ (B) $-\sin 5x$ (C) $\sin 5x$ (D) $\cos 5x$

e. If $\vec{A} = 2\hat{i} + 2\hat{j} - \hat{k}, \vec{B} = 6\hat{i} - 3\hat{j} + 2\hat{k},$ then the unit vector perpendicular to both $\vec{A}$ and $\vec{B}$ is

(A) $\hat{i} + 10\hat{j} + 18\hat{k} / 5\sqrt{17}$ (B) $\hat{i} - 10\hat{j} + 18\hat{k} / 5\sqrt{17}$

(C) $-\hat{i} - 10\hat{j} - 18\hat{k} / 5\sqrt{17}$ (D) $\hat{i} - 10\hat{j} - 18\hat{k} / 5\sqrt{17}$

f. The vectors A and B are perpendicular to each other if and only if

(A) $A \times B = 0$ (B) $A \bullet B = 0$ (C) $A \times B \neq 0$ (D) $A \bullet B \neq 0$

g. If $x + iy = \sqrt{2} + 3i$ then $x^2 + y$ is equal to

(A) 7 (B) 5 (C) 13 (D) $\sqrt{2} + 3$

h. $\lim_{n\to\infty}\left(\frac{n}{n-1}\right)^2$ is equal to

(A) 1 (B) ∞ (C) 2 (D) 0

i. If value of $L\{F(t)\} = f(s)$, then $L\{(\cosh at)F(t)\}$ is equal to

(A) $\frac{1}{2}[f(s-a) - f(s+a)]$ (B) $\frac{1}{2}[f(s-a) + f(s+a)]$

(C) $-\frac{1}{2}[f(s-a) - f(s+a)]$ (D) $-\frac{1}{2}[f(s+a) + f(s-a)]$

j. $L^{-1}\left\{\frac{1}{(s-a)^n}\right\}$ is equal to

(A) $\frac{e^{at}t^{n-1}}{n!}$ (B) $\frac{e^{at}t^n}{(n-1)!}$ (C) $\frac{e^{at}t^{n-1}}{(n-1)!}$ (D) $\frac{e^{at}t^n}{n!}$

ANSWER ANY FIVE QUESTINS OUT OF EIGHT QUESTIONS

EACH QUESTION CARRIES 16 MARKS

Q.2 a. Verify Lagrange's mean value theorem for the function $f(x) = x^3 - 3x - 1$ in $(-11/7, 13/7)$. **(8)**

b. Evaluate $\lim_{x\to 0}\frac{2\sin x - \sin 2x}{x^3}$ **(8)**

Q.3 a. Evaluate $\int_0^{\frac{\pi}{2}} \cos^9 x dx$ **(8)**

b. Find the area lying between the parabola $y = 4x - x^2$ and the line $y = x$. **(8)**

Q.4 a. If z is a complex number, then prove that:

(*i*) $|z|^2 = |\overline{z}|^2 = z\overline{z}$ (*ii*) $Amp(z) + Amp(\overline{z}) = 0$ **(8)**

b. If $x = \cos\alpha + i\sin\alpha$, $y = \cos\beta + i\sin\beta$, prove that $\frac{x-y}{x+y} = i\tan\left(\frac{\alpha-\beta}{2}\right)$ **(8)**

Q.5 a. If $\vec{A} = \hat{i} + 2\hat{j} - 3\hat{k}, \vec{B} = 3\hat{i} - \hat{j} + 2\hat{k}$. Find the angle between $2\vec{A} + \vec{B}$ and $\vec{A} + 2\vec{B}$. **(8)**

b. Evaluate $\vec{A} \bullet \left(\vec{B} \times \vec{C}\right)$, *if* $\vec{A} = 2\hat{i} - 3\hat{j}, \vec{B} = \hat{i} + \hat{j} + \hat{k}$ and $\vec{C} = 3\hat{i} - \hat{k}$. **(8)**

Q.6 a. Solve the differential equation $\dfrac{d^2y}{dx^2} + 3\dfrac{dy}{dx} + 2y = 4\sin^2 2x$. **(8)**

b. Solve the differential equation $\dfrac{d^2y}{dx^2} + 4y = x^2 + \cos 2x$. **(8)**

Q.7 a. Discuss the convergence of the series $1 + \dfrac{2!}{2^2} + \dfrac{3!}{3^3} + \dfrac{4!}{4^4} \ldots\infty$ **(8)**

b. Test the convergence of the series $\sum \dfrac{(n+1)^n x^n}{n^{n+1}}$. **(8)**

Q.8 a. Find the Laplace transform of $[t^2 \sin^2 3t]$. **(8)**

b. Find the Laplace transform of $\left[\dfrac{1 - \cos 2t}{t^2}\right]$. **(8)**

Q.9 a. Apply Convolution theorem to evaluate $L^{-1}\left[\dfrac{s^2}{(s^4 - 16)}\right]$. **(8)**

b. Using Laplace transform solve the differential equation

$y'' - 2y' + y = e^t$, given that $y(0) = 2$ and $y(0) = -1$. **(8)**

June 2018

Engineering Mathematics II

DE 105/DC 105 **ET/CS (New Scheme)**

Time: 3 Hours **Max. Marks: 100**

NOTE: There are 9 Questions in all

- **Question 1 is compulsory and carries 20 marks. Answer to Q.1 must be written in the space provided for it in the answer book supplied and nowhere else.**
- **The answer sheet for the Q.1 will be collected by the invigilator after 45 minutes of the commencement of the examination.**
- **Out of the remaining EIGHT Questions answer any FIVE Questions. Each question carries 16 marks.**
- **Any required data not explicitly given, may be suitably assumed and stated.**

Q.1 Choose the correct or the best alternative in the following: (2 × 10)

a. If $f(x) = x^3 + x$, $a = 0$, $b = 1$, and $\dfrac{f(b) - f(a)}{b - a} = f(c), a < c < b$ then 'c' is equal to:

(A) $-\dfrac{1}{\sqrt{3}}$ (B) $\sqrt{3}$ (C) $-\sqrt{3}$ (D) $\dfrac{1}{\sqrt{3}}$

b. The value of $\int_0^{\pi/2} \cos^7 x\, dx$ is equal to:

(A) $\dfrac{16}{35}$ (B) $\dfrac{8}{35}$ (C) $-\dfrac{16}{35}$ (D) $\dfrac{32}{35}$

c. If $z_1 = (2 + 3i)$ and $z_2 = (3 - 2i)$ be two complex numbers, then $\left(\dfrac{z_1}{z_2}\right)$ is equal to

(A) $\dfrac{12i}{13}$ (B) $-i$ (C) i (D) $\dfrac{-12i}{13}$

d. If $\vec{a} = (2\hat{i} + \hat{j} + \hat{k})$ and $\vec{b} = (\hat{i} - 2\hat{j} + 2\hat{k})$ be two vectors, then unit vector perpendicular to both $\vec{a}$ and $\vec{b}$ is equal to:

(A) $\left[\dfrac{4\hat{i} - 3\hat{j} + 5\hat{k}}{5\sqrt{2}}\right]$ (B) $\left[\dfrac{4\hat{i} + 3\hat{j} - 5\hat{k}}{5\sqrt{2}}\right]$ (C) $\left[\dfrac{4\hat{i} - 3\hat{j} - 5\hat{k}}{5\sqrt{2}}\right]$ (D) $\left[-\dfrac{4\hat{i} - 3\hat{j} - 5\hat{k}}{5\sqrt{2}}\right]$

e. The solution of differential equation $(D^2 + 4)\, y = \cos 2x$ is:

(A) $y = c_1 \cos 2x + c_2 \sin 2x + \dfrac{x}{4} \cos 2x$ (B) $y = c_1 \cos 2x + c_2 \sin 2x + \dfrac{x}{4} \sin 2x$

(C) $y = c_1 \cos 2x + c_2 \sin 2x - \frac{x}{4} \sin 2x$ (D) $y = c_1 \cos 2x + c_2 \sin 2x - \frac{x}{4} \cos 2x$

f. The series:

$1 + \frac{1}{1!} + \frac{1}{2!} + \frac{1}{3!} + \ldots\ldots + \frac{1}{(n-1)!} + \ldots$ is

(A) convergent (B) divergent
(C) oscillatory (D) conditional divergence

g. The Laplace transform of $(e^t \cdot \sin 4t)$ is:

(A) $\frac{4}{s^2 - 2s + 15}$ (B) $\frac{4}{s^2 - 2s + 17}$ (C) $\frac{-4}{s^2 + 2s + 17}$ (D) $\frac{4}{s^2 + 2s - 15}$

h. If $\vec{a} = (\hat{i} - 2\hat{j})$; $\vec{b} = (\hat{i} + 2\hat{j} - 4\hat{k})$; and $\vec{c} = (2\hat{i} - 3\hat{j})$ then value of $\vec{a} \cdot (\vec{b} \times \vec{c})$ is equal to:

(A) – 4 (B) – 3 (C) 3 (D) 4

i. The inverse Laplace transform of $\left[\frac{1}{s^2 - 5s + 6}\right]$ is equal to:

(A) $(e^{3t} + e^{2t})$ (B) $(-e^{3t} - e^{2t})$ (C) $(-e^{3t} + e^{2t})$ (D) $(-e^{2t} + e^{3t})$

j. For $n \in N$ and $i = \sqrt{-1}$, $\left(\frac{\cos nx + i \sin nx}{\cos nx - i \sin nx}\right)$ is equal to:

(A) $\cos 2nx + i \sin 2nx$ (B) $\cos 2nx - i \sin 2nx$
(B) $\sin 2nx + i \cos 2nx$ (C) $\sin 2nx - i \cos 2nx$

ANSWER ANY FIVE QUESTINS OUT OF EIGHT QUESTIONS

EACH QUESTION CARRIES 16 MARKS

Q.2 a. Expand $\log(1 + e^x)$ in ascending powers of 'x' as far as the term containing x^4 by using Maclaurin's theorem. **(8)**

b. Evaluate: $\lim_{x \to 0} \left[\frac{1}{x} - \frac{1}{x^2} \log(1 + x)\right]$ **(8)**

Q.3 a. Evaluate $\int_0^{\pi/2} \sin^3 x \cos^4 x \, dx$ by using "Reduction Formula". **(8)**

b. The area bounded by the parabola $y^2 = 4x$ and the straight line $4x - 3y + 2 = 0$ is rotated about y-axis. Find the volume of solid so formed. **(8)**

Q.4 a. If 'n' is a positive integer, then show that: **(8)**

$$\left[\frac{1 + \cos\theta + i \sin\theta}{1 + \cos\theta - i \sin\theta}\right] = \cos n\theta + i \sin n\theta$$

b. A coil of resistance of 4 ohms and inductive resistance of 42 ohms is connected in parallel with a resistance of 15 ohms and capacitive reactance of 18 ohms. This parallel circuit is connected across 220 volts mains as shown in figure. Find (*i*) current taken by each circuit (*ii*) total current. **(8)**

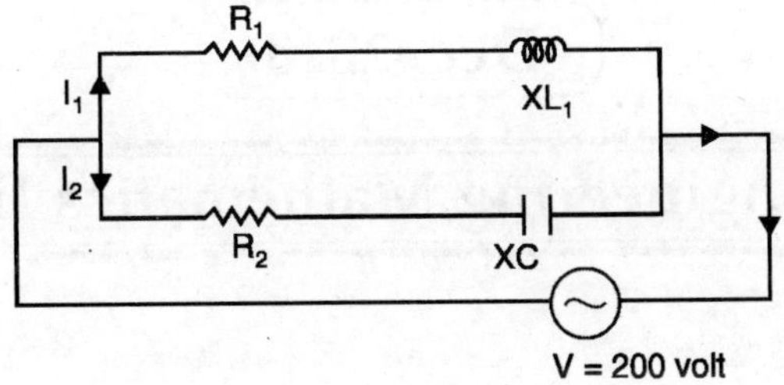

Q.5 a. If $\vec{a} = a_1\hat{i} + a_2\hat{j} + a_3\hat{k}$; $\vec{b} = b_1\hat{i} + b_2\hat{j} + b_3\hat{k}$ and $\vec{c} = c_1\hat{i} + c_2\hat{j} + c_3\hat{k}$, then show that:

$$\vec{a}\cdot\left(\vec{b}\times\vec{c}\right) = \begin{Bmatrix} a_1 & a_2 & a_3 \\ b_1 & b_2 & b_3 \\ c_1 & c_2 & c_3 \end{Bmatrix}$$ **(8)**

b. The point of application of the force $\vec{F} = 5\hat{i} + 10\hat{j} + \hat{k}$ is displaced from the point A (2, 1, 3) to the point B (4, 0, – 5), find work done by the force $\vec{F}$. **(8)**

Q.6 a. Solve differential equation: $D^2y - 2Dy + 3y = x + \cos x$ **(8)**

b. An inductance of 2 henries and a resistance of 20 ohms are connected in series with an e.m.f. 100 sin 150t. If the current is zero, when t = 0, find the current at the end of 0.01 sec. **(8)**

Q.7 a. Test the convergence of the series **(8)**

$$\sum_{n=1}^{\infty}\left[\sqrt{\left(\frac{n}{n+1}\right)}\right]x^n$$

b. Test the convergence of the series $\frac{1}{1.3} + \frac{1}{3.5} + \frac{1}{5.7} + \ldots\ldots$ **(8)**

Q.8 Find Laplace transform of following: **(8+8)**

a. $f(t) = \sin^2 t$

b. $f(t) = \left(\frac{\sin t}{t}\right)$

Q.9 a. Find inverse Laplace transform of $\left[\frac{s}{(s+3)^2 + 4}\right]$ **(8)**

b. Using Laplace transform, find solution of initial value problem:
$y'' + 9y = 6\cos 3x$; $y(0) = 2$, $y'(0) = 0$ **(8)**

Dec. 2018

Engineering Mathematics II

DE 105/DC 105 **ET/CS (New Scheme)**

Time: 3 Hours **Max. Marks: 100**

NOTE: There are 9 Questions In all

- **Question 1 is compulsory and carries 20 marks. Answer to Q.1 must be written in the space provided for it in the answer book supplied and nowhere else.**
- **The answer sheet for the Q.1 will be collected by the invigilator after 45 minutes of the commencement of the examination.**
- **Out of the remaining EIGHT Questions answer any FIVE Questions. Each question carries 16 marks.**
- **Any required data not explicitly given, may be suitably assumed and stated.**

Q.1 Choose the correct or the best alternative in the following: **(2 ×10)**

a. The value of $\lim\limits_{x\to 0} \dfrac{\log x}{\cot x}$ is

(A) 2 (B) 3 (C) 1 (D) 0

b. The value of $\int_0^{\pi/2} \sin^6 x\, dx$ is

(A) $3\pi/12$ (B) $5\pi/32$ (C) $7\pi/13$ (D) $\pi/12$

c. If $z = 1 + i$ then z^2 is

(A) $5i$ (B) $4i$ (C) $2i$ (D) $3i$

d. If $\vec{A} = 2\hat{i} + \hat{j} + \hat{k}$, $\vec{B} = 3\hat{i} + 2\hat{j} + \hat{k}$, then $|\vec{A} \times \vec{B}|$ is

(A) $\sqrt{178}$ (B) $\sqrt{179}$ (C) $\sqrt{168}$ (D) none of these

e. The series $\sqrt{1/4} + \sqrt{2/6} + \sqrt{3/8} + \sqrt{n/2(n+1)}$ is

(A) divergent (B) convergent (C) oscillating (D) none of these

f. The solution of the differential equation $\dfrac{d^2y}{dx^2} - 8\dfrac{dy}{dx} + 15y = 0$ is

(A) $y = c_1 e^{3x} + c_2 e^{5x}$ (B) $y = c_1 e^{3x} + c_2 e^{-5x}$

(C) $y = c_1 e^{3x} + c_2 e^{4x}$ (D) $y = c_1 e^{4x} + c_2 e^{5x}$

g. The complementary function of $D^2x + 4Dx + 5y = 0$

(A) $e^{-2x} (A \cos x + B \sin x)$ (B) $e^{+2x} (A \cos x + B \sin x)$

(C) $e^{-4x} (A \cos x + B \sin x)$ (D) $e^{-5x} (A \cos x - B \sin x)$

h. The Laplace transform of $1/\sqrt{t}$ is

(A) $\sqrt{\pi}/4s$ (B) $\sqrt{\pi}/2s$ (C) $\sqrt{\pi}/s$ (D) $-\sqrt{\pi}/s$

i. The value of $L^{-1}\left\{\frac{1}{\sqrt{s}}\right\}$ is

(A) $-1/\sqrt{\pi t}$ (B) $2/\sqrt{\pi t}$ (C) $1/\sqrt{\pi t}$ (D) $3/\sqrt{\pi t}$

j. The particular integral of $(D^2 - 2D + 4)\ y = e^x \cos x$ is

(A) $\frac{1}{2}e^{-x}\cos x$ (B) $\frac{1}{2}e^{x}\cos x$ (C) $\frac{1}{2}e^{-x}\sin x$ (D) $\frac{1}{2}e^{x}\sin x$

ANSWER ANY FIVE QUESTINS OUT OF EIGHT QUESTIONS

EACH QUESTION CARRIES 16 MARKS

Q.2 a. Expand $\tan x$ by Maclaurin's series upto the torn containing x^4. **(8)**

b. Evaluate $\lim\limits_{x\to 0}\frac{\sqrt{1+\sin x}-\sqrt{1-\sin x}}{x}$ **(8)**

Q.3 a. Evaluate $\int_0^{\frac{\pi}{6}} \cos^5\theta\, d\theta$ **(8)**

b. Find the area of the surface revolution formed by revolving the curve $r = 2a\cos\theta$ about the initial line. **(8)**

Q.4 a. Express $(1 + \cos\theta + i\sin\theta)$ in modulus – arguments form. **(8)**

b. The admittance and current are given the complex number $7 + 5j$ and $17 - 16j$ respectively. Find the voltage of the current. **(8)**

Q.5 a. Show that the vector $2\hat{i} + \hat{j} + \hat{k}, \hat{i} - 3\hat{j} - 5\hat{k}$ and $3\hat{i} - 4\hat{j} - 4\hat{k}$ form the sides of a right angled-triangle. **(8)**

b. A rigid body is rotating with angular velocity 2 radians/sec about an axis OR where R is $2\hat{i} + 2\hat{j} + \hat{k}$ and O is the origin. Find the vector of the point $3\hat{i} + 2\hat{j} - \hat{k}$ on the body. **(8)**

Q.6 a. Solve $\frac{d^2y}{dx^2} - 4y = (1+e^x)^2$ **(8)**

b. A voltage E e^{-at} is applied at $t = 0$ to an LR circuit. Find the current at any time t. **(8)**

Q.7 a. Test for the absolute convergence of the series $\sum\limits_{n=2}^{\infty}\frac{(-1)^n}{n(\log n)^2}$. **(8)**

b. Test for the convergence of the series $1+\frac{x}{2}+\frac{x^2}{5}+\frac{x^3}{10}+\ldots+\frac{x^n}{n^2+1}$. **(8)**

Q.8 a. Find the Laplace transform of $f(t) = \sin at \sin bt$. **(8)**

b. Find the Laplace transform of $\frac{1-\cos t}{t^2}$ **(8)**

Q.9 a. Find the inverse Laplace transform of $\log\left(\frac{s+1}{s-1}\right)$. **(8)**

b. Using convolution theorem find $L^{-1}\left\{\frac{s}{(s^2+a^2)^2}\right\}$ **(8)**

June 2018

Applied Mathematics II

2K7-BSN-01

Time: 3 Hours **Max. Marks: 100**

NOTE:

- **Part 'A' may be attempted in first 6 pages of Answer Sheet.**
- **Part 'B' in rest of the Sheets of Answer Sheet.**
- **Answers may be given in English or Hindi.**

PART A

Q.1 Attempt any 10 parts: **(10 × 2 = 20)**

a. Express $\frac{1}{2}+\frac{\sqrt{3}}{2}i$ in the polar form.

b. Evaluate: $\lim_{x\to 0}\frac{1-\cos x}{x^2}$.

c. Differentiate $\cos\sqrt{x}$ with respect to x.

d. Find $\frac{d}{dx}\left(x(\log x)^2\right)$, if $x > 0$.

e. Find the slope of the Normal to the curve $y = 4x^2$ at $x = \frac{1}{2\sqrt{2}}$.

f. Find the points at which the function $f(x) = x^3 + 6x^2 - 15x + 5$ has a maxima or minima.

g. Evaluate $\int\frac{dx}{1-\sin x}$.

h. Evaluate $\int_{-\pi/4}^{\pi/4}\text{cosec}^2\, x\, dx$.

i. The distance moved by a particle travelling in a straight line in t seconds is given by $s = 48\, t - t^3$. Find the time taken by the particle to come to rest.

j. If $\int e^x(\sin x+\cos x)dx = e^x f(x)+c$, find $f(x)$.

k. If $u=\tan^{-1}\left(\frac{x}{y}\right)$, write the value of $\frac{\partial u}{\partial y}$ at (1, 2).

l. Find the order and degree of the differential equation $\left(1+\left(\frac{dy}{dx}\right)^2\right)^{1/2}=\frac{d^2y}{dx^2}$.

m. Draw the phasor diagram of the function $f(t) = 4 \sin(\omega t + 45°)$.

n. Write the integrating factor of the differential equation $\frac{dy}{dx}+y\cot x=\operatorname{cosec} x$.

Q.2 Attempt any 5 parts: **(5 × 4 = 20)**

a. Separate $\tan(x - iy)$ into real and imaginary parts.

b. Differentiate $\tan x$ with respect to x by first principles.

c. First the derivative of $\log\sqrt{\frac{(1-\cos x)}{(1+\cos x)}}$ with respect to x.

d. Find $\frac{dy}{dx}$, if $x=a\left(\cos t+\log\tan\frac{t}{2}\right)$, $y=a\sin t$.

e. Show that the maximum value of $\left(\frac{1}{x}\right)^x$ is $e^{1/e}$.

f. Evaluate $\int_0^1\sqrt{9+x^2}\,dx$.

g. Evaluate $\int_0^1\frac{xdx}{\sqrt{1+x^2}}$.

h. Solve the differential equation $\frac{dy}{dx}=e^{x-y}+x^2e^{-y}$.

PART B

Attempt any three questions: **(3 × 20 = 60)**

Q.3 a. Find the value of k if $\lim_{x\to 1}\frac{x^4-1}{x-1}=\lim_{x\to k}\frac{x^3-k^3}{x^2-k^2}$.

b. Find $\frac{dy}{dx}$, given $x^y + y^x = 1$.

Q.4 a. If $\sin y = x \sin(a + y)$ show that $\frac{dy}{dx}=\frac{\sin^2(a+y)}{\sin a}$.

b. Evaluate $\int\frac{\sqrt{(\log x)^2}}{x}dx$.

Q.5 a. Evaluate $\int\frac{dx}{1+x^4}$.

b. Evaluate $\int_0^{\pi/6}\sin^4 6\theta\cos^6 3\theta\,d\theta$ by Gamma function.

Q.6 a. Find the volume of the solid generated by the revolution of the area under the curves $x^2 = 4y$ and $y = 1$ about the y-axis.

b. Using the Simpson's rule, obtain the approximate value of the integral $\int_0^1 \log(1+x)\,dx$, taking 4 equal intervals.

Q.7 a. If $u = \sin^{-1}\left[\dfrac{x+y}{\sqrt{x}+\sqrt{y}}\right]$, prove that: $x^2\dfrac{\partial^2 u}{\partial x^2} + 2xy\dfrac{\partial^2 u}{\partial x \partial y} + y^2\dfrac{\partial^2 u}{\partial y^2} = -\dfrac{\sin u \cos^2 u}{4\cos^3 u}$.

b. Solve the differential equation: $ydy - (1 - x^2 - y^2)\,xdx = 0$.

Applied Mathematics II

3K-BSN-02A

Time: 3 Hours **Max. Marks: 100**

NOTE:
- **Part 'A' may be attended in first 5 pages of Answer Sheet.**
- **Part 'B' may be attended in rest of the Sheets of Answer Sheet.**
- **Answers may be given in English or Hindi.**

PART A

Q.1 Attempt any ten questions: **(10 × 2 = 20)**

a. Express (1 + 2i) in Eulerian form.

b. If $u = \cos^{-1}(x + y)$, then find $\frac{\partial u}{\partial x}$ and $\frac{\partial u}{\partial y}$.

c. Find $\frac{dy}{dx}$ for $y = x \cos y + y \cos x$.

d. Find the derivative of $y = e^{ax} \cos(bx + c)$.

e. Evaluate $\int \sin^2 x\, dx$.

f. Evaluate $\int_{-1}^{1} (x^5 - 5x^2 + 12)\, dx$.

g. Evaluate $\lim_{x \to 0} \frac{x^2 - \sin x}{x}$.

h. Find the integrating factor of the differential equation $\frac{dy}{dx} = x^5 - 4x^2 y$.

i. Find the complementary function of the differential equation $(D^2 + 2D + 1)\, y = \cos 3x$

j. For the curve $x^2 + y^2 = 2ax$, find $\frac{dy}{dx}$ at (– 1, 1).

k. If $\int_{-a}^{a} (9x^2)\, dx = 6,$ then find real values of a.

l. Find the derivative of $y = x^{\cos 2x}$.

m. Evaluate $\lim_{x \to 1} \left(\frac{x^3 - 1}{x - 1} \right)$

n. Find the derivative of $y = e^{\sin x}$.

Q.2 Attempt any five questions, Answer in brief: **(5 × 4 = 20)**

a. Find the modules and amplitude of $\frac{1 - 2i}{1 + 3i}$.

b. Solve $\frac{dy}{dx} = e^{2x + 3y}$

c. Find $\frac{du}{dt}$, when $u = x^2 + y^2, x = -at^3, y = -2at$.

d. If $x^y = e^{x-y}$ prove that $\frac{dy}{dx} = \frac{\log x}{(1+\log x)^2}$

e. Evaluate $\int \frac{dx}{x\sqrt{x-8}}$

f. Evaluate $\int (x^2 - 5x)^2\, dx$

g. Evaluate $\lim_{x\to 0} \frac{1-e^x}{x^2}$

h. Find the first principle, find the derivative of sin x with respect to x.

PART B

Attempt any three questions: **(3 × 20 = 60)**

Q.3 a. Evaluate $\int \frac{x^2 + 5x - 1}{\sqrt{x}}\, dx$

b. Evaluate $\int_0^6 \frac{dx}{1+x^2}$ by Simpson's 1/3 rule using six equal intervals.

Q.4 a. Evaluate $\int_0^\pi \frac{x^2 \cos x}{1+\sin^2 x}\, dx$.

b. If $u = \log(x^2 + y^2 + z^2)$, then prove that: $x\frac{\partial^2 u}{\partial y\, \partial z} = y\cdot\frac{\partial^2 u}{\partial z\, \partial x} = z\frac{\partial^2 u}{\partial x\, \partial z}$.

Q.5 a. Find the maximum and minimum value for the function $f(x) = 8x^5 - 15x^4 + 10x^2$.

b. Find the area bounded by the hyperbola $x^2 - y^2 = a^2$ between the straight lines $x = a$ and $x = 2a$.

Q.6 a. Find the area enclosed by $y = x(x-1)(x-2)$ and x-axis.

b. (1) Evaluate $\lim_{x\to 0} (\sin x)^{\tan x}$

(2) If $x = a\cos^3 t$ and $y = a\sin^3 t$, show that $\frac{dy}{dx} = -1$ at $t = \frac{\pi}{4}$.

Q.7 a. Solve the differential equation $\frac{d^2y}{dx^2} - 2\frac{dy}{dx} + y = \sin 2x + e^{2x}$.

b. If $u = f(r)$, where $r = \sqrt{x^2 + y^2}$, prove that $\frac{\partial^2 u}{\partial x^2} + \frac{\partial^2 u}{\partial y^2} = f''(r) + \frac{1}{r}f'(r)$.

Dec. 2018

Applied Mathematics II

2K5-BS-01

Time: 3 Hours **Max. Marks: 100**

NOTE:

- **Part 'A' may be attemted in first 5 pages of Answer Sheet.**
- **Part 'B' may be attempted in rest of the Sheets of Answer Sheet.**
- **Answers may be given in English or Hindi.**

PART A

Q.1 Attempt any ten parts: **(10×2=20)**

a. Evaluate $\int \frac{(\log x)^2}{x} dx$.

b. Evaluate $\int (3^x + 2^x)^2 \, dx$.

c. If $\int_0^a \frac{1}{4+x^2} dx = \frac{\pi}{8}$, find the value of a.

d. Find the value of the integral $\int_0^{\frac{\pi}{2}} \cos^2 x \sin x \, dx$.

e. Find the Laplace transform of 2^{3t}.

f. Find the inverse Laplace transform of $\frac{5s+12}{s^2}$.

g. If $\frac{a_0}{2} + \sum_{n=1}^{\infty} (a_n \cos nx + b_n \sin nx)$ is the fourier series expansion of function $f(x) = x$ defined on $[-\pi, \pi]$, find the value of a_0.

h. Write the period of the function $f(x) = \tan \pi x$.

i. If $A = \{x : x = 2n, n \in N\}$ and $B = \{x : x \text{ is a prime natural number}\}$, find $A \cap B$.

j. Let $U = \{1, 2, 3, 4, 5, 6, 7, 8, 9\}$ be the universal set and $A = \{1, 2, 3, 4\}$, $B = \{2, 4, 6, 8\}$ be two sets. Find $(A \cup B)'$ and $(A \cap B)$.

k. Find the Laplace transform of $\sin(2t + 3)$.

l. If $f(x) = \begin{cases} -1 & \text{for } -\pi \le x \le 0 \\ 1 & \text{for } 0 \le x \le \pi \end{cases}$, find the sum of the fourier series of $f(x)$ at $x = 0$.

m. Find the inverse Laplace transform of $\frac{1}{(S-1)^3}$

n. Evaluate $\int \frac{(\log x)\, dx}{x}$

Q.2 Attempt any five parts: **(5×4=20)**

a. Find the fourier series for the function $f(x) = x$ in the interval $[-\pi, \pi,]$.

b. Evaluate $\int_0^{\frac{\pi}{2}} \sin^4 x \cos^2 x\, dx$.

c. Find Laplace inverse of $\dfrac{1}{s^2 + 8s + 16}$.

d. In a group of 800 people, 550 can speak Hindi and 450 can speak English, how many can speak both Hindi and English?

e. Find the Laplace transform of $e^t \sin^2 t$.

f. Evaluate $\int 5^x dx$.

g. Prove that the total number of subsets of a set of n elements is 2^n.

h. Find Laplace transform of $(te^t \sin 4t)$.

PART B

Attempt any three questions: **(3×20=60)**

Q.3 a. Find the Fourier series expansion for the function $f(x)$ given by

$$f(x) = \begin{cases} \pi x, 0 \le x \le 1 \\ \pi(2 - x), 1 \le x \le 2 \end{cases}$$

b. Obtain the fourier series to represent the function $f(x) = |x|$ for $-\pi \le x \le \pi$ and hence deduce that

$$\frac{\pi^2}{8} = \frac{1}{1^2} + \frac{1}{3^2} + \frac{1}{5^2} + \frac{1}{7^2} + \ldots\ldots$$

Q.4 a. Find the equation of the waveform as in figure below.

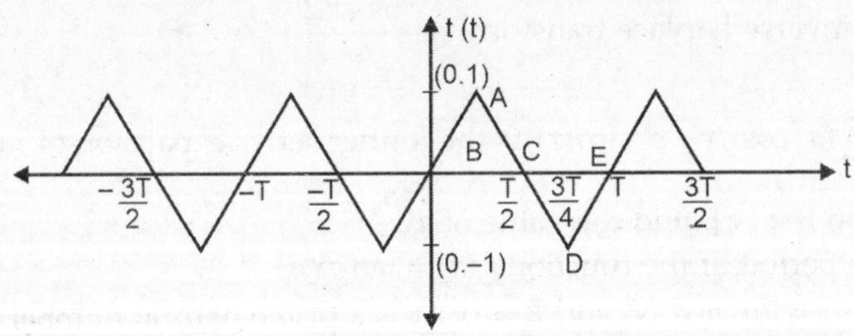

b. Find Laplace inverse of $\log\left(1 + \frac{1}{s^2}\right)$.

Q.5 a. If $A = \{x : x$ is a positive integer < 8 and a multiple of 3 or 5$\}$,

$B = \{x : x^3 - 6x^2 + 11x - 6 = 0\}$

$C = \{x : x$ is an even positive integer $< 7\}$

then show that (i) $A \cup (B \cup C) = (A \cup B) \cup C$ (ii) $A \cap (B \cap C) = (A \cap B) \cap C$.

b. Evaluate $\int \dfrac{x^2 + 1}{(x^2 + 2)(2x^2 + 1)} dx$.

Q.6 a. Calculate by Simpson's rule an approximate value of the integral $\int_0^1 \frac{1}{1+x} dx$ by dividing the interval [0, 1] into 10 equal parts. Hence, obtain an approximate value of $\log_e 2$.

b. Evaluate $\int_0^{\frac{\pi}{4}} \tan^3 x\, dx$.

Q.7 a. Find Laplace transform of $\left\{\frac{\sin at}{t}\right\}$.

b. Verify the truth of the following by Venn diagram:

(*i*) $A - B = A \cap B'$ (*ii*) $(A \cup B) \cap (A \cup B') = A$.

Applied Mathematics II

97-BSN-11

Time: 3 Hours **Max. Marks: 100**

NOTE:
- **Part 'A' may be attempted in first 6 pages of Answer Sheet.**
- **Part 'B' in rest of the Sheets of Answer Sheet.**
- **Answers may be given in English or Hindi.**

PART A

Q.1 The questions consists of 14 parts of 2 marks each. Attempted any 10 parts. **(10 × 2 = 20)**

a. If $f(x) = 3\cos 2x - \sin x$; find $f\left(\frac{5\pi}{6}\right)$

b. Evaluate $\lim\limits_{x\to 0}\frac{e^{2x}-1}{x}$

c. Find $\frac{d}{dx}(x|x|)$, if $x > 0$

d. Differentiate $\tan^{-1}\left(\frac{1-x}{1+x}\right)$ with respect to x.

e. Find the slope of the tangent to the curve $3x^2 + y = 4$ at the point $(-1, 1)$.

f. Find the integrating factor of the differential equation $\frac{dy}{dx} + y = 2$

g. Find $\int\frac{dx}{1+\tan^2 x}$

h. Find $\int\frac{\log x}{x}dx$

i. Fill in the blank: $\int\sqrt{9-x^2}\,dx = \frac{x}{2}\sqrt{9-x^2} + ... + \sin^{-1}\left(\frac{x}{3}\right) + C$

j. Evaluate $\int 3^{2\log_3 x}dx$

k. Find the degree of the homogeneous function $f(x, y) = y^2 \log\left(\frac{x}{y}\right)$

l. Find $\frac{dy}{dx}$ where $y = x^3 + 2\log_e x + 4\tan x$

m. Find the area bounded by the curve $y^2 = 2x$ and the line $x = 1$

n. Evaluate $\int_1^e \log x\,dx$

Q.2 Attempt any five questions: **(5 × 4 = 20)**

a. Evaluate $\lim_{x\to 0}\frac{\tan x-\sin x}{x^3}$

b. Differentiate e^x with respect to x using the first principles.

c. Find the percentage error in calculating the volume of a spherical ball if an error of 1% is made in measuring the radius of the ball.

d. Find $\int \tan^{-1}\left(\sqrt{\frac{1-\cos 2x}{1+\cos 2x}}\right)dx$

e. Evaluate $\int_1^2 \frac{1}{x(1+\log x)^2}dx$

f. Find the area lying in the first quadrant and bounded by the circle $x^2+y^2=4$, the line $x=\sqrt{3}\,y$ and x-axis.

g. Show that the function $y = A\sin nx + B\cos nx$ is a solution of the differential equation $\frac{d^2y}{dx^2}+n^2y=0.$

h. Solve the differential equation $(x^2-1)\frac{dy}{dx}+2xy=\frac{1}{x^2-1}.$

PART B

Attempt any three questions: **(3 × 20 = 60)**

Q.3 a. Differentiate $\sin^{-1}\left(\frac{2x}{1+y^2}\right)$ w.r.t. $\tan^{-1}\left(\frac{2x}{1-x^2}\right)$.

b. Find the equation of the tangent and normal to the curve $y = 5x^2 - 2x + 3$ at (1, 6).

Q.4 a. Find the maximum and minimum values of the function $x^3 + 8x^2 + 96x$.

b. From a differential equation by eliminating a and b from $y = ae^{3x} + be^x$.

Q.5 a. Evaluate $\int \frac{2x+1}{\sqrt{x^2+4x+3}}dx$

b. Evaluate $\int_0^1 \frac{(\tan^{-1}x)^3}{1+x^2}dx$

Q.6 a. Find the volume of the solid generated by revolving the ellipse $2x^2 + y^2 = 4$ about its minor axis.

b. Using trapezoidal rule, obtain an approximate value of the integral $\int_1^2 \frac{1}{x}dx$ taking 4 equal intervals. Hence obtain an approximate value of $\log_e 2$.

Q.7 a. Solve the differential equation: $x^2\frac{dy}{dx}=2xy+y^2$.

b. Solve: $\frac{dy}{dx}=\frac{x+2y-1}{x+2y+1}$.

May/June 2017

Applied Mathematics II

3K-BSN-2A

Time: 3 Hours **Max. Marks: 100**

NOTE:

- **Part 'A' may be attempted in first 6 pages of Answer Sheet.**
- **Part 'B' in rest of the Sheets of Answer Sheet.**
- **The questions paper consists of two parts, namely, Part A and Part B.**
- **Part A consists of two questions and Part B consists of 5 questions.**
- **Answers may be given in Hindi or English language.**

PART A

Q.1 Answer any 10 parts: **(2×10=20)**

a. Express $-1-i$ in polar and Eulerian forms.

b. Draw the phasor diagram of the function $f(t) = -6\cos(\omega t - 45°)$

c. If $f(x+3) = 2x^2 - 3x + 1$, find $f(x+1)$

d. Evaluate $\lim_{x\to 0}\frac{8^x - 2^x}{x}$.

e. Find the derivative of $\log|x|$ with respect to x.

f. If the curves $y = ae^x$ and $y = be^{-x}$ cut orthogonally, then find the value of ab

g. If the rate of change of area of a circle is equal to the rate of change of its diameter, then find its radius.

h. Evaluate $\int e^{-\log x}dx$.

i. Find $f(x)$ satisfying $\int e^x(\sec^2 x + \tan x)\,dx = e^x f(x) + C$.

j. If $\int_0^1 (3x^2 + 2x + k)\,dx = 0,$ find the value of of k.

j. Evaluate $\int_{-\pi/2}^{\pi/2} \sin(x)\,dx$.

l. Find the degree of the homogeneous function $f(x,y) = \frac{x+y}{\sqrt{x}+\sqrt{y}}$

m. Find the order and degree of the differential equation $\left[1+\left(\frac{dy}{dx}\right)^2\right] = \frac{d^2y}{dx^2}$

n. Find the integrating factor of the differential equation $x\frac{dy}{dx} - y = \sin x$

Q.2 Answer any 10 parts: **(2 × 10 = 20)**

a. If $x + iy = \sin(A + iB)$, prove that $\dfrac{x^2}{\sin^2 A} - \dfrac{y^2}{\cos^2 A} = 1$.

b. Evaluate $\lim\limits_{x\to 0} \dfrac{\tan x - \sin x}{x^3}$.

c. Find, from the first principle, the derivative of $\log_e x$, $(x > 0)$.

d. Differentiate $\tan^{-1}\left(\dfrac{\sqrt{1+x^2}-1}{x}\right)$ w.r.t. $\tan^{-1} x$.

e. Integrate: $\displaystyle\int \frac{1}{\sin(x-a)\sin(x-b)}\,dx$.

f. Find the area bounded by the parabola $y^2 = 4ax$ and its latus-rectum.

g. Find $\dfrac{dz}{dt}$ as a total derivative when $z = x^2 - y^2$, $\quad x = e^t \sin t$, $\quad y = e^t \cos t$

h. Solve the differential equation $x(1 + y^2)\,dx - y(1 + x^2)\,dy = 0$.

PART B

Q.3 Attempt any three questions: **(3×20=60)**

a. If $x\sqrt{1+y} + y\sqrt{1+x} = 0$, prove that $\dfrac{dy}{dx} = \dfrac{-1}{(x+1)^2}$.

b. A square piece of tin of side 24 cm is to be made into a box without top by cutting a square from each corner and folding up the flaps to form box. What should be the side of the square to be cut off so that the volume of the box is maximum? Also, find the maximum volume.

Q.4 a. For a positive constant a find $\dfrac{dy}{dx}$,

where $\qquad y = a^{t+\frac{1}{t}}$, and $x = \left(t + \dfrac{1}{t}\right)^a$

b. Find the equations of the tangent and normal to the curve

$$y = \frac{8a^3}{4a^2 + x^2} \text{ at the point where } x = 2a.$$

Q.5 a. Evaluate $\displaystyle\int_0^a x^4\sqrt{a^2 - x^2}\,dx$.

b. Apply Simpson's rule to find the approximate value of $\displaystyle\int_0^1 \frac{1}{1+x}\,dx$, taking 10 equal intervals. Hence, obtain an approximate value of log 2.

Q.6 a. Integrate: $\displaystyle\int \frac{1}{x\{6(\log x)^2 + 7\log x + 2\}}\,dx$.

b. Find the volume of the solid generated by revoluting the ellipse

$$\frac{x^2}{a^2} + \frac{y^2}{b^2} = 1$$

about the minor axis.

Q.7 a. Verify Euler's theorem for the function

$$u = \sin^{-1}\left(\frac{x}{y}\right) + \tan^{-1}\left(\frac{y}{x}\right).$$

b. Solve the differential equation

$$\frac{d^2x}{dt^2} - 3\frac{dx}{dt} + 2x = 4t + e^{3t}, \text{ given that } x = 1 \text{ and } \frac{dx}{dt} = -1 \text{ when } t = 0.$$

May/June 2017

Applied Mathematics II

3K4-MBM-5 **For Mechanical Branch**

Time: 3 Hours **Max. Marks: 100**

NOTE:

- **Part 'A' may be attempted in first 6 pages of Answer Sheet.**
- **Part 'B' in rest of the Sheets of Answer Sheet.**
- **Answers may be given in English or Hindi.**

PART A

Q.1 This question consists of 14 parts each of 2 marks. Attempt any 10 parts: (2×10=20)

a. If $f(x) = x + |x|$, find $f(3)$ and $f(-3)$.

b. If $\lim_{x\to 0}(1+ax)^{2/x} = e$, find the value of a.

c. If $y = e^{3\log x}$, find $\frac{dy}{dx}$.

d. Find the derivative of tan x' with respect to x.

e. Evaluate $\frac{d}{dx}(x|x|)$ at $x = -2$.

f. Determine the slope of the tangent to the curve $y^2 = 2x^3$ at the point (2, 4).

g. Integrate $\int \tan^2 x\,dx$.

h. Evaluate $\int_0^1 \frac{1}{\sqrt{1+x^2}}dx$.

i. If $f(x)$ is an odd function of x, find the value of $\int_{-a}^{+a} f(x)\,dx$.

j. Find the mean value of the function $f(x) = \sqrt{4-x}$ over [0, 3].

k. Evaluate $\Gamma\left(\frac{7}{2}\right)\cdot\Gamma\left(\frac{1}{2}\right)$, where the symbol 'Γ'denotes gamma.

l. If $u = f(x, y)$, where $x = \phi(t)$, $y = \psi(t)$, write the expression $\frac{du}{dt}$ as a total derivative.

m. Write the order and degree of the differential equation:

$$\left\{1+\left(\frac{dy}{dx}\right)^2\right\}^{1/4} = 2\sqrt{\frac{d^2y}{dx^2}}.$$

n. Find the integrating factor of the differential equation:

$$x\frac{dy}{dx} - y = 2x^3$$

Q.2 This question consists of 8 parts each of 4 marks. Attempt any 5 parts: (5×4 = 20)

a. Differentiate $\log_e x$ with respect to x from the first principles.

b. If $x = a(\theta - \sin\theta)$, $y = a(\sin\theta - \theta\cos\theta)$, find $\frac{dy}{dx}$.

c. Air is filled in a spherical balloon at the rate of 15 cm^3/sec. At what rate is its surface area increasing when the radius is 5 cm?

d. Find $\int \frac{xe^x}{(1+x)^2}dx$

e. Evaluate $\int_0^{\pi/2} \frac{1}{a^2\sin^2 x + b^2\cos^2 x}dx$

f. Solve the differential equation $\frac{dy}{dx} = \frac{y^2 - x^2}{2xy}$

g. If $u = \tan^{-1}\left(\frac{x^3 + y^3}{x - y}\right)$ prove that: $x\frac{\partial u}{\partial x} + y\frac{\partial u}{\partial y} = \sin 2u$

h. Solve $\frac{dy}{dx} = 1 + x + y + xy$

PART B

Part B consists of 5 questions each of 20 marks. Attempt any 3 questions: (3×20=60)

Q.3 a. Evaluate $\lim_{x\to 3}\frac{\sqrt{10-3x} - \sqrt{2x-5}}{\sqrt{x+1} - 2}$.

b. If $x^m y^n = (x + y)^{m+n}$, show that $\frac{dy}{dx} = \frac{y}{x}$.

Q.4 a. If $\sin y = x\sin(a + y)$, show that: $\frac{dy}{dx} = \frac{\sin^2(a+y)}{\sin a}$.

b. A square piece of tin of side 24 cm is to be made into a box without top by cutting a square from each corner and folding up the flaps to form a box. What should be the side of the square to be cut off so that the volume of the box is maximum? Also find this maximum volume.

Q.5 a. Evaluate $\int_0^a x^4\sqrt{a^2 - x^2}dx$.

b. Evaluate $\int_0^{\pi/2} \frac{\sin^5 x}{\sin^5 x + \cos^5 x}dx$.

Q.6 a. Find the area of the region bounded by the ellipse $\frac{x^2}{a^2} + \frac{y^2}{b^2} = 1$ using integration.

b. Apply Simpson's one-third rule, to obtain an approximate value of the definite integral $\int_0^{\pi} \frac{\sin x}{x}\,dx$, taking 4 equal intervals.

Q.7 a. Solve the differential equation:

$(1+x^2)\frac{dy}{dx}+2xy=4x^2$, subject to the initial condition $y(0) = 0$.

b. If $u = \log(x^3 + y^3 + z^3 - 3xyz)$, show that: $\left(\frac{\partial}{\partial x}+\frac{\partial}{\partial y}+\frac{\partial}{\partial z}\right)^2 u = -\frac{9}{(x+y+z)^2}$.

May/June 2017

Applied Mathematics II

3K-BSN-02

Time: 3 Hours **Max. Marks: 100**

NOTE:

- **Part 'A' may be attempted in first 6 pages of Answer Sheet.**
- **Part 'B' in rest of the Sheets of Answer Sheet.**
- **The questions paper consists of two parts, namely, Part A and Part B.**
- **A candidate has to attempt both parts.**
- **Part A consists of two questions and Part-B consists of 5 questions.**
- **Answers may be given in Hindi or English language.**

PART A

Q.1 Answer any 10 parts. Each part can be answered in either one word or one sentence or as per requirement of the questions. **(2×10=20)**

a. Write the modulus and argument of $1 - i$.

b. Evaluate $\lim_{x\to 0}\left(1+\frac{2}{x}\right)^{3/x}$.

c. Find $\frac{d}{dx}(x^2\,|x|)$ for $x < 0$.

d. If $y = \tan 3x^\circ$, find $\frac{dy}{dx}$.

e. Evaluate $\lim_{x\to 0}\frac{\sin 3x^\circ}{x}$.

f. If $f(x, y) = x^y$, find $\frac{\partial f}{\partial x}$ and $\frac{\partial f}{\partial y}$.

g. If the slope of the tangent to the curve $x = 1 - a\sin\theta$, $y = b\cos^2\theta$ at $\theta = \frac{\pi}{2}$ is 1, find the value of $a - 2b$.

h. Evaluate $\int \text{cosec}^2(2-3x)\,dx$.

i. If $\int_0^a 5x^4\,dx = 243$, find the value of a.

j. If $f(x)$ is a continuous function defined on $[a, b]$ write the expression for the root mean square (RMS) value of $f(x)$ over $[a, b]$.

k. Evaluate $\int \frac{\sec^2\sqrt{x}}{\sqrt{x}}\,dx$.

l. Solve the differential equation $2ydx + xdy = 0$.

m. Find the intergrating factor of the linear differential equation $\frac{dx}{dy} - \frac{x}{y} = 2y^2$.

n. Draw the phasor diagram of the function $f(t) = 12 \sin(\omega t + 45°)$.

Q.2 Answer any five parts: **(5×4=20)**

i. Prove that:

(a) $\cos h^2x - \sin h^2x = 1$

(b) $\cos h^2x + \sin h^2x = \cosh 2x$

ii. Evaluate $\lim_{x\to\infty}\left(\sqrt{x^2+x+1} - x\right)$.

iii. Differentiate $\sin^2x$ from first principles.

iv. Find the equation of the normal to the curve $y^2 = 4ax$ at $(at^2, 2at)$.

v. Evaluate $\int \sec^4 x \tan x\,dx$.

vi. Evaluate $\int_0^{\pi/2} \sin^2 x \cos^6 x\,dx$.

vii. If $u = \log(x^2 + xy + y^2)$, prove that $x\frac{\partial u}{\partial x} + y\frac{\partial u}{\partial y} = 2$.

viii. Solve the differential equation $\frac{d^2y}{dx^2} - 3\frac{dy}{dx} + 2y = 0$.

PART B

Q.3 Attempt any three questions: **(3×20=60)**

a. If $x \sin(a + y) + \sin a \cos(a + y) = 0$, prove that $\frac{dy}{dx} = \frac{\sin^2(a+y)}{\sin a}$.

b. A figure consists of a semi-circle with a rectangle on its diameter. Given the perimeter of the figure, find its dimensions in order that the area any may be maximum.

Q.4 Evaluate

a. $\int \frac{x^2+1}{(x^2+2)(2x^2+1)}\,dx$ b. $\int_0^a \frac{1}{(x^2+a^2)}\,dx$

Q.5 a. Find the area of the region included between the parabolas $y^2 = 4ax$ and $x^2 = 4ay$, where $a > 0$.

b. Use Simpson's rule to evaluate integral $\int_0^{\pi/2} \sqrt{\sin\theta}\, d\theta$, dividing the interval $\left[0, \frac{\pi}{2}\right]$ into 6 equal parts.

Q.6 a. If $y = A\cos(\log x) + B\sin(\log x)$, prove that $x^2 \frac{d^2y}{dx^2} + x\frac{dy}{dx} + y = 0$.

b. If $f(x, y)$ is a homogeneous function of degree n, prove that

(i) $x\frac{\partial f}{\partial x} + y\frac{\partial f}{\partial y} = nf$.

(ii) $x^2\frac{\partial^2 f}{\partial x^2} + 2xy\frac{\partial^2 f}{\partial x\,\partial y} + y^2\frac{\partial^2 f}{\partial y^2} = n(n-1)f$

Q.7 a. Solve the differential equation: $xdy - ydx = \sqrt{x^2 + y^2}\, dx$.

b. Find the volume of the solid generated by the revolution of the curve $y^2 = 25x$ between the ordinates $x = 1$ and $x = 2$ about x-axis.

May/June 2017

Applied Mathematics II

3K4-CBM-06

Time: 3 Hours **Max. Marks: 100**

NOTE:

- **Part 'A' may be attempted in first 6 pages of Answer Sheet.**
- **Part 'B' in rest of the Sheets of Answer Sheet.**
- **Answers may be given in English or Hindi.**

PART A

Q.1 This question of 14 parts each of 2 marks. Attempt any 10 parts: **(10×2=20)**

a. Write the domain of the function $f(x) = \dfrac{x^2 - 4}{x - 2}$.

b. Evaluate $\lim\limits_{x \to 0} \dfrac{e^{-x} - 1}{x}$.

c. If $f(x) = \sin 2x - \cos x$, determine the value of $f'\left(\dfrac{-\pi}{3}\right)$.

d. If $y = \tan^{-1}(\sqrt{x})$, where $0 < x < 1$, find $\dfrac{dy}{dx}$.

e. Find the slope of the normal to the curve $y = 4x^2$ at $x = \dfrac{1}{2\sqrt{2}}$.

f. If $f(x) = x^4 - 62x^2 + ax + 9$, attains a local extreme at $x = 1$, write the value of a.

g. Find $\int \dfrac{dx}{\sqrt{1 - \cos x}}$.

h. Evaluate $\int_0^\infty \int \dfrac{1}{3 + x^2}\, dx$.

i. If $\int \sqrt{9 - x^2}\, dx = \dfrac{x}{2}\sqrt{9 - x^2} + k \sin^{-1}\left(\dfrac{x}{3}\right) + C$, find the value of k.

j. Find the R.M.S. value of the function $f(x) = x\sqrt{x}$, where $1 \le x \le 3$.

k. Find the degree of the homoneous function $f(x, y) = x^2 \log\left(\dfrac{y}{x}\right)$.

l. If $z = \sin^{-1}\left(\dfrac{x}{y}\right)$, write the value of $\dfrac{\partial z}{\partial y}$ at (1, 2).

m. Write the order and degree of the differential equation:

$$\left\{1+\left(\frac{dy}{dx}\right)^2\right\}^{1/3}=2\sqrt{\frac{d^2y}{dx^2}}$$

n. Solve the differential equation $\frac{dy}{dx}=\frac{y}{x}$.

Q.2 This question consists of 8 parts each of 4 marks. Attempt any 5 parts: (5×4=20)

(a) Differentiate tan x with respect to x with the first principles.

(b) Find $\frac{dy}{dx}$, if $y=x^{x^x}$.

(c) An error of 1 percent is recorded in measuring the radius of a hemi-spherical bowl. Compute the percentage error caused in its volume.

(d) Find $\int\frac{dx}{3-2x-x^2}$.

(e) Evaluate $\int_0^a x^2\sqrt{a^2-x^2}\,dx$.

(f) Solve the differential equation: $\frac{dy}{dx}=\frac{\sqrt{x^2+y^2}+y}{x}, x\neq 0$.

(g) If $u=\frac{x^2+y^2}{x^2-xy}$, find the value of $x\frac{\partial u}{\partial x}+y\frac{\partial u}{\partial y}$.

(h) Solve $\frac{dy}{dx}=(1+x^2)(1+y^2)$.

PART B

Part B consists of 5 questions each of 20 marks. Attempt any 3 questions: (3×20=60)

Q.3 a. Evaluate $\lim_{x\to 3\sqrt{2}}\left[\frac{\sqrt{11-2x}-(3-\sqrt{2})}{x^2-18}\right]$.

b. Differential $\tan^{-1}\left(\frac{\sqrt{1+\sin x}-\sqrt{1-\sin x}}{\sqrt{1+\sin x}+\sqrt{1-\sin x}}\right)$ with respect to x.

Q.4 a. If $x\cos(a+y)+\cos a\sin(a+y)=0$, show that:

$$\frac{dy}{dx}=-\frac{\cos^2(a+y)}{\cos a}.$$

b. Show that the semi-vertical angle of a right circular cone of a given surface area and maximum volume is $\sin^{-1}\left(\frac{1}{3}\right)$.

Q.5 a. Find $\int\frac{x^2+x-1}{(x+1)^2(x+2)}dx$.

b. Evaluate $\int_0^{\pi/4} \log(1+\tan x)\,dx$.

Q.6 a. The area enclosed between the line $y = 2x$ and two ordinates $x = 1$ and $x = 2$ is revolved about the x-axis. Find the volume of the solid thus generated.

b. Using trapezoidal rule, obtain an approximate value of the integral $\int_1^2 \frac{1}{x}\,dx$ taking 8 equal intervals. Hence, obtain an approximate value of $\log_e 2$.

Q.7 a. Solve $(1 + y^2) + (x - e^{\tan^{-1} y}) \frac{dy}{dx} = 0$.

b. If $v = (x^2 + y^2 + z^2)^{-1/2}$, show that: $\frac{\partial^2 v}{\partial x^2} + \frac{\partial^2 v}{\partial y^2} + \frac{\partial^2 v}{\partial z^2} = 0$.

Index

I

L

M

N

O

P

U

V

W

Corrigendum

Pages with Corrections Incorporated

Page Nos.

5	6	13	14	20
29	47	72	182	189
194	239	634	753	837
897	923	944	1007	1019
1041	1136	1186	1203	1223
1228	1297	1367	1404	

EXERCISE 1.1

1. *Find the value of the following:*

(*i*) $(128)^{\frac{3}{7}}$ **Ans.** 8 (*ii*) $(243)^{-\frac{2}{5}}$ **Ans.** $\frac{1}{9}$

(*iii*) $\dfrac{1}{(216)^{-\frac{2}{3}}}$ **Ans.** 36 (*iv*) $\left(\dfrac{8}{27}\right)^{-\frac{4}{3}}$ **Ans.** $\frac{81}{16}$

2. *Solve the following:*

(*i*) $3^2 . 3^4$ **Ans.** 3^6 (*ii*) $4^2 . 8^3$ **Ans.** 2^{13}

(*iii*) $5^3 . 25^2$ **Ans.** 5^7 (*iv*) $49^2 . 7^3$ **Ans.** 7^7

(*v*) $(49)^{\frac{1}{2}} \cdot (16)^{\frac{1}{2}}$ **Ans.** 28 (*vi*) $(125)^{\frac{1}{3}} \cdot (25)^{\frac{1}{2}}$ **Ans.** 5^2

(*vii*) $(216)^{\frac{1}{3}} \cdot (36)^{\frac{1}{2}}$ **Ans.** 6^2 (*viii*) $(256)^{\frac{1}{4}} \cdot (32)^{\frac{1}{5}}$ **Ans.** 8

3. *Simplify:*

(*i*) $\dfrac{(49)^{\frac{1}{2}}}{(343)^{\frac{1}{3}}}$ **Ans.** 7^0 (*ii*) $\dfrac{(27)^2}{(81)^{\frac{1}{3}}}$ **Ans.** $3^{\frac{14}{3}}$

(*iii*) $\dfrac{(125)^{\frac{1}{3}}}{(5)^{\frac{4}{3}}}$ **Ans.** $5^{-\frac{1}{3}}$ (*iv*) $\dfrac{(121)^{\frac{1}{2}}}{(11)^{\frac{3}{2}}}$ **Ans.** $11^{-\frac{1}{2}}$

(*v*) $\dfrac{(16)^{\frac{1}{3}}}{(8)^{\frac{1}{2}}} \times \dfrac{(4)^{\frac{1}{3}}}{(32)^{\frac{1}{5}}}$ **Ans.** $2^{-\frac{1}{2}}$ (*vi*) $\dfrac{(6)^{\frac{1}{3}} \times (9)^{\frac{2}{5}}}{(2) \times (3)^{\frac{2}{3}}}$ **Ans.** $\dfrac{3^{\frac{7}{15}}}{2^{\frac{2}{3}}}$

(*vii*) $\dfrac{(15)^{\frac{2}{3}} (3)^{\frac{1}{2}}}{(5)^2 (9)^{\frac{1}{3}}}$ **Ans.** $\dfrac{3^{\frac{1}{2}}}{5^{\frac{4}{3}}}$ (*viii*) $\dfrac{(30)^{\frac{3}{4}} \cdot (2)^{\frac{1}{3}}}{(25)^{\frac{3}{8}} \cdot (6)^{\frac{13}{12}}}$ **Ans.** $3^{-\frac{1}{3}}$

4. *Simplify the following:*

(*i*) $\left\{(64)^{\frac{3}{5}}\right\}^2 \times (8^3)^{-2}$ **Ans.** 4

(*ii*) $\left(\dfrac{a^x}{a^y}\right)^{x+y} \times \left(\dfrac{a^y}{a^z}\right)^{y+z} \times \left(\dfrac{a^z}{a^x}\right)^{z+x}$ **Ans.** 1

(*iii*) $\left(\dfrac{x^b}{x^c}\right)^{\frac{1}{bc}} \times \left(\dfrac{x^c}{x^a}\right)^{\frac{1}{ca}} \times \left(\dfrac{x^a}{x^b}\right)^{\frac{1}{ab}}$ **Ans.** 1

(*iv*) $\left(\dfrac{x^a}{x^b}\right)^{a^2+ab+b^2} \times \left(\dfrac{x^b}{x^c}\right)^{b^2+bc+c^2} \times \left(\dfrac{x^c}{x^a}\right)^{c^2+ca+a^2}$ **Ans.** 1

(*v*) $\left[\dfrac{b+c}{x^{c-a}}\right]^{\frac{1}{a-b}} \cdot \left[\dfrac{c+a}{x^{a-b}}\right]^{\frac{1}{b-c}} \cdot \left[\dfrac{a+b}{x^{b-c}}\right]^{\frac{1}{c-a}}$ **Ans.** 1

(vi) $\left(x^{\frac{1}{2}} + y^{\frac{1}{2}}\right)\cdot\left(x^{\frac{1}{4}} + y^{\frac{1}{4}}\right)\cdot\left(x^{\frac{1}{4}} - y^{\frac{1}{4}}\right)$ **Ans.** $x - y$

(vii) $\dfrac{2^{m+3}\cdot 3^{2m-n}\cdot 5^{m+n+3}\cdot 6^{n+1}}{6^{m+1}\cdot 10^{n+3}\cdot 15^{m}}$ **Ans.** 1

(viii) $\dfrac{2^{n}+2^{n-1}}{2^{n+1}-2^{n}}$ **Ans.** $\dfrac{3}{2}$

(ix) $\dfrac{3^{a}\cdot 3^{a^2-a}}{3^{a+1}3^{a-1}}\times\left[\dfrac{(3^3)^{\frac{\alpha}{3}}}{3^2}\right]^{-\alpha}$ **Ans.** 1

(x) $\left[\dfrac{(a^2b^3)^{\frac{2}{3}}\cdot(a^{-2}\cdot b^{-2}c)^{\frac{3}{2}}}{(a\cdot b)^{-\frac{2}{3}}\cdot c^{\frac{1}{2}}}\right]$ **Ans.** $\dfrac{c^3}{a\cdot b}$

(xi) $\left\{(x^{p})^{1-\frac{1}{p}}\right\}^{p^2+p+1}$ **Ans.** x^{p^3-1}

(xii) $\dfrac{1}{1+x^{a-b}+x^{a-c}}+\dfrac{1}{1+x^{b-c}+x^{b-a}}+\dfrac{1}{1+x^{c-a}+x^{c-b}}$ **Ans.** 1

(xiii) Show that $\dfrac{y^{-1}}{x^{-1}+y^{-1}}+\dfrac{y^{-1}}{x^{-1}-y^{-1}}=\dfrac{2xy}{y^2-x^2}$.

(xiv) If $a = xy^{p-1}$, $b = xy^{q-1}$ and $c = xy^{r-1}$, prove that $a^{q-r}\,b^{r-p}\,c^{p-q} = 1$.

(xv) If $a = \sqrt[3]{3} + \left(\sqrt[3]{3}\right)^{-1}$, prove that: $3a^3 - 9a = 10$.

(xvi) If $m = a^x$, $n = a^y$ and $a^z = (m^y \,.\, n^x)^z$, show that: $x\,y\,z = 1$.

(xvii) If $a^x = b^y = c^z$ and $b^2 = ac$, then show that: $\dfrac{1}{x}+\dfrac{1}{z}=\dfrac{2}{y}$.

5. ***Solve the following equations:***

(i) $7^{x+7} = 49^{4x-7}$ **Ans.** $x = 3$

(ii) $2^{x+4} = 2^{x+3} + 4$ **Ans.** $x = -1$

(iii) $\left(\dfrac{a}{b}\right)^{4x-1} = \left(\dfrac{b}{a}\right)^{2x-5}$ **Ans.** $x = 1$

(iv) $9^{x-y} = 81$, $9^{x+y} = 729$ **Ans.** $x = \dfrac{5}{2}, y = \dfrac{1}{2}$

1.7 ALGEBRAIC IDENTITIES

1. $(a + b)^2 = a^2 + 2ab + b^2$
2. $(a - b)^2 = a^2 - 2ab + b^2$
3. $a^2 - b^2 = (a + b)(a - b)$
4. $(a + b + c)^2 = a^2 + b^2 + c^2 + 2ab + 2bc + 2ac$
5. $(a + b)^3 = a^3 + b^3 + 3ab(a + b)$
   $= a^3 + b^3 + 3a^2b + 3ab^2$
6. $(a - b)^3 = a^3 - b^3 - 3ab(a - b)$
   $= a^3 - b^3 - 3a^2b + 3ab^2$

EXERCISE 1.7

Write the following in the form of logarithms:

1. $2^6 = 64$ **Ans.** $\log_2 64 = 6$

2. $4^5 = 1024$ **Ans.** $\log_4 1024 = 5$

3. $3^4 = 81$ **Ans.** $\log_3 81 = 4$

4. $4^3 = 64$ **Ans.** $\log_4 64 = 3$

5. $7^2 = 49$ **Ans.** $\log_7 49 = 2$

6. $8^3 = 512$ **Ans.** $\log_8 512 = 3$

7. $9^{\frac{5}{2}} = 243$ **Ans.** $\log_9 243 = \frac{5}{2}$

8. $10^0 = 1$ **Ans.** $\log_{10} 1 = 0$

Express each of the following in exponential form:

9. $\log_5 1 = 0$ **Ans.** $5^0 = 1$

10. $\log_{10} 1000 = 3$ **Ans.** $10^3 = 1000$

11. $\log_4 64 = 3$ **Ans.** $4^3 = 64$

12. $\log_7 343 = 3$ **Ans.** $7^3 = 343$

13. $\log_{10} 0.001 = -3$ **Ans.** $10^{-3} = 0.001$

14. $\log_3 \frac{1}{9} = -2$ **Ans.** $3^{-2} = \frac{1}{9}$

15. $\log_8 4 = \frac{2}{3}$ **Ans.** $8^{\frac{2}{3}} = 4$

16. $\log_9 6561 = 4$ **Ans.** $9^4 = 6561$

Find the value of each of the following by the definition of logarithm:

17. $\log_2 16$ **Ans.** 4

18. $\log_2 \sqrt{32}$ **Ans.** $\frac{5}{2}$

19. $\log_{10} 10^5$ **Ans.** 5

20. $\log_n 1$ **Ans.** 0

1.9 LAWS OF LOGARITHM

In this section, we shall learn the following laws of logarithm. These laws hold for any base $a(a > 0$ and $a \neq 1)$.

(i) First Law (Product Law) $\boxed{\log_a(mn) = \log_a m + \log_a n}$.

The logarithm of the product of two numbers is equal to the sum of their logarithms with reference to the same base.

Proof: Let $\log_a m = x$ and $\log_a n = y$. Then,

$$\log_a m = x \Rightarrow a^x = m \quad \text{... (1)}$$

and $$\log_a n = y \Rightarrow a^y = n \quad \text{... (2) [By definition]}$$

On multiplication of (1) and (2), we get

$$\therefore \quad m \cdot n = a^x \cdot a^y \Rightarrow m \cdot n = a^{x+y} \quad \text{[By laws of indices]}$$

$$\Rightarrow \log_a mn = x + y \quad \text{... (3) [By definition of log]}$$

On putting values of x and y in (3), we get

$$\log_a mn = \log_a m + \log_a n$$

or $$\log 2 \times 3 = \log 2 + \log 3$$

(ii) Second Law (Quotient Law) $\boxed{\log_a\left(\frac{m}{n}\right) = \log_a m - \log_a n}$

The logarithm of quotient of two numbers is equal to the difference of logarithm of the numerator and the logarithm of the denominator.

Proof. Let $\log_a m = x$ and $\log_a n = y$. Then,

$$\log_a m = x \Rightarrow a^x = m \quad \text{... (1)}$$

and $$\log_a n = x \Rightarrow a^y = n \quad \text{... (2) [By definition of log]}$$

On dividing (1) by (2), we get

$$\therefore \quad \frac{a^x}{a^y} = \frac{m}{n} \quad \Rightarrow \quad \frac{m}{n} = a^{x-y} \qquad \text{[By laws of indices]}$$

$$\log_e\left(\frac{m}{n}\right) = x - y \qquad \text{... (3)}$$

On putting value of $x = \log_a m$ and $y = \log_a n$ in (3), we get

$$\log a\left(\frac{m}{n}\right) = \log_a m - \log_a n$$ **Proved.**

or $$\log\left(\frac{3}{4}\right) = \log 3 - \log 4$$

(iii) Third Law (Power Law) $\boxed{\log_a m^n = n\log_a m}$

The logarithm of a number raised to a power n is n times the logarithm of the number.

Proof. Let $\log_a m = x$. Then, $a^x = m$

Now, $a^x = m \Rightarrow (a^x)^n = m^n \Rightarrow a^{xn} = m^n$

$\Rightarrow \quad m^n = a^{nx} \Rightarrow \log_a m^n = nx$ [By definition of log]

$\Rightarrow \quad \log_a m^n = n \log_a m$ $(\because x = \log_a m)$

Hence, $\log_a m^n = n\log_a m$ **Proved.**

For example, $\log m^2 = 2\log m$

Example 25. *Find the value:* $\log_3 27\sqrt{729}$

Solution. We have,

$$\log_3 27\sqrt{729} = \log_3 27 + \log_3\sqrt{729} = \log_3(3)^3 + \log_3(729)^{\frac{1}{2}} \qquad [\because \sqrt{a} = a^{\frac{1}{2}}]$$

$$= 3\log_3 3 + \frac{1}{2}\log_3(729) \qquad [\because \log_a m^n = n\log_a m]$$

$$= 3(1) + \frac{1}{2}\log_3(3^6) = 3 + \frac{1}{2}\cdot 6\log_3 3$$

$$= 3 + \frac{1}{2}\cdot 6(1) = 3 + 3 = 6 \qquad [\because \log_3 3 = 1]$$

Example 26. *Show that:* $3\log 4 - 2\log 6 + \log(18)^{\frac{3}{2}} = \log(96\sqrt{2})$

Solution.

$$\text{L.H.S.} = 3\log 4 - 2\log 6 + \log(18)^{\frac{3}{2}}$$

$$= \log 4^3 - \log 6^2 + \log(18)^{\frac{3}{2}} = \log 64 - \log 36 + \log(2 \times 3^2)^{\frac{3}{2}}$$

$$= \log\frac{64 \times (2 \times 3^2)^{\frac{3}{2}}}{36} = \log\frac{64 \times 2^{\frac{3}{2}} \times 3^{2 \times \frac{3}{2}}}{36} = \log\frac{64 \times 2\sqrt{2} \times 27}{36}$$

$$= \log(16 \times 2\sqrt{2} \times 3) = \log(96\sqrt{2}) = \text{R.H.S.}$$ **Proved.**

Example 27. *Prove that:*

(*i*) $\log(1 + 2 + 3) = \log 1 + \log 2 + \log 3$ (*ii*) $\log\frac{a^2}{bc} + \log\frac{b^2}{ac} + \log\frac{c^2}{ab} = 0$

(*iii*) $\log_b a \times \log_c b \times \log_a c = 1$

2.4 GENERAL TERMS

T_1 represents the first term and T_n represents the nth term in the sequence $T_1, T_2, T_3, ..., T_n$. The nth term of a sequence is denoted by a_n or U_n or T_n is called a general term.

For example third term of the sequence 1, 7, 13, 19, 25, ..., is given by $T_3 = 13$.

Notes:1. The nth term of the sequence 2, 4, 6, ..., is given by $T_n = 2n$, where n is a natural number.

2. In the sequence of odd numbers 1, 3, 5, ..., the nth term is given by the formula $T_n = 2n - 1$, where n is a natural number.

3. A sequence can be regarded as a function, whose domain is a set of natural numbers.

4. A series is generally represented in a compact form called sigma with notation Σ.

$$\sum_{n=1}^{\infty} T_n = T_1 + T_2 + T_3 + ... + T_n$$

Example 1. *Write down the first five terms of each of the following sequences, whose nth terms are*

(*i*) $T_n = 2n + 5$ (*ii*) $T_n = (-1)^{n-1}(5)^{n+1}$

(*iii*) $T_n = \frac{n(n^2 + 5)}{4}$

Solution. (*i*) Here, $T_n = 2n + 5$ ...(1)

Substituting $n = 1, 2, 3, 4, 5$ in (1), we get

$$T_1 = 2 \times 1 + 5 = 2 + 5 = 7$$
$$T_2 = 2 \times 2 + 5 = 4 + 5 = 9$$
$$T_3 = 2 \times 3 + 5 = 6 + 5 = 11$$
$$T_4 = 2 \times 4 + 5 = 8 + 5 = 13$$
$$T_5 = 2 \times 5 + 5 = 10 + 5 = 15$$

Thus, the required terms are 7, 9, 11, 13, 15 **Ans.**

(*ii*) Here, $T_n = (-1)^{n-1}(5)^{n+1}$

Substituting $n = 1, 2, 3, 4, 5$ in (1), we get ...(1)

$$T_1 = (-1)^{1-1}(5)^{1+1} = 25$$
$$T_2 = (-1)^{2-1}(5)^{2+1} = -125$$
$$T_3 = (-1)^{3-1}(5)^{3+1} = 625$$
$$T_4 = (-1)^{4-1}(5)^{4+1} = -3125$$
$$T_5 = (-1)^{5-1}(5)^{5+1} = 15625$$

Thus, the required terms are 25, – 125, 625, – 3125, 15625 **Ans.**

(*iii*) Here, $T_n = \frac{n(n^2 + 5)}{4}$ ...(1)

Substituting $n = 1, 2, 3, 4, 5$ in (1), we get

$$T_1 = \frac{1(1^2 + 5)}{4} = \frac{6}{4} = \frac{3}{2}$$

Solution. Let the three parts of 69 be $a - d$, a and $a + d$.

Sum of three terms = 69 $\Rightarrow$ $(a - d) + a + (a + d) = 69$

$\Rightarrow$ $3a = 69$ $\Rightarrow$ $a = 23$ ... (1)

Product of two smaller parts = 483 $\Rightarrow$ $a(a - d) = 483$ [using Eq. (1)]

$\Rightarrow$ $23(23 - d) = 483$ $\Rightarrow$ $23 - d = 21$ $\Rightarrow$ $d = 23 - 21 = 2$

$$\text{Part I} = a - d = 23 - 2 = 21$$

$$\text{Part II} = a = 23$$

$$\text{Part III} = a + d = 23 + 2 = 25$$

Hence, three parts are 21, 23, 25. **Ans.**

Example 17. *Divide 32 into four parts which are in A.P., such that the product of extremes to the product of means is 7 : 15.*

Solution. Let the four parts be $(a - 3d)$, $(a - d)$, $(a + d)$, $(a + 3d)$. Then,

sum of four parts = 32

$\Rightarrow$ $(a - 3d) + (a - d) + (a + d) + (a + 3d) = 32$

$\Rightarrow$ $4a = 32$ $\Rightarrow$ $a = 8$

and $$\frac{(a-3d)(a+3d)}{(a-d)(a+d)} = \frac{7}{15}$$

$\Rightarrow$ $$\frac{a^2 - 9d^2}{a^2 - d^2} = \frac{7}{15} \Rightarrow \frac{64 - 9d^2}{64 - d^2} = \frac{7}{15}$$

$\Rightarrow$ $128d^2 = 512$ $\Rightarrow$ $d^2 = 4$ $\Rightarrow$ $d = \pm 2$

Hence, the required parts are 2, 6, 10, 14 **Ans.**

Example 18. *If the nth term of a progression is a linear expression in n, then show that it is an A.P.*

Solution. Let $T_n = an + b$, where a and b are constants.

Then $T_{n-1} = a(n - 1) + b$

$\therefore$ $T_n - T_{n-1} = (an + b) - [a(n - 1) + b]$

$= an - an + b + a - b$

$= a$, which is a constant.

Thus, the difference between any two consecutive terms of the given progression is constant.

Hence, the given progression is an A.P. **Proved.**

Example 19. *If m times the mth term of an A.P. is equal to n times its nth term, show that the (m + n)th term of the A.P. is zero.*

Solution. Let a be the first term and d the common difference, then

$mT_m = nT_n$ [Given]

$\Rightarrow$ $m[a + (m - 1)d] = n[a + (n - 1)d]$

$\Rightarrow$ $[(m^2 - n^2) - (m - n)]d = (n - m)a$

$\Rightarrow$ $[(m - n)(m + n) - (m - n)]d = (n - m)a$ $\Rightarrow$ $(m - n)(m + n - 1)d = (n - m)a$

$\Rightarrow$ $(m + n - 1)d = -a$ $\Rightarrow$ $a + (m + n - 1)d = 0$

$\Rightarrow$ $T_{m+n} = 0$ **Proved.**

2.9 ARITHMETIC MEAN BETWEEN ANY TWO GIVEN NUMBERS

Let a, b be any two numbers. Let A, be the arithmetic mean between a and b.

Therefore, a, A, b are in A.P.

$\Rightarrow \quad A - a = b - A$ [common difference]

$$\Rightarrow \quad 2A = a + b \Rightarrow A = \frac{a+b}{2}$$

$\therefore$ Arithmetic mean between two given numbers is equal to half their sum.

$$\boxed{\textbf{A.M.} = \frac{a+b}{2}}$$

2.10 *n*th A.M. BETWEEN TWO GIVEN NUMBERS

Let a, b be any two numbers. Let $A_1, A_2, A_3, ..., A_n$ be the n A.M. between a and b. Then, $a, A_1, A_2, ..., A_n, b$ are in A.P.

Here, b is the $(n + 2)$th term *i.e.*, $T_{n+2} = a + (n + 2 - 1)d \quad \Rightarrow \quad b = a + (n + 1)d$

$$\Rightarrow \quad d = \frac{b-a}{n+1}$$

Thus, the n arithmetic means are given below

$$A_1 = a + d = a + \frac{b-a}{n+1}$$

$$A_2 = a + 2d = a + 2\left(\frac{b-a}{n+1}\right)$$

$$A_3 = a + 3d = a + 3\left(\frac{b-a}{n+1}\right)$$

....

....

$$A_n = a + nd = a + n\left(\frac{b-a}{n+1}\right)$$

Example 43. *Find the A.M. between 11 and 89.*

Solution. A.M. $= \dfrac{11+89}{2} = \dfrac{100}{2} = 50$ **Ans.**

Example 44. *If* $\dfrac{a^n + b^n}{a^{n-1} + b^{n-1}}$ *is the A.M. between a and b, then find the value of n.*

Solution. We know that A.M. between a and b = $\dfrac{a+b}{2}$

$$\Rightarrow \quad \frac{a^n + b^n}{a^{n-1} + b^{n-1}} = \frac{a+b}{2} \text{ [Given]} \quad \Rightarrow \quad 2a^n + 2b^n = a^n + ab^{n-1} + a^{n-1}b + b^n$$

$$\Rightarrow \quad a^n - a^{n-1}b = ab^{n-1} - b^n \quad \Rightarrow \quad a^{n-1}(a-b) = b^{n-1}(a-b)$$

Dividing (2) by (1), we get

$$\frac{1-r^6}{1-r^3}=\frac{144}{16} \Rightarrow \frac{(1+r^3)(1-r^3)}{(1-r^3)}=9$$

$$\Rightarrow \quad 1+r^3=9 \Rightarrow r^3=8 \Rightarrow r=2$$

Putting $r = 2$ in (1), we get

$$\frac{a(1-2^3)}{1-2}=16 \Rightarrow 7a=16 \Rightarrow a=\frac{16}{7}$$

Here, $a=\frac{16}{7}$ and $r = 2$ (> 1). Therefore,

$$S_n=a\left(\frac{r^n-1}{r-1}\right) \Rightarrow S_n=\frac{16}{7}\left(\frac{2^n-1}{2-1}\right) \Rightarrow S_n=\frac{16}{7}(2^n-1) \quad \textbf{Ans.}$$

Example 25. *In a geometric progression, $\{a_n\}$, if $T_1 = 3$, $T_n = 96$ and $S_n = 189$. Find n.*

Solution. Here, $T_1 = a = 3,\quad T_n = 96,\quad S_n = 189$

We know that $T_n = ar^{n-1} = 96$ ... (1)

$$S_n=\frac{a(r^n-1)}{r-1}=\frac{ar^n-a}{r-1}$$

$$\Rightarrow \quad S_n=\frac{a(r^{n-1})r-a}{r-1} \Rightarrow S_n=\frac{96r-a}{r-1} \quad \text{[using (1)]}$$

$$189=\frac{96r-3}{(r-1)} \Rightarrow 189r-189=96r-3 \Rightarrow r=2, [\because S_n=189]$$

Now, $T_n = ar^{n-1} \Rightarrow 96 = 3 \times 2^{n-1}$

$$\Rightarrow \quad \frac{96}{3}=2^{n-1} \Rightarrow 2^{n-1}=32$$

$$\Rightarrow \quad 2^{n-1}=(2)^5 \Rightarrow n-1=5 \Rightarrow n=6 \quad \textbf{Ans.}$$

Example 26. *Sum the series:*

$$(x+y)+(x^2+xy+y^2)+(x^3+x^2y+xy^2+y^3)+\ldots \text{ up to } n \text{ terms.}$$

Solution. Required sum

$$S=(x+y)+(x^2+xy+y^2)+(x^3+x^2y+xy^2+y^3)+\ldots \text{ up to } n \text{ terms.}$$

Multiplying and dividing by $(x - y)$, we get

$$S=\frac{1}{x-y}\left[(x^2-y^2)+(x^3-y^3)+(x^4-y^4)+\ldots \text{ up to } n \text{ terms}\right]$$

$$\Rightarrow \quad S=\frac{1}{x-y}\left[(x^2+x^3+x^4+\ldots \text{ upto } n \text{ terms})-(y^2+y^3+y^4+\ldots \text{ upto } n \text{ terms}\right]$$

$$\Rightarrow \quad S=\frac{1}{x-y}\left[\frac{x^2(1-x^n)}{1-x}-\frac{y^2(1-y^n)}{1-y}\right] \quad \textbf{Ans.}$$

Example 27. *The sum of some terms of G.P. is 315 whose first term and the common ratio are 5 and 2 respectively. Find the last term and the number of terms.*

8. Find the 11th term from the end in the expansion of $\left(2x-\frac{1}{x^2}\right)^{25}$ **Ans.** ${}^{25}C_{15}.\frac{2^{10}}{x^{20}}$

9. In the expansion of $\left(x+\frac{1}{x}\right)^6$, find the third term from the end. **Ans.** $\frac{15}{x^2}$

10. Find the $(n + 1)$th term from the end in the expansion of $\left(x-\frac{1}{x}\right)^{3n}$ **Ans.** $\frac{(3n)!}{n!(2n)!}.\frac{1}{x^n}$

8.10 MIDDLE TERMS IN A BINOMIAL EXPANSION

Since the Binomial expansion of $(a + b)^n$ has $(n + 1)$ terms. Therefore,

(*i*) If n is even, then the middle term is $\left(\frac{n}{2}+1\right)$ th term.

(*ii*) If n is odd, then the middle terms are $\left(\frac{n+1}{2}\right)$ th and $\left(\frac{n+3}{2}\right)$ th terms

Type II. To find the middle term:

Example 15. *Find the middle term in the expansion of* $\left(\frac{x}{3}+9y\right)^{10}$

Solution. Here $n = 10$, which is an even number. So, $\left(\frac{10}{2}+1\right)$ th term, i.e. 6th term is the middle term.

Hence, the middle term $= T_6 = T_{5+1} = {}^{10}C_5\left(\frac{x}{3}\right)^{10-5}(9y)^5$

$$= {}^{10}C_5\left(\frac{x}{3}\right)^5(9y)^5$$

$$= \frac{10!}{5!(10-5)!}\times\frac{x^5\times y^5\times(9)^5}{3^5}$$

$$= 61236x^5y^5$$ **Ans.**

Example 16. *Find the middle terms in the expansion of* $\left(3-\frac{x^3}{6}\right)^7$ **[S.B.T.E. 2014]**

Solution. The general expression $= \left(3-\frac{x^3}{6}\right)^7$. Here $n = 7$, which is an odd number.

So, $\left(\frac{7+1}{2}\right)$ th and $\left(\frac{7+1}{2}+1\right)$ th, i.e. 4th and 5th terms are two middle terms.

Now, $T_4 = T_{3+1} = {}^7C_3(3)^{7-3}\left(-\frac{x^3}{6}\right)^3$

According to the question :

$$^nC_{r-1} : {}^nC_r : {}^nC_{r+1} = 1 : 7 : 42$$

$$\Rightarrow \quad \frac{^nC_{r-1}}{1} = \frac{^nC_r}{7} = \frac{^nC_{r+1}}{42} \qquad ...(1)$$

If $\quad \frac{^nC_{r-1}}{1} = \frac{^nC_r}{7} \Rightarrow 7\,{}^nC_{r-1} = {}^nC_r$

$$\Rightarrow \quad 7 \cdot \frac{n!}{(r-1)!\,(n-r+1)!} = \frac{n!}{r!\,(n-r)!}$$

$$\Rightarrow \quad \frac{7}{(r-1)!\,(n+1-r)(n-r)!} = \frac{1}{r(r-1)!\,(n-r)!}$$

$$\Rightarrow \quad \frac{7}{n+1-r} = \frac{1}{r}$$

$$\Rightarrow \quad 7r = n + 1 - r$$

$$\Rightarrow \quad n = 8r - 1 \qquad ...(2)$$

Again, if $\frac{^nC_r}{7} = \frac{^nC_{r+1}}{42}$

$$\Rightarrow \quad {}^{6n}C_r = {}^{6n}C_{r+1} \Rightarrow 6\frac{n!}{r!\,(n-r)!} = \frac{n!}{(r+1)!\,(n-r-1)!}$$

$$\Rightarrow \quad \frac{6}{r!(n-r)\,(n-r-1)!} = \frac{1}{(r+1)r!\,(n-r-1)!}$$

$$\Rightarrow \quad \frac{6}{n-r} = \frac{1}{r+1} \Rightarrow 6\,(r+1) = n - r$$

$$\Rightarrow \quad 6r + 6 = n - r \Rightarrow n = 7r + 6 \qquad ...(3)$$

From (2) and (3), we get

$$8r - 1 = 7r + 6 \Rightarrow r = 7$$

Putting $r = 7$ in (3), we get.

$$\Rightarrow \quad n = 7 \times 7 + 6 \Rightarrow n = 55$$ **Ans.**

Example 28. *The coefficients of $(r-1)^{th}$, r^{th} and $(r+1)^{th}$ terms in the expansion of $(x+1)^n$ are in the ratio 1 : 3 : 5. Find both n and r.*

Solution. According to the question,

coefficient of T_{r-1} : coefficient of T_r : coefficient of $T_{r-1} = 1 : 3 : 5$

$$^nC_{r-2} : {}^nC_{r-1} : {}^nC_r = 1 : 3 : 5$$

$$\frac{^nC_{r-2}}{1} = \frac{^nC_{r-1}}{3} = \frac{^nC_r}{5} \qquad ...(1)$$

If $\quad \frac{^nC_{r-2}}{1} = \frac{^nC_{r-1}}{3} \Rightarrow 3\,{}^nC_{r-2} = {}^nC_{r-1}$

$$\Rightarrow \quad 3\frac{n!}{(r-2)!\,(n-r+2)!} = \frac{n!}{(r-1)!\,(n-r+1)!}$$

Type VI. To find the expansion:

Example 32. *The first three terms in the expansion of a binomial are 1, 10 and 40. Find the expansion.*

Solution. (i) Let $$(1+x)^n = 1 + nx + \frac{n(n-1)}{2!}x^2 + \ldots$$

According to the problem,

$$nx = 10 \quad \ldots(1)$$

and $$\frac{n(n-1)}{2!}x^2 = 40 \quad \ldots(2)$$

Dividing (2) by the square of (1), we get

$$\frac{n(n-1)}{2!}x^2 \cdot \frac{1}{n^2x^2} = \frac{40}{100}$$

$$\Rightarrow \quad \frac{n-1}{2n} = \frac{2}{5} \Rightarrow 5n - 5 = 4n$$

$$\Rightarrow \quad 5n - 4n = 5 \Rightarrow n = 5$$

Putting $n = 5$ in (1), we get

$$5x = 10 \Rightarrow x = 2$$

Hence, the binomial is $(1 + 2)^5$ **Ans.**

Example 33. *Write first three terms of the expansion of* $\frac{2+x}{(3-2x)^2}$

Solution. We have,

$$\frac{2+x}{(3-2x)^2} = \frac{2+x}{9\left(1-\frac{2}{3}x\right)^2} = \frac{1}{9}(2+x)\left(1-\frac{2}{3}x\right)^{-2}$$

$$= \frac{1}{9}(2+x)\left[1+\frac{4}{3}x+\frac{(-2)(-3)}{2!}\left(-\frac{2}{3}x\right)^2+\ldots.\right]$$

$$= \frac{1}{9}(2+x)\left[1+\frac{4}{3}x+\frac{4}{3}x^2+\ldots.\right]$$

$$= \frac{1}{9}\left[2+\frac{11}{3}x+4x^2+\ldots.\right]$$

$$= \frac{2}{9}+\frac{11}{27}x+\frac{4}{9}x^2+\ldots.$$

Therefore, the first three terms in the expansion of $\frac{2+x}{(3-2x)^2}$ are $\frac{2}{9}, \frac{11}{27}x$ and $\frac{4}{9}x^2+\ldots$ **Ans.**

Choose the correct alternative:

11. The value of the determinant $\begin{vmatrix} 1 & \omega^3 & \omega^5 \\ \omega^3 & 1 & \omega^4 \\ \omega^5 & \omega^4 & 1 \end{vmatrix}$, where ω is an imaginary cube root of unity is

(*a*) $(1-\omega)^2$ (*b*) 3 (*c*) -3 (*d*) 4 **Ans.** (*b*)

12. The value of $\begin{vmatrix} x+a & x & x \\ x & x+a & x \\ x & x & x+a \end{vmatrix}$ is equal to

(*a*) $3a^2x$ (*b*) $a^2(3x-a)$ (*c*) $a^2(3x+a)$ (*d*) $3ax^2$ **Ans.** (*c*)

9.8 FACTOR THEOREM

If the elements of a determinant are polynomials in a variable x and if the substitution $x = a$ makes two rows (or columns) identical, then $(x - a)$ is a factor of the determinant.

When two rows are identical, the value of the determinant is zero. The expansion of the determinant being a polynomial in x vanish on putting $x = a$, then $x - a$ is its factor by the remainder theorem.

Example 24. *Show that* $\begin{vmatrix} 1 & 1 & 1 \\ x & y & z \\ x^2 & y^2 & z^2 \end{vmatrix} = (x-y)(y-z)(z-x)$ **[S.B.T.E. 2017, 2012]**

Solution. If we put $x = y$; $y = z$; $z = x$ then in each case two columns become identical and the determinant vanishes.

$\therefore$ $(x - y)$; $(y - z)$; $(z - x)$ are the factors.

Since the determinant is of third degree, the other factors can be numerical only say k.

$$\begin{vmatrix} 1 & 1 & 1 \\ x & y & z \\ x^2 & y^2 & z^2 \end{vmatrix} = k(x-y)(y-z)(z-x)$$

This leading term (product of the elements of the diagonal elements) in the given determinant is yz^2 and in the expansion

$$k(x-y)(y-z)(z-x), \text{ we get } kyz^2$$

Equating the coefficient of yz^2, we have

$$k = 1$$

Hence the expansion $= (x-y)(y-z)(z-x)$. **Proved.**

Example 25. *Show that* $\begin{vmatrix} x & x^2 & x^3 \\ y & y^2 & y^3 \\ z & z^2 & z^3 \end{vmatrix} = xyz(x-y)(y-z)(z-x)$ **[S.B.T.E. 2016, 2014, 2013]**

Solution. $$\begin{vmatrix} x & x^2 & x^3 \\ y & y^2 & y^3 \\ z & z^2 & z^3 \end{vmatrix} = xyz\begin{vmatrix} 1 & x & x^2 \\ 1 & y & y^2 \\ 1 & z & z^2 \end{vmatrix}$$

$$= xyz(x-y)(y-z)(z-x) \text{ (see example 24)}$$ **Proved.**

The value of the product changes if the order is non-cycle.

Note. $\vec{a} \cdot (\vec{b} \cdot \vec{c})$, $\vec{a} \times (\vec{b} \cdot \vec{c})$ are meaningless.

20.28 GEOMETRICAL INTERPRETATION OF TRIPLE PRODUCT

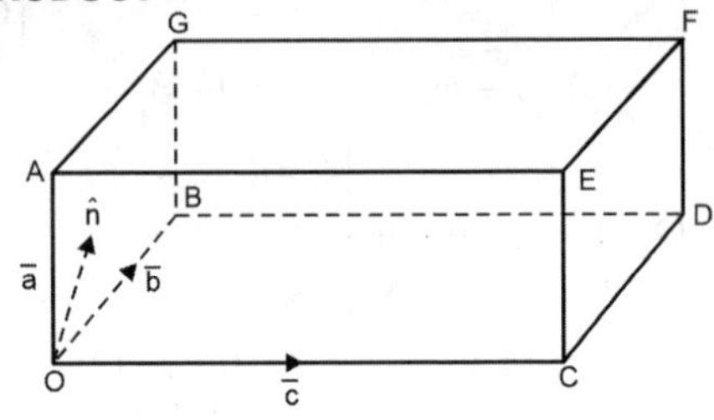

The scalar triple product $\vec{a} \cdot (\vec{b} \times \vec{c})$ represents the volume of the parallelepiped having $\vec{a}, \vec{b}, \vec{c}$ as its coterminus edges.

$\vec{a} \cdot (\vec{b} \times \vec{c}) = a \cdot$ Area of || gm $OBDC\ \hat{n}$

= Area of || gm $OBDC$ × perpendicular distance between the parallel faces $OBDC$ and $AEFG$.

$\boxed{\vec{a} \cdot (\vec{b} \times \vec{c}) = \text{Volume of the parallelepiped}}$

Note. (1) If $\vec{a} \cdot (\vec{b} \times \vec{c}) = 0$, then $\vec{a}, \vec{b}, \vec{c}$ are coplanar.

(2) $\boxed{\text{Volume of tetrahedron} = \frac{1}{6}(\vec{a}\ \vec{b}\ \vec{c}).}$

Example 53. *Find the volume of parallelepiped*

If $\vec{a} = -3\hat{i} + 7\hat{j} + 5\hat{k}$, $\vec{b} = -3\hat{i} + 7\hat{j} - 3\hat{k}$ *and* $\vec{c} = 7\hat{i} - 5\hat{j} - 3\hat{k}$ *are the three coterminus edges of the parallelopiped.*

Solution. Volume = $\vec{a} \cdot (\vec{b} \times \vec{c}) = \begin{vmatrix} -3 & 7 & 5 \\ -3 & 7 & -3 \\ 7 & -5 & -3 \end{vmatrix}$

$= -3(-21 - 15) - 7(9 + 21) + 5(15 - 49)$

$= 108 - 210 - 170 = -272$. cubic units. **Ans.**

EXERCISE 20.10

1. Find the volume of the parallelepiped with adjacent sides $\overrightarrow{OA} = 3\hat{i} - \hat{j}$, $\overrightarrow{OB} = \hat{j} + 2\hat{k}$, $\overrightarrow{OC} = \hat{i} + 5\hat{j} + 4\hat{k}$ extending from the origin of co-ordinates O. **Ans.** 20
2. Find the volume of the tetrahedron whose vertices are the points $A(2, -1, -3)$; $B(4, 1, 3)$; $C(3, 2, -1)$ and $D(1, 4, 2)$. **Ans.** $7\frac{1}{3}$
3. Find the volume of the tetrahedron whose vertices are the points $A(2, 1, 8)$; $B(3, 2, 9)$; $C(2, 1, 4)$ and $D(3, 3, 10)$. **Ans.** $\frac{2}{3}$
4. Prove that $[\vec{a} + \vec{b}, \vec{b} + \vec{c}, \vec{c} + \vec{a}] = 2[\vec{a}\ \vec{b}\ \vec{c}]$.

$$\Rightarrow \qquad b^2 = a^2 (2-1) \qquad (\because e = \sqrt{2})$$

$$b^2 = a^2 \quad \Rightarrow \quad b = a = 4\sqrt{2}$$

The equation is

$$\frac{x^2}{(4\sqrt{2})^2} - \frac{y^2}{(4\sqrt{2})^2} = 1 \quad \Rightarrow \quad \frac{x^2}{32} - \frac{y^2}{32} = 1$$

$$\therefore \qquad x^2 - y^2 = 32.$$ **Ans.**

Example 27. *Find the equation of the hyperbola whose vertices are (0, 0) (10, 0) and one of whose foci is (18, 0).*

Solution. Let the vertices be A' (0, 0) and A (10, 0).

Let $AA' = 2a \quad \therefore 2a = 10 \quad \Rightarrow \quad a = 5$

Let other focus be $(x, 0)$

Mid-point of foci SS' = Mid-point of vertices AA'.

x-coordinate of mid-point SS' = x-coordinate of mid-point of AA'.

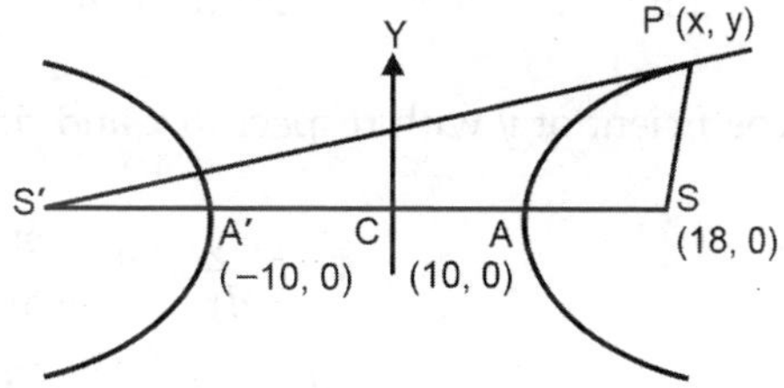

$$\frac{x+18}{2} = \frac{0+10}{2}$$

$$x + 18 = 10 \quad \Rightarrow \quad x = -8$$

$\therefore$ The coordinates of S' are (–8, 0).

Let $P(x, y)$ be any point on the hyperbola.

By the property of hyperbola

$$S'P - SP = 2a$$

$$\Rightarrow \quad \sqrt{(x+8)^2 + (y-0)^2} - \sqrt{(x-18)^2 + (y-0)^2} = 10$$

$$\Rightarrow \quad \sqrt{(x+8)^2 + y^2} = 10 + \sqrt{(x-18)^2 + y^2}$$

Squaring both sides, we have

$$(x+8)^2 + y^2 = (10)^2 + (x-18)^2 + y^2 + 20\sqrt{(x-18)^2 + y^2}$$

$$\Rightarrow \quad x^2 + 16x + 64 + y^2 = 100 + x^2 - 36x + 324 + y^2 + 20\sqrt{(x-18)^2 + y^2}$$

$$\Rightarrow \quad 52x - 360 = 20\sqrt{(x-18)^2 + y^2}$$

$$\Rightarrow \quad 13x - 90 = 5\sqrt{(x-18)^2 + y^2}$$

$$\Rightarrow \quad (13x - 90)^2 = 25[(x-18)^2 + y^2]$$

$$\Rightarrow \quad 169x^2 - 2340x + 8100 = 25(x^2 - 36x + 324 + y^2)$$

$$\Rightarrow \quad 169x^2 - 2340x + 8100 = 25x^2 - 900x + 25y^2 + 8100$$

$$\Rightarrow \quad 144x^2 - 1440x - 25y^2 = 0$$

$$\Rightarrow \quad \frac{(x-5)^2}{25} - \frac{y^2}{144} = 1$$ **Ans.**

which is the required equation.

Differentiation

28.1 DIFFERENTIAL FROM THE FIRST PRINCIPLE

The limit of incremental ratio, *i.e.* lim $\frac{\delta y}{\delta x}$ as δx approaches zero is called the differential coefficient of y with respect to x and denoted by $\frac{dy}{dx}$.

$$\frac{dy}{dx} = \lim_{\delta x \to 0} \frac{\delta y}{\delta x}$$

$$\frac{d}{dx} f(x) = \lim_{\delta x \to 0} \frac{f(x + \delta x) - f(x)}{\delta x}$$

Example 1. *Find the differential coefficient of $y = x^2$ at $x = 2$.*

Solution. We have, $y = x^2$

$x + \delta x$	$y + \delta y$	δx	δy	$\frac{\delta y}{\delta x}$
2.0	4.00	0	0	$\frac{0}{0}$
2.1	4.41	0.1	0.41	$\frac{0.41}{0.1} = 4.1$
2.01	4.0401	0.01	0.0401	$\frac{0.0401}{0.01} = 4.01$
2.001	4.004001	0.001	0.004001	$\frac{0.004001}{0.001} = 4.001$
2.0001	4.00040001	0.0001	0.00040001	$\frac{0.00040001}{0.0001} = 4.0001$
		$\downarrow$ 0	$\downarrow$ 0	$\downarrow$ 4

As $\delta x \to 0$, hence $\delta y \to 0$. But the value of $\frac{dy}{dx}$ remains constant.

Example 1. *Find the percentage error in the area of a rectangle when an error of + 1 percent is made in measuring its length and breath.*

Solution. Let A be the area and l, b the length and breath

$$\Rightarrow \quad A = lb$$

$$\Rightarrow \quad \log A = \log l + \log b$$

On differentiating we get

$$\Rightarrow \quad \frac{\delta A}{A} = \frac{\delta l}{l} + \frac{\delta b}{b}$$

$$\Rightarrow \quad 100\frac{\delta A}{A} = 100\frac{\delta l}{l} + 100\frac{\delta b}{b}$$

$$= +1 + 1$$

Percentage error in area $= +2$ **Ans.**

Example 2. *If the radius of a sphere is measured as 7 m with an error of 0.02 m, then find the approximate error in calculating its volume.*

Solution. Let the radius of sphere be $r = 7\text{m}$, $\delta r = 0.02$ m and

volume of sphere $V = \dfrac{4\pi r^3}{3}$

$$\Rightarrow \quad \frac{dV}{dr} = 4\pi r^2$$

Therefore,

$$\delta V = \left(\frac{dV}{dr}\right)\delta r = (4\pi r^2)\delta r$$

$$= [4\pi(7)^2]\,(0.02) = (196\pi)\,(0.02) = 3.92\pi \text{ m}^3$$

Hence, the approximate error in calculating the volume is 3.92π m^3. **Ans.**

Example 3. *Find the percentage error in calculating the volume of cubical box if an error of 1% is made in measuring the length of edges of the cube.* **[SBTE. 2016, 2015, 2014]**

Solution. Let Δx be small error in 'x' and Δv be the small error in 'v'. then,

$$\frac{\Delta x}{x} \times 100 = 1 \quad \text{[Given]}$$

$$\frac{dx}{x} \times 100 = 1 \quad ...(1)$$

$$v = x^3$$

$$\Rightarrow \quad \frac{dv}{dx} = 3x^2$$

$$\Rightarrow \quad dv = 3x^2 dx$$

$$\Rightarrow \quad \frac{dv}{v} = \frac{3x^2}{v} dx$$

$$\Rightarrow \quad \frac{dv}{v} = \frac{3x^2}{x^3} dx \quad [\because v = x^3]$$

$$\Rightarrow \quad \frac{dv}{v} \times 100 = 3\frac{dx}{x} \times 100 = 3 \times 1 \quad \text{[from (1)]}$$

Example 42. *The function $f(x) = a \log_e x + bx^2 + x$ has extreme values at $x = 1$ and $x = 2$. Find a and b.*

Solution. Here we have $f(x) = a \log_e x + bx^2 + x$

$$\Rightarrow \quad f'(x) = \frac{a}{x} + 2bx + 1$$

$f(x)$ has extreme values at $x = 1$ and $x = 2$

then $f'(1) = 0$ and $f'(2) = 0$

$$f'(1) = a + 2b + 1 = 0$$

or $$a + 2b = -1 \quad \text{... (1)}$$

$$\Rightarrow \quad f'(2) = \frac{a}{2} + 4b + 1 = 0$$

or $$a + 8b = -2 \quad \text{... (2)}$$

On solving (1) and (2), we get

$$a = \frac{-2}{3}, b = -\frac{1}{6}$$ **Ans.**

EXERCISE 29.4

Find the points at which the following functions have a maximum and minimum values, also find the point of inflexion.

1. $y = 9x^3 - 45x^2 + 48x + 11.$ **Ans.** Minimum at $x = \frac{8}{3}$, minimum value $= \frac{31}{3}$

Maximum at $x = \frac{2}{3}$, maximum value $= \frac{77}{3}$

2. $y = 11 - 12x + 6x^2 + x^3.$ **Ans.** Point of inflexion at $x = 2$

3. $y = x^5 - 3x^4 + 5$ **[B.T.E. Delhi, May 2008]**

Ans. At $x = \frac{12}{5}$, Minimum

4. $y = x^5 - 5x^4.$ **Ans.** At $x = 4$, Minimum

5. $y = x^5 - 5x^4 + 5x^3 - 1.$ **Ans.** At $x = 0$, Pt. of inflexion

At $x = 1$, Maximum

At $x = 3$, Minimum

6. $y = 3 \sin^2 x + 4 \cos^2 x.$ **[Diploma IETE, Dec. 2005]**

Ans. Maximum at $x = 0$, maximum value = 4, minimum at $x = \pi/2$, minimum value = 3

7. Find the maximum value of $f(x) = (x - 1)(x - 2)(x - 3)$

Ans. Minimum at $x = \frac{6+\sqrt{3}}{3}$, maximum at $x = \frac{6-\sqrt{3}}{3}$

8. Find the minimum value of $f(x) = \sin x$ in the interval $\pi \le x \le 2\pi$. **Ans.** $x = \frac{3\pi}{2}$

9. Show that $\frac{\log x}{x}$ has a maximum value at $x = e$.

10. The strength S of a rectangular beam which can be cut from a circular log of radius r is given by $S = kx(4r^2 - x^2)$, where k is a constant and x is one side, find the value of x for which S is maximum. **Ans.** $x = \frac{2}{\sqrt{3}}r$

$$\Rightarrow \qquad \rho_2 = \frac{\left(\frac{a^2}{\theta^2}+\frac{r^2}{\theta^2}\right)^{3/2}}{\frac{a^2}{\theta^2}+2\frac{r^2}{\theta^2}-\left(\frac{a}{\theta}\right)\left(\frac{2r}{\theta^2}\right)}$$

$$\Rightarrow \qquad \rho_2 \text{ at } (\theta=1) = \frac{(a^2+a^2)^{3/2}}{a^2+2a^2-2a^2} = \frac{(2a^2)^{3/2}}{a^2} = 2\sqrt{2}\,a$$

$$\therefore \qquad \rho_1:\rho_2 = \frac{2\sqrt{2}}{3}a : 2\sqrt{2}\,a = 1:3$$

Proved

29.11 RADIUS OF CURVATURE FOR PEDAL EQUATIONS

Pedal equation for a plane curve is a relative between r and p

We know that $\quad p = r\sin\varphi$

Differentiating w.r.t. 'r',

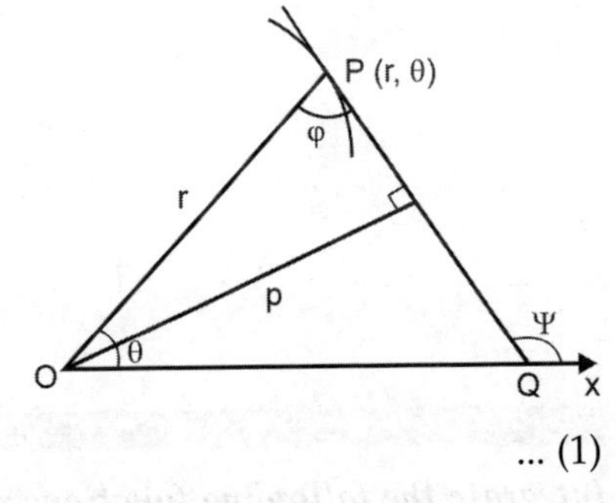

$$\Rightarrow \qquad \frac{dp}{dr} = r\cdot\cos\varphi\,\frac{d\varphi}{dr} + \sin\varphi$$

$$= r\cdot\frac{dr}{ds}\frac{d\varphi}{dr} + r\,\frac{d\theta}{ds}$$

$$= r\cdot\frac{d\varphi}{ds} + r\,\frac{d\theta}{ds}$$

$$\Rightarrow \qquad \frac{dp}{dr} = r\left(\frac{d\varphi}{ds}+\frac{d\theta}{ds}\right) \qquad \text{... (1)}$$

We also know that $\theta + \varphi = \psi$

$$\Rightarrow \qquad \frac{d\theta}{ds}+\frac{d\varphi}{ds} = \frac{d\psi}{ds} \qquad \text{... (2)}$$

Putting the value of $\frac{d\theta}{ds}+\frac{d\varphi}{ds}$ from (2) in (1), we get

$$\Rightarrow \qquad \frac{dp}{dr} = r\,\frac{d\psi}{ds} \quad \text{or} \quad \frac{ds}{d\psi} = r\,\frac{dr}{dp}$$

$$\Rightarrow \qquad \boldsymbol{\rho = r\,\frac{dr}{dp}}$$

Example 62. *Find the radius of curvature for a curve* $pa^m = r^{m+1}$.

Solution. $\quad pa^m = r^{m+1} \qquad \text{...(1)}$

Differentiating (1) w.r.t. 'r', we get

$$\Rightarrow \qquad \frac{dp}{dr}a^m = (m+1)\,r^m \quad \Rightarrow \quad \frac{dp}{dr} = (m+1)\,\frac{r^m}{a^m}$$

$$\Rightarrow \qquad \frac{dr}{dp} = \frac{a^m}{(m+1)\,r^m}$$

$$\therefore \qquad K = \frac{1}{3}$$ **Ans.**

Example 42. *Evaluate:* $\int e^x \frac{x^2+1}{(x+1)^2}\, dx$ **[SBTE. 2016, 2014]**

Solution. Let
$$I = \int e^x \frac{x^2+1}{(x+1)^2}\, dx$$

$$= \int e^x \frac{(x+1)^2 - 2x}{(x+1)^2}\, dx = \int e^x dx - 2\int e^x \frac{x}{(x+1)^2}\, dx$$

$$= e^x - 2\int e^x \left[\frac{(x+1)-1}{(x+1)^2}\right] dx$$

$$= e^x - 2\int e^x \cdot \frac{1}{x+1}\, dx + 2\int e^x \frac{1}{(x+1)^2}\, dx$$

$$= e^x - 2\cdot\frac{e^x}{x+1} - 2\int e^x \cdot \frac{1}{(x+1)^2}\, dx + 2\int e^x \cdot \frac{1}{(x+1)^2}\, dx$$

$$\therefore \qquad I = e^x - 2\cdot\frac{e^x}{x+1} + C$$ **Ans.**

EXERCISE 31.3

Integrate the following functions w.r.t. x.

1. $x\,(x+9)^{3/2}$ **Ans.** $\frac{2}{5}x\,(x+9)^{5/2} - \frac{4}{35}(x+9)^{7/2} + C$

2. $x \sin x$ **[SBTE. 2016, 2015]** **Ans.** $-\,x \cos x + \sin x + C$

3. $x \cos 2x$ **Ans.** $\frac{x \sin 2x}{2} + \frac{\cos 2x}{4} + C$

4. $x \cos^2 x$ **Ans.** $\frac{x^2}{4} + \frac{x}{4}\sin 2x + \frac{1}{8}\cos 2x + C$

5. $\frac{\quad}{\cos}$ **Ans.** $x \tan x + \log(\cos x) + C$

6. $x \sin x \cos x$ **Ans.** $\frac{1}{8}(\sin 2x - 2x \cos 2x) + C$

7. $x^2 \sin x \cos x$ **Ans.** $\frac{-x^2 \cos 2x}{4} + \frac{x \sin 2x}{2} + \frac{\cos 2x}{4} + C$

8. $\sec^3 x$ **[SBTE. 2014]** **Ans.** $\frac{1}{2}[\sec x \tan x + \log(\sec x + \tan x)] + C$

9. $x \log(1+x)$ **Ans.** $\frac{1}{2}(x^2-1)\log(1+x) - \frac{x^2}{4} + \frac{x}{2} + C$

Example 60. *Evaluate* $\int \sqrt{1+x-2x^2}\,dx$

Solution. $\int \sqrt{1+x-2x^2}\,dx = \sqrt{2}\int \sqrt{\frac{1}{2}+\frac{x}{2}-x^2}\,dx$

$$= \sqrt{2}\int \sqrt{\frac{1}{2}-\left(x^2-\frac{x}{2}\right)}dx$$

$$= \sqrt{2}\int \sqrt{\frac{1}{2}-\left(x-\frac{1}{4}\right)^2+\frac{1}{16}}\,dx$$

$$= \sqrt{2}\int \sqrt{\left(\frac{3}{4}\right)^2-\left(x-\frac{1}{4}\right)^2}\,dx$$

$$\left[\text{This is the form} = \int \sqrt{a^2-x^2}\,dx = \frac{x}{2}\sqrt{a^2-x^2}+\frac{a^2}{2}\sin^{-1}\left(\frac{x}{a}\right)\right]$$

$$\Rightarrow \quad \int \sqrt{1+x-2x^2}\,dx = \sqrt{2}\left[\frac{1}{2}\left(x-\frac{1}{4}\right)\sqrt{\frac{9}{16}-\left(x-\frac{1}{4}\right)^2}+\frac{1}{2}\cdot\frac{9}{16}\sin^{-1}\left(\frac{x-\frac{1}{4}}{\frac{3}{4}}\right)\right]$$

$$= \frac{1}{2}\left[x-\frac{1}{4}\right]\sqrt{1+x-2x^2}+\frac{9\sqrt{2}}{32}\sin^{-1}\left[\frac{4x-1}{3}\right]+C \qquad \textbf{Ans.}$$

Example 61. *Evaluate* $\int \frac{1}{\sqrt{1-x-x^2}}\,dx$ **[SBTE. 2015, 2014, 2010]**

Solution.

Let $$I = \int \frac{1}{\sqrt{1-x-x^2}}\,dx$$

$$= \int \frac{1}{\sqrt{-\left(x^2+x+\frac{1}{4}\right)+\frac{1}{4}+1}}\,dx$$

$$= \int \frac{1}{\sqrt{\frac{5}{4}-\left(x+\frac{1}{2}\right)^2}}\,dx = \int \frac{1}{\sqrt{\left(\frac{\sqrt{5}}{2}\right)^2-\left(x+\frac{1}{2}\right)^2}}\,dx$$

$$= \sin^{-1}\left(\frac{x+\frac{1}{2}}{\frac{\sqrt{5}}{2}}\right)+C$$

$\therefore$ $$I = \sin^{-1}\left(\frac{2x+1}{\sqrt{5}}\right)+C \qquad \textbf{Ans.}$$

Putting the value of $\int_0^{\frac{\pi}{2}} \sin^{n-2} x\,dx$ from (2) in (1), we get

$$\int_0^{\frac{\pi}{2}} \sin^n x\,dx = \frac{n-1}{n} \times \frac{n-3}{n-2} \int_0^{\frac{\pi}{2}} \sin^{n-4} x\,dx$$

Similarly, $$\int_0^{\frac{\pi}{2}} \sin^n x\,dx = \frac{n-1}{n} \times \frac{n-3}{n-2} \times \frac{n-5}{n-4} \int_0^{\frac{\pi}{2}} \sin^{n-6} x\,dx$$

and so on.

Case I. If n is even, then

$$\int_0^{\frac{\pi}{2}} \sin^n x\,dx = \frac{n-1}{n} \times \frac{n-3}{n-2} \times \frac{n-5}{n-4} \times \ldots \frac{1}{2} \times \int_0^{\frac{\pi}{2}} (\sin x)^0\,dx$$

$$= \frac{(n-1)(n-3)(n-5)\ldots 1}{n(n-2)(n-4)\ldots 2} \int_0^{\frac{\pi}{2}} 1 \cdot dx$$

$$= \frac{(n-1)(n-3)(n-5)\ldots 1}{n(n-2)(n-4)\ldots 2} [x]_0^{\frac{\pi}{2}}$$

$$\boxed{\int_0^{\frac{\pi}{2}} \sin^n x\,dx = \frac{(n-1)(n-3)(n-5)\ldots 1}{n(n-2)(n-4)\ldots 2} \times \frac{\pi}{2}}$$ **Ans.**

Case II. If n is odd, then

$$\int_0^{\frac{\pi}{2}} \sin^n x\,dx = \frac{n-1}{n} \times \frac{n-3}{n-2} \times \frac{n-5}{n-4} \times \ldots \times \frac{2}{3} \times \int_0^{\frac{\pi}{2}} \sin x\,dx$$

$$= \frac{(n-1)(n-3)(n-5)\ldots 2}{n(n-2)(n-4)\ldots 3} [-\cos x]_0^{\frac{\pi}{2}}$$

$$\boxed{\int_0^{\frac{\pi}{2}} \sin^n x\,dx = \frac{(n-1)(n-3)(n-5)\ldots 2}{n(n-2)(n-4)\ldots 3}}$$ **Ans.**

Similarly, $$\int_0^{\frac{\pi}{2}} \cos^n x\,dx = \frac{(n-1)(n-3)(n-5)\ldots 1}{n(n-2)(n-4)\ldots 2} \times \frac{\pi}{2} \text{ when } n \text{ is even}$$

$$\int_0^{\frac{\pi}{2}} \cos^n x\,dx = \frac{(n-1)(n-3)(n-5)\cdots \times 2}{n(n-2)(n-4)\ldots \times 3} \text{ when } n \text{ is odd.}$$

Example 12. *Evaluate:* (*i*) $\int_0^{\frac{\pi}{2}} \sin^7 x\,dx$ (*ii*) $\int_0^{\frac{\pi}{2}} \sin^6 x\,dx$

Solution. (*i*) $\int_0^{\frac{\pi}{2}} \sin^7 x\,dx = \frac{6 \times 4 \times 2}{7 \times 5 \times 3} = \frac{16}{35}$ **Ans.**

(*ii*) $\int_0^{\frac{\pi}{2}} \sin^6 x\,dx = \frac{5 \times 3 \times 1}{6 \times 4 \times 1} \times \frac{\pi}{2} = \frac{5\pi}{32}$ **Ans.**

Hence solution is $\int\limits_{y \text{ as constant}} M\,dx + \int (\text{terms of } N \text{ not containing } x)\,dy = C$

$$\Rightarrow \quad \int (2xy\cos x^2 - 2xy + 1)\,dx + \int 3\,dy = C$$

$$\Rightarrow \quad \int [y(2x\cos x^2) - y(2x) + 1]\,dx + 3\int dy = C$$

$$\Rightarrow \quad y\int 2x\cos x^2 dx - y\int 2x\,dx + \int 1\,dx + 3\int dy = C$$

Put $x^2 = t$ so that $2x\,dx = dt$

$$\Rightarrow \quad y\int \cos t\,dt - 2y\frac{x^2}{2} + x + 3y = C$$

$$\Rightarrow \quad y \sin t - x^2 y + x + 3y = C$$

$$\therefore \quad y \sin x^2 - x^2 y + x + 3y = C$$ **Ans.**

Example 59. *Solve:* $(1 + e^{x/y}) + e^{x/y}\left(1 - \frac{x}{y}\right)\frac{dy}{dx} = 0$

[Nagpur University, Summer 2008, AMIETE, June 2018]

Solution. We have,

$$\Rightarrow \quad \left(1 + e^{\frac{x}{y}}\right) + e^{\frac{x}{y}}\left(1 - e^{\frac{x}{y}}\right)\frac{dy}{dx} = 0 \Rightarrow \left(1 + e^{\frac{x}{y}}\right)dx + \left(e^{\frac{x}{y}} - e^{\frac{x}{y}}\frac{x}{y}\right)dy = 0$$

$$\Rightarrow \quad M = 1 + e^{\frac{x}{y}} \Rightarrow \frac{\partial M}{\partial y} = -\frac{x}{y^2}e^{\frac{x}{y}}$$

$$\Rightarrow \quad N = e^{\frac{x}{y}} - e^{\frac{x}{y}}\frac{x}{y} \Rightarrow \frac{\partial N}{\partial x} = \frac{1}{y}e^{\frac{x}{y}} - \frac{1}{y}e^{\frac{x}{y}} - \frac{x}{y^2}e^{\frac{x}{y}} = -\frac{x}{y^2}e^{\frac{x}{y}}$$

$$\Rightarrow \quad \frac{\partial M}{\partial y} = \frac{\partial N}{\partial x}$$

$\therefore$ Given equation is exact.

Its solution is $\int \left(1 + e^{\frac{x}{y}}\right)dx + \int (\text{terms of } N \text{ not containing } x)\,dy = C$

$$\Rightarrow \quad \int \left(1 + e^{\frac{x}{y}}\right)dx + \int 0\,dy = C \Rightarrow x + ye^{\frac{x}{y}} = C$$ **Ans.**

Example 60. *Solve:* $[1 + \log(xy)]dx + \left[1 + \frac{x}{y}\right]dy = 0$

Solution. $[1 + \log x + \log y]dx + \left[1 + \frac{x}{y}\right]dy = 0$

which is in the form $M\,dx + N\,dy = 0$

$$\Rightarrow \quad M = [1 + \log x + \log y] \quad \text{and} \quad N = 1 + \frac{x}{y}$$

Laplace and Inverse Laplace Transform

35.1 INTRODUCTION

Laplace transforms help in solving the differential equations with boundary values without finding the general solution and the values of the arbitrary constants.

35.2 LAPLACE TRANSFORM

Let $f(t)$ be function defined for all positive values of t, then

$$F(s) = \int_0^\infty e^{-st} f(t)\, dt$$

provided the integral exists, is called the **Laplace transform** of $f(t)$. It is denoted as

$$\boxed{L[f(t)] = F(s) = \int_0^\infty e^{-st} f(t)\, dt}$$

35.3 IMPORTANT FORMULAE

1. $L(1) = \dfrac{1}{s}$

2. $L(t^n) = \dfrac{n!}{s^{n+1}}$, then $n = 0, 1, 2, 3...$

3. $L\left(e^{at}\right) = \dfrac{1}{s-a}$ $\qquad (s > a)$

4. $L(\cosh at) = \dfrac{s}{s^2 - a^2}$ $\qquad (s^2 > a^2)$

5. $L(\sinh at) = \dfrac{a}{s^2 - a^2}$ $\qquad (s^2 > a^2)$

6. $L(\sin at) = \dfrac{a}{s^2 + a^2}$ $\qquad (s > 0)$

7. $L(\cos at) = \dfrac{s}{s^2 + a^2}$ $\qquad (s > 0)$

Proof of the above important formulae

$$\boxed{L(1) = \frac{1}{s}}$$

Example 18. *If* $f(t) = \dfrac{e^{at} - \cos bt}{t}$*, find the Laplace transform of* $f(t)$.

Solution.
$$f(t) = \frac{e^{at} - \cos bt}{t} = \frac{e^{at}}{t} - \frac{\cos bt}{t}$$

We know that,
$$L(e^{at} - \cos bt) = \left(\frac{1}{s-a} - \frac{s}{s^2+b^2}\right)$$

$$\Rightarrow \quad L\left(\frac{e^{at} - \cos bt}{t}\right) = \int_s^\infty \left(\frac{1}{s-a} - \frac{s}{s^2+b^2}\right) ds$$

$$= \left[\log(s-a) - \frac{1}{2}\log(s^2+b^2)\right]_s^\infty$$

$$= \left[\frac{2\log(s-a) - \log(s^2+b^2)}{2}\right]_s^\infty$$

$$= \frac{1}{2}\left[\log(s-a)^2 - \log(s^2+b^2)\right]_s^\infty$$

$$= \frac{1}{2}\left[\log\frac{(s-a)^2}{s^2+b^2}\right]_s^\infty = \frac{1}{2}\left[\log\left\{\frac{\left(1-\frac{a}{s}\right)^2}{1+\frac{b^2}{s^2}}\right\}\right]_s^\infty$$

$$= \frac{1}{2}\left[0 - \log\frac{\left(1-\frac{a}{s}\right)^2}{1+\frac{b^2}{s^2}}\right] = \frac{1}{2}\left[\log\frac{s^2+b^2}{(s-a)^2}\right] \quad \textbf{Ans.}$$

Example 19. *Find the Laplace transform of* $\dfrac{1-\cos t}{t^2}$. **[DIP IETE 2018, 2019]**

Solution.
$$L(1-\cos t) = L(1) - L(\cos t) = \frac{1}{s} - \frac{s}{s^2+1}$$

$$\Rightarrow \quad L\left[\frac{1-\cos t}{t}\right] = \int_s^\infty \left(\frac{1}{s} - \frac{s}{s^2+1}\right) ds = \left[\log s - \frac{1}{2}\log(s^2+1)\right]_s^\infty$$

$$= \frac{1}{2}\left[\log s^2 - \log(s^2+1)\right]_s^\infty = \frac{1}{2}\left[\log\left(\frac{s^2}{s^2+1}\right)\right]_s^\infty$$

$$= \frac{1}{2}\left[\log\frac{s^2}{s^2\left(1+\frac{1}{s^2}\right)}\right]_s^\infty = \frac{1}{2}\left[0 - \log\left(\frac{s^2}{s^2+1}\right)\right]$$

Example 41. *Find the inverse Laplace transform of the following :*

(i) $\dfrac{1}{s-2}$ (ii) $\dfrac{1}{s^2-9}$ (iii) $\dfrac{1}{s^2-16}$

(iv) $\dfrac{1}{s^2+25}$ (v) $\dfrac{s}{s^2+9}$ (vi) $\dfrac{1}{(s-2)^2+1}$

(vii) $\dfrac{s-1}{(s-1)^2+4}$ (viii) $\dfrac{1}{(s+3)^2-4}$ (ix) $\dfrac{s+2}{(s+2)^2-25}$

(x) $\dfrac{1}{2s-7}$

Solution.

(i) $$L^{-1}\left(\frac{1}{s-2}\right) = e^{2t}$$

(ii) $$L^{-1}\left(\frac{1}{s^2-9}\right) = L^{-1}\left[\frac{1}{3}\cdot\frac{3}{s^2-(3)^2}\right] = \frac{1}{3}\sinh 3t$$

(iii) $$L^{-1}\left(\frac{s}{s^2-16}\right) = L^{-1}\left[\frac{s}{s^2-(4)^2}\right] = \cosh 4t$$

(iv) $$L^{-1}\left(\frac{1}{s^2+25}\right) = \frac{1}{5}\frac{5}{s^2+(5)^2} = \frac{1}{5}\sin 5t$$

(v) $$L^{-1}\left(\frac{s}{s^2+9}\right) = \frac{s}{s^2+(3)^2} = \cos 3t$$

(vi) $$L^{-1}\left[\frac{1}{(s-2)^2+1}\right] = e^{2t}\sin t$$

(vii) $$L^{-1}\left[\frac{s-1}{(s-1)^2+4}\right] = e^{t}\cos 2t$$

(viii) $$L^{-1}\left[\frac{1}{(s+3)^2-4}\right] = \frac{1}{2}\frac{2}{(s+3)^2-(2)^2} = \frac{1}{2}e^{-3t}\sinh 2t$$

(ix) $$L^{-1}\left[\frac{s+2}{(s+2)^2-25}\right] = L^{-1}\left[\frac{(s+2)}{(s+2)^2-(5)^2}\right] = e^{-2t}\cosh 5t$$

(x) $$L^{-1}\left(\frac{1}{2s-7}\right) = \frac{1}{2}e^{\frac{7}{2}t} \qquad \left[\because\ L^{-1}F(as) = \frac{1}{a}f\left(\frac{t}{a}\right)\right]$$

Example 42. *Find Inverse Laplace transform of*

(a) $\left\{\dfrac{6}{2s-3} - \dfrac{3+4s}{9s^2-16} + \dfrac{8-6s}{16s^2+9}\right\}$ (b) $\dfrac{2s-5}{9s^2-25}$ (c) $\dfrac{s-2}{6s^2+20}$

Example 45. *Find the inverse Lapace transform of*

(i) $\dfrac{1}{(s+2)^5}$ (ii) $\dfrac{s}{s^2+4s+13}$ (iii) $\dfrac{1}{9s^2+6s+1}$

Solution.

(i) $$L^{-1}\left(\frac{1}{s^5}\right) = \frac{t^4}{4!}$$

then $$L^{-1}\left[\frac{1}{(s+2)^5}\right] = e^{-2t}.\frac{t^4}{4!}$$ **Ans.**

(ii) $$L^{-1}\left(\frac{s}{s^2+4s+13}\right) = L^{-1}\left[\frac{s+2-2}{(s+2)^2+(3)^2}\right]$$

$$= L^{-1}\left[\frac{s+2}{(s+2)^2+(3)^3}\right]-L^{-1}\left[\frac{2}{(s+2)^2+(3)^2}\right]$$

$$= e^{-2t}L^{-1}\left[\frac{s}{s^2+3^2}\right]-e^{-2t}L^{-1}\left[\frac{2}{3}\left(\frac{3}{s^2+3^2}\right)\right]$$

$$= e^{-2t}\cos 3t-\frac{2}{3}e^{-2t}\sin 3t$$ **Ans.**

(iii) $$L^{-1}\left(\frac{1}{9s^2+6s+1}\right) = L^{-1}\left[\frac{1}{(3s+1)^2}\right]=\frac{1}{9}L^{-1}\left[\frac{1}{\left(s+\frac{1}{3}\right)^2}\right]$$

$$= \frac{1}{9}e^{-\frac{t}{3}}L^{-1}\left(\frac{1}{s^2}\right)$$

$$= \frac{1}{9}e^{-\frac{t}{3}}t = \frac{te^{-\frac{t}{3}}}{9}$$ **Ans.**

EXERCISE 35.11

Obtain the inverse Laplace transform of the following :

1. $\dfrac{s+8}{s^2+4s+5}$ **Ans.** $e^{-2t}(\cos t+6\sin t)$

2. $\dfrac{s}{(s+3)^2+4}$ **Ans.** $e^{-3t}(\cos 2t-1.5\sin 2t)$

3. $\dfrac{s}{(s+7)^4}$ **Ans.** $e^{-7t}\dfrac{t^2}{6}(3-7t)$

4. $\dfrac{s+2}{s^2-2s-8}$ **Ans.** $e^{t}(\cos h\,3t+\sin h\,3t)$

9. Of the cigarette smoking population, 70% are men and 30% women, 10% of these men and 20% of these women smoke 'WILLS.' What is the probability that a person seen smoking a 'WILLS' will be a man?

Ans. $\frac{7}{13}$

10. A machine contains a component C that is vital to its operation. The reliability of component C is 80%. To improve the reliability of a machine, a similar component is used in parallel to form a system S. The machine will work provided that one of these components functions correctly. Calculate the reliability of the system S.

Ans. 96%

11. In a bolt factory, machines A, B and C manufacture 25%, 35% and 40% of the total output respectively. Of their outputs, 5%, 4% and 2% are defective bolts. A bolt is chosen at random and found to be defective. What is the probability that the bolt came from machine A ? B ? C ?

Ans. $\frac{25}{69}, \frac{28}{69}, \frac{16}{69}$

12. One bag contains four white and two black beads and another contains three of each colour. A bead is drawn from each bag. What is the probability that one is white and one is black?

Ans. $\frac{1}{2}$

13. The odds that a book will be favourably reviewed by three independent critics are 5 to 2, 4 to 3, 3 to 4 respectively. What is the probability that of the three reviews, a majority will be favourable?

Ans. $\frac{209}{343}$

14. Let E and F be independent events. The probability that both E and F happen is $\frac{1}{12}$ and the probability that neither E nor F happen is $\frac{1}{2}$. Then find $P(E)$ and $P(F)$.

Ans. $P(E) = \frac{1}{3}, P(F) = \frac{1}{4}$

15. Given a random variable whose range set is (1, 2) and whose probability is $f(1) = \frac{1}{4}$ and $f(2) = \frac{3}{4}$. Find the mean and variance of the distribution.

Ans. $\text{Mean} = \frac{7}{4}, \text{Var} = \frac{3}{16}$

16. A man takes a step forward with probability 0.4 and backward with probability 0.6. Find the probability that at the end of 11 steps, he is just one step away from the starting point.

Ans. 0.5263

17. What would be the expectation of the number of failures preceding the first success is an infinite series of independent trials with the constant probability of success p?

Hint: The probabilities of success in 1st, 2nd, 3rd trials respectively are p, qp, q^2p, q^3p, ...

The expected number of failures preceding the first success

$$E(x) = (0 \cdot p) + (1 \cdot qp) + (2 \cdot q^2p) + \dots \infty$$

$$= qp\,[1 + 2q + 3q^2 + \dots \infty] \text{ where } q < 1$$

$$= \frac{qp}{(1-q)^2} = \frac{qp}{p^2} = \frac{q}{p}$$

Applied Mathematics I

2K5-AS-02

Time: 3 Hours **Max. Marks: 100**

NOTE:
- **Part 'A' may be attempted in first 6 pages of Answer Sheet.**
- **Part 'B' in rest of the Sheets of Answer Sheet.**
- **Answers may be given in English or Hindi.**

PART A

Q.1 Attempt any 10 questions: **(10×2=20)**

a. Find the inverse of the matrix $A = \begin{bmatrix} 5 & -2 \\ 4 & -3 \end{bmatrix}$.

b. Find the product AB of the matrices $A = \begin{bmatrix} 1 & -2 \\ 3 & 1 \end{bmatrix}$ and $B = \begin{bmatrix} 2 & -2 \\ 1 & -4 \end{bmatrix}$.

c. Determine the intercepts on x and y-axis for the straight line $3x + 4y - 12 = 0$.

d. Find the equation of the circle centred at (2, –2) having radius 6.

e. Find the coordinates of the vertex and focus for the parabola $x^2 + 12y = 0$.

f. Find the slope of a line perpendicular to the line $2x + 5y + 1 = 0$.

g. If A is a square matrics such that $|A^T| = 8$, write the value of $|A^{-1}|$.

h. If $\begin{bmatrix} a & 14 \\ 1 & b \end{bmatrix} + 2\begin{bmatrix} 2 & -4 \\ 3 & 1 \end{bmatrix} = \begin{bmatrix} 8 & 5 \\ 7 & 7 \end{bmatrix}$, then find the values of a and b.

i. If $\vec{a} = 2\hat{i} + \hat{j} - \hat{k}$ and $\vec{b} = 2\hat{i} - 2\hat{j} + 3\hat{k}$, determine $|\vec{a} + \vec{b}|$

j. Find $5\vec{a} - \vec{b}$, if $\vec{a} = \hat{i} - 2\hat{j} + \hat{k}$ and $\vec{b} = 2\hat{i} + \hat{j} - 3\hat{k}$.

k. Find $\frac{d}{dx}(e^{ax-b})$.

l. Find $\frac{d}{dx}(\sin^2 x)$.

m. Evaluate $\lim_{x \to 2}(2x^2 - 4)$.

n. Find $\frac{d}{dx}(5x^3)$.

Q.2 Attempt any 5 questions: **(5×4=20)**

a. Convert the Cartesian coordinates $(-3, -\sqrt{3})$ in to the corresponding polar coordinates.

b. Find the equation of the line which passes through the point (2, –4) and parallel to the line $8x + 2y + 3 = 0$.

b. Find the volume of the solid generated by revoluting the ellipse

$$\frac{x^2}{a^2} + \frac{y^2}{b^2} = 1$$

about the minor axis.

Q.7 a. Verify Euler's theorem for the function

$$u = \sin^{-1}\left(\frac{x}{y}\right) + \tan^{-1}\left(\frac{y}{x}\right)$$

b. Solve the differential equation

$$\frac{d^2x}{dt^2} - 3\frac{dx}{dt} + 2x = 4t + e^{3t}, \text{ given that } x = 1 \text{ and } \frac{dx}{dt} = -1 \text{ when } t = 0.$$